P9-CMX-034

BIOLOGY

The Unity and Diversity of Life

STARR
TAGGART
EVERS
STARR

CENGAGE Learning

14TH EDITION

Australia • Brazil • Mexico • Singapore • United Kingdom • United States

Biology: The Unity and Diversity of Life,
Fourteenth Edition
Cecie Starr, Ralph Taggart, Christine Evers, Lisa Starr

Product Director: Mary Finch

Senior Product Team Manager: Yolanda Cossio

Senior Product Manager: Peggy Williams

Associate Content Developers: Kellie Petruzzelli, Casey Lozier

Product Assistant: Victor Luu

Media Developer: Lauren Oliveira

Senior Market Development Manager: Tom Ziolkowski

Content Project Manager: Harold Humphrey

Senior Art Directors: John Walker, Bethany Casey

Manufacturing Planner: Karen Hunt

Production Service: Grace Davidson & Associates

Photo Researcher: Cheryl DuBois, PreMedia Global

Text Researcher: Kristine Janssens, PreMedia Global

Copy Editor: Anita Wagner Heuftle

Illustrators: Lisa Starr, Gary Head, ScEYEnce Studios

Text Designer: Lisa Starr

Cover Designer: Bethany Casey

Cover and Title Page Image: © Pete Oxford/Minden Pictures

Butterflies sip the tears of a yellow-spotted river turtle sunning itself in Yasuní National Park, Ecuador. Turtle tears supply the butterflies with sodium, an essential nutrient missing from their flower nectar diet in the Amazon rainforest. Butterflies are almost never observed sipping turtle tears outside of this small region, which is famous for having one of the most diverse assortments of species in the world. Currently, oil drilling operations are destroying the forest and wildlife in the park.

Compositor: Lachina Publishing Services

Library of Congress Control Number: 2014944585

Student Edition:

ISBN-13: 978-1-305-07395-1

ISBN-10: 1-305-07395-9

Loose-leaf Edition:

ISBN-13: 978-1-305-25131-1

ISBN-10: 1-305-25131-8

Cengage Learning
20 Channel Center Street
Boston, MA 02210
USA

Printed in Canada
Print Number: 01 Print Year: 2014

Contents in Brief

Detailed Contents

INTRODUCTION

1 Invitation to Biology

UNIT I PRINCIPLES OF CELLULAR LIFE

2 Life's Chemical Basis

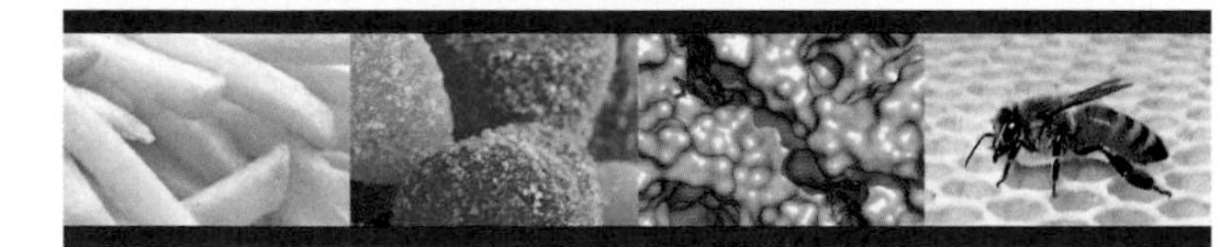

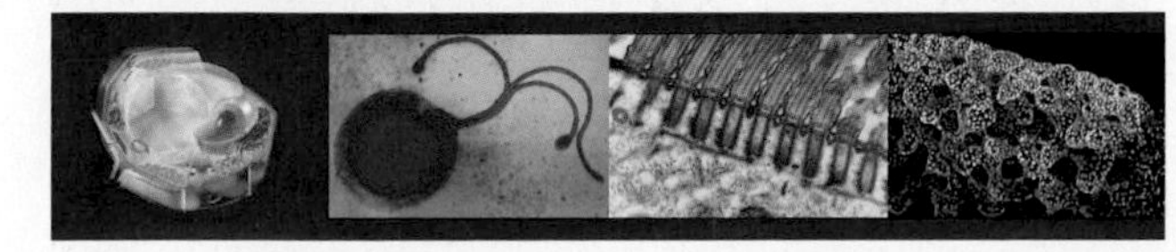

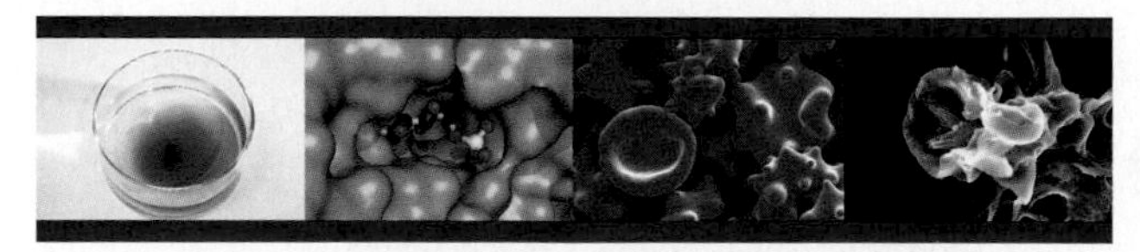

Detailed Contents (continued)

9 From DNA to Protein

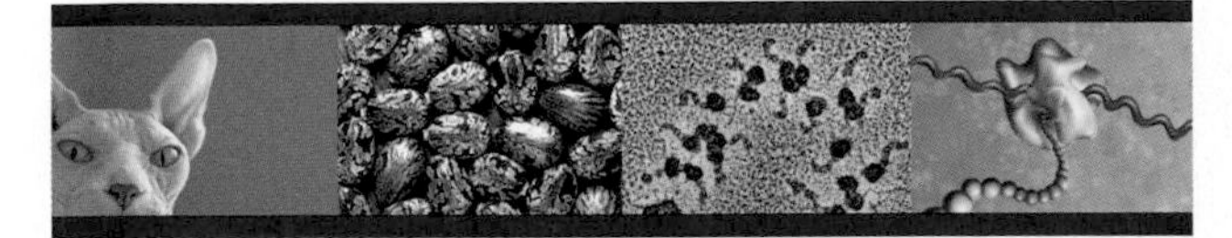

10 Control of Gene Expression

Detailed Contents (continued)

11 How Cells Reproduce

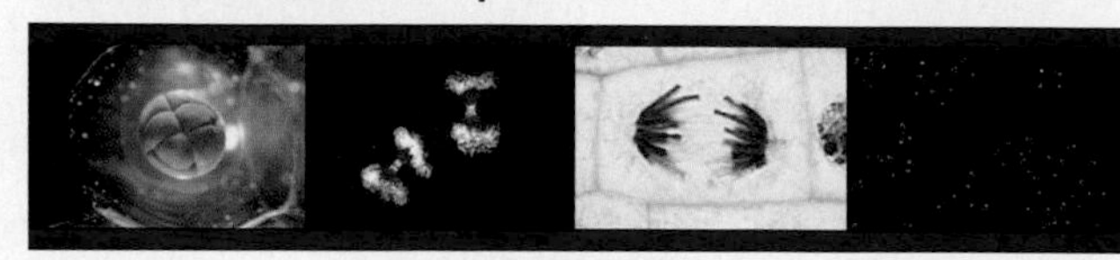

12 Meiosis and Sexual Reproduction

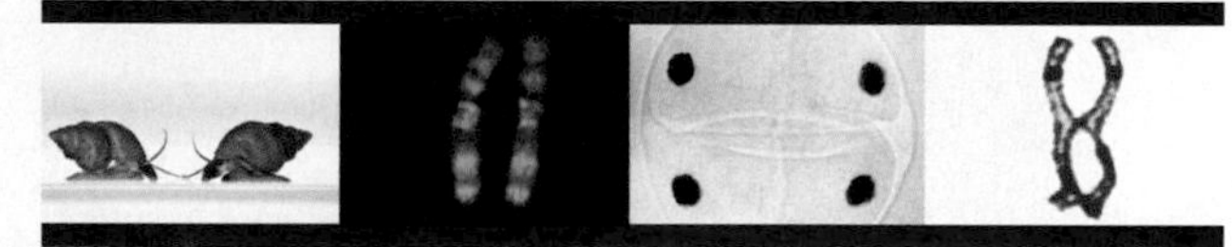

13 Observing Patterns in Inherited Traits

14 Chromosomes and Human Inheritance

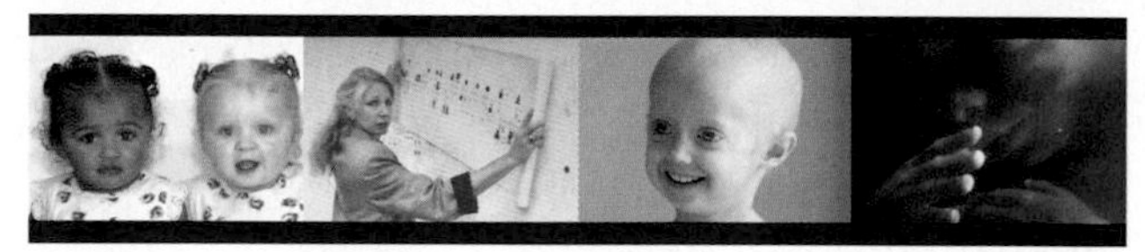

15 Studying and Manipulating Genomes

UNIT III PRINCIPLES OF EVOLUTION

16 Evidence of Evolution

17 Processes of Evolution

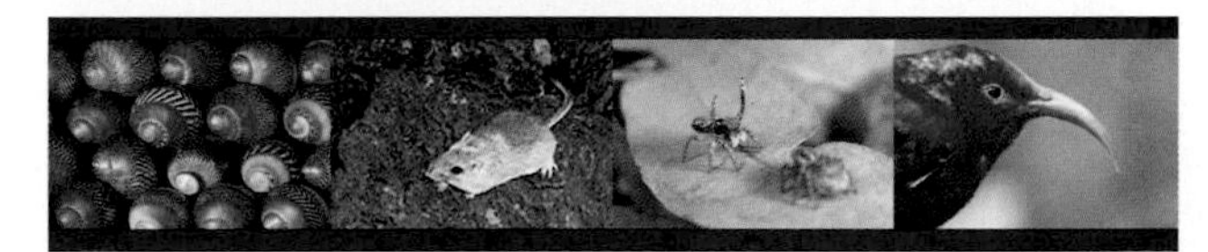

Detailed Contents (continued)

21 Protists—The Simplest Eukaryotes

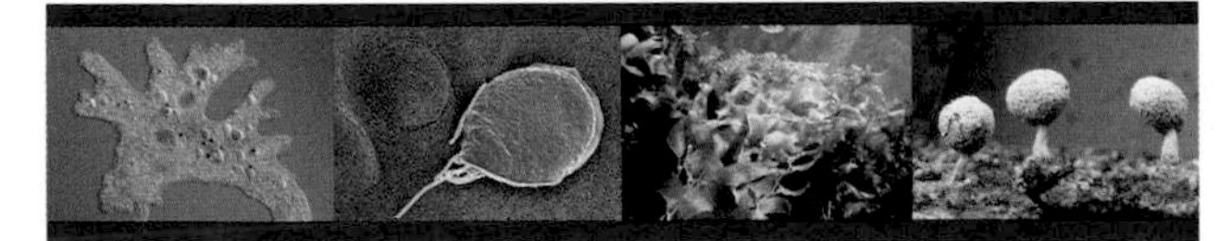

Detailed Contents (continued)

Detailed Contents (continued)

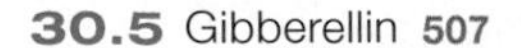

UNIT VI HOW ANIMALS WORK

31 Animal Tissues and Organ Systems

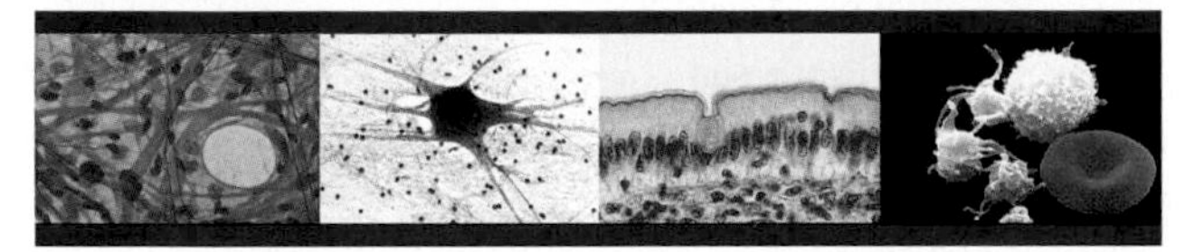

Detailed Contents (continued)

32 Neural Control

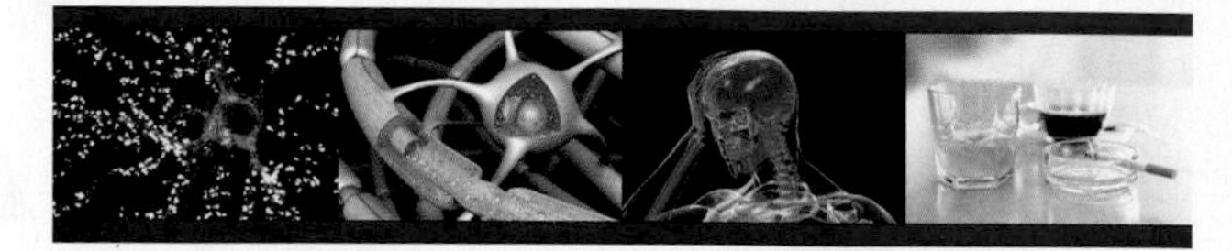

33 Sensory Perception

34 Endocrine Control

Detailed Contents (continued)

39 Digestion and Nutrition

40 Maintaining the Internal Environment

Detailed Contents (continued)

43 Animal Behavior

UNIT VII PRINCIPLES OF ECOLOGY

44 Population Ecology

45 Community Ecology

Detailed Contents (continued)

Preface

This edition of *Biology: The Unity and Diversity of Life* includes a wealth of new information reflecting recent discoveries in biology (details can be found in the *Power Bibliography*, which lists journal articles and other references used in the revision process; available upon request). Descriptions of current research, along with photos and videos of scientists who carry it out, underscore the concept that science is an ongoing endeavor carried out by a diverse community of people. Discussions include not only what was discovered, but also how the discoveries were made, how our understanding has changed over time, and what remains to be discovered. These discussions are provided in the context of a thorough, accessible introduction to well-established concepts and principles that underpin modern biology. Every topic is examined from an evolutionary perspective, emphasizing the connections between all forms of life.

Throughout the book, text and art have been revised to help students grasp difficult concepts. This edition also continues to focus on real world applications pertaining to the field of biology, including social issues arising from new research and developments. This edition covers in detail the many ways in which human activities are continuing to alter the environment and threaten both human health and Earth's biodiversity.

Changes to this Edition

Here are a few highlights of the revisions to this edition.

1 Invitation to Biology Renewed and updated emphasis on the relevance of new species discovery and the process of science.

2 Life's Chemical Basis New graphics illustrate elements and radioactive decay.

3 Molecules of Life New figure illustrates protein domains.

4 Cell Structure and Function New table summarizing cell theory; new photos of prokaryotes. Comparison of microscopy techniques updated using *Paramecium*. New figure shows food vacuoles in *Nassula*.

5 Ground Rules of Metabolism Temperature-dependent enzyme activity now illustrated with polymerases. New art and photos illustrate coenzymes, adhesion proteins, membrane trafficking, and energy transfer in redox reactions.

6 Where It Starts—Photosynthesis New photos illustrate phycobilins, stomata, adaptations of C4 plants, ice core sampling, smog in China. Light-dependent reactions art simplified.

7 How Cells Release Chemical Energy New photos illustrate mitochondrial disease and aerobic respiration.

8 DNA Structure and Function Concepts and illustrations of DNA hybridization and primers added to replication section. New photo of mutations caused by radiation at Chernobyl; new illustration of mutation.

9 From DNA to Protein Expanded material on the effects of mutation includes discussion of hairlessness in cats and a new micrograph of a sickled blood cell.

10 Gene Control New photos show transcription factors, X chromosome inactivation; new material explains evolution of lactose tolerance. New critical thinking question requires understanding of the effects of floral identity gene mutations.

11 How Cells Reproduce New photos illustrate mitosis, the mitotic spindle, and telomeres.

12 Meiosis and Sexual Reproduction New material on asexuality in mud snails and bdelloid rotifers. New micrograph shows multiple crossovers.

13 Observing Patterns in Inherited Traits New material about environmental effects on hemoglobin gene expression in *Daphnia*. New photos illustrate continuous variation.

14 Chromosomes and Human Inheritance Material on Tay-Sachs has been moved to this chapter as an illustration of autosomal recessive inheritance.

15 Studying and Manipulating Genomes Coverage of personal genetic testing updated with new medical applications, including the social impact of Angelina Jolie's response to her test. New photos of genetically modified animals. New "who's the daddy" critical thinking question offers students an opportunity to analyze a paternity test based on SNPs.

16 Evidence of Evolution New MRI showing coccyx illustrates a vestigial structure. Photos of 19th century naturalists added to emphasize the process of science that led to natural selection theory. Expanded coverage of fossil formation includes how banded iron formations provide evidence of the evolution of photosynthesis.

17 Processes of Evolution New opening essay on resistance to antibiotics as an outcome of agricultural overuse (warfarin material moved to illustrate directional selection). New art illustrates founder effect, and hypothetical example in text replaced with reduced diversity of *ABO* alleles in Native Americans. New art illustrates stasis in coelacanths.

18 Organizing Information About Species New material on DNA barcoding added to biochemical comparisons section. Data analysis activity revised to incorporate new data on honeycreeper ancestry.

19 Life's Origin and Early Evolution Added material about new discovery of 3.4-billion-year old fossil bacteria. New graphic illustrates endosymbiotic origin of mitochondria and chloroplasts.

20 Viruses, Bacteria, and Archaea Added information about Ebola and West Nile viruses, and newly discovered giant viruses.

21 Protists—The Simplest Eukaryotes New graphic depicts primary and secondary endosymbiosis. Added information about diatoms as a source of oil.

22 The Land Plants New essay about seed banks and the importance of sustain plant biodiversity.

23 Fungi More extensive coverage of fungal ecology; added information about white nose syndrome, a fungal disease of bats.

24 Animal Evolution—Invertebrates Updated information of medicines from invertebrates. New photos of terrestrial flatworm, plant-infecting roundworm.

25 Animal Evolution—Vertebrates Improved discussion of transition to land, with new illustration. Reorganized coverage of mammal evolution and diversity.

26 Human Evolution Updated to include latest discoveries about *Australopithecus sediba*, Denisovans, and Neanderthals.

27 Plant Tissues Carbon sequestration essay revised to include new data on wood production by old-growth redwoods. Reorganized to consolidate primary growth into its own section. Many new photos illustrate stem, leaf, and root structure. Material on fire scars added to section on dendroclimatology.

28 Plant Nutrition and Transport Illustration of Casparian strip integrated with new micrograph. Revisited section discusses phytoremediation at Ford's Rouge Center.

29 Life Cycles of Flowering Plants Updated material reflects current research on bee pollination behavior and colony collapse. New photos illustrate pollinators, fruit classification, asexual reproduction.

30 Communication Strategies in Plants Updates reflect ongoing major breakthroughs in the field of plant hormone function. New photos show apical dominance, effect of gibberellin, and abscission.

31 Animal Tissues and Organ Systems Added information about tissue regeneration in nonhuman animals; updated information about use of human and embryonic stem cells. Added information about blubber as a specialized adipose tissue.

32 Neural Control New opening essay about the effects of concussion on the brain. Reorganized coverage of psychoactive drugs. Added information about epidural anesthesia. Updated, improved coverage of memory.

33 Sensory Perception New opening essay about cochlear implants; revisited section discusses retinal implants, artificial limbs. Updated information about human sense of taste.

34 Endocrine Control Updated discussion of endocrine disruptors. New examples of pituitary gigantisms, dwarfism. Added information about role of melatonin in seasonal coat color changes.

35 Structural Support and Movement Added information about myostatin polymorphism in race horses to opening essay. New section discusses principles of animal location. Added information about boneless muscular organs such as the tongue.

36 Circulation More extensive coverage of plasma components. Discussion of genetics of blood types deleted. Improved coverage of and illustration of capillary exchange. Added information about blood pressure and jugular vein valves in giraffes.

37 Immunity Updated material on HIV/AIDS treatment strategies. New photos show T cell/APC interaction, skin as a surface barrier, a cytotoxic T cell killing a cancer cell, contact allergy, and victims of HIV.

38 Respiration Improved comparison of water and air as respiratory media with accompanying figure. Revised figure depicting first aid for choking victims to reflect latest guidelines. Discussion of human adaptation to high altitude now compares mechanisms in Tibetan and Andean populations.

39 Digestion and Nutrition New graphic depicting functional variations in animal dentition. New figure showing arrangement of organs that empty into the small intestine. Improved discussion of vitamin and mineral functions. New MRI illustrates how abdominal fat compresses internal organs. Added information about basal metabolic rate.

40 Maintaining the Internal Environment New subsection about climate-related adaptations in human populations.

41 Animal Reproductive Systems Coverage of intersex conditions dropped. Opening essay now discusses reproductive technology (IVF, egg banking); Revisited section discusses sperm banks. New section discusses location of animal gonads and the general mechanism of gamete formation. Reproductive function of human females now discussed before that of males; improved figure depicting the ovarian cycle.

42 Animal Development New opening essay about human birth defects, with a focus on cleft lip and palate. Improved photos illustrating apoptosis in digit development. Reorganized coverage of early human development. Added information about surgical delivery (cesarean section).

43 Animal Behavior Opening essay about effects of noise pollution on animal communication moved here and updated to reflect recent research. Revised discussion of the possible benefits of grouping.

44 Population Ecology Improved presentation of effects of predation on guppy life history. Revised, updated graphics.

45 Community Ecology Added information about and a photo of a brood parasite of ants. Added photo of the keystone species Pisaster.

46 Ecosystems More extensive discussion of aquifer depletion, salination; added information about ecological effects of over-allocation of river water. Updated discussion of the rise in atmospheric CO_2.

47 The Biosphere New opening essay about how winds and ocean currents distributed and are distributing material from the 2011 earthquake and tidal wave that affected Japan. Discussion of El Nino now a subsection within the chapter.

48 Human Impacts on the Biosphere New graphics of extinct animals: mastadon and dodo. Added information about and photo of endangered Florida lichen; added information about the Great Pacific Garbage Patch. Updated coverage of ozone depletion and effects of global climate change.

Student and Instructor Resources

Cengage Learning Testing Powered by Cognero is a flexible, online system that allows you to:

• author, edit, and manage test bank content from multiple Cengage Learning solutions
• create multiple test versions in an instant
• deliver tests from your LMS, your classroom or wherever you want

Instructor Companion Site Everything you need for your course in one place! This collection of book-specific lecture and class tools is available online via www.cengage.com/login. Access and download PowerPoint presentations, images, instructor's manual, videos, and more

Cooperative Learning Cooperative Learning: Making Connections in General Biology, 2nd Edition, authored by Mimi Bres and Arnold Weisshaar, is a collection of separate, ready-to-use, short cooperative activities that have broad application for first year biology courses. They fit perfectly with any style of instruction, whether in large lecture halls or flipped classrooms. The activities are designed to address a range of learning objectives such as reinforcing basic concepts, making connections between various chapters and topics, data analysis and graphing, developing problem solving skills, and mastering terminology. Since each activity is designed to stand alone, this collection can be used in a variety of courses and with any text.

MindTap A personalized, fully online digital learning platform of authoritative content, assignments, and services that engages students with interactivity while also offering instructors their choice in the configuration of coursework and enhancement of the curriculum via web-apps known as MindApps. MindApps range from ReadSpeaker (which reads the text out loud to students), to Kaltura (allowing you to insert inline video and audio into your curriculum). MindTap is well beyond an eBook, a homework solution or digital supplement, a resource center website, a course delivery platform, or a Learning Management System. It is the first in a new category —the Personal Learning Experience.

New for this edition! MindTap has an integrated Study Guide, expanded quizzing and application activities, and an integrated Test Bank.

Aplia for Biology The Aplia system helps students learn key concepts via Aplia's focused assignments and active learning opportunities that include randomized, automatically graded questions, exceptional text/art integration, and immediate feedback. Aplia has a full course management system that can be used independently or in conjunction with other course management systems such as MindTap, D2L, or Blackboard.

Acknowledgments

Writing, revising, and illustrating a biology textbook is a major undertaking for two full-time authors, but our efforts constitute only a small part of what is required to produce and distribute this one. We are truly fortunate to be part of a huge team of very talented people who are as committed as we are to creating and disseminating an exceptional science education product.

Biology is not dogma; paradigm shifts are a common outcome of the fantastic amount of research in the field. Ideas about what material should be taught and how best to present that material to students changes even from one year to the next. It is only with the ongoing input of our many academic reviewers and advisors (see opposite page) that we can continue to tailor this book to the needs of instructors and students while integrating new information and models. We continue to learn from and be inspired by these dedicated educators. A special thanks goes to Jose Panero for his extensive and detailed review for this edition.

On the production side of our team, the indispensable Grace Davidson orchestrated a continuous flow of files, photos, and illustrations while managing schedules, budgets, and whatever else happened to be on fire at the time. Grace, thank you as always for your patience and dedication. Thank you also to Cheryl DuBois, John Sarantakis, and Christine Myaskovsky for your help with photoresearch. Copyeditor Anita Hueftle and proofreader Kathy Dragolich, your valuable suggestions kept our text clear and concise.

Yolanda Cossio, thank you for continuing to support us and for encouraging our efforts to innovate and improve. Peggy Williams, we are as always grateful for your enthusiastic, thoughtful guidance, and for your many travels (and travails) on behalf of our books.

Thanks to Hal Humphrey our Cengage Production Manager, Tom Ziolkowski our Marketing Manager, Lauren Oliveira who creates our exciting technology package, Associate Content Developers Casey Lozier and Kellie Petruzzelli, and Product Assistant Victor Luu.

Lisa Starr and Christine Evers, May 2014

Influential Class Testers and Reviewers

Brenda Alston-Mills
North Carolina State University

Kevin Anderson
Arkansas State University - Beebe

Norris Armstrong
University of Georgia

Tasneem Ashraf
Coshise College

Dave Bachoon
Georgia College & State University

Neil R. Baker
The Ohio State University

Andrew Baldwin
Mesa Community College

David Bass
University of Central Oklahoma

Lisa Lynn Boggs
Southwestern Oklahoma State University

Gail Breen
University of Texas at Dallas

Marguerite "Peggy" Brickman
University of Georgia

David Brooks
East Central College

David William Bryan
Cincinnati State College

Lisa Bryant
Arkansas State University - Beebe

Katherine Buhrer
Tidewater Community College

Uriel Buitrago-Suarez
Harper College

Sharon King Bullock
Virginia Commonwealth University

John Capehart
University of Houston - Downtown

Daniel Ceccoli
American InterContinental University

Tom Clark
Indiana University South Bend

Heather Collins
Greenville Technical College

Deborah Dardis
Southeastern Louisiana University

Cynthia Lynn Dassler
The Ohio State University

Carole Davis
Kellogg Community College

Lewis E. Deaton
University of Louisiana - Lafayette

Jean Swaim DeSaix
University of North Carolina - Chapel Hill

(Joan) Lee Edwards
Greenville Technical College

Hamid M. Elhag
Clayton State University

Patrick Enderle
East Carolina University

Daniel J. Fairbanks
Brigham Young University

Amy Fenster
Virginia Western Community College

Kathy E. Ferrell
Greenville Technical College

Rosa Gambier
Suffok Community College - Ammerman

Tim D. Gaskin
Cuyahoga Community College - Metropolitan

Stephen J. Gould
Johns Hopkins University

Laine Gurley
Harper College

Marcella Hackney
Baton Rouge Community College

Gale R. Haigh
McNeese State University

John Hamilton
Gainesville State

Richard Hanke
Rose State Community College

Chris Haynes
Shelton St. Community College

Kendra M. Hill
South Dakota State University

Juliana Guillory Hinton
McNeese State University

W. Wyatt Hoback
University of Nebraska, Kearney

Kelly Hogan
University of North Carolina

Norma Hollebeke
Sinclair Community College

Robert Hunter
Trident Technical College

John Ireland
Jackson Community College

Thomas M. Justice
McLennan College

Timothy Owen Koneval
Laredo Community College

Sherry Krayesky
University of Louisiana - Lafayette

Dubear Kroening
University of Wisconsin - Fox Valley

Jerome Krueger
South Dakota State University

Jim Krupa
University of Kentucky

Mary Lynn LaMantia
Golden West College

Dale Lambert
Tarrant County College

Kevin T. Lampe
Bucks County Community College

Susanne W. Lindgren
Sacramento State University

Madeline Love
New River Community College

Dr. Kevin C. McGarry
Kaiser College - Melbourne

Ashley McGee
Alamo College

Jeanne Mitchell
Truman State University

Alice J. Monroe
St. Petersburg College - Clearwater

Brenda Moore
Truman State University

Erin L. G. Morrey
Georgia Perimeter College

Rajkumar "Raj" Nathaniel
Nicholls State University

Francine Natalie Norflus
Clayton State University

Harold Olivey
Indiana University Northwest

Alexander E. Olvido
Virginia State University

John C. Osterman
University of Nebraska, Lincoln

Jose L. Panero
University of Texas

Bob Patterson
North Carolina State University

Shelley Penrod
North Harris College

Carla Perry
Community College of Philadelphia

Mary A. (Molly) Perry
Kaiser College - Corporate

John S. Peters
College of Charleston

Carlie Phipps
SUNY IT

Michael Plotkin
Mt. San Jacinto College

Ron Porter
Penn State University

Karen Raines
Colorado State University

Larry A. Reichard
Metropolitan Community College - Maplewood

Jill D. Reid
Virginia Commonwealth University

Robert Reinswold
University of Northern Colorado

Ashley E. Rhodes
Kansas State University

David Rintoul
Kansas State University

Darryl Ritter
Northwest Florida State College

Amy Wolf Rollins
Clayton State University

Sydha Salihu
West Virginia University

Jon W. Sandridge
University of Nebraska

Robin Searles-Adenegan
Morgan State University

Erica Sharar
IVC; National University

Julie Shepker
Kaiser College - Melbourne

Rainy Shorey
Illinois Central College

Eric Sikorski
University of South Florida

Phoebe Smith
Suffolk County Community College

Robert (Bob) Speed
Wallace Junior College

Tony Stancampiano
Oklahoma City Community College

Jon R. Stoltzfus
Michigan State University

Peter Svensson
West Valley College

Jeffrey L. Travis
University at Albany

Nels H. Troelstrup, Jr.
South Dakota State University

Allen Adair Tubbs
Troy University

Will Unsell
University of Central Oklahoma

Rani Vajravelu
University of Central Florida

Jack Waber
West Chester University of Pennsylvania

Kathy Webb
Bucks County Community College

Amy Stinnett White
Virginia Western Community College

Virginia White
Riverside Community College

Robert S. Whyte
California University of Pennsylvania

Kathleen Lucy Wilsenn
University of Northern Colorado

Penni Jo Wilson
Cleveland State Community College

Robert Wise
University of Wisconsin Oshkosh

Michael L. Womack
Macon State College

Maury Wrightson
Germanna Community College

Mark L. Wygoda
McNeese State University

Lan Xu
South Dakota State University

Poksyn ("Grace") Yoon
Johnson and Wales University

Muriel Zimmermann
Chaffey College

1 Invitation to Biology

LEARNING ROADMAP

Whether or not you have studied biology, you already have an intuitive understanding of life on Earth because you are part of it. Every one of your experiences with the natural world—from the warmth of the sun on your skin to the love of your pet—contributes to that understanding.

THE SCIENCE OF NATURE

We can understand life by studying it at many levels, starting with atoms that are components of all matter, and extending to interactions of organisms with their environment.

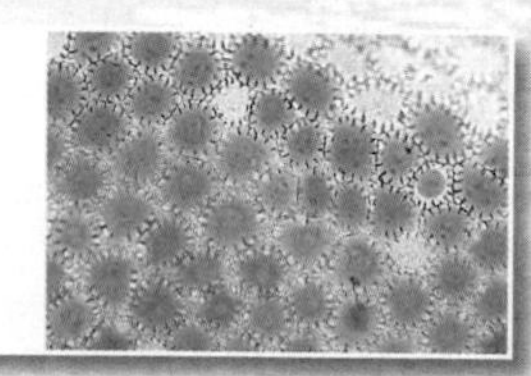

LIFE'S UNITY

All living things require ongoing inputs of energy and raw materials; all sense and respond to change; and all have DNA that guides their functioning.

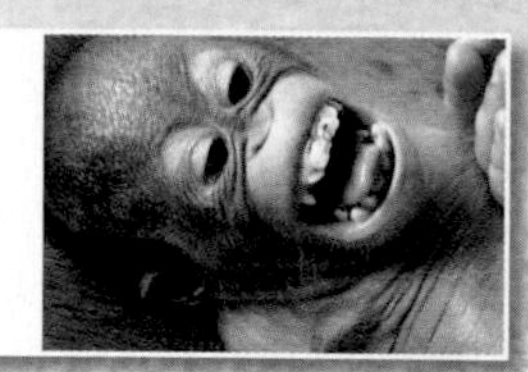

LIFE'S DIVERSITY

Observable characteristics vary tremendously among organisms. Various classification systems help us keep track of the differences.

THE NATURE OF SCIENCE

Carefully designing experiments helps researchers unravel cause-and-effect relationships in complex natural systems.

WHAT SCIENCE IS (AND WHAT IT IS NOT)

Science addresses only testable ideas about observable events and processes. It does not address the untestable, including beliefs and opinions.

This book parallels nature's levels of organization, from atoms to the biosphere. Learning about the structure and function of atoms and molecules will prime you to understand how living cells work. Learning about processes that keep a single cell alive can help you understand how multicelled organisms survive. Knowing what it takes for organisms to survive can help you see why and how they interact with one another and their environment.

1.1 The Secret Life of Earth

In this era of detailed satellite imagery and cell phone global positioning systems, could there possibly be any places left on Earth that humans have not yet explored? Actually, there are plenty of them. In 2005, for example, helicopters dropped a team of scientists into the middle of a vast and otherwise inaccessible cloud forest atop New Guinea's Foja Mountains. Within a few minutes, the explorers realized that their landing site, a dripping, moss-covered swamp, had been untouched by humans. Team member Bruce Beehler remarked, "Everywhere we looked, we saw amazing things we had never seen before. I was shouting. This trip was a once-in-a-lifetime series of shouting experiences."

How did the explorers know they had landed in uncharted territory? For one thing, the forest was filled with plants and animals previously unknown even to native peoples that have long inhabited other parts of the region. During the next month, the team members discovered many new species, including a rhododendron plant with flowers the size of a plate and a frog the size of a pea. They also came across hundreds of species that are on the brink of extinction in other parts of the world, and some that supposedly had been extinct for decades. The animals had never learned to be afraid of humans, so they could easily be approached. A few were discovered as they casually wandered through campsites (**FIGURE 1.1**).

New species are discovered all the time, often in places much more mundane than Indonesian cloud forests. How do we know what species a particular organism belongs to? What is a species, anyway, and why should discovering a new one matter to anyone other than a scientist? You will find the answers to such questions in this book. They are part of the scientific study of life, **biology**, which is one of many ways we humans try to make sense of the world around us.

Trying to understand the immense scope of life on Earth gives us some perspective on where we fit into it. For example, hundreds of new species are discovered every year, but about 20 species become extinct every minute in rain forests alone—and those are only the ones we know about. The current rate of extinctions is about 1,000 times faster than normal, and human activities are responsible for the acceleration. At this rate, we will never know about most of the species that are alive on Earth today. Does that matter? Biologists think so. Whether or not we are aware of it, humans are intimately connected with the world around us. Our activities are profoundly changing the entire fabric of life on Earth. These changes are, in turn, affecting us in ways we are only beginning to understand.

Ironically, the more we learn about the natural world, the more we realize we have yet to learn. But don't take our word for it. Find out what biologists know, and what they do not, and you will have a solid foundation upon which to base your own opinions about how humans fit into this world. By reading this book, you are choosing to learn about the human connection—your connection—with all life on Earth.

biology The scientific study of life.

FIGURE 1.1 Explorers found hundreds of rare species and dozens of new ones during recent survey expeditions to the Foja Mountain cloud forest (left). Right, Paul Oliver discovered this tree frog (*Litoria*) perched on a sack of rice during a particularly rainy campsite lunch. The explorers dubbed the new species "Pinocchio frog" after the Disney character because the male frog's long nose inflates and points upward during times of excitement.

1.2 Life Is More Than the Sum of Its Parts

✔ Biologists study life by thinking about it at different levels of organization.

✔ The quality of life emerges at the level of the cell.

❶ **Atoms**
Atoms are fundamental units of all substances, living or not. This image shows a model of a single atom.

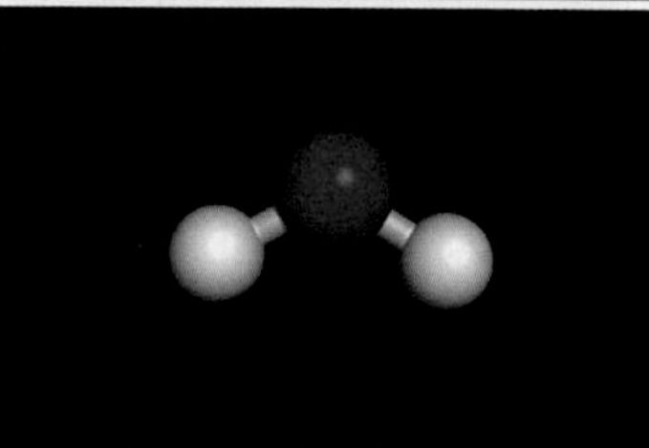

❷ **Molecule**
Atoms join other atoms in molecules. This is a model of a water molecule. The molecules special to life are much larger and more complex than water.

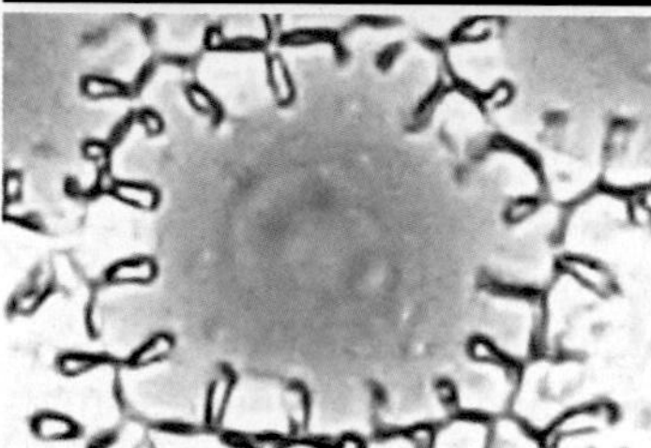

❸ **Cell**
The cell is the smallest unit of life. Some, like this plant cell, live and reproduce as part of a multicelled organism; others do so on their own.

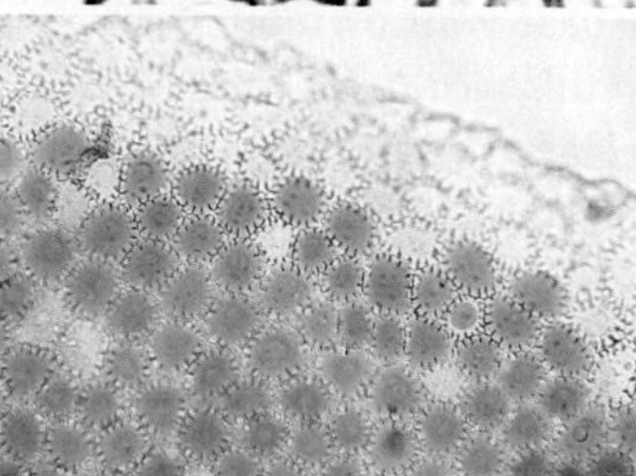

❹ **Tissue**
Organized array of cells that interact in a collective task. This is epidermal tissue on the outer surface of a flower petal.

❺ **Organ**
Structural unit of interacting tissues. Flowers are the reproductive organs of many plants.

❻ **Organ system**
A set of interacting organs. The shoot system of this poppy plant includes its aboveground parts: leaves, flowers, and stems.

FIGURE 1.2 ▸**Animated** Levels of life's organization.

What, exactly, is the property we call "life"? We may never actually come up with a good definition, because living things are too diverse, and they consist of the same basic components as nonliving things. When we try to define life, we end up with a list of properties that differentiate living from nonliving things. These properties often emerge from the interactions of basic components. To understand how that works, take a look at these groups of squares:

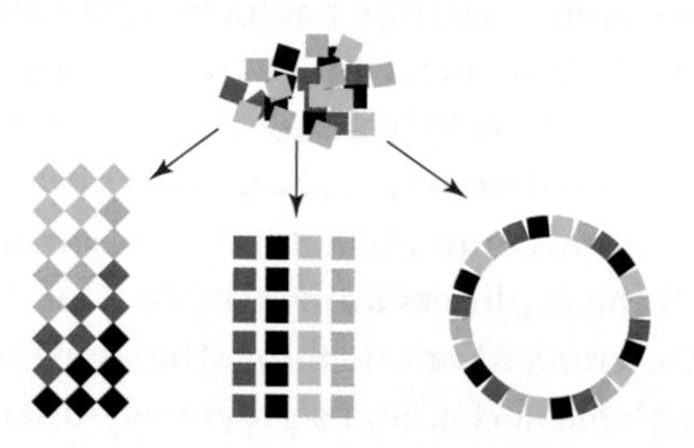

The property of "roundness" emerges when the component squares are organized one way, but not other ways. Characteristics of a system that do not appear in any of the system's components are called **emergent properties**. The idea that structures with emergent properties can be assembled from the same basic building blocks is a recurring theme in our world—and also in biology.

Life has successive levels of organization, with new emergent properties appearing at each level (FIGURE 1.2). This organization begins with interactions between **atoms**, which are fundamental building blocks of all substances ❶. Atoms bond together to form **molecules** ❷. There are no atoms unique to living things, but there are unique molecules. In today's natural world, only living things make the "molecules of life," which are lipids, proteins, DNA, RNA, and complex carbohydrates. The emergent property of "life" appears at the next level, when many molecules of life become organized as a cell ❸. A **cell** is the smallest unit of life. Cells survive and reproduce themselves using energy, raw materials, and information in their DNA.

Some cells live and reproduce independently. Others do so as part of a multicelled organism. An **organism** is an individual that consists of one or more cells. A poppy plant is an example of a multicelled organism ❼. In most multicelled organisms, cells are organized as tissues ❹. A **tissue** consists of specific types of cells organized in a particular pattern. The arrangement allows the cells to collectively perform a special function such as protection from injury (dermal tissue) or movement (muscle tissue). An **organ** is an organized array of tissues that collectively carry out a particular task or set of tasks ❺. For example, a flower is an organ of reproduction in plants; a heart, an organ that pumps blood in animals. An **organ system** is a set of organs and tissues that interact to keep the individual's body working properly ❻. Examples of organ systems include the aboveground parts of a plant (the shoot system), and the heart and blood vessels of an animal (the circulatory system).

A **population** is a group of interbreeding individuals of the same type, or species, living in a given area ❽. An example may be all California poppies living in California's Antelope Valley Poppy Reserve. At the next level, a **community** consists of all populations of all species in a given area. The Antelope Valley Reserve community includes California poppies and all other plants, animals, microorganisms, and so on ❾. Communities may be large or small, depending on the area defined.

The next level of organization is the **ecosystem**, which is a community interacting with its environment ❿. The most inclusive level, the **biosphere**, encompasses all regions of Earth's crust, waters, and atmosphere in which organisms live ⓫.

atom Fundamental building block of all matter.
biosphere All regions of Earth where organisms live.
cell Smallest unit of life.
community All populations of all species in a given area.
ecosystem A community interacting with its environment.
emergent property A characteristic of a system that does not appear in any of the system's component parts.
molecule Two or more atoms bonded together.
organ In multicelled organisms, a grouping of tissues engaged in a collective task.
organism Individual that consists of one or more cells.
organ system In multicelled organisms, set of organs engaged in a collective task that keeps the body functioning properly.
population Group of interbreeding individuals of the same species that live in a given area.
tissue In multicelled organisms, specialized cells organized in a pattern that allows them to perform a collective function.

TAKE-HOME MESSAGE 1.2

How do living things differ from nonliving things?

- ✔ All things, living or not, consist of the same building blocks: atoms. Atoms join as molecules.
- ✔ In today's natural world, only living things make lipids, proteins, DNA, RNA, and complex carbohydrates. The unique properties of life emerge as these molecules become organized into cells.
- ✔ Higher levels of life's organization include multicelled organisms, populations, communities, ecosystems, and the biosphere.
- ✔ Emergent properties occur at each successive level of life's organization.

❼ Multicelled organism Individual that consists of more than one cell. Cells of this California poppy plant are part of its two organ systems: aboveground shoots and belowground roots.

❽ Population Group of single-celled or multicelled individuals of a species in a given area. This population of California poppy plants is in California's Antelope Valley Poppy Reserve.

❾ Community All populations of all species in a specified area. These plants are part of the Antelope Valley Poppy Reserve community.

❿ Ecosystem A community interacting with its physical environment through the transfer of energy and materials. Sunlight and water sustain the natural community in the Antelope Valley.

⓫ Biosphere The sum of all ecosystems: every region of Earth's waters, crust, and atmosphere in which organisms live. No ecosystem in the biosphere is truly isolated from any other.

1.3 How Living Things Are Alike

✔ Continual inputs of energy and the cycling of materials maintain life's complex organization.

✔ Organisms sense and respond to change.

✔ All organisms use information in the DNA they inherited from their parent or parents to function.

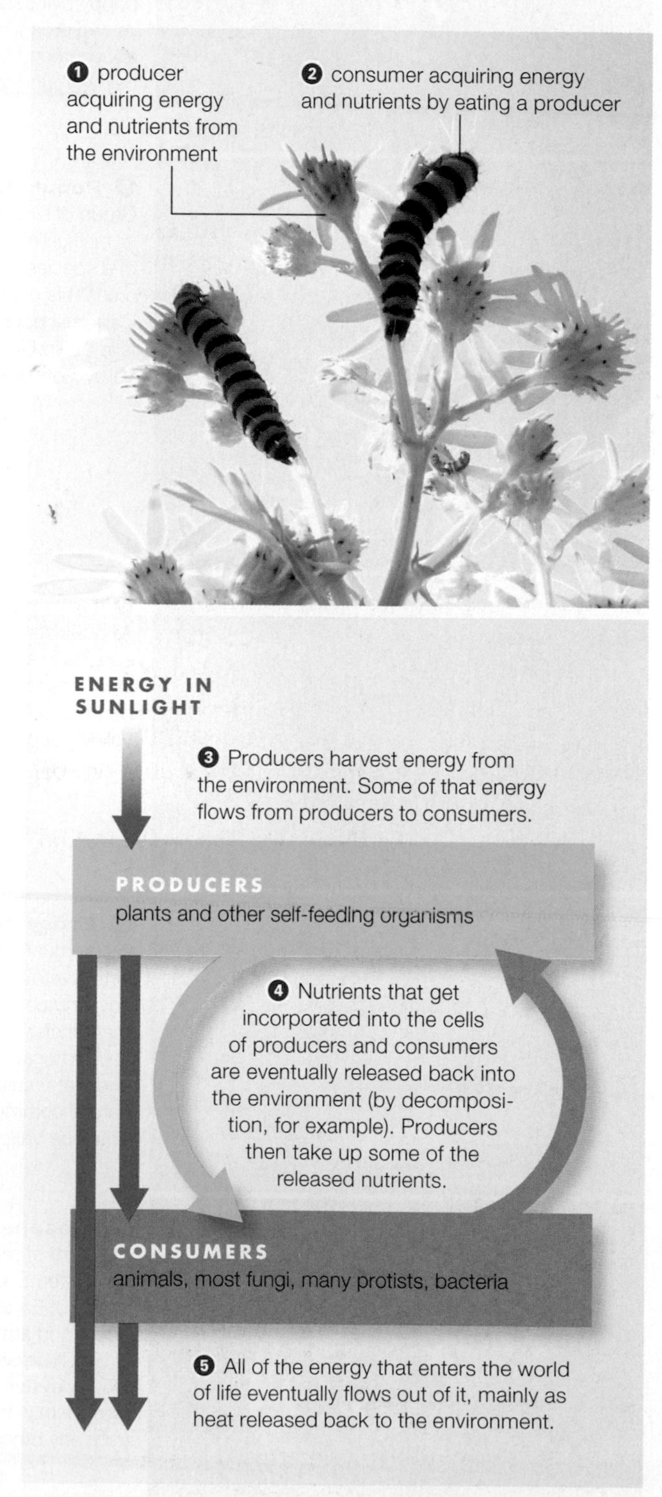

FIGURE 1.3 ▶**Animated** The one-way flow of energy and cycling of materials in the world of life.

Even though we cannot precisely define "life," we can intuitively understand what it means because all living things share a set of key features. All require ongoing inputs of energy and raw materials; all sense and respond to change; and all pass DNA to offspring.

Organisms Require Energy and Nutrients

Not all living things eat, but all require energy and nutrients on an ongoing basis. Both are essential to maintain the functioning of individual organisms and the organization of life. A **nutrient** is a substance that an organism needs for growth and survival but cannot make for itself.

Organisms spend a lot of time acquiring energy and nutrients (FIGURE 1.3). However, the source of energy and the type of nutrients acquired differ among organisms. These differences allow us to classify all living things into two categories: producers and consumers. **Producers** make their own food using energy and simple raw materials they obtain from nonbiological sources ❶. Plants are producers that use the energy of sunlight to make sugars from water and carbon dioxide (a gas in air), a process called **photosynthesis**. By contrast, **consumers** cannot make their own food. They obtain energy and nutrients by feeding on other organisms ❷. Animals are consumers. So are decomposers, which feed on the wastes or remains of other organisms. The leftovers from consumers' meals end up in the environment, where they serve as nutrients for producers. Said another way, nutrients cycle between producers and consumers.

Unlike nutrients, energy is not cycled. It flows through the world of life in one direction: from the environment ❸, through organisms ❹, and back to

consumer Organism that gets energy and nutrients by feeding on tissues, wastes, or remains of other organisms.
development Multistep process by which the first cell of a new multicelled organism gives rise to an adult.
DNA Deoxyribonucleic acid; carries hereditary information that guides development and other activities.
growth In multicelled species, an increase in the number, size, and volume of cells.
homeostasis Process in which an organism keeps its internal conditions within tolerable ranges by sensing and responding to change.
inheritance Transmission of DNA to offspring.
nutrient Substance that an organism needs for growth and survival but cannot make for itself.
photosynthesis Process by which producers use light energy to make sugars from carbon dioxide and water.
producer Organism that makes its own food using energy and nonbiological raw materials from the environment.
reproduction Processes by which parents produce offspring.

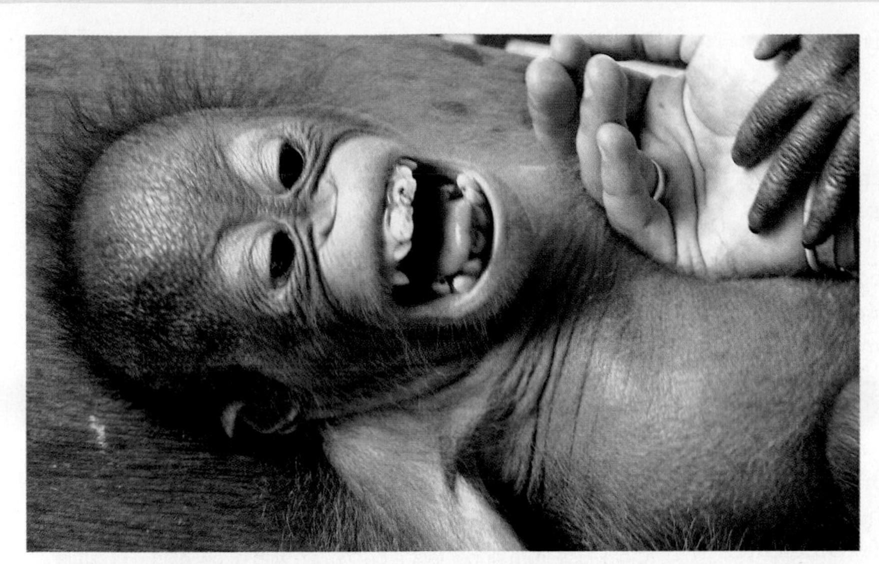

FIGURE 1.4 Organisms sense and respond to stimulation. This baby orangutan is laughing in response to being tickled. Apes and humans make different sounds when being tickled, but the airflow patterns are so similar that we can say apes really do laugh.

the environment ❺. This flow maintains the organization of every living cell and body, and it also influences how individuals interact with one another and their environment. The energy flow is one-way, because with each transfer, some energy escapes as heat, and cells cannot use heat as an energy source. Thus, energy that enters the world of life eventually leaves it (we return to this topic in Chapter 5).

Organisms Sense and Respond to Change

An organism cannot survive for very long in a changing environment unless it adapts to the changes. Thus, every living thing has the ability to sense and respond to change both inside and outside of itself (**FIGURE 1.4**). For example, after you eat, the sugars from your meal enter your bloodstream. The added sugars set in motion a series of events that causes cells throughout the body to take up sugar faster, so the sugar level in your blood quickly falls. This response keeps your blood sugar level within a certain range, which in turn helps keep your cells alive and your body functioning.

The fluid portion of your blood is a component of your internal environment, which is all of the body fluids outside of cells. That internal environment must be kept within certain ranges of temperature and other conditions, or the cells that make up your body will die. By sensing and adjusting to change, you and all other organisms keep conditions in the internal environment within a range that favors survival. **Homeostasis** is the name for this process, and it is one of the defining features of life.

Organisms Use DNA

With little variation, the same types of molecules perform the same basic functions in every organism. For example, information in an organism's **DNA** (deoxyribonucleic acid) guides ongoing functions that sustain the individual through its lifetime. Such functions include **development**: the process by which the first cell of a new individual gives rise to a multicelled adult; **growth**: increases in cell number, size, and volume; and **reproduction**: processes by which individuals produce offspring.

Individuals of every natural population are alike in certain aspects of their body form and behavior because their DNA is very similar: Orangutans look like orangutans and not like caterpillars because they inherited orangutan DNA, which differs from caterpillar DNA in the information it carries. **Inheritance** refers to the transmission of DNA to offspring. All organisms inherit their DNA from one or two parents.

DNA is the basis of similarities in form and function among organisms. However, the details of DNA molecules differ, and herein lies the source of life's diversity. Small variations in the details of DNA's structure give rise to differences among individuals, and also among types of organisms. As you will see in later chapters, these differences are the raw material of evolutionary processes.

TAKE-HOME MESSAGE 1.3

How are all living things alike?

✔ A one-way flow of energy and a cycling of nutrients sustain life's organization.

✔ Organisms sense and respond to conditions inside and outside themselves. They make adjustments that keep conditions in their internal environment within a range that favors cell survival, a process called homeostasis.

✔ All organisms use information in the DNA they inherited from their parent or parents to develop, grow, and reproduce. DNA is the basis of similarities and differences in form and function among organisms.

1.4 How Living Things Differ

✔ There is great variation in the details of appearance and other observable characteristics of living things.

Living things differ tremendously in their observable characteristics. Various classification schemes help us organize what we understand about the scope of this variation, which we call Earth's **biodiversity**.

For example, organisms can be grouped on the basis of whether they have a nucleus, which is a sac with two membranes that encloses and protects a cell's DNA. **Bacteria** (singular, bacterium) and **archaea** (singular, archaeon) are organisms whose DNA is *not* contained within a nucleus. All bacteria and archaea are single-celled, which means each organism consists of one cell (**FIGURE 1.5A,B**). Collectively, these organisms are the most diverse representatives of life. Different kinds are producers or consumers in nearly all regions of Earth. Some inhabit such extreme environments as frozen desert rocks, boiling sulfurous lakes, and nuclear reactor waste. The first cells on Earth may have faced similarly hostile environments.

Traditionally, organisms without a nucleus have been called **prokaryotes**, but this designation is now used only informally. This is because, despite the similar appearance of bacteria and archaea, the two types of cells are less related to one another than we once thought. Archaea turned out to be more closely related to **eukaryotes**, which are organisms whose DNA is contained within a nucleus. Some eukaryotes live as individual cells; others are multicelled (**FIGURE 1.5C**). Eukaryotic cells are typically larger and more complex than bacteria or archaea.

Structurally, **protists** are the simplest eukaryotes, but as a group they vary dramatically, from single-celled consumers to giant, multicelled producers.

Fungi (singular, fungus) are eukaryotic consumers that secrete substances to break down food externally, then absorb nutrients released by this process. Many fungi are decomposers. Most fungi, including those that form mushrooms, are multicellular. Fungi that live as single cells are called yeasts.

Plants are multicelled eukaryotes; the majority are photosynthetic producers that live on land. Besides feeding themselves, plants also serve as food for most other land-based organisms.

Animals are multicelled consumers that ingest tissues or juices of other organisms. Unlike fungi, animals break down food inside their body. They also develop through a series of stages that lead to the adult form. All animals actively move about during at least part of their lives.

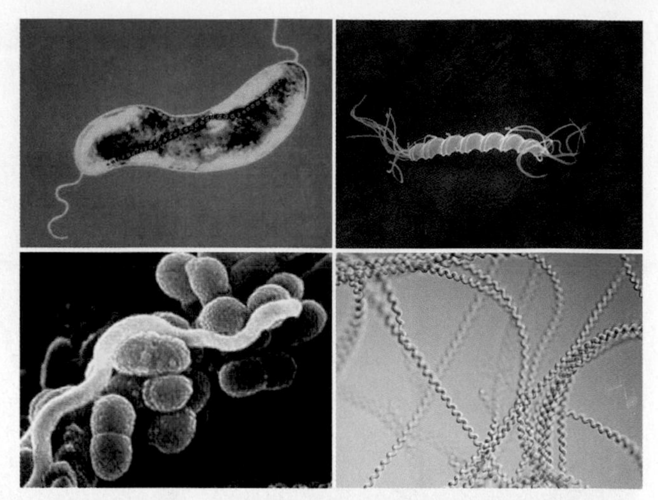

A Bacteria are the most numerous organisms on Earth. Clockwise from upper left, a bacterium with a row of iron crystals that acts like a tiny compass; a common resident of cat and dog stomachs; spiral cyanobacteria; types found in dental plaque.

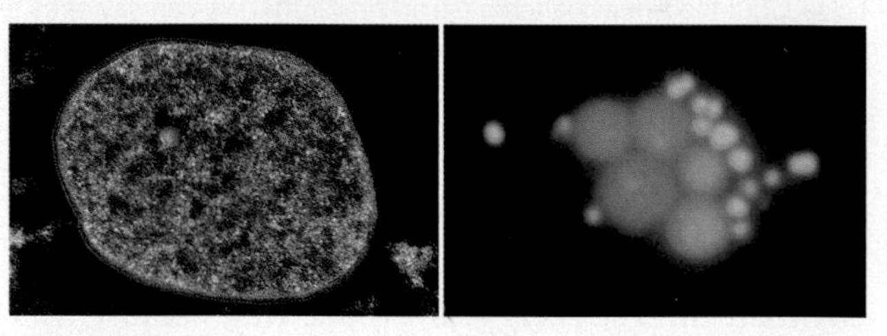

B Archaea resemble bacteria, but are more closely related to eukaryotes. Left, an archaeon that grows in sulfur hot springs. Right, two types of archaea from a seafloor hydrothermal vent.

FIGURE 1.5 ▶**Animated** A few representatives of life's diversity.

animal Multicelled consumer that develops through a series of stages and moves about during part or all of its life.
archaea Group of single-celled organisms that lack a nucleus but are more closely related to eukaryotes than to bacteria.
bacteria The most diverse and well-known group of single-celled organisms that lack a nucleus.
biodiversity Scope of variation among living organisms.
eukaryote Organism whose cells characteristically have a nucleus.
fungus Single-celled or multicelled eukaryotic consumer that breaks down material outside itself, then absorbs nutrients released from the breakdown.
plant A multicelled, typically photosynthetic producer.
prokaryote Single-celled organism without a nucleus.
protist Member of a diverse group of simple eukaryotes.

TAKE-HOME MESSAGE 1.4
How do organisms differ from one another?

✔ Organisms differ in their details; they show tremendous variation in observable characteristics.

✔ We divide Earth's biodiversity into broad groups based on traits such as having a nucleus or being multicellular.

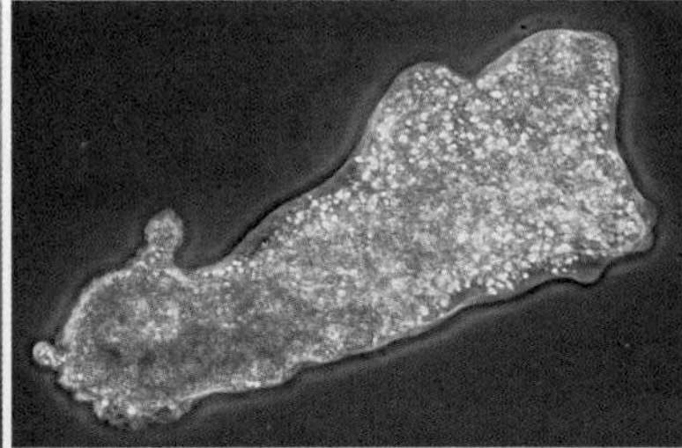

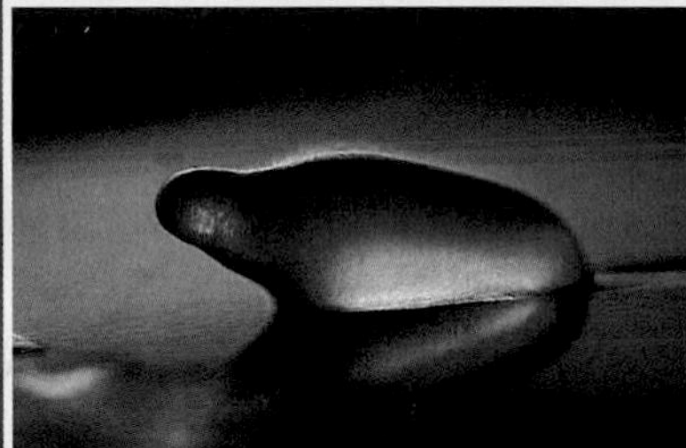

Protists are a group of extremely diverse eukaryotes that range from giant multicelled seaweeds to microscopic single cells.

Plants are multicelled eukaryotes, most of which are photosynthetic. Nearly all have roots, stems, and leaves.

Fungi are eukaryotic consumers that secrete substances to break down food outside their body. Most are multicelled (left), but some are single-celled (right).

Animals are multicelled eukaryotes that ingest tissues or juices of other organisms. All actively move about during at least part of their life.

C **Eukaryotes** are single-celled or multicelled organisms whose DNA is contained within a nucleus.

CREDITS: (5C) Protists: from left, Courtesy of Allen W. H. Bé and David A. Caron; top, M I Walker/Science Source; middle, © Carolina Biological Supply Company; bottom, Oliver Meckes/Science Source; top, Courtesy of Allen W. H. Bé and David A. Caron; bottom, © Emiliania Huxleyi photograph, Vita Pariente, scanning electron micrograph taken on a Jeol T330A instrument at Texas A&M University Electron Microscopy Center; Plants: left, © Jag.ca.Shutterstock.com; right, © Martin Ruegner/Radium Images/Jupiter Images; Fungi, left, © Robert C. Simpson/Nature Stock; right, By London Scientific Films/Oxford Scientific/Getty Images; Animals, from left, © Tom & Pat Leeson, Ardea London Ltd.; Thomas Eisner, Cornell University; © Martin Zimmerman, Science, 1961, 133:73-79, © AAAS.; © Pixtal/SuperStock.

1.5 Organizing Information About Species

✔ Each type of organism, or species, is given a unique name.
✔ We define and group species based on shared traits.

Each time we discover a new **species**, or unique kind of organism, we name it. **Taxonomy**, a system of naming and classifying species, began thousands of years ago, but naming species in a consistent way did not become a priority until the eighteenth century. At that time, European explorers who were just discovering the scope of life's diversity started having more and more trouble communicating with one another because species often had multiple names. For example, the dog rose (a plant native to Europe, Africa, and Asia) was alternately known as briar rose, witch's briar, herb patience, sweet briar, wild briar, dog briar, dog berry, briar hip, eglantine gall, hep tree, hip fruit, hip rose, hip tree, hop fruit, and hogseed—and those are only the English names! Species often had multiple scientific names too, in Latin that was descriptive but often cumbersome. The scientific name of the dog rose was *Rosa sylvestris inodora seu canina* (odorless woodland dog rose), and also *Rosa sylvestris alba cum rubore, folio glabro* (pinkish white woodland rose with smooth leaves).

An eighteenth-century naturalist, Carolus Linnaeus, standardized a naming system that we still use. By the Linnaean system, every species is given a unique two-part scientific name. The first part is the name of the **genus** (plural, genera), a group of species that share a unique set of features. The second part is the **specific epithet**. Together, the genus name and the specific epithet designate one species. Thus, the dog rose now has one official name, *Rosa canina*, that is recognized worldwide.

Genus and species names are always italicized. For example, *Panthera* is a genus of big cats. Lions belong to the species *Panthera leo*. Tigers belong to a different species in the same genus (*Panthera tigris*), and so do leopards (*P. pardus*). Note how the genus name may be abbreviated after it has been spelled out.

A Rose by Any Other Name . . .

The individuals of a species share a unique set of inherited characteristics, or **traits**. For example, giraffes normally have very long necks, brown spots on white coats, and so on. These are morphological traits (*morpho–* means form). Individuals of a species also share biochemical traits (they make and use the same molecules) and behavioral traits (they respond the same way to certain stimuli, as when hungry giraffes feed on tree leaves).

We can rank species into ever more inclusive categories based on traits. Each rank, or **taxon** (plural, taxa), is a group of organisms that share a unique set of traits. Each category above species—genus, family, order, class, phylum (plural, phyla), kingdom, and

	wild carrot	marijuana	apple	prickly rose	dog rose
domain	Eukarya	Eukarya	Eukarya	Eukarya	Eukarya
kingdom	Plantae	Plantae	Plantae	Plantae	Plantae
phylum	Magnoliophyta	Magnoliophyta	Magnoliophyta	Magnoliophyta	Magnoliophyta
class	Magnoliopsida	Magnoliopsida	Magnoliopsida	Magnoliopsida	Magnoliopsida
order	Apiales	Rosales	Rosales	Rosales	Rosales
family	Apiaceae	Cannabaceae	Rosaceae	Rosaceae	Rosaceae
genus	*Daucus*	*Cannabis*	*Malus*	*Rosa*	*Rosa*
species	*carota*	*sativa*	*domestica*	*acicularis*	*canina*

FIGURE 1.6 Taxonomic classification of five species that are related at different levels. Each species has been assigned to ever more inclusive groups, or taxa: in this case, from genus to domain.

Which of the plants shown here are in the same order?

Answer: Marijuana, apple, prickly rose, and dog rose

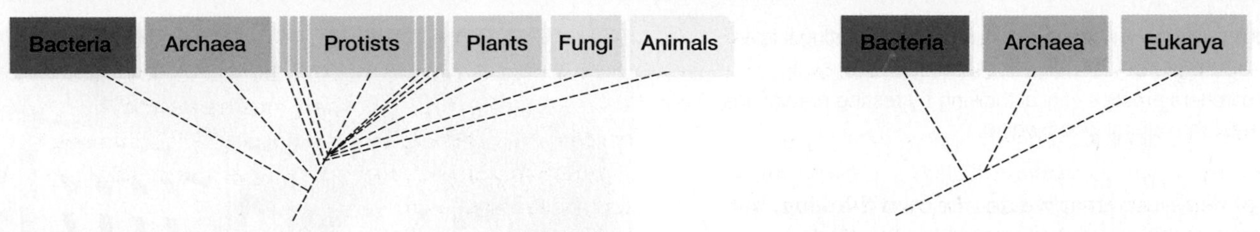

A Six-kingdom classification system. The protist kingdom includes the most ancient multicelled and all single-celled eukaryotes.

B Three-domain classification system. The Eukarya domain includes protists, plants, fungi, and animals.

FIGURE 1.7 ▶Animated Two ways to see the big picture of life. Lines in such diagrams indicate evolutionary connections. Compare **FIGURE 1.6**.

domain—consists of a group of the next lower taxon (**FIGURE 1.6**). Using this system, all life can be sorted into categories (**FIGURE 1.7** and **TABLE 1.1**).

It is easy to tell that orangutans and caterpillars are different species because they appear very different. Distinguishing between species that are more closely related may be much more challenging (**FIGURE 1.8**). In addition, traits shared by members of a species often vary a bit among individuals, as eye color does among people. How do we decide whether similar-looking organisms belong to the same species? The short answer to that question is that we rely on whatever information we have. Early naturalists studied anatomy and distribution—essentially the only methods available at the time—so species were named and classified according to what they looked like and where they lived. Today's biologists are able to compare traits that the early naturalists did not even know about, including biochemical ones.

The discovery of new information sometimes changes the way we distinguish a particular species or how we group it with others. For example, Linnaeus grouped plants by the number and arrangement of reproductive parts, a scheme that resulted in odd pairings such as castor-oil plants with pine trees. Having more information today, we place these plants in separate phyla.

Evolutionary biologist Ernst Mayr defined a species as one or more groups of individuals that potentially can interbreed, produce fertile offspring, and do not interbreed with other groups. This "biological species concept" is useful in many cases, but it is not universally applicable. For example, we may never know whether two widely separated populations could interbreed if they got together. As another example, populations often continue to interbreed even as they diverge, so the exact moment at which two populations become two species is often impossible to pinpoint. We return to speciation and how it occurs in Chapter 17, but for now it is important to remember that a "species" is a convenient but artificial construct of the human mind.

Table 1.1 All of Life in Three Domains

Bacteria	Single cells, no nucleus. Most ancient lineage.
Archaea	Single cells, no nucleus. Evolutionarily closer to eukaryotes than bacteria.
Eukarya	Eukaryotic cells (with a nucleus). Single-celled and multicelled species of protists, plants, fungi, and animals.

FIGURE 1.8 Four butterflies, two species: Which are which?

The top row shows two forms of the butterfly species *Heliconius melpomene*; the bottom row, two forms of *H. erato*.

H. melpomene and *H. erato* never cross-breed. Their alternate but similar patterns of coloration evolved as a shared warning signal to predatory birds that these butterflies taste terrible.

genus A group of species that share a unique set of traits; also the first part of a species name.
species Unique type of organism.
specific epithet Second part of a species name.
taxon A rank of organisms that share a unique set of traits.
taxonomy The science of naming and classifying species.
trait An inherited characteristic of an organism or species.

TAKE-HOME MESSAGE 1.5

How do we keep track of known species?

✔ Each type of organism, or species, is given a unique, two-part scientific name.

✔ Classification systems group species on the basis of shared, inherited traits.

1.6 The Science of Nature

✔ Judging the quality of information before accepting it is an active process called critical thinking.

✔ Researchers practice critical thinking by testing predictions about how the natural world works.

Most of us assume that we do our own thinking, but do we, really? You might be surprised to find out how often we let others think for us. Consider how a school's job (which is to impart as much information to students as quickly as possible) meshes perfectly with a student's job (which is to acquire as much knowledge as quickly as possible). In this rapid-fire exchange of information, it can be very easy to forget about the quality of what is being exchanged. Anytime you accept information without questioning it, you let someone else think for you.

Thinking About Thinking

Critical thinking is the deliberate process of judging the quality of information before accepting it. "Critical" comes from the Greek *kriticos* (discerning judgment). When you use critical thinking, you move beyond the content of new information to consider supporting evidence, bias, and alternative interpretations. How does the busy student manage this? Critical thinking does not necessarily require extra time, just a bit of extra awareness. There are many ways to do it. For example, you might ask yourself some of the following questions while you are learning something new:

What message am I being asked to accept?
Is the message based on facts or opinion?
Is there a different way to interpret the facts?
What biases might the presenter have?
How do my own biases affect what I'm learning?

Such questions are a way of being conscious about learning. They can help you decide whether to allow new information to guide your beliefs and actions.

The Scientific Method

Critical thinking is a big part of **science**, the systematic study of the observable world and how it works (**FIGURE 1.9**). A scientific line of inquiry usually begins with curiosity about something observable, such as, say, a decrease in the number of birds in a particular area. Typically, a scientist will read about what others have discovered before making a **hypothesis**, a testable explanation for a natural phenomenon. An example of a hypothesis would be, "The number of birds is decreasing because the number of cats is increasing." Making a hypothesis this way is an example of **inductive reasoning**, which means arriving at a conclusion based on one's observations. Inductive reasoning is the way we come up with new ideas about groups of objects or events.

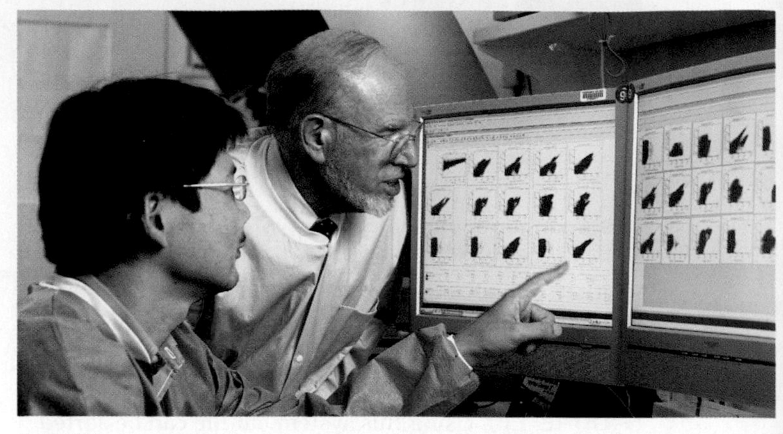

A Devising a vaccine for cancer.

FIGURE 1.9 Examples of research in the field of biology.

A **prediction**, or statement of some condition that should exist if the hypothesis is correct, comes next. Making predictions is called the if–then process, in which the "if" part is the hypothesis, and the "then" part is the prediction: *If* the number of birds is decreasing because the number of cats is increasing, *then* reducing the number of cats should stop the decline. Using a hypothesis to make a prediction is a form of **deductive reasoning**, the logical process of using a general premise to draw a conclusion about a specific case.

control group Group of individuals identical to an experimental group except for the independent variable under investigation.
critical thinking Evaluating information before accepting it.
data Experimental results.
deductive reasoning Using a general idea to make a conclusion about a specific case.
dependent variable In an experiment, a variable that is presumably affected by an independent variable being tested.
experiment A test designed to support or falsify a prediction.
experimental group In an experiment, a group of individuals who have a certain characteristic or receive a certain treatment.
hypothesis Testable explanation of a natural phenomenon.
independent variable Variable that is controlled by an experimenter in order to explore its relationship to a dependent variable.
inductive reasoning Drawing a conclusion based on observation.
model Analogous system used for testing hypotheses.
prediction Statement, based on a hypothesis, about a condition that should exist if the hypothesis is correct.
science Systematic study of the observable world.
scientific method Making, testing, and evaluating hypotheses.
variable In an experiment, a characteristic or event that differs among individuals or over time.

CREDIT: (9A) National Cancer Institute.

B Improving efficiency of biofuel production from agricultural waste.

C Studying the ecological benefits of weedy buffer zones on farms.

D Discovering medically active natural products made by marine animals.

Next, a researcher will test the prediction. Tests may be performed on a **model**, or analogous system, if working with an object or event directly is not possible. For example, animal diseases are often used as models of similar human diseases. Careful observations are one way to test predictions that flow from a hypothesis. So are **experiments**: tests designed to support or falsify a prediction. A typical experiment explores a cause-and-effect relationship using **variables,** which are characteristics or events that can differ among individuals or over time. An **independent variable** is defined or controlled by the person doing the experiment. A **dependent variable** is an observed result that is supposed to be influenced by the independent variable. For example, in an investigation of our hypothetical bird–cat relationship, an independent variable may be the presence or absence of cats. The dependent variable in this test would be the number of birds.

Biological systems are typically complex, with many interdependent variables. It can be difficult to study one variable separately from the rest. Thus, biology researchers often test two groups of individuals simultaneously. An **experimental group** is a set of individuals that have a certain characteristic or receive a certain treatment. This group is tested side by side with a **control group**, which is identical to the experimental group except for one independent variable: the characteristic or the treatment being tested. Any differences in experimental results between the two groups is likely to be an effect of changing the variable.

Test results—**data**—that are consistent with the prediction are evidence in support of the hypothesis. Data inconsistent with the prediction are evidence that the hypothesis is flawed and should be revised.

Table 1.2 The Scientific Method

1. Observe some aspect of nature.
2. Think of an explanation for your observation (in other words, form a hypothesis).
3. Test the hypothesis.
    a. Make a prediction based on the hypothesis.
    b. Test the prediction using experiments or surveys.
    c. Analyze the results of the tests (data).
4. Decide whether the results of the tests support your hypothesis or not (form a conclusion).
5. Report your results to the scientific community.

A necessary part of science is reporting one's results and conclusions in a standard way, such as in a peer-reviewed journal article. The communication gives other scientists an opportunity to evaluate the information for themselves, both by checking the conclusions drawn and by repeating the experiments. Forming a hypothesis based on observation, and then systematically testing and evaluating the hypothesis, are collectively called the **scientific method** (TABLE 1.2).

TAKE-HOME MESSAGE 1.6

How does science work?

✔ The scientific method consists of making, testing, and evaluating hypotheses. It is a way of critical thinking—systematically judging the quality of information before allowing it to guide one's beliefs and actions.

✔ Experiments measure how changing an independent variable affects a dependent variable.

1.7 Examples of Experiments in Biology

✔ Researchers unravel cause-and-effect relationships in complex natural processes by changing one variable at a time.

There are many different ways to do research, particularly in biology. Some biologists survey, which means they observe without making hypotheses. Some make hypotheses and leave the experimentation to others. When it comes to scientific experimentation, however, consistency is the rule. Researchers try to change one independent variable at a time, and see what happens to a dependent variable. To give you a sense of how biology experiments work, we summarize two published studies here.

Potato Chips and Stomachaches

In 1996 the U.S. Food and Drug Administration (FDA) approved Olestra®, a fat replacement manufactured from sugar and vegetable oil, as a food additive. Potato chips were the first Olestra-containing food product to be sold in the United States. Controversy about the chip additive soon raged. Many people complained of intestinal problems after eating the chips, and thought that the Olestra was at fault. Two years later, researchers at the Johns Hopkins University School of Medicine designed an experiment to test whether Olestra causes cramps.

The researchers predicted *if* Olestra causes cramps, *then* people who eat Olestra should be more likely to get cramps than people who do not eat it. To test the prediction, they used a Chicago theater as a "laboratory." They asked 1,100 people between the ages of thirteen and thirty-eight to watch a movie and eat their fill of potato chips. Each person received an unmarked bag containing 13 ounces of chips.

In this experiment, the individuals who received Olestra-laden potato chips constituted the experimental group, and individuals who received regular chips were the control group. The independent variable was the presence or absence of Olestra in the chips.

A few days after the movie, the researchers contacted all of the people and collected any reports of post-movie gastrointestinal problems. Of 563 people making up the experimental group, 89 (15.8 percent) complained about cramps. However, so did 93 of the 529 people (17.6 percent) making up the control group—who had eaten the regular chips.

People were about as likely to get cramps whether or not they ate chips made with Olestra. These results did not support the prediction, so the researchers concluded that eating Olestra does not cause cramps (FIGURE 1.10).

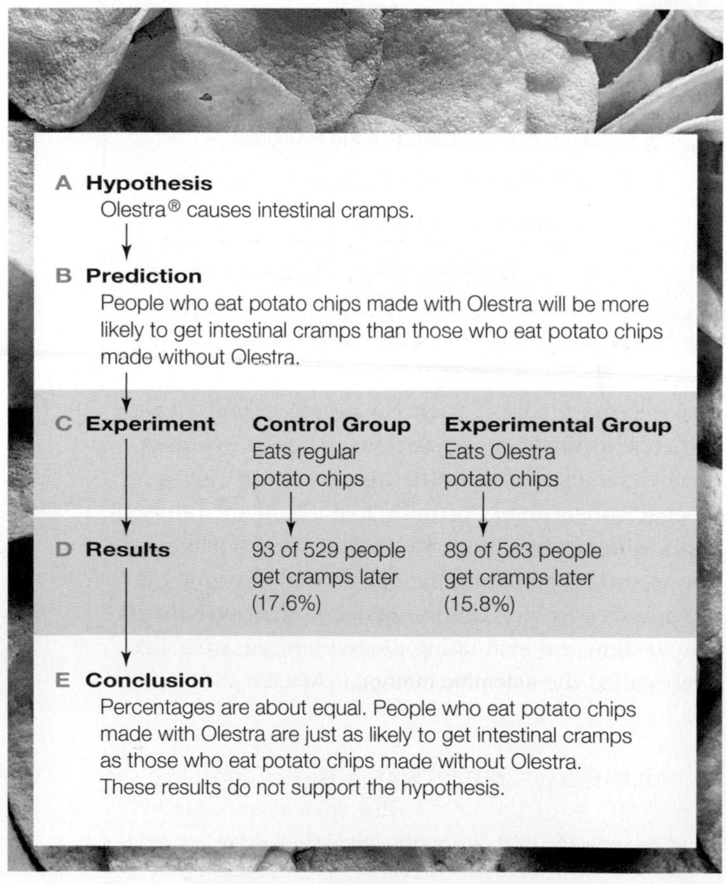

FIGURE 1.10 The steps in a scientific experiment to determine whether Olestra causes cramps. A report of this study was published in the *Journal of the American Medical Association* in January 1998.

What was the dependent variable in this experiment?

Answer: Whether or not a person got cramps

Butterflies and Birds

The peacock butterfly is a winged insect named for the large, colorful spots on its wings. In 2005, researchers reported the results of experiments investigating whether certain behaviors help peacock butterflies defend themselves against insect-eating birds. The study began with the observation that a resting peacock butterfly sits motionless, wings folded (FIGURE 1.11A). The dark underside of the wings provide appropriate camouflage. However, when the butterfly sees a predator approaching, it repeatedly flicks its wings open, exposing brilliant spots (FIGURE 1.11B). At the same time, it moves the hindwings in a way that produces a hissing sound and a series of clicks. Typically, a colorful, moving, noisy insect is very attractive to predatory birds, so the researchers were curious about why the peacock butterfly moves and makes noises only in the *presence* of a predator. After review-

A With wings folded, a resting peacock butterfly resembles a dead leaf.

B When a bird approaches, a butterfly repeatedly flicks its wings open. This behavior exposes brilliant spots and also produces hissing and clicking sounds.

C Researchers tested whether peacock butterfly wing flicking and hissing reduce predation by blue tits.

FIGURE 1.11 Testing the defensive value of peacock butterfly behaviors.

Researchers painted out the spots of some butterflies, cut the sound-making part of the wings on others, and did both to a third group; then exposed each butterfly to a hungry blue tit.

Results, listed below in **TABLE 1.3**, support the hypotheses that peacock butterfly spots and sounds can deter predatory birds.

What was the dependent variable in this series of experiments?

Answer: Getting eaten

Table 1.3 Results of Peacock Butterfly Experiment*

Wing Spots	Wing Sound	Total Number of Butterflies	Number Eaten	Number Survived
Spots	Sound	9	0	9 (100%)
No spots	Sound	10	5	5 (50%)
Spots	No sound	8	0	8 (100%)
No spots	No sound	10	8	2 (20%)

* *Proceedings of the Royal Society of London, Series B (2005) 272: 1203–1207.*

ing earlier studies, the scientists made two hypotheses to explain these behaviors:

1. The wing flicking may startle predatory birds because the peacock butterfly's wing spots resemble owl eyes, and anything that looks like owl eyes is known to startle birds.
2. The hissing and clicking sounds produced when the peacock butterfly moves its hindwings may be an additional defense that startles predatory birds.

The researchers used these hypotheses to make the following predictions:

1. If peacock butterflies startle predatory birds by exposing their brilliant wing spots, then individuals having wing spots will be less likely to get eaten by predatory birds than those lacking wing spots.
2. If peacock butterfly hisses and clicks deter predatory birds, then sound-producing individuals will be less likely to get eaten by predatory birds than silent individuals.

The next step was the experiment. The researchers used a marker to paint the wing spots of some butterflies black, and scissors to cut off the sound-making part of the hindwings of others. A third group had their wing spots painted and their hindwings cut. The researchers then put each butterfly into a large cage with a hungry blue tit (**FIGURE 1.11C**) and watched the pair for thirty minutes.

TABLE 1.3 lists the results of the experiment. All of the butterflies with unmodified wing spots survived, regardless of whether they made sounds. By contrast, only half of the butterflies that had spots painted out but could make sounds survived. Most of the silenced butterflies with painted-out spots were eaten quickly. The test results confirmed both predictions, so they support both hypotheses. Predatory birds are indeed deterred by peacock butterfly sounds, and even more so by wing spots.

TAKE-HOME MESSAGE 1.7

Why do biologists perform experiments?

✔ Natural processes are often very complex and influenced by many interacting variables.

✔ Experiments help researchers unravel causes of complex natural processes by focusing on the effects of changing a single variable.

1.8 Analyzing Experimental Results

✔ Checks and balances inherent in the scientific process help researchers to be objective about their observations.

✔ Science is, ideally, a self-correcting process because scientists present their work in a way that allows others to check it.

Sampling Error

Researchers can rarely observe all individuals of a group. For example, the explorers you read about in Section 1.1 did not—and could not—survey every uninhabited part of the Foja Mountains. The cloud forest itself cloaks more than 2 million acres, so surveying all of it would take unrealistic amounts of time and effort.

When researchers cannot directly observe all individuals of a population, all instances of an event, or some other aspect of nature, they may test or survey a subset. Results from the subset are then used to make generalizations about the whole. However, generalizing from a subset is risky because subsets are not necessarily representative of the whole. Consider the golden-mantled tree kangaroo, which was first discovered in 1993 on a single forested mountaintop in New Guinea. For more than a decade, the species was never seen outside of that habitat, which is getting smaller every year because of human activities. Thus, the golden-mantled tree kangaroo was considered to be one of the most endangered animals on the planet. Then, in 2005, the New Guinea explorers discovered that this kangaroo species is fairly common in the Foja Mountain cloud forest (**FIGURE 1.12**). As a result, biologists now believe its future is secure, at least for the moment.

FIGURE 1.12 Kris Helgen holds a golden-mantled tree kangaroo he found during the 2005 Foja Mountains survey. This kangaroo species is extremely rare in other areas, so it was thought to be critically endangered prior to the expedition.

Sampling error is a difference between results obtained from a subset, and results from the whole (**FIGURE 1.13A**). Sampling error may be unavoidable, but knowing how it can occur helps researchers design their experiments to minimize it. For example, sampling error can be a substantial problem with a

A Natalie chooses a random jelly bean from a jar. She is blindfolded, so she does not know that the jar contains 120 green and 280 black jelly beans.

The jar is hidden from Natalie's view before she removes her blindfold. She sees one green jelly bean in her hand and assumes that the jar must hold only green jelly beans. This assumption is incorrect: 30 percent of the jelly beans in the jar are green, and 70 percent are black. The small sample size has resulted in sampling error.

B Still blindfolded, Natalie randomly chooses 50 jelly beans from the jar. She ends up choosing 10 green and 40 black ones.

The larger sample leads Natalie to assume that one-fifth of the jar's jelly beans are green (20 percent) and four-fifths are black (80 percent). The larger sample more closely approximates the jar's actual green-to-black ratio of 30 percent to 70 percent.

The more times Natalie repeats the sampling, the greater the chance she has of guessing the actual ratio.

FIGURE 1.13 ▸Animated How sample size affects sampling error.

small subset, so experimenters try to start with a relatively large sample, and they repeat their experiments (FIGURE 1.13B). To understand why these practices reduce the risk of sampling error, think about flipping a coin. There are two possible outcomes of each flip: The coin lands heads up, or it lands tails up. Thus, the chance that the coin will land heads up is one in two (1/2), or 50 percent. However, when you flip a coin repeatedly, it often lands heads up, or tails up, several times in a row. With just 3 flips, the proportion of times that heads actually land up may not even be close to 50 percent. With 1,000 flips, however, the overall proportion of times the coin lands heads up is much more likely to approach 50 percent.

In cases such as flipping a coin, it is possible to calculate **probability**, which is the measure, expressed as a percentage, of the chance that a particular outcome will occur. That chance depends on the total number of possible outcomes. For instance, if 10 million people enter a drawing, each has the same probability of winning: 1 in 10 million, or (an extremely improbable) 0.00001 percent.

Analysis of experimental data often includes probability calculations. If a result is very unlikely to have occurred by chance alone, it is said to be **statistically significant**. In this context, the word "significant" does not refer to the result's importance. Rather, it means that a rigorous statistical analysis has shown a very low probability (usually 5 percent or less) of the result being skewed by sampling error.

Variation in data is often shown as error bars on a graph (FIGURE 1.14). Depending on the graph, error bars may indicate variation around an average for one sample set, or the difference between two sample sets.

Bias in Interpreting Results

Particularly when studying humans, changing a single variable apart from all others is not often possible. For example, remember that the people who participated in the Olestra experiment were chosen randomly. Thus, the study was not controlled for gender, age, weight, medications taken, and so on. Such variables may have influenced the results.

probability The chance that a particular outcome of an event will occur; depends on the total number of outcomes possible.
sampling error Difference between results derived from testing an entire group of events or individuals, and results derived from testing a subset of the group.
statistically significant Refers to a result that is statistically unlikely to have occurred by chance alone.

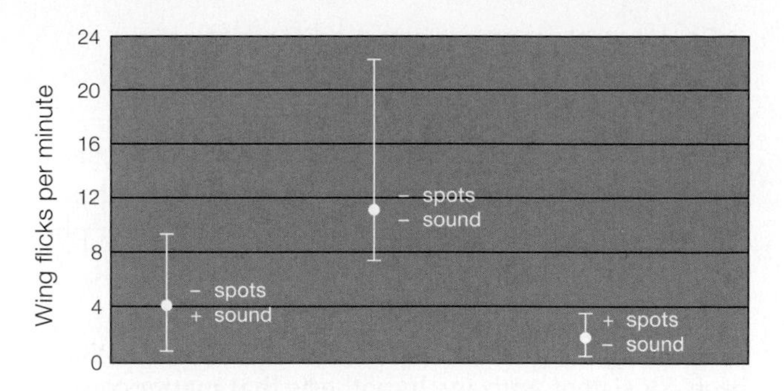

FIGURE 1.14 Example of error bars in a graph. This graph was adapted from the peacock butterfly research described in Section 1.7. The researchers recorded the number of times each butterfly flicked its wings in response to an attack by a bird.

The squares represent average frequency of wing flicking for each sample set of butterflies. The error bars that extend above and below the dots indicate the range of values—the sampling error.

What was the fastest rate at which a butterfly with no spots or sound flicked its wings?

Answer: 22 times per minute

Human beings are by nature subjective, and scientists are no exception. Researchers risk interpreting their results in terms of what they want to find out. That is why they often design experiments to yield quantitative results, which are counts or some other data that can be measured or gathered objectively. Such results minimize the potential for bias, and also give other scientists an opportunity to repeat the experiments and check the conclusions drawn from them.

This last point gets us back to the role of critical thinking in science. Scientists expect one another to recognize and put aside bias in order to test their hypotheses in ways that may prove them wrong. If one scientist does not, then others will, because exposing errors is just as useful as applauding insights. The scientific community consists of critically thinking people trying to poke holes in one another's ideas. Their collective efforts make science a self-correcting endeavor.

TAKE-HOME MESSAGE 1.8

How do scientists avoid potential pitfalls of sampling error and bias when doing research?

✔ Researchers minimize sampling error by using large sample sizes and by repeating their experiments.

✔ Probability calculations can show whether a result is likely to have occurred by chance alone.

✔ Science is a self-correcting process because it is carried out by a community of people systematically checking one another's work and conclusions.

1.9 The Nature of Science

✔ Scientific theories are our best descriptions of reality.
✔ Science is limited to the observable.

Suppose a hypothesis stands even after years of tests. It is consistent with all data ever gathered, and it has helped us make successful predictions about other phenomena. When a hypothesis meets these criteria, it is considered to be a **scientific theory** (TABLE 1.4).

To give an example, all observations to date have been consistent with the hypothesis that matter consists of atoms. Scientists no longer spend time testing this hypothesis for the compelling reason that, since we started looking 200 years ago, no one has discovered matter that doesn't consist of atoms. Thus, researchers use the hypothesis, now called atomic theory, to make other hypotheses about matter and the way it behaves.

Scientific theories are our best descriptions of reality. However, they can never be proven absolutely, because to do so would necessitate testing under every possible circumstance. For example, in order to prove atomic theory, the atomic composition of all matter in the universe would have to be checked—an impossible task even if someone wanted to try.

Like all hypotheses, a scientific theory can be disproven by a single observation or result that is inconsistent with it. For example, if someone discovers a form of matter that does not consist of atoms, atomic theory would have to be revised. The potentially falsifiable nature of scientific theories means that science has a built-in system of checks and balances. A theory is revised until no one can prove it to be incorrect. For example, the theory of evolution, which states that change occurs in a line of descent over time, still holds after a century of observations and testing. As with all other scientific theories, no one can be absolutely sure that it will hold under all possible conditions, but it has a very high probability of not being wrong. Few other theories have withstood as much scrutiny.

You may hear people apply the word "theory" to a speculative idea, as in the phrase "It's just a theory." This everyday usage of the word differs from the way it is used in science. Speculation is an opinion, belief, or personal conviction that is not necessarily supported by evidence. A scientific theory is different. By definition, it is supported by a large body of evidence, and it is consistent with all known data.

A scientific theory also differs from a **law of nature**, which describes a phenomenon that has been observed to occur in every circumstance without fail, but the scientific explanation for it is incomplete. The laws of thermodynamics, which describe energy, are examples. We know how energy behaves, but not exactly why it behaves the way it does.

The Limits of Science

Science helps us be objective about our observations in part because of its limitations. For example, science does not address many questions, such as "Why do I exist?" Answers to questions like this can only come from within, as an integration of the personal experiences and mental connections that shape our consciousness. This is not to say subjective answers have

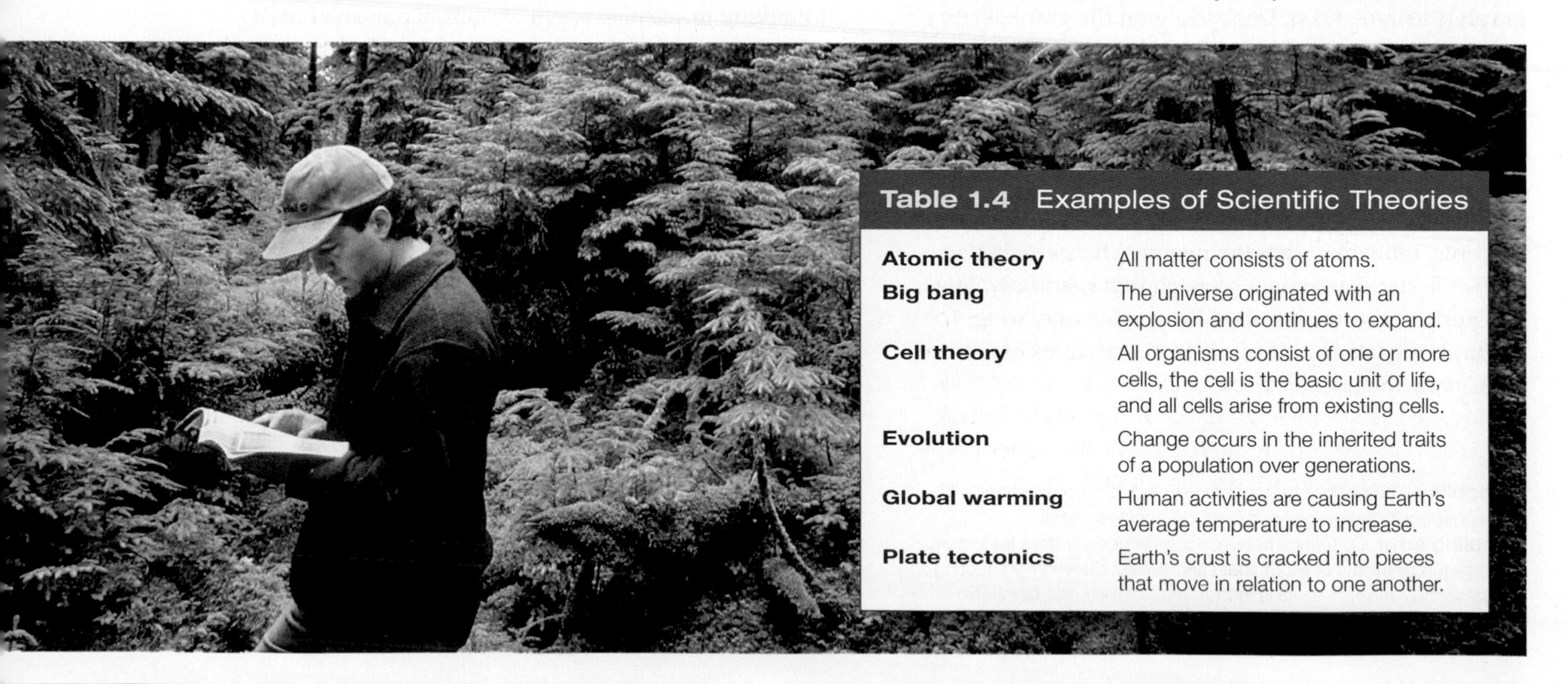

Table 1.4 Examples of Scientific Theories

Atomic theory	All matter consists of atoms.
Big bang	The universe originated with an explosion and continues to expand.
Cell theory	All organisms consist of one or more cells, the cell is the basic unit of life, and all cells arise from existing cells.
Evolution	Change occurs in the inherited traits of a population over generations.
Global warming	Human activities are causing Earth's average temperature to increase.
Plate tectonics	Earth's crust is cracked into pieces that move in relation to one another.

The Secret Life of Earth (revisited)

Of an estimated 100 billion species that have ever lived, at least 100 million are still with us. That number is only an estimate because we are still discovering them. For example, a mouse-sized opossum and a cat-sized rat turned up on a return trip to the Foja Mountains. Other recent surveys have revealed new species of dolphin, gecko, skink, and frog in Australia; a giant fish in Brazil; legless lizards in Southern California; a giant crayfish in Tennessee; a rat-eating plant in the Philippines; a sausage-sized millipede in Tanzania; and scores of plants and single-celled organisms. Most were discovered by biologists simply trying to find out what lives where.

Biologists discover thousands of new species per year (**FIGURE 1.15**). Each is a reminder that we do not yet know all of the organisms living on our own planet. We don't even know how many to look for. The vast information about the 1.8 million species we do know about changes so quickly that collating it has been impossible—until recently. A website titled the Encyclopedia of Life is intended to be an online reference source and database of species information maintained by collaborative effort. See its progress at www.eol.org.

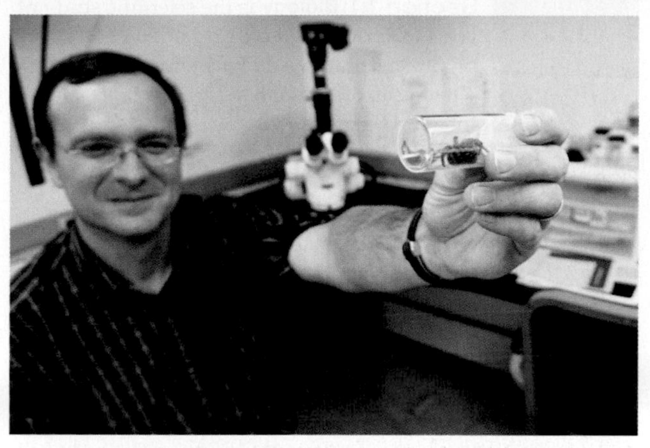

FIGURE 1.15 The discoverer of a new species typically has the honor of naming it. Dr. Jason Bond holds a new species of spider he discovered in California in 2008. Bond named the spider *Aptostichus stephencolberti*, after TV personality Stephen Colbert.

---

no value, because no human society can function for long unless its individuals share standards for making judgments, even if they are subjective. Moral, aesthetic, and philosophical standards vary from one society to the next, but all help people decide what is important and good. All give meaning to our lives.

Neither does science address the supernatural, which is anything "beyond nature." Science neither assumes nor denies the existence of supernatural phenomena, but controversy may arise when researchers discover a natural explanation for something that was thought to have none. Such controversy often occurs when a society's moral standards are interwoven with its understanding of nature. Consider Nicolaus Copernicus, who concluded in 1540 that Earth orbits the sun. Today that idea is generally accepted, but during Copernicus's time the prevailing belief system had Earth as the immovable center of the universe. In 1610, astronomer Galileo Galilei published evidence for the Copernican model of the solar system, an act that resulted in his imprisonment. He was publicly forced to recant his work, spent the rest of his life under house arrest, and was never allowed to publish again.

As Galileo's story illustrates, exploring a traditional view of the natural world from a scientific perspective is often misinterpreted as a violation of morality. As a group, scientists are no less moral than anyone else. However, they follow a particular set of rules that do not necessarily apply to others: Their work concerns only the natural world, and their ideas must be testable in ways others can repeat.

Science helps us communicate our experiences without bias. As such, it may be as close as we can get to a universal language. We are fairly sure, for example, that the law of gravity applies everywhere in the universe. Intelligent beings on a distant planet would likely understand the concept of gravity. We might well use gravity or another scientific concept to communicate with them, or anyone, anywhere. The point of science, however, is not to communicate with aliens. It is to find common ground here on Earth.

TAKE-HOME MESSAGE 1.9

Why does science work?

✔ Science is concerned only with testable ideas about observable aspects of the natural world.

✔ Because a scientific theory is thoroughly tested and revised until no one can prove it wrong, it is our best way of describing reality.

law of nature Generalization that describes a consistent natural phenomenon for which there is incomplete scientific explanation.
scientific theory Hypothesis that has not been disproven after many years of rigorous testing.

CREDITS: (in text) Tim Laman/National Geographic Stock; (15) Courtesy East Carolina University.

summary

Section 1.1 **Biology** is the scientific study of life. We know about only a fraction of the organisms that live on Earth, in part because we have explored only a fraction of its inhabited regions.

Section 1.2 Biologists think about life at different levels of organization, with **emergent properties** appearing at successive levels. All matter consists of **atoms**, which combine as **molecules**. **Organisms** are individuals that consist of one or more **cells**, the organizational level at which life emerges. Cells of larger multicelled organisms are organized as **tissues**, then **organs**, and **organ systems**. A **population** is a group of interbreeding individuals of a species in a given area; a **community** is all populations of all species in a given area. An **ecosystem** is a community interacting with its environment. The **biosphere** includes all regions of Earth that hold life.

Section 1.3 Life has underlying unity in that all living things have similar characteristics: (1) All organisms require energy and **nutrients** to sustain themselves. **Producers** harvest energy from the environment to make their own food by processes such as **photosynthesis**; **consumers** ingest other organisms, their wastes, or remains. (2) Organisms keep the conditions in their internal environment within ranges that their cells tolerate—a process called **homeostasis**. (3) **DNA** contains information that guides an organism's **growth**, **development**, and **reproduction**. The passage of DNA from parents to offspring is called **inheritance**.

Section 1.4 The many types of organisms that currently exist on Earth differ greatly in the details of their body form and function. **Biodiversity** is the sum of differences among living things. **Bacteria** and **archaea** are both **prokaryotes**, single-celled organisms whose DNA is not contained within a nucleus. The DNA of single-celled or multicelled **eukaryotes** (**protists**, **plants**, **fungi**, and **animals**) is contained within a nucleus.

Section 1.5 Each **species** is given a two-part name. The first part is the **genus** name. When combined with the **specific epithet**, it designates the particular species. With **taxonomy**, species are ranked into ever more inclusive **taxa** (genus, family, order, class, phylum, kingdom, domain) on the basis of shared inherited **traits**.

Section 1.6 **Critical thinking**, the self-directed act of judging the quality of information as one learns, is an important part of **science**. Generally, a researcher observes something in nature, uses **inductive reasoning** to form a **hypothesis** (testable explanation) for it, then uses **deductive reasoning** to make a testable **prediction** about what might occur if the hypothesis is correct. **Experiments** with **variables** may be performed on an **experimental group** as compared with a **control group**, and sometimes on **models**. A researcher typically changes an **independent variable**, then observes the effects of the change on a **dependent variable**. Conclusions are drawn from the resulting **data**. The **scientific method** consists of making, testing, and evaluating hypotheses, and sharing results with the scientific community.

Section 1.7 Biological systems are typically influenced by many interacting variables. Research approaches differ, but experiments are designed in a consistent way, in order to study a single cause-and-effect relationship in a complex natural system.

Section 1.8 Small sample size increases the potential for **sampling error** in experimental results. In such cases, a subset may be tested that is not representative of the whole. Researchers design experiments carefully to minimize sampling error and bias, and they use **probability** rules to check the **statistical significance** of their results. Science is ideally a self-correcting process because scientists check and test one another's ideas.

Section 1.9 Science helps us be objective about our observations because it is only concerned with testable ideas about observable aspects of nature. Opinion and belief have value in human culture, but they are not addressed by science. A **scientific theory** is a long-standing hypothesis that is useful for making predictions about other phenomena. It is our best way of describing reality. A **law of nature** is a phenomenon that occurs without fail, but has an incomplete scientific explanation.

self-quiz

Answers in Appendix VII

1. ______ are fundamental building blocks of all matter.
   a. Atoms
   b. Molecules
   c. Cells
   d. Organisms

2. The smallest unit of life is the ______ .
   a. atom
   b. molecule
   c. cell
   d. organism

3. Organisms require ______ and ______ to maintain themselves, grow, and reproduce.
   a. DNA; energy
   b. food; sunlight
   c. nutrients; energy
   d. DNA; cells

4. By sensing and responding to change, organisms keep conditions in the internal environment within ranges that cells can tolerate. This process is called ______ .

5. DNA ______ .
   a. guides form and function
   b. is the basis of traits
   c. is transmitted from parents to offspring
   d. all of the above

data analysis activities

Peacock Butterfly Predator Defenses The photographs below represent the experimental and control groups used in the peacock butterfly experiment discussed in Section 1.7. See if you can identify the experimental groups, and match them up with the relevant control group(s). *Hint:* Identify which variable is being tested in each group (each variable has a control).

A Wing spots painted out

B Wing spots visible; wings silenced

C Wing spots painted out; wings silenced

D Wings painted but spots visible

E Wings cut but not silenced

F Wings painted, spots visible; wings cut, not silenced

6. A process by which an organism produces offspring is called ______ .
   a. reproduction
   b. development
   c. homeostasis
   d. inheritance

7. ______ is the transmission of DNA to offspring.
   a. Reproduction
   b. Development
   c. Homeostasis
   d. Inheritance

8. A butterfly is a(n) ______ (choose all that apply).
   a. organism
   b. domain
   c. species
   d. eukaryote
   e. consumer
   f. producer
   g. prokaryote
   h. trait

9. ______ move around for at least part of their life.

10. A bacterium is ______ (choose all that apply).
    a. an organism
    b. single-celled
    c. an animal
    d. a eukaryote

11. Bacteria, Archaea, and Eukarya are three ______.

12. A control group is ______ .
    a. a set of individuals that have a certain characteristic or receive a certain treatment
    b. the standard against which an experimental group is compared
    c. the experiment that gives conclusive results

13. Fifteen randomly selected students are found to be taller than 6 feet. The researchers concluded that the average height of a student is greater than 6 feet. This is an example of ______ .
    a. experimental error
    b. sampling error
    c. a subjective opinion
    d. experimental bias

14. Science only addresses that which is ______ .
    a. alive
    b. observable
    c. variable
    d. indisputable

15. Match the terms with the most suitable description.

___ life	a. if–then statement
___ probability	b. unique type of organism
___ species	c. emerges with cells
___ hypothesis	d. testable explanation
___ prediction	e. measure of chance
___ producer	f. makes its own food

CENGAGE brain To access course materials, please visit www.cengagebrain.com.

critical thinking

1. A person is declared dead upon the irreversible ceasing of spontaneous body functions: brain activity, blood circulation, and respiration. Only about 1% of a body's cells have to die in order for all of these things to happen. How can a person be dead when 99% of his or her cells are alive?

2. Explain the difference between a one-celled organism and a single cell of a multicelled organism.

3. Why would you think twice about ordering from a restaurant menu that lists the specific epithet but not the genus name of its offerings? *Hint:* Look up *Homarus americanus, Ursus americanus, Bufo americanus, Lepus americanus, Necator americanus, Lysichiton americanus, Leucoagaricus americanus*, and *Nicrophorus americanus*.

4. Once there was a highly intelligent turkey that had nothing to do but reflect on the world's regularities. Morning always started out with the sky turning light, followed by the master's footsteps, which were always followed by the appearance of food. Other things varied, but food always followed footsteps. The sequence of events was so predictable that it eventually became the basis of the turkey's theory about the goodness of the world. One morning, after more than 100 confirmations of this theory, the turkey listened for the master's footsteps, heard them, and had its head chopped off. Any scientific theory is modified or discarded upon discovery of contradictory evidence. The absence of absolute certainty has led some people to conclude that "theories are irrelevant because they can change." If that is so, should we stop doing scientific research? Why or why not?

5. In 2005, researcher Woo-suk Hwang reported that he had made immortal stem cells from human patients. His research was hailed as a breakthrough for people affected by degenerative diseases, because stem cells may be used to repair a person's own damaged tissues. Hwang published his results in a peer-reviewed journal. In 2006, the journal retracted his paper after other scientists discovered that Hwang's group had faked their data.

Does the incident show that results of scientific studies cannot be trusted? Or does it confirm the usefulness of a scientific approach, because other scientists discovered and exposed the fraud?

2 Life's Chemical Basis

LEARNING ROADMAP

In this chapter, you will explore the first level of life's organization—atoms—as you encounter an example of how the same building blocks, arranged different ways, form different products (Section 1.2). You will also see one aspect of homeostasis, the process by which organisms keep themselves in a state that favors cell survival (Section 1.3).

ATOMS AND ELEMENTS

Atoms, the building blocks of all matter, differ in their numbers of protons, neutrons, and electrons. Atoms of an element have the same number of protons.

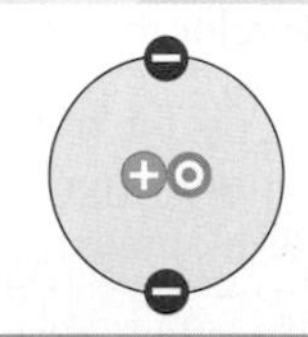

WHY ELECTRONS MATTER

Whether an atom interacts with other atoms depends on the number of electrons it has. An atom with an unequal number of electrons and protons is an ion.

ATOMS BOND

Atoms of many elements interact by acquiring, sharing, and giving up electrons. Interacting atoms may form ionic, covalent, or hydrogen bonds.

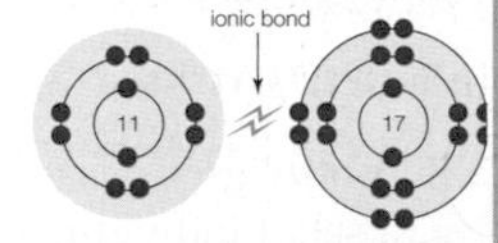

WATER

Hydrogen bonding among individual molecules gives water properties that make life possible: temperature stabilization, cohesion, and the ability to dissolve many other substances.

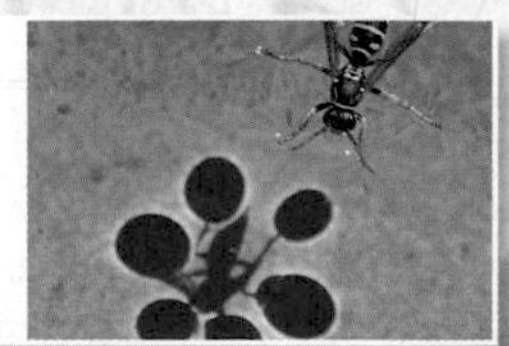

HYDROGEN POWER

Most of the chemistry of life occurs in a narrow range of pH, so most fluids inside organisms are buffered to stay within that range.

Electrons will come up again as you learn how energy drives metabolism, especially in photosynthesis (Chapter 6) and respiration (Chapter 7). Hydrogen bonding is critical for the molecules of life (3.4, 3.6–3.8); the properties of water, for membranes (Section 4.2 and Chapter 5), plant nutrition and transport (28.4), and temperature regulation (40.9). You will also see how radioisotopes are used to date rocks and fossils (16.6), and explore the dangers of free radicals (8.6) and acid rain (48.5).

2.1 Mercury Rising

Actor Jeremy Piven, best known for his Emmy-winning role on the television series *Entourage*, began starring in a Broadway play in 2008. He quit suddenly after two shows, citing medical problems. Piven explained that he was suffering from mercury poisoning caused by eating too much sushi. The play's producers and his co-actors were skeptical, and the playwright ridiculed Piven, saying he was leaving to pursue a career as a thermometer. However, mercury poisoning is no laughing matter.

Mercury is a naturally occurring toxic metal. Most of it is safely locked away in rocky minerals, but volcanic activity and other geologic processes release it into the atmosphere. So do human activities, especially burning coal (**FIGURE 2.1**). Airborne mercury can drift long distances before settling to Earth's surface, where microbes combine it with carbon to form a substance called methylmercury.

Unlike mercury alone, methylmercury easily crosses skin and mucous membranes. In water, it ends up in the tissues of aquatic organisms. All fish and shellfish contain it. Humans contain it too, mainly as a result of eating seafood.

When mercury enters the body, it damages the nervous system, brain, kidneys, and other organs. An average-sized adult who ingests as little as 200 micrograms of methylmercury may experience blurred vision, tremors, itching or burning sensations, and loss of coordination. Exposure to larger amounts can result in thought and memory impairment, coma, and death. Methylmercury in a pregnant woman's blood passes to her unborn child, along with a legacy of permanent developmental problems.

It takes months or even years for mercury to be cleared from the body, so the toxin can build up to high levels if even small amounts are ingested on a regular basis. That is why large predatory fish have a lot of mercury in their tissues. It is also why the U.S. Environmental Protection Agency recommends that adult humans ingest less than 0.1 microgram of mercury per kilogram of body weight per day. For an average-sized person, that limit works out to be about 7 micrograms per day, which is not a big amount if you eat seafood. A typical 6-ounce can of albacore tuna contains about 60 micrograms of mercury, and the occasional can has many times that amount. It does not matter if the fish is canned, grilled, or raw, because methylmercury is unaffected by cooking. Eat a medium-sized tuna steak, and you could be getting more than 700 micrograms of mercury along with it.

With this chapter, we turn to the first of life's levels of organization: atoms. Interactions between atoms make the molecules that sustain life, and also some that destroy it.

FIGURE 2.1 The Navajo Generating Station, a coal-fired power plant in Arizona. Emissions from such plants are by far the largest source of mercury pollution worldwide.

Mercury can accumulate to toxic levels in the tissues of tuna and other large predatory fish that we harvest for food.

CREDITS: (opposite) Francois Gohier/Science Source; (l) left, Ed Darack/Science Faction/SuperStock; top, Ventin/Shutterstock; right, Nanisimova/Shutterstock.

2.2 Start With Atoms

✔ Atomic structure gives rise to chemical properties of atoms.

✔ The number of protons in the atomic nucleus defines the element, and the number of neutrons defines the isotope.

Even though atoms are about 20 million times smaller than a grain of sand, they consist of even smaller subatomic particles. Positively charged **protons** (p^+) and uncharged **neutrons** occur in an atom's core, or **nucleus**. Negatively charged **electrons** (e^-) move around the nucleus (FIGURE 2.2A). **Charge** is an electrical property: Opposite charges attract, and like charges repel.

A typical atom has about the same number of electrons and protons. The negative charge of an electron is the same magnitude as the positive charge of a proton, so the two charges cancel one another. Thus, an atom with exactly the same number of electrons and protons carries no charge.

All atoms have protons. The number of protons in the nucleus is called the **atomic number**, and it determines the type of atom, or element. **Elements** are pure substances, each consisting only of atoms with the same number of protons in their nucleus (FIGURE 2.2B). For example, the element carbon has an atomic number of 6. All atoms with six protons in their nucleus are carbon atoms, no matter how many electrons or neutrons they have. Elemental carbon (the substance) consists only of carbon atoms, and all of those atoms have six protons.

Knowing the numbers of electrons, protons, and neutrons in atoms helps us predict how elements will behave. In 1869, chemist Dmitry Mendeleyev arranged the elements known at the time by their chemical properties. The arrangement, which he called the **periodic table**, turned out to be by atomic number, even though subatomic particles would not be discovered until the early 1900s. In the periodic table, each element is represented by a symbol that is typically an abbreviation of the element's Latin or Greek name (FIGURE 2.2C). For example, the symbol for lead, Pb, is short for its Latin name: *plumbum*. The word "plumbing" is related (ancient Romans made their water pipes with lead). Carbon's symbol, C, is from *carbo*, the Latin word for coal, which is mostly carbon.

A Atoms consist of electrons moving around a nucleus of protons and neutrons. Models such as this one do not show what atoms look like. Electrons move in defined, three-dimensional spaces about 10,000 times bigger than the nucleus.

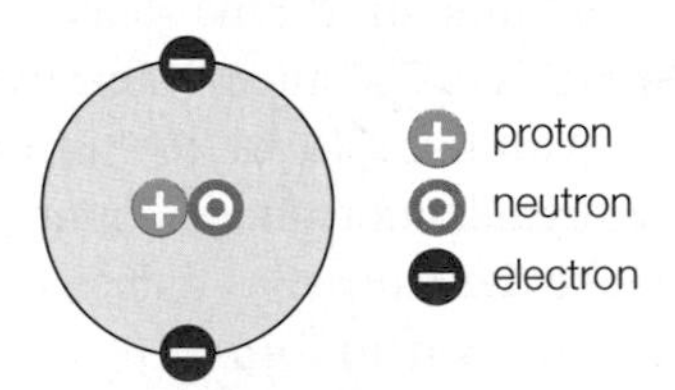

B Example of an element.

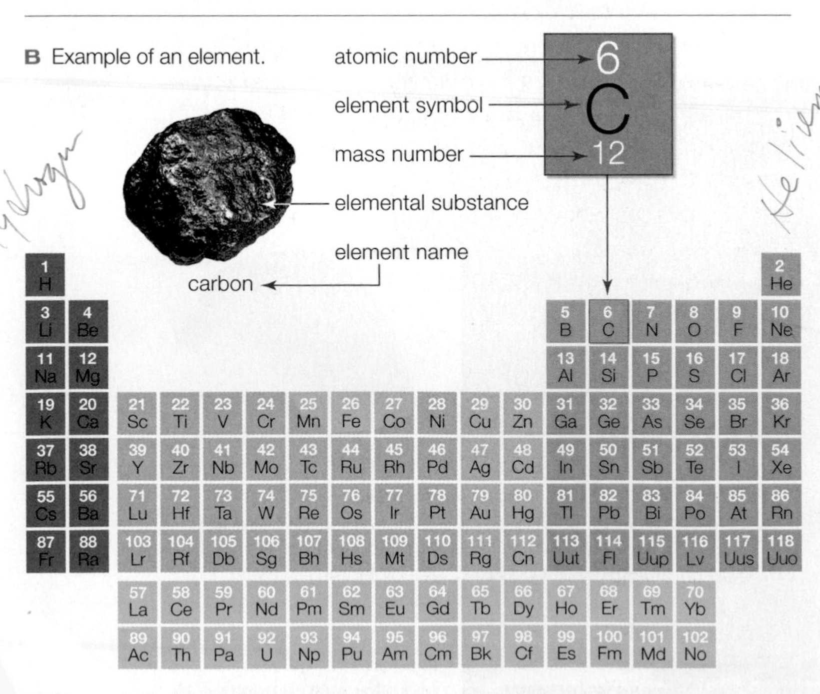

C The periodic table of the elements.

FIGURE 2.2 ▶**Animated** Atoms and elements.

Isotopes and Radioisotopes

All atoms of an element have the same number of protons, but they can differ in the number of other subatomic particles. Those that differ in the number of neutrons are called **isotopes**. The total number of neutrons and protons in the nucleus of an isotope is its **mass number**. Mass number is written as a superscript to the left of the element's symbol. For example, the most common isotope of hydrogen has one proton

atomic number Number of protons in the atomic nucleus; determines the element.
charge Electrical property. Opposite charges attract, and like charges repel.
electron Negatively charged subatomic particle.
element A pure substance that consists only of atoms with the same number of protons.
isotopes Forms of an element that differ in the number of neutrons their atoms carry.
mass number Of an isotope, the total number of protons and neutrons in the atomic nucleus.
neutron Uncharged subatomic particle in the atomic nucleus.
nucleus Core of an atom; occupied by protons and neutrons.
periodic table Tabular arrangement of all known elements by their atomic number.
proton Positively charged subatomic particle that occurs in the nucleus of all atoms.
radioactive decay Process by which atoms of a radioisotope emit energy and/or subatomic particles when their nucleus spontaneously breaks up.
radioisotope Isotope with an unstable nucleus.
tracer A substance that can be traced via its detectable component.

and no neutrons, so it is designated ^1H. Other isotopes include deuterium (^2H, one proton and one neutron), and tritium (^3H, one proton and two neutrons).

The most common isotope of carbon has six protons and six neutrons (^{12}C). Another naturally occurring carbon isotope has six protons and eight neutrons (^{14}C). Carbon 14 is an example of a **radioisotope**, or radioactive isotope. Atoms of a radioisotope have an unstable nucleus that breaks up spontaneously. As a nucleus breaks up, it emits radiation (subatomic particles, energy, or both), a process called **radioactive decay**. The atomic nucleus cannot be altered by ordinary means, so radioactive decay is unaffected by external factors such as temperature, pressure, or whether the atoms are part of molecules.

Each radioisotope decays at a predictable rate into predictable products. For example, when carbon 14 decays, one of its neutrons splits into a proton and an electron. The nucleus emits the electron as radiation. Thus, a carbon atom with eight neutrons and six protons (^{14}C) becomes a nitrogen atom, with seven neutrons and seven protons (^{14}N):

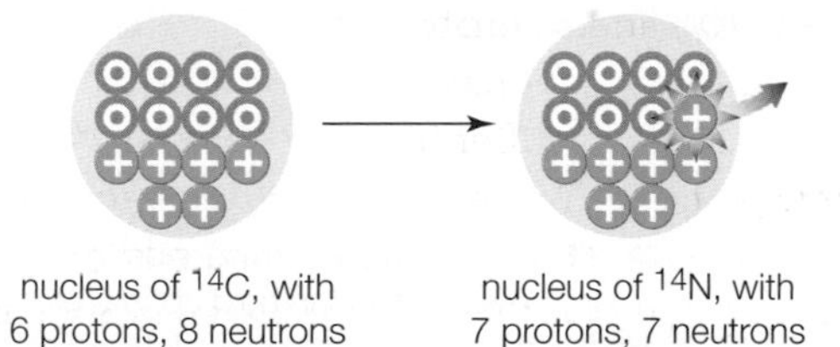

This process is so predictable that we can say with certainty that about half of the atoms in any sample of ^{14}C will be ^{14}N atoms after 5,730 years. The predictable rate of radioactive decay makes it possible for scientists to estimate the age of a rock or fossil by measuring its isotope content (we return to this topic in Section 16.6).

All isotopes of an element generally have the same chemical properties regardless of the number of neutrons in their atoms. This consistency means that the atoms of one isotope behave the same way inside organisms as atoms of another isotope. Thus, radioisotopes can be used as or in **tracers**, which are substances with a detectable component. For example, a molecule in which an atom (such as ^{12}C) has been replaced with a radioisotope (such as ^{14}C) can be used as a radioactive tracer. When delivered into a biological system such as a cell, body, or ecosystem, this tracer may be followed as it moves through the system with instruments that detect radiation.

Radioactive tracers are widely used in research. A famous example is a series of experiments carried out by Melvin Calvin and Andrew Benson. These researchers synthesized carbon dioxide with ^{14}C, then let green algae take up the radioactive gas. Using instruments that detect electrons emitted by the radioactive decay of ^{14}C, they tracked carbon through steps by which the algae—and all plants—make sugars.

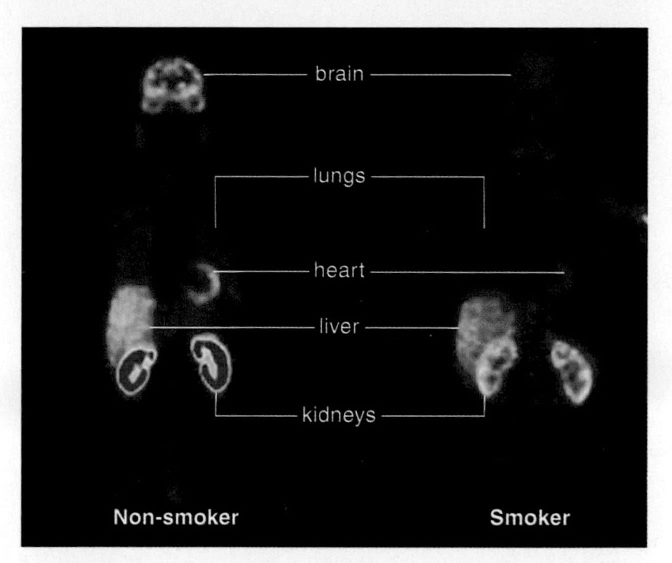

FIGURE 2.3 ▶**Animated** PET scans use radioactive tracers to form a digital image of a process in the body's interior. These two PET scans reveal the activity of a molecule called MAO-B in the body of a nonsmoker (left) and a smoker (right). The activity is color-coded from red (highest activity) to purple (lowest). Low MAO-B activity is associated with violence, impulsiveness, and other behavioral problems.

Radioisotopes have medical applications as well. For example, PET (short for positron-emission tomography) helps us "see" a functional process inside the body. By this procedure, a radioactive sugar or other tracer is injected into a patient. Inside the patient's body, cells with differing rates of activity take up the tracer at different rates. A scanner detects radioactive decay wherever the tracer is, then translates that data into an image (**FIGURE 2.3**).

TAKE-HOME MESSAGE 2.2

What are the basic building blocks of all matter?

✔ All matter consists of atoms, tiny particles that in turn consist of electrons moving around a nucleus (core). Protons and neutrons are components of the atomic nucleus.

✔ An elemental substance consists only of atoms with the same number of protons. Isotopes are forms of an element that have different numbers of neutrons.

✔ Unstable nuclei of radioisotopes emit radiation as they spontaneously break down (decay). Radioisotopes decay at a predictable rate to form predictable products.

2.3 Why Electrons Matter

✔ Whether an atom will interact with other atoms depends on how many electrons it has.

The more we learn about electrons, the weirder they seem. Consider that an electron has mass but no size, and its position in space is described as more of a smudge than a point. It carries energy, but only in incremental amounts (this concept will be important to remember when you learn how cells harvest and release energy). An electron gains energy only by absorbing the precise amount needed to boost it to the next energy level. Likewise, it loses energy only by emitting the exact difference between two energy levels.

A lot of electrons may be occupying the same atom. However, despite moving very fast—almost the speed of light—they never collide. Why not? For one reason, electrons in an atom occupy different orbitals, which are defined volumes of space around the atomic nucleus. To understand how orbitals work, imagine that an atom is a multilevel apartment building, with the nucleus in the basement. Each "floor" of the building corresponds to a certain energy level, and each has a certain number of "rooms" (orbitals) available for rent. Two electrons can occupy each room. Pairs of electrons populate rooms from the ground floor up; in other words, they fill orbitals from lower to higher energy levels. The farther an electron is from the nucleus in the basement, the greater its energy. An electron can move to a room on a higher floor if an energy input gives it a boost, but it immediately emits the extra energy and moves back down.

A **shell model** helps us visualize how electrons populate atoms (**FIGURE 2.4**). In this model, nested "shells" correspond to successively higher energy levels. Thus, each shell includes all of the rooms (orbitals) on one floor (energy level) of our atomic apartment building.

We draw a shell model of an atom by filling it with electrons (represented as balls or dots), from the innermost shell out, until there are as many electrons as the atom has protons. There is only one room on the first floor, one orbital at the lowest energy level. It fills up first. In hydrogen, the simplest atom, a single electron occupies that room (**FIGURE 2.4A**). Helium, with two protons, has two electrons that fill the room—and the first shell. In larger atoms, more electrons rent the second-floor rooms (**FIGURE 2.4B**). When the second floor fills, more electrons rent third-floor rooms (**FIGURE 2.4C**), and so on.

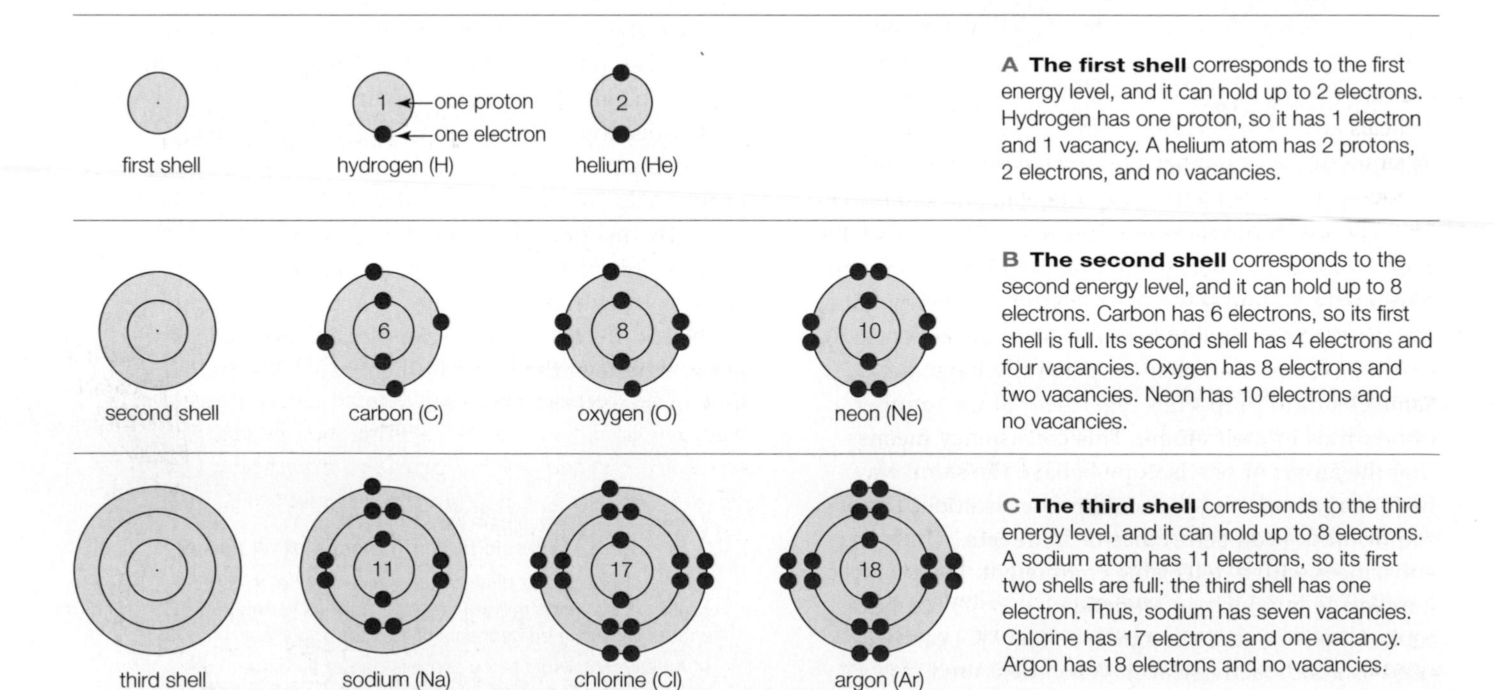

FIGURE 2.4 ▶Animated Shell models. Each circle (shell) represents one energy level. To make these models, we fill the shells with electrons from the innermost shell out, until there are as many electrons as the atom has protons. The number of protons in each model is indicated.

FIGURE IT OUT Which of these models have unpaired electrons in their outer shell?

Answer: Hydrogen, carbon, oxygen, sodium, and chlorine

About Vacancies

When an atom's outermost shell is filled with electrons, we say that it has no vacancies, and it is in its most stable state. Helium, neon, and argon are examples of elements with no vacancies. Atoms of these elements are chemically stable, which means they have very little tendency to interact with other atoms. Thus, these elements occur most frequently in nature as solitary atoms.

By contrast, when an atom's outermost shell has room for another electron, it has a vacancy. Atoms with vacancies tend to get rid of them by interacting with other atoms; in other words, they are chemically active. For example, the sodium atom (Na) depicted in **FIGURE 2.4C** has one electron in its outer (third) shell, which can hold eight. With seven vacancies, we can predict that this atom is chemically active.

In fact, this particular sodium atom is not just active, it is extremely so. Why? The shell model shows that a sodium atom has an unpaired electron, but in the real world, electrons really like to be in pairs when they occupy orbitals. Atoms that have unpaired electrons are called **free radicals**. With a few exceptions, free radicals are very unstable, easily forcing electrons upon other atoms or ripping electrons away from them. This property makes free radicals dangerous to life (we return to this topic in Section 5.6). A sodium atom with 11 electrons (a sodium radical) can easily evict the one unpaired electron, so that its second shell—which is full of electrons—becomes its outermost, and no vacancies remain. This is the atom's most stable state. The vast majority of sodium atoms on Earth are like this one, with 11 protons and 10 electrons.

Atoms with an unequal number of protons and electrons are called **ions**. Ions carry a net (or overall) charge. Sodium ions (Na^{+}) offer an example of how atoms gain a positive charge by losing an electron (**FIGURE 2.5A**). Other atoms gain a negative charge by accepting an electron. For example, an uncharged chlorine atom has 17 protons and 17 electrons. The outermost shell of this atom can hold eight electrons, but it has only seven. With one vacancy and one unpaired electron, we can predict—correctly—that this atom is chemically very active. An uncharged chlorine atom (a chlorine radical) easily fills its third shell by accepting an electron. When that happens, the atom becomes a chloride ion (Cl^{-}) with 17 protons, 18 electrons, and a net negative charge (**FIGURE 2.5B**).

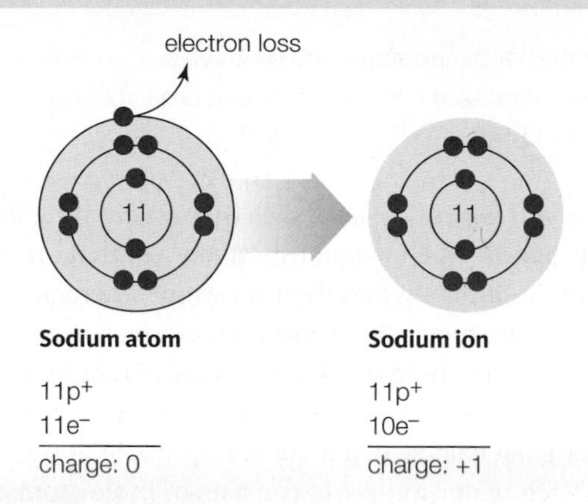

A A sodium atom (Na) becomes a positively charged sodium ion (Na^{+}) when it loses the single electron in its third shell. The atom's full second shell is now its outermost, so it has no vacancies.

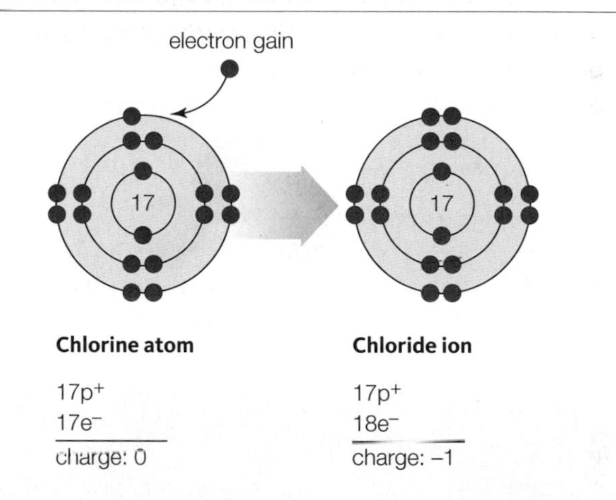

B A chlorine atom (Cl) becomes a negatively charged chloride ion (Cl^{-}) when it gains an electron and fills the vacancy in its third, outermost shell.

FIGURE 2.5 Ion formation. **FIGURE IT OUT** Does a chloride ion have an unpaired electron?

Answer: No

TAKE-HOME MESSAGE 2.3

Why do atoms interact?

- ✔ An atom's electrons are the basis of its chemical behavior.
- ✔ Shells represent all electron orbitals at one energy level in an atom. When the outermost shell is not full of electrons, the atom has a vacancy.
- ✔ Atoms with vacancies tend to interact with other atoms.

free radical Atom with an unpaired electron.
ion Charged atom.
shell model Model of electron distribution in an atom.

2.4 Chemical Bonds: From Atoms to Molecules

✔ Chemical bonds link atoms into molecules.

✔ The characteristics of a chemical bond arise from the properties of the atoms taking part in it.

An atom can get rid of vacancies by participating in a **chemical bond**, which is an attractive force that arises between two atoms when their electrons interact. When atoms interact, they often form molecules. A molecule consists of atoms held together in a particular number and arrangement by chemical bonds.

Water is an example of a substance made of molecules. Each water molecule consists of three atoms: two hydrogen atoms bonded to the same oxygen atom (FIGURE 2.6). Because a water molecule consists of two or more elements, it is called a **compound**. Other molecules, including molecular oxygen (a gas in air), have atoms of one element only.

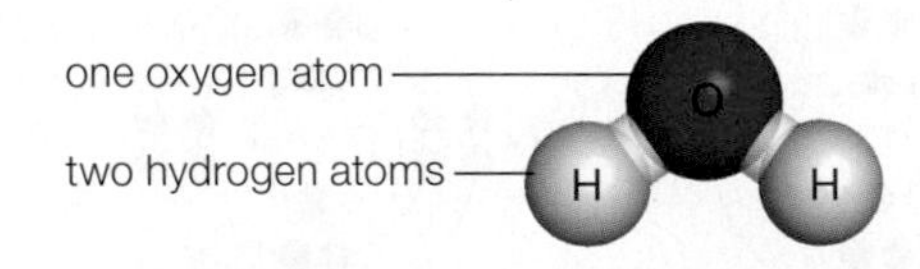

FIGURE 2.6 The water molecule. Each water molecule has two hydrogen atoms bonded to the same oxygen atom.

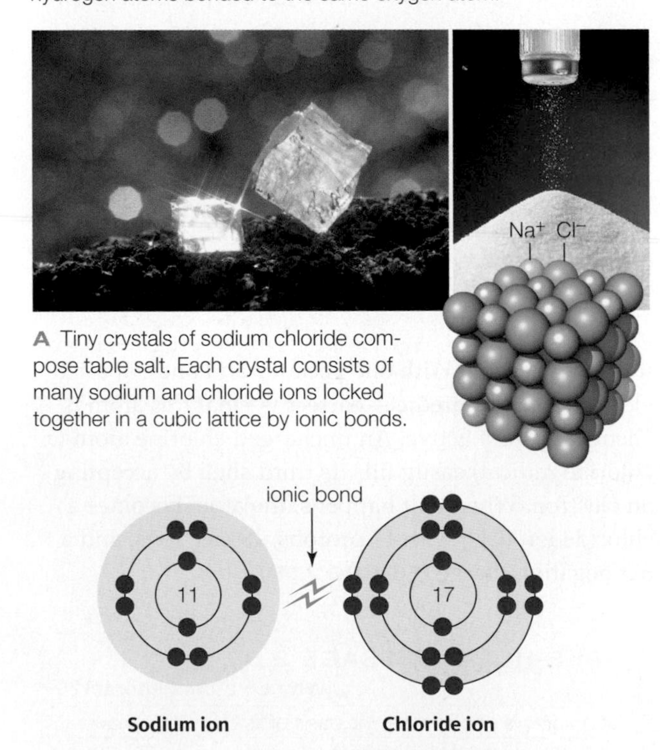

A Tiny crystals of sodium chloride compose table salt. Each crystal consists of many sodium and chloride ions locked together in a cubic lattice by ionic bonds.

B The strong mutual attraction of opposite charges holds a sodium ion and a chloride ion together in an ionic bond.

FIGURE 2.7 ▸Animated Ionic bonds in table salt, or NaCl.

The term "bond" applies to a continuous range of atomic interactions. However, we can categorize most bonds into distinct types based on their different properties. Which type forms depends on the atoms taking part in the molecule.

Ionic Bonds

Two ions may be held together by the mutual attraction of their opposite charges, an association called an **ionic bond**. Ionic bonds can be quite strong. Ionically bonded sodium and chloride ions make sodium chloride (NaCl), which we know as table salt; a crystal of this substance consists of a lattice of sodium and chloride ions interacting in ionic bonds (FIGURE 2.7A).

Ions retain their respective charges when participating in an ionic bond (FIGURE 2.7B). Thus, one "end" of an ionic bond has a positive charge, and the other "end" has a negative charge (right). Any separation of charge into distinct positive and negative regions is called **polarity**.

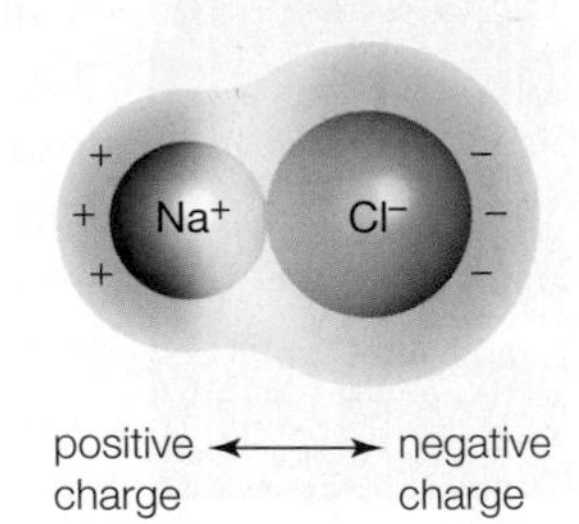

A sodium chloride molecule is polar because the chloride ion keeps a very strong hold on its extra electron. In other words, it is strongly electronegative. **Electronegativity** is a measure of an atom's ability to pull electrons away from another atom. Electronegativity is not the same as charge. Rather, an atom's electronegativity depends on its size, how many vacancies it has, and its interactions with other atoms.

An ionic bond is completely polar because the atoms participating in it have a very large difference in electronegativity. When atoms with a lower difference in electronegativity interact, they tend to form chemical bonds that are less polar than ionic bonds.

Covalent Bonds

In a **covalent bond**, two atoms share a pair of electrons, so that each atom's vacancy becomes partially filled (FIGURE 2.8). Sharing electrons links two atoms just as sharing earphones links two friends (left). Covalent bonds can be stronger than ionic bonds, but they are not always so.

TABLE 2.1 shows different ways of representing covalent bonds. In structural formulas, a line between two atoms represents a single covalent bond. For example, molecular hydrogen (H_2) has one covalent bond

Table 2.1 Ways of Representing Covalent Bonds in Molecules

Common name:	Water	Familiar term.
Chemical name:	Dihydrogen monoxide	Describes elemental composition.
Chemical formula:	H_2O	Indicates unvarying proportions of elements. Subscripts show number of atoms of an element per molecule. The absence of a subscript means one atom.
Structural formula:	H—O—H	Represents each covalent bond as a single line between atoms.
Structural model:	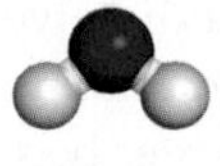	Shows relative sizes and positions of atoms in three dimensions.
Shell model:	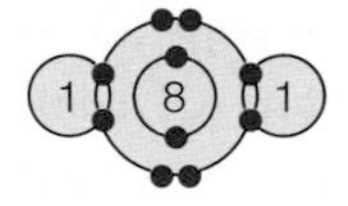	Shows how pairs of electrons are shared in covalent bonds.

MOLECULAR HYDROGEN (H—H)

Two hydrogen atoms, each with one proton, share two electrons in a nonpolar covalent bond.

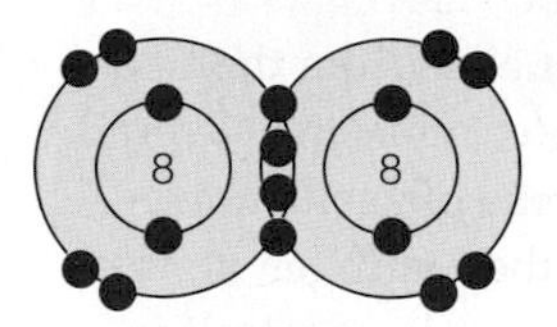

MOLECULAR OXYGEN (O=O)

Two oxygen atoms, each with eight protons, share four electrons in a double covalent bond.

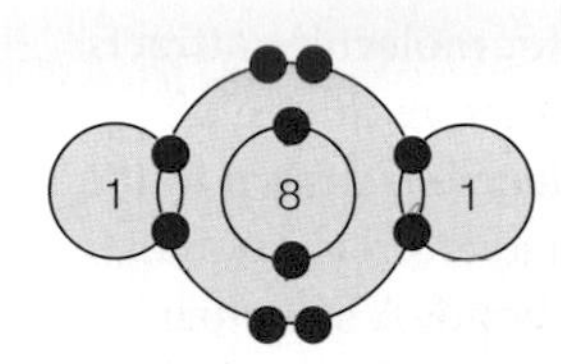

WATER (H—O—H)

Two hydrogen atoms share electrons with an oxygen atom in two covalent bonds. The bonds are polar because the oxygen exerts a greater pull on the shared electrons than the hydrogens do.

FIGURE 2.8 ▸**Animated** Covalent bonds, in which atoms fill vacancies by sharing electrons. Two electrons are shared in each covalent bond. When sharing is equal, the bond is nonpolar. When one atom exerts a greater pull on the electrons, the bond is polar.

between hydrogen atoms (H—H). Two, three, or even four covalent bonds may form between atoms when they share multiple pairs of electrons. For example, two atoms sharing two pairs of electrons are connected by two covalent bonds, which are represented by a double line between the atoms. A double bond links the two oxygen atoms in molecular oxygen (O=O). Three lines indicate a triple bond, in which two atoms share three pairs of electrons. A triple covalent bond links the two nitrogen atoms in molecular nitrogen (N≡N). Comparing bonds between the same two atoms: A triple bond is stronger than a double bond, which is stronger than a single bond.

Double and triple bonds are not distinguished from single bonds in structural models, which show the positions and relative sizes of the atoms in three dimensions. The bonds are shown as one stick connecting two balls, which represent atoms. Elements are usually coded by color:

carbon hydrogen oxygen nitrogen phosphorus

Atoms share electrons unequally in a polar covalent bond. A bond between an oxygen atom and a hydrogen atom in a water molecule is an example. One atom (the oxygen, in this case) is a bit more electronegative. It pulls the electrons a little more toward its side of the bond, so that atom bears a slight negative charge. The atom at the other end of the bond (the hydrogen) bears a slight positive charge. In most cases, covalent bonds in compounds are polar. By contrast, atoms participating in a nonpolar covalent bond share electrons equally, so there is no difference in charge between the two ends of the bond. The bonds in molecular hydrogen (H_2), oxygen (O_2), and nitrogen (N_2) are examples.

chemical bond An attractive force that arises between two atoms when their electrons interact.
compound Molecule that has atoms of more than one element.
covalent bond Type of chemical bond in which two atoms share a pair of electrons.
electronegativity Measure of the ability of an atom to pull electrons away from other atoms.
ionic bond Type of chemical bond in which a strong mutual attraction links ions of opposite charge.
polarity Separation of charge into positive and negative regions.

TAKE-HOME MESSAGE 2.4

How do atoms interact in chemical bonds?

✔ A chemical bond forms between atoms when their electrons interact. A chemical bond may be ionic or covalent depending on the atoms taking part in it.

✔ An ionic bond is a strong mutual attraction between two ions of opposite charge.

✔ Atoms share a pair of electrons in a covalent bond. When the atoms share electrons unequally, the bond is polar.

2.5 Hydrogen Bonds and Water

✔ The unique properties of liquid water arise because of the water molecule's polarity.

✔ Extensive hydrogen bonds form among water molecules.

Hydrogen Bonding in Water

Water has unique properties that arise from the two polar covalent bonds in each water molecule. Overall, the molecule has no charge, but the oxygen atom carries a slight negative charge; the hydrogen atoms, a slight positive charge. Thus, the molecule itself is polar (**FIGURE 2.9A**).

The polarity of individual water molecules attracts them to one another. The slight positive charge of a hydrogen atom in one water molecule is drawn to the slight negative charge of an oxygen atom in another, an interaction called a hydrogen bond. A **hydrogen bond** is an attraction between a covalently bonded hydrogen atom and another atom taking part in a separate polar covalent bond (**FIGURE 2.9B**). Like ionic bonds, hydrogen bonds form by the mutual attraction of opposite charges. However, unlike ionic bonds, hydrogen bonds do not make molecules out of atoms, so they are not chemical bonds.

A Polarity of the water molecule. Each of the hydrogen atoms in a water molecule bears a slight positive charge (represented by a blue overlay). The oxygen atom carries a slight negative charge (red overlay).

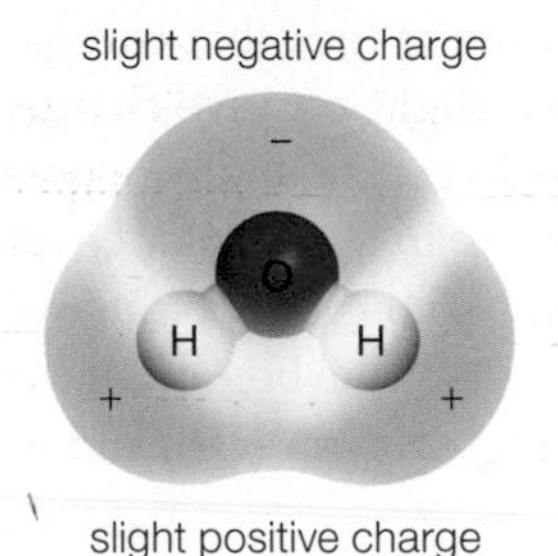

B A hydrogen bond is an attraction between a hydrogen atom and another atom taking part in a separate polar covalent bond.

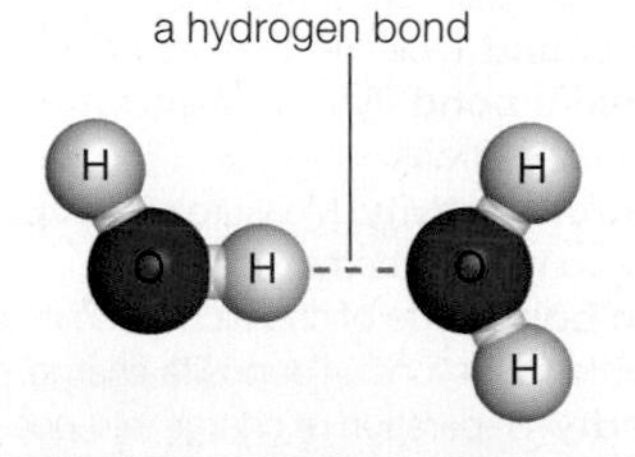

C The many hydrogen bonds that form among water molecules impart special properties to liquid water.

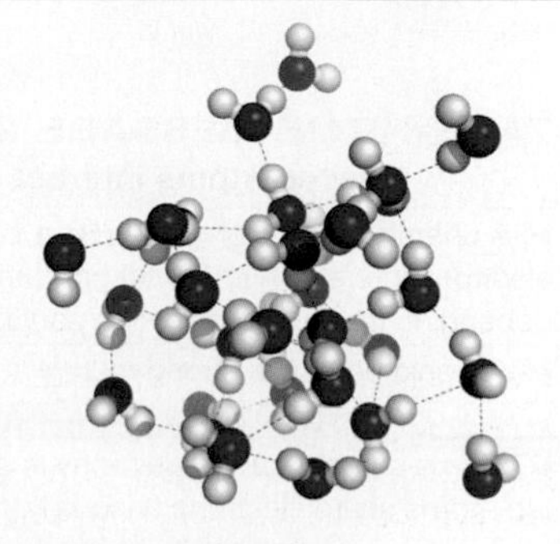

FIGURE 2.9 ▸**Animated** Hydrogen bonds and water.

Hydrogen bonds are on the weaker end of the spectrum of atomic interactions; they form and break much more easily than covalent or ionic bonds. Even so, many of them form, and collectively they are quite strong. As you will see, hydrogen bonds stabilize the characteristic structures of biological molecules such as DNA and proteins. They also form in tremendous numbers among water molecules (**FIGURE 2.9C**). Extensive hydrogen bonding among water molecules gives liquid water several special properties that make life possible.

Water's Special Properties

Water Is an Excellent Solvent The ability of water molecules to form hydrogen bonds make water an excellent **solvent**, which means that many other substances can dissolve in it. Substances that dissolve easily in water are **hydrophilic** (water-loving). Ionic solids such as sodium chloride (NaCl) dissolve in water because the slight positive charge on each hydrogen atom in a water molecule attracts negatively charged ions (Cl^-), and the slight negative charge on the oxygen atom attracts positively charged ions (Na^+). Hydrogen bonds among many water molecules are collectively stronger than an ionic bond between two ions, so the solid dissolves as water molecules tug the ions apart and surround each one (left).

Sodium chloride is called a **salt** because it releases ions other than H^+ and OH^- when it dissolves in water (more about this in the next section). When an ionic solid dissolves, its component ions disperse uniformly among molecules of the solvent, and it becomes a **solute**. A uniform mixture such as salt dissolved in water is called a **solution**. Chemical bonds do not form between molecules of solute and solvent, so the proportions of the two substances in a solution can vary. The amount of a solute that is dissolved in a given volume of fluid is its **concentration**.

Many nonionic solids also dissolve easily in water. Sugars are examples. Molecules of these substances have one or more polar covalent bonds, and atoms participating in a polar covalent bond can form hydrogen bonds with water molecules. Hydrogen bonding with water pulls individual molecules of the solid away from one another and keeps them apart. Unlike ionic solids, these substances retain their molecular integrity when they dissolve, which means they do not dissociate into atoms.

Water does not interact with **hydrophobic** (water-dreading) substances such as oils. Oils consist of nonpolar molecules, and hydrogen bonds do not form between nonpolar molecules and water. When you mix oil and water, the water breaks into small droplets, but quickly begins to cluster into larger drops as new hydrogen bonds form among its molecules. The bonding excludes molecules of oil and pushes them together into drops that rise to the surface of the water. The same interactions occur at the thin, oily membrane that separates the watery fluid inside cells from the watery fluid outside of them. As you will see in Chapter 3, such interactions give rise to the structure of cell membranes.

Water Has Cohesion Molecules of some substances resist separating from one another, and this resistance gives rise to a property called **cohesion**. Water has cohesion because hydrogen bonds collectively exert a continuous pull on its individual molecules. You can see cohesion in water as surface tension, which means that the surface of liquid water behaves a bit like a sheet of elastic (left).

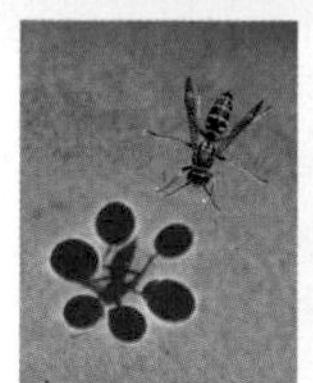

Cohesion plays a role in many processes that sustain multicelled bodies. Consider how water molecules constantly escape from the surface of liquid water as vapor, a process called **evaporation**. Evaporation is resisted by hydrogen bonding among water molecules. In other words, overcoming water's cohesion takes energy. Thus, evaporation sucks energy (in the form of heat) from liquid water, and this lowers the water's surface temperature. Evaporative water loss helps you and some other mammals cool off when you sweat in hot, dry weather. Sweat, which is about 99 percent water, cools the skin as it evaporates.

Cohesion works inside organisms, too. Consider how plants absorb water from soil as they grow. Water molecules evaporate from leaves, and replacements are pulled upward from roots. Cohesion makes it possible for columns of liquid water to rise from roots to leaves inside narrow pipelines of vascular tissue. In some trees, these pipelines extend hundreds of feet above the soil (Section 28.4 returns to this topic).

cohesion Property of a substance that arises from the tendency of its molecules to resist separating from one another.
concentration Amount of solute per unit volume of solution.
evaporation Transition of a liquid to a vapor.
hydrogen bond Attraction between a covalently bonded hydrogen atom and another atom taking part in a separate covalent bond.
hydrophilic Describes a substance that dissolves easily in water.
hydrophobic Describes a substance that resists dissolving in water.
salt Ionic compound that releases ions other than H^+ and OH^- when it dissolves in water.
solute A dissolved substance.
solution Uniform mixture of solute completely dissolved in solvent.
solvent Liquid in which other substances dissolve.
temperature Measure of molecular motion.

FIGURE 2.10 Hydrogen bonds lock water molecules in a rigid lattice in ice. The molecules in this lattice pack less densely than in liquid water, which is why ice floats on water. A covering of ice can insulate water underneath it, thus keeping aquatic organisms from freezing during cold winters.

Water Stabilizes Temperature All atoms jiggle nonstop, so the molecules they make up jiggle too. We measure the energy of this motion as degrees of **temperature**. Adding energy (in the form of heat, for example) makes the jiggling faster, so the temperature rises. Hydrogen bonding keeps water molecules from moving as much as they would otherwise, so it takes more heat to raise the temperature of water compared with other liquids. Temperature stability is an important part of homeostasis, because most of the molecules of life function properly only within a certain range of temperature.

Below 0°C (32°F), water molecules do not jiggle enough to break hydrogen bonds between them, and they become locked in the rigid, lattice-like bonding pattern of ice (**FIGURE 2.10**). Individual water molecules pack less densely in ice than they do in water, which is why ice floats on water. Sheets of ice that form on the surface of ponds, lakes, and streams can insulate the water under them from subfreezing air temperatures. Such "ice blankets" protect aquatic organisms during long, cold winters.

TAKE-HOME MESSAGE 2.5

What gives water the special properties that make life possible?

✔ Extensive hydrogen bonding among water molecules arises from the polarity of the individual molecules.

✔ Hydrogen bonding among water molecules imparts cohesion to liquid water, and gives it the ability to stabilize temperature and dissolve many substances.

2.6 Acids and Bases

✔ The number of hydrogen ions in a fluid is measured as pH.

✔ Most biological processes occur within a narrow range of pH, typically around pH 7.

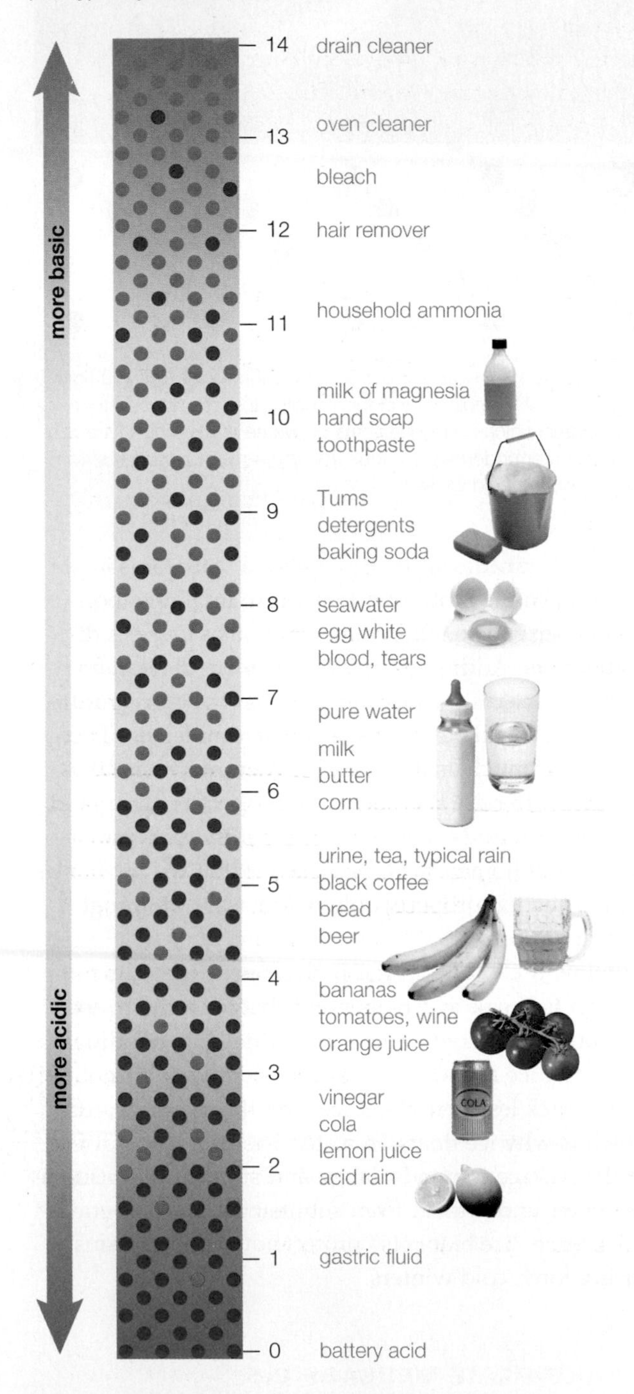

FIGURE 2.11 A pH scale. Here, red dots signify hydrogen ions (H^+) and gray dots signify hydroxyl ions (OH^-). Also shown are approximate pH values for some common solutions.

This pH scale ranges from 0 (most acidic) to 14 (most basic). A change of one unit on the scale corresponds to a tenfold change in the amount of H^+ ions.

FIGURE IT OUT What is the approximate pH of cola?

Answer: 2.5

A hydrogen atom, remember, is just a proton and an electron. When a hydrogen atom participates in a polar covalent bond with a more electronegative atom (such as oxygen), the electron is pulled away from the proton, just a bit. Hydrogen bonding in water tugs on that proton even more, so much that the proton can be pulled right off of the molecule. The electron stays with the rest of the molecule, which becomes negatively charged (ionic), and the proton becomes a hydrogen ion (H^+). For example, a water molecule that loses a proton becomes a hydroxyl ion (OH^-):

$$\underset{\text{water molecule}}{H_2O} \longrightarrow \underset{\text{hydroxide ion}}{OH^-} + \underset{\text{hydrogen ion}}{H^+}$$

The loss is more or less temporary, because these two ions easily get back together to form a water molecule:

$$\underset{\text{hydroxide ion}}{OH^-} + \underset{\text{hydrogen ion}}{H^+} \longrightarrow \underset{\text{water molecule}}{H_2O}$$

With other molecules, the loss of a hydrogen ion in water is essentially permanent. We use a value called **pH** to measure of the number of hydrogen ions floating around in a water-based fluid. In pure water, the number of H^+ ions is the same as the number of OH^- ions, and the pH is 7, or neutral. The higher the number of hydrogen ions, the lower the pH. A one-unit decrease in pH corresponds to a tenfold increase in the number of H^+ ions, and a one-unit increase corresponds to a tenfold decrease in the number of H^+ ions (**FIGURE 2.11**). One way to get a sense of the pH scale is to taste dissolved baking soda (pH 9), distilled water (pH 7), and lemon juice (pH 2).

An **acid** is a substance that gives up hydrogen ions in water. Acids can lower the pH of a solution and make it acidic (below pH 7). **Bases** accept hydrogen ions from water, so they can raise the pH of a solution and make it basic, or alkaline (above pH 7).

Strong acids ionize completely in water to give up all of their H^+ ions; weak acids give up only some of them. Hydrochloric acid (HCl) is an example of a strong acid. When HCl dissolves in water, all of the molecules give up hydrogen ions, leaving Cl^- ions behind. Hydrogen ions released from HCl makes gastric fluid inside your stomach very acidic (pH 1–2).

acid Substance that releases hydrogen ions in water.
base Substance that accepts hydrogen ions in water.
buffer Set of chemicals that can keep the pH of a solution stable by alternately donating and accepting ions that contribute to pH.
pH Measure of the number of hydrogen ions in a fluid.

Mercury Rising (revisited)

All ecosystems now have detectable effects of air pollution, but many of those effects are not as well understood as acid rain. We do know that the concentration of mercury in Earth's waters is rising, and is predicted to double within forty years. We also know that this rise is occurring as a consequence of human activities, which release more than 2,000 tons of mercury into the atmosphere every year.

All human bodies now have detectable amounts of mercury; the average adult living in the U.S. has about 4 micrograms of it circulating in his or her blood. Some comes from dental fillings, imported skin-bleaching cosmetics, and broken fluorescent lamps. However, most comes from dietary seafood: The more fish and shellfish you eat, the more mercury your body has. A diet that consists of a high proportion of seafood can result in a blood mercury content thirty times the average.

Carbonic acid forms when carbon dioxide gas dissolves in plasma, the fluid portion of human blood:

$$\underset{\text{carbon dioxide}}{CO_2} + \underset{\text{water molecule}}{H_2O} \longrightarrow \underset{\text{carbonic acid}}{H_2CO_3}$$

Carbonic acid is a weak acid, so only some of its molecules give up a hydrogen ion in water. When carbonic acid loses a hydrogen ion, it becomes an ionic molecule called bicarbonate:

$$\underset{\text{carbonic acid}}{H_2CO_3} \longrightarrow \underset{\text{hydrogen ion}}{H^+} + \underset{\text{bicarbonate}}{HCO_3^-}$$

Bicarbonate can act like a base by accepting a hydrogen ion. When it does, carbonic acid forms again:

$$\underset{\text{hydrogen ion}}{H^+} + \underset{\text{bicarbonate}}{HCO_3^-} \longrightarrow \underset{\text{carbonic acid}}{H_2CO_3}$$

Together, carbonic acid and bicarbonate constitute a buffer. A **buffer** is a set of chemicals that can keep the pH of a solution stable by alternately donating and accepting ions that contribute to pH. Consider how the pH of pure water rises when a base is added to it. This is because the base accepts hydrogen ions from the water, thus reducing the number of hydrogen ions floating around in it (and contributing to pH). By contrast, the carbonic acid–bicarbonate buffer system can keep the pH of blood plasma from rising when base is added. The added base causes carbonic acid to give up hydrogen ions (and become bicarbonate). These hydrogen ions replace the ones that the base removed from the solution. The same buffer system can also keep plasma pH from declining when acid is added. Hydrogen ions released by the acid combine with bicarbonate, so they do not contribute to pH. In both cases, the proportion of carbonic acid and bicarbonate molecules in plasma shifts, but the pH stays stable (typically between 7.3 and 7.5).

The addition of too much acid or base can overwhelm a buffer's capacity to stabilize pH. Such buffer failure can be catastrophic in a cell or body because most biological molecules function properly only within a narrow range of pH: Even a slight deviation from that range can halt cellular processes. Consider what happens when breathing is impaired suddenly. Carbon dioxide gas accumulates in tissues, and too much carbonic acid forms in plasma. If the excess acid reduces blood pH below 7.3, a dangerous level of unconsciousness called coma can be the outcome. By contrast, hyperventilation (sustained rapid breathing) causes the body to lose too much CO_2. The loss results in a rise in blood pH. If blood pH rises too much, prolonged muscle spasm (tetany) or coma may occur.

Burning fossil fuels such as coal releases sulfur and nitrogen compounds that affect the pH of rain and other forms of precipitation. These fluids are not buffered, so the addition of acids or bases has a dramatic effect. In places with a lot of fossil fuel emissions, the rain and fog can be more acidic than vinegar. The corrosive effect of this acid rain is visible in urban areas (left). Acid rain also drastically changes the pH of water in soil, lakes, and streams. Such changes can overwhelm the buffering capacity of fluids inside organisms that live in these environments, with lethal effects. Section 48.5 returns to the topic of acid rain.

TAKE-HOME MESSAGE 2.6

Why are hydrogen ions important in biological systems?

✔ The number of hydrogen ions in a fluid determines its pH. Most biological systems function properly only within a narrow range of pH.

✔ Acids release hydrogen ions in water; bases accept them.

✔ Buffers help keep pH stable. Inside organisms, they play a role in homeostasis.

summary

Section 2.1 Interactions between atoms make the molecules that sustain life, and also some that destroy it. Mercury in air pollution ends up in the bodies of fish, and in turn, in the bodies of humans.

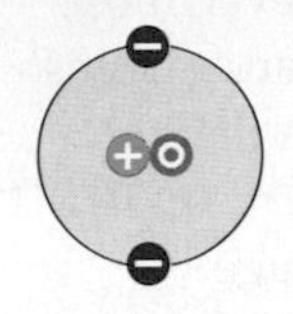

Section 2.2 Atoms consist of **electrons**, which carry a negative **charge**, moving about a **nucleus** of positively charged **protons** and uncharged **neutrons** (**TABLE 2.2**). The **periodic table** lists **elements** in order of **atomic number**. **Isotopes** of an element differ in the number of neutrons. The total number of protons and neutrons is the **mass number**. **Tracers** can be made with **radioisotopes**, which, by a process called **radioactive decay**, emit particles and energy when their nucleus spontaneously breaks up.

Section 2.3 Up to two electrons occupy each orbital (volume of space around a nucleus). Which orbital an electron occupies depends on its energy. A **shell model** represents successive energy levels as concentric circles. Atoms are in their most stable state when all of their shells are full, so they tend to get rid of vacancies. Many can do so by gaining or losing electrons, thereby becoming charged **ions**. Atoms with unpaired electrons are called **free radicals**. Most free radicals are highly chemically active, easily forcing electrons onto other atoms or pulling electrons away from them.

Section 2.4 A **chemical bond** unites two atoms in a molecule. A **compound** is a molecule that consists of two or more elements. Atoms form different types of bonds depending on their **electronegativity**. An **ionic bond** is a strong association between oppositely charged ions; it arises from the mutual attraction of opposite charges. **Polarity** is a separation of charge into positive and negative regions. Ionic bonds are completely polar. Atoms share a pair of electrons in a **covalent bond**, which is nonpolar if the sharing is equal, and polar if it is not.

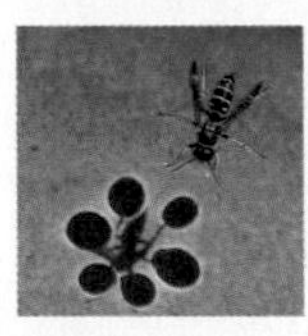

Section 2.5 Two polar covalent bonds give each water molecule an overall polarity. **Hydrogen bonds** that form among water molecules in tremendous numbers are the basis of water's unique life-sustaining properties: **cohesion**, resistance to **temperature** change, and the ability to act as a **solvent** that dissolves **salts** and other **solutes**. **Hydrophilic** substances dissolve easily in water to form **solutions**; **hydrophobic** substances do not. The amount of solute in a given volume of fluid is the solute's **concentration**. **Evaporation** is the transition of a liquid to vapor.

Section 2.6 **pH** is a measure of the number of hydrogen ions (H^+) in a liquid. At neutral pH (7), there are an equal number of H^+ and OH^- ions. **Acids** release hydrogen ions in water; **bases** accept them. A **buffer** can keep a solution within a consistent range of pH. Most cell and body fluids are buffered because most molecules of life work only within a narrow range of pH.

Table 2.2 Players in the Chemistry of Life

Atoms	Particles that are basic building blocks of all matter.
Proton (p^+)	Positively charged subatomic particle in the nucleus.
Electron (e^-)	Negatively charged subatomic particle that can occupy a defined volume of space (orbital) around the nucleus.
Neutron	Uncharged subatomic particle of the nucleus.
Element	Pure substance that consists entirely of atoms with the same, characteristic number of protons.
Isotopes	Atoms of an element that differ in the number of neutrons.
Radioisotope	Isotope with an unstable nucleus that emits radiation when it decays (breaks up).
Tracer	Substance with a detectable component (such as a radioisotope) used to track its movement or destination in a biological system.
Ion	Atom that carries a charge after it has gained or lost one or more electrons. A single proton without an electron is a hydrogen ion (H^+).
Molecule	Two or more atoms joined in a chemical bond.
Compound	Molecule of two or more different elements in unvarying proportions (for example, water: H_2O).
Solute	Substance dissolved in a solvent.
Hydrophilic	Refers to a substance that dissolves easily in water.
Hydrophobic	Refers to a substance that resists dissolving in water.
Acid	Compound that releases H^+ when dissolved in water.
Base	Compound that accepts H^+ when dissolved in water.
Salt	Ionic compound that releases ions other than H^+ or OH^- when dissolved in water.
Solvent	Substance that can dissolve other substances.
Buffer	Set of chemicals that can stabilize the pH of a fluid.

self-quiz

Answers in Appendix VII

1. What atom has only one proton?
   a. hydrogen
   b. an isotope
   c. a free radical
   d. a radioisotope

2. A molecule into which a radioisotope has been incorporated can be used as a(n) ________ .
   a. compound
   b. tracer
   c. salt
   d. acid

3. Which of the following statements is incorrect?
   a. Isotopes have the same atomic number and different mass numbers.
   b. Atoms have about the same number of electrons as protons.
   c. All ions are atoms.
   d. Free radicals are dangerous because they emit energy.

data analysis activities

Mercury Emissions by Continent By weight, coal does not contain much mercury, but we burn a lot of it. Several industries besides coal-fired power plants contribute substantially to atmospheric mercury pollution. **FIGURE 2.12** shows mercury emissions by industry from different regions of the world in 2006.

1. About how many tons of mercury were released?

2. Which industry tops the list of mercury emitters? Which industry is next on the list?

3. Which region emitted the most mercury from producing cement?

4. About how many tons of mercury were released from gold production in South America?

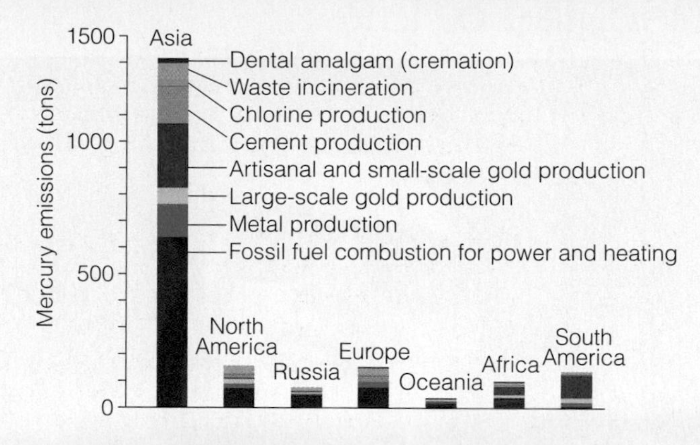

FIGURE 2.12 Global mercury emissions, 2006.

4. In the periodic table, symbols for the elements are arranged according to ________ .
   a. size
   b. charge
   c. mass number
   d. atomic number

5. An ion is an atom that has ________ .
   a. the same number of electrons and protons
   b. a different number of electrons and protons
   c. electrons, protons, and neutrons

6. The measure of an atom's ability to pull electrons away from another atom is called ________ .
   a. electronegativity
   b. charge
   c. polarity

7. The mutual attraction of opposite charges holds atoms together as molecules in a(n) ________ bond.
   a. ionic
   b. hydrogen
   c. polar covalent
   d. nonpolar covalent

8. Atoms share electrons unequally in a(n) ________ bond.
   a. ionic
   b. hydrogen
   c. polar covalent
   d. nonpolar covalent

9. A(n) ________ substance repels water.
   a. acidic
   b. basic
   c. hydrophobic
   d. polar

10. A salt does not release ________ in water.
   a. ions
   b. energy
   c. H^+

11. Hydrogen ions (H^+) are ________ .
   a. in blood
   b. protons
   c. indicated by a pH scale
   d. all of the above

12. When dissolved in water, a(n) ________ donates H^+; a(n) ________ accepts H^+.
   a. acid; base
   b. base; acid
   c. buffer; solute
   d. base; buffer

13. A ________ can help keep the pH of a solution stable.
   a. covalent bond
   b. hydrogen bond
   c. buffer
   d. pH

CENGAGE brain .com To access course materials, please visit www.cengagebrain.com.

14. A ________ is dissolved in a solvent.
   a. molecule
   b. solute
   c. salt

15. Match the terms with their most suitable description.

____ hydrophilic	a. protons > electrons
____ atomic number	b. number of protons in nucleus
____ hydrogen bonds	c. polar; dissolves easily in water
____ positive charge	d. collectively strong
____ temperature	e. protons < electrons
____ negative charge	f. measure of molecular motion

critical thinking

1. Alchemists were medieval scholars and philosophers who were the forerunners of modern-day chemists. Many spent their lives trying to transform lead (atomic number 82) into gold (atomic number 79). Explain why they never did succeed in that endeavor.

2. Draw a shell model of a lithium atom (Li), which has 3 protons. Predict whether the majority of lithium atoms on Earth are uncharged, positively charged, or negatively charged.

3. Polonium is a rare element with 33 radioisotopes. The most common one, ^{210}Po, has 82 protons and 128 neutrons. When ^{210}Po decays, it emits an alpha particle, which is a helium nucleus (2 protons and 2 neutrons). ^{210}Po decay is tricky to detect because alpha particles do not carry very much energy compared to other forms of radiation. They can be stopped by, for example, a sheet of paper or a few inches of air. This property is one reason why authorities failed to discover toxic amounts of ^{210}Po in the body of former KGB agent Alexander Litvinenko until after he died suddenly and mysteriously in 2006. What element does an atom of ^{210}Po change into after it emits an alpha particle?

4. Some undiluted acids are not as corrosive as when they are diluted with water. That is why lab workers are told to wipe off splashes with a towel before washing. Explain.

SOURCE: (12) Global Atmospheric Mercury Assessment: Sources, Emissions and Transport, United Nations Environmental Programme, Chemicals Branch. 2008

3 Molecules of Life

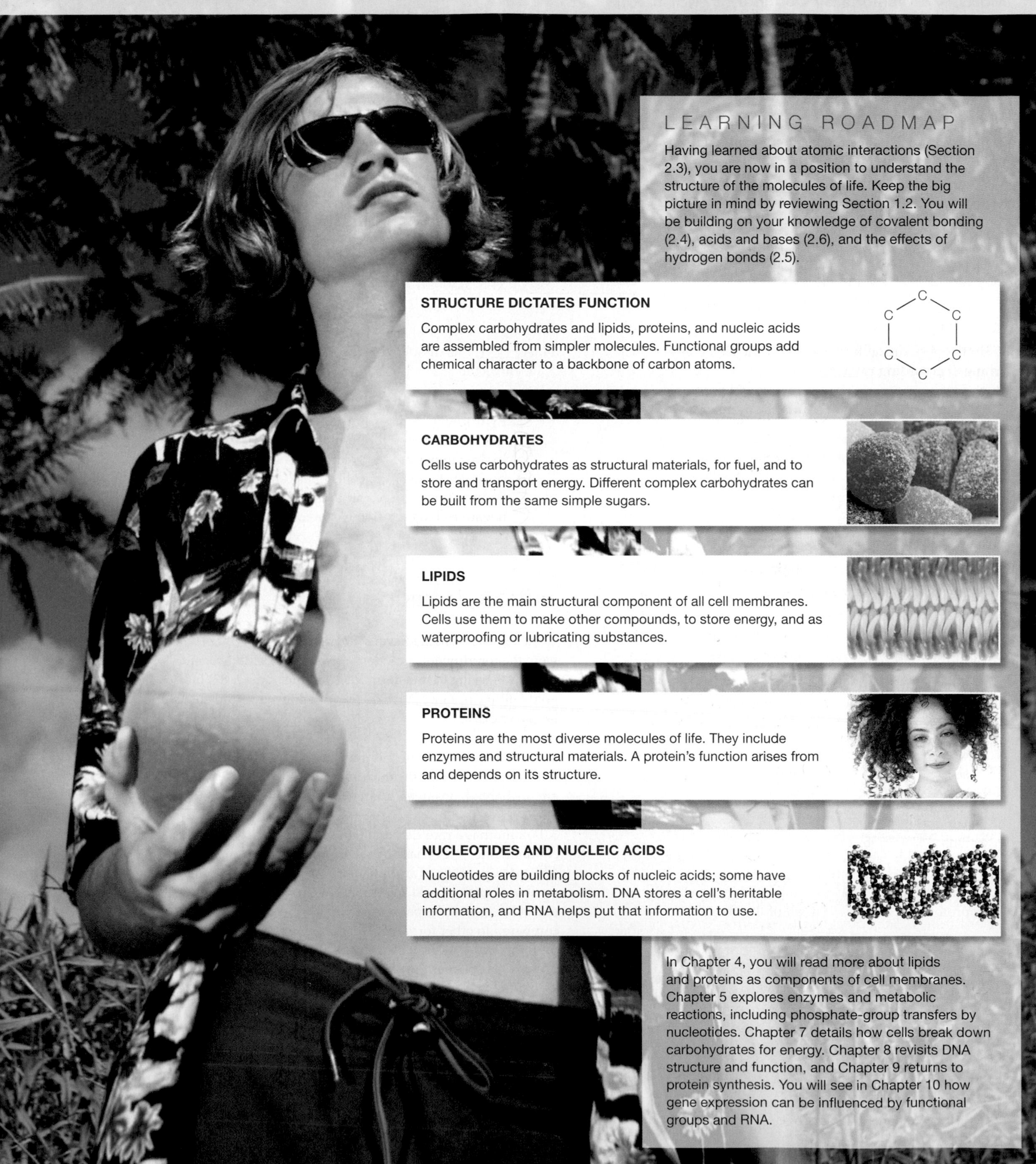

LEARNING ROADMAP

Having learned about atomic interactions (Section 2.3), you are now in a position to understand the structure of the molecules of life. Keep the big picture in mind by reviewing Section 1.2. You will be building on your knowledge of covalent bonding (2.4), acids and bases (2.6), and the effects of hydrogen bonds (2.5).

STRUCTURE DICTATES FUNCTION

Complex carbohydrates and lipids, proteins, and nucleic acids are assembled from simpler molecules. Functional groups add chemical character to a backbone of carbon atoms.

CARBOHYDRATES

Cells use carbohydrates as structural materials, for fuel, and to store and transport energy. Different complex carbohydrates can be built from the same simple sugars.

LIPIDS

Lipids are the main structural component of all cell membranes. Cells use them to make other compounds, to store energy, and as waterproofing or lubricating substances.

PROTEINS

Proteins are the most diverse molecules of life. They include enzymes and structural materials. A protein's function arises from and depends on its structure.

NUCLEOTIDES AND NUCLEIC ACIDS

Nucleotides are building blocks of nucleic acids; some have additional roles in metabolism. DNA stores a cell's heritable information, and RNA helps put that information to use.

In Chapter 4, you will read more about lipids and proteins as components of cell membranes. Chapter 5 explores enzymes and metabolic reactions, including phosphate-group transfers by nucleotides. Chapter 7 details how cells break down carbohydrates for energy. Chapter 8 revisits DNA structure and function, and Chapter 9 returns to protein synthesis. You will see in Chapter 10 how gene expression can be influenced by functional groups and RNA.

3.1 Fear of Frying

The human body requires only about a tablespoon of fat each day to stay healthy, but most people in developed countries eat far more than that. The average American eats about 70 pounds of fat per year, which may be part of the reason why the average American is overweight. Being overweight increases one's risk for many chronic illnesses. However, the total quantity of fat in the diet may have less impact on health than the types of fats. Fats are more than inert molecules that accumulate in strategic areas of our bodies. They are the main constituents of cell membranes, and as such they have powerful effects on cell function.

The typical fat molecule has three fatty acid tails, each a long chain of carbon atoms that can vary a bit in structure. Fats with a certain arrangement of hydrogen atoms around those carbon chains are called *trans* fats (**FIGURE 3.1**). Small amounts of *trans* fats occur naturally in red meat and dairy products. However, the main source of these fats in the American diet is an artificial food product called partially hydrogenated vegetable oil.

Hydrogenation is a manufacturing process that adds hydrogen atoms to oils in order to change them into solid fats. In 1908, Procter & Gamble Co. developed partially hydrogenated soybean oil as a substitute for the more expensive solid animal fats they had been using to make candles. However, the demand for candles began to wane as more households in the United States became wired for electricity, and P & G began to look for another way to sell its proprietary fat. Partially hydrogenated vegetable oil looks a lot like lard, so in 1911 the company began aggressively marketing it as a revolutionary new food: a solid cooking fat with a long shelf life, mild flavor, and lower cost than lard or butter.

By the mid-1950s, hydrogenated vegetable oil had become a major part of the American diet. At this writing, it can still be found in many manufactured and fast foods: stick margarines, ready-to-use frostings, french fries, cookies, crackers, cakes and pancakes, peanut butter, pies, doughnuts, muffins, chips, granola bars, breakfast bars, chocolate, microwave popcorn, pizzas, burritos, chicken nuggets, fish sticks, and so on.

For decades, hydrogenated vegetable oil was considered more healthy than animal fats because it was made from plants, but we now know otherwise. The *trans* fats in hydrogenated vegetable oils raise the level of cholesterol in our blood more than any other fat, and they directly alter the function of our arteries and veins. The effects of such changes are quite serious.

Eating as little as 2 grams per day (about 0.4 teaspoon) of hydrogenated vegetable oil measurably increases one's risk of atherosclerosis (hardening of the arteries), heart attack, and diabetes. A small serving of french fries made with hydrogenated vegetable oil contains about 5 grams of *trans* fat.

All organisms consist of the same kinds of molecules, but small differences in the way those molecules are put together can have big effects. With this concept, we introduce you to the chemistry of life. This is your chemistry. It makes you far more than the sum of your body's molecules.

FIGURE 3.1 *Trans* fats, an unhealthy food. Double bonds in the tail of most naturally occurring fatty acids are *cis*, which means that the two hydrogen atoms flanking the bond are on the same side of the carbon backbone. Hydrogenation creates abundant *trans* bonds, with hydrogen atoms on opposite sides of the tail.

3.2 Organic Molecules

✔ All of the molecules of life are built with carbon atoms.

✔ We use different models to highlight different aspects of the same molecule.

Carbon: The Stuff of Life

The same elements that make up a living body also occur in nonliving things, but their proportions differ. For example, compared to sand or seawater, a human body has a much larger proportion of carbon atoms. Why? Unlike sand or seawater, a body contains a lot of the molecules of life—complex carbohydrates and lipids, proteins, and nucleic acids—and these molecules consist of a high proportion of carbon atoms.

Compounds that consist primarily of carbon and hydrogen atoms are said to be **organic**. The term is a holdover from a time when these molecules were thought to be made only by living things, as opposed to the "inorganic" molecules that formed by nonliving processes. We now know that organic compounds were present on Earth long before organisms were, and we can also make them synthetically in laboratories.

Carbon's importance to life arises from its versatile bonding behavior. Carbon has four vacancies in its outer shell (Section 2.3), so it can form four covalent bonds with other atoms, including other carbon atoms. Many organic molecules have a backbone—a chain of carbon atoms—to which other atoms attach. The ends of a backbone may join to form a carbon ring structure (**FIGURE 3.2**). Carbon's ability to form chains and rings, and also to bond with many other elements, means that atoms of this element can be assembled into a wide variety of organic compounds.

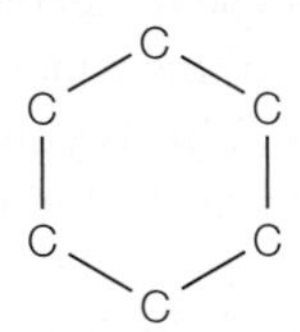

A Carbon's versatile bonding behavior allows it to form a variety of structures, including rings.

B Carbon rings form the framework of many sugars, starches, and fats, such as those found in doughnuts.

FIGURE 3.2 Carbon rings.

Modeling Organic Molecules

As you will see in the next few sections, the function of an organic molecule depends on its structure. Researchers routinely make models of organic molecules such as proteins in order to study (for example) surface properties, structure–function relationships, changes during synthesis or other biochemical processes, and molecular recognition. A molecule's structure can be modeled in various ways. The different models allow us to visualize different characteristics of the same molecule.

Structural formulas of organic molecules can be quite complex, even when the molecules are relatively small (**FIGURE 3.3A**). Thus, formulas of organic molecules are typically simplified. Hydrogen atoms and some of the bonds may not be shown, but are understood to exist where they should. Carbon ring structures such as the ones that occur in glucose and other sugars are often represented as polygons (**FIGURE 3.3B**). If no atom is shown at a corner or at the end of a bond, a carbon atom is implied there.

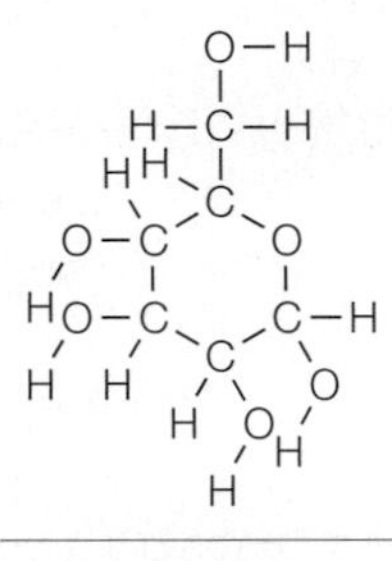

A A structural formula for an organic molecule—even a simple one—can be very complicated. The overall structure is obscured by detail.

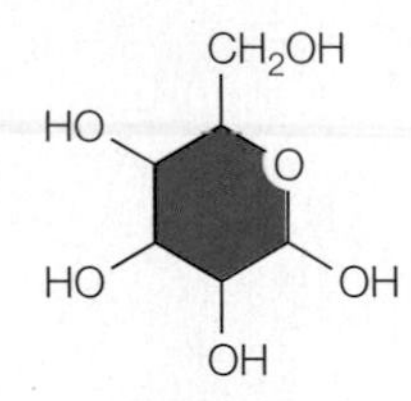

B Structural formulas of organic molecules are typically simplified by using polygons as symbols for rings, omitting some bonds and element labels.

C A ball-and-stick model is often used to show the arrangement of atoms and bonds in three dimensions.

D A space-filling model can be used to show a molecule's overall shape. Individual atoms are visible in this model.

FIGURE 3.3 Modeling an organic molecule. All of these models represent the same molecule: glucose.

organic Describes a compound that consists mainly of carbon and hydrogen atoms.

Ball-and-stick models show the positions of individual atoms in three dimensions (**FIGURE 3.3C**). Single, double, and triple covalent bonds are all shown as one stick connecting two balls, which represent atoms. Ball size reflects relative sizes of the atoms, and ball color indicates the element according to a standard code (Section 2.4).

Space-filling models represent atomic volume most accurately (**FIGURE 3.3D**). This type of model shows the overall shape of an organic molecule. Atoms in space-filling models may be color-coded by element using the same scheme as ball-and-stick models.

Many organic molecules are so large that ball-and-stick or space-filling models of them may be incomprehensible. **FIGURE 3.4** shows three different ways to represent hemoglobin, a large molecule that functions as the main oxygen carrier in your blood. Many interesting features of this molecule are not visible in the space-filling model (**FIGURE 3.4A**). Consider that a properly functioning hemoglobin molecule has embedded hemes, which are small carbon-ring structures with an iron atom at their center (Section 5.6 returns to hemes). The hemes are impossible to distinguish in a space-filling model of hemoglobin, but become visible when depicted as in **FIGURE 3.4B**, which shows a surface model. This model reveals the hemes (red sticks) within the molecule's crevices. Surface models are often used to highlight large-scale features such as charge distribution that can be difficult to distinguish in models depicting individual atoms. Other types of models further reduce visual complexity. Proteins and large nucleic acids are typically represented as ribbons that show only the carbon backbone. In a ribbon model of hemoglobin (**FIGURE 3.4C**), you can see that the molecule consists of four coiled protein components, each folded around a heme. Such structural details are clues about function: Oxygen binds at the hemes, so each hemoglobin molecule can carry up to four molecules of oxygen.

TAKE-HOME MESSAGE 3.2

How are all of the molecules of life alike?

✔ The molecules of life (carbohydrates, lipids, proteins, and nucleic acids) are organic, which means they consist mainly of carbon and hydrogen atoms.

✔ The structure of an organic molecule starts with a chain of carbon atoms (the backbone) that may form a ring.

✔ We use different models to represent different structural characteristics. Considering a molecule's structural features gives us insight into how it functions.

A The complexity of a space-filling model of hemoglobin obscures many interesting features of the molecule.

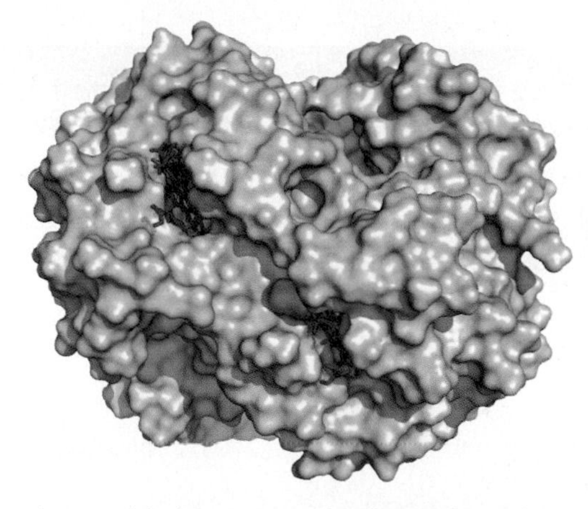

B A surface model of the same molecule reveals crevices and folds that are important for its function. Hemes, in red, are cradled in pockets of the molecule.

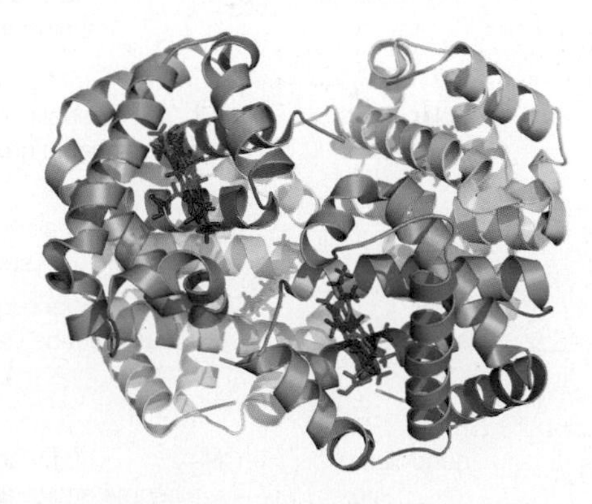

C A ribbon model of hemoglobin reveals all four hemes, also in red. The hemes are held in place by the coiled backbones of the molecule's four protein components.

FIGURE 3.4 Visualizing the structure of hemoglobin, the oxygen-transporting molecule in human blood. Models that show individual atoms usually depict them color-coded by element. Other models may be shown in various colors, depending on which features are being highlighted.

3.3 Molecules of Life—From Structure to Function

✔ How an organic molecule functions in a biological system begins with its structure.

Functional Groups

An organic molecule that consists only of hydrogen and carbon atoms is called a **hydrocarbon**. Methane, the simplest hydrocarbon, is one carbon atom bonded to four hydrogen atoms. Other organic molecules, including the molecules of life, have at least one functional group. A **functional group** is an atom (other than hydrogen) or small molecular group covalently bonded to a carbon atom of an organic compound. These groups impart chemical properties such as acidity or polarity.

The chemical behavior of the molecules of life arises mainly from the number, kind, and arrangement of their functional groups. **TABLE 3.1** lists some of the most common functional groups in these molecules. A hydroxyl group adds polar character to an organic compound, thus increasing its solubility in water. A methyl group adds nonpolar character, and may dampen the effect of a polar functional group. Methyl groups added to DNA act like an "off" switch for this molecule; acetyl groups act like an "on" switch (we return to this topic in Chapter 10). Acetyl groups also carry two carbons from one molecule to another in some metabolic reactions.

Aldehyde and ketone groups are part of simple sugars. Some sugars convert to a ring form when the highly reactive aldehyde group on one carbon of the backbone reacts with a hydroxyl group on another (**FIGURE 3.5**). Carboxyl groups make amino acids and fatty acids acidic; amine and amide groups make nucleotide bases basic.

When a phosphate group is transferred from one molecule to another, energy is transferred along with it. Bonds between sulfhydryl groups stabilize the structure of many proteins, including those that make up human hair. Heat and some kinds of chemicals can temporarily break sulfhydryl bonds, which is why we can curl straight hair and straighten curly hair.

Table 3.1
Some Functional Groups in Biological Molecules

Group	Structure	Character	Formula	Found In
acetyl	—C(=O)—CH_3	polar, acidic	$-COCH_3$	some proteins, coenzymes
aldehyde	—C—C(=O)—H	polar, reactive	—CHO	simple sugars
amide	—C(=O)—N<	weakly basic, stable, rigid	—C(O)N—	proteins, nucleotide bases
amine	—N(H)H	very basic	$-NH_2$	nucleotide bases, amino acids
carboxyl	—C—C(=O)—OH	very acidic	—COOH	fatty acids, amino acids
carbonyl	—C(=O)—	polar, reactive	—CO	alcohols, other functional groups
hydroxyl	—O—H	polar	—OH	alcohols, sugars
ketone	—C—C(=O)—C—	polar, acidic	—CO—	simple sugars, nucleotide bases
methyl	—C(H)(H)—H	nonpolar	$-CH_3$	fatty acids, some amino acids
sulfhydryl	—S—H	forms rigid disulfide bonds	—SH	cysteine, many cofactors
phosphate	—O—P(=O)(O⁻)—O⁻	polar, reactive	$-PO_4$	nucleotides, DNA, RNA phospholipids, proteins

What Cells Do to Organic Compounds

All biological systems are based on the same organic molecules, a similarity that is one of many legacies of life's common origin. However, the details of those molecules differ among organisms. Just as atoms

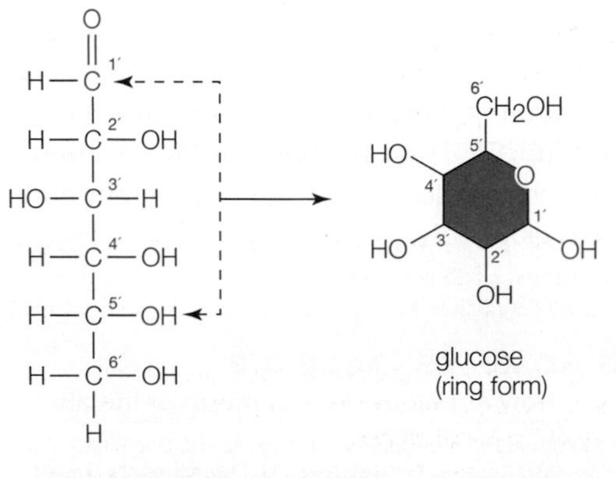

glucose (open-chain form)

glucose (ring form)

FIGURE 3.5 Glucose. This simple sugar converts from a straight-chain into a ring form when the aldehyde group (on carbon 1) reacts with a hydroxyl group (on carbon 5). In water, the cyclic structure is the more common one.

Note that the carbons in sugars such as glucose are numbered in a standard way: 1′, 2′, 3′, and so on.

A Metabolism refers to processes by which cells acquire and use energy as they make and break down molecules. Humans and other consumers break down the molecules in food. They use energy and raw materials from the breakdown to maintain themselves and to build new components.

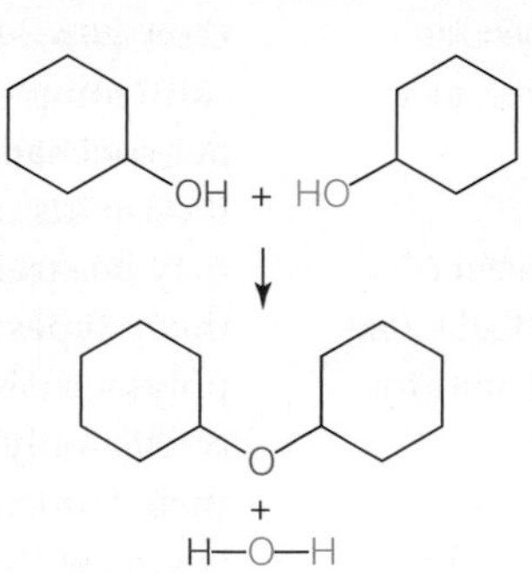

B Condensation. Cells build a large molecule from smaller ones by this reaction. An enzyme removes a hydroxyl group from one molecule and a hydrogen atom from another. A covalent bond forms between the two molecules; water also forms.

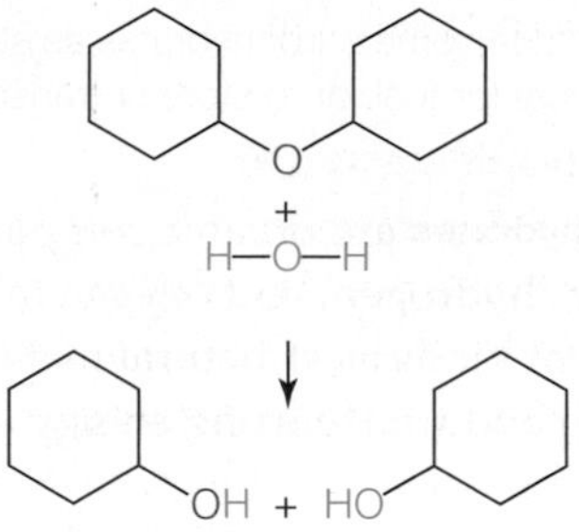

C Hydrolysis. Cells split a large molecule into smaller ones by this water-requiring reaction. An enzyme attaches a hydroxyl group and a hydrogen atom (both from water) at the cleavage site.

FIGURE 3.6 Metabolism. Two common reactions by which cells build and break down organic molecules are shown.

bonded in different numbers and arrangements form different molecules, simple organic building blocks bonded in different numbers and arrangements form different versions of the molecules of life. Cells assemble complex carbohydrates, lipids, proteins, and nucleic acids from small organic molecules. These small organic molecules—simple sugars, fatty acids, amino acids, and nucleotides—are called **monomers** when they are used as subunits of larger molecules. A molecule that consists of multiple monomers is called a **polymer**.

Cells build polymers from monomers, and break down polymers to release monomers. These and other processes of molecular change are called **reactions**. Cells constantly run reactions as they acquire and use energy to stay alive, grow, and reproduce—activities that are collectively called **metabolism** (**FIGURE 3.6A**).

Metabolism requires **enzymes**, which are organic molecules (usually proteins) that speed up reactions without being changed by them. Enzymes drive metabolic reactions in which large organic molecules are assembled from smaller ones. With **condensation**, an enzyme covalently bonds two molecules together. Water (H—O—H) forms as a product of condensation when a hydrogen atom (H—) from one of the molecules combines with a hydroxyl group (—OH) from the other molecule (**FIGURE 3.6B**). With **hydrolysis**, the reverse of condensation, an enzyme breaks apart a large organic molecule into smaller ones. During hydrolysis, a bond between two atoms breaks when a hydroxyl group gets attached to one of the atoms, and a hydrogen atom gets attached to the other (**FIGURE 3.6C**). The hydroxyl group and hydrogen atom come from a water molecule, so this reaction requires water.

We will revisit enzymes and metabolic reactions in Chapter 5. The remainder of this chapter introduces the different types of biological molecules and the monomers from which they are built.

condensation Chemical reaction in which an enzyme builds a large molecule from smaller subunits; water also forms.
enzyme Organic molecule that speeds up a reaction without being changed by it.
functional group An atom (other than hydrogen) or a small molecular group bonded to a carbon of an organic compound; imparts a specific chemical property.
hydrocarbon Compound or region of one that consists only of carbon and hydrogen atoms.
hydrolysis Water-requiring chemical reaction in which an enzyme breaks a molecule into smaller subunits.
metabolism All of the enzyme-mediated chemical reactions by which cells build and break down organic molecules.
monomers Molecules that are subunits of polymers.
polymer Molecule that consists of multiple monomers.
reaction Process of molecular change.

TAKE-HOME MESSAGE 3.3

How do organic molecules work in living systems?

- ✔ Functional groups of an organic molecule impart chemical characteristics to its hydrocarbon backbone. These groups contribute to the function of a biological molecule.
- ✔ An organic molecule's structure dictates its function in biological systems.
- ✔ All life is based on the same types of organic compounds: complex carbohydrates, lipids, proteins, and nucleic acids.
- ✔ By processes of metabolism, cells assemble the molecules of life from monomers. They also break apart polymers into component monomers.

3.4 Carbohydrates

✔ Carbohydrates are the most plentiful biological molecules.

✔ Cells use some carbohydrates as structural materials; they use others for fuel, or to store or transport energy.

Carbohydrates are organic compounds that consist of carbon, hydrogen, and oxygen in a 1:2:1 ratio. Cells use different kinds as structural materials, for fuel, and for storing and transporting energy.

Carbohydrates in Biological Systems

Simple Sugars "Saccharide" is from *sacchar*, a Greek word that means sugar. **Monosaccharides** (one sugar) are the simplest type of carbohydrate. These molecules have important biological roles. Common monosaccharides have a backbone of five or six carbon atoms, one carbonyl group (—C=O), and two or more hydroxyl groups (—OH). The functional groups, which are polar, make monosaccharides soluble (able to dissolve) in water. Thus, these molecules move easily through the water-based internal environments of all organisms.

Cells break the bonds of sugars to release energy that can be harnessed to power other reactions (we return to this important metabolic process in Chapter 7). Monosaccharides are also used as structural materials to build larger molecules, and as precursors, or parent molecules, that are remodeled into other molecules. For example, cells of plants and many animals make vitamin C from glucose, which is a monosaccharide. Human cells are unable to make vitamin C, so we need to get it from our food.

Short-Chain Carbohydrates Oligosaccharides are short chains of covalently bonded monosaccharides (*oligo–* means a few). **Disaccharides** consist of two sugar monomers. The lactose in milk, with one glucose and one galactose, is a disaccharide. Sucrose, the most plentiful sugar in nature, has a glucose and a fructose unit (FIGURE 3.7). Sucrose extracted from sugarcane or sugar beets is our table sugar. Oligosaccharides attached to lipids or proteins function in immunity.

Complex Carbohydrates Foods that we call "complex" carbohydrates consist mainly of **polysaccharides**, which are chains of hundreds or thousands of monosaccharide monomers. The chains may be straight or branched, and can have one or many types of monosaccharides. The most common polysaccharides are cellulose, starch, and glycogen. All consist only of glucose monomers, but as substances their properties are very different. Why? The answer begins with differences in patterns of covalent bonding that link their monomers.

Cellulose, the major structural material of plants, is the most abundant biological molecule on Earth. Hydrogen bonding locks its long, straight chains of covalently bonded glucose monomers into tight, sturdy bundles (FIGURE 3.8A). The bundles form tough fibers that act like reinforcing rods inside stems and other plant parts, helping these structures resist wind and other forms of mechanical stress. Cellulose is insoluble (it does not dissolve) in water, and it is not easily broken down. Some bacteria and fungi make enzymes that can break it apart into its component sugars, but humans and other mammals do not. Dietary fiber, or "roughage," usually refers to the cellulose in our vegetable foods. Bacteria that live in the guts of termites and grazers such as cattle and sheep help these animals digest the cellulose in plants.

In **starch**, a different covalent bonding pattern between glucose monomers makes a chain that coils up into a spiral (FIGURE 3.8B). Like cellulose, starch does not dissolve readily in water, but it is easier to break down than cellulose. These properties make the molecule ideal for storing sugars in the watery, enzyme-filled interior of plant cells. Most plant leaves make glucose during the day, and their cells store it by building starch. At night, hydrolysis enzymes break the bonds between starch's glucose monomers. The released glucose can be broken down immediately for energy, or converted to sucrose that is transported to other parts of the plant. Humans also have hydrolysis

glucose + fructose → sucrose + water

FIGURE 3.7 ▶Animated The synthesis of a sucrose molecule is an example of a condensation reaction. You are already familiar with sucrose—it is common table sugar.

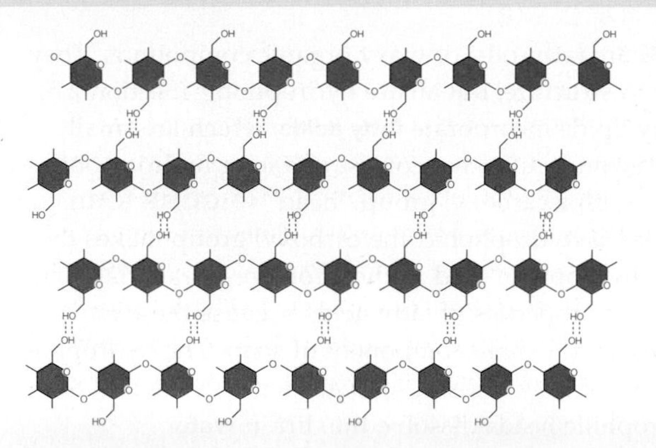

A Cellulose
Cellulose is the main structural component of plants. Above, in cellulose, chains of glucose monomers stretch side by side and hydrogen-bond at many —OH groups. The hydrogen bonds stabilize the chains in tight bundles that form long fibers. Few types of organisms can digest this tough, insoluble material.

B Starch
Starch is the main energy reserve in plants, which store it in their roots, stems, leaves, seeds, and fruits. Below, in starch, a series of glucose monomers form a chain that coils up.

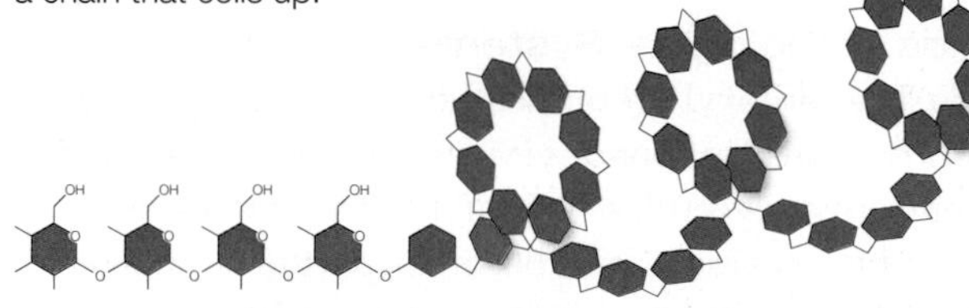

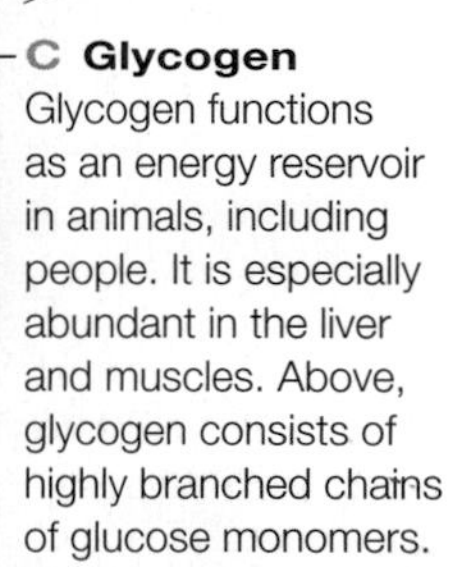

C Glycogen
Glycogen functions as an energy reservoir in animals, including people. It is especially abundant in the liver and muscles. Above, glycogen consists of highly branched chains of glucose monomers.

FIGURE 3.8 ▸**Animated** Three of the most common complex carbohydrates and their locations in a few organisms. Each polysaccharide consists only of glucose subunits, but different bonding patterns result in substances with very different properties.

enzymes that break down starch, so this carbohydrate is an important component of our food.

Animals store sugars in the form of **glycogen**, a polysaccharide that consists of highly branched chains of glucose monomers (**FIGURE 3.8C**). Muscle and liver cells contain most of the body's glycogen. When the blood sugar level falls, liver cells break down the glycogen, and the released glucose subunits enter the blood.

In chitin, a polysaccharide similar to cellulose, long, unbranching chains of nitrogen-containing monomers are linked by hydrogen bonds (**FIGURE 3.9**). As a structural material, chitin is durable, translucent, and flexible. It strengthens hard parts of many animals, including the outer cuticle of crustaceans, beetles, and ticks, and it reinforces the cell wall of many fungi.

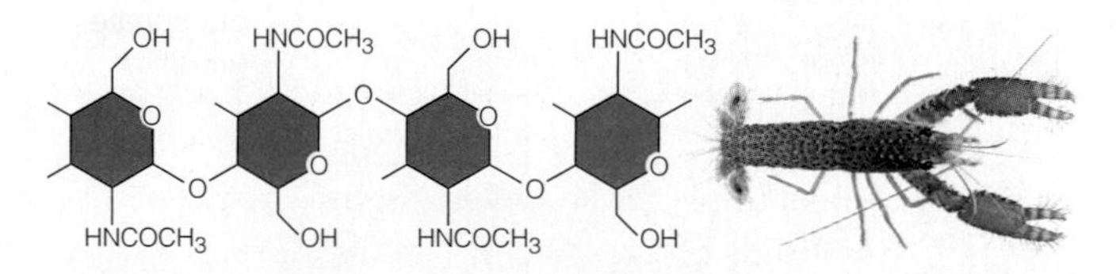

FIGURE 3.9 Chitin. This polysaccharide strengthens the hard parts of many small animals such as lobsters.

carbohydrate Molecule that consists primarily of carbon, hydrogen, and oxygen atoms in a 1:2:1 ratio.
cellulose Tough, insoluble polysaccharide that is the major structural material in plants.
disaccharide Polymer of two sugar subunits.
glycogen Polysaccharide; energy reservoir in animal cells.
monosaccharide Simple sugar; monomer of polysaccharides.
polysaccharide Polymer of many monosaccharides.
starch Polysaccharide; energy reservoir in plant cells.

TAKE-HOME MESSAGE 3.4
What is a carbohydrate?

✔ Cells use simple carbohydrates (sugars) for energy and to build other molecules.

✔ Glucose monomers, bonded different ways, form complex carbohydrates such as cellulose, starch, and glycogen.

3.5 Lipids

✔ Triglycerides, phospholipids, waxes, and steroids are lipids common in biological systems.

hydrophilic "head" (acidic carboxyl group)

hydrophobic "tail"

A stearic acid (saturated)

B linoleic acid (omega-6)

C linolenic acid (omega-3)

FIGURE 3.10 ▸**Animated** Fatty acids. **A** The tail of stearic acid is fully saturated with hydrogen atoms. **B** Linoleic acid, with two double bonds, is unsaturated. The first double bond occurs at the sixth carbon from the end, so linoleic acid is called an omega-6 fatty acid. Omega-6 and **C** omega-3 fatty acids are "essential fatty acids." Your body does not make them, so they must come from food.

Lipids are fatty, oily, or waxy organic compounds. They vary in structure, but all are hydrophobic (Section 2.5). Many lipids incorporate **fatty acids**, which are small organic molecules that consist of a long hydrocarbon "tail" with a carboxyl group "head" (**FIGURE 3.10**). The tail is hydrophobic; the carboxyl group makes the head hydrophilic (and acidic). You are already familiar with the properties of fatty acids because these molecules are the main component of soap. The hydrophobic tails of fatty acids in soap attract oily dirt, and the hydrophilic heads dissolve the dirt in water.

Saturated fatty acids have only single bonds linking the carbons in their tails. In other words, their carbon chains are fully saturated with hydrogen atoms (**FIGURE 3.10A**). Saturated fatty acid tails are flexible and they wiggle freely. Double bonds between carbons limit the flexibility of the tails of **unsaturated fatty acids** (**FIGURE 3.10B,C**). **FIGURE 3.1** shows how these bonds are *cis* or *trans*, depending on the way the hydrogens are arranged around them.

Lipids in Biological Systems

Fats The carboxyl group head of a fatty acid can easily form a covalent bond with another molecule. When it bonds to a glycerol, a type of alcohol, it loses its hydrophilic character and becomes part of a fat. **Fats** are lipids with one, two, or three fatty acids bonded to the same glycerol. A fat with three fatty acid tails is called a **triglyceride** (**FIGURE 3.11A**). Triglycerides are entirely hydrophobic, so they do not dissolve in water. Most "neutral" fats, such as butter and vegetable oils, are examples. Triglycerides are the most abundant and richest energy source in vertebrate bodies. Gram for gram, these fats store more energy than carbohydrates.

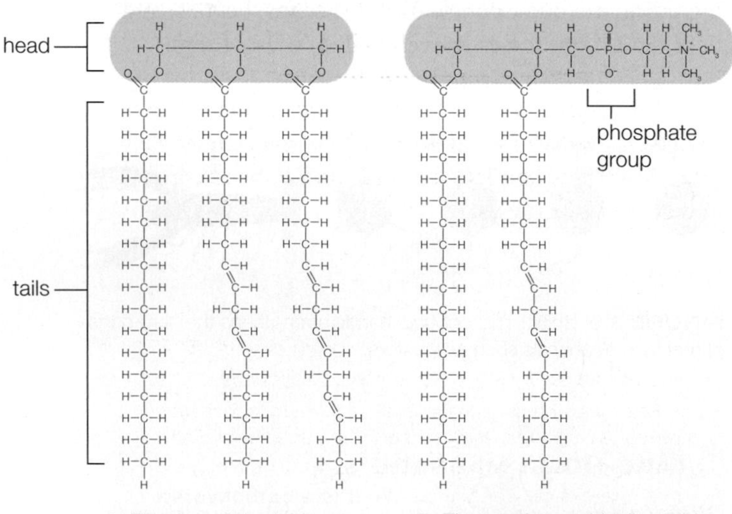

A The three fatty acid tails of a triglyceride are attached to a glycerol head.

B The two fatty acid tails of this phospholipid are attached to a phosphate-containing head.

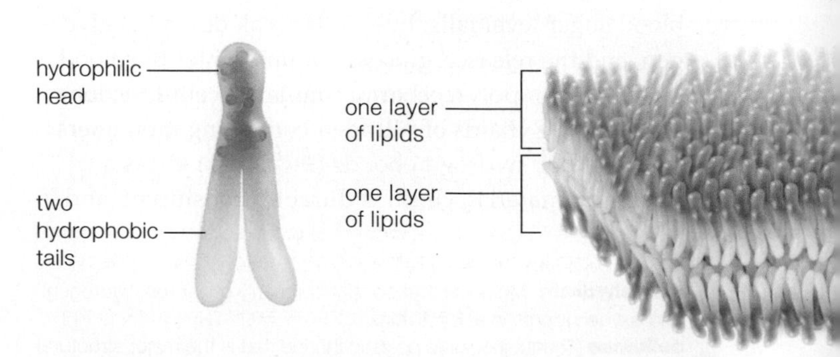

C A double layer of phospholipids—the lipid bilayer—is the structural foundation of all cell membranes. You will read more about the structure of cell membranes in Chapter 5.

FIGURE 3.11 ▸**Animated** Lipids with fatty acid tails. **FIGURE IT OUT** Is the triglyceride saturated or unsaturated?

Answer: Unsaturated

Butter, cream, and other high-fat animal products have a high proportion of saturated fats, which means they consist mainly of triglycerides with three saturated fatty acid tails. Saturated fats tend to be solid at room temperature because their floppy saturated tails can pack tightly. Most vegetable oils are unsaturated fats, which means they consist mainly of triglycerides with one or more unsaturated fatty acid tails. Each double bond in a fatty acid tail makes a rigid kink. Kinky tails do not pack tightly, so unsaturated fats are typically liquid at room temperature. The partially hydrogenated vegetable oils that you learned about in Section 3.1 are an exception. These fats are solid at room temperature because the special *trans* double bond keeps fatty acid tails straight, allowing the fat molecules to pack tightly just like saturated fats do.

Phospholipids A **phospholipid** has a phosphate-containing head and two long hydrocarbon tails that are typically derived from fatty acids (**FIGURE 3.11B**). The tails are hydrophobic, but the highly polar phosphate group makes the head hydrophilic. These opposing properties give rise to the basic structure of cell membranes, which consist mainly of phospholipids. In a cell membrane, phospholipids are arranged in two layers—a **lipid bilayer** (**FIGURE 3.11C**). The heads of one layer are dissolved in the cell's watery interior, and the heads of the other layer are dissolved in the cell's fluid surroundings. All of the hydrophobic tails are sandwiched between the hydrophilic heads. You will read more about the structure of cell membranes in Chapters 4 and 5.

Waxes A **wax** is a complex, varying mixture of lipids with long fatty acid tails bonded to long-chain alcohols or carbon rings. The molecules pack tightly, so waxes are firm and water-repellent. Plants secrete waxes onto their exposed surfaces to restrict water loss and keep out parasites and other pests. Other types of waxes protect, lubricate, and soften skin and hair. Waxes, together with fats and fatty acids, make feathers waterproof. Bees store honey and raise new generations of bees inside a honeycomb of secreted beeswax.

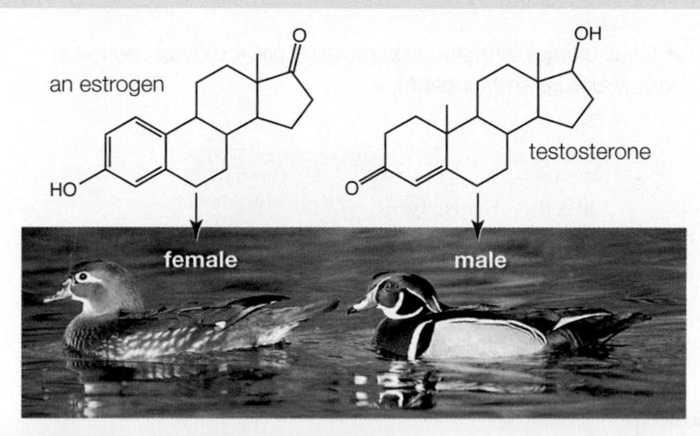

FIGURE 3.12 Steroids. Estrogen and testosterone are steroid hormones that govern reproduction and secondary sexual traits. The two hormones are the source of gender-specific traits in many species, including these wood ducks.

Steroids **Steroids** are lipids with no fatty acid tails; they have a rigid backbone that consists of twenty carbon atoms arranged in a characteristic pattern of four rings (**FIGURE 3.12**). These molecules serve varied and important physiological functions in plants, fungi, and animals. Functional groups attached to the rings define the type of steroid. Cholesterol, the most common steroid in animal tissue, is a precursor for many other molecules, including bile salts (which help digest fats), vitamin D (required to keep teeth and bones strong), and steroid hormones.

TAKE-HOME MESSAGE 3.5

What are lipids?

- ✔ Lipids are fatty, waxy, or oily organic compounds.
- ✔ Fats have one, two, or three fatty acid tails; triglyceride fats are an important energy reservoir in vertebrate animals.
- ✔ Phospholipids arranged in a lipid bilayer are the main component of cell membranes.
- ✔ Waxes have complex, varying structures. They are components of water-repelling and lubricating secretions.
- ✔ Steroids serve varied and important physiological roles in plants, fungi, and animals.

fat Lipid that consists of a glycerol molecule with one, two, or three fatty acid tails. Saturated fats have three saturated fatty acid tails. Unsaturated fats have one or more unsaturated fatty acid tails.
fatty acid Organic compound that consists of an acidic carboxyl group "head" and a long hydrocarbon "tail."
lipid Fatty, oily, or waxy organic compound.
lipid bilayer Double layer of lipids arranged tail-to-tail; structural foundation of cell membranes.
phospholipid A lipid with a phosphate group in its hydrophilic head, and two nonpolar fatty acid tails; main constituent of eukaryotic cell membranes.
saturated fatty acid Fatty acid with only single bonds linking the carbons in its tail.
steroid Type of lipid with four carbon rings and no fatty acid tails.
triglyceride A fat with three fatty acid tails.
unsaturated fatty acid Fatty acid with one or more carbon–carbon double bonds in its tail.
wax Water-repellent mixture of lipids with long fatty acid tails bonded to long-chain alcohols or carbon rings.

CREDITS: (12) top, © Cengage Learning; bottom, Tim Davis/Science Source; (in text) Kenneth Lorenzen.

3.6 Proteins

✔ Of all biological molecules, proteins are the most diverse in both structure and function.

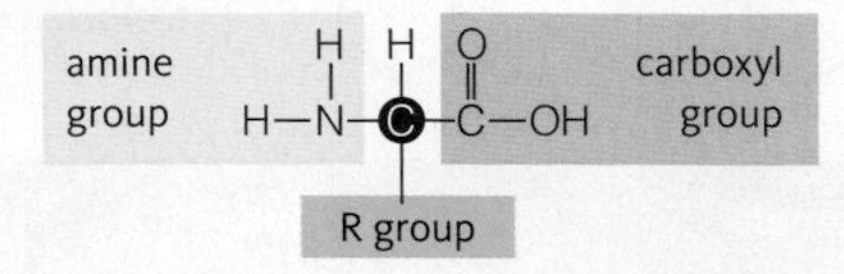

FIGURE 3.13 Generalized structure of an amino acid. See Appendix II for the complete structures of the twenty most common amino acids found in eukaryotic proteins.

Proteins participate in all processes that sustain life. Structural proteins support cell parts and, as part of tissues, multicelled bodies. Most enzymes that drive metabolic reactions are proteins. Proteins move substances, help cells communicate, and defend the body.

With a few exceptions, cells can make all of the thousands of different proteins they need from only twenty kinds of amino acid monomers. An **amino acid** is a small organic compound with an amine group (—NH_2), a carboxyl group (—COOH, the acid), and a side chain called an "R group" that defines the kind of amino acid. In most amino acids, all three groups are attached to the same carbon atom (**FIGURE 3.13**).

The covalent bond that links amino acids in a protein is called a **peptide bond**. During protein synthesis, a peptide bond forms between the carboxyl group of the first amino acid and the amine group of the second (**FIGURE 3.14** ❶). Another peptide bond links a third amino acid to the second, and so on (you will learn more about the details of protein synthesis in Chapter 9). A short chain of amino acids is called a **peptide**; as the chain lengthens, it becomes a **polypeptide**. **Proteins** consist of polypeptides that can be hundreds or even thousands of amino acids long.

The idea that structure dictates function is particularly appropriate as applied to proteins, because the diversity in biological activity among these molecules arises from differences in their three-dimensional shape. Protein structure begins with the linear series of amino acids composing a polypeptide chain ❷. The order of the amino acids in the chain, which is called primary structure, defines the type of protein. The molecule begins to take on shape during protein synthesis, when hydrogen bonds that form between amino acids cause the lengthening polypeptide chain to twist and fold. Hydrogen bonding holds sections of the polypeptide in loops, helices (coils), or flat sheets, and these patterns constitute the protein's secondary structure ❸. The primary structure of each type of protein is unique, but most proteins have similar patterns of secondary structure.

Much as an overly twisted rubber band coils back upon itself, hydrogen bonding also makes the loops, helices, and sheets of a protein fold up into even more compact domains (**FIGURE 3.15A**). These domains are called tertiary structure ❹. Tertiary structure is what makes a protein a working molecule. For example, the helices and loops in a globin chain fold up together to form a pocket that can hold a heme, which is a small compound essential to the finished protein's function. In other proteins, sheets, loops, and helices come together as complex structures that resemble barrels, propellers, sandwiches, and so on. Barrel domains often form tunnels through cell membranes, allowing small molecules to cross. Some proteins have barrel domains that rotate like motors in small molecular machines (**FIGURE 3.15B**). A protein may have several domains, each contributing a particular structural or functional property to the molecule.

FIGURE 3.14 ▶**Animated** Protein structure.

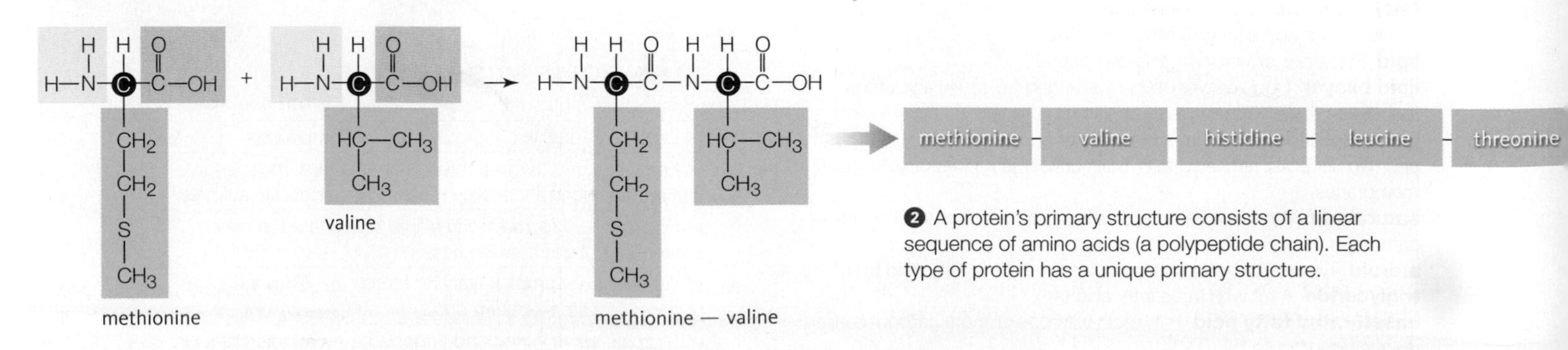

❷ A protein's primary structure consists of a linear sequence of amino acids (a polypeptide chain). Each type of protein has a unique primary structure.

❶ A condensation reaction joins the carboxyl group of one amino acid and the amine group of another to form a peptide bond. In this example, a peptide bond forms between the amino acids methionine and valine.

A In this protein, loops (green), coils (red), and a sheet (yellow) fold up together into a chemically active pocket. The pocket gives this protein the ability to transfer electrons from one molecule to another. Many other proteins have the same pocket structure.

B This barrel domain is part of a rotary mechanism in a larger protein. The protein functions as a molecular motor that pumps hydrogen ions through cell membranes.

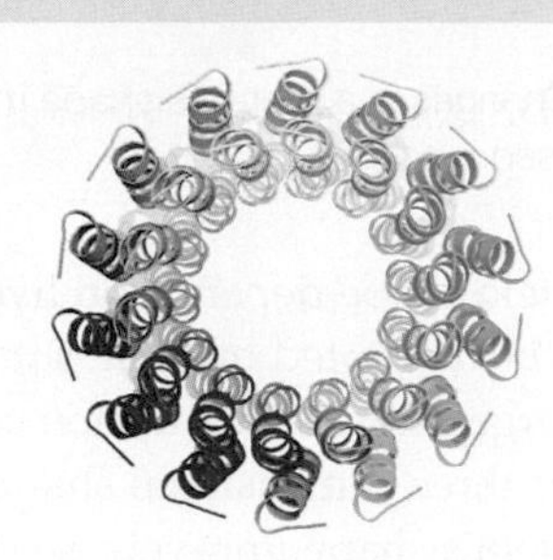

FIGURE 3.15 Examples of domains in proteins.

Many proteins also have quaternary structure, which means they consist of two or more polypeptide chains that are closely associated or covalently bonded together. Hemoglobin is like this ❺. So are most enzymes, which have multiple polypeptide chains that collectively form a roughly spherical shape.

Fibrous proteins aggregate by many thousands into much larger structures, with their polypeptide chains organized into strands or sheets. The keratin in your hair is an example ❻. Some fibrous proteins contribute to the structure and organization of cells and tissues. Others, such as the actin and myosin filaments in muscle cells, are part of the mechanisms that help cells, cell parts, and multicelled bodies move.

Carbohydrates, lipids, or both may get attached to a protein after synthesis. A protein with one or more oligosaccharides attached to it is called a glycoprotein. Molecules that allow a tissue or a body to recognize its own cells are glycoproteins, as are other molecules that help cells interact in immunity. A protein with one or more lipids attached to it is called a lipoprotein. Some lipoproteins are aggregate structures that consist of variable amounts and types of proteins and lipids (**FIGURE 3.16**).

amino acid Small organic compound that is a subunit of proteins. Consists of a carboxyl group, an amine group, and a characteristic side group (R), all typically bonded to the same carbon atom.
peptide Short chain of amino acids linked by peptide bonds.
peptide bond A bond between the amine group of one amino acid and the carboxyl group of another. Joins amino acids in proteins.
polypeptide Long chain of amino acids linked by peptide bonds.
protein Organic molecule that consists of one or more polypeptides.

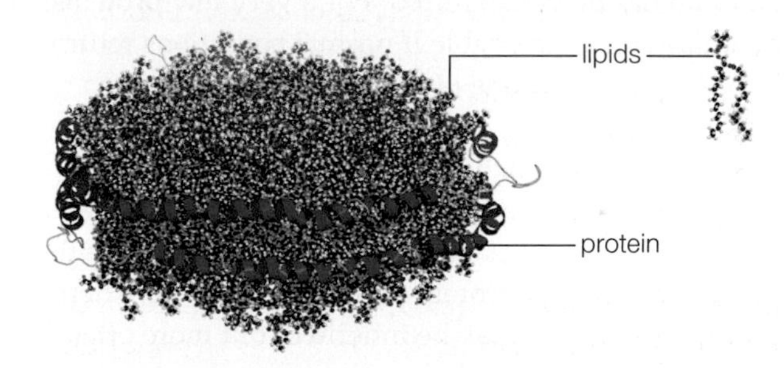

FIGURE 3.16 A lipoprotein particle. The one depicted here (HDL, which is often called "good" cholesterol) consists of thousands of lipids lassoed into a clump by two proteins.

TAKE-HOME MESSAGE 3.6

What is a protein?

✔ A protein is a chain of amino acids. The order of amino acids in a polypeptide chain dictates the type of protein.

✔ Polypeptide chains twist and fold into coils, sheets, and loops, which fold and pack further into functional domains.

✔ A protein's function arises from its shape.

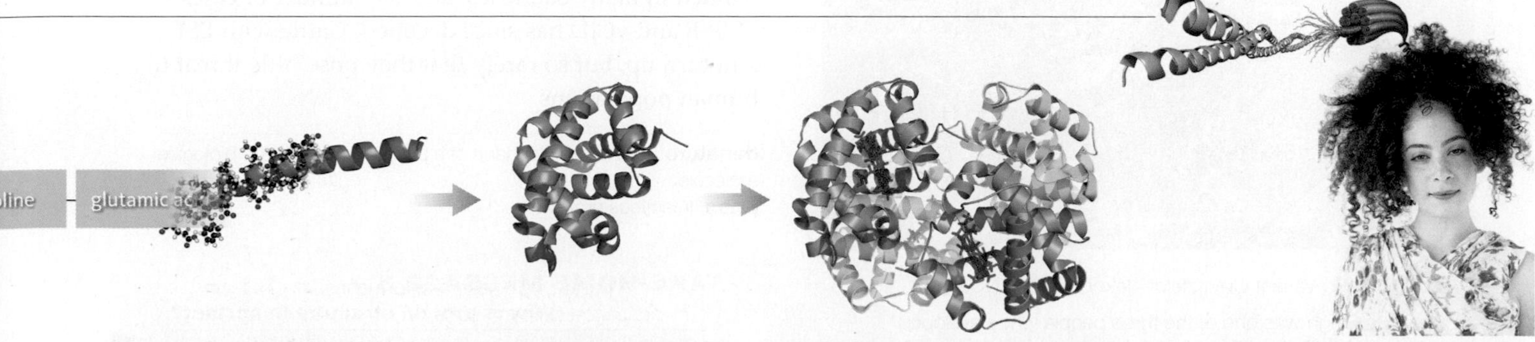

❸ Secondary structure arises as a polypeptide chain twists into a helix (coil), loop, or sheet held in place by hydrogen bonds.

❹ Tertiary structure arises when loops, helices, and sheets fold up into a domain. In this example, the helices of a globin chain form a pocket.

❺ Many proteins have two or more polypeptide chains (quaternary structure). Hemoglobin, shown here, consists of four globin chains (green and blue). Each globin pocket now holds a heme group (red).

❻ Some types of proteins aggregate into much larger structures. As an example, organized arrays of keratin, a fibrous protein, compose filaments that make up your hair.

CREDITS: (14) #3–5: 1BBB, A third quaternary structure of human hemoglobin A at 1.7-A resolution. Silva, M.M., Rogers, P.H., Arnone, A., Journal: (1992) *J.Biol.Chem.* 267: 17248–17256; #6: © JupiterImages Corporation. (15A, 16) Castrignanò T, De Meo PD, Cozzetto D, Talamo IG, Tramontano A. (2006). The PMDB Protein Model Database. Nucleic Acids Research, 34: D306-D309. (15B) pdb ID2W5J, Vollmar, M., Shlieper, D., Winn M., Buechner, C., Groth, G. "Structure of the C14 rotor ring of the proton translocating chloroplast ATP synthase." (2009) *J. Biol. Chem.* 284:18228.

3.7 Why Is Protein Structure So Important?

✔ Changes in a protein's shape may have drastic health consequences.

Protein shape depends on hydrogen bonding, which can be disrupted by heat, some salts, shifts in pH, or detergents. Such disruption can cause proteins to lose their three-dimensional shape, or **denature**. Once a protein's shape unravels, so does its function.

You can see denaturation in action when you cook an egg. A protein called albumin is a major component of egg white. Cooking does not disrupt the covalent bonds of albumin's primary structure, but it does destroy the hydrogen bonds that maintain the protein's shape. When a translucent egg white turns opaque, the albumin has been denatured. For a very few proteins, denaturation is reversible if normal conditions return, but albumin is not one of them. There is no way to uncook an egg.

Prion diseases such as mad cow disease (bovine spongiform encephalitis, or BSE) in cattle, Creutzfeldt–Jakob disease in humans, and scrapie in sheep, are the dire aftermath of a protein that changes shape. These infectious diseases may be inherited, but more often they arise spontaneously. All are characterized by relentless deterioration of mental and physical abilities that eventually causes death (**FIGURE 3.17A**).

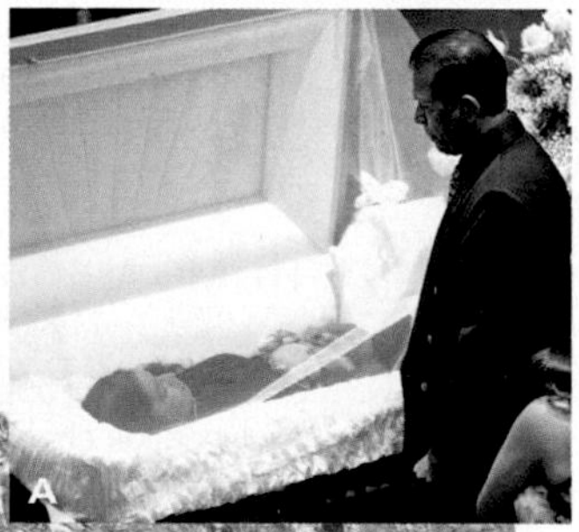

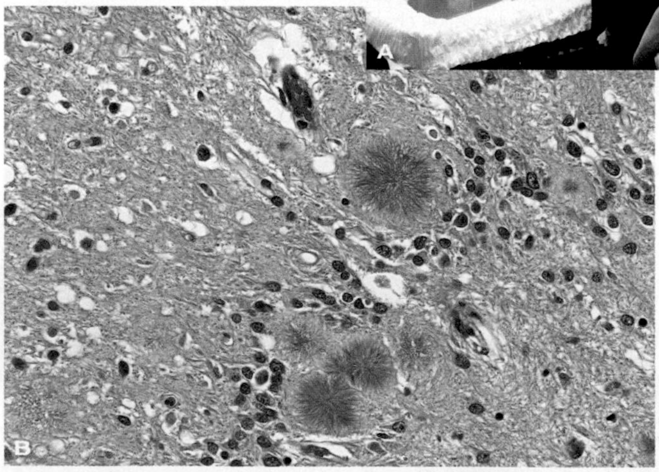

FIGURE 3.17 Variant Creutzfeldt–Jakob disease (vCJD).

A Charlene Singh was one of the three people who developed symptoms of vCJD disease while living in the United States. Like the others, Singh most likely contracted the disease elsewhere; she spent her childhood in Britain. Diagnosed in 2001, she died in 2004.

B Slice of brain tissue from a person with vCJD. Fibers of prion proteins (amyloid fibrils) radiating from several deposits are visible.

All prion diseases begin with a glycoprotein called PrPC that occurs normally in cell membranes of the mammalian body. This protein is especially abundant in brain cells, but we still know very little about what it does. Sometimes, a PrPC protein misfolds so that part of the molecule forms a sheet instead of a helix. One misfolded molecule should not pose much of a threat, but when this particular protein misfolds it becomes a **prion**, or infectious protein. The shape of a misfolded PrPC protein causes normally folded PrPC proteins to misfold too. Each protein that misfolds becomes infectious, so the number of prions increases exponentially.

The shape of misfolded PrPC proteins allows them to align tightly into long, thin, insoluble fibers, which are called amyloid fibrils. Amyloid fibrils grow in patches from their ends as more PrPC proteins misfold (**FIGURE 3.17B**). These patches disrupt normal brain function, causing symptoms such as confusion, memory loss, and lack of coordination. Holes form in the brain as its cells die. Eventually, the brain becomes so riddled with holes that it looks like a sponge.

In the mid-1980s, an epidemic of mad cow disease in Britain was followed by an outbreak of a new variant of Creutzfeldt–Jakob disease (vCJD) in humans. Researchers isolated a prion similar to the one in scrapie-infected sheep from cows with BSE, and also from humans affected by the new type of Creutzfeldt–Jakob disease. How did the prion get from sheep to cattle to people? Prions resist denaturation, so treatments such as cooking that inactivate other types of infectious agents have little effect on them. The cattle became infected by the prion after eating feed prepared from the remains of scrapie-infected sheep, and people became infected by eating beef from the infected cattle.

Two hundred people have died from vCJD since 1990. The use of animal parts in livestock feed is now banned in many countries, and the number of cases of BSE and vCJD has since declined. Cattle with BSE still turn up, but so rarely that they pose little threat to human populations.

denature To unravel the shape of a protein or other large biological molecule.
prion Infectious protein.

TAKE-HOME MESSAGE 3.7
Why is protein structure important?

✔ Protein shape can be unraveled by heat or other conditions that disrupt hydrogen bonding.

✔ A protein's function depends on its shape, so conditions that alter a protein's shape also alter its function.

3.8 Nucleic Acids

✔ DNA and RNA consist of nucleotides.

✔ Some nucleotides also have roles in metabolism.

Nucleotides are small organic molecules that function as energy carriers, enzyme helpers, chemical messengers, and subunits of DNA and RNA. Each consists of a monosaccharide ring bonded to a nitrogen-containing base and one, two, or three phosphate groups (**FIGURE 3.18**). The monosaccharide is a five-carbon sugar, either ribose or deoxyribose; the base, one of five compounds with a flat ring structure (we return to the structure of nucleotide bases in Sections 8.3 and 9.2). When the third phosphate group of a nucleotide is transferred to another molecule, energy is transferred along with it. You will read about such phosphate-group transfers and their important metabolic role in Chapter 5. The nucleotide **ATP** (adenosine triphosphate) serves an especially important role as an energy carrier in cells.

Nucleic acids are polymers, chains of nucleotides in which the sugar of one nucleotide is joined to the phosphate group of the next (**FIGURE 3.19A**). An example is **RNA**, or ribonucleic acid, named after the ribose sugar of its component nucleotides. An RNA molecule is a chain of four kinds of nucleotide monomers, one of which is ATP. RNA molecules carry out protein synthesis, which we discuss in detail in Chapter 9.

DNA, or deoxyribonucleic acid, is a nucleic acid named after the deoxyribose sugar of its component nucleotides. A DNA molecule consists of two chains of nucleotides twisted into a double helix (**FIGURE 3.19B**). Hydrogen bonds between the nucleotides hold the two chains together. Each cell starts life with DNA inherited from a parent cell. That DNA contains all of the information necessary to build a new cell and, in the case of multicelled organisms, an entire individual. The cell uses the order of nucleotide bases in DNA—the DNA sequence—to guide production of RNA and proteins. Parts of the sequence are identical or nearly so in all organisms, and parts are unique to a species or an individual (Chapter 8 returns to DNA structure and function).

ATP Adenosine triphosphate. Nucleotide that consists of an adenine base, a ribose sugar, and three phosphate groups; serves an important role as an energy carrier in cells.
DNA Deoxyribonucleic acid. Nucleic acid that consists of two chains of nucleotides twisted into a double helix; carries hereditary information.
nucleic acid Polymer of nucleotides; DNA or RNA.
nucleotide Monomer of nucleic acids; has a five-carbon sugar, a nitrogen-containing base, and one, two, or three phosphate groups.
RNA Ribonucleic acid. Single-stranded chain of nucleotides. Some types have roles in protein synthesis.

Fear of Frying (revisited)

Trans fatty acids are relatively rare in unprocessed foods, so it makes sense from an evolutionary standpoint that our bodies may not have enzymes to deal with them efficiently. The enzymes that hydrolyze *cis* fatty acids have difficulty breaking down *trans* fatty acids, a problem that may be a factor in the ill effects of *trans* fats. All prepackaged foods in the United States are now required to list *trans* fat content, but may be marked "zero grams of *trans* fats" even when a single serving contains up to half a gram.

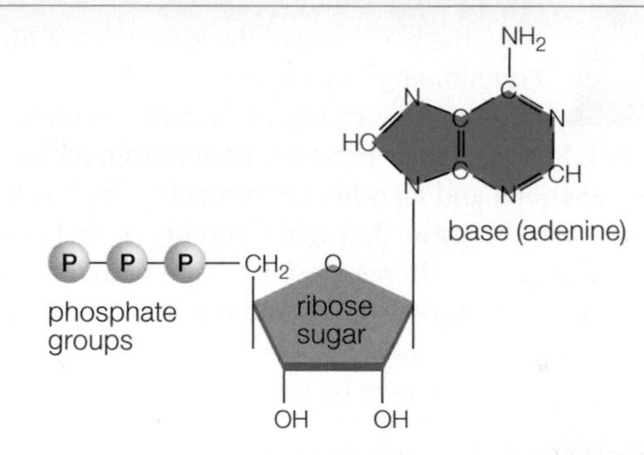

FIGURE 3.18 Example of a nucleotide: ATP. ATP is a monomer of RNA, and also a participant in many metabolic reactions.

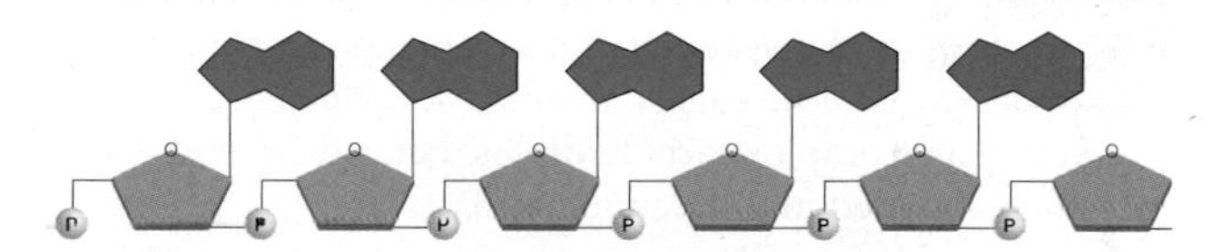

A A chain of nucleotides is a nucleic acid. The sugar of one nucleotide is covalently bonded to the phosphate group of the next, forming a sugar–phosphate backbone.

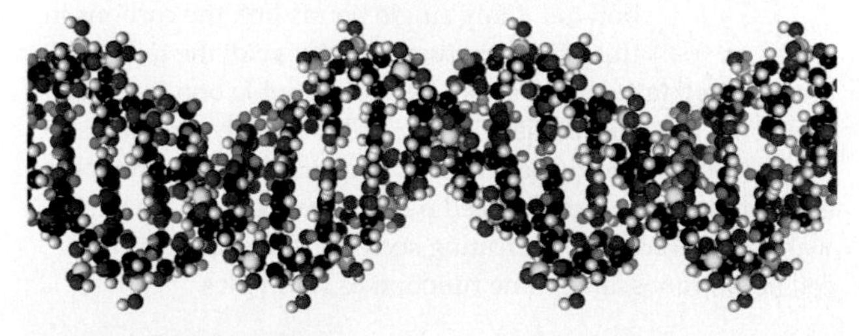

B DNA consists of two chains of nucleotides, twisted into a double helix. Hydrogen bonding maintains the three-dimensional structure of this nucleic acid.

FIGURE 3.19 Nucleic acid structure.

TAKE-HOME MESSAGE 3.8

What are nucleotides and nucleic acids?

✔ Nucleotides are monomers of nucleic acids. ATP has an important metabolic role as an energy carrier.

✔ The nucleic acid DNA holds information necessary to build cells and multicelled individuals.

✔ RNAs are nucleic acids that carry out protein synthesis.

summary

Section 3.1 All organisms consist of the same kinds of molecules. Seemingly small differences in the way those molecules are put together can have big effects inside a living organism.

Section 3.2 Molecules of life—complex carbohydrates and lipids, proteins, and nucleic acids—consist mainly of carbon and hydrogen atoms, so they are **organic**. Different models reveal different aspects of their structure.

Section 3.3 **Hydrocarbons** have only carbon and hydrogen atoms. Carbon chains or rings form the backbone of the molecules of life. **Functional groups** attached to the backbone influence the chemical character of these compounds, and thus their function. **Metabolism** includes chemical **reactions** and all other processes by which cells acquire and use energy as they make and break the bonds of organic compounds. In reactions such as **condensation**, **enzymes** build **polymers** from **monomers** of simple sugars, fatty acids, amino acids, and nucleotides. Reactions such as **hydrolysis** release monomers by breaking apart polymers.

Section 3.4 Cells use different kinds of **carbohydrates** for energy, and as structural materials. Enzymes assemble **disaccharides** and **polysaccharides** from **monosaccharide** (simple sugar) monomers. **Cellulose, glycogen**, and **starch** are complex carbohydrates that consist of glucose monomers bonded in different patterns.

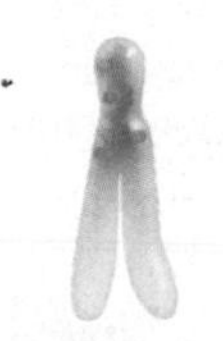

Section 3.5 **Lipids** are fatty, oily, or waxy compounds. All are nonpolar. A **fatty acid** is a lipid with an acidic head and a long hydrocarbon tail. Only single bonds link the carbons in the tail of a **saturated fatty acid**; the tail of an **unsaturated fatty acid** has one or more double bonds. **Fats** and some other lipids have fatty acid tails; **triglycerides** have three. A **lipid bilayer** (that consists primarily of **phospholipids**) is the basic structure of all cell membranes. **Waxes** are part of water-repellent and lubricating secretions. **Steroids** occur in cell membranes, and some function as hormones.

Section 3.6 Structurally and functionally, **proteins** are the most diverse molecules of life. The shape of a protein is the source of its function. Protein structure begins as a sequence of **amino acids** linked by **peptide bonds** into a **peptide**, then a **polypeptide** (primary structure). Polypeptides twist into loops, helices, and sheets (secondary structure) that can pack further into functional domains (tertiary structure). Many proteins, including most enzymes, consist of two or more polypeptides (quaternary structure). Fibrous proteins aggregate into much larger structures.

Section 3.7 A protein's structure dictates its function, so changes in a protein's structure may also alter its function.

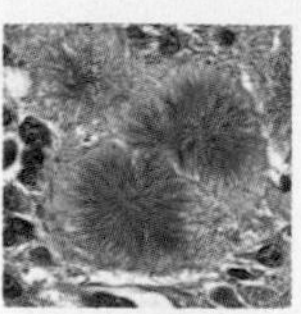

Hydrogen bonds that stabilize a protein's shape may be disrupted by shifts in pH or temperature, or exposure to detergent or some salts. If that happens, the protein unravels, or **denatures**, and so loses its function. **Prion** diseases are a fatal consequence of misfolded proteins.

Section 3.8 **Nucleotides** are small organic molecules that consist of a five-carbon sugar; a nitrogen-containing base; and one, two, or three phosphate groups. Nucleotides are monomers of **DNA** and **RNA**, which are **nucleic acids**. Some, especially **ATP**, have additional functions such as carrying energy. DNA encodes heritable information; RNAs carry out protein synthesis.

self-quiz

Answers in Appendix VII

1. Organic molecules consist of mainly of ______ atoms.
   a. carbon
   b. carbon and oxygen
   c. carbon and hydrogen
   d. carbon and nitrogen

2. Each carbon atom can bond with as many as ______ other atom(s).

3. ______ groups are the "acid" part of amino acids and fatty acids.
   a. Hydroxyl (—OH)
   b. Carboxyl (—COOH)
   c. Methyl (—CH_3)
   d. Phosphate (—PO_4)

4. ______ is a simple sugar (a monosaccharide).
   a. Glucose
   b. Sucrose
   c. Ribose
   d. Starch
   e. both a and c
   f. a, b, and c

5. Which three carbohydrates can be built using only glucose monomers?
   a. Starch, cellulose, and glycogen
   b. Glucose, sucrose, and ribose
   c. Cellulose, steroids, and polysaccharides
   d. Starch, chitin, and DNA
   e. Triglycerides, nucleic acids, and polypeptides

6. Unlike saturated fats, the fatty acid tails of unsaturated fats incorporate one or more ______ .
   a. phosphate groups
   b. glycerols
   c. double bonds
   d. single bonds

7. Is this statement true or false? Unlike saturated fats, all unsaturated fats are beneficial to health because their fatty acid tails kink and do not pack together.

8. Steroids are among the lipids with no ______ .
   a. double bonds
   b. fatty acid tails
   c. hydrogens
   d. carbons

9. Which of the following is a class of molecules that encompasses all of the other molecules listed?
   a. triglycerides
   b. fatty acids
   c. waxes
   d. steroids
   e. lipids
   f. phospholipids

Effects of Dietary Fats on Lipoprotein Levels Cholesterol that is made by the liver or that enters the body from food cannot dissolve in blood, so it is carried through the bloodstream by lipoproteins. Low-density lipoprotein (LDL) carries cholesterol to body tissues such as artery walls, where it can form deposits associated with cardiovascular disease. Thus, LDL is often called "bad" cholesterol. High-density lipoprotein (HDL, **FIGURE 3.16**) carries cholesterol away from tissues to the liver for disposal, so HDL is often called "good" cholesterol.

In 1990, Ronald Mensink and Martijn Katan published a study that tested the effects of different dietary fats on blood lipoprotein levels. Their results are shown in **FIGURE 3.20**.

1. In which group was the level of LDL ("bad" cholesterol) highest?

2. In which group was the level of HDL ("good" cholesterol) lowest?

3. An elevated risk of heart disease has been correlated with increasing LDL-to-HDL ratios. Which group had the highest LDL-to-HDL ratio?

4. Rank the three diets from best to worst according to their potential effect on heart disease.

	Main Dietary Fats			
	cis fatty acids	*trans* fatty acids	saturated fats	optimal level
LDL	103	117	121	<100
HDL	55	48	55	>40
ratio	1.87	2.44	2.2	<2

FIGURE 3.20 Effect of diet on lipoprotein levels. Researchers placed 59 men and women on a diet in which 10 percent of their daily energy intake consisted of *cis* fatty acids, *trans* fatty acids, or saturated fats.

Blood LDL and HDL levels were measured after three weeks on the diet; averaged results are shown in mg/dL (milligrams per deciliter of blood). All subjects were tested on each of the diets. The ratio of LDL to HDL is also shown.

10. ______ are to proteins as ______ are to nucleic acids.
   a. Sugars; lipids
   b. Sugars; proteins
   c. Amino acids; hydrogen bonds
   d. Amino acids; nucleotides

11. A denatured protein has lost its ______ .
   a. hydrogen bonds
   b. shape
   c. function
   d. all of the above

12. ______ consists of nucleotides.
   a. Ribose
   b. RNA
   c. DNA
   d. b and c

13. In the following list, identify the carbohydrate, the fatty acid, the amino acid, and the polypeptide:
   a. NH_2—CHR—COOH
   b. $C_6H_{12}O_6$
   c. $(methionine)_{20}$
   d. $CH_3(CH_2)_{16}COOH$

14. Match the molecules with the best description.

___ wax	a. protein primary structure
___ starch	b. an energy carrier
___ triglyceride	c. water-repellent secretions
___ DNA	d. carries heritable information
___ polypeptide	e. sugar storage in plants
___ ATP	f. richest energy source

15. Match each polymer with the component monomers.

___ protein	a. phosphate, fatty acids
___ phospholipid	b. amino acids, sugars
___ glycoprotein	c. glycerol, fatty acids
___ fat	d. nucleotides
___ nucleic acid	e. glucose only
___ wax	f. sugar, phosphate, base
___ nucleotide	g. amino acids
___ lipoprotein	h. glucose, fructose
___ sucrose	i. lipids, amino acids
___ glycogen	j. fatty acids, carbon rings

critical thinking

1. Lipoproteins are like suitcases that move cholesterol, fatty acid remnants, triglycerides, and phospholipids from one place to another in the body. Given what you know about the insolubility of lipids in water, which of the four kinds of lipids would you predict to be on the outside of a lipoprotein clump, bathed in the water-based fluid portion of blood?

2. In 1976, a team of chemists in the United Kingdom was developing new insecticides by modifying sugars with chlorine (Cl_2), phosgene (Cl_2CO), and other toxic gases. One young member of the team misunderstood his verbal instructions to "test" a new molecule. He thought he had been told to "taste" it. Luckily for him, the molecule was not toxic, but it was very sweet. It became the food additive sucralose.

Sucralose has three chlorine atoms substituted for three hydroxyl groups of sucrose (table sugar). It binds so strongly to the sweet-taste receptors on the tongue that the human brain perceives it as 600 times sweeter than sucrose. Sucralose was originally marketed as an artificial sweetener called Splenda®, but it is now available under several other brand names.

Researchers investigated whether the body recognizes sucralose as a carbohydrate by feeding sucralose labeled with ^{14}C to volunteers. Analysis of the radioactive molecules in the volunteers' urine and feces showed that 92.8 percent of the sucralose passed through the body without being altered. Many people are worried that the chlorine atoms impart toxicity to sucralose. How would you respond to that concern?

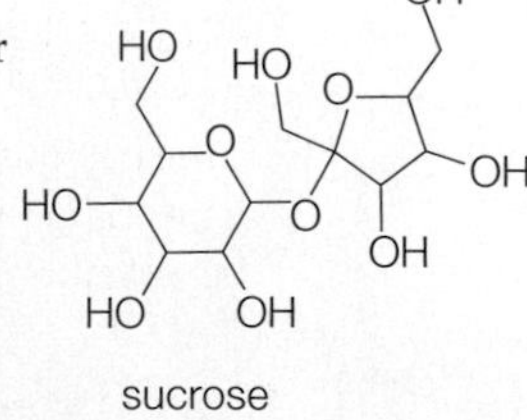

sucrose

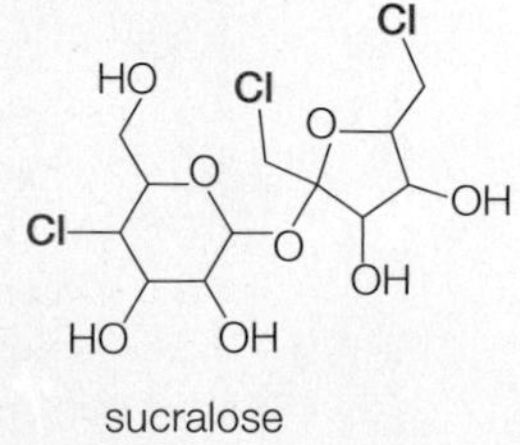

sucralose

CREDITS: (20) Source, Mensink RP, Katan MB, "Effect of dietary trans fatty acids on high-density and low-density lipoprotein cholesterol levels in healthy subjects." NEJM 323(7):439–45; (in text) © Cengage Learning.

4 Cell Structure

LEARNING ROADMAP

Reflect on life's levels of organization (Section 1.2). In this chapter, you will see how properties of lipids (3.5) give rise to cell membranes, consider the location of DNA (3.8) and the sites of carbohydrate metabolism (3.3, 3.4), and expand your understanding of the roles of proteins (3.6, 3.7). You will also revisit tracers (2.2) and the philosophy of science (1.9).

COMPONENTS OF ALL CELLS

Every cell has a plasma membrane separating its interior from the exterior environment. A cell's interior contains cytoplasm, DNA, and other structures.

PROKARYOTIC CELLS

Archaea and bacteria have no nucleus. In general, they are smaller and structurally more simple than eukaryotic cells. As a group, they are by far the most numerous and diverse organisms.

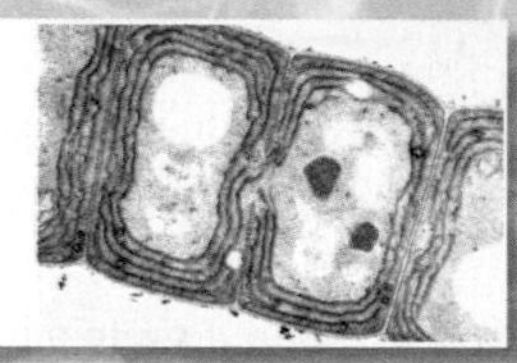

EUKARYOTIC CELLS

Protists, plants, fungi, and animals are eukaryotes. Cells of these organisms differ in internal components and surface specializations, but all start out life with a nucleus.

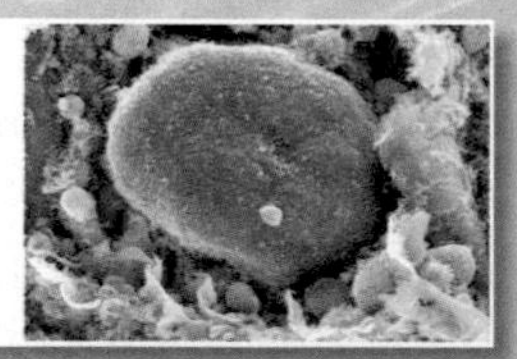

ORGANELLES OF EUKARYOTES

Membranes around eukaryotic organelles maintain internal environments that allow these structures to carry out specialized functions within a cell.

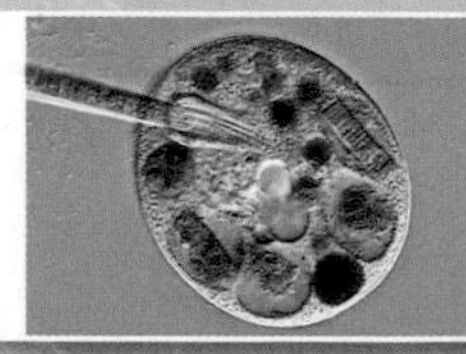

OTHER CELL COMPONENTS

Cytoskeletal elements organize and move cells and cell components. Cells secrete protective and structural materials, and connect to one another via cell junctions.

Chapter 5 explores cell function, and Chapters 6-8 detail individual metabolic processes. Some cellular structures are required for cell division (Chapter 11). Human genetic disorders return in Chapter 14. Chapter 20 details prokaryotes; Chapter 21, protists. Cell structures introduced in this chapter return in context of the physiology of plants (Chapters 27–30) and animals (Chapters 31–42). You will see some medical consequences of biofilms in Section 37.3.

4.1 Food for Thought

Cell for cell, microorganisms that live in and on a human body outnumber the person's own cells by about ten to one. Most are bacteria that live in the digestive tract, but these cells are not just stowaways. Gut bacteria help with digestion, make vitamins that mammals cannot, prevent the growth of dangerous germs, and shape the immune system.

One of the most common intestinal bacteria of warm-blooded animals (including humans) is *Escherichia coli*. Most of the hundreds of types, or strains, of *E. coli*, are helpful, but a few make a toxic protein that can severely damage the lining of the intestine. After ingesting as few as ten cells of a toxic strain, a person may become ill with severe cramps and bloody diarrhea that lasts up to ten days. In some people, complications of infection result in kidney failure, blindness, paralysis, and death. Each year, about 265,000 people in the United States become infected with toxin-producing *E. coli*.

Strains of *E. coli* that are toxic to people live in the intestines of other animals—mainly cattle, deer, goats, and sheep—apparently without sickening them. Humans are exposed to the bacteria when they come into contact with feces of animals that harbor them, for example, by eating contaminated ground beef. During slaughter, meat can come into contact with feces. Bacteria in the feces stick to the meat, then get thoroughly mixed into it during the grinding process. Unless contaminated meat is cooked to at least 71°C (160°F), live bacteria will enter the digestive tract of whoever eats it.

People also become infected with toxic *E. coli* by eating fresh fruits and vegetables that have contacted animal feces. Washing produce with water does not remove all of the bacteria because they are sticky (**FIGURE 4.1**). In 2012, more than 4,000 people in Europe were sickened after eating sprouts, and 49 of them died. The outbreak was traced to a single shipment of contaminated sprout seeds from Egypt.

The impact of such outbreaks, which occur with unfortunate regularity, extends beyond casualties. The contaminated sprouts cost growers in the European Union at least $600 million in lost sales. In 2011 alone, the United States Department of Agriculture (USDA) recalled 36.7 million pounds of ground meat products contaminated with toxic bacteria, at a cost in the billions of dollars. Such costs are eventually passed to taxpayers and consumers.

Food growers and processors are implementing new procedures intended to reduce the number and scope of these outbreaks. Meat and produce are being tested for some bacteria before sale, and improved documentation should allow a source of contamination to be pinpointed more quickly.

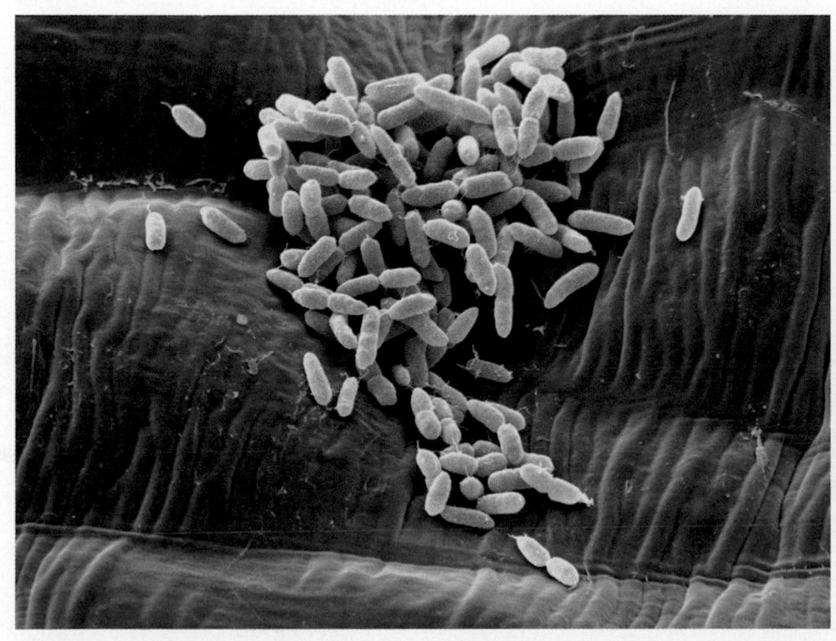

FIGURE 4.1 *Escherichia coli* cells sticking to the surface of a lettuce leaf. Some strains of this bacteria can cause a serious intestinal illness when they contaminate human food (right).

4.2 What Is a Cell?

✔ All cells have a plasma membrane and cytoplasm, and all start out life with DNA.

Cell Theory

Before microscopes were invented, no one knew that cells existed because nearly all are invisible to the naked eye. By 1665, Antoni van Leeuwenhoek had constructed an early microscope that revealed tiny organisms in rainwater, insects, fabric, sperm, feces, and other samples. In scrapings of tartar from his teeth, Leeuwenhoek saw "many very small animalcules, the motions of which were very pleasing to behold." He (incorrectly) assumed that movement defines life, but (correctly) concluded that the moving animalcules he saw were alive. Robert Hooke, a contemporary of Leeuwenhoek, observed cork under a microscope and discovered it to consist of "a great many little Boxes." He called the tiny compartments cells, after the small chambers that monks lived in.

Today we know that a cell carries out metabolism and homeostasis, and reproduces either on its own or as part of a larger organism. By this definition, each cell is alive even if it is part of a multicelled body, and all living organisms consist of one or more cells. We also know that cells reproduce by dividing, so it follows that all existing ctells must have arisen by division of other cells (later chapters discuss the processes by which cells divide). As a cell divides, it passes its hereditary material—its DNA—to offspring. Taken together, these generalizations constitute the **cell theory**, which is one of the foundations of modern biology (TABLE 4.1).

Table 4.1 The Cell Theory

1. Every living organism consists of one or more cells.
2. The cell is the structural and functional unit of all organisms. A cell is the smallest unit of life, individually alive even as part of a multicelled organism.
3. All living cells arise by division of preexisting cells.
4. Cells contain hereditary material, which they pass to their offspring when they divide.

Components of All Cells

Cells vary in shape and in what they do, but all have a plasma membrane, cytoplasm, and DNA (FIGURE 4.2).

Plasma Membrane A cell's **plasma membrane** is its outermost, and this membrane separates the cell's contents from the external environment. Like all other cell membranes, a plasma membrane is selectively permeable, which means that only certain materials can cross it. Thus, a plasma membrane controls exchanges between the cell and its environment. Like all cell membranes, a plasma membrane consists mainly of a phospholipid bilayer (Section 3.5). Many different proteins embedded in this bilayer or attached to one of its surfaces carry out various tasks (FIGURE 4.3). For example, some types of membrane proteins form channels through the bilayer; others pump substances across it. You will see in Chapter 5 how a membrane's function arises from its composite structure.

Cytoplasm The plasma membrane encloses a jellylike mixture of water, sugars, ions, and proteins "called **cytoplasm**. A major part of a cell's metabolism

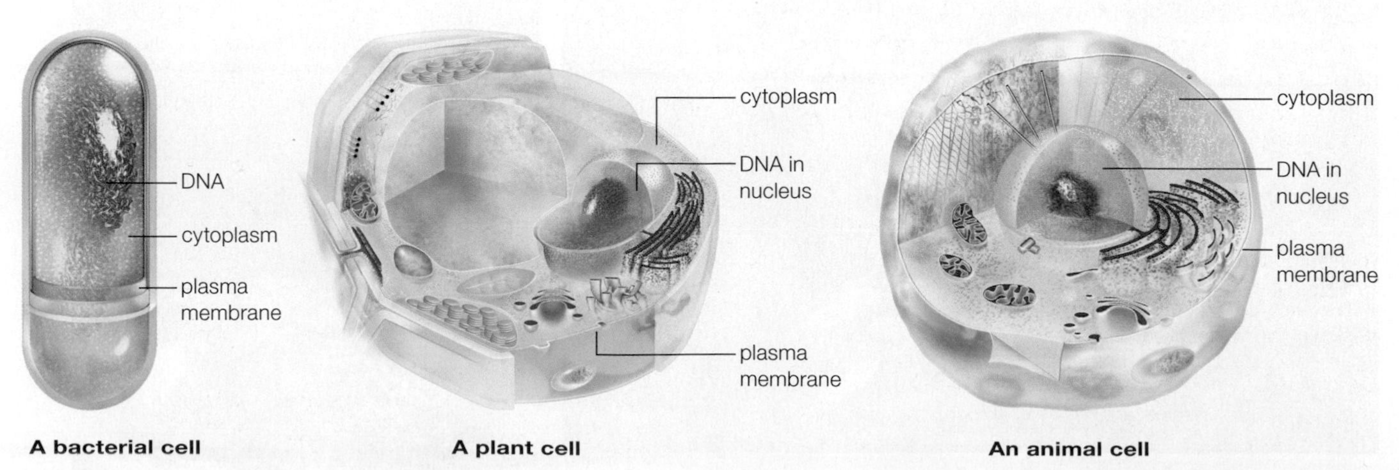

FIGURE 4.2 ▶**Animated** Overview of the general organization of a cell. All cells start out life with a plasma membrane, cytoplasm, and DNA. Archaea are similar to bacteria in overall structure; both are typically much smaller than eukaryotic cells. If the cells depicted here had been drawn to the same scale, the bacterium would be about this big:

occurs in the cytoplasm, and the cell's other internal components, including organelles, are suspended in it. **Organelles** are structures that carry out special functions inside a cell. Membrane-enclosed organelles allow a cell to compartmentalize substances and activities.

DNA Every cell starts out life with DNA. In nearly all bacteria and archaea, that DNA is suspended directly in cytoplasm. By contrast, the DNA of a eukaryotic cell is contained in a **nucleus** (plural, nuclei), an organelle with a double membrane. All protists, fungi, plants, and animals are eukaryotes. Some of these organisms are independent, free-living cells; others consist of many cells working together as a body.

Constraints on Cell Size

A cell exchanges substances with its environment at a rate that keeps pace with its metabolism. These exchanges occur across the plasma membrane, which can handle only so many exchanges at a time. The rate of exchange across a plasma membrane depends on its surface area: The bigger it is, the more substances can cross it during a given interval. Thus, cell size is limited by a physical relationship called the **surface-to-volume ratio**. By this ratio, an object's volume increases with the cube of its diameter, but its surface area increases only with the square.

Apply the surface-to-volume ratio to a round cell. As FIGURE 4.4 shows, when a cell expands in diameter, its volume increases faster than its surface area does. Imagine that a round cell expands until it is four times its original diameter. The volume of the cell has increased 64 times (4^3), but its surface area has increased only 16 times (4^2). Each unit of plasma membrane must now handle exchanges with four times as much cytoplasm ($64 \div 16 = 4$). If the cell gets too big, the inward flow of nutrients and the outward flow of wastes across that membrane will not be fast enough to keep the cell alive.

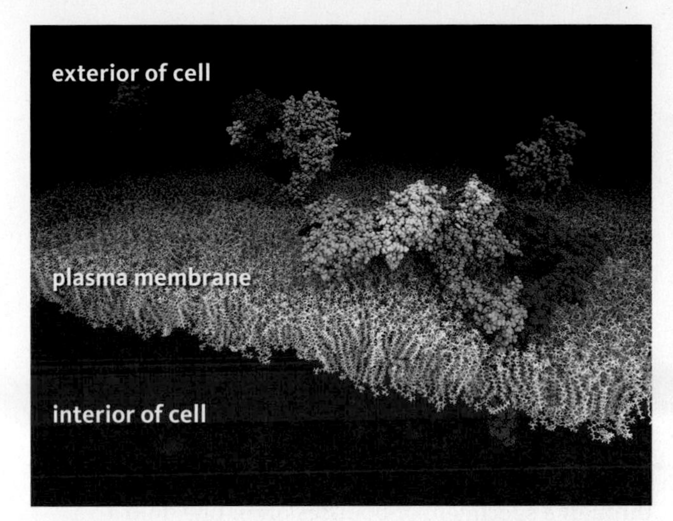

FIGURE 4.3 A plasma membrane separates a cell from its external environment. Proteins (in color) embedded in the lipid bilayer (gray) carry out special membrane functions. Chapter 5 returns to membrane structure and function.

Surface-to-volume limits also affect the form of colonial organisms and multicelled bodies. For example, small cells attach end to end to form strandlike algae, so each can interact directly with the environment. Some muscle cells in your thighs run the length of your upper leg, but each is thin, so it exchanges substances efficiently with fluids in the surrounding tissue.

Diameter (cm)	2	3	6
Surface area (cm^2)	12.6	28.2	113
Volume (cm^3)	4.2	14.1	113
Surface-to-volume ratio	3:1	2:1	1:1

FIGURE 4.4 Three examples of the surface-to-volume ratio. This physical relationship between increases in volume and surface area constrains cell size and shape.

cell theory Theory that all organisms consist of one or more cells, which are the basic unit of life; all cells come from division of preexisting cells; and all cells pass hereditary material to offspring.
cytoplasm Jellylike mixture of water and solutes enclosed by a cell's plasma membrane.
nucleus Of a eukaryotic cell, organelle with a double membrane that holds the cell's DNA.
organelle Structure that carries out a specialized metabolic function inside a cell.
plasma membrane A cell's outermost membrane.
surface-to-volume ratio A relationship in which the volume of an object increases with the cube of the diameter, and the surface area increases with the square.

TAKE-HOME MESSAGE 4.2

How are all cells alike?

✔ Observations of cells led to the cell theory: All organisms consist of one or more cells; the cell is the smallest unit of life; each new cell arises from another cell; and a cell passes hereditary material to its offspring.

✔ All cells start life with a plasma membrane, cytoplasm, and a region of DNA, which, in eukaryotic cells only, is enclosed within a nucleus.

✔ The surface-to-volume ratio limits cell size and influences cell shape.

4.3 How Do We See Cells?

✔ We use different types of microscopes to study different aspects of cells and their parts.

Most cells are 10–20 micrometers in diameter, about fifty times smaller than the unaided human eye can perceive (**FIGURE 4.5**). One micrometer (μm) is one-thousandth of a millimeter, which is one-thousandth of a meter (**TABLE 4.2**). We use microscopes to observe cells and other objects in the micrometer range of size.

In a light microscope, visible light illuminates a sample. As you will learn in Chapter 6, all light travels in waves. This property of light causes it to bend when passing through a curved glass lens. Inside a light microscope, such lenses focus light that passes through a specimen, or bounces off of one, into a magnified image (**FIGURE 4.6A**). Microscopes that use polarized light can yield images in which the edges of some structures appear in three-dimensional relief (**FIGURE 4.6B**). Photographs of images enlarged with a microscope are called micrographs; those taken with visible light are called light micrographs (LM).

Most cells are nearly transparent, so their internal details may not be visible unless they are first stained, or exposed to dyes that only some cell parts soak up. Parts that absorb the most dye appear darkest. Staining results in an increase in contrast (the difference between light and dark) that allows us to see a greater range of detail.

Researchers often use light-emitting tracers (Section 2.2) to pinpoint the location of a particular molecule of interest within a cell. When illuminated with laser light, these tracers fluoresce (emit light), and an image of the emitted light can be captured with a fluorescence microscope (**FIGURE 4.6C**). Such images are called fluorescence micrographs.

Table 4.2 Equivalent Units of Length

Unit		Equivalent: Meter	Equivalent: Inch
centimeter	cm	1/100	0.4
millimeter	mm	1/1000	0.04
micrometer	μm	1/1,000,000	0.00004
nanometer	nm	1/1,000,000,000	0.00000004
meter	m	100 cm 1,000 mm 1,000,000 μm 1,000,000,000 nm	

Structures smaller than about 200 nanometers across appear blurry under light microscopes because the wavelength of light—the distance from the peak of one wave to the peak behind it—limits the resolving power of even the best light microscope. To observe objects of this size range clearly, we would have to switch to an electron microscope. Electrons travel in wavelengths much shorter than those of visible light, so these microscopes can resolve details thousands of times smaller than light microscopes do.

There are two types of electron microscope; both use magnetic fields as lenses to focus a beam of electrons onto a sample. A transmission electron microscope

FIGURE 4.5 ▸**Animated** Relative sizes. Below, the diameter of most cells is in the range of 1 to 100 micrometers. **TABLE 4.2** shows conversions among units of length; also see Units of Measure, Appendix VI.

Which one is smallest: a protein, a lipid, or a water molecule?

Answer: A water molecule

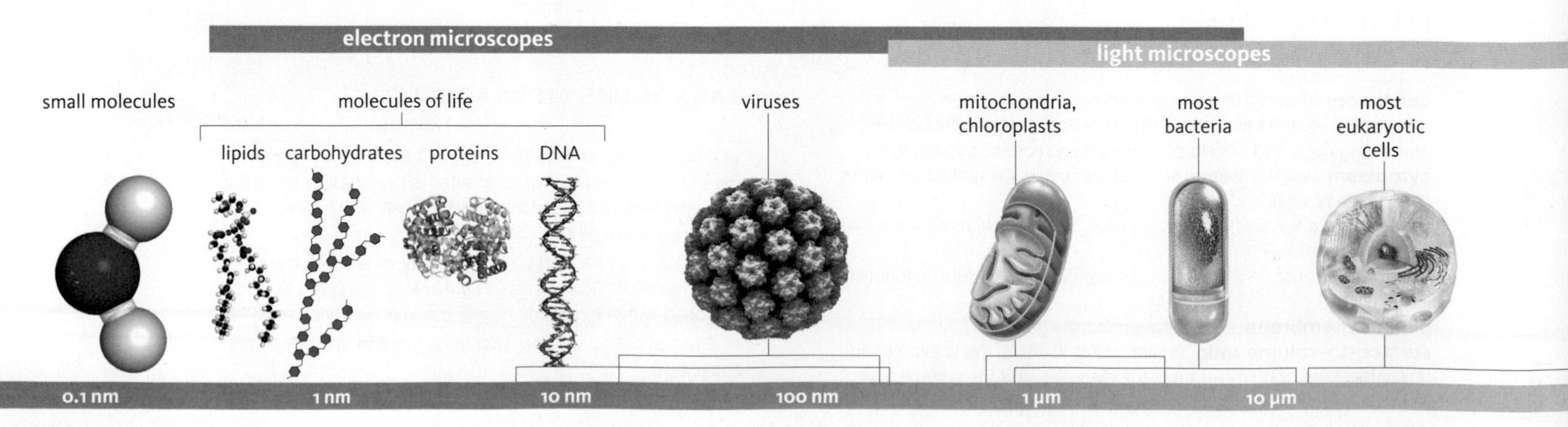

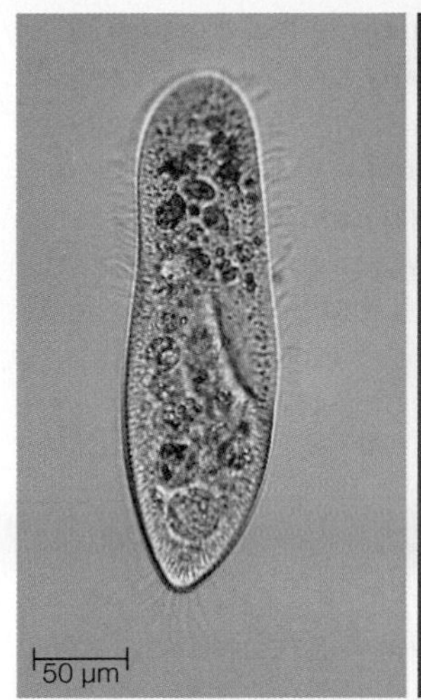

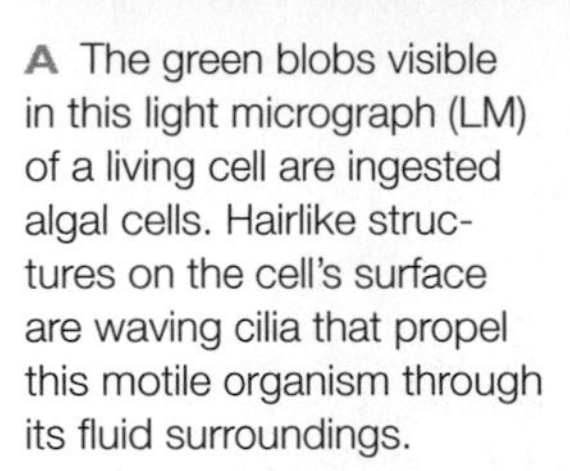

A The green blobs visible in this light micrograph (LM) of a living cell are ingested algal cells. Hairlike structures on the cell's surface are waving cilia that propel this motile organism through its fluid surroundings.

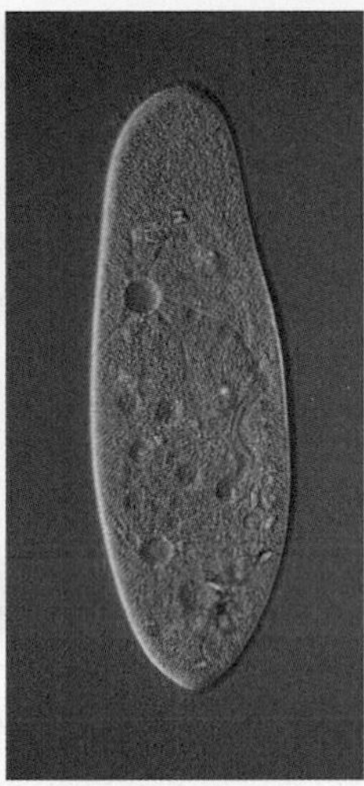

B A light micrograph taken with polarized light shows edges in relief. This technique reveals ingested algal cells, and some internal structures not visible in A.

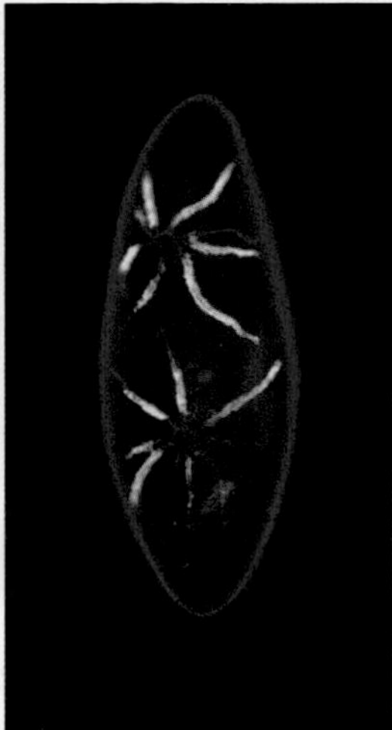

C In this fluorescence micrograph, yellow pinpoints the location of a particular protein in the membrane of organelles called contractile vacuoles. These organelles are also visible in B.

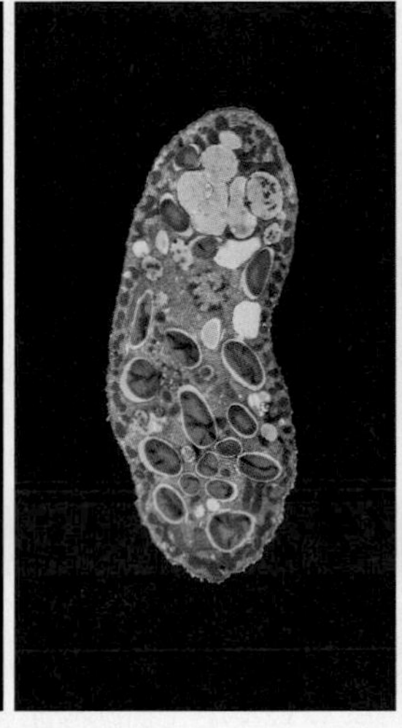

D A colorized transmission electron micrograph (TEM) reveals several types of internal structures in a plane (slice). Ingested algal cells are being broken down inside food vacuoles.

E A scanning electron micrograph (SEM) shows details of the cell's surface, including the thick coat of cilia. The cell ingests its food via the indentation (also visible in A).

FIGURE 4.6 Different microscopy techniques reveal different characteristics of the same type of organism, a protist (*Paramecium*).

What are the approximate dimensions of a *Paramecium*?

Answer: Approximately 250 μm long and 50 μm wide

directs electrons through a thin specimen, and the specimen's internal details appear as shadows in the resulting image, which is called a transmission electron micrograph, or TEM (**FIGURE 4.6D**). A scanning electron microscope directs a beam of electrons back and forth across the surface of a specimen that has been coated with a thin layer of gold or other metal. The irradiated metal emits electrons and x-rays, which are converted into an image (a scanning electron micrograph, or SEM) of the surface (**FIGURE 4.6E**). SEMs and TEMs are always black and white; colored versions have been digitally altered to highlight specific details.

TAKE-HOME MESSAGE 4.3

How do we see cells?

- ✔ Most cells are visible only with the help of microscopes.
- ✔ We use different microscopes and techniques to reveal different aspects of cell structure.

human eye (no microscope)

frog eggs

small animals

largest organisms

100 μm — 1 mm — 1 cm — 10 cm — 1 m — 10 m — 100 m

4.4 Introducing Prokaryotes

✔ Prokaryotic cells (bacteria and archaea) have no nucleus.

All bacteria and archaea are single-celled organisms, although individual cells of many species cluster in filaments or colonies (**FIGURE 4.7**). Outwardly, cells of the two groups appear so similar that archaea were once presumed to be an unusual group of bacteria. Both were classified as prokaryotes, a word that means "before the nucleus." By 1977, it had become clear that archaea are more closely related to eukaryotes than to bacteria, so they were given their own separate domain. The term "prokaryote" is now an informal designation only.

Bacteria and archaea are the smallest and most metabolically diverse forms of life that we know about. Chapter 20 revisits them in more detail; here we present an overview of structures shared by both groups (**FIGURE 4.8**).

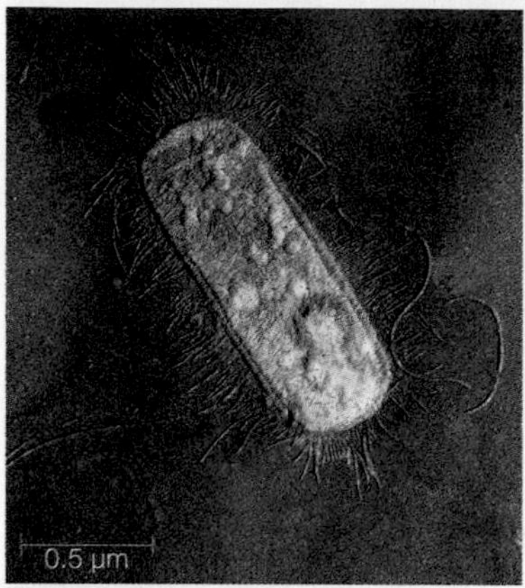

A *Escherichia coli*, a common bacterial inhabitant of human intestines. Short, hair-like structures are pili; longer ones are flagella.

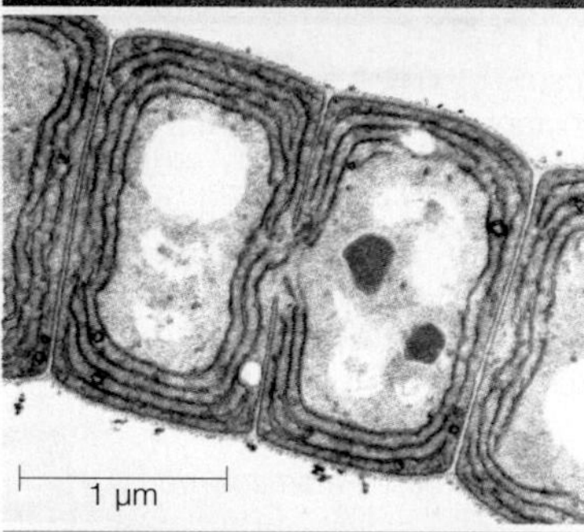

B *Oscillatoria*, a type of cyanobacteria that forms long filaments. Like other members of this ancient lineage, *Oscillatoria* has internal membranes (in green) where photosynthesis occurs. The multi-sided structures (pink) are carboxysomes, protein-enclosed organelles that assist photosynthesis.

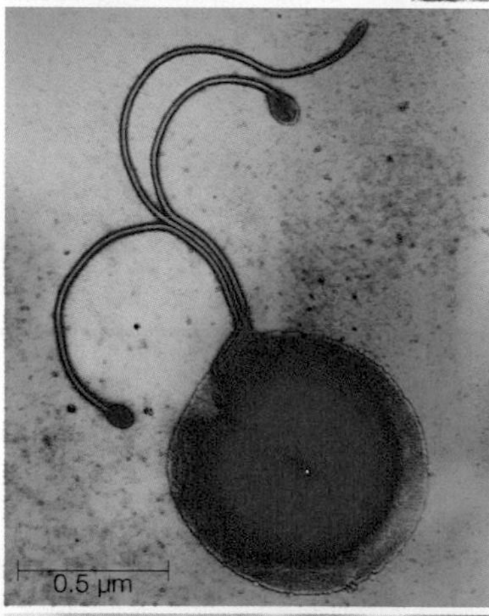

C *Helicobacter pylori*, a bacterium that can cause stomach ulcers when it infects the lining of the stomach. In unfavorable conditions, this species takes on a ball-shaped form (shown) that may offer the cells protection from environmental challenges such as antibiotic treatment.

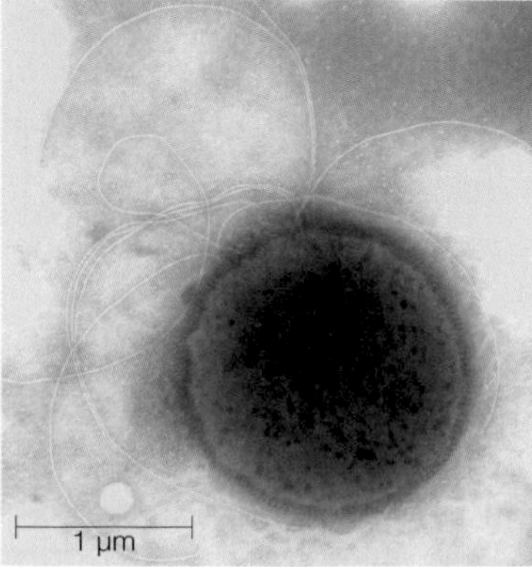

D *Thermococcus gammatolerans*, an archaeon discovered at a deep-sea hydrothermal vent, where it lives under extreme conditions of salt, temperature, and pressure. It is by far the most radiation-resistant organism ever discovered, capable of withstanding thousands of times more radiation than humans can.

FIGURE 4.7 Some representatives of bacteria (**A–C**) and an archaeon (**D**).

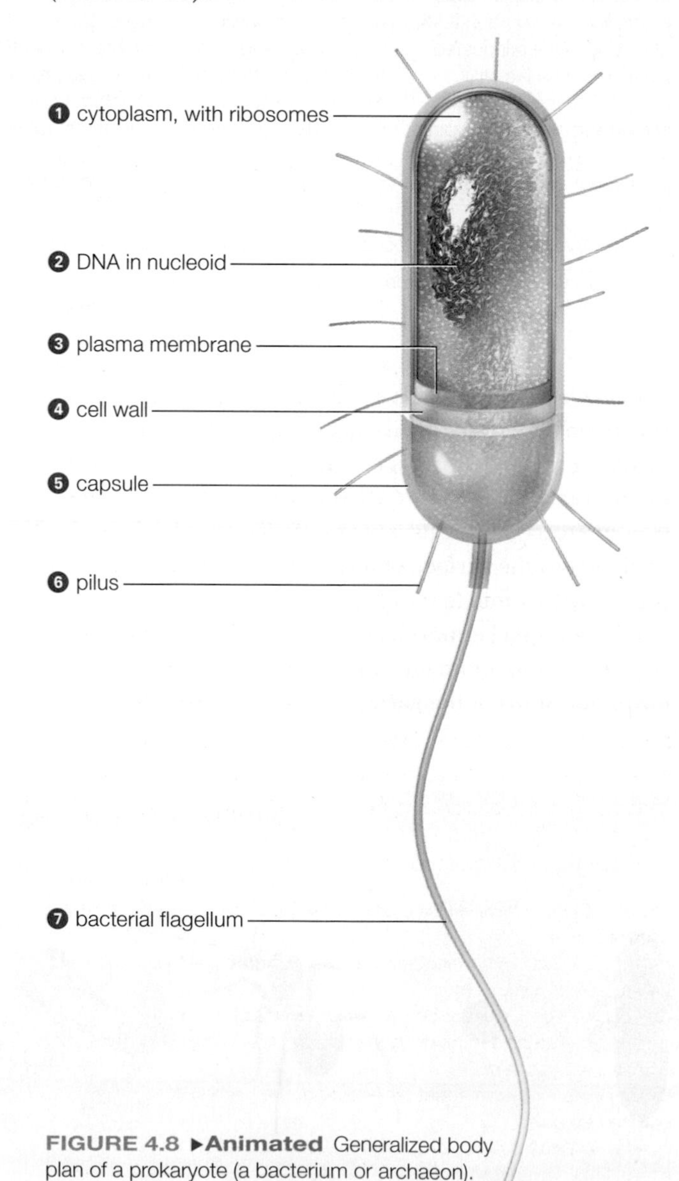

FIGURE 4.8 ▸Animated Generalized body plan of a prokaryote (a bacterium or archaeon).

Compared with eukaryotic cells, prokaryotes have little in the way of internal framework, but they do have protein filaments under the plasma membrane that reinforce the cell's shape and act as scaffolding for internal structures. The cytoplasm of these cells ❶ contains many **ribosomes** (organelles upon which polypeptides are assembled), and in some species, additional organelles. Cytoplasm also contains **plasmids**, which are small circles of DNA that carry a few genes (units of inheritance). Plasmid genes can provide advantages such as resistance to antibiotics. The cell's remaining genes typically occur on one large circular molecule of DNA located in an irregularly shaped region of cytoplasm called the **nucleoid** ❷. In a few species, the nucleoid is enclosed by a membrane. Other internal membranes carry out special metabolic processes such as photosynthesis in some prokaryotes (FIGURE 4.7B).

Like all cells, bacteria and archaea have a plasma membrane ❸. In nearly all prokaryotes, a rigid **cell wall** ❹ surrounding the plasma membrane protects the cell and supports its shape. Most archaeal cell walls consist of proteins; most bacterial cell walls consist of peptides and polysaccharides. Both types are permeable to water, so dissolved substances easily cross.

Polysaccharides form a slime layer or capsule ❺ around the wall of many types of bacteria. These sticky structures help the cells adhere to various surfaces, and they also offer some protection against predators.

Protein filaments called **pili** (singular, pilus) ❻ project from the surface of some prokaryotes. Pili help these cells move across or cling to surfaces. One kind, a "sex" pilus, attaches to another bacterium and then shortens. The attached cell is reeled in, and DNA is transferred from one cell to the other. Many types of bacteria and archaea also have one or more flagella projecting from their surface ❼. **Flagella** (singular, flagellum) are long, slender cellular structures used for motion. A prokaryotic flagellum rotates like a propeller that drives the cell through fluid habitats.

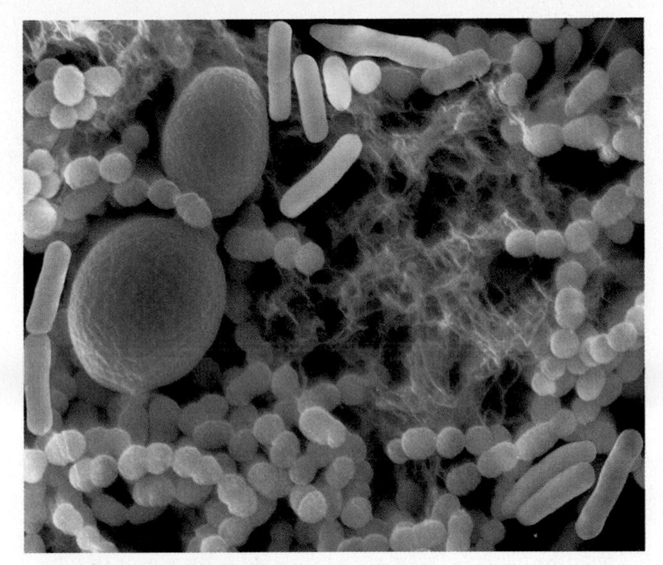

FIGURE 4.9 Oral bacteria in dental plaque, a biofilm. This micrograph shows two species of bacteria (tan, green) and a yeast (red) sticking to one another and to teeth via a gluelike mass of shared, secreted polysaccharides (pink). Other secretions of these organisms cause cavities and periodontal disease.

Biofilms

Bacterial cells often live so close together that an entire community shares a layer of secreted polysaccharides and proteins. A communal living arrangement in which single-celled organisms occupy a shared mass of slime is called a **biofilm**. A biofilm is often attached to a solid surface, and may include bacteria, algae, fungi, protists, and/or archaea. Participating in a biofilm allows the cells to linger in a favorable spot rather than be swept away by fluid currents, and to reap the benefits of living communally. For example, rigid or netlike secretions of some species serve as permanent scaffolding for others; species that break down toxic chemicals allow more sensitive ones to thrive in habitats that they could not withstand on their own; and waste products of some serve as raw materials for others. Later chapters discuss medical implications of biofilms, including the dental plaque that forms on teeth (FIGURE 4.9).

biofilm Community of microorganisms living within a shared mass of secreted slime.
cell wall Rigid but permeable structure that surrounds the plasma membrane of some cells.
flagellum Long, slender cellular structure used for motility.
nucleoid Of a bacterium or archaeon, region of cytoplasm where the DNA is concentrated.
pilus A protein filament that projects from the surface of some prokaryotic cells.
plasmid Small circle of DNA in some bacteria and archaea.
ribosome Organelle of protein synthesis.

TAKE-HOME MESSAGE 4.4

How are bacteria and archaea alike?

✔ Bacteria and archaea do not have a nucleus. Most kinds have a cell wall around their plasma membrane. The permeable wall reinforces and imparts shape to the cell body.

✔ The structure of bacteria and archaea is relatively simple, but as a group these organisms are the most diverse forms of life.

4.5 Introducing Eukaryotic Cells

✔ Eukaryotic cells carry out much of their metabolism inside organelles enclosed by membranes.

In addition to the nucleus, a typical eukaryotic cell has many other membrane-enclosed organelles, including endoplasmic reticulum, Golgi bodies, and at least one mitochondrion (**TABLE 4.3** and **FIGURES 4.10** and **4.11**). An enclosing membrane allows an organelle to regulate the types and amounts of substances that enter and exit. Through this type of control, an organelle maintains a special internal environment that allows it to carry out a particular function—for example, isolating toxic or sensitive substances from the rest of the cell, moving substances through cytoplasm, maintaining fluid balance, or providing a favorable environment for a special process.

The remaining sections of this chapter detail the structures and functions of organelles typical of eukaryotic cells.

TAKE-HOME MESSAGE 4.5

What do all eukaryotic cells have in common?

✔ All eukaryotic cells start life with a nucleus and other membrane-enclosed organelles.

Table 4.3
Some Organelles in Eukaryotic Cells

Organelles with membranes	
Nucleus	Protects, controls access to DNA
Endoplasmic reticulum (ER)	Makes, modifies polypeptides and lipids; other tasks
Golgi body	Modifies and sorts polypeptides and lipids
Vesicle	Transports, stores, or breaks down substances
Mitochondrion	Makes ATP by glucose breakdown
Chloroplast	Makes sugars (in plants, some protists)
Lysosome	Intracellular digestion
Peroxisome	Breaks down fatty acids, amino acids, toxins
Vacuole	Storage, breaks down food or waste
Organelles without membranes	
Ribosomes	Assembles polypeptides
Centriole	Anchors cytoskeleton
Other components	
Cytoskeleton	Contributes to cell shape, internal organization, movement

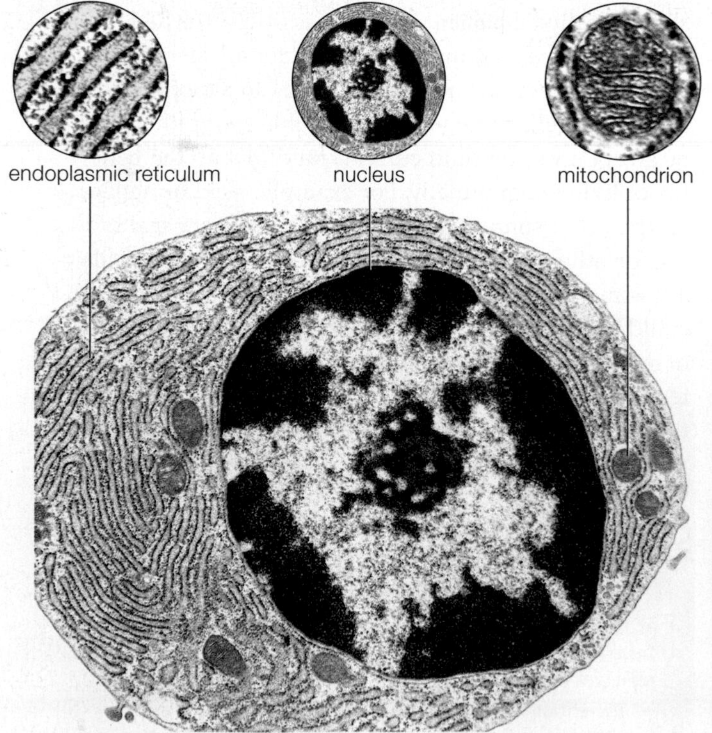

An animal cell (a white blood cell of a guinea pig)

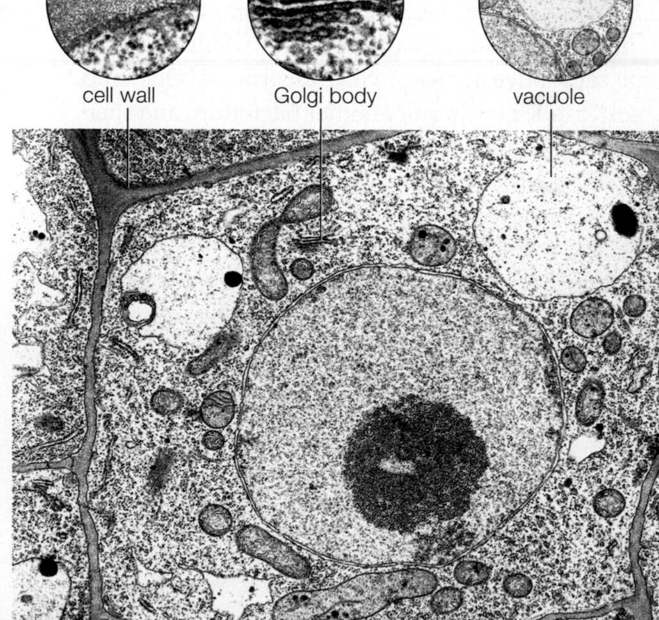

A plant cell (from a root of thale cress)

FIGURE 4.10 Transmission electron micrographs of two eukaryotic cells.

A Typical plant cell components.

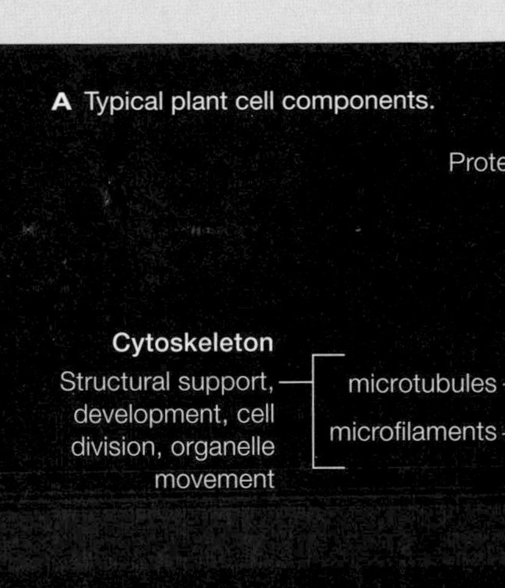

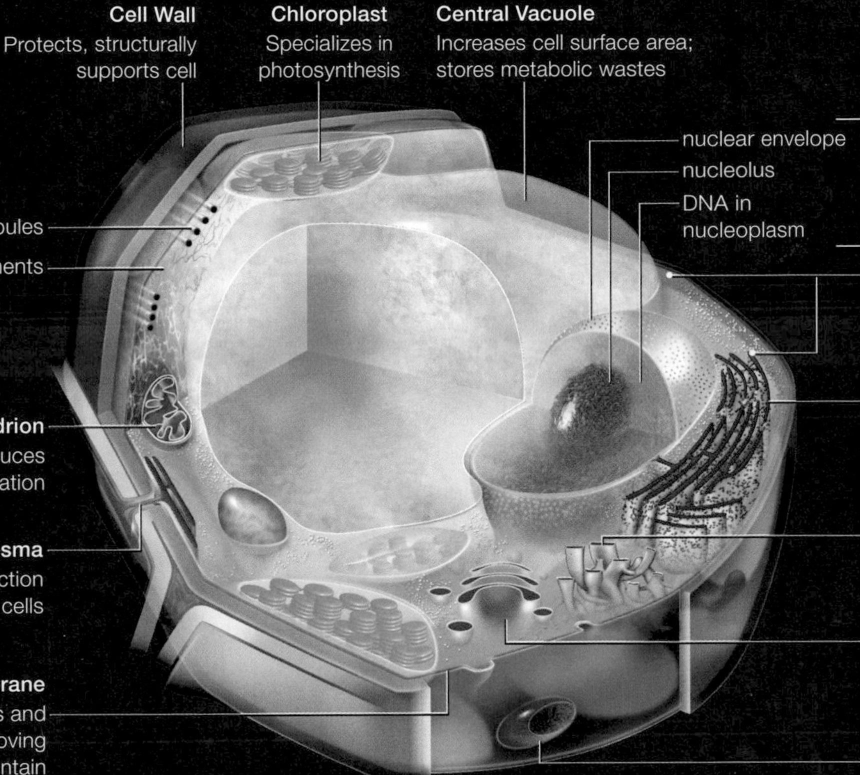

B Typical animal cell components.

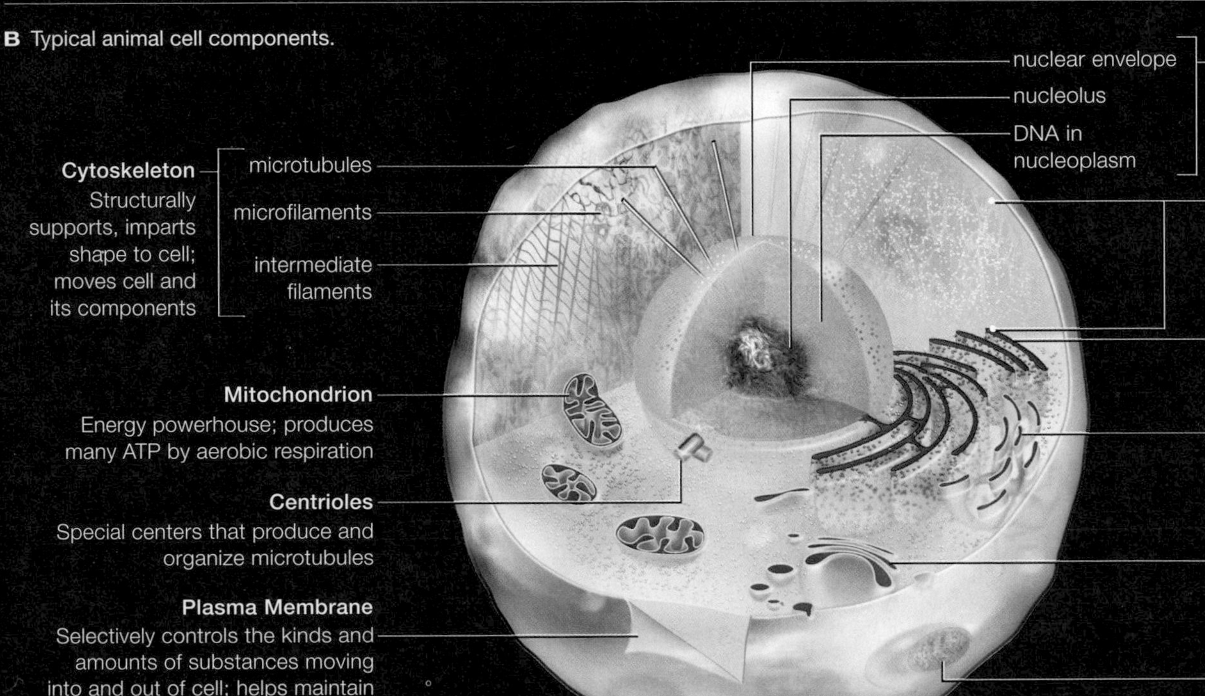

FIGURE 4.11 Animated Organelles and structures typical of **A** plant cells and **B** animal cells.

4.6 The Nucleus

✔ A nucleus keeps a eukaryotic cell's DNA away from potentially damaging reactions in the cytoplasm.

✔ Pores in the nuclear envelope selectively restrict access to the cell's DNA.

The cell nucleus serves two important functions. First, it keeps the cell's genetic material—DNA—safe from metabolic processes that might damage it. Isolated in its own compartment, DNA stays separated from the bustling activity of the cytoplasm. Second, a nucleus controls the passage of certain molecules across its membrane. **FIGURE 4.12** shows the components of the nucleus. **TABLE 4.4** lists their functions. Let's zoom in on the individual components.

Table 4.4 Components of the Nucleus

Chromatin	DNA and associated proteins in a cell nucleus
Nucleoplasm	Semifluid interior portion of the nucleus
Nuclear envelope	Double membrane with nuclear pores that control which substances enter and exit the nucleus
Nucleolus	Dense region of proteins and nucleic acid where ribosomal subunits are being produced

Chromatin

As molecules go, DNA is gigantic. Unraveled and stretched out, the DNA in the nucleus of a single human cell would be about 2 meters (6–1/2 feet) long. All of that DNA fits into a nucleus only six microns in diameter. How? Proteins associate with and organize each DNA molecule so it can pack tightly—and precisely—into the nucleus (Chapter 8 returns to this topic). All of the DNA in a cell's nucleus, together with associated proteins, is collectively called **chromatin**. Chromatin is suspended in **nucleoplasm**, a viscous fluid similar to cytoplasm, that fills the nucleus.

The Nuclear Envelope

The special membrane that encloses the nucleus is called the **nuclear envelope**. It consists of two phospholipid bilayers, folded around an intermembrane space 20–40 nm wide (**FIGURE 4.13**). The outer lipid bilayer is continuous with the lipid bilayer of endoplasmic reticulum, and like rough endoplasmic reticulum it is studded with ribosomes (more about endoplasmic reticulum in the next section). The inner lipid bilayer

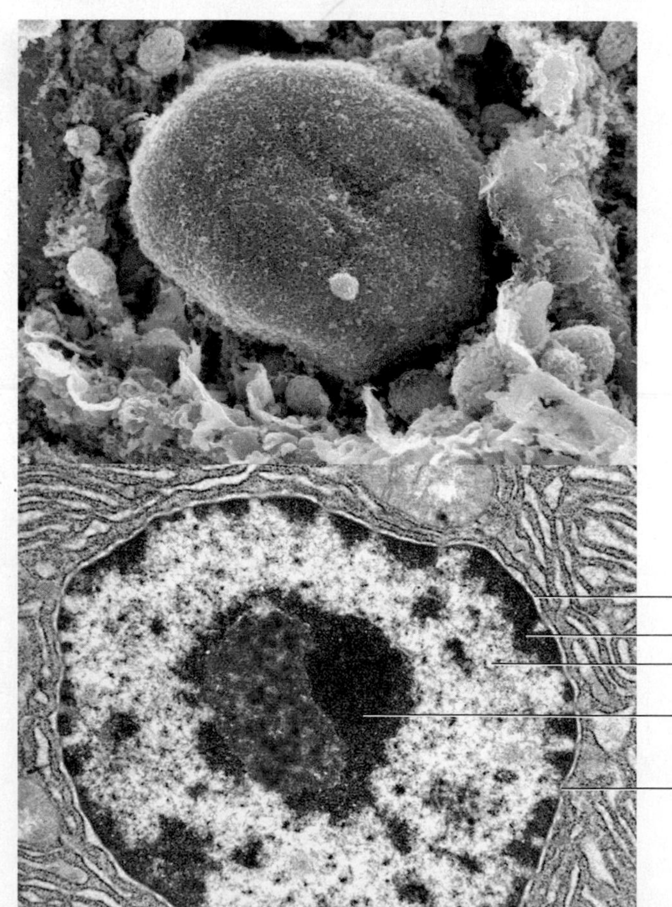

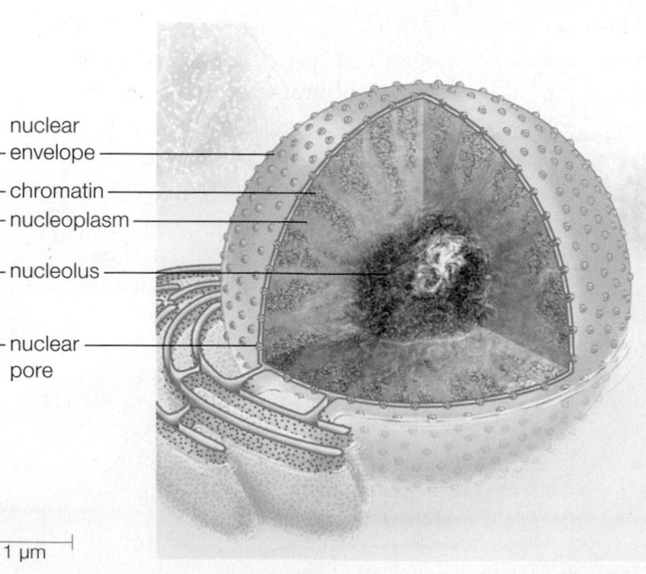

FIGURE 4.12 ▸Animated The cell nucleus. The SEM shows the nucleus of a liver cell (in pink); the TEM, the nucleus of a pancreas cell.

of animal cell nuclei is covered and supported by the nuclear lamina, a dense mesh of fibrous proteins (Section 4.10 returns to these proteins). Membrane proteins connect the lamina to the nuclear envelope.

The two lipid bilayers of a nuclear envelope connect at nuclear pores. Some bacteria have membranes around their DNA, but we do not consider the bacteria to have nuclei because there are no pores in these membranes. By contrast, a eukaryotic cell's nucleus has thousands of nuclear pores, each composed of hundreds of membrane proteins. A nuclear pore forms a hole in the envelope that can be widened or constricted by conformational changes in these proteins.

As you will see in Chapter 5, large molecules, including RNA and proteins, cannot cross lipid bilayers on their own. Nuclear pores function as gateways for these molecules to enter and exit a nucleus. Protein synthesis offers an example of why this movement is important. Protein synthesis occurs in cytoplasm, and it requires the participation of many molecules of RNA. RNA is produced in the nucleus. Thus, RNA molecules must move from nucleus to cytoplasm, and they do so through nuclear pores. Proteins that carry out RNA synthesis must move in the opposite direction, because this process occurs in the nucleus. A cell can regulate the amounts and types of proteins it makes at a given time by selectively restricting the passage of certain molecules through nuclear pores. Chapter 9 returns to the details of protein synthesis; and Chapter 10, to controls over this process.

The Nucleolus

Depending on a cell's metabolic state, its nucleus contains one or more nucleoli. A **nucleolus** (plural, nucleoli) is an irregularly shaped region, dense with proteins and nucleic acids, where subunits of ribosomes are being produced. The subunits pass through nuclear pores into the cytoplasm, where they join and become active in protein synthesis (Section 9.4 returns to ribosome structure). New research is revealing additional roles for nucleoli, for example in cell division, cell death, and responses to cellular stress.

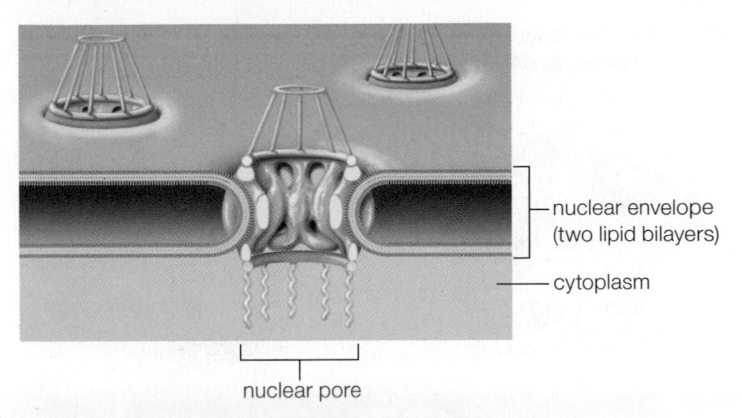

A The nuclear envelope consists of two lipid bilayers that connect at nuclear pores. Each pore is a complex structure formed by hundreds of proteins; each forms a hole in the envelope that can change in diameter. Part of a pore is a basket-shaped, multifunctional scaffold that can, for example, bind to chromatin. Cytoplasmic fibrils guide molecules through the pore.

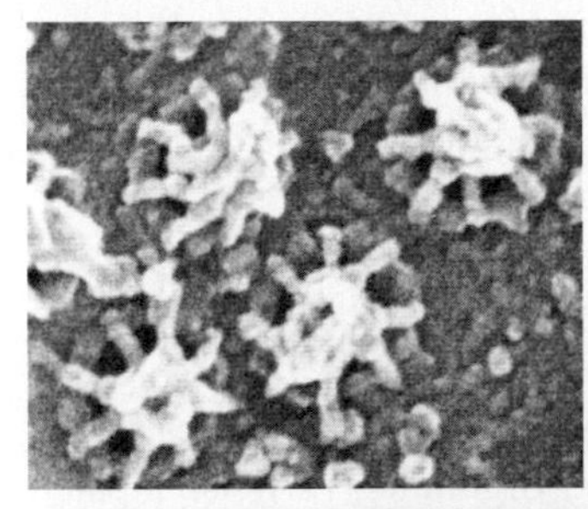

B The proteins that compose a nuclear pore allow only certain molecules to cross the nuclear membrane. The "baskets" on the inside of the nuclear envelope are visible in this SEM.

100 nm

nucleoplasm
chromatin
nuclear envelope
cytoplasm
10 nm
nuclear pore

C This TEM shows a pore in the nuclear envelope of a bone marrow cell nucleus. The identity of the "plug" seen in the center of this and many other nuclear pores is unknown; it may be a molecule caught in transit.

FIGURE 4.13 Structure of the nuclear membrane.

TAKE-HOME MESSAGE 4.6
What is the function of the cell nucleus?

✔ A nucleus protects and controls access to a eukaryotic cell's DNA.

✔ The nuclear envelope is a double lipid bilayer. Proteins embedded in the bilayer form pores that control the passage of molecules between the nucleus and cytoplasm.

chromatin Collective term for all of the DNA and associated proteins in a cell nucleus.
nuclear envelope A double membrane that constitutes the outer boundary of the nucleus. Nuclear pores in the membrane control the entry and exit of large molecules.
nucleolus In a cell nucleus, a dense, irregularly shaped region where ribosomal subunits are being produced.
nucleoplasm Viscous fluid enclosed by the nuclear envelope.

4.7 The Endomembrane System

✔ The endomembrane system is a set of organelles that makes, modifies, and transports proteins and lipids.

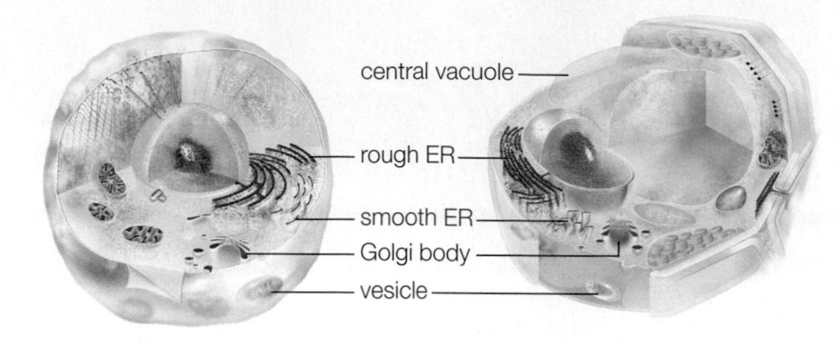

The **endomembrane system** (above) is a series of interacting organelles between the nucleus and the plasma membrane (**FIGURE 4.14**). Its main function is to make lipids, enzymes, and proteins for insertion into the cell's membranes or secretion to the external environment. The endomembrane system also destroys toxins, recycles wastes, and has other special functions. Components of the system vary among different types of cells, but here we present an overview of the most common ones.

Small, membrane-enclosed sacs called **vesicles** ❶ form by budding from other organelles or when a patch of plasma membrane sinks into the cytoplasm. Many types carry substances from one organelle to another, or to and from the plasma membrane. Some are a bit like trash cans that collect and dispose of waste, debris, or toxins. Enzymes in vesicles called **peroxisomes** break down fatty acids, amino acids, and poisons such as alcohol. They also break down hydrogen peroxide, a toxic by-product of fatty acid breakdown. **Lysosomes** are vesicles that take part in intracellular digestion. They contain powerful enzymes that break down cellular debris and wastes (carbohydrates, proteins, nucleic acids, and lipids). In cells such as amoebas and white blood cells, ingested bacteria, cell parts, and other particles are delivered to lysosomes for breakdown.

Vacuoles are sacs that form by the fusion of multiple vesicles. Many isolate or break down waste, debris, toxins, or food (**FIGURE 4.15**). Plant cells have a large **central vacuole**, in which amino acids, sugars, ions, wastes, and toxins accumulate. Fluid pressure in a central vacuole keeps plant cells plump, so stems, leaves, and other plant parts stay firm.

Endoplasmic reticulum (**ER**) comprises an interconnected system of tubes and flattened sacs. The ER membrane is an extension of the outer lipid bilayer of the nuclear envelope, so the space it encloses is continuous with the intermembrane space of the nuclear envelope. Two kinds of ER, rough and smooth, are named for their appearance. Thousands of ribosomes that attach to the outer surface of rough ER give this organelle its "rough" appearance. These ribosomes make polypeptides that thread into the interior of the ER as they are assembled ❷. Inside the ER, the polypeptide chains fold and take on their tertiary structure, and many assemble with other polypeptide chains (Section 3.6). Cells that make, store, and secrete proteins have a lot of rough ER. For example, ER-rich cells in the pancreas make digestive enzymes that they secrete into the small intestine.

Some proteins made in rough ER become part of its membrane. Others migrate through the ER compartment to smooth ER. Smooth ER has no ribosomes, so it

FIGURE 4.14 ▶**Animated** Some interactions among components of the endomembrane system.

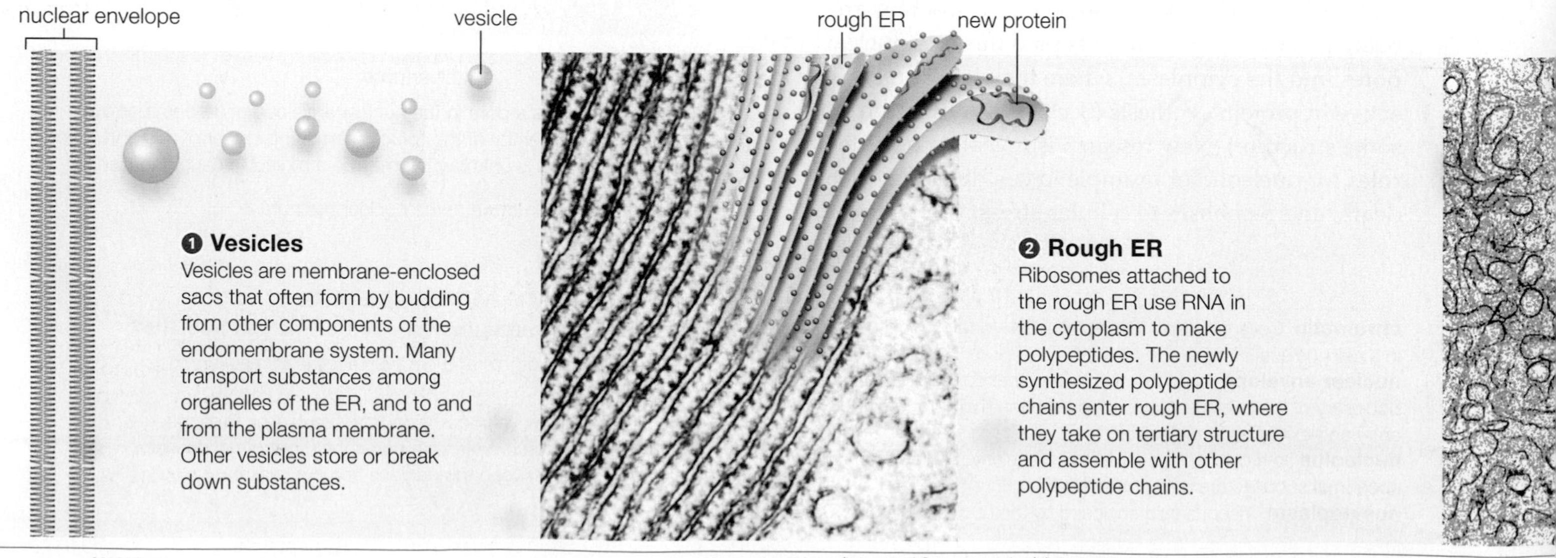

does not make its own proteins ❸. Some proteins that arrive in smooth ER are immediately packaged into vesicles for delivery elsewhere. Others are enzymes that stay and become part of the smooth ER. Smooth ER enzymes break down carbohydrates, fatty acids, and some drugs and poisons; and also make lipids for the cell's membranes.

A **Golgi body** has a folded membrane that often looks like a stack of pancakes ❹. Enzymes inside of Golgi bodies put finishing touches on proteins and lipids that have been delivered from ER. These enzymes attach phosphate groups or carbohydrates, and cleave certain proteins. The finished products (such as membrane proteins and proteins for secretion, and other enzymes) are sorted and packaged in new vesicles. Some of the vesicles deliver their cargo to the plasma membrane; others become lysosomes.

central vacuole Large fluid-filled organelle in many plant cells.
endomembrane system Series of interacting organelles (endoplasmic reticulum, Golgi bodies, vesicles) between nucleus and plasma membrane; produces lipids, proteins.
endoplasmic reticulum (ER) Membrane-enclosed organelle that is a continuous system of sacs and tubes extending from the nuclear envelope. Smooth ER makes lipids and breaks down carbohydrates and fatty acids; ribosomes on the surface of rough ER make proteins.
Golgi body Membrane-enclosed organelle that modifies proteins and lipids, then packages the finished products into vesicles.
lysosome Enzyme-filled vesicle that breaks down cellular wastes and debris.
peroxisome Enzyme-filled vesicle that breaks down amino acids, fatty acids, and toxic substances.
vacuole A membrane-enclosed organelle filled with fluid; isolates or disposes of waste, debris, or toxic materials.
vesicle Small, membrane-enclosed organelle; different kinds store, transport, or break down their contents.

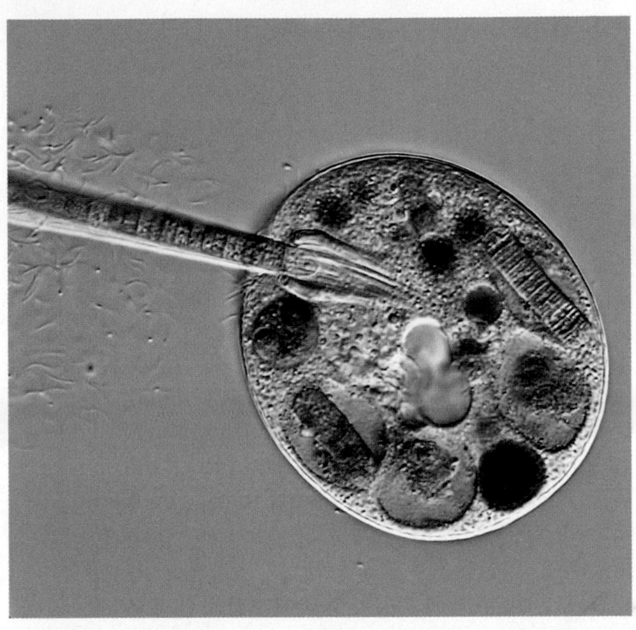

FIGURE 4.15 An example of vacuole function. The protist *Nassula* (round cell) uses a distinctive oral basket to feed on strands of photosynthetic algae. Ingested cells are packaged in food vacuoles that change color (green to purple to brown to gold) as chlorophyll molecules inside them get broken down.

TAKE-HOME MESSAGE 4.7
What is the endomembrane system?

✔ The endomembrane system includes rough and smooth endoplasmic reticulum, vesicles, and Golgi bodies.

✔ Rough ER produces enzymes, membrane proteins, and secreted proteins. Smooth ER produces lipids and breaks down carbohydrates, fatty acids, and toxins.

✔ Golgi bodies modify and ship new proteins and lipids.

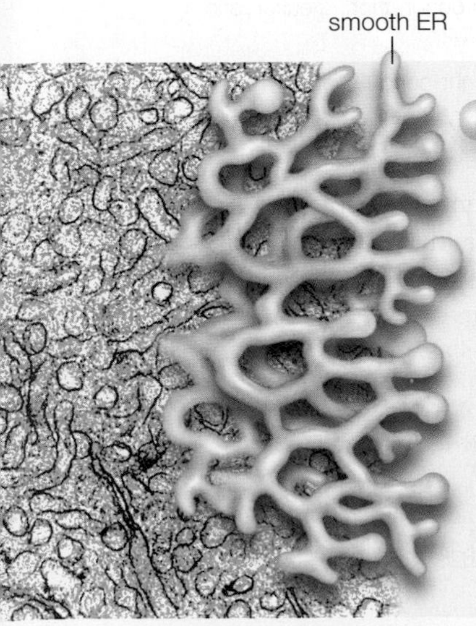

❸ Smooth ER
Proteins migrate through the interior of the rough ER, and end up in the smooth ER. Some stay in smooth ER, as enzymes that assemble lipids and break down carbohydrates, wastes, and toxins. Other proteins are packaged in vesicles for transport to Golgi bodies.

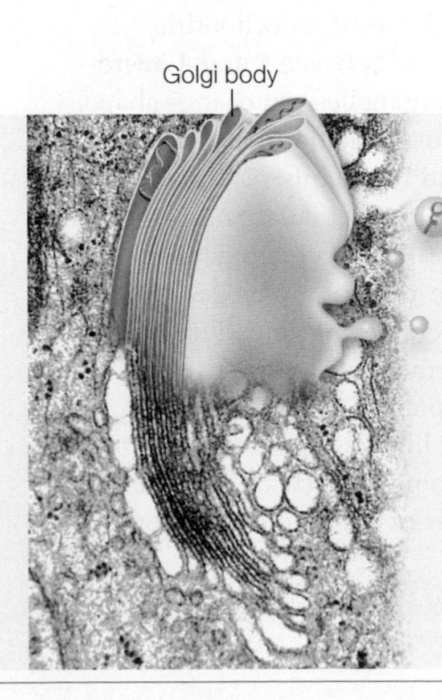

❹ Golgi Body
Proteins and lipids arriving in vesicles are modified into final form, sorted, and repackaged into new vesicles. Some of these vesicles ferry proteins to the plasma membrane for secretion or insertion into the lipid bilayer. Others become lysosomes.

4.8 Mitochondria

✔ Eukaryotic cells make most of their ATP in mitochondria.

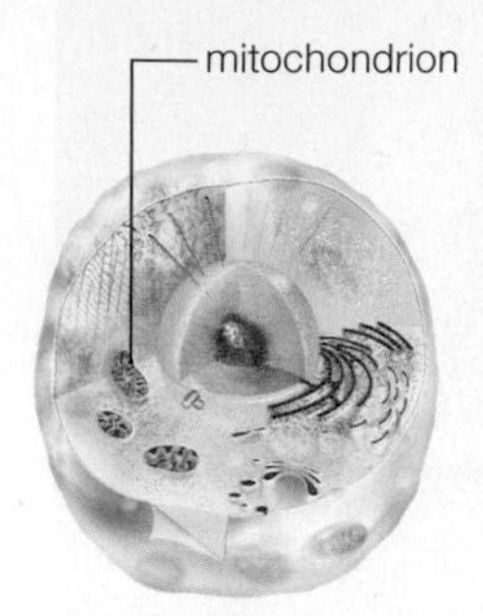

As you will see in Chapter 5, biologists think of the nucleotide ATP as a type of cellular currency because it carries energy between reactions. Cells require a lot of ATP. The most efficient way they can produce it is by aerobic respiration, a series of oxygen-requiring reactions that harvests the energy in sugars by breaking their bonds. In eukaryotes, aerobic respiration occurs inside organelles called **mitochondria** (singular, mitochondrion). With each breath, you are taking in oxygen mainly for the mitochondria in your trillions of aerobically respiring cells.

The structure of a mitochondrion is specialized for carrying out reactions of aerobic respiration. Each mitochondrion has two membranes, one highly folded inside the other (**FIGURE 4.16**). This arrangement creates two compartments: an outer one (between the two membranes), and an inner one (inside the inner membrane). During aerobic respiration, hydrogen ions accumulate between the two membranes. The buildup causes the ions to flow across the inner mitochondrial membrane, and this flow drives ATP formation (we return to aerobic respiration in Chapter 7).

Nearly all eukaryotic cells (including plant cells) have mitochondria, but the number varies by the type of cell and by the organism. For example, single-celled organisms such as yeast often have only one mitochondrion, but human skeletal muscle cells have a thousand or more. In general, cells that have the highest demand for energy tend to have the most mitochondria.

Typical mitochondria are between 1 and 4 micrometers in length. These organelles can change shape, split in two, branch, or fuse together. They resemble bacteria in size, form, and biochemistry. Mitochondria have their own DNA, which is circular and otherwise similar to bacterial DNA. They divide independently of the cell, and have their own ribosomes. Such clues led to a theory that mitochondria evolved from aerobic bacteria that took up permanent residence inside a host cell (we return to this topic in Section 19.6).

Some eukaryotes that live in oxygen-free environments have hydrogenosomes, which are modified mitochondria that produce hydrogen in addition to ATP. Like mitochondria, hydrogenosomes have two membranes. Unlike mitochondria, they have no DNA, so they cannot divide independently of the cell.

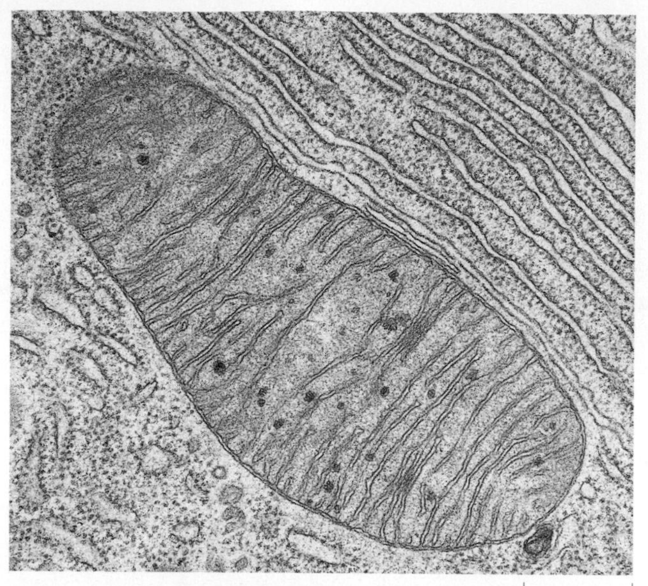

A Mitochondrion in a cell from bat pancreas.

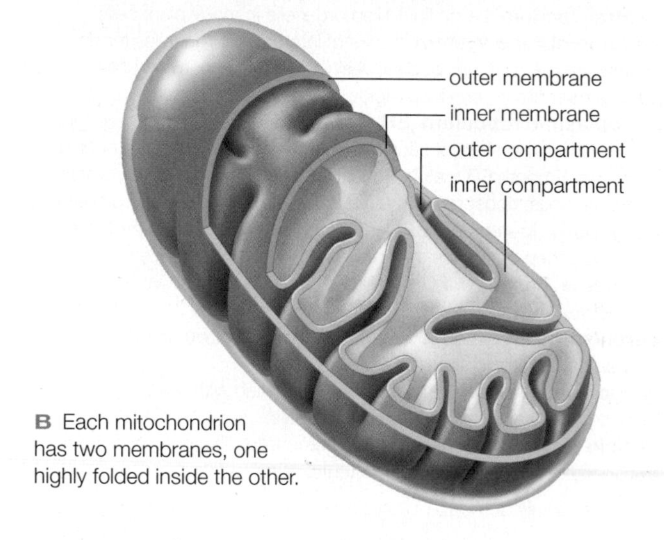

B Each mitochondrion has two membranes, one highly folded inside the other.

FIGURE 4.16 ▸**Animated** The mitochondrion, a eukaryotic organelle that specializes in producing ATP.

What organelle is visible in the upper right-hand corner of the TEM?

Answer: Rough ER

mitochondrion Double-membraned organelle that produces ATP by aerobic respiration in eukaryotes.

TAKE-HOME MESSAGE 4.8

What do mitochondria do?

✔ Mitochondria are eukaryotic organelles specialized to produce ATP by aerobic respiration, an oxygen-requiring series of reactions that breaks down carbohydrates.

4.9 Chloroplasts and Other Plastids

✔ Plastids function in storage and photosynthesis in plants and some types of algae.

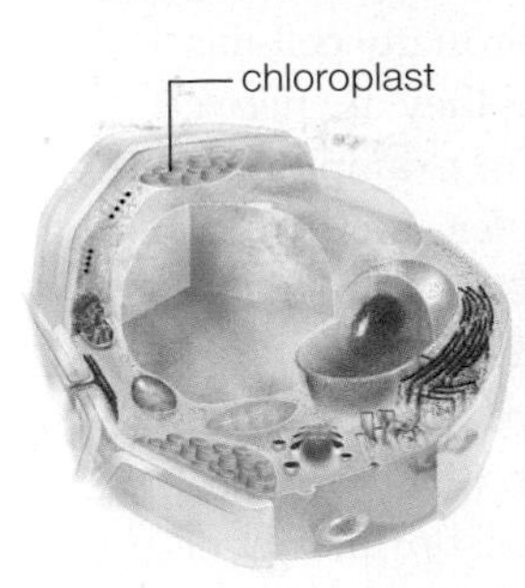

Plastids are double-membraned organelles that function in photosynthesis, storage, or pigmentation in plant and algal cells. Photosynthetic cells of plants and many protists contain **chloroplasts**, which are plastids specialized for photosynthesis (FIGURE 4.17). Most chloroplasts are oval or disk-shaped. Each has two outer membranes enclosing a semifluid interior, the stroma, that contains enzymes and the chloroplast's own DNA. In the stroma, a third, highly folded membrane forms a single, continuous compartment. Photosynthesis occurs at this inner membrane.

The innermost membrane of a chloroplast incorporates many pigments, including a green one called chlorophyll (the abundance of chlorophyll in plant cell chloroplasts is the reason most plants are green). During photosynthesis, these pigments capture energy from sunlight, and pass it to other molecules that require energy to make ATP. The resulting ATP is used inside the stroma to build sugars from carbon dioxide and water. (Chapter 6 returns to details of these processes.) In many ways, chloroplasts resemble the photosynthetic bacteria from which they evolved.

Chromoplasts are plastids that make and store pigments other than chlorophylls. They often contain red or orange carotenoids that color flowers, leaves, roots, and fruits (FIGURE 4.18). Chromoplasts are related to chloroplasts, and the two types of plastids are interconvertible. For example, as fruits such as tomatoes ripen, green chloroplasts in their cells are converted to red chromoplasts, so the color of the fruit changes.

Amyloplasts are unpigmented plastids that make and store starch grains. They are notably abundant in cells of stems, tubers (underground stems), fruits, and seeds. Like chromoplasts, amyloplasts are related to chloroplasts, and one type can change into the other. Starch-packed amyloplasts are dense and heavy compared to cytoplasm; in some plant cells, they function as gravity-sensing organelles (we return to this topic in Section 30.8).

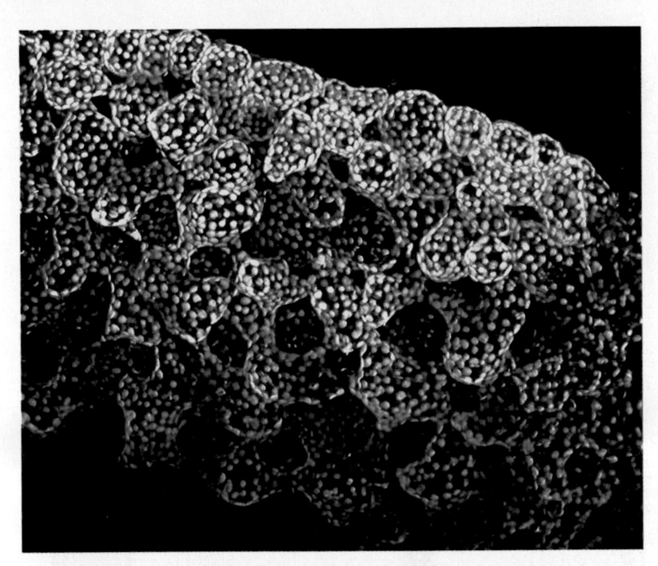

A Chloroplast-packed cells make up a leaf of a flowering plant.

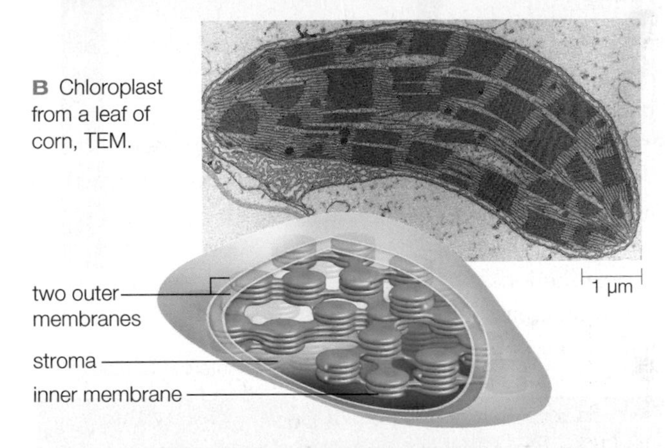

B Chloroplast from a leaf of corn, TEM.

C Each chloroplast has two outer membranes. Photosynthesis occurs at a much-folded inner membrane.

FIGURE 4.17 ▶**Animated** The chloroplast.

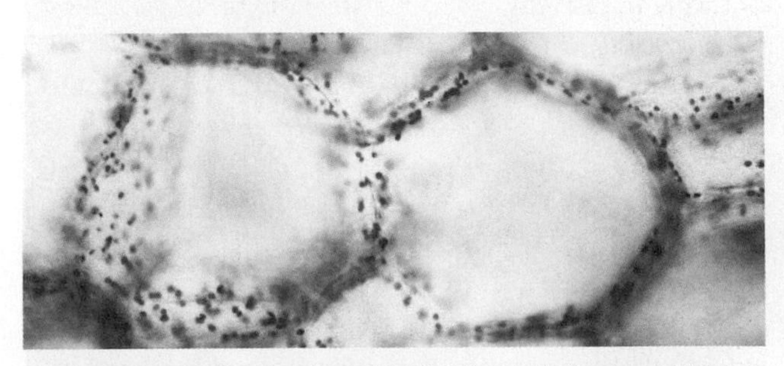

FIGURE 4.18 Chromoplasts in cells of a red bell pepper. The color of the fruit arises from these plastids.

TAKE-HOME MESSAGE 4.9

What are plastids?

✔ Plastids occur in plants and some protists; they function in photosynthesis, storage, and pigmentation.

✔ Chloroplasts are plastids that carry out photosynthesis.

chloroplast Organelle of photosynthesis in the cells of plants and photosynthetic protists.

plastid Double-membraned organelle that functions in photosynthesis, pigmentation, or storage in plants and algal cells; for example, a chloroplast, chromoplast, or amyloplast.

4.10 The Cytoskeleton

✔ A cytoskeleton supports a eukaryotic cell and helps it move.

Between the nucleus and plasma membrane of all eukaryotic cells is a system of interconnected protein filaments collectively called the **cytoskeleton**. Elements of the cytoskeleton reinforce, organize, and move cell structures, and often the whole cell. Some are permanent; others form only at certain times.

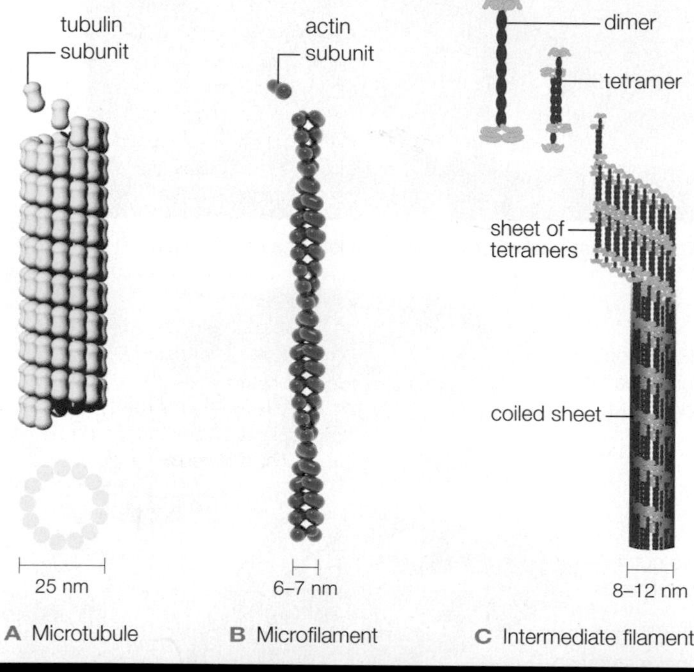

A Microtubule **B** Microfilament **C** Intermediate filament

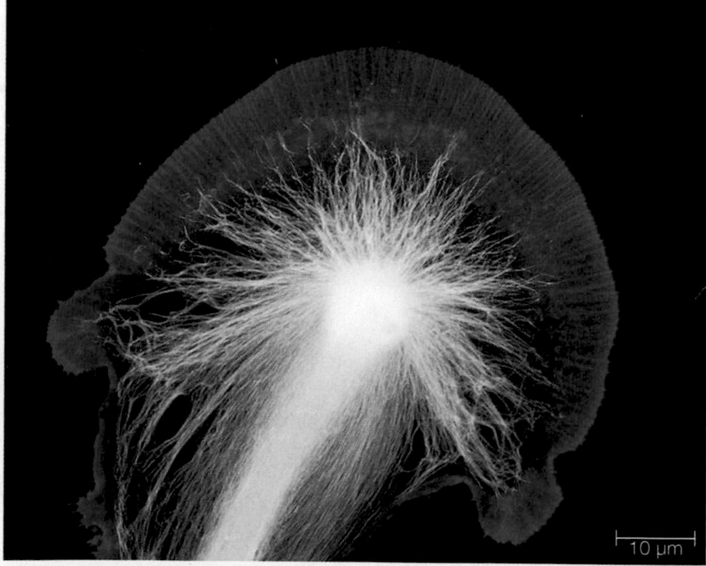

D A fluorescence micrograph shows microtubules (yellow) and microfilaments (blue) in the growing end of a nerve cell. These cytoskeletal elements support and guide the cell's lengthening in a particular direction.

FIGURE 4.19 ▸Animated Cytoskeletal elements.

Microtubules are long, hollow cylinders that consist of subunits of the protein tubulin (**FIGURE 4.19A**). They form a dynamic scaffolding for many cellular processes, rapidly assembling when they are needed, disassembling when they are not. For example, before a eukaryotic cell divides, microtubules assemble, separate the cell's duplicated DNA molecules, then disassemble. As another example, microtubules that form in the growing end of a young nerve cell support its lengthening in a particular direction (**FIGURE 4.19D**).

Microfilaments are fibers that consist primarily of subunits of a protein called actin (**FIGURE 4.19B**). These fine fibers strengthen or change the shape of eukaryotic cells, and have a critical function in cell migration, movement, and contraction. Crosslinked, bundled, or gel-like arrays of them make up the **cell cortex**, a reinforcing mesh under the plasma membrane. Microfilaments also connect plasma membrane proteins to other proteins inside the cell.

Intermediate filaments are the most stable elements of the cytoskeleton, forming a framework that lends structure and resilience to cells and tissues in multicelled organisms. Several types of intermediate filaments are assembled from different proteins (**FIGURE 4.19C**). For example, intermediate filaments that make up your hair consist of keratin, a fibrous protein (Section 3.6). Intermediate filaments that consist of lamins, another type of fibrous protein, form the nuclear lamina of animal cells. In addition to providing structural support, nuclear lamins are part of mechanisms that regulate DNA replication and other processes that take place inside the nucleus.

Motor proteins that associate with cytoskeletal elements move cell parts when energized by a phosphate-group transfer from ATP (Section 3.8). A cell is like a bustling train station, with molecules and structures being moved continuously throughout its interior. Motor proteins are like freight trains, dragging cellular cargo along tracks of microtubules and microfilaments (**FIGURE 4.20**). The motor protein myosin interacts with microfilaments to bring about muscle cell contraction. Another motor protein, dynein, interacts with microtubules to bring about movement of flagella and cilia in eukaryotes. Eukaryotic flagella whip back and forth to propel motile cells such as sperm through fluid (**FIGURE 4.21A**). **Cilia** (singular, cilium) are short, hair-

FIGURE 4.20 ▸Animated Motor proteins. Here, kinesin (tan) drags a pink vesicle as it inches along a microtubule.

like structures that project from the surface of some cells. The coordinated waving of many cilia propels some cells through fluid, and stirs fluid around other cells that are stationary. The waving movement of eukaryotic flagella and cilia, which differs from the propeller-like rotation of prokaryotic flagella, arises from their internal architecture. Microtubules extend lengthwise through these structures, in what is called a 9+2 array (**FIGURE 4.21B**). The array consists of nine pairs of microtubules ringing another pair in the center. The microtubules grow from a barrel-shaped organelle, the **centriole**, which remains below the finished array as part of a **basal body** (**FIGURE 4.21C**).

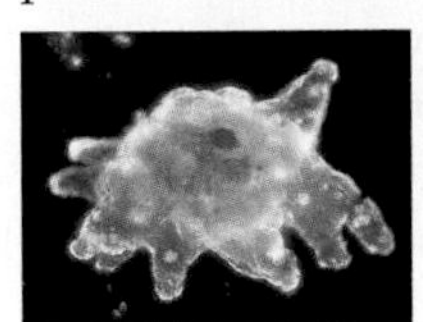

Some eukaryotic cells, including the amoeba at left, form **pseudopods**, or "false feet." As these temporary, irregular lobes bulge outward, they move the cell and engulf a target such as prey. Elongating microfilaments force the lobe to advance in a steady direction. Motor proteins attached to the microfilaments drag the plasma membrane along with them.

basal body Organelle that develops from a centriole.
cell cortex Reinforcing mesh of cytoskeletal elements under a plasma membrane.
centriole Barrel-shaped organelle from which microtubules grow.
cilium Short, movable structure that projects from the plasma membrane of some eukaryotic cells.
cytoskeleton Dynamic framework of protein filaments that support, organize, and move eukaryotic cells and their internal structures.
intermediate filament Stable cytoskeletal element that structurally supports cells and tissues.
microfilament Reinforcing cytoskeletal element that functions in cell movement; a fiber of actin subunits.
microtubule Cytoskeletal element involved in cellular movement; hollow filament of tubulin subunits.
motor protein Type of energy-using protein that interacts with cytoskeletal elements to move the cell's parts or the whole cell.
pseudopod A temporary protrusion that helps some eukaryotic cells move and engulf prey.

TAKE-HOME MESSAGE 4.10

What is a cytoskeleton?

✔ A cytoskeleton of protein filaments is the basis of eukaryotic cell shape, internal structure, and movement.

✔ Microtubules organize eukaryotic cells and help move their parts. Networks of microfilaments reinforce cell shape and function in movement. Intermediate filaments strengthen and maintain the shape of cell membranes and tissues, and form external structures such as hair.

✔ When energized by ATP, motor proteins move along tracks of microtubules and microfilaments. As part of cilia, flagella, and pseudopods, they can move the whole cell.

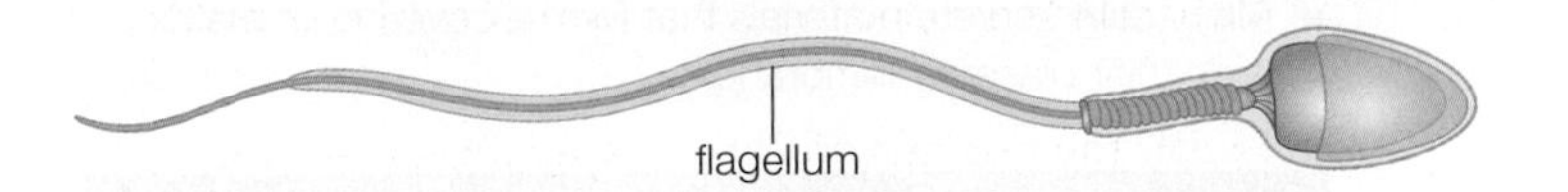

A Flagella propel motile eukaryotic cells such as sperm through fluid surroundings. The waving motion of eukaryotic flagella and cilia arises from microtubules that run lengthwise through them, in a 9+2 array.

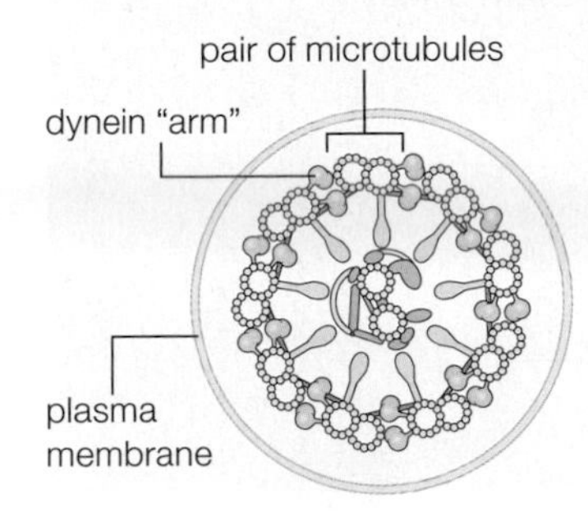

B A 9+2 array consists of a ring of nine pairs of microtubules plus one pair at their core. Stabilizing spokes and linking elements connect the microtubules and keep them aligned in this pattern. Projecting from each pair of microtubules are "arms" of the motor protein dynein.

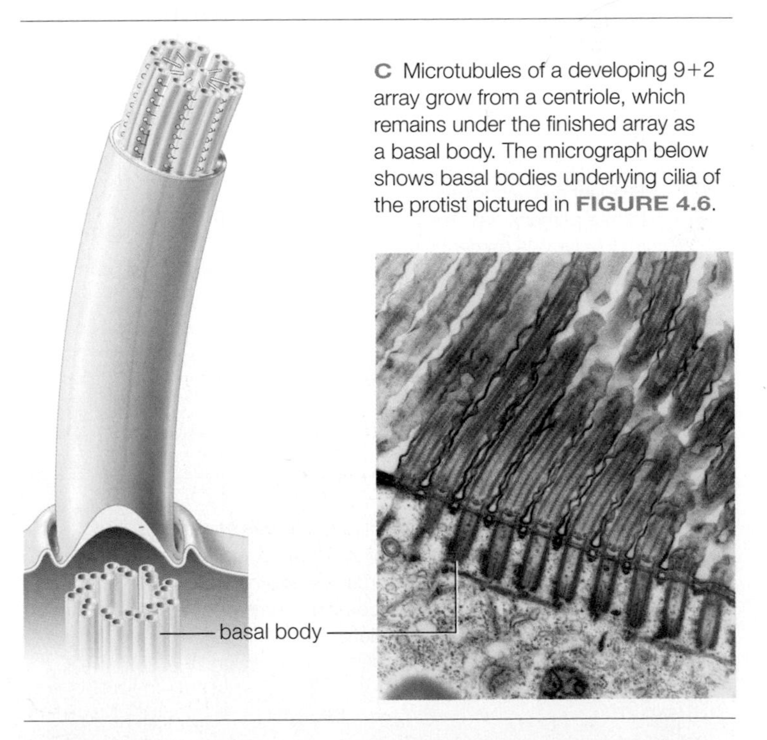

C Microtubules of a developing 9+2 array grow from a centriole, which remains under the finished array as a basal body. The micrograph below shows basal bodies underlying cilia of the protist pictured in **FIGURE 4.6**.

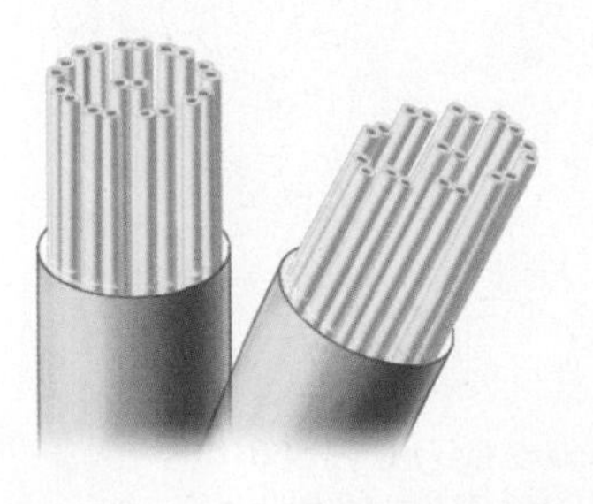

D Phosphate-group transfers from ATP cause the dynein arms in a 9+2 array to repeatedly bind the adjacent pair of microtubules, bend, and then disengage. The dynein arms "walk" along the microtubules, so adjacent microtubule pairs slide past one another. The short, sliding strokes of the dynein arms occur in a coordinated sequence around the ring, down the length of the microtubules. The movement causes the entire structure to bend.

FIGURE 4.21 ▶**Animated** How eukaryotic flagella and cilia move.

4.11 Cell Surface Specializations

✔ Many cells secrete materials that form a covering or matrix outside their plasma membrane.

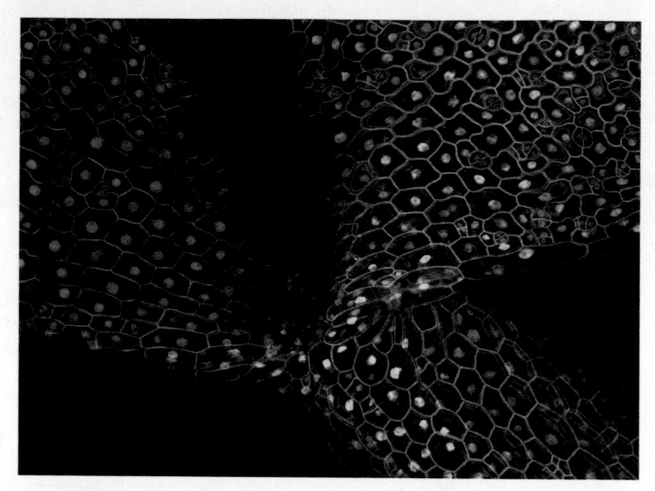

A Nuclei appear orange, and cell walls are green in this fluorescent micrograph of an *Arabidopsis* seedling. The walls, which are rigid and permeable, protect but do not isolate the cells.

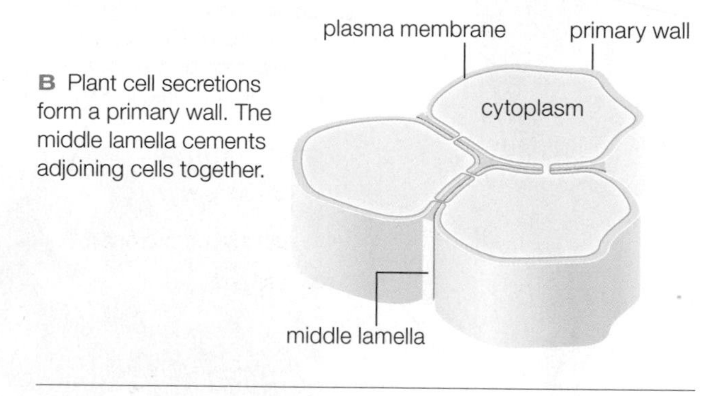

B Plant cell secretions form a primary wall. The middle lamella cements adjoining cells together.

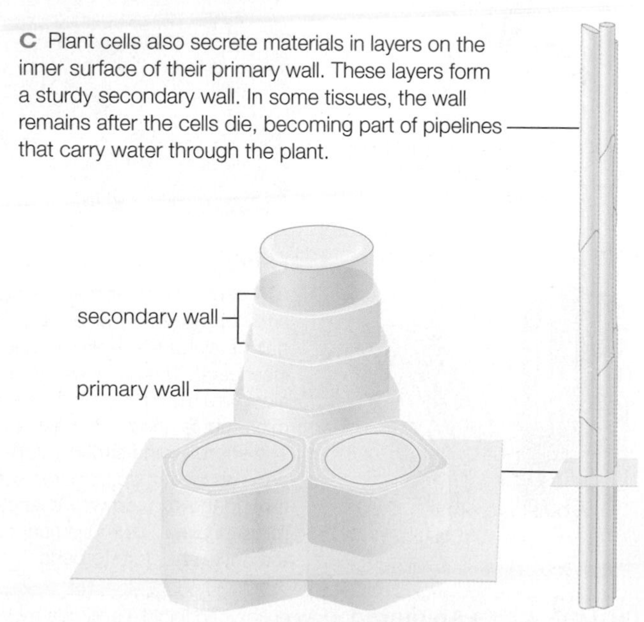

C Plant cells also secrete materials in layers on the inner surface of their primary wall. These layers form a sturdy secondary wall. In some tissues, the wall remains after the cells die, becoming part of pipelines that carry water through the plant.

FIGURE 4.22 ▶Animated Plant cell walls.

Cell Matrices

Many cells secrete an **extracellular matrix (ECM)**, a complex mixture of molecules that often includes polysaccharides and fibrous proteins. The composition and function of ECM vary by cell type.

A cell wall is an example of ECM. You learned in Section 4.4 that many prokaryotes have cell walls. Plants have them too (FIGURE 4.22A), as do fungi and some protists. The composition of the wall differs among these groups, but in all cases it supports and protects the cell. Like a prokaryotic cell wall, a eukaryotic cell wall is porous: Water and solutes easily cross it on the way to and from the plasma membrane.

In plants, the cell wall forms as a young cell secretes pectin and other polysaccharides onto the outer surface of its plasma membrane. The sticky coating is shared between adjacent cells, and it cements them together. Each cell then forms a **primary wall** by secreting strands of cellulose into the coating. Some pectin remains as the middle lamella, a sticky layer in between the primary walls of abutting plant cells (FIGURE 4.22B).

Being thin and pliable, a primary wall allows a growing plant cell to enlarge and change shape. In some plants, mature cells secrete material onto the primary wall's inner surface. These deposits form a firm **secondary wall** (FIGURE 4.22C). One of the materials deposited is **lignin**, an organic compound that makes up as much as 25 percent of the secondary wall of cells in older stems and roots. Lignified plant parts are stronger, more waterproof, and less susceptible to plant-attacking organisms than younger tissues.

Animal cells have no walls, but some types secrete an extracellular matrix called basement membrane. Despite the name, basement membrane is not a cell membrane because it does not consist of a lipid bilayer. Rather, it is a sheet of fibrous material that structurally supports and organizes tissues, and it has roles in cell signaling. Bone is an ECM composed mostly of the fibrous protein collagen, and hardened by deposits of calcium and phosphorus.

A **cuticle** is a type of ECM secreted by cells at a body surface. In plants, a cuticle of waxes and proteins helps stems and leaves fend off insects and retain water. Crabs, spiders, and other arthropods have a cuticle that consists mainly of chitin (Section 3.4).

Cell Junctions

In multicelled species, cells can interact with one another and their surroundings by way of cell junctions. **Cell junctions** are structures that connect a cell directly to other cells or to its environment. Cells send and receive substances and signals through some junctions. Other junctions help cells recognize and stick to each other and to ECM.

Three types of cell junctions are common in animal tissues (**FIGURE 4.23A**). In tissues that line body surfaces and internal cavities, rows of **tight junctions** fasten the plasma membranes of adjacent cells. These junctions prevent body fluids from seeping between the cells. For example, the lining of the stomach is leakproof because tight junctions seal its cells together. These junctions keep gastric fluid, which contains acid and destructive enzymes, safely inside the stomach. If a bacterial infection damages the stomach lining, gastric fluid leaks into and damages the underlying layers. A painful peptic ulcer results.

Adhering junctions fasten cells to one another and to basement membrane. These junctions make a tissue quite strong because they connect to cytoskeletal elements inside the cells. Contractile tissues (such as heart muscle) have a lot of adhering junctions, as do tissues subject to abrasion or stretching (such as skin).

Gap junctions are closable channels that connect the cytoplasm of adjoining animal cells. When open, they permit water, ions, and small molecules to pass directly from the cytoplasm of one cell to another. These channels allow entire regions of cells to respond to a single stimulus. Heart muscle and other tissues in which the cells perform a coordinated action have many gap junctions.

In plants, **plasmodesmata** (singular, plasmodesma) are open channels that connect the cytoplasm of adjoining cells. These cell junctions extend across the cell walls (**FIGURE 4.23B**). Like gap junctions, plasmodesmata allow substances to flow quickly from cell to cell.

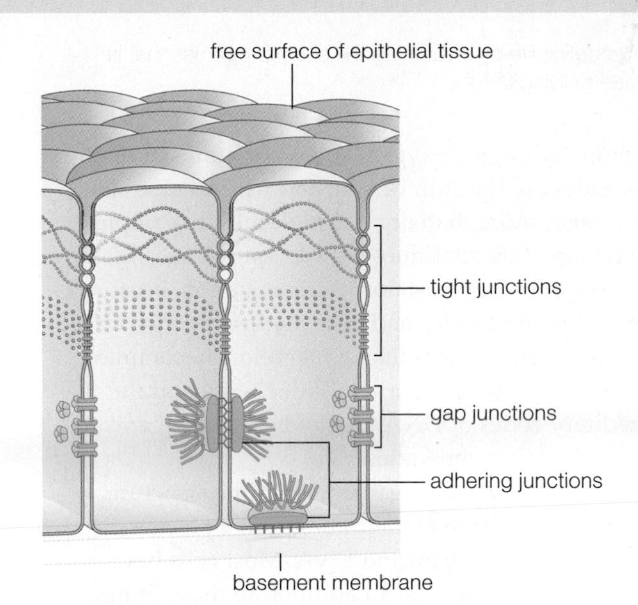

A Three types of cell junctions in animal tissues.

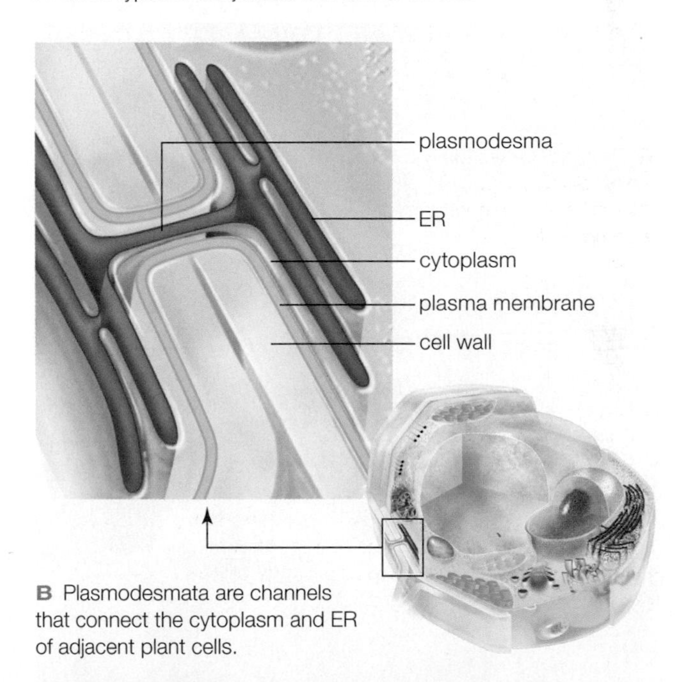

B Plasmodesmata are channels that connect the cytoplasm and ER of adjacent plant cells.

FIGURE 4.23 ▶**Animated** Cell junctions.

adhering junction Cell junction that fastens an animal cell to another cell, or to basement membrane. Connects to cytoskeletal elements inside the cell.
cell junction Structure that connects a cell to another cell or to extracellular matrix.
cuticle Secreted covering at a body surface.
extracellular matrix (ECM) Complex mixture of cell secretions; its composition and function vary by cell type.
gap junction Cell junction that forms a closable channel across the plasma membranes of adjoining animal cells.
lignin Material that strengthens plant cell walls.
plasmodesma Cell junction that forms an open channel between the cytoplasm of adjacent plant cells.
primary wall The first cell wall of young plant cells.
secondary wall Lignin-reinforced wall that forms inside the primary wall of a plant cell.
tight junction Cell junction that fastens together the plasma membrane of adjacent animal cells; collectively prevent fluids from leaking between the cells.

TAKE-HOME MESSAGE 4.11

What structures form on the outside of eukaryotic cells?

✔ Many cells secrete an extracellular matrix (ECM). ECM varies in composition and function depending on cell type.

✔ Plant cells, fungi, and some protists have a cell wall around the plasma membrane. Animal cells have no wall.

✔ Cell junctions structurally and functionally connect cells in tissues. In animal tissues, cell junctions also connect cells with basement membrane.

4.12 The Nature of Life

✔ We define life by describing the set of properties that is unique to living things.

Carbon, hydrogen, oxygen, and other atoms of organic molecules are the stuff of you, and us, and all of life. Yet it takes more than organic molecules to complete the picture. Life continues only as long as an ongoing flow of energy sustains its organization, because assembling molecules and cells requires energy. Life is no more and no less than a marvelously complex system for prolonging order. With energy and the hereditary codes of DNA, matter becomes organized, generation after generation.

In this chapter, you learned about the structure of cells, which have at their minimum a plasma membrane, cytoplasm, and DNA. Most cells have many other components in addition to these things.

We often use differences in cellular components—the presence or absence of a particular organelle, for example—to categorize life's diversity. What about life's commonality? The cell is the smallest unit with the properties of life, but what is it, exactly, that makes a cell, or an organism that consists of them, alive?

According to evolutionary biologist Gerald Joyce, the simplest definition of life might well be "that which is squishy." He says, "Life, after all, is protoplasmic and cellular. It is made up of cells and organic stuff and is undeniably squishy."

Defining life more unambiguously than "squishy" is challenging, if not impossible. Even deciding what sets the living apart from the nonliving can be tricky. For example, living things have a high proportion of the organic molecules of life, but so do the remains of dead organisms in seams of coal. Living things use energy to reproduce themselves, but computer viruses, which are arguably not alive, can do that too.

So how do biologists, who study life as a profession, define it? The short answer is that their best definition is a long list of descriptions that collectively apply to living things, and not to nonliving things. You already know about two of these properties:

1. They make and use the organic molecules of life.
2. They consist of one or more cells.

The remainder of this book details the others properties of life:

3. Living things engage in self-sustaining biological processes such as metabolism and homeostasis.
4. They change over their lifetime, for example by growing, maturing, and aging.
5. They use DNA as their hereditary material when they reproduce.
6. They have the collective capacity to change over successive generations, for example by adapting to environmental pressures.

TAKE-HOME MESSAGE 4.12

What, exactly, is life?

✔ We describe the characteristic of "life" in terms of properties that are collectively unique to living things.

✔ Organisms make and use the organic molecules of life. DNA is their hereditary material.

✔ In living things, the molecules of life are organized as one or more cells that engage in self-sustaining biological processes.

✔ Living things change over lifetimes, and also over successive generations.

Food for Thought (revisited)

One way to ensure food safety involves sterilization, which kills toxic *E. coli* and other bacteria. In the U.S., recalled, contaminated ground beef is often cooked or otherwise sterilized, then processed into ready-to-eat products such as canned chili. Raw beef trimmings, which have a high risk of contact with fecal matter during the butchering process, are effectively sterilized when sprayed with ammonia. Ground to a paste and formed into pellets or blocks, the resulting product is termed "lean finely textured beef" or "boneless lean beef trimmings." Although this product cannot be sold directly to consumers, lean finely textured beef has been and continues to be routinely used as a filler in prepared food products such as hamburger patties, fresh ground beef, hot dogs, lunch meats, sausages, frozen entrees, canned foods, and other items sold to quick service restaurants, hotel and restaurant chains, institutions, and school lunch programs.

A series of news reports in 2012 provoked public outrage at the widespread, unlabeled use of lean finely textured beef, pejoratively nicknamed pink slime. Meat industry organizations and the USDA agree that lean finely textured beef, appetizing or not, is perfectly safe to eat because it has been sterilized—any live bacteria in it have been killed. However, the public controversy prompted some companies to discontinue use of ground beef containing the filler product.

summary

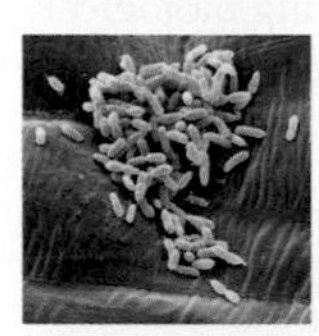

Section 4.1 Bacteria are found in all parts of the biosphere, including the human body. A few types can cause disease. Contamination of food with disease-causing bacteria can result in food poisoning that is sometimes fatal.

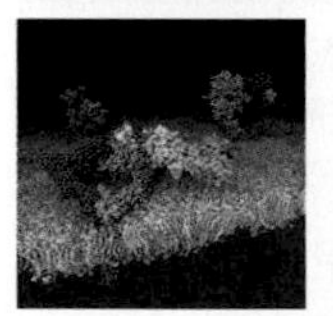

Section 4.2 By the **cell theory**, all organisms consist of one or more cells; the cell is the smallest unit of life; each new cell arises from another, preexisting cell; and a cell passes hereditary material to its offspring. All cells start out life with **cytoplasm**, DNA, and a **plasma membrane** that controls the types and kinds of substances that cross it. Most have many additional components. A eukaryotic cell's DNA is contained within a **nucleus**, which is a membrane-enclosed **organelle**. All cell membranes, including the plasma membrane and organelle membranes, consist mainly of phospholipids organized as a lipid bilayer.

A cell's surface area increases with the square of its diameter, while its volume increases with the cube. This **surface-to-volume ratio** limits cell size and influences cell shape.

Section 4.3 Most cells are far too small to see with the naked eye, so we use microscopes to observe them. Different types of microscopes and techniques reveal different internal and external details of cells.

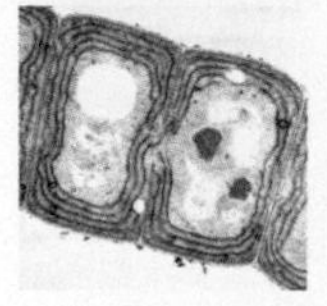

Section 4.4 Bacteria and archaea, informally grouped as prokaryotes, are the most diverse forms of life that we know about. These single-celled organisms have no nucleus, but they do have a **nucleoid** and **ribosomes**. Many also have a protective, rigid **cell wall** and a sticky capsule, and some have motile structures (**flagella**) and other projections (**pili**). There are often **plasmids** in addition to the single circular molecule of DNA. Bacteria and other microbial organisms may live together in a shared mass of slime as a **biofilm**.

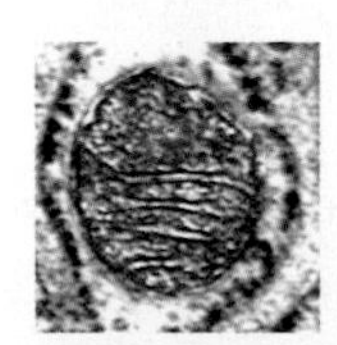

Section 4.5 Protists, fungi, plants, and animals are eukaryotic. Cells of these organisms start out life with a nucleus, and typically many other membrane-enclosed organelles. Organelles compartmentalize tasks and substances that are sensitive or dangerous to the rest of the cell.

Section 4.6 A nucleus protects and controls access to a eukaryotic cell's DNA. A double membrane studded with pores constitutes the **nuclear envelope**. The pores serve as gateways for molecules passing into and out of the nucleus. Inside the nuclear envelope, **chromatin** is suspended in viscous **nucleoplasm**. In the nucleus, ribosome subunits are produced in dense, irregularly shaped **nucleoli**.

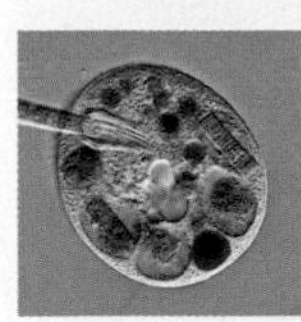

Section 4.7 The **endomembrane system** is a series of organelles (endoplasmic reticulum, Golgi bodies, vesicles) that interact mainly to make lipids, enzymes, and proteins for insertion into membranes or secretion. **Endoplasmic reticulum (ER)** is a continuous system of sacs and tubes extending from the nuclear envelope. Ribosome-studded rough ER makes proteins; smooth ER makes lipids and breaks down carbohydrates and fatty acids. **Golgi bodies** modify proteins and lipids before sorting them into vesicles.

CREDITS: (in text) left, JupiterImages Corporation; right, mrivserg/Shutterstock.com.

Different types of **vesicles** store, break down, or transport substances through the cell. Enzymes in **peroxisomes** break down substances such as amino acids, fatty acids, and toxins. **Lysosomes** contain enzymes that break down cellular wastes and debris. Fluid-filled **vacuoles** store or break down waste, food, and toxins. Fluid pressure inside a **central vacuole** keeps plant cells plump, which in turn keeps plant parts firm.

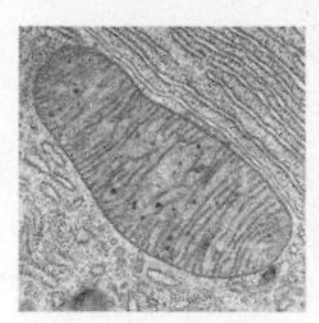

Section 4.8 Double-membraned organelles called **mitochondria** specialize in making many ATP by breaking down organic compounds in the oxygen-requiring pathway of aerobic respiration.

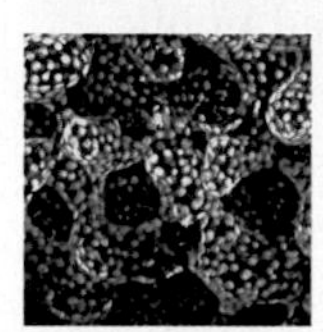

Section 4.9 Different types of **plastids** are specialized for photosynthesis or storage. In eukaryotes, photosynthesis takes place inside **chloroplasts**. Pigment-filled chromoplasts and starch-filled amyloplasts are used for storage; many of these plastids serve additional roles.

Section 4.10 Elements of a **cytoskeleton** reinforce, organize, and move cell structures and often the cell. Cytoskeletal elements include **microtubules, microfilaments,** and **intermediate filaments**. Interactions between ATP-driven **motor proteins** and hollow, dynamically assembled microtubules bring about the movement of cell parts. A microfilament mesh called the **cell cortex** reinforces plasma membranes. Elongating microfilaments bring about movement of **pseudopods**. Intermediate filaments lend structural support to cells and tissues, and they help support the nuclear membrane. **Centrioles** give rise to a special 9+2 array of microtubules inside **cilia** and eukaryotic flagella, then remain beneath these motile structures as **basal bodies**.

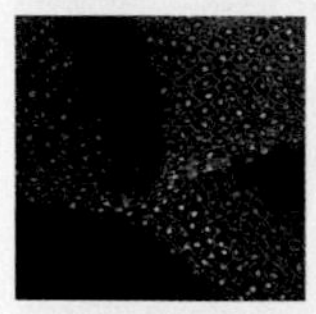

Section 4.11 A secreted mixture of materials forms **extracellular matrix (ECM)** that has different functions depending on the cell type. In animals, a secreted basement membrane supports and organizes cells in tissues. Among the eukaryotes, plant cells, fungi, and many protists secrete a cell wall around their plasma membrane. Older plant cells secrete a rigid, **lignin**-containing **secondary wall** inside their pliable **primary wall**. Many eukaryotic cell types also secrete a protective **cuticle**. **Plasmodesmata** are open **cell junctions** that connect the cytoplasm of adjacent plant cells. In animals, **gap junctions** are closable channels between adjacent cells. **Adhering junctions** that connect to cytoskeletal elements fasten cells to one another and to basement membrane. **Tight junctions** form a waterproof seal between cells.

Section 4.12 Differences among cell components (TABLE 4.5) allow us to categorize life, but not to define it. We can describe the quality of "life" as a set of properties that are collectively unique to living things. Living things consist of cells that engage in self-sustaining biological processes, pass their hereditary material (DNA) to offspring, and have the capacity to change.

Table 4.5 Comparing Components of Prokaryotic and Eukaryotic Cells

			Eukaryotes			
Cell Component	**Example(s) of Function**	**Prokaryotes**	**Protists**	**Fungi**	**Plants**	**Animals**
Cell wall	Protection, structural support	+	+	+	+	–
Plasma membrane	Control of substances moving into and out of cell	+	+	+	+	+
Nucleus	Physical separation of DNA from cytoplasm	–	+	+	+	+
Nucleolus	Assembly of ribosome subunits	–	+	+	+	+
DNA	Encoding of hereditary information	+	+	+	+	+
RNA	Protein synthesis	+	+	+	+	+
Ribosome	Protein synthesis	+	+	+	+	+
Endoplasmic reticulum	Protein, lipid synthesis; carbohydrate, fatty acid breakdown	–	+	+	+	+
Golgi body	Final modification of proteins, lipids	–	+	+	+	+
Lysosome	Intracellular digestion	–	+	+	+	+
Peroxisome	Breakdown of fatty acids, amino acids, and toxins	–	+	+	+	+
Mitochondrion	Production of ATP by aerobic respiration	–	+	+	+	+
Hydrogenosome	Anaerobic production of ATP	–	+	+	–	+
Chloroplast	Photosynthesis; starch storage	–	+	–	+	–
Central vacuole	Increasing cell surface area; storage	–	–	+	+	–
Flagellum	Locomotion through fluid surroundings	+	+	+	+	+
Cilium	Locomotion through fluid surroundings; movement of surrounding fluid	+	+	–	+	+
Cytoskeleton	Physical reinforcement; internal organization; movement	+	+	+	+	+

+ found in at least some species; – not found in any species.

data analysis activities

Abnormal Motor Proteins Cause Kartagener Syndrome An abnormal form of a motor protein called dynein causes Kartagener syndrome, a genetic disorder characterized by chronic sinus and lung infections. Biofilms form in the thick mucus that collects in the airways and the resulting bacterial activities and inflammation damage tissues.

Affected men can produce sperm but are infertile. Some have become fathers after a doctor injects their sperm cells directly into eggs. Review **FIGURE 4.21** and **FIGURE 4.24**, then explain how abnormal dynein could cause the observed effects.

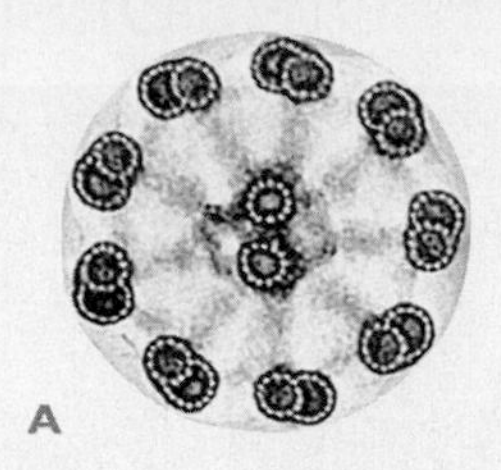

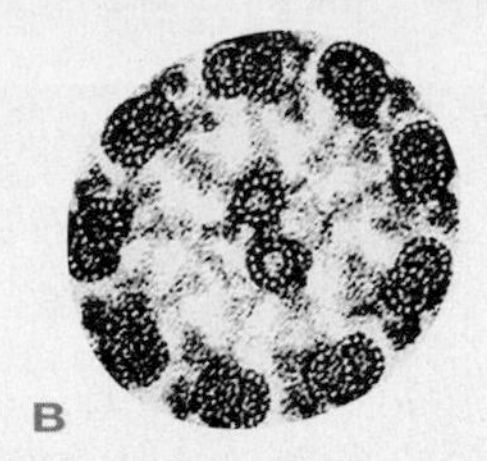

FIGURE 4.24 Cross-section of the flagellum of a sperm cell from **A** a human male affected by Kartagener syndrome and **B** an unaffected male.

self-quiz

Answers in Appendix VII

1. All cells have these three things in common: ________ .
   a. cytoplasm, DNA, and organelles with membranes
   b. a plasma membrane, DNA, and a nucleus
   c. cytoplasm, DNA, and a plasma membrane
   d. a cell wall, cytoplasm, and DNA

2. Name one major principle of the cell theory.

3. The surface-to-volume ratio ________ .
   a. does not apply to prokaryotic cells
   b. constrains cell size
   c. is part of the cell theory
   d. b and c

4. True or false? Some protists start out life with no nucleus.

5. Cell membranes consist mainly of ______ and ______ .
   a. lipids; carbohydrates
   b. phospholipids; protein
   c. lipids; carbohydrates
   d. phospholipids; ECM

6. True or false? Ribosomes are only found in eukaryotes.

7. Unlike eukaryotic cells, prokaryotic cells ________ .
   a. have no plasma membrane
   b. have RNA but not DNA
   c. have no nucleus
   d. a and c

8. Enzymes contained in ________ break down worn-out organelles, bacteria, and other particles.
   a. lysosomes
   b. amyloplasts
   c. endoplasmic reticulum
   d. peroxisomes

9. Put the following structures in order according to the pathway of a secreted protein:
   a. plasma membrane
   b. Golgi bodies
   c. endoplasmic reticulum
   d. post-Golgi vesicles

10. The main function of the endomembrane system is:
    a. building and modifying proteins and lipids
    b. isolating DNA from toxic substances
    c. secreting extracellular matrix onto the cell surface
    d. producing ATP by aerobic respiration

11. True or false? The plasma membrane is the outermost component of all cells. Explain.

12. Cell junctions called ________ connect the cytoplasm of plant cells.

13. Which of the following organelles contains no DNA?
    a. nucleus
    b. Golgi body
    c. mitochondrion
    d. chloroplast

14. No animal cell has a ________ .
    a. plasma membrane
    b. flagellum
    c. lysosome
    d. cell wall

15. Match each cell component with a function.

___ mitochondrion	a. protein synthesis
___ chloroplast	b. connection
___ cell junction	c. stores starch
___ smooth ER	d. breaks down toxins
___ Golgi body	e. sorts and ships
___ rough ER	f. assembles lipids
___ peroxisome	g. photosynthesis
___ amyloplast	h. ATP production
___ flagellum	i. movement

CENGAGE brain To access course materials, please visit www.cengagebrain.com.

critical thinking

1. In a classic episode of *Star Trek*, a gigantic amoeba engulfs an entire starship. Spock blows the cell to bits before it can reproduce. Think of at least one inaccuracy that a biologist would identify in this scenario.

2. In plants, the cell wall forms as a young plant cell secretes polysaccharides onto the outer surface of its plasma membrane. Being thin and pliable, this primary wall allows the cell to enlarge and change shape. In mature woody plants, cells in some tissues deposit material onto the primary wall's inner surface. Why doesn't this secondary wall form on the outer surface of the primary wall?

3. Which structures can you identify in the organism below? Is it prokaryotic or eukaryotic? How can you tell?

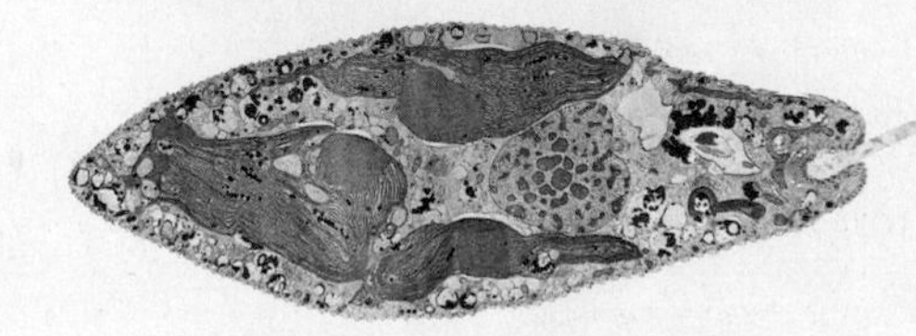

CREDITS: (24) From "Tissue & Cell", Vol. 27, pp.421–427, Courtesy of Bjorn Afzelius, Stockholm University; (in text CT) P. L. Walne and J. H. Arnott, Planta, 77:325-354, 1967.

5 Ground Rules of Metabolism

LEARNING ROADMAP

In this chapter, you will gain insight into the one-way flow of energy (Section 1.3) through the world of life (1.2). You will learn more about energy, including the laws of nature (1.9) that describe it, and heat (2.5). This chapter also revisits the structure and function of atoms (2.3), molecules (2.4, 3.2, 3.3–3.6), and cells (4.6, 4.7, 4.10, 4.11).

ENERGY FLOW

Each time energy is transferred, some of it disperses. An organism can sustain its life only as long as it continues to harvest energy from the environment.

HOW ENZYMES WORK

Enzymes increase the rate of chemical reactions. They are assisted by cofactors, and affected by temperature, salinity, pH, and other environmental factors.

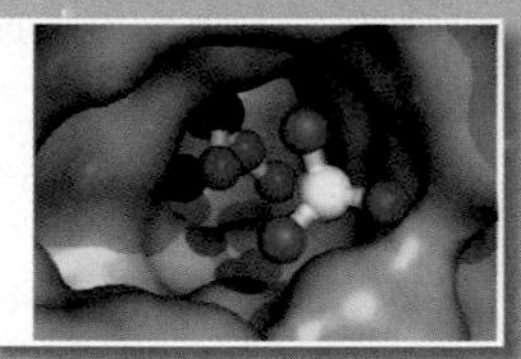

THE NATURE OF METABOLISM

Sequences of enzyme-mediated reactions build, remodel, and break down organic molecules. Controls that govern steps in these pathways quickly shift cell activities.

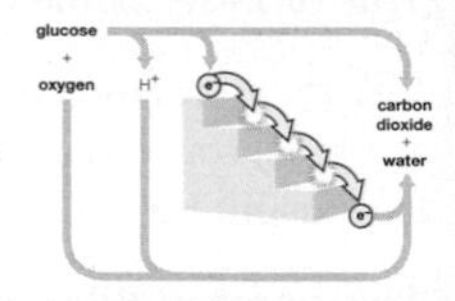

MOVEMENT OF FLUIDS

Gradients drive the directional movements of substances across membranes. Water tends to diffuse across a cell membrane to a region of higher solute concentration.

MEMBRANE TRANSPORT

Transport proteins control solute concentrations in cells and organelles by helping substances move across membranes. Substances also move across cell membranes inside vesicles.

The next two chapters detail metabolic pathways of photosynthesis (Chapter 6) and aerobic respiration (Chapter 7). You will reencounter metabolism in the context of cancer (Chapter 10) and inherited disease (Chapter 14); and membrane proteins in processes of immunity (Chapter 37). Later chapters return to membrane transport and calcium ions in cell signaling (Chapters 30, 32, and 35), and enzymes in digestion (Chapter 39).

5.1 A Toast to Alcohol Dehydrogenase

Most college students are under the legal drinking age, but alcohol abuse continues to be the most serious drug problem on college campuses throughout the United States. Before you drink, consider what you are consuming. All alcoholic drinks—beer, wine, hard liquor—contain the same psychoactive ingredient: ethanol. Ethanol molecules move quickly from the stomach and small intestine into the bloodstream. Almost all of the ethanol ends up in the liver, a large organ in the abdomen. Liver cells have impressive numbers of enzymes. One of them, ADH (alcohol dehydrogenase), helps break down ethanol and other toxic compounds (**FIGURE 5.1**).

If you put more ethanol into your body than your enzymes can deal with, then you will damage it. Ethanol and its breakdown products harm liver cells, so the more a person drinks, the fewer liver cells are left to do the breaking down. Ethanol also interferes with normal processes of metabolism. For example, oxygen that would ordinarily take part in breaking down fatty acids is diverted to breaking down ethanol. As a result, fats tend to accumulate as large globules in the tissues of heavy drinkers.

Long-term heavy drinking causes alcoholic hepatitis, a disease characterized by inflammation and destruction of liver tissue. It also causes cirrhosis, a condition in which the liver becomes so scarred, hardened, and filled with fat that it loses its function. (The term cirrhosis is from the Greek *kirros*, meaning "orange-colored," after the abnormal skin color of people with the disease.) The liver is the largest gland in the human body, and it has many important functions. In addition to breaking down fats and toxins, it helps regulate the body's blood sugar level, and it makes proteins that are essential for blood clotting, immune function, and maintaining the solute balance of body fluids. Loss of these functions can be deadly.

Heavy drinking is dangerous in the short term too. Tens of thousands of undergraduate students have been polled about their drinking habits in recent surveys. More than half of them reported that they regularly drink five or more alcoholic beverages within a two-hour period—a self-destructive behavior called binge drinking. Consuming large amounts of alcohol in a brief period of time does far more than damage one's liver. Aside from the related 500,000 injuries from accidents, the 600,000 assaults by intoxicated students, 100,000 cases of date rape, and 400,000 incidences of unprotected sex among students, binge drinking is responsible for killing or causing the death of more than 1,700 college students every year.

With this sobering example, we invite you to learn about how and why your cells break down organic compounds, including toxic molecules such as ethanol.

FIGURE 5.1 Alcohol dehydrogenase. This enzyme helps the body break down toxic alcohols such as ethanol, making it possible for humans to drink beer, wine, and other alcoholic beverages. The photo shows a tailgate party at a Notre Dame–Alabama football game. During 2012 alone, Indiana State police arrested 138 Notre Dame students for underage drinking at tailgate parties.

5.2 Energy in the World of Life

✔ Sustaining life's organization requires ongoing energy inputs.

Energy Disperses

Energy is formally defined as the capacity to do work, but this definition is not very satisfying. Even brilliant physicists who study energy cannot say exactly what it is. However, we do have an intuitive understanding of energy just by thinking about familiar forms of it, such as light, heat, electricity, and motion. We also understand intuitively that one form of energy can be converted to another. Think about how a lightbulb changes electricity into light, or how an automobile changes the chemical energy of gasoline into the energy of motion, which is also called **kinetic energy**.

The formal study of heat and other forms of energy is thermodynamics (*therm* is a Greek word for heat; *dynam* means energy). By making careful measurements, thermodynamics researchers discovered that the total amount of energy before and after every conversion is always the same. In other words, energy cannot be created or destroyed—a phenomenon that is the **first law of thermodynamics**. Remember, a law of nature describes something that occurs without fail, but our explanation of why it occurs is incomplete (Section 1.9).

Energy also tends to spread out, or disperse, until no part of a system holds more than another part. In a kitchen, for example, heat always flows from a hot pan to cool air until the temperature of both is the same. We never see cool air raising the temperature of a hot pan. **Entropy** is a measure of how much the energy of a particular system has become dispersed. We can use the hot pan in a cool kitchen as an example of a system. As heat flows from the pan into the air, the entropy of the system increases (**FIGURE 5.2**). Entropy continues to increase until the heat is evenly distributed throughout the kitchen, and there is no longer a net (or overall) flow of heat from one area to another. Our system has now reached its maximum entropy with respect to heat. The tendency of entropy to increase is the **second law of thermodynamics**. This is the formal way of saying that energy tends to spread out spontaneously.

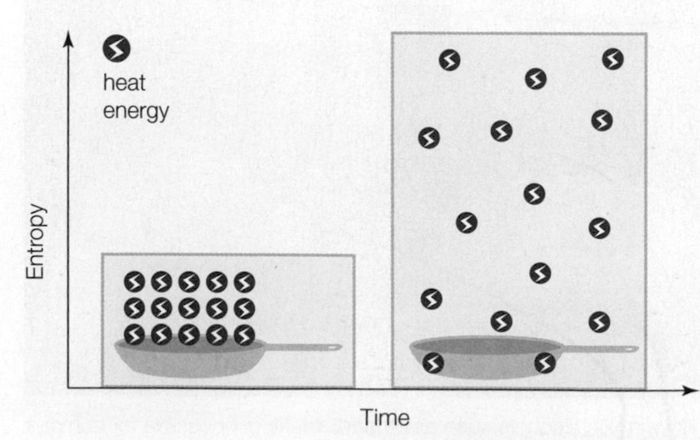

FIGURE 5.2 Entropy. Entropy tends to increase, but the total amount of energy in any system always stays the same.

FIGURE 5.3 It takes more than 10,000 pounds of soybeans and corn to raise a 1,000-pound steer. Where do the other 9,000 pounds go? About half of the steer's food is indigestible and passes right through it. The animal's body breaks down molecules in the remaining half to access energy stored in chemical bonds. Only about 15 percent of that energy goes toward building body mass. The rest is lost during energy conversions, as metabolic heat.

Biologists use the concept of entropy as it applies to chemical bonding, because energy flow in living things occurs mainly by the making and breaking of chemical bonds. How is entropy related to chemical bonding? Think about it just in terms of motion. Two unbound atoms can vibrate, spin, and rotate in every direction, so they are at high entropy with respect to motion. A covalent bond between the atoms restricts their movement, so they are able to move in fewer ways than they did before bonding. Thus, the entropy of two atoms decreases when a bond forms between them. Such entropy changes are part of the reason why some reactions occur spontaneously and others require an energy input, as you will see in the next section.

Energy's One-Way Flow

Work occurs as a result of energy transfers. Consider how it takes work to push a box across a floor. In this case, a body (you) transfers energy to another body (the box) to make it move. Similarly, a plant cell works to make sugars. Inside the cell, one set of molecules harvests energy from light, then transfers it to another set of molecules. The second set of molecules uses the energy to build the sugars from carbon dioxide and water. This particular energy transfer involves the conversion of light energy to chemical energy. Most other types of cellular work occur by the transfer of chemical energy from one molecule to another.

energy The capacity to do work.
entropy Measure of how much the energy of a system is dispersed.
first law of thermodynamics Energy cannot be created or destroyed.
kinetic energy The energy of motion.
potential energy Stored energy.
second law of thermodynamics Energy tends to disperse spontaneously.

As you learn about such processes, remember that every time energy is transferred, a bit of it disperses. Energy lost from a transfer is usually in the form of heat. As a simple example, a typical incandescent lightbulb converts only about 5 percent of the energy of electricity into light. The remaining 95 percent of the energy ends up as heat that disperses from the bulb.

Dispersed heat is not useful for doing work, and it is not easily converted to a more useful form of energy (such as electricity). Because some energy in every transfer disperses as heat, and heat is not useful for doing work, we can say that the total amount of energy in the universe available for doing work is always decreasing.

Is life an exception to this inevitable flow? An organized body is hardly dispersed. Energy becomes concentrated in each new organism as the molecules of life organize into cells. Even so, living things constantly use energy—to grow, to move, to acquire nutrients, to reproduce, and so on—and some energy is lost in every one of these processes (**FIGURE 5.3**). Unless those losses are replenished with energy from another source, the complex organization of life will end.

The energy that fuels most life on Earth comes from the sun. That energy flows through producers such as plants, then consumers such as animals (**FIGURE 5.4**). During this journey, the energy is transferred many times. With each transfer, some energy escapes as heat until, eventually, all of it is permanently dispersed. However, the second law of thermodynamics does not say how quickly the dispersal has to happen. Energy's spontaneous dispersal is resisted by chemical bonds. The energy in chemical bonds is a type of **potential energy**, which is energy stored in the position or arrangement of objects in a system (**FIGURE 5.5**). Think of all the bonds in the countless molecules that make up your skin, heart, liver, fluids, and other body parts. Those bonds hold the molecules, and you, together—at least for the time being.

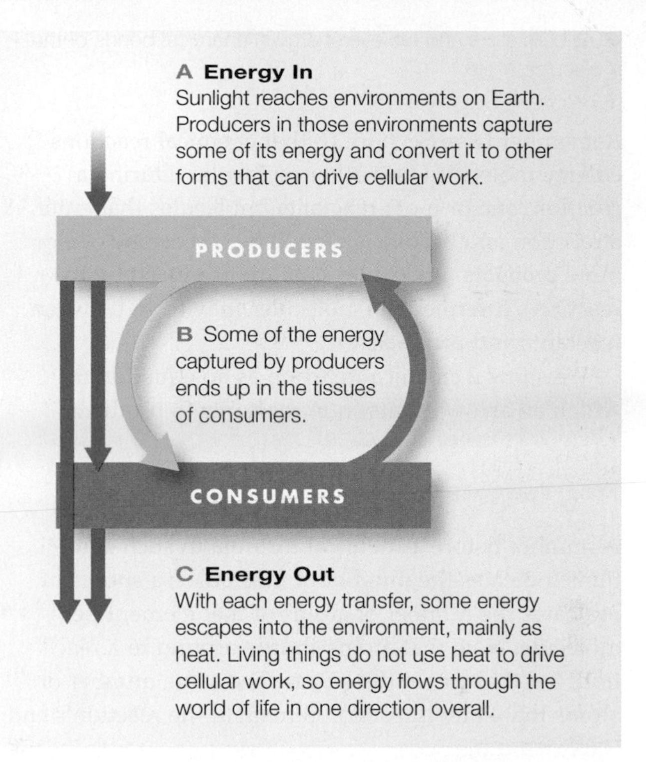

FIGURE 5.4 ▶**Animated** Energy flows from the environment into living organisms, then back to the environment. The flow drives a cycling of materials among producers and consumers.

FIGURE 5.5 Illustration of potential energy. By opposing gravity's downward pull, the rope attached to the rock keeps the man from falling. Similarly, a chemical bond attaches two atoms and keeps them from flying apart.

TAKE-HOME MESSAGE 5.2

What is energy?

- ✔ Energy, which is the capacity to do work, cannot be created or destroyed.
- ✔ Energy disperses spontaneously.
- ✔ Energy can be transferred between systems or converted from one form to another, but some is lost (as heat, typically) during every such exchange.
- ✔ Sustaining life's organization requires ongoing energy inputs to counter energy loss. Organisms stay alive by replenishing themselves with energy they harvest from someplace else.

5.3 Energy in the Molecules of Life

✔ All cells store and retrieve energy in chemical bonds of the molecules of life.

Remember from Section 3.3 that chemical reactions change molecules into other molecules. During a reaction, one or more **reactants** (molecules that enter a reaction and become changed by it) become one or more **products** (molecules that are produced by the reaction). Intermediate molecules may form between reactants and products.

We show a chemical reaction as an equation in which an arrow points from reactants to products:

$$\underset{\text{(hydrogen)}}{2H_2} + \underset{\text{(oxygen)}}{O_2} \longrightarrow \underset{\text{(water)}}{2H_2O}$$

A number before a chemical formula in such equations indicates the number of molecules; a subscript indicates the number of atoms of that element per molecule. Note that atoms shuffle around in a reaction, but they never disappear: The same number of atoms that enter a reaction remain at the reaction's end (**FIGURE 5.6**).

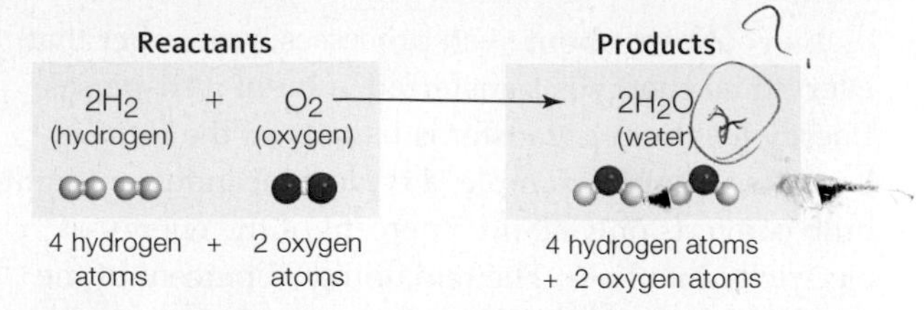

FIGURE 5.6 ▸**Animated** Chemical bookkeeping. In equations that represent chemical reactions, reactants are written to the left of an arrow that points to the products. A number before a formula indicates the number of molecules. The same number of atoms that enter the reaction remain at its end.

Chemical Bond Energy

Every chemical bond holds a certain amount of energy. That is the amount of energy required to break the bond, and it is also the amount of energy released when the bond forms. The particular amount of energy held by a bond depends on which elements are taking part in it. For example, two covalent bonds—one between an oxygen and a hydrogen atom in a water molecule, the other between two oxygen atoms in molecular oxygen (O_2)—both hold energy, but different amounts of it.

Bond energy and entropy both contribute to a molecule's free energy, which is the amount of energy that is available ("free") to do work. In most reactions, the free energy of reactants differs from the free energy of products. If the reactants have less free energy than the products, the reaction will not proceed without a net energy input. Such reactions are **endergonic**, which means "energy in" (**FIGURE 5.7A**). If the reactants have more free energy than the products, the reaction will end with a net release of energy. Such reactions are **exergonic**, which means "energy out" (**FIGURE 5.7B**).

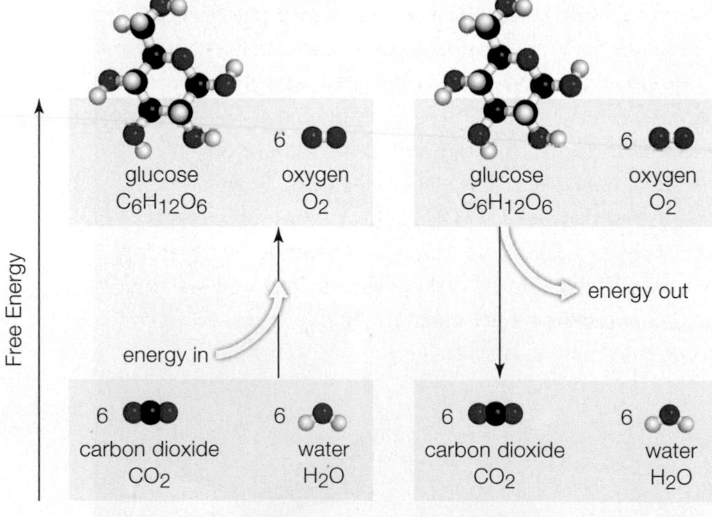

A Endergonic reactions convert molecules with lower free energy to molecules with higher free energy, so they require a net energy input in order to proceed.

B Exergonic reactions convert molecules with higher free energy to molecules with lower free energy, so they end with an energy release.

FIGURE 5.7 ▸**Animated** The ins and outs of energy in chemical reactions. **FIGURE IT OUT** Which law of thermodynamics explains energy inputs and outputs in chemical reactions?

Answer: The first law

Why Earth Does Not Go Up in Flames

The molecules of life release energy when they combine with oxygen. Think of how a spark ignites tinder-dry wood in a campfire. Wood is mostly cellulose, which consists of long chains of repeating glucose monomers (Section 3.4). A spark starts a reaction that converts cellulose (in wood) and oxygen (in air) to water and carbon dioxide. The reaction is highly exergonic, which means it releases a lot of energy—enough to initiate the same reaction with other cellulose and oxygen molecules. That is why wood keeps burning after it has been lit.

Earth is rich in oxygen—and in potential exergonic reactions. Why doesn't it burst into flames? Luckily, chemical bonds do not break without at least a small input of energy, even in an energy-releasing reaction. We call this input activation energy. **Activation energy**, the minimum amount of energy required to get a chemical reaction started, is a bit like a hill that reactants must climb before they can coast down the other side to become products (**FIGURE 5.8**).

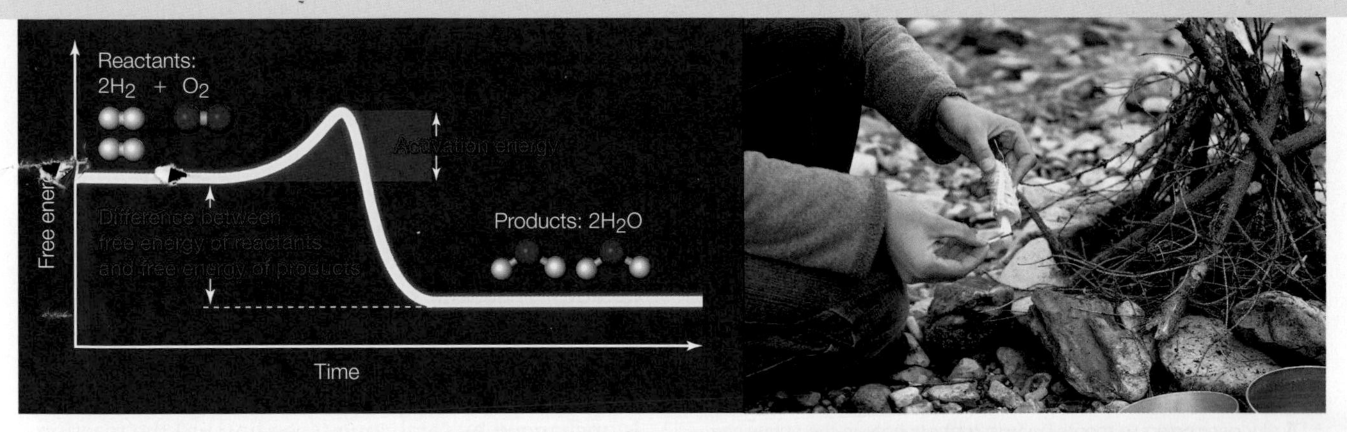

FIGURE 5.8 ▸Animated Activation energy. Most reactions will not begin without an input of activation energy, which is shown in the graph as a bump in a free energy hill. Reactants in this example have more energy than the products. Activation energy keeps this and other exergonic reactions, including the combustion of cellulose in wood, from starting spontaneously.

Both endergonic and exergonic reactions have activation energy, but the amount varies with the reaction. Consider guncotton (nitrocellulose), a highly explosive derivative of cellulose. Christian Schönbein accidentally discovered a way to make it when he used his wife's cotton apron to wipe up a nitric acid spill on his kitchen table, then hung it up to dry next to the oven. The apron exploded. Being a chemist in the 1800s, Schönbein immediately thought of marketing guncotton as a firearm explosive, but it proved to be too unstable to manufacture. So little activation energy is needed to make guncotton react with oxygen that it tends to explode unexpectedly. Several manufacturing plants burned to the ground before guncotton was abandoned for use as a firearm explosive. The substitute? Gunpowder, which has a higher activation energy for a reaction with oxygen.

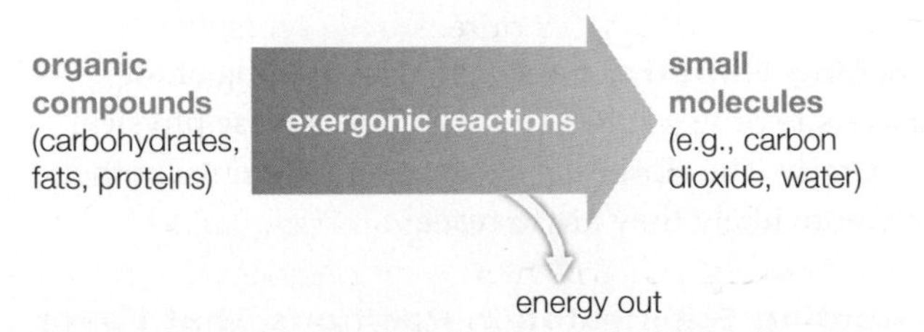

A Cells run endergonic reactions that store energy in the bonds of organic compounds.

organic compounds (carbohydrates, fats, proteins) → **exergonic reactions** → small molecules (e.g., carbon dioxide, water); energy out

B Cells run exergonic reactions that retrieve energy stored in the bonds of organic compounds.

FIGURE 5.9 How cells store and retrieve free energy.

activation energy Minimum amount of energy required to start a chemical reaction.
endergonic Describes a reaction that requires a net input of free energy to proceed.
exergonic Describes a reaction that ends with a net release of free energy.
product A molecule that is produced by a reaction.
reactant A molecule that enters a reaction and is changed by participating in it.

Energy In, Energy Out

Cells store energy by running endergonic reactions that build organic compounds (**FIGURE 5.9A**). For example, light energy drives the overall reactions of photosynthesis, which produce sugars such as glucose from carbon dioxide and water. Unlike light, glucose can be stored in a cell. Cells harvest energy by running exergonic reactions that break the bonds of organic compounds (**FIGURE 5.9B**). Most cells do this when they carry out the overall reactions of aerobic respiration, which releases the energy of glucose by breaking the bonds between its carbon atoms. You will see in the next few sections how cells use energy released from some reactions to drive others.

TAKE-HOME MESSAGE 5.3
How do cells use energy?

✔ Endergonic reactions will not run without a net input of energy. Exergonic reactions end with a net release of energy.

✔ Both endergonic and exergonic reactions require an input of activation energy to begin.

✔ Cells store energy in chemical bonds by running endergonic reactions that build organic compounds. To release this stored energy, they run exergonic reactions that break the bonds.

5.4 How Enzymes Work

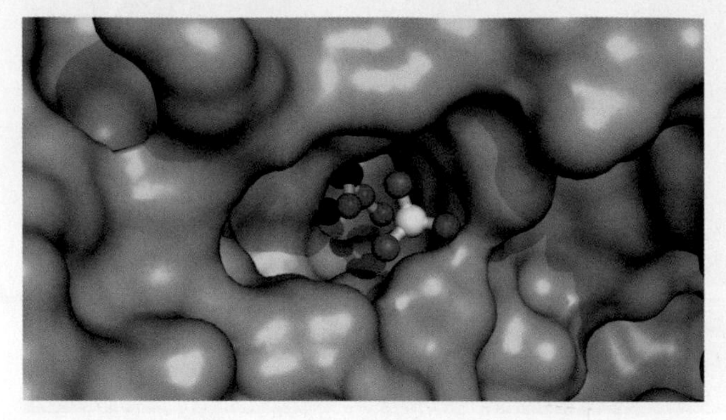

A A glucose molecule meets up with a phosphate in the active site of a hexokinase enzyme.

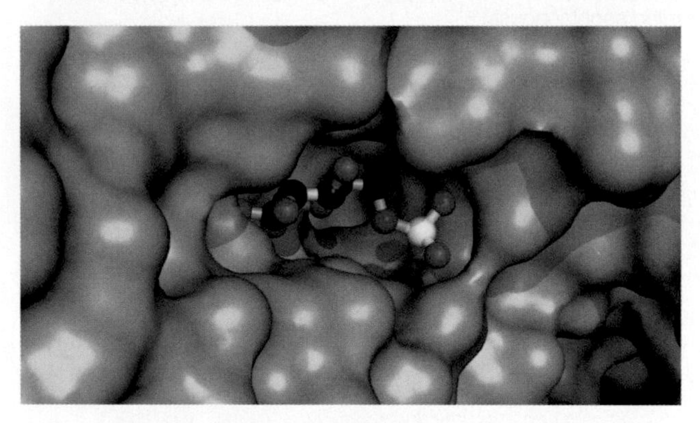

B The enzyme has catalyzed the reaction between glucose and phosphate. The product of this reaction, glucose-6-phosphate, is shown leaving the active site.

FIGURE 5.10 Example of an active site. This one is in hexokinase, an enzyme that adds a phosphate group to glucose and other six-carbon sugars.

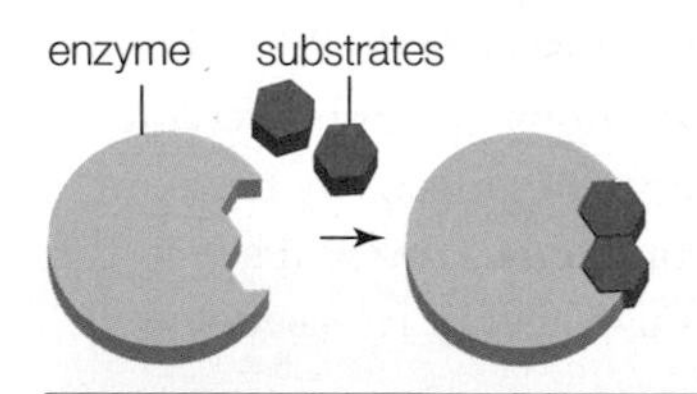

A An active site binds substrates that are complementary in shape, size, polarity, and charge.

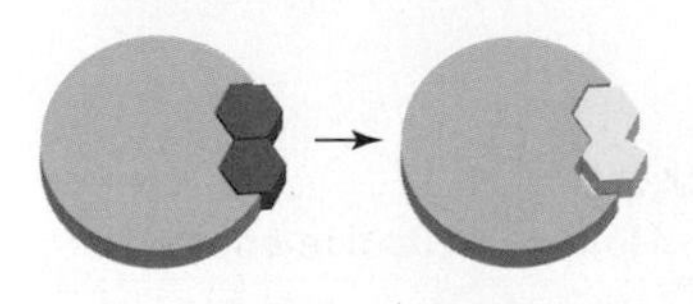

B The binding squeezes substrates together, influences their charge, or causes some change that lowers activation energy, so the reaction proceeds.

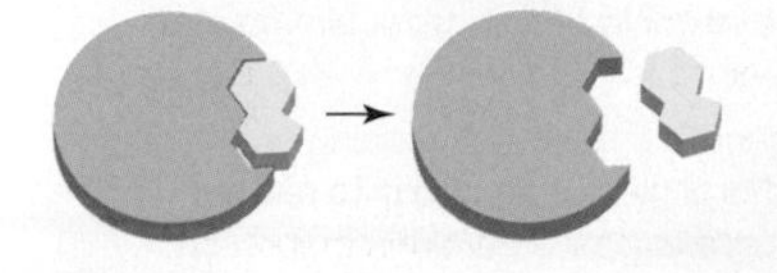

C The product leaves the active site after the reaction is finished. The enzyme is unchanged, so it can work again.

FIGURE 5.11 How an active site works.

✔ Enzymes make specific reactions occur much faster than they would on their own.

The Need for Speed

Metabolism requires enzymes. Why? Consider that a molecule of glucose can break down to carbon dioxide and water on its own, but the process might take decades. That same conversion takes just seconds inside your cells. Enzymes make the difference. In a process called **catalysis**, an enzyme makes a reaction run much faster than it would on its own. The enzyme is unchanged by catalyzing (speeding up) the reaction, so it can work again and again.

Most enzymes are proteins, but some are RNAs. Each kind of enzyme interacts only with specific reactants, or **substrates**, and alters them in a specific way. Such specificity occurs because an enzyme's polypeptide chains fold up into one or more **active sites**, which are pockets where substrates bind and where reactions proceed (FIGURE 5.10). An active site is complementary in shape, size, polarity, and charge to the enzyme's substrate (FIGURE 5.11). This fit is the reason why each enzyme acts in a specific way on a specific substrate.

The Transition State

When we talk about activation energy, we are really talking about the energy required to bring reactant bonds to their breaking point. At that point, which is called the **transition state**, the reaction can run without any additional energy input. Enzymes bring on the transition state by lowering activation energy (FIGURE 5.12). They do so by the following four mechanisms.

Forcing Substrates Together Binding at an active site brings substrates together in close physical proximity. The closer the substrates are to one another, the more likely they are to react.

Orienting Substrates in Positions That Favor Reaction Substrate molecules in a solution collide from random directions. By contrast, binding at an active site positions substrates optimally for reaction.

Inducing a Fit Between Enzyme and Substrate By the **induced-fit model**, an enzyme's active site is not quite complementary to its substrate. Interacting with a substrate molecule causes the enzyme to change shape so that the fit between them improves. The improved fit may result in a stronger bond between enzyme and substrate.

CREDITS: (10) PDB ID: 1GZX; Paoli, M., Liddington, R., Tame, J., Wilkinson, A., Dodson, G., Crystal structure of T state hemoglobin with oxygen bound at all four haems. *J.Mol.Bio.*, v256, pp. 775–792, 1996; (11) © Cengage Learning.

Excluding Water Metabolism occurs in water-based fluids, but water molecules can interfere with certain reactions. The active sites of some enzymes repel water, and keep it away from the reactions.

Enzyme Activity

Environmental factors such as pH, temperature, salt, and pressure influence an enzyme's shape, and so influence its function (Sections 3.6 and 3.7). Each enzyme works best in a particular range of conditions that reflect the environment in which it evolved.

Consider pepsin, a digestive enzyme that works best at low pH (**FIGURE 5.13A**). Pepsin begins the process of protein digestion in the very acidic environment of the stomach (pH 2). During digestion, the stomach's contents pass into the small intestine, where the pH rises to about 7.5. Pepsin denatures (unfolds) above pH 5.5, so this enzyme becomes inactive in the small intestine. Here, protein digestion continues with the assistance of trypsin, an enzyme that functions well at the higher pH.

Adding heat boosts free energy, which is why the jiggling motion of atoms and molecules (Section 2.5) increases with temperature. The greater the free energy of reactants, the closer they are to reaching activation energy. Thus, the rate of an enzymatic reaction typically increases with temperature—but only up to a point. An enzyme denatures above a characteristic temperature. Then, the reaction rate falls sharply as the shape of the enzyme changes and it stops working (**FIGURE 5.13B**). Body temperatures above 42°C (107.6°F) adversely affect the function of many of your enzymes, which is why such severe fevers are dangerous.

The activity of many enzymes is also influenced by the amount of salt in the surrounding fluid. Too little salt, and polar parts of the enzyme attract one another so strongly that the enzyme's shape changes. Too much salt interferes with the hydrogen bonds that hold the enzyme in its characteristic shape, so the enzyme denatures (Section 3.7).

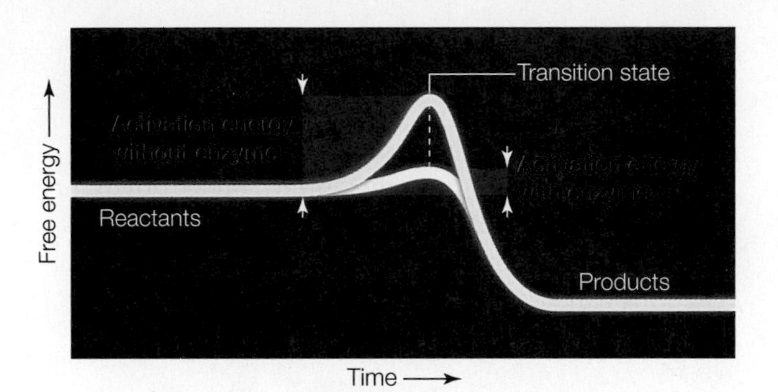

FIGURE 5.12 The transition state. An enzyme enhances the rate of a reaction by lowering activation energy.

FIGURE IT OUT Is this reaction endergonic or exergonic?

Answer: Exergonic

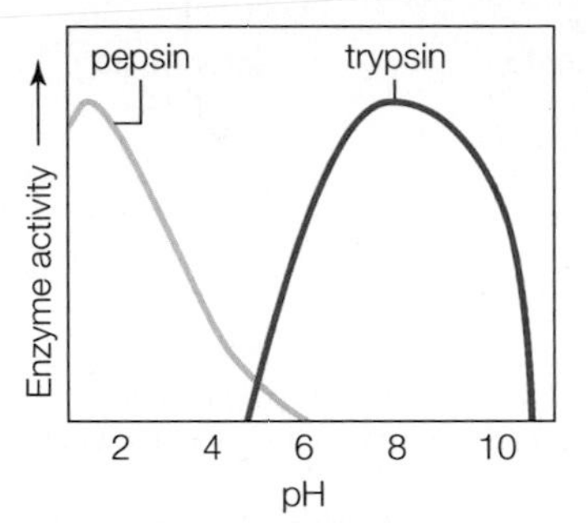

A The pH-dependent activity of two digestive enzymes, pepsin and trypsin. Pepsin acts in the stomach, where the normal pH is 2. Trypsin acts in the small intestine, where the normal pH is around 7.5.

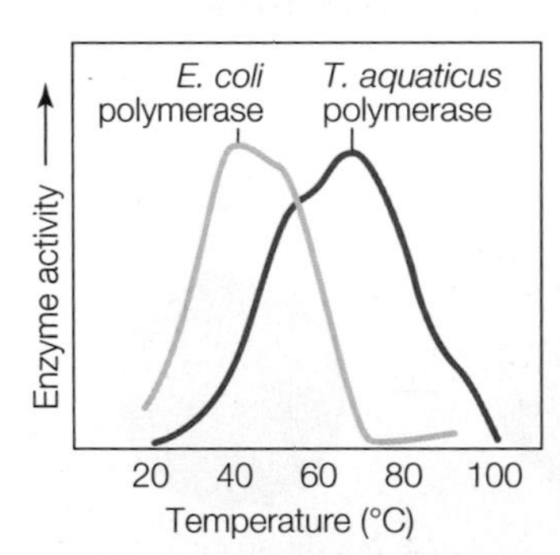

B Comparison of temperature-dependent activity of a DNA synthesis enzyme from two species of bacteria: *E. coli*, which inhabits the human gut (normally 37°C); and *Thermus aquaticus*, which lives in hot springs around 70°C.

FIGURE 5.13 Each enzyme works best within a characteristic range of conditions—generally, the same conditions that occur in the environment in which the enzyme normally functions.

FIGURE IT OUT At what temperature does the *E. coli* DNA polymerase work fastest?

Answer: About 37°C

TAKE-HOME MESSAGE 5.4

How do enzymes work?

- ✔ Enzymes greatly enhance the rate of specific reactions.
- ✔ Binding at an enzyme's active site causes a substrate to reach its transition state. In this state, the substrate's bonds are at the breaking point, and the reaction can run spontaneously to completion.
- ✔ Each enzyme works best within a certain range of environmental conditions that include temperature, pH, pressure, and salt concentration.

active site Pocket in an enzyme where substrates bind and a chemical reaction occurs.
catalysis The acceleration of a chemical reaction by a molecule that is unchanged by participating in the reaction.
induced-fit model Substrate binding to an active site improves the fit between the two.
substrate Of an enzyme, a reactant that is specifically acted upon by the enzyme.
transition state Point during a reaction at which substrate bonds will break and the reaction will run spontaneously to completion.

✔ ATP, enzymes, and other molecules interact in organized pathways of metabolism.

Metabolism includes all activities by which cells acquire and use energy as they build, break down, or remodel organic molecules (Section 3.3). Such activities often occur stepwise, in a series of enzymatic reactions called a **metabolic pathway**. Some metabolic pathways are linear, meaning that the reactions run straight from reactant to product (**FIGURE 5.14A**). Other reactions are cyclic. In a cyclic pathway, the last step regenerates a reactant required for the first step (**FIGURE 5.14B**). Both linear and cyclic pathways are common in cells; both can involve thousands of molecules and be quite complex. Later chapters detail the steps in some important pathways.

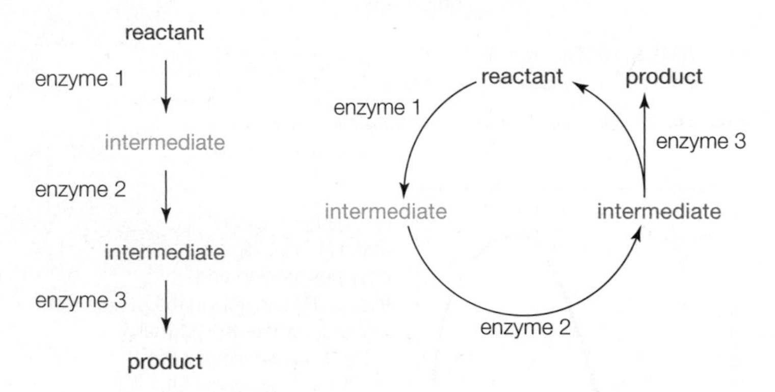

A A linear pathway runs straight from reactant to product.

B The last step of a cyclic pathway regenerates a reactant for the first step.

FIGURE 5.14 ▶Animated Linear and cyclic metabolic pathways.

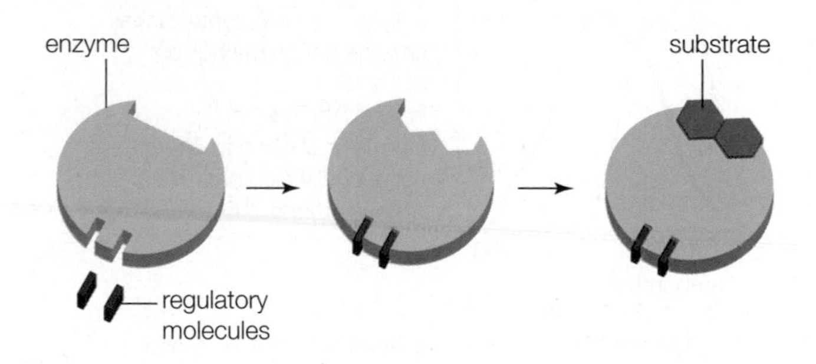

FIGURE 5.15 ▶Animated Allosteric regulation, in which regulatory molecules bind to a region of an enzyme that is not the active site. The binding changes the shape of the enzyme, and thus alters its activity.

FIGURE IT OUT Does the binding of regulatory molecules help or hinder this enzyme's function?

Answer: It helps.

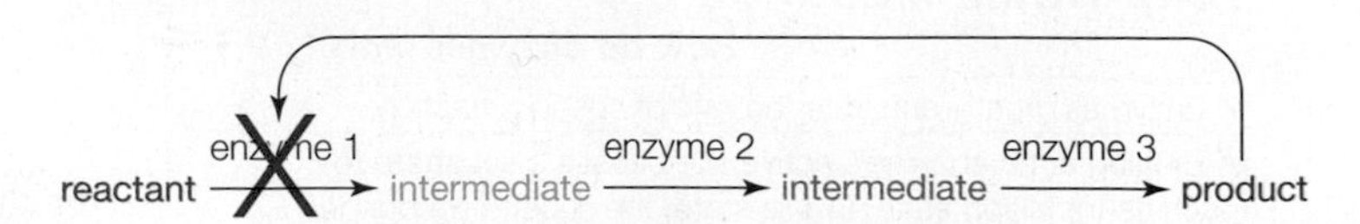

FIGURE 5.16 Feedback inhibition. In this example, three different enzymes act in sequence to convert a substrate to a product. The product inhibits the activity of the first enzyme.

FIGURE IT OUT Is this metabolic pathway cyclic or linear?

Answer: Linear

Controls Over Metabolism

Cells conserve energy and resources by making only what they require at any given moment—no more, no less. Several mechanisms help a cell maintain, raise, or lower its production of thousands of different substances. Consider that reactions do not only run from reactants to products. Many also run in reverse at the same time, with some of the products being converted back to reactants. The rates of the forward and reverse reactions often depend on the concentrations of reactants and products: A high concentration of reactants pushes the reaction in the forward direction, and a high concentration of products pushes it in the reverse direction.

Other mechanisms more actively regulate enzymatic reactions. Certain substances—regulatory molecules or ions—can influence enzyme activity. In some cases, the regulatory substance activates or inhibits an enzyme by binding directly to the active site. In other cases, the regulatory substance binds outside of the active site, a mechanism called **allosteric regulation** (*allo–* means other; *–steric* means structure). Binding of an allosteric regulator alters the shape of the enzyme in a way that enhances or inhibits its function (**FIGURE 5.15**).

Regulation of a single enzyme can affect an entire metabolic pathway. For example, the end product of a series of enzymatic reactions often inhibits the activity of one of the enzymes in the series (**FIGURE 5.16**). This type of regulatory mechanism, in which a change that results from an activity decreases or stops the activity, is called **feedback inhibition**.

Electron Transfers

The bonds of organic molecules hold a lot of energy that can be released in a reaction with oxygen. Burning is one type of reaction with oxygen, and it releases the energy of organic molecules all at once—explosively (**FIGURE 5.17A**). Cells use oxygen to break the bonds of organic molecules, but they have no way to harvest the explosive burst of energy that occurs during burning. Instead, they break the molecules apart in pathways that release the energy in small, manage-

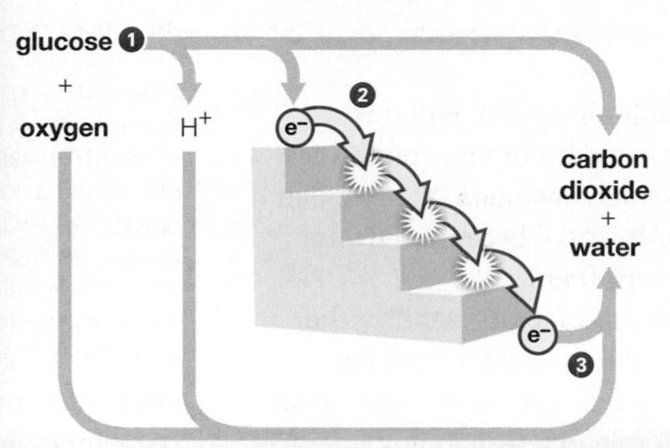

A Left, glucose in a metal spoon reacts (burns) with oxygen inside a glass jar. Energy in the form of light and heat is released all at once as CO_2 and water form.

B In cells, the same overall reaction occurs in a stepwise fashion that involves an electron transfer chain, represented here by a staircase. Energy is released in amounts that cells are able to use.

1 An input of activation energy splits glucose into carbon dioxide, electrons, and hydrogen ions (H^+).

2 Electrons lose energy as they move through an electron transfer chain. Energy released by electrons is harnessed for cellular work.

3 Electrons, hydrogen ions, and oxygen combine to form water.

FIGURE 5.17 ▸**Animated** Comparing uncontrolled (**A**) and controlled (**B**) energy release.

able steps. Most of these steps are oxidation–reduction reactions, or redox reactions for short. A typical **redox reaction** is an electron transfer, in which one molecule accepts electrons (thereby becoming reduced) from another molecule (which becomes oxidized when that happens). To remember what reduced means, think of how the negative charge of an electron "reduces" the charge of a recipient molecule.

Energy is often transferred during redox reactions (**FIGURE 5.18**). In the next two chapters, you will learn about the importance of these energy transfers in electron transfer chains. An **electron transfer chain** is a series of membrane-bound enzymes and other molecules that give up and accept electrons in turn. Electrons are at a higher energy level (Section 2.3) when they enter a chain than when they leave. Energy given off by an electron as it drops to a lower energy level is harvested by molecules of the electron transfer chain to do cellular work (**FIGURE 5.17B**). Electron transfer chains are part of photosynthesis and aerobic respiration. Energy released at certain steps in those chains helps drive the synthesis of ATP.

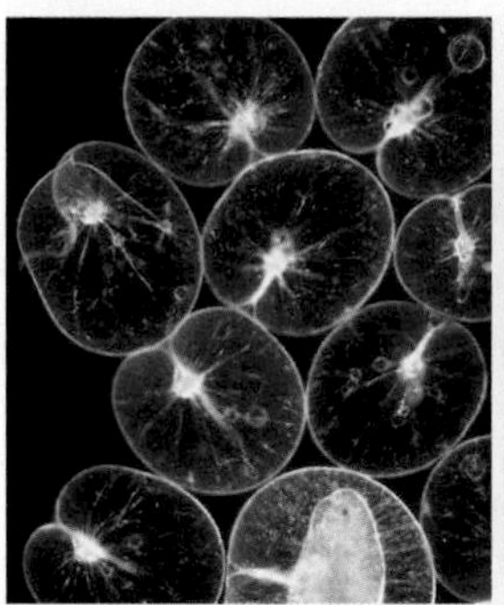

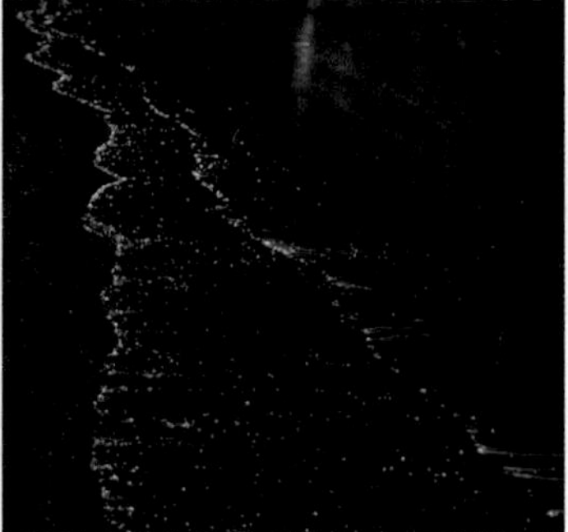

FIGURE 5.18 Visible evidence of energy transferred during a redox reaction: a glowing protist, *Noctiluca scintillans* (left). The metabolic pathway that produces the blue glow involves an enzyme, luciferase, and its substrate, luciferin. It runs when the cells are mechanically stimulated, as by waves (right) or an attack by a protist-eating predator.

The pathway, summarized below, includes a redox reaction in which luciferin becomes oxidized, and oxygen becomes reduced:

$$\text{luciferin} + 2H^+ + O_2 \xrightarrow{\text{luciferase}} \text{luciferin}{=}O + H_2O + \textbf{light}$$

Light given off by a living organism is called bioluminescence.

TAKE-HOME MESSAGE 5.5

What is a metabolic pathway?

✔ A metabolic pathway is a stepwise series of enzyme-mediated reactions.

✔ Cells conserve energy and resources by producing only what they need at a given time. Such metabolic control can arise from mechanisms (such as regulatory molecule binding to an enzyme) that start, stop, or alter the rate of a single reaction. Other mechanisms (such as feedback inhibition) influence an entire pathway.

✔ Many metabolic pathways involve electron transfers. Electron transfer chains are sites of energy exchange.

allosteric regulation Control of enzyme activity by a regulatory molecule or ion that binds to a region outside the enzyme's active site.
electron transfer chain Array of membrane-bound enzymes and other molecules that accept and give up electrons in sequence, thus releasing the energy of the electrons in steps.
feedback inhibition Regulatory mechanism in which a change that results from some activity decreases or stops the activity.
metabolic pathway Series of enzyme-mediated reactions by which cells build, remodel, or break down an organic molecule.
redox reaction Oxidation–reduction reaction; typically, one molecule accepts electrons (it becomes reduced) from another molecule (which becomes oxidized).

CREDITS: (17) left, Martyn F. Chillmaid/Science Source; right, © Cengage Learning; (18) left, FLPA/Alamy; right, Travelart/Alamy; (in text) © Cengage Learning.

5.6 Cofactors in Metabolic Pathways

✔ Cofactors help enzymes work.

✔ Energy in ATP drives many endergonic reactions.

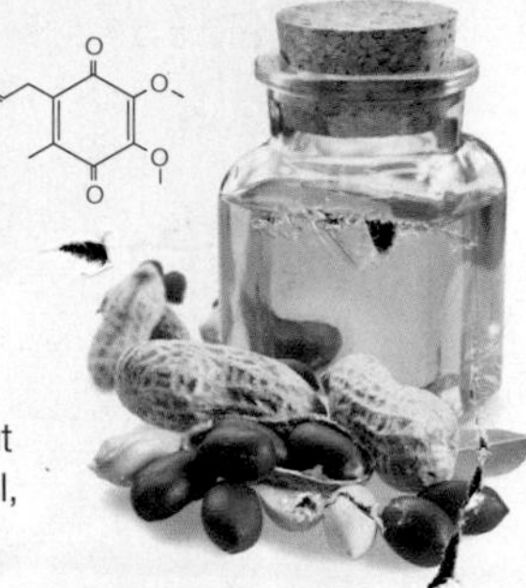

FIGURE 5.19 Example of a coenzyme. Coenzyme Q_{10} (above) is an essential part of the ATP-making machinery in your mitochondria. It carries electrons between enzymes of electron transfer chains during aerobic respiration. Your body makes it, but some foods—particularly red meats, soy oil, and peanuts—are rich dietary sources.

Most enzymes cannot function properly without assistance from metal ions or small organic molecules. Such enzyme helpers are called **cofactors**. Many dietary vitamins and minerals are essential because they are cofactors or are precursors for them.

Some metal ions that act as cofactors stabilize the structure of an enzyme, in which case the enzyme denatures if the ions are removed. In other cases, metal cofactors play a functional role in a reaction by interacting with electrons in nearby atoms. Atoms of metal elements readily lose or gain electrons, so a metal cofactor can help bring on the transition state by donating electrons, accepting them, or simply tugging on them.

Organic cofactors are called **coenzymes** (**TABLE 5.1** and **FIGURE 5.19**). Coenzymes carry chemical groups, atoms, or electrons from one reaction to another, and often into or out of organelles. Unlike enzymes, many coenzymes are modified by taking part in a reaction. They are regenerated in separate reactions. Consider NAD^+ (nicotinamide adenine dinucleotide), a coenzyme derived from niacin (vitamin B_3). NAD^+ can accept electrons and hydrogen atoms, thereby becoming reduced to NADH. When electrons and hydrogen atoms are removed from NADH (an oxidation reaction), NAD^+ forms again:

NAD^+ + electrons + H^+ → **NADH** → NAD^+ + electrons + H^+

Table 5.1 Some Common Coenzymes

Coenzyme	Example of Function
ATP	Transfers energy with a phosphate group
NAD, NAD^+	Carries electrons during glycolysis
NADP, NADPH	Carries electrons, hydrogen atoms during photosynthesis
FAD, FADH, $FADH_2$	Carries electrons during aerobic respiration
Coenzyme A (CoA)	Carries acetyl group ($COCH_3$) during glycolysis
Coenzyme Q_{10}	Carries electrons in electron transfer chains of aerobic respiration
Heme	Accepts and donates electrons
Ascorbic acid	Carries electrons during peroxide breakdown (in lysosomes)
Biotin (vitamin B_7)	Carries CO_2 during fatty acid synthesis

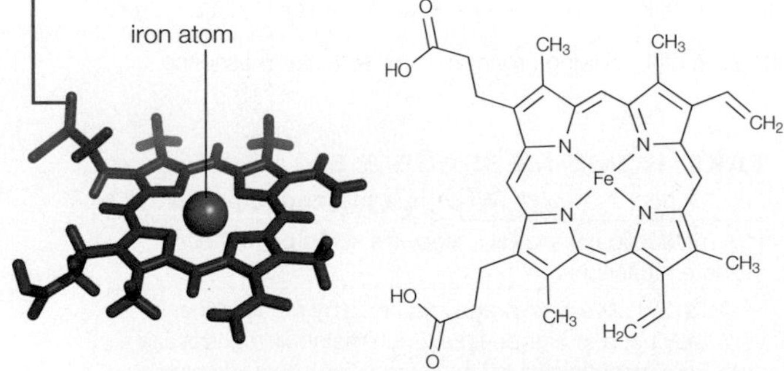

FIGURE 5.20 ▸Animated Heme, modeled two ways. This organic molecule is part of the active site in many enzymes (such as catalase). In other contexts, it carries oxygen (e.g., in hemoglobin), or electrons (e.g., in molecules of electron transfer chains).

FIGURE IT OUT Is heme a cofactor or a coenzyme?

Answer: It is both.

In some reactions, cofactors participate as separate molecules. In others, they stay tightly bound to the enzyme. Catalase, an enzyme of peroxisomes, has four tightly bound cofactors called hemes. A heme is a small organic compound with an iron atom at its center (**FIGURE 5.20**). Catalase's substrate is hydrogen peroxide (H_2O_2), a highly reactive molecule that forms during some normal metabolic reactions. Hydrogen peroxide is dangerous because it can easily oxidize and destroy the organic molecules of life, or form free radicals that do. Catalase neutralizes this threat. Catalase's active site holds a hydrogen peroxide molecule close to a heme. Two H_2O_2 molecules alternately oxidize and then reduce the heme's iron atom, an interaction that causes the molecules to break down and form water.

Substances such as catalase that interfere with the oxidation of other molecules are called **antioxidants**. Antioxidants are essential to health because they limit the amount of damage that cells sustain as a result of oxidation by free radicals or other molecules. Oxidative damage is associated with many diseases, including cancer, diabetes, atherosclerosis, stroke, and neurodegenerative problems such as Alzheimer's disease.

ATP—A Special Coenzyme

In cells, the nucleotide ATP (adenosine triphosphate, Section 3.8) functions as a cofactor in many reactions. Bonds between phosphate groups hold a lot of energy compared to other bonds. ATP has two of these bonds holding its three phosphate groups together (**FIGURE 5.21A**). When a phosphate group is transferred to or from a nucleotide, energy is transferred along with it. Thus, the nucleotide can receive energy from an exergonic reaction, and it can contribute energy to an endergonic one. ATP is such an important currency in the energy economy of cells that we use a cartoon coin to symbolize it.

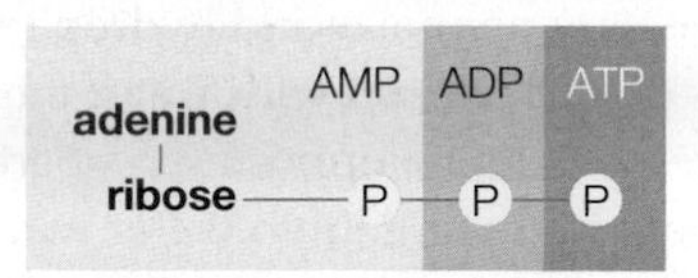

A reaction in which a phosphate group is transferred from one molecule to another is called a **phosphorylation**. ADP (adenosine diphosphate) forms when an enzyme transfers a phosphate group from ATP to another molecule (**FIGURE 5.21B**). Cells constantly run this reaction in order to drive a variety of endergonic reactions. Thus, they must constantly replenish their stockpile of ATP—by running exergonic reactions that phosphorylate ADP. The cycle of using and replenishing ATP is called the **ATP/ADP cycle** (**FIGURE 5.21C**).

The ATP/ADP cycle couples endergonic reactions with exergonic ones (**FIGURE 5.22**). As you will see in Chapter 7, cells harvest energy from organic compounds by running metabolic pathways that break them down. Energy that cells harvest in these pathways is not released to the environment, but rather stored in the high-energy phosphate bonds of ATP molecules and in electrons carried by reduced coenzymes. Both the ATP and the reduced coenzymes that form in these pathways can be used to drive many of the endergonic reactions that a cell runs.

antioxidant Substance that prevents oxidation of other molecules.
ATP/ADP cycle Process by which cells regenerate ATP. ADP forms when ATP loses a phosphate group, then ATP forms again as ADP gains a phosphate group.
coenzyme An organic cofactor.
cofactor A coenzyme or metal ion that associates with an enzyme and is necessary for its function.
phosphorylation A phosphate-group transfer.

TAKE-HOME MESSAGE 5.6

How do cofactors work?

- ✔ Cofactors associate with enzymes and assist their function.
- ✔ Many coenzymes carry chemical groups, atoms, or electrons from one reaction to another.
- ✔ The formation of ATP from ADP is an endergonic reaction. ADP forms again when a phosphate group is transferred from ATP to another molecule.
- ✔ When a phosphate group is transferred from ATP to another molecule, energy is transferred along with it. This energy drives cellular work.

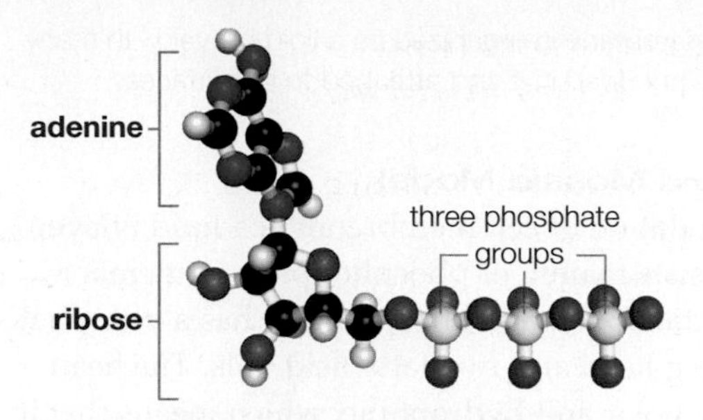

A ATP. Bonds between its phosphate groups hold a lot of energy.

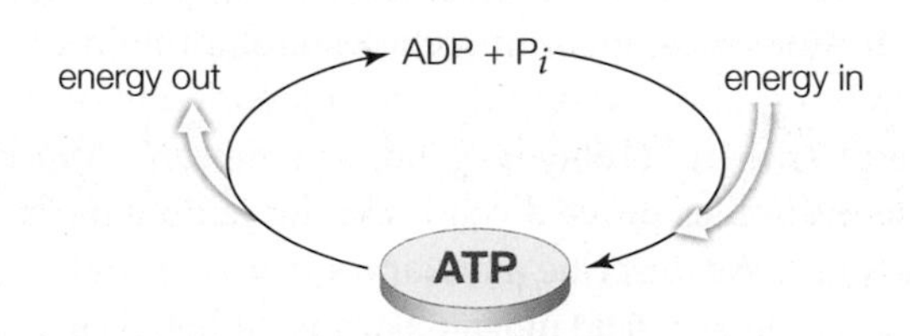

B After ATP loses one phosphate group, the nucleotide is ADP (adenosine diphosphate); after losing two, it is AMP (adenosine monophosphate).

C The ATP/ADP cycle. ADP forms in a reaction that removes a phosphate group from ATP (P_i is an abbreviation for phosphate group). Energy released in this reaction drives other reactions that are the stuff of cellular work. ATP forms again in reactions that phosphorylate ADP.

FIGURE 5.21 ATP, an important energy currency in metabolism.

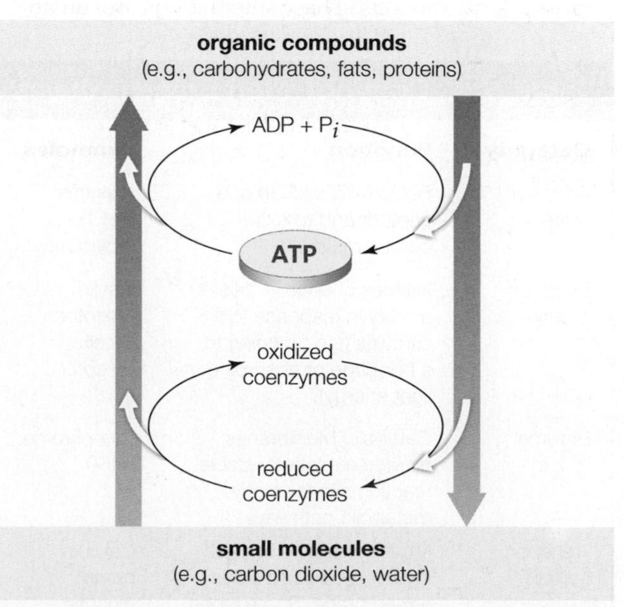

FIGURE 5.22 How ATP and coenzymes couple endergonic reactions with exergonic reactions. Yellow arrows indicate energy flow. Compare with **FIGURES 5.9** and **5.21C**.

5.7 A Closer Look at Cell Membranes

✔ A cell membrane is organized as a lipid bilayer with many proteins embedded in it and attached to its surfaces.

The Fluid Mosaic Model

The foundation of cell membranes is a lipid bilayer that consists mainly of phospholipids. Remember from Section 3.5 that a phospholipid has a phosphate-containing head and two fatty acid tails. The head is highly polar and hydrophilic, which means that it interacts with water molecules. The long hydrocarbon tails are very nonpolar and hydrophobic, so they do not interact with water molecules. As a result of these opposing properties, phospholipids swirled into water will spontaneously organize themselves into lipid bilayer sheets or bubbles (left), with hydrophobic tails together, hydrophilic heads facing the watery surroundings (**FIGURE 5.23A**).

fluid

Other molecules, including cholesterol, proteins, glycoproteins, and glycolipids, are embedded in or attached to the lipid bilayer of a cell membrane. Many of these molecules move around the membrane more or less freely. We describe a eukaryotic or bacterial cell membrane as a **fluid mosaic** because it behaves like a two-dimensional liquid of mixed composition. The "mosaic" part of the name comes from the many different types of molecules in the membrane. A cell membrane is fluid because its phospholipids are not chemically bonded to one another; they stay organized in a bilayer as a result of collective hydrophobic and hydrophilic attractions. These interactions are, on an individual basis, relatively weak. Thus, individual phospholipids in the bilayer drift sideways and spin around their long axis, and their tails wiggle.

Table 5.2 Common Membrane Proteins

Category	Function	Examples
Adhesion protein	Helps cells stick to one another and to extracellular matrix.	Integrins; MHC molecules
Receptor protein	Initiates change in a cell's activity in response to a stimulus (e.g., binding to a hormone or absorbing light energy).	Insulin receptor; B cell receptor
Enzyme	Catalysis. Membranes provide a relatively stable reaction site for many metabolic pathways.	Cytochrome P450
Transport protein	Moves or allows specific ions or molecules across a membrane. Some types require an energy input, as from ATP.	Calcium pump; glucose transporter

A In a watery fluid, phospholipids spontaneously line up into two layers: the hydrophobic tails cluster together, and the hydrophilic heads face outward, toward the fluid. This lipid bilayer forms the framework of all cell membranes. Many types of proteins intermingle among the lipids; a few that are typical of plasma membranes are shown opposite.

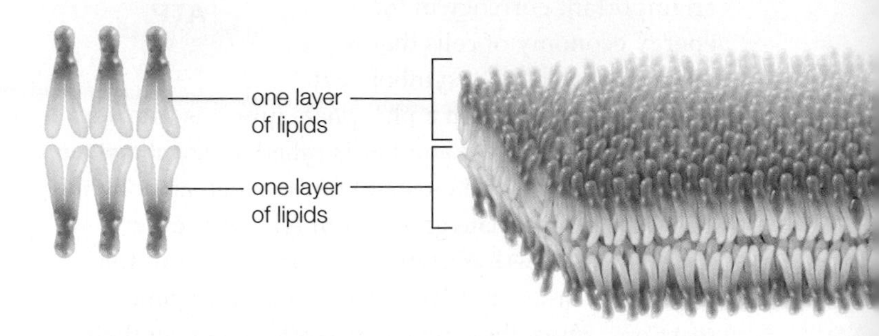

FIGURE 5.23 Cell membrane structure.

Organization of phospholipids in cell membranes (**A**) and examples of common membrane proteins (**B–E**). For clarity, these proteins are often modeled as blobs or geometric shapes; their structure can be extremely complex.

A cell membrane's properties vary depending on the types and proportions of molecules composing it. For example, membrane fluidity decreases with increasing cholesterol content. A membrane's fluidity also depends on the length and saturation of its phospholipids' fatty acid tails (Section 3.5).

Archaea do not even use fatty acids to build their phospholipids. Instead, they use molecules with reactive side chains, so the tails of archaeal phospholipids form covalent bonds with one another. As a result of this rigid crosslinking, archaeal phospholipids do not drift, spin, or wiggle in a bilayer. Thus, membranes of archaea are stiffer than those of bacteria or eukaryotes, a characteristic that may help these cells survive in extreme habitats.

Proteins Add Function

Many types of proteins are associated with a cell membrane (**TABLE 5.2**). These proteins can be assigned to one of two categories, depending on the way they are attached to the lipid bilayer. Integral membrane proteins are permanently anchored in the membrane by one or more domains sunk deeply into the lipid bilayer's hydrophobic core. Integral proteins that span the entire bilayer are called transmembrane proteins. By contrast, a peripheral membrane protein temporar-

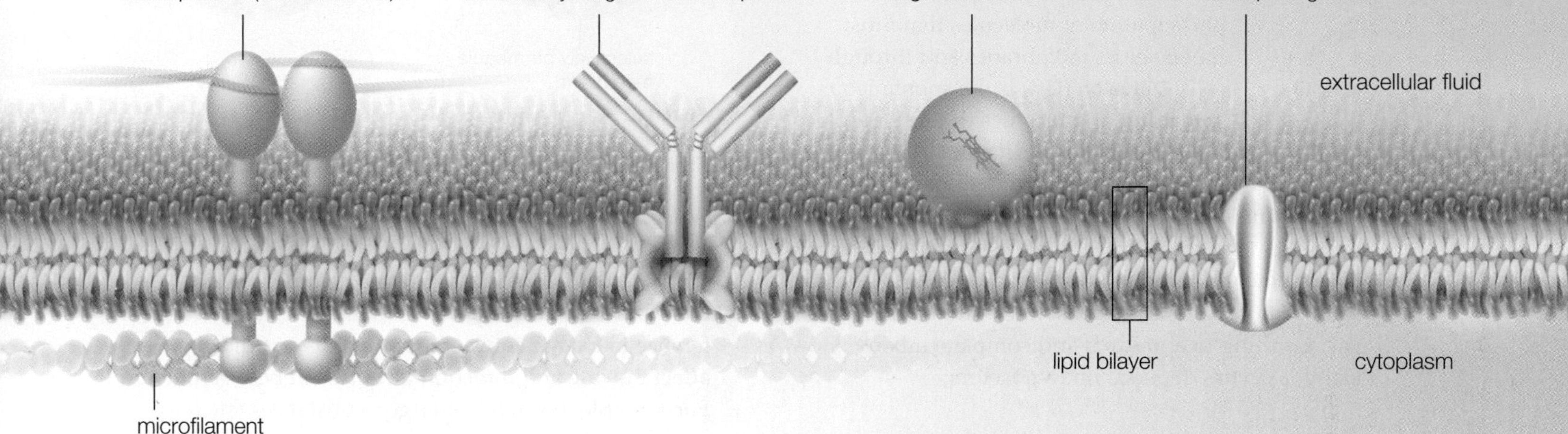

ily attaches to one of the lipid bilayer's surfaces by interacting with lipid heads or an integral protein.

A cell membrane physically separates an external environment from an internal one, but that is not its only task. Each kind of protein in a membrane imparts a specific function to it. Thus, different cell membranes can have different functions depending on which proteins are associated with them. A plasma membrane incorporates certain proteins that no internal cell membrane has, so it carries out functions that no other membrane does. For example, adhesion proteins occur only on plasma membranes. **Adhesion proteins** fasten cells to one another, or connect extracellular matrix outside the cell to cytoskeletal elements inside of it (**FIGURE 5.23B**). This arrangement strengthens a tissue, and can constrain certain membrane proteins to an upper or lower surface of the cell. Adhesion proteins are the sticky components of adhering and tight junctions in animal tissues (Section 4.11). Many adhesion proteins also have important roles in cell signaling, helping cells sense and respond to external conditions.

Plasma membranes and some internal membranes incorporate **receptor proteins**, which trigger a change in the cell's activities in response to a stimulus (**FIGURE 5.23C**). Each type of receptor protein receives a particular stimulus, for example absorbing light at a certain wavelength, or binding to a certain hormone. Each receptor protein also triggers a specific response inside the cell, which may involve metabolism, movement, division, or even cell death.

All cell membranes incorporate enzymes (**FIGURE 5.23D**). A lipid bilayer provides a scaffold for enzymes that work in series, for example in electron transfer chains. Some membrane enzymes act on other proteins or lipids that are part of the lipid bilayer. All membranes also have **transport proteins**, which move specific substances across the bilayer (**FIGURE 5.23E**). These proteins are important because, as you will see in the next section, lipid bilayers are impermeable to most substances, including ions and polar molecules.

adhesion protein Protein that helps cells stick together in animal tissues. Some types form adhering junctions and tight junctions.
fluid mosaic Model of a cell membrane as a two-dimensional fluid of mixed composition.
receptor protein Membrane protein that triggers a change in cell activity in response to a stimulus such as binding a certain substance.
transport protein Protein that passively or actively assists specific ions or molecules across a membrane.

TAKE-HOME MESSAGE 5.7

What is a cell membrane?

✔ The foundation of almost all cell membranes is the lipid bilayer—two layers of lipids (mainly phospholipids), with tails sandwiched between heads.

✔ Proteins embedded in or attached to a lipid bilayer add specific functions to each cell membrane.

5.8 Diffusion and Membranes

✔ Ions and molecules tend to move spontaneously from regions of higher to lower concentration.

✔ Water diffuses across cell membranes by osmosis.

a drop of pink dye diffusing in water

Metabolic pathways require the participation of molecules that must move across membranes and through cells. **Diffusion** (left) is the spontaneous spreading of molecules or ions, and it is an essential way in which substances move into, through, and out of cells. An atom or molecule is always jiggling, and this internal movement causes it to randomly bounce off of nearby objects, including other atoms or molecules. Rebounds from such collisions propel solutes through a liquid or gas, resulting in a gradual and complete mixing. How fast this occurs depends on five factors:

Molecular Size It takes more energy to move a large object than it does to move a small one, so small molecules diffuse more quickly than large ones.

Temperature Atoms and molecules jiggle faster at higher temperature, so they collide more often. Thus, diffusion occurs more quickly at higher temperatures.

Concentration A difference in solute concentration (Section 2.6) between adjacent regions of solution is called a concentration gradient. Solutes tend to diffuse "down" their concentration gradient, from a region of higher concentration to one of lower concentration. Why? Consider that moving objects (such as molecules) collide more often as they get more crowded. Thus, during a given interval, more molecules get bumped out of a region of higher concentration than get bumped into it.

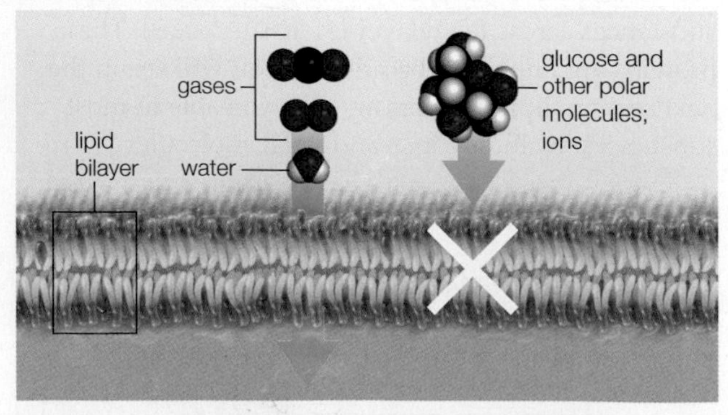

FIGURE 5.24 Selective permeability of lipid bilayers. Hydrophobic molecules, gases, and water molecules can cross a lipid bilayer on their own. Ions in particular and most polar molecules, including glucose, cannot.

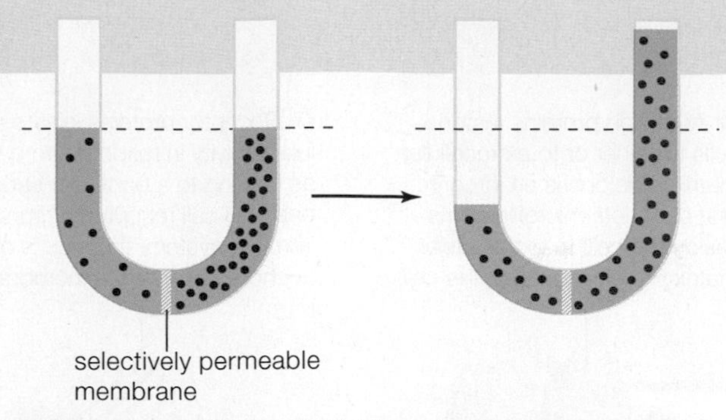

FIGURE 5.25 ▶Animated Osmosis. Water moves across a selectively permeable membrane that separates two fluids of differing solute concentration. The fluid volume changes in the two compartments as water diffuses across the membrane from the hypotonic solution to the hypertonic one.

Charge Each ion or charged molecule in a fluid contributes to the fluid's overall electric charge. A difference in charge between two regions of fluid can affect the rate and direction of diffusion between them. For example, positively charged substances (such as sodium ions) will tend to diffuse toward a region with an overall negative charge.

Pressure The rate of diffusion may be affected by a difference in pressure between two adjoining regions. Pressure squeezes objects—including atoms and molecules—closer together. Atoms and molecules that are more crowded collide and rebound more frequently. Thus, diffusion occurs faster at higher pressures.

Semipermeable Membranes

Lipid bilayers are selectively permeable (Section 4.2); water, gases, and hydrophobic molecules can cross them, but ions and most polar molecules cannot (**FIGURE 5.24**). When two fluids with different solute concentrations are separated by a selectively permeable membrane, water will diffuse across the membrane. The direction of water movement depends on the relative solute concentration of the two fluids, which we describe in terms of tonicity. Fluids that are **isotonic** have the same overall solute concentration. If the overall solute concentrations of the two fluids differ, the fluid with the lower concentration of solutes is said to be **hypotonic** (*hypo*–, under). The other one, with the higher solute concentration, is **hypertonic** (*hyper*–, over).

When a selectively permeable membrane separates two fluids that are not isotonic, water will move across the membrane from the hypotonic fluid into the hypertonic one (**FIGURE 5.25**). The diffusion will continue until the two fluids are isotonic, or until pressure against the hypertonic fluid counters it. The movement of water across membranes is so important in biology that it is given a special name: **osmosis**.

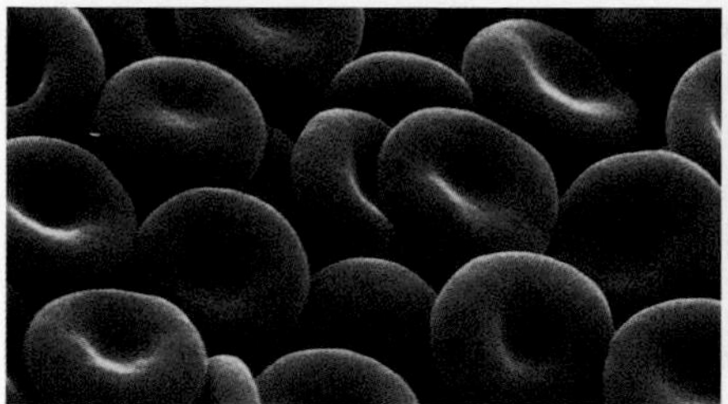

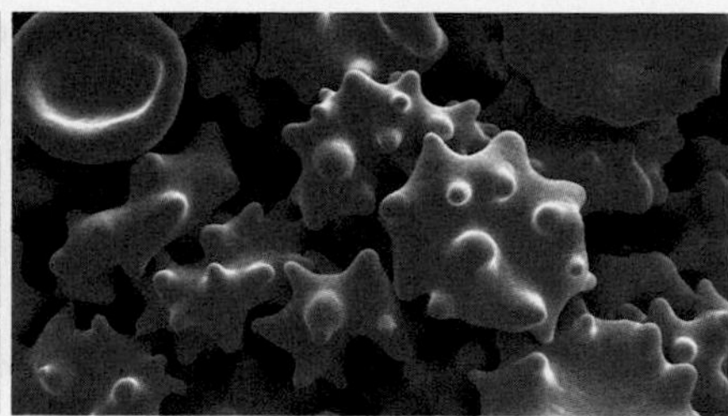

A Red blood cells in an isotonic solution (such as the fluid portion of blood) have a normal, indented disk shape.

B Water diffuses out of red blood cells immersed in a hypertonic solution, so they shrivel up.

C Water diffuses into red blood cells immersed in a hypotonic solution, so they swell up. Some of these have burst.

FIGURE 5.26 Effects of tonicity in human red blood cells. These cells have no mechanism to compensate for differences in solute concentration between cytoplasm and extracellular fluid.

If a cell's cytoplasm becomes hypertonic with respect to the fluid outside of its plasma membrane, water will diffuse into the cell. If the cytoplasm becomes hypotonic, water will diffuse out. In either case, the solute concentration of the cytoplasm may change. If it changes enough, the cell's enzymes will stop working, with lethal results. Many cells have built-in mechanisms that compensate for differences in solute concentration between cytoplasm and extracellular (external) fluid. In cells with no such mechanism, the volume—and solute concentration—of cytoplasm changes when water diffuses into or out of the cell (**FIGURE 5.26**).

Turgor

The rigid cell walls of plants and many protists, fungi, and bacteria can resist an increase in the volume of cytoplasm even in hypotonic environments. In the case of plant cells, cytoplasm usually contains more solutes than soil water does. Thus, water usually diffuses from soil into a plant—but only up to a point. Stiff walls keep plant cells from expanding very much, so an inflow of water causes pressure to build up inside them. Pressure that a fluid exerts against a structure that contains it is called **turgor**. When enough pressure builds up inside a plant cell, water stops diffusing into its cytoplasm. The amount of turgor that is enough to stop osmosis is called **osmotic pressure**.

Osmotic pressure keeps walled cells plump, just as high air pressure inside a tire keeps it inflated. A young land plant can resist gravity to stay erect because its cells are plump with cytoplasm (**FIGURE 5.27A**). When soil dries out, it loses water, so the concentration of solutes increases in it. If soil water becomes hypertonic with respect to cytoplasm, water will start diffusing out of the plant's cells, causing their cytoplasm to shrink (**FIGURE 5.27B**). As turgor inside the cells decreases, the plant wilts.

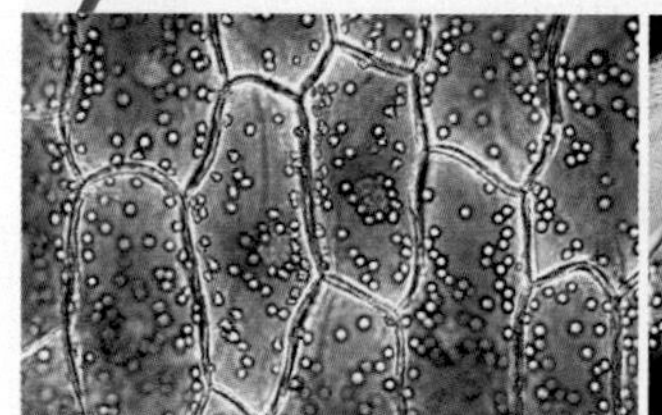

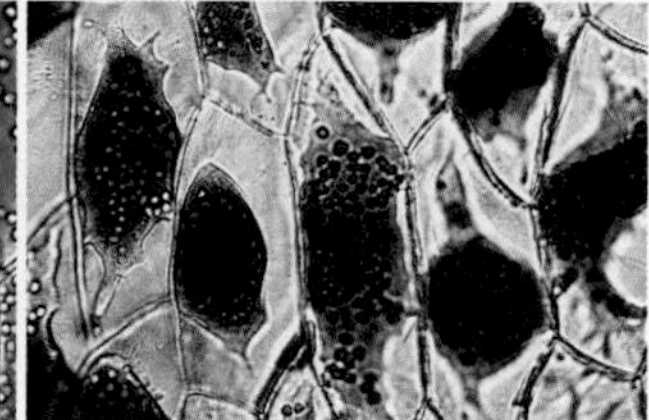

A Osmotic pressure keeps plant parts erect. These cells in an iris petal are plump with cytoplasm.

B Cells from a wilted iris petal. The cytoplasm shrank, and the plasma membrane moved away from the wall.

FIGURE 5.27 Turgor, as illustrated in cells of iris petals.

TAKE-HOME MESSAGE 5.8

What influences the movement of solutes?

✔ Solutes tend to diffuse into an adjoining region of fluid in which they are not as concentrated. The steepness of a concentration gradient as well as temperature, molecular size, charge, and pressure affect the rate of diffusion.

✔ When two fluids of different solute concentration are separated by a selectively permeable membrane, water diffuses from the hypotonic to the hypertonic fluid. This movement, osmosis, is opposed by turgor.

diffusion Spontaneous spreading of molecules or ions.
hypertonic Describes a fluid that has a high solute concentration relative to another fluid separated by a semipermeable membrane.
hypotonic Describes a fluid that has a low solute concentration relative to another fluid separated by a semipermeable membrane.
isotonic Describes two fluids with identical solute concentrations and separated by a semipermeable membrane.
osmosis Diffusion of water across a selectively permeable membrane; occurs in response to a difference in solute concentration between the fluids on either side of the membrane.
osmotic pressure Amount of turgor that prevents osmosis into cytoplasm or other hypertonic fluid.
turgor Pressure that a fluid exerts against a structure that contains it.

CREDITS: (26A) Annie Cavanagh/Wellcome Images; (26B, C) CMSP/Getty Images; (27A, B) Claude Nuridsany & Marie Perennou/Science Source; (27 inset) © Evgenyi/Shutterstock.com.

5.9 Membrane Transport Mechanisms

✔ Many types of molecules and ions can cross a lipid bilayer only with the help of transport proteins.

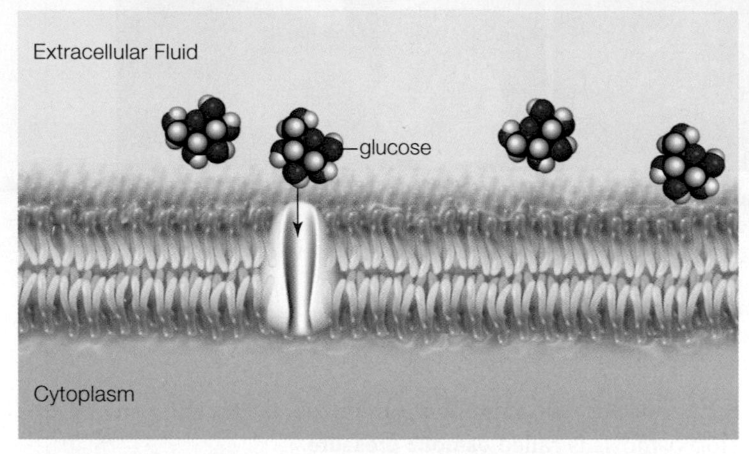

A A glucose molecule (here, in extracellular fluid) binds to a glucose transporter (gray) in the plasma membrane.

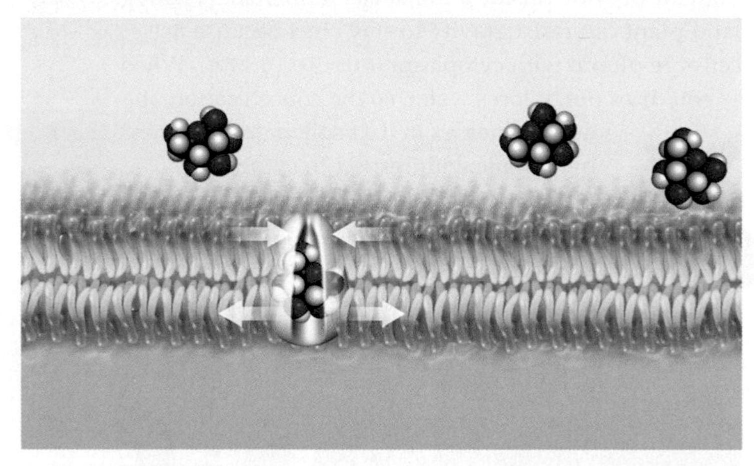

B Binding causes the transport protein to change shape.

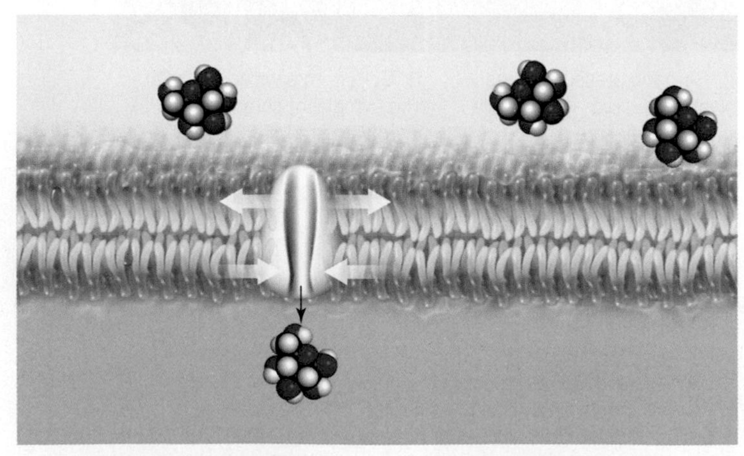

C The transport protein releases the glucose on the other side of the membrane (here, in cytoplasm) and resumes its original shape.

FIGURE 5.28 An example of facilitated diffusion.

FIGURE IT OUT In this example, which fluid is hypotonic: extracellular fluid or the cytoplasm?

Answer: Cytoplasm

Transport Protein Specificity

Substances that cannot diffuse directly through lipid bilayers—ions in particular—cross cell membranes only with the help of transport proteins. Each type of transport protein allows a specific substance to cross: Calcium pumps pump only calcium ions; glucose transporters transport only glucose; and so on. This specificity is an important part of homeostasis. For example, the composition of cytoplasm depends on the movement of particular solutes across the plasma membrane, which in turn depends on the transporters embedded in it. Glucose is an important source of energy for most cells, so they normally take up as much as they can from extracellular fluid. They do so with the help of glucose transporters in the plasma membrane. As soon as a molecule of glucose enters cytoplasm, an enzyme (hexokinase) phosphorylates it. Phosphorylation traps the molecule inside the cell because the transporters are specific for glucose, not phosphorylated glucose. Thus, phosphorylation prevents the molecule from moving back through the transport protein and leaving the cell.

Facilitated Diffusion

Osmosis is an example of **passive transport**, a membrane-crossing mechanism that requires no energy input. The diffusion of solutes through transport proteins is another example. In this case, the movement of the solute (and the direction of its movement) is driven entirely by the solute's concentration gradient. Some transport proteins form pores: permanently open channels through a membrane. Other channels are gated, which means they open and close in response to a stimulus such as a shift in electric charge or binding to a signaling molecule.

With a passive transport mechanism called **facilitated diffusion**, a solute binds to a transport protein, which then changes shape so the solute is released to the other side of the membrane. A glucose transporter is an example of a transport protein that works in facilitated diffusion (**FIGURE 5.28**). This protein changes shape when it binds to a molecule of glucose. The shape change moves the glucose to the opposite side of the membrane, where it detaches from the transport protein. Then, the glucose transporter reverts to its original shape.

active transport Energy-requiring mechanism in which a transport protein pumps a solute across a cell membrane against the solute's concentration gradient.
facilitated diffusion Passive transport mechanism in which a solute follows its concentration gradient across a membrane by moving through a transport protein.
passive transport Membrane-crossing mechanism that requires no energy input.

Active Transport

Maintaining a solute's concentration often means transporting the solute against its gradient, to the side of the membrane where it is more concentrated. This takes energy. In **active transport**, a transport protein uses energy to pump a solute against its gradient across a cell membrane. Typically, an energy input (for example, in the form of a phosphate-group transfer from ATP) changes the shape of an active transport protein. The shape change causes the protein to release a bound solute to the other side of the membrane.

A calcium pump moves calcium ions across cell membranes by active transport (**FIGURE 5.29**). Calcium ions act as potent messengers inside cells, and they also affect the activity of many enzymes. Thus, their concentration in cytoplasm is tightly regulated. Calcium pumps in the plasma membrane of all eukaryotic cells can keep the concentration of calcium ions in cytoplasm thousands of times lower than it is in extracellular fluid.

Another example of active transport involves sodium–potassium pumps (**FIGURE 5.30**). Nearly all cells in your body have these transport proteins. Sodium ions in cytoplasm diffuse into the pump's open channel and bind to its interior. A phosphate-group transfer from ATP causes the pump to change shape so that its channel opens to extracellular fluid, where it releases the sodium ions. Then, potassium ions from extracellular fluid diffuse into the channel and bind to its interior. The transporter releases the phosphate group and reverts to its original shape. The channel opens to the cytoplasm, where it releases the potassium ions.

Bear in mind that the membranes of all cells, not just those of animals, have active transport proteins. In plants, for example, active transport proteins in the plasma membranes of leaf cells pump sucrose into tubes that thread throughout the plant body.

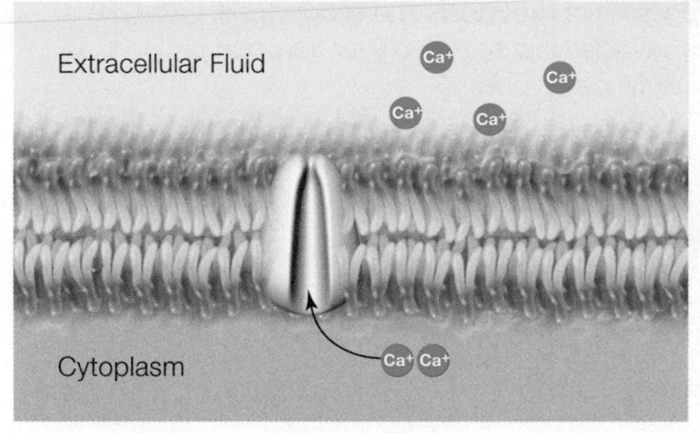

A Two calcium ions (blue) bind to the transport protein (gray).

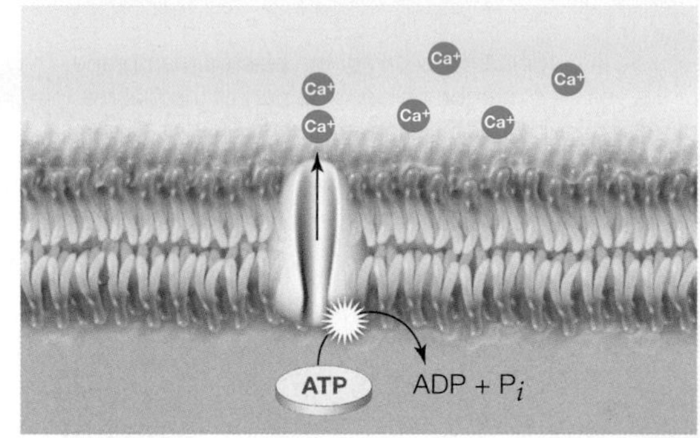

B A phosphate group from ATP causes the protein to change shape so that the calcium ions are ejected to the opposite side of the membrane.

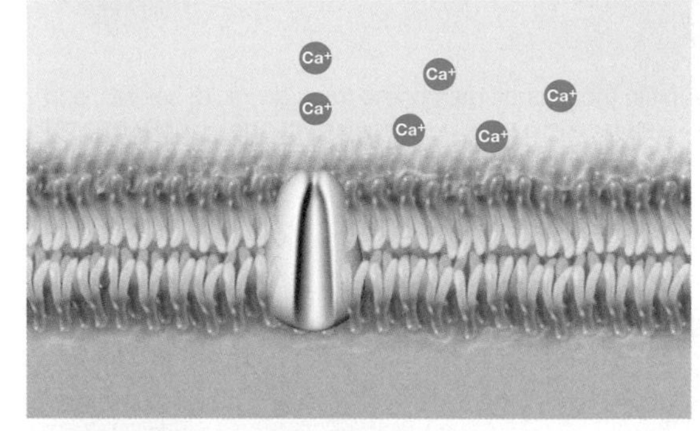

C After it loses the calcium ions, the transport protein resumes its original shape.

FIGURE 5.29 Active transport of calcium ions.

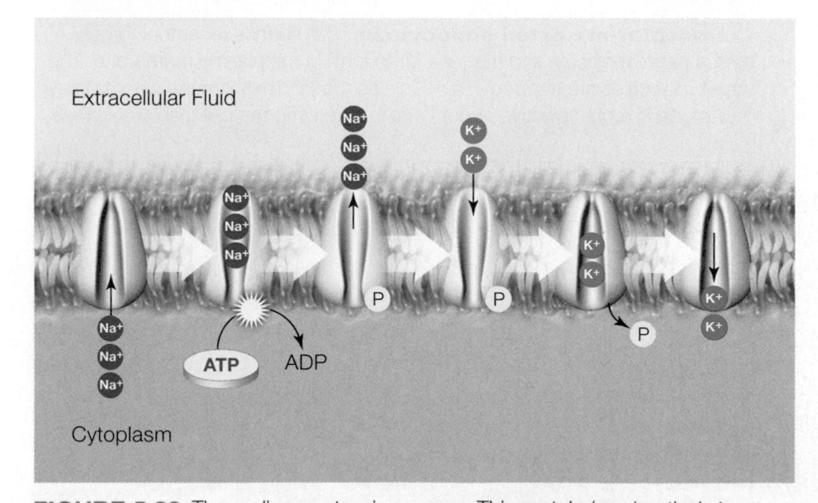

FIGURE 5.30 The sodium–potassium pump. This protein (gray) actively transports sodium ions (Na^+) from cytoplasm to extracellular fluid, and potassium ions (K^+) in the other direction. The transfer of a phosphate group (P) from ATP provides energy required for transporting the ions against their concentration gradient.

TAKE-HOME MESSAGE 5.9

How do solutes that cannot diffuse through lipid bilayers cross cell membranes?

✔ Transport proteins move specific ions or molecules across a cell membrane. The types and amounts of substances that cross a membrane depend on the transport proteins embedded in it.

✔ In facilitated diffusion (a type of passive transport), a solute binds to a transport protein that releases it on the opposite side of the membrane. The movement is driven by the solute's concentration gradient.

✔ In active transport, a transport protein pumps a solute across a membrane against its concentration gradient. The movement requires energy, as from ATP.

5.10 Membrane Trafficking

✔ By processes of exocytosis and endocytosis, cells take in and expel particles that are too big for transport proteins, as well as substances in bulk.

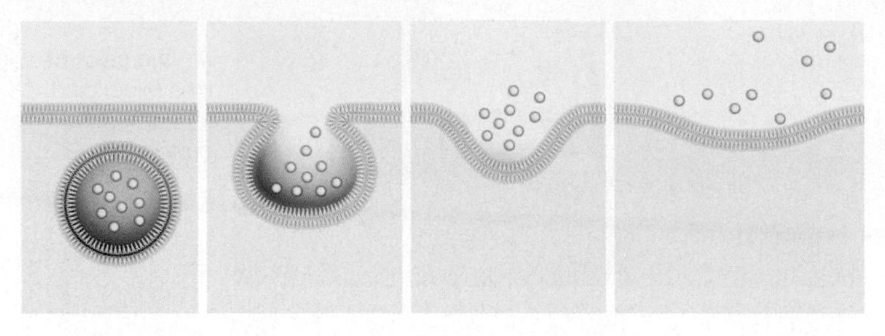

A **Exocytosis**. A vesicle in cytoplasm fuses with the plasma membrane. Lipids and proteins of the vesicle's membrane become part of the plasma membrane as its contents are expelled to the environment.

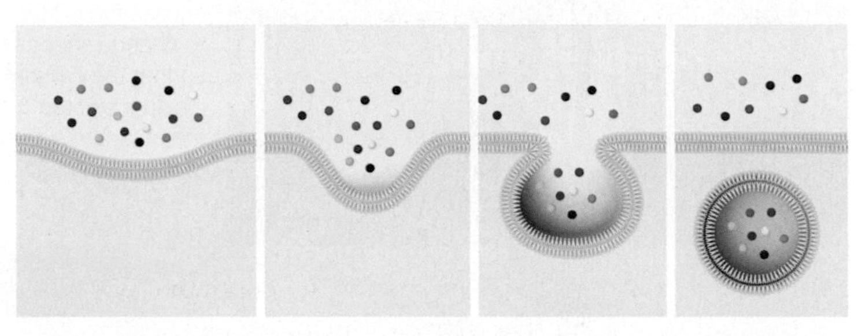

B **Pinocytosis**. A pit in the plasma membrane traps any fluid, solutes, and particles near the cell's surface in a vesicle as it sinks into the cytoplasm.

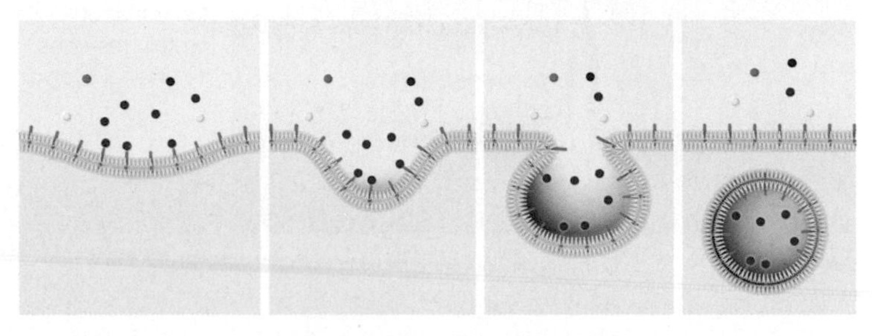

C **Receptor-mediated endocytosis**. Cell surface receptors (green) bind a target molecule and trigger a pit to form in the plasma membrane. The target molecules are trapped in a vesicle as the pit sinks into the cytoplasm. This mode is more selective about what is taken into the cell than pinocytosis.

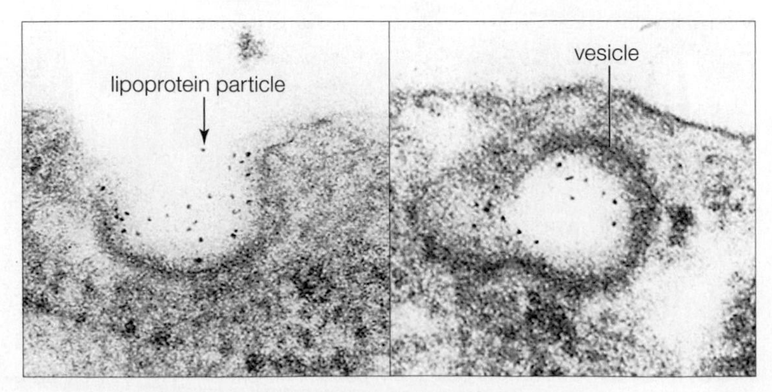

D Receptor-mediated endocytosis of lipoprotein particles.

FIGURE 5.31 Exocytosis and endocytosis.

Endocytosis and Exocytosis

Think back on the fluid mosaic structure of a lipid bilayer. When a membrane is disrupted, the fatty acid tails of the phospholipids in the bilayer become exposed to their watery surroundings. Remember, in water, phospholipids spontaneously rearrange themselves so that their nonpolar tails stay together. Thus, a membrane tends to seal itself after a disruption. Vesicles form the same way. When a patch of membrane bulges into the cytoplasm, the hydrophobic tails of the lipids in the bilayer are repelled by the watery fluid on both sides. The fluid "pushes" the phospholipid tails together, which helps round off the bud as a vesicle, and also seals the rupture in the membrane.

Vesicles are constantly carrying materials to and from a cell's plasma membrane. This movement typically requires ATP because it involves motor proteins that drag the vesicles along cytoskeletal elements. We describe the movement based on where and how the vesicle originates, and where it goes.

By **exocytosis**, a vesicle in the cytoplasm moves to the cell's surface and fuses with the plasma membrane (**FIGURE 5.31A**). As the exocytic vesicle loses its identity, its contents are released to the surroundings.

There are several pathways of **endocytosis**, but all take up substances in bulk near the cell's surface (as opposed to one molecule or ion at a time via transport proteins). **Pinocytosis** is an endocytic pathway that brings a drop of extracellular fluid (along with solutes and particles suspended in it) into the cell (**FIGURE 5.31B**). With this pathway, a small patch of plasma membrane balloons inward and then pinches off as it sinks into the cytoplasm. The membrane patch becomes the outer boundary of a vesicle.

Receptor-mediated endocytosis is more selective than pinocytosis about what it brings into the cell (**FIGURE 5.31C**). With this pathway, molecules of a hormone, vitamin, mineral, or another substance bind to receptors on the plasma membrane. The binding triggers a shallow pit to form in the membrane, just under the receptors. The pit sinks into the cytoplasm and traps the target substance in a vesicle as it closes back on itself. LDL and other lipoproteins (Section 3.6) enter cells this way (**FIGURE 5.31D**).

Phagocytosis (which means "cell eating") is a type of receptor-mediated endocytosis in which motile cells engulf microorganisms, cellular debris, or other large particles. Many single-celled protists such as amoebas feed by phagocytosis. Some of your white blood cells use phagocytosis to engulf viruses and bacteria, cancerous body cells, and other threats to health.

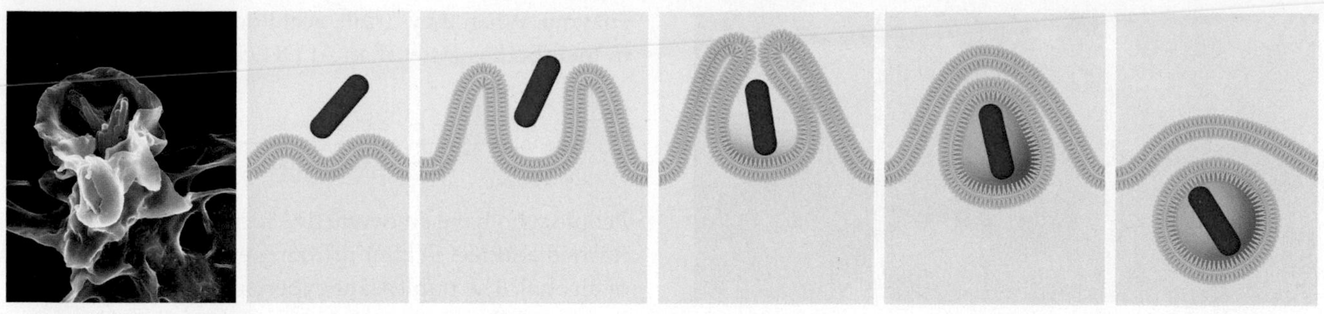

FIGURE 5.32 An example of phagocytosis. The SEM on the left shows a phagocytic white blood cell's pseudopods (extending lobes of cytoplasm) surrounding *Tuberculosis* bacteria (in red). The artwork shows how plasma membrane above the bulging lobes fuses and forms a vesicle. Once inside the cytoplasm, the endocytic vesicle will fuse with a lysosome. Enzymes delivered by the lysosome will break down the contents of the vesicle.

Phagocytosis begins when receptor proteins bind to a particular target. The binding causes microfilaments to assemble in a mesh under the plasma membrane. The microfilaments then contract, forcing a lobe of membrane-enclosed cytoplasm to bulge outward as a pseudopod (Section 4.10). Pseudopods that merge around a target trap it inside a vesicle that sinks into the cytoplasm (**FIGURE 5.32**). Material taken in by phagocytosis is typically digested by lysosomal enzymes, and the resulting molecular bits may be recycled by the cell, or expelled by exocytosis.

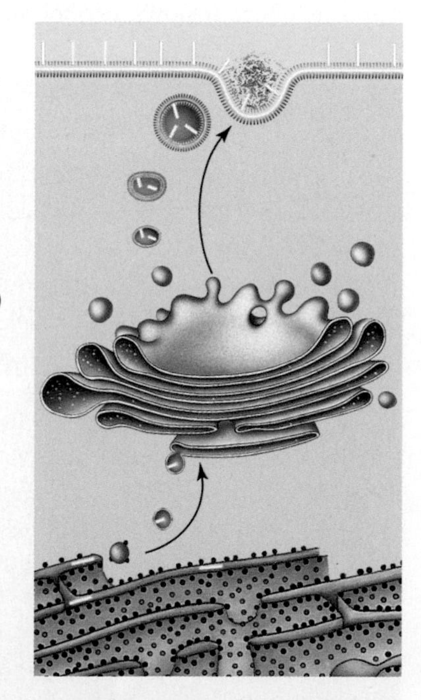

FIGURE 5.33 How membrane proteins become oriented to the inside or the outside of a cell. Proteins of the plasma membrane are assembled in the ER, and finished inside Golgi bodies. The proteins (shown in white) become part of vesicle membranes that bud from the Golgi. The membrane proteins automatically become oriented in the proper direction when the vesicles fuse with the plasma membrane.

FIGURE IT OUT What process does the upper arrow represent?

Answer: Exocytosis

Recycling Membrane

The composition of a plasma membrane begins in the endoplasmic reticulum (ER). There, membrane proteins and lipids are made and modified, and both become part of vesicles that transport them to Golgi bodies for final modification. New plasma membrane forms when the finished proteins and lipids are repackaged as vesicles that travel to the plasma membrane and fuse with it.

FIGURE 5.33 shows what happens when an exocytic vesicle fuses with the plasma membrane. Membrane proteins are oriented toward the interior of a vesicle that buds from a Golgi body, so after the vesicle fuses with the plasma membrane, the proteins face the extracellular environment.

As long as a cell is alive, exocytosis and endocytosis continually replace and withdraw patches of its plasma membrane. If the cell is not enlarging, the total area of the plasma membrane remains more or less constant. Membrane lost as a result of endocytosis is replaced by membrane arriving as exocytic vesicles.

endocytosis Process by which a cell takes in a small amount of extracellular fluid (and its contents) by the ballooning inward of the plasma membrane.
exocytosis Process by which a cell expels a vesicle's contents to extracellular fluid.
phagocytosis "Cell eating"; an endocytic pathway by which a cell engulfs large particles such as microbes or cellular debris.
pinocytosis Endocytic pathway by which fluid and materials in bulk are brought into the cell.

TAKE-HOME MESSAGE 5.10

How do large particles and bulk substances move into and out of cells?

- ✔ Exocytosis and endocytosis move materials in bulk across plasma membranes.
- ✔ In exocytosis, a cytoplasmic vesicle fuses with the plasma membrane and releases its contents to the cell's exterior.
- ✔ In endocytosis, a patch of plasma membrane sinks inward and forms a vesicle in the cytoplasm.
- ✔ Some cells can engulf large particles by phagocytosis.

CREDITS: (32) left, Science Source; right, © Cengage Learning; (33) © Cengage Learning.

A Toast to Alcohol Dehydrogenase (revisited)

In most organisms, the main function of the enzyme alcohol dehydrogenase (ADH) is to detoxify the tiny quantities of alcohols that form in some metabolic pathways. In animals, the enzyme also detoxifies small amounts of alcohols made by gut-inhabiting bacteria, and those in foods such as ripe fruit.

ADH in the human body converts ethanol to acetaldehyde, an organic molecule even more toxic than ethanol and the most likely source of various hangover symptoms:

$$\underset{\text{ethanol}}{H_3C{-}CH_2{-}OH} + NAD^+ \xrightarrow{\text{ADH}} \underset{\text{acetaldehyde}}{H_3C{-}C(=O){-}H} + NADH$$

A different enzyme, aldehyde dehydrogenase (ALDH), very quickly converts the toxic acetaldehyde to nontoxic acetate:

$$\underset{\text{acetaldehyde}}{H_3C{-}C(=O){-}H} + NAD^+ \xrightarrow{\text{ALDH}} \underset{\text{acetate}}{H_3C{-}C(=O){-}O^-} + NADH + H^+$$

Both ADH and ALDH use the coenzyme NAD^+ to accept electrons and hydrogen atoms. Thus, the overall pathway of ethanol metabolism in humans is:

$$\text{ethanol} \xrightarrow[NAD^+ \rightarrow NADH]{\text{ADH}} \text{acetaldehyde} \xrightarrow[NAD^+ \rightarrow NADH]{\text{ALDH}} \text{acetate}$$

In the average healthy adult human, this metabolic pathway can detoxify between 7 and 14 grams of ethanol per hour. The average alcoholic beverage contains between 10 and 20 grams of ethanol, which is why having more than one drink in any two-hour interval may result in a hangover.

A person's ability to metabolize ethanol in alcoholic drinks depends on the amount and activity of ADH they make. Some people have an overactive form of the enzyme. When they drink, acetaldehyde accumulates in their bodies faster than ALDH can detoxify it:

$$\text{ethanol} \xrightarrow{\text{ADH}} \text{acetaldehyde, acetaldehyde, acetaldehyde} \xrightarrow{\text{ALDH}} \text{acetate}$$

People who have an overactive form of ADH become flushed and feel ill after drinking even a small amount of alcohol. The unpleasant experience may be part of the reason that these people are statistically unlikely to become alcoholics.

Having an underactive form of ALDH also results in an accumulation of acetaldehyde after drinking:

$$\text{ethanol} \xrightarrow{\text{ADH}} \text{acetaldehyde, acetaldehyde, acetaldehyde} \;\not\longrightarrow\; \text{acetate}$$

Underactive ALDH is associated with the same protection from alcoholism as overactive ADH. Both types of variant enzymes are common in people of Asian descent. For this reason, the alcohol flushing reaction is informally called "Asian flush."

Having an underactive ADH enzyme has the opposite effect. It slows alcohol metabolism, so

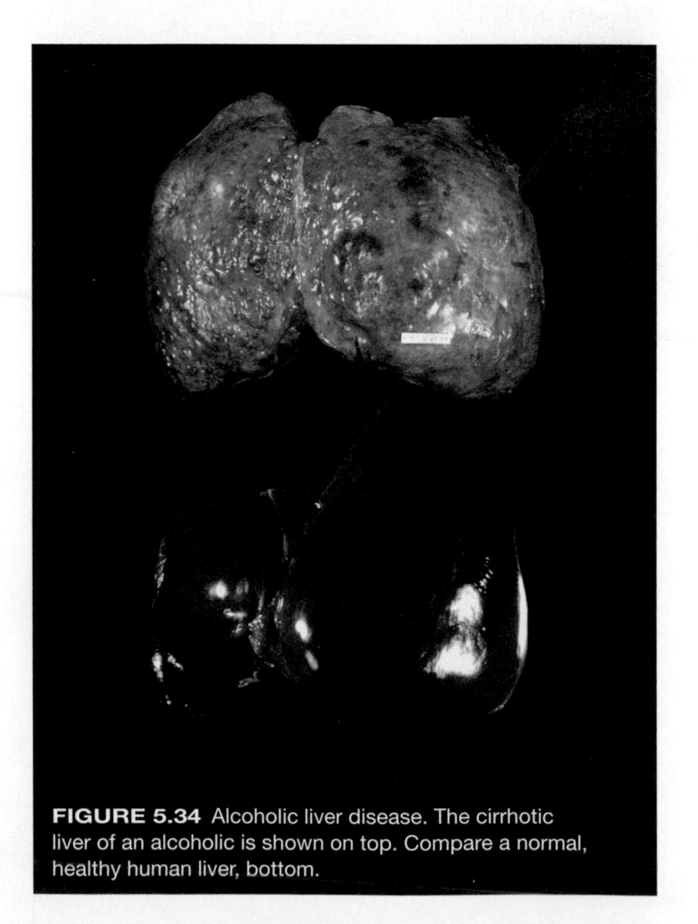

FIGURE 5.34 Alcoholic liver disease. The cirrhotic liver of an alcoholic is shown on top. Compare a normal, healthy human liver, bottom.

people with a low level of ADH activity may not feel the ill effects of drinking alcoholic beverages as much as others do. When these people drink, they have a tendency to become alcoholics. Alcoholism is characterized by compulsive, uncontrolled drinking that damages the individual's health and social relationships. One-quarter of undergraduate students who binge also have other signs of alcoholism.

Alcoholics will continue to drink despite the knowledge that doing so has tremendous negative consequences. In the United States, alcohol abuse is the leading cause of cirrhosis of the liver. A cirrhotic liver is so scarred, hardened, and filled with fat that it no longer functions properly (**FIGURE 5.34**). It no longer produces the protein albumin, so the solute balance of body fluids is disrupted, and the legs and abdomen swell with watery fluid. It can no longer remove drugs and other toxins from the blood, so they accumulate in the brain—which impairs mental functioning and alters personality. Restricted blood flow through the liver causes veins to enlarge and rupture, so internal bleeding is a risk. The damage to the body results in a heightened susceptibility to diabetes and liver cancer. Once cirrhosis has been diagnosed, a person has about a 50 percent chance of dying within 10 years (**FIGURE 5.35**).

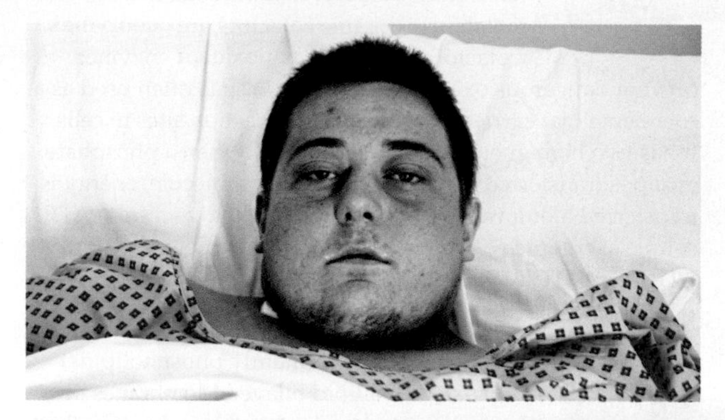

FIGURE 5.35 Gary Reinbach, who died at the age of 22 from alcoholic liver disease shortly after this photograph was taken, in 2009. The odd color of his skin is a symptom of cirrhosis.

Transplantation is a last-resort treatment for a failed liver, but there are not enough liver donors for everyone who needs a transplant. Reinbach was refused a transplant that may have saved his life because he had not abstained from drinking for the prior 6 months.

summary

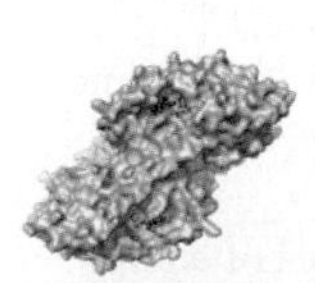

Section 5.1 Currently the most serious drug problem on college campuses is binge drinking, which is often a symptom of alcoholism. Drinking more alcohol than the body's enzymes can detoxify can be lethal in both the short term and the long term.

Section 5.2 **Energy** is the capacity to do work. One form of energy (such as **potential energy**) can be converted to another (such as **kinetic energy**). Energy cannot be created or destroyed (**first law of thermodynamics**), and it tends to disperse spontaneously (**second law of thermodynamics**). **Entropy** is a measure of how much the energy of a system is dispersed. A bit disperses at each energy transfer, usually in the form of heat. Living things maintain their organization only as long as they harvest energy from someplace else. Energy flows in one direction through the biosphere, starting mainly from the sun, then into and out of ecosystems. Producers and then consumers use the captured energy to assemble, rearrange, and break down organic molecules that cycle among organisms in an ecosystem.

Section 5.3 Cells store and retrieve energy by making and breaking chemical bonds in reactions that convert **reactants** to **products**. **Endergonic** reactions require a net input of energy to proceed. **Exergonic** reactions end with a net release of energy. **Activation energy** is the minimum energy required to start a reaction.

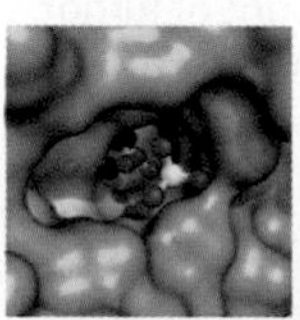

Section 5.4 Enzymes greatly enhance the rate of reactions without being changed by them, a process called **catalysis**. They lower a reaction's activation energy by boosting local concentrations of **substrates**, orienting substrates in positions that favor reaction, inducing the fit between a substrate and the enzyme's **active site** (**induced-fit model**), or excluding water. These mechanisms bring on the substrate's **transition state**. Each type of enzyme works best within a characteristic range of conditions, including temperature, salt concentration, and pH.

Section 5.5 Cells build, convert, and dispose of substances in **metabolic pathways**, which are sequences of enzyme-mediated reactions. Regulating metabolic pathways allows a cell to conserve energy and resources by making only what it needs at a given time. With **allosteric regulation**, a regulatory molecule or ion alters the activity of an enzyme by binding to it in a region other than the active site. The products of some metabolic pathways inhibit their own production, a regulatory mechanism called **feedback inhibition**. **Redox** (oxidation–reduction) **reactions** in **electron transfer chains** allow cells to harvest energy in small, manageable steps.

CREDIT: (35) Stuart Clark/The Sunday Times/nisyndication.

Section 5.6 Most enzymes require assistance from **cofactors**. Some cofactors are metal ions; organic cofactors are **coenzymes**. Cofactors help some **antioxidant** enzymes prevent dangerous oxidation reactions. ATP is often used as a coenzyme that carries energy between reaction sites in cells. It has two high-energy phosphate bonds. When a phosphate group is transferred from ATP to another molecule, energy is transferred along with it. **Phosphorylations** to and from ATP couple exergonic with endergonic reactions. Cells regenerate ATP in the **ATP/ADP cycle**.

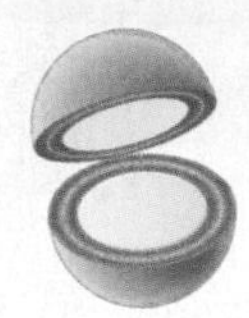

Section 5.7 A cell membrane is a mosaic of proteins and lipids (mainly phospholipids) organized as a lipid bilayer. Membranes of bacteria and eukaryotic cells can be described as a **fluid mosaic**; membranes of archaea are not fluid. Proteins transiently or permanently associated with a membrane carry out most membrane functions. All cell membranes have enzymes, and all have **transport proteins** that help substances move across the membrane. Plasma membranes also incorporate **adhesion proteins** that lock cells together in tissues. Plasma membranes and some internal membranes have **receptor proteins** that trigger a change in cell activities in response to a specific stimulus.

Section 5.8 The rate of **diffusion** is influenced by temperature, solute size, and regional differences in concentration, charge, and pressure. Gases, water, and nonpolar molecules can diffuse across a lipid bilayer. Most other molecules, and ions in particular, cannot.

Osmosis is the diffusion of water across a selectively permeable membrane, from a **hypotonic** fluid toward a **hypertonic** fluid. There is no net movement of water between **isotonic** solutions. **Osmotic pressure** is the amount of **turgor** (fluid pressure against a cell membrane or wall) sufficient to halt osmosis.

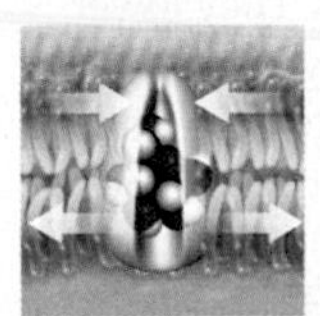

Section 5.9 Ions and most polar molecules can cross cell membranes only with the help of a transport protein. With **facilitated diffusion**, a solute follows its concentration gradient across a membrane through a transport protein. Facilitated diffusion is a type of **passive transport** (no energy input is required). With **active transport**, a transport protein uses energy to pump a solute across a membrane against its concentration gradient. A phosphate-group transfer from ATP often supplies the necessary energy for active transport.

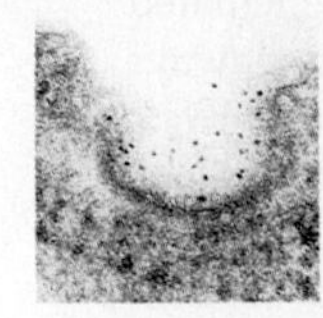

Section 5.10 Substances in bulk and large particles are moved across plasma membranes by processes of **exocytosis** and **endocytosis**. With exocytosis, a cytoplasmic vesicle fuses with the plasma membrane, and its contents are released to the outside of the cell. **Pinocytosis** is an endocytic pathway in which a patch of plasma membrane balloons into the cell, and forms a vesicle that sinks into the cytoplasm. Some cells engulf large particles such as prey or cell debris by the endocytic pathway of **phagocytosis**.

self-quiz

Answers in Appendix VII

1. Which of the following statements is *not* correct?
   a. Energy cannot be created or destroyed.
   b. Energy cannot change from one form to another.
   c. Energy tends to disperse spontaneously.

2. _______ is life's primary source of energy.
   a. Food
   b. Water
   c. Sunlight
   d. ATP

3. Entropy _______ .
   a. disperses
   b. is a measure of disorder
   c. always increases, overall
   d. b and c

4. If we liken a chemical reaction to an energy hill, then a(n) _______ reaction is, overall, an uphill run.
   a. endergonic
   b. exergonic
   c. catalytic
   d. both a and c

5. If we liken a chemical reaction to an energy hill, then activation energy is like _______ .
   a. a burst of speed
   b. coasting downhill
   c. a bump at the top of the hill
   d. putting on the brakes

6. _______ are always changed by participating in a reaction. (Choose all that are correct.)
   a. Enzymes
   b. Cofactors
   c. Reactants
   d. Coenzymes

7. Name one environmental factor that typically influences enzyme function.

8. Which of the following statements is *not* correct?
   a. Metabolic pathways build or break down the organic molecules of life.
   b. All metabolic pathways generate heat.
   c. Electron transfer chains are important sites of energy exchange in many metabolic pathways.
   d. All metabolic pathways require ATP.

9. A molecule that donates electrons becomes _______, and the one that accepts electrons becomes _______ .
   a. reduced; oxidized
   b. ionic; electrified
   c. oxidized; reduced
   d. electrified; ionic

10. All antioxidants _______ .
    a. prevent other molecules from being oxidized
    b. are coenzymes
    c. balance charge
    d. deoxidize free radicals

11. Solutes tend to diffuse from a region where they are _______ (more/less) concentrated to another where they are _______ (more/less) concentrated.

12. _______ cannot easily diffuse across a lipid bilayer.
    a. Water
    b. Gases
    c. Ions
    d. all of the above

13. A transport protein requires ATP to pump sodium ions across a membrane. This is a case of _______ .
    a. passive transport
    b. active transport
    c. facilitated diffusion
    d. a and c

data analysis activities

One Tough Bug The genus *Ferroplasma* consists of a few species of acid-loving archaea. One species, *F. acidarmanus*, was discovered to be the main constituent of slime streamers (a type of biofilm) deep inside an abandoned California copper mine (**FIGURE 5.36**). These cells use an ancient energy-harvesting pathway that combines oxygen with iron-sulfur compounds in minerals such as pyrite. Oxidizing these minerals dissolves them, so groundwater that seeps into the mine ends up with extremely high concentrations of metal ions such as copper, zinc, cadmium, and arsenic. The reaction also produces sulfuric acid, which lowers the pH of the water around the cells to zero.

F. acidarmanus cells maintain their internal pH at a cozy 5.0 despite living in an environment similar to hot battery acid. Thus, researchers investigating *Ferroplasma* were surprised to discover that most of the cells' enzymes function best at very low pH (**FIGURE 5.37**).

1. What does the dashed line signify?

2. Of the four enzymes profiled in the graph, how many function optimally at a pH lower than 5? How many retain significant function at pH 5?

3. What is the optimal pH for *Ferroplasma* carboxylesterase?

FIGURE 5.36 Deep inside one of the most toxic sites in the United States: Iron Mountain Mine, in California. The water in this stream, which is about 1 meter (3 feet) wide in this view, is hot (around 40°C, or 104°F), heavily laden with arsenic and other toxic metals, and has a pH of zero. The slime streamers growing in it are a biofilm dominated by a species of archaea, *Ferroplasma acidarmanus*.

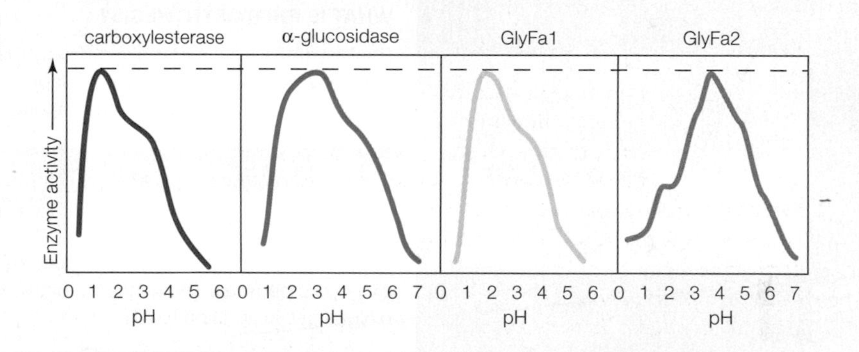

FIGURE 5.37 pH anomaly of *Ferroplasma acidarmanus* enzymes. The graphs (right) show the pH activity profiles of four enzymes isolated from *Ferroplasma*. Researchers had expected these enzymes to function best at the cells' cytoplasmic pH (5.0).

14. Immerse a human red blood cell in a hypotonic solution, and water ________ .
    a. diffuses into the cell
    b. diffuses out of the cell
    c. shows no net movement
    d. moves in by endocytosis

15. Vesicles are part of ________ .
    a. endocytosis
    b. exocytosis
    c. phagocytosis
    d. all of the above

16. Match each term with its most suitable description.

___ reactant	a. assists enzymes
___ phagocytosis	b. forms at reaction's end
___ first law of thermodynamics	c. enters a reaction
___ product	d. requires energy input
___ cofactor	e. one cell engulfs another
___ concentration gradient	f. energy cannot be created or destroyed
___ passive transport	g. basis of diffusion
___ active transport	h. no energy input required
___ redox reaction	i. phospholipids + water
___ cyclic pathway	j. goes in circles
___ lipid bilayer	k. electron exchange

critical thinking

1. Beginning physics students are often taught the basic concepts of thermodynamics with two phrases: First, you can never win. Second, you can never break even. Explain.

2. How is diffusion similar to entropy?

3. How do you think a cell regulates the amount of glucose it brings into its cytoplasm from the extracellular fluid?

4. The enzyme trypsin is sold as a dietary enzyme supplement. Explain what happens to trypsin taken with food.

5. Catalase combines two hydrogen peroxide molecules ($H_2O_2 + H_2O_2$) to make two molecules of water. A gas also forms. What is the gas?

CREDITS: (36) Katrina J. Edwards; (37) From Golyshina et al., *Environmental Microbiology*, 8(3): 416–425, © 2006 John Wiley and Sons. Used with permission of the publisher.

6 Where It Starts—Photosynthesis

LEARNING ROADMAP

This chapter explores the main metabolic pathways (Sections 5.5, 5.6) by which organisms harvest energy from the sun (5.2, 5.3). We revisit experimental design (1.6), electrons and energy levels (2.3), bonds (2.4), carbohydrates (3.4), plastids (4.9), plant cell specializations (4.7, 4.11), membrane proteins (5.7), and concentration gradients (5.8).

THE RAINBOW CATCHERS

The main flow of energy through the biosphere starts when photosynthetic pigments absorb light. In plants and other eukaryotes, these pigments occur in chloroplasts.

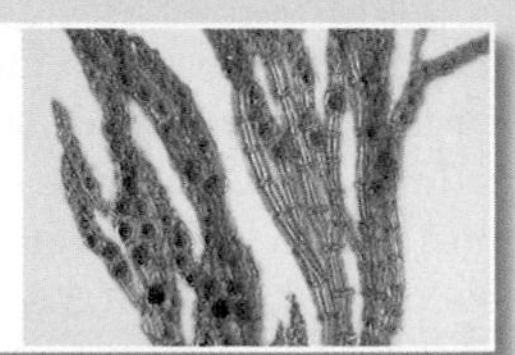

WHAT IS PHOTOSYNTHESIS?

Photosynthesis is a metabolic pathway that occurs in two stages. Light energy harvested in the first stage is used to make molecules that power sugar formation in the second.

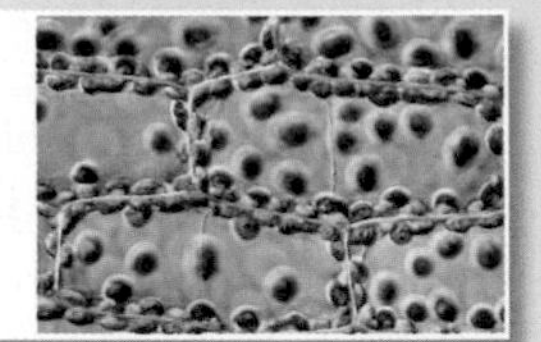

MAKING ATP AND NADPH

The light-dependent reactions produce ATP by either a noncyclic or a cyclic pathway. The noncyclic pathway produces NADPH and oxygen gas in addition to ATP.

MAKING SUGARS

In the second stage of photosynthesis, sugars are assembled from CO_2. The reactions run on ATP and NADPH—molecules that formed in the first stage of photosynthesis.

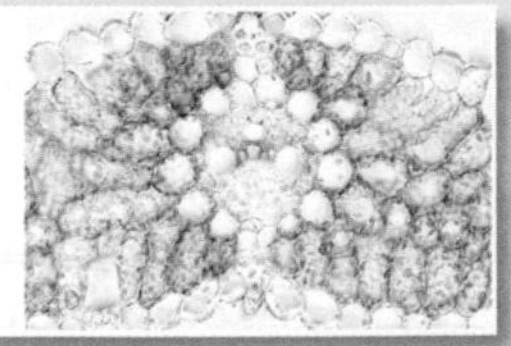

ALTERNATE PATHWAYS

Metabolic pathways are shaped by evolution. Variations in photosynthetic pathways are evolutionary adaptations that allow plants to thrive in a variety of environments.

You will see in Chapter 7 how molecules originally assembled by photosynthesizers are disassembled to harvest energy stored in their bonds. Chapter 22 returns to evolutionary adaptations of plants; and Chapters 27–30, to plant structure and function. In Chapter 46, you will see how photosynthetic organisms sustain almost all life on Earth and how carbon cycles through the biosphere. Chapter 48 returns to human impacts on the biosphere.

6.1 Biofuels

Today, the expression "food is fuel" is not just about eating. With fossil fuel prices soaring, there is an increasing demand for biofuels, which are oils, gases, or alcohols made from organic matter that is not fossilized. Much of the material currently used for biofuel production in the United States consists of food crops—mainly corn, soybeans, and sugarcane. Growing these crops in large quantities is typically expensive and damaging to the environment, and using them to make biofuel competes with our food supply.

How did we end up competing with our vehicles for food? We both run on the same fuel: energy that plants have stored in chemical bonds. Fossil fuels such as coal and natural gas are the remains of ancient swamp forests that decayed and compacted over millions of years. These fuels consist mainly of molecules originally assembled by ancient plants. By contrast, biofuels—and foods—consist mainly of molecules originally assembled by modern plants.

Autotrophs are organisms that make their own food by harvesting energy directly from the environment (*auto-* means self; *-troph* refers to nourishment). All organisms need carbon; autotrophs obtain it from inorganic molecules such as carbon dioxide (CO_2). Plants and most other autotrophs make their food by the metabolic pathway of photosynthesis (Section 1.3). During this pathway, the energy of sunlight is used to drive the assembly of carbohydrates—sugars—from carbon dioxide and water.

Heterotrophs get their carbon by breaking down organic molecules assembled by other organisms (*hetero-* means other). Heterotrophs are an ecosystem's consumers. We and almost all other heterotrophs sustain ourselves by extracting energy from organic molecules that photosynthesizers make. Thus, photosynthesis feeds most life on Earth.

A lot of energy is locked up in the chemical bonds of molecules made by plants. That energy can fuel heterotrophs, as when an animal cell powers ATP synthesis by breaking the bonds of sugars (a topic detailed in the next chapter). It can also fuel our cars, which run on energy released by burning biofuels or fossil fuels. Both processes are fundamentally the same: They release energy by breaking the bonds of organic molecules. Both use oxygen to break those bonds, and both produce carbon dioxide.

Corn and other food crops are rich in oils, starches, and sugars that can be easily converted to biofuels. The starch in corn kernels, for example, can be enzymatically broken down to glucose, which is converted to ethanol by heterotrophic bacteria or yeast. Making biofuels from other types of plant matter requires additional steps, because these materials contain a higher proportion of cellulose. Breaking down this tough, insoluble carbohydrate to its glucose monomers adds a lot of cost to the biofuel product. Researchers are currently working on cost-effective ways to break down the abundant cellulose in fast-growing weeds such as switchgrass (**FIGURE 6.1**), and agricultural wastes such as wood chips, wheat straw, cotton stalks, and rice hulls.

autotroph Organism that makes its own food using energy from the environment and carbon from inorganic molecules such as CO_2.
heterotroph Organism that obtains carbon from organic compounds assembled by other organisms.

FIGURE 6.1 Making biofuels. Left, Ratna Sharma and Mari Chinn are researching ways to reduce the cost of producing biofuel from renewable sources such as wild grasses and agricultural wastes. Right, switchgrass (*Panicum virgatum*) growing wild in a North American prairie.

6.2 Sunlight as an Energy Source

✔ Photosynthetic organisms use pigments to capture the energy of sunlight.

Properties of Light

Photosynthesizers make their own food by converting light energy to chemical energy. In order to understand how that happens, you have to know a little about the nature of light. Light is electromagnetic radiation, a type of energy that moves through space in waves, a bit like waves move across an ocean. The distance between the crests of two successive waves is called **wavelength**, and it is measured in nanometers (nm).

Light that humans can see is a small part of the spectrum of electromagnetic radiation emitted by the sun (**FIGURE 6.2A**). Visible light travels in wavelengths between 380 and 750 nm, and this is the main form of energy that drives photosynthesis. Our eyes perceive all of these wavelengths combined as white light, and particular wavelengths in this range as different colors. White light separates into its component colors when it passes through a prism, or raindrops that act as tiny prisms (**FIGURE 6.3**). A prism bends longer wavelengths more than it bends shorter ones, so a rainbow of colors forms.

Light travels in waves, but it is also organized in packets of energy called photons. A photon's energy and its wavelength are related, so all photons traveling at the same wavelength carry the same amount of energy. Photons that carry the least amount of energy travel in longer wavelengths; those that carry the most energy travel in shorter wavelengths (**FIGURE 6.2B**).

FIGURE 6.3 A rainbow. Sunlight passing through raindrops separates into its component colors.

Pigments: The Rainbow Catchers

Photosynthesizers use pigments to capture light. A **pigment** is an organic molecule that selectively absorbs light of specific wavelengths. Wavelengths of light that are not absorbed are reflected, and that reflected light gives each pigment its characteristic color.

Chlorophyll *a* is the most common photosynthetic pigment in plants and photosynthetic protists. It also occurs in some bacteria. Chlorophyll *a* absorbs violet, red, and orange light, and it reflects green light, so it appears green to us. Accessory pigments, including other chlorophylls, collectively harvest a wide range of additional light wavelengths for photosynthesis (**FIGURE 6.4**).

A pigment molecule is a bit like an antenna specialized for receiving light. It has a light-trapping region, in

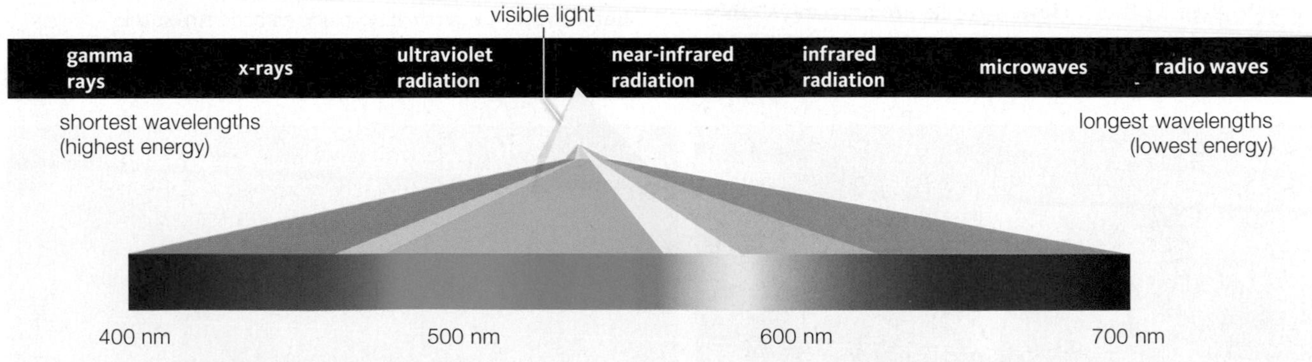

A Electromagnetic radiation moves through space in waves that we measure in nanometers (nm). Visible light makes up a very small part of this energy. Raindrops or a prism can separate visible light's different wavelengths, which we see as different colors. About 25 million nanometers are equal to 1 inch.

B Light is organized as packets of energy called photons. The shorter a photon's wavelength, the greater its energy.

FIGURE 6.2 Properties of light.

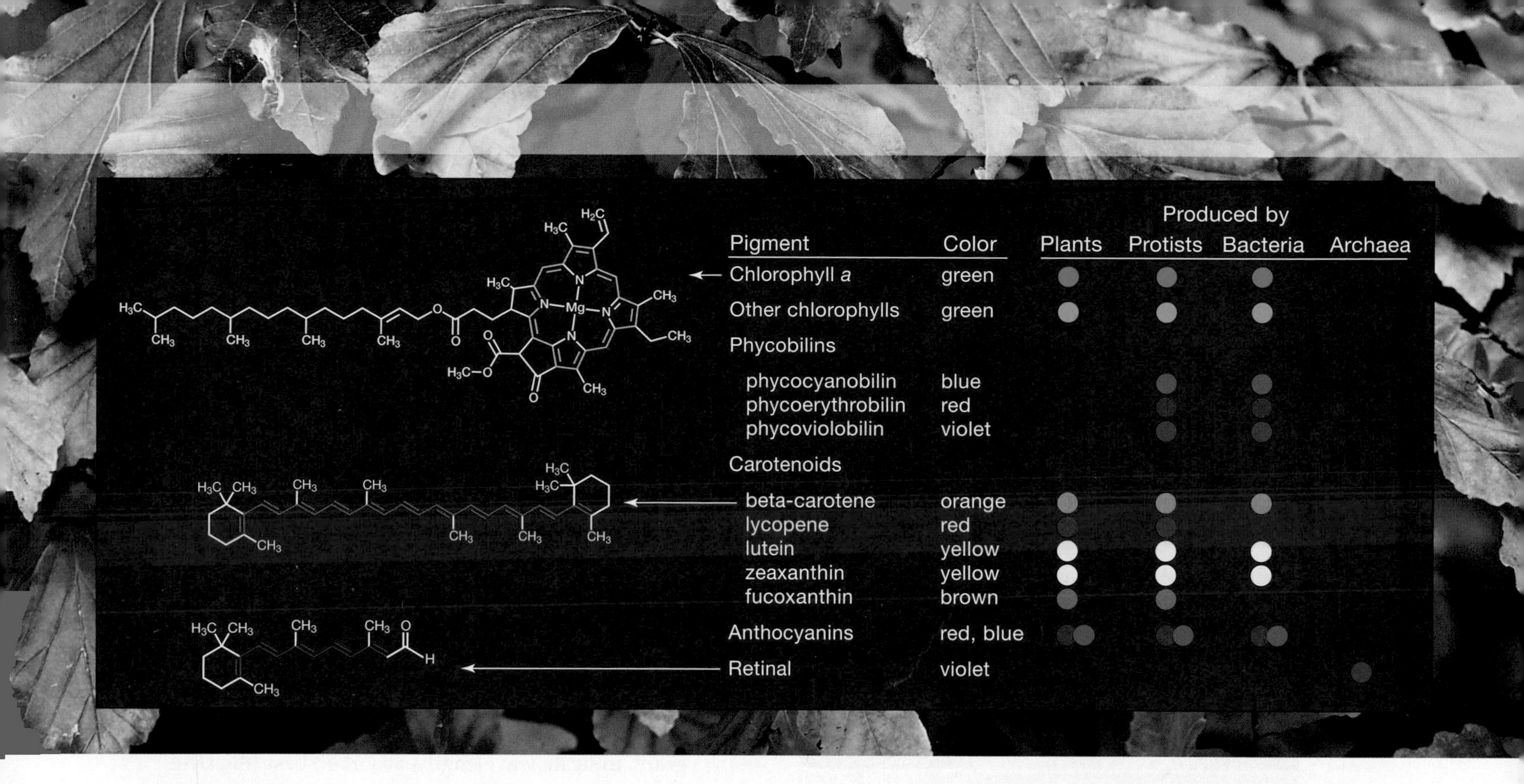

FIGURE 6.4 Examples of photosynthetic pigments. Photosynthetic pigments can collectively absorb almost all visible light wavelengths. Left, the light-catching part of a pigment (shown in color) is the region in which single bonds alternate with double bonds. These and many other pigments (including heme, Section 5.6) are derived from evolutionary remodeling of the same compound. Animals convert dietary beta-carotene into a similar pigment (retinal) that is the basis of vision.

which single bonds alternate with double bonds. Electrons populating the atoms of this region easily absorb a photon—but not just any photon. Only a photon with exactly enough energy to boost an electron to a higher energy level is absorbed (Section 2.3). This is why a pigment absorbs light of only certain wavelengths.

An excited electron (one that has been boosted to a higher energy level) quickly emits its extra energy and returns to a lower energy level. As you will see, photosynthetic cells capture energy emitted from an electron returning to a lower energy level.

Most photosynthetic organisms use a combination of pigments to capture light for photosynthesis—and often for additional purposes. Many accessory pigments are antioxidants that protect cells from the damaging effects of ultraviolet (UV) light in the sun's rays (Section 5.6). Appealing colors attract animals to ripening fruit or pollinators to flowers. You may already be familiar with some of these molecules. Carrots, for example, are orange because they contain beta-carotene (β-carotene); roses are red and violets are blue because of their anthocyanin content.

chlorophyll *a* Main photosynthetic pigment in plants.
pigment An organic molecule that can absorb light of certain wavelengths.
wavelength Distance between the crests of two successive waves.

In green plants, chlorophylls are usually so abundant that they mask the colors of the other pigments. Plants that change color during autumn are preparing for a period of dormancy; they conserve resources by moving nutrients from tender parts that would be damaged by winter cold (such as leaves) to protected parts (such as roots). Chlorophylls are not needed during dormancy, so they are disassembled and their components recycled. Yellow and orange accessory pigments are also recycled, but not as quickly as chlorophylls. Their colors begin to show as the chlorophyll content declines in leaves. Anthocyanin synthesis also increases in some plants, adding red and purple tones to turning leaf colors. (Chapter 29 returns to the topic of dormancy in plants.)

TAKE-HOME MESSAGE 6.2

How do photosynthesizers absorb light?

✔ The sun emits electromagnetic radiation (light). Visible light is the main form of energy that drives photosynthesis.

✔ Light travels in waves and is organized as photons. We see different wavelengths of visible light as different colors.

✔ Pigments absorb light at specific wavelengths. Photosynthetic species use pigments such as chlorophyll *a* to harvest the energy of light for photosynthesis.

6.3 Exploring the Rainbow

✔ Photosynthetic pigments work together to harvest light of different wavelengths.

In 1882, botanist Theodor Engelmann designed a series of experiments to test his hypothesis that the color of light affects the rate of photosynthesis. It had long been known that photosynthesis releases oxygen, so Engelmann used oxygen emission as an indirect measure of photosynthetic activity. He directed a spectrum of light across individual strands of green algae suspended in water (**FIGURE 6.5A**). Oxygen-sensing equipment had not yet been invented, so Engelmann used motile, oxygen-requiring bacteria to show him where the oxygen concentration in the water was highest. The bacteria moved through the water and gathered mainly where blue and red light fell across the algal cells (**FIGURE 6.5B**). Engelmann concluded that photosynthetic cells illuminated by light of these colors were releasing the most oxygen—a sign that blue and red light are the best for driving photosynthesis in these algal cells.

Today we can directly measure the efficiency at which a photosynthetic pigment absorbs different wavelengths of light. A graph that shows this efficiency is called an absorption spectrum. Peaks in the graph indicate wavelengths absorbed best (**FIGURE 6.5C**). Engelmann's results (the distribution of bacteria around the algal cells) represent the combined spectra of all the photosynthetic pigments present in this alga.

The combination of pigments used for photosynthesis differs among species. Why? Photosynthesizers are adapted to the environment in which they evolved, and light that reaches different environments varies in its proportions of wavelengths. Consider that seawater absorbs green and blue-green light less efficiently than other colors. Thus, more green and blue-green light penetrates deep ocean water. Algae that can live far below the sea surface (such as *Polysiphonia*, below) tend to be rich in pigments—mainly phycobilins—that absorb green and blue-green light.

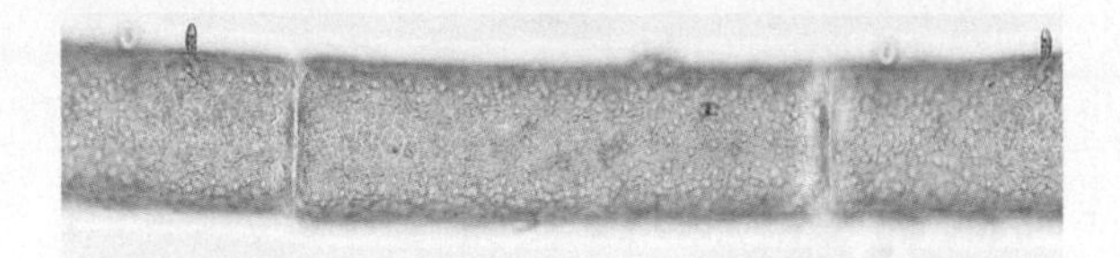

A Each cell in a strand of *Cladophora* is filled with a single chloroplast. Theodor Engelmann used this and other species of green algae in a series of experiments to determine whether some colors of light are better for photosynthesis than others.

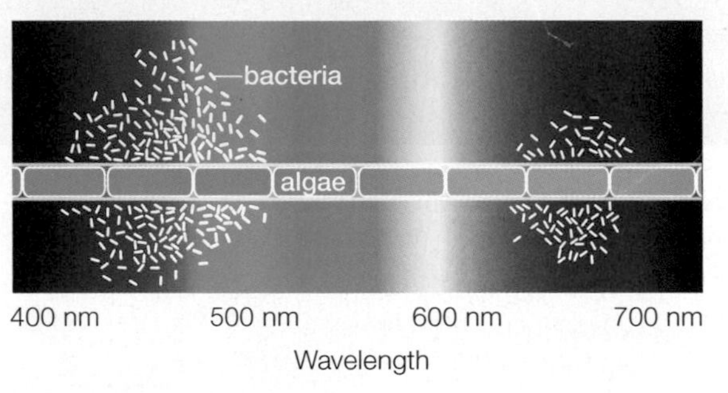

B Engelmann directed light through a prism so that bands of colors crossed a water droplet on a microscope slide. The water held a strand of photosynthetic algae, and also oxygen-requiring bacteria. The bacteria swarmed around the algal cells that were releasing the most oxygen—the ones most actively engaged in photosynthesis. Those cells were under blue and red light.

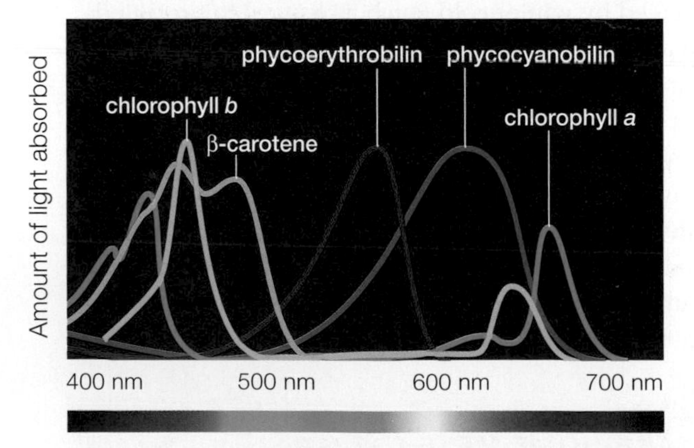

C Absorption spectra of chlorophylls *a* and *b*, β-carotene, and two phycobilins reveal the efficiency with which these pigments absorb different wavelengths of visible light. Line color indicates the characteristic color of each pigment.

FIGURE 6.5 ▸Animated Discovery that photosynthesis is driven by particular wavelengths of light.

Of the five pigments represented in **C**, which three are the main photosynthetic pigments in *Cladophora*?

Answer: Chlorophyll *a*, chlorophyll *b*, and β-carotene

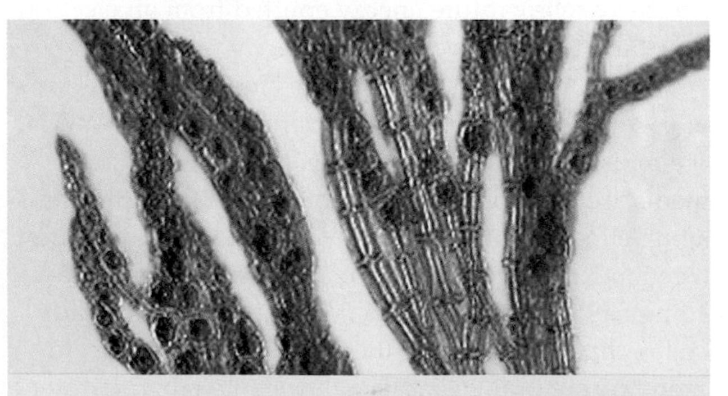

TAKE-HOME MESSAGE 6.3

Why do cells use multiple pigments for photosynthesis?

✔ A combination of pigments allows a photosynthetic cell to efficiently capture the wavelenghths of light most abundant in the habitat in which it evolved.

6.4 Overview of Photosynthesis

✔ In eukaryotes, photosynthesis takes place in chloroplasts.
✔ Photosynthesis occurs in two stages.

All life is sustained by inputs of energy, but not all forms of energy can sustain life. Sunlight, for example, is abundant here on Earth, but it cannot be used to directly power protein synthesis or other energy-requiring reactions that keep organisms alive. Photosynthesis converts the energy of light into the energy of chemical bonds. Unlike light, chemical energy can power the reactions of life, and it can be stored for later use.

In plants and other photosynthetic eukaryotes, photosynthesis takes place in chloroplasts (Section 4.9). Plant chloroplasts have two outer membranes, and they are filled with a thick, cytoplasm-like fluid called **stroma** (FIGURE 6.6). Suspended in the stroma are the chloroplast's own DNA, some ribosomes, and an inner, much-folded **thylakoid membrane**. The folds of a thylakoid membrane typically form stacks of interconnected disks called thylakoids. The space enclosed by the thylakoid membrane is a single, continuous compartment.

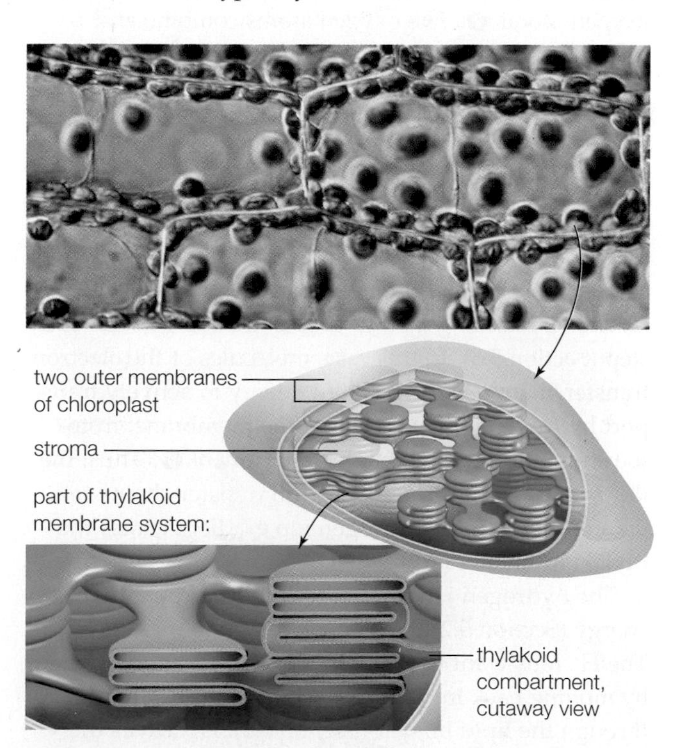

FIGURE 6.6 ▶Animated Zooming in on a chloroplast, the site of photosynthesis in a plant cell. The micrograph shows chloroplasts in cells of a moss leaf.

light-dependent reactions First stage of photosynthesis; convert light energy to chemical energy.
light-independent reactions Second stage of photosynthesis; use ATP and NADPH to assemble sugars from water and CO_2.
stroma Cytoplasm-like fluid between the thylakoid membrane and the two outer membranes of a chloroplast.
thylakoid membrane A chloroplast's highly folded inner membrane system; forms a continuous compartment in the stroma.

Photosynthesis is often summarized by this equation:

$$CO_2 + \text{water} \xrightarrow{\text{light energy}} \text{sugars} + O_2$$

CO_2 (carbon dioxide) and O_2 (oxygen) are gases abundant in air. Keep in mind that photosynthesis is not a single reaction. Rather, it is a metabolic pathway (Section 5.5) comprising many reactions that occur in two stages. Molecules embedded in the thylakoid membrane carry out the reactions of the first stage, which are driven by light and thus called the **light-dependent reactions**. The "photo" in photosynthesis means light, and it refers to the conversion of light energy to the chemical bond energy of ATP during this stage. In addition to making ATP, the main light-dependent pathway in chloroplasts splits water molecules and releases O_2. Hydrogen ions and electrons from the water molecules end up in the coenzyme NADPH:

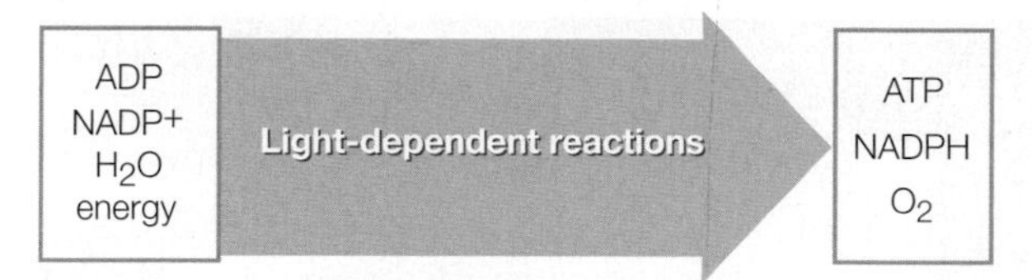

The "synthesis" part of photosynthesis refers to the reactions of the second stage, which build sugars from CO_2 and water. These sugar-building reactions run in the stroma. They are collectively called the **light-independent reactions** because light energy does not power them. Instead, they run on energy delivered by NADPH and ATP that formed during the first stage:

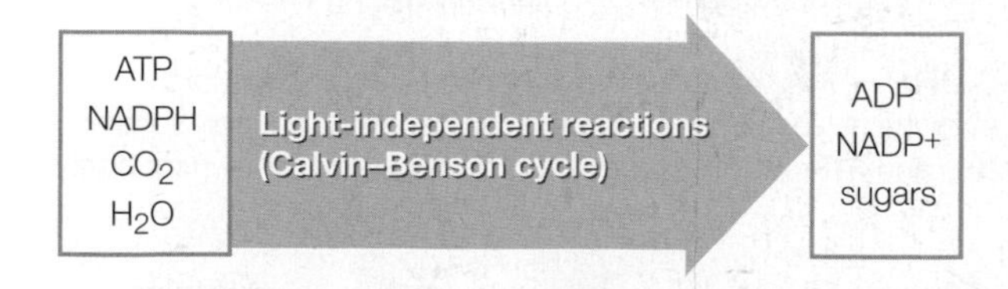

TAKE-HOME MESSAGE 6.4
What happens during photosynthesis?

✔ In eukaryotic cells, the first stage of photosynthesis occurs at the thylakoid membrane of chloroplasts. During these light-dependent reactions, light energy drives the formation of ATP and NADPH.

✔ In eukaryotic cells, the second stage of photosynthesis occurs in the stroma of chloroplasts. During these light-independent reactions, ATP and NADPH drive the synthesis of sugars from water and carbon dioxide.

6.5 Light-Dependent Reactions

✔ The reactions of the first stage of photosynthesis convert the energy of light to the energy of chemical bonds.

A chloroplast's thylakoid membrane contains millions of light-harvesting complexes, which are circular arrays of chlorophylls, various accessory pigments, and proteins (**FIGURE 6.7**). When a chlorophyll or accessory pigment in a light-harvesting complex absorbs light, one of its electrons jumps to a higher energy level (shell, Section 2.3). The electron quickly drops back down to a lower shell by emitting its extra energy. Light-harvesting complexes hold onto that emitted energy by passing it back and forth, a bit like volleyball players pass a ball among team members. The reactions of photosynthesis begin when energy being passed around the thylakoid membrane reaches a photosystem. A **photosystem** is a group of hundreds of chlorophylls, accessory pigments, and other molecules that work as a unit to begin the reactions of photosynthesis.

FIGURE 6.7 A view of some components of the thylakoid membrane as seen from the stroma. Molecules of electron transfer chains and ATP synthases are also present, but not shown for clarity.

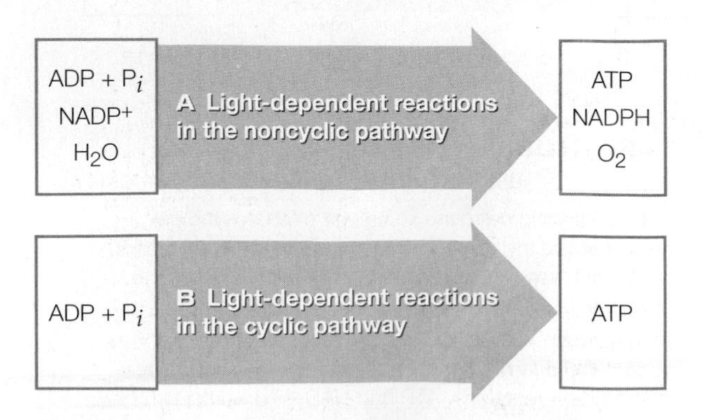

FIGURE 6.8 Summary of the inputs and outputs of the two pathways of light-dependent reactions.

The Noncyclic Pathway

Thylakoid membranes have two kinds of photosystems, type I and type II, that were named in the order of their discovery. In cyanobacteria, plants, and all photosynthetic protists, both photosystem types work together in the noncyclic pathway of photosynthesis (**FIGURE 6.8A**). This pathway begins when energy being passed among light-harvesting complexes reaches a photosystem II (**FIGURE 6.9**). At the center of each photosystem are two very closely associated chlorophyll *a* molecules (a "special pair"). When a photosystem absorbs energy, electrons are ejected from its special pair ❶.

A photosystem can lose only a few electrons before it must be restocked with more. Where do replacements come from? Photosystem II gets more electrons by removing them from water molecules in the thylakoid compartment. This reaction causes the water molecules to dissociate into hydrogen ions and oxygen atoms ❷. The oxygen atoms combine and diffuse out of the cell as oxygen gas (O_2). This and any other process by which a molecule is broken apart by light energy is called **photolysis**.

The actual conversion of light energy to chemical energy occurs when electrons ejected from photosystem II enter an electron transfer chain in the thylakoid membrane ❸. Remember that electron transfer chains can harvest the energy of electrons in a series of redox reactions, releasing a bit of their extra energy with each step (Section 5.5). In this case, molecules of the electron transfer chain use the released energy to actively transport hydrogen ions (H^+) across the membrane, from the stroma to the thylakoid compartment ❹. Thus, the flow of electrons through electron transfer chains sets up and maintains a hydrogen ion gradient across the thylakoid membrane.

The hydrogen ion gradient is a type of potential energy (Section 5.2) that can be tapped to make ATP. The H^+ ions want to follow their concentration gradient by moving back into the stroma, but ions cannot diffuse through the lipid bilayer (Section 5.8). H^+ leaves the thylakoid compartment only by flowing through proteins called ATP synthases embedded in the thylakoid membrane ❼. An ATP synthase is both a transport protein and an enzyme. When hydrogen ions flow through its interior, the protein phosphorylates ADP, so ATP forms in the stroma ❽. The process by which the flow of electrons through electron transfer chains drives ATP formation is called **electron transfer phosphorylation.**

After the electrons have moved through the first electron transfer chain, they are accepted by a photo-

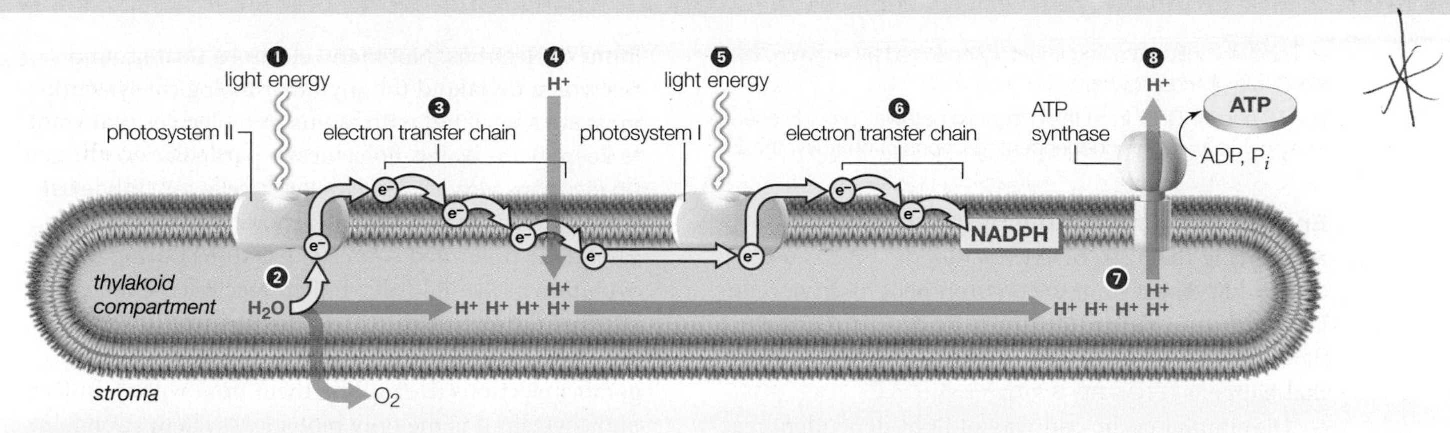

❶ Light energy ejects electrons from a photosystem II.

❷ The photosystem pulls replacement electrons from water molecules, which then break apart into oxygen and hydrogen ions. The oxygen leaves the cell as O_2.

❸ The electrons enter an electron transfer chain in the thylakoid membrane.

❹ Energy lost by the electrons as they move through the chain is used to actively transport hydrogen ions from the stroma into the thylakoid compartment. A hydrogen ion gradient forms across the thylakoid membrane.

❺ Light energy ejects electrons from a photosystem I. Replacement electrons come from an electron transfer chain.

❻ The ejected electrons move through a second electron transfer chain, then combine with $NADP^+$ and H^+, so NADPH forms.

❼ Hydrogen ions in the thylakoid compartment follow their gradient across the thylakoid membrane by flowing through the interior of ATP synthases.

❽ Hydrogen ion flow causes ATP synthases to phosphorylate ADP, so ATP forms in the stroma.

FIGURE 6.9 ▸Animated Light-dependent reactions, noncyclic pathway. ATP and oxygen gas are produced in this pathway. Electrons that travel through two different electron transfer chains end up in NADPH.

system I. When this photosystem absorbs light energy, its special pair of chlorophylls emits electrons ❺. These electrons enter a second, different electron transfer chain. At the end of this chain, the coenzyme $NADP^+$ accepts the electrons along with H^+, so NADPH forms ❻:

$$NADP^+ + 2e^- + H^+ \longrightarrow \textbf{NADPH}$$

The Cyclic Pathway

As you will see shortly, ATP and NADPH produced in the light-dependent reactions are used to make sugars. On its own, the noncyclic pathway does not yield enough ATP to balance NADPH use in sugar production pathways. The cyclic pathway produces additional ATP for this purpose (**FIGURE 6.8B**).

In the cyclic pathway, electrons that are ejected from photosystem I enter an electron transfer chain, and then return to photosystem I. As in the noncyclic pathway, the electron transfer chain uses electron energy to move hydrogen ions into the thylakoid compartment, and the resulting hydrogen ion gradient drives ATP formation. However, the cyclic pathway does not produce NADPH or oxygen gas.

The cyclic pathway allows light-dependent reactions to continue when the noncyclic pathway stops, for example under intense illumination. Light energy in excess of what can be used for photosynthesis can result in the formation of dangerous free radicals (Section 2.3). A light-induced structural change in photosystem II prevents this from happening. The photosystem stops initiating the noncyclic pathway, and traps excess energy instead. At such times, the cyclic pathway predominates.

electron transfer phosphorylation Process in which electron flow through electron transfer chains sets up a hydrogen ion gradient that drives ATP formation.
photolysis Process by which light energy breaks down a molecule.
photosystem Cluster of pigments and proteins that converts light energy to chemical energy in photosynthesis.

TAKE-HOME MESSAGE 6.5

How do the light-dependent reactions of photosynthesis work?

✔ Photosynthetic pigments in the thylakoid membrane transfer the energy of light to photosystems, which eject electrons that enter electron transfer chains.

✔ In both noncyclic and cyclic pathways, the flow of electrons through the transfer chains sets up hydrogen ion gradients that drive ATP formation.

✔ The noncyclic pathway uses two photosystems. Water molecules are split, oxygen is released, and electrons end up in NADPH.

✔ The cyclic pathway uses only photosystem I. No NADPH forms, and no oxygen is released.

6.6 The Light-Independent Reactions

✔ The chloroplast is a sugar factory operated by enzymes of the Calvin–Benson cycle.

✔ ATP and NADPH from the noncyclic pathway provide energy that powers the light-independent reactions of photosynthesis.

Energy Flow in Photosynthesis

A recurring theme in biology is that organisms use energy harvested from the environment to drive cellular processes. Energy flow in the noncyclic pathway of light-dependent reactions is a classic example of how that happens (**FIGURE 6.10**).

The simpler cyclic pathway of light-dependent reactions was the first one to evolve. Later, the photosynthetic machinery in some organisms evolved so that photosystem II became part of it. Of the two photosystems, photosystem I is the stronger reducing agent, which means it more easily gives up electrons. Those electrons either cycle back to it (in the cyclic pathway), or they end up in NADPH (in the noncyclic pathway). Running the noncyclic pathway requires a continuous input of electrons. Not many electrons float around freely in a thylakoid (or any other biological system); most are associated with atoms or molecules that want to keep them. Water molecules in particular do not give up electrons very easily. However, cells can only exist where there is water, so water molecules provided an essentially unlimited source of electrons during the evolution of the light-dependent reactions. Thus, it is perhaps not suprising that, of the two photosystems, photosystem II is the stronger oxidizer: It is best at gaining electrons (by pulling them from water). In fact, photosystem II is the only biological system strong enough to oxidize water molecules.

Light-Independent Reactions

NADPH is a powerful reducing agent (electron donor), and a lot of it is required to run the second stage of photosynthesis. You learned in Section 6.4 that reactions of this stage are light-independent because light energy does not power them. Energy that drives these

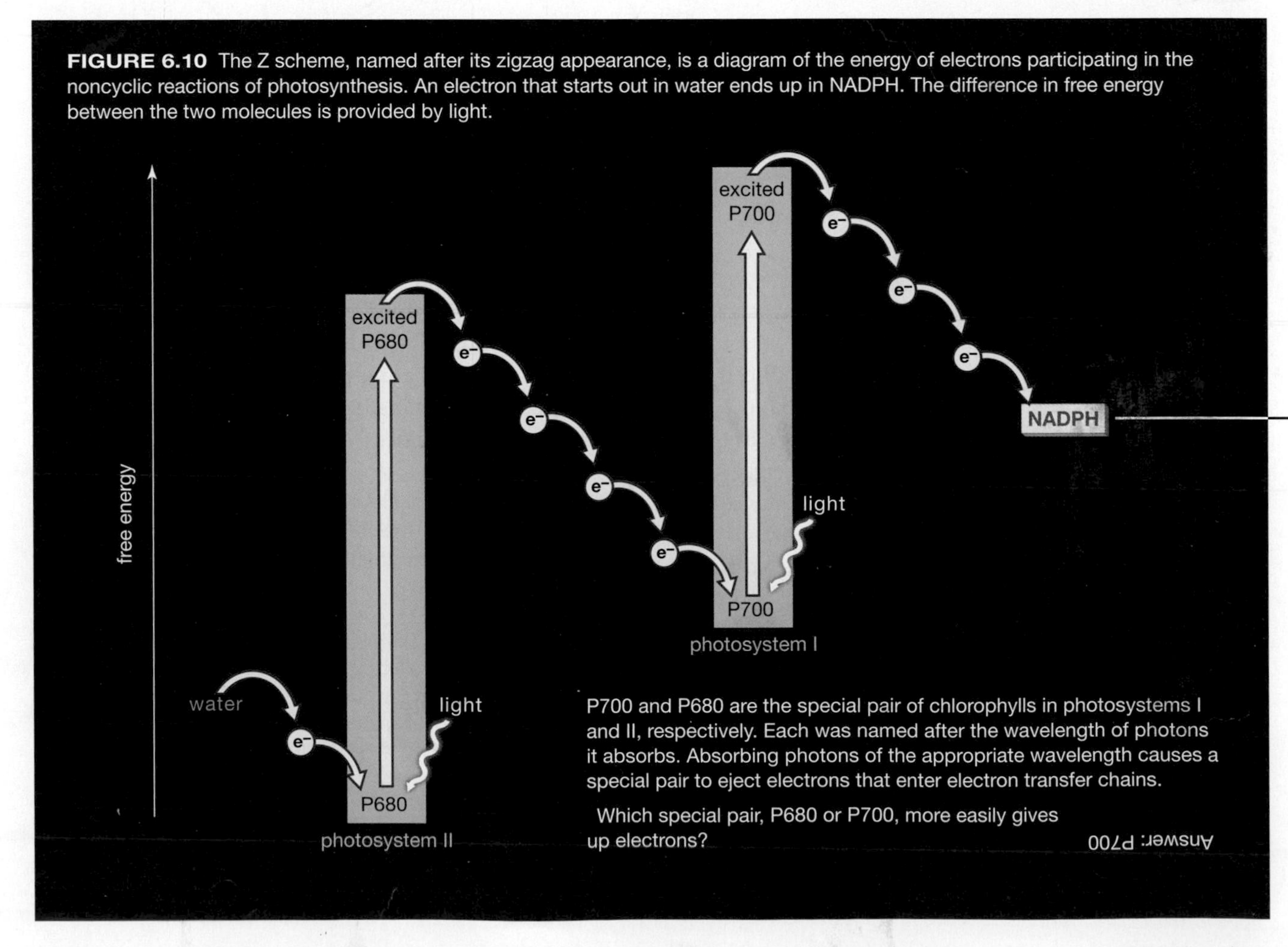

FIGURE 6.10 The Z scheme, named after its zigzag appearance, is a diagram of the energy of electrons participating in the noncyclic reactions of photosynthesis. An electron that starts out in water ends up in NADPH. The difference in free energy between the two molecules is provided by light.

reactions is provided by phosphate-group transfers from ATP, and electrons from NADPH.

The light-independent reactions, which are collectively called the **Calvin–Benson cycle**, produce sugars in the stroma of chloroplasts (**FIGURE 6.11**). This cyclic pathway uses carbon atoms from CO_2 to build carbon backbones of the sugar molecules. Extracting carbon atoms from an inorganic source (such as CO_2) and incorporating them into an organic molecule is a process called **carbon fixation**.

In most plants, photosynthetic protists, and some bacteria, the enzyme **rubisco** fixes carbon by attaching CO_2 to RuBP (ribulose bisphosphate), a five-carbon molecule ❶. The six-carbon intermediate that forms by this reaction is unstable, so it splits right away into two three-carbon molecules of PGA (phosphoglycerate). Each PGA receives a phosphate group from ATP, and hydrogen and electrons from NADPH ❷. Thus, ATP energy and the reducing power of NADPH convert each molecule of PGA into a molecule of PGAL (phosphoglyceraldehyde), a phosphorylated sugar.

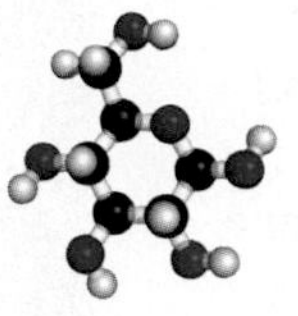

In later reactions, two or more of the three-carbon PGAL molecules can be combined and rearranged to form larger carbohydrates. Glucose has six carbon atoms (left). To make one glucose molecule, six CO_2 must be attached to six RuBP molecules, so twelve PGAL form. Two PGAL combine to form one glucose molecule ❸. The ten remaining PGAL regenerate the starting compound of the cycle, RuBP ❹.

Plants can break down the glucose they make in the Calvin–Benson cycle to access the energy stored in its bonds (we return to this process in Chapter 7). However, most of the glucose is converted at once to sucrose or starch by other pathways that conclude the light-independent reactions. Excess glucose is stored as starch grains in chloroplast stroma. When sugars are needed in other parts of the plant, the starch is broken down and sugar monomers are exported from the cell.

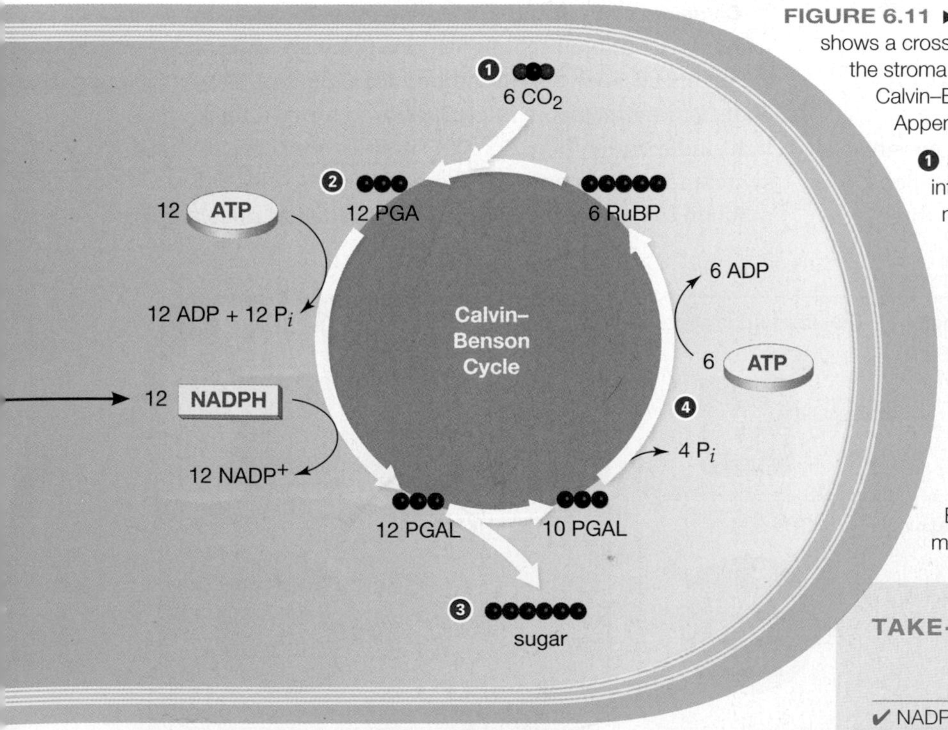

FIGURE 6.11 ▶**Animated** The Calvin–Benson cycle. This sketch shows a cross-section of a chloroplast with the reactions cycling in the stroma. The steps shown are a summary of six cycles of the Calvin–Benson reactions. Black balls signify carbon atoms. Appendix III details the reaction steps.

❶ Six CO_2 diffuse into a photosynthetic cell, and then into a chloroplast. Rubisco attaches each to a RuBP molecule. The resulting intermediates split, so twelve molecules of PGA form.

❷ Each PGA molecule gets a phosphate group from ATP, plus hydrogen and electrons from NADPH, so twelve PGAL form. ❸ Two PGAL may combine to form one six-carbon sugar (such as glucose).

❹ The remaining ten PGAL receive phosphate groups from ATP. The transfer primes them for endergonic reactions that regenerate the 6 RuBP.

How many times does the Calvin–Benson cycle need to run in order to produce one molecule of glucose?

Answer: Six

TAKE-HOME MESSAGE 6.6

How do the light-independent reactions of photosynthesis work?

✔ NADPH and ATP produced by the light-dependent reactions power the light-independent reactions of the Calvin–Benson cycle.

✔ The Calvin–Benson cycle uses atoms of hydrogen (from NADPH), and carbon and oxygen (from CO_2) to build sugars.

✔ Incorporating carbon atoms from an inorganic source (such as CO_2) into an organic molecule is called carbon fixation.

Calvin–Benson cycle Cyclic carbon-fixing pathway that builds sugars from CO_2; light-independent reactions of photosynthesis.
carbon fixation Process by which carbon from an inorganic source such as carbon dioxide gets incorporated into an organic molecule.
rubisco Ribulose bisphosphate carboxylase. Carbon-fixing enzyme of the Calvin–Benson cycle.

6.7 Adaptations: Alternative Carbon-Fixing Pathways

✔ Carbon fixation sustains all life on Earth, but its details differ among species.

Most plants have a thin, waterproof cuticle that limits evaporative water loss from their aboveground parts. Gases cannot diffuse across the cuticle, but oxygen produced by the light-dependent reactions must escape the plant, and carbon dioxide needed for the Calvin–Benson cycle must enter it. Thus, the surfaces of leaves and stems are studded with tiny, closable gaps called **stomata** (**FIGURE 6.12**). When stomata are open, CO_2 diffuses from the air into photosynthetic tissues, and O_2 diffuses out of the tissues into the air. Stomata close to conserve water on hot, dry days. When that happens, gas exchange comes to a halt.

FIGURE 6.12 Stomata on the surface of a leaf. When these tiny pores are open, gases are exchanged between the plant's internal tissues and air. Stomata close to conserve water on hot, dry days.

Both stages of photosynthesis run during the day. With stomata closed, the O_2 level in the plant's tissues rises, and the CO_2 level declines. This outcome can reduce the efficiency of sugar production because both gases are substrates of rubisco, and they compete for its active site. Rubisco initiates the Calvin–Benson cycle by attaching CO_2 to RuBP. It also initiates a pathway called **photorespiration**, by attaching O_2 to RuBP. Photorespiration is an inefficient way to produce sugars because it requires more ATP and NADPH than the Calvin–Benson cycle, and it produces CO_2 (**FIGURE 6.13A**).

The detrimental effects of photorespiration are greatest in **C3 plants**, which fix carbon by the Calvin–Benson cycle only (they are called C3 plants because a three-carbon molecule, PGA, is the first stable intermediate to form in their light-independent reactions). C3 plants have no way to compensate for a decline in CO_2 level when stomata close during the day, so sugar production becomes less and less efficient with rising daytime temperature. These plants compensate for photorespiration by making a lot of rubisco: It is the most abundant protein on Earth.

C4 plants compensate for rubisco's inefficiency by using an additional set of reactions (they are called C4 plants because a four-carbon molecule, oxaloacetate, is the first stable intermediate to form in their light-independent reactions). Examples are corn, switchgrass, and bamboo. These plants also close stomata on dry days, but their sugar production does not

FIGURE 6.13 ▶**Animated** Adaptations of C4 and CAM plants minimize photorespiration. Compare **FIGURE 6.14**.

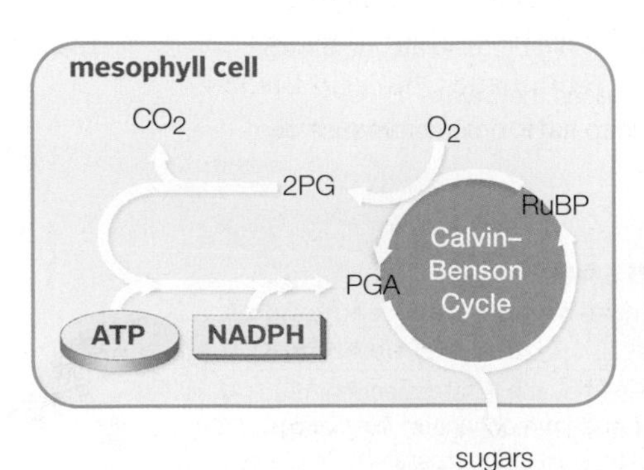

A Photorespiration. When stomata close during photosynthesis, O_2 accumulates in tissues and CO_2 declines. Rubisco initiates more photorespiration, shunting resources away from the sugar-producing reactions of the Calvin–Benson cycle.

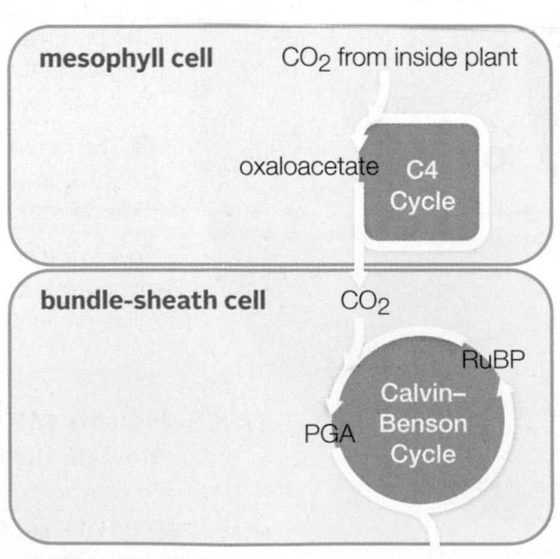

B In C4 plants, oxygen also builds up in leaves when stomata close during photosynthesis. An additional pathway in these plants keeps the CO_2 concentration high enough in bundle-sheath cells to prevent photorespiration.

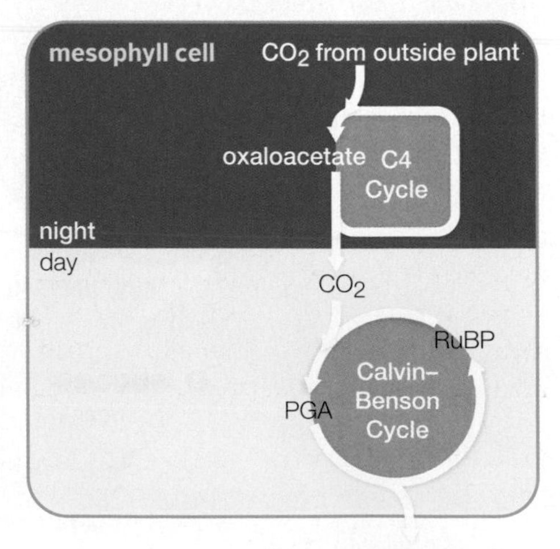

C CAM plants open stomata and fix carbon using a C4 pathway at night. When the plant's stomata are closed during the day, the organic compound made during the night is converted to CO_2 that enters the Calvin–Benson cycle.

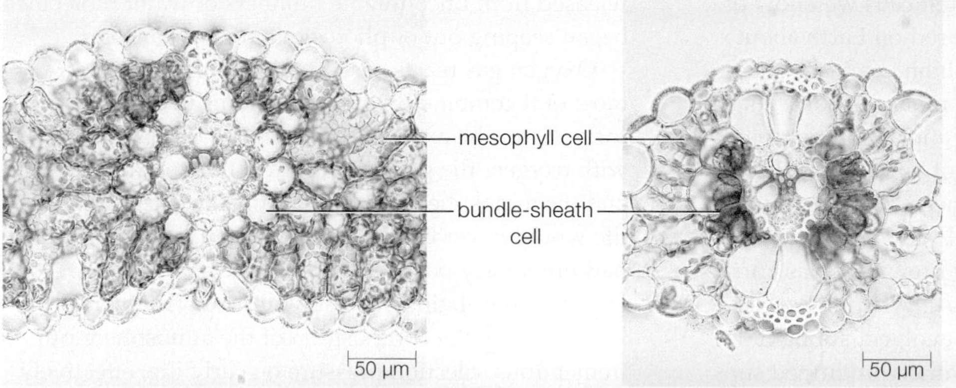

A In a leaf of a C3 plant (here, barley), chloroplasts—the sites of carbon fixation—occur mainly in mesophyll cells. Bundle sheath cells ringing the leaf vein are mostly empty of chloroplasts.

Photorespiration in C3 plants reduces the efficiency of sugar synthesis on hot, dry days.

B In a leaf of a C4 plant (millet, left), carbon is fixed the first time in mesophyll cells, which are near air spaces in the leaf. Carbon fixation occurs for the second time in chloroplast-stuffed bundle-sheath cells.

The photo on the right shows crabgrass "weeds" overgrowing a lawn. Crabgrasses, which are C4 plants, thrive in hot, dry summers, when they easily outcompete Kentucky bluegrass and other fine-leaved C3 grasses commonly planted in residential lawns.

FIGURE 6.14 ▸Animated Comparing the sites of carbon fixation in C3 plants and C4 plants. Micrographs show cross sections. In both types of plants, bundle-sheath cells ring the leaf veins. In C4 plants only, a second carbon fixation occurs in these cells. The adaptation maintains a high CO_2/O_2 ratio near rubisco, minimizing photorespiration when stomata close on hot, dry days.

decline. C4 plants fix carbon twice, in two kinds of cells (**FIGURE 6.13B** and **FIGURE 6.14**). The first set of reactions occurs in mesophyll cells, where carbon is fixed by an enzyme that does not use oxygen even when the carbon dioxide level is low. The product of this C4 pathway, a four-carbon acid, is transported to bundle-sheath cells. There, it is converted to CO_2, and rubisco fixes carbon for the second time as the CO_2 enters the Calvin–Benson cycle.

Bundle-sheath cells of C4 plants have chloroplasts that carry out light-dependent reactions, but only in the cyclic pathway. No oxygen is released, so the O_2 level near rubisco stays low. This, along with the high CO_2 level provided by the C4 reactions, minimizes photorespiration, so sugar production stays efficient in these plants even in hot, dry weather.

Succulents, cacti, and other **CAM plants** use a carbon-fixing pathway that allows them to minimize photorespiration even in desert regions with extremely high daytime temperatures. CAM stands for crassulacean acid metabolism, after the Crassulaceae family of plants in which this pathway was first studied. Like C4 plants, CAM plants fix carbon twice, but the reactions occur at different times rather than in different cells (**FIGURE 6.13C**). Stomata on a CAM plant open only at night, when typically lower temperatures minimize evaporative water loss. The plants use a C4 pathway to fix carbon from CO_2 in the air at this time. The product of this pathway, a four-carbon acid, is stored in the cell's central vacuole. When the stomata close the next day, the acid moves out of the vacuole and becomes broken down to CO_2, which is fixed for the second time when it enters the Calvin–Benson cycle. The high CO_2 level provided by the acid breakdown minimizes photorespiration in these plants.

C3 plant Type of plant that uses only the Calvin–Benson cycle to fix carbon.
C4 plant Type of plant that minimizes photorespiration by fixing carbon twice, in two cell types.
CAM plant Type of C4 plant that minimizes photorespiration by fixing carbon twice, at different times of day.
photorespiration Pathway initiated when rubisco attaches oxygen instead of carbon dioxide to RuBP (ribulose bisphosphate).
stomata Gaps that open on plant surfaces; allow water vapor and gases to diffuse across the epidermis.

TAKE-HOME MESSAGE 6.7

How do carbon-fixing reactions vary?

- ✔ When stomata close on hot, dry days, they also prevent the exchange of gases between plant tissues and the air.
- ✔ Rubisco can initiate photorespiration by attaching oxygen (instead of carbon dioxide) to RuBP. Photorespiration reduces the efficiency of sugar production, especially in C3 plants.
- ✔ Plants adapted to hot, dry conditions limit photorespiration by fixing carbon twice. C4 plants separate the two sets of reactions in space; CAM plants separate them in time.

CREDITS: (14A) Masahiro Yamada, Michio Kawasaki, Tatsuo Sugiyama, Hiroshi Miyake, Mitsutaka Taniguchi; Differential Positioning of C4 Mesophyll and Bundle Sheath Chloroplasts: Aggregative Movement of C4 Mesophyll Chloroplasts in Response to Environmental Stresses: Plant and Cell Physiology; (2009) 50(10): 1736-1749; (14B) left, Eri Maai, Shouu Shimada, Masahiro Yamada, Tatsuo Sugiyama, Hiroshi Miyake, Mitsutaka Taniguchi; The avoidance and aggregative movements of mesophyll chloroplasts in C4 monocots in response to blue light and abscisic acid; Journal of Experimental Botany, doi:10.1093/jxb/err008, by permission of Oxford University Press; right, Image courtesy msuturfweeds.net.

The first cells we know of appeared on Earth about 3.4 billion years ago. Like some modern prokaryotes, these ancient organisms did not tap into sunlight: They extracted the energy they needed from simple molecules such as methane and hydrogen sulfide. Both gases were plentiful in the nasty brew that was Earth's early atmosphere (**FIGURE 6.15A**). When the cyclic pathway of photosynthesis first evolved, sunlight offered cells that used it an essentially unlimited supply of energy. Shortly afterward, this pathway became modified. The new noncyclic pathway split water molecules into hydrogen and oxygen. Cells that used the pathway were very successful. Oxygen gas (O_2) released from uncountable numbers of water molecules began seeping out of photosynthetic prokaryotes.

Oxygen gas reacts easily with metals, so at first, most of it combined with metal atoms in exposed rocks. After the exposed minerals became saturated with oxygen, the gas began to accumulate in the ocean and the atmosphere. From that time on, the world of life would never be the same. Molecular oxygen, which had previously been very rare in the atmosphere, began accumulating. As you will see in Chapter 7, the change in the composition of the atmosphere put tremendous selection pressure on early life, effectively spurring the evolution of aerobic respiration.

Earth's atmosphere is changing again, and this time, the level of carbon dioxide is increasing. To understand why, think about your own body. A human body is about 9.5 percent carbon by weight, which means that you contain an enormous number of carbon atoms. Where did all of those carbon atoms come from?

You and other heterotrophs, remember, ingest tissues of other organisms to get carbon. Thus, your body is built from organic compounds obtained from other organisms. The carbon atoms in those compounds may have passed through other heterotrophs before you ate them, but at some point they were part of photosynthetic organisms (**FIGURE 6.15B**). Plants and almost all other photosynthesizers in the human food chain obtain their carbon from carbon dioxide. Your carbon atoms—and those of most other organisms that live on land—were recently part of Earth's atmosphere, in molecules of CO_2.

Photosynthesis removes carbon dioxide from the atmosphere, and fixes its carbon atoms in organic compounds. When you and other organisms break down organic compounds for energy, carbon atoms are released in the form of CO_2, which then reenters the atmosphere. Since photosynthesis evolved, these two processes have constituted a more or less balanced cycle of the biosphere. You will learn more about the carbon cycle in Section 46.7. For now, know that the amount of carbon dioxide that photosynthesis removes from the atmosphere is roughly the same amount that organisms release back into it. At least it was, until humans came along.

As early as 8,000 years ago, humans began burning forests to clear land for agriculture. When trees and other plants burn, most of the carbon locked in their tissues is released into the atmosphere as carbon dioxide. Fires that occur naturally release carbon dioxide the same way. Today, we burn a lot more than our ancestors ever did. In addition to wood, we are burn-

A An artist's view of Earth's early atmosphere, which was abundant in gases such as methane, sulfur, ammonia, and chlorine.

B Today, photosynthesis is now the main pathway by which energy and carbon enter the web of life. The plants in this orchard are producing oxygen and carbon-rich parts (apples) at the Jerzy Boyz farm in Chelan, Washington.

FIGURE 6.15 Then and now—a view of how our atmosphere was irrevocably altered by photosynthesis.

ing fossil fuels—coal, petroleum, and natural gas—to satisfy our greater and greater demands for energy. Fossil fuels are the organic remains of ancient organisms. When we burn fossil fuels, we release the carbon that has been sequestered in their organic molecules for hundreds of millions of years, mainly as CO_2 that reenters the atmosphere.

Our extensive use of fossil fuels has put Earth's atmospheric cycle of carbon dioxide out of balance: We are adding far more CO_2 to the atmosphere than photosynthetic organisms are removing from it. The resulting imbalance is fueling global climate change (we return to this topic in Section 46.8). In 2010 alone, human activities released 10 billion tons of CO_2, an increase of 5.9 percent over 2009, and 49 percent over 1990. Most of this CO_2 comes from burning fossil fuels (**FIGURE 6.16**). How do we know? Researchers can determine how long ago the carbon atoms in a sample of CO_2 were part of a living organism by measuring the ratio of different carbon isotopes in it (you will read more about radioisotope dating techniques in Section 16.6). The results are correlated with global statistics on the extraction, refining, and trade of fossil fuels.

Tiny pockets of Earth's ancient atmosphere remain in Antarctica, preserved in snow and ice that have been accumulating in layers, year after year, for millions of years (**FIGURE 6.17**). Air and dust trapped in each layer reveal the composition of the atmosphere that prevailed when the layer formed. These layers tell us that the atmospheric CO_2 level was relatively stable for about 10,000 years before the industrial revolution began in the mid-1800s. Since then, the CO_2 level has been steadily rising. Today, the atmospheric CO_2 level is higher than it has been for *15 million years*.

Such alarming statistics are why researchers are scrambling to find a way to make cost-effective biofuels. Unlike fossil fuels, biofuels are a renewable source of energy: We can always make more of them simply by growing more plants. Also unlike fossil fuels, biofuels do not contribute to global climate change, because growing plant matter for fuel recycles carbon that is already in the atmosphere.

FIGURE 6.16 Carbon dioxide is an invisible component of the smog shrouding Lianyungang, China, in December 2013. Most air pollution—here and worldwide—comes from fossil fuel combustion.

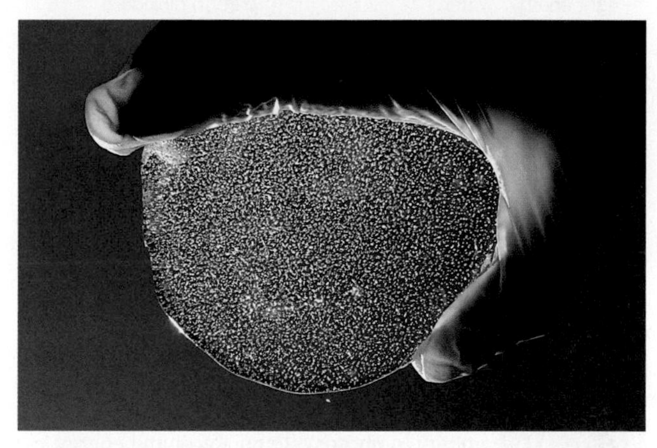

FIGURE 6.17 A slice of ancient history. Air bubbles trapped in Antarctic ice core slices such as this one are samples of Earth's atmosphere as it was when the ice formed. The deeper the slice, the older the air in the bubbles.

summary

Section 6.1 Plants and other **autotrophs** make their own food using energy from the environment and carbon from inorganic sources such as CO_2. By metabolic pathways of photosynthesis, plants and most other autotrophs capture the energy of light and use it to build sugars from water and carbon dioxide. **Heterotrophs** get carbon from molecules that other organisms have already assembled; most also get energy from these molecules.

Earth's early atmosphere held very little free oxygen. When the noncyclic pathway of photosynthesis evolved, oxygen released by organisms that used it permanently changed the atmosphere. Photosynthesis removes CO_2 from the atmosphere, and the metabolic activity of most organisms puts it back. This global cycle has been balanced for millions of years. Human activites, especially burning fossil fuels, are disrupting the cycle by adding extra CO_2 to the atmosphere. The resulting imbalance is fueling global climate change.

Sections 6.2, 6.3 Visible light is a very small part of the spectrum of electromagnetic energy radiating from the sun. That energy travels in waves, and it is organized as photons. A photon's wavelength is related to its energy. Wavelengths that we can see—visible light—drive photosynthesis, which begins when photons are absorbed by photosynthetic pigments. A **pigment** absorbs light of particular **wavelengths** only; wavelengths not captured are reflected as its characteristic color. The main photosynthetic pigment, **chlorophyll *a***, absorbs violet and red light, so it appears green. Accessory pigments absorb additional wavelengths.

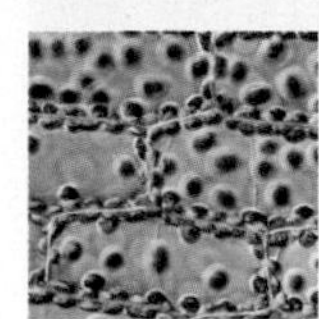

Section 6.4 In chloroplasts, the **light-dependent reactions** of photosynthesis occur at a much-folded **thylakoid membrane**. The membrane forms a compartment in the chloroplast's interior (**stroma**), in which the **light-independent reactions** occur. An overview is shown below:

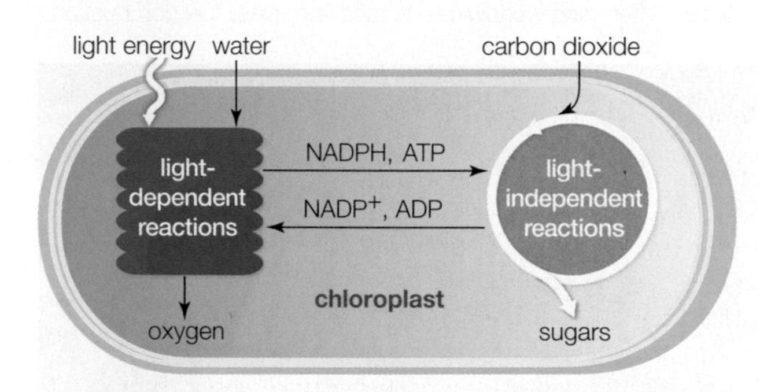

Section 6.5 In the light-dependent reactions, light-harvesting complexes in the thylakoid membrane absorb photons and pass the energy to **photosystems**. Receiving energy causes photosystems to release electrons.

In the noncyclic pathway, electrons released from photosystem II flow through an electron transfer chain, then to photosystem I. Photosystem II replaces lost electrons by pulling them from water, which then splits into H^+ and O_2 (an example of **photolysis**). Electrons released from photosystem I end up in NADPH.

In the cyclic pathway, electrons released from photosystem I enter an electron transfer chain, then cycle back to photosystem I. NADPH does not form and O_2 is not released.

ATP forms by **electron transfer phosphorylation** in both pathways. Electrons flowing through electron transfer chains cause hydrogen ions to accumulate in the thylakoid compartment. The hydrogen ions follow their gradient back across the membrane through ATP synthases, driving ATP synthesis.

Section 6.6 **Carbon fixation** occurs as part of the light-independent reactions of photosynthesis. Inside the stroma of chloroplasts, the enzyme **rubisco** attaches CO_2 to RuBP to start the **Calvin–Benson cycle**. This cyclic pathway builds the carbon backbones of sugars using carbon atoms from CO_2. It is driven by phosphate-group transfers from ATP and electrons from NADPH. Both molecules form during the light-dependent reactions.

Section 6.7 On hot, dry days, a plant conserves water by closing **stomata**, so carbon dioxide for the light-independent reactions cannot enter it, and oxygen produced by the light-dependent reactions cannot leave. The resulting high O_2/CO_2 ratio in the plant's tissues can shift sugar production toward **photorespiration**. This inefficient pathway limits the growth of **C3 plants** in hot, dry climates. Other types of plants minimize photorespiration by fixing carbon twice, thus keeping the CO_2 level high near rubisco. **C4 plants** carry out the two sets of reactions in different cell types; **CAM plants** carry them out at different times.

self-quiz

Answers in Appendix VII

1. A cat eats a bird, which ate a caterpillar that chewed on a weed. Which organisms are autotrophs? Which ones are heterotrophs?

2. Plants use ________ as an energy source to drive photosynthesis.
   a. sunlight c. O_2
   b. hydrogen ions d. CO_2

3. Most of the carbon dioxide that plants use for photosynthesis comes from ________ .
   a. glucose c. rainwater
   b. the atmosphere d. photolysis

4. Which of the following statements is incorrect?
   a. Pigments absorb light of certain wavelengths only.
   b. Many accessory pigments are multipurpose molecules.
   c. Chlorophyll is green because it absorbs green light.

5. In plants and other photosynthetic eukaryotes, the light-dependent reactions proceed in/at the ________ .
   a. thylakoid membrane c. stroma
   b. plasma membrane d. cytoplasm

6. When a photosystem absorbs light, ________ .
   a. sugar phosphates are produced
   b. electrons are transferred to ATP
   c. RuBP accepts electrons
   d. electrons are ejected from its special pair

7. In the light-dependent reactions, ________ .
   a. carbon dioxide is fixed d. CO_2 accepts electrons
   b. ATP forms e. b and c
   c. sugars form f. a and c

8. What accumulates inside the thylakoid compartment of chloroplasts during the light-dependent reactions?
   a. sugars c. O_2
   b. hydrogen ions d. CO_2

9. The atoms in the molecular oxygen released during photosynthesis come from ________ molecules.
   a. glucose c. water
   b. CO_2 d. O_2

10. Light-independent reactions in plants proceed in/at the ________ of chloroplasts.
   a. thylakoid membrane c. stroma
   b. plasma membrane d. cytoplasm

data analysis activities

Energy Efficiency of Biofuel Production Most of the plant material currently used for biofuel production in the United States consists of food crops—mainly corn, soybeans, and sugarcane. In 2006, David Tilman and his colleagues published the results of a 10-year study comparing the net energy output of various biofuels. The researchers grew a mixture of native perennial grasses without irrigation, fertilizer, pesticides, or herbicides, in sandy soil that was so depleted by intensive agriculture that it had been abandoned. They measured the usable energy in biofuels made from the grasses, and also from corn and soy, then measured the energy it took to grow and produce biofuel from each kind of crop (**FIGURE 6.18**).

1. About how much energy did ethanol produced from one hectare of corn yield? How much energy did it take to grow the corn and make that ethanol?

2. Which of the biofuels tested had the highest ratio of energy output to energy input?

3. Which of the three crops would require the least amount of land to produce a given amount of biofuel energy?

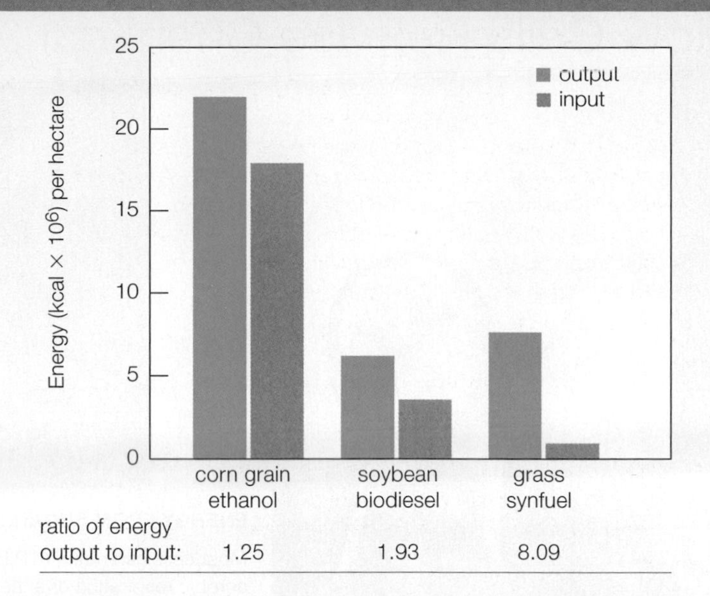

FIGURE 6.18 Energy inputs and outputs of biofuels made from three different crops. One hectare is about 2.5 acres.

11. The Calvin–Benson cycle starts when ________ .
    a. light is available
    b. carbon dioxide is attached to RuBP
    c. electrons leave a photosystem II

12. Which of the following substances does *not* participate in the Calvin–Benson cycle?
    a. ATP
    b. NADPH
    c. RuBP
    d. PGAL
    e. O_2
    f. CO_2

13. Closed stomata ________ .
    a. limit gas exchange
    b. permit water loss
    c. restrict photosynthesis
    d. absorb light

14. In C3 plants, ________ makes sugar production inefficient when stomata close during the day.
    a. photosynthesis
    b. photolysis
    c. photorespiration
    d. carbon fixation

15. Match each with its most suitable description.

___ PGAL formation	a. absorbs light
___ CO_2 fixation	b. converts light to chemical energy
___ photolysis	c. self-feeder
___ ATP forms; NADPH does not	d. electrons cycle back to photosystem I
___ photorespiration	e. problem in C3 plants
___ photosynthesis	f. ATP, NADPH required
___ pigment	g. water molecules split
___ autotroph	h. rubisco function

CENGAGE brain .com To access course materials, please visit www.cengagebrain.com.

critical thinking

1. About 200 years ago, Jan Baptista van Helmont wanted to know where growing plants get the materials necessary for increases in size. He planted a tree seedling weighing 5 pounds in a barrel filled with 200 pounds of soil and then watered the tree regularly. After five years, the tree weighed 169 pounds, 3 ounces, and the soil weighed 199 pounds, 14 ounces. Because the tree had gained so much weight and the soil had lost so little, he concluded that the tree had gained all of its additional weight by absorbing the water he had added to the barrel, but of course he was incorrect. What really happened?

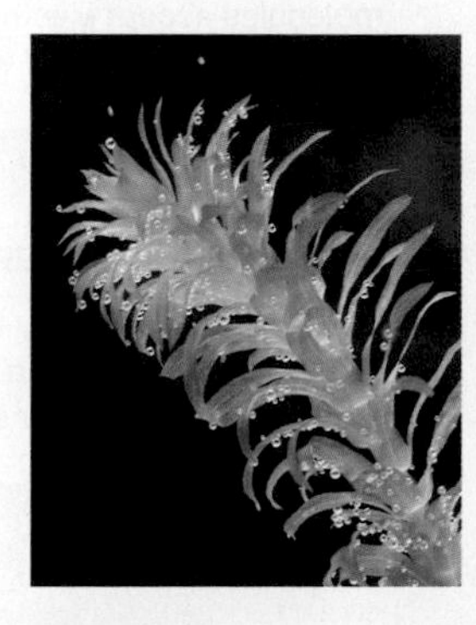

2. While gazing into an aquarium, you see bubbles coming from an aquatic plant (left). What are the bubbles?

3. A C3 plant absorbs a carbon radioisotope (as part of $^{14}CO_2$). In which compound does the labeled carbon appear first? Which compound forms first if a C4 plant absorbs the same radioisotope?

4. As you learned in this chapter, cell membranes are required for electron transfer phosphorylation. Thylakoid membranes in chloroplasts serve this purpose in photosynthetic eukaryotes. Prokaryotic cells do not have this organelle, but many are photosynthesizers. How do you think they carry out the light-dependent reactions, given that they have no chloroplasts?

7 How Cells Release Chemical Energy

This chapter focuses on metabolic pathways (Section 5.5) that harvest energy (5.2) stored in sugars (3.4). Some reactions (3.3) occur in mitochondria (4.8). You will revisit free radicals (2.3), lipids (3.5), proteins (3.6), electron transfer chains (5.5, 6.5), coenzymes (5.6, 6.6), membrane transport (5.8, 5.9), and photosynthesis (6.4).

ENERGY FROM SUGARS

Most cells can make ATP by breaking down sugars in either aerobic respiration or anaerobic fermentation pathways. Aerobic respiration yields the most ATP.

GLYCOLYSIS

Aerobic respiration and fermentation start in the cytoplasm with glycolysis, a pathway that splits glucose into two pyruvate molecules and yields two ATP.

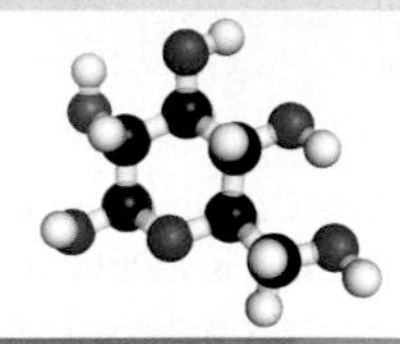

AEROBIC RESPIRATION

Eukaryotes break down pyruvate to CO_2 in mitochondria. Many coenzymes are reduced; these deliver electrons and hydrogen ions to electron transfer chains that drive ATP formation.

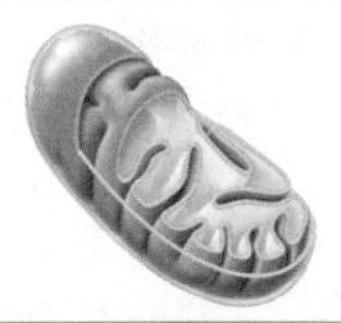

FERMENTATION

Fermentation occurs entirely in cytoplasm, where organic molecules accept electrons from pyruvate. The net yield of ATP is small compared with that from aerobic respiration.

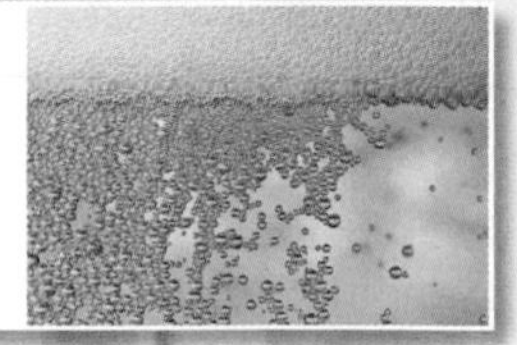

ENERGY FROM OTHER MOLECULES

Cells also make ATP by breaking down molecules other than sugars. Dietary lipids and proteins are converted to substrates of glycolysis or another step in the aerobic respiration pathway.

In Section 11.6, you will see examples of how metabolic pathways are disrupted in cancer cells. Chapter 14 explains why metabolic disorders can be inherited. We return to plant adaptations for photosynthesis in Chapter 27, muscle function in Chapter 35, how the body acquires oxygen for respiration in Chapter 38, and digestion and nutrition in Chapter 39.

7.1 Risky Business

Before photosynthesis evolved, molecular oxygen had been a very small component of Earth's atmosphere. In what may have been the earliest case of catastrophic pollution, the new abundance of this gas exerted tremendous pressure on all life at the time. Why? Then, like now, enzymes that require metal cofactors were a critical part of metabolism. Oxygen reacts with metal cofactors, and free radicals (Section 2.3) form during those reactions. Free radicals damage biological molecules, so they are dangerous to life. Most cells had no way to cope with them, and so were wiped out everywhere except deep water, muddy sediments, and other **anaerobic** (oxygen-free) places.

A very few cells were able to survive in **aerobic** (oxygen-containing) places. By lucky circumstance, these cells made antioxidants that could detoxify or prevent the formation of free radicals. As these organisms evolved, their antioxidant molecules became incorporated into new metabolic pathways. One of the new pathways, aerobic respiration, put the reactive properties of oxygen to use. In modern eukaryotic cells, most of the aerobic respiration pathway takes place inside mitochondria (Section 4.8). An internal folded membrane system allows these organelles to make ATP very efficiently. Electron transfer chains in this membrane set up hydrogen ion gradients that power ATP synthesis. Oxygen molecules accept electrons at the end of these chains.

ATP participates in almost all cellular reactions, so a cell benefits from making a lot of it. However, aerobic respiration is a dangerous occupation. When an oxygen molecule (O_2) accepts electrons from an electron transfer chain, it dissociates into oxygen atoms. Most of the atoms immediately combine with hydrogen ions and end up in water molecules. Occasionally, however, an oxygen atom escapes this final reaction. The atom has an unpaired electron, so it is a free radical.

Mitochondria cannot detoxify free radicals, so they rely on antioxidant enzymes and vitamins in the cell's cytoplasm to do it for them. The system works well, at least most of the time. However, a genetic disorder or an encounter with a toxin or pathogen can tip the normal cellular balance of aerobic respiration and free radical formation. Free radicals accumulate and destroy first the function of mitochondria, then the cell. The resulting tissue damage is called oxidative stress.

At least 83 proteins are directly involved in mitochondrial electron transfer chains. A defect in any one of them—or in any of the thousands of other proteins used by mitochondria—can wreak havoc in the body. New research is showing that oxidative stress caused by mitochondrial malfunction is also involved in many other illnesses, including cancer, hypertension, Alzheimer's and Parkinson's diseases, and even aging. Hundreds of incurable genetic disorders are associated with mitochondrial defects (**FIGURE 7.1**), and more are being discovered all the time. Nerve cells, which require a lot of ATP, are particularly affected. Symptoms of these disorders range from mild to major progressive neurological deficits, blindness, deafness, diabetes, strokes, seizures, gastrointestinal malfunction, and disabling muscle weakness.

aerobic Involving or occurring in the presence of oxygen.
anaerobic Occurring in the absence of oxygen.

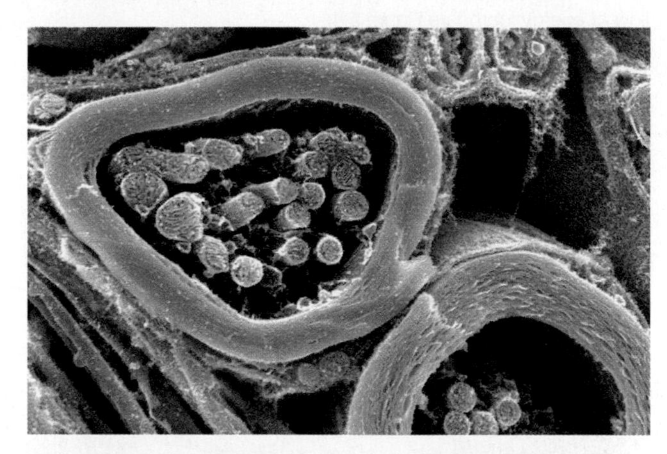

A A cross-section shows mitochondria (yellow) inside nerve cells. These cells are particularly affected by mitochondrial malfunction.

B "Tom does not look sick, but inside his organs are all getting badly damaged," says Martine Martin, pictured here with her eight-year-old son. Tom, who was born with a mitochondrial disease, eats with the help of a machine, suffers intense pain, and will soon be blind. Despite intensive medical intervention, he is not expected to reach his teens.

FIGURE 7.1 Mitochondria, powerhouses of all eukaryotic cells. When they malfunction, the lights go off in cellular businesses.

7.2 Overview of Carbohydrate Breakdown Pathways

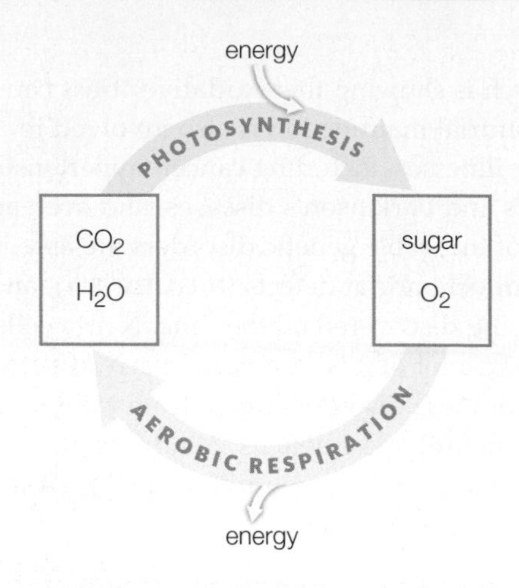

FIGURE 7.2 The global connection between photosynthesis and aerobic respiration. Note the cycling of materials, and the one-way flow of energy (compare Figure 5.4).

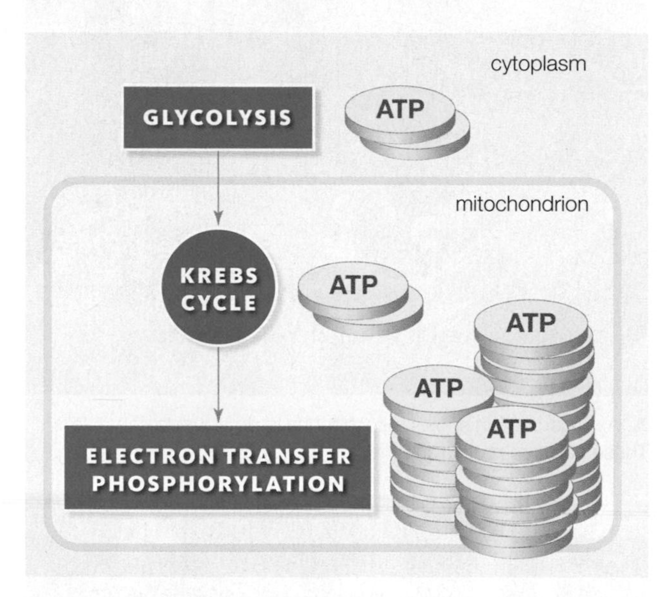

A Aerobic respiration.

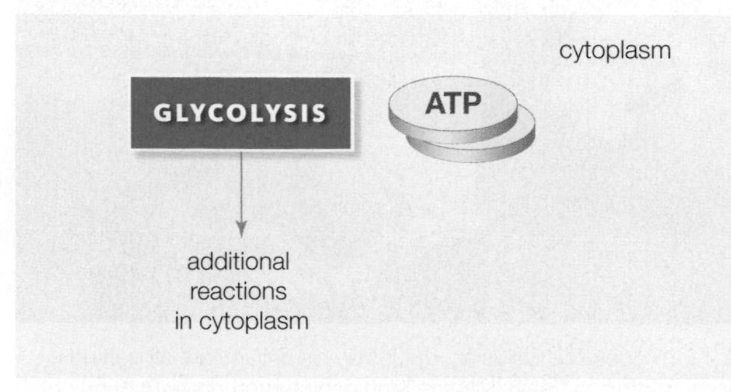

B Fermentation.

FIGURE 7.3 ▶Animated Comparison of aerobic respiration and fermentation.

FIGURE IT OUT Which pathway produces more ATP?

Answer: Aerobic respiration

✔ Most organisms, including photosynthetic ones, make ATP by breaking down sugars.

Photosynthetic organisms capture energy from the sun and store it in the form of sugars (Sections 6.5 and 6.6). They and most other organisms use energy stored in sugars to run various endergonic reactions of metabolism that sustain life. However, sugars rarely participate in such reactions, so how do cells harness their energy? In order to use the energy stored in sugars, cells must first transfer it to molecules—ATP in particular—that do participate in energy-requiring reactions. Cells break the bonds between carbon atoms of a sugar molecule, and use energy released as these bonds break to drive ATP synthesis.

In nearly all eukaryotes and some bacteria, the main carbohydrate-breakdown pathway is **aerobic respiration**. Remember, the bonds of organic molecules hold a lot of energy that can be released in a reaction with oxygen (Section 5.5). Aerobic respiration harvests energy from sugars by completely breaking apart their carbon backbones, bond by bond, and it uses oxygen to do this. The pathway's products—carbon dioxide and water—are the raw materials used by the vast majority of photosynthetic organisms to build the sugars in the first place. With this connection, the cycling of carbon, hydrogen, and oxygen through living things comes full circle through the biosphere (**FIGURE 7.2**).

The following equation summarizes the overall pathway of aerobic respiration:

$$\underset{\text{glucose}}{C_6H_{12}O_6} + \underset{\text{oxygen}}{6O_2} \longrightarrow \underset{\text{carbon dioxide}}{6CO_2} + \underset{\text{water}}{6H_2O}$$

The reactions in the pathway occur in three stages (**FIGURE 7.3A**). The first stage, glycolysis, is a linear pathway that takes place in cytoplasm. Glycolysis begins the breakdown of one sugar molecule for a net yield of 2 ATP. In eukaryotes, the next two stages take place in mitochondria. In the second stage, the Krebs cycle completes the breakdown of the sugar molecule to CO_2. This cyclic pathway produces 2 ATP and reduces many coenzymes. In the third stage, electron transfer phosphorylation (Section 6.5), the coenzymes reduced during glycolysis and the Krebs cycle deliver electrons and hydrogen ions to electron transfer chains. Energy released by electrons as they move through the chains drives the formation of as many as 32 ATP. At the end of the electron transfer chains, water forms when oxygen accepts hydrogen ions and electrons. In exactly the reverse of the photolysis reaction that

FIGURE 7.4 Like you, a whale breathes air to provide its cells with a fresh supply of oxygen for aerobic respiration. Carbon dioxide released from aerobically respiring cells leaves the body in each exhalation.

splits water during the noncyclic, light-dependent reactions of photosynthesis, oxygen combines with electrons and hydrogen ions to form water:

$$O_2 + 4e^- + 4H^+ \longrightarrow 2H_2O$$

Aerobic respiration, which means "taking a breath of air," refers to this pathway's requirement for oxygen as the final acceptor of electrons.

Fermentation refers to sugar breakdown pathways that produce ATP and do not require oxygen (FIGURE 7.3B). Like aerobic respiration, fermentation begins with glycolysis in cytoplasm. Unlike aerobic respiration, fermentation includes no electron transfer chains, and, in all organisms that use the pathway, it takes place entirely in cytoplasm. The reactions that conclude it produce no additional ATP, and an organic molecule (not oxygen) accepts electrons.

aerobic respiration Oxygen-requiring metabolic pathway that breaks down sugars to produce ATP.
fermentation A metabolic pathway that breaks down sugars to produce ATP and does not require oxygen.

The breakdown of a sugar molecule by fermentation yields only 2 ATP, but the pathway is efficient enough to sustain many single-celled species. It also helps cells of multicelled species produce ATP under anaerobic conditions, but aerobic respiration is a much more efficient way of harvesting energy from carbohydrates. You and other large, multicelled organisms could not live without its higher yield (FIGURE 7.4).

TAKE-HOME MESSAGE 7.2

How do cells access the chemical energy in sugars?

- ✔ Most cells can make ATP by breaking down sugars, in aerobic respiration, fermentation, or both.
- ✔ Aerobic respiration and fermentation begin in cytoplasm.
- ✔ Fermentation does not require oxygen and ends in cytoplasm.
- ✔ Aerobic respiration requires oxygen and, in eukaryotes, it ends in mitochondria. This pathway yields much more ATP per sugar molecule than fermentation does.

7.3 Glycolysis—Sugar Breakdown Begins

✔ The reactions of glycolysis convert one molecule of glucose (or other six-carbon sugar) to two molecules of pyruvate.

✔ Glycolysis reactions use 2 ATP and produce 4 ATP, so the net yield is 2 ATP per molecule of sugar.

Glycolysis is a series of reactions that begin the sugar breakdown pathways of both aerobic respiration and fermentation. The word "glycolysis" comes from two Greek words: *glyk-*, sweet, and *-lysis*, loosening; it refers to the release of chemical energy from sugars. The reactions of glycolysis, which occur with some variation in the cytoplasm of almost all cells, convert one molecule of a six-carbon sugar (such as glucose) into two molecules of **pyruvate**, an organic compound with a three-carbon backbone:

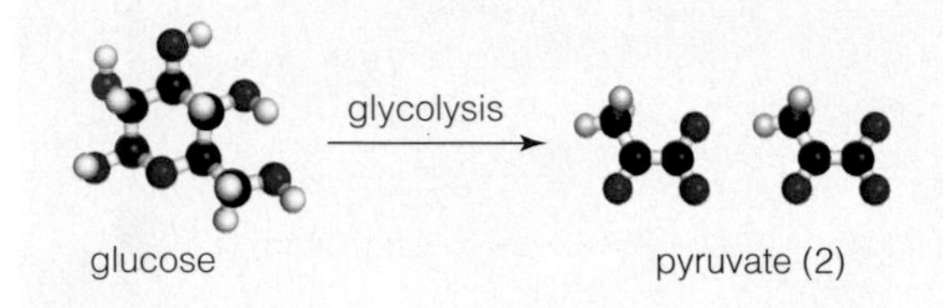

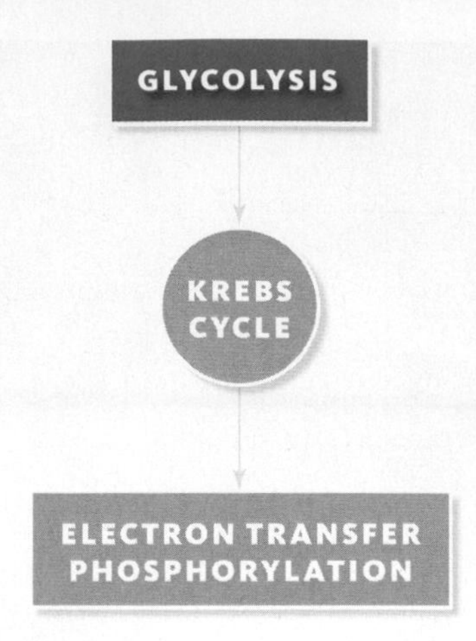

FIGURE 7.5 ▸**Animated** Glycolysis. This first stage of sugar breakdown starts and ends in the cytoplasm of all cells. Opposite, for clarity, we track only the six carbon atoms (black balls) that enter the reactions as part of glucose.

Cells invest two ATP to start glycolysis, so the net yield from one glucose molecule is two ATP. Two NADH also form, and two pyruvate molecules are the end products. Appendix III has more details for interested students.

Other six-carbon sugars (such as galactose and fructose) can enter glycolysis, but we focus here on glucose, for clarity.

Glycolysis begins when a molecule of glucose enters a cell through a glucose transporter, a passive transport protein that you encountered in Section 5.9. The cell invests two ATP in the endergonic reactions that begin the pathway (**FIGURE 7.5**). In the first reaction, a phosphate group is transferred from ATP to the glucose, thus forming glucose-6-phosphate ❶. A model of hexokinase, the enzyme that catalyzes this reaction, is pictured in Section 5.4.

Glycolysis continues as the glucose-6-phosphate accepts a phosphate group from another ATP ❷, then splits to form two PGAL (phosphoglyceraldehyde). Remember from Section 6.6 that this phosphorylated sugar also forms during the Calvin–Benson cycle.

In the next reaction, each PGAL receives a second phosphate group, and each gives up two electrons and a hydrogen ion. Two molecules of PGA (phosphoglycerate) form as products of this reaction ❸. The electrons and hydrogen ions are accepted by two NAD^+, which thereby become reduced to NADH. Aerobic respiration's final stage requires this NADH, as does fermentation (Sections 7.5 and 7.6 detail the final stages of these pathways).

Next, a phosphate group is transferred from each PGA to ADP, so two ATP form ❹. The direct transfer of a phosphate group from a substrate to ADP is called **substrate-level phosphorylation**. Substrate-level phosphorylation is a completely different process from electron transfer phosphorylation, which is the way ATP forms during the light-dependent reactions of photosynthesis (Section 6.5).

Glycolysis ends with the formation of two more ATP by substrate-level phosphorylation ❺. Remember, two ATP were invested to begin the reactions of glycolysis. A total of four ATP form, so the net yield is two ATP per molecule of glucose ❻. The pathway also produces two three-carbon pyruvate molecules. Pyruvate is a substrate for the second-stage reactions of aerobic respiration, and also for fermentation reactions (Sections 7.4 and 7.6 return to this topic).

TAKE-HOME MESSAGE 7.3

What happens during glycolysis?

✔ Glycolysis is the first stage of sugar breakdown in both aerobic respiration and fermentation.

✔ The reactions of glycolysis occur in the cytoplasm.

✔ Glycolysis converts one molecule of glucose to two molecules of pyruvate, with a net energy yield of two ATP. Two NADH also form.

glycolysis Set of reactions in which a six-carbon sugar (such as glucose) is converted to two pyruvate for a net yield of two ATP.
pyruvate Three-carbon end product of glycolysis.
substrate-level phosphorylation The formation of ATP by the direct transfer of a phosphate group from a substrate to ADP.

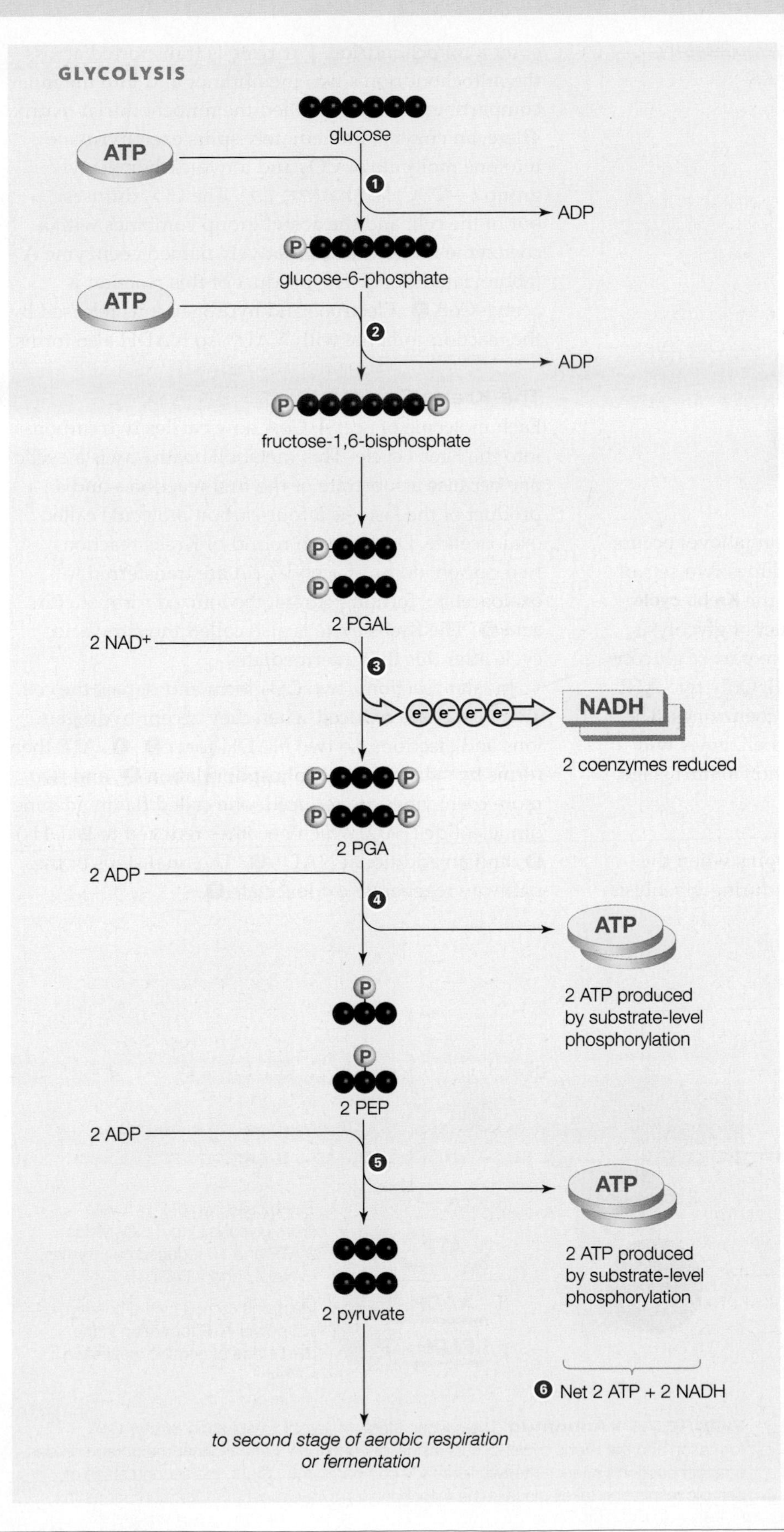

ATP-Requiring Steps

❶ An enzyme transfers a phosphate group from ATP to glucose, forming glucose-6-phosphate. (You learned about this enzyme, hexokinase, in **FIGURE 5.10** and Section 5.4.)

❷ A phosphate group from a second ATP is transferred to the glucose-6-phosphate. The resulting molecule is unstable, and it splits into two three-carbon molecules. The molecules are interconvertible, so we will call them both PGAL (phosphoglyceraldehyde).

Two ATP have now been invested in the reactions.

ATP-Generating Steps

❸ An enzyme attaches a phosphate to each PGAL, so two PGA (phosphoglycerate) form. Two electrons and a hydrogen ion (not shown) from each PGAL are accepted by NAD^+, so two NADH form.

❹ An enzyme transfers a phosphate group from each PGA to ADP, forming two ATP and two intermediate molecules (PEP).

The original energy investment of two ATP has now been recovered.

❺ An enzyme transfers a phosphate group from each PEP to ADP, forming two more ATP and two molecules of pyruvate.

❻ Summing up, glycolysis yields two NADH, two ATP (net), and two pyruvate for each glucose molecule.

Depending on the type of cell and environmental conditions, the pyruvate may enter the second stage of aerobic respiration or it may be used in other ways, such as in fermentation.

7.4 Second Stage of Aerobic Respiration

✔ The second stage of aerobic respiration completes the breakdown of glucose that began in glycolysis.

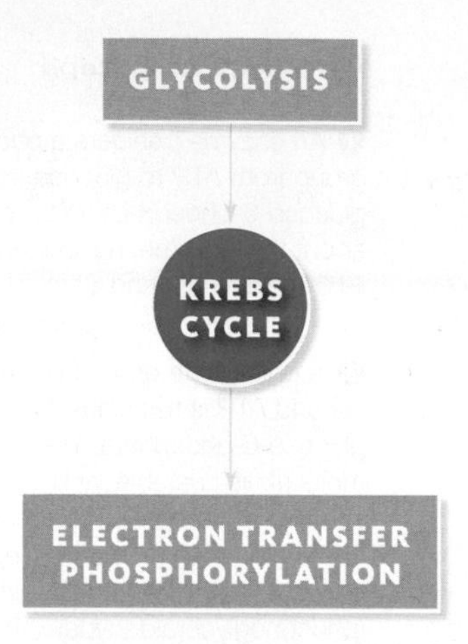

The second stage of aerobic respiration (above) occurs in mitochondria (**FIGURE 7.6**). It includes two sets of reactions, acetyl–CoA formation and the **Krebs cycle**, that break down pyruvate, the product of glycolysis. All of the carbon atoms that were once part of glucose end up in CO_2, which departs the cell. Only two ATP form, but the reactions reduce many coenzymes. The energy of electrons carried by these coenzymes will drive the reactions of aerobic respiration's third stage.

Acetyl–CoA Formation

Aerobic respiration's second stage begins when the two pyruvate molecules that formed during glycolysis enter a mitochondrion. Pyruvate is transported across the mitochondrion's two membranes and into the inner compartment, which is called the mitochondrial matrix. There, an enzyme immediately splits each pyruvate into one molecule of CO_2 and a two-carbon acetyl group (—$COCH_3$, **FIGURE 7.7**). The CO_2 diffuses out of the cell, and the acetyl group combines with a coenzyme rather unimaginatively named coenzyme A (abbreviated CoA). The product of this reaction is acetyl–CoA ❶. Electrons and hydrogen ions released by the reaction combine with NAD^+, so NADH also forms.

The Krebs Cycle

Each molecule of acetyl–CoA now carries two carbons into the Krebs cycle. This metabolic pathway is a cyclic one because a substrate of the first reaction—and a product of the last—is a four-carbon molecule called oxaloacetate. During each round of Krebs reactions, two carbon atoms of acetyl–CoA are transferred to oxaloacetate, forming citrate, the ionized form of citric acid ❷. The Krebs cycle is also called the citric acid cycle after this first intermediate.

In later reactions, two CO_2 form and depart the cell. Two NAD^+ are reduced when they accept hydrogen ions and electrons, so two NADH form ❸, ❹. ATP then forms by substrate-level phosphorylation ❺, and two more coenzymes are reduced: one called flavin adenine dinucleotide (FAD, which becomes reduced to $FADH_2$) ❻, and an additional NAD^+ ❼. The final steps of the pathway regenerate oxaloacetate ❽.

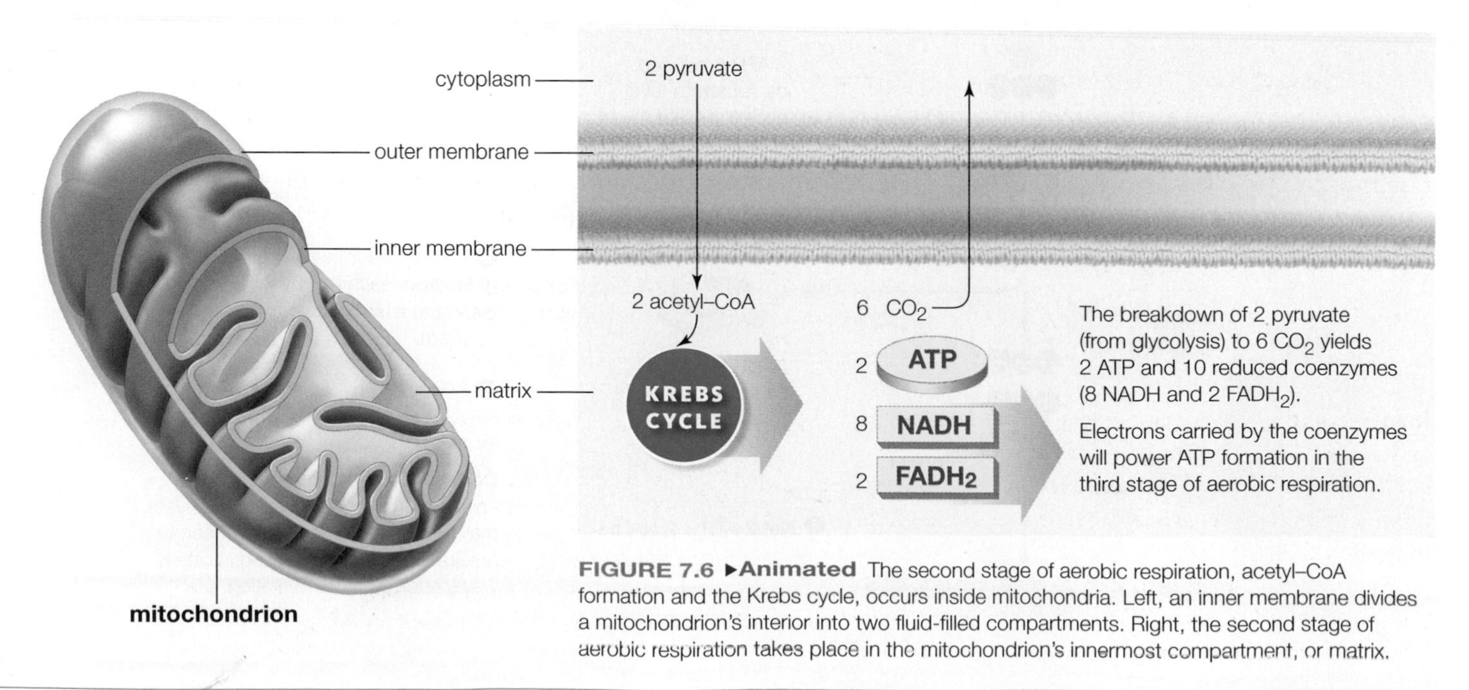

FIGURE 7.6 ▸**Animated** The second stage of aerobic respiration, acetyl–CoA formation and the Krebs cycle, occurs inside mitochondria. Left, an inner membrane divides a mitochondrion's interior into two fluid-filled compartments. Right, the second stage of aerobic respiration takes place in the mitochondrion's innermost compartment, or matrix.

Acetyl–CoA Formation and the Krebs Cycle

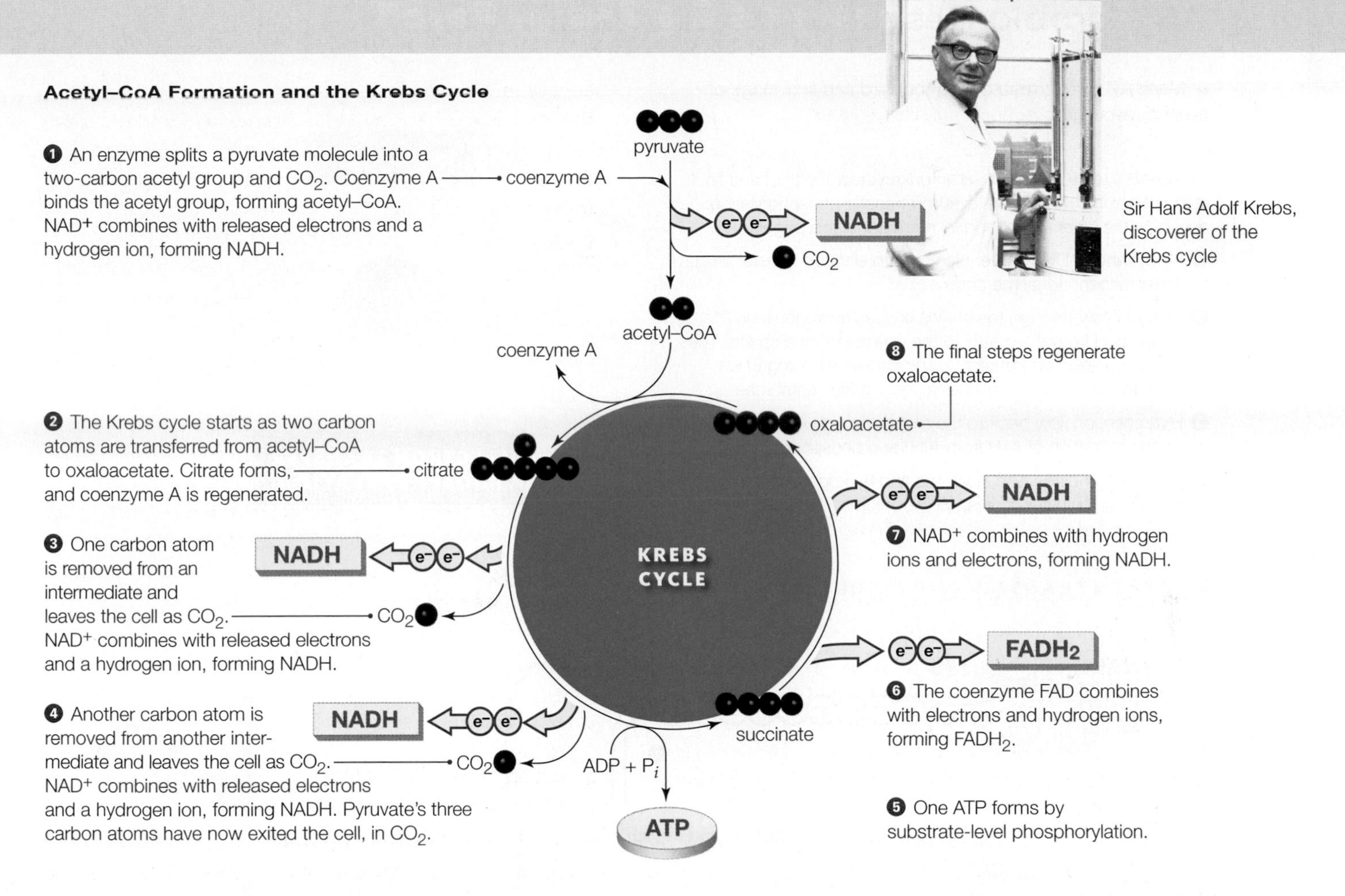

FIGURE 7.7 Acetyl–CoA formation and the Krebs cycle.

It takes two cycles of Krebs reactions to break down two pyruvate molecules that formed during glycolysis of one glucose molecule. After two cycles, all six carbons that entered glycolysis in one glucose molecule have left the cell, in six CO_2. Electrons and hydrogen ions are released as each carbon is removed from the backbone of intermediate molecules; ten coenzymes will carry them to the third and final stage of aerobic respiration. Not all reactions are shown; Appendix III has more details.

After two cycles of Krebs reactions, the two carbon atoms carried by each acetyl–CoA end up in CO_2. Thus, the combined second-stage reactions of aerobic respiration break down two pyruvate to six CO_2:

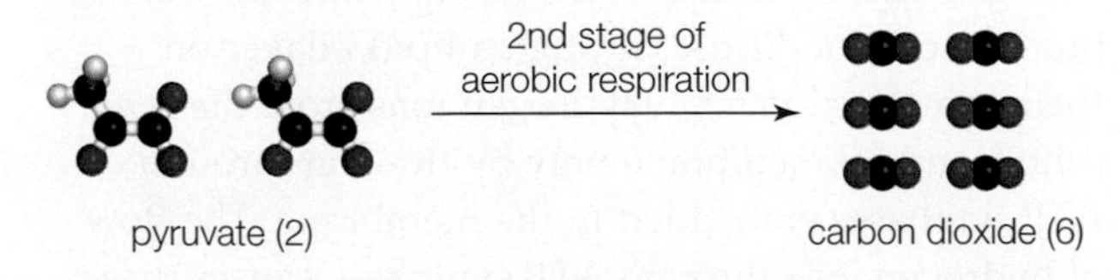

Remember, the two pyruvate were a product of glycolysis. So, at this point in aerobic respiration, the carbon backbone of one glucose molecule has been broken down completely, its six carbon atoms having exited the cell in CO_2.

Two ATP that form during the second stage add to the small net yield (two ATP) of glycolysis. However, this stage reduces ten coenzymes—eight NAD^+ and two FAD. Add in the two NAD^+ that were reduced in glycolysis, and the full breakdown of each glucose molecule has a big potential payoff. Twelve reduced coenzymes will deliver electrons—and the energy they carry—to the third stage of aerobic respiration.

Krebs cycle Cyclic pathway that, along with acetyl–CoA formation, breaks down pyruvate to carbon dioxide during aerobic respiration.

TAKE-HOME MESSAGE 7.4

What happens during the second stage of aerobic respiration?

✔ The second stage of aerobic respiration, acetyl–CoA formation and the Krebs cycle, occurs in the inner compartment (matrix) of mitochondria.

✔ The second-stage reactions convert the two pyruvate that formed in glycolysis to six CO_2. Two ATP form, and ten coenzymes (eight NAD^+ and two FAD) are reduced.

7.5 Aerobic Respiration's Big Energy Payoff

✔ Many ATP are formed during the third and final stage of aerobic respiration.

FIGURE 7.8 ▸**Animated** In eukaryotes, the third and final stage of aerobic respiration, electron transfer phosphorylation, occurs at the inner mitochondrial membrane (this page).

❶ NADH and $FADH_2$ deliver electrons to electron transfer chains in the inner mitochondrial membrane.

❷ Electron flow through the chains causes hydrogen ions (H^+) to be pumped from the matrix to the intermembrane space. The activity of the electron transfer chains causes a hydrogen ion gradient to form across the inner mitochondrial membrane.

❸ Hydrogen ion flow back to the matrix through ATP synthases drives the formation of ATP from ADP and phosphate (P_i).

❹ Oxygen combines with electrons and hydrogen ions at the end of the electron transfer chains, so water forms.

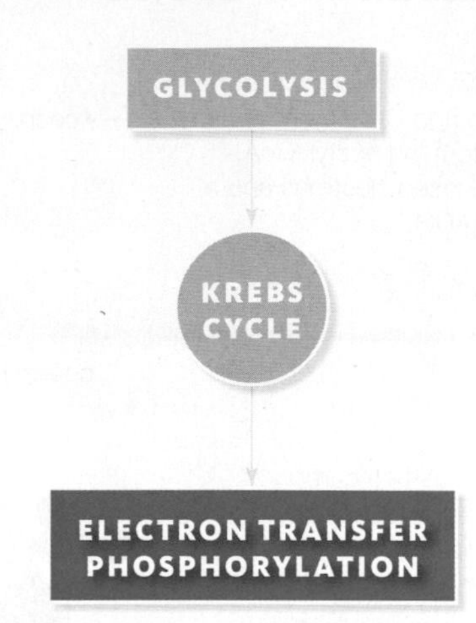

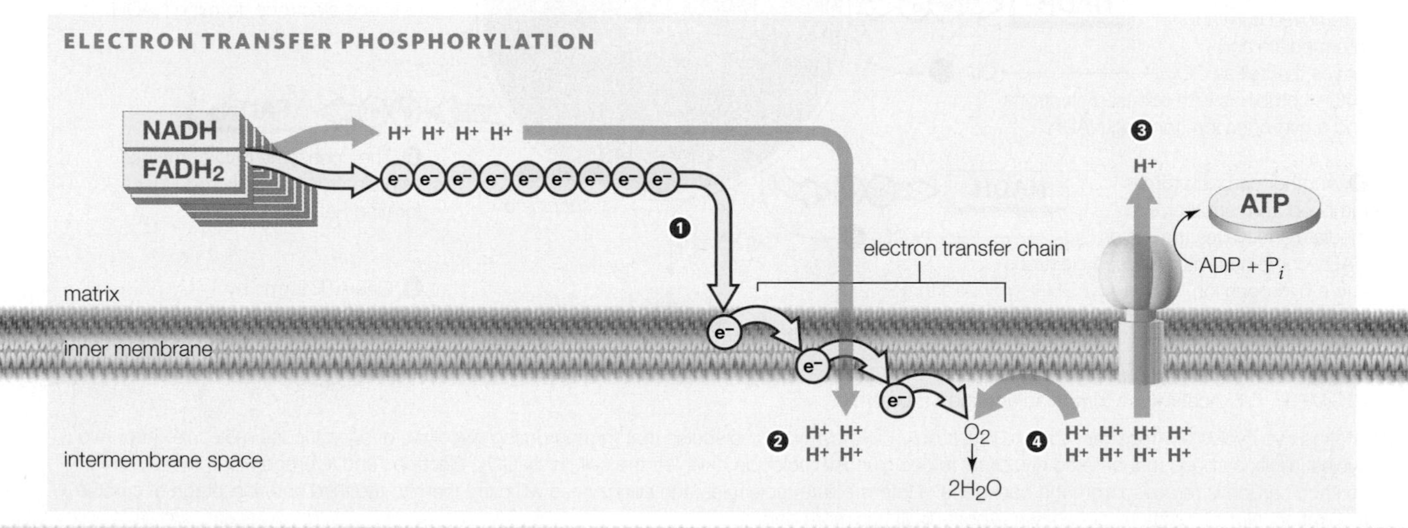

In eukaryotes, the third stage of aerobic respiration, electron transfer phosphorylation, occurs at the inner mitochondrial membrane (**FIGURE 7.8**). The reactions of electron transfer phosphorylation begin with the coenzymes NADH and $FADH_2$, which became reduced in the first two stages of aerobic respiration. These coenzymes donate their cargo of electrons and hydrogen ions to electron transfer chains embedded in the inner mitochondrial membrane ❶.

As the electrons move through the chains, they give up energy little by little (Section 5.5). Some molecules of the electron transfer chains harness that energy to actively transport hydrogen ions across the inner membrane, from the matrix to the intermembrane space ❷. The accumulating hydrogen ions form a gradient across the inner mitochondrial membrane. This gradient attracts the ions back toward the matrix, but ions cannot diffuse through a lipid bilayer on their own (Section 5.8). Hydrogen ions cross the inner mitochondrial membrane only by flowing through ATP synthases embedded in the membrane. The flow of hydrogen ions through ATP synthases causes these proteins to attach phosphate groups to ADP, so ATP forms ❸.

Oxygen accepts electrons at the end of the mitochondrial electron transfer chains ❹. When oxygen accepts electrons, it combines with H^+ to form water, which is a product of the third-stage reactions.

For each glucose molecule that enters aerobic respiration, four ATP form in the first- and second-stage

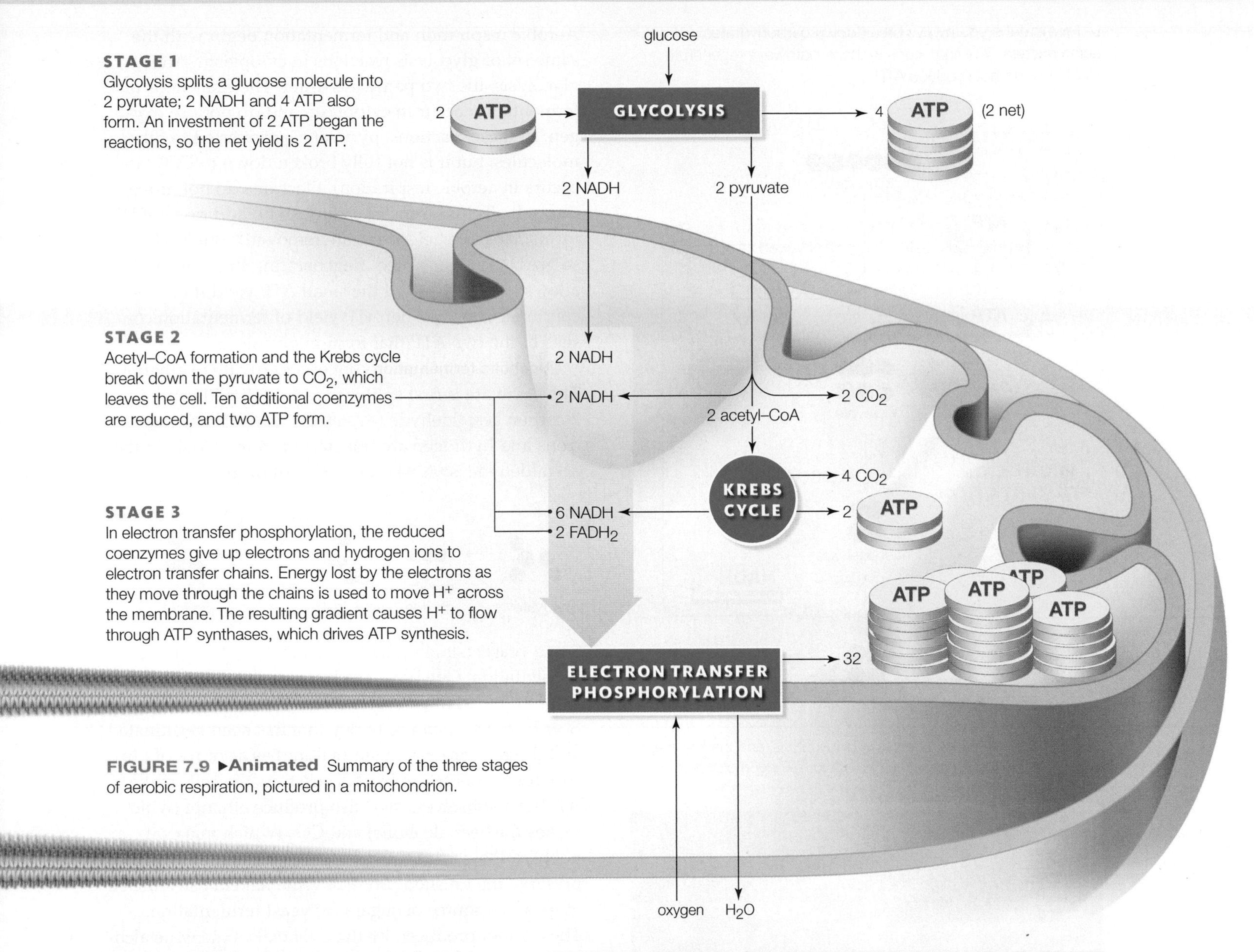

FIGURE 7.9 ▶**Animated** Summary of the three stages of aerobic respiration, pictured in a mitochondrion.

reactions. The twelve coenzymes reduced in these two stages deliver enough H^+ and electrons to fuel the synthesis of about thirty-two additional ATP in the third stage. Thus, the breakdown of one glucose molecule typically yields thirty-six ATP (**FIGURE 7.9**). The ATP yield varies depending on cell type. For example, aerobic respiration in brain and skeletal muscle cells yields thirty-eight ATP per glucose.

Remember that some energy dissipates with every transfer (Section 5.2). Even though aerobic respiration is a very efficient way of retrieving energy from carbohydrates, about 60 percent of the energy harvested in this pathway disperses as metabolic heat.

TAKE-HOME MESSAGE 7.5

What happens during the third stage of aerobic respiration?

✔ In aerobic respiration's third stage, electron transfer phosphorylation, energy released by electrons moving through electron transfer chains is ultimately captured in the attachment of phosphate to ADP.

✔ Coenzymes that were reduced in the first and second stages deliver electrons and hydrogen ions to electron transfer chains in the inner mitochondrial membrane.

✔ Energy released by electrons as they pass through electron transfer chains is used to pump H^+ from the mitochondrial matrix to the intermembrane space. The H^+ gradient that forms across the inner mitochondrial membrane drives the flow of hydrogen ions through ATP synthases, which results in ATP formation.

✔ About thirty-two ATP form during the third-stage reactions, so a typical net yield of all three stages is thirty-six ATP per glucose.

7.6 Fermentation

✔ Fermentation pathways break down carbohydrates without using oxygen. The final steps in these pathways regenerate NAD^+ but do not produce ATP.

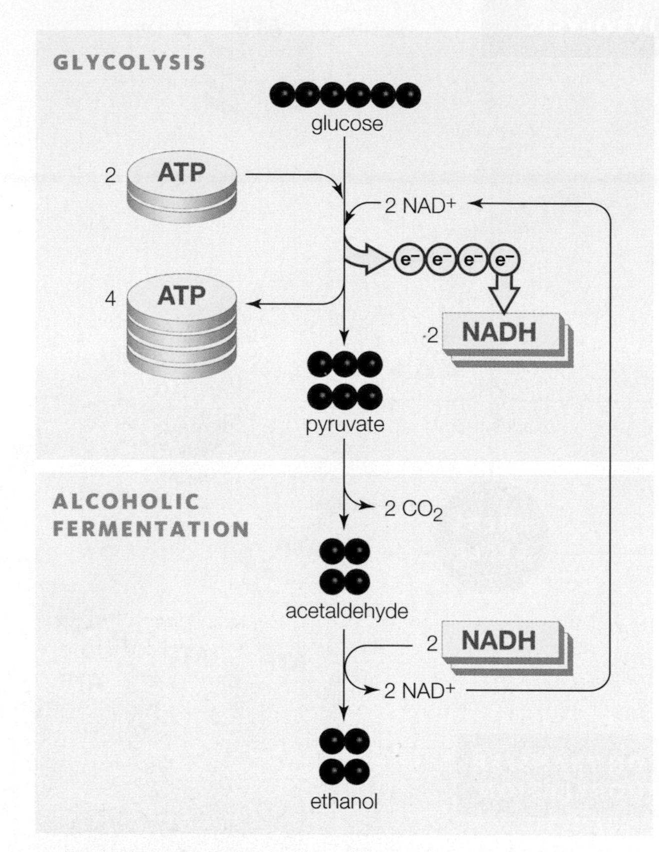

A Alcoholic fermentation begins with glycolysis, and the final steps regenerate NAD^+. The reactions yield two ATP per molecule of glucose (from glycolysis).

B *Saccharomyces cerevisiae* yeast (top). One product of alcoholic fermentation in these cells (ethanol) makes beer alcoholic; another (CO_2) makes it bubbly. Holes in bread are pockets where CO_2 released by fermenting yeast cells accumulated in the dough.

FIGURE 7.10 Alcoholic fermentation.

Aerobic respiration and fermentation begin with the same set of glycolysis reactions in cytoplasm. After glycolysis, the two pathways differ. The final steps of fermentation occur in cytoplasm and do not use oxygen. In these reactions, pyruvate is converted to other molecules, but it is not fully broken down to CO_2 (as occurs in aerobic respiration). Electrons do not move through electron transfer chains, so no additional ATP forms. However, electrons are removed from NADH, so NAD^+ is regenerated. Regenerating this coenzyme allows glycolysis—and the small ATP yield it offers—to continue. Thus, the net ATP yield of fermentation consists of the two ATP that form in glycolysis.

Alcoholic fermentation converts pyruvate to ethanol. The pyruvate is first split into carbon dioxide and 2-carbon acetaldehyde (FIGURE 7.10A). Then, electrons and hydrogen are transferred from NADH to the acetaldehyde, so NAD^+ and ethanol form:

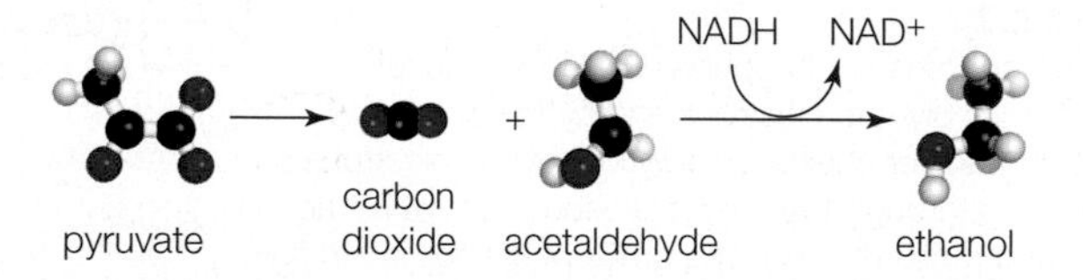

Some yeasts (single-celled fungi) carry out alcoholic fermentation. One type, *Saccharomyces cerevisiae*, helps us produce beer, wine, and bread (FIGURE 7.10B). Beer brewers often use barley that has been germinated and dried (a process called malting) as a source of glucose for fermentation by this yeast. As the cells make ATP for themselves, they also produce ethanol (which makes the beer alcoholic) and CO_2 (which makes it bubbly). Flowers of the hop plant add flavor and help preserve the finished product. Winemakers use crushed grapes as a source of sugars for yeast fermentation. The ethanol produced by the cells makes the wine alcoholic, and the CO_2 is allowed to escape to the air.

To make bread, flour is kneaded with water, yeast, and sometimes other ingredients. Flour contains a protein (gluten) and a disaccharide (maltose). Kneading causes the gluten to form polymers in long, interconnected strands that make the resulting dough stretchy and resilient. The yeast cells in the dough first break down the maltose into its two glucose subunits, then use the released sugars for alcoholic fermentation. CO_2 they produce accumulates in bubbles that are trapped by the mesh of gluten strands. As the bubbles expand, they cause the dough to rise. The ethanol product of fermentation evaporates during baking.

In **lactate fermentation**, the electrons and hydrogen ions carried by NADH are transferred directly to pyruvate (**FIGURE 7.11A**). This reaction converts pyruvate to three-carbon lactate (the ionized form of lactic acid), and also converts NADH to NAD^+:

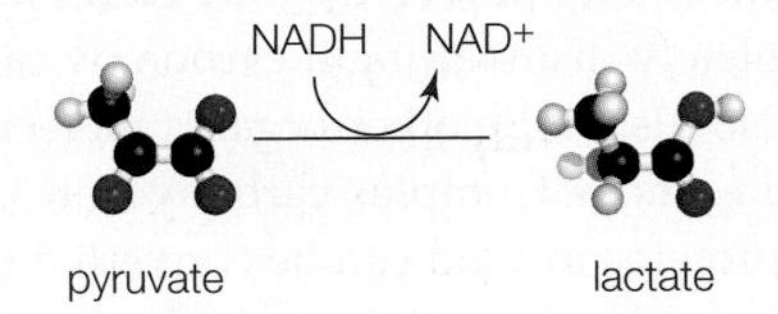

We use lactate fermentation by beneficial bacteria to prepare many foods. Yogurt, for example, is made by allowing species such as *Lactobacillus bulgaricus* and *Streptococcus thermophilus* to grow in milk. Milk contains a disaccharide (lactose) and a protein (casein). The bacterial cells first break down the lactose into its sugar subunits, then use the released glucose for lactate fermentation. Lactate reduces the pH of the milk, which imparts tartness and causes the casein to form a gel.

Cells in animal skeletal muscles are fused as long fibers that carry out aerobic respiration, lactate fermentation, or both. Red fibers have many mitochondria and produce ATP mainly by aerobic respiration. These fibers sustain prolonged activity. They are red because they contain myoglobin, a protein that stores oxygen for aerobic respiration (**FIGURE 7.11B**). White muscle fibers contain few mitochondria and no myoglobin; they make most of their ATP by lactate fermentation. This pathway makes ATP quickly, so it is useful for quick, strenuous bursts of activity (**FIGURE 7.11C**). The low ATP yield does not support prolonged activity.

Most animal muscles are a mixture of white and red fibers, but the proportions vary. For example, great sprinters tend to have more white fibers in their leg muscles; great marathon runners have more red fibers. Chickens cannot fly far because their flight muscles consist mostly of white fibers (thus, the "white" breast meat). A chicken most often walks or runs. Its leg muscles consist mostly of red muscle fibers, the "dark meat." Section 35.7 returns to skeletal muscle fibers.

alcoholic fermentation Anaerobic sugar breakdown pathway that produces ATP, CO_2, and ethanol.
lactate fermentation Anaerobic sugar breakdown pathway that produces ATP and lactate.

TAKE-HOME MESSAGE 7.6

What is fermentation?

✔ Prokaryotes and eukaryotes use fermentation pathways, which are anaerobic, to produce ATP by breaking down carbohydrates. Fermentation's small ATP yield (two per molecule of glucose) occurs by glycolysis.

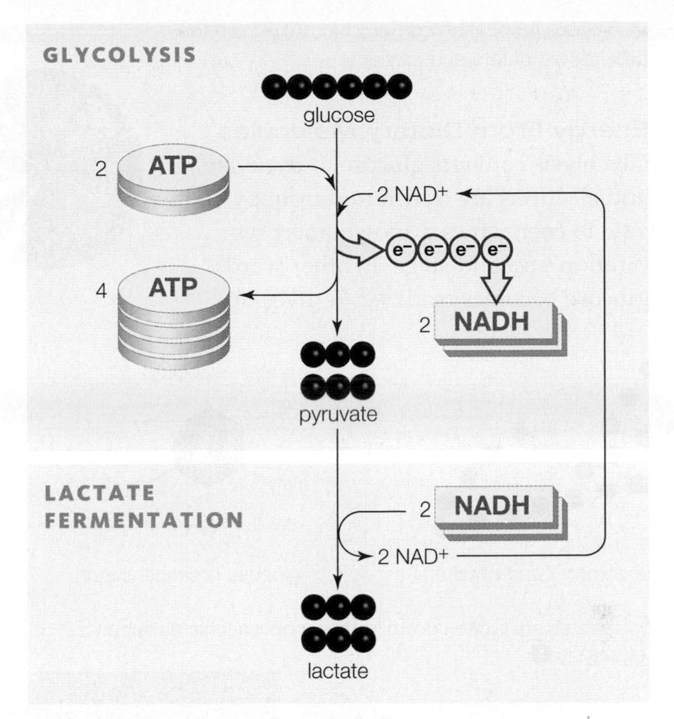

A Lactate fermentation begins with glycolysis, and the final steps regenerate NAD^+. The reactions yield two ATP per molecule of glucose (from glycolysis).

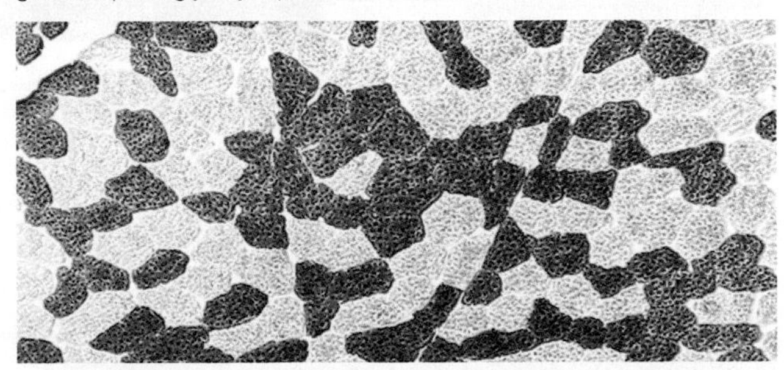

B Lactate fermentation occurs in white muscle fibers, visible in this cross-section of human thigh muscle. The red fibers, which make ATP by aerobic respiration, sustain endurance activities.

C Intense activity such as sprinting quickly depletes oxygen in muscles. Under anaerobic conditions, ATP is produced mainly by lactate fermentation in white muscle fibers. Fermentation does not make enough ATP to sustain this type of activity for long.

FIGURE 7.11 ▶**Animated** Lactate fermentation.

7.7 Alternative Energy Sources in Food

✔ Aerobic respiration can produce ATP from the breakdown of fats and proteins.

Energy From Dietary Molecules

Glycolysis converts glucose to pyruvate, and electrons are transferred from pyruvate to coenzymes during aerobic respiration's second stage. In other words, glucose becomes oxidized (it gives up electrons) and coenzymes become reduced (they accept electrons). Oxidizing an organic molecule can break the covalent bonds of its carbon backbone. Aerobic respiration produces a lot of ATP by fully oxidizing glucose, completely dismantling it carbon by carbon.

Cells also dismantle other organic molecules by oxidizing them. Complex carbohydrates, fats, and proteins in food can be converted to molecules that enter glycolysis or the Krebs cycle (**FIGURE 7.12**). As in glucose metabolism, many coenzymes are reduced, and the energy of the electrons they carry ultimately drives the synthesis of ATP in electron transfer phosphorylation.

Complex Carbohydrates In humans and other mammals, the digestive system breaks down starch and other complex carbohydrates to monosaccharides (**FIGURE 7.12A**). The sugars are taken up by cells and

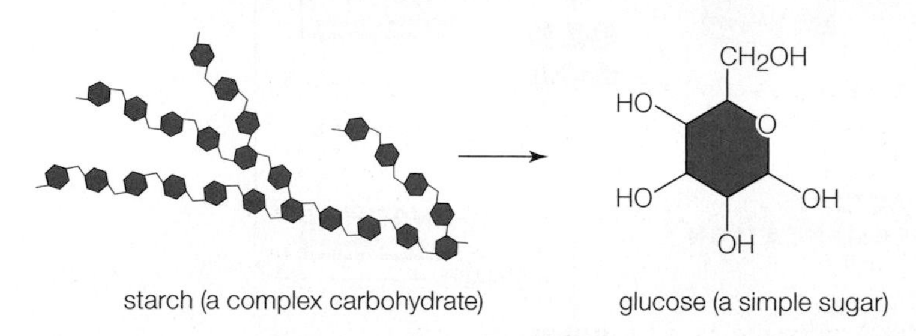

A Complex carbohydrates are broken down to their monosaccharide subunits, which can enter glycolysis ❶.

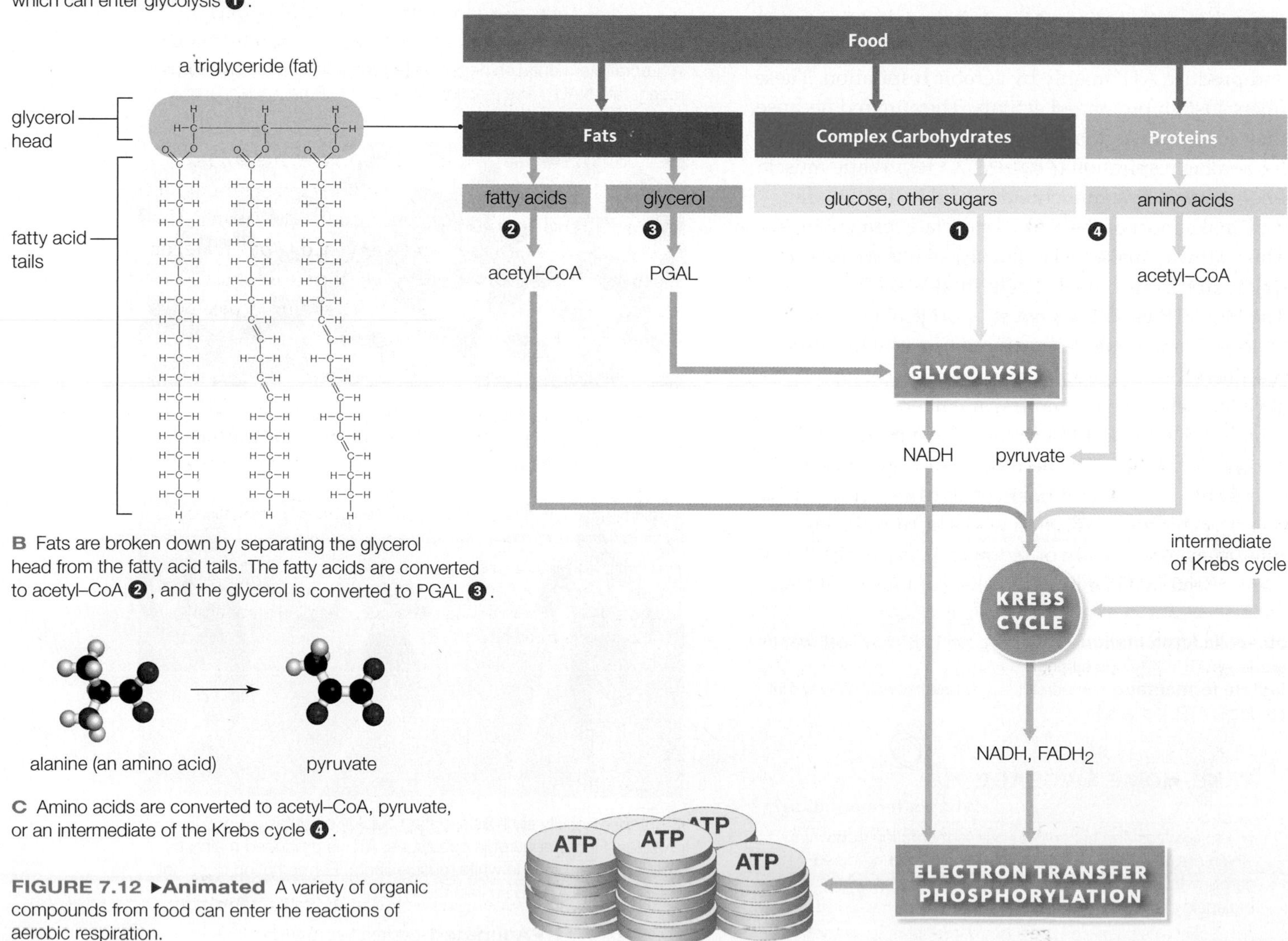

B Fats are broken down by separating the glycerol head from the fatty acid tails. The fatty acids are converted to acetyl–CoA ❷, and the glycerol is converted to PGAL ❸.

C Amino acids are converted to acetyl–CoA, pyruvate, or an intermediate of the Krebs cycle ❹.

FIGURE 7.12 ▸Animated A variety of organic compounds from food can enter the reactions of aerobic respiration.

Risky Business (revisited)

Mitochondrial diseases are devastating, incurable, and heritable. Mitochondria, remember, contain their own DNA (Section 4.8). A child inherits them from one parent only: the mother. For a variety of reasons, some women's eggs contain mostly defective mitochondria; children born to these women have a higher than normal risk of mitochondrial disease. Thanks to efforts by Martine Martin and many others, the United Kingdom recently approved a new reproductive technology, three-person IVF, intended to help these women have healthy babies. With this procedure, nuclear DNA from the mother and father is inserted into a donor egg from a woman with healthy mitochondria. The resulting embryo will be implanted into the mother, much as normal IVF works. A baby born from this procedure will have three parents: Its cells will have nuclear DNA inherited from the mother and father, and mitochondria inherited from the donor.

---

converted to glucose-6-phosphate for glycolysis ❶. When a cell produces more ATP than it uses, ATP accumulates in the cytoplasm. The high concentration of ATP causes glucose-6-phosphate to be diverted away from glycolysis and into a pathway that builds glycogen (Section 3.4). Liver and muscle cells especially favor the conversion of glucose to glycogen, and these cells contain the body's largest stores of it. Between meals, the liver maintains the glucose level in blood by converting the stored glycogen to glucose.

Fats A fat molecule has a glycerol head and one, two, or three fatty acid tails (Section 3.5). Cells dismantle these molecules by first breaking the bonds that connect the fatty acid tails to the glycerol head (FIGURE 7.12B). Nearly all cells in the body can oxidize the released fatty acids by splitting their long backbones into two-carbon fragments. These fragments are converted to acetyl–CoA, which can enter the Krebs cycle ❷. Glycerol released by fat breakdown is converted by liver cells to PGAL, an intermediate of glycolysis ❸.

On a per carbon basis, fats are a richer source of energy than carbohydrates. Carbohydrate backbones have many oxygen atoms, so they are partially oxidized. A fat's long fatty acid tails are hydrocarbon chains that typically have no oxygen atoms bonded to them, so they have a longer way to go to become oxidized—more reactions are required to fully break them down. Coenzymes accept electrons in these oxidation reactions. The more reduced coenzymes that form, the more electrons can be delivered to the ATP-forming machinery of electron transfer phosphorylation.

What happens if you eat too many carbohydrates? When the blood level of glucose gets too high, acetyl–CoA is diverted away from the Krebs cycle and into a pathway that makes fatty acids. That is why excess dietary carbohydrate ends up as fat.

Proteins Enzymes in the digestive system split dietary proteins into their amino acid subunits, which are absorbed into the bloodstream. Cells use the amino acids to build proteins or other molecules. When you eat more protein than your body needs for this purpose, the amino acids are broken down. The amino (—NH_3^+) group is removed, and it becomes ammonia (NH_3), a waste product that is eliminated in urine. The carbon backbone is split, and acetyl–CoA, pyruvate, or an intermediate of the Krebs cycle forms, depending on the amino acid (FIGURE 7.12C). These molecules enter aerobic respiration's second stage ❹.

TAKE-HOME MESSAGE 7.7

Can the body break down any organic molecule for energy?

✔ Oxidizing organic molecules can break their carbon backbones, releasing electrons whose energy can be harnessed to drive ATP formation in aerobic respiration.

✔ First the digestive system and then individual cells convert molecules in food (fats, complex carbohydrates, and proteins) into substrates of glycolysis or aerobic respiration's second-stage reactions.

summary

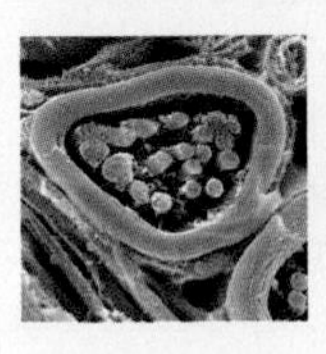

Section 7.1 Photosynthesis changed the composition of Earth's atmosphere, with profound effects on life's evolution. Organisms that could not tolerate the increased atmospheric oxygen persisted only in **anaerobic** habitats. The evolution of antioxidants allowed organisms to thrive under **aerobic** conditions.

Free radicals that form during aerobic respiration are detoxified by antioxidant molecules in the cell's cytoplasm. Missing molecules or heritable defects in mitochondrial electron transfer chain components can cause a buildup of free radicals that damage the cell—and the individual. The resulting oxidative stress plays a role in many illnesses.

Section 7.2 Most organisms can make ATP by breaking down sugars, either by fermentation or by aerobic respiration. Both pathways begin in cytoplasm. **Aerobic respiration** requires oxygen and, in eukaryotes, ends inside mitochondria. This pathway includes electron transfer chains, and ATP forms by electron transfer phosphorylation. **Fermentation** pathways run entirely in cytoplasm and do not require oxygen. Aerobic respiration yields much more ATP per glucose molecule than fermentation.

Section 7.3 **Glycolysis**, the first stage of aerobic respiration and fermentation pathways, occurs in cytoplasm. During glycolysis, enzymes use two ATP to convert one molecule of glucose or another six-carbon sugar to two molecules of three-carbon **pyruvate**. Electrons and hydrogen ions are transferred to two NAD^+, which are thereby reduced to NADH. Four ATP also form by **substrate-level phosphorylation**.

Section 7.4 In eukaryotes, aerobic respiration continues in mitochondria. The second stage of aerobic respiration, acetyl–CoA formation and the **Krebs cycle**, takes place in the inner compartment (matrix) of the mitochondrion. The first steps convert the two pyruvate from glycolysis to two acetyl–CoA and two CO_2. The acetyl–CoA delivers carbon atoms to the Krebs cycle. Then, electrons and hydrogen ions are transferred to NAD^+ and FAD, which are thereby reduced to NADH and $FADH_2$. ATP forms by substrate-level phosphorylation.

Two cycles of Krebs reactions break down the two pyruvate from glycolysis. At this stage of aerobic respiration, the glucose molecule that entered glycolysis has been dismantled completely: All of its carbon atoms have exited the cell in CO_2.

Section 7.5 In the third stage of aerobic respiration, electron transfer phosphorylation, the many coenzymes reduced in the first two stages now deliver electrons and hydrogen ions to electron transfer chains in the inner mitochondrial membrane. Energy lost by electrons moving through the chains is harnessed to move H^+ from the matrix to the intermembrane space. The resulting hydrogen ion gradient across the inner membrane drives these ions through ATP synthases, which in turn drives ATP synthesis. Oxygen accepts electrons at the end of the chains and combines with hydrogen ions, so water forms. The ATP yield of aerobic respiration varies, but typically it is about thirty-six ATP per glucose.

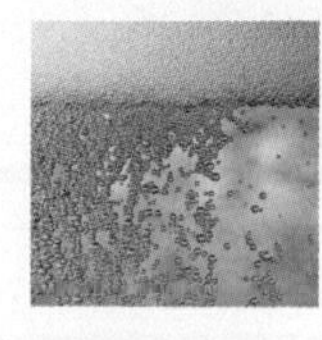

Section 7.6 Anaerobic fermentation pathways begin with glycolysis, and they run in the cytoplasm. The electron acceptor at the end of these reactions is an organic molecule such as acetaldehyde (in **alcoholic fermentation**) or pyruvate (in **lactate fermentation**). The final steps of fermentation regenerate NAD^+, which is required for glycolysis to continue, but they produce no ATP. Thus, the breakdown of one glucose molecule yields only the two ATP from glycolysis. The small yield is enough to sustain single-celled organisms, for example in beneficial microbes that we use in production of foods such as bread and yogurt. Fermentation also supplements metabolism of multicelled eukaryotes, for example in mammalian white skeletal muscle fibers that use lactate fermentation to support intense bursts of activity.

Section 7.7 Oxidizing an organic molecule can break its carbon backbone. Aerobic respiration fully oxidizes glucose, dismantling its backbone carbon by carbon. Each carbon removed releases electrons that drive ATP formation in electron transfer phosphorylation. Organic molecules other than sugars are also broken down (oxidized) for energy. In humans and other mammals, first the digestive system and then individual cells convert fats, proteins, and complex carbohydrates in food to molecules that are substrates of glycolysis or the second-stage reactions of aerobic respiration.

self-quiz

Answers in Appendix VII

1. True or false? Unlike animals, which make many ATP by aerobic respiration, plants make all of their ATP by photosynthesis.

2. Glycolysis starts and ends in the ________ .
   a. nucleus
   b. mitochondrion
   c. plasma membrane
   d. cytoplasm

3. Which of the following metabolic pathways require(s) molecular oxygen (O_2)?
   a. aerobic respiration
   b. lactate fermentation
   c. alcoholic fermentation
   d. all of the above

4. Which molecule does not form during glycolysis?
   a. NADH b. pyruvate c. $FADH_2$ d. ATP

5. In eukaryotes, aerobic respiration is completed in the ________ .
   a. nucleus
   b. mitochondrion
   c. plasma membrane
   d. cytoplasm

6. Which of the following reaction pathways is *not* part of the second stage of aerobic respiration?
   a. electron transfer phosphorylation
   b. acetyl–CoA formation
   c. Krebs cycle
   d. glycolysis
   e. a and d

7. After Krebs reactions run through ________ cycle(s), one glucose molecule has been completely broken down to CO_2.
   a. one b. two c. three d. six

8. In the third stage of aerobic respiration, ________ is the final acceptor of electrons.
   a. water b. hydrogen c. oxygen d. NADH

9. ________ accepts electrons in alcoholic fermentation.
   a. Oxygen
   b. Pyruvate
   c. Acetaldehyde
   d. Sulfate

data analysis activities

Mitochondrial Abnormalities in Tetralogy of Fallot Tetralogy of Fallot (TF) is a genetic disorder in which heart malformations result in abnormal blood circulation, so oxygen does not reach body cells as it should. With insufficient oxygen to accept electrons at the end of miotchondrial electron transfer chains, too many free radicals form. This damages the mitochondria—and the cells. In 2004, Sarah Kuruvilla studied mitochondria in the heart muscle of TF patients. Some of her results are shown in **FIGURE 7.13**.

1. In this study, which abnormality was most strongly associated with TF?

2. What percentage of the TF patients had mitochondria that were abnormal in size?

3. Can you make any correlations between blood oxygen content and mitochondrial abnormalities in these patients?

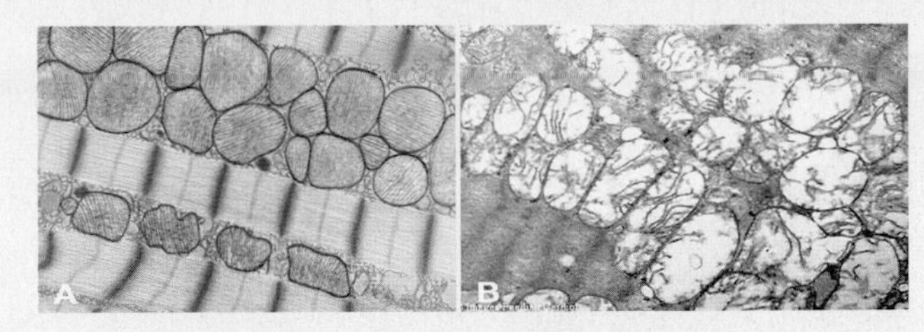

Patient (age)	SPO_2 (%)	Mitochondrial Abnormalities in TF: Number	Shape	Size	Broken
1 (5)	55	+	+	–	–
2 (3)	69	+	+	–	–
3 (22)	72	+	+	–	–
4 (2)	74	+	+	–	–
5 (3)	76	+	+	–	+
6 (2.5)	78	+	+	–	+
7 (1)	79	+	+	–	–
8 (12)	80	+	–	+	–
9 (4)	80	+	+	–	–
10 (8)	83	+	–	+	–
11 (20)	85	+	+	–	–
12 (2.5)	89	+	–	+	–

C

FIGURE 7.13 Mitochondrial changes in tetralogy of Fallot (TF).

(**A**) Normal heart muscle. Many mitochondria between the fibers provide muscle cells with ATP for contraction. (**B**) Heart muscle from a person with TF has swollen, broken mitochondria.

(**C**) Types of mitochondrial abnormalities in TF patients. SPO_2 is oxygen saturation of the blood. A normal value of SPO_2 is 96%. Abnormalities are marked +.

10. Fermentation pathways make no more ATP beyond the small yield from glycolysis. The remaining reactions serve to regenerate ________ .
  a. FAD
  b. NAD^+
  c. glucose
  d. oxygen

11. Most of the energy that is released by the full breakdown of glucose to CO_2 and water ends up in ________ .
  a. NADH
  b. ATP
  c. heat
  d. electrons

12. Your body cells can break down ________ as a source of energy to fuel ATP production.
  a. fatty acids
  b. glycerol
  c. amino acids
  d. all of the above

13. Which of the following is *not* produced by an animal muscle cell operating under anaerobic conditions?
  a. heat
  b. pyruvate
  c. NAD^+
  d. ATP
  e. lactate
  f. all are produced

14. Match the reactions with the events.

  ____ glycolysis
  ____ fermentation
  ____ Krebs cycle
  ____ electron transfer phosphorylation

  a. ATP, NADH, $FADH_2$, and CO_2 form
  b. glucose to two pyruvate
  c. NAD^+ regenerated, little ATP
  d. H^+ flow via ATP synthases

15. Match the term with the best description.

  ____ mitochondrial matrix
  ____ pyruvate
  ____ NAD^+
  ____ mitochondrion
  ____ NADH
  ____ anaerobic

  a. needed for glycolysis
  b. inner space
  c. makes many ATP
  d. product of glycolysis
  e. reduced coenzyme
  f. no oxygen required

critical thinking

1. The higher the altitude, the lower the oxygen level in air. Climbers of very tall mountains risk altitude sickness, a condition characterized by shortness of breath, weakness, dizziness, and confusion. The early symptoms of cyanide poisoning are the same as those for altitude sickness. Cyanide binds tightly to cytochrome *c* oxidase, a protein complex that is the last component of mitochondrial electron transfer chains. Cytochrome *c* oxidase with bound cyanide can no longer transfer electrons. Explain why cyanide poisoning starts with the same symptoms as altitude sickness.

2. As you learned, membranes impermeable to hydrogen ions are required for electron transfer phosphorylation. Membranes in mitochondria serve this purpose in eukaryotes. Bacteria do not have this organelle, but they can make ATP by electron transfer phosphorylation. How do you think they do it, given that they have no mitochondria?

3. The bar-tailed godwit is a type of shorebird that makes an annual migration from Alaska to New Zealand and back. The birds make each 11,500-kilometer (7,145-mile) trip by flying over the Pacific Ocean in about nine days, depending on weather, wind speed, and direction of travel. One bird was observed to make the entire journey uninterrupted, a feat that is comparable to a human running a nonstop seven-day marathon at 70 kilometers per hour (43.5 miles per hour). Would you expect the flight (breast) muscles of bar-tailed godwits to be light or dark colored? Explain your answer.

CREDITS: (13A) Steve Gschmeissner/Science Source; (13B) © Images Paediatr Cardiol; (13C) © Cengage Learning.

LEARNING ROADMAP

Radioisotope tracers (Section 2.2) were used in research that led to the discovery that DNA (3.8), not protein (3.6), is the hereditary material of all organisms (1.3). This chapter revisits free radicals (2.3), the cell nucleus (4.6), and metabolism (5.4–5.6). Your knowledge of carbohydrate ring numbering (3.3) will help you understand DNA replication.

DISCOVERY OF DNA'S FUNCTION

The work of many scientists over nearly a century led to the discovery that DNA, not protein, stores hereditary information in all living things.

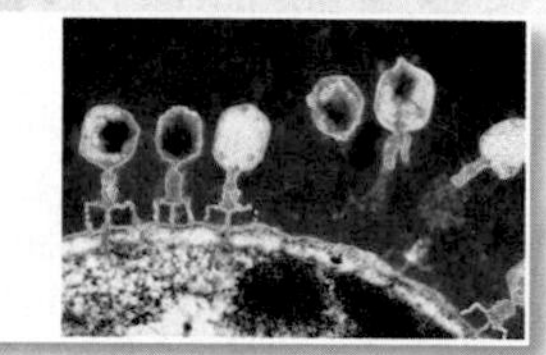

STRUCTURE OF DNA MOLECULES

A DNA molecule consists of two long chains of nucleotides coiled into a double helix. The order of the four types of nucleotides in a chain differs among individuals and among species.

CHROMOSOMES

The DNA of eukaryotes is divided among a characteristic number of chromosomes. A living cell's chromosomes contain all of the information necessary to build a new individual.

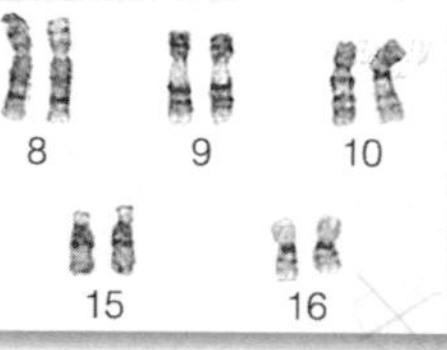

DNA REPLICATION

Before a cell divides, it copies its DNA so both descendant cells will inherit a full complement of chromosomes. Replication of each DNA molecule produces two duplicates.

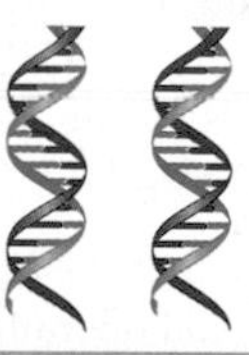

MUTATIONS

DNA damage by environmental agents can cause replication errors. Newly forming DNA is monitored for errors, most of which are corrected. Uncorrected errors become mutations.

You will revisit DNA structure and function many times, particularly when you learn about how genetic information is converted into parts of a cell (Chapter 9), and how cells control that process (Chapter 10). Chromosome structure will turn up again in the context of cell division in Chapters 11 and 12, and in human inheritance and disease in Chapter 14. Viruses such as bacteriophage will be explained in more detail in Chapter 20, and stem cells in Chapter 31.

8.1 A Hero Dog's Golden Clones

On September 11, 2001, Constable James Symington drove his search dog Trakr from Nova Scotia to Manhattan. Within hours of arriving, the dog led rescuers to the area where the final survivor of the World Trade Center attacks was buried. She had been clinging to life, pinned under rubble from the building where she had worked. Symington and Trakr helped with the search and rescue efforts for three days nonstop, until Trakr collapsed from smoke and chemical inhalation, burns, and exhaustion (**FIGURE 8.1**).

Trakr survived the ordeal, but later lost the use of his limbs from a degenerative neurological disease probably linked to toxic smoke exposure at Ground Zero. The hero dog died in April 2009, but his DNA lives on in his genetic copies—his **clones**. Symington's essay about Trakr's superior nature and abilities as a search and rescue dog won the Golden Clone Giveaway, a contest to find the world's most clone-worthy dog. Trakr's DNA was shipped to Korea, where it was inserted into donor dog eggs, which were then implanted into surrogate mother dogs. Five puppies, all clones of Trakr, were delivered to Symington in July 2009.

Many other adult mammals have been cloned besides Trakr, but the technique is still unpredictable. Depending on the species, few implanted embryos may survive until birth. Of the clones that do survive, many have serious health problems. Why the difficulty? Even though all cells of an individual inherit the same DNA, an adult cell uses only a fraction of it compared with an embryonic cell. To make a clone from an adult cell, researchers must reprogram its DNA to function like the DNA of an egg. They are getting better at it. The research continues because the potential benefits are enormous. Already, cells of cloned human embryos are helping researchers unravel the molecular mechanisms of human genetic diseases. Replacement tissues and organs for people with incurable diseases are being generated from cloned human cells. Endangered animals might be saved from extinction; extinct animals may be brought back.

Perfecting methods for cloning animals brings us closer to the possibility of cloning humans, both technically and ethically. For example, if cloning a lost cat for a grieving pet owner is acceptable, why would it not be acceptable to clone a lost child for a grieving parent? Different people have very different answers to such questions, so controversy over cloning continues even as the techniques improve.

clone Genetically identical copy of an organism.

FIGURE 8.1 James Symington and his dog Trakr at Ground Zero, September 2001.

8.2 The Discovery of DNA's Function

✔ Investigations that led to our understanding that DNA is the molecule of inheritance reveal how science advances.

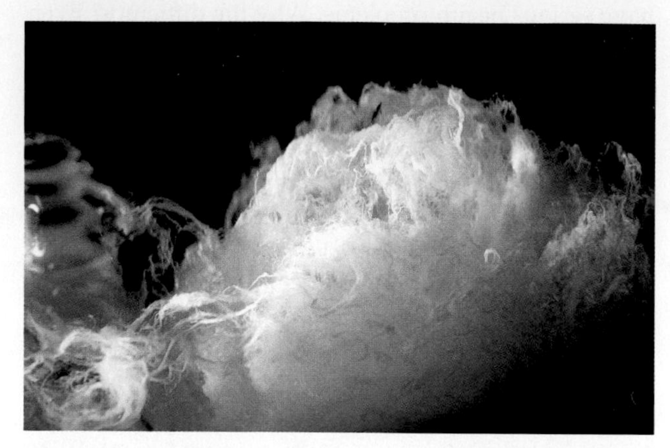

FIGURE 8.2 DNA, the substance, extracted from human cells.

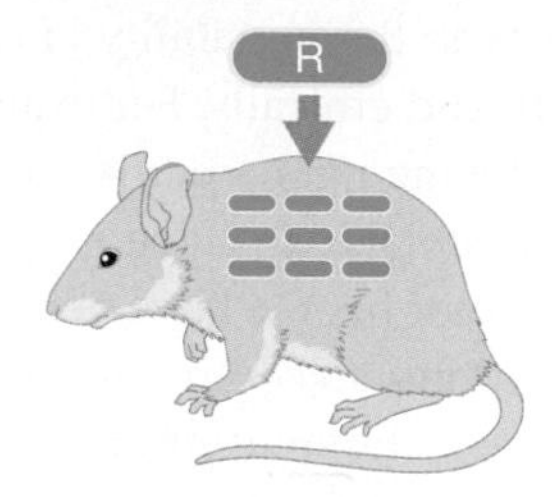

A Griffith's first experiment showed that R cells were harmless. When injected into mice, the bacteria multiplied, but the mice remained healthy.

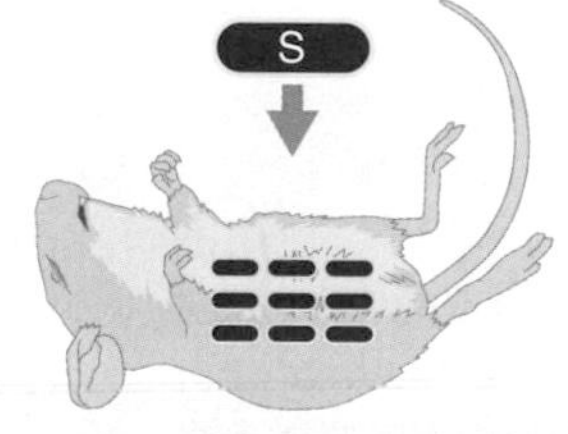

B The second experiment showed that an injection of S cells caused mice to develop fatal pneumonia. Their blood contained live S cells.

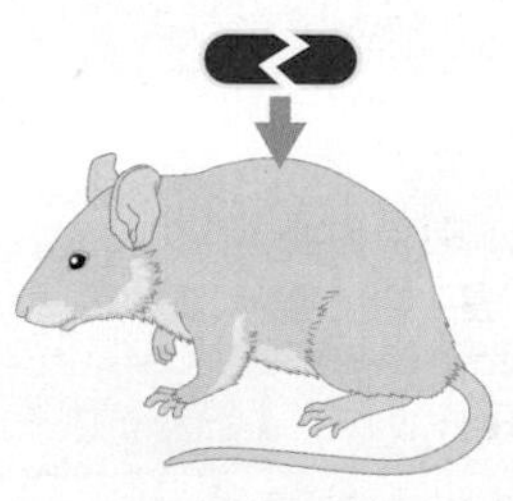

C For a third experiment, Griffith killed S cells with heat before injecting them into mice. The mice remained healthy, indicating that the heat-killed S cells were harmless.

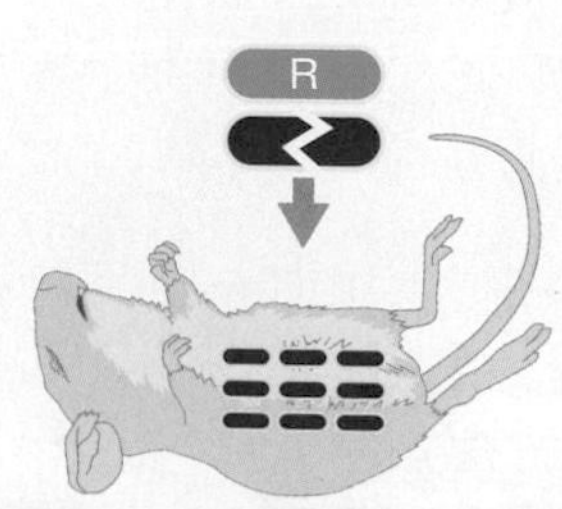

D In his fourth experiment, Griffith injected a mixture of heat-killed S cells and live R cells. To his surprise, the mice became fatally ill, and their blood contained live S cells.

FIGURE 8.3 ▶**Animated** Fred Griffith's experiments with two strains (R and S) of *Streptococcus pneumoniae* bacteria.

The substance we now call DNA (**FIGURE 8.2**) was first described in 1869 by Johannes Miescher, a chemist who extracted it from cell nuclei. Miescher determined that DNA is not a protein, and that it is rich in nitrogen and phosphorus, but he never learned its function.

Sixty years after Miescher's work, Frederick Griffith unexpectedly found a clue about DNA's function. Griffith was studying pneumonia-causing bacteria in the hope of creating a vaccine. He isolated two types, or strains, of the bacteria. One was harmless (R); the other lethal (S). Griffith used R and S cells in a series of experiments testing their ability to cause pneumonia in mice (**FIGURE 8.3**). He discovered that heat destroyed the ability of lethal S bacteria to cause pneumonia, but it did not destroy their hereditary material, including whatever specified "kill mice." That material could be transferred from dead S cells to live R cells, which put it to use. The transformation was permanent and heritable: Even after hundreds of generations, descendants of transformed R cells retained the ability to kill mice.

What substance had caused this transformation? In 1940, Oswald Avery and Maclyn McCarty set out to identify this substance, which they called the "transforming principle." The team extracted lipids, proteins, and nucleic acids from S cells, then used a process of elimination to see which component would cause the transformation. Treating the extract with lipid- and protein-destroying enzymes did not prevent it from transforming R cells, so the transforming principle could not be lipid or protein. Avery and McCarty realized that the substance they were seeking must be nucleic acid—DNA or RNA. DNA-degrading enzymes destroyed the extract's ability to transform cells, but RNA-degrading enzymes did not. Thus, DNA had to be the transforming principle.

The result surprised Avery and McCarty, who, along with most other scientists, had assumed that proteins were the material of heredity. After all, traits are diverse, and proteins are the most diverse of all biological molecules. The two scientists were so skeptical that they published their results only after they had convinced themselves, by years of painstaking experimentation, that DNA was indeed hereditary material. They were also careful to point out that they had not proven DNA was the *only* hereditary material.

Avery and McCarty's tantalizing results prompted a stampede of other scientists into the field of DNA research. The resulting explosion of discovery confirmed the molecule's role as carrier of hereditary information. Key in this advance was the realization that any molecule—DNA or otherwise—had to have certain

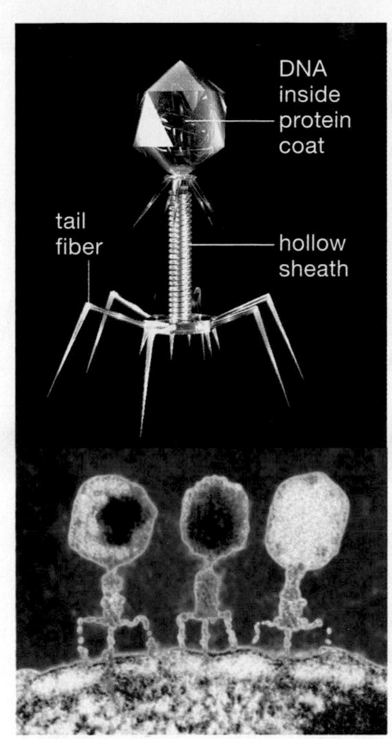

A Top, a model of a bacteriophage. Bottom, micrograph of three viruses injecting DNA into an *E. coli* cell.

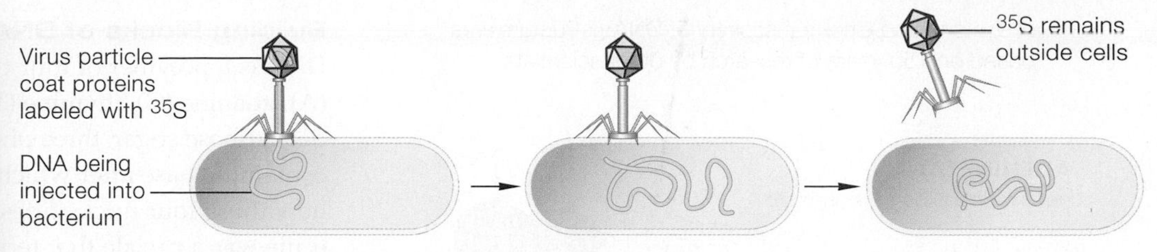

B In one experiment, bacteriophage were labeled with a radioisotope of sulfur (^{35}S), a process that makes their protein components radioactive. The labeled viruses were mixed with bacteria long enough for infection to occur, and then the mixture was whirled in a kitchen blender. Blending dislodged viral parts that remained on the outside of the bacteria. Afterward, most of the radioactive sulfur was detected outside the bacterial cells. The viruses had not injected protein into the bacteria.

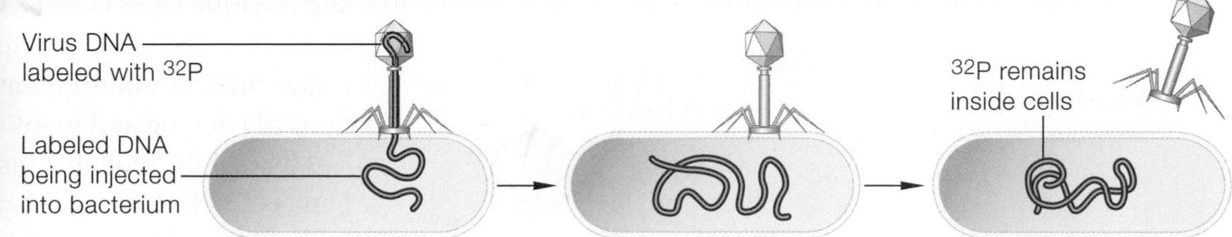

C In another experiment, bacteriophage were labeled with a radioisotope of phosphorus (^{32}P), which makes their DNA radioactive. The labeled viruses were allowed to infect bacteria. After the external viral parts were dislodged from the bacteria, the radioactive phosphorus was detected mainly inside the bacterial cells. The viruses had injected DNA into the cells—evidence that DNA is the genetic material of this virus.

FIGURE 8.4 ▸**Animated** The Hershey–Chase experiments. Alfred Hershey and Martha Chase carried out experiments to determine the composition of the hereditary material that bacteriophage inject into bacteria. The experiments were based on the knowledge that proteins contain more sulfur (S) than phosphorus (P), and DNA contains more phosphorus than sulfur.

properties in order to function as hereditary material. First, a full complement of hereditary information must be transmitted along with the molecule; second, each cell of a given species should contain the same amount of it; third, because the molecule functions as a genetic bridge between generations, it has to be exempt from major change; and fourth, it must be capable of encoding the almost unimaginably huge amount of information required to build a new individual.

In the late 1940s, Alfred Hershey and Martha Chase proved that DNA, and not protein, satisfies the first property of a hereditary molecule: It transmits a full complement of hereditary information. Hershey and Chase specialized in working with **bacteriophage**, a type of virus that infects bacteria (**FIGURE 8.4A**). Like all viruses, these infectious particles carry information about how to make new viruses in their hereditary material. After a virus injects a cell with this material, the cell starts making new virus particles. Hershey and Chase carried out an elegant series of experiments proving that the material a bacteriophage injects into bacteria is DNA, not protein (**FIGURE 8.4B,C**).

bacteriophage Virus that infects bacteria.

The second property expected of a hereditary molecule was pinned on DNA by André Boivin and Roger Vendrely, who meticulously measured the amount of DNA in cell nuclei from a number of species. In 1948, they proved that body cells of any individual of a species contain precisely the same amount of DNA. Daniel Mazia's laboratory discovered that the protein and RNA content of cells varies over time, but not the DNA content, demonstrating that DNA is not involved in metabolism (and proving DNA has the third property expected of a hereditary molecule). The fourth property—that a hereditary molecule must somehow encode a huge amount of information—would be proven along with the elucidation of DNA's structure, a topic we continue in the next section.

TAKE-HOME MESSAGE 8.2

How was the function of DNA discovered?

✔ DNA, the molecule of inheritance, was first discovered in the late 1800s. Its role as the carrier of hereditary information was uncovered over many years, as scientists built upon one another's discoveries.

✔ Watson and Crick's discovery of DNA's structure was based on 150 years of research by other scientists.

ADENINE (A)
deoxyadenosine triphosphate

BASE
SUGAR

GUANINE (G)
deoxyguanosine triphosphate

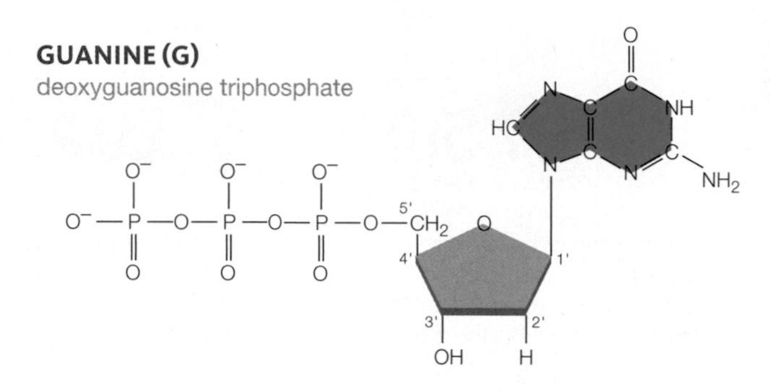

THYMINE (T)
deoxythymidine triphosphate

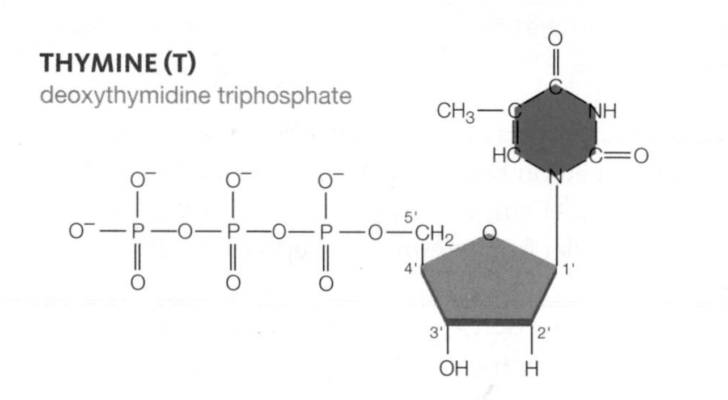

CYTOSINE (C)
deoxycytidine triphosphate

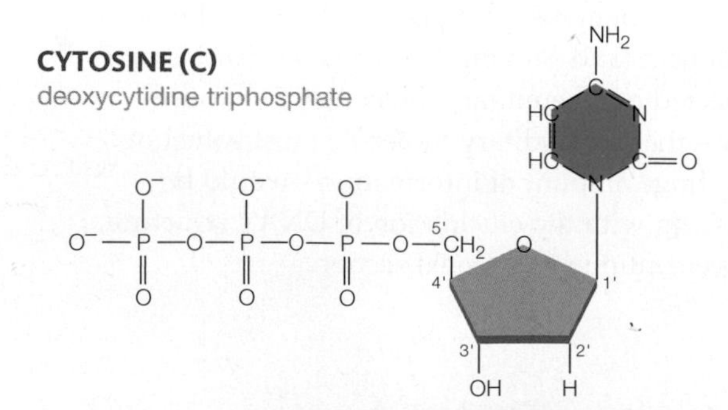

FIGURE 8.5 The four nucleotides in DNA. Each has three phosphate groups, a deoxyribose sugar (orange), and a nitrogen-containing base (blue) after which it is named. Adenine and guanine bases are purines; thymine and cytosine, pyrimidines. Biochemist Phoebus Levene identified the structure of these bases and how they are connected in nucleotides in the early 1900s. Levene worked with DNA for almost 40 years.

Building Blocks of DNA

DNA is a polymer of four types of nucleotides—adenine (A), guanine (G), thymine (T), and cytosine (C). Each has a deoxyribose sugar, three phosphate groups, and a nitrogen-containing base after which it is named (**FIGURE 8.5**). Just how those four nucleotides are arranged in a DNA molecule was a puzzle that took over 50 years to solve.

Clues about DNA's structure started coming together around 1950, when Erwin Chargaff (one of many researchers investigating its function) made two important discoveries about the molecule. First, the amounts of thymine and adenine are identical, as are the amounts of cytosine and guanine (A = T and G = C). We call this discovery Chargaff's first rule. Chargaff's second discovery, or rule, is that the DNA of different species differs in the proportions of adenine and guanine.

Meanwhile, biologist James Watson and biophysicist Francis Crick had been sharing ideas about the structure of DNA. The helical (coiled) pattern of secondary structure that occurs in many proteins (Section 3.6) had just been discovered, and Watson and Crick suspected that the DNA molecule was also a helix. The two spent many hours arguing about the size, shape, and bonding requirements of the four DNA nucleotides. They pestered chemists to help them identify bonds they might have overlooked, fiddled with cardboard cutouts, and made models from scraps of metal connected by suitably angled "bonds" of wire.

Biochemist Rosalind Franklin had also been working on the structure of DNA. Like Crick, Franklin specialized in x-ray crystallography, a technique in which x-rays are directed through a purified and crystallized substance. Atoms in the substance's molecules scatter the x-rays in a pattern that can be captured as an image. Researchers can use the pattern to calculate the size, shape, and spacing between any repeating elements of the molecules—all of which are details of molecular structure. As molecules go, DNA is gigantic, and it was difficult to crystallize given the techniques of the time. Franklin made the first clear x-ray diffraction image of DNA as it occurs in cells (left). From the information in that image, she calculated that DNA is very long compared to its 2-nanometer diameter. She also identified a repeating pattern every 0.34 nanometer along its length, and another every 3.4 nanometers.

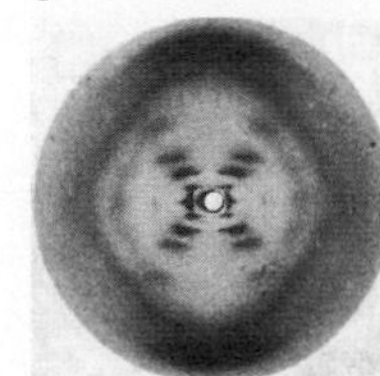

Franklin's image and data came to the attention of Watson and Crick, who now had all the information they needed to build the first accurate model of DNA (**FIGURE 8.6**). A DNA molecule has two chains (strands) of nucleotides running in opposite directions and coiled into a double helix (**FIGURE 8.7**). Bonds between the deoxyribose of one nucleotide and the phosphate of the next form the sugar–phosphate backbone of each chain. Hydrogen bonds between the internally positioned bases hold the two strands together. Only two kinds of base pairings form: A to T, and G to C, and this explains the first of Chargaff's rules. Most scientists had assumed (incorrectly) that the bases had to be on the outside of the helix, because they would be more accessible to DNA-copying enzymes that way. You will see in Section 8.5 how DNA replication enzymes access the bases on the inside of the double helix.

DNA's Base Sequence

A small piece of DNA from a tulip, a human, or any other organism might be:

T G A G G A C T C C T C
A C T C C T G A G G A G
one base pair

Notice how the two strands of DNA match. They are complementary—the base of each nucleotide on one strand pairs with a suitable partner base on the other. This base-pairing pattern (A to T, G to C) is the same in all molecules of DNA. How can just two kinds of base pairings give rise to the incredible diversity of traits we see among living things? Even though DNA is composed of only four nucleotides, the *order* in which one nucleotide follows the next along a strand—the **DNA sequence**—varies tremendously among species (which explains Chargaff's second rule). DNA molecules can be hundreds of millions of nucleotides long, so their sequence can encode a massive amount of information (we return to the nature of that information in the next chapter). DNA sequence variation is the basis of traits that define species and distinguish individuals. Thus DNA, the molecule of inheritance in every cell, is the basis of life's unity. Variations in its sequence are the foundation of life's diversity.

DNA sequence Order of nucleotides in a strand of DNA.

TAKE-HOME MESSAGE 8.3

What is the structure of DNA?

✔ A DNA molecule consists of two nucleotide chains (strands) running in opposite directions and coiled into a double helix. Internally positioned nucleotide bases hydrogen-bond between the two strands. A pairs with T, and G with C.

✔ The sequence of bases along a DNA strand varies among species and among individuals. This variation is the basis of life's diversity.

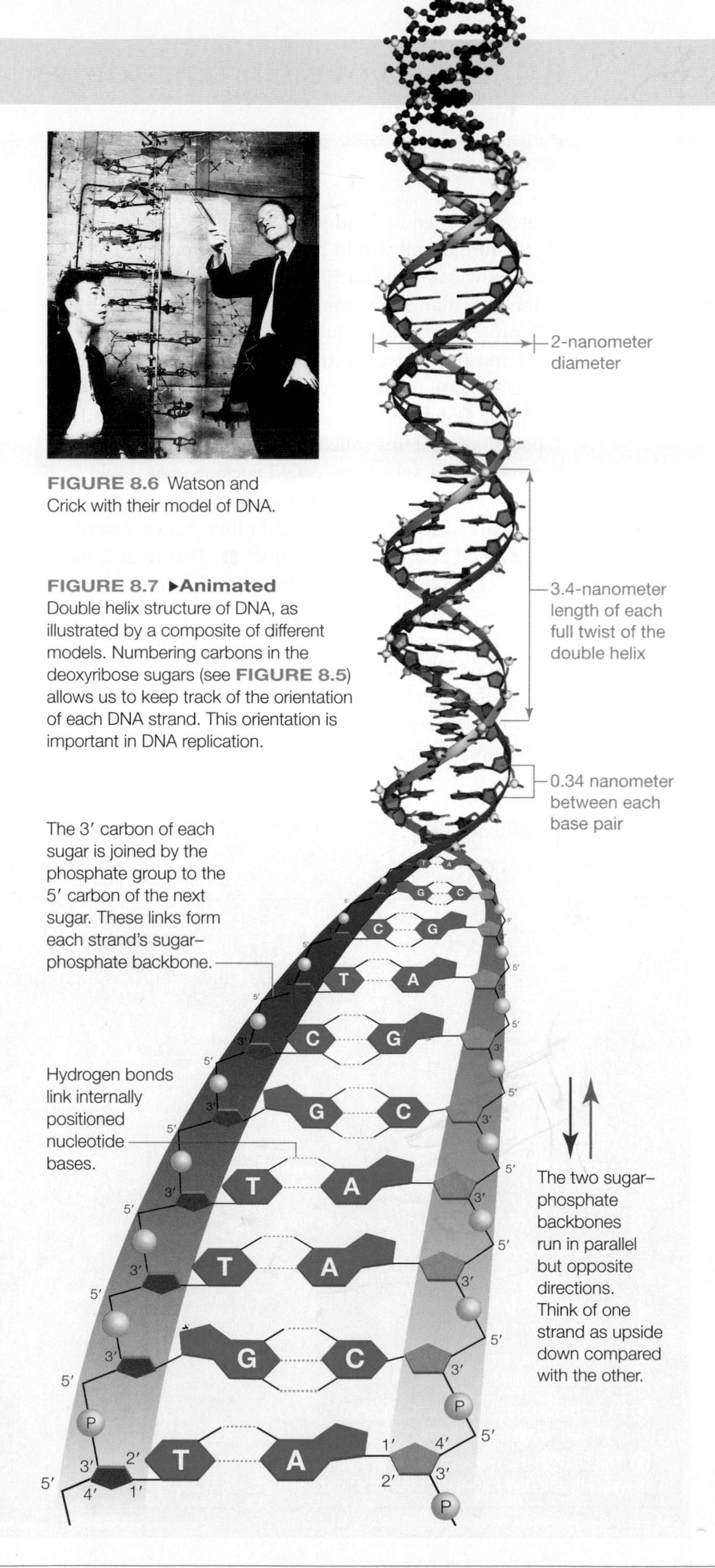

FIGURE 8.6 Watson and Crick with their model of DNA.

FIGURE 8.7 ▶Animated Double helix structure of DNA, as illustrated by a composite of different models. Numbering carbons in the deoxyribose sugars (see **FIGURE 8.5**) allows us to keep track of the orientation of each DNA strand. This orientation is important in DNA replication.

8.4 Eukaryotic Chromosomes

✔ In cells, DNA and associated proteins are organized as chromosomes.

Stretched out end to end, the DNA molecules in a single human cell would be about 2 meters (6.5 feet) long. How can that much DNA cram into a nucleus that is less than 10 micrometers in diameter? Inside a cell, proteins that associate with each DNA molecule twist and pack it into a structure called a **chromosome** (**FIGURE 8.8**). In a eukaryotic cell, for example, a DNA molecule ❶ wraps twice at regular intervals around "spools" of proteins called **histones** ❷. These DNA–histone spools, which are called **nucleosomes**, look like beads on a string in micrographs (**FIGURE 8.9**). Interactions among histones and other proteins twist the spooled DNA into a tight fiber ❸. This fiber coils, and then it coils again into a hollow cylinder a bit like an old-style telephone cord ❹.

During most of the cell's life, each chromosome consists of one DNA molecule. When the cell prepares to divide, it duplicates its chromosomes by DNA replication (more about this process in the next section). After replication, each chromosome consists of two DNA molecules, or **sister chromatids**, attached to one another at a constricted region called the **centromere**:

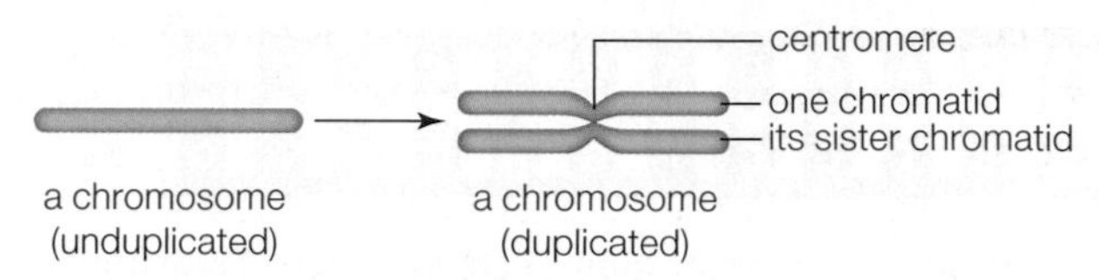

The chromosomes condense into their familiar "X" shapes ❺ just before the cell divides.

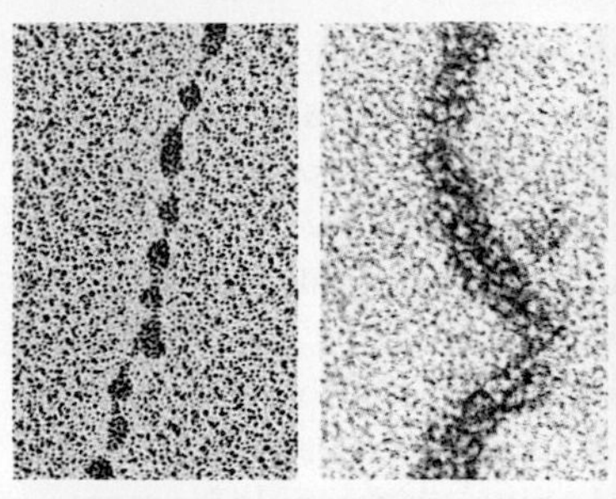
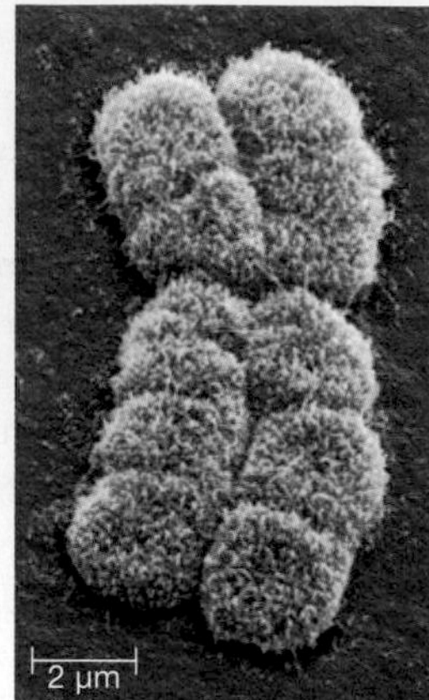

FIGURE 8.9 Chromosome packing. Left, beads-on-a string appearance of DNA–histone spools. Middle, coiled coils of a DNA fiber. Right, a duplicated chromosome just before cell division.

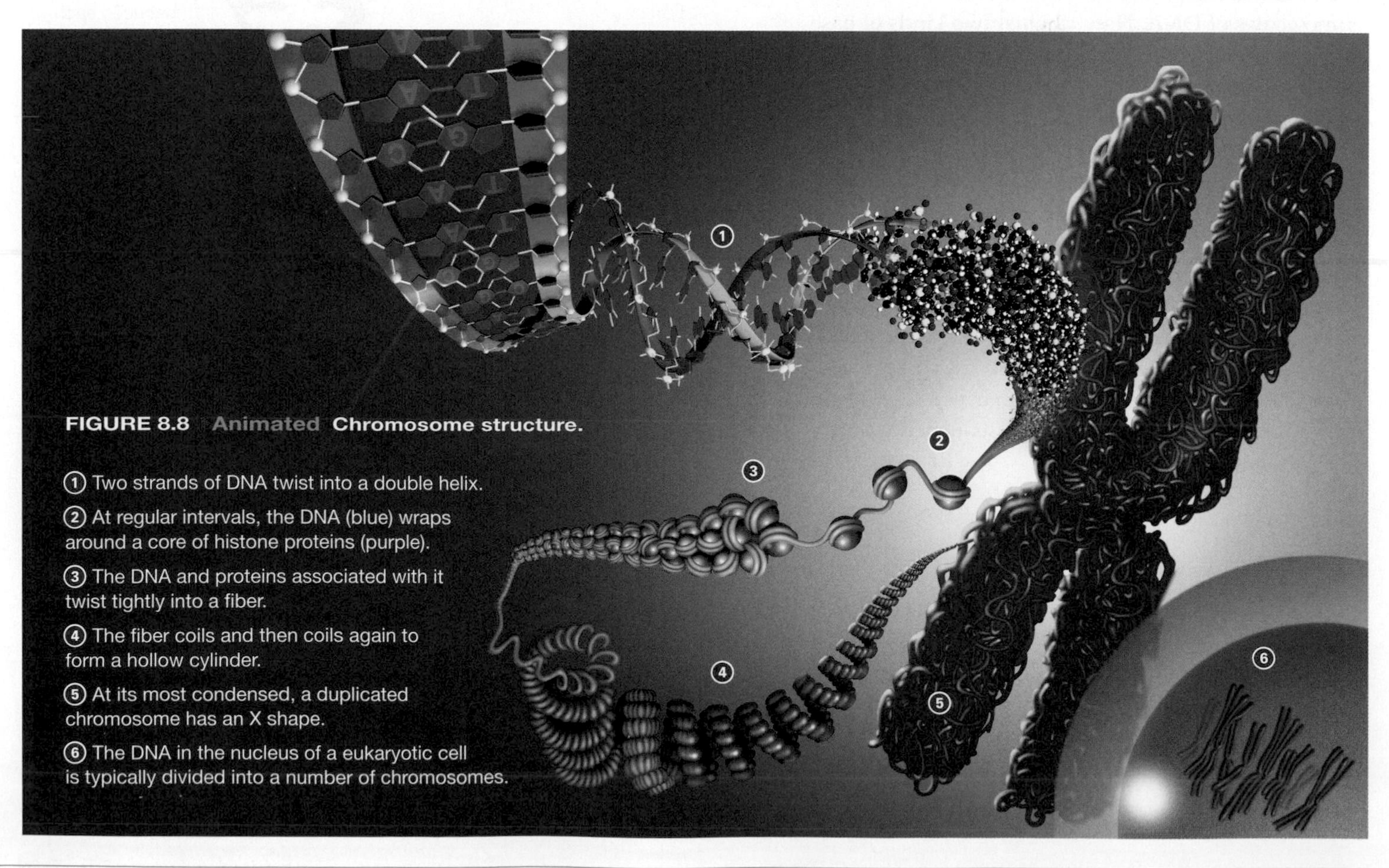

FIGURE 8.8 Animated **Chromosome structure.**

① Two strands of DNA twist into a double helix.

② At regular intervals, the DNA (blue) wraps around a core of histone proteins (purple).

③ The DNA and proteins associated with it twist tightly into a fiber.

④ The fiber coils and then coils again to form a hollow cylinder.

⑤ At its most condensed, a duplicated chromosome has an X shape.

⑥ The DNA in the nucleus of a eukaryotic cell is typically divided into a number of chromosomes.

Chromosome Number and Type

The DNA of a eukaryotic cell is divided among some number of chromosomes that differ in length and shape ❻. That number is called the **chromosome number**, and it is a characteristic of the species. For example, the chromosome number of humans is 46, so our cells have 46 chromosomes.

Actually, human body cells have two sets of 23 chromosomes—two of each type. Having two sets of chromosomes means these cells are **diploid**, or $2n$. An image of an individual's diploid set of chromosomes is called a **karyotype** (**FIGURE 8.10**). To create a karyotype, cells taken from the individual are treated to make the chromosomes condense, and then stained so the chromosomes can be distinguished under a microscope. A micrograph of a single cell is digitally rearranged so the images of the chromosomes are lined up by centromere location, and arranged according to size, shape, and length.

In a human body cell, all but one pair of chromosomes are **autosomes**, which are the same in both females and males. The two autosomes of a pair have the same length, shape, and centromere location. They also hold information about the same traits. Think of them as two sets of books on how to build a house. Your father gave you one set. Your mother had her own ideas about wiring, plumbing, and so on. She gave you an alternate set that says slightly different things about many of those tasks.

Members of a pair of **sex chromosomes** differ between females and males, and the differences determine an individual's sex. The sex chromosomes of humans are called X and Y. The body cells of typical human females have two X chromosomes (XX, **FIGURE 8.10A**); those of typical human males have one X and one Y chromosome (XY). This pattern—XX females and XY males—is the rule among fruit flies, mammals, and many other animals, but there are other patterns (**FIGURE 8.10B**). Female butterflies, moths, birds, and certain fishes have two nonidentical sex chromosomes; the two sex chromosomes of males are identical. Environmental factors (not chromosomes) determine sex in some species of invertebrates, turtles, and frogs. As an example, the temperature of the sand in which sea turtle eggs are buried determines the sex of the hatchlings.

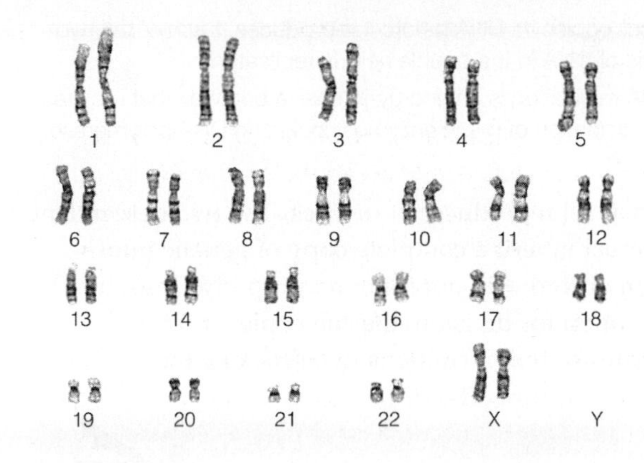

A Karyotype of a female human, with identical sex chromosomes (XX).

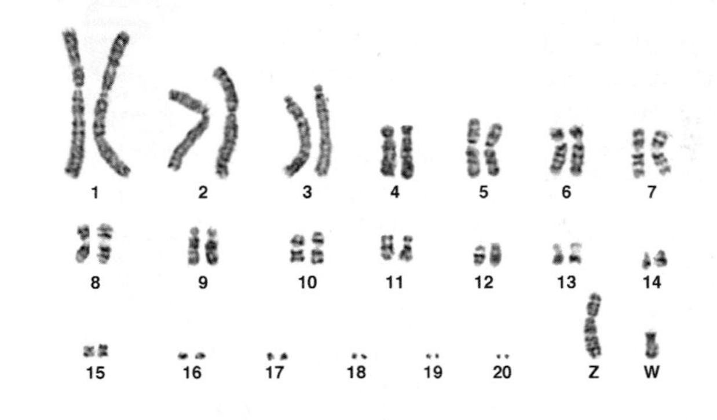

B Karyotype of a female chicken, with nonidentical sex chromosomes (ZW).

FIGURE 8.10 ▸**Animated** Karyotypes.

autosome A chromosome that is the same in males and females.
centromere Of a duplicated eukaryotic chromosome, constricted region where sister chromatids attach to each other.
chromosome A molecule of DNA together with associated proteins; carries part or all of a cell's genetic information.
chromosome number The total number of chromosomes in a cell of a given species.
diploid Having two of each type of chromosome characteristic of the species ($2n$).
histone Type of protein that associates with DNA and structurally organizes eukaryotic chromosomes.
karyotype Image of an individual's set of chromosomes arranged by size, length, shape, and centromere location.
nucleosome A length of DNA wound twice around a spool of histone proteins.
sex chromosome Member of a pair of chromosomes that differs between males and females.
sister chromatids The two attached DNA molecules of a duplicated eukaryotic chromosome.

TAKE-HOME MESSAGE 8.4
What is a chromosome?

- ✔ A chromosome is a molecule of DNA together with associated proteins that organize it and allow it to pack tightly.
- ✔ A eukaryotic cell's DNA is divided among a characteristic number of chromosomes, which differ in length and shape.
- ✔ Members of a pair of sex chromosomes differ between males and females. Chromosomes that are the same in both sexes are called autosomes.

CREDITS: (10A) © University of Washington Department of Pathology; (10B) With kind permission from Springer Science+Business Media: *Chromosome Research*, Volume 17, Number 1, 99 133, DOI: 10 1007/s10577-009-9021-6; *Avian comparative genomics reciprocal chromosome painting between domestic chicken (Gallus gallus) and the stone curlew (Burkinus oedicnemus, Charadriiformes)—An atypical species with low diploid number*, Wenhui Nie, Patricia C.M. O'Brien, Bee L. Ng. Biyuan Fu, Vitaly Volobouev, Nigel P. Carter, Malcolm A. Ferguson-Smith and Fengtang Yang; fig 2a.

8.5 DNA Replication

✔ A cell copies its DNA before it reproduces. Each of the two strands of DNA in the double helix is replicated.

✔ DNA replication is an energy-intensive pathway that requires the participation of many enzymes, including DNA polymerase.

When a cell reproduces, it divides. The two descendant cells must inherit a complete copy of genetic information or they will not function properly. Thus, in preparation for division, the cell copies its chromosomes so that it contains two sets: one for each of its future offspring.

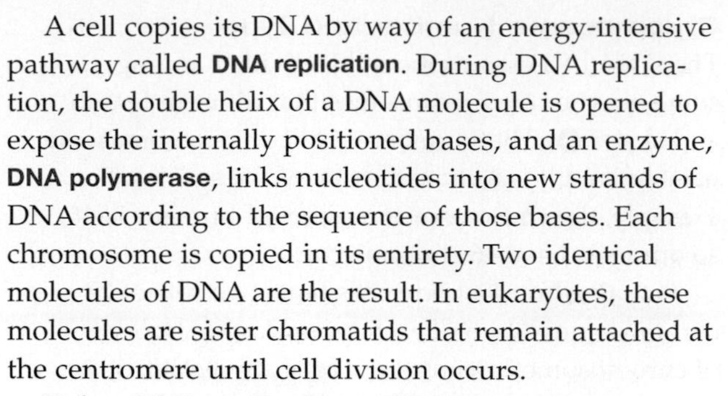

❶ As replication begins, many initiator proteins attach to the DNA at certain sites in the chromosome. Eukaryotic chromosomes have many of these origins of replication; DNA replication proceeds more or less simultaneously at all of them.

❷ Enzymes recruited by the initiator proteins begin to unwind the two strands of DNA from one another.

❸ Primers base-paired with the exposed single DNA strands serve as initiation sites for DNA synthesis.

❹ Starting at primers, DNA polymerases (green boxes) assemble new strands of DNA from nucleotides, using the parent strands as templates.

❺ DNA ligase seals any gaps that remain between bases of the "new" DNA, so a continuous strand forms.

❻ Each parental DNA strand (blue) serves as a template for assembly of a new strand of DNA (magenta). Both strands of the double helix serve as templates, so two double-stranded DNA molecules result. One strand of each is parental (old), and the other is new, so DNA replication is said to be semiconservative.

initiator proteins

topoisomerase (untwists the double helix)

helicase (breaks hydrogen bonds between bases)

primer

DNA polymerase

nucleotide

DNA ligase

FIGURE 8.11 ▶Animated DNA replication, in which a double-stranded molecule of DNA is copied in its entirety. Green arrows show the direction of synthesis for each strand. The Y-shaped structure of a DNA molecule undergoing replication is called a replication fork.

A cell copies its DNA by way of an energy-intensive pathway called **DNA replication**. During DNA replication, the double helix of a DNA molecule is opened to expose the internally positioned bases, and an enzyme, **DNA polymerase**, links nucleotides into new strands of DNA according to the sequence of those bases. Each chromosome is copied in its entirety. Two identical molecules of DNA are the result. In eukaryotes, these molecules are sister chromatids that remain attached at the centromere until cell division occurs.

Before DNA replication, a chromosome has one molecule of DNA—one double helix (**FIGURE 8.11**). As replication begins, proteins called initiators bind to certain sequences of nucleotides in the DNA ❶. Initiator proteins allow other molecules to bind, including enzymes that pry apart the two DNA strands: one (topoisomerase) that untwists the double helix, and another (helicase) that breaks the hydrogen bonds holding the double helix together. The two DNA strands begin to unwind from one another ❷. Another enzyme (primase) then starts making **primers**, which are short, single strands of nucleotides that serve as attachment points for DNA polymerase. The nucleotide bases of a primer can form hydrogen bonds with the exposed bases of a single strand of DNA ❸. Thus, a primer can base-pair with a complementary strand of DNA:

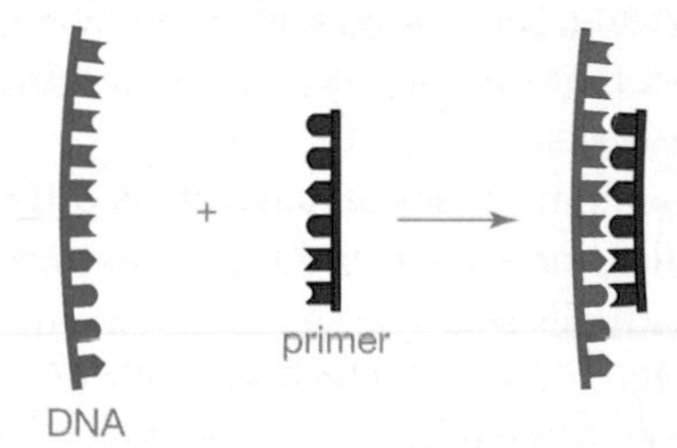

The establishment of base pairing between two strands of DNA (or DNA and RNA) is called **nucleic acid hybridization**. Hybridization is spontaneous and is driven entirely by hydrogen bonding.

DNA polymerases attach to the hybridized primers and begin DNA synthesis. As a DNA polymerase moves along a strand, it uses the sequence of exposed nucleotide bases as a template, or guide, to assemble a new strand of DNA from free nucleotides ❹. Each nucleotide provides energy for its own attachment. The bonds between a nucleotide's phosphate groups hold a lot of energy (Section 5.6). Two of three phosphate groups are removed when a nucleotide is added to a DNA strand. Breaking those bonds releases enough energy to drive the attachment.

A DNA polymerase follows base-pairing rules: It adds a T to the end of the new DNA strand when it reaches an A in the template strand; it adds a G when it reaches a C; and so on. Thus, the DNA sequence of each new strand is complementary to the template (parental) strand. The enzyme **DNA ligase** seals any gaps, so the new DNA strands are continuous ❺. Both of the two strands of the parent molecule are copied at the same time. As each new DNA strand lengthens, it winds up with the template strand into a double helix. So, after replication, two double-stranded molecules of DNA have formed ❻. One strand of each molecule is conserved (parental), and the other is new; hence the name of the process, **semiconservative replication**. Both double-stranded molecules produced by DNA replication are duplicates of the parent molecule.

Numbering the carbons of the deoxyribose sugars in nucleotides allows us to keep track of the orientation of strands in a DNA double helix (see FIGURES 8.5 and 8.7). Each strand has two ends. The last carbon atom on one end of the strand is a 5′ (5 prime) carbon of a sugar; the last carbon atom on the other end is a 3′ (three prime) carbon of a sugar:

5′ ——— 3′
3′ ——— 5′

DNA polymerase can attach a nucleotide only to a 3′ end. Thus, during DNA replication, only one of two new strands of DNA can be constructed in a single piece (FIGURE 8.12). Synthesis of the other strand occurs in segments that must be joined by DNA ligase where they meet up. This is why we say that DNA synthesis proceeds only in the 5′ to 3′ direction.

DNA ligase Enzyme that seals gaps in double-stranded DNA.
DNA polymerase DNA replication enzyme. Uses a DNA template to assemble a complementary strand of DNA.
DNA replication Process by which a cell duplicates its DNA before it divides.
nucleic acid hybridization Spontaneous establishment of base-pairing between two nucleic acid strands.
primer Short, single strand of DNA that base-pairs with a specific DNA sequence.
semiconservative replication Describes the process of DNA replication, which produces two copies of a DNA molecule: one strand of each copy is new, and the other is a strand of the original DNA.

TAKE-HOME MESSAGE 8.5

How does a cell copy its DNA?

✔ During replication of a molecule of DNA, each strand of its double helix serves as a template for synthesis of a new, complementary strand of DNA.

✔ Replication of a molecule of DNA produces two double helices that are duplicates of the parent molecule. One strand of each is parental; the other is new.

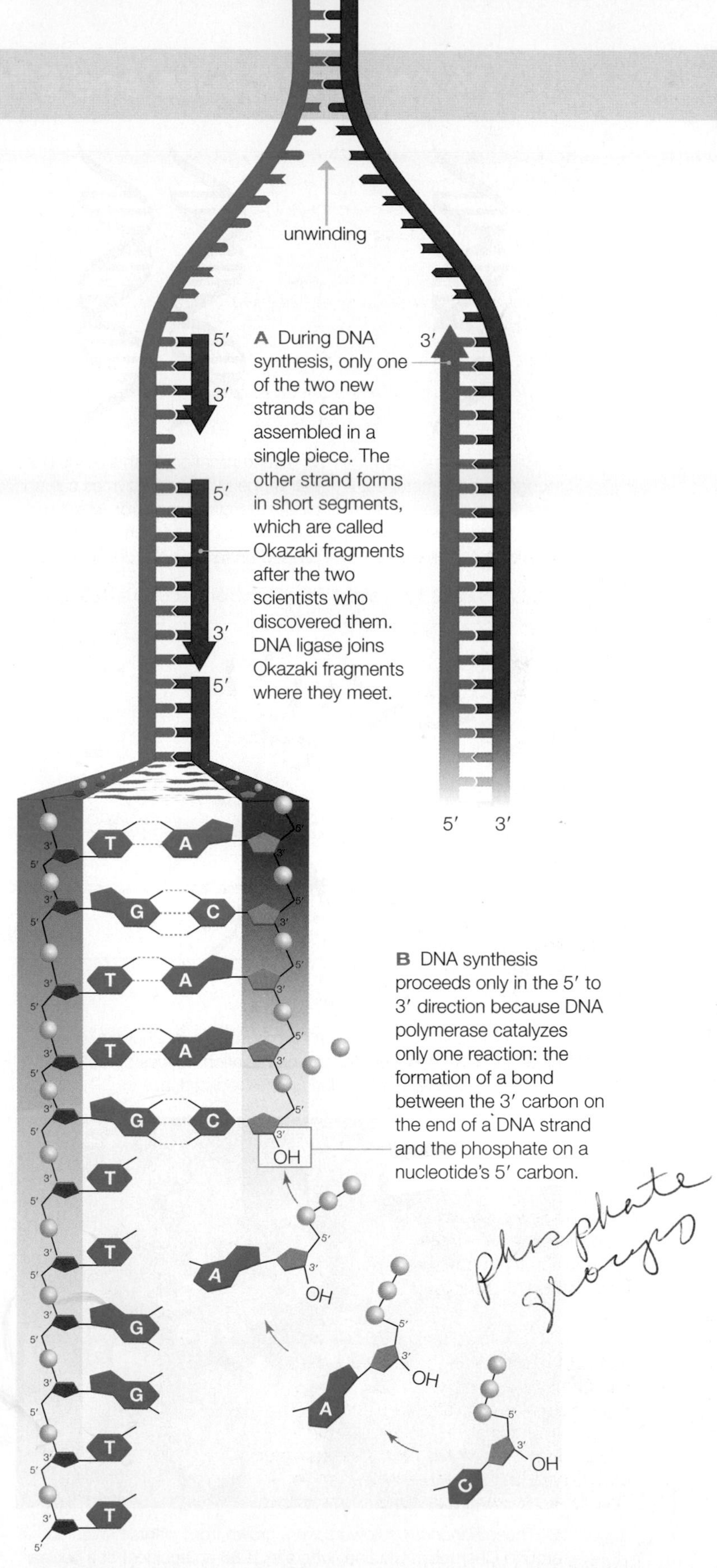

FIGURE 8.12 ▸**Animated** Discontinuous synthesis of DNA. This close-up of a replication fork shows that only one new DNA strand is assembled continuously.

FIGURE IT OUT What do the yellow balls represent? Answer: Phosphate groups

8.6 Mutations: Cause and Effect

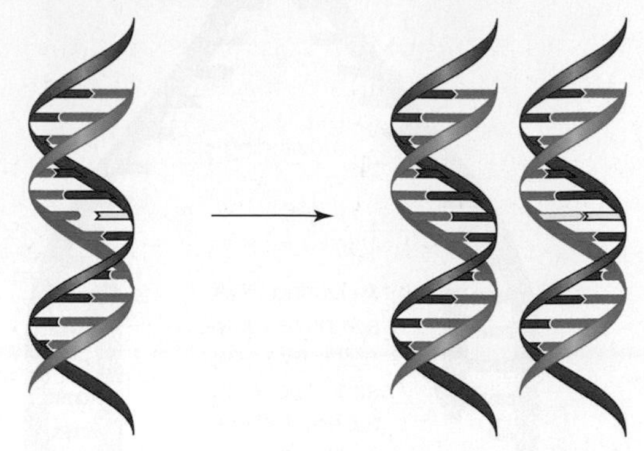

A Repair enzymes can recognize a mismatched base (yellow), but sometimes fail to correct it before DNA replication.

B After replication, both strands base-pair properly. Repair enzymes can no longer recognize the error, which has now become a mutation that will be passed on to the cell's descendants.

FIGURE 8.13 How a replication error can become a mutation.

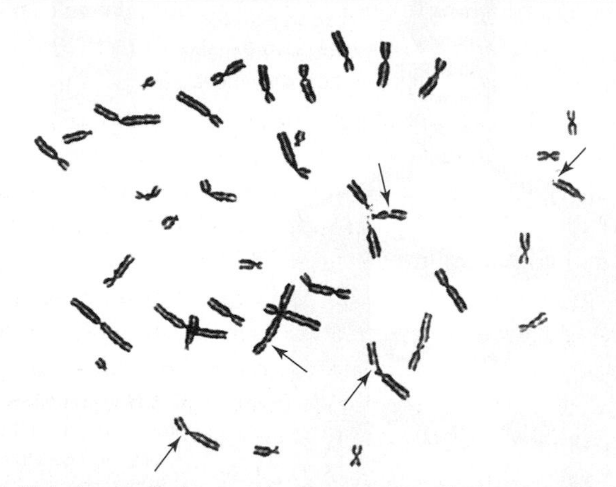

A Major breaks (red arrows) in chromosomes of a human white blood cell after exposure to ionizing radiation. Pieces of broken chromosomes can get lost during DNA replication.

B These *Ranunculus* flowers were grown from plants harvested around Chernobyl, Ukraine, where in 1986 an accident at a nuclear power plant released huge amounts of radiation. A normal flower is shown for comparison, in the inset.

FIGURE 8.14 Exposure to ionizing radiation causes mutations.

✔ Mutations, which are permanent changes in the DNA sequence of a chromosome, can be passed to descendants.

Replication Errors

Sometimes, a new DNA strand is not exactly complementary to its parent strand. A nucleotide may get lost during DNA replication, or an extra one slips in. Occasionally, the wrong nucleotide is added. Most of these replication errors occur simply because DNA polymerases work very fast. Mistakes are inevitable, and some DNA polymerases make a lot of them. Luckily, most DNA polymerases also proofread their work. They can correct a mismatch by reversing the synthesis reaction to remove the mispaired nucleotide, then resuming synthesis in the forward direction.

Replication errors also occur after the cell's DNA gets broken or otherwise damaged, because DNA polymerases do not copy damaged DNA very well. In most cases, repair enzymes and other proteins remove and replace damaged or mismatched bases in DNA before replication begins. When proofreading and repair mechanisms fail, an error becomes a **mutation**, a permanent change in the DNA sequence of a cell's chromosome(s). Repair enzymes cannot fix a mutation after the altered strand has been replicated, because they do not recognize correctly paired bases (**FIGURE 8.13**). Thus, a mutation is passed to one of the cell's offspring and all of its descendants.

Mutations can form in any type of cell. Those that occur during egg or sperm formation can be inherited by offspring, and in fact each human child is born with an average of 36 new ones. Mutations that alter DNA's instructions may have a harmful or lethal outcome; most cancers begin with them (we return to this topic in Section 11.6). However, not all mutations are dangerous. As you will see in Chapter 17, they give rise to the variation in traits that is the raw material of evolution.

Agents of DNA Damage

Electromagnetic energy with a wavelength shorter than 320 nanometers, including x-rays, most ultraviolet (UV) light, and gamma rays, has enough energy to knock electrons out of atoms. Exposure to such ionizing radiation damages DNA, breaking it into pieces that get lost during replication (**FIGURE 8.14A**). Ionizing radiation can also cause covalent bonds to form between bases on opposite strands of the double helix, an outcome that permanently blocks DNA replication. (We consider cancer-causing effects of such cell cycle interruptions in Chapter 11.) The nucleotide bases themselves can be irreparably damaged by ionizing radiation. Repair enzymes remove bases damaged in this way, but they leave an empty space in the double helix or even a strand break. Any of these events can result in mutations (**FIGURE 8.14B**).

UV light in the range of 320 to 380 nanometers does not have enough energy to knock electrons out of atoms. However, it has enough energy to open up the double bond in the ring of a cytosine or thymine base. The open ring can form a covalent bond with the ring of an adjacent pyrimidine (left), forming a dimer that kinks the DNA strand. A specific set of proteins can recognize, remove, and replace pyrimidine dimers before replication begins. Mutations arise if they do not, because DNA polymerase tends to copy kinked DNA incorrectly. Mutations that arise as a result of pyrimidine dimers are the cause of most skin cancers. Exposing unprotected skin to sunlight increases the risk of cancer because UV wavelengths in the light cause dimers to form. For every second a skin cell spends in the sun, 50 to 100 pyrimidine dimers form in its DNA.

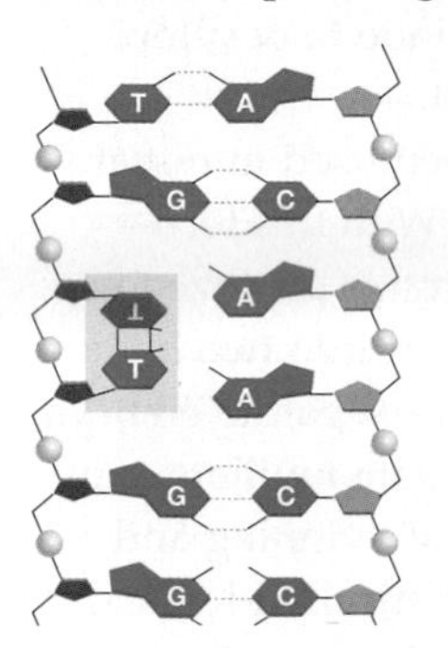

a thymine dimer

Exposure to some chemicals also causes mutations. For instance, several of the fifty-five or more cancer-causing chemicals in tobacco smoke transfer methyl groups (—CH_3) to the nucleotide bases in DNA. Nucleotides altered in this way do not base-pair correctly. The body converts other chemicals in the smoke to compounds that bind irreversibly to DNA. In both cases, the resulting replication errors can lead to mutation. Cigarette smoke also contains free radicals, which inflict the same damage on DNA as ionizing radiation.

Rosalind Franklin, X-Rays, and Cancer

Rosalind Franklin arrived at King's College, London, in 1951. The expert x-ray crystallographer had been told she would be the only one in her department working on the structure of DNA, so she did not know that Maurice Wilkins was already doing the same thing just down the hall. No one had told Wilkins about Franklin's assignment; he assumed she was a technician hired to do his x-ray crystallography work. And so a clash began. Wilkins thought Franklin displayed an appalling lack of deference that technicians of the era usually accorded researchers. To Franklin, Wilkins seemed prickly and oddly overinterested in her work.

mutation Permanent change in the DNA sequence of a chromosome.

Wilkins and Franklin had been given identical samples of DNA. Franklin's meticulous work with hers yielded the first clear x-ray diffraction image of DNA as it occurs in cells. She gave a presentation on her work in 1952. DNA, she said, had two chains twisted into a double helix, with a backbone of phosphate groups on the outside, and bases arranged in an unknown way on the inside. She had calculated DNA's diameter, the distance between its chains and between its bases, the angle of the helix, and the number of bases in each coil. Crick, with his crystallography background, would have recognized the significance of the work—if he had been there. Watson was in the audience but he was not a crystallographer, and he did not understand the implications of Franklin's data.

Franklin started to write a research paper on her findings. Meanwhile, and perhaps without her knowledge, Watson reviewed Franklin's x-ray diffraction image with Wilkins, and Watson and Crick read a report detailing Franklin's unpublished data. Crick, who had more experience with molecular modeling than Franklin, immediately understood what the image and the data meant. Watson and Crick used that information to build their model of DNA.

On April 25, 1953, Franklin's paper appeared third in a series of articles about the structure of DNA in the journal *Nature*. It supported with solid experimental evidence Watson and Crick's theoretical model, which appeared in the first article of the series.

Rosalind Franklin (left) died in 1958, at the age of 37, of ovarian cancer probably caused by extensive exposure to x-rays during her work. At the time, the link between x-rays, mutations, and cancer was not understood. Because the Nobel Prize is not given posthumously, Franklin did not share in the 1962 honor that went to Watson, Crick, and Wilkins for the discovery of the structure of DNA.

TAKE-HOME MESSAGE 8.6

What are mutations?

✔ Proofreading and repair mechanisms usually maintain the integrity of a cell's genetic information by correcting mispaired bases and fixing damaged DNA before replication.

✔ Mismatched or damaged nucleotides that are not repaired can become mutations—permanent changes in the DNA sequence of a chromosome.

✔ DNA damage by environmental agents such as UV light, chemicals, and free radicals can result in mutations, because damaged DNA is not replicated very well.

8.7 Cloning Adult Animals

✔ Various reproductive interventions produce genetically identical individuals.

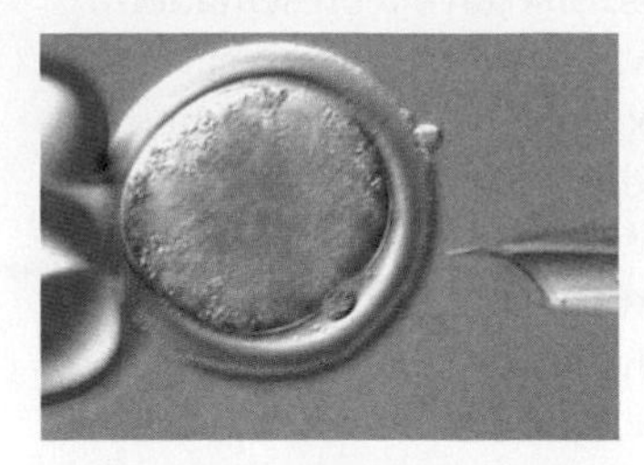

A A cow's egg is held in place by suction through a hollow glass tube called a micropipette. DNA is identified by a purple stain.

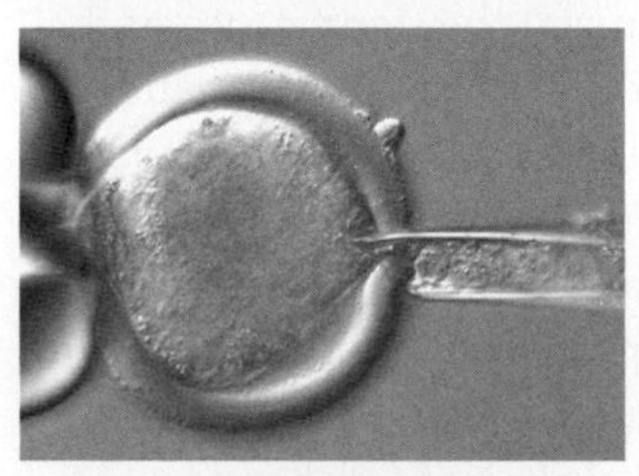

B Another micropipette punctures the egg and sucks out the DNA. All that remains inside the egg's plasma membrane is cytoplasm.

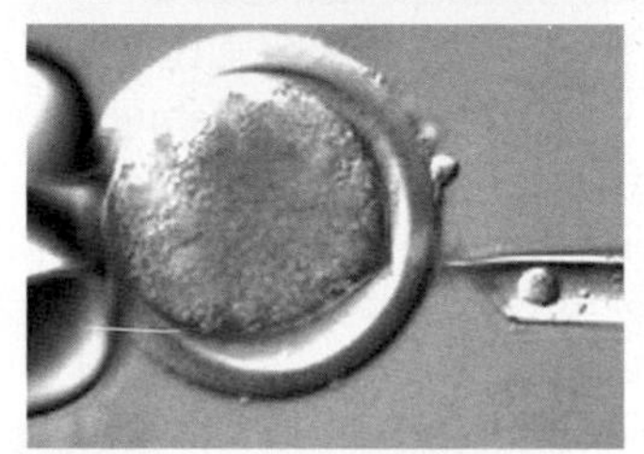

C A new micropipette prepares to enter the egg at the puncture site. The pipette contains a cell grown from the skin of a donor animal.

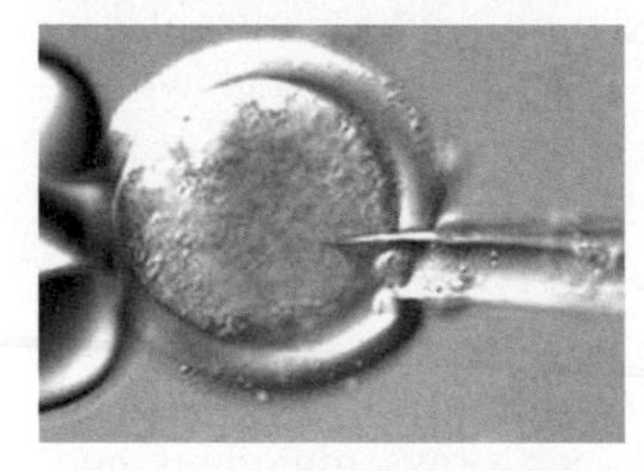

D The micropipette enters the egg and delivers the skin cell to a region between the cytoplasm and the plasma membrane.

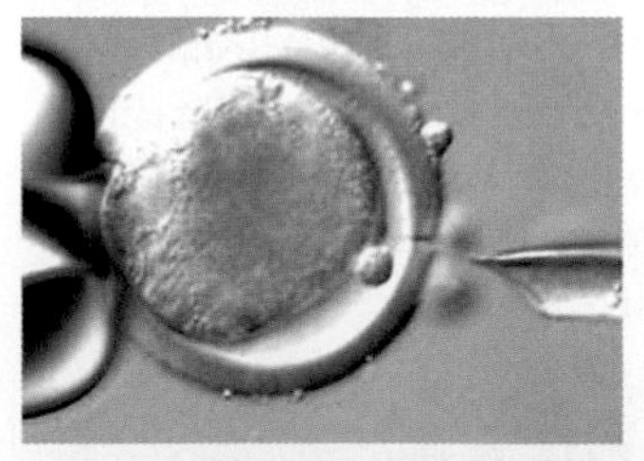

E After the pipette is withdrawn, the donor's skin cell is visible next to the cytoplasm of the egg. The transfer is now complete.

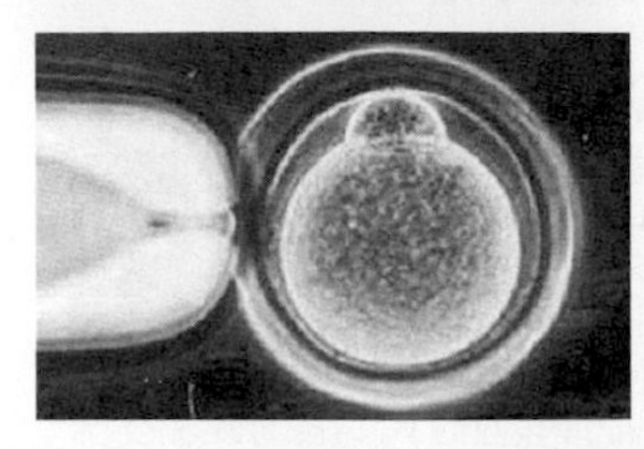

F An electric current causes the foreign cell to fuse with and deposit its nucleus into the cytoplasm of the egg. The egg begins to divide, and an embryo forms.

FIGURE 8.15 ▶Animated Somatic cell nuclear transfer, using cattle cells. This series of micrographs was taken by scientists at Cyagra, a company that specializes in cloning livestock.

The word "cloning" means making an identical copy of something, and it can refer to deliberate interventions in reproduction that produce an exact genetic copy of an organism. Genetically identical organisms occur all the time in nature, arising most often by the process of asexual reproduction (which we discuss in Chapter 11). Embryo splitting, another natural process, results in identical twins. The first few divisions of a fertilized egg form a ball of cells that sometimes splits spontaneously. If both halves of the ball continue to develop independently, identical twins result.

Artificial embryo splitting has been used in research and animal husbandry for decades. With this technique, a tiny ball of cells is grown from a fertilized egg in a laboratory. The ball is teased apart into two halves, each of which goes on to develop as a separate embryo. The embryos are implanted in surrogate mothers, who give birth to identical twins. Artificial twinning and any other technology that yields genetically identical individuals is called **reproductive cloning**.

Twins get their DNA from two parents that typically differ in their DNA sequence. Thus, although twins produced by embryo splitting are identical to one another, they are not identical to either parent. Animal breeders who want an exact copy of a specific individual may turn to a cloning method that starts with a somatic cell taken from an adult organism (a somatic cell is a body cell, as opposed to a reproductive cell; *soma* is a Greek word for body). All cells descended from a fertilized egg inherit the same DNA. Thus, the DNA in each living cell of an individual is like a master blueprint that contains enough information to build an entirely new individual.

Even though a somatic cell contains all the DNA needed to produce a new individual, it will not automatically start dividing and form an embryo. The cell must first be tricked into rewinding its developmental clock. During development, cells in an embryo start using different subsets of their DNA. As they do, the cells become different in form and function, a process called **differentiation**. Differentiation is usually a one-way path in animal cells. Once a cell has become specialized, all of its descendant cells will be specialized the same way.

By the time a liver cell, muscle cell, or other differentiated cell forms, most of its DNA has been turned off, and is no longer used. To clone an adult, scientists transform one of its differentiated cells into an undifferentiated cell by turning the unused DNA back on. One way to do this is **somatic cell nuclear transfer (SCNT)**, a laboratory procedure in which an unfertil-

CREDIT: (15) Courtesy of Cyagra, Inc., www.cyagra.com.

A Hero Dog's Golden Clones (revisited)

Today, Trakr's clones (right) are search and rescue dogs for Team Trakr Foundation, an international humanitarian organization that Symington established in 2010. The ability to clone dogs is a recent development, but the technique is not. SCNT (the technique used to produce Trakr's clones) first made headlines in 1997, when Scottish geneticist Ian Wilmut and his team produced a clone from the udder cell of an adult sheep. The cloned lamb was named Dolly. At first, Dolly looked and acted like a normal sheep. However, she died early, suffering from health problems that were most likely an outcome of being a clone.

SCNT has also been used to clone mice, rats, rabbits, pigs, cattle, goats, sheep, horses, mules, deer, cats, a camel, a ferret, a monkey, and a wolf. Until recently, many of the clones were unusually overweight or had enlarged organs. Cloned mice developed lung and liver problems, and almost all died prematurely. Cloned pigs tended to limp and have heart problems; some developed without a tail or, even worse, an anus. SCNT technology has improved so much in recent years that such problems are much less common in animals cloned today.

---

ized egg's nucleus is replaced with the nucleus of a donor's somatic cell (**FIGURE 8.15**). If all goes well, the egg's cytoplasm reprograms the transplanted DNA to direct the development of an embryo, which is then implanted into a surrogate mother. The animal born to the surrogate is the donor's clone, genetically identical with the donor of the nucleus.

SCNT is now a common practice among people who breed prized livestock. Among other benefits, many more offspring can be produced in a given time frame by cloning than by traditional breeding methods. Cloned animals have the same championship features as their DNA donors (**FIGURE 8.16**). Offspring can also be produced from a donor animal that is castrated or even dead.

As the techniques become routine, cloning humans is no longer only within the realm of science fiction. SCNT is already being used to produce human embryos for medical purposes, a practice called **therapeutic cloning**. Undifferentiated (stem) cells taken from the cloned human embryos are used to treat human patients and to study human diseases. For example, embryos created using cells from people with genetic heart defects are allowing researchers to study how the defect causes developing heart cells to malfunction. Such research may ultimately lead to treatments for people who suffer from diseases that are otherwise incurable. (We return to the topic of stem cells and their potential medical benefits in Chapter 31.) Human cloning is not the intent of the research, but if it were, SCNT would indeed be the first step toward that end.

FIGURE 8.16 Champion Holstein dairy cow Nelson's Estimate Liz (right) and her clone, Nelson's Estimate Liz II (left). Liz II was produced by somatic cell nuclear transfer in 2003. She had already begun to win championships by the time she was one year old.

TAKE-HOME MESSAGE 8.7

How are clones of adult animals produced?

✔ Reproductive cloning technologies produce genetically identical individuals.

✔ The DNA inside a living cell contains all the information necessary to build a new individual.

✔ In somatic cell nuclear transfer (SCNT), the nucleus of a donor's somatic cell is transferred to an egg with no nucleus. The donor DNA directs the cell to develop into an embryo that is implanted into a surrogate. The individual born to the surrogate is a clone of the adult donor.

differentiation Process by which cells become specialized during development; occurs as different cells in an embryo begin to use different subsets of their DNA.
reproductive cloning Any of several technologies that produce genetically identical individuals.
somatic cell nuclear transfer (SCNT) Reproductive cloning method in which the DNA of an adult donor's body cell is transferred into an unfertilized egg.
therapeutic cloning The use of SCNT to produce human embryos for research purposes.

summary

Section 8.1 Making **clones**, or exact genetic copies, of adult animals is now a common practice. The techniques are improving, but their outcome is still unpredictable. The practice continues to raise ethical questions, particularly about cloning human cells.

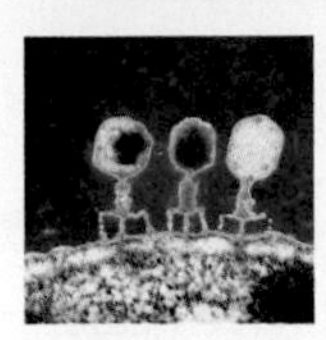

Section 8.2 Eighty years of experimentation with cells and **bacteriophage** offered solid evidence that deoxyribonucleic acid (DNA), not protein, is the hereditary material of all life.

Section 8.3 Each DNA nucleotide has a five-carbon sugar (deoxyribose), three phosphate groups, and one of four nitrogen-containing bases after which the nucleotide is named: adenine, thymine, guanine, or cytosine. DNA is a polymer that consists of two strands of these nucleotides coiled into a double helix. Hydrogen bonding between the internally positioned bases holds the strands together. The bases pair in a consistent way: adenine with thymine (A–T), and guanine with cytosine (G–C). The order of bases along a strand of DNA—the **DNA sequence**—varies among species and among individuals, and this variation is the basis of life's diversity.

Section 8.4 The DNA of eukaryotes is typically divided among a number of **chromosomes** that differ in length and shape. In eukaryotic chromosomes, the DNA wraps around **histones** to form **nucleosomes**. When duplicated, a eukaryotic chromosome consists of two **sister chromatids** attached at a **centromere**.

Diploid cells have two of each type of chromosome. **Chromosome number** is the total number of chromosomes in a cell of a given species. A human body cell has twenty-three pairs of chromosomes. Members of a pair of **sex chromosomes** differ among males and females. Chromosomes that are the same in males and females are **autosomes**. Autosomes of a pair have the same length, shape, and centromere location. A **karyotype** is an image of an individual's complete set of chromosomes.

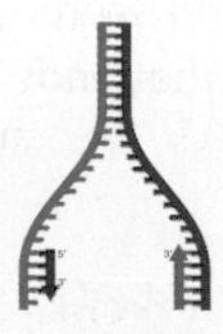

Section 8.5 **DNA replication** is the energy-intensive metabolic pathway in which a cell copies its chromosomes. For each molecule of DNA that is copied, two DNA molecules are produced; each is a duplicate of the parent. One strand of each molecule is new, and the other is parental; thus the name **semiconservative replication**.

During DNA replication, enzymes unwind the double helix. **Primers** base-pair with the exposed single strands of DNA, a process called **nucleic acid hybridization**. Starting at the primers, **DNA polymerase** enzymes use each strand as a template to assemble new, complementary strands of DNA from free nucleotides. Synthesis of one strand necessarily occurs discontinuously. **DNA ligase** seals any gaps to form continuous strands.

Section 8.6 Proofreading by DNA polymerases corrects most DNA replication errors as they occur. DNA damage by environmental agents, including ionizing and nonionizing radiation, free radicals, and some chemicals, can lead to replication errors because DNA polymerase does not copy damaged DNA very well. Most DNA damage is repaired before replication begins. Uncorrected replication errors become **mutations**: permanent changes in the nucleotide sequence of a cell's DNA. Cancer begins with mutations, but not all mutations are harmful.

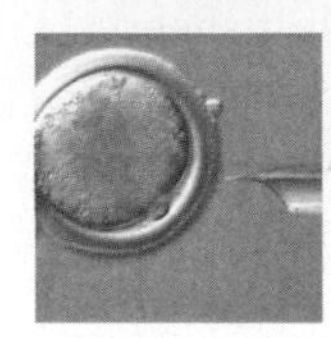

Section 8.7 **Somatic cell nuclear transfer (SCNT)** and other types of **reproductive cloning** technologies produce genetically identical individuals (clones). SCNT using human cells is called **therapeutic cloning**. The DNA in each living cell contains all the information necessary to build a new individual. During development, cells of an embryo become specialized as they begin to use different subsets of their DNA (a process called **differentiation**).

self-quiz

Answers in Appendix VII

1. Which is *not* a nucleotide base in DNA?
   a. adenine
   b. guanine
   c. glutamine
   d. thymine
   e. cytosine
   f. All are in DNA.

2. What are the base-pairing rules for DNA?
   a. A–G, T–C
   b. A–C, T–G
   c. A–T, G–C

3. Variation in ______ is the basis of variation in traits.
   a. karyotype
   b. the DNA sequence
   c. the double helix
   d. chromosome number

4. One species' DNA differs from others in its ______ .
   a. nucleotides
   b. DNA sequence
   c. sugar–phosphate backbone
   d. all of the above

5. In eukaryotic chromosomes, DNA wraps around ______ .
   a. histone proteins
   b. nucleosomes
   c. centromeres
   d. none of the above

6. Chromosome number ______ .
   a. refers to a particular chromosome in a cell
   b. is a characteristic feature of a species
   c. is the number of autosomes in cells of a given type

7. Human body cells are diploid, which means ______ .
   a. they are complete
   b. they have two sets of chromosomes
   c. they contain sex chromosomes

8. When DNA replication begins, ______ .
   a. the two DNA strands unwind from each other
   b. the two DNA strands condense for base transfers
   c. old strands move to find new strands

9. DNA replication requires ______ .
   a. DNA polymerase
   b. nucleotides
   c. primers
   d. all are required

data analysis activities

Hershey–Chase Experiments The graph shown in **FIGURE 8.17** is reproduced from an original publication by Hershey and Chase. The data are from the experiments described in Section 8.2, in which bacteriophage DNA and protein were labeled with radioactive tracers and allowed to infect bacteria. The virus–bacteria mixtures were then whirled in a blender to dislodge any viral components attached to the exterior of the bacteria. Afterward, radioactivity from the tracers was measured.

1. Before blending, what percentage of each isotope, ^{35}S and ^{32}P, was extracellular (outside the bacteria)?
2. After 4 minutes in the blender, what percentage of each isotope was extracellular?
3. How did the researchers know that the radioisotopes in the fluid came from outside of the bacterial cells and not from bacteria that had been broken apart by whirling in the blender?
4. The extracellular concentration of which isotope increased the most with blending?
5. Do these results imply that viruses inject DNA or protein into bacteria? Why or why not?

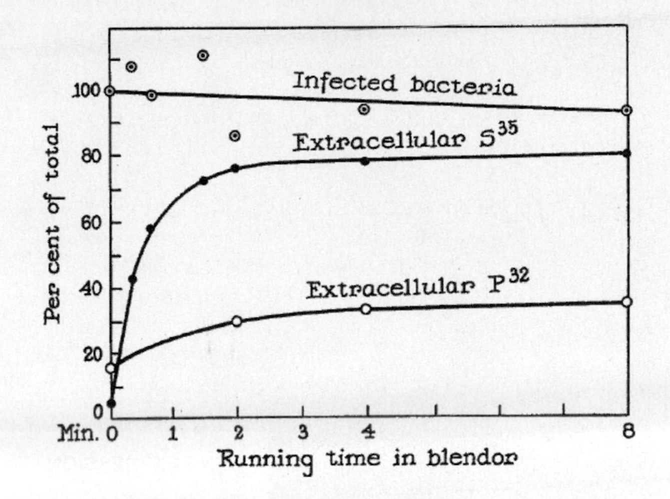

FIGURE 8.17 Detail of Alfred Hershey and Martha Chase's 1952 publication describing their experiments with bacteriophage. "Infected bacteria" refers to the percentage of bacteria that survived the blender.

10. Energy that drives the attachment of a nucleotide to the end of a growing strand of DNA comes from ________ .
    a. ATP c. the nucleotide
    b. DNA polymerase d. a and c

11. The phrase "5′ to 3′" refers to the ________ .
    a. timing of DNA replication
    b. directionality of DNA synthesis
    c. number of phosphate groups

12. After DNA replication, a eukaryotic chromosome ______ .
    a. consists of two sister chromatids
    b. has a characteristic X shape
    c. is constricted at the centromere
    d. all of the above

13. All mutations ________ .
    a. cause cancer c. are caused by radiation
    b. lead to evolution d. change the DNA sequence

14. ________ is an example of reproductive cloning.
    a. Somatic cell nuclear transfer (SCNT)
    b. Multiple offspring from the same pregnancy
    c. Artificial embryo splitting
    d. a and c

15. Match the terms appropriately.

___ bacteriophage	a. nitrogen-containing base, sugar, phosphate group(s)
___ clone	b. copy of an organism
___ nucleotide	c. does not determine sex
___ diploid	d. injects DNA
___ DNA ligase	e. seals breaks in a DNA strand
___ DNA polymerase	f. can cause cancer
___ autosome	g. two chromosomes of each type
___ mutation	h. adds nucleotides to a growing DNA strand

critical thinking

1. Show the complementary strand of DNA that forms on this template DNA fragment during replication:

   5′—GGTTTCTTCAAGAGA—3′

2. Woolly mammoths have been extinct for about 10,000 years, but we often find their well-preserved remains in Siberian permafrost. Research groups are now planning to use SCNT to resurrect these huge elephant-like mammals. No mammoth eggs have been recovered so far, so elephant eggs would be used instead. An elephant would also be the surrogate mother for the resulting embryo. The researchers may try a modified SCNT technique used to clone a mouse that had been dead and frozen for sixteen years. Ice crystals that form during freezing break up cell membranes, so cells from the frozen mouse were in bad shape. Their DNA was transferred into donor mouse eggs, and cells from the resulting embryos were fused with mouse stem cells. Four healthy clones were born from the hybrid embryos. What are some of the pros and cons of cloning an extinct animal?

3. Xeroderma pigmentosum is an inherited disorder characterized by rapid formation of many skin sores that develop into cancers. All forms of radiation trigger these symptoms, including fluorescent light, which contains UV light in the range of 320 to 400 nm. In most affected individuals, at least one of nine particular proteins is missing or defective. What is the collective function of these proteins?

CREDIT: (17) © *The Journal of General Physiology*, (36(1), Sept. 20, 1952.

9 From DNA to Protein

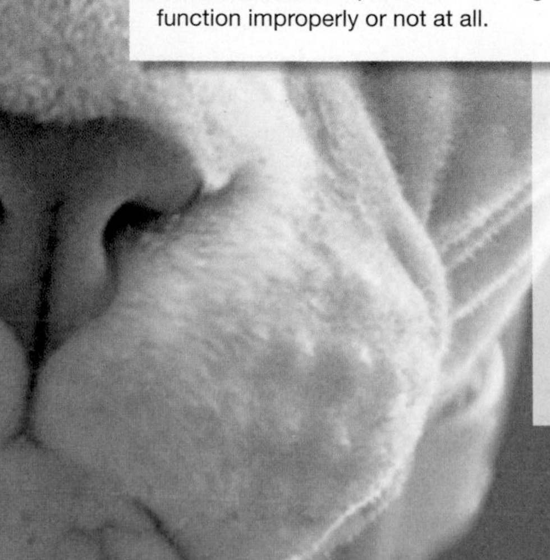

LEARNING ROADMAP

Your knowledge of base pairing (Section 8.3) and chromosomes (8.4) will help you understand how cells use nucleic acids (3.8) to build proteins (3.6). This chapter revisits hydrophobicity (2.5), hemoglobin (3.2), pathogenic bacteria (4.1), organelles (4.5, 4.7), free radicals and cofactors (5.6), enzyme function (3.3, 5.4), DNA replication (8.5), and mutations (8.6).

GENE EXPRESSION

The information encoded in DNA occurs in subsets called genes. Converting genetic information to a protein product involves RNA, and it occurs in two stages: transcription and translation.

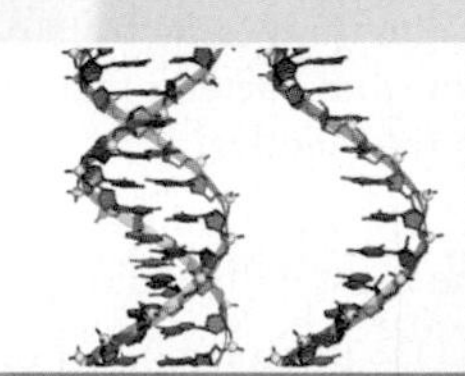

DNA TO RNA: TRANSCRIPTION

During transcription, a gene region in one strand of DNA serves as a template for assembling a strand of RNA. In eukaryotes, a new RNA is modified before leaving the nucleus.

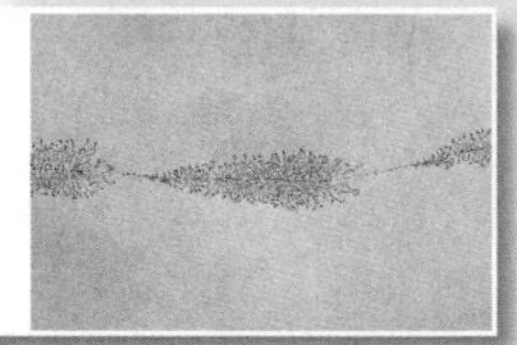

RNA

A messenger RNA carries a gene's protein-building instructions as a string of three-nucleotide codons. Transfer RNA and ribosomal RNA translate those instructions into a protein.

RNA TO PROTEIN: TRANSLATION

During translation, amino acids are assembled into a polypeptide in the order determined by the sequence of codons in an mRNA.

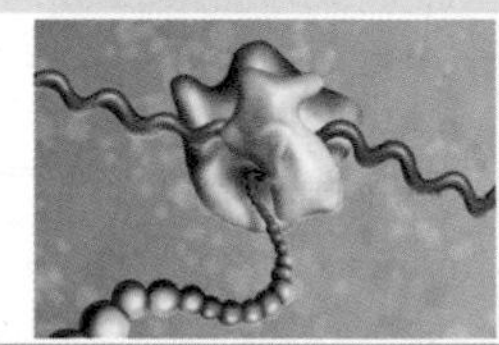

ALTERED PROTEINS

Mutations that change a gene's DNA sequence alter the instructions it encodes. A protein built using altered instructions may function improperly or not at all.

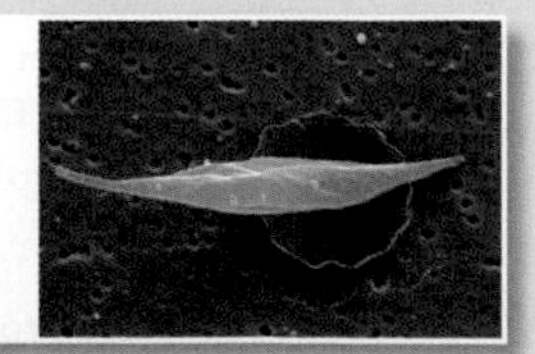

What you learn in this chapter about genes will be the foundation for concepts of gene expression (Chapter 10), inheritance (Chapters 13 and 14), and genetic engineering (Chapter 15). Chapters 16 and 17 will show you how mutations are the raw material of natural selection and other processes of evolution. You will also revisit hemoglobin, sickle-cell anemia, and the circulatory system in Chapter 36, and immunity in Chapter 37.

9.1 Ricin, RIP

A dose of ricin as small as a few grains of salt can kill an adult human, and there is no antidote. Ricin is a protein that deters beetles, birds, mammals, and other animals from eating seeds of the castor-oil plant (*Ricinus communis*), which grows wild in tropical regions and is widely cultivated. Castor-oil seeds are the source of castor oil, an ingredient in plastics, cosmetics, paints, soaps, polishes, and many other items. After the oil is extracted from the seeds, the ricin typically is discarded along with the leftover seed pulp.

Ricin's lethal effects were known as long ago as 1888, but using it as a weapon is now banned by most countries under the Geneva Protocol. However, controlling the production of ricin is impossible, because no special skills or equipment are required to manufacture the toxin from easily obtained raw materials. Thus, ricin appears periodically in the news as a tool of criminals. For example, a Texas actress sent ricin-laced letters to President Obama and the mayor of New York City in 2013. Perhaps the most famous example occurred in 1978, at the height of the Cold War when defectors from countries under Russian control were targets for assassination. Bulgarian journalist Georgi Markov had defected to England and was working for the BBC. As he made his way to a bus stop on a London street, an assassin used a modified umbrella to fire a tiny, ricin-laced ball into Markov's leg. Markov died in agony three days later.

Ricin is called a ribosome-inactivating protein (RIP) because it inactivates ribosomes, the organelles that assemble amino acids into proteins. Other RIPs are made by some bacteria, mushrooms, algae, and many plants (including food crops such as tomatoes, barley, and spinach). Most of these proteins are not particularly toxic in humans because they do not cross intact cell membranes very well. Those that do, including ricin, have a domain that binds tightly to plasma membrane glycolipids or glycoproteins (**FIGURE 9.1**). Binding causes the cell to take up the RIP by endocytosis (Section 5.10). Once inside the cell, the second domain of the RIP—an enzyme—begins to inactivate ribosomes. One molecule of ricin can inactivate more than 1,000 ribosomes per minute. If enough ribosomes are affected, protein synthesis grinds to a halt. Proteins are critical to all life processes, so cells that cannot make them die quickly.

Fortunately, few people actually encounter ricin. Other toxic RIPs are more prevalent. Bracelets made from beautiful seeds were recalled from stores in 2011 after a botanist recognized the seeds as jequirity beans.

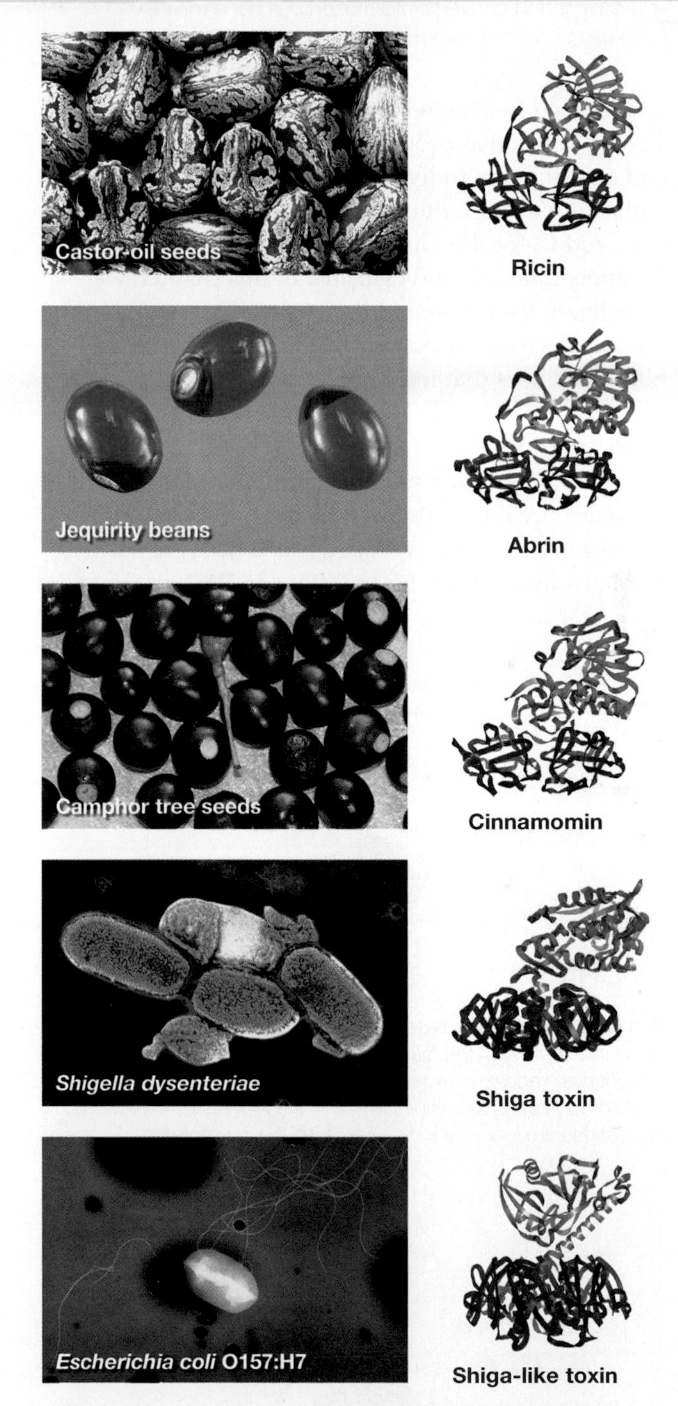

FIGURE 9.1 Lethal lineup: a few toxic ribosome-inactivating proteins (RIPs) and their sources. One of the two chains of a toxic RIP (brown) helps the molecule cross a cell's plasma membrane; the other (gold) is an enzyme that inactivates ribosomes.

These beans contain abrin, an RIP even more toxic than ricin. Shiga toxin made by *Shigella dysenteriae* bacteria causes dysentery. Some strains of *E. coli* bacteria make Shiga-like toxin, an RIP that is the source of intestinal illness (Section 4.1).

9.2 DNA, RNA, and Gene Expression

✔ Transcription converts information in a gene to RNA; translation converts information in an mRNA to protein.

You learned in Chapter 8 that chromosomes are like a set of books that provide instructions for building and operating an individual. You already know the alphabet used to write those books: the four letters A, T, G, and C, for the four nucleotides in DNA—adenine, thymine, guanine, and cytosine. In this chapter, we investigate the nature of information represented by the sequence of nucleotides in a DNA molecule, and how a cell uses that information.

DNA to RNA

Information encoded within a chromosome's DNA sequence occurs in hundreds or thousands of units called genes. The DNA sequence of a **gene** encodes (contains instructions for building) an RNA or protein product. Converting the information encoded by a gene into a product starts with RNA synthesis, or transcription. During **transcription**, enzymes use the gene's DNA sequence as a template to assemble a strand of RNA:

$$\textbf{DNA} \xrightarrow{\textit{transcription}} \textbf{RNA}$$

Most of the RNA inside cells occurs as a single strand that is similar in structure to a single strand of DNA. For example, both are chains of four kinds of nucleotides. Like a DNA nucleotide, an RNA nucleotide has three phosphate groups, a sugar, and one of four bases. However, the sugar in an RNA nucleotide is a ribose, which differs just a bit from deoxyribose, the sugar in a DNA nucleotide (**FIGURE 9.2**). Three bases (adenine, cytosine, and guanine) occur in both RNA and DNA nucleotides, but the fourth base differs between the two molecules (**FIGURE 9.3**). In DNA, the fourth base is thymine (T); in RNA, it is uracil (U).

Despite these small differences in structure, DNA and RNA have very different functions. DNA's important but only role is to store a cell's genetic information. By contrast, a cell makes several kinds of RNAs, each with a different function. MicroRNAs are important in gene control, which is the subject of the next chapter. Three other types of RNA have roles in protein synthesis. **Ribosomal RNA (rRNA)** is the main component of ribosomes (Section 4.4), which assemble amino acids into polypeptide chains (Section 3.6). **Transfer RNA (tRNA)** delivers amino acids to ribosomes, one by one, in the order specified by a **messenger RNA (mRNA)**.

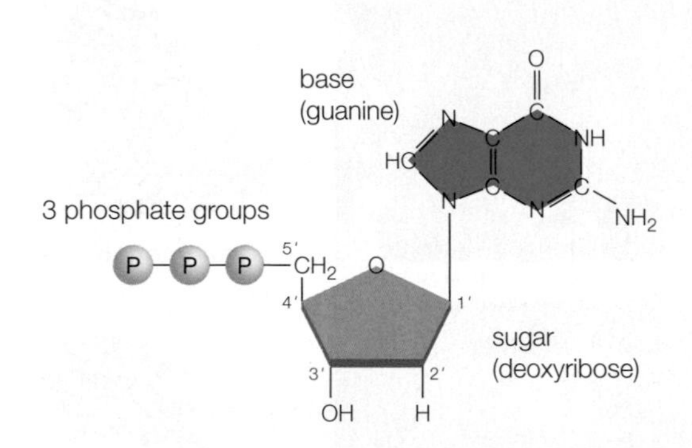

A The DNA nucleotide guanine (G), or deoxyguanosine triphosphate, one of the four nucleotides in DNA. The other nucleotides—adenine, thymine, and cytosine—differ only in their component bases (blue). Three of the four bases in RNA nucleotides are identical to the bases in DNA nucleotides.

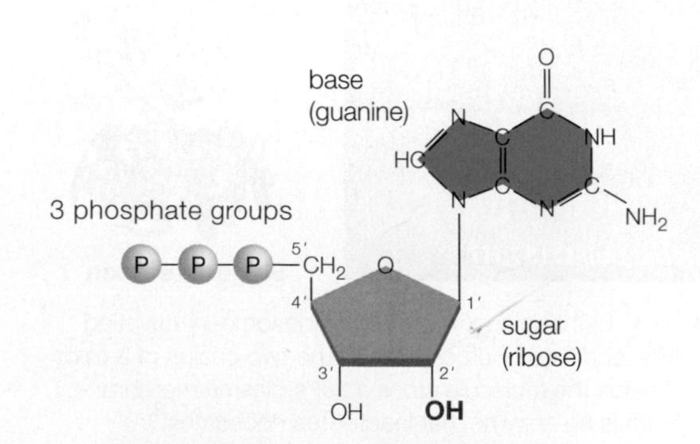

B The RNA nucleotide guanine (G), or guanosine triphosphate. The only difference between the DNA and RNA versions of guanine (or adenine, or cytosine) is that RNA has a hydroxyl group (shown in red) at the 2′ carbon of the sugar.

FIGURE 9.2 ▶**Animated** Comparing nucleotides of DNA and RNA.

RNA to Protein

Messenger RNA was named for its function as the "messenger" between DNA and protein. An mRNA carries a protein-building message that is encoded by sets of three nucleotides, "genetic words" that occur one after another along its length. Like the words of a sentence, a series of these genetic words can form a meaningful parcel of information—in this case, the sequence of amino acids of a protein.

By the process of **translation**, the protein-building information in an mRNA is decoded (translated) into a sequence of amino acids. The result is a polypeptide chain that twists and folds into a protein:

$$\textbf{mRNA} \xrightarrow{\textit{translation}} \textbf{protein}$$

Transcription and translation are both part of **gene expression**, the multistep process by which information

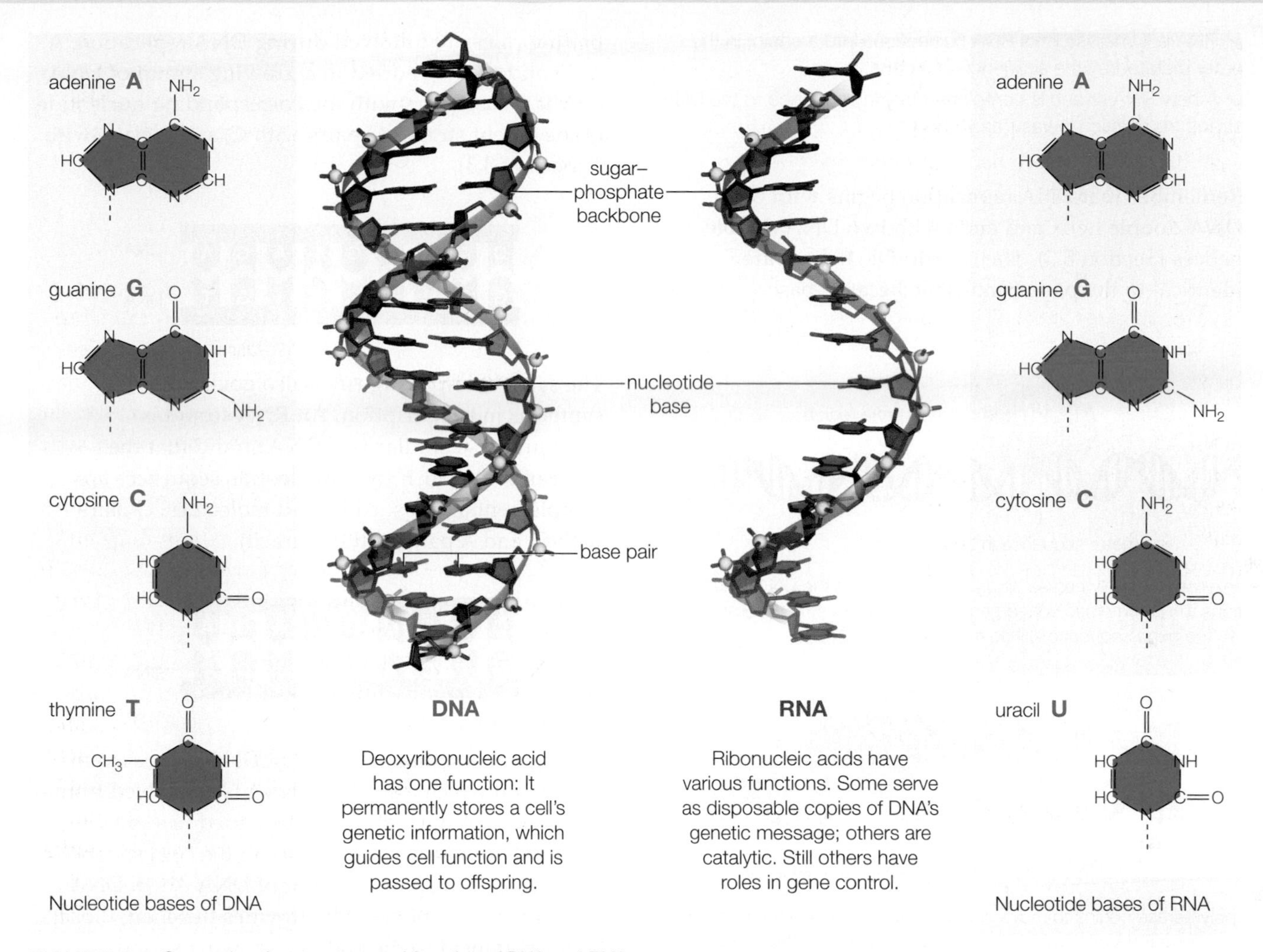

FIGURE 9.3 Comparing the structure and function of DNA and RNA.

in a gene guides the assembly of an RNA or protein product. Expression of genes that encode RNA products (such as tRNA and rRNA) involves transcription. Expression of genes that encode protein products involves both transcription and translation:

DNA —*transcription*→ **mRNA** —*translation*→ **protein**

gene A part of a chromosome that encodes an RNA or protein product in its DNA sequence.
gene expression Process by which the information in a gene guides assembly of an RNA or protein product.
messenger RNA (mRNA) RNA that has a protein-building message.
ribosomal RNA (rRNA) RNA that becomes part of ribosomes.
transcription Process by which enzymes assemble an RNA using the nucleotide sequence of a gene as a template.
transfer RNA (tRNA) RNA that delivers amino acids to a ribosome during translation.
translation Process by which a polypeptide chain is assembled from amino acids in the order specified by an mRNA.

The DNA sequence of a cell's chromosome(s) contains all the information it needs to make the molecules of life. Each gene encodes an RNA, and RNAs interact to assemble proteins from amino acids (Section 3.6). Proteins (enzymes, in particular) assemble lipids and carbohydrates, replicate DNA, make RNA, and perform many other functions that keep the cell alive.

TAKE-HOME MESSAGE 9.2

What is the nature of the information carried by a DNA sequence?

- ✔ Information in a DNA sequence occurs in units that are called genes.
- ✔ A cell uses the information encoded in a gene to make an RNA or protein product, a process called gene expression.
- ✔ The DNA sequence of a gene is transcribed into RNA.
- ✔ Information carried by a messenger RNA (mRNA) is translated into a protein.

9.3 Transcription: DNA to RNA

✔ RNA polymerase links RNA nucleotides into a chain, in the order dictated by the sequence of a gene.

✔ A new RNA strand is complementary in sequence to the DNA strand from which it was transcribed.

Remember that DNA replication begins with one DNA double helix and ends with two DNA double helices (Section 8.5). The two double helices are identical to the parent molecule because base-pairing rules are followed during DNA replication. A nucleotide can be added to a growing strand of DNA only if it base-pairs with the corresponding nucleotide of the parent strand: G pairs with C, and A pairs with T (Section 8.3):

The same base-pairing rules also govern RNA synthesis in transcription. An RNA strand is structurally so similar to a DNA strand that the two can base-pair if their nucleotide sequences are complementary. In such hybrid molecules, G pairs with C, and A pairs with U (uracil):

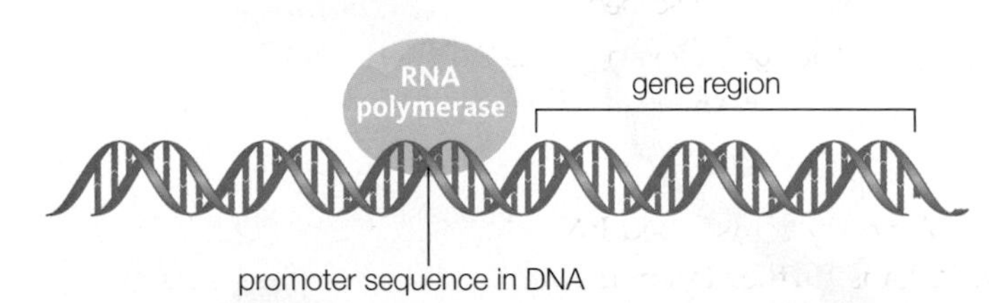

❶ The enzyme RNA polymerase binds to a promoter in the DNA. The binding positions the polymerase near a gene. Only the DNA strand complementary to the gene sequence will be translated into RNA.

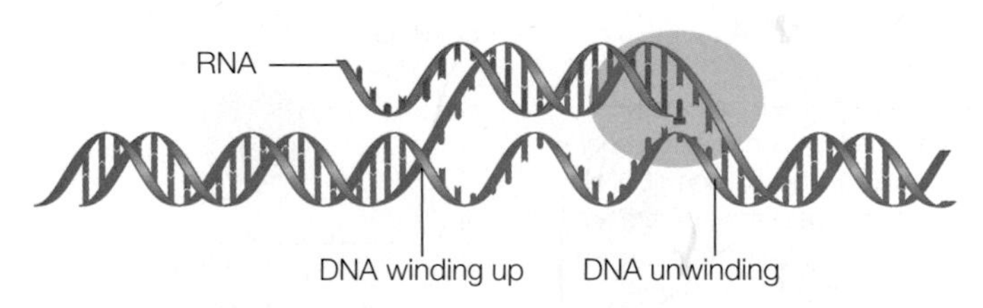

❷ RNA polymerase begins to move along the gene and unwind the DNA. As it does, it links RNA nucleotides in the order specified by the sequence of the complementary (noncoding) DNA strand. The DNA winds up again after the polymerase passes.

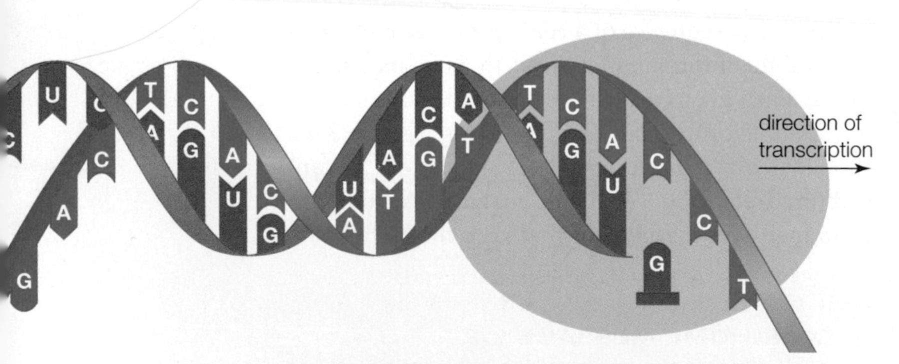

❸ Zooming in on the site of transcription, we see that RNA polymerase covalently bonds successive nucleotides into a new strand of RNA. The new RNA is complementary in sequence to the template DNA strand, so it is an RNA copy of the gene.

FIGURE 9.4 ▶Animated Transcription. By this process, a strand of RNA is assembled from nucleotides according to a template: a gene region in DNA. **FIGURE IT OUT** After the guanine, what is the next nucleotide that will be added to this growing strand of RNA?

Answer: Another guanine (G)

During transcription, a strand of DNA acts as a template upon which a strand of RNA is assembled from nucleotides. A nucleotide can be added to a growing RNA only if it is complementary to the corresponding nucleotide of the parent strand of DNA. As in DNA replication, each nucleotide provides the energy for its own attachment to the end of a growing strand.

Transcription is also similar to DNA replication in that one strand of a nucleic acid serves as a template for synthesis of another. However, in contrast with DNA replication, only part of one DNA strand, not the whole molecule, is used as a template for transcription. The enzyme **RNA polymerase**, not DNA polymerase, adds nucleotides to the end of a growing RNA. Also, transcription produces a single strand of RNA, not two DNA double helices.

In eukaryotic cells, transcription occurs in the nucleus; in prokaryotes, it occurs in cytoplasm. In both cases, the process begins at a regulatory site called a **promoter** (FIGURE 9.4 ❶), a short DNA sequence close to a gene and upstream from it (in the 5′ direction). A promoter is recognized by DNA-binding proteins that in turn bind RNA polymerase. After the polymerase binds, it moves along the DNA toward the gene and over it, unwinding the double helix just a bit so it can "read" the base sequence of the noncoding strand (the strand complementary to the gene) ❷. As it does, the polymerase joins free RNA nucleotides into a chain,

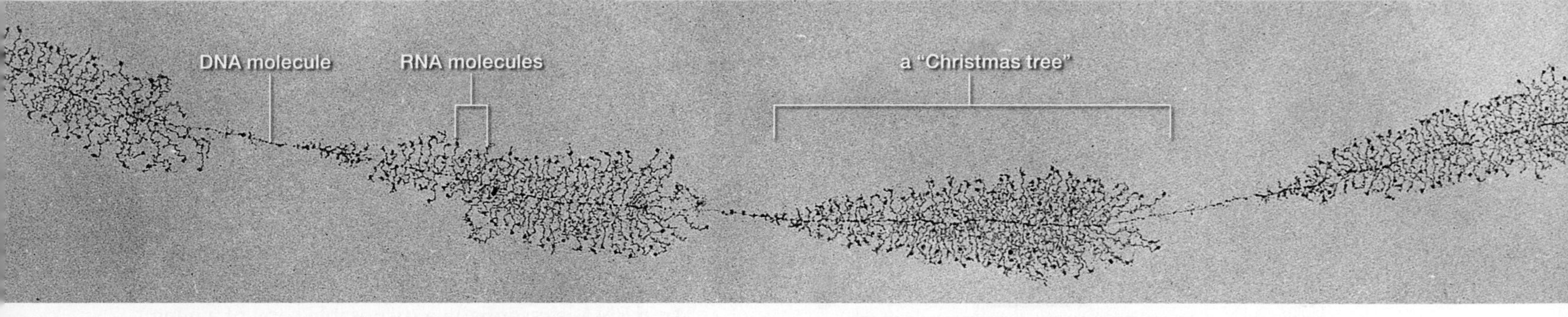

FIGURE 9.5 Typically, many RNA polymerases simultaneously transcribe the same gene, producing a structure often called a "Christmas tree" after its shape. Here, four genes next to one another on the same chromosome are being transcribed.

FIGURE IT OUT Are the polymerases transcribing this DNA molecule moving from left to right or from right to left?

Answer: Left to right (the RNAs get longer as the polymerases move along the DNA)

in the order dictated by that base sequence. As in DNA replication, the synthesis is directional: An RNA polymerase adds nucleotides only to the 3′ end of the growing strand of RNA.

When the polymerase reaches the end of the gene region, the DNA and the new RNA are released. RNA polymerase follows base-pairing rules, so the new RNA is complementary in base sequence to the DNA strand that served as its template ❸. It is an RNA copy of a gene, the same way that a paper transcript of a conversation carries the same information in a different format. Typically, many polymerases transcribe a particular gene region at the same time, so many new RNA strands can be produced quickly (**FIGURE 9.5**).

Post-Transcriptional Modifications

Just as a dressmaker may snip off loose threads or add bows to a dress before it leaves the shop, so do eukaryotic cells tailor their RNA before it leaves the nucleus. Consider that most eukaryotic genes contain intervening sequences called **introns**. Intron sequences are removed in chunks from a newly transcribed RNA before it leaves the nucleus. Sequences that remain in the RNA after this process are called **exons** (**FIGURE 9.6**). Exons can be rearranged and spliced together in different combinations—a process called **alternative splicing**—so one gene may encode two or more versions of the same product.

alternative splicing Post-transcriptional RNA modification process in which some exons are removed or joined in different combinations.
exon Nucleotide sequence that remains in an RNA after post-transcriptional modification.
intron Nucleotide sequence that intervenes between exons and is removed during post-transcriptional modification.
promoter DNA sequence that is a site where transcription begins.
RNA polymerase Enzyme that carries out transcription.

A newly transcribed RNA that will become an mRNA is further tailored after splicing. Enzymes attach a modified guanine "cap" to the 5′ end; later, this cap will help the finished mRNA bind to a ribosome. Between 50 and 300 adenines are also added to the 3′ end of a new mRNA. This poly-A tail is a signal that allows an mRNA to be exported from the nucleus, and as you will see in Chapter 10, it helps regulate the timing and duration of the mRNA's translation.

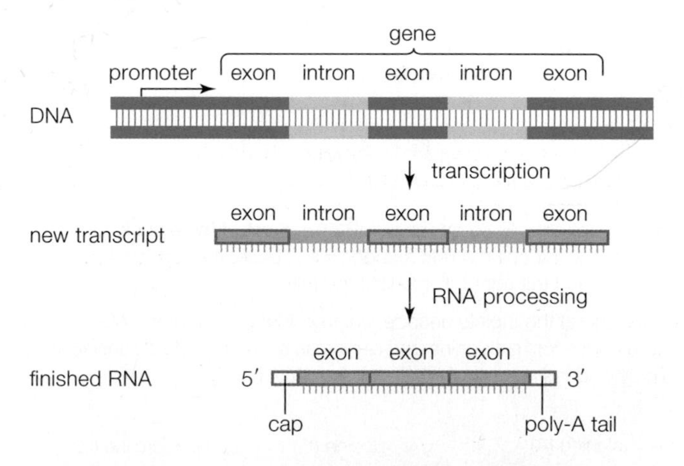

FIGURE 9.6 ▶**Animated** Post-transcriptional modification of RNA. Introns are removed and exons spliced together. Messenger RNAs also get a poly-A tail and modified guanine "cap."

TAKE-HOME MESSAGE 9.3

How is RNA assembled?

✔ Transcription produces an RNA from a gene.

✔ RNA polymerase uses a gene region in a chromosome as a template to assemble a strand of RNA. The new strand is an RNA copy of the gene from which it was transcribed.

✔ Post-transcriptional modification of RNA occurs in the nucleus of eukaryotes.

CREDITS: (5) O. L. Miller; (6) © Cengage Learning.

9.4 RNA and the Genetic Code

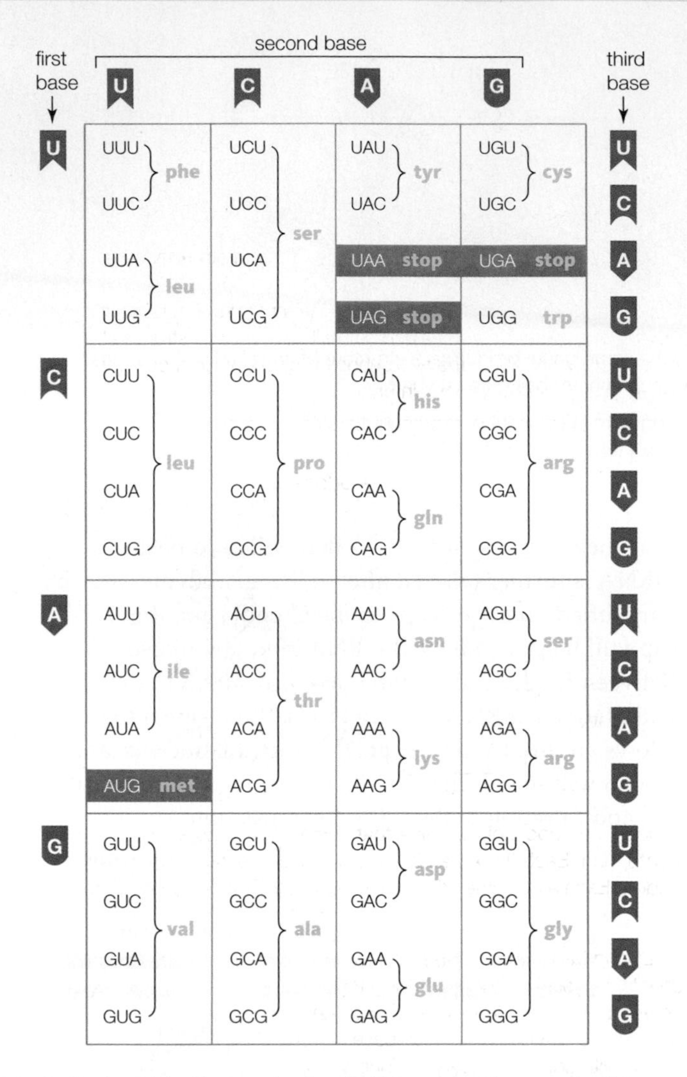

A Codon table. Each codon in mRNA is a set of three nucleotide bases. The left column lists a codon's first base, the top row lists the second, and the right column lists the third.

Sixty-one of the triplets encode amino acids; one of those, AUG, both codes for methionine and serves as a signal to start translation. Three codons are signals that stop translation.

ala alanine (A)	gly glycine (G)	pro proline (P)
arg arginine (R)	his histidine (H)	ser serine (S)
asn asparagine (N)	ile isoleucine (I)	thr threonine (T)
asp aspartic acid (D)	leu leucine (L)	trp tryptophan (W)
cys cysteine (C)	lys lysine (K)	tyr tyrosine (Y)
glu glutamic acid (E)	met methionine (M)	val valine (V)
gln glutamine (Q)	phe phenylalanine (F)	

B Names and abbreviations of the 20 naturally occurring amino acids specified by the genetic code (**A**).

FIGURE 9.7 The genetic code.

FIGURE IT OUT Which codons specify the amino acid lysine?

Answer: AAA and AAG

✔ Nucleotide base triplets in an mRNA encode a protein-building message. Ribosomal RNA and transfer RNA translate that message into a polypeptide chain.

DNA stores heritable information about proteins, but making those proteins requires messenger RNA (mRNA), transfer RNA (tRNA), and ribosomal RNA (rRNA). The three types of RNA interact to translate DNA's information into a protein.

An mRNA is essentially a disposable copy of a gene. Its job is to carry the gene's protein-building information to the other two types of RNA during translation. That protein-building information consists of a linear sequence of genetic "words" spelled with an alphabet of the four nucleotide bases A, C, G, and U. Each of the genetic "words" carried by an mRNA is three bases long, and each is a code—a **codon**—for a particular amino acid. With four possible bases in each of the three positions of a codon, there are a total of sixty-four (or 4^3) mRNA codons. Collectively, the sixty-four codons constitute the **genetic code** (**FIGURE 9.7**). The sequence of bases in a triplet determines which amino acid the codon specifies. For instance, the codon UUU codes for the amino acid phenylalanine (phe), and UUA codes for leucine (leu).

Codons occur one after another along the length of an mRNA. When an mRNA is translated, the order of its codons determines the order of amino acids in the resulting polypeptide. Thus, the DNA sequence of a gene is transcribed into the nucleotide sequence of an mRNA, which is in turn translated into an amino acid sequence (**FIGURE 9.8**).

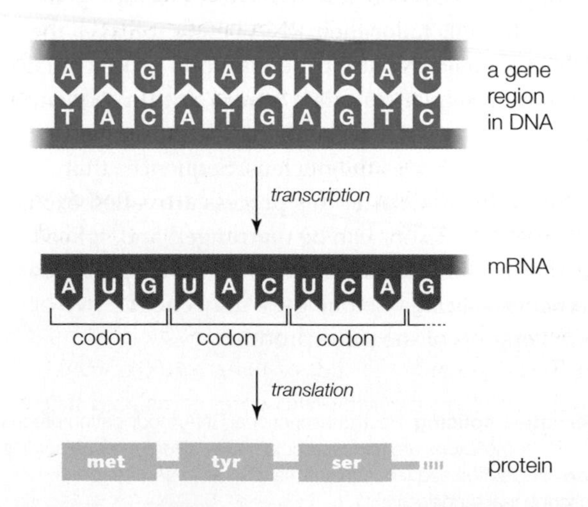

FIGURE 9.8 Example of the correspondence between DNA, RNA, and protein. A gene region in a strand of chromosomal DNA is transcribed into an mRNA, and the codons of the mRNA specify a chain of amino acids—a protein.

With a few exceptions, twenty naturally occurring amino acids are encoded by the sixty-four codons in the genetic code. Some amino acids are specified by more than one codon. For instance, the amino acid tyrosine (tyr) is specified by two codons: UAU and UAC. Other codons signal the beginning and end of a protein-coding sequence. The first AUG in an mRNA usually serves as the signal to start translation. AUG is the codon for methionine, so methionine is always the first amino acid in new polypeptides of such organisms. The codons UAA, UAG, and UGA do not specify an amino acid. These are signals that stop translation, so they are called stop codons. A stop codon marks the end of the protein-coding sequence in an mRNA.

The genetic code is highly conserved, which means that most organisms use the same code and probably always have. Bacteria, archaea, and some protists have a few codons that differ from the eukaryotic code, as do mitochondria and chloroplasts—a clue that led to a theory of how these two organelles evolved (we return to this topic in Section 19.6).

Ribosomes interact with transfer RNAs (tRNAs) to translate the sequence of codons in an mRNA into a polypeptide. A ribosome has two subunits, one large and one small (**FIGURE 9.9**). Both subunits consist mainly of rRNA, with some associated structural proteins. During translation, a large and a small ribosomal subunit converge as an intact ribosome on an mRNA. Ribosomal RNA is one example of RNA with enzymatic activity: During translation, rRNA catalyzes formation of peptide bonds between amino acids.

Amino acids are delivered to ribosomes by tRNAs. Each tRNA has two attachment sites. The first is an **anticodon**, which is a triplet of nucleotides that base-pairs with an mRNA codon (**FIGURE 9.10A**). The other attachment site binds to an amino acid—the one specified by the codon. Transfer RNAs with different anticodons carry different amino acids.

During translation, tRNAs deliver amino acids to a ribosome, one after the next in the order specified by the codons in an mRNA (**FIGURE 9.10B**). As the amino acids are delivered, the ribosome joins them via peptide bonds into a new polypeptide (Section 3.6). Thus, the order of codons in an mRNA—DNA's protein-building message—becomes translated into a new protein.

anticodon In a tRNA, set of three nucleotides that base-pairs with an mRNA codon.
codon In an mRNA, a nucleotide base triplet that codes for an amino acid or stop signal during translation.
genetic code Complete set of sixty-four mRNA codons.

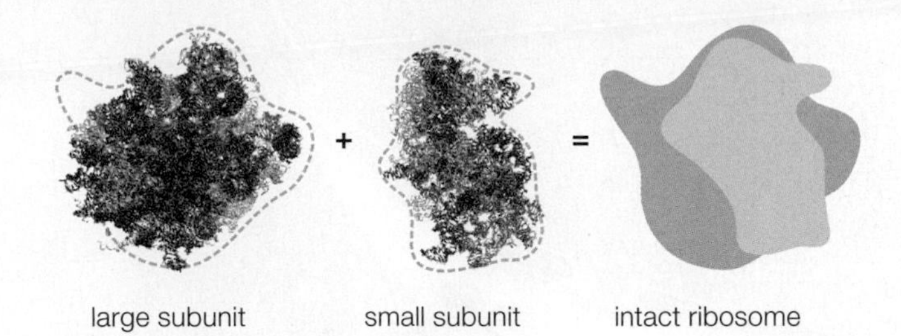

FIGURE 9.9 ▸**Animated** Ribosome structure. Each intact ribosome consists of a large and a small subunit. The structural protein components of the two subunits are shown in green; the catalytic rRNA components, in brown.

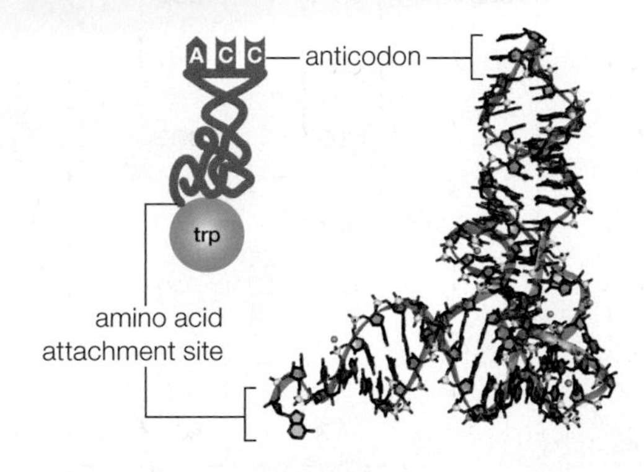

A Icon and model of the tRNA that carries the amino acid tryptophan. Each tRNA's anticodon is complementary to an mRNA codon. Each also carries the amino acid specified by that codon.

B During translation, tRNAs dock at an intact ribosome (for clarity, only the small subunit is shown, in tan). Here, the anticodons of two tRNAs have base-paired with complementary codons on an mRNA (red). The amino acids they carry are not shown, for clarity.

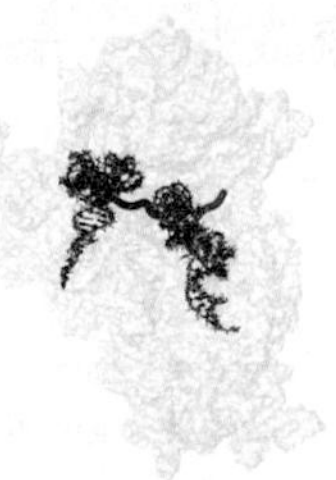

FIGURE 9.10 tRNA structure.

TAKE-HOME MESSAGE 9.4

What roles do mRNA, tRNA, and rRNA play during translation?

✔ The sequence of nucleotide triplets (codons) in an mRNA encodes a gene's protein-building message.

✔ The genetic code consists of sixty-four codons. Three are signals that stop translation; the remaining codons specify an amino acid. In most mRNAs, the first occurrence of the codon that specifies methionine is a signal to begin translation.

✔ Ribosomes, which consist of two subunits of rRNA and proteins, assemble amino acids into polypeptide chains.

✔ A tRNA has an anticodon complementary to an mRNA codon, and a binding site for the amino acid specified by that codon. During translation, tRNAs deliver amino acids to ribosomes.

9.5 Translation: RNA to Protein

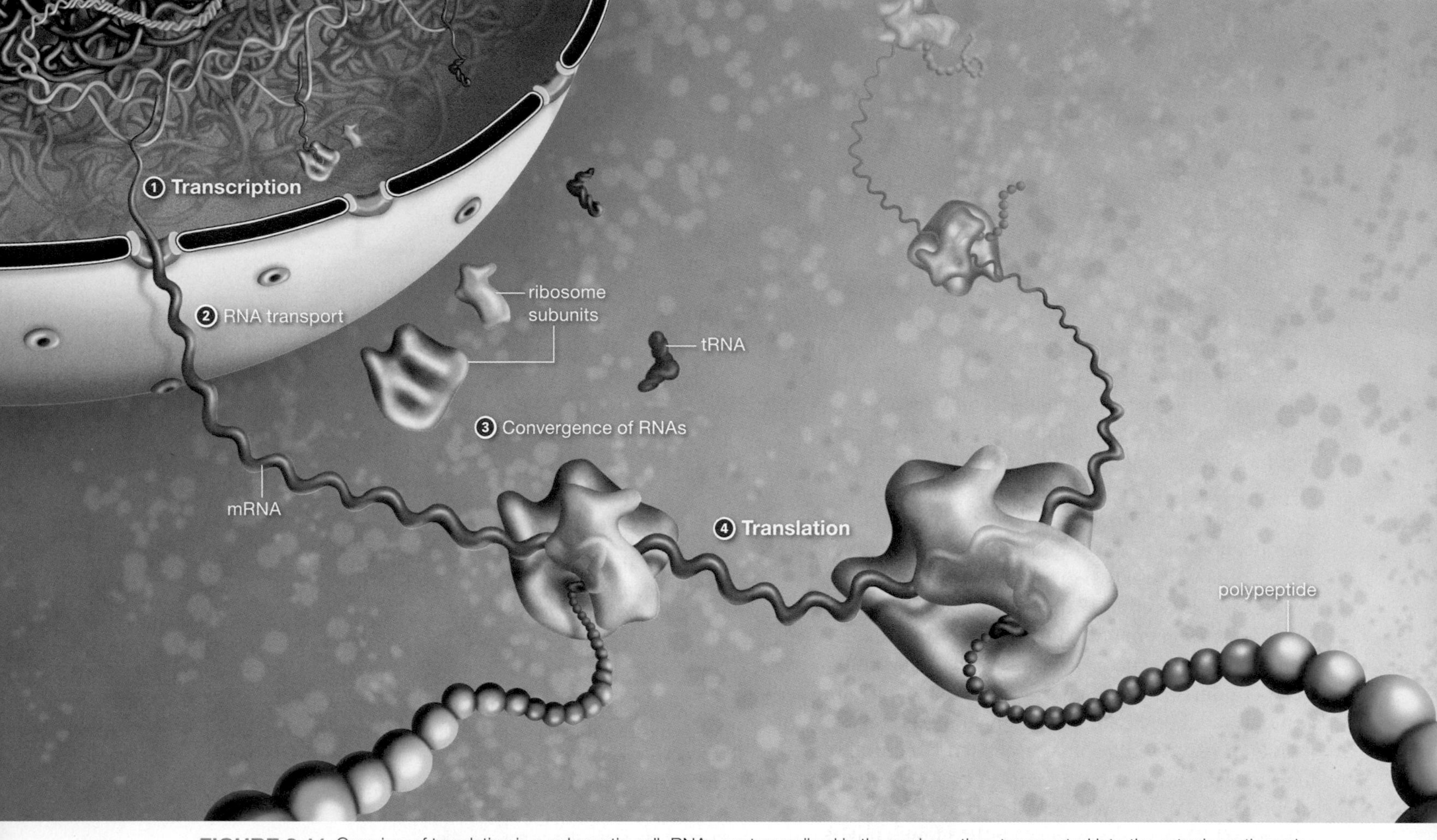

FIGURE 9.11 Overview of translation in a eukaryotic cell. RNAs are transcribed in the nucleus, then transported into the cytoplasm through nuclear pores. Translation begins when ribosomal subunits and tRNA converge on an mRNA in cytoplasm.

✔ Translation converts the information carried by an mRNA into a new polypeptide chain.

✔ The order of codons in an mRNA determines the order of amino acids in the polypeptide chain translated from it.

Translation, the second part of protein synthesis, proceeds in three stages: initiation, elongation, and termination. In all cells, translation occurs in cytoplasm. Cytoplasm contains many free amino acids, tRNAs, and ribosomal subunits available to participate in protein synthesis. In a eukaryotic cell, RNAs that carry out transcription are produced in the nucleus (**FIGURE 9.11** ❶), then transported through nuclear pores into the cytoplasm ❷. Translation is initiated when ribosomal subunits and tRNAs converge on an mRNA. First, a small ribosomal subunit binds to an mRNA, and the anticodon of a special tRNA called an initiator base-pairs with the mRNA's first AUG codon. Then, a large ribosomal subunit joins the small subunit ❸.

The complex of molecules is now ready to carry out protein synthesis. In the elongation stage, the intact ribosome moves along the mRNA and assembles a polypeptide chain ❹. **FIGURE 9.12** shows how this works. Initiator tRNAs carry methionine, so the first amino acid of the new polypeptide chain is a methionine. Another tRNA joins the complex when its anticodon base-pairs with the second codon in the mRNA ❺. This tRNA brings with it the second amino acid. The ribosome then catalyzes formation of a peptide bond between the first two amino acids ❻.

As the ribosome moves to the next codon, it releases the first tRNA. Another tRNA brings the third amino acid to the complex as its anticodon base-pairs with the third codon of the mRNA ❼. The ribosome catalyzes the formation of a peptide bond between the second and third amino acids ❽.

The second tRNA is released and the ribosome moves to the next codon. Another tRNA brings the fourth amino acid to the complex as its anticodon base-pairs with the fourth codon of the mRNA ➒. The ribosome catalyzes the formation of a peptide bond between the third and fourth amino acids.

The new polypeptide chain continues to elongate as amino acids are delivered by successive tRNAs. Translation terminates when the ribosome reaches a stop codon in the mRNA ➓. The mRNA and the polypeptide detach from the ribosome, and the ribosomal subunits separate from each other. Translation is now complete. The new polypeptide joins the pool of proteins in cytoplasm, or it enters rough ER (Section 4.7).

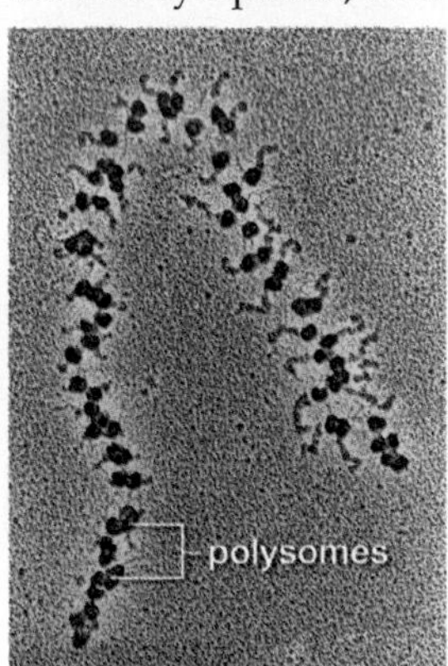

Many ribosomes may simultaneously translate the same mRNA, in which case they are called polysomes (left). In bacteria and archaea, transcription and translation both occur in the cytoplasm, and these processes are closely linked in time and space. Translation begins before transcription ends, so a transcription "Christmas tree" is often decorated with polysome "balls."

Given that many polypeptides can be translated from one mRNA, why would any cell also make many copies of an mRNA? Compared with DNA, RNA is not very stable. An mRNA may last only a few minutes in cytoplasm before enzymes disassemble it. The fast turnover allows cells to adjust their protein synthesis quickly in response to changing needs.

Translation is energy intensive. Most of that energy is provided in the form of phosphate-group transfers from the RNA nucleotide GTP (shown in **FIGURE 9.2B**) to molecules that help the ribosome move from one codon to the next along the mRNA.

TAKE-HOME MESSAGE 9.5

How is mRNA translated into protein?

✔ Translation is an energy-requiring process that converts a sequence of codons in an mRNA into a polypeptide.

✔ During initiation, an mRNA joins with an initiator tRNA and two ribosomal subunits.

✔ During elongation, amino acids are delivered to the ribosome by tRNAs in the order dictated by successive mRNA codons. As amino acids arrive, the ribosome joins each to the end of the polypeptide.

✔ Termination occurs when the ribosome encounters a stop codon in the mRNA. The mRNA and the polypeptide are released, and the ribosome disassembles.

➎ Ribosomal subunits and an initiator tRNA converge on an mRNA. A second tRNA binds to the second codon.

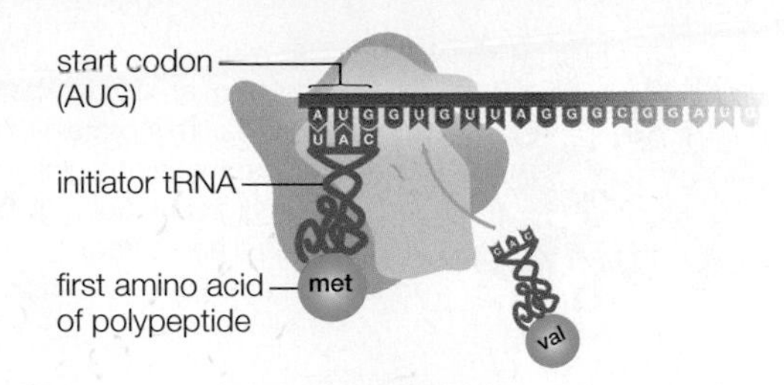

➏ A peptide bond forms between the first two amino acids.

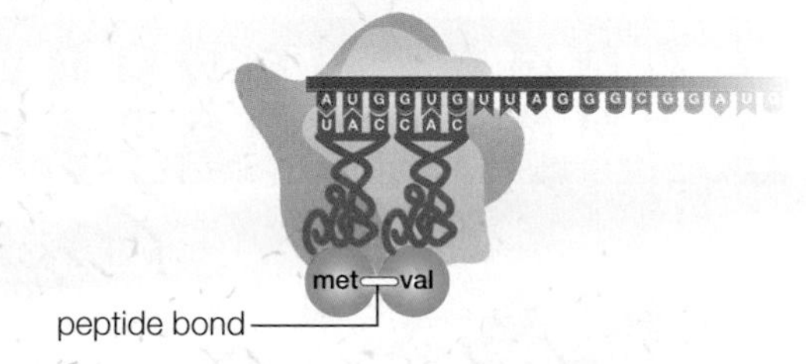

➐ The first tRNA is released and the ribosome moves to the next codon. A third tRNA binds to the third codon.

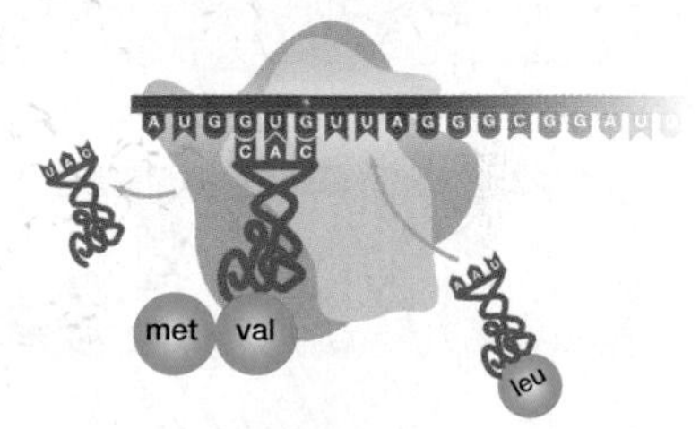

➑ A peptide bond forms between the second and third amino acids.

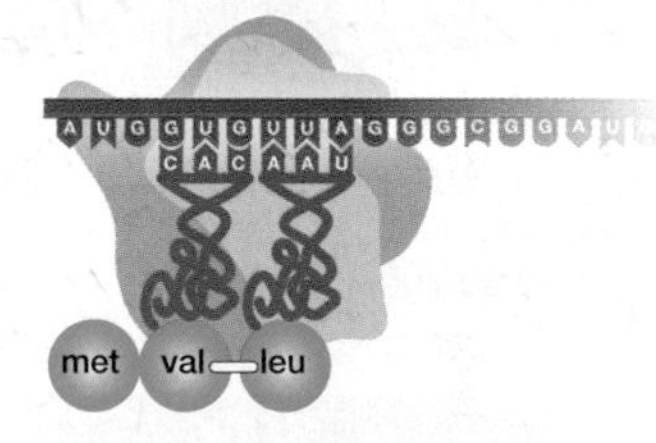

➒ The second tRNA is released and the ribosome moves to the next codon. A fourth tRNA brings the next amino acid to be added to the polypeptide chain.

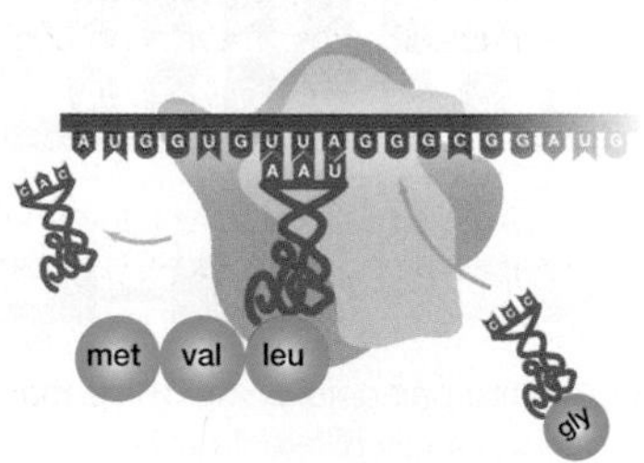

➓ The process repeats until the ribosome encounters a stop codon. Then, the new polypeptide is released and the ribosomal subunits separate.

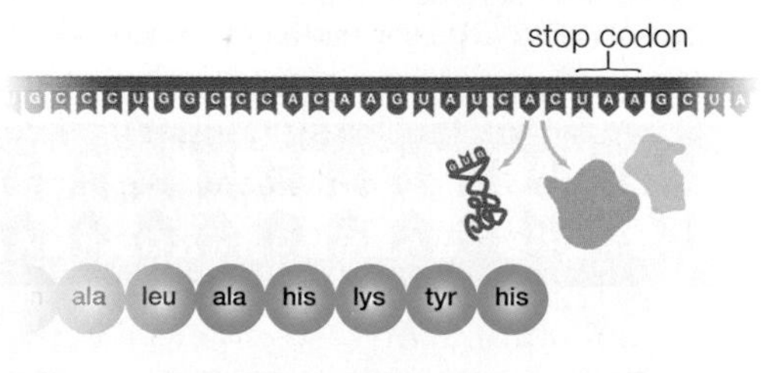

FIGURE 9.12 ▸**Animated** Translation. Translation begins when ribosomal subunits and an initiator tRNA converge on an mRNA. Then, tRNAs deliver amino acids in the order dictated by successive codons in the mRNA. The ribosome links the amino acids together as it moves along the mRNA, so a polypeptide forms and elongates. Translation ends when the ribosome reaches a stop codon.

9.6 Mutated Genes and Their Protein Products

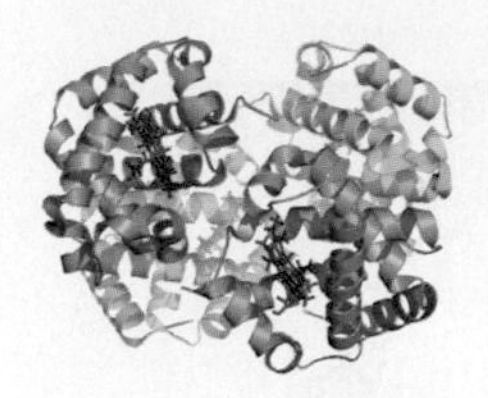

A Hemoglobin, an oxygen-binding protein in red blood cells. This protein consists of four polypeptides: two alpha globins (blue) and two beta globins (green). Each globin has a pocket that cradles a heme (red). Oxygen molecules bind to the iron atom at the center of each heme.

B Part of the DNA (blue), mRNA (brown), and amino acid sequence of human beta globin. Numbers indicate nucleotide position in the mRNA.

C A base-pair substitution replaces a thymine with an adenine. When the altered mRNA is translated, valine replaces glutamic acid as the sixth amino acid. Hemoglobin with this form of beta globin is called HbS, or sickle hemoglobin.

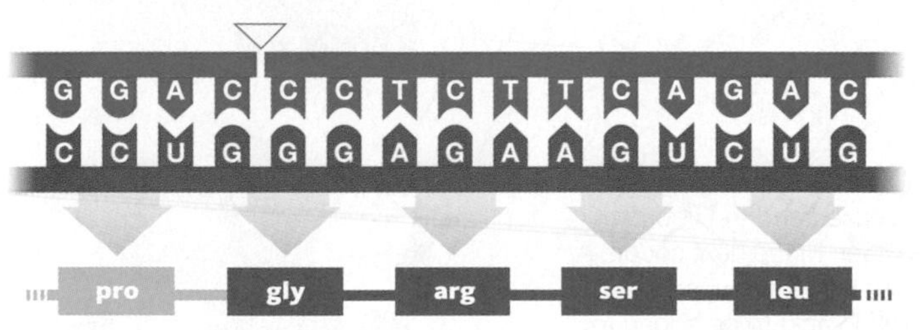

D A base-pair deletion shifts the reading frame for the rest of the mRNA, so a completely different protein product forms. The mutation shown results in a defective beta globin. The outcome is beta thalassemia, a genetic disorder in which a person has an abnormally low amount of hemoglobin.

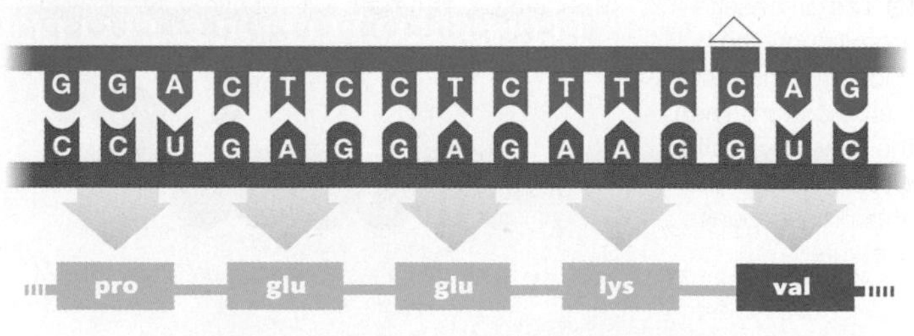

E An insertion of one nucleotide causes the reading frame for the rest of the mRNA to shift. The protein translated from this mRNA is too short and does not assemble correctly into hemoglobin molecules. As in D, the outcome is beta thalassemia.

FIGURE 9.13 ▸Animated Examples of mutations.

✔ If the nucleotide sequence of a gene changes, it may result in an altered gene product, with harmful effects.

Mutations, remember, are permanent changes in the DNA sequence of a chromosome (Section 8.6). A mutation in which one nucleotide and its partner are replaced by a different base pair is called a **base-pair substitution**. Other mutations involve the loss of one or more base pairs (a **deletion**) or the addition of extra base pairs (an **insertion**).

Mutations are relatively uncommon events in a normal cell. Consider that the chromosomes in a diploid human cell collectively consist of about 6.5 billion nucleotides, any of which may become mutated each time that cell divides. On average, about 175 nucleotides do change during DNA replication. However, only about 3 percent of the cell's DNA encodes protein products, so there is a low probability that any of those mutations will be in a protein-coding region.

When a mutation does occur in a protein-coding region, the redundancy of the genetic code offers the cell a margin of safety. For example, a mutation that changes a CCU codon to CCC may have no further effect, because both of these codons specify proline. Other mutations may change an amino acid in a protein, or result in a premature stop codon that shortens it.

Mutations that alter a protein can have drastic effects on an organism. Consider the effects of such mutations on hemoglobin, an oxygen-transporting protein in your red blood cells (Section 3.2). Hemoglobin's structure allows it to bind and release oxygen. In adult humans, a hemoglobin molecule consists of four polypeptides called globins: two alpha globins and two beta globins (FIGURE 9.13A). Each globin folds around a heme, a cofactor with an iron atom at its center (Section 5.6). Oxygen molecules bind to hemoglobin at those iron atoms.

Mutations in the genes for alpha or beta globin cause a condition called anemia, in which a person's blood is deficient in red blood cells or in hemoglobin. Both outcomes limit the blood's ability to carry oxygen, and the resulting symptoms can range from mild to life-threatening.

Sickle-cell anemia, a type of anemia that is most common in people of African ancestry, arises because of a base-pair substitution in the beta globin gene. The substitution causes the body to produce a version of beta globin in which the sixth amino acid is valine instead of glutamic acid (FIGURE 9.13B,C). Hemoglobin assembled with this altered beta globin chain is called sickle hemoglobin, or HbS.

Unlike glutamic acid, which carries a negative charge, valine carries no charge. As a result of that one base-pair substitution, a tiny patch of the beta globin polypeptide that is normally hydrophilic becomes hydrophobic. This change slightly alters hemoglobin behavior. Under certain conditions, HbS molecules stick together and form large, rodlike clumps. Red blood cells that contain the clumps become distorted into a crescent (sickle) shape (**FIGURE 9.14**). Sickled cells clog tiny blood vessels, thus disrupting blood circulation throughout the body. Over time, repeated episodes of sickling can damage organs and eventually cause death.

A different type of anemia, beta thalassemia, is caused by the deletion of the twentieth nucleotide in the coding region of the beta globin gene (**FIGURE 9.13D**). Like many other deletions, this one causes the reading frame of the mRNA codons to shift. A frameshift usually has drastic consequences because it garbles the genetic message, just as incorrectly grouping a series of letters garbles the meaning of a sentence:

The fat cat ate the sad rat
T hef atc ata tet hes adr at

The frameshift caused by the beta globin deletion results in a polypeptide that differs drastically from normal beta globin in amino acid sequence and in length. This outcome is the source of the anemia. Beta thalassemia can also be caused by insertion mutations, which, like deletions, often result in frameshifts (**FIGURE 9.13E**).

Not all mutations that affect protein structure disrupt codons for amino acids. DNA also contains special nucleotide sequences that influence the expression of nearby genes (we return to this topic in the next chapter). A promoter is one example; an intron–exon splice site is another. Consider a mutation that causes the hairless appearance of sphynx cats (**FIGURE 9.15**). In this case, a base-pair substitution disrupts an intron–exon splice site in a gene for keratin, a fibrous protein (Section 3.6). The intron is not correctly removed during post-transcriptional processing. The altered protein translated from the resulting mRNA cannot properly assemble into filaments that make up hair. Cats with this mutation still make hair, but it falls out before it gets very long.

base-pair substitution Type of mutation in which a single base pair changes.
deletion Mutation in which one or more nucleotides are lost.
insertion Mutation in which one or more nucleotides become inserted into DNA.

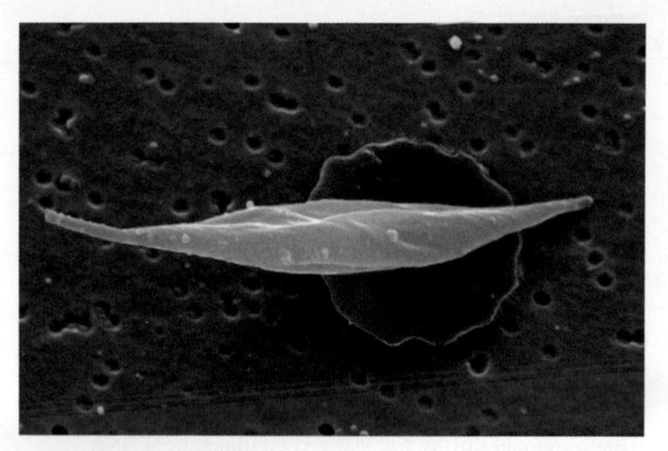

FIGURE 9.14 ▸Animated A sickled red blood cell compared with a normal one. A single base-pair substitution gives rise to an abnormal beta globin chain that, when assembled in hemoglobin molecules, forms HbS. The sixth amino acid in these abnormal beta globin chains is valine, not glutamic acid. In the body, the difference causes HbS molecules to form rod-shaped clumps that distort normally round blood cells (red) into the sickle shape (tan) characteristic of sickle-cell anemia. Sickled cells clog small blood vessels.

FIGURE 9.15 The hairless appearance of a sphynx cat arises from a single base-pair mutation in a gene for keratin, a fibrous protein that makes up hair. The altered keratin that results from the mutation does not assemble correctly into filaments. Sphynx cats are not truly hairless; they produce hair, but it is easily dislodged.

TAKE-HOME MESSAGE 9.6

What happens after a gene becomes mutated?

✔ Mutations that result in an altered protein can have drastic consequences.

✔ A base-pair substitution may change an amino acid in a protein, or it may introduce a premature stop codon.

✔ Frameshifts that occur after an insertion or deletion can change an mRNA's codon reading frame, thus garbling its protein-building instructions.

CREDITS: (14) EM Unit, UCL Medical School, Royal Free Campus/Wellcome Images; (15) Glennis Siverson/National Geographic Creative.

summary

Section 9.1 The ability to make proteins is critical to all life processes. Ribosome-inactivating proteins (RIPs) have an enzyme domain that alters ribosomes. Ricin and other toxic RIPs have an additional protein domain that triggers endocytosis. Once in cytoplasm, the enzyme domain destroys the cell's ability to make proteins.

Section 9.2 Information encoded by the nucleotide sequence of DNA occurs in subsets called **genes**. **Gene expression** is the conversion of information in a gene to an RNA or protein product. RNA is produced during **transcription**. **Ribosomal RNA (rRNA)** and **transfer RNA (tRNA)** interact during **translation** of a **messenger RNA (mRNA)** into a protein product.

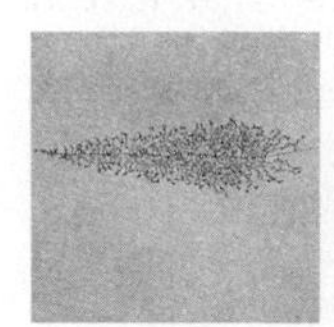

Section 9.3 During transcription, **RNA polymerase** binds to a **promoter** near a gene region, then links RNA nucleotides in the order dictated by the nucleotide base sequence of the noncoding DNA strand. The resulting RNA strand is an RNA copy of the gene.

In eukaryotes, mRNA is modified before it leaves the nucleus. **Intron** sequences are removed, and the remaining **exon** sequences may be rearranged and spliced in different combinations, a process called **alternative splicing**. New mRNAs also receive a cap and poly-A tail.

Section 9.4 An mRNA's protein-building information consists of a series of **codons**. Sixty-four codons constitute the **genetic code**. Three codons are signals that terminate translation. The remaining codons specify a particular amino acid; one of those also serves as a signal to start translation. Some amino acids are specified by multiple codons. Each tRNA has an **anticodon** that can base-pair with a codon, and it binds to the amino acid specified by that codon. Proteins and enzymatic rRNA make up the two subunits of ribosomes.

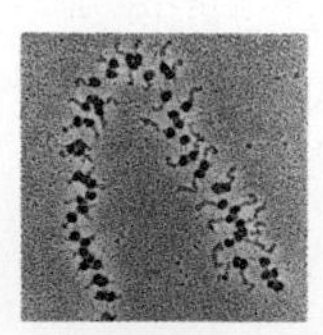

Section 9.5 During translation, codons in an mRNA direct synthesis of a polypeptide. First, the mRNA, an initiator tRNA, and two ribosomal subunits converge. Next, amino acids are delivered by tRNAs in the order specified by the codons in the mRNA. The intact ribosome catalyzes formation of a peptide bond between the successive amino acids, so a polypeptide forms. Translation ends when the ribosome encounters a stop codon in the mRNA.

Section 9.6 **Insertions, deletions,** and **base-pair substitutions** are mutations. A mutation that changes a gene's product may have harmful effects. In humans, an example is sickle-cell anemia, a disorder caused by a single base-pair substitution in the gene for the beta globin chain of hemoglobin.

Ricin, RIP (revisited)

RIPs remove a particular adenine base from one of the rRNAs in the ribosome's large subunit. The adenine is part of a binding site for proteins involved in GTP-requiring steps of elongation. After the base has been removed, the ribosome can no longer bind to these proteins, and elongation stops.

Despite their toxicity, the main function of RIPs may not be destroying ribosomes. Many are part of plant immune systems, but it is their antiviral and anticancer activity that has researchers abuzz. Plants that make RIPs have been used as traditional medicines for many centuries; now, Western scientists are investigating RIPs as drugs to combat HIV and cancer. For example, researchers who design drugs for cancer therapy have modified ricin's glycolipid-binding domain to recognize plasma membrane proteins (Section 5.7) especially abundant in cancer cells. The modified ricin preferentially enters—and kills—cancer cells. Ricin's toxic enzyme has also been attached to an antibody that can find cancer cells in a person's body. The intent of both strategies: to assassinate the cancer cells without harming normal ones.

self-quiz

Answers in Appendix VII

1. A chromosome contains many different gene regions that are transcribed into different ________ .
   a. proteins
   b. polypeptides
   c. RNAs
   d. a and b

2. A binding site for RNA polymerase is called a ________ .
   a. gene
   b. promoter
   c. codon
   d. protein

3. An RNA molecule is typically ________ ; a DNA molecule is typically ________ .
   a. single-stranded; double-stranded
   b. double-stranded; single-stranded

4. RNAs form by ________ ; proteins form by ________ .
   a. replication; translation
   b. translation; transcription
   c. transcription; translation
   d. replication; transcription

5. The main function of a DNA molecule is to ________ .
   a. store heritable information
   b. carry a translatable message
   c. form peptide bonds between amino acids

6. The main function of an mRNA molecule is to ________ .
   a. store heritable information
   b. carry a translatable message
   c. form peptide bonds between amino acids

7. Most codons specify a(n) ________ .
   a. protein
   b. mRNA
   c. amino acid

CREDITS: (in text revisited) Vaughan Fleming/Science Source.

data analysis activities

RIPs as Cancer Drugs Researchers are taking a page from the structure–function relationship of RIPs in their quest for cancer treatments. The most toxic RIPs, remember, have one domain that interferes with ribosomes, and another that carries them into cells. Melissa Cheung and her colleagues incorporated a peptide that binds to skin cancer cells into the enzymatic part of an RIP, the *E. coli* Shiga-like toxin. The researchers created a new RIP that specifically kills skin cancer cells, which are notoriously resistant to established therapies. Some of their results are shown in **FIGURE 9.16**.

1. Which cells had the greatest response to an increase in concentration of the engineered RIP?
2. At what concentration of RIP did all of the different kinds of cells survive?
3. Which cells survived best at 10^{-6} grams per liter RIP?
4. Why are some of the data points linked by curved lines?

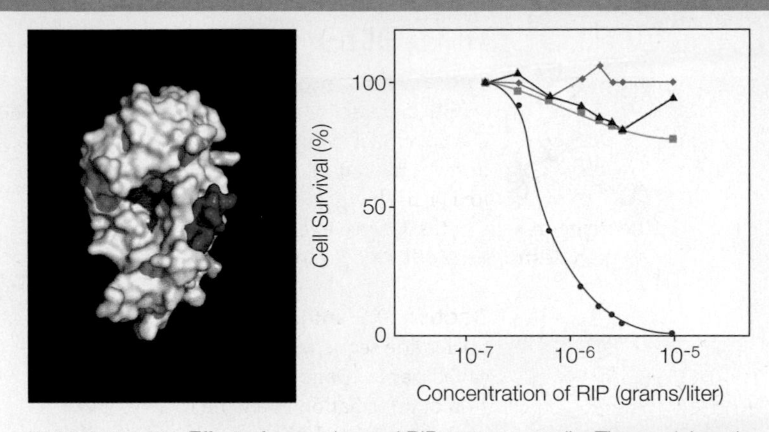

FIGURE 9.16 Effect of an engineered RIP on cancer cells. The model on the left shows the enzyme portion of *E. coli* Shiga-like toxin engineered to carry a small sequence of amino acids (in blue) that targets skin cancer cells. (Red indicates the active site.) The graph (right) shows the effect of this engineered RIP on human cancer cells of the skin (●); breast (◆); liver (▲); and prostate (■).

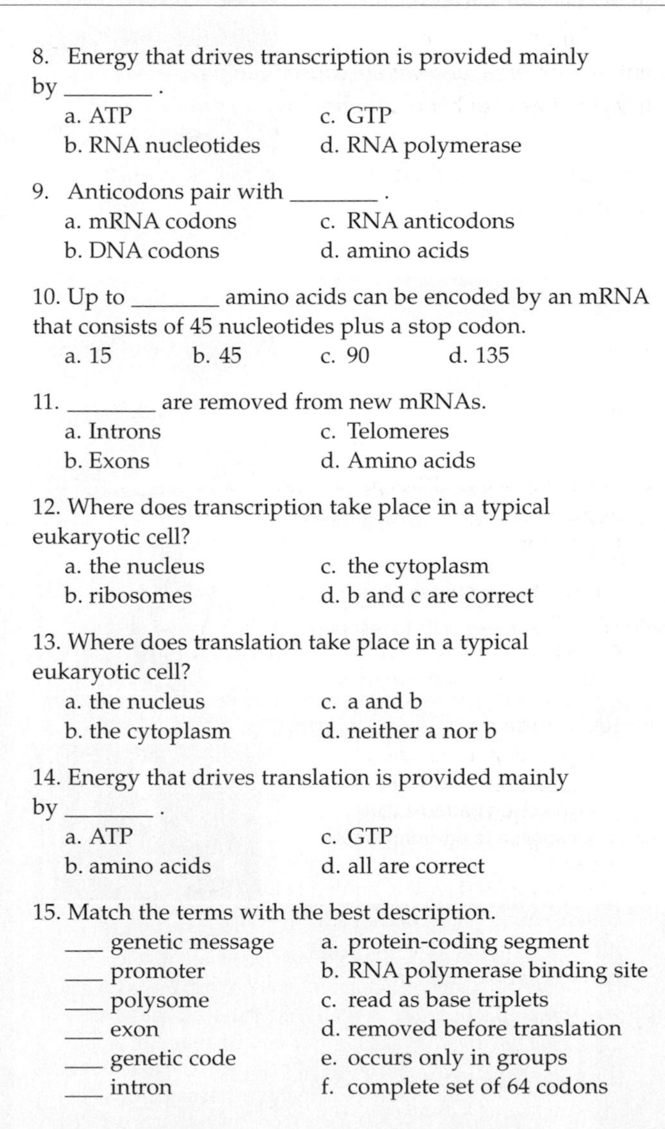

8. Energy that drives transcription is provided mainly by ________ .
   a. ATP c. GTP
   b. RNA nucleotides d. RNA polymerase

9. Anticodons pair with ________ .
   a. mRNA codons c. RNA anticodons
   b. DNA codons d. amino acids

10. Up to ________ amino acids can be encoded by an mRNA that consists of 45 nucleotides plus a stop codon.
    a. 15 b. 45 c. 90 d. 135

11. ________ are removed from new mRNAs.
    a. Introns c. Telomeres
    b. Exons d. Amino acids

12. Where does transcription take place in a typical eukaryotic cell?
    a. the nucleus c. the cytoplasm
    b. ribosomes d. b and c are correct

13. Where does translation take place in a typical eukaryotic cell?
    a. the nucleus c. a and b
    b. the cytoplasm d. neither a nor b

14. Energy that drives translation is provided mainly by ________ .
    a. ATP c. GTP
    b. amino acids d. all are correct

15. Match the terms with the best description.
    ___ genetic message a. protein-coding segment
    ___ promoter b. RNA polymerase binding site
    ___ polysome c. read as base triplets
    ___ exon d. removed before translation
    ___ genetic code e. occurs only in groups
    ___ intron f. complete set of 64 codons

critical thinking

1. Researchers are designing and testing antisense drugs as therapies for a variety of diseases, including cancer, AIDS, diabetes, and muscular dystrophy. The drugs are also being tested to fight infection by deadly viruses such as Ebola. Antisense drugs consist of short mRNA strands that are complementary in base sequence to mRNAs that form during the progression of disease. Speculate on how these drugs work.

2. An anticodon has the sequence GCG. What amino acid does this tRNA carry? What would be the effect of a mutation that changed the C of the anticodon to a G?

3. Each position of a codon can be occupied by one of four (4) nucleotides. What is the minimum number of nucleotides per codon necessary to specify all 20 of the amino acids that are found in proteins?

4. Refer to **FIGURE 9.7**, then translate the following mRNA nucleotide sequence into an amino acid sequence, starting at the first base:

5′—UGUCAUGCUCGUCUUGAAUCUUGUGAUGC UCGUUGGAUUAAUUGU—3′

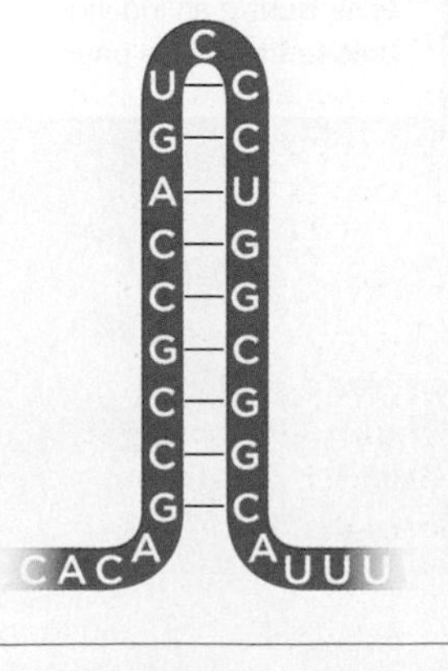

5. Translate the sequence of bases in the previous question, starting at the second base.

6. Bacteria use the same stop codons as eukaryotes. However, bacterial transcription is also terminated in places where the mRNA folds back on itself to form a hairpin-looped structure like the one shown on the left. How do you think that this structure stops transcription?

CREDITS: (16) Cheung et al. *Molecular Cancer*, 9:28, 2010; (in text CT#6) © Cengage Learning.

10 Control of Gene Expression

LEARNING ROADMAP

This chapter explores metabolism (Sections 5.5, 5.6) in the context of gene expression (9.2). You will be applying what you know about DNA (8.2, 8.4, 8.5), mutations (8.6, 9.6), and differentiation (8.7), as well as transcription (9.3) and translation (9.5). You will also revisit functional groups (3.3), carbohydrates (3.4), glycolysis (7.3), and fermentation (7.6).

MECHANISMS OF GENE CONTROL

Every step of gene expression is regulated. This control is critical for development, and it allows individual cells to respond to changes in external and internal conditions.

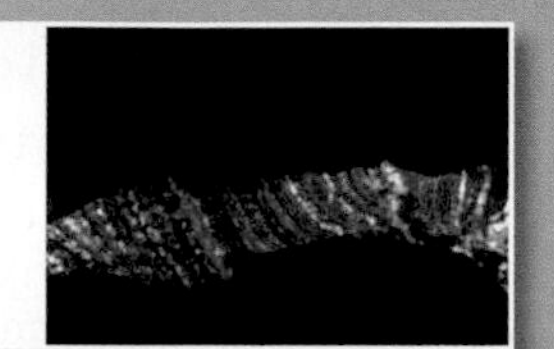

MASTER GENES

During development, the orderly, localized expression of master genes gives rise to the body plan of complex multicelled animals.

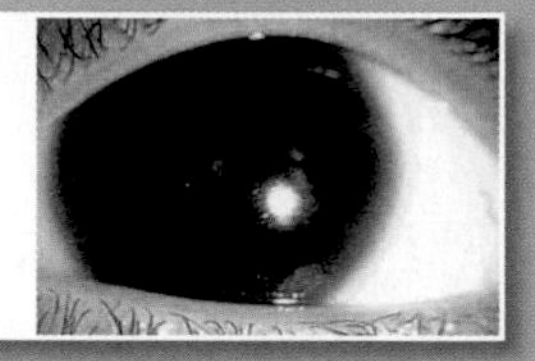

GENE CONTROL IN EUKARYOTES

Examples of gene control in eukaryotes include X chromosome inactivation and male sex determination in mammals, and flower formation in plants.

GENE CONTROL IN PROKARYOTES

Most gene control in prokaryotes occurs at the level of transcription. Fast adjustment of transcription allows these cells to respond quickly to changes in external conditions.

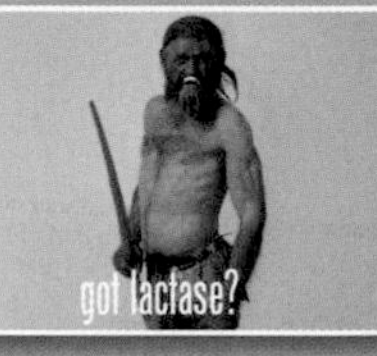

EPIGENETICS

New research is revealing how gene expression patterns that arise during an individual's lifetime in response to environmental pressures can be passed to descendants.

Chapter 11 discusses the cell cycle and how it goes awry in cancer; Chapter 12, the genetic basis of reproduction; Chapter 13, inheritance; Chapter 14, human inheritance patterns; and Chapter 30, plant gene controls. The study of genomes is explained in Section 15.5, and evolutionary processes involving mutation will be discussed more fully in Chapter 17. Section 19.4 discusses the hypothesis that RNA, not DNA, was genetic material in the distant past.

10.1 Between You and Eternity

You are in college, your whole life ahead of you. Your risk of developing cancer is as remote as old age, an abstract statistic that is easy to forget. "There is a moment when everything changes—when the width of two fingers can suddenly be the total distance between you and eternity." Robin Shoulla wrote those words after being diagnosed with breast cancer. She was seventeen years old.

At an age when most young women are thinking about school, friends, parties, and potential careers, Robin was dealing with radical mastectomy: the removal of a breast, all lymph nodes under the arm, and skeletal muscles in the chest wall under the breast. She was pleading with her oncologist not to use her jugular vein for chemotherapy and wondering if she would survive to see the next year (**FIGURE 10.1**).

Robin's ordeal became part of a statistic, one of more than 200,000 new cases of breast cancer diagnosed in the United States each year. About 5,700 of those cases occur in women and men under thirty-four years of age.

Every second, millions of cells in your skin, bone marrow, gut lining, liver, and elsewhere are dividing and replacing their worn-out, dead, and dying predecessors. They do not divide at random; in normal cells, growth and division is tightly regulated. When this control fails, cancer is the outcome.

Cancer is a process in which abnormally growing and dividing cells disrupt body tissues. Mechanisms that normally keep cells from getting overcrowded in tissues are lost, so cancer cell populations may reach extremely high densities. Unless chemotherapy, surgery, or another procedure eradicates them, cancer cells can put an individual on a painful road to death. In developed countries, cancers cause 15 to 20 percent of all human deaths.

Cancer typically begins with a mutation in a gene whose product is part of a system that controls cell growth and division. Such controls govern when and how fast specific genes are transcribed and translated. A cancer-causing mutation may be inherited, or it may arise after birth, as when DNA becomes damaged by environmental agents. If the mutation alters the gene's protein product so that it no longer works properly, one level of control over the cell's growth and division has been lost. You will be considering the impact of gene controls in chapters throughout this book, and also in some chapters of your life.

Robin Shoulla survived. Radical mastectomy is rarely performed today, but it was her only option. Now, seventeen years later, she has what she calls a normal life: career, husband, children. Her goal as a cancer survivor: "To grow very old with gray hair and spreading hips, smiling."

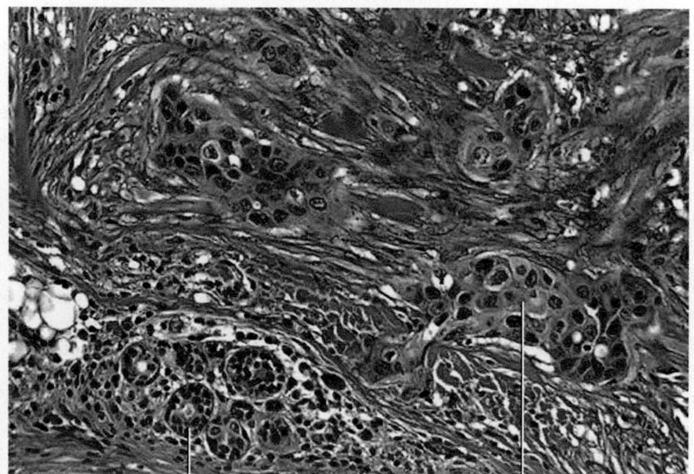

FIGURE 10.1 A case of breast cancer. Left, Robin Shoulla. Right, this light micrograph shows irregular clusters of cancer cells that have infiltrated milk ducts in human breast tissue. Diagnostic tests revealed abnormal cells such as these in Robin's body when she was just seventeen years old.

10.2 Switching Genes On and Off

✔ Controls over gene expression govern the kinds and amounts of substances that are present in a cell at any given time.

A typical cell in your body uses only about 10 percent of its genes at a time. Some of the active genes affect structures and metabolic pathways common to all cells; others are expressed only by certain subsets of cells. For example, most body cells express genes that encode the enzymes of glycolysis, but only immature red blood cells express globin genes. Differentiation (Section 8.7) occurs as different cell lineages begin to express different subsets of their genes during development. Which genes a cell uses determines the molecules it will produce, which in turn determines what kind of cell it will be. Thus, control over gene expression is necessary for proper development of complex, multicelled bodies. It also allows individual cells to respond appropriately to changes in their internal and external environments.

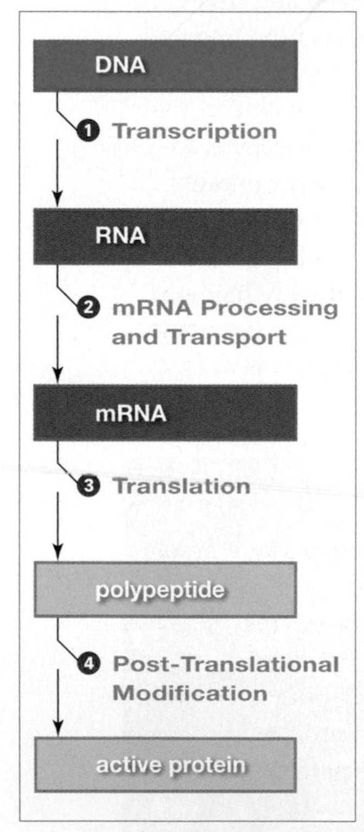

FIGURE 10.2 ▶**Animated** Points of control over gene expression.

Gene Expression Control

The "switches" that turn a gene on or off are molecules or processes that affect individual steps of its expression (**FIGURE 10.2**).

❶ **Transcription** Proteins called **transcription factors** affect whether and how fast a gene is transcribed by binding directly to the DNA. In eukaryotes especially, transcription is often governed by multiple transcription factors; overlapping and opposing effects of these proteins give the cell a nuanced level of control over RNA production.

Transcription factors called **repressors** shut off transcription or slow it down, either by preventing RNA polymerase from accessing a promoter or by impeding its progress along the DNA strand. Some repressors work by binding directly to the promoter. Others bind to a silencer—a region in the DNA that may be thousands of base pairs away from the gene. **Activators** are transcription factors that help RNA polymerase bind to a promoter, so they speed up transcription. Some eukaryotic activators work by binding to DNA sequences called **enhancers**, which, like silencers, may be far away from the gene

enhancer

FIGURE 10.3 Hypothetical part of a chromosome that contains a gene. Molecules that affect the rate of transcription of the gene bind at promoter (yellow) or enhancer (green) sequences.

they affect (**FIGURE 10.3**). An enhancer operating on a distant gene can inappropriately affect a nearby gene; insulators prevent this from occurring. An insulator is a region of DNA that, upon binding a transcription factor, can block the effect of an enhancer on a neighboring gene (**FIGURE 10.4**).

Chromatin structure also affects transcription. The DNA of eukaryotic cells and some archaea wraps around histones (Section 8.4). In these cells, only the regions of DNA that have been unwound from histones are accessible to RNA polymerase. Modifications to histone proteins change the way they interact with DNA wrapped around them, thus affecting transcription. Some modifications make histones release their grip on DNA; others make them tighten it. For example, adding acetyl groups (—$COCH_3$) to a histone loosens the DNA, so enzymes that acetylate histones allow transcription to proceed. Conversely, adding methyl groups (—CH_3) to a histone tightens DNA, so enzymes that methylate histones shut down transcription.

In some specialized eukaryotic cells, a high level of gene expression can be achieved by copying DNA repeatedly without cell division. The result is a polytene chromosome consisting of hundreds or thousands of side-by-side copies of the same DNA molecule (the chromosome shown in **FIGURE 10.4** is polytene). Transcription of one gene occurs simultaneously on all of the DNA strands, quickly producing a lot of mRNA.

❷ **mRNA Processing and Transport** As you know, transcription in eukaryotes occurs in the nucleus, and translation occurs in cytoplasm (Section 9.5). A newly transcribed mRNA can pass through pores of a nuclear envelope only after it has been processed appropriately—spliced, capped, and finished with a poly-A tail. Mechanisms that delay these post-transcriptional modifications also delay the mRNA's appearance in cytoplasm for translation.

Control over post-transcriptional modification can also affect the form of a protein. Consider RNA transcribed from the gene for fibronectin, a protein produced by cells of vertebrate animals. Two cell types splice this RNA alternatively, so they produce different mRNAs—and different forms of fibronectin. Liver cells produce a soluble form that circulates in blood plasma.

promoter | exon1 | intron | exon2 | enhancer

transcription start site

transcription end

Fibroblasts produce an insoluble form that is a major component of extracellular matrix (Section 4.11).

In eukaryotic cells, most mRNAs are delivered to organelles or specific regions of cytoplasm, thus allowing proteins to be produced close to where they are being used. In an egg, cytoplasmic localization of mRNA is crucial for proper development of the future embryo. How does localization occur? A short nucleotide sequence near an mRNA's poly-A tail is like a zip code that specifies a particular destination. Proteins that attach to the zip code drag an mRNA along cytoskeletal elements to its destination. Other proteins influence localization by interacting with mRNA-binding proteins. mRNA localization also occurs in prokaryotes, but the mechanism is not yet understood.

3 Translation Production of the many molecules that participate in translation is a major point of control in eukaryotic cells. An mRNA's sequence also affects translation. For example, proteins that bind to an mRNA's zip code region prevent transcription from occurring before the mRNA is delivered to its final destination. As another example, consider that the longer an mRNA lasts, the more protein can be made from it. Enzymes begin disassembling an mRNA as soon as it arrives in cytoplasm. The fast turnover allows a cell to adjust protein synthesis quickly. How long an mRNA persists depends on its base sequence, the length of its poly-A tail, and which proteins are attached to it.

In eukaryotes, translation of a particular mRNA can be shut down by microRNAs, which are tiny bits of noncoding RNA. A microRNA is complementary in sequence to part of an mRNA, and when the two show up together in cytoplasm they base-pair to form a small double-stranded region of RNA. By a process called RNA interference, any double-stranded RNA is cut up into small bits that are taken up by special enzyme complexes. These complexes then destroy every RNA in a cell that can base-pair with the bits. Thus, expression of a microRNA results in the destruction of all mRNA complementary to it.

Double-stranded RNA is also a factor in control over translation in prokaryotes. For example, bacteria can shut off translation of a particular mRNA by expressing an antisense RNA (one that is complementary in sequence to the mRNA). The two molecules hybridize (Section 8.5) to form a double-stranded RNA that ribosomes cannot translate. As another example, some bacterial mRNAs can loop back on themselves to form a small double-stranded region. Translation only occurs when this structure is unraveled, for example by exposure to heat.

4 Post-Translational Modification Many newly synthesized polypeptide chains must be modified before they become functional. For example, some enzymes become active only after they have been phosphorylated (Section 5.6). Such post-translational modifications inhibit, activate, or stabilize many molecules, including enzymes that participate in transcription and translation.

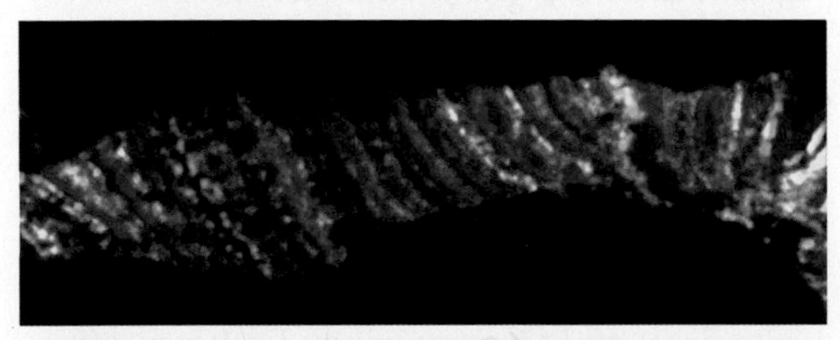

FIGURE 10.4 Fluorescence micrograph showing part of a chromosome in a fruit fly salivary gland cell. DNA appears blue. Red and green show the locations of two transcription factors bound to insulator sequences (yellow shows where these two colors overlap). The transcription factors are inhibiting the effect of nearby enhancers.

activator Regulatory protein that increases the rate of transcription when it binds to a promoter or enhancer.
enhancer Binding site in DNA for proteins that enhance the rate of transcription.
repressor Regulatory protein that blocks transcription.
transcription factor Regulatory protein that influences transcription; e.g., an activator or repressor.

TAKE-HOME MESSAGE 10.2

What is gene expression control?

✔ Gene expression can be switched on or off, or speeded or slowed, by molecules and processes that operate at each step.

✔ These gene expression controls allow cells to respond appropriately to environmental change. They are also critical for differentiation and development in multicelled eukaryotes.

CREDIT: (4) *Journal of Bioscience*, Volume 36, Number 3, August 2011, Indian Academy of Sciences, Springer.

10.3 Master Genes

✔ Cascades of gene expression govern the development of a complex, multicelled body.

As an animal embryo develops, its cells differentiate and form tissues, organs, and body parts. The entire process of development is driven by cascades of master gene expression. The products of **master genes** affect the expression of many other genes. Expression of a master gene causes other genes to be expressed, which in turn cause other genes to be expressed, and so on.

The orchestration of gene expression during animal development begins when different maternal mRNAs localize to different regions of cytoplasm in an egg as it forms. These mRNAs are translated only after the egg is fertilized. Then, their protein products diffuse away, forming gradients that span the developing embryo. The position of a cell within the embryo determines how much of these proteins it is exposed to. This in turn determines which master genes it turns on. The products of those master genes also form in gradients that diffuse away from cells expressing them. Still other master genes are transcribed depending on where a cell falls within these gradients, and so on. Eventually, the products of master genes cause undifferentiated cells to differentiate, and specialized structures form in specific regions of the embryo (**FIGURE 10.5**).

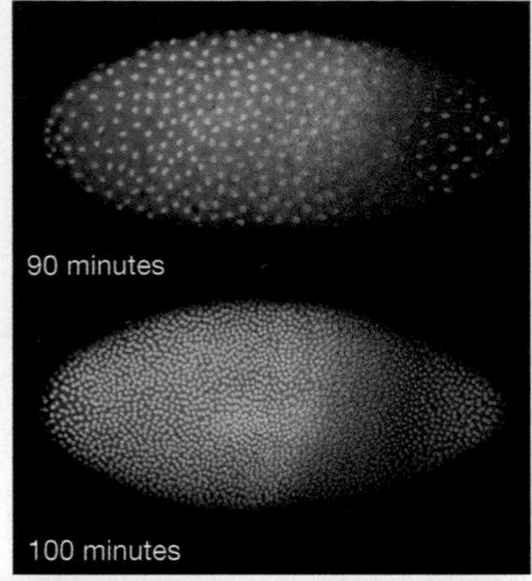

A 100 minutes after fertilization, the master gene *even-skipped* is expressed (in red) only where two transcription factors (green and blue) overlap. The transcription factors are the protein products of maternal mRNAs.

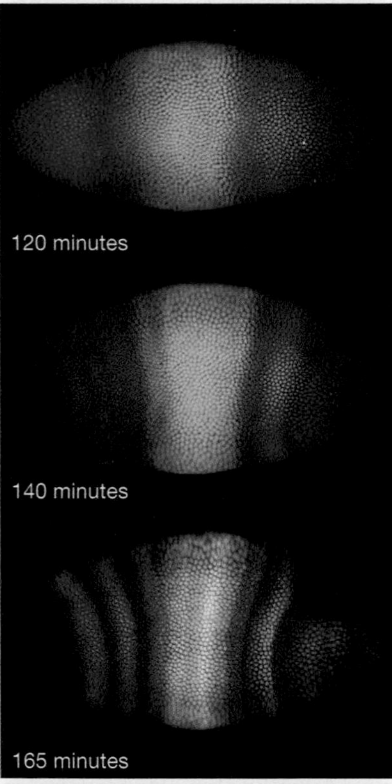

B By 165 minutes after fertilization, the products of several master genes, including the two shown here in green and blue, have confined the expression of *even-skipped* (red) to seven stripes. (Pink and yellow areas are regions in which red fluorescence has overlapped with blue or green.)

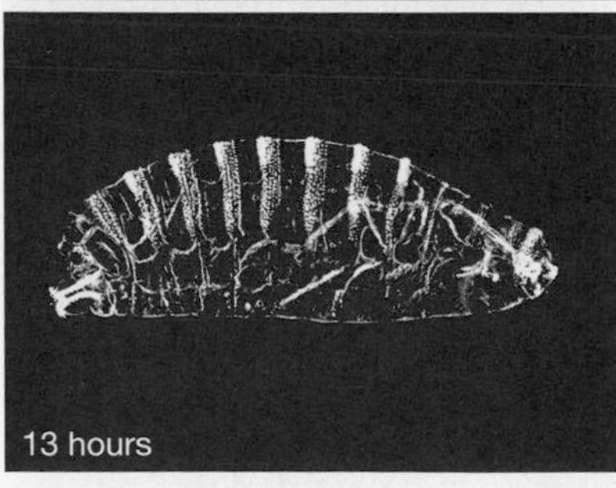

C One day later, seven segments have developed. The position of the segments corresponds to the position of the *even-skipped* stripes.

FIGURE 10.5 How gene expression control makes a fly, as illuminated by the formation of segments in a *Drosophila* embryo.

Expression of different master genes is shown by different colors in fluorescence microscopy images of whole embryos at successive stages of development (time after fertilization is indicated). Bright dots are individual nuclei.

Homeotic Genes

A **homeotic gene** is a type of master gene whose expression directs the formation of a specific body part such as an eye, leg, or wing. Animal homeotic genes encode transcription factors with a homeodomain, which is a region of about sixty amino acids that can bind directly to a promoter or some other sequence of nucleotides in a chromosome (**FIGURE 10.6A**).

The function of many homeotic genes has been discovered by deliberately manipulating their expression. In an experiment called a **knockout**, researchers inactivate a gene by introducing a mutation that prevents its expression, or by deleting it entirely. A knockout organism (one that has had a gene knocked out) may differ from normal individuals, and the differences are clues to the function of the missing gene product.

Homeotic genes are often named for what happens when a mutation alters their function. For example, fruit flies with a mutation that affects their *antennapedia* gene (*ped* means foot) have legs in place of antennae (**FIGURE 10.6B**). The *dunce* gene is required for learning and memory. *Wingless*, *wrinkled*, and *minibrain* are self-explanatory. *Tinman* is necessary for development of a heart. Flies with a mutated *groucho* gene have extra

groucho mutation

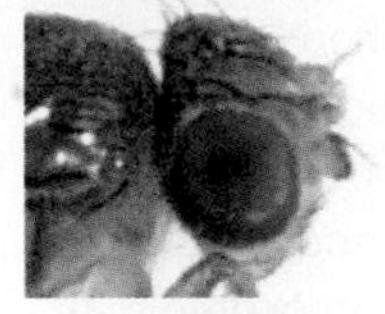

normal fly head

bristles that resemble bushy eyebrows (left). Flies with a mutated *eyeless* gene lack eyes (**FIGURE 10.6C**). One gene was named *toll*, after what its German discoverer exclaimed upon seeing the disastrous effects of the mutation (*toll* is German slang that means "cool!").

Homeotic genes control development by the same mechanisms in all multicelled eukaryotes, and many are interchangeable among different species. Thus, we can infer that they evolved in the most ancient eukaryotic cells. Homeodomains often differ among species only in conservative substitutions (one amino acid has replaced another with similar chemical properties). Consider the *eyeless* gene. Eyes form in embryonic fruit flies wherever this gene is expressed, which is normally in tissues of the head only. If the *eyeless* gene is expressed in another part of the developing embryo, eyes form there too (**FIGURE 10.6D**).

Humans, squids, mice, fishes, and many other animals have a gene called *PAX6*, which is very similar in DNA sequence to the *eyeless* gene of flies. In humans, mutations in *PAX6* cause eye disorders such as aniridia, in which the irises are underdeveloped or missing (**FIGURE 10.6E**). If a functional *PAX6* gene from a human or a mouse is inserted into a fly, it has the same effect as the *eyeless* gene: An eye forms wherever it is expressed. (Because *PAX6* is just a switch, the eye that forms is a fly eye, not a human or mouse eye.) The same principle applies in reverse: The *eyeless* gene from flies switches on eye formation in frogs. Such studies are evidence of shared ancestry among these evolutionarily distant animals.

homeotic gene Type of master gene with a homeodomain; its expression directs formation of a specific body part in development.
knockout An experiment in which a gene is deliberately inactivated in a living organism; also, an organism that has a knocked-out gene.
master gene Gene encoding a product that affects the expression of many other genes.

TAKE-HOME MESSAGE 10.3

How do genes control development in animals?

- ✔ Animal development is orchestrated by cascades of master gene expression in embryos.
- ✔ The expression of homeotic genes during development directs the formation of specific body parts.
- ✔ Homeotic genes that function in similar ways in evolutionarily distant animals are evidence of shared ancestry.

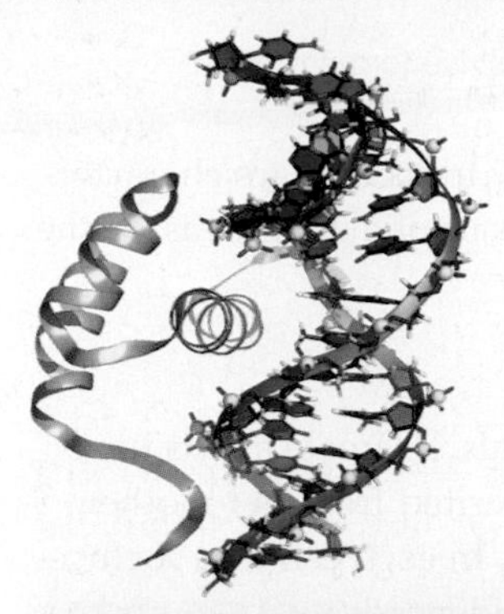

A A model of the protein product (in gold) of the homeotic gene *antennapedia* attached to a promoter. The homeodomain is the region that binds to the DNA. Expression of *antennapedia* in embryonic tissues of the insect thorax causes legs to form.

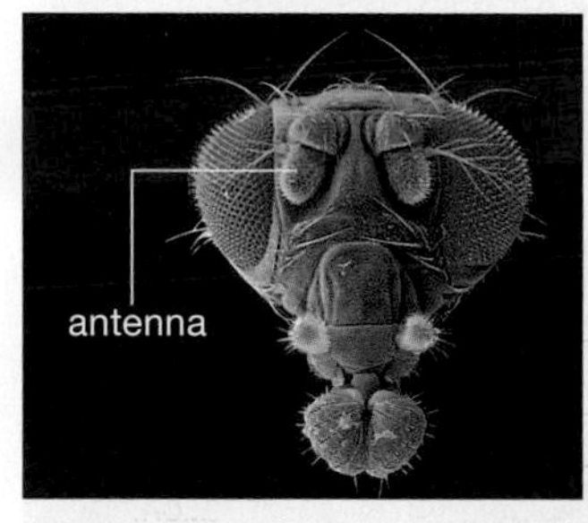

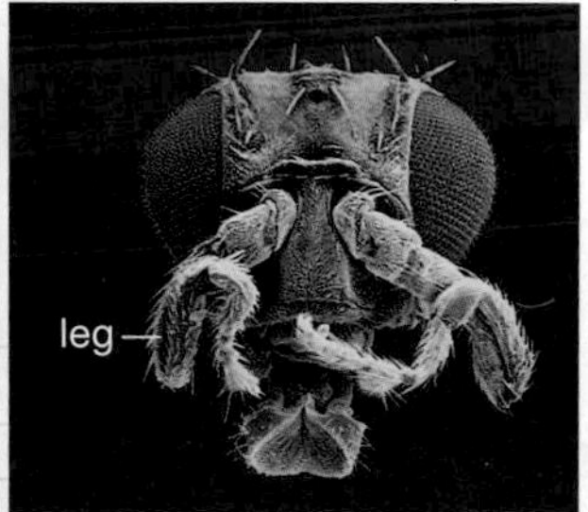

B The head of a normal fruit fly (left) has two antennae. Right, a mutation that triggers expression of the *antennapedia* gene in embryonic tissues of the head causes legs to form instead of antennae.

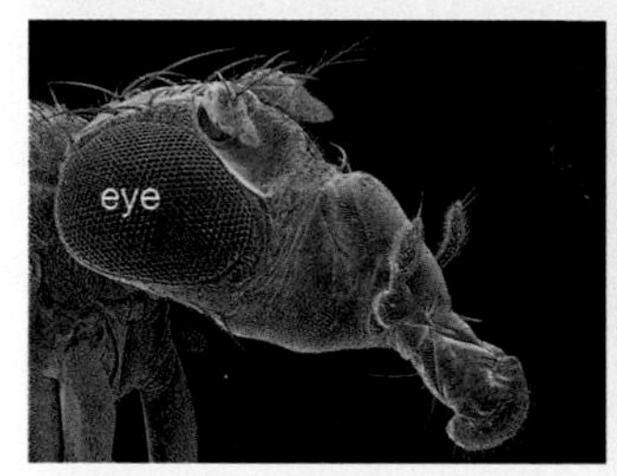

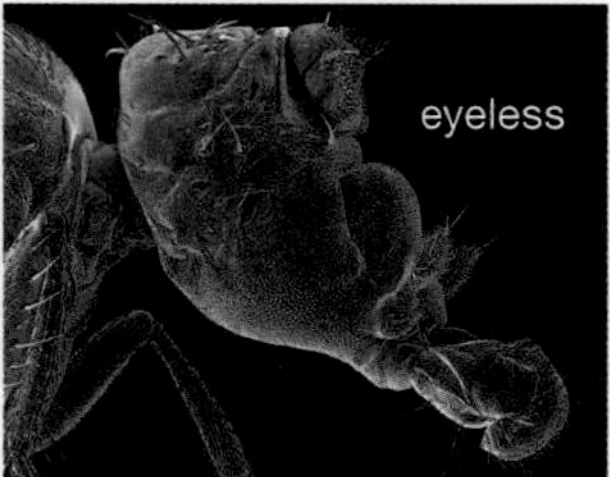

C A normal fruit fly (left) has large, round eyes. A fruit fly with a mutation in its *eyeless* gene (right) develops without eyes.

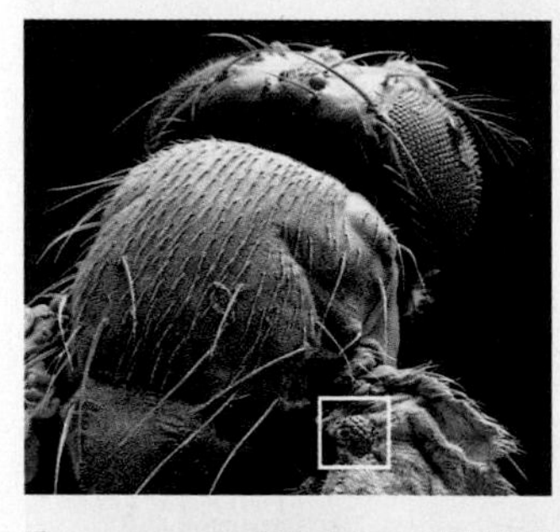

D Eyes form wherever the *eyeless* gene is expressed in fly embryos—here, on the head and also on the wing.

The *PAX6* gene of humans, mice, squids, and some other animals is so similar to *eyeless* that it similarly triggers eye development in fruit flies.

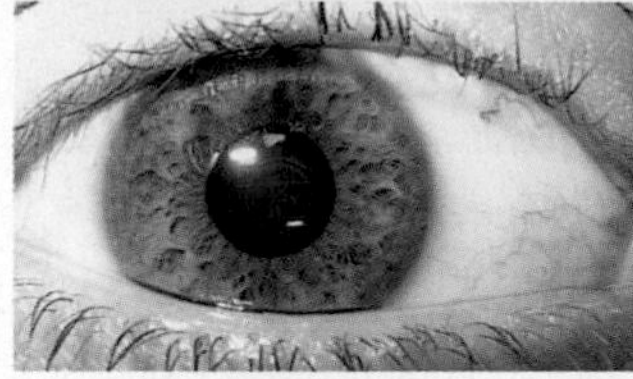

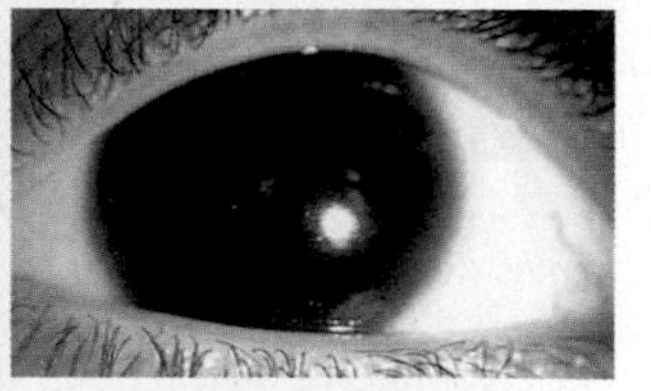

E A normal human eye has a colored iris surrounding the pupil (dark area where light enters). Mutations in *PAX6* cause eyes to develop without an iris, a condition called aniridia.

FIGURE 10.6 Some effects of homeotic gene mutations.

10.4 Examples of Gene Control in Eukaryotes

✔ Selective gene expression gives rise to many traits.

Gene control influences many traits that are characteristic of humans and other eukaryotic organisms, as the following examples illustrate.

X Marks the Spot

In humans and other mammals, a female's cells have two X chromosomes, one inherited from her mother, the other one from her father. In each cell, one X chromosome is always tightly condensed (FIGURE 10.7A). We call the condensed X chromosomes **Barr bodies**, after Murray Barr, who discovered them. Condensation inhibits transcription, so most of the genes on a Barr body are not expressed. This **X chromosome inactivation** ensures that only one of the two X chromosomes in a female's cells is active, thus equalizing expression of X chromosome genes between the sexes—a mechanism called **dosage compensation**. The body cells of male mammals (XY) have one set of X chromosome genes. Body cells of female mammals (XX) have two sets, but female embryos do not develop properly when both sets are expressed.

X chromosome inactivation occurs when an embryo is a ball of about 200 cells. In humans and most other mammals, it occurs independently in every cell of a female embryo. Which X chromosome condenses is random: The paternal or maternal X chromosome may get inactivated in any cell. Once the selection is made in a cell, it is permanent. All of that cell's descendants make the same selection as they continue dividing and forming tissues. As a result of random inactivation of maternal and paternal X chromosomes, an adult female mammal is a "mosaic" for the expression of X chromosome genes. She has patches of tissue in which genes of the maternal X chromosome are expressed, and patches in which genes of the paternal X chromosome are expressed (FIGURE 10.7B).

How does just one of two X chromosomes get inactivated? An X chromosome gene called *XIST* is transcribed on only one of the two X chromosomes. The gene's product, a long noncoding RNA, sticks to the chromosome that expresses the gene. The RNA coats the chromosome, and by an unknown mechanism causes it to condense into a Barr body. Thus, transcription of the *XIST* gene keeps the chromosome from transcribing other genes. The other chromosome does not express *XIST*, so it does not get coated with RNA; its genes remain available for transcription. How a cell "chooses" which X chromosome will express *XIST* is still unknown.

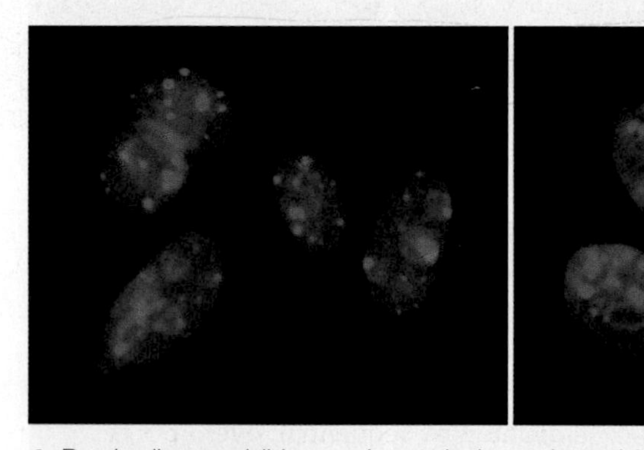

A Barr bodies are visible as red spots in the nucleus of the four XX cells on the left. Compare the nucleus of two XY cells to the right.

B When this calico cat was an embryo, one of the two X chromosomes was inactivated in each of her cells. The descendants of the cells formed her adult body, which is a mosaic for expression of X chromosome genes. Black fur arises in patches where genes on the X chromosome inherited from one parent are expressed; orange fur arises in patches where genes on the X chromosome inherited from the other parent are expressed.

FIGURE 10.7 ▶**Animated** X chromosome inactivation.

Male Sex Determination in Humans

Only a few of the 1,113 genes on the human X chromosome are associated with traits such as body fat distribution that differ between males and females. Most X chromosome genes govern nonsexual traits such as blood clotting and color perception. Such genes are expressed in both males and females. Males, remember, also inherit one X chromosome.

The human Y chromosome carries only 128 genes, but one of them is *SRY*—the master gene for male sex determination in mammals. Its expression in XY embryos triggers the formation of testes, which are male reproductive organs. Some of the cells in testes make testosterone, a sex hormone that controls the emergence of male secondary sexual traits such as

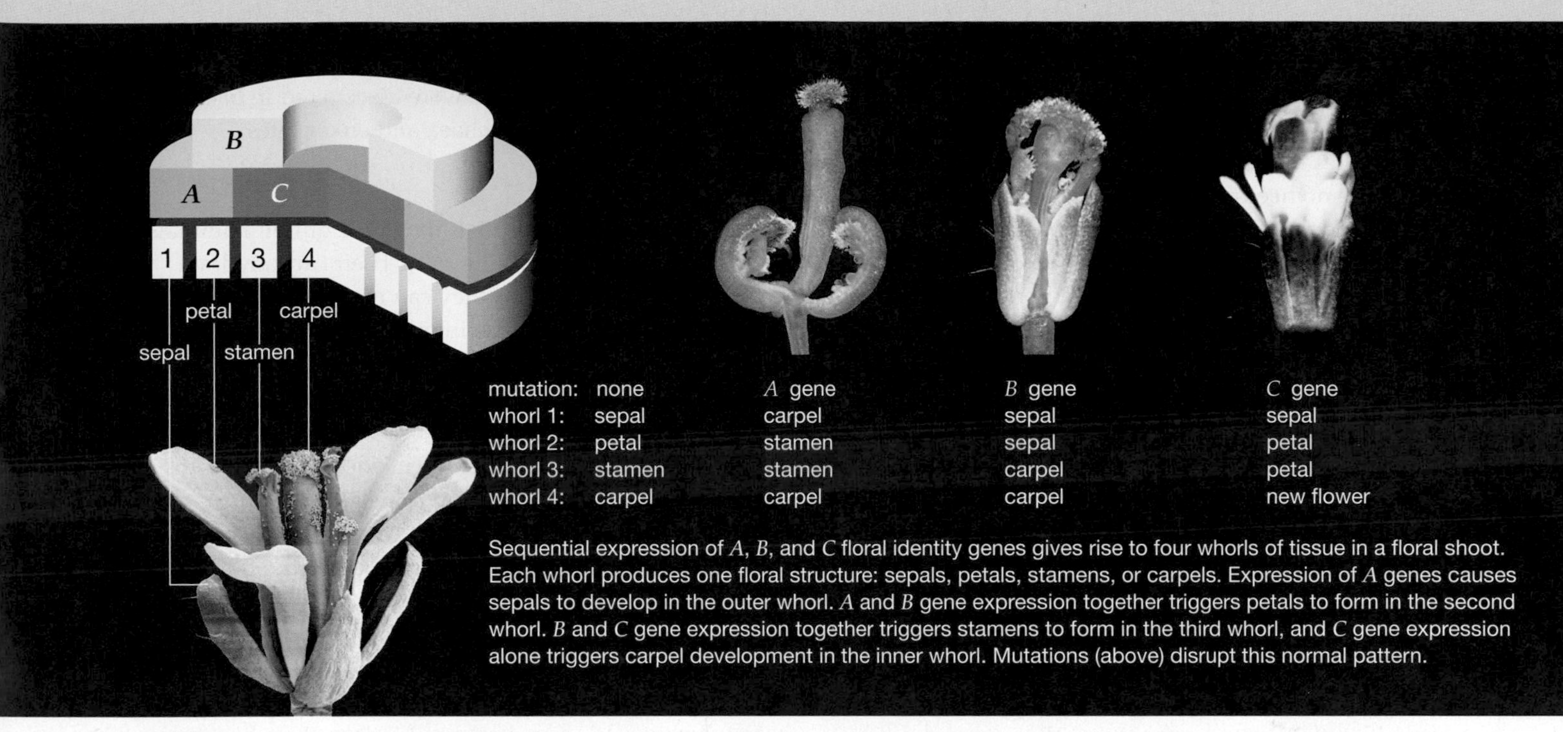

FIGURE 10.8 ▶**Animated** Control of flower formation, as revealed by mutations in *Arabidopsis thaliana*.

facial hair, increased musculature, and deepened voice. We know that *SRY* is the master gene that controls emergence of male sexual traits because mutations in this gene cause XY individuals to develop external genitalia that appear female. An XX embryo has no Y chromosome, no *SRY* gene, and much less testosterone, so primary female reproductive organs (ovaries) form instead of testes. Ovaries make estrogens and other sex hormones that will govern the development of female secondary sexual traits, such as enlarged, functional breasts, and fat deposits around the hips and thighs.

Flower Formation

In flowering plants, populations of cells in a shoot tip may give rise to a flower instead of leaves. Studies of mutations in thale cress, *Arabidopsis thaliana*, revealed the gene control behind this switch in development. Transcription factors produced by three sets of floral identity genes (called *A*, *B*, and *C*) guide the process. These genes are switched on by environmental cues such as seasonal changes in the length of night, as you will see in Section 30.9.

Barr body Inactivated X chromosome in a cell of a female mammal. The other X chromosome is active.
dosage compensation Mechanism in which X chromosome inactivation equalizes gene expression between males and females.
X chromosome inactivation Developmental shutdown of one of the two X chromosomes in the cells of female mammals.

When a flower forms at the tip of a shoot, differentiating cells form whorls of tissue, one over the other like layers of an onion. Each whorl produces one type of floral structure—sepals, petals, stamens, or carpels. This pattern is dictated by sequential, overlapping expression of the *ABC* genes (**FIGURE 10.8**). The *A* genes are switched on first in a shoot tip, and their products trigger events that cause the outer whorl to form and give rise to sepals. *B* genes switch on before *A* genes turn off. Together, *A* and *B* gene products cause the second whorl to form and produce petals. Next, *A* genes turn off and the *C* gene switches on (there is only one *C* gene). Together, the products of the *B* and *C* genes trigger formation of the third whorl, which makes stamens. Finally, the *B* gene turns off, and the *C* gene product on its own gives rise to the fourth, inner whorl, which produces the carpel.

TAKE-HOME MESSAGE 10.4

How do genes control development in animals?

- ✔ X chromosome inactivation balances expression of X chromosome genes between female (XX) and male (XY) mammals.
- ✔ *SRY* gene expression triggers the development of male traits in mammals.
- ✔ In plants, expression of *ABC* floral identity genes governs development of the specialized parts of a flower.

CREDITS: (8) top left, © Cengage Learning; bottom left, Juergen Berger, Max Planck Institute for Developmental Biology, Tuebingen, Germany; (8A, B gene) © Jose Luis Riechmann; (8C gene) Image by Marty Yanofsky.

10.5 Examples of Gene Control in Prokaryotes

✔ Bacteria control gene expression mainly by adjusting the rate of transcription.

Prokaryotes do not undergo development, so these cells have no need for master genes. However, they do respond to environmental fluctuations by adjusting gene expression, mainly at the level of transcription. For example, when a preferred nutrient becomes available, a bacterium begins transcribing genes whose products allow the cell to use the nutrient. When the nutrient is no longer available, transcription of those genes stops. Thus, the cell does not waste energy and resources producing gene products that are not needed.

In bacteria, genes that are used together often occur together on the chromosome, one after the other. One promoter precedes the genes, so all are transcribed together into a single RNA strand. Thus, their transcription can be controlled in a single step that typically involves repressor binding to an **operator**, which is a type of silencer sequence in DNA. A group of genes together with a promoter and one or more operators that control their transcription are collectively called an **operon**. Operons were discovered in bacteria, but they also occur in archaea and eukaryotes.

The *lac* Operon

Escherichia coli bacteria that live in the gut of mammals dine on nutrients traveling past. Their carbohydrate of choice is glucose, but they can make use of other sugars such as the lactose in milk. An operon called *lac* allows *E. coli* cells to metabolize lactose (**FIGURE 10.9**). The *lac* operon includes three genes and a promoter flanked by two operators ❶.

One gene in the *lac* operon encodes an active transport protein (Section 5.7) that brings lactose across the plasma membrane into the cell. Another encodes an enzyme (β-galactosidase) that breaks the bond between lactose's two monosaccharide monomers, glucose and galactose. The third gene encodes an enzyme whose function is still being investigated. Bacteria make these three proteins only when lactose is present. When lactose is not present, a repressor binds to the two opera-

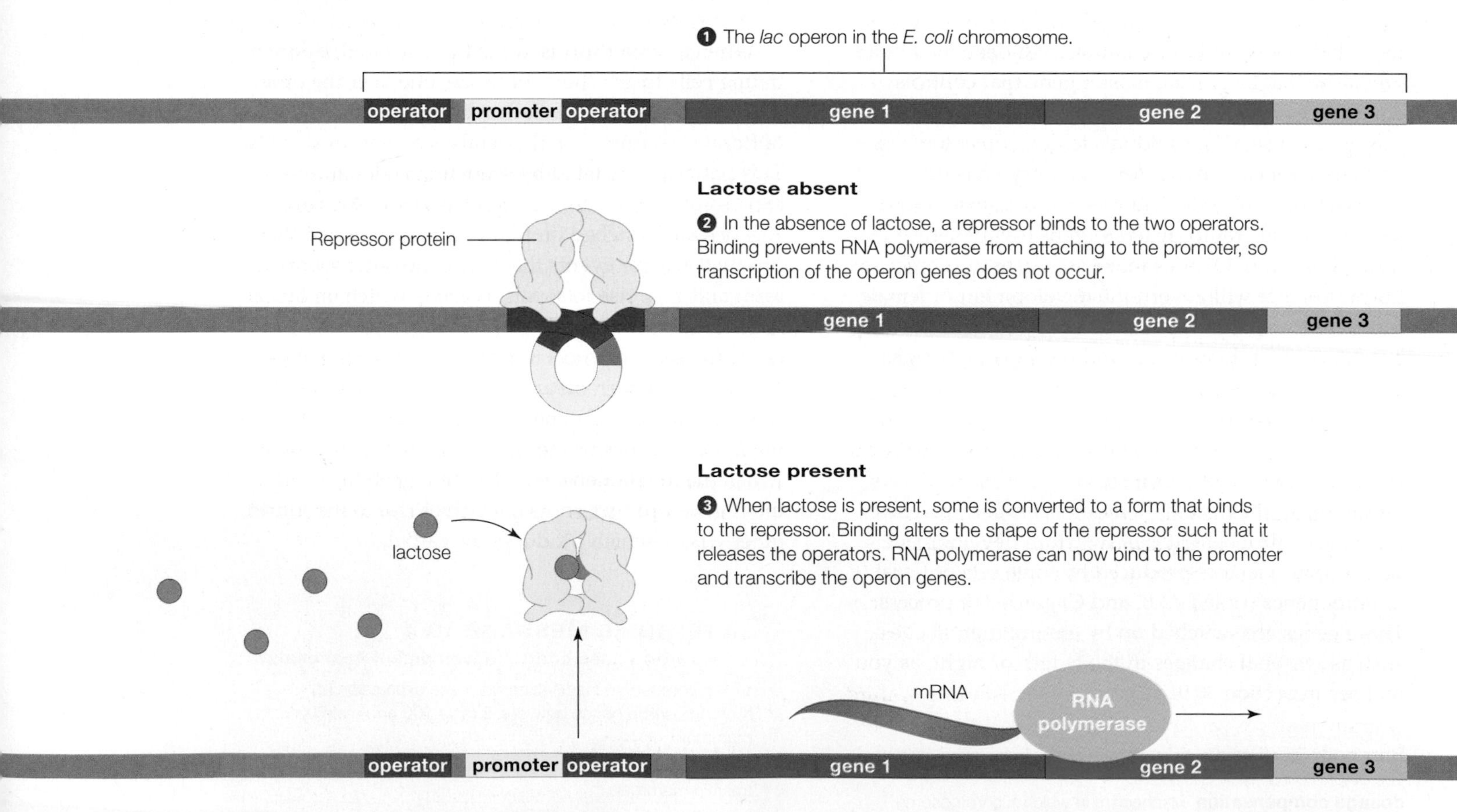

FIGURE 10.9 ▶**Animated** Example of gene control in bacteria: the lactose operon on a bacterial chromosome. The operon consists of a promoter flanked by two operators, and three genes for lactose-metabolizing enzymes. **FIGURE IT OUT** What portion of the operon binds RNA polymerase when lactose is present?

Answer: The promoter

tors and twists the region of DNA with the promoter into a loop ❷. RNA polymerase cannot bind to the twisted-up promoter, so the *lac* operon's genes cannot be transcribed. When lactose is present, some of it is converted to another sugar that binds to the repressor and changes its shape. The altered repressor releases the operators and the looped DNA unwinds. The promoter is now accessible to RNA polymerase, and transcription of lactose-metabolizing genes begins ❸.

In bacteria, glucose metabolism requires fewer enzymes than lactose metabolism does, so it requires less energy and fewer resources. Accordingly, when both lactose and glucose are present, the cells will use up all of the available glucose before switching to lactose metabolism. How does a cell ignore the presence of one sugar while it uses another? Another level of gene control over the *lac* operon regulates this metabolic switch, but how it works is still being debated.

Lactose Intolerance

Like *E. coli*, humans and other mammals break down lactose, but most do so only when young. An individual's ability to digest lactose ends at a species-specific age. In the majority of humans worldwide, this switch occurs at about age five, when transcription of the gene for lactase (a type of β-galactosidase) slows. The resulting decline in production of this enzyme causes a common condition known as lactose intolerance.

Cells in the intestinal lining secrete lactase into the small intestine, where the enzyme cleaves lactose into its glucose and galactose monomers. These sugars are absorbed directly by the small intestine, but lactose and other disaccharides are not. Thus, when lactase production slows, lactose passes undigested through the small intestine. The lactose ends up in the large intestine, which hosts huge numbers of *E. coli* and a variety of other bacteria. These resident organisms respond to the presence of lactose by switching on their *lac* operons. Carbon dioxide, methane, hydrogen, and other gaseous products of their various fermentation reactions accumulate quickly in the large intestine, distending its wall and causing pain. Other products of their metabolism disrupt the solute–water balance inside the large intestine, and diarrhea results.

Not everybody is lactose intolerant. About one-third of human adults carry a mutation that allows them to digest milk; this mutation is more common in some

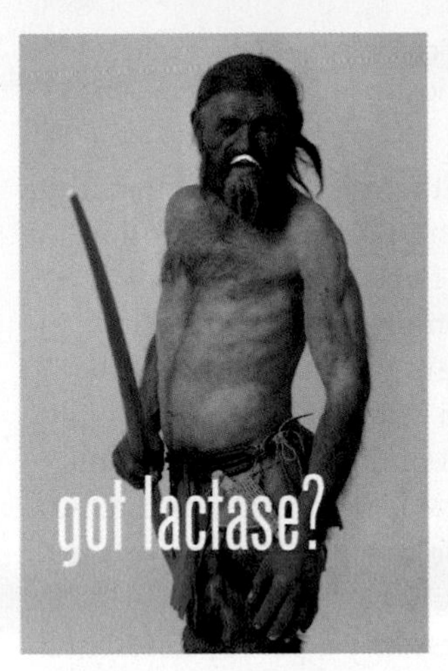

populations than in others. Recent analyses of DNA from well-preserved skeletons shows that the vast majority of adult humans living about 8,000 years ago in Europe were lactose intolerant. Around that time, a mutation appeared in the DNA of prehistoric people inhabiting a region between what is now central Europe and the Balkans. This mutation allowed its bearers to continue digesting milk as adults, and it spread rapidly to the rest of the continent along with the practice of dairy farming. Today, most adults of northern and central European ancestry are able to digest milk because they carry this mutation, a single base-pair substitution in an enhancer that controls the lactase gene promoter. Other mutations in the same enhancer arose independently in North Africa, southern Asia, and the Middle East. Some people descended from these populations can continue to digest milk as adults.

Riboswitches

mRNAs that regulate their own translation are common in bacteria. These mRNAs have small sequences called riboswitches that bind to a metal ion, cofactor, or metabolic product. The binding causes a conformational change in the mRNA that affects its translation. Consider what happens when bacteria make vitamin B_{12}. The enzymes involved in synthesis of this vitamin are produced from mRNAs that have riboswitches. In this case, the riboswitches bind to vitamin B_{12}. Binding changes the shape of the mRNA so that ribosomes can no longer attach to it, so production of B_{12}-making enzymes stops. This example also illustrates feedback inhibition (Section 5.5).

TAKE-HOME MESSAGE 10.5

What are some outcomes of gene control in prokaryotes?

✔ The bacterial *lac* operon allows expression of lactose-metabolizing proteins only when lactose is present.

✔ Most adult humans do not produce the enzyme that breaks down lactose. When undigested lactose enters the large intestine, resident bacteria give rise to symptoms of lactose intolerance.

operator Part of an operon; a DNA binding site for a repressor.
operon Group of genes together with a promoter–operator DNA sequence that controls their transcription.

10.6 Epigenetics

✔ Methylations and other modifications that accumulate in DNA during an individual's lifetime can be passed to offspring.

You learned in Section 10.2 that the addition of methyl groups to histone proteins suppresses transcription. Direct methylation of DNA nucleotides also suppresses transcription, often more permanently than histone modifications. Once a particular nucleotide has become methylated in a cell's DNA, it will usually stay methylated in the DNA of the cell's descendants. Methylation and other heritable modifications to DNA that affect its function but do not alter the nucleotide sequence are said to be **epigenetic**.

DNA methylation is necessarily a part of differentiation, so it begins very early in embryonic development: Genes actively expressed in a zygote (the first cell of a new individual) become silenced as their promoters get methylated. This silencing is the basis of selective gene expression that drives differentiation. During development, and also during the remainder of the individual's life, each cell's DNA continues to acquire methylations. Between 3 and 6 percent of the DNA has been methylated in a normal, differentiated body cell.

In eukaryotes, methyl groups are usually added to a cytosine that is followed by a guanine (**FIGURE 10.10**), but which of these cytosines are methylated varies by the individual. This is because methylation is influenced by environmental factors. For instance, humans conceived during a famine end up with an unusually low number of methyl groups attached to the nucleotides of certain genes. The product of one of those genes is a hormone that fosters prenatal growth and development. The resulting increase in expression of this gene may offer a survival advantage in a poor nutritional environment.

Methyl groups are also added to nucleotides by chance during DNA replication, so cells that divide a lot tend to have more methyl groups in their DNA than inactive cells. Free radicals and toxic chemicals add more methyl groups. These and other factors that influence DNA methylation can have multigenerational effects. When an organism reproduces, it passes its DNA to offspring. Methylation of parental DNA is normally "reset" in the zygote, with new methyl groups being added and old ones being removed. This reprogramming does not remove all of the parental methyl groups, however, so methylations acquired during an individual's lifetime can be passed to future offspring.

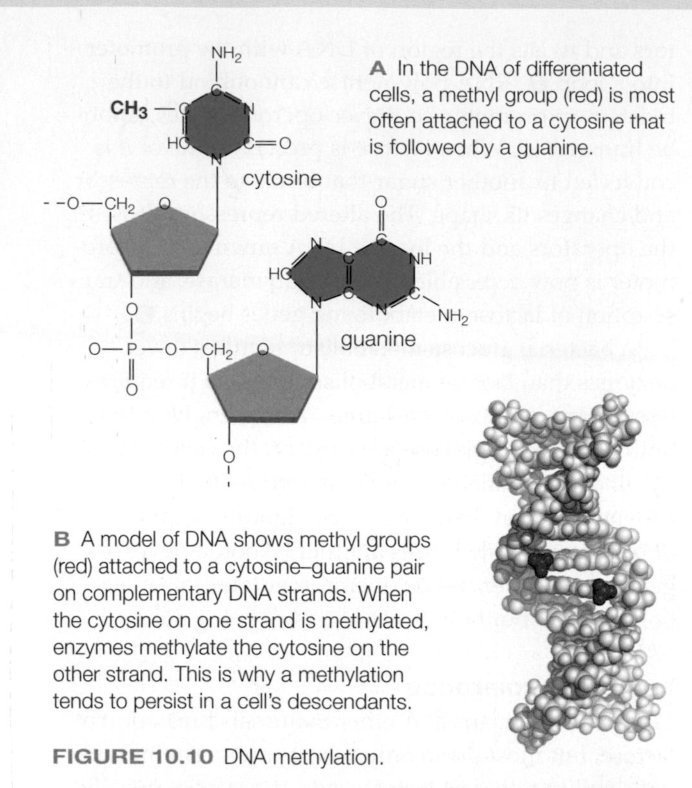

A In the DNA of differentiated cells, a methyl group (red) is most often attached to a cytosine that is followed by a guanine.

B A model of DNA shows methyl groups (red) attached to a cytosine–guanine pair on complementary DNA strands. When the cytosine on one strand is methylated, enzymes methylate the cytosine on the other strand. This is why a methylation tends to persist in a cell's descendants.

FIGURE 10.10 DNA methylation.

FIGURE 10.11 An epigenetic effect. Grandsons of men who endured a famine when they were boys tend to live about 32 years longer than grandsons of men who ate well during the same winter.

epigenetic Refers to heritable changes in gene expression that are not the result of changes in DNA sequence.

Between You and Eternity (revisited)

An effective cancer treatment must eliminate cancerous cells from a person's body. However, cancer cells are body cells, so drugs and other therapies that kill them also kill normal body cells. There is often a fine line between eliminating cancer from a patient's body, and killing the patient.

As you will see in Chapter 11, a normal cell has layers upon layers of gene expression controls and fail-safe mechanisms that determine when division occurs and when it does not. This finely tuned system becomes unbalanced in a cancer cell, so that some genes are expressed at a higher level than they should be, and some are expressed at a lower level.

For decades, we have been studying controls over cell division and how they go awry in cancer, because understanding how a cancer cell differs from a normal cell at a molecular level offers our best chance for developing a cure. In the early 1980s, researchers discovered that mutations in some genes predispose individuals to develop certain kinds of cancer. These genes are tumor suppressors, so named because tumors are more likely to occur when these genes mutate. Two examples are *BRCA1* and *BRCA2*: A mutated version of one or both of these genes is often found in breast and ovarian cancer cells. Because mutations in genes such as *BRCA* can be inherited, cancer is not only a disease of the elderly, as Robin Shoulla's story illustrates. Robin is one of the unlucky people who carry mutations in both of her *BRCA1* and *BRCA2* genes. A woman who carries one of three particularly dangerous *BRCA* mutations has an 80 percent chance of developing breast cancer before the age of seventy.

BRCA genes are master genes whose protein products help maintain the structure and number of chromosomes in a dividing cell. The multiple functions of these proteins are still being unraveled. We do know they participate directly in DNA repair (Section 8.6), so any mutations that alter this function also alter the cell's capacity to repair damaged DNA. Other mutations are likely to accumulate, and that sets the stage for cancer.

The products of *BRCA* genes also bind to receptors for the hormones estrogen and progesterone, which are abundant on cells of breast and ovarian tissues. Binding suppresses transcription of growth factor genes in these cells. Among other effects, growth factors stimulate cells to divide during normal, cyclic renewals of breast and ovarian tissues. When a mutation alters a *BRCA* gene so that its product cannot bind to hormone receptors, the cells overproduce growth factors. Cell division goes out of control, and tissue growth becomes disorganized. In other words, cancer develops (Section 11.6 returns to this topic).

Researchers discovered that the RNA product of the *XIST* gene (Section 10.4) localizes abnormally in breast cancer cells. In those cells, both X chromosomes are active. It makes sense that two active X chromosomes would have something to do with abnormal gene expression, but why the RNA product of an unmutated *XIST* gene does not localize properly in cancer cells remains a mystery.

Mutations in the *BRCA1* gene may be part of the answer. The researchers found that the protein product of the *BRCA1* gene physically associates with the RNA product of the *XIST* gene. They were able to restore proper XIST RNA localization—and proper X chromosome inactivation—by restoring the function of the *BRCA1* gene product in breast cancer cells.

Inheritance of epigenetic modifications can adapt offspring to an environmental challenge much more quickly than evolution (we return to evolutionary processes in Chapter 17). Epigenetic modifications are not considered to be evolutionary because the underlying DNA sequence does not change. Even so, they may persist for generations after an environmental challenge has faded. For example, a recent study showed that grandsons of boys who endured a winter of famine (**FIGURE 10.11**) tend to outlive—by far—grandsons of boys who overate at the same age. The effect is presumed to be due to epigenetic modification because these results were corrected for socioeconomic and genetic factors. In a similar study, nine-year-old boys whose fathers smoked cigarettes before age eleven were overweight compared with boys whose fathers did not smoke in childhood.

TAKE-HOME MESSAGE 10.6

Can gene expression patterns be inherited?

✔ Epigenetic modifications of chromosomal DNA, including DNA methylations acquired during an individual's lifetime, can be passed to offspring.

CREDIT: (in text) Courtesy of Robin Shoulla and Young Survival Coalition.

summary

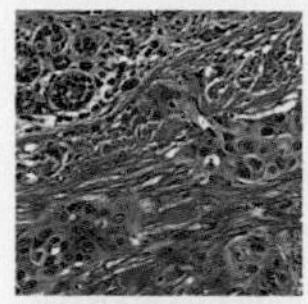

Section 10.1 A complex interplay of controls over gene expression is a critical part of normal functioning of cells in a multicelled body. Cancer typically begins with mutations in master genes that govern cell division.

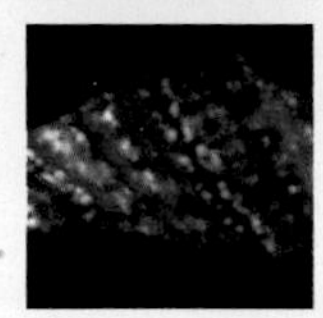

Section 10.2 Which genes a cell uses depends on the type of organism, the type of cell, conditions inside and outside the cell, and, in complex multicelled species, the organism's stage of development.

Gene expression control is necessary for differentiation during development in multicelled eukaryotes, and it also allows individual cells to respond appropriately to changes in their internal and external environments.

Different molecules and processes govern every step between transcription of a gene and delivery of the gene's product to its final destination. **Transcription factors** such as **activators** and **repressors** influence transcription by binding to chromosomal DNA at sites such as promoters, **enhancers**, and silencers.

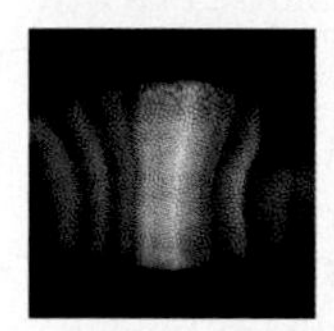

Section 10.3 Control over gene expression drives embryonic development of complex, multicelled animal bodies. Various **master genes** are expressed locally in different parts of an embryo as it develops. Their products, which form gradients as they diffuse through the embryo, affect expression of other master genes, which in turn affect the expression of others, and so on. Cells differentiate according to where they fall in these gradients. Eventually, master gene expression induces the expression of **homeotic genes**, the products of which govern the development of body parts. The function of many homeotic genes was revealed by **knockouts** in fruit flies.

Section 10.4 In cells of female mammals, one of the two X chromosomes is condensed as a **Barr body**, rendering most of its genes permanently inaccessible. By the theory of **dosage compensation**, this **X chromosome inactivation** balances gene expression between the sexes. The inactivation occurs because the *XIST* gene's RNA product shuts down the one X chromosome that transcribes it. The *SRY* gene determines male sex in humans.

Overlapping expression of master genes guides expression of flower formation in plants. Cells differentiate and form sepals, petals, stamens, or carpels depending on which floral identity gene products they are exposed to.

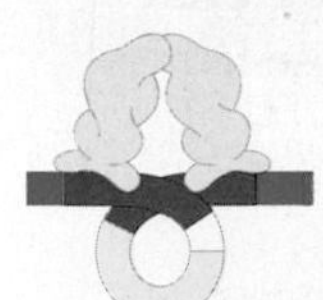

Section 10.5 Prokaryotes, being single-celled, do not undergo development. Most of their gene control reversibly adjusts transcription rates in response to environmental conditions, especially nutrient availability. The *lac* **operon** governs expression of three genes, the three products of which allow a bacterial cell to metabolize lactose. Two **operators** that flank the promoter are binding sites for a repressor that blocks transcription. Binding to an mRNA's riboswitch affects its translation.

Section 10.6 **Epigenetic** refers to heritable modifications of DNA that affect gene expression but do not involve changes to the DNA sequence. DNA methylations and other epigenetic modifications acquired during an individual's lifetime can persist for generations.

self-quiz

Answers in Appendix VII

1. The expression of a gene may depend on ______ .
   a. the type of organism
   b. environmental conditions
   c. the type of cell
   d. all of the above

2. Gene expression in multicelled eukaryotic organisms changes in response to ______ .
   a. extracellular conditions
   b. master gene products
   c. operons
   d. a and b

3. Binding of ______ to ______ in DNA can increase the rate of transcription of specific genes.
   a. activators; repressors
   b. activators; enhancers
   c. repressors; operators
   d. both a and b

4. Proteins that influence RNA synthesis by binding directly to DNA are called ______ .
   a. promoters
   b. transcription factors
   c. operators
   d. enhancers

5. In eukaryotes, control over gene expression occurs at the level of ______ .
   a. transcription
   b. RNA processing
   c. RNA transport
   d. mRNA degradation
   e. translation
   f. protein modification
   g. a through e
   h. all of the above

6. Muscle cells differ from bone cells because ______ .
   a. they carry different genes
   b. they use different genes
   c. both a and b

7. Control over gene expression drives ______ in complex, multicelled eukaryotes.
   a. transcription factors
   b. nutrient availability
   c. development
   d. all of the above

8. Homeotic gene products ______ .
   a. flank a bacterial operon
   b. map out the overall body plan in embryos
   c. control the formation of specific body parts

9. A gene that is knocked out is ______ .
   a. deleted
   b. inactivated
   c. expressed
   d. either a or b

10. Which of the following includes all of the others?
    a. homeotic genes
    b. master genes
    c. *SRY* gene
    d. *PAX6*

11. The expression of *ABC* genes ______ .
    a. occurs in layers of an onion
    b. controls flower formation
    c. causes mutations in flowers

Effect of Paternal Grandmother's Food Supply on Infant Mortality Researchers are investigating long-reaching epigenetic effects of starvation, in part because historical data on periods of famine are widely available.

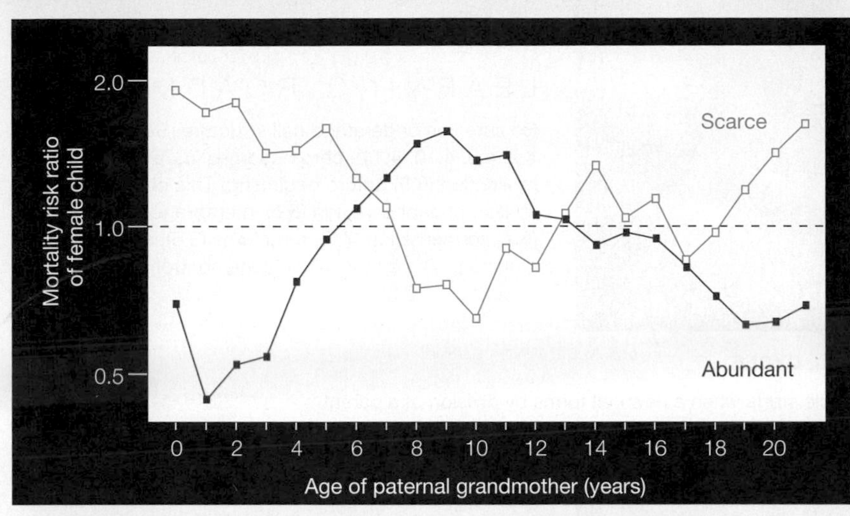

FIGURE 10.12 Graph showing the relative risk of early death of a female child, correlated with the age at which her paternal grandmother experienced a winter with a food supply that was scarce (blue) or abundant (red) during childhood. The dotted line represents no difference in risk of mortality. A value above the line means an increased risk; one below the line indicates a reduced risk.

Before the industrial revolution, a failed harvest in one autumn typically led to severe food shortages the following winter. A retrospective study has correlated female infant mortality at certain ages with the abundance of food during the paternal grandmother's childhood. **FIGURE 10.12** shows some of the results of this study.

1. Compare the mortality of girls whose paternal grandmothers ate well at age 2 with that of those who experienced famine at the same age. Which girl was more likely to die early? How much more likely was she to die?

2. Children have a period of slow growth around age 9. What trend in the data can you see around that age?

3. There was no correlation between early death of a male child and eating habits of his paternal grandmother, but there was a strong correlation with the eating habits of his paternal grandfather. What does this tell you about the probable location of epigenetic changes that gave rise to these data?

12. During X chromosome inactivation, ________ .
   a. female cells shut down
   b. RNA coats a chromosome
   c. pigments form
   d. both a and b

13. A cell with a Barr body is ________ .
   a. a bacterium
   b. a sex cell
   c. from a female mammal
   d. infected by the Barr virus

14. Operons ________ .
   a. only occur in bacteria
   b. have multiple genes
   c. involve selective gene expression

15. Match the terms with the most suitable description.

___ *SRY* gene	a. makes a man out of you
___ operator	b. binding site for repressor
___ Barr body	c. may be epigenetic
___ differentiation	d. inactivated X chromosome
___ mRNA zip code	e. controls multiple genes
___ methylation	f. localization mechanism
___ *eyeless*	g. works by binding DNA
___ homeodomain	h. required for eye formation
___ operon	i. driven by gene expression control

critical thinking

1. Why are some genes expressed and some not?

2. Do the same types of gene control operate in bacterial cells and eukaryotic cells?

3. Almost all calico cats (one is pictured in **FIGURE 10.7B**) are female. Why?

4. The photos above show flowers from *Arabidopsis* plants. One plant is wild-type (unmutated); the other carries a mutation in one of its *ABC* floral identity genes. This mutation causes sepals and petals to form instead of stamens and carpels. Refer to Figure 10.8 to decide which gene (*A*, *B*, or *C*) has been inactivated by the mutation.

CREDITS: (12) Pembrey et al. *European Journal of Human Genetics* (2006) 14, 159–166; (in text CT #4) left, © Jose Luis Riechmann; right, Science Source.

11 How Cells Reproduce

LEARNING ROADMAP

Be sure you understand cell structure (Sections 4.2, 4.6, 4.10, 4.11), chromosomes (8.4), and DNA replication (8.5) before beginning. This chapter also revisits phosphorylation (5.6), membrane proteins (5.7), fermentation (7.6), mutations (8.6), animal cloning (8.7), cancer (10.1), gene control (10.2), and knockouts (10.3).

THE CELL CYCLE

A cell cycle starts when a new cell forms by division of a parent cell, and ends when the cell completes its own division. Built-in checkpoints control the timing and rate of the cycle.

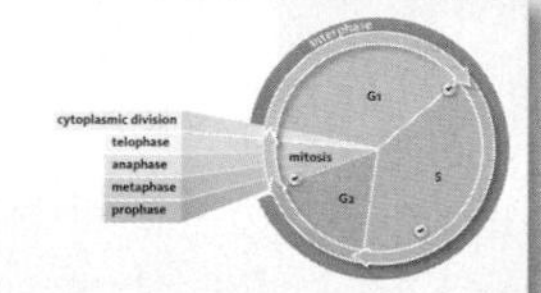

MITOSIS

Mitosis, a mechanism by which a cell's nucleus divides, maintains the chromosome number. Four sequential stages parcel the cell's duplicated chromosomes into two new nuclei.

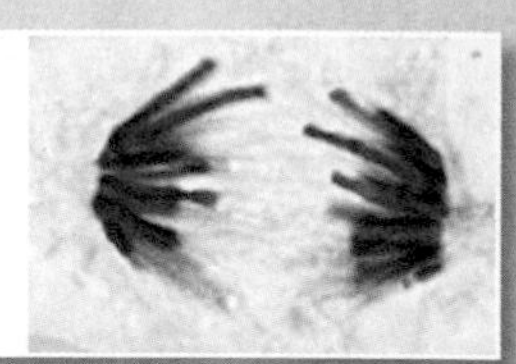

CYTOPLASMIC DIVISION

After nuclear division, the cytoplasm may divide, so one nucleus ends up in each of two new cells. Cytoplasmic division proceeds by different mechanisms in animal and plant cells.

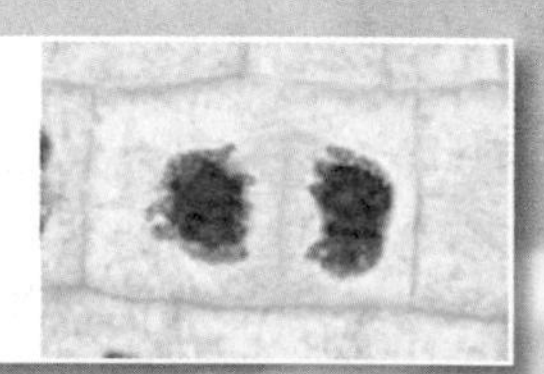

MITOTIC CLOCKS

Built into eukaryotic chromosomes are DNA sequences that protect the cell's genetic information. Degradation of these sequences is associated with cell death and aging.

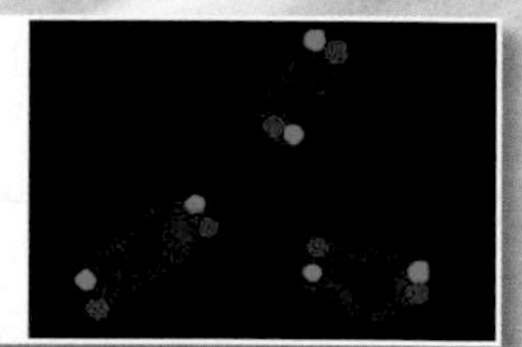

THE CELL CYCLE GONE AWRY

On rare occasions, cell cycle checkpoint mechanisms fail, and cell division becomes uncontrollable. Tumor formation and cancer are outcomes.

Mitosis is the basis of reproduction in single-celled eukaryotic organisms (Chapters 21 and 23). In multicelled eukaryotes, it has a role in reproduction and development (Chapters 29 and 42), as well as growth and tissue repair (Chapter 31). We compare mitosis with meiosis in Chapter 12. The HPV virus causes a sexually transmitted disease (Section 41.9) and cervical cancer (Section 37.1). Other cancers are discussed in Chapters 31, 32, and 34.

11.1 Henrietta's Immortal Cells

Each human starts out as a fertilized egg. By the time of birth, that cell has given rise to about a trillion cells, all organized as a human body. Even in an adult, billions of cells divide every day as new cells replace worn-out ones. However, despite the ability of human cells to continue dividing as part of a body, they tend to divide a limited number of times and die within weeks when grown in the laboratory.

Researchers had been trying to coax human cells to keep dividing outside of the body as early as the mid-1800s. Immortal cell lineages—cell lines—would allow the researchers to study human diseases (and potential cures for them) without experimenting on people. The quest to create a human cell line continued unsuccessfully until 1951. By this time, George and Margaret Gey had been trying to culture human cells for nearly thirty years. Then their assistant, Mary Kubicek, prepared a new sample of human cancer cells. Mary named the cells HeLa, after the first and last names of the patient from whom the cells had been taken. The HeLa cells began to divide, again and again. The cells were astonishingly vigorous, quickly coating the inside of their test tube and consuming their nutrient broth. Four days later, there were so many cells that the researchers had to transfer them to more tubes. The cell populations increased at a phenomenal rate. The cells were dividing every twenty-four hours and coating the inside of the tubes within days.

Sadly, cancer cells in the patient were dividing just as fast. Only six months after she had been diagnosed with cervical cancer, malignant cells had invaded tissues throughout her body. Two months after that, Henrietta Lacks, a young African American woman from Baltimore, was dead.

Although Henrietta passed away, her cells lived on in the Geys' laboratory (**FIGURE 11.1**). The Geys were able to grow poliovirus in HeLa cells, a practice that enabled them to determine which strains of the virus cause polio. That work was a critical step in the development of polio vaccines, which have since saved millions of lives.

Henrietta Lacks was just thirty-one, a wife and mother of five, when runaway cell divisions of cancer killed her. Her cells, however, are still dividing, again and again, more than fifty years after she died. Frozen away in tiny tubes and packed in Styrofoam boxes, HeLa cells continue to be shipped among laboratories all over the world. They are still widely used to investigate cancer, viral growth, protein synthesis, the effects of radiation, and countless other processes important in medicine and research. HeLa cells helped several researchers win Nobel Prizes, and some even traveled into space for experiments on satellites.

Understanding why cancer cells are immortal—and why we are not—begins with understanding the structures and mechanisms that cells use to divide.

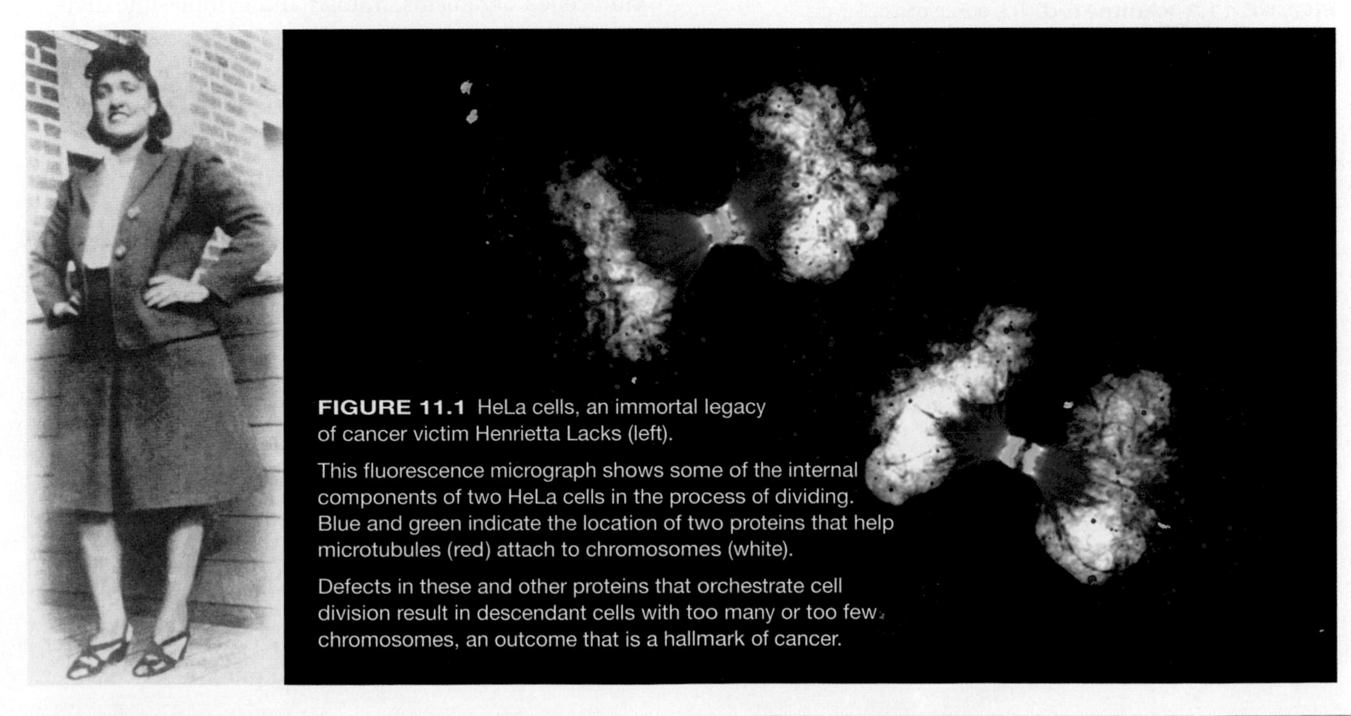

FIGURE 11.1 HeLa cells, an immortal legacy of cancer victim Henrietta Lacks (left).

This fluorescence micrograph shows some of the internal components of two HeLa cells in the process of dividing. Blue and green indicate the location of two proteins that help microtubules (red) attach to chromosomes (white).

Defects in these and other proteins that orchestrate cell division result in descendant cells with too many or too few chromosomes, an outcome that is a hallmark of cancer.

11.2 Multiplication by Division

✔ A cell's life occurs in a recognizable series of intervals and events collectively called the cell cycle.

✔ Division of a eukaryotic cell typically occurs in two steps: nuclear division followed by cytoplasmic division.

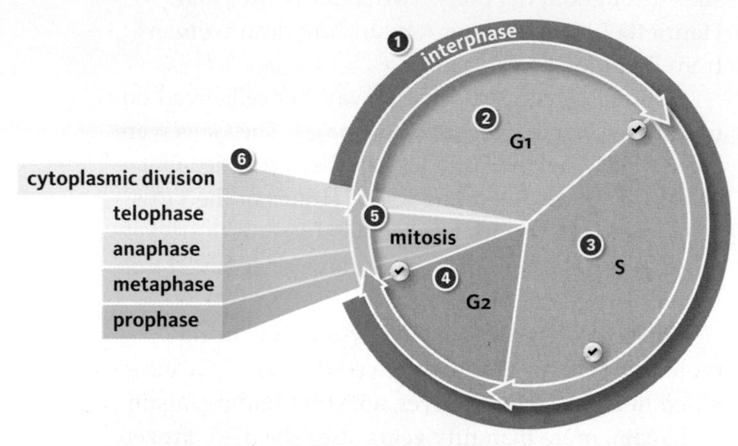

❶ A cell spends most of its life in interphase, which includes three stages: G1, S, and G2.

❷ G1 is the phase of growth before DNA replication. The cell's chromosomes are unduplicated.

❸ S is the phase of synthesis, during which the cell makes copies of its chromosome(s) by DNA replication.

❹ G2 is the phase after DNA replication and before mitosis. The cell prepares to divide during this stage.

❺ The nucleus divides during mitosis, the four stages of which are detailed in the next section.

❻ After mitosis, the cytoplasm may divide. Each descendant cell begins the cycle anew, in interphase.

✔ Built-in checkpoints stop the cycle from proceeding until certain conditions are met.

FIGURE 11.2 ▸Animated The eukaryotic cell cycle. The length of the intervals differs among cells. G1, S, and G2 are part of interphase.

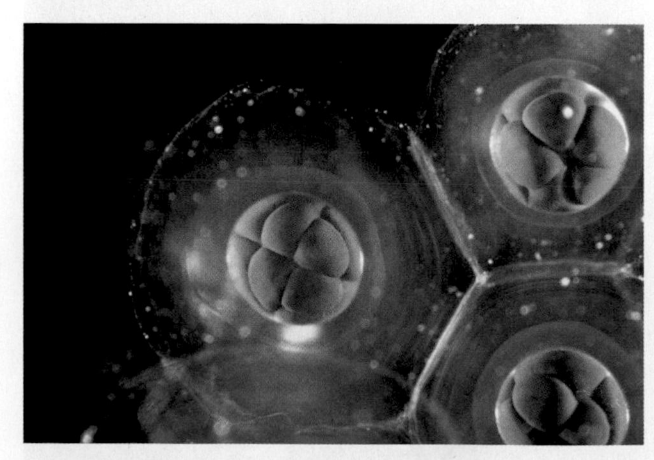

FIGURE 11.3 A multicelled eukaryote develops by repeated cell divisions. This photo shows early frog embryos, each a product of three mitotic divisions of one fertilized egg.

FIGURE IT OUT Each of these embryos has how many cells?

Answer: Eight

A life cycle is a sequence of recognizable stages that occur during an organism's lifetime, from the first cell of the new individual until its death. Multicelled organisms and free-living cells have life cycles, but what about cells that make up a multicelled body? Biologists consider such cells to be individually alive, each with its own lifetime. A cell's life passes through a series of recognizable intervals and events collectively called the **cell cycle** (**FIGURE 11.2**).

A typical cell spends most of its life in the interval called interphase ❶. During **interphase**, the cell increases its mass, roughly doubles the number of its cytoplasmic components, and replicates its DNA in preparation for division. Interphase comprises three major stages: G1, S, and G2. G1 and G2 were named "Gap" phases because outwardly they seem to be periods of inactivity, but they are not. Most cells going about their metabolic business are in G1 ❷. Cells preparing to divide enter S ❸, the phase of DNA synthesis, when they copy their chromosomes by DNA replication (Section 8.5). During G2 ❹, cells make the proteins that will drive the division process.

The remainder of the cell cycle consists of the division process itself. When the cell divides, both of its two cellular offspring end up with a blob of cytoplasm and some DNA. Each of the offspring of a eukaryotic cell inherits its DNA packaged inside a nucleus. Thus, a eukaryotic cell's nucleus has to divide before its cytoplasm does. **Mitosis** is a nuclear division mechanism that maintains the chromosome number ❺. In multicelled organisms, mitosis and cytoplasmic division ❻ are the basis of increases in body size and tissue remodeling during development (**FIGURE 11.3**), as well as ongoing replacements of damaged or dead cells. Mitosis and cytoplasmic division are also part of **asexual reproduction**, a reproductive mode by which offspring are produced by one parent only. Some multicelled eukaryotes and many single-celled ones use this mode of reproduction. (Prokaryotes do not have a nucleus and do not undergo mitosis. We discuss their reproduction in Section 20.6.)

When a cell divides by mitosis, it produces two descendant cells, each with the chromosome number of the parent. However, if only the total number of chromosomes mattered, then one of the descendant cells might get, say, two pairs of chromosome 22 and no chromosome 9. A cell cannot function properly without a full complement of DNA, which means it needs to have *a copy of each* chromosome. Thus, the two cells produced by mitosis have the same number and types of chromosomes as the parent.

Your body's cells are diploid, which means their nuclei contain pairs of chromosomes—two of each type (Section 8.4). One chromosome of a pair was inherited from your father; the other, from your mother. Except for a pairing of nonidentical sex chromosomes (XY) in males, the chromosomes of each pair are homologous. **Homologous chromosomes** have the same length, shape, and genes (*hom–* means alike).

FIGURE 11.4 shows how homologous chromosomes are distributed into descendant cells when a diploid cell divides by mitosis. When a cell is in G1, each of its chromosomes consists of one double-stranded DNA molecule ❷. The cell replicates its DNA in S, so by G2, each of its chromosomes consists of two double-stranded DNA molecules ❹. These molecules stay attached to one another at the centromere as sister chromatids until mitosis is almost over, and then they are pulled apart and packaged into two separate nuclei (the next section details this process).

When sister chromatids are pulled apart, each becomes an individual chromosome that consists of one double-stranded DNA molecule. Thus, each of the two new nuclei that form in mitosis contains a full complement of (unduplicated) chromosomes. When the cytoplasm divides, these nuclei are packaged into separate cells ❻. Each new cell starts the cell cycle over again in G1 of interphase.

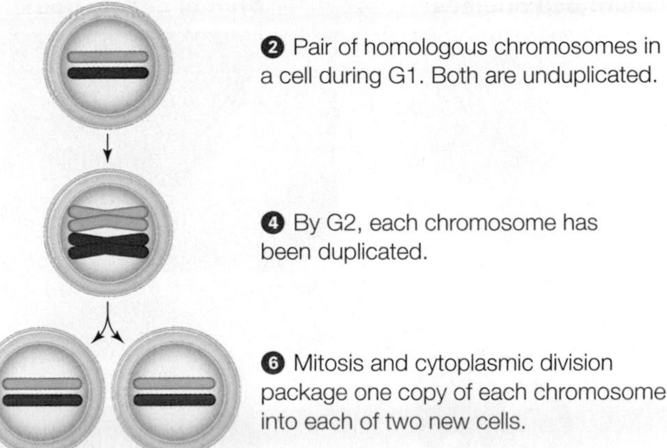

FIGURE 11.4 How mitosis maintains chromosome number in a diploid cell. For clarity, only one pair of homologous chromosomes is shown. The maternal chromosome is shown in pink, the paternal one in blue. Numbered balls indicate cell cycle stages in **FIGURE 11.2**.

Controls Over the Cell Cycle

When a cell divides—and when it does not—is determined by mechanisms of gene expression control (Section 10.2). Like the accelerator of a car, some of these mechanisms cause the cell cycle to advance. Others are like brakes, preventing the cycle from proceeding. In the adult body, brakes on the cell cycle normally keep the vast majority of cells in G1. Most of your nerve cells, skeletal muscle cells, heart muscle cells, and fat-storing cells have been in G1 since you were born, for example.

Control over the cell cycle also ensures that a dividing cell's descendants receive intact copies of its chromosomes. Built-in checkpoints ensure the cell's DNA has been copied completely, that it is not damaged, and even that enough nutrients are available to support division. Protein products of "checkpoint genes" interact to carry out this control process. For example, at a checkpoint that operates in S, these molecules put the brakes on the cycle if the cell's chromosomes have been damaged during DNA replication (Section 8.6). Checkpoint proteins that function as sensors recognize damaged DNA and bind to it. Upon binding, they trigger other events that stall the cell cycle, and also enhance expression of genes involved in DNA repair. After the problem has been corrected, the brakes are lifted and the cell cycle proceeds. If the problem remains uncorrected, other checkpoint proteins may initiate a series of events that eventually cause the cell to self-destruct (you will read more about cell suicide, or apoptosis, in Section 42.4).

asexual reproduction Reproductive mode of eukaryotes by which offspring arise from a single parent only.
cell cycle The collective series of intervals and events of a cell's life, from the time it forms until it divides.
homologous chromosomes Chromosomes with the same length, shape, and genes.
interphase In a eukaryotic cell cycle, the interval during which a cell grows, roughly doubles the number of its cytoplasmic components, and replicates its DNA in preparation for division.
mitosis Nuclear division mechanism that maintains the chromosome number.

TAKE-HOME MESSAGE 11.2
How do eukaryotic cells reproduce?

- ✔ A cell cycle is the sequence of events and stages through which a cell passes during its lifetime. The cell cycle includes interphase, mitosis, and cytoplasmic division.
- ✔ A eukaryotic cell reproduces by division: nucleus first, then cytoplasm. Each descendant cell receives a set of chromosomes and some cytoplasm.
- ✔ When a nucleus divides by mitosis, each new nucleus has the same chromosome number as the parent cell.
- ✔ Gene expression controls advance, delay, or block the cell cycle in response to internal and external conditions. Checkpoints built into the cell cycle allow problems to be corrected before the cycle proceeds.

11.3 A Closer Look at Mitosis

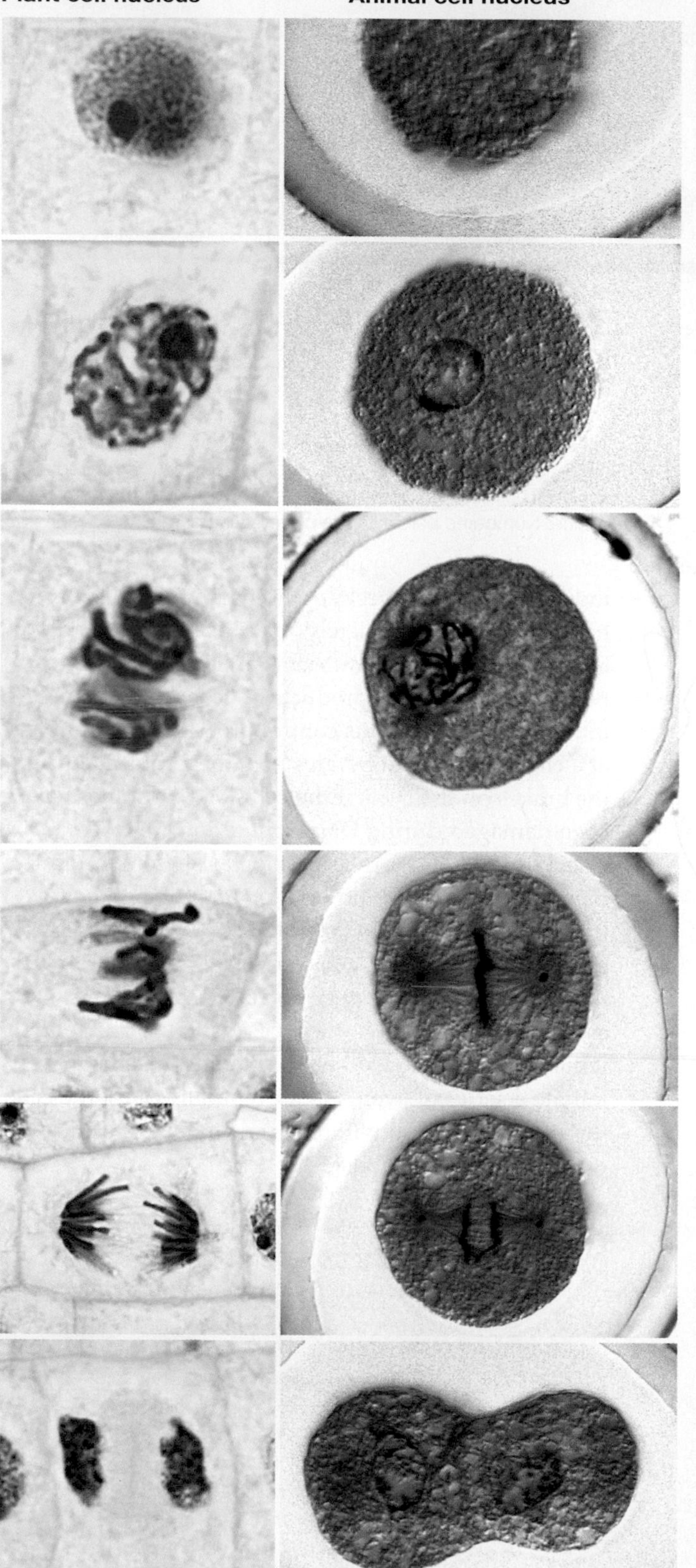

❶ Interphase
Interphase cells are shown for comparison, but interphase is not part of mitosis. The nuclear envelope is intact.

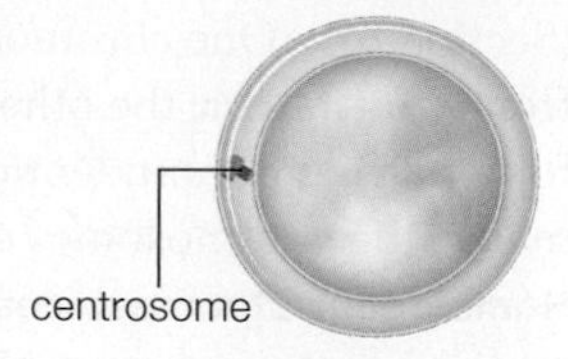

❷ Early Prophase
Mitosis begins. Transcription stops, and the DNA begins to appear grainy as it starts to condense. The nuclear envelope begins to break up and the centrosome gets duplicated.

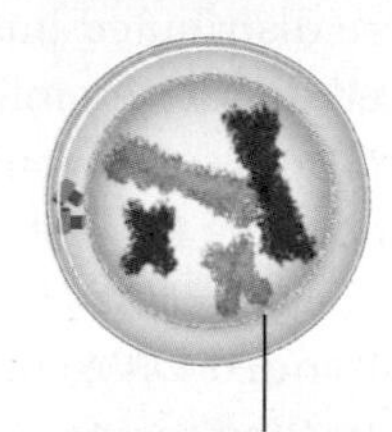

❸ Prophase
The duplicated chromosomes become visible as they condense. One of the two centrosomes moves to the opposite side of the cell as the nuclear envelope breaks up completely. Spindle microtubules assemble and bind to chromosomes at the centromere.

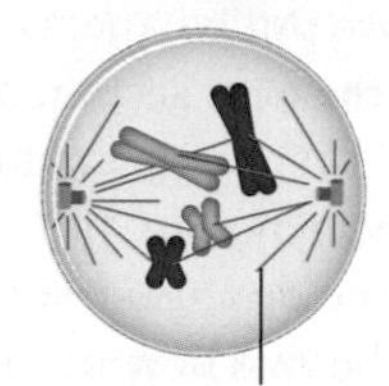

❹ Metaphase
All of the chromosomes are aligned midway between the spindle poles.

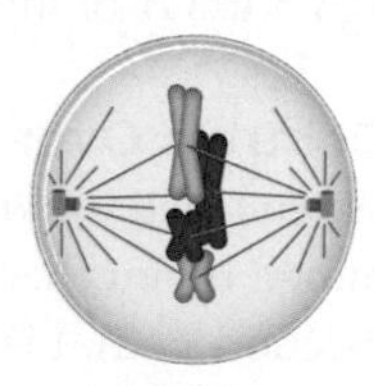

❺ Anaphase
Spindle microtubules separate the sister chromatids and move them toward opposite spindle poles. Each sister chromatid has now become an individual, unduplicated chromosome.

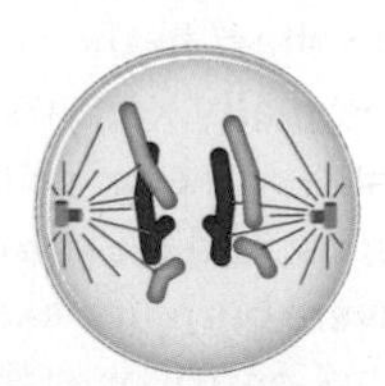

❻ Telophase
The chromosomes reach opposite sides of the cell and loosen up. Mitosis ends when a new nuclear envelope forms around each cluster of chromosomes.

FIGURE 11.5 ▸Animated Mitosis. Micrographs show nuclei of plant cells (onion root, left), and animal cells (fertilized eggs of a roundworm, right). A diploid (2*n*) animal cell with two chromosome pairs is illustrated.

CREDITS: (5) left, Michael Clayton/University of Wisconsin, Department of Botany; right, ISM/Phototake; far right, © Cengage Learning.

✔ The four main stages of mitosis are prophase, metaphase, anaphase, and telophase.

During interphase, a cell's chromosomes are loosened from histones to allow transcription and DNA replication. Loosening spreads out the chromosomes, so they are not easily visible under a light microscope (**FIGURE 11.5** ❶). In preparation for nuclear division, the chromosomes begin to pack tightly ❷. Transcription and DNA replication stop as the chromosomes condense into their most compact "X" forms (Section 8.4). Tight condensation keeps the chromosomes from getting tangled and breaking during nuclear division.

Most animal cells have a structure called a centrosome, which typically consists of a pair of centrioles (Section 4.10) surrounded by a region of dense cytoplasm. The centrosome gets duplicated as mitosis begins. During **prophase**, the first stage of mitosis, the chromosomes condense so much that they become visible under a light microscope ❸. "Mitosis" is from *mitos*, the Greek word for thread, after the threadlike appearance of chromosomes during nuclear division. If the cell has centrosomes, one of them now moves to the opposite side of the cell. Microtubules begin to assemble and lengthen from the centrosomes (or from other structures in cells that have no centrosomes). The lengthening microtubules form a **spindle**, which is a temporary structure for moving chromosomes (**FIGURE 11.6**). The area from which the spindle originates on each side of the cell is called a spindle pole.

Spindle microtubules penetrate the nuclear region as the nuclear envelope breaks up. Some of the microtubules stop lengthening when they reach the middle of the cell. Others lengthen until they reach a chromosome and attach to it at the centromere. By the end of prophase, one sister chromatid of each chromosome has become attached to microtubules extending from one spindle pole, and the other sister chromatid has become attached to microtubules extending from the other spindle pole.

The opposing sets of microtubules then begin a tug-of-war by adding and losing tubulin subunits. As the microtubules lengthen and shorten, they push and pull the chromosomes. When all the microtubules are the same length, the chromosomes are aligned midway between spindle poles ❹. The alignment marks **metaphase** (from *meta*, the ancient Greek word for between).

During **anaphase**, the spindle pulls the sister chromatids of each duplicated chromosome apart and moves them toward opposite spindle poles ❺. Each DNA molecule has now become a separate chromosome.

Telophase begins when one set of chromosomes reaches each spindle pole ❻. Each set consists of the same number and kinds of chromosomes as the parent cell nucleus had: two of each type of chromosome, if the parent cell was diploid. A new nuclear envelope forms around each set of chromosomes as they loosen up again. At this point, telophase—and mitosis—are over.

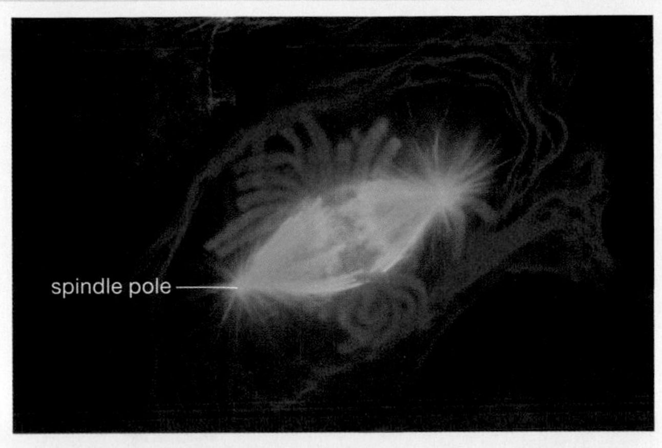

FIGURE 11.6 The spindle. In this dividing lung cell of a salamander, microtubules (green) have extended from two centrosomes to form the spindle, which has attached to and aligned the chromosomes (blue) midway between its two poles. Red shows actin microfilaments.

FIGURE IT OUT This cell is in which stage of mitosis?

Answer: Metaphase

anaphase Stage of mitosis during which sister chromatids separate and move toward opposite spindle poles.
metaphase Stage of mitosis at which all chromosomes are aligned midway between spindle poles.
prophase Stage of mitosis during which chromosomes condense and become attached to a newly forming spindle.
spindle Temporary structure that moves chromosomes during nuclear division; consists of microtubules.
telophase Stage of mitosis during which chromosomes arrive at opposite spindle poles and decondense, and two new nuclei form.

TAKE-HOME MESSAGE 11.3

What is the sequence of events in mitosis?

✔ DNA replication occurs before mitosis. As mitosis begins, each chromosome consists of two DNA molecules attached as sister chromatids.

✔ In prophase, the chromosomes condense and a spindle forms. Spindle microtubules attach to the chromosomes as the nuclear envelope breaks up.

✔ At metaphase, the spindle has aligned all of the (still duplicated) chromosomes in the middle of the cell.

✔ In anaphase, sister chromatids separate and move toward opposite spindle poles. Each DNA molecule is now an individual chromosome.

✔ In telophase, two clusters of chromosomes reach opposite spindle poles. A new nuclear envelope forms around each cluster, so two new nuclei form.

11.4 Cytokinesis: Division of Cytoplasm

✔ The mechanism by which cytoplasm divides differs between plant and animal cells.

In most eukaryotes, the cell cytoplasm divides between late anaphase and the end of telophase, so two cells form, each with their own nucleus. The mechanism of cytoplasmic division, which is called **cytokinesis**, differs between plants and animals.

Typical animal cells pinch themselves in two after nuclear division ends (**FIGURE 11.7**). How? The spindle begins to disassemble during telophase ❶. The cell cortex, which is the mesh of cytoskeletal elements just under the plasma membrane (Section 4.10), includes a band of actin and myosin filaments that wraps around the cell, midway between the former spindle poles. The band is called a contractile ring because it contracts when its component proteins are energized by phosphate-group transfers from ATP. When the ring contracts, it drags the attached plasma membrane inward ❷. The sinking plasma membrane becomes visible on the outside of the cell as an indentation between the former spindle poles ❸. The indentation, which is called a **cleavage furrow**, deepens until the cytoplasm (and the cell) is pinched in two ❹. Each of the two cells formed by this division has its own nucleus and some of the parent cell's cytoplasm, and each is enclosed by a plasma membrane.

Dividing plant cells face a particular challenge because a stiff cell wall surrounds their plasma membrane (Section 4.11). Accordingly, plant cells have their own mechanism of cytokinesis. By the end of anaphase, a set of short microtubules has formed on either side of the future plane of division. These microtubules guide vesicles from Golgi bodies and the cell surface to the division plane ❺. These vesicles provide material for a new cell membrane and wall. After mitosis, the vesicles fuse into a disk-shaped **cell plate** ❻. The plate expands at its edges until it reaches the plasma membrane and attaches to it, thus partitioning the cytoplasm ❼. In time, the cell plate will develop into two new cell walls, so each of the descendant cells will be enclosed by its own plasma membrane and wall ❽.

Animal cell cytokinesis

❶ In a dividing animal cell, the spindle disassembles as mitosis ends.

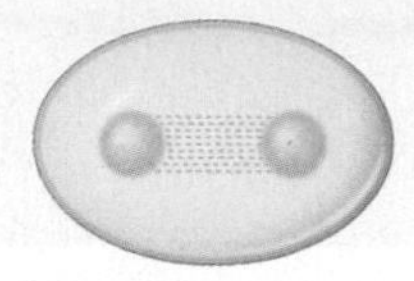

❷ At the midpoint of the former spindle, a ring of actin and myosin filaments attached to the plasma membrane contracts.

❸ This contractile ring pulls the cell surface inward, forming a cleavage furrow as it shrinks.

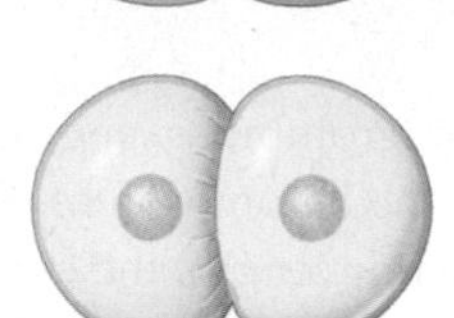

❹ The ring contracts until it pinches the cell in two.

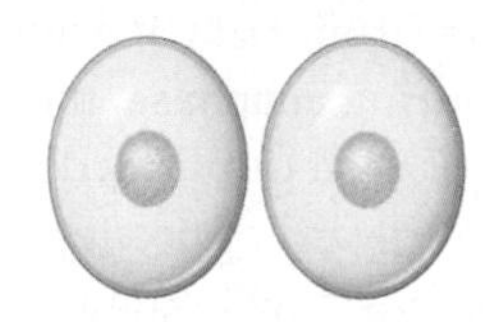

Plant cell cytokinesis

❺ In a dividing plant cell, vesicles cluster at the future plane of division before mitosis ends.

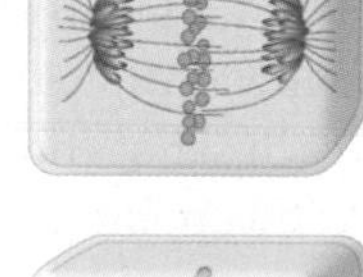

❻ The vesicles fuse with each other, forming a cell plate along the plane of division.

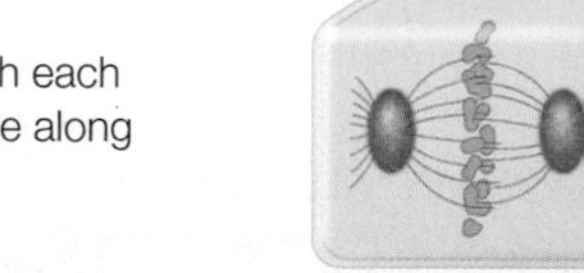

❼ The cell plate expands outward along the plane of division. When it reaches and attaches to the plasma membrane, it partitions the cytoplasm.

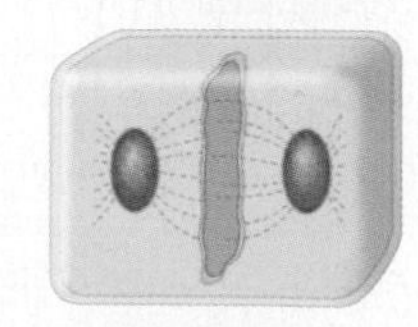

❽ The cell plate matures as two new cell walls. These walls join with the parent cell wall, so each descendant cell becomes enclosed by its own wall.

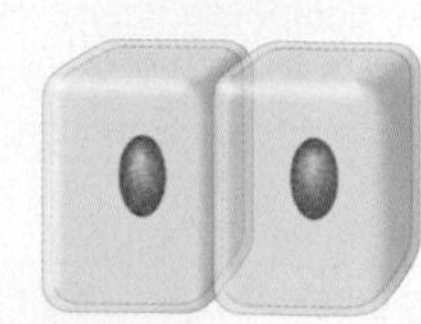

FIGURE 11.7 ▶**Animated** Cytoplasmic division of animal cells (top) and plant cells (bottom).

cell plate A disk-shaped structure that forms during cytokinesis in a plant cell; matures as a cross-wall between the two new nuclei.
cleavage furrow In a dividing animal cell, the indentation where cytoplasmic division will occur.
cytokinesis Cytoplasmic division.

TAKE-HOME MESSAGE 11.4
How do eukaryotic cells divide?

✔ In most eukaryotes, the cell cytoplasm divides between late anaphase and the end of telophase. Two descendant cells form, each with its own nucleus.

✔ In animal cell cytokinesis, a contractile ring pinches the cytoplasm in two. In plant cell cytokinesis, a cell plate that forms in the middle of the cell partitions the cytoplasm when it reaches and connects to the parent cell wall.

11.5 Marking Time With Telomeres

✔ Telomeres protect eukaryotic chromosomes from losing genetic information at their ends.

Remember that Dolly, the first clone produced by somatic cell nuclear transfer (SCNT, Section 8.7), died early. The life expectancy of a sheep is normally about 10 to 12 years. By the time Dolly was five, however, she was as fat and arthritic as a twelve-year-old sheep. The following year, she contracted a lung disease typical of much older sheep, and had to be euthanized.

Dolly's early demise may have been the result of abnormally short telomeres. **Telomeres** are noncoding DNA sequences that occur at the ends of eukaryotic chromosomes (**FIGURE 11.8**). Vertebrate telomeres consist of a short DNA sequence, 5′-TTAGGG-3′, repeated perhaps thousands of times. These "junk" repeats provide a buffer against the loss of more valuable DNA internal to the chromosomes.

A telomere buffer is particularly important because, under normal circumstances, a eukaryotic chromosome shortens by about 100 nucleotides with each DNA replication. When a cell's offspring receive chromosomes with too-short telomeres, checkpoint gene products halt the cell cycle, and the descendant cells die shortly thereafter. Most body cells can divide only a certain number of times before this happens. This cell division limit may be a fail-safe mechanism in case a cell loses control over the cell cycle and begins to divide again and again. A limit on the number of divisions keeps such cells from overrunning the body (an outcome that, as you will see in the next section, has dangerous consequences to health).

The cell division limit varies by species, and it may be part of the mechanism that sets an organism's life span. Dolly's DNA came from the nucleus of a mammary gland cell donated by an adult sheep. When Dolly was only two years old, her telomeres were as short as those of a six-year-old sheep—the exact age of the adult animal that had been her genetic donor.

A few normal cells in an adult retain the ability to divide indefinitely. Their descendants replace cell lineages that eventually die out when they reach their division limit. These cells are called stem cells, and they are immortal because they continue to make an enzyme called telomerase. Telomerase reverses the telomere shortening that normally occurs after DNA replication.

Mice that have had their telomerase enzyme knocked out age prematurely. Their tissues degenerate much more quickly than those of normal mice, and their life expectancy declines to about half that of a normal mouse. When one of these knockout mice is close to the end of its shortened life span, rescuing the function of its telomerase enzyme results in lengthened telomeres. The rescued mouse also regains vitality: Decrepit tissue in its brain and other organs repairs itself and begins to function normally, and the once-geriatric individual even begins to reproduce again.

Researchers are careful to point out that shortening telomeres may be an effect of aging rather than a cause. Also, while telomerase holds therapeutic promise for rejuvenating aged tissues, it can also be dangerous. Cancer cells—including the HeLa cells you learned about in Section 11.1—characteristically express high levels of this molecule, which is why, like stem cells, they can divide indefinitely.

telomere Noncoding, repetitive DNA sequence at the end of chromosomes; protects the coding sequences from degradation.

FIGURE 11.8 ▸Animated Telomeres. The bright dots at the end of each DNA strand in these duplicated chromosomes are telomeres.

TAKE-HOME MESSAGE 11.5

What is the function of telomeres?

✔ Telomeres are noncoding sequences at the end of eukaryotic chromosomes. These extra sequences prevent loss of genetic information.

✔ Telomeres shorten with every cell division in normal body cells. When they are too short, the cell stops dividing and dies. Thus, telomeres are associated with aging.

✔ On rare occasions, controls over cell division are lost and a neoplasm forms.

✔ Cancer develops as cells of a neoplasm become malignant.

Sometimes a checkpoint gene mutates so that its protein product no longer works properly. In other cases, the controls that regulate its expression fail, and a cell makes too much or too little of its product. When enough checkpoint mechanisms fail, a cell loses control over its cell cycle. Interphase may be skipped, so division occurs over and over with no resting period. Signaling mechanisms that cause abnormal cells to die may stop working. The problem is compounded because checkpoint malfunctions are passed along to the cell's descendants, which form a **neoplasm**, an accumulation of cells that lost control over how they grow and divide.

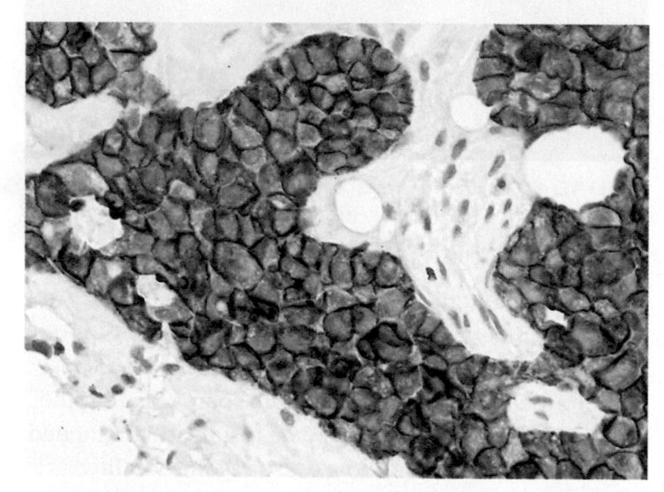

FIGURE 11.9 Effects of an oncogene. In this section of human breast tissue, a brown-colored tracer shows the active form of the EGF receptor. Normal cells are lighter in color. The dark cells have an overactive EGF receptor that is constantly stimulating mitosis; these cells have formed a neoplasm.

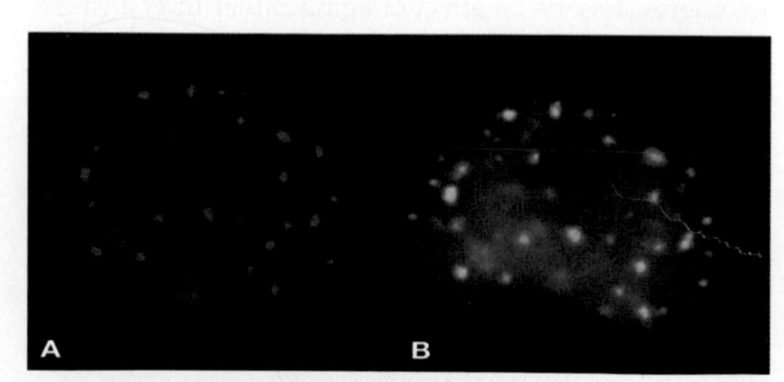

FIGURE 11.10 Checkpoint genes in action.

Radiation damaged the DNA inside this nucleus. (**A**) Red dots show the location of the *BRCA1* gene product. (**B**) Green dots pinpoint the location of the product of a gene called *53BP1*.

Both proteins have clustered around the same chromosome breaks in the same nucleus; both function to recruit DNA repair enzymes. The integrated action of these and other checkpoint gene products blocks mitosis until the DNA breaks are fixed.

A neoplasm that forms a lump in the body is called a **tumor**, but the two terms are sometimes used interchangeably. Once a tumor-causing mutation has occurred, the gene it affects is called an oncogene. An **oncogene** is any gene that can transform a normal cell into a tumor cell (Greek *onkos*, or bulging mass). Oncogene mutations in reproductive cells can be passed to offspring, which is a reason that some types of tumors run in families.

Genes encoding proteins that promote mitosis are called **proto-oncogenes** because mutations can turn them into oncogenes. One proto-oncogene, *EGFR*, encodes a plasma membrane receptor for EGF (epidermal growth factor). EGF and other **growth factors** are molecules that stimulate a cell to divide and differentiate. When the EGF receptor binds to EGF, it becomes activated and triggers the cell to begin mitosis. When EGF is not present, the receptor is in an inactive form and does not trigger mitosis. Tumor-causing mutations in *EGFR* change the receptor so that it stimulates mitosis even in the absence of EGF. Most neoplasms carry mutations resulting in an overactivity or overabundance of this particular receptor (**FIGURE 11.9**).

Checkpoint gene products that inhibit mitosis are called tumor suppressors because tumors form when they are missing. The products of the *BRCA1* and *BRCA2* genes that you learned about in Chapter 10 are examples of tumor suppressors. These proteins regulate, among other things, the expression of DNA repair enzymes (**FIGURE 11.10**). Tumor cells often have mutations in their *BRCA* genes.

Viruses such as HPV (human papillomavirus) cause a cell to make proteins that interfere with its own tumor suppressors. Infection with HPV causes skin growths called warts, and some kinds are associated with neoplasms that form on the cervix.

Cancer

Benign neoplasms such as warts are not usually dangerous. They grow very slowly, and their cells retain the plasma membrane adhesion proteins that keep them properly anchored to the other cells in their home tissue (**FIGURE 11.11** ❶).

A malignant neoplasm is one that gets progressively worse, and is dangerous to health. Malignant cells typically display the following three characteristics:

First, like cells of all neoplasms, malignant cells grow and divide abnormally. Controls that usually keep cells from getting overcrowded in tissues are lost in malignant cells, so their populations may reach extremely high densities with cell division occurring very rapidly. The number of small blood vessels that transport blood to the growing cell mass also increases abnormally.

Second, the cytoplasm and plasma membrane of malignant cells are altered. Both are indications of cellular malfunction. The cytoskeleton may be shrunken, disorganized, or both. Malignant cells typically have an abnormal chromosome number, with some chromosomes present in multiple copies, and others missing or damaged. The balance of metabolism is often shifted, as in an amplified reliance on ATP formation by fermentation rather than aerobic respiration.

Altered or missing proteins impair the function of the plasma membrane of malignant cells. For example, these cells do not stay anchored properly in tissues because their plasma membrane adhesion proteins are defective or missing ❷. Malignant cells can slip easily into and out of vessels of the circulatory and lymphatic systems ❸. By migrating through these vessels, the cells can establish neoplasms elsewhere in the body ❹. The process in which malignant cells break loose from their home tissue and invade other parts of the body is called **metastasis**. Metastasis is the third hallmark of malignant cells.

The disease called **cancer** occurs when the abnormally dividing cells of a malignant neoplasm disrupt body tissues, both physically and metabolically. Unless chemotherapy, surgery, or another procedure eliminates malignant cells from the body, they can put an individual on a painful road to death. Each year, cancer causes 15 to 20 percent of all human deaths in developed countries. The good news is that mutations in multiple checkpoint genes are required to transform a normal cell into a malignant one, and such mutations may take a lifetime to accumulate. Lifestyle choices such as not smoking and avoiding exposure of unprotected skin to sunlight can reduce one's risk of acquiring mutations in the first place. Some neoplasms can be detected with periodic screening such as gynecology or dermatology exams (**FIGURE 11.12**). If detected early enough, many types of malignant neoplasms can be removed before metastasis occurs.

cancer Disease that occurs when a malignant neoplasm physically and metabolically disrupts body tissues.
growth factor Molecule that stimulates mitosis and differentiation.
metastasis The process in which malignant cells spread from one part of the body to another.
neoplasm An accumulation of abnormally dividing cells.
oncogene Gene that helps transform a normal cell into a tumor cell.
proto-oncogene Gene that, by mutation, can become an oncogene.
tumor A neoplasm that forms a lump.

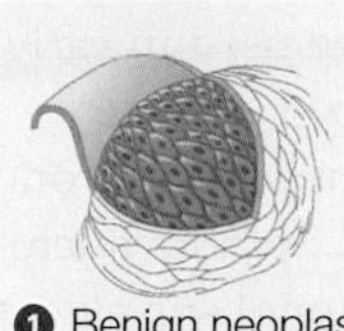

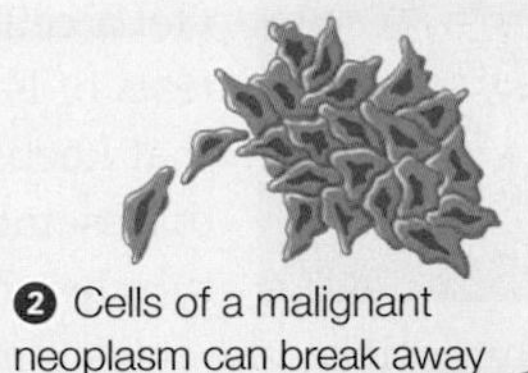

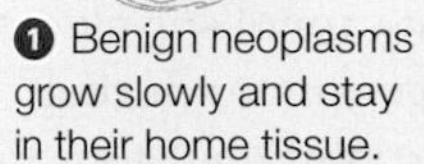

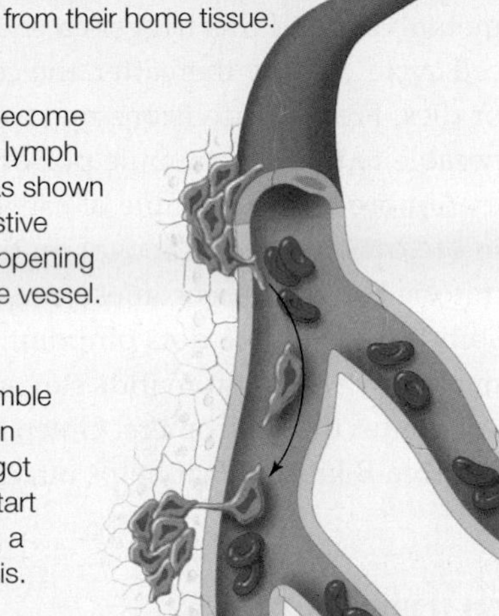

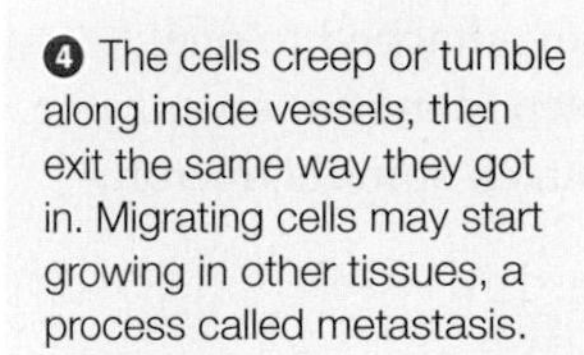

FIGURE 11.11 ▶**Animated** Neoplasms and malignancy.

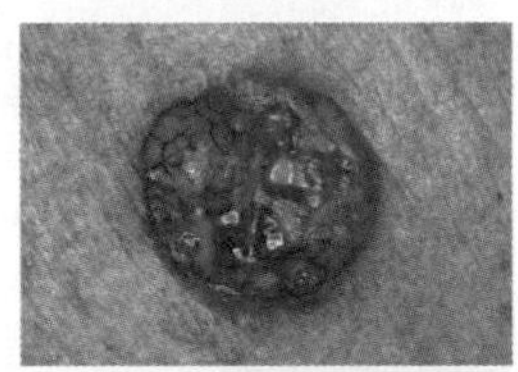

A Basal cell carcinoma is the most common type of skin cancer. This slow-growing, raised lump may be uncolored, reddish-brown, or black.

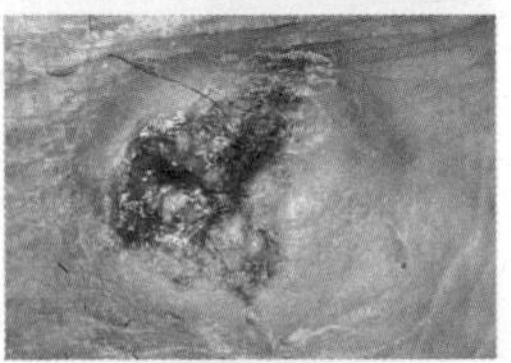

B Squamous cell carcinoma is the second most common form of skin cancer. This pink growth, firm to the touch, grows under the skin's surface.

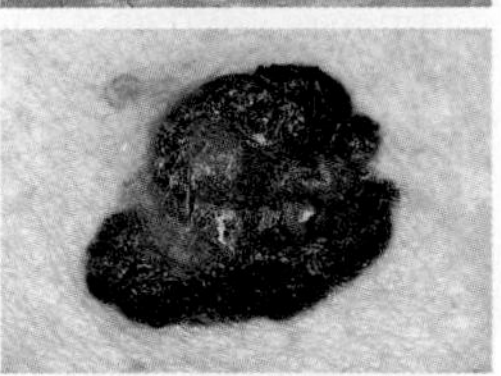

C Melanoma spreads fastest. Cells form dark, encrusted lumps that may itch or bleed easily.

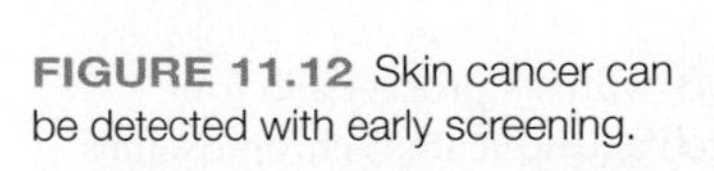

FIGURE 11.12 Skin cancer can be detected with early screening.

TAKE-HOME MESSAGE 11.6

What is cancer?

- ✔ Neoplasms form when cells lose control over their cell cycle and begin dividing abnormally.
- ✔ Mutations in multiple checkpoint genes can give rise to a malignant neoplasm that gets progressively worse.
- ✔ Cancer is a disease that occurs when the abnormally dividing cells of a malignant neoplasm physically and metabolically disrupt body tissues.
- ✔ Although some mutations are inherited, lifestyle choices and early intervention can reduce one's risk of cancer.

Henrietta's Immortal Cells (revisited)

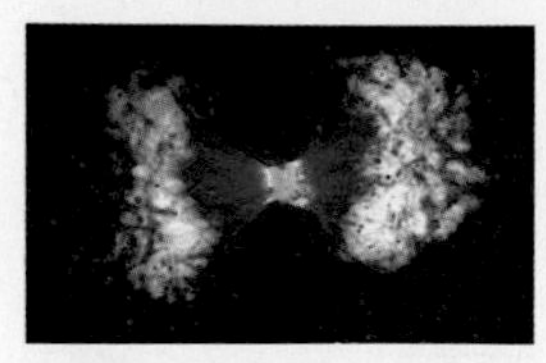

HeLa cells were used in early tests of Paclitaxel, a drug that keeps microtubules from disassembling. Spindle microtubules that cannot shrink cannot properly position the cell's chromosomes during metaphase, and this triggers a checkpoint that stops the cell cycle. Shortly thereafter, the cell either exits mitosis or dies. Frequent divisions make cancer cells more vulnerable to this microtubule poison than normal cells.

A more recent example of cancer research is shown in **FIGURE 11.1** (and above). In this micrograph of mitotic HeLa cells, chromosomes appear white and the spindle is red. Blue dots pinpoint a protein (INCENP) that helps sister chromatids stay attached to one another at the centromere. Green identifies an enzyme (Aurora B kinase) that helps attach spindle microtubules to centromeres. At this stage of telophase, these two proteins should be closely associated midway between the two clusters of chromosomes. The abnormal distribution means that the chromosomes are not properly attached to the spindle.

Defects in Aurora B kinase or its expression result in unequal distribution of chromosomes into descendant cells. Researchers recently correlated overexpression of Aurora B in cancer cells with shortened patient survival rates. Thus, drugs that inhibit Aurora B function are now being tested as potential cancer therapies.

Ongoing research with HeLa cells may one day allow researchers to identify drugs that target and destroy malignant cells or stop them from dividing. The research is far too late to have saved Henrietta Lacks, but it may one day yield drugs that put the brakes on cancer.

summary

Section 11.1 An immortal line of human cells (HeLa) is a legacy of cancer victim Henrietta Lacks. Researchers all over the world continue to work with these cells as they try to unravel the mechanisms of cancer.

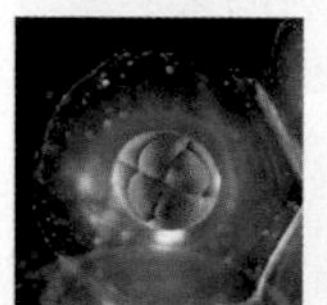

Section 11.2 A **cell cycle** includes all the recognizable stages and events of a cell's lifetime; it starts when a new cell forms, and ends when the cell reproduces. Most of a cell's activities, including DNA replication that copies its **homologous chromosomes**, occur during **interphase**. A eukaryotic cell reproduces by dividing: nucleus first, then cytoplasm. **Mitosis** is a mechanism of nuclear division that maintains the chromosome number. It is the basis of growth, cell replacements, and tissue repair in multicelled species, and **asexual reproduction** in many species.

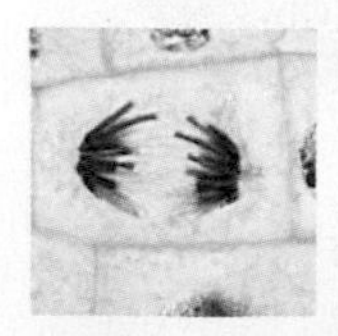

Section 11.3 Mitosis proceeds in four stages. The cell's (duplicated) chromosomes condensense during **prophase**. Microtubules assemble and form a **spindle**, and the nuclear envelope breaks up. Some microtubules that extend from one spindle pole attach to one chromatid of each chromosome; some that extend from the opposite spindle pole attach to its sister chromatid. These microtubules drag each chromosome toward the center of the cell.

At **metaphase**, all chromosomes are aligned at the spindle's midpoint.

During **anaphase**, the sister chromatids of each chromosome detach from each other, and the spindle microtubules move them toward opposite spindle poles.

During **telophase**, a complete set of chromosomes reaches each spindle pole. A nuclear envelope forms around each cluster. Two new nuclei, each with the parental chromosome number, are the result.

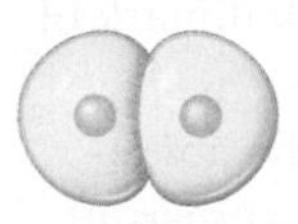

Section 11.4 **Cytokinesis** typically follows nuclear division. In animal cells, a contractile ring of microfilaments pulls the plasma membrane inward, forming a **cleavage furrow** that pinches the cytoplasm in two. In plant cells, vesicles guided by microtubules to the future plane of division merge as a **cell plate**. The plate expands and fuses with the parent cell wall, thus becoming a cross-wall that partitions the cytoplasm.

Section 11.5 **Telomeres** that protect the ends of eukaryotic chromosomes shorten with every DNA replication. Cells that inherit too-short telomeres die, and in most cells this limits the number of divisions that can occur.

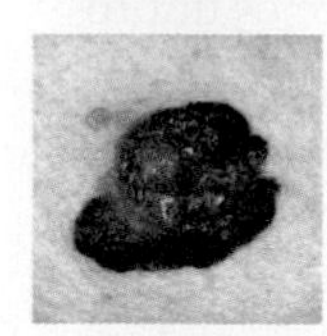

Section 11.6 The products of checkpoint genes work together to control the cell cycle. These molecules monitor the integrity of the cell's DNA, and can pause the cycle until breaks or other problems are fixed. When checkpoint mechanisms fail, a cell loses control over its cell cycle, and the cell's descendants form a **neoplasm**. Neoplasms may form lumps called **tumors**.

Genes encoding **growth factor** receptors are examples of **proto-oncogenes**, which means mutations can turn them into tumor-causing **oncogenes**. Mutations in multiple checkpoint genes can transform benign neoplasms into malignant ones. Cells of malignant neoplasms can break loose from their home tissues and colonize other parts of the body, a process called **metastasis**. **Cancer** occurs when malignant neoplasms physically and metabolically disrupt normal body tissues.

data analysis activities

HeLa Cells Are a Genetic Mess HeLa cells can vary in chromosome number. Defects in proteins that orchestrate cell division result in descendant cells with too many or too few chromosomes, an outcome that is one of the hallmarks of cancer cells. The panel of chromosomes on the right, originally published in 1989, shows all of the chromosomes in a single metaphase HeLa cell.

1. What is the chromosome number of this HeLa cell?

2. How many extra chromosomes does this cell have, compared to a normal human body cell?

3. Can you tell that this cell came from a female? How?

self-quiz

Answers in Appendix VII

1. Mitosis and cytoplasmic division function in ______ .
   a. asexual reproduction of single-celled eukaryotes
   b. growth and tissue repair in multicelled species
   c. asexual reproduction in prokaryotes
   d. both a and b
   e. all of the above

2. A duplicated chromosome has how many chromatids?

3. Homologous chromosomes ______ .
   a. carry the same genes
   b. are the same shape
   c. are the same length
   d. all of the above

4. Most cells spend the majority of their lives in ______ .
   a. prophase
   b. metaphase
   c. anaphase
   d. telophase
   e. interphase
   f. a and d

5. The spindle attaches to chromosomes at the ______ .
   a. centriole
   b. contractile ring
   c. centromere
   d. centrosome

6. Only ______ is not a stage of mitosis.
   a. prophase
   b. metaphase
   c. interphase
   d. anaphase

7. In intervals of interphase, G stands for ______ .
   a. gap b. growth c. Gey d. gene

8. Interphase is the part of the cell cycle when ______ .
   a. a cell ceases to function
   b. a cell forms its spindle apparatus
   c. a cell grows and duplicates its DNA
   d. mitosis proceeds

9. After mitosis, the chromosome number of a descendant cell is ______ the parent cell's.
   a. the same as
   b. one-half of
   c. rearranged compared to
   d. double that of

10. A plant cell divides by the process of ______ .
    a. telekinesis
    b. nuclear division
    c. fission
    d. cytokinesis

11. *BRCA1* and *BRCA2* ______ .
    a. are checkpoint genes
    b. are proto-oncogenes
    c. encode tumor suppressors
    d. all of the above

12. ______ are characteristic of cancer.
    a. Malignant cells b. Neoplasms c. Tumors

13. Match each term with its best description.

___ cell plate	a. lump of cells
___ spindle	b. made of microfilaments
___ tumor	c. divides plant cells
___ cleavage furrow	d. organize(s) the spindle
___ contractile ring	e. dangerous metastatic cells
___ cancer	f. made of microtubules
___ centrosomes	g. indentation
___ telomere	h. shortens with age

14. Match each stage with the events listed.

___ metaphase	a. sister chromatids move apart
___ prophase	b. chromosomes condense
___ telophase	c. new nuclei form
___ interphase	d. all chromosomes are aligned midway between spindle poles
___ anaphase	e. DNA replication
___ cytokinesis	f. cytoplasmic division

critical thinking

1. When a cell reproduces by mitosis and cytoplasmic division, does its life end?

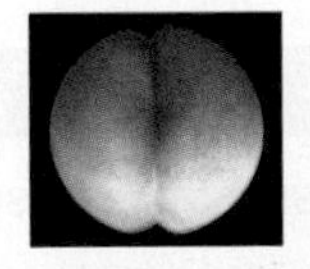

2. The eukaryotic cell in the photo on the left is in the process of cytoplasmic division. Is this cell from a plant or an animal? How do you know?

3. Exposure to radioisotopes or other sources of radiation can damage DNA. Humans exposed to high levels of radiation face a condition called radiation poisoning. Why do you think that exposure to radiation is used as a therapy to treat some kinds of cancers?

4. Suppose you have a way to measure the amount of DNA in one cell during the cell cycle. You first measure the amount at the G1 phase. At what points in the rest of the cycle will you see a change in the amount of DNA per cell?

CENGAGE brain To access course materials, please visit www.cengagebrain.com.

CREDITS: (in text) Data Analysis Activities, Courtesy of © Dr. Thomas Ried, NIH and the American Association for Cancer Research; (CT #2) © Michel Delarue © ISM/Phototake.

12 Meiosis and Sexual Reproduction

LEARNING ROADMAP

This chapter will draw on your knowledge of eukaryotic chromosomes (Section 11.2), DNA replication (8.5), genes (9.2), cytoplasmic division (11.4), and cell cycle controls (11.6) as we compare meiosis with mitosis (11.3). You will also revisit clones (8.1, 8.7), and the effects of mutation (9.6).

ALLELES AND SEXUAL REPRODUCTION

Genes that vary a bit in sequence as alleles are the basis of variation in traits. In sexual reproduction, offspring inherit genes from two parents who usually differ in some number of alleles.

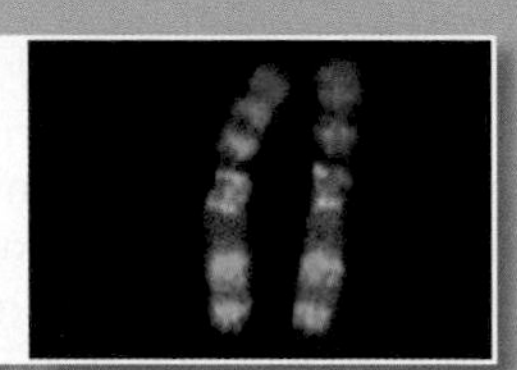

MEIOSIS IN THE LIFE CYCLE

Meiosis is a nuclear division process that reduces the chromosome number. It occurs only in cells that play a role in sexual reproduction in eukaryotes.

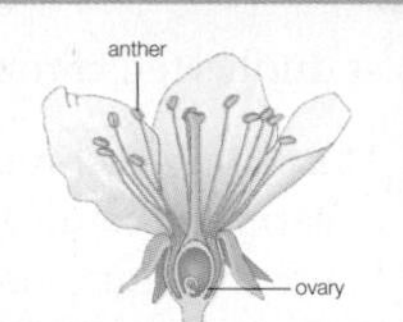

STAGES OF MEIOSIS

The chromosome number becomes reduced by the two nuclear divisions of meiosis. During this process, the chromosomes are sorted into four new nuclei.

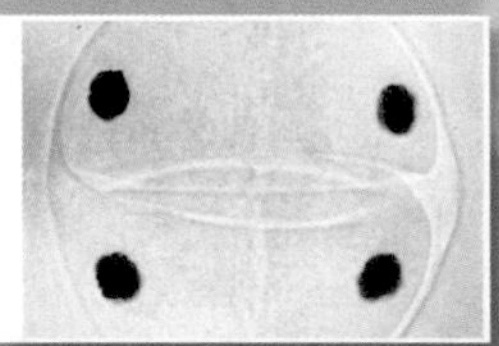

SHUFFLING PARENTAL DNA

During meiosis, homologous chromosomes swap segments, then are randomly sorted into separate nuclei. Both processes lead to novel combinations of alleles among offspring.

MITOSIS AND MEIOSIS COMPARED

Similarities between mitosis and meiosis suggest meiosis originated by evolutionary remodeling of mechanisms that already existed for mitosis and for repairing damaged DNA.

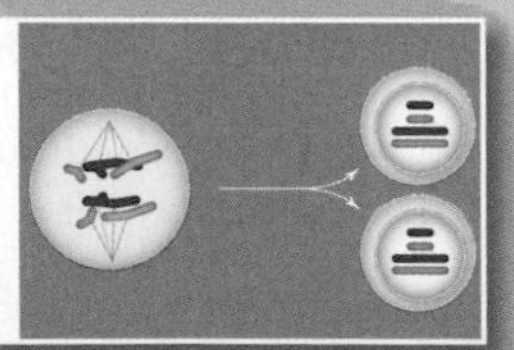

Meiosis is the basis of sexual reproduction, a topic covered in detail in Chapters 29 (plants) and 41 (animals). The variation in traits among individuals of sexually reproducing species arises from allele differences, a concept we revisit in context of inheritance in Chapters 13 and 14, and natural selection in Chapter 17.

12.1 Why Sex?

A few species reproduce asexually, which means one individual gives rise to offspring that are identical to itself and to one another. By contrast, **sexual reproduction** involves two individuals and mixes their genetic material (**FIGURE 12.1**). Almost all species reproduce sexually. If the function of reproduction is the perpetuation of one's genes, then an asexual reproducer would seem to win the evolutionary race. When it reproduces, it passes all of its genes to every one of its offspring. Only about half of a sexual reproducer's genes are passed to each offspring. Yet most eukaryotes reproduce sexually, at least some of the time. Why?

Consider that all offspring of an asexual reproducer are clones: They have the same traits as their parent. Consistency is a good thing if an organism lives in a favorable, unchanging environment. Traits that help it survive and reproduce do the same for its descendants. However, most environments are constantly changing. All offspring of an asexual reproducer are identical to one another and to the parent, so all are equally vulnerable to a change that is unfavorable. In changing environments, sexual reproducers have the evolutionary edge. Their offspring vary in the details of their traits. Some may have a particular combination of traits that suits them perfectly to a change in their environment. As a group, their diversity offers them a better chance of surviving this challenge than clones.

Consider the interaction between a predatory species and its prey. In each generation, prey individuals with traits that allow them to hide from, fend off, or escape the predator will leave the most offspring. However, the predator is constantly changing too: In each generation, individuals best able to find, capture, and overcome prey leave the most descendants. Thus, predators and prey are locked in a constant race, with each genetic improvement in one species countered by a genetic improvement in the other. This idea is called the Red Queen hypothesis, a reference to Lewis Carroll's book *Through the Looking Glass*. In the book, the Queen of Hearts tells Alice, "It takes all the running you can do, to keep in the same place." Environmental challenges that constantly change favor sexual reproduction, because the genetic diversity it fosters is evolutionarily advantageous in a changing environment.

Perhaps the most important advantage of sexual reproduction involves the inevitable occurrence of harmful mutations. A population of sexual reproducers has a better chance of weathering the effects of such mutations. With asexual reproduction, individuals bearing a harmful mutation necessarily pass it to all of their offspring. This outcome would be rare in sexual reproduction, because each offspring of a sexual union has a 50 percent chance of inheriting a parent's mutation. Thus, all else being equal, harmful mutations accumulate in an asexually reproducing population more quickly than in a sexually reproducing one.

sexual reproduction Reproductive mode by which offspring arise from two parents and inherit genes from both.

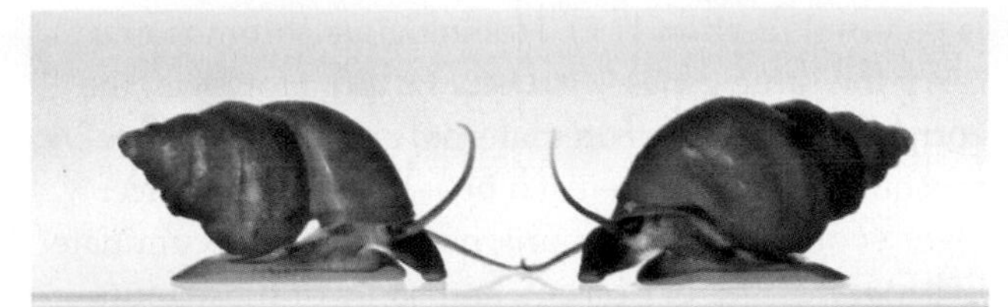

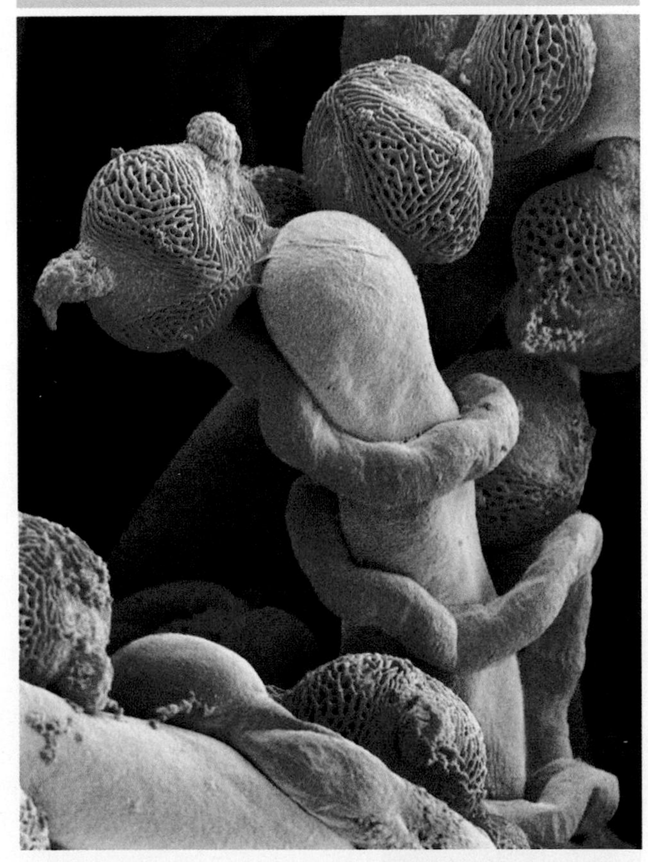

FIGURE 12.1 Moments in the stages of sexual reproduction. Sex mixes up the genetic material of two individuals.

Top, New Zealand mud snails can reproduce on their own (asexually) or with a partner (sexually). Natural populations of this species vary in their proportion of sexual and asexual individuals.

Bottom, flowering plants can reproduce asexually or sexually. The micrograph shows sexual reproduction, in which pollen grains (orange) germinate on flower carpels (yellow). Pollen tubes with male gametes inside grow from the grains down into tissues of the ovary, which house the flower's female gametes.

12.2 Meiosis in Sexual Reproduction

✔ Small differences in shared genes are the basis of differences in shared traits.

✔ Sexual reproduction mixes up alleles from two parents.

✔ Meiosis, the basis of sexual reproduction, occurs in eukaryotic cells set aside for reproduction.

Introducing Alleles

Your **somatic** (body) cells and those of many other sexually reproducing eukaryotes are diploid, which means they contain pairs of chromosomes. One chromosome of each homologous pair is maternal, and the other is paternal (Section 11.2). Homologous chromosomes carry the same genes (**FIGURE 12.2A**). However, the corresponding genes on maternal and paternal chromosomes often vary—just a bit—in DNA sequence. Over evolutionary time, unique mutations accumulate in separate lines of descent, and some of those mutations occur in genes. Thus, the DNA sequence of any gene may differ from the corresponding gene on the homologous chromosome (**FIGURE 12.2B**). Different forms of the same gene are called **alleles**.

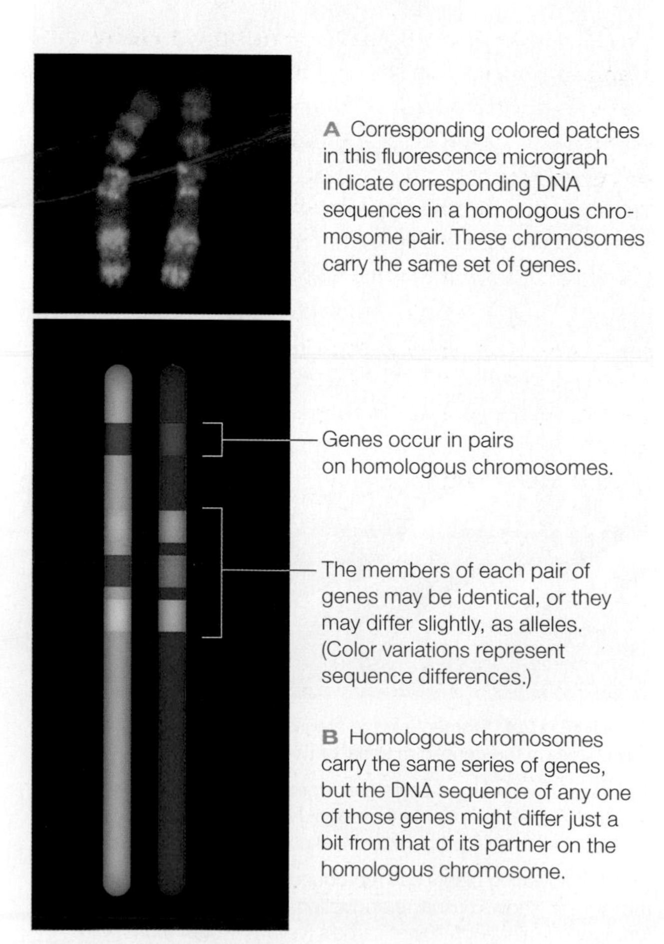

A Corresponding colored patches in this fluorescence micrograph indicate corresponding DNA sequences in a homologous chromosome pair. These chromosomes carry the same set of genes.

B Homologous chromosomes carry the same series of genes, but the DNA sequence of any one of those genes might differ just a bit from that of its partner on the homologous chromosome.

FIGURE 12.2 ▸**Animated** Genes on chromosomes. Different forms of a gene are called alleles.

Alleles may encode slightly different forms of a gene's product, and such differences influence the details of traits shared by a species. Consider that one of the approximately 20,000 genes in human chromosomes encodes beta globin, a subunit of hemoglobin (Section 9.6). Like most human genes, the beta globin gene has multiple alleles—more than 700 in this case. A few beta globin alleles cause sickle-cell anemia, several cause beta thalassemia, and so on. Allele differences among individuals are one reason that the members of a sexually reproducing species are not identical. Offspring of sexual reproducers vary in the combinations of alleles they inherit; thus, they vary in their combinations of traits.

Meiosis Halves the Chromosome Number

Sexual reproduction involves the fusion of mature reproductive cells—**gametes**—from two parents. Gametes have a single set of chromosomes, so they are **haploid** (n): Their chromosome number is half of the diploid ($2n$) number (Section 8.4). **Meiosis**, the nuclear division mechanism that halves the chromosome number, is necessary for gamete formation. Meiosis also gives rise to new combinations of parental alleles.

Gametes arise by division of **germ cells**, which are immature reproductive cells that form in organs set aside for reproduction (**FIGURE 12.3**). Animals and plants make gametes somewhat differently. In animals, meiosis in diploid germ cells gives rise to eggs (female gametes) or sperm (male gametes). In plants, haploid germ cells form by meiosis. Gametes form when these cells divide by mitosis.

The first part of meiosis is similar to mitosis. A cell duplicates its DNA before either nuclear division process begins. As in mitosis, a spindle forms, and its microtubules move the duplicated chromosomes to opposite spindle poles. However, meiosis sorts the chromosomes into new nuclei not once, but twice, so it results in the formation of four haploid nuclei. The two consecutive nuclear divisions are called meiosis I and meiosis II:

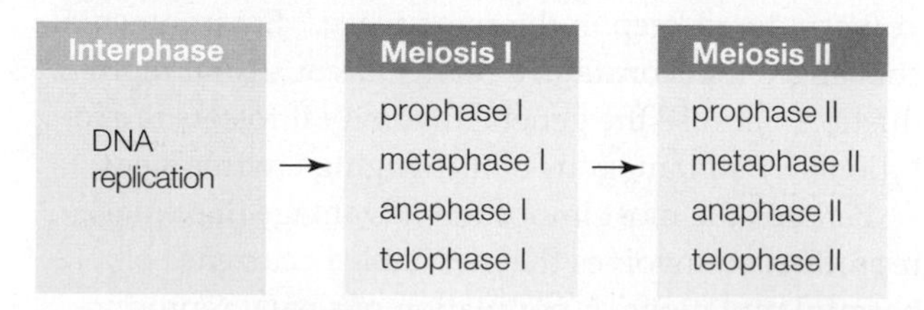

In some cells, meiosis II occurs immediately after meiosis I. In others, a period of protein synthesis—but no DNA replication—intervenes between the divisions.

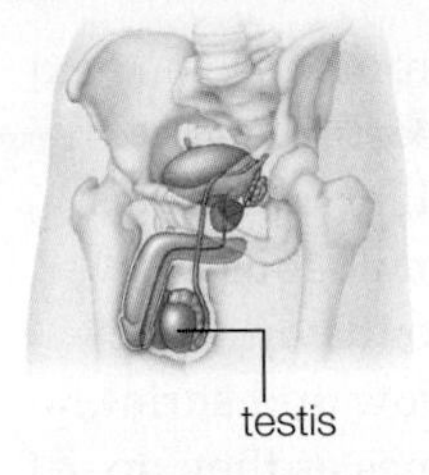

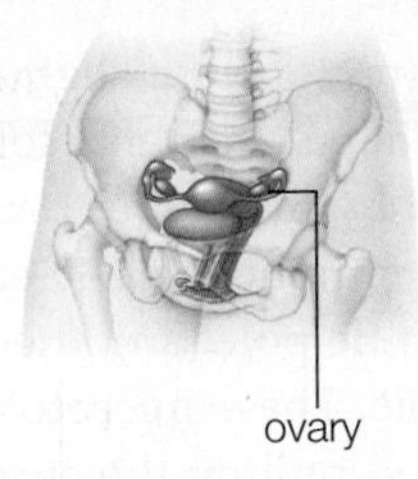

A Reproductive organs of humans. Meiosis in germ cells inside testes and ovaries produces gametes (sperm and eggs).

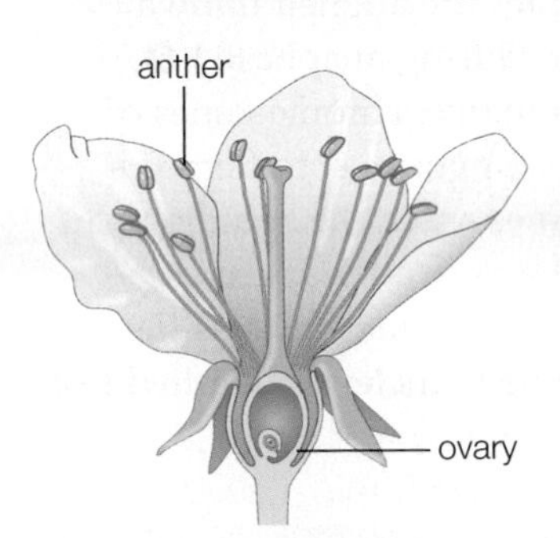

B Reproductive organs of a flowering plant. Meiosis produces haploid germ cells inside anthers and ovaries. These cells divide by mitosis to give rise to gametes (sperm and eggs).

FIGURE 12.3 ▶Animated Examples of reproductive organs in (A) animals and (B) plants.

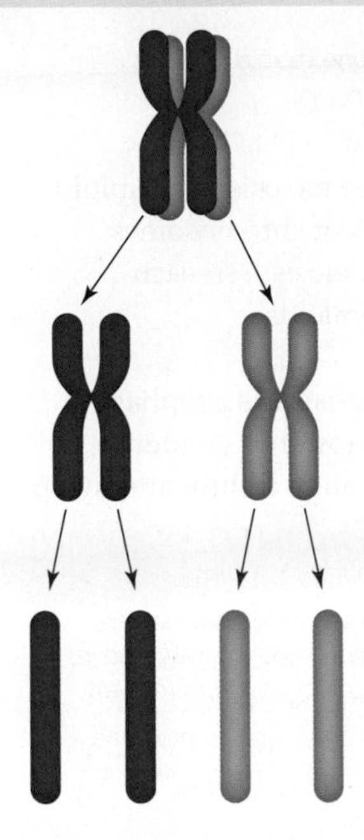

FIGURE 12.4 How meiosis halves the chromosome number.

During meiosis I, every duplicated chromosome aligns with its homologous partner (**FIGURE 12.4** ❶). Then the homologous chromosomes are pulled away from one another and packaged into separate nuclei ❷. Each of the two new nuclei is haploid (n)—it has one copy of each chromosome—so the chromosome number is now half that of the parent.

The chromosomes are still duplicated (the sister chromatids remain attached to one another). During meiosis II, sister chromatids are pulled apart, and each becomes an individual, unduplicated chromosome ❸. The chromosomes are sorted into four new nuclei. Each new nucleus still has one copy of each chromosome, so it is haploid (n).

Thus, meiosis partitions the chromosomes of one diploid nucleus ($2n$) into four haploid (n) nuclei. The next section zooms in on the details of this process.

Fertilization Restores Chromosome Number

Haploid gametes form by meiosis. The diploid chromosome number is restored at **fertilization**, when two haploid gametes fuse to form a **zygote**, the first cell of a new individual. Thus, meiosis halves the chromosome number, and fertilization restores it.

If meiosis did not precede fertilization, then the chromosome number would double with every generation. As you will see in Chapter 14, chromosome number changes can have drastic consequences for health, particularly in animals. An individual's set of chromosomes is like a fine-tuned blueprint that must be followed exactly, page by page, in order to build a body that functions normally.

alleles Forms of a gene with slightly different DNA sequences; may encode slightly different versions of the gene's product.
fertilization Fusion of two gametes to form a zygote.
gamete Mature, haploid reproductive cell; e.g., an egg or a sperm.
germ cell Immature reproductive cell that gives rise to haploid gametes when it divides.
haploid Having one of each type of chromosome characteristic of the species.
meiosis Nuclear division process that halves the chromosome number. Basis of sexual reproduction.
somatic Relating to the body.
zygote Diploid cell that forms when two gametes fuse; the first cell of a new individual.

TAKE-HOME MESSAGE 12.2

Why is meiosis necessary for sexual reproduction?

✔ Paired genes on homologous chromosomes may vary in DNA sequence as alleles. Alleles arise by mutation.

✔ Alleles give rise to differences in shared traits. Offspring of sexual reproducers inherit new combinations of parental alleles—thus new combinations of traits.

✔ The nuclear division process of meiosis is the basis of sexual reproduction in plants and animals.

✔ Meiosis halves the diploid ($2n$) chromosome number, to the haploid number (n), for forthcoming gametes.

✔ When two gametes fuse at fertilization, the diploid chromosome number is restored in the resulting zygote.

12.3 Visual Tour of Meiosis

✔ During meiosis, chromosomes of one diploid nucleus become distributed into four haploid nuclei.

FIGURE 12.5 shows the stages of meiosis in a diploid (2*n*) cell, which contains two sets of chromosomes. DNA replication occurs before meiosis I, so each chromosome has two sister chromatids.

Meiosis I The first stage of meiosis I is prophase I ❶. During this phase, the chromosomes condense, and homologous chromosomes align tightly and swap segments (more about segment-swapping in the next section). The nuclear envelope breaks up. A spindle forms, and by the end of prophase I, microtubules attach one chromosome of each homologous pair to one spindle pole, and the other to the opposite spindle pole. These microtubules grow and shrink, pushing and pulling the chromosomes as they do. At metaphase I ❷, all of the microtubules are the same length, and the chromosomes are aligned midway between the spindle poles. During anaphase I ❸, the spindle pulls the homologous chromosomes of each pair apart and toward opposite spindle poles. The two sets of chromosomes reach the spindle poles during telophase I ❹, and a new nuclear envelope forms around each cluster of chromosomes as the DNA loosens up. The two new nuclei are haploid (*n*);

FIGURE 12.5 ▸Animated Meiosis. Two pairs of chromosomes are illustrated in a diploid (2*n*) cell. Homologous chromosomes are indicated in blue and pink. Micrographs show meiosis in a lily plant cell (*Lilium regale*).

FIGURE IT OUT During which phase of meiosis does the chromosome number become reduced?

Answer: Anaphase I

MEIOSIS I One diploid nucleus to two haploid nuclei

① **Prophase I**
Homologous chromosomes condense, pair up, and swap segments. Spindle microtubules attach to them as the nuclear envelope breaks up.

② **Metaphase I**
Homologous chromosome pairs are aligned between spindle poles. Spindle microtubules attach the two chromosomes of each pair to opposite spindle poles.

③ **Anaphase I**
All of the homologous chromosomes separate and begin heading toward the spindle poles.

④ **Telophase I**
A complete set of chromosomes clusters at both ends of the cell. A nuclear envelope forms around each set, so two haploid (*n*) nuclei form.

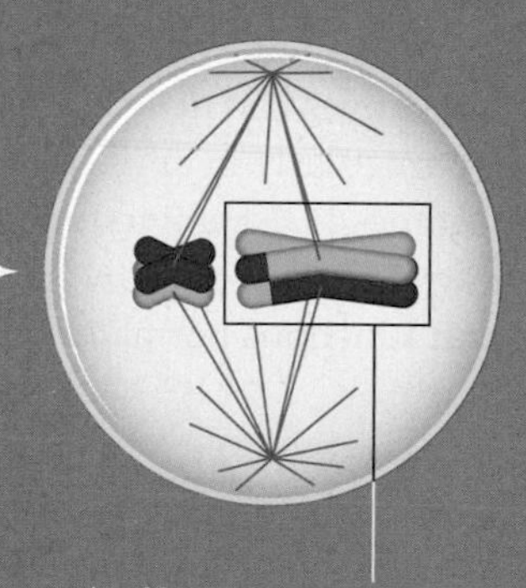

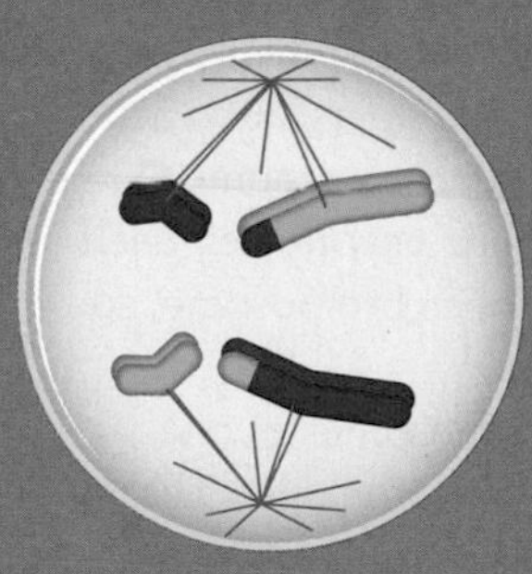

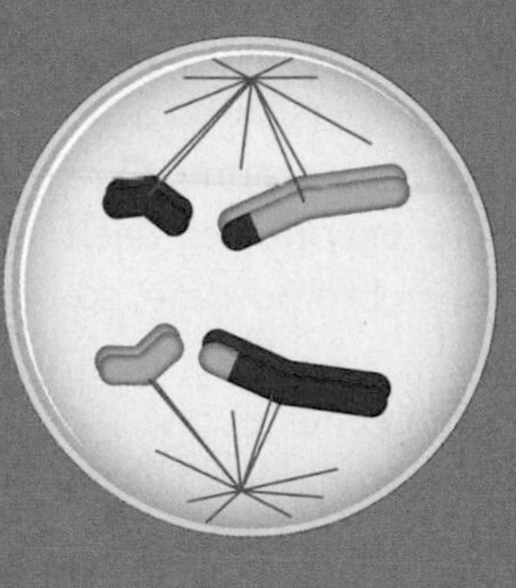

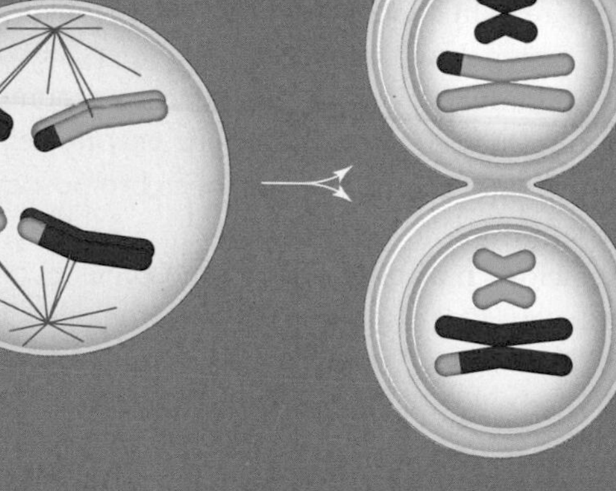

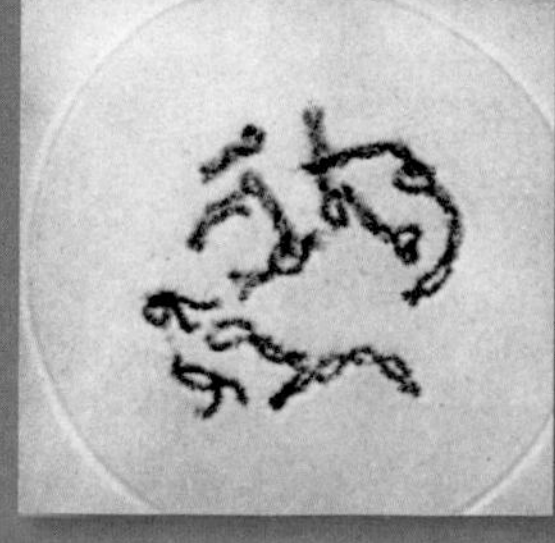

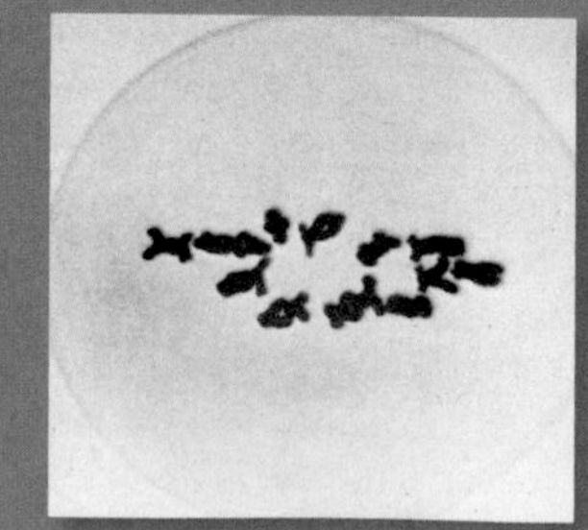

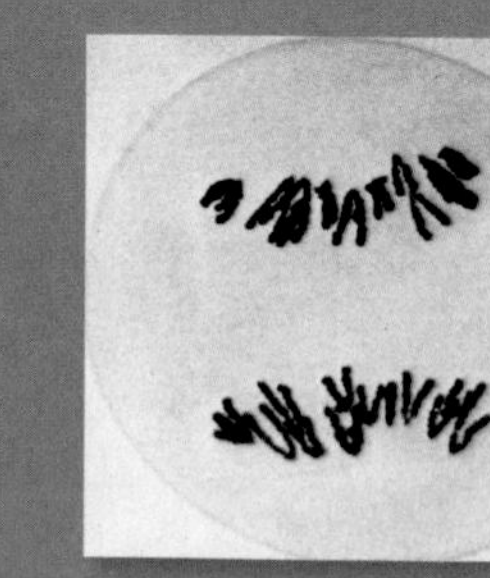

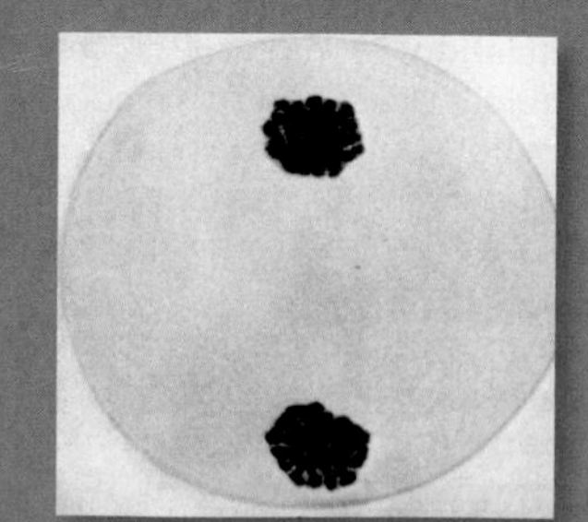

each contains one set of chromosomes. The cytoplasm often divides at this point. Each chromosome is still duplicated (it consists of two sister chromatids).

Meiosis may pause at this point, but no DNA replication occurs before meiosis II.

Meiosis II Meiosis II proceeds simultaneously in both nuclei that formed in meiosis I. During prophase II ❺, the chromosomes condense and the nuclear envelope breaks up. A new spindle forms. By the end of prophase II, spindle microtubules attach each chromatid to one spindle pole, and its sister chromatid to the opposite spindle pole. These microtubules push and pull the chromosomes, aligning them midway between spindle poles at metaphase II ❻. During anaphase II ❼, the spindle microtubules pull the sister chromatids apart and toward opposite spindle poles. Each chromosome is now unduplicated (it consists of one molecule of DNA). During telophase II ❽, these chromosomes reach the spindle poles. New nuclear envelopes form around the four clusters of chromosomes as the DNA loosens up. Each of the four nuclei that form are haploid (n), with one set of (unduplicated) chromosomes. The cytoplasm often divides at this point.

TAKE-HOME MESSAGE 12.3

What happens to a cell during meiosis?

✔ During meiosis, the nucleus of a diploid ($2n$) cell divides twice. Four haploid (n) nuclei form, each with a full set of chromosomes—one of each type.

MEIOSIS II Two haploid nuclei to four haploid nuclei

⑤ **Prophase II**
The chromosomes condense. Spindle microtubules attach to each sister chromatid as the nuclear envelope breaks up.

⑥ **Metaphase II**
The (still duplicated) chromosomes are aligned midway between spindle poles.

⑦ **Anaphase II**
Sister chromatids separate. The now unduplicated chromosomes head to the spindle poles.

⑧ **Telophase II**
A complete set of chromosomes clusters at both ends of the cell. A new nuclear envelope forms around each set, so four haploid (n) nuclei form.

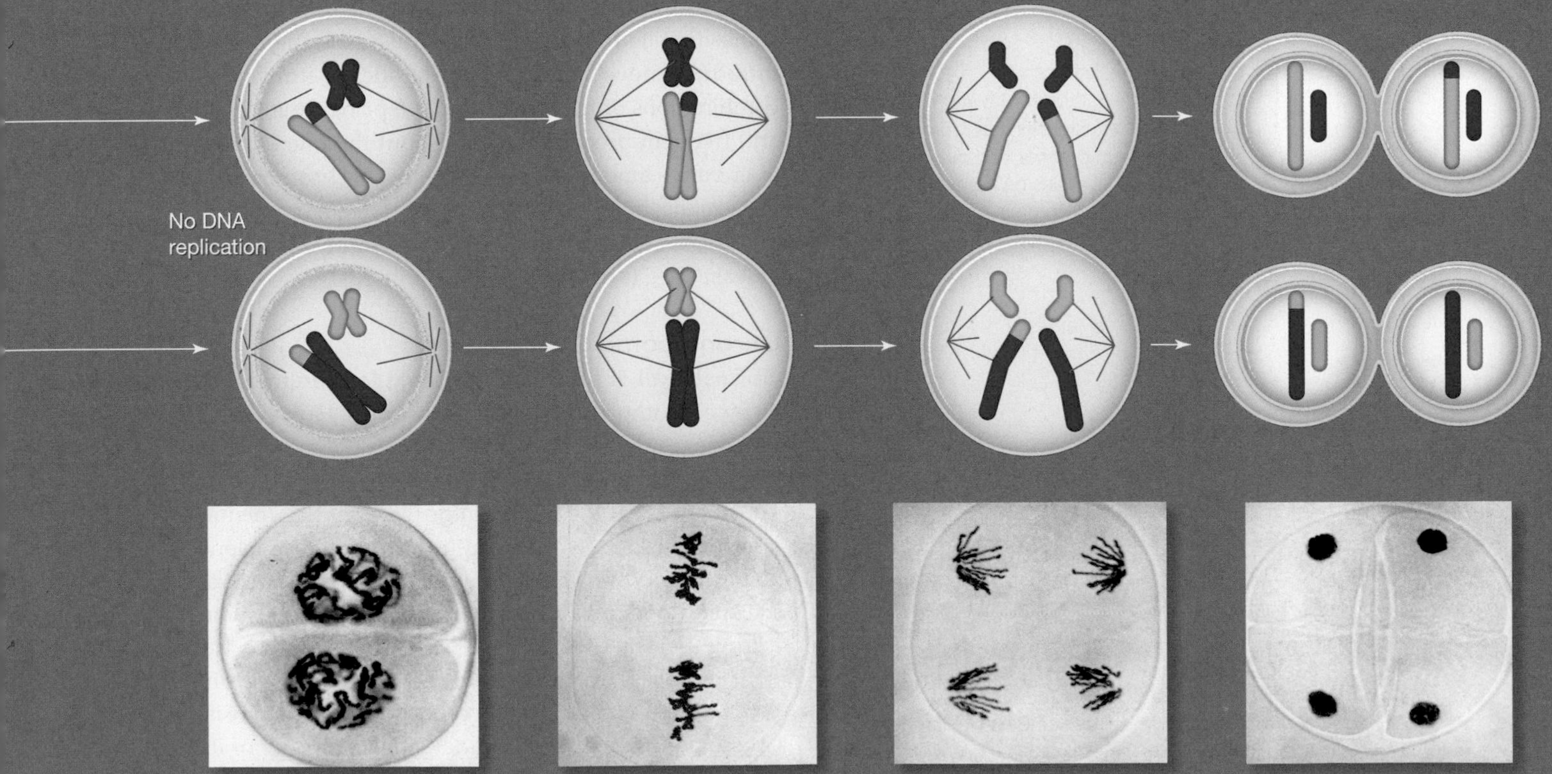

No DNA replication

✔ Crossovers and the random sorting of chromosomes into gametes result in new combinations of traits among offspring of sexual reproducers.

The previous section mentioned briefly that duplicated chromosomes swap segments with their homologous partners during prophase I. It also showed how spindle microtubules align and then separate homologous chromosomes during anaphase I. These events, along with fertilization, contribute to the variation in combinations of traits among the offspring of sexually reproducing species.

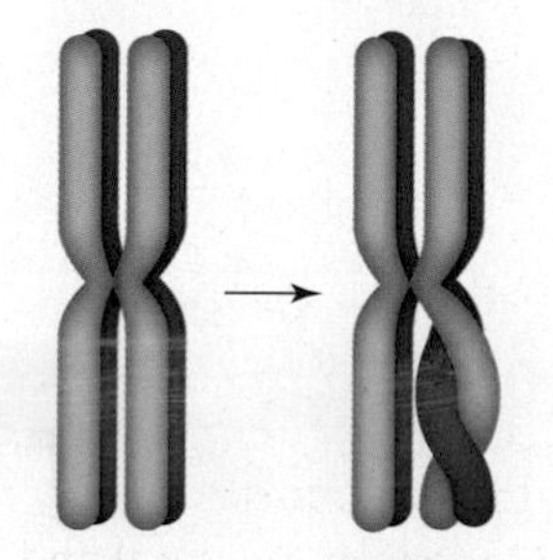

A Here, we focus on only two of the many genes on a chromosome. In this example, one gene has alleles *A* and *a*; the other has alleles *B* and *b*.

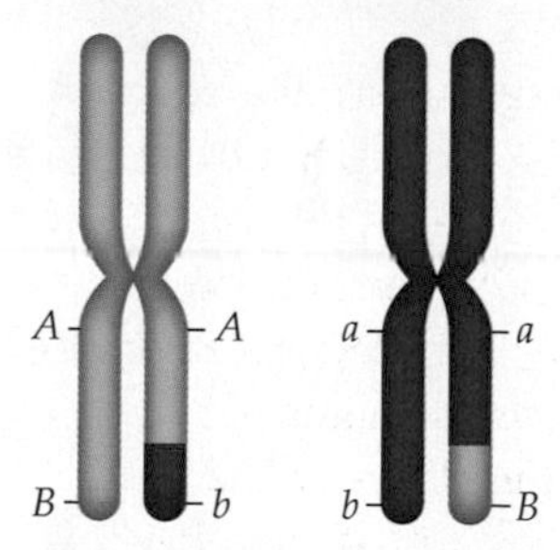

B Close contact between homologous chromosomes promotes crossing over between nonsister chromatids. Paternal and maternal chromatids exchange corresponding pieces.

C Crossing over mixes up paternal and maternal alleles on homologous chromosomes.

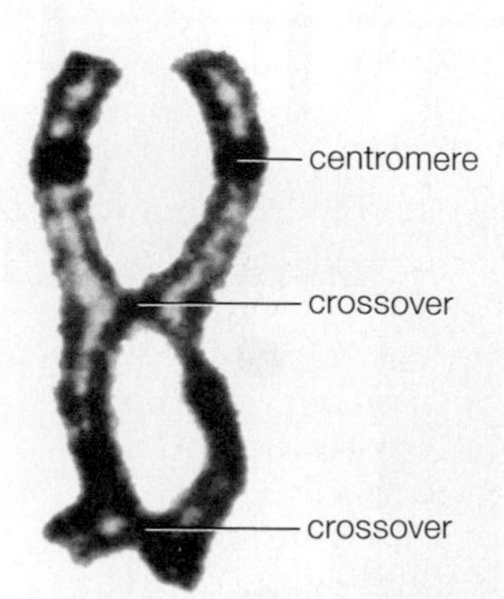

D Each pair of homologous chromosomes can cross over multiple times. This is a normal and common process of meiosis.

FIGURE 12.6 ▸Animated Crossing over. Blue signifies a paternal chromosome, and pink, its maternal homologue. For clarity, we show only one pair of homologous chromosomes.

Crossing Over

Early in prophase I of meiosis, all chromosomes in the cell condense. When they do, each chromosome is drawn close to its homologous partner, so that the chromatids align along their length:

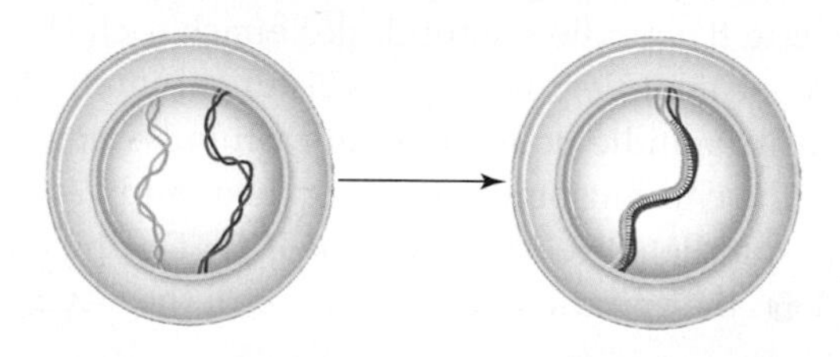

This tight, parallel orientation favors **crossing over**, a process by which a chromosome and its homologous partner exchange corresponding pieces of DNA during meiosis (**FIGURE 12.6A–C**). Homologous chromosomes may swap any segment of DNA along their length, although crossovers tend to occur more frequently in certain regions.

Swapping segments of DNA shuffles alleles between homologous chromosomes. It breaks up the particular combinations of alleles that occurred on the parental chromosomes, and makes new ones on the chromosomes that end up in gametes. Thus, crossing over introduces novel combinations of alleles—thus new combinations of traits—among offspring. It is a normal and frequent process in meiosis, but the rate of crossing over varies among species and among chromosomes. In humans, between 46 and 95 crossovers occur per meiosis, so on average each chromosome crosses over at least once (**FIGURE 12.6D**).

Chromosome Segregation

Normally, all of the new nuclei that form in meiosis I receive a complete set of chromosomes. However, whether a new nucleus ends up with the maternal or paternal version of a chromosome is entirely random. The chance that the maternal or the paternal version of any chromosome will end up in a particular nucleus is 50 percent. Why? The answer has to do with the way the spindle segregates the homologous chromosomes during meiosis I.

The process of chromosome segregation begins in prophase I. Imagine a germ cell undergoing meiosis. Crossovers have already made genetic mosaics of its chromosomes, but for simplicity let's put crossing over

CREDITS: (6A–C, in text) © Cengage Learning; (6D) © James Kezer, Courtesy of Dr. Sessions.

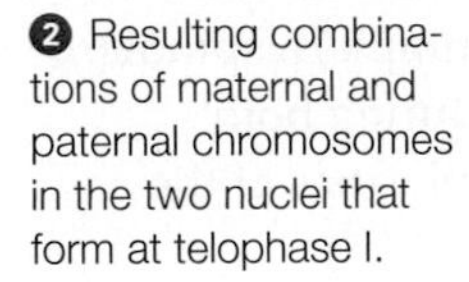

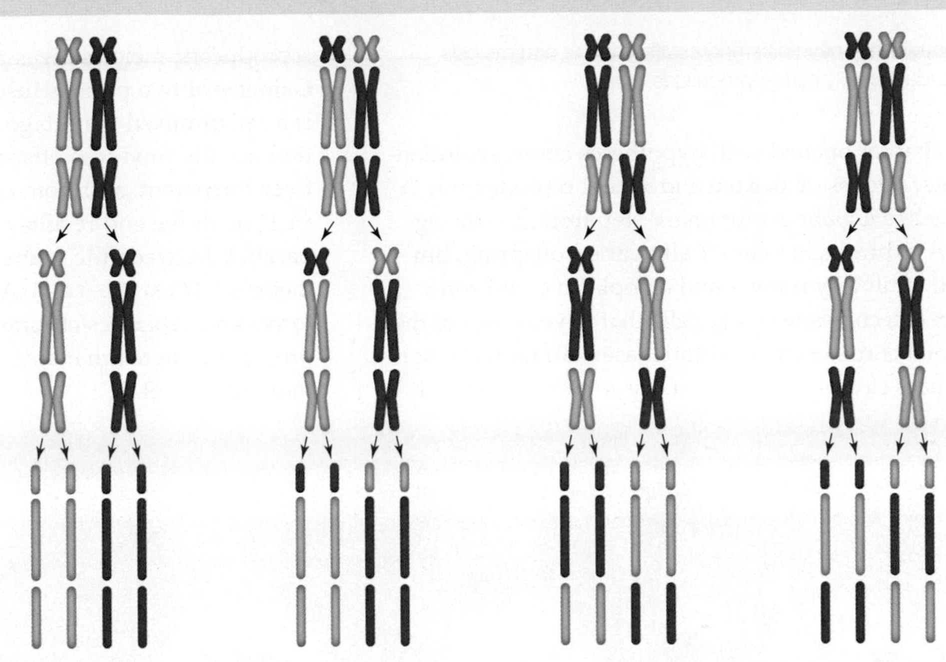

FIGURE 12.7 ▸**Animated** Hypothetical segregation of three pairs of chromosomes in meiosis I. Maternal chromosomes are pink; paternal, blue. Which chromosome of each pair gets packaged into which of the two new nuclei that form at telophase I is random. For simplicity, no crossing over occurs in this example, so all sister chromatids are identical.

aside for a moment. Just call the chromosomes inherited from the mother the maternal ones, and the ones from the father the paternal ones.

During prophase I, microtubules fasten the cell's chromosomes to the spindle poles. It is very unlikely that all of the maternal chromosomes will be attached to one pole, and all of the paternal chromosomes will be attached to the other. Microtubules extending from a spindle pole bind to the centromere of the first chromosome they contact, regardless of whether it is maternal or paternal. Though each homologous partner becomes attached to the opposite spindle pole, there is no pattern to the attachment of the maternal or paternal chromosomes to a particular pole.

Now imagine that the germ cell has three pairs of chromosomes (**FIGURE 12.7**). By metaphase I, those three pairs of maternal and paternal chromosomes have been divided up between the two spindle poles in one of four ways ❶. In anaphase I, homologous chromosomes separate and are pulled toward opposite spindle poles. In telophase I, a new nucleus forms around the chromosomes that cluster at each spindle pole. Each nucleus contains one of eight possible combinations of maternal and paternal chromosomes ❷.

In telophase II, each of the two nuclei divides and gives rise to two new haploid nuclei. The two new nuclei are identical because no crossing over occurred in our hypothetical example, so all of the sister chromatids were identical. Thus, at the end of meiosis in this cell, two (2) spindle poles have divvied up three (3) chromosome pairs. The resulting four nuclei have one of eight (2^3) possible combinations of maternal and paternal chromosomes ❸.

Cells that give rise to human gametes have twenty-three pairs of homologous chromosomes, not three. Each time a human germ cell undergoes meiosis, the four gametes that form end up with one of 8,388,608 (or 2^{23}) possible combinations of homologous chromosomes. That number does not even take into account crossing over, which mixes up the alleles on maternal and paternal chromosomes, or fusion with another gamete at fertilization.

crossing over Process by which homologous chromosomes exchange corresponding segments during prophase I of meiosis.

TAKE-HOME MESSAGE 12.4

How does meiosis give rise to new combinations of parental alleles?

✔ Crossing over—recombination between nonsister chromatids of homologous chromosomes—occurs during prophase I.

✔ Homologous chromosomes can get attached to either spindle pole in prophase I, so each chromosome of a homologous pair can end up in either of the two new nuclei.

✔ Crossing over and random sorting of chromosomes into gametes give rise to new combinations of alleles—thus new combinations of traits—among offspring of sexual reproducers.

12.5 Mitosis and Meiosis—An Ancestral Connection?

✔ Though they have different results, mitosis and meiosis are fundamentally similar processes.

This chapter opened with hypotheses about evolutionary advantages of asexual and sexual reproduction. It seems like a giant evolutionary step from producing clones to producing genetically varied offspring, but was it really? By mitosis and cytoplasmic division, one cell becomes two new cells that have copies of the parental chromosomes. Mitotic (asexual) reproduction produces clones of the parent. By contrast, in sexual reproducers, meiosis gives rise to haploid gametes. Gametes of two parents fuse to form a zygote, which is a cell of mixed parentage. Meiotic (sexual) reproduction usually produces offspring that differ genetically from the parent, and from one another.

Though the end results differ, there are striking parallels between the four stages of mitosis and meiosis II (**FIGURE 12.8**). As one example, a spindle forms and separates chromosomes during both processes. There are many more similarities at the molecular level.

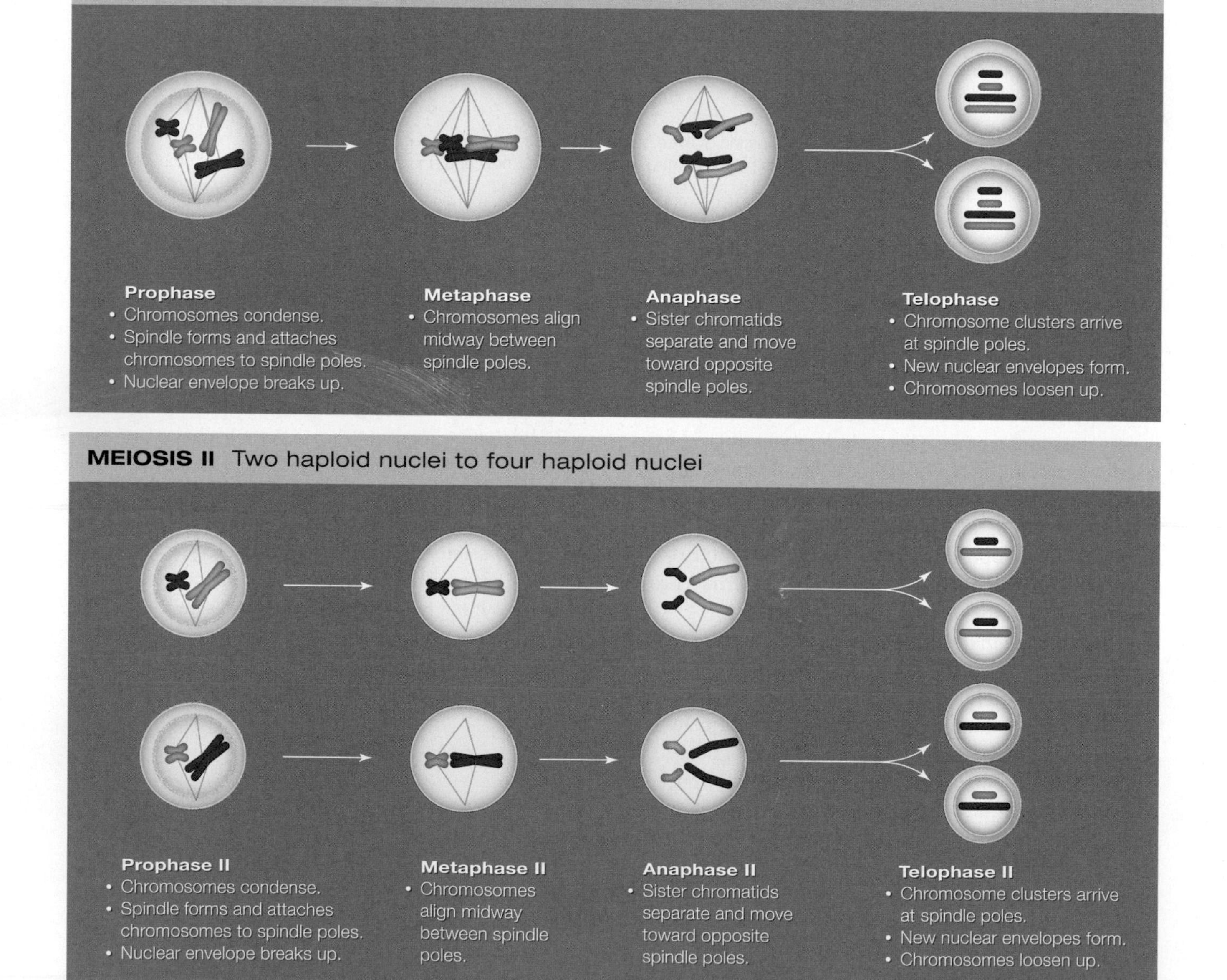

FIGURE 12.8 ▸**Animated** Comparing meiosis II with mitosis.

Why Sex? (revisited)

Why do males exist? No male has ever been found among the tiny freshwater creatures called bdelloid rotifers (FIGURE 12.10). Females have been reproducing for 80 million years solely through cloning themselves. Bdelloids are one of the few animal groups to have completely abandoned sex.

Compared to sex, asexual reproduction is often seen as a poor long-term strategy because it lacks crossing over—the chromosomal shuffling that brings about genetic diversity thought to give species an adaptive edge in the face of new challenges. Bdelloids have contradicted this theory by being very successful; over 360 species are alive today.

A newly discovered ability may help to explain the success of the bdelloids despite their rejection of sex. These rotifers can apparently import genes from bacteria, fungi, protists, and even plants. If the main advantage of sex is that it promotes genetic diversity, why worry about it when you have the genes of entire kingdoms available to you?

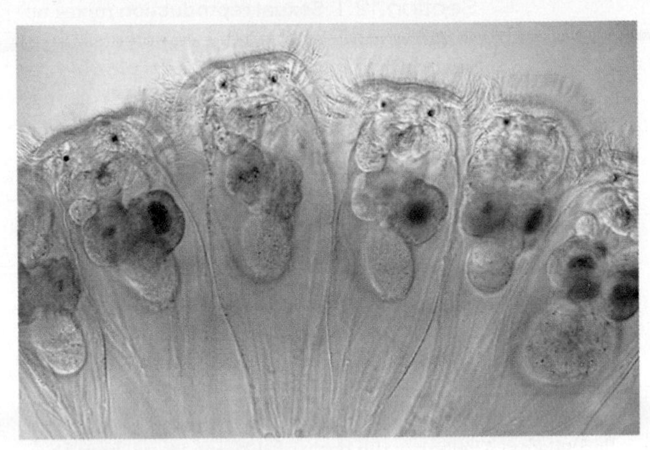

FIGURE 12.10 A common bdelloid rotifer (*Philodina rosea*). All of these tiny animals are female.

The direct swapping of genetic material is incredibly rare in animals, but bdelloids are bringing in external genes to an extent completely unheard of in complex organisms. Each rotifer is a genetic mosaic whose DNA spans almost all the major kingdoms of life: About 10 percent of its active genes have been pilfered from other organisms.

Long ago, the molecular machinery of mitosis may have been remodeled into meiosis. Evidence for this hypothesis includes a host of shared molecules, including the products of the *BRCA* genes (Section 11.6) that are made by all modern eukaryotes. By monitoring and fixing problems with the DNA—such as damaged or mismatched bases (Section 8.6)—these molecules actively maintain the integrity of a cell's chromosomes, particularly during DNA replication and mitosis. Many of the same molecules help homologous chromosomes cross over in prophase I of meiosis (FIGURE 12.9). As another example, consider that the same regulatory molecules involved in checkpoints of mitosis are also involved in checkpoints of meiosis.

During anaphase of mitosis, sister chromatids are pulled apart. What would happen if the connections between the sisters did not break? Each duplicated chromosome would be pulled to one or the other spindle pole—which is exactly what happens during anaphase I of meiosis. The shared molecules and mechanisms imply a shared evolutionary history; sexual reproduction probably originated with mutations that affected processes of mitosis. As you will see in later chapters, the remodeling of existing processes into new ones is a common evolutionary theme.

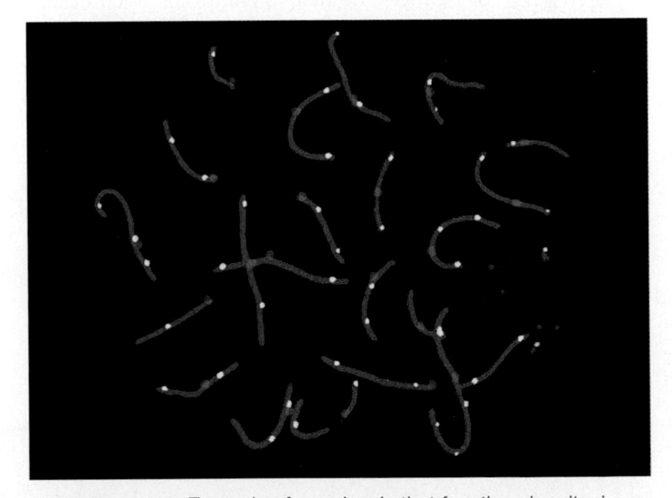

FIGURE 12.9 Example of a molecule that functions in mitosis and meiosis. This fluorescence micrograph shows homologous chromosome pairs (red) in the nucleus of a human cell during prophase I of meiosis. Centromeres are blue. Yellow pinpoints the location of a protein called MLH1 assisting with crossovers. MLH1 also helps repair mismatched bases during mitosis.

TAKE-HOME MESSAGE 12.5

Are the processes of mitosis and meiosis related?

✔ Meiosis may have evolved by the remodeling of existing mechanisms of mitosis.

summary

Section 12.1 **Sexual reproduction** mixes up the genetic information of two parents. The offspring of sexual reproducers typically vary in shared, inherited traits. Particularly in environments where pressures change rapidly, this variation can offer an evolutionary advantage over genetically identical offspring produced by asexual reproduction.

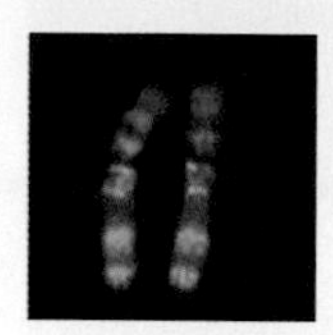

Section 12.2 Sexual reproduction produces offspring whose **somatic** cells contain pairs of chromosomes, one of each homologous pair from the mother and the other from the father. The two chromosomes of a homologous pair carry the same genes. Paired genes on homologous chromosomes may vary in DNA sequence, in which case they are called **alleles**. Alleles are the basis of differences in shared, heritable traits. They arise by mutation.

Meiosis, the basis of sexual reproduction in eukaryotes, is a nuclear division mechanism that halves the chromosome number for forthcoming **gametes**. It occurs only in cells that are involved in sexual reproduction. **Haploid** (n) gametes are mature reproductive cells that form from division of **germ cells**. The fusion of two haploid gametes during **fertilization** restores the diploid parental chromosome number in the **zygote**, the first cell of the new individual.

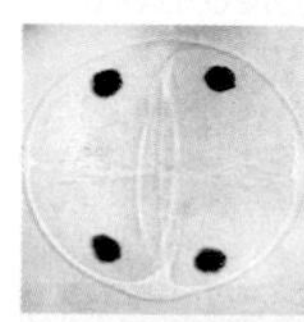

Section 12.3 DNA replication occurs before meiosis begins, so each chromosome consists of two molecules of DNA (sister chromatids). The process of meiosis comprises two nuclear divisions, meiosis I and meiosis II.

The first nuclear division (meiosis I) begins when the chromosomes condense and align tightly with their homologous partners during prophase I. Microtubules then extend from the spindle poles, penetrate the nuclear region, and attach to one or the other chromosome of each homologous pair. At metaphase I, all chromosomes are lined up midway between the spindle poles. During anaphase I, homologous chromosomes separate and move to opposite spindle poles. Two nuclear envelopes form around the two sets of chromosomes during telophase I. The cytoplasm may divide at this point. There may be a resting period before meiosis resumes, but DNA replication does not occur.

The second nuclear division (meiosis II) occurs simultaneously in both haploid nuclei that formed during meiosis I. The chromosomes are still duplicated; each still consists of two sister chromatids. The chromosomes condense during prophase II, and align in metaphase II. The sister chromatids of each chromosome are pulled apart from each other during anaphase II, so at the end of meiosis each chromosome consists of one molecule of DNA. By the end of telophase II, four haploid nuclei have typically formed, each with a complete set of (unduplicated) chromosomes.

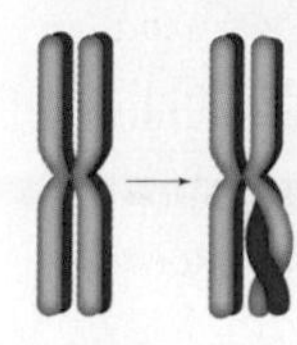

Section 12.4 Meiosis shuffles parental alleles, so offspring inherit nonparental combinations of them. During prophase I, homologous chromosomes exchange corresponding segments. This **crossing over** mixes up the alleles on maternal and paternal chromosomes, thus giving rise to combinations of alleles not present in either parental chromosome. Novel combinations of alleles are the basis of novel combinations of traits among offspring of sexual reproducing organisms. Meiosis also contributes to variation in traits by randomly segragating maternal and paternal chromosomes into gametes. Microtubules can attach the maternal or the paternal chromosome of each pair to one or the other spindle pole. Either chromosome may end up in any new nucleus, and in any gamete.

Section 12.5 The process of meiosis resembles that of mitosis, and may have evolved from it. Many of the same molecules function the same way in both processes. For example, a spindle forms, moves and sorts chromosomes, and disassembles during both mitosis and meiosis.

self-quiz

Answers in Appendix VII

1. One evolutionary advantage of sexual over asexual reproduction may be that it produces ______.
   a. more offspring per individual
   b. more variation among offspring
   c. healthier offspring

2. Meiosis is a necessary part of sexual reproduction because it ______.
   a. divides two nuclei into four new nuclei
   b. reduces the chromosome number for gametes
   c. produces clones that can cross over

3. Meiosis ______.
   a. occurs in all eukaryotes
   b. supports growth and tissue repair in multicelled species
   c. gives rise to genetic diversity among offspring
   d. is part of the life cycle of all cells

4. Sexual reproduction in animals requires ______.
   a. meiosis
   b. fertilization
   c. germ cells
   d. all of the above

5. Meiosis ______ the parental chromosome number.
   a. doubles
   b. halves
   c. maintains
   d. mixes up

6. Dogs have a diploid chromosome number of 78. How many chromosomes do their gametes have?
   a. 39
   b. 78
   c. 156
   d. 234

7. The cell in the diagram to the right is in anaphase I, not anaphase II. I know this because ______.

8. Which of the following cells can undergo meiosis?
   a. the diploid body cells of an animal
   b. cells set aside for reproduction in eukaryotes
   c. haploid gametes
   d. all of the above

data analysis activities

BPA and Abnormal Meiosis In 1998, researchers at Case Western University were studying meiosis in mouse oocytes when they saw an unexpected and dramatic increase in abnormal meiosis events (**FIGURE 12.11**). The improper segregation of chromosomes during meiosis is one of the main causes of human genetic disorders, which we will discuss in Chapter 14.

The researchers discovered that the spike in meiotic abnormalities began immediately after the mouse facility started washing the animals' plastic cages and water bottles in a new, alkaline detergent. The detergent had damaged the plastic, which as a result was leaching bisphenol A (BPA). BPA is a synthetic chemical that mimics estrogen, the main female sex hormone in animals. BPA is still widely used to manufacture polycarbonate plastic items (such as water bottles) and epoxies (such as the coating on the inside of metal cans of food).

1. What percentage of mouse oocytes displayed abnormalities of meiosis with no exposure to damaged caging?
2. Which group of mice had the most meiotic abnormalities in their oocytes?
3. What is abnormal about metaphase I as it is occurring in the oocytes shown in **FIGURE 12.11B**, **C**, and **D**?

Caging materials	Total number of oocytes	Abnormalities
Control: New cages with glass bottles	271	5 (1.8%)
Damaged cages with glass bottles		
Mild damage	401	35 (8.7%)
Severe damage	149	30 (20.1%)
Damaged bottles	197	53 (26.9%)
Damaged cages with damaged bottles	58	24 (41.4%)

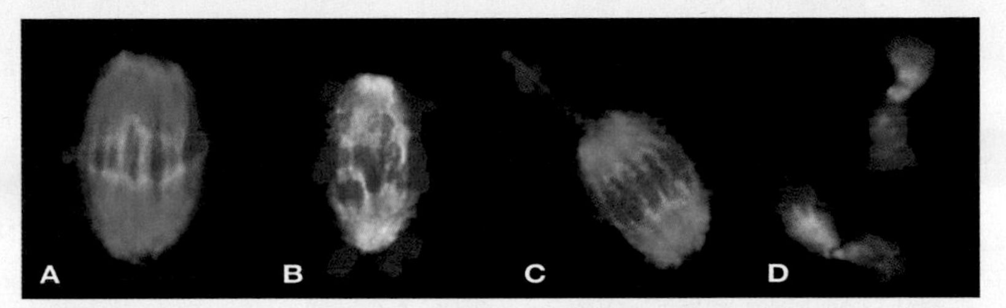

FIGURE 12.11 Meiotic abnormalities associated with exposure to damaged plastic caging.

Fluorescent micrographs show nuclei of single mouse oocytes in metaphase I. (**A**) Normal metaphase; (**B–D**) examples of abnormal metaphase. Chromosomes are stained red; spindle fibers, green.

9. The cell pictured to the right is in which stage of nuclear division?
   a. anaphase
   b. anaphase I
   c. anaphase II
   d. none of the above

10. Crossing over mixes up ________ .
    a. chromosomes
    b. alleles
    c. zygotes
    d. gametes

11. Crossing over happens during which phase of meiosis?
    a. prophase I
    b. prophase II
    c. anaphase I
    d. anaphase II

12. Which of the following is one of the very important differences between mitosis and meiosis?
    a. Chromosomes align midway between spindle poles only in meiosis.
    b. Homologous chromosomes pair up only in meiosis.
    c. DNA is replicated only in mitosis.
    d. Sister chromatids separate only in meiosis.
    e. Interphase occurs only in mitosis.

13. Match each term with the best description.

___ interphase	a. different forms of a gene
___ metaphase I	b. useful for varied offspring
___ alleles	c. none between meiosis I and meiosis II
___ zygotes	d. chromosome lineup
___ gametes	e. haploid
___ males	f. form at fertilization
___ prophase I	g. mash-up time

14. ________ contributes to variation in traits among the offspring of sexual reproducers.
    a. Crossing over
    b. Random attachment of chromosomes to spindle poles
    c. Fertilization
    d. both a and b
    e. all are factors

CENGAGE brain To access course materials, please visit www.cengagebrain.com.

critical thinking

1. The diploid chromosome number for the body cells of a frog is 26. What would that number be after three generations if meiosis did not occur before gamete formation?

2. In your own words, explain why sexual reproduction tends to give rise to greater genetic diversity among offspring in fewer generations than asexual reproduction.

3. Different populations of the tiny freshwater snails pictured in **FIGURE 12.1** reproduce sexually, or asexually. Individuals of the sexual populations are diploid; those in asexual populations are triploid ($3n$, having three sets of chromosomes). Huge populations of asexual snails are disrupting ecosystems worldwide.

Fertilizers and detergents contain a lot of phosphorus. So does DNA. Explain why you might expect to find more sexual snail populations in an unpolluted river, and more asexual ones in a river polluted by agricultural and urban runoff.

4. Make a simple sketch of meiosis in a cell with a diploid chromosome number of 4. Now try it when the chromosome number is 3.

13 Observing Patterns in Inherited Traits

LEARNING ROADMAP

You may want to review traits (Section 1.5), chromosomes (8.4), genes and gene expression (9.2), sexual reproduction (12.1), alleles (12.2), and meiosis (12.3, 12.4). This chapter revisits probability and sampling error (1.8), laws of nature (1.9), protein structure (3.6), pigments (6.2), clones (8.7), gene expression control (10.2, 11.6), and epigenetics (10.6).

WHERE MODERN GENETICS STARTED

Gregor Mendel discovered that inherited traits are specified in units. The units, which are distributed into gametes in predictable patterns, were later identified as genes.

MONOHYBRID CROSSES

Tracking inheritance patterns of single traits led to the discovery that during meiosis, pairs of genes on homologous chromosomes separate and end up in different gametes.

DIHYBRID CROSSES

Tracking inheritance patterns of two unrelated traits led to the discovery that in most cases, genes of a pair segregate into gametes independently of other gene pairs.

VARIATIONS ON MENDEL'S THEME

An allele may be partly dominant over a nonidentical partner, or codominant with it. Multiple genes may influence a trait; some genes influence many traits.

COMPLEX VARIATIONS IN TRAITS

Environmental factors can alter the expression of genes that influence a trait. Many traits appear in a continuous range of forms.

We return to human skin color in Section 14.1, and human genetic disorders in the rest of Chapter 14. Chapter 17 explores the interplay between genes and the environment in an evolutionary context. Signaling pathways that affect gene expression comprise stimulus, perception, and response (Section 34.2). Chapter 29 details flowering plant reproduction. Neurons and how they work are the topic of Chapter 32, with neurological disorders in Section 32.7.

13.1 Menacing Mucus

In 1988, researchers discovered a gene that, when mutated, causes cystic fibrosis (CF). Cystic fibrosis is the most common fatal genetic disorder in the United States. The gene, *CFTR*, encodes an active transport protein that moves chloride ions out of epithelial cells. Sheets of epithelial cells form the skin and also line the passageways and ducts of the lungs, liver, pancreas, intestines, and reproductive system. When the CFTR protein pumps chloride ions out of these cells, water follows the ions by osmosis. The two-step process maintains a thin film of water on the surface of epithelial cell sheets. Mucus slides easily over the wet sheets of cells.

The mutation most commonly associated with cystic fibrosis is a 3-base-pair deletion in the *CFTR* gene. The mutation is called *ΔF508* because it encodes a CFTR protein missing the phenylalanine (F) that is normally the 508th amino acid (Δ means deleted). This defect interferes with membrane trafficking of the newly synthesized CFTR protein. Normally, a new CFTR polypeptide moves from endoplasmic reticulum (ER) to a Golgi body, which packages it in vesicles routed to the plasma membrane. CFTR polypeptides with the *ΔF508* deletion misfold in a tiny domain, and this change is recognized by a quality control system in the endoplasmic reticulum. Most of the altered proteins are left stranded in the ER; the few that make it to the plasma membrane are quickly taken back into the cell by endocytosis and destroyed.

Epithelial cell membranes that lack the CFTR protein cannot transport chloride ions. Too few chloride ions leave these cells. Not enough water leaves them either, so the surfaces of epithelial cell sheets are too dry. Mucus that normally slips through the body's tubes sticks to their walls instead. Thick globs of mucus accumulate and clog passageways and ducts throughout the body. Breathing becomes difficult as the mucus obstructs the smaller airways of the lungs. Digestive problems arise as ducts that lead to the gut become clogged with mucus. Males are typically infertile because their sperm flow is hampered.

CFTR also helps alert the immune system to the presence of disease-causing bacteria in the lungs. The CFTR protein functions as a receptor: It binds directly to bacteria and triggers endocytosis. Endocytosis of bacteria into epithelial cells lining the respiratory tract initiates an immune response. When the cells lack CFTR, this early alert system fails, so bacteria have time to multiply before being detected by the immune system. Thus, chronic bacterial infections of the lungs are a hallmark of cystic fibrosis. Antibiotics help control infections, but there is no cure for the disorder. Most affected people die before age thirty, when their tormented lungs fail (**FIGURE 13.1**).

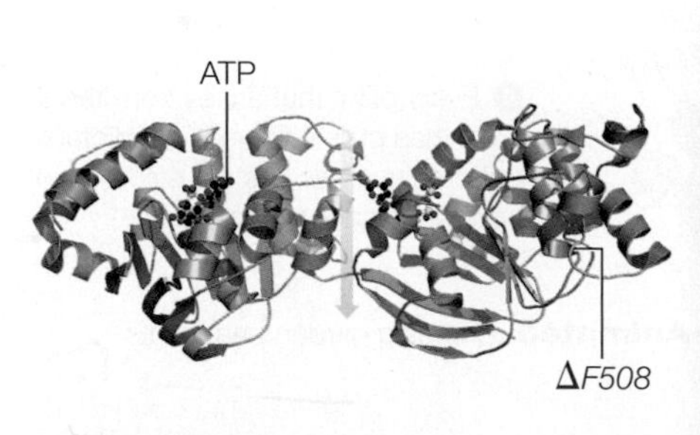

FIGURE 13.1 Cystic fibrosis.

Above, model of the CFTR protein. The parts shown here are a pair of ATP-driven motors that widen or narrow a channel (gray arrow) across the plasma membrane. The tiny part of the protein that is deleted in most people with cystic fibrosis is shown in green.

Right, a few of the many young victims of cystic fibrosis, which occurs most often in people of northern European ancestry. At least one young person dies every day in the United States from complications of this incurable disease.

CREDITS: (opposite) Jean M. Labat/Ardea London; (1) art, © Cengage Learning; photos, top row from left, Courtesy of © The Cody Dieruf Benefit Foundation, www.breathinisbelievin.org; Courtesy of © Bobby Brooks and The Family of Jeff Baird; Courtesy of © Steve & Ellison Widener and Breathe Hope, breathehope.tamu.edu; bottom row from left, Courtesy of The Family of Benjamin Hill, reprinted with permission of © Chappell/Marathonfoto; Courtesy of © The Family of Savannah Brooke Snider; Courtesy of © the family of Brandon Herriott.

13.2 Mendel, Pea Plants, and Inheritance Patterns

✔ Recurring patterns of inheritance offer observable evidence of how heredity works.

In the nineteenth century, people thought that hereditary material must be some type of fluid, with fluids from both parents blending at fertilization like milk into coffee. However, the idea of "blending inheritance" failed to explain what people could see with their own eyes. Children sometimes have traits such as freckles that do not appear in either parent. A cross between a black horse and a white one does not produce gray offspring.

The naturalist Charles Darwin doubted the idea of blending inheritance, but he could not come up with an alternative hypothesis even though inheritance was central to his theory of natural selection. (We return to Darwin and this theory in Chapter 16.) At the time, no one knew that hereditary information is divided into discrete units (genes, Section 9.2), an insight that is critical to understanding how traits are inherited. However, even before Darwin presented his theory, someone had been gathering data that would support it. Gregor Mendel (left), an Austrian monk, had been carefully breeding thousands of pea plants. By keeping detailed records of how traits passed from one generation to the next, Mendel had been collecting evidence of how inheritance works.

Mendel's Experiments

Mendel cultivated the garden pea plant (**FIGURE 13.2**). This species is naturally self-fertilizing, which means its flowers produce male and female gametes ❶ that form viable embryos when they meet up. In order to study inheritance, Mendel had to carry out controlled matings between individuals with specific traits, then observe and document the traits of their offspring. To prevent an individual pea plant from self-fertilizing, Mendel removed the pollen-bearing parts (anthers) from its flowers. He then cross-fertilized the flowers by brushing their egg-bearing parts (carpels) with pollen from another plant ❷. He collected seeds ❸ from the cross-fertilized individual, and recorded the traits of the new pea plants that grew from them ❹.

Many of Mendel's experiments started with plants that "bred true" for particular traits such as white flowers or purple flowers. Breeding true for a trait means that, new mutations aside, all offspring have the same form of the trait as the parent(s), generation after generation. For example, all offspring of two pea plants that breed true for white flowers also have white flowers. As you will see in the next section, Mendel cross-fertilized pea plants that breed true for different forms of a trait, and discovered that the traits of the offspring often appear in predictable patterns. Mendel's meticulous work tracking pea plant traits led him to conclude (correctly) that hereditary information passes from one generation to the next in discrete units.

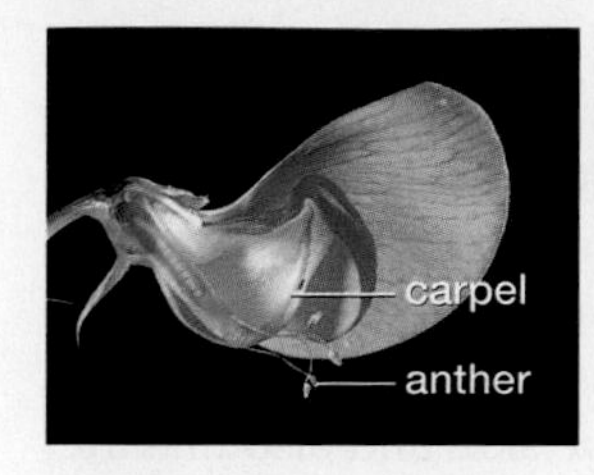

❶ In the flowers of garden pea plants, pollen grains that form in anthers produce male gametes; female gametes form in carpels.

❷ Experimenters control the transfer of hereditary material from one pea plant to another by snipping off a flower's pollen-producing anthers (to prevent it from self-fertilizing), then brushing pollen from another flower onto its egg-producing carpel.

In this example, pollen from a plant with purple flowers is brushed onto the carpel of a white-flowered plant.

❸ Later, seeds develop inside pods of the cross-fertilized plant. An embryo in each seed develops into a mature pea plant.

❹ Every plant that arises from the cross has purple flowers. Predictable patterns such as this are evidence of how inheritance works.

FIGURE 13.2 ▸Animated Breeding garden pea plants.

Inheritance in Modern Terms

DNA was not proven to be hereditary material until the 1950s (Section 8.2), but Mendel discovered its units, which we now call genes, almost a century before then. Today, we know that individuals of a species share certain traits because their chromosomes carry the same genes.

Each gene occurs at a specific location, or **locus** (plural, loci), on a particular chromosome (**FIGURE 13.3**). The somatic cells of humans and other animals are diploid, so their nuclei contain pairs of genes, on pairs of homologous chromosomes. In most cases, both genes of a pair are expressed. Genes at the same locus on a pair of homologous chromosomes may be identical, or they may vary as alleles (Section 12.2).

Organisms breed true for a trait because they carry identical alleles of genes governing that trait. An individual with two identical alleles of a gene is **homozygous** for the allele. By contrast, an individual with two different alleles of a gene is **heterozygous** (*hetero–* means mixed). A **hybrid** is a heterozygous individual, such as an offspring of a cross or mating between individuals that breed true for different forms of a trait.

When we say that an individual is homozygous or heterozygous, we are discussing its **genotype**, the particular set of alleles it carries. Genotype is the basis of **phenotype**, which refers to the individual's observable traits. "White-flowered" and "purple-flowered" are examples of pea plant phenotypes that arise from differences in genotype.

The phenotype of a heterozygous individual depends on how the products of its two different alleles interact. In many cases, the effect of one allele influences the effect of the other, and the outcome of this interaction is reflected in the individual's phenotype. An allele is **dominant** when its effect masks that of a **recessive** allele paired with it. Usually, a dominant allele is represented by an italic capital letter such as *A*; a recessive allele, with a lowercase italic letter such as *a*. Consider the purple- and white-flowered pea plants that Mendel studied. In these plants, the allele that specifies purple flowers (let's call it *P*) is dominant over the allele that specifies white flowers (*p*). Thus, a pea plant homozygous for the dominant allele (*PP*) has purple flowers; one homozygous for the recessive allele (*pp*) has white flowers (**FIGURE 13.4**). A heterozygous plant (*Pp*) has purple flowers.

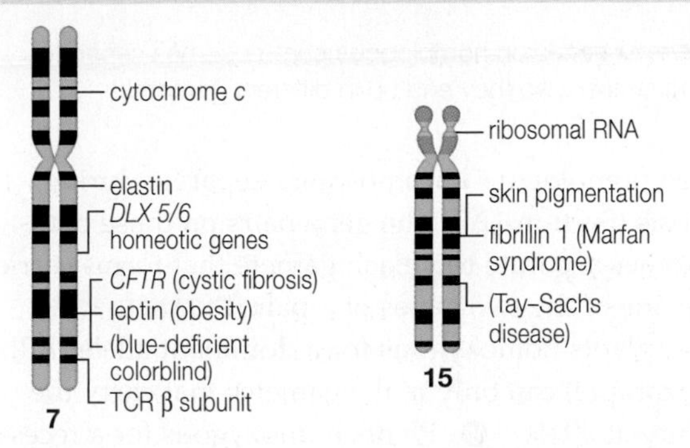

FIGURE 13.3 Loci of a few human genes. Genetic disorders arising from mutations in the genes are shown in parentheses. The number or letter below a chromosome is its name; characteristic banding patterns appear after staining. Appendix IV has a similar map of all 23 human chromosomes.

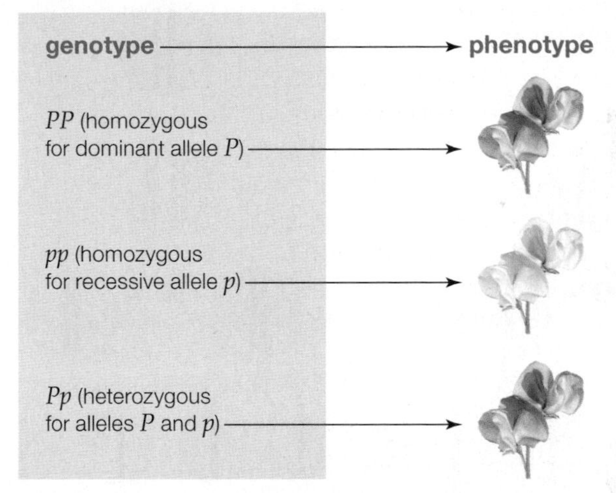

FIGURE 13.4 Genotype gives rise to phenotype. In this example, the dominant allele *P* specifies purple flowers; the recessive allele *p*, white flowers. **FIGURE IT OUT** Which individual is a hybrid?

Answer: The heterozygous one

TAKE-HOME MESSAGE 13.2

How do alleles contribute to traits?

✔ Gregor Mendel indirectly discovered the role of alleles in inheritance by carefully breeding pea plants and tracking traits of their offspring.

✔ Genotype refers to the particular set of alleles that an individual carries. Genotype is the basis of phenotype, which refers to the individual's observable traits.

✔ A homozygous individual has two identical alleles of a gene. A heterozygous individual has two nonidentical alleles.

✔ A dominant allele masks the effect of a recessive allele paired with it in a heterozygous individual.

dominant Refers to an allele that masks the effect of a recessive allele paired with it in heterozygous individuals.
genotype The particular set of alleles that is carried by an individual's chromosomes.
heterozygous Having two different alleles of a gene.
homozygous Having identical alleles of a gene.
hybrid A heterozygous individual.
locus Location of a gene on a chromosome.
phenotype An individual's observable traits.
recessive Refers to an allele with an effect that is masked by a dominant allele on the homologous chromosome.

13.3 Mendel's Law of Segregation

✔ Pairs of genes on homologous chromosomes separate during meiosis, so they end up in different gametes.

When homologous chromosomes separate during meiosis (Section 12.3), the gene pairs on those chromosomes separate too. Each gamete that forms carries only one of the two genes of a pair (**FIGURE 13.5**). Thus, plants homozygous for a dominant allele (*PP*, for example) can only make gametes that carry the dominant allele *P* ❶. Plants homozygous for a recessive allele (*pp*) can only make gametes that carry the recessive allele *p* ❷. If these homozygous plants are crossed (*PP* × *pp*), only one outcome is possible: A gamete carrying allele *P* meets up with a gamete carrying allele *p* ❸. All offspring of this cross will have both alleles—they will be heterozygous (*Pp*). A grid called a **Punnett square** is helpful for predicting the outcomes of such crosses (**FIGURE 13.6**).

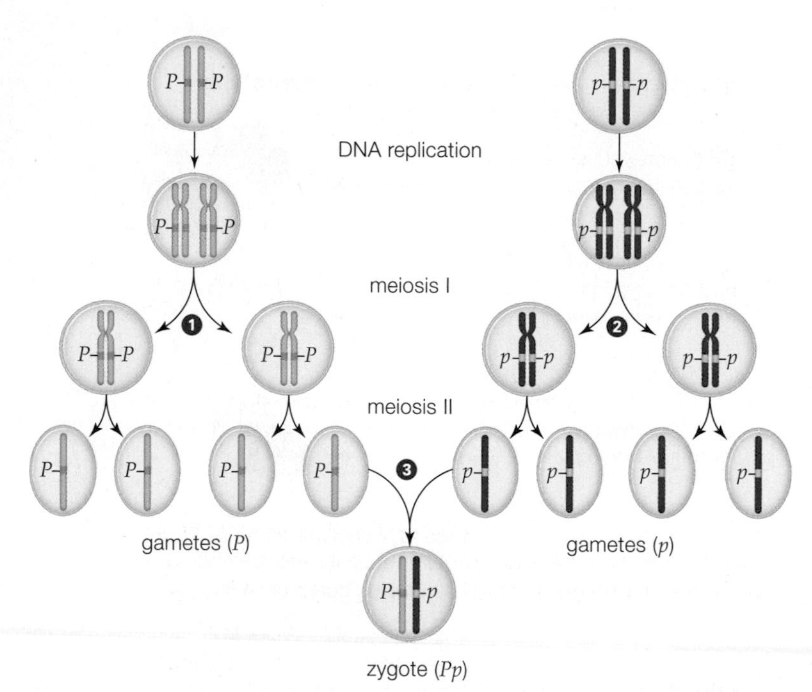

FIGURE 13.5 ▸Animated Segregation of genes into gametes. Homologous chromosomes separate during meiosis, so the pairs of genes they carry separate too. Each of the resulting gametes carries one of the two members of each gene pair. For clarity, only one set of chromosomes is illustrated.

❶ All gametes made by a parent homozygous for a dominant allele carry that allele.

❷ All gametes made by a parent homozygous for a recessive allele carry that allele.

❸ If these two parents are crossed, the union of any of their gametes at fertilization produces a zygote with both alleles. All offspring of this cross will be heterozygous.

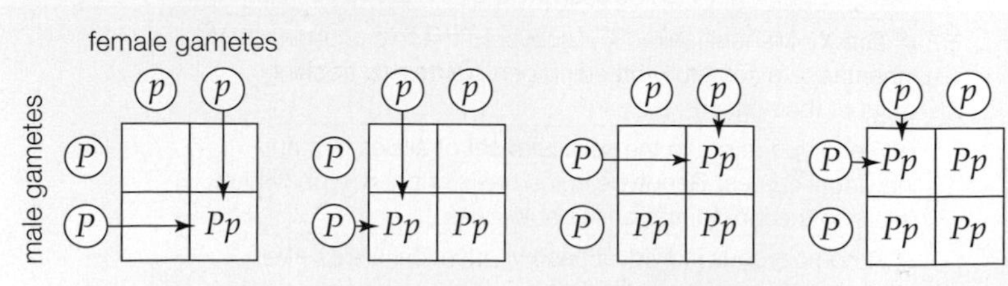

FIGURE 13.6 Making a Punnett square. Parental gametes are listed in circles on the top and left sides of a grid. Each square is filled with the combination of alleles that would result if the gametes in the corresponding row and column met up.

Our example illustrated a pattern so predictable that it can be used as evidence of a dominance relationship between alleles. In a **testcross**, an individual that has a dominant trait (but an unknown genotype) is crossed with an individual known to be homozygous for the recessive allele. The pattern of traits among the offspring of the cross can reveal whether the tested individual is heterozygous or homozygous. If all of the offspring of the testcross have the dominant trait (as occurred in our example above), then the parent with the unknown genotype is homozygous for the dominant allele. If some of the offspring have the recessive trait, then the parent is heterozygous.

Dominance relationships between alleles determine the phenotypic outcome of a **monohybrid cross**, in which individuals identically heterozygous for one gene (they have the same two alleles: *Pp*, for example)

Table 13.1
Mendel's Seven Pea Plant Traits

Trait	Dominant Form	Recessive Form
Seed Shape	Round	Wrinkled
Seed Color	Yellow	Green
Pod Texture	Smooth	Wrinkled
Pod Color	Green	Yellow
Flower Color	Purple	White
Flower Position	Along Stem	At Tip
Stem Length	Tall	Short

are bred together or self-fertilized ($Pp \times Pp$). The frequency at which the phenotype associated with each allele appears among offspring depends on whether one of the alleles is dominant over the other.

To do a monohybrid cross, we would start with two individuals that breed true for two distinct forms of a trait. In garden pea plants, flower color (purple or white) is one example of a trait with two distinct forms, but there are many others. Mendel investigated seven of them: stem length (tall or short), seed color (yellow or green), pod texture (smooth or wrinkled), and so on (**TABLE 13.1**). A cross between individuals that breed true for different forms of the trait yields offspring identically heterozygous for the alleles that govern the trait. A cross between these F_1 (first generation) hybrids is the monohybrid cross. The frequency at which the two traits appear in the F_2 (second generation) offspring offers information about a dominance relationship between the alleles. F is an abbreviation for filial, which means offspring.

A cross between two purple-flowered heterozygous individuals (Pp) is an example of a monohybrid cross. Each of these plants can make two types of gametes: ones that carry a P allele, and ones that carry a p allele (**FIGURE 13.7A**). So, in a monohybrid cross between two Pp plants ($Pp \times Pp$), the two types of gametes can meet up in four possible ways at fertilization:

Possible Event	Outcome
Sperm P meets egg P ⟶	zygote genotype is PP
Sperm P meets egg p ⟶	zygote genotype is Pp
Sperm p meets egg P ⟶	zygote genotype is Pp
Sperm p meets egg p ⟶	zygote genotype is pp

Three out of four possible outcomes of this cross include at least one copy of the dominant allele P. Each time fertilization occurs, there are 3 chances in 4 that the resulting offspring will inherit a P allele, and have purple flowers. There is 1 chance in 4 that it will inherit two recessive p alleles, and have white flowers. Thus, the probability that a particular offspring of this cross will have purple or white flowers is 3 purple to 1 white, which we represent as a ratio of 3:1 (**FIGURE 13.7B**). The 3:1 pattern is an indication that purple and white flower color are specified by alleles with a clear dominance relationship: Purple is dominant; white, recessive. If the probability of an individual inheriting a particular genotype is difficult to imagine, think about it in terms of many offspring: In this example, there will be roughly three purple-flowered plants for every white-flowered one.

The phenotype ratios in the F_2 offspring of Mendel's monohybrid crosses were all close to 3:1. These results became the basis of his **law of segregation**, which we state here in modern terms: Diploid cells carry pairs of genes, on pairs of homologous chromosomes. The two genes of each pair are separated from each other during meiosis, so they end up in different gametes.

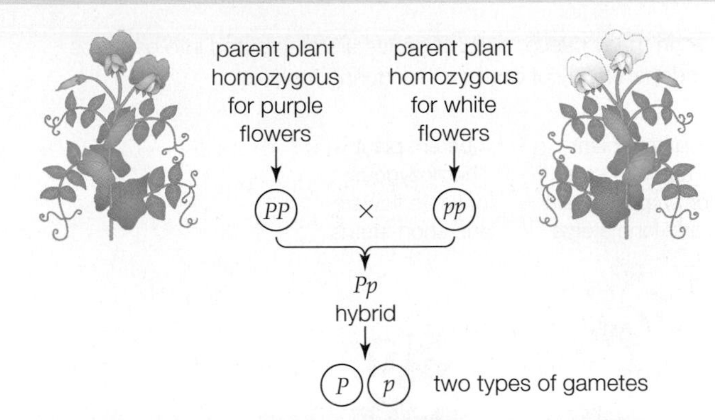

A All of the F_1 offspring of a cross between two plants that breed true for different forms of a trait are identically heterozygous (Pp). These offspring make two types of gametes: P and p.

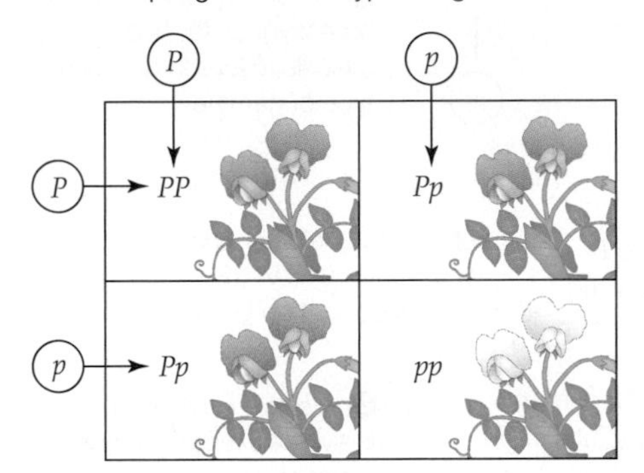

B A monohybrid cross is a cross between these F_1 offspring. In this example, the phenotype ratio in F_2 offspring is 3:1 (3 purple to 1 white).

FIGURE 13.7 ▶**Animated** An example of a monohybrid cross.
FIGURE IT OUT In this example, how many possible genotypes are there in the F_2 generation?
Answer: Three: PP, Pp, and pp

law of segregation The two members of each pair of genes on homologous chromosomes end up in different gametes during meiosis.
monohybrid cross Cross between two individuals identically heterozygous for one gene; for example, $Aa \times Aa$.
Punnett square Diagram used to predict the genotypic and phenotypic outcomes of a cross.
testcross Method of determining genotype of an individual with a dominant phenotype: a cross between the individual and another individual known to be homozygous recessive.

TAKE-HOME MESSAGE 13.3

How are alleles distributed into gametes?

✔ Homologous chromosomes carry pairs of genes. The two genes of each pair are separated from each other during meiosis, so they end up in different gametes.

13.4 Mendel's Law of Independent Assortment

✔ In many cases, pairs of genes are distributed into gametes independently of one another during meiosis.

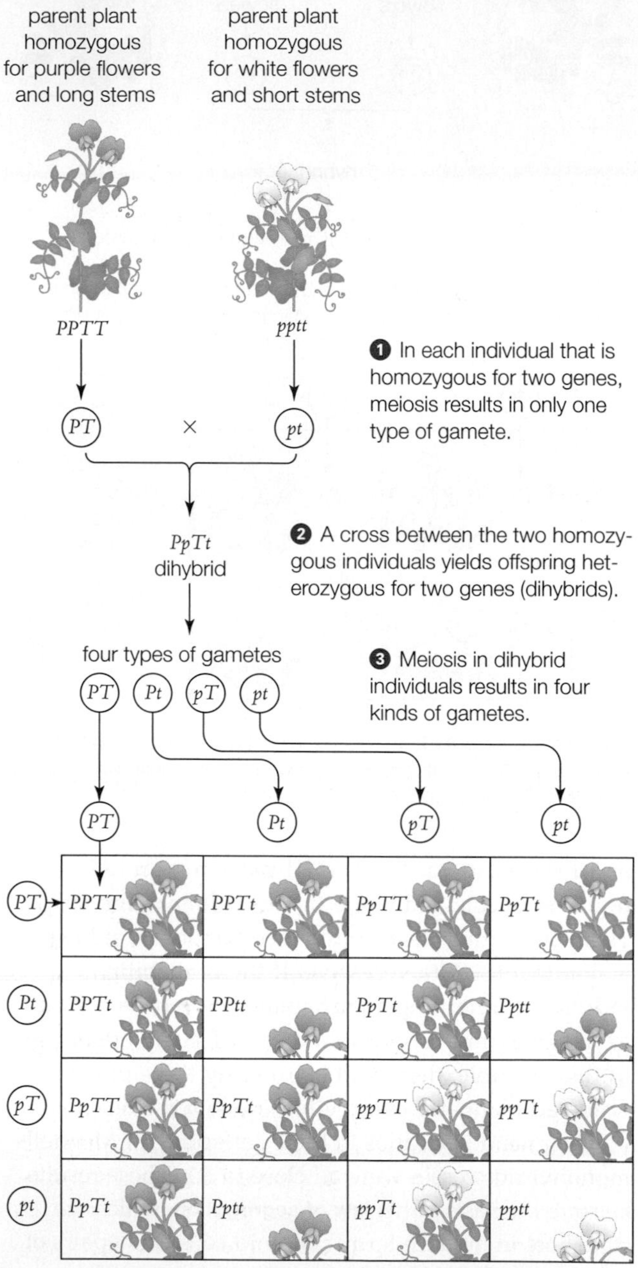

parent plant homozygous for purple flowers and long stems

parent plant homozygous for white flowers and short stems

PPTT | *pptt*

PT × *pt*

❶ In each individual that is homozygous for two genes, meiosis results in only one type of gamete.

PpTt dihybrid

❷ A cross between the two homozygous individuals yields offspring heterozygous for two genes (dihybrids).

four types of gametes

PT *Pt* *pT* *pt*

❸ Meiosis in dihybrid individuals results in four kinds of gametes.

	PT	*Pt*	*pT*	*pt*
PT	*PPTT*	*PPTt*	*PpTT*	*PpTt*
Pt	*PPTt*	*PPtt*	*PpTt*	*Pptt*
pT	*PpTT*	*PpTt*	*ppTT*	*ppTt*
pt	*PpTt*	*Pptt*	*ppTt*	*pptt*

❹ If two of the dihybrid individuals are crossed, the four types of gametes can meet up in 16 possible ways. Of 16 possible offspring genotypes, 9 will result in plants that are purple-flowered and tall; 3, purple-flowered and short; 3, white-flowered and tall; and 1, white-flowered and short. Thus, the ratio of phenotypes is 9:3:3:1.

FIGURE 13.8 ▶Animated A dihybrid cross between plants that differ in flower color and plant height. *P* and *p* stand for dominant and recessive alleles for flower color; *T* and *t*, dominant and recessive alleles for height.

FIGURE IT OUT What do the flowers inside the boxes represent?

Answer: Phenotypes of the F_2 offspring

A monohybrid cross allows us to study a dominance relationship between alleles of one gene. What about alleles of two genes? Dihybrids are individuals that have two alleles at two loci (*AaBb*, for example). In a **dihybrid cross**, individuals identically heterozygous for two genes are crossed (*AaBb* × *AaBb*). As with a monohybrid cross, the pattern of traits seen among the offspring of the cross depends on the dominance relationships between the alleles.

Let's use a gene for flower color (*P*, purple; *p*, white) and one for plant height (*T*, tall; *t*, short) in an example. **FIGURE 13.8** shows a dihybrid cross starting with one parent plant that breeds true for purple flowers and tall stems (*PPTT*), and one that breeds true for white flowers and short stems (*pptt*). The *PPTT* plant only makes gametes with the dominant alleles (*PT*); the *pptt* plant only makes gametes with the recessive alleles (*pt*) ❶. So, all offspring from a cross between these parent plants (*PPTT* × *pptt*) will be dihybrids (*PpTt*) with purple flowers and tall stems ❷.

Four combinations of alleles are possible in the gametes of *PpTt* dihybrids ❸. If two *PpTt* plants are crossed (a dihybrid cross, *PpTt* × *PpTt*), the four types of gametes can combine in sixteen possible ways at fertilization ❹. Nine of the sixteen genotypes would give rise to tall plants with purple flowers; three, to short plants with purple flowers; three, to tall plants with white flowers; and one, to short plants with white flowers. Thus, the ratio of phenotypes among the offspring of this dihybrid cross would be 9:3:3:1.

Mendel discovered the 9:3:3:1 ratio of phenotypes among the offspring of his dihybrid crosses, but he had no idea what it meant. He could only say that "units" specifying one trait (such as flower color) are inherited independently of "units" specifying other traits (such as plant height). In time, Mendel's hypothesis became known as the **law of independent assortment**, which we state here in modern terms: During meiosis, the two genes of a pair tend to be sorted into gametes independently of how other gene pairs are sorted into gametes.

Mendel published his results in 1866, but apparently his work was read by few and understood by no one at the time. In 1871 he was promoted, and his pioneering experiments ended. When he died in 1884, he did not know that his work with pea plants would be the starting point for modern genetics.

The Contribution of Crossovers

How two pairs of genes get distributed into gametes depends partly on whether the genes are on the same chromosome. When homologous chromosomes sepa-

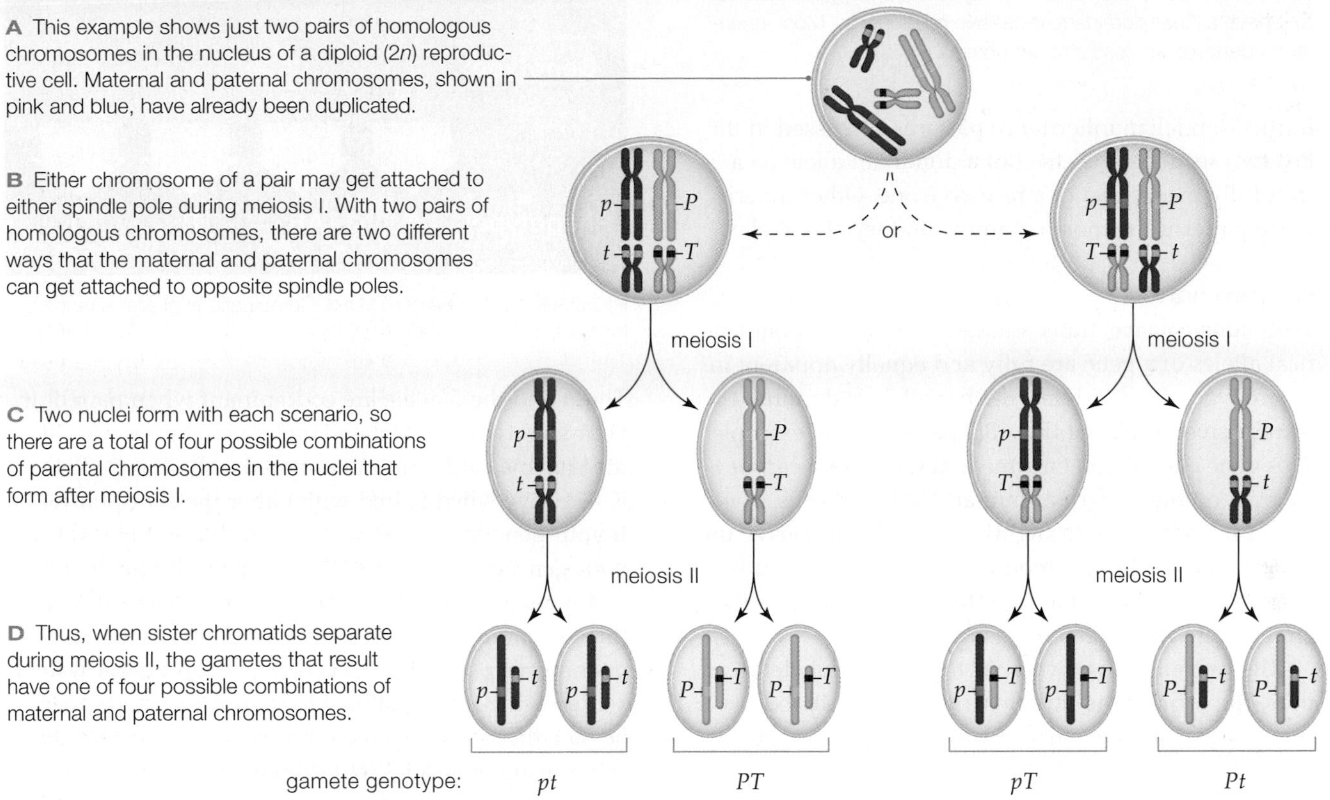

A This example shows just two pairs of homologous chromosomes in the nucleus of a diploid (2*n*) reproductive cell. Maternal and paternal chromosomes, shown in pink and blue, have already been duplicated.

B Either chromosome of a pair may get attached to either spindle pole during meiosis I. With two pairs of homologous chromosomes, there are two different ways that the maternal and paternal chromosomes can get attached to opposite spindle poles.

C Two nuclei form with each scenario, so there are a total of four possible combinations of parental chromosomes in the nuclei that form after meiosis I.

D Thus, when sister chromatids separate during meiosis II, the gametes that result have one of four possible combinations of maternal and paternal chromosomes.

or

meiosis I

meiosis II

gamete genotype: *pt* *PT* *pT* *Pt*

FIGURE 13.9 ▸Animated Independent assortment of genes on different chromosomes. Genes that are far apart on the same chromosome usually assort independently too, because crossovers typically separate them.

rate during meiosis, either member of a gene pair can end up in either of the two new nuclei that form. This random assortment happens independently for each pair of homologous chromosomes in the cell. Thus, genes on one chromosome pair assort into gametes independently of genes on the other chromosome pairs (**FIGURE 13.9**).

Pea plants have seven chromosomes. Mendel studied seven pea genes, and all of them assorted into gametes independently of one another. Was he lucky enough to choose one gene on each of those chromosomes? As it turns out, some of the genes Mendel studied *are* on the same chromosome. These genes are far enough apart that crossing over occurs between them very frequently—so frequently that they tend to assort into gametes independently, just as if they were on different chromosomes. By contrast, genes that are very close together on a chromosome usually do not assort independently into gametes, because crossing over does not happen between them very often. Thus, gametes usually end up with parental combinations of alleles of these genes.

Genes that do not assort independently into gametes are said to be linked. A **linkage group** comprises all genes on a chromosome. Peas have 7 different chromosomes, so they have 7 linkage groups. Humans have 23 different chromosomes, so they have 23 linkage groups.

dihybrid cross Cross between two individuals identically heterozygous for two genes; for example *AaBb* × *AaBb*.
law of independent assortment During meiosis, members of a pair of genes on homologous chromosomes tend to be distributed into gametes independently of other gene pairs.
linkage group All of the genes on a chromosome.

TAKE-HOME MESSAGE 13.4

How are gene pairs distributed into gametes?

✔ During meiosis, gene pairs on homologous chromosomes tend to be distributed into gametes independently of how other gene pairs are distributed.

✔ Independent assortment of genes on the same chromosome depends on proximity. Genes that are closer together get separated less frequently by crossovers, so gametes often receive parental combinations of alleles of these genes.

13.5 Beyond Simple Dominance

✔ Mendel studied traits with distinct forms arising from alleles that have a clear dominant–recessive relationship. Most inheritance patterns are less straightforward.

In the Mendelian inheritance patterns discussed in the last two sections, the effect of a dominant allele on a trait fully masks that of a recessive one. Other inheritance patterns are more common, and more complex.

Codominance

With **codominance**, traits associated with two nonidentical alleles of a gene are fully and equally apparent in heterozygous individuals; neither allele is dominant or recessive. Alleles of the *ABO* gene offer an example. This gene encodes an enzyme that modifies a carbohydrate on the surface of human red blood cells. The *A* and *B* alleles encode slightly different versions of this enzyme, which in turn modify the carbohydrate differently. The *O* allele has a mutation that prevents its enzyme product from becoming active at all.

The two alleles you carry at the *ABO* locus determine the form of the carbohydrate on your blood cells, so they are the basis of your ABO blood type.

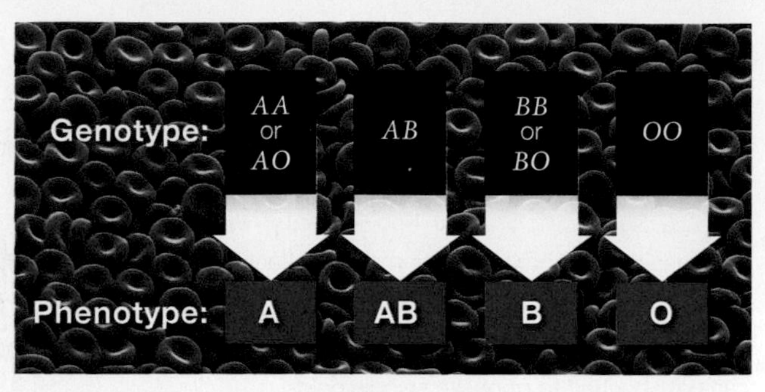

FIGURE 13.10 ▶**Animated** Combinations of alleles that are the basis of human ABO blood type.

The *A* and the *B* allele are codominant when paired. If your genotype is *AB*, then you have both versions of the enzyme, and your blood type is AB. The *O* allele is recessive when paired with either the *A* or *B* allele. If your genotype is *AA* or *AO*, your blood type is A. If your genotype is *BB* or *BO*, it is type B. If you are *OO*, it is type O (**FIGURE 13.10**). A gene such as *ABO*, with three or more alleles persisting at relatively high frequency in a population, is called a **multiple allele system**.

Receiving incompatible blood in a transfusion can be dangerous because the immune system attacks any cell bearing molecules that do not occur in one's own body. An immune system attack causes red blood cells to clump or burst, with potentially fatal results. Almost everyone makes the type O carbohydrate, so type O blood does not usually trigger an immune response in transfusion recipients. People with type O blood are called universal donors because they can donate blood to anyone. However, because their body is unfamiliar with the carbohydrates made by people with type A or B blood, they can receive type O blood only. People with type AB blood can receive a transfusion of any blood type, so they are called universal recipients.

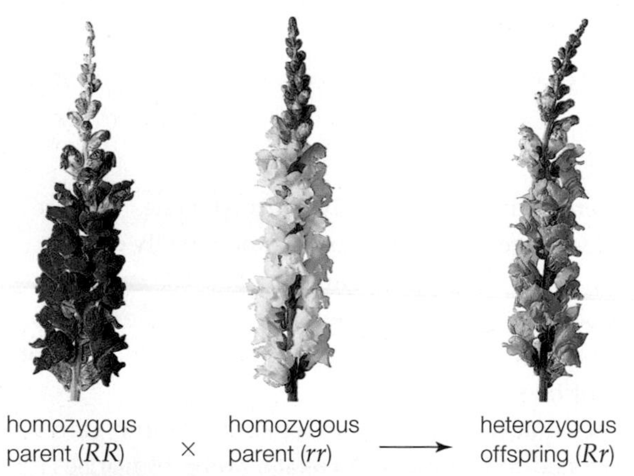

A Cross a red-flowered with a white-flowered snapdragon plant, and all of the offspring will have pink flowers.

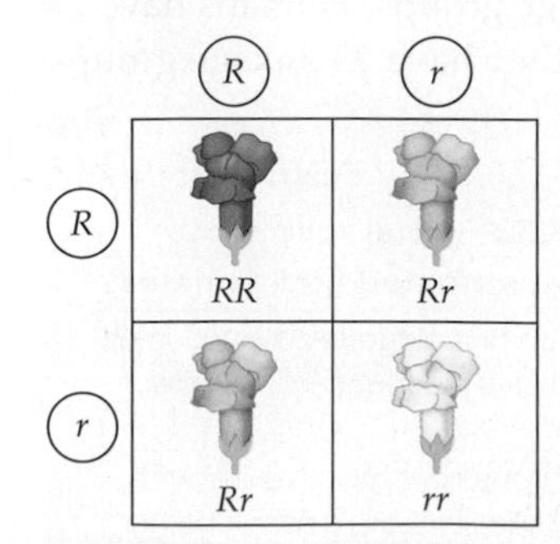

B If two of the pink-flowered snapdragons are crossed, the phenotypes of their offspring will occur in a 1:2:1 ratio.

FIGURE 13.11 ▶**Animated** Incomplete dominance in heterozygous (pink) snapdragons. One allele (*R*) results in the production of a red pigment; the other (*r*) results in no pigment.

Incomplete Dominance

With **incomplete dominance**, one allele is not fully dominant over the other, so the heterozygous phenotype is an intermediate blend of the two homozygous phenotypes. A gene that affects flower color in snapdragons is an example (**FIGURE 13.11**). One allele of the gene (*R*) encodes an enzyme that makes a red pigment. The enzyme encoded by a mutated allele (*r*) cannot make any pigment. Plants homozygous for the *R* allele (*RR*) make a lot of red pigment, so they have red flowers. Plants homozygous for the *r* allele (*rr*) make no pigment, so their flowers are white. Heterozygous plants (*Rr*) make only enough red pigment to tint their flowers pink. A cross between two heterozygous plants yields red-, pink-, and white-flowered offspring in a 1:2:1 ratio.

FIGURE 13.12 ▸Animated An example of epistasis. Interactions among products of two gene pairs affect coat color in Labrador retrievers. Dogs with alleles *E* and *B* have black fur. Those with an *E* and two recessive *b* alleles have brown fur. Dogs homozygous for the recessive *e* allele have yellow fur.

	EB	*Eb*	*eB*	*eb*
EB	*EEBB*	*EEBb*	*EeBB*	*EeBb*
Eb	*EEBb*	*EEbb*	*EeBb*	*Eebb*
eB	*EeBB*	*EeBb*	*eeBB*	*eeBb*
eb	*EeBb*	*Eebb*	*eeBb*	*eebb*

Epistasis

Some traits are affected by multiple genes, an effect called polygenic inheritance or **epistasis**. Coat color in Labrador retriever dogs, which depends on pigments called melanins, is an example of a trait affected by multiple genes (**FIGURE 13.12**). A dark brown melanin gives rise to brown or black fur; a reddish melanin, to yellow fur. Melanin synthesis and deposition in fur requires several genes. The product of one gene (*TYRP1*) helps make the brown melanin. A dominant allele (*B*) of this gene results in a higher production of brown melanin than the recessive allele (*b*). A different gene (*MC1R*) affects which type of melanin is produced. A dominant allele (*E*) of this gene triggers production of the brown melanin; its recessive partner (*e*) carries a mutation that results in production of the reddish form. Dogs homozygous for the *e* allele are yellow because they make only the reddish melanin.

Pleiotropy

In many cases, a single gene influences multiple traits, an effect called **pleiotropy**. Mutations that affect the gene's product or its expression affect all of the traits at once. Many complex genetic disorders, including sickle-cell anemia (Section 9.6) and cystic fibrosis, are caused by mutations in single genes. Another example, Marfan syndrome, is a result of mutations that affect fibrillin. Long fibers of this protein impart elasticity to tissues of the heart, skin, blood vessels, tendons, and other body parts. Mutations can cause tissues to form with defective fibrillin or none at all. The largest blood vessel leading from the heart, the aorta, is particularly affected. The aorta's thick wall is not as elastic as it should be, and it eventually stretches and becomes leaky. Thinned and weakened, the aorta can rupture during exercise—an abruptly fatal outcome.

About 1 in 5,000 people have Marfan syndrome, and there is no cure. Its effects—and risks—are manageable with early diagnosis, but symptoms are easily missed. Affected people are tall and loose-jointed, but there are plenty of tall, loose-jointed people who do not have the syndrome. Thus, people with Marfan syndrome may die suddenly and early without ever knowing they had the disorder (**FIGURE 13.13**).

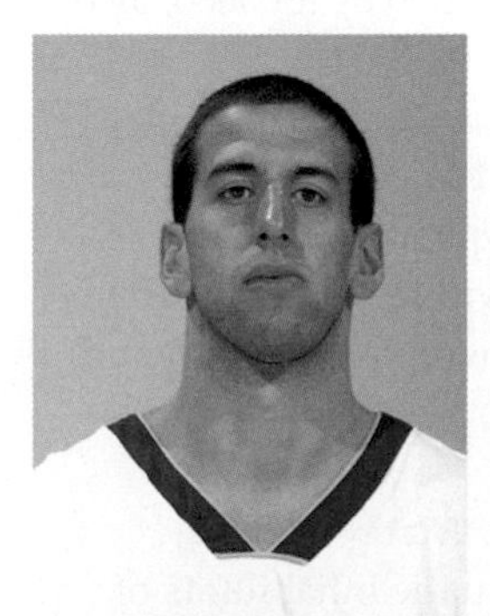

FIGURE 13.13 A heartbreaker: Marfan syndrome.

In 2006, 21-year-old basketball star Haris Charalambous collapsed and died suddenly during warm-up exercises. An autopsy revealed that his aorta had burst, an effect of the Marfan syndrome that Charalambous did not realize he had.

Assistant trainer Brian Jones says, "Haris was just the nicest, funniest kid in the world. With his size, he was sort of lovably goofy. He was everybody's best friend."

TAKE-HOME MESSAGE 13.5

Are all genes inherited in a Mendelian pattern?

✔ Some alleles are not dominant or recessive when paired.

✔ With incomplete dominance, one allele is not fully dominant over another, so the heterozygous phenotype is an intermediate blend of the two homozygous phenotypes.

✔ In codominance, two alleles have full and separate effects, so the phenotype of a heterozygous individual comprises both homozygous phenotypes.

✔ In some cases, one gene influences multiple traits. In other cases, multiple genes influence the same trait.

codominance Effect in which the full and separate phenotypic effects of two alleles are apparent in heterozygous individuals.
epistasis Polygenic inheritance, in which a trait is influenced by multiple genes.
incomplete dominance Effect in which one allele is not fully dominant over another, so the heterozygous phenotype is an intermediate blend between the two homozygous phenotypes.
multiple allele system Gene for which three or more alleles persist in a population at relatively high frequency.
pleiotropy Effect in which a single gene affects multiple traits.

CREDITS: (12) left, © Cengage Learning; right, Susan Schmitz/Shutterstock; (13) Courtesy of The Family of Haris Charalambous and the University of Toledo.

13.6 Nature and Nurture

✔ Variations in traits are not always the result of differences in alleles. Many traits are also influenced by environmental factors.

The phrase "nature versus nurture" refers to a centuries-old debate about whether human behavioral traits arise from one's genetics (nature) or from environmental factors (nurture). It turns out that both play a role. The environment affects the expression of many genes, which in turn affects phenotype. We can summarize this thinking with an equation:

genotype + environment ⟶ phenotype

Epigenetics research is revealing that the environment has an even greater contribution to this equation than most biologists had suspected (Section 10.6).

Environmental cues initiate cell-signaling pathways that trigger changes in gene expression (you will learn more about such pathways in later chapters). Some of these cell-signaling pathways methylate or demethylate particular regions of DNA, so they suppress or enhance gene expression in those regions (Section 10.2). In humans, research has shown that DNA methylation patterns can be permanently and heritably affected by diet, stress, and exercise, and also by exposure to drugs and toxins such as tobacco, alcohol, arsenic, and asbestos.

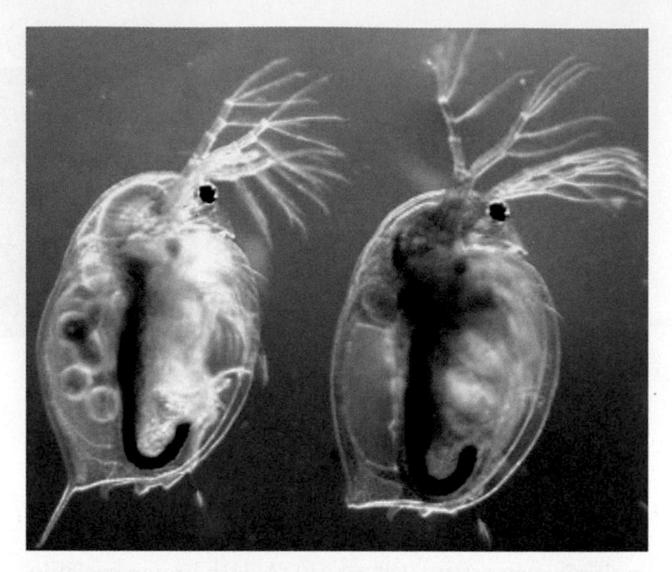

FIGURE 13.14 An environmental effect on phenotype of the water flea (*Daphnia*). Under low-oxygen conditions, water fleas can switch on their hemoglobin genes. Increased production of this protein enhances the individual's ability to take up oxygen from the water. The flea on the left has been living in water with a normal oxygen content; the one on the right, in water with a low oxygen concentration. The red pigment is hemoglobin.

Some Environmental Effects

Mechanisms that adjust phenotype in response to external cues are part of an individual's normal ability to adapt to a changing environment, as the following examples illustrate.

Alternative Phenotypes in Water Fleas

Water fleas (*Daphnia*) are tiny aquatic inhabitants of seasonal ponds, ditches, and other standing bodies of fresh water. Conditions such as temperature, oxygen content, and salinity vary dramatically over time and in different areas of these habitats. For example, water at the top of a still pond is typically warmer than water at the bottom, and it also contains more light and dissolved oxygen. Individual water fleas acclimate to such environmental differences by adjusting their gene expression. The adjustment provides an appropriate set of proteins to maintain cellular function in current environmental conditions.

Consider that *Daphnia* has a lot of genes—far more than other animals—and the abundance offers a striking plasticity of phenotype. For example, eleven *Daphnia* genes encode hemoglobin subunits; seven are differentially expressed depending on the temperature and oxygen content of the water. A flea survives low oxygen conditions by turning on synthesis of these genes—and turning red (**FIGURE 13.14**). The newly-produced hemoglobin improves the individual's ability to absorb oxygen from the water.

Other environmental factors also affect water flea phenotype. The presence of insect predators causes water fleas to form a protective pointy helmet and lengthened tail spine, for example. Individual water fleas also switch between asexual and sexual modes of reproduction. During early spring, food and space are typically abundant, and competition for these resources is scarce. Under these conditions, water fleas reproduce rapidly by asexual means, giving birth to large numbers of female offspring that quickly fill the ponds. Later in the season, competition intensifies as the pond water becomes warmer, saltier, and more crowded. Then, some of the water fleas start giving birth to males, and the population begins to reproduce sexually. The increased genetic diversity of sexually produced offspring may offer an advantage in the more challenging environment.

Seasonal Changes in Coat Color Seasonal changes in temperature and the length of day affect the production of melanin and other pigments that color the skin and fur of many animals. These species have different color phases in different seasons (**FIGURE 13.15A**). Hormonal signals triggered by the seasonal changes cause fur to be shed, and new fur grows back with different types and amounts of pigments deposited in it. The resulting change in phenotype provides these animals with seasonally appropriate camouflage from predators.

Effect of Altitude on Yarrow In plants, a flexible phenotype gives immobile individuals an ability to thrive in diverse habitats. For example, genetically identical yarrow plants grow to different heights at different altitudes (**FIGURE 13.15B**). More challenging temperature, soil, and water conditions are typically encountered at higher altitudes. Differences in altitude are also correlated with changes in the reproductive mode of yarrow: Plants at higher altitude tend to reproduce asexually, and those at lower altitude tend to reproduce sexually.

Psychiatric Disorders Researchers recently discovered that mutations in four human gene regions are associated with five psychiatric disorders: autism, depression, schizophrenia, bipolar disorder, and attention deficit hyperactivity disorder (ADHD). However, there must be environmental components to these disorders too, because one person with the mutations might get one type of disorder, while a relative with the same mutations might get another: two different results from the same genetic underpinnings. Moreover, the majority of people who carry these mutations never end up with a psychiatric disorder.

Recent discoveries in animal models are beginning to unravel some of the mechanisms by which environment can influence mental state in humans. For example, we now know that learning and memory are associated with dynamic and rapid DNA modifications in brain cells. Mood is, too. Stress-induced depression causes methylation-based silencing of a particular nerve growth factor gene; some antidepressants work by reversing this methylation. As another example, rats whose mothers are not very nurturing end up anxious and having a reduced resilience for stress as adults. The difference between these rats and ones who had nurturing maternal care is traceable to epigenetic DNA modifications that result in a lower than normal level of another nerve growth factor. Drugs can reverse these modifications—and their effects.

We do not yet know all of the genes that influence human mental state, but the implication of such research is that future treatments for many psychiatric disorders will involve deliberate modification of methylation patterns in an individual's DNA.

A The color of the snowshoe hare's fur varies by season. In summer, the fur is brown (left); in winter, white (right). Both forms offer seasonally appropriate camouflage from predators.

B The height of a mature yarrow plant (*Achillea millefolium*) depends on the elevation at which it grows.

FIGURE 13.15 ▶Animated Examples of environmental effects on phenotype.

TAKE-HOME MESSAGE 13.6

Does an individual's environment affect its phenotype?

✔ The environment influences gene expression, and therefore can alter phenotype.

✔ Cell-signaling pathways link environmental cues with changes in gene expression.

13.7 Complex Variation in Traits

✓ Individuals of most species vary in some of their shared traits. In some cases, the variation occurs in a continuous range.

FIGURE 13.16 Face length varies continuously in dogs. A gene with 12 alleles influences this trait; all arose by the spontaneous insertion of short tandem repeats. The longer the alleles, the longer the face.

A To see if human height varies continuously, male biology students at the University of Florida were divided into categories of one-inch increments in height and counted.

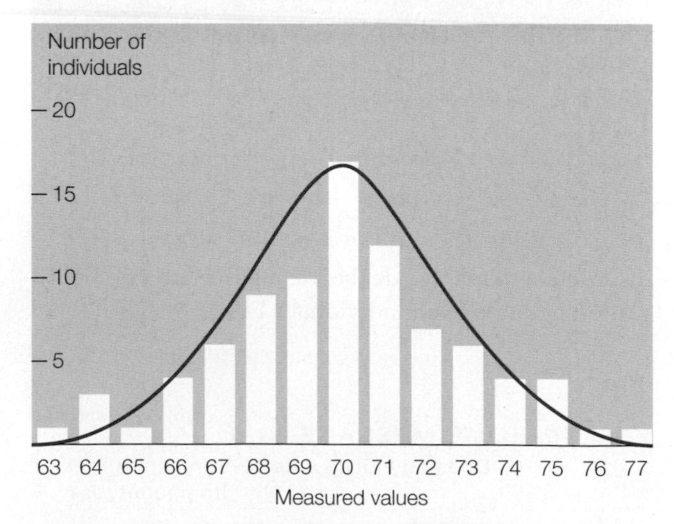

B Graphing the data that resulted from the experiment in (**A**) produces a bell-shaped curve, which is an indication that height does vary continuously in humans.

FIGURE 13.17 ▶**Animated** Continuous variation.

The pea plant phenotypes that Mendel studied appeared in two or three forms, which made them easy to track through generations. However, many other traits do not appear in distinct forms. Such traits are often the result of complex genetic interactions—multiple genes, multiple alleles, or both—with added environmental influences. Tracking traits with complex variation presents a special challenge, which is why the genetic basis of many of them has not yet been completely unraveled.

Continuous Variation

Some traits occur in a range of small differences that is called **continuous variation**. Continuous variation can be an outcome of epistasis, in which multiple genes affect a single trait. The more genes that influence a trait, the more continuous is its variation. Traits that arise from genes with a lot of alleles may also vary continuously. Some genes have regions of DNA in which a series of 2 to 6 nucleotides is repeated hundreds or thousands of times in a row. These **short tandem repeats** can spontaneously expand or contract very quickly compared with the typical rate of mutation, and the resulting changes in the gene's DNA sequence may be preserved as alleles. For example, short tandem repeats have given rise to 12 alleles of a homeotic gene that influences the length of the face in dogs, with longer repeats associated with longer faces (**FIGURE 13.16**).

How do we know if a particular trait varies continuously? Let's use a human trait, height, in an example. First, the total range of phenotypes is divided into measurable categories—inches, in this case (**FIGURE 13.17A**). Next, the individuals in each category are counted; these counts reveal the relative frequencies of phenotypes across the range of values. Finally, this data is plotted as a bar chart (**FIGURE 13.17B**). A graph line around the top of the bars shows the distribution of values for the trait. If the line is a bell-shaped curve, or **bell curve**, then the trait varies continuously.

Human skin color varies continuously, as does human eye color (**FIGURE 13.18**). The colored part of the eye is a doughnut-shaped structure called the iris. Iris color, like skin color, results from interactions

bell curve Bell-shaped curve; typically results from graphing frequency versus distribution for a trait that varies continuously.
continuous variation Range of small differences in a shared trait.
short tandem repeat In chromosomal DNA, sequences of a few nucleotides repeated multiple times in a row.

Menacing Mucus (revisited)

The *ΔF508* allele that causes cystic fibrosis is at least 50,000 years old and very common: In some populations, 1 in 25 people are heterozygous for it. Why does the allele persist if it is so harmful?

The *ΔF508* allele is eventually lethal in homozygous individuals, but not in those who are heterozygous. It is codominant with the normal allele. Heterozygous individuals typically have no symptoms of cystic fibrosis because their cells have plasma membranes with enough CFTR to transport chloride ions normally.

Researchers think that the *ΔF508* allele has persisted because it offers heterozygous individuals an advantage in surviving certain deadly infectious diseases. CFTR's receptor function is an essential part of the immune response to bacteria in the respiratory tract. However, the same function allows bacteria to enter cells of the gastrointestinal tract, where they can be deadly. Thus, people who carry *ΔF508* are probably less susceptible to dangerous bacterial diseases that begin in the intestinal tract.

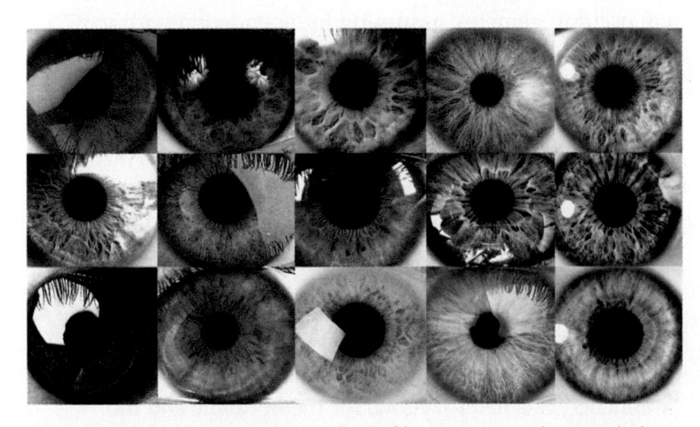

FIGURE 13.18 A small sampling of human eye color, a trait that varies continuously.

among several gene products that make and distribute melanins (an example of epistasis). The more melanin deposited in the iris, the less light is reflected from it. Dark irises have dense melanin deposits that absorb almost all light, and reflect almost none. Green and blue irises have the least amount of melanin, so they reflect the most light.

TAKE-HOME MESSAGE 13.7

Do all traits occur in distinct forms?

✔ The more genes and other factors that influence a trait, the more continuous is its range of variation.

summary

Section 13.1 Symptoms of cystic fibrosis are pleiotropic effects of mutations in the *CFTR* gene. The allele associated with most cases persists at high frequency despite its devastating effects in homozygous people. Carrying the allele may offer heterozygous individuals protection from dangerous gastrointestinal tract infections.

Section 13.2 Gregor Mendel indirectly discovered the role of alleles in inheritance by breeding pea plants and carefully tracking traits of the offspring over many generations.

Each gene occurs at a **locus**, or location, on a chromosome. Individuals with identical alleles are **homozygous** for the allele. **Heterozygous** individuals, or **hybrids**, have two nonidentical alleles.

A **dominant** allele masks the effect of a **recessive** allele partnered with it on the homologous chromosome. **Genotype** (an individual's particular set of alleles) gives rise to **phenotype** (an individual's observable traits).

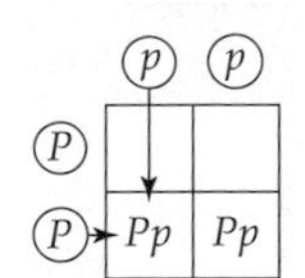

Section 13.3 Crossing two individuals that breed true for different forms of a trait yields identically heterozygous offspring. A cross between such offspring is called a **monohybrid cross**. The frequency at which traits appear in offspring of a **testcross** can reveal the genotype of an individual with a dominant phenotype. **Punnett squares** are useful for determining the probability of offspring genotype and phenotype.

Mendel's monohybrid cross results led him to formulate his **law of segregation**, which we state here in modern terms: Diploid cells have pairs of genes on homologous chromosomes. The two genes of a pair become separated from each other during meiosis, so they end up in different gametes.

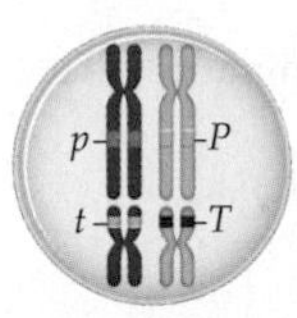

Section 13.4 Crossing two individuals that breed true for different forms of two traits yields F_1 offspring identically heterozygous for alleles governing those traits. A cross between such offspring is a **dihybrid cross**. The frequency at which the two traits appear in F_2 offspring can reveal dominance relationships between alleles associated with those traits. Mendel's dihybrid cross results led to his **law of independent assortment**, which we state here in modern terms: Pairs of genes on homologous chromosomes tend to sort into gametes independently of other gene pairs during meiosis. Crossovers can break up **linkage groups**.

Section 13.5 With **incomplete dominance**, the phenotype of heterozygous individuals is an intermediate blend of the two homozygous phenotypes. With **codominant** alleles, heterozygous individuals have both homozygous phenotypes. Codominance may occur in **multiple allele systems** such as the one underlying ABO blood typing. With **epistasis**, two or more genes affect the same trait. With **pleiotropy**, one gene affects two or more traits.

Section 13.6 An individual's phenotype is influenced by environmental factors. Environmental cues alter gene expression by way of cell signaling pathways that ultimately affect gene expression control.

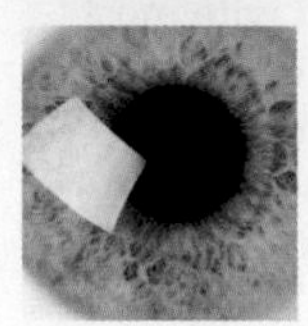

Section 13.7 A trait that is influenced by multiple genes often occurs in a range of small increments of phenotype called **continuous variation**. Continuous variation typically occurs as a **bell curve** in the range of values. Multiple alleles such as those that arise in regions of **short tandem repeats** can give rise to continuous variation.

self-quiz

Answers in Appendix VII

1. A heterozygous individual has a ________ for a trait being studied.
   a. pair of identical alleles
   b. pair of nonidentical alleles
   c. haploid condition, in genetic terms

2. An organism's observable traits constitute its _______ .
   a. phenotype c. genotype
   b. variation d. pedigree

3. In genetics, independent assortment means ________ .
   a. genes of a pair end up in different gametes.
   b. gene pairs separate independently of other gene pairs

4. The second-generation offspring of a cross between individuals who are homozygous for different alleles of a gene are called the ________ .
   a. F_1 generation c. hybrid generation
   b. F_2 generation d. none of the above

5. The F_1 offspring of the cross *AA* × *aa* are ________ .
   a. all *AA* c. all *Aa*
   b. all *aa* d. half are *AA*; half are *aa*

6. Refer to question 5. Assuming complete dominance, the F_2 generation will show a phenotypic ratio of ______ .
   a. 3:1 b. 9:1 c. 1:2:1 d. 9:3:3:1

7. A testcross is a way to determine ________ .
   a. phenotype b. genotype c. both a and b

8. Assuming complete dominance, crosses between two dihybrid F_1 pea plants, which are offspring from a cross *AABB* × *aabb*, result in F_2 phenotype ratios of ________ .
   a. 1:2:1 b. 3:1 c. 1:1:1:1 d. 9:3:3:1

9. The probability of a crossover occurring between two genes on the same chromosome ________ .
   a. is unrelated to the distance between them
   b. decreases with increasing distance between them
   c. increases with the distance between them

10. True or false? All traits are inherited in a Mendelian pattern.

11. A gene that affects three traits is ________ .
    a. epistatic c. pleiotropic
    b. a multiple allele system d. dominant

12. The phenotype of individuals heterozygous for ________ alleles comprises both homozygous phenotypes.
    a. epistatic c. pleiotropic
    b. codominant d. hybrid

13. ________ in a trait is indicated by a bell curve.
    a. Epigenetic effects c. Incomplete dominance
    b. Pleiotropy d. Continuous variation

14. Match the terms with the best description.
    ___ dihybrid cross a. *bb*
    ___ monohybrid cross b. *AaBb* × *AaBb*
    ___ homozygous condition c. *Aa*
    ___ heterozygous condition d. *Aa* × *Aa*

genetics problems

Answers in Appendix VII

1. Assuming that independent assortment occurs during meiosis, what type(s) of gametes will form in individuals with the following genotypes?
   a. *AABB* b. *AaBB* c. *Aabb* d. *AaBb*

2. Refer to problem 1. Determine the frequencies of each genotype among offspring from the following matings:
   a. *AABB* × *aaBB* c. *AaBb* × *aabb*
   b. *AaBB* × *AABb* d. *AaBb* × *AaBb*

3. Refer to problem 2. Assume a third gene has alleles *C* and *c*. For each genotype listed, what allele combinations will occur in gametes, assuming independent assortment?
   a. *AABBCC* c. *AaBBCc*
   b. *AaBBcc* d. *AaBbCc*

4. Heterozygous individuals perpetuate some alleles that have lethal effects in homozygous individuals. A mutated allele (M^L) associated with taillessness in Manx cats is an example (left). Cats homozygous for this allele (M^LM^L) typically die before birth due to severe spinal cord defects. In a case of incomplete dominance, cats heterozygous for the M^L allele and the normal, unmutated allele (*M*) have a short, stumpy tail or none at all. Two M^LM cats mate. What is the probability that any one of their surviving kittens will be heterozygous?

5. Suppose you identify a new gene in mice. One of its alleles specifies white fur, another specifies brown. You want to see if these alleles are inherited in a Mendelian pattern, or with incomplete dominance. What crosses would give you the answer?

6. Mendel crossed a true-breeding pea plant with green pods and a true-breeding pea plant with yellow pods. All the F_1 plants had green pods. Which color is recessive?

7. Several alleles affect traits of roses, such as plant form and bud shape. Alleles of one gene govern whether a plant will be a climber (dominant) or shrubby (recessive). All F_1 offspring from a cross between a true-breeding climber and a shrubby plant are climbers. If an F_1 plant is crossed with a shrubby plant, about 50 percent of the offspring will be shrubby; 50 percent will be climbers. Using symbols *A* and *a*

data analysis activities

Carrying the Cystic Fibrosis Allele Offers Protection from Typhoid Fever Epithelial cells that lack the CFTR protein cannot take up bacteria by endocytosis. Endocytosis is an important part of the respiratory tract's immune defenses against common *Pseudomonas* bacteria, which is why *Pseudomonas* infections of the lungs are a chronic problem in cystic fibrosis patients. Endocytosis is also the way that *Salmonella typhi* bacteria (shown at right) enter cells of the gastrointestinal tract, where internalization of this bacteria can result in typhoid fever.

Typhoid fever is a common worldwide disease. Its symptoms include extreme fever and diarrhea, and the resulting dehydration causes delirium that may last several weeks. If untreated, it kills up to 30 percent of those infected. Around 600,000 people, most of whom are children, die annually from typhoid fever.

In 1998, Gerald Pier and his colleagues compared the uptake of *S. typhi* by different types of epithelial cells: those homozygous for the normal allele, and those heterozygous for the Δ*F508* allele associated with CF. (Cells that are homozygous for the mutation do not take up any *S. typhi* bacteria.) Some of the results are shown in **FIGURE 13.19**.

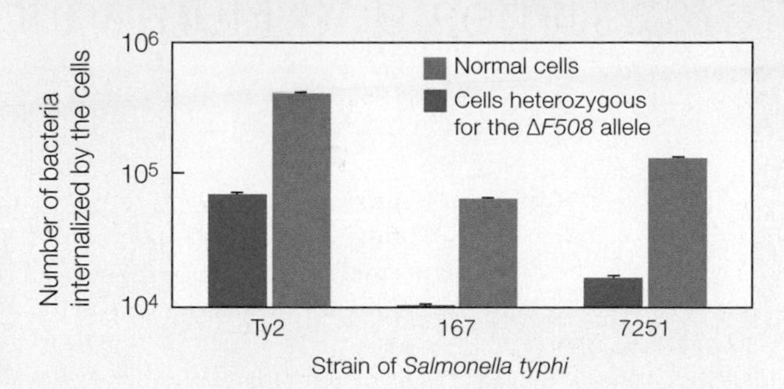

FIGURE 13.19 Effect of the Δ*F508* mutation on the uptake of three different strains of *Salmonella typhi* bacteria by epithelial cells.

1. Regarding the Ty2 strain of *S. typhi*, about how many more bacteria were able to enter normal cells (those heterozygous for the normal allele) than cells heterozygous for the Δ*F508* allele?

2. Which strain of bacteria entered normal epithelial cells most easily?

3. Entry of all three *S. typhi* strains into the heterozygous epithelial cells was inhibited. Is it possible to tell from this graph which strain was most inhibited?

---

for the dominant and recessive alleles, make a Punnett-square diagram of the expected genotypes and phenotypes in the cross between the F_1 offspring and the shrubby plant.

8. Mutations in the *TYR* gene may render its enzyme product—tyrosinase—nonfunctional. Individuals homozygous for such mutations cannot make the pigment melanin. Albinism, the absence of melanin, results. Humans and many other organisms can have this phenotype (left). Mutated tyrosinase alleles are recessive when paired with the normal allele in heterozygous individuals. In the following situations, what are the probable genotypes of the father, the mother, and their children?
   a. Both parents have normal phenotypes; some of their children have the albino phenotype and others are unaffected.
   b. Both parents and children have the albino phenotype.
   c. The mother and three children are unaffected; the father and one child have the albino phenotype.

9. In sweet pea plants, an allele for purple flowers (*P*) is dominant when paired with a recessive allele for red flowers (*p*). An allele for long pollen grains (*L*) is dominant when paired with a recessive allele for round pollen grains (*l*). Bateson and Punnett crossed a plant having purple flowers and long pollen grains with one having white flowers and round pollen grains. All F_1 offspring have purple flowers and long pollen grains. Among the F_2 generation, the researchers observed the following phenotypes:
   296 purple flowers/long pollen grains
   19 purple flowers/round pollen grains
   27 red flowers/long pollen grains
   85 red flowers/round pollen grains
   What is the best explanation for these results?

10. Red-flowering snapdragons are homozygous for allele R^1. White-flowering snapdragons are homozygous for a different allele (R^2). Heterozygous plants (R^1R^2) bear pink flowers. What phenotypes should appear among first-generation offspring of the crosses listed? What are the expected proportions for each phenotype?
   a. $R^1R^1 \times R^1R^2$
   b. $R^1R^1 \times R^2R^2$
   c. $R^1R^2 \times R^1R^2$
   d. $R^1R^2 \times R^2R^2$
   (Note that alleles inherited in a pattern of incomplete dominance are designated by superscript numerals, as shown, rather than by upper- and lowercase letters.)

11. A single allele gives rise to the HbS form of hemoglobin. Individuals who are homozygous for the allele (*HbS*/*HbS*) develop sickle-cell anemia (Section 9.6). Heterozygous individuals (*HbA*/*HbS*) have few symptoms. A couple who are both heterozygous for the *HbS* allele plan to have children. For each of the pregnancies, state the probability that they will have a child who is:
   a. homozygous for the *HbS* allele
   b. homozygous for the normal allele (*HbA*)
   c. heterozygous: *HbA*/*HbS*

14 Chromosomes and Human Inheritance

LEARNING ROADMAP

This chapter revisits proteins (Section 3.6); cell components (4.6, 4.8, 4.10, 4.11, 5.7); metabolism (5.5); pigments (6.2); chromosomes (8.4); DNA replication and repair (8.5); mutations (8.6); gene expression (9.2, 9.3) and control (10.4, 10.6); telomeres (11.5); oncogenes (11.6); meiosis (12.3); and inheritance patterns (13.2, 13.5, and 13.7).

TRACKING TRAITS IN HUMANS

Inheritance patterns in humans are revealed by following traits through generations of a family. Tracked traits are often genetic abnormalities or syndromes associated with a genetic disorder.

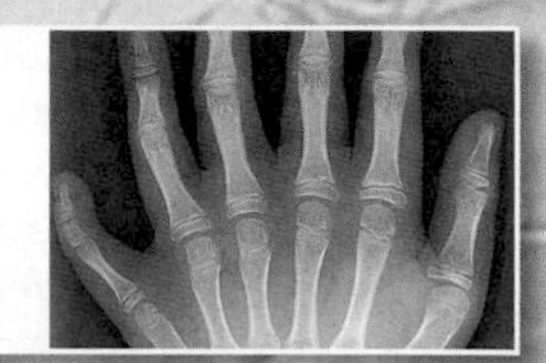

AUTOSOMAL INHERITANCE

Traits associated with dominant alleles on autosomes appear in every generation. Traits associated with recessive alleles on autosomes can skip generations.

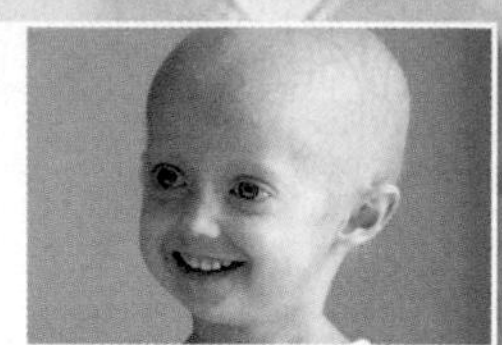

SEX-LINKED INHERITANCE

Traits associated with alleles on the X chromosome tend to affect more men than women. Men cannot pass such alleles to a son; carrier mothers bridge affected generations.

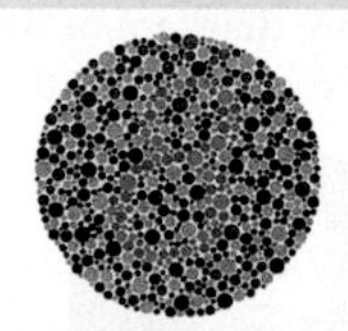

CHROMOSOME CHANGES

Some genetic disorders arise after large-scale change in chromosome structure. With few exceptions, a change in the number of autosomes is fatal in humans.

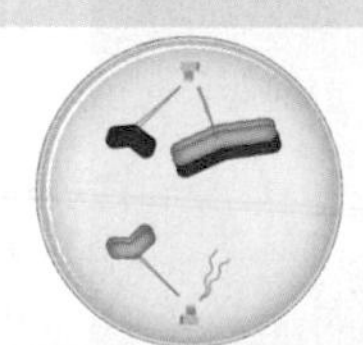

GENETIC TESTING

Genetic testing provides information about the risk of passing a harmful allele to offspring. Prenatal testing can reveal a genetic abnormality or disorder in a developing fetus.

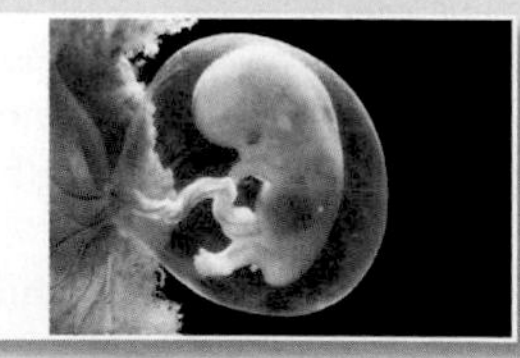

Genetic disorders are discussed in later chapters, in the context of the systems that they affect. Chapter 15 returns to human chromosomes as part of genomics and genetic engineering. Chapter 17 explores evolutionary adaptations and factors that influence the frequency of alleles in a population. The cells and other structural components of human skin are covered in detail in Section 31.8. Chapters 41 and 42 return to human reproduction and development.

14.1 Shades of Skin

The color of human skin begins with melanosomes, which are organelles that make melanin pigments. Most people have about the same number of melanosomes in their skin cells. Variations in skin color arise from differences in the size, shape, and cellular distribution of melanosomes in the skin, as well as in the kinds and amounts of melanins they make.

Human skin color variation may have evolved as a balance between vitamin production and protection against harmful ultraviolet (UV) radiation in the sun's rays. Dark skin is beneficial under the intense sunlight of African savannas where humans first evolved. Melanin is a natural sunscreen: It prevents UV radiation from breaking down folate, a vitamin essential for normal sperm formation and embryonic development.

Early human groups that migrated to regions with cold climates were exposed to less sunlight. In these regions, lighter skin color is beneficial. Why? UV radiation stimulates skin cells to make a molecule the body converts to vitamin D. Where sunlight exposure is minimal, UV radiation is less of a risk than vitamin D deficiency, which has serious health consequences for developing fetuses and children. People with dark, UV-shielding skin have a high risk of this deficiency in regions with long, dark winters.

Skin color, like most other human traits, has a genetic basis; at least 100 gene products are involved in pigmentation. The evolution of regional variations in human skin color began with mutations in these genes. Consider a gene on chromosome 15, *SLC24A5*, that encodes a transport protein in melanosome membranes. Nearly all people of African, Native American, or east Asian descent carry the same allele of this gene. Between 6,000 and 10,000 years ago, a mutation gave rise to a different allele. The mutation, a single base-pair substitution (Section 9.6), changed the 111th amino acid of the transport protein from alanine to threonine. The change results in less melanin—and lighter skin color—than the original African allele does. Today, nearly all people of European descent are homozygous for the mutated allele.

A person of mixed ethnicity may make gametes that contain different combinations of alleles for dark and light skin. It is fairly rare that one of those gametes contains all of the alleles for dark skin, or all of the alleles for light skin, but it happens (**FIGURE 14.1**). Skin color is only one of many human traits that vary as a result of single nucleotide mutations. The small scale of such changes offers a reminder that all of us share the genetic legacy of common ancestry.

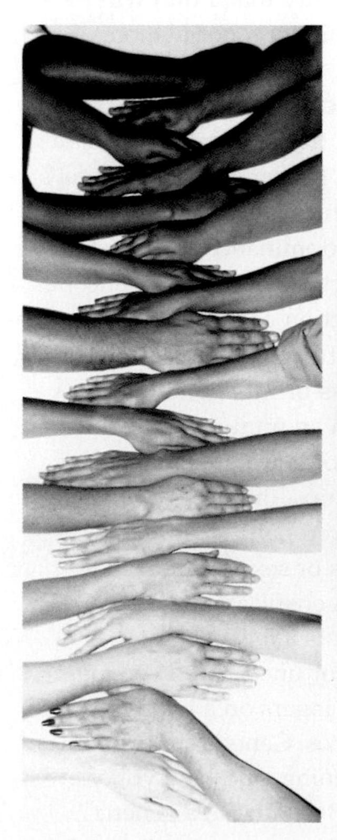

FIGURE 14.1 Variation in human skin color (left) begins with differences in alleles inherited from parents. Above, twins Kian and Remee with their parents. Both of the children's grandmothers are of European descent, and have pale skin. Both of their grandfathers are of African descent, and have dark skin. The twins inherited different alleles of some genes that affect skin color from their parents, who, given the appearance of their children, must be heterozygous for those alleles.

14.2 Human Chromosomes

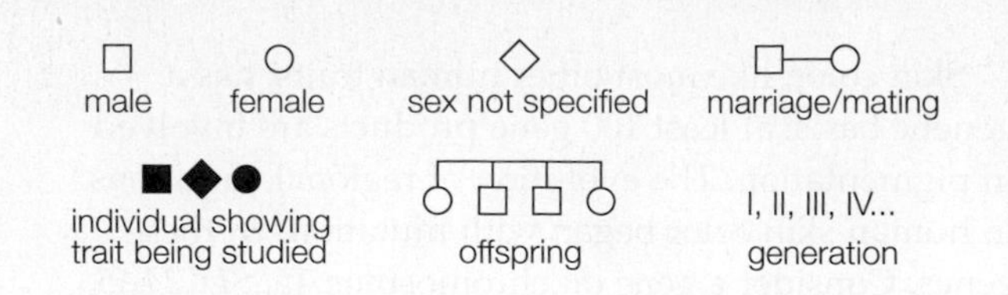

A Standard symbols used in pedigrees.

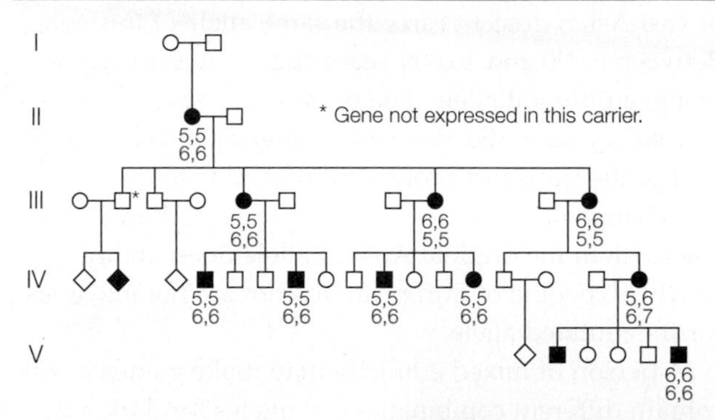

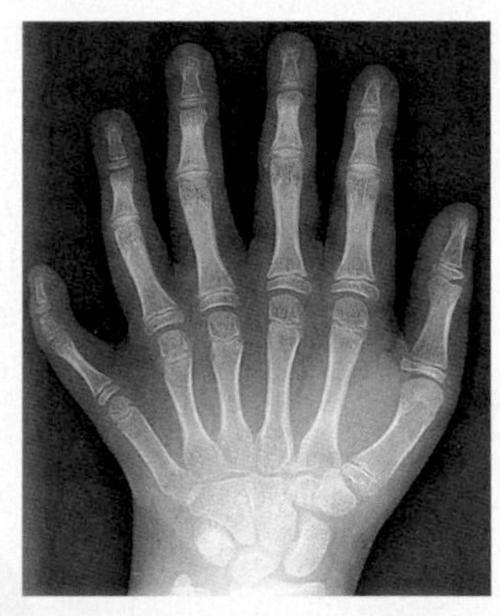

B Above, a pedigree for polydactyly, which is characterized by extra fingers (right), toes, or both. The black numbers signify the number of fingers on each hand; the red numbers signify the number of toes on each foot. Polydactyly that appears as part of a syndrome (such as Ellis–van Creveld syndrome) can be inherited in an autosomal recessive pattern. It also occurs on its own, in which case it is typically inherited in an autosomal dominant pattern.

C For more than 30 years, researcher Nancy Wexler has studied the genetic basis of Huntington's disease, an inherited disorder that causes progressive degeneration of the nervous system. Wexler and her team constructed an extended family tree for nearly 10,000 Venezuelans. Their analysis of relationships among unaffected and affected individuals revealed that a dominant allele on human chromosome 4 is the culprit. Wexler has a special interest in the disorder: It runs in her family.

FIGURE 14.2 Pedigrees.

✔ Geneticists study inheritance patterns in humans by tracking the appearance of genetic disorders and abnormalities through generations of families.

✔ Charting these genetic connections with pedigrees can reveal patterns of inheritance for certain traits.

Some organisms, including pea plants and fruit flies, are ideal for genetic studies. They have relatively few chromosomes, they reproduce quickly under controlled conditions, and breeding them poses few ethical problems. It does not take long to follow a trait through many generations. Humans, however, are a different story. Unlike flies grown in laboratories, we humans live under variable conditions, in different places, and we live as long as the geneticists who study our inheritance patterns. Most of us select our own mates and reproduce if and when we want to. Our families tend to be on the small side, so sampling error (Section 1.8) is a major factor in studying them.

Because of these and other challenges, geneticists often use historical records to track traits through many generations of a family. They use standardized charts called **pedigrees** to illustrate the phenotypes of family members and genetic connections among them (**FIGURE 14.2**). Analysis of a pedigree can reveal whether a trait is associated with a dominant or recessive allele, and whether the allele is on an autosome or a sex chromosome. Pedigree analysis also allows geneticists to determine the probability that a trait will recur in future generations of a family or a population.

Types of Genetic Variation

Some easily observed human traits follow Mendelian inheritance patterns. Like the flower color of Mendel's pea plants, these traits are controlled by a single gene with alleles that have a clear dominance relationship. Consider the *MC1R* gene (Section 13.5), which encodes a protein that triggers production of the brownish melanin. Mutations can result in a defective protein; an allele with one of these loss-of-function mutations is recessive when paired with an unmutated allele. A person who is homozygous for a mutated allele does not make the brownish melanin—only the reddish type—so this individual has red hair.

Single genes on autosomes or sex chromosomes also govern more than 6,000 genetic abnormalities and disorders. **TABLE 14.1** lists a few examples. A genetic abnormality is a rare or uncommon version of a trait, such as having six fingers on a hand or having a web between two toes. Genetic abnormalities are not inherently life-threatening, and how you view them is a matter of opinion. By contrast, a genetic disorder sooner or later causes medical problems that may be severe. A genetic disorder is often characterized by a specific set of symptoms (a syndrome). In general, much more research focuses on genetic disorders than on other human traits, because what we learn helps us develop treatments for affected people.

The next two sections of this chapter focus on inheritance patterns of human single-gene disorders, which collectively affect about 1 in 200 people. Keep in mind that these inheritance patterns are the least common kind. Most human traits, including skin color, are polygenic (influenced by multiple genes, Section 13.5), and some have epigenetic contributions or causes (Section 10.6). Environmental effects (Section 13.6) make these traits even harder to study. Many genetic disorders, including diabetes, asthma, obesity, cancers, heart disease, and multiple sclerosis, are inherited in patterns so complex that our understanding of the genetics behind them remains incomplete despite intense research. For example, mutations associated with an increased risk of autism (a developmental disorder) have been found on almost every chromosome, but most people who carry these mutations do not have autism. Appendix IV shows a map of human chromosomes with the locations of some alleles known to play a role in genetic disorders and other human traits.

Alleles that give rise to severe genetic disorders are generally rare in populations because they compromise the health and reproductive ability of their bearers. Why do they persist? Mutations periodically reintroduce them. In some cases, a codominant allele offers a survival advantage in a particular environment. You learned about one example, the $\Delta F508$ allele that causes cystic fibrosis, in Chapter 13: People heterozygous for this allele are protected from infection by bacteria that cause typhoid fever. You will see additional examples in later chapters.

pedigree Chart of family connections that shows the appearance of a trait through generations.

TAKE-HOME MESSAGE 14.2

How do we study inheritance patterns in humans?

✔ Human inheritance patterns are often studied by tracking genetic abnormalities or disorders through family trees.

✔ A genetic disorder is an inherited condition that causes medical problems. A genetic abnormality is a rare but harmless version of an inherited trait.

✔ A few genetic disorders are governed by single genes inherited in a Mendelian fashion. Most human traits are polygenic, and some have epigenetic contributions.

Table 14.1 Examples of Genetic Abnormalities and Disorders in Humans

Disorder/Abnormality	Main Symptoms
Autosomal dominant inheritance pattern	
Achondroplasia	One form of dwarfism
Aniridia	Defects of the eyes
Camptodactyly	Rigid, bent fingers
Familial hypercholesterolemia	High cholesterol level; clogged arteries
Huntington's disease	Degeneration of the nervous system
Marfan syndrome	Abnormal or missing connective tissue
Polydactyly	Extra fingers, toes, or both
Progeria	Drastic premature aging
Neurofibromatosis	Tumors of nervous system, skin
Autosomal recessive inheritance pattern	
Albinism	Absence of pigmentation
Hereditary methemoglobinemia	Blue skin coloration
Cystic fibrosis	Difficulty breathing; chronic lung infections
Ellis–van Creveld syndrome	Dwarfism, heart defects, polydactyly
Fanconi anemia	Physical abnormalities, marrow failure
Galactosemia	Brain, liver, eye damage
Hereditary hemochromatosis	Joints, organs damaged by iron overload
Phenylketonuria (PKU)	Mental impairment
Sickle-cell anemia	Anemia, pain, swelling, frequent infections
Tay–Sachs disease	Deterioration of mental and physical abilities; early death
X-linked recessive inheritance pattern	
Androgen insensitivity syndrome	XY individual but having some female traits; sterility
Red–green color blindness	Inability to distinguish red from green
Hemophilia	Impaired blood clotting ability
Muscular dystrophies	Progressive loss of muscle function
X-linked anhidrotic dysplasia	Mosaic skin (patches with or without sweat glands); other ill effects
X-linked dominant inheritance pattern	
Fragile X syndrome	Intellectual, emotional disability
Incontinentia pigmenti	Abnormalities of skin, hair, teeth, nails, eyes; neurological problems
Changes in chromosome number	
Down syndrome	Mental impairment; heart defects
Turner syndrome (XO)	Sterility; abnormal ovaries, sexual traits
Klinefelter syndrome	Sterility; mild mental impairment
XXX syndrome	Minimal abnormalities
XYY condition	Mild mental impairment or no effect
Changes in chromosome structure	
Chronic myelogenous leukemia (CML)	Overproduction of white blood cells; organ malfunctions
Cri-du-chat syndrome	Mental impairment; abnormal larynx

✔ An allele is inherited in an autosomal dominant pattern if the trait it specifies appears in heterozygous people.

✔ An allele is inherited in an autosomal recessive pattern if the trait it specifies appears only in homozygous people.

The Autosomal Dominant Pattern

A trait associated with a dominant allele on an autosome appears in people who are heterozygous for it as well as those who are homozygous. Such traits appear in every generation of a family, and they occur with equal frequency in both sexes. When one parent is heterozygous, and the other is homozygous for the recessive allele, each of their children has a 50 percent chance of inheriting the dominant allele and having the associated trait (**FIGURE 14.3A**).

Achondroplasia A form of hereditary dwarfism called achondroplasia offers an example of an autosomal dominant disorder (one caused by a dominant allele on an autosome). Mutations associated with achondroplasia occur in a gene for a growth factor receptor. The mutations cause the receptor, which normally slows bone development, to be overly active. About 1 in 10,000 people is heterozygous for one of these mutations. As adults, affected people are, on average, about 4 feet 4 inches (1.3 meters) tall, with arms and legs that are short relative to torso size (**FIGURE 14.3B**). An allele that causes achondroplasia can be passed to children because its expression does not interfere with reproduction, at least in heterozygous people. The homozygous condition results in severe skeletal malformations that cause early death.

Huntington's Disease Alleles associated with Huntington's disease are also inherited in an autosomal dominant pattern. Mutations that cause this disorder alter a gene for a cytoplasmic protein whose function is still unknown. The mutations are insertions caused by expansion of a short tandem repeat (Section 13.7), in which the same three nucleotides become repeated many times in the gene's sequence. The oversized protein product of the altered gene gets chopped into pieces inside nerve cells of the brain. The pieces accumulate in cytoplasm as large clumps that eventually prevent the cells from functioning properly. Brain cells involved in movement, thinking, and emotion are particularly affected. Dramatic, involuntary jerking and writhing movements that are symptoms of the most common form of Huntington's appear after age thirty. Affected people die during their forties or fifties. With this and other late-onset disorders, people may reproduce before symptoms appear, so the allele can be passed unknowingly to children.

Hutchinson-Gilford Progeria Hutchinson–Gilford progeria is an autosomal dominant disorder characterized by drastically accelerated aging. It is usually caused by a mutation that affects lamin A, a

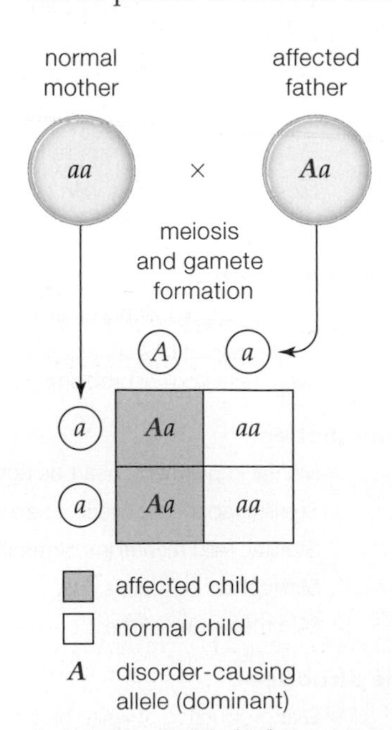

A A dominant allele (red) on an autosome affects all heterozygous people.

B Achondroplasia affects Ivy Broadhead (left), her brother, father, and grandfather.

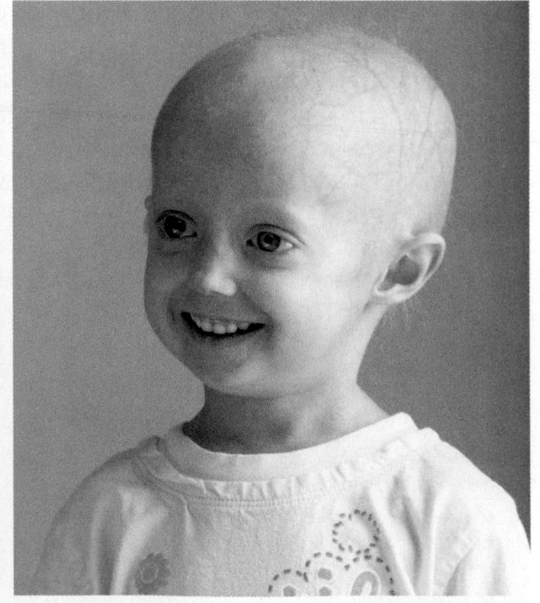

C Five-year-old Megan is already showing symptoms of Hutchinson–Gilford progeria.

FIGURE 14.3 ▸**Animated** Autosomal dominant inheritance.

fibrous protein component of intermediate filaments of the nuclear lamina (Section 4.10). The mutation, a base-pair substitution, adds a signal for an alternative splice site (Section 9.3). The resulting protein is defective, and so is the nuclear lamina. In cells that carry this mutation, the nucleus is grossly abnormal, with improperly assembled nuclear pore complexes and membrane proteins localized to the wrong side of the nuclear envelope. The function of the nucleus as protector of chromosomes and gateway for transcription is severely impaired, and DNA damage accumulates quickly. The effects are pleiotropic. Outward symptoms begin to appear before age two, as skin that should be plump and resilient starts to thin, muscles weaken, and bones soften. Premature baldness is inevitable (**FIGURE 14.3C**). Most people with the disorder die in their early teens as a result of a stroke or heart attack brought on by hardened arteries, a condition typical of advanced age. Progeria does not run in families because affected people do not live long enough to reproduce.

The Autosomal Recessive Pattern

A recessive allele on an autosome is expressed only in homozygous individuals, so traits associated with the allele tend to skip generations. Both sexes are equally affected. Heterozygous individuals are called carriers because they have the allele but not the trait. Any child of two carriers has a 25 percent chance of inheriting the allele from both parents—and developing the trait (**FIGURE 14.4A**).

Tay–Sachs Disease Alleles associated with Tay–Sachs disease are inherited in an autosomal recessive pattern. In the general population, about 1 in 300 people is a carrier for one of these alleles, but the incidence is ten times higher in some groups, such as Jews of eastern European descent. The gene altered in Tay–Sachs encodes a lysosomal enzyme responsible for breaking down a particular type of lipid. Mutations result in an enzyme that misfolds and becomes destroyed, so cells make the lipid but cannot break it down. Typically, newborns homozygous for a Tay–Sachs allele seem normal, but within three to six months they become irritable, listless, and may have seizures as the lipid accumulates in their nerve cells. Blindness, deafness, and paralysis follow. Affected children usually die by age five (**FIGURE 14.4B**).

Albinism Albinism, a phenotype characterized by an abnormally low level of the pigment melanin, is also inherited in an autosomal recessive pattern. Mutations associated with albinism affect proteins involved in melanin synthesis. Skin, hair, or eye pigmentation may be reduced or missing. In the most dramatic form, the skin is very white and does not tan, and the hair is white. The irises of the eyes appear red because the lack of pigment allows underlying blood vessels to show through. Melanin also plays a role in the retina, so vision problems are typical. In skin, melanin acts as a sunscreen; without it, the skin is defenseless against UV radiation. Thus, people with the albino phenotype have a very high risk of skin cancer.

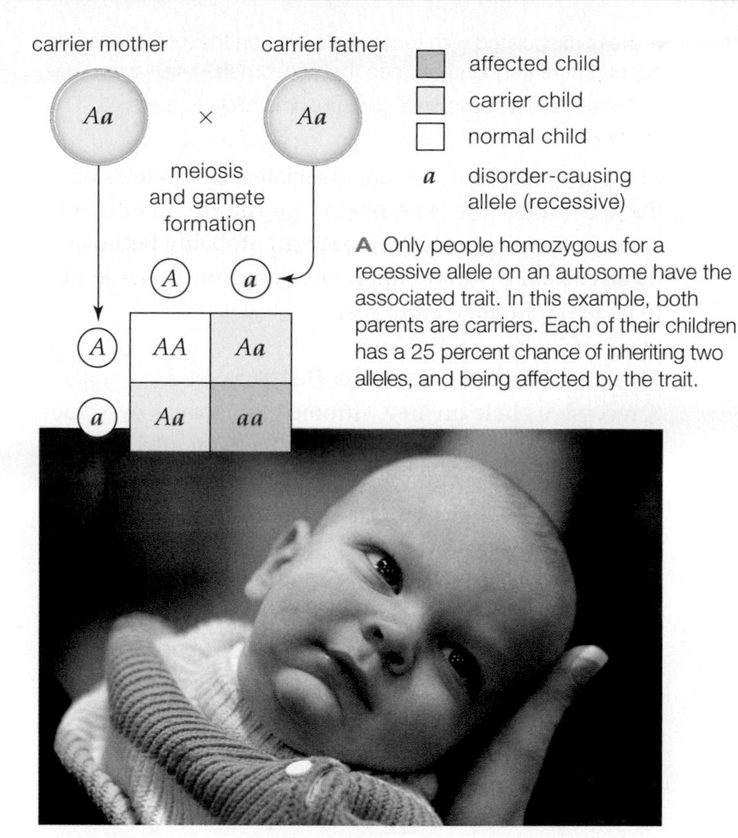

A Only people homozygous for a recessive allele on an autosome have the associated trait. In this example, both parents are carriers. Each of their children has a 25 percent chance of inheriting two alleles, and being affected by the trait.

B Conner Hopf was diagnosed with Tay–Sachs disease at age 7½ months. He died before his second birthday.

FIGURE 14.4 ▶Animated Autosomal recessive inheritance.

TAKE-HOME MESSAGE 14.3

How do we know when a trait is affected by an allele on an autosome?

✔ With an autosomal dominant inheritance pattern, anyone with the allele, homozygous or heterozygous, has the associated trait. The trait typically appears in every generation.

✔ With an autosomal recessive inheritance pattern, only persons who are homozygous for an allele have the associated trait. The trait tends to skip generations.

14.4 Examples of X-Linked Inheritance Patterns

✔ Traits associated with recessive alleles on the X chromosome appear more frequently in men than in women.

✔ A man cannot pass an X chromosome allele to a son.

Many genetic disorders are associated with alleles on the X chromosome (**FIGURE 14.5**). Almost all of them are inherited in a recessive pattern, probably because those caused by dominant X chromosome alleles tend to be lethal in male embryos.

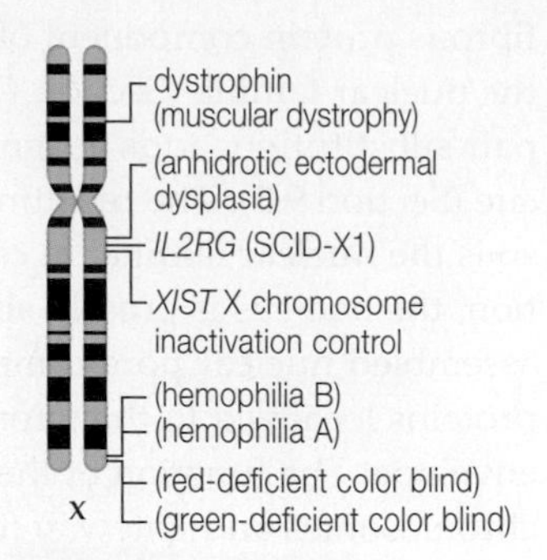

FIGURE 14.5 The human X chromosome. This chromosome carries about 2,000 genes—almost 10 percent of the total. Most X chromosome alleles that cause genetic disorders are inherited in a recessive pattern. A few disorders are listed (in parentheses).

The X-Linked Recessive Pattern

A recessive allele on an X chromosome leaves two clues when it causes a genetic disorder. First, an affected father never passes the disorder to a son, because all children who inherit their father's X chromosome are female (**FIGURE 14.6A**). Thus, a heterozygous female is always the bridge between an affected male and his affected grandson. Second, the disorder appears in males more often than in females. This is because all males who carry the allele have the disorder, but not all heterozygous females do. Remember that one of the two X chromosomes in each cell of a female is inactivated as a Barr body (Section 10.4). As a result, only about half of a heterozygous female's cells express the recessive allele. The other half of her cells express the dominant, normal allele that she carries on her other X chromosome, and this expression can mask the phenotypic effects of the recessive allele.

Red–Green Color Blindness Color blindness refers to a range of conditions in which an individual cannot distinguish among colors in the spectrum of visible light. These conditions are typically inherited in an X-linked recessive pattern, because most of the genes involved in color vision are on the X chromosome.

Humans can sense the differences among 150 colors, and this perception depends on pigment-containing receptors in the eyes. Mutations that result in altered or missing receptors affect color vision. For example, people who have red–green color blindness see fewer than 25 colors because receptors that respond to the red

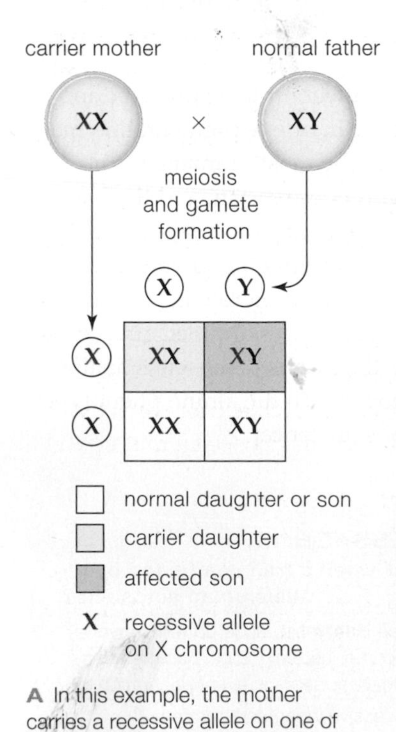

A In this example, the mother carries a recessive allele on one of her two X chromosomes (red).

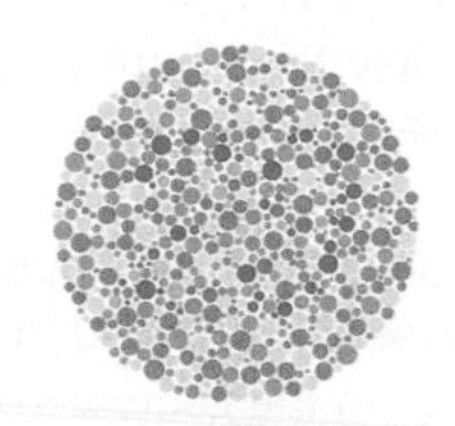

You may have one form of red–green color blindness if you see a 7 in this circle instead of a 29.

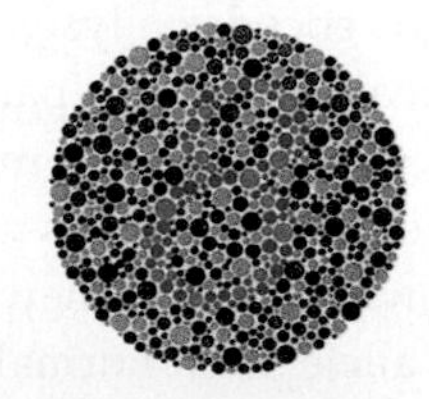

You may have another form of red–green color blindness if you see a 3 instead of an 8 in this circle.

B A view of color blindness. The photo on the left shows how a person with red–green color blindness sees the photo on the right. The perception of blues and yellows is normal; red and green appear similar. The circle diagrams are part of a standardized test for color blindness. A set of 38 of these diagrams is commonly used to diagnose deficiencies in color perception.

FIGURE 14.6 ▶**Animated** X-linked recessive inheritance.

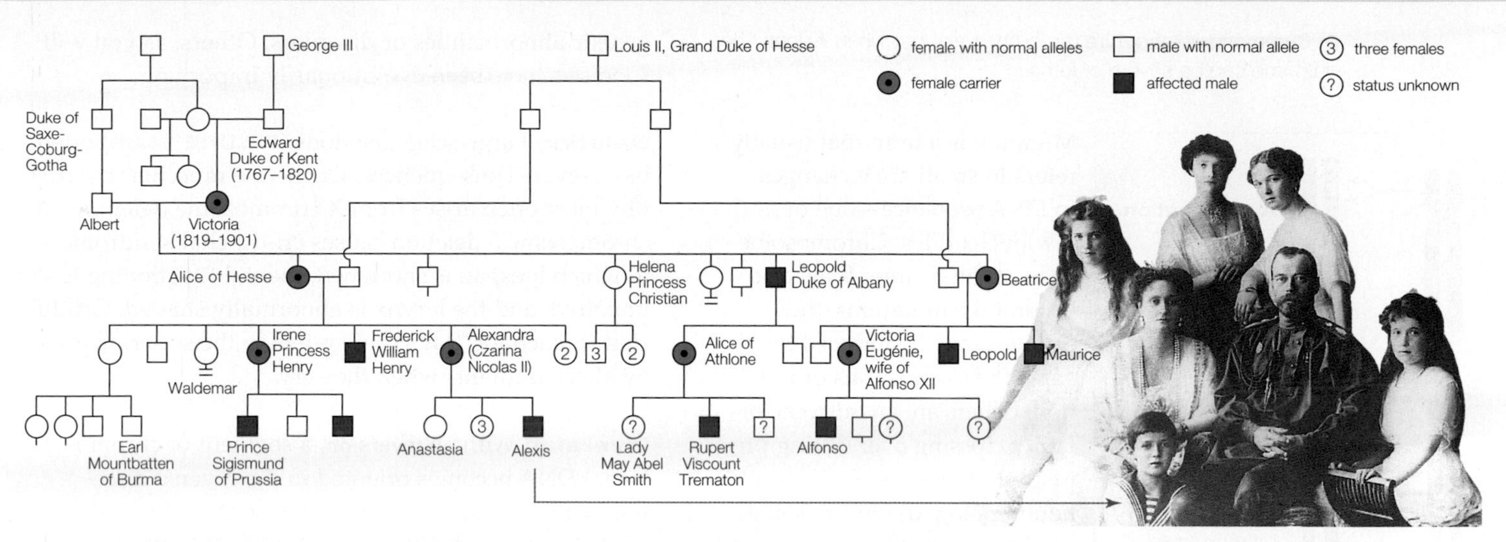

FIGURE 14.7 ▸Animated A classic case of X-linked recessive inheritance: a partial pedigree of the descendants of Queen Victoria of England. At one time, the recessive X-linked allele that resulted in hemophilia was present in eighteen of Victoria's sixty-nine descendants, who sometimes intermarried. Of the Russian royal family members shown in the photo, the mother (Alexandra Czarina Nicolas II) was a carrier.

FIGURE IT OUT How many of Alexis's siblings were affected by hemophilia?

Answer: None

and green wavelengths of light are weakened or absent (**FIGURE 14.6B**). Some confuse red and green; others see green as gray.

Duchenne Muscular Dystrophy A genetic disorder called Duchenne muscular dystrophy (DMD) is characterized by progressive muscle degeneration. It is caused by mutations in the X chromosome gene for dystrophin, a cytoskeletal protein that links actin microfilaments in cytoplasm to a complex of proteins in the plasma membrane. This complex structurally and functionally links the cell to extracellular matrix. When dystrophin is absent, the entire protein complex is unstable. Muscle cells, which are subject to stretching, are particularly affected. Their plasma membrane is easily damaged, and they become flooded with calcium ions. Eventually, the cells die and become replaced by fat cells and connective tissue.

DMD affects about 1 in 3,500 people; almost all are boys. Symptoms begin between ages three and seven. Anti-inflammatory drugs can slow the progression of DMD, but there is no cure. When an affected boy is about twelve years old, he will begin to use a wheelchair and his heart will start to fail. Even with the best care, he will probably die before the age of thirty, from a heart disorder or respiratory failure (suffocation).

Hemophilia Hemophilias are genetic disorders in which the blood does not clot properly. Most of us have a blood clotting mechanism that quickly stops bleeding from minor injuries. That mechanism involves two proteins, clotting factor VIII and IX, both products of X chromosome genes. Mutations in these two genes cause two type of hemophilia (A and B, respectively). Males who carry one of these mutations have prolonged bleeding, as do homozygous females (heterozygous females make enough clotting protein to have a clotting time that is close to normal). Affected people bruise easily, but internal bleeding is their most serious problem. Repeated bleeding inside the joints disfigures them and causes chronic arthritis.

In the nineteenth century, the incidence of hemophilia A was relatively high in royal families of Europe and Russia, probably because the common practice of inbreeding kept the allele in their family trees (**FIGURE 14.7**). Today, about 1 in 7,500 people in the general population is affected. That number may be rising because the disorder is now treatable. More affected people now live long enough to transmit a mutated allele to children.

TAKE-HOME MESSAGE 14.4

How do we know when a trait is affected by an allele on an X chromosome?

✔ Men who have an X-linked allele have the associated trait, but not all heterozygous women do. Thus, the trait appears more often in men.

✔ Men transmit an X-linked allele to their daughters, but not to their sons.

14.5 Heritable Changes in Chromosome Structure

✔ Chromosome structure rarely changes, but when it does, the outcome can be severe or lethal.

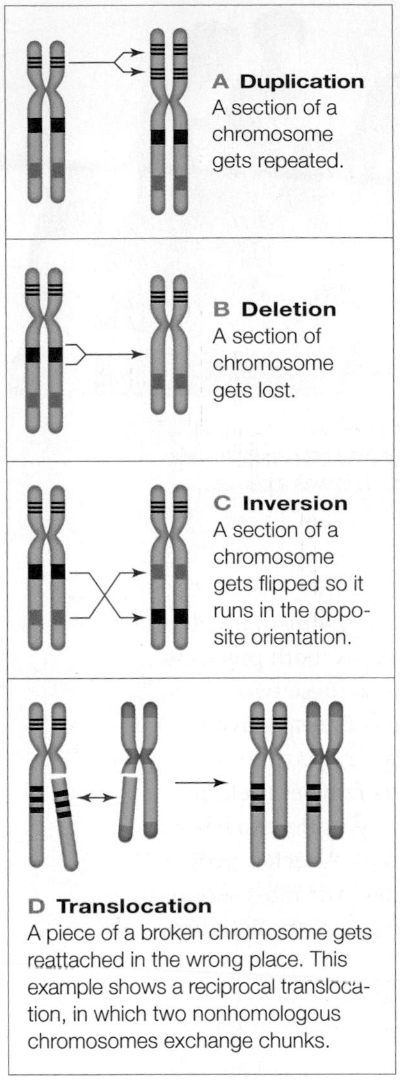

A Duplication
A section of a chromosome gets repeated.

B Deletion
A section of chromosome gets lost.

C Inversion
A section of a chromosome gets flipped so it runs in the opposite orientation.

D Translocation
A piece of a broken chromosome gets reattached in the wrong place. This example shows a reciprocal translocation, in which two nonhomologous chromosomes exchange chunks.

FIGURE 14.8 ▶**Animated** Major changes in chromosome structure.

Mutation is a term that usually refers to small-scale changes in DNA sequence—one or a few nucleotides. Chromosome changes on a larger scale also occur. Like mutations, these changes may be induced by exposure to chemicals or radiation. Others are an outcome of faulty crossing over during prophase I of meiosis. For example, nonhomologous chromosomes sometimes align and swap segments at spots where the DNA sequence is similar. Homologous chromosomes also may misalign along their length. In both cases, crossing over results in the exchange of segments that are not equivalent. The activity of transposable elements also alters chromosome structure. A **transposable element** is a segment of DNA hundreds or thousands of nucleotides long that can move spontaneously within or between chromosomes. Repeated DNA sequences at the ends allow the element to "jump" during mitosis or meiosis. Transposable elements are common in the DNA of all species; about 45 percent of human DNA consists of them and their evolutionary remnants.

Types of Chromosomal Change

Large-scale changes in chromosome structure can be categorized into several groups (**FIGURE 14.8**). In most cases, these changes drastically affect health; about half of all miscarriages are due to chromosome abnormalities of the developing embryo.

Duplication Even normal chromosomes have DNA sequences that are repeated two or more times. These repetitions are called **duplications** (**FIGURE 14.8A**). Some newly occurring duplications, such as the expansion mutations that cause Huntington's disease, cause genetic abnormalities or disorders. Others, as you will soon see, have been evolutionarily important.

Deletion Large-scale deletions (**FIGURE 14.8B**) often have severe consequences. Duchenne muscular dystrophy most often arises from X chromosome deletions. A chromosome 5 deletion causes cri-du-chat syndrome, in which lifespan is shortened, mental functioning is impaired, and the larynx is abnormally shaped. Cri-du-chat (French for "cat's cry") refers to the sound made by affected infants when they cry.

Inversion With an **inversion**, a segment of chromosomal DNA becomes oriented in the reverse direction, with no loss of nucleotides (**FIGURE 14.8C**). An inversion may not affect a carrier's health if it does not interrupt a gene or gene control region, because the individual's cells still contain their full complement of genetic material. However, fertility may be compromised because a chromosome with an inversion does not pair properly with its homologous partner during meiosis. Crossovers may occur between the mispaired chromosomes, producing other chromosome abnormalities that reduce the viability of forthcoming embryos. People who carry an inversion may not know about it until they are diagnosed with infertility and their karyotype is checked.

Translocation If a chromosome breaks, the broken part may become attached to a different chromosome, or to a different part of the same one. This type of structural change is called a **translocation**. Most translocations are reciprocal, in which two nonhomologous chromosomes exchange broken parts (**FIGURE 14.8D**). A reciprocal translocation between chromosomes 8 and 14 is the usual cause of Burkitt's lymphoma, an aggressive cancer of the immune system. The translocation moves a proto-oncogene to a region that is vigorously transcribed in immune cells, with the result being uncontrolled cell divisions that are characteristic of cancer (Section 11.6).

Many other reciprocal translocations have no adverse effects on health, but, like inversions, they can compromise fertility. During meiosis, translocated chromosomes pair abnormally and segregate improperly; about half of the resulting gametes carry major duplications or deletions. If one of these gametes unites with a normal gamete at fertilization, the resulting embryo almost always dies. As with inversions, people who carry a translocation may not know about it until they have difficulty with fertility.

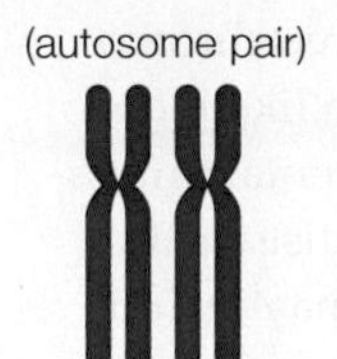

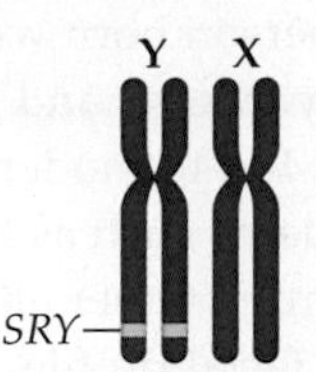

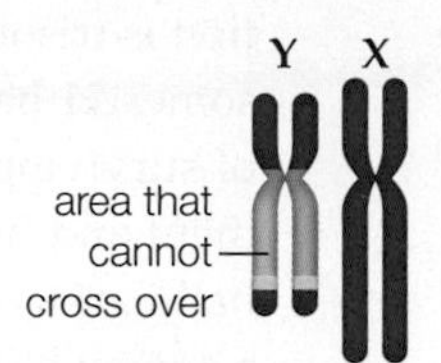

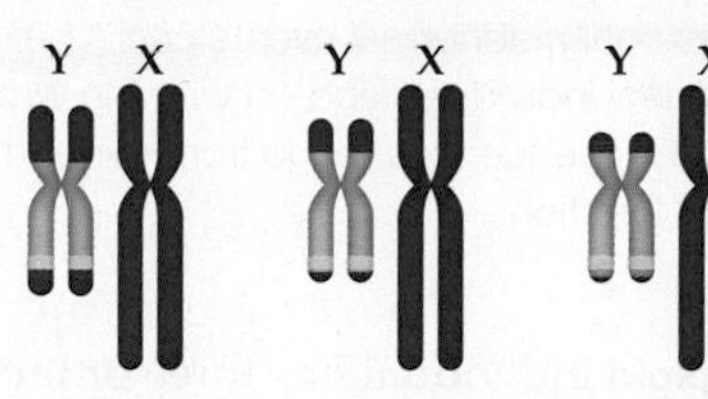

A Before 350 mya, sex was determined by temperature, not by chromosome differences.

B The *SRY* gene begins to evolve 350 mya. The DNA sequences of the chromosomes diverge as other mutations accumulate.

C By 320–240 mya, the DNA sequences of the chromosomes are so different that the pair can no longer cross over in one region. The Y chromosome begins to shorten.

D Three more times, the pair stops crossing over in yet another region. Each time, the DNA sequences of the chromosomes diverge, and the Y chromosome shortens. Today, the pair crosses over only at a small region near the ends.

FIGURE 14.9 Evolution of the Y chromosome. Today, the *SRY* gene determines male sex. Homologous regions of the chromosomes are shown in pink; mya, million years ago. Monotremes are egg-laying mammals; marsupials are pouched mammals.

Chromosome Changes in Evolution

There is evidence of major structural alterations in the chromosomes of all known species. For example, duplications have often allowed a copy of a gene to mutate while the original carried out its unaltered function. The multiple and strikingly similar globin chain genes of mammals apparently evolved by this process. Globin chains, remember, associate to form molecules of hemoglobin (Section 9.6). Two identical genes for the alpha chain—and five other slightly different versions of it—form a cluster on chromosome 16. The gene for the beta chain clusters with four other slightly different versions on chromosome 11.

As another example, X and Y chromosomes were once homologous autosomes in ancient, reptilelike ancestors of mammals (**FIGURE 14.9**). Ambient temperature probably determined the gender of those organisms, as it still does in turtles and some other modern reptiles. About 350 million years ago, a gene on one of the two homologous chromosomes mutated. The mutation, which interfered with crossing over during meiosis, was the beginning of the male sex determination gene *SRY* (Section 10.4). A reduced frequency of crossovers allowed the chromosomes to diverge around the changed region as mutations began to accumulate separately in the two chromosomes. Over evolutionary time, the chromosomes became so different that they no longer crossed over at all in the changed region, so they diverged even more. Today, the Y chromosome is much smaller than the X, and is homologous with it only in a tiny part. The Y crosses over mainly with itself—by translocating duplicated regions of its own DNA.

Some chromosome structure changes contributed to differences among closely related organisms, such as apes and humans. Human somatic cells have twenty-three pairs of chromosomes, but cells of chimpanzees, gorillas, and orangutans have twenty-four. Thirteen human chromosomes are almost identical with chimpanzee chromosomes. Nine more are similar, except for some inversions. One human chromosome matches up with two in chimpanzees and the other great apes (**FIGURE 14.10**). During human evolution, two chromosomes evidently fused end to end and formed our chromosome 2. How do we know? The region where the fusion occurred contains remnants of a telomere (Section 11.5).

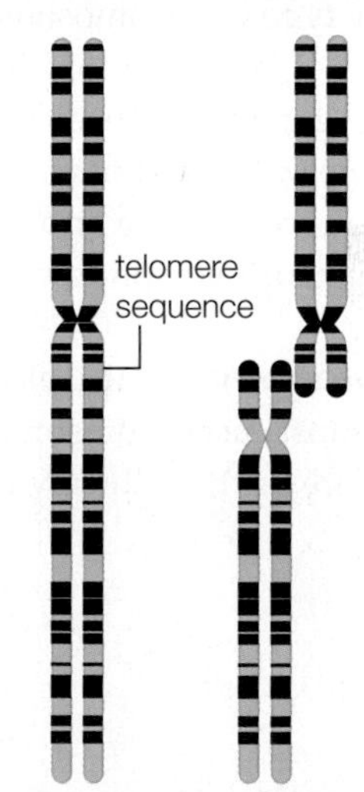

FIGURE 14.10 Human chromosome 2 compared with chimpanzee chromosomes 2A and 2B.

duplication Repeated section of a chromosome.
inversion Structural rearrangement of a chromosome in which part of the DNA becomes oriented in the reverse direction.
translocation Structural change of a chromosome in which a broken piece gets reattached in the wrong location.
transposable element Segment of DNA that can move spontaneously within or between chromosomes.

TAKE-HOME MESSAGE 14.5

How does chromosome structure change?

✔ A segment of a chromosome may be duplicated, deleted, inverted, or translocated. Any of these changes are usually harmful or lethal, but may be conserved in the rare circumstance that it has a neutral or beneficial effect.

✔ Occasionally, abnormal events occur before or during meiosis, and new individuals end up with the wrong chromosome number. Consequences range from minor to lethal changes in form and function.

A **polyploid** individual has three or more complete sets of chromosomes. About 70 percent of flowering plant species are polyploid, as are some insects, fishes, and other animals—but not humans. In our species, inheriting more than two full sets of chromosomes is invariably fatal, although some somatic cells are normally polyploid in adult tissues.

An **aneuploid** individual has too many or too few copies of a particular chromosome. Less than 1 percent of children are born with a diploid chromosome number that differs from the normal 46. Changes in chromosome number are usually an outcome of **nondisjunction**, the failure of chromosomes to separate properly during mitosis or meiosis. Nondisjunction during meiosis (FIGURE 14.11) can affect chromosome number at fertilization. For example, if a normal gamete (n) fuses with a gamete that has an extra chromosome ($n+1$), the resulting zygote will have three copies of one type of chromosome and two of every other type ($2n+1$), an aneuploid condition called trisomy. If a normal gamete (n) fuses with a gamete missing a chromosome ($n-1$), the new individual will have one copy of one chromosome and two of every other type ($2n-1$), an aneuploid condition called monosomy.

Autosomal Aneuploidy and Down Syndrome

In most cases, autosomal aneuploidy in humans is fatal before birth or shortly thereafter. An important exception is trisomy 21. A person born with three chromosomes 21 has Down syndrome and a high likelihood of surviving infancy. Mild to moderate mental impairment and health problems such as heart disease are hallmarks of this disorder. Other effects may include a somewhat flattened facial profile, a fold of skin that starts at the inner corner of each eyelid, white spots on the iris (FIGURE 14.12), and one deep crease (instead of two shallow creases) across each palm. The skeleton grows and develops abnormally, so older children have short body parts, loose joints, and misaligned bones of the fingers, toes, and hips. Muscles and reflexes are weak, and motor skills such as speech develop slowly. With medical care, affected individuals live about fifty-five years. Early training can help these individuals learn to care for themselves and to take part in normal activities. Down syndrome occurs in about 1 of 700 births, and the risk increases with maternal age.

Sex Chromosome Aneuploidy

Nondisjunction also causes alterations in the number of X and Y chromosomes, with a frequency of about 1 in 400 live births. Most often, such alterations lead to mild difficulties in learning and impaired motor skills such as a speech delay. These problems may be very subtle.

Turner Syndrome Individuals with Turner syndrome have an X chromosome and no corresponding X or Y chromosome (XO). The syndrome is thought to arise most frequently as an outcome of inheriting an unstable Y chromosome from the father. The zygote starts out being genetically male, with an X and a Y chromosome. Sometime during early development,

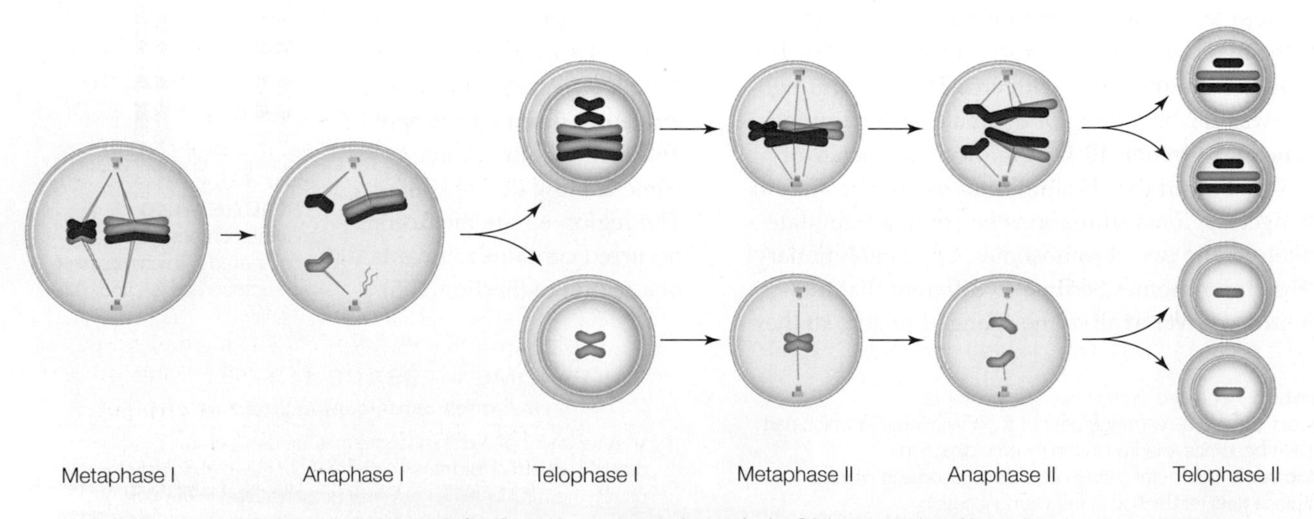

FIGURE 14.11 ▸Animated An example of nondisjunction during meiosis. Of the two pairs of homologous chromosomes shown here, one fails to separate during anaphase I. The chromosome number is altered in the resulting gametes.

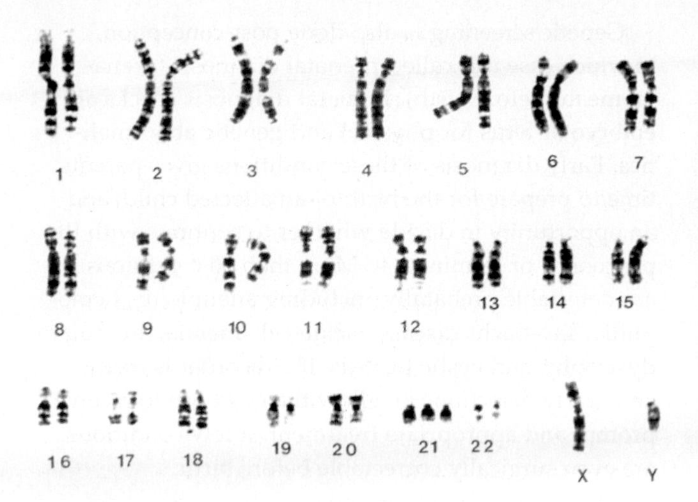

A Example of a Down syndrome genotype.

B Example of a Down syndrome phenotype. Excess tissue deposits on the iris give rise to a ring of starlike white speckles, a lovely effect of the chromosome number change that causes Down syndrome.

FIGURE 14.12 Down syndrome. **FIGURE IT OUT** Is the karyotype from an individual who is male or female? Answer: Male (XY)

the Y chromosome breaks up and is lost, so the embryo continues to develop as a female.

There are fewer people affected by Turner syndrome than other chromosome abnormalities: Only about 1 in 2,500 newborn girls has it. XO individuals grow up well proportioned but short, with an average height of 4 feet 8 inches (1.4 meters). Their ovaries do not develop properly, so they do not make enough sex hormones to become sexually mature and do not develop secondary sexual traits such as enlarged breasts.

XXX Syndrome A female may inherit multiple X chromosomes, a condition called XXX syndrome. This syndrome occurs in about 1 of 1,000 births. As with Down syndrome, the risk increases with maternal age. Only one X chromosome is typically active in female cells, so having extra X chromosomes usually does not cause physical or medical problems, but mild mental impairment may occur.

Klinefelter Syndrome About 1 out of every 500 males has an extra X chromosome (XXY). The resulting disorder, Klinefelter syndrome, becomes apparent at puberty. As adults, XXY males tend to be overweight and tall, with mild mental impairment. They make more estrogen and less testosterone than normal males. This hormone imbalance causes affected men to have small testes and a small prostate gland, a low sperm count, sparse facial and body hair, a high-pitched voice, and enlarged breasts. Testosterone injections during puberty can minimize these traits.

XYY Syndrome About 1 in 1,000 males is born with an extra Y chromosome (XYY), a result of nondisjunction of the Y chromosome during sperm formation. Adults tend to be taller than average and have mild mental impairment, but most are otherwise normal. XYY men were once thought to be predisposed to live a life of crime. This misguided view was based on sampling error (too few cases in narrowly chosen groups such as prison inmates) and bias (the researchers who gathered the karyotypes also took the personal histories of the participants). That view has since been disproven: Men with XYY syndrome are only slightly more likely to be convicted for crimes than unaffected men. Researchers believe this slight increase can be explained by poor socioeconomic conditions related to the effects of the syndrome.

aneuploid Having too many or too few copies of a particular chromosome.
nondisjunction Failure of sister chromatids or homologous chromosomes to separate during nuclear division.
polyploid Having three or more of each type of chromosome characteristic of the species.

TAKE-HOME MESSAGE 14.6

What are the effects of chromosome number changes in humans?

- ✔ Polyploidy is fatal in humans, but not in flowering plants and some other organisms.
- ✔ Aneuploidy can arise from nondisjunction during meiosis. In humans, most cases of aneuploidy are associated with some degree of mental impairment.

14.7 Genetic Screening

✔ Our understanding of human inheritance can provide prospective parents with information about the health of their future children.

Studying human inheritance patterns has given us many insights into how genetic disorders arise and progress, and how to treat them. Some disorders can be detected early enough to start countermeasures before symptoms develop. For this reason, most hospitals in the United States now screen newborns for mutations that cause phenylketonuria, or PKU. The mutations affect an enzyme that converts one amino acid (phenylalanine) to another (tyrosine). Without this enzyme, the body becomes deficient in tyrosine, and phenylalanine accumulates to high levels. The imbalance inhibits protein synthesis in the brain, which in turn results in severe neurological symptoms. Restricting all intake of phenylalanine can slow the progression of PKU, so routine early screening has resulted in fewer individuals suffering from the symptoms of the disorder.

The probability that a child will inherit a genetic disorder can be estimated by testing prospective parents for alleles known to be associated with genetic disorders. Karyotypes and pedigrees are also useful in this type of screening, which can help the parents make decisions about family planning.

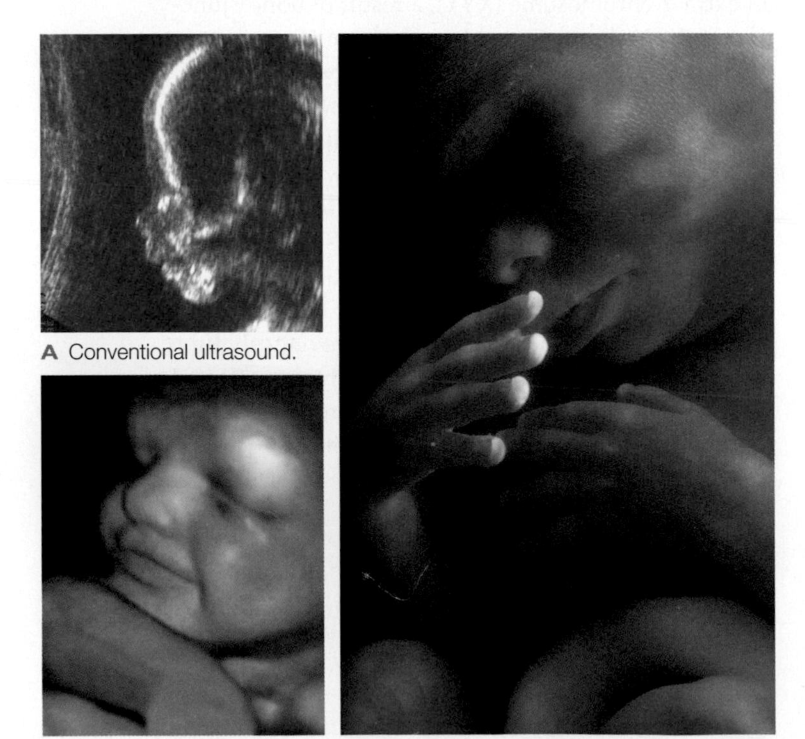

A Conventional ultrasound.

B 4D ultrasound.

C Fetoscopy.

FIGURE 14.13 Three ways of imaging a developing human fetus.

Genetic screening is also done post-conception, in which case it is called prenatal diagnosis (prenatal means before birth). Prenatal diagnosis checks an embryo or fetus for physical and genetic abnormalities. Early diagnosis of these conditions gives parents time to prepare for the birth of an affected child, and an opportunity to decide whether to continue with the pregnancy or terminate it. More than 30 conditions are detectable prenatally, including aneuploidy, hemophilia, Tay–Sachs disease, sickle-cell anemia, muscular dystrophy, and cystic fibrosis. If a disorder is treatable, early detection can allow the newborn to receive prompt and appropriate treatment. A few conditions are even surgically correctable before birth.

As an example of how prenatal diagnosis works, consider a woman who becomes pregnant at age thirty-five. Her doctor will probably perform a procedure called obstetric sonography, in which ultrasound waves directed across the woman's abdomen form images of the fetus's limbs and internal organs (**FIGURE 14.13A,B**). If the images reveal a physical defect that may be the result of a genetic disorder, a more invasive technique such as fetoscopy would be recommended for further diagnosis. With fetoscopy, sound waves pulsed from inside the mother's uterus yield images much higher in resolution than ultrasound (**FIGURE 14.13C**). Samples of tissue or blood are often taken at the same time, and some corrective surgeries can be performed.

Human genetics studies show that our thirty-five-year-old woman has about a 1 in 80 chance that her baby will be born with a chromosomal abnormality, a risk more than six times greater than when she was twenty years old. Thus, even if no abnormalities are detected by ultrasound, she probably will be offered an additional diagnostic procedure, amniocentesis, in which a small sample of fluid is drawn from the amniotic sac enclosing the fetus (**FIGURE 14.14**). The fluid contains cells shed by the fetus, and those cells can be tested for genetic disorders. Chorionic villus sampling (CVS) can be performed earlier than amniocentesis. With this technique, a few cells from the chorion are removed and tested for genetic disorders. (The chorion is a membrane that surrounds the amniotic sac and helps form the placenta, an organ that allows substances to be exchanged between mother and embryo.)

An invasive procedure often carries a risk to the fetus. The risks vary by the procedure. Amniocentesis has improved so much that, in the hands of a skilled physician, the procedure no longer increases the risk of miscarriage. CVS occasionally disrupts the placenta's

Shades of Skin (revisited)

Individuals of European descent share several alleles that influence skin pigmentation with individuals of east Asian descent. However, most people of east Asian descent carry a particular mutation in their *OCA2* gene—a single base-pair substitution in which an adenine changed to a cytosine—that results in lightened skin color. The product of the *OCA2* gene is a protein of unknown function, but it is named after the condition that occurs when the protein is missing: oculocutaneous albinism type II.

The *OCA2* mutation that lightens east Asian skin is uncommon in people of European ancestry. The *SLC24A5* allele that lightens European skin is uncommon in people of east Asian ancestry. Taken together, the distribution of these alleles suggests that (1) an African population with dark skin was ancestral to both east Asians and Europeans, and (2) east Asian and European populations separated before their pigmentation genes mutated and their skin color changed.

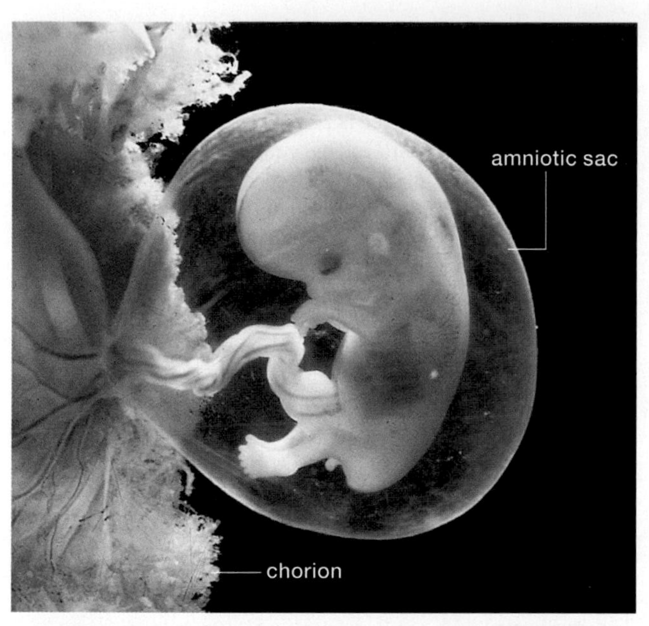

FIGURE 14.14 An 8-week-old fetus. With amniocentesis, a tiny bit of the fluid inside the amniotic sac is removed, and fetal cells that have been shed into the fluid are tested for genetic disorders. Chorionic villus sampling tests cells of the chorion, which is part of the placenta.

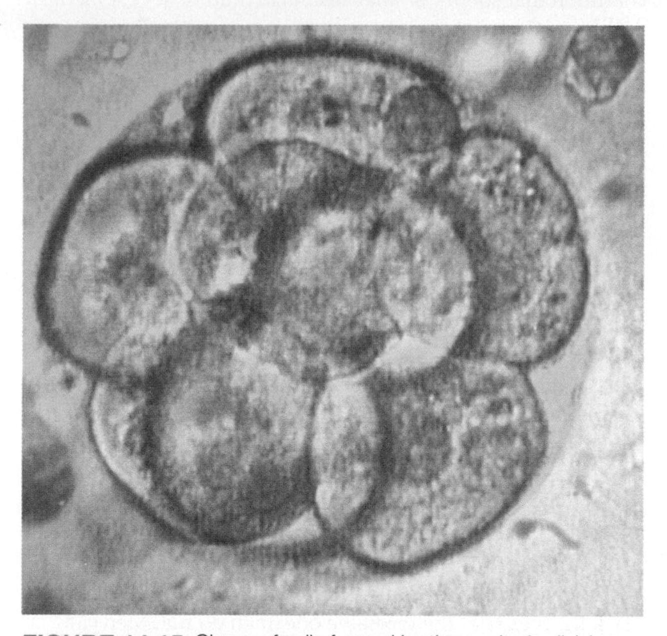

FIGURE 14.15 Clump of cells formed by three mitotic divisions after *in vitro* fertilization. All eight of the cells are identical; one can be removed for genetic analysis to determine whether the embryo carries any genetic defects. The remaining cells can continue development to form a viable embryo.

development and thus causes underdeveloped or missing fingers and toes in 0.3 percent of newborns. Fetoscopy raises the miscarriage risk by a whopping 2 to 10 percent.

Couples who discover they are at high risk of having a child with a genetic disorder may opt for reproductive interventions such as *in vitro* fertilization. With this procedure, sperm and eggs taken from prospective parents are mixed in a test tube. If an egg becomes fertilized, the resulting zygote will begin to divide. In about forty-eight hours, it will have become an embryo that consists of a ball of eight cells (**FIGURE 14.15**). All of the cells in this ball have the same genes, but none has yet committed to being specialized one way or another. Doctors can remove one of these undifferentiated cells and analyze its genes, a procedure called preimplantation diagnosis. The withdrawn cell will not be missed. If the embryo has no detectable genetic defects, it is inserted into the woman's uterus to continue developing. Many of the resulting "test-tube babies" are born in good health.

TAKE-HOME MESSAGE 14.7

How do we use what we know about human inheritance?

✔ Studying inheritance patterns for genetic disorders has helped researchers develop treatments for some of them.

✔ Genetic testing can provide prospective parents with information about the health of their future children.

summary

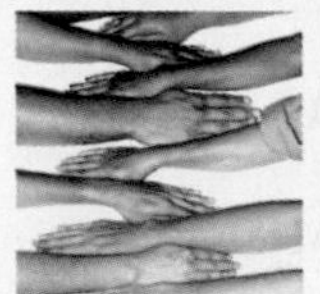

Section 14.1 Like most other human traits, skin color has a genetic basis. Minor differences in the alleles that govern melanin production and the size, shape, and distribution of melanosomes affect skin color. Skin color differences probably evolved as a balance between vitamin production and protection against harmful UV radiation.

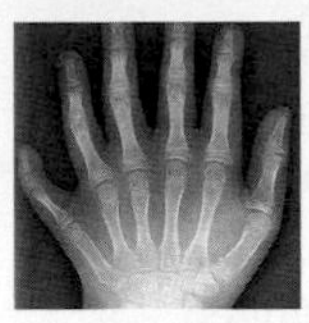

Section 14.2 Geneticists study inheritance patterns in humans by tracking genetic disorders and abnormalities through generations of families. A genetic abnormality is an uncommon version of a heritable trait that does not result in medical problems. A genetic disorder is a heritable condition that sooner or later results in mild or severe medical problems. Geneticists make **pedigrees** to reveal inheritance patterns for alleles that can be predictably associated with specific phenotypes.

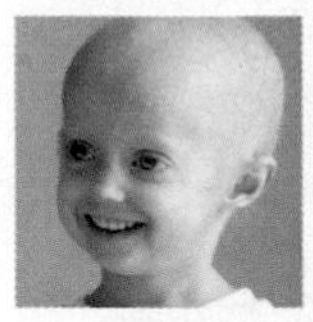

Section 14.3 An allele is inherited in an autosomal dominant pattern if the trait it specifies appears in everyone who carries it, and both sexes are affected with equal frequency. Such traits appear in every generation of families that have the allele. An allele is inherited in an autosomal recessive pattern if the trait it specifies appears only in homozygous people. Such traits also appear in both sexes equally, but they can skip generations.

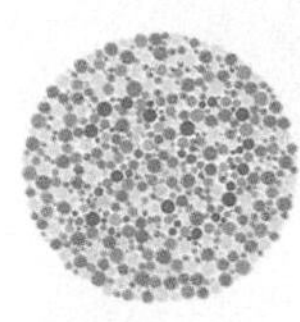

Section 14.4 An allele is inherited in an X-linked pattern when it occurs on the X chromosome. Most X-linked disorders are inherited in a recessive pattern, and these tend to appear in men more often than in women. Heterozygous women have a dominant, normal allele that can mask the effects of the recessive one; men do not. Men can transmit an X-linked allele to their daughters, but not to their sons. Only a woman can pass an X-linked allele to a son.

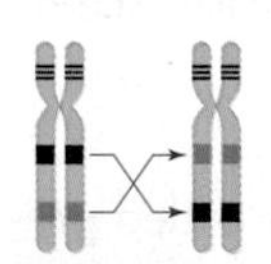

Section 14.5 Faulty crossovers and the activity of **transposable elements** can give rise to major changes in chromosome structure, including **duplications**, **inversions**, and **translocations**. Some of these changes are harmful or lethal in humans; others affect fertility. Even so, major structural changes have accumulated in the chromosomes of all species over evolutionary time.

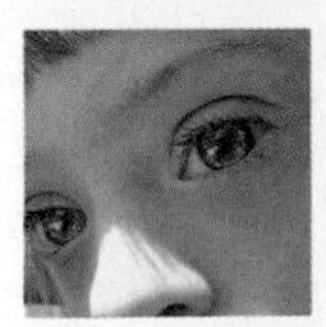

Section 14.6 Occasionally, abnormal events occur before or during meiosis, and new individuals end up with the wrong chromosome number. Consequences of such changes range from minor to lethal alterations in form and function. Chromosome number change is usually an outcome of **nondisjunction**, in which chromosomes fail to separate properly during nuclear division. **Polyploid** individuals have three or more of each type of chromosome. Polyploidy is lethal in humans, but not in flowering plants and some insects, fishes, and other animals. **Aneuploid** individuals have too many or too few copies of a chromosome. In humans, most cases of autosomal aneuploidy are lethal. Trisomy 21, which causes Down syndrome, is an exception. A change in the number of sex chromosomes usually results in some degree of impairment in learning and motor skills.

Section 14.7 Prospective parents can use genetic screening to estimate their risk of transmitting a harmful allele to offspring. The procedure involves analysis of parental pedigrees and genotype by a genetic counselor. Amniocentesis and other methods of prenatal genetic testing can reveal a genetic disorder before birth.

self-quiz

Answers in Appendix VII

1. Constructing a pedigree is particularly useful when studying inheritance patterns in organisms that ________ .
   a. produce many offspring per generation
   b. produce few offspring per generation
   c. have a very large chromosome number
   d. reproduce asexually
   e. have a fast life cycle

2. Pedigree analysis is necessary when studying human inheritance patterns because ________ .
   a. humans have more than 20,000 genes
   b. of ethical problems with experimenting on humans
   c. inheritance in humans is more complicated than it is in other organisms
   d. genetic disorders occur only in humans
   e. all of the above

3. A recognized set of symptoms that characterize a genetic disorder is a(n) ________ .
   a. syndrome b. disease c. abnormality

4. If one parent is heterozygous for a dominant allele on an autosome and the other parent does not carry the allele, any child of theirs has a ________ chance of being heterozygous.
   a. 25 percent c. 75 percent
   b. 50 percent d. no chance; it will die

5. True or false? A son can inherit an X-linked recessive allele from his father.

6. A trait that is present in a male child but not in either of his parents is characteristic of ________ inheritance.
   a. autosomal dominant
   b. autosomal recessive
   c. X-linked recessive
   d. It is impossible to answer this question without more information.

7. Color blindness is a case of ________ inheritance.
   a. autosomal dominant
   b. autosomal recessive
   c. X-linked dominant
   d. X-linked recessive

8. A female child inherits one X chromosome from her mother and one from her father. What sex chromosome does a male child inherit from each of his parents?

data analysis activities

Skin Color Survey of Native Peoples In 2000, researchers measured the average amount of UV radiation received in more than fifty regions of the world, and correlated it with the average skin reflectance of people native to those regions (reflectance is a way to measure the amount of melanin pigment in skin). Some of the results of this study are shown in **FIGURE 14.16**.

1. Which country receives the most UV radiation? The least?

2. The people native to which country have the darkest skin? The lightest?

3. According to these data, how does the skin color of indigenous peoples correlate with the amount of UV radiation incident in their native regions?

Country	Skin Reflectance	UVMED
Australia	19.30	335.55
Kenya	32.40	354.21
India	44.60	219.65
Cambodia	54.00	310.28
Japan	55.42	130.87
Afghanistan	55.70	249.98
China	59.17	204.57
Ireland	65.00	52.92
Germany	66.90	69.29
Netherlands	67.37	62.58

FIGURE 14.16 Skin color of indigenous peoples and regional incident UV radiation. Skin reflectance measures how much light of 685-nanometer wavelength is reflected from skin; UVMED is the annual average UV radiation received at Earth's surface.

9. Alleles for Tay–Sachs disease are inherited in an autosomal recessive pattern. Why would two parents with a normal phenotype have a child with Tay–Sachs?
   a. Both parents are homozygous for a Tay–Sachs allele.
   b. Both parents are heterozygous for a Tay–Sachs allele.
   c. A new mutation gave rise to Tay–Sachs in the child.
   d. b or c

10. The *SRY* gene gives rise to the male phenotype in humans (Sections 10.4 and 14.5). What do you think the inheritance pattern of *SRY* alleles is called?

11. Nondisjunction may occur during ________ .
   a. mitosis   c. fertilization
   b. meiosis   d. both a and b

12. Nondisjunction can result in ________ .
   a. duplications   c. crossing over
   b. aneuploidy   d. pleiotropy

13. True or false? An individual may inherit three or more of each type of chromosome characteristic of the species, a condition called polyploidy.

14. Klinefelter syndrome (XXY) can be easily diagnosed by ________ .
   a. pedigree analysis   c. karyotyping
   b. aneuploidy   d. phenotypic treatment

15. Match the chromosome terms appropriately.

___ polyploidy	a. symptoms of a genetic disorder
___ deletion	b. chromosomal mashup
___ aneuploidy	c. extra sets of chromosomes
___ translocation	d. gets around
___ syndrome	e. a chromosome segment lost
___ transposable element	f. one extra chromosome

genetics problems

Answers in Appendix VII

1. Duchenne muscular dystrophy (DMD), which is inherited in an X-linked recessive pattern, occurs almost exclusively in males. Suggest why.

CENGAGE brain To access course materials, please visit www.cengagebrain.com.

2. Does the phenotype indicated by the red circles and squares in this pedigree show an inheritance pattern that is autosomal dominant, autosomal recessive, or X-linked?

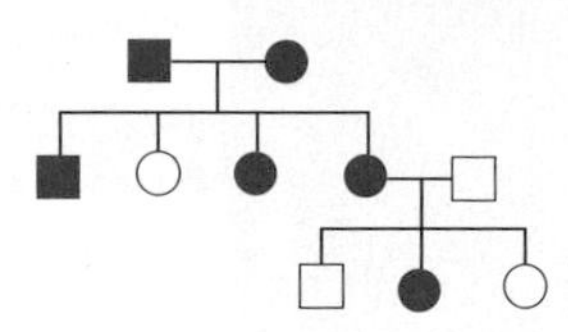

3. Human females have two X chromosomes (XX); males have one X and one Y chromosome (XY).
   a. With respect to X-linked alleles, how many different types of gametes can a male produce?
   b. If a female is homozygous for an X-linked allele, how many types of gametes can she produce with respect to that allele?
   c. If a female is heterozygous for an X-linked allele, how many types of gametes can she produce with respect to that allele?

4. A mutated allele responsible for Marfan syndrome (Section 13.5) is inherited in an autosomal dominant pattern. What is the chance that any child will inherit it if one parent does not carry the allele and the other is heterozygous for it?

5. Somatic cells of individuals with Down syndrome usually have an extra chromosome 21; they contain forty-seven chromosomes.
   a. At which stage(s) of meiosis could nondisjunction alter the chromosome number?
   b. A few individuals with Down syndrome have forty-six chromosomes: two normal-appearing chromosomes 21, and a longer-than-normal chromosome 14. Speculate on how this chromosome abnormality may arise.

6. Mutations in the genes for clotting factor VIII and IX cause hemophilia A and B, respectively. A woman may be heterozygous for mutations in both genes, with a mutated factor VIII allele on one X chromosome, and a mutated factor IX allele on the other. All of her sons should have either hemophilia A or B. However, on rare occasions, one of these women gives birth to a son who does not have hemophilia, and his one X chromosome does not have either mutated allele. Explain how this boy's X chromosome probably arises.

15 Studying and Manipulating Genomes

LEARNING ROADMAP

This chapter builds on your understanding of DNA (Sections 8.3–8.5, 13.2, 14.5, 14.7). Clones (8.1), gene expression (9.2, 9.3), and knockouts (10.3) turn up in the context of human traits (13.7) and genetic disorders (Chapter 14). You will revisit tracers (2.2), triglycerides (3.5), denaturation (3.7), β-carotene (6.2), the lac operon (10.5), and cancer (11.6).

DNA CLONING

Researchers make recombinant DNA by cutting and pasting together DNA from different species. Plasmids and other vectors can carry foreign DNA into host cells.

FINDING NEEDLES IN HAYSTACKS

Genetic engineering, the directed modification of an organism's genes, relies on laboratory techniques for isolating and identifying targeted fragments of DNA.

DNA SEQUENCING

Sequencing reveals the linear order of nucleotides in DNA. Comparing genomes offers insights into human genes and evolution. DNA sequence can be used to identify individuals.

GENETIC ENGINEERING

Genetic engineering is now a routine part of research and industrial applications. Genetically modified organisms are used to produce food, medicines, and other products.

GENE THERAPY

The directed modification of human DNA continues to be tested in medical applications. It also continues to raise ethical questions about modifying the human genome.

Evolutionary biology relies heavily on DNA sequence comparisons (Section 18.4). The escape of transgenic genes into the environment is an example of gene flow (17.8). In nature, plasmids transfer genes among bacteria (20.5). Genetic engineering returns in the context of phytoremediation (28.1) and vaccines (37.12). Research using engineered animals is explained in relevant chapters. Heat-loving bacteria return in Section 20.7.

15.1 Personal Genetic Testing

About 99 percent of your DNA is exactly the same as everyone else's. The shared part is what makes you human; the differences make you a unique member of the species. If you compared your DNA with your neighbor's, about 2.97 billion nucleotides of the two sequences would be identical; the remaining 30 million nonidentical nucleotides are sprinkled throughout your chromosomes. The sprinkling is not entirely random because some regions of DNA vary less than others. These conserved regions are of particular interest because they are the ones most likely to have an essential function. If a conserved sequence does vary among people, the variation tends to be in single nucleotides at a particular location. A base-pair substitution that is carried by a measurable percentage of a population, usually above 1 percent, is called a **single-nucleotide polymorphism**, or **SNP** (pronounced "snip").

Alleles of most genes differ by single nucleotides, and differences in alleles are the basis of the variation in human traits that makes each individual unique (Section 12.2). Thus, SNPs account for many of the differences in the way humans look, and they also have a lot to do with differences in the way our bodies work—how we age, respond to drugs, weather assaults by pathogens and toxins, and so on.

Consider the lipoprotein particles that carry fats and cholesterol through our bloodstreams (Section 3.6). These particles consist of variable amounts and types of lipids and proteins. One of these proteins is called apolipoprotein E, and it is encoded by the *APOE* gene. About one in four people carries an allele of this gene, *E4*, that has a cytosine instead of the more common thymine at a particular location in its sequence. The gene with this SNP encodes an apolipoprotein E with one amino acid substitution, an arginine instead of a cysteine in position 112. How this change affects the function of the protein is not yet clear, but we do know that having the *E4* allele increases one's risk of developing Alzheimer's disease later in life, particularly in people homozygous for it.

At this writing, about 73 million SNPs in human DNA have been identified, and that number grows every day. A few companies now offer to determine some of the SNPs you carry. The companies extract your DNA from the cells in a few drops of spit, then analyze it for SNPs. Such personalized genetic testing is now revolutionizing medicine, for example by allowing physicians to determine a patient's ability to respond to certain drugs before treatment begins. Cancer treatments are being tailored to fit the genetic makeup of individual patients and their tumor cells. People who discover they carry SNPs associated with a heightened risk of a medical condition are being encouraged to make lifestyle changes that could delay the condition's onset or prevent it entirely; preventive medical treatments based on these SNPs are becoming more common—and more mainstream (**FIGURE 15.1**).

single-nucleotide polymorphism (SNP) One-nucleotide DNA sequence variation carried by a measurable percentage of a population.

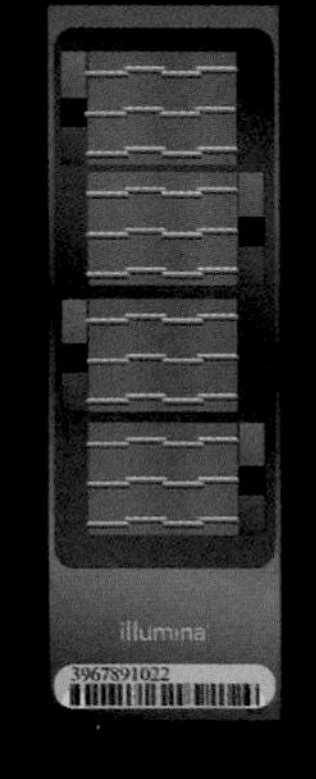

Only about 1 percent of the 3 billion bases in a person's DNA are unique to the individual. Personal genetic testing companies use chips like this one to analyze their customers' chromosomes for SNPs. This chip reveals which versions of 1,140,419 SNPs occur in the DNA of four individuals at a time.

FIGURE 15.1 Celebrity Angelina Jolie chose preventive treatment after genetic testing showed she had a very high risk of breast cancer. She carries a *BRCA1* mutation associated with an 87% lifetime risk of developing breast cancer. Even though Jolie did not yet have cancer, she underwent a double mastectomy, thereby reducing her risk of breast cancer to 5%.

15.2 Cloning DNA

✔ Researchers cut up DNA from different sources, then paste the resulting fragments together.

✔ Cloning vectors can carry foreign DNA into host cells.

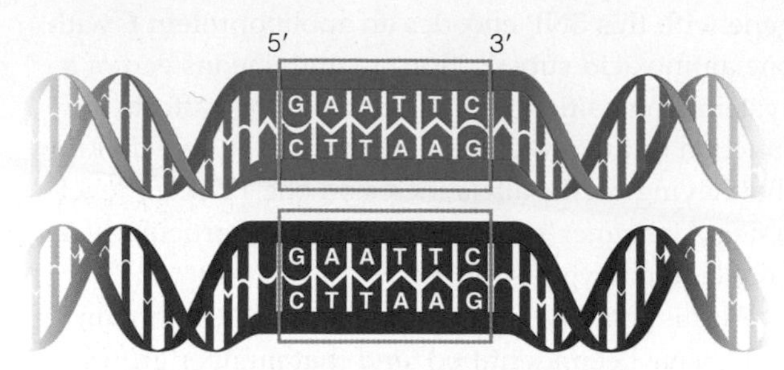

❶ The restriction enzyme *Eco*RI (named after the *E. coli* bacteria from which it was isolated) recognizes a specific base sequence (GAATTC) in DNA from two different sources.

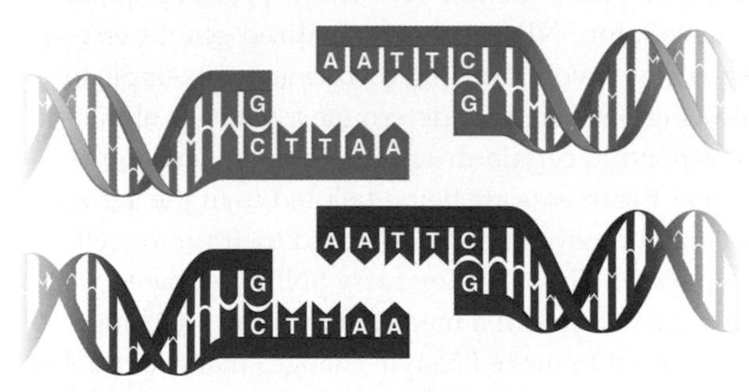

❷ The enzyme cuts the DNA into fragments. *Eco*RI leaves single-stranded tails ("sticky ends") where it cuts DNA.

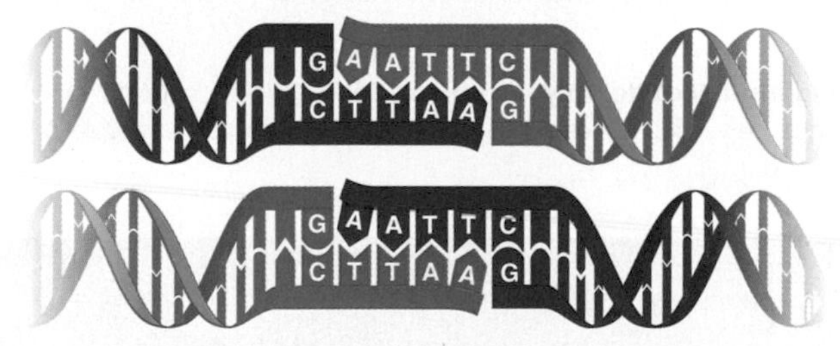

❸ When the DNA fragments from the two sources are mixed together, matching sticky ends base-pair with each other.

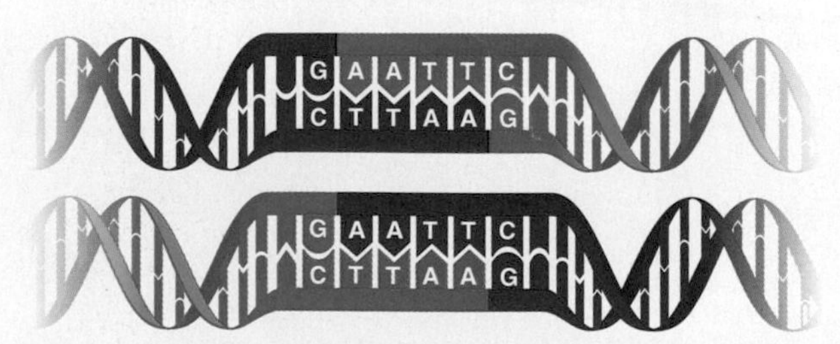

❹ DNA ligase joins the base-paired DNA fragments to produce molecules of recombinant DNA.

FIGURE 15.2 ▸**Animated** Making recombinant DNA.

FIGURE IT OUT Why did the enzyme cut both strands of DNA?

Answer: Because the recognition sequence occurs on both strands.

Cut and Paste

In the 1950s, excitement over the discovery of DNA's structure (Section 8.3) gave way to frustration: No one could determine the order of nucleotides in a molecule of DNA. Identifying a single base among thousands or millions of others turned out to be a huge technical hurdle. Research in a seemingly unrelated field yielded a solution when Werner Arber, Hamilton Smith, and their coworkers discovered how some bacteria resist infection by bacteriophage (Section 8.2). These bacteria have enzymes that chop up any injected viral DNA before it has a chance to integrate into the bacterial chromosome. The enzymes restrict viral growth; hence their name, restriction enzymes. A **restriction enzyme** cuts DNA wherever a specific nucleotide sequence occurs (**FIGURE 15.2** ❶).

The discovery of restriction enzymes allowed researchers to cut chromosomal DNA into manageable chunks. It also allowed them to combine DNA fragments from different organisms. How? Many restriction enzymes leave single-stranded tails on DNA fragments ❷. Researchers realized that complementary tails will base-pair, regardless of the source of DNA ❸. The tails are called "sticky ends" because two DNA fragments stick together when their matching tails base-pair. The enzyme DNA ligase (Section 8.5) can be used to seal the gaps between base-paired sticky ends, so continuous DNA strands form ❹. Thus, using appropriate restriction enzymes and DNA ligase, researchers can cut and paste DNA from different sources. The result, a hybrid molecule that consists of genetic material from two or more organisms, is called **recombinant DNA**.

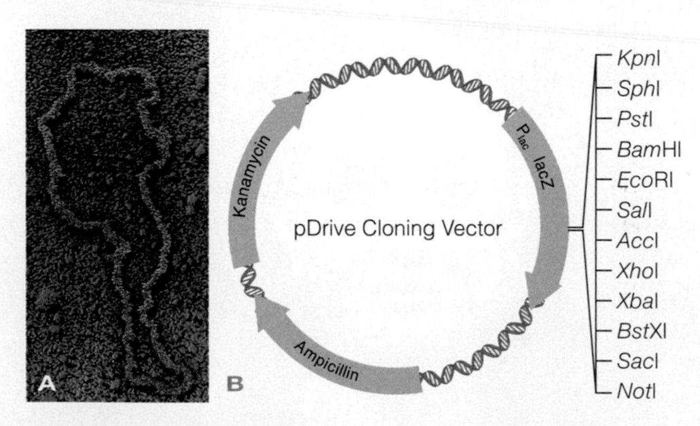

FIGURE 15.3 Plasmid cloning vectors. (**A**) Micrograph of a plasmid. (**B**) Commercial plasmid cloning vector. Restriction enzyme recognition sequences are indicated (right) by the name of the enzyme that cuts them. Researchers insert foreign DNA into the vector at these sequences. Bacterial genes (gold) help them identify host cells that take up a vector with inserted DNA. This vector carries two antibiotic resistance genes and the *lac* operon (Section 10.5).

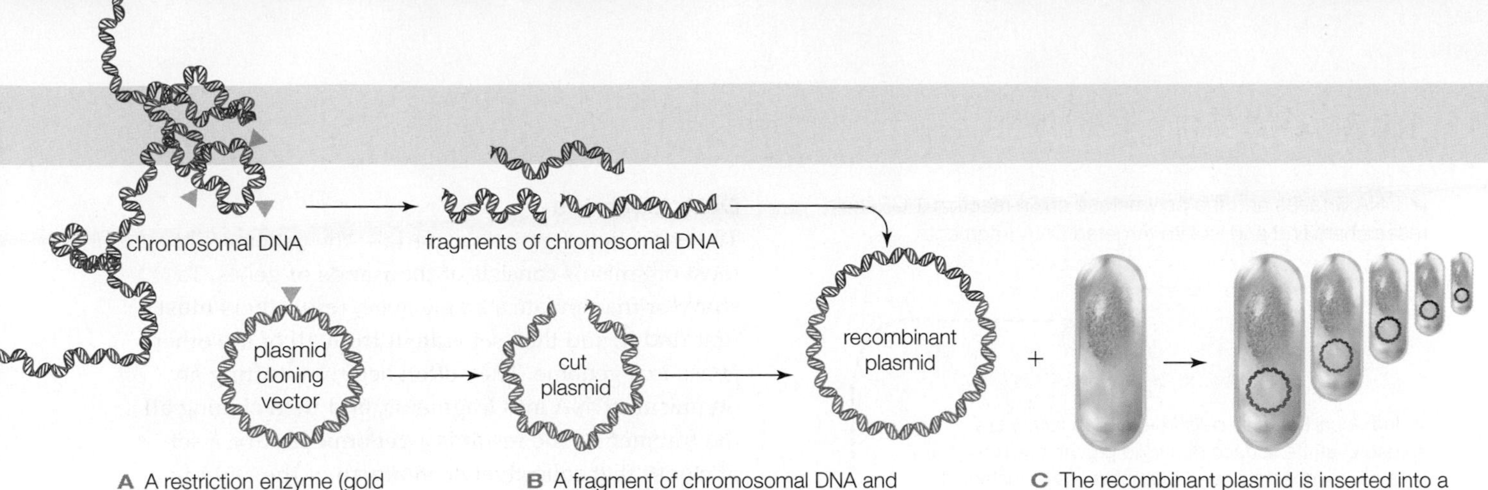

A A restriction enzyme (gold triangles) cuts a specific nucleotide sequence in chromosomal DNA and in a plasmid cloning vector.

B A fragment of chromosomal DNA and the cut plasmid base-pair at their sticky ends. DNA ligase joins the two pieces of DNA, so a recombinant plasmid forms.

C The recombinant plasmid is inserted into a host bacterial cell. When the cell reproduces, it copies the plasmid along with its chromosome. Each descendant cell receives a plasmid.

FIGURE 15.4 ▶Animated An example of cloning. Here, a fragment of chromosomal DNA is inserted into a plasmid.

Making recombinant DNA is the first step in **DNA cloning**, a set of laboratory methods that uses living cells to mass-produce specific DNA fragments. Researchers clone a fragment of DNA by inserting it into a **cloning vector**, which is a molecule that can carry foreign DNA into host cells. Bacterial plasmids (Section 4.4) may be used as cloning vectors (**FIGURE 15.3**). A bacterium copies all of its DNA before it divides, so its offspring inherit plasmids along with chromosomes. If a plasmid carries a fragment of foreign DNA, that fragment gets copied and distributed to descendant cells along with the plasmid DNA (**FIGURE 15.4**).

A host cell into which a cloning vector has been inserted can be grown in the laboratory (cultured) to yield a huge population of genetically identical cells, or clones (Section 8.7). Each clone contains a copy of the vector and the inserted DNA fragment. The hosted DNA fragment can be harvested in large quantities from the clones.

cDNA Cloning

Remember from Section 9.3 that eukaryotic DNA contains introns. Unless you are a eukaryotic cell, it is not very easy to determine which parts of eukaryotic DNA encode gene products. Thus, researchers who study gene expression in eukaryotes often start with mature mRNA. Post-transcriptional processing removes introns from an mRNA, so just the coding sequence remains.

An mRNA cannot be cut with restriction enzymes or pasted with DNA ligase, because these enzymes work only on double-stranded DNA. Thus, cloning with mRNA requires **reverse transcriptase**, a replication enzyme that uses an RNA template to assemble a strand of complementary DNA, or **cDNA**:

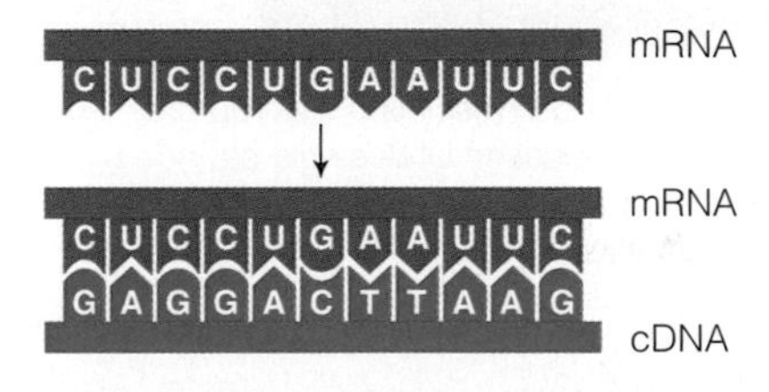

DNA polymerase is used to copy the cDNA into a second strand of DNA. The outcome is a double-stranded DNA version of the original mRNA:

Like any other double-stranded DNA, this fragment may be cut with restriction enzymes and pasted into a cloning vector using DNA ligase.

cDNA Complementary strand of DNA synthesized from an RNA template by the enzyme reverse transcriptase.
cloning vector A DNA molecule that can accept foreign DNA and be replicated inside a host cell.
DNA cloning Set of methods that uses living cells to mass-produce targeted DNA fragments.
recombinant DNA A DNA molecule that contains genetic material from more than one organism.
restriction enzyme Type of enzyme that cuts DNA at a specific nucleotide sequence.
reverse transcriptase An enzyme that uses mRNA as a template to make a strand of cDNA.

TAKE-HOME MESSAGE 15.2

What is DNA cloning?

✔ DNA cloning uses living cells to mass-produce targeted DNA fragments. Restriction enzymes cut DNA into fragments, then DNA ligase seals the fragments into cloning vectors. Recombinant DNA molecules result.

✔ A cloning vector that holds foreign DNA can be introduced into a living cell. When the host cell divides, it gives rise to huge populations of genetically identical cells (clones), each with a copy of the foreign DNA.

15.3 Isolating Genes

✔ DNA libraries and the polymerase chain reaction (PCR) help researchers find and isolate targeted DNA fragments.

A Individual bacterial cells from a DNA library are spread over the surface of a solid growth medium. The cells divide repeatedly and form colonies—clusters of millions of genetically identical descendant cells.

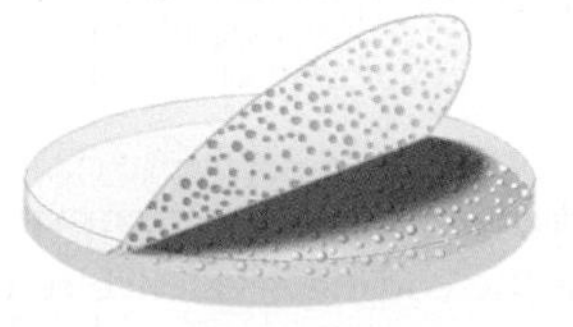

B Special paper is pressed onto the surface of the growth medium. Some cells from each colony stick to the paper.

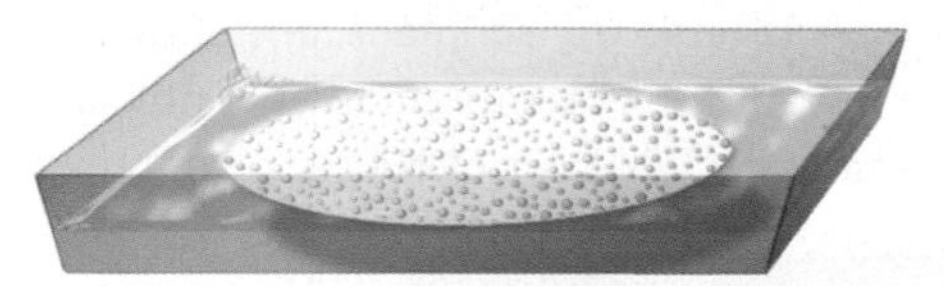

C The paper is soaked in a solution that ruptures the cells and makes the released DNA single-stranded. The DNA clings to the paper in spots mirroring the distribution of colonies.

D A radioactive probe is added to the liquid bathing the paper. The probe hybridizes with any spot of DNA that contains a complementary sequence.

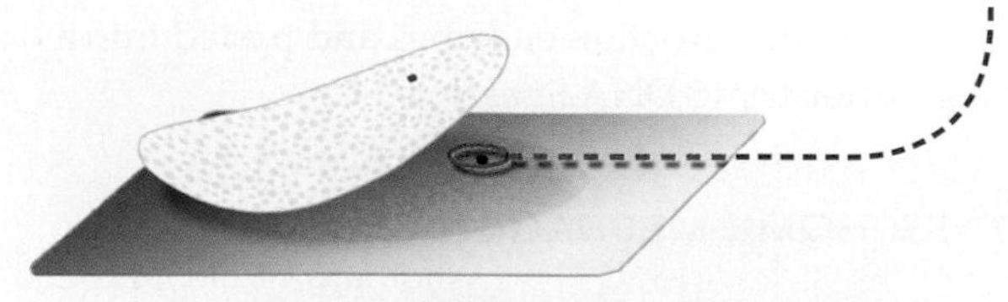

E The paper is pressed against x-ray film. The radioactive probe darkens the film in a spot where it has hybridized. The spot's position is compared to the positions of the original bacterial colonies. Cells from the colony that corresponds to the spot are cultured, and their DNA is harvested.

FIGURE 15.5 ▶**Animated** Nucleic acid hybridization. In this example, a radioactive probe helps identify a colony of bacteria that contain a targeted fragment of DNA.

DNA Libraries

The entire set of genetic material—the **genome**—of most organisms consists of thousands of genes. To study or manipulate a single gene, researchers must first find it, and then separate it from all of the other genes in a genome. They often begin by cutting an organism's DNA into fragments, and then cloning all the fragments. The result is a genomic library, a set of clones that collectively contain all of the DNA in a genome. Researchers may also harvest mRNA, make cDNA copies of it, and then clone the cDNA. The resulting cDNA library represents only those genes being expressed at the time the mRNA was harvested.

Genomic and cDNA libraries are **DNA libraries**, sets of cells that host various cloned DNA fragments. In such libraries, a clone that contains a targeted DNA fragment of interest is mixed up with thousands or millions of others that do not—a needle in a genetic haystack. One way to find that clone among the others involves the use of a **probe**, which is a fragment of DNA or RNA labeled with a tracer (Section 2.2).

For example, to find a targeted gene, researchers may use radioactive nucleotides to synthesize a short strand of DNA complementary in sequence to a similar gene. Because the nucleotide sequences of the probe and the gene are complementary, the two can hybridize. (Remember from Section 8.5 that nucleic acid hybridization is the establishment of base pairing between nucleic acid strands.) When the probe is mixed with DNA from a library, it will hybridize with the gene, but not with other DNA (**FIGURE 15.5**). Researchers can pinpoint a cell that hosts the gene by detecting the label on the probe. That cell is isolated and cultured, and DNA can be extracted in bulk from the cultured cells for research or other purposes.

PCR

The **polymerase chain reaction** (**PCR**) is a technique used to mass-produce copies of a particular section of DNA without having to clone it in living cells (**FIGURE 15.6**). The reaction can transform a needle in a haystack—that one-in-a-million fragment of DNA—into a huge stack of needles with a little hay in it.

The starting material for PCR is any sample of DNA with at least one molecule of a targeted sequence. It might be extracted from a mixture of 10 million different clones, a sperm, a hair left at a crime scene, or a mummy—essentially any sample that has DNA in it.

The PCR reaction is similar to DNA replication (Section 8.5). It requires two primers; each base-pairs with one end of the section of DNA to be amplified,

FIGURE 15.6 ▸Animated Two rounds of PCR. Each cycle of this reaction can double the number of copies of a targeted sequence of DNA. Thirty cycles can make a billion copies.

targeted section

❶ DNA template (blue) is mixed with primers (pink), nucleotides, and heat-tolerant *Taq* DNA polymerase.

❷ When the mixture is heated, the double-stranded DNA separates into single strands. When the mixture is cooled, some of the primers base-pair with the DNA at opposite ends of the targeted sequence.

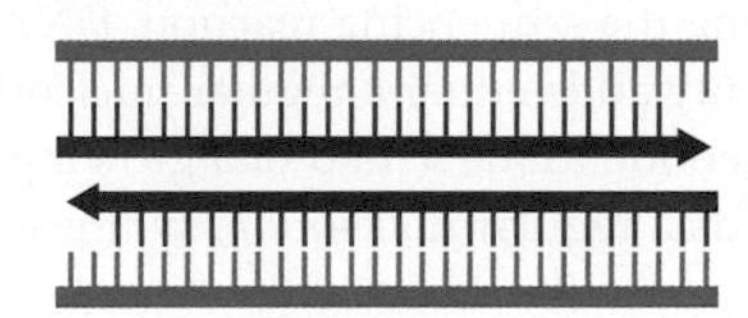

❸ *Taq* polymerase begins DNA synthesis at the primers, so it produces complementary strands of the targeted DNA sequence.

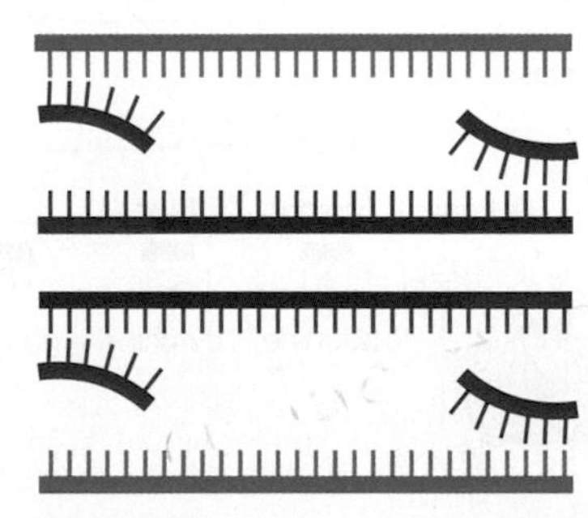

❹ The mixture is heated again, so all double-stranded DNA separates into single strands. When it is cooled, primers base-pair with the targeted sequence in the original template DNA and in the new DNA strands.

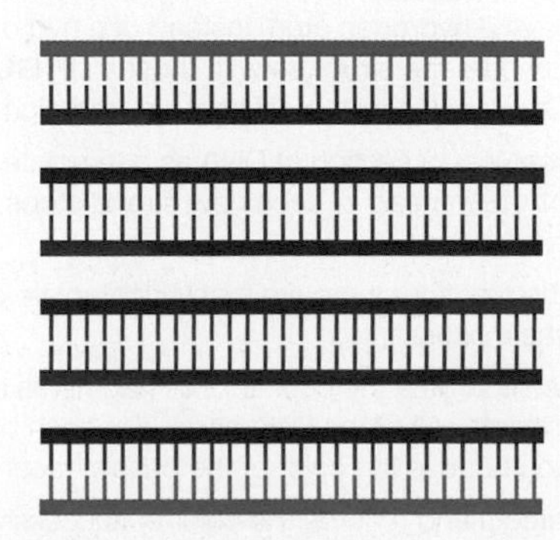

❺ Each cycle of heating and cooling can double the number of copies of the targeted DNA section.

or mass-produced ❶. Researchers mix these primers with the starting (template) DNA, nucleotides, and DNA polymerase, then expose the reaction mixture to repeated cycles of high and low temperatures. A few seconds at high temperature disrupts the hydrogen bonds that hold the two strands of a DNA double helix together (Section 8.3), so every molecule of DNA unwinds and becomes single-stranded. As the temperature of the reaction mixture is lowered, the single DNA strands hybridize with the primers ❷.

The DNA polymerases of most organisms denature at the high temperature required to separate DNA strands. The kind that is used in PCR reactions, *Taq* polymerase, is from *Thermus aquaticus*. This bacterial species lives in hot springs and hydrothermal vents, so its DNA polymerase necessarily tolerates heat. *Taq* polymerase, like other DNA polymerases, recognizes hybridized primers as places to start DNA synthesis ❸. Synthesis proceeds along the template strand until the temperature rises and the DNA separates into single strands ❹. The newly synthesized DNA is a copy of the targeted section. When the mixture is cooled, the primers rehybridize, and DNA synthesis begins again. Each cycle of heating and cooling takes only a few minutes, but it can double the number of copies of the targeted section of DNA ❺. Thirty PCR cycles may amplify that number a billionfold.

DNA library Collection of cells that host different fragments of foreign DNA, often representing an organism's entire genome.
genome An organism's complete set of genetic material.
polymerase chain reaction (PCR) Method that rapidly generates many copies of a specific section of DNA.
probe Short fragment of DNA designed to hybridize with a nucleotide sequence of interest and labeled with a tracer.

TAKE-HOME MESSAGE 15.3

How do researchers study one gene in the context of many?

✔ A DNA library can be made from cDNA or genomic DNA. Probes are used to identify one clone that hosts a targeted DNA fragment among many other clones in a DNA library.

✔ PCR quickly mass-produces copies of a targeted section of DNA.

15.4 DNA Sequencing

✔ DNA sequencing reveals the order of nucleotide bases in a section of DNA.

Once a fragment of DNA has been isolated (for example by cloning or PCR), researchers can use a technique called **sequencing** to determine the order of nucleotides in it. The most common method uses DNA polymerase (Section 8.5). This enzyme is mixed with a primer, nucleotides, and the DNA to be sequenced (the template). Starting at the primer, the polymerase joins the nucleotides into a new strand of DNA, in the order dictated by the sequence of the template (**FIGURE 15.7**).

The sequencing reaction mixture includes all four kinds of nucleotides, and also four kinds of dideoxynucleotides—DNA nucleotides that lack the hydroxyl group on their 3′ carbon ❶. Each dideoxynucleotide base (A, C, G, or T) is labeled with a different colored pigment. During the sequencing reaction, DNA polymerase randomly adds either a regular nucleotide or a dideoxynucleotide to the 3′ end of a growing DNA strand. If it adds a regular nucleotide, synthesis can continue because the 3′ carbon of the strand will have a hydroxyl group on it. If it adds a dideoxynucleotide, the 3′ carbon will not have a hydroxyl group, so synthesis of the strand ends there ❷. (Remember that DNA synthesis proceeds only in the 5′ to 3′ direction.)

The reaction produces millions of DNA fragments of different lengths—incomplete, complementary copies of the starting DNA ❸. Each fragment of a given length ends with the same dideoxynucleotide base. For example, if the tenth base in the template DNA was thymine, then any newly synthesized fragment that is 10 nucleotides long ends with an adenine.

The DNA fragments are then separated by length. Using a technique called **electrophoresis**, an electric field pulls the fragments through a semisolid gel. Fragments of different sizes move through the gel at different rates. The shorter the fragment, the faster it moves, because shorter fragments slip through the tangled molecules of the gel faster than longer fragments do. All fragments of the same length move through the gel at the same speed, so they gather into bands. Because all fragments in a given band are the same length, all have the same dideoxynucleotide base at

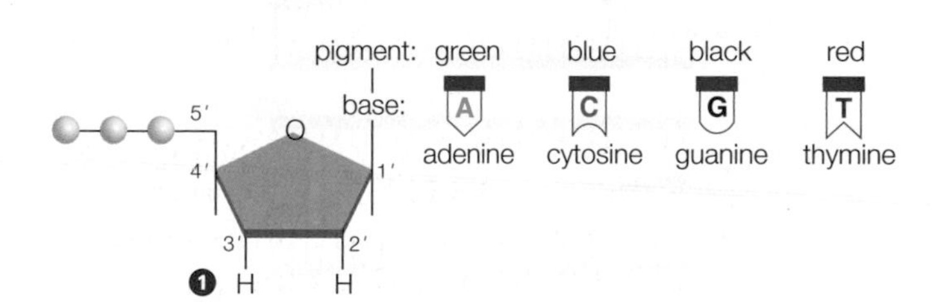

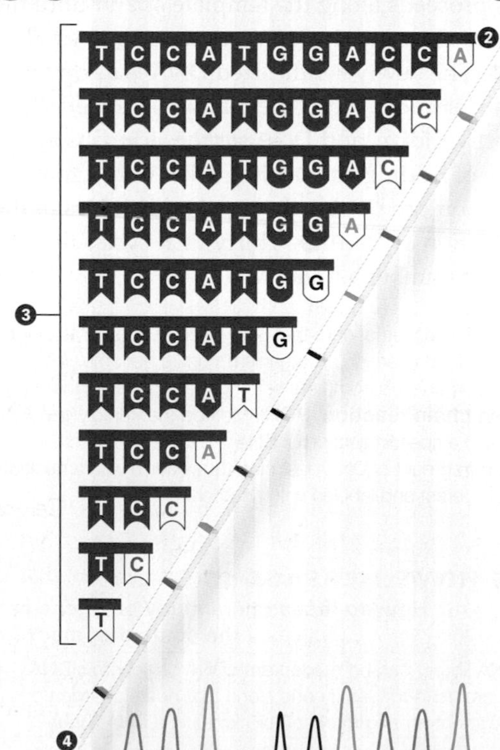

FIGURE 15.7 ▸**Animated** A method of sequencing DNA, in which DNA polymerase is used to incompletely replicate a section of DNA.

❶ This sequencing method depends on dideoxynucleotides, which are nucleotides that have a hydrogen atom instead of a hydroxyl group on their 3′ carbon (compare the structure with those in **FIGURE 9.2**). Each base (G, A, T, or C) is labeled with a different colored pigment.

❷ DNA polymerase uses a section of DNA as a template to synthesize new strands of DNA. Synthesis of each new strand stops when a dideoxynucleotide is added.

❸ At the end of the reaction, there are many incomplete copies of the template DNA in the mixture.

❹ Electrophoresis separates the copied DNA fragments into bands according to their length. All of the DNA strands in each band end with the same base, so each band is the color of the base's tracer pigment.

❺ A computer detects and records the color of successive bands on the gel (see **FIGURE 15.8** for an example). The order of colors of the bands represents the sequence of the template DNA.

their ends, and the pigment labels now impart distinct colors to the bands ❹. Each color designates one of the four bases, so the order of colored bands in the gel represents the DNA sequence ❺.

The Human Genome Project

The sequencing method we have just described was invented in 1975. Ten years later, it had become so routine that scientists began to consider sequencing the entire human genome—all 3 billion nucleotides. Proponents of the idea said it could provide huge payoffs for medicine and research. Opponents said this daunting task would divert attention and funding from more urgent research. It would require 50 years to sequence the human genome given the techniques of the time. However, the techniques continued to improve rapidly, and with each improvement more nucleotides could be sequenced in less time. Automated (robotic) DNA sequencing and PCR had just been invented. Both were still too cumbersome and expensive to be useful in routine applications, but they would not be so for long. Waiting for faster, cheaper technologies seemed the most efficient way to sequence the genome, but just how fast did they need to be before the project should begin?

A few privately owned companies decided not to wait, and started sequencing. One of them intended to determine the genome sequence in order to patent it. The idea of patenting the human genome provoked widespread outrage, but it also spurred commitments in the public sector. In 1988, the National Institutes of Health (NIH) essentially took over the project by hiring James Watson (of DNA structure fame) to head an official Human Genome Project, and providing $200 million per year to fund it. A partnership formed between the NIH and international institutions that were sequencing different parts of the genome. Watson set aside 3 percent of the funding for studies of ethical and social issues arising from the work. He later resigned over a patent disagreement, and geneticist Francis Collins took his place.

Amid ongoing squabbles over patent issues, Celera Genomics formed in 1998. With biologist Craig Venter at its helm, the company intended to commercialize human genetic information. Celera invented faster techniques for sequencing genomic DNA, because the first to have the complete sequence had a legal basis for patenting it. The competition motivated the international partnership to accelerate its efforts. Then, in 2000, U.S. President Bill Clinton and British Prime Minister Tony Blair jointly declared that the sequence of the human genome could not be patented. Celera kept sequencing anyway, and, in 2001, the competing governmental and corporate teams published about 90 percent of the sequence. In 2003, fifty years after the discovery of the structure of DNA, the sequence of the human genome was officially completed (**FIGURE 15.8**). The technology has improved so much that a human genome can now be sequenced in a few hours (**FIGURE 15.9**).

electrophoresis Technique that separates DNA fragments by size.
sequencing Method of determining the order of nucleotides in DNA.

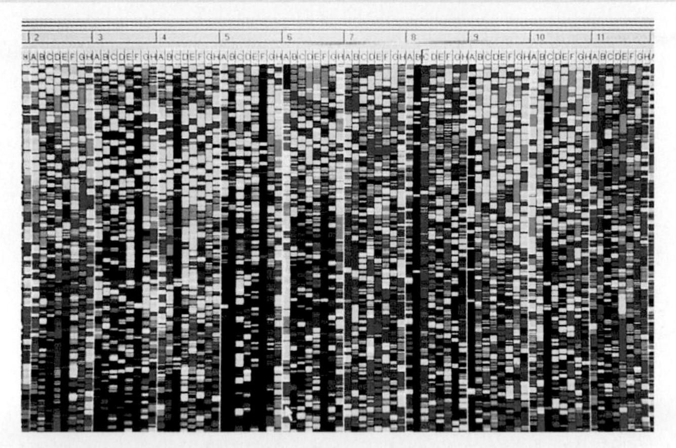

FIGURE 15.8 Human DNA sequence data. The order of colors in each vertical lane reveals one part of the DNA sequence.

FIGURE 15.9 Today's automated DNA sequencing machines can sequence an individual's entire genome in 2–4 hours, for a cost of about $1,000.

TAKE-HOME MESSAGE 15.4

How is DNA sequence determined?

✔ In the most common method of DNA sequencing, a strand of DNA is partially replicated. Electrophoresis is used to separate the resulting fragments by length.

✔ Improved sequencing techniques and worldwide efforts allowed the human genome sequence to be determined.

CREDITS: (8) Patrick Landmann/Science Source; (9) © Michelle McLoughlin/Reuters/Corbis.

15.5 Genomics

✔ Comparing the human genome sequence with that of other species is helping us understand how the human body works.

✔ Unique sequences of genomic DNA can be used to distinguish an individual from all others.

Despite our ability to determine the sequence of an individual's genome, it will be a long time before we understand all the information coded within that sequence. The human genome contains a massive amount of seemingly cryptic data. We can decipher some of this data by comparing genomes of different species, the premise being that all organisms are descended from shared ancestors, so all genomes are related to some extent. We see evidence of such genetic relationships simply by comparing the raw sequence data, which, in some regions, is extremely similar across many species (**FIGURE 15.10**).

The study of genomes is called **genomics**. This broad field encompasses whole-genome comparisons, structural analysis of gene products, and surveys of small-scale variations in sequence. Genomics is providing powerful insights into evolution. For example, comparing primate genomes revealed a change in chromosome structure that occurred during the evolution of our species (Section 14.5). Comparing genomes also showed us that structural changes in chromosomes are not entirely random; some specific alterations are much more common than others. This is because chromosomes that break tend to do so in particular spots. Human, mouse, rat, cow, pig, dog, cat, and horse chromosomes have undergone several translocations at these breakage hot spots during evolution. In humans, chromosome abnormalities that contribute to the progression of cancer also occur at the very same breakage hot spots.

Comparing the coding regions of genomes offers medical benefits. We have learned the function of many human genes by studying their counterpart genes in other species. For instance, researchers comparing human and mouse genomes discovered a human version of a mouse gene, *APOA5*, that encodes a lipoprotein. Mice with an *APOA5* knockout have four times the normal level of triglycerides in their blood. The researchers then looked for—and found—a correlation between *APOA5* mutations and high triglyceride levels in humans. High triglycerides are a risk factor for coronary artery disease.

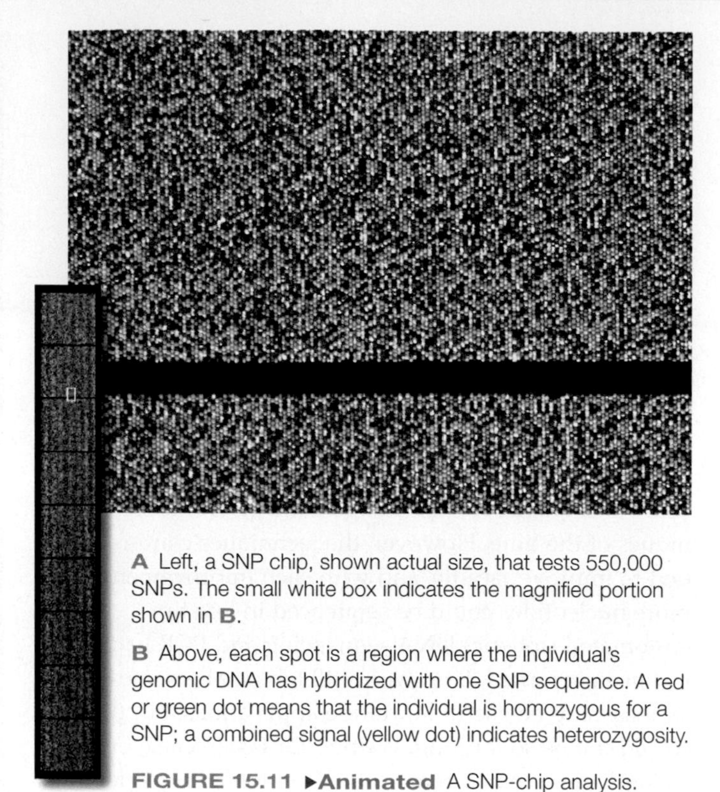

A Left, a SNP chip, shown actual size, that tests 550,000 SNPs. The small white box indicates the magnified portion shown in **B**.

B Above, each spot is a region where the individual's genomic DNA has hybridized with one SNP sequence. A red or green dot means that the individual is homozygous for a SNP; a combined signal (yellow dot) indicates heterozygosity.

FIGURE 15.11 ▶**Animated** A SNP-chip analysis.

DNA Profiling

As you learned in Section 15.1, about 99% of the human genome sequence is identical in every member of the species. The differences you carry in your DNA make you unique. In fact, those differences are so unique that they can be used to identify you. Identifying an individual by his or her DNA is a method called **DNA profiling**. One DNA profiling method uses SNP-chips (**FIGURE 15.11**). A SNP-chip

```
758 GATAATCCTGTTTTGAACAAAAGGTCAAATTGCTGAATAGAAA-GTCTTGATTAACTAAAAGATGTACAAAGTGGAATTA 836  Human
752 GATAATCCTGTTTTGAACAAAAGGTCAAATTGCTGAATAGAAA-GTCTTGATTAACTAAAAGATGTACAAAGTGGAATTA 830  Mouse
751 GATAATCCTGTTTTGAACAAAAGGTCAAATTGCTGAATAGAAA-GTCTTGATTAACTAAAAGATGTACAAAGTGGAATTA 829  Rat
754 GATAATCCTGTTTTGAACAAAAGGTCAAATTGCTGAATAGAAA-GTCTTGATTAACTAAAAGATGTACAAAGTGGAATTA 832  Dog
782 GATAATCCTGTTTTGAACAAAAGGTCAAATTGCTGAATAGAAA-GTCTTGATTAACTAAAAGATGTACAAAGTGGAATTA 860  Chicken
758 GATAATCCTGTTTTGAACAAAAGGTCAAATTGCTGAATAGAAA-GTCTTGATTAAGTAAAAGATGTACAAAGTGGAATTA 836  Frog
823 GATAATCCTGTTTTGAACAAAAGGTCAGATTGCTGAATAGAAAAGGCTTGATTAAAGCAGAGATGTACAAAGTGGACGCA 902  Zebrafish
763 GATAATCCTGTTTTGAACAAAAGGTCAAATTGTTGAATAGAGACGCTTTGATAAAGCGGAGGAGGTACAAAGTGGGACC- 841  Pufferfish
```

FIGURE 15.10 Genomic DNA alignment. This is a region of the gene for a DNA polymerase. Nucleotides that differ from those in the human sequence are highlighted. The chance that any two of these sequences would randomly match is about 1 in 10^{46}.

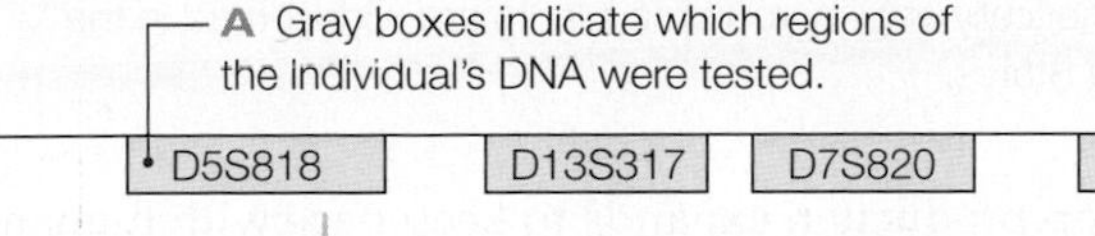

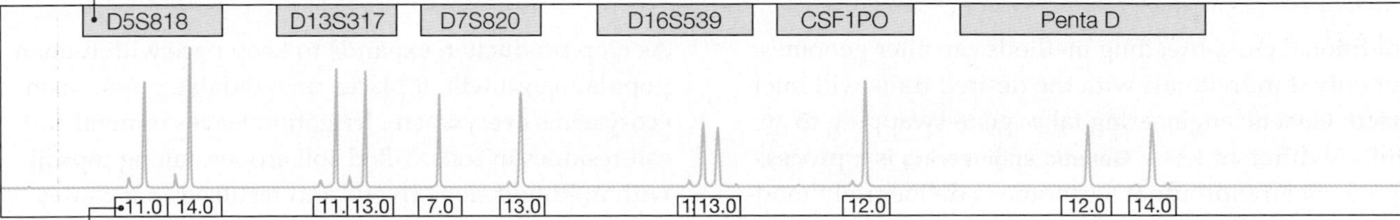

B The number of repeats is shown in a box below each peak. A peak's location on the x-axis corresponds to the length of the DNA fragment amplified (a measure of the number of repeats). Peak size reflects the amount of DNA.

FIGURE 15.12 ▶**Animated** An individual's (partial) short tandem repeat profile. Remember, human body cells are diploid. Double peaks appear on a profile when the two members of a chromosome pair carry a different number of repeats.

is a tiny glass plate with microscopic spots of DNA stamped on it. The DNA sample in each spot is a short, synthetic single strand with a unique SNP sequence. When an individual's genomic DNA is washed over a SNP-chip, it hybridizes only with DNA spots that have a matching SNP sequence. Probes reveal where the genomic DNA has hybridized—and which of the SNPs are carried by the individual.

Another method of DNA profiling involves analysis of short tandem repeats in an individual's chromosomes (Section 13.7). Short tandem repeats usually occur in the same location in human chromosomes, but the number of times a sequence is repeated in each location differs among individuals. For example, one person's DNA may have fifteen repeats of the nucleotides TTTTC at a certain spot on one chromosome. Another person's DNA may have four repeats of this sequence in the same location. Short tandem repeats slip spontaneously into DNA during replication, and their numbers grow or shrink over generations. Unless two people are identical twins, the chance that they have identical short tandem repeats in even three regions of DNA is 1 in a quintillion (10^{18}), which is far more than the number of people who have ever lived. Thus, an individual's array of short tandem repeats is, for all practical purposes, unique.

Analyzing a person's short tandem repeats begins with PCR, which is used to copy ten to thirteen particular regions of chromosomal DNA known to have repeats. The lengths of the copied DNA fragments differ among most individuals, because the number of tandem repeats in those regions also differs. Thus, electrophoresis can be used to reveal an individual's unique array of short tandem repeats (**FIGURE 15.12**).

Short tandem repeat analysis will soon be replaced by full genome sequencing, but for now it continues to be a common DNA profiling method. Geneticists compare short tandem repeats on Y chromosomes to determine relationships among male relatives, and to trace an individual's ethnic heritage. They also track mutations that accumulate in populations over time by comparing DNA profiles of living humans with those of ancient ones. Such studies are allowing us to reconstruct population dispersals that happened in the ancient past.

Short tandem repeat profiles are routinely used to resolve kinship disputes, and as evidence in criminal cases. Within the context of a criminal or forensic investigation, DNA profiling is called DNA fingerprinting. As of January 2014, the database of DNA fingerprints maintained by the Federal Bureau of Investigation (the FBI) contained the short tandem repeat profiles of 10.7 million convicted offenders, and had been used in more than 200,000 criminal investigations. DNA fingerprinting is also used to identify human remains, including the individuals who died in the World Trade Center on September 11, 2001.

TAKE-HOME MESSAGE 15.5

How do we use what researchers discover about genomes?

✔ Analysis of the human genome sequence is yielding new information about our genes and how they work.

✔ DNA profiling identifies individuals by the unique parts of their DNA.

DNA profiling Identifying an individual by analyzing the unique parts of his or her DNA.
genomics The study of genomes.

15.6 Genetic Engineering

✔ Bacteria and yeast are the most common genetically engineered organisms.

Traditional cross-breeding methods can alter genomes, but only if individuals with the desired traits will interbreed. Genetic engineering takes gene-swapping to an entirely different level. **Genetic engineering** is a process by which an individual's genome is deliberately modified. A gene from one species may be transferred to another to produce an organism that is **transgenic**, or a gene may be altered and reinserted into an individual of the same species. Both methods yield a **genetically modified organism**, or **GMO**.

The most common GMOs are yeast and bacteria (**FIGURE 15.13**). Both types of cells have the metabolic machinery to make complex organic molecules, and they are easily modified to produce, for example, medically important proteins. People with diabetes were among the first beneficiaries of such organisms. Insulin for their injections was once extracted from animals, but it provoked an allergic reaction in some people. Human insulin, which does not provoke allergic reactions, has been produced by transgenic *E. coli* since 1982. Slight modifications of the gene have yielded fast-acting and slow-release forms of human insulin.

Genetically modified microorganisms also make proteins used in foods. For example, enzymes produced by modified microorganisms improve the taste and clarity of beer and fruit juice, slow bread staling, or modify certain fats. Cheese is traditionally made with an enzyme, chymosin, extracted from calf stomachs. Today, almost all cheese is made with calf chymosin produced by transgenic fungi.

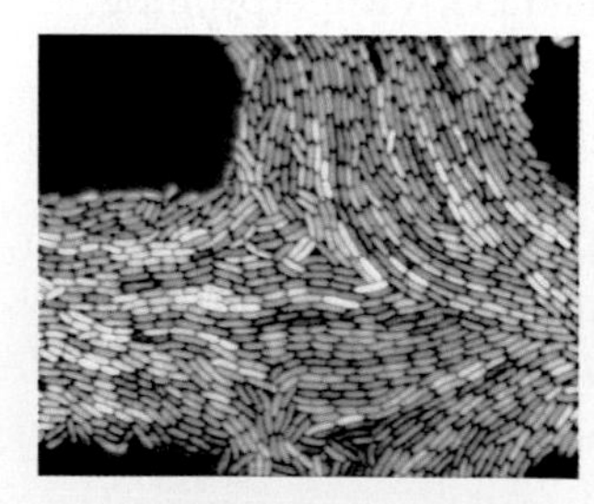

FIGURE 15.13 *E. coli* bacteria transgenic for a fluorescent jellyfish protein. Variation in fluorescence among the genetically identical cells reveals differences in gene expression that may help us understand why some bacteria become dangerously resistant to antibiotics, and others do not.

TAKE-HOME MESSAGE 15.6

What is genetic engineering?

✔ Genetic engineering is the deliberate alteration of an individual's genome, and it results in a genetically modified organism (GMO).

✔ A transgenic organism carries a gene from a different species. Transgenic bacteria and yeast are used in research, medicine, and industry.

15.7 Designer Plants

✔ Genetically engineered crop plants are widespread in the United States.

As crop production expands to keep pace with human population growth, it places unavoidable pressure on ecosystems everywhere. Irrigation leaves mineral and salt residues in soils. Tilled soil erodes, taking topsoil with it. Runoff clogs rivers, and fertilizer in it causes algae to grow so fast that fish suffocate. Pesticides can be harmful to humans and other animals, including beneficial insects such as bees.

Pressured to produce more food at lower cost and with less damage to the environment, many farmers have begun to rely on genetically modified crop plants. Genes can be introduced into plant cells by way of electric or chemical shocks, by blasting them with microscopic DNA-coated pellets, or by using *Agrobacterium tumefaciens* bacteria. *A. tumefaciens* carries a plasmid with genes that cause tumors to form on infected plants; hence the name Ti plasmid (for Tumor-inducing). Researchers replace the tumor-inducing genes with foreign or modified genes, then use the plasmid as a vector to deliver the desired genes into plant cells. Whole plants can be grown from plant cells that integrate a recombinant plasmid into their chromosomes (**FIGURE 15.14**).

Many genetically modified crops carry genes that impart resistance to devastating plant diseases. Others offer improved yields. GMO crops such as Bt corn and soy help farmers use smaller amounts of toxic pesticides. Organic farmers often spray their crops with spores of Bt (*Bacillus thuringiensis*), a bacterial species that makes a protein toxic only to some insect larvae. Researchers transferred the gene encoding the Bt protein into plants. The engineered plants produce the Bt protein, but otherwise they are essentially identical with unmodified plants. Larvae die shortly after eating their first and only GMO meal. Farmers can use much less pesticide on crops that make their own (**FIGURE 15.15**).

Transgenic crop plants are also being developed for impoverished regions of the world. Genes that confer drought tolerance, insect resistance, and enhanced nutritional value are being introduced into plants such as corn, rice, beans, sugarcane, cassava, cowpeas, banana, and wheat. The resulting GMO crops may help

genetic engineering Process by which deliberate changes are introduced into an individual's genome.
genetically modified organism (GMO) Organism whose genome has been modified by genetic engineering.
transgenic Refers to a genetically modified organism that carries a gene from a different species.

people in these regions who rely mainly on agriculture for food and income.

Genetic modifications can make food plants more nutritious. For example, rice plants have been engineered to make β-carotene, an orange photosynthetic pigment that is remodeled by cells of the small intestine into vitamin A. These rice plants carry two genes in the β-carotene synthesis pathway: one from corn, the other from bacteria. One cup of the engineered rice seeds—grains of Golden Rice—has enough β-carotene to satisfy a child's daily need for vitamin A.

The USDA Animal and Plant Health Inspection Service (APHIS) regulates the introduction of GMOs into the environment. At this writing, APHIS has deregulated ninety-two crop plants, which means the plants are approved for unrestricted use in the United States. Worldwide, more than 330 million acres are currently planted in GMO crops, the majority of which are corn, sorghum, cotton, soy, canola, and alfalfa engineered for resistance to the herbicide glyphosate. Rather than tilling the soil to control weeds, farmers can spray their fields with glyphosate, which kills the weeds but not the engineered crops.

Crops genetically engineered to resist glyphosate have been used in conjunction with the herbicide since the mid 1970s. Genes that confer glyphosate resistance are now appearing in weeds and other wild plants, as well as in nonengineered crops—which means that recombinant DNA can (and does) escape into the environment. Glyphosate resistance genes are probably being transferred from transgenic plants to nontransgenic ones via pollen carried by wind or insects.

Many people are opposed to any GMO. Some worry that our ability to tinker with genetics has surpassed our ability to understand the impact of the tinkering. Controversy raised by GMO use invites you to read the research and form your own opinions. The alternative is to be swayed by media hype (the term "Frankenfood," for instance), or by reports from possibly biased sources (such as herbicide manufacturers).

TAKE-HOME MESSAGE 15.7

Are genetically modified plants used as commercial crops?

✔ Genetically modified crop plants can help farmers be more productive while reducing overall costs.

✔ The widespread use of GMO crops has had unintended environmental effects. Herbicide resistant weeds are now common, and recombinant genes have spread to wild plants and non-GMO crops.

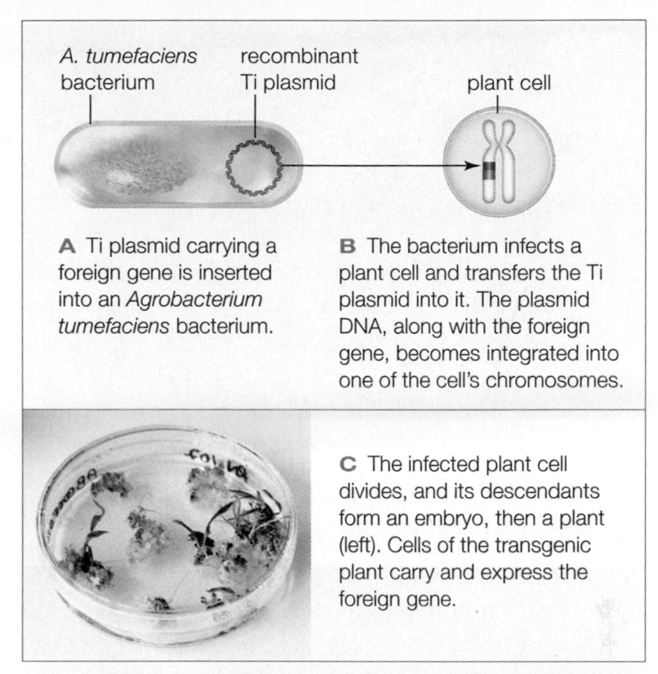

A Ti plasmid carrying a foreign gene is inserted into an *Agrobacterium tumefaciens* bacterium.

B The bacterium infects a plant cell and transfers the Ti plasmid into it. The plasmid DNA, along with the foreign gene, becomes integrated into one of the cell's chromosomes.

C The infected plant cell divides, and its descendants form an embryo, then a plant (left). Cells of the transgenic plant carry and express the foreign gene.

FIGURE 15.14 **▸Animated** Using the Ti plasmid to make a transgenic plant.

FIGURE 15.15 Farmers can use much less pesticide on crops that make their own. The genetically modified plants that produced the row of corn on the top carry a gene from the bacteria *Bacillus thuringiensis* (Bt) that conferred insect resistance. Compare the corn from unmodified plants, bottom. No pesticides were used on either crop.

CREDITS: (14A, B) © Cengage Learning; (14C) Pascal Goetgheluck/Science Source; (15) The Bt and Non-Bt corn photos were taken as part of field trial conducted on the main campus of Tennessee State University at the Institute of Agricultural and Environmental Research. The work was supported by a competitive grant from the CSREES, USDA titled *Southern Agricultural Biotechnology Consortium for Underserved Communities*, (2000–2005). Dr. Fisseha Tegegne and Dr. Ahmad Aziz served as Principal and Co-principal Investigators respectively to conduct the portion of the study in the State of Tennessee.

15.8 Biotech Barnyards

✔ Genetically engineered animals are invaluable in medical research and in other applications.

Traditional cross-breeding can produce animals so unusual that transgenic animals may appear mundane by comparison. Consider featherless chickens (right) that were bred to survive in deserts where cooling systems are not an option. Cross-breeding is a form of genetic manipulation, but many transgenic animals would probably never have occurred without laboratory intervention.

The first genetically modified animals were mice. Today, such mice are commonplace, and they are invaluable in research (**FIGURE 15.16A**). For example, we have discovered the function of human genes (including the *APOA5* gene discussed in Section 15.5) by inactivating their counterparts in mice. Genetically modified mice are also used as models of human diseases. For example, researchers inactivated the molecules involved in the control of glucose metabolism, one by one. Studying the effects of the knockouts in mice has resulted in much of our current understanding of how diabetes works in humans.

Genetically modified animals other than mice are also useful in research (**FIGURE 15.16B**), and some make molecules that have medical and industrial applications (**FIGURE 15.16C**). Various transgenic goats produce proteins used to treat cystic fibrosis, heart attacks, blood clotting disorders, and even nerve gas exposure. Milk from goats transgenic for lysozyme, an antibacterial protein in human milk, may protect infants and children in developing countries from acute diarrheal disease. Goats transgenic for a spider silk gene produce the silk protein in their milk; researchers can spin this protein into nanofibers that have medical and electronics applications. Rabbits make human interleukin-2, a protein that triggers immune cells to divide and is used as a cancer drug.

We also engineer food animals. Genetic engineering has given us pigs with heart-healthy fat and environmentally friendly low-phosphate feces, muscle-bound trout, chickens that do not transmit bird flu, and cows that do not get mad cow disease, among other examples. Many people think that genetically engineering livestock is unconscionable. Others see it as an extension of thousands of years of acceptable animal husbandry practices. The techniques have changed, but not the intent: We humans continue to have an interest

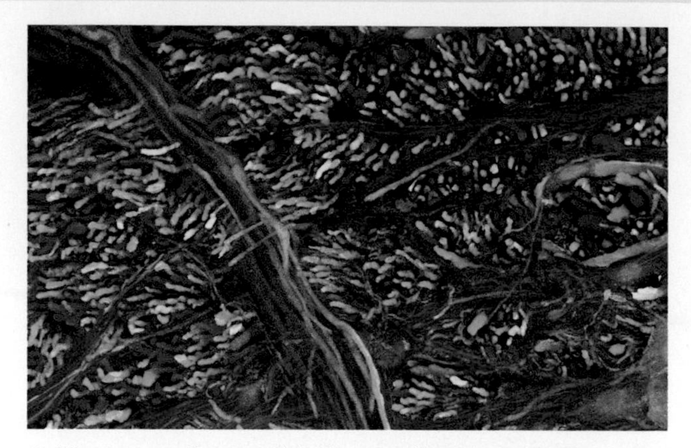

A Mice transgenic for multiple pigments ("brainbow mice") are allowing researchers to map the complex neural circuitry of the brain. Individual nerve cells in the brain stem of a brainbow mouse are visible in this fluorescence micrograph.

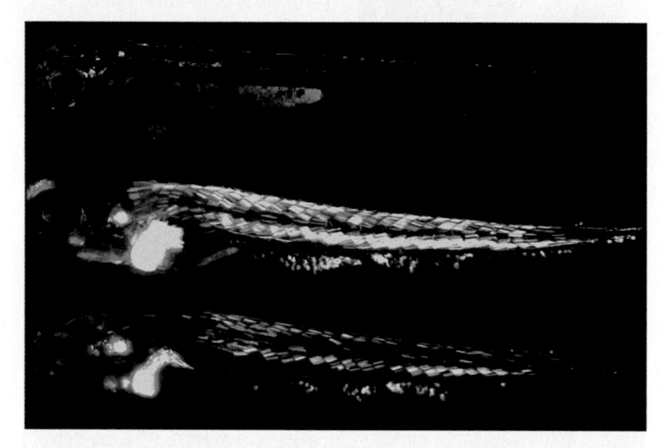

B Zebrafish engineered to glow in places where BPA, an endocrine-disrupting chemical, is present. The fish are literally illuminating where this pollutant acts in the body—and helping researchers discover what it does when it gets there.

C Transgenic goats produce human antithrombin, an anticlotting protein. Antithrombin harvested from their milk is used as a drug during surgery or childbirth to prevent blood clotting in people with hereditary antithrombin deficiency. This genetic disorder carries a high risk of life-threatening clots.

FIGURE 15.16 Examples of genetically modified animals.

in improving our livestock. Either way, tinkering with the genes of animals raises a host of ethical dilemmas. Consider animals genetically modified to carry mutations associated with human diseases—multiple sclerosis, cystic fibrosis, diabetes, cancer, or Huntington's disease, for example. Researchers study these animals in order to understand the diseases, and to test potential treatments, without experimenting on humans. However, the modified animals often suffer the same terrible symptoms of the condition as humans do.

Knockouts and Organ Factories

Millions of people have a heart, kidney, or other organ that has been damaged and cannot heal. In any given year, more than 80,000 of them are on waiting lists for an organ transplant in the United States alone. Human donors are in such short supply that illegal organ trafficking is now a common problem.

Pigs are considered a potential source of organs for transplantation, because pig and human organs are about the same in both size and function. However, the human immune system battles anything it recognizes as nonself. It rejects a pig organ at once, because it recognizes proteins and carbohydrates on the plasma membrane of pig cells. Within a few hours, blood coagulates inside the organ's vessels and dooms the transplant. Drugs can suppress the immune response, but they also render organ recipients particularly vulnerable to infection. Researchers have produced genetically modified pigs that lack the offending molecules on their cells. The human immune system may not reject tissues or organs transplanted from these pigs.

Pig-to-human transplants are an example of **xenotransplantation**, the transfer of an organ from one species into another. Critics of xenotransplantation are concerned that, among other things, such transplants would invite pig viruses to infect humans, perhaps with catastrophic results. Their concerns are not unfounded: All human pandemics have been caused by animal viruses that adapted to replicate in people.

xenotransplantation Transplantation of an organ from one species into another.

TAKE-HOME MESSAGE 15.8
Why do we genetically engineer animals?

✔ Genetic engineering creates animals that would be impossible to produce using traditional cross-breeding methods.

✔ Most engineered animals are used for research and medical applications.

✔ The technique also offers a way to improve livestock.

15.9 Safety Issues

✔ The first transfer of foreign DNA into bacteria ignited an ongoing debate about potential dangers of transgenic organisms that enter the environment.

When James Watson and Francis Crick presented their model of DNA in 1953, they ignited a global blaze of optimism. The very book of life seemed to be open for scrutiny, but in reality, no one could read it. New techniques would have to be invented before that book would be readable. Twenty years later, Paul Berg and his coworkers discovered how to make recombinant organisms by fusing DNA from two species of bacteria, providing the tools to be able to study DNA sequence in detail. They began to clone DNA from many different organisms. The technique of genetic engineering was born, and suddenly everyone was worried about it. Researchers knew that DNA itself was not toxic, but they could not predict with certainty what would happen each time they fused genetic material from different organisms. Would they accidentally make a new, dangerous form of life by fusing DNA of two harmless organisms? What if an engineered organism escaped the laboratory and transformed other organisms?

In a remarkably quick act of self-regulation, scientists reached a consensus on new safety guidelines for DNA research. Adopted at once by the NIH, these guidelines included precautions for laboratory procedures. They covered the design and use of host organisms that could survive only under a narrow range of conditions inside the laboratory. Researchers stopped using DNA from pathogenic or toxic organisms for recombinant DNA experiments until proper containment facilities were developed.

Today, all genetic engineering should be done under these laboratory guidelines, but the rules are not a guarantee of safety. We are still learning about escaped GMOs and their effects, and enforcement is a problem. For example, the expense of deregulating a GMO is prohibitive for endeavors in the public sector. Thus, most commercial GMOs were produced by large, private companies—the same ones that typically wield tremendous political influence over the very government agencies charged with regulating them.

TAKE-HOME MESSAGE 15.9
Is genetic engineering safe?

✔ Guidelines for DNA research have been in place for decades in the United States and other countries. Researchers are expected to comply, but the guidelines are not a guarantee of safety.

15.10 Genetically Modified Humans

✔ We as a society continue to work our way through the ethical implications of applying new DNA technologies.

✔ The manipulation of individual genomes continues even as we are weighing the risks and benefits of this research.

Gene Therapy

We know of more than 15,000 serious genetic disorders. Collectively, they cause 20 to 30 percent of infant deaths each year, and account for half of all mentally impaired patients and a fourth of all hospital admissions. They also contribute to many age-related disorders, including cancer, Parkinson's disease, and diabetes. Drugs and other treatments can minimize the symptoms of some genetic disorders, but gene therapy is the only cure. **Gene therapy** is the transfer of recombinant DNA into an individual's body cells, with the intent to correct a genetic defect or treat a disease. Typically, the transfer inserts an unmutated gene into the individual's chromosomes.

DNA can be introduced into human cells in many ways, for example by direct injection, electrical pulses, lipid clusters, nanoparticles, or genetically engineered viruses. Viruses have molecular machinery that delivers their genomes into cells they infect. Those used as vectors have DNA that splices itself into the infected cells' chromosomes, along with foreign DNA that has been inserted into it.

Human gene therapy is a compelling reason to embrace genetic engineering research. It is now being tested as a treatment for AIDS, muscular dystrophy, heart attack, sickle-cell anemia, cystic fibrosis, hemophilia A, Parkinson's disease, Alzheimer's disease, several types of cancer, and inherited diseases of the eye, the ear, and the immune system.

People have already benefited from gene therapy. Consider Rhys Evans (**FIGURE 15.17**), who was born with SCID-X1, a severe genetic disorder that stems from a mutated allele of the *IL2RG* gene. The gene encodes a receptor for an immune signaling molecule. Without treatment, people affected by this disorder can survive only in germ-free isolation tents because they cannot fight infections (a diminished life in a sterile isolation tent was the source of the term "bubble boy"). In the late 1990s, researchers used a genetically engineered virus to insert unmutated copies of *IL2RG* into cells taken from the bone marrow of twenty boys with SCID-X1. Each child's modified cells were infused back into his bone marrow. Within months of their treatment, eighteen of the boys left their isolation tents for good. Rhys was one of them. Gene therapy had permanently repaired their immune systems.

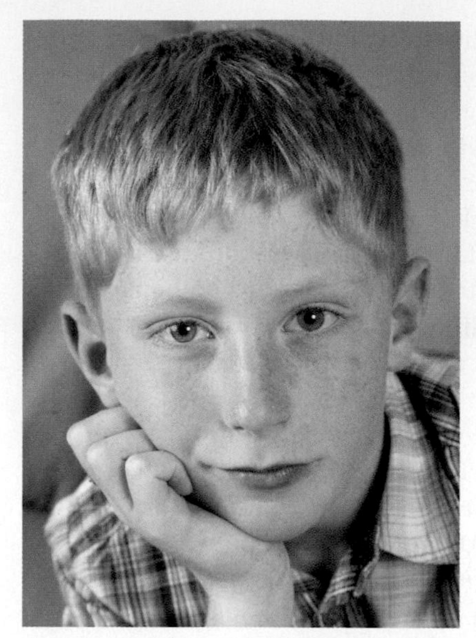

FIGURE 15.17 Rhys Evans, shown here at age 10, was born with SCID-X1. His immune system has been permanently repaired by gene therapy.

Recently, gene therapy has been used to treat acute lymphoblastic leukemia, a typically fatal cancer of bone marrow cells. A viral vector was used to insert a gene into immune cells extracted from patients. When the engineered cells were reintroduced into the patients' bodies, the inserted gene directed the destruction of the cancer cells. The therapy worked astonishingly well: In one patient, all traces of the leukemia vanished in eight days.

Despite the successes, manipulating a gene within the context of a living individual is unpredictable even when we know its sequence and location on a chromosome. No one, for example, can predict with absolute certainty where a virus-injected gene will become integrated into a chromosome. Its insertion might disrupt other genes. If it interrupts a gene that is part of the controls over cell division, then cancer might be the outcome. Consider that five of the twenty boys first treated with gene therapy for SCID-X1 developed a type of bone marrow cancer called leukemia, and one of them died. Developers of the gene therapy had wrongly predicted that cancer related to it would be rare. Research now implicates the very gene targeted for repair, especially when combined with the virus that delivered it. The viral DNA preferentially inserted itself into the children's chromosomes at a site near a proto-oncogene (Section 11.6). The insertion activated the gene by triggering its transcription, and that is how the leukemia began. Since that time, researchers used PCR to detect viral integration sites and improve the design of the vector. The development of more efficient and specific viral vectors has reduced the risk associated with all types of gene therapy.

Personal Genetic Testing (revisited)

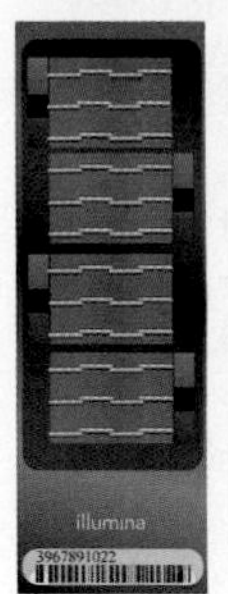

Results of a personal genetic test may include estimated risks of developing conditions associated with your particular set of SNPs. For example, the test will probably determine whether you are homozygous for certain alleles of the *MC1R* gene (Section 13.5). If you are, the company's report may tell you that you have red hair. Few SNPs have such a clear effect, however. Most human traits are polygenic, and many are also influenced by environmental factors (Section 13.6). Thus, although a DNA test can reliably determine the SNPs in an individual's genome, it cannot reliably predict the effect of those SNPs on the individual.

For example, if you carry one *E4* allele of the *APOE* gene, a DNA testing company cannot tell you whether you will develop Alzheimer's disease later in life. However, it may report your lifetime risk of developing the disease, which is about 29 percent, as compared with about 9 percent for someone who has no *E4* allele.

What, exactly, does a 29 percent lifetime risk of Alzheimer's mean? The number is a probability statistic; it means, on average, 29 of every 100 people who have the *E4* allele eventually get the disease. However, a risk is just that. Not everyone who has the allele develops Alzheimer's, and not everyone who develops the disease has the allele. Other unknown factors, including epigenetic modifications of DNA, contribute to the disease. We still have a limited understanding of how genes contribute to many health conditions, particularly age-related ones such as Alzheimer's. Geneticists believe that it will be at least five to ten more years before genotyping can be used to accurately predict an individual's future health problems.

Eugenics

The idea of selecting the most desirable human traits, **eugenics**, is an old one. It has been used as a justification for some of the most horrific episodes in human history, including the genocide of 6 million Jews during World War II; thus, it continues to be a hotly debated social issue. For example, using gene therapy to cure human genetic disorders seems like an acceptable goal to most people, but imagine taking this idea a bit further. Would it also be acceptable to engineer the genome of an individual who is within a normal range of phenotype in order to modify a particular trait? Researchers have already produced mice that have improved memory, enhanced learning ability, bigger muscles, and longer lives. Why not people?

Given the pace of genetics research, the debate is no longer about how we would engineer desirable traits, but how we would choose the traits that are desirable. Realistically, cures for many severe but rare genetic disorders will not be found, because the financial return would not cover the cost of the research. Eugenics, however, may be profitable. How much would potential parents pay to be sure that their child will be tall or blue-eyed, with breathtaking strength or intelligence? What about a treatment that can help you lose that extra weight, and keep it off permanently? The gray area between interesting and abhorrent can be very different depending on who is asked. In a survey conducted in the United States, more than 40 percent of those interviewed said they would be fine with using gene therapy to make smarter and cuter babies. In one poll of British parents, 10 percent would use it to keep a child from growing up to be homosexual, and 18 percent would be willing to use it to keep a child from being aggressive.

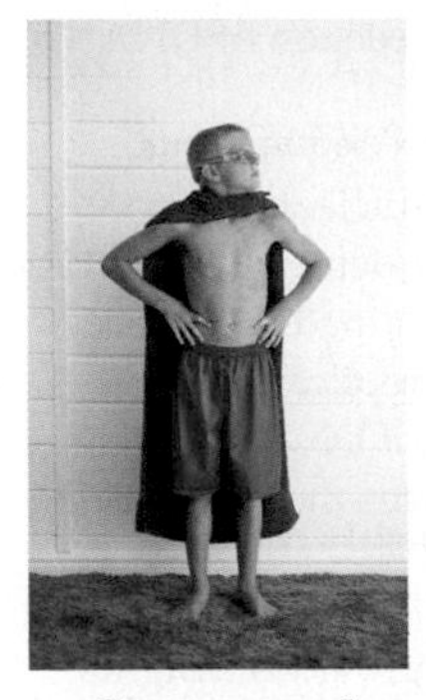

Some people are concerned that gene therapy puts us on a slippery slope that may result in irreversible damage to ourselves and to the biosphere. We as a society may not have the wisdom to know how to stop once we set foot on that slope; one is reminded of our peculiar human tendency to leap before we look. And yet, something about the human experience allows us to dream of such things as wings of our own making, a capacity that carried us into space. In this brave new world, the questions before you are these: What do we stand to lose if serious risks are not taken? And, do we have the right to impose the potential consequences on people who would choose not to take those risks?

TAKE-HOME MESSAGE 15.10
Can people be genetically modified?

✔ Genes can be transferred into a person's cells to correct a genetic defect or treat a disease. However, the outcome of altering a person's genome can be unpredictable.

eugenics Idea of deliberately improving the genetic qualities of the human race.
gene therapy Treating a genetic defect or disorder by transferring a normal or modified gene into the affected individual.

summary

Section 15.1 Personal genetic testing involves identifying a person's unique array of **single-nucleotide polymorphisms (SNPs)**. This type of test is revolutionizing the way medicine is practiced.

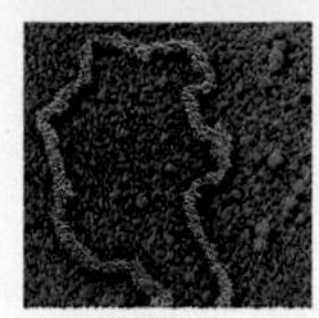

Section 15.2 In **DNA cloning**, researchers use **restriction enzymes** to cut DNA into pieces, and then use DNA ligase to splice the fragments into plasmids or other **cloning vectors**. The resulting molecules of **recombinant DNA** are inserted into host cells such as bacteria. Division of host cells produces huge populations of genetically identical descendant cells (clones), each with a copy of the cloned DNA fragment. The enzyme **reverse transcriptase** is used to transcribe RNA into **cDNA** for cloning.

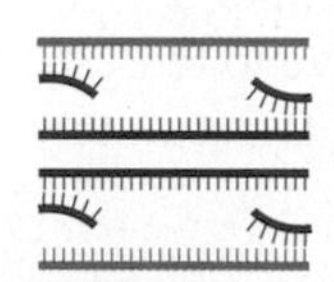

Section 15.3 A **DNA library** is a collection of cells that host different fragments of DNA. A genomic library represents an organism's entire **genome**. Researchers can use **probes** to identify cells in a library that carry a targeted fragment of DNA. The **polymerase chain reaction (PCR)** uses primers and a heat-resistant DNA polymerase to rapidly increase the number of copies of a targeted section of DNA.

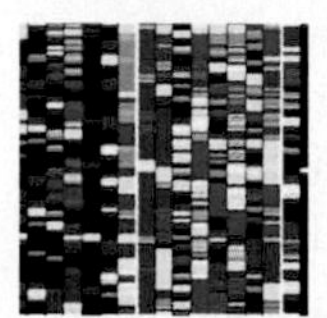

Section 15.4 Advances in **sequencing**, which reveals the order of nucleotides in DNA, allowed the DNA sequence of the entire human genome to be determined. DNA polymerase is used to partially replicate a DNA template. The reaction produces a mixture of DNA fragments of different lengths; **electrophoresis** separates the fragments by length into bands.

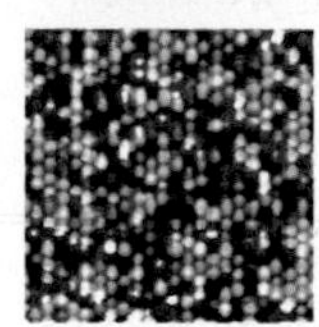

Section 15.5 **Genomics** provides insights into the function of the human genome. Similarities between genomes of different organisms are evidence of evolutionary relationships, and can be used as a predictive tool in research. **DNA profiling** identifies a person by the unique parts of his or her DNA. An example is the determination of an individual's array of short tandem repeats or single-nucleotide polymorphisms. Within the context of a criminal investigation, a DNA profile is called a DNA fingerprint.

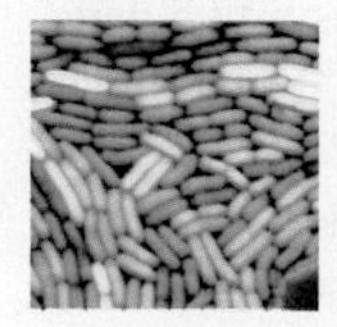

Sections 15.6–15.9 Recombinant DNA technology is the basis of **genetic engineering**, the directed modification of an organism's genetic makeup with the intent to change its phenotype. A gene from one species is inserted into an individual of a different species to make a **transgenic** organism, or a gene is modified and reinserted into an individual of the same species. The result of either process is a **genetically modified organism (GMO)**.

Bacteria and yeast, the most common genetically engineered organisms, produce proteins that have medical value. Most transgenic crop plants, which are now in widespread use worldwide, were created to help farmers produce food more efficiently. Engineered animals, which are mainly used for research and medical applications, may one day provide organs and tissues for **xenotransplantation** into humans.

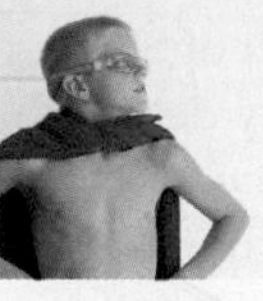

Section 15.10 With **gene therapy**, a gene is transferred into body cells to correct a genetic defect or treat a disease. Potential benefits of genetically modifying humans must be weighed against potential risks. The practice raises ethical issues such as whether **eugenics** is desirable in some circumstances.

self-quiz

Answers in Appendix VII

1. ______ cut(s) DNA molecules at specific sites.
   a. DNA polymerase
   b. DNA probes
   c. Restriction enzymes
   d. Reverse transcriptase

2. A ______ is a molecule that can be used to carry a fragment of DNA into a host organism.
   a. cloning vector
   b. chromosome
   c. GMO
   d. cDNA

3. Reverse transcriptase assembles a(n) ______ on a(n) ______ template.
   a. mRNA; DNA
   b. cDNA; mRNA
   c. DNA; ribosome
   d. protein; mRNA

4. For each species, all ______ in the complete set of chromosomes is the ______.
   a. genomes; library
   b. DNA; genome
   c. mRNA; start of cDNA
   d. cDNA; start of mRNA

5. A set of cells that host various DNA fragments collectively representing an organism's entire set of genetic information is a ______.
   a. genome
   b. clone
   c. genomic library
   d. GMO

6. ______ is a technique to determine the order of nucleotide bases in a fragment of DNA.
   a. PCR
   b. Sequencing
   c. Electrophoresis
   d. Nucleic acid hybridization

7. Fragments of DNA can be separated by electrophoresis according to ______.
   a. sequence
   b. length
   c. species

8. PCR can be used ______.
   a. to increase the number of specific DNA fragments
   b. in DNA fingerprinting
   c. to modify a human genome
   d. a and b are correct

9. An individual's set of unique ______ can be used in DNA profiling.
   a. DNA sequences
   b. short tandem repeats
   c. SNPs
   d. all of the above

10. A transgenic organism ______.
    a. carries a gene from another species
    b. has been genetically modified
    c. both a and b

data analysis activities

Enhanced Spatial Learning in Mice With an Autism Mutation Autism is a neurobiological disorder with symptoms that include impaired social interactions and stereotyped patterns of behavior. Around 10 percent of autistic people have an extraordinary skill or talent such as greatly enhanced memory.

Mutations in neuroligin 3, an adhesion protein that connects brain cells to one another, have been associated with autism. One mutation changes amino acid 451 from arginine to cysteine. In 2007, Katsuhiko Tabuchi and his colleagues genetically modified mice to carry the same arginine-to-cysteine substitution in their neuroligin 3. Mice with the mutation had impaired social behavior.

To test spatial learning ability, the mice were placed in a water maze: a deep pool of warm water in which a platform is submerged a few millimeters below the surface. The platform is not visible to swimming mice. Mice do not particularly enjoy swimming, so they locate a hidden platform as fast as they can. When tested again, they can remember its location by checking visual cues around the edge of the pool. How quickly they remember the platform's location is a measure of spatial learning ability (**FIGURE 15.18**).

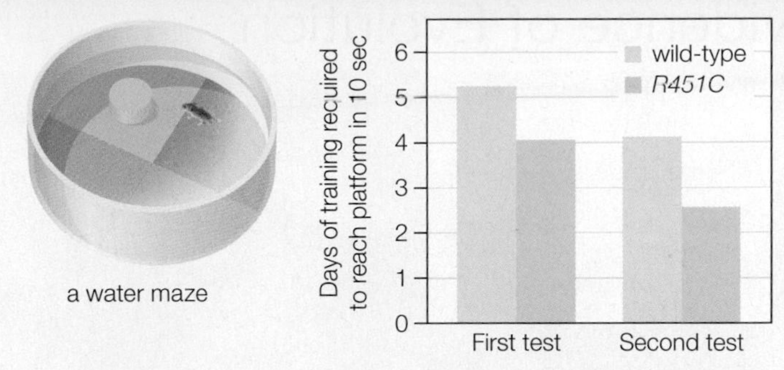

FIGURE 15.18 Spatial learning ability in mice with a mutation in neuroligin 3 (*R451C*), compared with unmodified (wild-type) mice.

1. In the first test, how many days did it take unmodified mice to learn to find the location of the hidden platform within 10 seconds?

2. Did the modified or the unmodified mice learn the location of the platform faster in the first test?

3. Which mice learned faster the second time around?

4. Which mice showed the greatest improvement in memory between the first and the second test?

11. True or false? A transgenic organism can pass a foreign gene to offspring.

12. Which of the following can be used to carry foreign DNA into host cells? Choose all correct answers.
   a. RNA
   b. viruses
   c. PCR
   d. plasmids
   e. lipid clusters
   f. blasts of pellets
   g. xenotransplantation
   h. nanoparticles

13. True or false? Some humans are genetically modified.

14. Match the method with the appropriate enzyme.

___ PCR	a. *Taq* polymerase
___ cutting DNA	b. DNA ligase
___ cDNA synthesis	c. reverse transcriptase
___ DNA sequencing	d. restriction enzyme
___ pasting DNA	e. DNA polymerase (not *Taq*)

15. Match each term with the most suitable description.

___ DNA profile	a. GMO with a foreign gene
___ Ti plasmid	b. alleles commonly have them
___ eugenics	c. a person's unique collection of short tandem repeats
___ SNPs	d. selecting "desirable" traits
___ transgenic	e. genetically modified
___ GMO	f. used in plant gene transfers

critical thinking

1. In 1918, an influenza pandemic that originated with avian flu killed 50 million people. Researchers isolated samples of that virus from bodies of infected people preserved in Alaskan permafrost since 1918. From the samples, they sequenced the viral genome, then reconstructed the virus. The reconstructed virus is 39,000 times more infectious than modern influenza strains, and 100 percent lethal in mice.

Understanding how this virus works can help us defend ourselves against other strains that may arise. For example, discovering what makes it so infectious and deadly would help us design more effective vaccines. Critics of the research are concerned: If the virus escapes the containment facilities (even though it has not done so yet), it might cause another pandemic. Worse, the published DNA sequence and methods to make the virus could be used for criminal purposes. Do you think this research makes us more or less safe?

2. The results of a paternity test using short tandem repeats are listed in the table below. Who's the daddy? How sure are you?

	Mother	Baby	Alleged Father #1	Alleged Father #2
CSF1PO	15, 17	17, 23	23, 27	17, 15
FGA	9, 9	9, 9	9, 12	9, 12
THO1	29, 29	29, 27	27, 28	29, 28
TPOX	14, 18	18, 20	15, 20	17, 22
VWA	14, 14	14, 14	14, 14	14, 16
D3S1358	11, 14	14, 16	12, 16	14, 20
D5S818	11, 13	10, 13	8, 10	18, 18
D7S820	7, 13	13, 13	13, 19	13, 13
D8S1179	13, 13	13, 15	12, 15	10, 12
D13S317	12, 12	10, 12	8, 10	12, 17
D16S539	12, 14	14, 12	14, 14	18, 25
D18S51	5, 6	6, 22	22, 6	5, 22
D21S11	15, 17	17, 22	15, 22	22, 22

16 Evidence of Evolution

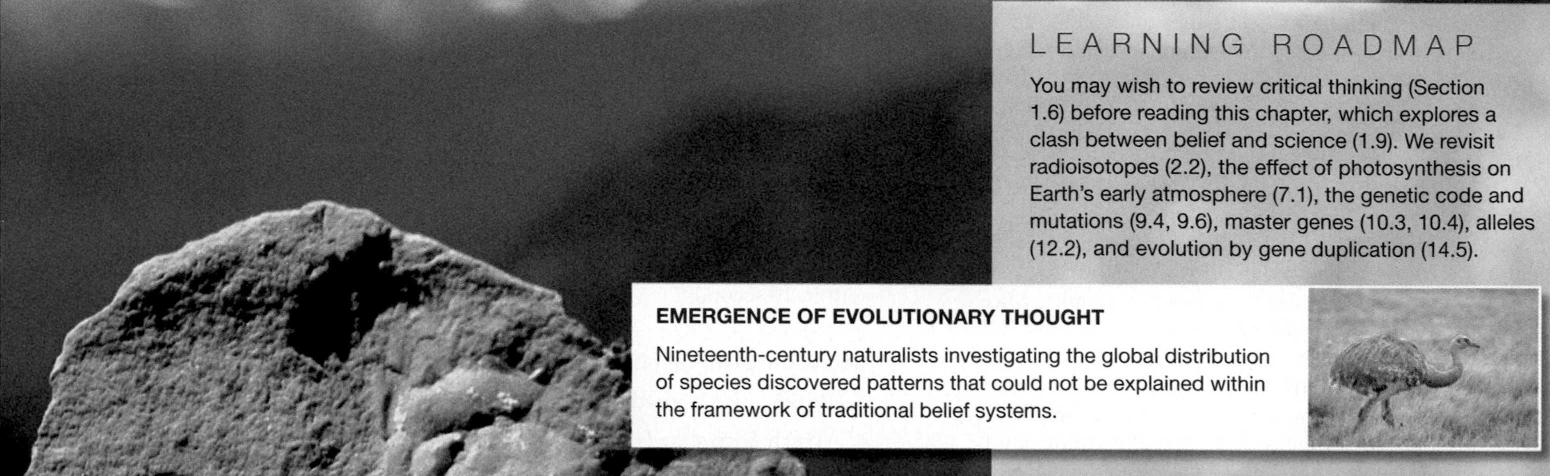

LEARNING ROADMAP

You may wish to review critical thinking (Section 1.6) before reading this chapter, which explores a clash between belief and science (1.9). We revisit radioisotopes (2.2), the effect of photosynthesis on Earth's early atmosphere (7.1), the genetic code and mutations (9.4, 9.6), master genes (10.3, 10.4), alleles (12.2), and evolution by gene duplication (14.5).

EMERGENCE OF EVOLUTIONARY THOUGHT

Nineteenth-century naturalists investigating the global distribution of species discovered patterns that could not be explained within the framework of traditional belief systems.

A THEORY TAKES FORM

Evidence of evolution, or change in lines of descent, led Charles Darwin and Alfred Wallace to develop a theory of how traits that define each species change over time.

EVIDENCE FROM FOSSILS

The fossil record provides physical evidence of past changes in many lines of descent. We use the property of radioisotope decay to determine the age of rocks and fossils.

INFLUENTIAL GEOLOGIC FORCES

Over millions of years, slow movements of Earth's outer crust have affected land and oceans. The changes have profoundly influenced life's evolution.

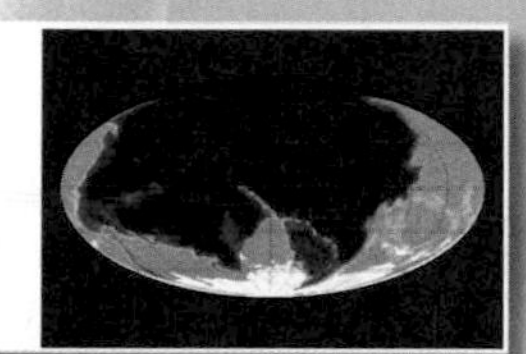

THE GEOLOGIC TIME SCALE

By studying rock layers and fossils in them, we can correlate geologic and evolutionary events. The correlation helps explain the distribution of species, past and present.

The next two chapters continue the theme of evolutionary processes, including how natural selection works (17.4–17.7), how the movement of tectonic plates can affect evolution (Section 17.10), and what comparative morphology (18.3) and DNA sequence comparisons (18.4) can tell us about shared evolutionary history. A continent's climate is affected by its position on the globe, as you will see in Chapter 47.

16.1 Reflections of a Distant Past

How do you think about time? Perhaps you can conceive of a few hundred years of human events, maybe a few thousand, but how about a few million? Envisioning the very distant past requires an intellectual leap from the familiar to the unknown. One way to make that leap involves, surprisingly, asteroids. Asteroids are small planets hurtling through space. They range in size from 1 to 1,500 kilometers (roughly 0.5 to 1,000 miles) across. Millions of them orbit the sun between Mars and Jupiter—cold, stony leftovers from the formation of our solar system. Asteroids are difficult to see even with the best telescopes, because they do not emit light. Many cross Earth's orbit, but most of those pass us by before we know about them. Some have not passed us at all.

Consider the mile-wide Barringer Crater in Arizona (**FIGURE 16.1A**). An asteroid 45 meters (150 feet) wide made this impressive pockmark in the desert sandstone when it slammed into Earth 50,000 years ago with an impact 150 times more powerful than the bomb that leveled Hiroshima. No humans were in North America at the time of the impact. If there were no witnesses, how is it possible to know anything about what happened? We often reconstruct history by studying physical evidence of past events. Geologists were able to infer the most probable cause of the Barringer Crater by analyzing tons of meteorites, melted sand, and other rocky clues at the site.

Similar evidence points to even larger asteroid impacts in the more distant past. For example, fossil hunters have long known about a mass extinction, or permanent loss of major groups of organisms, that occurred 66 million years ago. The event is marked by an unusual, worldwide formation of sedimentary rock (**FIGURE 16.1B**). This formation is called the K–Pg boundary sequence (it was formerly known as the K–T boundary). There are plenty of dinosaur fossils below this formation. Above it, in rock layers that were deposited more recently, there are no dinosaur fossils, anywhere. A gigantic impact crater—274 kilometers (170 miles) across and 1 kilometer (3,000 feet) deep—off the coast of what is now the Yucatán Peninsula dates to about 66 million years ago. Coincidence? Many scientists say no. The asteroid that made the Yucatán crater had to be at least 20 kilometers (12 miles) wide when it hit. The impact of an asteroid that size would have been *40 million times* more powerful than the one that made the Barringer Crater. The scientists infer that the impact caused a global catastrophe of sufficient scale to wipe out the dinosaurs.

You are about to make an intellectual leap through time to consider events that were not known even a few centuries ago. We invite you to launch yourself from this premise: Natural phenomena that occurred in the past can be explained by the very same physical, chemical, and biological processes that operate today. That premise is the foundation for scientific research into the history of life. The research represents a shift from experience to inference—from the known to what can only be surmised—and it gives us astonishing glimpses into the distant past.

A What made the Barringer Crater in Arizona? Rocky evidence points to a 300,000-ton asteroid that collided with Earth 50,000 years ago.

B The K–Pg boundary sequence, an unusual, worldwide formation of sedimentary rock that formed 66 million years ago. This formation marks an abrupt transition in the fossil record that implies a mass extinction. The red pocketknife gives an idea of scale.

FIGURE 16.1 Evidence to inference.

16.2 Early Beliefs, Confounding Discoveries

✔ Belief systems are influenced by the extent of our understanding of the natural world. Those that are inconsistent with systematic observations tend to change over time.

About 2,300 years ago, the Greek philosopher Aristotle described nature as a continuum of organization, from lifeless matter through complex plants and animals. Aristotle's work greatly influenced later European thinkers, who adopted his view of nature and modified it in light of their own beliefs. By the fourteenth century, Europeans generally believed that a "great chain of being" extended from the lowest form (snakes), up through humans, to spiritual beings. Each link in the chain was a species, and each was said to have been forged at the same time, in one place, and in a perfect state. The chain itself was complete and continuous. Because everything that needed to exist already did, there was no room for change.

In the 1800s, European naturalists embarked on globe-spanning survey expeditions and brought back tens of thousands of plants and animals from Asia, Africa, North and South America, and the Pacific Islands. Each newly discovered species was carefully catalogued as another link in the chain of being. The explorers began to see patterns in where species live and similarities in body plans, and had started to think about the natural forces that shape life. These explorers were pioneers in **biogeography**, the study of patterns in the geographic distribution of species and communities. Some of the patterns raised questions that could not be answered within the framework of prevailing belief systems. For example, globe-trotting explorers had discovered plants and animals living in extremely isolated places. The isolated species looked suspiciously similar to species living on the other side of impassable mountain ranges, or across vast expanses of open ocean. Consider the emu, rhea, and ostrich, three types of bird native to three different continents (FIGURE 16.2). These birds share a set of unusual features. All are very large, with long, muscular legs and necks. All are also flightless, sprinting about in flat, open grasslands about the same distance from the equator. Alfred Wallace, an explorer particularly interested in the geographical distribution of animals, thought that the shared traits might mean that the birds descended from a common ancestor (and he was correct), but he had no idea how they could have ended up on different continents.

Naturalists of the time also had trouble classifying organisms that are very similar in some features, but different in others. For example, both plants shown in FIGURE 16.3 live in hot deserts where water is seasonally scarce. Both have rows of sharp spines that deter herbivores, and both store water in their thick, fleshy

FIGURE 16.3 Similar-looking, unrelated species. On the left, an African milk barrel cactus (*Euphorbia horrida*), native to the Great Karoo desert of South Africa. On the right, saguaro cactus (*Carnegiea gigantea*), native to the Sonoran Desert of Arizona.

A Emu, native to Australia

B Rhea, native to South America

C Ostrich, native to Africa

FIGURE 16.2 Similar-looking, related animals native to distant geographic realms. These birds are unlike most others in several unusual features, including long, muscular legs and an inability to fly. All are native to open grassland regions about the same distance from the equator.

stems. However, their reproductive parts are very different, so these plants cannot be (and are not) as closely related as their outward appearance might suggest.

Observations such as these are examples of **comparative morphology**, the study of anatomical patterns: similarities and differences among the body plans of organisms. Today, comparative morphology is only one branch of taxonomy (Section 1.5), but in the nineteenth century it was the only way to distinguish species. In some cases, comparative morphology revealed anatomical details (body parts with no apparent function, for example) that added to the mounting confusion. If every species had been created in a perfect state, then why were there useless parts such as wings in birds that do not fly, eyes in moles that are blind, or remnants of a tail in humans (**FIGURE 16.4**)?

Fossils were puzzling too. A **fossil** is physical evidence—remains or traces—of an organism that lived in the ancient past. Geologists mapping rock formations exposed by erosion or quarrying had discovered identical sequences of rock layers in different parts of the world. Deeper layers held fossils of simple marine life. Layers above those held similar but more complex fossils (**FIGURE 16.5**). In higher layers, fossils that were similar but even more complex resembled modern species. What did these sequences mean? Fossils of many animals unlike any living ones were also being unearthed. If these animals had been perfect at the time of creation, then why had they become extinct?

Taken as a whole, the accumulating discoveries from biogeography, comparative morphology, and geology did not fit with prevailing beliefs of the nineteenth century. If species had not been created in a perfect state (and extinct species, fossil sequences, and "useless" body parts implied that they had not), then perhaps species had indeed changed over time.

biogeography Study of patterns in the geographic distribution of species and communities.
comparative morphology The scientific study of similarities and differences in body plans.
fossil Physical evidence of an organism that lived in the ancient past.

TAKE-HOME MESSAGE 16.2

Why did observations of nature change our thinking in the nineteenth century?

✔ Increasingly extensive observations of nature in the nineteenth century did not fit with prevailing belief systems.

✔ Cumulative findings from biogeography, comparative morphology, and geology led naturalists to question traditional ways of interpreting the natural world.

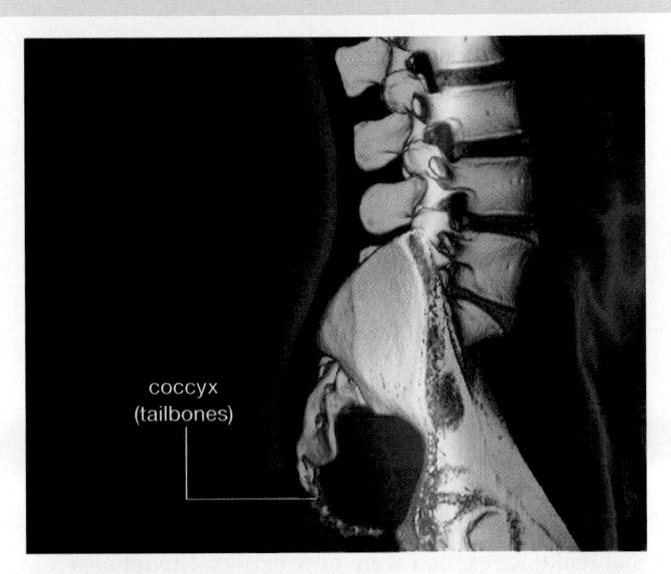

FIGURE 16.4 A vestigial structure: human tailbones. Nineteenth-century naturalists were well aware of—but had trouble explaining—body structures such as human tailbones that had apparently lost most or all function.

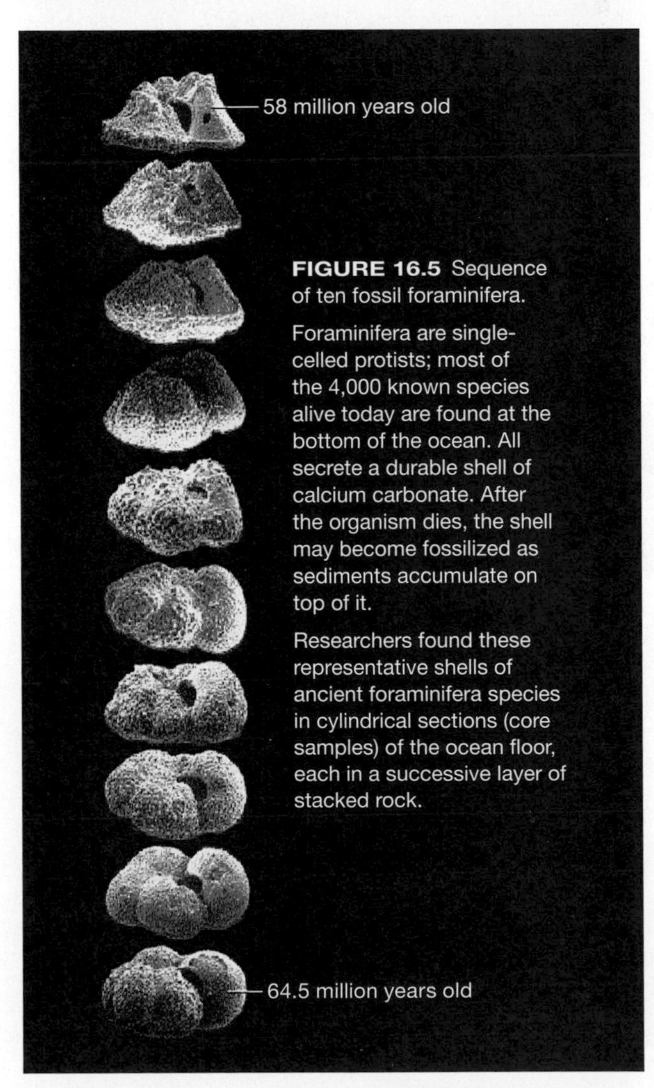

FIGURE 16.5 Sequence of ten fossil foraminifera.

Foraminifera are single-celled protists; most of the 4,000 known species alive today are found at the bottom of the ocean. All secrete a durable shell of calcium carbonate. After the organism dies, the shell may become fossilized as sediments accumulate on top of it.

Researchers found these representative shells of ancient foraminifera species in cylindrical sections (core samples) of the ocean floor, each in a successive layer of stacked rock.

16.3 A Flurry of New Theories

✔ In the 1800s, many naturalists realized that life on Earth had changed over time, and began to think about what could have caused the changes.

Squeezing New Evidence Into Old Beliefs

In the nineteenth century, naturalists were faced with increasing evidence that life on Earth, and even Earth

itself, had changed over time. Around 1800, Georges Cuvier (left), an expert in zoology and paleontology, was trying to make sense of the new information. He had observed abrupt changes in the fossil record, and knew that many fossil species seemed to have no living counterparts. Given this evidence, he proposed an idea startling for the time: Many species that had once existed were now extinct. Cuvier also knew about evidence that Earth's surface had changed. For example, he had seen fossilized seashells in rocks at the tops of mountains far from modern seas. Like most others of his time, he assumed Earth's age to be in the thousands, not billions, of years. He reasoned that geologic forces unlike any known at the time would have been necessary to raise seafloors to mountaintops in this short time span. Catastrophic geological events would have caused extinctions, after which surviving species repopulated the planet. Cuvier's idea came to be known as **catastrophism**. We now know it is incorrect; geologic processes have not changed over time.

Another naturalist, Jean-Baptiste Lamarck, was thinking about processes that might drive **evolution**, or change in a line of descent. A line of descent is also

called a **lineage**. Lamarck (left) thought that a species gradually improved over generations because of an inherent drive toward perfection, up the chain of being. The drive directed an unknown "fluida" into body parts needing change. By Lamarck's hypothesis, environmental pressures cause an internal requirement for change in an individual's body, and the resulting change is inherited by offspring.

Try using Lamarck's hypothesis to explain why a giraffe's neck is very long. We might predict that some short-necked ancestor of the modern giraffe stretched its neck to browse on leaves beyond the reach of other animals. The stretches may have even made its neck a bit longer. By Lamarck's hypothesis, that animal's offspring would inherit a longer neck. The modern giraffe would have been the result of many generations that strained to reach ever loftier leaves. Lamarck was correct in thinking that environmental factors affect a species' traits, but his understanding of how inheritance works was incomplete.

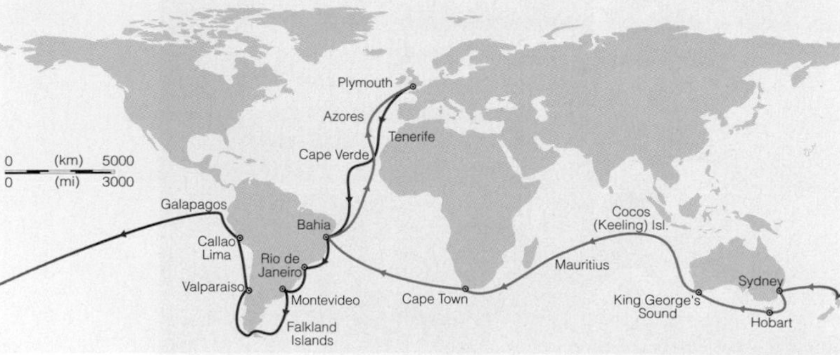

FIGURE 16.6 Voyage of the HMS *Beagle*. With Darwin aboard as ship's naturalist, the vessel (top) originally set sail to map the coast of South America, but ended up circumnavigating the globe (bottom). The path of the voyage is shown from red to blue. Darwin's detailed observations of the geology, fossils, plants, and animals he encountered on this expedition changed the way he thought about evolution.

Darwin and the HMS *Beagle*

Lamarck's ideas about evolution influenced the thinking of Charles Darwin, who, at the age of 22, joined a survey expedition to South America on the ship *Beagle*. Since he was eight years old, Darwin had wanted to hunt, fish, collect shells, or watch insects and birds—anything but sit in school. After a failed attempt to study medicine in college, he earned a degree in theol-

catastrophism Now-abandoned hypothesis that catastrophic geologic forces unlike those of the present day shaped Earth's surface.
evolution Change in a line of descent.
lineage Line of descent.
theory of uniformity Idea that gradual repetitive processes occurring over long time spans shaped Earth's surface.

ogy from Cambridge. During his studies, Darwin had spent most of his time with faculty members and other students who embraced natural history.

The *Beagle* set sail for South America in December 1831 (**FIGURE 16.6**). The young man who had hated school and had no formal training in science quickly became an enthusiastic naturalist. During the *Beagle*'s five-year voyage, Darwin found many unusual fossils, and saw diverse species living in environments that ranged from the sandy shores of remote islands to plains high in the Andes. Along the way, he read the

first volume of a new and popular book, Charles Lyell's *Principles of Geology*. Lyell (left) was a proponent of what became known as the **theory of uniformity**, the idea that gradual, repetitive change had shaped Earth. For many years, geologists had been chipping away at the sandstones, limestones, and other types of rocks that form from accumulated sediments at the bottom of lakes, rivers, and oceans. These rocks held evidence that gradual processes of geologic change operating in the present were the same ones that operated in the distant past.

By the theory of uniformity, strange catastrophes were not necessary to explain Earth's surface. Gradual, everyday geologic processes such as erosion by wind and water could have sculpted Earth's current landscape over great spans of time. This theory challenged the prevailing belief in Europe that Earth was 6,000 years old. According to traditional scholars, people had recorded everything that happened in those 6,000 years—and in all that time, no one had mentioned seeing a species evolve. However, by Lyell's calculations, it must have taken millions of years to sculpt Earth's surface. Darwin's exposure to Lyell's ideas gave him insights into the geologic history of the regions he would encounter on his journey. Was millions of years enough time for species to evolve? Darwin thought that it was (**FIGURE 16.7**).

FIGURE 16.7 Charles Darwin and part of a page from his 1836 notes on the "Transmutation of Species."

The text reads as follows: "Let a pair be introduced and increase slowly, from many enemies, so as often to intermarry who will dare say what result. According to this view animals on separate islands ought to become different if kept long enough apart with slightly differing circumstances."

TAKE-HOME MESSAGE 16.3

How did new evidence change the way people in the 19th century thought about the history of life?

✔ In the 1800s, fossils and other evidence led some naturalists to propose that Earth and the species on it had changed over time. The naturalists also began to reconsider the age of Earth.

✔ Darwin's detailed observations of nature during a five-year voyage around the world changed his ideas about how evolution occurs.

CREDITS: (7) top, Painting by George Richmond; bottom, Cambridge University Library.

16.4 Darwin, Wallace, and Natural Selection

✔ Darwin's observations of species in different parts of the world helped him understand a driving force of evolution.

Among the thousands of specimens Darwin collected on his voyage and sent to England were fossil glyptodons from Argentina. These armored mammals are extinct, but they have many traits in common with modern armadillos (**FIGURE 16.8**). For example, armadillos live only in places where glyptodons once lived. Like glyptodons, armadillos have helmets and protective shells that consist of unusual bony plates. Could the shared traits mean that glyptodons were ancient relatives of armadillos? If so, perhaps traits of their common ancestor had changed in the line of descent that led to armadillos. But why would such changes have occurred?

A Fossil of a glyptodon, an automobile-sized mammal that existed from 2 million to 15,000 years ago.

B A modern armadillo, about a foot long.

FIGURE 16.8 Ancient relatives: glyptodon and armadillo.

Even though these animals are widely separated in time, they share a restricted distribution and unusual traits, including a shell and helmet of keratin-covered bony plates—a material similar to crocodile and lizard skin. (The fossil in **A** is missing its helmet.) Their unique shared traits were a clue that helped Darwin develop the theory of evolution by natural selection.

A Key Insight—Variation in Traits

After Darwin returned to England, he pondered his notes and fossils, and read an essay by one of his contemporaries, economist Thomas Malthus (left). Malthus had correlated increases in the size of human populations with episodes of famine, disease, and war. He proposed the idea that humans run out of food, living space, and other resources because they tend to reproduce beyond the capacity of their environment to sustain them. When that happens, the individuals of a population must either compete with one another for the limited resources, or develop new technologies to increase productivity. Darwin realized that Malthus's ideas had wider application: All populations, not just human ones, must have the capacity to produce more individuals than their environment can support.

Reflecting on his journey, Darwin started thinking about how individuals of a species often vary a bit in the details of shared traits such as size, coloration, and so on. He saw such variation among finch species on isolated islands of the Galápagos archipelago. This island chain is separated from South America by 900 kilometers (550 miles) of open ocean, so most species living on the islands did not have the opportunity for interbreeding with mainland populations. The Galápagos island finches resembled finch species in South America, but many had unique traits that suited their particular island habitat.

Darwin was familiar with dramatic variations in traits that selective breeding could produce in pigeons, dogs, and horses. He recognized that a natural environment could similarly select forms of traits that make individuals of a population suited to it. It dawned on Darwin that having a particular form of a shared trait might give an individual an advantage over competing members of its species. In any population, some individuals have forms of shared traits that make them better suited to their environment than others. In other words, individuals of a natural population vary in fitness. Today, we define **fitness** as the degree of adaptation to a specific environment, and measure it by relative genetic contribution to future generations. A form of a heritable trait that enhances an individual's fitness is called an evolutionary **adaptation**, or **adaptive trait**. Over many generations, individuals that have adaptive traits tend to survive longer and reproduce more than their less fit rivals. Darwin understood that this process, which he called **natural selection**, could be a mechanism by which evolution occurs. If an individual has a form of a trait that makes it better suited to an environment, then it is better able to survive. If an individual is better able to survive, then it has

FIGURE 16.9 Alfred Wallace, codiscoverer of natural selection.

Table 16.1 Principles of Natural Selection

Observations About Populations

- ✔ Natural populations have an inherent capacity to increase in size over time.
- ✔ As a population expands, resources that are used by its individuals (such as food and living space) eventually become limited.
- ✔ When resources are limited, individuals of a population compete for them.

Observations About Genetics

- ✔ Individuals of a species share certain traits.
- ✔ Individuals of a natural population vary in the details of those shared traits.
- ✔ Shared traits have a heritable basis, in genes. Slightly different versions of those genes (alleles) give rise to variation in shared traits.

Inferences

- ✔ A certain form of a shared trait may make its bearer better able to survive.
- ✔ Individuals of a population that are better able to survive tend to leave more offspring.
- ✔ Thus, an allele associated with an adaptive trait tends to become more common in a population over time.

a better chance of living long enough to produce offspring. If individuals with an adaptive form of a trait produce more offspring than those that do not, then the frequency of that form will tend to increase in the population over successive generations. TABLE 16.1 summarizes this reasoning in modern terms.

Great Minds Think Alike

Darwin wrote out his ideas about natural selection, but let ten years pass without publishing them. In the meantime, Alfred Wallace (FIGURE 16.9), who had been studying wildlife in the Amazon basin and the Malay Archipelago, wrote an essay and sent it to Darwin for advice. Wallace's essay outlined evolution by natural selection—the very same theory as Darwin's. Wallace had written earlier letters to Darwin and Lyell about patterns in the geographic distribution of species, and had come to the same conclusion.

In 1858, just weeks after Darwin received Wallace's essay, the theory of evolution by natural selection was presented at a scientific meeting. Both Darwin and Wallace were credited as authors. Wallace was still in the field and knew nothing about the meeting, which Darwin did not attend. The next year, Darwin published *On the Origin of Species*, which laid out detailed evidence in support of the theory. Many people had already accepted the idea of descent with modification (evolution). However, there was a fierce debate over the idea that natural selection drives evolution. Decades would pass before experimental evidence from the field of genetics led to its widespread acceptance as a theory by the scientific community.

As you will see in the remainder of this unit, the theory of evolution by natural selection is supported by and helps explain the fossil record as well as similarities and differences in the form, function, and biochemistry of living things.

TAKE-HOME MESSAGE 16.4

What is natural selection?

- ✔ Natural selection is a process that drives evolutionary change: Individuals of a population survive and reproduce with differing success depending on the details of their shared, heritable traits.
- ✔ Traits favored in a particular environment are adaptive.

adaptation (adaptive trait) A form of a heritable trait that enhances an individual's fitness in a particular environment.
fitness Degree of adaptation to an environment, as measured by an individual's relative genetic contribution to future generations.
natural selection Differential survival and reproduction of individuals of a population based on differences in shared, heritable traits.

16.5 Fossils: Evidence of Ancient Life

✔ The fossil record holds clues to life's evolution, but it will always be incomplete.

A Fossil skeleton of an ichthyosaur that lived about 200 million years ago. These marine reptiles were about the same size as modern porpoises, breathed air like them, and probably swam as fast, but the two groups are not closely related.

B Extinct wasp encased in amber, which is ancient tree sap. This 9-mm-long insect lived about 20 million years ago.

C Fossilized leaf from a 260-million-year-old *Glossopteris*, a type of plant called a seed fern.

D Fossilized footprints of a theropod, a name that means "beast foot." This group of carnivorous dinosaurs, which includes the familiar *Tyrannosaurus rex*, arose about 250 million years ago.

E Coprolite (fossilized feces). Fossilized food remains and parasitic worms inside coprolites offer clues about the diet and health of extinct species. A foxlike animal excreted this one.

FIGURE 16.10 Examples of fossils.

Even before Darwin's time, fossils were recognized as stone-hard evidence of earlier forms of life. Most fossils consist of mineralized bones, teeth, shells, seeds, spores, or other body parts (**FIGURE 16.10A–C**). Trace fossils such as footprints and other impressions, nests, burrows, trails, eggshells, or feces are evidence of an organism's activities (**FIGURE 16.10D,E**).

The process of fossilization typically begins when an organism or its traces become covered by sediments, mud, or ash. Groundwater then seeps into the remains, filling spaces around and inside of them. Minerals dissolved in the water gradually replace minerals in bones and other hard tissues. Mineral particles that crystallize and settle out of the groundwater inside cavities and impressions form detailed imprints of internal and external structures. Sediments that slowly accumulate on top of the site exert increasing pressure, and, after a very long time, extreme pressure transforms the mineralized remains into rock.

Most fossils are found in layers of sedimentary rock (**FIGURE 16.11**). Sedimentary rocks form as rivers wash silt, sand, volcanic ash, and other materials from land to sea. Mineral particles in the materials settle on seafloors in horizontal layers that vary in thickness and composition. After hundreds of millions of years, the layers of sediments become compacted into layers of rock. Even though most sedimentary rock forms at the bottom of a sea, geologic processes can tilt the rock and lift it far above sea level, where the layers may become exposed by the erosive forces of water and wind.

Biologists study sedimentary rock formations in order to understand life's historical context. Features of the formations can provide information about conditions in the environment in which they formed. Consider banded iron, a unique formation named after its distinctive striped appearance (left). Huge deposits of this sedimentary rock are the source of most iron we mine for steel today, but they also hold a record of how the evolution of the noncyclic pathway of photosynthesis changed the chemistry of Earth. Banded iron started forming about 2.4 billion years ago, right after photosynthesis evolved (Section 7.1). At that time, Earth's atmosphere and ocean contained very little oxygen, so almost all of the iron on Earth was in a reduced form (Section 5.5). Reduced iron dissolves in water, and ocean water contained a lot of it. Oxygen released into the ocean by early photosynthetic bacteria quickly combined with the dissolved iron. The resulting oxidized iron compounds are completely insoluble in water, and

they began to rain down on the ocean floor in massive quantities. These compounds accumulated in sediments that would eventually become compacted into banded iron formations. This process continued for about 600 million years. After that, ocean water no longer contained very much dissolved iron, and oxygen gas bubbling out of it had oxidized the iron in rocks exposed to the atmosphere.

The Fossil Record

We have fossils for more than 250,000 known species. Considering the current range of biodiversity, there must have been many millions more, but we will never know all of them. Why not? The odds are against finding evidence of an extinct species, because fossils are relatively rare. Typically, when an organism dies, its remains are obliterated quickly by scavengers. Organic materials decompose in the presence of moisture and oxygen, so remains that escape scavenging endure only if they dry out, freeze, or become encased in an air-excluding material such as sap, tar, or mud. Remains that do become fossilized are usually crushed or scattered by erosion and other geologic assaults.

In order for us to know about an extinct species that existed long ago, we have to find a fossil of it. At least one specimen had to be buried before it decomposed or something ate it. The burial site had to escape destructive geologic events, and end up in a place that we can find today. Most ancient species had no hard parts to fossilize, so we do not find much evidence of them. For example, there are many fossils of bony fishes and mollusks with hard shells, but few fossils of the jellyfishes and soft worms that were probably much more common. Also think about relative numbers of organisms. Fungal spores and pollen grains are typically released by the billions. By contrast, the earliest humans lived in small bands and few of their offspring survived. The odds of finding even one fossilized human bone are much smaller than the odds of finding a fossilized fungal spore. Finally, imagine two species, one that existed only briefly and the other for billions of years. Which is more likely to be represented in the fossil record?

TAKE-HOME MESSAGE 16.5

What are fossils?

✔ Fossils are evidence of organisms that lived in the remote past, a stone-hard historical record of life.

✔ The fossil record will never be complete because fossils are relatively rare. It is slanted toward hard-bodied species that lived in large populations and persisted for a long time.

A Two types of sedimentary rock. Sandstones (left) consist of compacted grains of sand or minerals; shales (right) consist of ancient compacted clay or mud.

B Cindy Looy and Mark Sephton climb the walls of Butterloch Canyon, Italy, to look for fossilized fungal spores in a 251-million-year-old layer of sedimentary rock.

C A fossilized trilobite (an ancient marine relative of centipedes) found in a shale formation by fossil hunters in Yoho National Park, British Columbia.

FIGURE 16.11 Fossils are most often found in layered sedimentary rock. This type of rock forms over hundreds of millions of years, often at the bottom of a sea. Geologic processes can tilt the rock and lift it far above sea level, where the layers become exposed by the erosive forces of water and wind.

16.6 Filling In Pieces of the Puzzle

✔ Radiometric dating reveals the age of rocks and fossils.

✔ New fossil discoveries are continually filling gaps in our understanding of the ancient history of many lineages.

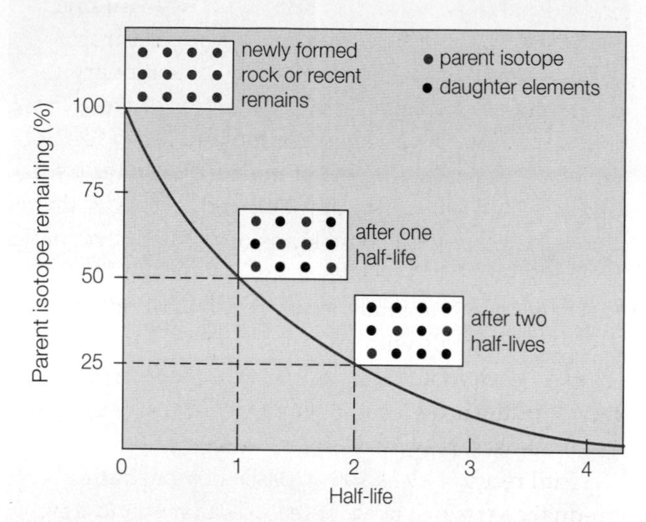

FIGURE 16.12 ▶Animated Half-life. **FIGURE IT OUT** How much of any radioisotope remains after two half-lives have passed?

Answer: 25 percent

A Long ago, ^{14}C and ^{12}C were incorporated into the tissues of a nautilus. The carbon atoms were part of organic molecules in the animal's food. ^{12}C is stable and ^{14}C decays, but the proportion of the two isotopes in the nautilus's tissues remained the same. Why? The nautilus continued to gain both types of carbon atoms in the same proportions from its food.

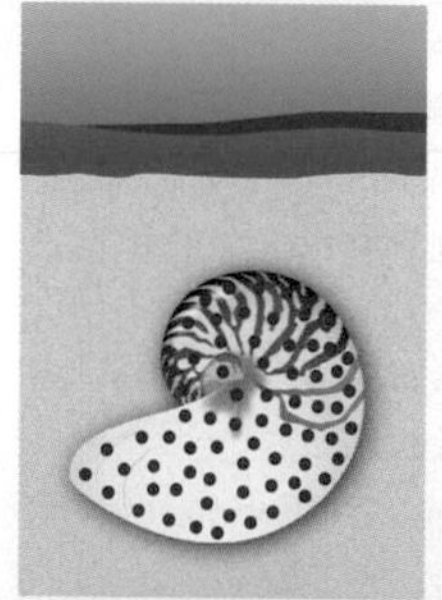

B The nautilus stopped eating when it died, so its body stopped gaining carbon. The ^{12}C atoms in its tissues were stable, but the ^{14}C atoms (represented as red dots) were decaying into nitrogen atoms. Thus, over time, the amount of ^{14}C decreased relative to the amount of ^{12}C. After 5,730 years, half of the ^{14}C had decayed; after another 5,730 years, half of what was left had decayed, and so on.

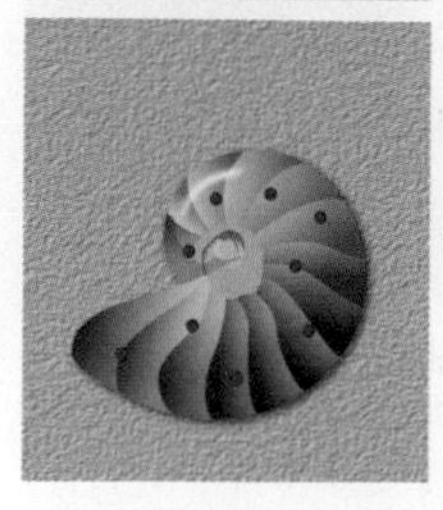

C Fossil hunters discover the fossil and measure its content of ^{14}C and ^{12}C. They use the ratio of these isotopes to calculate how many half-lives have passed since the organism died. For example, if its ^{14}C to ^{12}C ratio is one-eighth of the ratio in living organisms, then three half-lives $(1/2)^3$ must have passed since it died. Three half-lives of ^{14}C is 17,190 years.

FIGURE 16.13 ▶Animated Example of how radiometric dating can be used to find the age of a carbon-containing fossil. Carbon 14 (^{14}C) is a radioisotope of carbon that decays into nitrogen. It forms in the atmosphere and combines with oxygen to become CO_2, which enters food chains by way of photosynthesis.

Radiometric Dating

Remember from Section 2.2 that a radioisotope is a form of an element with an unstable nucleus. Atoms of a radioisotope become atoms of other elements—daughter elements—as their nucleus disintegrates. This radioactive decay is not influenced by temperature, pressure, chemical bonding state, or moisture; it is influenced only by time. Thus, like the ticking of a perfect clock, each type of radioisotope decays at a constant rate. The time it takes for half of the atoms in a sample of radioisotope to decay is called a **half-life** (FIGURE 16.12).

Half-life is a characteristic of each radioisotope. For example, radioactive uranium 238 decays into thorium 234, which decays into something else, and so on until it becomes lead 206. The half-life of the decay of uranium 238 to lead 206 is 4.5 billion years.

The predictability of radioactive decay can be used to find the age of a volcanic rock (the date it solidified). Rock forms from magma, which is a hot, molten material deep under Earth's surface. Because magma is fluid, atoms swirl and mix in it. When the material cools, for example after reaching the surface as lava, it hardens and becomes rock. Minerals crystallize in the rock as it hardens. Each kind of mineral has a characteristic structure and composition. For example, the mineral called zircon (left) consists mainly of orderly arrays of zirconium silicate molecules ($ZrSiO_4$). Some of the molecules in a newly formed zircon crystal have uranium atoms substituted for zirconium atoms, but never lead atoms. However, uranium decays into lead at a predictable rate. Thus, over time, uranium atoms disappear from a zircon crystal, and lead atoms accumulate in it. The ratio of uranium atoms to lead atoms in a zircon crystal can be measured precisely. That ratio can be used to calculate how long ago the crystal formed (its age).

zircon

We have just described **radiometric dating**, a method that can reveal the age of a material by measuring its content and proportions of a radioisotope and daughter elements. The oldest known terrestrial rock, a tiny zircon crystal from the Jack Hills in Western Australia, formed 4.4 billion years ago. More recent fossils that still contain carbon can be radiometrically dated by measuring their content of carbon 14 (FIGURE 16.13). Most of the ^{14}C in a fossil will have decayed after about 60,000 years. The age of fossils older than that can be estimated by dating volcanic rock in lava flows above and below the fossil-containing layer.

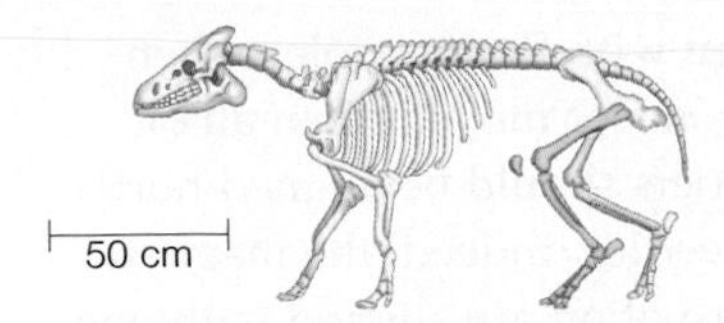

A *Elomeryx*, a small terrestrial animal that lived about 30 million years ago. This is a member of the same artiodactyl group (even-toed hooved mammals) that gave rise to modern representatives, including hippopotamuses. *Elomeryx* is thought to resemble a 60-million-year-old ancestor that it shares with whales.

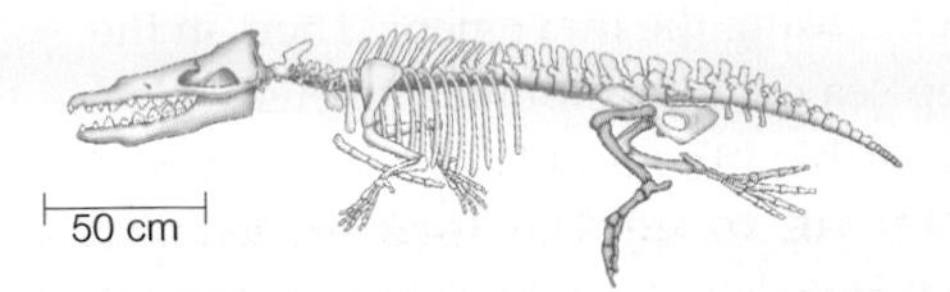

B *Rodhocetus kasrani*, an ancient whale that lived about 47 million years ago. Its distinctive ankle bones are evidence of a close evolutionary connection to artiodactyls. Artiodactyls are defined by the unique "double-pulley" shape of the bone (right) that forms the lower part of their ankle joint.

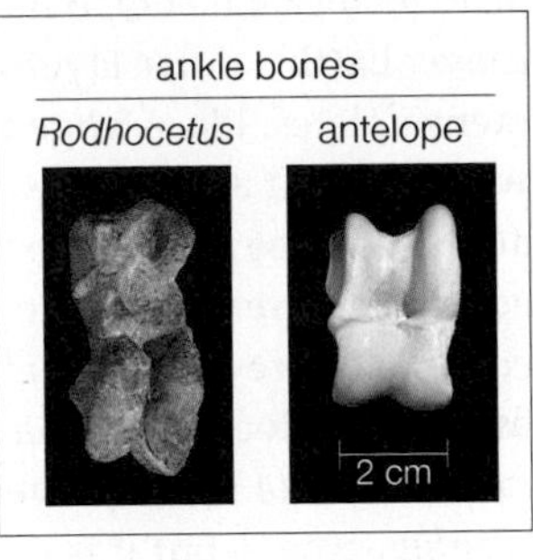

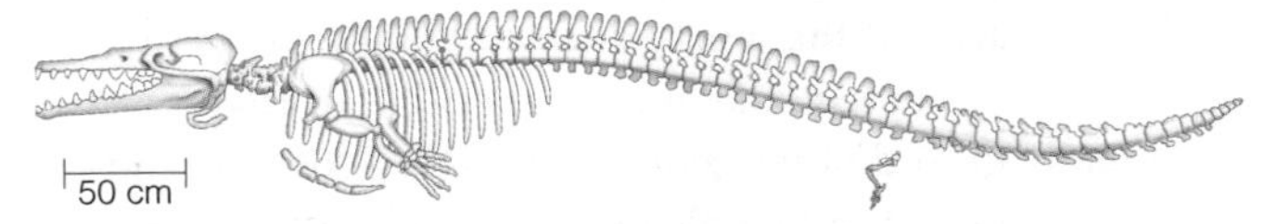

C *Dorudon atrox*, an ancient whale that lived about 37 million years ago. Its tiny, artiodactyl-like ankle bones were much too small to have supported the weight of its huge body on land, so this mammal had to be fully aquatic.

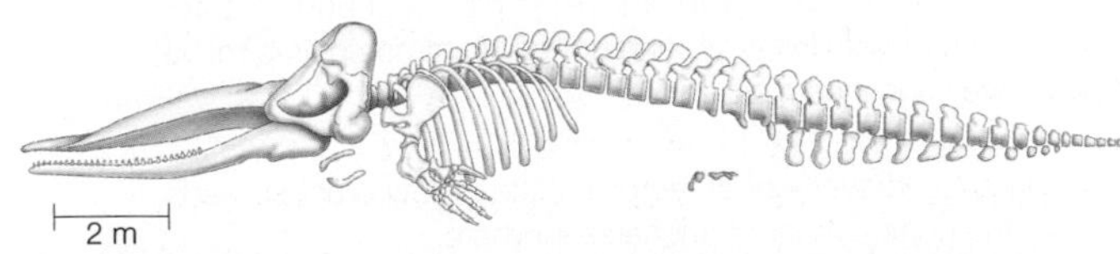

D Modern cetaceans such as the sperm whale have remnants of a pelvis and leg, but no ankle bones.

FIGURE 16.14 Comparison of cetacean skeletal features. The ancestor of whales was an artiodactyl that walked on land. Over millions of years, the lineage transitioned from life on land to life in water, and as it did, bones of the hindlimb (highlighted in blue) became smaller.

Rodhocetus and *Dorudon* are long-lost relatives of modern whales, not direct ancestors. Both were offshoots of the ancient-artiodactyl-to-modern-whale lineage during its transition from life on land to life in water.

half-life Characteristic time it takes for half of a quantity of a radioisotope to decay.
radiometric dating Method of estimating the age of a rock or a fossil based on the predictability of radioactive decay.

Missing Links

The discovery of intermediate forms of cetaceans (an order of animals that includes whales, dolphins, and porpoises) offers an example of how fossil finds and radiometric dating can be used to reconstruct evolutionary history. For some time, evolutionary biologists had thought that the ancestors of modern cetaceans walked on land, then took up life in the water. Evidence in support of this idea includes a set of distinctive features of the skull and lower jaw that cetaceans share with some kinds of ancient carnivorous land animals. DNA sequence comparisons indicate that the ancient land animals were probably artiodactyls, hooved mammals with an even number of toes (two or four) on each foot (**FIGURE 16.14A**). Modern representatives of the artiodactyl lineage include camels, hippopotamuses, pigs, deer, sheep, and cows.

Until recently, we had no fossils demonstrating gradual changes in skeletal features that would have accompanied a transition of whale lineages from terrestrial to aquatic life. Researchers knew there were intermediate forms because they had found a representative fossil skull of an ancient whalelike animal, but without a complete skeleton the rest of the story remained speculative. Then, in 2000, Philip Gingerich and his colleagues unearthed complete fossilized skeletons of two ancient whales: *Rodhocetus kasrani* (**FIGURE 16.14B**) excavated from a 47-million-year-old rock formation in Pakistan, and *Dorudon atrox* (**FIGURE 16.14C**), from 37-million-year-old rock in Egypt. Both fossil skeletons had whalelike skull bones, as well as intact ankle bones. The ankle bones of both fossils have distinctive features in common with those of extinct and modern artiodactyls. Modern cetaceans do not have even a remnant of an ankle bone (**FIGURE 16.14D**).

The proportions of limbs, skull, neck, and thorax indicate *Rodhocetus* swam with its feet, not its tail. Like modern whales, the 5-meter (16-foot) *Dorudon* was clearly a fully aquatic tail-swimmer: The entire hindlimb was only about 12 centimeters (5 inches) long, much too small to have supported the animal's tremendous body out of water.

TAKE-HOME MESSAGE 16.6

How is the age of rocks and fossils determined?

✔ The predictability of radioisotope decay can be used to estimate the age of rock layers and fossils in them.

✔ Radiometric dating helps evolutionary biologists retrace changes in ancient lineages.

16.7 Drifting Continents, Changing Seas

✔ Over billions of years, movements of Earth's outer layer of rock have changed the land, atmosphere, and oceans.

Wind, water, and other forces continuously sculpt Earth's surface, but they are only part of a much bigger picture of geological change. For example, all continents that exist today were once part of one huge supercontinent—**Pangea**—that split into fragments and drifted apart. The idea that continents move around, originally called continental drift, was proposed in the early 1900s to explain why the Atlantic coasts of South America and Africa seem to "fit" like jigsaw puzzle pieces, and why the same types of fossils occur in identical rock formations on both sides of the Atlantic Ocean. It also explained why the magnetic poles of gigantic rock formations point in different directions on different continents. As magma solidifies into rock, some iron-rich minerals in it become magnetic, and their magnetic poles align with Earth's poles when they do. If the continents never moved, then all of these ancient rocky magnets should be aligned north-to-south, like compass needles. Indeed, the magnetic poles of rocks in each formation are aligned with one another, but the alignment is not always north-to-south. Either Earth's magnetic poles veer dramatically from their north–south axis, or the continents wander.

The idea that continents move was initially greeted with skepticism because there was no known mechanism capable of causing the movement. Then, in the late 1950s, deep-sea explorers found immense ridges and trenches stretching thousands of kilometers across the seafloor (**FIGURE 16.15**). The discovery led to the **plate tectonics theory**, which explains how continents move: Earth's outer layer of rock is cracked into immense plates, like a huge cracked eggshell. Magma welling up at an undersea ridge ❶ or continental rift at one edge of a plate pushes old rock at the opposite edge into a trench ❷. The movement is like that of a colossal conveyor belt that transports continents on top of it to new locations. The plates move no more than 10 centimeters (4 inches) a year—about half as fast as your toenails grow—but it is enough to carry a continent all the way around the world after 40 million years or so.

Evidence of tectonic movement is all around us, in faults ❸ and other geological features of our landscapes. For example, volcanic island chains (archipela-

The San Andreas Fault, extending 800 miles in California, marks the boundary between two tectonic plates.

FIGURE 16.15 Plate tectonics. Huge pieces of Earth's outer layer of rock slowly drift apart and collide. As these plates move, they convey continents around the globe.

❶ At oceanic ridges, plumes of magma (red) welling up from Earth's interior drive the movement of tectonic plates. New crust spreads outward as it forms on the surface, forcing adjacent tectonic plates away from the ridge and into trenches elsewhere.

❷ At trenches, the advancing edge of one plate plows under an adjacent plate and buckles it.

❸ Faults are ruptures in Earth's crust where plates meet. The diagram shows a rift fault, in which plates move apart. The photo above shows a strike-slip fault, in which two abutting plates slip against one another in opposite directions.

❹ Plumes of magma rupture a tectonic plate at what are called "hot spots." The Hawaiian Islands have been forming from magma that continues to erupt at a hot spot under the Pacific Plate. This and other tectonic plates are shown in Appendix V.

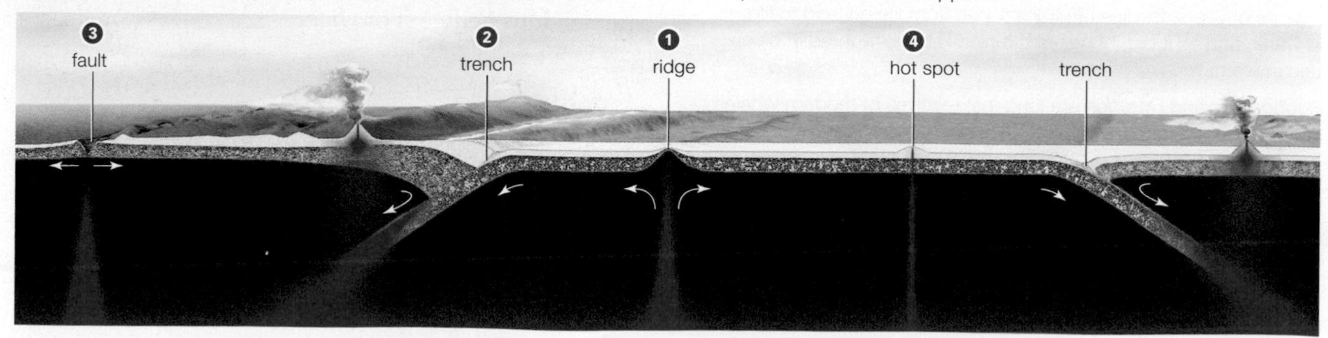

CREDITS: (15) top, Courtesy of SanAndreasFault.org; bottom, © Cengage Learning.

gos) form as a plate moves across an undersea hot spot. These hot spots are places where a plume of magma ruptures a tectonic plate ❹.

The fossil record also provides evidence in support of plate tectonics. Consider an unusual rock formation that exists in a huge belt across Africa. The sequence of rock layers in this formation is so complex that it is quite unlikely to have formed more than once, but identical sequences of layers also occur in huge belts that span India, South America, Madagascar, Australia, and Antarctica. Across all of these continents, the layers are the same ages. They also hold fossils found nowhere else, including imprints of the seed fern *Glossopteris* (pictured in **FIGURE 16.10C**). The most probable explanation for these observations is that the layered rock formed in one long belt on a single continent, which later broke up.

We now know that at least five times since Earth's outer layer of rock solidified 4.55 billion years ago, supercontinents formed and split up again. One called **Gondwana** formed about 500 million years ago. Over the next 230 million years, this supercontinent wandered across the South Pole, then drifted north until it merged with other landmasses to form Pangea (**FIGURE 16.16**). Most of the landmasses currently in the Southern Hemisphere as well as India and Arabia were once part of Gondwana. Some modern species, including the birds pictured in **FIGURE 16.2**, live only in these places.

Geologic changes brought on by plate tectonics have had a profound impact on life. For example, colliding continents have physically separated organisms living in oceans, and brought together those that had been living apart on land. As continents broke up, they separated organisms living on land, and brought together ones that had been living in separate oceans. These events have been a major driving force of evolution, as you will see in the next chapter.

Gondwana Supercontinent that existed before Pangea, more than 500 million years ago.
Pangea Supercontinent that formed about 270 million years ago.
plate tectonics theory Theory that Earth's outer layer of rock is cracked into plates, the slow movement of which conveys continents to new locations over geologic time.

TAKE-HOME MESSAGE 16.7
How has Earth changed over geologic time?

✔ Over geologic time, movements of Earth's crust have caused dramatic changes in continents and oceans. These changes profoundly influenced the course of life's evolution.

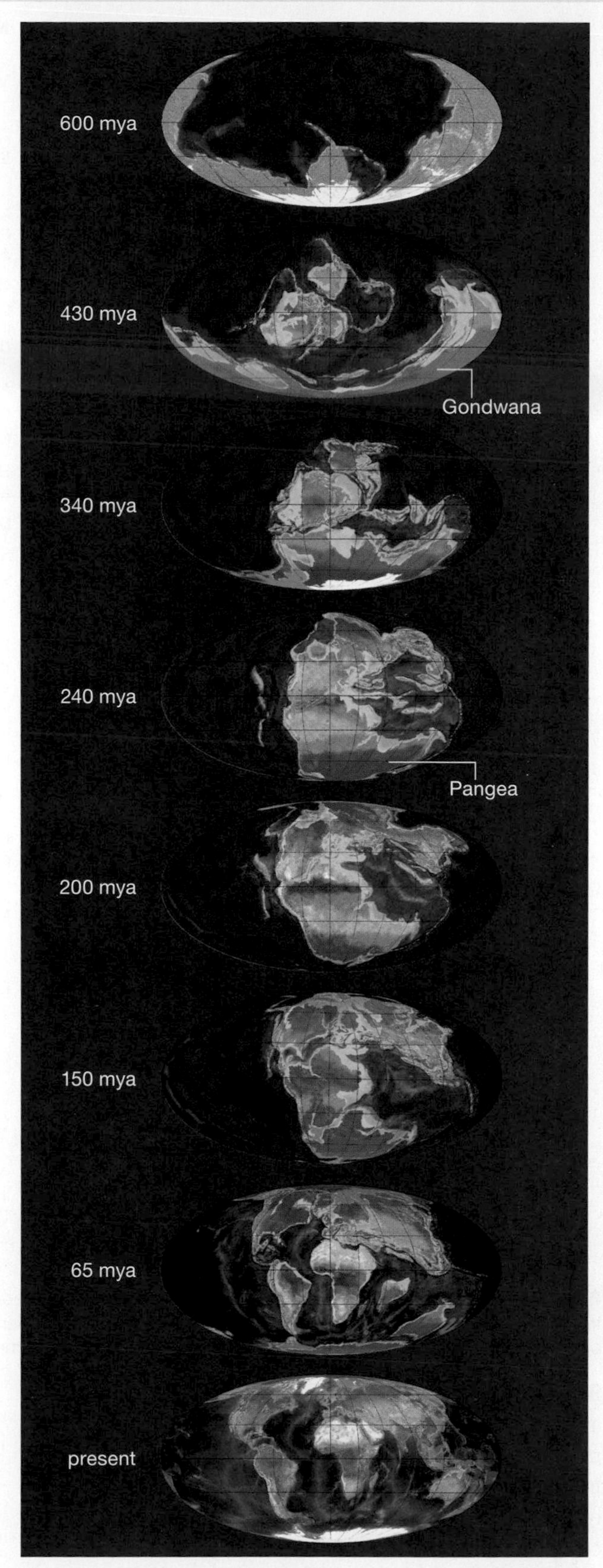

FIGURE 16.16 A series of reconstructions of the drifting continents. mya: million years ago.

16.8 Putting Time Into Perspective

Eon	Era	Period	Epoch	mya	Major Geologic and Biological Events
Phanerozoic	Cenozoic	Quaternary	Holocene	0.01	Modern humans evolve. Major extinction event is now under way.
			Pleistocene	2.6	
		Neogene	Pliocene	5.3	Tropics, subtropics extend poleward. Climate cools; dry woodlands and grasslands emerge. Adaptive radiations of mammals, insects, birds.
			Miocene	23.0	
		Paleogene	Oligocene	33.9	
			Eocene	56.0	
			Paleocene	66.0 ◀	Major extinction event
	Mesozoic	Cretaceous	Upper		Flowering plants diversify; sharks evolve. All dinosaurs and many marine organisms disappear at the end of this epoch.
				100.5	
			Lower		Climate very warm. Dinosaurs continue to dominate. Important modern insect groups appear (bees, butterflies, termites, ants, and herbivorous insects including aphids and grasshoppers). Flowering plants originate and become dominant land plants.
				145.0	
		Jurassic			Age of dinosaurs. Lush vegetation; abundant gymnosperms and ferns. Birds appear. Pangea breaks up.
				201.3 ◀	Major extinction event
		Triassic			Recovery from the major extinction at end of Permian. Many new groups appear, including turtles, dinosaurs, pterosaurs, and mammals.
				252 ◀	Major extinction event
	Paleozoic	Permian			Supercontinent Pangea and world ocean form. Adaptive radiation of conifers. Cycads and ginkgos appear. Relatively dry climate leads to drought-adapted gymnosperms and insects such as beetles and flies.
				299	
		Carboniferous			High atmospheric oxygen level fosters giant arthropods. Spore-releasing plants dominate. Age of great lycophyte trees; vast coal forests form. Ears evolve in amphibians; penises evolve in early reptiles (vaginas evolve later, in mammals only).
				359 ◀	Major extinction event
		Devonian			Land tetrapods appear. Explosion of plant diversity leads to tree forms, forests, and many new plant groups including lycophytes, ferns with complex leaves, seed plants.
				419	
		Silurian			Radiations of marine invertebrates. First appearances of land fungi, vascular plants, bony fishes, and perhaps terrestrial animals (millipedes, spiders).
				443 ◀	Major extinction event
		Ordovician			Major period for first appearances. The first land plants, fishes, and reef-forming corals appear. Gondwana moves toward the South Pole and becomes frigid.
				485	
		Cambrian			Earth thaws. Explosion of animal diversity. Most major groups of animals appear (in the oceans). Trilobites and shelled organisms evolve.
				541	
Precambrian	Proterozoic				Oxygen accumulates in atmosphere. Origin of aerobic metabolism. Origin of eukaryotic cells, then protists, fungi, plants, animals. Evidence that Earth mostly freezes over in a series of global ice ages between 750 and 600 mya.
				2,500	
	Archean and earlier				3,800–2,500 mya. Origin of bacteria and archaea. 4,600–3,800 mya. Origin of Earth's crust, first atmosphere, first seas. Chemical, molecular evolution leads to origin of life (from protocells to anaerobic single cells).

FIGURE 16.17 The geologic time scale (above) correlated with sedimentary rock exposed by erosion in the Grand Canyon (opposite). Red triangles mark times of great mass extinctions. "First appearance" refers to appearance in the fossil record, not necessarily the first appearance on Earth. mya: million years ago. Dates are from the International Commission on Stratigraphy, 2014.

Similar sequences of sedimentary rock layers occur around the world. Transitions between the layers mark boundaries between great intervals of time in the **geologic time scale**, which is a chronology of Earth's history (**FIGURE 16.17**). Each layer's composition offers clues about conditions on Earth during the time the layer was deposited. Fossils in the layers are a record of life during that period of time.

geologic time scale Chronology of Earth's history.

TAKE-HOME MESSAGE 16.8

What is the geologic time scale?

✔ The geologic time scale correlates geological and evolutionary events of the ancient past.

Each rock layer has a composition and set of fossils that reflect events during its deposition. For example, Coconino Sandstone, which stretches from California to Montana, is mainly weathered sand. Ripple marks and reptile tracks are the only fossils in it. Many think it is the remains of a vast sand desert, similar to the modern Sahara.

FIGURE IT OUT: Which formation is marked with the ?? Answer: Tapeats sandstone

Reflections of a Distant Past (revisited)

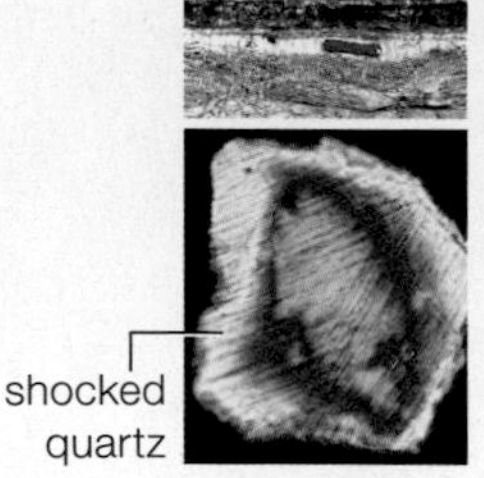

The K–Pg boundary sequence is unusually rich in iridium, an element rare on Earth's surface but common in asteroids. After researchers discovered the iridium, they looked for evidence of an asteroid impact massive enough to cover the entire Earth with extraterrestrial debris. In the Yucatán Peninsula, they found a crater so big that no one had realized it was a crater before. The K–Pg boundary sequence also contains shocked quartz and small glass spheres called tektites—rocks that form when quartz or sand (respectively) undergoes a sudden, violent application of extreme pressure. As far as we know, the only processes on Earth that produce shocked quartz and tektites are atomic bomb explosions and meteorite impacts.

summary

Section 16.1 Events of the ancient past can be explained by the same physical, chemical, and biological processes that operate today. An asteroid impact may have caused a mass extinction 66 million years ago.

Section 16.2 Expeditions by nineteenth-century naturalists yielded increasingly detailed observations of nature. Geology, **biogeography**, and **comparative morphology** of organisms and their **fossils** led to new ways of thinking about the natural world.

Section 16.3 Prevailing belief systems often influence interpretation of the cause of natural events. Nineteenth-century European naturalists proposed **catastrophism** and the **theory of uniformity** in their attempts to reconcile traditional beliefs with physical evidence of **evolution**, or change in a **lineage** over time.

Section 16.4 Humans select desirable traits in animals by selective breeding. Charles Darwin and Alfred Wallace independently came up with a theory of how environments also select traits, stated here in modern terms: A population tends to grow until it exhausts environmental resources. As that happens, competition for those resources intensifies among the population's members. Individuals with forms of shared, heritable traits that give them an advantage in this competition tend to produce more offspring. Thus, **adaptive traits** (**adaptations**) that impart greater **fitness** to an individual become more common in a population over generations. The process in which environmental pressures result in the differential survival and reproduction of individuals of a population is called **natural selection**. It is one of the processes that drives evolution.

Section 16.5 Fossils are typically found in stacked layers of sedimentary rock. Younger fossils usually occur in layers deposited more recently, on top of older fossils in older layers. Fossils of many organisms are relatively scarce, so the fossil record will always be incomplete.

Section 16.6 A radioisotope's characteristic **half-life** can be used to determine the age of rocks and fossils. This technique, **radiometric dating**, helps us understand the ancient history of many lineages.

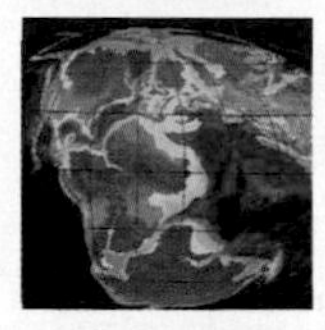

Section 16.7 According to the **plate tectonics theory**, Earth's crust is cracked into giant plates that convey landmasses to new positions as they move. Earth's landmasses have periodically converged as supercontinents such as **Gondwana** and **Pangea**.

Section 16.8 Transitions in the fossil record are the boundaries of great intervals of the **geologic time scale**, a chronology of Earth's history that correlates geologic and evolutionary events.

self-quiz

Answers in Appendix VII

1. The number of species on an island depends on the size of the island and its distance from a mainland. This statement would most likely be made by ________ .
   a. an explorer
   b. a biogeographer
   c. a geologist
   d. a philosopher

2. The bones of a bird's wing are similar to the bones in a bat's wing. This observation is an example of ________ .
   a. uniformity
   b. evolution
   c. comparative morphology
   d. a lineage

3. Evolution ________ .
   a. is natural selection
   b. is change in a line of descent
   c. can occur by natural selection
   d. b and c are correct

4. A trait is adaptive if it ________ .
   a. arises by mutation
   b. increases fitness
   c. is passed to offspring
   d. occurs in fossils

5. In which type of rock are you most likely to find a fossil?
   a. basalt, a dark, fine-grained volcanic rock
   b. limestone, composed of sedimented calcium carbonate
   c. slate, a volcanically melted and cooled shale
   d. granite, which forms by crystallization of magma cooling below Earth's surface

CREDITS: (in text revisited) left, © David A. Kring, NASA/Univ. Arizona Space Imagery Center; right, U.S. Geological Survey.

data analysis activities

Discovery of Iridium in the K–Pg Boundary Sequence In the late 1970s, geologist Walter Alvarez was investigating the composition of the K–Pg boundary sequence in different parts of the world. He asked his father, Nobel Prize–winning physicist Luis Alvarez, to help him analyze the elemental composition of the layer. The Alvarezes and their colleagues tested the K–Pg boundary sequence in Italy and Denmark. They discovered that it contains a much higher iridium content than the surrounding rock layers. Some of their results are shown in **FIGURE 16.18**.

Iridium belongs to a group of elements (Appendix I) that are much more abundant in asteroids and other solar system materials than they are in Earth's crust. The Alvarez group concluded that the K–Pg boundary sequence must have originated with extraterrestrial material. They calculated that an asteroid 14 kilometers (8.7 miles) in diameter would contain enough iridium to account for the extra iridium in the K–Pg boundary sequence.

1. What was the iridium content of the K–Pg boundary sequence?
2. How much higher was the iridium content of the boundary sequence than the sample taken 0.7 meter above the sequence?

Sample Depth	Average Abundance of Iridium (ppb)
+ 2.7 m	< 0.3
+ 1.2 m	< 0.3
+ 0.7 m	0.36
boundary layer	41.6
– 0.5 m	0.25
– 5.4 m	0.30

FIGURE 16.18 Abundance of iridium in and near the K–Pg boundary sequence in Stevns Klint, Denmark. Many rock samples taken from above, below, and at the boundary were tested for iridium content. Depths are given as meters above or below the boundary.

The iridium content of an average Earth rock is 0.4 parts per billion (ppb) of iridium. An average meteorite contains about 550 parts per billion of iridium.

The photo shows Luis and Walter Alvarez with a section of the boundary sequence.

6. Which of the following is a fossil?
   a. an insect encased in 10-million-year-old tree sap
   b. a woolly mammoth frozen in Arctic permafrost for the last 50,000 years
   c. mineral-hardened remains of a whalelike animal found in an Egyptian desert
   d. an impression of a plant leaf in a rock
   e. all of the above can be considered fossils

7. If the half-life of a radioisotope is 20,000 years, then a sample in which three-quarters of that radioisotope has decayed is ______ years old.
   a. 15,000 b. 26,667 c. 30,000 d. 40,000

8. Did Pangea or Gondwana form first?

9. The dinosaurs died out ______ million years ago.

10. On the geologic time scale, life originated in the ______ .
   a. Archean
   b. Proterozoic
   c. Phanerozoic
   d. Cambrian

11. Match the terms with the most suitable description.

___ fitness	a. measured by reproductive success
___ fossils	b. geologic change occurs continuously
___ natural selection	c. geologic change occurs in unusual major events
___ half-life	d. evidence of life in distant past
___ catastrophism	e. survival of the fittest
___ uniformity	f. characteristic of a radioisotope

12. Forces that cause geologic change include ______ (select all that are correct).
   a. erosion
   b. natural selection
   c. volcanic activity
   d. tectonic plate movement
   e. wind
   f. meteorite impacts

critical thinking

1. Radiometric dating does not measure the age of an individual atom. It is a measure of the age of a quantity of atoms—a statistic. As with any statistical measure, its values may deviate around an average (see sampling error, Section 1.8). Imagine that one sample of rock is dated ten different ways. Nine of the tests yield an age close to 225,000 years. One test yields an age of 3.2 million years. Do the nine consistent results imply that the one that deviates is incorrect, or does the one odd result invalidate the nine that are consistent?

2. If you think of geologic time spans as minutes, life's history might be plotted on a clock such as the one shown below. According to this clock, the most recent epoch started in the last 0.1 second before noon. Where does that put you?

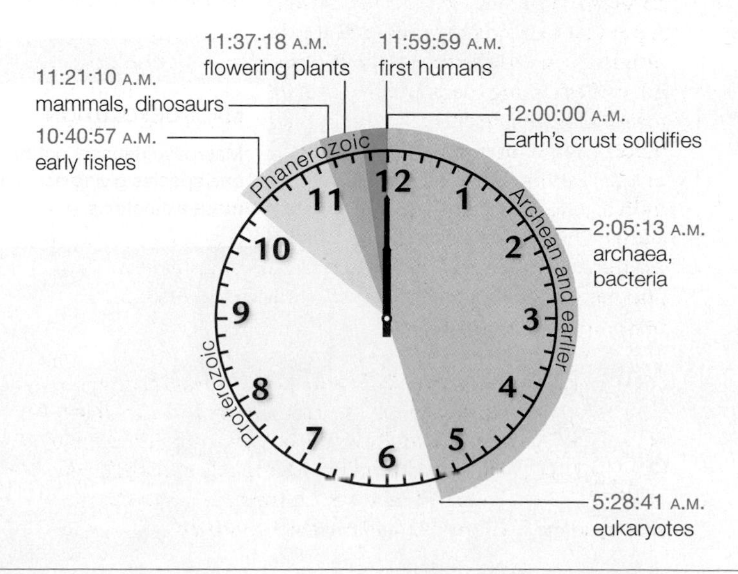

CREDITS: (18) left, Lawrence Berkeley National Laboratory; right, © Cengage Learning; (CT #2) © Cengage Learning.

17 Processes of Evolution

LEARNING ROADMAP

A review of species (Section 1.5), mutations (9.6), sexual reproduction (12.1–12.3), inheritance (13.2, 13.3), traits (13.5–13.7, 14.2), and natural selection (16.4) will be helpful. We revisit experiments (1.6, 1.8), homeotic genes (10.3), polyploidy (14.6), genetic screening (14.7), transgenic plants (15.7), plate tectonics (16.7), and the geologic time scale (16.8).

MICROEVOLUTION

Members of a population inherit different alleles, which are the basis of differences in phenotype. An allele may increase or decrease in frequency in a population, a change called microevolution.

PATTERNS OF NATURAL SELECTION

Natural selection is one of the processes that drive microevolution. Depending on the population and its environment, natural selection can shift or maintain a range of variation in a heritable trait.

OTHER PROCESSES OF MICROEVOLUTION

With genetic drift, change can occur in a line of descent by chance alone. Gene flow counters the evolutionary effects of mutation, natural selection, and genetic drift.

HOW SPECIES ARISE

Speciation typically starts after gene flow ends. Microevolution leads to genetic divergences, which are reinforced as mechanisms evolve that prevent interbreeding.

MACROEVOLUTION

Macroevolutionary patterns include the origin of major groups, one species giving rise to many, two species evolving jointly, and mass extinctions.

Chapter 18 explores the techniques we use to keep track of species and evolutionary patterns. Later chapters return to polyploid plants (Section 29.9), the genetic basis of behavior (Section 43.2), mating behaviors (Section 43.7), examples of natural selection at work in populations (Section 44.6), coevolved species (45.6), and competition and other interactions between species (Chapter 45).

17.1 Superbug Farms

Scarlet fever, tuberculosis, and pneumonia once caused one-fourth of the annual deaths in the United States. Since the 1940s, we have been relying on antibiotics to fight these and other dangerous bacterial diseases. We have also been using them in other, less dire circumstances. For an as-yet-unknown reason, antibiotics promote growth in cattle, pigs, poultry, and even fish. The agricultural industry uses a lot of antibiotics, mainly for this purpose. In 2011, 13.7 million kilograms (about 30 million pounds) of antibiotics were used for agriculture in the U.S.—more than four times the amount used to treat people in the same year.

A natural population of bacteria is diverse, and it can evolve astonishingly fast. Consider how each cell division is an opportunity for mutation (Section 8.6). The common intestinal bacteria *E. coli* can divide every 17 minutes, so even if a population starts out as clones, its cells diversify quickly. In addition, bacteria share DNA even among different species, and this adds even more genetic diversity to their populations.

Genetic diversity is an advantage in a changing environment (Section 12.1). When a natural population of bacteria is exposed to a selection pressure such as an antibiotic, some cells in the population are likely to survive because they carry an allele that offers an advantage—antibiotic resistance, in this case. As susceptible cells die and the survivors reproduce, the frequency of antibiotic-resistance alleles in the population increases. A typical two-week course of treatment with antibiotics can exert selection pressure on over a thousand generations of bacteria. The pressure drives genetic change in bacterial populations so they become composed mainly of antibiotic resistant cells. Thus, the practice of treating livestock with growth-promoting antibiotics essentially guarantees the production of antibiotic-resistant bacterial populations (**FIGURE 17.1**).

FIGURE 17.1 The vast majority of chickens raised for meat in the United States spend their lives in gigantic flocks that crowd huge buildings like this one. Growth-promoting antibiotics are given to the entire flock in food, a practice that pressures normal bacterial populations to become antibiotic resistant.

Farms where antibiotics are used to promote growth are hot spots for the spread of resistant bacteria to humans. Veterinarians and other people who work with the animals on these farms tend to carry more antibiotic-resistant bacteria in their bodies. So do neighbors who live within a mile. The bacteria spread much farther than the farm, however. Bacteria on an animal's skin or in its digestive tract can easily contaminate its meat during slaughter (Section 4.1), and contaminated meat ends up in restaurant and home kitchens. A 2013 investigation found "worrisome" amounts of bacteria in 97% of the chicken meat in stores across the United States. About half of the samples tested were contaminated with superbugs—bacteria that are resistant to multiple antibiotics—and one in ten contained multiple superbug species. An earlier study found antibiotic-resistant bacteria in more than half of supermarket ground beef and pork chops, and in over 80 percent of ground turkey. Bacteria can be killed by the heat of cooking, but it is almost impossible to prevent them from spreading from contaminated meat to kitchen surfaces—and to people—during the process.

We have only a limited number of antibiotic drugs, and developing new ones is much slower than bacterial evolution. As resistant bacteria become more common, the number of antibiotics that can be used to effectively treat infections in humans dwindles. Using a particular antibiotic only in animals, or only in humans, is not a solution to this problem; there are only a few mechanisms by which these drugs kill bacteria, so resistance to one antibiotic often confers resistance to others. For example, bacteria that are resistant to flavomycin (a phosphoglycolipid antibiotic used only in animals) also resist vancomycin (a glycopeptide antibiotic used only in humans). Superbugs resistant to most currently available antibiotics are turning up at an alarming rate.

All of this amounts to bad news. An infection with antibiotic-resistant bacteria tends to be longer, more severe, and more likely to be deadly than one more easily treatable with antibiotics. Superbugs cause more than 2 million cases of serious illness each year in the United states alone; they outright kill 23,000 of these people. Many, many more die because the infection complicates another, preexisting illness.

17.2 Individuals Don't Evolve, Populations Do

✔ Mutations in individuals are the original source of new alleles in a population's pool of genetic resources.

✔ A change in an allele's frequency in a population is called microevolution.

Table 17.1 Some Sources of Variation in Shared Traits

Genetic Event	Effect
Mutation	Original source of new alleles
Crossing over at meiosis I	Introduces new combinations of alleles into chromosomes
Independent assortment at meiosis I	Mixes maternal and paternal chromosomes
Fertilization	Combines alleles from two parents
Changes in chromosome number or structure	Often dramatic changes in structure and function

Alleles in Populations

Section 1.2 introduced a **population** as a group of interbreeding individuals of the same species in some specified area. The individuals of a population (and a species) share certain features. For example, giraffes normally have long necks, brown spots on white coats, and so on. These are examples of morphological traits (*morpho–* means form). Individuals of a species also share physiological traits, such as metabolic activities. They also respond the same way to certain stimuli, as when hungry giraffes feed on tree leaves. These are behavioral traits.

Members of a population have the same traits because they have the same genes. However, almost every shared trait varies a bit among individuals of a population (FIGURE 17.2). Alleles of the shared genes are the basis of this variation. Many traits have two or more distinct forms, or morphs. A trait with only two forms is dimorphic (*di–* means two). Purple and white flower color in the pea plants that Gregor Mendel studied is an example of a dimorphic trait (Section 13.3). Dimorphic flower color occurs in this case because the interaction of two alleles with a clear dominance relationship gives rise to the trait. Traits with more than two distinct forms are polymorphic (*poly–*, many). ABO blood type in humans, which is determined by the codominant alleles of the *ABO* gene, is an example (Section 13.5). The genetic basis of traits that vary continuously among the individuals of a population is typically quite complex (Sections 13.5–13.7). Any or all of the genes that influence such traits may have multiple alleles.

In earlier chapters, you learned about genetic events that contribute to the variation in shared traits we see among individuals of a population (TABLE 17.1). Mutation is the original source of new alleles. Other events shuffle alleles into different combinations, and what a shuffle that is! There are $10^{116,446,000}$ possible combinations of human alleles. Not even 10^{10} people are living today. Unless you have an identical twin, it is extremely unlikely that another person with your precise genetic makeup has ever lived, or ever will.

An Evolutionary View of Mutations

Being the original source of new alleles, mutations are worth another look, this time in the context of their impact on populations. We cannot predict when or in which individual a particular gene will mutate. We can, however, predict the average mutation rate of a species, which is the probability that a mutation will occur in a given interval. In the human species, that rate is about 2.2×10^{-9} mutations per base pair per year. In other words, about 70 nucleotides in the human genome sequence change every decade.

In humans at least, most mutations are neutral. A **neutral mutation** changes the DNA sequence of a chromosome, but the alteration has no effect on survival or reproduction—it neither helps nor hurts the individual. For instance, if you carry a mutation that keeps your earlobes attached to your head instead of swinging freely, attached earlobes should not in itself stop you from surviving and reproducing as well as anybody else. So, natural selection would not affect the frequency of this mutation in the human population.

Some mutations give rise to structural, functional, or behavioral alterations that reduce an individual's chances of surviving and reproducing. Even one biochemical change may be devastating. For instance, the skin, bones, tendons, lungs, blood vessels, and other vertebrate organs incorporate the protein collagen. If one of the genes for collagen mutates in a way that changes the protein's function, the entire body may be affected. A mutation such as this can change phenotype so drastically that it results in death, in which case it is a **lethal mutation**.

Occasionally, a change in the environment favors a mutation that had previously been neutral or even somewhat harmful. Even if a beneficial mutation bestows only a slight advantage, its frequency tends to increase in a population over time. This is because

FIGURE 17.2 Sampling morphological variation among zigzag Nerite snails (left) and humans (right). Variation in shared traits among individuals is mainly an outcome of variations in alleles that influence those traits.

natural selection operates on traits with a genetic basis. With natural selection, remember, environmental pressures result in an increase in the frequency of an adaptive form of a trait in a population over generations (Section 16.4). Mutations have been altering genomes for billions of years, and they continue to do so. Cumulatively, mutations have given rise to Earth's staggering biodiversity. Think about it: The reason you do not look like an avocado or an earthworm or even your next-door neighbor began with mutations that occurred in different lines of descent.

Allele Frequencies

Together, all the alleles of all the genes of a population make up a pool of genetic resources—a **gene pool**. Members of a population breed with one another more often than they breed with members of other populations, so their gene pool is more or less isolated. **Allele frequency** refers to the abundance of a particular allele among the individuals of a population, expressed as a fraction of the total number of alleles. Any change in an allele's frequency in the gene pool of a population (or a species) is called **microevolution**.

Microevolution is always occurring in natural populations because, as you will see in the next section, processes that drive it are always operating in nature. The remaining sections of this chapter explore some of the processes and evolutionary effects of mutation, natural selection, genetic drift, and gene flow. As you learn about these patterns, remember an important point: Evolution is not purposeful; it simply fills the nooks and crannies of opportunity.

allele frequency Abundance of a particular allele among members of a population.
gene pool All the alleles of all the genes in a population; a pool of genetic resources.
lethal mutation Mutation that alters phenotype so drastically that it causes death.
microevolution Change in an allele's frequency in a population.
neutral mutation A mutation that has no effect on survival or reproduction.
population A group of organisms of the same species who live in a specific location and breed with one another more often than they breed with members of other populations.

TAKE-HOME MESSAGE 17.2
What is microevolution?

✔ Individuals of a natural population share morphological, physiological, and behavioral traits characteristic of the species. Alleles are the main basis of differences in the details of those shared traits.

✔ All alleles of all individuals in a population make up the population's gene pool. An allele's abundance in the gene pool is called its allele frequency.

✔ Microevolution is change in allele frequency. It is always occurring in natural populations because processes that drive it are always operating.

17.3 Genetic Equilibrium

✔ Natural populations are always evolving.

✔ Researchers trace evolution within a population by tracking deviations from a baseline of genetic equilibrium.

Early in the twentieth century, Godfrey Hardy (a mathematician) and Wilhelm Weinberg (a physician) independently applied the rules of probability to population genetics. Both realized that, under certain theoretical conditions, allele frequencies in a sexually reproducing population's gene pool would remain stable from one generation to the next. The population would stay in this stable state, called **genetic equilibrium**, as long as all of the following five conditions are met:

1. Mutations never occur.
2. The population is infinitely large.
3. The population is isolated from all other populations (no individual enters or leaves).
4. Mating is random.
5. All individuals survive and produce the same number of offspring.

As you can imagine, all five of these conditions never occur in nature, so natural populations are never in genetic equilibrium.

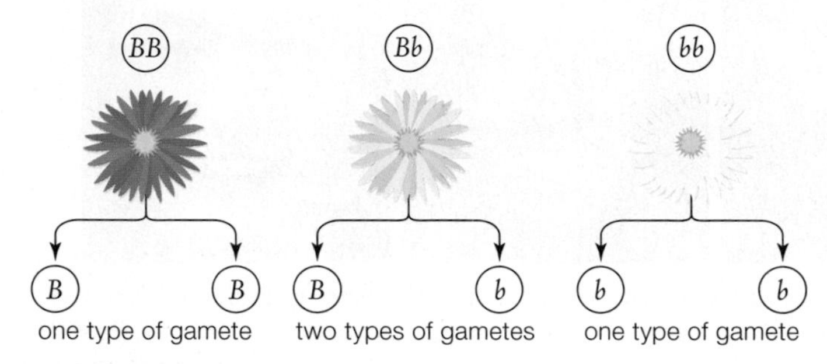

A In this two-allele system, *B* specifies dark blue flowers; *b*, white. Plants that are homozygous (*BB* or *bb*) make one kind of gamete. Heterozygous plants (*Bb*) have light blue flowers and make two kinds of gametes (*B* and *b*).

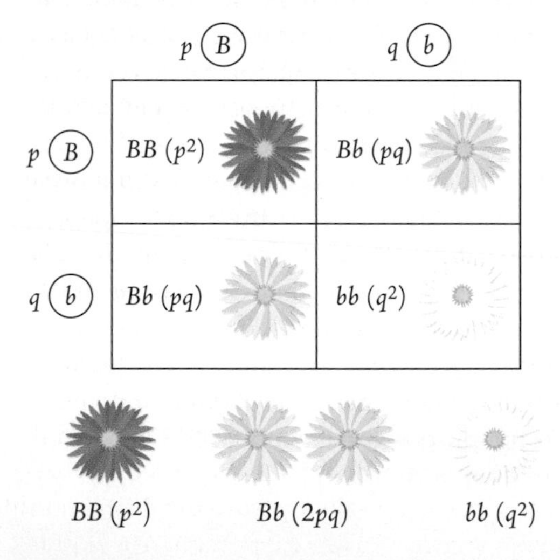

B Say p is the proportion of *B* alleles in the gene pool, and q is the proportion of *b* alleles. This Punnett square shows that in each generation of a randomly mating population, the predicted proportion of offspring that will inherit two *B* alleles is $p \times p$, or p^2. Likewise, the proportion that will inherit both alleles is $2pq$, and the proportion that will inherit two *b* alleles is q^2.

FIGURE 17.3 ▸Animated Hardy–Weinberg calculations. In this example, two alleles show incomplete dominance over flower color.

FIGURE IT OUT If 1/4 of this population has dark blue flowers and 1/4 has white flowers, what proportion of the next generation will have light blue flowers (assuming genetic equilibrium)?

Answer: Half of the gametes have a *B* allele; the other half have a *b* allele. Both p and q = 0.5, so $2pq$ = 50 percent.

Applying the Hardy–Weinberg Law

The concept of genetic equilibrium under ideal conditions is called the Hardy–Weinberg law. To see how it works, consider a hypothetical gene that encodes a blue pigment in daisies. A plant homozygous for one allele (*BB*) has dark blue flowers. A plant homozygous for the other allele (*bb*) has white flowers. These two alleles are inherited in a pattern of incomplete dominance, so a heterozygous plant (*Bb*) has medium-blue flowers (**FIGURE 17.3A**).

Start with the concept that allele frequencies always add up to one. For a gene with two alleles, the following equation is true:

$$p + q = 1.0$$

where p is the frequency of one allele in the population, and q is the frequency of the other. These alleles assort into different gametes during meiosis (Section 13.3), and then meet up at fertilization. If our hypothetical population of plants mate at random, the fraction of offspring that inherit two *B* alleles (*BB*) is $p \times p$, or p^2; the fraction that inherit two *b* alleles (*bb*) is q^2; and the fraction that inherit one *B* allele and one *b* allele (*Bb*) is $2pq$ (**FIGURE 17.3B**). Note that the frequencies of the three genotypes, whatever they may be, add up to 1.0:

$$p^2 + 2pq + q^2 = 1.0$$

Imagine that a population of daisies consists of 1,000 plants: 490 homozygous (*BB*), 420 heterozygous (*Bb*), and 90 homozygous (*bb*), and each of these individuals makes just two gametes. The *BB* individuals make 980 gametes, all with the *B* allele. The *Bb* individuals make 840 gametes, half (420) with the *B* allele. Thus, the frequency of the *B* allele among the pool of gametes is:

$$B\ (p) = \frac{980 + 420}{2{,}000 \text{ alleles}} = \frac{1{,}400}{2{,}000} = 0.7$$

The *bb* individuals make 180 gametes, all with the *b* allele. The other half of the 840 gametes made by the heterozygous individuals also have the b allele. Thus, the frequency of the b allele among the population's pool of gametes is:

$$b\ (q) = \frac{180 + 420}{2{,}000 \text{ alleles}} = \frac{600}{2{,}000} = 0.3$$

With $p = 0.7$ and $q = 0.3$, the proportion of genotypes among the next generation of individuals should be:

$$\begin{array}{llcccc} BB & (p^2) & = & (0.7)^2 & = & 0.49 \\ Bb & (2pq) & = & 2\,(0.7 \times 0.3) & = & 0.42 \\ bb & (q^2) & = & (0.3)^2 & = & 0.09 \end{array}$$

These proportions are the same as the ones in the parent population. As long as the five conditions required for genetic equilibrium are met, traits specified by the alleles should show up in the same proportions in each generation. If they do not, the population is evolving.

Real-World Situations

Genetic equilibrium is often used as a benchmark. For example, researchers used it to determine the carrier frequency of an allele that causes a genetic disorder called hereditary hemochromatosis (HH). Individuals affected by HH absorb too much iron from their food, and this causes liver problems, fatigue, and arthritis. The allele is inherited in an autosomal recessive pattern, so carriers show no symptoms. The researchers found the allele's frequency among people of Irish ancestry to be 14 percent. If $q = 0.14$, then p, the frequency of the normal allele, must be 0.86. Thus, the carrier frequency, $2pq$, was calculated to be 0.24 (24 percent of the population).

As another example, consider the *BRCA* genes, mutations in which are linked to breast cancer (Section 15.1). A deviation from predicted allele frequencies suggested that *BRCA* mutations have effects even before birth, so researchers investigated the frequency of mutated alleles of these genes among newborn girls. They found fewer individuals homozygous for these alleles than expected, based on the number of heterozygous individuals. Thus, in homozygous form, *BRCA* mutations impair the survival of female embryos.

genetic equilibrium Theoretical state in which an allele's frequency never changes in a population's gene pool.

TAKE-HOME MESSAGE 17.3

How do we know when a population is evolving?

✔ Researchers measure genetic change by comparing it with a theoretical baseline of genetic equilibrium.

17.4 Patterns of Natural Selection

✔ Natural selection occurs in different patterns depending on the organisms involved and their environment.

The rest of this chapter explores the mechanisms and effects of natural selection and other processes that drive evolution. Remember from Section 16.4 that natural selection is a process in which environmental pressures result in the differential survival and reproduction of individuals of a population based on their shared, heritable traits. It influences the frequency of alleles in a population by operating on traits with a genetic basis.

We observe different patterns of natural selection. In some cases, individuals with a trait at one extreme of a range of variation are selected against, and forms at the other extreme are adaptive. We call this directional selection. With stabilizing selection, midrange forms of a trait are adaptive, and extremes are selected against. With disruptive selection, forms at the extremes of the range of variation are adaptive, and intermediate forms are selected against.

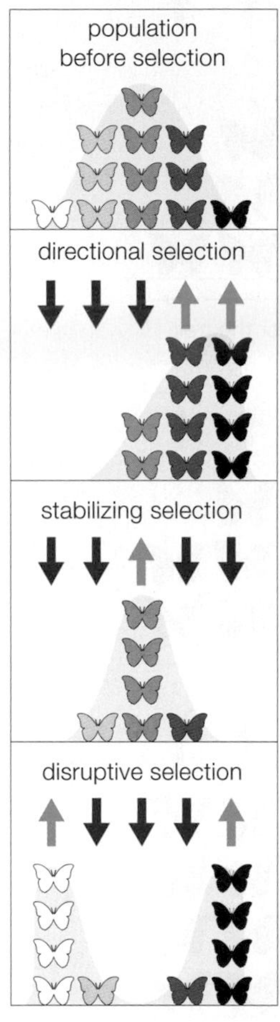

FIGURE 17.4 Overview of three modes of natural selection.

These modes of natural selection, which **FIGURE 17.4** summarizes, are discussed in the next two sections.

Section 17.7 explores sexual selection, a mode of natural selection that operates on a population by influencing mating success. This section also discusses balanced polymorphism, a particular case in which natural selection maintains a relatively high frequency of multiple alleles in a population. Natural selection and other processes of evolution can alter a population so much that it becomes a new species. We discuss mechanisms of speciation in the final sections.

TAKE-HOME MESSAGE 17.4

How do allele frequencies change?

✔ Natural selection, one of the most influential processes of evolution, can increase, decrease, or maintain the frequency of specific traits in a population.

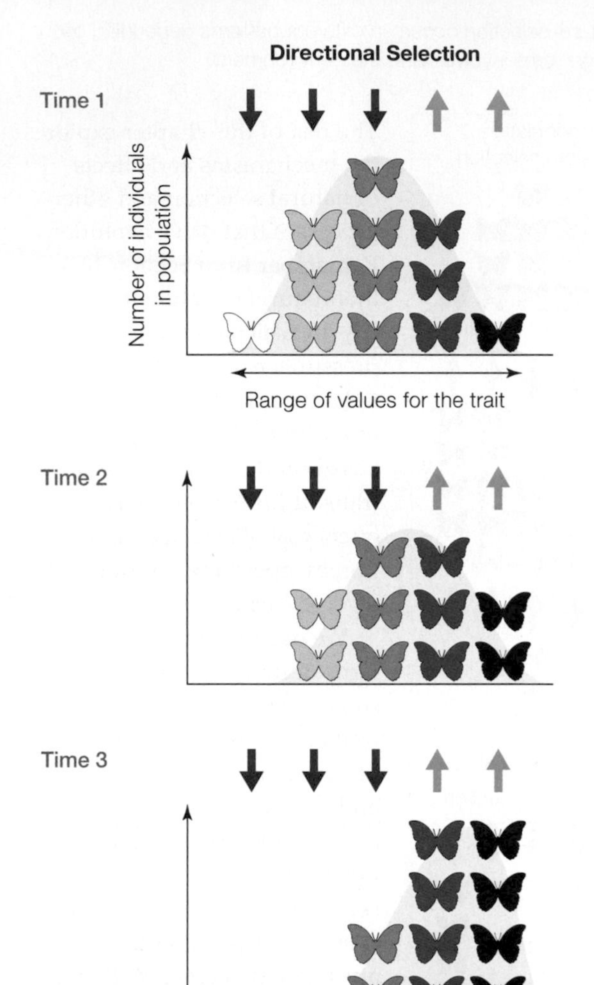

FIGURE 17.5 ▸Animated With directional selection, a form of a trait at one end of a range of variation is adaptive. Bell-shaped curves indicate continuous variation. Red arrows indicate which forms are being selected against; green, forms that are adaptive.

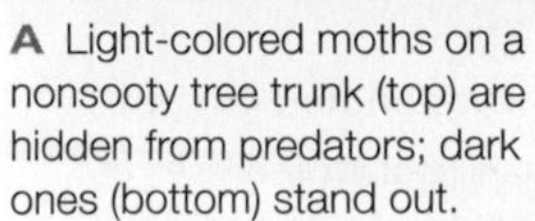

A Light-colored moths on a nonsooty tree trunk (top) are hidden from predators; dark ones (bottom) stand out.

B In places where soot darkens tree trunks, the dark color (bottom) provides more camouflage than the light color (top).

FIGURE 17.6 ▸Animated Adaptive value of two color forms of the peppered moth.

✔ Changing environmental conditions can result in a directional shift in an allele's frequency.

Directional selection shifts allele frequencies in a consistent direction, so forms at one end of a range of phenotypic variation become more common over time (**FIGURE 17.5**). Antibiotic use that fosters resistant bacterial populations is one example of directional selection. Additional examples follow.

Examples of Directional Selection

The Peppered Moth A well-documented case of directional selection involves coloration changes in peppered moths. These moths feed and mate at night, then rest on trees during the day. In preindustrial England, the vast majority of peppered moths were white with black speckles, and a small number were much darker. At this time, the air was clean, and light-gray lichens grew on the trunks and branches of most trees. When light-colored moths rested on lichen-covered trees, they were well camouflaged, whereas darker moths were not (**FIGURE 17.6A**). By the 1850s, the industrial revolution had begun, and smoke emitted by coal-burning factories was killing the lichens. Dark moths, which were better camouflaged on lichen-free, soot-darkened trees, had become more common (**FIGURE 17.6B**).

Scientists suspected that predation by birds was the selective pressure that shaped moth coloration, and in the 1950s, H. B. Kettlewell set out to test this hypothesis. He bred dark and light moths in captivity, marked them for easy identification, then released them in several areas. His team recaptured more of the dark moths in the polluted areas and more light ones in the less polluted ones. The researchers also observed predatory birds eating more light-colored moths in soot-darkened forests, and more dark-colored moths in cleaner, lichen-rich forests. Dark-colored moths were clearly at a selective advantage in industrialized areas.

Pollution controls went into effect in 1952. As a result of improved environmental standards, tree trunks gradually became free of soot, and lichens made a comeback. Kettlewell observed that moth phenotypes shifted too: Wherever pollution decreased, the frequency of dark moths decreased as well. Recent research has confirmed Kettlewell's results implicating birds as selective agents of peppered moth coloration. It has also shown that coloration in peppered moths is determined by a single gene. Individuals with a dominant allele of this gene are dark; those homozygous for a recessive allele are light.

CREDITS: (5) © Cengage Learning; (6) J. A. Bishop, L. M. Cook.

Rock Pocket Mice Directional selection also affects the color of rock pocket mice in Arizona's Sonoran Desert. Rock pocket mice are small mammals that spend the day sleeping in underground burrows, emerging at night to forage for seeds. Light brown granite dominates their environment, but there are also patches of dark basalt: the remains of ancient lava flows. Most of the mice in populations that inhabit the dark rock have dark gray coats (**FIGURE 17.7A**). Most of the mice in populations that inhabit the light brown rock have light brown coats (**FIGURE 17.7B**). The difference arises because mice that match the rock color in each habitat are camouflaged from their natural predators. Night-flying owls more easily see mice that do not match the rocks, and they preferentially eliminate easily seen mice from each population. Thus, in both habitats, selective predation has resulted in a directional shift in the frequency of alleles that affect coat color.

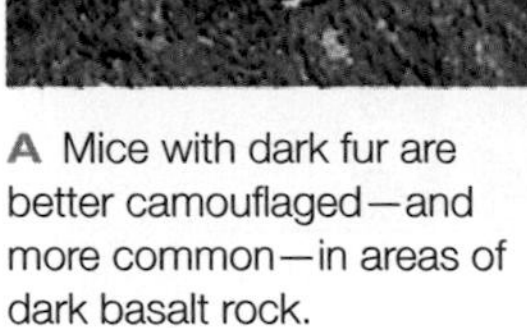

A Mice with dark fur are better camouflaged—and more common—in areas of dark basalt rock.

B Mice with light fur are better camouflaged—and more common—in areas dominated by light-colored granite.

FIGURE 17.7 Directional selection in the rock pocket mouse (*Chaetodipus intermedius*). Predators preferentially eliminate individuals with coat colors that do not match their surroundings in the Sonoran Desert.

Warfarin Resistance in Rats Rats thrive in urban centers, where garbage is plentiful and natural predators are not. Part of their success stems from an ability to reproduce very quickly: Rat populations can expand within weeks to match the amount of garbage available for them to eat. For decades, people have been fighting back with poisons. Baits laced with warfarin, an organic compound that interferes with blood clotting, became popular in the 1950s. Rats that ate the poisoned baits died within days after bleeding internally or losing blood through cuts or scrapes. Warfarin was extremely effective, and its impact on harmless species was much lower than that of other rat poisons. It quickly became the rat poison of choice. By 1980, however, about 10 percent of rats in urban areas were resistant to warfarin. What happened?

Warfarin interferes with blood clotting because it inhibits the function of an enzyme called VKORC1. This enzyme regenerates vitamin K, which participates as a cofactor in the post-translational modification of blood clotting factors (Section 14.4). When vitamin K is not regenerated, the clotting factors are not properly processed, and clotting cannot occur. Rats resistant to warfarin have a mutated version of the *VKORC1* gene; the enzyme encoded by this allele is insensitive to warfarin.

"What happened" was evolution by natural selection. Rats with the normal allele died after eating warfarin; the lucky ones with a mutated allele survived and passed it to their offspring. The rat populations recovered quickly, and a higher proportion of individuals in the next generation carried a mutation. With each onslaught of warfarin, the frequency of the mutation in rat populations increased. Exposure to warfarin had exerted directional selection.

The mutation that results in warfarin resistance also reduces the activity of the VKORC1 enzyme, so rats that have it require a lot of extra vitamin K. However, being vitamin K deficient is not so bad when compared with being dead from rat poison. In the absence of warfarin, though, rats with the allele are at a serious disadvantage because they cannot easily obtain enough vitamin K from their diet to sustain normal blood clotting and bone formation. Thus, the frequency of a warfarin resistance allele in a rat population declines quickly after warfarin exposure ends.

directional selection Mode of natural selection in which phenotypes at one end of a range of variation are favored.

TAKE-HOME MESSAGE 17.5

What is directional selection?

✔ With directional selection, a range of variation in a trait shifts in a consistent direction. The frequency of alleles underlying the trait also shifts directionally.

17.6 Stabilizing and Disruptive Selection

✔ Stabilizing selection is a mode of natural selection in which an intermediate phenotype is adaptive.

✔ Disruptive selection is a mode of natural selection in which extreme forms of a trait are adaptive.

Natural selection does not always result in a directional shift in a population's range of phenotypes. In some cases, environmental pressures favor a midrange form of a trait; in others, a midrange form is eliminated and the most extreme forms are adaptive.

Stabilizing Selection

With **stabilizing selection**, an intermediate form of a trait is favored, and extreme forms are selected against (**FIGURE 17.8**).

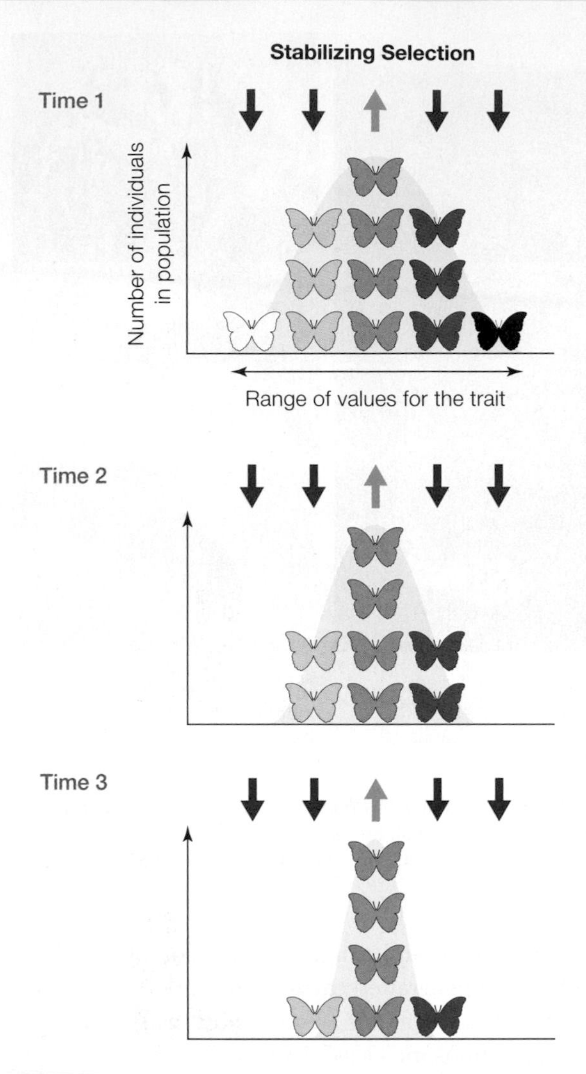

FIGURE 17.8 **▶Animated** With stabilizing selection, extreme forms of a trait are eliminated, and an intermediate form is maintained. Red arrows indicate which forms are being selected against; green, the form that is adaptive. Compare the data set from a field experiment in **FIGURE 17.9**.

Consider how environmental pressures maintain an intermediate body mass in populations of sociable weavers (**FIGURE 17.9**). These birds live in the African savanna, where they build large communal nests. Their body mass has a genetic basis. Between 1993 and 2000, Rita Covas and her colleagues investigated selection pressures that operate on sociable weaver body mass by capturing and weighing thousands of birds before and after the breeding seasons. The results of this study indicated that optimal body mass in sociable weavers is a trade-off between the risks of starvation and predation. Birds that carry less fat are more likely to starve than fatter birds. However, birds that carry more fat spend more time eating, which in this species means foraging in open areas where they are easily accessible to predators. Fatter birds are also more attractive to predators, and not as agile when escaping. Thus, predators are agents of selection that eliminate the fattest individuals. Birds of intermediate weight have

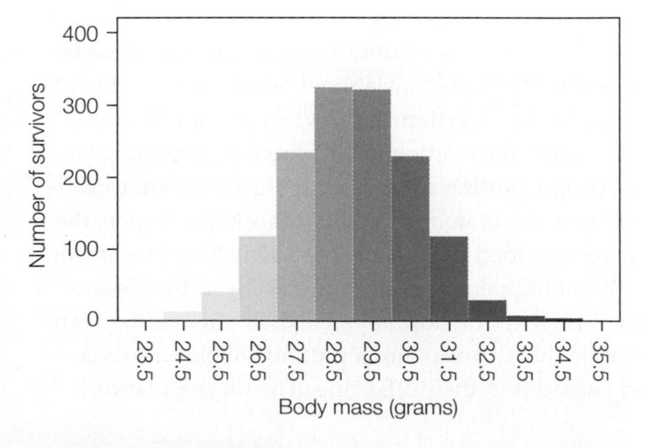

FIGURE 17.9 Stabilizing selection in sociable weavers (left). Graph (right) shows the number of birds (out of 977) that survived a breeding season. **FIGURE IT OUT** According to these data, what is the optimal weight of a sociable weaver?

Answer: About 29 grams

the selective advantage, and they make up the bulk of sociable weaver populations.

Disruptive Selection

With **disruptive selection**, forms of a trait at both ends of a range of variation are favored, and intermediate forms are selected against (**FIGURE 17.10**).

The black-bellied seedcracker is a colorful finch species native to Cameroon, Africa. In these birds, there is a genetic basis for bill size. The bill of a typical black-bellied seedcracker, male or female, is either 12 millimeters wide, or wider than 15 millimeters (**FIGURE 17.11**). Birds with a bill size between 12 and 15 millimeters are uncommon. It is as if every human adult were 4 feet or 6 feet tall, with no one of intermediate height. Seedcrackers with the large and small bill forms inhabit the same geographic range, and they breed randomly with respect to bill size.

The dimorphism in bill size of seedcrackers arises from (and is maintained by) environmental factors that affect feeding performance. The finches feed mainly on the seeds of two types of sedge, which is a grass-like plant. One sedge produces hard seeds; the other produces soft seeds. Small-billed birds are better at opening the soft seeds, but large-billed birds are better at cracking the hard ones. Both hard and soft sedge seeds are abundant during Cameroon's semiannual wet seasons. At these times, all seedcrackers feed on both seed types. The seeds become scarce during the region's dry seasons. As competition for food intensifies, each bird focuses on eating the seeds that it opens most efficiently: Small-billed birds feed mainly on soft seeds, and large-billed birds feed mainly on hard seeds. Birds with intermediate-sized bills cannot open either type of seed as efficiently as the other birds, so they are less likely to survive the dry seasons.

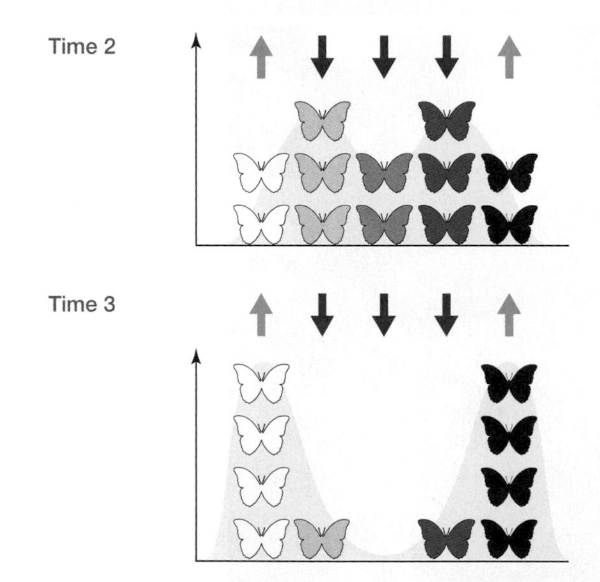

FIGURE 17.10 ▶Animated With disruptive selection, a midrange form of a trait is eliminated, and extreme forms are maintained. Red arrows indicate which form is being selected against; green, the forms that are adaptive.

FIGURE 17.11 ▶Animated Disruptive selection in African seedcracker populations maintains a distinct dimorphism in bill size.

Competition for scarce food during dry seasons favors birds with bills that are either 12 millimeters wide (left) or 15 to 20 millimeters wide (right). Birds with bills of intermediate size are selected against.

disruptive selection Mode of natural selection in which traits at the extremes of a range of variation are adaptive, and intermediate forms are not.

stabilizing selection Mode of natural selection in which an intermediate form of a trait is adaptive, and extreme forms are not.

TAKE-HOME MESSAGE 17.6

In what modes of natural selection are intermediate or extreme forms of traits adaptive?

✔ With stabilizing selection, an intermediate phenotype is adaptive, and extreme forms are selected against.

✔ With disruptive selection, an intermediate form of a trait is selected against, and extreme phenotypes are adaptive.

✔ Some adaptive traits help individuals secure mates.

✔ Any mode of natural selection may maintain multiple alleles in a population.

A Male elephant seals engaged in combat. Males of this species typically compete for access to clusters of females.

B A male bird of paradise engaged in a flashy courtship display has caught the eye (and, perhaps, the sexual interest) of a female. Female birds of paradise are choosy; a male mates with any female that accepts him.

C Female stalk-eyed flies prefer to mate with males that have the longest eyestalks, a trait that provides no known selective advantage other than sexual attractiveness.

FIGURE 17.12 Sexual selection in action.

Survival of the Sexiest

Not all evolution is driven by selection for traits that enhance survival. Competition for mates is another selective pressure that can shape form and behavior. Consider dimorphisms among males and females of some sexually reproducing species (a trait that differs among males and females is called a **sexual dimorphism**). Individuals of one sex are more colorful, larger, or more aggressive than individuals of the other sex. These traits can seem puzzling because they take energy and time away from activities that enhance survival, and some actually hinder an individual's ability to survive. Why, then, do they persist? The answer is **sexual selection**, in which the evolutionary winners outreproduce others of a population because they are better at securing mates. With this mode of natural selection, the most adaptive forms of a trait are those that help individuals defeat rivals for mates, or are most attractive to the opposite sex.

For example, the females of some species cluster in defensible groups when they are sexually receptive, and males compete for sole access to the groups. Competition for the ready-made harems favors brawny, combative males (FIGURE 17.12A).

Males or females that are choosy about mates act as selective agents on their own species. The females of some species shop for a mate among males that display species-specific cues such as a highly specialized appearance or courtship behavior (FIGURE 17.12B). The cues often include flashy body parts or movements, traits that tend to attract predators and in some cases are a physical hindrance. However, to a female member of the species, a flashy male's survival despite his obvious handicap may imply health and vigor, two traits that are likely to improve her chances of bearing healthy, vigorous offspring. Selected males pass alleles for their attractive traits to the next generation of males, and females pass alleles that influence mate preference to the next generation of females. Highly exaggerated traits can be the evolutionary outcome (FIGURE 17.12C).

Maintaining Multiple Alleles

Any mode of natural selection may keep two or more alleles of a gene circulating at relatively high frequency in a population's gene pool, a state called **balanced polymorphism**. For example, sexual selection maintains multiple alleles that govern eye color in populations of

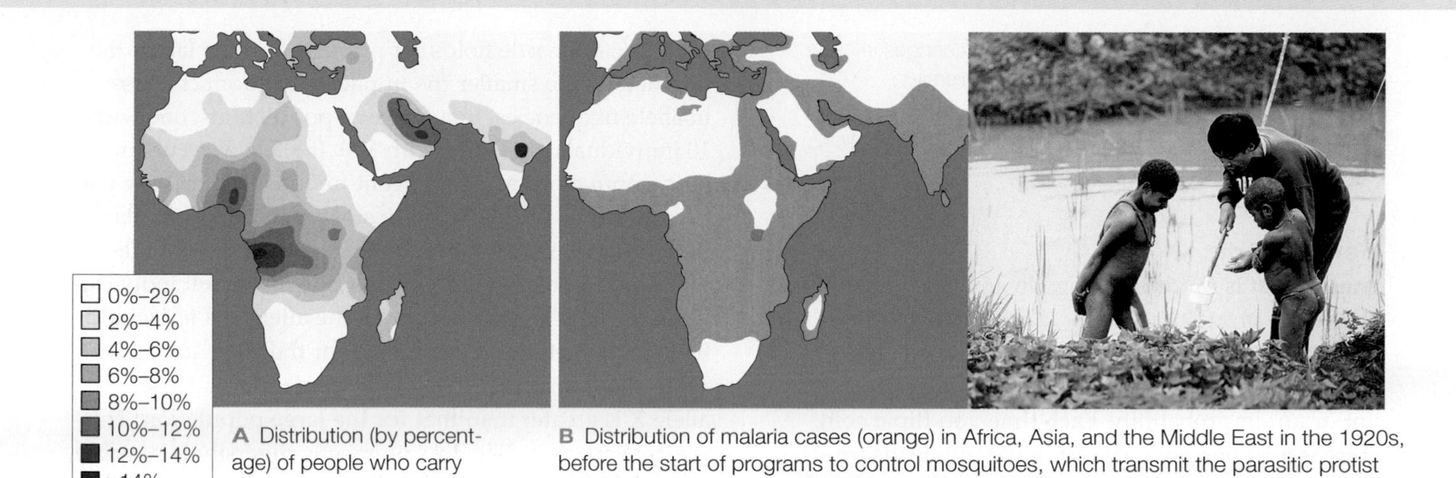

A Distribution (by percentage) of people who carry the sickle-cell allele.

B Distribution of malaria cases (orange) in Africa, Asia, and the Middle East in the 1920s, before the start of programs to control mosquitoes, which transmit the parasitic protist that causes the disease. Notice the correlation with the distribution of the sickle-cell allele in **A**. The photo shows a physician searching for mosquito larvae in Southeast Asia.

FIGURE 17.13 Malaria and sickle-cell anemia.

Drosophila fruit flies. Female flies prefer to mate with rare white-eyed males, until the white-eyed males become more common than red-eyed males, at which point the red-eyed flies are again preferred. This is also an example of **frequency-dependent selection**, in which the adaptive value of a particular form of a trait depends on its frequency in a population.

Balanced polymorphism can also arise in environments that favor heterozygous individuals. Consider the gene that encodes the beta globin chain of hemoglobin. *HbA* is the normal allele; the codominant *HbS* allele carries a mutation that causes sickle-cell anemia (Section 9.6). Even with medical care, about 15 percent of individuals homozygous for the *HbS* allele die by age 18 from complications of the disorder.

Despite being so harmful, the *HbS* allele persists at very high frequency among the human populations in tropical and subtropical regions of Asia, Africa, and the Middle East (**FIGURE 17.13A**). Why? Populations with the highest frequency of the *HbS* allele also have the highest incidence of malaria (**FIGURE 17.13B**). Mosquitoes transmit *Plasmodium*, the parasitic protist that causes malaria, to human hosts (more about this in Section 21.7). *Plasmodium* multiplies in the liver and then in red blood cells. The cells rupture and release new parasites during recurring bouts of severe illness.

balanced polymorphism Maintenance of two or more alleles of a gene at high frequency in a population.
frequency-dependent selection Natural selection in which a trait's adaptive value depends on its frequency in a population.
sexual dimorphism Difference in appearance between males and females of a species.
sexual selection Mode of natural selection in which some individuals outreproduce others of a population because they are better at securing mates.

People who make both normal and sickle hemoglobin are more likely to survive malaria than people who make only normal hemoglobin. In *HbA*/*HbS* heterozygous individuals, *Plasmodium*-infected red blood cells sometimes sickle. The abnormal shape brings the cells to the attention of the immune system, which destroys them along with the parasites they harbor. By contrast, *Plasmodium*-infected red blood cells of individuals homozygous for the *HbA* allele do not sickle, so the parasite may remain hidden from the immune system.

In areas where malaria is common, the persistence of the *HbS* allele is a matter of relative evils. Malaria and sickle-cell anemia are both potentially deadly. Heterozygous individuals may not be completely healthy, but they do have a better chance of surviving malaria than people homozygous for the normal allele (*HbA*/*HbA*). With or without malaria, people who have both alleles (*HbA*/*HbS*) are more likely to live long enough to reproduce than individuals homozygous for the sickle allele (*HbS*/*HbS*). The result is that nearly one-third of people living in the most malaria-ridden regions of the world carry the *HbS* allele.

TAKE-HOME MESSAGE 17.7

How does natural selection maintain diversity?

✔ With sexual selection, adaptive forms of a trait are those that give an individual an advantage in securing mates.

✔ Sexual selection can reinforce phenotypic differences between males and females, and sometimes it results in exaggerated traits.

✔ Balanced polymorphism can be an outcome of frequency-dependent selection, or of environmental pressures that favor heterozygous individuals.

✔ Especially in small populations, random changes in allele frequencies can lead to a loss of genetic diversity.

✔ Interbreeding among populations can change or stabilize allele frequencies, as can individuals (along with their alleles) moving from one population to another.

Genetic Drift

Genetic drift is random change in an allele's frequency over time, brought about by chance alone. We explain genetic drift in terms of probability (the chance that some event will occur, Section 1.8). Sample size is important in probability. Each time you flip a coin, there is a 50 percent chance it will land heads up. With 10 flips, the proportion of times heads actually land up may be very far from 50 percent. With 1,000 flips, that proportion is more likely to be near 50 percent. The same rule holds for populations: the larger the population, the smaller the impact of random changes in allele frequencies. Imagine two populations, one with 10 individuals, the other with 100. If allele *X* occurs in both populations at a 10 percent frequency, then only one person carries the allele in the small population. If that individual dies without reproducing, then the population's gene pool will lose allele *X*. However, ten individuals in the large population carry the allele. All ten would have to die without reproducing for the allele to be lost. Thus, the chance that the small population will lose allele *X* is greater than that for the large population. This is a general effect: The loss of genetic diversity is possible in all populations, but it is more likely to occur in small ones (**FIGURE 17.14**). When all individuals of a population are homozygous for an allele, we say that the allele is **fixed**. The frequency of a fixed allele will not change unless a new mutation occurs, or an individual bearing another allele enters the population.

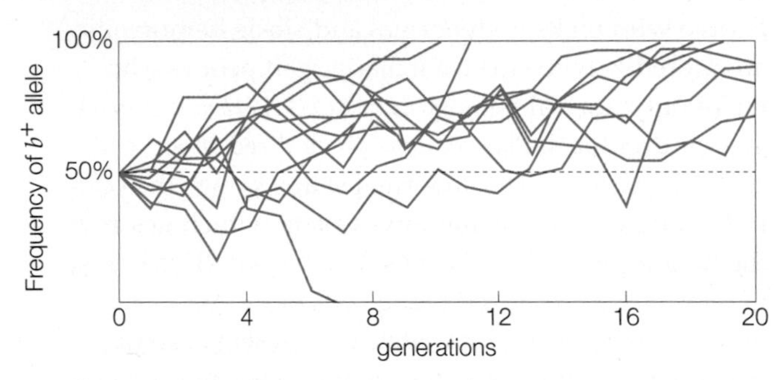

A The size of these populations was maintained at 10 breeding individuals. Allele b^+ was lost in one population (one graph line ends at 0).

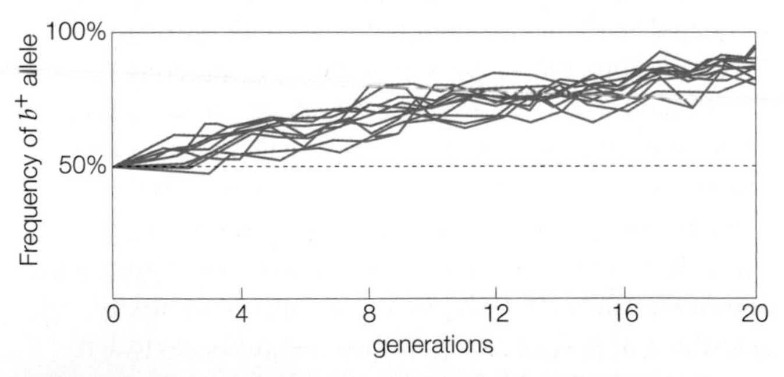

B The size of these populations was maintained at 100 individuals. Drift in these populations was less than in the small populations in **A**.

FIGURE 17.14 ▶**Animated** Genetic drift experiment in flour beetles (*Tribolium castaneum*), shown left on a flake of cereal.

Beetles heterozygous for alleles b^+ and b were maintained in populations of (**A**) 10 individuals or (**B**) 100 individuals for 20 generations. Graph lines in **B** are smoother than in **A**, indicating that drift was greatest in the sets of 10 beetles and least in the sets of 100.

Notice that the average frequency of allele b^+ rose at the same rate in both groups, an indication that natural selection was at work too: Allele b^+ was weakly favored. **FIGURE IT OUT** In how many populations did allele b^+ become fixed?

Answer: Six

Bottlenecks and the Founder Effect

A drastic reduction in population size, which is called a **bottleneck**, can greatly reduce genetic diversity. For example, northern elephant seals (shown in **FIGURE 17.12A**) underwent a bottleneck during the late 1890s, when hunting reduced their population size to about twenty individuals. Hunting restrictions have since allowed the population to recover, but genetic diversity among its members has been greatly reduced. The bottleneck and subsequent genetic drift eliminated many alleles that had previously been present in the population.

A loss of genetic diversity can also occur when a small group of individuals establishes a new population. If the founding group is not representative of the original population in terms of allele frequencies, then the new population will not be representative of it either. This outcome is called the **founder effect** (**FIGURE 17.15A**). Consider that all three *ABO* alleles for blood type (Section 13.5) are common in most human populations. Native Americans are an exception, with the majority of individuals being homozygous for the *O* allele. Native Americans are descendants of early humans who migrated from Asia

bottleneck Reduction in population size so severe that it reduces genetic diversity.
fixed Refers to an allele for which all members of a population are homozygous.
founder effect After a small group of individuals founds a new population, allele frequencies in the new population differ from those in the original population.
gene flow The movement of alleles into and out of a population.
genetic drift Change in allele frequency due to chance alone.
inbreeding Mating among close relatives.

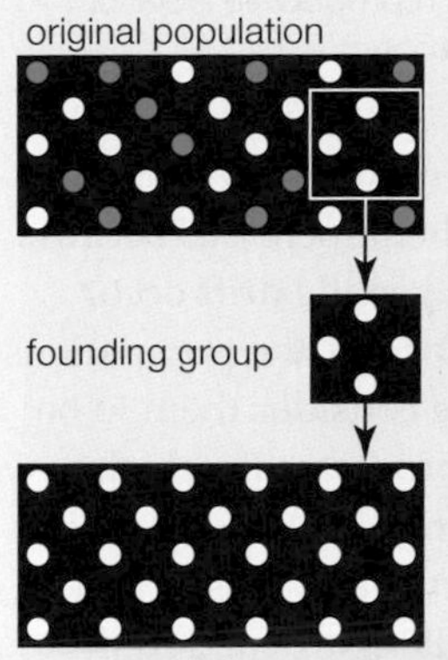

A The founder effect: a group that founds a new population is not representative of the original population, so allele frequencies differ between the new and the old populations.

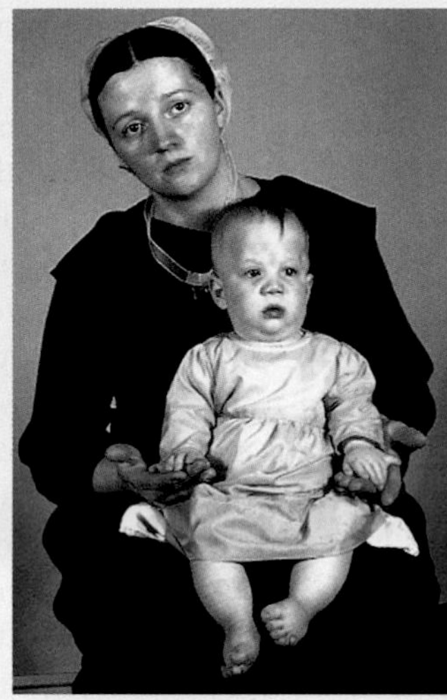

B A high frequency of an allele that causes Ellis–van Creveld syndrome among the Lancaster Amish began with the founder effect.

FIGURE 17.15 The founder effect and one outcome.

between 14,000 and 21,000 years ago, across a narrow land bridge that once connected Siberia and Alaska. Analysis of DNA from ancient skeletal remains reveals that most early Americans were also homozygous for the *O* allele. Modern Siberian populations have all three alleles. Thus, the first humans in the Americas were probably members of a small group that had reduced genetic diversity compared with the general population.

Founding populations are often necessarily inbred. **Inbreeding** is mating between close relatives. Closely related individuals tend to share more alleles than nonrelatives do, so inbred populations often have unusually high numbers of individuals homozygous for recessive alleles, some of which are harmful. This outcome is minimized in human populations that discourage or forbid incest (mating between parents and children or between siblings).

The Old Order Amish in Lancaster County, Pennsylvania, offer an example of the effects of inbreeding. Amish people marry only within their community. Intermarriage with other groups is not permitted, and no "outsiders" are allowed to join the community. As a result, Amish communities are moderately inbred, and many of their individuals are homozygous for harmful recessive alleles. The Lancaster community has an unusually high frequency of a recessive allele that causes Ellis–van Creveld syndrome, a genetic disorder characterized by dwarfism, polydactyly, and heart defects, among other symptoms. This allele has been traced to a man and his wife, two of a group of 400 Amish people who immigrated to the United States in the mid-1700s. As a result of the founder effect and inbreeding since then, about 1 of 8 people in the Lancaster community is now heterozygous for the allele, and 1 in 200 is homozygous for it (**FIGURE 17.15B**).

Gene Flow

Individuals of natural populations tend to mate or breed most frequently with other members of their own population. However, not all populations of a species are completely isolated from one another, and nearby populations may occasionally interbreed. Also, individuals sometimes leave one population and join another. **Gene flow**, the movement of alleles between populations, occurs in both cases. Gene flow can change or stabilize allele frequencies, thus countering the evolutionary effects of mutation, natural selection, and genetic drift.

Gene flow is typical among populations of animals, but it also occurs in less mobile organisms. Consider

the acorns that jays disperse when they gather nuts for the winter (left). Every fall, these birds visit acorn-bearing oak trees repeatedly, then bury the acorns in the soil of territories as much as a mile away. The jays transfer acorns (and the alleles carried by these seeds) among populations of oak trees that may otherwise be genetically isolated. Gene flow also occurs when wind or an animal transfers pollen from one plant to another, often over great distances (more about this in Chapter 29). Many opponents of genetic engineering cite gene flow from transgenic crop plants into wild populations via pollen transfer. For example, engineered genes that confer resistance to herbicides and Bt (Section 15.7) are now commonly found in weeds and unmodified crop plants. Long-term effects of this gene flow are currently unknown.

TAKE-HOME MESSAGE 17.8

Other than natural selection, what mechanisms affect allele frequencies?

✔ Genetic drift can reduce a population's genetic diversity. Its effect is greatest in small populations.

✔ A population's genetic diversity may be lowered after a bottleneck, or as a result of the founder effect.

✔ Gene flow tends to oppose the evolutionary effects of mutation, natural selection, and genetic drift in a population.

17.9 Reproductive Isolation

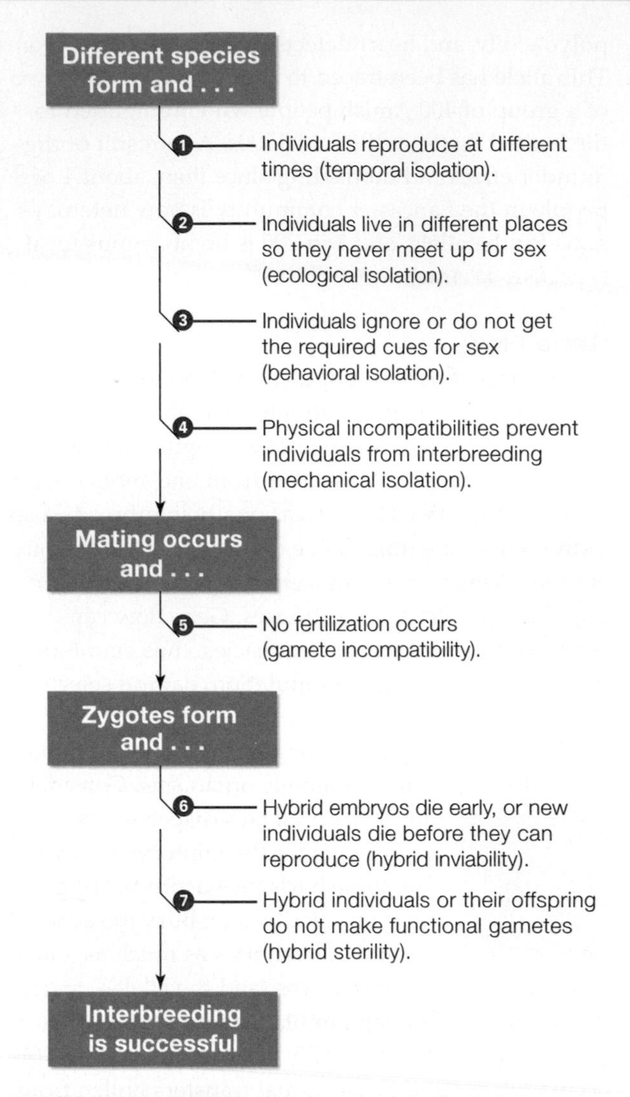

FIGURE 17.16 ▶**Animated** How reproductive isolation prevents interbreeding.

FIGURE 17.17 Behavioral isolation. A male peacock spider (*Maratus volans*) approaches a female, signaling his intent to mate with her by raising and waving colorful flaps, and gesturing his legs in time with abdominal vibrations. If his species-specific courtship display fails to impress her, she will kill him.

✔ Speciation differs in its details, but reproductive isolation mechanisms are always part of the process.

When two populations do not interbreed, the number of genetic differences between them increases because mutation, natural selection, and genetic drift occur independently in each one. Over time, the populations may become so different that we consider them to be different species. Evolutionary processes in which new species arise are called **speciation**.

Evolution is a dynamic, extravagant, messy, and ongoing process that can be challenging for people who like categories. Speciation offers a perfect example, because it rarely occurs at a precise moment in time: Individuals often continue to interbreed even as populations are diverging, and populations that have already diverged may come together and interbreed again.

Every time speciation happens, it happens in a unique way, and each species is a product of its own unique evolutionary history. However, there are recurring patterns. For example, reproductive isolation is always part of speciation. **Reproductive isolation,** the end of gene flow between populations, is part of the process by which sexually reproducing species attain and maintain their separate identities. Mechanisms that prevent successful interbreeding reinforce differences between diverging populations (**FIGURE 17.16**).

❶ Some closely related species cannot interbreed because the timing of their reproduction differs (an effect called temporal isolation). Consider the periodical cicada (left).

Larvae of these insects feed on roots as they mature underground, then the adults emerge to reproduce. Three cicada species reproduce every 17 years. Each has a sibling species with nearly identical form and behavior, except that the siblings emerge on a 13-year cycle instead of a 17-year cycle. Sibling species have the potential to interbreed, but they can only get together once every 221 years!

❷ Adaptation to different microenvironments may prevent closely related species from interbreeding (ecological isolation). For example, two species of manzanita, a plant native to the Sierra Nevada mountain range, rarely hybridize. One species that lives on dry, rocky hillsides is better adapted for conserving water. The other, less drought-adapted species lives on lower slopes where water stress is not as intense. The physical separation makes cross-pollination unlikely.

❸ Differences in behavior can prevent mating between related animal species (behavioral isolation). For example, males and females of many animal species engage in courtship displays before sex (**FIGURE 17.17**). In a typical pattern, the female recognizes the sounds and movements of a male of her species as an overture to sex; females of different species do not.

❹ The size or shape of an individual's reproductive parts may prevent it from mating with members of closely related species (mechanical isolation). For example, plants called black sage and white sage grow in the same areas, but hybrids rarely form because the flowers of these two related species have become specialized for different pollinator species (**FIGURE 17.18**).

❺ Even if gametes of different species do meet up, they often have molecular incompatibilities that prevent a zygote from forming. For example, the molecular signals that trigger pollen germination in flowering plants are species-specific (we return to pollen germination and other aspects of flowering plant reproduction in Section 29.4). Gamete incompatibility may be the primary speciation route among animals that release their eggs and free-swimming sperm into water.

❻ Genetic changes are the basis of divergences in form, function, and behavior. Even chromosomes of species that diverged relatively recently may be different enough that a hybrid zygote inherits extra or missing genes, or genes with incompatible products—outcomes that typically disrupt development. Hybrids that do survive embryonic development often have reduced fitness. For example, hybrid offspring of lions and tigers have more health problems and a shorter life expectancy than individuals of either parent species.

❼ Some interspecies crosses produce robust but sterile hybrid offspring. For example, mating between a female horse (64 chromosomes) and a male donkey (62 chromosomes) produces a mule. Mules are healthy, but their 63 chromosomes cannot pair up evenly during meiosis, so this animal makes few viable gametes. If hybrids are fertile, their offspring usually have lower and lower fitness with each successive generation. Incompatible nuclear and mitochondrial DNA may be the cause (mitochondrial DNA is inherited from the mother only).

reproductive isolation The end of gene flow between populations.
speciation Evolutionary process in which new species arise.

TAKE-HOME MESSAGE 17.9

How do species attain and maintain separate identities?

✔ Speciation is an evolutionary process in which new species form. It varies in its details and duration, but reproductive isolation is always a part of speciation.

A Black sage is pollinated mainly by small insects.

B The flowers of black sage are too delicate to support larger insects. Big insects access the nectar of small sage flowers only by piercing from the outside, as this carpenter bee is doing. When they do so, they avoid touching the flower's reproductive parts.

C The reproductive parts (anthers and stigma) of white sage flowers are too far away from the petals to be brushed by honeybees, so honeybees are not efficient pollinators of this species. White sage is pollinated mainly by larger bees and hawkmoths, which brush the flower's stigma and anthers as they pry apart the petals to access nectar.

FIGURE 17.18 Mechanical isolation in sage.

CREDITS: (18A) Courtesy of Dr. James French; (18B) Courtesy of © Ron Brinkmann, www.flickr.com/photos/ronbrinkmann; (18C) © David Goodin.

17.10 Allopatric Speciation

✔ In allopatric speciation, a physical barrier arises and ends gene flow between populations.

Genetic changes that lead to a new species can begin with physical separation between populations. With **allopatric speciation**, a physical barrier arises and separates two populations, ending gene flow between them (*allo–* means different; *patria*, fatherland). Then, reproductive isolating mechanisms evolve that prevent interbreeding even if the diverging populations meet again.

Gene flow between populations separated by distance is often inconsistent. Whether a geographic barrier can completely block that gene flow depends on how the species travels (such as by swimming, walking, or flying), and how it reproduces (for example, by internal fertilization or by pollen dispersal).

A geographic barrier can arise in an instant, or over an eon. The Great Wall of China is an example of a barrier that arose abruptly. As it was being built, the wall interrupted gene flow among nearby populations of insect-pollinated plants; DNA sequence comparisons show that trees, shrubs, and herbs on either side of the wall are diverging genetically. Geographic isolation usually occurs much more slowly. For example, it took millions of years of tectonic plate movements (Section 16.7) to bring the two continents of North and South America close enough to collide. The land bridge where the two continents now connect is called the Isthmus of Panama. When this isthmus formed about 4 million years ago, it cut off the flow of water—and gene flow among populations of aquatic organisms—as it separated one large ocean into what are now the Pacific and Atlantic Oceans (**FIGURE 17.19**).

Speciation in Archipelagos

New species rarely form on island chains such as the Florida Keys that are in close proximity to a mainland. Being close to a mainland means gene flow is essentially unimpeded between island and mainland populations. By contrast, allopatric speciation is common on archipelagos (island chains) such as the Hawaiian and Galápagos Islands. These islands are so geographically isolated that, for most species, no gene flow occurs between island and mainland populations.

The Hawaiian archipelago includes 19 islands and more than 100 atolls stretching 1,500 miles in the Pacific Ocean. These islands are the product of hot spots on the ocean floor (Section 16.7). Because they were the tops of volcanoes, we can assume that their fiery surfaces were initially barren and inhospitable to life. Later, winds and currents carried individuals of mainland species to them. The individuals reproduced, and their descendants established populations. The lack of gene flow with mainland populations allowed the island populations to diverge. Today, thousands of species are unique to this island chain.

Consider Hawaiian honeycreepers, birds that are descendants of Asian finches that arrived on the

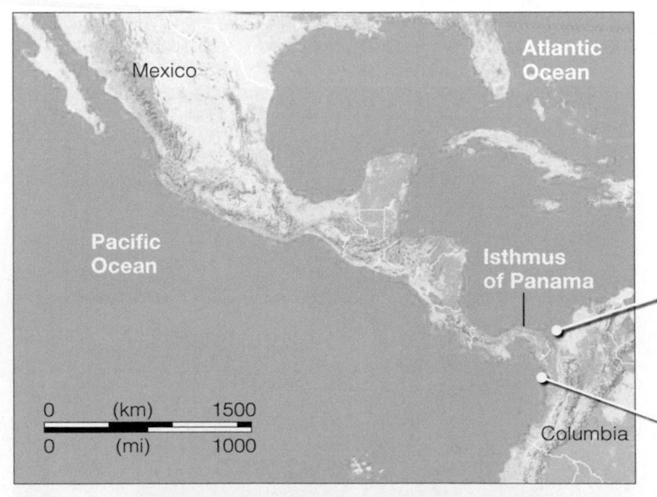

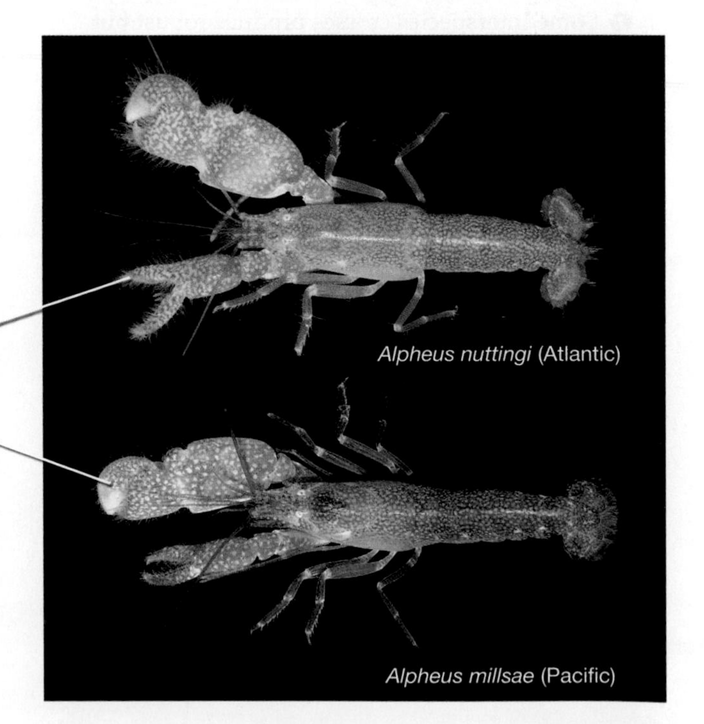

FIGURE 17.19 Example of allopatric speciation. When the Isthmus of Panama formed, it cut off gene flow among ocean-dwelling populations of snapping shrimp. Today, shrimp on opposite sides of the isthmus might be able to interbreed were it not for behavioral isolation: Instead of mating when they are brought together, they snap their claws at one another aggressively. The photos show two of the many closely related species that live on opposite sides of the isthmus.

FIGURE 17.20 ▶Animated
Example of allopatric speciation on an archipelago. The ancestor of all Hawaiian honeycreepers was a rosefinch species (*Carpodacus*, left) from southern Asia. About 5.8 million years ago, a population of these finches somehow managed to fly thousands of miles across the open ocean to the Hawaiian archipelago (right). The expanse of ocean prevented gene flow between mainland populations and the island colonizers, which subsequently diverged into many honeycreeper species.

Hawaiian archipelago

islands at least 5.8 million years ago (**FIGURE 17.20**). A buffet of fruits, seeds, nectars, tasty insects, and the near absence of competitors and predators allowed the finch's descendants to thrive. In the absence of gene flow, the island finch population diverged from the ancestral mainland species. Further divergences occurred as new islands arose; habitats on the landmasses of the archipelago vary dramatically—from lava beds, rain forests, and grasslands to dry woodlands and snow-capped peaks. Selection pressures differ within and between these habitats. The cumulative result of all these divergences is a spectacular array of honeycreeper species.

allopatric speciation Speciation pattern in which a physical barrier arises and ends gene flow between populations.

TAKE-HOME MESSAGE 17.10

What is allopatric speciation?

✔ A physical barrier that intervenes between populations of a species prevents gene flow among them. When gene flow ends, genetic divergences give rise to new species. This pattern is called allopatric speciation.

17.11 Other Speciation Models

✔ Populations sometimes speciate even without a physical barrier that bars gene flow between them.

Sympatric Speciation

In **sympatric speciation**, populations inhabiting the same geographic region speciate in the absence of a physical barrier between them (*sym*– means together). Sympatric speciation can occur in a single generation when the chromosome number multiplies. Polyploidy (having three or more sets of chromosomes, Section 14.6) typically arises when an abnormal nuclear division during meiosis or mitosis doubles the chromosome number. For example, if the nucleus of a somatic cell in a flowering plant fails to divide during mitosis, the resulting cell—which is polyploid—may proliferate and give rise to shoots and flowers. If the flowers can self-fertilize, a new polyploid species may be the result. Common bread wheat originated after related species hybridized, and then the chromosome number of the hybrid offspring doubled (FIGURE 17.21).

Sympatric speciation can also occur with no change in chromosome number. The mechanically isolated sage plants you learned about in Section 17.9 speciated with no physical barrier to gene flow. As another example, more than 500 species of cichlid fishes arose by sympatric speciation in the shallow waters of Lake Victoria. This large freshwater lake sits isolated from river inflow on an elevated plain in Africa's Great Rift Valley. Since Lake Victoria formed about 400,000 years ago, it has dried up three times. DNA sequence comparisons indicate that almost all of the cichlid species in this lake arose since the last dry spell, which was 12,400 years ago. How could hundreds of species arise so quickly? In this case, the answer begins with differences in the color of ambient light in different parts of the lake. The light in the lake's shallower, clear water is mainly blue; light that penetrates the deeper, muddier water is mainly red. The cichlid species vary in color

FIGURE 17.22 Red fish, blue fish: Males of four closely related species of cichlid native to Lake Victoria, Africa. Hundreds of cichlid species arose by sympatric speciation in this lake. Mutations that affect female cichlids' perception of the color of ambient light in deeper or shallower regions of the lake also affect their choice of mates. Female cichlids prefer to mate with brightly colored males of their own species.

FIGURE IT OUT Which form of natural selection is driving sympatric speciation in these cichlids? Answer: Sexual selection

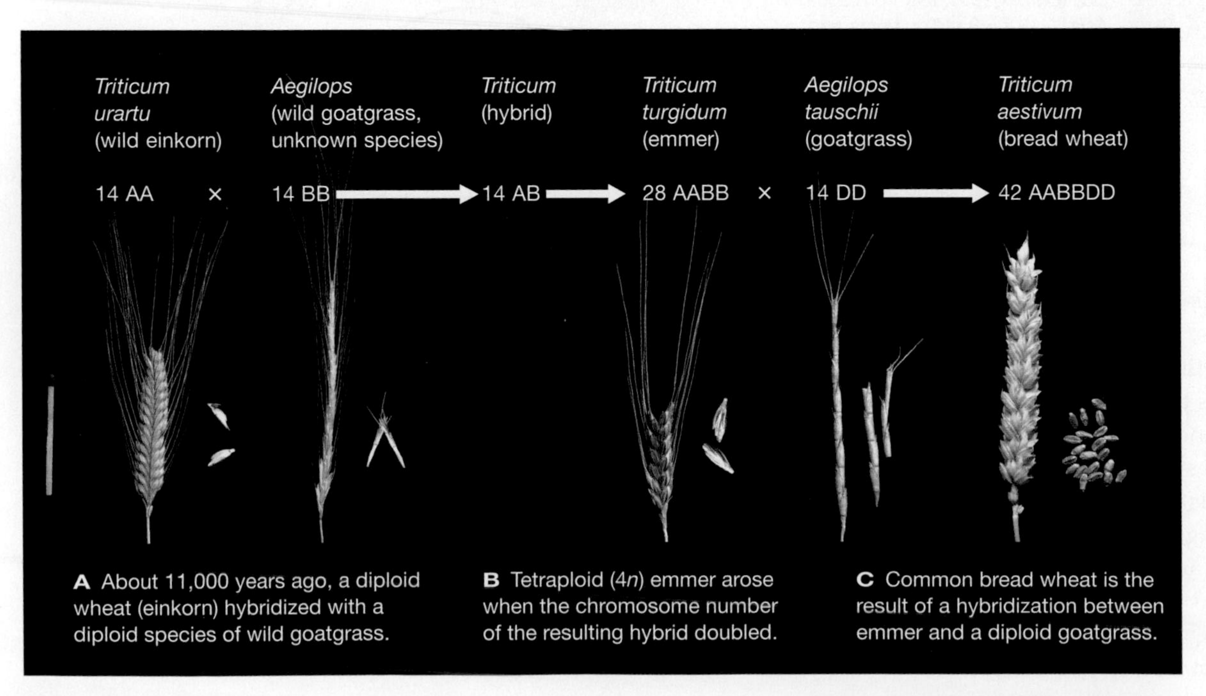

FIGURE 17.21 ▸**Animated** Sympatric speciation in wheat. The wheat genome, which consists of seven chromosomes, occurs in slightly different forms called A, B, C, D, and so on. Many wheat species are polyploid, carrying more than two copies of the genome. For example, modern bread wheat (*Triticum aestivum*) is hexaploid, with six copies of the wheat genome: two each of genomes A, B, and D (or 42 AABBDD).

A Giant velvet walking worm, *Tasmanipatus barretti*.

B Blind velvet walking worm, *T. anophthalmus*.

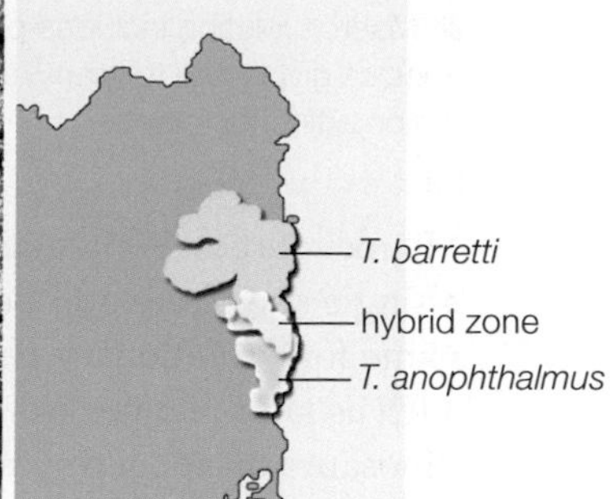

C The habitats of the worms overlap in a hybrid zone on the island of Tasmania.

FIGURE 17.23 Example of parapatric speciation: velvet walking worms in Tasmania.

(**FIGURE 17.22**). Outside of captivity, female cichlids rarely mate with males of other species. Given a choice, they prefer to mate with brightly colored males of their own species. Their preference has a genetic basis, in alleles that encode light-sensitive pigments of the retina (part of the eye). Retinal pigments made by species that live mainly in shallow areas of the lake are more sensitive to blue light. The males of these species are also the bluest. Retinal pigments made by species that prefer deeper areas of the lake are more sensitive to red light. Males of these species are redder. In other words, the colors that a female cichlid sees best are the same colors displayed by males of her species. Thus, mutations that affect color perception are likely to affect a female's choice of mates. Such mutations are probably the way sympatric speciation occurs in these fishes.

Sympatric speciation has also occurred in greenish warblers of central Asia (*Phylloscopus trochiloides*). A chain of populations of this bird encircles the Tibetan plateau (left). Adjacent populations of greenish warblers interbreed easily, except for the two populations at the ends of the chain. These two populations overlap in northern Siberia, but their individuals do not interbreed because they do not recognize one another's songs (an example of behavioral isolation). Small genetic differences between adjacent populations have added up to major differences between the two populations at the ends of the chain. Greenish warbler populations that make up the chain are collectively called a ring species. Ring species present one of those paradoxes for people who like neat categories: Gene flow occurs continuously all around the chain, but the two populations at the ends of the chain are clearly different species. Where should we draw the line that divides those two species?

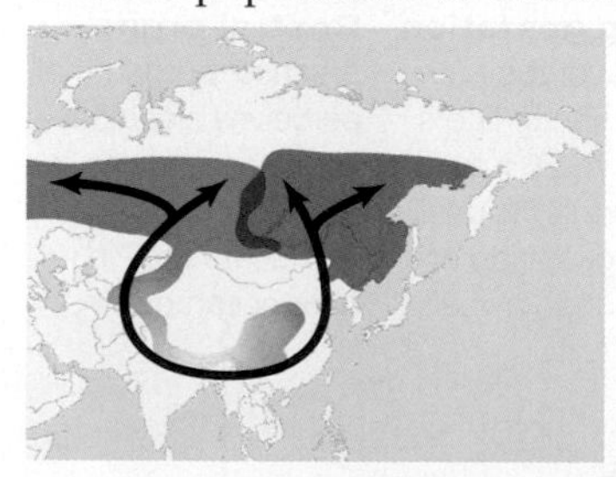

parapatric speciation Populations inhabiting different areas speciate while in contact along a common border.
sympatric speciation Divergence within a population leads to speciation; occurs in the absence of a physical barrier to gene flow.

Parapatric Speciation

With **parapatric speciation**, adjacent populations speciate despite being in contact across a common border. Divergences spurred by local selection pressures are reinforced because hybrids that form in the contact zone are less fit than individuals on either side of it.

Consider velvet walking worms, which resemble caterpillars but may be more related to spiders: They are predatory, and shoot streams of glue from their head to entangle insect prey. Two rare species of velvet walking worm are native to the island of Tasmania (**FIGURE 17.23**). The giant velvet walking worm and the blind velvet walking worm can interbreed, but they only do so in a tiny area where their habitats overlap. Hybrid offspring are sterile, which may be the main reason the two species are maintaining separate identities in the absence of a physical barrier between their adjacent populations.

TAKE-HOME MESSAGE 17.11

Does speciation occur in the absence of a physical barrier to gene flow?

✔ With sympatric speciation, divergence within a population leads to new species that inhabit the same geographical area, with no physical barrier to gene flow.

✔ With parapatric speciation, populations maintaining contact along a common border evolve into distinct species.

CREDITS: (23A, B) Courtesy of Dr. Robert Mesibov; (23C, in text) © Cengage Learning.

17.12 Macroevolution

✔ Macroevolution includes patterns of change such as one species giving rise to many, the origin of major groups, and major extinction events.

Microevolution is change in allele frequencies within a single species or population. **Macroevolution** is our name for evolutionary patterns on a larger scale: trends such as land plants evolving from green algae, the dinosaurs disappearing in a mass extinction, a burst of divergences from a single species, and so on.

The simplest macroevolutionary pattern is **stasis**, in which little change occurs over a very long period of time. Consider coelacanths, an order of ancient lobe-finned fish that had been assumed extinct for at least 70 million years until a fisherman caught one in 1938. In form and other aspects, modern coelacanth species are similar to fossil specimens hundreds of millions of years old (**FIGURE 17.24**).

Major evolutionary novelties often stem from the adaptation of an existing structure for a completely new purpose. This macroevolutionary pattern is called **exaptation**. For example, the feathers that allow modern birds to fly are derived from feathers that first evolved in some dinosaurs. Those dinosaurs could not have used their feathers for flight, but they probably did use them for insulation. Thus, we say that flight feathers in birds evolved by exaptation from insulating feathers in dinosaurs.

By current estimates, more than 99 percent of all species that ever lived are now **extinct**, which means they no longer have living members. In addition to continuing small-scale extinctions, the fossil record indicates that there have been more than twenty mass extinctions, which are simultaneous losses of many lineages. These include five catastrophic events in which the majority of species on Earth disappeared (Section 16.8).

With **adaptive radiation**, one lineage rapidly diversifies into several new species. An adaptive radiation typically occurs after a population colonizes a new environment that has a variety of different habitats and few or no competitors. Speciation occurs along with adaptation to the different habitats. The Hawaiian honeycreepers arose this way, as did the Lake Victoria cichlids. Adaptive radiation may also occur after a key innovation evolves. A **key innovation** is a new trait that allows its bearer to exploit a habitat more efficiently or in a novel way. The evolution of lungs offers an example, because lungs were a key innovation that opened the way for an adaptive radiation of vertebrates on land. A geologic or climatic event that eliminates some species from a habitat can spur adaptive radiation; species that survive the event then have access to resources from which they had previously been excluded. This is the way mammals were able to undergo an adaptive radiation after the dinosaurs disappeared.

The process by which close ecological interactions between two species cause them to evolve jointly is called **coevolution**. One species acts as an agent of selection on the other, and each adapts to changes in the other. Over evolutionary time, the two species may become so interdependent that they can no longer survive without one another.

Notochord This tough, elastic tube, which is partially hollow and filled with fluid, is ancestral to the spinal cord.

Lobed fins These fleshy fins retain a few of the ancestral bones that gave rise to legs and arms in other lineages.

Long gestation Coelacanths give birth to litters of up to 26 fully developed "pups" after gestation of more than a year.

Rostral organ A sensory organ that perceives electrical impulses in water, it probably helps the fish locate prey in dark ocean depths.

FIGURE 17.24 An example of stasis. Photos (left) compare a 320-million-year-old coelacanth fossil found in Montana with a live coelacanth. The diagram (right) shows a few of the coelacanth's unusual ancestral features that have been lost in almost all other fish lineages over evolutionary time.

CREDITS: (24) top left, Courtesy of The Virtual Fossil Museum, www.fossilmuseum.net; bottom left, AlessandroZocc/Shutterstock.com; right, Raul Martin Domingo/National Geographic Creative.

Superbug Farms (revisited)

Antibiotic-resistant bacteria bred inside people or treated animals end up in the environment, where they can easily spread to other individuals. These bacteria have plagued hospitals for years, and now we find them everywhere. They are common in day-care centers, schools, gyms, prisons, and other places where people are in close contact. We also find them in our pets, and in wildlife such as crows, rabbits, mongooses, foxes, sharks, rodents, reptiles, birds, frogs, whales, chimpanzees, penguins, and insects such as moths and houseflies. They have even been found in beach sand, coastal waters, and Antarctic seawater.

Relationships between coevolved species can be quite intricate. Consider the large blue butterfly (*Maculinea arion*), a parasite of ants. After hatching, the butterfly larvae (caterpillars) feed on wild thyme flowers and then drop to the ground. An ant that finds a caterpillar strokes it, which makes the caterpillar exude honey. The ant eats the honey and continues to stroke the caterpillar, which secretes more honey. This interaction continues for about an hour, until the caterpillar suddenly hunches itself up (**FIGURE 17.25**). The ant then picks up the caterpillar and carries it back to its nest, where, in most cases, other ants kill it—except if the ants are of the species *Myrmica sabuleti*. The caterpillar secretes the same chemicals as *Myrmica sabuleti* larvae, and makes the same sounds as their queen—behaviors that deceive the ants into adopting the caterpillar and treating it better than their own larvae. The adopted caterpillar feeds on ant larvae for about 10 months, then undergoes metamorphosis, changing into a butterfly that emerges from the ground to mate. Eggs are deposited on wild thyme near another *M. sabuleti* nest, and the cycle starts anew. This relationship between ant and butterfly is typical of coevolved relationships in that it is extremely specific. Any increase in the ants' ability to identify a caterpillar in their nest selects for caterpillars that better deceive the ants, which in turn select for ants that can better identify the caterpillars. Each species exerts directional selection on the other.

FIGURE 17.25 A *Myrmica sabuleti* ant preparing to carry a hunched-up *Maculinea arion* caterpillar back to its nest. If adopted by the ant colony, the caterpillar will eat ant larvae until it matures.

Evolutionary Theory

Biologists do not doubt that macroevolution occurs, but many disagree about how it occurs. However we choose to categorize evolutionary processes, the very same genetic change may be at the root of all evolution—fast or slow, large-scale or small-scale. Dramatic jumps in morphology, if they are not artifacts of gaps in the fossil record, may be the result of mutations in homeotic or other regulatory genes. Macroevolution may include more processes than microevolution, or it may not. It may be an accumulation of many microevolutionary events, or it may be an entirely different process. Evolutionary biologists may disagree about these and other hypotheses, but all of them are trying to explain the same thing: how all species are related by descent from common ancestors.

TAKE-HOME MESSAGE 17.12

What is macroevolution?

✔ Macroevolution comprises large-scale patterns of evolutionary change such as adaptive radiation, the origin of major groups, and mass extinctions.

---

adaptive radiation A burst of genetic divergences from a lineage gives rise to many new species.
coevolution The joint evolution of two closely interacting species; each species is a selective agent for traits of the other.
exaptation Evolutionary adaptation of an existing structure for a completely new purpose.
extinct Refers to a species that no longer has living members.
key innovation An evolutionary adaptation that gives its bearer the opportunity to exploit a particular environment much more efficiently or in a new way.
macroevolution Large-scale evolutionary patterns and trends.
stasis Evolutionary pattern in which little or no change occurs over long spans of time.

CREDITS: (25) © Jeremy Thomas/Natural Visions; (in text revisited) Bob Nichols/USDA photo.

summary

Section 17.1 Our overuse of antibiotics exerts directional selection favoring resistant bacterial populations, which are now common in the environment. We are running out of effective antibiotics to use as human drugs.

Sections 17.2, 17.3 All alleles of all genes in a **population** constitute a **gene pool.** Mutations may be **neutral, lethal,** or adaptive. **Microevolution** is change in **allele frequency** of a population. Deviations from **genetic equilibrium** indicate that a population is evolving.

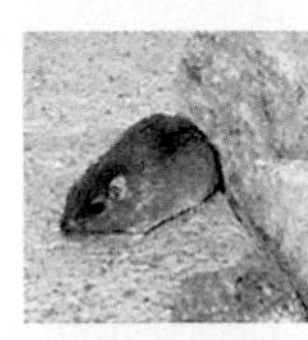

Sections 17.4–17.6 In **directional selection,** a phenotype at one end of a range of variation is adaptive. An intermediate form of a trait is adaptive in **stabilizing selection**; extreme forms are adaptive in **disruptive selection.**

Section 17.7 **Sexual dimorphism** is a potential outcome of **sexual selection,** a mode of natural selection in which adaptive traits are those that make their bearers better at securing mates. **Frequency-dependent selection** or any other mode of natural selection can give rise to a **balanced polymorphism.**

Section 17.8 **Genetic drift,** which is most pronounced in small or **inbreeding** populations, can cause alleles to become **fixed.** The **founder effect** may occur after an evolutionary **bottleneck. Gene flow** can counter the effects of mutation, natural selection, and genetic drift.

Section 17.9 The details of **speciation** differ every time it occurs, but **reproductive isolation,** the end of gene flow between populations, is always a part of the process (**TABLE 17.2**). The moment at which two populations become separate species is often impossible to pinpoint.

Table 17.2 Comparison of Speciation Models

	Allopatric	Parapatric	Sympatric
Original population(s)			
Initiating event:	physical barrier arises	selection pressures differ	genetic change
Reproductive isolation occurs			
New species arises:	in isolation	in contact along common border	within existing population

Section 17.10 In **allopatric speciation,** a geographic barrier arises and interrupts gene flow between populations. After gene flow ends, genetic divergences occur independently in each population, and this can result in separate species.

Section 17.11 Speciation can occur in the absence of a barrier to gene flow. **Sympatric speciation** occurs by divergence within a population. Polyploid species of many plants (and a few animals) have originated this way. With **parapatric speciation,** populations in physical contact along a common border speciate.

Section 17.12 **Macroevolution** refers to large-scale patterns of evolution. With **stasis,** little or no change occurs over long spans of time. In **exaptation,** a lineage uses a structure for a different purpose than its ancestor. A **key innovation** can result in an **adaptive radiation. Coevolution** occurs when two species act as agents of selection upon one another. A lineage with no more living members is **extinct.**

self-quiz

Answers in Appendix VII

1. _______ is the original source of new alleles.
   a. Mutation
   b. Natural selection
   c. Genetic drift
   d. Gene flow
   e. All are original sources of new alleles

2. Which is required for evolution to occur in a population?
   a. random mating
   b. selection pressure
   c. gene flow
   d. none of the above

3. Match the modes of natural selection with their best descriptions.
   ___ stabilizing — a. eliminates extreme forms of a trait
   ___ disruptive — b. eliminates midrange form of a trait

4. Sexual selection frequently influences aspects of body form and can lead to _______ .
   a. a sexual dimorphism
   b. male aggression
   c. exaggerated traits
   d. all of the above

5. The persistence of sickle-cell anemia in a population with a high incidence of malaria is a case of _______ .
   a. bottlenecking
   b. inbreeding
   c. balanced polymorphism
   d. the founder effect
   e. frequency-dependent selection

6. _______ tends to keep populations of a species similar to one another.
   a. Genetic drift
   b. Gene flow
   c. Mutation
   d. Natural selection

7. The theory of natural selection does not explain _______ .
   a. genetic drift
   b. the founder effect
   c. gene flow
   d. how mutations arise
   e. inheritance
   f. any of the above

data analysis activities

Resistance to Rodenticides in Wild Rat Populations Beginning in 1990, rat infestations in northwestern Germany started to intensify despite continuing use of rat poisons. In 2000, Michael H. Kohn and his colleagues analyzed the genetics of wild rat populations around Münster. For part of their research, they trapped wild rats in five towns, and tested those rats for resistance to warfarin and the more recently developed poison bromadiolone. The results are shown in **FIGURE 17.26**.

1. In which of the five towns were most of the rats susceptible to warfarin?

2. Which town had the highest percentage of poison-resistant wild rats?

3. What percentage of rats in Olfen were resistant to warfarin?

4. In which town do you think the application of bromadiolone was most intensive?

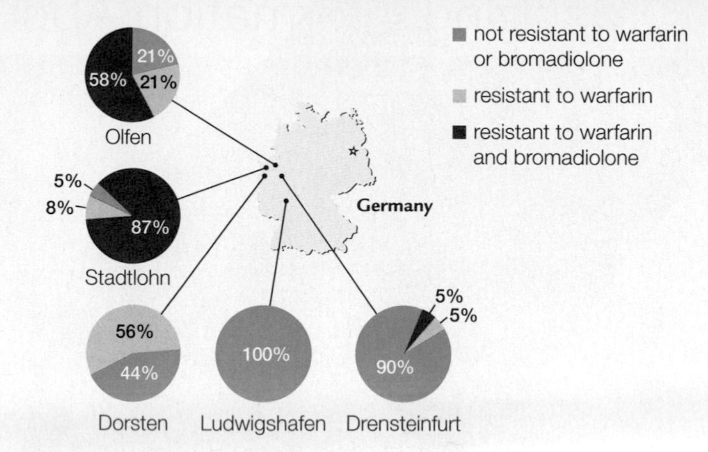

FIGURE 17.26 Resistance to rat poisons in wild populations of rats in Germany, 2000.

8. Which of the following is *not* part of how we define a species?
   a. Its individuals appear different from other species.
   b. It is reproductively isolated from other species.
   c. Its populations can interbreed.
   d. Fertile offspring are produced.

9. Sex in many birds is typically preceded by an elaborate courtship dance. If a male's movements are unrecognized by the female, she will not mate with him. This is an example of ________ .
   a. reproductive isolation
   b. behavioral isolation
   c. sexual selection
   d. all of the above

10. The difference between sympatric and parapatric speciation is ________ .
    a. parapatric speciation occurs only in worms
    b. sympatric speciation requires a barrier to gene flow
    c. the extent of overlap in range
    d. reproductive isolation does not occur

11. A fire devastates all trees in a wide swath of forest. Populations of a species of tree-dwelling frog on either side of the burned area diverge to become separate species. This is an example of ________ .
    a. allopatric speciation
    b. parapatric speciation
    c. sympatric speciation
    d. adaptive radiation

12. Match the evolution concepts.

___ gene flow	a. can lead to interdependent species
___ sexual selection	b. changes in a population's allele frequencies due to chance alone
___ mutation	c. alleles enter or leave a population
___ genetic drift	d. adaptive traits make their bearers better at securing mates
___ coevolution	e. original source of new alleles
___ adaptive radiation	f. burst of divergences from one lineage into many

13. Change in allele frequency of a population is called ________ .
    a. macroevolution
    b. adaptive radiation
    c. inbreeding
    d. microevolution

critical thinking

1. Species have been traditionally characterized as "primitive" and "advanced." For example, mosses were considered to be primitive, and flowering plants advanced; crocodiles were primitive and mammals were advanced. Why do most biologists of today think it is incorrect to refer to any modern species as primitive?

2. Rama the cama, a llama–camel hybrid, was born in 1997. The idea was to breed an animal that has the camel's strength and endurance, and the llama's gentle disposition. However, instead of being large, strong, and sweet, Rama is smaller than expected and has a camel's short temper. The breeders plan to mate him with Kamilah, a female cama. What potential problems with this mating should the breeders anticipate?

3. Two species of antelope, one from Africa, the other from Asia, are put into the same enclosure in a zoo. To the zookeeper's surprise, individuals of the different species begin to mate and produce healthy, hybrid baby antelopes. Explain why a biologist might not view these offspring as evidence that the two species of antelope are in fact one.

4. Some people think that many of our uniquely human traits arose by sexual selection. Over thousands of years, women attracted to charming, witty men perhaps prompted the development of human intellect beyond what was necessary for mere survival. Men attracted to women with juvenile features may have shifted the species as a whole to be less hairy and softer featured than any of our simian relatives. Can you think of a way to test these hypotheses?

18 Organizing Information About Species

LEARNING ROADMAP

This chapter adds the concept of evolution (Sections 16.3 and 16.4) to taxonomy (1.5). Before starting, you should review DNA sequences (8.3) and sequencing (15.4); the genetic code (9.4); master genes (10.3, 10.4); genomics (15.5), neutral mutations (17.2), genetic equilibrium (17.3); gene flow (17.8); and speciation (17.9).

PHYLOGENY

Evolutionary biologists can reconstruct the evolutionary history of a group of organisms by identifying shared, heritable traits that evolved in a common ancestor.

COMPARING BODY FORM

Similar body parts in different lineages may indicate descent from a shared ancestor, or they may have evolved independently in response to similar environmental pressures.

COMPARING BIOCHEMISTRY

Neutral changes tend to accumulate at a fairly constant rate in DNA. Molecular comparisons help us discover and confirm relationships among species and lineages.

COMPARING DEVELOPMENT

Patterns of development have a basis in master genes conserved over evolutionary time. Lineages with more recent common ancestry often develop in similar ways.

APPLICATIONS OF PHYLOGENY

Understanding a group's evolutionary history can help us protect endangered species. Applied to agents of infectious disease, it can also reveal large-scale patterns of transmission.

You will revisit cladistics throughout Unit IV as you learn about life's diversity. Classification of birds and reptiles is explained in Chapter 25. Chapter 44 returns to population ecology, the study of which is important for understanding human impacts on the biosphere (Chapter 48). Our efforts to mitigate that impact include conservation efforts aimed at maintaining biodiversity (Chapter 48) in ecosystems (Chapter 46). Animal development is detailed in Chapter 42.

18.1 Bye Bye Birdie

Some finch species migrate far outside of their normal range when food becomes scarce in a preferred overwintering spot, traveling in flocks of thousands or even tens of thousands of individuals. About 5.8 million years ago in southern Asia, one of these migratory flocks was caught up in the winds of a huge storm. The birds—rosefinches—were blown at least seven thousand miles (11,000 kilometers) across the open ocean to the islands of the Hawaiian archipelago. Enough individuals survived the journey to found a new population. The birds' arrival had been preceded by insects and plants, but no predators, and their descendants thrived. Isolation from mainland finch populations allowed the island colonizers to diverge in adaptive radiations that gave rise to the Hawaiian honeycreepers (Section 17.10).

The first Polynesians arrived on the islands sometime before 1000 A.D.; Europeans followed in 1778. Hawaii's rich ecosystem was hospitable to the newcomers and their domestic animals and crops. Escaped livestock ate and trampled rain forest plants that had provided the honeycreepers with food and shelter. Entire forests were cleared to grow imported crops, and plants that escaped cultivation began to crowd out native plants. Mosquitoes accidentally introduced in 1826 spread diseases such as avian malaria from imported chickens to native bird species. Stowaway rats ate their way through populations of native birds and their eggs. Mongooses deliberately imported to eat the rats preferred to eat birds and bird eggs.

The isolation that had allowed honeycreepers to arise by adaptive radiation also made them vulnerable to extinction. Divergence from the ancestral species had led to the loss of unnecessary traits such as defenses against mainland predators and diseases. Traits that had previously been adaptive—such as a long, curved beak matching the flower of a particular plant—became hindrances when habitats suddenly changed or disappeared. Thus, at least 43 Hawaiian honeycreeper species that had thrived on the islands before humans arrived were extinct by 1778. Conservation efforts began in the 1960s, but another 43 species have since disappeared.

Today, the few remaining Hawaiian honeycreepers are still pressured by established populations of nonnative species of plants and animals. Rising global temperatures are also allowing mosquitoes to invade high-altitude habitats that had previously been too cold for the insects, so honeycreeper species remaining in these habitats are now succumbing to mosquito-borne diseases. Of the 18 remaining honeycreeper species, only two are not in danger of extinction (**FIGURE 18.1**).

A A Palila (*Loxioides bailleui*) feeds on the seeds of the mamane plant, which are toxic to most other birds. The one remaining Palila population is declining because mamane plants are being trampled by cows and gnawed to death by goats and sheep. About 2,176 Palila remained in 2012.

B The unusual skewed bill of the Akekee (*Loxops caeruleirostris*) allows this bird to easily pry open buds that harbor insects. Avian malaria carried by mosquitoes to higher altitudes is decimating the last population of this species. In 2008, about 3,111 remained.

C This male Poouli (*Melamprosops phaeosoma*)—rare, old, and missing an eye—died in 2004 from avian malaria. There were two other Poouli alive at the time, but neither has been seen since then.

FIGURE 18.1 Three honeycreeper species: going, going, gone.

CREDITS: (opposite) John Steiner/Smithsonian Institution; (1A) © Eric VanderWerf/Pacific Rim Photos; (1B) © Courtesy of © Lucas Behnke; (1C) Bill Sparklin/Ashley Dayer.

18.2 Phylogeny

✔ Evolutionary history can be reconstructed by studying shared, heritable traits.

Table 18.1 Examples of Characters

	Bird	Bat	Lion
Warm-blooded	Y	Y	Y
Hair	N	Y	Y
Milk	N	Y	Y
Teeth	N	Y	Y
Wings	Y	Y	N
Feathers	Y	N	N

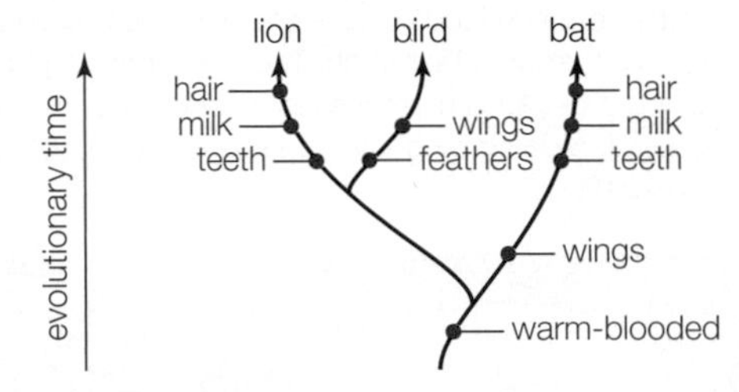

A If the bird and lion are most closely related, the derived traits would have evolved ten times in total.

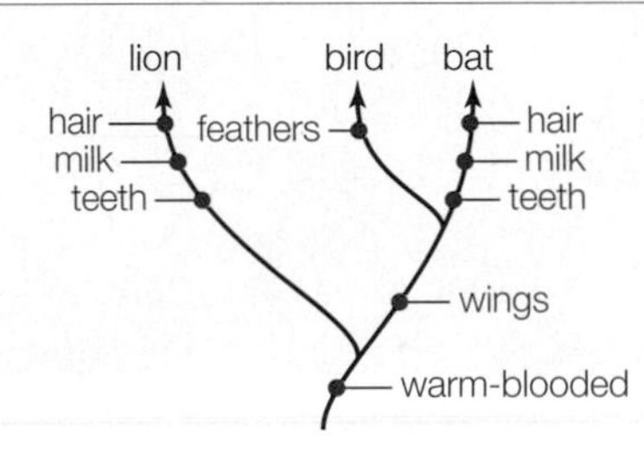

B If the bird and bat are most closely related, the derived traits would have evolved nine times in total.

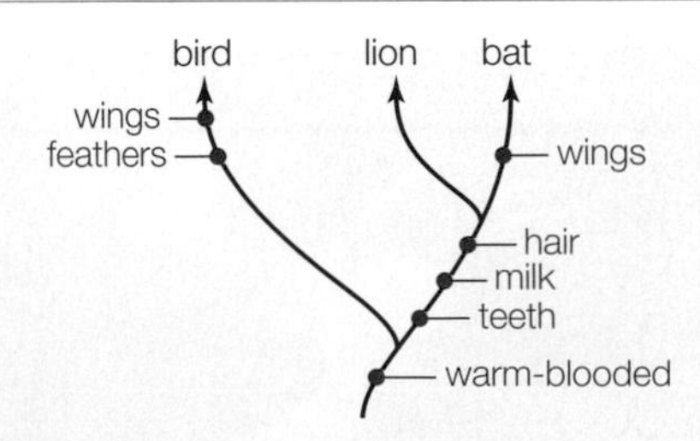

C If the lion and bat are most closely related, the derived traits would have evolved seven times in total.

FIGURE 18.2 An example of cladistics, using parsimony analysis with the characters listed in **TABLE 18.1**.

A, **B**, and **C** show the three possible evolutionary pathways that could connect birds, bats, and lions; red indicates the evolution of a derived trait. The pathway most likely to be correct (**C**) is the simplest—the one in which the derived traits would have had to evolve the fewest number of times in total.

Classifying life's tremendous diversity into a series of taxonomic ranks (Section 1.5) is a useful endeavor, in the same way that it is useful to organize a telephone book or contact list in alphabetical order. Today, biologists try to classify organisms according to evolutionary relationships among them. Thus, they focus on reconstructing **phylogeny**, the evolutionary history of a species or a group of species. Phylogeny is a kind of genealogy that follows a lineage's evolutionary relationships through time.

Humans were not around to witness the evolution of most species, but we can use evidence to understand events in the past (Section 16.1). Consider how each species bears traces of its own unique evolutionary history in its characters. A **character** is a quantifiable, heritable trait such as the number of segments in a backbone, the nucleotide sequence of ribosomal RNA, or the presence of wings (**TABLE 18.1**). Traditional classification schemes group organisms based on shared characters: Birds have feathers, cacti have spines, and so on. Such schemes do not necessarily reflect evolutionary history because species that are not closely related may appear very similar (Section 16.2). By contrast, evolutionary biology tries to fit each species into a bigger picture of evolution: Every living thing is related if you just go back far enough in time. Instead of grouping organisms by shared characters, evolutionary biologists try to pinpoint what makes the organisms share the characters in the first place: a common ancestor. They determine common ancestry by looking for derived traits. A **derived trait** is a character present in a group under consideration, but not in any of the group's ancestors.

A group whose members share one or more defining derived traits is called a **clade**. By definition, a clade is a **monophyletic group**: one that consists of an ancestor (in which a derived trait evolved) together with any and all of its descendants.

Each species is a clade. Many higher taxonomic rankings are also equivalent to clades—flowering plants, for example, are both a phylum and a clade—but some are not. For example, the traditional Linnaean class Reptilia ("reptiles") includes crocodiles, alligators, tuataras, snakes, lizards, turtles, and tortoises. While it is convenient to classify these animals together, they would not constitute a clade unless birds are also included, as you will see in Chapter 25.

It is the recent nature of a derived trait that defines a clade. Consider how alligators look a lot more like lizards than birds. In this case, the similarity in appearance does indicate shared ancestry, but it is a more distant relationship than alligators have with birds. Evolutionary biologists discovered that alligators and birds share a more recent common ancestor than alligators and lizards do. Derived traits—a gizzard and a four-chambered heart—evolved in the lineage that gave rise to alligators and birds, but not in the one that gave rise to lizards.

The remaining sections of this chapter explore some of the character comparisons that evolutionary biologists use to group organisms into clades. Remember that evolutionary history does not change because of events in the present: A species' ancestry remains the same no matter how it evolves. However, as with traditional taxonomy, we can make mistakes grouping organisms into clades if the information we have is incomplete. A clade is necessarily a hypothesis, and which organisms it includes may change when new discoveries are made. As with all hypotheses, the more data that support a cladistic grouping, the less likely it is to require revision.

Cladistics

In the big picture of evolution, all clades are interconnected; an evolutionary biologist's job is to figure out where the connections are. Making hypotheses about evolutionary relationships among clades is called **cladistics**. One way of doing this involves the logical rule of simplicity: When there are several possible ways that a group of clades can be connected, the simplest evolutionary pathway is probably the correct one. By comparing all of the possible connections among the clades, we can identify the simplest: the one in which the defining derived traits evolved the fewest number of times (**FIGURE 18.2**). The process of finding the simplest pathway is called parsimony analysis.

The result of a cladistic analysis is an **evolutionary tree**—a diagram of evolutionary connections—called a cladogram. A **cladogram** visually summarizes a hypothesis about how a group of clades are related (**FIGURE 18.3**). Data from an outgroup (a species not closely related to any member of the group under study) may be included in order to "root" the tree. Each line in a cladogram represents a lineage, which may branch into two lineages at a node.

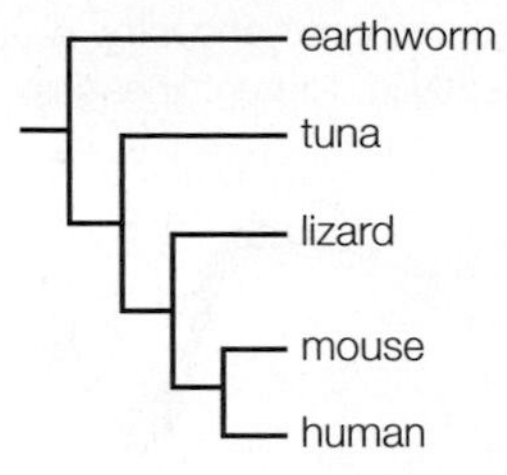

A Evolutionary connections among clades are represented as lines on a cladogram. Sister groups emerge from a node, which represents a common ancestor.

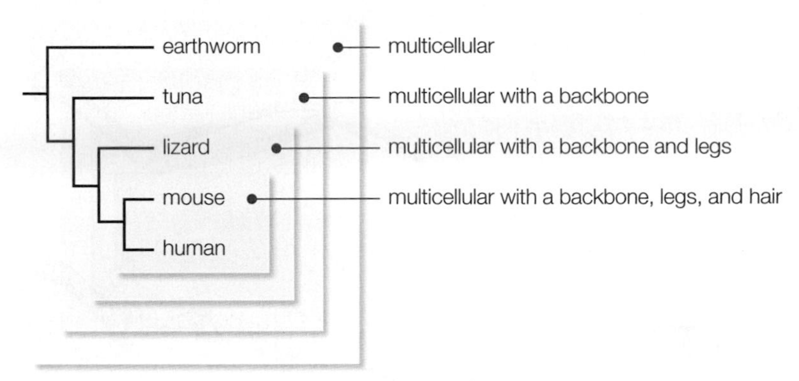

B A cladogram can be viewed as "sets within sets" of derived traits. Each set (an ancestor together with all of its descendants) is a clade.

FIGURE 18.3 ▶**Animated** An example of a cladogram.

The node represents a common ancestor of two lineages. Every branch of a cladogram is a clade; the two lineages that emerge from a node on a cladogram are called **sister groups**.

character Quantifiable, heritable characteristic or trait.
clade A group whose members share one or more defining derived traits.
cladistics Making hypotheses about evolutionary relationships among clades.
cladogram Evolutionary tree diagram that summarizes hypothesized relationships among a group of clades.
derived trait A novel trait present in a clade but not in any of the clade's ancestors.
evolutionary tree Diagram showing evolutionary connections.
monophyletic group An ancestor in which a derived trait evolved, together with all of its descendants.
phylogeny Evolutionary history of a species or group of species.
sister groups The two lineages that emerge from a node on a cladogram.

TAKE-HOME MESSAGE 18.2

Why do we study evolutionary history?

✔ Evolutionary biologists study phylogeny in order to understand how all species are connected by shared ancestry.

✔ A clade is a monophyletic group whose members share one or more derived traits. Cladistics is a method of making hypotheses about evolutionary relationships among clades.

✔ Cladograms and other evolutionary tree diagrams are based on our understanding of the evolutionary history of a group of organisms.

18.3 Comparing Form and Function

✔ Physical similarities are often evidence of shared ancestry, but sometimes a trait evolves independently in different lineages.

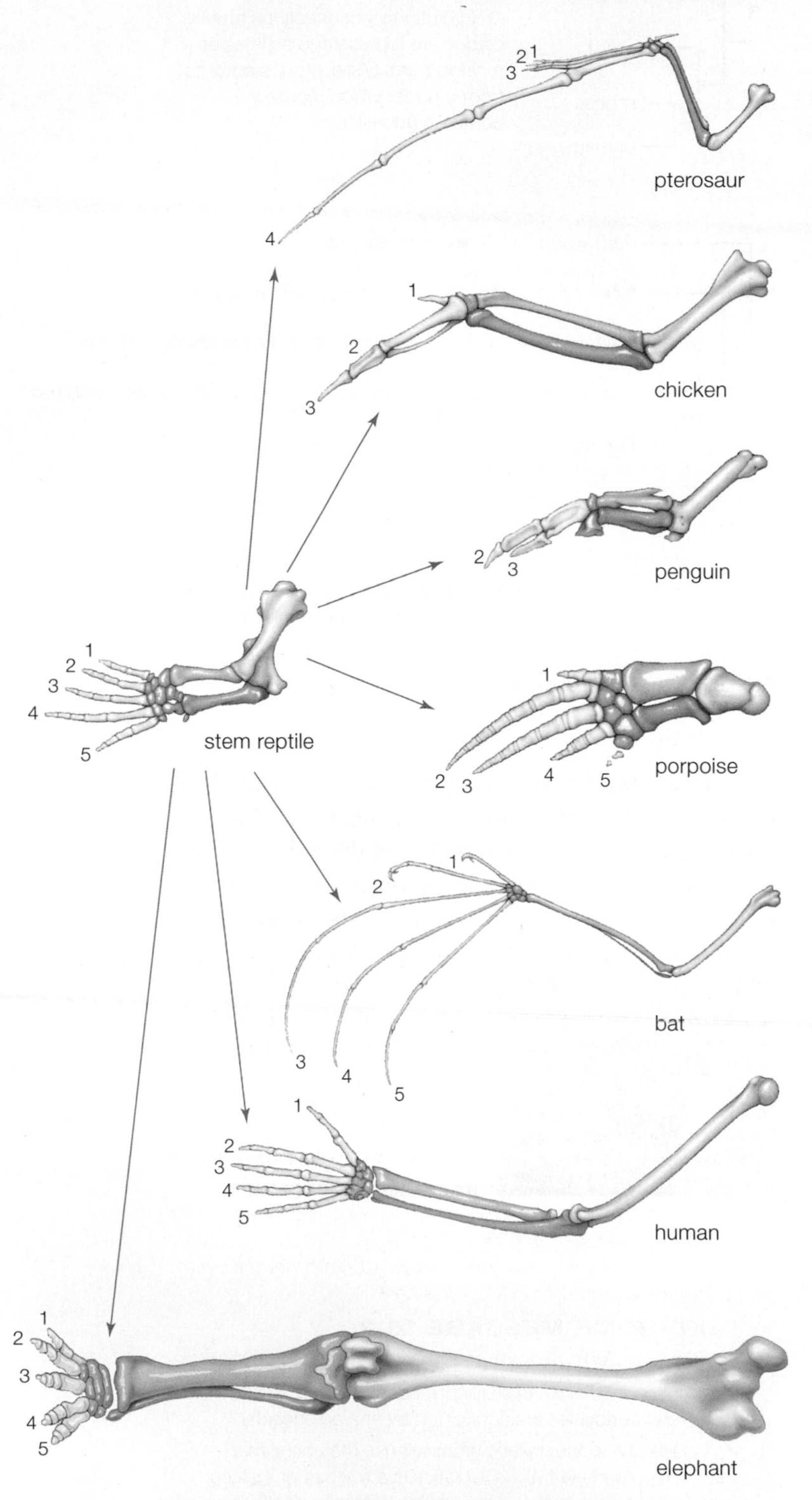

FIGURE 18.4 ▸**Animated** Morphological divergence among vertebrate forelimbs, starting with the bones of an ancient stem reptile. The number and position of many skeletal elements were preserved when these diverse forms evolved; notice the bones of the forearms. Certain bones were lost over time in some of the lineages (compare the digits numbered 1 through 5). Drawings are not to scale.

To biologists, remember, evolution means change in a line of descent. How do they reconstruct evolutionary events that occurred in the ancient past? Evolutionary biologists are a bit like detectives, using clues to piece together a history that they did not witness in person. Fossils provide some clues. The body form and function of organisms that are alive today provide others.

Morphological Divergence

Body parts that appear similar in separate lineages because they evolved in a common ancestor are called **homologous structures** (*hom–* means "the same"). Homologous structures may be used for different purposes in different groups, but the very same genes direct their development.

A body part that outwardly appears very different in separate lineages may be homologous in underlying form. Vertebrate forelimbs, for instance, vary in size, shape, and function. However, they are alike in the structure and positioning of bony elements, and in internal patterns of nerves, blood vessels, and muscles.

You learned in Chapter 17 how populations that are not interbreeding diverge genetically; these divergences give rise to changes in body form. Change from the body form of a common ancestor is an evolutionary pattern called **morphological divergence**. Consider the limb bones of vertebrate animals. Fossil evidence suggests that many modern vertebrates are descended from a family of ancient "stem reptiles" that crouched low to the ground on five-toed limbs. Descendants of this ancestral group diversified over millions of years, and eventually gave rise to modern reptiles, birds, and mammals; a few lineages that had become adapted to walking on land even returned to aquatic living. As these lineages diversified, their five-toed limbs became adapted for many different purposes (**FIGURE 18.4**). Limbs became modified for flight in extinct reptiles called pterosaurs and in bats and most birds. In penguins and porpoises, they are now flippers useful for swimming. Forelimbs of humans are arms and hands with four fingers and an opposable thumb. Among elephants, the limbs are now strong and pillarlike, capable of supporting a great deal of weight. Limbs degenerated to nubs in pythons and boa constrictors, and they disappeared entirely in other snakes.

Morphological Convergence

Body parts that appear similar in different species are not always homologous; they sometimes evolve independently in lineages subject to the same environmental pressures. The independent evolution of similar body parts in different lineages is **morphological convergence**. Structures that are similar as a result of morphological convergence are called **analogous structures**. Analogous structures look alike but did not evolve in a shared ancestor; they evolved independently after the lineages diverged.

For example, bird, bat, and insect wings all perform the same function, which is flight. However, several clues tell us that the wing surfaces are not homologous. All of the wings are adapted to the same physical constraints that govern flight, but each is adapted in a different way. In the case of birds and bats, the limbs themselves are homologous, but the adaptations that make those limbs useful for flight differ. The surface of a bat wing is a thin, membranous extension of the animal's skin. By contrast, the surface of a bird wing is a sweep of feathers, which are specialized structures derived from skin. Insect wings differ even more. An insect wing forms as a saclike extension of the body wall. Except at forked veins, the sac flattens and fuses into a thin membrane. The sturdy, chitin-reinforced veins structurally support the wing. Unique adaptations for flight are evidence that wing surfaces of birds, bats, and insects are analogous structures that evolved after the ancestors of these modern groups diverged (**FIGURE 18.5**).

As another example of morphological convergence, consider the similar external structures of American cacti and African euphorbias (see **FIGURE 16.3**). These structures adapt the plants to similarly harsh desert environments where rain is scarce. Distinctive accordion-like pleats allow the plant body to swell with water when rain does come. Water stored in the plants' tissues allows them to survive long dry periods. As the stored water is used, the plant body shrinks, and the folded pleats provide it with some shade in an environment that typically has none. Despite these similarities, a closer look reveals many differences that indicate the two types of plants are not closely related. For example, cactus spines have a simple fibrous structure; they are modified leaves that arise from dimples on the plant's surface. Euphorbia spines project smoothly from the plant surface, and they are not modified leaves: In many species the spines are actually dried flower stalks (left).

FIGURE 18.5 Morphological convergence in animals. The surfaces of an insect wing, a bat wing, and a bird wing are analogous structures. The diagram shows how the evolution of wings (red dots) occurred independently in the three separate lineages.

analogous structures Similar body structures that evolved separately in different lineages.
homologous structures Body structures that are similar in different lineages because they evolved in a common ancestor.
morphological convergence Evolutionary pattern in which similar body parts evolve separately in different lineages.
morphological divergence Evolutionary pattern in which a body part of an ancestor changes in its descendants.

TAKE-HOME MESSAGE 18.3

What evidence does evolution leave in body form?

✔ In a pattern of morphological divergence, body parts are often modified differently in different lines of descent. Such parts are called homologous structures.

✔ In a pattern of morphological convergence, body parts that appear alike evolved independently in different lineages. Such parts are called analogous structures.

18.4 Comparing Biochemistry

✔ Evolution leaves clues in biochemistry. In general, more closely related species share more biochemical similarities.

Over time, inevitable mutations change a genome's DNA sequence. Most of these mutations are neutral. Neutral mutations have no effect on an individual's survival or reproduction, so we can assume they accumulate at a constant rate. For example, a nucleotide substitution that changes one codon from AAA to AAG in a gene's protein-coding region would probably not affect the protein product, because both codons specify lysine (Section 9.4). In other cases, a neutral mutation can change the amino acid sequence but not the function of a protein product.

The accumulation of neutral mutations in the DNA of a lineage can be likened to the predictable ticks of a **molecular clock**. Turn the hands of such a clock back, so the ticks wind back through the past, and the last tick will be the time when the lineage embarked on its own unique evolutionary road. To calibrate the molecular clock, the number of differences between genomes can be correlated with the timing of morphological changes observed in the fossil record.

Mutations alter the DNA of a lineage independently of all other lineages. The more recently two lineages diverged, the less time there has been for unique neutral mutations to accumulate in the DNA of each one. That is why the genomes of closely related species tend to be more similar than those of distantly related ones—a general rule that can be used to estimate relative times of divergence.

DNA and Protein Sequence Comparisons

Similarities in the amino acid sequence of a shared protein, or the nucleotide sequence of a shared gene, can be evidence of an evolutionary relationship. Biochemical comparisons like these are often used together with morphological comparisons in phylogenetic analyses, in order to provide data for hypotheses about shared ancestry.

In general, species that are more closely related tend to have more similar proteins. Two species with very few similar proteins probably have not shared an ancestor for a long time—long enough for many mutations to have accumulated in the DNA of their separate lineages. Evolutionary biologists can compare a protein's sequence among several species, and use the number of amino acid differences as one measure of relative relatedness (**FIGURE 18.6**).

The amino acids that differ are also clues. For example, a leucine to isoleucine change may not affect the function of a protein very much, because both amino acids are nonpolar, and both are about the same size. Such changes are called conservative amino acid substitutions. By contrast, the substitution of a lysine (which is basic) for an aspartic acid (which is acidic) may dramatically change the character of a protein. Such nonconservative substitutions—as well as deletions and insertions—often affect phenotype. Most mutations that affect phenotype are selected against, but occasionally one proves adaptive. Thus, the longer it has been since two lineages diverged, the more nonconservative amino acid substitutions we are likely to see when comparing their proteins.

Among species that diverged relatively recently, many proteins have identical amino acid sequences. Nucleotide sequence differences may be instructive in such cases. Even if the amino acid sequence of a protein is identical among species, the nucleotide sequence of the gene that encodes the protein may differ because of redundancies in the genetic code.

The DNA from nuclei, mitochondria, or chloroplasts can be used in nucleotide comparisons. Mitochondrial DNA accumulates mutations faster than nuclear DNA,

```
    honeycreepers (10) ...CRDVQFGWLIRNLHANGASFFFICIYLHIGRGIYYGSYLNK--ETWNIGVILLLTLMATAFVGYVLPWGQMSFWG...
          song sparrow ...CRDVQFGWLIRNLHANGASFFFICIYLHIGRGIYYGSYLNK--ETWNVGIILLLALMATAFVGYVLPWGQMSFWG...
    Gough Island finch ...CRDVQFGWLIRNIHANGASFFFICIYLHIGRGLYYGSYLYK--ETWNVGVILLLTLMATAFVGYVLPWGQMSFWG...
            deer mouse ...CRDVNYGWLIRYMHANGASMFFICLFLHVGRGMYYGSYTFT--ETWNIGIVLLFAVMATAFMGYVLPWGQMSFWG...
    Asiatic black bear ...CRDVHYGWIIRYMHANGASMFFICLFMHVGRGLYYGSYLLS--ETWNIGIILLFTVMATAFMGYVLPWGQMSFWG...
        bogue (a fish) ...CRDVNYGWLIRNLHANGASFFFICIYLHIGRGLYYGSYLYK--ETWNIGVVLLLLVMGTAFVGYVLPWGQMSFWG...
                 human ...TRDVNYGWIIRYLHANGASMFFICLFLHIGRGLYYGSFLYS--ETWNIGIILLLATMATAFMGYVLPWGQMSFWG...
thale cress (a plant) ...MRDVEGGWLLRYMHANGASMFLIVVYLHIFRGLYHASYSSPREFVWCLGVVIFLLMIVTAFIGYVLPWGQMSFWG...
          baboon louse ...ETDVMNGWMVRSIHANGASWFFIMLYSHIFRGLWVSSFTQP--LVWLSGVIILFLSMATAFLGYVLPWGQMSFWG...
          baker's yeast ...MRDVHNGYILRYLHANGASFFFMVMFMHMAKGLYYGSYRSPRVTLWNVGVIIFTLTIATAFLGYCCVYGQMSHWG...
```

FIGURE 18.6 Example of a protein comparison. Here, part of the amino acid sequence of mitochondrial cytochrome *b* from 20 species is aligned. This protein is a crucial component of mitochondrial electron transfer chains. The honeycreeper sequence is identical in ten species of honeycreeper; amino acids that differ in the other species are shown in red. Dashes are gaps in the alignment.

FIGURE IT OUT In this comparison, which species is the most closely related to honeycreepers?

Answer: The song sparrow

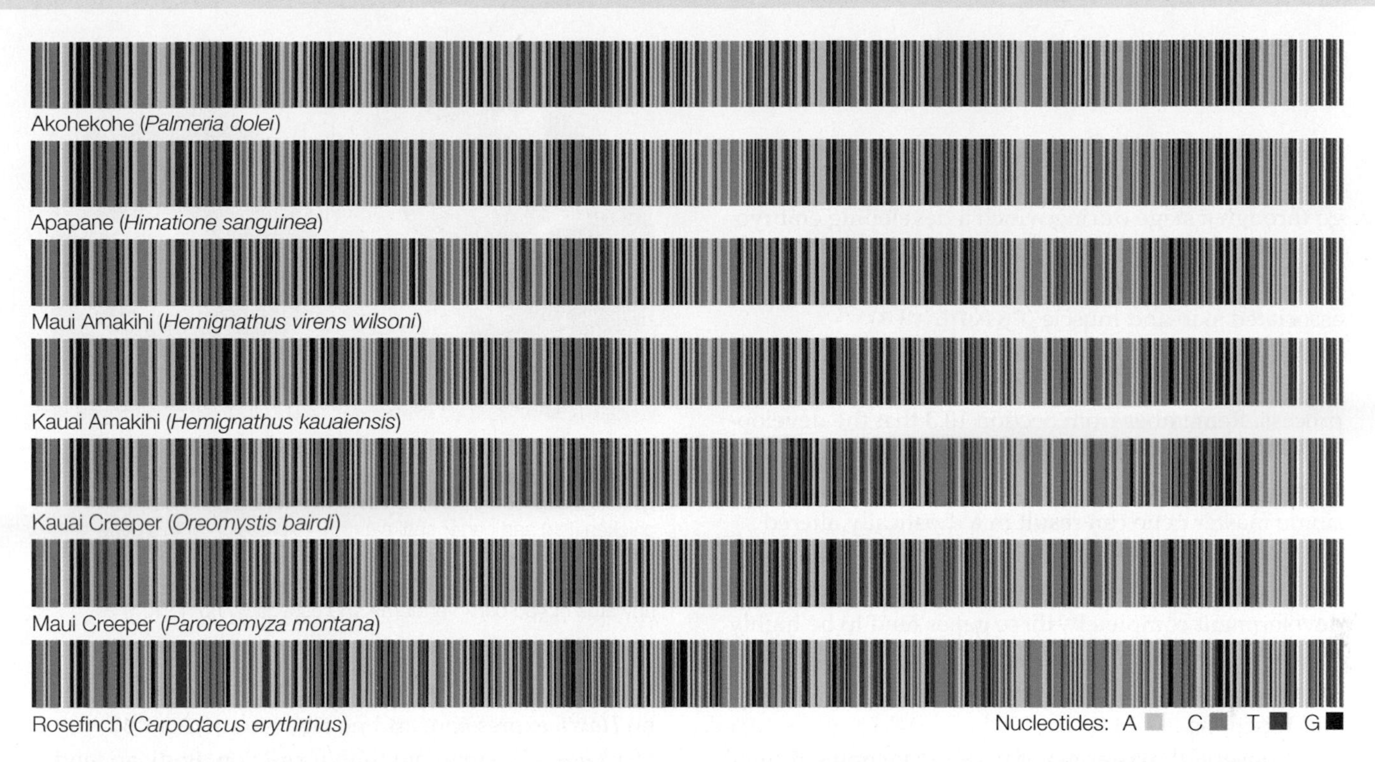

FIGURE 18.7 DNA barcoding. This example allows a quick visual comparison of 591 nucleotides of a mitochondrial gene called *ND2* (NADH dehydrogenase subunit 2) from 7 Hawaiian honeycreeper species and a rosefinch. The Akohekohe and Apapane are sister species, as are the Amakihis, and the Creepers.

so it can even be used to compare different individuals of the same sexually reproducing animal species. In most animals, mitochondria are inherited intact from a single parent (usually the mother). They also contain their own DNA, and they reproduce by dividing. Thus, in most cases, differences in mitochondrial DNA sequences between maternally related individuals are due to mutations, not genetic recombination events.

As you will see in the next section, some essential genes are highly conserved, which means their DNA sequences have changed very little or not at all over evolutionary time. Other genes are not conserved, and these underly differences that define species. DNA sequence differences are the basis of a method called DNA barcoding, which is used to identify an individual as belonging to a particular species. A standard region of the individual's DNA is amplified using PCR (Section 15.3) and then sequenced. The sequence is illustrated as a barcode in which each base is represented by a different color (**FIGURE 18.7**). The technique is used when other methods of identification are difficult or impossible. For example, researchers can use DNA barcoding of feces to identify components of an endangered animal's diet.

Getting useful information from comparing DNA requires a lot more data than comparing proteins. This is because coincidental homologies are statistically more likely to occur with DNA comparisons—there are only four nucleotides in DNA versus twenty amino acids in proteins. However, DNA sequencing has become so fast that there is a lot of data available to compare. Genomics studies with such data have shown us (for example) that about 88 percent of the mouse genome sequence is identical with the human genome, as is 73 percent of the zebrafish genome, 47 percent of the fruit fly genome, and 25 percent of the rice genome.

molecular clock Technique that uses molecular change to estimate how long ago two lineages diverged.

TAKE-HOME MESSAGE 18.4

Why do DNA or protein similarities reflect evolutionary history?

✔ Mutations change the nucleotide sequence of a lineage's DNA over time.

✔ Lineages that diverged long ago generally have more differences between their DNA (and their proteins) than do lineages that diverged more recently.

18.5 Comparing Patterns of Animal Development

✔ Similar patterns of embryonic development are an outcome of highly conserved master genes.

In general, the more closely related animals are, the more similar is their development. For example, all vertebrates go through a stage during which a developing embryo has four limb buds, a tail, and a series of somites—divisions of the body that give rise to the backbone and associated skin and muscle (**FIGURE 18.8**).

Animals have similar patterns of embryonic development because the very same master genes direct the process. Remember from Section 10.3 that the development of an embryo into a body is orchestrated by layer after layer of master gene expression. The failure of any single master gene can result in a drastically altered body plan, typically with devastating consequences. Because a mutation in a master gene typically unravels development completely, these genes tend to be highly conserved. Even among lineages that diverged a very long time ago, such genes often retain similar sequences and functions.

If conserved master genes direct development in all vertebrate lineages, how do the adult forms end up so different? Part of the answer is that there are differences in the onset, rate, or completion of early steps in development. These differences are brought about by variations in master gene expression patterns. Consider homeotic genes called *Hox*. Like other homeotic genes, *Hox* gene expression helps sculpt details of the body's form during embryonic development. Vertebrate animals have multiple sets of the same ten *Hox* genes that occur in insects and other arthropods. You have already read about one of these genes, *antennapedia*, which determines the identity of the thorax (the body part with legs) in fruit flies. One vertebrate version of *antennapedia* is called *Hoxc6*, and it determines the identity of the back (as opposed to the neck or tail). Expression of the *Hoxc6* gene causes ribs to develop on a vertebra.

FIGURE 18.9 How differences in body form can arise from differences in master gene expression. Expression of the *Hoxc6* gene is indicated by purple stain in two vertebrate embryos, chicken (left) and garter snake (right). Expression of this gene causes a vertebra to develop ribs. Chickens have 7 vertebrae in their back and 14 to 17 vertebrae in their neck; snakes have upwards of 450 back vertebrae and essentially no neck.

Vertebrae of the neck and tail normally develop with no *Hoxc6* expression, and no ribs (**FIGURE 18.9**). *Hox* genes also regulate limb formation. Body appendages as diverse as crab legs, beetle legs, sea star arms, butterfly wings, fish fins, and mouse feet start out as clusters of cells that bud from the surface of the embryo. The buds form wherever a homeotic gene called *Dlx* is expressed. *Dlx* encodes a transcription factor that signals clusters of embryonic cells to "stick out from the body" and give rise to an appendage. *Hox* genes suppress *Dlx* expression in all parts of an embryo that will not have appendages.

TAKE-HOME MESSAGE 18.5

Why do animals develop in similar ways?

✔ Similarities in patterns of development are the result of master genes that have been conserved over evolutionary time.

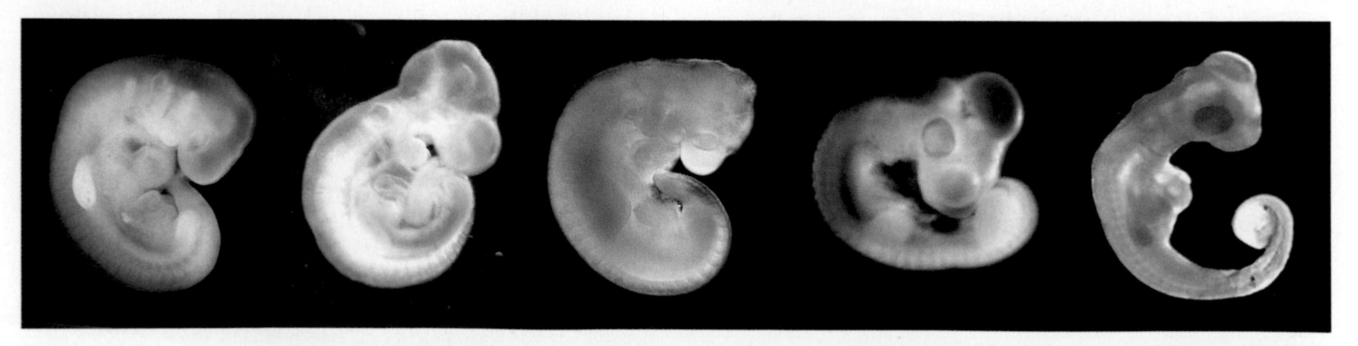

FIGURE 18.8 Visual comparison of vertebrate embryos. All vertebrates go through an embryonic stage in which they have four limb buds, a tail, and divisions called somites along their back. From left to right: human, mouse, bat, chicken, alligator.

18.6 Applications of Phylogeny Research

✔ We use phylogeny to understand how to preserve species that exist today.

Studies of phylogeny reveal how species relate to one another and to species that are now extinct. In doing so, they inform our understanding of how shared ancestry interconnects all species—including our own.

Conservation Biology

The story of the Hawaiian honeycreepers is a dramatic illustration of how evolution works. It also shows how finding ancestral connections can help species that are still living. As more and more honeycreeper species become extinct, the group's reservoir of genetic diversity dwindles. The lowered diversity means the group as a whole is less resilient to change, and more likely to suffer catastrophic losses.

Deciphering their phylogeny can tell us which honeycreeper species are most different from the others—and those are the ones most valuable in terms of preserving the group's genetic diversity. Such research allows us to concentrate our resources and conservation efforts on those species whose extinction would mean a greater loss to biodiversity. For example, we now know the Poouli (pictured in **FIGURE 18.1C**) to be the most distant relative in the Hawaiian honeycreeper family. Unfortunately, the knowledge came too late; the Poouli is probably extinct. Its extinction means the loss of a large part of evolutionary history of the group: One of the longest branches of the honeycreeper family tree is gone forever.

Cladistics analyses are also used to correlate past evolutionary divergences with behavior and dispersal patterns of existing populations. Such studies are useful in conservation efforts. For example, a decline in antelope populations in African savannas is at least partly due to competition with domestic cattle. A cladistic analysis of mitochondrial DNA sequences suggested that current populations of blue wildebeest (**FIGURE 18.10**) are genetically less similar than they should be, based on other antelope groups of similar age. Combined with behavioral and geographic data, the analysis helped conservation biologists realize that a patchy distribution of preferred food plants is preventing gene flow among blue wildebeest populations. The absence of gene flow can lead to a catastrophic loss of genetic diversity in populations under pressure. Restoring appropriate grasses in intervening, unoccupied areas of savanna would allow isolated wildebeest populations to reconnect.

FIGURE 18.10 A blue wildebeest in Africa. Conservation biologists discovered that a patchy availability of preferred food was hampering gene flow among wildebeest populations. The biologists recommended restoring grasses in some areas that had been cleared, to re-establish gene flow among isolated wildebeests.

Medical Research

Researchers often study the evolution of viruses and other infectious agents by grouping them into clades based on biochemical characters. Even though viruses are not alive, they can mutate every time they infect a host, so their genetic material changes over time. Consider the H5N1 strain of influenza (flu) virus, which infects birds and other animals. H5N1 has a very high mortality rate in humans, but human-to-human transmission has been rare to date. However, the virus replicates in pigs without causing symptoms. Pigs transmit the virus to other pigs—and apparently to humans too. A phylogenetic analysis of H5N1 isolated from pigs showed that the virus "jumped" from birds to pigs at least three times since 2005, and that one of the isolates had acquired the potential to be transmitted among humans. An increased understanding of how this virus adapts to new hosts is helping researchers design more effective vaccines for it.

TAKE-HOME MESSAGE 18.6

How can we use what we learn about evolutionary history?

✔ Phylogeny research is yielding an ever more specific and accurate picture of how all life is related by shared ancestry.

✔ Among other applications, phylogeny research can help us to prioritize efforts to preserve endangered species, and to understand the spread of infectious diseases.

CREDITS: (in text) John Steiner/Smithsonian Institution; (10) Alan Lucas/Shutterstock.

Bye Bye Birdie (revisited)

In 2004, researchers captured one of the three remaining Pooulis, with the intent of starting a captive breeding program before the species became extinct. They were unable to capture a female to mate with this male before he died in captivity a month later. Cells from this last bird were frozen, and may be used in the future for cloning. However, with no parents left to demonstrate the species' natural behavior to chicks, cloned birds would probably never be able to establish themselves as a natural population.

summary

Section 18.1 The Hawaiian honeycreepers are highly specialized. Their specialization makes them particularly vulnerable to extinction as a result of habitat loss and exotic species introductions that followed human colonization of the Hawaiian Islands.

Section 18.2 Evolutionary biologists reconstruct evolutionary history (**phylogeny**) by comparing physical, behavioral, and biochemical traits, or **characters**, among species. A **clade** is a **monophyletic group** that consists of an ancestor in which one or more **derived traits** evolved, together with all of its descendants.

Making hypotheses about the evolutionary history of a group of clades is called **cladistics**. **Evolutionary tree** diagrams are based on the premise that all organisms are connected by shared ancestry. In a **cladogram**, each line represents a lineage. A lineage branches into two **sister groups** at a node, which represents a shared ancestor.

Section 18.3 Comparative morphology is one way to study evolutionary connections among lineages. **Homologous structures** are similar body parts that became modified differently in different lineages (a pattern called **morphological divergence**). Such parts are evidence of a common ancestor. **Analogous structures** are body parts that look alike in different lineages but did not evolve in a common ancestor. Rather, they evolved separately after the lineages diverged, a pattern called **morphological convergence**.

Section 18.4 We can discover and clarify evolutionary relationships through comparisons of DNA and protein sequences. In general, these sequences are more similar among lineages that diverged more recently. Neutral mutations accumulate in DNA at a predictable rate; like the ticks of a **molecular clock**, they help researchers estimate how long ago lineages diverged.

Section 18.5 Master genes that affect development tend to be highly conserved, so similarities in patterns of embryonic development reflect shared ancestry that can be evolutionarily ancient. Mutations that alter the timing of master gene expression can alter the rate or onset of development, and thus result in different adult body forms.

Section 18.6 Reconstructing phylogeny is part of our efforts to preserve endangered species. Phylogeny is also used for studying the spread of viruses and other agents of infectious diseases.

self-quiz

Answers in Appendix VII

1. In cladistics, the only taxon that is always correct as a clade is the _______ .
   a. genus
   b. family
   c. species
   d. kingdom

2. In evolutionary trees, each node represents a(n) _______ .
   a. single lineage
   b. extinction
   c. point of divergence
   d. adaptive radiation

3. A clade is _______ .
   a. defined by a derived trait
   b. a monophyletic group
   c. a hypothesis
   d. all of the above

4. Cladistics _______ .
   a. may involve parsimony analysis
   b. is based on derived traits
   c. both of the above are correct

5. In cladograms, sister groups are _______ .
   a. inbred
   b. the same age
   c. represented by nodes
   d. in the same family

6. Through _______ , a body part of an ancestor is modified differently in different lines of descent.
   a. homologous evolution
   b. morphological convergence
   c. adaptive divergence
   d. morphological divergence

7. Homologous structures among major groups of organisms may differ in _______ .
   a. size
   b. shape
   c. function
   d. all of the above

8. Neutral mutations are those that do not affect _______ .
   a. amino acid sequence
   b. nucleotide sequence
   c. the chances of survival
   d. all of the above

9. Mitochondrial DNA sequences are often used in cladistic comparisons of _______ .
   a. different species
   b. individuals of the same species
   c. different taxa

CREDIT: (in text revisited) Bill Sparklin/Ashley Dayer.

data analysis activities

Hawaiian Honeycreeper Phylogeny The Poouli (*Melamprosops phaeosoma*) was discovered in 1973 by a group of students from the University of Hawaii. Its membership in the Hawaiian honeycreeper clade was (until recently) controversial, mainly because its appearance and behavior are so different from other living honeycreepers. It particularly lacked the "old tent" odor characteristic of other honeycreepers.

In 2011, Heather Lerner and her colleagues deciphered phylogeny of the 19 Hawaiian honeycreepers that were not yet officially declared to be extinct at the time, including the Poouli. The researchers sequenced mitochondrial and nuclear DNA samples taken from the honeycreepers, and also from 28 other birds (outgroups). Phylogenetic analysis of these data firmly establishes the Poouli as a member of the clade, and also reveals the Eurasian rosefinch as the clade's closest relative (**FIGURE 18.11**).

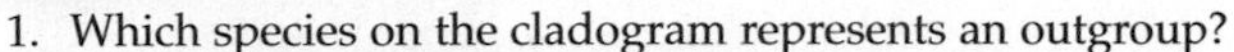

1. Which species on the cladogram represents an outgroup?

2. Which species is most closely related to the Apapane (*Himatione sanguinea*)?

3. What is the sister group of the Akikiki (*Oreomystis bairdi*)?

4. Which species is more closely related to the Palila (*Loxioides bailleui*): the Iiwi (*Vestiaria coccinea*) or the Maui Alauahio (*Paroreomyza montana*)?

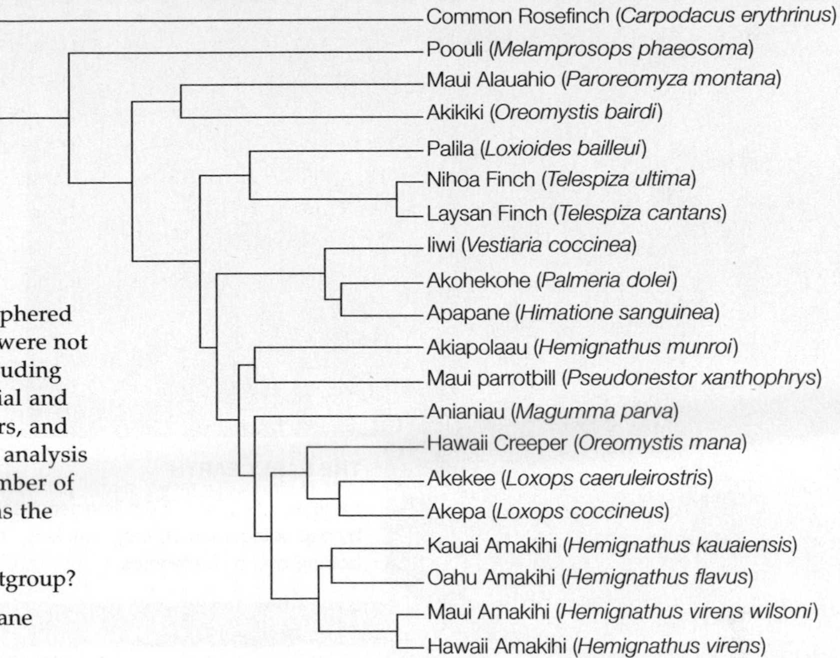

FIGURE 18.11 Hawaiian honeycreeper phylogeny. This cladogram was constructed using sequence comparisons of mitochondrial DNA (whole genome), and 13 nuclear DNA loci of 19 Hawaiian honeycreepers and 28 other finch species.

10. Molecular clocks are based on comparisons of the number of ________ mutations between species.
   a. lethal
   b. neutral
   c. conservative
   d. nonconservative

11. True or false? DNA barcoding can identify an individual as belonging to a particular species.

12. A mutation that alters the embryonic expression pattern of a ________ may lead to major differences in the adult form.
   a. derived trait
   b. master gene
   c. homologous structure
   d. all of the above

13. All of the following data types can be used as evidence of shared ancestry except similarities in ________ .
   a. amino acid sequences
   b. DNA sequences
   c. fossil morphologies
   d. embryonic development
   e. form due to convergence
   f. all are appropriate

14. True or false? Phylogeny helps us study the spread of viruses through human populations.

15. Match the terms with the most suitable description.

___ phylogeny	a. novel character
___ cladogram	b. evolutionary history
___ homeotic genes	c. human arm and bird wing
___ homologous structures	d. similar across diverse taxa
	e. measures neutral mutations
___ molecular clock	f. insect wing and bird wing
___ analogous structures	g. evolutionary tree
___ derived trait	

critical thinking

1. In the late 1800s, a biologist studying animal embryos coined the phrase "ontogeny recapitulates phylogeny," meaning that the physical development of an animal embryo (ontogeny) seemed to retrace the changing form of the species during its evolutionary history (phylogeny). Why would embryonic development retrace evolutionary steps?

2. The photos shown below illustrate a case of synpolydactyly. This genetic abnormality is characterized by webbing between partially or completely duplicated fingers or toes. The same mutations that give rise to the human phenotype also give rise to a similar phenotype in mice. In which family of genes do you think these mutations occur?

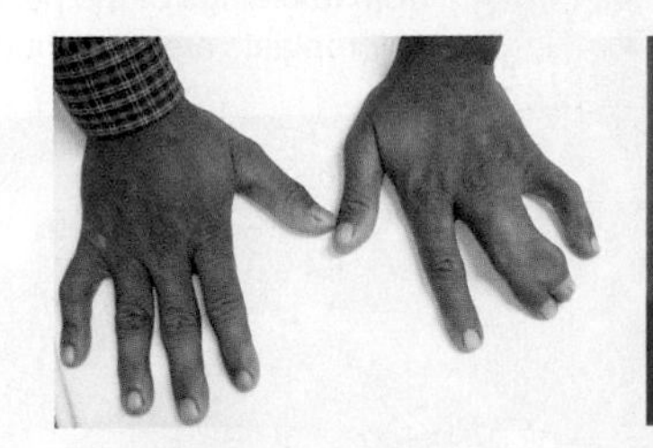

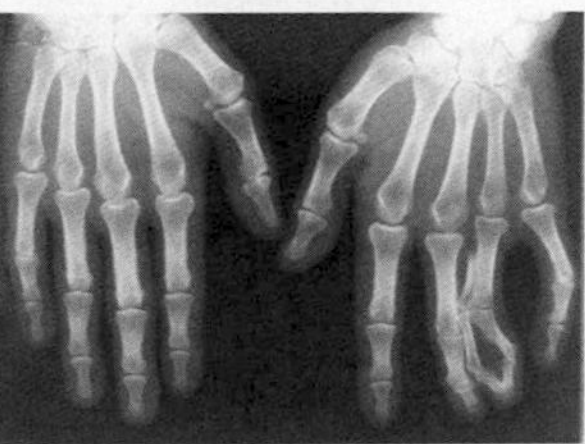

CREDITS: (11) © Cengage Learning; (in text CT #3) Courtesy of Dr. Sajid Malik, from www.biomedcentral.com/1471-2350/8/78.

19 Life's Origin and Early Evolution

LEARNING ROADMAP

This chapter explains how the molecular subunits of life introduced in Section 3.3 could have originated and assembled to form the first cells. It considers the evolution of eukaryotic traits (Sections 4.5–4.12) and draws on your understanding of fossils and the geologic time scale (16.5 and 16.8), aerobic respiration (7.2), and photosynthesis (6.4).

THE EARLY EARTH

Earth formed about 4.6 billion years ago from material released by exploding stars. At first, it lacked oxygen and was constantly bombarded by meteorites.

BUILDING BLOCKS OF LIFE

Simulations show how organic monomers could have formed on the early Earth. Such compounds also form in space and could have been delivered by meteorites.

ORIGIN OF CELLS

Experiments provide insight into how cells could have arisen from nonliving material through known physical and chemical processes.

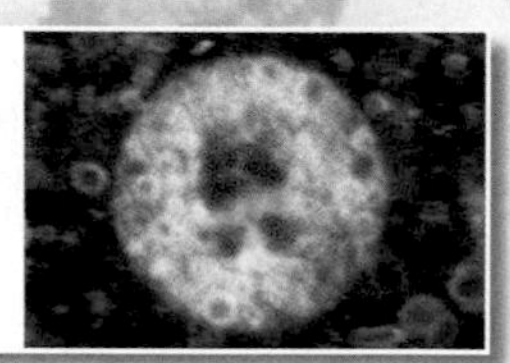

EVOLUTION OF EARLY LIFE

The first cells were probably anaerobic and prokaryotic. Evolution of oxygen-producing photosynthesis altered Earth's atmosphere, creating selection pressure that favored aerobic organisms.

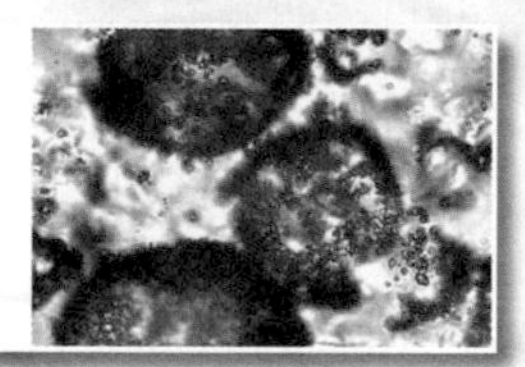

ORIGINS OF EUKARYOTIC ORGANELLES

The nucleus and endomembrane system are evolutionarily derived from infoldings of the plasma membrane. Mitochondria and chloroplasts are thought to be descendants of bacteria.

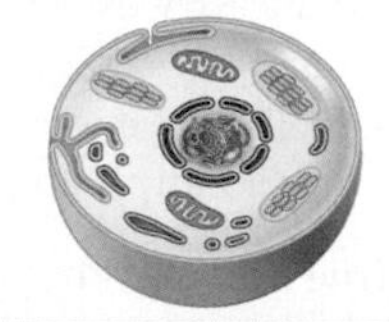

We continue our discussion of the features of bacterial and archaeal cells in Sections 20.5–20.8. You will learn much more about the features of simple eukaryotes when we discuss the protists in Chapter 21. We return to the topic of the ozone layer in Section 48.6 as we discuss the damage done by air pollutants.

19.1 Looking for Life

We live in a vast universe that we have only begun to explore. So far, we know of only one planet—Earth—that has life. In addition, biochemical, genetic, and metabolic similarities among Earth's species imply that all evolved from a common ancestor that lived billions of years ago. What properties of the ancient Earth allowed life to arise, survive, and diversify? Could similar processes occur on other planets? These are some of the questions posed by **astrobiology**, the study of life's origins and distribution in the universe.

Astrobiologists study Earth's extreme habitats to determine the range of conditions that can support life. They have found species that withstand extraordinary levels of temperature, pH, salinity, and pressure. For example, some bacteria live deep in the soil of Chile's Atacama Desert, the driest place on Earth (**FIGURE 19.1**). Other bacteria thrive at high pressure and temperature 3 kilometers (almost 2 miles) beneath the soil surface in Virginia. There is even a biofilm (Section 4.4) under Antarctica's western ice sheet.

Understanding how these species withstand such extreme conditions on Earth informs the search for life elsewhere. For example, the details of metabolism vary, but in all known species—even the extraordinary ones—metabolic reactions occur in or at the boundary of water-based solutions. Thus, liquid water is considered an essential requirement for life as we know it. Excitement over the discovery of water on Mars, our closest planetary neighbor, stems from this assumption. A robotic lander found water frozen in the soil of Mars, and geological formations suggest water flowed across the planet's surface and pooled in giant lakes billions of years ago. Other features suggest a small amount of water may still flow seasonally in some areas.

If there is any life on Mars today, it is likely to be underground. Unlike Earth, Mars does not have an **ozone layer**: an atmospheric layer with a high concentration of ozone (O_3). Earth's ozone layer serves as a natural sunscreen, preventing most ultraviolet (UV) radiation emitted by the sun from reaching the planet's surface. Because Mars has no ozone layer, its surface receives intense UV radiation. As explained in Section 8.6, UV radiation damages DNA. Thus, the UV radiation that reaches Mars probably sterilizes the upper layer of the Martian soil.

Suppose scientists do find evidence that microbial life exists or existed on Mars, or on some other planet. Why would it matter? The discovery of extraterrestrial microbes would lend credence to the idea that nonhuman intelligent life exists somewhere else in the universe. The more places microbial life exists, the more likely it is that complex, intelligent life evolved on planets other than Earth.

This chapter is your introduction to a slice through time. We begin with Earth's formation and move on to life's chemical origins and the evolution of traits present in modern eukaryotes. The picture we paint here sets the stage for the next unit, which takes you along lines of descent to the present range of biodiversity.

astrobiology The scientific study of life's origin and distribution in the universe.

ozone layer High atmospheric layer rich in ozone; prevents most ultraviolet radiation in sunlight from reaching Earth's surface.

FIGURE 19.1 Looking for life in a part of Chile's Atacama Desert, where rain falls only once every few decades and the soil evolved without plant life. Scientist Jay Quade of the University of Arizona was a member of a team that found bacteria living 30 centimeters beneath the surface of this arid desert soil.

CREDITS: (opposite) Michael Aw/Lonely Planet Images/Getty Images; (1) Photo by Julio Betancourt/U.S. Geological Survey. See Drees, K.P. et al. 2006. Bacterial community structure of soils in a hyperarid region of the Atacama Desert. Applied and Environmental Microbiology 72, 7902-7908 and Quade, J et al, 2007, Soils at the hyperarid margin: the isotopic composition of soil carbonate from the Atacama Desert. Geochimica et Cosmochima Acta 71, 3772-3795

19.2 The Early Earth

FIGURE 19.2 Artist's depiction of our sun surrounded by a cloud of dust, debris, and gases. Earth and other planets formed from the material in this cloud.

✔ Modern chemistry and physics are the bases for scientific hypotheses about early events in Earth's history.

Origin of the Universe and Earth

Studies of the modern universe allow astronomers and physicists to propose and test ideas about its origin. According to the **big bang theory**, the universe began in a single instant, about 13 to 15 billion years ago. In that instant, all existing matter and energy suddenly appeared and exploded outward from a single point. Simple elements such as hydrogen and helium formed within minutes. Then, over millions of years, gravity drew the gases together and they condensed to form giant stars.

Explosions of these early stars scattered heavier elements from which today's galaxies formed. About 5 billion years ago, a cloud of dust and rocks (asteroids) orbited the star we now call our sun (**FIGURE 19.2**). The asteroids collided and merged into bigger asteroids. The heavier these pre-planetary objects became, the more gravitational pull they exerted, and the more material they gathered. By about 4.6 billion years ago, this gradual buildup of materials had formed Earth and the other planets of our solar system.

Conditions on the Early Earth

Planet formation did not remove all the material orbiting the sun, so the early Earth received a constant hail of meteorites and was struck by many asteroids. This extraterrestrial material, along with substances released by frequent volcanic eruptions, provided components of Earth's land, seas, and atmosphere.

The composition of Earth's early atmosphere remains a matter of debate, but geologic evidence suggests our planet started out with little or no free oxygen (O_2). Had O_2 been present early on, we would see evidence of iron oxidation (rust formation) in Earth's most ancient rocks. However, these rocks show no sign of such oxidation. The apparent lack of O_2 interests scientists because it would have facilitated some proposed steps on the path to life. Had O_2 been present, oxidation reactions would have broken apart small organic compounds as quickly as they formed.

Liquid water is essential to life as we know it because molecules that carry out metabolic reactions have to be dissolved in water. At first, Earth's surface was molten rock, so all water was in the form of vapor. However, evidence from ancient rocks indicates that by 4.3 billion years ago, Earth had cooled enough for water to pool on its surface.

big bang theory Well-supported hypothesis that the universe originated by a nearly instant distribution of matter through space.

TAKE-HOME MESSAGE 19.2

What were conditions like on the early Earth?

✔ During Earth's early years, meteorites pummeled the planet's surface and volcanic eruptions were common.

✔ The early atmosphere probably had no oxygen gas.

✔ As Earth cooled, liquid water began to pool on its surface.

CREDIT: (2) Painting by William K. Hartmann.

19.3 Organic Monomers Form

✔ All living things are made from the same organic subunits: amino acids, fatty acids, nucleotides, and simple sugars.

FIGURE 19.4 A hydrothermal vent on the seafloor. Mineral-rich water heated by geothermal energy streams out, into cold ocean water. As the water cools, dissolved minerals come out of solution and form a chimney-like structure around the vent.

Organic Molecules From Inorganic Precursors

Until the early 1900s, chemists thought that organic molecules possessed a special "vital force" and that only living organisms could make them. Then, in 1925, a chemist synthesized urea, an organic molecule abundant in urine. Later, another chemist synthesized alanine, an amino acid. These synthesis reactions showed organic molecules could be formed by nonliving processes.

Sources of Life's First Building Blocks

Today, there are three main hypotheses regarding the mechanism by which organic monomers appeared on the early Earth. Keep in mind that these mechanisms are not mutually exclusive. All three may have operated simultaneously and contributed to an accumulation of simple organic compounds in Earth's early seas.

Lightning-Fueled Atmospheric Reactions
In 1953, Stanley Miller and Harold Urey tested the hypothesis that lightning could have powered synthesis reactions in Earth's early atmosphere. To simulate this process, they filled a reaction chamber with methane, ammonia, and hydrogen gas, and zapped it with sparks from electrodes (**FIGURE 19.3**). Within a week, a variety of organic molecules formed, including amino acids that are common in living things. Our understanding of the composition of Earth's early atmosphere has changed since that time, but more recent experimental simulations using gas mixtures that more accurately represent the early atmosphere also produced amino acids.

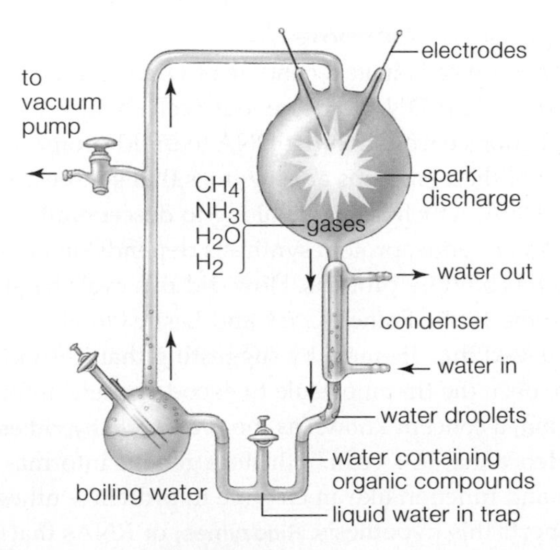

FIGURE 19.3 ▶Animated Diagram of an apparatus used to test whether organic compounds could have formed spontaneously on the early Earth. Water vapor, hydrogen gas (H_2), methane (CH_4), and ammonia (NH_3) simulated Earth's early atmosphere. Sparks from an electrode simulated lightning.

FIGURE IT OUT What was the source of the nitrogen for the amino acids formed in this apparatus?

Answer: Ammonia

Reactions at Hydrothermal Vents By another hypothesis, synthesis of life's building blocks occurred at deep-sea hydrothermal vents. A **hydrothermal vent** is like an underwater geyser, a place where mineral-rich water heated by geothermal energy streams out through a rocky opening in the seafloor (**FIGURE 19.4**).

Delivery From Space Modern-day meteorites that fall to Earth sometimes contain amino acids, sugars, and nucleotide bases and these compounds (or precursors of them) have been discovered in gas clouds surrounding nearby stars. Thus it is possible that some of the many meteorites that fell on the early Earth carried organic monomers that had formed in outer space.

hydrothermal vent Underwater opening from which mineral-rich water heated by geothermal energy streams out.

TAKE-HOME MESSAGE 19.3

What was the source of the small organic molecules required to build the first life?

✔ Small organic molecules that serve as the building blocks for living things can be formed by nonliving mechanisms. For example, amino acids form in reaction chambers that simulate conditions on the early Earth. They are also present in some meteorites.

19.4 From Polymers to Protocells

✔ We will never know for sure how the first cells came to be, but we can investigate the possible steps on the road to life.

Properties of Cells

In addition to sharing the same molecular components, all cells have a plasma membrane with a lipid bilayer. They have a genome of DNA that enzymes transcribe into RNA, and ribosomes that translate RNA into proteins. All cells replicate, and they pass on copies of their genetic material to their descendants. The many similarities in structure, metabolism, and replication processes among all life provide evidence of descent from a common cellular ancestor.

No one was around to witness the origin of life on Earth, and time has erased all traces of the earliest cells. However, scientists can still investigate this first chapter in life's history by making hypotheses about how life began, and testing those hypotheses experimentally. Their work has shown that cells could have arisen in a stepwise process beginning with inorganic materials (**FIGURE 19.5**). Each step on this hypothetical road to life can be explained by familiar chemical and physical mechanisms that still occur today.

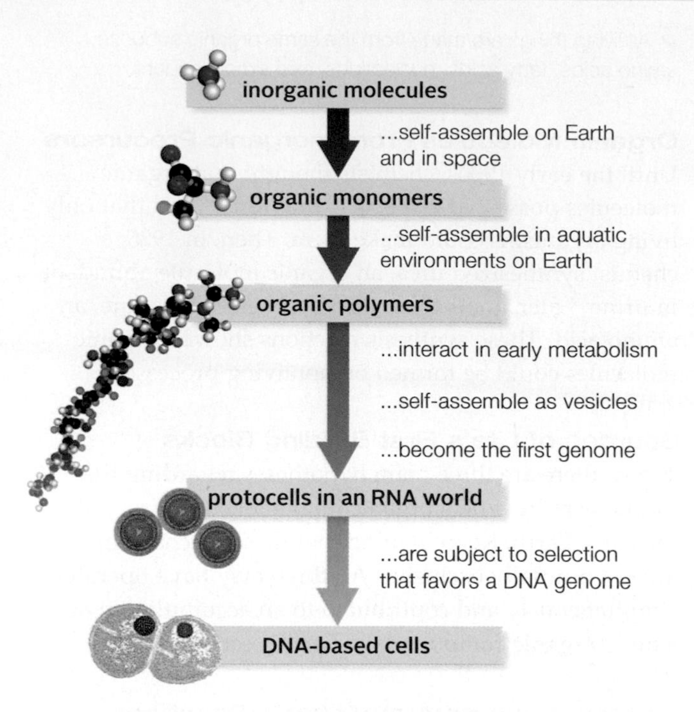

FIGURE 19.5 ▸Animated Proposed sequence for the evolution of cells. Scientists carry out experiments and simulations that test the feasibility of individual steps.

Origin of Metabolism

Modern cells take up organic monomers, concentrate them, and assemble them into organic polymers. Before there were cells, a nonbiological process that concentrated organic subunits would have increased the chance of polymer formation. By one hypothesis, this process occurred on clay-rich tidal flats. Clay particles have a slight negative charge, so positively charged molecules (such as organic subunits dissolved in seawater) stick to them. Such binding would have concentrated the subunits. Evaporation at low tide would have continued this process, and energy from the sun might have triggered polymerization. Amino acids do form short chains under simulated tidal flat conditions.

The **iron–sulfur world hypothesis**, proposed by Günter Wächtershäuser, holds that life originated at deep-sea hydrothermal vents. The porous rocks that form around these vents are rich in iron sulfides, compounds that easily donate electrons to inorganic gases dissolved in the hot seawater spewing from the vents. Accepting electrons causes the gases (such as carbon dioxide and hydrogen cyanide) to react and form organic molecules (such as pyruvate). The organic molecules stick to the iron sulfur compounds and accumulate inside tiny, cell-sized chambers of the rocks, where they can undergo further reactions as they become concentrated. Long ago, such molecules became catalytic: early versions of enzymes that carry out modern metabolic reactions. The iron sulfur cofactors required by all modern organisms may be a legacy of life's rocky beginnings at hydrothermal vents.

Origin of the Genome

All modern cells have a genome of DNA. They pass copies of their DNA to descendant cells, which use instructions encoded in the DNA to build proteins. Some of these proteins are enzymes that synthesize new DNA, which is passed along to descendant cells, and so on. Thus, protein synthesis depends on DNA, which is built by proteins. How did this cycle begin?

In the 1960s, Francis Crick and Leslie Orgel addressed this dilemma by suggesting that RNA may have been the first molecule to encode genetic information, a concept known as the **RNA world hypothesis**. Evidence that RNA can both store genetic information and function like an enzyme in protein synthesis, supports this hypothesis. **Ribozymes**, or RNAs that function as enzymes, are common in living cells. For example, the rRNA in ribosomes speeds formation of peptide bonds during protein synthesis. Other ribozymes cut noncoding bits (introns) out of newly formed mRNAs. Researchers have also produced self-replicating ribozymes that assemble free nucleotides.

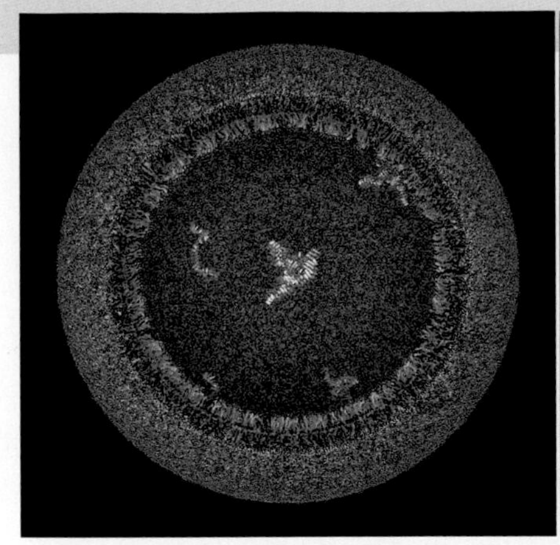

A Illustration of a laboratory-produced protocell with a bilayer membrane of fatty acids and strands of RNA inside.

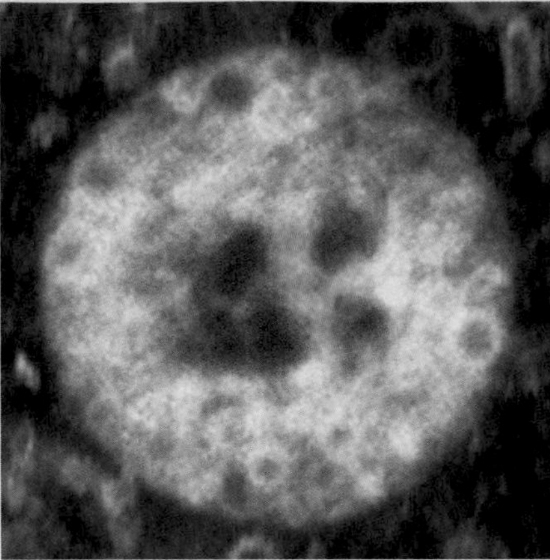

B Laboratory-formed protocell consisting of RNA-coated clay (red) surrounded by fatty acids and alcohols.

C Field-testing a hypothesis about protocell formation. David Deamer pours a mix of small organic molecules and phosphates into a hot acidic pool in Russia.

FIGURE 19.6 ▶**Animated** Protocells. Scientists test hypotheses about protocell formation through laboratory simulations and field experiments.

If the earliest self-replicating genetic systems were RNA-based, then why do all organisms now have a genome of DNA? The structure of DNA may hold the answer. Compared to a double-stranded DNA molecule, single-stranded RNA breaks more easily. Thus, a switch from RNA to DNA would make larger, more stable genomes possible.

Origin of the Plasma Membrane

Self-replicating molecules and products of other early synthetic reactions would have floated away from one another unless something enclosed them. In modern cells, a plasma membrane serves this function. If the first reactions took place in tiny rock chambers, the rock would have acted as a boundary. Over time, lipids produced by reactions inside a chamber could have accumulated and lined the chamber wall, forming a protocell. A **protocell** is a membrane-enclosed collection of interacting molecules that can take up material and replicate. Protocells are thought to be the ancestors of cellular life.

Experiments by Jack Szostak and others have shown that protocells can form even without rock chambers. **FIGURE 19.6A** illustrates one type of protocell that Szostak investigates. **FIGURE 19.6B** is a photo of a protocell that formed in his laboratory. This synthetic protocell consists of a membrane of lipid bilayer enclosing strands of RNA. The protocell can "grow" by adding fatty acids to its membrane and nucleotides to its RNA; mechanical force can make it "divide."

David Deamer studies protocell formation both in the laboratory and in the field. In the lab, he has shown that the small organic molecules carried to Earth on meteorites can react with minerals and seawater to form vesicles with a bilayer membrane. However, Deamer has yet to locate a natural environment that facilitates the same process. In one experiment, he added a mix of organic subunits to the acidic waters of a clay-rich volcanic pool in Russia (**FIGURE 19.6C**). The organic subunits bound tightly to the clay, but no vesicle-like structures formed. Deamer concluded that hot acidic waters of volcanic springs do not provide the right conditions for protocell formation. He continues to carry out experiments to determine what naturally occurring conditions would favor this process.

iron–sulfur world hypothesis Hypothesis that the metabolic reactions that led to the first cells took place on the porous surface of iron–sulfide-rich rocks at hydrothermal vents.
protocell Membranous sac that contains interacting organic molecules; hypothesized to have formed prior to the earliest life-forms.
ribozyme RNA that functions as an enzyme.
RNA world hypothesis Hypothesis that RNA served as the genetic information of early life.

TAKE-HOME MESSAGE 19.4

What have existing life and simulations revealed about the steps that led to the first cells?

✔ All living cells carry out metabolic reactions, are enclosed within a plasma membrane, and can replicate themselves.

✔ Metabolic reactions may have begun when molecules became concentrated on clay particles or inside tiny rock chambers near hydrothermal vents.

✔ RNA can serve as an enzyme, as well as a genome. An RNA world may have preceded evolution of DNA-based genomes.

✔ Vesicle-like structures with outer membranes form spontaneously when certain organic molecules are mixed with water.

✔ Fossils and molecular comparisons among modern organisms inform us about the early history of life.

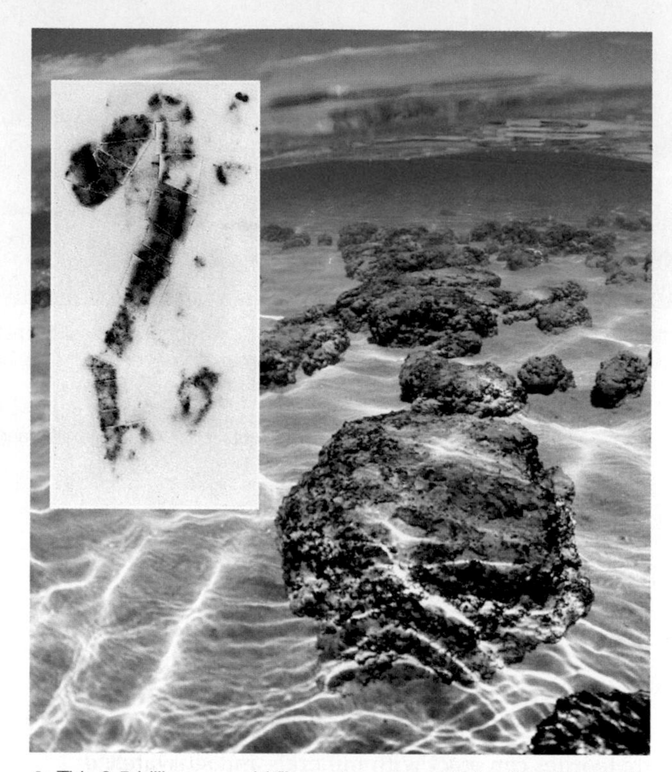

A This 3.5-billion-year-old filament may be a chain of fossil bacteria from an ancient stromatolite. The underlying photo shows modern stromatolites in Australia's Shark Bay. Each stromatolite consists of living photosynthetic bacteria atop the remains of countless earlier generations of cells along with the sediment that they trapped.

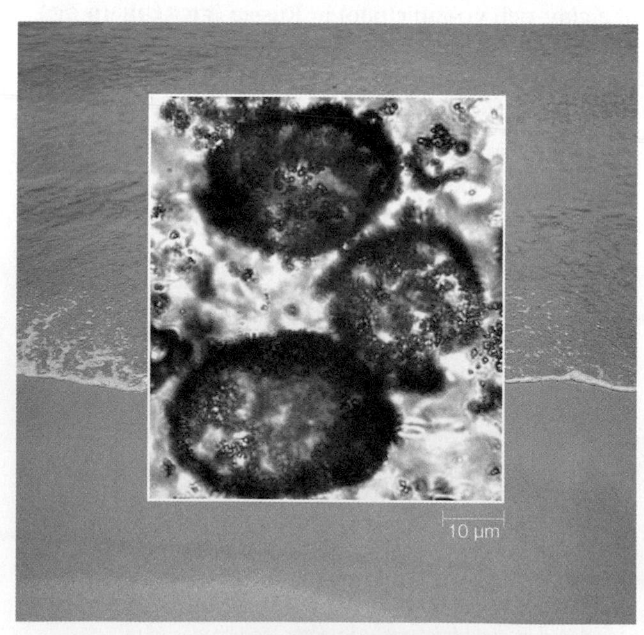

B These 3.4-billion-year-old spheres may be fossil bacteria that lived among sand grains of an ancient shore.

FIGURE 19.7 The oldest proposed bacterial microfossils.

The Common Ancestor of All Life

The processes described in Section 19.4 may have produced cellular life more than once. If so, all but one of those early cell lineages became extinct. Analysis of modern genomes tells us that all living species are descended from the same cell, and that this cell may have lived as early as 4 billion years ago.

Given what scientists know about relationships among modern species, most assume that this common ancestor was prokaryotic, meaning it did not have a nucleus (Section 1.4). Oxygen was scarce on the early Earth, so the ancestral cell must also have been anaerobic (capable of living without oxygen).

Looking for Evidence of Early Life

Finding and identifying signs of early cells poses a challenge. Cells are microscopic and most have no hard parts to fossilize. In addition, few ancient rocks that could hold early fossils still exist. Tectonic plate movements have destroyed nearly all rocks older than about 4 billion years, and most slightly younger rocks have been heated or undergone other processes that destroy traces of biological material. To add to the difficulty, structures formed by nonbiological mechanisms sometimes resemble fossils. To avoid mistakenly accepting such material as a genuine fossil, scientists constantly reanalyze purported fossil finds and they often question one another's conclusions.

The Oldest Fossil Cells

The divergence that separated the two prokaryotic domains, Bacteria and Archaea, occurred very early in the history of life, and no fossils from before this divergence have been discovered. At present, the oldest proposed cell microfossils (microscopic fossils) are filaments from 3.5-billion-year-old rocks in Western Australia. The filaments resemble chains of modern photosynthetic bacteria, and the rocks in which they occur are thought to be remains of ancient stromatolites (**FIGURE 19.7A**).

A **stromatolite** is a mounded, layered structure that forms in shallow sunlit water when a mat of photosynthetic bacteria traps minerals and sediment. The stromatolite increases in size over time as new layers form over the old. When sediment covers the living bacteria atop the stromatolite, new bacterial cells grow over that sediment, then trap more sediment to form a new layer. Scientists can observe this process in modern stromatolites such as those in Australia's Shark Bay. Stromatolites in this bay began growing about 2,000 years ago and now stand up to 1.5 meters high.

CREDITS: (7A) background, Michael Aw/Lonely Planet Images/Getty Images; inset, Courtesy of John Fuerst, University of Queensland. Originally published in *Archives of Microbiology* vol 175, p 413–29 (Lindsay MR, Webb RI, Strous M, Jetten MS, Butler MK, Forde RJ, Fuerst JA. Cell compartmentalisation in planctomycetes: Novel types of structural organization for the bacterial cell. *Arch. Microbiol.* 2001 Jun, 175(6): 413–29); (7B) background, tratong/Shutterstock.com; inset, Courtesy of David Wacey.

Stromatolites reached their peak abundance about 1.25 billion years ago, when they were common worldwide.

Another set of proposed microfossils, also from Western Australia, dates to 3.4 billion years ago. The fossils have a spherical shape and occur in rocks composed of sand grains from an ancient beach (**FIGURE 19.7B**). The presence of pyrite (an iron–sulfide mineral) in these fossils suggests they were similar to the sulfur bacteria common in modern mud flats. Sulfur bacteria use sulfur as the final electron acceptor in their energy-producing pathway, and they produce pyrite as a by-product of this process.

Changes in the Air

Many types of bacteria carry out photosynthesis, but only one group, the cyanobacteria, do so by an oxygen-producing pathway. Evolution of oxygen-producing photosynthesis in cyanobacteria had a dramatic effect on early life. By about 2.5 billion years ago, oxygen released by these bacteria had begun to accumulate in Earth's air and seas creating a new, global selection pressure. The oxygen was toxic to many species that had evolved in its absence. Inside cells, it reacts with metal ions, forming free radicals that can damage DNA and other essential cell components (Sections 2.3 and 5.6). Species unable to detoxify the free radicals either went extinct or became restricted to the low-oxygen environments that remained. By contrast, species that happened to have metabolic machinery capable of detoxifying the free radicals thrived. Some of these survivors even began to use some of this machinery for aerobic respiration, an energy-releasing pathway in which oxygen serves as the final electron acceptor (Section 7.5).

The increase in atmospheric oxygen also led to the formation of the ozone layer, a region of the upper atmosphere that contains a high concentration of ozone gas (O_3). Before the ozone layer formed, life could exist only in water, which shielded organisms from incoming UV radiation. Without an ozone layer to screen out some of this radiation, life could not have moved onto land.

Rise of the Eukaryotes

Nuclei are not often preserved during fossilization, but other traits provide evidence that a fossilized cell was eukaryotic. Eukaryotic cells are generally larger than prokaryotic ones. A cell wall with complex patterns, spines, or spikes probably belonged to a eukaryote. Researchers also look for biomarkers specific to eukaryotes. A **biomarker** is a substance that occurs only or predominantly in cells of a specific type. For example, certain steroids are found only in eukaryotes, so traces of these steroids are biomarkers for this group.

Steroids found in ancient rocks suggest eukaryotes may have arisen as early as 2.7 billion years ago. However, the oldest microfossils that most scientists agree are fossil eukaryotes date to about 1.8 billion years ago. The first eukaryotes were protists, and the oldest eukaryotic fossils that we can assign to a modern group are a type of red algae (**FIGURE 19.8**).

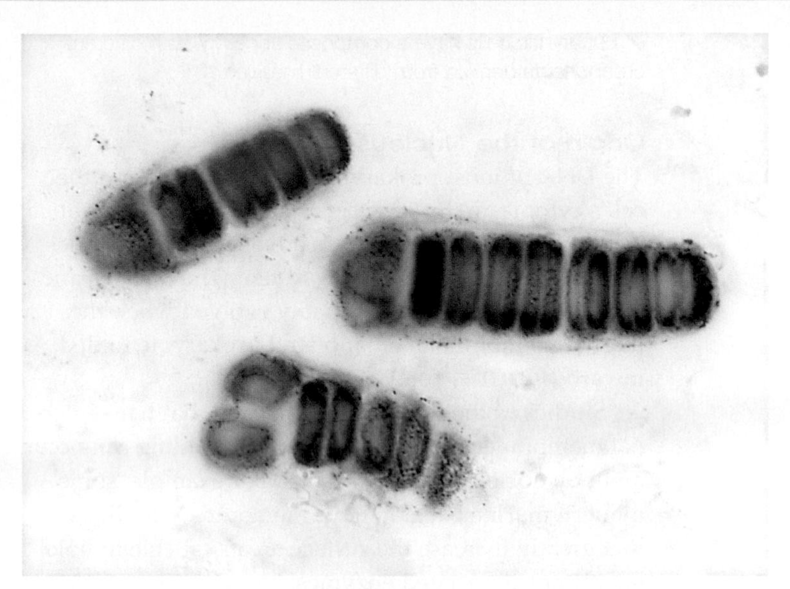

FIGURE 19.8 Fossil red algae (*Bangiomorpha pubescens*) that lived 1.2 billion years ago. Protists such as these algae were the earliest eukaryotes.

TAKE-HOME MESSAGE 19.5

What was early life like and how did it change Earth?

✔ The cellular ancestor of all modern life arose by 3–4 billion years ago.

✔ The first cells were most likely anaerobic prokaryotes.

✔ By 2.5 billion years ago, oxygen released by photosynthetic cyanobacteria had begun to accumulate. The rise in oxygen concentration resulted in formation of the protective ozone layer and favored organisms capable of carrying out aerobic respiration.

✔ The first eukaryotic organisms may have arisen as early as 2.7 billion years ago. They were protists.

biomarker Molecule produced only by a specific type of cell; its presence indicates the presence of that cell.
stromatolite Dome-shaped structure composed of layers of bacterial cells, their secretions, and sediments.

19.6 Origin and Evolution of Eukaryotes

✔ Eukaryotic cells have a composite ancestry, with different components derived from different lineages.

Origin of the Nucleus

The DNA of most prokaryotes lies unenclosed in the cell's cytoplasm. By contrast, the DNA of a eukaryotic cell is always enclosed within a nucleus that is associated with an endomembrane system. The nucleus and endomembrane system probably evolved when the plasma membrane of an ancestral prokaryote folded inward (**FIGURE 19.9**).

Studies of the few types of bacteria that have internal membranes illustrates that such infolding can occur and how it can be advantageous. For example, some modern marine bacteria have membrane infoldings that greatly increase the surface area available to hold membrane-associated enzymes.

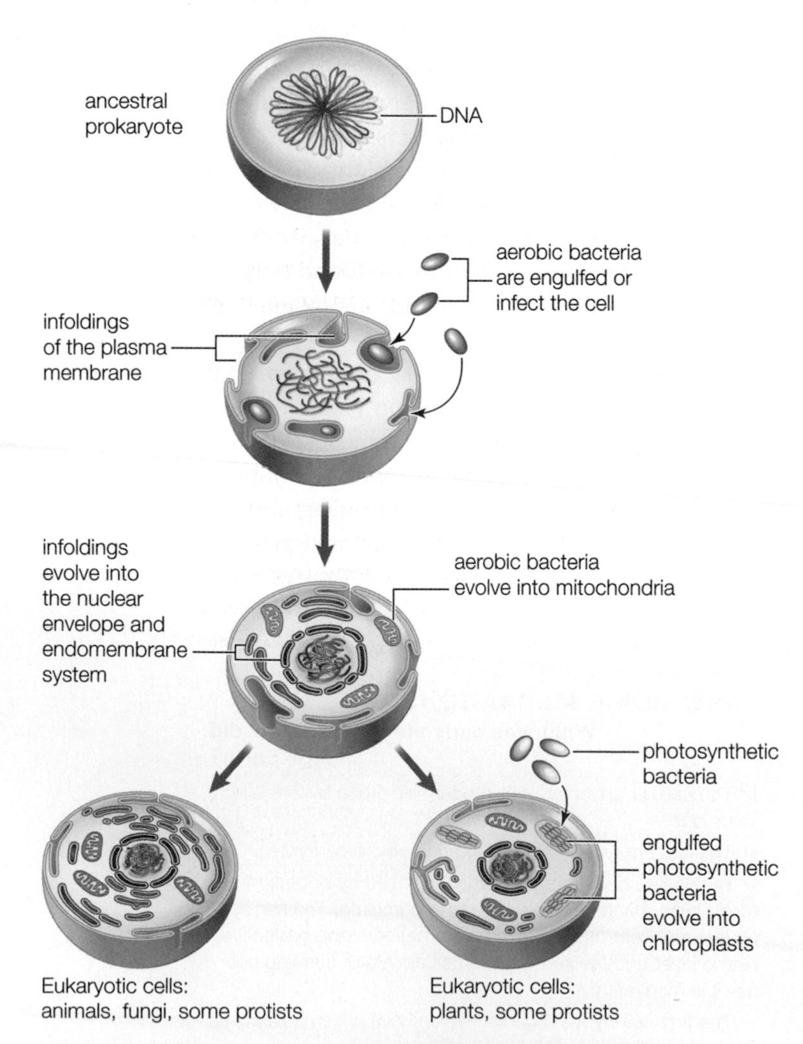

FIGURE 19.9 ▶**Animated** Proposed steps in the evolution of some eukaryotic organelles.

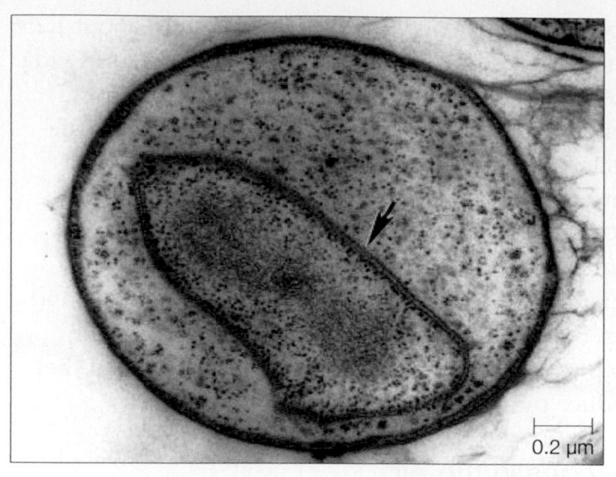

FIGURE 19.10 Nucleus-like structure in a prokaryote. DNA of *Gemmata obscuriglobus*, a species of bacteria, is enclosed by a double lipid bilayer membrane (indicated by the arrow).

Internal membranes also protect a genome from physical or biological threats. Consider *Gemmata obscuriglobus*, a bacterial species that withstands high levels of mutation-causing radiation. It is one of the few bacteria that has a membrane around its DNA (**FIGURE 19.10**). Like a eukaryotic nuclear envelope (Section 4.6), this membrane consists of two lipid bilayers. However, the membrane lacks nuclear pores or their equivalent, so it is not a true nuclear envelope. Researchers attribute the radiation resistance of *G. obscuriglobus* to the tighter packing, and therefore higher shielding, of its DNA within the membrane-enclosed compartment. DNA of other bacteria is more vulnerable to radiation because it is more spread out.

Origin of Mitochondria and Chloroplasts

Mitochondria and chloroplasts resemble bacteria in their size and shape, and they replicate independently of the cell that holds them. Like bacteria, they have their own DNA in the form of a single circular chromosome. They also have at least two outer membranes, with the innermost membrane structurally similar to a bacterial plasma membrane.

Recognition of these similarities led to the formulation of the **endosymbiont hypothesis**, which states that mitochondria and chloroplasts evolved from bacteria. (Endosymbiosis means "living inside" and refers to a relationship in which one organism lives inside another.) Nearly all eukaryotic lineages have mitochondria or mitochondria-like organelles, but only some have chloroplasts. Thus, biologists postulate that the two types of organelles were acquired independently in the sequence illustrated in **FIGURE 19.9**. In both

CREDITS: (9) From Russell/Wolfe/Hertz/Starr, *Biology*, 2e, © 2011 Cengage Learning Inc.; (10) Courtesy of John Fuerst, University of Queensland. Originally published in *Archives of Microbiology* vol 175, p 413–29 (Lindsay MR, Webb RI, Strous M, Jetten MS, Butler MK, Forde RJ, Fuerst JA. Cell compartmentalisation in planctomycetes: Novel types of structural organization for the bacterial cell. *Arch. Microbiol.* 2001 Jun, 175(6): 413–29).

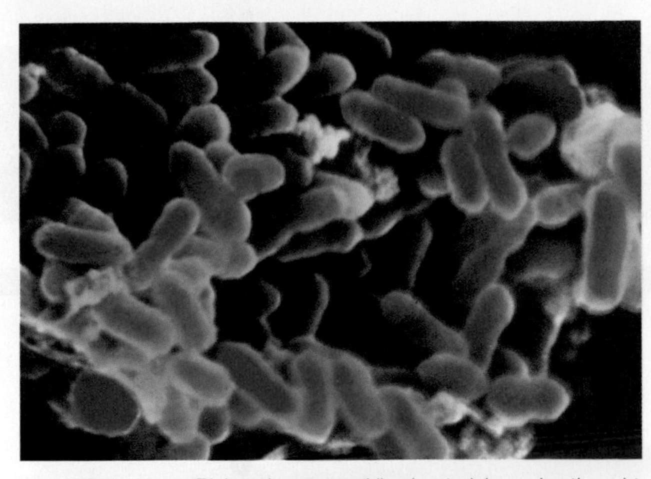

FIGURE 19.11 *Rickettsia prowazekii*, a bacterial species thought to be a close relative of the bacteria ancestral to mitochondria. It causes the disease typhus in humans.

cases, the process began when bacteria were taken up by or invaded a host cell, then replicated inside it. When the host cell divided, it passed some "guest" cells, referred to as endosymbionts, along to its offspring. As the two species lived together over many generations, genes carried by both partners were free to mutate. A gene could lose its function in one partner if the duplicate gene carried by the other partner still worked. Eventually, the host and endosymbiont lost enough duplicated genes to become incapable of living independently. At that point, the endosymbionts had evolved into organelles.

Given the evidence that mitochondria and chloroplasts evolved from bacteria, scientists are now investigating which modern bacteria are the closest relatives of these organelles. In such investigations, metabolic and genetic similarities between organelles and specific bacterial groups are considered to be evidence of shared ancestry.

So far, two groups of bacteria have been identified as close relatives of mitochondria. One group is the rickettsias, which are tiny bacteria that invade eukaryotic cells and replicate inside them. *Rickettsia prowazekii*, which causes the disease typhus, has a genome very similar to that of some mitochondria (**FIGURE 19.11**). The other potential mitochondrial relative is a type of free-living bacteria that floats in the ocean's surface waters. These marine bacteria have a genome similar to both rickettsias and mitochondria. Scientists do not know exactly how the two modern bacterial groups relate to one another and to mitochondria. By one hypothesis, all three—mitochondria, rickettsias, and the modern marine bacteria—descended from an ancient free-living marine species.

The ancestry of chloroplasts is better understood. Cyanobacteria are the only modern bacteria that carry out photosynthesis by the oxygen-producing pathway, as chloroplasts do. Thus, ancient cyanobacteria are thought to have given rise to chloroplasts.

The endosymbiont hypothesis assumes that cells can enter and live inside other cells. It also assumes that such a relationship can, over time, become essential to the partners. Studies of modern-day cell partnerships lend support to both assumptions. One such study was carried out by microbiologist Kwang Jeon. In 1966, Jeon was studying *Amoeba proteus*, a species of single-celled protist. By accident, his amoebas became infected by a rod-shaped bacterium. Most infected amoebas died right away. A few, however, survived and reproduced despite their infection. Intrigued, Jeon maintained these infected cultures to see what would happen. Five years later, the descendant amoebas were host to many bacterial cells, yet they seemed healthy. In fact, when these amoebas were treated with bacteria-killing drugs that usually do not harm amoebas, they died. Apparently the amoebas had come to require the bacteria for some life-sustaining function. Further investigations revealed that the amoebas had lost the ability to make an essential enzyme. They now depended on their bacterial partners to make that enzyme for them.

endosymbiont hypothesis Hypothesis that mitochondria and chloroplasts evolved from bacteria.

Eukaryotic Divergence

However they arose, early eukaryotic cells had a nucleus, endomembrane system, mitochondria, and—in certain lineages—chloroplasts. These cells were the first protists. Over time, their many descendants came to include the modern protist lineages, as well as plants, fungi, and animals. The next section provides a time frame for these pivotal evolutionary events.

TAKE-HOME MESSAGE 19.6

How might the nucleus and other eukaryotic organelles have evolved?

- ✔ A nucleus and other organelles are defining features of eukaryotic cells.
- ✔ The nucleus and endomembrane system could have arisen through modification of infoldings of the plasma membrane.
- ✔ Mitochondria and chloroplasts most likely descended from bacteria.

CREDIT: (11) CNRI/Science Source.

19.7 Time Line for Life's Origin and Evolution

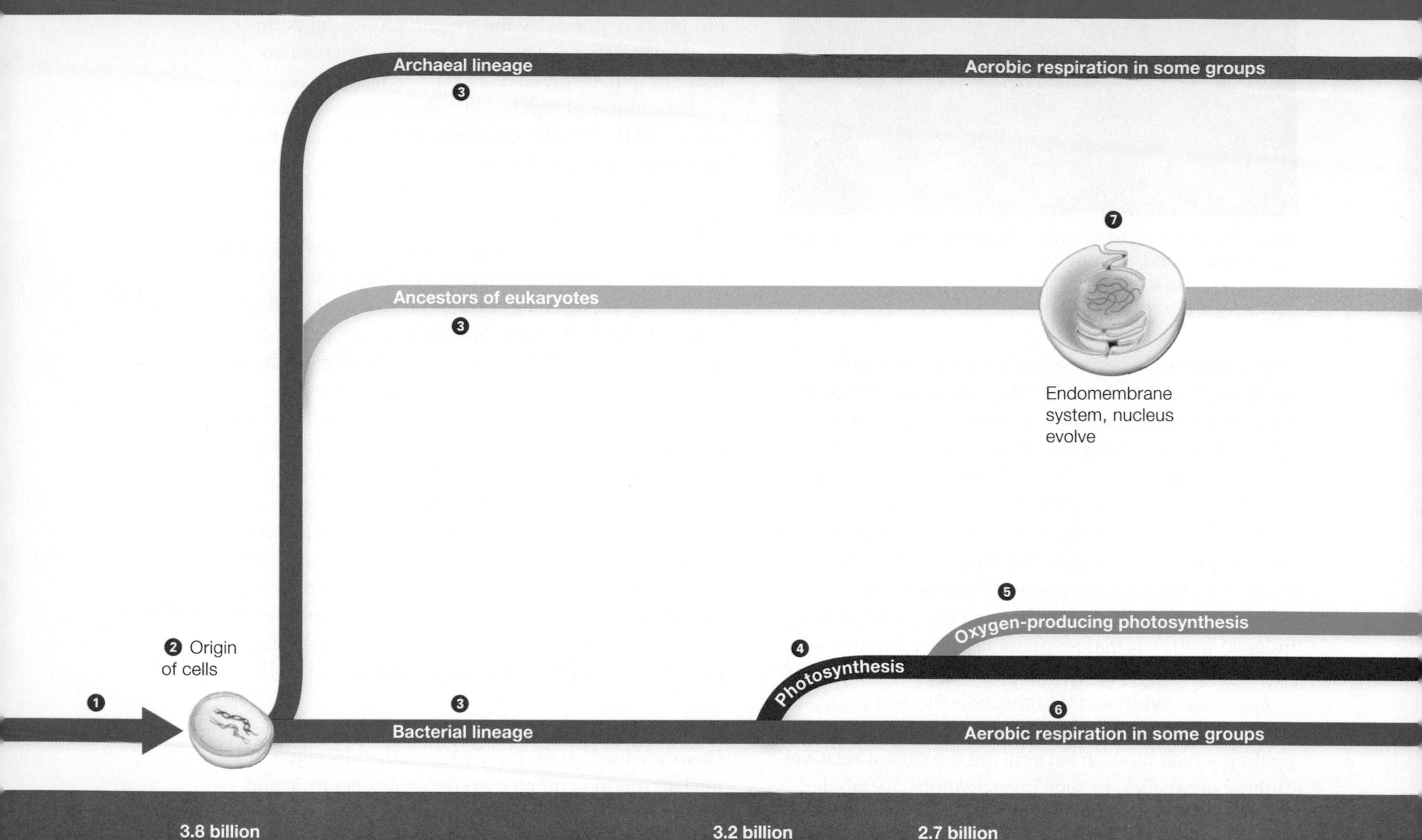

Building Blocks Form

❶ Lipids, proteins, nucleic acids, and complex carbohydrates formed from the simple organic compounds present on early Earth.

Origin of Cells

❷ The first cells did not have a nucleus or other organelles. Oxygen was scarce, so the first cells made ATP by anaerobic pathways.

Domains Diverge

❸ An early divergence separated bacteria from the common ancestor of archaeal and eukaryotic cells. Not long after that, archaeal and eukaryotic cells diverged.

Photosynthesis and Aerobic Respiration Evolve

❹ Photosynthetic pathways that did not produce oxygen evolved in some bacterial lineages.

❺ Oxygen-producing photosynthesis evolved in a branch from this lineage, and oxygen began to accumulate.

❻ Aerobic respiration became the predominant metabolic pathway in some bacteria and archaea.

Endomembrane System and Nucleus Evolve

❼ Cell sizes and the amount of genetic information continued to expand in ancestors of what would become the eukaryotic cells. The endomembrane system, including the nuclear envelope, arose through the modification of cell membranes.

FIGURE 19.12 ▸Animated Milestones in the history of life, based on the most widely accepted hypotheses. As you read the next unit on life's past and present diversity, refer to this visual overview. It can serve as a simple reminder of the evolutionary connections among all groups of organisms. Time line not to scale.

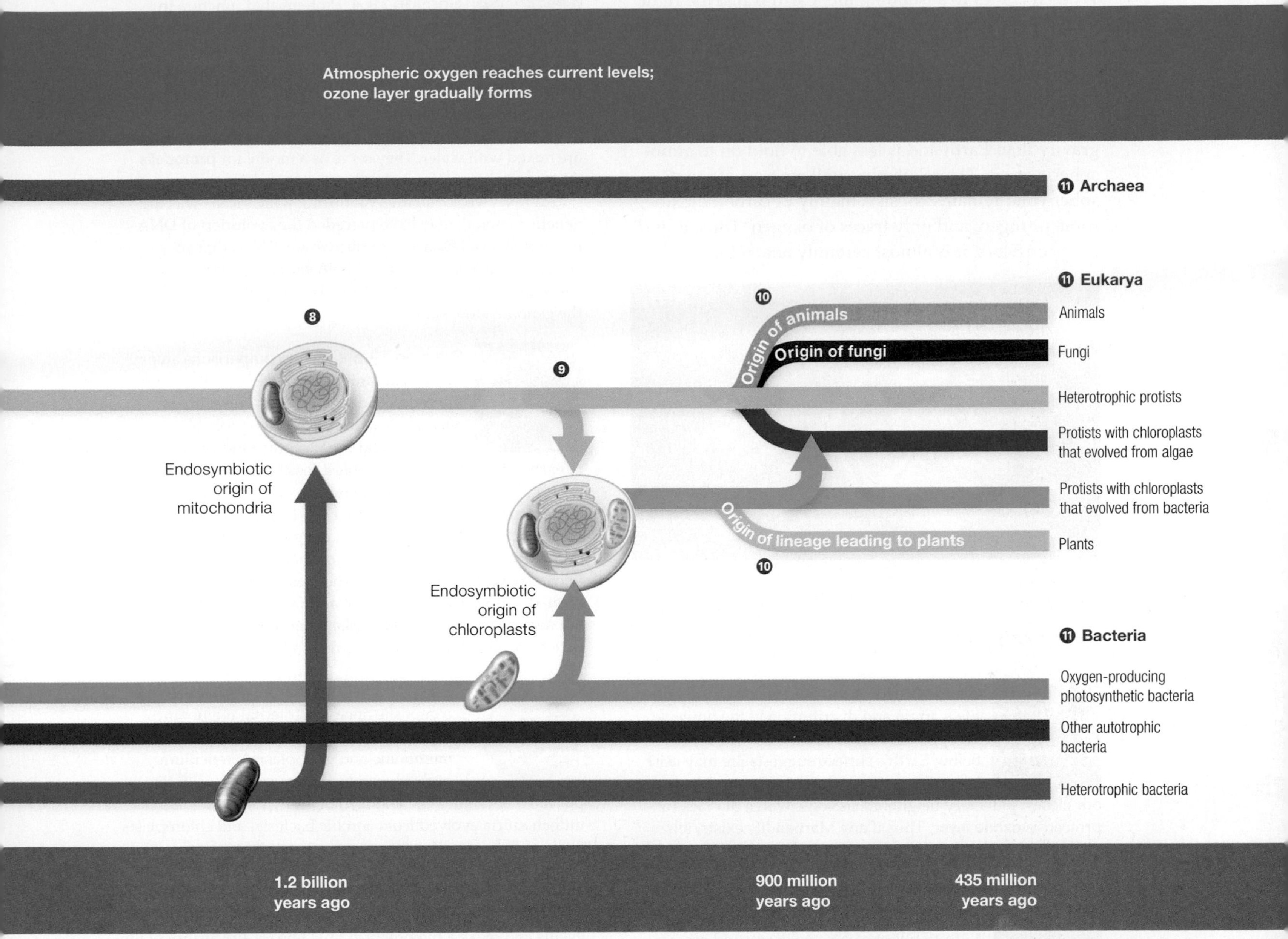

Atmospheric oxygen reaches current levels; ozone layer gradually forms

⓫ Archaea

⓫ Eukarya

Animals

Fungi

Heterotrophic protists

Protists with chloroplasts that evolved from algae

Protists with chloroplasts that evolved from bacteria

Plants

⓫ Bacteria

Oxygen-producing photosynthetic bacteria

Other autotrophic bacteria

Heterotrophic bacteria

❽ Endosymbiotic origin of mitochondria

❾ Endosymbiotic origin of chloroplasts

❿ Origin of animals

Origin of fungi

❿ Origin of lineage leading to plants

1.2 billion years ago

900 million years ago

435 million years ago

Origin of Mitochondria

❽ Aerobic bacteria entered and lived inside a eukaryotic cell. Over many generations, descendants of these bacteria evolved into mitochondria.

Origin of Chloroplasts

❾ An oxygen-producing, photosynthetic bacterial cell entered a eukaryotic cell. Over generations, bacterial descendants evolved into chloroplasts.

Origins of Fungi, Animals, and Plants

❿ By 900 million years ago, representatives of all major lineages—including fungi, animals, and the algae that would give rise to plants—had evolved in the seas.

Modern Life

⓫ Modern organisms are related by descent, so all share certain traits. However, each lineage also has characteristic traits that evolved in response to the unique selective pressures it experienced.

FIGURE IT OUT From which lineage are mitochondria descended, bacteria or archaea?

Answer: Bacteria

Looking for Life (revisited)

When it comes to sustaining life, Earth is just the right size. If the planet were much smaller, it would not exert enough gravitational pull to keep atmospheric gases from drifting off into space. The photo below shows the relative sizes of Earth and Mars. As you can see, Mars is only about half the size of Earth. Mars has less gravity than Earth and is less able to hold on to atmospheric gases. The relatively small amount of atmosphere that remains consists mainly of carbon dioxide, some nitrogen, and only traces of oxygen. Thus, if life exists on Mars, it is almost certainly anaerobic.

summary

Section 19.1 **Astrobiology** is the study of life's origin and distribution in the universe. The presence of cells in deserts and deep below Earth's surface suggests life may exist in similar settings on other planets. Mars, our closest planetary neighbor, has some water, but lacks a protective **ozone layer**. Thus if any Martian life exists, it is likely deep in the soil.

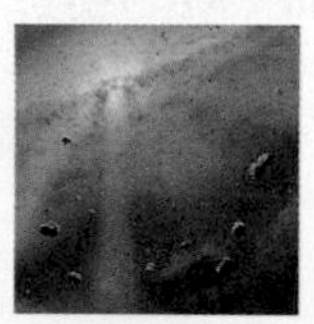

Section 19.2 According to the **big bang theory**, the universe formed in an instant 13 to 15 billion years ago. Earth and other planets formed about 4.6 billion years ago from material released by explosions of giant stars. Early in Earth's existence, the planet had little or no free oxygen, and it received a constant hail of meteorites. Oxygen is highly reactive, so a lack of it would have facilitated assembly of simple organic compounds. Earth's surface was initially molten, but by 4.3 billion years ago it had cooled enough for life-sustaining water to pool on its surface.

Section 19.3 Laboratory simulations provide indirect evidence that organic monomers could have formed by lightning-fueled reactions in Earth's early atmosphere or in the hot, mineral-rich water around **hydrothermal vents**. Examination of meteorites shows that such compounds form in deep space and could have been transported to Earth by meteorites.

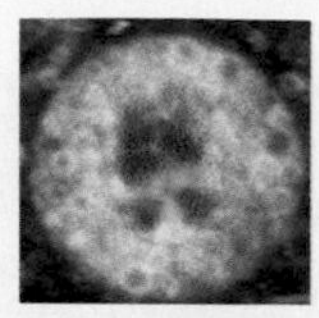

Section 19.4 Proteins that function in metabolic pathways might have first formed when amino acids stuck to clay, then bonded under the heat of the sun. The **iron–sulfur world hypothesis** postulates that metabolism began on the surface of rocks at hydrothermal vents.

Membrane-like structures and vesicles form when lipids are mixed with water. They serve as a model for **protocells**, which may have preceded cells.

An RNA world, an interval during which RNA was the genetic material, may have preceded the evolution of DNA-based systems. Discovery of **ribozymes**, RNAs that act as enzymes, lends support to the **RNA world hypothesis**. A subsequent switch from RNA to DNA would have made the genome more stable.

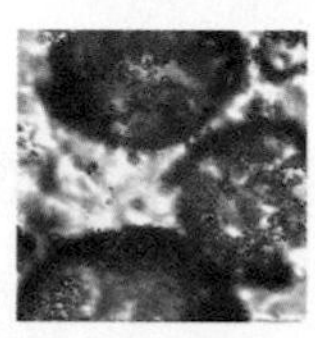

Section 19.5 Genetic comparisons among modern organisms indicate that all modern life is descended from the same cellular ancestor. The first cells were probably anaerobic and prokaryotic. Finding fossils of early cells is difficult. The oldest fossil evidence of cells is of photosynthetic bacteria that formed dome-shaped **stromatolites** and of bacteria that used sulfur as an electron acceptor. By 2.5 billion years ago, oxygen released as a by-product of photosynthesis by cyanobacteria had begun to change Earth's atmosphere. The increased oxygen level created a protective ozone layer, and favored cells that carried out aerobic respiration. **Biomarkers** and fossils of eukaryotes date back more than 2 billion years.

Sections 19.6, 19.7 Eukaryotes have a composite ancestry. Internal membranes of eukaryotic cells such as a nuclear membrane and endoplasmic reticulum probably evolved from infoldings of the plasma membrane. The **endosymbiont hypothesis** holds that mitochondria evolved from aerobic bacteria, and chloroplasts evolved from cyanobacteria.

Protists were the first eukaryotes. Over time, various protist lineages gave rise to plants, fungi, and animals. Evidence from many sources allows us to reconstruct the order of events and make a hypothetical time line for the history of life.

self-quiz

Answers in Appendix VII

1. According to the big bang theory, ________ .
   a. the universe expanded out from a single point
   b. Earth and our sun formed simultaneously
   c. carbon and oxygen were the first elements to form
   d. all of the above

2. An abundance of ________ in Earth's early atmosphere would have interfered with assembly of organic compounds.
   a. carbon dioxide
   b. ammonia
   c. water
   d. oxygen

data analysis activities

The Changing Earth Studies of ancient rocks and fossils can reveal changes that have taken place during Earth's existence. **FIGURE 19.13** shows how asteroid impacts and the composition of the atmosphere are thought to have changed over time. Use this figure and information in the chapter to answer the following questions.

1. Which occurred first, a decline in asteroid impacts, or a rise in the atmospheric level of oxygen (O_2)?
2. How do modern levels of carbon dioxide (CO_2) and O_2 compare to those at the time when the first cells arose?
3. Which is now more abundant, oxygen or carbon dioxide?
4. What do you think accounts for the rise in CO_2 at the far right of the graph?

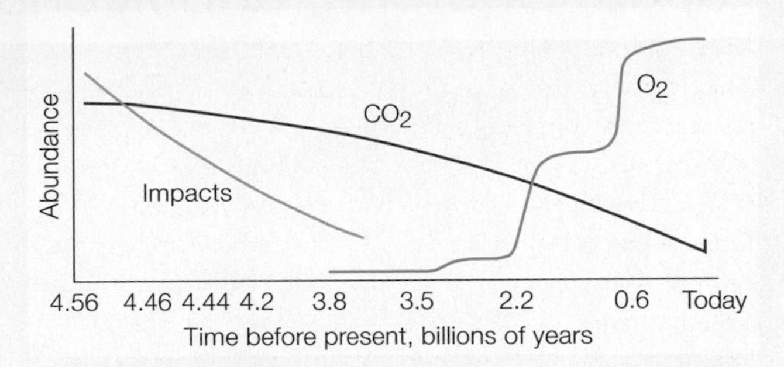

FIGURE 19.13 Changes in the abundance of asteroid impacts (green), atmospheric carbon dioxide concentration (pink), and atmospheric oxygen (blue) over geologic time.

3. Stanley Miller's experiment demonstrated that ________.
   a. Earth is more than 4 billion years old
   b. under some conditions, amino acids can assemble spontaneously
   c. oxygen is necessary for all life
   d. DNA is less stable than RNA

4. According to one hypothesis, negatively charged clay particles played a role in early ________.
   a. protein formation
   b. DNA replication
   c. photosynthesis
   d. oxygen declines

5. The prevalence of ________ in living organisms is taken as support for the idea that life arose near deep-sea vents.
   a. mitochondria
   b. iron–sulfide cofactors
   c. DNA
   d. a plasma membrane

6. An RNA that functions as an enzyme is a ________.
   a. protein
   b. protocell
   c. ribosome
   d. ribozyme

7. Among prokaryotes, only the cyanobacteria ________.
   a. live near hydrothermal vents
   b. produce oxygen during photosynthesis
   c. cannot tolerate oxygen
   d. have a nucleus-like structure

8. The evolution of ________ resulted in an increase in the levels of atmospheric oxygen.
   a. DNA-based genomes
   b. aerobic respiration
   c. sexual reproduction
   d. photosynthesis that releases oxygen

9. Bacteria that cause the disease typhus are close relatives of bacteria that evolved into ________.
   a. protists
   b. protocells
   c. chloroplasts
   d. mitochondria

10. Infoldings of the plasma membrane into the cytoplasm of some ancestral cells may have evolved into the ________.
    a. nuclear envelope
    b. ER membranes
    c. primary cell wall
    d. both a and b

11. An ________ is a relationship in which one organism lives inside another.

12. A stromatolite is a structure ________.
    a. produced by endosymbiosis
    b. that formed only on the early Earth
    c. consisting of layered bacteria and sediment
    d. that expels hot water from deep in the Earth

13. The first eukaryotes were ________.
    a. fungi
    b. plants
    c. protists
    d. animals

14. Evidence that Mars ________ suggests that it may have supported or still supports life.
    a. has an ozone layer
    b. has water
    c. is about the same size as Earth
    d. all of the above

15. Chronologically arrange the evolutionary events, with 1 being the earliest and 6 the most recent.

___ 1	a. onset of oxygen-releasing pathway of photosynthesis
___ 2	b. origin of mitochondria
___ 3	c. origin of protocells
___ 4	d. emergence of the first eukaryotes
___ 5	e. origin of chloroplasts
___ 6	f. the big bang

critical thinking

1. Researchers looking for fossils of the earliest life-forms face many hurdles. For example, few sedimentary rocks date back more than 3 billion years. Review what you learned about plate tectonics (Section 16.7). Then, explain why so few remaining samples of these early rocks remain.

2. The astronomer Carl Sagan once said, "We are made of star stuff." Explain why this is true of all life on Earth.

3. How did the evolution of oxygen-releasing photosynthesis in cyanobacteria increase the likelihood that mitochondria would one day evolve?

20 Viruses, Bacteria, and Archaea

LEARNING ROADMAP

Section 1.4 gave you an early glimpse of the bacteria and archaea. Section 4.4 began our discussion of their structure and Section 19.5 put these groups into an evolutionary time frame. Section 17.1 focused on antibiotic resistance in bacteria. This chapter also discusses bacteriophages, a type of virus that will be familiar from Section 8.2.

VIRUSES AND VIROIDS

A virus is a noncellular infectious particle made of nucleic acid and protein. It must infect a living cell to replicate. A viroid is an even simpler infectious particle that consists solely of RNA.

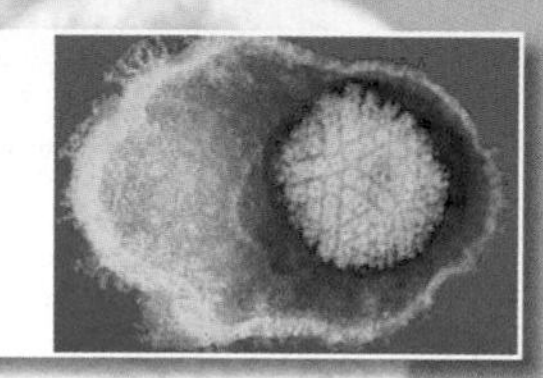

VIRAL REPLICATION AND RECOMBINATION

Viruses infect and replicate in all organisms, and some cause human disease. When multiple types of virus infect a cell, their genes can recombine to create a new type of virus.

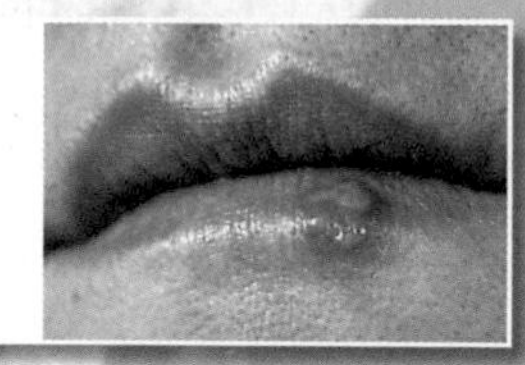

TWO LINEAGES OF SIMPLE CELLS

Bacteria and archaea are two distinct lineages of asexually reproducing, structurally simple cells that do not have a nucleus. They can exchange genetic material among existing individuals.

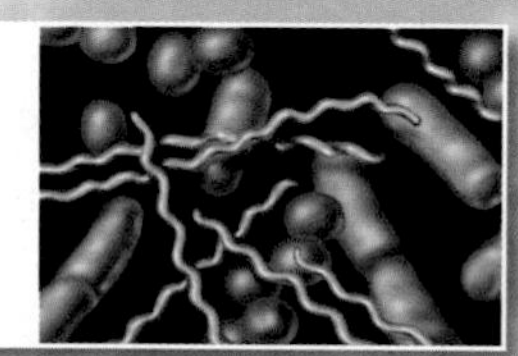

BACTERIA

Bacteria play many essential ecological roles. They put oxygen into the air, make nitrogen available to plants, and act as decomposers. A minority cause disease in humans.

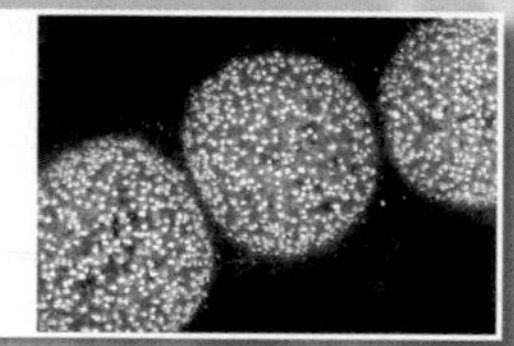

ARCHAEA

Archaea live in some astonishingly hostile environments such as hot springs and pools of brine. They also live alongside bacteria in soil and in the animal gut.

You will learn more about the effects of viral and bacterial pathogens in chapters that discuss physiology. For example, Section 37.11 looks at the immune effects of AIDS, and Section 41.9 discusses sexually transmitted diseases. You will also learn about the effects of the bacteria in your gut (39.1), and the bacteria that partner with plants (28.3). Bacteria play an integral role in food webs (46.3) and biogeochemical cycles (46.5, 46.9).

20.1 Evolution of a Disease

Billions of years before there were fungi, plants, or animals, Earth's seas were home to two groups of microscopic organisms: bacteria and archaea. These single-celled organisms do not have a nucleus or other typical eukaryotic organelles. Viruses are simpler still, with no chromosomes or metabolic machinery. By many definitions, viruses are not even alive. Despite their simplicity, viruses can evolve because like living organisms they have a genome that can mutate.

In recent years, scientists have learned quite a bit about the origin and evolution of **HIV (human immunodeficiency virus)**. This virus causes the emerging disease AIDS (acquired immune deficiency syndrome). An **emerging disease** is a disease that is relatively new to humans or has newly expanded its range.

HIV was first isolated in the early 1980s. Since then, gene sequence comparisons have shown that the most common strain (HIV-1) evolved from a type of simian immunodeficiency virus (SIV) that infects chimpanzees in west central Africa. A recent study investigated the health effects of SIV in a wild chimpanzee population that has been studied by primatologist Jane Goodall and others for many years. Analyzing DNA in chimpanzee feces allowed researchers to determine whether individual animals were infected with SIV (FIGURE 20.1). This information was then combined with observational data from the field. The resulting data showed that, in this population, SIV reduces fitness. SIV-infected chimpanzees die earlier than unaffected animals and leave fewer offspring.

How did SIV make the jump from chimpanzees into people? In some African communities, people commonly hunt nonhuman primates for food. Humans are exposed to SIV during butchery of an infected animal, an intimate and bloody process during which the virus can enter the butcher's body through a cut. Researchers believe SIV mutated and became HIV inside a person exposed this way.

The earliest known evidence of HIV infection comes from two tissue samples stored at a hospital in west central Africa. One is a blood sample taken from a man in 1959. The other is a woman's lymph node that was removed in 1960. The viral gene sequences from the two samples differ a bit, which implies that HIV had already been around and mutating by the time these two people became infected. Given the known mutation rate for HIV, researchers estimate that HIV first infected humans in the early 1900s.

FIGURE 20.1 Analyzing wild chimpanzee feces for SIV, the virus from which HIV evolved. Rebecca Rudicell was part of a team that is studying the effects of SIV in chimpanzees.

Gene sequence comparisons have also allowed researchers to trace the movement of the virus out of Africa. One recent study concluded that HIV-1 was carried from Africa to Haiti in the mid-1960s. The virus diversified in Haiti and acquired distinctive mutations not seen in Africa. By 1969, HIV-1 with these mutations had been introduced to the United States. It may have arrived in an infected individual or in infected blood. Once there, it spread quietly until AIDS was identified as a threat in 1981.

Today, more than 20 million people worldwide have died from AIDS. About 30 million are currently infected with HIV. The virus infects and replicates inside white blood cells that are essential to immune responses. Eventually, the infected white blood cells die, destroying the body's ability to defend itself. As a result, disease-causing organisms run rampant, causing symptoms of AIDS and health problems that can be fatal. Section 37.11 discusses in detail how AIDS affects the immune system.

Knowing about HIV's ancestry may help us develop new weapons against the virus. For example, although SIV does harm chimpanzees, the effects are nowhere near as devastating as untreated HIV in humans. Determining how the chimpanzee immune system fights against SIV may provide insights that we can put to use in our own battle against AIDS.

emerging disease Disease that is relatively new to a species, or has recently expanded its range.
HIV (human immunodeficiency virus) Retrovirus that causes the disease AIDS.

20.2 Viruses and Viroids

✔ A virus consists of nucleic acid and protein. It is smaller than any cell and has no metabolic machinery of its own.

In the late 1800s, scientists studying stunted tobacco plants discovered a new kind of disease-causing agent, or **pathogen**. It was so small that it passed through screens that filtered out bacteria, and it could not be seen with a light microscope. The scientists called this unseen infectious entity a virus, a term that means "poison" in Latin.

Today, we define a **virus** as a noncellular, infectious particle that can replicate only in a living cell. A virus is an obligate intracellular parasite. It does not have ribosomes or other metabolic machinery and it cannot make ATP. To replicate, the virus must insert its genetic material into a cell of a specific type of organism. We call the organism that a pathogen infects its "host."

Viral Structure

Most viruses are so small (about 25 to 300 nanometers in diameter) that they can only be seen with an electron microscope. However, some recently discovered viruses, appropriately referred to as giant viruses, can be larger than most bacteria.

A free viral particle, or virion, always includes a viral genome enclosed within a protein coat, or capsid. The viral genome may be RNA or DNA, and it may be single-stranded or double-stranded. The capsid may also enclose one or more viral enzymes.

The capsid consists of protein subunits that assemble in a repeating pattern to produce a helical or many-sided (polyhedral) shape. A capsid protects the viral genetic material and facilitates its delivery into a host cell. In all viruses, some components of the viral coat bind to proteins at the surface of a host cell.

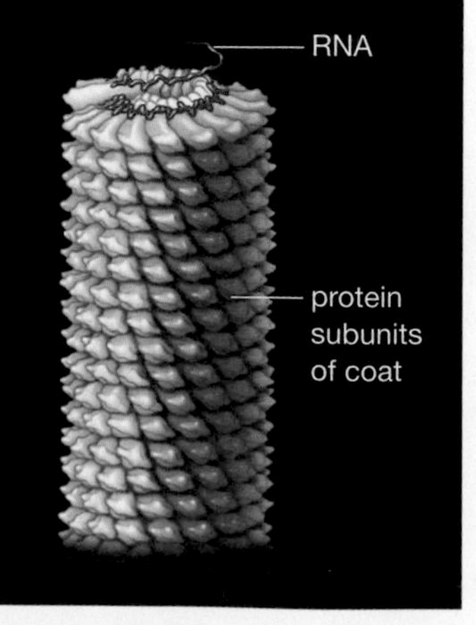

FIGURE 20.2 Tobacco mosaic virus, a virus that infects tobacco (above) and related plants. The helical arrangement of the capsid subunits (right) gives the virus a rodlike structure. The genome is single-stranded RNA.

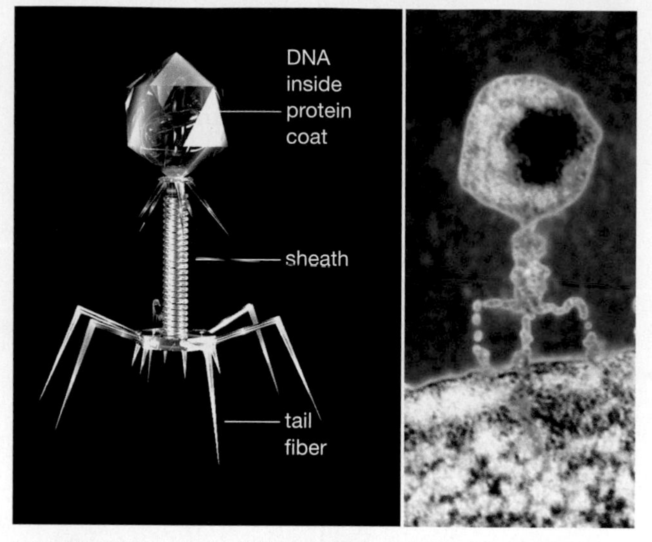

FIGURE 20.3 Model (left) and electron micrograph (right) of a bacteriophage, a bacteria-infecting virus with a complex structure.

Like many plant viruses, the tobacco mosaic virus has a helical structure. Its coat proteins assemble in a tight helix around its genetic material, a single strand of RNA (FIGURE 20.2). Plant viruses typically enter their host through a wound made by an insect, pruning, or another mechanical injury. Symptoms of viral infection typically include stunting and curling, yellowing, or spotting of leaves.

Bacteriophages, viruses that infect bacteria, have a complex structure (FIGURE 20.3). Their headlike capsid encloses the viral DNA. Other protein components of the virus allow it to pierce a bacterial cell wall and inject DNA into the cell. Hershey and Chase used a type of bacteriophage called lambda to identify DNA as the genetic material of all organisms (Section 8.2).

Adenoviruses, the most frequent cause of the common cold, have a 20-sided, polyhedral capsid with a distinctive protein spike at each corner (FIGURE 20.4A). They are nonenveloped ("naked") viruses, meaning that the capsid is their outermost layer.

By contrast, the capsid of most animal viruses is enclosed within an "envelope," a layer of membrane derived from the host cell in which the virus assembled. Depending on the virus, the envelope may be derived from either the plasma membrane or the nuclear membrane of the host cell. Herpesvirus is an example of an enveloped DNA virus. Like an adenovirus, a herpesvirus has a 20-sided capsid, but it is surrounded by an envelope made of fragments of a host cell's nuclear membrane (FIGURE 20.4B).

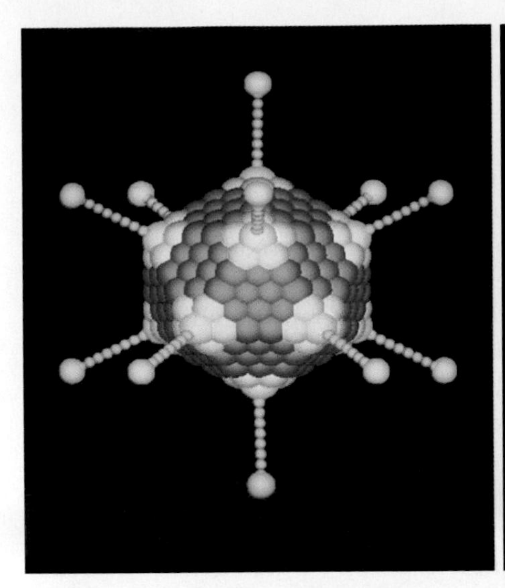

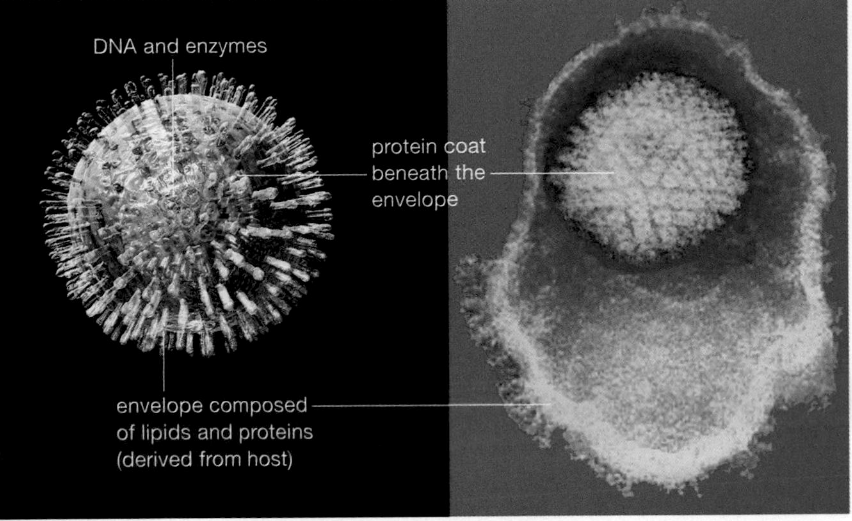

A Model of an adenovirus, a polyhedral virus. Protein subunits form a 20-sided polyhedron around double-stranded DNA.

B Model (left) and electron micrograph (right) of a herpesvirus, an enveloped virus. The envelope is derived from the nuclear membrane of the cell in which the virus assembled. In the micrograph, the envelope is peeled back to reveal the protein coat beneath.

FIGURE 20.4 ▶**Animated** Two animal viruses.

Origin and Ecological Role of Viruses

We do not know how viruses are related to cellular life. The fact that they can replicate only inside cells suggests that they may have evolved from cells, perhaps being derived from bits of DNA or RNA that escaped a plasma membrane. Alternatively, viruses may be remnants of a time before cells. This could explain why some viral genomes consist largely of genes that have no counterparts in modern cells.

Viruses infect and replicate in all organisms, no matter how simple or complex. A viral infection often decreases a host's ability to survive and reproduce, so viruses affect ecological interactions among species throughout the biosphere.

Some viruses assist humans through their effects on other species. For example, we benefit when baculoviruses infect and kill caterpillars that feed on crop species or when bacteriophages kill bacteria that cause food poisoning. On the other hand, viruses can have devastating economic effects when they infect livestock or agriculturally important plants. In recent years, outbreaks of influenza among pigs and chickens have led to the slaughter of hundreds of thousands of animals. Viruses also harm us directly by impairing human health (a topic we return to in Section 20.4).

bacteriophage Virus that infects bacteria.
pathogen Disease-causing agent.
viroid Small, noncoding, infectious RNA.
virus Noncellular, infectious particle of protein and nucleic acid; replicates only in a host cell.

Viroids

Viroids are small RNAs that cause disease in many commercially valuable plants, including potatoes, tomatoes, citrus, apples, coconuts, avocados, and chrysanthemums. They were discovered in the 1970s by the plant pathologist Theodor Diener. He named the previously unknown pathogen a viroid because it seemed like a stripped-down version of a virus.

All viroids are circular, single-stranded RNAs. They are remarkably small, with fewer than 400 nucleotides. By comparison, even the smallest viral genome has thousands of nucleotides. Unlike the genetic material of a virus, viroid RNA does not encode proteins. The viroid replicates with the help of a plant enzyme (RNA polymerase) that normally transcribes the plant's DNA to RNA. Viroids cause disease by interfering with normal gene expression in plant cells, thus stunting the plant and deforming its parts.

TAKE-HOME MESSAGE 20.2

What are viruses and viroids?

- ✔ Viruses are noncellular infectious particles that consist of nucleic acid surrounded by capsid proteins and, in some types, an envelope derived from membrane of its host.
- ✔ A virus is an obligate intracellular parasite. It possesses no metabolic machinery of its own and can multiply only inside living cells.
- ✔ Viroids are infectious, noncoding RNAs that cause some plant diseases.

CREDITS: (4A) Dr. Richard Feldmann/National Cancer Institute; (4B) left, Russell Knightly/Science Source, Inc.; right, Dr. Linda Stannard, UCT/Science Source.

20.3 Viral Replication

✔ A viral infection is like a cellular hijacking; viral genes take over a host cell's machinery and direct it to synthesize viral components that can self-assemble as new viral particles.

Overview of Viral Replication

Details of viral replication processes vary, but all involve the steps outlined in **TABLE 20.1**. A virus cannot move itself toward a host, so infection begins with a chance encounter. During attachment, viral proteins bind receptor proteins on the surface of a host cell. The virus's genetic material enters the host cell and takes over its genetic machinery. Under the virus's influence, the cell puts aside its normal tasks and turns to replicating and expressing viral genes. When viral proteins and nucleic acid come into contact, they self-assemble as new virions. The virus either buds from the host cell or escapes when the host cell lyses (breaks open).

Table 20.1 Steps in Most Viral Replication Cycles

1. **Attachment**	Proteins on viral particle chemically recognize and lock onto specific receptors at the host cell surface.
2. **Penetration**	Either the viral particle or its genetic material crosses the plasma membrane of a host cell and enters the cytoplasm.
3. **Replication and synthesis**	Viral DNA or RNA directs host to make viral nucleic acids and viral proteins.
4. **Assembly**	Viral components self-assemble as new viral particles.
5. **Release**	The new viral particles are released from the cell.

Bacteriophage Replication

Bacteriophages replicate by two pathways. Both begin when a bacteriophage attaches to a bacterial cell and injects its DNA (**FIGURE 20.5A**). In the **lytic pathway**, viral genes are expressed immediately ❶. The infected host first produces viral components that self-assemble as virus particles. Then, a viral-encoded enzyme breaks down the bacterial cell wall, triggering a process called lysis in which the cell disintegrates and dies.

In the **lysogenic pathway**, viral DNA injected into a bacterial cell becomes integrated into the host cell's genome ❷. Viral genes are not expressed, so this cell remains healthy. When the host cell reproduces, viral DNA is copied and passed on along with the host's genome. Like miniature time bombs, viral DNA inside these descendant cells awaits a signal to enter the lytic pathway (**FIGURE 20.5B4**).

Some bacteriophages always replicate by the lytic pathway. They kill their host cell quickly and are not passed from one bacterial generation to the next. Others embark upon either the lytic or lysogenic pathway, depending on conditions in the host cell.

Replication of HIV

HIV is an enveloped virus that replicates inside human white blood cells (**FIGURE 20.6**). It attaches to a cell via a glycoprotein that extends through its envelope ❶. The glycoprotein binds two different proteins on the host cell. After attachment, the viral envelope fuses

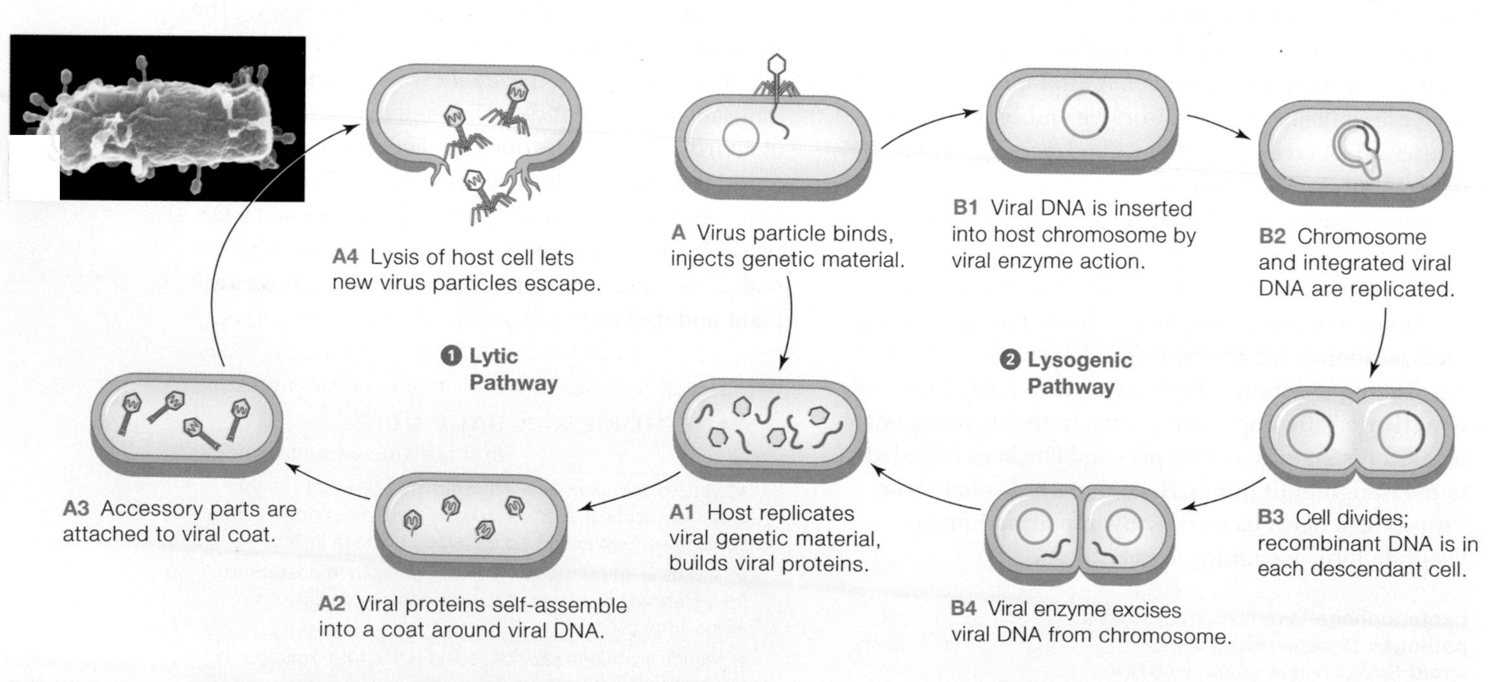

FIGURE 20.5 ▸**Animated** Pathways in the replication cycle of a bacteriophage.
FIGURE IT OUT What is the blue circle in A?
Answer: Bacterial chromosome

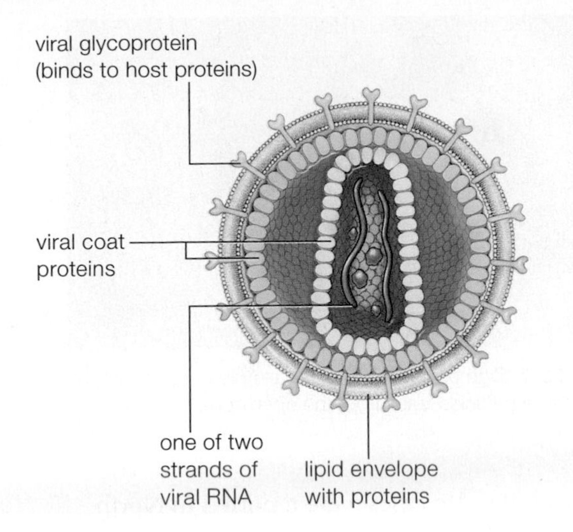

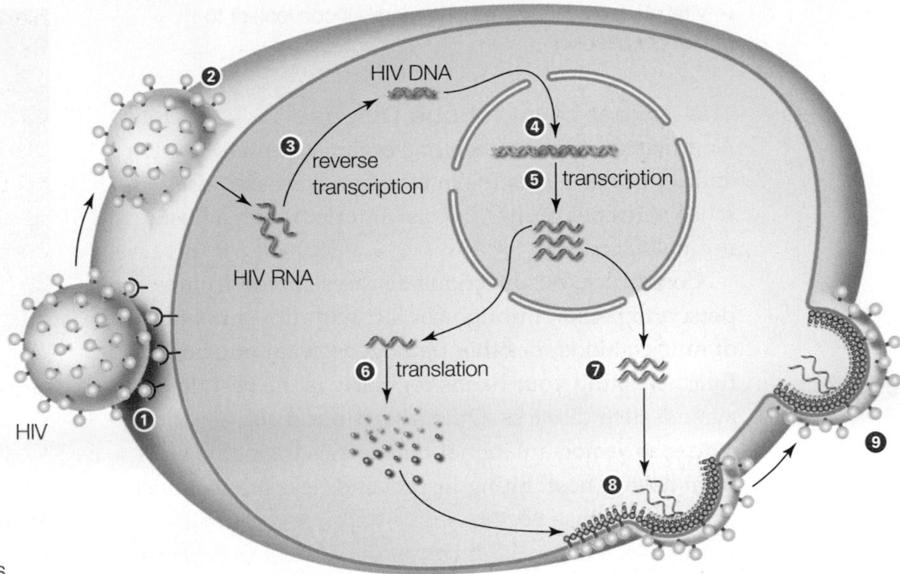

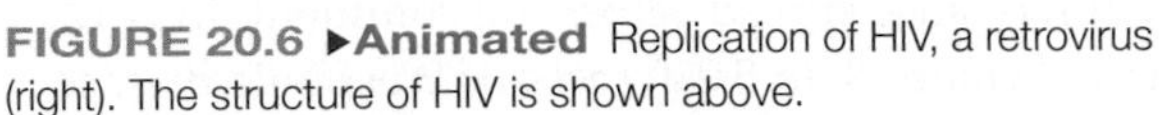

FIGURE 20.6 ▸Animated Replication of HIV, a retrovirus (right). The structure of HIV is shown above.

❶ Viral glycoprotein spikes bind to proteins at the surface of a white blood cell.

❷ The viral envelope fuses with the host cell plasma membrane, allowing viral RNA and enzymes to enter the cell.

❸ Viral reverse transcriptase uses viral RNA to make double-stranded viral DNA.

❹ Viral DNA enters the nucleus and becomes integrated into the host genome.

❺ Transcription produces viral RNA.

❻ Some viral RNA is translated to produce viral proteins.

❼ Other viral RNA forms the new viral genome.

❽ Viral proteins and viral RNA self-assemble at the host plasma membrane.

❾ New virus buds from the host cell, with an envelope of host plasma membrane.

with the blood cell's plasma membrane, releasing viral enzymes and RNA into the cell ❷.

HIV is a **retrovirus**, a virus whose RNA genome is used as a template to produce double-stranded DNA inside a host cell. This conversion is carried out by the viral enzyme reverse transcriptase ❸. DNA produced by reverse transcriptase enters the nucleus together with a viral enzyme called integrase. Integrase inserts the DNA into one of the host's chromosomes ❹. Once integrated, viral DNA is replicated and transcribed along with the host genome ❺.

Some of the resulting viral RNA is translated into viral proteins ❻ and some becomes the genetic material of new HIV ❼. HIV particles self-assemble at the plasma membrane ❽. As the virus buds from the host cell, some of the host's plasma membrane becomes the viral envelope ❾. Each new virus can then infect another white blood cell. New HIV-infected cells are also produced when an infected cell replicates.

Drugs designed to fight HIV take aim at steps in viral replication. Some interfere with the way HIV binds to a host cell. Others impair the viral reverse transcriptase. Integrase inhibitors prevent viral DNA from integrating into a human chromosome. Protease inhibitors prevent the processing of newly translated polypeptides into mature viral proteins. These antiviral drugs lower the number of HIV particles, so a person stays healthier. Less HIV in body fluids also reduces the risk of passing the virus to others. However, no drug has yet been found that can eliminate the virus entirely, all have unpleasant side effects, and all must be taken for life.

lysogenic pathway Bacteriophage replication mechanism in which viral DNA becomes integrated into the host's chromosome and is passed to the host's descendants.
lytic pathway Bacteriophage replication mechanism in which a virus replicates in its host and kills it quickly.
retrovirus Virus whose RNA is used as a template to produce to double-stranded viral DNA within a host cell.

TAKE-HOME MESSAGE 20.3

How do viruses replicate?

✔ A virus binds to a specific type of host cell, and viral genetic material enters the cell. Viral genes direct the production of viral components that then self-assemble as new viral particles.

20.4 Viruses as Human Pathogens

✔ Viral diseases range from the merely inconvenient to potentially deadly.

The Threat of Infectious Disease

An infection occurs when one organism enters another and replicates inside it. An infectious disease arises when activities of the "guests" interfere with a host's normal functions.

Communicable infectious diseases spread from person to person through contact with tiny amounts of mucus, blood, or other pathogen-containing body fluid. Washing your hands regularly is the best defense against such diseases. Other infectious diseases require a **disease vector**, an animal that carries the pathogen from host to host. Biting insects and ticks are the most common disease vectors.

Common Viral Diseases

Most viral diseases cause mild symptoms and trouble us only briefly. For example, some adenoviruses infect the membranes of our upper respiratory system and cause common colds. Others colonize the lining of our gut and cause a brief bout of vomiting and diarrhea.

A few viral diseases can be more persistent. Various types of herpesviruses cause cold sores, genital herpes, mononucleosis, or chicken pox. Typically the initial infection causes symptoms that subside quickly. However, the virus remains in the body in a latent state, and can reawaken later on. After a person has been infected by herpes simplex virus 1 (HSV-1), the virus can lie latent in nerve cells for years. When activated, the virus replicates and causes painful "cold sores" on the edge of the lips (**FIGURE 20.7**). Similarly, the virus responsible for a childhood case of chicken pox can later cause the disease shingles.

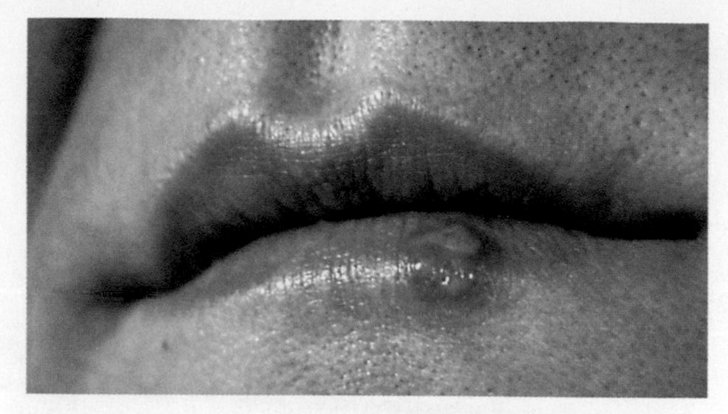

FIGURE 20.7 Sign of an active herpes simplex virus 1 infection. Fluid rich in viral particles leaks from the open sore.

Measles, mumps, rubella (German measles), and chicken pox are viral diseases of childhood that, until recently, were common worldwide. Today, most children in developed countries have been vaccinated against these illnesses. Administering a vaccine primes the body to fight off a specific pathogen, a process explained in detail in Section 37.12.

A few types of viral infections increase the risk of cancer. Infection by certain strains of sexually transmitted human papillomaviruses (HPV) can cause cancer of the cervix, anus, mouth, or throat. Similarly, infection by some hepatitis viruses raises the risk of liver cancer.

Emerging Viral Diseases

Like AIDS, West Nile disease is an emerging viral disease. It was first identified in humans in 1981 in the West Nile region. No cases were reported in North America until 1999, when the virus turned up in New York. It has since spread across the continent. In 2012, there were more than 5,000 cases of West Nile disease in the United States and 286 deaths. Children and the elderly are most likely to die after infection.

Mosquitoes are the vector for West Nile virus and birds serve as its main reservoir. A reservoir for a human disease is an animal that hosts the disease-causing pathogen, but does not transmit the pathogen to humans. Animal reservoirs can be a source of virus for new human infections.

Ebola virus causes the emerging disease Ebola hemorrhagic fever. Symptoms of this incurable disease include fever, muscle pain, and massive internal and external bleeding. The death rate is close to 90 percent. The Ebola virus, which was first isolated in 1976, is highly contagious in humans. So far, outbreaks of Ebola have occurred only in Africa, where fruit bats are the main reservoir. The recent discovery that Asian fruit bats also harbor the virus raises concerns that human infections could become more widespread.

Viral Mutation and Reassortment

Like living organisms, viruses have genomes that can be altered by mutation. RNA viruses such as HIV and influenza virus mutate especially quickly. The viral reverse transcriptase makes frequent replication errors. These errors remain uncorrected because the host's proofreading and repair mechanisms evolved to fix errors of transcription and do not operate during reverse transcription.

To keep up with ongoing mutations in influenza viruses, scientists create a new flu shot every year. The flu shot is a vaccine designed to thwart the newly mutated influenza viruses that scientists predict are most likely to pose a threat during the upcoming flu

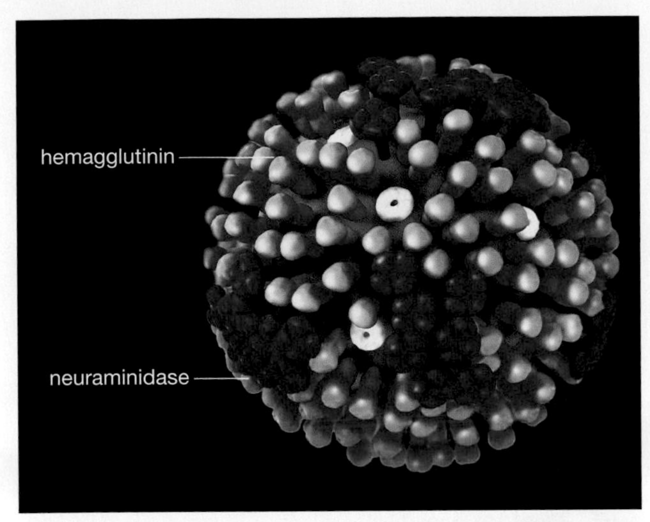

FIGURE 20.8 Influenza virus. Subtypes such as H1N1 or H5N1 are defined by the structure of viral proteins—hemagglutinin (H) and neuraminidase (N)—that extend through the outer envelope.

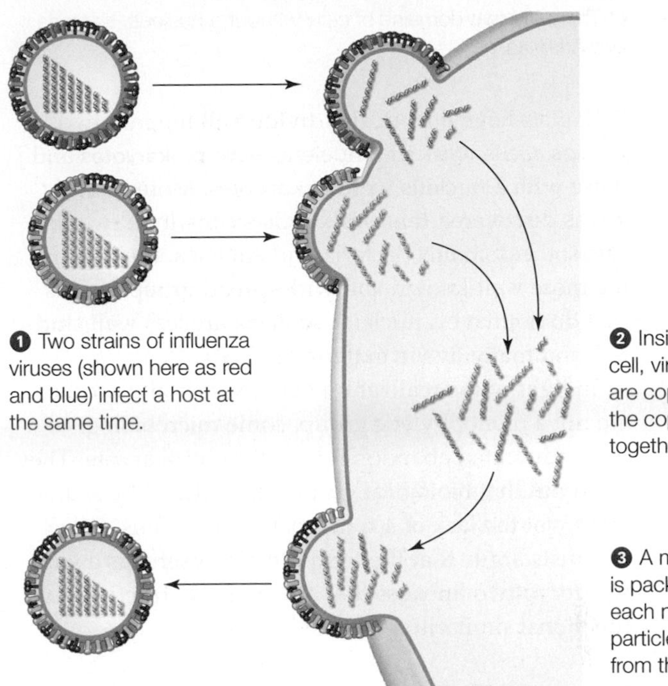

FIGURE 20.9 Viral reassortment. When a host cell is infected by two viruses of the same type, copies of viral genes reassort to form new combinations.

season. Unfortunately, determining which flu strains will be circulating is not an exact science. Even after a flu shot, a person remains susceptible to a virus that differs from the virus types targeted by the vaccine.

Influenza subtypes are named for the structure of two proteins at the viral surface (**FIGURE 20.8**). One protein, hemagglutinin (H), is a glycoprotein that allows the virus to bind to a host cell. The other is an enzyme, neuraminidase (N), that helps new viral particles exit an infected cell.

In April 2009, a new version of the H1N1 subtype of influenza appeared unexpectedly. Although the media referred to this virus as "swine flu," it actually had a composite genome, with genes from a human flu virus, bird flu virus, and two different swine flu viruses. Such novel genomes arise as a result of **viral reassortment**, the swapping of genes between viruses that infect a host at the same time (**FIGURE 20.9**).

The 2009 H1N1 virus was discovered when it caused an epidemic in Mexico. An **epidemic** is an outbreak of disease in a limited region. Within months, there was a **pandemic**, an outbreak of disease that simultaneously affects people throughout the world. Fortunately, governments quickly released reserves of antiviral drugs to treat those infected. These drugs interfere with neuraminidase function, and so impair the virus's ability to infect cells. A vaccine was created to prevent new infections and the World Health Organization declared the H1N1 pandemic over in August 2010.

Another strain of influenza, influenza H5N1, is a bird flu that occasionally infects people who have direct contact with birds. When the virus does infect people, the death rate is disturbingly high. From 2003 to 2013, the World Health Organization received reports of 641 human cases of influenza H5N1, mainly in Asia. Of these, 380 (about 60 percent) were fatal. Fortunately, person-to-person transmission of the H5N1 virus remains exceedingly rare.

Health officials continue to carefully monitor H5N1 and H1N1 influenza. Either virus could mutate, and their coexistence raises the possibility of a potentially disastrous gene exchange. If H1N1 picked up genes from avian H5N1, the result could be a flu virus that is both easily transmissible and deadly.

disease vector An animal that transmits a pathogen from one host to the next.
epidemic Disease outbreak that occurs in a limited region.
pandemic Disease outbreak with cases worldwide.
viral reassortment Two related viruses infect the same individual simultaneously and swap genes.

TAKE-HOME MESSAGE 20.4

How do viruses affect human health?

- ✔ Viruses cause many diseases, most short-lived and relatively mild, but some that are deadly.
- ✔ Viral pathogens can change by mutation or reassortment.

20.5 Shared Traits of Prokaryotes

✔ There are two domains of cells without a nucleus: Bacteria and Archaea.

Biologists have historically divided all life into two groups. Cells without a nucleus were **prokaryotes** and those with a nucleus were eukaryotes. More recently it was discovered that prokaryotes constitute two lineages: the domains Bacteria and Archaea. **Bacteria** are the more well-known and widespread group of cells that do not have a nucleus. **Archaea** are less well studied, and many live in extreme habitats.

In light of the realization that bacteria and archaea are not a monophyletic group, some microbiologists have advocated abandoning the term prokaryote. They point out that biological groups are defined by shared traits, not the lack of a trait, such as a nucleus. Other scientists argue that the term remains useful as a way to refer to two lineages that share many structural and functional similarities (TABLE 20.2).

Structural Traits

A typical bacterial or archaeal cell cannot be seen without a light microscope. Three cell shapes are common in both groups: rods, spheres, and spirals (FIGURE 20.10A). Rod-shaped cells are called bacilli (singular, bacillus), spherical cells are cocci (singular, coccus), and spiral cells are called spirilla (singular, spirillum).

Most bacteria and archaea secrete a semirigid, porous cell wall around their plasma membrane (FIGURE 20.10B). Bacteria may also have a slime layer or capsule around the cell wall. Slime helps a cell stick to surfaces. A capsule helps some pathogenic bacteria evade the immune defenses of their vertebrate hosts.

Bacteria and archaea typically have a single chromosome. This circle of double-stranded DNA resides in a cytoplasmic region called the **nucleoid**. There is no nuclear envelope as in eukaryotes, although at least one bacterial species has a membrane surrounding its DNA (Section 19.6). Membranes of some bacteria fold inward, but no bacteria or archaea have an endoplasmic reticulum or Golgi bodies.

Table 20.2 Traits Common to Bacteria and Archaea

1. No nuclear envelope; chromosome in nucleoid
2. Generally a single chromosome (a circular DNA molecule); many species also contain plasmids
3. Cell wall (in most species)
4. Ribosomes distributed in the cytoplasm
5. Asexual reproduction by binary fission
6. Capacity for gene exchange among cells by way of conjugation, transduction, and transformation

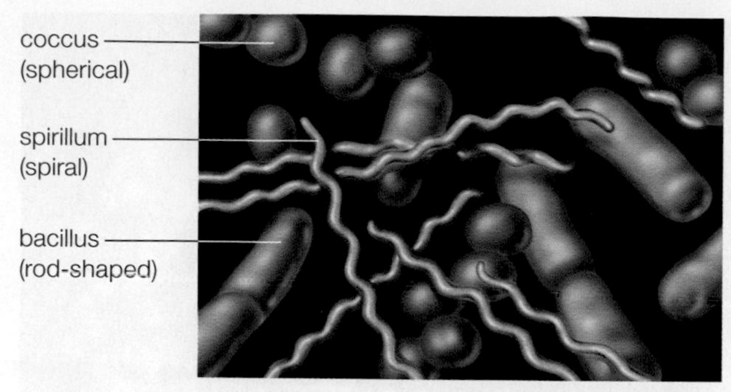

A Three cell shapes common among prokaryotic cells.

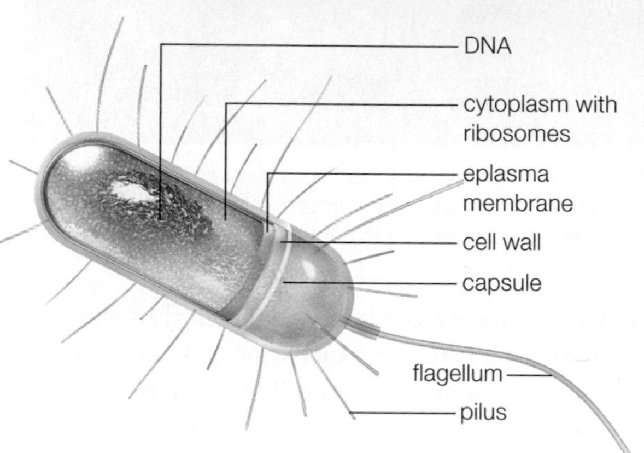

B Features of a typical bacterial cell.

FIGURE 20.10 ▶Animated Prokaryotic cell structure.

Many bacteria and archaea have a flagellum. However, unlike the eukaryotic flagellum it does not contain microtubules and it does not bend side to side. Rather, the flagellum rotates like a propeller.

Hairlike filaments called **pili** (singular, pilus) may also extend from the cell surface. Some cells use pili to stick to surfaces. Others glide along by using their pili as grappling hooks. A pilus extends out to a surface, sticks to it, then shortens, drawing the cell forward. Another type of retractable pilus draws cells together for gene exchanges, as described below.

Reproduction and Gene Transfers

Bacteria and archaea usually reproduce by **binary fission**, a type of asexual reproduction (FIGURE 20.11). The process begins when the cell replicates its single chromosome, which is attached to the inside of the plasma membrane ❶. The DNA replica attaches to the membrane adjacent to the parent molecule. Addition of membrane and wall material elongates the cell and moves the two DNA molecules apart ❷. Next,

❶ A bacterium has one circular chromosome that attaches to the inside of the plasma membrane.

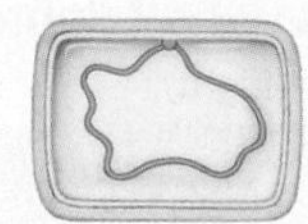

❷ The cell duplicates its chromosome, attaches the copy beside the original, and adds membrane and wall material between them.

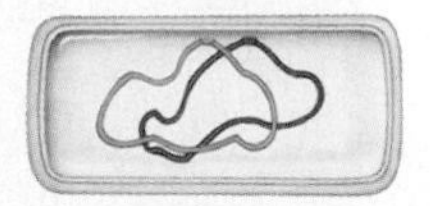

❸ When the cell has almost doubled in size, new membrane and wall are deposited across its midsection.

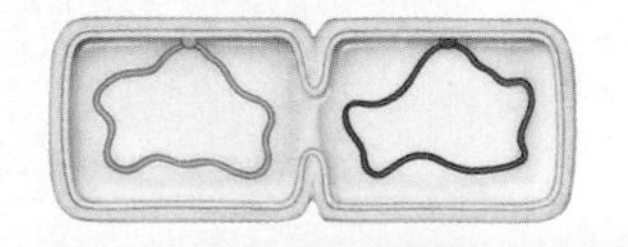

❹ Two genetically identical cells result.

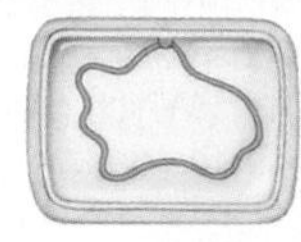

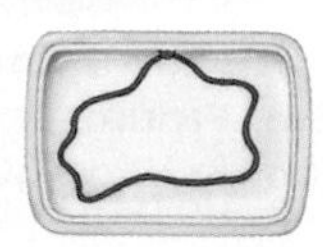

FIGURE 20.11 ▶**Animated** Binary fission, the reproductive mode of bacteria and archaea.

❶ Conjugation begins when a cell with a specific kind of plasmid (red) extends a sex pilus to a cell lacking this plasmid.

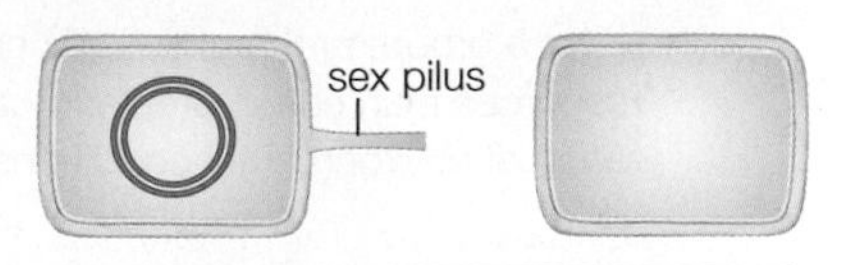

❷ The pilus attaches to the potential recipient cell, then shortens, drawing the cells together so their cytoplasm can connect.

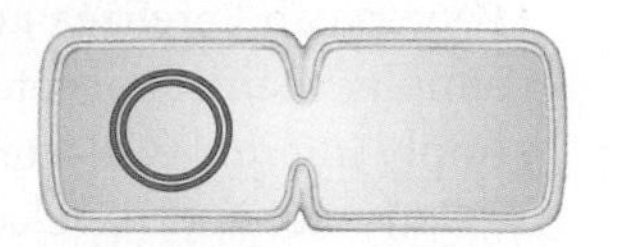

❸ A single strand of plasmid DNA breaks and moves into the recipient cell. Each cell will duplicate its single strand of plasmid DNA.

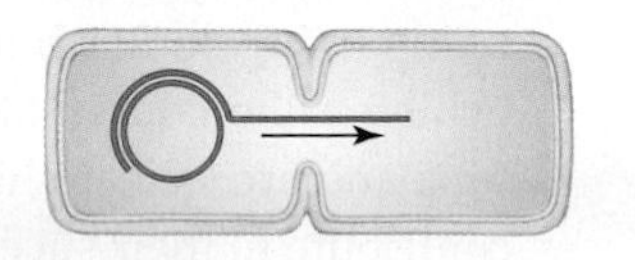

❹ The result is two cells, each with a plasmid and the capacity to donate it to other cells.

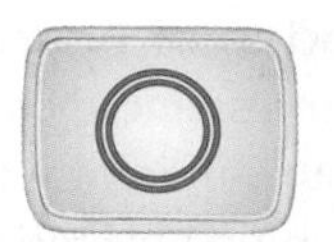

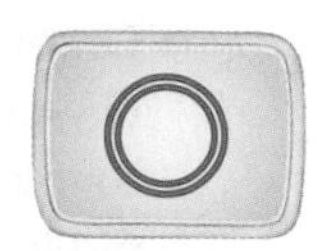

FIGURE 20.12 ▶**Animated** Conjugation, a type of horizontal gene transfer. For clarity, the chromosome is not shown.

membrane and cell wall material is deposited across the cell's midsection ❸. This material partitions the cell, eventually yielding two genetically identical descendant cells ❹.

In addition to inheriting DNA "vertically" from a parent cell, bacteria and archaea engage in **horizontal gene transfers**, in which an individual acquires genes by a mechanism other than inheritance. The gene donor can be a cell of the same species or a different species. One type of horizontal transfer, **conjugation**, involves movmement of genes on a plasmid (**FIGURE 20.12**). A **plasmid** is a small circle of double-stranded DNA separate from the chromosome. During conjugation, a plasmid-bearing cell extends a sex pilus out to a plasmid-free cell ❶ and draws it close ❷. Then, a single strand of the plasmid DNA is moved from the donor to the recipient cell ❸. Both cells create a complementary strand of plasmid DNA, so each ends up with a functional plasmid ❹. In another type of horizontal transfer, called **transduction**, bacteriophages move genes between cells. The virus picks up a bit of DNA from one host cell, then transfers the DNA to its next host. Bacteria and archaea also acquire new genes by taking up DNA from their environment, a process called **transformation**.

The ability of bacteria to acquire new genes has important implications for public health. Gene transfers increase the rate at which a gene spreads through a population of bacteria, thus enhancing the population's ability to respond to a selection pressure such as that arising from antibiotic use.

archaea More recently discovered and less well-known lineage of single-celled organisms without a nucleus.
bacteria More diverse and well-known lineage of single-celled organisms without a nucleus.
binary fission Cell reproduction process of bacteria and archaea.
conjugation Mechanism of horizontal gene transfer in which one bacterial or archaeal cell passes a plasmid to another.
horizontal gene transfer Transfer of genetic material by a mechanism other than inheritance from a parent or parents.
nucleoid DNA-containing region of a bacterial or archaeal cell.
pilus Protein filament that projects from the surface of some bacterial and archaeal cells.
plasmid Of many bacteria and archaea, a small ring of DNA replicated independently of the chromosome.
prokaryote Member of one of two single-celled lineages (bacteria and archaea) that do not have a nucleus; a bacterium or archaeon.
transduction In bacteria and archaea, a mechanism of horizontal gene transfer in which a bacteriophage transfers DNA between cells.
transformation In bacteria and archaea, a type of horizontal gene transfer in which DNA is taken up from the environment.

TAKE-HOME MESSAGE 20.5

What structural and functional features do bacteria and archaea share?

✔ In both lineages, cells are typically walled and a single chromosome is not enclosed by a nucleus.

✔ Both lineages reproduce by binary fission. They can swap genes by conjugation and other mechanisms of horizontal gene transfer.

20.6 Factors in the Success of Prokaryotes

✔ As a group, prokaryotic cells can utilize a wider range of resources than eukaryotes and can tolerate a much broader range of environmental conditions.

Bacteria and archaea are smaller and structurally simpler than eukaryotes, but this simplicity does not imply inferiority. Bacteria and archaea existed before eukaryotes and have coexisted with them for more than 2 billion years. The number of bacterial cells currently living on Earth has been estimated at 5 million trillion trillion. From an evolutionary perspective, bacteria and archaea are highly successful. Several factors contribute to their success.

CARBON SOURCE	ENERGY SOURCE: Light	ENERGY SOURCE: Chemicals
Inorganic source such as CO_2	**Photoautotrophs** bacteria, archaea, photosynthetic protists, plants	**Chemoautotrophs** bacteria, archaea
Organic source such as glucose	**Photoheterotrophs** bacteria, archaea	**Chemoheterotrophs** bacteria, archaea, fungi, animals, nonphotosynthetic protists

FIGURE 20.13 ▸**Animated** Modes of nutrition in bacteria and archaea. **FIGURE IT OUT** Which group of organisms can build their own food from CO_2 in the dark? Answer: Chemoautotrophs

Diverse Modes of Nutrition

Organisms harvest energy and nutrients from the environment in four different ways. All of these nutritional modes occur among bacteria, archaea, or both (**FIGURE 20.13**). In addition, some bacteria and archaea can switch from one metabolic mode to another.

As you learned in Section 6.1, autotrophs build their own food using carbon dioxide (CO_2) as their carbon source. There are two subgroups of autotrophs: those that obtain energy from light, and those that obtain energy from chemicals.

Photoautotrophs are photosynthetic. They use the energy of light to assemble organic compounds from CO_2 and water. Many bacteria are photoautotrophs, as are plants and photosynthetic protists. As Section 19.6 explained, the ancestors of eukaryotic chloroplasts are cyanobacteria, a type of photosynthetic bacteria.

Chemoautotrophs obtain energy by oxidizing (removing electrons from) inorganic molecules such as hydrogen sulfide or methane. They use energy released by this process to build organic compounds from CO_2. Chemoautotrophic bacteria and archaea are the main producers in dark environments such as the seafloor. We know of no eukaryotic chemoautotroph.

Heterotrophs cannot tap into inorganic sources of carbon, so they obtain carbon by taking up organic molecules from their environment. **Photoheterotrophs** harvest energy from light, and carbon from alcohols, fatty acids, or other small organic molecules. Heliobacteria that live in the soils of rice paddies are an example. No eukaryotic photoheterotrophs are known.

Chemoheterotrophs obtain energy and carbon by breaking down carbohydrates, lipids, and proteins. Most bacteria and some archaea are chemoheterotrophs, as are fungi, animals, and nonphotosynthetic protists. All pathogenic bacteria are chemoheterotrophs that extract organic compounds they need from their host. Other chemoheterotrophic bacteria live in or on a host without causing any harm, and some even provide benefits. For example, chemoheterotrophic bacteria that live in the gut of many animals assist in digestion and may produce nutrients such as vitamins that the host requires. Still other bacterial chemoheterotrophs serve as decomposers. **Decomposers** break down complex organic molecules in wastes and remains into inorganic components that plants can take up and use. Decomposers also break down pesticides and pollutants, thus improving the environment for other organisms.

Aerobes and Anaerobes

With rare exceptions, eukaryotic organisms rely on aerobic respiration (Section 7.2) and thus need oxygen. By contrast, many bacteria and most archaea are anaerobes. Some can tolerate an oxygen-free environment, but are not harmed by oxygen. Others are obligate anaerobes, meaning oxygen either slows their growth or kills them outright. Such cells are harmed by oxygen because oxidation reactions damage their biological molecules and, unlike aerobic cells, they have no enzymes that can repair such damage. We find obligate anaerobes in aquatic sediments and the animal gut. They also can infect deep wounds. *Clostridium tetani*, a species of bacteria that causes the disease tetanus when it infects wounds, is an example.

Dormant Resting Structures

C. tetani also has another characteristic that contributes to the success of many prokaryotes—an ability to enter a state of suspended animation when conditions do not favor growth. Many bacteria and some archaea respond to adverse conditions by forming a dormant resting structure. Depending on the group and how the structure forms, it may be called a spore or a cyst. *C. tetani* and related bacteria form an **endospore**, an especially resilient resting structure consisting of a stripped-down bacterial cell within a thick protective covering (**FIGURE 20.14**). The covering includes a coat of enzyme-resistant proteins with an underlying layer of peptidoglycan: a polymer of sugars and amino acids found in the walls of some bacteria. Unlike metabolically active cells, endospores withstand heating, freezing, drying out, and exposure to ultraviolet radiation.

Endospores can survive in a dormant state for many years. Consider that scientists have extracted bacterial endospores from the gut of a bee that had been encased in amber (fossilized tree sap) for at least 20 million years. When given nutrients and moisture, the endospores germinated; the cells within the spores became active once again.

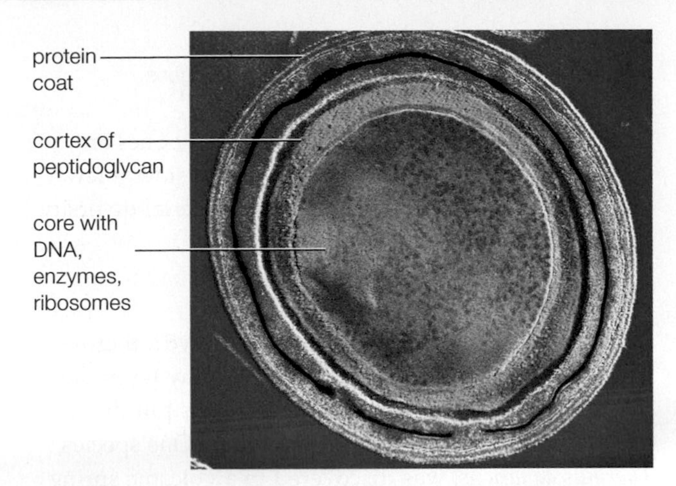

FIGURE 20.14 Endospore of *Clostridium tetani* (colorized micrograph). Endospore formation allows cells to survive in a dormant state when environmental conditions do not favor growth.

Investigating Species Diversity

In eukaryotes, a species is defined on the basis of the ability of its members to mate and produce fertile offspring (Section 1.5). However, this biological species concept is not easily applied to bacteria and archaea, which reproduce only asexually. In these groups, a species is defined as a group of individuals that share an ancestor and have a high degree of similarity in many independently inherited traits.

Historically, classification of bacteria was based on numerical taxonomy. By this process, an unidentified cell is compared against a known group on the basis of shape, metabolism, and properties of the cell wall. The more traits the cell shares with the known group, the closer is their inferred relatedness. This approach works best for cells that can be grown in the laboratory, where their responses to different conditions can be carefully observed. However, many bacteria and archaea do not grow in the lab.

The relatively new field of metagenomics assesses microbial diversity by sequence analysis of DNA in samples collected directly from an environment. Metagenomic studies often reveal a remarkable degree of species diversity. For example, air samples collected in two cities in Texas contained about 1,800 different kinds of bacteria. Seawater and soil samples from a variety of habitats have also shown similar diversity, with a high proportion of previously unknown and unnamed species.

The Human Microbiome Project is an ongoing metagenomic study of the microorganisms that live in or on the human body. Already, a survey of bacteria that live on the skin of the inner elbow turned up nearly 200 species. Another study revealed that more than a thousand species of bacteria and a few species of archaea can live in the human gut.

TAKE-HOME MESSAGE 20.6

Why do biologists consider prokaryotes successful?

- ✔ Both bacteria and archaea have survived for billions of years and continue to coexist beside the eukaryotes.
- ✔ Bacteria and archaea are Earth's most abundant organisms. We are only beginning to appreciate their enormous species diversity.
- ✔ Collectively, prokaryotes can live in a wider range of habitats than eukaryotes because they are more diverse in terms of their metabolism. Prokaryotes utilize all four modes of nutrition and may be aerobic or anaerobic.

chemoautotroph Organism that uses carbon dioxide as its carbon source and obtains energy by oxidizing inorganic molecules.
chemoheterotroph Organism that obtains both energy and carbon by breaking down organic compounds.
decomposer Organism that breaks organic wastes and remains down into their inorganic subunits.
endospore Spore (resting structure) formed by some soil bacteria; contains a dormant cell and is highly resistant to adverse conditions.
photoautotroph Organism that obtains carbon from carbon dioxide and energy from light.
photoheterotroph Organism that obtains carbon from organic compounds and its energy from light.

20.7 A Sample of Bacterial Diversity

✔ Most bacteria play important ecological roles.
✔ A small minority of bacteria are human pathogens.

There are many bacterial lineages and new ones are constantly being discovered. Here we consider a few major groups to provide insight into bacterial diversity and ecology.

Heat Lovers

If life emerged in thermal pools or near hydrothermal vents, the modern heat-loving bacteria may resemble those early cells. Biochemical comparisons put them near the base of the bacterial family tree. One species, *Thermus aquaticus*, was discovered in a volcanic spring in Yellowstone National Park. A heat-stable DNA polymerase first isolated from *T. aquaticus* is used to make copies of DNA by the polymerase chain reaction (PCR). This method of copying DNA is widely used in biotechnology (Section 15.3).

Oxygen-Producing Cyanobacteria

Photosynthesis evolved in many bacterial lineages, but only **cyanobacteria** utilize the noncyclic pathway and release free oxygen. When cyanobacteria incorporate the carbon from carbon dioxide into an organic compound, we say that they fix carbon. Some cyanobacteria also carry out **nitrogen fixation**, meaning they incorporate nitrogen from the air into ammonia (NH_3).

Nitrogen fixation is an important ecological service provided only by bacteria. Photosynthetic eukaryotes need nitrogen, but they cannot use the gaseous form ($N{\equiv}N$) because they do not have an enzyme that can break the molecule's triple bond. They can, however, take up ammonia released by nitrogen-fixing bacteria.

Some cyanobacteria partner with fungi to form lichens (which we discuss in Section 23.7) and others grow on the surface of soils, but most are aquatic. Aquatic cyanobacteria grow as single cells or as filaments of cells arranged end to end (**FIGURE 20.15A**). When conditions become unfavorable for growth, some filamentous cyanobacteria produce thick-walled resting cells. These cells are easily dispersed and remain dormant until environmental conditions improve.

Highly Diverse Proteobacteria

Proteobacteria are the most diverse bacterial group. One member, *Thiomargarita namibiensis*, is the largest bacterium known and can be seen without a microscope (**FIGURE 20.15B**). It gets energy by stripping electrons from sulfur that it stores in a giant vacuole.

Many chemoheterotrophic proteobacteria live inside other organisms. Rickettsias, the intercellular parasites that are relatives of mitochondria (Section 19.6), are an example. So are members of the genus *Rhizobium*, which live inside roots of legumes such as peas and beans. The bacteria receive sugars from the plant, and in turn provide it with ammonia from nitrogen they fix.

Helicobacter pylori commonly lives in the human stomach, where it sometimes causes ulcers. Prolonged infection can also cause stomach cancer.

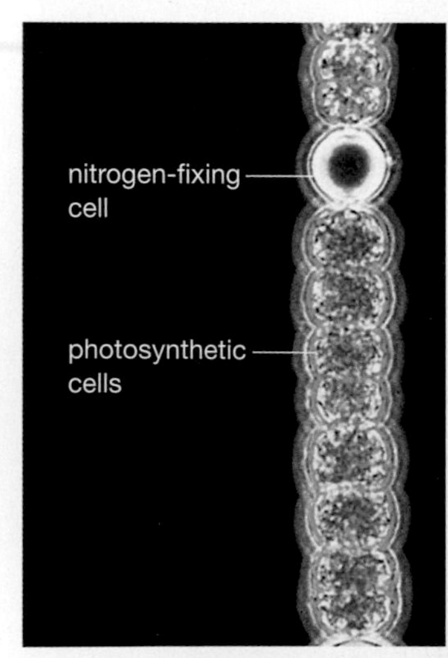

A *Anabaena*, a type of aquatic cyanobacterium. It carries out oxygen-producing photosynthesis and fixes nitrogen.

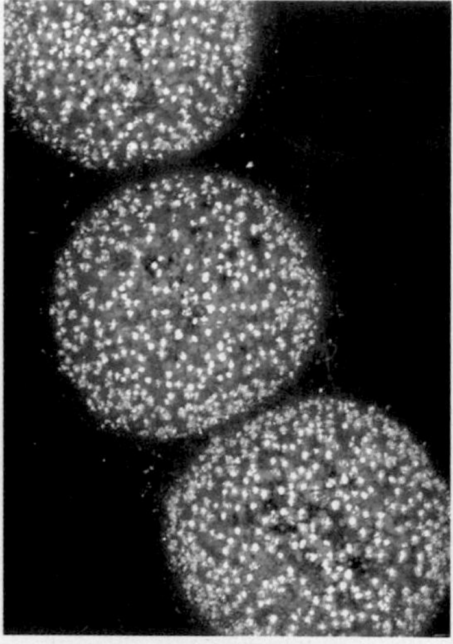

B *Thiomargarita namibiensis*, the biggest bacterium known. It lives in sea sediments and takes up and stores sulfates (white dots).

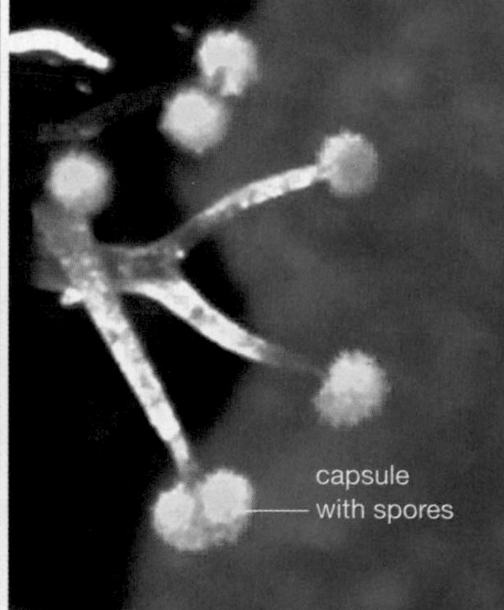

C *Chondromyces crocatus*, a myxobacterium, hunts other soil bacteria. When food runs out, thousands of cells form a fruiting body with spores at its tip.

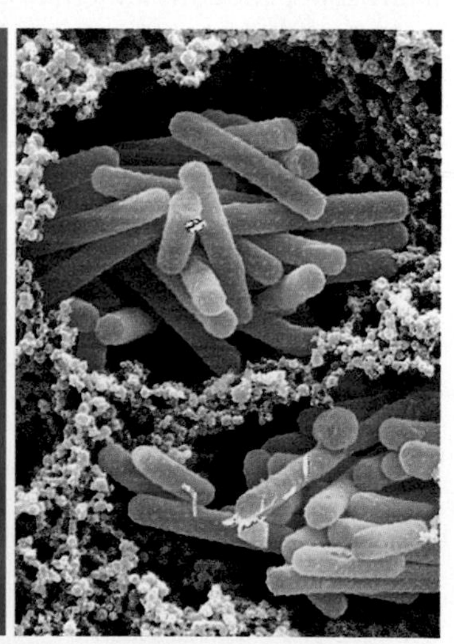

D *Lactobacillus* ferments sugars and produces lactate. The cells shown here were used to turn milk into yogurt. Other lactobacilli are important as decomposers.

FIGURE 20.15 A sampling of bacterial diversity. Most bacteria do not cause disease.

The most studied prokaryote is the proteobacterial species *Escherichia coli*, which normally lives in the mammalian gut. Researchers often investigate genetic and metabolic processes in *E. coli* because it is easily grown in laboratories. *E. coli* is also used in industrial biotechnology. Recombinant *E. coli* now make hormones and other proteins for medical use.

When biotechnologists want to alter a plant's genome, they may turn to *Agrobacterium*. These soil proteobacteria have a plasmid that gives them the capacity to infect plants and cause a tumor. As Section 15.5 explained, scientists produce recombinant plants by inserting genes into the tumor-inducing plasmid, then infecting a plant with recombinant bacteria.

The proteobacteria known as slime bacteria, or myxobacteria, are notable for their collective behavior. These tiny hunters live in soil, where they glide about as a swarm and eat other bacteria. When food dwindles, hundreds of thousands of cells form a multicelled fruiting body with spores at its tips (**FIGURE 20.15C**). Each spore contains an inactive cell. Spores released into the environment remain dormant until conditions favor growth. Then they germinate (become active).

Thick-Walled, Gram-Positive Bacteria

Gram-positive bacteria have thick cell walls that stain purple when prepared for microscopy by a process known as Gram staining. Most are chemoheterotrophs and many serve as decomposers. Actinomycetes are Gram-positive decomposers that grow through soil as long, branching chains of cells. Their presence gives freshly exposed soil its distinctive "earthy" smell. Many antibiotics, including streptomycin and vancomycin, were first isolated from actinomycetes. The actinomycetes make these bacteria-killing compounds to eliminate their competitors. They are not themselves harmed by the antibiotic they release.

Mycobacteria are rod-shaped cells related to the actinomycetes. One species, *Mycobacterium tuberculosis*, causes tuberculosis, a respiratory disease that kills more than a million people each year.

Soil also contains members of the Gram-positive genus *Lactobacillus*. These cells ferment sugars and produce lactate. Lactobacilli sometimes spoil milk, but they are also used to produce sour foods such as sauerkraut and yogurt (**FIGURE 20.15D**). Streptococci, which are close relatives of lactobacilli, include species that cause strep throat and the skin disease impetigo.

Gram-positive bacteria also include the endospore-forming soil bacteria. Various members of this group cause the muscle-paralyzing disease tetanus, the deadly food poisoning known as botulism, and the respiratory disorder anthrax.

A The spirochete that causes Lyme disease. Infected ticks spread the disease to humans.

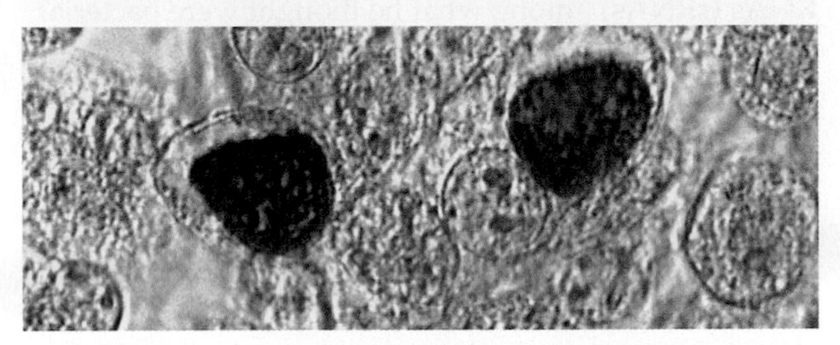

B A culture of human cells with two cells colored dark by an abundance of chlamydia living as parasites inside them.

FIGURE 20.16 Two bacterial pathogens of humans.

The Tiny Spirochetes and Chlamydias

Spirochetes and chlamydias are two unrelated lineages of very small bacteria. They can barely be seen with a light microscope. **Spirochetes** are shaped like a stretched-out spring (**FIGURE 20.16A**). Some live in the cattle gut and help their host break down cellulose. Others are aquatic decomposers and some fix nitrogen. One spirochete causes the sexually transmitted disease syphilis, and another causes Lyme disease, which is spread by ticks. **Chlamydias** live inside cells of vertebrates (**FIGURE 20.16B**). One species, *Chlamydia trachomatis*, is a cause of sexually transmitted disease.

chlamydias Bacteria that are intracellular parasites of vertebrates.
cyanobacteria Oxygen-producing photosynthetic bacteria.
Gram-positive bacteria Lineage of thick-walled bacteria that are colored purple by Gram staining.
nitrogen fixation Incorporation of nitrogen gas into ammonia.
proteobacteria Most diverse bacterial lineage; includes species that carry out photosynthesis, fix nitrogen. Some cause disease.
spirochetes Lineage of bacteria shaped like a stretched-out spring.

TAKE-HOME MESSAGE 20.7

What ecological roles do bacteria play?

- ✔ Bacteria benefit other organisms by putting oxygen into the air, fixing nitrogen, and acting as decomposers.
- ✔ We use bacteria in scientific research, biotechnology, and food production.
- ✔ Some bacteria are human pathogens.

CREDITS: (16A) Stem Jems/Science Source; (16B) CDC/Dr. E. Arum and Dr. N. Jacobs.

20.8 Archaea

✔ Archaea, the more recently discovered prokaryotic lineage, are found in some very inhospitable places.

Discovery of the Third Domain

The distinctive features of archaea first came to light in the 1970s. Carl Woese was comparing the ribosomal RNAs (rRNAs) among what he thought were bacterial species to find out how they related to one another. He discovered that some species fell into a distinct group. Their rRNA gene sequences positioned them between all other bacteria and the eukaryotes. On the basis of this evidence, Woese proposed the three-domain classification system (Section 1.5).

Woese's ideas were initially greeted with skepticism, but as years went by, evidence in support of them mounted. Archaea differ from bacteria in the composition of their cell wall and plasma membrane. Like eukaryotes, archaea organize their DNA around histone proteins, which bacteria do not have. The first sequencing of an archaeal genome provided the definitive evidence that archaea and bacteria are distinct lineages—most of this archaeon's genes have no counterpart in bacteria.

Woese has compared the discovery of archaea to the discovery of a new continent, which he and others are now exploring.

Here, There, Everywhere

Many archaea thrive in seemingly hostile habitats. The first archaeon to have its genome sequenced was discovered near a hydrothermal vent on the seafloor. It is an **extreme thermophile**, an organism that grows only at a very high temperature. Some archaea that live near hydrothermal vents can grow even at 110°C (230°F). Heat-loving archaea also live in volcanically heated geysers and hot springs (**FIGURE 20.17A**).

Other archaea are among the **extreme halophiles**, organisms that live in highly salty water. Salt-loving archaea live in the Dead Sea, the Great Salt Lake, and many smaller brine-filled lakes (**FIGURE 20.17B**). One extreme halophile, *Halobacterium*, has gas-filled vesicles that keep it afloat in well-lit surface waters. Its plasma membrane contains a unique protein, a purple pigment called bacteriorhodopsin. When excited by light, this protein pumps protons (H^+) out of the cell, against their gradient. The H^+ flows back into the cell through ATP synthases, thus driving the formation of ATP. A similar process drives ATP formation during photosynthesis (Section 6.5). However, *Halobacterium* does not use energy stored in ATP energy to fix carbon dioxide as photosynthetic organisms do. It is a photohetero-

A Thermally heated waters. Pigmented archaea color the rocks in waters of this hot spring in Nevada.

B Highly salty waters. Pigmented extreme halophiles such as color the brine in this California lake.

C The gut of many animals. Cows belch frequently to release the methane produced by archaea in their stomach.

FIGURE 20.17 Examples of archaeal habitats.

Evolution of a Disease (revisited)

Infectious diseases that are immediately fatal are rare. This is fortunate and not surprising. Evolutionarily speaking, the pathogens that leave the most descendants win. Think of a person infected with HIV as a factory that makes and distributes the virus (**FIGURE 20.18**). When the host dies, this facility shuts down.

Of course, hosts also evolve in response to disease. A disease with a high mortality rate acts as a selective agent, favoring individuals capable of resisting infection or surviving in spite of it. For example, about 10 percent of people of European ancestry have a mutation that lessens the likelihood of infection by HIV. The mutation is absent in American Indian, east Asian, and African populations.

People with this protective mutation currently enjoy a selective advantage as a result of the AIDS epidemic. However, the protective mutation did not become prevalent in Europeans as a result of AIDS. Studies of ancient remains tell us it has been in the northern European gene pool for thousands of years. Like all mutations, it arose randomly. It probably increased to its current frequency in Europe because it provided protection against one of the many historical epidemics that occurred there. Its current selective advantage is simply a matter of luck.

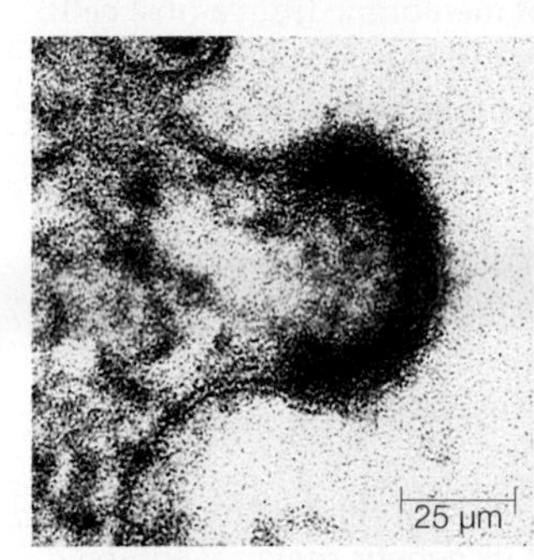

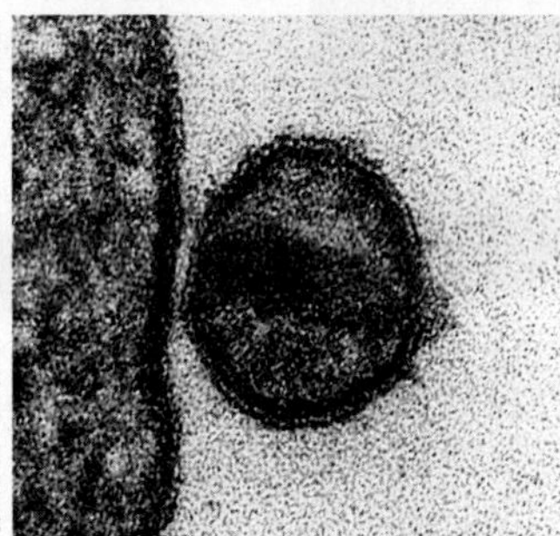

FIGURE 20.18 Micrographs of a new HIV particle budding from an infected white blood cell.

---

troph that obtains carbon by taking up small organic molecules from its environment.

Many archaea, including some extreme thermophiles and extreme halophiles, are **methanogens**, or methane makers. These chemoautotrophs form ATP by pulling electrons from hydrogen gas or acetate. The reactions produce methane (CH_4), an odorless gas. Methanogenic archaea are strict anaerobes, meaning they cannot live in the presence of oxygen. They abound in sewage, marsh sediments, and the animal gut (**FIGURE 20.17C**). Cattle have methanogens in their stomach and release methane gas by belching.

Some people release methane too. About a third of the human population has significant numbers of methanogens in their intestine, so their flatulence (farts) contains methane. Archaea have also been discovered in the human mouth and vagina. Although no archaea are known to cause disease on their own, some types can contribute to ill health. For example, certain methanogens that can live in the human mouth encourage gum disease. By taking up hydrogen, the archaea make the mouth more hospitable to the pathogenic bacteria that cause this disease.

Similarly, some methanogenic archaea live in the human gut, and their abundance may affect our weight. Hydrogen uptake by methanogens improves the environment for bacteria that break down complex carbohydrates. As a result, the more methanogens in your gut, the more calories you can extract from food. Some studies have found a correlation between an abundance of gut methanogens and obesity.

Researchers continue to investigate the distribution and diversity of archaea, and to delve into the evolutionary relationships among them. They are finding that archaea live alongside bacteria nearly everywhere. In deep, dark ocean waters, archaea are the most abundant cells. Two major lineages have been identified. Most of the extreme thermophiles belong to the lineage Crenarchaeota, and most methanogenic and salt-loving archaea belong to the lineage Euryarchaeota.

extreme halophile Organism adapted to life in a highly salty environment.
extreme thermophile Organism adapted to life in a very high-temperature environment.
methanogen Organism that produces methane gas as a metabolic by-product.

TAKE-HOME MESSAGE 20.8

What are archaea?

- ✔ Archaea are single cells without a nucleus that are closer to eukaryotes than to bacteria. Many live in very hot or very salty habitats, but there are archaea nearly everywhere.
- ✔ No archaea are known to be human pathogens, although some may foster the presence of pathogenic bacteria.

CREDIT: (18) Micrographs Z. Salahuddin, National Institutes of Health.

summary

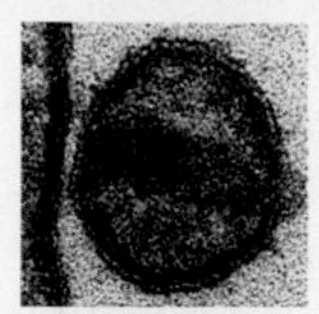

Section 20.1 AIDS is an **emerging disease** caused by **HIV**. The oldest evidence of this virus comes from central Africa, where many chimpanzees are infected by a related virus (SIV). Analysis of viral genes has allowed researchers to trace the spread of HIV.

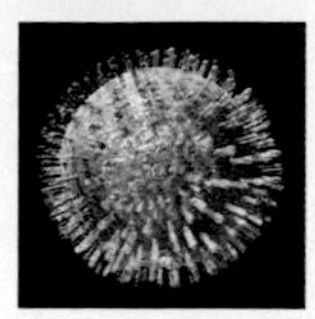

Section 20.2 A **virus** consists of protein, nucleic acid, and, in some cases, a viral envelope (a bit of membrane from a host cell). A virus replicates inside a cell of a specific host type. For example, **bacteriophages** infect bacteria. Some viruses are **pathogens** that cause human disease. Viruses may have evolved before cells, or they may be descended from cells or their components.

Viroids are infectious particles consisting only of a small circle of RNA that does not encode any proteins. They enter plants through a wound and cause disease.

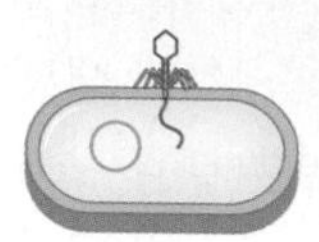

Section 20.3 To replicate, a virus attaches to a host cell and its genetic material enters the cell. Viral genes and enzymes direct host machinery to replicate the viral genome and make viral proteins. These components self-assemble as viral particles, which are then released.

Bacteriophages may multiply by a **lytic pathway**, in which the new viral particles are made fast and released by lysis, or by a **lysogenic pathway**, in which viral DNA becomes part of the host chromosome and is passed on to descendant cells.

HIV is an enveloped **retrovirus**. The viral enzyme reverse transcriptase uses HIV RNA as its template to make DNA that gets inserted into a chromosome of the host cell.

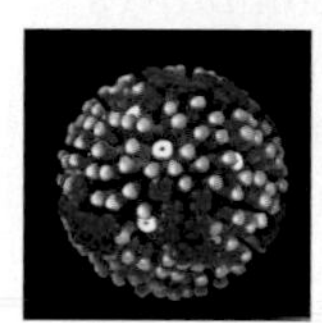

Section 20.4 Viral diseases may be spread by contact with a viral particle or delivered by a **disease vector** such as a tick. Most viral diseases such as common colds cause symptoms only briefly. Some viruses persist in the body and reawaken after a latent period. Viral genes mutate and viruses can swap genes in a host, a process called **viral reassortment**. An **epidemic** is an outbreak of disease in only a limited region. A **pandemic** is a worldwide outbreak. Like AIDS, West Nile virus and Ebola are emerging diseases.

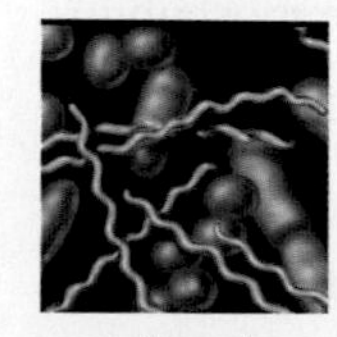

Section 20.5 The organisms known as **prokaryotes** actually constitute two distinct lineages: **bacteria** and **archaea**. Unlike eukaryotes, these lineages do not typically have a nucleus or endomembrane system, and they do not reproduce sexually. Members of both groups typically have cell walls. Many have surface projections such as flagella and **pili**. The chromosome is a circular molecule of DNA that resides in the cytoplasm, in a region called the **nucleoid**. There may also be one or more **plasmids**, circles of DNA that are separate from the chromosome and carry a few genes. Reproduction occurs by **binary fission**, a type of asexual reproduction.

Three types of **horizontal gene transfer** move genes between existing prokaryotic cells. **Conjugation** transfers a plasmid from one cell to another. Virus-assisted transfer of genes is **transduction**. With **transformation**, cells take up DNA from the environment.

Section 20.6 Bacteria and archaea are abundant, and, as a group, metabolically diverse. Some are aerobic and others cannot tolerate oxygen. **Photoautotrophs** carry out photosynthesis. **Photoheterotrophs** capture light, but get carbon from organic molecules rather than CO_2. **Chemoautotrophs** build food from CO_2 using energy from inorganic substances. **Chemoheterotrophs** such as **decomposers** get energy and carbon from organic molecules. Some bacteria and archaea escape adverse conditions by becoming inactive, as when some bacteria form **endospores**.

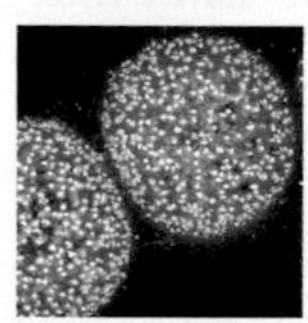

Section 20.7 Bacteria are the more known and diverse lineage of prokaryotic organisms. Many are ecologically important. **Cyanobacteria** produce oxygen as a by-product of photosynthesis. They and other bacteria carry out **nitrogen fixation**, the incorporation of nitrogen from nitrogen gas into ammonia. **Proteobacteria** and **Gram-positive bacteria** are the most diverse bacterial lineages. Both include some human pathogens. Spiral-shaped cells called **spirochetes** and intracellular parasites called **chlamydias** can also cause disease.

Section 20.8 Archaea are the more recently discovered lineage of prokaryotes. Comparisons of structure, function, and genetic sequences position them in a separate domain, between eukaryotes and bacteria. Many live in extreme environments. **Extreme halophiles** live in very salty waters and **extreme thermophiles** live at very high temperatures. Some archaea are photoheterotrophs with a unique purple protein that captures light energy. Most, including the **methanogens**, are chemoautotrophs. Some archaea live in our bodies, but none are considered pathogens.

self-quiz

Answers in Appendix VII

1. DNA or RNA may be the genetic material of ________ .
   a. a bacterium b. a viroid c. a virus d. an archaeon
2. A viroid consists entirely of ________ .
   a. DNA b. RNA c. protein d. lipids
3. Bacteriophages can kill their host quickly by ________ .
   a. binary fission c. a lysogenic pathway
   b. a lytic pathway d. both b and c
4. The genetic material of HIV is ________ .
   a. DNA b. RNA c. protein d. lipids
5. ________ do not reproduce sexually.
   a. Eukaryotes b. Bacteria c. Archea d. b and c
6. One cell transfers a plasmid to another by ________ .
   a. binary fission c. conjugation
   b. transformation d. the lytic pathway

data analysis activities

Adapting to Host Defenses Surface proteins called HLAs allow white blood cells to detect HIV particles and fight an infection. In a recent study, scientists tested whether HIV is adapting to this host defense. They did so by looking at the frequency of a specific mutation (I135X) in HIV. This "escape mutation" helps the virus avoid detection by a version of the HLA protein (HLA-B*51) that is common in some regions of the world, but not in others. **FIGURE 20.19** shows the percentage of HIV-positive people who had HIV with the I135X mutation. Data were collected at medical centers from several parts of the world.

1. What percentage of people with HLA-B*51 in Vancouver had HIV with the escape mutation for this protein?

2. Overall, are people with HLA-B*51 more or less likely than those with other HLAs to have virus with the mutation?

3. Do these results support the hypothesis that the virus is adapting to host defenses?

4. Japan has a high frequency of HLA-B*51; about half the population has it. How might this explain the high frequency of the I135X mutation in Japanese with other HLAs?

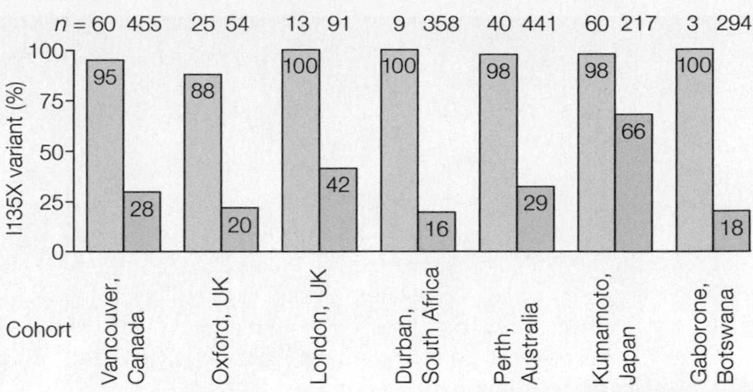
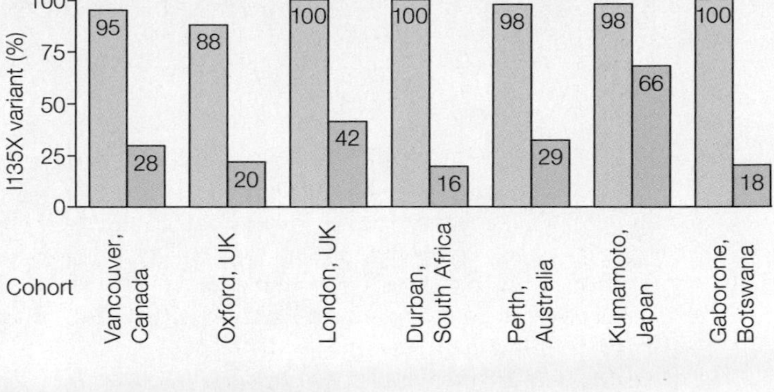

FIGURE 20.19 Regional variation in the frequency of the I135X escape mutation among HIV-positive people. For each region, pink bars represent the percentage of people whose blood cells have HLA-B*51, and thus cannot detect I135X mutants. Blue bars represent the percentage of people with other versions of the HLA protein. These people have blood cells that can detect and fight HIV even if it has the I135X mutation.

7. All ________ are oxygen-releasing photoautotrophs.
   a. spirochetes c. cyanobacteria
   b. chlamydias d. proteobacteria

8. Bacteria that serve as decomposers are ________ .
   a. photoautotrophs c. chemoautotrophs
   b. photoheterotrophs d. chemoheterotrophs

9. Some ________ can cause cancer in humans.
   a. viruses c. viroids
   b. bacteria d. a and b

10. Formation of a(n) ________ allows some soil bacteria to survive adverse conditions.
   a. pilus c. endospore
   b. nucleoid d. plasmid

11. ________ reproduce by binary fission.
   a. Viruses c. Bacteria
   b. Archaea d. both b and c

12. A plasmid is a circle of ________ .
   a. RNA b. DNA c. either RNA or DNA

13. Which of these diseases is caused by bacteria?
   a. flu b. AIDS c. measles d. syphilis

14. A worldwide outbreak of a disease is a(n) ________ .

15. Match the terms with their most suitable description.

___ methanogen	a. infectious RNA
___ nucleoid	b. has RNA genome, protein coat
___ retrovirus	c. draws cells together
___ plasmid	d. releases methane gas
___ extreme halophile	e. region with DNA
	f. circle of nonchromosomal DNA
___ viroid	g. salt lover
___ sex pilus	

critical thinking

1. If a cut or scrape becomes infected, *Staphylococcus aureus* is probably the culprit (right). These bacteria often live on the skin and they can cause a problem if they get into a wound. Most "staph" infections can be cured with the antibiotic methicillin. Unfortunately, methicillin-resistant *S. aureus* (MRSA) is on the rise. Previously, antibiotic-resistant staph infections occurred mainly in hospitals and nursing homes. Now they are breaking out in schools and health clubs. The bacteria are transmitted by contact with an infected person or something that person has touched, as by sharing towels and razors. The gene conferring methicillin resistance is on a plasmid. Explain why a gene that is on a plasmid can spread more quickly than one on the bacterial chromosome.

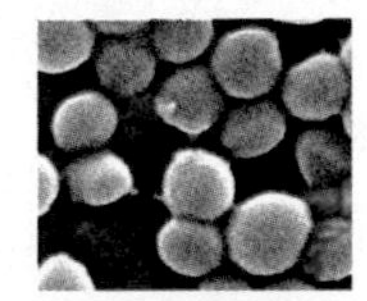

S. aureus

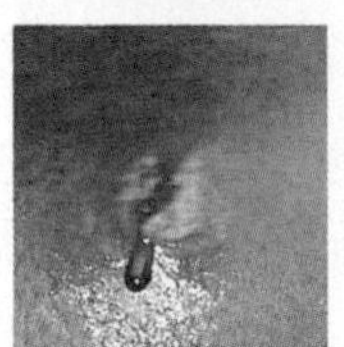

Wound infected by MRSA

2. The adenoviruses that cause colds do not have a lipid envelope and they tend to remain infectious outside the body for longer than enveloped viruses. "Naked" viruses are also less likely to be rendered harmless by soap and water. Why might enveloped viruses be less hardy than naked viruses?

3. The antibiotic penicillin acts by interfering with the production of new bacterial cell walls. Cells treated with penicillin do not die immediately, but they cannot reproduce. Explain how penicillin halts binary fission.

4. Review the description of Fred Griffith's experiments with *Streptococcus pneumoniae* in Section 8.2. Using your knowledge of bacterial biology, explain the process by which the harmless bacteria became dangerous after being exposed to components of harmful bacterial cells.

21 Protists—The Simplest Eukaryotes

LEARNING ROADMAP

This chapter is devoted to the protists, a eukaryotic group introduced in Section 1.4. Section 4.5 described traits of eukaryotic cells and Section 19.6 explored eukaryotic origins. The chapter's discussion of relationships among protist groups will draw on your understanding of cladistics, a topic presented in detail in Section 18.2.

MULTIPLE UNRELATED LINEAGES

Diverse lineages of eukaryotic organisms that are not fungi, plants, or animals are referred to as protists. They include single cells and multicelled organisms, autotrophs and heterotrophs.

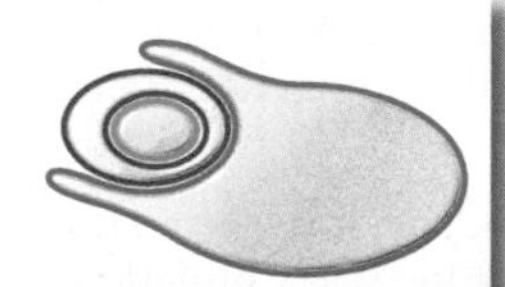

SINGLE-CELLED HETEROTROPHS

Protists that live mainly as single cells include the flagellated protozoans, shelled cells called foraminifera and radiolaria, and the alveolates (ciliates, dinoflagellates, and apicomplexans).

BROWN ALGAE AND RELATIVES

Brown algae are multicelled seaweeds. They are not close relatives of red or green algae, but rather belong to the same lineage as single-celled diatoms and colorless water molds.

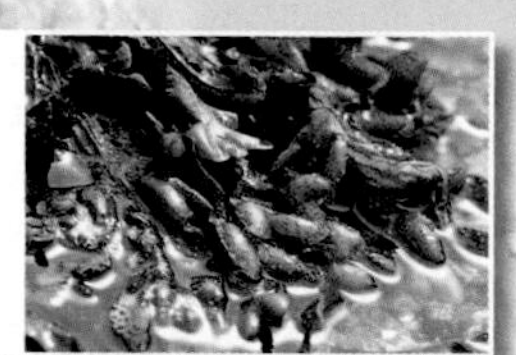

RELATIVES OF LAND PLANTS

Red algae and green algae are photosynthetic groups that both include single cells as well as multicelled seaweeds. One lineage of green algae includes the closest living relatives of land plants.

RELATIVES OF FUNGI AND ANIMALS

Animals and fungi share a common ancestor with the protist group that includes amoebas and slime molds. Choanoflagellates are the sister group to animals.

We explore the connection between green algae and land plants in more detail when we discuss plant evolution (Sections 22.2 and 22.3). Similarly, the choanoflagellates figure in our discussion of animal origins (Section 24.3). You will be reminded of the photosynthetic protists when we consider primary productivity (Section 46.4) and the carbon cycle (Section 46.7), and of flagellated protozoans when we look at sexually transmitted diseases (Section 41.9).

21.1 Malaria: From Tutankhamun to Today

Protists are eukaryotes that are not fungi, plants, or animals. As a group, they are sometimes described as "simple" eukaryotes. As you will see, protists are structurally less complex than plants or animals; however, like these more familiar organisms, they are well adapted to their environment.

The vast majority of protists are free-living. They play important ecological roles as producers or as tiny predators on microorganisms. A few groups include important agents of human disease. Most notably, the single-celled protist *Plasmodium* causes malaria, a disease that kills more than 500,000 people each year.

Plasmodium is transmitted from person to person by mosquitoes. The parasite invades and multiplies inside human liver cells and oxygen-carrying red blood cells. Infection destroys red blood cells (**FIGURE 21.1A**), resulting in fatigue and weakness. When blood cells infected by *Plasmodium* get into the brain, they can cause blindness, seizures, coma, and death.

Like AIDS, malaria has an evolutionary connection to a primate disease. The *Plasmodium* species that causes most cases of human malaria descended from a parasite that infects gorillas in western Africa. Unlike AIDS, malaria has been affecting human populations for millennia. For example, researchers reported finding bits of *Plasmodium* DNA in the 3,300-year-old mummified remains of the Egyptian pharaoh Tutankhamun, also known as "King Tut" (**FIGURE 21.1B**).

As Section 17.7 explained, mortality from malaria has driven up the frequency of the hemoglobin allele associated with sickle-cell anemia (HbS) in some human populations. Having the HbS allele reduces the risk of dying from malaria.

Natural selection also acts on *Plasmodium*. In recent years, the protist has become resistant to several widely used antimalarial drugs. Over a longer time scale, *Plasmodium* has evolved an astonishing capacity to affect the behavior of its hosts. Compared to a mosquito free of *Plasmodium*, an infected mosquito is more likely to feed several times a night, and thus more likely to bite several people. The parasite also alters the odor of infected humans, making them especially appetizing to hungry mosquitoes. By manipulating both its insect and human hosts, the protist maximizes its own reproductive success.

Malaria was common in the southern United States until the 1940s, when crews drained swamps and ponds where mosquitoes breed, and sprayed the insecticide DDT inside millions of homes. Today, nearly all cases of malaria in the United States occur in people who contracted the disease elsewhere.

Malaria remains a threat in many tropical regions. It is endemic (constantly present) in parts of Mexico, Central America, and South America, as well as Asia and in the Pacific Islands. However, the disease takes its greatest toll in Africa. One African child dies of this disease every minute.

protist General term for member of one of the eukaryotic lineages that is not a fungus, animal, or plant.

FIGURE 21.1 An ancient disease. **A** *Plasmodium*, a protist pathogen that causes malaria. The graphic depicts an infected red blood cell rupturing and releasing new *Plasmodium* cells (blue). **B** Some of the oldest evidence of malaria in humans comes from the mummified remains of Tutankhamun that were found in this elaborate tomb.

21.2 A Collection of Lineages

✔ Protists include many lineages of mostly single-celled eukaryotes, some only distantly related to one another.

Classification and Phylogeny

No single trait is unique to the protists, so this collection of organisms is not a clade, or monophyletic group (Section 18.2). Many protist groups are only distantly related to one another. In fact, some are more closely related to plants, fungi, or animals than to other protists. **FIGURE 21.2** shows how the lineages discussed in this book fit into the eukaryotic family tree. Keep in mind that there are many additional protist lineages.

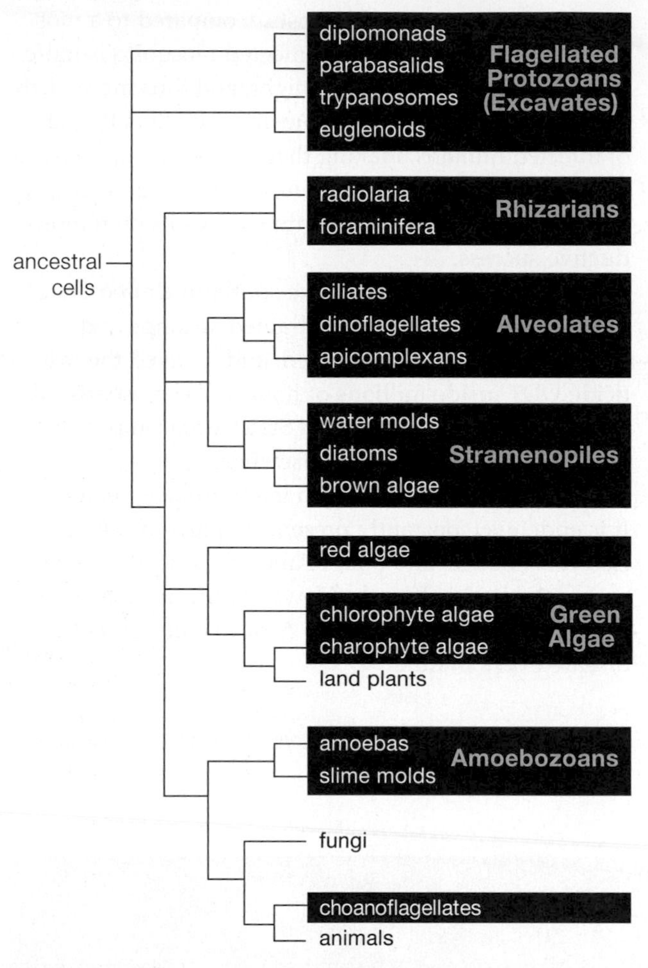

FIGURE 21.2 Proposed phylogenetic tree for eukaryotic groups discussed in this book. Protist groups are indicated by black boxes.

FIGURE IT OUT Are land plants more closely related to the red algae or the brown algae?

Answer: Red algae

Cells, Colonies, and Multicellular Organisms

Most protist lineages include only single-celled species (**TABLE 21.1**). However, colonial protists exist, and multicellularity evolved independently in several lineages. Cells of a **colonial organism** live together and behave in an integrated fashion, but still remain self-sufficient. Each retains the traits required to survive and reproduce on its own. By contrast, the cells of a **multicellular organism** have a division of labor and rely on one another for survival.

Modes of Nutrition

Though typically small, protists have a huge ecological impact. Heterotrophic protists decompose organic material, prey on smaller organisms such as bacteria, or live inside the bodies of larger species. Photosynthetic protists use the same oxygen-producing photosynthetic pathway as cyanobacteria, and their activity produces much of the oxygen we breathe. They also take up carbon dioxide and build organic compounds, thus serving as the base for aquatic food chains.

All protist chloroplasts evolved by endosymbiosis (**FIGURE 21.3**). **Primary endosymbiosis** occurs when a bacterium enters a cell and its descendants evolve into an organelle ❶. Red algae, green algae, and land plants share an ancestor in which chloroplasts evolved by primary endosymbiosis. In these lineages, chloroplasts have two membranes, one from the engulfed bacterium, and one from the vacuole in which it was engulfed. **Secondary endosymbiosis** occurs when a photosynthetic protist taken up by a heterotrophic protist evolves into a chloroplast ❷. In this case, the chloroplast often has more than two membranes. For example, some red algae evolved into chloroplasts in other protists and those chloroplasts have four membranes.

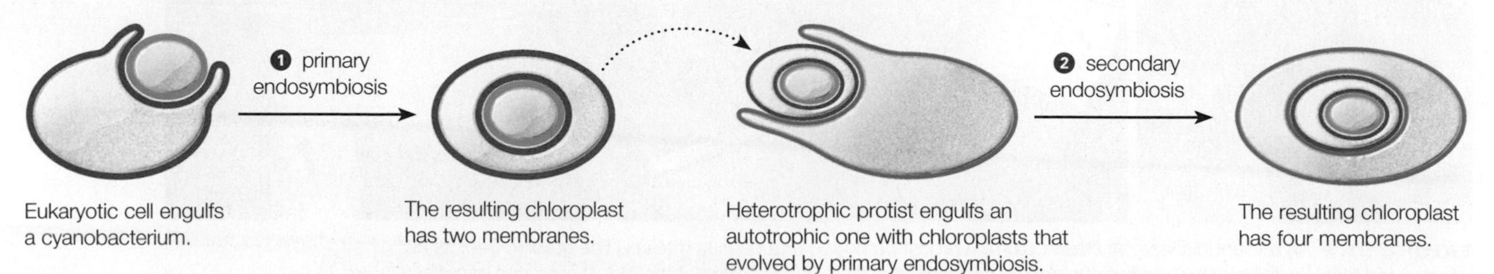

FIGURE 21.3 Evolution of chloroplasts by primary and secondary endosymbiosis.

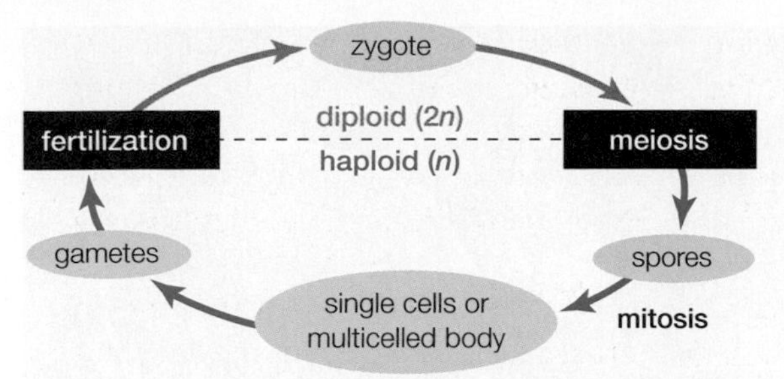

A Haploid-dominant cycle; the zygote is the only diploid cell.

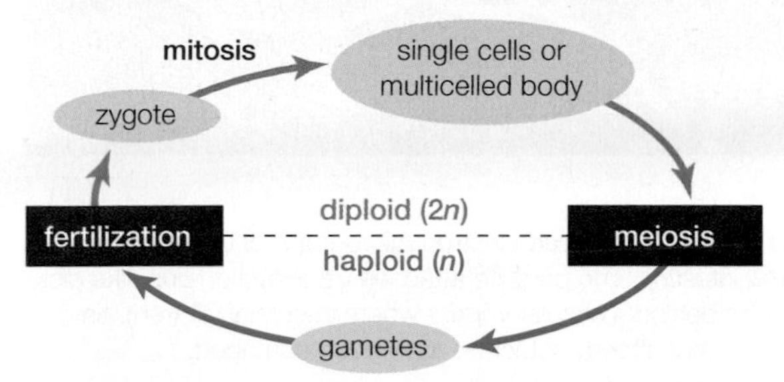

B Diploid-dominant cycle; gametes are the only haploid cells.

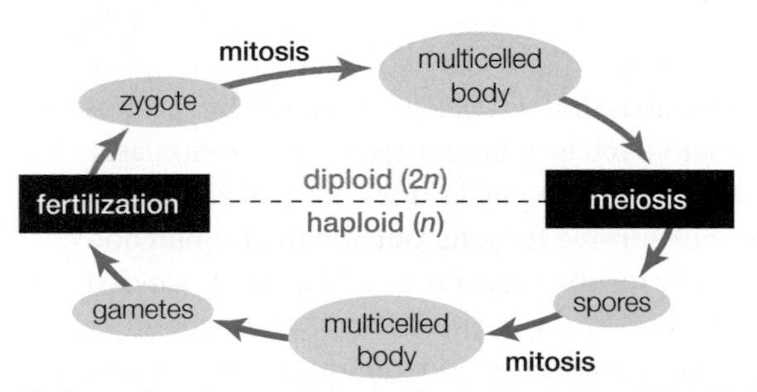

C Alternation of generations; multicelled haploid and diploid stages.

FIGURE 21.4 Three types of life cycle in sexually reproducing eukaryotes. Collectively, protists utilize all three.

Diverse Life Cycles

Most protists reproduce asexually when conditions favor growth, but switch to sexual reproduction when they do not. Some protists have a cycle dominated by haploid forms (**FIGURE 21.4A**). Other protists, like all animals, have a cycle in which diploid forms dominate (**FIGURE 21.4B**). Some multicellular algae (and all plants) have an **alternation of generations**, in which both haploid and diploid bodies form (**FIGURE 21.4C**).

alternation of generations Of land plants and some algae, a life cycle in which both haploid and diploid multicelled bodies form.
colonial organism Organism composed of many integrated cells, each capable of living and reproducing on its own.
multicellular organism Organism composed of interdependent cells that vary in their structure and function.
primary endosymbiosis Evolution of an organelle from bacteria that entered a host cell and lived inside it.
secondary endosymbiosis Evolution of an chloroplast from a protist that itself contains chloroplasts that arose by primary endosymbiosis.

Table 21.1 Characteristics of Some Protist Groups

Protist Group	Organization	Nutritional Mode
Flagellated Protozoans		
Diplomonads	Single cell	Heterotrophs; free-living or parasites of animals
Parabasalids	Single cell	Heterotrophs; free-living or parasites of animals
Kinetoplastids	Single cell	Heterotrophs; mostly parasites, some free-living
Euglenoids	Single cell	Free-living heterotrophs; autotrophs and mixotrophs, most free-living
Radiolaria	Single cell	Free-living heterotrophs
Foraminifera	Single cell	Free-living heterotrophs
Alveolates		
Ciliates	Single cell	Heterotrophs; most free-living, some parasites of animals
Dinoflagellates	Single cell	Autotrophs; mixotrophs; free-living or parasitic heterotrophs
Apicomplexans (sporozoans)	Single cell	Heterotrophs; all parasites of animals
Stramenopiles		
Water molds	Single cell or multicelled	Heterotrophs; free-living or parasites
Diatoms	Single cell	Mostly autotrophs; a few heterotrophs or mixotrophs
Brown algae	Multicelled	Autotrophs
Red Algae	Most multicelled	Autotrophs
Green Algae	Single cell, colonial, or multicelled	Autotrophs
Amoebozoans		
Amoebas	Single cell	Heterotrophs; most free-living, a few parasites of animals
Slime molds	Single-celled and aggregated stages	Heterotrophs
Choanoflagellates	Single cell or colony	Heterotrophs

TAKE-HOME MESSAGE 21.2

What are protists?

✔ Protists do not have a specific defining trait; they are a collection of many eukaryotic lineages rather than a clade.

✔ Most protists are single-celled, but there are multicelled and colonial species. Life cycles vary as does nutrition; some species are autotrophs, others are heterotrophs.

21.3 Flagellated Protozoans

✔ Flagellated protozoans are single-celled species that swim in lakes, seas, and the body fluids of animals.

Single-celled heterotrophic protists are commonly referred to as protozoans. **Flagellated protozoans** are single, unwalled cells with flagella. All are entirely or mostly heterotrophic. The life cycle is dominated by haploid cells that reproduce by mitosis. A **pellicle**, consisting of the plasma membrane and a thin layer of elastic proteins just beneath it, protects the cell and allows it to retain a particular shape.

Anaerobic Flagellates

Diplomonads and parabasalids have multiple flagella and are among the few protists adapted to oxygen-poor waters. Instead of typical mitochondria, they have **hydrogenosomes**, organelles that make ATP by an anaerobic pathway and release hydrogen gas (H_2) as a by-product. Hydrogenosomes adapt organisms to anaerobic aquatic habitats. Free-living diplomonads and parabasalids thrive deep in seas and lakes. Others live in animal hosts; they may be harmful, helpful, or have no effect on their host's health.

The diplomonad *Giardia lamblia* infects mammals, attaching to their intestinal lining and sucking up nutrients (**FIGURE 21.5**). In humans, it causes the disease giardiasis. Symptoms of giardiasis can include cramps, nausea, and severe diarrhea that sometimes persist for weeks. Infected people and animals excrete cysts of *G. lamblia* in their feces. A protistan cyst is a dormant cell enclosed within a thick protective wall. Ingesting even a few *G. lamblia* cysts in contaminated water can result in an infection.

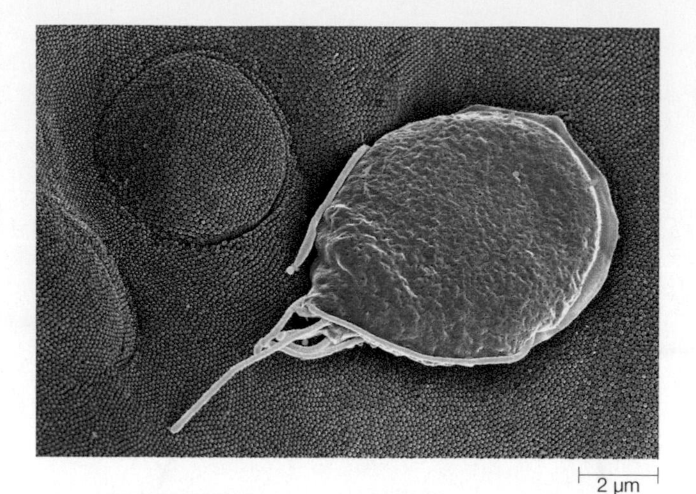

FIGURE 21.5 Colorized electron micrograph of *Giardia lamblia* in the small intestine. The parasite attaches via a suction cup–like disk that leaves behind a circular imprint where intestinal villi (red), tiny projections that absorb nutrients, have been damaged.

The parabasalid *Trichomonas vaginalis* infects human reproductive tracts, where it causes trichomoniasis (**FIGURE 21.6**). *T. vaginalis* does not make cysts, so it does not survive for long outside the human body. Fortunately for the parasite, sexual intercourse delivers it directly into hosts. In the United States, about 6 million people are infected. In women, symptoms include vaginal soreness, itching, and a yellowish discharge. Infected males typically show no symptoms. Untreated infections damage the urinary tract, cause infertility, and increase risk of HIV infection. A dose of an antiprotozoal drug provides a quick cure.

FIGURE 21.6 Colorized electron micrograph of *Trichomonas vaginalis*, the parabasalid that causes trichomoniasis. The parasite can be seen in vaginal secretions examined under a light microscope, and can be identified by its twitching movements.

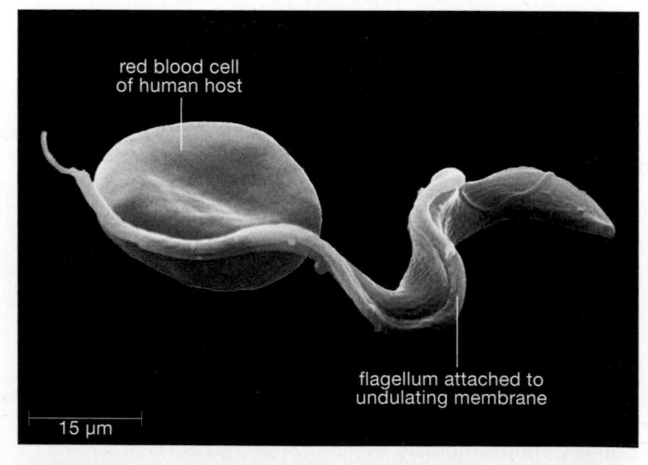

FIGURE 21.7 Colorized electron micrograph of the kinetoplastid *Trypanosoma brucei*. A bite of an infected tsetse fly delivers this parasite into the blood. It lives in the fluid portion of the blood (the plasma) and absorbs nutrients across its body wall.

CREDITS: (5) CDC/Dr. Stan Erlandsen; (6) David M. Phillips/The Population Council/Science Source; (7) Oliver Meckes/Science Source.

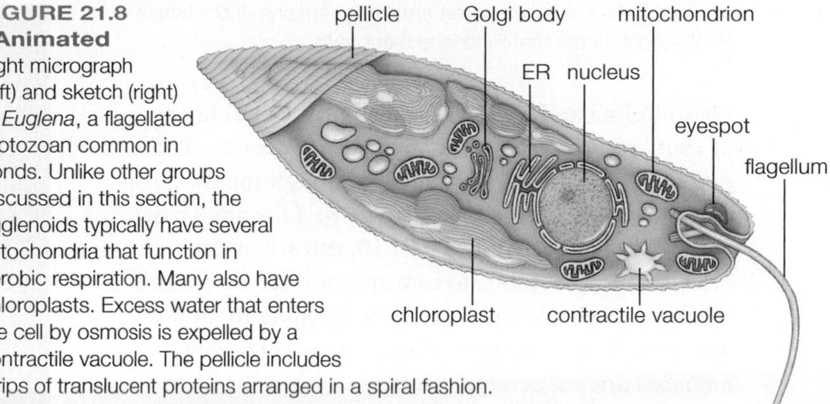

FIGURE 21.8 ▸**Animated** Light micrograph (left) and sketch (right) of *Euglena*, a flagellated protozoan common in ponds. Unlike other groups discussed in this section, the euglenoids typically have several mitochondria that function in aerobic respiration. Many also have chloroplasts. Excess water that enters the cell by osmosis is expelled by a contractile vacuole. The pellicle includes strips of translucent proteins arranged in a spiral fashion.

Trypanosomes and Other Kinetoplastids

Kinetoplastids are flagellated protozoans that have a single large mitochondrion. Inside the mitochondrion, near the base of the flagellum, is a clump of DNA. The clump is the kinetoplast for which the group is named.

Some kinetoplastids prey on bacteria in fresh water and seas, but **trypanosomes**, the largest subgroup, are parasitic; they live in plants and animals. Trypanosomes are long, tapered cells. Their flagellum emerges from the posterior, then runs forward along the cell body within a membrane (**FIGURE 21.7**). The membrane moves in an undulating (wavelike) manner.

Biting insects act as vectors for many trypanosomes that parasitize humans. (Again, a vector is an insect or other animal that carries a pathogen between hosts.)

Tsetse flies spread *Trypanosoma brucei*, which causes African trypanosomiasis, commonly known as African sleeping sickness. Infected people are drowsy during daytime and often cannot sleep at night. If untreated, the infection is fatal. Tsetse flies that spread *T. brucei* occur only in sub-Saharan Africa.

Bloodsucking bugs transmit *Trypanosoma cruzi*, the cause of Chagas disease. Untreated infection can harm the heart and digestive organs. Chagas disease is prevalent in parts of Central and South America, and occurs at low frequency in the southern United States.

Large influxes of immigrants from South and Central America into the United States raised concerns about *T. cruzi* contamination of the blood supply. As a result, blood banks now test donated blood for this parasite.

Euglenoids

Euglenoids are flagellated protists closely related to kinetoplastids. Most live in fresh water and none are human pathogens. Many prey on bacteria or on other protists, a few are parasites of larger organisms, and about a third have chloroplasts (**FIGURE 21.8**). The structure of euglenoid chloroplasts and the pigments inside them indicate that these organelles evolved from a green alga by secondary endosymbiosis.

Photosynthetic euglenoids detect light with a light-sensitive organelle (an eyespot) near the base of their long flagellum. They typically revert to heterotrophic nutrition if light levels decline, or conditions for photosynthesis become otherwise unfavorable.

A euglenoid is hypertonic (Section 5.8) relative to fresh water. As in other freshwater protists, one or more **contractile vacuoles** counter the tendency of water to diffuse into the cell. Excess water collects in contractile vacuoles, which contract and expel it to the outside.

contractile vacuole In freshwater protists, an organelle that collects and expels excess water.
euglenoid Flagellated protozoan with multiple mitochondria; may be heterotrophic or have chloroplasts descended from a green alga.
flagellated protozoan Protist belonging to an entirely or mostly heterotrophic lineage with no cell wall and one or more flagella.
hydrogenosome Organelle that produces ATP and hydrogen gas by an anaerobic pathway; evolved from mitochondria.
pellicle Outer layer of plasma membrane and elastic proteins that protects and gives shape to many unwalled, single-celled protists.
trypanosome Parasitic flagellate with a single mitochondrion and a membrane-encased flagellum.

TAKE-HOME MESSAGE 21.3

What are flagellated protozoans?

✔ Flagellated protozoans are single-celled protists with one or more flagella. They are typically heterotrophic and reproduce asexually by mitosis.

✔ Diplomonads and parabasalids have adapted to life in oxygen-poor waters. Some members of these groups commonly infect humans and cause disease.

✔ Trypanosomes also include human pathogens that are transmitted by insects. Their relatives, the euglenoids, do not infect humans. Most prey on bacteria, and some have chloroplasts that evolved from green algae.

CREDITS: (8) left, © Kage Mikrofotografie/Phototake; right, © Cengage Learning.

21.4 Foraminifera and Radiolarians

✔ Limestone, chalk, and chert are rocky remains of countless single shelled cells that lived in ancient seas.

Foraminifera and radiolarians are two related lineages of primarily heterotrophic single-celled protists. Both groups secrete a sieve-like shell, although the composition of the shell differs between them. Like amoebas, which we discuss in Section 21.10, rhizarians use cytoplasmic extensions to capture prey. However, this mechanism of feeding evolved independently in the two groups; genetic comparisons show rhizarians and amoebas are not close relatives.

Foraminifera secrete a calcium carbonate ($CaCO_3$) shell. Including its shell, an individual foraminiferal cell can be as big as a grain of sand. Most species live on the seafloor, where they probe the water and sediments for prey. Others are members of the marine plankton (**FIGURE 21.9**). **Plankton** is a community of mostly microscopic organisms that drift or swim in open waters. Planktonic foraminifera often have smaller photosynthetic protists such as diatoms or algae living on or in them.

By taking up carbon dioxide (CO_2) and incorporating it into their calcium carbonate shells, foraminifera help the sea to absorb more carbon dioxide from the air. Lowering the concentration of CO_2 in the air is important because excess atmospheric CO_2 is one cause of global climate change.

Radiolaria secrete a glassy silica (SiO_2) shell (**FIGURE 21.10**) and have two cytoplasmic layers. The inner layer holds all the typical eukaryotic organelles and the outer one contains numerous gas-filled vacuoles that keep the cell afloat. Radiolaria are planktonic and they are most abundant in nutrient-rich tropical waters.

Foraminifera and radiolaria have lived and died in the oceans for more than 500 million years, so remains of countless cells have fallen to the seafloor. Over time, geologic processes transformed some accumulations of foraminiferal shells into chalk and limestone, two types of calcium carbonate–rich sedimentary rock. The giant blocks of limestone used to build the great pyramids of Egypt consist largely of the shells of foraminifera that fell to the seafloor about 50 million years ago. Similarly, geologic processes transformed silica-rich deposits of radiolarian shells into a rock called chert.

FIGURE 21.9 A planktonic foraminiferan with algal cells (yellow dots) living on its cytoplasmic extensions.

FIGURE 21.10 A planktonic radiolarian.

TAKE-HOME MESSAGE 21.4

What are foraminifera and radiolaria?

✔ Foraminifera and radiolaria are related lineages of heterotrophic, single cells that live mainly in seawater.

✔ Foraminifera make a shell of calcium carbonate and most live on the seafloor.

✔ Radiolaria have a glassy silica shell; most are planktonic.

✔ Deposits of foraminiferal shells that fell to the seafloor have become limestone or chalk. Similarly, silica-rich remains of radiolaria have become chert.

foraminifera Heterotrophic single-celled protists with a porous calcium carbonate shell and long cytoplasmic extensions.
plankton Community of tiny drifting or swimming organisms.
radiolaria Heterotrophic single-celled protists with a porous silica shell and long cytoplasmic extensions.

CREDITS: (9) Courtesy of Allen W. H. Bé and David A. Caron; (10) © Franz Neidl.

21.5 Ciliates

✔ Ciliated cells hunt bacteria, other protists, and one another in freshwater habitats and the oceans.

Three groups of single-celled protists—ciliates, dinoflagellates, and apicomplexans—are **alveolates**, members of a lineage that have sacs of unknown function under their plasma membrane. Alveolus means "sac."

Ciliates, or ciliated protozoans, are highly diverse heterotrophs, with about 8,000 species. They occur just about anywhere there is water, and most are predators (**FIGURE 21.11**) that engulf other microorganisms.

Paramecium is a freshwater ciliate (**FIGURE 21.12**). Cilia that cover its entire surface function in feeding and locomotion. They sweep water laden with bacteria, algae, and other food particles into an oral groove at the cell surface, and then to the gullet. Enzyme-filled vesicles digest food in the gullet. Like other ciliates, *Paramecium* has organelles called trichocysts beneath its pellicle (**FIGURE 21.12B**). A trichocyst holds a protein thread that can be expelled to help the cell capture prey or fend off a predator.

A few types of ciliates have adapted to life in the animal gut. Some help cattle, sheep, and related grazers break down the cellulose in plant material. Others help termites digest wood. Only one ciliate, *Balantidium coli*, is a known human pathogen. It can live in the gut of both humans and pigs. Human infections arise when dormant cysts of the ciliate that were shed in feces contaminate food or water.

All ciliates reproduce asexually, and some types also reproduce sexually. A ciliate has a macronucleus that controls daily function, and one or more small micronuclei. During sexual reproduction, two cells get together and their micronuclei undergo meiosis. The cells swap haploid micronuclei, then use their new combination of micronuclei to form a macronucleus with a mixed genome. After the cells separate, each divides and passes a macronucleus with a mix of genes to its descendant cells.

alveolate Member of a protist lineage having small sacs beneath the plasma membrane; dinoflagellate, ciliate, or apicomplexan.
ciliate Single-celled, heterotrophic protist with many cilia.

FIGURE 21.11 *Didinium*, a barrel-shaped ciliate with tufts of cilia, catching (left) and engulfing (right) *Paramecium*, another ciliate.

TAKE-HOME MESSAGE 21.5

What are ciliates?

✔ Ciliates are heterotrophic single cells that move about and feed with the help of cilia. Most are free-living predators, but some live inside animals.

✔ Ciliates, along with dinoflagellates and apicomplexans, are alveolates, a group characterized by tiny sacs (alveoli) beneath the plasma membrane.

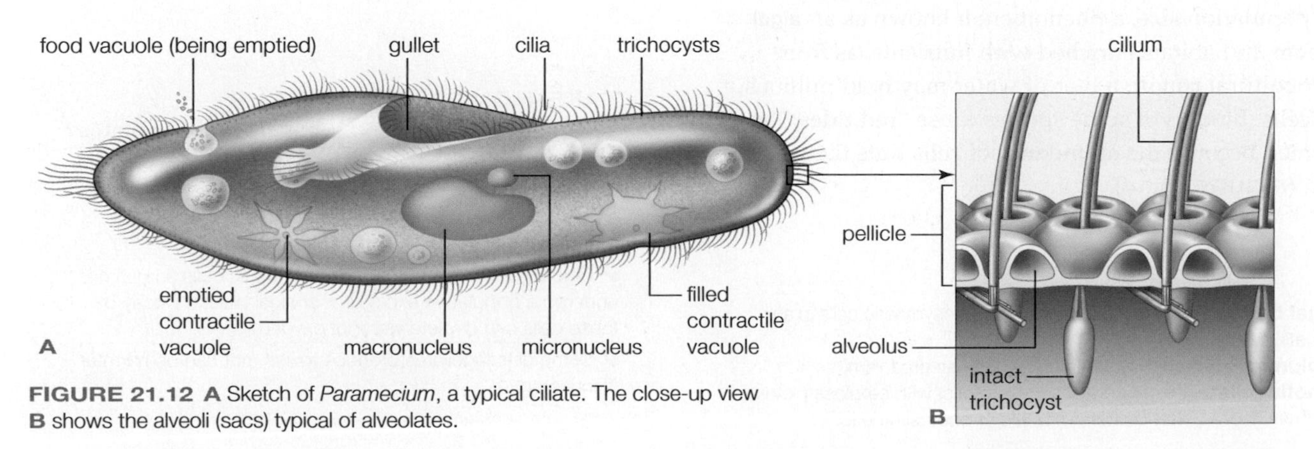

FIGURE 21.12 A Sketch of *Paramecium*, a typical ciliate. The close-up view **B** shows the alveoli (sacs) typical of alveolates.

CREDITS: (11) Gary W. Grimes and Steven L'Hernault; (12A, B) © Cengage Learning.

21.6 Dinoflagellates

✔ Dinoflagellates are single-celled heterotrophs and autotrophs. Most whirl about in the seas. Some live inside corals.

The name **dinoflagellate** means "whirling flagellate." These single-celled protists typically have two flagella; one extends out from the base of the cell, and the other wraps around the cell's middle like a belt (**FIGURE 21.13**). Combined action of these two flagella causes the cell to rotate as it moves forward. Dinoflagellates are alveolates, and most deposit cellulose in the alveoli beneath their plasma membrane. The cellulose accumulates as thick but porous plates.

Most dinoflagellates are photosynthetic and have chloroplasts descended from red algae. A few photosynthetic dinoflagellate species are themselves endosymbionts inside animals, such as reef-building corals. A coral is an invertebrate animal, and its relationship with its dinoflagellates is mutually beneficial. The protists supply the coral with oxygen needed for aerobic respiration and with food (photosynthetically produced sugar). The coral provides the protists with nutrients, shelter, and the carbon dioxide necessary for photosynthesis. Without its dinoflagellate partners, a reef-building coral will die.

The vast majority of dinoflagellates are part of the marine plankton, and they are especially abundant in warm climates. In tropical seas, dinoflagellates are the most common source of **bioluminescence**, light produced by a living organism. When disturbed, dinoflagellates produce a blue or blue-green glow by an oxidation–reduction reaction (Section 5.5). The light may protect the protist by startling a predator that was about to eat it. By another hypothesis, the flash of light acts like a car alarm. It attracts the attention of other organisms, including predators that pursue the would-be dinoflagellate eaters.

Free-living photosynthetic dinoflagellates or other photosynthetic cells sometimes undergo great increases in population size, a phenomenon known as an **algal bloom**. In habitats enriched with nutrients, as from agricultural runoff, a liter of water may hold millions of cells. Blooms of some species cause "red tides," so named because the abundance of cells tints the water red (**FIGURE 21.14**).

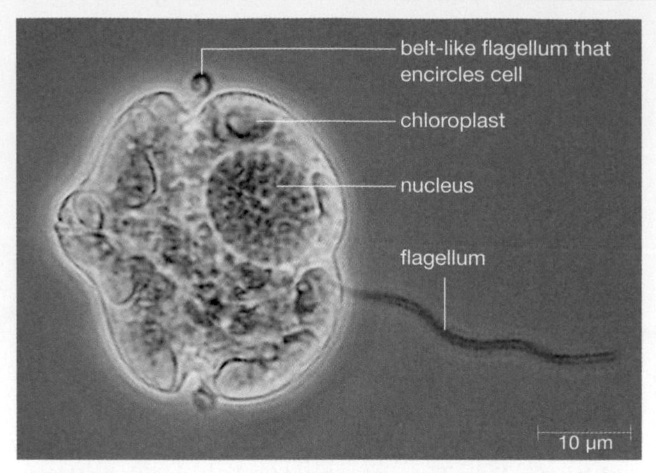

FIGURE 21.13 A photosynthetic dinoflagellate (*Karenia brevis*).

FIGURE 21.14 Red tide. A population explosion of dinoflagellates tints waters off Florida's Gulf Coast red.

Algal blooms can pose a threat to other organisms. Aerobic bacteria that feed on algal remains can use up the oxygen in the water, so that aquatic animals suffocate. Some dinoflagellates also produce toxins. For example, *Karenia brevis* produces a toxin that binds to transport proteins in the plasma membrane of nerve cells. Eat shellfish contaminated with this toxin and you might end up dizzy and nauseated from neurotoxic shellfish poisoning. Symptoms usually develop hours after the meal and persist for a few days.

TAKE-HOME MESSAGE 21.6
What are dinoflagellates?

✔ Dinoflagellates are a group of mostly marine single-celled alveolate protists. Some are heterotrophs. Others are photosynthetic members of plankton or symbionts in reef-building corals.

✔ Dinoflagellates and other photosynthetic cells sometimes undergo a population explosion, or algal bloom. Decay of these cells can deplete water of oxygen.

✔ Some dinoflagellates produce toxins that can be harmful if ingested.

algal bloom Population explosion of photosynthetic cells in an aquatic habitat.
bioluminescence Production of light by an organism.
dinoflagellate Single-celled, aquatic protist with cellulose plates and two flagella; may be heterotrophic or photosynthetic.

21.7 Apicomplexans

✔ As noted in Section 21.1, malaria is a major cause of human death. Here we take a closer look at the protist that causes malaria and at another pathogenic relative.

Apicomplexans are parasitic alveolates that spend part of their life cycle inside cells of their hosts, which include both vertebrate and invertebrate animals. Their name refers to a complex of microtubules at their apical (top) end that they use to enter a host cell.

Malaria is the most studied apicomplexan disease. FIGURE 21.15 shows the life cycle of *Plasmodium*, the agent of malaria. A female *Anopheles* mosquito transmits a motile infective cell (a sporozoite) to a vertebrate such as a human ❶. The sporozoite travels to liver cells, where it reproduces asexually ❷. Some of the resulting cells (merozoites) enter red blood cells and liver cells, where they divide asexually ❸. Others develop into immature gametes, or gametocytes ❹. A mosquito that sucks blood from an infected person takes up gametocytes that later mature as gametes in its gut ❺. When two gametes fuse, the resulting zygote develops into a new sporozoite ❻. Sporozoites migrate to the insect's salivary glands, where they await transfer to a new vertebrate host.

Fever and weakness usually start a week or two after a bite, when infected liver cells rupture, releasing merozoites and cellular debris into the bloodstream. After the initial episode, symptoms may subside. However, continued infection harms organs throughout the body and—if the infection remains untreated—eventually kills the host.

Another apicomplexan disease, toxoplasmosis, is caused by *Toxoplasma gondii*. Many infected people have no apparent symptoms. However, an infection can be fatal in immune-suppressed people, such as those with AIDS. A *T. gondii* infection during pregnancy can cause neurological birth defects in offspring.

T. gondii infects cattle, sheep, pigs, and poultry, and most people become infected by eating cysts in undercooked meat. Cats that prey on rodents and birds are another source of infection. Keeping a cat indoors and feeding it only commercially prepared food minimizes risk that it will become infected. Pregnant women and people with a weakened immune system should avoid contact with cat feces, which can contain infective *T. gondii* cysts.

apicomplexan Single-celled alveolate protist that lives as a parasite inside animal cells; some cause malaria or toxoplasmosis.

TAKE-HOME MESSAGE 21.7

What diseases are caused by apicomplexans?

✔ Apicomplexans cause the deadly disease malaria, as well as toxoplasmosis.

✔ Mosquitoes transmit malaria. People get toxoplasmosis when they ingest cysts (resting cells) of the parasite.

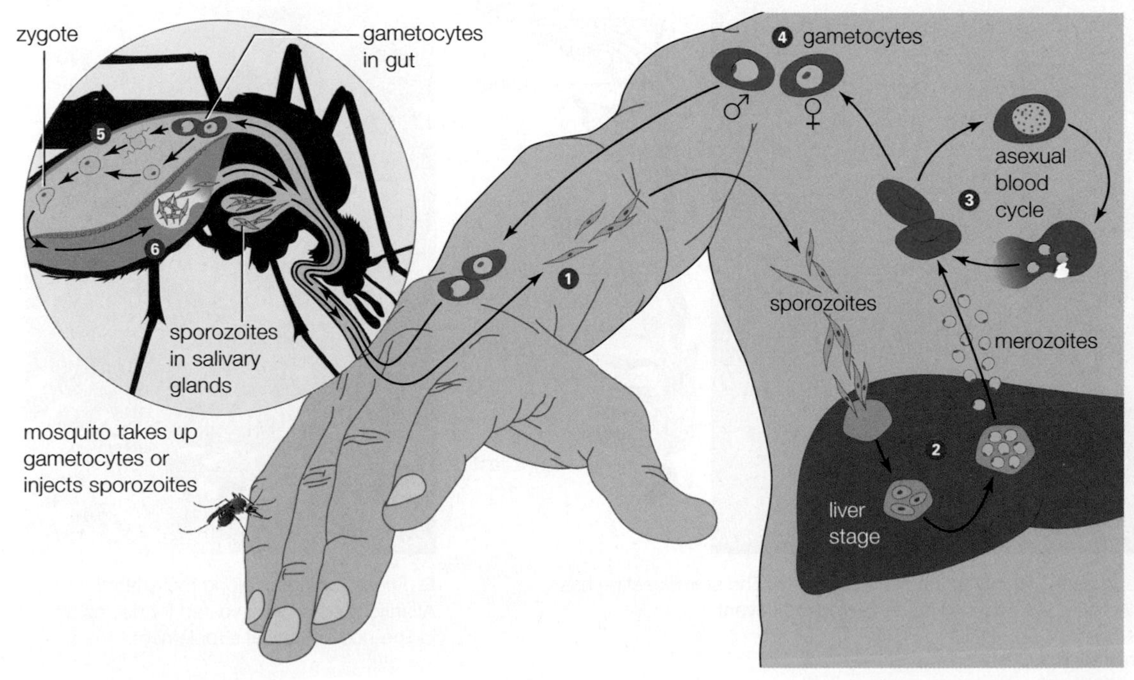

FIGURE 21.15 ▶**Animated** Life cycle of a *Plasmodium* species that causes malaria.

❶ A mosquito bites a human. Haploid sporozoites enter blood, which carries them to the liver.

❷ Sporozoites reproduce asexually in liver cells, then mature into merozoites. Merozoites leave the liver and infect red blood cells.

❸ Merozoites reproduce asexually in some red blood cells.

❹ In other red blood cells, merozoites become gametocytes.

❺ A mosquito bites and sucks blood from the infected person. Gametocytes in blood enter its gut and mature into gametes, which fuse to form diploid zygotes.

❻ Meiosis of zygotes produces cells that develop into sporozoites. The sporozoites migrate to the mosquito's salivary glands.

21.8 Stramenopiles

✔ Colorless filamentous heterotrophs, photosynthetic single cells, and huge seaweeds belong to the stramenopile lineage.

We turn now to another major protist lineage known as the **stramenopiles**. The group name means "straw-haired," and it refers to an anterior flagellum with short hairlike extensions that forms during the life cycle of most members of the lineage.

Most stramenopiles are autotrophs, with chloroplasts that contain a brown accessory pigment called fucoxanthin, along with chlorophylls *a* and *c*. The chloroplasts have four membranes and are thought to have evolved from a red alga by secondary endosymbiosis.

Diatoms

Diatoms are diploid, photosynthetic, single-celled or colonial protists that secrete a protective silica shell (**FIGURE 21.16**). The shell consists of two overlapping parts that fit one inside the other, like a shoe box and its lid. Many diatoms have a cylindrical shape, but others are triangular, square, or needlelike.

Diatoms live in lakes, seas, and damp soils. They are among the most abundant members of the phytoplankton in temperate and polar waters. When marine diatoms die, their shells accumulate on the seafloor. In places where accumulations of diatom shells have been lifted onto land by geological processes, people mine the resulting "diatomaceous earth." This silica-rich material is used in filters and cleaners, and as an insecticide that is nontoxic to vertebrates. Diatomaceous earth kills crawling insects by nicking their outer covering, causing them to dry out and die.

Diatom cells contain a large amount of oil. Oil is less dense than water, and its presence helps these photosynthetic cells stay afloat in sunlit waters. The oil also serves as a store of energy. In some places,

FIGURE 21.16 An assortment of silica diatom shells. Each held a single photosynthetic cell.

A Giant kelp (above and right). A holdfast anchors it in place. The stemlike stipe has leaflike blades. Gas-filled bladders make stipes and blades buoyant.

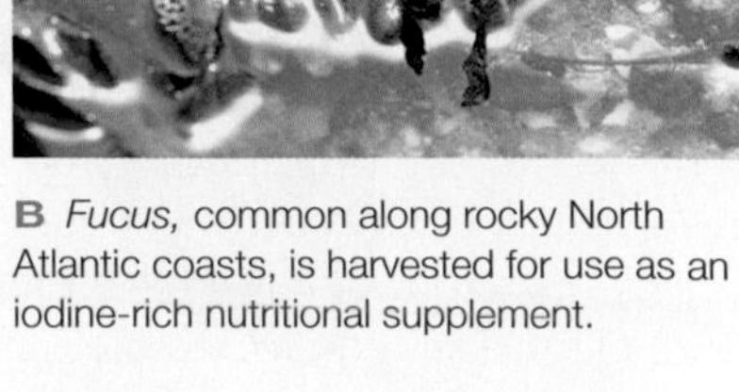

B *Fucus*, common along rocky North Atlantic coasts, is harvested for use as an iodine-rich nutritional supplement.

FIGURE 21.17 Brown algae.

CREDITS: (16) M. I. Walker/Getty Images; (17A) left, © Lewis Trusty/Animals Animals; right, © Cengage Learning; (17B) Philippe Garo/Science Source.

deposits of ancient diatom oil have been transformed into petroleum, which we extract to produce gasoline. Researchers are currently investigating the use of diatoms to produce biofuels.

Brown Algae

Brown algae are multicelled protists that live mainly in cool coastal waters. Although some brown algae appear plantlike, this similarity in form is not evidence of shared ancestry. Brown algae evolved from a different single-celled ancestor than the lineage that includes the red algae, green algae, and land plants.

About 1,500 species of brown algae range in size from microscopic filaments to kelps that stand 30 meters (100 feet) tall. Giant kelps form forestlike stands in coastal waters of the Pacific Northwest (**FIGURE 21.17A**). Like trees in a forest, these kelps shelter a wide variety of other organisms.

An alternation of generations occurs during the kelp life cycle. A large, spore-bearing body forms in the longer, diploid stage of the cycle. Microscopic gamete-forming bodies form during the shorter, haploid stage.

Brown algae have several commercial uses. Some species are harvested for use as food or as nutritional supplements (**FIGURE 21.17B**). Kelp is also the source of alginates; these polysaccharides are used to thicken foods, beverages, cosmetics, and body lotions.

Water Molds

The 500 species of **water molds**, or oomycetes, are heterotrophs that form a mesh of nutrient-absorbing filaments. They were once classified with the fungi, because both groups share a similar growth habit. However, water mold filaments consist of diploid cells that have cell walls made of cellulose. By contrast, fungal filaments are haploid and have cell walls of chitin.

Most water molds decompose organic matter in aquatic habitats, but some are aquatic parasites. For example, *Saprolegnia* often infects fish in aquariums, fish farms, and hatcheries (**FIGURE 21.18**). Other water molds live in damp places on land, or in plants.

Phytophthora means "plant destroyer," and this genus of water molds lives up to its name. In the mid-1800s, one species caused an outbreak of disease that destroyed Ireland's potato crop and resulted in widespread famine. Today, *Phytophthora* species cause an estimated 5 billion dollars in crop losses every year. Forests in California and Oregon are currently under attack by *Phytophthora ramorum*, a previously unknown species of water mold (**FIGURE 21.19**).

FIGURE 21.18 The water mold *Saprolegnia* growing on a fish.

FIGURE 21.19 In California, an oak forest with many trees dead as a result of sudden oak death, a disease caused by the pathogenic water mold *Phytophthora ramorum*. The photo at the right shows the trunk of an infected oak.

brown alga Multicelled marine protist with a brown accessory pigment (fucoxanthin) in its chloroplasts.
diatom Single-celled photosynthetic protist with a brown accessory pigment (fucoxanthin) and a two-part silica shell.
stramenopiles Protist lineage that includes the photosynthetic diatoms and brown algae, as well as the heterotrophic water molds. Some members of the group have a hairy flagellum.
water mold Heterotrophic protist that grows as nutrient-absorbing filaments.

TAKE-HOME MESSAGE 21.8
What are stramenopiles?

✔ Stramenopiles include single-celled diatoms and multicelled brown algae, which are both autotrophs.

✔ Heterotrophic water molds also belong to this group.

CREDITS: (18) Heather Angel; (19) top, © Susan Frankel, USDA-FS; bottom, Dr. Pavel Svihra.

21.9 Red Algae and Green Algae

✔ Red algae and green algae are two related groups of primarily aquatic photosynthetic protists.

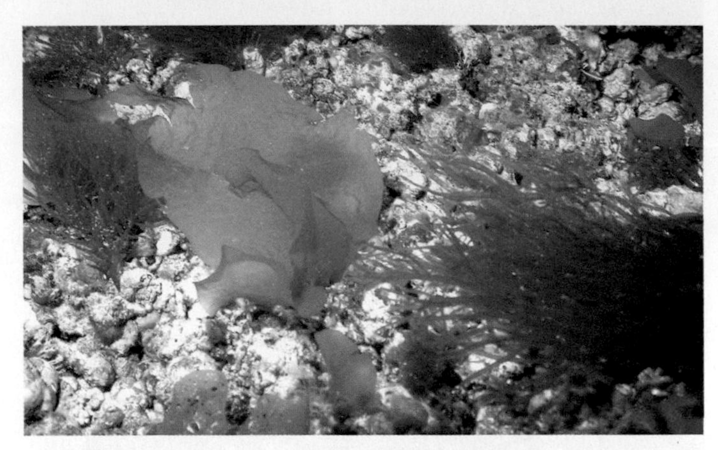

FIGURE 21.20 Branching and sheetlike red algae growing 75 meters (250 ft) beneath the sea surface in the Gulf of Mexico.

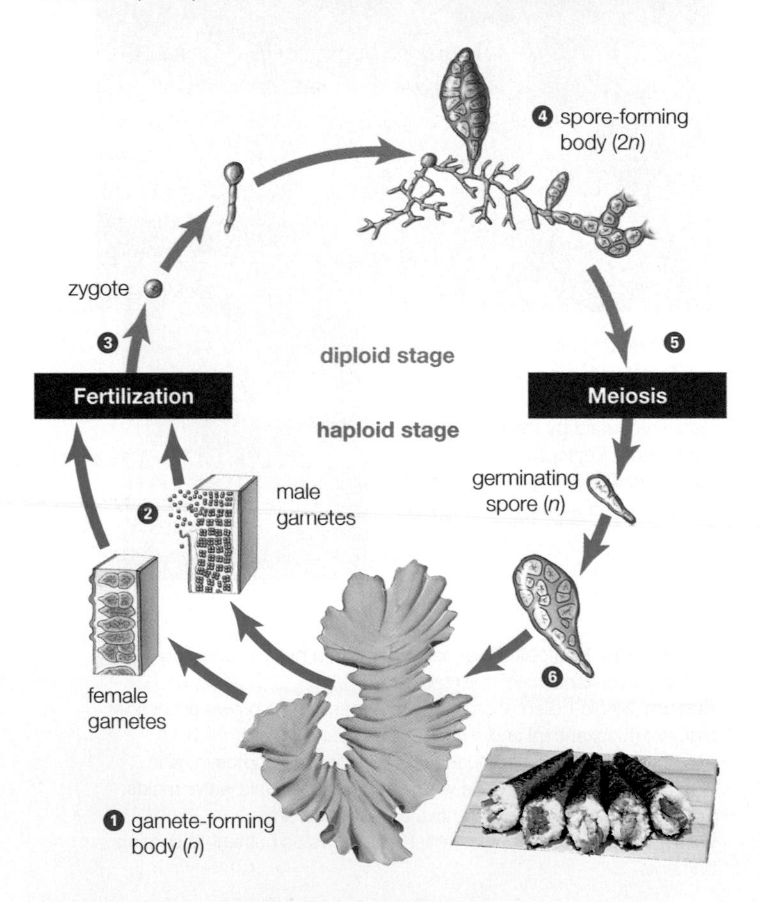

FIGURE 21.21 ▸**Animated** Life cycle of a red alga (*Porphyra*), commonly known as nori.

❶ The haploid gamete-forming body is sheetlike.
❷ Gametes form at its edges.
❸ Fertilization produces a diploid zygote.
❹ The zygote develops into a diploid spore-forming body.
❺ Haploid spores form by meiosis and are released.
❻ Spores germinate and develop into a new gamete-forming body.

Red algae, green algae, and land plants belong to a clade of organisms (Archaeplastida) whose cellulose-walled cells contain chloroplasts with two membranes. Most scientists think all members of this group share a common ancestor in which chloroplasts evolved from cyanobacteria by primary endosymbiosis. We discuss protist members of the archaeplastid clade here, then turn to land plants in the next chapter.

Red Algae

Red algae (clade Rhodophyta) are photosynthetic, typically multicelled protists that live in clear, warm seas. Most of the 6,000 or so species grow as thin sheets or in a branching pattern (**FIGURE 21.20**).

Chloroplasts of red algae contain chlorophyll *a* and red accessory pigments called phycobilins. Phycobilins absorb blue-green light, which penetrates deeper into water than light of other wavelengths. Thus, red algae can thrive at greater depths than other algae.

Agar and carrageenan are two gelatinous polysaccharides extracted from red algae for use as thickeners or stabilizers in foods, cosmetics, and personal care products. Agar is also widely used in microbiology. When mixed with appropriate nutrients, agar can be used to form a semisolid culture medium for growing bacteria or fungi.

Nori, the edible seaweed used to wrap sushi, is a red alga (*Porphyra*). Like many multicelled algae and all plants, *Porphyra* has an alternation of generations (**FIGURE 21.21**). The haploid gamete-forming body is a sheetlike seaweed ❶ that forms gametes by mitosis ❷. When two gametes join at fertilization, they form a diploid zygote ❸ that grows and develops into the diploid spore-forming body ❹. The spore-forming body is a tiny, branching filament that produces haploid spores by meiosis ❺. The spore germinates, then grows and develops into a new gamete-forming body to complete the cycle ❻.

Green Algae

Green algae is the informal name for about 7,000 photosynthetic species ranging in size from microscopic cells to multicelled filamentous or branching forms more than a meter long. Like land plants, and unlike other protists, they have chloroplasts that contain chlorophylls *a* and *b*.

Chlorophyte algae (clade Chlorophyta) are the most diverse lineage of green algae. They include both freshwater and marine species. Some single-celled chlorophytes live in the soil or grow as a film on surfaces, including ice. Others partner with a fungus and form

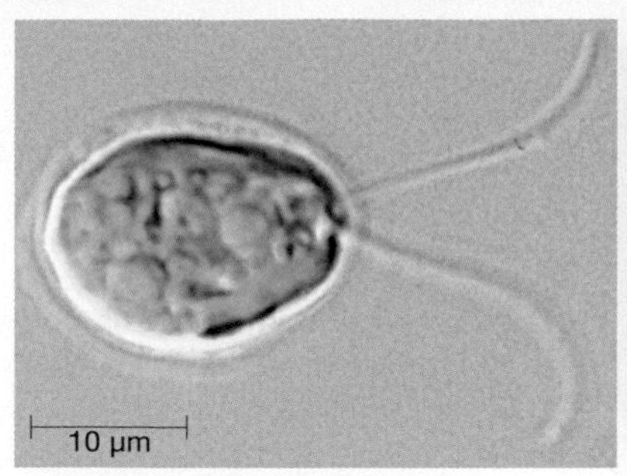

A Single-celled organism. *Chlamydomonas*, a freshwater alga, uses its two flagella to swim. An eyespot allows it to detect light.

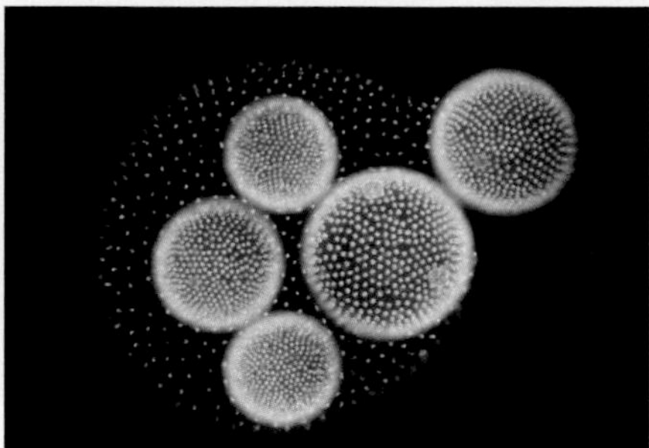

B Colony. *Volvox* colonies consist of flagellated cells connected by thin strands of cytoplasm. New colonies form inside a parent colony and are released.

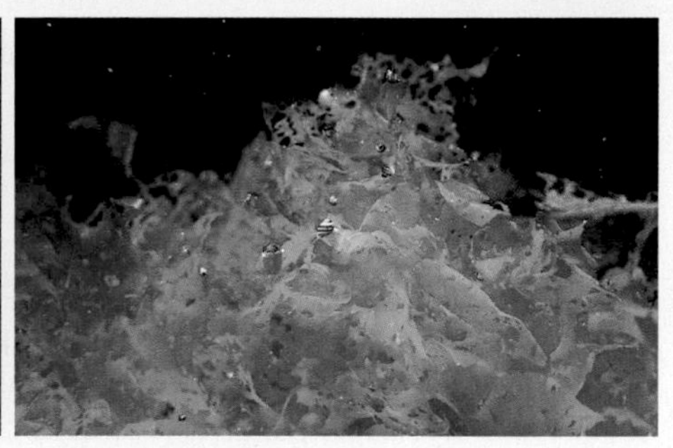

C Multicellular organism. Long, thin sheets of sea lettuce (*Ulva*) are common along coasts where the water movement from waves is minimal.

FIGURE 21.22 Three levels of organization among chlorophytes, the most diverse lineage of green algae.

a composite organism known as a lichen. However, most single-celled chlorophytes live in fresh water. Melvin Calvin used one of them, *Chlorella*, to study the light-independent reactions of photosynthesis (Section 2.2). Today, *Chlorella* is grown as a health food. It is also being used in experimental production of biofuel.

Chlamydomonas is a single-celled green alga common in ponds. Flagellated haploid cells (**FIGURE 21.22A**) reproduce asexually when conditions favor growth. If nutrients become scarce, gametes form by mitosis and fuse to form a diploid zygote with a thick wall. When conditions improve, the zygote undergoes meiosis, yielding four haploid, flagellated cells.

Volvox is a colonial freshwater species. Hundreds to thousands of flagellated cells that resemble those of *Chlamydomonas* are joined together by thin cytoplasmic strands to form a whirling, spherical colony (**FIGURE 21.22B**). Daughter colonies form inside the parental sphere, which eventually ruptures and releases them.

Other freshwater chlorophytes form long filaments. Theodor Engelmann used one of these, *Cladophora*, in his studies of photosynthesis (Section 6.3).

Many "seaweeds" are chlorophytes. Wispy sheets of *Ulva* cling to coastal rocks worldwide (**FIGURE 21.22C**). In some species, sheets grow longer than your arm, but are less than 40 microns thick. *Ulva* is commonly known as sea lettuce and people harvest it for use in salads and soups.

Land plants are thought to be most closely related to the second lineage of green algae, the charophyte algae, with which they share unique structural and functional features (**FIGURE 21.23**). Like plants, and unlike other green algae, charophyte cells divide their cytoplasm by cell plate formation (Section 11.4) and have plasmodesmata (cytoplasmic connections between neighboring cells, Section 4.11).

FIGURE 21.23 The charophyte alga *Coleochaete orbicularis*, which grows as a multicellular disk attached to a surface.

green alga Common term for one of the single-celled, colonial, or multicelled photosynthetic protists that has chloroplasts containing chlorophylls *a* and *b*; a chlorophyte or charophyte alga.

red alga Photosynthetic protist; typically multicelled, with chloroplasts containing red accessory pigments (phycobilins).

TAKE-HOME MESSAGE 21.9

What are red algae and green algae?

- ✔ Red algae and green algae are descendants of a common ancestor in which bacteria evolved into chloroplasts.
- ✔ Red algae are mostly multicellular and marine. They have red accessory pigments that allow them to live at greater depths than other algae.
- ✔ Green algae live in many habitats and may be single-celled, colonial, or multicelled. One subgroup, charophyte algae, includes the closest living relatives of land plants.

21.10 Amoebozoans and Choanoflagellates

✔ Amoebozoans and choanoflagellates are heterotrophic cells with close links to fungi and animals.

Amoebozoans

Amoebozoans do not have a cell wall or a stiff pellicle, so they continually change shape. A compact blob of a cell can extend lobes of cytoplasm called pseudopods (Section 4.10) to move about and to capture food.

Amoebas live as single cells (**FIGURE 21.24**). Most amoebas are predators in freshwater habitats, but a few live inside animals, and some cause human disease. Each year, about 50 million people suffer from amebic dysentery after drinking water contaminated by *Entamoeba histolytica* cysts. Inadequate sterilization of contact lenses or swimming with such lenses on can result in an eye infection by *Acanthamoeba*, which is common in soil, standing water, and even tap water.

Slime molds are sometimes described as "social amoebas." There are two types, cellular slime molds and plasmodial slime molds. **Cellular slime molds** spend the majority of their existence as individual haploid amoeboid (amoeba-like) cells. *Dictyostelium discoideum* is an example (**FIGURE 21.25**). Each cell eats bacteria and reproduces by mitosis ❶. When food runs out, thousands of cells aggregate to form a multicelled mass ❷. Environmental gradients in light and moisture induce the mass to crawl along as a cohesive unit often referred to as a "slug" ❸. When the slug reaches a suitable spot, its component cells differentiate to form a fruiting body. Some cells become a stalk, and others become spores atop it ❹. When a spore germinates, it releases a cell that starts the life cycle anew ❺.

Plasmodial slime molds spend most of their life cycle as a multinucleated mass called a plasmodium. The plasmodium forms when a diploid cell divides its nucleus repeatedly by mitosis, but does not undergo cytoplasmic division. The resulting mass can be as big as a dinner plate. It streams out along the for-

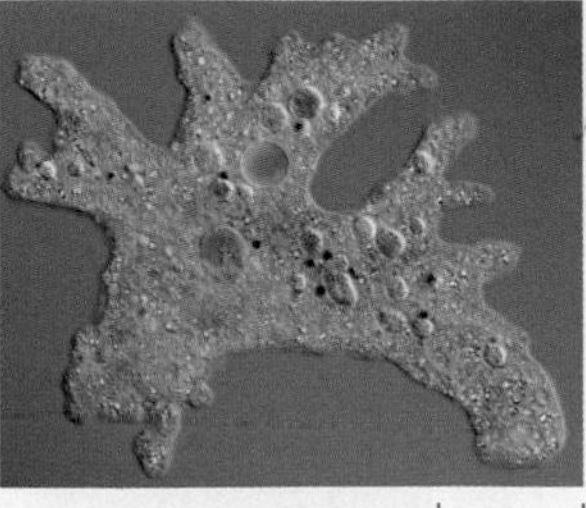

FIGURE 21.24 *Amoeba proteus*, a free-living freshwater amoeba. It feeds or shifts position by extending lobes of cytoplasm (pseudopods). This individual's food vacuoles contain green algae that it engulfed.

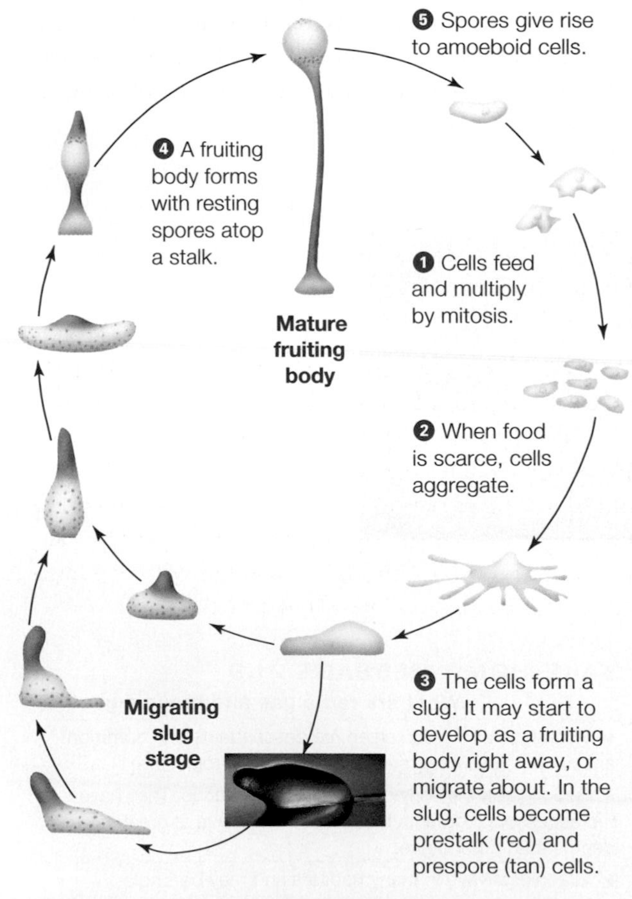

FIGURE 21.25 ▸**Animated** Life cycle of *Dictyostelium discoideum*, a cellular slime mold. During aggregation, the cells secrete and respond to cyclic AMP.

A Multinucleated mass (a plasmodium) streaming across a log.

B Fruiting bodies, with spores.

FIGURE 21.26 Plasmodial slime mold (*Physarum*).

Malaria: From Tutankhamun to Today (revisited)

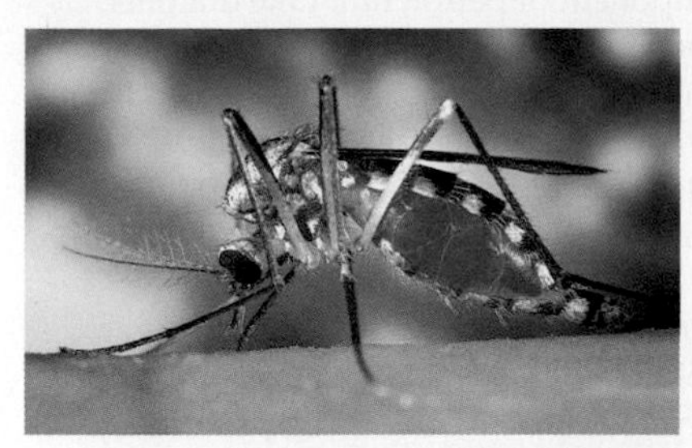

Biting people is a risky business. Each time a mosquito bites, it runs the risk of being detected and swatted. To facilitate speedy feedings and quick getaways, mosquitos inject their host with apyrase, an enzyme that impairs blood clotting. An injection of apyrase before a meal helps ensure that blood will flow smoothly up the insect's proboscis, rather than turning chunky and clogging it.

After *Plasmodium* (the apicomplexan that causes malaria) has replicated in a mosquito's gut, it moves to the insect's salivary glands. Here, it inhibits production of the anticlotting enzyme. A mosquito with infectious *Plasmodium* in its saliva has to bite more often because blood clots as it feeds. This is bad for the mosquito, but good for *Plasmodium*. A mosquito that is forced to eat many small meals, rather than a single large one, is more likely to deliver sporozoites to many new hosts.

Other apicomplexan parasites also alter host behavior to their own benefit. Consider the parasite that causes toxoplasmosis, *Toxoplasma gondii*. It reproduces asexually in rodents, but must enter a cat to complete its life cycle. Parasite-induced changes in the rat's behavior facilitate the parasite's movement from rat to cat. Unlike healthy rats, those infected by *T. gondii* do not avoid cat-scented areas but rather show a preference for such areas. Exactly how the parasite alters the rodent's behavior is not clear, but we do know that *T. gondii* infects the amygdala, the region of the brain that governs fear and anxiety.

---

est floor feeding on microbes and organic matter (**FIGURE 21.26A**). When food supplies dwindle, a plasmodium develops into spore-bearing fruiting bodies (**FIGURE 21.26B**). Later, after the spores disperse, a cell will emerge from each spore.

FIGURE 21.27 Choanoflagellates. The colony at the right is made up of single cells of the type illustrated at the left.

Choanoflagellates

Aquatic, heterotrophic protists called **choanoflagellates** are the closest known protistan relatives of animals. A choanoflagellate cell has a flagellum surrounded by a "collar" of threadlike projections (**FIGURE 21.27**). Movement of the flagellum sets up a current that draws food-laden water through the collar.

Most choanoflagellates live as single cells, but some form colonies (**FIGURE 21.27**). Colonies arise by mitosis, when descendant cells do not separate after division. Instead, cells stick together with the help of adhesion proteins. Choanoflagellate adhesion proteins are similar to those of animals, and even solitary choanoflagellates have these proteins. By one hypothesis, the common ancestor of animals and choanoflagellates was a single-celled protist with adhesion proteins that helped it catch prey. Later, these proteins were put to use helping choanoflagellates stick together to form multicelled colonies. Later still, the proteins allowed animal cells to adhere to one another in multicelled bodies. This is an example of exaptation, an evolutionary process by which a trait that evolves with one function later takes on another (Section 17.12).

---

amoeba Single-celled, unwalled protist that extends pseudopods to move and to capture prey.
amoebozoans Lineage of heterotrophic, unwalled protists that live in soils and water; include amoebas and slime molds.
cellular slime mold Soil-dwelling protist that feeds as solitary cells but congregates under adverse conditions to form a cohesive unit that develops into a fruiting body.
choanoflagellates Heterotrophic protists thought to be the sister group of animals; collared cells strain food from water.
plasmodial slime mold Soil-dwelling protist that feeds as a multinucleated mass. Develops into a fruiting body under adverse conditions.

TAKE-HOME MESSAGE 21.10

What are amoebozoans and choanoflagellates?

✔ Amoebozoans are a recently recognized lineage that includes amoebas and slime molds. All are unwalled heterotrophs that feed and move by extending cytoplasmic extensions (pseudopods).

✔ Choanoflagellates are flagellated, heterotrophic cells that are considered to be close relatives of the animals.

summary

Section 21.1 Protists include many eukaryotic lineages, some only distantly related to one another. *Plasmodium*, the protist that causes malaria, is an example of a parasitic protist with important health effects. Malaria has afflicted humans for thousands of years. The parasite alters traits of its human and mosquito hosts in ways that enhance its reproduction.

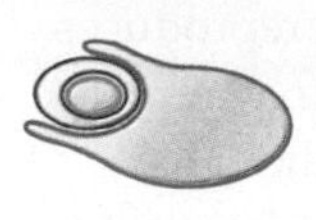

Section 21.2 Most protists live as single cells, but some are **colonial organisms** or **multicellular organisms**. Autotrophic protists have chloroplasts derived from bacteria (**primary endosymbiosis**) or other protists (**secondary endosymbiosis**). Life cycles vary: In some, diploid cells dominate, in others only the zygote is diploid, and in still others there is an **alternation of generations**.

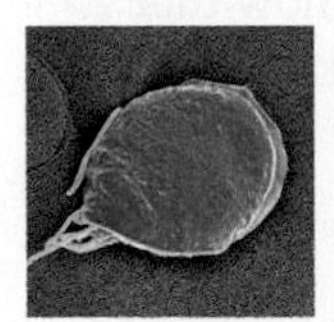

Section 21.3 Flagellated protozoans are single-celled and mostly or entirely heterotrophic. They have no cell wall, but an elastic **pellicle** gives them a distinct shape. Diplomonads and parabasalids are anaerobic and have **hydrogenosomes** instead of mitochondria. Both groups include species that infect humans.

Trypanosomes are parasites with a single giant mitochondrion and an undulating membrane. Some infect humans and cause disease. The related **euglenoids** live mainly in fresh water; a **contractile vacuole** rids them of excess water. Some have chloroplasts derived from a green alga.

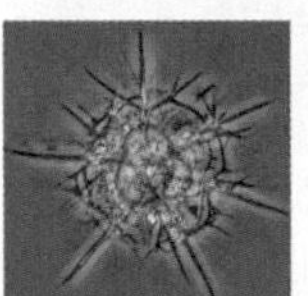

Section 21.4 Foraminifera and **radiolaria** are single-celled heterotrophs with a secreted shell. Foraminifera tend to live on the seafloor, and radiolaria drift as **plankton**. Remains of both accumulate in abundance on the seafloor. Deposits of foraminiferal shells can be transformed over time into limestone or chalk. Similarly, radiolarian shells can become chert.

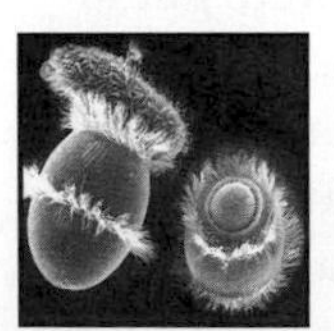

Sections 21.5–21.7 Alveolates are single cells characterized by tiny sacs (alveoli) beneath the plasma membrane. **Ciliates** are aquatic heterotrophs with many cilia. Most are freshwater predators. Some live in the gut of animals. Ciliates reproduce asexually and can also engage in sexual reproduction by exchanging micronuclei. **Dinoflagellates** are cellulose-covered aquatic heterotrophs and autotrophs. Some are capable of **bioluminescence**. In nutrient-enriched water, dinoflagellates and other photosynthetic protists undergo population explosions known as **algal blooms**. **Apicomplexans** such as the organisms that cause malaria and toxoplasmosis are parasites that spend part of their time inside host cells.

Section 21.8 Stramenopiles are named after their flagellum, which has hairlike filaments. **Diatoms** are photosynthetic single cells with a two-part silica shell. Deposits of ancient diatoms are mined as diatomaceous earth. Diatom oils are

a component of petroleum. Like diatoms, **brown algae** such as giant kelps contain the pigment fucoxanthin. Algins extracted from brown algae are used as thickeners. **Water molds** are heterotrophic stramenopiles that grow as filaments. Some are important plant pathogens.

Section 21.9 The red algae, green algae, and plants share a common ancestor in which chloroplasts evolved by primary endosymbiosis.

Red algae are multicelled. They can survive in deeper water than most photoautotrophs because their chloroplasts contain phycobilins. Red algae are the source of carrageenan, agar, and the nori used to wrap sushi.

Green algae are mostly aquatic autotrophs. They include single-celled, colonial, and multicelled groups. Like plants, green algae have chlorophylls *a* and *b*, and they store excess sugars as starch. Charophyte algae, the green algal lineage that includes the closest relatives of plants, have additional structural and functional similarities to the plants.

Section 21.10 Amoebozoans are heterotrophic cells that do not have a cell wall or pellicle. They move and feed by forming pseudopods. **Amoebas** live as solitary cells in aquatic habitats and the animal gut.

There are two lineages of slime molds: **plasmodial slime molds**, which feed as a multinucleated mass, and **cellular slime molds**, which feed as individual amoeba-like cells. When food becomes scarce, cellular slime molds aggregate and move about as a cohesive unit. Both types of slime molds form fruiting bodies that produce resting spores.

Choanoflagellates are the closest protist relatives of animals. Cells have a flagellum surrounded by threadlike projections that filter food from the water. They may live on their own or in colonies of cells attached by adhesion proteins.

self-quiz

Answers in Appendix VII

1. Multicelled haploid and diploid bodies form in protists that have a(n) ________ .
   a. haploid-dominant life cycle
   b. diploid-dominant life cycle
   c. alternation of generations

2. The ________ of diplomonads and parabasalids produce ATP by an anaerobic pathway.
   a. trypanosomes
   b. hydrogenosomes
   c. microvilli
   d. flagella

3. Radiolaria and diatoms have a shell of ________ .
   a. cellulose
   b. silica
   c. calcium carbonate
   d. chitin

4. Which of the following might you find in seawater?
   a. an apicomplexan
   b. a cellular slime mold
   c. a dinoflagellate
   d. a euglenoid
   e. all of the above

data analysis activities

Summoning Mosquitoes Parasites sometimes alter their host's behavior in a way that increases their chances of transmission to another host.

Dr. Jacob Koella and his associates hypothesized that *Plasmodium* might benefit by making its human host more attractive to hungry mosquitoes when gametocytes are available in the host's blood. Gametocytes taken up by the mosquito will mature into gametes and mate inside its gut.

To test their hypothesis, the researchers recorded the response of mosquitoes to the odor of *Plasmodium*-infected children and uninfected children over the course of 12 trials on 12 separate days. **FIGURE 21.28** shows their results.

1. On average, which group of children was most attractive to mosquitoes?

2. Which group of children averaged the fewest mosquitoes attracted?

3. What percentage of the total number of mosquitoes were attracted to the most attractive group?

4. Did the data support Dr. Koella's hypothesis?

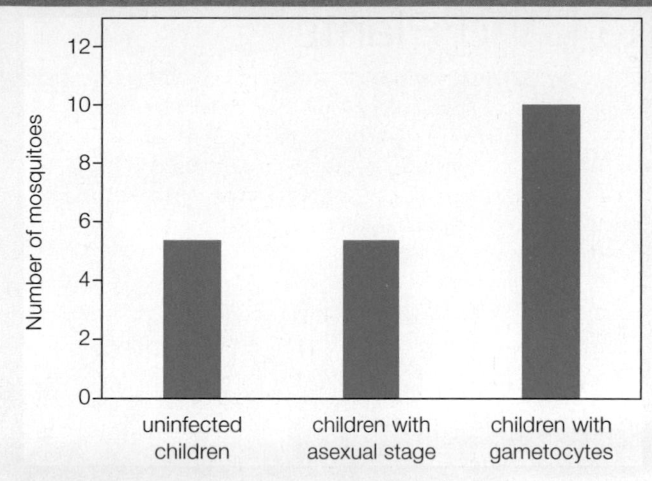

FIGURE 21.28 Number of mosquitoes (out of 100) attracted to uninfected children, children harboring the asexual stage of *Plasmodium*, and children with gametocytes in their blood. The bars show the average number of mosquitoes attracted to that category of child over the course of 12 separate trials.

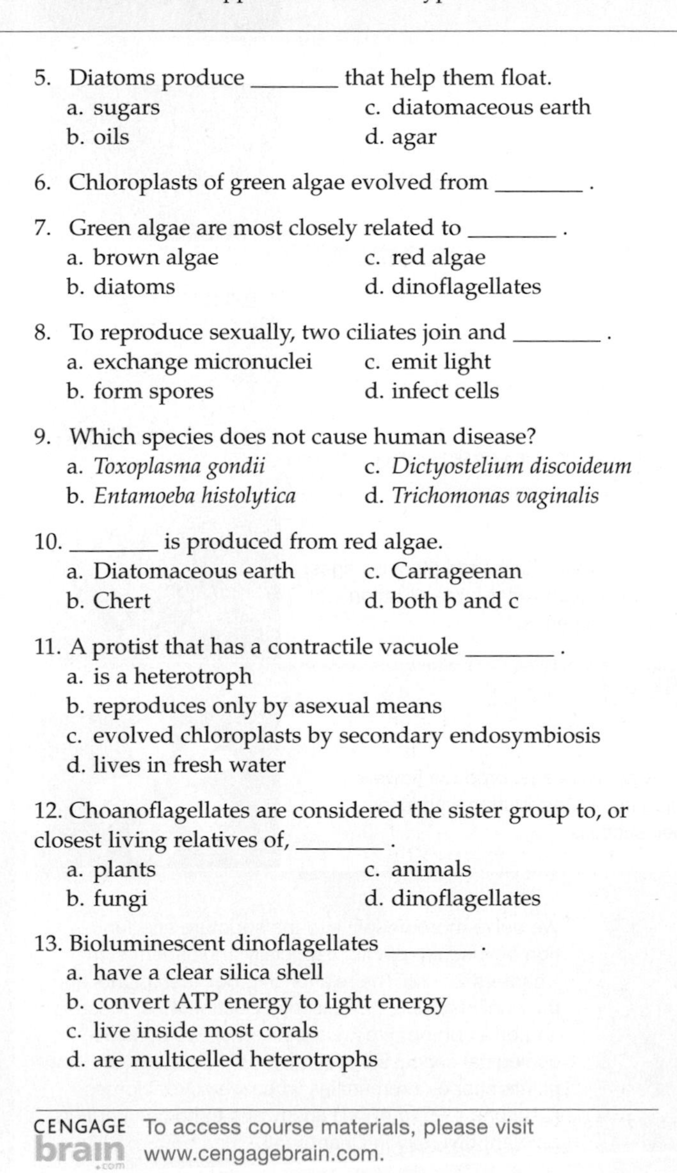

5. Diatoms produce _______ that help them float.
   a. sugars
   b. oils
   c. diatomaceous earth
   d. agar

6. Chloroplasts of green algae evolved from _______ .

7. Green algae are most closely related to _______ .
   a. brown algae
   b. diatoms
   c. red algae
   d. dinoflagellates

8. To reproduce sexually, two ciliates join and _______ .
   a. exchange micronuclei
   b. form spores
   c. emit light
   d. infect cells

9. Which species does not cause human disease?
   a. *Toxoplasma gondii*
   b. *Entamoeba histolytica*
   c. *Dictyostelium discoideum*
   d. *Trichomonas vaginalis*

10. _______ is produced from red algae.
   a. Diatomaceous earth
   b. Chert
   c. Carrageenan
   d. both b and c

11. A protist that has a contractile vacuole _______ .
   a. is a heterotroph
   b. reproduces only by asexual means
   c. evolved chloroplasts by secondary endosymbiosis
   d. lives in fresh water

12. Choanoflagellates are considered the sister group to, or closest living relatives of, _______ .
   a. plants
   b. fungi
   c. animals
   d. dinoflagellates

13. Bioluminescent dinoflagellates _______ .
   a. have a clear silica shell
   b. convert ATP energy to light energy
   c. live inside most corals
   d. are multicelled heterotrophs

14. Match each term with its most suitable description.

___ limestone	a. extracted from kelps
___ chert	b. dried red alga
___ algin	c. remains of radiolaria
___ nori	d. extracted from red algae
___ agar	e. remains of foraminifera

15. Match each term with its most suitable description.

___ diplomonad	a. protist population explosion
___ apicomplexan	b. silica-shelled producer
___ algal bloom	c. cells with "collared" flagellum
___ diatom	d. no mitochondria, anaerobic
___ brown alga	e. closest relative of land plants
___ red alga	f. multicelled, with fucoxanthin
___ green alga	g. one type causes malaria
___ choanoflagellate	h. deep dweller with phycobilins

critical thinking

1. Which groups of protists would you be most likely to find as fossils? Why?

2. Diatoms are the main producers in polar seas. Like other producers, they require dissolved nitrogen and phosphorus to grow. Unlike most producers, their reproduction can be limited by the lack of dissolved silica. Explain why a lack of silica would interfere with diatom reproduction.

3. The parasite *Toxoplasma gondii* is known to alter rodent behavior. This parasite also infects humans, and some researchers suspect that infection may have psychological effects in our species as well. People with the mental illness schizophrenia are more likely than unaffected people to be infected by *T. gondii*. Can you suggest a possible explanation for this finding, other than *T. gondii* increasing a person's risk of becoming schizophrenic?

22 The Land Plants

LEARNING ROADMAP

This chapter continues the story of how plants evolved from green algae (Section 21.9). Like some algae, plants have an alternation of generations (21.2, 21.9). Understanding methods of classification (18.2) will help you comprehend how plants are now grouped.

MILESTONES IN PLANT EVOLUTION

Plants evolved from a green alga. Structural and developmental adaptations allowed plants to colonize land, and to move into increasingly drier habitats.

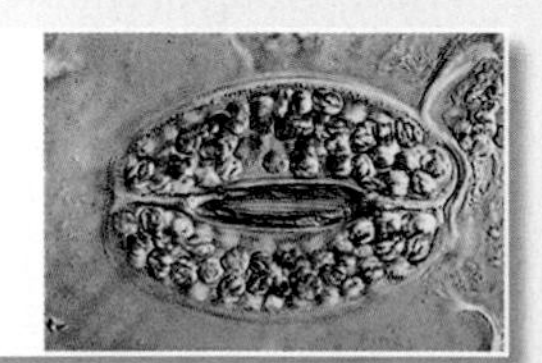

NONVASCULAR PLANTS

In hornworts, liverworts, and mosses, a gamete-producing body dominates the life cycle, and sperm reach the eggs by swimming through droplets or films of water.

SEEDLESS VASCULAR PLANTS

In ferns and related plants, a large spore-producing body with vascular tissues dominates the life cycle. As with bryophytes, sperm swim through water to reach eggs.

GYMNOSPERMS

Gymnosperms are the older of the two existing seed plant lineages. They make pollen and thus do not require water for fertilization. Conifers are the most diverse gymnosperms.

ANGIOSPERMS

Angiosperms, the most diverse plant lineage, produce flowers and disperse their seeds inside fruits. Coevolution with insect pollinators contributed to their success.

We delve more deeply into the structure and function of vascular plants, especially angiosperms, in Chapters 27–29. The relationship between plants and their pollinators is discussed in Section 45.3. Mosses and other bryophytes reappear when we consider ecological succession (Section 45.8). We discuss how plants shape communities when we cover biomes (Sections 47.5 to 47.11) and return to human impacts on plant diversity in Chapter 48.

22.1 Saving Seeds

Plant diversity is declining. We know of threats to about 12,000 wild plant species, and the actual number of threatened plants is no doubt much higher. As a result, many valuable sources of food, medicine, and other products could disappear before we ever learn their value.

A decline in the diversity of cultivated plants raises additional concerns. In the past, food crops were locale specific. People developed and planted crop varieties that did well in their region, and they saved seeds from their crops to plant the following year. Now, many traditional varieties of crop plants are disappearing as farmers worldwide turn to the same few large companies for seeds. Different varieties of plants are resistant to different diseases, so widespread planting of one variety increases the risk that a disease could decimate the global supply of a crop. The more varieties of a crop we plant, the more likely it is that some will resist a particular pathogen. Sustaining the wild relatives of crop plants provides a form of insurance. Such plants represent a reservoir of genetic diversity that plant breeders can draw upon to meet future challenges.

One way to ensure the survival of potentially useful plants is by storing their seeds in seed banks. There are now more than 1,500 seed banks around the world. The most ambitious of these, the Svalbard Global Seed Vault, was built in 2008 on a Norwegian island 700 miles from the North Pole. This so-called doomsday vault serves as a backup for other seed banks (**FIGURE 22.1**). Here, deep inside a mountain, seeds are stored in a permanently chilled, earthquake-free zone 400 feet above sea level. The location was chosen to ensure that the seeds will remain high and dry even if global climate change causes the polar ice caps to melt. The vault also has an advanced security system and has been engineered to withstand any nearby explosions.

The Svalbard vault now holds the world's most diverse collection of seeds, with more than 750,000 samples and material from nearly every nation. The United States has stored seeds of its native crops such as chile peppers from New Mexico, as well as seeds that American scientists collected elsewhere in the world. Under the deep-freeze conditions in Svalbard, the seeds are expected to survive about 100 years.

FIGURE 22.1 The Svalbard Global Seed Vault. To learn more about the vault and the seeds stored inside it, visit www.croptrust.org.

CREDITS: (opposite) Dr. Annkatrin Rose, Appalachian State University; (1) Jim Richardson/National Geographic Creative.

22.2 Plant Ancestry and Diversity

✔ Land plants evolved from a green alga. An adaptive radiation that began 500 million years ago produced diverse groups.

From Algal Ancestors to Embryophytes

Plants are multicelled, typically photosynthetic eukaryotes adapted to life on land. They are close relatives of red algae and green algae, and the charophyte algae are considered their sister group:

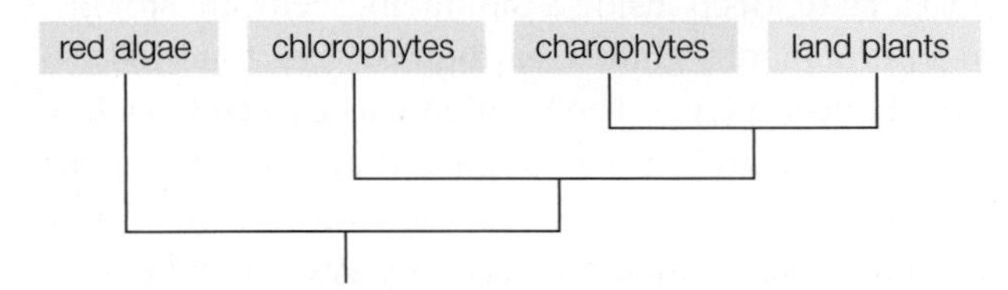

Like all green algae, plants have cell walls made of cellulose, chloroplasts with chlorophylls *a* and *b*, and they store sugars as starch. Shared traits that unite the plants and the charophyte algae include the mechanism by which a new cell wall forms after cell division and the arrangement of flagella on sperm (among those species that have flagellated sperm).

A developmental trait defines the clade of land plants. They are called **embryophytes** (meaning "embryo-bearing plants"), because their embryos form within a chamber of parental tissues and receive nourishment from the parent during early development. In some multicelled charophytes the zygote forms in a chamber on the parental body, but the zygote is released into the environment and the embryo develops without further parental assistance.

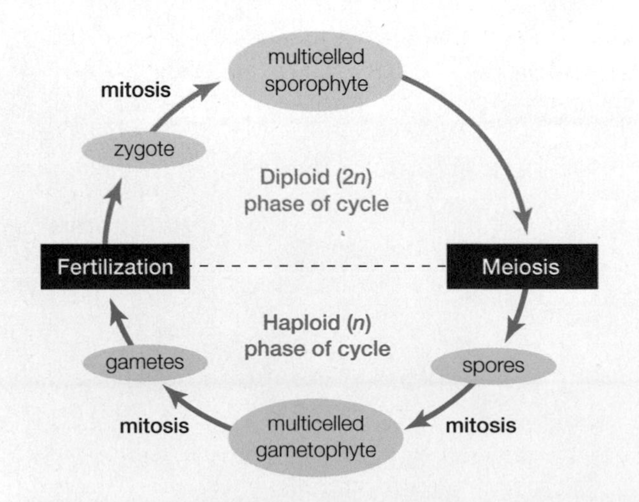

FIGURE 22.2 Generalized life cycle for land plants. A life cycle in which a multicelled haploid generation alternates with a multicelled diploid generation is called an alternation of generations.

Table 22.1 Diversity of Modern Land Plants

Bryophytes	
Liverworts	9,000 species
Mosses	15,000 species
Hornworts	100 species
Seedless Vascular Plants	
Lycophytes	1,100 species
Whisk ferns	7 species
Horsetails	25 species
Ferns	12,000 species
Gymnosperms	
Cycads	130 species
Ginkgos	1 species
Conifers	600 species
Gnetophytes	70 species
Angiosperms (Flowering Plants)	
Basal groups (e.g., magnoliids)	9,200 species
Monocots	80,000 species
Eudicots	>180,000 species

An Adaptive Radiation on Land

The first plants probably evolved by about 500 million years ago. By that time, enormous numbers of photosynthetic cells had come and gone. Oxygen released by photosynthetic bacteria and protists had altered the composition of the atmosphere. High above Earth, the sun's energy had converted some of this oxygen into a dense ozone layer, which screened out much incoming ultraviolet radiation. UV radiation is a mutagen, so the decrease in the amount reaching Earth's surface made land much more hospitable to life.

Spores that date to about 475 million years ago are generally considered the earliest fossil evidence of land plants. The spores were discovered in Argentina but, at the time of their release, this area was in the eastern part of the supercontinent Gondwana (Section 16.7).

A plant spore is a haploid cell with a thick wall. Like some algae, plants have an alternation of generations, a life cycle in which a diploid generation alternates with a haploid one (**FIGURE 22.2**). During the diploid generation, a multicellular **sporophyte** produces spores by meiosis. When these spores germinate (become active), they undergo mitosis and grow into the next generation, a multicelled haploid **gametophyte** that produces gametes by mitosis. When male and female gametes meet up, fertilization occurs and a diploid zygote forms. The zygote grows and develops into a new sporophyte.

In their structure, the oldest fossil spores resemble spores of modern liverworts. Liverworts, mosses, and

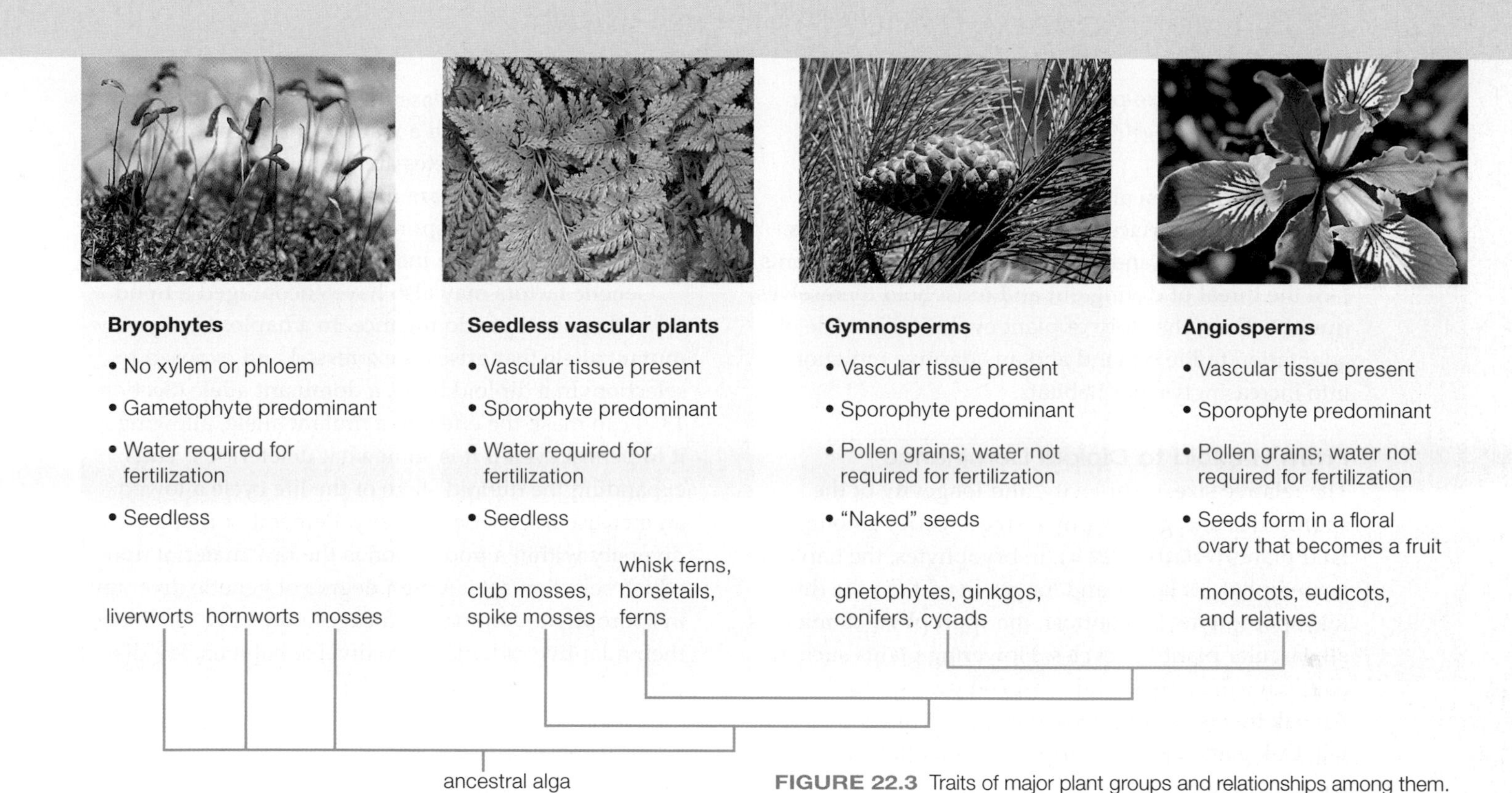

FIGURE 22.3 Traits of major plant groups and relationships among them.

hornworts are three early plant lineages informally referred to as **bryophytes** (**FIGURE 22.3** and **TABLE 22.1**). In all bryophytes, the gametophyte is larger and longer-lived than the sporophyte.

The first vascular plants evolved about 430-million years ago. In **vascular plants**, the sporophyte is larger and longer-lived than the gametophyte, and it has specialized internal pipelines that transport water and sugars. Because bryophytes do not have these specialized tissues, they are sometimes described as nonvascular plants.

Seed plants evolved about 385 million years ago, about the same time the first insects evolved. **Seed plants** are vascular plants that hold on to their spores and disperse by releasing seeds. Seed plants are also the only plants that produce pollen. As the next section explains, producing seeds and pollen allowed seed plants to expand into dry habitats that seedless plants could not tolerate.

Two lineages of seed plants survived to the present: gymnosperms and angiosperms. The gymnosperms are the more ancient of the two groups. Modern conifers such as pines, firs, and spruces are gymnosperms.

The flowering plants, or angiosperms, arose from a gymnosperm ancestor in the late Jurassic or early Cretaceous. In less than 40 million years, they would become the dominant plants in most habitats.

bryophyte Member of an early-evolving plant lineage with a gametophyte-dominant life cycle; a moss, liverwort, or hornwort.
embryophytes Land plants; clade of multicelled, photosynthetic species that protect and nourish the embryo on the parental body.
gametophyte A haploid, multicelled body that produces gametes in the life cycle of land plants and some algae.
plants Lineage of multicelled, typically photosynthetic eukaryotes adapted to life on land.
seed plant Plant that produces seeds and pollen; an angiosperm or gymnosperm.
sporophyte Diploid, spore-producing body in the life cycle of land plants and some algae.
vascular plant Plant having specialized tissues (xylem and phloem) that transport water and sugar within the plant body.

TAKE-HOME MESSAGE 22.2

What traits define plants and the various plant lineages?

✔ Land plants are embryophytes; their embryo forms in a chamber on the parental body and is nourished by the parent.

✔ Three early-evolving plant lineages are known collectively as bryophytes. These nonvascular plants disperse by releasing spores.

✔ Vascular plants have internal pipelines that transport water and sugars. Some produce spores; others produce seeds.

✔ Gymnosperms and angiosperms are two modern lineages of seed plants. Angiosperms (flowering plants) are the most diverse plant lineage, and the most recently evolved.

22.3 Evolutionary Trends Among Plants

✔ Over time, the spore-producing bodies of plants became larger, more complex, and better adapted to dry habitats.

A multicelled green alga absorbs the water it needs across its body surface. Water also buoys algal parts, helping the alga stand upright. In contrast, land plants face the threat of drying out and must hold themselves upright. Thus, the story of plant evolution is a tale of adaptation to life on land and an adaptive radiation into increasingly drier habitats.

From Haploid to Diploid Dominance

The relative size, complexity, and longevity of the sporophyte and gametophyte stages varies among the land plants (**FIGURE 22.4**). In bryophytes, the haploid gametophyte is larger and longer-lived than the diploid sporophyte. By contrast, the sporophyte dominates all vascular plant life cycles. Flowering plants such as oaks have the largest and most complex sporophytes. An oak tree is a sporophyte that stands many meters tall. Oak gametophytes form inside oak flowers and consist of only a few cells.

What drove the trend toward gametophyte reduction and sporophyte enlargement? Early on, structural differences between spores and gametes were probably a factor. Plants that release their spores into the environment encase them in a waterproof, decay-resistant wall. By contrast, gametes are unwalled cells. Under dry conditions, spores are more likely to survive than gametes, so increased spore production provides a greater advantage than increased gamete production.

Genetic factors may also have encouraged a trend toward sporophyte dominance. In a haploid body, any mutant allele that arises is expressed and exposed to selection. In a diploid body, a dominant allele (Section 13.2) can mask the effect of a mutant allele, allowing it to persist even if it is somewhat deleterious. Thus, expanding the diploid stage of the life cycle allowed an increase in genetic diversity. Remember that genetic diversity within a population is the raw material upon which selection acts. A high degree of genetic diversity in sporophyte-dominated lineages may have facilitated their adaptive radiation into diverse habitats. We discuss some of their relevant adaptations below.

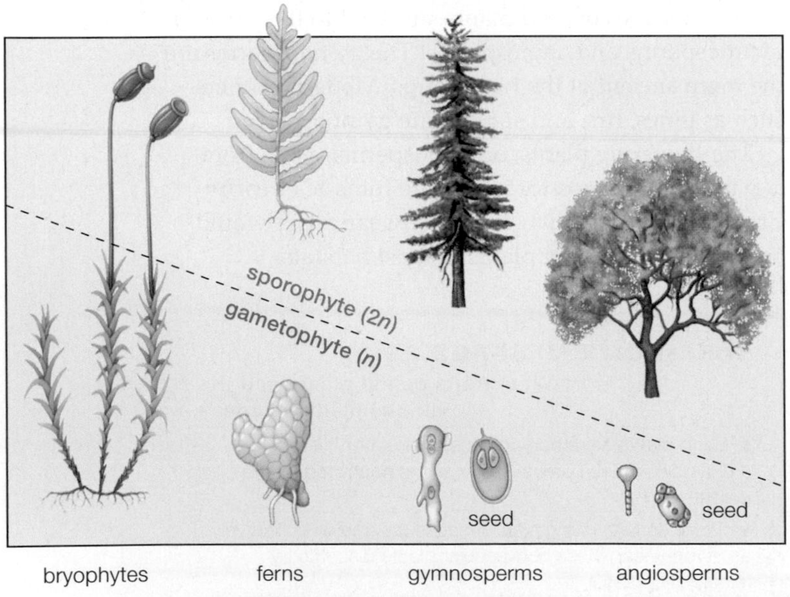

FIGURE 22.4 ▶Animated Evolutionary trend in plant life cycles. Mosses and other bryophytes put the most energy into making gametophytes. Later-evolving groups invested increasingly in making sporophytes.

FIGURE IT OUT A pine tree is a gymnosperm. Which is the larger, more prominent phase in its life cycle, the sporophyte or gametophyte?

Answer: Sporophyte

Structural Adaptations

Like algae, bryophytes absorb all the water and dissolved minerals they require across their body surface. They also lose water by evaporation across this surface. In some bryophytes and all vascular plants, the sporophyte secretes a waxy covering, or **cuticle**, that helps reduce evaporative water loss (**FIGURE 22.5A**). Closable pores called **stomata** (singular, stoma) extend across the cuticle (**FIGURE 22.5B**). Stomata open to allow gas exchange for photosynthesis, or close to conserve water.

Early plants had threadlike structures to hold them in place, but these structures did not deliver water and dissolved minerals to the rest of the plant body, as roots do. Moving substances from roots to other body regions requires **vascular tissues**, a system of internal pipelines (**FIGURE 22.6**). **Xylem** is a vascular tissue that distributes water and mineral ions. **Phloem** is a vascular tissue that distributes sugars made in photosynthetic cells. Most modern plants have xylem and phloem, and thus are vascular plants (tracheophytes).

Vascular tissues not only distribute material, they also provide structural support. An organic compound called **lignin** (Section 4.11) stiffens the walls of xylem cells. Evolution of lignin-stiffened tissue allowed vascular plants to stand taller than bryophytes and to branch, giving them an advantage in spore dispersal.

Most vascular plants also have leaves. Leaves are flattened, aboveground organs that increase the surface area available for intercepting sunlight and for gas exchange. They contain veins of vascular tissue.

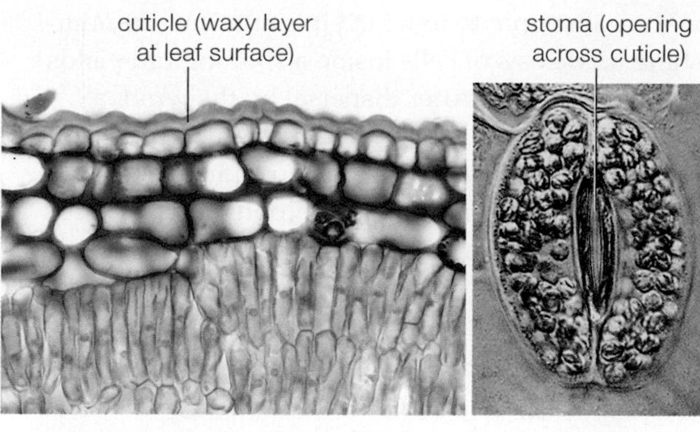

A Light micrograph showing waxy cuticle (stained pink) at a leaf surface. **B** One stoma.

FIGURE 22.5 Water-conserving adaptations. A secreted cuticle reduces evaporation. Stomata are openings across the cuticle.

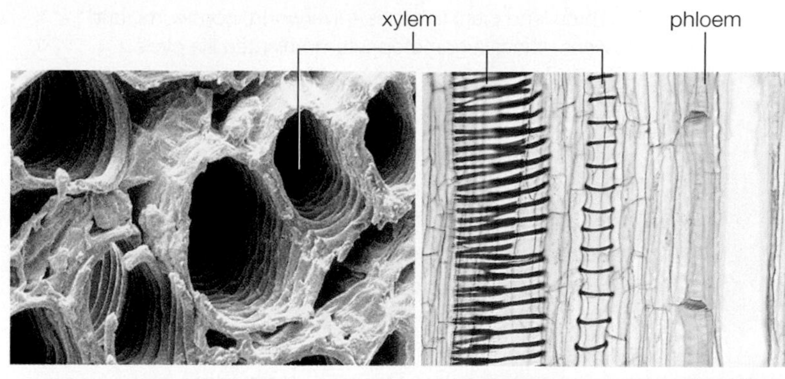

A Colorized micrograph of xylem (hollow tubes) in cross-section. **B** Light micrograph of vascular tissue of a squash stem with lignin stained red.

FIGURE 22.6 Vascular tissues. Xylem have walls stiffened with lignin and carry water. Phloem carries dissolved sugars produced by photosynthesis.

Pollen and Seeds

Novel reproductive traits gave seed-bearing vascular plants a competitive edge. All bryophytes, and some vascular plants such as the ferns, release sperm that must swim through the environment to eggs. Only seed-bearing vascular plants release pollen grains. A **pollen grain** is a walled, immature gametophyte that will give rise to the male gametes. After pollen grains are released, they travel to female gametophytes on the wind or on the bodies of animals, most often insects. An ability to produce pollen gave seed plants an ability to reproduce even in dry environments.

Bryophytes and seedless vascular plants release spores (**FIGURE 22.7A**), but seed plants protect spores within their tissues and release seeds. A **seed** consists of an embryo sporophyte and some nutritive tissue enclosed within a waterproof seed coat. Many seeds have features that facilitate their dispersal away from the parent plant. Angiosperms, or flowering plants, disperse the seeds inside a fruit that forms from floral tissue (**FIGURE 22.7B**). The great majority of modern plants are angiosperms.

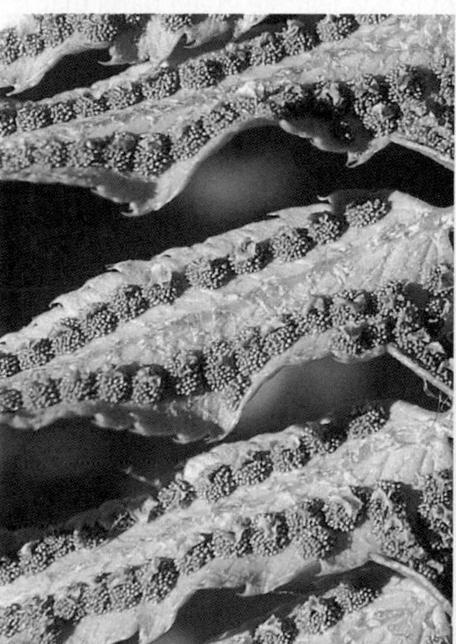

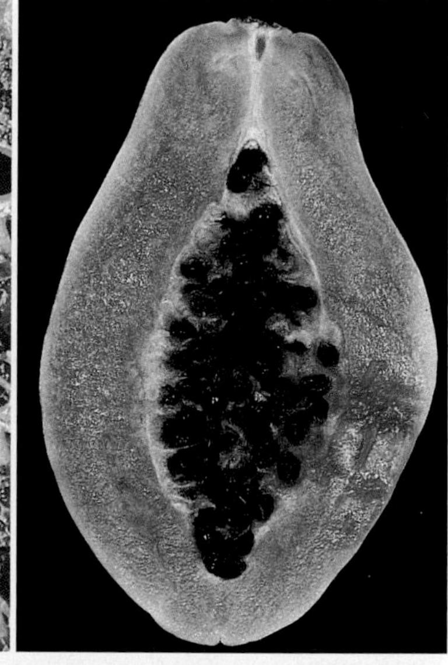

A Ferns disperse by releasing spores that form on the underside of leaves. **B** Flowering plants such as papayas hold on to their spores and disperse by releasing seeds enclosed in fruit.

FIGURE 22.7 Mechanisms of dispersal.

TAKE-HOME MESSAGE 22.3

What adaptations contributed to plant diversification?

✔ Plant life cycles shifted from a gametophyte-dominated cycle in bryophytes to a sporophyte-dominated cycle in vascular plants.

✔ Life on land favored water-conserving features such as a cuticle. In vascular plants, a system of vascular tissue—xylem and phloem—distributes material through the leaves, stems, and roots of sporophytes.

✔ Bryophytes and seedless vascular plants release spores. Only seed plants release embryos inside protective seeds. Only in the flowering plants do seeds form inside floral tissue that later develops into a fruit.

cuticle Secreted covering at a body surface.
lignin Material that stiffens cell walls of vascular plants.
phloem In vascular plants, tissue that distributes photosynthetically produced sugars through the plant body.
pollen grain Male gametophyte of a seed plant.
seed Embryo sporophyte of a seed plant packaged with nutritive tissue inside a protective coat.
stomata Closable gaps defined by guard cells on plant surfaces; when open, they allow water vapor and gases to diffuse across the epidermis.
vascular tissue In vascular plants, tissue (xylem and phloem) that distributes water and nutrients through the plant body.
xylem In vascular plants, tissue that distributes water and dissolved minerals through the plant body.

✔ Three land plant lineages—liverworts, hornworts, and mosses—have a gametophyte-dominated life cycle.

Three lineages—liverworts, hornworts, and mosses—are informally referred to as bryophytes, and also as nonvascular plants. Some mosses have internal pipelines, but none of the bryophytes have true vascular tissue reinforced by lignin, so few are more than 20 centimeters (8 inches) tall.

Mosses

Mosses are the most diverse and familiar group of bryophytes. We will use the life cycle of the moss *Polytrichum* as our example of a bryophyte life cycle (**FIGURE 22.8**). Like all bryophytes, this moss has a gametophyte-dominated life cycle.

A moss gametophyte has leaflike green parts that grow from a central stalk ❶. Threadlike **rhizoids** hold the gametophyte in place. The moss sporophyte is a **sporangium** (spore-producing structure) on a stalk ❷. The sporophyte is not photosynthetic and so depends on the gametophyte to which it is attached for nourishment. Meiosis of cells inside a sporangium yields haploid spores ❸. After dispersal by the wind, a spore germinates and grows into a gametophyte. Multicellular **gametangia** (gamete-producing structures) develop in or on the gametophyte. The moss we are using as our example has separate sexes, with each gametophyte producing eggs or sperm ❹. Other bryophytes are bisexual. In either case, rain triggers the release of flagellated sperm that swim through a film of water to eggs ❺. Some moss gametangia facilitate sperm movement by attracting mites and crawling insects that unknowingly pick up sperm and move them to adjacent plants. Fertilization inside the egg chamber produces a zygote ❻ that develops into a new sporophyte ❼. Mosses also reproduce asexually by fragmentation when a bit of gametophyte breaks off and develops into a new plant.

The 350 or so species of peat moss (*Sphagnum*) are the most economically important bryophytes. Peat mosses are the dominant plants in **peat bogs** that cover

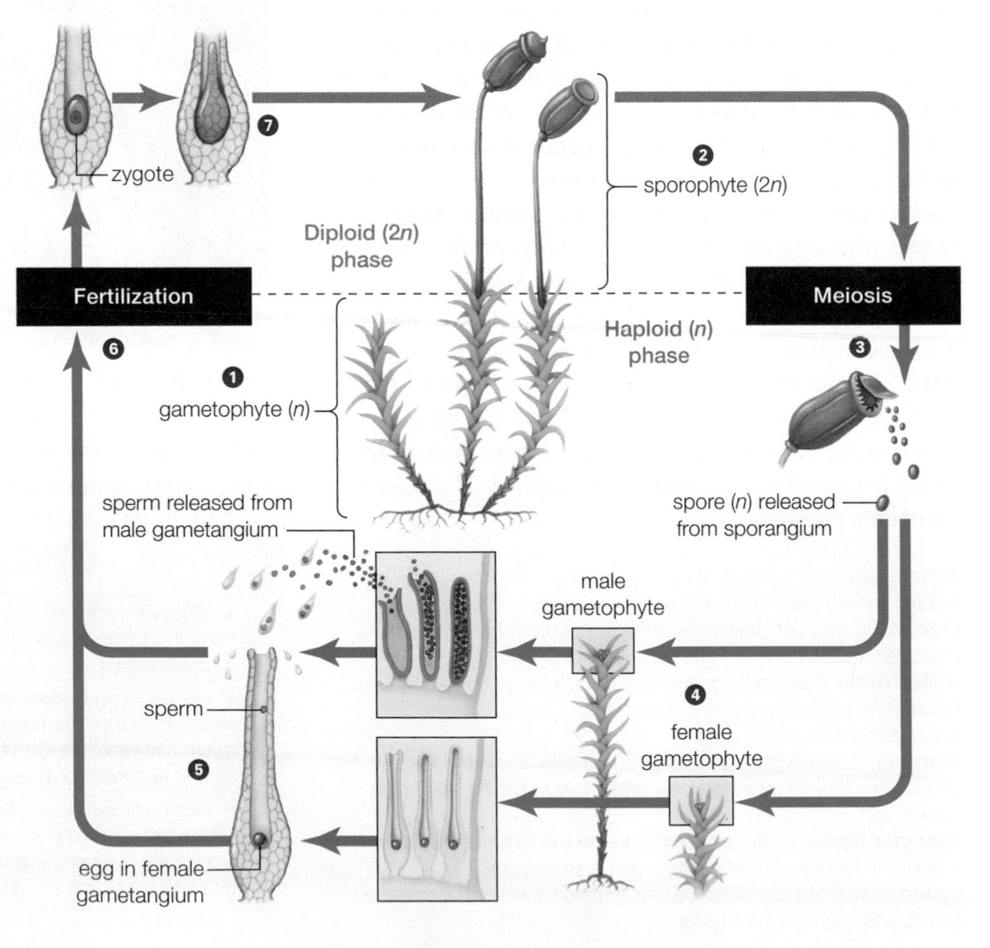

FIGURE 22.8 ▸**Animated** Life cycle of a moss (*Polytrichum*).

❶ The leafy green part of a moss is the haploid gametophyte.

❷ The diploid sporophyte has a stalk and a capsule (sporangium). It is not photosynthetic.

❸ Haploid spores form by meiosis in the capsule, are released, and drift with the winds.

❹ Spores germinate and develop into male or female gametophytes with gametangia that produce eggs or sperm by mitosis.

❺ Sperm swim to eggs.

❻ Fertilization produces a zygote.

❼ The zygote grows and develops into a new sporophyte while remaining attached to and nourished by the female gametophyte.

FIGURE 22.9 Peat bog in Ireland. These boys are cutting blocks of peat and stacking them to dry for use as fuel.

FIGURE 22.10 Liverwort (*Marchantia*). Tiny sporophytes with yellow capsules form on umbrella-shaped, female gametangia.

hundreds of millions of acres in high-latitude regions of Europe, Asia, and North America. Many peat bogs have persisted for thousands of years, and layer upon layer of plant remains have become compressed as a carbon-rich material called peat. Blocks of peat are cut, dried, and burned as fuel, especially in Ireland (**FIGURE 22.9**). Freshly harvested peat moss is also an important commercial product. The moss is dried and added to planting mixes to help soil retain moisture.

Liverworts

Liverworts may be the most ancient of the surviving plant lineages. The oldest known fossils of land plants are spores that resemble those of modern liverworts. In addition, genetic comparisons put liverworts near the base of the plant family tree.

Some liverwort gametophytes look leafy and others are flattened sheets. In the widespread liverwort genus *Marchantia*, eggs and sperm form on separate plants. The gametangia are elevated above the main gametophyte body on stalks (**FIGURE 22.10**). Members of this genus also reproduce asexually by producing small clumps of cells in cups on the gametophyte surface.

Hornworts

A pointy, hornlike sporophyte that can be several centimeters tall gives the hornworts their common name (**FIGURE 22.11**). The base of the sporophyte is embedded in gametophyte tissues, and spores form in an upright sporangium, or capsule. When spores mature, the tip of the capsule splits, releasing them. The sporophyte grows continually from its base, so it can make and release spores over an extended period.

The sporophyte has chloroplasts and, in some cases, can survive even after the death of the gametophyte. These traits and genetic similarities suggest that hornworts are probably the sister group to vascular plants.

FIGURE 22.11 Hornwort. Photosynthetic hornlike sporophytes grow from a flattened gametophyte body.

TAKE-HOME MESSAGE 22.4

What are bryophytes?

✔ Bryophyte is the common name for three lineages of plants: mosses, liverworts, and hornworts. All are low-growing with no lignin-reinforced vascular tissues. All have flagellated sperm that require a film of water to swim to eggs, and all disperse by releasing spores.

✔ Bryophytes are unique among land plants in having a life cycle in which the gametophyte is the dominant generation. The sporophyte remains attached to the gametophyte even when mature.

gametangium Gamete-producing organ of a plant.
peat bog High-latitude community dominated by *Sphagnum* moss.
rhizoid Threadlike structure that anchors a bryophyte.
sporangium Of plants and fungi, a structure in which haploid spores form by meiosis.

22.5 Seedless Vascular Plants

✔ A sporophyte with lignified vascular tissue is the dominant phase in the life cycle of the seedless vascular plants.

Some mosses have internal pipelines that transport fluid within their body. However, only vascular plants have lignin-strengthened vascular tissue with xylem and phloem. This innovation allowed the evolution of larger, branching sporophytes, which are the predominant generation in all vascular plants.

Seedless vascular plants are the oldest vascular plant lineages. Like bryophytes, they have flagellated sperm that swim to eggs, and they disperse by releasing spores directly into the environment.

Two lineages of seedless vascular plants survived to the present. Lycophytes include plants that are commonly known as club mosses and spike mosses, but are not mosses. Monilophytes include whisk ferns, horsetails, and ferns.

Lycophytes and monilophytes diverged from a common ancestor before leaves and roots had evolved, and each developed these features in a different way. For example, lycophytes form spores along the sides of branches. Their leaves have one unbranched vein and probably evolved from a lateral sporangium. In contrast, monilophytes have spores at branch tips. Their leaves have branching veins, so they probably evolved from a branching network of stems.

FIGURE 22.12 Club moss (*Lycopodium*) with yellow stotobili.

Club Mosses

Most of the 1,200 modern lycophytes are club mosses. Members of the club moss genus *Lycopodium* often grow on the floor of temperate forests (**FIGURE 22.12**). Their sporophytes resemble miniature pine trees and are sometimes called ground pines. The plant has a horizontal stem, or **rhizome**, that runs along the ground. Roots and upright stems with tiny leaves grow from the rhizome. When a plant is several years old, it produces a strobilus seasonally. A **strobilus** (plural, strobili) is a cone-shaped, spore-producing structure composed of modified leaves.

Lycopodium is gathered from the wild for use in wreaths and bouquets, and an extract of the plant is marketed as an herbal medicine. The spores, which are covered with a waxy coating that ignites easily, are sold as "flash powder" for creating special effects.

FIGURE 22.13 Whisk fern with sporangia at tips of short lateral branches.

Whisk Ferns and Horsetails

Whisk ferns (*Psilotum*) are native to the southeastern United States. Their sporophytes have underground rhizomes and photosynthetic stems that appear leafless (**FIGURE 22.13**). Spores form in fused sporangia at the tips of short branches. Florists often use branches of whisk ferns in mixed bouquets.

The 25 *Equisetum* species include plants commonly known as horsetails and rushes. Their sporophytes have rhizomes and hollow stems with tiny nonphotosynthetic leaves at the joints. Photosynthesis occurs in stems and leaflike branches (**FIGURE 22.14A**). Deposits of silica in the stem support the plant and give stems a sandpapery texture that helps fend off herbivores. Before scouring powders and pads were widely available, people used stems of *Equisetum* species as pot scrubbers, thus the common name "scouring rush."

Depending on the species, strobili form either at tips of photosynthetic stems or on specialized reproductive stems without chlorophyll (**FIGURE 22.14B**).

Ferns

With 12,000 or so species, ferns are the most diverse and familiar seedless vascular plants. Most live in the tropics. A typical sporophyte has fronds (leaves) and roots that grow from a rhizome (**FIGURE 22.15** ❶).

A Photosynthetic shoots. The long, needlelike structures are not leaves, but branches.

B Nonphotosynthetic shoot with a strobilus at its tip.

FIGURE 22.14 Horsetails (*Equisetum*).

FIGURE 22.15 ▸**Animated** Life cycle of a fern.

❶ The familiar leafy form is the diploid sporophyte.

❷ Meiosis in cells on the underside of fronds (leaves) produces haploid spores.

❸ After spores are released, they germinate and grow into tiny gametophytes that produce eggs and sperm.

❹ Sperm swim to eggs and fertilize them, forming a zygote.

❺ The sporophyte begins its development attached to the gametophyte, but it continues to grow and live independently after the gametophyte dies.

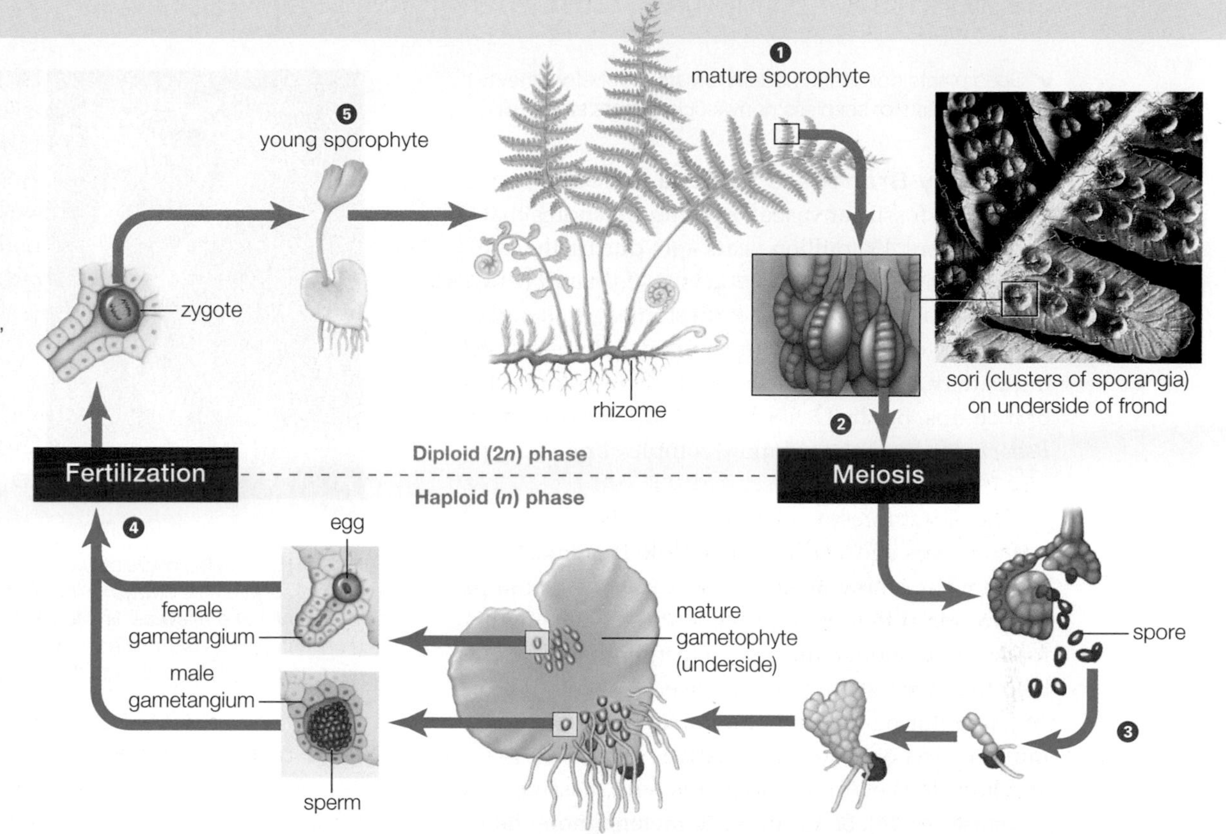

Fern spores form in **sori** (singular, sorus), which are clusters of capsules (sporangia) that develop on the lower surface of fronds ❷. When these capsules pop open, spores disperse on the wind.

After a spore germinates, it grows into a photosynthetic, heart-shaped gametophyte just a few centimeters wide ❸. Eggs and sperm form in chambers (gametangia) on the underside of the gametophyte. Rain stimulates release of sperm, which swim through a film of water to reach and fertilize an egg ❹. The resulting zygote develops into a new sporophyte ❺, and its parental gametophyte dies.

Fern sporophytes vary in size and structure. The floating fern *Azolla* has fronds only 1 millimeter long. Chambers in its fronds shelter nitrogen-fixing bacteria, so Southeast Asian farmers grow this fern in rice fields as an alternative to chemical fertilizers. Many ferns are **epiphytes**, plants that attach to and grow on another plant but do not withdraw nutrients from it. The largest modern nonvascular plants are tree ferns (**FIGURE 22.16**). Some stand 20 meters (65 feet) high.

FIGURE 22.16 Tree ferns in Australia. The trunks are rhizomes.

TAKE-HOME MESSAGE 22.5

What are seedless vascular plants?

✔ Club mosses and relatives belong to one seedless vascular lineage (lycophytes). Ferns, horsetails, and whisk ferns belong to the other (monilophytes).

✔ A sporophyte with vascular tissues (xylem and phloem) dominates the seedless vascular plant life cycle, dispersal occurs by release of spores.

epiphyte Plant that grows on another plant but does not harm it.
rhizome Stem that grows horizontally along or under the ground.
sorus Cluster of spore-producing capsules on a fern leaf.
strobilus (strobili) Of some nonflowering plants such as horsetails and cycads, a cluster of spore-producing structures.

CREDITS: (15) photo, A. & E. Bomford/ Ardea, London; art, © Cengage Learning; (16) © Klein Hubert/Peter Arnold, Inc.

22.6 History of the Vascular Plants

✔ Seed plants dominate modern forests, but before they evolved, forests of seedless nonvascular plants stood tall.

From Tiny Branchers to Coal Forests

The oldest fossils of vascular plants are spores that date to about 450 million years ago, during the late Ordovician period (**FIGURE 22.17**). *Cooksonia* (**FIGURE 22.18A**) may be one of the earliest lineages. It stood only a few centimeters high and had a simple branching pattern, and no leaves or roots. Spores formed at branch tips. By the Devonian period, plants such as *Psilophyton* had evolved more complex branching pattern (**FIGURE 22.18B**).

The oldest forest we know about existed about 385 million years ago, during the middle Devonian, at a site in what is now upstate New York. Fossil stumps and fronds at this site indicate that the plants in this forest stood about 8 meters (26 feet) high and resembled tree ferns in their appearance.

Later, during the Carboniferous period (359–299 million years ago), ancient relatives of club mosses and horsetails evolved into giants with massive stems (**FIGURE 22.19**). Some stood 40 meters (more than 130 feet) high. After forests of these plants formed, climates changed, and the sea level rose and fell many times. When the waters receded, the forests flourished. After the sea moved back in, submerged trees became buried in sediments that protected them from decomposition. Layers of sediments accumulated one on top of the other. Their weight squeezed the water out of the saturated, undecayed remains, and the compaction generated heat. Over time, pressure and heat transformed the compacted organic remains into **coal**.

It took millions of years of photosynthesis, burial, and compaction to form coal. When you hear about annual production of coal or other fossil fuels, keep in mind that we do not really "produce" these materials, we only extract them. This is why fossil fuels are said to be nonrenewable sources of energy.

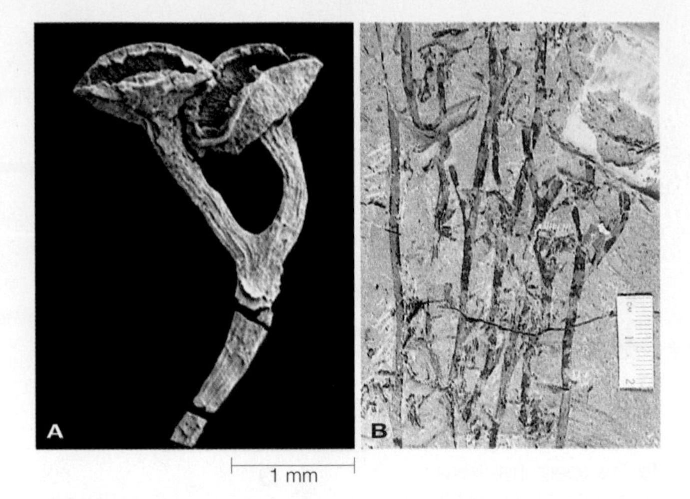

FIGURE 22.18 Fossils of two early seedless vascular plants. **A** *Cooksonia* stems always divided into two equal branches. This plant stood a few centimeters tall. **B** *Psilophyton* shows a more complex growth pattern. It branched unequally with a main stem and smaller branches to the side.

Rise of the Seed Plants

The first gymnosperms evolved from a seedless ancestor late in the Devonian. Representatives of some modern gymnosperm lineages (cycads and conifers) first appeared in the late Carboniferous. Angiosperms, or flowering plants, arose by about 120 million years ago, during the Cretaceous, when dinosaurs reigned (**FIGURE 22.20**).

Reproductive traits of seed-bearing plants gave them a selective advantage over earlier lineages. Gametophytes of seedless vascular plants develop in the environment. By contrast, gametophytes of seed plants develop within the protection of a sporophyte body (**FIGURE 22.21**). Sporangia called **pollen sacs** produce microspores. **Microspores** develop into sperm-making gametophytes (pollen grains). Sporangia called **ovules** produce megaspores. Egg-producing gametophytes develop from **megaspores**. A sporophyte releases pollen grains, but holds on to its eggs. Wind

Ordovician	Silurian	Devonian	Carboniferous	Permian	Triassic / Jurassic	Cretaceous / Tertiary
Origin of first land plants (bryophytes) by 475 mya.	Origin of seedless vascular plants.	Bryophytes and seedless vascular plants diversify. Seed plants arise by 385 mya.	Lycophytes and horsetails diversify. Cycads and conifers arise late in this period.	Ginkgos appear. Horsetails and lycophytes decline in diversity.	Diversification of conifers; by the end of the Jurassic they are the dominant trees. Cycad, ginkgos, tree ferns remain common.	Flowering plants appear in the early Cretaceous, undergo adaptive radiation, and become dominant.

488 443 416 359 299 251 200 146 66

Millions of years ago (mya)

FIGURE 22.17 A time line for major events in the evolution of plants.

FIGURE 22.19 Painting of a Carboniferous coal forest. An understory of ferns is shaded by tree-sized relatives of modern horsetails and club mosses.

FIGURE 22.20 Early angiosperms such as magnolias (foreground) evolved while dinosaurs walked the earth.

or animals can deliver pollen from one seed plant to the ovule of another, a process known as **pollination**. Because the sperm of seed plants do not need to swim through a film of water to reach eggs, these plants can reproduce even during dry times.

After pollination and fertilization, an ovule develops into a seed. Releasing seeds puts seed plants at an advantage over spore-bearing plants. A seed contains a multicelled embryo sporophyte and stored food that the embryo can draw on during early development. By contrast, seedless plants release single-celled spores without stored food.

Structural traits also gave seed plants an advantage over seedless lineages. Some seed plants undergo secondary growth (growth in diameter) and produce wood. Wood is lignin-stiffened tissue that strengthens and protects older stems and roots. The giant nonvascular plants that lived in Carboniferous forests did undergo secondary growth. However, because their trunks were softer and more flexible than those of woody seed plants, they could not grow as tall.

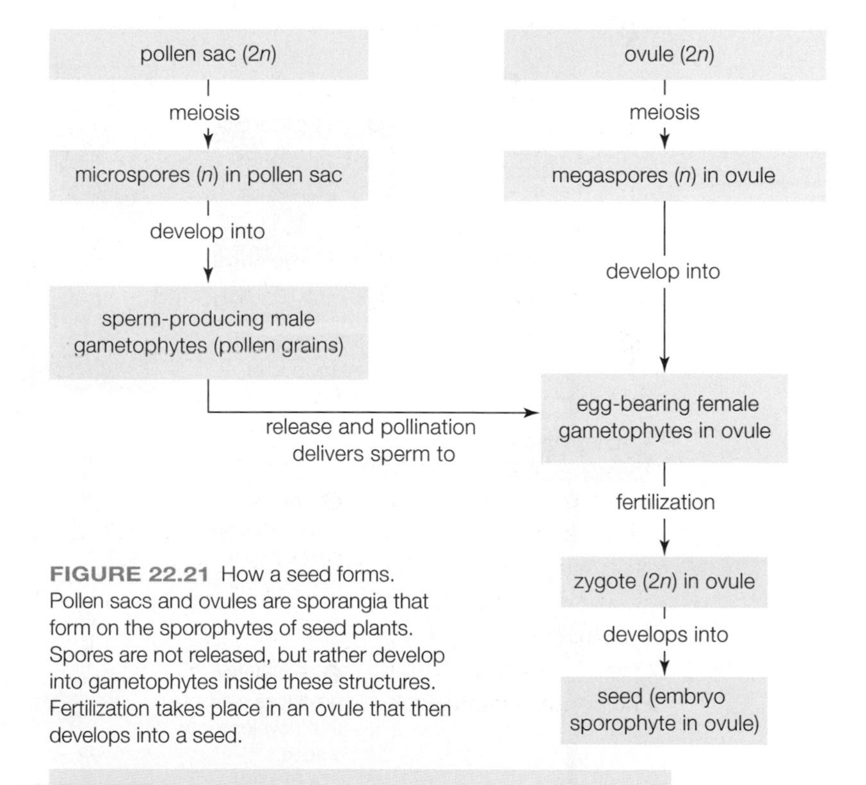

FIGURE 22.21 How a seed forms. Pollen sacs and ovules are sporangia that form on the sporophytes of seed plants. Spores are not released, but rather develop into gametophytes inside these structures. Fertilization takes place in an ovule that then develops into a seed.

coal Fossil fuel formed over millions of years by compaction and heating of plant remains.
megaspore Haploid spore formed in ovule of seed plants; develops into an egg-producing gametophyte.
microspore Haploid spore formed in pollen sacs of seed plants; develops into a sperm-producing gametophyte (a pollen grain).
ovule Of seed plants, sporangium that produces megaspores that develop into egg-producing gametophytes.
pollen sac Of seed plants, sporangium where microspores form and develop into pollen grains.
pollination Arrival of a pollen grain on the egg-bearing part of a seed plant.

TAKE-HOME MESSAGE 22.6

What are the major events in the evolutionary history of vascular plants?

✔ Seedless vascular plants arose and became widespread during the Carboniferous. Coal is the remains of some of these ancient plants.

✔ The first seed plants evolved in the late Devonian. Because they produced pollen and seeds, they were able to live in drier places than other plants. An ability to produce wood allowed some seed plants to grow very tall.

22.7 Gymnosperms

✔ Gymnosperms are seed-bearing plants that produce their seeds on the surface of modified leaves.

✔ Gymnosperms do not make flowers or fruit.

Gymnosperms are vascular seed plants that produce seeds on the surface of ovules. Their seeds are said to be "naked," because unlike those of angiosperms they are not inside a fruit. (*Gymnos* means naked; *sperma* is seed.) However, many gymnosperms enclose their seeds in a fleshy or papery covering.

Conifers

The 600 or so species of **conifers** are trees and shrubs with woody cones. Conifers typically have needlelike or scalelike leaves with a thick cuticle. They tend to be more resistant to drought and cold than flowering plants and are the main plants in high-latitude forests of the Northern Hemisphere. Most conifers are evergreen: They shed some leaves year round but always retain enough to remain green. Conifers include the tallest trees in the Northern Hemisphere (redwoods), and the most abundant (pines). They also include the longest-lived plants; some bristlecone pines are more than 4,000 years old.

We mulch our gardens with fir bark, use oils from cedar in cleaning products, and eat the seeds, or "pine nuts," of some pines. Pines also provide lumber for building homes and furniture. Some pines make a sticky resin that deters insects from boring into them. We use this resin to make turpentine, a paint solvent.

Pines have a life cycle typical of conifers (FIGURE 22.22). The pine tree is a sporophyte that produces spores on specialized strobili called cones. In most pine species, each tree makes only one type of cone: either small, soft pollen cones or large, woody, ovulate cones. Both have scales arranged around a central axis.

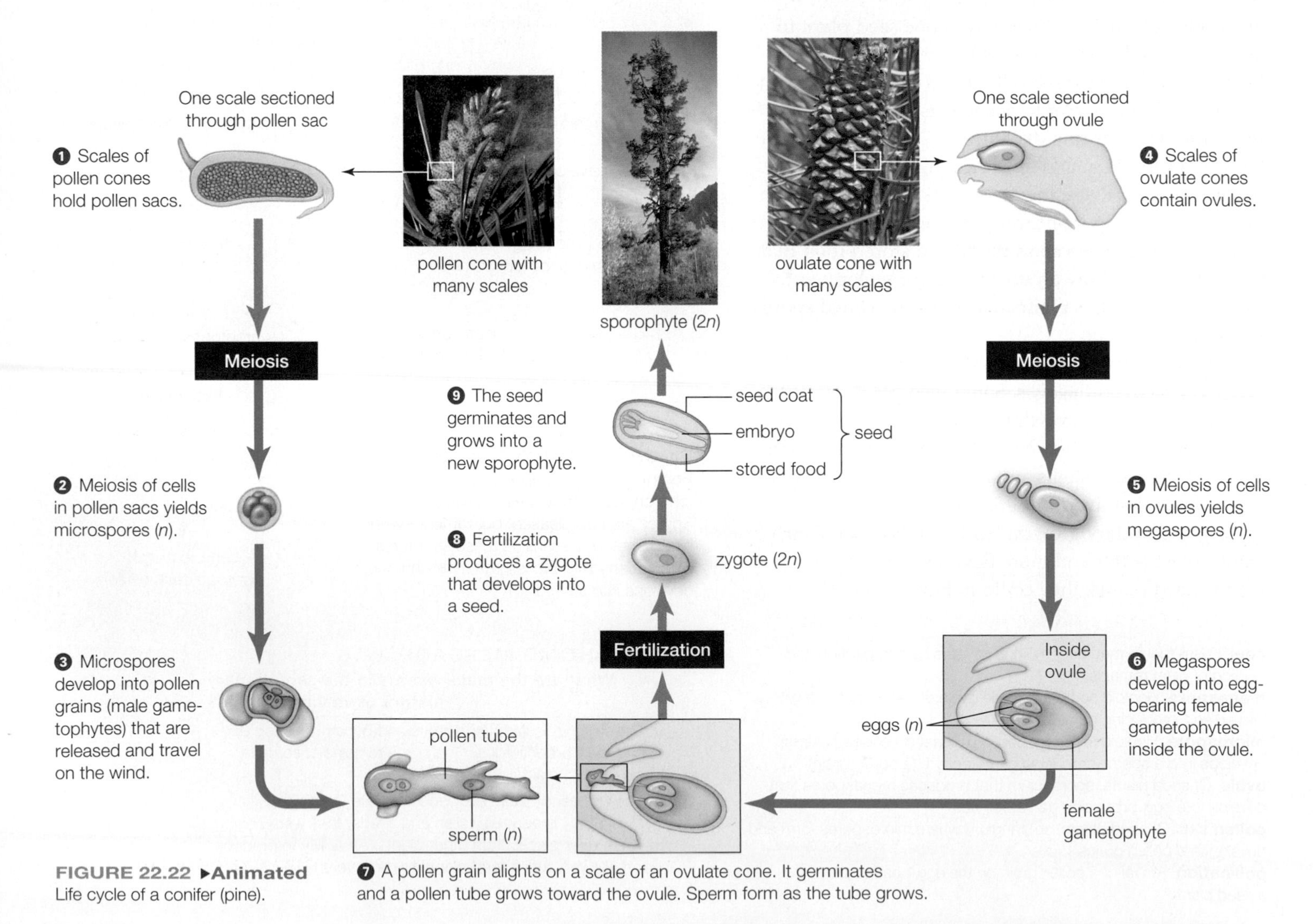

FIGURE 22.22 ▸**Animated** Life cycle of a conifer (pine).

Each scale of a pollen cone contains pollen sacs ❶. Microspores that form by meiosis in these sacs ❷ develop into pollen grains ❸. Pollen cones release these tiny grains to drift with the winds.

Each scale of an ovulate cone, holds a pair of ovules ❹. Meiosis of diploid cells in an ovule produces megaspores ❺ that develop into egg-producing female gametophytes ❻.

An unpollinated ovule exudes a sugary "pollination droplet" that captures windborne pollen. The pollen grain then germinates, and a pollen tube grows through the ovule tissue to deliver sperm to the egg ❼. Fertilization produces a zygote ❽. The zygote develops into an embryo sporophyte that, along with tissues of the ovule, becomes a seed. The seed is released, germinates, then develops into a new sporophyte ❾.

Lesser-Known Lineages

Diversity of cycads and ginkgos peaked in the time of the dinosaurs. Today, they are the only seed plants with flagellated sperm. Sperm emerge from pollen grains, then swim in fluid released by the plant's ovule.

The 130 modern **cycads** are native to dry tropics and subtropics. Many resemble palms (FIGURE 22.23A) but the two groups are not close relatives.

The single living **ginkgo**, *Ginkgo biloba*, is native to China (FIGURE 22.23B). It is deciduous, meaning it loses all its leaves at once seasonally. The ginkgo's pretty fan-shaped leaves and resistance to insects, disease, and air pollution make it a popular tree along city streets. Female trees produce fleshy plum-sized seeds that emit a strong unpleasant odor, so male trees are preferred for urban landscaping.

conifer Gymnosperm with nonmotile sperm and woody cones; for example, a pine.
cycad Tropical or subtropical gymnosperm with flagellated sperm, palmlike leaves, and fleshy seeds.
ginkgo Deciduous gymnosperm with flagellated sperm, fan-shaped leaves, and fleshy seeds.
gnetophyte Shrubby or vinelike gymnosperm, with nonmotile sperm; for example, *Ephedra*.
gymnosperm Seed plant that does not make flowers or fruits; for example, a conifer.

TAKE-HOME MESSAGE 22.7
What are gymnosperms?

✔ Gymnosperms include conifers, cycads, ginkgos, and gnetophytes.

✔ These seed-bearing vascular plants produce seeds on strobili. In the case of conifers, the seed-bearing strobilus is a woody cone.

Gnetophytes include tropical trees, desert shrubs, and leathery vines. *Ephedra*, a twiggy shrub (FIGURE 22.23C), produces ephedrine, a chemical sold as an herbal stimulant and weight loss aid. *Welwitschia* lives in Africa's Namib desert (FIGURE 22.23D). It has a taproot, woody stem, and two long, straplike leaves that split lengthwise repeatedly, giving the plant a shaggy appearance.

A Cycad (*Cycas*) growing in the understory of a Cambodian forest. Seeds are visible near its base.

B Fan-shaped leaves and fleshy seeds of *Ginkgo biloba*, the maidenhair tree.

C Pollen cones of *Ephedra*, a shrubby gnetophye.

D *Welwitschia*, a gnetophyte, with seed cones and two long, wide leaves.

FIGURE 22.23 Gymnosperm diversity.

CREDITS: (23A, D) Fletcher and Baylis/Science Source; (23B) Georgia Silvera Seamans, localecology.org; (23C) Christine Evers.

22.8 Angiosperms—The Flowering Plants

✔ Angiosperms are the most diverse plant lineage and the only plants that make flowers and fruits.

✔ Coevolution with animals enhanced angiosperm success.

Angiosperms are vascular seed plants that make flowers and fruits. A **flower** is a specialized reproductive shoot that consists of modified leaves arranged in three concentric whorls (**FIGURE 22.24**). Sepals form the outermost whorl; petals, the middle whorl; pollen-bearing **stamens**, the inner whorl. The innermost part of the flower is the **carpel**. The **ovary**, a chamber at the base of the carpel, contains one or more ovules where eggs form. After fertilization, an ovule matures into a seed, and the ovary becomes the **fruit**. The name "angiosperm" refers to the development of seeds inside an ovary. (*Angio*– means enclosed chamber; *sperma*, seed.)

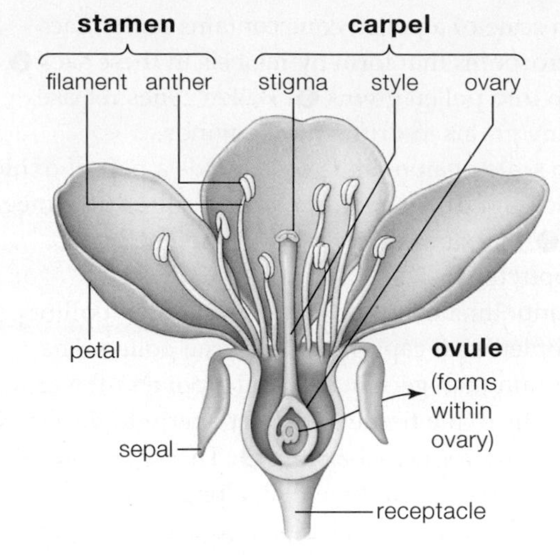

FIGURE 22.24 Floral components.

FIGURE 22.25 ▸**Animated** Flowering plant life cycle.

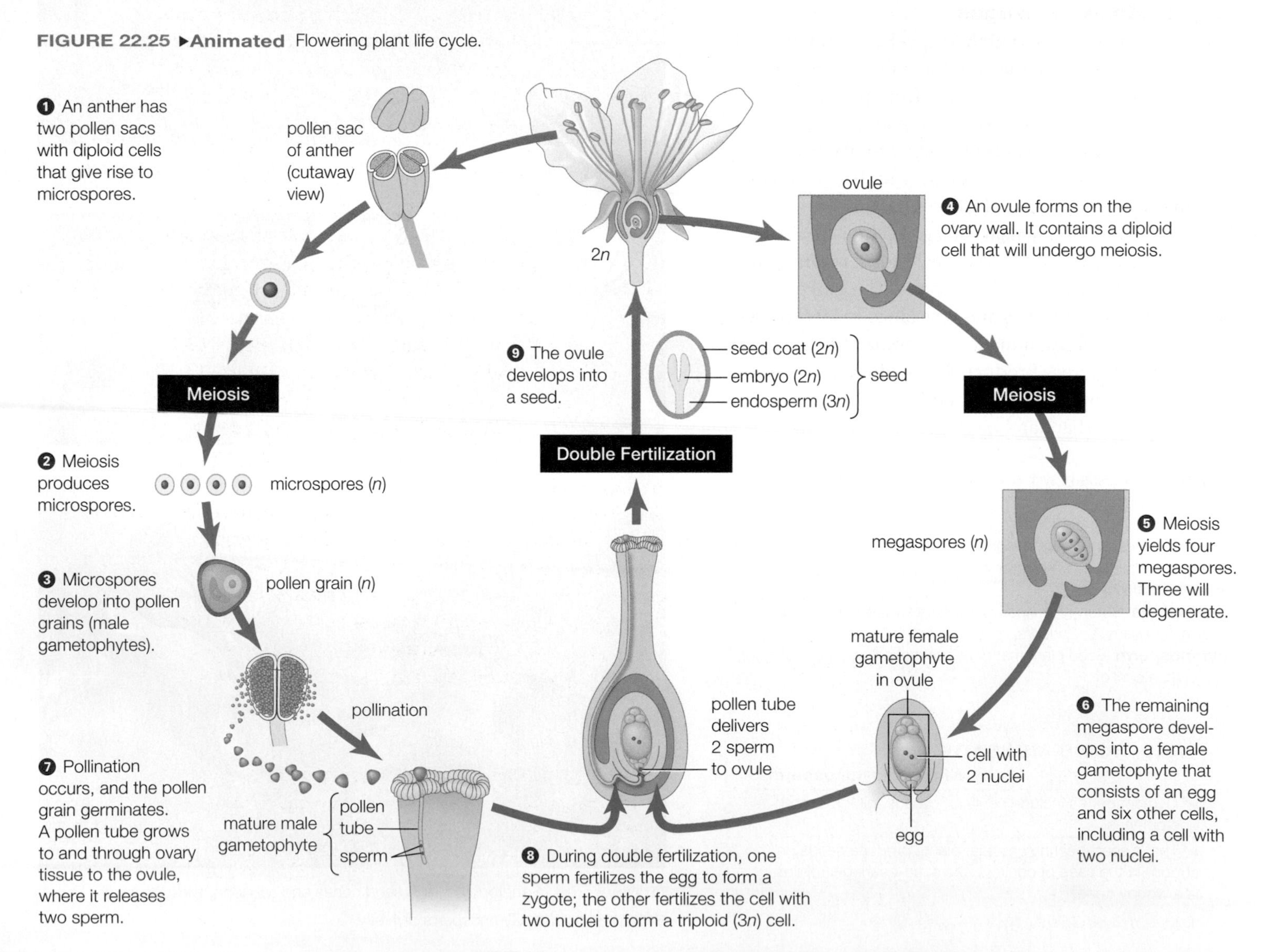

The Angiosperm Life Cycle

FIGURE 22.25 shows a generalized life cycle for a flowering plant. The upper portion of a stamen is an anther that holds two pollen sacs. Inside those sacs are diploid cells ❶ that give rise to microspores by meiosis ❷. The microspores develop into pollen grains (immature male gametophytes) ❸. A flowering plant's ovules form on the wall of an ovary at the base of a carpel ❹. Meiosis of a cell in an ovule yields four haploid megaspores ❺. Only one of these megaspore develops into a female gametophyte, which consists of a haploid egg, a cell with two nuclei, and a few additional cells ❻.

Pollination occurs when a pollen grain arrives on a receptive stigma, the uppermost part of the carpel ❼. The pollen grain germinates, and a pollen tube grows through the style (the structure that elevates the stigma) to the ovary at the base of the carpel. Two nonflagellated sperm form inside the tube as it grows.

Double fertilization occurs when a pollen tube delivers the two sperm into the ovule ❽. One sperm fertilizes the egg to create a zygote. The other sperm fuses with the cell that has two nuclei, forming a triploid ($3n$) cell. After double fertilization, the ovule matures into a seed ❾. The zygote develops into an embryo sporophyte and the triploid cell develops into **endosperm**, a nutritious tissue that will serve as a source of food for the developing embryo.

Factors Contributing to Angiosperm Success

Today, nearly 90 percent of all plant species are flowering plants. Several factors gave angiosperms a selective advantage over gymnosperms.

Accelerated Life Cycle Compared to gymnosperms, most angiosperms have a shorter life cycle. A dandelion or grass can grow from a seed, mature, and produce seeds of its own within a month or so. In contrast, gymnosperms tend to take years to mature. Producing and dispersing seeds quickly helps angiosperms expand their range faster than gymnosperms.

A Bee transferring pollen.

B Bird dispersing seeds.

FIGURE 22.26 Animal assistants to flowering plants.

Animal-Pollinated Flowers Evolution of flowers gave angiosperms an edge by facilitating animal-assisted pollination. After seed plants evolved, some insects began feeding on protein-rich pollen. The plants lost a bit of pollen, but benefited when the insects inadvertently transferred pollen from one plant to another of the same species. An animal that aids in pollination by transferring pollen between plants is a **pollinator**. Most pollinators are insects (FIGURE 22.26A), but birds, bats, and other animals also fulfill this role.

Over time, many flowering plants coevolved with pollinators. **Coevolution** refers to the joint evolution of two or more species as a result of their close ecological interactions (Section 17.12). Producing sugary nectar encouraged more pollinator visits, which improved pollination rates and enhanced seed production. Conspicuous petals and distinctive scents that attracted pollinators provided a selective advantage. At the same time, selection favored pollinators that searched out nectar-rich flowers and accessed the nectar reward.

Enhanced Seed Dispersal Fruit production also contributed to angiosperm success. Fleshy, sugary fruits attract animals that carry the fruits (and their enclosed seeds) away from a parent plant (FIGURE 22.26B). Other fruits have shapes that help them catch the wind or stick to fur. As a group, gymnosperms have fewer structural adaptations for seed dispersal.

TAKE-HOME MESSAGE 22.8

What are the features of angiosperms?

✔ Angiosperms produce seeds inside the ovaries of flowers. After pollination, an ovary becomes a fruit.

✔ Angiosperms are the most successful plants. Short life cycles, coevolution with insect pollinators, and diverse fruit structures enhanced their success.

angiosperms Most diverse seed plant lineage. Only group that makes flowers and fruits.
carpel Of flowering plants, a reproductive structure that produces female gametophytes; consists of a stigma, a style, and an ovary.
coevolution The joint evolution of two closely interacting species; each species is a selective agent for traits of the other.
double fertilization In flowering plants only, one sperm fertilizes an egg to produce a zygote and another fertilizes a diploid cell to create a triploid cell that will develop into endosperm.
endosperm Triploid ($3n$) nutritive tissue in an angiosperm seed.
flower Specialized reproductive shoot of a flowering plant.
fruit Mature ovary of a flowering plant; encloses a seed or seeds.
ovary In flowering plants, the enlarged base of a carpel, inside which one or more ovules form and eggs are fertilized.
pollinator Animal that moves pollen, thus facilitating pollination.
stamen Pollen-producing organ of flowering plant.

22.9 Angiosperm Diversity and Importance

✔ Most flowering plants are monocots or eudicots. These two groups of plants are the main source of food worldwide.

Angiosperm Lineages

Gene comparisons have identified the oldest angiosperm lineages that include modern representatives. Among living groups, *Amborella*, a small shrub native to New Caledonia, has the most ancient ancestry. Like gymnosperms, and unlike all other angiosperms, it has only one type of xylem. Water lilies (FIGURE 22.27A) and star anise are also modern representatives of ancient plant lineages.

Genetic divergences gave rise to other groups that became dominant: magnoliids, eudicots (true dicots), and monocots. Among the 9,200 or so magnoliids are magnolias (FIGURE 22.27B) and avocados. The 80,000 or so **monocots** include orchids, palms, lilies, grasses, and irises (FIGURE 22.27C). The approximately 170,000 **eudicots** (FIGURE 22.27D) include familiar broadleaf plants such as tomatoes, cabbages, roses, and poppies, as well as flowering shrubs, trees, and cacti.

Monocots and eudicots derive their group names from the number of seed leaves, or cotyledons, in the embryo. Monocots have one cotyledon and parallel leaf veins; eudicots have two cotyledons and branching veins. The two groups also differ in the arrangement of their vascular tissues, number of flower petals, and other traits. Some eudicots undergo secondary growth and become woody, but no monocots produce true wood. We discuss the differences between monocots and dicots in greater detail in Section 27.2.

A Water lily (basal angiosperm)

B Magnolia (magnoliid)

C Iris (monocot)

D Chickweed (eudicot)

FIGURE 22.27 Representatives of four angiosperm lineages. The vast majority of angiosperms are monocots or eudicots.

Ecological and Economic Importance

It would be nearly impossible to overestimate the importance of angiosperms. As the most abundant plants in most habitats, they provide food and shelter for a variety of animals. Animals feed on angiosperm roots, shoots, and fruits, and they sip floral nectar.

Angiosperms provide nearly all human food, either directly or as food for livestock. Cereal grains are the most widely planted crops. All are grasses. Like other monocots, cereals store nutrients in their endosperm mainly as starch. In the United States, corn is the top cereal crop. Worldwide, rice feeds the greatest number of people. Other widely grown cereal crops include rye, oats, barley, sorghum, and wheat (FIGURE 22.28A).

The widespread use of cereal crops is partially a consequence of their varied carbon-fixing pathways (Section 6.7). C3 grasses such as rice, wheat, oats, and barley grow well under cool conditions. C4 grasses such as corn and sorghum thrive in hot, dry climates. Thus, people throughout the world can grow a cereal grain that is well suited to their climate.

Soybeans, lentils, peas, and peanuts are among the legumes, the second most important source of human food. Legumes are eudicots, and their endosperm stores nutrients as proteins and oils, as well as carbohydrates. Legumes can be paired with grains to provide all the amino acids that a human body needs to synthesize proteins.

Humans enliven their diet with a variety of plant parts. We dine on leaves of lettuce and spinach, stems of asparagus and celery, immature floral shoots of broccoli and cauliflower, and fleshy fruits of tomatoes and blueberries. Stamens of crocus flowers provide the spice saffron, and cinnamon is the grated bark of a tropical tree. Maple syrup is fluid tapped from a tree's xylem and boiled down to a syrupy consistency.

Angiosperms are the source of fabrics such as linen, ramie, hemp, burlap, and cotton (FIGURE 22.28B). The fruit of a cotton plant, the cotton boll, is nearly pure cellulose. Fibers from the leaves of agave are used to make sisal rugs.

Oils extracted from seeds of eudicots such as rapeseed and hemp are used in detergents, skin care products, and as industrial lubricants and fuel.

Saving Seeds (revisited)

The seed bank at Svalbard is maintained at a temperature just below freezing. Most flowering plants make seeds that remain viable for long periods after being dried and frozen at this temperature. However, seeds of some plants, most often tropical species, are usually short-lived when frozen under standard conditions. To preserve such species, botanists turn to cryopreservation, which is the storage of biological material at an ultralow temperature. Typically, the seeds are stored in tanks of liquid nitrogen at –190°C (–320°F), as shown in **FIGURE 22.29**. Cryopreservation can also be used to preserve spores of seedless plants and vegetative bits of plants that typically reproduce mainly by asexual mechanisms such as fragmentation.

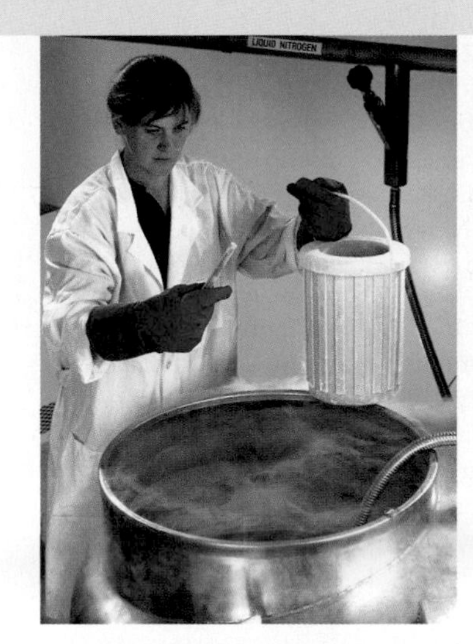

FIGURE 22.29 A botanist lowers a container of seeds into a vat of liquid nitrogen for long-term storage. Cryopreservation has the advantage of killing most plant pathogens in the plant material being preserved.

Eudicot woods are "hardwoods," as opposed to gymnosperm "softwoods." Furniture and flooring are often made from oak or other hardwoods. Hardwoods are also preferred as firewood.

We use secondary metabolites of plants as medicines and mood-altering drugs. A **secondary metabolite** is a compound with no known metabolic role in the organism that makes it. Many plant secondary metabolites are defenses against grazing animals. Aspirin is derived from a compound made by willows, and digitalis used to strengthen heartbeats is from foxglove plants. Caffeine from coffee beans and nicotine from tobacco leaves are widely used stimulants. Marijuana (**FIGURE 22.28C**) is one of the United States' most valuable cash crops. Worldwide, cultivation of opium poppies (the source of heroin) and coca (the source of cocaine) have wide-reaching health, economic, and political effects.

eudicots Most diverse lineage of angiosperms; members have two seed leaves, branching leaf veins.
monocots Highly diverse angiosperm lineage; includes plants such as grasses that have one seed leaf and parallel veins.
secondary metabolite Molecule that is produced by an organism but does not play any known role in its metabolism. Some serve as defense against predation.

TAKE-HOME MESSAGE 22.9

What are the major angiosperm lineages and how are they important?

✔ Most flowering plants are monocots or eudicots.

✔ The grains that serve as staples of the human diet are monocots. Many other crop plants are eudicots. We use the secondary metabolites of some flowering plants as medicines or mood-altering drugs.

A Mechanized harvesting of wheat.

B A field of cotton ready for harvest.

C Marijuana found growing illegally in Oregon.

FIGURE 22.28 Angiosperms as crops.

CREDITS: (28A) Photo USDA; (28B) Photo by Scott Bauer, USDA/ARS; (28C) Courtesy of Linn County, Oregon Sheriff's Office; (29) Photo by Scott Bauer/USDA Agricultural Research Service.

summary

Section 22.1 Sustaining many varieties of crop plants and their wild relatives ensures that plant breeders will have a reservoir of genetic diversity to tap into if widely planted varieties fail. Seed banks can help us maintain a wide variety of potentially valuable plant species. Seeds and other plant parts can be frozen for long-term storage.

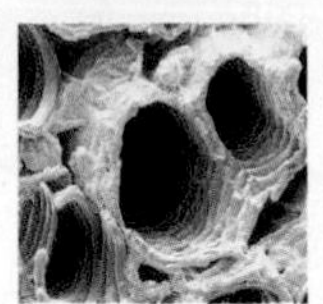

Sections 22.2, 22.3 **Embryophytes**, more commonly referred to as **plants**, are close relatives of charophyte green algae. Nearly all are photoautotrophs. Life cycles have changed as new lineages evolved. A **gametophyte** dominates the life cycles of the oldest lineages, the **bryophytes**. A **sporophyte** dominates the life cycle of **vascular plants**.

Features that contributed to success on land include **cuticle** and **stomata** that minimize water loss, **xylem** and **phloem** (**vascular tissues**), and **lignin** in cell walls. Evolution of **pollen grains** allowed seed plants to reproduce without standing water. **Seed plants** protect their embryo sporophytes in **seeds**.

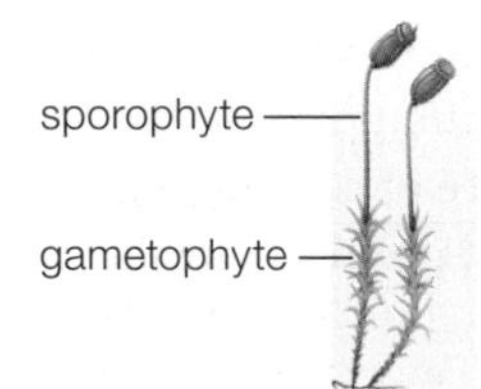

Section 22.4 Three lineages (mosses, liverworts, and hornworts) are collectively referred to as bryophytes. Mosses are the most diverse bryophytes and the main plants in **peat bogs**.

Bryophytes are nonvascular plants (no xylem or phloem). The sporophyte, which produces spores in a **sporangium**, is attached to and nourished by the gametophyte. **Rhizoids** attach a gametophyte to soil or another surface. **Gametangia** in or on the surface of the gametophyte produce gametes. Sperm are flagellated and swim to eggs through a film of water.

Section 22.5 Club mosses belong to one lineage of seedless vascular plants; horsetails, whisk ferns, and true ferns belong to another. In both lineages, the sporophyte dominates the life cycle. Roots and aboveground stems grow from **rhizomes**. Spore-bearing structures include the **strobili** of horsetails and the **sori** of ferns. Many ferns live as **epiphytes** attached to another plant. Sperm swim through water to reach eggs.

Section 22.6 The earliest vascular plants were tiny and had a simple branching pattern. By the Carboniferous, swamp forests were dominated by giant lycophytes. The remains of these forests became **coal**.

Seed plants, which began to diversify during the Carboniferous, do not release spores. Their sporophytes have **pollen sacs**, where **microspores** form and develop into pollen grains (male gametophytes). **Megaspores** form inside **ovules** and develop into female gametophytes. **Pollination** unites the egg and sperm of a seed plant. A seed is a mature ovule. It includes nutritive tissue and a tough seed coat that protects the embryo sporophyte inside the seed from harsh conditions.

Section 22.7 **Gymnosperms** include the **conifers** and the lesser-known **cycads**, **ginkgos**, and **gnetophytes**. Most conifers are evergreen trees, and the group is an important source of lumber. Conifers produce soft, pollen-bearing cones as well as woody, ovulate cones. They are wind pollinated.

Sections 22.8, 22.9 **Angiosperms** are the most diverse plants. They alone produce **flowers**, which are modified shoots. The **stamens** of a flower produce pollen. An **ovary** at the base of a **carpel** holds one or more ovules. After pollination and **double fertilization**, the flower's ovary becomes a **fruit** that contains one or more seeds. A flowering plant seed includes an embryo sporophyte and **endosperm**, a nutritious tissue.

Factors that contributed to angiosperm success include a short life cycle, **coevolution** with **pollinators**, and a variety of mechanisms for dispersing fruits.

As the dominant plants in most land habitats, flowering plants are ecologically important, as well as essential to human existence. The two major lineages are **monocots** and **eudicots**. The starch-rich endosperm of monocot seeds such as wheat and rice makes them staples of human diets throughout the world. Angiosperms also supply us with vegetables, spices, fiber, furniture, oils, medicines, and mood-altering drugs. Many useful plant compounds are **secondary metabolites**, compounds that probably help defend the plant against predation.

self-quiz

Answers in Appendix VII

1. The first land plants were ________ .
   a. gnetophytes
   b. gymnosperms
   c. bryophytes
   d. lycophytes

2. Lignin is not found in stems of ________ .
   a. mosses
   b. ferns
   c. monocots
   d. a and b

3. A waxy cuticle helps land plants ________ .
   a. conserve water
   b. take up carbon dioxide
   c. reproduce
   d. stand upright

4. True or false? Ferns produce seeds inside sori.

5. ________ attach mosses to soil.
   a. Rhizoids
   b. Rhizomes
   c. Roots
   d. Strobili

6. Bryophytes alone have a relatively large ________ and an attached, dependent ________ .
   a. sporophyte; gametophyte
   b. gametophyte; sporophyte

7. Club mosses, horsetails, and ferns are ________ plants.
   a. multicelled aquatic
   b. nonvascular seed
   c. seedless vascular
   d. seed-bearing vascular

8. Coal consists primarily of compressed remains of the ________ that dominated Carboniferous swamp forests.
   a. seedless vascular plants
   b. conifers
   c. flowering plants
   d. hornworts

data analysis activities

Insect-Assisted Fertilization in Moss Moss sperm can swim, but plant ecologist Nils Cronberg suspected that they sometimes hitch a ride on crawling insects or mites (tiny animals related to spiders). To test this hypothesis he carried out an experiment. He placed patches of male and female moss gametophytes in dishes, either next to one another or with water-absorbing plaster between them. The plaster prevented sperm from swimming between plants. He then looked at how the presence or absence of insects affected the number of sporophytes formed. **FIGURE 22.30** shows his results.

1. Why is sporophyte formation a good way to determine if fertilization occurred?

2. How close did the male and female patches have to be for sporophytes to form in the absence of insects?

3. Does this study support the hypothesis that insects and mites aid moss fertilization?

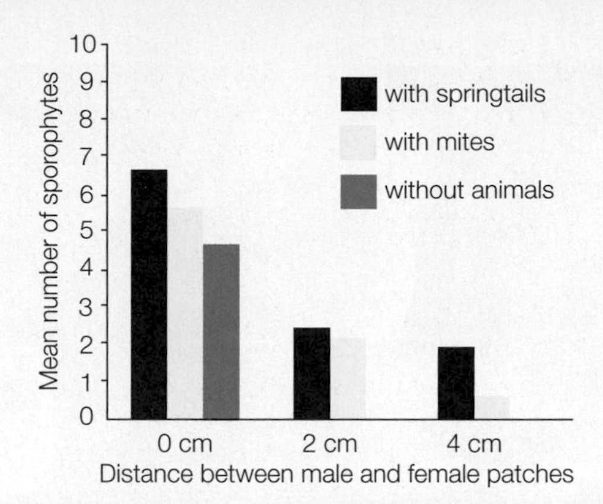

FIGURE 22.30 Sporophyte production in female moss patches with and without either crawling insects (springtails) or mites. No sporophytes formed in the animal-free dishes when moss patches were 2 or 4 centimeters apart.

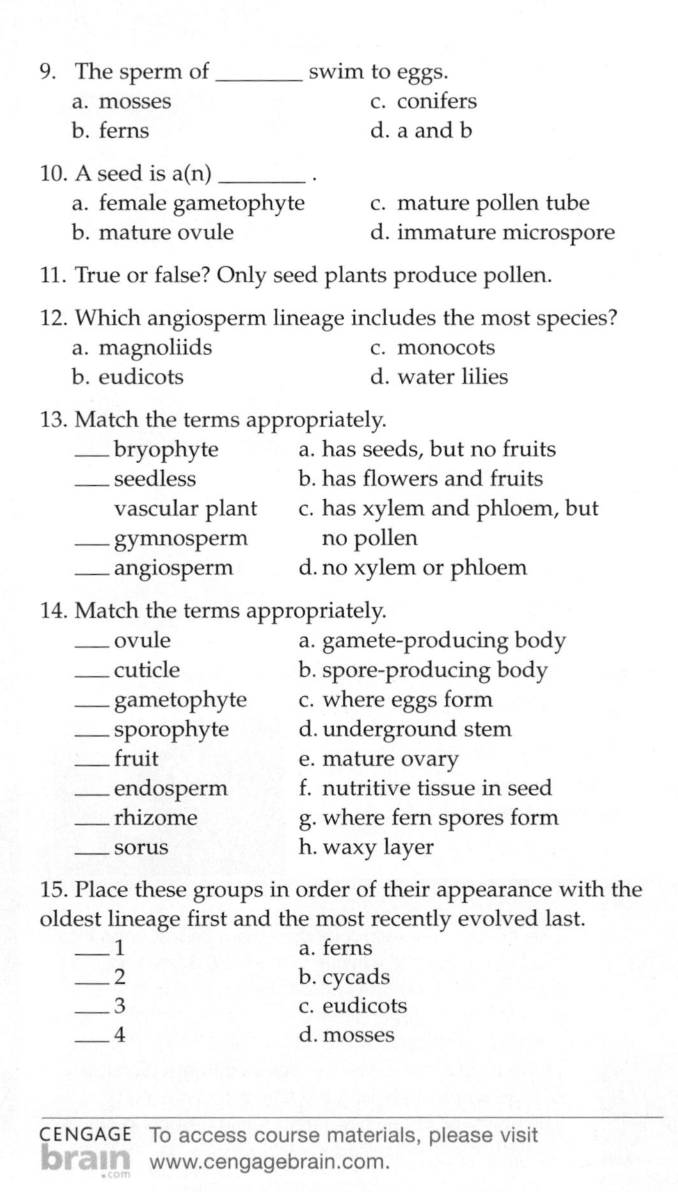

9. The sperm of ________ swim to eggs.
   a. mosses c. conifers
   b. ferns d. a and b

10. A seed is a(n) ________ .
    a. female gametophyte c. mature pollen tube
    b. mature ovule d. immature microspore

11. True or false? Only seed plants produce pollen.

12. Which angiosperm lineage includes the most species?
    a. magnoliids c. monocots
    b. eudicots d. water lilies

13. Match the terms appropriately.

___ bryophyte	a. has seeds, but no fruits
___ seedless vascular plant	b. has flowers and fruits
___ gymnosperm	c. has xylem and phloem, but no pollen
___ angiosperm	d. no xylem or phloem

14. Match the terms appropriately.

___ ovule	a. gamete-producing body
___ cuticle	b. spore-producing body
___ gametophyte	c. where eggs form
___ sporophyte	d. underground stem
___ fruit	e. mature ovary
___ endosperm	f. nutritive tissue in seed
___ rhizome	g. where fern spores form
___ sorus	h. waxy layer

15. Place these groups in order of their appearance with the oldest lineage first and the most recently evolved last.

___ 1	a. ferns
___ 2	b. cycads
___ 3	c. eudicots
___ 4	d. mosses

critical thinking

1. Early botanists admired ferns but found their life cycle perplexing. In the 1700s, they learned to propagate ferns by sowing what appeared to be tiny dustlike "seeds" from the undersides of fronds. Despite many attempts, the scientists could not find the pollen source, which they assumed must stimulate these "seeds" to develop. Imagine you could write to one of these botanists. Compose a note that would clear up the confusion.

2. With the exception of the bryophytes, the dominant stage in land plants is the diploid sporophyte. By one hypothesis, diploid dominance was favored because it allowed a greater level of genetic diversity. Suppose that a recessive mutation arises. It is mildly disadvantageous now, but it will be useful in some future environment. Explain why such a mutation would be more likely to persist in a plant with a sporophyte-dominated life cycle, such as a fern, than in a moss.

3. A pine tree produces far more pollen than an apple or cherry tree produces. Explain why a fruit tree does not need to make as much pollen as a pine to ensure pollination.

4. Compared to vascular plants, bryophytes are more easily harmed by environmental contaminants and acid rain. Thus the diversity of these plants is sometimes used to assess pollution levels. What is it about the structure or function of a moss that would make it more susceptible to damage from air pollution than a horsetail or a cactus?

5. Lignin and vascular tissue first evolved in relatives of club moss, and some extinct species stood 40 meters (130 feet) high. Explain how the evolution of vascular tissues and lignin would have allowed a dramatic increase in plant height. How might being tall give one plant species a competitive advantage over another?

CREDIT: (30) *Science*, 1 September 2006: Vol 313 no 5791 p.1255 DOI:10.1126/scince.1128707.

23 Fungi

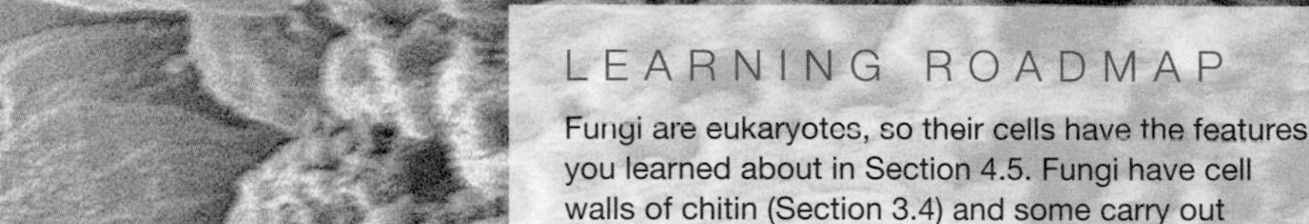

LEARNING ROADMAP

Fungi are eukaryotes, so their cells have the features you learned about in Section 4.5. Fungi have cell walls of chitin (Section 3.4) and some carry out fermentation (7.6). We discussed the photosynthetic components of lichens earlier (20.7 and 21.9). Here we look at the fungal component.

TRAITS AND CLASSIFICATION

Fungi are heterotrophs that secrete digestive enzymes onto organic matter, then absorb the released nutrients. Fungi reproduce sexually and asexually by producing spores.

THE OLDEST LINEAGES

Chytrids are an ancient lineage of flagellated fungi. The zygote fungi include diverse molds. Some relatives of zygote fungi live as parasites in animal cells. Others partner with plant roots.

SAC FUNGI AND CLUB FUNGI

Many sac fungi and club fungi make complex spore-bearing structures such as mushrooms. Meiosis in cells on these structures produces spores.

LIVING TOGETHER

Many fungi live on, in, or with other species. Some benefit plants by enhancing the plant's ability to take up nutrients. Others form lichens by living with algae or cyanobacteria.

FUNGAL PATHOGENS AND TOXINS

A minority of fungi are parasites. Some of these species cause disease in humans. Fungi also make toxins that can be deadly when eaten.

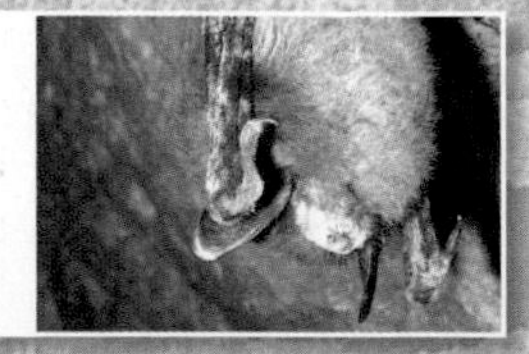

This chapter considers fungus–plant interactions from the fungal point of view. In Section 28.3, we focus on how these interactions affect plant nutrition. Fungal pathogens of humans come up once again when we discuss symptoms of AIDS (Section 37.11). The role of fungi as decomposers is covered in our discussion of food webs (46.3) and the nitrogen cycle (46.9). The ability of fungal hyphae to retain carbon in soil is discussed in Section 46.7.

23.1 High-Flying Fungi

Fungi are not known for their mobility. You probably don't think of mushrooms and their relatives as world travelers, but some do get around. Fungi produce microscopic spores that can lodge in crevices on tiny dust particles. When winds carry these particles aloft, spores go along for the ride. Dustborne fungal spores can travel long distances riding the winds that swirl high above Earth's surface.

Dust storms in North African deserts sometimes lift fungus-laden particles more than 4.5 kilometers (3 miles) above the desert floor. Winds carry this dust out over the Atlantic Ocean, and sometimes completely across it. Most fungal spores that hitchhike on African dust are harmless, but occasionally spores of fungi that cause plant disease make the journey. For example, winds of a 1978 cyclone introduced sugar cane rust (a fungal disease) from Cameroon to the Dominican Republic. Similarly, winds probably carried coffee rust fungus from Angola to Brazil in 1980.

Today, an African outbreak of an old fungal foe has agricultural officials around the world on edge. The fungus that arouses their concern, *Puccinia graminis*, causes wheat stem rust disease. Like other rust fungi, *P. graminis* is an obligate plant parasite, meaning it can grow and reproduce only in living plant tissue. An infection begins when a spore lands on the leaf of a wheat plant. The spore germinates, and a fungal filament enters the plant through a stoma. As fungal filaments grow through the plant's tissues, they suck up photosynthetic sugars that the plant would normally use to meet its own needs. As a result, an infected plant is stunted and produces little or no wheat. About a week after a plant becomes infected, tens of thousands of rust-colored spores appear on its stem (**FIGURE 23.1**). Each spore can disperse and infect a new plant.

Wheat stem rust disease routinely decimated wheat crops worldwide until the 1960s, when strains resistant to *P. graminis* became available. The fungus-resistant strains were the product of a plant breeding program headed by Norman Borlaug. The great importance of Borlaug's work was highlighted in 1970, when he received the Nobel Peace Prize. He was honored for his role in preventing food shortages that can contribute to global instability.

Worldwide use of rust-resistant wheats provided a respite from outbreaks of wheat stem rust for decades. Then, in 1999, a new strain of wheat stem rust (called Ug99) was discovered in Uganda, a country in eastern Africa. Ug99 had mutations that allowed it to infect most wheat varieties that were previously considered resistant to wheat stem rust.

Since then, windblown spores of Ug99 have dispersed to Kenya, Ethiopia, South Africa, and Sudan, crossed the Red Sea to Yemen, and from there crossed the Persian Gulf to Iran. India, the world's second largest wheat producer, is expected to be affected soon. Most likely, winds will eventually distribute Ug99 throughout the world. Fungicides can help minimize the damage but are too expensive for farmers in many developing nations.

With one of the world's most important crops at risk, plant scientists set to work. They are working to develop new varieties of wheat with genes that will provide resistance against Ug99. Some Ug99-resistant wheats have already been developed, but multiplying enough seed and deploying it to farmers who currently grow susceptible varieties is a challenge. Farmers save their own wheat seed and resow it, so the resistant wheat is not of commercial interest to seed companies that have the capacity to distribute it.

FIGURE 23.1 Wheat stem, with rust-colored fungal sporangia (spore-producing structures) on its surface. The inset micrograph shows spores (red) escaping from the sporangia.

CREDITS: (opposite) Eye of Science/Science Source; (l) background, Photo by Yue Jin/USDA; inset, Courtesy of Charles Good, Ohio State University at Lima.

FIGURE 23.2 Multicelled fungi.

✔ Fungi are heterotrophs that obtain nutrition by extracellular digestion, and disperse by producing spores.

Structure and Function

Like plants, **fungi** have walled cells, spend their lives fixed in place, and produce haploid spores by meiosis. However, fungi are more closely related to animals than to plants. Like animals, fungi are heterotrophs that store excess sugars as glycogen. Most fungi are decomposers that feed on organic wastes and remains. A lesser number live on or in other living organisms.

Fungal digestive enzymes break down many sturdy structural proteins that animal digestive enzymes cannot. For example, fungi digest cellulose and lignin in plant cell walls, as well as keratin, the main protein in animal skin, claws, hair, and fur. The capacity to digest these tough structural materials makes fungi important both as decomposers and as parasites.

Some fungi live as single cells, and are commonly called yeasts. However, most fungi are multicelled. Molds, mildews, and mushrooms are familiar examples of multicelled fungi (**FIGURE 23.2**). These fungi grow as a **mycelium** (plural, mycelia), a network of microscopic interwoven filaments. Each filament, or **hypha** (plural, hyphae), is a strand of walled cells arranged end to end (**FIGURE 23.3**). Fungal cell walls consist primarily of chitin, a polysaccharide also found in the body covering of crabs and insects (Section 3.4).

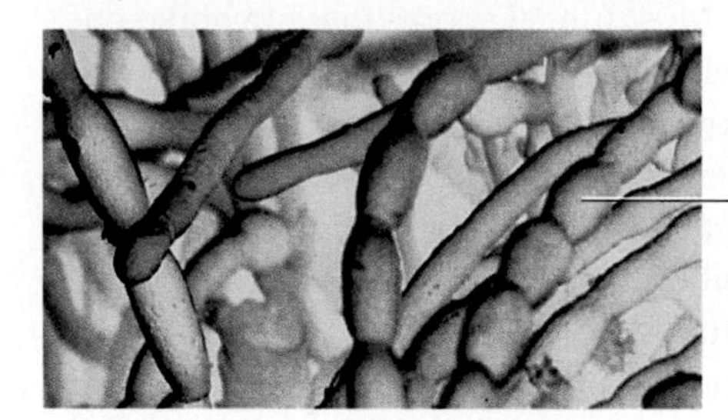

FIGURE 23.3 A mycelium, a mass of threadlike hyphae. Each hypha is a strand of cells attached one to the other.

The structure of fungal hyphae varies among groups. In the oldest fungal lineages, which include chytrids, zygote fungi (zygomycetes), and glomeromycetes, cells of a hypha do not have cross-walls between them. Each hypha is a long tube full of cytoplasm and nuclei. Hyphae divided into compartments by cross-walls, or septae (singular, septa), evolved in the common ancestor of the sac fungi and club fungi. These

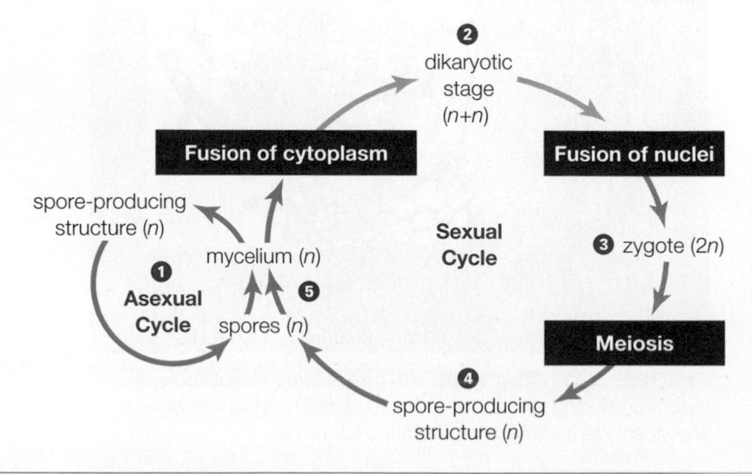

FIGURE 23.4 ▶**Animated** Generalized life cycle of a fungus.

❶ Asexual reproduction occurs when a haploid mycelium produces spores by mitosis at the tips of specialized hyphae.

❷ Sexual reproduction begins when haploid hyphae of two individuals meet and cells at their tips fuse. This cytoplasmic fusion produces a dikaryotic cell (a cell with two nuclei).

❸ Fusion of nuclei in a dikaryotic cell creates a diploid zygote.

❹ The zygote undergoes meiosis and produces a haploid spore-bearing structure.

❺ Haploid spores form by mitosis and germinate to form a new haploid mycelium.

are the two most diverse fungal lineages, and septate hyphae contributed to their success.

When cross-walls are present, they are porous, so materials can flow between adjacent cells:

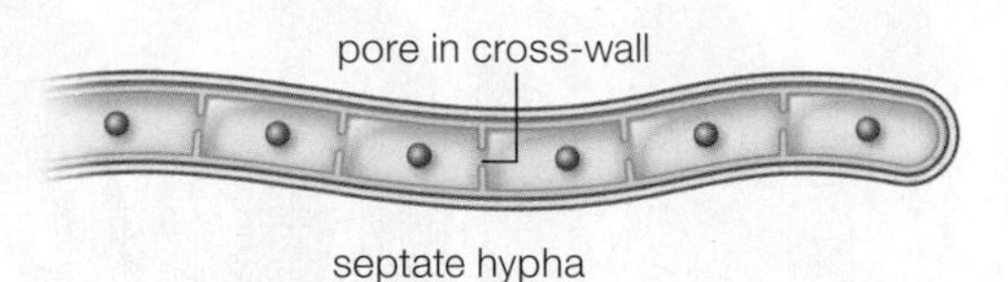

As a result, a fungus can share nutrients or water taken up in one part of its mycelium with cells in other parts of the fungal body. The presence of cross-walls made hyphae sturdier, allowing the evolution of larger, more elaborate spore-producing bodies, also known as fruiting bodies. Fungi with septate hyphae also are more resistant to drying out than those with nonseptate hyphae. Thus sac fungi and club fungi can live in drier environments than other fungi.

Life Cycles

Fungi produce spores both sexually and asexually (**FIGURE 23.4**). During asexual reproduction, multicelled fungi form spores by mitosis at the tips of specialized hyphae ❶. Sexual reproduction begins when two hyphae meet and the cytoplasm of cells at their tips fuses to produce a dikaryotic cell ❷. **Dikaryotic** means having two genetically distinct types of nuclei ($n+n$). The fungal zygote forms when the two nuclei inside a dikaryotic cell fuse ❸. The resulting diploid zygote undergoes meiosis and gives rise to a structure that produces haploid spores by mitosis ❹. Those spores germinate, releasing cells that divide by mitosis to form a new haploid mycelium ❺.

dikaryotic Having two genetically different nuclei in a cell ($n+n$).
fungus Eukaryotic heterotroph with cell walls of chitin; obtains nutrients by extracellular digestion and absorption.
hypha Component of a fungal mycelium; a filament made up of cells arranged end to end.
mycelium Mass of threadlike filaments (hyphae) that make up the body of a multicelled fungus.

TAKE-HOME MESSAGE 23.2
What are characteristics of fungi?

✔ Fungi are heterotrophs that secrete digestive enzymes to break down organic material, then absorb the released nutrients. Some live as single cells; others grow as a multicelled mycelium. All have cell walls with chitin.

✔ Fungi disperse by producing spores, which may be produced either by mitosis or meiosis.

23.3 Flagellated Fungi

✔ Chytrids are the only modern fungi with a life cycle that includes flagellated cells.

Chytrids (Chytridiomycota) include the only fungal lineages that form flagellated spores through mitosis. The group name, which comes from the Greek word *chytridion*, means "little pot." It refers to the vessel-like sporangium in which the spores form (**FIGURE 23.5A**). Like animal sperm, chytrid zoospores have a single flagellum at the rear of the cell that pushes the cell forward. Some chytrid species live as single cells and others form hyphae. When hyphae do form, there are no cross-walls between the cells.

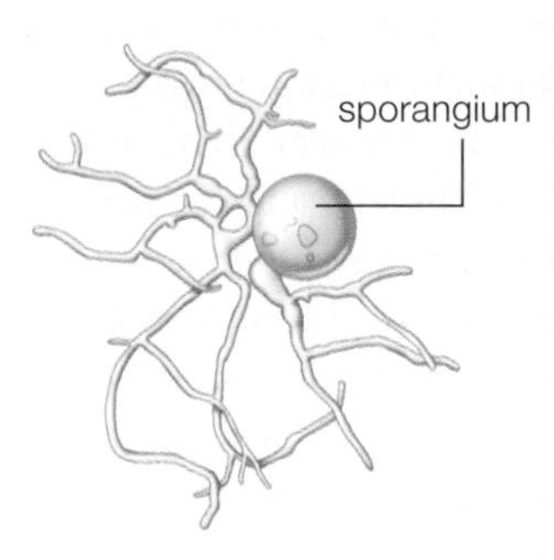

A Structure of a hypha-forming chytrid. Flagellated zoospores form inside the sporangium.

B Frogs that have been killed by chytridiomycosis, a disease caused by a parasitic chytrid.

FIGURE 23.5 Chytrids.

Chytrids thrive in a variety of habitats. Many are decomposers in freshwater, nearshore coastal waters, or in soil. Others live in the gut of mammalian grazers such as sheep and cattle, where they assist their hosts by breaking down cellulose. Still other are parasites; chytrid parasites infect protists, plants, and animals.

Amphibian populations worldwide are currently threatened by the spread of a parasitic chytrid that causes chytridiomycosis (**FIGURE 23.5B**). The chytrid infects the animal's skin, causing it to thicken. Thickened skin prevents an amphibian from absorbing water properly, so it eventually dies of dehydration.

chytrid Fungus that produces flagellated spores.

TAKE-HOME MESSAGE 23.3
What are chytrids?

✔ Chytrids are fungi that produce flagellated spores. Most are decomposers in soil, but some live in the animal gut, and some are parasites. The spread of a parasitic species that kills amphibians is a matter of concern.

23.4 Zygote Fungi and Related Groups

✔ Zygote fungi form a branching haploid mycelium on organic material, and inside living plants and animals.

Zygote Fungi

The 1,100 or so species of **zygote fungi** (zygomycetes) live in damp places. Many are molds, meaning they typically grow over or through organic matter as a mass of asexually reproducing hyphae. As noted in Section 23.2, the hyphae of zygote fungi consist of haploid cells without septae between them.

Black bread mold (*Rhizopus stolinifera*) has a life cycle typical of zygote fungi (FIGURE 23.6). As long as food is plentiful, it grows as a haploid mycelium and produces spores by mitosis ❶. If the food supply dwindles and a compatible sexual partner is nearby, hyphae develop special side branches (gametangia) ❷. Zygote fungi do not have walls between their cells, so many haploid nuclei from within a hypha can flow into each gametangium. When gametangia of two individuals come into contact, their walls break down, and their cytoplasm fuses. The result is an immature zygospore containing multiple nuclei from each parent ❸. As the zygospore matures, haploid nuclei from the two parents pair up and form diploid nuclei. A mature zygospore contains multiple diploid nuclei and has a thick, protective wall ❹. It is the only diploid stage in the zygote fungus life cycle. When the zygospore germinates, a hypha emerges and cells at its tip undergo meiosis to produce haploid spores ❺.

Many *Rhizopus* species spoil foods, but others are used to produce tempeh, a high-protein fermented soy product with a chewy, meatlike consistency. Hyphae of the fungus bind the partially fermented soybeans together into a dense, sliceable form.

FIGURE 23.7 Spore-bearing structures of *Pilobolus*. The name means "hat-thrower." The dark "hats" are sporangia (spore sacs).

Another zygote fungus, *Pilobolus*, is notable for its interesting mechanism of spore dispersal. Grazing animals ingest *Pilobolus* spores along with grass. The spores pass through the animal's gut and are deposited in feces. They germinate and grow into a mycelium that produces spore-bearing hyphae (FIGURE 23.7). At the tip of each hypha, a tiny dark sporangium perches atop a large fluid-filled vesicle. When fluid pressure builds up in the vesicle, it bursts, hurling the sporangium as far as 2 meters (about 6.5 feet) away. Animals ingest the sporangium and the process begins again.

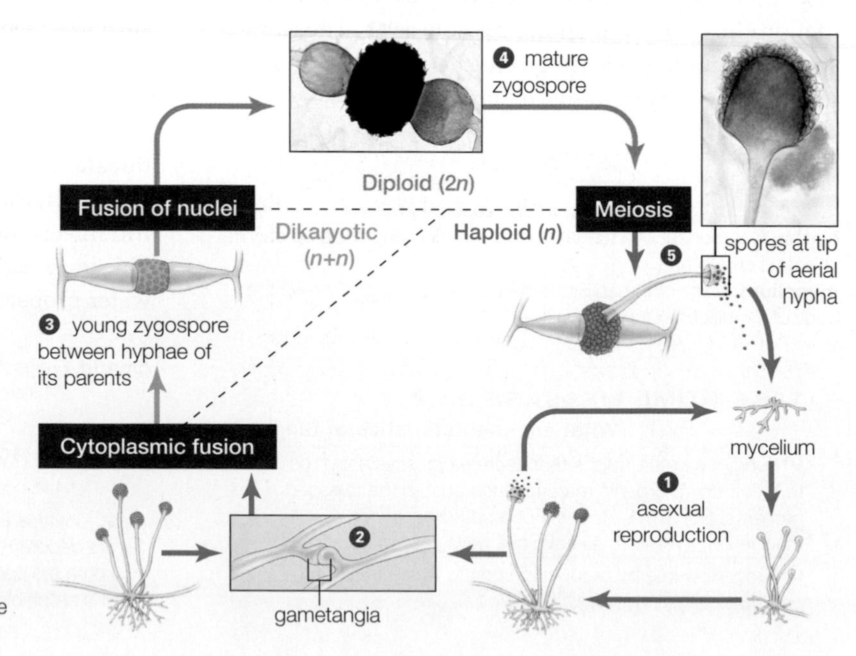

FIGURE 23.6 ▶Animated Life cycle of black bread mold, a zygote fungus.

❶ As long as food is plentiful, a haploid mycelium grows in size and produces spores by mitosis on specialized hyphae.

❷ When nutrients are limited and hyphae of two compatible individuals come into close proximity, they produce branches (gametangia) that grow toward one another. As these branches grow, haploid nuclei stream into them and accumulate at their tips.

❸ Cytoplasmic fusion of the gametangia produces a zygospore that contains many haploid nuclei from each parent.

❹ Nuclei within the zygospore pair up and fuse to produce a mature zygospore with many diploid nuclei.

❺ The zygospore germinates, and an aerial hypha emerges. Meiosis of cells within this hypha gives rise to haploid spores, which are released from its tip.

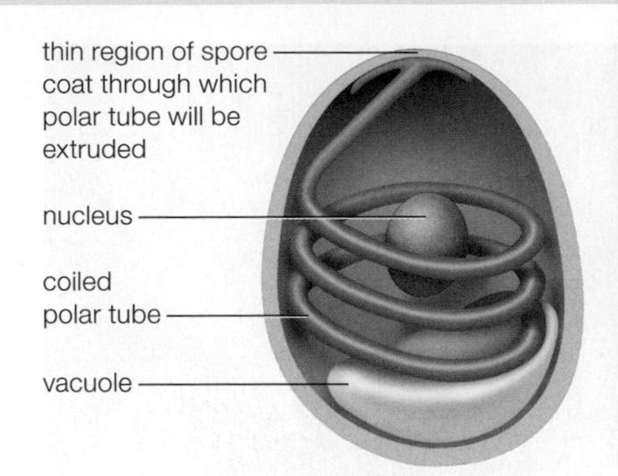

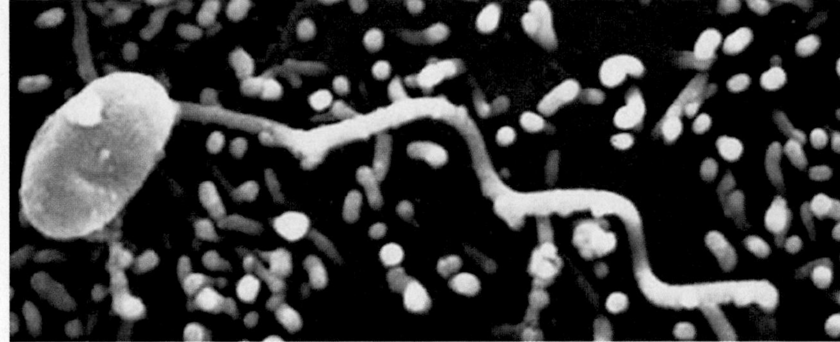

A Diagram of a microsporidian spore. **B** Microsporidium with its polar tube extruded (scanning electron micrograph).

FIGURE 23.8 A microsporidium. Microsporidia are single-celled intracellular parasites of animals.

Microsporidia—Intracellular Parasites

Microsporidia (singular, microsporidium) include about 1,300 named species of single-celled intracellular parasites. Most infect animals, with fish and insects being the most common hosts. The relationship of microsporidians to other fungi remains unclear, although some genetic similarities suggest that they may be closer to zygote fungi than to other fungal groups.

A microsporidium spore has a tough coat that allows it to survive adverse conditions for years. Beneath the coat, a long tube lies coiled in the cytoplasm (**FIGURE 23.8**). During infection, the tube uncoils and perforates a host cell. The microsporidium's nucleus and cytoplasm then flow through the tube into the host.

Some microsporidia infect the human gut, but they generally cause symptoms only in people with a weakened immune system. Microsporidia of the genus *Nosema* infect bees, and the spread of these parasites is contributing to an ongoing loss of honeybee colonies.

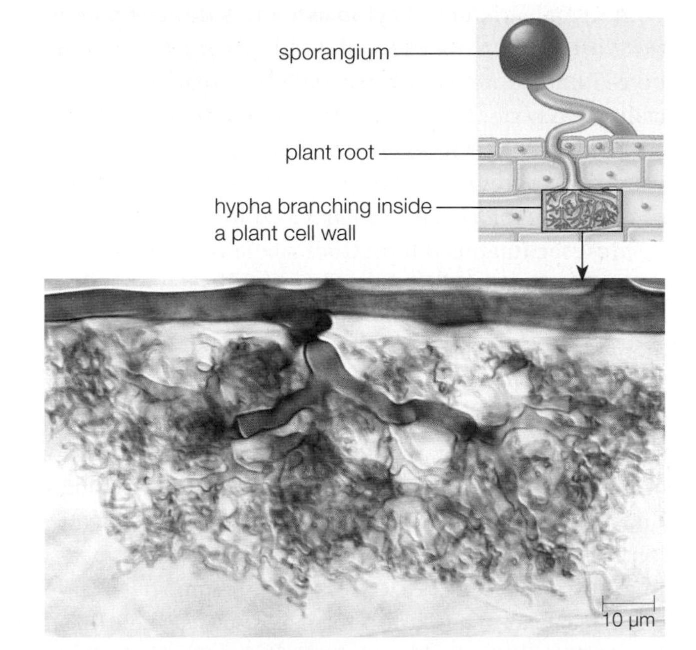

FIGURE 23.9 Glomeromycete fungus. Specialized hyphae of these fungi enter and branch inside the cells of a plant root.

Glomeromycetes—Partners of Plants

Glomeromycetes were previously placed with zygote fungi, but are now considered a separate lineage. All 150 or so species take part in a plant root–fungus partnership called a **mycorrhiza** (plural, mycorrhizae). A glomeromycete hypha grows into a root and branches inside the wall of a root cell (**FIGURE 23.9**). Having a fungal roommate does not harm a root cell; the fungus shares nutrients it takes up from the soil with its host. We return to fungus–plant associations in Section 23.7.

glomeromycete Fungus with hyphae that grow inside the wall of a plant root cell.
microsporidium Single-celled spore-forming fungus that is an intracellular animal parasite.
mycorrhiza Mutually beneficial partnership between a fungus and a plant root.
zygote fungus Fungus that forms a zygospore during sexual reproduction.

TAKE-HOME MESSAGE 23.4

What are zygote fungi and their relatives?

✔ Zygote fungi form a thick-walled diploid spore when they reproduce sexually. Some spoil food or cause disease.

✔ Microsporidia are single-celled parasites that invade an animal cell by way of a polar tube.

✔ Glomeromycetes are mycorrhizal fungi; they partner with plant roots.

23.5 Sac Fungi—Ascomycetes

✔ Sac fungi are the most diverse fungal group. There are single-celled and multicelled forms.

Sac fungi, or ascomycetes, are the most diverse fungal lineage, with more than 64,000 known species. Some are yeasts (single-celled), but most are multicelled. Hyphae of multicelled species have cross-walls at regular intervals.

Life Cycles

Sac fungi commonly reproduce by asexual mechanisms. Yeasts often reproduce asexually by **budding**, a process in which mitosis is followed by unequal cytoplasmic division. During budding, a descendant cell with a small amount of cytoplasm buds from its parent (FIGURE 23.10A). Sac fungal molds grow as a haploid mycelium and reproduce asexually by means of specialized hyphae. These hyphae produce spores by mitosis in cells at their tips (FIGURE 23.10B). Mitotically produced spores of sac fungi are called conidiospores or conidia. *Conidia* is the Greek word for dust.

Most sac fungi can reproduce sexually. When they do, spores form within a saclike cell called an ascus (plural, asci). In single-celled species, the ascus stands alone. In multicelled species, the ascus often forms on a fruiting body called an ascocarp. In ascocarp-forming fungi, sexual reproduction begins when the haploid hyphae of two compatible individuals meet and cells at their tips undergo cytoplasmic fusion. The resulting dikaryotic cell divides by mitosis to produce dikaryotic hyphae. These hyphae intertwine with haploid hyphae to form an ascocarp (FIGURE 23.10C). Some of dikaryotic hyphae in the ascocarp form an ascus at their tip. Inside these asci, nuclear fusion followed by meiosis produces four haploid cells. Generally, these cells then divide by mitosis to yield eight haploid spores.

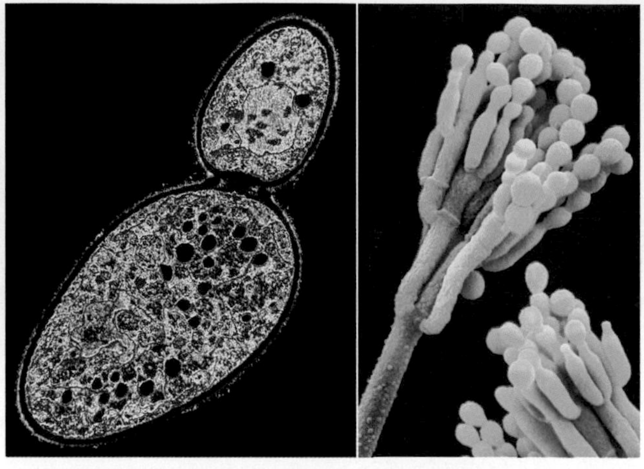

A Asexual reproduction by budding in a yeast (*Saccharomyces*). B Conidia (asexually produced spores) of the mold *Eupenicillium*.

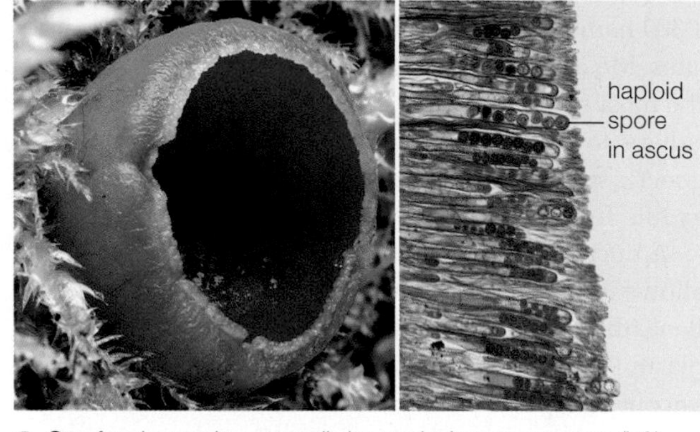

C Cup fungi reproduce sexually by producing an ascocarp (left). Spores (right) form by meiosis of cells on the cup's concave surface.

FIGURE 23.10 Modes of reproduction in sac fungi.

Diversity

Sac fungi have a variety of ecological roles. Most sac fungi live in the soil and serve as decomposers. Other soil sac fungi form mycorrhizae with plant roots. Sac fungi are also the fungi most commonly found in lichens. Pathogenic sac fungi include powdery mildews that grow on plant leaves and the fungi that commonly cause human yeast infections.

Human Uses of Sac Fungi

Of all fungal groups, sac fungi have the greatest number of commercial uses. Morels (FIGURE 23.11A) and truffles (FIGURE 23.11B) are examples of edible ascocarps. Truffles form underground. When spores mature, the fungus gives off an odor like that emitted by an amorous male pig. Female pigs that catch a whiff of this scent disperse truffle spores as they root through the soil in search of a seemingly subterranean suitor. Dogs can also be trained to snuffle out truffles. The search can be highly lucrative. In 2006, a single 1.5-kilogram (about 3-pound) Italian truffle sold for $160,000.

Some sac fungi are used to produce foods and beverages (FIGURE 23.12). One *Aspergillus* species ferments soybeans and wheat in the production of soy sauce; another makes citric acid used as a preservative and to flavor soft drinks. *Penicillium roqueforti* adds tangy blue veins to cheeses such as Roquefort. A packet of baker's yeast holds spores of the sac fungus *Saccharomyces cerevisiae*. When these spores germinate in bread dough, they ferment sugar and produce carbon dioxide that causes the dough to rise. A different strain of *S. cerevisiae* helps produce beer and wine.

A Morels. Asci line pits on the upper part of the fruiting body.

B Truffles. These ascocarps form underground; asci are inside.

FIGURE 23.11 Two types of edible ascocarp.

FIGURE 23.12 Products made with the assistance of sac fungi. Yeasts carry out fermentation necessary to make bread and wine. The blue material in "blue cheese" is hyphae of a sac fungal mold.

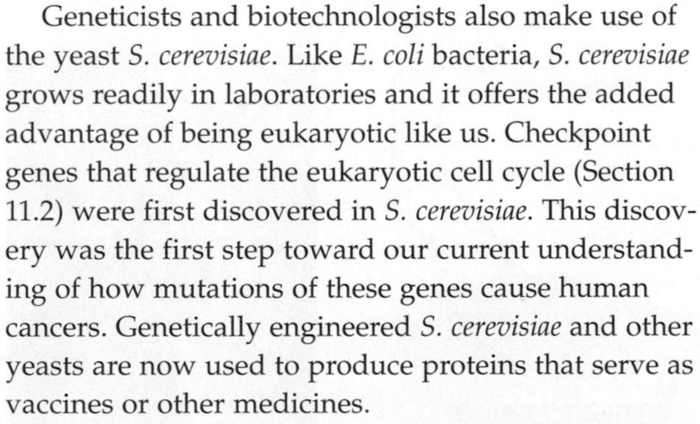

Geneticists and biotechnologists also make use of the yeast *S. cerevisiae*. Like *E. coli* bacteria, *S. cerevisiae* grows readily in laboratories and it offers the added advantage of being eukaryotic like us. Checkpoint genes that regulate the eukaryotic cell cycle (Section 11.2) were first discovered in *S. cerevisiae*. This discovery was the first step toward our current understanding of how mutations of these genes cause human cancers. Genetically engineered *S. cerevisiae* and other yeasts are now used to produce proteins that serve as vaccines or other medicines.

Sac fungi have provided us with some useful medications. Most famously, the initial source of the antibiotic penicillin was the soil fungus *Penicillium chrysogenum*. Another antibiotic, cephalosporin, was first isolated from *Cephalosporium*. Statins from *Aspergillus* help lower cholesterol levels, and cyclosporin from *Trichoderma* helps prevent the rejection of transplanted organs.

budding Mechanism of asexual reproduction by which a small cell forms on a parent, then is released.
sac fungi Most diverse group of fungi; sexual reproduction produces spores inside a saclike structure (an ascus).

TAKE-HOME MESSAGE 23.5

What are sac fungi?

✔ Sac fungi are the most diverse group of fungi. Some are yeasts, but most are multicelled.

✔ Yeasts reproduce asexually by budding and multicelled species by forming conidia. Sac fungi that reproduce sexually typically form spores inside an ascus.

✔ We use sac fungi as sources of food, beverages, and drugs. Sac fungal yeasts are important in biotechnology and scientific research.

23.6 Club Fungi—Basidiomycetes

✔ Club fungi make the largest and most elaborate fruiting bodies; some familiar mushrooms are examples.

Life Cycle

Club fungi are fungi that form spores inside club-shaped cells during sexual reproduction. Most often the spores form on a spore-bearing structure (a basidiocarp) composed of dikaryotic hyphae. Basidocarp-producing club fungi do not produce spores asexually.

The button mushroom often seen sliced atop pizzas and in salad bars is a familiar club fungus (**FIGURE 23.13**). Most of the time, its septate hyphae grow through soil and feed by decomposing organic material. When hyphae of two mushroom-forming club fungi meet, they fuse and form a dikaryotic mycelium ❶. This mycelium continues to grow through the soil unseen and can reach great size.

Embryonic mushrooms develop on the mycelium. When rains come, hyphae soak up water, and these tiny mushrooms expand and break through the soil surface ❷. Producing fruiting bodies after a rain ensures that spores will disperse when conditions in the environment are most likely to favor their germination. Spores must take up water to germinate.

The underside of a mushroom's cap has thin tissue sheets (gills) fringed with club-shaped, dikaryotic cells ❸. Fusion of the nuclei in a dikaryotic cell forms a diploid zygote ❹. The zygote undergoes meiosis, forming four haploid spores ❺. After dispersal, these spores germinate and the life cycle begins again ❻.

Ecology

There are more than 32,000 club fungus species. They include all commercially grown mushrooms. Many club fungi are capable of breaking down lignin, thus making this group important decomposers and parasites of woody plants. For example, the shelf fungus shown in **FIGURE 23.14A** is feeding on its host tree. Hyphae that extend into the tree's heartwood weaken it, and will eventually kill it.

Jelly fungi (**FIGURE 23.14B**) form gelatinous fruiting body on logs and tree trunks, but do not feed on wood. Rather, they steal nutrients from the mycelia of other fungi that do feed on wood.

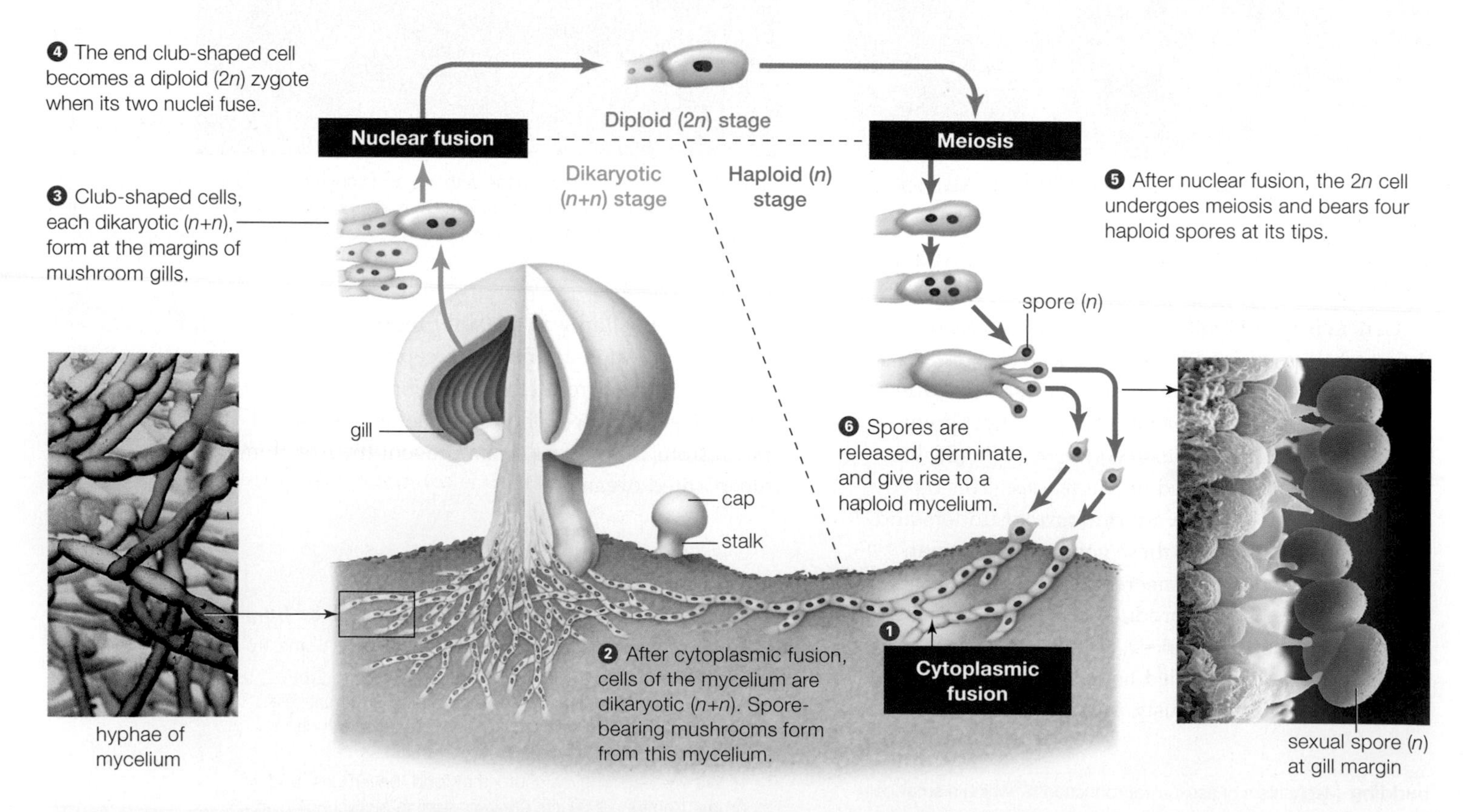

FIGURE 23.13 ▸**Animated** Life cycle of a club fungus having two mating strains of hyphae.

FIGURE IT OUT What are the blue and red dots in this figure?

Answer: Genetically different nuclei

FIGURE 23.14 Club fungus fruiting bodies (basidiocarps). Spores are produced on or, in the case of puffballs, inside the basidiocarp.

Chanterelles (FIGURE 23.14C) are edible forest mushrooms with a vaselike shape. They partner with tree roots as mycorrhizae.

Puffballs produce some of the largest fruiting bodies (FIGURE 23.14D). These decomposers can be as large as a meter across. When young, they have a white, edible flesh. When mature, their covering turns dark and splits, allowing spores to escape.

Stinkhorns have a phallic shape and an aroma like rotting meat or feces (FIGURE 23.14E). Flies and beetles attracted by the scent alight on the spore-laden tip of the fungus and distribute spores to new locations.

Many club fungi make toxins that fend off predators, including humans. Each year thousands of people become ill after eating poisonous mushrooms that they mistook for edible ones. For instance, untrained foragers sometimes mistake a young death cap mushroom (*Amanita phalloides*) for an edible puffball. Only when the basidiocarp matures does the distinctive cap form (FIGURE 23.14F). Eating *Amanita* species can cause nausea and abdominal cramping, followed by liver and kidney failure, and even death. Other club fungal toxins can alter mood and cause hallucinations. LSD is a compound that was first isolated from a club fungus. So-called magic mushrooms make the mind-altering chemical psilocybin.

TAKE-HOME MESSAGE 23.6

What are club fungi?

✔ Club fungi are a group of mostly multicelled fungi. Multicelled species spend most of their life cycle as a dikaryotic mycelium and have the largest, most complex spore-bearing structures of all fungi.

✔ Club fungi include all cultivated mushrooms and are the only fungi that break down lignin. Club fungi are important in forests both as decomposers and as pathogens.

club fungi Fungi that have septate hyphae and produce spores by meiosis in club-shaped cells.

23.7 Ecological Roles of Fungi

✔ Fungi directly affect other species as mutualistic partners and as pathogens.

Fungi as Decomposers

Fungi provide an important ecological service by breaking down complex compounds in organic wastes and remains. When a fungus secretes digestive enzymes onto these materials, some soluble nutrients escape into nearby soil or water. Plants and other producers can then take up these substances to meet their own needs. Bacteria also serve as decomposers, but they tend to grow mainly on surfaces. By contrast, fungal hyphae extend deep into a dead log or another bulky food source and break it down from the inside.

Fungal Partnerships

A **lichen** is a composite organism consisting of a sac fungus and cyanobacteria or green algae. The fungus makes up most of a lichen's mass (**FIGURE 23.15**) Photosynthetic cells that provide the fungus with sugars are held in place by hyphae. If these cells are cyanobacteria, they can also provide fixed nitrogen.

A lichen may be a **mutualism**, an interaction that benefits both partners. However, in some cases the fungus may be parasitically exploiting captive photosynthetic cells that would do better on their own.

Lichens disperse by fragmentation or by releasing special structures that contain cells of both partners. In addition, the fungus can release spores. The fungus that germinates from a spore can form a new lichen only if it alights near an appropriate photosynthetic cell. This is not as unlikely as it may seem; the required algae and bacteria are common as free-living cells.

Lichens colonize places too hostile for most organisms, such as newly exposed bedrock. They break down rock by releasing acids and by holding water that freezes and thaws. When soil conditions improve, plants move in and take root. Long ago, lichens may have preceded plants onto land.

Today, an estimated 80 percent of the vascular plants form mycorrhizae with glomeromycete fungi. Hyphae of these fungi enter root cells and branch in the space between the cell wall and the plasma membrane. Some plants form mycorrhizae with sac fungi and club fungi. In this case, fungal hyphae surround a root and grow in between its cells.

Hyphae of all mycorrhizal fungi functionally increase the absorptive surface area of their plant partner. Hyphae are thinner than even the smallest roots and can grow more easily between soil particles. The fungus shares water and nutrients taken up by its hyphae with root cells. In return, the plant supplies sugar to the fungus. It is a beneficial trade; many plants do poorly without mycorrhizae (**FIGURE 23.16**).

Fungal partners also enhance the nutrition of some animals. Chytrid fungi that live in the stomachs of grazing hoofed mammals such as cattle, deer, and moose aid their hosts by breaking down otherwise indigestible cellulose. Similarly, fungal partners of some ants and beetles serve as an external digestive system. Leaf-cutter ants gather bits of leaf to sustain

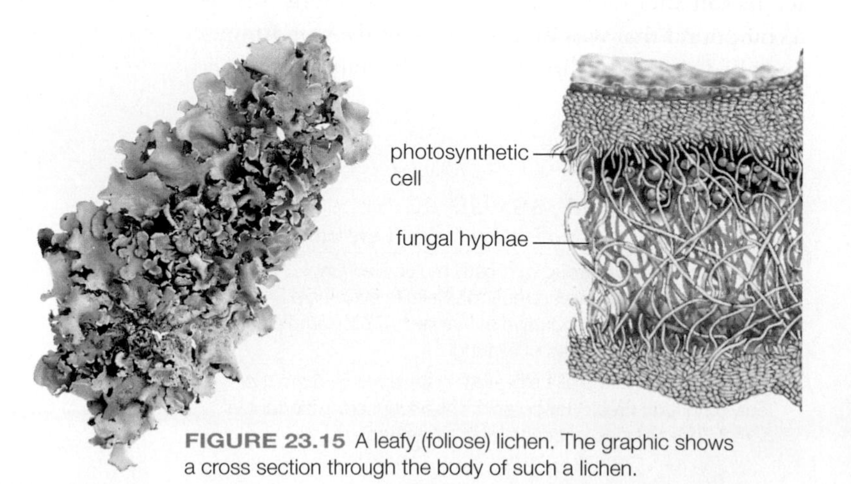

FIGURE 23.15 A leafy (foliose) lichen. The graphic shows a cross section through the body of such a lichen.

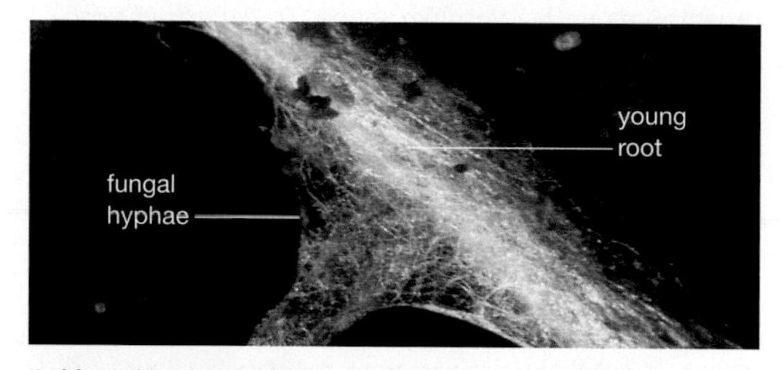

A Mycorrhiza formed by a fungus and root of a hemlock tree.

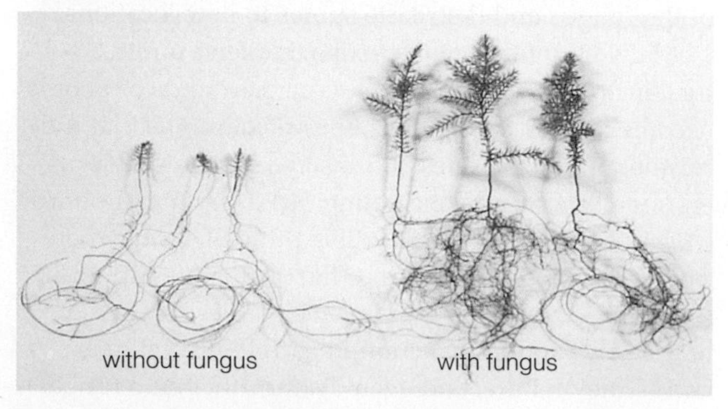

B Six-month-old juniper seedlings grown with or without mycorrhizal fungi in sterilized, phosphorus-poor soil.

FIGURE 23.16 Mycorrhiza.

FIGURE 23.17 Bat with white nose syndrome. Bats with this disease fly when they should be hibernating, lose weight, and have fuzzy white filaments of the sac fungus *Pseudogymnoascus destructans* on their wings, ears, and muzzle.

High-Flying Fungi (revisited)

Plant breeders have produced some new wheat varieties resistant to the pathogen Ug99 by inducing mutations through radiation, then selecting for Ug99 resistance among offspring of the radiated plants. They have also investigated the source of Ug99 resistance in wild relatives of wheat. Recently, they isolated Ug99 resistance genes from two such relatives, einkorn wheat and a wild goatgrass. The newly isolated genes do not occur in any of the cultivated bread wheats. However, it may be possible to introduce one or both of them into cultivated strains and thus produce transgenic Ug99-resistant wheats suited to cultivation.

the fungus that they cultivate in their nest. The ants cannot digest leaves, but they do eat the fungus.

Fungal Parasites and Pathogens

Many sac fungi and club fungi are plant parasites. Powdery mildews (sac fungi) and rusts and smuts (club fungi) grow only in living plants. Their hyphae extend into cells of stems and leaves, where they suck up photosynthetically produced sugars. The resulting loss of nutrients stunts the plant, prevents it from producing seeds, and may eventually kill it. However, a plant usually does not die before the fungus has produced spores on the surface of its infected parts.

Other pathogenic fungi produce toxins that kill plant tissues, then feed on the resulting remains. The club fungus *Armillaria* causes root rot by infecting trees and woody shrubs in forests worldwide. Once an infected tree dies, the fungus decomposes the stumps and logs left behind. In one Oregon forest, the mycelium of a single honey mushroom (*A. ostoyae*) extends across nearly 4 square miles (10 km^2). It has been growing for an estimated 2,400 years and is one of the world's largest organisms.

Many more fungi infect plants than animals. Among animals, those that do not maintain a high body temperature are most vulnerable to fungal infections. Hundreds of fungal species infect insects.

In mammals, fungal infections are usually not fatal. White nose syndrome, an emerging fungal disease affecting North American bats, is an exception (FIGURE 23.17). The disease was first described in New York in 2006. By early 2013, it had been discovered in 21 states and four Canadian provinces, and had killed millions of North American bats. Scientists think the sac fungus that causes the disease was introduced to North America from Europe, where the bats have evolved resistance to it and are not sickened by it.

Most human fungal infections involve body surfaces. Infected areas become raised, red, and itchy, as when a fungus infects skin between the toes and on the sole of the foot, causing "athlete's foot." Fungal vaginitis (a vaginal yeast infection) occurs when a yeast (*Candida*) that normally lives in the vagina in low numbers undergoes a population explosion. Fungi also cause skin infections misleadingly known as "ringworm." No worm is involved. Rather, a ring-shaped lesion forms as fungal hyphae grow outward from the initial infection site. Life-threatening fungal infections in humans usually occur only in people whose immune response is weak as a result of other factors.

TAKE-HOME MESSAGE 23.7

What ecological roles do fungi play?

✔ Fungi decompose materials, thus releasing nutrients that producers can take up and use.

✔ Fungi form mutually beneficial partnerships with plants, cyanobacteria, and green algae. Some ants farm fungus as a source of food.

✔ Some fungi are pathogens that invade the tissues of plants and animals. Human fungal infections are usually not life-threatening in otherwise healthy people.

lichen Composite organism consisting of a fungus and a single-celled alga or a cyanobacterium.
mutualism Mutually beneficial relationship between two species.

CREDITS: (17) USFWS/Science Source; (in text) Volosina/Shutterstock.com.

summary

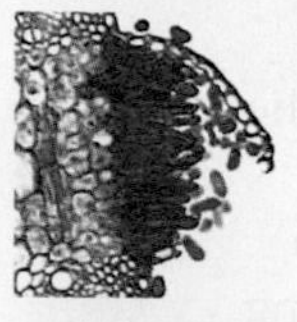

Section 23.1 Fungi disperse by releasing spores. Winds can distribute spores far from their point of origin. Windblown spores of a new strain of wheat stem rust fungus threatens global food supplies.

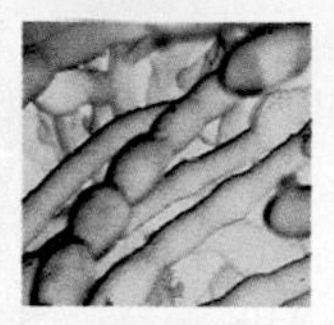

Section 23.2 **Fungi** are heterotrophs that secrete digestive enzymes onto organic matter and absorb released nutrients. Most feed on organic wastes or remains (they are decomposers) but some live in or on other organisms. Fungi are more closely related to animals than to plants. They include single-celled yeasts and multicelled species. In the multicelled species, spores germinate and give rise to filaments called **hyphae**. The filaments typically grow as an extensive mesh called a **mycelium**. Depending on the group, a mycelium may be haploid or **dikaryotic** ($n+n$).

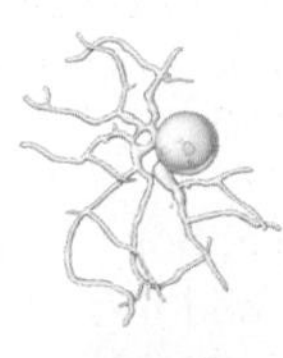

Section 23.3 **Chytrids** are an ancient group of fungi and the only modern fungi that produce flagellated spores. One parasitic species is the cause of an emerging disease that threatens many amphibian populations.

Section 23.4 Hyphae of **zygote fungi**, which include molds that spoil bread, are continuous tubes with no cross-walls (septae). A thick-walled, diploid zygospore forms during sexual reproduction. Meiosis of cells inside the zygospore produces haploid spores that germinate and produce a haploid mycelium. Mycelia also produce asexual spores.

Microsporidia are possible relatives of zygote fungi and live inside animal cells, sometimes causing disease. **Glomeromycetes**, also relatives of zygote fungi, partner with plant roots in a relationship called a **mycorrhiza.**

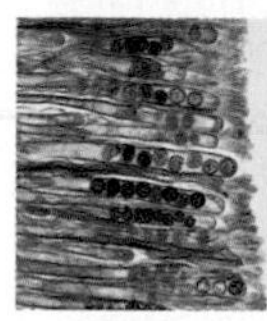

Section 23.5 **Sac fungi** are the most diverse group of fungi. They include single-celled yeasts and multicelled species that have hyphae with cross-walls. Yeasts reproduce asexually by **budding**. Other sac fungi grow as haploid hyphae and produce asexual spores called conidia. Sexual spores are produced in asci. In multicelled species, these saclike structures form on an ascocarp consisting of intertwined haploid and dikaryotic hyphae. Sac fungi are economically important as food and as sources of medicines.

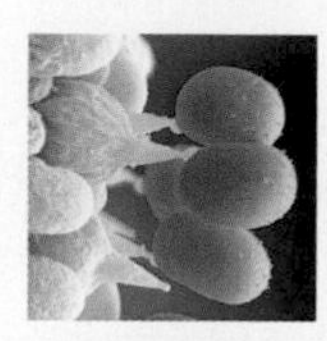

Section 23.6 The mostly multicelled **club fungi** have hyphae with cross-walls. This group produces the largest and most complex fruiting bodies. Club fungi that break down the lignin in wood are important decomposers and pathogens in forests.

Typically, a dikaryotic mycelium dominates the life cycle. It grows by mitosis and, in some species, extends through a vast volume of soil. When conditions favor reproduction, a basidiocarp, also made up of dikaryotic hyphae, develops. Haploid spores form by meiosis at the tips of club-shaped cells in the basidiocarp.

A mushroom is a familiar basidiocarp, but club fungus basidiocarps come in a wide variety of shapes and sizes. Some produce toxins that can sicken or kill people.

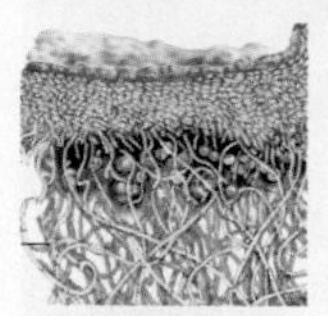

Section 23.7 Many fungi spend their life closely associated with another species.

A **lichen** is a composite organism that usually consists of a sac fungus and one or more photoautotrophs, such as green algae or cyanobacteria. The fungus, which makes up the bulk of the lichen, obtains nutrients from its photosynthetic partner. Lichens fix nitrogen, and they contribute to the breakdown of rocks to soil.

In a mycorrhiza, hyphae surround or penetrate a root and supplement the plant's surface area for absorbing water and nutrients. The fungus shares some absorbed nutrients with the plant and gets sugars in return, so the relationship is usually considered a **mutualism**.

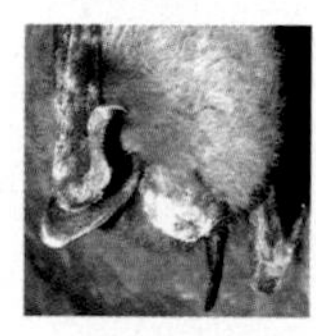

Fungal infections are more common in plants than in animals. In humans, most fungal infections occur at a body surface such as the skin, mouth, or vaginal lining. Such infections seldom pose a severe threat unless the immune system is impaired.

self-quiz

Answers in Appendix VII

1. All fungi ________ .
   a. are multicelled
   b. form flagellated spores
   c. are heterotrophs
   d. all of the above

2. Most fungi obtain nutrients from ________ .
   a. nonliving organic matter
   b. living plants
   c. living animals
   d. photosynthesis

3. In ________ , a hypha has no cross-walls.
   a. zygote fungi
   b. sac fungi
   c. club fungi
   d. all of the above

4. The yeasts whose fermentation reactions produce the carbon dioxide that makes bread rise are a type of ________ .
   a. chytrid
   b. zygote fungus
   c. sac fungus
   d. club fungus

5. In most ________ , an extensive dikaryotic mycelium is the longest-lived phase of the life cycle.
   a. chytrids
   b. zygote fungi
   c. sac fungi
   d. club fungi

6. The mycelium of a multicelled fungus is a mesh of filaments, each called a ________ .
   a. septa
   b. hypha
   c. spore

7. A mushroom is ________ .
   a. a fungal digestive organ
   b. the only part of the fungal body made of hyphae
   c. a reproductive structure that releases sexual spores
   d. made of haploid hyphae

8. Spores released from a mushroom's gills are ________ .
   a. flagellated
   b. produced by mitosis
   c. dikaryotic
   d. haploid

data analysis activities

Fighting a Forest Fungus The honey mushroom, *Armillaria ostoyae*, is a parasite of living trees; it withdraws nutrients from them. If an infected tree dies, the fungus continues to dine on its remains. Hyphae grow out from the roots of infected trees and and from dead stumps. If these hyphae contact roots of a healthy tree, they can invade and cause a new infection.

Canadian forest pathologists hypothesized that removing stumps after logging could help prevent tree deaths. To test this hypothesis, they carried out an experiment. In half of a forest they removed stumps after logging. In a control area, they left stumps behind. For more than 20 years, they recorded tree deaths and whether *A. ostoyae* caused them. **FIGURE 23.18** shows the results.

1. Which tree species was most often killed by *A. ostoyae* in control forests? Which was least affected by the fungus?

2. For the most affected species, what percentage of deaths did *A. ostoyae* cause in control and in experimental forests?

3. Looking at the overall results, do the data support the hypothesis? Does stump removal reduce effects of *A. ostoyae*?

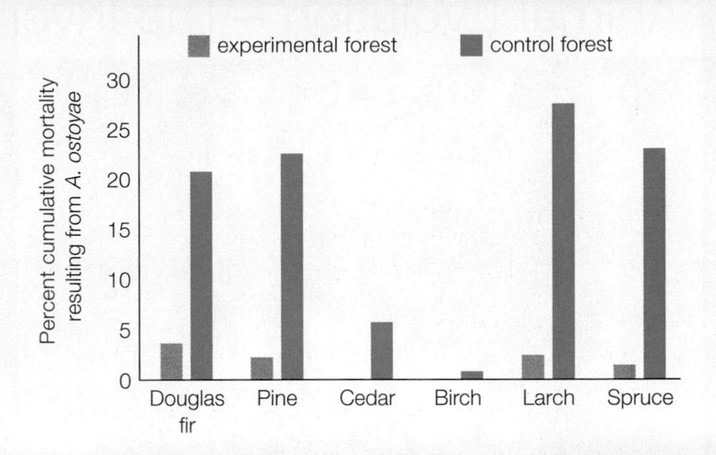

FIGURE 23.18 Results of a long-term study of how logging practices affect tree deaths caused by the fungus *A. ostoyae*. In the experimental portion of the forest, whole trees—including stumps—were removed (brown bars). The control portion of the forest was logged conventionally, with stumps left behind (blue bars).

9. _______ are fungi that produce flagellated spores.
   a. Chytrids c. Zygote fungi
   b. Sac fungi d. Club fungi

10. Nitrogen-fixing cyanobacteria often partner with a fungus to form a _______ .
   a. mycelium c. mycorrhiza
   b. lichen d. fruiting body

11. _______ are fungi that live as intracellular parasites.
   a. Glomeromycetes c. Microsporidia
   b. Chytrids d. Club fungi

12. Fungal infections are most common in _______ .
   a. plants c. mammals
   b. insects d. birds

13. All _______ form partnerships with plant roots.
   a. Glomeromycetes c. Microsporidia
   b. Chytrids d. Club fungi

14. Budding is a mechanism of _______ in yeasts.
   a. extracellular digestion c. defense
   b. asexual reproduction d. toxin production

15. Match the terms appropriately.

___ club fungus	a. first discovered in a sac fungus
___ chitin	b. component of fungal cell walls
___ penicillin	c. partnership between a fungus and one or more photoautotrophs
___ sac fungus	d. basidocarp producer
___ zygote fungus	e. fungus–plant partnership
___ lichen	f. forms sexual spores in an ascus
___ mycorrhiza	g. bread mold is an example
___ hypha	h. fungal skin disease
___ ringworm	i. strand of many cells arranged end to end

critical thinking

1. Researchers working in a Brazilian rain forest recently discovered eight species of bioluminescent mushrooms at a single site. The mushrooms continually emit a faint glow that, although undetectable in daylight, makes them visible at night. Suggest a mechanism by which glowing in the dark could benefit a mushroom. Why do you think so many species with this unusual trait live in the same region?

2. Fungi play an important role in the breakdown of cellulose. What type of chemical reaction do you think this breakdown entails? Which simple sugar is the final product?

3. Health professionals refer to fungal skin diseases as "tineas" and name them according to the region affected (**TABLE 23.1**). Fungal skin diseases are persistent, in part because fungi can penetrate deeper layers of skin than can ointments and creams. There are fewer antifungal drugs than antibacterial ones, and antifungals often have more severe side effects. Reflect on the evolutionary relationships among bacteria, fungi, and humans. Why it is harder to fight fungi than bacteria?

Table 23.1 Fungal Diseases of Skin

Disease	Infected Body Parts
Tinea corporis (ringworm)	Trunk, limbs
Tinea pedis (athlete's foot)	Feet, toes
Tinea capitis	Scalp, eyebrows, eyelashes
Tinea cruris (jock itch)	Groin, perianal area
Tinea barbae	Bearded areas
Tinea unguium	Toenails, fingernails

CREDITS: (18) After graph from www.pfc.forestry.ca; (Table 23.1) © Cengage Learning.

24 Animal Evolution—The Invertebrates

LEARNING ROADMAP

Our discussion of animals explains their adaptive traits (Section 16.4) and the role of exaptation (17.12). We draw on earlier discussions of homeotic genes (10.3), comparative genomics (15.5), and patterns of development (18.5).

ANIMAL TRAITS AND EVOLUTION

Animals are multicelled heterotrophs that move about. Early animals were small and structurally simple. More complex multilayered body plans and specialized parts evolved later.

STRUCTURALLY SIMPLE GROUPS

Sponges have no body symmetry or tissues. They filter food from the water. The radially symmetrical cnidarians such as jellies have two tissue layers and unique stinging cells that help capture prey.

INTRODUCING BILATERAL ANIMALS

Flatworms are bilateral and have organs but do not have a body cavity. Annelid worms and the mollusks have a lined body cavity, and both develop from the same type of larva.

BILATERAL ANIMALS THAT MOLT

Roundworms and arthropods molt as they grow. Arthropods, the most diverse animal group, include aquatic crustaceans as well as land-dwelling insects and spiders.

ON THE ROAD TO VERTEBRATES

Echinoderms are on the same branch of the animal family tree as vertebrates. They are invertebrates with bilateral ancestors, but adults have a decidedly radial body plan.

Section 25.2, introduces invertebrate chordates, the closest relatives of vertebrates. We return to insects as pollinators in Sections 29.1 and 29.3. We also look at invertebrate nervous systems (32.2), vision (33.5) and hormonal control of arthropod molting (34.13). We study solute-regulating organs of invertebrates in Section 40.2, as well as their skeletal systems (35.3), circulation (36.2), respiration (38.3), and development (42.2). Social insects are the focus of Section 43.9.

24.1 Medicines From the Sea

Animal life began in the sea, and the oceans remain the greatest repository of animal diversity. In the oceans, as on land, the majority of animals are **invertebrates**, meaning they do not have a backbone. Only about 5 percent of animals have a backbone and thus are considered vertebrates. Invertebrates evolved long before vertebrates, and they remain the most diverse and numerous animals. The longevity and diversity of invertebrate lineages attests to how well they have adapted to their environment.

Many marine invertebrates produce secondary metabolites (Section 22.9) that help them survive. These compounds may protect an animal from predators, help it fend off pathogens, or assist in the capture of prey. Some such compounds also have effects in the human body, and so can be useful as medicines.

Consider the venom that some fish-eating cone snails inject into their prey to subdue it (**FIGURE 24.1**). The snail's venom anesthetizes and paralyzes a fish, thus preventing it from struggling with and possibly harming the snail as it is captured and consumed. Human nerves and fish nerves use the same chemical communication signals, so compounds used by a cone snail to drug fish can also affect the function of the human nervous system. A person stung by a fish-eating cone snail may become numb at the site of injection, suffer from temporary paralysis, or even die.

A synthetic version of one peptide in cone snail venom is now used as a pain reliever. The drug, ziconotide (Prialt), is injected to suppress severe pain that cannot be controlled by other means. Other peptides isolated from cone snail venom are being tested as treatments for epilepsy, diabetes, or cancer.

Compounds derived from other marine invertebrates are also already in use. AZT (azidothymidine), the first drug successfully used to treat AIDS, is a synthetic derivative of a molecule first discovered in a sponge. Other compounds from sponges can be used to treat infections caused by herpesviruses. Gorgonians, which are relatives of sea anemones, are the source of a variety of anti-inflammatory compounds, one of which is used in an "anti-aging" face cream.

invertebrate Animal that does not have a backbone.

FIGURE 24.1 One of the fish-eating cone snails with a sedated fish in its mouth.

24.2 Animal Traits and Body Plans

✔ Like plants, all animals are multicelled. Like fungi, they are consumers. Unlike either group, animals can move.

What Is an Animal?

Animals are multicelled heterotrophs that take food into their body, where they digest it and absorb the released nutrients. An animal body consists of a few to hundreds of types of unwalled cells. These cells become specialized as an animal develops from an embryo (an individual in the earliest stage of development) to an adult. Most animals reproduce sexually, some reproduce asexually, and some do both. All are motile (move from place to place) during part or all of their life.

Variation in Animal Body Plans

FIGURE 24.2 shows an evolutionary tree for the major animal groups covered in this book, and we will use it as a guide to discuss evolutionary trends. All animals are multicellular ❶ and constitute the clade Metazoa. The earliest animals were probably aggregations of cells, and this level of organization persists in sponges. However, most modern animals have cells organized as tissues ❷.

Tissue organization begins in animal embryos. Embryos of jellies and other cnidarians have two tissue layers: an outer ectoderm and an inner endoderm. In other modern animals, embryonic cells typically rearrange themselves to form a middle tissue layer called mesoderm (**FIGURE 24.3**). Evolution of a three-layer embryo allowed an important increase in structural complexity. Most internal organs in animals develop from embryonic mesoderm.

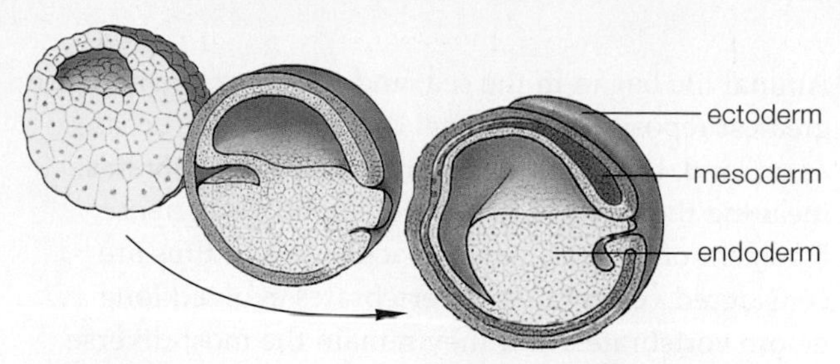

FIGURE 24.3 How a three-layer animal embryo forms through cell movements. Most animals have a three-layer embryo.

Animals with the simplest structural organization, such as sponges, are asymmetrical; their body cannot be divided into two halves that are mirror images. Jellies, sea anemones, and other cnidarians have **radial symmetry**: Body parts are repeated around a central axis, like spokes of a wheel ❸. Radial animals usually attach to an underwater surface or drift along. A radial body plan allows them to capture food that can arrive from any direction. Animals with a three-layer body plan typically have **bilateral symmetry**: The body's left and right halves are mirror images ❹. Such lineages typically undergo **cephalization**, an evolutionary process whereby many nerve cells and sensory structures

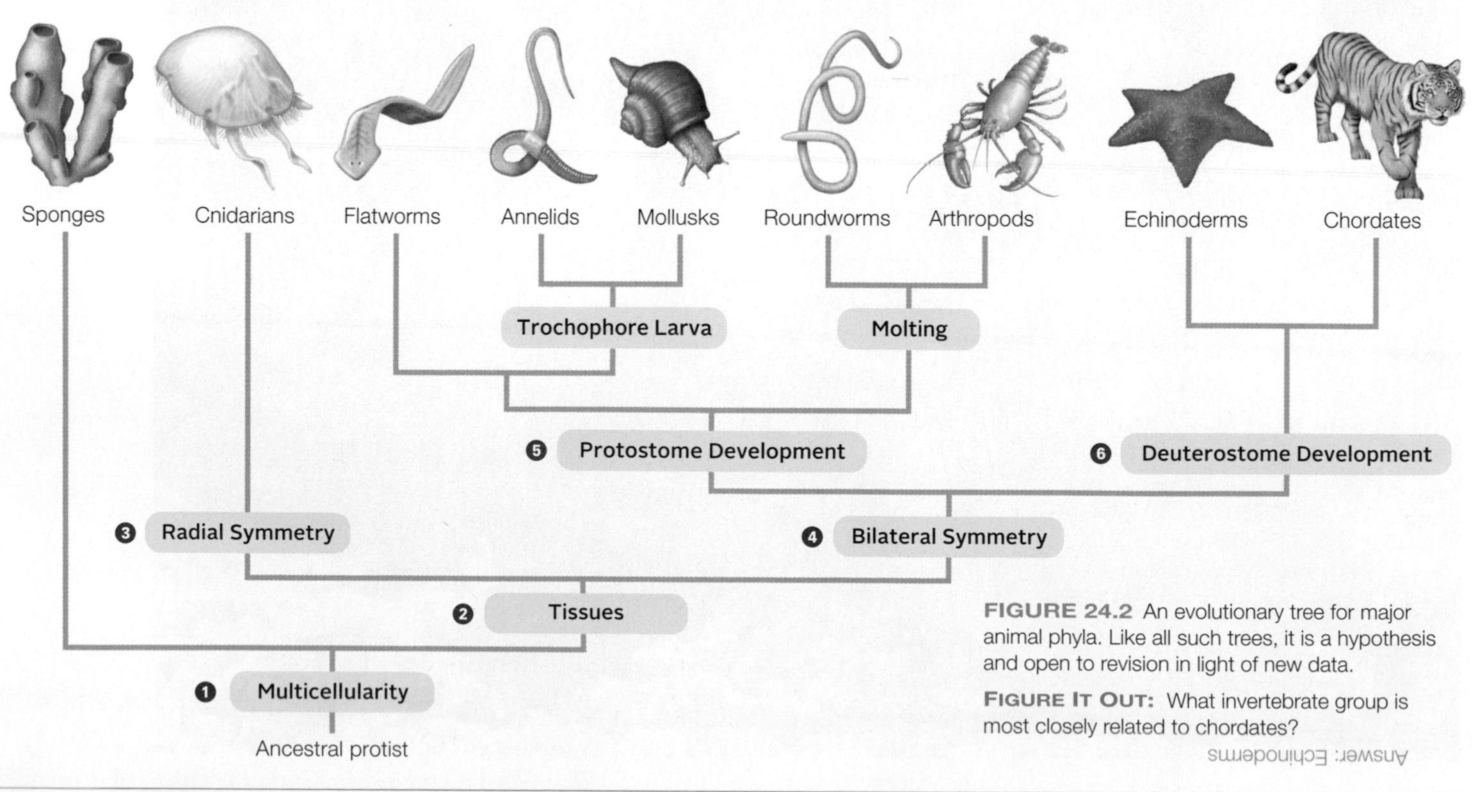

FIGURE 24.2 An evolutionary tree for major animal phyla. Like all such trees, it is a hypothesis and open to revision in light of new data.

FIGURE IT OUT: What invertebrate group is most closely related to chordates?

Answer: Echinoderms

become concentrated at the front of the body. These structures help the animal find food or avoid threats as it moves headfirst through its environment.

The two lineages of bilateral animals are defined in part by developmental differences. In **protostomes**, the first opening that appears on an embryo becomes a mouth ❺. *Proto–* means first and *stoma* means opening. Most bilateral invertebrates are protostomes. In **deuterostomes**, the mouth develops from the second embryonic opening ❻. Deuterostomes include some invertebrates, and all vertebrates.

Animals also differ in how they digest food. In sponges, digestion is intracellular. Cnidarians and flatworms digest food inside a saclike gut called a **gastrovascular cavity**. Food enters this cavity through the same opening that expels wastes. The cavity also functions in gas exchange. Most bilateral animals have a tubular gut, or **complete digestive tract**, with an opening at either end. Parts of the tube specialize in taking in food, digesting it, absorbing nutrients, or compacting waste. Unlike a gastrovascular cavity, a tubular gut carries out all of these tasks simultaneously.

A mass of tissues and organs surrounds a flatworm's gut (FIGURE 24.4A). Most other animals have a "tube within a tube" body plan. Their gut runs through a fluid-filled body cavity. In roundworms, this cavity is partially lined with tissue derived from mesoderm, and is called a **pseudocoelom** (FIGURE 24.4B). More typically, bilateral animals have a **coelom**, a body cavity fully lined with tissue derived from mesoderm (FIGURE 24.4C). Sheets of tissue called mesentery suspend the gut in the center of a coelom. Coelomic fluid cushions the gut and keeps it from being distorted by body movements. The fluid also helps distribute material through the body, and in some animals it plays a role in locomotion.

animals Multicelled heterotrophs with unwalled cells; are motile during part or all of the life cycle.
bilateral symmetry Having paired structures so the right and left halves are mirror images.
cephalization Evolutionary process whereby nerve and sensory cells become concentrated in an animal's front end.
coelom Of many animals, a body cavity that surrounds the gut and is lined with tissue derived from mesoderm.
complete digestive tract Tubular gut.
gastrovascular cavity Of some invertebrates, a saclike cavity that functions in digestion and in gas exchange.
deuterostomes Lineage of bilateral animals in which the second opening on the embryo surface develops into a mouth.
protostomes Lineage of bilateral animals in which the first opening on the embryo surface develops into a mouth.
pseudocoelom Incompletely lined body cavity of some animals.
radial symmetry Having parts arranged around a central axis, like spokes around a wheel.

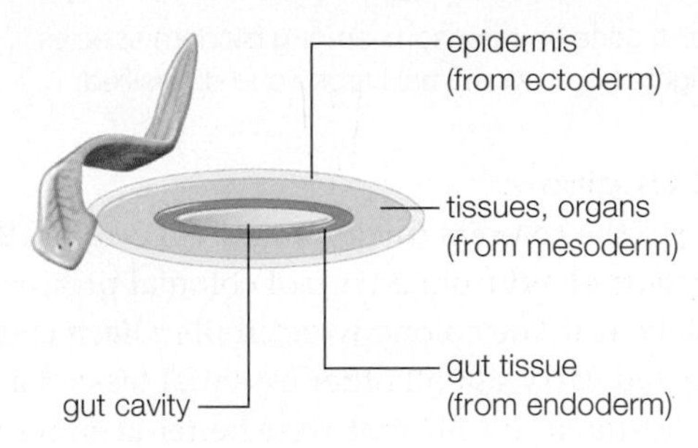

A Flatworms have no body cavity except their gut.

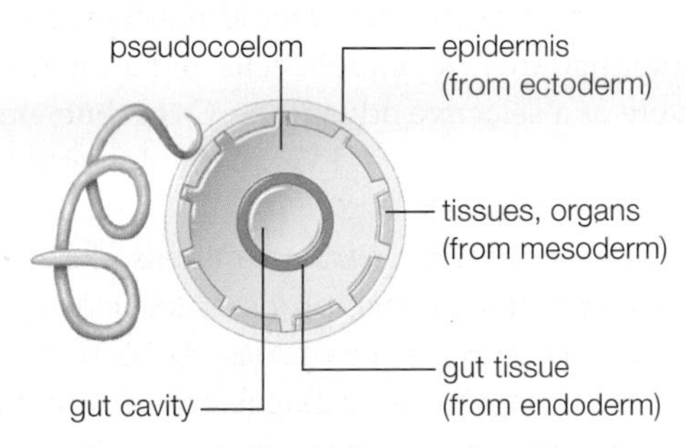

B Roundworms have a fluid-filled pseudocoelom.

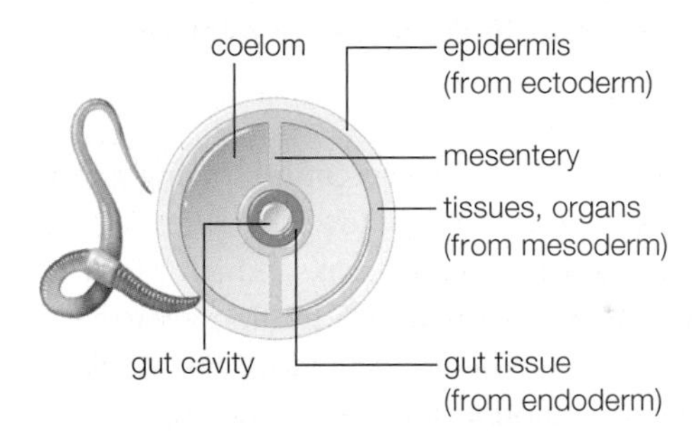

C Annelids have a fluid-filled coelom with sheets of tissue (mesentery) that hold the gut in place.

FIGURE 24.4 ▸Animated Variations in body plans among three bilateral invertebrates. Graphics are schematics of a cross-section through the body. Tissue layers are shown, but their relative width is not to scale.

TAKE-HOME MESSAGE 24.2

What are animals?

✔ Animals are multicelled heterotrophs with bodies made of unwalled cells. Animals are motile for part or all of their life.

✔ Sponges have an asymmetrical body plan. Later, radially symmetrical and then bilaterally symmetrical bodies evolved.

✔ Most animals with bilateral symmetry have a fluid-filled coelom.

✔ The two lineages of bilaterally symmetrical animals, protostomes and deuterostomes, differ in the details of their development, but both develop from an embryo that has three tissue layers.

24.3 Animal Origins and Adaptive Radiation

✔ Fossils and gene comparisons among modern species provide insights into how animals arose and diversified.

Colonial Origins

According to the colonial theory of animal origins, the first animals evolved from a type of colonial protist. At first, all cells in the colony were similar. Each could reproduce and carry out all other essential tasks. Later, mutations resulted in cells that were better at some tasks but did not carry out others. Perhaps these cells captured food more efficiently, but did not reproduce. Colonies that had interdependent cells and a division of labor were at a selective advantage. Over time, new specialized cell types evolved. Eventually this process produced the first animal.

What was the protist ancestral to animals like? Choanoflagellates, the modern protists most closely related to animals, provide some clues. As Section 21.10 explained, choanoflagellates are flagellated cells that live either as single cells or as a colony of genetically identical cells. In their structure, choanoflagellate cells closely resemble some cells in the bodies of modern sponges.

FIGURE 24.5 Fossil animals. Top, *Spriggina*, an Ediacaran, lived about 570 million years ago. It was about 3 centimeters (1 inch) long. By one hypothesis, it was a soft-bodied ancestor of arthropods, such as trilobites (bottom) that arose during the Cambrian.

Early Animals

Chemicals discovered in rocks found during an oil-drilling project in Oman provide the earliest evidence of animals. The sedimentary rocks, which were laid down as seafloor more than 635 million years ago, contain a biomarker characteristic of sponges. As Section 19.5 explained, a biomarker is a distinctive molecule produced only by a particular type of cell or lineage. Similarly aged rocks from Australia contain what appear to be fossils of sponge bodies.

A collection of 570-million-year-old fossil organisms first discovered in Australia's Ediacara Hills provides evidence of an early animal diversification. These fossils, collectively known as Ediacarans, include a variety of soft-bodied organisms. Many of these fossils appear to be marine invertebrates, and some are thought to be early representatives of modern invertebrate groups (**FIGURE 24.5**). Others belong to groups that have no surviving members.

The Cambrian Explosion

Animals underwent a dramatic adaptive radiation during the Cambrian (542–488 million years ago). By the end of this period, all major animal lineages were present in the seas. What caused this Cambrian explosion in diversity? Environmental factors probably played a role. During this period, the global climate warmed and the oxygen concentration rose in the seas. Both factors made the environment more hospitable to animal life. With rare exceptions, animals require oxygen. In modern seas, highly oxygenated waters support a greater species diversity than cooler ones. Also during the Cambrian, the supercontinent Gondwana (Section 16.7) underwent a dramatic rotation. Movement of this landmass could have isolated populations, thus increasing the likelihood that allopatric speciation events would occur.

Biological factors could also have encouraged diversification. After predatory animals arose, evolution of novel prey defenses such as protective hard parts would have been favored. Duplications and divergence of homeotic genes (Section 10.3) would have facilitated diversification. Changes in these genes have dramatic effects on body plans. Some mutations produced adaptive traits that better allowed animals to more easily escape predators or to survive in novel habitats.

TAKE-HOME MESSAGE 24.3

What do we know about animal origins and diversification?

✔ Animals probably evolved from a colonial protist that resembled modern choanoflagellates.

✔ The earliest evidence of animals is sponge biomarkers that date back more than 635 million years.

✔ Some representatives of modern animal lineages are found among the Ediacarans. A great adaptive radiation of animals took place during the Cambrian period.

24.4 Sponges

✔ Sponges have no tissues or organs. They filter food from water that they draw through pores in their body.

Sponges (phylum Porifera) are aquatic animals that do not have tissues or organs. Most of the more than 5,000 species live in tropical seas. Some are as small as a fingertip, whereas others stand meters tall. An asymmetrical vaselike or columnar shape is most common, but some grow as a thin crust.

An adult sponge is a **sessile animal**, meaning it lives attached to a surface. The body has many pores (FIGURE 24.6). Flat, nonflagellated cells cover the body's outer surface, flagellated collar cells line its inner surface, and a jellylike extracellular matrix fills the space in between. Amoeba-like sponge cells live in the matrix. Many sponges also have cells that secrete fibrous proteins or glassy silica spikes. These materials structurally support the body, and help fend off predators. Some protein-rich sponges are harvested from the sea, dried, cleaned, and bleached. Their rubbery protein remains (right) are used for bathing and cleaning.

The typical sponge is a **suspension feeder**, meaning it filters its food from the surrounding water. The movement of flagella on collar cells draws food-laden water through pores in a sponge's body wall. The collar cells filter food from the water and engulf it by phagocytosis. Digestion is intracellular. The amoeba-like cells in the matrix receive breakdown products of digestion from collar cells, and distribute these nutrients to other cells in the sponge body.

Most sponges are **hermaphrodites**, meaning each individual can produce both eggs and sperm. Typically, a sponge releases sperm into the water (FIGURE 24.7) but holds on to its eggs. Fertilization produces a zygote that develops into a ciliated larva. A **larva** (plural, larvae) is a young, sexually immature animal with a body form that differs from that of an adult. Sponge larvae swim briefly, then settle and develop into adults. Many sponges can also reproduce asexually when small buds or fragments break away and grow into new sponges.

Some freshwater sponges can survive dry conditions by producing gemmules, which are tiny clumps of resting cells encased in a hardened coat. Gemmules are dispersed by the wind. Those that land in a hospitable habitat become active and grow into new sponges.

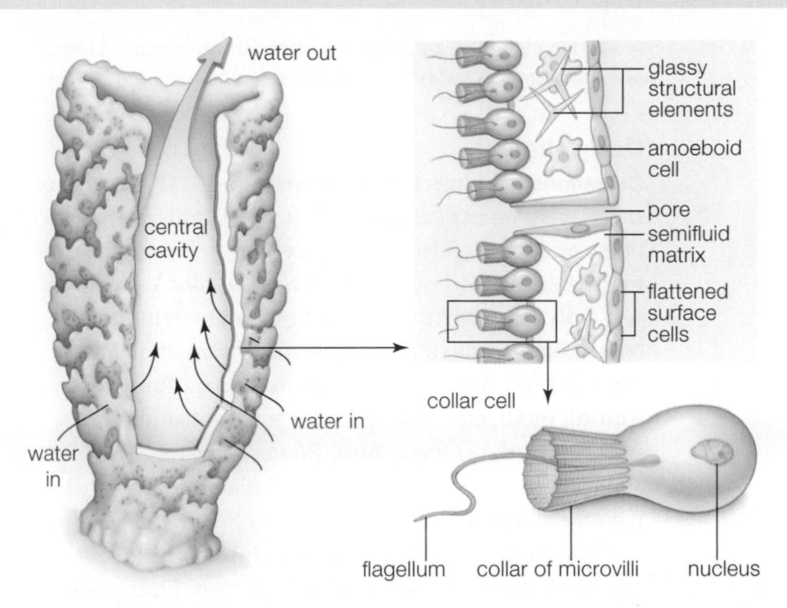

FIGURE 24.6 ▶Animated Adult sponge, with a porous, asymmetrical body.

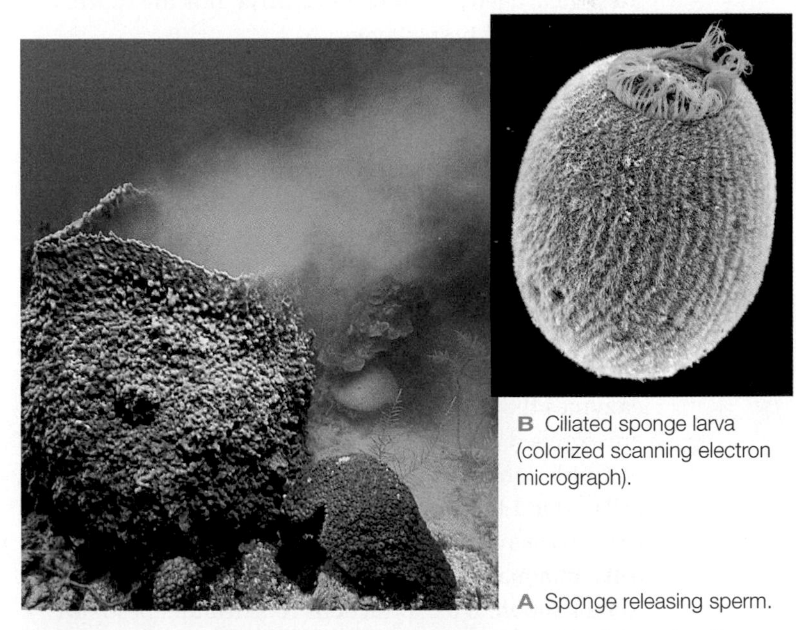

B Ciliated sponge larva (colorized scanning electron micrograph).

A Sponge releasing sperm.

FIGURE 24.7 Sexual reproduction in sponges. Sponges are hermaphrodites.

TAKE-HOME MESSAGE 24.4

What are sponges?

✔ Sponges are mostly marine animals that have an asymmetrical body without any tissues or organs.

✔ Ciliated larvae swim briefly, but adults are sessile filter feeders that strain bits of food from water flowing through their pores.

hermaphrodite Animal that can produce both eggs and sperm.
larva An immature animal with a body form that differs from the adult form. Develops during the life cycle of some animals.
sessile animal Animal that lives fixed in place on some surface.
sponge Aquatic invertebrate that has no body symmetry, tissues, or organs; feeds by filtering food from the water.
suspension feeder Animal that filters food from water around it.

24.5 Cidarians—Predators With Stinging Cells

✔ Cnidarians are radial aquatic animals with two tissue layers. They capture food with their tentacles.

Body Plans

Cnidarians (phylum Cnidaria) include 10,000 species of radially symmetrical animals such as corals, sea anemones, and jellies (also called jellyfishes). Nearly all are marine. There are two cnidarian body plans—medusa and polyp (**FIGURE 24.8**). In both, a tentacle-ringed orifice opens onto a gastrovascular cavity that functions in both digestion and gas exchange. A **medusa** (plural, medusae) is shaped like a bell or umbrella, with a mouth on the ventral (lower) surface. Most swim or drift about. A **polyp** is tubular, and one end usually attaches to a surface.

Cnidarians have an embryo with two tissue layers. Their outer epidermis develops from embryonic ectoderm, and the inner gastrodermis from endoderm. Mesoglea, a jellylike secreted matrix, fills the space between these tissue layers.

Cnidarians are predators that use their tentacles to sting and capture prey. The name Cnidaria is from *cnidos*, the Greek word for nettle, a kind of stinging plant. The tentacles have cells called **cnidocytes** with specialized organelles (nematocysts) that act like a jack-in-the-box (**FIGURE 24.9**). When something brushes against the cell, a coiled thread pops out and entangles prey or sticks a venomous barb into it. The captured prey is pushed through the mouth, into the saclike gastrovascular cavity. Gland cells of the gastrodermis secrete enzymes that digest the prey, then digestive remains are expelled through the mouth.

Cnidarians are brainless, but interconnecting nerve cells extend through their tissues, forming a **nerve net**. Body parts move when nerve cells signal contractile cells. In a manner analogous to squeezing a water-filled balloon, the contractions redistribute mesoglea or water trapped in the gastrovascular cavity. A fluid-filled cavity or cellular mass on which contractile cells exert force is a **hydrostatic skeleton**.

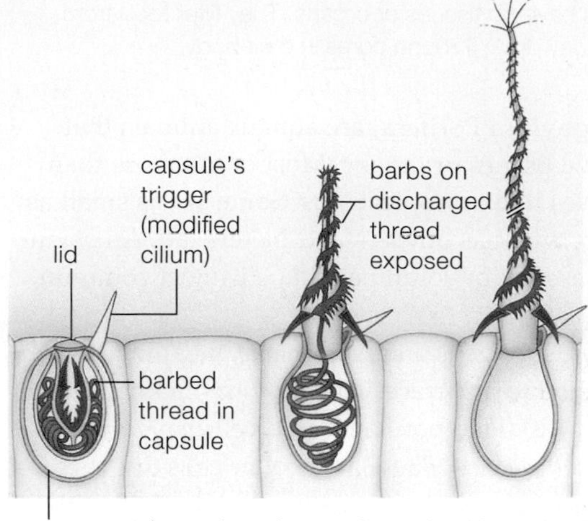

FIGURE 24.9 ▶**Animated** Example of cnidocyte action. Mechanical stimulation causes the thread coiled up inside the capsule to spring out and penetrate prey.

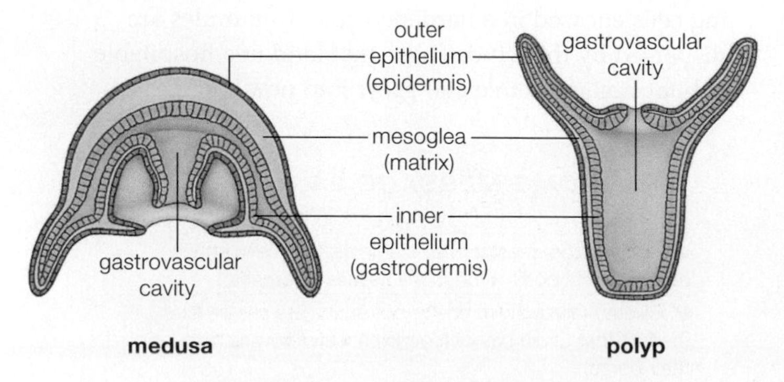

FIGURE 24.8 ▶**Animated** The two cnidarian body plans.

Diversity and Life Cycles

There are four cnidarian classes: hydrozoans, anthozoans, cubozoans, and scyphozoans.

Some marine hydrozoans have a life cycle that includes polyp, medusa, and larval stages (**FIGURE 24.10**). A cnidarian larva, called a planula, is ciliated and bilateral. *Hydra*, another hydrozoan, is a freshwater predatory polyp about 20 millimeters (3/4 inch) tall (**FIGURE 24.11A**). There is no medusa stage, and reproduction usually occurs asexually by budding.

Anthozoans such as corals and sea anemones also do not have a medusa stage (**FIGURE 24.11B**). Gametes form on polyps. Coral reefs consist of colonies of polyps that enclose themselves in a framework of secreted calcium carbonate. Photosynthetic dinoflagellates (Section 21.6) live in the polyp's tissues and provide it with sugars.

Cubozoans and scyphozoans have a medusa-dominated life cycle. Cubozoans, commonly called box jellies, are active swimmers with structurally complex eyes, complete with a lens. Box jellies that live in the Indo-Pacific ocean produce a powerful venom, and their sting can be deadly to humans (**FIGURE 24.11C**).

Scyphozoans include most jellies that commonly wash up on beaches. Some are harvested and dried for use as food, especially in Asia. The Portuguese man-of-war (*Physalia*) is a large colonial scyphozoan that consists of many individuals (**FIGURE 24.11D**).

FIGURE 24.10 ▶**Animated**
Life cycle of *Obelia*, a hydrozoan that has alternating medusa (left) and polyp generations.

1. A medusa makes and releases eggs or sperm. Gametes combine and a zygote forms.
2. The zygote develops into a ciliated bilateral larva.
3. The larva settles and develops into a polyp.
4. The polyp grows and reproduces asexually, eventually producing a branching colony.
5. Some branches of the colony are specialized for capturing and consuming prey.
6. Other branches produce and release medusae that begin the sexual phase of the life cycle again.

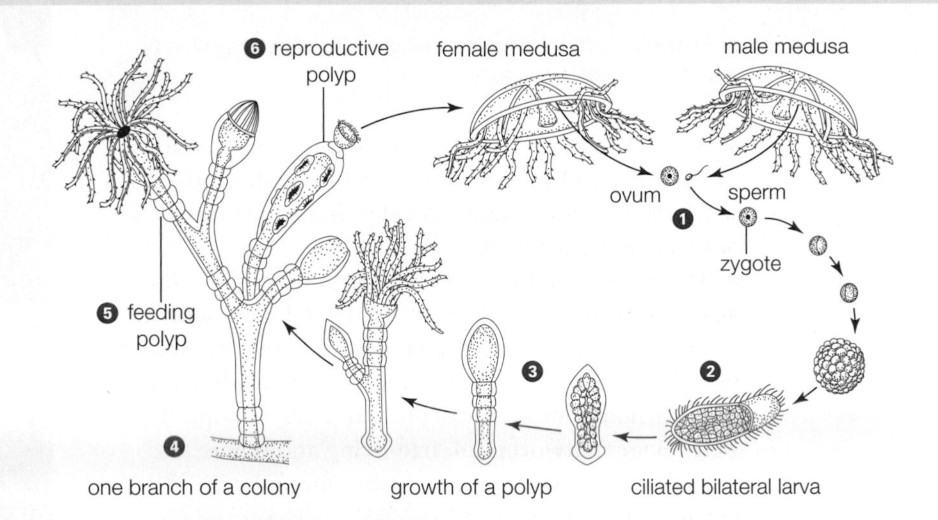

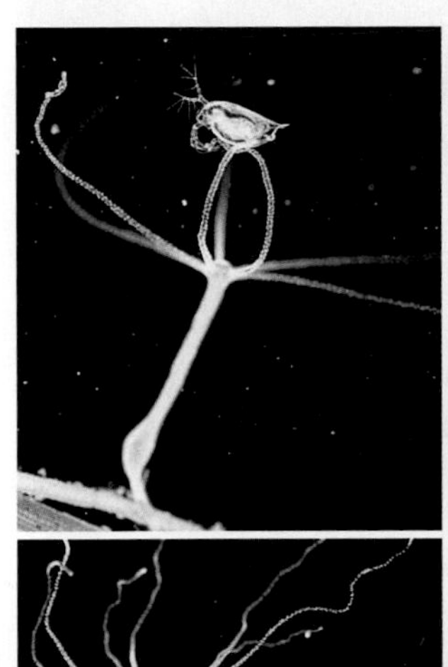

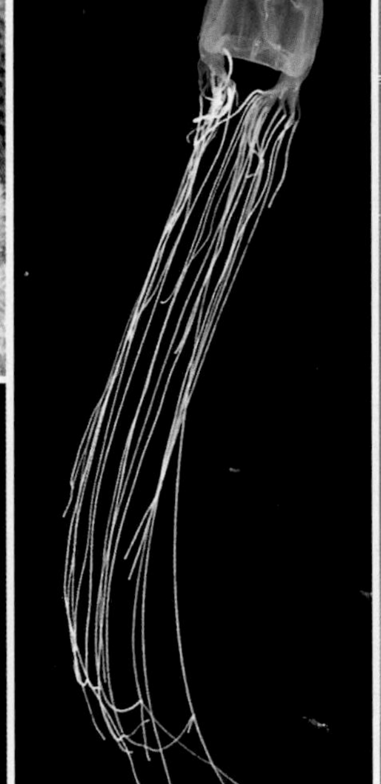

A Hydrozoan. *Hydra*, a freshwater species, capturing and digesting a water flea.

B Anthozoans. A reef-building coral with polyps extended for feeding (top), and a sea anemone (bottom).

C Cubozoan. The box jelly *Chironex* makes a toxin that can kill a person.

D Scyphozoan. The Portuguese man-of-war (*Physalia*) is a colony. The purplish-blue, air-filled float is a modified polyp.

FIGURE 24.11 Representatives of the four cnidarian classes.

TAKE-HOME MESSAGE 24.5

What are cnidarians?

✔ Cnidarians are radial, mostly marine predators with two tissue layers. There are two body plans: medusa and polyp. In both, tentacles with cnidocytes surround a mouth that opens onto a gastrovascular cavity. A nerve net interacts with the hydrostatic skeleton to allow movement.

✔ Cnidarians include sea anemones, jellies, and corals. Reef-building corals form reefs by secreting calcium carbonate.

cnidarian Radially symmetrical invertebrate that has tentacles with stinging cells (cnidocytes).
cnidocyte Stinging cell unique to cnidarians.
hydrostatic skeleton Of soft-bodied invertebrates, a fluid-filled chamber that muscles exert force against, redistributing the fluid.
medusa A bell-like cnidarian body fringed with tentacles.
nerve net A mesh of nerve cells with no central control organ.
polyp A pillarlike cnidarian body topped with tentacles.

CREDITS: (10) © Cengage Learning; (11A) Kim Taylor/Bruce Coleman, Ltd.; (11B) top, © Expeditieteam Aldabra, Foto Natura/Minden Pictures; bottom, Ethan Daniels/Shutterstock; (11C) Courtesy of Dr. William H. Hamner; (11D) A.N.T./Science Source.

24.6 Flatworms—Simple Organ Systems

✔ Flatworms are the simplest animals with bilateral symmetry and a three-layer embryo.

With this section, we begin our survey of protostomes, one of the two lineages of bilaterally symmetrical animals. All of these animals develop from an embryo with three tissue layers.

Flatworms (phylum Platyhelminthes) are the simplest protostomes. They have a flattened body with an array of organ systems, but no body cavity other than a gastrovascular cavity. Like cnidarians, they rely entirely on diffusion to move nutrients and gases through their body. Some flatworms are free-living and others are parasites. Nearly all are hermaphrodites.

FIGURE 24.12 A marine flatworm. Brightly colored flatworms are common inhabitants of coral reef ecosystems.

FIGURE 24.13 A land-dwelling flatworm (*Bipalium*). It is a predator that hunts other invertebrates in moist leaf litter.

Free-Living Flatworms

Most free-living flatworms live in tropical seas, and many of these are brilliantly colored (FIGURE 24.12). A lesser number live in fresh water, and a few live in damp places on land (FIGURE 24.13). Most free-living flatworms glide along, propelled by the movement of cilia that cover the body surface. Some marine species can also swim with an undulating motion.

FIGURE 24.14 shows the structure of a planarian, a type of free-living flatworm common in ponds. A planarian uses a muscular tube called a **pharynx** on its ventral surface to suck food into a highly branched gastrovascular cavity (FIGURE 24.14A). Nutrients and oxygen diffuse from the fine branches to all body cells.

A planarian's head has chemical receptors and eyespots that detect light. These sensory structures send messages to a simple brain consisting of paired ganglia. A **ganglion** is a cluster of nerve cell bodies. Bundles of nerve fibers called **nerve cords** extend from the head and run the length of the body (FIGURE 24.14B).

A planarian's body fluid has a higher solute concentration than the fresh water around it, so water tends to move into the body by osmosis. A system of tubes regulates internal water and solute levels by driving excess fluids out through a pore at the body surface (FIGURE 24.14C).

FIGURE 24.14 ▸**Animated** Organ systems of a planarian, one of the flatworms.

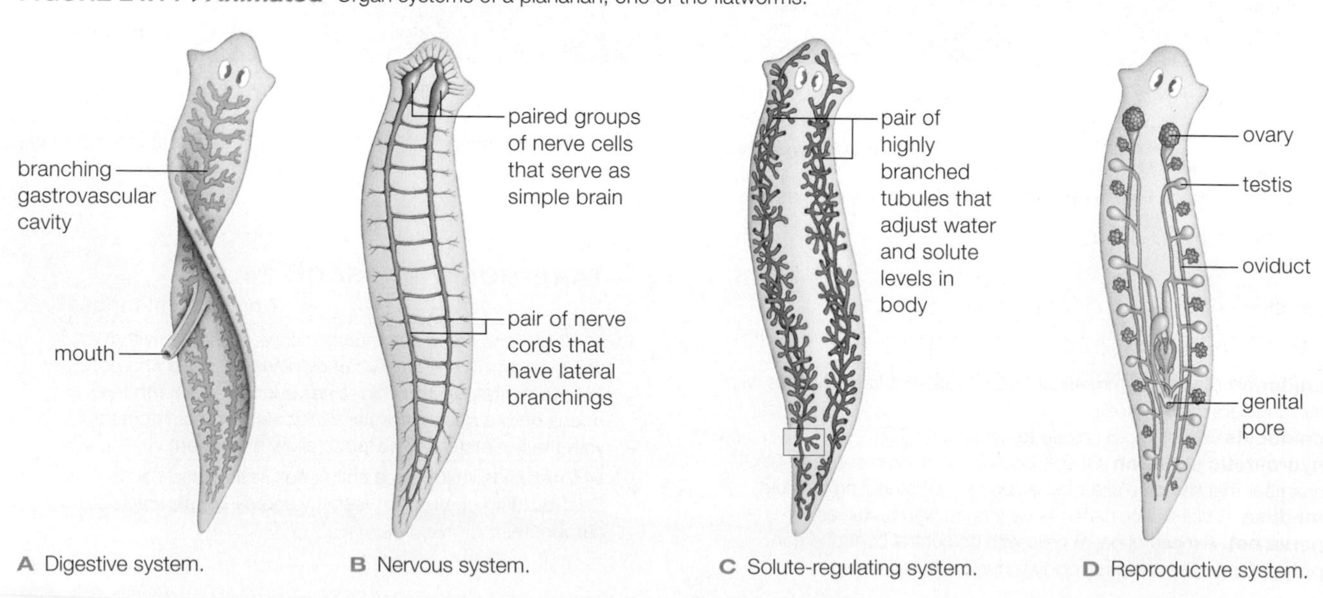

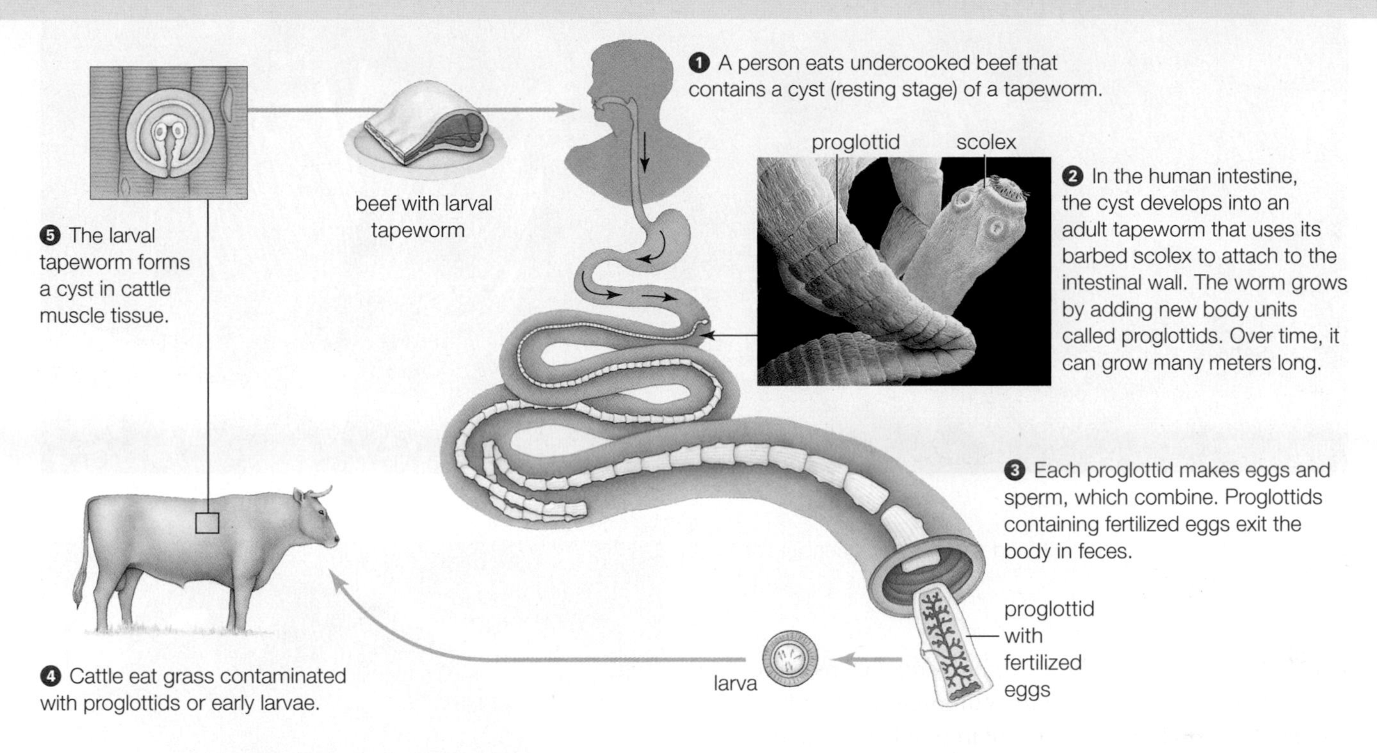

FIGURE 24.15 ▶**Animated** Life cycle of a beef tapeworm.

A planarian is a hermaphrodite (**FIGURE 24.14D**), but it cannot fertilize its own eggs. Freshwater planarians typically swap sperm. By contrast, some marine flatworms battle over who will assume the male role. In a behavior described as "penis fencing," each flatworm attempts to stab its penis into a partner's body and squirt in some sperm, while fending off its partner's attempts to do the same.

Planarians also reproduce asexually and some have an amazing capacity for regeneration. During asexual reproduction, the body splits in two near the middle, then each piece regrows the missing parts. This capacity for regrowth is also put to use when a planarian is injured, as by a predator. A planarian can regenerate even when a large portion of its body is gone.

Parasitic Flatworms

Flukes and tapeworms are parasitic flatworms whose life cycle often involves multiple hosts. Typically, larvae reproduce asexually in one or more intermediate hosts before developing into adults. Adults reproduce sexually in a definitive (final) host. For example, aquatic snails are the intermediate host for blood fluke larvae (*Schistosoma*), but the adults can only reproduce sexually inside a mammal, such as a human. Infection by the blood fluke causes schistosomiasis, a disease that affects about 200 million people. Most cases occur in Southeast Asia and northern Africa.

Tapeworms are parasites that live and reproduce in the vertebrate gut. The head has a scolex, a structure with hooks or suckers that allow the worm to attach to the gut wall. Behind the head are body units called proglottids. Unlike planarians and flukes, tapeworms do not have a gastrovascular cavity. Instead, the worm absorbs nutrients across its body wall. **FIGURE 24.15** shows the life cycle of a beef tapeworm, for which humans are the definitive host.

flatworm Acoelomate, unsegmented worm; for example, a planarian or tapeworm.
ganglion Cluster of nerve cell bodies.
nerve cord Bundle of nerve fibers that runs the length of the body in many invertebrates.
pharynx Tube connecting the mouth to the digestive tract.

TAKE-HOME MESSAGE 24.6
What are flatworms?

✔ Flatworms are bilateral, acoelomate animals with a gastrovascular cavity, a cephalized nervous system, and a system for regulating water and solutes.

✔ Planarians live in fresh water, and some flatworms live in moist places on land. Other free-living flatworms are marine.

✔ Flukes and tapeworms are parasitic flatworms. Both groups include species that infect humans.

A The sandworm, an active predator.

B Feather duster worm, a suspension feeder.

FIGURE 24.16 Polychaete worms.

✔ An annelid body is coelomate and segmented; it consists of many repeated units.

Annelids (phylum Annelida) are bilateral worms with a coelom and conspicuous segmentation, inside and out. Segmentation refers to a body plan in which similar units are repeated one after the other along a body's main axis. A tubular gut extends through all segments. Annelids also have a **closed circulatory system**, in which blood flows through a continuous system of vessels. Exchanges between blood and tissues take place across a vessel wall.

Two annelid subgroups, polychaetes and oligochaetes, are named for the chitin-reinforced bristles called chaetae on their segments. Polychaetes have many bristles and oligochaetes have few. (*Poly–* means many; *oligo–* means few.) Members of a third subgroup, the leeches, lack bristles entirely.

The Marine Polychaetes

The best-known polychaetes are sandworms (*Nereis*) (**FIGURE 24.16A**). They are often sold as bait for saltwater fishing. These active predators use their chitin-strengthened jaws to capture other soft-bodied invertebrates. Each segment has pair of paddlelike appendages called parapodia that help the worm burrow in sediments and pursue prey.

Other polychaetes have modifications of this basic body plan. Fan worms and feather duster worms live inside a tube of secreted mucus and sand grains. The head protrudes from the tube, and its elaborate tentacles capture food (**FIGURE 24.16B**). Modified parapodia are used to draw water into the tube.

A larva with a band of cilia above its mouth (a trochophore larva), forms during polychaete development. Mollusks have the same type of larva, and this developmental similarity is taken as evidence of a close relationship between the annelid and mollusk lineages.

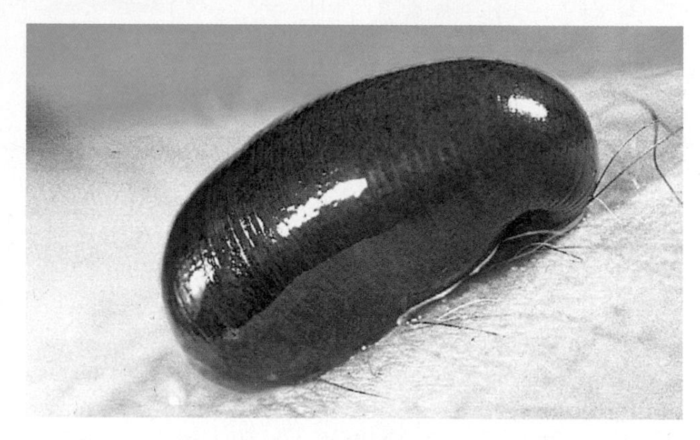
FIGURE 24.17 A leech feeding on human blood.

Leeches—Bloodsuckers and Others

Leeches occur in the ocean and damp habitats on land, but are most common in fresh water. Their body lacks conspicuous bristles and has a sucker at either end. Many leeches are scavengers or predators of small invertebrates. Others attach to a vertebrate, pierce its skin, and suck blood (**FIGURE 24.17**). Their saliva has a protein that keeps blood from clotting while the leech feeds. For this reason, doctors who reattach a severed finger or ear often apply leeches to the reattached part. As they feed, the leeches prevent unwanted clots from forming inside blood vessels of this part.

The Earthworm—An Oligochaete

Oligochaetes include marine and freshwater species, but the land-dwelling earthworms are most familiar. In the United States, most earthworms observed in backyards, bait shops, and compost heaps are introduced species native to Europe. However, some undisturbed habitats retain native earthworms,

CREDITS: (16A) Darlyne A. Murawski, National Geographic Creative; (16B) © Jon Kenfield/Bruce Coleman Ltd.; (17) J. A.L. Cooke/Oxford Scientific Films.

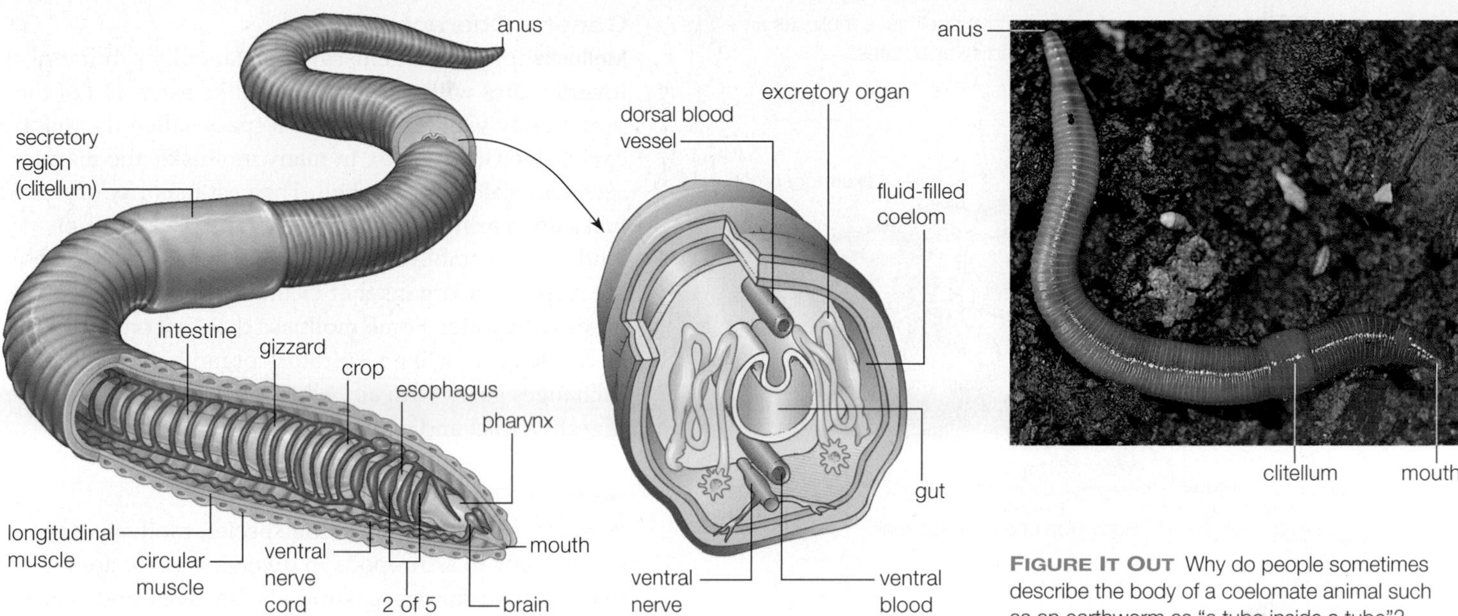

FIGURE 24.18 ▶**Animated** Earthworm body plan. Each segment contains a coelomic chamber full of organs. A gut, ventral nerve cord, and dorsal and ventral blood vessels run through all coelomic chambers.

FIGURE IT OUT Why do people sometimes describe the body of a coelomate animal such as an earthworm as "a tube inside a tube"?

Answer: One tube, the gut, is inside another tube, the body wall. The fluid-filled coelom lies between tubes.

including a species in Oregon that can reach more than a meter in length.

A cuticle of secreted proteins coats an earthworm's body (**FIGURE 24.18**). Visible grooves at the body surface correspond to internal partitions. Gas exchange occurs across the body surface, and the circulatory system helps distribute oxygen. Multiple hearts in the anterior of the worm keep the blood moving. A fluid-filled coelom runs the length of the body and is divided into chambers, one per segment. A tubular gut extends through all coelomic chambers.

Earthworms are scavengers that eat their way through the soil and digest organic debris. The worms improve soil by loosening its particles and by excreting tiny bits of organic matter that decomposers can easily break down. Excreted earthworm "castings" are sold as a natural fertilizer.

Most body segments have a pair of excretory organs (nephridia) that regulate the solute composition and volume of coelomic fluid. These organs collect coelomic fluid, adjust its composition, and expel waste through a pore in the body wall.

A simple brain connects to a pair of nerve cords that run the length of the body. The brain coordinates locomotion and receives sensory input, such as information from light-detecting cells in the worm's body wall

An earthworm has two sets of muscles. Longitudinal muscles parallel the body's long axis, and circular muscles ring the body. The worm's coelomic fluid is a hydrostatic skeleton. Contraction of muscles puts pressure on fluid trapped inside body segments, causing them to change shape. When a segment's longitudinal muscles contract, the segment gets shorter and fatter. When circular muscles contract, a segment gets longer and thinner. Coordinated waves of contraction that run along the body propel the worm through soil.

Earthworms are hermaphrodites, but cannot fertilize themselves. During mating, a secretory organ (the clitellum) produces mucus that glues two worms together while they swap sperm. Later, the same organ secretes a silky case that will protect fertilized eggs in the soil.

TAKE-HOME MESSAGE 24.7

What are annelids?

✔ Annelids are bilateral, coelomate, and segmented. They include earthworms, marine polychaete worms, and leeches. All have a tubular gut and a closed circulatory system.

✔ Larval similarities imply a close relationship between annelids and mollusks.

annelid Segmented worm with a coelom, complete digestive system, and closed circulatory system.
closed circulatory system System in which blood flows to and from a heart or hearts through a continuous series of vessels.

CREDITS: (18) art, After Solomon, 8th edition, p624, figure 29-4; photo, Christine Evers.

24.8 Mollusks—Animals With a Mantle

✔ The ability to secrete a protective shell gave mollusks an advantage over other soft-bodied invertebrates.

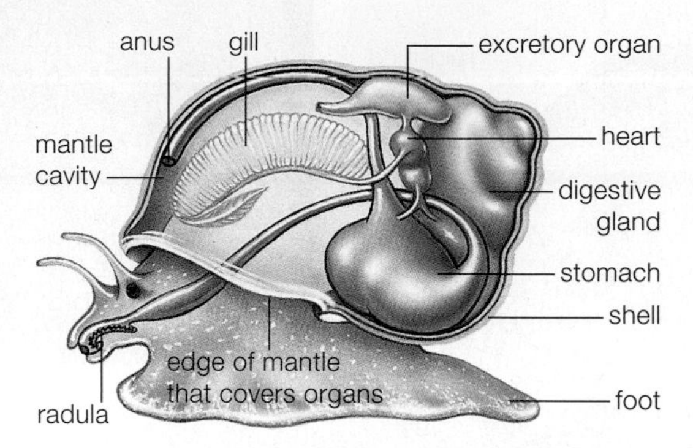

FIGURE 24.19 Body plan of an aquatic snail.

A Chitons (overlapping plates).

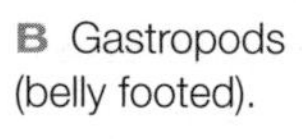

B Gastropods (belly footed).

C Bivalves (two-part shell).

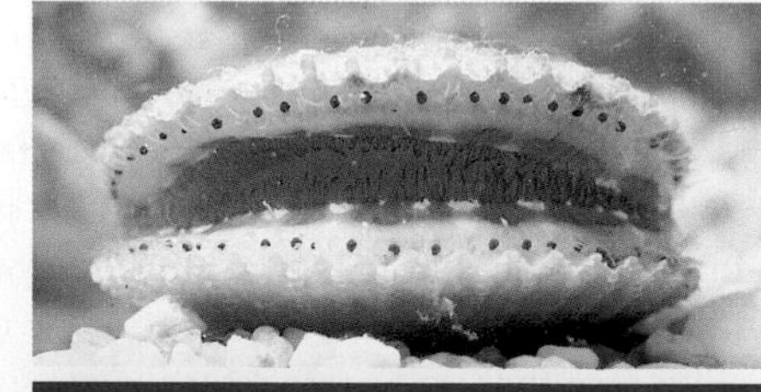

D Cephalopods (jet-propelled).

FIGURE 24.20 Representatives of the four mollusk classes.

General Characteristics

Mollusks (phylum Mollusca) are bilaterally symmetrical invertebrates with a **mantle**, a skirtlike extension of the upper body wall that encloses a space called the mantle cavity (**FIGURE 24.19**). In many mollusks, the mantle secretes a calcium-rich shell. The vast majority of mollusks are marine, but some live in fresh water or on land. Aquatic mollusks have one or more **gills**, which are respiratory organs that facilitate the exchange of gases with water. Some mollusks that live on land have a **lung**, a saclike respiratory organ in which blood exchanges gases with air. All mollusks have a complete digestive tract and a reduced coelom.

Mollusk Diversity

With more than 100,000 living species, mollusks are second only to arthropods in diversity. There are four main classes: chitons, gastropods, bivalves, and cephalopods (**FIGURE 24.20**).

Chitons are probably the most similar to ancestral mollusks. All are marine and have a dorsal shell that consists of eight plates (**FIGURE 24.20A**). Chitons cling to rocks and scrape up algae using their **radula**, which is a tonguelike organ hardened with chitin.

With more than 60,000 species of snails and slugs, the gastropods are the most diverse mollusks. Their name means "belly foot." Most species glide about on the broad muscular foot that makes up most of the lower body mass (**FIGURES 24.19** and **24.20B**). A gastropod shell, when present, is one-piece and often coiled.

Gastropods have a distinct head that usually has eyes and sensory tentacles. In many aquatic species, a part of the mantle forms an inhalant siphon, a tube through which water is drawn into the mantle cavity. The cone snails discussed in Section 24.1 use their siphon to sniff out prey. Cone snails are predatory, harpooning prey with a modified radula, but most other gastropods are herbivores.

Gastropods include the only terrestrial mollusks. In land-dwelling snails and slugs, a lung replaces the gill. Glands on the foot continually secrete mucus that protects the animal as it moves across dry, abrasive surfaces. Most mollusks have separate sexes, but land dwellers are typically hermaphrodites. Unlike other mollusks, which produce a swimming larva, these groups develop directly into adults.

Bivalves include many of the mollusks that end up on our dinner plates, including mussels, oysters, clams, and scallops (**FIGURE 24.20C**). All bivalves have a hinged, two-part shell. Powerful adductor muscles connect the two parts (**FIGURE 24.21**). Contraction

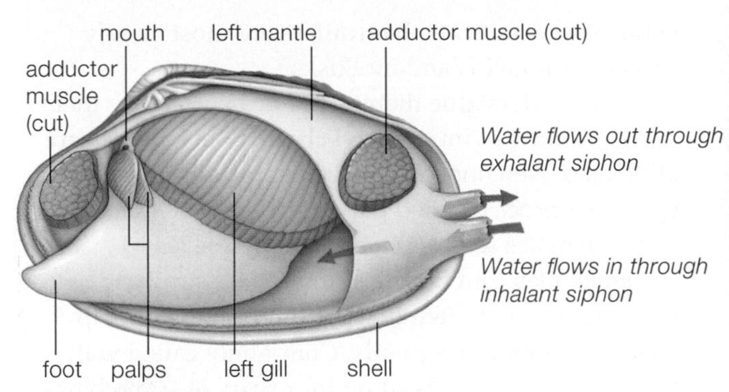

FIGURE 24.21 Body plan of a clam, a bivalve.

A Chambered nautilus, one of six living nautilus species. The shell consists of many gas-filled chambers. The numerous tentacles do not have suckers.

B Octopus. Its eight arms are covered with many individually controlled suckers that allow it to hold on to objects, and to adhere to or walk along surfaces.

FIGURE 24.22 ▸**Animated** Examples of cephalopods.

of these muscles pulls the shell shut, enclosing the body and protecting it from predation or drying out. A bivalve has a reduced head, but eyes arrayed around the edge of its mantle alert it to danger.

Bivalves have a large triangular foot that most use to burrow. A clam burrows beneath sand and extends its siphons into the water. Like other bivalves, it lacks a radula. It feeds by drawing water into its mantle cavity and trapping bits of food in mucus on its gills. Waving cilia directs particle-laden mucus to a pair of fleshy extensions called palps that sort out particles and sweep food into the mouth.

Cephalopods include squids (**FIGURE 24.20D**), nautiluses (**FIGURE 24.22A**), octopuses (**FIGURE 24.22B**), and cuttlefish. Cephalopod means "head-footed," and their foot has been modified into arms and/or tentacles that extend from the head. All cephalopods are predators and most have beaklike, biting mouthparts in addition to a radula. Cephalopods move by jet propulsion. They draw water into the mantle cavity, then force it out through a funnel-shaped siphon.

Five hundred million years ago, large cephalopods with a long, conelike shell were top predators in the seas. Modern nautiluses have a coiled external shell, but other cephalopods have a reduced shell or none at all. Competition with jawed fishes, which evolved 400 million years ago, may have favored a shift in body form. Cephalopods with the smallest shell could be fastest and most agile. A speedier lifestyle required other changes as well. Noncephalopod mollusks have an **open circulatory system**, in which fluid leaves vessels and seeps among tissues before returning to the heart. Only cephalopods have a closed circulatory system that enhances flow of oxygen to muscle tissues. Competition with fishes also favored improved eyesight. Like vertebrates, cephalopods have eyes with a lens that focuses light.

Cephalopods include the fastest (squids), biggest (giant squid), and smartest (octopuses) invertebrates. Of all invertebrates, octopuses have the largest brain relative to body size, and the most complex behavior.

gill Respiratory organ that facilitates gas exchanges with water.
lung Saclike respiratory organ in which blood exchanges gases with the air.
mantle Of mollusks, extension of the body wall.
mollusk Invertebrate with a reduced coelom and a mantle, includes chitons, bivalves, gastropods, and cephalopods.
open circulatory system Circulatory system in which fluid leaves vessels and mingles with tissue fluid before returning to the heart.
radula Of many mollusks, a tonguelike organ hardened with chitin.

TAKE-HOME MESSAGE 24.8

What are mollusks?

✔ Mollusks are invertebrates with a bilateral body plan, a reduced coelom, and a mantle that drapes over their internal organs. In most species, the mantle secretes a protective hardened shell.

✔ Most mollusks are aquatic, but some gastropods have adapted to life on land. In addition to gastropods, mollusks include chitons, bivalves, and cephalopods.

24.9 Rotifers and Tardigrades—Tiny and Tough

✔ Rotifers and tardigrades are two phyla of minuscule animals that can withstand extreme environmental conditions.

The 2,150 species of **rotifers** (phylum Rotifera) live in fresh water and in damp land habitats. Most are less than one millimeter long. The group name, Latin for "wheel bearer," refers to the constantly moving cilia on the head, which look like turning wheels (**FIGURE 24.23**). Movement of the cilia directs food into the mouth. Rotifers have excretory organs (protonephridia) and a complete digestive tract, but no circulatory or respiratory organs. Digestive and excretory organs are located in a pseudocoelom.

Traditionally, rotifers and roundworms were grouped together as pseudocoelomates. However, gene comparisons indicate that rotifers are most closely related to annelids and mollusks.

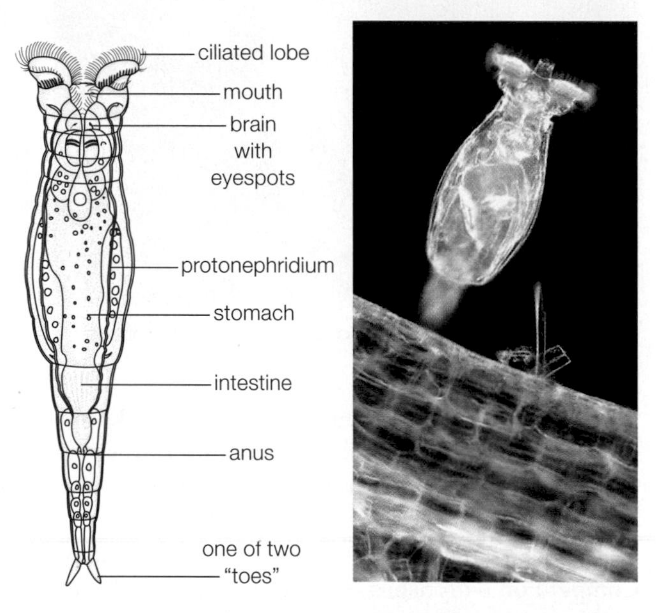

FIGURE 24.23 Rotifer body plan (left) and light micrograph (right).

Some rotifers glue themselves to a surface by their toes, but most swim or crawl about. Some species are all female. New individuals develop from unfertilized eggs by a process called parthenogenesis. Other species produce males seasonally or have two sexes.

Tardigrades (phylum Tardigrada) are similarly tiny animals that often live beside rotifers in damp moss and temporary ponds. Commonly called water bears, they waddle about on four pairs of stubby legs (**FIGURE 24.24**). "Tardigrada" means slow walker.

Most of the 950 or so named tardigrade species suck juices from plants or algae. Some, including the one in **FIGURE 24.24**, are predators. The digestive system is complete and there are excretory organs, but no circulatory or respiratory organs. The coelom is reduced.

Like roundworms and arthropods, tardigrades have an external body covering that they molt (shed periodically) as they grow. Tardigrades may be related to arthropods and roundworms, but relationships among these groups are poorly understood.

Tardigrades and rotifers living in habitats that often dry up completely have evolved a remarkable ability. They survive dry periods by entering a sort of suspended animation. As the habitat dries up, sugar replaces water in their tissues and metabolism slows to a nearly nonexistent pace. In tardigrades, the water content of the body can drop to 1 percent of normal.

Dormant tardigrades withstand extraordinary heat and cold. They have survived for a few days at –200°C (–328°F) and a few minutes at 151°C (304°F). Also, a tardigrade can remain dormant for years, then revive when placed in water. For all of these reasons, tardigrades are often said to be the toughest animals.

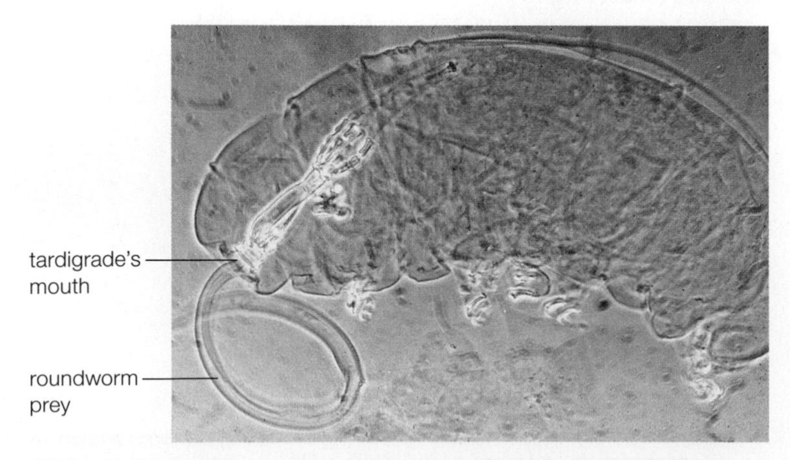

FIGURE 24.24 ▸**Animated** Light micrograph of a tardigrade eating a roundworm.

rotifer Tiny bilateral, pseudocoelomate animal with a ciliated head.
tardigrade A tiny coelomate animal with four pairs of legs; in a dormant state, it can survive extremely adverse conditions.

TAKE-HOME MESSAGE 24.9
What are rotifers and tardigrades?

✔ Rotifers and tardigrades (water bears) are tiny bilateral animals. Most live in damp habitats or fresh water. An ability to enter a dormant state allows some to survive in an environment that often dries out.

✔ Rotifers have a pseudocoelom, but genetic comparisons indicate they are closest to annelids and mollusks.

✔ Tardigrades have a coelom and molt; they are probably relatives of the roundworms and insects.

24.10 Roundworms—Unsegmented Worms That Molt

✔ Most roundworms act as decomposers in soil and water, but some infect us, our pets, our livestock, or our crops.

Roundworms, or nematodes (phylum Nematoda), are unsegmented worms that have a pseudocoelom and a cuticle-covered, cylindrical body (**FIGURE 24.25**). The collagen-rich cuticle is periodically molted as the worm grows. Roundworms have a complete digestive tract, excretory organs, and a nervous system, but no circulatory or respiratory organs.

Nearly 20,000 named roundworm species live in the seas, in fresh water, in damp soil, and inside other animals. Most are free-living decomposers less than a millimeter long. Parasitic species tend to be larger.

Like the fruit fly, the soil roundworm *Caenorhabditis elegans* is frequently used in scientific studies. It serves as a model for developmental processes because it has the same tissue types as more complex organisms, but it is transparent, has less than 1,000 body cells, reproduces fast, and has a small genome.

Several kinds of parasitic roundworms infect humans. The intestinal parasite *Ascaris lumbricoides* (**FIGURE 24.26A**) currently infects more than 1 billion people. Most of those affected live in developing tropical nations, but occasional infections occur in rural parts of the American Southeast. Pinworms (*Enterobius vermicularis*) infect children worldwide. The small worms, about the size of a staple, live in the rectum. Mosquito-transmitted roundworms cause a disfiguring tropical disease called lymphatic filariasis. The worms travel in the body's lymph vessels and destroy the vessels' valves so lymph pools in the lower limbs (**FIGURE 24.26B**). Elephantiasis, the common name for this disease, refers to fluid-filled, "elephant-like" legs.

Parasitic roundworms also infect our livestock and pets. A roundworm that lives in pigs can also infect humans who eat undercooked pork, causing the disease trichinosis. Dogs are susceptible to infection by heartworms, which are roundworms transmitted by mosquitoes. In cats, roundworm infections usually occur after the cat feeds on an infected rodent that serves as an intermediate host.

Many crop plants are susceptible to nematode parasites. In some cases, the worms suck on plant roots and in others they actually enter the plant (**FIGURE 24.26C**). Either way, the infection stunts the plant's growth and lowers crop yields.

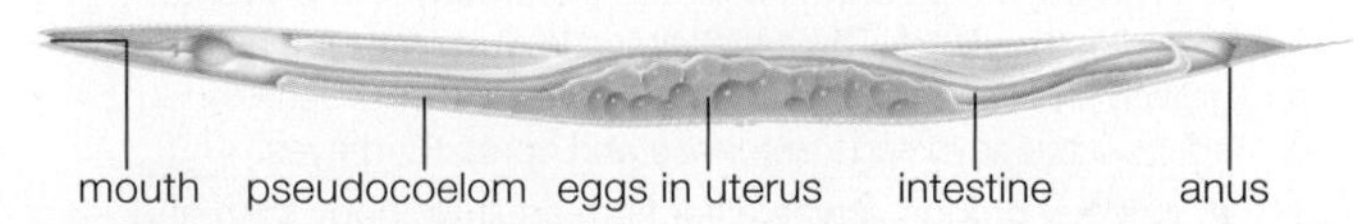

FIGURE 24.25 Body plan of a free-living roundworm (*Caenorhabditis elegans*).

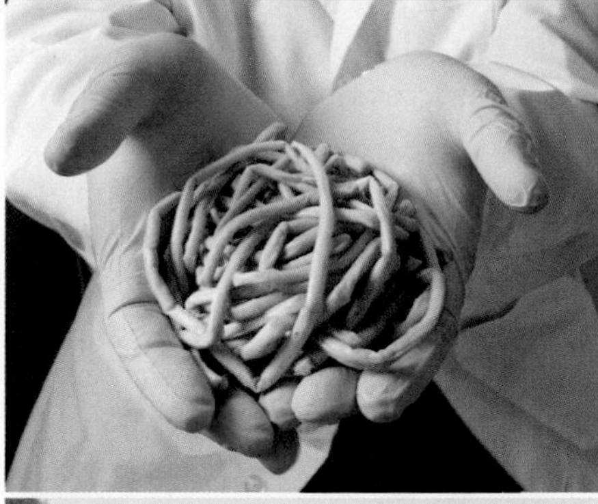

A Intestinal parasite, *Ascaris*, passed by a child.

B The grossly enlarged leg of the man on the left is a sign of lymphatic filariasis.

C Plant-infecting roundworm entering a root.

FIGURE 24.26 Parasitic roundworms.

roundworm Unsegmented pseudocoelomate worm with a cuticle that is molted periodically as the animal grows.

TAKE-HOME MESSAGE 24.10

What are roundworms?

✔ Roundworms are unsegmented, pseudocoelomate worms with a cuticle that they molt as they grow.

✔ Most roundworms are decomposers, but some are parasites. One free-living species is often used by scientists in studies development.

CREDITS: (25) © Cengage Learning; (26A) CDC/ Henry Bishop; (26B) Courtesy of © Emily Howard Staub and The Carter Center; (26C) William Wergin and Richard Sayre. Colorized by Stephen Ausmus.

24.11 Arthropods—Molting Animals With Jointed Legs

✔ Arthropods are the most diverse invertebrate group. A variety of features contribute to their success.

Arthropods (phylum Arthropoda) are bilateral, with a reduced coelom. They have a hard, jointed external skeleton, a complete digestive tract, an open circulatory system, and respiratory and excretory organs. Sexes are usually separate, although hermaphrodites occur in some groups.

One lineage, the trilobites, was abundant in seas until the Permian, when they went extinct. Modern groups include myriapods such as centipedes, crustaceans such as crabs, and insects. Here we begin by thinking about the key adaptations that contribute to arthropod success.

FIGURE 24.27 An insect molting its old exoskeleton. Notice the jointed legs, antennae, and compound eyes.

Key Arthropod Adaptations

Hardened Exoskeleton Arthropods secrete a cuticle composed of chitin (Section 3.4), proteins, and waxes. It functions as an **exoskeleton**, a hard, external skeleton that gives the body shape, offers protection from predators, and serves as a framework to which muscles can attach. When some groups invaded land, the exoskeleton also helped them conserve water. Although the arthropod exoskeleton is hard, it does not restrict growth, because—like the roundworms—arthropods molt their cuticle after each growth spurt (**FIGURE 24.27**). Hormones that regulate molting trigger a new cuticle to form beneath the old one. The new cuticle remains soft until the old one is shed.

Jointed Appendages "Arthropod" means jointed leg. If an arthropod's cuticle were uniformly hard and thick like a plaster cast, it would prevent movement. An arthropod's cuticle thins at regions where two hard body parts meet. When muscles that span one of these joints contract, they cause the cuticle to bend, which in turn causes the parts on either side of the joint to move relative to one another.

Highly Modified Segments In early arthropods, the body segments were distinct, and all appendages were similar. In many of their descendants, the segments became fused into structural units such as a head, a thorax (midsection), and an abdomen (hind section). Appendages on some segments became modified for special tasks. For example, in insects, thin extensions of the wall of some segments evolved as wings.

Sensory Specializations Most arthropods have one or more pairs of eyes. Insects and crustaceans have **compound eyes**, which consist of many individual units, each with a lens. Such eyes excel at detecting movement. With the exception of chelicerates, most arthropods also have paired **antennae** that can detect touch and waterborne or airborne chemicals.

Specialized Developmental Stages The body plan of many arthropods changes during the life cycle. In many groups, individuals undergo **metamorphosis**: Tissues get remodeled as larvae develop into adults. For example, crab larvae float in the plankton, but adults are bottom-feeders. Each stage is specialized for a different lifestyle. Having different bodies prevents adults and juveniles from competing for resources.

antenna Of some arthropods, sensory structure on the head that detects touch and odors.
arthropod Invertebrate with jointed legs and a hard exoskeleton that is periodically molted; for example, an insect or crustacean.
compound eye Eye that consists of many individual units; each with a lens; excels at detecting movement.
exoskeleton Of some invertebrates, hard external parts that muscles attach to and move.
metamorphosis Remodeling of body form during the transition from larva to adult.

TAKE-HOME MESSAGE 24.11

What are arthropods and what factors contribute to their success?

✔ Arthropods are the most diverse animal phylum. In addition to modern groups, they include the extinct trilobites.

✔ A hardened exoskeleton protects the body and prevents water loss on land. The exoskeleton is molted as the animal grows, and it is thin at joints to allow movement. Sensory specializations include antennae and compound eyes.

✔ In many groups, larvae differ from adults in body form and utilize different resources.

24.12 Chelicerates—Spiders and Their Relatives

✔ Chelicerates include spiders, scorpions, and ticks, as well as the marine horseshoe crabs.

Chelicerates have a body with two regions, a cephalothorax (fused head and thorax) and an abdomen. Walking legs attach to the cephalothorax. The head has eyes, but no antennae. Paired feeding appendages near the mouth, called chelicerae, give the group its name.

Four species of horseshoe crabs are the only marine chelicerates and members of the oldest surviving arthropod lineage. Horseshoe crabs are bottom feeders that eat clams and worms. A horseshoe-shaped shield covers the cephalothorax, and the last segment has evolved into a long spine (**FIGURE 24.28**).

Arachnids include spiders, scorpions, ticks, and mites (**FIGURE 24.29**). Nearly all live on land. All have four pairs of walking legs, and a pair of appendages (pedipalps) between the chelicerae and the front legs.

Scorpions and spiders are venomous predators. Spider chelicerae are fanglike and have venom glands. Of 38,000 spider species, about 30 produce venom harmful to humans. Most spiders benefit us by eating insect pests. Some weave prey-catching webs made of silk that is exuded from glands on their abdomen. Others such as tarantulas (**FIGURE 24.29A**) are active hunters. Scorpions hunt prey and dispense venom through a stinger on their last segment (**FIGURE 24.29B**). Their pedipalps have evolved into claws.

Ticks are parasites of vertebrates (**FIGURE 24.29C**). They pierce a host's skin with their chelicerae, and then suck blood. Some transmit bacteria that cause Lyme disease or other diseases.

Mites are the smallest arachnids; most are less than a millimeter long. Dust mites (**FIGURE 24.29D**) are scavengers that can cause problems for people who are allergic to their feces. Other mites parasitize plants or animals. Some larval mites, commonly called chiggers, attach to human skin at hair follicles and sip our tissue fluids. Mites that burrow beneath the skin cause scabies in humans and mange in dogs.

arachnids Land-dwelling chelicerate arthropods with four pairs of walking legs; spiders, scorpions, mites, and ticks.
chelicerates Arthropod subgroup with specialized feeding structures (chelicerae) and no antennae.

TAKE-HOME MESSAGE 24.12
What are chelicerates?

✔ Chelicerates include horseshoe crabs and arachnids. The horseshoe crabs are an ancient marine lineage. Arachnids are land dwellers with eight walking legs and no antennae.

FIGURE 24.28 Horseshoe crab. There are five pairs of walking legs (visible in the photo of the underside at right), and the final segment on the abdomen has been modified as a long spine that serves as a rudder. Horseshoe crabs do not produce venom.

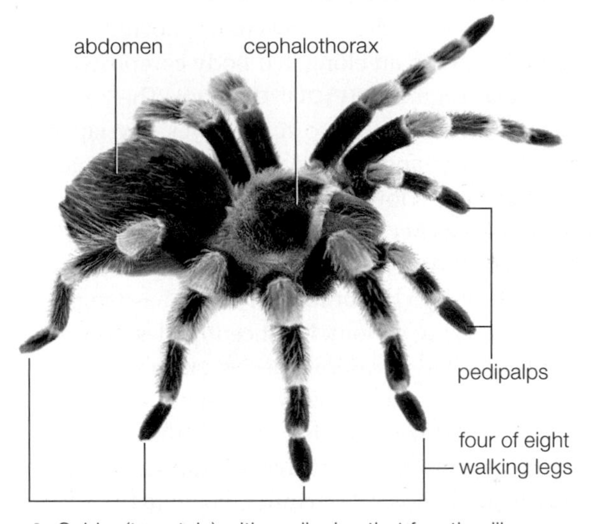

A Spider (tarantula) with pedipalps that function like arms.

B Scorpion with clawlike pedipalps. The last segment has a venomous stinger.

C Female tick at the tip of a blade of grass waiting for a host to walk by.

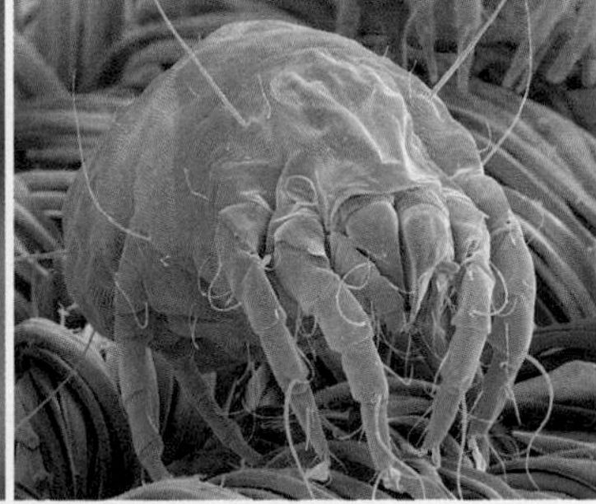

D Dust mite, a scavenger (colorized scanning electron micrograph).

FIGURE 24.29 Arachnids. All have chelicerae, pedipalps, and eight walking legs.

24.13 Myriapods

A Centipede, a speedy predator.

B Millipede, a scavenger of decaying plant material.

FIGURE 24.30 ▶**Animated** Myriapods.

✔ Centipedes and millipedes use their many legs to walk about on land, hunting prey or scavenging.

Centipedes and millipedes are **myriapods**, nocturnal ground dwellers with an elongated body composed of many similar segments (**FIGURE 24.30**). The head has a pair of antennae and two simple eyes. Myriapod means "many feet," and aptly describes these animals.

Centipedes have a low-slung, flattened body with a single pair of legs per segment, for a total of 30 to 50 (**FIGURE 24.30A**). They are fast-moving predators. Their first pair of legs has become modified as fangs that inject paralyzing venom. Most centipedes prey on insects, but some large tropical species capture and eat small vertebrates, including lizards, rodents, and birds.

Millipedes are slower-moving animals that feed on decaying vegetation. They have a cylindrical body with a cuticle hardened by calcium carbonate (**FIGURE 24.30B**). As a result of segment fusion during development, adults have two pairs of legs per segment. Most have a few hundred pairs of legs. Because millipedes are not predators, they do not produce venom. However, they do make and release toxic compounds as a defense. When threatened, a millipede curls up and secretes a foul-smelling, unpalatable fluid.

Myriapods have a long history. The earliest known fossil of a land-dwelling, air-breathing animal is of a myriapod that lived 428 million years ago. Another extinct myriapod is one of the largest known arthropods. It lived during the Carboniferous and reached more than 2.5 meters (8 feet) in length.

myriapod Land-dwelling arthropod with two antennae and an elongated body with many segments; a millipede or centipede.

TAKE-HOME MESSAGE 24.13

What are myriapods?

✔ Myriapods are land-dwelling arthropods that have two antennae and an abundance of body segments. Centipedes are predators and millipedes are scavengers.

24.14 Crustaceans

✔ Most marine arthropods are crustaceans. Their amazing diversity and abundance are reflected in their nickname "the insects of the seas."

Crustaceans are a diverse group of mostly marine arthropods that have two pairs of antennae. As in chelicerates, the body has two distinctive regions, a cephalothorax and an abdomen. In many crustaceans, the exoskeleton is stiffened in some regions by incorporation of calcium carbonate.

You are probably familiar with some of the decapod crustaceans. This group of bottom-feeding scavengers includes the lobsters, crayfish, crabs, and shrimps. Decapods typically have five pairs of walking legs. In some lobsters, crayfish, and crabs, the first pair of legs has become modified into claws (**FIGURE 24.31**).

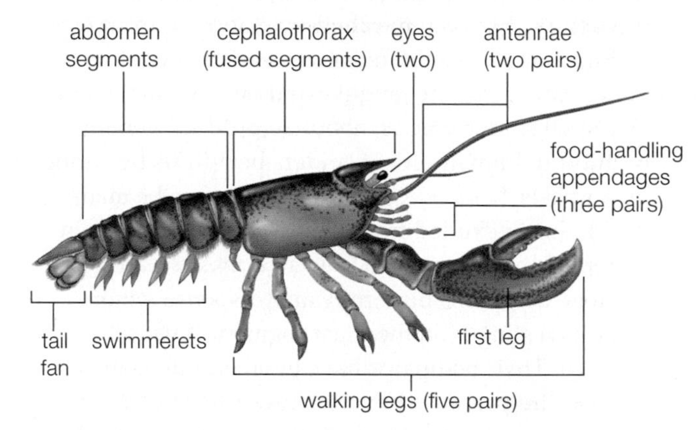

FIGURE 24.31 ▶**Animated** Body plan of a lobster (*Homarus americanus*).

An aquatic crustacean starts life as a microscopic, planktonic larva that typically bears little resemblance to the adult animal. The transition from larval form to adult form takes place over the course of several molts. For example, larvae of Dungeness crabs harvested along the Pacific coast of North America live as members of the plankton for months and molt repeatedly before taking on their adult form (**FIGURE 24.32**).

Krill and copepods are small swimmers that feed on plankton. They in turn serve as a major food source for larger animals. Krill (euphausiids) are relatives of the decapods and have a shrimplike body up to 6 centimeters long (**FIGURE 24.33A**). They swim through cool marine waters worldwide. Krill have historically been so abundant that a 30-ton blue whale could sustain itself largely on the krill that it filtered from the water. Recent overharvesting of krill from North Pacific and Antarctic waters has caused steep population declines and may endanger the animals that depend on this

FIGURE 24.32 ▶**Animated** Development of the Dungeness crab (*Cancer magister*). The crabs live for 8 to 10 years.

❶ After a female crab has mated, she fertilizes eggs with stored sperm, then holds them under her abdomen for 2–3 months until they hatch.

❷ Within an hour of hatching, an early-stage planktonic larva (called a zoea) develops.

❸ The early-stage larva grows and molts repeatedly, altering its form only slightly.

❹ The fifth molt of the early-stage larva produces a late-stage planktonic larva with an altered body form (a megalops).

❺ The late-stage larva molts to become a bottom-dwelling juvenile crab. Growth and repeated molting will produce a sexually mature adult in about 2 years.

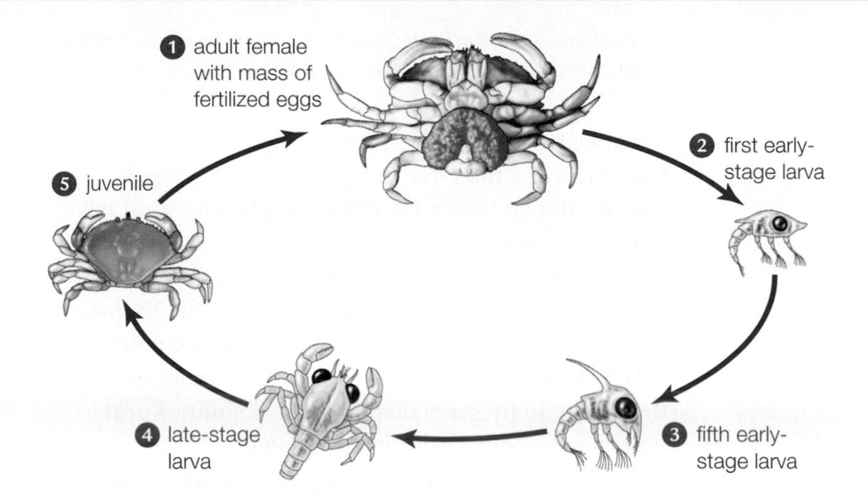

resource. Humans harvest krill to feed farm-raised salmon and as a source of omega-3 fatty acids.

Free-swimming copepods abound in seas and lakes (**FIGURE 24.33B**). They are probably the most numerous of all crustaceans. Adults are typically about 1 to 2 millimeters long. Some copepod species affect human health by serving as the intermediate host for bacteria that cause cholera.

Larval barnacles swim, but adults secrete a thick calcified shell and live fixed, head down, in place. They filter food from the water with feathery legs (**FIGURE 24.33C**). Some barnacles are notable for the length of their penis, which can be eight times that of their body.

Isopods are a mostly marine group, but you are probably familiar with the species that have adapted to life on land. Sow bugs and pill bugs are often found beneath flowerpots or rotting logs. They require a damp habitat where they can feed on organic debris or on soft, young plant parts. Some can be pests of agricultural crops such as soybeans. The species commonly known as pill bugs defend themselves from threats by rolling into a ball (right).

crustaceans Mostly marine arthropod group with two pairs of antennae and a calcium-stiffened exoskeleton.

TAKE-HOME MESSAGE 24.14
What are crustaceans?

✔ Crustaceans are mostly marine arthropods that have two pairs of antennae and an exoskeleton that includes calcium. They include bottom-feeding decapods, planktonic copepods and krill, filter-feeding barnacles, and even some species that have adapted to life on land.

A Antarctic krill (*Euphausia superba*). Individuals can be up to 6 centimeters long.

B Female copepod with eggs. **C** Barnacle extending feathery legs.

FIGURE 24.33 Crustaceans that feed on plankton.

CREDITS: (in text) © Chris Howey/Shutterstock.com; (32) © Cengage Learning; (33A) © David Tipling/Photographer's Choice/Getty Images; (33B) Herve Chaumeton/Agence Nature; (33C) Peter Parks/Image Quest Marine.

24.15 Insects—Diverse and Abundant

✔ The insects are the most abundant land arthropods, and the most diverse animal class. They have six legs.

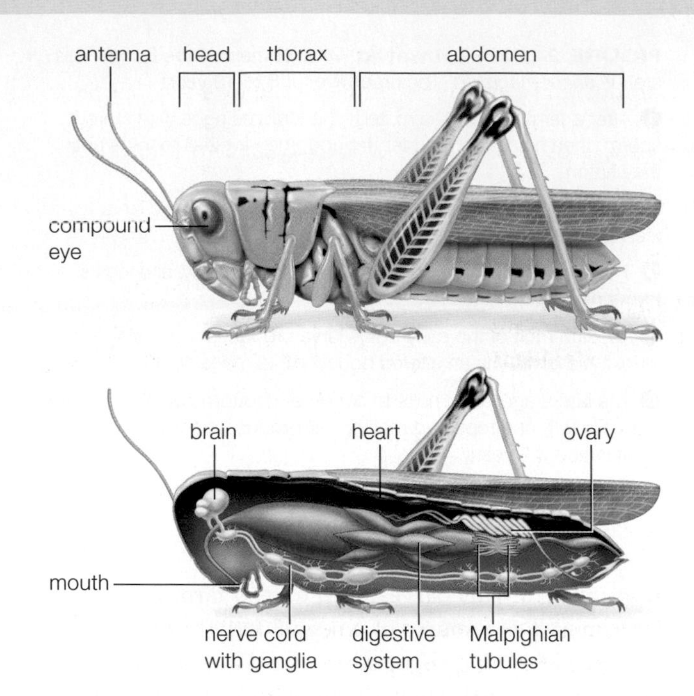

FIGURE 24.34 Body plan of a female grasshopper.

Characteristic Features

Insects have a three-part body plan, with a head, thorax, and abdomen (**FIGURE 24.34**). The head has one pair of antennae and two compound eyes. Near the mouth are jawlike mandibles and other feeding appendages. Three pairs of legs attach to the thorax. In some groups, the thorax also has one or two pairs of wings. Insects are the only winged invertebrates, and the ability to fly gives them dispersal abilities unrivaled among other land invertebrates.

The group is overwhelmingly terrestrial. A respiratory system consisting of tracheal tubes carries air from openings at the outer surface to deep inside the body. An insect abdomen contains digestive organs, sex organs, and water-conserving excretory organs called **Malpighian tubules**.

Until recently, insects were thought to be close relatives of myriapods. Both groups have a single pair of antennae and unbranched legs. Then, new gene comparisons made scientists rethink the connections. The currently favored hypothesis holds that insects are most closely related to crustaceans. Specifically, insects are thought to be descended from freshwater crustaceans. If this is correct, then insects are the crustaceans of the land.

The earliest insects were wingless, ground-dwelling scavengers that did not undergo metamorphosis. A few modern insects such as bristletails and silverfish retain this type of body form and development. When they hatch from an egg, they look like a tiny adult, and they simply grow bigger with each molt.

Most groups of modern insects have wings and undergo metamorphosis. In groups that undergo incomplete metamorphosis, an egg hatches into a nymph that differs somewhat in form from the adult. It gradually achieves adult form over the course of several molts. Cockroaches, grasshoppers, and dragonflies develop in this manner. A dragonfly lives as an aquatic nymph for one to three years, molting several times, before emerging from the water and undergoing a final molt to the winged adult form.

With complete metamorphosis, a larva grows and molts without altering its form, then undergoes pupation; it becomes a pupa. A pupa is a nonfeeding body in which larval tissues are remodeled into the adult form (**FIGURE 24.35**). Members of the four most diverse insect orders have wings and undergo complete metamorphosis. There are about 150,000 species of flies (Diptera), and at least as many beetles (Coleoptera). The order Hymenoptera includes 130,000 species of wasps, ants, and bees. Moths and butterflies are Lepidoptera, a group of about 120,000 species. As a comparison, consider that there are about 4,500 species of mammals.

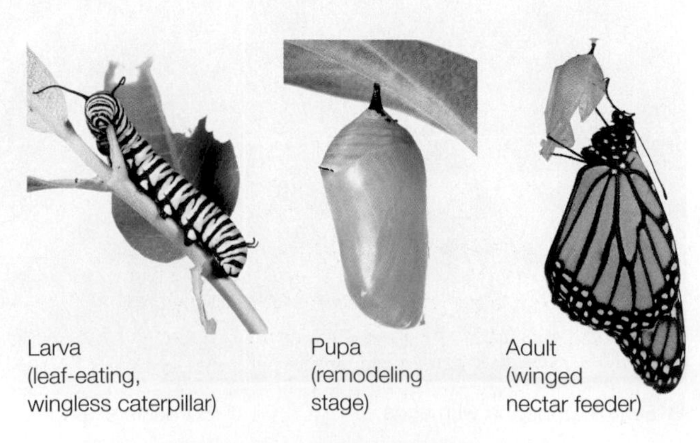

FIGURE 24.35 Complete metamorphosis in a butterfly.

The Importance of Insects

With more than a million species, insects are the most diverse arthropod group, and they are breathtakingly abundant. By one estimate, ants alone make up about 10 percent of the world's terrestrial animal biomass (the total weight of all living land animals). Given their diversity and numbers, it would be difficult to overestimate the importance of insects.

Ecological Services Most flowering plants are pollinated by members of one of the four most diverse insect orders (**FIGURE 24.36A**). Other insect orders contain few or no pollinators. By one hypothesis, such interactions between insect groups and flowering plants contributed to an increased rate of speciation in both groups.

Insects are also important as food for wildlife. Most songbirds nourish their nestlings on a diet consisting largely of insects. Migratory songbirds often travel long distances to nest and raise young in areas where insect abundance is seasonally high. Aquatic larvae of insects such as dragonflies, mayflies, and mosquitoes serve as food for trout and other freshwater fish. Amphibians and reptiles feed mainly on insects. Even humans eat insects. In many cultures, they are considered a tasty source of protein.

Insects dispose of wastes and remains. Flies and beetles are quick to discover an animal corpse or a pile of feces (**FIGURE 24.36B**). They lay their eggs in or on this organic material, and the larvae that hatch devour it. By their actions, these insects keep organic wastes and remains from accumulating, and help distribute nutrients through the ecosystem.

A Butterfly (a lepidopteran) serving as a pollinator.

B A dung beetle collecting feces that will nourish its offspring.

FIGURE 24.36 Helpful insects.

Competitors for Crops Insects are our main competitors for food and other plant products. About a quarter to a third of all crops grown in the United States are lost to insects. Also, in an age of global trade and travel, we have more than just homegrown pests to worry about. Consider the Mediterranean fruit fly, or Med fly (**FIGURE 24.37A**). Med flies lay eggs in citrus and other fruits, as well as many vegetables. Damage done to plants and fruits by larvae of the Medfly can cut crop yield in half. Medflies are not native to the United States, and there is an ongoing inspection program for imported produce, but some Medflies still slip in. So far, eradication efforts have been successful, but they have cost hundreds of millions of dollars.

A Medfly, a pest of citrus.

B Bedbug, parasite of humans.

FIGURE 24.37 Harmful insects.

Parasites and Pathogens Some insects spread human diseases. Mosquitoes are probably the most important vectors. They transmit malaria, which causes more than 500,000 deaths each year (Section 21.1), as well as other diseases. Bites of other insects also spread pathogens. Biting flies transmit African sleeping sickness; biting bugs (Order, Hemiptera) spread Chagas disease (Section 21.3). Fleas that bite rats and then bite humans can transmit deadly bubonic plague. Body lice transmit typhus.

As far as we know, bedbugs (**FIGURE 24.37B**) do not transmit disease, although their bites can itch. In addition, a bedbug infestation causes psychological stress and can have a severe negative economic impact on a commercial establishment.

insect Six-legged arthropod with two antennae and two compound eyes. Member of the most diverse class of animals.
Malpighian tubule Water-conserving excretory organ of insects.

TAKE-HOME MESSAGE 24.15

What are insects?

✔ Insects are six-legged arthropods adapted to life on land. They probably evolved from a crustacean ancestor.

✔ Tracheal tubes allow insects to breathe air. Malpighian tubules help conserve water while eliminating wastes. In some groups, wings evolved as extensions from the body wall.

✔ Insects are both numerous and diverse. They are food for other animals, pollinate plants, and scavenge wastes and remains. They also compete with us for crops, parasitize us, and infect us with diseases.

CREDITS: (36A) Marcia Straub/Shutterstock; (36B) Gregory G. Dimijian, M.D./Science Source; (37A) Photo by Scott Bauer/USDA; (37B) CDC/Piotr Naskrecki.

24.16 The Spiny-Skinned Echinoderms

✔ Echinoderms begin life as bilateral larvae and develop into spiny-skinned, radial adults. All are marine.

The Protostome–Deuterostome Split

In Section 24.2 we introduced the two major lineages of animals, protostomes and deuterostomes. Here we begin our survey of deuterostome lineages. Echinoderms are the largest group of invertebrate deuterostomes. We will discuss other invertebrate deuterostomes and the vertebrates (also deuterostomes) in the next chapter.

Echinoderm Characteristics and Body Plan

Echinoderms (phylum Echinodermata) include about 6,000 marine invertebrates. Their name means "spiny-skinned" and refers to interlocking spines and plates of calcium carbonate embedded in their skin. Adults are coelomate animals and most have a radial body, with five parts (or multiples of five) around a central axis. The larvae are bilateral, which suggests that the ancestor of echinoderms was a bilateral animal.

Sea stars (also called starfish) are the most familiar echinoderms, and we will use them to illustrate the general body plan (**FIGURE 24.38**). Sea stars do not have a brain, but they do have a nerve net. A nerve ring surrounds the mouth, and nerves that branch from this ring extend into the arms. Eyespots at the tips of arms detect light and movement.

A typical sea star is an active predator that moves about on tiny, fluid-filled tube feet. Tube feet are part of a **water–vascular system** unique to echinoderms. The system includes a central ring and fluid-filled canals that extend into each arm. Side canals deliver coelomic fluid into muscular ampullae, which are tiny bulbs that function like the bulb on a medicine dropper. Contraction of an ampulla forces fluid into the attached tube foot, extending the foot. A sea star glides along as coordinated contraction and relaxation of the ampullae redistribute fluid among hundreds of tube feet.

Sea stars typically feed on bivalve mollusks. A sea star's mouth is on its lower surface. To feed, the animal slides its stomach out through its mouth and into a bivalve's shell. The stomach secretes acid and enzymes that kill the mollusk and begin the process of digestion. Partially digested food is taken into the stomach, then digestion is completed with the aid of digestive glands in the arms.

There are no specialized respiratory or circulatory organs. Gas exchange occurs by diffusion across the tube feet and tiny skin projections at the body surface. There are also no specialized solute-regulating organs.

Sexes are separate. The arms hold sexual organs that release eggs or sperm into the water. Fertilization produces an embryo that develops into a ciliated, bilateral larva. The larva swims briefly, then undergoes metamorphosis into the adult form.

Sea stars and other echinoderms have a remarkable ability to regenerate lost body parts. If a sea star is cut into pieces, any portion with some of the central disk can regrow the missing body parts.

Echinoderm Diversity

Brittle stars have a central disk and highly flexible arms that move in a snakelike way (**FIGURE 24.39A**). Brittle stars are the most diverse and abundant echinoderms, but are less familiar than sea stars because they gener-

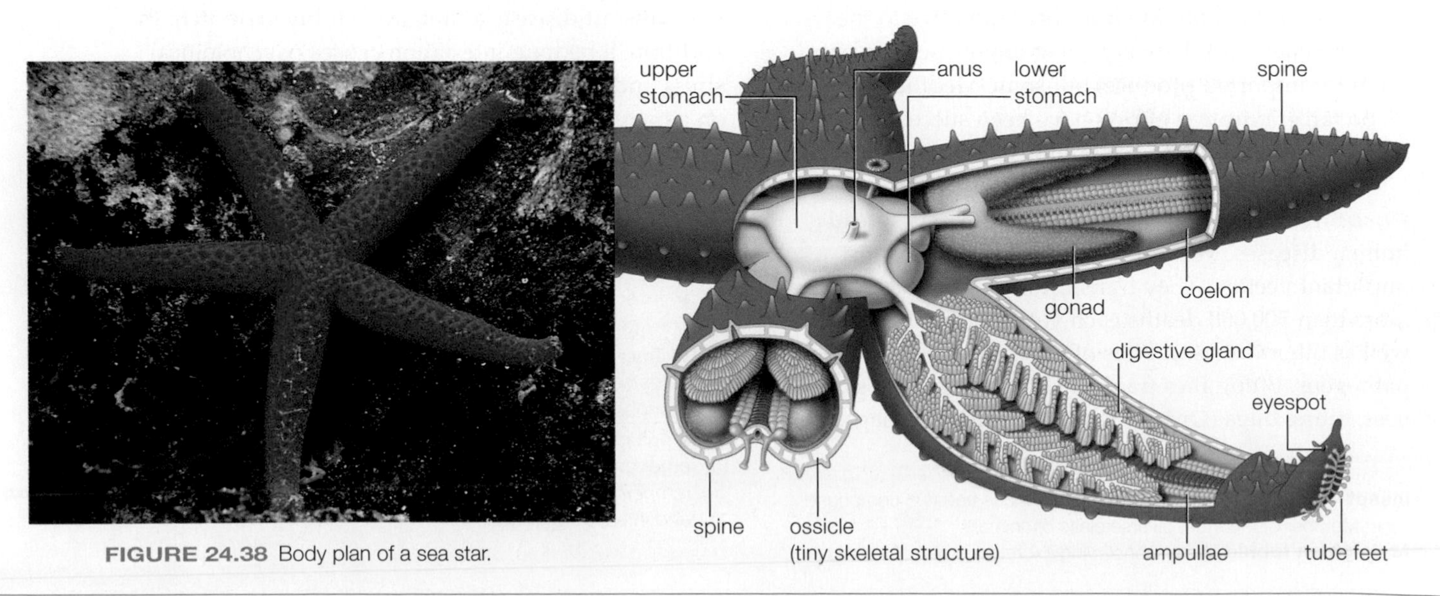

FIGURE 24.38 Body plan of a sea star.

CREDITS: (38) left, Herve Chaumeton/Agence Nature; right, © Cengage Learning.

Medicines From the Sea (revisited)

Many invertebrates make chemicals that inhibit the growth of bacteria or protozans. To find new drugs with these properties, researchers extract compounds from invertebrates, then test the ability of the compounds to kill pathogens cultured in the laboratory. The effect of each compound is also tested on cultured human cells, because an ideal candidate drug does no harm to human cells. Chemicals are also tested for their ability to kill cancer cells, while sparing normal ones.

Finding a compound that may have medicinal value is only the first step in drug development. For a compound to be used in clinical tests, researchers must obtain a sufficient amount of it. This can be difficult because many compounds of interest occur only at very low concentrations in animals. Consider the drug Eribulin (Halhaven), now used to fight some metastatic breast cancers. Sponges make the compound on which it is modeled, but only in tiny amounts. Obtaining enough of the compound to test its efficiency as a cancer drug (300 milligrams) required processing more than one metric ton of sponges.

Generally, once the structure of a useful compound has been determined, chemists find a way to produce it or a compound with a similar structure and properties. Eribulin is a variant of the sponge compound and it is synthesized, not extracted from sponges.

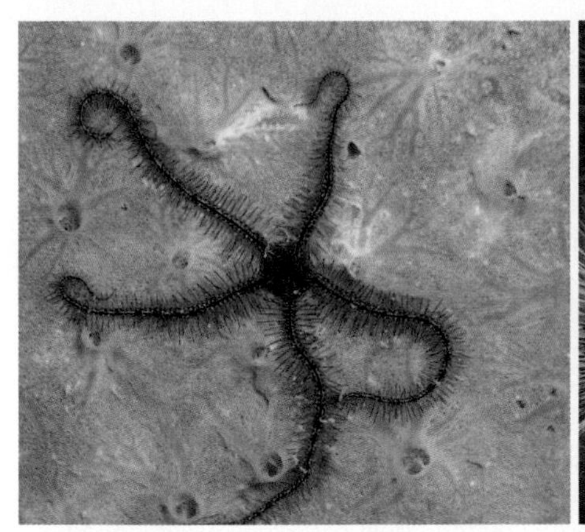

A Brittle star

B Sea urchin

C Sea cucumber

FIGURE 24.39 Echinoderm diversity.

ally live in deeper water. Most brittle stars are scavengers on the seafloor.

In sea urchins, calcium carbonate plates form a stiff, rounded cover from which spines protrude (**FIGURE 24.39B**). The spines provide protection and are used in movement. Some urchins graze on algae. Others act as scavengers or prey on invertebrates.

In sea cucumbers, hardened parts have been reduced to microscopic plates embedded in a soft body. Most species have a wormlike body and, like earthworms, they eat their way through sediments and digest any organic material (**FIGURE 24.39C**). Some deep-sea species can swim with the aid of capelike body extensions.

Lacking spines or sharp plates, sea cucumbers have alternative defenses. When threatened, they expel a sticky mass of specialized threads or internal organs out through their anus. If this maneuver successfully distracts the predator, the sea cucumber escapes, and its missing parts grow back.

Sea urchin roe (eggs) and sea cucumbers are popular foods in Asia, and overharvesting for this market threatens some species in both groups. Farming these animals may provide a way to meet growing consumer demand while protecting natural ecosystems.

echinoderms Invertebrates with hardened plates and spines embedded in the skin or body, and a water-vascular system.
water–vascular system Of echinoderms, a system of fluid-filled tubes and tube feet that function in locomotion.

TAKE-HOME MESSAGE 24.16

What are echinoderms?

✔ Echinoderms are deuterostome invertebrates that have a radial body as adults. They are brainless and have a unique water-vascular system that functions in locomotion.

CREDITS: (39A) Stubblefield Photography/Shutterstock.com; (39B) © Derek Holzapfel/photos.com; (39C) Andrew David, NOAA/NMFS/SEFSC Panama City, Lance Horn, UNCW/NURC - Phantom II ROV operator.

summary

Section 24.1 **Invertebrates** (animals with no backbone) are the most diverse animal group. Some compounds that invertebrates produce to fight off infections or for use in capturing prey can be used as drugs to treat infections or cancer in humans.

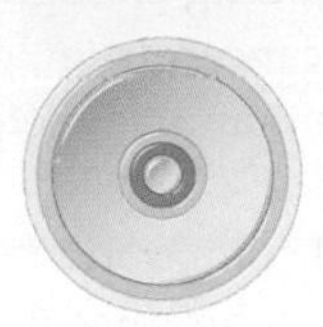

Sections 24.2, 24.3 **Animals** are multicelled heterotrophs with unwalled cells. **TABLE 24.1** summarizes the traits of groups covered in this chapter. Some have no body symmetry; others have **radial symmetry**, like a wheel. Most have **bilateral symmetry** and **cephalization**, a concentration of nerves and sensory structures at the head end. The two lineages of bilateral animals, **protostomes** and **deuterostomes**, develop from a three-layered embryo. Some animals digest food in a saclike **gastrovascular cavity**, others in a tubular **complete digestive tract**. A digestive tract is usually inside a fluid-filled cavity (a **coelom** or a **pseudocoelom**).

Animals most likely evolved from a colonial species similar to choanoflagellates, a type of protist. The oldest animal body fossils (of Ediacarans) date back about 600 million years. A great adaptive radiation during the Cambrian gave rise to most modern lineages.

Section 24.4 **Sponges** are asymmetrical and do not have tissues. They are **suspension feeders** and **hermaphrodites**: Each makes eggs and sperm. Adults are **sessile animals**, but the ciliated **larvae** swim.

Section 24.5 **Cnidarians** have two radially symmetrical body forms: **medusa** and **polyp**. Both have tentacles with **cnidocytes** that help them catch prey. Both also have two tissues with a jellylike layer between them that functions as a **hydrostatic skeleton**. A **nerve net** controls movements, and a gastrovascular cavity functions in both respiration and digestion.

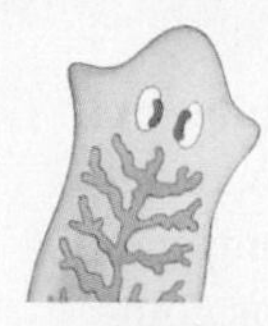

Section 24.6 **Flatworms** are bilateral acoelomate worms with organ systems. They include free-living species and parasitic tapeworms and flukes. **Nerve cords** connect to **ganglia** in the head that serve as a control center. In planarians, a **pharynx** on the lower surface sucks up food and delivers it to the gastrovascular cavity.

Section 24.7 **Annelids** are segmented worms and leeches. They have a **closed circulatory system**, in which blood is always within blood vessels or a heart. Nephridia regulate body fluid composition. Like mollusks, annelids have a trochophore larva.

Section 24.8 **Mollusks** include chitons, gastropods, bivalves, and cephalopods. Their **mantle** is an extension of the body wall. Most have a food-scraping **radula**, and use **gills** or (in land dwellers) a **lung** to breathe. All except the cephalopods have an **open circulatory system**.

Sections 24.9, 24.10 **Rotifers** and **tardigrades** are tiny animals of damp or aquatic habitats. Both groups can dry out and survive long periods of adverse conditions. The **roundworms** (nematodes) have an unsegmented body, a cuticle that is molted, a complete gut, and a false coelom. Some are parasites of humans.

Sections 24.11–24.15 **Arthropods**, the largest phylum of animals, have a jointed **exoskeleton**. Most have sensory **antennae**; insects and crustaceans have **compound eyes**. Development often includes **metamorphosis**, a change in body form.

Chelicerates include marine horseshoe crabs and the eight-legged, land-dwelling **arachnids**. **Crustaceans** have an exoskeleton hardened with calcium, and most are marine.

Table 24.1 Comparison of Invertebrate Groups Surveyed in This Chapter

Animal Phylum	Representative Groups	Living Species	Organization	Symmetry	Digestion	Body Circulation
Porifera	Barrel sponges, encrusting sponges	8,000	Connected cells	None	Intracellular	Diffusion
Cnidaria	Sea anemones, jellyfishes, corals	11,000	2 tissue layers	Radial	Saclike gut	Diffusion
Platyhelminthes	Planarians, tapeworms, flukes	15,000	3 tissue layers	Bilateral	Saclike gut	Diffusion
Annelida	Polychaetes, earthworms, leeches	15,000	3 tissue layers	Bilateral	Complete gut	Closed system
Mollusca	Snails, slugs, clams, octopuses	110,000	3 tissue layers	Bilateral	Complete gut	Open in most, closed in cephalopods
Rotifera	Rotifers	2,150	3 tissue layers	Bilateral	Complete gut	Diffusion
Tardigrada	Water bears	950	3 tissue layers	Bilateral	Complete gut	Diffusion
Nematoda	Pinworms, hookworms	20,000	3 tissue layers	Bilateral	Complete gut	Diffusion
Arthropoda	Spiders, crabs, insects	>1,000,000	3 tissue layers	Bilateral	Complete gut	Open system
Echinodermata	Sea stars, sea urchins, sea cucumbers	6,000	3 tissue layers	Larvae bilateral; adults radial	Complete gut	Open system

data analysis activities

Sustainable Use of Horseshoe Crabs Horseshoe crab blood clots immediately upon exposure to bacterial toxins, so it can be used to test injectable drugs for the presence of dangerous bacteria. To keep horseshoe crab populations stable, blood is extracted from captured animals, which are then returned to the wild. Concerns about the survival of animals after bleeding led researchers to do an experiment. They compared survival of animals captured and maintained in a tank with that of animals captured, bled, and kept in a similar tank. **FIGURE 24.40** shows the results.

1. In which trial did the most control crabs die? In which did the most bled crabs die?

2. Looking at the overall results, how did the mortality of the two groups differ?

3. Based on these results, would you conclude that bleeding harms horseshoe crabs more than capture alone does?

	Control Animals		Bled Animals	
Trial	Number of crabs	Number that died	Number of crabs	Number that died
1	10	0	10	0
2	10	0	10	3
3	30	0	30	0
4	30	0	30	0
5	30	1	30	6
6	30	0	30	0
7	30	0	30	2
8	30	0	30	5
Total	200	1	200	16

FIGURE 24.40 Mortality of young male horseshoe crabs kept in tanks during the 2 weeks after their capture. Blood was taken from half the animals on the day of their capture. Control animals were handled, but not bled. This procedure was repeated 8 times with different sets of horseshoe crabs.

Myriapods include centipedes and millipedes. **Insects**, the most diverse arthropods, include the only winged invertebrates. Tracheal tubes and **Malpighian tubules** adapt them to life on land. Insects pollinate plants, dispose of wastes, and serve as food, but some eat crops or transmit disease.

Section 24.16 **Echinoderms** are deuterostomes. Their skin is hardened with bits of calcium carbonate. A **water–vascular system** with tube feet helps most glide about. Adults are radial with some bilateral features, and the larvae are bilateral.

self-quiz

Answers in Appendix VII

1. True or false? Animal cells do not have walls.

2. A body cavity fully lined with tissue derived from mesoderm is called a(n) ________ .

3. Flatworms, annelids, and roundworms are all ________ .
   a. protostomes c. deuterostomes
   b. vertebrates d. radial

4. The oldest body fossil of a land animal is of a ________ .
   a. sponge c. cnidarian
   b. roundworm d. myriapod

5. ________ function(s) in the movement of cnidarians.
   a. A hydrostatic skeleton c. Cnidocytes
   b. Tube feet d. Malpighian tubules

6. Earthworms are most closely related to ________ .
   a. tapeworms c. planarians
   b. roundworms d. leeches

7. Annelids and ________ have a closed circulatory system.
   a. insects c. flatworms
   b. cephalopods d. sea stars

8. Which invertebrate phylum includes the most species?
   a. mollusks c. arthropods
   b. roundworms d. flatworms

9. Did Ediacarans evolve before or after the Cambrian?

10. A slug is a land-dwelling ________ .
    a. arthropod c. cephalopod
    b. gastropod d. crustacean

11. A barnacle is a shelled ________ .
    a. arthropod c. cephalopod
    b. gastropod d. crustacean

12. ________ include the only winged invertebrates.
    a. Cnidarians c. Rotifers
    b. Echinoderms d. Arthropods

13. The ________ and ________ have similar larvae.
    a. cnidarians/arthropods c. annelids/mollusks
    b. echinoderms/rotifers d. flatworms/roundworms

14. Match the organisms with their descriptions.
    ___ mollusks a. complete gut, pseudocoelom
    ___ echinoderms b. nematocyst producers
    ___ sponges c. simplest organ systems
    ___ cnidarians d. no tissues, filters out food
    ___ flatworms e. jointed exoskeleton
    ___ roundworms f. mantle over body mass
    ___ annelids g. segmented worms
    ___ arthropods h. tube feet, spiny skin

critical thinking

1. A massive die-off of lobsters in the Long Island Sound was blamed on pesticides sprayed to control mosquitoes that carry West Nile virus. Why might a chemical designed to kill insects also harm lobsters but have no effect on sea stars?

2. A rise in atmospheric CO_2 increases ocean acidity, making it harder for animals to build calcium carbonate parts. List some animals that high ocean acidity could harm in this way.

CREDIT: (40) Data adapted from Walls, E., Berkson J., *Fish. Bull.* 101:457-459 (2003).

25 Animal Evolution—The Chordates

LEARNING ROADMAP

This chapter introduces the remaining lineages of deuterostomes, a group defined in Section 24.2. We refer to discussions of fossils (16.5), plate tectonics (16.7), the geologic time scale (16.8), and adaptive radiation (17.12). Understanding cladistics (18.2) and how comparative studies help determine relationships (18.3–18.5) will be extremely useful.

CHARACTERISTICS OF CHORDATES

Distinctive embryonic traits characterize chordates, a group that includes two lineages of marine invertebrates, as well as all animals with a backbone (vertebrates).

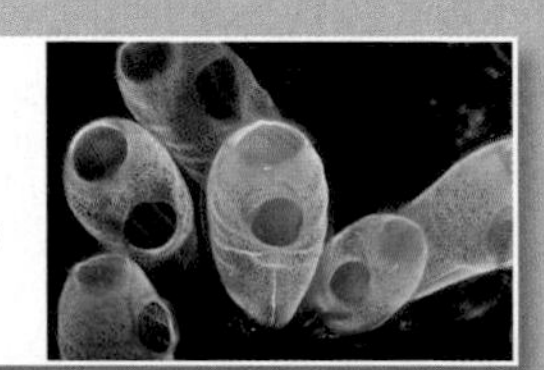

THE FISHES

Fishes were the first vertebrates and remain the most diverse members of this group. Early fishes were jawless. Evolution of jaws and paired fins opened the way to a great adaptive radiation.

TRANSITION FROM WATER TO LAND

One group of fishes gave rise to aquatic tetrapods (four-legged walkers). Amphibians are tetrapods that typically live on land and breathe air, but require fresh water to breed.

THE REPTILES

Reptiles have a scaly body and are amniotes: vertebrates with eggs that develop on land. Birds are reptiles with feathers and an ability to adjust their body temperature through heat production.

THE MAMMALS

Mammals are amniotes that produce milk and have hair. Some lay eggs, but young of the most diverse group are nourished by their mother as they develop inside her body.

The next chapter continues the story of mammals, with a closer look at the primates. We compare vertebrate brains in Section 32.10, endoskeletons in 35.3, and circulatory systems in 36.2. Fish gills and bird lungs are covered in Section 38.4. Section 39.2 compares the digestive systems of various vertebrates, and 40.2 looks at urinary systems. Sections 40.8 and 40.9 cover vertebrate regulation of body temperature, and 42.9 describes the development and function of the placenta.

25.1 Very Early Birds

In Darwin's time, acceptance of his theory of evolution by natural selection was hindered, in part, by an apparent absence of transitional fossils. Skeptics wondered, if new species evolve from existing ones, where are the fossils that represent these transitions? In fact, one of these "missing links" was unearthed by workers at a limestone quarry in Germany just one year after Darwin's *On the Origin of Species* was published.

That fossil, about the size of a large crow, looked like a small dinosaur. It had a long, bony tail, three clawed fingers on each forelimb, and a heavy jaw with short, spiky teeth, but it also had feathers (**FIGURE 25.1A**). The fossil species was named *Archaeopteryx* (meaning ancient winged one). To date, eight fossilized members of this species have been unearthed. Radiometric dating indicates that they lived about 150 million years ago.

Archaeopteryx was the first early bird fossil discovered but many others are now known. *Confuciusornis sanctus*, a fossil species discovered in China, had a beak and a short tailbone like that of a modern bird (**FIGURE 25.1B**). However, unlike wings of nearly all modern birds, those of *C. sanctus* have claws at their tips.

Another fossil takes us back even farther in time and supports the hypothesis that birds descended from dinosaurs. In 1994, a farmer in China discovered a fossil of a small dinosaur with short forelimbs and a long tail (**FIGURE 25.1C**). Unlike most dinosaurs, this one was covered with tiny filaments that resemble downy feathers of modern birds. Researchers named the farmer's fuzzy find *Sinosauropteryx prima*, meaning first Chinese feathered dragon. Given its shape and lack of long feathers, *S. prima* was certainly unable to fly. Its fuzzy feathers probably functioned as insulation, as they do in modern birds.

Piecing together the evolutionary history of animals, like that of other groups, relies on evidence of past events that no one witnessed. Fossils and biochemical comparisons of living organisms provide clues to the past. However, interpretations of any historical event can differ when evidence of it is incomplete. For this reason, evolutionary biologists often disagree about the relationships among lineages, both modern and extinct. These disagreements do not call into question the fundamental idea that all animal lineages arose by descent with modification from a common ancestor. Rather, the disputes—which involve details such as the timing of a divergence, or a taxonomic grouping—are part of the process of science. Any hypothesis is revised when data that is inconsistent with it is discovered. Given that the fossil record will always be incomplete, it is not surprising that new discoveries can prompt major revisions in our understanding of evolutionary history.

A *Archaeopteryx*, an early bird with a long, bony tail, teeth, and clawed digits on its wings.

B *Confuciusornis sanctus*, an early bird with a beak, short tail, and clawed digits on its wings. Males had long tail feathers.

C *Sinosauropteryx prima*, a feathered dinosaur.

FIGURE 25.1 Feathered fossil species. The *Archaeopteryx* fossil was discovered in Germany. The reconstructions of *Confuciusornis* and *Sinosauropteryx* are based on fossils unearthed in China.

25.2 Chordate Traits and Evolutionary Trends

✔ Chordates, distinguished by their embryonic traits, include two lineages of marine invertebrates, as well as the vertebrates.

Chordate Characteristics

The previous chapter ended with the echinoderms, a deuterostome lineage. The other major deuterostome lineage is the **chordates** (phylum Chordata), a group of bilaterally symmetrical, coelomate animals, with a complete digestive system and a closed circulatory system. Four traits unique to chordate embryos define the lineage:

1. A **notochord**, a rod of stiff but flexible connective tissue, extends the length of the body and supports it.
2. A hollow nerve cord parallels the notochord and runs along the dorsal (upper) surface.
3. Gill slits (narrow openings) extend across the wall of the pharynx (the throat region).
4. A muscular tail extends beyond the anus.

Depending on the group, some, none, or all of the defining chordate traits persist in the adult.

Most of the 50,000 or so chordates are **vertebrates**, animals with a backbone. The group also includes two lineages of marine invertebrates.

Invertebrate Chordates

Lancelets (subphylum Cephalochordata) have a fish-shaped body (**FIGURE 25.2**). An adult is about 5 centimeters (2 inches) long, and it retains all four characteristic chordate traits. The nerve cord extends into the head, where there is a simple brain. An eyespot at the end of the nerve cord detects light, but there are no paired sensory organs as in fishes. Lancelets spend most of their time buried up to the mouth in sediments. Waving of cilia inside the pharynx moves water into the pharynx and out through gill slits. The gill slits filter food particles out of the water.

In **tunicates** (subphylum Urochordata), only larvae have all the typical chordate traits (**FIGURE 25.3A**). The larva does not feed. Almost immediately after it hatches, metamorphosis transforms it into a barrel-shaped adult that secretes a "tunic" of polysaccharides (**FIGURE 25.3B**). Adults feed by sucking in water through a tube, capturing food on their pharynx, then expelling the filtered water through another tube. Some species are sessile as adults, but others drift or swim.

Overview of Chordate Evolution

Until recently, lancelets were considered the closest invertebrate relatives of vertebrates. An adult lance-

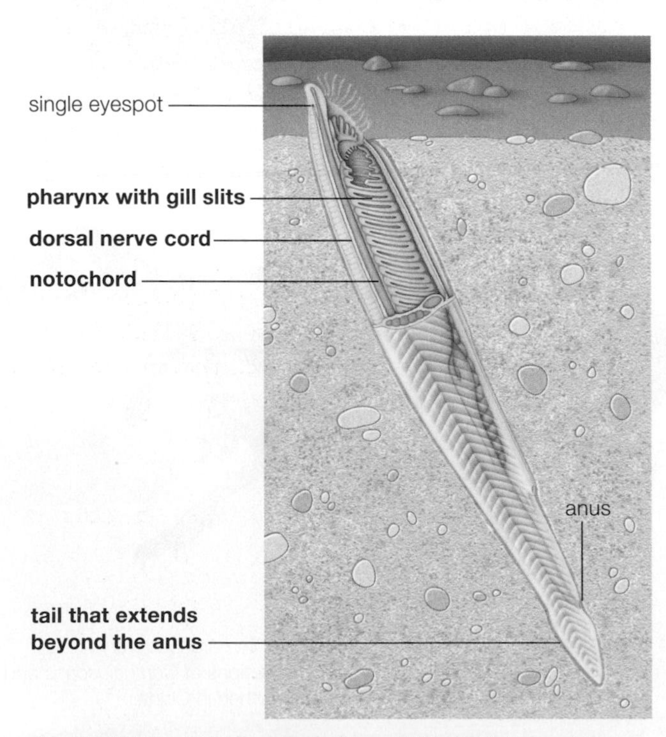

FIGURE 25.2 ▸Animated A lancelet. Lancelets can swim, but spend most of their buried tail down in sediment. Both larvae and adults have all four chordate traits (bold labels).

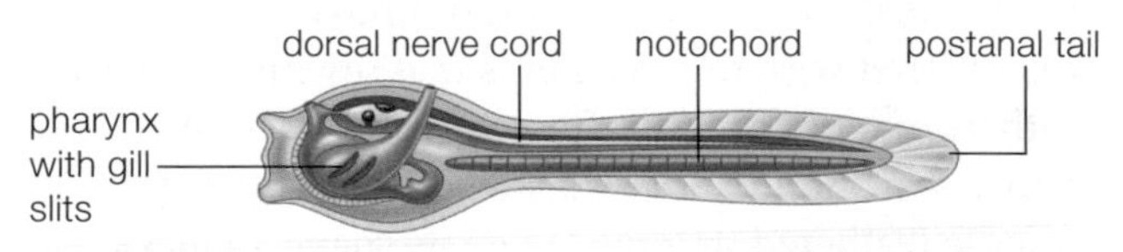

A Free-swimming tunicate larva with all the defining chordate traits.

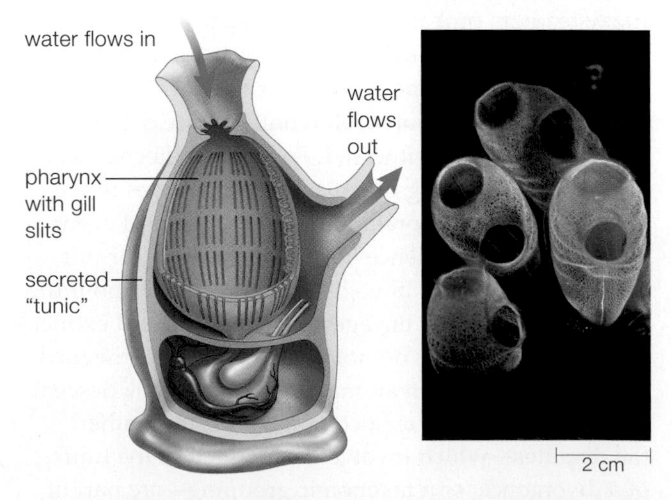

B Adult tunicate. The only defining chordate trait it retains is the pharynx with gill slits. The species in the photo is sessile as an adult.

FIGURE 25.3 ▸Animated Tunicates.

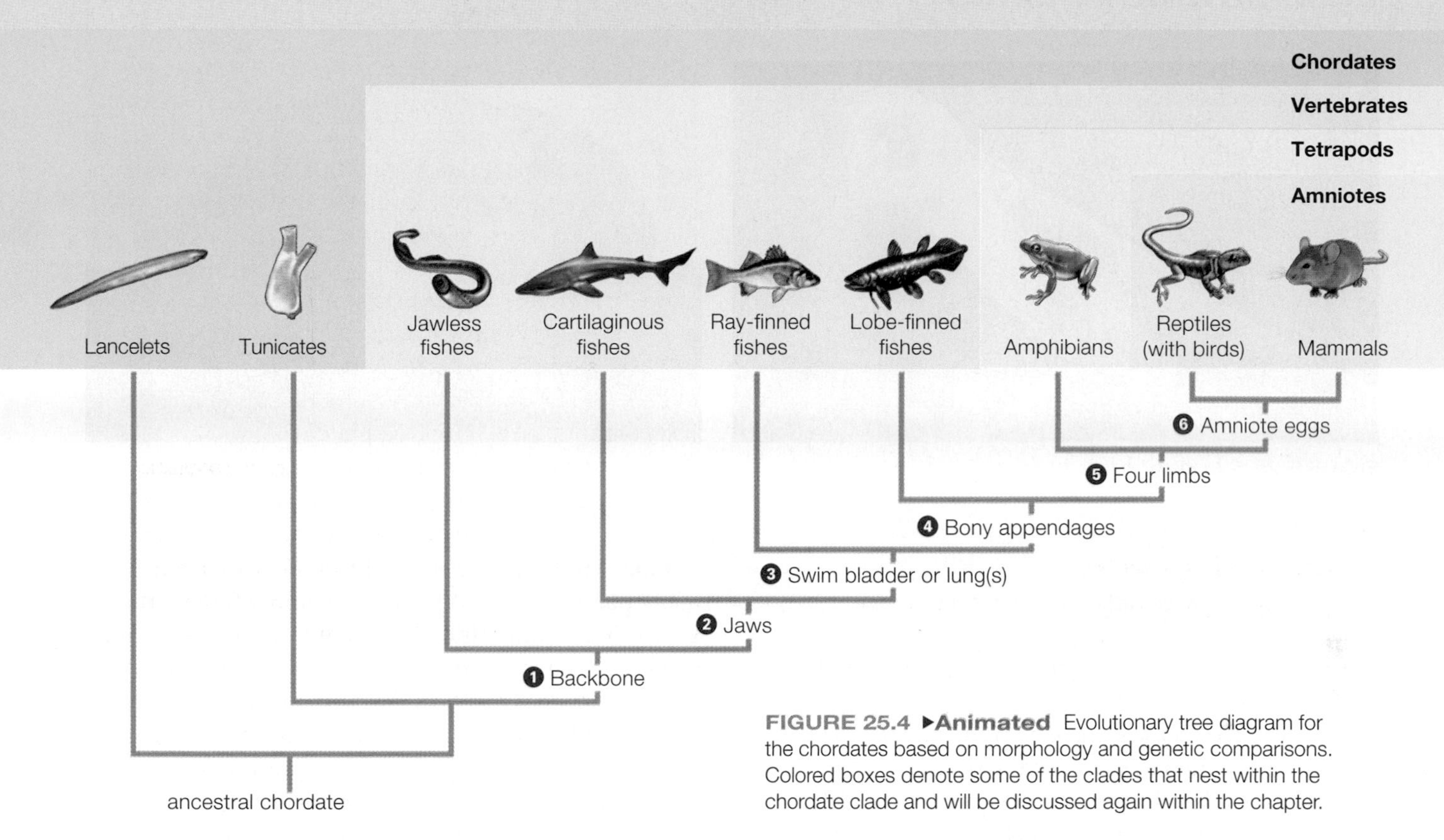

FIGURE 25.4 ▸**Animated** Evolutionary tree diagram for the chordates based on morphology and genetic comparisons. Colored boxes denote some of the clades that nest within the chordate clade and will be discussed again within the chapter.

let looks more like a fish than an adult tunicate does, but morphological traits can be deceiving. Studies of developmental processes and gene sequences revealed that tunicates are the invertebrate lineage most closely related to the vertebrates (**FIGURE 25.4**).

Most chordates have a backbone and thus are vertebrates ❶. The backbone and other skeletal elements are components of the vertebrate **endoskeleton**, or internal skeleton. A vertebrate endoskeleton consists of living cells, so it grows with the animal and does not have to be molted as an exoskeleton does.

The first vertebrates were fishes that sucked up or scraped up food. Later, hinged skeletal elements called jaws evolved ❷. Jaws allowed their bearers to exploit new strategies for feeding. The vast majority of fishes and other modern vertebrates have jaws.

Later evolutionary modifications allowed animals to move onto land. In one group of fishes, two outpouchings on the side of the gut wall evolved into lungs: moist, internal sacs that enhance gas exchange with the air ❸. Fins with bones inside them evolved in a subgroup of these fishes ❹. These fins would later evolve into limbs of the four-legged walkers, or **tetrapods** ❺.

Early tetrapods spent time on land, but laid their eggs in water. Later, eggs that enclosed an embryo within a series of waterproof membranes evolved in one lineage. These specialized eggs allowed animals known as **amniotes** to become the most diverse group of vertebrates on land ❻.

amniote Vertebrate with eggs that enclose the embryo within waterproof membranes.
chordate Animal with an embryo that has a notochord, dorsal nerve cord, pharyngeal gill slits, and a tail that extends beyond the anus. For example, a lancelet or a vertebrate.
endoskeleton Internal skeleton made up of hardened components such as bones.
lancelet Invertebrate chordate with a fishlike shape; retains all the defining embryonic chordate traits into adulthood.
notochord Stiff rod of connective tissue that runs the length of the body in chordate larvae or embryos.
tetrapod Vertebrate with four legs, or a descendant thereof.
tunicate Marine invertebrate chordate; a fish-shaped, swimming larva, undergoes metamorphosis into a barrel-shaped, sessile adult.
vertebrate Animal with a backbone.

TAKE-HOME MESSAGE 25.2

What traits define the major subgroups of chordates?

✔ All chordate embryos have a notochord, a dorsal tubular nerve cord, a pharynx with gill slits in its wall, and a tail that extends past the anus. There are two groups of invertebrate chordates: lancelets and tunicates.

✔ Most chordates have a backbone and so are vertebrates. Limbs evolved in one lineage that later colonized the land. Amniotes, a tetrapod subgroup with specialized eggs, are the predominant vertebrates on land.

25.3 Jawless Fishes

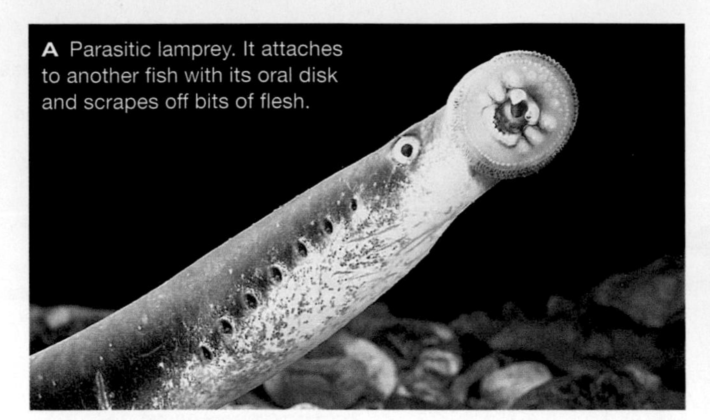

A Parasitic lamprey. It attaches to another fish with its oral disk and scrapes off bits of flesh.

FIGURE 25.5 Modern jawless fishes.

B Hagfish. It feeds on worms and scavenges on the seafloor.

✔ The first fishes were jawless.
✔ Two groups of jawless fishes survived to the present.

We begin our survey of vertebrate diversity with the fishes. **Fishes** were the first vertebrate lineages to evolve, and they remain the most fully aquatic. Nearly all rely on gills for oxygen exchange.

The earliest fossil fishes date to about 530 million years ago, during the late Cambrian period. The fossil animals had a tapered body a few centimeters long, and a head with a pair of eyes, but no jaws. Their skeleton consisted of cartilage, the same rubbery tissue that supports your ears and nose.

By about 480 million years ago, jawless fishes called ostracoderms had evolved and begun to diversify. Body size remained small; most were only a few centimeters long. Ostracoderm means "shelled skin" and refers to bony external plates that covered the head or, in some cases, the entire body. The plates probably helped the fishes fend off predatory invertebrates such as sea scorpions. Ostracoderms became extinct after jawed fishes arose.

Two lineages of jawless fishes (lampreys and hagfishes) survived to the present. Both groups have a cylindrical body about a meter long and a skeleton composed of cartilage. Both also lack the scales and paired fins typical of jawed fishes. Their gill slits are not covered and are visible at the body surface.

We know from fossils that lampreys date back at least 350 million years. The 50 or so modern species all breed in fresh water, although some spend most of their adult life in the sea. Unlike most fishes, lampreys undergo metamorphosis. Their larvae resemble larval tunicates or adult lancelets.

Some lamprey species do not eat as adults, but many parasitize fish. A parasitic lamprey (**FIGURE 25.5A**) has an oral disk with toothlike structures made of keratin, the main protein in your nails and hair. A lamprey attaches to a host fish with its oral disk, then scrapes off bits of flesh. Fish rarely survive this attack. In the early 1900s, parasitic Atlantic lampreys entered the Great Lakes via newly built canals and decimated native fish populations. Fishery managers now lower lamprey numbers with dams, nets, and poisons.

The 60 or so species of flexible-bodied hagfishes are marine bottom-feeders (**FIGURE 25.5B**). Hagfishes have poor eyesight and use sensory tentacles near their mouth to locate worms and carcasses. Their mouth has dental plates covered with sharp barbs of keratin. The barbed plates are used to grab and pull apart food. A frightened hagfish secretes a compound that combines with water to form a gelatinous slime. Slime deters most predators but has not kept humans from harvesting hagfish. Most belts, wallets, and other products labeled as "eelskin" are actually made of hagfish skin.

Relationships among lampreys, hagfishes, and jawed fishes have long inspired debate. Recent genetic comparisons indicate that hagfishes and lampreys constitute a monophyletic group. The name of this group, Cyclostomata, means "round mouthed."

TAKE-HOME MESSAGE 25.3

What are jawless fishes?

✔ Jawless fishes are gilled, aquatic vertebrates with a cartilage skeleton. They do not have jaws or scales.

✔ Lampreys and hagfishes have hard mouthparts made of keratin. Lampreys undergo metamorphosis and those that feed as adults parasitize other fishes. Hagfishes are marine; they scavenge and eat soft-bodied invertertebrates.

fish Gilled aquatic vertebrate that is not a tetrapod.

CREDITS: (5A) Heather Angel; (5B) NHPA/SuperStock.

25.4 Evolution of Jawed Fishes

✔ The evolution of jaws and fins opened the way to a great diversification of fishes.

The first jawed vertebrates (subphylum Gnathostomata) evolved during the late Silurian period, by about 420 million years ago. Jaws evolved from gill arches, which are skeletal elements that support fish gills (**FIGURE 25.6**). Fishes with jaws were at an advantage over jawless fishes. Jaws help a fish catch and kill prey, and also allow it to tear large prey into easy-to-swallow chunks.

Jawed fishes were also the first animals with scales and paired fins. **Scales** are hard, flattened structures that grow from and often cover the skin. Fins are flattened appendages used to propel and steer a body while swimming. Most jawed fishes have movable paired fins.

The Devonian period (416–359 million years ago) is called the "Age of Fishes," and placoderms were the most numerous and diverse vertebrates in Devonian seas. They became extinct at the end of this period. About 200 different species of placoderms have been identified from fossils. Placoderm means "tablet skin" and refers to bony armor that covered the animal's head and neck (**FIGURE 25.7**). A typical placoderm did not have teeth, but sharp bony plates performed the same function. Placoderms grew larger than the jawless fishes that preceded them, and some were enormous. *Dunkleosteus*, which once inhabited a shallow sea in what is now Ohio, grew as long as a bus. Placoderms are the earliest vertebrates for which we have fossil evidence of internal fertilization and development. Scientists recently discovered a fossilized female that apparently died while giving birth.

Another group of early jawed fish lineages is collectively referred to as acanthodians (spiny fins). Acanthodians arose at about the same time as placoderms, but were smaller (only centimeters long), less diverse, and did not have bony armor. As a result, they left fewer fossils than placoderms, so we know less about them. Acanthodians became extinct at the end of the Permian period.

Tiny fossilized scales reveal that small sharks also swam in Devonian seas. We discuss sharks and the other jawed fish lineages that survived to the present in detail in the next section.

scales Flattened structures that grow from and sometimes cover the skin in some groups of vertebrates.

FIGURE 25.7 Artist's depiction of *Dunkleosteus*, a bus-sized placoderm that lived in the seas of what is now Ohio during the Devonian. These early jawed fishes had bony coverings on their head and bony plates that functioned like teeth. Like most jawed fishes, they had movable paired fins.

TAKE-HOME MESSAGE 25.4

What were the first jawed fishes like?

✔ Jaws evolved during the Silurian period by modification of the first pair of gill arches in a jawless ancestor. Jawed fishes were also the first vertebrates with paired fins.

✔ Placoderms were an early group of jawed fishes that had bony armor on their head and neck. Some grew to great size. The acanthodian lineages were smaller and lacked bony armor.

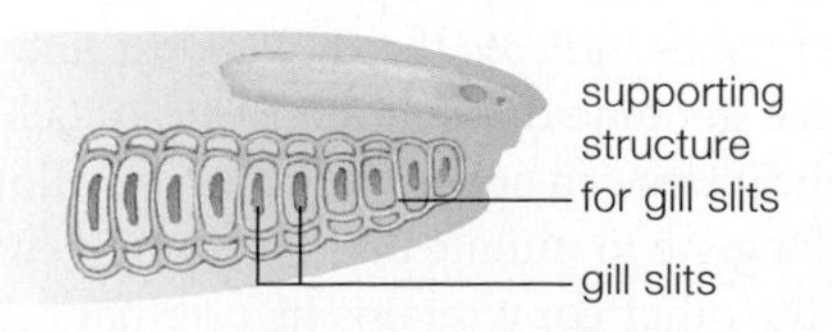

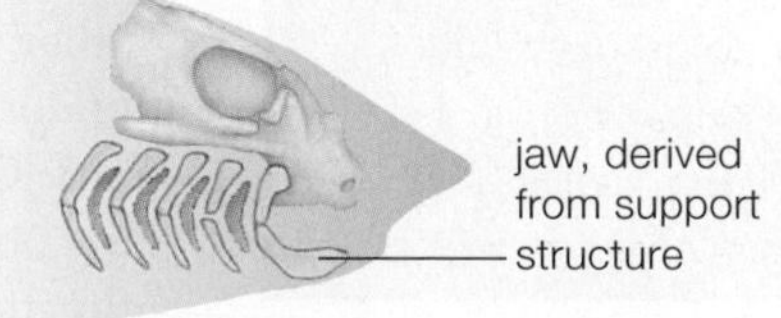

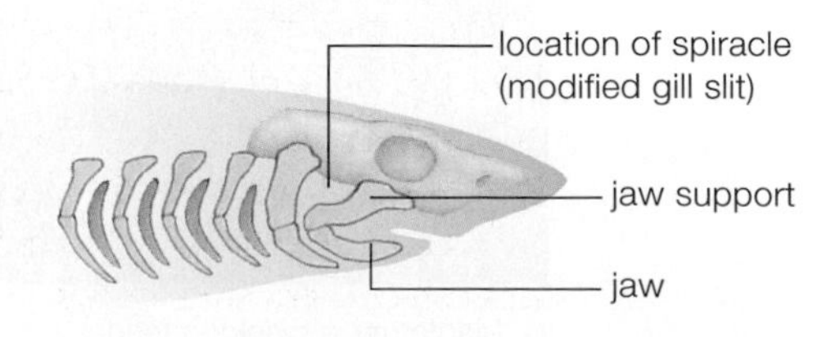

FIGURE 25.6 ▶**Animated** Proposed steps in the evolution of jaws.

25.5 Modern Jawed Fishes

✔ There are two living lineages of jawed fishes, cartilaginous fishes and bony fishes.

FIGURE 25.8 shows the one hypothesis for relationships among living jawed vertebrates. Cartilaginous fishes may comprise a monophyletic group, but the bony fishes certainly do not.

Cartilaginous Fishes

Cartilaginous fishes are a mostly marine group of jawed fishes with a cartilage skeleton. The five to seven pairs of gill slits are uncovered at the body surface in most species. Jaws have teeth that grow in rows and are continually shed and replaced. Sexes are separate and fixed for life. Eggs typically develop in an egg case inside the mother's body. When the young are ready to hatch, the egg case ruptures and they are released into the environment. Less commonly, females release egg cases containing developing embryos. In most groups of cartilaginous fishes, a single opening on the ventral surface, called a **cloaca**, functions in reproduction and also serves as the exit for digestive and urinary waste. All living jawless fishes also have a cloaca.

Most of the 850 species of cartilaginous fishes are sharks or rays. The most well-known sharks are speedy predators that chase down and tear apart prey (**FIGURE 25.9A**). Others are bottom-feeders that suck up invertebrates and act as scavengers. Still others strain plankton from the seawater. The largest living fish, the whale shark, feeds in this manner. It can weigh several tons.

Rays have a flattened body with large pectoral fins. Manta rays glide through warm seas and feed by filtering out plankton (**FIGURE 25.9B**). Stingrays are bottom-feeders. Their barbed tail has a venom gland that serves as a defense against predators.

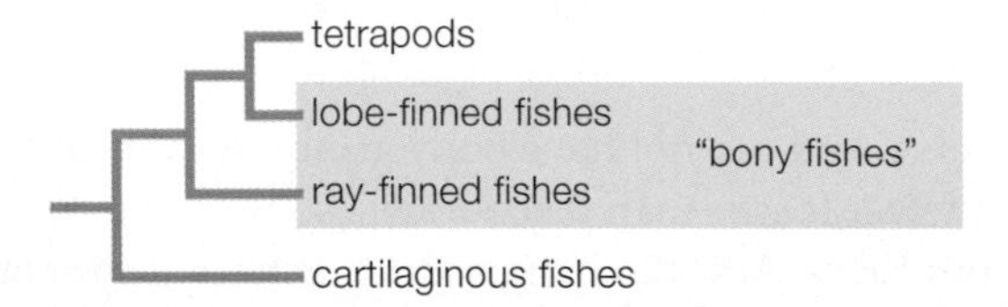

FIGURE 25.8 Evolutionary tree for modern jawed vertebrates. Note that bony fishes (blue box) are not a clade.

A Galápagos shark, a streamlined, fast-moving predator.

B Manta ray, a plankton feeder.

FIGURE 25.9 Cartilaginous fishes. Note the visible gill slits.

Bony Fishes

Modern **bony fishes** include members of two lineages: ray-finned fishes and lobe-finned fishes. In both groups, bone replaces cartilage in some or most of the adult skeleton and gill slits are hidden beneath a gill cover. Bony fishes also have a gas-filled organ or organs derived from outpouchings of their pharynx.

Ray-Finned Fishes Thin, membranous fins with flexible fin supports derived from skin are the defining trait of **ray-finned fishes** (**FIGURE 25.10A**). In most members of this lineage the gas-filled organ is a **swim bladder**, a sac whose volume of gas can be varied to adjust buoyancy (**FIGURE 25.10B**). Instead of a cloaca, a typical ray-finned fish has separate openings for the urinary, digestive, and reproductive systems.

With about 24,000 living species, ray-finned fishes are the most diverse group of modern fishes. Sturgeons are modern representatives of one ancient ray-finned lineage. Humans harvest eggs from some sturgeons for use as caviar. Gars, predatory fish with an elongated body, are members of another ray-finned lineage.

A third ray-finned lineage, the teleosts, includes 99 percent of ray-finned fishes, and about half of all vertebrates. Comparisons of the teleost genome with that of other ray-finned fishes indicate that the teleost lineage underwent whole-genome duplication early in its history. Gene duplications can speed evolution by allowing one copy of a gene to mutate and take on a new function while the other copy retains its original role. The diversification of copied genes is thought to have facilitated an adaptive radiation of this group.

A Goldfish (a type of carp). Note the flexible fins supported by thin rays. As in most bony fishes, a bony cover hides the gills.

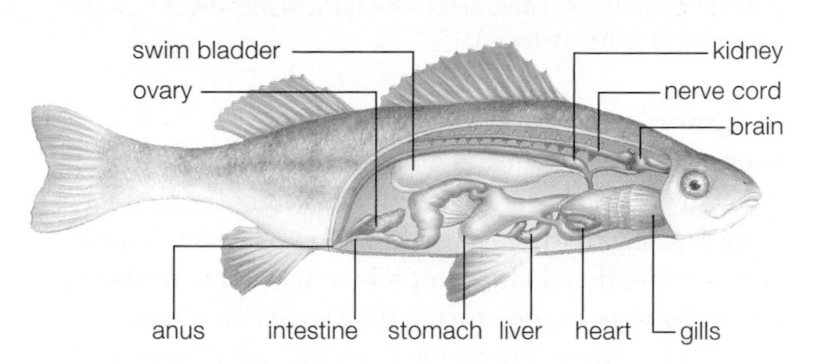

B Anatomy of a perch. The swim bladder allows the fish to adjust its buoyancy (its tendency to float).

FIGURE 25.10 ▸Animated Ray-finned fishes.

Teleosts have a vast array of body forms and the group includes most fishes that humans harvest as food, including anchovies, salmon, sardines, bass, swordfish, trout, tuna, halibut, carp, and cod. Teleosts also have a wider variety of reproductive patterns than other fishes. In some species, individuals are hermaphrodites that simultaneously produce eggs and sperm. In others, individuals change sex during the course of their lifetime. In still others, sexes are separate and fixed for life.

Lobe-Finned Fishes Thick, fleshy fins with supporting bones inside them characterize **lobe-finned fishes**. There are two lineages, the marine coelacanths and the freshwater lungfishes. In both, sexes are separate and fixed. Until 1938, coelacanths were known only from fossils, which showed these fish were abundant from the Devonian through the Cretaceous period. In 1938, a living coelacanth was found in the Indian Ocean. We now know that there are several modern species (**FIGURE 25.11A**).

As their common name implies, the six species of lungfishes (**FIGURE 25.11B**) have lungs—air-filled sacs with an associated network of tiny blood vessels. A lungfish gulps air to fill its lungs, then oxygen diffuses from the lungs into its blood. Similarities between fin bones of lobe-finned fishes and limb bones of tetrapods indicate tetrapods descended from a lobe-finned ancestor. Genome comparisons indicate that, of the two lobe-finned groups, tetrapods are closest to lungfishes.

A Coelacanth.

B Lungfish.

FIGURE 25.11 Lobe-finned fishes. These fishes have fleshy pelvic and pectoral fins supported by sturdy bones. Lungfishes have been observed to "walk" underwater. They thrust their pelvic fins against the bottom to propel their body forward.

TAKE-HOME MESSAGE 25.5

What are jawed fishes?

✔ Jawed fishes include cartilaginous fishes and bony fishes. Cartilaginous fishes have a skeleton of cartilage. Bony fishes have a skeleton made primarily of bone.

✔ The ray-finned lineage of bony fishes is the most diverse group of vertebrates.

✔ Lobe-finned fishes include lungfishes—the closest living relatives of the tetrapods—and coelocanths.

bony fish Fish whose skeleton includes bone; a ray-finned or lobe-finned fish.
cartilaginous fish Jawed fish with a cartilage skeleton.
cloaca In some vertebrates, a body opening through which both wastes and gametes exit.
lobe-finned fish Bony fish with fleshy fins supported by bones.
ray-finned fish Bony fish with fin supports derived from skin.
swim bladder Adjustable flotation sac of some bony fish.

25.6 Amphibians—First Tetrapods on Land

✔ Amphibians spend part of their life on land, but most still return to water to breed.

Adapting to Life on Land

Amphibians are scaleless, land-dwelling vertebrates that typically breed in water. Amphibians were the first tetrapods. Recently discovered fossil footprints in Poland demonstrate that a large amphibian walked here about 395 million years ago, during the Devonian. The animal that left these footprints was about 2.5 meters (8 feet) long.

Fossils show how fishes adapted to swimming gave rise to tetrapods that walk (FIGURE 25.12). Bones of a lobe-finned fish's pectoral fins and pelvic fins are homologous with those of an amphibian's front and hind limbs. During the transition to land, these bones became larger and better able to bear weight. Ribs enlarged and a distinct neck emerged, allowing the head to move independently of the rest of the body.

The transition to land was not simply a matter of skeletal changes. Lungs became larger and more complex. Division of the previously two-chambered heart into three chambers allowed blood to flow in two circuits, one to the body and one to those increasingly important lungs. Changes to the inner ear improved detection of airborne sounds. Evolution of eyelids prevented delicate eye tissues from drying out.

What drove the move onto land? An ability to spend time out of water would have been favored in seasonally dry places. In addition, it would have allowed escape from aquatic predators and access to a new food source—insects—which also arose in the Devonian.

FIGURE 25.13 Shasta salamander, with equal-sized forelimbs and hind limbs.

Modern Amphibians

Modern amphibians include salamanders, caecilians, frogs, and toads. All are carnivores as adults, preying mainly on insects and worms.

The 535 species of salamanders and related newts live mainly in North America, Europe, and Asia. In body form, they are the modern group most like early tetrapods. Forelimbs and back limbs are of similar size and there is a long tail (FIGURE 25.13). As salamanders walk, their body bends from side to side, like the body of a swimming fish. Larval salamanders look like small versions of adults, except for the presence of gills. Typically, they lose their gills and develop lungs

FIGURE 25.12 Fossil species from the late Devonian illustrate how a fish body became adapted for life on land.

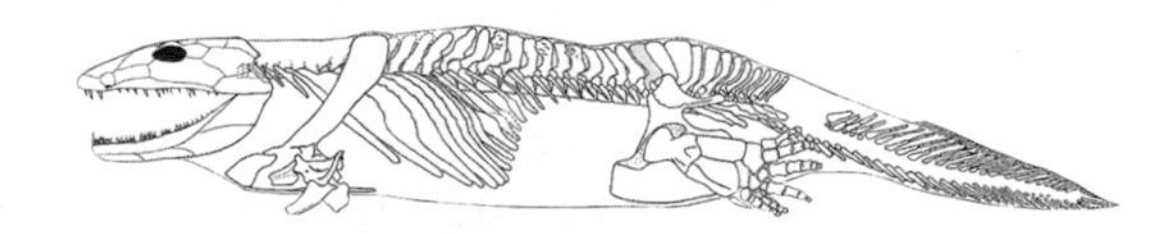

③ Early amphibian (*Icthyostega*) with well-developed ribs, and thick limbs with distinct digits.

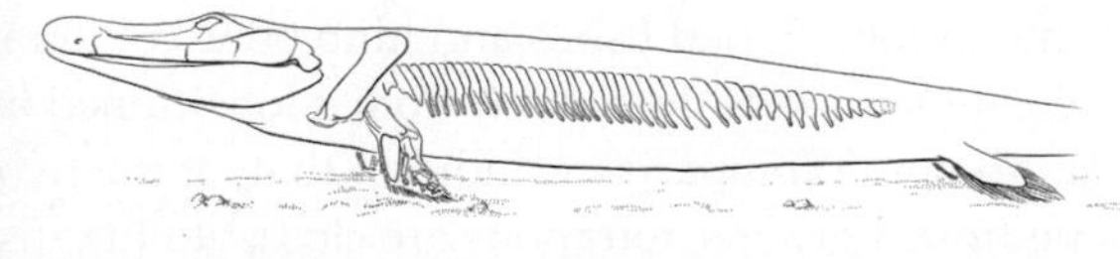

② Fish (*Tiktaalik*) with sturdier weight-bearing pectoral fins, wristlike bones, and enlarged ribs.

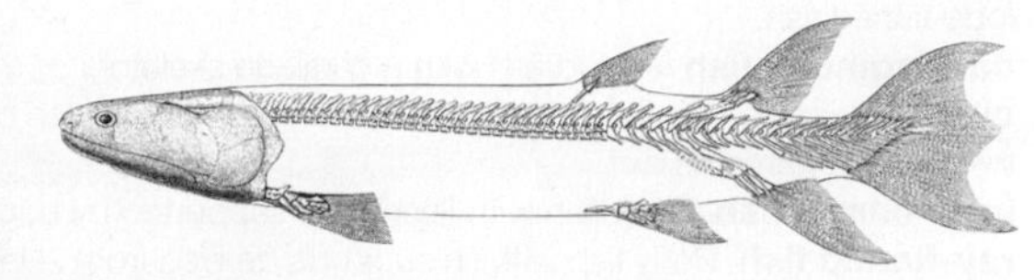

① Fish (*Eusthenopteron*) with bony fins.

A Long, muscular hind limbs allow an adult frog to make spectacular leaps. Frogs typically have smooth skin and spend much of their time in water.

as they mature. However, some salamanders (axolotls) retain gills even as adults. Others salamander species lose their gills but do not develop lungs; gas exchange occurs across the skin.

Caecilians are amphibians closely related to salamanders. They live in the tropics and have a wormlike form that adapts them to their burrowing life. There are about 165 species, all limbless and blind.

Frogs and toads belong to the most diverse amphibian order, with more than 5,000 species. The order's name, Anura, means "without a tail" and adults are always tailless. The long, muscular hind limbs of an adult frog allow it to swim, hop, and make spectacular leaps (**FIGURE 25.14A**). The much smaller forelimbs help to absorb the impact of landings. Frogs usually have a smooth, thin skin and live in a moist environment. Toads are better adapted to dry conditions. They have thick, bumpy skin (**FIGURE 25.14B**). Compared with frogs, toads have a stubbier body and their hind legs are proportionately shorter. Toads can hop, but more often they walk.

Both frogs and toads undergo metamorphosis, during which a gilled, tailed larva (**FIGURE 25.14C**) transforms itself into an adult with lungs and no tail.

B Toads have rougher, thicker skin than frogs and typically spend less time in water. The flattened disk visible behind this cane toad's eye is its eardrum.

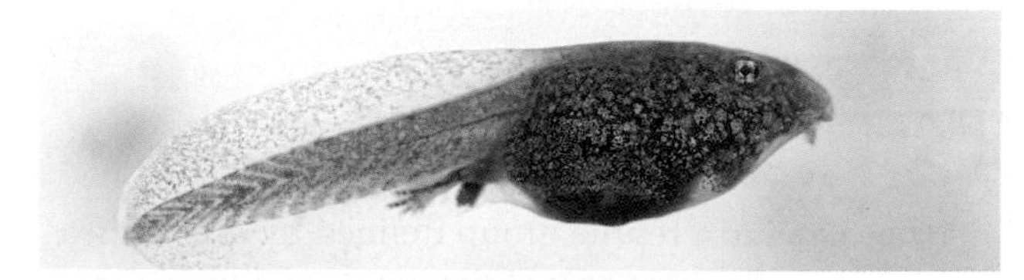
C Aquatic larva (a tadpole), with gills and a tail.

FIGURE 25.14 Anurans: frogs and toads.

Declining Diversity

We are in the midst of an alarming decline in amphibian numbers. Population reductions are best documented in North America and Europe, but similar declines are occurring worldwide. An amphibian's thin, scaleless skin makes it relatively easy for parasites, pathogens, and pollutants to enter the body. Section 23.3 described the effects of an introduced chytrid fungus on frogs. Section 34.1 explains how agricultural chemicals can disrupt amphibian hormone production. Habitat loss is another important threat. In many places, people have filled in low-lying areas that once collected water from seasonal rains. Such seasonal pools are important breeding sites for amphibians.

amphibian Tetrapod with a three-chambered heart and scaleless skin; typically develops in water, then lives on land as an air-breathing carnivore.

TAKE-HOME MESSAGE 25.6

What are amphibians?

✔ Amphibians are carnivorous vertebrates that typically live on land but breed in water. The most diverse group includes the frogs and toads, which undergo metamorphosis from gilled larvae. Salamanders and the closely related caecilians are less diverse lineages.

✔ An amphibian's scaleless, permeable skin makes it vulnerable to pollutants, and its requirement for water in which to breed makes it sensitive to habitat alteration.

25.7 Amniote Evolution

✔ Amniotes are vertebrates that have adapted to a life lived entirely on land.

Amniotes branched off from an amphibian ancestor during the Carboniferous. A variety of traits adapt them to life in dry places. They have lungs throughout their life, and their skin is rich in keratin, a protein that makes it waterproof. A pair of well-developed kidneys help conserve water, and fertilization usually takes place in the female's body. Sexes are typically separate and fixed for life. Amniotes produce eggs in which an embryo develops bathed in fluid, so amniotes can develop on dry land (**FIGURE 25.15**). Membranes within the egg function in gas exchange, nutrition, and waste removal.

An early branching of the amniote lineage separated ancestors of mammals from the common ancestor of all modern **reptiles**. You probably do not think of birds as reptiles, but the reptile clade includes turtles, lizards, snakes, crocodilians, and birds:

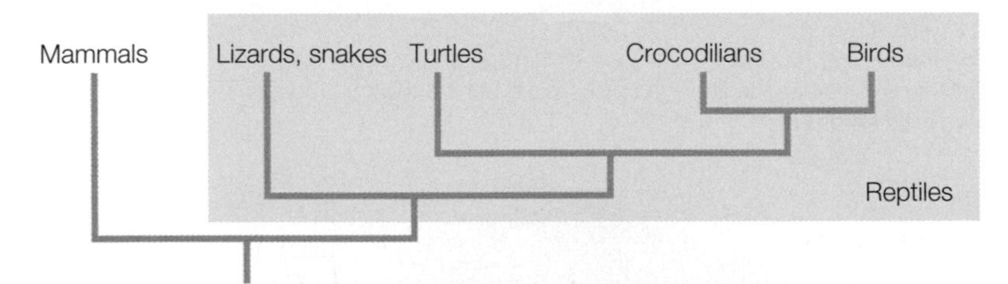

Dinosaurs are a reptile group defined by skeletal features such as the anatomy of the pelvis and hips. One dinosaur lineage, the theropods, included many feathered species. Like modern birds and mammals, these may have been **endotherms**, which maintain their body temperature by adjusting metabolic heat production. All modern nonbird reptiles are **ectotherms**, animals whose temperature varies with that of their environment.

FIGURE 25.15 Snakes hatching from amniote eggs.

Birds branched off from a theropod dinosaur lineage during the Jurassic (**FIGURE 25.16**), and are the only surviving descendants of dinosaurs. All dinosaurs became extinct by the end of the Cretaceous, probably as a result of an asteroid impact (Section 16.1).

dinosaur Group of reptiles that include the ancestors of birds; became extinct at the end of the Cretaceous.
ectotherm Animals whose body temperature varies with that of its environment.
endotherm Animal that maintains its temperature by adjusting its production of metabolic heat; for example, a bird or mammal.
reptile Amniote subgroup that includes lizards, snakes, turtles, crocodilians, and birds.

TAKE-HOME MESSAGE 25.7

What are amniotes?

✔ Amniotes are animals that produce eggs in which the young can develop away from water. They have waterproof skin and highly efficient kidneys.

✔ An early divergence separated the ancestors of mammals from the ancestors of reptiles—a group in which biologists include turtles, lizards, snakes, crocodilians, and birds.

FIGURE 25.16 Painting of a Jurassic scene. In the foreground, the early bird *Archaeopteryx* glides along. Behind the birds, a meat-eating dinosaur sizes up a larger plant-eating one. At the far right, an early mammal surveys the scene from its perch on a tree.

FIGURE IT OUT Which of the animals mentioned above do biologists consider amniotes? Which do they group as reptiles?

Answer: All are amniotes. The dinosaurs and birds are reptiles.

25.8 Nonbird Reptiles

✔ Reptiles have a scale-covered body. Most have four limbs of approximately equal size, but the snakes are limbless.

Together, lizards and snakes constitute the most diverse group of modern reptiles, with about 9,000 species. Members of this group are covered with overlapping scales and they periodically shed their skin. Iguanas are herbivores, but other lizards are predators. The largest lizard, the Komodo dragon (**FIGURE 25.17A**), grows up to 3 meters (10 feet) long. After inflicting a venomous bite, it trails its prey for hours or days until the poisoned animal collapses.

Snakes first evolved during the Cretaceous, from short-legged, long-bodied lizards. Some modern snakes have bony remnants of hind limbs, but most lack limb bones entirely. All snakes are predators and all have teeth, but only some have fangs. Rattlesnakes and other fanged snakes subdue prey with a venom they produce in modified salivary (saliva-producing) glands. Other snakes are constrictors that suffocate a prey animal by wrapping around it so tightly that it cannot expand its chest to inhale.

The 300 or so species of turtles have a bony, keratin-covered shell attached to their skeleton (**FIGURE 25.17B**). We can see from fossils how turtles have evolved. One 200-million-year-old fossil turtle found in China has a protective plate derived from expanded ribs on its belly side, but no shell on its back. The fossil turtle also has teeth. By contrast, modern turtles are toothless. As in birds, a layer of keratin covers their jaws and forms a horny beak. Most turtles that live in the sea feed on invertebrates such as sponges or jellies; others feed mainly on sea grass. Freshwater turtles prey on fishes and invertebrates. Land-dwelling turtles, which are commonly called tortoises, feed on plants.

Crocodilians—crocodiles, alligators, and caimans—are stealthy predators with a long snout and many sharp peglike teeth (**FIGURE 25.17C**). They spend much of their time in water, where a long, powerful tail propels them when they swim. Crocodilians are the closest living relatives of birds and, like many birds, they are highly vocal and engage in complex parental behavior. During courtship, males and females grunt and bellow. After a female mates, she digs a nest, lays eggs, then buries and guards them. When the young are ready to hatch they call, and their mother helps them dig their way to the surface.

Crocodilians and birds also share another trait: a four-chambered heart. In lizards, snakes, and turtles (as in amphibians) the heart has three chambers. In one of these chambers, oxygen-rich and oxygen-poor blood mix a bit. In a four-chambered heart, oxygen-poor blood from body tissues never mixes with oxygen-rich from the lungs.

A Komodo dragon, the largest lizard.

B Turtle in a defensive posture.

C Crocodile with its fish prey.

FIGURE 25.17 ▶**Animated** Examples of nonbird reptiles.

TAKE-HOME MESSAGE 25.8

What are the traits of nonbird reptiles?

✔ Lizards and snakes have skin covered with overlapping scales. Most are predators.

✔ Turtles have a bony, keratin-covered shell and a horny keratin beak.

✔ Crocodilians are aquatic predators with a four-chambered heart and a long snout with peglike teeth.

CREDITS: (17A) Soren Egeberg Photography/Shutterstock.com; (17B) Joel Sartore/National Geographic Creative; (17C) Johan Swanepoel/Shutterstock.com.

25.9 Birds—The Feathered Ones

FIGURE 25.18 An owl in flight.

✔ In one group of dinosaurs, the scales became modified as feathers. Birds are modern descendants of this group.

General Characteristics

Birds, the only living animals with feathers, first arose during the Jurassic and are thought to be descended from feathered dinosaurs. Feathers are filamentous keratin structures derived from scales.

Bird feathers have a variety of functions. Birds are endotherms, and downy feathers help them maintain their temperature. Feathers slow the loss of metabolic heat in cool environments, and prevent heat gain in hot ones. Feathers also shed water and thus help to keep a bird's skin dry. In many birds, colorful plumage plays a role in courtship. Feather color can result from the way that keratin reflects light or from deposition of pigments derived from the diet. For example, dietary carotenoids deposited in feathers make flamingos pink and American cardinals red.

Feathers also play a role in flight (**FIGURE 25.18**). Like humans, birds stand upright on their hind limbs, and their wings are homologous with our arms. Each wing is covered with long flight feathers that extend outward and increase its surface area. Flight feathers give the wing a shape that helps lift the bird as air passes over it.

Flight muscles run from the keel (a bony extension) of the large breastbone (sternum) to bones of the upper limb (**FIGURE 25.19**). Contraction of one set of flight muscles produces a powerful downstroke that lifts the bird. A less powerful set of muscles contracts to raise the wing.

Most birds are surprisingly lightweight, and this feature helps them become and remain airborne. Air cavities inside a bird's bones keep its body weight low, as does the lack of a bladder (an organ that stores urinary waste in many other vertebrates). In addition, birds do not have teeth, which are heavy. Instead, their jaws are covered with lightweight keratin to form a beak.

Keeping muscles supplied with ATP that fuels their contractions requires plenty of oxygen for aerobic respiration (Section 7.2). A unique system of air sacs keeps air flowing continually through a bird's lungs, maintaining a high oxygen concentration even in active muscles. Also, as previously noted, birds (like crocodilians) have a four-chambered heart that pumps blood in two fully separated circuits.

Flying requires good eyesight and a great deal of coordination. Compared to a lizard of similar body mass, a bird has larger brain and much larger eyes.

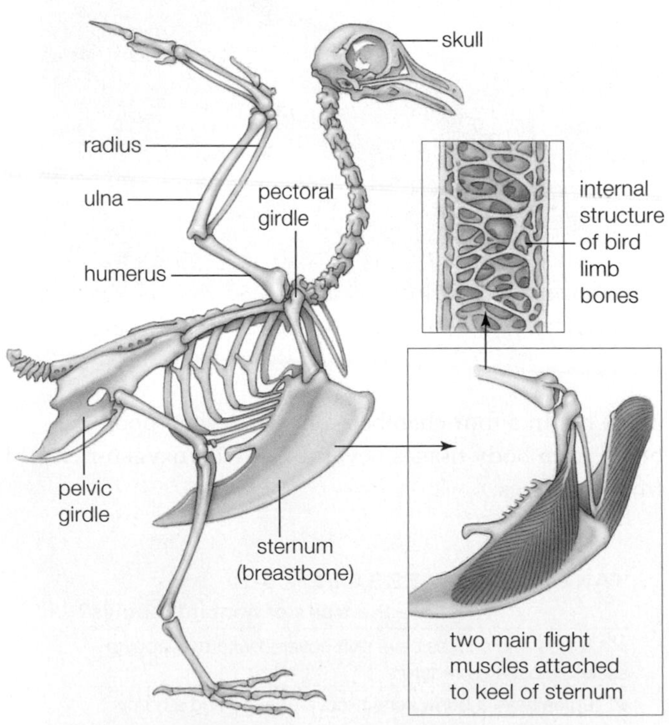

FIGURE 25.19 Bird skeleton. A bird's skeleton is made up of lightweight bones with internal air pockets. A wing is a modified forelimb (see **FIGURE 18.5**). Powerful flight muscles attach to a large breastbone, or sternum.

Each eye is protected within a circle of bone that holds it fixed in place. Thus, a bird cannot move its eyes to look to the side, as you can. Instead, the bird must turn its highly flexible neck.

As in other reptiles, fertilization is internal. However, most male birds do not have a penis. To inseminate a female, a male presses his cloaca against hers, a maneuver poetically described as a cloacal kiss (**FIGURE 25.20**). After fertilization occurs, the female lays a fertilized egg that has the characteristic amniote membranes (**FIGURE 25.21**). Nutrients from the egg's yolk and water from the egg white (albumen) sustain the developing embryo. Like some turtles and all crocodilians, birds encase their eggs within a rigid shell hardened by calcium carbonate. The egg must be kept warm in order for the embryo it contains to develop. In nearly all bird species, one or both parents incubate the eggs until they hatch. Parental responsibilities typically continue even after eggs hatch (**FIGURE 25.22**).

Avian Diversity

More than half of the approximately 10,000 living bird species belong to the order of perching birds (Passiformes). This group includes familiar backyard birds such as sparrows, crows, jays, starlings, swallows, finches, robins, warblers, orioles, and cardinals. The next most diverse order includes 450 species, most of them hummingbirds. Hummingbirds are agile fliers and the only birds capable of flying backward.

Many birds, including some hummingbirds and perching birds, make a seasonal migration; they fly from one region to another in response to a seasonal change, such as a shift in day length. Migration typically involves spring travel to a breeding site where insects are abundant in the summer, then an autumn flight to an overwintering site. One shorebird monitored by researchers flew from Alaska to New Zealand, a distance of 11,500 kilometers (7,145 miles), without stopping to feed or rest.

At the other extreme, penguins and ratite birds such as ostriches cannot fly. Penguins live along coasts of the Southern Hemisphere. They flap their wings to propel themselves through water. The ostrich, the largest living bird, is native to Africa. It can weigh up to 150 kilograms (330 pounds). Other ratite birds include Australia's emus and cassowaries, New Zealand's kiwis, and South America's rheas. Unlike other birds, ratites have a flat sternum; there is no keel.

bird Common name for a member of the only surviving reptile lineage with feathers.

FIGURE 25.20 Mating house sparrows. A male bird does not have a penis. To inseminate his partner, he must balance on her back and bend his body so his cloaca meets hers.

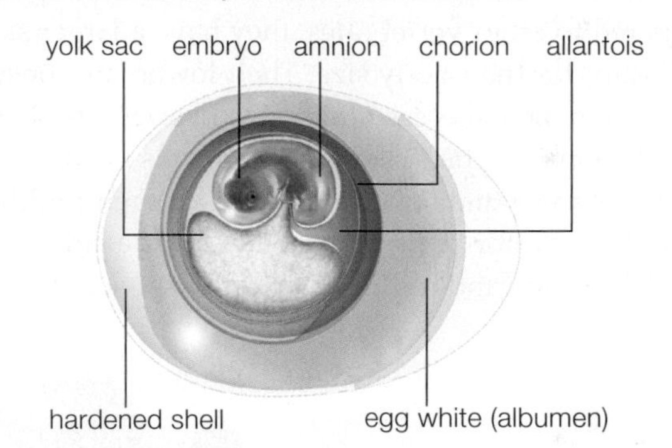

FIGURE 25.21 ▶**Animated** Bird's egg. The egg has a calcium-hardened shell that encloses the embryo and four characteristic amniote membranes (yolk sac, amnion, chorion, and allantois).

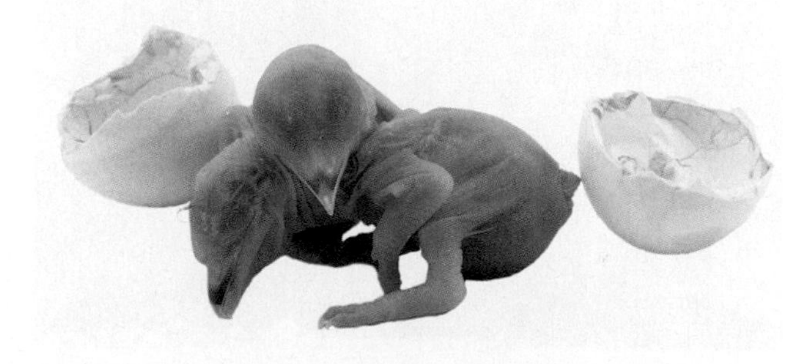

FIGURE 25.22 Newly hatched parrots. Unlike other reptiles, many birds hatch in a relatively undeveloped state and require extensive parental care before they can live on their own.

TAKE-HOME MESSAGE 25.9

What are birds?

✔ Birds are the only living animals with feathers. They evolved from dinosaurs. Adaptations for flight include lightweight bones; air sacs that increase the efficiency of respiration; and a four-chambered heart that keeps blood moving rapidly.

25.10 Mammals—Milk Makers

✔ Mammals scurried about while dinosaurs dominated, then underwent an adaptive radiation once dinosaurs were gone.

Mammalian Traits

Mammals are animals in which females nourish their offspring with milk that they secrete from mammary glands. The group name is derived from the Latin *mamma*, meaning breast. Mammals have hair or fur that, like feathers, is made of the protein keratin (**FIGURE 25.23A**). Mammals are endotherms, and a coat of fur or head of hair helps them maintain their body temperature. Mammals also include the only animals that sweat, although not all mammals do so. The mammalian heart is four-chambered, and gas exchange occurs in a pair of well-developed lungs.

Mammals have distinctive skeletal and dental traits. Compared to other vertebrates, they have a larger skull (and brain) for their body size. Their lower jaw consists of a single bone, whereas other jawed vertebrates have multiple bones in the lower jaw. Mammals are also the only vertebrates that have three bones in their middle ear. In the ancestor of mammals, there was a single middle ear bone and multiple lower jaw bones, as in reptiles. Over time, two of the jaw bones moved and became modified to serve as bones of the middle ear.

As a group, mammals have four different types of teeth, each with a distinctive shape (**FIGURE 25.23B**). Incisors are used to gnaw, canines tear and rip flesh, and premolars and molars grind and crush hard foods. Not all mammals have teeth of all four types, but most have some combination. In other vertebrates, an individual's teeth may differ in size, but they are all the same shape. Having a variety of tooth shapes gives mammals an ability to eat a wider variety of foods than other vertebrates.

A Mammary glands and hair or fur. A young grizzly bear sucks milk from its mother's nipples.

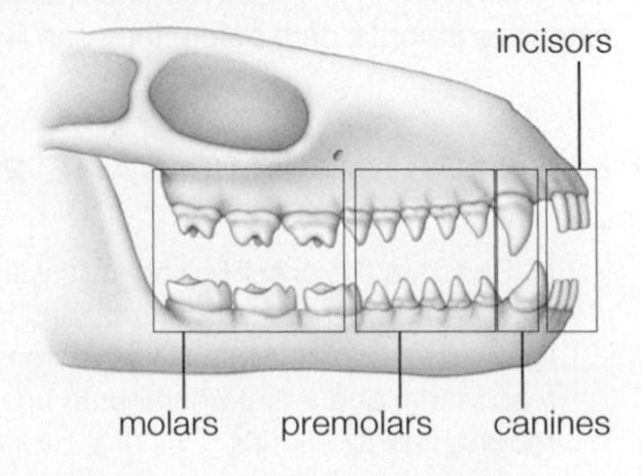

B Multiple types of teeth and a single bone in the lower jaw.

FIGURE 25.23 ▶**Animated** Distinctive mammalian traits.

Mammalian Origins and Diversification

During the Carboniferous, the amniote lineage split in two. One of the resulting branches would give rise to the groups we now define as reptiles (including the birds). The other branch, the synapsids, would give rise to the mammals.

Therapsids, the synapsid subgroup to which mammals belong, evolved and underwent an adaptive radiation during the late Permian. By the end of this period, they had become the dominant animals on land. Then, disaster struck. The end of the Permian period (250 million years ago) marks the most extensive extinction event known; 70 percent of species that lived on land disappeared. Most synapsids perished at this time, but a group of therapsids called cynodonts was among the survivors. Cynodonts means "dog toothed," and these animals had canines, incisors, and molars. Most likely they were endotherms and had insulating fur or hair. Mammals evolved from a cynodont ancestor during the Jurassic.

Three mammal lineages survived to the present: monotremes, marsupials, and placental mammals. An early divergence separated the lineage leading to monotremes from that leading to marsupial and placental mammals:

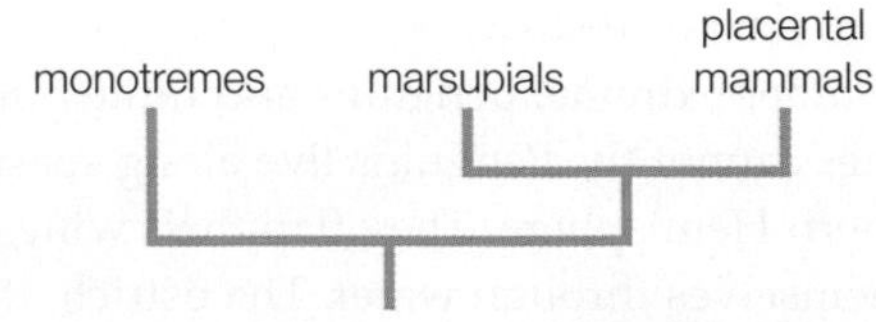

Monotremes, or egg-laying mammals, are the oldest surviving mammal lineage. Female monotremes lay and incubate eggs that have a leathery shell like that of lizards. Offspring hatch in a relatively undeveloped state—tiny, hairless, and blind. Young cling to the mother or are held in a skin fold on her belly. Milk oozes from openings on the mother's skin; monotremes do not have nipples.

A Mammals in a Wyoming forest, about 63 milion years ago. With the exception of the marsupial on the tree branch, all animals shown are members of extinct mammalian lineages.

B The giant placental mammal, *Indricotherium*, weighed as much as 20 tons and stood 5.5 meters (18 feet) tall at the shoulder. It lived in Asia between 20 and 35 million years ago.

FIGURE 25.24 After the dinosaurs; early results of the mammalian adaptive radiation.

Marsupials are pouched mammals. Young marsupials develop for a brief period inside their mother's body, where they nourished by egg yolk and nutrients that diffuse from maternal tissues. They are born while still blind and tiny, when their limbs have just begun to develop. After birth, they must use these stubby limbs to crawl along their mother's body to a permanent pouch on her ventral surface. Once inside the pouch, they attach to a nipple, suckle, and grow.

In **placental mammals**, young inside a mother's body are nourished by means of a placenta. The **placenta** is an organ that forms during pregnancy and allows material to diffuse between maternal and embryonic bloodstreams. A placenta transfers nutrients more efficiently than diffusion does, so placental embryos can grow faster than those of other mammals. After birth, young placental mammals suckle milk from nipples on their mother's ventral surface. Placental mammals have separate openings for the urinary, reproductive, and excretory systems, unlike the monotremes and marsupials, which have a cloaca.

Representatives of all three mammalian lineages lived alongside the dinosaurs, and many mammals perished 65 million years ago in the mass extinction (the K-Pg extinction) that resulted in the dinosaurs' demise (Section 16.1). In the aftermath of this catastrophic event, mammals underwent a great adaptive radiation. **FIGURE 25.24** shows some early results of this diversification and the section that follows describes the diversity of modern mammals.

mammal Animal with hair or fur; females feed young with milk secreted by mammary glands.
marsupial Mammal in which young are born at an early stage and complete development in a pouch on the mother's surface.
monotreme Egg-laying mammal.
placenta Of placental mammals, organ that forms during pregnancy and allows diffusion of substances between the maternal and embryonic bloodstreams.
placental mammal Mammal in which a mother and her embryo exchange materials by means of an organ called the placenta.

TAKE-HOME MESSAGE 25.10

What are mammals and how did they evolve?

✔ Mammals are animals that nourish young with milk and have hair or fur. Their four kinds of teeth allow them to eat many different kinds of foods.

✔ Mammals evolved during the Jurassic and underwent an adaptive radiation after dinosaurs died out in a mass extinction at the end of the Cretaceous.

✔ Three lineages of mammals have survived to the present: monotremes, marsupials, and placental mammals.

25.11 Modern Mammalian Diversity

✔ Mammals successfully established themselves on every continent and in the seas.

Egg-Laying Monotremes

Montremes are the least diverse group of modern mammals, with only five species. Four are echidnas, also known as spiny anteaters (**FIGURE 25.25A**) that live in Australia or New Guinea. The fifth species is the platypus, a semiaquatic Australian animal with a beaverlike tail, a ducklike bill, and webbed feet (**FIGURE 25.25B**). Platypuses burrow into riverbanks using claws exposed when they retract the web on their feet. Both males and females have spurs on their hind feet. The male's spurs produce venom, making platypuses the only known venomous mammals.

Marsupials

About 300 species of marsupials live in Australia and on nearby islands. They include the plant-eating kangaroos (**FIGURE 25.26A**) and koalas, and the Tasmanian devil, a carnivore the size of a small dog.

About 100 marsupial species live in South or Central America. Most are opossums, as are North America's only surviving native marsupials (**FIGURE 25.26B**).

Placental Mammals

With more than 5,000 species, placental mammals are the dominant mammals in most land habitats, and the only mammals that live in the seas. **FIGURE 25.27** shows examples of major, currently accepted orders.

About 40 percent of all placental mammals are classified as rodents (order Rodentia). They include rats, mice, hamsters, squirrels, beavers, porcupines, and guinea pigs. Short generation times and a diverse diet probably contributed to the group's success. All rodents have teeth specialized for gnawing. A rodent's incisors grow continually, to prevent them from being worn down to nubs.

About 1,200 species of bats constitute the second most diverse order (Chiroptera). Most are nocturnal (active at night). Bats are the only mammals capable of sustained flight. Their wings consist of membranes

A Echidna (spiny anteater).

B Platypus with its young.

FIGURE 25.25 Monotremes.

A Kangaroo with a juvenile in its pouch.

B North American oppossum with its young.

FIGURE 25.26 Marsupials.

CREDITS: (25A) D. & V. Blagden/ANT Photo Library; (25B) Jean Phillipe Varin/Jacana/Science Source; (26A) © iStockphoto.com/Craig Dingle; (26B) Jack Dermid.

Very Early Birds (revisited)

How do we know what color a feathered dinosaur or bird was? Some fossil feathers such as those of the dinosaur *Sinosauropteryx* contain microscopic granules of the type that contain pigment in modern bird feathers. In birds, different shaped granules contain different shades of melanin, a pigment that comes in red to brown shades. By studying the distribution and shape of pigment granules in *Sinosauropteryx* fossils, scientists concluded this animal had a reddish color and a tail with alternating red and white bands.

---

of skin stretched between bones. Old World bats are larger and eat fruit, whereas New World bats are smaller and most are insect eaters. Insect-eating bats locate their prey by echolocation. They emit high-pitched sounds, then listen for echoes that arise when sound waves hit prey and reflected back.

Moles and shrews make up the next most diverse order (Soricomorpha). Most moles are adapted to a burrowing way of life, with strong forelimbs for tunneling, reduced eyes, and no external ear flaps. They eat earthworms and insects. Shrews resemble mice with spiky teeth, but are not close relatives of rodents. Moles and shrews were historically grouped with other small insect eaters such as hedgehogs in an order (Insectivora) that was dismantled when gene comparisons showed its members were not closely related.

Carnivora means meat eater, and members of this order have enlarged canine teeth suited to tearing at flesh. They include cats, dogs, wolves, bears, foxes, and weasels, as well as marine pinnipeds (fin-footed animals) such as seals, sea lions, and walruses.

Whales and dolphins (Cetacea) have adapted to life in the sea. Section 16.6 described some fossils representing the Cetacean transition from land dwellers to marine mammals. As far as we know, the blue whale, which can be 30 meters (100 feet) long, is the largest mammal that ever lived. Like bats, some cetaceans have a capacity for echolocation, which they use to locate prey and in navigation.

Cetaceans are close relatives of even-toed mammals (Artiodactyla). Most large mammalian grazers such as cattle, deer, and goats are members of this order, as are pigs and hippos. Horses, zebras, and rhinos are grazers too, but they belong to the odd-toed order (Perissodactyla). Both orders of grazing mammals subsist on a diet of plant material that they digest with the assistance of microbes that live in their gut.

Humans are members of the order Primates. The next chapter explores the unique adaptations of this group, and the history of the human lineage.

FIGURE 25.27 Representatives of major orders of placental mammals.

TAKE-HOME MESSAGE 25.11

What are modern mammals like?

✔ Most modern mammals are placental mammals, with rodents and bats being the most diverse orders. Most mammals live on land, but some placental species have adapted to life in the seas.

summary

Section 25.1 Fossils provide evidence of evolutionary transitions between major animal groups. For example, fossils show that feathers, a defining trait of modern birds, evolved in a dinosaur ancestor of birds. Fossils also document how birds lost their long tail and teeth, and evolved a beak.

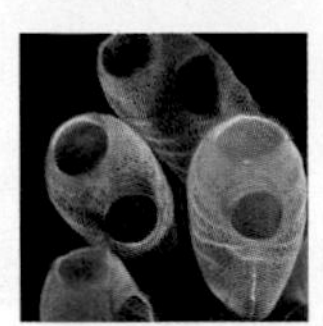

Section 25.2 Four embryonic features define the **chordates**: a **notochord**, a dorsal hollow nerve cord, a pharynx with gill slits, and a tail that extends past the anus. Depending on the group, these features may or may not persist in adults. Two groups of marine invertebrates, **tunicates** and **lancelets**, are chordates, but most chordates are **vertebrates** and have a backbone as part of their **endoskeleton**. Early vertebrates that swam in the seas gave rise to **tetrapods** that walked on land. **Amniotes** are a tetrapod subgroup that adapted to a life spent entirely on land.

Section 25.3 **Fishes** are gilled aquatic vertebrates. The earliest fishes such as the ostracoderms were jawless. Modern jawless fishes—lampreys and hagfishes—have an endoskeleton of cartilage.

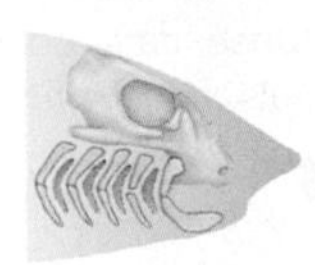

Section 25.4 Jaws evolved through a modification of gill-supporting bones in a jawless ancestor. Jawed fishes were the first vertebrates to have **scales** and paired fins. Placoderms, a now-extinct group of armored jawed fishes, dominated seas during the Devonian period.

Section 25.5 **Cartilaginous fishes** such as sharks and rays are jawed fishes with a cartilage skeleton. Products of their reproductive, digestive, and urinary systems exit the body by way of a single opening called the **cloaca**.

Bony fishes have a skeleton that consists of bone as well as cartilage. A bony cover overlays their gills and there are usually separate openings serving the reproductive, digestive, and urinary systems. **Ray-finned fishes** are bony fishes with thin, flexible fins and a **swim bladder**. They include teleosts, the most diverse lineage of modern fishes. A whole-genome duplication may have contributed to teleost diversity. **Lobe-finned fishes** have thick fins with bony supports. Coelacanths and lungfishes are in this group. Lungfishes have lungs and are the sister group to tetrapods. They use their lobed fins to walk underwater.

Section 25.6 **Amphibians** are scaleless, carnivorous tetrapods that live on land, but lay eggs in water. The evolutionary transition from an aquatic life to one largely lived on land required changes to the skeleton and to other organ systems. The heart became three-chambered and the lungs more efficient. Frogs and toads, the most diverse amphibians, undergo metamorphosis from a gilled aquatic larva with a tail to a tailless adult with lungs. Salamanders retain their tail into adulthood.

Section 25.7 Amniotes are tetrapods adapted to life on land by waterproof skin, highly efficient kidneys, and special eggs that allow embryos to develop away from water. **Reptiles** include the now extinct **dinosaurs** and their bird descendants. Some dinosaurs may have been **endotherms**, which regulate their temperature as birds and mammals do. All modern nonbird reptiles are **ectotherms**.

Section 25.8 Lizards and snakes constitute the most diverse lineage of modern reptiles. Most are predators and some are venomous. Turtles are reptiles with a keratin-covered shell and beak. Crocodilians are predators that spend much of their time in water. Unlike other nonbird reptiles, crocodilians have a four-chambered heart that prevents oxygen-poor and oxygen-rich blood from mixing.

Section 25.9 **Birds** are the only living animals with feathers. Like their closest living relatives, the crocodilians, they have a four-chambered heart. Most bird have a body that is highly adapted for flight. Front limbs are wings that are moved by muscles attached to a keeled sternum. Bones are lightweight and a system of air sacs keeps air flowing continually through the lungs. Bird eggs must be incubated to develop properly, and many birds also provide additional care to the young.

Perching birds are the most diverse lineage. Some birds migrate seasonally, typically moving between a breeding area and an area where they spend the winter. Penguins and ratites are two lineages of flightless birds.

Sections 25.10, 25.11 **Mammals** nourish young with milk secreted by their mammary glands. They also have fur or hair, and more than one kind of tooth. Three lineages are egg-laying mammals (**monotremes**), pouched mammals (**marsupials**), and **placental mammals**, the most diverse group. A **placenta** is an organ that facilitates exchange of substances between the embryonic and maternal blood.

Placental mammals develop faster than other mammals and are born at a later stage of development. They have become the dominant mammal group in most environments. The three most diverse placental lineages are rodents, bats, and moles and shrews.

self-quiz

Answers in Appendix VII

1. List the four distinguishing chordate traits.
2. Which of these traits are retained by an adult lancelet?
3. Vertebrate jaw bones evolved from ________ .
   a. gill supports b. ribs c. scales d. ear bones
4. Both cartilaginous and bony fishes have ________ .
   a. jaws
   b. a bony skeleton
   c. lungs
   d. a swim bladder
   e. a four-chambered heart
   f. all of the above

data analysis activities

Deathly Lamprey Repellent Predation by sea lampreys on native fishes in the Great Lakes is an ongoing problem. To help solve it, Michael Wagner and his team test methods of repelling lampreys. They carried out an experiment to investigate reports that sea lampreys detect the scent of lamprey carcasses and tend to avoid them. The researchers made alcohol-based lamprey carcass extracts, then observed what happened when lampreys were put in tanks and exposed to either this extract or to alcohol alone. **FIGURE 25.28** shows their results.

1. Why was it necessary to test the response of lampreys to the scent of alcohol alone?

2. What was the lowest proportion on lampreys on the scented side of the tank when the scent was alcohol? When the scent was alcohol-based carcass extract?

3. Do the results indicate that lampreys detect and avoid the scent of dead lampreys?

4. Do the results show that lampreys cannot detect alcohol?

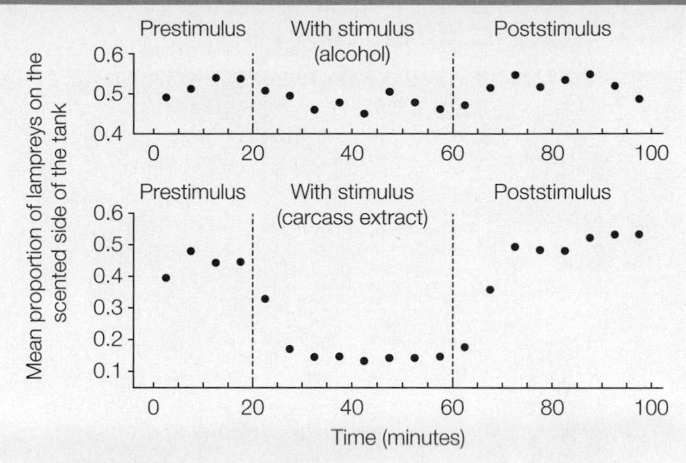

FIGURE 25.28 Mean proportion of lampreys on the scented side of the test tank during 8 trials with 10 lampreys. Lamprey placed in the tank for 20 minutes before exposure to alcohol or carcass extract, and remained there for 40 minutes after exposure. The upper graph shows results with alcohol as the stimulus; the lower shows the results with carcass extracts. Bars indicate standard error.

5. Tetrapods evolved from ________ .
   a. sharks
   b. teleosts
   c. lobe-finned fishes
   d. placoderms

6. Turtles, lizards, and birds belong to one major lineage of amniotes, and ________ belong to another.
   a. sharks
   b. frogs and toads
   c. mammals
   d. salamanders

7. Reptiles are adapted to life on land by ________ .
   a. waterproof skin
   b. internal fertilization
   c. efficient kidneys
   d. amniote eggs
   e. both a and c
   f. all of the above

8. The closest living relatives of hagfishes are ________ .
   a. lampreys b. placoderms c. snakes d. lizards

9. Among living animals, only birds have ________ .
   a. a cloaca
   b. a four-chamber heart
   c. feathers
   d. amniote eggs

10. Feathers and hair consist mainly of ________ .
   a. chitin
   b. cartilage
   c. keratin
   d. lignin

11. Unlike *Archaeopteryx*, modern birds have ________ .
   a. a long bony tail
   b. a toothless beak
   c. a two-chambered heart
   d. feathers

12. ________ eggs are typically released into the water.
   a. Frog
   b. Turtle
   c. Shark
   d. all of the above

13. Match each structure with its description.

___ cloaca	a. adjusts buoyancy
___ swim bladder	b. nourishes embryo
___ fur	c. supports body
___ pectoral fins	d. multipurpose exit from body
___ endoskeleton	e. homologous to forelimbs
___ placenta	f. insulates body

14. Match the organisms with the appropriate description.

___ lancelets	a. pouched mammals
___ lampreys	b. most diverse mammal lineage
___ amphibians	c. feathered amniotes
___ lizards	d. egg-laying mammals
___ birds	e. extinct jawed fishes
___ sharks	f. ectothermic amniotes
___ monotremes	g. cartilaginous fishes
___ marsupials	h. first land tetrapods
___ placoderms	i. living jawless fishes
___ placental mammals	j. invertebrate chordates

15. Arrange the groups in order in which they evolved.

___ 1 (earliest)	a. Jawless fishes
___ 2	b. Birds
___ 3	c. Dinosaurs
___ 4	d. Amphibians
___ 5 (most recent)	e. Jawed fishes

critical thinking

1. In 1798, a stuffed platypus specimen was delivered to the British Museum. Reports that it laid eggs created much confusion. To modern biologists, a platypus is clearly a mammal. It has fur and the females produce milk. Young animals have typical mammalian teeth that are replaced by hardened pads of the "bill" as the animal matures. Why do you think modern biologists can more easily accept that a mammal can have some reptilelike traits than scientists who were considering this animal in the late 1700s?

2. Controlling for body size, would you expect a flightless bird to produce eggs that are larger than, smaller than, or the same size as eggs produced by a bird that flies? Why?

3. Why is it more difficult to determine the sex of a newly hatched canary than a newborn puppy?

CREDIT: (28) Based on C. Michael Wagner, Eric M. Stroud, and Trevor D. Meckley, "A deathly odor suggests a new sustainable tool for controlling a costly invasive species," Published at www.nrcresearchpress.com/cjfas, July 2011, J2011-0010.

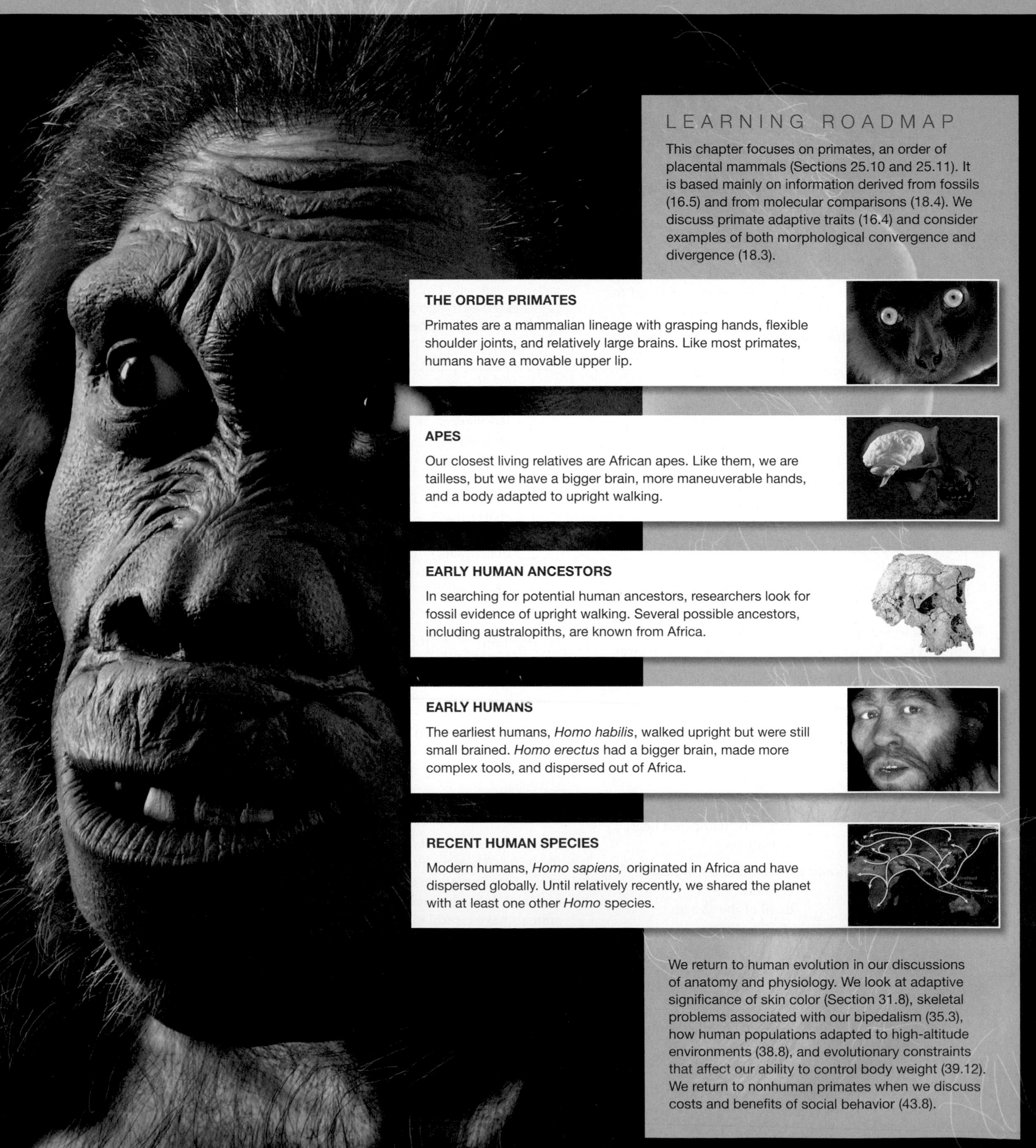

LEARNING ROADMAP

This chapter focuses on primates, an order of placental mammals (Sections 25.10 and 25.11). It is based mainly on information derived from fossils (16.5) and from molecular comparisons (18.4). We discuss primate adaptive traits (16.4) and consider examples of both morphological convergence and divergence (18.3).

THE ORDER PRIMATES

Primates are a mammalian lineage with grasping hands, flexible shoulder joints, and relatively large brains. Like most primates, humans have a movable upper lip.

APES

Our closest living relatives are African apes. Like them, we are tailless, but we have a bigger brain, more maneuverable hands, and a body adapted to upright walking.

EARLY HUMAN ANCESTORS

In searching for potential human ancestors, researchers look for fossil evidence of upright walking. Several possible ancestors, including australopiths, are known from Africa.

EARLY HUMANS

The earliest humans, *Homo habilis*, walked upright but were still small brained. *Homo erectus* had a bigger brain, made more complex tools, and dispersed out of Africa.

RECENT HUMAN SPECIES

Modern humans, *Homo sapiens,* originated in Africa and have dispersed globally. Until relatively recently, we shared the planet with at least one other *Homo* species.

We return to human evolution in our discussions of anatomy and physiology. We look at adaptive significance of skin color (Section 31.8), skeletal problems associated with our bipedalism (35.3), how human populations adapted to high-altitude environments (38.8), and evolutionary constraints that affect our ability to control body weight (39.12). We return to nonhuman primates when we discuss costs and benefits of social behavior (43.8).

26.1 A Bit of a Neanderthal

Paleoanthropology, the scientific study of prehistoric humans and their relatives, was born in the mid-1800s, when scientists found humanlike fossils in Germany's Neander Valley. The scientists postulated that the fossils were of an extinct human relative, which they named *Homo neanderthalensis*. At the time the fossils were discovered, their age was unknown. We now know that they date to about 40,000 years ago.

Since the discovery of those first Neanderthal bones, many other fossils with similar features have been unearthed, including a few nearly complete skeletons. Analysis of these many fossil finds confirmed that Neanderthals were a distinct ancient species, and evolutionary biologists now consider Neanderthals our closest extinct relatives.

Compared with modern humans (*Homo sapiens*), Neanderthals had a shorter, stockier build, with thicker bones and bulkier muscles. A reconstruction of a Neanderthal male, based on material from multiple fossils, stands about 164 centimeters (5 feet 4 inches) tall (**FIGURE 26.1**). Neanderthals had a braincase that was longer and lower than that of modern humans, but their brain was as big as ours or bigger. Their face had pronounced brow ridges and a large nose with widely spaced nostrils.

We know from fossils that Neanderthals lived in the Middle East, Europe, and in central Asia as far east as Siberia. The last known population lived in seaside caves in Gibraltar until perhaps as recently as 28,000 years ago.

In some regions, Neanderthals existed side by side with our own species (*Homo sapiens*) for thousands of years. People have long speculated about whether the two species met and mated. Scientists are less interested in the logistics of such matings than in whether they produced fertile offspring whose descendants survive to this day. In other words, they want to know whether Neanderthals contributed in any substantive way to the modern human gene pool.

Advances in our ability to extract, amplify, and sequence DNA now allow us to answer this question. In 2010, an international team of scientists headed by Svante Pääbo sequenced DNA extracted from Neanderthal fossils discovered in Croatia, Russia, Spain, and Germany. The sequence data they obtained represents much of the Neanderthal genome. Pääbo and his team compared this genome with homologous portions of genomes from five modern human populations and from chimpanzees. The results revealed that many of us do, in fact, have a Neanderthal in our family tree. Genomes of people from France, China, and Papua New Guinea share with the Neanderthal genome certain mutations that are not present in people from Africa or in chimpanzees. This suggests that the human–Neanderthal matings which left their mark on our gene pool took place in the Middle East after *H. sapiens* began venturing out of Africa, but before they dispersed to Europe, Asia, and elsewhere.

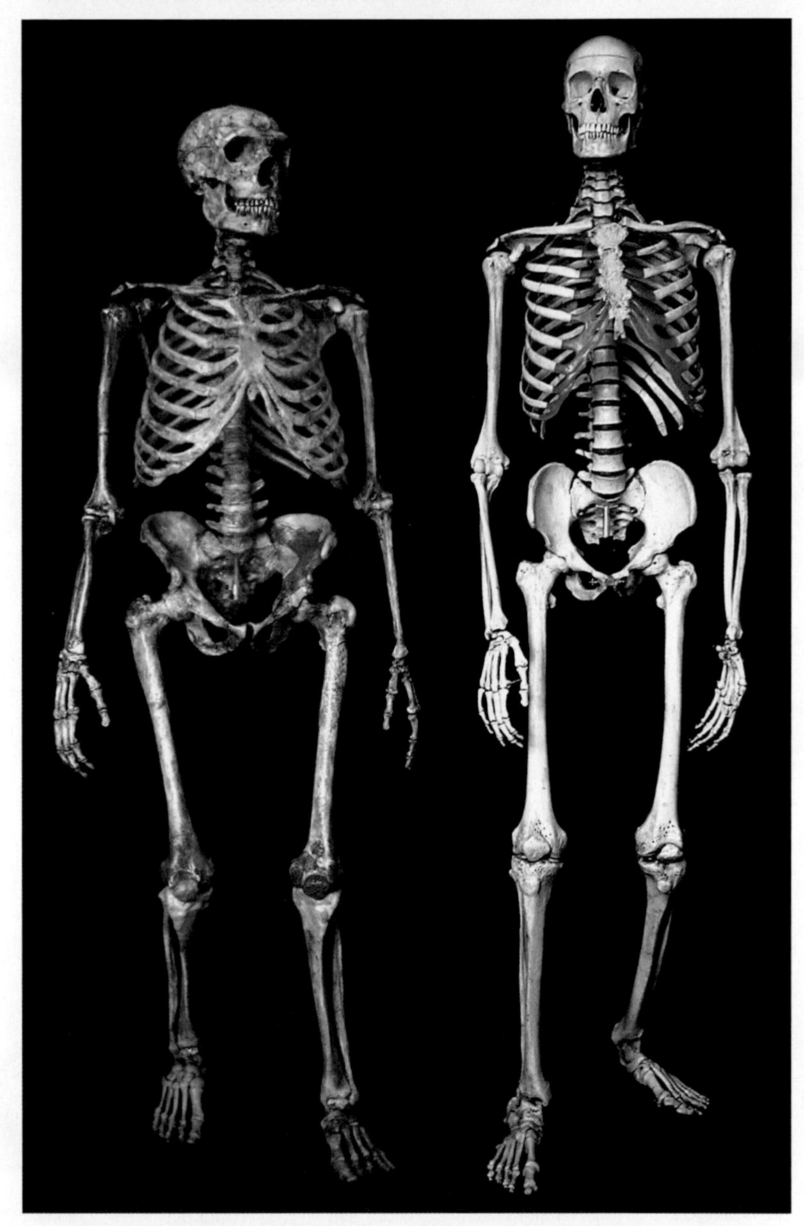

Homo neanderthalensis *Homo sapiens*

FIGURE 26.1 Skeletal anatomy of Neanderthal male compared with a modern human male. The Neanderthal skeleton is a reconstruction based on fossils of multiple males. Each color denotes a different fossil.

26.2 Primates: Our Order

✔ We share our five-digit grasping hands, forward-facing eyes, and a movable upper lip with other primate species.

Primate Characteristics

Primates are an order of placental mammals that includes humans, apes, monkeys, and close relatives. **FIGURE 26.2** shows relationships among living groups. Primates first evolved in warm forests, and many traits characteristic of the group adapt them to life among the branches. Primate shoulders have an extensive range of motion that facilitates climbing. Unlike most mammals, a primate can extend its arms out to its sides, reach above its head, and rotate its forearm at the elbow. With the exception of humans, all living primates have both hands and feet capable of grasping. A typical mammal has claws, but in primates the tips of fingers and toes have touch-sensitive pads protected by flattened nails.

Unlike the eyes of most mammals, which are widely spaced and set toward the side of the skull, primate eyes tend to be at the front of the head. As a result, both eyes view the same area, each from a slightly different vantage point. The brain integrates the differing signals it receives from the two eyes to produce a three-dimensional mental image. A primate's excellent depth perception adapts it to a life spent leaping or swinging from limb to limb.

Compared to other mammals, primates have a large brain for their body size. The regions of the brain devoted to vision and to information processing are expanded, and the area devoted to smell is reduced. Primates have a varied diet, and their teeth reflect this lack of specialization; they have all four types of mammalian teeth (Section 25.10).

Most primates spend their life in a social group that includes adults of both sexes. Females usually give birth to only one or two young at a time and provide care for an extended period after birth.

Origins and Early Branchings

Primates most likely arose before the demise of the dinosaurs, some time between 85 and 66 million years ago. Their sister group (the group that is their closest living relatives) is the colugos. Colugos are nocturnal mammals that live in the forests of Southeast Asia and glide from tree to tree like flying squirrels.

An early divergence among primates created two suborders: the wet-nosed primates (Strepsirrhini) and the dry-nosed primates (Haplorhini). Modern representatives of the wet-nosed suborder include lemurs of Madagascar, lorises of India and Asia, and galagos of Africa. Wet-nosed primates have a typical mammalian nose. Like a dog, they have nostrils set in an area of continually moist skin, and their cleft upper

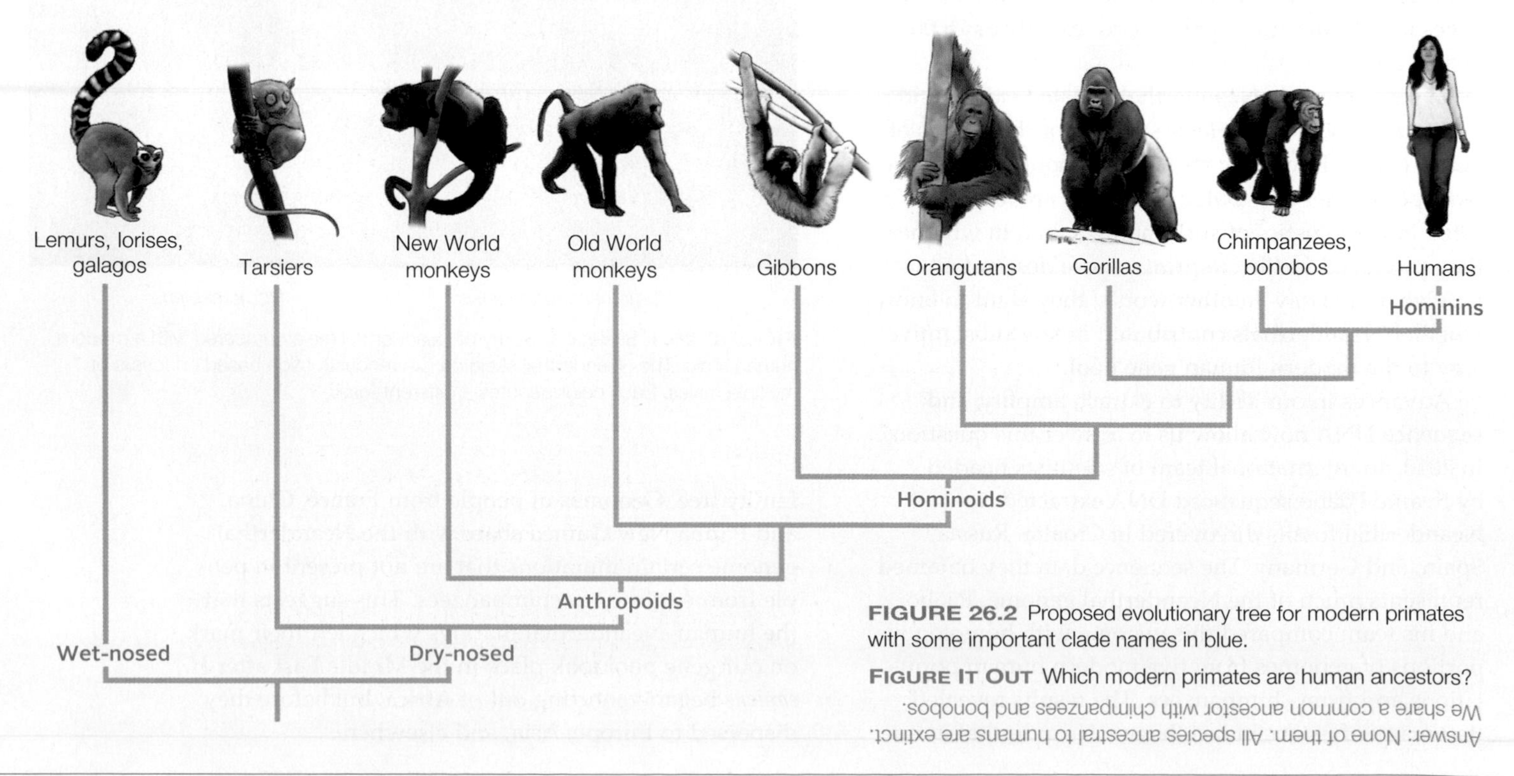

FIGURE 26.2 Proposed evolutionary tree for modern primates with some important clade names in blue.

FIGURE IT OUT Which modern primates are human ancestors?

Answer: None of them. All species ancestral to humans are extinct. We share a common ancestor with chimpanzees and bonobos.

A Lemur, with a wet nose and cleft upper lip.

B Tarsier, with a dry nose and uncleft upper lip.

C Squirrel monkey (New World monkey), with a flat face and prehensile tail.

D Baboon (Old World monkey), with a long nose and short tail.

FIGURE 26.3 Examples of nonhominoid primates.

lip attaches tightly to the underlying gum (**FIGURE 26.3A**). A wet nose aids in detecting scents.

Humans and most other modern primates belong to the dry-nosed suborder. Their nostrils are set in an area of dry, hairy skin. The upper lip is not cleft, and its attachment to the underlying gum is greatly reduced. A movable upper lip allows a wider range of facial expressions and vocalizations.

Tarsiers, the oldest surviving dry-nosed lineage (**FIGURE 26.3B**), are small, nocturnal insect eaters that live on southern Asian islands. Tarsiers were traditionally grouped with lemurs as prosimians (*pro–*, before, *simian*, monkey) because both groups have claws on some digits. However, DNA comparisons show tarsiers are more closely related to monkeys than to lemurs.

The **anthropoid** lineage includes monkeys, apes, and humans; "anthropoid" means humanlike. Nearly all anthropoids are diurnal (active during the day) and have good eyesight, including color vision. Most feed primarily on plant material. The group is thought to have originated either in Africa or Asia.

Modern New World monkeys climb through forests of Central and South America in search of fruits. They have a flat face and a nose with widely separated nostrils that open to the side. A long tail helps them maintain balance. In some species, the tail is prehensile, meaning it can grasp things (**FIGURE 26.3C**).

New World monkeys arrived in South America from western Africa about 35 million years ago. Ancestors of South American rodents such as guinea pigs arrived from Africa at about the same time. Both groups of animals probably crossed the Atlantic on rafts of vegetation. This journey may not have been as daunting as it sounds, because the continents were closer than they are today. Also, until 40 million years ago, sea level fluctuations periodically exposed islands that would have been convenient way stations. Crossing the ocean probably took many generations, with animals colonizing one island after another.

Old World monkeys live in Africa, the Middle East, and Asia. They tend to be larger than New World monkeys and have a longer nose with close set, downward facing nostrils. Some are tree-climbing forest dwellers. Others, such as baboons (**FIGURE 26.3D**), spend most of their time on the ground in grasslands and deserts. Not all Old World monkeys have a tail, but in those that do, it is usually short and never prehensile.

Apes and humans belong to the hominoid lineage. **Hominoids** are tailless primates with an upright posture and relatively large brains.

TAKE-HOME MESSAGE 26.2

What are primates?

✔ Primates include lemurs, tarsiers, monkeys, apes, and humans. Most primates are tree-dwellers. Traits such as a flexible shoulder joint and grasping hands with nails are adaptations to a climbing lifestyle.

✔ Compared to other mammals, primates have a relatively large brain with more area devoted to vision and less to smell.

✔ There are two main subgroups. Lemurs and their relatives belong to the subgroup with a wet nose and a fixed upper lip. Tarsiers, monkeys, apes, and humans belong to the subgroup with a dry nose and a movable upper lip.

anthropoids Primate lineage that includes monkeys, apes, and humans.
hominoids Tailless primate lineage that includes apes and humans.
primates Mammalian order that includes lemurs, tarsiers, monkeys, apes, and humans.

26.3 The Apes

✔ A tailless body and an upright posture evolved in the common ancestor of humans and other apes.

Hominoid Origins and Divergences

Hominoids are not descended from any existing group of monkeys, but they share a common ancestor with the Old World monkeys. That common ancestor may have lived as early as 30 million years ago. By 20 million years ago, during the Miocene, a variety of hominoid species now assigned to the genus *Proconsul* lived in the then lush forests of northeast Africa. Like modern apes, they had flattened molars and were probably tailless; however, they had monkey-sized brain and a body adapted to walking along on all four limbs.

As the Miocene continued, Africa became hotter and drier. Seasonally dry woodlands and savannas replaced the forests where *Proconsul* had thrived, and new species of apes evolved. Some of these species dispersed to Europe and Asia, where they are known from fossils. By one hypothesis, modern African apes descend from an ape lineage that diversified in Europe before some members returned to Africa. Alternatively, modern African apes may have evolved in Africa, and the fossil European apes may have no modern descendants.

Modern Apes

About 15 species of small apes called gibbons inhabit Southeast Asian forests. They use their elongated arms and permanently curved fingers to swing from limb to limb (**FIGURE 26.4A**). Gibbons live in family groups with similarly sized males and females. Depending on the species, adults weigh 7 to 14 kilograms (15 to 30 lb).

Gibbons are sometimes referred to as "lesser apes," in comparison with the larger apes, or "great apes." Unlike gibbons, all great apes are sexually dimorphic; males are substantially larger than females.

The forest-dwelling orangutan (*Pongo pygmaeus*) of Sumatra and Borneo is the only surviving Asian great ape. Like gibbons, orangutans are tree dwellers, but they live solitary lives and climb slowly using all four limbs (**FIGURE 26.4B**). Males are twice as big as females and can weigh 80 to 90 kg (176 to 198 lb).

All African great apes (gorillas, chimpanzees, and bonobos) are native to central Africa, live in social groups, and spend most of their time on the ground. When walking, they lean forward and support their weight on their knuckles (**FIGURE 26.4C**). Gorillas (*Gorilla gorilla*), the largest living primates, live in forests and feed mainly on leaves.

Chimpanzees (*Pan troglodytes*) (**FIGURE 26.4D**) and the bonobos, or pygmy chimpanzees (*Pan paniscus*), are our closest living relatives. The chimpanzee/bonobo lineage and the lineage leading to humans diverged about 13 million years ago. Chimpanzees and bonobos both live in large social groups. Both species feed mainly on fruit, but also prey on insects and small mammals, including monkeys. The two species differ in their social behavior, with chimpanzees engaging in more intraspecific aggression and bonobos spending more time in nonreproductive sexual acts.

A Gibbon B Orangutan C Gorilla D Chimpanzee

FIGURE 26.4 Modern ape diversity. A lesser ape (A) and three of the great apes (B–D).

A Human–Ape Comparison

Humans and great apes share a common ancestor, but these groups differ in many aspects. Compared to a chimpanzee, a human has a flatter face, a smaller jaw with reduced canine teeth, and a larger brain (FIGURE 26.5A). Our bodies are similarly sized, but our brains are three times bigger.

We also walk differently. When gorillas and chimpanzees walk, they lean forward and support their weight on their knuckles. By contrast, humans walk upright. Habitual upright walking is called **bipedalism**.

Evolution of bipedalism involved many skeletal changes (FIGURE 26.6). A knuckle-walking ape has a C-shaped backbone, flat feet capable of grasping, and a spinal cord that enters near the rear of skull. By contrast, the human backbone has an S-shaped curve that keeps our head centered over our feet, which have a pronounced arch and a nonopposable big toe. Our spinal cord enters at the skull's base (FIGURE 26.5B).

Human hands also differ from those of apes. A human thumb is longer, stronger, and more maneuverable than that of a chimpanzee. Many monkeys and apes wrap their hand around objects in a power grip, but only humans routinely use a precision grip, pinching together the tips of the thumb and forefinger for fine manipulation of objects:

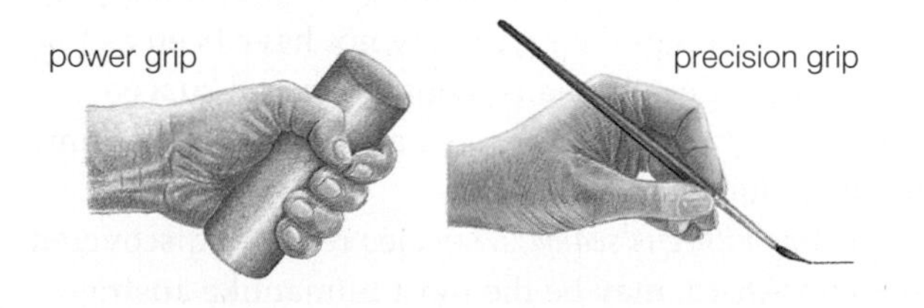

Humans are the only living primates without a thick coat of body hair. We have about the same number of body hairs as other primates, but ours are shorter and finer, giving our skin a more naked appearance. The decrease in hairiness was probably driven by a need to stay cool. Sweat evaporates more quickly from bare skin than from under thick hair, and an improved cooling ability would have reduced the risk of overheating while walking or running across hot, sunny grasslands.

The many differences between humans and apes did not arise all at once. Rather, different traits changed at different rates. As you will see in the next sections, our ancestors walked upright for many millions of years before their brain size began to increase.

bipedal Adapted to habitually walking upright.

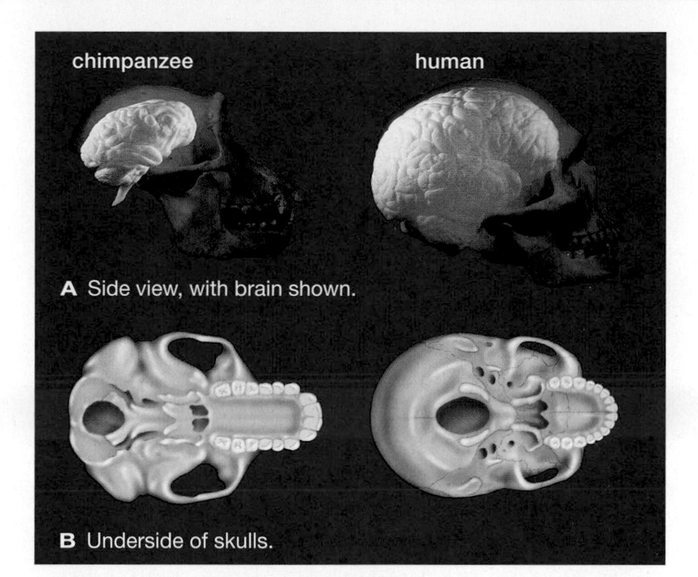

A Side view, with brain shown.

B Underside of skulls.

FIGURE 26.5 Comparison of chimpanzee and human skulls.

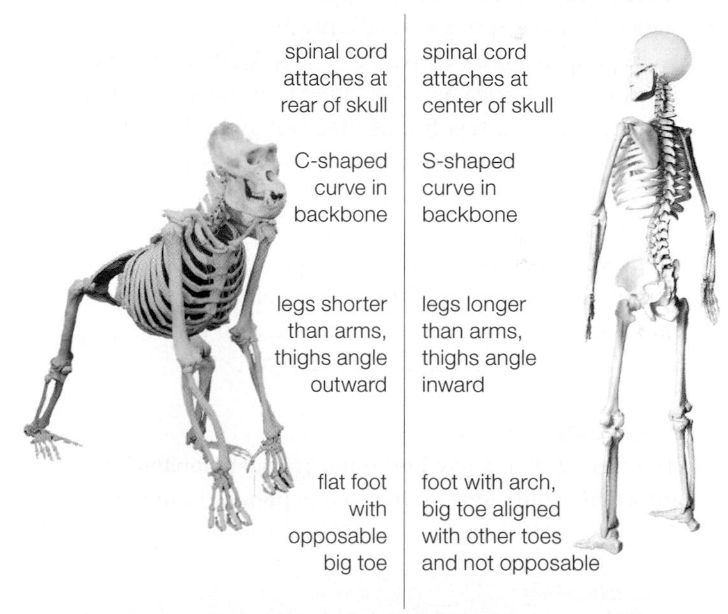

FIGURE 26.6 ▸Animated Some skeletal differences between a knuckle-walking gorilla (left) and a bipedal human (right).

TAKE-HOME MESSAGE 26.3

Which human traits are typical of hominoids and which are novel?

✔ Hominoids share a common ancestor with Old World monkeys, but they are generally larger than monkeys, lack a tail, and have an upright stance. Modern hominoids include Asian apes, African apes, and humans.

✔ Upright walking (bipedalism) evolved in the lineage leading to humans. This lineage also shows trends toward increased brain size, increased maneuverability of the thumb, and decreased body hair.

CREDITS: (5A) Kenneth Garrett/National Geographic Image Collection; (5B, in text) © Cengage Learning; (6) left, Bone Clones, www.boneclones.com; right, Gary Head.

26.4 Rise of the Hominins

✔ Humans are upright walkers with big brains, but bipedalism first evolved in our relatively small-brained ancestors.

Early Proposed Hominins

Hominins include humans and all extinct species more closely related to them than to any other group (FIGURE 26.7). Bipedalism is a defining hominin trait, so researchers interested in human history look for fossil evidence that a species walked upright.

The earliest proposed hominin, *Sahelanthropus tchadensis*, lived about 7 million years ago in western Africa (Chad). Only a few fossils have been found, but one is a skull, and it has the opening for the spinal cord positioned as in modern upright walkers. Like modern humans and unlike chimpanzees, *S. tchadensis* had a flat face and small canine teeth (FIGURE 26.8).

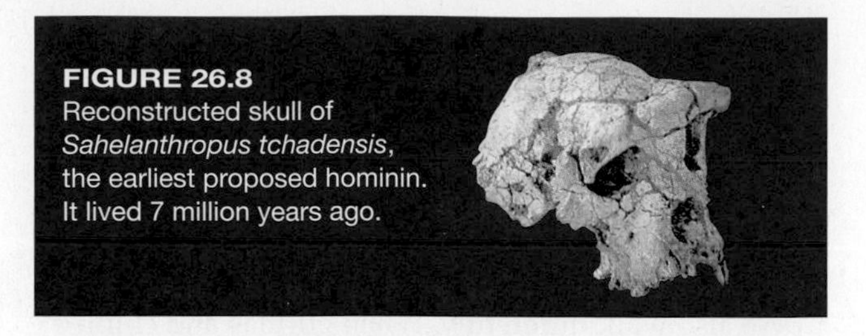

FIGURE 26.8 Reconstructed skull of *Sahelanthropus tchadensis*, the earliest proposed hominin. It lived 7 million years ago.

Another proposed early hominin, *Orrorin tugenensis*, lived about 6 million years ago in East Africa. Two sturdy fossilized femurs (thighbones) are cited as evidence that this species stood upright.

Two species of chimpanzee-sized *Ardipithecus* that lived in East Africa are also likely hominins. *A. kadabba* is known from a few fossils that date from 5.8 to 5.2 million years ago. *A. ramidus* left many fossils, including a nearly complete female skeleton known as Ardi. *A. ramidus* had smaller teeth than a chimpanzee, and its pelvis was more humanlike than apelike. It walked upright on the ground, but long arms, curved fingers, and a splayed big toe indicate that it also walked on all fours along branches (FIGURE 26.9A).

Australopiths

The best known early hominins are **australopiths**, an informal group comprising two genera of hominins that lived in Africa from about 4 million to 1.2 million years ago. *Australopithecus* species were small boned. Their fossil history reveals trends toward smaller teeth and improvements in the ability to walk upright, but little increase in brain size.

Fossil footprints in Tanzania document the passage of a bipedal species 3.6 million years ago. We can tell from the prints that the walkers' feet had a pronounced arch and a big toe in line with the other toes—both adaptations to upright walking. The prints were probably made by *Australopithecus afarensis*, an australopith that lived in Tanzania and other parts of eastern Africa from 3.9 to 3 million years ago. A partial *A. afarensis* skeleton known as Lucy (FIGURE 26.9B), is the best known representative of this species.

A. afarensis males stood about 1.5 to 1.8 meters (5 to 5.5 feet) tall, and females 1 meter (3 feet) tall. Members of this species had a pelvis and legs suited to upright walking, although their gait may not have been as fluid as that of modern humans. Long arms and curved fingers indicate that *A. afarensis* also spent a lot of time climbing among tree branches.

Australopithecus sediba, a species recently discovered in South Africa, may be the most humanlike australopith discovered to date (FIGURE 26.9C). *A. sediba*,

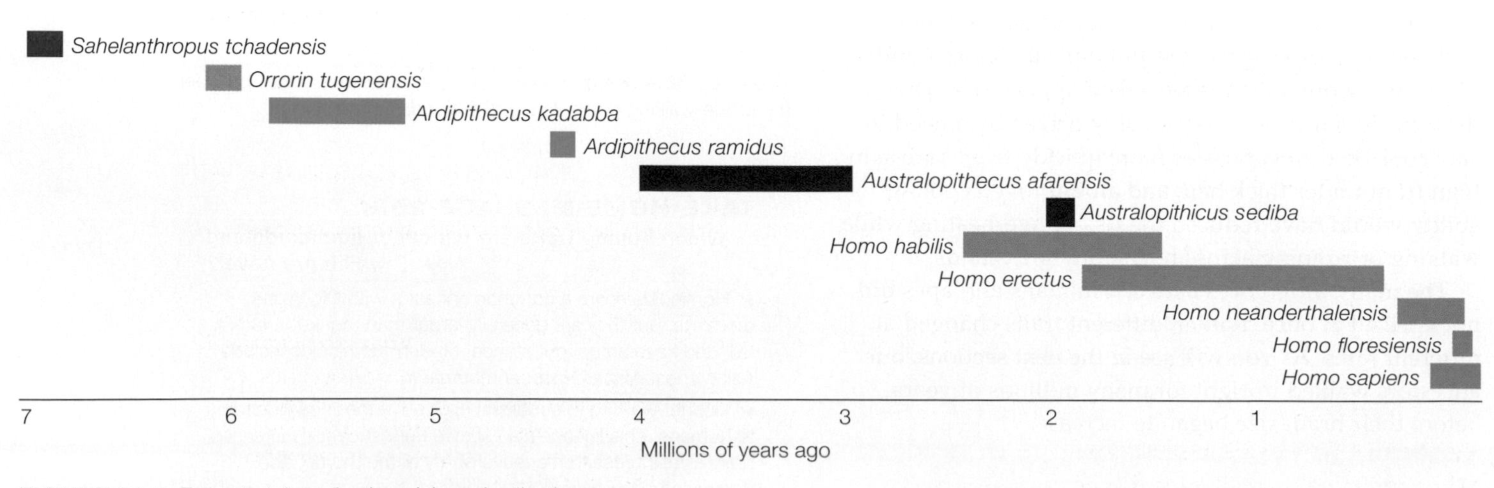

FIGURE 26.7 Estimated dates for the origin and extinction of a sampling of hominins and proposed hominins.

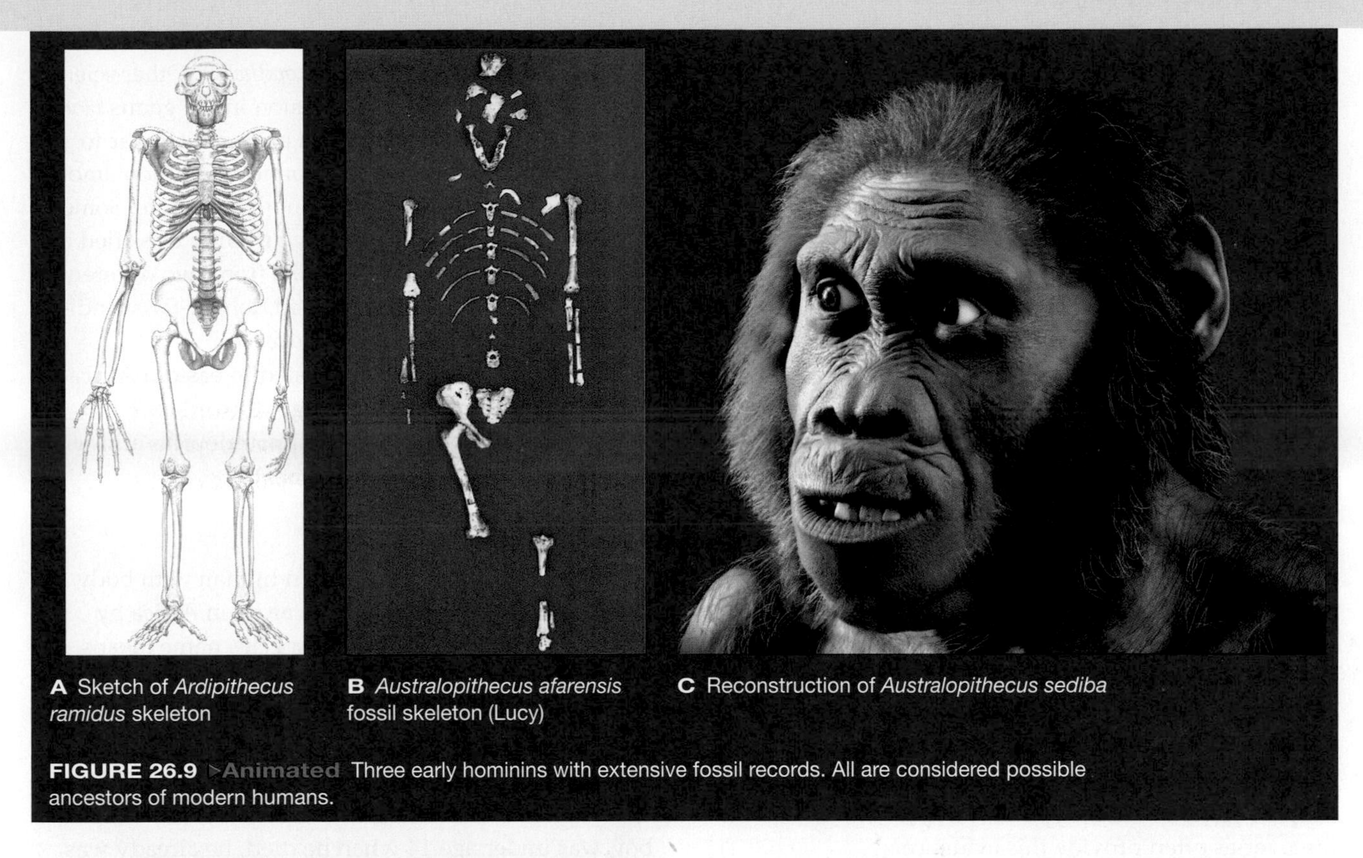

A Sketch of *Ardipithecus ramidus* skeleton

B *Australopithecus afarensis* fossil skeleton (Lucy)

C Reconstruction of *Australopithecus sediba*

FIGURE 26.9 ▷Animated Three early hominins with extensive fossil records. All are considered possible ancestors of modern humans.

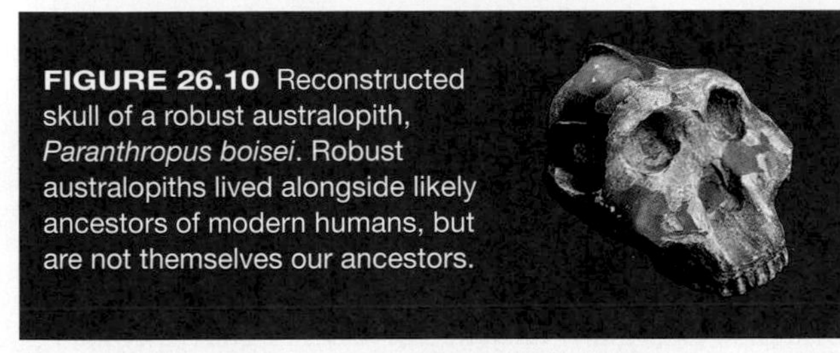

FIGURE 26.10 Reconstructed skull of a robust australopith, *Paranthropus boisei*. Robust australopiths lived alongside likely ancestors of modern humans, but are not themselves our ancestors.

which lived about 2 million years ago had remarkably humanlike hands; they were probably capable of a precision grip. In addition, the *A. sediba* brain, although chimpanzee-sized and largely apelike, has a humanlike frontal region.

A. afarensis and *A. sediba* are considered possible ancestors of modern humans, but the hominin family tree is bushy, with many branches that certainly did not lead to modern species. The species known as robust australopiths were on one of these dead-end branches. These heavy-bodied species had a skull with a pronounced crest, to which large jaw muscles attached (**FIGURE 26.10**).

Factors Favoring Bipedalism

What advantages could have driven an evolutionary shift from walking on all fours to bipedalism? A trend toward a drier and hotter climate was probably a factor. Hominins originated at a time when Africa's lush rain forests were giving way to woodlands interspersed with grassy plains. In this altered habitat, an ability to move efficiently across open ground between trees would have been favored, and upright walking can be energy efficient. Human walkers use less energy than chimpanzees who walk on all fours. Bipedalism also keeps the body cooler. A bipedal animal gains less heat from the ground than a four-legged walker and intercepts less warming sunlight. In addition, an upright stance makes it easier to scan the horizon for predators, and hands freed from locomotion can be used to gather and carry food and other items.

australopith Informal name for chimpanzee-sized hominins that lived in Africa from 4 million to 1.2 million years ago.

hominins Modern humans and their closest extinct relatives.

TAKE-HOME MESSAGE 26.4

What are hominins?

✔ Hominins include modern humans and extinct members of their lineage. Possible hominin fossils date back as far as 7 million years ago. Early hominins had small brains. Australopiths are a diverse group of African hominins that probably include human ancestors.

CREDITS: (9A) HO/Reuters/Corbis; (9B) Dr. Donald Johanson, Institute of Human Origins; (9C) Courtesy of © John Gurche; (10) Pascal Goetgheluck/Science Source.

26.5 Early Humans

✔ The genus *Homo* arose in Africa, and early humans dispersed from there to Europe and Asia.

Classifying Fossils—Lumpers and Splitters

When it comes to classifying fossil species, evolutionary biologists fall into two general camps. The "lumpers" look for similarities among fossils and tend to assign fossils to relatively few species. Lumpers support this approach by pointing out that members of a single species can vary widely in their traits. For example, modern humans share many traits, but vary greatly in their height. By contrast, "splitters" focus on differences between fossils and often name new species on the basis of subtle differences. In support of their approach, splitters point out that many modern species differ only slightly in their morphology. For example, chimpanzees and bonobos appear very similar, although bonobos are a bit smaller.

From a scientific standpoint, both approaches are equally valid. Both lumpers and splitters formulate hypotheses about how a particular fossil relates to other known fossils. They then seek evidence that will support or refute their hypotheses. Additional fossil discoveries often provide this evidence.

In this section, we discuss early fossils of our own genus: *Homo*. Lumpers assign these fossils to two species; splitters divide them into four or more.

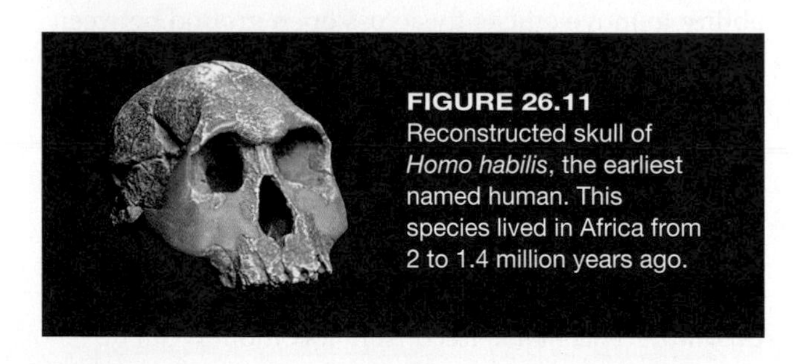

FIGURE 26.11 Reconstructed skull of *Homo habilis*, the earliest named human. This species lived in Africa from 2 to 1.4 million years ago.

Homo habilis

Humans are members of the genus *Homo*. The oldest named member of this species is ***Homo habilis*** (FIGURE 26.11). Fossils of *H. habilis* date from 2.3 million years to 1.4 million years before the present. The species was named based on a fossil discovered in Kenya in 1964. At that time, tool production was considered a diagnostic trait of the genus *Homo*. The name *Homo habilis* means "handy man," and is a reference to the stone tools found near the fossil.

Classification of *H. habilis* continues to inspire debate. *H. habilis* resembles some australopiths in body proportions and brain size, and some scientists think it should be classified in *Australopithecus*. Other scientists argue for the species' inclusion in the genus *Homo*, pointing out that the hands and arms are similar to those of modern humans. Anatomical variation among the fossils has also resulted in another dispute. Some splitters contend that the fossils currently classified as *H. habilis* actually include remains from two different species. They place larger-brained, longer-faced individuals in the species *Homo rudolfensis*.

Homo habilis/rudolfensis persisted in eastern Africa until at least 1.4 million years ago. It overlapped in time and range both with some australopiths and with *Homo erectus*, which we turn to next.

Homo erectus

Homo erectus, the earliest known human with body proportions similar to our own, arose in Africa by about 2 million years ago. The species name means "upright man," and like us, *H. erectus* stood on legs that were longer than its arms. The most complete *H. erectus* fossil found thus far is a skeleton of a young male who lived about 1.5 million years ago in Kenya. Although this individual, informally known as Turkana boy, was under age 14 when he died, he already was 1.60 meters (5 feet 2 inches) tall and his brain was twice the size of a chimpanzee's. Footprints from the same region and time suggest that *H. erectus* had a gait like that of modern humans (FIGURE 26.12A).

As far as we know, *H. erectus* was the first hominin to venture out of Africa. By 1.75 million years ago, a population had become established in what is now Dmanisi, Georgia (FIGURE 26.12B). *H. erectus* colonized Indonesia by 1.6 million years ago, and China by 1.15 million years ago (FIGURE 26.12C).

As with *Homo habilis*, some scientists split *Homo erectus* fossils into two species. The splitters reserve the name *H. erectus* for fossils in Asia, and refer to the similar African fossils as *H. ergaster* (working man).

Early Culture

Culture is a set of learned behaviors that are passed from one individual to another, and from one generation to the next. The earliest hominin cultural trait for which we have fossil evidence is toolmaking. Distinctive scrape marks on bones provide indirect evidence that hominins used sharp stone edges to scrape meat from bones as early as 3.4 million years ago. Given the age of the bones, these early tool users were most likely australopiths.

The first stone tools were sharp flakes chipped from larger rocks. Toolmaking improved about 1.4 million

A 1.5-million-year-old footprints from Kenya indicate a smooth gait.

B 1.75-million-year-old remains from the Republic of Georgia.

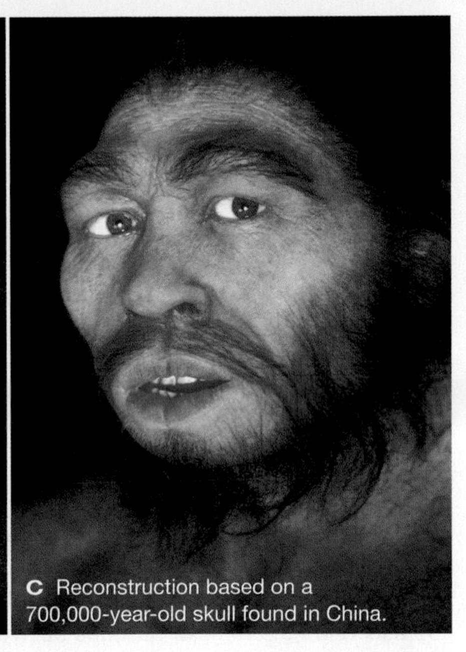
C Reconstruction based on a 700,000-year-old skull found in China.

FIGURE 26.12 *Homo erectus*, the first hominin for which we have evidence outside of Africa.

years ago, when African *H. erectus* began to use a succession of carefully placed strikes to sculpt tools in a variety of shapes (**FIGURE 26.13**).

Marks on fossilized bones of prey animals indicate that *H. erectus* used stone tools to cut up scavenged animal carcasses, to scrape meat from bones, and to extract marrow. Some researchers have suggested that increased meat consumption may have been a necessary prerequisite for the evolution of a larger brain. Meat is a more concentrated source of energy and protein than plant foods, so it can better fuel development and maintenance of a large brain.

Chimpanzees use gestures and calls to communicate, and early humans probably did the same. However, it is unlikely that *H. habilis* or *H. erectus* had a spoken language. Evolution of a capacity for speech involved changes in the brain, remodeling of the larynx (voice box), and an improved ability to control air flow through the vocal tract. To date, scientists have not found fossil evidence of these necessary anatomical modifications in either early species of *Homo*.

FIGURE 26.13 Ancient stone tools discovered in Tanzania.

culture Learned behaviors transmitted between individuals and down through generations.
Homo erectus Early human species that arose in Africa by about 2 million years ago; some populations migrated to other regions.
Homo habilis Earliest named human species; lived in Africa from about 2.3 to 1.4 million years ago.
human Living or extinct member of the genus *Homo*.

TAKE-HOME MESSAGE 26.5
What traits characterized early humans?

✔ *Homo habilis*, the most ancient human species known, lived in Africa. It was similar in many respects to australopiths, but its hands and arms more closely resemble those of modern humans.

✔ *Homo erectus* is known from fossils found Africa, Europe, and Asia. It had a significantly larger brain than earlier hominins and a gait similar to modern humans.

✔ Early humans made stone tools and used them to cut up animal carcasses. However, production and use of stone tools may have begun among the australopiths.

CREDITS: (12A) Professor Matthew Bennett, Bournemouth University/Science Source; (12B) Kenneth Garrett/National Geographic Creative; (12C) Philippe Plailly & Atelier Daynes/Science Source; (13) John Reader/Science Source.

✔ Our status as the only living members of the genus *Homo* came about relatively recently.

Homo erectus is considered the most likely ancestor of our own species, Neanderthals, and some other more recently discovered hominins.

Origin and Dispersal of *Homo Sapiens*

Anatomically modern humans are assigned to ***Homo sapiens***. Compared to *H. erectus*, we have a higher, rounder skull, a larger brain, and a flatter face with smaller jawbones and teeth. We also have a chin, which is a protruding area of thickened bone in the middle of our lower jawbone.

Much evidence suggests our species arose in Africa. To date, the oldest *H. sapiens* fossils discovered are two partial male skulls from Ethiopia, in East Africa. Known as Omo I and Omo II, they date to 195,000 years ago. Fossil remains of two adult males and a child who lived 160,000 years ago were also unearthed in this region. Other fossils show that *H. sapiens* reached South Africa by 115,000 years ago.

Modern Africans are more genetically diverse than people of any other region. This diversity indicates that our species has existed here for a very long time—long enough to accumulate a very large number of random mutations compared with other populations. Furthermore, most genetic variations seen in people native to regions outside of Africa are subsets of the variation found in Africa. This is evidence that founder effects (Section 17.8) occurred after some people left Africa to colonize the rest of the world.

Our species expanded its range gradually, as small groups ventured away from their homelands. As they did, unique mutations arose in different lineages and were passed to descendants. Today, the distribution of specific mutations among different ethnic groups provides evidence of routes taken by the ancient travelers. By mapping the frequency of maternal and paternal genetic markers in modern peoples, geneticists have created a picture of when and where ancient *Homo sapiens* moved around the world (**FIGURE 26.14**).

Other Recent *Homo* Species

Neanderthals ***Homo neanderthalensis***, the Neanderthals, are our closest relatives. The human and Neanderthal lineages diverged between 400,000 and 500,000 years ago, although as Section 26.1 explained, some interbreeding occurred about 60,000 years ago. Neanderthal fossils have been found in the Middle East, Europe, and Asia.

The common conception of Neanderthals as brutish cavemen with poor posture arose from an early reconstruction based on a fossil of an individual deformed by arthritis. More recent reconstructions reveal Neanderthals were shorter than modern humans, but stood upright. They lived in regions where winters are cold, and a short, stocky body minimized the surface

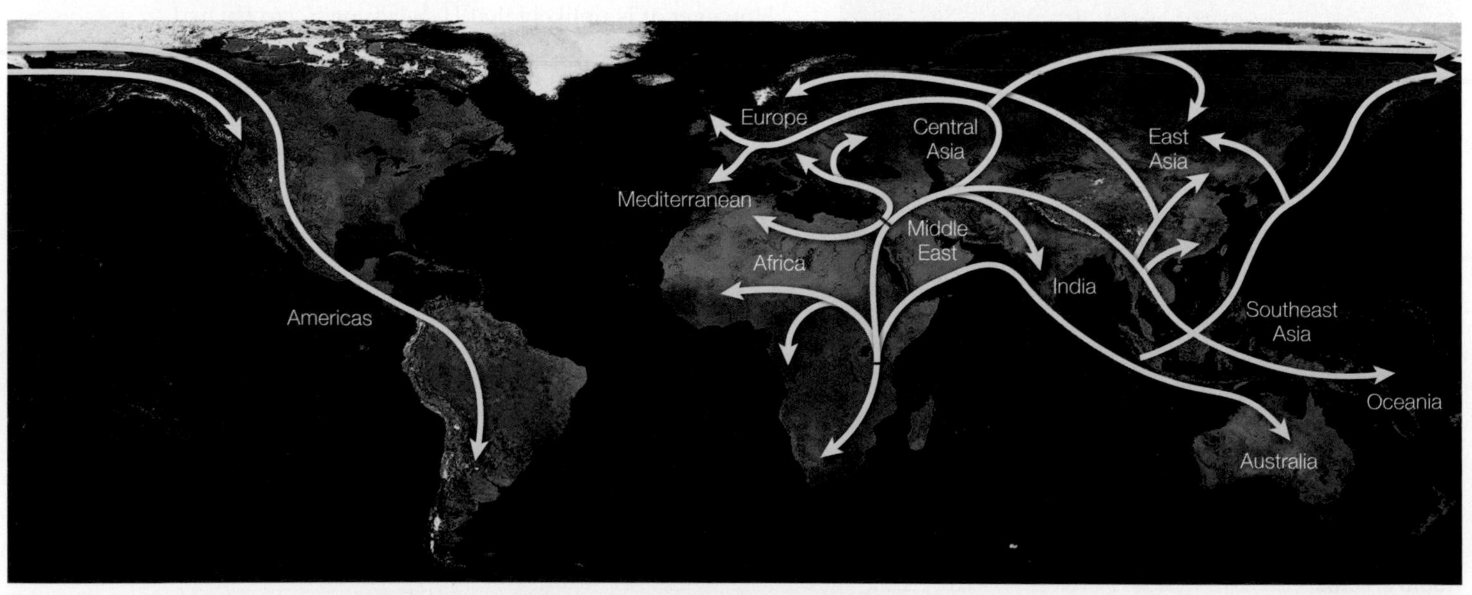

FIGURE 26.14 Human migration routes. People crossed from the Horn of Africa onto the Arabian Peninsula, then moved along the coast to India, reaching Southeast Asia and Australia by 50,000 years ago. Later, another group traveled through the Middle East into south central Asia. Offshoots from this lineage peopled North Asia and Europe. About 15,000 years ago, people crossed a temporary land bridge from Siberia to North America. Their descendants spread to the tip of South America.

CREDITS: (14) photo, NASA; data, National Geographic.

A Bit of a Neanderthal (revisited)

Until relatively recently, it was thought that humans had simply replaced earlier hominins. The more recent genetic evidence of interbreeding has led to what is known as the leaky replacement model (**FIGURE 26.15**). As previously noted, the lineages leading to modern humans and to Neanderthals parted ways about 500,000 years ago ❶. Then, about 60,000 years ago, as modern humans ancestral to non-African populations left Africa, they interbred with Neanderthals ❷. Another hybridization took place about 40,000 years ago, when humans moving through Asia on their way to Melanesia (Papua New Guinea and adjacent islands) interbred with Denisovans ❸. As a result of this later event, people of Melanesian descent carry genetic markers of both Neanderthals and Denisovans.

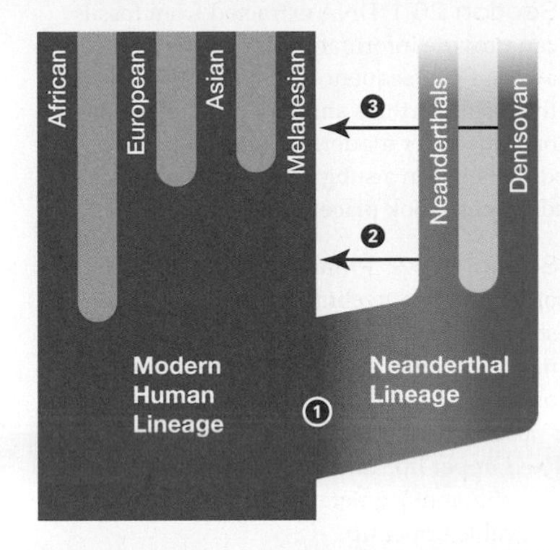

FIGURE 26.15 The leaky replacement model for human evolution. Red arrows denote instances of interbreeding.

area available for heat loss. Modern Arctic peoples have a similar body shape. Maintaining body temperature in a cool environment burns a lot of energy and requires a lot of calories.

Neanderthals had a braincase that was longer and lower than that of modern humans, but their brain was as big as ours or bigger. Fossils of Neanderthal individuals who survived despite disabilities such as the loss of a limb testify to a compassionate and cooperative social structure. Some simple burials suggest possible symbolic thought, and several lines of evidence suggest Neanderthals were able to speak.

The most recent evidence of Neanderthals dates to 28,000 years ago. Neanderthals may have been outcompeted by newly arrived *H. sapiens* or killed by diseases these migrants brought with them. Climate changes and volcanic eruptions may have contributed to their demise by altering the abundance of the animals they hunted. Meat made up the bulk of the Neanderthal diet, so they would have been more affected by such declines than *H. sapiens*, who had a more varied diet.

Denisovans Denisovans are a newly discovered lineage. Their existence came to light in 2010, when researchers found a 50,000-year-old fossil pinky finger in Denisova cave in Siberia. Extraction and analysis of DNA from this fossil revealed it was from a young female who was more closely related to Neanderthals than to modern humans, but was not a typical Neanderthal. More recently, 400,000-year-old fossils from a cave in Spain were found to have a Denisovan-like DNA. The relationship between the Spanish individuals and Siberian Denisovans remains an open question. At this writing, Denisovans have not been assigned to a new species; they are considered a subspecies of the Neanderthal lineage.

Flores Hominins In 2003, scientists discovered 18,000-year-old hominin fossils on the Indonesian island of Flores. The fossil individuals were only a meter tall, with a heavy brow and a small brain. Scientists who found the fossils assigned them to a new species, *Homo floresiensis*, which they believe descended from *Homo erectus*. Some other scientists who have examined the fossils agree with this assessment. Others think the fossils could be *H. erectus* or *H. sapiens* who had a genetic or nutritional disorder. Further study of existing fossils, a search for more fossils, and possibly DNA analysis will allow testing of these hypotheses.

TAKE-HOME MESSAGE 26.6

How are the most recent hominins related?

✔ *H. erectus* is considered an ancestor of both modern humans and Neanderthals. The human and Neanderthal lineages parted ways about 500,000 years ago, although some interbreeding occurred later as *H. sapiens* dispersed.

✔ Denisovans are a genetically distinct group of Neanderthals whose existence was discovered recently when DNA from a fossil was sequenced.

✔ *Homo floresiensis* is a recently proposed species from Indonesia that may also be descended from *H. erectus*.

Homo sapiens Modern humans; only surviving *Homo* species.
Homo neanderthalensis Neanderthals. Extinct hominins that lived in the Middle East, Europe, and Asia; the closest relatives of modern humans.

CREDIT: (15) © Cengage Learning.

summary

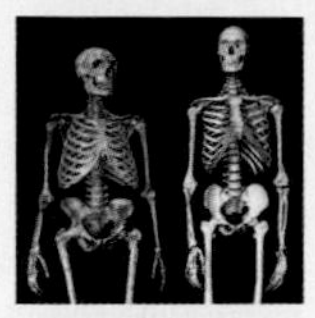

Section 26.1 DNA extracted from fossils can provide information about our extinct relatives. DNA sequence comparisons indicate that Neanderthals and modern humans interbred just after modern humans left Africa. A later interbreeding between a subgroup of Neanderthals (Denisovans) and humans took place in Asia.

Section 26.2 **Primates** are a mammalian order adapted to climbing. They have flexible shoulder joints, grasping hands tipped by nails, and good depth perception. They rely on vision more than smell. Lemurs and their relatives are wet-nosed primates that have a typical mammalian nose and a fixed upper lip. Tarsiers, monkeys, apes, and humans belong to a different primate lineage; they are dry-nosed and have movable upper lip.

Anthropoids include monkeys, apes, and humans. The **hominoids** (apes and humans) do not have tails. Hominoids are more closely related to Old World monkeys than to New World monkeys.

Section 26.3 Fossils show that *Proconsul*, an early ape, was living in Africa by about 20 million years ago. Fossil apes have also been found in Europe.

Modern apes include the gibbons and orangutans of Asia, and the gorillas, chimpanzees, and bonobos of Africa. Chimpanzees and bonobos are our closest living relatives. Both live in social groups and are sexually dimorphic. Our lineage and the chimpanzee/bonobo lineage diverged from a common ancestor about 8 million years ago.

Proconsul walked on all fours, but gorillas, chimpanzees, and bonobos are knuckle walkers. Humans are **bipedal**, meaning they habitually walk and stand upright. Human skeletons have adaptations related to bipedalism. The position of the hole that attaches our spinal cord to our brain is at the base of our skull, we have a foot with an arch and a nonopposable big toe, and our backbone has an S-shaped curve.

Humans also have a much larger brain and more flexible hands than any living ape.

Section 26.4 **Hominins** include modern humans and their extinct bipedal relatives. The earliest proposed hominins date to 6 or 7 million years ago and are known from fossil fragments found in Africa. *Ardipithecus* lived in Africa 5.8 to 5.2 million years ago and left many fossils, including an almost complete skeleton of a female. Her skeletal anatomy suggests that she walked upright when on the ground, but also walked on all fours along branches.

Footprints dating to 3.6 million years ago are evidence of the passage of a fully bipedal species. It was most likely one of the **australopiths**, a group of hominins that lived in Africa from 4 to 1.2 million years ago and left many fossils. Some are probable ancestors of humans. Australopiths were upright walkers, but their brains were chimpanzee-sized.

Bipedalism evolved at a time when the African climate was warming and grasslands were replacing forests. Walking upright increases the efficiency of movement on the ground, keeps a body cooler than four-legged walking, and makes it easier to carry food or other material.

Section 26.5 **Humans** are members of the genus *Homo*. The earliest named human species, ***Homo habilis***, appeared in Africa by 2.3 million years ago and survived until at least 1.4 million years ago. *H. habilis* had an australopith-like build and a small brain.

Homo erectus evolved by 1.9 million years ago. Compared to *H. habilis*, they had a significantly larger brain. They also had longer legs than arms, as do modern humans. *H. erectus* dispersed out of Africa and fossils of this species have been found in Europe, China, and Indonesia.

Stone tool production is an aspect of **culture**, a set of learned behaviors passed down from one generation to the next. Evidence of stone tool use dates back to 3.4 million years ago, suggesting it probably began among the australopiths. Hominins used tools to scrape meat from animal bones. By one hypothesis, increased meat consumption provided energy necessary to develop larger bodies and brains. *H. habilis* and *H. erectus* probably did not have a spoken language.

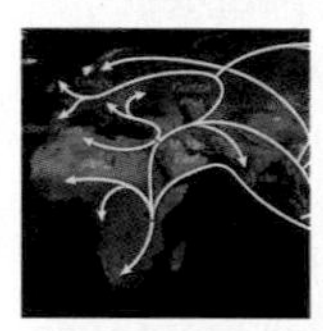

Section 26.6 The earliest fossils of our own species, ***Homo sapiens***, are from central Africa and date to 195,000 years ago. They are distinguished by a larger brain, flatter face, smaller jaw, and a bony protrusion on the lower jaw (a chin). By 115,000 years ago, *H. sapiens* had migrated south into South Africa. Migrations that began about 60,000 years ago spread our species across the globe. The genomes of people native to different regions differ, in part, as a result of founder effects and accumulation of unique mutations that occurred during this migration.

The lineages leading to ***Homo neanderthalensis*** and *H. sapiens* diverged about 500,000 years ago. Neanderthals dispersed to the Middle East, Europe, and Asia, then died out by 28,000 years ago. They may have had a language. Sequencing of DNA from a Siberian bone led to the discovery of Denisovans, a previously unknown type of Neanderthal.

Fossils of short, small-brained hominins that lived 18,000 years ago on an island in Indonesia have been named as a new species, *Homo floresiensis*. However, not all scientists agree that the fossils belong to a new species of *Homo*.

self-quiz

Answers in Appendix VII

1. New World monkeys ________ .
   a. lack a tail
   b. are bipedal
   c. live only in Africa
   d. are dry-nosed primates
   e. are human ancestors
   f. all of the above

2. The closest relatives of bonobos are ________ .
   a. chimpanzees
   b. humans
   c. tarsiers
   d. Old World monkeys

3. An S-shaped backbone is an adaptation to ________ .
   a. tool use
   b. climbing trees
   c. bipedalism
   d. carnivory

data analysis activities

Neanderthal Hair Color The *MC1R* gene regulates pigmentation in humans (Sections 14.1 and 15.1 revisited), so loss-of-function mutations in this gene affect hair and skin color. A person with two mutated alleles for this gene makes more of the reddish melanin than the brownish melanin, resulting in red hair and pale skin. DNA extracted from two Neanderthal fossils contains a mutated *MC1R* allele that has not yet been found in humans. To see how the Neanderthal mutation affects the function of the *MC1R* gene, Carles Lalueza-Fox and her team introduced the allele into cultured monkey cells (**FIGURE 26.16**).

1. How did *MCR1* activity in monkey cells with the mutant allele differ from that in cells with the normal allele?

2. What does this imply about the mutation's effect on Neanderthal hair color?

3. What purpose do the cells with the gene for green fluorescent protein serve in this experiment?

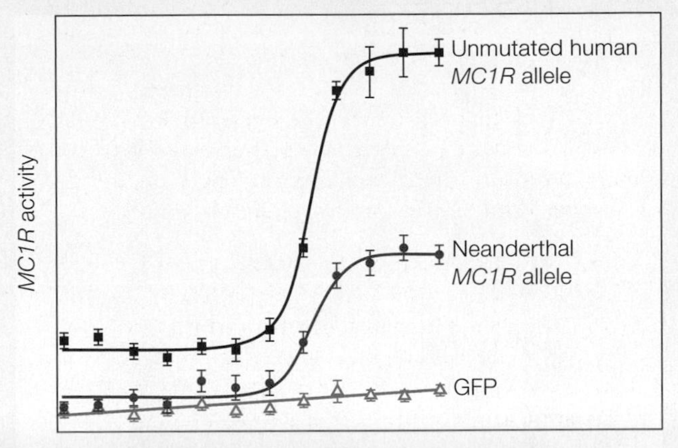

FIGURE 26.16 *MC1R* activity in monkey cells transgenic for an unmutated *MC1R* gene, the Neanderthal *MC1R* allele, or the gene for green fluorescent protein (GFP). GFP is not related to *MC1R*.

4. Like most mammals, a(n) ________ has a wet nose.
   a. New World monkey
   b. Old World monkey
   c. chimpanzee
   d. lemur

5. The 3.6-million-year-old footprints left by bipedal walkers in Tanzania were probably made by ________.
   a. australopiths
   b. Neanderthals
   c. modern humans
   d. *Homo erectus*

6. The position where a spinal cord enters the skull provides evidence about whether a fossil species ________.
   a. was nocturnal
   b. was carnivorous
   c. walked upright
   d. all of the above

7. Australopiths are ________.
   a. extinct
   b. placental mammals
   c. hominoids
   d. all of the above

8. The oldest *Homo* fossils found outside of Africa are assigned to the species ________.
   a. *H. sapiens*
   b. *H. habilis*
   c. *H. erectus*
   d. *H. floresiensis*

9. A prominent chin is typical of ________.
   a. *Homo sapiens*
   b. *Homo habilis*
   c. *Homo erectus*
   d. *Homo floresiensis*

10. Compared to modern humans, Neanderthals had a ________ brain and were ________.
   a. smaller, taller
   b. smaller, shorter
   c. similar-sized, taller
   d. similar-sized, shorter

11. The oldest *Homo sapiens* fossils were found in ________.
   a. the Middle East
   b. Africa
   c. Indonesia
   d. Europe

12. ________ probably had a spoken language.
   a. *Ardipithecus ramidus*
   b. *Homo habilis*
   c. *Australopithecus sediba*
   d. *Homo neanderthalensis*

13. Modern people native to ________ have a higher degree of genetic diversity than people native to other regions.
   a. Europe
   b. Asia
   c. Africa
   d. North America

14. Match each group with its description.
   ___ hominins — a. first to be found outside Africa
   ___ australopiths — b. modern humans
   ___ *Homo erectus* — c. includes all other groups listed
   ___ *Homo sapiens* — d. nonhuman but bipedal
   ___ *Homo habilis* — e. short-statured, from Indonesia
   ___ *Homo floresiensis* — f. most ancient human species

15. Place the events in order.
   ___ 1 (earliest) — a. Neanderthals disappear
   ___ 2 — b. monkeys reach the New World
   ___ 3 — c. early apes colonize Europe
   ___ 4 — d. *Homo erectus* leaves Africa
   ___ 5 — e. divergence of lineages leading to humans and to chimpanzees
   ___ 6 — f. divergence of wet-nosed and dry-nosed primates
   ___ 7 (most recent) — g. *Homo sapiens* evolve in Africa

critical thinking

1. Male aggression is rare in bonobo society and common in chimpanzee society. Various authors have argued that either one species or the other should be considered a model for "natural" human behavior. Explain why, from the standpoint of relatedness, there is no reason to think that one of these species is a better model for human behavior than the other.

2. Consider the human dispersal pattern described and illustrated in **FIGURE 26.14**. Given what you know about the founder effect, would you expect populations native to South America to be more or less genetically diverse than those native to North America? Explain your reasoning.

CREDIT: (16) © Cengage Learning.

27 Plant Tissues

LEARNING ROADMAP

This chapter builds on discussions of plant structure and life cycles in Sections 22.3 and 22.8. It explores the anatomy of flowering plants in the context of life's organization (1.2), and revisits carbohydrates (3.4), plant cell specializations (4.9, 4.11, 6.7), membrane permeability (5.8), and differentiation (10.2).

TYPES OF PLANT TISSUES

Most flowering plants have shoots and roots. All plant parts consist of the same tissues, but the arrangement of the tissues differs between the two main groups, monocots and eudicots.

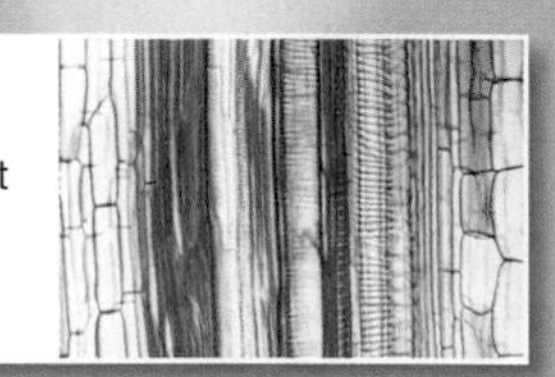

STEM STRUCTURE

Stems support the plant body. A specialized structure allows some stems to have additional functions such as reproduction, water storage, and nutrient storage.

LEAF STRUCTURE

Leaves are plant parts that specialize in intercepting sunlight and exchanging gases for photosynthesis. Eudicot leaves in particular vary in shape.

ROOT STRUCTURE

Roots provide a large surface area for absorbing water and nutrients from soil, and often they anchor the plant. Each has a central column of vascular tissue.

PLANT GROWTH

Young plant parts grow by lengthening at their tips; older plant parts grow by thickening. Tree rings hold evidence of environmental conditions that prevailed when they formed.

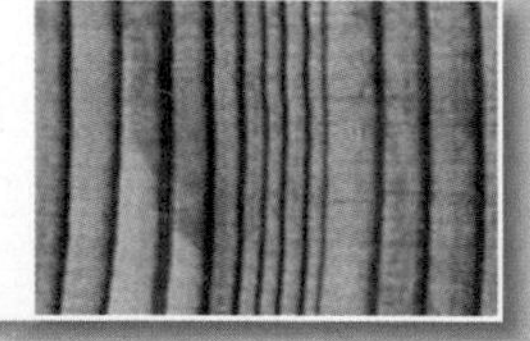

In Chapter 28, you will learn more about the function of plant structures, including how roots take up water from soil, how fluids move through vascular tissue, and how epidermal tissue conserves water. Chapter 29 details plant reproduction. Plants form the basis of almost all food webs (Section 46.3), so they are a critical factor in communities (45.8) and the carbon cycle (46.7). We return to greenhouse gases and climate change in Section 46.8.

27.1 Carbon Sequestration

A 2,000-year-old giant sequoia tree is just cranking out wood (**FIGURE 27.1**): Researchers recently discovered that growth does not slow with age in long-lived trees; rather, more and more wood is produced until the tree dies. This finding is very relevant for carbon offsets, which are financial instruments designed to reduce global emissions of carbon dioxide and other carbon-based gases. Consider that plants are specialized to absorb carbon dioxide (CO_2) from the air and lock it in their tissues via photosynthesis. In an actively growing plant, photosynthetically produced sugars are continually remodeled into other carbon-containing compounds that end up in the plant's tissues. Biologists calculated that a single 3,200-year-old giant sequoia in California's Sequoia National Park holds more than 1,500 cubic meters (54,000 cubic feet) of wood and bark. A lot of carbon is locked up in that one old tree!

Carbon in sturdy plant tissues such as wood can stay out of the atmosphere for many thousands of years. After a tree dies, its tissues decompose, and the carbon in them is re-released to the atmosphere. However, plant matter decomposes more slowly than other organic materials, because compounds such as lignin and cellulose that waterproof and reinforce plant parts are relatively stable.

You learned in Chapter 6 that humans release a lot of carbon dioxide by burning fossil fuels and other plant-derived materials. As a result of these activities, the amount of CO_2 in the atmosphere is increasing exponentially, with unintended and potentially catastrophic effects on Earth's climate. Companies and individuals buy carbon offsets to "offset" activities that release CO_2 and other carbon-containing greenhouse gases. The funds are then used to support projects aimed at reducing emissions of the gases, or activities that remove them from the atmosphere. Some carbon offsets finance endeavors such as increasing plant density in existing forests, and replanting deforested areas. Efforts like these typically cost less than other methods of removing carbon from the atmosphere, in part because they take advantage of the ability of plants to do this naturally. Forests worldwide absorb about 3 billion tons of CO_2 from the atmosphere every year, about one-third of the amount released by human activities.

FIGURE 27.1 A giant sequoia (*Sequoiadendron giganteum*) in California's Sequoia National Park makes a lot of wood.

27.2 The Plant Body

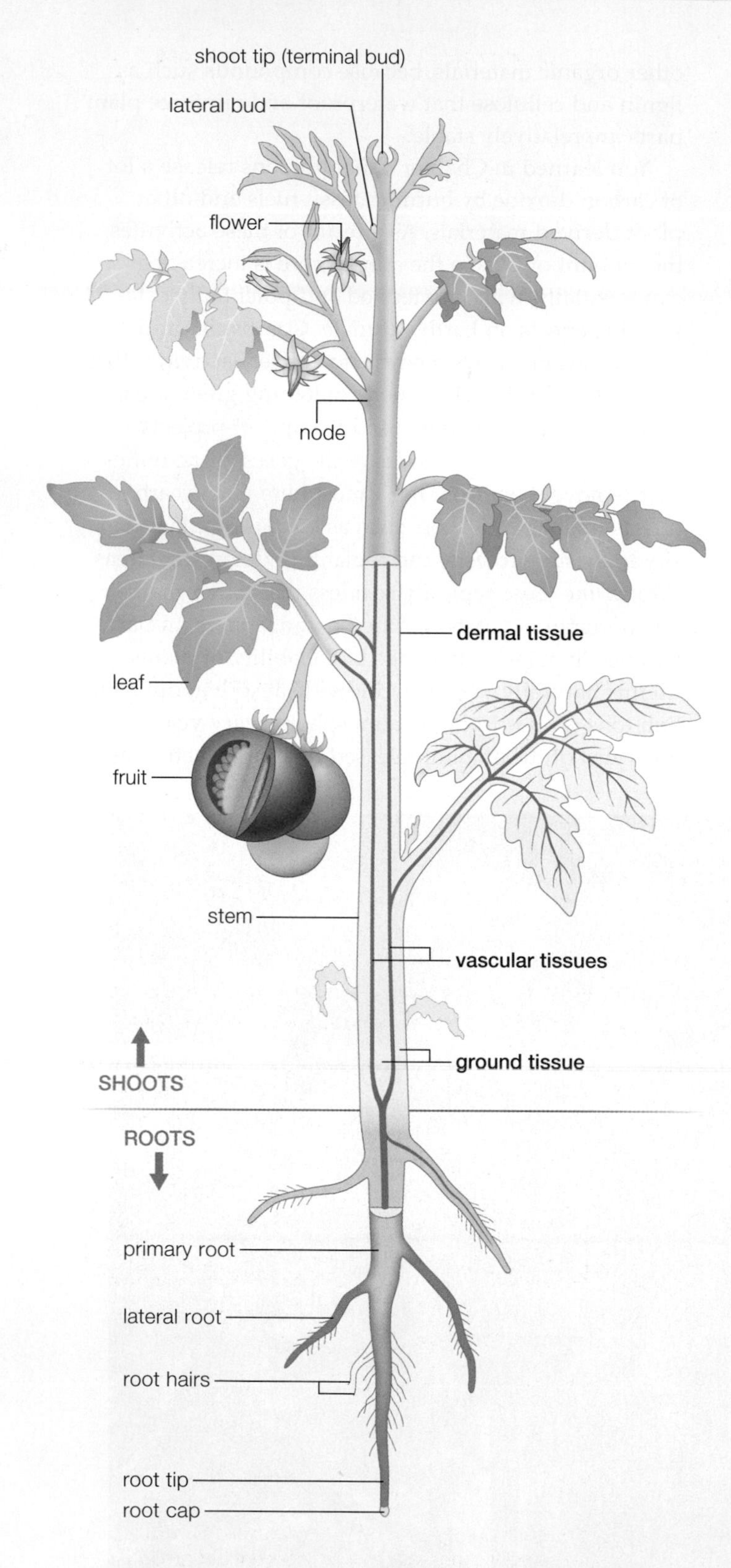

FIGURE 27.2 ▶Animated Body plan of a tomato plant. Dermal tissue covers its surfaces. Vascular tissues (purple) conduct water and solutes. They thread through ground tissue that makes up most of the plant body.

✔ Roots and shoots compose the plant body.

✔ In vascular plants, roots and shoots consist mainly of ground tissue with vascular tissue threading through it. Dermal tissue covers exposed surfaces of the plant.

✔ Monocots and eudicots differ in tissue organization.

With more than 260,000 species (and counting), flowering plants dominate the plant kingdom. Magnoliids, eudicots (true dicots), and monocots (Section 22.9) are the major angiosperm groups. In this chapter, we focus mainly on the structure of eudicots and monocots. Eudicots include flowering shrubs and trees, vines, and many nonwoody plants such as tomatoes and dandelions. Lilies, orchids, grasses, and palms are examples of monocots.

Like cells of most other multicelled organisms, those in plants are organized as tissues, organs, and organ systems (Section 1.2). A vascular plant's body consists of two organ systems: shoots and roots (**FIGURE 27.2**). Most shoots are above the ground and most roots are below it, but as you will see, there are many exceptions to this general rule.

A shoot system includes stems, leaves, and reproductive organs such as flowers. Stems, the structural framework of the shoot system, support the plant's growth above soil and (in many cases) below it. Leaves are specialized to intercept sunlight for photosynthetic production of sugars. Roots absorb water and minerals dissolved in it, and they often serve to anchor the plant in soil. They are as essential as a plant's shoots, and often at least as extensive.

Stems, leaves, flowers, roots—all of these plant parts are composed of ground, vascular, and dermal tissues (**FIGURE 27.3**). **Ground tissues** constitute the bulk of the plant, and include the cells specialized for photosynthesis and for storage. Cells of **vascular tissues** form pipelines that thread through ground tissue and distribute water and nutrients to all parts of the plant body. **Dermal tissues** cover and protect the plant's exposed surfaces. All three types of plant tissues arise during lengthening of young roots and shoots.

Monocots and eudicots have the same types of cells and tissues composing their parts, but the two lineages differ in many aspects of their organization and hence in their structure (**FIGURE 27.4**). The names of the groups refer to one such difference, the number of

cotyledon Seed leaf; embryonic leaf of a flowering plant.
dermal tissues Tissues that cover and protect the plant body.
ground tissues Tissues that make up the bulk of the plant body.
vascular tissues Tissues that distribute water and nutrients through a vascular plant body.

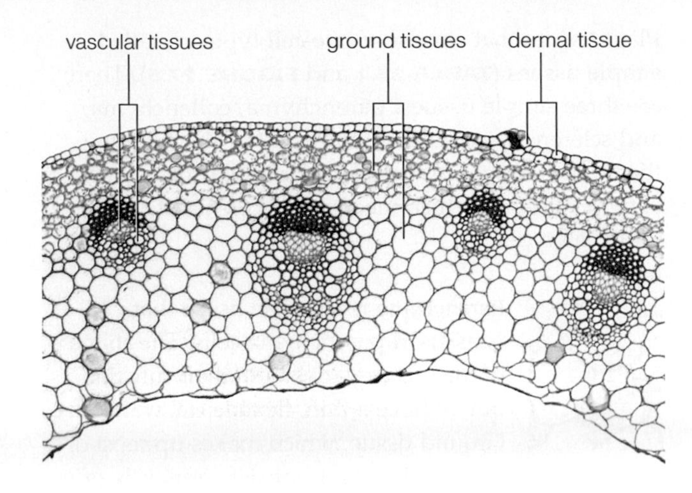

A Vascular, ground, and dermal tissues make up a stem of buttercup (*Ranunculus*).

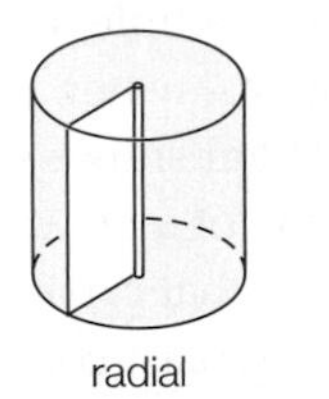

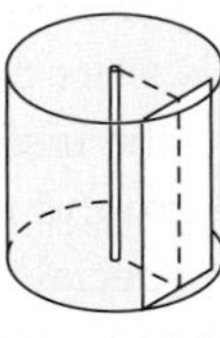

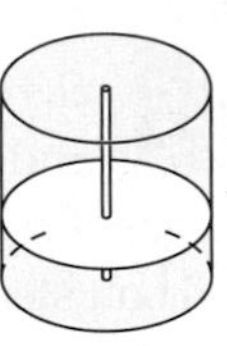

B To simplify interpretation of micrographs, plant parts are typically cut along standard planes. A longitudinal cut along a radius gives a radial section. A cut at right angles to the radius gives a tangential section. A cut perpendicular to the long axis gives a transverse (cross) section.

FIGURE 27.3 Plant organs consist of three types of tissue: dermal, ground, and vascular tissues.

FIGURE IT OUT Along which plane was the section in **A** cut?

Answer: Transverse

seed leaves, or **cotyledons**, in their embryos. In monocots, the embryo has a single cotyledon; in eudicots, it has two (Chapter 29 returns to development in plants). As you will see, the tissue organization of shoots and roots also differs among eudicots and monocots.

TAKE-HOME MESSAGE 27.2

[illegible]

✔ Flowering plants typically have aboveground shoots and belowground roots. Both consist of ground, vascular, and dermal tissue systems.

✔ Ground tissues make up most of a plant. Vascular tissues that thread through ground tissue distribute water and nutrients. Dermal tissues cover and protect plant surfaces.

✔ Eudicots and monocots have the same types of tissues, but differ somewhat in their pattern of tissue organization.

Eudicots	**Monocots**

In seeds, two cotyledons (seed leaves of embryo)	In seeds, one cotyledon (seed leaf of embryo)

Flower parts in fours or fives (or multiples of four or five)	Flower parts in threes (or multiples of three)
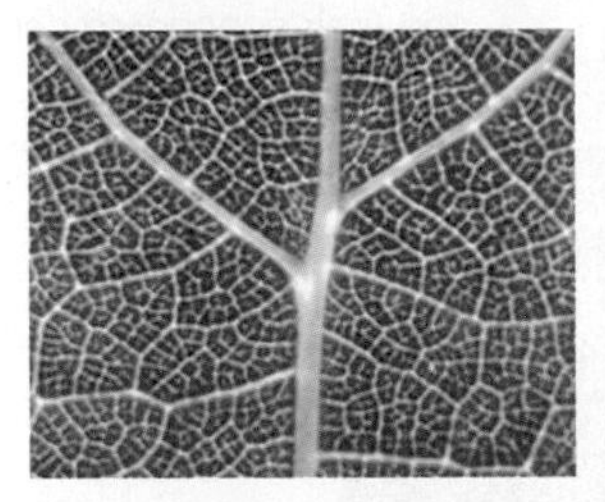	
Leaf veins usually forming a netlike array	Leaf veins usually running parallel with one another
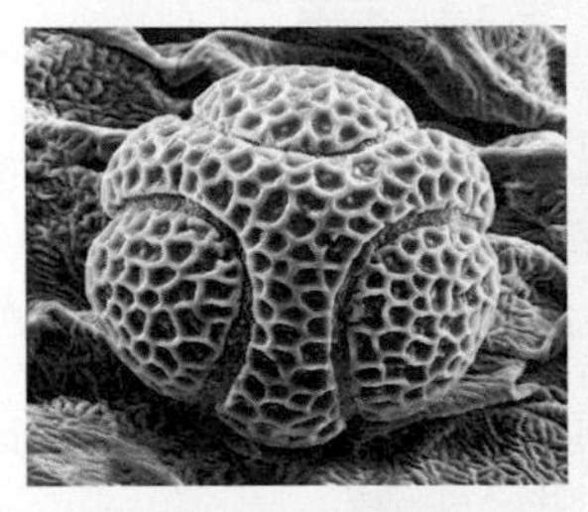	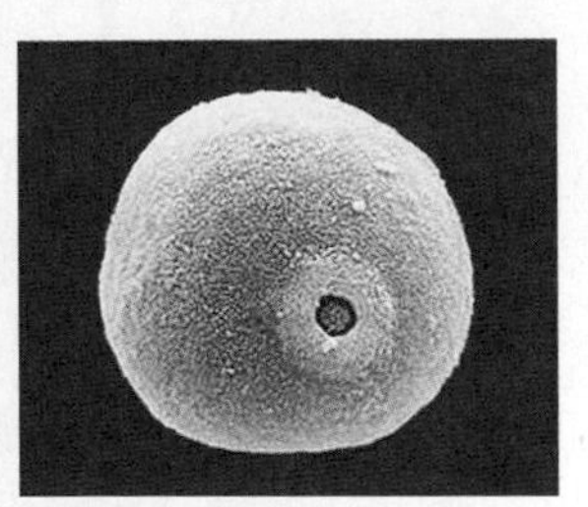
Pollen grains with three pores or furrows	Pollen grains with one pore or furrow
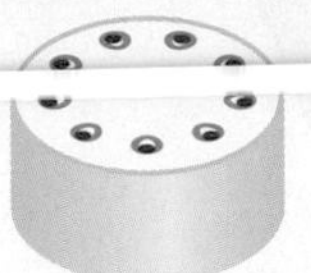	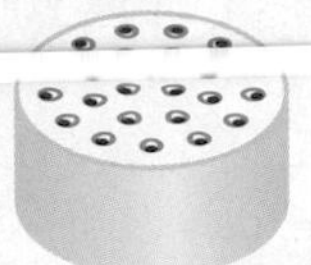
Vascular bundles organized in a ring in ground tissue of stem	Vascular bundles throughout ground tissue of stem

FIGURE 27.4 ▶**Animated** Some of the structural differences between eudicots and monocots.

27.3 Plant Tissues

✔ In plants, simple tissues consist primarily of one type of cell; complex tissues have two or more cell types.

Table 27.1 Simple and Complex Plant Tissues

Tissue	Main Components	Main Functions
Simple Tissues		
Parenchyma	Parenchyma cells	Photosynthesis, storage, secretion, tissue repair
Collenchyma	Collenchyma cells	Pliable structural support
Sclerenchyma	Fibers or sclereids	Structural support
Complex Tissues		
Epidermis (dermal)	Epidermal cells and their secretions	Secretion of cuticle; protection; control of gas exchange and water loss
Periderm (dermal)	Cork cambium; cork cells; parenchyma	Forms protective cover on older stems, roots
Xylem (vascular)	Tracheids; vessel elements; parenchyma cells; sclerenchyma cells	Water-conducting tubes; structural support
Phloem (vascular)	Sieve elements, parenchyma cells; sclerenchyma cells	Sugar-conducting tubes and their supporting cells

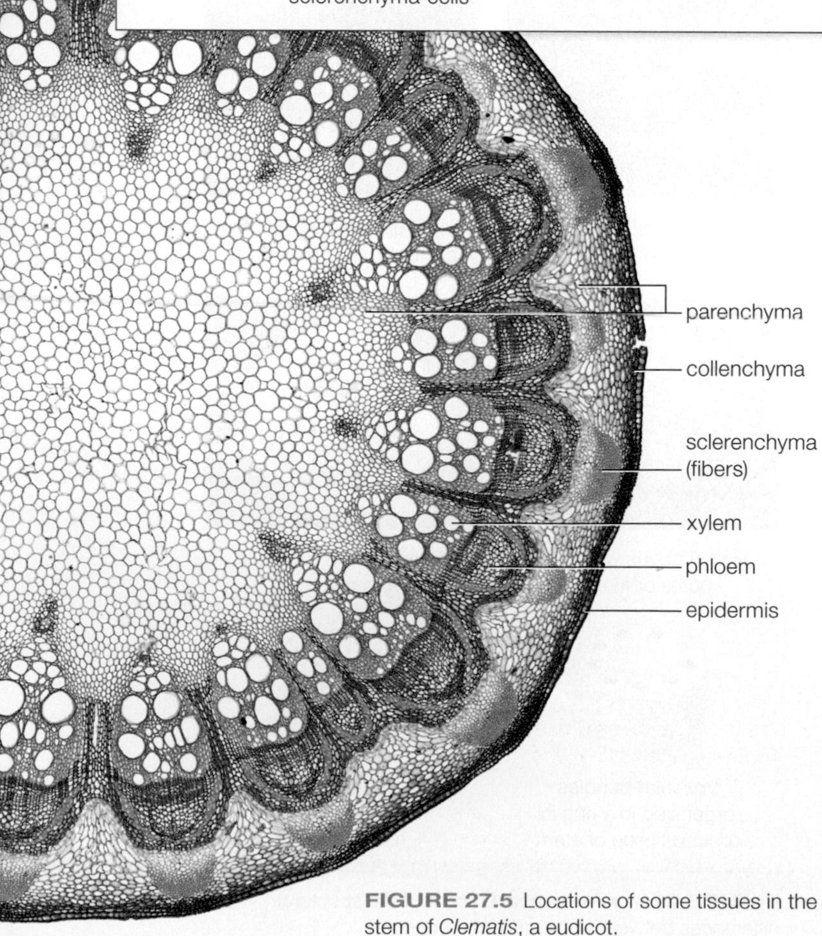

FIGURE 27.5 Locations of some tissues in the stem of *Clematis*, a eudicot.

Plant tissues that consist of one cell type are called simple tissues (**TABLE 27.1** and **FIGURE 27.5**). There are three simple tissues: parenchyma, collenchyma, and sclerenchyma. Each is named after the cell type that makes it up. Epidermal and vascular tissues are complex tissues, which means they consist of two or more cell types.

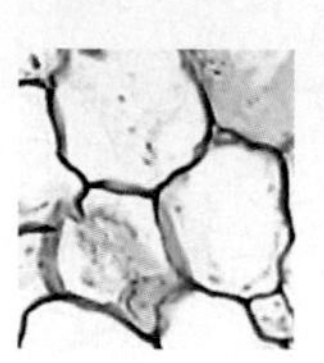

Parenchyma is a simple tissue that consists of parenchyma cells. The shape of these cells varies with their function, but all have a thin, flexible cell wall (left). Ground tissue, which makes up most of the soft internal parts of a plant, consists primarily of parenchyma cells.

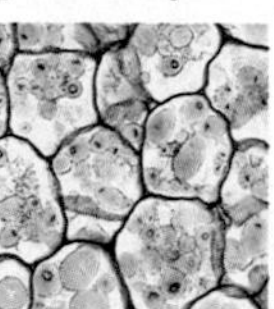

Parenchyma has specialized roles that vary with its location in the plant. In stems and roots, for example, parenchyma cells store proteins, water, oils, and starch (the inset at left shows starch-packed organelles in the ground tissue of a potato). In leaves and soft stems, the special tissue that carries out

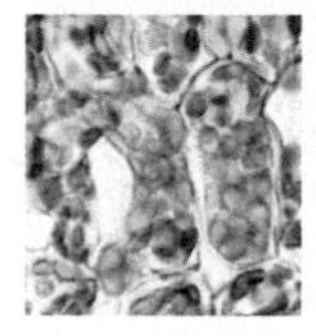

photosynthesis is a chloroplast-containing parenchyma called **mesophyll** (left). Other functions of parenchyma include structural support, nectar secretion, and gas exchange in stems and leaves.

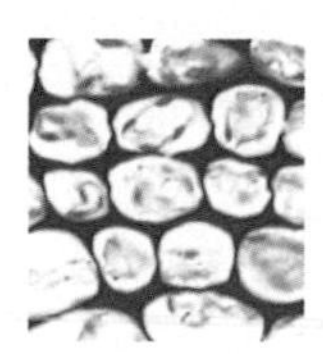

Collenchyma (left) is a simple tissue that supports rapidly growing plant parts such as young stems and leaf stalks. Collenchyma cells stay alive in mature tissue. A complex polysaccharide called pectin imparts flexibility to these cells' primary wall, which is thickened unevenly where three or more of the cells abut one another.

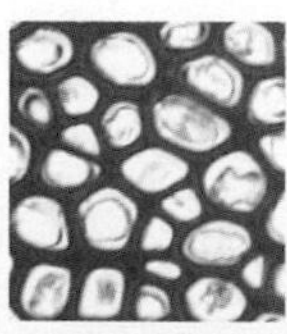

Variably shaped cells of **sclerenchyma** die when they mature, but their thick cell walls remain (left). The walls, which contain a high proportion of durable compounds such as cellulose and lignin, lend sturdiness to plant parts and help them resist stretching and compression. Fibers are long, tapered sclerenchyma cells; they occur in bundles that support and protect vascular tissues in stems and leaves. We

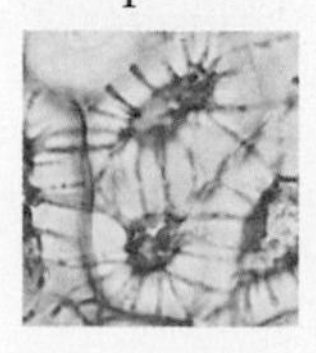

harvest fibers of some plants to make cloth, rope, paper, and other commercial products. Sclereids (left) are sclerenchyma cells that strengthen hard seed coats, and they make pear flesh gritty.

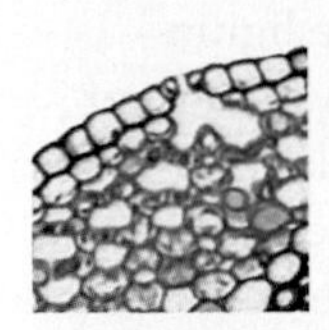

The first dermal tissue to form on a plant is **epidermis**, which, in most species, consists of one layer of cells on the plant's outer surface (left). Epidermal cells secrete waxy substances on their outward-facing cell walls. The deposits form a waterproof cuticle that helps the plant conserve water and repel pathogens. The epidermis of leaves and young stems includes specialized cells and, often, hairs and other epidermal cell outgrowths. Special paired epidermal cells form stomata. Plants control the diffusion of water vapor and gases across epidermis by opening and closing these tiny gaps (Section 6.7). In older stems and roots, a complex dermal tissue called periderm (left) replaces epidermis.

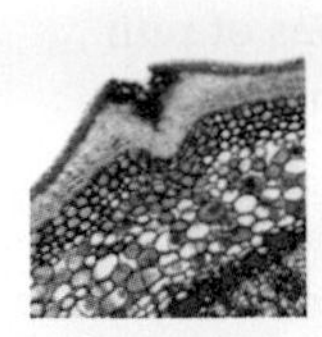

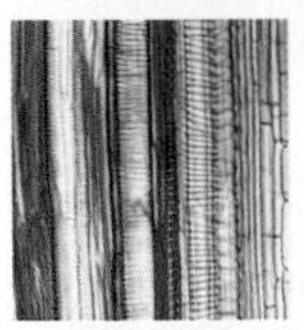

Pipelines of complex vascular tissues (left) distribute water and nutrients through all parts of the plant body. The two types of vascular tissue, xylem and phloem, are typically bundled with sclerenchyma fibers in young plant parts. Both vascular tissues are composed of elongated conducting tubes.

Xylem, the vascular tissue that conducts water and mineral ions, consists of two types of cells: vessel elements (**FIGURE 27.6A**) and tracheids (**FIGURE 27.6B**). Both types of cells are dead in mature tissue, but their stiff, waterproof walls remain. These interconnected walls form tubes (left) that conduct water and also lend structural support to the plant. Water can move vertically through the tubes and laterally between them (Section 28.4 returns to xylem).

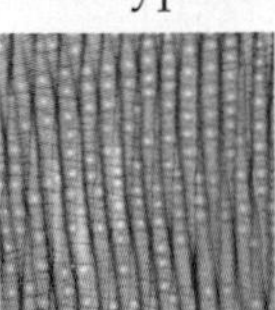

Phloem, the vascular tissue that conducts sugars and other organic solutes, consists of cells called sieve elements and their associated companion cells. Both types of cells are alive in mature tissue. Sieve elements connect end to end at sieve plates (left), forming sieve tubes that conduct sugars to all parts of the plant (**FIGURE 27.6C**). Companion cells are a type of parenchyma. A companion cell provides each sieve element with metabolic support, and also transfers sugars into it (Section 28.6 returns to phloem).

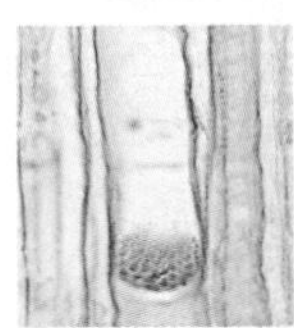

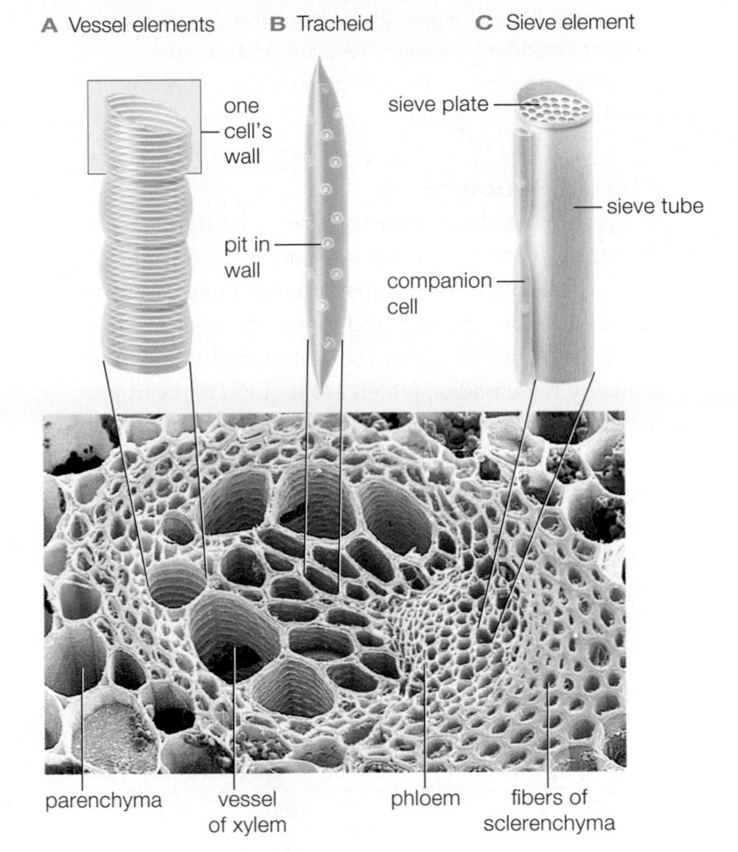

FIGURE 27.6 ▸**Animated** Vascular tissues. Parenchyma cells and fibers of sclerenchyma are also visible in the micrograph.

FIGURE IT OUT What are the green structures inside the parenchyma cells?

Answer: Chloroplasts

collenchyma In plants, simple tissue composed of living cells with unevenly thickened walls; provides flexible support.
epidermis Dermal tissue; outermost layer of a young plant.
mesophyll Photosynthetic parenchyma.
parenchyma In plants, simple tissue composed of living cells that have different functions depending on location; main component of ground tissue.
phloem Complex vascular tissue of plants; its living sieve elements compose sieve tubes that distribute sugars.
sclerenchyma In plants, simple tissue composed of cells that die when they are mature; their lignin-reinforced cell walls remain and structurally support plant parts. Includes fibers, sclereids.
xylem Complex vascular tissue of plants; its dead tracheids and vessel elements distribute water and mineral ions.

TAKE-HOME MESSAGE 27.3

What are the main types of plant tissues?

✔ Dermal tissues (epidermis and periderm) cover and protect plant surfaces.

✔ Cells of parenchyma have diverse roles, including photosynthesis. Collenchyma and sclerenchyma support and strengthen plant parts.

✔ Vascular tissues distribute water and solutes through ground tissue. In xylem, water and ions flow through tubes of dead tracheids and vessel elements. In phloem, sugars flow through sieve tubes that consist of living cells.

27.4 Stems

✔ The organization of ground, vascular, and dermal tissues inside stems differs between monocots and eudicots.

✔ Primary growth in a stem arises by divisions of apical meristem cells in shoot tips.

Internal Structure

Stems form the basic structure for growth of a flowering plant, providing support and keeping leaves positioned for photosynthesis. They can grow above or below the soil, and many species have stems specialized for storage or asexual reproduction. Stems typically have **nodes**, which are regions that can give rise to new shoots or roots.

Inside stems, xylem and phloem are organized as long, multistranded **vascular bundles**. The main function of vascular bundles is to conduct water, ions, and nutrients between different parts of the plant. Some components of the bundles—fibers and the lignin-reinforced walls of tracheids—also play an important role in supporting upright stems of land plants.

Vascular bundles extend through the ground tissue of all stems and leaves, but the arrangement of the bundles inside these plant parts differs between monocots and eudicots. The vascular bundles of monocot stems are typically distributed throughout the ground tissue (**FIGURE 27.7A**). By contrast, all of the vascular bundles inside a typical eudicot stem are arranged in a characteristic ring (**FIGURE 27.7B**). The ring divides the stem's ground tissue into distinct regions of pith (inside the ring) and cortex (outside the ring).

Variations on a Stem

Many plants have modified stem structures that function in storage and reproduction.

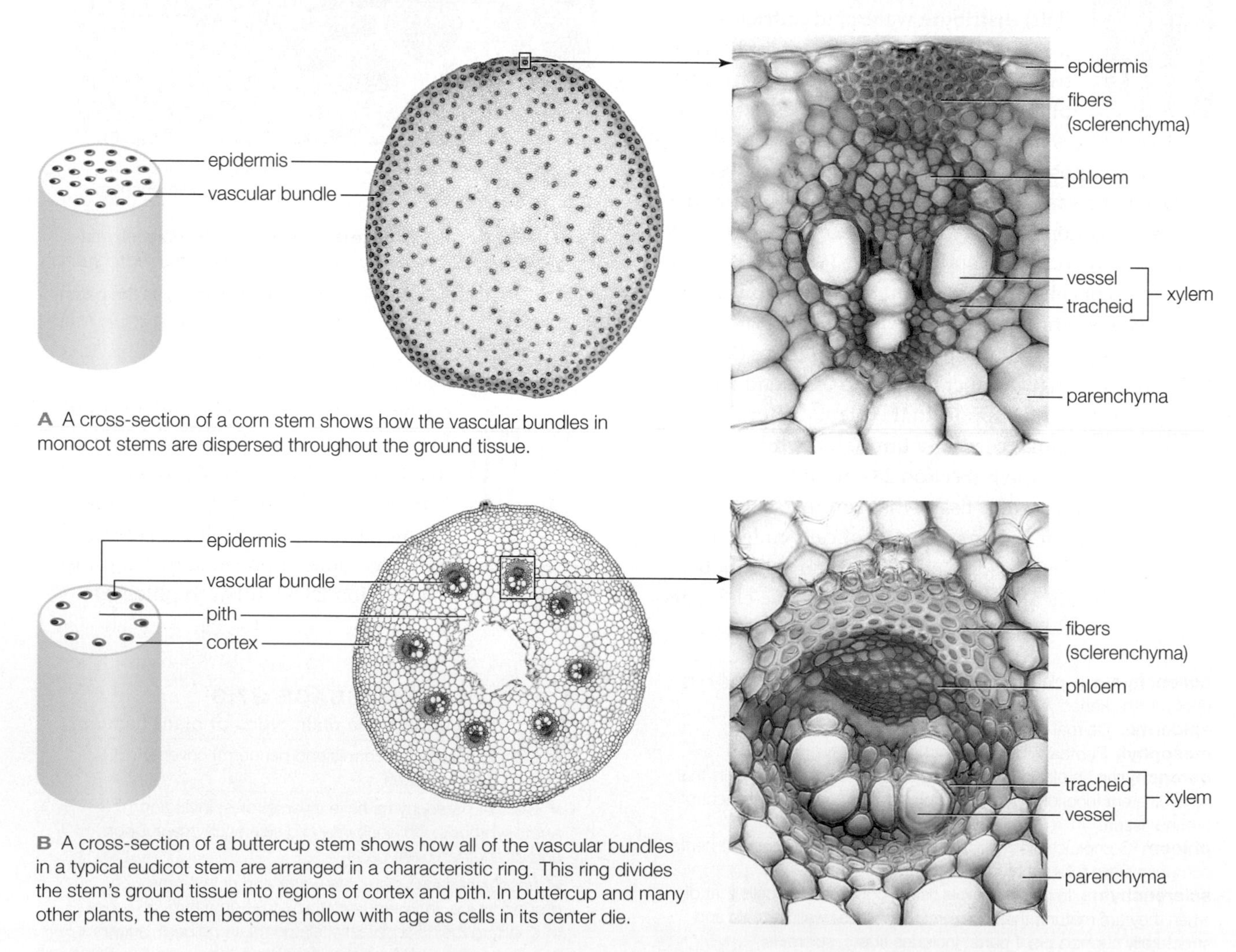

A A cross-section of a corn stem shows how the vascular bundles in monocot stems are dispersed throughout the ground tissue.

B A cross-section of a buttercup stem shows how all of the vascular bundles in a typical eudicot stem are arranged in a characteristic ring. This ring divides the stem's ground tissue into regions of cortex and pith. In buttercup and many other plants, the stem becomes hollow with age as cells in its center die.

FIGURE 27.7 Comparing the structure of a young shoot from (**A**) a monocot and (**B**) a eudicot. Cells appear different colors in the sections because they have been stained with different dyes that bind to certain carbohydrates and lignin.

Stolons Stolons are stems that branch from the main stem of the plant and grow horizontally along the surface of the ground or just under it. Stolons may look like roots, but they have nodes (roots do not have nodes). Roots and leafy shoots that sprout from the nodes develop into new plants. Stolons are commonly called runners because in many plants they "run" along the surface of the soil. The strawberry plant (*Fragaria*) in the photo above is reproducing asexually by sending out runners.

Bulbs A bulb is a short section of underground stem encased by overlapping layers of thickened, modified

leaves called scales. The photo (left) shows clearly visible scales surrounding the stem at the center of an onion, which is the bulb of an *Allium cepa* plant. The scales develop from a basal plate at the base of the bulb, as do roots. Bulb scales contain starch and other substances that a plant holds in reserve during times when conditions in the environment are unfavorable for growth. When favorable conditions return, the plant then uses these stored substances to sustain rapid growth. The dry, paperlike outer scale of an onion and many other bulbs serves as a protective covering.

Corms A corm is a short, thickened underground stem. A corm is like a bulb in that it stores nutrients for times when conditions in the environment are unfavorable for growth, and roots grow from its basal plate. Unlike a bulb, however, a corm is solid rather than layered, and it has nodes. Taro, also known as arrowroot, is the corm of *Colocasia esculenta* plants (above).

node A region of stem where new shoots and roots can form.
vascular bundle In a stem or leaf, multistranded bundle formed by xylem, phloem, and sclerenchyma fibers.

Rhizomes Ginger, irises, and some grasses have rhizomes, which are fleshy stems that typically grow under the soil and parallel to its surface. A rhizome is the main stem of the plant, and it also serves as the plant's primary storage tissue. Shoots that sprout from nodes grow aboveground for photosynthesis and flowering. The photo on the left shows shoots sprouting from rhizomes of turmeric plants (*Curcuma longa*).

Stem Tubers Stem tubers are thick, fleshy storage structures that form on the stolons or rhizomes of some plant species. Most are underground and temporary. Like corms, stem tubers have nodes from which new shoots and roots can sprout. Unlike corms, they do not have a basal plate. The photo (left) shows how potatoes, which are stem tubers, grow on stolons of *Solanum tuberosum* plants. The "eyes" of a potato are nodes.

Cladodes Many types of cacti and other succulents have cladodes, which are flattened, photosynthetic stems that store water. New plants form at the nodes. The cladodes of some plants appear leaflike, but most are unmistakably fleshy. The photo (above) shows a spiky cladode of a prickly pear plant (*Opuntia*).

TAKE-HOME MESSAGE 27.4

How are plant tissues organized inside stems?

✔ The arrangement of vascular bundles, which are multistranded bundles of vascular tissue, differs between eudicot and monocot stems.

✔ Many plants have modified stems that function in storage and reproduction. Stolons, rhizomes, bulbs, corms, stem tubers, and cladodes are examples.

CREDITS: (in text left column) top, © Anthony Tripodi, www.thecompostbin.com; middle, © iStockphoto.com/mjutabor; bottom, © Eric Sueyoshi/Audrey Magazine; (in text right column) top, © Dinodia Photo Library/Botanica/Getty Images; middle, Chase Studio/Science Source; bottom, © Chris Hellier/Corbis.

27.5 Leaves

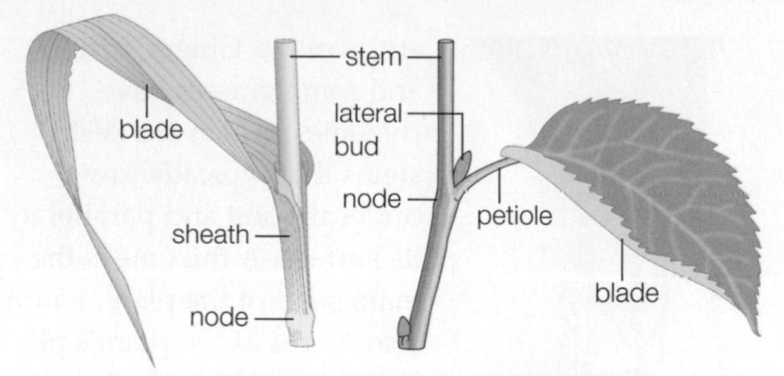

A Typical leaves of monocots (left) and eudicots (right).

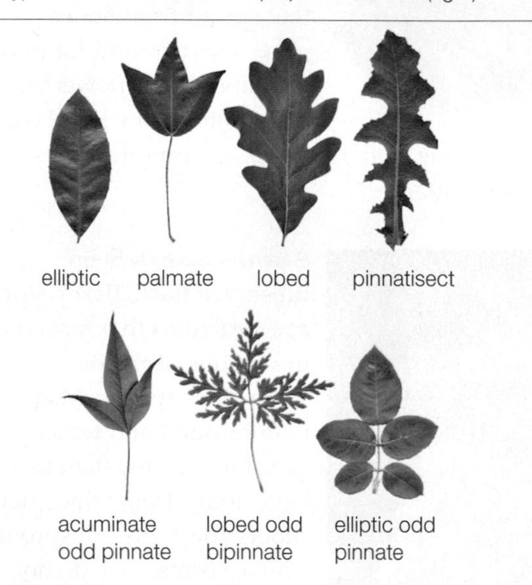

B Examples of common leaf forms in eudicots. Top row: simple leaves; bottom row, compound leaves.

C Example of specialized leaf form. Carnivorous plants of the genus *Nepenthes* grow in nitrogen-poor soil. They secrete acids and protein-digesting enzymes into fluid in a cup-shaped modified leaf. The enzymes release nitrogen from small animals (such as insects, frogs, and mice) that are attracted to odors from the fluid and then drown in it.

D Example of leaf epidermis specialization: trichomes, which are outgrowths of epidermal cells. Glandular types such as the ones on the surface of this marijuana leaf secrete substances that deter plant-eating animals. Marijuana trichomes produce a chemical (tetrahydrocannabinol, or THC) that has a psychoactive effect in humans.

FIGURE 27.8 Leaf structure.

✔ Most leaves are metabolic factories where photosynthetic cells churn out sugars. They vary in size, shape, surface specializations, and internal structure.

Leaves are the main organs of photosynthesis in most flowering plant species. They also function in gas exchange, and they are the major site of evaporative water loss. Typical leaves are thin, with a high surface-to-volume ratio. Leaves of most monocots are also long and narrow, with the base of the leaf wrapping around the stem and forming a sheath around it (**FIGURE 27.8A**). In most eudicots, a short stalk called a petiole attaches the leaf to a stem. Leaf structure varies widely among eudicot species (**FIGURE 27.8B**). Simple leaves are undivided, although the margins of many are lobed, serrated, or otherwise contoured. Compound leaves are divided into leaflets. Some eudicot leaves have a highly specialized form that permits a special function (**FIGURE 27.8C**).

Internal Structure

FIGURE 27.9 illustrates the internal structure of a typical eudicot leaf. The bulk of the leaf consists of mesophyll, which is photosynthetic parenchyma with air spaces between cells. Eudicot leaves are typically oriented perpendicular to the sun's rays, and have two layers of mesophyll. The uppermost layer is palisade mesophyll ❶. Cells in this layer have more chloroplasts than cells of the spongy mesophyll layer below. Cells of spongy mesophyll are more rounded than those of palisade mesophyll, and have larger air spaces between them ❷.

The vascular bundles of leaves are called **leaf veins**. Inside each vein, strands of xylem transport water and dissolved mineral ions to photosynthetic cells, while strands of phloem transport products of photosynthesis (sugars) away from them ❸. Layers of sclerenchyma around the vascular tissue stiffen a vein, and also provide support for the leaf's softer tissues. In most eudicot leaves, large veins branch into a network of minor veins.

The outermost tissue of a leaf is a sheet of epidermis one cell thick ❹. This surface tissue may be smooth, sticky, or slimy. Outgrowths of epidermal cells can form hairs, scales, spikes, hooks, and other special structures (**FIGURE 27.8D**).

Epidermal cells secrete a translucent, waxy cuticle that slows water loss. A leaf's upper surface, which typically receives the most direct sunlight, may have a thicker cuticle than the lower surface, which tends to be shaded. The lower surface usually has more

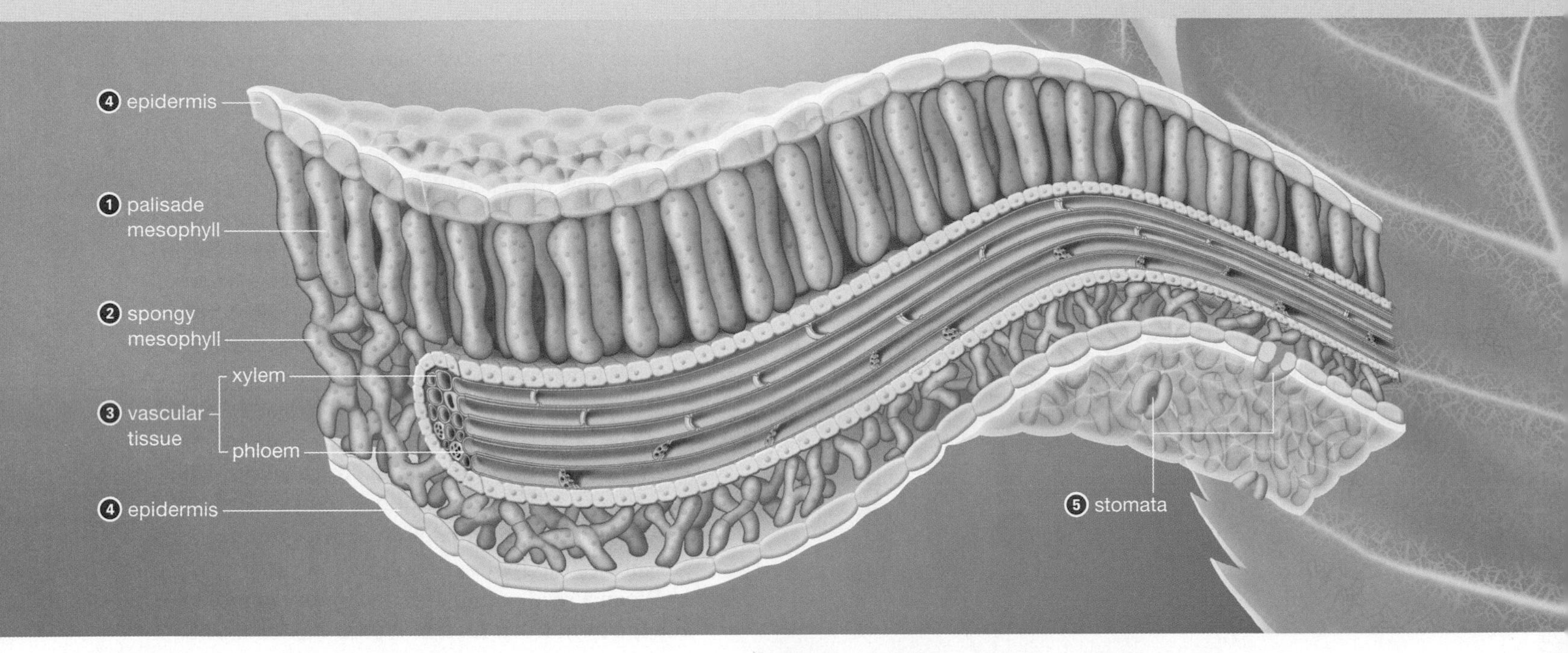

FIGURE 27.9 Anatomy of a eudicot leaf.

The bulk of the leaf is mesophyll, a type of photosynthetic parenchyma. Eudicot leaves often have two distinct layers of this tissue: palisade mesophyll with elongated cells ❶, and spongy mesophyll with rounded cells ❷.

❸ Inside leaf veins, vascular bundles of xylem (blue) and phloem (pink) transport materials to and from photosynthetic cells.

❹ The leaf surface is epidermis with a secreted layer of cuticle.

❺ Gas exchanges between air inside and outside of the leaf occur at stomata.

stomata ❺ (Section 6.7). Guard cells on either side of each stoma may contain chloroplasts, but these are the only photosynthetic cells in leaf epidermis. As you will see in Section 28.5, changes in the shape of guard cells close stomata to prevent water loss, or open stomata to allow gases to cross the epidermis. Carbon dioxide needed for photosynthesis enters a leaf through stomata, then diffuses through air spaces to mesophyll cells. Oxygen released by photosynthesis diffuses in the opposite direction.

Monocot and eudicot leaves have a somewhat different internal structure. Blades of grass and other monocot leaves that grow vertically can intercept light from all directions. Such leaves typically have a single layer of mesophyll (**FIGURE 27.10**). Unlike a eudicot's branching veins, monocot veins are similar in length and run parallel with the leaf's long axis (as shown in **FIGURE 27.4**).

leaf vein A vascular bundle in a leaf.

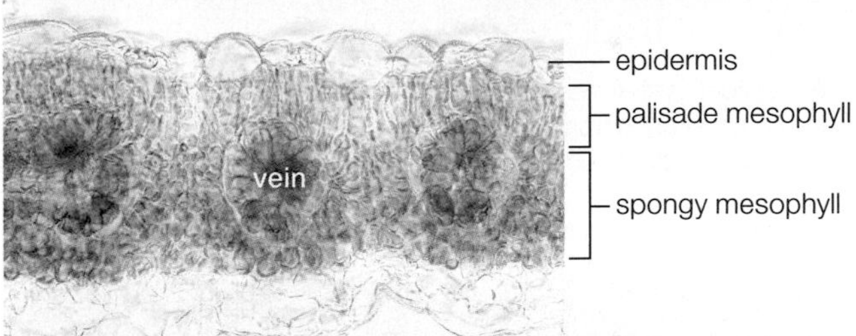

A Leaf section of coastal plain yellowtops (*Flaveria bidentis*), a eudicot.

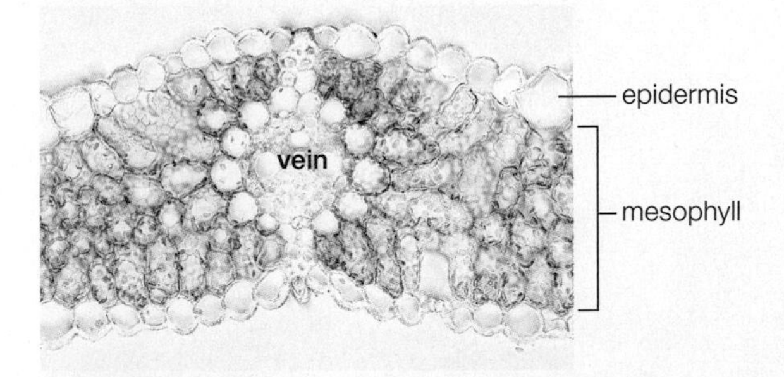

B Leaf section of barley (*Hordeum vulgare*), a monocot.

FIGURE 27.10 Comparing the arrangement of cells and tissues in monocot and eudicot leaves.

TAKE-HOME MESSAGE 27.5

How does a leaf's structure contribute to its function?

✔ Leaves are structurally adapted to intercept sunlight and exchange gases for photosynthesis.

✔ Leaf components include veins (bundles of vascular tissue), mesophyll (photosynthetic cells), and cuticle-secreting epidermis.

27.6 Roots

✔ Roots provide large surface area for absorbing water and dissolved mineral ions from soil.

A Fibrous root systems have similar-sized adventitious roots that branch from nodes on the stem. Grasses and other typical monocots have a fibrous root system.

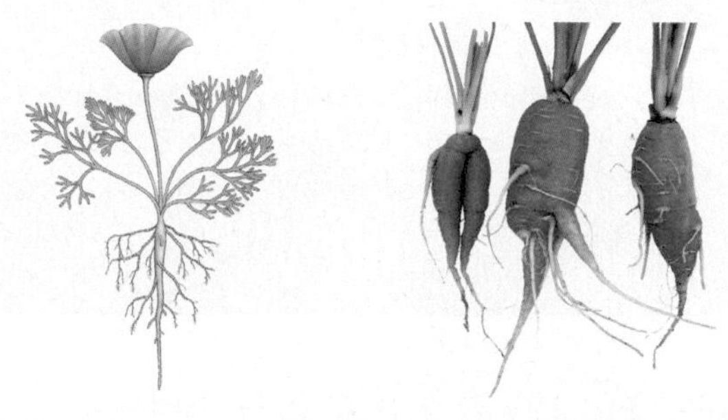

B Taproot systems consist of a large main root together with lateral roots that branch from it. Carrots are the taproots of one variety of *Daucus carota*, a eudicot.

FIGURE 27.11 ▶**Animated** Root systems.

FIGURE 27.12 Root hairs on a primary root emerging from a cabbage seed.

The main function of a root is to take up water and mineral ions from soil. Roots also anchor a plant in soil, and in many species they store nutrients. The roots of a typical plant are as extensive as its shoots, and often more so. For example, the root system of a single young rye plant, if laid out as a sheet, would cover about 600 square meters, or about 6,500 square feet! An extensive root system provides the plant with a very large surface area for absorbing soil water.

Roots branch from existing roots. They can also form on stems, leaves, or other structures, in which case they are called adventitious roots. Adventitious roots that arise from nodes on above-ground stems are called prop roots.

When a seed germinates, the first structure to emerge from it is a root. In typical monocots, this primary root is quickly replaced with a mat of adventitious roots that arise from nodes on the developing stem. All of the roots in the resulting **fibrous root system** are about equal in diameter—there is no dominant, central root (**FIGURE 27.11A**). Roots in a fibrous system may or may not be branched.

In typical eudicots, the primary root that emerges from a seed thickens and grows longer to become a taproot. A taproot typically gives rise to other, lateral-branching roots as it lengthens, but even so it remains the largest, dominant root. A eudicot taproot together with its smaller lateral root branchings constitutes a **taproot system** (**FIGURE 27.11B**). In some eudicots, the taproot is eventually replaced by adventitious roots, so a taproot system becomes a fibrous root system.

Dermal tissue on the surface of roots is the plant's absorptive interface with soil. Epidermal cells on young roots often send out very thin extensions called **root hairs** (**FIGURE 27.12**). These structures are tiny, but a huge number of them form. Collectively, they greatly increase the root's surface area.

Internal Structure

Water and substances dissolved in it enter a root by crossing epidermal cells and seeping into parenchyma that forms the root's region of cortex (**FIGURE 27.13**). The water moves from cell to cell of the cortex until it reaches a column of vascular tissue called the **vascular cylinder**, or **stele**, that runs lengthwise through the center of each root. The outer boundary of the vascular cylinder is a layer of cells called **endodermis**. Cells of the endodermis help control which solutes are taken

CREDITS: (11A) right, Don Nichols/Getty Images; (11B) right, Caitlin Winner/Getty Images; (11A, B) left, © Cengage Learning; (12) Nigel Cattlin/Science Source.

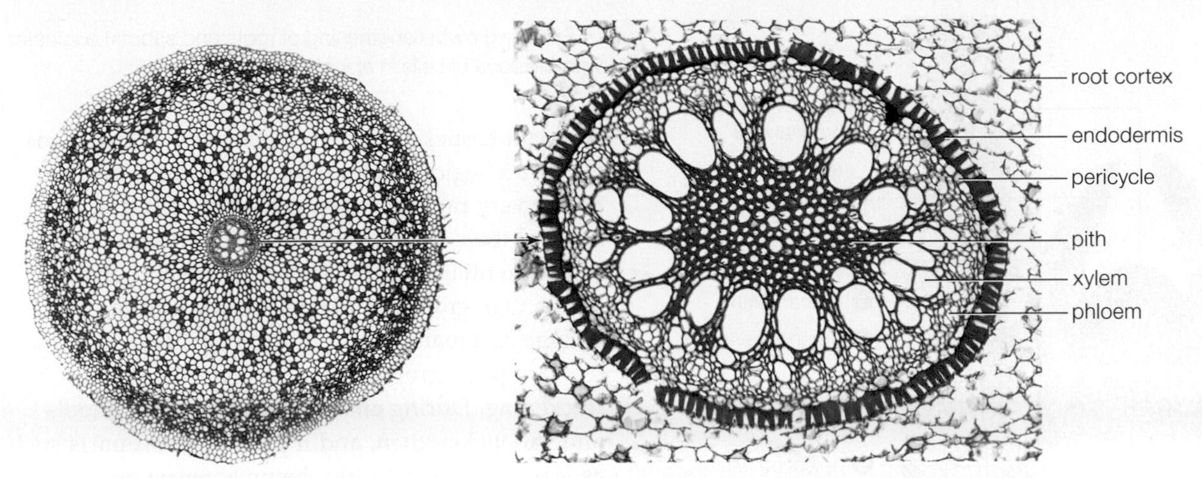

A Vascular cylinder in a root of iris (*Iris*), a monocot. In this and other typical monocots, a substantial portion of the vascular cylinder is pith.

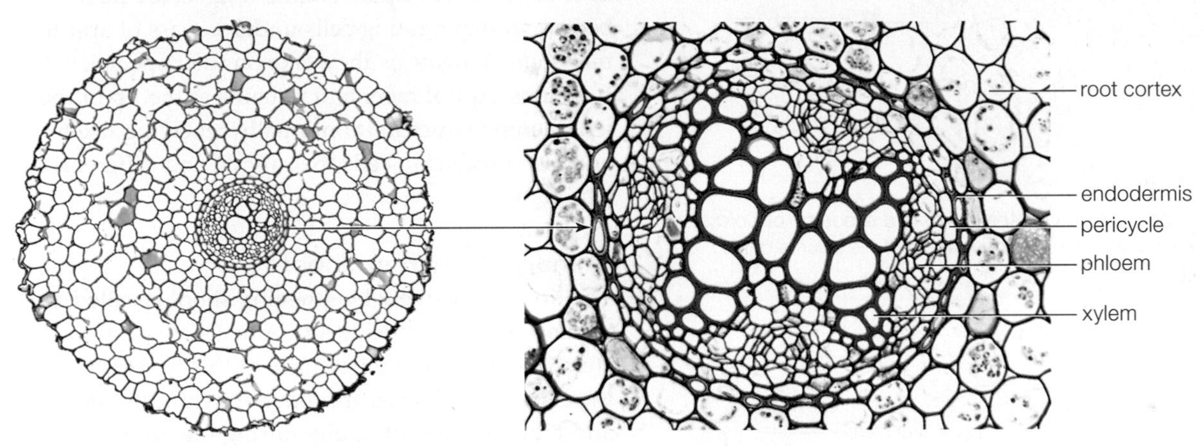

B Vascular cylinder in a root of buttercup (*Ranunculus*), a eudicot. In this and other typical eudicots, the vascular cylinder consists mainly of vascular tissue (xylem and phloem).

FIGURE 27.13 Zooming in on the arrangement of tissues inside monocot and eudicot vascular cylinders.

into the plant's vascular system (Section 28.3 returns to this topic). Just inside the endodermis is **pericycle**, a layer of specialized parenchyma one or two cells thick. These cells retain the capacity to divide and differentiate into other cell types, so they can give rise to several different structures. Lateral roots arise from divisions of pericycle cells.

In a typical monocot root, the vascular cylinder divides ground tissue into regions of pith (inside the ring) and cortex (outside the ring). By contrast, the root of a typical eudicot root has cortex, but the vascular cylinder contains very little pith or none at all. In these plants, the bulk of the vascular cylinder is vascular tissue—xylem and phloem.

endodermis Outer layer of the vascular cylinder in a plant root; sheet of cells just outside the pericycle.
fibrous root system Root system composed of an extensive mass of similar-sized adventitious roots; typical of monocots.
pericycle Layer of cells just inside root endodermis; can give rise to lateral roots.
root hairs Hairlike, absorptive extensions of a root epidermis cell; form on young roots.
taproot system In eudicots, an enlarged primary root together with all of the lateral roots that branch from it.
vascular cylinder (stele) Of a root, central column that contains vascular tissue, endodermis, pericycle, and other supporting cells.

TAKE-HOME MESSAGE 27.6

What is the basic structure of plant roots?

✔ Roots absorb water and dissolved minerals from soil for distribution to the rest of the plant.

✔ Taproot systems consist of an enlarged primary root and lateral roots that branch from it. Fibrous root systems consist of adventitious roots that replace the primary root.

✔ A vascular cylinder that runs lengthwise through each root contains pipelines of xylem and phloem.

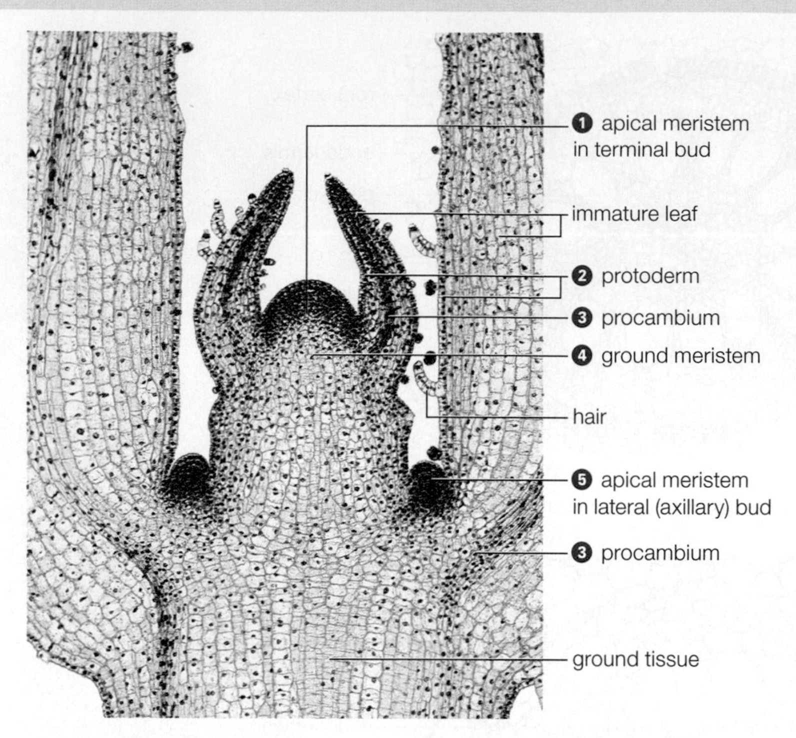

FIGURE 27.14 A longitudinal cut through the center of a shoot top of *Coleus*, a eudicot. Cell nuclei are stained red (nuclei take up most of the volume of the smaller cells in the darker regions).

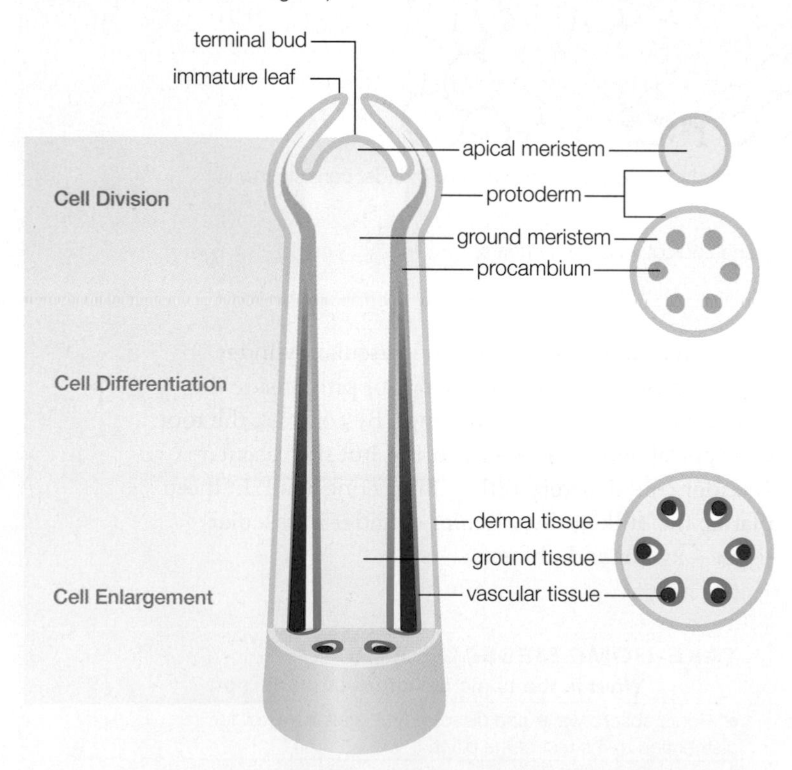

FIGURE 27.15 Primary growth in a shoot. New tissues arise by divisions of undifferentiated cells in shoot apical meristem. Some of the dividing cells are pushed to the periphery of the apical meristem, where they begin to differentiate as protoderm, procambium, and ground meristem. Cells in these primary meristem tissues divide and fully differentiate into dermal, vascular, and ground tissues, respectively. Regions of cell division, differentiation, and enlargement are indicated.

✔ Primary growth (lengthening of roots and shoots) originates with divisions of cells in apical meristems.

All plant tissues arise from the activity of **meristems**, which are regions of undifferentiated cells that can divide very rapidly. When meristem cells divide, some of their descendants remain meristematic (undifferentiated); others differentiate and give rise to vascular, ground, and epidermal tissues as they enlarge and mature.

Plant parts grow both by lengthening and by thickening. During **primary growth**, a plant's roots and shoots lengthen, and it produces soft parts such as leaves. These activities begin at **apical meristems** in shoot and root tips. Lengthening occurs mainly because differentiating cells under a mass of apical meristem elongate as they take on their specialized functions. Apical meristem remains at the tip of the lengthening structure, continually renewing itself and also producing cells that divide and differentiate behind it.

Primary Growth in Shoots

The tip of an actively lengthening shoot is called a **terminal bud** (**FIGURE 27.14**). A mass of apical meristem lies just below the surface of this bud ❶. Cells at the center of the mass are completely undifferentiated, and they divide continually during the growing season. The divisions push some of the cells toward the periphery of the mass. Cells that are displaced away from the center of the meristem begin to differentiate. Depending on its location in relationship to the bud's surface and to other cells, the cell will become a component of one of three primary meristem tissues. Cells that reach the bud's outer surface form protoderm ❷; those in its interior form procambium ❸ and ground meristem ❹. Further divisions and differentiation of cells in these primary meristems give rise to mature tissues: protoderm, to dermal tissue; procambium, to vascular tissue; ground meristem, to ground tissue (**FIGURE 27.15**).

As the shoot lengthens, masses of tissue near the sides of the apical meristem in the terminal bud bulge

apical meristem Meristem in the tip of a shoot or root; gives rise to primary growth (lengthening).
lateral bud Axillary bud; forms in a leaf axil.
meristem Zone of undifferentiated plant cells; all plant growth arises from divisions of meristem cells.
primary growth Lengthening of young shoots and roots; originates at apical meristems.
terminal bud Tip of an actively growing shoot; contains apical meristem.

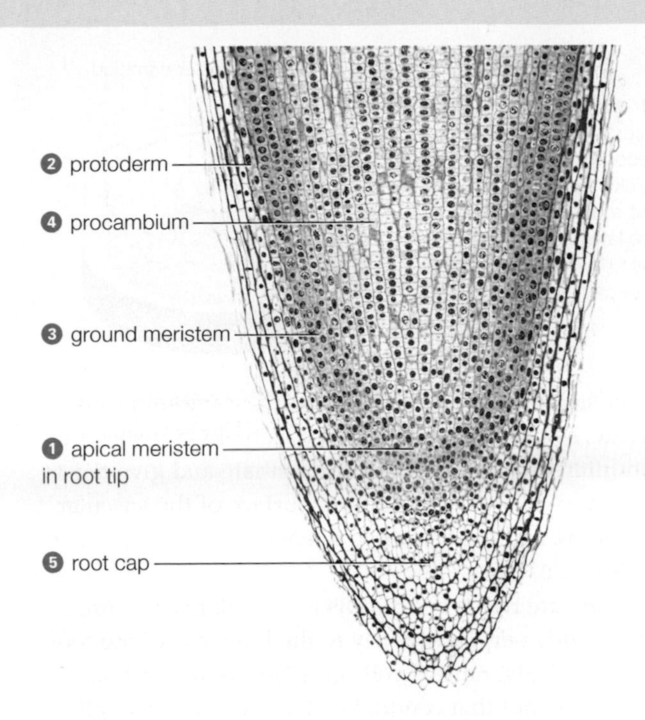

A A longitudinal cut through the center of a root tip of onion (*Allium*), a monocot. Labels indicate where procambium is giving rise to the vascular cylinder; protoderm, to the epidermis; ground meristem, to the root cortex.

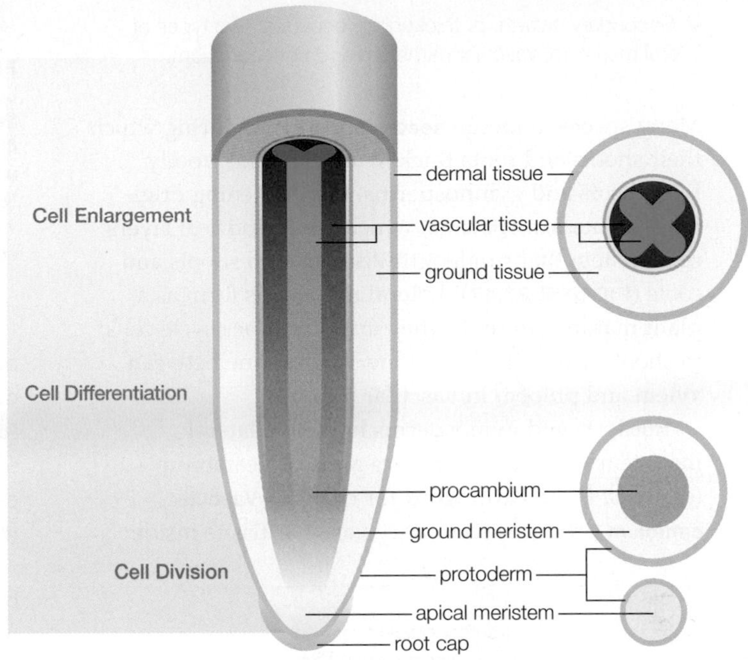

B Dividing cells of root apical meristem give rise to protoderm, ground meristem, and procambium, which differentiate into dermal tissue, ground tissue, and vascular tissue. Regions of cell division, differentiation, and enlargement are indicated.

FIGURE 27.16 Primary growth in a root. New tissues arise by divisions of undifferentiated cells in root apical meristem. Some of the dividing cells are pushed to the periphery of the apical meristem, where they begin to differentiate as protoderm, procambium, and ground meristem. Cells in these primary meristem tissues divide and fully differentiate into dermal, vascular, and ground tissues, respectively.

out and then develop into leaves. Leaves that develop nearest the tip of a lateral bud may develop into bud scales, which are tiny, modified leaves that protect the bud as the shoot lengthens behind it. Other leaves form and mature in orderly tiers, one after the next, along the lengthening stem. As they do, **lateral buds** ❺ (also called axillary buds) form in the leaf axils, which are places where the leaves attach to the stem. Lateral buds also contain apical meristem cells. Hormonal signals can trigger these cells to divide and give rise to a shoot—either a stem, a leaf, or a flower—that branches laterally from the main stem.

Primary Growth in Roots

As in stems, primary growth in roots originates at apical meristems (**FIGURE 27.16**). Cells in the apical meristem of a root tip divide repeatedly ❶, giving rise to primary meristem tissues: protoderm ❷, ground meristem ❸, and procambium ❹. Cells in these three tissues differentiate and enlarge to form dermal, ground, and vascular tissues, respectively. Some protoderm cells give rise to a root cap ❺, a dome-shaped mass of cells that protects the soft, young root as it grows through soil. Cells of the root cap are shed continually as the root lengthens.

Unlike shoots, roots have no nodes. Lateral roots that form on an existing root originate with pericycle cells in the root's vascular cylinder. Pericycle cells divide in a direction perpendicular to the longitudinal axis of the root. Further differentiation of these cells gives rise to a lateral root. A lateral root arising from pericycle in a root of corn (*Zea mays*) is shown at right.

TAKE-HOME MESSAGE 27.7

What is primary growth?

✔ Primary growth (lengthening) occurs at apical meristems in shoot and root tips.

CREDITS: (16A) © Ed Reschke/Photolibrary/Getty Images; (16B) © Cengage Learning; (in text) Michael Clayton/University of Wisconsin, Department of Botany.

27.8 Secondary Growth

✔ Secondary growth, or thickening, occurs at two types of lateral meristem, vascular cambium and cork cambium.

Many species undergo **secondary growth**, during which their shoots and roots thicken and become woody. In eudicots and gymnosperms, the thickening originates at **lateral meristems**, which are cylindrical layers of meristem that run lengthwise through shoots and roots (**FIGURE 27.17**). Lateral meristems form as a plant matures. In roots, they arise from pericycle cells; in shoots, from tiny bits of meristem tissue between xylem and phloem in vascular bundles.

Eudicots and gymosperms have two lateral meristems: vascular cambium and cork cambium (*cambium* is the Latin word for change). **Vascular cambium** produces secondary vascular tissue inside older stems and roots. When vascular cambium cells divide, some of their descendants remain meristematic (undifferentiated). The rest differentiate and give rise to secondary phloem on the outer surface of the vascular cambium, and to secondary xylem on the inner surface (**FIGURE 27.18**).

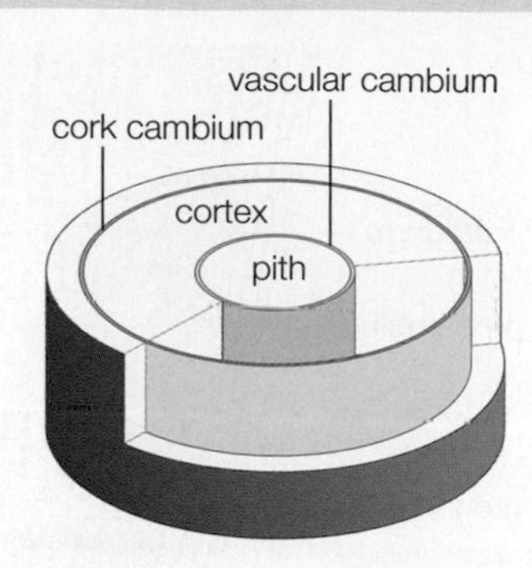

FIGURE 27.17 Lateral meristems. Secondary growth (thickening of older stems and roots) originates at two lateral meristems, vascular cambium and cork cambium.

There are two types of cells in vascular cambium; both divide perpendicularly to the long axis of the root or shoot. Long, narrow cells give rise to the long vascular pipelines that conduct water and solutes lengthwise through the entire structure. Small, rounded cells give rise to "rays" of parenchyma, radially oriented like spokes of a bicycle wheel. Secondary xylem and phloem in these rays conduct water and solutes laterally through the structure.

Thin-walled, living parenchyma cells and sieve tubes of secondary phloem lie in a narrow zone outside the vascular cambium. This secondary phloem often has bands of thick-walled reinforcing sclerenchyma fibers interspersed through it.

Secondary xylem that has accumulated inside a cylinder of vascular cambium is called **wood**. Oak, hickory, and other long-lived angiosperm species are called hardwoods because their secondary xylem

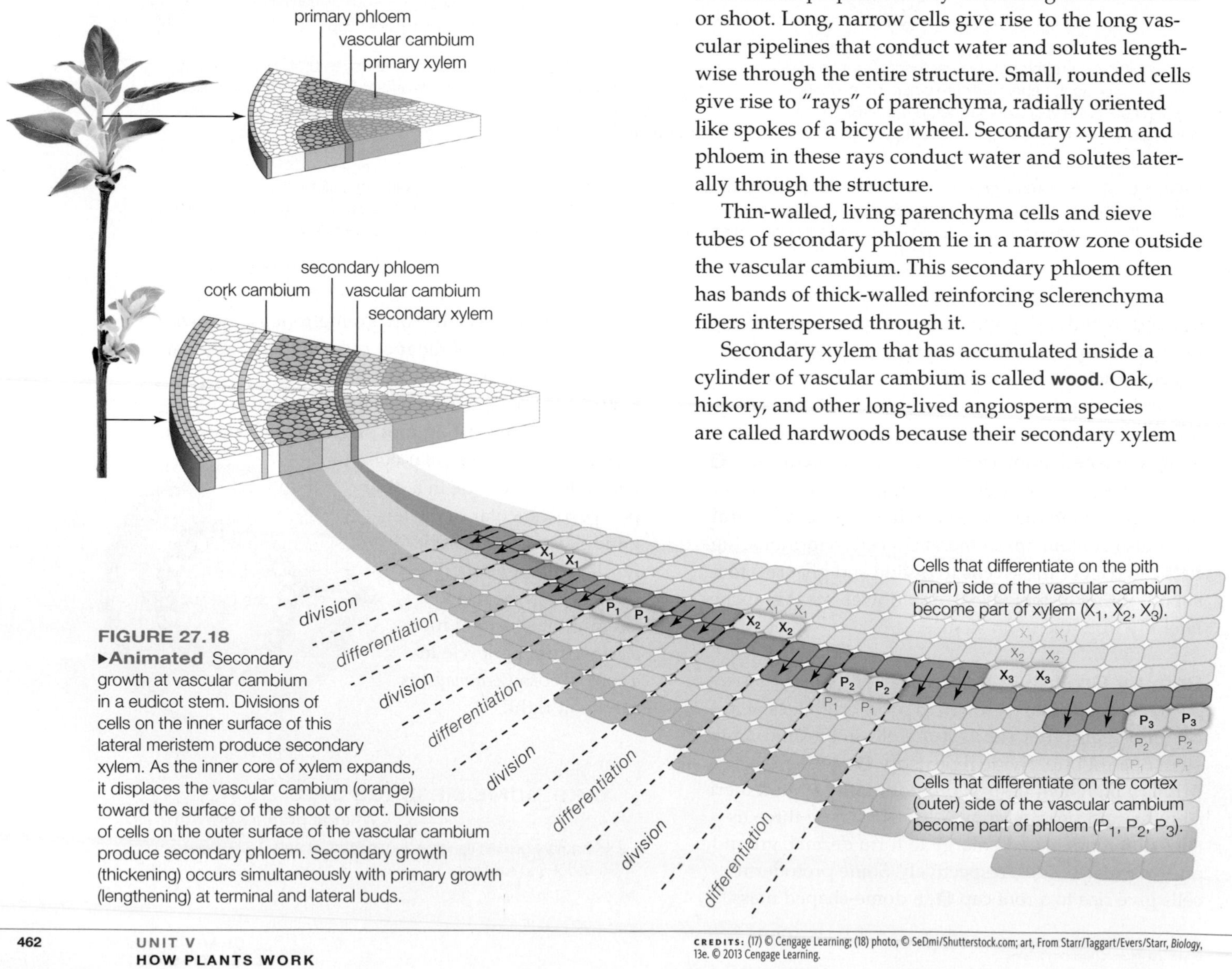

FIGURE 27.18 ▶**Animated** Secondary growth at vascular cambium in a eudicot stem. Divisions of cells on the inner surface of this lateral meristem produce secondary xylem. As the inner core of xylem expands, it displaces the vascular cambium (orange) toward the surface of the shoot or root. Divisions of cells on the outer surface of the vascular cambium produce secondary phloem. Secondary growth (thickening) occurs simultaneously with primary growth (lengthening) at terminal and lateral buds.

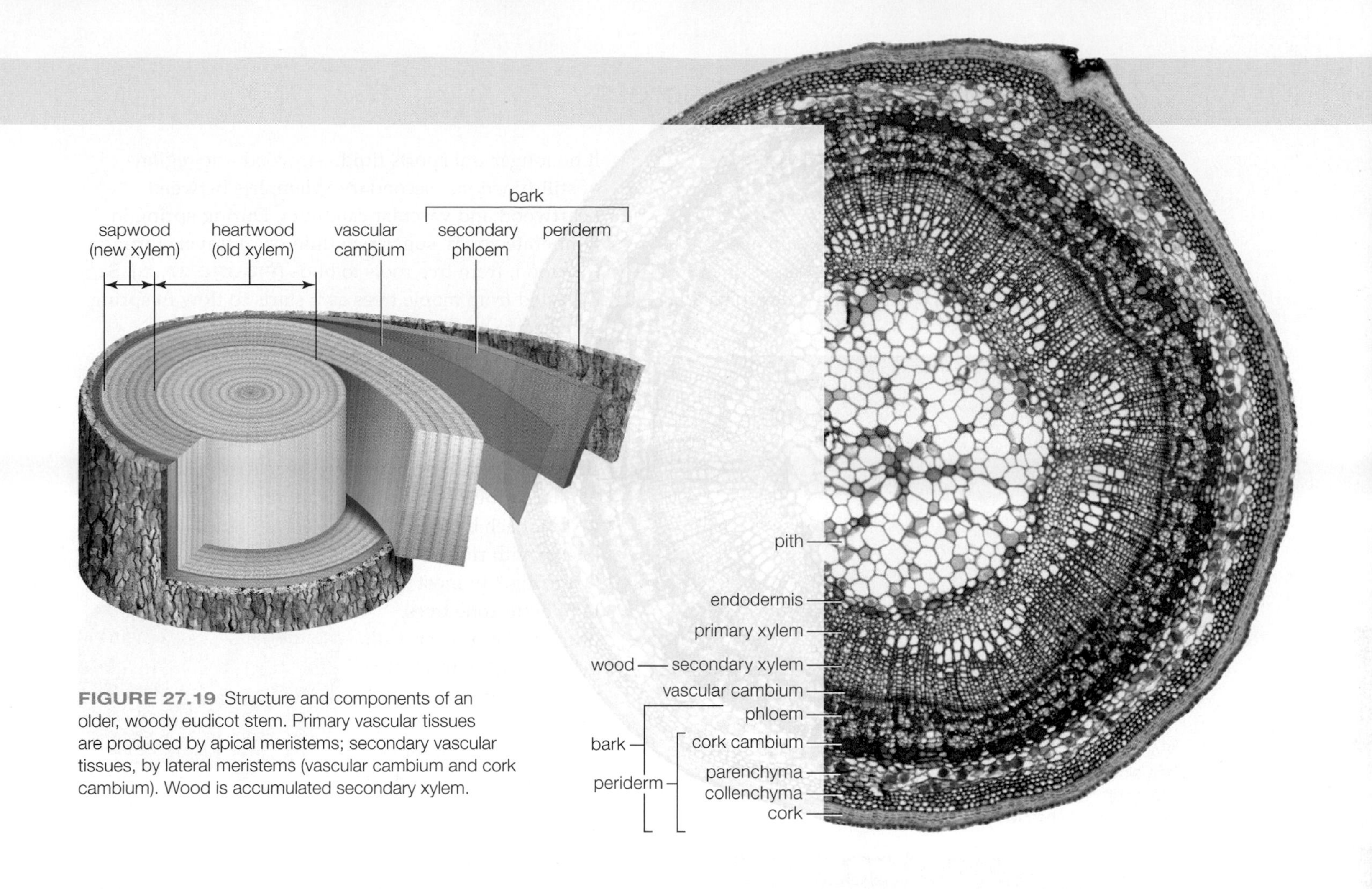

FIGURE 27.19 Structure and components of an older, woody eudicot stem. Primary vascular tissues are produced by apical meristems; secondary vascular tissues, by lateral meristems (vascular cambium and cork cambium). Wood is accumulated secondary xylem.

(wood) consists of tracheids, vessels, and fibers. By contrast, the secondary xylem of softwoods such as pines, cypress, and other conifers has tracheids but no vessels or fibers. The terms hardwood and softwood are somewhat misleading because wood hardness varies greatly in both groups: The wood of some gymnosperms is much heavier and more dense than the wood of some angiosperms.

As wood accumulates in a root or shoot, it displaces the vascular cambium toward the outer surface of the structure. Displaced cells of the vascular cambium divide in a widening circle, so this meristem tissue keeps its cylindrical form. Another lateral meristem, **cork cambium**, eventually forms outside the cylinder of vascular cambium. Cork cambium produces **cork**, a tissue that consists of densely packed dead cells with thickened, waxy walls. Cork protects, insulates, and waterproofs the surface of a stem or root. It is part of **periderm**, a dermal tissue that replaces epidermis on the surfaces of older stems and roots. Periderm comprises cork, cork cambium, and every other tissue between these two. **Bark** is the informal term for periderm and all other living and dead tissues outside the cylinder of vascular cambium (**FIGURE 27.19**).

bark In woody plants, informal term for all living and dead tissues outside of the vascular cambium.
cork Tissue that waterproofs, insulates, and protects the surfaces of woody stems and roots.
cork cambium Lateral meristem that produces cork.
lateral meristem Vascular cambium or cork cambium; cylindrical sheet of meristem that runs lengthwise through shoots and roots; gives rise to secondary growth (thickening).
periderm Plant dermal tissue that replaces epidermis during secondary growth of eudicots and gymnosperms.
secondary growth Thickening of older stems and roots; originates at lateral meristems.
vascular cambium Lateral meristem that produces secondary xylem and phloem.
wood Accumulated secondary xylem.

TAKE-HOME MESSAGE 27.8
What is secondary growth?

✔ Secondary growth, which thickens older roots and shoots, occurs at lateral meristems (vascular cambium and cork cambium) in many eudicots and gymnosperms.

✔ Wood is accumulated secondary xylem. Bark comprises all living and dead tissue outside of the vascular cambium.

27.9 Tree Rings and Old Secrets

✔ The relative thicknesses of a tree's rings hold clues to environmental conditions during its lifetime.

Wood's structure and function change with age. Over time, the secondary xylem at the center of an older woody stem or root can become plugged so that it no longer transports fluid. Sapwood, the region of still-functional secondary xylem, lies between heartwood and vascular cambium. During spring in temperate zones, sugar-rich fluid (sap) travels through sapwood, from tree roots to buds (**FIGURE 27.20**). Sap collected from maple trees as it starts to flow in spring is boiled to make maple syrup.

FIGURE 27.20 Sap oozes from xylem in a freshly cut pine log. Sapwood is secondary xylem that is still functional. As a tree ages, the oldest xylem at the center of its trunk may no longer be able to transport fluid because its tubes have become plugged. This older xylem is called heartwood.

In trees that are dormant in winter, vascular cambium gives rise to large-diameter, thin-walled xylem cells (early wood) during spring. Late wood, with small-diameter, thick-walled xylem cells, forms during summer. Thus, a cross-section reveals rings that consist of alternating bands of early and late wood (**FIGURE 27.21**). Each band is one growth ring, or "tree ring." In most temperate zone trees, one ring forms each year (right). In tropical regions where weather does not vary seasonally, trees grow at the same rate all year and do not have rings.

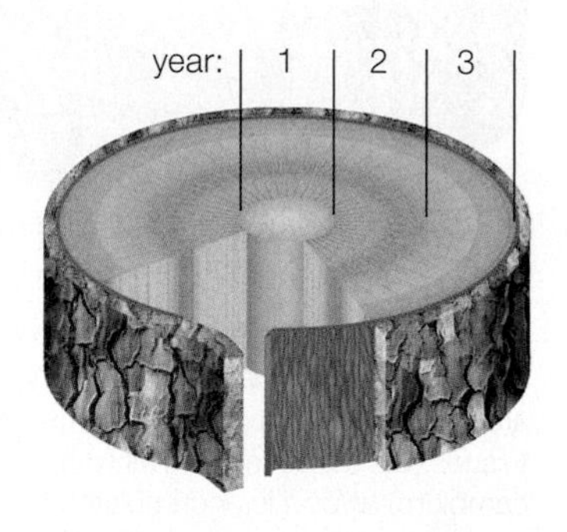

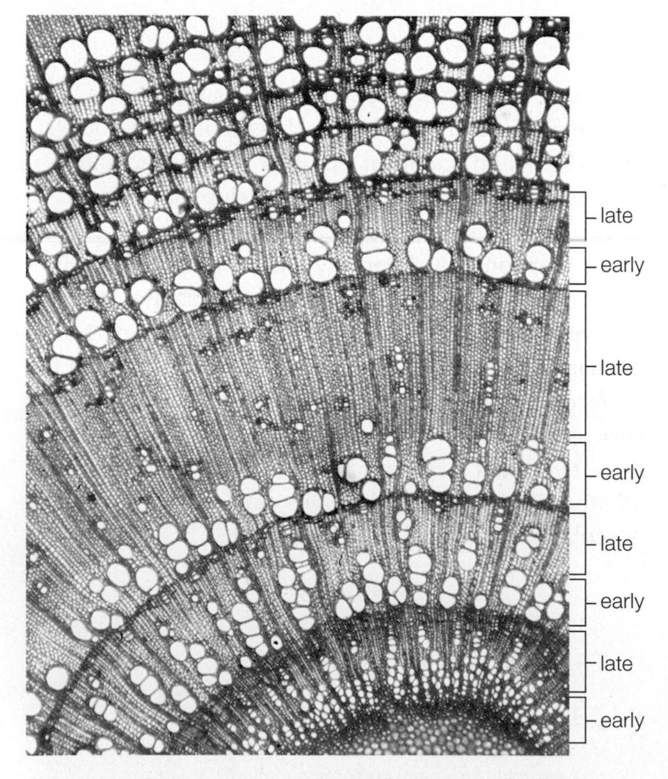

FIGURE 27.21 Early and late wood in an ash tree (*Fraxinus*). Early wood forms during warm, wet springs. Late wood forms during warm, dry summers. Each band of early and late wood is one growth ring, or "tree ring."

Tree rings are often used to estimate average annual rainfall; to date archaeological ruins; to gather evidence of wildfires, floods, landslides, and glacier movements; and to study the ecology and effects of parasitic insect populations. How can tree rings provide all of this information? Some tree species, such as redwoods, giant sequoias, and bristlecone pines, add wood over many centuries, one ring per year. Count an old tree's rings, and you have an idea of its age. If you know the year in which the tree was cut, you can determine when a particular ring formed by counting rings backward from the outer edge. Thickness, density, composition, and other features of the rings offer clues about environmental conditions that prevailed during the years they formed.

Consider how more fires tend to occur in warmer, drier conditions, so an increased frequency of fires can be evidence of a period of drought. Very old trees often bear the scars of many forest fires in their rings (**FIGURE 27.22A**). In 2010, researchers used the rings and fire scars of ancient trees in California's Sequoia National Park to reconstruct a 3,000-year history of the region's climate. They discovered, for example, a dramatic increase in the number of forest fires between 800 A.D. and 1300 A.D. This 500-year period coincided with the Medieval Warm Period, an anomaly in climate that had previously been documented in some other

Carbon Sequestration (revisited)

Carbon offsets that support reforestation are typically purchased well in advance of their benefit because most trees grow very slowly. It can be many decades before a tree sequesters a substantial amount of carbon in its secondary growth. Critics of carbon offsets point out that carbon sequestration in forests is also impermanent: There is no guarantee that reforested trees will survive to maturity, or that the carbon they have sequestered will not be released again, for example by fire or by logging.

parts of the world. Taken together, the evidence suggests this warming period was a global climate pattern that caused worldwide drought.

For another example of how we use information held in tree rings to reconstruct past events, consider the history of early English settlers of the United States. In 1587, the first set of settlers arrived at what is now Roanoke Island, off the coast of North Carolina. Supply ships that returned in 1589 found the island abandoned; searches of the mainland coast failed to turn up the missing colonists. In 1607, a second set of English settlers arrived in the New World, this time at Jamestown, Virginia. These colonists survived, but just barely; more than 40 percent of them died in the summer of 1610 alone.

Differences in the thicknesses of tree rings from nearby bald cypress trees that had been growing at the time of the Roanoke and Jamestown colonies revealed that both sets of settlers had terrible timing (**FIGURE 27.22B**). The Roanoke settlers arrived just in time for the worst drought in 800 years, and nearly a decade of severe drought struck Jamestown. We know that the corn crop of the Jamestown colony failed, so similar drought-related crop failures probably occurred at Roanoke. The Jamestown settlers also would have had difficulty finding fresh water, because they had established their colony at the head of an estuary. When the river levels dropped during the drought, the settlers' drinking water would have mixed with ocean water and become salty. Piecing together these bits of evidence gives us an idea of what life must have been like for the early settlers.

TAKE-HOME MESSAGE 27.9

What are tree rings?

✔ Tree rings are bands of early and late wood laid down one per year by temperate zone trees. Each band reflects conditions in the environment during the time it formed.

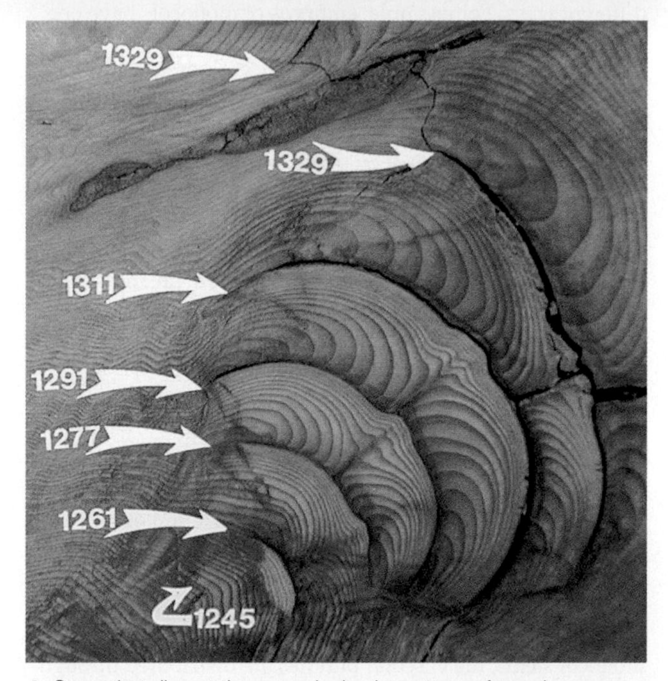

A Scars that disrupt the normal, circular pattern of tree rings are evidence of past fires. This photo shows fire scars in the wood of an ancient giant sequoia from California's Sequoia National Park; numbers indicate the year a particular ring was laid down.

B Relative thickness of tree rings can be used to estimate data such as average annual rainfall long before records of climate were kept. This section of a bald cypress tree (*Taxodium distichum*) was living near English colonists when they first settled in North America. Narrower annual rings mark years of severe drought.

FIGURE 27.22 Examples of data gleaned from tree rings.

CREDITS: (22A) photo by Tom Swetnam; (22B) David W. Stahle, Department of Geosciences, University of Arkansas; (in text revisited) Michael Nichols/National Geographic Creative.

summary

Section 27.1 By the process of photosynthesis, plants naturally remove carbon from the atmosphere and incorporate it into their tissues. Carbon that is locked in molecules of wood and other durable plant tissues can stay out of the atmosphere for centuries.

Section 27.2 Most flowering plants have belowground roots and aboveground shoots, including stems, leaves, and flowers. **Ground tissue** makes up the bulk of a plant, and **dermal tissues** protect its surfaces. **Vascular tissues** conduct water and nutrients to all parts of the plant. Monocots and eudicots have the same tissues organized in different ways. For example, embryos of monocots have one **cotyledon**; those of eudicots have two.

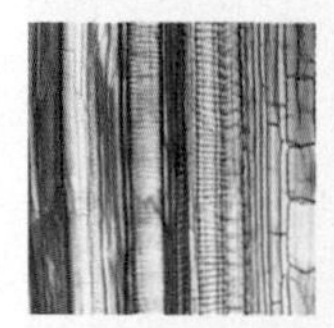

Section 27.3 **Parenchyma**, **collenchyma**, and **sclerenchyma** are simple tissues; each consists of only one type of cell. **Mesophyll** is photosynthetic parenchyma. Living cells in collenchyma have sturdy, flexible walls that support fast-growing plant parts. Cells in sclerenchyma die at maturity, but their lignin-reinforced walls remain and support the plant. Stomata open across **epidermis**, a dermal tissue that covers soft plant parts.

In vascular tissue, water and dissolved minerals flow through vessels of **xylem**, and sugars travel through vessels of **phloem**.

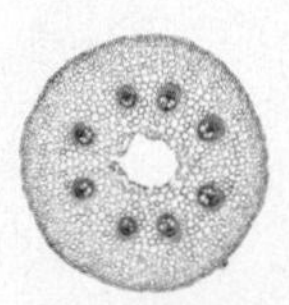

Section 27.4 **Vascular bundles** extending through stems conduct water and nutrients between different parts of the plant, and also help structurally support the plant body. In most eudicot stems, vascular bundles form a ring that divides the ground tissue into cortex and pith. In monocot stems, the vascular bundles are distributed throughout the ground tissue. New shoots and roots form at **nodes** on stems. Stem specializations such as rhizomes, corms, stem tubers, bulbs, cladodes, and stolons are adaptations for storage or reproduction in many types of plants.

Section 27.5 Leaves, which are specialized for photosynthesis, contain mesophyll and vascular bundles (**leaf veins**) between upper and lower epidermis. Eudicots typically have two layers of mesophyll; monocots do not. Water vapor and gases cross cuticle-covered epidermis at stomata.

Section 27.6 Roots absorb water and mineral ions for the entire plant. Inside each is a **vascular cylinder (stele)**. The outer boundary of the vascular cylinder is a layer of **endodermis**. **Root hairs** increase the surface area of roots. Many monocots have a **fibrous root system** that consists of similar-sized adventitious roots. Most eudicots have a **taproot system**—an enlarged primary root with its lateral root branchings. Lateral roots arise from divisions of **pericycle** cells inside the root vascular cylinder.

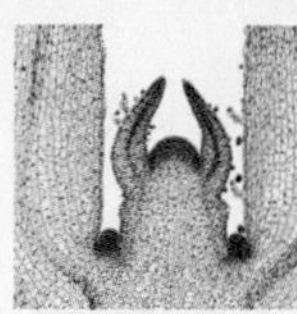

Section 27.7 All plant tissues originate at **meristems**, which are regions of undifferentiated cells that retain their ability to divide. **Primary growth** (lengthening) arises at **apical meristems** in **terminal buds** and **lateral buds** in the tips of shoots and roots.

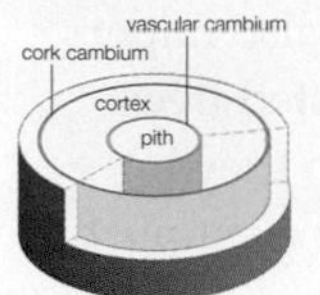

Section 27.8 **Secondary growth** (thickening) arises at **lateral meristems (vascular cambium** and **cork cambium)** in older stems and roots. Vascular cambium produces secondary xylem (**wood**) on its inner surface, and secondary phloem on its outer surface. Cork cambium gives rise to **cork**, which is part of **periderm**. **Bark** is all tissue outside of the vascular cambium of a woody plant.

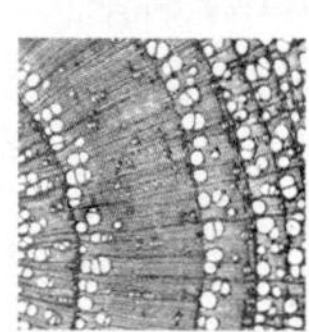

Section 27.9 In many trees, one ring forms during each growing season. Tree rings hold information about environmental conditions that prevailed while the rings were forming. For example, the relative thicknesses of the rings reflect the relative availability of water.

self-quiz

Answers in Appendix VII

1. In plants, fibers are a type of ________ cell.
   a. parenchyma
   b. sclerenchyma
   c. collenchyma
   d. mesophyll

2. Ground tissue consists mainly of ________ .
   a. waxes and cellulose
   b. lignified cell walls
   c. parenchyma cells
   d. cork but not bark

3. Which of the following cell types remain alive in mature tissue? Choose all that apply.
   a. companion cells
   b. sieve elements
   c. tracheids
   d. vessel elements

4. All of the vascular bundles inside a typical ________ are arranged in a ring.
   a. monocot stem
   b. eudicot stem
   c. monocot root
   d. eudicot root

5. True or false? Lateral roots form at nodes on roots.

6. Epidermis and periderm are ________ tissues.
   a. ground
   b. vascular
   c. dermal

7. A vascular bundle in a leaf is called ________ .
   a. xylem
   b. mesophyll
   c. a vein

8. Typically, vascular tissue is organized as ________ in stems and as ________ in roots.
   a. multiple vascular bundles; one vascular cylinder
   b. one vascular bundle; multiple vascular cylinders
   c. one vascular cylinder; multiple vascular bundles
   d. multiple vascular cylinders; one vascular bundle

9. An onion is a ________ (choose all that apply).
   a. root
   b. stem
   c. bulb
   d. corm

Tree Rings and Droughts El Malpais National Monument, in west central New Mexico, has pockets of vegetation that have been surrounded by lava fields for about 3,000 years, so they have escaped wildfires, grazing animals, agricultural activity, and logging. Henri Grissino-Mayer generated a 2,129-year annual precipitation record using tree ring data from living and dead trees in this park (**FIGURE 27.23**).

1. Around 770 A.D., the Mayan civilization began to suffer a massive population loss, particularly in the southern lowlands of Mesoamerica. The El Malpais tree ring data show a drought during that time. Was it more or less severe than the "dust bowl" drought?

2. One of the worst population catastrophes ever recorded occurred in Mesoamerica between 1519 and 1600 A.D., when around 22 million people native to the region died. Which period between 137 B.C. and 1992 had the most severe drought? How long did that drought last?

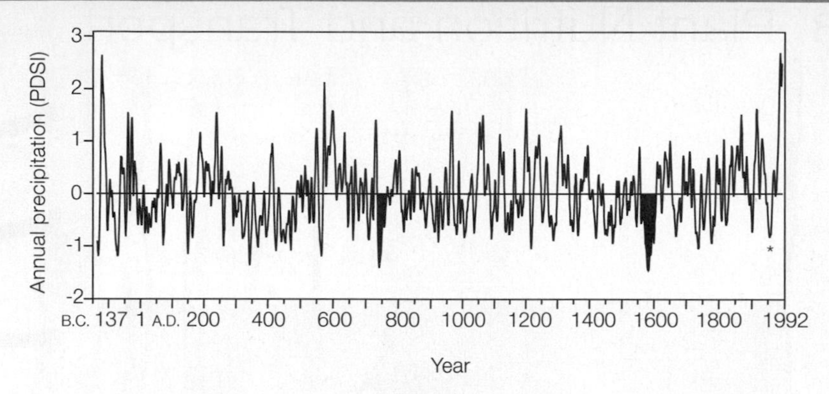

FIGURE 27.23 A 2,129-year annual precipitation record inferred from compiled tree ring data in El Malpais National Monument, New Mexico. Data were averaged over 10-year intervals; graph correlates with other indicators of rainfall collected in all parts of North America. PDSI, Palmer Drought Severity Index: 0, normal rainfall; increasing numbers mean increasing excess of rainfall; decreasing numbers mean increasing severity of drought.

* A severe drought contributed to a series of catastrophic dust storms that turned the midwestern United States into a "dust bowl" between 1933 and 1939.

10. In a(n) ________ , the primary root is typically the largest.
   a. lateral meristem
   b. adventitious root system
   c. fibrous root system
   d. taproot system

11. Root hairs ________ .
   a. conduct water from cortex to aboveground shoots
   b. increase the root's surface area for absorption
   c. anchor the plant in soil

12. Roots and shoots lengthen through activity at ________ .
   a. apical meristems
   b. lateral meristems
   c. vascular cambium
   d. cork cambium

13. The activity of lateral meristems ________ older roots and stems.
   a. lengthens
   b. thickens
   c. both a and b

14. Tree rings occur because ________ .
   a. there are droughts during the time the rings form
   b. environmental conditions influence xylem cell size
   c. heartwood alternates with sapwood
   d. periderm replaces epidermis

15. Match the plant parts with the best description.

___ vascular cambium	a. ground tissue
___ mesophyll	b. stem structure
___ wood	c. a lateral meristem
___ cortex	d. photosynthetic parenchyma
___ potato	e. secondary xylem
___ parallel veins	f. between epidermal cells
___ stoma	g. characteristic of monocot leaves

critical thinking

1. Oscar and Lucinda meet in a tropical rain forest and fall in love, and he carves their initials into the bark of a tree. They never do get together, though. Ten years later, still heartbroken, Oscar searches for the tree. Given what you know about primary and secondary growth, will he find the carved initials higher relative to ground level? If he goes berserk and chops down the tree, what kinds of growth rings will he see?

2. Was the transverse section shown at right taken from a stem or a root? Monocot or eudicot?

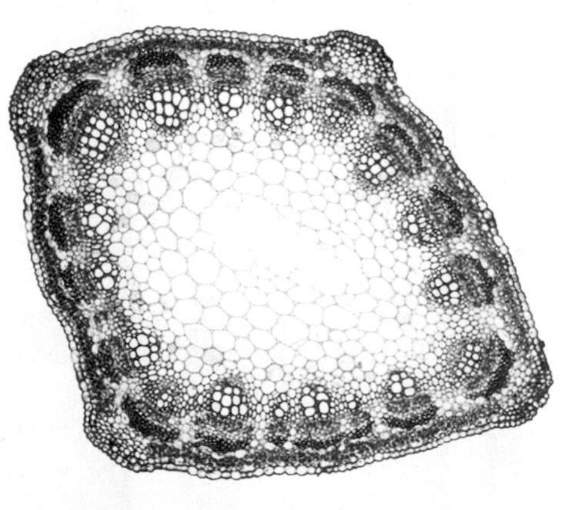

3. Aboveground plant surfaces are typically covered with a waxy cuticle. Why do roots lack this protective coating?

4. Why do eudicot trees tend to be wider at the base than at the top?

5. Is the plant with the yellow flower below a eudicot or a monocot? What about the plant with the purple flower?

CREDITS: (23) Data by Henri D. Grissino-Mayer; (in text) CT #2, © Mike Clayton/University of Wisconsin Department of Botany; CT #5, Edward S. Ross.

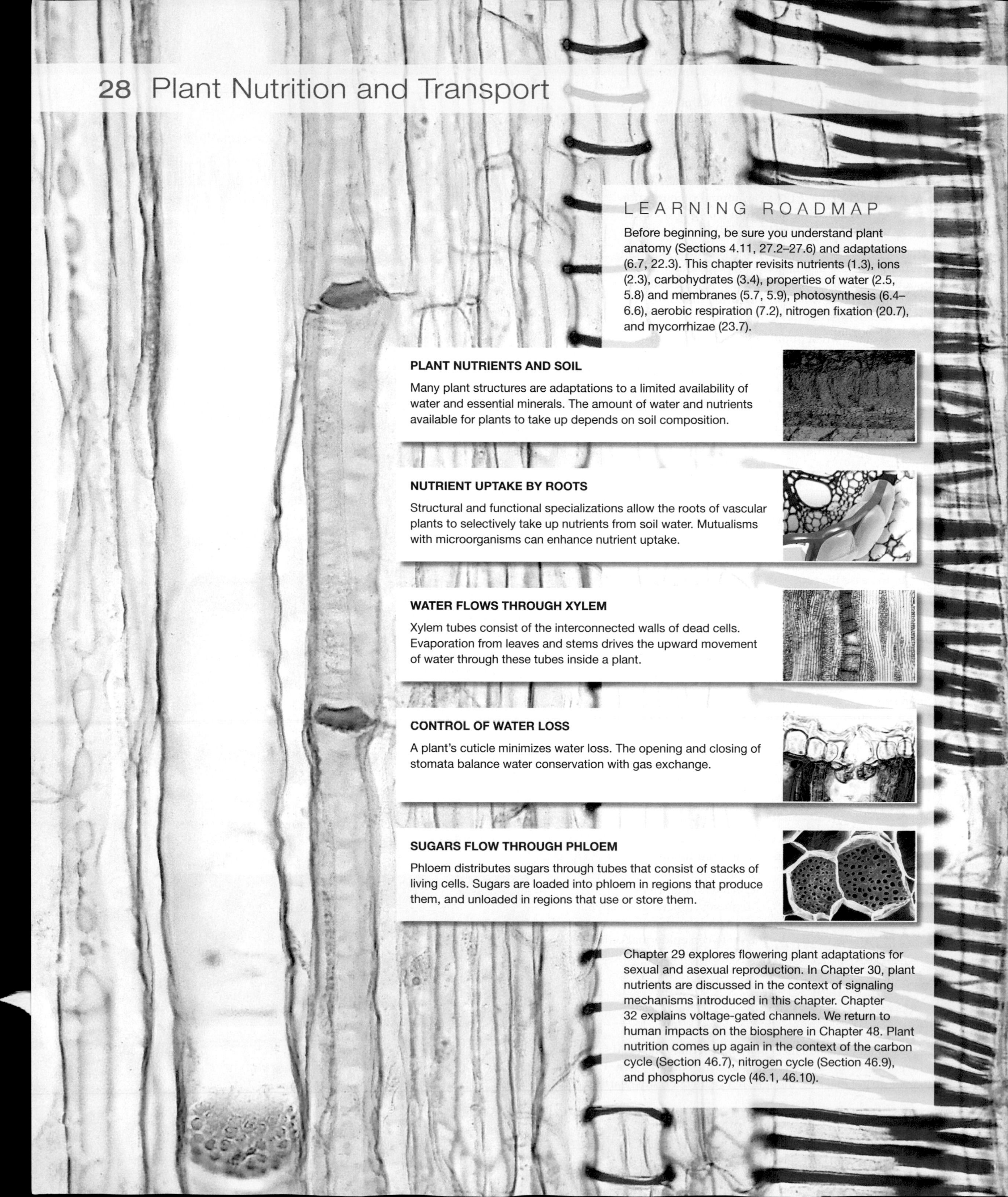

28 Plant Nutrition and Transport

LEARNING ROADMAP

Before beginning, be sure you understand plant anatomy (Sections 4.11, 27.2–27.6) and adaptations (6.7, 22.3). This chapter revisits nutrients (1.3), ions (2.3), carbohydrates (3.4), properties of water (2.5, 5.8) and membranes (5.7, 5.9), photosynthesis (6.4–6.6), aerobic respiration (7.2), nitrogen fixation (20.7), and mycorrhizae (23.7).

PLANT NUTRIENTS AND SOIL

Many plant structures are adaptations to a limited availability of water and essential minerals. The amount of water and nutrients available for plants to take up depends on soil composition.

NUTRIENT UPTAKE BY ROOTS

Structural and functional specializations allow the roots of vascular plants to selectively take up nutrients from soil water. Mutualisms with microorganisms can enhance nutrient uptake.

WATER FLOWS THROUGH XYLEM

Xylem tubes consist of the interconnected walls of dead cells. Evaporation from leaves and stems drives the upward movement of water through these tubes inside a plant.

CONTROL OF WATER LOSS

A plant's cuticle minimizes water loss. The opening and closing of stomata balance water conservation with gas exchange.

SUGARS FLOW THROUGH PHLOEM

Phloem distributes sugars through tubes that consist of stacks of living cells. Sugars are loaded into phloem in regions that produce them, and unloaded in regions that use or store them.

Chapter 29 explores flowering plant adaptations for sexual and asexual reproduction. In Chapter 30, plant nutrients are discussed in the context of signaling mechanisms introduced in this chapter. Chapter 32 explains voltage-gated channels. We return to human impacts on the biosphere in Chapter 48. Plant nutrition comes up again in the context of the carbon cycle (Section 46.7), nitrogen cycle (Section 46.9), and phosphorus cycle (46.1, 46.10).

28.1 Leafy Cleanup

From 1940 until the 1970s, the United States Army tested and disposed of weapons at J-Field, Aberdeen Proving Ground in Maryland (**FIGURE 28.1**). Obsolete chemical weapons and explosives were burned in open pits, together with plastics, solvents, and other wastes. Lead, arsenic, mercury, and other metals heavily contaminated the soil and groundwater. So did highly toxic organic compounds, including TCE (trichloroethylene). TCE damages the nervous system, lungs, and liver, and exposure to large amounts can be fatal. Today, the toxic groundwater is seeping toward nearby marshes and the Chesapeake Bay.

To protect the bay and clean up the soil, the Army and the Environmental Protection Agency turned to phytoremediation: the use of plants to take up and concentrate or degrade environmental contaminants. They planted poplar trees (*Populus trichocarpa*) that cleanse groundwater by removing TCE and other organic pollutants from it.

Like other vascular plants, poplar trees take up soil water through their roots. Along with the water come nutrients and chemical contaminants, including TCE. Although TCE is toxic to animals, it does not harm the trees. The poplars break down some of the toxin, but they release most of it into the atmosphere. Airborne TCE is the lesser of two evils: It breaks down much more quickly in air than it does in groundwater.

Analyses at J-Field since the poplars were planted show that the phytoremediation strategy is working splendidly. By 2001, the trees had removed 60 pounds of TCE from the soil and groundwater. In addition, they had helped foster communities of microorganisms that were also breaking down toxins. Researchers estimate that the trees will have removed the majority of contaminants from J-Field by 2030.

With metal pollutants, the best phytoremediation strategies use plants that take up toxins and store them in aboveground tissues. The toxin-laden plant parts can then be harvested for safe disposal. This strategy is being tested in Ukraine, where in 1986 an accident at the Chernobyl nuclear power plant released huge amounts of radioactive material that still heavily contaminates the region. Transgenic sunflowers planted on rafts in heavily contaminated ponds quickly take up radioactive cesium (a metal) from the water.

Phytoremediation uses the ability of plants to take up toxic chemicals dissolved in soil water. Plants have this capacity not for our benefit, but rather to meet their own needs for water and nutrients. Scientists who know about plant physiology are finding ways to put plants to work as leafy cleanup crews. Compared to other methods of cleaning up toxic waste sites, phytoremediation is usually less expensive—and it is more appealing to neighbors.

FIGURE 28.1 An example of phytoremediation. The soil at J-Field became heavily contaminated by chemical weapons and explosives that were burned and dumped at the site (inset photo) for about 30 years. The photo in the background shows J-Field today.

CREDITS: (opposite) © M.I. Walker/Wellcome Images; (l) inset, © OPSEC Control Number #4 077-A-4; background, Billy Wrobel, 2004.

28.2 Plant Nutrients and Availability in Soil

✔ Plants require elemental nutrients in soil, water, and air.
✔ Soil type affects the growth of plants.

Plants require sixteen elemental nutrients (**TABLE 28.1**). Nine are macronutrients, meaning they are required in amounts above 0.5 percent of a plant's dry weight. Carbon, oxygen, and hydrogen are macronutrients, as are nitrogen, phosphorus, and sulfur (components of proteins and nucleic acids), potassium and calcium (which affect processes such as cell signaling), and magnesium (a component of chlorophyll). Chlorine, iron, boron, manganese, zinc, copper, and molybdenum are micronutrients, meaning they make up traces of the plant's dry weight. Some micronutrients serve as enzyme cofactors (Section 5.6).

Properties of Soil

Carbon, oxygen, and hydrogen atoms are abundantly available in carbon dioxide and water. Plants get the other elements they need when their roots take up minerals dissolved in soil water. Soils vary in their nutrient content and other properties, so some support plant growth better than others. Soil consists mainly of mineral particles—sand, silt, and clay—that form by the weathering of rocks. Sand grains are about one millimeter in diameter. Silt particles are hundreds or thousands of times smaller than sand grains, and clay particles are even smaller. Clay particles enhance the quality of soils because they have a negative charge that attracts positively charged mineral ions in soil water. Thus, clay-rich soil retains dissolved nutrients that might otherwise trickle past roots too quickly to be absorbed. Sand and silt are also necessary for plant growth because they intervene between tiny particles of clay. Soils with too much clay pack so tightly that they exclude air—and the oxygen in it. Cells in a plant's roots, like cells in aboveground parts, require oxygen for aerobic respiration.

Soils with the best oxygen and water penetration are **loams**, which have roughly equal proportions of sand, silt, and clay. Most plants do best in loams that contain between 10 and 20 percent **humus**, which is decomposing organic material—fallen leaves, feces, and so on. Humus affects plant growth because it releases nutrients, and its negatively charged organic acids can trap positively charged mineral ions in soil water. Humus also swells and shrinks as it absorbs and releases water, and these changes in size aerate soil by opening spaces for air to penetrate. Humus tends to accumulate in waterlogged soils because water excludes air, and organic matter breaks down much more slowly in the absence of oxygen. Soils in swamps, bogs, and other perpetually wet areas often contain more than 90 percent humus. Very few types of plants can grow in these soils.

Table 28.1 Plant Nutrients and Their Main Functions

Macronutrient	Main Functions
Carbon, Hydrogen, Oxygen	Raw materials for photosynthesis. Main components of biological molecules (nucleic acids, proteins, fats, carbohydrates), cofactors, and organic signaling molecules such as hormones
Nitrogen	Important component of proteins (including enzymes), nucleic acids, cofactors, photosynthetic pigments
Potassium	Involved in enzyme activation, pH, opening stomata; also contributes to water–solute balance,* which in turn influences osmosis and turgor
Calcium	Stabilizes structure of cell membranes and cell walls; important in cell signaling, mitosis, stomata function
Magnesium	Functional component of chlorophyll, many enzymes and cofactors; stabilizes structure of cell membranes, cell walls
Phosphorus	Important component of phospholipids and nucleic acids (particularly ATP and other coenzymes)
Sulfur	Component of proteins, some cofactors

Micronutrient	Main Functions
Chlorine	Photosystem II cofactor; signaling; solute and electrochemical balances
Iron	Component of chlorophyll and many enzymes; particularly important in electron transfers
Boron	Cross-links polysaccharides in cell walls; also links cell wall polysaccharides to surface of plasma membrane
Manganese	Component of photosystem II that allows it to catalyze photolysis during light reactions of photosynthesis; also functional component of many coenzymes
Zinc	Cofactor for many enzymes; component of many proteins, including auxin receptors and transcription factors
Copper	Cofactor in many proteins and enzymes involved in electron transfer chains (including plastocyanin of light-dependent reactions), lignin synthesis, detoxification of free radicals
Molybdenum	Cofactor required for nitrogen fixation and for abscisic acid synthesis

** All mineral elements contribute to water–solute balances.*

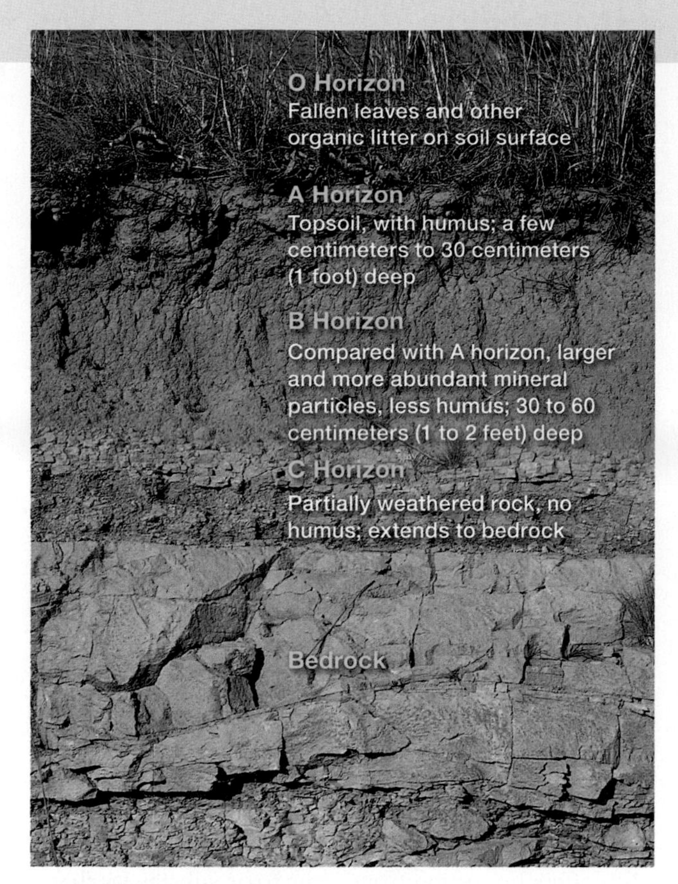

FIGURE 28.2 An example of soil horizons.

FIGURE 28.3 Runaway erosion in Providence Canyon, Georgia. European settlers who arrived in 1800 plowed the land straight up and down the hills. The furrows made excellent conduits for rainwater, which proceeded to carve out deep crevices that made even better rainwater conduits. The area became useless for farming by 1850. It now consists of about 445 hectares (1,100 acres) of deep canyons that continue to expand at the rate of about 2 meters (6 feet) per year.

How Soils Change

Soils develop over thousands of years. They exist in different stages of development in different regions. Most form in layers, or horizons, that are distinct in color and other properties (**FIGURE 28.2**). Identifying the layers helps us compare soils in different places. For instance, the A horizon, which is **topsoil**, contains the greatest amount of organic matter, so the roots of most plants grow most densely in this layer. Grasslands typically have a deep layer of topsoil; tropical forests do not.

Minerals, salts, and other molecules dissolve in water as it filters through soil. **Leaching** is the process by which water removes soil nutrients and carries them away. Leaching is fastest in sandy soils, which do not bind nutrients as well as clay soils.

Soil erosion is a loss of soil under the force of wind and water. Poor farming practices can also lead to erosion (**FIGURE 28.3**). Each year, about 25 billion metric tons of topsoil erode from croplands in the midwestern United States. The topsoil enters the Mississippi River, which then dumps it into the Gulf of Mexico. Nutrient loss because of this erosion affects not only plants that grow in the region, but also humans and other organisms that depend on the plants for survival.

humus Decaying organic matter in soil.
leaching Process by which water moving through soil removes nutrients from it.
loam Soil with roughly equal amounts of sand, silt, and clay.
soil erosion Loss of soil under the force of wind and water.
topsoil Uppermost soil layer; contains the most organic matter and nutrients for plant growth.

TAKE-HOME MESSAGE 28.2

Where do plants get nutrients they require?

✔ Plants require sixteen elements. All are available in water, air, or soil.

✔ Soil consists mainly of mineral particles: sand, silt, and clay. Clay retains positively charged mineral ions in soil water.

✔ Humus aerates soil, and like clay it retains positively charged mineral ions in soil water.

✔ Most plants grow best in loams (soils with equal proportions of sand, silt, and clay) that contain 10 to 20 percent humus.

✔ Leaching and erosion remove nutrients from soil.

CREDITS: (2) William Ferguson; (3) Photo courtesy of Stephanie G. Harvey, Georgia Southwestern State University.

28.3 Root Adaptations for Nutrient Uptake

✔ Root specializations such as root hairs, mycorrhizae, and nodules help plants absorb water and nutrients.

The Function of Endodermis

Water moves from soil, through a root's epidermis and cortex, to the vascular cylinder (**FIGURE 28.4**). Osmosis drives this movement, because fluid in the plant typically contains more solutes than soil water (Section 5.8). After soil water enters the vascular cylinder, xylem distributes it to the rest of the plant.

Most of the soil water that enters a root moves through cell walls ❶. Cell walls are permeable to water and ions, and in plants they are shared between adjacent cells (Section 4.11). The walls of tightly packed root cells form a pathway for water to move between epidermis and vascular cylinder. Thus, soil water can diffuse from epidermis, through the cortex, to the vascular cylinder without ever entering a cell.

Although soil water can reach the vascular cylinder by diffusing through cell walls, it cannot enter the cylinder the same way. Why not? A vascular cylinder is separated from root cortex by its endodermis, a tissue that consists of a single layer of tightly packed endodermal cells (Section 27.6). These cells secrete a waxy substance into their walls wherever they abut. The substance forms a **Casparian strip**, which is a waterproof band between the plasma membranes of endodermal cells (**FIGURE 28.5**). A Casparian strip prevents water from diffusing through endodermal cell walls ❷. Thus, soil water can enter a vascular cylinder only by passing through the cytoplasm of an endodermal cell.

Water can enter the cytoplasm of an endodermal cell directly, by diffusing across its plasma membrane. Water can also enter the cytoplasm of an endodermal cell indirectly, by first diffusing across the plasma membrane of a cell in the root cortex, then through plasmodesmata (Section 4.11). Because these cell junctions connect abutting plant cells, water can diffuse through them, one cell to the next, until it enters an endodermal cell. Once in endodermal cell cytoplasm, water molecules move into xylem of the vascular cylinder by diffusing through plasmodesmata of adjacent cells or across their plasma membranes.

Water can diffuse directly across lipid bilayers, but ions cannot (Section 5.7). Mineral ions in soil water enter or exit a cell only through transport proteins in its plasma membrane. Thus, transport proteins in root cell membranes control the types and amounts of ions that move from soil water into the plant body. This mechanism offers protection from some toxic substances that may be in soil water. In addition, most mineral ions important for plant nutrition are more concentrated

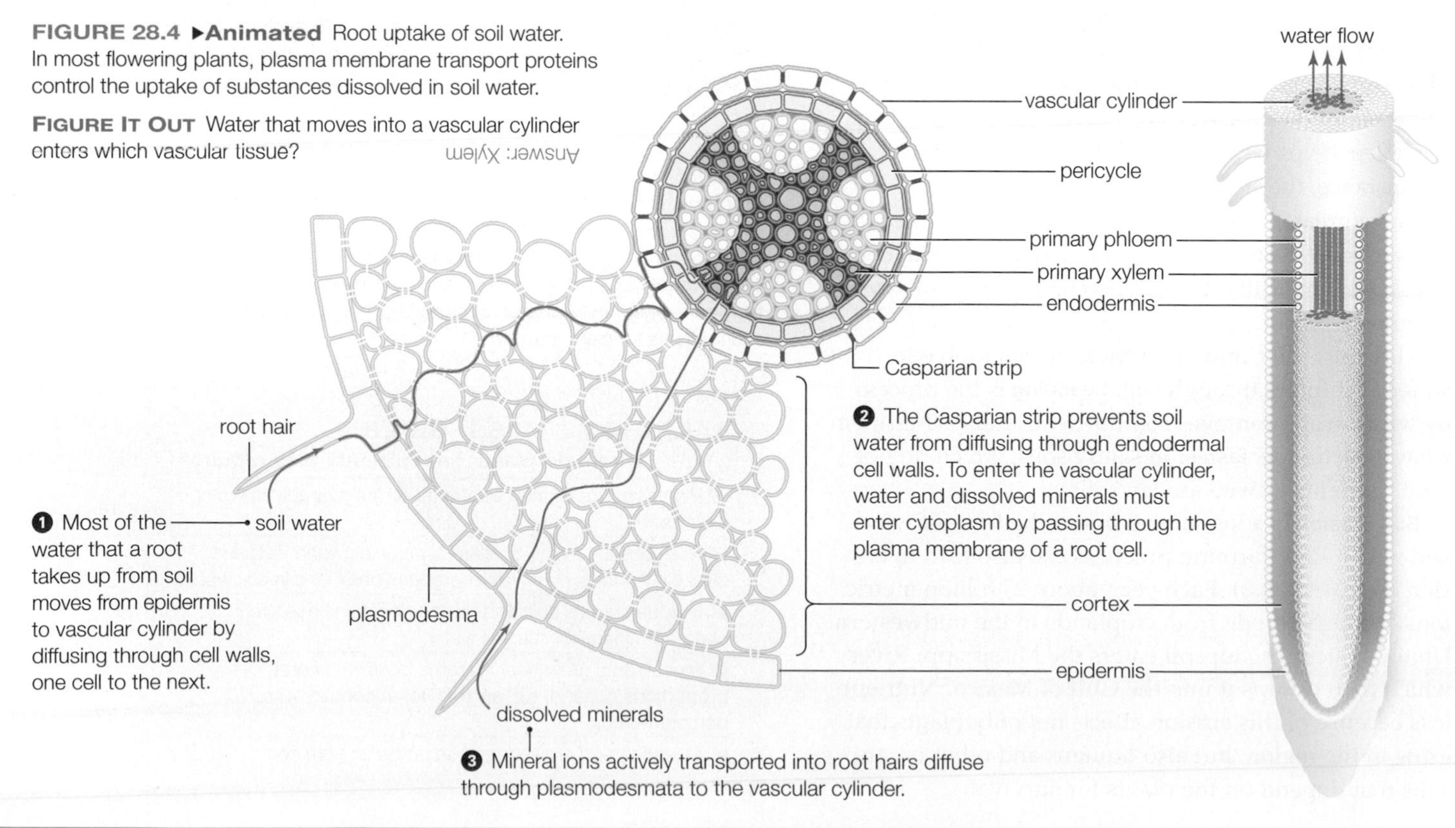

FIGURE 28.4 ▶**Animated** Root uptake of soil water. In most flowering plants, plasma membrane transport proteins control the uptake of substances dissolved in soil water.

FIGURE IT OUT Water that moves into a vascular cylinder enters which vascular tissue?

Answer: Xylem

in the root than in soil, so they will not spontaneously diffuse into the root. These nutrients must be actively transported into root cells. The transport occurs mainly at the membrane of root hairs ❸. Once in cell cytoplasm, mineral ions can diffuse from cell to cell through plasmodesmata until they reach pericycle. Transport proteins in the membranes of pericycle cells load mineral ions into xylem of the vascular cylinder.

Mutualisms

Most flowering plants take part in mycorrhizae and other mutualisms that provide nutritional benefit. As Section 23.7 explained, a mycorrhiza is a mutually beneficial interaction between a root and a fungus that grows on or in it. Filaments of the fungus (hyphae) form a velvety cloak around roots or penetrate their cells. Collectively, the hyphae have a large surface area, so they absorb mineral ions from a larger volume of soil than roots alone. The fungus takes up some sugars and nitrogen-rich compounds from root cells. In return, the root cells get some scarce minerals that the fungus is better able to absorb.

Some plant species form mutualisms with nitrogen-fixing *Rhizobium* bacteria (Section 20.7). The roots of these plants release certain compounds into the soil that are recognized by compatible bacteria. The bacteria respond by releasing signaling molecules that, in turn, trigger the roots to grow around and encapsulate them inside swellings called **root nodules** (FIGURE 28.6). The association is beneficial to both parties. The plants require a lot of nitrogen, but cannot use nitrogen gas ($N{\equiv}N$, or N_2) that is abundant in air. The bacteria in root nodules fix this gas to ammonia (NH_3), which is a form of nitrogen that the plant can use. In return for this valuable nutrient, the plant provides an oxygen-free environment for the anaerobic bacteria, and shares its photosynthetically produced sugars with them.

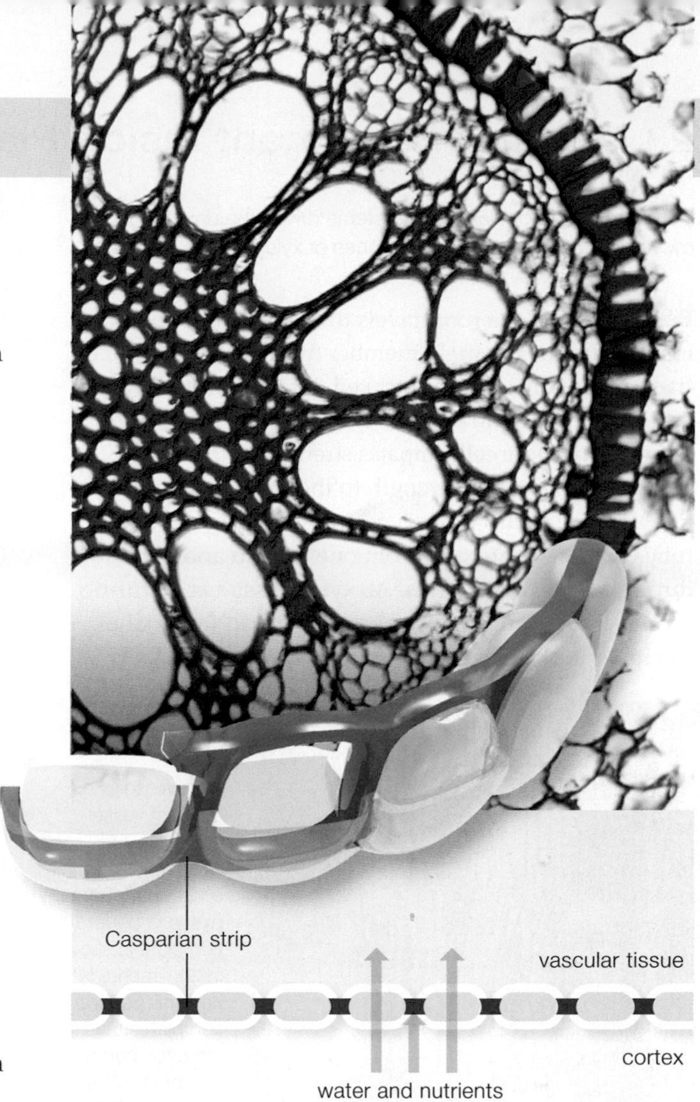

FIGURE 28.5 ▸**Animated** The Casparian strip.

The micrograph shows a vascular cylinder in the root of an iris plant. Parenchyma cells that make up endodermis are specialized to secrete a waxy substance into their walls wherever they touch. The secretions, which form a Casparian strip, prevent water from diffusing through endodermal cell walls to enter the vascular cylinder.

FIGURE 28.6 Soybean plants in nitrogen-poor soil show how root nodules (right) affect growth. Only the darker green plants growing in the rows on the right were infected with nitrogen-fixing *Rhizobium* bacteria.

Casparian strip Waxy, waterproof band that seals abutting cell walls of root endodermal cells. Helps the plant control the amounts and types of substances that it takes up from soil water.
root nodules Swellings of some plant roots that contain mutualistic nitrogen-fixing bacteria.

TAKE-HOME MESSAGE 28.3

How do plant roots absorb water and nutrients?

✔ Root hairs and mutualisms greatly enhance a root's ability to take up water and nutrients.

✔ Transport proteins in root cell plasma membranes control the uptake of mineral ions into the vascular cylinder.

CREDITS: (5) photo, Dr. Keith Wheeler/Science Photo Library; art, © Cengage Learning; (6) left, NifTAL Project, Univ. of Hawaii, Maui; right, Ninjatacoshell.

28.4 Water Movement Inside Plants

✔ Evaporation from leaves and stems drives the upward movement of water through pipelines of xylem inside a plant.

Water that enters a root travels to the rest of the plant inside tubes of xylem. Remember from Section 27.3 that these tubes consist of the stacked, interconnected walls of dead cells. Lignin deposited in rings or spirals into the walls of these cells imparts strength to the tubes, which in turn impart strength to the stem or other organ they service. Water flows lengthwise through xylem tubes, and also laterally, from one tube to another, through their pitted walls. As xylem tissue is maturing, its still-living cells deposit secondary wall material on the inner surface of their primary wall (Section 4.11).

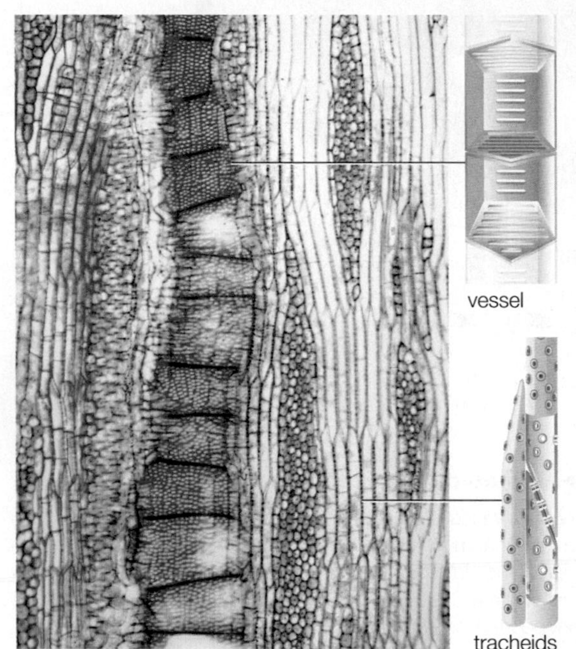

A A vessel is a stack of vessel elements that connect end to end at perforation plates (a perforation plate is shown in **B**). Tracheids have no perforation plates, and are often narrower and pointier than vessel elements. The micrograph shows a section of Japanese wisteria (*Wisteria floribunda*).

B Water moves through interconnected pits in the secondary walls of dead vessel elements and tracheids. The pits are often bordered by thickened cell wall deposits. Micrograph: Japanese holly (*Ilex crenata*).

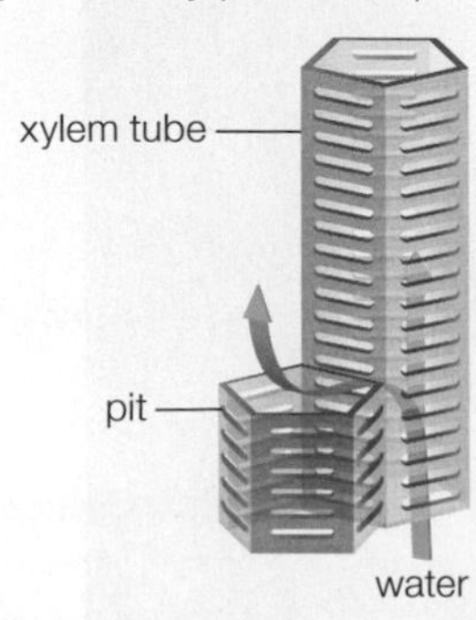

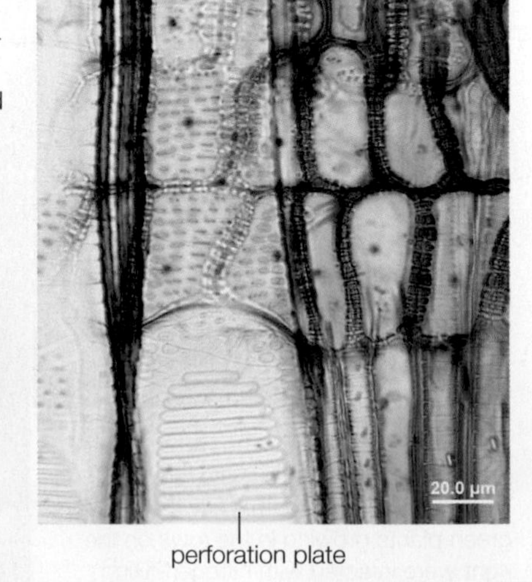

FIGURE 28.7 Pipelines of xylem.

The secondary wall forms around plasmodesmata, but not over them, so the cytoplasm of adjacent cells stays connected. Just before the cells die, their primary wall breaks down. Small holes (pits) remain in the secondary walls where plasmodesmata had once connected the living cells. In mature xylem, water can flow laterally through these pits, between adjacent tubes. Pits are often bordered, which means their edges are thickened by pectin-containing secondary wall material. Pectins shrink when dry, and swell when wet. When water flows through a bordered pit, the border swells and eventually plugs the hole. Thus, water tends to flow toward the thirstiest regions of the plant, where bordered pits are dry and open.

Angiosperms have two types of xylem tubes, each composed of one cell type: vessel elements or tracheids (**FIGURE 28.7**). Vessels, the main water-conducting tubes in angiosperms, consist of **vessel elements** stacked end to end. Vessel elements evolved more recently than tracheids, and probably contributed to the evolutionary success of the group (most gymnosperms have no vessels). In maturing xylem tissue, stacked vessel elements digest very large holes in their end walls (where they meet) just before they die. The resulting perforation plates allow water to flow freely through vessel elements in a stack. Some perforation plates have ladderlike bars that break up air bubbles in water flowing through larger tubes.

Tracheids, like vessel elements, die in stacks to form water-conducting tubes in xylem. Unlike vessel elements, tracheids have no perforation plates; their ends are typically pointed and closed. Pits in the narrowed ends of stacked tracheids match up. Water flows vertically through the tube by moving through these matched pits. Being narrower than vessel elements, tracheids are more resistant to vertical compression. Thus, in addition to conducting water, tracheids have a substantial role in structural support.

Cohesion–Tension Theory

How does water in xylem move all the way from roots to leaves that may be far above soil? Tracheids and vessel elements that compose xylem tubes are dead, so these cells cannot be expending any energy to pump water upward against gravity. Rather, the upward movement of water in vascular plants occurs as a consequence of evaporation and cohesion (Section 2.5). By the **cohesion–tension theory**, water is pulled upward through a plant by air's drying power, which creates a continuous negative pressure called tension. The tension extends from leaves at the tips of shoots to roots that may be hundreds of feet below. Consider how almost all of the water that a plant takes up is lost by evaporation, typically from stomata on the plant's leaves and stems (**FIGURE 28.8** ❶). The evaporation of water from aboveground plant parts is called **transpiration**. Transpiration's effect on water inside a plant is a bit like what happens when you suck a drink through a straw. Transpiration exerts negative pressure (it pulls) on water. Because water molecules are connected by hydrogen bonds, a pull on one tugs all of them. Thus, negative pressure (tension) created by transpiration pulls on entire columns of water that fill xylem tubes ❷. The tension extends all the way from leaves that may be hundreds of feet in the air, down through stems, into young roots where water is being taken up from soil ❸. Water is pulled upward in continuous columns because xylem tubes are narrow, and water moving through a narrow conduit (such as a straw or a xylem tube) resists breaking into droplets. This phenomenon is partly an effect of water's cohesion, and partly because water molecules are attracted to hydrophilic materials (such as cellulose in the walls of xylem) making up the conduit.

Transpiration drives almost all of the upward movement of water through a vascular plant, but many metabolic pathways also contribute to the negative pressure that sucks water through xylem. For example, the noncyclic reactions of photosynthesis (Section 6.5) split water molecules into oxygen and hydrogen ions. Thus, cells carrying out photosynthesis must receive a constant supply of water molecules, and these are delivered by xylem.

cohesion–tension theory Explanation of how transpiration creates a tension that pulls a cohesive column of water upward through xylem, from roots to shoots.
tracheids Tapered cells of xylem that die when mature; their interconnected, pitted walls remain and form water-conducting tubes.
transpiration Evaporation of water from aboveground plant parts.
vessel elements Of xylem, cells that form in stacks and die when mature; their pitted walls remain to form water-conducting tubes. Each vessel consists of a stack of vessel elements that meet end to end at perforation plates.

TAKE-HOME MESSAGE 28.4

What makes water move inside plants?

✔ In plants, water moves inside continuous tubes of xylem that consist of interconnected, pitted secondary walls of dead cells—tracheids and vessel elements.

✔ Transpiration, the evaporation of water from aboveground plant parts, pulls on cohesive columns of water inside xylem tubes. This tension pulls water upward through the plant, from roots to shoots.

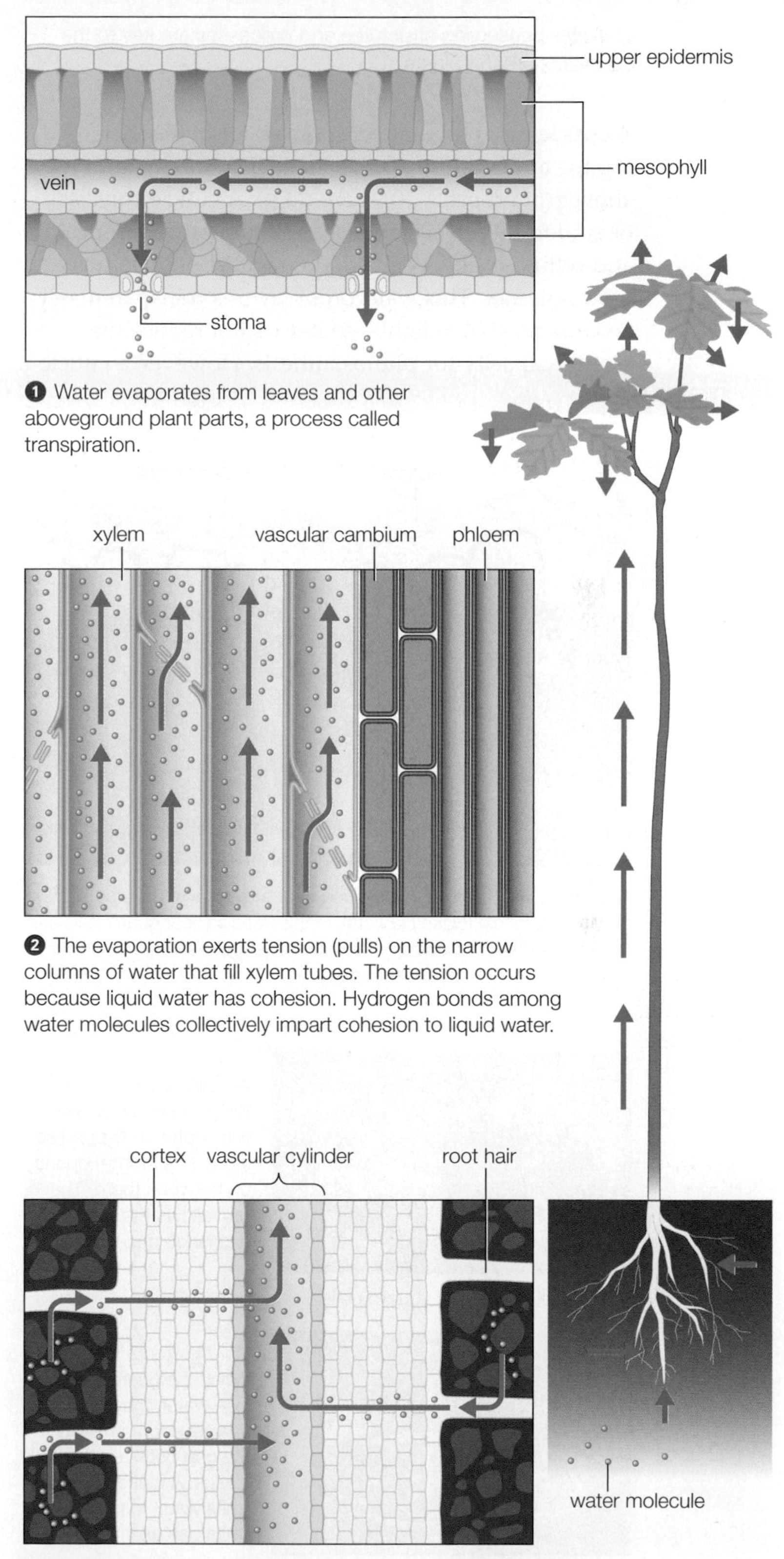

❶ Water evaporates from leaves and other aboveground plant parts, a process called transpiration.

❷ The evaporation exerts tension (pulls) on the narrow columns of water that fill xylem tubes. The tension occurs because liquid water has cohesion. Hydrogen bonds among water molecules collectively impart cohesion to liquid water.

❸ The tension inside xylem tubes extends from leaves to roots, where water molecules are being taken up from the soil.

FIGURE 28.8 Cohesion–tension theory of water transport in vascular plants.

28.5 Water-Conserving Adaptations of Stems and Leaves

✔ Water-conserving structures and processes are key to the survival of all land plants.

A cuticle helps a plant conserve water by restricting the amount of water vapor that diffuses out of its aboveground parts (**FIGURE 28.9**). A cuticle consists of epidermal cell secretions: a mixture of waxes, pectin, and cellulose fibers embedded in cutin, an insoluble lipid polymer. This waterproof layer is transparent—it absorbs no visible light—so it does not reduce the plant's capacity for photosynthesis. However, a cuticle does absorb light of less than 400 nm, so it offers protection from UV wavelengths in sunlight.

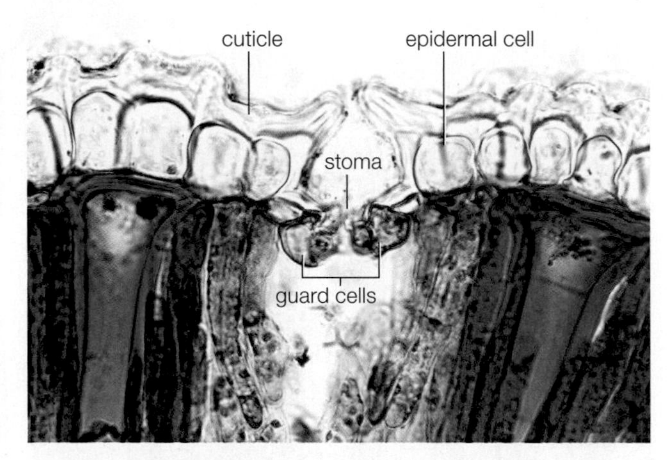

FIGURE 28.9 Water-conserving structures in a transverse section of pincushion leaf (*Hakea laurina*). In this species, leaf guard cells are recessed beneath a ledge of cuticle. The ledge creates a small cup that traps moist air above the stoma.

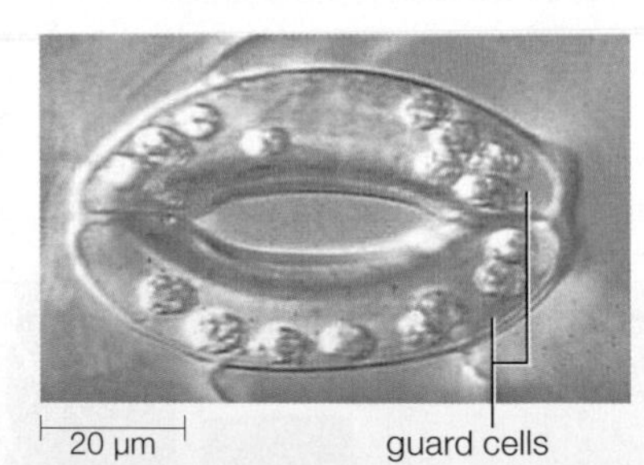

A This stoma is open. When guard cells swell with water, they bend so a gap (the stoma) opens up between them. The gap allows the plant to exchange gases with air.

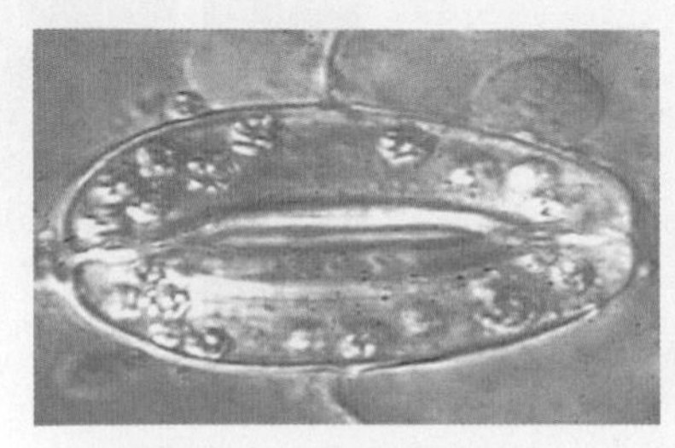

B This stoma is closed. When the guard cells lose water, they collapse against each other so the gap between them closes. A closed stoma limits water loss. It also limits gas exchange.

FIGURE 28.10 Stomata. Whether a stoma is open or closed depends on how much water is plumping up guard cells. The round structures inside the cells are chloroplasts. Guard cells are the only type of plant epidermal cell with these organelles.

A cuticle stops most water loss from aboveground plant parts, but only when stomata are closed. A pair of specialized epidermal cells defines each stoma (stoma is the singular form of stomata). When these two **guard cells** swell with water, they bend slightly so a gap (the stoma) forms between them (**FIGURE 28.10A**). When the guard cells lose water, they collapse against one another, so the stoma between them closes (**FIGURE 28.10A**). Open stomata allow water—a lot of it—to exit the plant. Even under conditions of high humidity, the interior of a leaf or stem contains more water vapor than the environment. Thus, water vapor diffuses out of a stoma whenever it is open. Indentations, ledges, or other structures that reduce air movement next to a stoma decrease the rate of evaporation from it (**FIGURE 28.9**). Nonetheless, in land plants, around 95 percent of the water taken up from soil is lost by transpiration from open stomata.

Water is important for land plants, so why do they let almost all of it evaporate away? A cuticle is not only waterproof, it is also gasproof. When a plant's stomata are closed, it cannot exchange enough carbon dioxide and oxygen with the air to support critical metabolic processes (such as the sugar-building reactions of the Calvin–Benson cycle in C3 plants, Section 6.7). Opening and closing stomata allow a plant to balance its need for water with its need to exchange gases. A lot more water is lost through open stomata than gases are gained, however, and this is the reason for the fantastic amount of water loss by transpiration. During the day, each open stoma loses about 400 molecules of water for every molecule of carbon dioxide it takes in.

How Stomata Work

Stomata open or close based on an integration of cues that include humidity, light intensity, the level of carbon dioxide inside the leaf, and hormonal signals from other parts of the plant. These cues trigger changes in osmotic pressure (Section 5.8) that affect turgor in guard cells. For example, when soil water becomes scarce, root cells release abscisic acid (ABA), a plant hormone. ABA travels through the plant's vascular system to leaves and stems, where it binds to receptors on guard cell plasma membranes. ABA binding to these receptors causes the cell to release nitric oxide (NO), a gas that functions as a hormone. The NO activates

guard cell One of a pair of cells that define a stoma across the epidermis of a leaf or other plant part.

membrane transport proteins that allow calcium ions to enter guard cell cytoplasm (**FIGURE 28.11** ❶). Cells use these ions for signaling purposes. In the presence of ABA, an increase in cytoplasmic calcium ion concentration triggers transport proteins to pump negatively charged malate (an organic acid) and chloride (Cl^-) out of the cell ❷. As a result, the overall charge of guard cell cytoplasm increases, and the overall charge of extracellular fluid decreases. A difference in charge is called voltage; when the voltage across guard cell plasma membranes changes, gated transport proteins in the membranes open (Section 5.9). The open gates allow potassium ions (K^+) to follow the charge gradient and exit the cells ❸. Water follows the solutes by osmosis, and the stomata close as the guard cells lose turgor and collapse against one another ❹.

A different type of voltage change across guard cell membranes triggers stomata to open (**FIGURE 28.12**). At sunrise, for example, the stomata of C3 and C4 plants open in response to light absorbed by receptor proteins called phototropins. Light-activated phototropins initiate events that phosphorylate ATP synthases in guard cell membranes. Remember from Section 6.5 that hydrogen ion flow through these transport proteins drives the formation of ATP during the light reactions of photosynthesis. In this case, phosphorylation drives the reverse reaction, in which ATP synthases actively pump hydrogen ions (H^+) out of guard cell cytoplasm ❶. The outflow of positively charged hydrogen ions decreases the overall charge of the cytoplasm, and increases the overall charge of extracellular fluid. The resulting voltage change across guard cell plasma membranes causes gated transport proteins in them to open. Potassium ions follow the charge gradient and flow into the cell through the opened gates ❷. Water follows the ions by osmosis. Turgor in the guard cells increases as they plump up with water, and the gap between them opens ❸. Carbon dioxide in air diffuses into the plant's tissues, and photosynthesis begins.

TAKE-HOME MESSAGE 28.5

How do stomata balance water loss with gas exchange?

✔ A cuticle minimizes water loss from a plant's aboveground parts when stomata are closed. Open stomata allow water loss, but also allow gas exchange required for metabolism.

✔ Environmental and internal cues trigger stomata to open or close. A stoma opens when the guard cells that define it become plump with water. It closes when the cells lose water and collapse against each other.

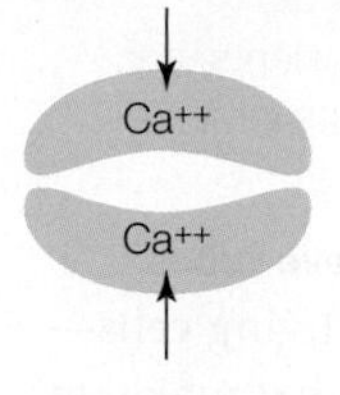

❶ ABA binding to its receptor on a guard cell membrane triggers the release of nitric oxide (NO). The NO activates calcium ion transport proteins, so these ions enter guard cell cytoplasm.

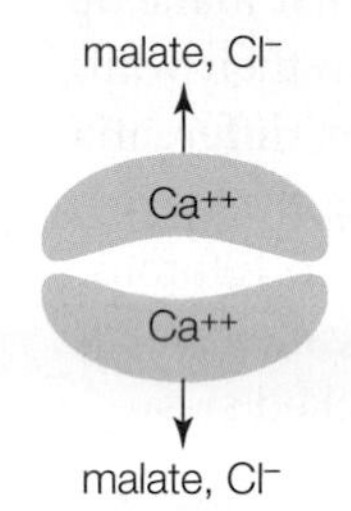

❷ The influx of calcium ions activates transport proteins that pump negatively charged malate and chloride ions out of the cells. Thus, the overall charge of cytoplasm increases, and the overall charge of extracellular fluid decreases.

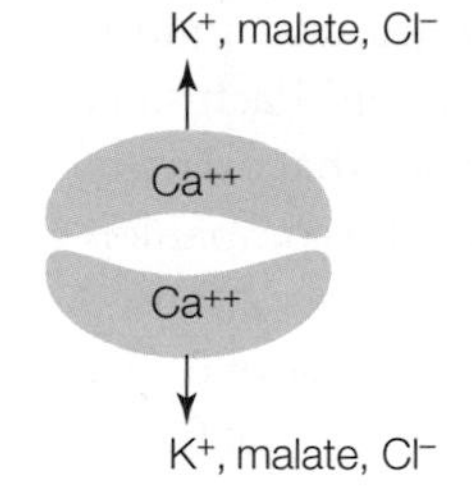

❸ The resulting voltage change across guard cell plasma membranes opens gated transport proteins in them. Potassium ions then exit the cells through the open gates.

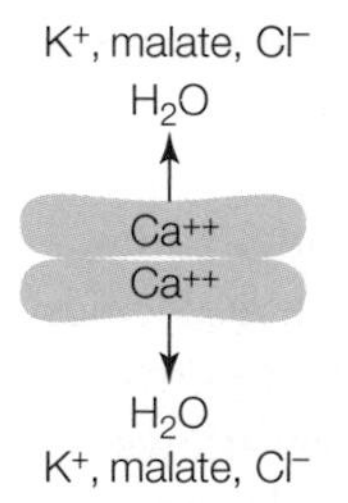

❹ Water follows the solutes by osmosis. The guard cells lose turgor and collapse against one another, so the stoma closes.

FIGURE 28.11 Abscisic acid (ABA) triggers a decrease in osmotic pressure that closes stomata.

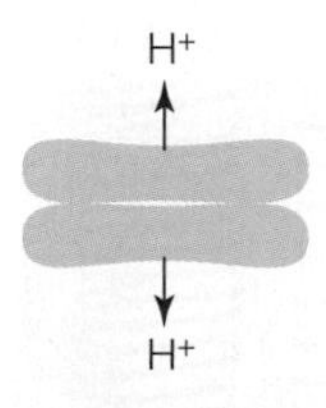

❶ Light-triggered phosphorylation of ATP synthases causes these transport proteins to pump hydrogen ions (H^+) out of guard cells.

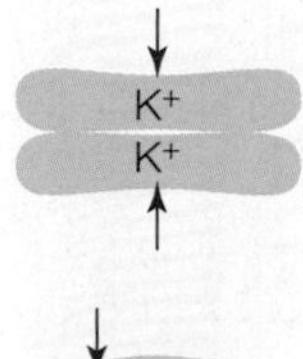

❷ The resulting change in voltage across guard cell plasma membranes opens gated transport proteins in them. Potassium ions enter the cell through the open gates.

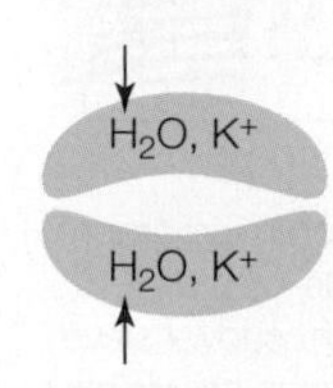

❸ Water follows the potassium ions by osmosis. Turgor increases inside the guard cells, and the stoma opens.

FIGURE 28.12 Light triggers an increase in osmotic pressure that opens stomata.

28.6 Movement of Organic Compounds in Plants

✔ Xylem distributes water and minerals through plants, and phloem distributes the organic products of photosynthesis.

Conducting tubes of phloem are called **sieve tubes** (Section 27.3). Each sieve tube consists of living cells—a stack of **sieve elements**. Unlike the cells that make up xylem tubes, sieve elements have no pits in their walls; fluid flows through their ends only. During differentiation of phloem, plasmodesmata connecting the sieve elements in a stack enlarge, sometimes up to 100-fold. The resulting pitted end walls, which are called sieve plates after their appearance, separate stacked sieve elements in mature tissue (**FIGURE 28.13**).

A maturing sieve element loses most of its organelles, including the nucleus, Golgi bodies, and almost all of its ribosomes. How does it stay alive? Each sieve element has an associated **companion cell** that arises by division of the same parenchyma cell. The companion cell retains its nucleus and other components. Many plasmodesmata connect the cytoplasm of the two cells, so the companion cell can provide all metabolic functions necessary to sustain its paired sieve element.

A plant produces sugar molecules during photosynthesis. Some of these molecules are used by or stored in the cells that make them; the rest are conducted to other parts of the plant inside sieve tubes. The movement of sugars and other organic molecules through phloem is called **translocation**. Inside sieve tubes, fluid rich in sugars—mainly sucrose—flows from a **source** (a region of plant tissue where sugars are being produced or released from storage) to a **sink** (a region where sugars are being broken down for energy, remodeled into other compounds, or stored for later use). Photosynthetic tissues are typical source regions; tissues of developing shoots and fruits are typical sink regions.

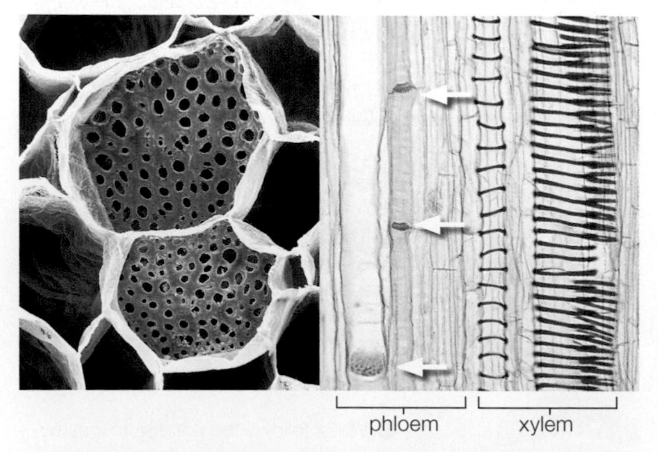

FIGURE 28.13 Sieve plates. The scanning electron micrograph on the left shows sieve plates on the ends of two side-by-side sieve-tube elements. The light micrograph on the right shows sieve plates (arrows) in two columns of phloem. Rings of lignin are visible as red stripes in xylem vessel members (older vessels stretch longitudinally as cells in newer tissue lengthen).

Why does sugar-rich fluid flow from source to sink? A pressure gradient between the two regions drives the movement (**FIGURE 28.14**). Consider how photosynthesizing mesophyll cells in a leaf produce far more sugar than they use for their own metabolism. Their activity sets up the leaf as a source region. Excess sugar molecules diffuse from photosynthesizing mesophyll cells into adjacent companion cells through plasmodesmata. Depending on the species and the source region, sugar molecules may also be pumped into a companion cell through active transport proteins in its membrane. Either way, the sugar molecules then diffuse from the companion cells to their associated sieve elements through other plasmodesmata ❶. This sugar loading increases the solute concentration of cytoplasm in a sieve element so that it becomes hypertonic with respect to the surrounding cells. Water follows the sugar by osmosis, moving into the sieve element from the surrounding cells ❷. The rigid cell walls of a mature sieve element cannot expand very much, so the influx of water raises the osmotic pressure (turgor) inside the cell. This pressure can be very high—up to five times higher than that of an automobile tire. The high fluid pressure pushes the sugar-rich cytoplasm from one sieve element to the next, toward a sink region where the osmotic pressure is lower ❸. Pressure inside a sieve tube decreases at a sink because sugars leave the tube in this region. Sugar molecules move into companion cells (either through plasmodesmata or by active transport), and then diffuse into sink cells through plasmodesmata. Water follows, again by osmosis ❹, so turgor inside sieve elements decreases in these regions. This explanation of how osmotic pressure pushes sugar-rich fluid inside a sieve tube from source to sink is called the **pressure flow theory**.

companion cell In phloem, specialized parenchyma cell that provides metabolic support to its partnered sieve element.
pressure flow theory Explanation of how a difference in turgor between sieve elements in source and sink regions pushes sugar-rich fluid through a sieve tube.
sieve elements Living cells that compose sugar-conducting sieve tubes of phloem. Each sieve tube consists of a stack of sieve elements that meet end to end at sieve plates.
sieve tube Sugar-conducting tube of phloem; consists of stacked sieve elements.
sink Region of plant tissue where sugars are being used.
source Region of a plant tissue where sugars are being produced or released from storage.
translocation Movement of organic compounds through phloem.

CREDITS: (13) left, © J.C. Revy/ISM/Phototake; right, © M.I. Walker/Wellcome Images.

companion cell — sieve element

Source

sugars

1. At a source region, sugars move into a companion cell, then into a sieve element.

water

2. The increase in solute concentration causes fluid in the sieve element to become hypertonic with respect to the surrounding cells. Water moves by osmosis from these cells into the sieve element, thus increasing pressure inside of it (turgor).

flow

3. The high pressure pushes the fluid through the sieve tube, toward a sink region where internal pressure is lower.

Sink

sugars

water

4. At a sink region, sugars move from sieve elements into sink cells. Water follows by osmosis.

FIGURE 28.14 ▶**Animated** Translocation of organic compounds in phloem from source to sink. Water molecules are represented by blue balls; sugar molecules, by red balls. Each sieve tube is a stack of sieve elements that meet end to end at perforation plates. Sugars move into and out of a sieve element via its associated companion cell.

TAKE-HOME MESSAGE 28.6

How do sugars move through phloem?

✔ A sieve tube consists of stacked sieve elements separated by porous end walls (sieve plates). Each sieve element is connected by plasmodesmata to a companion cell.

✔ At source regions, sugars move into companion cells, then into sieve tubes. Sugars exit sieve tubes at sink regions.

✔ A pressure gradient pushes sugar-rich fluid inside a sieve tube from a source to a sink.

Leafy Cleanup (revisited)

At Ford Motor Company's Rouge Center in Dearborn, Michigan, decades of steelmaking left soil contaminated with highly carcinogenic compounds known as polycyclic aromatic hydrocarbons, or PAHs. In 2000, the automaker funded development of a phytoremediation system that also boosted an initiative to restore the facility's native wildlife habitat (**FIGURE 28.15**). Unlike the phytoremediation strategy at J-Field, Ford's is based on native plants. Today, the Rouge Center's plantings are attracting insects, birds, and other wildlife while aggressively accelerating the natural degradation process of toxins. "If left undisturbed, it would take decades or centuries for these contaminants to naturally decompose," said researcher Clayton Rugh. "What our research is indicating is that we can achieve at least 50 percent degradation in three to five years—and that's at the very least. Some [species] seem to be approaching 70 percent in just three growing seasons."

FIGURE 28.15 Top, Ford Motor Company's Rouge Center in 1927. Bottom, the Rouge Center today. This 10-acre green roof over a truck manufacturing plant is the heart of a phytoremediation system that cleanses contaminated storm water as it moves across the site.

summary

Section 28.1 The ability of plants to take up substances from soil water is the basis for phytoremediation, which is a method that uses plants to remove pollutants from a contaminated area.

Section 28.2 Plant growth requires steady sources of elemental nutrients. Oxygen, carbon, and hydrogen atoms are abundant in air and water; nitrogen, phosphorus, sulfur, and other elements are available in soil.

The availability of water and mineral ions in a particular soil depends on its proportions of sand, silt, and clay, and also on its **humus** content. **Loams** have roughly equal proportions of sand, silt, and clay. **Leaching** and **soil erosion** deplete nutrients from soil, particularly **topsoils**.

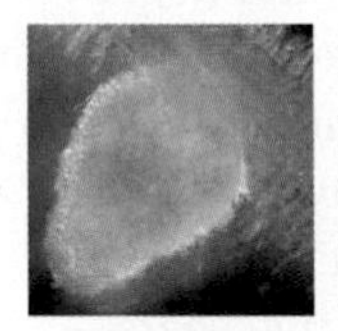

Section 28.3 Transport proteins in root cell plasma membranes control the plant's uptake of substances in soil water. Endodermal cells that form the vascular cylinder's outer layer deposit a **Casparian strip** into their abutting walls. This waxy, waterproof band prevents soil water from diffusing around endodermal cells into root xylem. Substances such as ions in soil water must pass through membrane transport proteins of an endodermal cell (or other root cell). Once in cytoplasm, the ions can diffuse through plasmodesmata to pericycle cells, which load them into xylem.

Many plants form mutualisms with microorganisms. Fungi associate with young roots in mycorrhizae, which enhance the plant's ability to absorb mineral ions from soil. Nitrogen fixation by bacteria in **root nodules** gives a plant extra nitrogen. In both cases, the microorganisms receive some sugars made by the plant.

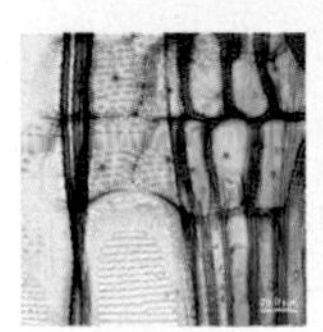

Section 28.4 Water and dissolved mineral ions flow through xylem from roots to shoot tips. Xylem tubes consist of stacks of dead cells: **tracheids** and **vessel elements**. Water moves through the perforated and lignin-reinforced secondary walls of these cells.

The **cohesion–tension theory** explains how water moves upward through xylem: **Transpiration** (the evaporation from aboveground plant parts, mainly at stomata) pulls water upward. This pull (tension) extends from leaves to roots because of water's cohesion inside the narrow tubes of xylem. Cohesion also keeps the water from breaking into droplets, so it moves upward in continuous columns.

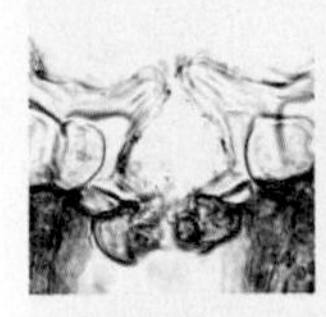

Section 28.5 A cuticle helps a plant conserve water; stomata help it balance water conservation with gas exchange required for metabolism. A stoma, which is a gap between two **guard cells**, may be surrounded by an indentation, protrusions, or other specializations that reduce airflow around it.

Environmental and internal signals cause stomata to open or close. The signals trigger guard cells to pump ions into or out of their cytoplasm; water follows the ions by osmosis. Water moving into guard cells plumps them, which opens the stoma between them. Water diffusing out of the cells causes them to collapse against each other, so the stoma closes.

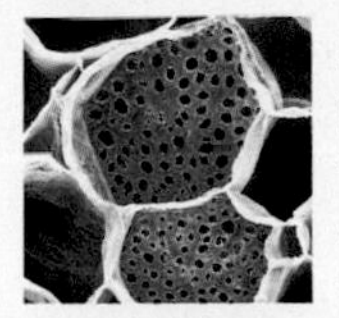

Section 28.6 Sugars move through a plant by **translocation** in phloem's **sieve tubes**, which consist of stacked **sieve elements** separated by perforated sieve plates. By the **pressure flow theory**, the movement of sugar-rich fluid through a sieve tube is driven by a pressure gradient between **source** and **sink**. **Companion cells** load sugars into sieve elements at sources.

self-quiz

Answers in Appendix VII

1. Carbon, hydrogen, and oxygen are ________ for plants.
   a. macronutrients
   b. micronutrients
   c. trace elements
   d. required elements
   e. both a and b
   f. both a and d

2. Decomposing organic matter in soil is called ________ .
   a. clay
   b. humus
   c. silt
   d. sand

3. A vascular cylinder consists of ________ .
   a. exodermis
   b. endodermis
   c. root cortex
   d. xylem and phloem
   e. b and d
   f. all of the above

4. A ________ strip between abutting endodermal cell walls forces water and solutes to move through these cells rather than around them.
   a. cutin b. Casparian c. cohesion d. cellulose

5. The nutrition of some plants is enhanced by a mutually beneficial association between a root and a fungus. The association is known as a ________ .
   a. root nodule
   b. mycorrhiza
   c. root hair
   d. hypha

6. Water evaporation from plant parts is called ________ .
   a. translocation
   b. respiration
   c. transpiration
   d. tension

7. Water transport from roots to leaves occurs by ________ .
   a. a pressure gradient inside sieve tubes
   b. different solutes at source and sink regions
   c. the pumping force of xylem vessels
   d. transpiration, tension, and cohesion of water

8. Tracheids are part of ________ .
   a. cortex
   b. mesophyll
   c. phloem
   d. xylem

9. Sieve tubes are part of ________ .
   a. cortex
   b. mesophyll
   c. phloem
   d. xylem

10. With stomata closed, a waterproof cuticle ________ .
    a. minimizes water loss through plant surfaces
    b. inhibits gas exchange between the plant and the air
    c. both a and b

data analysis activities

TCE Uptake by Transgenic Plants Plants used for phytoremediation take up organic pollutants, then transport the chemicals to plant tissues, where they are stored or broken down. Researchers are now designing transgenic plants with enhanced ability to take up or break down toxins. In 2007, Sharon Doty and her colleagues published the results of their efforts to design plants for phytoremediation of soil and air containing organic solvents. The researchers used *Agrobacterium tumefaciens* (Section 15.7) to deliver a mammalian gene into poplar plants. The gene encodes cytochrome P450, an enzyme involved in the breakdown of a range of organic molecules, including solvents such as TCE. **FIGURE 28.16** shows data from one test on the resulting transgenic plants.

1. How many transgenic plants did the researchers test?
2. In which group did the researchers see the slowest rate of TCE uptake? The fastest?
3. On day 6, what was the difference between the TCE content of air around planted transgenic plants and that around vector control plants?
4. Assuming no other experiments were done, what two explanations are there for the results of this experiment? What other control might the researchers have used?

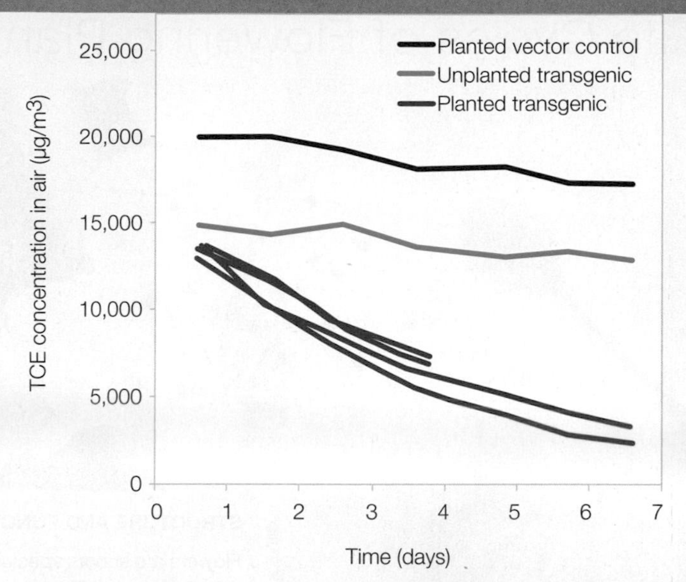

FIGURE 28.16 TCE uptake from air by transgenic poplar plants. Individual potted plants were kept in separate sealed containers with an initial level of TCE (trichloroethylene) around 15,000 micrograms per cubic meter of air. Samples of the air in the containers were taken daily and measured for TCE content. Controls included a tree transgenic for a Ti plasmid with no cytochrome P450 in it (vector control), and a bare-root transgenic tree (one that was not planted in soil).

11. When guard cells swell, ________ .
    a. transpiration ceases
    b. stomata close
    c. stomata open
    d. root cells die

12. Stomata open in response to light when ________ .
    a. ions flow into guard cell cytoplasm
    b. ions flow out of guard cell cytoplasm
    c. water evaporates out of guard cells

13. In phloem, organic compounds flow through ______ .
    a. collenchyma cells
    b. sieve tubes
    c. vessels
    d. tracheids

14. Sugar transport from leaves to roots occurs by ________ .
    a. a pressure gradient inside sieve tubes
    b. different solutes at source and sink regions
    c. the pumping force of xylem vessels
    d. transpiration, tension, and cohesion of water

15. Match the concepts of plant nutrition and transport.

___ stomata	a. separates cells in tubes of phloem
___ sieve plate	b. takes up soil water and nutrients
___ sink	c. balance water loss with gas exchange
___ root system	d. cohesion in xylem tubes
___ hydrogen bonds	e. sugars unloaded from sieve tubes
___ xylem	f. distributes sugars through the plant body
___ phloem	g. separates cells in tubes of xylem
___ perforation plate	h. distributes water through the plant body

critical thinking

1. Nitrogen deficiency stunts plant growth; leaves yellow and then die. Why does nitrogen deficiency causes these symptoms? *Hint:* think about which biological molecules incorporate nitrogen atoms.

2. You just returned home from a three-day vacation. Your severely wilted plants tell you they were not watered before you left. Use the cohesion–tension theory of water transport to explain what happened to them.

3. When you dig up a plant to move it from one spot to another, the plant is more likely to survive if some of the soil around the roots is transferred to the new location along with the plant. Make a hypothesis that explains this observation.

4. If a plant's stomata are made to stay open at all times, or closed at all times, it will die. Why?

5. What are the structures shown in the micrographs in **FIGURE 28.17**? In which plant tissue(s) can these structures be found?

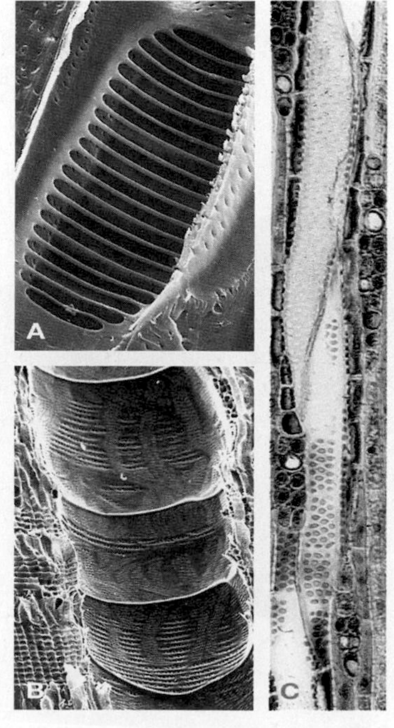

FIGURE 28.17 Name the mystery structures.

CREDITS: (16) © Cengage Learning; (17A–B) H. A. Core, W. A. Cote and A. C. Day, *Wood Structure and Identification*, 2nd Ed., Syracuse University Press, 1979; (17C) Alison W. Roberts, University of Rhode Island.

29 Life Cycles of Flowering Plants

LEARNING ROADMAP

This chapter revisits what you have learned about plants (Sections 4.11, 17.9, 17.11, 17.12, 22.2, 22.3, 22.6, 22.8, 22.9, 27.2–27.4, 27.7), molecules of life (3.3, 3.4), membranes (5.7), photosynthesis (6.4), respiration (7.5), cloning (8.7), gene control (10.2, 10.4), reproduction (12.2, 12.3), inheritance (13.2, 14.6), radiometric dating (16.6), and phylogeny (18.2).

STRUCTURE AND FUNCTION OF FLOWERS

Flowers are shoots specialized for reproduction. Their parts are modified leaves. Flowers of many plants are specialized to attract and reward coevolved animal pollinators.

GAMETES AND FERTILIZATION

Male and female gametophytes develop inside the reproductive parts of flowers. Pollination is followed by double fertilization. As in animals, signals are key to sex.

SEEDS AND FRUIT

After fertilization, an ovule matures into a seed. As this occurs, tissues of the ovary and often other parts of the flower mature into a fruit, which functions in seed dispersal.

GROWTH AND DEVELOPMENT

Plant development includes germination, root and shoot development, flowering, fruit formation, and dormancy. All have a genetic basis; all are triggered by environmental cues.

ASEXUAL REPRODUCTION IN PLANTS

Many species of plants can undergo vegetative reproduction. Humans take advantage of this natural tendency by propagating plants asexually for agriculture and research.

Chapter 30 details how plant development (including germination and cyclic patterns of growth) is mediated by signaling molecules and triggered by environmental cues. Chapter 44 discusses limits on population growth. Competitive interactions among species in a community are explained in Chapter 45, and Chapter 47 returns to forests and other biomes. Human impacts on the biosphere are further explored in Chapter 48.

29.1 Plight of the Honeybee

In the fall of 2006, commercial beekeepers worldwide began to notice something was amiss in their honeybee hives. All of the adult worker bees were missing from the hives, but there were no bee corpses to be found. The bees had suddenly (and quite unusually) abandoned their queen and many living larvae. The few young workers that remained were reluctant to feed on abundant pollen and honey stored in their own hive, or in other abandoned hives.

A third of the colonies did not survive the following winter. By spring, the phenomenon had a name: colony collapse disorder (CCD). Since then, honeybee populations have continued to decline. In 2011, thirty percent of bee colonies in the United States and twenty percent of those in Europe collapsed.

CCD is a problem that affects more than just bees and honey lovers. Nearly all of our food crops are the fruits of angiosperms (Section 22.9). Like other seed-bearing plants, angiosperms produce pollen (Section 22.6). Many angiosperm species depend on honeybees and other **pollinators** to carry pollen from one plant to another (**FIGURE 29.1A**). Fruits do not form from unpollinated flowers, and in some species they only form when flowers receive pollen from a different plant of the same species. Even species with flowers that can self-pollinate tend to make bigger fruits and more of them when they are cross-pollinated (**FIGURE 29.1B**).

Many types of insects pollinate plants, but honeybees are especially efficient pollinators of a wide variety of plant species. They are also the only ones that tolerate living in man-made hives that are loaded onto trucks and transported to wherever crops require pollination. The potential loss of their portable pollination service is a huge threat to our agricultural economy. Currently, about half of the world's leading crops—about $212 billion worth—depend on insect pollinators, mainly bees.

CCD is currently in the spotlight because it directly affects our food supply. However, populations of bumblebees and other insect pollinators are also dwindling. Habitat loss may be the main factor, but diseases and pesticides that harm honeybees also harm other invertebrate pollinators.

Flowering plants rose to dominance in part because they coevolved with animal pollinators. Most flowers are specialized to attract and be pollinated by a specific type (or even a specific species) of animal. Those adaptations become handicaps if coevolved pollinator populations decline. Wild animal species that depend on the plants for fruits and seeds will also be affected. Recognizing the prevalence and importance of these interactions is our first step toward finding workable ways to protect them.

pollinator An organism that moves pollen from one plant to another.

A Honeybees are efficient pollinators of a variety of flowers, including those of berry plants (*Rubus*).

B Raspberry flowers can pollinate themselves, but the fruit that forms from a self-fertilized flower is of lower quality than that of a cross-pollinated flower.

The two raspberries on the left formed from self-pollinated flowers. The one on the right formed from an insect-pollinated flower.

FIGURE 29.1 An example of the importance of insect pollinators.

29.2 Reproductive Structures

✔ Specialized reproductive structures called flowers consist of whorls of modified leaves.

Flowers are the specialized reproductive structures of angiosperms. The form and composition of flowers vary dramatically among species, but all develop at the tips of special reproductive shoots (**FIGURE 29.2A**).

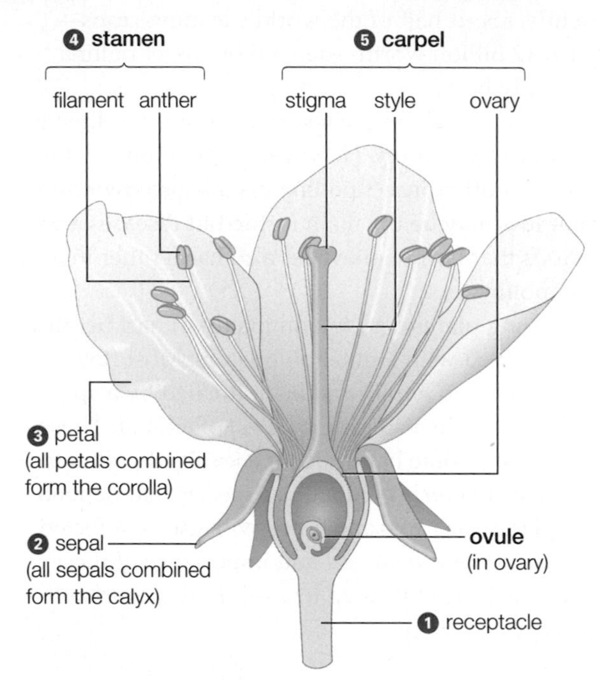

A The structures that produce male gametophytes are stamens, which consist of pollen-bearing anthers atop slender filaments. The structure that produces female gametophytes is the carpel, which consists of ovary, stigma, and style.

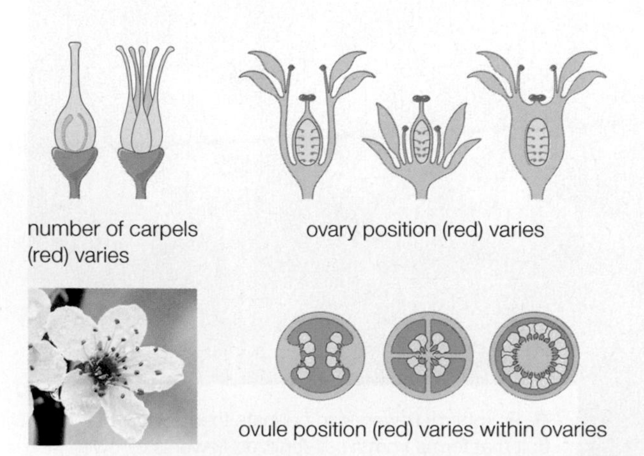

B Above, a typical cherry flower has five petals, several stamens, and one carpel. Within the carpel, one ovule forms inside one ovary. Other species have different patterns.

FIGURE 29.2 ▸**Animated** Anatomy of a typical flower: cherry (*Prunus avium*).

A flower's parts are modified leaves that develop from a thickened region of stem called a receptacle ❶. A typical flower has four spirals or rings (whorls) of modified leaves. As the flower forms, expression of floral identity genes (Section 10.4) determines the fate of each whorl. Typically, the outermost whorl is a **calyx**, which is a ring of leaflike **sepals** ❷. Sepals of most species are photosynthetic and inconspicuous; they serve to protect the flower's more delicate parts. The whorl just inside the calyx is the **corolla** (from the Latin *corona*, or crown), which is a ring of **petals** ❸. Petals are typically the largest and most brightly colored parts of a flower.

Inside the corolla is a whorl of **stamens**, the reproductive organs that produce the plant's male gametophytes (remember from Section 22.2 that gametophytes are the structures that produce gametes). In most flowers, a stamen consists of a thin filament with an **anther** at its tip ❹. A typical anther contains four pouches called pollen sacs. Pollen grains, which are immature male gametophytes (Section 22.3), form inside pollen sacs.

The flower's innermost whorl consists of modified leaves folded and fused into one or more carpels. A **carpel** is the reproductive organ that produces the plant's female gametophytes ❺. The flowers of some species have one carpel; those of other species have several (**FIGURE 29.2B**). A carpel—or a compound structure that consists of multiple fused carpels—is commonly called a pistil. The upper region of a carpel is a sticky or hairy **stigma** that is specialized to receive pollen grains. Typically, the stigma sits on top of a slender stalk called a style. The lower, swollen region of a carpel is the **ovary**, which contains one or more ovules. An **ovule** is a structure that bulges from the inside of the ovary wall; a female gametophyte forms inside it.

anther Part of the stamen that produces pollen grains.
calyx A flower's outer, protective whorl of sepals.
carpel Floral reproductive organ that consists of an ovary, stigma, and often a style.
corolla A flower's whorl of petals; forms within sepals and encloses reproductive organs.
flower Specialized reproductive structure of a flowering plant.
ovary In flowering plants, the enlarged base of a carpel, inside which one or more ovules form.
ovule Of a seed-bearing plant, structure in which a female gametophyte forms.
petal Unit of a flower's corolla; often showy and conspicuous.
sepal Unit of a flower's calyx; typically photosynthetic and inconspicuous.
stamen Floral reproductive organ that consists of an anther and, in most species, a filament.
stigma Upper part of a carpel; adapted to receive pollen.

A The solitary flower of the lady's slipper orchid (*Paphiopedilum*) is irregular.

B A flower of hyacinth (*Hyacinthus orientalis*) is an elongated inflorescence.

C A daisy (*Gerbera jamesonii*) is a composite flower with many individual florets.

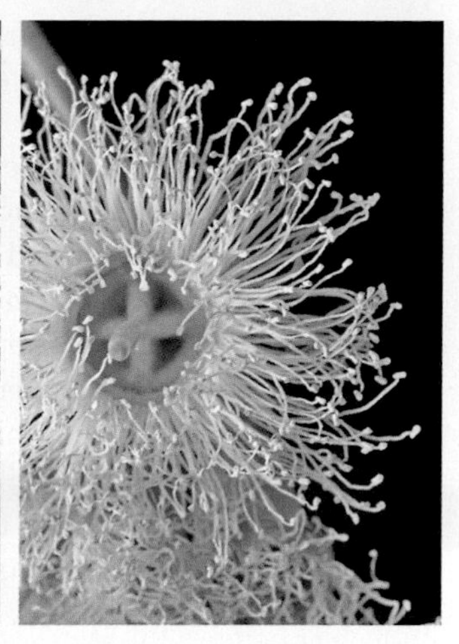

D A flower of eucalyptus (*Eucalyptus robusta*) has no petals, so it is incomplete.

FIGURE 29.3 Examples of structural variation in flowers.

Diversity of Floral Structure

Master gene mutations give rise to dramatic variations in flower structure, and we see many such variations in the range of diversity of flowering plants. For example, the number of floral parts differs between monocots and eudicots (Section 27.2). Flowers may form as solitary blossoms (**FIGURE 29.3A**) or in clusters called inflorescences (**FIGURE 29.3B**). A composite flower is an inflorescence of many flowers ("florets") grouped as a single head (**FIGURE 29.3C**). Cherry blossoms and other regular flowers are symmetric around their center axis, which means if the flower were cut like a pie, the pieces would be roughly identical. Irregular flowers are not radially symmetric (the flower shown in **FIGURE 29.3A** is an example).

A cherry blossom has all four sets of modified leaves (sepals, petals, stamens, and carpels), so it is called a "complete" flower. An "incomplete" flower lacks one or more of these structures (**FIGURE 29.3D**). Cherry blossoms are also "perfect" flowers, which means they have both stamens and carpels. Perfect flowers can be cross-pollinated (fertilized by pollen from other plants), or they can self-pollinate. Self-pollination can be adaptive in situations where plants are widely spaced, such as in newly colonized areas. However, offspring that arise from self-pollinated flowers are often less vigorous than those of cross-pollinated plants. Accordingly, floral adaptations of many plant species encourage or even require cross-pollination. For example, a perfect flower may release pollen only after its stigma is no longer receptive to being fertilized. "Imperfect" flowers, which lack stamens or carpels (**FIGURE 29.4**), cannot self-fertilize. In some species, stamen-bearing and carpel-bearing flowers form on different plants, so fertilization only occurs by cross-pollination.

FIGURE 29.4 Imperfect flowers: in some begonias, female blossoms (with no stamens, left) form on the same plants as male blossoms (with no carpels, right).

TAKE-HOME MESSAGE 29.2

What are flowers?

✔ Flowers are the reproductive structures of angiosperms. The different parts of a flower (sepals, petals, stamens, and carpels) are modified leaves. Flowers vary in structure.

✔ The parts of flowers that produce male gametophytes are stamens, which typically consist of a filament with an anther at the tip. Pollen forms inside the anthers.

✔ The parts of flowers that produce female gametophytes are carpels, which typically consist of stigma, style, and ovary. The gametophytes form in an ovule inside the ovary.

29.3 Flowers and Their Pollinators

✔ Flower traits often reflect a dependence on a coevolved animal pollinator.

Pollination, the arrival of pollen on a receptive stigma, is essential to sexual reproduction in flowering plants. Pollen is typically transferred from anther to stigma by a **pollination vector**, which can be an animal pollinator or an environmental agent such as wind.

Fragrance and other flower traits not directly related to reproduction are evolutionary adaptations that attract coevolved pollinators. An animal attracted to a particular flower often picks up pollen on a visit, then inadvertently transfers it to another flower on a subsequent visit. The more specific the attraction, the more efficient the transfer of pollen among plants of the same species. Thus, a flower's traits typically reflect the sensory abilities and preferences of its coevolved pollinator. In an example of morphological convergence (Section 18.3), the flowers of angiosperm species with the same type of pollinator (bats, for example, or birds) often share a set of traits called a pollination syndrome (**TABLE 29.1**). Consider that bees have a keen sense of smell, and they can see UV light, so bee-pollinated flowers tend to be fragrant, with a bull's-eye pattern of UV-reflecting pigments in their petals (**FIGURE 29.5**). Bee-pollinated flowers also attract and communicate with individual bees via electric fields. Birds, which rely more on their excellent sense of vision than their relatively poor sense of smell, are attracted to bright red, unscented flowers (**FIGURE 29.6A**). Night-flying bats have a good sense of smell but poor vision, so bat-pollinated flowers tend to be large and light-colored, with a strong nighttime fragrance (**FIGURE 29.6B**). Not all flowers smell sweet; odors like dung or rotting flesh beckon beetles and flies.

FIGURE 29.5 UV-reflecting patterns in flower petals. The image on the left shows a primrose flower as we see it. The image on the right shows the same flower photographed with a special filter that allows us to see reflected UV light, here represented by the color red. The bull's-eye pattern guides bees, which can see UV light, directly to the flower's reproductive parts.

Rewards offered by a flower reinforce a pollinator's memory of a visit, and encourage the animal to seek out other flowers of the same species. Many flowers exude a sweet fluid called **nectar** that is prized by many pollinators. Nectar is the only food for most adult butterflies, and it is the food of choice for hummingbirds. Honeybees convert nectar to honey, which helps feed the bees and their larvae through the winter. Bees also collect protein-rich pollen for food. Many beetles feed primarily on pollen. Pollen is the only reward for a pollinator of roses, poppies, and other flowers that produce no nectar. Some flowers offer other, less typical rewards (**FIGURE 29.6C**).

Table 29.1 Pollination Syndromes

Flower Trait:	Animal Pollination Vector: Bats	Bees	Beetles	Birds	Butterflies	Flies	Moths
Color:	Dull white, green, purple	Bright white, yellow, blue, UV	Dull white or green	Scarlet, orange, red, white	Bright, such as red, purple	Pale, dull, dark brown or purple	Pale/dull red, pink, purple, white
Odor:	Strong, musty, emitted at night	Fresh, mild, pleasant	None to strong	None	Faint, fresh	Putrid	Strong, sweet, emitted at night
Nectar:	Abundant, hidden	Usually	Sometimes, not hidden	Ample, deeply hidden	Ample, deeply hidden	Usually absent	Ample, deeply hidden
Pollen:	Ample	Limited, often sticky, scented	Ample	Modest	Limited	Modest	Limited
Shape:	Regular, bowl-shaped, closed during the day	Shallow with landing pad; tubular	Large, bowl-shaped	Large funnel-shaped cups, strong perch	Narrow tube with spur; wide landing pad	Shallow, funnel-shaped or trap-like and complex	Regular; tube-shaped with no lip
Examples:	Banana, agave	Larkspur, violet	Magnolia, dogwood	Fuschia, hibiscus	Phlox	Skunk cabbage, philodendron	Tobacco, lily, some cacti

A Birds. Pollen accumulates on the feathers of a greater double-collared sunbird as it sips nectar from a bottlebrush flower.

B Mammals. A lesser long-tongued fruit bat sips nectar from (and pollinates) flowers of wild banana.

C Insects. Female burnet moths perch on purple blossoms—preferably pincushion flowers—when they are ready to mate. The visual combination attracts male moths; the activity of the moths pollinates the flowers.

FIGURE 29.6 Examples of animals pollinating flowers.

Specializations of some flowers prevent pollination by all but one type of pollinator. For example, nectar or pollen at the bottom of a long floral tube may be accessible only to an insect with a matching feeding device. In some plants, stamens adapted to brush against a pollinator's body or lob pollen onto it will function only when triggered by that pollinator (**FIGURE 29.7**). Such relationships are to both species' mutual advantage: A flower that captivates the attention of an animal has a pollinator that spends its time seeking (and pollinating) only those flowers. Pollinators also benefit when they receive an exclusive supply of the flower's reward.

Grasses and other plants pollinated by wind do not benefit by attracting animals, so they do not waste resources making scented, brilliantly colored, or nectar-laden flowers. Wind-pollinated flowers are typically small, nonfragrant, and green, with large stigmas and insignificant petals or sepals. Wind disperses pollen randomly throughout the environment, so these flowers typically make and release pollen grains by the billions, insurance in numbers that some of their pollen will reach a receptive stigma.

FIGURE 29.7 An example of floral specialization for a specific coevolved pollinator. The zebra orchid (*Caladenia cairnsiana*) mimics the scent of a female wasp. Male wasps follow the scent to the flower, then try to copulate with and lift the dark red mass of tissue on the lip. The wasp's movements trigger the lip to tilt upward, which brushes the wasp's back against the flower's stigma and pollen.

nectar Sweet fluid exuded by some flowers that attracts animal pollinators.
pollination The arrival of pollen on a receptive stigma.
pollination vector Environmental agent that moves pollen grains from one plant to another.

TAKE-HOME MESSAGE 29.3

What is the purpose of the nonreproductive traits of flowers?

✔ Most flowering plants coevolved with animal pollinators. The shape, pattern, color, and fragrance of a flower attract the plant's coevolved pollinator.

✔ Pollinators are often rewarded for visiting a flower, for example by receiving pollen or nectar from it.

CREDITS: (6A) © Dietmar Nill/Foto Natura/Minden Pictures/Corbis; (6B) © Ch'ien Lee/Minden Pictures/Corbis; (6C) © David Goodin; (7) John Alcock, Arizona State University.

29.4 A New Generation Begins

✔ In flowering plants, fertilization has two outcomes: It results in a zygote, and it is the start of endosperm.

The life cycle of flowering plants (**FIGURE 29.8**) is dominated by the sporophyte, which is a diploid spore-producing plant body that grows by mitotic cell divisions of a fertilized egg. Spores that form by meiosis inside a sporophyte's flowers develop into haploid gametophytes, which produce gametes.

The production of female gametes begins when a mass of tissue—the ovule—starts growing on the inner wall of an ovary (**FIGURE 29.9** ❶). One cell in the middle of the mass undergoes meiosis and cytoplasmic division, forming four haploid **megaspores** ❷. Three of the four megaspores typically disintegrate. The remaining megaspore undergoes three rounds of mitosis without cytoplasmic division, thus producing a single cell that has eight haploid nuclei ❸. The cytoplasm of this cell divides unevenly, forming a seven-celled female gametophyte ❹. The gametophyte is enclosed and protected by layers of cells, or integuments, that developed from the outer layers of the ovule. One of the cells in the gametophyte, the endosperm mother cell, has two nuclei ($n + n$). Another cell is the egg.

The production of male gametes begins as masses of diploid, spore-producing cells form by mitosis inside anthers. Walls typically develop around the masses, so four pollen sacs form ❺. Each cell inside the pollen sacs undergoes meiosis and cytoplasmic division to form four haploid **microspores** ❻. Mitosis and differentiation of a microspore produce a pollen grain ❼. A pollen grain consists of two cells, one inside the cytoplasm of the other, enclosed by a durable coat. The coat will protect the cells inside on their journey to meet an egg. After the pollen grains form, they enter **dormancy**, a period of suspended metabolism, before being released from the anther when the pollen sacs split open ❽.

A pollen grain that lands on a receptive stigma **germinates**, which means it resumes metabolic activity after dormancy. When a pollen grain germinates, its outer cell develops into a tubular outgrowth called a pollen tube ❾. Its inner cell undergoes mitosis to produce two sperm cells (the male gametes) within the pollen tube. A pollen tube together with its contents of male gametes constitutes the mature male gametophyte. The pollen tube elongates at its tip, carrying the two sperm cells down through the tissues of the style and ovary toward the ovule.

A pollen tube that reaches and penetrates the female gametophyte releases the two sperm cells ❿. Flowering plants undergo **double fertilization**, in which one of the sperm cells delivered by the pollen tube fuses with (fertilizes) the egg and forms a diploid zygote; the other sperm cell fuses with the endosperm mother cell, forming a triploid ($3n$) cell. This cell gives rise to triploid **endosperm**, a nutritious tissue that forms only in seeds of flowering plants. When the seed sprouts, endosperm will sustain rapid growth of the seedling until its new leaves begin photosynthesis.

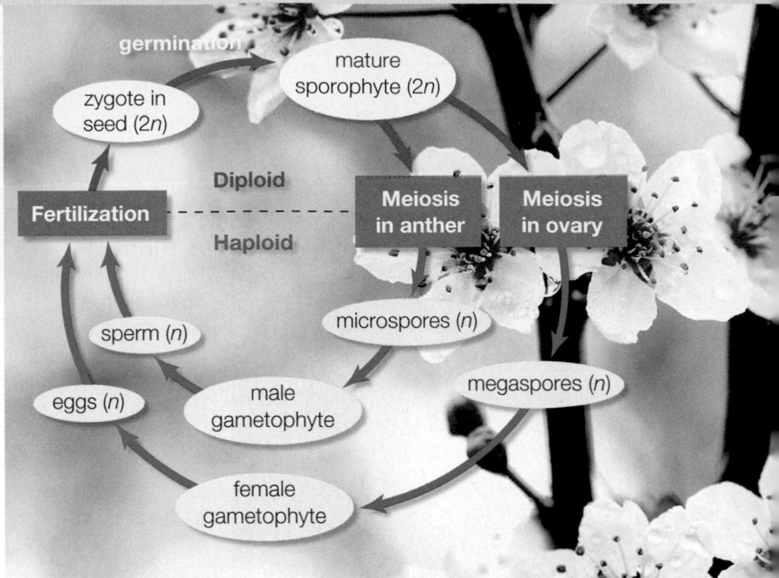

FIGURE 29.8 ▸**Animated** Overview of the life cycle of a typical flowering plant (compare **FIGURE 29.9**, opposite).

dormancy Period of temporarily suspended metabolism.
double fertilization Mode of fertilization in flowering plants in which one sperm cell fuses with the egg, and a second sperm cell fuses with the endosperm mother cell.
endosperm Nutritive tissue in the seeds of flowering plants.
germinate To resume metabolic activity after dormancy.
megaspore Of seed plants, haploid spore that forms in an ovule and gives rise to an egg-producing gametophyte.
microspore Of seed plants, haploid spore that gives rise to a pollen grain.

TAKE-HOME MESSAGE 29.4

How does sexual reproduction occur in flowering plants?

✔ A pollen grain that germinates on a stigma develops into the male gametophyte, which consists of a pollen tube and two sperm cells. The pollen tube grows into the carpel, enters an ovule, and releases the sperm cells.

✔ Double fertilization occurs when one of the sperm cells delivered by the pollen tube fuses with the egg, and the other fuses with the endosperm mother cell.

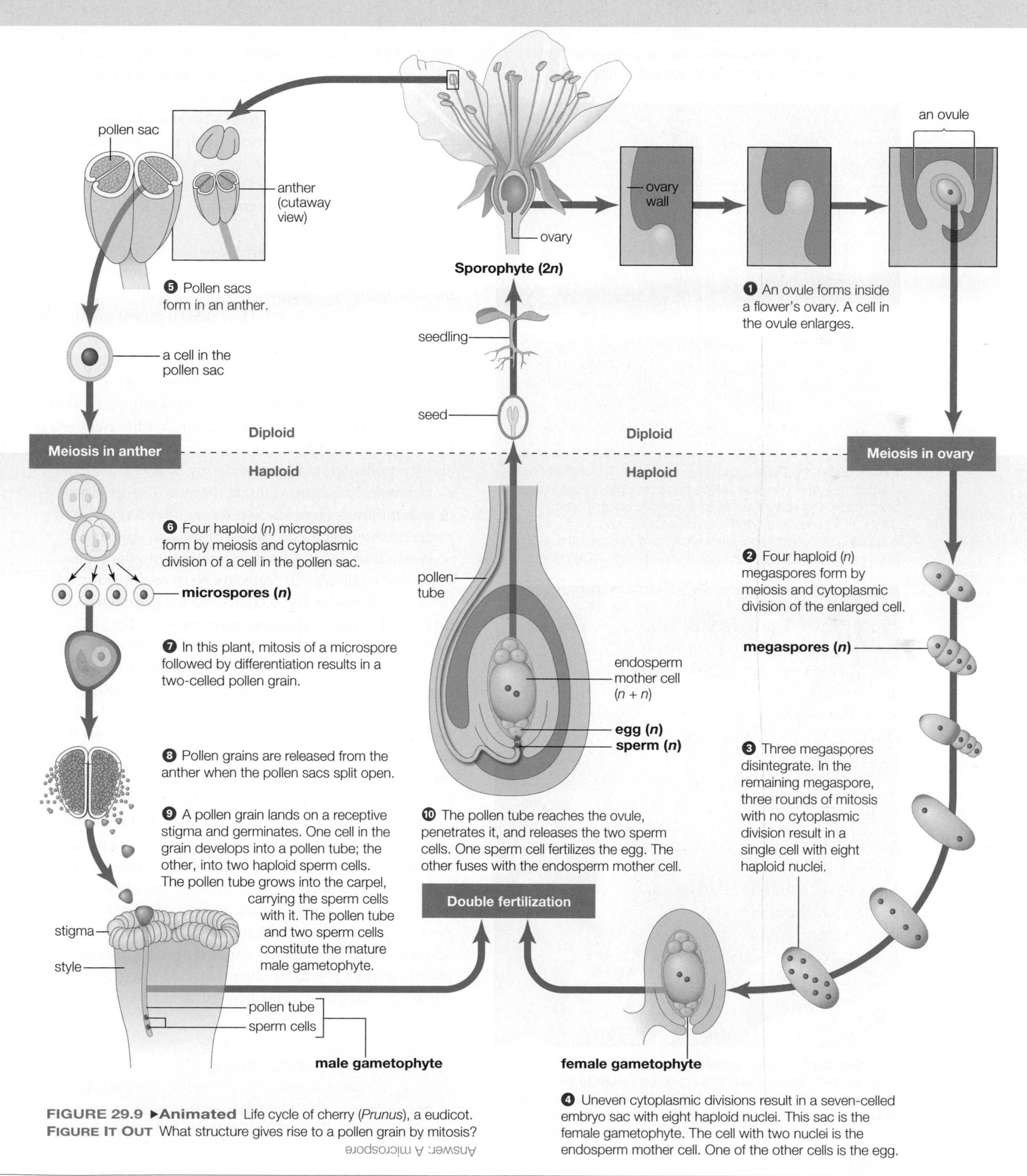

FIGURE 29.9 ▸**Animated** Life cycle of cherry (*Prunus*), a eudicot.
FIGURE IT OUT What structure gives rise to a pollen grain by mitosis?
Answer: A microspore

29.5 Flower Sex

✔ Species-specific interactions between pollen grain and stigma govern pollen germination and pollen tube growth.

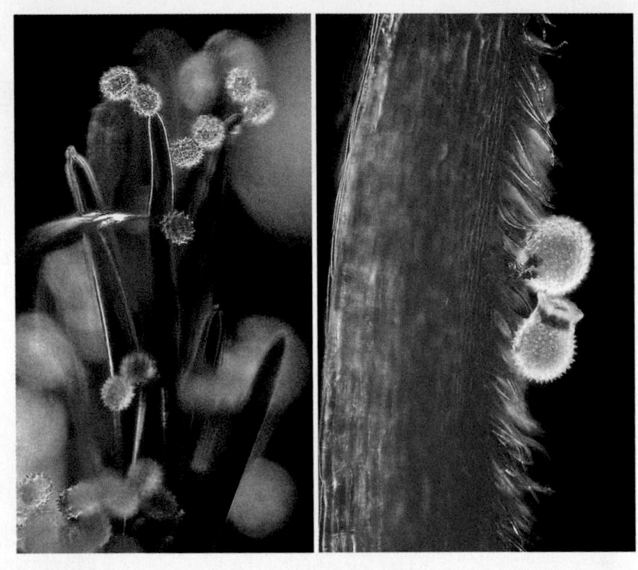

FIGURE 29.10 Pollen on stigmas of blanket flower (*Gaillardia grandiflora*, left) and mallow (*Malva neglecta*, right). A pollen grain's size, shape, and texture are species-specific adaptations to dispersal mechanisms, stigma specializations, and environmental conditions. All are morphological characters that can be useful in phylogenetic studies of both living and ancient flowering plants.

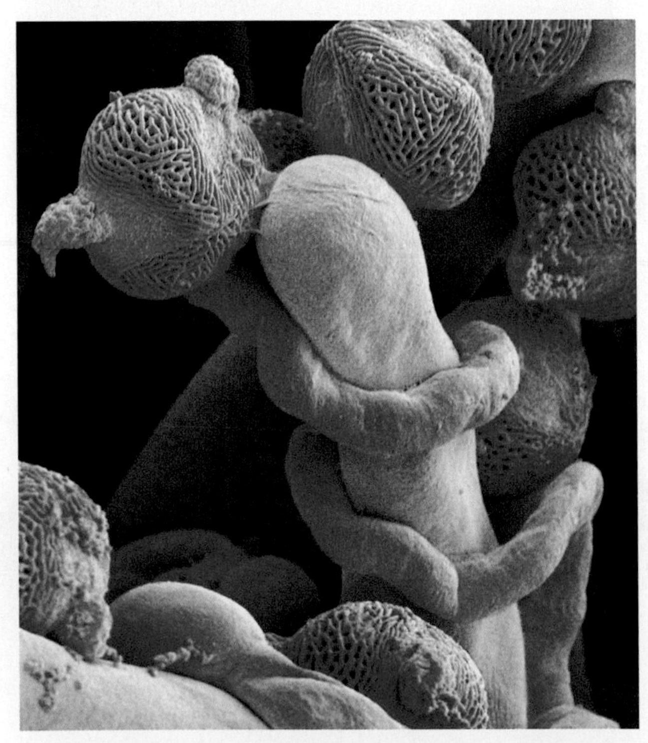

FIGURE 29.11 Adhesion proteins and signaling molecules guide a pollen tube's growth through carpel tissues to the egg. This electron micrograph shows pollen tubes growing from pollen grains (orange) that germinated on stigmas (yellow) of prairie gentian (*Gentiana*). Many pollen tubes may grow down into a carpel, but usually only one penetrates the female gametophyte.

The main function of a pollen grain's durable coat is to protect the two dormant cells inside of it on what may be a long, turbulent ride to a stigma (**FIGURE 29.10**). Pollen grains make terrific fossils because the outer layer of the coat consists primarily of sporopollenin, an extremely hard mixture of long-chain fatty acids and other organic molecules. This polymer is so resistant to enzymatic and chemical degradation that its molecular structure is still unknown.

A pollen grain's coat folds up to conserve water in dry environments, and swells in moist ones, but the cells inside of it will not germinate until the grain reaches a receptive stigma. How does a pollen grain "know" when it has arrived on an appropriate stigma? Sex in plants, like sex in animals, involves an interplay of signals. In this case, the signaling begins when recognition proteins on epidermal cells of a stigma bind to molecules in the coat of a pollen grain. Within minutes, lipids and proteins in the coat diffuse onto the stigma, and the pollen grain becomes tightly bound via adhesion proteins in stigma cell membranes. The specificity of recognition protein binding means that a stigma can preferentially hold on to pollen of its own species.

Next, fluid diffuses from the stigma into the pollen grain and stimulates the cells inside to resume metabolism. A pollen tube that contains the male gametes grows out of one of the furrows or pores in the pollen's coat (**FIGURE 29.11**). How does a microscopic pollen tube that grows through centimeters of tissue find its way to a single cell deep inside of the carpel? Cell signaling is involved in this process too. Adhesion proteins and signaling molecules in the style direct the deposition of membrane and wall material at the tube's tip, so the pollen tube grows in the appropriate direction. Two cells flanking the egg in the female gametophyte secrete a polypeptide (aptly named LURE) that guides the pollen tube's growth from the bottom of the style to the egg. These signals are species-specific: Pollen tubes do not recognize signals from eggs of other species. In some species, the signals also keep a flower's pollen from fertilizing its own stigma. In these species, only pollen from another flower (or another plant) can give rise to a pollen tube that recognizes the female gametophyte's chemical guidance.

TAKE-HOME MESSAGE 29.5

What constitutes sex in flowering plants?

✔ Species-specific molecular signals stimulate pollen germination and guide pollen tube growth to the egg.

29.6 Seed Formation

✔ After fertilization, mitotic cell divisions transform a zygote into an embryo sporophyte encased in a seed.

In flowering plants, double fertilization produces a zygote and a triploid (3*n*) cell; both immediately begin to divide by mitosis. The zygote develops into an embryo sporophyte, and the triploid cell develops into endosperm (**FIGURE 29.12A–C**). When the embryo approaches maturity, the integuments of the ovule separate from the ovary wall and become layers of a protective seed coat. The embryo sporophyte, its reserves of food, and the seed coat have now become a mature ovule, a self-contained package called a **seed** (**FIGURE 29.12D**). The seed may enter a period of dormancy until it receives signals that conditions in the environment are appropriate for germination.

As a seed is forming inside an ovule, the parent plant transfers nutrients to it. These nutrients accumulate in endosperm mainly as starch with some lipids and proteins. In typical monocots, nutrient reserves stay put in endosperm as the seed matures. By contrast, typical eudicot embryos transfer nutrients in endosperm to cotyledons as the seed matures, so most of the nutrients in a mature eudicot seed are stored in the two enlarged cotyledons.

The nutrients in endosperm and cotyledons nourish seedling sporophytes. They also nourish humans and other animals. For example, we cultivate many cereals—monocot grasses such as rice, wheat, rye, oats, and barley—for their nutritious seeds. The embryo (the germ) contains most of the seed's protein and vitamins, and the seed coat (the bran) contains most of the minerals and fiber. Milling removes the bran and germ, leaving only the starch-packed endosperm. Maize, or corn, is the most widely grown cereal crop. Popcorn pops because the moist endosperm steams when heated; pressure builds inside the seed until it bursts.

We also cultivate many eudicots for their seeds. For example, beans, lentils, and peas are eudicot seeds valued for their high protein and starch content; coffee and cacao seeds contain desirable stimulants.

seed Embryo sporophyte of a seed plant packaged with nutritive tissue inside a protective coat.

TAKE-HOME MESSAGE 29.6

What is a seed?

✔ After double fertilization, the zygote develops into an embryo, and the endosperm becomes enriched with nutrients.

✔ A seed is a mature ovule that consists of an embryo sporophyte, its food reserves, and a seed coat.

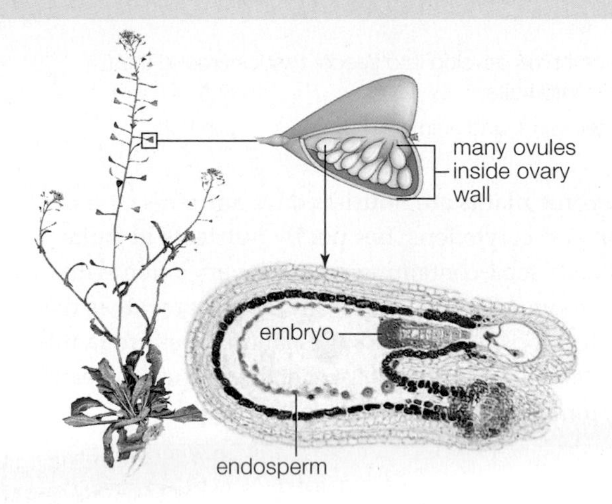

A After fertilization, a *Capsella* flower's ovary develops into a fruit, and an embryo forms inside each of the ovary's ovules.

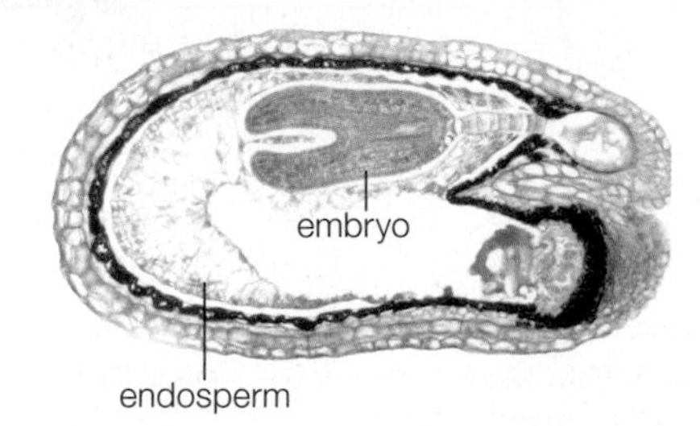

B The embryo is heart-shaped when its two cotyledons start forming. Endosperm tissue expands as the parent plant transfers nutrients into it.

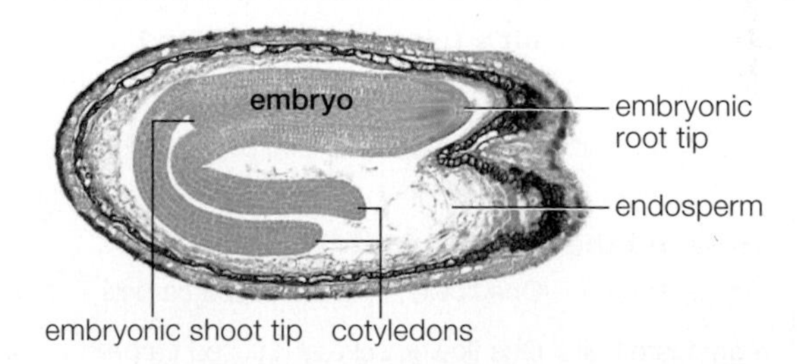

C In eudicots like *Capsella*, nutrients are transferred from endosperm into two cotyledons as the embryo matures. The developing embryo becomes shaped like a torpedo when the enlarging cotyledons bend.

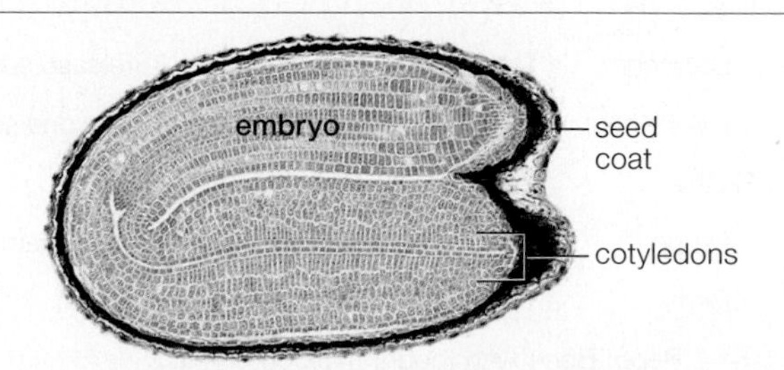

D A tough seed coat forms around the mature embryo and its two enlarged cotyledons.

FIGURE 29.12 ▸**Animated** Embryonic development of shepherd's purse (*Capsella*), a eudicot.

29.7 Fruits

✔ As embryos develop into seeds, tissues around them develop into fruits.

✔ Water, wind, and animals disperse seeds in fruits.

Embryonic plants are nourished by nutrients in endosperm and cotyledons, but not by nutrients in fruits. A **fruit** is a seed-containing mature ovary, often with fleshy tissues that develop from the ovary wall as the seed develops. Apples, oranges, and grapes are familiar fruits, but so are many "vegetables" such as beans, peas, tomatoes, grains, eggplant, and squash.

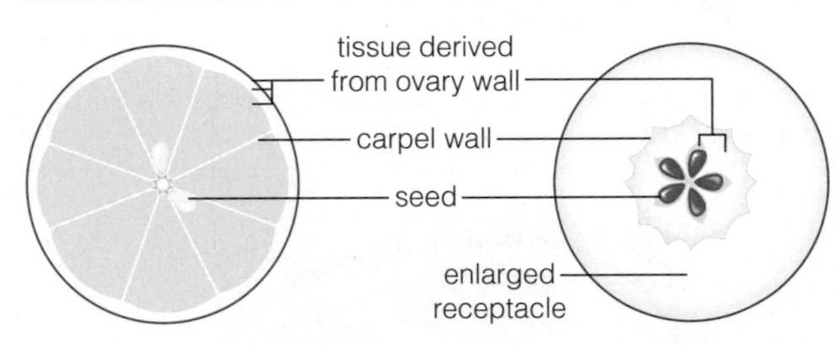

A The tissues of an orange develop from the ovary wall.

B The flesh of an apple is an enlarged receptacle.

FIGURE 29.13 Parts of a fruit develop from parts of a flower.
FIGURE IT OUT How many carpels were there in the flower that gave rise to the orange in A?

Answer: Eight

Table 29.2
Three Ways To Classify Fruits

What is the fruit's tissue composition?	
True fruit	Only ovarian wall and its contents
Accessory fruit	Ovary and other floral parts, such as receptacle
How did the fruit originate?	
Simple fruit	One flower, single or fused carpels
Aggregate fruit	One flower, several unfused carpels; becomes cluster of several fruits
Multiple fruit	Individually pollinated flowers fuse
Is the fruit dry or fleshy?	
Dry	
Dehiscent	Dry fruit wall splits on seam to release seeds
Indehiscent	Dry fruit wall does not split; usually one seed
Fleshy	
Drupe	Fleshy fruit, one seed enclosed by a hard pit
Berry	Fleshy fruit, often many seeds, no pit
Pepo: Berry with tough, thick outer rind	
Hesperidium: Berry with leathery rind, partioned sections	
Pome	Fleshy receptacle tissue surrounds core with seeds

A cherry is a true fruit, a simple fruit, and a fleshy drupe.

A blackberry is an accessory fruit, and an aggregate of many small, fleshy drupes.

A pineapple is an accessory fruit, and a multiple fruit formed by the fusion of many small, fleshy berries.

A pea pod is a true fruit, a simple fruit, and a dehiscent dry fruit.

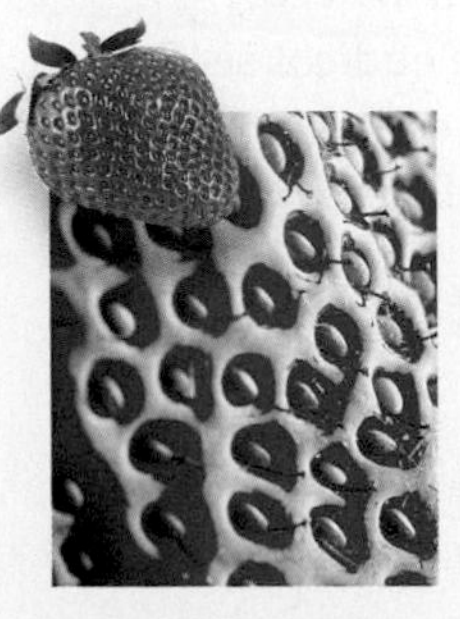

A strawberry is an accessory fruit, and an aggregate of many individual dry indehiscent fruits.

Fruits may be categorized by the composition of their tissues, how they originate, and whether they are fleshy (juicy) or dry (**TABLE 29.2**). True fruits such as oranges consist only of the ovary wall and its contents (**FIGURE 29.13A**). Other floral parts, such as the petals, sepals, stamens, or receptacle, expand along with the developing ovary as an accessory fruit. An apple is an accessory fruit; most of its flesh is an enlarged receptacle (**FIGURE 29.13B**).

Simple fruits are derived from one ovary or a few fused ovaries. Cherries, pea pods, acorns, and shepherd's purse are examples. By contrast, aggregate fruits such as blackberries are derived from separate ovaries of one flower that mature as a cluster. A multiple fruit such as a pineapple or a fig develops as a unit from several individually-pollinated flowers.

Dry fruits can be dehiscent or indehiscent. The wall of a dehiscent fruit splits along a seam to release the seeds inside when they are mature. Pea pods are examples. By contrast, a mature indehiscent fruit may have a seam, but its wall does not split open; its (typically single) seed is dispersed inside the intact fruit. Acorns and grains are dry indehiscent fruits, as are the fruits of sunflowers, maples, and strawberries. Strawberries are not berries and their fruits are not juicy. The red flesh of a strawberry is an enlarged receptacle, with individual dry indehiscent fruits on its surface.

Cherries, almonds, and olives are fleshy fruits, as are individual fruits of blackberries and other *Rubus* species. Grapes, tomatoes, and citrus fruits are berries—fleshy fruits produced from one ovary—as are pumpkins, watermelons, and cucumbers. Apples and pears are

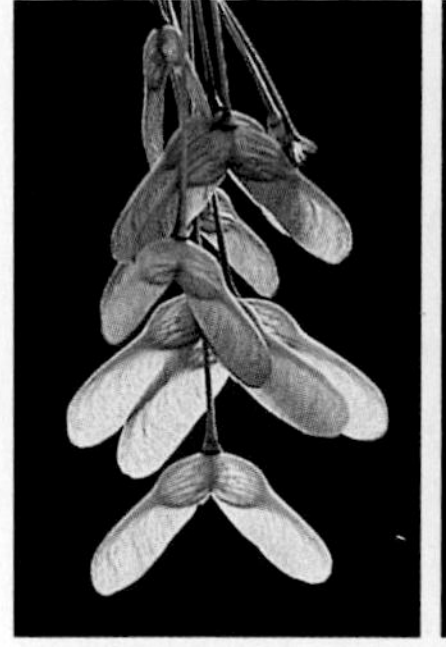

A Wind-dispersed fruits. Left, dry outgrowths of the ovary wall of a maple fruit form "wings" that catch the wind and spin the seeds away from the parent tree. Right, wind that lifts the hairy modified sepals of a dandelion fruit may carry the attached seed miles away from the parent plant.

B Water-dispersed fruits. Left, fruits of sedges native to American marshlands have seeds encased in a bladderlike envelope that floats. Right, buoyant fruits of the coconut palm have tough, waterproof husks. They can float for thousands of miles in seawater.

C Animal-dispersed fruits. Left, curved spines attach cocklebur fruits to the fur of animals (and clothing of humans) that brush past it. Right, the red, fleshy fruits of hawthorn plants are an important food source for cedar waxwings, which disperse the fruits' seeds in feces.

D The dry, dehiscent fruits of California poppy (*Eschscholzia californica*) and some other species spread their own seeds when they split open suddenly along a center seam. The movement propels the seeds through the air, away from the parent plant.

FIGURE 29.14 Examples of adaptations that aid fruit dispersal.

pomes: fruits in which fleshy tissues derived from the receptacle enclose a core derived from the ovary wall.

The function of a fruit is to protect and disperse seeds. Dispersal increases reproductive success by minimizing competition for resources among parent and offspring. Just as flower structure is adapted to certain pollination vectors, so are fruits adapted to certain dispersal vectors: environmental factors such as wind or water, or mobile organisms such as birds or insects. These adaptations are reflected in a fruit's form. Fruits dispersed by wind tend to be lightweight with breeze-catching specializations (**FIGURE 29.14A**). Fruits dispersed by water have water-repellent outer layers, and they float (**FIGURE 29.14B**). The fruits of many plants have specializations that facilitate dispersal by animals (**FIGURE 29.14C**). Some have hooks or spines that stick to the feathers, feet, fur, or clothing of more mobile species. Colorful, fleshy, or fragrant fruits attract birds and mammals that disperse seeds. The animal may eat the fruit and discard the seeds, or eat the seeds along with the fruit. Abrasion of the seed coat by teeth or by digestive enzymes in an animal's gut can help the seed germinate after it departs in feces. A few types of fruit have mechanical specializations that spread their seeds even without use of a dispersal vector (**FIGURE 29.14D**).

fruit Mature ovary of a flowering plant, often with accessory parts; encloses a seed or seeds.

TAKE-HOME MESSAGE 29.7

What is a fruit?

- ✔ A fruit is a mature ovary, with or without accessory tissues that develop from other parts of the flower.
- ✔ We can categorize a fruit in terms of how it originated, its composition, and whether it is dry or fleshy.
- ✔ A fruit protects and disperses seeds. Fruit specializations are adaptations to particular dispersal vectors.

CREDITS: (14A) left, R. Carr; right, © JupiterImages Corporation; (14B) left, © T. M. Jones; right, Alex James Bramwell/Shutterstock.com; (14C) left, © Robert H. Mohlenbrock © USDA-NRCS PLANTS Database; right, © Vishnevskiy Vasily/Shutterstock.com; (14D) © Trudi Davidoff, www.WinterSown.org.

29.8 Early Development

✔ Species-specific patterns of early plant development are triggered by environmental cues.

An embryonic plant complete with shoot and root apical meristems forms as part of a seed (**FIGURE 29.15**). The embryonic shoot (the plumule) is separated from the embryonic root (radicle) by a section of stem called hypocotyl. As the seed matures, the embryo may dry out and enter a period of dormancy. A dormant embryo can rest in its protective seed coat for many years before it resumes metabolic activity and germinates.

Breaking Dormancy

Seed dormancy is a climate-specific adaptation that allows germination to occur when conditions in the environment are most likely to support the growth of a seedling. For example, seeds of many annual plants native to cold winter regions are dispersed in autumn. If the seeds germinated immediately, the tender seedlings would not survive the coming winter. Instead, the seeds remain dormant until spring, when milder temperatures and longer day length favor the growth of seedling sporophytes. By contrast, the weather in regions near the equator does not vary by season. Seeds of most plants native to such regions do not enter dormancy; they can germinate as soon as they become mature.

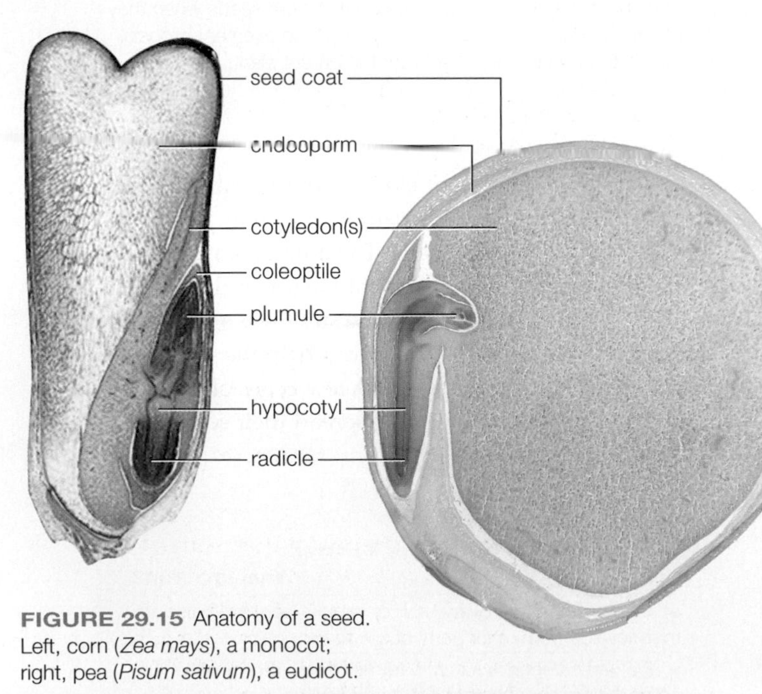

FIGURE 29.15 Anatomy of a seed. Left, corn (*Zea mays*), a monocot; right, pea (*Pisum sativum*), a eudicot.

As dormancy ends, cell divisions resume mainly at apical meristems of the plumule (the embryonic shoot) and radicle (the embryonic root). Plumule and radicle are separated by hypocotyl, a section of embryonic stem. In monocot grasses such as corn, the plumule is protected by a sheathlike coleoptile.

Other than the presence of water, the triggers for germination differ by species. Some seed coats are so dense that they must be abraded or broken (by being chewed, for example) before water can even enter the seed. Seeds of many cool-climate plants require exposure to freezing temperatures; those of some lettuce species, to bright light. In seeds of some species native to regions that have periodic wildfires, germination is inhibited by light and enhanced by smoke; in others, germination does not occur unless the seeds have been previously burned. Such requirements are evolutionary adaptations to life in a particular environment. All maximize a seedling's chance of survival.

The process of germination begins with water seeping into a seed. Water causes the seed's internal tissues to swell, forcing the seed coat to rupture. After the seed's coat breaks, oxygen diffuses into the tissues. The water also provokes a series of events that activate hydrolysis enzymes, which begin to break down stored starches into sugar subunits (we return to this topic in Section 30.5). Cells in the embryo's apical meristems (Section 27.7) use the sugars and the oxygen to run aerobic respiration. Energy released from the sugar breakdown fuels rapid divisions of the meristem cells, and the embryonic plant begins to grow. Germination ends when the first part of the embryo—the embryonic root, or radicle—breaks out of the seed coat.

After Germination

The pattern of early growth that occurs after germination varies. For example, in corn and other monocots, a rigid sheath called a **coleoptile** surrounds and protects the plumule (**FIGURE 29.16**). In a typical monocot pattern of development, germination is followed by the emergence of the radicle and coleoptile from the seed coat ❶. The radicle grows down into the soil, becoming a primary root ❷ as the coleoptile grows upward thorugh the soil ❸. When the coleoptile reaches the surface of the soil, it stops growing. The plumule develops into the primary (first) shoot as it emerges from the coleoptile ❹. Leaves form on the shoot and begin photosynthesis ❺. The embryo's single cotyledon, which typically stays beneath the soil, functions mainly to transfer endosperm breakdown products to the seedling.

In eudicot seedlings, the primary shoot is not protected by a coleoptile (**FIGURE 29.17**). In a typical eudicot pattern of development, the radicle emerges from

coleoptile In monocots, a rigid sheath that protects the plumule (embryonic shoot).

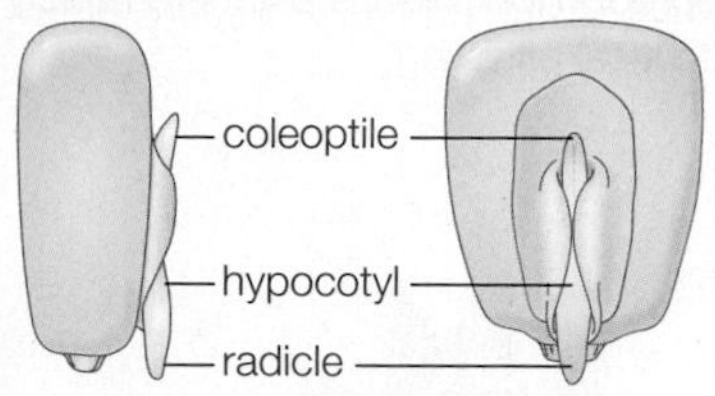

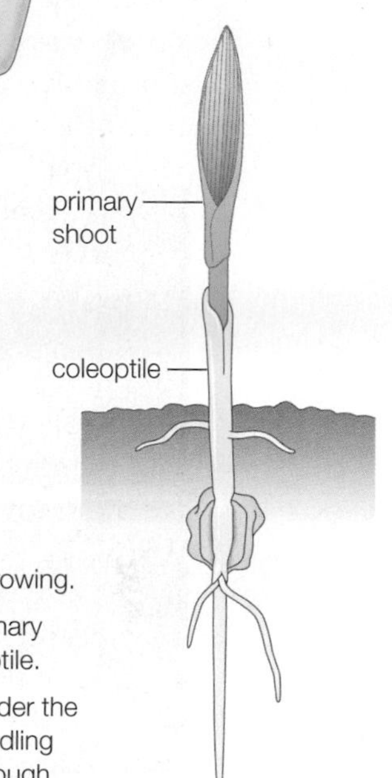

FIGURE 29.16 ▶**Animated** Early growth of corn (*Zea mays*), a typical monocot.

❶ As a corn grain (seed) germinates, its radicle and coleoptile emerge from the seed coat.

❷ The radicle develops into a primary root.

❸ The coleoptile grows upward and opens a channel through the soil to the surface, where it stops growing.

❹ The plumule develops into a primary shoot that emerges from the coleoptile.

❺ The single cotyledon remains under the soil, transferring nutrients to the seedling until its new leaves can produce enough sugars by photosynthesis.

FIGURE 29.17 ▶**Animated** Early growth of the common bean, a typical eudicot.

❶ The seed coat splits and the radicle emerges.

❷ The hypocotyl emerges from the seed and bends in the shape of a hook. As the stem lengthens, the bent hypocotyl drags the two cotyledons upward toward the surface of the soil.

❸ Exposure to sunlight causes the hypocotyl to straighten.

❹ Primary leaves emerge from between the cotyledons and begin photosynthesis.

❺ Cotyledons typically undergo a period of photosynthesis, then wither and fall off the stem.

the seed ❶, followed by the hypocotyl. The hypocotyl bends into a hook shape as it lengthens ❷. The bent hypocotyl pulls the cotyledons upward through the soil until it reaches the surface. There, exposure to light causes the hypocotyl to straighten ❸. Primary leaves emerge from between the cotyledons and begin photosynthesis ❹. The cotyledons typically undergo a period of photosynthesis before shriveling ❺. Eventually, the cotyledons fall off the lengthening stem, and the young plant's new leaves produce all of its food.

TAKE-HOME MESSAGE 29.8

What happens during early plant development?

✔ Seed dormancy and germination requirements are evolutionary adaptations to life in a particular climate.

✔ After a seed germinates, nutrients stored in endosperm (in monocots) or cotyledons (in eudicots) support the seedling's growth until new leaves begin photosynthesis.

✔ A coleoptile protects the embryonic shoot of many monocot species.

29.9 Asexual Reproduction of Flowering Plants

✔ Asexual reproduction permits plants to rapidly produce genetically identical offspring.

A Potatoes are tubers that grow on stolons. New roots and shoots sprout from their nodes, which we call "eyes." The new plants are genetically identical with the parent.

B Tiny new plants form at nodes on cladodes of many succulents such as this Mother of Thousands plant (*Bryophyllum daigremontianum*). Propagation occurs when the new plants break off the parent stem and fall to soil below, where they take root.

FIGURE 29.18 Examples of asexual reproduction in plants.

FIGURE 29.19 Apple trees do not breed true for fruit traits: color, flavor, size, sweetness, and texture. These are fruits of 21 wild trees.

Most flowering plants can reproduce asexually by **vegetative reproduction**, in which new roots and shoots grow from extensions or pieces of a parent plant. Each new plant is a clone, a genetic replica of its parent. For example, new roots and shoots can sprout from nodes on modified stems (**FIGURE 29.18**), and sometimes from pericycle in roots. Entire forests of quaking aspen trees are actually stands of clones that grew from root suckers, which are adventitious shoots that sprout from shallow lateral roots. Suckers appear after aboveground parts of the aspens are damaged or removed. One stand of quaking aspens in Utah consists of about 47,000 shoots and stretches for 107 acres. Such clones are as close as any organism gets to being immortal. One of the oldest known plants is a clone: The one and only population of King's holly, which consists of several hundred stems along a river gully in Tasmania. Radiometric dating of the plant's fossilized leaf litter shows that the clone is at least 43,600 years old.

Agricultural Applications

For thousands of years, we humans have been taking advantage of the natural capacity of many plants to reproduce asexually. For example, almost all houseplants, woody ornamentals, and orchard trees are clones that have been grown from stem fragments (cuttings) of a parent plant. Propagating some plants from cuttings may be as simple as jamming a broken stem into the soil. This method uses the plant's ability to form new roots and shoots from nodes on a stem. Other plants must be grafted. Grafting means inducing a cutting to fuse with and become supported by another plant. Often, the stem of a desired plant is grafted onto the roots of another.

Propagating a plant from cuttings ensures that offspring will have the same desirable traits as the parent. Consider familiar orchard apples, which are descendants of a wild species native to central Asia. Domesticated apple varieties are grafted because species in this genus do not breed true for fruit traits we value—color, flavor, size, sweetness, and texture (**FIGURE 29.19**). Most apple trees grown from seed produce unpalatable fruit. In the early 1800s, John Chapman (known as Johnny Appleseed) grew millions of apple trees from seed in the midwestern United States. He sold the trees to homesteading settlers, who made hard cider from the otherwise inedible apples. About one of every hundred trees produced sweet, tasty fruit; its lucky owner would graft the tree and patent it. An estimated 16,000 varieties of edible apples were discovered during this era. Few of them remain;

Plight of the Honeybee (revisited)

A single honeybee visits hundreds, sometimes thousands, of flowers a day to find nectar and pollen. The insect uses "scent memory" to remember the location of these rewards. Then it finds its way back to the hive, navigating distances up to 8 kilometers (5 miles) to communicate the location of the flowers to other bees. Bees' ability to rapidly learn, remember, and communicate with each other has made them highly efficient pollinators.

Long-term exposure to low levels of pesticides might impair a bee's ability to carry out its pollen mission. Many researchers suspect that synthetic pesticides called neonicotinoids are an important factor in colony collapse disorder. Neonicotinoids are now the most widely used insecticides in the United States, and they are still being used in other countries—about 2.5 million acres of croplands in total. The chemicals are systemic, which means they are taken up by all tissues of a plant treated with them, including the nectar and pollen that honeybees collect from flowers as they forage.

Neonicotinoids mimic the neurotoxic (nerve-killing) effect of nicotine, a natural insecticide made by plants such as tobacco, and they impair a bee's learning and memory. "Honeybees learn to associate floral colors and scents with the quality of food rewards," says Geraldine Wright, a neuroscientist at Newcastle University in England. Neonicotinoids affect the neurons involved in these behaviors, so bees exposed to these pesticides are likely to have difficulty communicating with other members of the colony. Neonicotinoids also increase bees' vulnerability to infection in amounts too tiny to detect in the insects themselves. Even dewdrops that form on neonicotinoid-treated plants apparently contain enough of the insecticide to affect bees.

---

only 15 varieties now account for 90 percent of the apples sold in U.S. grocery stores. All are clones of Chapman's original trees, and all are still grafted.

Grafting is also used to increase the hardiness of a desirable plant. In 1862, the plant louse *Phylloxera* was accidentally introduced into France via imported American grapevines. European grapevines had little resistance to this tiny insect, which attacks and kills the root systems of the vines. By 1900, *Phylloxera* had destroyed two-thirds of the vineyards in Europe, thus devastating the wine-making industry for decades. Today, French vintners routinely graft their prized grapevines onto the roots of *Phylloxera*-resistant American vines.

With **tissue culture propagation**, individual cells (typically from meristem) can be coaxed to divide and form embryos in a laboratory. This technique yields millions of genetically identical offspring from a single parent plant. It is frequently used in research aimed at improving food crops, and also to propagate rare ornamental plants such as orchids.

In some flowering plant species, fruits may form even in the absence of fertilization. Fruits that develop from unfertilized flowers may have underdeveloped seeds or none at all. Plants with mutations that cause ovules or embryos to abort during development bear seedless fruit; such plants are necessarily sterile and must be propagated by grafting. Seedless grapes and navel oranges are like this. All commercially produced bananas are seedless because the plants that bear them are triploid ($3n$). During meiosis, the three chromosomes cannot be divided equally between the two spindle poles, so gametes do not form and neither do seeds. Grafting bananas and other monocots is notoriously difficult (if not impossible), so seedless banana plants are propagated by cutting off and replanting shoots that sprout from their corms (Section 27.4).

Polyploid plants are common in nature (Sections 14.6 and 17.11), but it may take a long time for one to arise spontaneously. Plant breeders often use a microtubule poison called colchicine to artificially increase the frequency of polyploidy. Tetraploid ($4n$) plants produced by colchicine treatment are then crossed with diploid ($2n$) individuals. The resulting offspring are triploid and sterile: They make seedless fruit on their own or after pollination (but not fertilization) by a diploid plant. Seedless watermelons are produced commercially this way.

tissue culture propagation Laboratory method in which individual plant cells (typically from meristem) are induced to form embryos.
vegetative reproduction Growth of new roots and shoots from extensions or fragments of a parent plant; form of asexual reproduction in plants.

TAKE-HOME MESSAGE 29.9

How do plants reproduce asexually?

✔ Many plants propagate themselves when new shoots grow from a parent plant or pieces of it. Offspring of such asexual reproduction are clones.

✔ Humans propagate plants asexually for agricultural or research purposes, for example by grafting or tissue culture.

summary

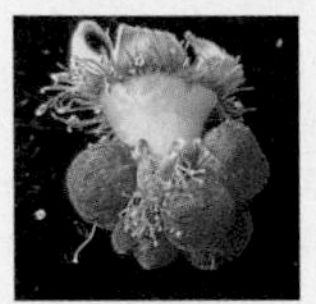

Section 29.1 Colony collapse disorder (CCD) is killing honeybees. Declines in populations of bees and other **pollinators** negatively affect plant populations as well as other animal species that depend on the plants, including humans. Widely used neonicotinoid pesticides may contribute to CCD.

Section 29.2 **Flowers** consist of whorls of modified leaves at the ends of specialized branches of angiosperms. A **calyx** of **sepals** surrounds a **corolla** of **petals**, which in turn surround **stamens** and **carpels**. A carpel consists of a **stigma**, often a style, and an **ovary** inside which one or more **ovules** develop. The female gametophyte forms inside an ovule. A stamen consists of an **anther** on a thin filament. Anthers produce pollen grains.

Section 29.3 **Pollination** is the arrival of pollen on a receptive stigma. A flower's shape, pattern, color, and fragrance typically reflect an evolutionary relationship with a particular **pollination vector**, often a coevolved animal. Coevolved pollinators receive **nectar**, pollen, or another reward for visiting a flower.

Sections 29.4, 29.5 Meiosis of diploid cells inside pollen sacs of anthers produces haploid **microspores**. Each microspore develops into a pollen grain that is released from a pollen sac after a period of **dormancy**.

Meiosis and cytoplasmic division of a cell in an ovule produce four **megaspores**, one of which gives rise to the female gametophyte. One of the seven cells of the gametophyte is the egg; another is the endosperm mother cell. An interplay of species-specific molecular signals trigger a pollen grain to **germinate** on a receptive stigma and form a pollen tube that contains two sperm cells. Other molecular signals guide pollen tube growth through tissues of the carpel to the egg. In **double fertilization**, one of the sperm cells in the pollen tube fertilizes the egg, forming a zygote; the other fuses with the endosperm mother cell and gives rise to triploid **endosperm**.

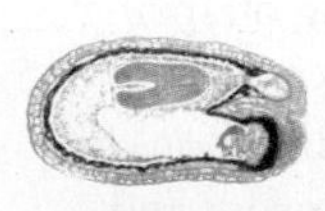

Section 29.6 As a zygote develops into an embryo, endosperm collects nutrients from the parent plant, and the ovule's protective layers develop into a seed coat. A **seed** is a mature ovule: an embryo sporophyte and endosperm enclosed within a seed coat. Nutrients in endosperm or cotyledons make seeds a nutritious food source.

Section 29.7 As an embryo sporophyte develops, the ovary wall and sometimes other tissues mature into a **fruit** that encloses the seeds. Fruit specializations are adaptations to seed dispersal by specific vectors such as wind, water, or animals. A fruit can be categorized by tissue of origin, composition, and whether it is dry or fleshy.

Section 29.8 Seeds often undergo a period of dormancy that does not end until species-specific environmental cues trigger germination. The radicle emerges from the seed coat at the end of germination; other patterns of early development vary. For example, monocot plumules are sheathed by a **coleoptile**; eudicot hypocotyls form a hook that pulls cotyledons up through soil.

Section 29.9 Many types of flowering plants produce clonal offspring by **vegetative reproduction**. Some are propagated commercially by grafting; the common laboratory technique of **tissue culture propagation** is used to propagate some valuable ornamentals.

self-quiz

Answers in Appendix VII

1. _______ is the arrival of pollen on a receptive stigma.

2. An animal pollinator may receive _______ when it visits a flower of a coevolved plant (choose all that apply).
   a. pollen c. pesticides
   b. nectar d. fruit

3. In flowers, a _______ contains one or more ovaries.
   a. pollen sac c. receptacle
   b. carpel d. sepal

4. In flowers, the structures that produce male and female gametophytes are called _______ and _______.
   a. pollen grains; flowers c. anthers; stigma
   b. stamens; carpels d. megaspores; microspores

5. Meiosis of cells in pollen sacs forms haploid _______.
   a. megaspores c. stamens
   b. microspores d. sporophytes

6. True or false? All flowers are pollinated by bees.

7. The three main parts of a mature seed are _______.
   a. pollen grain, egg, and seed coat
   b. embryo, endosperm, and seed coat
   c. megaspores, microspores, and ovule

8. The seed coat forms from the _______.
   a. integuments c. endosperm
   b. coleoptile d. sepals

9. Seeds are mature _______; fruits are mature _______.
   a. ovaries; ovules c. ovules; ovaries
   b. ovules; stamens d. stamens; ovaries

10. Dixie Bee prepares a plate of fruits for a party and cuts open a cantaloupe (*Cucumis melo*). A tough outer rind and soft fleshy tissue enclose many seeds (left). Knowing her friends will ask her what kind of fleshy fruit this is, she panics, runs to her biology book, and opens it to Table 29.2 in Section 29.7. What does she find out?

11. Cotyledons develop as part of ________ .
  a. carpels c. embryo sporophytes
  b. accessory fruits d. flowers

12. Exposure to ________ can trigger seed germination.
  a. light c. smoke
  b. cold d. all can be triggers

13. A new plant forms from a stem that broke off of the parent plant. This is an example of ________ .
  a. nodal cloning c. asexual reproduction
  b. exocytosis d. tissue culture propagation

14. Banana plants produce seedless fruit because they are ________ .
  a. triploid c. propagated by grafting
  b. monocots d. treated with colchicine

15. Match the terms with the most suitable description.

___ ovule	a. pollen tube together with its contents
___ receptacle	b. consists of seven cells, one with two nuclei
___ double fertilization	c. after fertilization, develops into a seed
___ anther	d. embryonic shoot
___ plumule	e. pollen sacs inside
___ mature female gametophyte	f. swollen stem; base of flower
___ mature male gametophyte	g. formation of zygote and first cell of endosperm

critical thinking

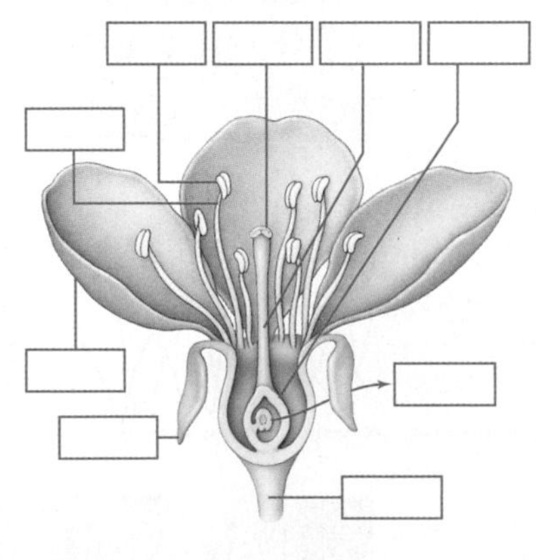

1. Label the parts of the flower shown at left.

2. Is the seedling shown on the right a monocot or eudicot? How can you tell?

3. All but one species of large-billed birds native to New Zealand's tropical forests are now extinct. Numbers of the surviving species, the kereru, are declining rapidly due to the habitat loss, poaching, predation, and interspecies competition that wiped out the other native birds. The kereru remains the sole dispersing agent for several native trees that produce big seeds and fruits. One tree, the puriri (*Vitex lucens*), is New Zealand's most valued hardwood. Explain, in terms of natural selection, why we might expect to see no new puriri trees in New Zealand.

data analysis activities

Who's the Pollinator? *Massonia depressa* is a low-growing succulent plant native to the desert of South Africa. The dull-colored flowers of this monocot develop at ground level, have tiny petals, emit a yeasty aroma, and produce a thick, jellylike nectar. These features led researchers to suspect that desert rodents such as gerbils pollinate this plant. To test their hypothesis, the researchers trapped rodents in areas where *M. depressa* grows and checked them for pollen. They also put some plants in wire cages that excluded mammals, but not insects, to see whether fruits and seeds would form in the absence of rodents. The results are shown in **FIGURE 29.20**.

1. How many of the 13 captured rodents showed some evidence of pollen from *M. depressa*?

2. Would this evidence alone be sufficient to conclude that rodents are the main pollinators for this plant?

3. How did the average number of seeds produced by caged plants compare with that of control plants?

4. Do these data support the hypothesis that rodents are required for pollination of *M. depressa*? Why or why not?

A The dull, petalless, ground-level flowers of *Massonia depressa* are accessible to rodents, who push their heads through the stamens to reach the nectar at the bottom of floral cups. Note the pollen on the gerbil's snout.

Type of rodent	Number caught	# with pollen on snout	# with pollen in feces
Namaqua rock rat	4	3	2
Cape spiny mouse	3	2	2
Hairy-footed gerbil	4	2	4
Cape short-eared gerbil	1	0	1
African pygmy mouse	1	0	0

B Evidence of visits to *M. depressa* by rodents.

	Mammals allowed access to plants	Mammals excluded from plants
Percent of plants that set fruit	30.4	4.3
Average number of fruits per plant	1.39	0.47
Average number of seeds per plant	20.0	1.95

C Fruit and seed production of *M. depressa* with and without visits by mammals. Mammals were excluded from plants by wire cages with openings large enough for insects to pass through. Twenty-three plants were tested in each group.

FIGURE 29.20 Testing pollination of *M. depressa* by rodents.

CREDITS: (CT #1) © Cengage Learning; (CT #2) © Jubal Harshaw/Shutterstock.com; (20A) © Steven D. Johnson; (20B, C) © Cengage Learning.

30 Communication Strategies in Plants

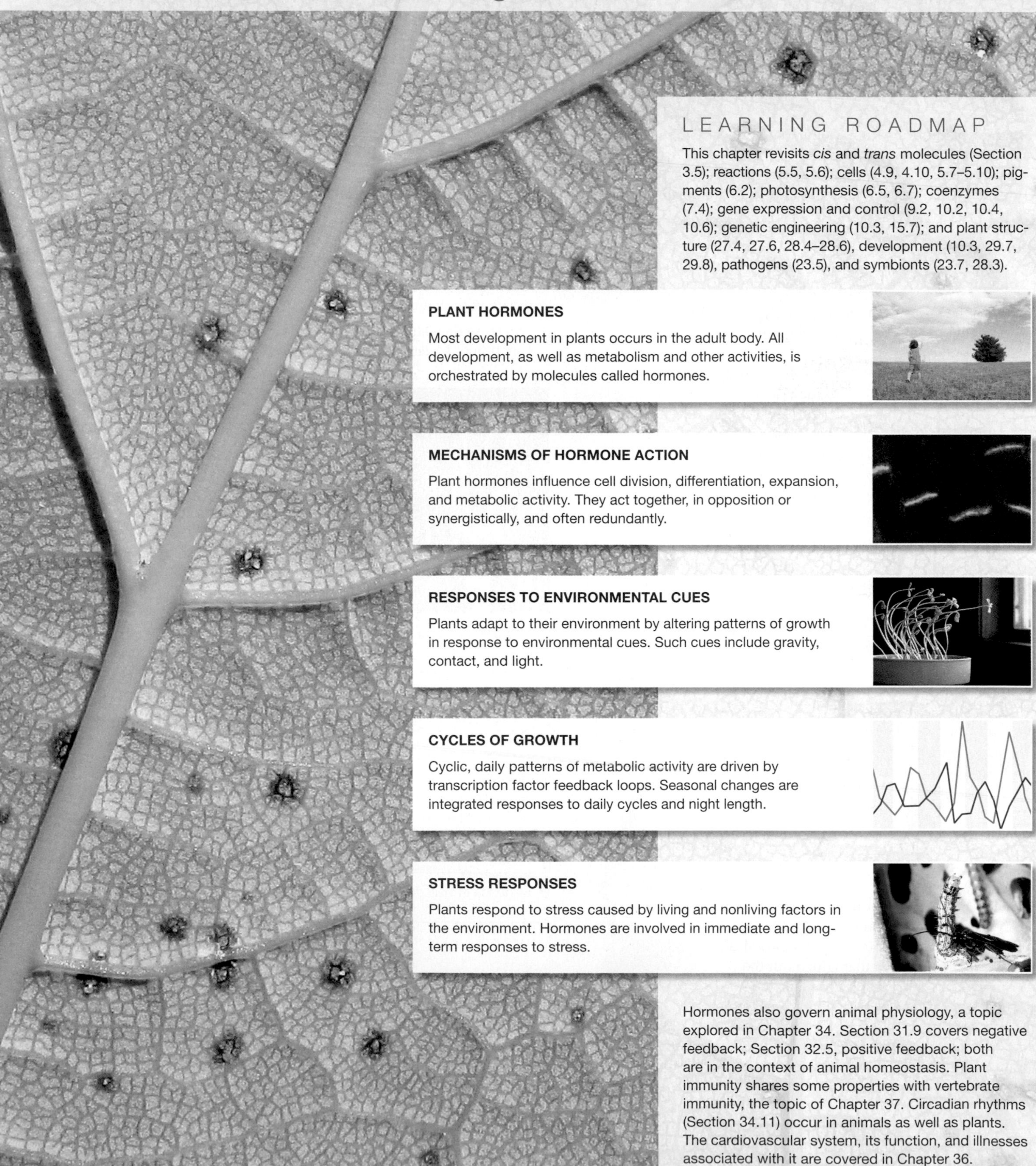

LEARNING ROADMAP

This chapter revisits *cis* and *trans* molecules (Section 3.5); reactions (5.5, 5.6); cells (4.9, 4.10, 5.7–5.10); pigments (6.2); photosynthesis (6.5, 6.7); coenzymes (7.4); gene expression and control (9.2, 10.2, 10.4, 10.6); genetic engineering (10.3, 15.7); and plant structure (27.4, 27.6, 28.4–28.6), development (10.3, 29.7, 29.8), pathogens (23.5), and symbionts (23.7, 28.3).

PLANT HORMONES

Most development in plants occurs in the adult body. All development, as well as metabolism and other activities, is orchestrated by molecules called hormones.

MECHANISMS OF HORMONE ACTION

Plant hormones influence cell division, differentiation, expansion, and metabolic activity. They act together, in opposition or synergistically, and often redundantly.

RESPONSES TO ENVIRONMENTAL CUES

Plants adapt to their environment by altering patterns of growth in response to environmental cues. Such cues include gravity, contact, and light.

CYCLES OF GROWTH

Cyclic, daily patterns of metabolic activity are driven by transcription factor feedback loops. Seasonal changes are integrated responses to daily cycles and night length.

STRESS RESPONSES

Plants respond to stress caused by living and nonliving factors in the environment. Hormones are involved in immediate and long-term responses to stress.

Hormones also govern animal physiology, a topic explored in Chapter 34. Section 31.9 covers negative feedback; Section 32.5, positive feedback; both are in the context of animal homeostasis. Plant immunity shares some properties with vertebrate immunity, the topic of Chapter 37. Circadian rhythms (Section 34.11) occur in animals as well as plants. The cardiovascular system, its function, and illnesses associated with it are covered in Chapter 36.

30.1 Prescription: Chocolate

For several centuries, the small islands of the San Blas Archipelago just off the coast of Panama have been inhabited by the Kuna, a tribe of indigenous South Americans. The Kuna have been untroubled by the malaria, yellow fever, and dengue fever that afflict their mainland cousins, primarily because no disease-carrying mosquitoes can survive on their windswept islands. The Kuna also have almost no incidence of the hypertension (high blood pressure and its associated cardiovascular problems) that plagues about one-quarter of people in mainland populations.

Scientists studying the Kuna hypothesized that their resistance to hypertension had a genetic basis, so in 1990, they began to search for an allele unique to the tribe that affects blood pressure. At the time, the human genome was being sequenced (Section 15.4), and genomic comparisons were becoming a routine avenue of research. However, a decade later, researchers still had not found the allele, and began to look for other correlations. They discovered that genetics really had nothing to do with the low incidence of hypertension in the Kuna. All else being equal, the most striking difference between island-dwelling populations and their mainland relatives is that the islanders consume an unusually large amount of cocoa. The Kuna tend to drink cocoa instead of water, and they live in a hot tropical climate so they drink a lot of it—a minimum of five cups per day.

Cocoa is made from cacao beans, which are the seeds of the *Theobroma cacao* tree (**FIGURE 30.1**). The seeds and young leaves of this tree have a particularly high content of flavonoids. Flavonoids are secondary metabolites: compounds that are not required for the immediate survival of the organism that makes them (Section 22.9). A particular flavonoid called epicatechin is probably responsible for the unusual absence of heart disease among the Kuna.

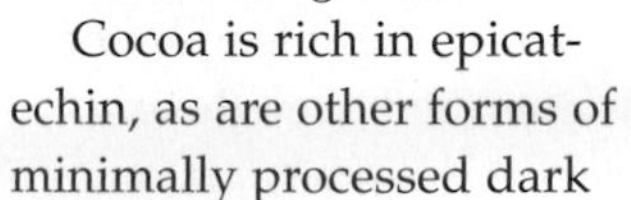

Cocoa is rich in epicatechin, as are other forms of minimally processed dark chocolate. Epicatechin influences a wide range of biological activities in the human body. This compound has a demonstrably protective effect against oxidative tissue damage that typically occurs after a stroke or a heart attack: A dose of epicatechin can reduce the amount of heart muscle damage caused by a heart attack by 52 percent; and reduce the brain damage that a stroke causes by a 32 percent. As an added benefit, epicatechin enhances memory and kills cancer cells.

Plants do not make this seemingly miraculous compound for our benefit. Neither do they make caffeine, resveratrol, curcumin (in coffee, red wine, and turmeric, respectively), or any other medically relevant chemical, for us. Epicatechin serves a role in plant immunity. Pigments, scents, and other chemicals that attract pollinators are also secondary metabolites, as are compounds that deter herbivores and pathogens, attract symbionts, or inhibit the growth of competing individuals. These chemicals are all part of the means by which plants interact with their environments.

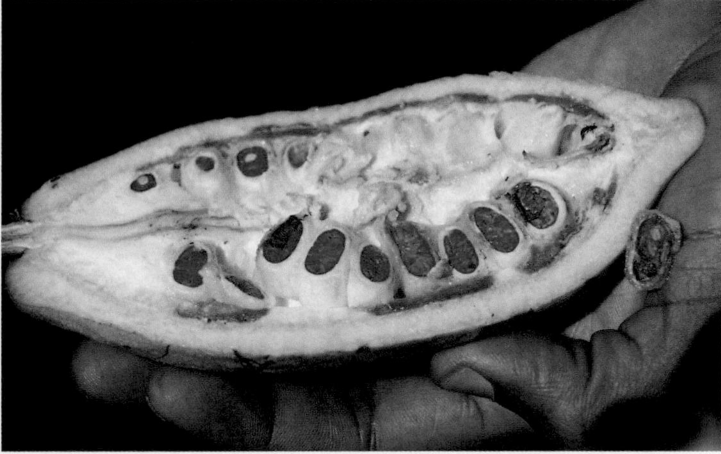

FIGURE 30.1 The cacao tree, *Theobroma cacao*. Left, the reddish pods are cacao fruits. Cocoa and other forms of chocolate are made from the dried, ground seeds of this plant (right). The seeds are particularly rich in flavonoids, especially epicatechin.

30.2 Introduction to Plant Hormones

✔ Plant development and function depend on cell-to-cell communication.

✔ Plant cells communicate via chemical signals.

In typical animals, most development occurs before adulthood. By contrast, most development in typical plants occurs at maturity (FIGURE 30.2). Plants, unlike animals, cannot move about to avoid unfavorable conditions. Instead, each plant adapts to its environment. It does so by adjusting development in response to cues such as temperature, gravity, night length, the availability of water and nutrients, and even the presence of pathogens or herbivores.

Developmental flexibility in plants depends on extensive coordination among individual cells. Cells in different tissues and even in different parts of a plant coordinate their activities by communicating with one another. As an example, a leaf being chewed by a caterpillar can signal other parts of the plant to produce appropriate caterpillar-deterring chemicals.

Chemical Signaling in Plants

Much of the cell-to-cell communication within plants involves hormones. A **plant hormone** is an extracellular signaling molecule that exerts its effect at very low concentrations. That effect may involve development of a plant part; the direction and rate of growth; immunity and other defense responses; circadian rhythms; the timing and duration of flowering; fruit and seed formation; aging; and initiating and breaking dormancy, to give some examples. In plants, one process is rarely controlled by one hormone. Rather, each process is usually influenced by multiple hormones in a complex network of signaling.

A hormonal signal is detected by receptor proteins (Section 5.7), with each type of receptor binding to a particular hormone. When a receptor binds to a hormone, it changes in a way that triggers a series of events inside the cell. These events often include DNA modifications that alter gene expression (Section 10.2).

A cell's response to a hormone depends on the receptors it has, and, in plants, it varies with the concentration of the hormone. The response also depends on the cell's integration of hormonal signals: One plant hormone can enhance or oppose another's effects on the same cell. Many hormones inhibit their own expression, a mechanism called negative feedback that is explored in more detail in Chapter 31. Positive feedback loops in which a hormone promotes its own transcription are part of daily cycles of activity.

TABLE 30.1 gives a preview of some major plant hormones and a few of their known effects. Those effects often vary by tissue. For example, brassinosteroids affect expression of genes governing anther and pollen development in cells of a developing flower. In leaves, the same hormones are part of different signaling pathways that cause cells to die in response to pathogen infection. Recent breakthroughs have

FIGURE 30.2 Plant versus animal development. In humans and other animals, most development occurs prior to adulthood. By contrast, a tree grows and adds new branches throughout its lifetime.

revealed many of the details of such signaling pathways, which we explore in the following sections.

Hormones operate in both plants and animals, but the structure, origin, and effects of these molecules differ between the two groups. For example, animals have organs specialized for hormone secretion; plants do not. All cells in a plant have the ability to make and release hormones. Animal hormones are defined by their ability to elicit an effect in a distant tissue. Similarly, a plant hormone may be released in a region of the body that is very far from cells in a targeted tissue, such as when cells in an actively growing root release cytokinins that keep shoot tips in a concurrent mode of active growth. However, some plant hormones can have very localized effects, such as when ethylene released from a cell in a ripening fruit affects the releasing cell as well as its neighbors.

Some molecules that act as hormones in plants act as hormones in animals, and vice versa. For example, plants make phytoestrogens that are structurally similar to estrogens, which are female sex hormones in mammals. The two types of compounds have similar hormonal effects in mammals, but phytoestrogens have nothing to do with sexual reproduction in plants. Rather, they function as part of plant defense mechanisms against fungal attack. The difference does not necessarily stem from differences in the structure of the hormones, but rather in the way the receptors for these compounds work in cells that bear them.

Unlike animals, plants do not have a circulatory system that moves hormones through the body. The movement of most plant hormones from source to target occurs by diffusion (for example through plasmodesmata between adjacent cells), and by fluid transport through xylem or phloem to more distant areas of the plant body. A few plant hormones move from cell to cell via active transport proteins in plasma membranes.

plant hormone Extracellular signaling molecule of plants that exerts its effect at very low concentration.

TAKE-HOME MESSAGE 30.2

What is a plant hormone?

✔ Plant hormones are signaling molecules that coordinate activities among cells in different parts of the plant body.

✔ Hormones are involved in all aspects of growth, development, and function in plants. They often work together, with synergistic or opposing effects on cells.

✔ Cells that bear receptors for a hormone—and thus can respond to it—may be in the same tissue as the hormone-releasing cell, or in another region of the plant body.

Table 30.1 Major Plant Hormones

Hormone	Examples of Effects
Abscisic acid (ABA)	Closes stomata in times of stress
	Inhibits shoot growth
	Inhibits seed germination
	Involved in environmental stress responses
	Involved in chloroplast movement
Auxin	Coordinates effects of other plant hormones during development of shoots and roots
	Role in apical dominance
	Stimulates cell expansion
	Role in abscission
	Mediates tropisms
	Stimulates division of meristem cells
Brassinosteroid	Regulates development of anthers and pollen
	Role in tissue defense
	Role in fruit ripening
Cytokinin	Stimulates differentiation of cells in root apical meristem
	Stimulates division of cells in shoot apical meristem
	Inhibits formation of lateral roots
Ethylene	Involved in breaking dormancy
	Stimulates flower organ development
	Stimulates fruit ripening
	Stimulates abscission
	Involved in stress responses
Gibberellin	Stimulates cell division, elongation in stems
	Mobilizes food reserves in germinating seeds
	Stimulates flowering in some plants
Jasmonic acid	Induces expression of genes used in defense
	Involved in thigmotropic response
Nitric oxide	Abiotic and biotic stress defense responses
	Enhances germination
	Involved in responses to contact
Salicylic acid	Activates systemic acquired resistance
Strigolactone	Opposes auxin's effects
	Role in apical dominance

30.3 Auxin: The Master Growth Hormone

✔ Auxin coordinates the effects of other plant hormones.

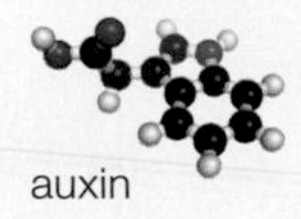

There are a few naturally occurring auxins, but the one that occurs most frequently in plants is IAA (indole-3-acetic acid), a small molecule derived from the amino acid tryptophan. In most cases, the term **auxin** refers to IAA. Auxin was first discovered for its ability to promote growth in plants (**FIGURE 30.3**), and its name is derived from the Greek word for growth. However, it has additional effects. Auxin plays a critical role in all aspects of plant development, starting with the first division of the zygote. It is involved in polarity and tissue patterning in the embryo, formation of plant parts (primary leaves, shoot tips, stems, and roots), differentiation of vascular tissues, formation of lateral roots (and adventitious roots in some species), and, as you will see in the later sections, shaping the plant body in response to environmental stimuli.

Auxin exerts many of its effects by influencing the levels of other plant hormones. For example, auxin produced in a shoot's tip supports growth in the shoot's stem and leaves. In stems, it induces synthesis of gibberellin, a plant hormone that stimulates growth; in leaves, it causes breakdown of another plant hormone, cyokinin, that inhibits cell division. Auxin has a different effect in roots, where it interacts with the hormone ethylene to inhibit lengthening of the root and to stimulate root hair growth.

Researchers are still working out the mechanisms of many of auxin's effects, but they do understand one way in which it promotes growth directly. During primary growth, auxin causes young cells to expand by increasing the activity of transport proteins that pump hydrogen ions from cytoplasm into the cell wall. The resulting increase in acidity softens the wall. Turgor, the pressure exerted by fluid inside the softened wall, enlarges the cell irreversibly.

Synthetic auxins have multiple uses. Application of auxin to a cut stem causes roots to develop, so this hormone is often sold as a rooting compound. Auxin can also be used as an herbicide. Applied to leaves, it causes uncontrolled cell division in the plant's apical meristems. The resulting growth is unsustainable: Stems curl as they become too long to be supported properly, leaves wither as the plant's resources are diverted to inappropriate growth, and the plant dies. Agent Orange, a defoliant used extensively during the Vietnam War, consisted of a combination of two synthetic auxins. One of those auxins, called 2,4-D, is now widely used as an herbicide on lawns and in cornfields, where it kills eudicots but not monocots.

Polar Transport

Auxin is present in almost all plant tissues, but is unevenly distributed through them. It is made mainly in shoot apical meristems and in young leaves, and transported elsewhere.

Auxin produced in shoot tips is loaded into phloem and travels to roots, where it is unloaded into root cells. A different mechanism dominates its movement over shorter distances. Auxin is pumped into and out of cells by active transport proteins called influx carriers and efflux carriers, respectively. Efflux carriers, unlike most other membrane proteins, are not always distributed evenly around a cell's plasma membrane. When efflux carriers in adjacent cells "point" in the same direction, they direct the flow of auxin through a tissue (**FIGURE 30.4A**). In an actively lengthening shoot, for example, efflux carriers are positioned on the side of the cell closest to the base of the stem. Thus,

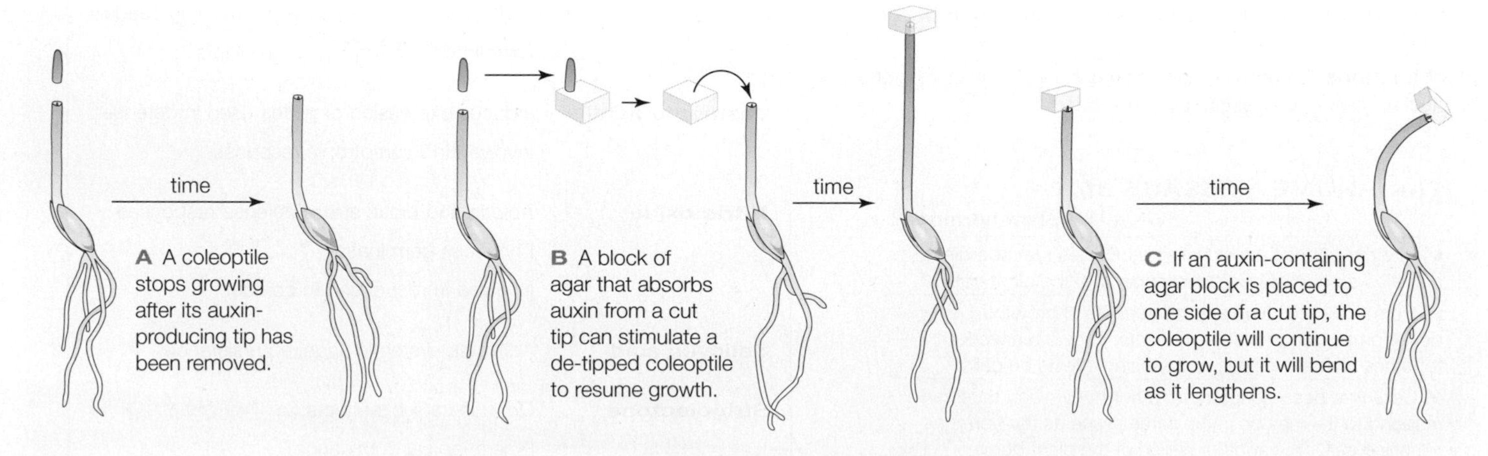

FIGURE 30.3 ▸**Animated** Experiments showing that a coleoptile lengthens in response to auxin produced in its tip.

auxin flows from the shoot's apical meristem toward the base of the stem (**FIGURE 30.4B**). Inside young root tips, the efflux carriers direct the flow of auxin in the opposite direction, from the root's tip toward the root–shoot interface (**FIGURE 30.4C**). Such polar transport, which is unique among plant hormones, establishes auxin concentration gradients.

Auxin affects the expression of other hormones, so auxin gradients set up localized patterns of hormone production that can vary from one cell layer to the next. The patterns change over time because efflux carriers are continually recycled by membrane trafficking (Section 5.10 and **FIGURE 30.4D**). This active process allows auxin-directed plant development to be flexible and responsive. Consider how efflux carriers help balance the growth of a plant's apical and lateral buds. When a shoot is lengthening at the tip, its lateral buds are usually dormant, an effect called **apical dominance**. In the shoot's tip, auxin produced by apical meristem is traveling through efflux carriers, and the flow through the stem causes it to lengthen. In the dormant lateral buds, auxin produced by apical meristem is not traveling, because cells in these buds have few efflux carriers in their membranes.

If a shoot's tip breaks off, its lateral buds begin to grow, an effect exploited by gardeners who pinch off shoot tips to make a plant bushier. When a shoot loses its tip, it also loses its source of auxin. The decrease in auxin stops the stem from lengthening. It also slows the production of another hormone, strigolactone, in roots. Strigolactone is transported from roots to all of the plant's shoots, where it interrupts trafficking of auxin efflux carriers so fewer of them get inserted into plasma membranes. Thus, strigolactone has a dampening effect on auxin transport—and as a result, it inhibits shoot lengthening. Cells that have a lot of efflux carriers (such as those in lengthening shoot tips) can still transport auxin even when the strigolactone level is high; cells with few efflux carriers (such as those in dormant lateral buds) cannot. When the strigolactone level declines, as occurs after a shoot's tip is removed, cells in dormant lateral buds begin to acquire efflux carriers in their membranes, auxin begins to be transported through them, and the buds lengthen.

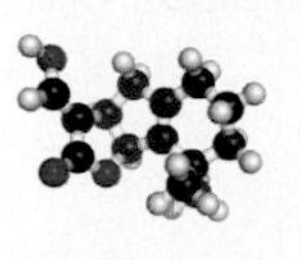
strigolactone

apical dominance Effect in which a lengthening shoot tip inhibits the growth of lateral buds.
auxin Plant hormone that causes cell enlargement; also has a central role in growth by coordinating the effects of other hormones. Indole-3-acetic acid (IAA) is the most common auxin.

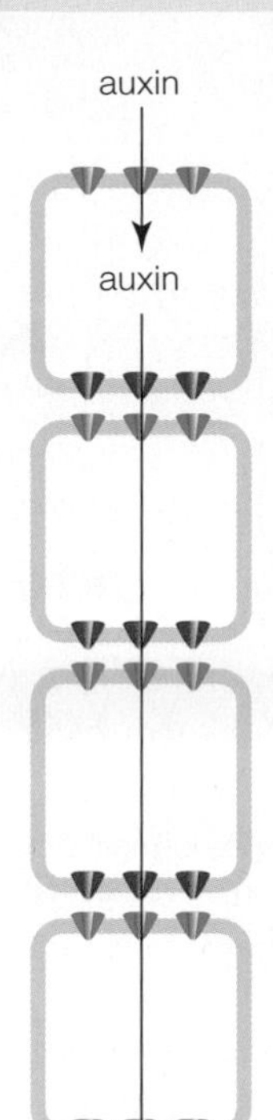

A Left, influx carriers (blue) actively transport auxin into the cell; efflux carriers (red) actively transport it out. Efflux carriers positioned asymmetrically in the membranes direct the flow of auxin in a particular direction.

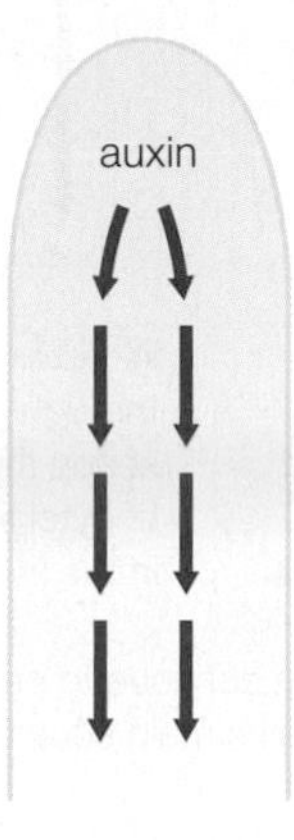

B Auxin efflux carriers in cells of a lengthening shoot tip direct auxin downward through the stem.

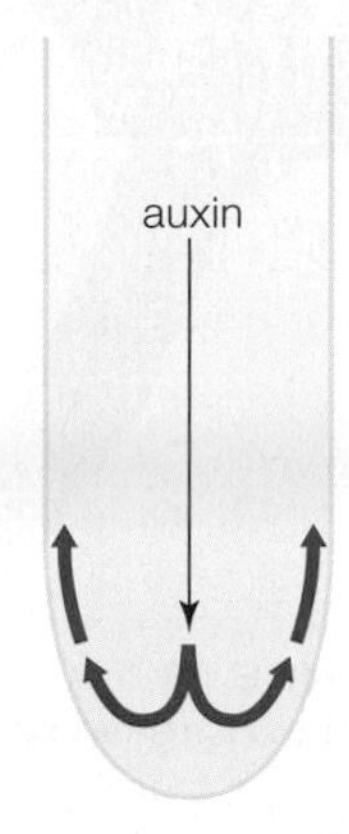

C Auxin enters a root tip in vascular tissue. Efflux carriers direct its flow through epidermal cells in the root tip.

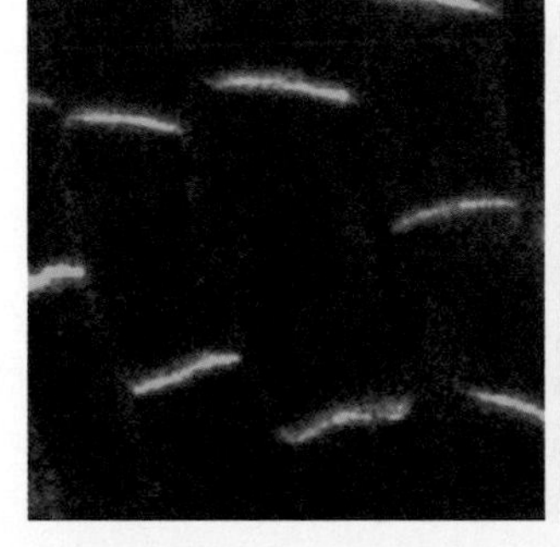
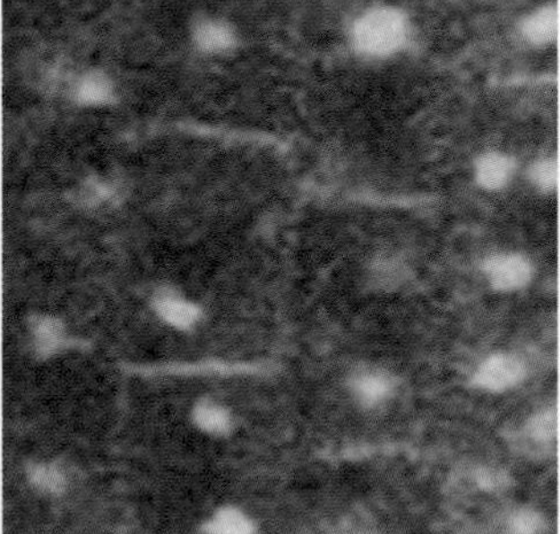

D Green fluorescence marks the location of auxin efflux carriers in these micrographs of *Arabidopsis* roots. Efflux carriers that cluster at one end of a cell membrane (left) direct the flow of auxin through a tissue. The polar positioning of these carriers in plant cell membranes is dynamic: In cells treated with a chemical that interrupts membrane trafficking between ER and Golgi bodies, efflux carriers accumulate quickly in cytoplasmic vesicles (right).

FIGURE 30.4 Polar transport of auxin is driven by asymmetrical positioning of plasma membrane active transport proteins called efflux carriers.

TAKE-HOME MESSAGE 30.3

What are the main effects of auxin in plants?

✔ Auxin directly promotes cell enlargement during primary growth. It also coordinates the effects of other plant hormones involved in growth and development.

✔ A dynamically regulated polar distribution system sets up auxin concentration gradients across a plant's tissues.

✔ Auxin gradients affect growth and development in localized patterns that can change over time.

30.4 Cytokinin

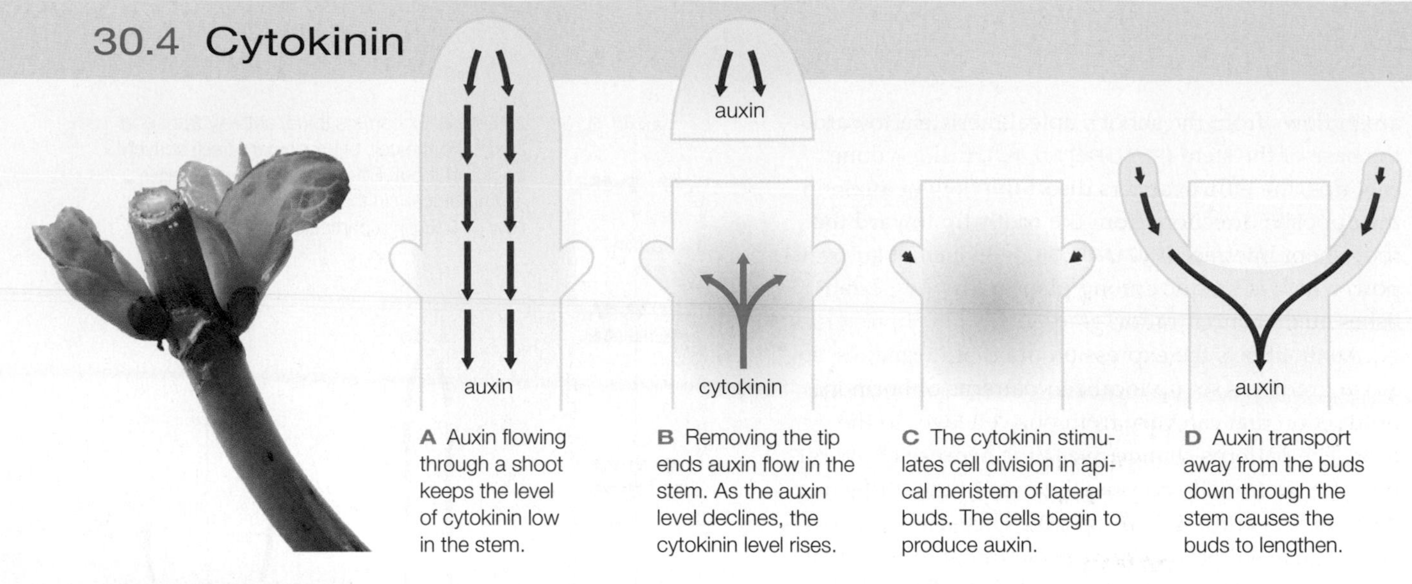

FIGURE 30.5 A shoot can continue to grow despite losing its tip: Interaction of auxin and cytokinin in the release of apical dominance. The loss of a shoot's tip ends its main supply of auxin—a signal that breaks dormancy in lateral buds.

✔ Cytokinin stimulates cell divisions in shoot apical meristem, and cell differentiation in root apical meristem.

A **cytokinin** is one of a group of plant hormones derived from the nucleotide adenine. Some cytokinin is synthesized locally in stems, but most is produced in roots and transported to shoots in xylem. Perhaps the best-known effect of cytokinin is that it stimulates growth of lateral buds. Apical dominance varies by species, and it is controlled by environmental cues as much as developmental programs. However, in general, lateral buds tend to remain dormant unless a shoot gets decapitated. Auxin maintains apical dominance mainly by regulating the production and transport of cytokinin in a stem, and by promoting expression of an enzyme that breaks it down. If a shoot's tip breaks off, its source of auxin disappears, so the cytokinin level rises in the stem (**FIGURE 30.5A,B**). The cytokinin moves into lateral buds and stimulates cell divisions of apical meristem inside them (**FIGURE 30.5C**). These newly active meristem cells produce auxin, which is then transported away from the bud tips and down the stem. This auxin flow causes the lateral buds to lengthen (**FIGURE 30.5D**).

Cytokinin and auxin influence one another's expression, a homeostatic mechanism that dynamically regulates their relative concentrations. They also influence the same cells. In shoot tips, for example, the two hormones act on the same set of receptors in shoot apical meristem cells to support cell division and to prevent differentiation. In most other contexts, cytokinin opposes auxin's effects. For example, in root tips, the two hormones work in opposition to maintain the balance of differentiating and undifferentiated cells in root apical meristem. In this context, auxin supports division of undifferentiated meristem cells, and cytokinin signals the cells to differentiate.

Cytokinin also opposes auxin's effect on lateral root formation, which requires an auxin gradient. Lateral roots grow from pericycle cells (Section 27.6). Only a few pericycle cells give rise to lateral roots, however. This is because, in roots, the level of cytokinin is normally high compared with auxin. (Most cytokinin is produced in roots; most auxin is produced in shoots.) Cytokinin inhibits production of auxin efflux carriers and their insertion into the membrane. Thus, in most pericycle cells (where the cytokinin to auxin ratio is high), an auxin gradient does not form, and neither does a lateral root. Lateral root formation is initiated by pericycle cells in which the auxin to cytokinin ratio increases. The increase results in efflux carriers being inserted into the cells' plasma membrane. The carriers set up an auxin gradient, and a lateral root forms.

cytokinin Plant hormone that promotes cell division in shoot apical meristem and cell differentiation in root apical meristem. Often opposes auxin's effects.

TAKE-HOME MESSAGE 30.4

What are the main effects of cytokinin in plants?

✔ Cytokinin stimulates cell divisions in shoot apical meristem, and cell differentiation in root apical meristem.

✔ Cyokinin and auxin act together and often antagonistically. The cytokinin–auxin balance controls cell division and differentiation in shoot and root apical meristem.

30.5 Gibberellin

✔ Gibberellins cause stems to elongate and seeds to germinate.

In 1926, researcher E. Kurosawa was studying what Japanese call *bakane*, or "foolish seedling" disease. The stems of rice plants infected with a fungus, *Gibberella fujikuroi*, grew twice the length of uninfected plants. Kurosawa discovered that he could induce the lengthening experimentally by applying extracts of the fungus to seedlings. Many years later, other researchers purified the substance from fungal extracts that brought about stem lengthening in plants (**FIGURE 30.6**). They named it gibberellin, after the fungus.

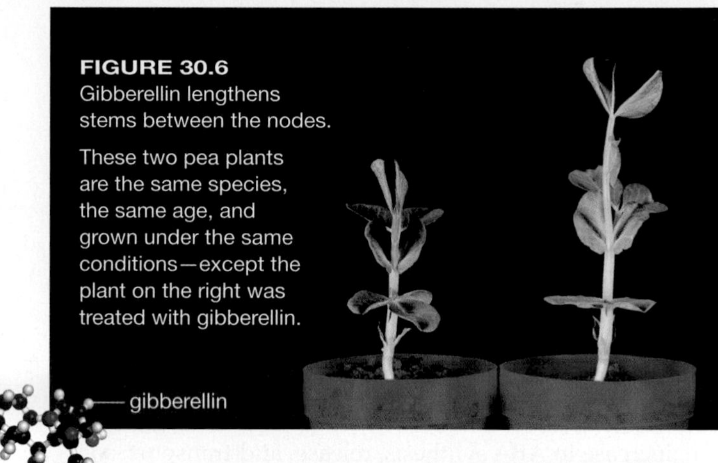

FIGURE 30.6 Gibberellin lengthens stems between the nodes.

These two pea plants are the same species, the same age, and grown under the same conditions—except the plant on the right was treated with gibberellin.

A **gibberellin**, as we now know, is a plant hormone that promotes growth in all flowering plants, among other functions. It causes a stem to lengthen between the nodes by inducing cell division and elongation. Along with auxin, gibberellin stimulates expansion along the long axis of a plant organ. Thus, it increases the length of the organ, and of the plant itself. The short stature of Mendel's dwarf pea plants (Section 13.3) is the result of a mutation that reduces the rate of gibberellin synthesis. A defective gibberellin receptor can result in a dwarf phenotype too. Gibberellin is also involved in slowing the aging of leaves and fruits, breaking dormancy and germination in seeds, and, in some plants, flowering.

Gibberellin made by cells in young leaves and root tips is transported through phloem to the rest of the plant. Seeds also make this hormone. Seedless grapes tend to be smaller than seeded varieties because their undeveloped seeds do not produce normal amounts of gibberellin. Farmers spray their seedless grape plants with synthetic gibberellin, which increases the size of the resulting fruit.

Gibberellin works by inhibiting inhibitors, thus removing the brakes on some cellular processes. Binding to a gibberellin receptor sets in motion the destruction of transcription repressors (Section 10.2) in the nucleus. The genes that these proteins repress are still being studied, but their products have overlapping functions involving cell proliferation and expansion, fertility, and germination.

For example, during germination of a barley seed, absorbed water causes cells of the embryo to release gibberellin (**FIGURE 30.7**). The hormone diffuses into the aleurone, a protein-rich layer of cells surrounding the endosperm. The presence of gibberellin causes these cells to destroy a transcriptional repressor. The result is transcription of the gene for amylase—the enzyme that breaks the bonds between starch's glucose monomers (Section 3.4). The cells start producing amylase and releasing it into the endosperm's starchy interior, where it breaks down stored starch molecules. The embryo then begins to use the released glucose for aerobic respiration, which fuels rapid cell divisions of meristem cells in the radicle and plumule.

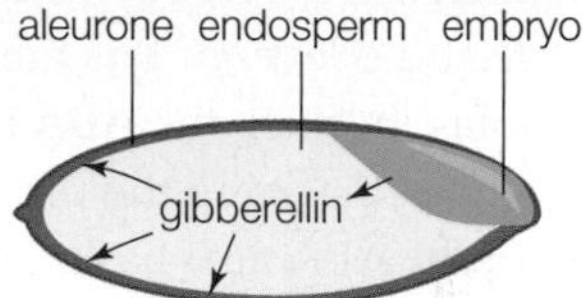

A Absorbed water causes cells of a barley embryo to release gibberellin, which diffuses through the seed into the aleurone layer of the endosperm.

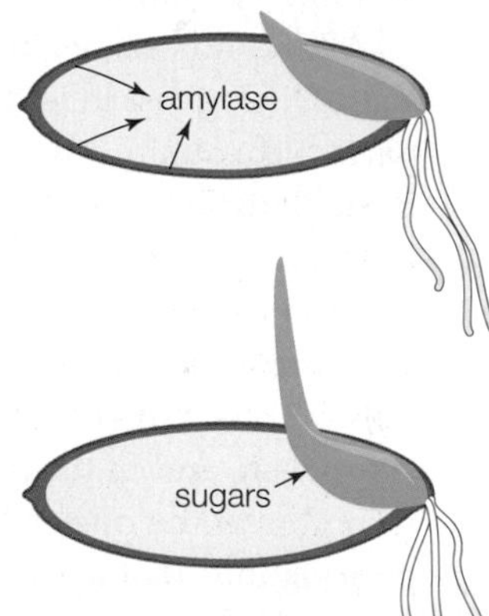

B Gibberellin causes cells of the aleurone layer to produce amylase. This enzyme diffuses into the starch-packed endosperm.

C The amylase breaks down starch into glucose. The glucose diffuses into the embryo and is used for aerobic respiration. Energy released by the reactions of aerobic respiration fuels meristem cell divisions in the embryo.

FIGURE 30.7 Gibberellin and germination, illustrated in a seed of barley (*Hordeum vulgare*), a monocot.

TAKE-HOME MESSAGE 30.5

What are the main effects of gibberellin in plants?

✔ Gibberellin causes lengthening in plants by stimulating cell division and elongation along the long axis of stems and other organs. Among other effects, it also helps mobilize nutrients in a seed during germination.

gibberellin Plant hormone that induces stem elongation and helps seeds break dormancy, among other effects.

30.6 Abscisic Acid

✔ Abscisic acid mediates germination, inhibits growth, and it is part of protective responses to stress.

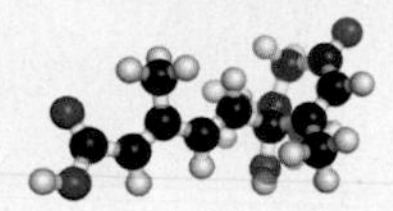

abscisic acid (ABA)

Abscisic acid (**ABA**) is a plant hormone that was named because its discoverers thought it mainly mediated **abscission**, the process by which a plant sheds leaves or other parts (we return to this topic in Section 30.9). However, ABA was later discovered to have a much greater role in plant stress responses. Temperature extremes, lack of water, and other environmental stresses trigger an increase in ABA synthesis, release, and transport, so that the level of this hormone rises in the plant's tissues. In turn, the ABA increase induces expression of genes that help the plant survive the adverse conditions. ABA also has an important role in embryo maturation, seed and pollen germination, and fruit ripening; and like cytokinin it suppresses lateral root formation.

ABA synthesis begins in chloroplasts, so its concentration is highest in leaves and other photosynthetic parts. Like auxin, ABA can exit a cell only by moving through active transport proteins in its plasma membrane, but ABA movement is not polar as is auxin's. Once in vascular tissue, ABA can move through xylem or phloem to all parts of the plant.

ABA receptors occur on the plasma membrane, in cytoplasm, and in the nucleus. When ABA binds to its receptors in the nucleus, it triggers a cascade of phosphorylations that activate hundreds of transcription factors. These transcription factors govern the expression of thousands of genes—around 10 percent of the total number in plants and a much larger proportion than any other plant hormone. In general, genes whose expression is repressed by ABA are involved in growth, such as those that encode components of ribosomes, chloroplasts, and cell walls; those whose expression is enhanced by ABA are involved in metabolism, stress responses, and embryonic development.

Consider how ABA encourages cells to produce NADPH oxidase, an enzyme associated with plasma membranes. This enzyme transfers electrons from NADPH inside the cell to oxygen molecules outside the cell—a redox reaction (Section 5.5) that produces oxygen free radicals and hydrogen peroxide. These reactive substances very quickly activate another plasma membrane enzyme, which, in turn, produces a burst of nitric oxide gas (NO). The NO burst is a critical part of plant stress responses (more about how this works in Section 30.10). A burst of NO also triggers production of an enzyme that breaks down ABA—a feedback loop that dynamically controls ABA signaling.

FIGURE 30.8 ABA prevents seed germination. Part of the ABA signaling pathway in this *Arabidopsis* plant has been knocked out. Its seeds germinated before they had a chance to disperse.

ABA that accumulates in a seed as it forms, matures, and dries out prevents the seed from germinating too early (**FIGURE 30.8**). It exerts this effect mainly by inhibiting expression of genes involved in cell wall loosening and expansion—both critical processes for growth of an embryonic plant. ABA also inhibits expression of genes involved in gibberellin synthesis. Thus, a seed cannot germinate until its ABA level declines. Exactly how this decline is triggered is unknown, but as the ABA level falls, cells of the embryo start expressing genes that result in cell wall softening and enlargement. Transcriptional control over gibberellin synthesis genes is also lifted. Hydrogen peroxide (produced by the activity of NADPH oxidase) enhances expression of these genes, so gibberellin is produced, and germination begins as enlarging cells of the embryo start using carbohydrates in endosperm.

TAKE-HOME MESSAGE 30.6

What are some effects of abscisic acid in plants?

✔ Abscisic acid plays a major role in stress responses.

✔ ABA also inhibits germination and growth, and has a role in fruit ripening and embryonic development.

abscisic acid (ABA) Plant hormone involved in stress responses; inhibits germination.

abscission Process by which plant parts are shed.

30.7 Ethylene

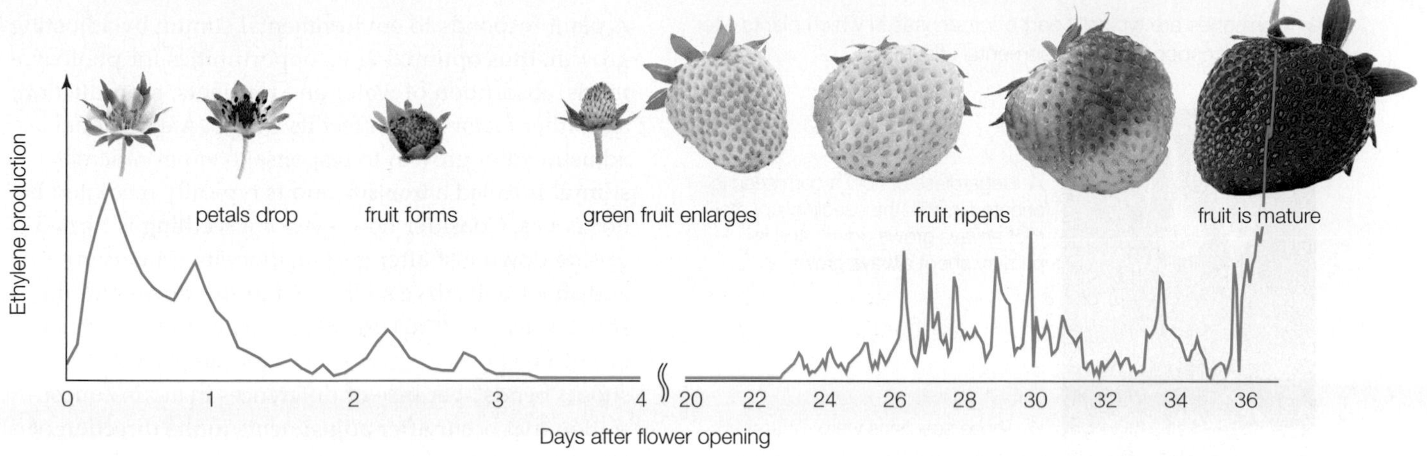

FIGURE 30.9 Ethylene production during strawberry formation and ripening. Oscillations are normal daily cycles (see Section 30.9).

✔ Ethylene is a gas involved in fine-tuning growth responses. It also triggers intermittent processes, including fruit ripening.

The plant hormone **ethylene** is part of regulatory pathways that govern a wide range of metabolic and developmental processes in plants, including germination, growth, abscission, ripening, and stress responses. Ethylene is a gas that is soluble in water, and it freely crosses lipid bilayers. Cells in all parts of a plant can produce this hormone from methionine and ATP.

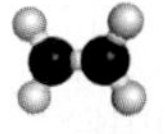

Two enzymes are involved in ethylene synthesis, and each occurs in multiple versions encoded by slightly different genes. Expression of some of these genes is inhibited by ethylene (a negative feedback loop); expression of the others is enhanced by ethylene (a positive feedback loop). Negative feedback loops produce a basal level of ethylene that helps fine-tune ongoing metabolic and developmental processes such as growth and cell expansion. For example, in roots, ethylene produced in a negative feedback loop enhances the transcription of genes involved in producing auxin and its efflux carriers (remember, auxin inhibits lengthening in roots). Positive feedback loops produce large amounts of ethylene required for intermittent processes such as germination, defense responses, abscission, and ripening (**FIGURE 30.9**). In one positive feedback loop, ethylene synthesis in flower petals increases until it provokes abscission, so the petals drop. Another feedback loop involves fruit ripening. ABA accumulates in a fruit as it forms and enlarges, eventually triggering production of ethylene in a positive feedback loop. The resulting ethylene burst enhances transcription of genes whose products have several effects associated with ripening. Chloroplasts are converted to chromoplasts (Section 4.9) as their (green) chlorophylls break down, and carotenoids such as (red) lycopene and (orange and yellow) beta-carotenes accumulate inside the organelles. Firm cell walls and middle lamellae break down. Starch and organic acids are converted to sugars, and aromatic molecules are produced. The resulting color change, softening, and increased palatability attract animals that can disperse the fruit's seeds.

Because ethylene is a gas, it can diffuse from one fruit to initiate ripening in another. Humans use this property to artificially ripen several types of fruit. Hard, unripe fruit can be transported long distances with less damage than soft, ripe fruit. Upon arrival at its final destination, the fruit is exposed to synthetic ethylene, which jump-starts the positive feedback ripening loop.

One of the ripening genes whose expression is induced by ethylene encodes an enzyme that breaks down pectin in plant cell walls. In 1994, the FDA approved "Flavr Savr" tomatoes, which had been genetically engineered to underproduce this enzyme. The tomatoes could be ripened artificially with ethylene, but the delayed softening doubled their shelf life.

ethylene Gaseous plant hormone involved in regulating growth and cell expansion. Participates in germination, abscission, ripening, and stress responses.

TAKE-HOME MESSAGE 30.7

What are some effects of ethylene in plants?

✔ Ethylene produced in negative feedback loops participates in ongoing metabolic and developmental processes.

✔ Ethylene produced in positive feedback loops is involved in intermittent processes such as abscission, fruit ripening, and defense responses.

✔ Hormones are typically part of processes in which plants alter growth in response to environmental stimuli.

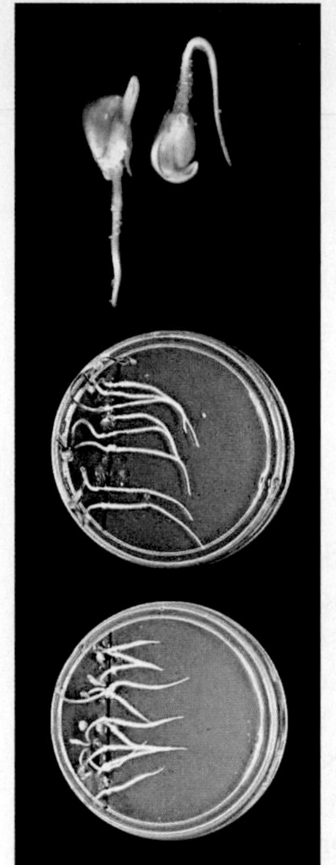

A Regardless of how a corn seed is oriented in soil, the seedling's primary root always grows down, and its primary shoot always grows up.

B These seedlings were rotated 90° counterclockwise after they germinated. The plant adjusts to the change by redistributing auxin, and the direction of growth shifts as a result.

C In the presence of auxin transport inhibitors, seedlings do not adjust their direction of growth after a 90° counterclockwise rotation. Mutations in genes that encode auxin efflux carriers have the same effect.

FIGURE 30.10 Gravitropism.

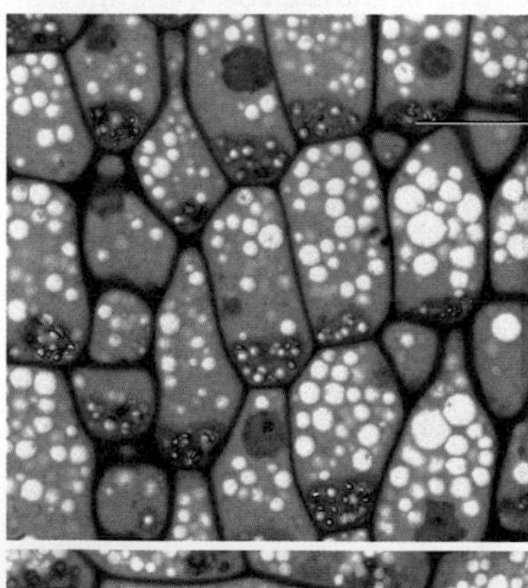

A This micrograph shows heavy, starch-packed statoliths settled on the bottom of gravity-sensing cells in a corn root cap.

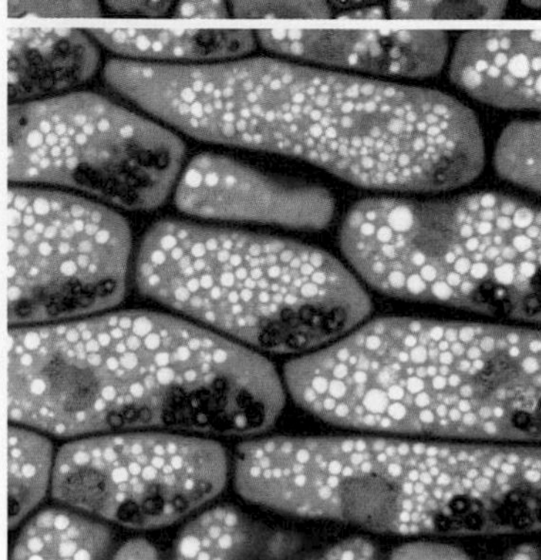

B This micrograph was taken ten minutes after the root in A was rotated 90°. The statoliths are already settling to the new "bottom" of the cells.

FIGURE 30.11 Gravity and statoliths.

FIGURE IT OUT In which direction was this root rotated?

Answer: Counterclockwise

A plant responds to environmental stimuli by adjusting growth, thus optimizing its opportunities for photosynthesis, absorption of water and nutrients, reproduction, and other factors that affect its fitness. A directional adjustment of growth in response to environmental stimuli is called a **tropism**, and is typically mediated by hormones. Consider how, even if a seedling is turned upside down just after germination, its primary root and shoot will curve so the root grows down and the shoot grows up (**FIGURE 30.10**). This and any other growth response to gravity is a **gravitropism**. A root or shoot "bends" because of differences in auxin concentration that occur after adjustments in the direction of auxin transport. In shoots, auxin enhances cell elongation, so auxin that accumulates on one side of a shoot causes the shoot to bend away from that side. Auxin has the opposite effect in roots: Auxin that accumulates on one side of a growing root causes the root to bend toward that side.

Gravity-sensing mechanisms of many organisms are based on organelles called **statoliths**. Plant statoliths are amyloplasts (Section 4.9) stuffed with dense grains of starch. They occur in root cap cells, and also in specialized cells at the edge of vascular tissues in the stem. (In shoots, chloroplasts may also function as gravity-sensing organelles because they specifically accumulate starch.) Starch grains are heavier than cytoplasm, so statoliths tend to sink to the lowest region of the cell, wherever that is (**FIGURE 30.11**). A shift in statolith position causes auxin efflux carriers to be redistributed to the down-facing side of the root or stem. The mechanism by which gravity influences the redistribution of efflux carriers is currently unknown, but it may involve a reorganization of actin filaments that connect amyloplasts to the cell's cytoskeleton.

A **phototropism** is a growth response that orients parts of a plant in a direction influenced by light. For example, an elongating stem curves toward a light

source (left), thus maximizing the amount of light intercepted by its photosynthetic cells. In shoots, phototropism occurs in response to blue light absorbed by nonphotosynthetic pigments called phototropins (these pigments are also part of a mechanism by which stomata open at sunrise, Section 28.5). In a shoot tip or coleoptile, light-energized phototropins repress transcription of a gene whose product causes auxin efflux carriers to be distributed evenly around the cell membrane. Thus, a cell exposed to a directional light source ends up with efflux carriers positioned mainly on its shady

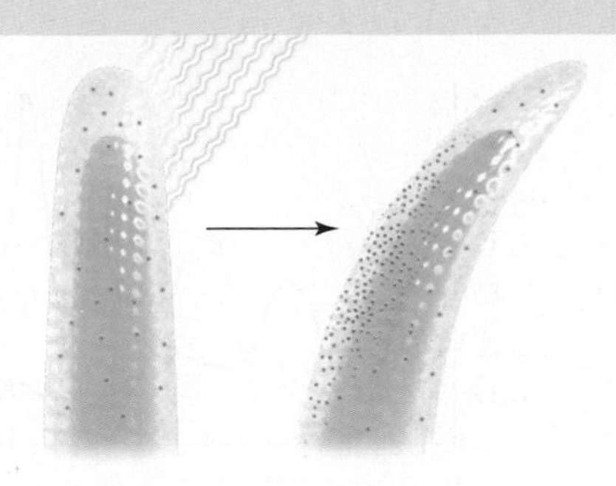

FIGURE 30.12 Phototropism. Auxin-mediated differences in cell elongation between two sides of a coleoptile-protected primary shoot induce bending toward light. Red dots signify auxin molecules.

side. The polarization directs auxin flow toward the shaded side of the structure, causing cells on that side to elongate more than cells on the illuminated side (**FIGURE 30.12**). The difference causes the entire structure to bend toward the light as it lengthens.

In another example of phototropism, chloroplasts are dragged from one position to another along actin filaments of the cytoskeleton. In low-intensity light, chloroplasts are directed toward leaf surfaces and oriented perpendicular to the light source, thus maximizing their exposure to light for photosynthesis. In high-intensity light, chloroplasts move in the opposite direction—toward internal tissues of the leaf where they are more shielded, and oriented parallel to the light (**FIGURE 30.13**). Moving away from high intensity light minimizes damage that can occur as a result of excess electrons accumulating in electron transfer chains of photosynthetic light reactions (Section 6.5). Blue light drives the movement in both cases.

Leaves or flowers of some plants change position in response to the changing angle of the sun throughout the day, a response called heliotropism (from Greek *helios*, sun). The mechanism that drives heliotropism is not understood, but it may be similar to phototropism because it has been observed to coincide with differential elongation of cells in stems, and it also occurs in response to blue light.

A plant's contact with an object may cause a change in the direction of its growth, a response called **thigmotropism** (Greek *thigma* means touch). We see thigmotropism when a vine's tendril grows around a wire (left). The mechanism that underlies the response is not well understood, but an immediate increase in cytoplasmic concentration of calcium that accompanies contact is likely to play a role. Plants (and animals) have plasma membrane transport proteins that flood cytoplasm with calcium ions upon mechanical disturbance of the membrane. Several gene products that can sense calcium ions are involved, as well as nitric oxide and another plant hormone called jasmonic acid. These molecules trigger

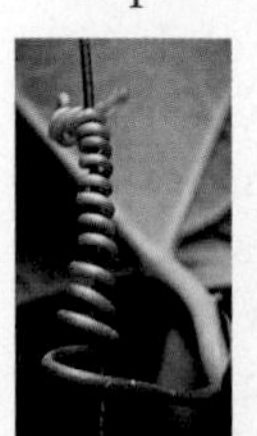

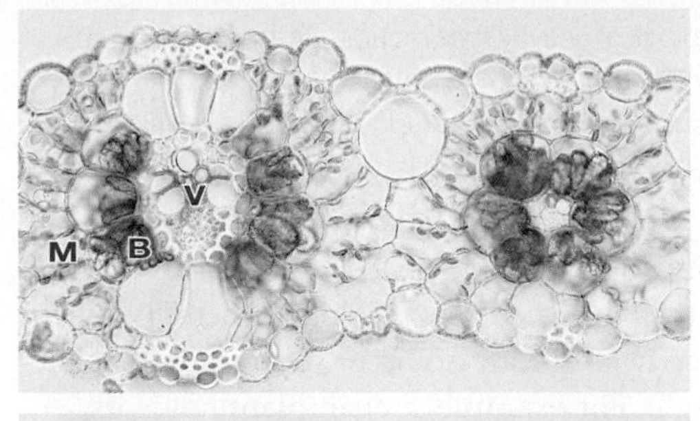

A Before light exposure.

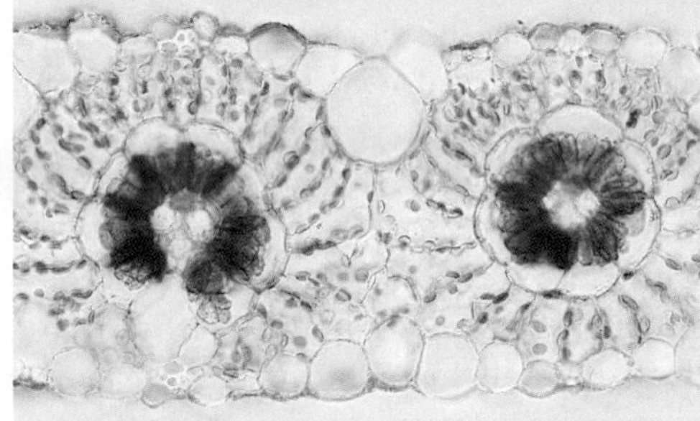

B After 2 hours of intense light exposure.

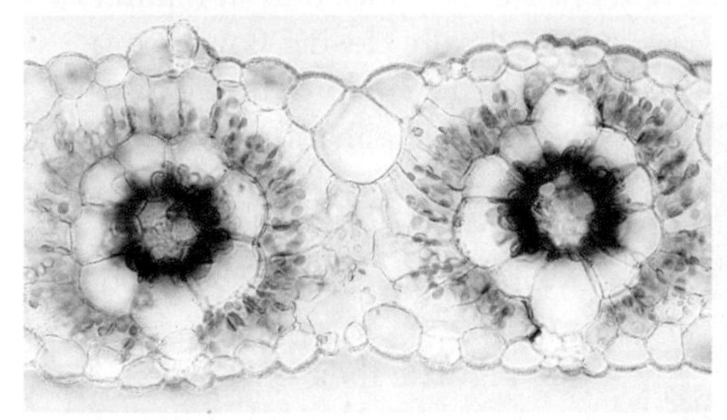

C After 16 hours of intense light exposure.

100 µm

FIGURE 30.13 Movement of chloroplasts in response to intense light, as seen in leaf sections of millet, a C4 plant. The chloroplasts respond by moving away from the external surfaces of the leaf, toward internal tissues that are more shielded from the light. Mesophyll (**M**), bundle sheath cells (**B**), and vascular bundles (**V**).

unequal growth rates of cells on opposite sides of the shoot top, which causes the shoot to coil as it grows. A similar mechanism causes roots to grow away from contact, so they "feel" their way around rocks and other impassable objects in the soil.

gravitropism Plant growth in a direction influenced by gravity.
phototropism Plant growth in a direction influenced by light.
statolith Amyloplast involved in sensing gravity.
thigmotropism Plant growth in a direction influenced by contact with a solid object.
tropism In plants, directional growth response to an environmental stimulus.

TAKE-HOME MESSAGE 30.8

How do environmental cues trigger growth responses in plants?

✔ Plants adjust the direction and rate of growth in response to environmental stimuli such as gravity, light, contact, and touch. Hormones are typically part of these responses.

30.9 Sensing Recurring Environmental Change

✔ Shifts in biological activity that recur in 24-hour cycles are mediated by cyclic shifts in gene expression.

✔ Seasonal shifts in night length trigger seasonal shifts in development in many plants.

Daily Change

A **circadian rhythm** is a cycle of biological activity that repeats every twenty-four hours or so (circadian means "about a day"). For example, a bean plant holds its leaves horizontally during the day but folds them close to its stem at night. Bean plants exposed to constant light or darkness for a few days will continue to move their leaves in and out of the "sleep" position at the time of sunrise and sunset (**FIGURE 30.14**). Similar mechanisms cause flowers of some plants to open only at certain times of day. For example, the flowers of plants pollinated mainly by night-flying bats open, secrete nectar, and release fragrance only at night. Periodically closing flowers protects delicate reproductive parts when the likelihood of pollination is lowest.

Circadian rhythms are driven by interconnected feedback loops involving transcription factors that regulate their own expression (typically, by the methylation and demethylation of histones, Section 10.2). These feedback loops give rise to cycles in the levels of more than 30 percent of a plant cell's mRNAs (**FIGURE 30.15**).

Cyclic shifts in gene expression underlie cyclic shifts in metabolism, for example between daytime starch-building reactions and nighttime starch-consuming reactions. Components of rubisco, photosystem II, ATP synthase, and other proteins used in photosynthesis are among the many gene products produced during the day and broken down at night. Similarly, many molecules used at night are broken down during the day.

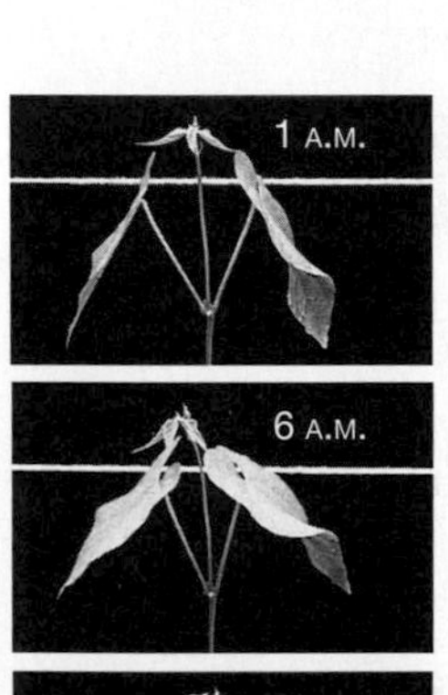

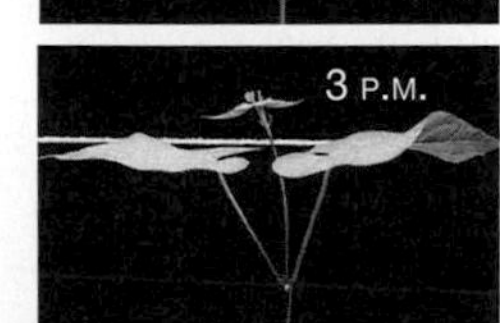

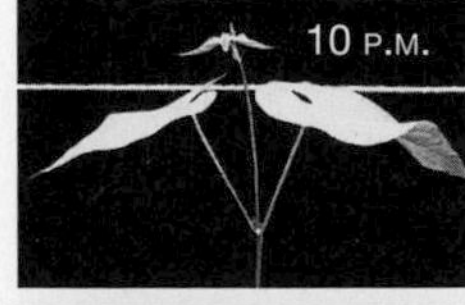

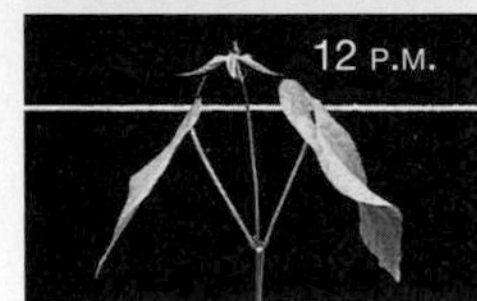

FIGURE 30.14 ▶**Animated** Rhythmic leaf movements by a young bean plant (*Phaseolus*). Physiologist Frank Salisbury kept this plant in darkness for twenty-four hours. Despite the lack of light cues, the leaves kept on folding and unfolding at sunrise (6 A.M.) and sunset (6 P.M.).

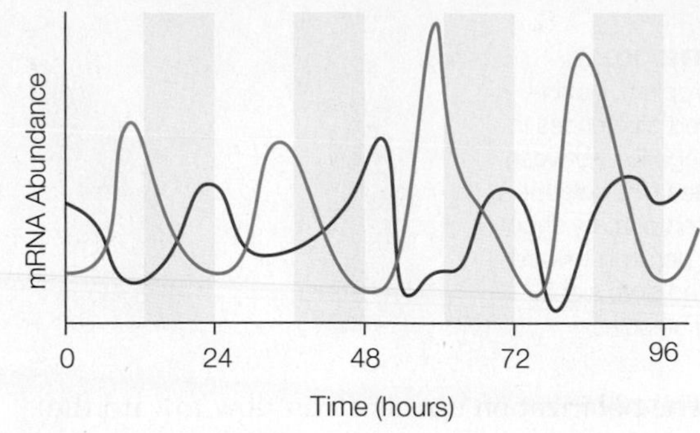

FIGURE 30.15 Circadian cycles of gene expression for some genes involved in temperature stress responses. Plants were kept in constant light; gray bars show nighttime hours. Molecules that help the plants tolerate cold nights or hot days are produced just before they are actually needed. Blue line: averaged expression of 46 cold tolerance genes; red line: averaged expression of 30 warm tolerance genes.

FIGURE IT OUT At what time of day is expression of cold tolerance genes lowest?

Answer: Early morning

At least six types of pigments provide light input into internal circadian "clocks," but the best known are phytochromes and cryptochromes. **Phytochromes** absorb red light; light with a wavelength of 660 nanometers (red) causes their structure to change from an inactive to an active form. Light with a wavelength of 730 nanometers (far-red, which predominates in shade) changes the molecule back to its inactive form. Cryptochromes absorb blue and UV light, thereby becoming activated and, interestingly, magnetic.

Seasonal Change

Except at the equator, the length of day varies with the season. Days are longer in summer than in winter, and the difference increases with latitude. These seasonal changes in light availability trigger seasonally appropriate responses in plants. **Photoperiodism** refers to an organism's response to changes in the length of day relative to night.

Flowering is a photoperiodic response in many species. Such species are termed long-day or short-day plants depending on the season in which they flower. These terms are somewhat misleading, as the main trigger for flowering is the length of night, not the length of day (**FIGURE 30.16**). Irises and other long-day plants flower only when the hours of darkness fall below a critical value, typically in summer. Chrysanthemums and other short-day plants flower only when the hours of darkness are greater than some critical value. Flowering is not photoperiodic in sunflowers, tomatoes, roses, and other day-neutral plants.

Some plants flower in spring only after exposure to a prolonged period of cold in the preceding winter, a response called **vernalization**. Researchers suspect that plant cells perceive temperature via their plasma membrane, because the lipid composition and calcium ion permeability of plasma membranes vary with temperature. Regardless of the mechanism of detection, a "cold" signal influences gene expression in these plant species.

Yearly cycles of abscission and dormancy are photoperiodic responses too. Plants that drop their leaves before dormancy are typically native to regions too dry or too cold for optimal growth during part of the year. For example, deciduous trees of the northeastern United States lose their leaves in autumn. The trees remain dormant during the months of harsh winter weather that would otherwise damage leaves and buds. Growth resumes in spring, when milder conditions return. On the opposite side of the world, many tree species native to tropical monsoon forests of south Asia lose their leaves during the summer dry season, which is between November and May. Although the region receives a lot of rain annually, almost none of it falls during this period of the year. Dormancy offers the trees a way to survive the extended droughtlike conditions. New growth appears at the beginning of June, just in time to be supported by the ample water of monsoon rains.

Seasonal abscission of leaves (and fruits) is mediated by ethylene. Let's use a horse chestnut tree as an example. In this species, most root and shoot growth occurs between spring and early summer, from April to July. By midsummer, the tree is producing fruits and seeds. In August, the growing season is coming to a close, and nutrients are being routed to stems and roots for storage during the forthcoming period of dormancy. Ripe fruits (**FIGURE 30.17A**) and seeds release ethylene that diffuses into nearby twigs, petioles, and fruit stalks. The ethylene triggers cells in these zones to produce wall-digesting enzymes. The cells bulge as their walls soften, and separate from one another as the extracellular matrix that cements them together dissolves. Tissue in the abscission zone weakens, and the structure above it drops. A scar often remains where the structure had been attached (**FIGURE 30.17B**).

circadian rhythm A biological activity that is repeated about every 24 hours.
photoperiodism Biological response to seasonal changes in the relative lengths of day and night.
phytochrome A light-sensitive pigment that helps set plant circadian rhythms based on length of night.
vernalization Stimulation of flowering in spring by prolonged exposure to low temperature in winter.

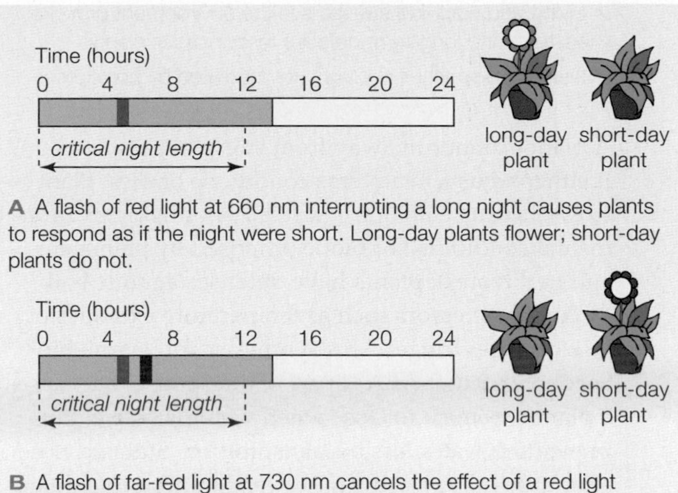

A A flash of red light at 660 nm interrupting a long night causes plants to respond as if the night were short. Long-day plants flower; short-day plants do not.

B A flash of far-red light at 730 nm cancels the effect of a red light flash. Short-day plants flower; long-day plants do not.

FIGURE 30.16 ▸Animated Experiments showing that long- or short-day plants flower in response to night length. Each horizontal bar represents 24 hours. Blue indicates night length; yellow, day length.

FIGURE IT OUT Which type of pigment detected the light flashes in these two experiments?
Answer: A phytochrome

A Ripening chestnuts emit ethylene that stimulates abscission of the fruits and nearby leaves.

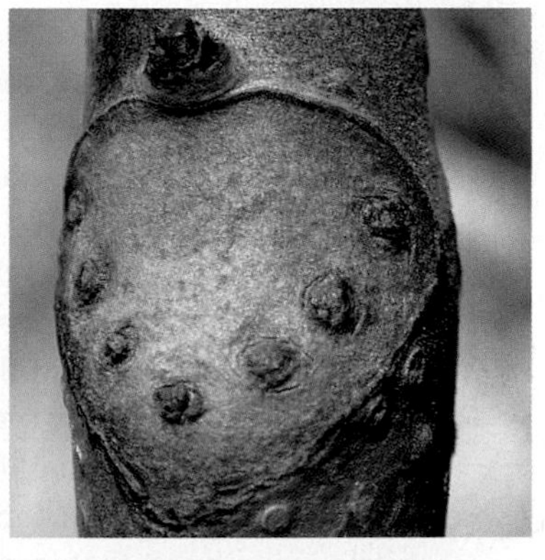

B Horse chestnut trees take their name from abscission zones that leave horseshoe-shaped scars in this species.

FIGURE 30.17 Abscission as part of the normal life cycle of deciduous trees such as chestnuts.

TAKE-HOME MESSAGE 30.9

How do plants sense and respond to recurring environmental change?

✔ Plants respond to recurring cues from the environment with recurring cycles of activity. Pigments provide input into circadian cycles.

✔ The main environmental cue for flowering is the length of night relative to the length of day, which varies seasonally in most places. Low winter temperatures stimulate spring flowering in some plant species.

✔ The timing of seasonal abscission and dormancy is an evolutionary adaptation to recurring periods of environmental conditions that do not favor growth.

30.10 Responses to Stress

✔ Living and nonliving stressors in the environment provoke short-term and long-term defense responses in plants.

✔ Defense responses in plants are mediated by hormones.

A plant cannot run away from stressful situations: It either adjusts to adverse conditions or dies. Plant stressors can be abiotic (caused by nonliving environmental conditions) or biotic (imposed by pathogens and herbivores); plants have defenses against both.

Abiotic stressors such as temperature extremes or lack of water trigger ABA synthesis. You learned in Section 28.5 that ABA is part of a response that causes a plant's stomata to close when water is scarce, thus preventing water loss by transpiration. Stomata close after transport proteins move calcium ions into guard cells; a nitric oxide burst produced in response to ABA activates these proteins.

ABA and nitric oxide also participate in biotic stress responses. Consider that receptors on plant cells recognize molecules specific to microbial pathogens, including flagellin, a protein component of bacterial flagella. Flagellin binding to one of these receptors triggers a burst of ethylene synthesis. When ethylene is not present, ethylene receptors block transcription of a set of genes involved in defense. When ethylene is present, it binds to its receptors and locks them in an inactive form that marks them for destruction. Thus, ethylene synthesis lifts the brakes on production of molecules used in defense. The cell starts producing more receptors that detect bacterial flagella—a positive feedback loop that sensitizes the plant to the presence of more bacteria. Any additional bacteria that bind to the accumulated flagellin receptors trigger an ABA-mediated nitric oxide burst that immediately closes stomata. Closing stomata defends the plant against bacterial invasion because bacteria cannot penetrate plant epidermis: They can enter plant tissues only through wounds or open stomata.

Mutualistic bacteria and fungi avoid triggering a plant's defense responses by exchanging chemical signals with it. For example, remember that *Rhizobium* bacteria in soil are attracted to compounds released from root cells (Section 28.3). Those compounds are flavonoids, a type of secondary metabolite you learned about in Section 30.1. *Rhizobium* bacteria respond to the plant flavonoids by producing molecules required for living in symbiosis with the plant. One of these molecules is secreted into the soil, and it is recognized by root cells as a signal to form a nodule.

When a pathogen penetrates a plant's epidermis, it triggers a large surge of hydrogen peroxide and nitric oxide that causes cells in the infected region to commit suicide. This "hypersensitive" response can prevent a pathogen from spreading to other parts of the plant, because it often kills the pathogen along with the infected tissue (**FIGURE 30.18**).

FIGURE 30.18 Evidence of a hypersensitive response in a leaf. Brown spots are dead tissue where germinating fungi penetrated the leaf's epidermis, triggering a release of hydrogen peroxide and nitric oxide that caused plant cell suicide.

A hypersensitive response is often ineffective against pathogens that gain nutrients by killing cells outright. Some sac fungi, for example, use toxins to kill their plant hosts, then absorb nutrients released from their decomposing tissues. A different, long-term defense response, **systemic acquired resistance**, increases resistance of the whole plant body to attack by a wide range of fungi and other pathogens, as well as tolerance to abiotic stressors. Systemic acquired resistance begins when an infected tissue releases an as yet unknown signal that travels to other parts of the plant, where it triggers cells to produce a plant hormone called salicylic acid. Salicylic acid increases transcription of hundreds of genes whose products confer general hardiness to the plant.

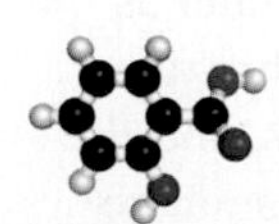

salicylic acid

One of the molecules produced in response to salicylic acid is epicatechin; this flavonoid prevents hardening of "pegs" that develop on germinating spores of many fungi. Fungi use these pegs to punch through plant epidermal tissue. Flavonoids related to epicatechin inhibit fatty acid synthesis in bacteria. Resveratrol produced by grapes and other plants has antifungal activ-

Prescription: Chocolate (revisited)

Epicatechin has now been shown to have a beneficial effect on a variety of illnesses, including hypertension and other cardiovascular disorders, Parkinson's disease, obesity, and even acne. Clinical trials involving chocolate consumption as an intervention in these disorders are now under way.

Epicatechin probably works its magic in humans because it activates an enzyme, nitric oxide synthase, in endothelial cells making up the walls of blood vessels. This enzyme makes nitric oxide, which in turn widens the blood vessels by relaxing muscles in their walls. In mammals, nitric oxide is part of hormonal systems that regulate blood vessel width, and it is also necessary for proper function of the endothelial cells that produce it. Dysfunction of these cells is linked to cardiovascular disease, a primary cause of illness and death in the United States and in many other countries. Thus, increasing nitric oxide production in blood vessel cells is a major area of medical research.

Humans have no stomata to close, and plants have no blood vessels, but nitric oxide operates as a hormone on both structures. In fact, NADPH oxidase, the plasma membrane enzyme that produces nitric oxide in plants, also occurs in human white blood cells, where it produces a nitric oxide burst in response to detection of bacterial flagellin inside the body. This response is mediated by abscisic acid, just as it is in plants—and all animals.

A Saliva of a tobacco budworm (*Heliothis virescens*) chewing on a leaf of a tobacco plant (*Nicotiana*) triggers the plant to emit a combination of 11 volatile secondary metabolites.

B Red-tailed wasps (*Cardiochiles nigriceps*), which parasitize tobacco budworms, recognize the unique chemical signature emitted by the plant. They follow the trail of chemicals back to the source.

C A wasp that finds a budworm attacks it and deposits an egg inside of it. When the egg hatches, a larva emerges and begins to eat the budworm, which eventually dies.

FIGURE 30.19 Interspecific plant defenses. Plant cells release a particular combination of volatile secondary metabolites in response to being chewed by a particular insect species. The chemical signature attracts wasps that parasitize the insect.

ity. Other gene products include lignin, which increases the strength of plant cell walls; compounds that break down structural components of fungal cell walls; antioxidants; and so on: The chemicals differ by species, but all help the plant resist biotic and abiotic stressors.

Wounding of a leaf, such as occurs when an insect chews on it, triggers the production of ABA, hydrogen peroxide, ethylene, and jasmonic acid. Jasmonic acid in turn increases transcription of genes whose products include secondary metabolites that are volatile (released as a vapor). These molecules are detected by wasps that parasitize insect herbivores (**FIGURE 30.19**). Volatile secondary metabolites released in response to herbivory are also detected by neighboring plants, which respond by increasing their own production of ethylene and jasmonic acid.

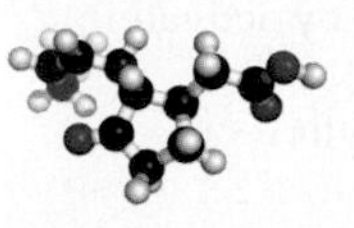

jasmonic acid

systemic acquired resistance In plants, inducible whole-body resistance to a wide range of pathogens and abiotic stressors.

TAKE-HOME MESSAGE 30.10

How do plants respond to stress?

- ✔ Abscisic acid is involved in responses to abiotic stressors such as temperature extremes or lack of water.
- ✔ Detection of plant pathogens can trigger stomatal closure or cell death.
- ✔ Systemic acquired resistance triggered by pathogen attacks increases a plant's ability to withstand biotic and abiotic stresses.

summary

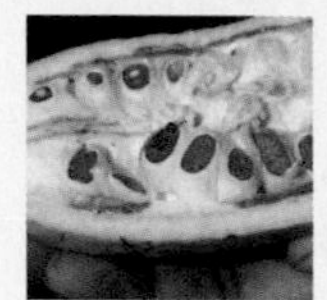

Section 30.1 Secondary metabolites are not required for immediate survival of the organism that produces them. Some plant secondary metabolites attract pollinators or symbionts, or function in defense. A few of these compounds, including a number of flavonoids, are beneficial to human health.

Section 30.2 Plants continue to develop through their lifetime. **Plant hormones** promote or arrest development in regions of a plant by stimulating or inhibiting cell division, differentiation, or enlargement. Some have roles in occasional responses such as pathogen defense. Hormones work together or in opposition, and many have different effects in different regions of the plant.

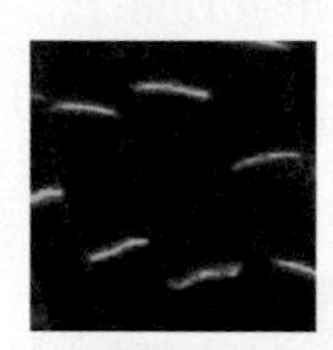

Section 30.3 **Auxin** promotes lengthening and also coordinates the effects of other hormones involved in growth. A unique transport system distributes auxin directionally. The polar distribution sets up auxin gradients that affect the growth and development of plant parts, such as when auxin gradients maintain **apical dominance** in a growing shoot tip.

Section 30.4 **Cytokinin** promotes cell division in shoot apical meristem, and differentiation in root apical meristem. This hormone acts together with auxin, often antagonistically, to balance growth with development in shoot and root tips.

Section 30.5 **Gibberellin** lengthens stems between nodes by inducing cell division and elongation. It also stimulates production of enzymes that break down endosperm during seed germination.

Section 30.6 **Abscisic acid** (**ABA**) plays a part in **abscission**, but has a greater role in other processes, especially responses to stress. ABA influences expression of thousands of genes, with effects that include the suppression of seed germination and growth. ABA also participates in embryonic development and fruit ripening.

Section 30.7 **Ethylene** is produced in negative and positive feedback loops. The negative feedback loops help regulate ongoing metabolism and development; the positive feedback loops trigger special processes such as abscission, ripening, and defense responses.

Section 30.8 A **tropism** is an adjustment in the direction of plant growth in response to environmental cues such as gravity (**gravitropism**), light (**phototropism**), or contact (**thigmotropism**). **Statoliths** play a part in gravitropism.

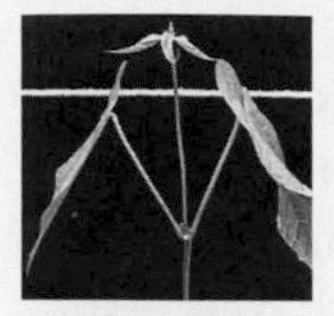

Section 30.9 **Circadian rhythms** are driven by gene expression feedback loops that have input from nonphotosynthetic pigments such as **phytochromes**. **Photoperiodism** is a response to change in the length of day relative to night. Seasonal cycles of abscission and dormancy are adaptations to seasonal changes in environmental conditions. Some plants require prolonged exposure to cold before they can flower (**vernalization**).

Section 30.10 Plants use hormones to respond to abiotic and biotic stresses. ABA is part of the stress response that closes stomata when water is scarce. Nitric oxide is part of a hypersensitive defense response that closes stomata or kills the cell in response to pathogen detection. Pathogen-triggered **systemic acquired resistance** increases the plant's resilience to biotic and abiotic stress in general.

self-quiz

Answers in Appendix VII

1. Plant hormones _______ .
   a. often have multiple, overlapping effects
   b. are active in developing plant embryos
   c. are active in adult plants
   d. may have different effects in different tissues
   e. all of the above

2. _______ is the hormone in most rooting compounds.
   a. Gibberellin
   b. Auxin
   c. Cytokinin
   d. ABA

3. Which of the following statements is false?
   a. Auxin and gibberellin promote stem elongation.
   b. Cytokinin promotes cell division in shoot tips.
   c. Abscisic acid promotes water loss and dormancy.
   d. Ethylene promotes fruit ripening and abscission.

4. Which combination of hormones is best for promoting lateral root growth?
   a. A high auxin level and low ABA and cytokinin levels
   b. A low auxin level and high ABA and cytokinin levels
   c. A high level of auxin, ABA, and cytokinin

5. Ethylene differs from other hormones in that it _______ .
   a. is a gas
   b. participates in defense
   c. operates during ripening
   d. none of the above

6. Heliotropism may be related to _______ .
   a. phototropism
   b. gravitropism
   c. photoperiodism
   d. a and c

7. Sunlight resets biological clocks in plants by activating and inactivating _______ .
   a. phototropins
   b. phytochromes
   c. cryptochromes
   d. chlorophylls
   e. b and c
   f. all of the above

8. In some plants, flowering is a _______ response.
   a. phototropic
   b. gravitropic
   c. photoperiodic
   d. thigmotropic

data analysis activities

Volatile Secondary Metabolites in Plant Stress Responses In 2007, researchers Casey Delphia, Mark Mescher, and Consuelo De Moraes (pictured at left) published a study on the production of different volatile chemicals by tobacco plants in response to predation by two types of insects: western flower thrips and tobacco budworms. Their results are shown in **FIGURE 30.20**.

1. Which treatment elicited the greatest production of volatiles?

2. Which volatile chemical was produced in the greatest amount? What was the stimulus?

3. Which one of the chemicals tested is most likely produced by tobacco plants in a nonspecific response to predation?

4. Are any chemicals produced in response to predation by budworms, but not in response to predation by thrips?

Volatile Compound Produced	Treatment					
	C	T	W	WT	HV	HVT
Myrcene	0	0	0	0	17	22
β-Ocimene	0	433	15	121	4,299	5,315
Linalool	0	0	0	0	125	178
Indole	0	0	0	0	74	142
Nicotine	0	0	233	160	390	538
β-Elemene	0	0	0	0	90	102
β-Caryophyllene	0	100	40	124	3,704	6,166
α-Humulene	0	0	0	0	123	209
Sesquiterpene	0	7	0	0	219	268
α-Farnesene	0	15	0	0	293	457
Caryophyllene oxide	0	0	0	0	89	166
Total	0	555	288	405	9,423	13,563

FIGURE 30.20 Volatile (airborne) compounds produced by tobacco plants (*Nicotiana tabacum*) in response to predation by different insects. Plants were untreated (C), attacked by thrips (T), mechanically wounded (W), mechanically wounded and attacked by thrips (WT), attacked by budworms (HV), or attacked by budworms and thrips (HVT). Values are nanograms/day.

9. Stomata close in response to _______ .
   a. ABA
   b. nitric oxide
   c. bacteria
   d. all of the above

10. Match the response with its main trigger.

___ phototropism	a. contact with an object
___ gravitropism	b. blue light
___ thigmotropism	c. a long period of cold
___ photoperiodism	d. gravity
___ vernalization	e. sun position
___ heliotropism	f. night length

11. Match the hormone with the characteristic.

___ ethylene	a. efflux carriers set up gradients
___ cytokinin	b. produced in positive and negative feedback loops
___ auxin	c. affects expression of more genes than any other hormone
___ gibberellin	d. works antagonistically to auxin in apical dominance
___ ABA	e. big in stems
___ nitric oxide	f. active in bursts

12. Match the hormone with the observation.

___ ethylene	a. Your cabbage plants bolt (they form elongated flowering stalks).
___ cytokinin	b. The potted plant in your room is leaning toward the window.
___ auxin	c. The last of your apples is getting really mushy.
___ gibberellin	d. The seeds of your roommate's marijuana plant do not germinate no matter what he does to them.
___ ABA	e. Lateral buds are sprouting.
___ nitric oxide	f. Your lettuce plants develop brown spots on their leaves.

CENGAGE brain To access course materials, please visit www.cengagebrain.com.

critical thinking

1. Professional gardeners often soak seeds in hydrogen peroxide before planting them. Why?

2. Photosynthesis sustains plant growth, and inputs of sunlight sustain photosynthesis. Why, then, do seedlings that germinate in a fully darkened room grow taller than seedlings of the same species that germinate in full sun?

3. Belgian scientists discovered that certain mutations in common wall cress (*Arabidopsis thaliana*) cause excess auxin production. Predict the impact on the plant's phenotype.

4. Cattle in industrial dairy farms are typically given rBST, or recombinant bovine somatotropin. This animal hormone increases a cow's milk production, but there is concern that such hormones may have unforeseen effects on milk-drinking humans. There is no similar concern about plant hormones in the plant foods we eat. Why not?

5. The oat coleoptiles below have been modified: either cut or placed in a light-blocking tube. Which ones will still bend toward a directional light source?

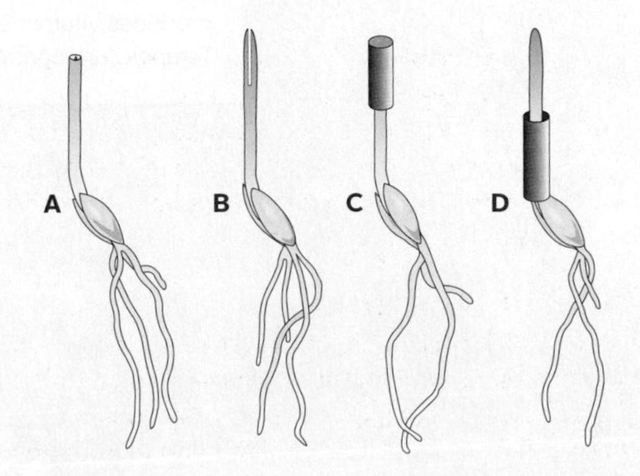

31 Animal Tissues and Organ Systems

LEARNING ROADMAP

With this chapter, we begin our survey of the tissues and organ systems (Section 1.2) in animals. The chapter expands on the nature of animal body plans (24.2) and trends in vertebrate evolution (25.2). We also revisit the topic of cancer (8.6,11.6) and look again at the evolution of human skin color (14.1).

ANIMAL ORGANIZATION

Most animals have cells organized as tissues, organs, and organ systems. The components function in concert to maintain conditions in the body's internal environment.

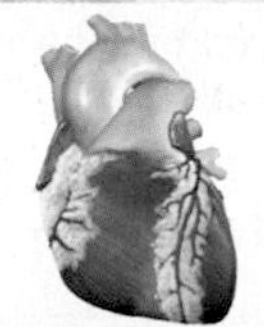

EPITHELIAL AND CONNECTIVE TISSUES

Epithelial tissue covers the body's surface and lines its internal tubes. Connective tissue underlies epithelial tissue and supports and connects body parts.

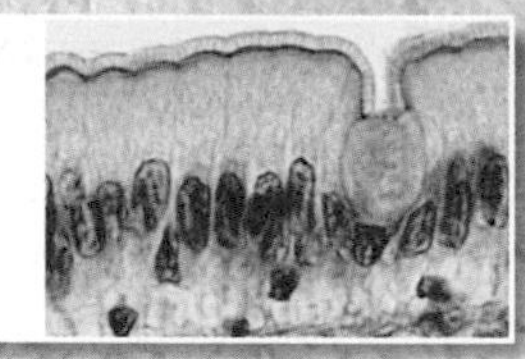

MUSCLE AND NERVOUS TISSUE

Muscle tissue consists of cells that contract in response to signals from nervous tissue. Nervous tissue also receives and integrates information from inside and outside the body.

ORGAN SYSTEMS

Vertebrates have a coelom, and many organs reside in body cavities derived from it. Interactions among organ systems sustain life.

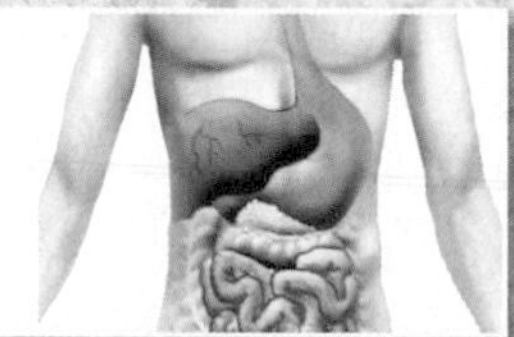

EXAMPLE OF AN ORGAN SYSTEM

Skin is an organ system that protects the body, conserves water, produces vitamin D, and helps maintain body temperature. Temperature control is an example of negative feedback.

The nervous system is the focus of Chapters 32 and 33. Chapter 34 describes the function of endocrine glands. Chapter 35 explains how muscles contract and interact with the skeleton. Chapters 36 and 37 consider the transport and immune functions of blood. Chapters 38, 39, and 40 describe how you take in essential substances and eliminate metabolic wastes. Finally, Chapters 41 and 42 describe organs involved in reproduction and how a body develops.

31.1 Stem Cells—It's All About Potential

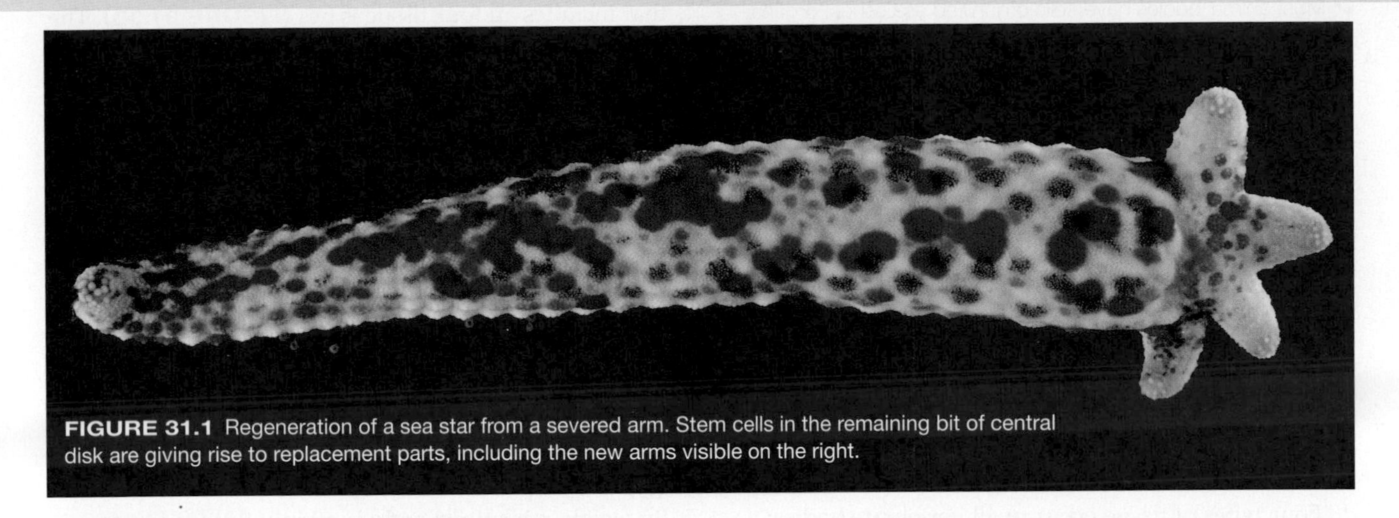

FIGURE 31.1 Regeneration of a sea star from a severed arm. Stem cells in the remaining bit of central disk are giving rise to replacement parts, including the new arms visible on the right.

Animals commonly replace tissues lost to injury. Invertebrates have the greatest capacity for regeneration. For example, some sea stars can regrow their entire body from a single arm and a bit of the central disk (**FIGURE 31.1**). No vertebrate can grow a body from a limb, but some salamanders can regrow a limb, and many salamanders and lizards replace a lost tail. Mammals do not replace limbs or tails, although like other animals they can replace skin and blood cells.

In all animals, stem cells are the key to producing new or replacement tissues. **Stem cells** (**FIGURE 31.2**) are unspecialized cells that can either divide to produce more stem cells ❶ or differentiate into one of the specialized cells that characterize specific body parts ❷. All cells in an animal body "stem" from stem cells.

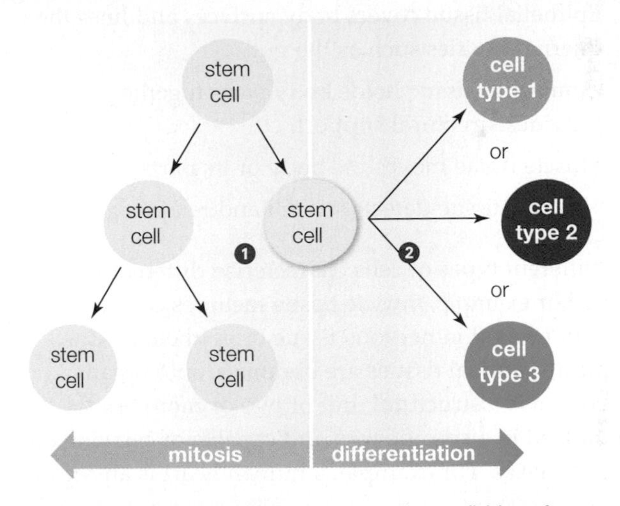

FIGURE 31.2 Stem cells. Each stem cell can divide to form two new stem cells, or differentiate into a specialized cell.

Stem cells vary in their potential to form a new individual or new tissues. A fertilized human egg and the cells produced by its first four divisions are "totipotent," meaning they can develop into a new individual if placed in a womb. Later divisions produce embryonic stem cells that are "pluripotent." A pluripotent cell does not have the ability to develop into a new individual, but it can give rise to any of the cell types in a body.

Adults have some stem cells, but these cells are "unipotent," meaning they can differentiate as one type of cell only. Some adult stem cells produce new skin cells; others produce new blood cells. However, few stem cells in an adult can become muscle cells or nerve cells. As a result, heart muscle lost to a heart attack, leg muscles destroyed by muscular dystrophy, or nerve cells destroyed by Parkinson's disease are not replaced.

Growing embryonic stem cells in culture and directing their differentiation into nerve or muscle cells could allow us to provide replacement tissues for people who need them. The first clinical studies (tests in humans) involving tissue derived from human embryonic stem cells have begun only recently. At this writing, one clinical trial is currently underway in the United States. This trial is testing whether transplanted tissue derived from embryonic stem cells can reverse a common form of blindness. Researchers have also begun preclinical research aimed at using embryonic stem cells in the treatment of heart disease, Alzheimer's disease, sickle-cell anemia, osteoporosis, muscular dystrophy, multiple sclerosis, Huntington's disease, and other conditions.

stem cell Cell that can divide to produce two stem cells or differentiate into some or all specific cell types.

CREDITS: (opposite) Eye of Science/Science Source; (1) Francois Michonneau/FLMNH; (2) © Cengage Learning.

31.2 Organization of Animal Bodies

✔ Most animal bodies have cells organized as tissues, organs, and organ systems.

✔ Physical constraints and evolutionary history are reflected in the structure and function of body parts.

Levels of Organization

In all animals, development produces a body with multiple cell types (FIGURE 31.3A). An adult human has about 200 different kinds of cells. In most animals, cells of different types are organized in tissues often anchored by extracellular matrix (FIGURE 31.3B). Cell junctions (Section 4.11) connect the cells of animal tissues to one another. These junctions fasten cells in place, and some that function in communication also allow cells to coordinate activities.

Four types of tissue occur in all vertebrate bodies:

1. Epithelial tissue covers body surfaces and lines the internal cavities such as the gut.
2. Connective tissue holds body parts together and provides structural support.
3. Muscle tissue moves the body or its parts.
4. Nervous tissue detects stimuli and relays signals.

Different types of cells characterize different tissues. For example, muscle tissue includes contractile cells not found in nervous tissue or epithelial tissue. Typically, animal tissues are organized into organs. An organ is a structural unit of two or more tissues organized in a specific way and capable of carrying out specific tasks. For example, a human heart is an organ that includes all four tissue types (FIGURE 31.3C). The heart's wall consists mainly of cardiac muscle tissue. A sheath of connective tissue covers the muscle, and internal chambers are lined with epithelial tissue. The heart receives signals via its nervous tissue.

In organ systems, two or more organs and other components interact physically, chemically, or both in a common task. For example, in the vertebrate circulatory system, the force generated by a beating heart (an organ) moves blood (a tissue) through blood vessels (organs), thereby transporting gases and solutes to and from all body cells (FIGURE 31.3D). Multiple organ systems sustain the organism (FIGURE 31.3E).

The Internal Environment

By weight, an animal body is mainly fluid: a water-based solution of salts, proteins, and other solutes. The bulk of this body fluid is intracellular, which means it is inside cells. The remainder is extracellular. **Extracellular fluid** is the environment in which body cells live. It bathes cells and provides them with the substances they require to stay alive. It also carries away cellular wastes. In vertebrates, extracellular fluid consists mainly of **interstitial fluid** (the fluid in spaces between cells) and plasma, the fluid portion of the blood (FIGURE 31.4).

Cells survive only if the fluid surrounding them remains within a narrow range of solute concentration and temperature. Maintaining conditions of the cell's environment within this range is an important aspect of homeostasis (Section 1.3).

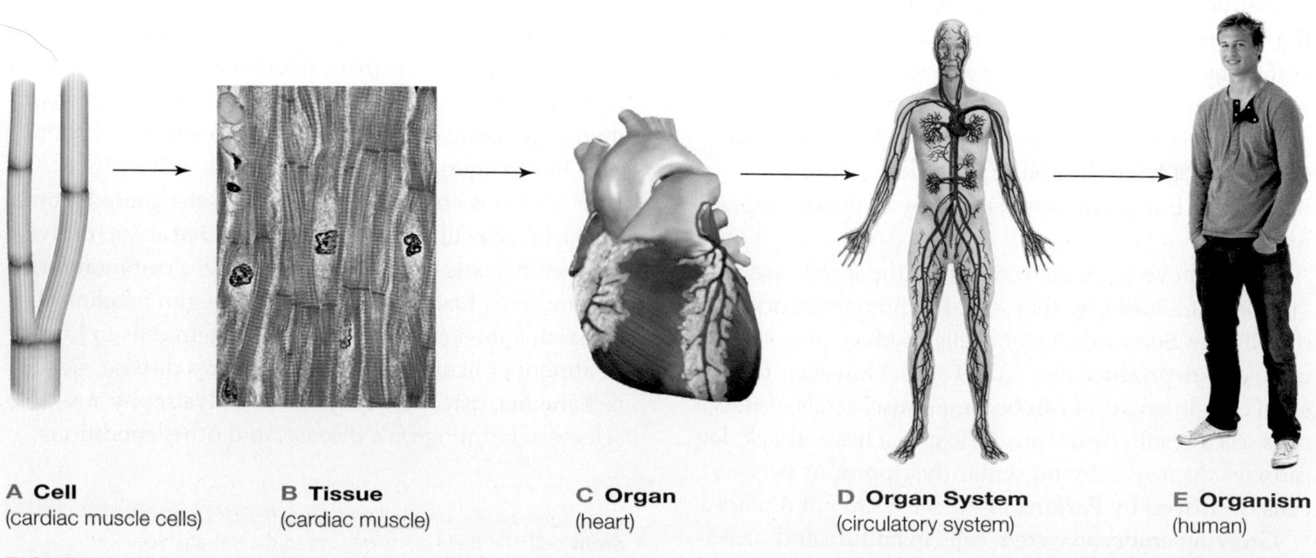

FIGURE 31.3 Levels of organization in a vertebrate (human) body.

Evolution of Animal Structure

An animal's structural traits (its anatomy) evolve in concert with its functional traits (its physiology). Both types of traits are genetically determined and vary among individuals. Genes for those traits that best help individuals survive and reproduce in their environment are passed on preferentially to the next generation. Over many generations, anatomical and structural traits become optimized in ways that reflect their function in a specific environment.

Physical constraints affect the evolution of body structure. For example, the rate of diffusion is a physical constraint that has influenced body size and shape. Substances dissolved in extracellular fluid diffuse through it, but only so fast. Gases, nutrients, and wastes cannot diffuse quickly enough through a large or thick body to keep up with cellular metabolism. Thus, mechanisms that speed the distribution of materials evolved along with increases in body size. In vertebrates, a circulatory system serves this purpose. The system includes a network of extensively branched blood vessels that extends through the body (**FIGURE 31.5**). Every living cell is close enough to a blood vessel to exchange substances with it by diffusion.

As another example of physical constraints that have affected body form, consider the challenges that vertebrates faced as they shifted from aquatic life to life on land (Section 25.6). Gases can only enter or leave an animal's body by diffusing across a moist surface. In an aquatic organism, the surrounding water both delivers oxygen and moistens the respiratory surface. By contrast, a land animal must extract oxygen from air, which can dry a respiratory surface. The evolution of lungs allowed vertebrates to maintain a moist respiratory surface inside their body. Cells inside the lung secrete the fluid that keeps this surface moist.

Lungs are not modified fish gills. Rather, lungs evolved from outpouchings of the gut in fishes ancestral to land vertebrates. As this example illustrates, evolution by natural selection often modifies existing tissues or organs. There is evidence of evolutionary compromise in the anatomy and physiology of many animals. For example, as a legacy of the lungs' ancestral connection to the gut, the human throat connects to both the digestive tract and respiratory tract. As a result of this dual connection, food sometimes goes where air should, and a person chokes. It would be safer if food and air entered the body through separate passageways. However, because evolution modifies existing structures, it does not necessarily produce the most optimal body plan.

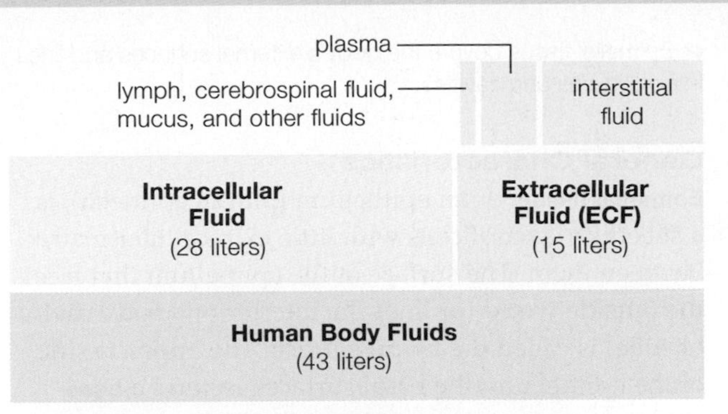

FIGURE 31.4 Distribution of fluids in a human body.

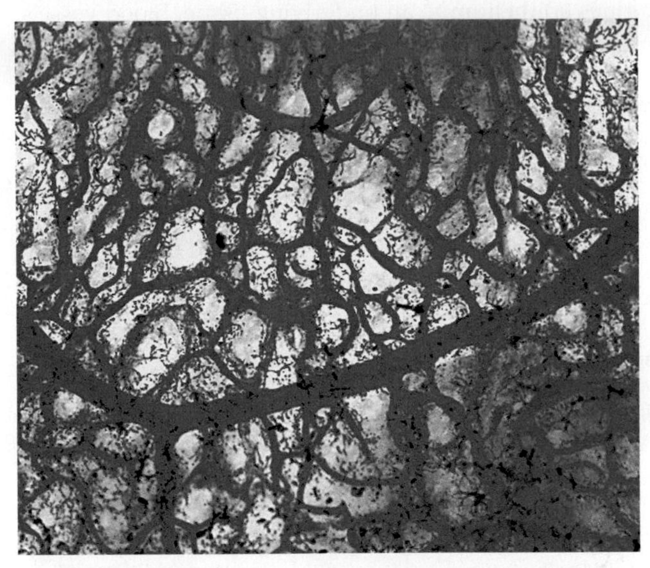

FIGURE 31.5 Branching blood vessels. The vessels deliver oxygen to within close proximity of all cells in a human body.

TAKE-HOME MESSAGE 31.2
How are animal bodies organized?

✔ In most animals, cells are organized as tissues. Each tissue consists of cells of a specific type that cooperate in carrying out a particular task. Tissues are organized into organs, which in turn are components of organ systems.

✔ The animal body consists mainly of fluid. Most of this fluid is inside cells. The fluid outside of cells (extracellular fluid) is the body's internal environment. Maintaining the solute concentration and temperature of this fluid is an important facet of homeostasis.

✔ Many anatomical traits evolved as solutions to physical challenges. However, these solutions are sometimes imperfect because evolution modifies existing structures.

extracellular fluid Of a multicelled organism, body fluid that is not inside cells; serves as the body's internal environment.
interstitial fluid Of a multicelled organism, body fluid in spaces between cells.

31.3 Epithelial Tissue

✔ Epithelial tissue covers the body's external surfaces and lines internal tubes and cavities.

General Characteristics

Epithelial tissue, or an epithelium (plural, epithelia), is a sheetlike layer of cells with little extracellular matrix between them. The surface of the epithelium that faces the outside world (or lines the interior of a body cavity or tube) is called the apical surface. The opposite side of the epithelium (the basal surface) secretes a **basement membrane**, a layer of noncellular material that glues epithelium to an underlying tissue. Blood vessels do not run through an epithelium, so nutrients reach cells by diffusing from vessels in an adjacent tissue.

Most of what you see when you look in a mirror—skin, hair, and nails—is epithelial tissue or structures derived from it. The visible portion of a hair, fingernail, beak, or feather consists of the remains of specialized epithelial cells that produced and stored large amounts of the fibrous protein keratin while they were alive.

Simple squamous epithelium
- Lines blood vessels, the heart, and air sacs of lungs
- Allows substances to cross by diffusion

Simple cuboidal epithelium
- Lines kidney tubules, ducts of some glands, reproductive tract
- Functions in absorption and secretion, movement of materials

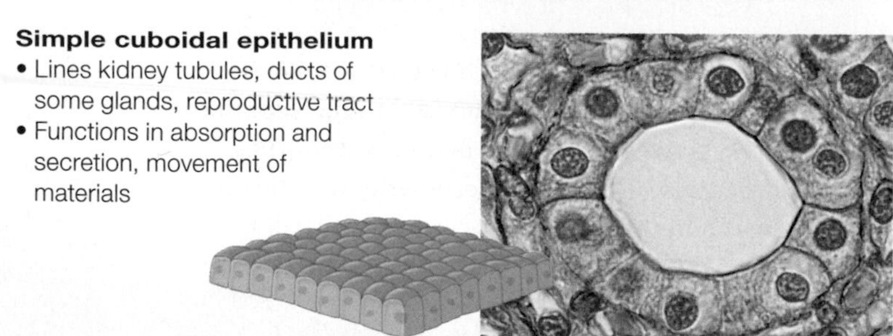

Simple columnar epithelium
- Lines some airways, parts of the gut
- Functions in absorption and secretion, protection

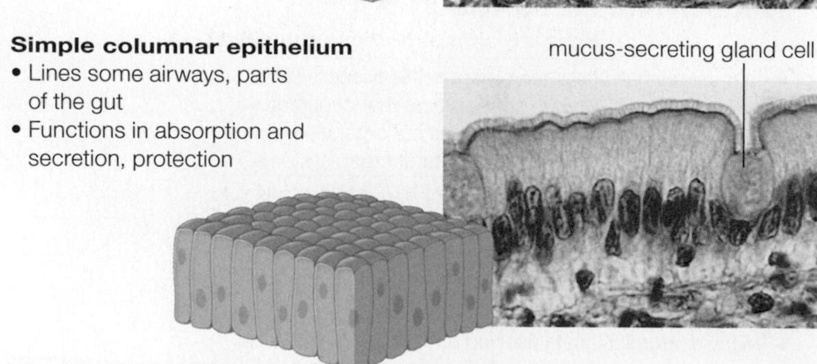

FIGURE 31.6 ▸Animated Micrographs and drawings of three types of simple epithelia, with examples of their functions and locations.

Variations in Structure and Function

Epithelial cells may be arranged as a single layer or multiple layers. A simple epithelium is a single cell layer thick, whereas a stratified epithelium includes multiple layers of cells.

Cells of an epithelium are typically described by their shape. Cells in squamous epithelium are flattened or scalelike. (*Squama* is the Latin word for scale.) Cells of cuboidal epithelium are short cylinders that look like cubes when viewed in cross-section. Cells in columnar epithelium are taller than they are wide. **FIGURE 31.6** shows the three types of simple epithelium and describes their functions.

Simple squamous epithelium facilitates the exchange of materials. It is the thinnest type of epithelium, and gases and nutrients diffuse across it easily. This type of epithelium lines blood vessels and the inner surface of the lungs. By contrast, stratified squamous epithelium is thick and serves a protective function. For example, it makes up the outermost layer of human skin.

Cells of cuboidal and columnar epithelium function in movement, absorption, or secretion of substances. Those that move substances along the surface of an epithelium have cilia at their apical surface. For example, ciliated epithelial cells line the tubes (oviducts) that convey a human egg from the ovary where it formed toward the uterus (the womb).

In some epithelia, cells have fingerlike extensions called **microvilli** at their free surface. Microvilli are typically shorter than cilia, do not move, and have an internal framework of actin filaments rather than microtubules. Microvilli increase the surface area across which substances can be detected by, absorbed into, or secreted from a cell.

Three types of intercellular junctions connect cells in animal tissues (Section 4.11). One type, the tight junction, occurs only in epithelial tissue. Tight junctions connect the plasma membranes of adjacent cells so securely that fluids cannot seep between the cells. An epithelium with cells connected by tight junctions keeps fluid contained within a particular body compartment from seeping into underlying tissue. For example, tight junctions join the epithelial cells in the lining of the gut. The junctions allow the gut epithelium to function as a selective barrier. Substances in the gut can enter the body's internal environment only by controlled movement into (and then out of) cells of the gut epithelium.

Epithelial tissues subject to mechanical stress such as skin epithelium have many adhering junctions. These junctions function like buttons that hold a shirt

CREDITS: (6) left, From Russell/Wolfe/Hertz/Starr. *Biology*, 1e. © 2008 Cengage Learning, Inc.; right from top, Ray Simmons/Science Source; Ed Reschke/Peter Arnold, Inc.; © Don W. Fawcett.

FIGURE 31.7 Examples of an exocrine and an endocrine gland.

FIGURE IT OUT Which type of gland is ductless?

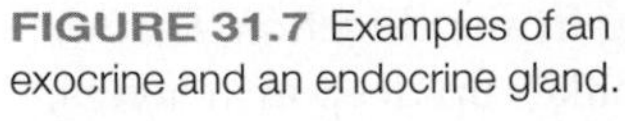

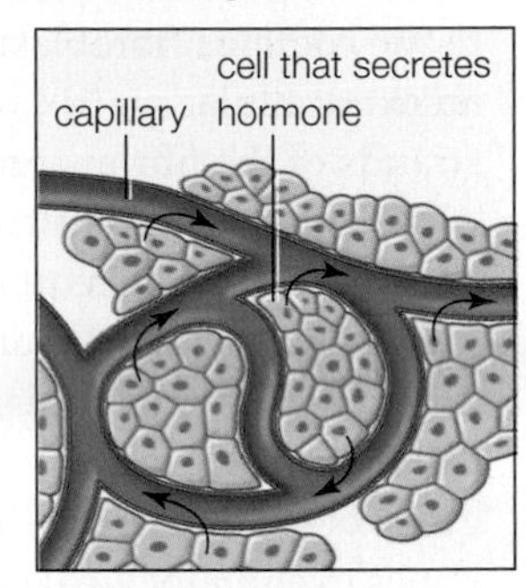

closed. They connect the plasma membranes of cells at distinct points but do not form a seal between them.

Epithelial Cell Secretions

Specialized epithelial cells called **gland cells** secrete substances that function outside the cell. In most animals, multicelled glands release substances onto the skin, or into a body cavity or fluid. There are two main types of glands (**FIGURE 31.7**). **Exocrine glands** have ducts or tubes that deliver their secretions onto an internal or external surface. Exocrine secretions include mucus, saliva, tears, digestive enzymes, earwax, and breast milk. **Endocrine glands** do not have ducts. They release signaling molecules called hormones into a body fluid. Most commonly, hormones are secreted into interstitial fluid near small blood vessels (capillaries). They enter the blood, which distributes them through the body.

Carcinomas—Epithelial Cell Cancers

Adult animals make few new muscle cells or nerve cells, but they constantly renew their epithelial tissue. For example, each day you lose skin cells and make new ones to replace them. An adult sheds about 0.7 kilogram (1.5 pounds) of skin each year. Similarly, the lining of your intestine is replaced every four to six days. The many cell divisions required to replace those cells provide lots of opportunities for DNA replication errors that can lead to cancer-causing mutations. As a result, epithelium is the animal tissue most likely to become cancerous.

An epithelial cell cancer is called a carcinoma. Nearly all skin cancers are carcinomas. Most breast cancers are carcinomas of epithelial cells that line the milk ducts or of the breast's glandular epithelium. Similarly, most lung cancers arise in cells of the lung's epithelial lining.

basement membrane Secreted material that attaches epithelium to an underlying tissue.
endocrine gland Ductless gland that secretes hormones into a body fluid.
epithelial tissue Sheetlike animal tissue that covers outer body surfaces and lines internal tubes and cavities.
exocrine gland Gland that secretes milk, sweat, saliva, or some other substance through a duct.
gland cell Secretory epithelial cell.
microvilli Thin projections from the plasma membrane of some epithelial cells; increase the cell's surface area.

TAKE-HOME MESSAGE 31.3

What are the functions of epithelial tissue?

- ✔ Epithelia are sheetlike tissues that line the body's surfaces, including its cavities, ducts, and tubes. They function in protection, absorption, and secretion. Some epithelia have cilia or microvilli at their apical surface.
- ✔ Specialized epithelial cells that produce large amounts of keratin are the source of hair, nails, hooves, and feathers.
- ✔ Glands are secretory organs derived from epithelium. Exocrine glands secrete material through a duct onto a body surface or into a body cavity. Endocrine glands secrete hormones that enter the blood.
- ✔ Epithelial tissues undergo continual turnover and are the most frequent site for cancers.

31.4 Connective Tissues

✔ Connective tissues connect body parts and provide structural and functional support to other body tissues.

General Characteristics

Connective tissues consist of cells scattered in an extracellular matrix of their own secretions. Most connective tissue contains fibroblasts, which are cells that secrete an extracellular matrix with polysaccharides and strands of the fibrous proteins collagen and elastin.

Collagen is the most abundant protein in an animal body. Its synthesis requires vitamin C, and a human deficiency in this vitamin causes the nutritional disorder known as scurvy. Early in this disorder, connective tissue surrounding tiny blood vessels breaks down, so gums bleed and bruiselike spots appear on the skin. Bones eventually weaken. Collagen is the main component of scar tissue, so a person with scurvy cannot heal cuts and scrapes. If untreated, scurvy can be fatal.

Types of Connective Tissue

Loose and dense connective tissue have the same components (fibroblasts and a matrix with elastin and collagen fibers) but in different proportions. Loose connective tissue is the most abundant tissue in vertebrates. Its fibroblasts, collagen fibers, and elastin fibers are widely scattered in a gel-like matrix (FIGURE 31.8A). Loose connective tissue underlies most epithelia, and it surrounds nerves and blood vessels. It also holds abdominal organs in place and stores fluid.

In dense, irregular connective tissue, fibroblasts lie among randomly arranged collagen fibers (FIGURE 31.8B). The random arrangement of fibers allows the tissue to withstand stretching in any direction. Dense, irregular connective tissue makes up deep skin layers, underlies the lining of the gut, and forms a protective capsule around the kidneys and the testes.

Dense, regular connective tissue consists of fibroblasts in orderly rows between parallel, tightly packed bundles of collagen fibers (FIGURE 31.8C). This parallel organization maximizes the strength of a tissue that is typically pulled in a single direction, parallel to the fibers. Dense, regular connective tissue is the main tissue in tendons and ligaments. Tendons connect skeletal muscle to bones. Ligaments attach one bone to another.

Adipose tissue, cartilage, bone tissue, and blood are specialized connective tissues.

Adipose tissue consists mainly of cells that make and store triglycerides (Section 3.5). These cells have little matrix between them, and a fat-containing vacuole fills their interior (FIGURE 31.8D). Tiny blood vessels that run between the cells carry fatty acids to or from the tissue. Adipose tissue is the body's main energy reservoir, and it cushions and protects body parts. It also serves as insulation. A special type of adipose tissue called blubber helps marine mammals retain heat (FIGURE 31.9).

Cartilage and bone are the main components of vertebrate skeletal systems. **Cartilage** has a matrix of collagen fibers and rubbery, compression-resistant glycoproteins. Living cartilage cells secrete the matrix that surrounds them (FIGURE 31.8E). Some fish such as sharks have a cartilage skeleton. In bony fishes and tet-

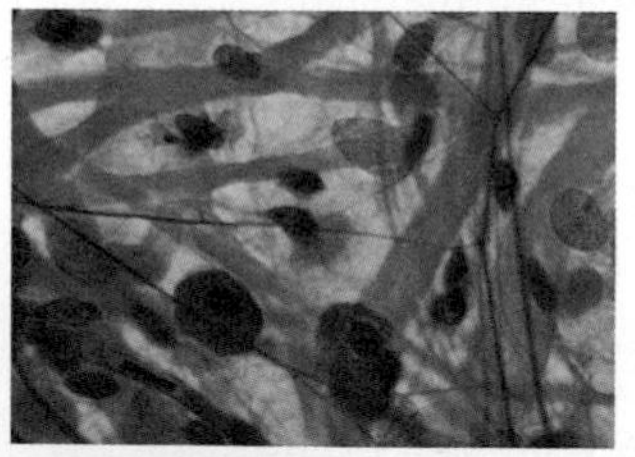

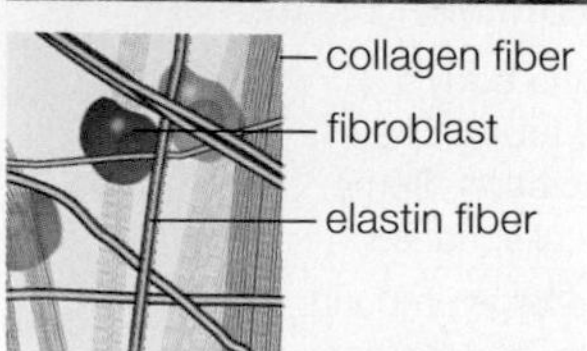

A Loose connective tissue
- Underlies most epithelia
- Provides elastic support and serves as a fluid reservoir

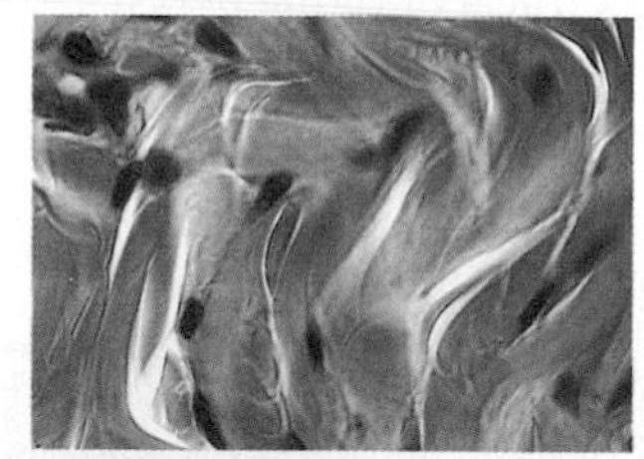

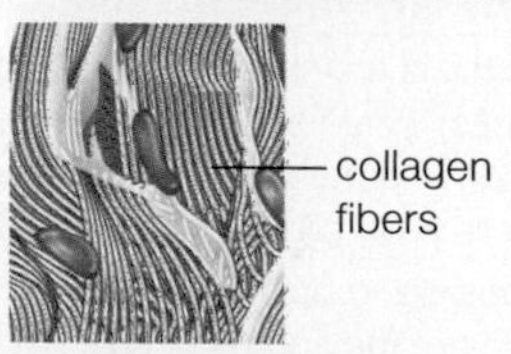

B Dense, irregular connective tissue
- In deep skin layers, around intestine, and in kidney capsule
- Binds parts, provides support and protection

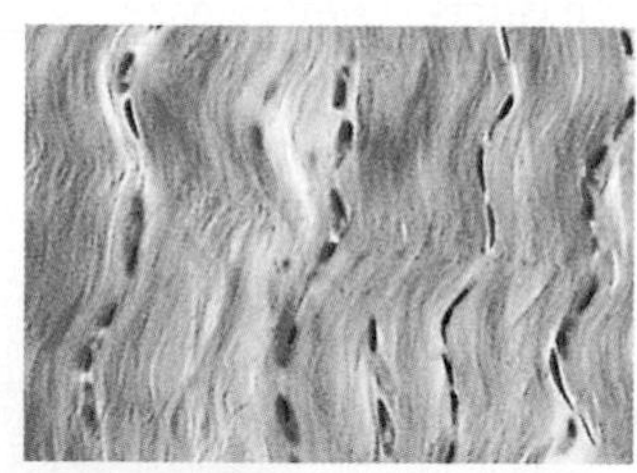

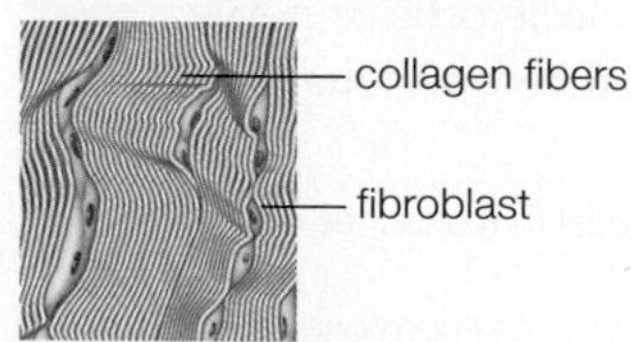

C Dense, regular connective tissue
- In tendons connecting muscle to bone and ligaments that attach bone to bone
- Provides stretchable attachment between body parts

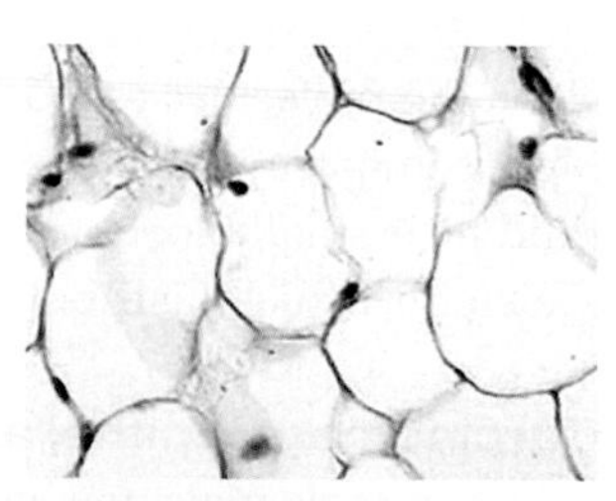

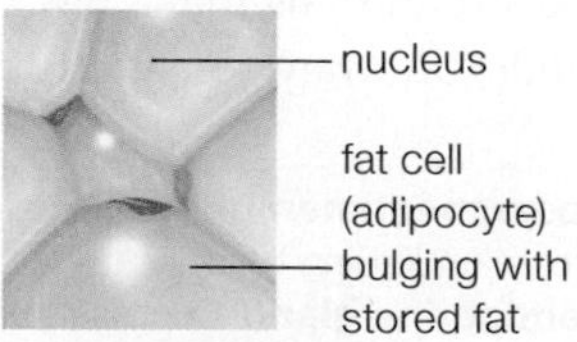

D Adipose tissue
- Underlies skin and occurs around heart and kidneys
- Stores energy, provides insulation, cushions and protects some body parts

FIGURE 31.8 ▸**Animated**
Connective tissue structure and function.

rapods, cartilage forms during development, then bone replaces most of it. After birth, cartilage still supports your nose, throat, and outer ears. It covers the ends of bones and reduces friction where they meet. Cartilage also acts as a shock absorber between segments of the backbone. Blood vessels do not extend through cartilage, as they do in other connective tissues.

Bone tissue is a connective tissue in which bone cells secrete and are surrounded by a collagen-rich matrix hardened by calcium and phosphorus (**FIGURE 31.8F**). Bone tissue is the main tissue of bones.

Blood (**FIGURE 31.8G**), the fluid in a circulatory system, is considered a connective tissue because its cellular components (red blood cells, white blood cells, and platelets) descend from stem cells in bone. Red blood cells transport oxygen. White blood cells defend the body against pathogens. Platelets help blood clot. Blood cells and platelets drift along in the plasma, a fluid extracellular matrix consisting of water with dissolved proteins, nutrients, gases, and other substances.

adipose tissue Connective tissue that specializes in fat storage.
blood Circulatory fluid; in vertebrates it is a fluid connective tissue consisting of plasma, red blood cells, white blood cells, and platelets.
bone tissue Connective tissue consisting of cells surrounded by a mineral-hardened matrix of their own secretions.
cartilage Connective tissue consisting of cells surrounded by a rubbery matrix of their own secretions.
connective tissue Animal tissue with an extensive extracellular matrix; structurally and functionally supports other tissues.

FIGURE 31.9 A walrus with a thick, insulating coat of blubber beneath its skin. Even when the external temperature dips far below freezing, a walrus's core body temperature is about the same as yours. Fat is less dense than water, so a coat of blubber also helps a walrus float. In addition, blubber includes many collagen and elastin fibers not found in other adipose tissue. The tendency of blubber to spring back when compressed reduces the energy the walrus expends while swimming.

TAKE-HOME MESSAGE 31.4

What are connective tissues?

✔ Soft connective tissues underlie epithelia, form capsules around organs, and connect muscle to bones or bones to one another.

✔ A vertebrate skeleton consists of two connective tissues: rubbery cartilage and mineral-hardened bone. Blood is a connective tissue because blood cells form in bone. The cells are carried by plasma, the fluid portion of the blood.

✔ Adipose tissue is a specialized connective tissue that stores fat.

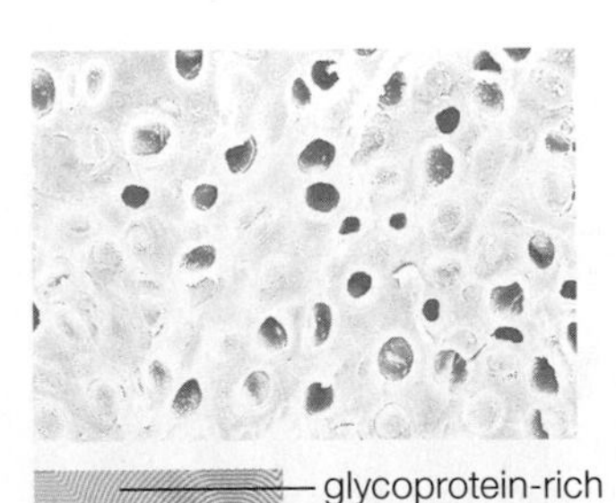

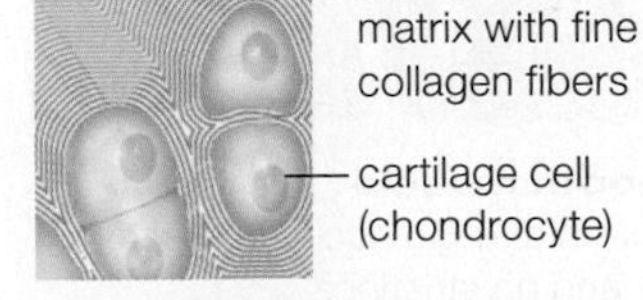

E Cartilage
- Internal framework of nose, ears, airways; covers ends of bones
- Supports soft tissues, cushions bone ends at joints, provides a low-friction surface for joint movements

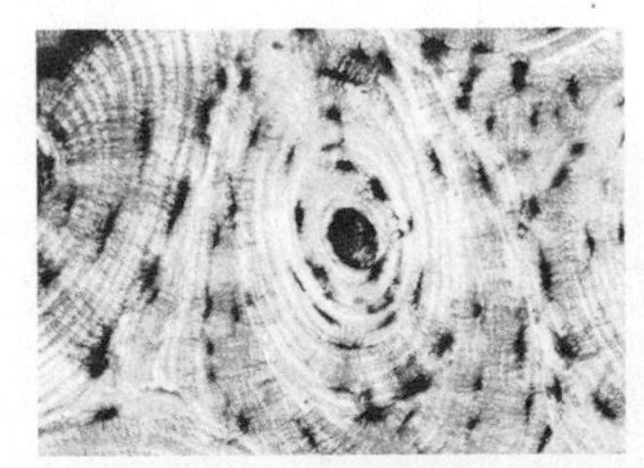

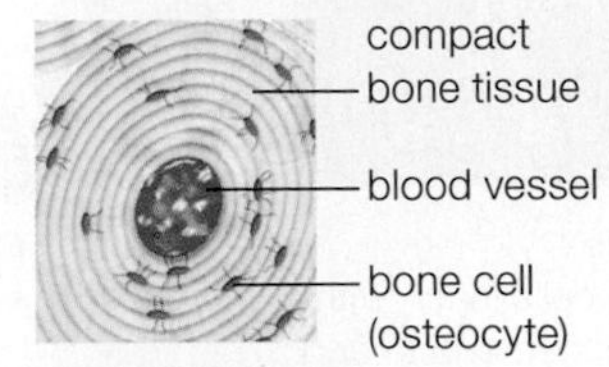

F Bone tissue
- Makes up the bulk of most vertebrate skeletons
- Provides rigid support, attachment site for muscles, protects internal organs, stores minerals, produces blood cells

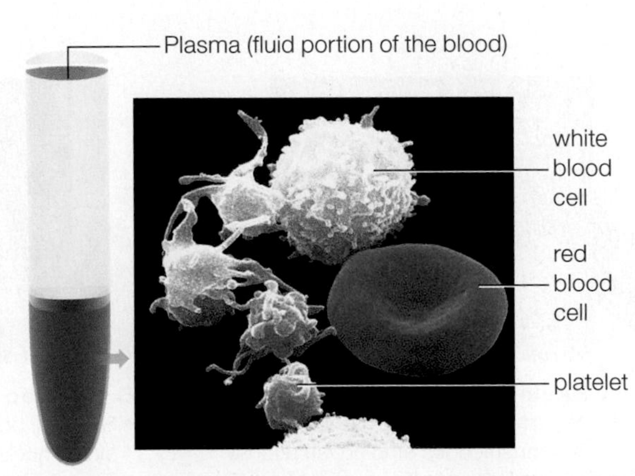

G Blood
- Flows through blood vessels, and the heart
- Distributes essential gases, nutrients to cells; removes wastes from them

31.5 Muscle Tissues

✔ Muscle moves bodies or propels materials through them.

Cells of muscle tissues contract (shorten) in response to signals from nervous tissue. ATP provides the energy that fuels the contractions. We discuss the mechanism of muscle contraction in detail in Section 35.7.

Skeletal Muscle Tissue

Skeletal muscle tissue pulls on bones to move body parts. It consists of parallel arrays of long, cylindrical cells called muscle fibers, which have a striated, or striped, appearance (**FIGURE 31.10A**). Muscle fibers are multinucleated and form by cell fusion during embryonic development. Skeletal muscle contracts reflexively, as when you pull your hand away after touching a hot object. More often, its contraction is deliberate, as when you reach for something. Thus, skeletal muscle is commonly described as "voluntary" muscle. By contrast, cardiac muscle and smooth muscle are said to be "involuntary" muscle because people cannot deliberately make these tissues contract.

Along with the liver, skeletal muscle is a major site for glycogen storage. Metabolic activity in skeletal muscles is the major source of body heat.

Cardiac Muscle Tissue

Cardiac muscle tissue occurs only in the heart wall (**FIGURE 31.10B**). Like skeletal muscle tissue, it has a striated appearance. Cardiac muscle consists of branching cells, each with a single nucleus, attached end to end by adhering junctions. The junctions hold the cells together during forceful contractions. Signals to contract pass swiftly from cell to cell at gap junctions that connect the cells along their length. The rapid flow of signals ensures that all cells in cardiac muscle tissue contract as a unit. Compared to other muscle tissues, cardiac muscle has a far greater supply of mitochondria to keep the continually beating heart supplied with ATP.

Smooth Muscle Tissue

We find layers of **smooth muscle tissue** in the wall of some blood vessels and soft internal organs, such as the stomach, uterus, and bladder. This tissue's unbranched cells contain a nucleus at their center and are tapered at both ends (**FIGURE 31.10C**). Contractile units are not arranged in an orderly repeating fashion, so smooth muscle tissue does not appear striated. Smooth muscle contracts more slowly than skeletal muscle, but its contractions can be sustained longer. These contractions propel material through the gut, alter the diameter of blood vessels, and adjust how much light enters the pupil of the eye.

cardiac muscle tissue Muscle of the heart wall.
skeletal muscle tissue Muscle that pulls on bones and moves body parts; under voluntary control.
smooth muscle tissue Muscle that lines blood vessels and forms the wall of hollow organs.

TAKE-HOME MESSAGE 31.5

What is muscle tissue?

✔ Muscle tissue consists of cells that contract in response to nervous signals. Contraction requires ATP.

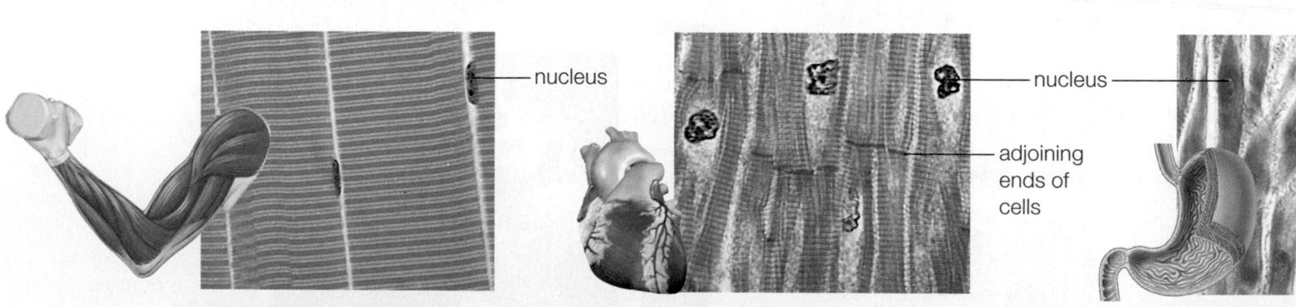

A Skeletal muscle
- Long, multinucleated, cylindrical cells with conspicuous striping (striations)
- Pulls on bones to bring about movement, maintain posture
- Reflex activated, but also under voluntary control

B Cardiac muscle
- Striated, branching cells (each with a single nucleus) attached end to end
- Found only in the heart wall
- Contraction is not under voluntary control

C Smooth muscle
- Cells with a single nucleus, tapered ends, and no striations
- Found in the walls of arteries, the digestive tract, the reproductive tract, the bladder, and other organs
- Contraction is not under voluntary control

FIGURE 31.10 ▶**Animated** Three types of muscle tissue.

CREDITS: (10A–C left) © Cengage Learning; (10A–B right) Ed Reschke; (10C right) Biophoto Associates/Science Source.

31.6 Nervous Tissue

✔ Nervous tissue detects changes in the internal or external environment, integrates information, and controls the activity of muscle and glands.

Nervous tissue consists of neurons and cell that support them. **Neurons** are the signaling cells in nervous tissue. Each neuron has a cell body, a region that contains the nucleus and other organelles. Cytoplasmic extensions that project from the cell body allow the cell to receive and send electrochemical signals (FIGURE 31.11).

When a neuron receives sufficient stimulation, an electrical signal travels along its plasma membrane to the ends of specialized cytoplasmic extensions. Arrival of this signal causes the release of chemical signaling molecules from these endings. The released molecules diffuse across a small gap to an adjacent neuron, muscle fiber, or gland cell, and alter that cell's behavior.

Your nervous system has more than 100 billion neurons. There are three types. Sensory neurons are excited by specific stimuli, such as light or pressure. Interneurons receive and integrate sensory information; these neurons allow the brain to store information and coordinate responses to stimuli. In vertebrates, interneurons occur mainly in the brain and spinal cord. Motor neurons relay commands from the brain and spinal cord to glands and muscle cells (FIGURE 31.12).

Neuroglial cells, also called neuroglia, keep neurons positioned where they should be, and provide them with nutrients. Neuroglial cells also wrap around the signal-sending cytoplasmic extensions of most motor neurons. They act as insulation, speeding the rate at which an electrical signal travels along the plasma membrane of the neuron.

nervous tissue Animal tissue composed of neurons and supporting cells; detects stimuli and controls responses to them.
neuroglial cell Cell that supports and assists neurons.
neuron One of the cells that make up communication lines of a nervous system; transmits electrical signals along its plasma membrane and communicates with other cells through chemical messages.

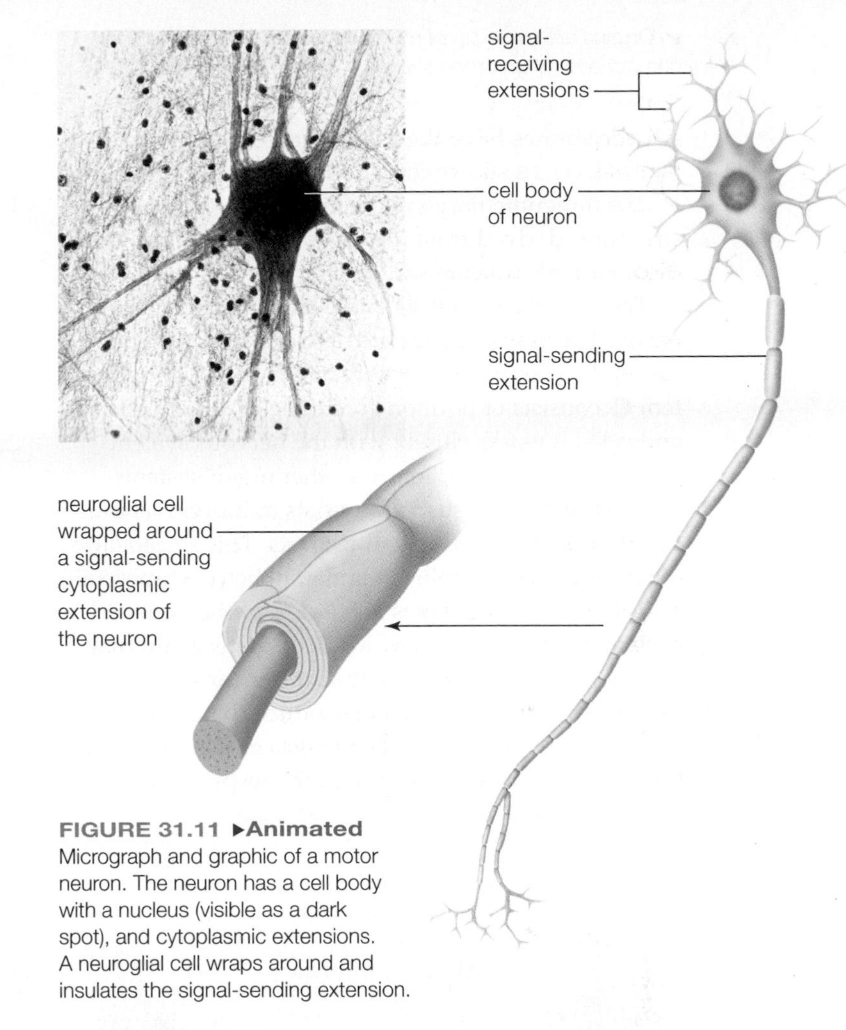

FIGURE 31.11 ▶**Animated** Micrograph and graphic of a motor neuron. The neuron has a cell body with a nucleus (visible as a dark spot), and cytoplasmic extensions. A neuroglial cell wraps around and insulates the signal-sending extension.

TAKE-HOME MESSAGE 31.6

What is nervous tissue?

✔ Nervous tissue consists of neurons and the cells that support them. Supporting cells in nervous tissue are referred to as neuroglial cells, or neuroglia.

✔ A neuron relays electrical signals along its length and communicates with other cells by chemical signals. Some neurons detect stimuli; others integrate information; others relay commands to glands and muscle cells.

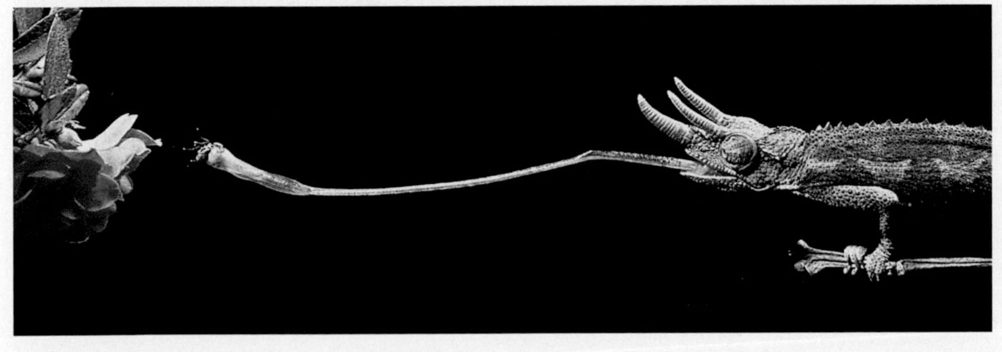

FIGURE 31.12 Example of a coordinated interaction between skeletal muscle tissue and nervous tissue.

Sensory neurons in the lizard's eyes relay information about the position of a fly to interneurons in the lizard's brain.

Signals from interneurons in the lizard's brain flow to motor neurons, which in turn send stimulatory signals to the muscle fibers of the lizard's long, coiled-up tongue. The tongue uncoils swiftly and precisely to reach the very spot where the fly is perched.

31.7 Organ Systems

✔ Organs are made up of multiple types of tissues and are components of an organ system.

All vertebrates have the same array of organ systems. **FIGURE 31.13** shows these systems in a human body.

The integumentary system ❶ includes skin and structures derived from it such as hair and nails. We discuss skin's functions in detail in the next section.

The nervous system ❷ is the body's main control center. The brain, spinal cord, and nerves are part of this system, as are sensory organs. The endocrine system ❸ consists of hormone-secreting endocrine glands and cells. It works closely with the nervous system and, like that system, controls other organ systems.

The muscular system ❹ consists of individual muscles that move the body and its parts. This system also plays an important role in regulating body temperature by generating heat. Bones are organs of the skeletal system, the body's framework ❺. It protects internal organs, serves as a point of attachment for skeletal muscles, stores minerals, and produces blood cells.

The circulatory system ❻ consists of the heart and blood vessels. It cooperates with the respiratory, digestive, and urinary systems in delivering oxygen and nutrients to cells and clearing away wastes (**FIGURE 31.14**). It also helps regulate body temperature by conveying heat from the body's warm core to skin. The lymphatic system ❼ has vessels that move fluid (lymph) from tissues to blood. Organs that help protect the body from pathogens are part of this system.

food, water intake
oxygen inhaled
Digestive System
Respiratory System
carbon dioxide exhaled
nutrients, water, solutes
oxygen
carbon dioxide
Circulatory System
Urinary System
water, solutes
excretion of food residues
transport of materials to and from cells
elimination of soluble wastes, excess water, and salts

FIGURE 31.14 Organ system interactions that keep the body supplied with essential substances and eliminate unwanted wastes.

FIGURE 31.13 Human organ systems.

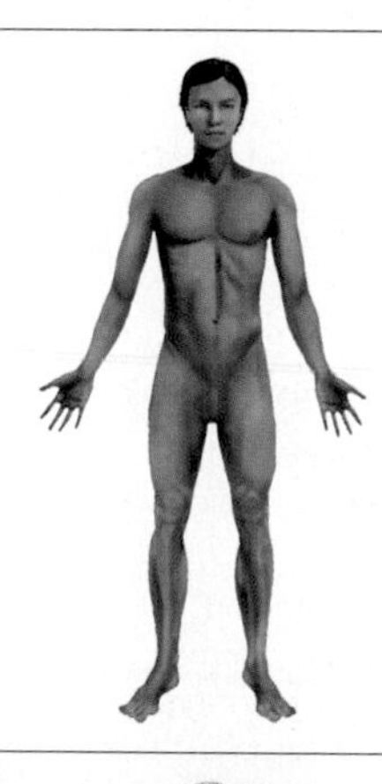

❶ Integumentary System

Protects body from injury, dehydration, pathogens; moderates temperature; excretes some wastes; detects external stimuli.

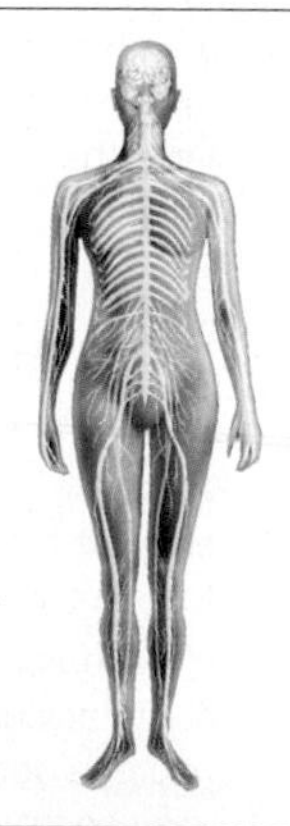

❷ Nervous System

Detects external and internal stimuli; coordinates responses to stimuli; integrates organ system activities.

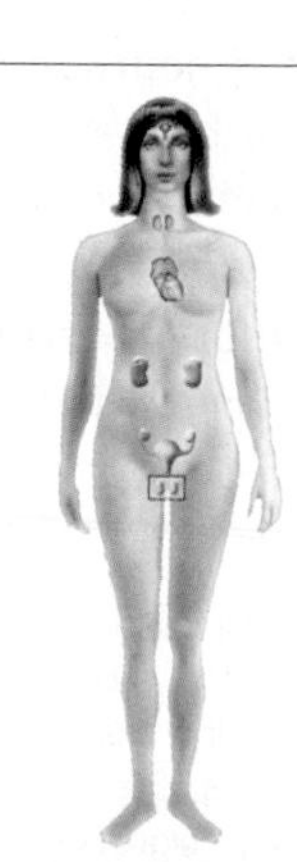

❸ Endocrine System

Secretes hormones that control activity of other organ systems. (Male testes added.)

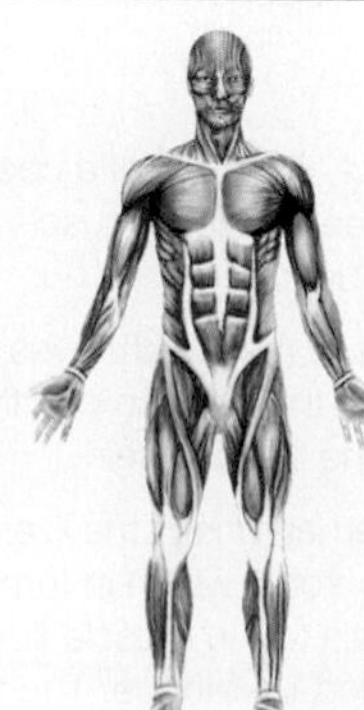

❹ Muscular System

Moves the body and its parts; maintains posture; produces heat to maintain body temperature.

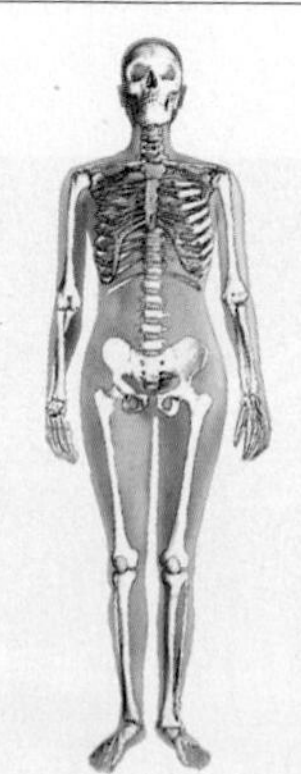

❺ Skeletal System

Supports and protects body parts; site of muscle attachment; produces red blood cells; stores minerals.

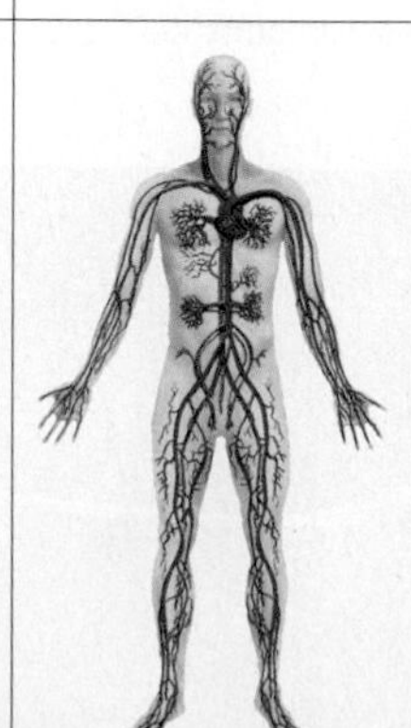

❻ Circulatory System

Distributes materials and heat throughout the body; helps maintain pH.

The respiratory system ❽ includes two lungs and the airways leading to them. It delivers oxygen from air to the blood, and expels carbon dioxide from blood into the air.

The digestive system takes in food, breaks it down, delivers nutrients to the blood, and eliminates undigested wastes ❾. It includes organs of the gut (esophagus, stomach, intestine), as well as glandular organs such as the liver and pancreas, which supply substances that function in digestion.

The urinary system consists of kidneys (organs that filter blood and create urine), the bladder, and ducts that deliver urine to the body surface for excretion ❿. It removes wastes from blood, and adjusts blood volume and solute composition. Reproductive systems of both sexes include gamete-making organs (ovaries or testes) and ducts through which gametes travel ⓫.

Some organs are components of more than one organ system. For example, the pancreas is an endocrine gland and a digestive organ.

Many organs reside within a body cavity (**FIGURE 31.15**). A cranial cavity in the head holds the brain, and a spinal cavity in the back holds the spinal cord. Like other vertebrates, humans have a coelom (Section 24.2). The diaphragm, a sheet of smooth muscle, divides a human coelom into two cavities. The upper (thoracic) cavity holds the heart and lungs. Digestive organs, including the stomach, intestines, and liver, are in the lower cavity's abdominal region. The pelvic region of the lower cavity holds the bladder and reproductive organs.

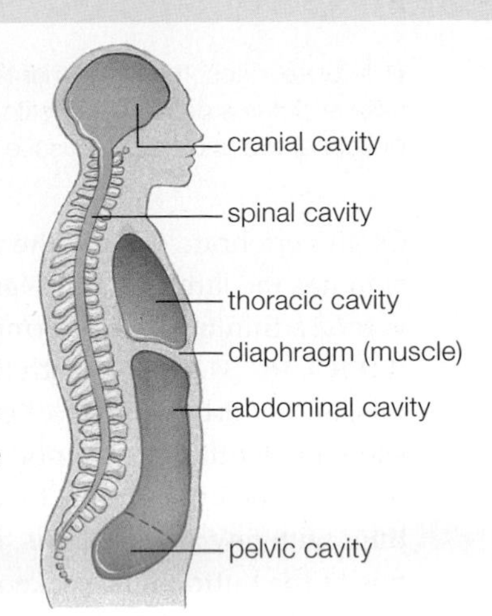

FIGURE 31.15 ▸Animated Body cavities that hold organs.
FIGURE IT OUT Which organs lie in body cavities that are not part of the coelom?
Answer: The spinal cord and brain

TAKE-HOME MESSAGE 31.7

What are organs and organ systems?

✔ Organs consist of multiple tissues and are themselves components of organ systems. Cooperative action of organ systems sustains the body.

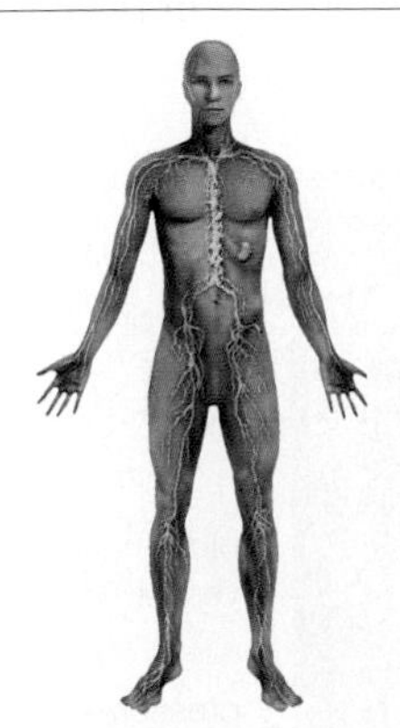

❼ Lymphatic System

Collects and returns tissue fluid to the blood; defends the body against infection, cancers.

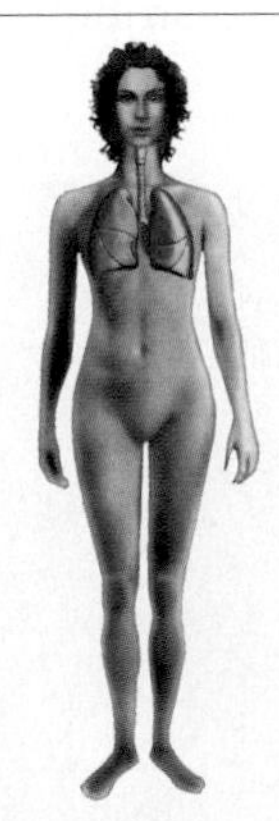

❽ Respiratory System

Takes in the oxygen for aerobic respiration; expels carbon dioxide released by this pathway.

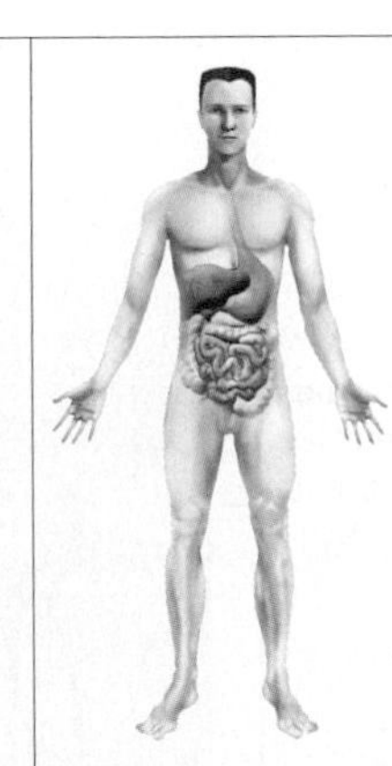

❾ Digestive System

Takes in food and water; breaks food down and absorbs needed nutrients, then eliminates food residues.

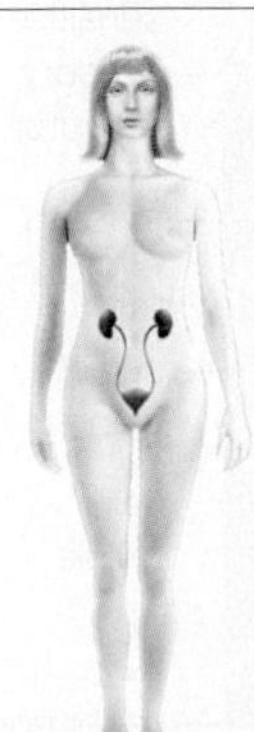

❿ Urinary System

Maintains volume and composition of blood; excretes excess fluid and wastes.

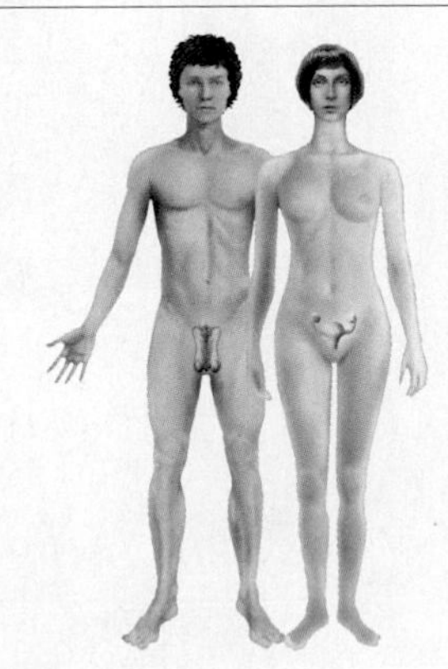

⓫ Reproductive System

Female: Produces eggs; nourishes and protects developing offspring.
Male: Produces sperm and transfers them to a female.

31.8 Human Integumentary System

✔ In vertebrates, the integumentary system consists of skin, structures derived from skin, and an underlying layer of connective and adipose tissue.

Of all vertebrate organs, the outer body covering called skin has the largest surface area. Skin consists of two layers, a thin upper epidermis and the dermis beneath it (**FIGURE 31.16**). Beneath the dermis is the hypodermis, an underlying layer of connective and adipose tissue. The depth of the hypodermis varies among body regions. The hypodermis beneath the skin of eyelids is thin, with few adipose cells. By contrast, the hypodermis of the buttocks is thickened by many adipose cells.

Vertebrate skin has many functions. It contains sensory receptors that keep the brain informed of external conditions. It serves as a barrier to keep out pathogens and it helps control internal temperature. In land vertebrates, skin also helps conserve water. In humans, reactions that produce vitamin D occur in the skin.

Structure of Human Skin

Epidermis is a stratified squamous epithelium with an abundance of adhering junctions and no extracellular matrix. Human epidermis consists mainly of keratinocytes, epithelial cells that synthesize the waterproofing protein keratin. Keratinocytes in deep epidermal layers continually divide by mitosis, so new cells displace the older ones toward the skin's surface. As cells move outward, they become flattened, lose their nucleus, and die. Dead keratinocytes accumulate at the skin surface, forming a tough protective layer that resists penetration by pathogens and helps the body conserve water.

FIGURE 31.17 Vitiligo. Lee Thomas, an African American television reporter, has vitiligo. The death of melanocytes has turned his hands white and produced white blotches on his face and arms.

Melanocytes, another type of epidermal cell, make pigments called melanins and donate them to keratinocytes. Variations in skin color arise from differences in the distribution and activity of melanocytes, and in the

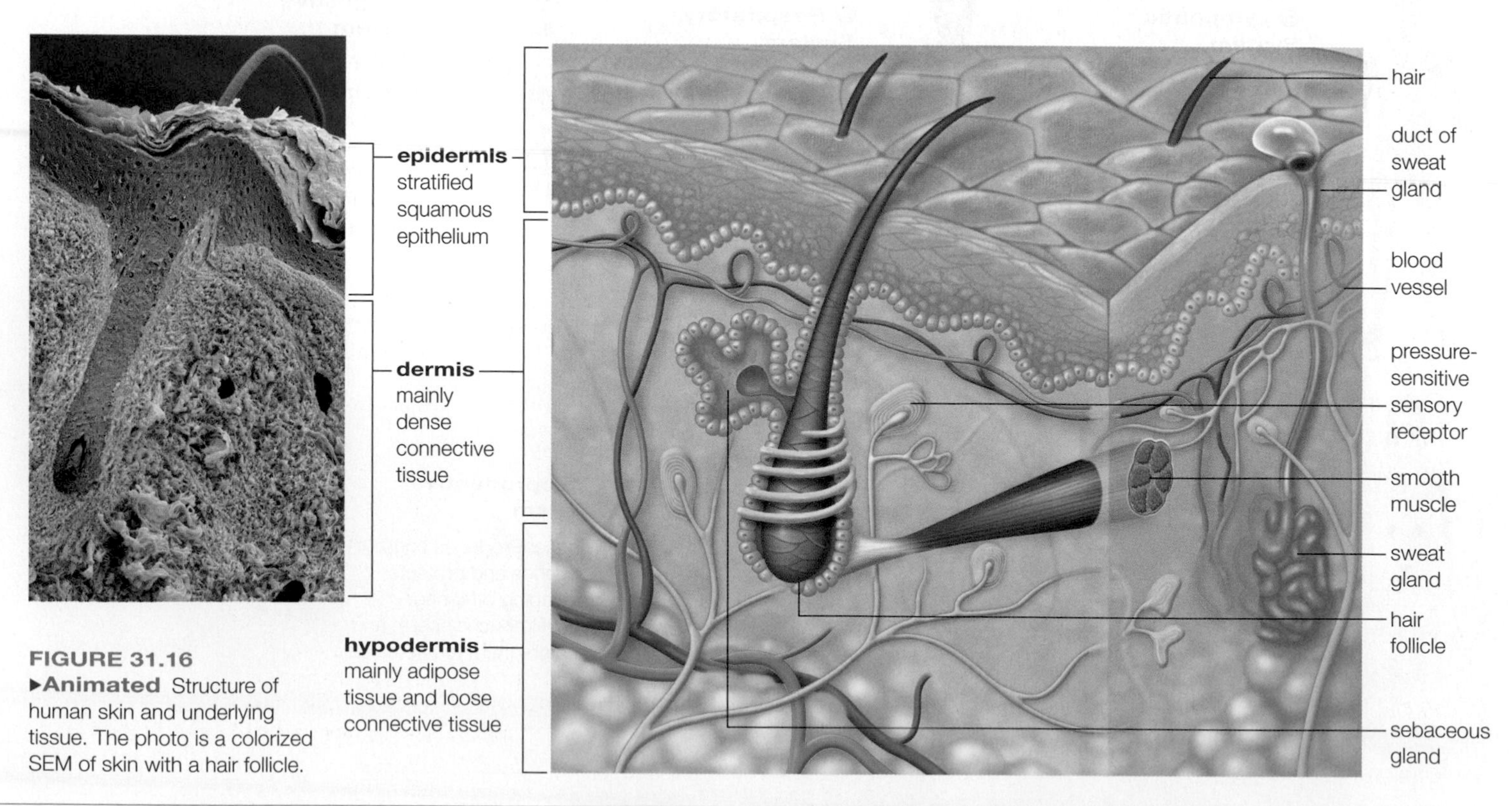

FIGURE 31.16 ▶**Animated** Structure of human skin and underlying tissue. The photo is a colorized SEM of skin with a hair follicle.

type of melanin they produce (Section 14.1). One melanin is brown to black. Another is red to yellow. The importance of melanocyte activity can be seen with vitiligo, a skin disorder in which the destruction of these cells results in light patches of skin (**FIGURE 31.17**).

Melanin functions as a sunscreen, absorbing ultraviolet (UV) radiation that could damage DNA and other biological molecules. When skin is exposed to sunlight, melanocytes produce more of the brownish-black melanin, resulting in a protective "tan."

Dermis consists primarily of dense connective tissue with stretchy elastin fibers and supportive collagen fibers. Blood vessels, lymph vessels, and sensory receptors weave through the dermis. Dermis is much thicker than epidermis, and more resistant to tearing. Leather is animal dermis that has been treated with chemical preservatives.

Sweat glands, sebaceous glands, and hair follicles are pockets of epidermal cells that migrated into the dermis during early development. Sweat is mostly water. As it evaporates, it cools the skin surface and helps keep the body from overheating. Sebaceous glands produce sebum, an oily mix of triglycerides, fatty acids, and other lipids. Sebum helps keep skin and hair soft. It also has antimicrobial properties.

The portion of a hair that you can see consists of the remains of cells that began life deep in the dermis, at the base of the hair follicle. Keratinocytes at a follicle's base divide every 24 to 72 hours, making them among the fastest-dividing cells in the body. As a result of these divisions, newly formed keratinocytes are continually forced away from the follicle's base. As these cells are pushed along through a sheath that extends to the skin's surface, they take up pigment from melanocytes and they produce and store keratin. Once filled with keratin, they die—before ever reaching the skin's surface. The threadlike stack of keratin-rich cell remains that eventually protrudes above the skin surface is what we call a shaft of hair.

Hair can "stand up" because smooth muscle attaches to a sheath that surrounds the base of each hair shaft. When this muscle reflexively contracts in response to cold or fright, hair is pulled upright.

Evolution and Human Skin

Compared to other primates, humans have far more sweat glands and shorter, finer body hairs (**FIGURE 31.18**). According to one hypothesis, an increase in sweat glands and a loss of body hair occurred in concert with the evolution of bipedalism. During brisk walking or running, the metabolic activity in skeletal muscles produces heat that raises the body temperature. Presumably, when our bipedal ancestors first made the transition to life in the open (rather than in trees), individuals with finer hair and more sweat glands were at an advantage because they were less likely to overheat.

FIGURE 31.18 Primate skin. Humans have less body hair and more sweat glands than other primates such as chimpanzees.

When young, our closest primate relatives, chimpanzees and bonobos, have pink skin and a covering of long, black body hair. Our forest-dwelling ancestors probably had similarly pink skin and dark hair. Thus, loss of body hair that facilitated cooling would have created a new selective challenge—an increased exposure to potentially damaging sunlight. The dark skin now observed in all African populations is considered an adaptation to this challenge. Later, as Section 14.1 explained, some populations of humans dispersed from Africa to higher latitudes where sunlight was less intense. Lighter skin color, which is adaptive in such habitats, became prevalent in these populations.

dermis Deep layer of skin that consists of connective tissue with nerves and blood vessels running through it.
epidermis Outermost tissue layer; in animals, the epithelial layer of skin.

TAKE-HOME MESSAGE 31.8

What are the functions of the integumentary system?

✔ The integumentary system consists of skin, derivatives of skin such as hair, and underlying connective tissue.

✔ Skin has sensory receptors that inform the brain about the environment. It also serves as a barrier against pathogens, produces vitamin D, and functions in temperature regulation.

31.9 Negative Feedback in Homeostasis

✔ A negative feedback system involving multiple organ systems allows the body to maintain its internal temperature.

In vertebrates, homeostasis involves interactions among sensory receptors, the brain, and muscles and glands. A **sensory receptor** is a cell or cell component that detects a specific stimulus. Sensory receptors involved in homeostasis function like watchmen that monitor the body for changes. Information from sensory receptors throughout the body flows to the brain. The brain evaluates incoming information, then signals effectors—muscles and glands—to take the necessary actions to keep the body functioning.

Homeostasis often involves **negative feedback**, a process in which a change causes a response that reverses the change. An air conditioner with a thermostat is a familiar nonbiological example of a negative feedback system. A person sets the air conditioner to a desired temperature. When the temperature rises above this preset point, a sensor in the air conditioner detects the change and turns the unit on. When the temperature declines to the desired level, the thermostat detects this change and turns off the air conditioner.

A negative feedback mechanism also keeps your internal temperature near 37°C (98.6°F). Consider what happens when you exercise on a hot day (**FIGURE 31.19**). Muscle activity generates heat, and your internal temperature rises. Sensory receptors in the skin detect the increase and signal the brain, which sends signals that bring about a response. Blood flow shifts, so more blood from the body's hot interior flows to the skin. The shift maximizes the amount of heat given off to the surrounding air. At the same time, sweat glands increase their output. Evaporation of sweat helps cool the body surface. Breathing quickens and deepens, speeding the transfer of heat from the blood in your lungs to the air. Hormonal changes make you feel more sluggish. As your activity level slows and your rate of heat loss increases, your temperature falls.

Sensory receptors also notify the brain when body temperature declines. The brain responds by sending signals that divert blood flow away from the skin and tighten smooth muscles attached to hairs so hairs stand up. With prolonged cold, the brain commands skeletal muscles to contract ten to twenty times a second. This shivering increases heat production by muscles. When you warm up, blood returns to your skin, you stop shivering, and your goose bumps subside.

Through the process of negative feedback, the body can prevent large variations in external temperature from causing similarly large changes inside the body.

negative feedback A change causes a response that reverses the change; important mechanism of homeostasis.

sensory receptor Cell or cell component that detects a specific type of stimulus.

TAKE-HOME MESSAGE 31.9

What is the role of negative feedback in homeostasis?

✔ Negative feedback prevents dramatic changes in internal conditions. Sensory receptors detect changes and signal the brain, which signals muscles and glands. The signals cause a response that reverses the initial change.

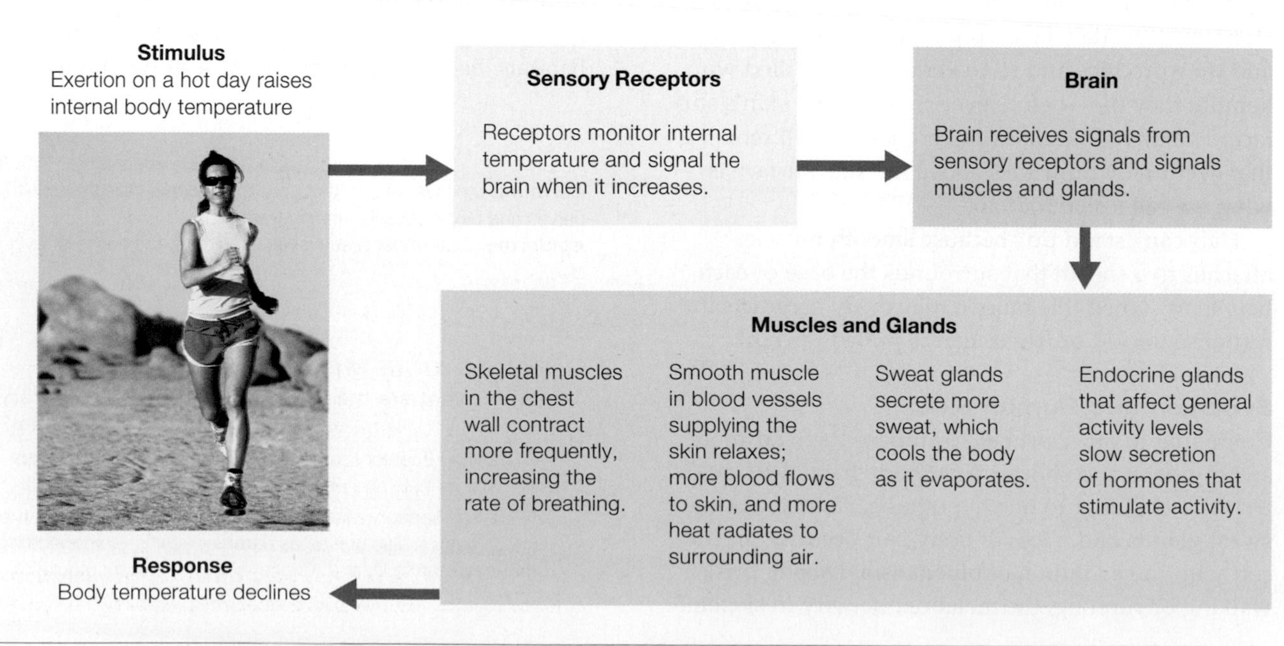

FIGURE 31.19 ▸**Animated** Negative feedback mechanism that reduces body temperature when it rises.

Stem Cells—It's All About Potential (revisited)

Human embryonic stem cells used in research are cultured cells derived from very early stage embryos (**FIGURE 31.20**). These embryos were produced by *in-vitro* fertilization (IVF), but are no longer needed. Typically, several embryos are produced during IVF in case the procedure needs to be repeated. If the resulting pregnancy is successful, the parents may donate their extra embryos for use in research, or have them destroyed. Either way, the embryos die, and some people find this death morally troubling.

With this concern in mind, some researchers have been looking into ways to remove cells from early embryos without destroying the embryo. Others are inducing adult cells to behave as if they were embryonic stem cells. Nonembryonic cells that have been altered so they behave like their embryonic counterparts are called **induced pluripotent stem cells (IPSCs)**.

The first IPSCs were produced using a virus to insert genes into fibroblasts. Expression of the inserted genes caused the cells to dedifferentiate (to reverse the process of differentiation that gave them their specific character). However, IPSCs produced in this manner are considered unsuitable for clinical use because their genome has been permanently altered by the gene insertion. The alteration could cause cells to behave in unexpected ways and perhaps become cancerous.

Researchers have since developed techniques that can bring about dedifferentiation without permanently altering the genome. For example, fibroblasts from skin can be turned into IPSCs by introducing synthetic modified RNAs (rather than genes). The fibroblasts take up the RNAs by endocytosis, then translate them into proteins that bring about dedifferentiation.

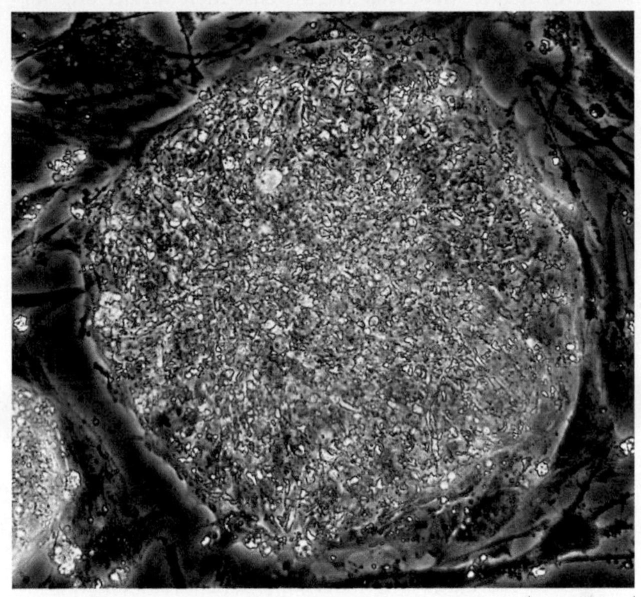

FIGURE 31.20 A cluster of human embryonic stem cells growing in culture at the University of Pittsburgh.

The first clinical experiment using specialized cells derived from IPSCs has just begun. The cells are being used to treat people who have common type of degenerative blindness.

induced pluripotent stem cells (IPSCs) Adult cells that have dedifferentiated so they behave like embryonic stem cells.

summary

Section 31.1 All cells in an animal body are derived from **stem cells**, which can either divide or differentiate into a specialized cell type. The first divisions of a fertilized egg yield totipotent cells that can form any tissue or develop into a new individual. Later embryos have pluripotent cells that can still form any tissue. After birth, cells are less versatile. Researchers hope to use embryonic stem cells to produce new cells of types that are not normally replaced in adults. **Induced pluripotent stem cells** derived from adult cells may also be used for this purpose.

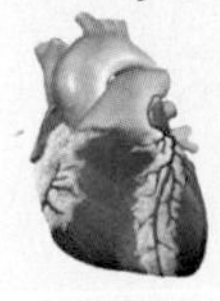

Section 31.2 Most animals have four types of tissues organized as organs and organ systems. An organ system consists of two or more organs that interact chemically, physically, or both in tasks that help keep individual cells as well as the whole body functioning smoothly. **Extracellular fluid** serves as the body's internal environment. In humans, extracellular fluid consists mainly of **interstitial fluid** and plasma. Animal structure is influenced both by physical constraints and by evolutionary history.

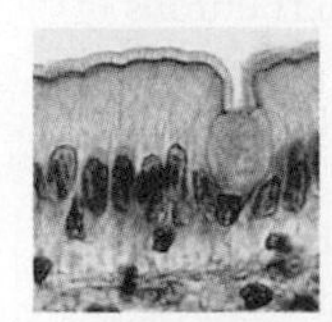

Section 31.3 **Epithelial tissue** covers the body surface and lines its internal tubes and cavities. Epithelial cells have little extracellular matrix between them. An epithelium has a free apical surface. Its basal surface secretes a **basement membrane** that attaches it to underlying connective tissue. Tight junctions occur only in epithelial cells. The gut is lined by an epithelium with tight junctions. Gut epithelium functions as a selective barrier by controlling

which substances move into the body's internal environment. Some ciliated epithelial cells move materials across their surface. Others have **microvilli** that increase their surface area for absorption or secretion. Hair, fur, and nails are keratin-rich remains of specialized epithelial cells.

Gland cells are epithelial cells whose secretions act outside the cell. Ductless **endocrine glands** secrete hormones into blood. **Exocrine glands** secrete products such as milk or saliva through ducts.

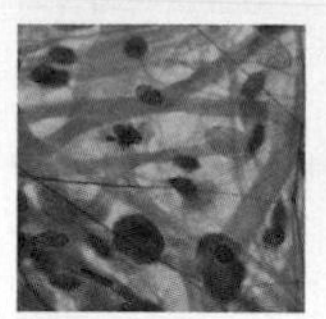

Section 31.4 **Connective tissues** "connect" tissues to one another, both functionally and structurally. Different types bind, organize, support, strengthen, protect, and insulate other tissues. All consist of cells in a secreted matrix. Soft connective tissue underlies skin, holds internal organs in place, and connects muscle to bone, or bones to one another. All soft connective tissues have the same components (fibroblasts and a matrix with elastin and collagen fibers) but in different proportions. Loose connective tissue holds internal organs in place. Ligaments and tendons consist of dense connective tissue.

Rubbery **cartilage** and mineral-hardened **bone tissue** are components of the skeleton. Fat stored in **adipose tissue** is the body's main energy reservoir. **Blood** consists of fluid plasma, cells, and platelets. It is considered a connective tissue because blood cells and platelets arise from stem cells in bone.

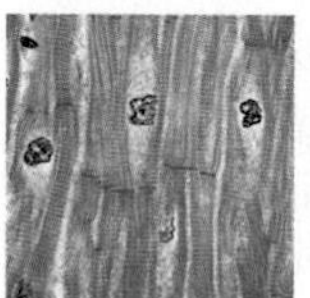

Section 31.5 Muscle tissues contract and move a body or its parts. Muscle contraction is a response to signals from the nervous system and is fueled by ATP.

Skeletal muscle consists of long fibers with multiple nuclei and has a striated (striped) appearance. Skeletal muscles, which pull on bones, are under voluntary control. They have large stores of glycogen. The metabolic reactions carried out by skeletal muscle are the main source of body heat.

Cardiac muscle, found only in the heart wall, consists of branching cells and has a striated appearance. Gap junctions allow signals to travel fast between the cells.

Smooth muscle tissue is found in the walls of tubular organs and some blood vessels. Its cells taper at both ends and are not striated.

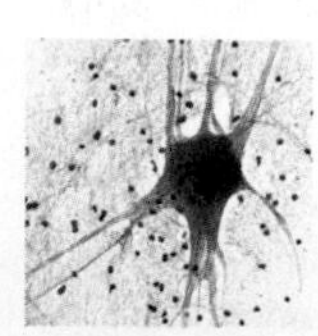

Section 31.6 **Nervous tissue** makes up the communication lines through the body. It consists of **neurons** that send and receive electrochemical signals, and **neuroglial cells** that support them. A neuron has a central cell body and long cytoplasmic extensions that send and receive signals. Sensory neurons detect information, interneurons integrate and assess information about internal and external conditions, and motor neurons command muscles and glands.

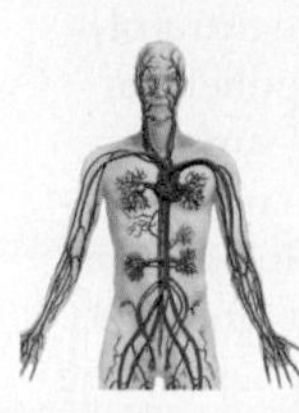

Section 31.7 Vertebrates are coelomate and many of their internal organs reside inside a body cavity derived from the coelom. All vertebrates have the same set of organ systems. These systems cooperate in maintaining the health of the body as a whole.

Section 31.8 The integumentary system consists of skin and structures such as hair that are derived from it. It functions in temperature control, detection of shifts in external conditions, vitamin production, and defense against pathogens. The outermost layer of skin, the **epidermis**, is a stratified squamous epithelium consisting mainly of keratinocytes. Melanocytes produce the melanin that gives skin its color and serves as a natural sunblock. The deeper **dermis** consists mainly of dense connective tissue and contains blood vessels, nerves, and muscles. Underlying the dermis is the hypodermis, a layer of connective tissue and adipose cells.

Sweat glands and hair follicles are collections of epidermal cells that descended into the dermis during development. Compared to our closest primate relatives, we have more sweat glands and finer, shorter body hair. These traits helped our early ancestors in Africa stay cool when walking and running under hot conditions. The reduction in body hair was accompanied by an increase in melanin deposition that protected the skin against sunlight. Later, when some human populations moved to regions with less sunlight, their skin color reverted to a lighter state.

Section 31.9 Homeostasis requires **sensory receptors** that detect changes, an integrating center (the brain), and effectors (muscles and glands) that bring about responses. **Negative feedback** often plays a role in homeostasis: A change causes the body to respond in a way that reverses the change.

self-quiz

Answers in Appendix VII

1. _______ tissues are sheetlike with one free surface.
   a. Epithelial
   b. Muscle
   c. Nervous
   d. Connective

2. _______ keep fluid from leaking between cells.
   a. Tight junctions
   b. Adhering junctions
   c. Gap junctions
   d. all of the above

3. Exocrine glands are specialized _______ tissue.
   a. epithelial
   b. muscle
   c. nervous
   d. connective

4. A rubbery secreted matrix of glycoproteins and collagen surrounds living cells in _______ .
   a. bone
   b. cartilage
   c. adipose tissue
   d. blood

5. Blood cells develop from stem cells in _______ .
   a. epidermis
   b. dermis
   c. cartilage
   d. bone

6. Your body's main energy reservoir is _______ .
   a. glycogen stored in cardiac muscle
   b. lipids stored in adipose tissue
   c. starch stored in skeletal muscle
   d. phosphorus stored in bone

7. Cytoplasmic extensions of _______ send and receive chemical messages.
   a. neuroglial cells
   b. neurons
   c. fibroblasts
   d. melanocytes

data analysis activities

Cultured Skin for Healing Wounds Diabetes is a disorder in which the blood sugar level is not properly controlled. Among other effects, it reduces blood flow to the lower legs and feet. As a result, about 3 million diabetes patients have ulcers (open wounds that do not heal) on their feet. Each year, about 80,000 require amputations.

Several companies provide cultured cell products designed to promote the healing of diabetic foot ulcers. **FIGURE 31.21** shows the results of a clinical experiment that tested the effect of one such cultured skin product versus standard treatment for diabetic foot wounds. Patients were randomly assigned to either the experimental treatment group or the control group and their progress was monitored for 12 weeks.

1. What percentage of wounds had healed at 8 weeks when treated the standard way? When treated with cultured skin?
2. What percentage of wounds had healed at 12 weeks when treated the standard way? When treated with cultured skin?
3. How early was the healing difference between the control and treatment groups obvious?

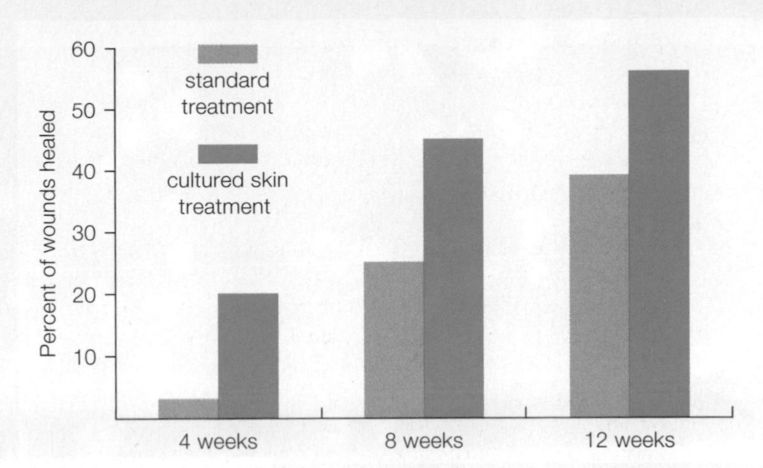

FIGURE 31.21 Results of a multicenter study of the effects of standard treatment versus use of a cultured cell product for diabetic foot ulcers. Bars show the percentage of foot ulcers that had completely healed.

8. _______ muscle pulls on bones and _______ muscle regulates the diameter of blood vessels.
   a. Skeletal/cardiac
   b. Smooth/cardiac
   c. Skeletal/smooth
   d. Smooth/skeletal

9. Straps of dense, regular connective tissue _______ .
   a. connect muscles to bones
   b. produce blood cells
   c. underlie the skin
   d. lack fibroblasts

10. _______ increase the surface area of some epithelial cells.
   a. Microfilaments
   b. Microvilli
   c. Gap junctions
   d. Adhering junctions

11. Tears are an _______ secretion released by specialized _______ tissue cells.
   a. endocrine; epithelial
   b. endocrine; connective
   c. exocrine; epithelial
   d. exocrine; connective

12. Cancers most commonly arise in _______ tissue.
   a. epithelial
   b. muscle
   c. nervous
   d. connective

13. The most abundant protein in your body is _______ .
   a. melanin
   b. actin
   c. collagen
   d. keratin

14. Match each term with the most suitable description.

___ exocrine gland	a. signaling cell in nervous tissue
___ endocrine gland	b. secretion through duct
___ fibroblast	c. collagen-producing cell
___ melanocyte	d. contraction is involuntary
___ neuron	e. pigment-producing cell
___ smooth muscle	f. main source of metabolic heat
___ skeletal muscle	g. main cell type in epidermis
___ blood	h. fluid connective tissue
___ keratinocyte	i. includes interstitial fluid, lymph
___ extracellular fluid	j. secretes hormones

15. With negative feedback, detection of a change brings about a response that _______ the change.
   a. reverses
   b. accelerates
   c. has no effect on
   d. mimics

CENGAGE brain To access course materials, please visit www.cengagebrain.com.

critical thinking

1. IPSCs are nearly identical to human embryonic stem cells in terms of gene expression, but there may be other ways in which they are not equivalent. For example, the telomeres of IPSCs often vary in length, with many IPSCs cells having telomeres shorter than those of embryonic cells. How might shortened telomeres affect the life-span of IPSCs or of differentiated cells derived from them?

2. Radiation and chemotherapy drugs preferentially kill cells that divide frequently, most notably cancer cells. These cancer treatments also cause hair to fall out. Why?

3. Each level of biological organization has emergent properties that arise from the interaction of its component parts. For example, cells have a capacity for inheritance that molecules making up the cell do not. What are some emergent properties of specific types of tissues?

4. The micrograph to the right shows cells from the lining of an airway leading to the lungs. The gold cells are ciliated and the darker brown ones secrete mucus. What type of tissue is this? How can you tell?

32 Neural Control

LEARNING ROADMAP

The nervous signals discussed in this chapter involve receptor proteins (Section 5.7) and transport mechanisms (5.9, 5.10). They depend on ion gradients, which are a type of potential energy (5.2). We also reconsider health-related topics such as cancer (11.6) and alcohol abuse (5.1).

NERVOUS SYSTEMS

In cnidarians, neurons interconnect as a nerve net. Most other animals have a bilateral nervous system with a brain or ganglia at the animals's anterior end.

HOW NEURONS WORK

A shift in the electric charge across a neuron membrane conveys information along a neuron. Neurons communicate with one another and with other cells via chemical signals.

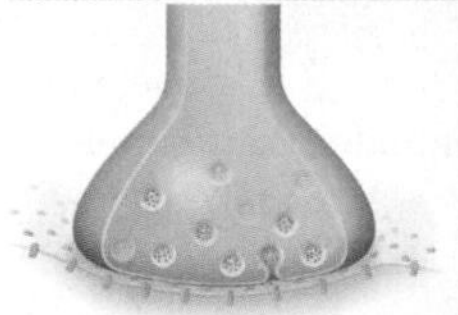

VERTEBRATE NERVOUS SYSTEM

The brain and spinal cord constitute the central nervous system. The peripheral nervous system includes nerves that connect the brain and spinal cord to the rest of the body.

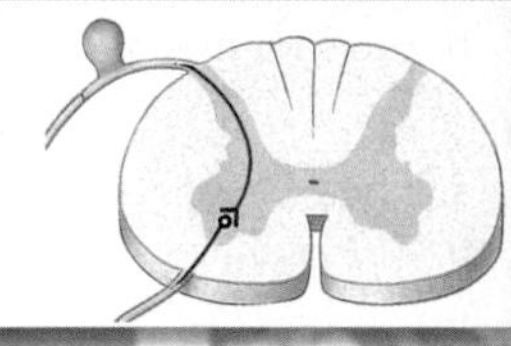

ABOUT THE BRAIN

A human brain includes evolutionarily ancient tissues that govern housekeeping tasks. The more recently expanded cerebral cortex provides our capacity for analytical thought and language.

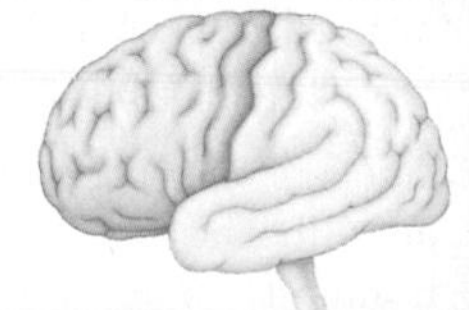

THE NEUROGLIAL SUPPORT SYSTEM

A variety of cells collectively known as neuroglia assist neurons. Neuroglia greatly outnumber neurons in the brain and play a role in some neurological disorders.

We continue our survey of the nervous system with the next chapter, which focuses on the function of sensory neurons and the integration of sensory input in the brain. We discuss the interactions between the nervous and endocrine systems in Chapter 34. We also consider nervous control of skeletal muscle contraction (Chapter 35), heart rate and blood vessel dilation (Chapter 36), breathing (Chapter 38), urination (Chapter 40), and sexual arousal (Chapter 41).

32.1 Impacts of Concussions

The human brain is surprisingly delicate. It is only about as firm as Jell-O or warm butter. Head impacts or rapid stops can cause this soft, squishy organ to slam against the inside of the bony skull, causing a mild traumatic brain injury called a concussion. Symptoms of a concussion can include confusion, dizziness, blurred vision, increased sensitivity to light, headache, impaired short-term memory, difficulty concentrating, irritability, nausea, altered sleep patterns, and a temporary loss of consciousness.

The Centers for Disease Control estimates that somewhere between 1.4 and 3.8 million concussions occur each year in the United States. Participating in contact sports raises the risk, but concussions also occur as a result of car accidents, workplace accidents, or simple falls. Military personnel deployed to combat zones often suffer concussions as a result of exposure to explosions. One recent study of U.S. soldiers who had served in Iraq found that about 15 percent had experienced symptoms of a concussion.

Usually, the brain resumes normal function within 7 to 10 days after a concussion. However, repeated head injuries can cause chronic traumatic encephalopathy (CTE), a neurodegenerative disorder that can involve memory loss, emotional problems, depression, suicidal impulses, and dementia.

Concussions are an occupational hazard for professional football players (**FIGURE 32.1**). Physicians for the National Football League routinely diagnose between 100 and 150 concussions per season, and these injuries sometimes have long-term consequences. Consider the case of Dave Duerson, a 50-year-old former player for the Chicago Bears. Duerson committed suicide in 2011 by shooting himself in the chest. He left behind a request that his brain be donated to a brain bank (a collection of brain tissue) devoted to the study of sport-related brain injuries. Examination of his brain revealed the extensive degenerative brain damage characteristic of CTE. Retired professional football players are three times more likely than the general population to die of neurodegenerative diseases. They are also more likely to be diagnosed with clinical depression, with the incidence of this disorder rising with the number of concussions sustained.

FIGURE 32.1 Jake Long of the St. Louis Rams football team is examined after suffering a concussion.

CREDITS: (opposite) Courtesy of © Riken Brain Science Institute; (l) Michael Zagaris/Getty Images Sport.

32.2 Evolution of Nervous Systems

✔ Interacting neurons give animals a capacity to respond to stimuli in the environment and inside their body.

For an animal body to function as an integrated whole, its cells must communicate with one another. Intercellular communication involves three steps. First, a signaling molecule released by one cell reversibly binds to a receptor on or in another cell. Second, the signal is transduced, meaning it is converted into a form that has an effect in the signal-receiving cell. Third, the signal-receiving cell changes in some way in response to the signal.

Two organ systems—the nervous system and the endocrine system—facilitate long-distance communication among cells of an animal body. The endocrine system coordinates activities by way of molecules called hormones that travel in the blood. We discuss this system in detail in Chapter 34. Here, we focus on the nervous system.

As Section 31.6 explained, neurons make up the communication lines of nervous systems. Neurons transmit electrical signals along their plasma membrane and also send chemical messages to other cells. In most animals, neuroglial cells support the neurons.

Nerve Net

Animals with radial symmetry have a mesh of interconnected neurons called a **nerve net**. Information flows in all directions among cells of the nerve net, and there is no centralized, controlling organ that functions like a brain. Sea anemones and other cnidarians (Section 24.5) are the simplest animals with a nerve net (**FIGURE 32.2A**). By causing cells in the body wall to contract, a cnidarian nerve net can alter the size of the mouth, change the body's shape, or shift the position of the tentacles.

Echinoderms (Section 24.16) are also radial, and they too have a nerve net. However, the echinoderm nerve net is a bit more complex than a cnidarian's because it includes nerves. A **nerve** is a bundle of neuron fibers (cytoplasmic extensions) wrapped in connective tissue. In sea stars, a ring of nerves surrounds the mouth and a nerve extends from this ring into each arm.

Getting a Head

Most animals have bilateral symmetry, with paired organs arranged on either side of the main body axis (Section 24.2). Evolution of bilateral body plans was accompanied by **cephalization**: an evolutionary process in which neurons that detect and process information about the external environment became concentrated at the body's anterior, or head end.

Planarian flatworms are bilateral animals with a simple nervous system. A pair of ganglia in the head serves as an integrating center (**FIGURE 32.2B**). Each **ganglion** (plural, ganglia) is a cluster of neuron cell bodies. (A cell body is the part of the neuron that holds the cell's nucleus and most other organelles.) A planarian's ganglia receive signals from eyespots and chemical-detecting cells on the head. The ganglia connect to a pair of nerve cords that run along the animal's ventral (lower) surface. A **nerve cord** consists of nerve fibers of many neurons, and it runs the length of a body. Nerves branch from the nerve cord and cross the body.

Annelids and arthropods have paired nerve cords that run along the ventral surface and connect to a simple brain (**FIGURE 32.2C**). A **brain** is a central control organ of a nervous system. It receives and integrates

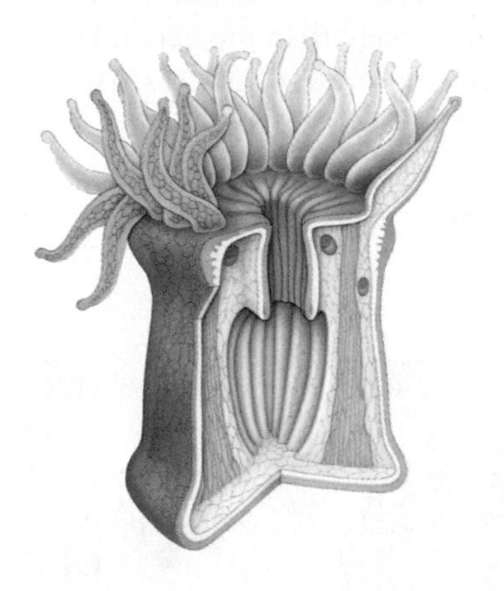

A Sea anemones have a nerve net (purple). There is no central organizing center in a nerve net.

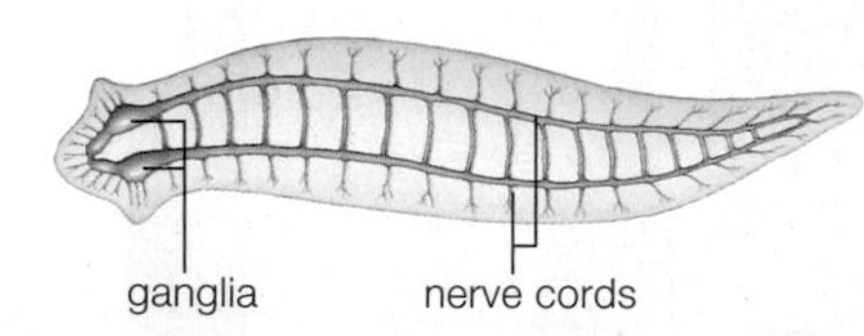

B Planarian flatworms have a pair of nerve cords and paired ganglia that serve as a brain.

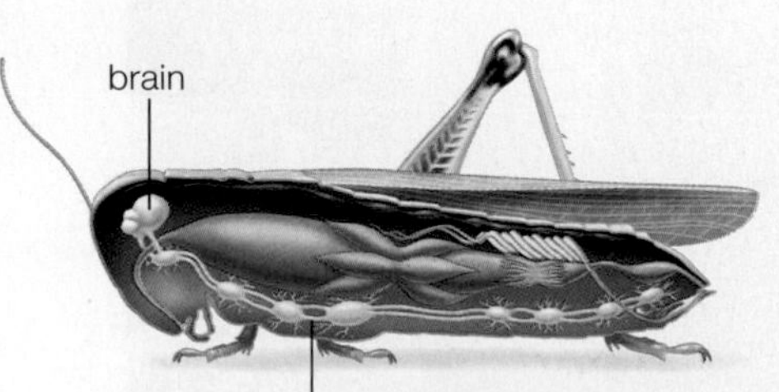

C Insects have paired ventral nerve cords with a ganglion in each segment. The cords connect to a simple brain.

FIGURE 32.2 ▸Animated Examples of invertebrate nervous systems. **FIGURE IT OUT** Which of these organisms has/have a cephalized nervous system? Answer: The planarian and grasshopper

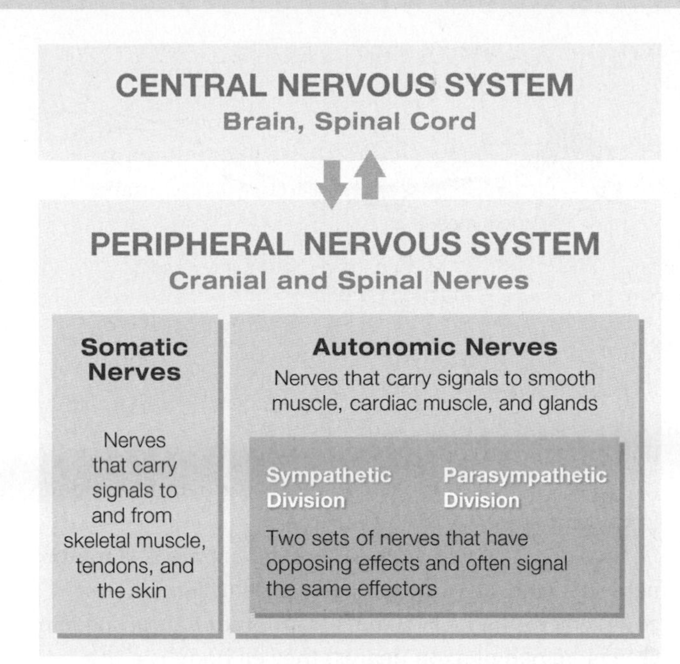

FIGURE 32.3 Functional divisions of vertebrate nervous systems. The spinal cord and brain are its central portion. The peripheral nervous system includes spinal nerves, cranial nerves, and their branches, which extend through the rest of the body. Peripheral nerves carry signals to and from the central nervous system.

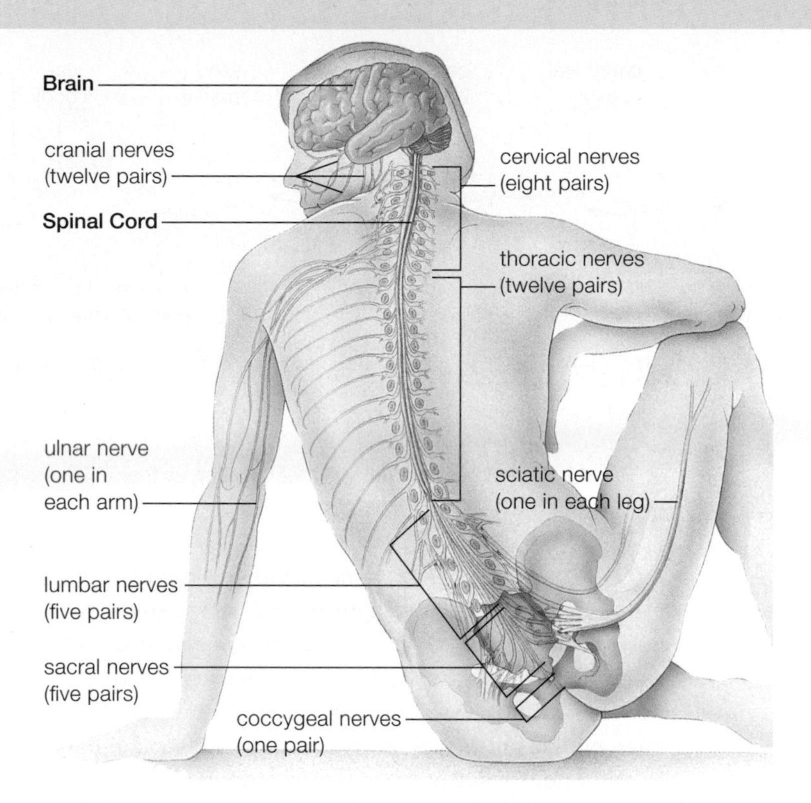

FIGURE 32.4 Some of the major nerves of the human nervous system.

sensory information, regulates internal processes, and sends out signals that bring about movements.

Organization of the mollusk nervous system varies. Clams and other bivalves are brainless; widely separated ganglia serve as local control centers. On the other hand, cephalopods such as octopuses have the largest and most complex nervous system of any invertebrate group. An octopus has a proportionately larger brain than many fishes or reptiles.

The Vertebrate Nervous System

A dorsal nerve cord (a nerve cord that runs along the back) is one defining feature of chordate embryos (Section 25.2). In vertebrates, the dorsal nerve cord evolved into a brain and spinal cord, which together constitute the animal's **central nervous system** (**FIGURE 32.3**). Nerves that extend from the central nervous system through the rest of the body are the **peripheral nervous system**.

The human peripheral nervous system consists of 12 cranial nerves that connect to the brain and 31 spinal nerves that connect to the spinal cord (**FIGURE 32.4**). Most cranial nerves, and all spinal nerves, carry signals both to and from the central nervous system. Consider the sciatic nerve, which runs from the spinal cord, through the buttock, and down the leg. When someone touches your thigh, this nerve carries signals from receptors in the skin to the spinal cord. When you move your leg, the same nerve carries commands for movement from the spinal cord to muscles in the leg.

brain Central control organ in a nervous system.
central nervous system Brain and spinal cord.
cephalization Evolutionary process whereby sensory structures and nerve cells accumulate at the anterior end of a bilateral body.
ganglion Cluster of neuron cell bodies.
nerve Neuron fibers bundled inside a sheath of connective tissue.
nerve cord Bundle of nerve fibers running the length of a body.
nerve net Of cnidarians, a mesh of interacting neurons with no central control organ.
peripheral nervous system Nerves that extend through the body and carry signals to and from the central nervous system.

TAKE-HOME MESSAGE 32.2
What are the features of animal nervous systems?

- ✔ Cnidarians and echinoderms have a simple nervous system—a nerve net with no central integrating organ.
- ✔ Flatworms have paired ganglia that serve as an integrating center. Some other invertebrates have brains.
- ✔ Bilateral invertebrates usually have a pair of ventral nerve cords. In contrast, chordates have one dorsal nerve cord.
- ✔ The vertebrate nervous system includes a well-developed brain, a spinal cord, and peripheral nerves.

32.3 Neurons—The Great Communicators

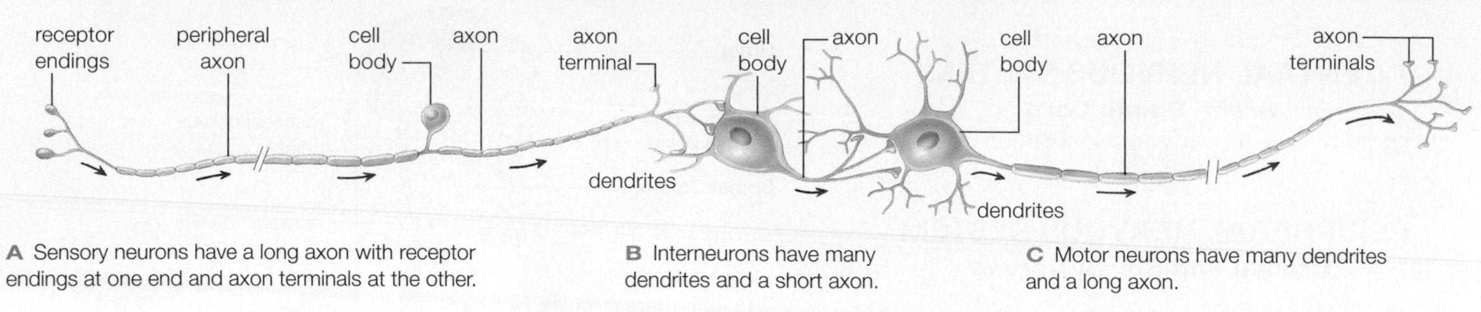

A Sensory neurons have a long axon with receptor endings at one end and axon terminals at the other.

B Interneurons have many dendrites and a short axon.

C Motor neurons have many dendrites and a long axon.

FIGURE 32.5 The three types of neurons. Arrows indicate the direction of information flow.

✔ The structure of neurons reflects their function as the communicating cells of nervous systems.

In vertebrate nervous systems, information typically flows from sensory neurons, to interneurons, to motor neurons (**FIGURE 32.5**). **Sensory neurons** detect a stimulus such as light or touch. When activated by this stimulus, they usually signal an interneuron. **Interneurons** both receive signals from and send signals to other neurons. An interneuron may signal another interneuron or a motor neuron. **Motor neurons** control muscles and glands.

Neuron structure varies, but all neurons have a central cell body with a nucleus and other organelles. All neurons also have one **axon**, a cytoplasmic extension that transmits electrical signals along its length and releases chemical signals at its terminals. Sensory neurons do not receive signals from other cells, so the axon is their only cytoplasmic extension (**FIGURE 32.5A**). One end of this axon has receptor endings that detect a specific stimulus; the other has signal-sending axon terminals. Interneurons and motor neurons both receive and send signals, so they have dendrites as well as an axon. **Dendrites** are cytoplasmic extensions that receive chemical signals from other neurons and convert them to electrical signals. Most vertebrate interneurons are located entirely in the brain or spinal cord. They have a short axon (**FIGURE 32.5B**). Motor neurons have a cell body in the brain or spinal cord, and an axon that extends out to the muscle or gland they control (**FIGURE 32.5C**). Axons of some motor neurons extend from your spinal cord to your toes.

FIGURE 32.6 shows the functional zones of a motor neuron. Dendrites and cell body are the signal input zone, where chemical signals from other neurons arrive ❶. The region of axon nearest the cell body is a trigger zone ❷. Excitation of this region triggers flow of electrical signals along the conducting zone of the axon ❸. Axon terminals are the output zone: They release signaling molecules that influence other cells ❹.

axon Of a neuron, a cytoplasmic extension that transmits electrical signals along its length and secretes chemical signals at its endings.
dendrite Of a neuron, a cytoplasmic extension that receives chemical signals sent by other neurons and converts them to electrical signals.
interneuron Neuron that both receives signals from and sends signals to other neurons. Located mainly in the brain and spinal cord.
motor neuron Neuron that controls a muscle or gland.
sensory neuron Neuron that is activated when its receptor endings detect a specific stimulus, such as light or pressure.

TAKE-HOME MESSAGE 32.3

How does a neuron's structure affect its function?

✔ All neurons have a cell body with organelles and an axon that sends chemical signals to other cells.

✔ Interneurons and motor neurons have signal-receiving dendrites. A sensory neuron has no dendrites. One end of its axon has receptor endings that detect a specific stimulus.

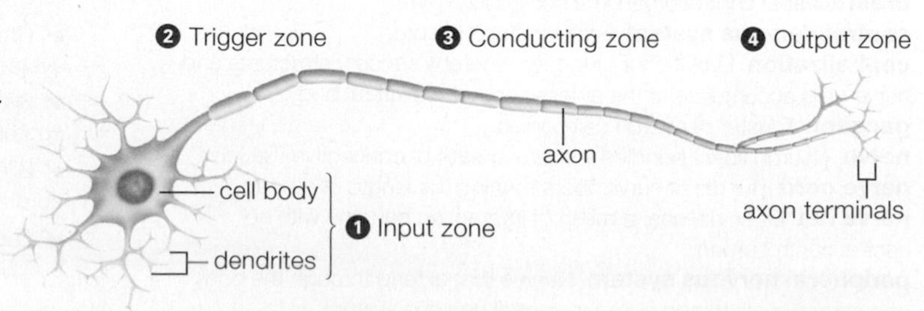

FIGURE 32.6 ▸Animated Functional zones of a motor neuron.

❶ Dendrites and the cell body function as an input zone. They receive chemical signals from other neurons and convert them to electrical signals.

❷ The stimulation spreads to a trigger zone at the start of an axon.

❸ If the electrical signals are sufficiently strong, they are conducted along the axon to its terminal endings.

❹ Axon terminals are a signal output zone. They release chemical signals that influence another cell. In the case of a motor neuron the signals affect a muscle cell.

32.4 Membrane Potential

✔ Transport proteins set up concentration and electrical gradients across the neuron cell membrane.

All animal cells have an electric gradient across their plasma membrane; cytoplasmic fluid near this membrane contains more negatively charged ions and proteins than extracellular fluid does. As in a battery, the separation of charge constitutes potential energy (Section 5.2). We describe potential energy in terms of voltage, and the voltage across a cell's membrane is called **membrane potential**. Researchers measure the membrane potential across a neuron's plasma membrane by inserting one electrode into a neuron and another into the interstitial fluid just outside of it.

The membrane potential of a neuron that is not being stimulated is the **resting potential**. This potential is usually about –70 millivolts (a millivolt [mV] is one-thousandth of a volt). The negative sign indicates the charge of the neuron's cytoplasm is more negative than that of the adjacent interstitial fluid.

What causes this potential? For one thing, cytoplasm contains many negatively charged proteins that are not present in the interstitial fluid. The proteins cannot cross the neuron's plasma membrane and so remain trapped within the cell. The distribution of positively charged potassium ions (K^+) and positively charged sodium ions (Na^+) also influence resting potential. These ions move in and out of the neuron with the assistance of transport proteins (Section 5.7).

Sodium–potassium pumps (a type of active transporter) move two potassium ions into the cell for every three sodium ions they move out (**FIGURE 32.7A**). By moving more positively charged ions out of the cell than into it, these pumps increase the electrical gradient across the membrane. The pumps also create concentration gradients for sodium and potassium ions across the membrane.

Sodium ions pumped out of the cell cannot easily cross the plasma membrane of a resting neuron.

membrane potential Voltage difference across a cell membrane.
resting potential Voltage difference across the plasma membrane of an excitable cell that is not receiving stimulation.

TAKE-HOME MESSAGE 32.4

What is a neuron's membrane potential?

✔ Membrane potential is the voltage or charge difference across a neuron's plasma membrane. The charge difference arises from differences in the distribution of ions and charged proteins across the membrane.

However, some of the potassium ions pumped into the neuron do exit it through passive transport proteins in the membrane (**FIGURE 32.7B**). This movement of positively charged potassium down its concentration gradient and out of the cell also contributes to the charge difference across the membrane.

In summary, the cytoplasm of a resting neuron contains more negatively charged proteins, fewer sodium ions (Na^+), and more potassium ions (K^+) than interstitial fluid. We illustrate the differences below, with the white minus sign representing negatively charged proteins:

interstitial fluid 150 Na^+ 5 K^+
plasma membrane
cytoplasm 15 Na^+ 150 K^+ 65 −

In addition to sodium–potassium pumps and the continually open potassium channels, some parts of a neuron membrane have voltage-gated channels for sodium and for potassium (**FIGURE 32.7C**). These channels are passive transport proteins whose shape varies depending on the membrane potential. When the neuron is at resting potential, part of the protein functions like a gate that prevents movement of ions through a channel in the protein. A shift in membrane potential can cause the gate to open, allowing ions to follow their gradients and flow through the channel, across the membrane. As you will learn in the next section, operation of voltage-gated channels is integral to a neuron's ability to transmit signals.

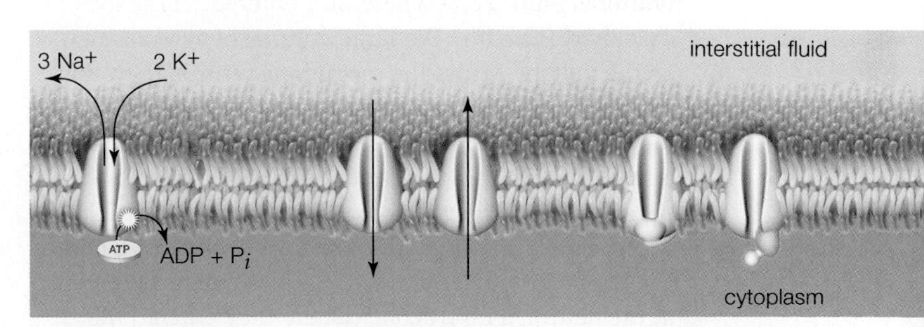

A Sodium–potassium cotransporters actively transport three Na^+ out of a neuron for every two K^+ they pump in. See Figure 5.30.

B Passive transporters allow K^+ ions to leak across the plasma membrane, following its concentration gradient.

C Voltage-gated channels for Na^+ or K^+ are closed in a neuron at rest (left), but open if the voltage changes (right).

FIGURE 32.7 ▶**Animated** Three types of transport proteins in a neuron membrane. Voltage-gated channels are only in the membrane of the axon.

32.5 The Action Potential

✔ Movement of sodium and potassium ions through gated channels causes a brief reversal of the membrane potential.

All cells have a membrane potential, but only neurons and muscle cells are "excitable," meaning that when stimulated, they will undergo an **action potential**—a brief reversal in the polarity of the electric gradient across the plasma membrane. If you plot the change in membrane potential during an action potential, the resulting graph will look like the one in **FIGURE 32.8**. During an action potential, membrane potential shoots up from its resting value (–70 mV) into positive territory. A positive membrane potential means the interior of the cell has more positively charged ions than interstitial fluid. The action potential peaks at +30 mV, then declines once again to resting potential. The whole process takes only a few milliseconds.

Graded Potentials and Reaching Threshold

Changes in a cell's membrane potential occur when ions move across its plasma membrane. In sensory neurons, a stimulus such as light or touch results in ion movement. Tap your wrist, and the pressure stimulates receptor endings of sensory neurons in your skin. The pressure deforms the plasma membrane at the input zone of these neurons, allowing some sodium ions to slip across it. The ion flow causes a local shift in the voltage difference across the neuron's membrane. The shift is a graded potential; "graded" means the size of the shift varies. In an interneuron or motor neuron, local graded potentials arise in response to arrival of signaling molecules released by another neuron.

When only a few sodium ions cross a neuron's membrane, they cause a small graded potential at the neuron's input zone where they entered. These ions quickly diffuse into the large volume of cytoplasm, so the voltage across the membrane returns to resting potential fast. By analogy, think about what would happen if you put a drop of ink into the water at the edge of a swimming pool. The ink would quickly dissipate into the large volume of water and be invisible.

When a larger number of sodium ions cross a neuron's membrane, they cause a larger graded potential at the neuron's input zone (where they entered the cell). The ions diffuse away from the input zone but do not dissipate into cytoplasm as quickly. If enough ions reach the neuron's trigger zone, they significantly shift the membrane potential there. The plasma membrane in the trigger zone has many voltage-gated sodium channels and potassium channels that remain closed when the neuron is at rest ❶. Diffusion of sodium ions into the trigger zone can shift the membrane potential there to a threshold level. When the membrane reaches this **threshold potential**, voltage-gated sodium channels open and an action potential gets under way.

An All-or-Nothing Spike

Remember from the previous section that transport proteins set up and maintain electrical and concentration gradients across a neuron's membrane. When voltage-gated sodium channels in the membrane open, they allow Na^+ ions to follow these gradients. The ions flow from interstitial fluid into the neuron ❷. This influx has an effect called **positive feedback**, in which a response intensifies the conditions that caused it to occur. In this case, the influx of sodium ions changes the membrane potential even more, which causes even more sodium channels to open, and so on:

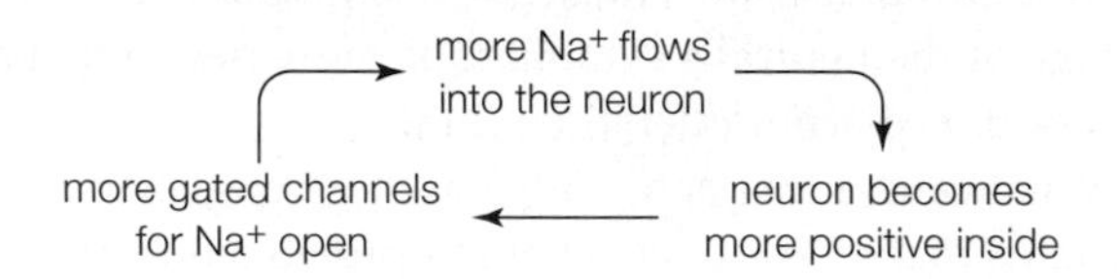

The positive feedback causes an accelerating rise in membrane potential, and an action potential results. An action potential is an all-or-nothing event, which means that once threshold potential is reached, an action potential always occurs. After sodium channels begin opening, the size of the stimulus that brought the membrane to threshold potential becomes irrelevant. The positive feedback loop is driven by Na^+ ions entering the neuron from interstitial fluid, not their diffusion in cytoplasm from the cell's input zone.

Every action potential peaks at +30 mV; no more, no less. Membrane potential cannot rise higher because, above a certain voltage, gates on voltage-gated Na^+ channels shut. About the same time, the voltage-gated potassium channels open ❸. Opening of these channels allows K^+ to diffuse down its concentration and electrical gradients out of the neuron.

This ion flow reduces membrane potential, which in turn causes the potassium channels to close ❹. By the time this occurs, so many K+ ions have exited the cell that the membrane potential is a bit below resting potential. Rapid diffusion of K^+ from the adjacent cyto-

action potential Brief reversal of the voltage difference across the plasma membrane of a neuron or muscle cell.
positive feedback A response intensifies the conditions that caused its occurrence.
threshold potential Neuron membrane potential at which gated sodium channels open, causing an action potential to occur.

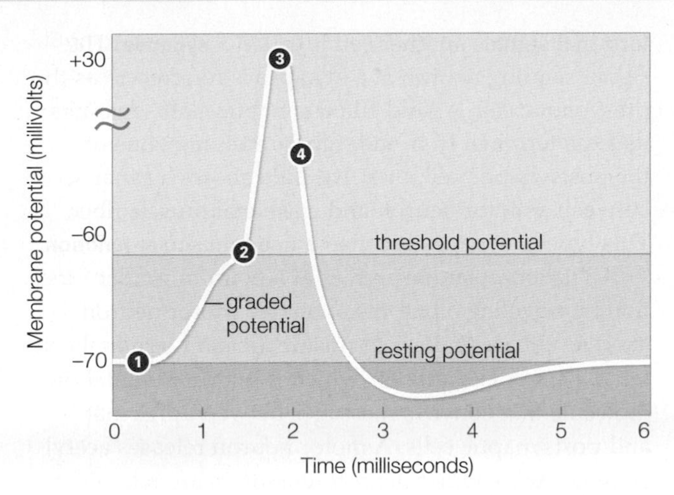

FIGURE 32.8 ▶Animated Action potential. Above, a plot of a neuron's membrane potential over time. The spike is the action potential—a brief reversal in the polarity of the electric gradient across the membrane. Numbers correlate with numbered graphics (right), which show the events at one region of an axon membrane.

plasm restores the K^+ concentration, and membrane potential returns to its resting value.

Propagation of an Action Potential

An action potential occurs only in a small area of membrane, but it propagates itself from trigger zone to axon terminals. When sodium channels open, some Na^+ diffuses through cytoplasm, raising the membrane potential in adjoining regions to threshold. An action potential cannot move backward, because after a sodium channel closes, it cannot open for a brief period. However, sodium channels in regions farther along the axon can and do open as membrane in these regions reaches threshold potential. As sodium channels open in one region after the next, the action potential moves quickly and steadily toward the axon terminals. The signal moves without weakening because, as previously noted, all action potentials are equal in magnitude.

> **TAKE-HOME MESSAGE 32.5**
> **What happens during an action potential?**
>
> ✔ An action potential begins in the neuron's trigger zone when a stimulus causes gated sodium channels to open, allowing sodium to flow into the neuron.
>
> ✔ The resulting reversal in the polarity across the membrane causes gated potassium channels to open, allowing potassium ions to flow out of the neuron.
>
> ✔ Action potentials are propagated along an axon. They move in one direction only—toward axon terminals—because after gated sodium channels close, they cannot reopen immediately.

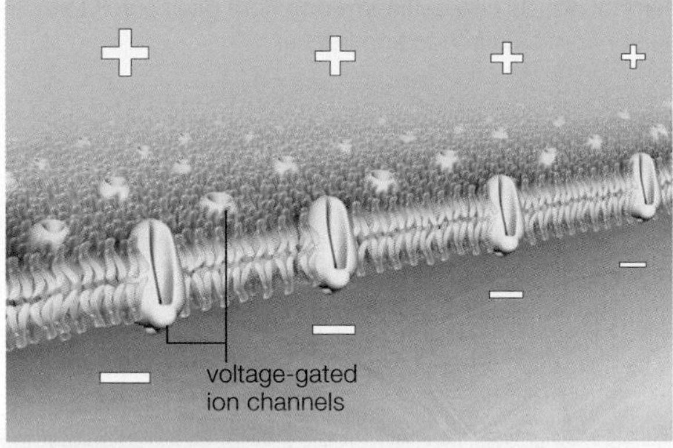

❶ Neuron at rest. Gated channels for Na^+ or K^+ are closed.

High Na^+ in interstitial fluid; high K^+ inside neuron.

Interstitial fluid has more positively charged ions than neuron cytoplasm does, making the membrane potential negative.

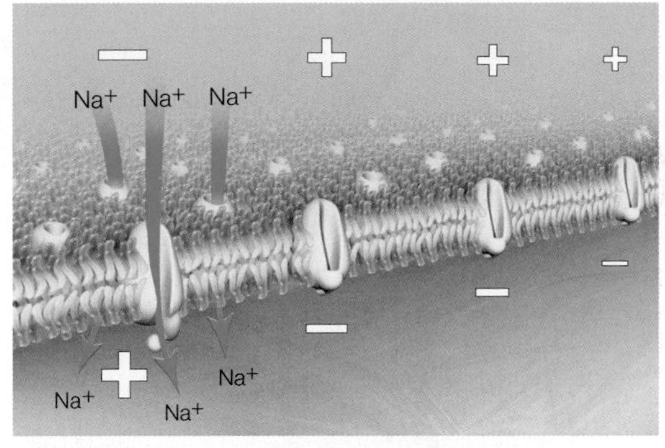

❷ Threshold is reached and Na^+ channels open.

Na^+ diffuses down its concentration gradient into the neuron.

As a result of the Na^+ influx, the neuron cytoplasm now has more positive charges than the interstitial fluid.

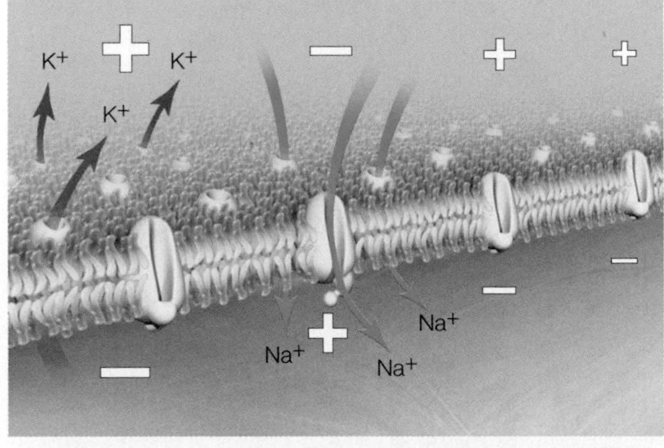

❸ High membrane potential makes Na^+ channels close and K^+ channels open.

K^+ diffuses down its concentration gradient out of the neuron, making neuron cytoplasm once again more negative than the interstitial fluid.

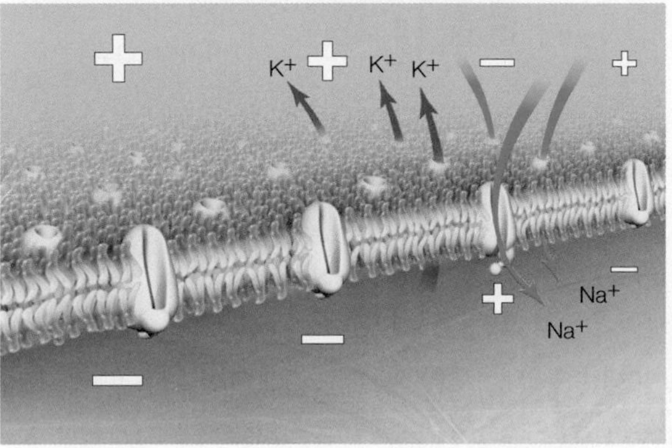

❹ K^+ channels close.

Diffusion of Na^+ to regions farther along the axon propagates the action potential.

Once closed, Na^+ channels are briefly inactivated, preventing action potentials from moving "backward."

✔ Chemical signals convey information from one neuron to another, or from a neuron to another cell.

The Synapse

What happens when an action potential arrives at an axon terminal? Membrane voltage changes cannot pass directly from one cell to another. Instead, chemicals relay the message. The region where an axon terminal signals another cell is called a **synapse**. The signal-sending neuron at a synapse is referred to as the presynaptic cell. A fluid-filled synaptic cleft separates its axon terminal from the signal-receiving zone of the postsynaptic cell. **FIGURE 32.9** shows a synapse between a motor neuron and a skeletal muscle fiber. This type of synapse is called a **neuromuscular junction**.

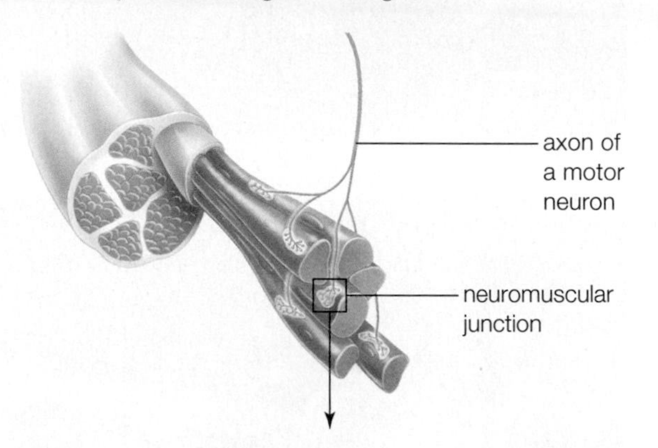

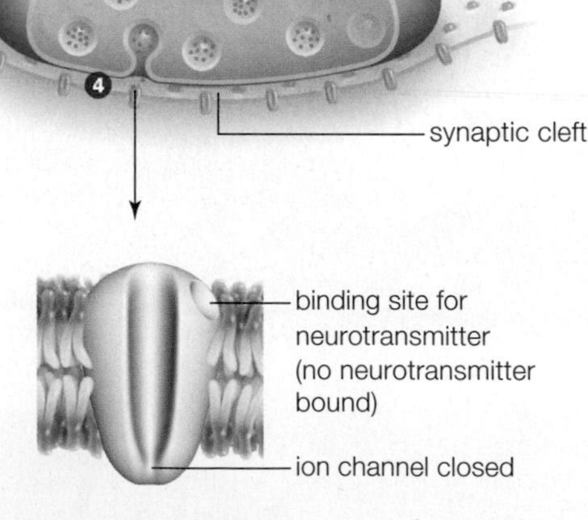

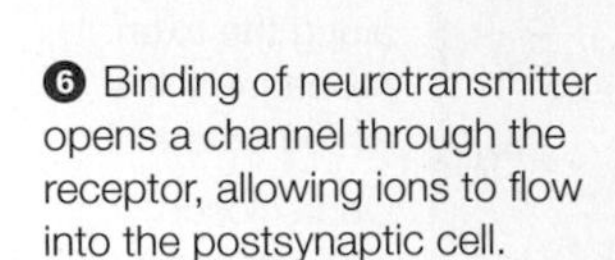

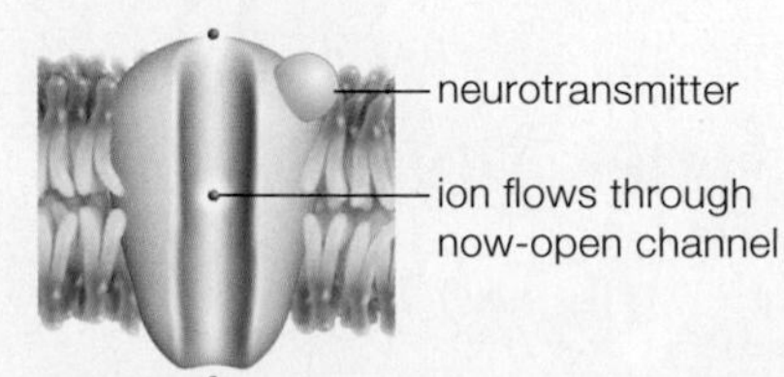

FIGURE 32.9 ▶Animated Cell–cell communication at a synapse. This example depicts what occurs at a neuromuscular junction—a synapse between an axon terminal of a motor neuron and a muscle fiber that the neuron controls.

An action potential arrives at a neuromuscular junction by traveling along the axon of a motor neuron to axon terminals ❶. Vesicles inside an axon terminal contain a **neurotransmitter** ❷, which is a type of signaling molecule that conveys messages between presynaptic and postsynaptic cells. A motor neuron releases acetylcholine (ACh). Other neurotransmitters are released by other types of neurons.

The plasma membrane of an axon terminal has voltage-gated calcium channels. When the neuron is resting, these channels are closed, and calcium pumps actively transport calcium ions (Ca^{++}) out of the cell. Thus, the cytoplasm of a resting neuron has fewer calcium ions than interstitial fluid. When an action potential arrives at an axon terminal, the calcium channels open. Calcium ions can then follow their gradient and diffuse from interstitial fluid into the axon terminal ❸. The concentration of calcium ions in the neuron's cytoplasm increases. This increase stimulates exocytosis, so neurotransmitter-filled vesicles release their contents into the synaptic cleft ❹.

The plasma membrane of a postsynaptic cell has receptors that bind neurotransmitter ❺. At a neuromuscular junction, the receptors are transport proteins that change shape when they bind to ACh. The change opens a channel through the protein. Sodium ions flow through the opening into the muscle cell ❻. Like a neuron, a muscle cell is excitable, meaning it can undergo an action potential. An increase in a muscle cell's cytoplasmic sodium ion concentration drives the membrane potential toward threshold. Once threshold is reached, action potentials stimulate muscle contraction by a process described in detail in Section 35.8.

Cleaning the Cleft

After neurotransmitter molecules do their work, they must be removed from synaptic clefts to make way for new signals. Active transporters pump some of the neurotransmitter back into presynaptic cells or into nearby neuroglial cells. In addition, postsynaptic cells have enzymes in their membrane that break down neurotransmitter. For example, the membranes of muscle cells at a neuromuscular junction contain acetylcholinesterase, an enzyme that breaks down ACh. Nerve gases such as sarin exert their deadly effects by binding to acetylcholinesterase and preventing it from breaking down ACh. The resulting accumulation of ACh in synaptic clefts causes muscle paralysis, confusion, headaches, and, when the dosage is high enough, death.

Synaptic Integration

A postsynaptic cell receives messages from many neurons at the same time (**FIGURE 32.10**). Some interneurons in the brain are on the receiving end of synapses with 10,000 neurons! Depending on the neurotransmitter released at a synapse and the type of receptor on the postsynaptic cell, an incoming signal may be excitatory or inhibitory. A signal that opens sodium gates has an excitatory effect because it pushes membrane potential closer to threshold. A signal that opens potassium gates has an inhibitory effect because it moves membrane potential farther away from threshold.

Incoming synaptic signals can amplify, dampen, or cancel one another's effects. All signals arriving at a neuron's input zone at the same time are summed, an effect called **synaptic integration** (**FIGURE 32.11**). An action potential will not occur in a postsynaptic neuron unless the summed signal is sufficient to drive the membrane potential to threshold level.

Competing signals cause the membrane potential at the postsynaptic cell's input zone to rise and fall. When excitatory signals outweigh inhibitory ones, ions diffuse from the input zone into the trigger zone and drive the postsynaptic cell to threshold. Gated sodium channels swing open, and an action potential occurs.

Neurons also integrate signals that arrive in quick succession from a single presynaptic cell. An ongoing stimulus can trigger a series of action potentials in a presynaptic cell, which will bombard a postsynaptic cell with waves of neurotransmitter.

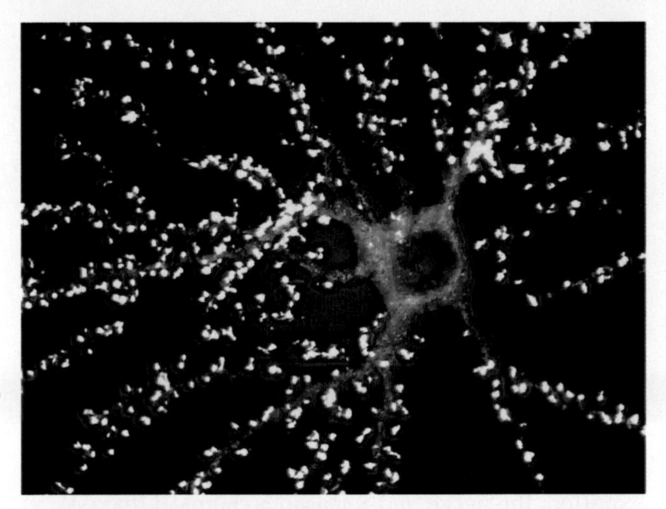

FIGURE 32.10 Synaptic density. This interneuron is stained with a yellow fluorescent dye that indicates the many locations where other neurons synapse on this cell.

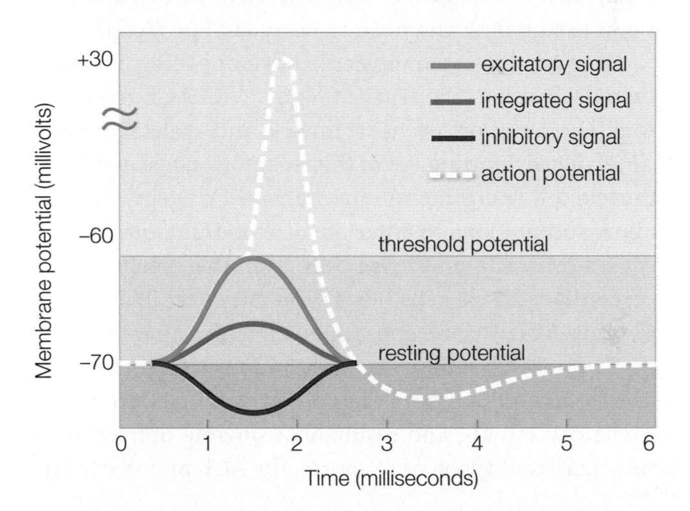

FIGURE 32.11 Synaptic integration. All excitatory and inhibitory signals arriving at a postsynaptic neuron's input zone at the same time are summed. The colored lines on the graph show the postsynaptic cell's response to an excitatory signal (green), to an inhibitory signal (red) and to both at once (blue). In this example, the excitatory signal would have been sufficient to trigger an action potential (white), but the inhibitory signal cancelled some of its effects.

FIGURE IT OUT Which colored line would you expect to see if a neurotransmitter caused an influx of potassium ions into the postsynaptic cell?

Answer: The red line. Influx of potassium drives the neuron away from threshold.

> **TAKE-HOME MESSAGE 32.6**
>
> **How does information pass between cells at a synapse?**
>
> ✔ An action potential travels along an axon to the axon's terminals, where it stimulates exocytosis of vesicles containing a neurotransmitter.
>
> ✔ Neurotransmitters are chemical signals that can have excitatory or inhibitory effects on a postsynaptic cell. Synaptic integration is the summation of all excitatory and inhibitory signals arriving at a postsynaptic cell's input zone at the same time.
>
> ✔ For a synapse to function properly, neurotransmitter in the synaptic cleft must be removed or inactivated after it has served its purpose.

neuromuscular junction Synapse between a motor neuron and the muscle it controls.
neurotransmitter Chemical signal released by axon terminals of a neuron; it binds to receptors on a postsynaptic cell.
synapse Region where a neuron's axon terminals transmit chemical signals to another cell.
synaptic integration The summation of excitatory and inhibitory signals by a postsynaptic cell.

32.7 Neurons' Chemical Signals

✔ There are a variety of neurotransmitters. Neurological disorders and psychoactive drugs interfere with their action.

In the early 1920s, Austrian scientist Otto Loewi was investigating what controls heart rate. He surgically removed a frog heart—with the nerve that controls its rate of beating still attached—and put it in saline solution. The heart continued to beat and, when Loewi stimulated the nerve, heartbeat slowed a bit.

Loewi suspected stimulation of the nerve caused release of a chemical signal. To test this hypothesis, he put two frog hearts into a saline-filled chamber and stimulated the nerve connected to one of them. Both hearts began to beat more slowly. As Loewi had expected, the nerve had released a chemical that not only affected the attached heart, but also diffused through the liquid and slowed the beating of the second heart. Later, another scientist, Henry Dale, identified the signaling chemical as acetylcholine (ACh).

How can the same molecule have opposing effects on muscle cells? The answer begins with ACh receptors. The receptors on heart muscle and skeletal muscle differ. Upon binding ACh, the receptor on a skeletal muscle cell opens membrane transport proteins that allow sodium ions to enter the cell. Sodium ion influx causes contraction in these cells. The ACh receptor on cardiac muscle cells has a different effect. Upon binding ACh, this receptor opens membrane transport proteins that allow potassium ions to exit the cell. Potassium ion efflux has a dampening effect on excitatory signals, and results in a slowing of the heart rate. Different types of receptors for ACh also occur on neurons of the central nervous system, where this neurotransmitter affects mood and memory.

FIGURE 32.12 Battling Parkinson's disease. This neurological disorder affects former heavyweight champion Muhammad Ali, actor Michael J. Fox, and about half a million other people in the United States. Concussion increases risk of the disease.

Table 32.1 Examples of Neurotransmitters and Their Effects

Neurotransmitter	Examples of Effects
Acetylcholine (ACh)	Induces skeletal muscle contraction; slows cardiac muscle contraction; affects mood and memory
Epinephrine and norepinephrine	Speed heart rate; dilate the pupils and airways; slow gut contractions; increase anxiety
Glutamate	Activates neurons; has roles in memory and learning
GABA	Inhibits neurons within the brain; influences motor control and anxiety
Dopamine	Involved in reward-based motivation and learning, motor control
Serotonin	Elevates mood; role in memory

The human body produces many neurotransmitters, each with a different function (**TABLE 32.1**). Norepinephrine and epinephrine (commonly known as adrenaline) are released in response to stress or excitement. You'll learn more about them in the next section.

Glutamate is the main excitatory neurotransmitter in the brain. With the genetic disorder Huntington's disease (Section 14.2), impaired glutamate reuptake causes overexcitation that damages neurons.

GABA (gamma aminobutyric acid) is the main inhibitory neurotransmitter in the brain. Tranquilizers such as Xanax (alprazolam) and Valium (diazepam) alleviate anxiety by enhancing effects of GABA.

Dopamine-secreting neurons in the brain govern motor control. Damage to dopamine-secreting neurons in the area governing motor control results in Parkinson's disease (**FIGURE 32.12**). Tremors are an early symptom. Later, the sense of balance becomes impaired, and voluntary movement, including speech, becomes difficult. Because the symptoms arise from a dopamine shortage, patients can be treated with a drug (levodopa) that the body converts to dopamine.

Dopamine also plays a role in attention and reward-based learning. A lower than normal dopamine level occurs with attention deficit hyperactivity disorder (ADHD). Affected people can have trouble concentrating and controlling impulses. Drugs used to treat

ADHD increase dopamine availability in the brain. For example, Ritalin (methylphenidate) acts by preventing the reuptake of dopamine at a synapse.

Serotonin also acts in the brain, where it influences mood and memory. Low levels of serotonin result in depression—a persistent feeling of sadness and inability to experience pleasure. The most widely prescribed antidepressants, including Prozac (fluoxetine) and Paxil (paroxetine), increase the level of serotonin by blocking transport proteins that return it to axon terminals.

In addition to neurotransmitters, some neurons release **neuromodulators**, molecules that influence the effects of neurotransmitters. One neuromodulator, substance P, enhances pain perception. Neuromodulators called endorphins are natural painkillers secreted by neurons in response to strenuous activity or injury. Endorphins also are released when people laugh, reach orgasm, or get a comforting hug or a relaxing massage.

Chemically Disrupting Signaling

Psychoactive drugs are chemicals that enter the brain and affect signal transmission at synapses. Some, including antidepressants, are taken to restore normal function. Others are taken to alleviate pain, relieve stress, or simply for pleasure (FIGURE 32.13).

Psychoactive drugs operate by a variety of mechanisms. Some mimic natural chemicals. They have a structure similar to a neurotransmitter or neuromodulator so they bind to and activate receptors for these molecules. Nicotine binds to and activates brain receptors for ACh. Narcotic analgesics, such as morphine, codeine, heroin, and oxycodone, bind to and activate receptors for natural painkillers. THC, the active ingredient in marijuana, binds to and activates receptors for neuromodulators called endocannabinoids. By doing so, it indirectly alters levels of dopamine, serotonin, norepinephrine, and GABA.

Binding of a drug to a receptor does not always activate it. For example, caffeine makes us alert by binding to and preventing normal action of receptors for adenosine, a neurotransmitter that causes drowsiness.

Other psychoactive drugs encourage or inhibit the release of neurotransmitter from a presynaptic cell. Ethanol does both. It slows motor response by inhibiting ACh release and makes us feel drowsy by encouraging adenosine release. It also stimulates the release of endorphins and GABA, resulting in a brief euphoria followed by depression. Still other drugs interfere with reuptake of neurotransmitter from the synaptic cleft. For example, cocaine interferes with reuptake of several neurotransmitters, including dopamine.

FIGURE 32.13 Commonly used psychoactive drugs: ethanol in beer, caffeine in coffee, and nicotine in tobacco.

Drug addiction has many causes, but dopamine plays an important role in creating dependency. A surge of dopamine is released in the brain during behaviors such as eating, which enhances survival, or engaging in sex, which enhances reproduction. This release helps individuals learn to repeat beneficial behaviors. Drugs that trigger dopamine release or prevent its reuptake hijack this ancient learning pathway. Drug users inadvertently teach themselves that the drug is essential to their well-being.

Continual use of a drug often results in tolerance, meaning the effectiveness of a given dose of the drug decreases over time. Although two beers can make a nondrinker feel drunk, they will have little effect on an alcoholic. Tolerance arises because repeated administration of the drug alters the user's body. Mechanisms of tolerance vary. Tolerance to a drug that causes the release of a particular neurotransmitter, for example, may arise from decreased synthesis of the neurotransmitter or its receptors. Tolerance can also involve changes in the rate at which neurotransmitter is cleared from the synaptic cleft.

neuromodulator Chemical signal that is released by a neuron and affects the response of neurons to a neurotransmitter.

TAKE-HOME MESSAGE 32.7

What are the effects of neurotransmitters?

✔ Neurotransmitters vary widely in their effects. There are a variety of different types, and the same type may elicit different responses in different cells.

✔ Psychoactive drugs act by influencing communication at synapses. Many are addictive, and continued use of them can alter the function of the user's synapses.

CREDIT: (13) Szasz-Fabian Jozsef/Shutterstock.com.

32.8 The Peripheral Nervous System

✔ Peripheral nerves, which extend throughout the body, carry signals to and from the central nervous system.

Nerve Structure

In humans, the peripheral nervous system includes 31 pairs of spinal nerves that connect to the spinal cord and 12 pairs of cranial nerves that connect directly to the brain. Each peripheral nerve consists of axons of many neurons bundled together inside a connective tissue sheath (**FIGURE 32.14A**). All spinal nerves include axons from both sensory and motor neurons. Cranial nerves may include axons of motor neurons, axons of sensory neurons, or axons of both sensory and motor neurons. Interneurons, remember, are not part of the peripheral nervous system.

In most vertebrate nerves, axons associate with neuroglial cells that wrap them in **myelin**, a fatty material that functions like insulation around a wire. In the peripheral nervous system, Schwann cells myelinate axons (**FIGURE 32.14B**). A myelinated axon has voltage-gated channels for sodium and potassium only at gaps (called nodes) in the myelin wrapping. As a result, disturbances associated with an action potential spread through cytoplasm until they reach a node. When voltage-gated channels at a node open, the voltage difference reverses abruptly. By jumping from node to node, an action potential can move as fast as 120 meters per second in a myelinated axon. In unmyelinated axons, the maximum speed is about 10 meters per second.

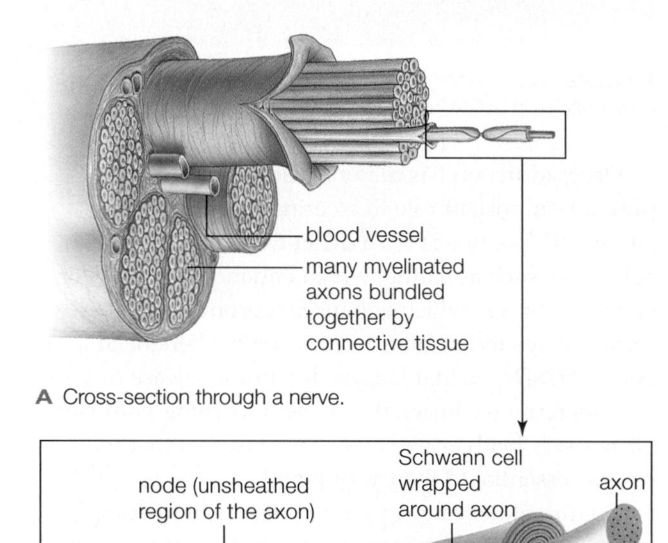

A Cross-section through a nerve.

B Myelinated axon of a peripheral nerve. Neuroglial cells called Schwann cells wrap around the axon one after another, creating a discontinuous covering called a myelin sheath.

FIGURE 32.14 ▸**Animated** Structure of a peripheral nerve.

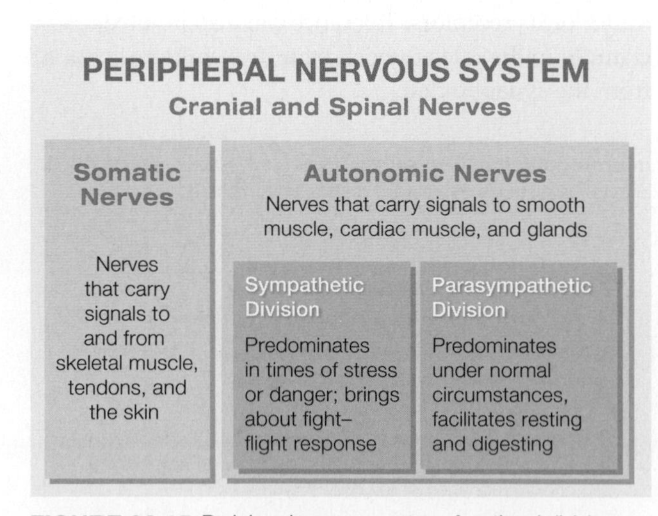

FIGURE 32.15 Peripheral nervous system, functional divisions.

Functional Divisions

There are two functional divisions of the peripheral nervous system (**FIGURE 32.15**). **Somatic nerves** carry signals from sensory neurons that monitor the external environment to the central nervous system, and carries commands for voluntary movements to skeletal muscles. The somatic nervous system allows you both to feel someone tickling your toes and to wiggle your toes in response.

Autonomic nerves relay signals from the central nervous system to smooth muscle, cardiac muscle, and glands. They also relay signals about internal conditions from sensory receptors in internal organs to the central nervous system. Signals that travel along autonomic nerves both adjust your heart rate and tell your brain when your blood pressure drops too low.

Nerves of the autonomic system signal muscles and glands by a two-neuron pathway. The axon of the first neuron (the preganglionic neuron) extends from a cell body in the brain or spinal cord to a ganglion (a cluster of cell bodies) outside the central nervous system. In the ganglion, axon terminals of the preganglionic neuron synapse on postganglionic neurons. Axons of the postganglionic neurons extend out to and synapse on an organ or gland.

There are two categories of autonomic nerves: sympathetic and parasympathetic. Both service most organs and they work antagonistically, meaning the signals from one type oppose signals from the other (**FIGURE 32.16**). **Sympathetic neurons** are most active in times of stress, excitement, and danger. **Parasympathetic neurons** are most active in times of relaxation. Signals

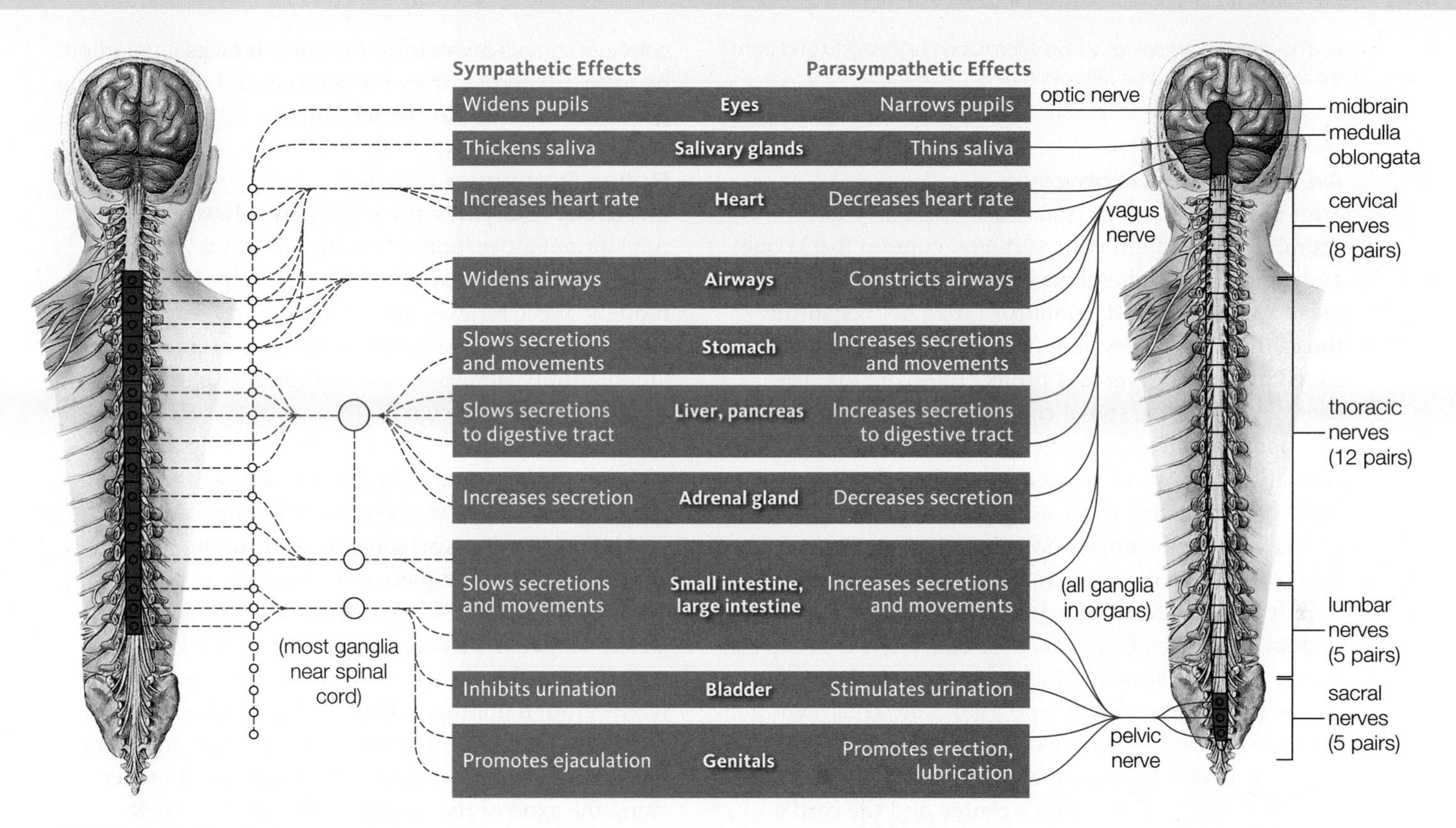

FIGURE 32.16 ▶**Animated** Autonomic nerves and their effects. Autonomic signals travel to organs by a two-neuron path. The first neuron has its cell body in the brain or spinal region (indicated in red in the two body sketches). This neuron synapses on a second neuron at a ganglion. Sympathetic ganglia are close to the spinal cord. Parasympathetic ganglia are in or near the organs they affect. Axons of the second neuron in the pathway synapse with the organ. **FIGURE IT OUT** What effect does stimulation of the vagus nerve (a parasympathetic nerve) have on the heart?

Answer: It decreases heart rate.

from these neurons promote housekeeping tasks such as digestion and urine formation.

When you are startled or frightened, parasympathetic neurons become less active and sympathetic ones become more active. Signals from sympathetic neurons trigger the adrenal glands to secrete epinephrine (adrenaline), which raises your heart rate and blood pressure, and makes you sweat more and breathe faster. These arousal effects prime the body to fight or make a fast getaway—an effect called the **fight–flight response**.

autonomic nerves Nerves that relay signals to and from internal organs and to glands.
fight–flight response Response to danger or excitement. Activity of parasympathetic neurons declines, sympathetic signals increase, and adrenal glands secrete epinephrine.
myelin Lipid-rich material made by some neuroglia; wraps around axons of some neurons; speeds propagation of action potentials.
parasympathetic neurons Neurons of the autonomic system that encourage "housekeeping" tasks in the body.
somatic nerves Nerves that control skeletal muscle and relay signals from sensory neurons in joints and skin to the central nervous system.
sympathetic neurons Neurons of the autonomic system that prepare the body for danger or excitement.

Opposing signals from sympathetic and parasympathetic neurons govern most organs. For instance, both act on smooth muscle cells in the gut wall. As sympathetic neurons release norepinephrine at synapses with these cells, parasympathetic neurons release ACh at other synapses with the same cells. One signal tells the gut to slow its contractions; the other calls for increased activity. Synaptic integration determines the outcome of the conflicting commands.

TAKE-HOME MESSAGE 32.8

What is the peripheral nervous system?

✔ The peripheral nervous system consists of nerves that extend through the body and relay signals to and from the central nervous system.

✔ Somatic nerves of the peripheral system control skeletal muscle and convey information about the external environment to the central nervous system.

✔ Autonomic nerves carries information to and from smooth muscle and cardiac muscle, and to glands. Signals from the two autonomic divisions—sympathetic and parasympathetic—have opposing effects.

32.9 The Spinal Cord

✔ The spinal cord serves as an information highway to and from the brain, and also as a reflex center.
✔ Spinal reflexes do not involve the brain.

An Information Highway

Your **spinal cord**, which is about as thick as your thumb, runs through your vertebral column (backbone) and connects peripheral nerves with the brain (**FIGURE 32.17**). The brain and spinal cord together constitute the central nervous system (CNS). Three membranes called **meninges** cover and protect these organs. The central canal of the spinal cord and the spaces between the meninges are filled with **cerebrospinal fluid**. This clear fluid consists of water and small molecules filtered out of the blood.

The bulk of the brain and spinal cord consists of two tissues: white matter and gray matter. **White matter** is mainly myelin-sheathed axons. In the central nervous system, a bundle of such axons is called a tract, rather than a nerve. **Gray matter** includes neuron cell bodies, dendrites, axon terminals, and glial cells. Thus, synapses of the central nervous system are located within the gray matter. The spinal cord's gray matter fills an H-shaped region in the cord's center, and the cord's white matter surrounds it.

Each spinal nerve has two nerve roots that connect it to the spinal cord. A nerve's dorsal root carries sensory signals to the spinal cord; it has a ganglion with cell bodies of sensory neurons. The nerve's ventral root conveys signals away from the cord. It lacks a ganglion because cell bodies of motor neurons and preganglionic autonomic neurons are in the spinal cord.

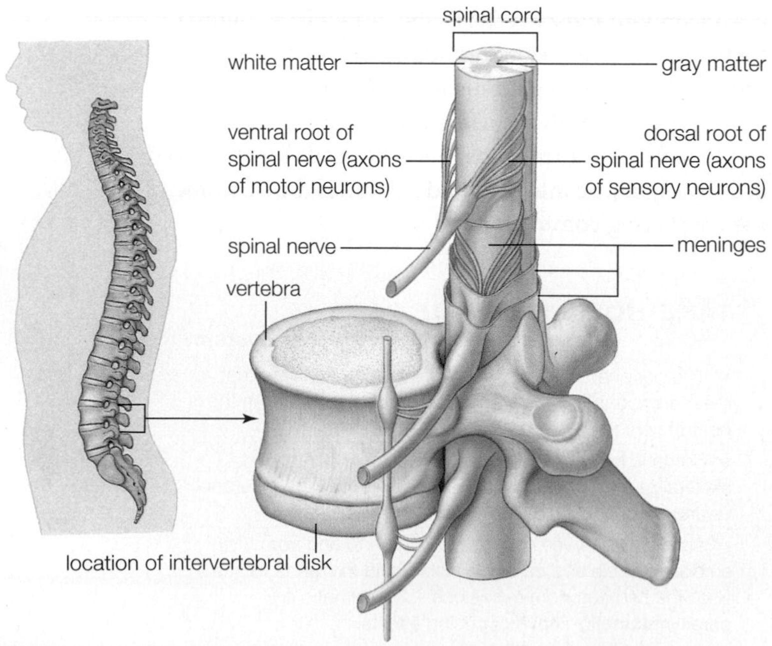

FIGURE 32.17 ▸Animated Location and organization of the spinal cord.

Reflex Pathways

The spinal cord plays a role in many reflexes. A **reflex** is an automatic response to a stimulus; it is a movement or other response that does not require conscious thought. Basic reflexes do not require any learning. They occur when the spinal cord or the brain stem automatically signals muscles to contract in response to a certain sensory signal.

The stretch reflex, which causes a muscle to contract after some force stretches it, is an example of a spinal reflex (**FIGURE 32.18**). Consider what happens when you hold a bowl as someone drops fruit into it ❶. The sudden addition of a piece of fruit increases the weight of the bowl, causing the biceps muscle in your upper arm to lengthen. Passive lengthening of the biceps excites a sensory receptor called a muscle spindle that wraps around a muscle. The muscle spindle is part of a sensory neuron, and its excitation triggers an action potential in that neuron. The action potential moves along the axon of the neuron to the spinal cord ❷.

In the spinal cord, the axon of the sensory neuron synapses on dendrites of one of the motor neurons that controls the biceps ❸. Binding of neurotransmitter released by the sensory neuron to receptors on the motor neuron causes the motor neuron to undergo an action potential. This action potential travels along the motor neuron's axon, which extends from the spinal cord to the biceps ❹. When the action potential reaches the terminal endings of the motor neuron, its arrival causes release ACh at the neuromuscular junction ❺. In response to this neurontransmitter, the biceps contracts and steadies the arm against the additional weight ❻. The entire process occurs without your having to think of it.

Another spinal reflex, the withdrawal reflex, helps prevent burns. Touch a hot surface and action potentials flow to the spinal cord. Unlike the stretch reflex, the withdrawal response involves a spinal interneuron. A heat-detecting sensory neuron alerts the spinal interneuron, which relays the signal to motor neurons. Before you know it, your biceps contracts, pulling your hand away from the potentially damaging heat.

Interrupted Spinal Signaling

An injury that interrupts nerve fibers running through the spinal cord can cause a permanent loss of sensation and paralysis. Symptoms depend on what portion of

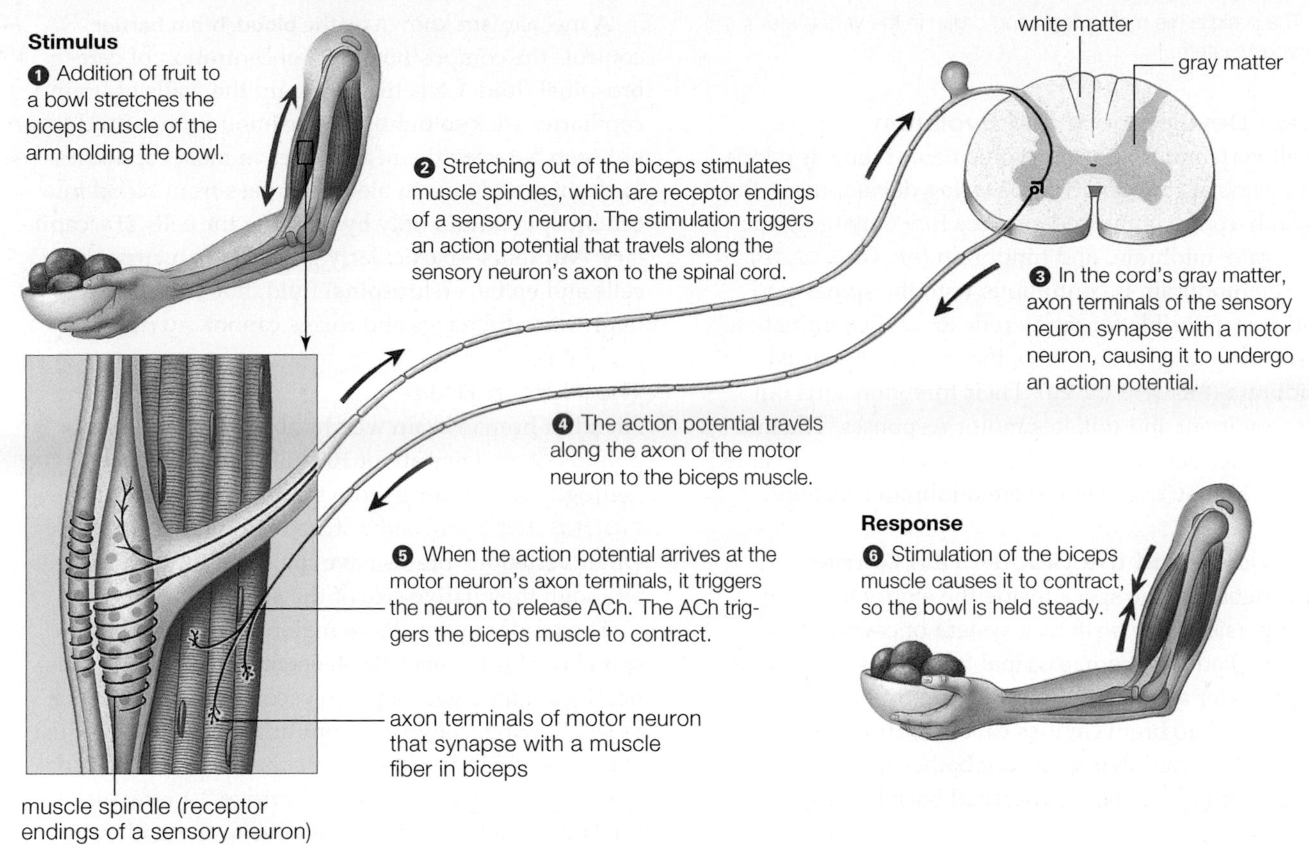

FIGURE 32.18 ▶**Animated** Stretch reflex, a spinal reflex.

FIGURE IT OUT Where is the cell body of the motor neuron that causes the biceps to contract located, in the cord's white matter or gray matter?

Answer: In the cord's gray matter.

the cord is damaged. Nerves carrying signals to and from the upper body are higher in the cord than nerves that service the lower body. An injury to the cord in the lower back often paralyzes the legs. An injury to higher cord regions can paralyze all limbs, as well as muscles involved in breathing. In the United States, more than a million people now live with paralysis sustained as a result of a spinal cord injury.

Doctors sometimes temporarily halt transmission of signals through the spinal cord to provide pain relief to a specific region of the body. Consider the most common way of lessening pain during childbirth—epidural anesthesia. During this procedure, a physician injects painkiller in the space between the meninges and spinal cord. The drug is administered in a region where it will partially numb the body from the waist down, without otherwise impairing perception or interfering with motor function.

cerebrospinal fluid Clear fluid that surrounds the brain and spinal cord and fills cavities within the brain.
gray matter Brain and spinal cord tissue that includes neuron cell bodies, dendrites, and axon terminals, as well as glial cells.
meninges The three membranes that enclose the brain and spinal cord.
reflex Automatic response that occurs without conscious thought or learning.
spinal cord Portion of central nervous system that connects peripheral nerves with the brain.
white matter Tissue of brain and spinal cord that consists of myelinated axons.

TAKE-HOME MESSAGE 32.9

What are the functions of the spinal cord?

✔ Tracts of the spinal cord relay information between peripheral nerves and the brain. Myelinated axons involved in these pathways make up the bulk of the cord's white matter. Synapses occur in the gray matter.

✔ The spinal cord also has a role in some simple reflexes, automatic responses that occur without conscious thought or learning. Signals from sensory neurons enter the cord through the dorsal root of spinal nerves. Commands for responses go out along the ventral root of these nerves.

32.10 The Vertebrate Brain

✔ The brain is the main integrating organ in the vertebrate nervous system.

Brain Development and Evolution

In all vertebrates, the embryonic neural tube develops into a spinal cord and brain. During development, the brain becomes organized as three functional regions: forebrain, midbrain, and hindbrain (**FIGURE 32.19**).

The hindbrain is continuous with the spinal cord and is responsible for many reflexes and coordination. Fishes and amphibians have the most pronounced midbrain (**FIGURE 32.20**). Their forebrain sorts out sensory input and initiates motor responses. The midbrain is reduced in birds and mammals; their expanded forebrain took over what were midbrain functions.

Ventricles and the Blood–Brain Barrier

In vertebrates, the space inside the embryonic neural tube persists after birth as a system of cavities and canals filled with cerebrospinal fluid. This fluid forms when water and small molecules are filtered out of the blood into brain cavities called ventricles (**FIGURE 32.21**). The fluid that seeps out bathes the brain and spinal cord before being absorbed back into the blood.

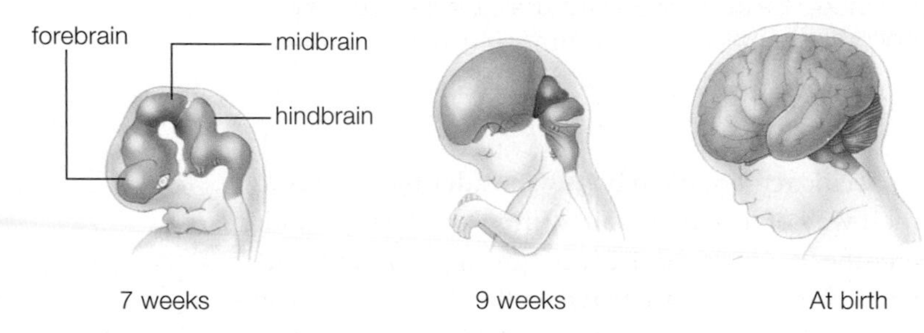

FIGURE 32.19 Human brain development. At 7 weeks, there is a hollow neural tube with regions that will develop into the forebrain, midbrain, and hindbrain.

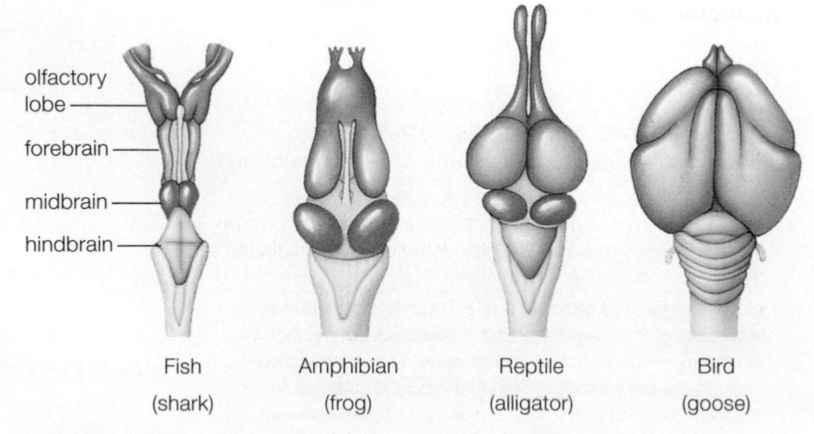

FIGURE 32.20 Vertebrate brains, dorsal views. Sketches are not to scale.

A mechanism known as the **blood–brain barrier** controls the composition and concentration of cerebrospinal fluid. Cells that make up the walls of brain capillaries stick so tightly to one another that fluid cannot seep between them, as it does in most capillaries. Molecules and ions in blood can pass from blood into cerebrospinal fluid only by crossing the cells of a capillary. Nutrients—particularly glucose—can cross these cells and enter cerebrospinal fluid, but wastes and many harmful drugs and toxins cannot.

The Human Brain

An adult human brain weighs about 1,400 grams, or 3 pounds. It contains about 100 billion interneurons, with neuroglia constituting more than half of its volume. **FIGURE 32.22** shows the structure of a human brain. Other vertebrate brains have the same functional areas, although the relative size of the areas varies.

The hindbrain has three regions. Connected to the spinal cord is the **medulla oblongata**, which influences heartbeat and breathing. It also controls reflexes such as swallowing, coughing, vomiting, and sneezing. Just above the medulla oblongata is the **pons**, which also affects breathing. Pons means "bridge," a reference to tracts that extend through the pons to the midbrain and functionally connect these regions. The plum-sized **cerebellum** at the rear of the skull controls posture, coordinates voluntary movements, and plays a role in learning new motor skills. The cerebellum is densely packed with neurons, having as many as all other brain regions combined.

The pons, medulla, and midbrain are collectively referred to as the **brain stem**. These are the most evolutionarily ancient parts of the brain. In humans, the midbrain is the smallest of these three brain regions. It has roles in reward-based learning and in voluntary movement. Parkinson's disease arises as a result of damage to or death of dopamine-producing neurons in one region of the midbrain.

The forebrain contains the **cerebrum**, the largest region of the human brain. The cerebrum receives sensory signals, integrates them, and initiates skeletal muscle movements. A fissure divides it into right and left hemispheres connected by a thick band of tissue (the corpus callosum) that relays signals between them. Each hemisphere has a cerebral cortex, an outer layer of gray matter. The cerebral cortex is responsible for memory, emotions, language, and abstract thought (more about this in the next section).

The **thalamus** is a two-lobed structure that sorts sensory signals and sends them to the proper region of the cerebral cortex. It also influences sleep and wakefulness. In the rare genetic disorder called fatal familial insomnia, deterioration of the thalamus causes an inability to sleep, followed by a coma, then death.

The **hypothalamus** ("under the thalamus") is the center for homeostatic control. It receives signals that convey information about the state of the body, and responds by sending out signals that affect thirst, appetite, sex drive, and body temperature. It also interacts with the adjacent pituitary gland as a control center for the endocrine system. We discuss endocrine functions of the hypothalamus in Chapter 34.

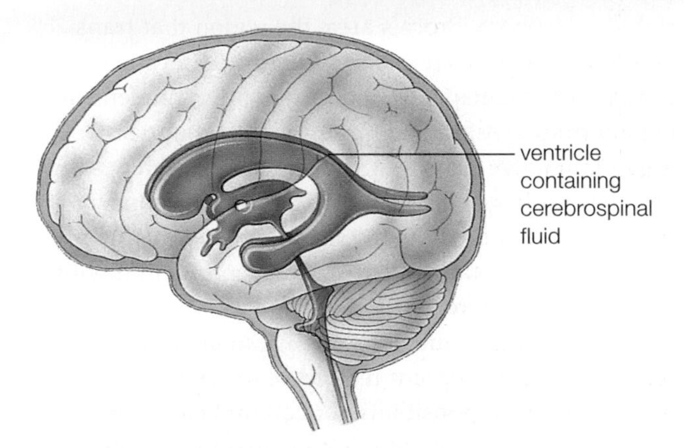

FIGURE 32.21 Cerebrospinal fluid. This clear fluid, shown here in blue, forms inside ventricles (chambers) within the brain.

blood–brain barrier Protective mechanism that controls the composition and concentration of cerebrospinal fluid.
brain stem The most evolutionarily ancient region of the vertebrate brain; includes the pons, medulla, and midbrain.
cerebellum Hindbrain region responsible for posture and for coordinating voluntary movements.
cerebrum Forebrain region that acts in language, abstract thought.
hypothalamus Forebrain region that controls processes related to homeostasis; control center for endocrine functions.
medulla oblongata Hindbrain region that controls breathing rhythm and reflexes such as coughing and vomiting.
pons Hindbrain region that influences breathing and serves as a functional bridge to the adjacent midbrain.
thalamus Forebrain region that relays signals to the cerebral cortex.

TAKE-HOME MESSAGE 32.10

How does the vertebrate brain develop, and what are its functional regions?

✔ The vertebrate brain develops from a hollow neural tube, the interior of which persists in adults as a system of cavities and canals filled with cerebrospinal fluid.

✔ Tissue of the embryonic neural tube develops into the hindbrain, forebrain, and midbrain. The hindbrain controls reflexes and coordination. The unique capacities of humans arise in regions of their enlarged forebrain.

FIGURE 32.22 ▶Animated Anatomy of the human brain.

A Right, view of a normal brain from above, showing the two cerebral hemispheres divided by a deep fissure. The meninges have been removed from the right hemisphere.

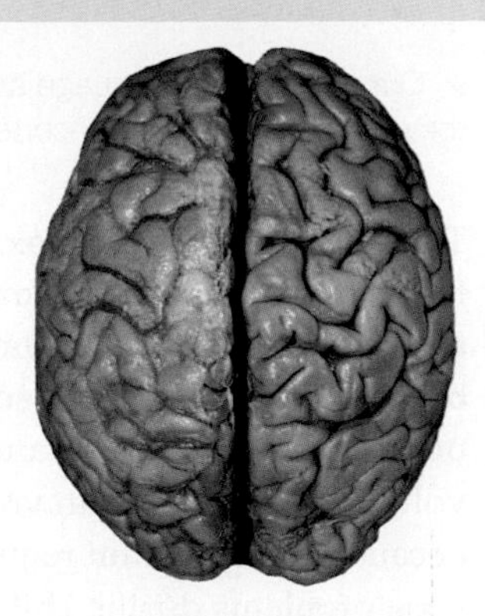

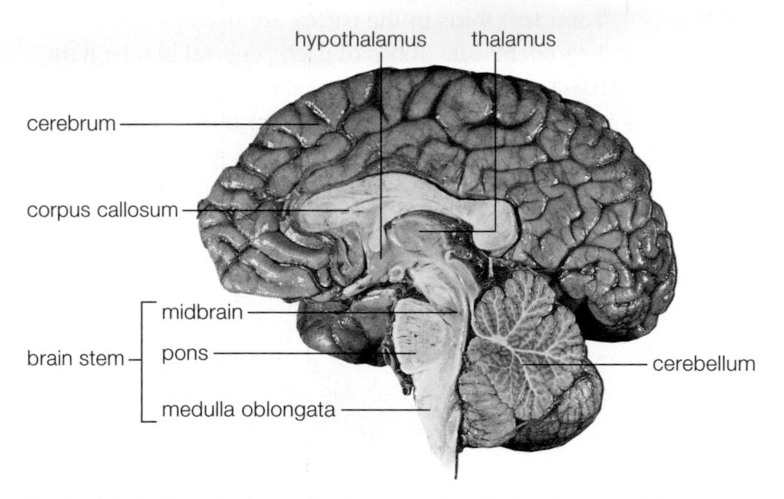

B The right half of a brain that has been sectioned along the central fissure.

Forebrain	
Cerebrum	Localizes, processes sensory inputs; initiates, controls skeletal muscle activity; governs memory, emotions, and, in humans, abstract thought
Thalamus	Relays sensory signals to and from cerebral cortex; has a role in memory
Hypothalamus	Functions in homeostatic control. Adjusts volume, composition, temperature of internal environment; governs behaviors that ensure homeostasis (e.g., thirst, hunger)
Midbrain	Relays sensory input to forebrain, affects coordination
Hindbrain	
Pons	Bridges cerebrum and cerebellum; also connects spinal cord with midbrain. With the medulla oblongata, controls rate and depth of respiration
Cerebellum	Coordinates motor activity for moving limbs and maintaining posture, and for spatial orientation
Medulla oblongata	Relays signals between spinal cord and pons; functions in reflexes that affect heart rate, blood vessel diameter, and respiratory rate. Also involved in vomiting, coughing, other reflexive functions

C Major components of the brain and their functions.

CREDITS: (21) © Cengage Learning; (22A) Living Art Enterprises/Science Source; (22B) C. Yokochi and J. Rohen, Photographic Anatomy of the Human Body, 2nd Ed., Igaku-Shoin, Ltd., 1979; (22C) © Cengage Learning.

32.11 The Human Cerebral Cortex

✔ Our capacity for language and abstract reasoning arises from the activity of the cerebral cortex.

The human **cerebral cortex**, the outermost portion of the cerebrum, is a 2-millimeter-thick layer of gray matter. Over the course of human evolution, this layer has become increasingly folded. Folds allowed expansion of our gray matter with a minimal increase in brain volume. (Keeping brain volume in check is important because a larger brain requires a larger skull, which can pose problems during childbirth.)

Prominent folds in the cortex are used as landmarks to define the lobes of each cerebral hemisphere (**FIGURE 32.23**). The cortex of the frontal lobe allows us to make reasoned choices, concentrate on tasks, plan for the future, and behave appropriately in social situations. During the 1950s, the frontal lobes of more than 20,000 people were deliberately damaged in a surgical procedure called frontal lobotomy. Lobotomy was used to treat mental illness, personality disorders, and even headaches. The procedure made patients calmer, but also blunted their emotions and impaired their ability to plan, concentrate, and behave appropriately.

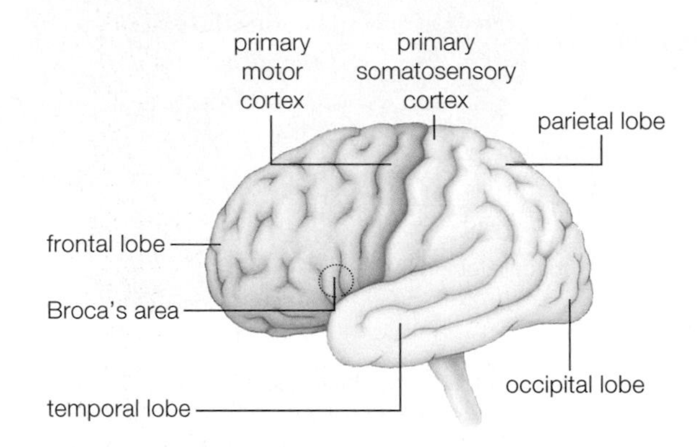

FIGURE 32.23 Four lobes of the cerebrum, indicated by color, and the location of some functional regions of the cortex.

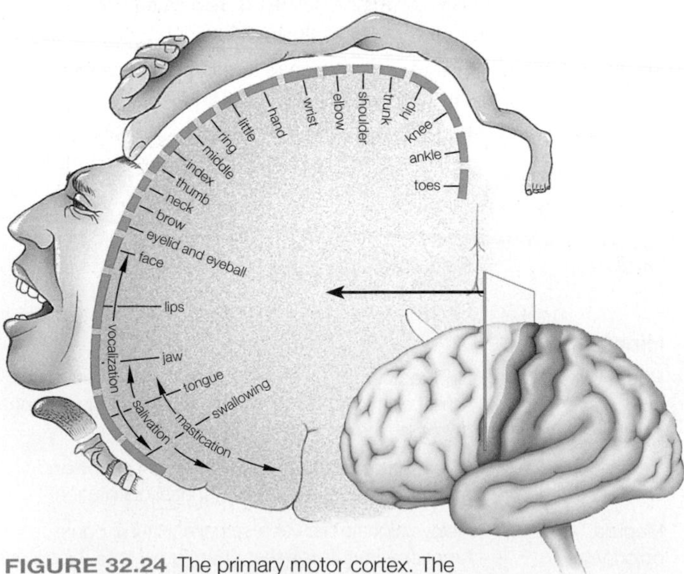

FIGURE 32.24 The primary motor cortex. The graphic shows a slice through this region with body parts draped over the part of the cortex that controls them. The most closely controlled body parts appear disproportionately large.

The two cerebral hemispheres differ somewhat in their function. In most people, the cortex of the left hemisphere plays the major role in movement and in language, whereas mathematical tasks, such as adding two numbers, typically activate cortical areas of the right hemisphere. Broca's area, the region that translates thoughts into speech, is in the cortex of the left frontal lobe. Damage to Broca's area often prevents a person from speaking, but does not affect understanding of language. Note that, despite their differences, both hemispheres are flexible. If a stroke or injury damages one side of the brain, the other hemisphere can take on new tasks. People can even function with a single hemisphere.

Although most functions involve many brain regions, we can pinpoint regions of the cortex that have primary responsibility for certain tasks. The main region involved in voluntary movement is the **primary motor cortex** at the rear of each frontal lobe. Neurons of this region are laid out like a map of the body, with body parts capable of fine movements taking up the greatest area (**FIGURE 32.24**). At the front of the parietal lobe is the **primary somatosensory cortex**, which receives sensory input from the skin and joints. A primary visual cortex at the rear of each occipital lobe receives signals from eyes. A primary auditory cortex in the temporal lobe acts in hearing.

Nearly all sensory and motor signals cross over on their way to and from the brain. Thus, commands to move parts on the body's left side originate in the right hemisphere, and visual signals from the left eye end up in the right hemisphere's occipital lobe.

cerebral cortex Outer gray matter layer of the cerebrum.
primary motor cortex Region of brain's frontal lobes that governs voluntary movements.
primary somatosensory cortex Region of the brain's parietal lobes that receives sensory information from skin and joints.

TAKE-HOME MESSAGE 32.11

What are the functions of the cerebral cortex?

✔ The cerebral cortex functions in motor activity, sensory perception, reasoning, and language and speech.

32.12 Emotion and Memory

✔ A collection of structures in the midbrain plays an important role in emotion and memory.

The Emotional Brain

The **limbic system** is a collection structures that encircle the upper part of the brain stem (**FIGURE 32.25**). This system governs emotions, assists in memory, and correlates organ activities with self-gratifying behavior such as eating and sex. The limbic system is sometimes described as our emotional-visceral brain, to contrast it with the cerebral cortex, which can often override the limbic system's "gut reactions."

Exactly how the structures of the limbic system give rise to different emotions is poorly understood, but we do know a bit about how its components function. For example, the hypothalamus summons up the physiological changes that accompany emotions. Signals from the hypothalamus make our heart pound and our palms sweat when we are fearful. The cingulate gyrus helps us suppress negative emotions. The adjacent, almond-shaped amygdala becomes active when we are fearful or perceive fear in others. The amygdala is often overactive in people with panic disorders.

Making Memories

At a cellular level, memory formation involves altering the number and placement of synaptic connections among existing neurons. The cerebral cortex receives information continually, but only a fraction of it is stored as a memory. Memory forms in stages. Short-term memory lasts seconds to hours. This type of memory holds a few bits of information: a set of numbers, words of a sentence, and so forth. In long-term memory, larger chunks of information can be stored more or less permanently.

Different types of memories are stored and brought to mind by different mechanisms. Repetition of motor tasks creates skill memories that can be highly persistent. Learn to ride a bicycle, dribble a basketball, or play the piano, and you are unlikely to lose that skill. Forming skill memories involves neurons in the cerebellum, which coordinates motor activity. By contrast, declarative memory stores facts and impressions. It helps you remember how a lemon smells, that a quarter is worth more than a dime, where your classes are held, and how to find you way home.

hippocampus Brain region essential to formation of declarative memories.
limbic system Group of structures deep in the brain that function in expression of emotion.

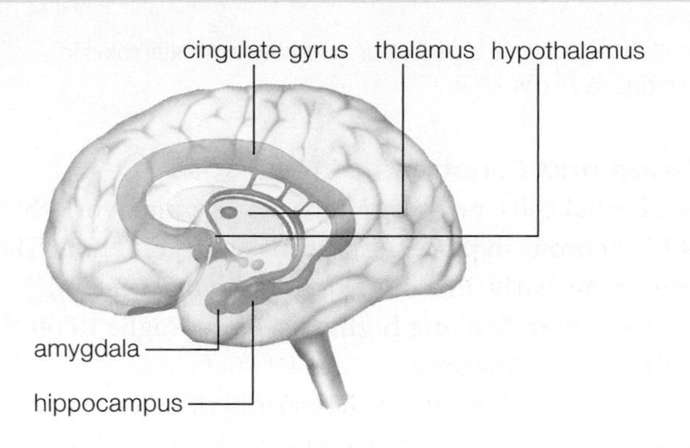

FIGURE 32.25 Components of the limbic system.

The **hippocampus** (plural, hippocampi), a structure adjacent to the amygdala, plays an essential role in the formation of declarative memories. Interactions between the amygdala and the hippocampus give rise to learned fears. The role of the hippocampus in memory was first discovered in the 1950s after a man known as HM had both his hippocampi surgically removed to treat his seizures. The surgery did alleviate HM's seizures, but also destroyed his ability to form new memories. Five minutes after meeting a person, HM was unable to remember that they had ever met. He still retained memories of pre-surgery events, indicating that the hippocampus is not the site for long-term memory storage. Additional evidence of the hippocampus's role in memory comes from people with Alzheimer's disease. The hippocampus is one of the first brain regions destroyed by this disorder. Like HM, people in early-stage Alzheimer's usually have impaired short-term memory, but retain their memories of long-ago events.

The hippocampus is one of the very few brain regions where new neurons continue to arise during adulthood. It it is not yet clear how producing new neurons assists in memory formation. By one hypothesis, continued neuron turnover helps make new memories by disrupting connections among existing cells.

TAKE-HOME MESSAGE 32.12

What is the cerebral cortex?

✔ The brain structures collectively described as the limbic system affect emotions and contribute to memory.

✔ Forming memories involves alterations to synaptic connections. Skill memory and declarative memory involve different regions of the brain. The hippocampus is essential to formation of new declarative memories.

32.13 Neuroglia—The Neurons' Support Staff

✔ A variety of cells called neuroglia play essential roles in nervous systems.

Types and Functions of Neuroglia

Neuroglial cells, or neuroglia, act as a framework that holds neurons in place; *glia* means "glue" in Latin. Their role begins early. In a developing nervous system, neurons migrate along highways of neuroglia to reach their final destination.

The main neuroglia of the peripheral nervous system are Schwann cells. As you have learned, these cells produce the myelin that wraps around and insulates peripheral nerves.

In the central nervous system, neuroglia outnumber neurons 10 to 1. An adult brain has four main types of neuroglial cells: oligodendrocytes, ependymal cells, microglia, and astrocytes.

Oligodendrocytes are the functional equivalent of Schwann cells; they make myelin sheaths that insulate axons in the central nervous system. However, the two types of cells differ in some respects. Each Schwann cell services a single axon, whereas an oligodendrocyte can myelinate multiple axons. The oligodendroctye has an octopus-like structure, with a central cell body from which long, tentacle-like cytoplasmic processes extend out to wrap around axons (**FIGURE 32.26**). If the axon of a peripheral nerve is cut, Schwann cells help it regrow. Oligodendrocytes, on the other hand, do not promote regrowth of damaged axons. This is one reason why axons of a severed finger can grow back, but axons cut by a spinal cord injury cannot.

FIGURE 32.26 Oligogodendrocyte. These neuroglia cells provide the myelin wrapping for axons in the brain and spinal cord.

Multiple sclerosis (MS) is an autoimmune disorder that arises when white blood cells mistakenly attack and destroy the myelin sheaths of oligodendrocytes. Noninsulating scar tissue replaces myelin in the brain and spinal cord. As this happens, the speed of action potential propagation along affected axons declines. The result is progressive weakness and fatigue, impaired balance, and vision problems. MS cannot be cured, although some treatments slow myelin loss.

MS typically appears in young adults, with women affected twice as often as men. The trigger for MS is unknown, although genetics certainly plays a role. In the general population, the risk of developing MS is about 1 in 1,000, but a person whose identical twin has MS has a risk of 1 in 4. The fact that one twin can be affected while the other is not indicates that environmental factors also influence the disease.

Ependymal cells line the brain's fluid-filled cavities (ventricles) and the spinal cord's central canal. Waving cilia on the surface of ependymal cells keep cerebrospinal fluid flowing in a consistent direction.

Microglia are, as the name implies, the smallest neuroglial cells. They continually survey the brain. If brain tissue is injured or infected, microglia become active; they move about, engulfing dead cells and cellular debris. They also produce chemical signals that alert the immune system to threats.

Star-shaped astrocytes are the most abundant cells in the brain. Scientists have only recently begun to understand their diverse and essential roles. Astrocytes that wrap around blood vessels supplying the brain stimulate formation of the blood–brain barrier. Astrocytes also play a role at synapses, where they take up neurotransmitters released by neurons. In addition, they assist in immune defense, release lactate that fuels the activity of neurons, and synthesize nerve growth factor. Neurons are stopped in G1 of the cell cycle (Section 11.2) and cannot divide, but nerve growth factor encourages a neuron to form new synapses.

Brain Tumors

Because neurons are mature, differentiated cells that cannot divide, they do not give rise to tumors. However, uncontrolled division of cells that give rise to neuroglial cells sometimes causes a tumor called a glioma. Tumors can also arise from epithelial cells in the meninges or endocrine glands of the brain, such as the pituitary. In addition, arrival of metastatic cancer cells from elsewhere in the body can cause a tumor.

CREDIT: (26) 3D Clinic/Getty Images.

Impacts of Concussions (revisited)

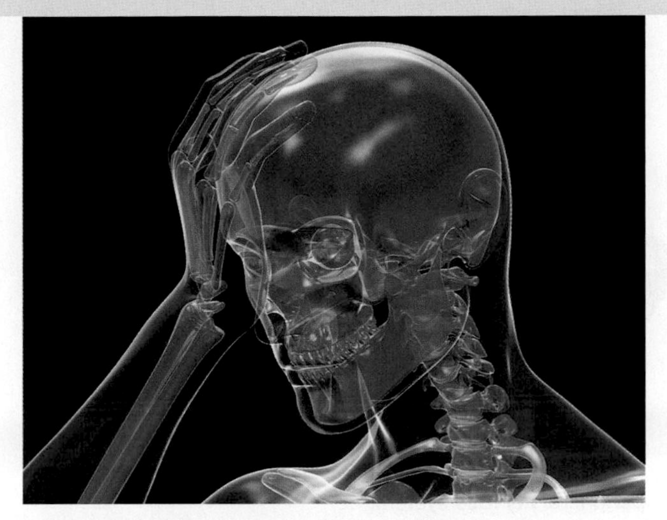

The brain fits pretty tightly into the skull, but there is a small cerebrospinal fluid-filled gap between the two. Thus, the brain can move a bit within the skull. An impact or sudden stop can cause enough movement to tear, bruise, and stretch brain tissue, causing concussion. Axons become twisted and torn, impairing their ability to signal. Tiny blood vessels are torn too, compromising the blood–brain barrier and reducing blood flow to the brain.

The mechanical stress also results in spontaneous action potentials, causing neurotransmitter from the brain's neurons to flood the cerebrospinal fluid. Glutamate, the most abundant neurotransmitter in the brain, has an excitatory effect on neurons. When it pours into the cerebrospinal fluid, the brain's neurons undergo additional spontaneous action potentials. Collectively, these events disrupt normal ion concentration gradients across neuron plasma membranes.

Energy metabolism is also disrupted. In response to alteration of the normal concentration gradients, sodium–potassium pumps and other active transport pumps go to work overtime, and fueling their activity uses up a neuron's ATP. Even under ordinary circumstances, a neuron's mitochondria operate near their peak capacity, so they cannot increase their ATP output to help out. Thus, to meet the increased demand for ATP, the neuron must step up its rate of glycolysis and lactate fermentation. The resulting accumulation of lactate (an acid) further impairs neuron function.

With all of these problems, it is no surprise that the brain does not function normally for a period after a concussion. There is no treatment for concussion, other than physical and mental rest. In most cases, the brain heals itself within about 10 days. During recovery, it is especially important to avoid additional head trauma. Even a seemingly slight injury to the healing brain can result in increased swelling that can lead to permanent paralysis and, in some cases, death. The threat of this "second impact syndrome" makes it all the more important that an initial concussion be properly diagnosed. If a concussion is suspected, a person should halt physical activities and see a doctor as soon as possible.

Most tumors that originate in the brain are not cancer. However, even a benign (noncancerous) tumor can pose a serious threat. Benign tumors do not metastasize (Section 11.6), but growth of a tumor within the confined space of the skull can put pressure on surrounding nervous tissue and damage neurons.

Some people are concerned that the radio waves emitted by cell phones could cause brain tumors. Epidemiological studies have not shown any increased brain tumor incidence related to cell phone use. However, cell phones are a relatively recent invention, their use has increased since they first became available, and brain tumors can take years to develop. Thus some doctors recommend the use of a headset to keep the radiation-emitting part of the phone from coming in close proximity to the brain.

TAKE-HOME MESSAGE 32.13

What are the functions of neuroglia and how do they affect health?

✔ Neuroglial cells make up the bulk of the brain. They provide a framework for neurons, insulate neuron axons, assist neurons metabolically, and protect the brain from injury and disease.

✔ Because neuroglia have essential roles in assisting neurons, diseases that impair neuroglia impair the function of the nervous system.

✔ Unlike neurons, most types of neuroglia continue to divide. Thus, neuroglia can be a source of brain tumors.

summary

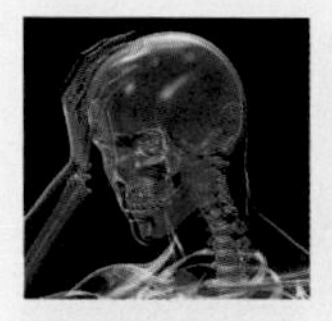

Section 32.1 A concussion is a type of traumatic brain injury that typically disrupts brain function for about 10 days. Repeated blows to the head can cause irreversible damage and result in chronic traumatic encephalopathy (CTE).

Section 32.2 Neurons are electrically excitable cells that signal other cells by means of chemicals. Cnidarians have a **nerve net**. Most other animals have a bilateral nervous system with **cephalization**, which means they have paired **ganglia** or a **brain** at the head end. Chordates have a dorsal **nerve cord**. A vertebrate **central nervous system** (CNS) consists of the brain and spinal cord. **Nerves** that run through the body and connect to the CNS constitute the **peripheral nervous system**.

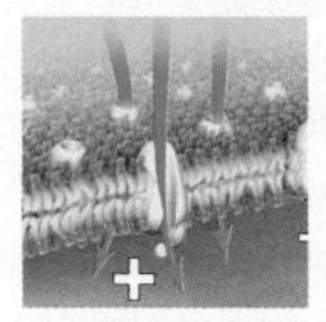

Sections 32.3–32.5 **Sensory neurons** detect stimuli. **Interneurons** relay signals between neurons. **Motor neurons** signal muscles and glands. A neuron's **dendrites** receive signals and its **axon** transmits signals. **Membrane potential** is the voltage across a plasma membrane. At **resting potential**, the interior of the neuron is more negative than interstitial fluid. An **action potential** occurs when a region of membrane reaches **threshold potential**, causing voltage-gated sodium channels to open. Inward flow of sodium causes more of these gates to open—an example of **positive feedback**. The resulting change in potential causes voltage-gated potassium channels to open, allowing potassium to exit the neuron. The outward flow of potassium and diffusion of ions within the axon restores resting potential. All action potentials are the same size and travel away from the cell body and toward the axon terminals.

Section 32.6 Neurons send chemical signals to cells at **synapses**. A motor neuron communicates with a muscle fiber at a type of synapse called a **neuromuscular junction**. Arrival of an action potential at a presynaptic cell's axon terminal triggers release of a **neurotransmitter**. Neurotransmitter diffuses to receptors on a postsynaptic cell and binds to them. A postsynaptic cell often receives signals from many presynaptic cells; its response is determined by **synaptic integration** of all of these signals.

Section 32.7 Different kinds of neurons produce different neurotransmitters. Receptors for a particular neurotransmitter also vary, so the same neurotransmitter can elicit different effects in different cells. Some neurons produce **neuromodulators** that can alter the effects of neurotransmitters.

Some neurological disorders involve production of too little or too much of a neurotransmitter. Psychoactive drugs interfere with signaling at synapses. Dopamine plays a role in reward-based learning, and drug addiction taps into this learning pathway.

Section 32.8 Nerves are bundles of axons wrapped in connective tissue. Most axons have associated neural glial cells that wrap them in a discontinuous sheath of **myelin**. This insulating material increases the speed of action potential propagation.

The peripheral nervous system is functionally divided into **somatic nerves**, which control skeletal muscles, and **autonomic nerves**, which control internal organs and glands.

There are two types of neurons in the autonomic system. In times of stress or danger, **sympathetic neurons** are the most active, as when they cause a **fight–flight response**. During less stressful times, **parasympathetic neurons** are more active. Organs receive signals from both types of autonomic neurons.

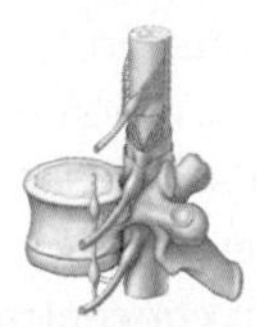

Section 32.9 Like the brain, the **spinal cord** consists of **white matter** (myelinated axons) and **gray matter** (cell bodies, dendrites, axon terminals, and neuroglia). The spinal cord and brain are enclosed by **meninges** and cushioned by **cerebrospinal fluid**. Spinal reflexes involve peripheral nerves and the spinal cord. A **reflex** is an automatic response to stimulation; it does not require conscious thought or learning.

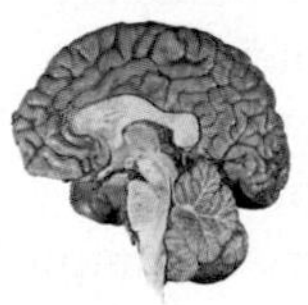

Sections 32.10–32.12 A vertebrate embryo's neural tube develops into the spinal cord and brain. Evolutionarily, the **brain stem** is the oldest brain tissue. It includes the **pons** and **medulla oblongata**, which control reflexes involved in breathing and other essential tasks. The **cerebellum** acts in motor control. The **thalamus** and **hypothalamus** function in homeostasis. A **blood–brain barrier** protects the brain from many harmful chemicals.

The corpus callosum connects the two halves of the **cerebrum**. The **cerebral cortex**, the most recently evolved part of the brain, governs complex functions. The **primary somatosensory cortex** receives sensory input from the skin and joints. The **primary motor cortex** controls voluntary movement. The **limbic system** governs emotion and its **hippocampus** is essential to memory.

Section 32.13 Neuroglial cells make up the bulk of the brain. They insulate axons of the CNS, produce growth factors that encourage synapse formation, contribute to the blood–brain barrier, and carry out other supportive tasks. Uncontrolled divisions of neuroglia can cause brain tumors, and damage to some neuroglia causes neurological disorders such as multiple sclerosis.

self-quiz

Answers in Appendix VII

1. ________ relay messages from the brain and spinal cord to muscles and glands.
   a. Motor neurons
   b. Sensory neurons
   c. Interneurons
   d. Neuroglia

data analysis activities

Effect of MDMA on Neural Development Animal studies are often used to assess effects of prenatal exposure to illicit drugs. For example, Jack Lipton used rats to study the behavioral effect of prenatal exposure to MDMA, the active ingredient in the drug commonly called Ecstasy. He injected female rats with either MDMA or saline solution when they were 14 to 20 days pregnant. This is the period when their offsprings' brains were forming. When those offspring were 21 days old, Lipton tested their ability to adjust to a new environment. He placed each young rat in a new cage and used a photobeam system to record how much each rat moved around before settling down. FIGURE 32.27 shows his results.

1. Which rats moved around most (caused the most photobeam breaks) during the first 5 minutes in a new cage, those prenatally exposed to MDMA or the controls?

2. How many photobeam breaks did the MDMA-exposed rats make during their second 5 minutes in the new cage?

3. Which rats moved around most during the last 5 minutes?

4. Does this study support the hypothesis that MDMA affects a developing rat's brain?

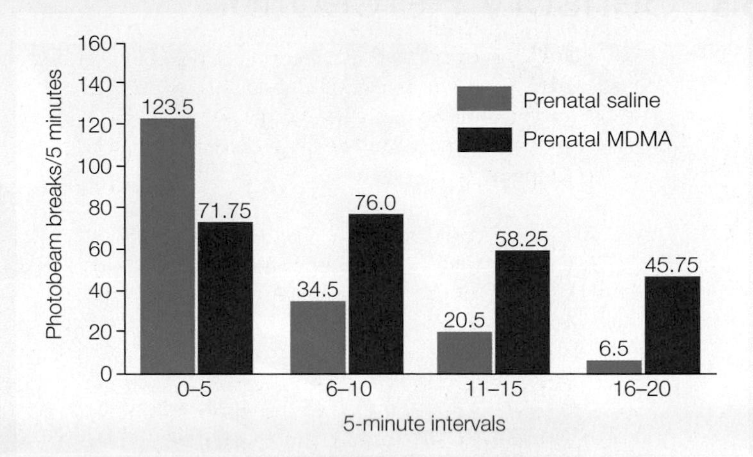

FIGURE 32.27 Effect of prenatal exposure to MDMA on activity levels of 21-day-old rats placed in a new cage. Movements were detected when the rat interrupted a photobeam. Rats were monitored at 5-minute intervals for a total of 20 minutes. Blue bars are average numbers of photobeam breaks for rats whose mothers received saline; red bars are for rats whose mothers received MDMA.

2. When a neuron is at rest, ________ .
   a. it is at threshold potential
   b. voltage-gated sodium channels are open
   c. sodium–potassium pumps are operating
   d. both a and c

3. Action potentials occur when ________ .
   a. a neuron receives adequate stimulation
   b. more and more sodium gates open
   c. sodium–potassium pumps kick into action
   d. both a and b

4. True or false? Action potentials vary in their size.

5. Neurotransmitters are released by ________ .
   a. axon terminals c. dendrites
   b. a neuron cell body d. the myelin sheath

6. What chemical is released by axon terminals of a motor neuron at a neuromuscular junction?
   a. ACh b. serotonin c. dopamine d. epinephrine

7. Which neurotransmitter is important in reward-based learning and drug addiction?
   a. ACh b. serotonin c. dopamine d. epinephrine

8. Skeletal muscles are controlled by ________ .
   a. sympathetic neurons c. somatic nerves
   b. parasympathetic neurons d. both a and b

9. When you sit quietly on the couch and read, output from ________ neurons prevails.
   a. sympathetic b. parasympathetic

10. Dorsal root ganglia contain ________ of sensory neurons.
   a. cell bodies c. dendrites
   b. axon terminals d. all of the above

11. Which of the following are not in the brain?
   a. Schwann cells b. astrocytes c. microglia

12. Neurons are arrested in ________ of the cell cycle.
   a. G1 b. anaphase c. prophase d. S phase

13. ________ plays a role in emotion and memory.
   a. The brain stem c. The limbic system
   b. Broca's area d. The somatosensory cortex

14. Commands to move your right arm start in the ________.
   a. left frontal lobe c. right temporal lobe
   b. right occipital lobe d. left parietal lobe

15. Match each item with its description.

___ gray matter	a. start of brain, spinal cord
___ neurotransmitter	b. connects the hemispheres
___ pons	c. protects brain and spinal cord from some toxins
___ corpus callosum	d. type of signaling molecule
___ cerebral cortex	e. brain's myelin makers
___ neural tube	f. brain stem structure
___ oligodendrocytes	g. controls language, reasoning
___ blood–brain barrier	h. cell bodies and dendrites

critical thinking

1. Some survivors of disastrous events develop post-traumatic stress disorder (PTSD). Symptoms include nightmares about the experience and suddenly feeling as if the event is recurring. Brain-imaging studies of people with PTSD showed that their hippocampus was shrunken and their amygdala unusually active. Given these changes, what other brain functions might be disrupted in PTSD?

2. In human newborns, especially premature ones, the blood–brain barrier is not yet fully developed. Why is this one reason to pay careful attention to the diet of infants?

33 Sensory Perception

LEARNING ROADMAP

This chapter explains how the cone snail venom discussed in Section 24.1 prevents pain. It draws on your knowledge of action potentials (32.5), neuromodulators (32.7), and the human brain (32.10–32.12). We also refer back to morphological convergence (18.3) and other phenomena of vertebrate evolution (25.6).

HOW SENSORY PATHWAYS WORK

Information from sensory receptors becomes encoded in the number and frequency of action potentials sent to the brain along particular nerve pathways.

SOMATIC AND VISCERAL SENSES

Somatic sensations are easily localized to skin or joints. Visceral sensations arise in soft organs and are less easily pinpointed.

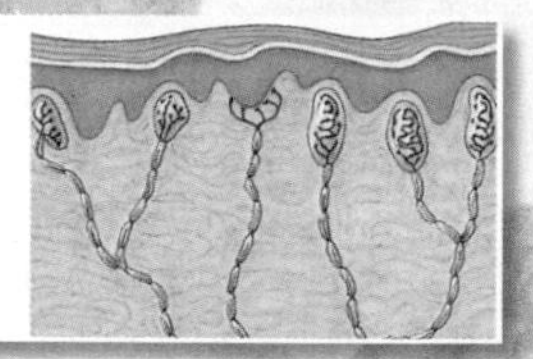

CHEMICAL SENSES

Sensations of smell and taste arise when chemoreceptors bathed in fluid bind molecules of specific substances in the fluid.

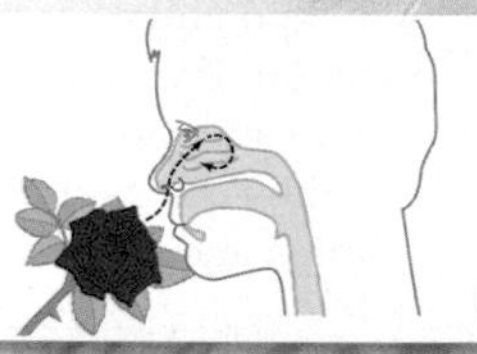

VISION

Vision begins with the activation of photoreceptors. Vertebrates have an eye that operates like a film camera. Their sensory pathway starts at the retina and ends in the visual cortex.

HEARING AND BALANCE

Mechanoreceptors in the ear function in hearing and balance. Sound waves excite receptors in the ear's cochlea, and body movements excite receptors in the vestibular apparatus.

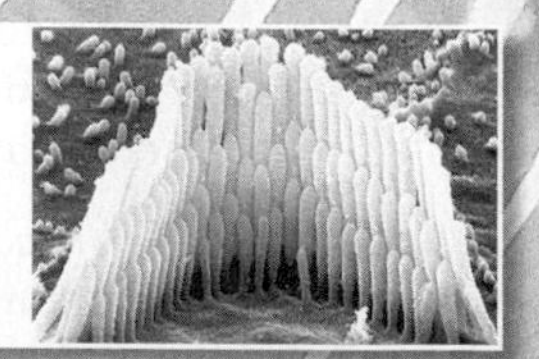

We return to the function of sensory receptors in later chapters. Section 40.5 describes the role of osmoreceptors in thirst, Section 36.7 discusses the role of pressure receptors in maintaining blood pressure, and Section 40.9 discusses the role of thermoreceptors in thermoregulation. The evolution of visual, auditory, and chemical communication signals are detailed in Section 43.6.

33.1 Bionic Senses

Like other parts of the nervous system, the sensory portion transmits information by way of electrochemical signals. As a result, electrical devices that artificially stimulate sensory neurons can be used to produce sensations somewhat similar to those evoked by normal sensory input.

Cochlear implants were the first such devices to reach the market. They became widely available in the early 1980s and their use has increased ever since. More than 220,000 people with impaired hearing or deafness now have a cochlear implant that helps them hear and understand sounds.

Aiden Kenny (**FIGURE 33.1**), who was born deaf, received his first cochlear implants when he was 10 months old. Surgeons inserted thin wires deep inside his ear and implanted electrodes in his cochlea, the location of the sensory receptors that function in hearing. A microphone behind each of Aiden's ears picks up sounds and sends them to an processor on his skull. The processor then sends signals wirelessly to the electrodes inside his cochlea. The resulting action of these electrodes triggers action potentials that travel from the cochlea along auditory nerves to the brain.

Since cochlear implants first became available, there have been multiple improvements in their design and function. Early implants provided only single-channel input of sound, whereas modern devices are multichannel. Advances in computer technology have increased the speed and accuracy of the conversion of sounds into electrical impulses. All components of the devices have also become smaller.

Despite the many improvements, sound perception with an implant is still not the same as normal hearing, and people's ability to comprehend spoken language using the implant varies. Aiden received his cochlea implant when he was very young, so his ability to recognize spoken language and to speak is expected to be quite good.

Congenitally deaf individuals who receive a cochlear implant later in life typically have a less successful result than those who received implants as young children. This is because a brain that does not receive auditory input reorganizes itself. Brain regions that would normally function in hearing do not simply sit idle, but rather are recruited to function in vision. Thus, deaf people often have more acute vision than those with normal hearing. After this reorganization has occurred early in life, it is largely irreversible. The brain of a deaf adult who receives a cochlear implant cannot re-reorganize itself to begin processing auditory information in an area now devoted to other senses.

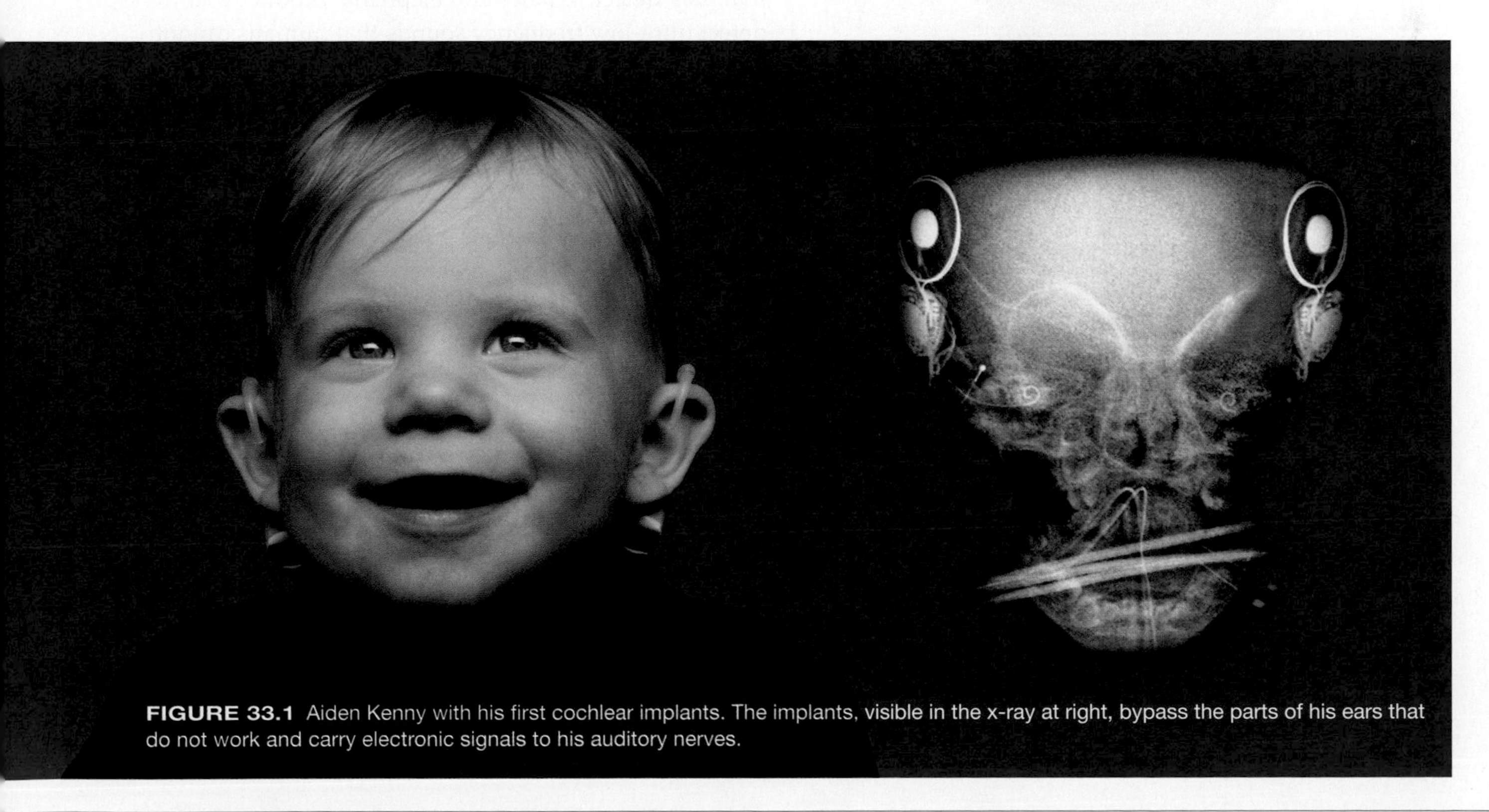

FIGURE 33.1 Aiden Kenny with his first cochlear implants. The implants, visible in the x-ray at right, bypass the parts of his ears that do not work and carry electronic signals to his auditory nerves.

CREDITS: (opposite) © Can Stock Photo, Inc./bgammache; (1) left, Mark Thiessen/National Geographic Creative; right, Johns Hopkins Hospital/National Geographic Creative.

33.2 Overview of Sensory Pathways

A Mechanoreceptors in a vampire bat's ear allow the bat to detect high-pitched sound waves (ultrasound). Thermoreceptors in the skin of the bat's nose help it locate veins filled with warm blood.

B Chemoreceptors embedded in the skin of a snail's two pairs of tentacles allow the snail to detect chemicals associated with food. Eyes at the tips of the two long tentacles detect light.

C Pigeons are among the animals that can detect variations in Earth's magnetic field and use this information to navigate.

FIGURE 33.2 Sensory diversity.

✔ An animal's sensory receptors determine what features of the environment it can detect and respond to.

The sensory portion of a vertebrate nervous system includes sensory neurons that become excited in the presence of specific stimuli, nerves that carry information about the stimulus to the brain, and brain regions that process this information. In the context of sensory systems, "stimulus" refers to an aspect of the environment that has the capacity to excite a sensory neuron.

Sensory Diversity

Detection of stimuli begins with excitation of sensory receptors. A sensory receptor can be either a sensory neuron or a specialized epithelial cell that responds to a stimulus by exciting a sensory neuron. In both cases, detection of a stimulus by a sensory receptor results in excitation of a sensory neuron.

We classify sensory receptors by the types of stimuli they respond to. Five types occur in most animals:

Mechanoreceptors respond to mechanical energy. Some detect shifts in a body's position or acceleration, others respond to touch or to stretching of a muscle, and still others respond to vibrations caused by pressure waves. Sound is a type of pressure wave, so auditory receptors are a type of mechanoreceptor. Auditory receptors of different animals vary in the frequency that they detect. Whales and elephants produce and detect ultra-low frequency sounds that humans cannot hear. Bats emit and respond to sounds too high for us to perceive (FIGURE 33.2A).

Thermoreceptors respond to a specific temperature or to a temperature change. All animals tolerate only a limited range of temperatures, and thermoreceptors help them avoid conditions outside their range of tolerance. Thermoreceptors can also help predatory animals locate the warm bodies of their prey. Some snakes have thermoreceptors on their head that allow them to detect any nearby rodents. Similarly, vampire bats use thermoreceptors on their nose to find blood vessels of their prey; a vessel filled with blood is warmer than the skin that surrounds it.

Chemoreceptors detect specific molecules in an environment; they function in the senses of taste and smell. Nearly all animals have chemoreceptors that help them locate nutrients and avoid ingesting poisons (FIGURE 33.2B). Other types of chemoreceptors monitor the internal environment. For example, chemoreceptors in some arteries monitor the level of CO_2 in your blood.

Photoreceptors detect light energy. Humans detect only visible light, but insects and some other animals, including rodents, also respond to ultraviolet light. Some flowers have UV-absorbing pigments arranged in patterns that are invisible to us, but draw the attention of the insects that pollinate them.

Pain receptors, also called nociceptors, detect tissue damage. They have a protective function and are often involved in reflexes that minimize further harm.

Other types of sensory receptors are less widespread among animals. For example, only some aquatic animals have receptors that can detect electrical signals in water. Such receptors help some fish, amphibians, and even platypuses detect electrical signals produced by the nerves and hearts of their prey. As another example, animals such as pigeons, sea turtles, and honeybees have magnetoreceptors that aid their navigation by allowing them to detect Earth's magnetic field (**FIGURE 33.2C**).

From Sensing to Sensation to Perception

In animals that have a brain, processing of sensory signals gives rise to **sensation**: awareness of a stimulus.

When a sensory receptor is stimulated sufficiently, it causes an action potential in a sensory neuron. Action potentials, remember, are always the same size (Section 32.5). The brain gathers additional information about a stimulus from three types of data: (1) the nerve pathway being triggered, (2) the number of axons in the pathway that are undergoing action potentials, and (3) the number of axons recruited by the stimulus.

An animal's brain interprets action potentials on the basis of where they originate. This is why you may "see stars" if you press on your eyes in a dark room. The pressure accidentally causes action potentials to travel along an optic nerve to the brain, which interprets all signals from this nerve as "light."

A strong stimulus causes a receptor to generate action potentials more often and longer than a weak signal does. The same receptors are stimulated by a whisper and a whoop. Your brain interprets the difference by variations in the frequency of the incoming

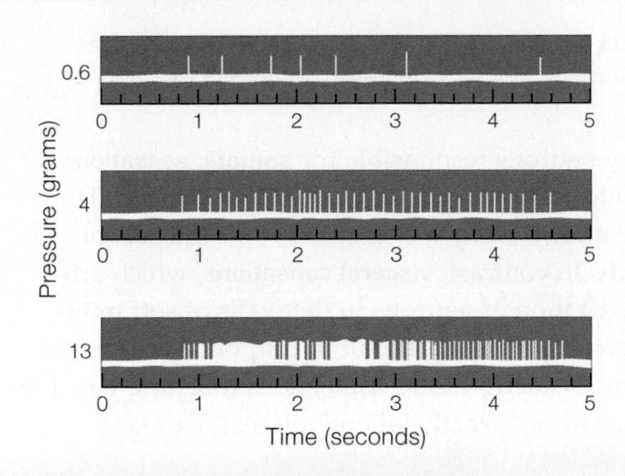

FIGURE 33.3 ▶**Animated** Action potential frequency in a pressure-sensitive mechanoreceptor in the skin. A rod was pressed against skin with varying amounts of pressure. Vertical bars above each thick horizontal line represent action potentials. The stronger the pressure, the more action potentials (white bars) occur per second.

signals (**FIGURE 33.3**). In addition, a strong stimulus recruits more sensory receptors, compared with a weak stimulus. A gentle tap on the arm activates fewer receptors than a slap.

Stimulus duration also affects how the stimulus is interpreted. In **sensory adaptation**, sensory neurons stop generating action potentials (or make fewer of them) despite continued stimulation. Walk into a house where an apple pie is in the oven and you will notice the sweet scent of baking apples immediately. Then, within a few minutes, the scent seems to lessen. The odor does not actually change in intensity, but chemoreceptors in your nose adapt to it

Sensory perception arises when the brain assigns meaning to sensory signals. Consider what happens when you watch a person walking away from you. As the distance between the two of you increases, the image of the person on your eye becomes smaller and smaller. You perceive this change in sensation as evidence of increasing distance between you and the person, rather than a sign that the person is shrinking.

TAKE-HOME MESSAGE 33.2

How do animals detect and process sensory stimuli?

✔ Sensory neurons undergo action potentials in response to specific stimuli. Different kinds of sensory receptors respond to different types of stimuli.

✔ Action potentials are all the same size, but which axons are responding, how many are responding, and the frequency of action potentials provide the brain with information about stimulus location and strength.

chemoreceptor Sensory receptor that responds to the presence of a specific chemical.
mechanoreceptor Sensory receptor that responds to pressure, position, or acceleration.
pain receptor Sensory receptor that responds to tissue damage.
photoreceptor Sensory receptor that responds to light.
sensation Awareness of a stimulus.
sensory adaptation Slowing or halting of a sensory response to an ongoing stimulus.
sensory perception The meaning a brain derives from a sensation.
thermoreceptor Temperature-sensitive sensory receptor.

33.3 Somatic and Visceral Sensations

✔ Signals from receptors in the skin, joints, muscles, and internal organs flow through the spinal cord to the brain.

Sensory neurons responsible for **somatic sensations** are located in skin, muscle, tendons, and joints. These sensations are easily localized to a specific part of the body. In contrast, **visceral sensations**, which arise from excitation of neurons in the walls of soft internal organs, are often difficult to pinpoint. It is easy to determine exactly where someone is touching you, but less easy to say exactly where you feel nausea.

The Somatosensory Cortex

Signals from the sensory neurons involved in somatic sensation travel along axons to the spinal cord, then along tracts in the spinal cord to the brain. The signals end up in the somatosensory cortex, a part of the cerebral cortex. Like the motor cortex (Section 32.11), the somatosensory cortex has neurons arrayed like a map of the body (**FIGURE 33.4**).

As an example of the types of receptors that signal the somatosensory cortex, consider those in the human skin (**FIGURE 33.5**). Receptor endings of sensory neurons are either surrounded by some sort of capsule or free (unenclosed) in the tissue. Some of the free nerve endings in the dermis coil around the roots of hair follicles and can detect even the slightest movement of the hair. Other free nerve endings detect temperature changes or tissue damage. Free nerve endings in skeletal muscles, tendons, joints, and walls of internal organs give rise to sensations that range from itching, to a dull ache, to sharp pain.

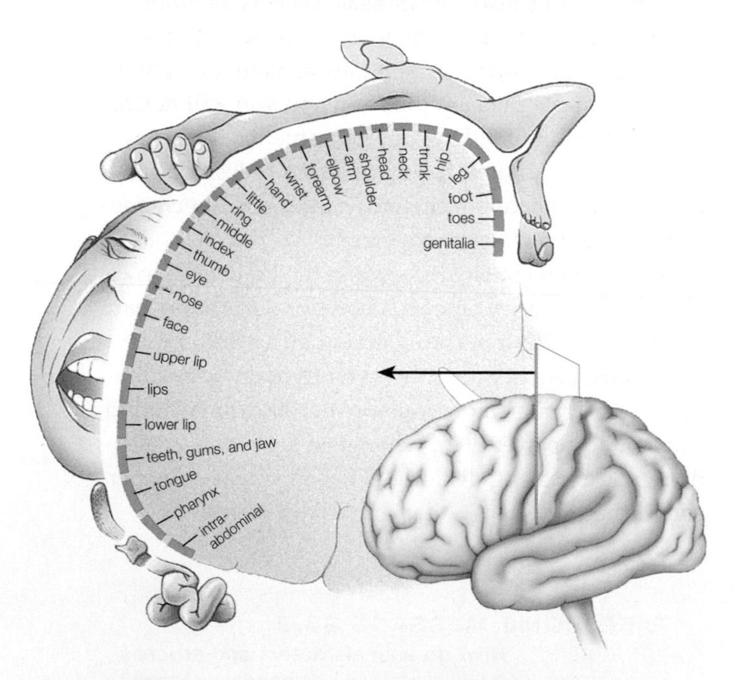

FIGURE 33.4 A map showing where the different body regions are represented in the human primary somatosensory cortex. This brain region is a narrow strip of the cerebral cortex (shown here in yellow) that runs from the top of the head to just above each ear.

Body parts that are disproportionately large in the "body" mapped onto this brain correspond to body regions with the most sensory receptors, such as the fingertips, face, and lips. Body parts that have relatively fewer sensory neurons, such as thighs, are disproportionately small. Compare Figure 32.23.

Concentric layers of epithelial tissue wrap around the receptor endings of many sensory neurons in the skin. These encapsulated receptors are named for the scientists who first described them. Meissner's corpuscles and Pacinian corpuscles detect touch and pressure in hairless skin regions such as fingertips, palms, and the soles of the feet. Meissner's corpuscles in the upper dermis are small and respond to light touches. Pacinian corpuscles respond to more pressure. They are larger, deeper in the dermis, and also occur near joints and in the wall of some organs. Ruffini endings are encapsulated receptors that are found in skin throughout the body and detect its stretching. Compared to Meissner's and Pacinian corpuscles, Ruffini endings adapt to continued stimulation more slowly. Thus, if you hold a stone in your hand, Ruffini endings inform your brain the stone is still there even after other receptors have stopped responding. Still another encapsulated receptor, called the bulb of Krause, responds to touch and to cold. Bulbs of Krause occur in skin throughout the body but are especially abundant in skin of the lips, tongue, nipples, and genitals.

Signals from mechanoreceptors that detect stretching of tendons and muscles are also conveyed to the somatosensory cortex. This information helps your brain keep track of where your body parts are located in space. The muscle spindle fibers that take part in the stretch reflex are one such type of receptor. The more a muscle stretches, the more frequently stretch receptors fire. Nearly all skeletal muscles and smooth muscles contain muscle spindles.

Pain

Pain is the perception of a tissue injury. Somatic pain begins with signals from pain receptors in skin, skeletal muscles, joints, and tendons. Visceral pain is associated with organs inside body cavities. It occurs as a response to a smooth muscle spasm, inadequate blood flow to an organ, overstretching of a hollow organ such as the stomach, and other abnormal conditions.

Injured or distressed body cells release local signaling molecules such as histamine and prostaglandins.

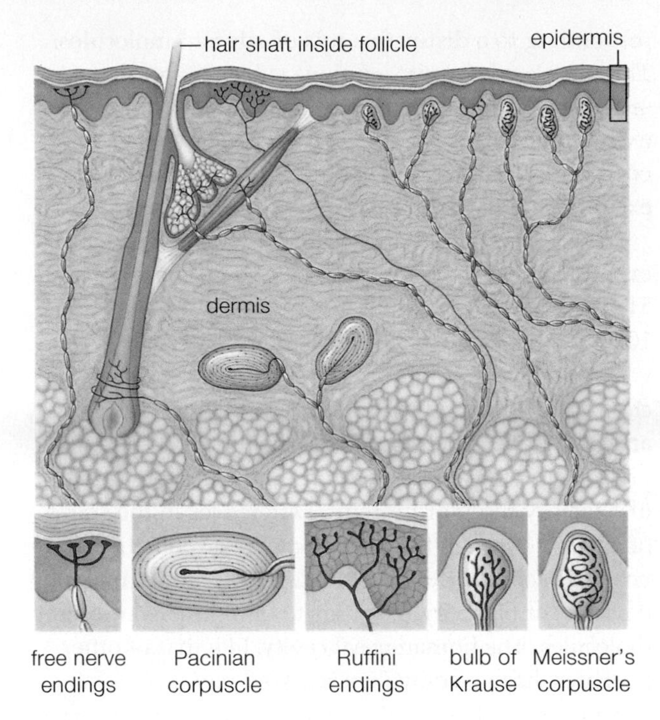

FIGURE 33.5 ▸**Animated** Sensory receptors in human skin.

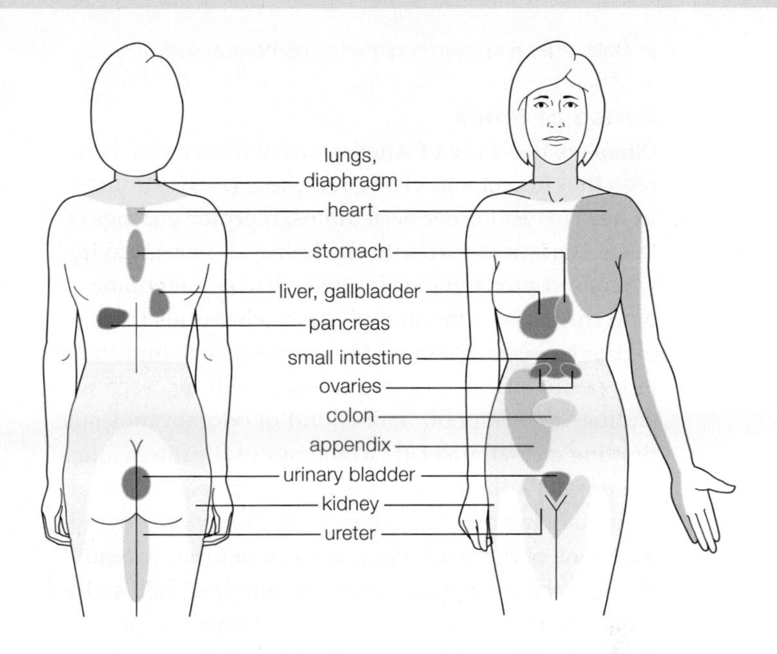

FIGURE 33.6 Sites of referred pain. Colored regions indicate the area that the brain interprets as affected when specific internal organs are actually distressed.

These molecules stimulate neighboring pain receptors, increasing the likelihood that the receptors will send signals along their axons to the spinal cord. In the spinal cord, the axons synapse with interneurons that relay signals about pain to the somatosensory cortex.

The neuromodulator substance P enhances pain perception by making spinal interneurons more likely to send signals about pain on to the sensory cortex. People with fibromyalgia, a disorder characterized by chronic pain in muscles and joints throughout the body, tend to have an elevated level of substance P. The pain-reducing effect of endorphins arises from their ability to slow release of substance P.

Pain-relieving drugs (analgesics) interfere with pain perception. Aspirin reduces pain by slowing prostaglandin production. Synthetic opiates such as morphine mimic the activity of endorphins. The drug ziconotide, a chemical first discovered in the venom of a cone snail (Section 24.1), blocks calcium channels in axon terminals of pain receptor neurons. Blocking these channels prevents the neurons from releasing neurotransmitter and inhibits transmission of pain signals.

The brain sometimes mistakenly interprets signals from receptors in internal organs as if they were from receptors in the skin or joints. The result is referred pain. The classic example is a pain that radiates from the chest across the shoulder and down the left arm during a heart attack (**FIGURE 33.6**). The arm is not affected, so why does it hurt? Referred pain occurs because each part of the spinal cord receives sensory input from both skin and internal organs. Skin encounters more painful stimuli than internal organs, so pain signals from skin flow more frequently along the neural pathway to the brain. The brain tends to attribute signals that arrive along a particular pathway to their most common source, even if they originate elsewhere.

TAKE-HOME MESSAGE 33.3

How do somatic and visceral sensations arise?

✔ Somatic sensations originate at sensory receptors in skin, skeletal muscle, and joints. They travel along sensory neuron axons to the spinal cord, then to the somatosensory cortex.

✔ Visceral sensations originate with the stimulation of sensory neurons in the walls of organs. These signals are relayed to the spinal cord, and then to the brain.

✔ Pain is the sensation associated with tissue damage. Referred pain occurs because the brain sometimes misinterprets visceral pain as if it were caused by a problem in the skin or a joint.

pain Perception of tissue injury.
somatic sensations Sensations such as touch and pain that arise when sensory neurons in skin, muscle, or joints are activated.
visceral sensations Sensations that arise when sensory neurons associated with organs inside body cavities are activated.

33.4 Chemical Senses

✔ Both smell and taste begin with chemoreceptors.

Sense of Smell

Olfaction, the sense of smell, starts with sensory neurons that function as chemoreceptors (**FIGURE 33.7**). In humans and other vertebrates, receptor endings of these neurons extend into the lining of a nasal cavity, where receptor proteins in their plasma membrane bind dissolved odorant molecules (chemicals that excite our sense of smell) ❶. Humans have hundreds of types of chemoreceptive sensory neurons, each with endings that respond to one kind of odorant molecule. Binding of that molecule to the receptor protein alters the protein's shape, and sets in motion reactions that culminate in an action potential in the sensory neuron.

Axons of these olfactory sensory neurons extend through the base of the skull and into the brain's olfactory bulb ❷, where they synapse on interneurons. Each interneuron receives signals from many sensory neurons that all detect the same odorant molecule ❸. In response to excitatory signals from these cells, the interneurons send signals to other brain regions such as the limbic system and the cerebral cortex ❹.

An average person can discriminate between and remember about 10,000 different odors. When you recognize an odor, such as the scent of a rose, you are responding to a distinctive mix of odorant molecules. These molecules excite a unique subset of your nose's sensory neurons, thus triggering a unique pattern of excitation in the cerebral cortex. Through past experience, your brain has learned to associate this pattern of excitatory signals with its source.

Vertebrates vary in their ability to detect odors. For example, dogs are known for their acute sense of smell. They detect some odorant molecules at concentrations 10,000 to 100,000 times lower than humans can. A variety of factors contribute to dogs' greater olfactory capacity. Adjusting for body size, a dog has a far larger area of nasal epithelium. This epithelium has both a denser array of sensory neurons and a greater variety of olfactory receptor proteins. Differences in the speed of air flow through the nose also play a role. A dog's nose has a recessed area where air flow slows, making it easier for olfactory receptors to bind inhaled odorant molecules. The human nasal cavity, like that of other primates, has no equivalent recessed region.

Sense of Taste

Like olfactory receptors, **taste receptors** are chemoreceptors that detect chemicals dissolved in fluid, but they differ in their function, location, and structure. Taste receptors help animals locate food and avoid

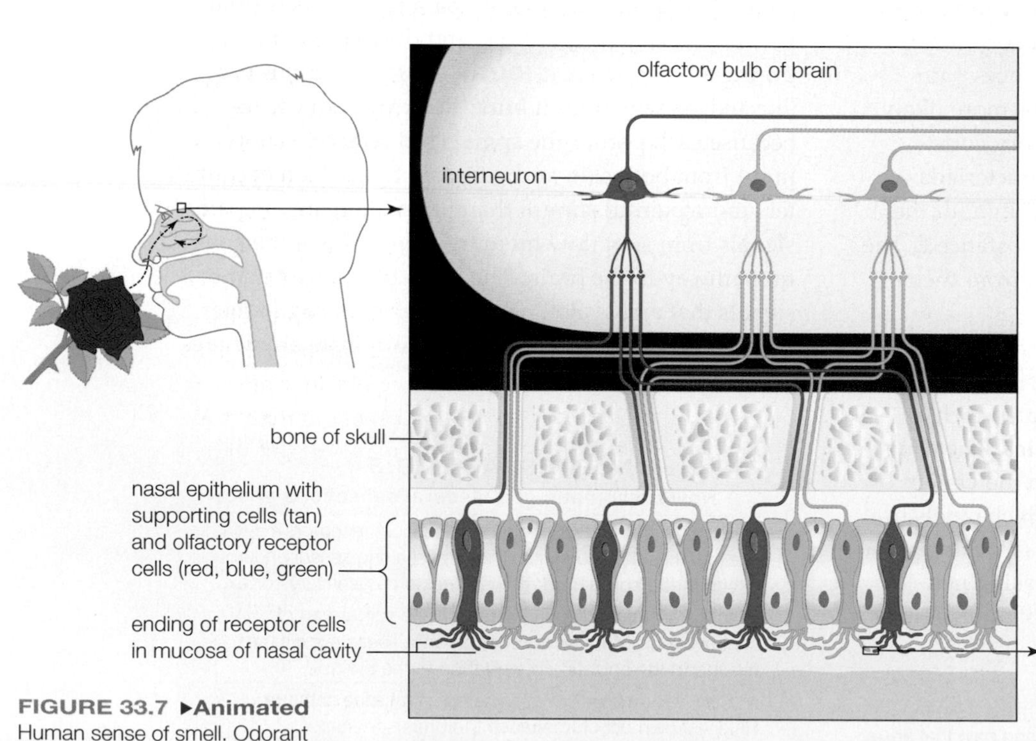

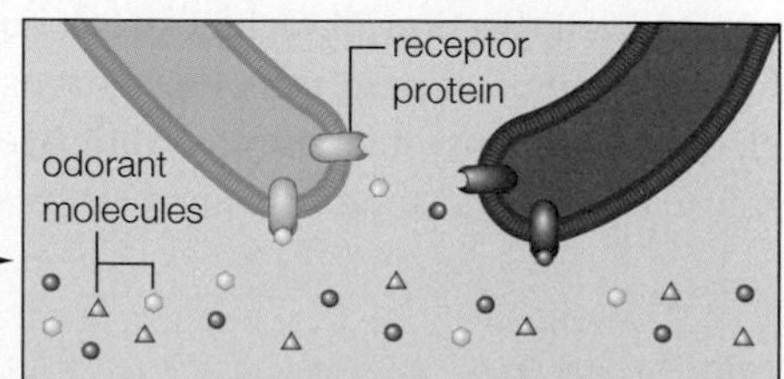

❶ Inhaled odorant molecules bind to receptor proteins on chemoreceptive sensory neurons in the nasal cavity. A receptor protein binds only one type of odorant molecule, and each cell has only one type of receptor. This art shows three types of receptor cells (coded red, green, and blue), but there are hundreds.

FIGURE 33.7 ▶Animated Human sense of smell. Odorant molecules excite a sensory neuron when they bind to chemoreceptors in its plasma membrane.

poisons. An octopus "tastes" substances with receptors in suckers on its tentacles. A fly "tastes" using receptors in its antennae and feet.

Humans have taste receptors in taste buds on the lining of their mouth and the upper surface of the tongue. Taste buds are located in specialized epithelial structures, or papillae, that are visible as raised bumps (**FIGURE 33.8**). Each taste bud contains taste receptor cells (specialized epithelial cells) and neurons. Receptor-covered microvilli of the taste receptor cells extend out through a pore, and thus come in contact with food molecules in saliva.

The sensation of taste begins with the binding of a molecule to receptor proteins on the microvilli of a taste receptor cell. Binding of the molecule excites this cell, which in turn excites a sensory neuron within the taste bud, which relays an action potential along its axon to the brain.

In humans, perceived taste arises from a combination of signals produced by different types of taste receptors. We have receptors that respond to *sweetness* (elicited by glucose and the other simple sugars), *sourness* (acids), *saltiness* (sodium chloride or other salts), *bitterness* (plant toxins, including alkaloids), *umami* (elicited by amino acids such as glutamate, which is found in cheese and aged meat), and—as recently discovered—*fattiness* (fatty acids). All mammals generally have the same types of taste receptors, although cats, dolphins, and some other carnivores have lost the ability to detect sweetness.

Each taste receptor cell is most sensitive to one of the six primary tastes, but taste buds in all regions of the tongue include all six types of cells. Thus, contrary to popular belief, the ability to detect specific tastes does not map to specific regions of the tongue.

Variation in the ability to taste certain bitter compounds arises from genetic differences in the structure of taste receptor proteins. Consider a receptor protein that binds to a synthetic molecule called phenylthiocarbamide (PTC). This protein is encoded by a gene that has two common alleles. A dominant allele encodes a receptor protein that triggers an action potential when it binds PTC. People with this allele perceive PTC as somewhat to very bitter. The receptor encoded by the recessive allele does not trigger an action potential when it binds PTC. People homozygous for this allele find PTC difficult to detect. PTC is not found in nature, but similar molecules occur in broccoli, cabbage, and related vegetables. People who can taste PTC are more likely to find these vegetables unpalatably bitter.

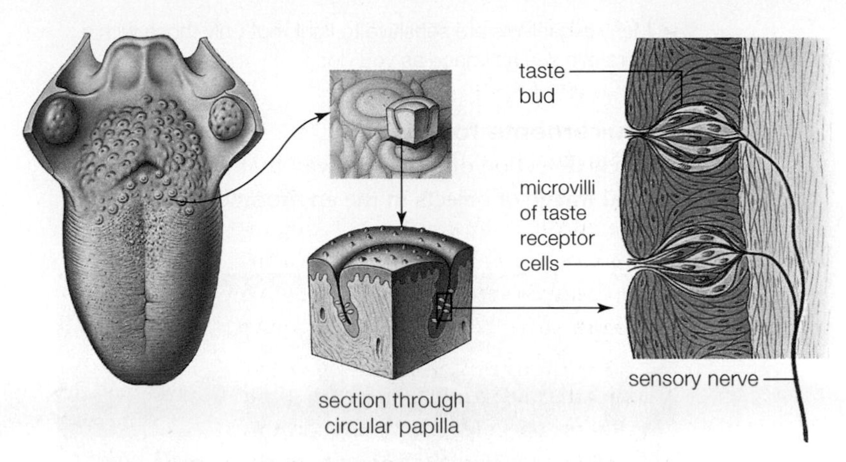

FIGURE 33.8 ▸**Animated** Taste receptors in the human tongue. Taste buds are clusters of receptor cells and supporting cells inside special epithelial papillae. One type, a circular papilla, is shown in the section here. The tongue has about 5,000 taste buds, each enclosing as many as 150 taste receptor cells.

Pheromones—Chemical Messages

Many animals release and detect chemicals that function in communication. A **pheromone** is a signaling molecule that is secreted by one individual and affects the behavior of other members of its species. For example, male silk moths have antennae that detect a sex pheromone secreted by female moths. Male moths will follow the pheromone and fly upwind toward its source.

Reptiles and most mammals make pheromones, and they have a **vomeronasal organ**—a collection of sensory neurons in the nasal cavity that bind and respond to pheromones. Humans and our closest primate relatives have a reduced version of this organ. Whether humans make and respond to pheromones remains a matter of debate. We discuss pheromones and animal communication in more detail in Chapter 43.

TAKE-HOME MESSAGE 33.4

What are the features of the chemical senses?

- ✔ Smell and taste involve stimulation of chemoreceptors by the binding of specific molecules.
- ✔ Humans have six types of taste receptors and hundreds of types of olfactory receptors.
- ✔ Many kinds of animals make and detect pheromones, chemicals that function in intraspecific communication.

olfaction The sense of smell.
pheromone Chemical that serves as a communication signal among members of an animal species.
taste receptors Chemoreceptors involved in the sense of taste.
vomeronasal organ Pheromone-detecting organ of some vertebrates.

33.5 Diversity of Visual Systems

✔ Many organisms are sensitive to light, but only those with a camera eye see an image as you do.

Requirements for Vision

Vision is detection of light in a way that provides a mental image of objects in the environment. It requires eyes and a brain with the capacity to interpret visual stimuli. Image perception arises when the brain integrates signals regarding shapes, brightness, positions, and movement of visual stimuli.

Eyes are sensory organs that hold photoreceptors. Pigment molecules inside the photoreceptors absorb light energy. That energy is converted to signals in the form of action potentials that travel to the brain.

Certain invertebrates, including earthworms, do not have eyes, but they do have photoreceptors dispersed under the epidermis or clustered in parts of it.

Scallops have as many as 60 eyes arrayed along the border of their mantle (**FIGURE 33.9**). Each eye has a **lens**, a transparent disk-shaped structure that bends light rays from any point in the visual field so they converge on photoreceptors. Despite its many eyes, a scallop cannot form a mental image of its surroundings because it has no brain to integrate visual information.

FIGURE 33.9 A scallop with many blue eyes along the edge of its mantle. The eyes are sensitive to movement and changes in light.

Insects have **compound eyes** consisting of many separate units called ommatidia, each with its own lens (**FIGURE 33.10**). The brain constructs an image based on the intensity of light detected by the different units. Compound eyes do not provide the clearest vision, but they are highly sensitive to movement.

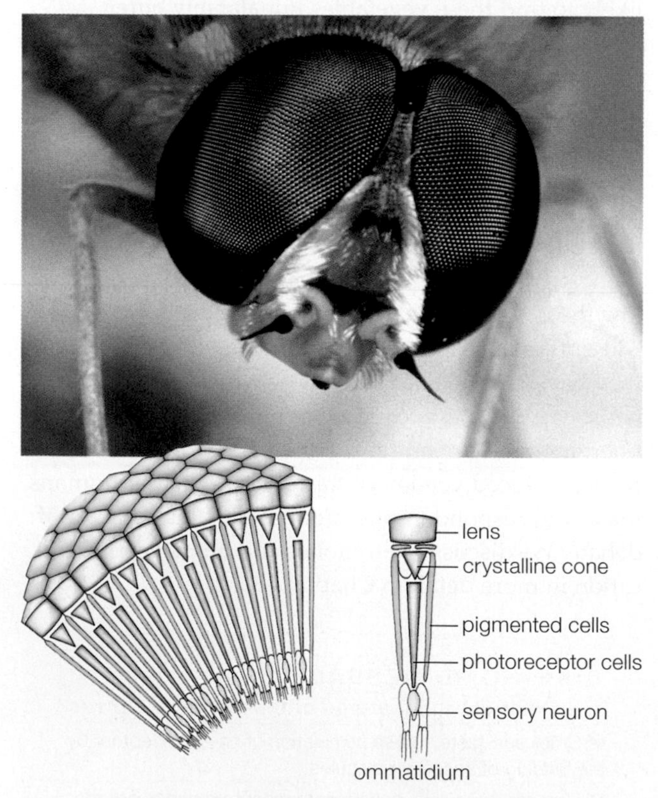

FIGURE 33.10 The compound eye of a deerfly, with many densely packed, identical units called ommatidia. Each ommatidium has a lens that focuses light on photoreceptor cells. Although the mosaic image produced by such an eye is fuzzy, the eye is very good at detecting movement.

Cephalopod mollusks such as squids and octopuses have the most complex eyes of any invertebrate (**FIGURE 33.11**). Their **camera eyes** have an adjustable opening that allows light to enter a dark chamber. Each eye has a single lens that focuses incoming light onto a **retina**, a tissue densely packed with photoreceptors. The retina of a camera eye is analogous to the light-sensitive film used in a traditional film camera. Signals from the photoreceptors in each eye travel along one of the two optic tracts to the brain. Compared to compound eyes, camera eyes can produce a more sharply defined and detailed image.

Vertebrates also have camera eyes, and because they are distant relatives of cephalopod mollusks, camera eyes are presumed to have evolved independently in the two lineages. This is an example of morphological convergence (Section 18.3).

Many animals have eyes on either side of their head, an arrangement that maximizes the area they can see. Predators, including owls, tend to have two forward-facing eyes (**FIGURE 33.12**). Primates also have eyes at the front of their head. Having eyes that face forward enables depth perception, because the same visual field can be surveyed simultaneously from two slightly different positions. The brain compares the overlapping information it receives to determine how far apart objects in the field are.

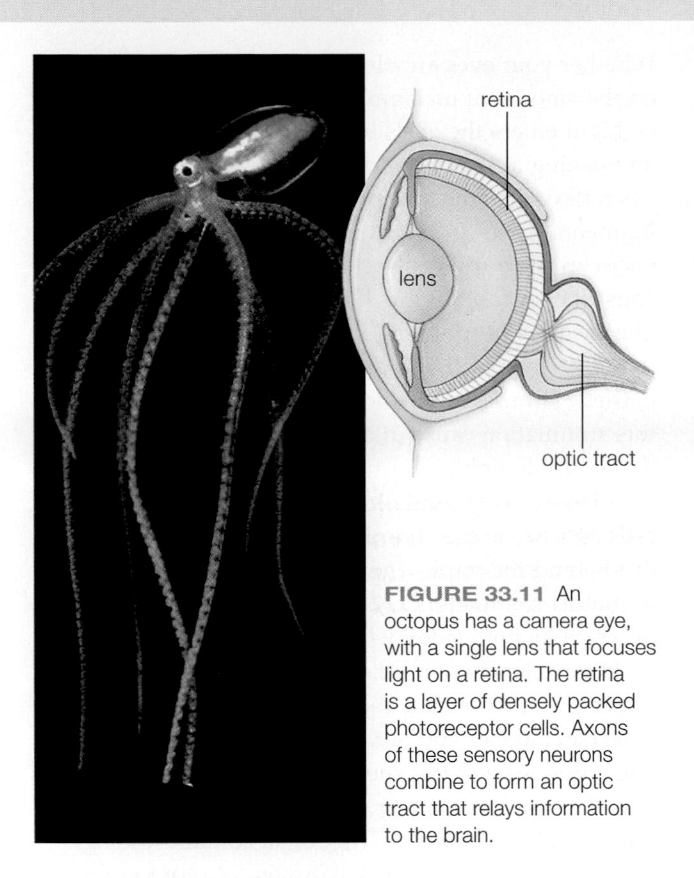

FIGURE 33.11 An octopus has a camera eye, with a single lens that focuses light on a retina. The retina is a layer of densely packed photoreceptor cells. Axons of these sensory neurons combine to form an optic tract that relays information to the brain.

FIGURE 33.12 The eyes of owls face forward, and photoreceptors are concentrated near the top of the inner eyeball. Owls mainly look down for prey. When on the ground, they must turn their heads almost upside down to see something above their head.

Animals that are active primarily at night often have large eyes relative to their body size. A large eye captures more available light than a smaller one. Light-reflecting layers also evolved in the eyes of many animals that are active under low-light conditions. When light shines on the eyes of these animals, the reflection from this layer makes their eyes appear to glow, a phenomenon known as "eyeshine" (**FIGURE 33.13**). Sharks, crocodiles, canines, felines, dolphins, deer, and rodents are among the many animals with reflective eyes. The location and structure of the reflective material varies among groups, indicating that this adaptation arose independently in the different lineages. Humans and other dry-nosed primates (Section 26.2) do not have a reflective layer in their eyes.

FIGURE 33.13 Eyeshine of a black-footed ferret. A reflective layer in this nocturnal predator's eyes enhances its night vision.

camera eye Eye with an adjustable opening and a single lens that focuses light on a retina.
compound eye Eye with many units, each having its own lens.
eye Sensory organ that incorporates a dense array of photoreceptors.
lens Disk-shaped structure that bends light rays so they fall on an eye's photoreceptors.
retina In an eye, a layer densely packed with photoreceptors.
vision Perception of visual stimuli based on light focused on a retina and image formation in the brain.

TAKE-HOME MESSAGE 33.5
How do animal visual systems differ?

✔ Some animals such as earthworms have photoreceptors that detect light, but do not form any sort of image.

✔ Other animals, including insects, have compound eyes. A compound eye has many individual units, each with its own lens. This eye produces a fuzzy, mosaic image, but its highly sensitive to movement.

✔ A camera eye with an adjustable opening and a lens that focuses light on a photoreceptor-rich retina provides a highly detailed image. Such eyes evolved independently in cephalopod mollusks and vertebrates.

33.6 A Closer Look at the Human Eye

✔ Protective structures surround the human eye, which is multilayered, with a light-bending cornea, a focusing lens, and a photoreceptor-rich retina.

Anatomy of the Eye

Each human eyeball sits inside a protective, cuplike, bony cavity called the orbit. Skeletal muscles that run from the rear of the eye to the bones of the orbit move the eyeball up and down or side to side.

Eyelids, eyelashes, and tears all help protect the eye's delicate tissues. Periodic blinking is a reflex that spreads a film of tears over the eyeball's exposed surface. Tears are secreted by exocrine glands in the eyelids and consist of water, lipids, salts, and proteins. Among the proteins are enzymes that break down bacterial cell walls and thus help prevent eye infections.

The **conjunctiva**, a protective mucous membrane, lines the inner surface of the eyelids and folds back to cover most of the eye's outer surface. Conjunctivitis, commonly called pinkeye, is an inflammation of this membrane; it is most often caused by a viral or bacterial infection.

The eyeball is spherical, with a three-layered structure (FIGURE 33.14). A dense, white, fibrous **sclera** covers most of the eye's outer surface. The front portion of the eye is covered by a **cornea** composed of transparent proteins called crystallins.

The eye's middle layer includes the choroid, iris, and ciliary body. The blood vessel–rich **choroid** is darkened by the brownish pigment melanin. Its dark color prevents light reflection within the eyeball. Attached to the choroid, and suspended behind the cornea, is a muscular, doughnut-shaped **iris**. It too has melanin. Whether your eyes are blue, brown, or green depends on the amount of melanin in your iris.

Light enters the eye's interior through the **pupil**, an opening at the center of the iris. Muscles of the iris reflexively adjust pupil diameter in response to light conditions. In bright light, an iris muscle that encircles the pupil contracts, causing the pupil to constrict (shrink). In low light, a different, spokelike iris muscle contracts, causing the pupil to dilate (widen). The pupil also dilates in response to sympathetic stimulation, which is why drugs that mimic this stimulation cause dilated pupils and increased sensitivity to light.

A ciliary body consisting of muscle and secretory cells attaches to the choroid. It holds the lens in place just behind the pupil. The stretchable, transparent lens is about 1 centimeter (1/2 inch) in diameter and bulges outward on both sides.

The eye has two internal chambers. The ciliary body produces a fluid called aqueous humor that fills the anterior chamber and bathes the iris and lens. A jellylike vitreous body fills the larger chamber behind the lens. The innermost layer of the eye, the retina, is at the back of this chamber. The retina contains the light-detecting photoreceptors. We discuss photoreceptor function in the next section.

The cornea and lens both bend incoming light. When light reflected from objects in the environment passes through these structures, it converges on the retina and forms an image of the objects. This image is upside down and a mirror image of the real world (FIGURE 33.15). The brain makes the necessary adjustments so you perceive the correct orientation.

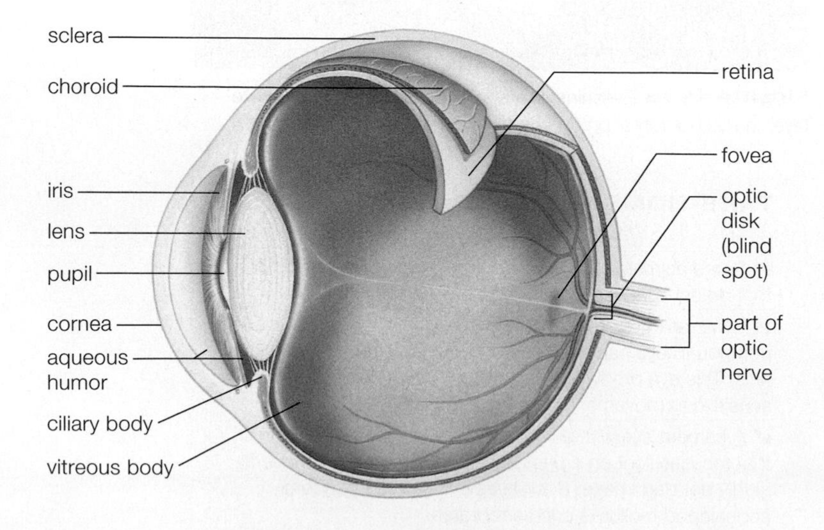

Wall of eyeball (three layers)	
Outer layer	*Sclera:* Protects eyeball
	Cornea: Helps the lens focus light
Middle layer	*Choroid:* Its blood vessels nutritionally support wall cells; its pigments lessen light scattering
	Iris: Smooth muscle that adjusts pupil size.
	Pupil: Opening that allows light into eye.
	Ciliary body: Its muscles control the lens shape; its fine fibers hold lens in place
	Start of optic nerve: Carries signals to brain
Inner layer	*Retina:* Absorbs, transduces light energy
Interior of eyeball	
Lens	Focuses light on photoreceptors
Aqueous humor	Transmits light, maintains fluid pressure
Vitreous body	Transmits light, supports lens and eyeball

FIGURE 33.14 ▶Animated Components and structure of the human eye.

FIGURE 33.15 Pattern of retinal stimulation in the human eye. The curved, transparent cornea changes the trajectory of light rays that enter the eye. As a result, light rays that fall on the retina produce a pattern that is upside down and inverted left to right.

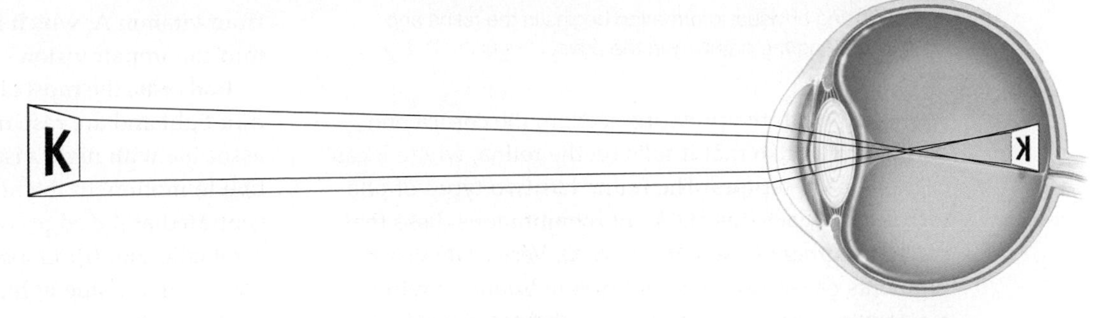

Focusing Mechanisms

With **visual accommodation**, the shape or position of a lens is adjusted so that incoming light forms an image on the retina, not in front of it or behind it. Without these adjustments, only objects at a fixed distance would be in focus. Objects closer or farther away would appear fuzzy.

In fishes and amphibians, the lens of an eye can be shifted forward or back, but its shape does not change. Extending or decreasing the distance between the lens and retina keeps light focused on the retina.

In mammals and other amniotes, a ring-shaped **ciliary muscle** (part of the ciliary body) adjusts the shape of the lens. This muscle encircles the lens and attaches to it by fine fibers. When the ciliary muscle relaxes, these fibers are taut, and the lens is under tension and flattened (FIGURE 33.16A). When the ciliary muscle contracts, fibers attached to the lens slacken, allowing the lens to become rounder (FIGURE 33.16B).

The curvature of the lens determines the extent to which it bends light rays. A flat lens focuses light from a distant object onto the retina; the lens must be rounder to focus light from nearby objects. When you read a book, ciliary muscle contracts and fibers that connect this muscle to the lens slacken. The decreased tension on the lens allows it to round up enough to focus light from the page onto your retina. Gaze into the distance and ciliary muscle around the lens relaxes, allowing the lens to flatten. Continual viewing of a close object, such as a computer screen or book, keeps ciliary muscle contracted. To reduce the strain on these muscles, take breaks and focus on more distant objects.

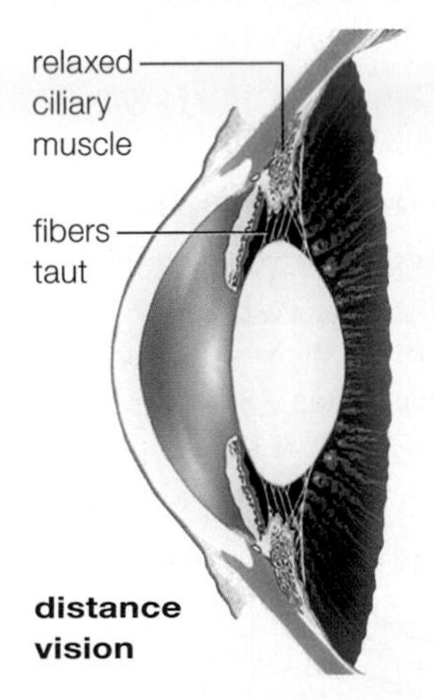

A Relaxed ciliary muscle pulls fibers taut; the lens is stretched into a flatter shape that focuses light from a distant object on the retina.

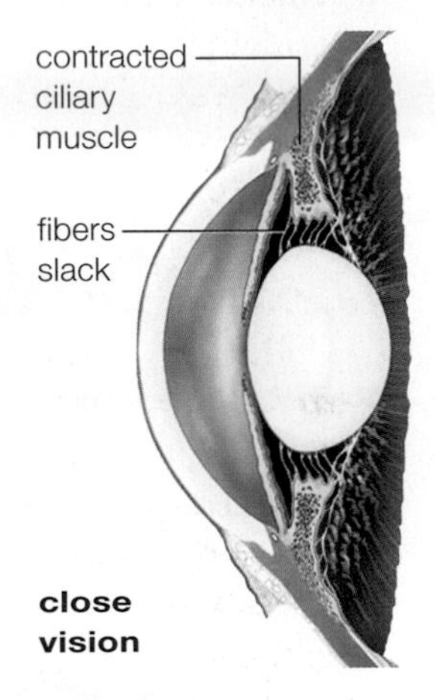

B Contracted ciliary muscle allows fibers to slacken; the lens rounds up and focuses light from a close object on the retina.

FIGURE 33.16 How the eye varies its focus. The lens is encircled by ciliary muscle. Elastic fibers attach the muscle to the lens. The shape of the lens is adjusted by contracting or relaxing the ciliary muscle, which increases or decreases the tension on the fibers, and thus changes the shape of the lens.

FIGURE IT OUT The thicker a lens, the more it bends light. Does the lens bend light more with distance vision or close vision? Answer: Close vision

TAKE-HOME MESSAGE 33.6

How is the structure of the human eye related to its function?

- ✔ The eye consists of delicate tissues surrounded by a bony orbit and constantly bathed in infection-fighting tears.
- ✔ The cornea at the front of the eye bends light rays entering the eye's interior through the pupil. The diameter of the pupil changes depending on the amount of available light.
- ✔ Behind the pupil, the lens focuses light on the retina, the eye's innermost photoreceptor-containing layer. Muscle contraction and relaxation alter the shape of the lens to focus light from near or distant objects.

choroid A blood vessel–rich layer of the middle eye. It is darkened by the brownish pigment melanin to prevent light scattering.
ciliary muscle A ring-shaped muscle of the eye that encircles the lens and attaches to it by short fibers.
conjunctiva Mucous membrane that lines the inner surface of the eyelids and folds back to cover the eye's sclera.
cornea Clear, protective covering at the front of the vertebrate eye.
iris Circular muscle that adjusts the shape of the pupil to regulate how much light enters the eye.
pupil Adjustable opening that allows light into a camera eye.
sclera Fibrous white layer that covers most of the eyeball.
visual accommodation Process of adjusting lens shape or position so that light from an object falls on the retina.

33.7 Light Reception and Visual Processing

✔ Processing of visual information begins in the retina and continues along the pathway in the brain.

As explained in the previous section, the cornea and lens bend light so that it falls on the retina, where it can excite photoreceptors. The retina has two types of photoreceptors. Each has stacks of membranous disks that contain pigment (FIGURE 33.17A). Vertebrate visual pigments consist of an opsin protein bound to retinal, a cofactor. Retinal (shown in FIGURE 6.4) is derived from vitamin A, which is why a deficiency in this vitamin can impair vision.

Rod cells, the most abundant photoreceptors, detect dim light and are responsible for the coarse image we associate with night vision. They are also highly sensitive to motion. In the human eye, rods tend to be concentrated at the edges of the retina. All rods have the same pigment (rhodopsin), which is excited by exposure to green-blue light.

Cone cells provide acute daytime vision and allow us to detect colors. There are three types of cone cells, each with a slightly different form of the cone pigment (iodopsin). One cone pigment absorbs mainly red light, another absorbs mainly blue, and a third absorbs green. Normal human color vision requires all three. The **fovea**, a pit in the central region of the retina, has the greatest density of cone cells. With normal vision, images focused on the retina fall mainly on the fovea.

Signal integration and processing begin in the retina, which is multilayered (FIGURE 33.17B). When a rod cell or cone cell is excited, it signals a neuron in the layer above. Neurons in this layer respond to patterns of photoreceptor activation by sending signals to ganglion cells. Bundled axons of these ganglion cells constitute the optic nerve (FIGURE 33.17C). The region of the retina through which the optic nerve exits the eye lacks photoreceptors. It cannot respond to light and thus is a "blind spot." We all have a blind spot in each eye, but do not notice it because the information missed by one eye is provided to the brain by the other.

Signals from the right visual field of each eye travel along an optic nerve to the brain's left hemisphere. Signals from the left visual field travel to the right hemisphere. Each optic nerve ends in a brain region (the lateral geniculate nucleus) that processes signals. From here, signals are conveyed to the visual cortex, where the final integration produces visual sensations.

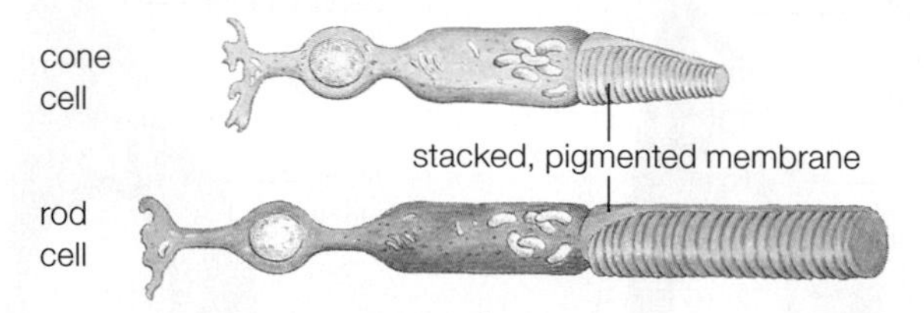

A The two types of photoreceptors in the retina.

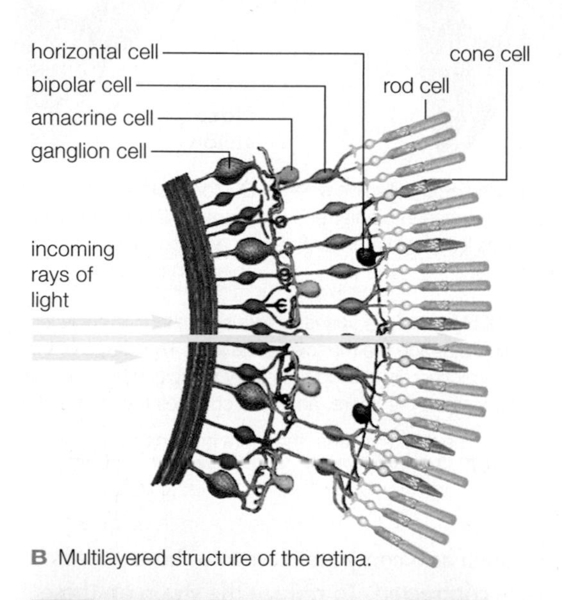

B Multilayered structure of the retina.

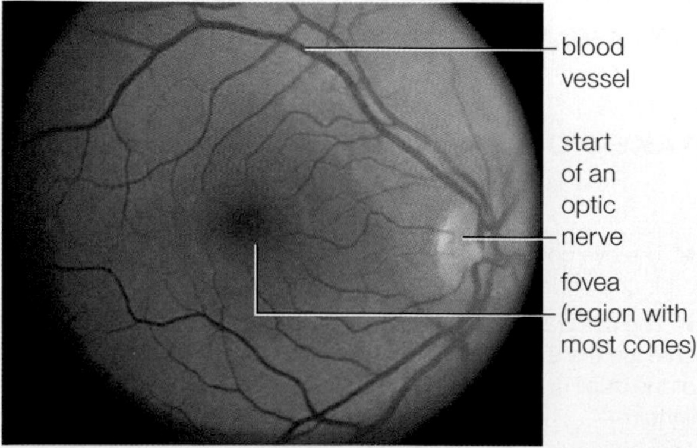

C Magnified view of the retina as seen through the pupil.

FIGURE 33.17 ▶**Animated** Structure of the retina. The two types of photoreceptors, rods and cones, lie at the very rear of the retina, beneath layers of signal-processing neurons.

cone cell Photoreceptor that provides sharp vision and allows detection of color.
fovea Retinal region where cone cells are most concentrated.
rod cell Photoreceptor that is active in dim light; provides coarse perception of image and detects motion.

TAKE-HOME MESSAGE 33.7

How do we detect and process visual information?

✔ When stimulated by light, rod cells and cone cells send signals to neurons in the layer of retina above them. These neurons process signals, then send messages to the brain.

✔ The visual cortex integrates signals from the eyes and gives rise to visual sensations.

CREDITS: (17A, B) Based on www.occipita.cfa.cmu.edu; (17C) Courtesy of Dr. Bryan Jones, University of Utah School of Medicine.

33.8 Visual Disorders

✔ A variety of disorders can impair vision.

Sometimes one or more types of cones are missing or impaired. The result is color blindness. With red–green color blindness, an X-linked recessive trait (Section 14.4), it is difficult to distinguish between red and green. Like other X-linked traits, red–green color blindness shows up most often in males. In people of European descent, about 8 percent of males and 0.5 percent of females are affected. The trait is about half as common in Africans and Asians.

A variety of problems can prevent light rays from converging on the retina as they should. With astigmatism, vision is blurred at all distances by an unevenly curved cornea. With nearsightedness, distant objects are out of focus. This can occur if the distance from the cornea to the retina is longer than normal (**FIGURE 33.18A**), or when ciliary muscles contract too much. With farsightedness, close objects are out of focus, either because the distance from the cornea to the retina is unusually short (**FIGURE 33.18B**) or ciliary muscles are too weak. Regardless of the cause, light rays from nearby objects get focused behind the retina.

Glasses, contact lenses, or surgery can correct most focusing problems. Glasses and contact lenses bend light before it reaches the eye. Laser surgery (LASIK) reshapes the cornea. Typically, LASIK can eliminate the need for glasses during most activities, although some older adults still need reading glasses. However, chronic eye irritation is a common complication.

As people age, changes in the structure of proteins in their lens often affect vision. The lens becomes less flexible, so most people over age forty have somewhat impaired near vision. Other changes to these proteins can cloud the lens, a condition called a cataract. Smoking, steroid use, and some diseases such as diabetes promote cataract formation, as does excessive exposure to ultraviolet radiation. Typically, both eyes are affected. At first, a cataract scatters light and blurs vision (**FIGURE 33.19A**). Eventually, the lens may become fully opaque, causing blindness. Worldwide, about 16 million people remain blinded by age-related cataracts. Cataract surgery, a common procedure in developed countries, can restore normal vision by replacing a clouded lens with a clear plastic implant.

In the United States, the leading cause of blindness among older adults is age-related macular degeneration (AMD). The macula, the part of the retina around and including the fovea, is essential to clear vision. Destruction of photoreceptors in the macula clouds the center of the visual field more than the periphery (**FIGURE 33.19B**). Damage caused by AMD usually cannot be reversed, but drug injections and laser therapy can slow its progression. A treatment using cells derived from embryonic stem cells is also being tested.

Glaucoma results when too much aqueous humor builds up inside the eyeball. The increased fluid pressure damages blood vessels and ganglion cells. It can also interfere with peripheral vision and visual processing. Although glaucoma is most often diagnosed in old age, conditions that give rise to the disorder start to develop long before symptoms arise. Screening for glaucoma in younger people allows doctors to detect increased fluid pressure inside the eye before the damage to ganglion cells becomes severe. They can then manage the disorder with medication, surgery, or both.

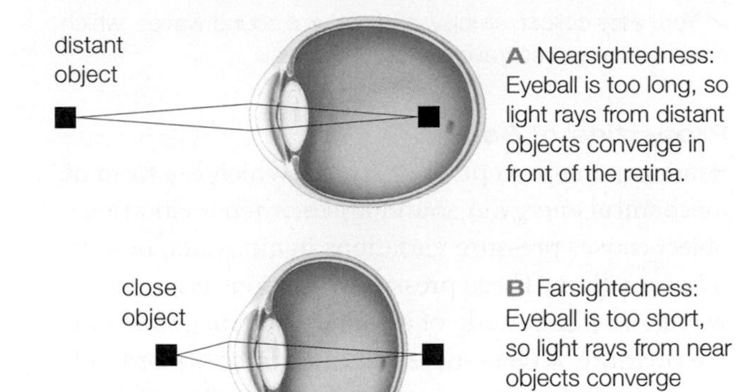

FIGURE 33.18 Focusing problems caused by abnormal eye shape.

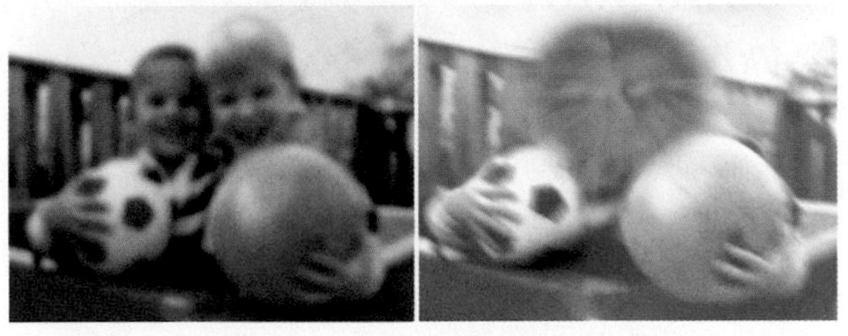

A With cataracts. **B** With macular degeneration.

FIGURE 33.19 Photos simulating vision with visual disorders.

TAKE-HOME MESSAGE 33.8

What causes common visual disorders?

✔ Problems with cone cells result in color blindness.

✔ With nearsightedness and farsightness, light rays are not properly focused on the retina. A lens that has become inflexible causes age-related farsightedness.

✔ Other age-related vision disorders include cataracts (clouding of the lens), macular degeneration (loss of photoreceptors), and glaucoma (excessive aqueous humor).

CREDITS: (18) © Cengage Learning; (19) National Eye Institute, U.S. National Institute of Health.

33.9 Vertebrate Hearing

✔ Your ears collect, amplify, and sort out sound waves, which are pressure waves traveling through the air.

Properties of Sound

Hearing is the perception of sound, which is a form of mechanical energy. A sound arises when a vibrating object causes pressure variations in air, water, or some other medium. These pressure variations occur as waves. The amplitude of a sound—the magnitude of the pressure waves—determines its intensity or loudness. The frequency of a sound—the number of waves per second—determines pitch (**FIGURE 33.20**). The more waves per second, the higher the frequency.

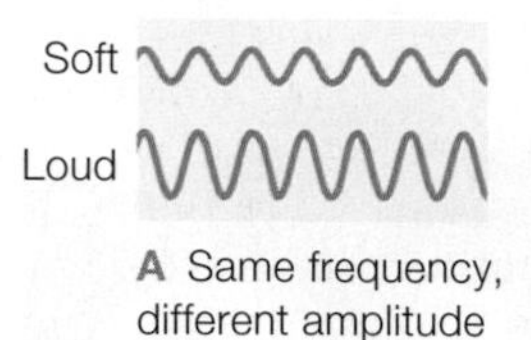

A Same frequency, different amplitude

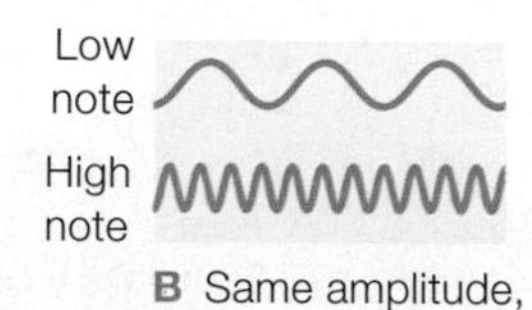

B Same amplitude, different frequency

FIGURE 33.20 Sound waves.

The Vertebrate Ear

Water readily transfers vibrations to body tissues, so fishes do not require elaborate ears to detect sounds. When vertebrates left water for land, their capacity to collect and amplify vibrations evolved in response to a new environmental challenge: The transfer of sound waves to body tissues is less efficient in air than in water. Certain features of mammalian ears are evolutionary adaptations that increase the efficiency of this transfer (**FIGURE 33.21**). For example, unlike amphibians and reptiles, most mammals have an **outer ear** that funnels sound inward ❶. A skin-covered flap of cartilage, called the pinna, projects from the side of the head. The pinna collects sound waves and directs them into the auditory canal, an air-filled passage that connects to the middle ear.

The **middle ear** amplifies and transmits sound waves to the inner ear. An **eardrum**, or tympanic membrane, first evolved in amphibians, in which it is visible as a round area on each side of the head. Sound waves cause an eardrum to vibrate. Behind the eardrum, an air-filled cavity holds three small bones known as the hammer, anvil, and stirrup ❷. The bones transmit the force of sound waves from the eardrum onto the surface of the oval window, an elastic membrane that is the boundary between the middle and inner ear.

The **inner ear** includes structures involved in hearing and in balance. We discuss the sense of balance in the next section; here, we focus on the role of the sound-detecting **cochlea**, a pea-sized, fluid-filled structure that resembles a coiled snail shell (the Greek *koklias* means snail). Membranes divide the interior of the cochlea into three fluid-filled ducts ❸. When sound waves make the three tiny bones of the middle ear vibrate, the stirrup pushes against the oval window. With each pulse of the sound wave, the oval window bows inward. This movement creates a pressure wave in the fluid inside the cochlea. These waves travel through the fluid, causing the lower wall of the cochlear duct (the basilar membrane) to vibrate up and down. The **organ of Corti**, an acoustical organ with arrays of hair cells, sits on top of this membrane ❹. Each hair cell is a mechanoreceptor with a tuft of modified cilia at one end. The cilia project into a tectorial membrane that drapes over them. When the basilar membrane moves, it pushes the cilia against the tectorial membrane and bends them ❺. The bending causes the hair cells to undergo action potentials that travel along an auditory nerve to the brain.

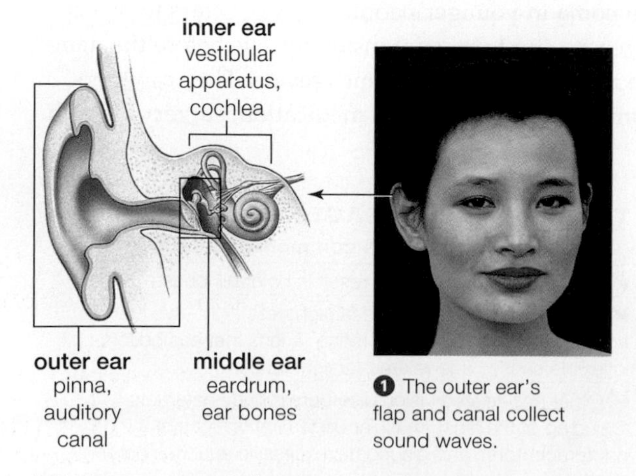

❶ The outer ear's flap and canal collect sound waves.

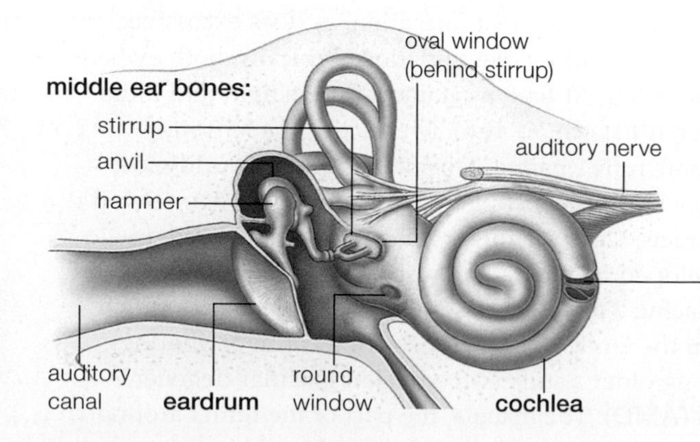

❷ The eardrum and middle ear bones amplify sound.

FIGURE 33.21 ▸**Animated** Anatomy of the ear and how we hear.

The number of hair cells that fire and the frequency of their signals inform the brain about the volume of a sound. The louder the sound, the more action potentials flow along the auditory nerve to the brain.

The brain can determine the pitch of a sound by assessing which part of the basilar membrane is vibrating most. The basilar membrane is not uniform along its length. It is stiff and narrow near the oval window, and broader and more flexible deeper into the coil. High-pitched sounds make the stiff, narrow, closer-in part of the basilar membrane vibrate most. Low-pitched sounds cause vibrations mainly in the wide, flexible part close to the membrane's tip. More vibrations make more hair cells in that region fire.

cochlea Coiled, fluid-filled structure in the inner ear that holds the sound-detecting organ of Corti.
eardrum The membrane that vibrates in response to pressure waves (sounds), thus transmitting vibrations to the bones of the middle ear.
hearing Perception of sound.
inner ear Fluid-filled vestibular apparatus and cochlea.
middle ear Eardrum and the tiny bones that transfer sound to the inner ear.
organ of Corti An acoustical organ in the cochlea that transduces mechanical energy of pressure waves into action potentials.
outer ear External ear and the air-filled auditory canal.

TAKE-HOME MESSAGE 33.9

How do vertebrates hear?

✔ The transition to life on land involved adaptations that improved the ability to detect airborne sounds.

✔ Like most mammals, humans have an outer ear that collects sound waves and directs them to the middle ear. In the middle ear, vibrations of the eardrum are transmitted via movement of small bones. These bones set up pressure waves in fluid inside the inner ear's cochlea.

✔ The organ of Corti inside the cochlea has hair cells that convert pressure waves into action potentials that travel along the auditory nerve to the brain.

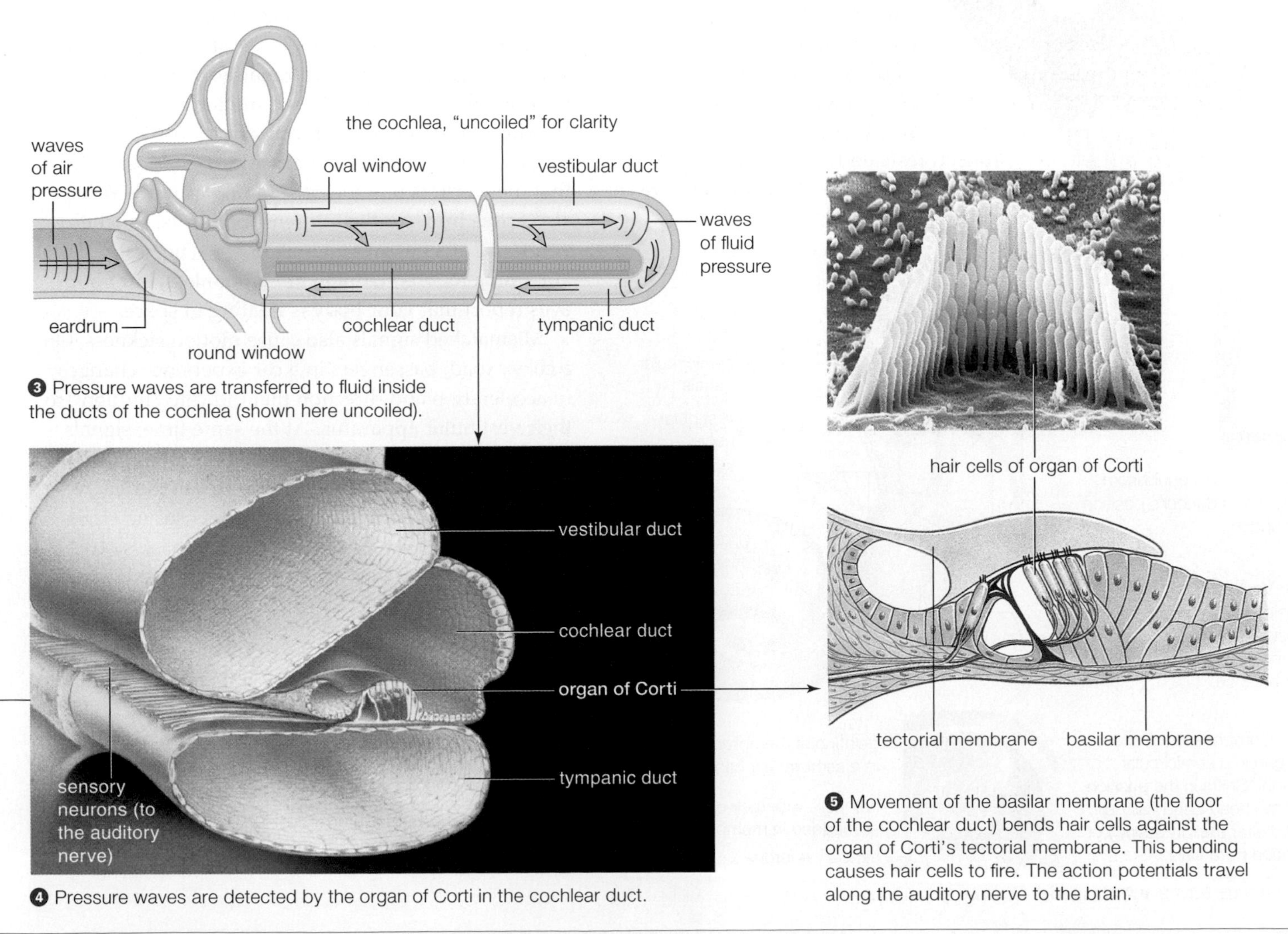

❸ Pressure waves are transferred to fluid inside the ducts of the cochlea (shown here uncoiled).

❹ Pressure waves are detected by the organ of Corti in the cochlear duct.

❺ Movement of the basilar membrane (the floor of the cochlear duct) bends hair cells against the organ of Corti's tectorial membrane. This bending causes hair cells to fire. The action potentials travel along the auditory nerve to the brain.

33.10 Organs of Equilibrium

✔ Organs inside your inner ear are essential to maintaining posture and a sense of balance.

✔ Somatic sensory receptors also contribute to balance.

Organs of equilibrium monitor the body's position and motion (**FIGURE 33.22**). Each vertebrate ear includes these organs inside a fluid-filled sensory structure called the **vestibular apparatus**. The organs are located in three semicircular canals, and in two sacs, called the saccule and utricle (**FIGURE 33.22B**).

Like the organ of Corti, organs of the vestibular apparatus have hair cells. Fluid pressure inside the canals and sacs makes the cilia bend. The mechanical energy of this bending deforms the hair cell plasma membrane just enough to let ions slip across and stimulate an action potential. A vestibular nerve carries the sensory input to the brain.

The three semicircular canals are oriented at right angles to one another, so rotation of the head in any combination of directions—front/back, up/down, or left/right—moves the fluid inside them. An organ of equilibrium rests on the bulging base of each canal. The cilia of its hair cells are embedded in a jellylike mass (**FIGURE 33.22C**). When fluid moves in the canal, it pushes the mass and generates pressure required for initiating action potentials.

The brain receives signals from semicircular canals on both sides of the head. By comparing the number and frequency of action potentials coming from both sides, the brain senses dynamic equilibrium: the angular movement and rotation of the head. Among other things, your sense of dynamic equilibrium allows you to keep your eyes locked on an object even when you swivel your head or nod.

Organs in the saccule and utricle act in the sense of static equilibrium. These organs help the brain keep track of the head's position and how fast it is moving in a straight line. They also help keep the head upright and maintain posture. Inside the saccule and utricle, a jellylike layer weighted with calcite overlies the mechanoreceptors (hair cells). When you tilt your head, or start or stop moving, the weighted mass shifts, bending hair cells and altering their rate of action potentials.

The brain also takes into account information from the eyes, and from receptors in the skin, muscles, and joints. Integration of this information provides awareness of the body's position and motion in space.

A stroke, an inner ear infection, or loose particles in the semicircular canals can cause vertigo, a sensation that the world is moving or spinning around. Vertigo also arises from conflicting sensory inputs, as when you stand at a height and look down. The vestibular apparatus reports that you are motionless, but your eyes report that your body is floating in space.

Mismatched signals also cause motion sickness. On a curvy road, passengers in a car experience changes in acceleration and direction that indicate "motion" to their vestibular apparatus. At the same time, signals from their eyes about objects inside the car tell their brain that the body is at rest. Driving can minimize motion sickness because the driver focuses on sights outside the car such as scenery rushing past, so the visual signals are consistent with vestibular signals.

semicircular canals
vestibular nerve
saccule
utricle
gelatinous membrane in a semicircular canal
hair cells with their cilia embedded in membrane
sensory neurons

A Organs of equilibrium monitor a dancer's position in space.

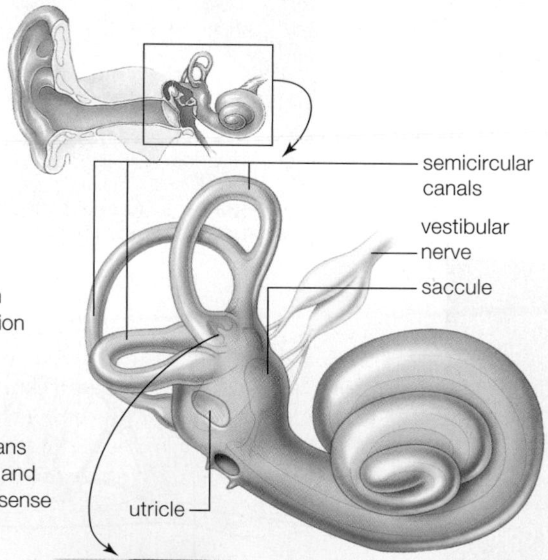

B Vestibular apparatus inside a human ear. Organs inside its fluid-filled sacs and canals contribute to the sense of balance.

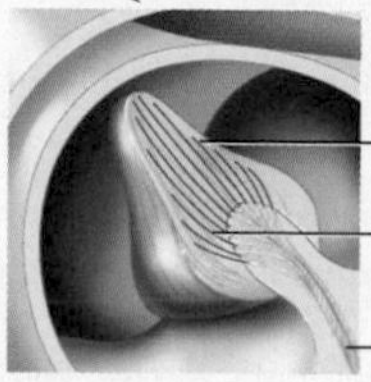

C Components of one organ in a semicircular canal. Shifts in the position of the head bend hair cells and alter their frequency of action potentials.

FIGURE 33.22 ▶**Animated** Organs of equilibrium.

organs of equilibrium Sensory organs that respond to body position and motion.
vestibular apparatus System of fluid-filled sacs and canals in the inner ear; contains the organs of equilibrium.

TAKE-HOME MESSAGE 33.10

What gives us our sense of balance?

✔ Mechanoreceptors in the fluid-filled vestibular apparatus of the inner ear detect the body's position in space, and when we start or stop moving.

Bionic Senses (revisited)

Cochlear implants allow people to hear despite a damaged or defective cochlea. Similarly, retinal implants assist people whose photoreceptors have been damaged. These implants, sometimes described as artificial retinas, are a relatively recent invention. The first such device to be approved, the Argus II retinal prosthesis system, became available for sale in Europe in 2011 and in the United States in 2013. It is currently approved for use only in adults with retinosis pigmentosa, a rare, inherited disease in which degeneration of the retina causes blindness. However, the Argus II or similar devices may eventually prove helpful to patients blinded by macular degeneration or other disorders.

The Argus II system consists of a pair of glasses with a video camera that captures a scene and sends information to a processing unit mounted on the glasses. The processor converts the visual information to instructions that are sent wirelessly to an electrode-bearing implant on the surface of the retina. In response to the wireless signal, the electrodes in the implant electrically excite the optic nerve, triggering action potentials that travel to the brain. Other systems that work in a similar manner are also being tested.

As with cochlear implants, the Argus II retinal implant cannot completely restore normal function. A previously blind person who uses this device can perceive differences between dark and light areas and even recognize letters, but only if they are quite large (**FIGURE 33.23**). All vision is in black and white.

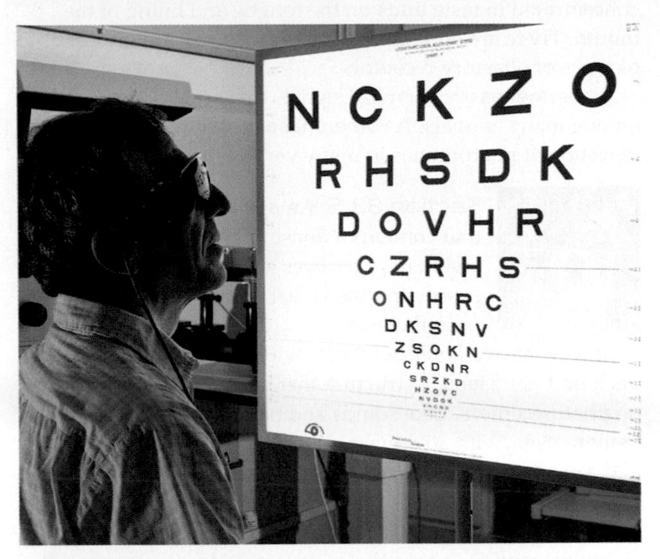

FIGURE 33.23 Vision testing in a patient with the Argus II retinal implant. The previously blind patient can now read the largest letters on this chart.

However, researchers expect updates to the device, development of competing device, and improved processing mechanisms to lead to improved function.

Researchers are also developing prosthetic hands that will give amputees not only an ability to grip and hold materials, but also restore some sensation of touch. The prostheses convert information about pressure to electrical signals, which cause action potentials to travel along nerves to the somatosensory cortex.

summary

Section 33.1 Devices that electrically stimulate nerves carrying sensory information to the brain can be used to restore some degree of hearing, sight, or touch to people in whom these senses are impaired.

Section 33.2 The types of sensory receptors that an animal has determine the types of stimuli it detects and can respond to. Stimulation of a sensory receptor causes a sensory neuron to undergo an action potential. **Mechanoreceptors** respond to mechanical energy such as touch. **Pain receptors** respond to tissue damage. **Thermoreceptors** are sensitive to temperature. **Chemoreceptors** respond when a dissolved chemical binds. **Photoreceptors** respond to light.

The brain evaluates action potentials from sensory receptors based on which nerves deliver them, their frequency, and the number of axons firing in any given interval. Continued stimulation of a receptor may lead to a diminished response (**sensory adaptation**). **Sensation** is detection of a stimulus, whereas **sensory perception** involves assigning meaning to some sensation.

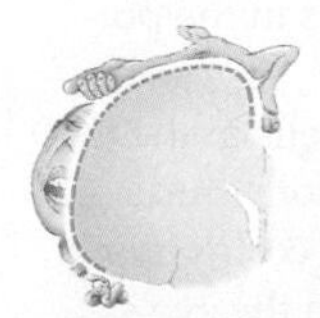

Section 33.3 **Somatic sensations** are easy to pinpoint and arise from receptors in the skin, joints, tendons, and skeletal muscles. Signals from these receptors flow to the somatosensory cortex, an area of the cerebral cortex where interneurons are organized like maps of individual parts of the body surface. **Visceral sensations** originate from receptors in the walls of soft organs and are less easily pinpointed. **Pain** is the perception of tissue damage. With referred pain, the brain mistakenly attributes signals that come from an internal organ to the skin or muscles.

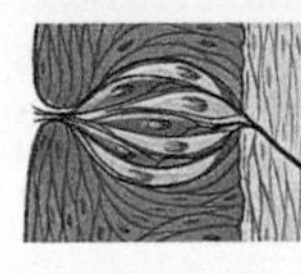

Section 33.4 The senses of taste and **olfaction** (smell) involve molecules binding to chemoreceptors. Binding triggers action potentials, and these signals are relayed to the

cerebral cortex and limbic system. In vertebrates, olfactory receptors line the nasal passages; and **taste receptors** are concentrated in taste buds on the tongue and lining of the mouth. There are six classes of taste receptors, but hundreds of types of olfactory receptors.

Pheromones are chemical signals that act as social cues among many animals. A **vomeronasal organ** functions in detection of pheromones in many vertebrates.

Section 33.5 An **eye** is a sensory organ that contains a dense array of photoreceptors. **Vision** requires eyes and a brain capable of processing the visual information that arrives from the eyes.

Insects have a **compound eye**, with many individual units. Each unit has a **lens**, a structure that bends light so that it falls on photoreceptors. Like squids and octopuses, humans have **camera eyes**, with an adjustable opening that lets in light, and a single lens that focuses images on a **retina** with a dense array of photoreceptors.

The arrangement and structure of vertebrate eyes vary. Depth perception arises when two forward-facing eyes send the brain information about the same viewed area. Large eyes with an internal light-reflecting layer adapt some animals to low light conditions.

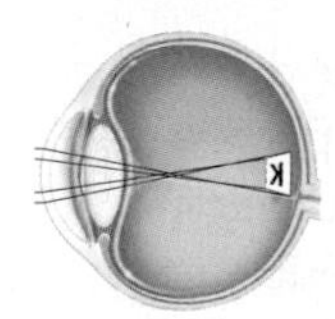

Section 33.6 A human eye sits in a bony orbit and is protected by eyelids lined by the **conjunctiva**. This membrane also covers the **sclera**, or white of the eye. The clear, curved **cornea** at the front of the eye bends incoming light. Light enters the eye's interior through the **pupil**, an adjustable opening in the center of the muscular, doughnut-shaped **iris**. Light that enters the eye is focused into an image on the retina. The retina sits on a pigmented **choroid** that minimizes reflections inside the eye.

With **visual accommodation**, the **ciliary muscle** adjusts the shape of the lens so that light reflected from a near or distant object focuses on the retina.

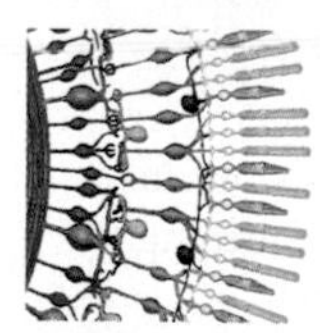

Sections 33.7, 33.8 Humans have two types of photoreceptors. **Rod cells** detect dim light and are important in coarse vision and peripheral vision. **Cone cells** detect bright light and colors, and they provide a sharp image. The greatest concentration of cone cells is in the portion of the retina called the **fovea**.

Rod cells and cone cells interact with other cells in the retina that begin processing visual information before sending it to the brain. Visual signals travel to the cerebral cortex along two optic nerves. There are no photoreceptors in the eye's blind spot, the area where the optic nerve begins.

Common vision disorders result from defective, degenerated or missing photoreceptors, misshapen eyes or parts thereof, a clouded lens, or excess aqueous humor.

Section 33.9 **Hearing** is the perception of sound, which is a form of mechanical energy. Sound waves are pressure waves. We perceive variations in their amplitude as differences in loudness, and variations in wave frequency as differences in pitch.

Human ears have three functional regions. The pinna of the **outer ear** collects sound waves. The **middle ear** contains the **eardrum** and a set of tiny bones that amplify sound waves and transmit them to the inner ear. In the **inner ear**, pressure waves elicit action potentials inside a **cochlea**. This coiled structure with fluid-filled ducts holds the mechanoreceptors responsible for hearing in its **organ of Corti**.

Pressure waves traveling through the fluid inside the cochlea bend hair cells of the organ of Corti. The brain gauges the loudness of a sound by the number of action potentials it elicits. It determines a sound's pitch by which part of the cochlea's basilar membrane the signals arrive from.

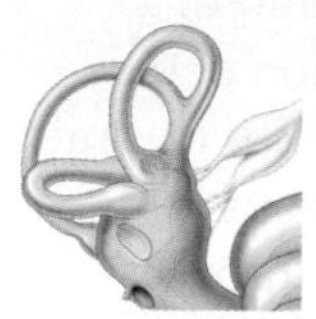

Section 33.10 **Organs of equilibrium** detect effects of gravity and acceleration to determine the body's position and motion. The **vestibular apparatus** is a system of fluid-filled sacs and canals in the inner ear. The sense of dynamic equilibrium arises when body movements cause shifts in the fluid, which causes cilia of hair cells to bend. Static equilibrium depends on signals from hair cells that lie beneath a weighted, jellylike mass. A shift in head position or a sudden stop or start shifts the mass, bends the hair cells, and alters the rate at which they undergo action potentials.

self-quiz

Answers in Appendix VII

1. The pain of heartburn is an example of a ________ .
   a. somatic sensation
   b. visceral sensation
   c. sensory adaptation
   d. spinal reflex

2. ________ is defined as a decrease in the response to an ongoing stimulus.
   a. Perception
   b. Visual accommodation
   c. Sensory adaptation
   d. Somatic sensation

3. Which is a somatic sensation?
   a. taste
   b. smell
   c. touch
   d. hearing
   e. a through c
   f. all of the above

4. Chemoreceptors play a role in the sense of ________ .
   a. taste
   b. smell
   c. touch
   d. hearing
   e. both a and b
   f. all of the above

5. In the ________ , neurons are arranged like maps that correspond to different parts of the body surface.
   a. retina
   b. somatosensory cortex
   c. basilar membrane
   d. occipital lobe

6. Mechanoreceptors in the ________ send signals to the brain about the body's position relative to gravity.
   a. eye b. ear c. tongue d. nose

7. The middle ear functions in ________ .
   a. detecting shifts in body position
   b. transmitting sound waves
   c. sorting sound waves out by frequency
   d. all of the above

8. Substance P ________ .
   a. increases pain-related signals
   b. is a natural painkiller
   c. is the active ingredient in aspirin

data analysis activities

Occupational Hearing Loss Frequent exposure to loud noise of a particular pitch can cause loss of hair cells in the part of the cochlea that responds to that pitch. People who work with or around noisy machinery are at risk for such frequency-specific hearing loss. Taking precautions such as using ear plugs to reduce sound exposure is important. Noise-induced hearing loss can be prevented, but once it occurs it is irreversible because dead or damaged hair cells are not replaced.

FIGURE 33.24 shows the threshold decibel levels at which sounds of different frequencies can be detected by an average 25-year-old carpenter, a 50-year-old carpenter, and a 50-year-old who has not been exposed to on-the-job noise. Sound frequencies are given in hertz (cycles per second). The more cycles per second, the higher the pitch.

1. Which sound frequency was most easily detected by all three people?

2. How loud did a 1,000-hertz sound have to be for the 50-year-old carpenter to detect it?

3. Which of the three people had the best hearing in the range of 4,000 to 6,000 hertz? Which had the worst?

4. Based on these data, would you conclude that the hearing decline in the 50-year-old carpenter was caused by age or by job-related noise exposure?

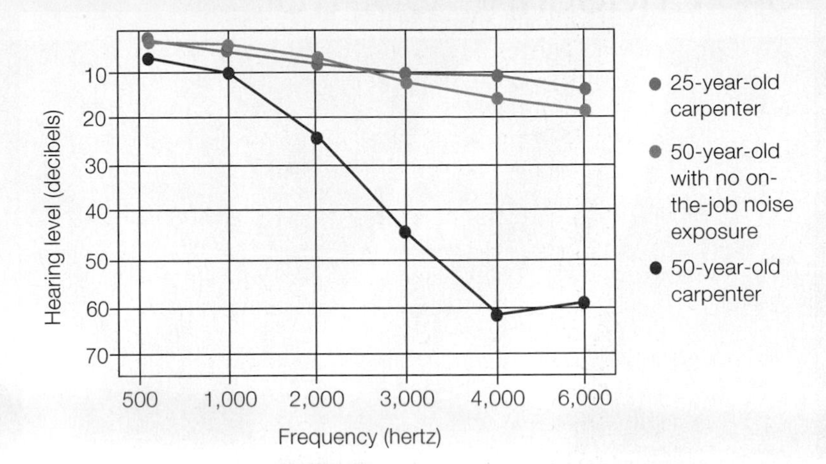

FIGURE 33.24 Effects of age and occupational noise exposure. The graph shows the threshold hearing capacities (in decibels) for sounds of different frequencies (given in hertz) in a 25-year-old carpenter (blue), a 50-year-old carpenter (red), and a 50-year-old who did not have any on-the-job noise exposure (brown).

9. The organ of Corti contains receptors that signal in response to ________ .
   a. heat b. sound c. light d. pheromones

10. Night vision begins with stimulation of ________ .
   a. hair cells b. rod cells c. cone cells d. neuroglia

11. Visual accommodation involves adjustment to the shape or position of the ________ .
   a. conjunctiva b. retina c. orbit d. lens

12. When you view a close object your lens gets ________ .
   a. flatter b. rounder c. darker d. cloudier

13. Defective or missing ________ cause color blindness.
   a. hair cells b. rod cells c. cone cells d. neuroglia

14. ________ causes the pupil to widen.
   a. Low light b. Bright light

15. Match each structure with its description.

___ cataract	a. protects eyeball
___ cochlea	b. transmits vibration to bone
___ eardrum	c. functions in balance
___ lens	d. detects pheromones
___ sclera	e. interferes with vision
___ fovea	f. contains chemoreceptors
___ taste bud	g. focuses rays of light
___ Pacinian corpuscle	h. has the most cones
___ pinna	i. collects sound waves
___ vestibular apparatus	j. sorts out sound waves
___ vomeronasal organ	k. detects touch

critical thinking

1. Like other nocturnal carnivores, the ferret shown in **FIGURE 33.13** has light-reflecting material in its choroid. Explain why the presence of reflective material in this layer of the eye maximizes the degree to which light excites photoreceptors. Explain also why having reflective material in this layer causes the perceived image to be somewhat blurry.

2. A compound extracted from the leaves of the shrub *Stevia rebaudiana* is a natural sugar substitute. The compound is 300 times sweeter than sugar, but has a slight bitter aftertaste. Given what you know about taste receptors, explain how a compound can be perceived as both sweet and bitter.

3. Most bats eat insects or fruit. Vampire bats, however, suck blood from birds or mammals. Like some snakes, and unlike any other mammals, vampire bats have thermoreceptors that can detect body heat given off by prey. Is the heat-detecting ability of vampire bats and snakes a homologous trait or an analogous one? Explain your reasoning.

4. If a mammal injures its leg, the resulting pain discourages the animal from putting too much weight on the leg while it is healing. An injured insect shows no such shielding response when its leg is injured. Some have cited the lack of such a shielding response as evidence that insects do not feel pain. On the other hand, insects do produce substances similar to one of our natural painkillers. Is the presence of these compounds in insects sufficient evidence to conclude that they do perceive pain as we do?

34 Endocrine Control

LEARNING ROADMAP

This chapter focuses on endocrine glands, which were introduced in Section 31.3. Knowing the properties of steroids (3.5), proteins (3.6), and the plasma membrane (5.7) will help you understand hormone action. Among other topics, you will learn how hormones affect human glucose metabolism (7.7) and arthropod molting (24.11).

VERTEBRATE ENDOCRINE SYSTEM

Nearly all vertebrates have an endocrine system that includes the same hormone-producing structures. The endocrine system interacts closely with the nervous system.

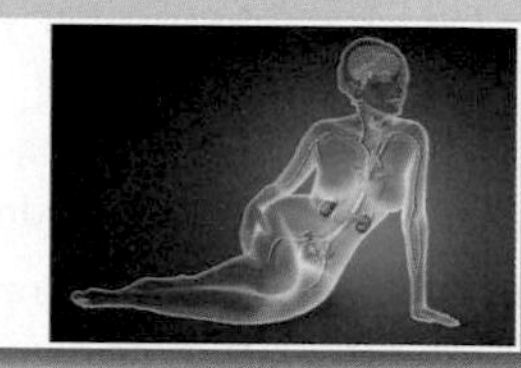

HORMONE ACTION

A hormone travels in the blood and binds receptors on target cells. Receptor activation converts the hormonal signal to a form that elicits a response in the target cell.

A MASTER REGULATORY CENTER

In vertebrates, two glands deep in the forebrain (the hypothalamus and pituitary gland) interact structurally and functionally in the control of many other glands.

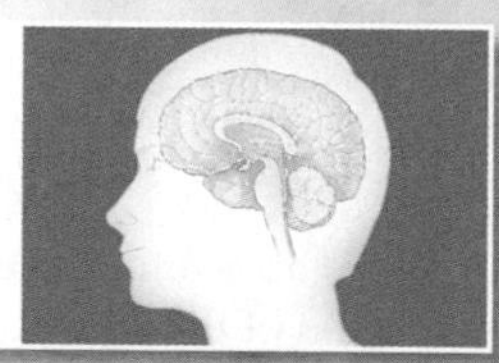

OTHER CONTROLLING FACTORS

Some glands alter their secretion of hormones in response to internal changes such as a shift in blood glucose level, or to external factors such as changes in day length.

INVERTEBRATE HORMONES

Hormones control molting and other events in invertebrate life cycles. Vertebrate hormones and receptors for them first evolved in ancestral lineages of invertebrates.

Later chapters discuss the effect of hormones on vertebrate muscle mass (Section 35.1), bone turnover (35.4), adjusting to high altitude (38.8), appetite (Section 39.5) and urine formation (40.5). Sex hormones have a central role in our discussion of gamete formation and reproduction (Chapter 41), and we look at hormonal effects on animal behavior in Section 43.2. We return to hormone disruptors in our discussion of pollutants in Chapter 48.

34.1 Hormones in the Balance

We live in a world awash in synthetic chemicals. We drink from plastic bottles, wear clothing made of synthetic fabrics, slather ourselves with synthetic skin care products, and eat food treated with synthetic pesticides. Huge numbers of man-made compounds are used to make computers and other electronic gadgets. Synthetic chemicals enter our bodies when we ingest them, inhale them, or absorb them across our skin.

We have learned by sad experience that some synthetic chemicals threaten animal and human health. For example, years after we started using them, we realized that DDT (a pesticide) and PCBs (used in electronic products, caulking, and solvents) are endocrine disruptors. **Endocrine disruptors** are molecules that interfere with the function of the endocrine system. DDT was banned in 1972 and PCBs in 1979. However, both chemicals were in wide use for years and are highly stable, so they still persist in the environment.

Other endocrine disruptors remain in use. Atrazine, an herbicide, is among them. This chemical has been widely used for more than forty years. In the United States, 76 million pounds of atrazine are sprayed each year, mostly to kill weeds in cornfields. Each year, some runs off from fields and contaminates water.

Living in atrazine-contaminated water can disrupt endocrine function in aquatic animals and diminish their reproductive capacity. Exposing genetically male frog tadpoles to atrazine alters their sexual development. As adults, the altered frogs have a lower than normal level of the male sex hormone testosterone, smaller testes, and a reduced sperm count. Atrazine even causes some males to develop ovaries in addition to testes, or in place of them (**FIGURE 34.1**).

Atrazine affects freshwater fish too. Exposing fish embryos to atrazine levels comparable to those in runoff from atrazine-treated fields doubles the proportion that develop into females. In addition, female minnows who breed in atrazine-contaminated water produce fewer eggs than those that breed in pure fresh water.

Atrazine is usually applied to fields in spring, so peak atrazine runoff often occurs when fish and frogs are mating and when their young are developing. If fish and frogs seems of little concern, consider this: All vertebrates have similar hormone-secreting glands and endocrine systems. Thus, chemicals that adversely affect one are likely to affect others in a similar manner.

In 2013, Syngenta, the company that makes atrazine, began paying out a $105 million settlement to community water suppliers across the United States. The payments are meant to reimburse these communities for costs incurred when they had to filter atrazine out of their drinking water supply. Runoff of atrazine from agricultural fields in these communities had resulted in water with an atrazine level above that considered safe by regulatory agencies.

endocrine disruptor Chemical that interferes with function of the endocrine system.

FIGURE 34.1 One effect of hormone disruption. This photo shows two frogs mating. The one on the bottom is a genetic male that has been turned into an egg-laying female by exposure to atrazine during development.

CREDITS: (opposite) Mark Hamblin/2020 VISION/naturepl.com; (1) Courtesy of Prof. Tyrone Hayes, UC Berkeley.

34.2 The Vertebrate Endocrine System

✔ Animal cells communicate with one another by way of a variety of short-range and long-range chemical signals.

Mechanisms of Intercellular Signaling

Cells of an animal body constantly signal one another. Gap junctions allow signaling substances to move directly between the cytoplasm of adjacent cells. Other cell–cell communication involves signaling molecules that are secreted into interstitial fluid (the fluid between cells). These molecules exert effects only when they bind to a receptor on or inside another cell. Any cell with receptors that bind and respond to a specific signaling molecule is a "target" of that molecule.

Secreted signaling molecules exert their effect by a three-step process. The molecules bind to a target cell's receptor, the receptor transduces the signal (changes it into a form that affects target cell behavior), and the target cell responds:

Some secreted signaling molecules diffuse a short distance through interstitial fluid and bind to nearby cells. For example, most neurons secrete neurotransmitters into the synaptic cleft that separates them from their target—a postsynaptic cell. Many types of cells produce and secrete **local signaling molecules**, which reach nearby targets by diffusion and so act over only a limited distance. Prostaglandins are an example of a local signaling molecule. When released by injured cells, they activate nearby pain receptors.

Animal hormones are long-range communication molecules that travel in blood. Remember from Section 31.3 that hormones are secreted by gland cells. Secreted hormones diffuse through interstitial fluid to nearby capillaries, then enter the bloodstream by seeping through openings in the walls of these blood vessels. Compared to neurotransmitters or local signaling molecules, hormones last longer, travel farther, and exert their effects on a greater number of cells.

Discovery of Hormones

Hormones were first discovered in the early 1900s by physiologists William Bayliss and Ernest Starling. They were studying how the secretion of pancreatic juices is regulated. Bayliss and Starling knew that food mixes with acid in the stomach, and that when this acidic mix reaches the small intestine, it stimulates the pancreas to secrete a bicarbonate buffer. However, they did not know how the message that triggered bicarbonate secretion reached the pancreas. To find out how the small intestine communicated with the pancreas, the scientists did an experiment. They surgically altered a laboratory animal, cutting all nerves that carry signals to and from its small intestine. Even after these nerves had been cut, the animal's small intestine responded to the presence of acid by secreting bicarbonate. This result indicated that communication between the intestine and the pancreas does not involve nerves.

Starling and Bayliss hypothesized that the small intestine produces a signal that travels in the blood. To test this idea, they exposed small intestinal cells to acid, then made an extract of those cells. Injecting this extract into the bloodstream of another animal caused its pancreas to secrete bicarbonate. The result confirmed the hypothesis: Exposure to acid causes the small intestine to release a chemical signal into the blood. This bloodborne substance triggers the pancreas to secrete bicarbonate into the gut.

The signaling substance discovered by Starling and Bayliss is now called secretin. Identifying its mode of action supported a hypothesis that dated back centuries: Blood carries internal secretions that influence the activities of the body's organs.

Starling coined the term "hormone" (the Greek *hormon* means to set in motion), which we now use to describe glandular secretions such as secretin. Later researchers identified additional hormones and their sources. Endocrine glands and other structures that secrete hormones are referred to collectively as the vertebrate **endocrine system**. FIGURE 34.2 shows the main glands of a human endocrine system. Many internal organs such as the small intestine and heart have cells that produce and release hormones as well.

Conversely, some of the major endocrine glands also have functions unrelated to hormone secretion. For example, the pancreas secretes hormones into the blood, but also secretes digestive enzymes into the small intestine. The gonads (ovaries and testes) produce sex hormones and also make gametes.

The discovery of hormones and subsequent studies of endocrine function have had important medical implications. They have allowed treatment of endocrine disorders, such as diabetes, as well as the invention of hormonal methods of contraception, a topic we discuss in detail in Chapter 41.

Neuroendocrine Interactions

Both neurons and endocrine cells develop from an embryo's ectodermal layer. Portions of the endocrine system and nervous system are so closely linked that

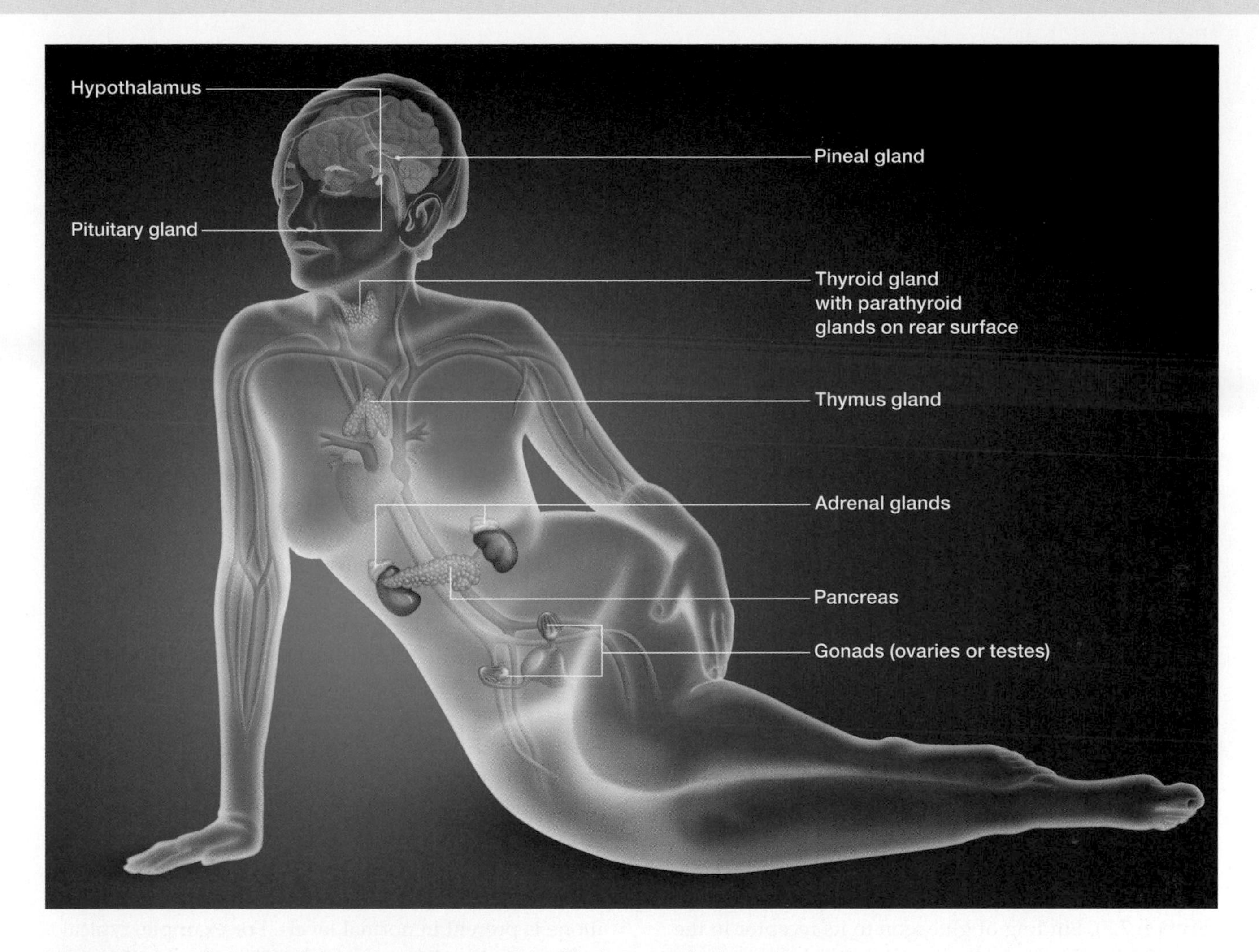

FIGURE 34.2 ▸**Animated** Main glandular components of the human endocrine system.

scientists sometimes refer to them collectively as the neuroendocrine system. Both endocrine glands and neurons receive signals from the hypothalamus, a command center in the forebrain (Section 32.10). Most organs respond both to hormones and to signals from the nervous system.

Hormones influence brain development, both before and after birth. Hormones can also affect nervous processes such as sleep–wake cycles, emotion, mood, and memory. Conversely, the nervous system regulates hormone secretion. For example, in a stressful situation, sympathetic nervous stimulation triggers an increase in the secretion of some hormones, and a decrease in the secretion of other hormones.

animal hormone Intercellular signaling molecule secreted by an endocrine gland or cell; travels in the blood to target cells.
endocrine system System of hormone-producing glands and secretory cells of a vertebrate body.
local signaling molecule Chemical signal that diffuses through interstitial fluid and affects nearby cells in an animal body.

TAKE-HOME MESSAGE 34.2

How do cells of an animal body communicate with one another?

✔ In all animals, cells release molecules that influence other cells. Each type of signal acts on all target cells that have receptors for it. Hormones are intercellular signaling molecules that travel in the bloodstream.

✔ Collectively, hormone-secreting glands and cells make up an endocrine system.

✔ Integrated interactions between the nervous system and nearly all endocrine glands coordinate many different functions for the body as a whole.

34.3 The Nature of Hormone Action

✔ For a hormone to have an effect, it must bind to receptors on or inside a target cell.

Categories of Hormones

There are two major categories of hormones, those derived from amino acids and those derived from cholesterol. **Amino acid–derived hormones** include amine hormones (modified amino acids), peptide hormones (short chains of amino acids), and protein hormones (longer chains of amino acids). All are typically polar, so they dissolve easily in blood, which is mostly water. Like other polar molecules, protein and peptide hormones cannot diffuse across a lipid bilayer. These hormones always bind to receptors at the plasma membrane of a target cell.

When an amino acid–derived hormone binds to a receptor in the plasma membrane, a second messenger transmits the signal into the cell (**FIGURE 34.3A**). A **second messenger** is a molecule that forms inside a cell in response to an external signal, and it triggers a change in cellular activities. Formation of the second messenger sets in motion a chain of events that bring about the target cell's response.

In many cases, the cascade of reactions that result from second messenger formation culminates in activation of an enzyme already present in the cytoplasm. Consider glucagon, a protein hormone that is secreted by your pancreas and targets cells in your liver. The liver contains the body's main store of glycogen (Section 7.7). Binding of glucagon to its receptor in the plasma membrane of a liver cell sets in motion a chain of reactions. The end result of this chain is activation of an enzyme that breaks down glycogen, releasing glucose. As you will learn in Section 34.8, glucagon helps regulate the concentration of glucose in your blood.

Endocrine cells are themselves the targets of some hormones. In this case, second messenger formation affects the target cell's secretion of a different hormone. **Inhibiting hormones** discourage hormone secretion by their target endocrine cells. **Releasing hormones** encourage hormone secretion by their target endocrine cells. For example, the pituitary gland in your brain produces a releasing hormone that targets cells of the thyroid gland in your neck. Binding of the releasing hormone to its receptors triggers events that result in exocytosis of vesicles filled with thyroid hormone.

Effects of second messenger formation sometimes extend into the nucleus, bringing about changes in gene expression. However, protein and peptide hormones cannot enter a nucleus and interact directly with molecules that affect gene expression.

By contrast, steroid hormones can directly affect gene expression. A **steroid hormone** is derived from cholesterol and is nonpolar; it can dissolve in lipids, but not in water. Being lipid soluble, it can diffuse into a target cell directly across the plasma membrane.

A steroid hormone typically binds to its receptor inside the cell, forming a hormone–receptor complex (**FIGURE 34.3B**). This complex functions in the nucleus, where it binds to a promoter in the target cell's DNA. Recall that a promoter is a region where RNA polymerase binds (Section 9.2). Depending on the hormone, the receptor, and the cell, binding of the complex to a promoter can increase or decrease the rate of transcription of a nearby gene or set of genes.

Once secreted into the blood, hormones remain active for a limited period. Some are taken into and broken down in the kidney or the liver. Others are taken into the cell they affect and broken down there. The amount of time it takes to clear a hormone from the blood varies. Hormones derived from amino acids are typically cleared from the blood more quickly than steroid hormones.

Receptor Function and Diversity

A cell can only respond to a hormone for which it has appropriate and functional receptors. All hormone receptors are proteins. Mutations that result in missing or defective receptors, or alter their rate of production, also affect the response to a hormone—even if the hormone is present in normal levels. For example, typical male genitals will not form in an XY embryo without testosterone, which is a steroid hormone (Section 10.4). XY individuals who have total androgen insensitivity syndrome make and secrete testosterone, but they carry a mutation that affects their testosterone receptors. Without functional receptors, it is as if testosterone is not present. As a result, testes form during embryonic development but do not descend into the scrotum, and the genitals appear female at birth. Such individuals are often raised as females. Their condition becomes apparent at puberty when, lacking a uterus, they do not begin to menstruate.

Variations in receptor structure also affect responses to hormones. Different tissues have receptor proteins that respond in different ways to binding the same hormone. For example, ADH (antidiuretic hormone) acts on kidney cells and affects urine formation. ADH is sometimes referred to as vasopressin, because it also binds to receptors in the wall of blood vessels and causes the vessels to narrow. In many mammals, ADH helps maintain blood pressure. ADH also binds to

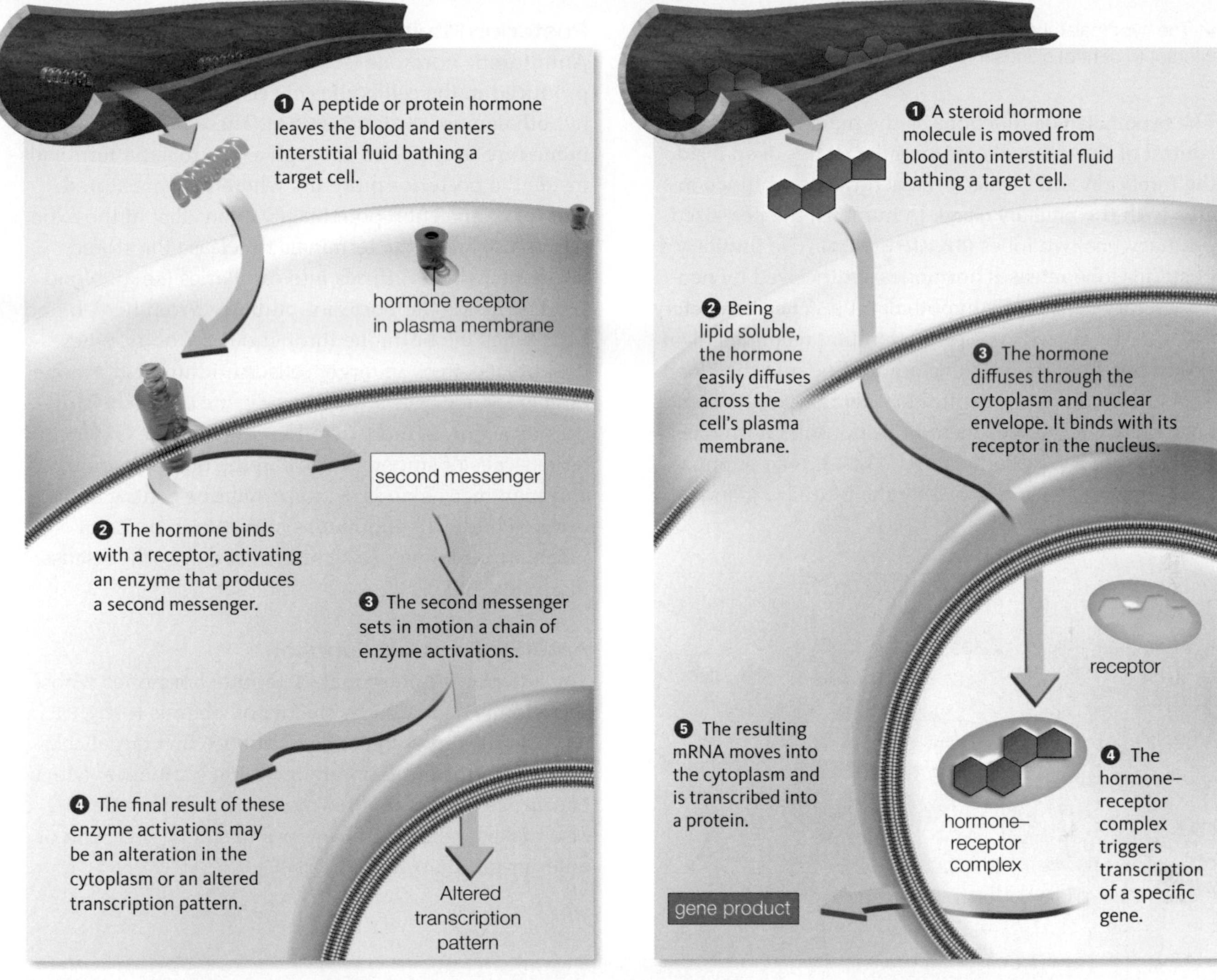

A Protein and peptide hormones act by way of a second messenger. **B** Steroid hormones can enter and act inside a target cell.

FIGURE 34.3 ▶**Animated** Examples of how hormones transduce signals.

FIGURE IT OUT Why does protein hormone action require a second messenger?

Answer: Protein hormones do not enter target cells.

brain cells and affects sexual and social behavior, as we will discuss in Section 43.2. This enormous diversity of responses to a single hormone is an outcome of variations in the structure of ADH receptors. In each kind of cell, a different kind of receptor summons up a different cellular response.

amino acid–derived hormone An amine (modified amino acid), peptide, or protein that functions as a hormone.
inhibiting hormone Hormone that deters release of a different hormone by its target endocrine cells.
releasing hormone Hormone that stimulates release of a different hormone by its target endocrine cells.
second messenger Molecule that forms inside a cell when a hormone binds to a receptor in the plasma membrane; sets in motion reactions that alter activity inside the cell.
steroid hormone Lipid-soluble hormone derived from cholesterol.

TAKE-HOME MESSAGE 34.3

How do hormones exert their effects on target cells?

✔ Hormones exert their effects by binding to protein receptors, either inside a cell or at the plasma membrane.

✔ Peptide and protein hormones cannot enter cells; they bind to a receptor at the plasma membrane. Often they trigger formation of a second messenger, a molecule that relays a signal into the cell.

✔ Steroid hormones can enter a cell and they often act in the nucleus, where they alter the expression of specific genes.

✔ Variations in receptor structure among cells types allow the same hormone to have different effects on different cells.

34.4 The Hypothalamus and Pituitary Gland

✔ The hypothalamus and pituitary gland deep inside the brain interact to control glands throughout the body.

The **hypothalamus** functions as the main center for control of the internal environment. It lies deep inside the forebrain and connects, structurally and functionally, with the **pituitary gland**. In humans, the pea-sized pituitary has two lobes (**FIGURE 34.4**). The pituitary's posterior lobe releases hormones synthesized by neurosecretory cells in the hypothalamus. A **neurosecretory cell** is a specialized type of neuron that responds to an action potential by releasing a hormone into the blood. The anterior lobe of the pituitary synthesizes its own hormones but releases them in response to hormones produced in the hypothalamus. **TABLE 34.1** summarizes the hormones released by the pituitary gland.

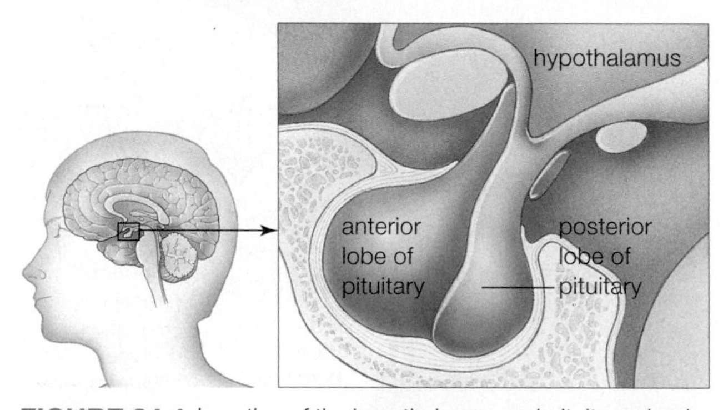

FIGURE 34.4 Location of the hypothalamus and pituitary gland.

Posterior Pituitary Function

Antidiuretic hormone (ADH) and oxytocin (OT) are produced in the cell bodies of secretory neurons of the hypothalamus (**FIGURE 34.5A**). These peptide hormones are transported through axons to axon terminals inside the posterior pituitary, where they are stored. Arrival of an action potential (Section 32.5) at the axon terminals causes the terminals to release the stored hormone, which diffuses into capillaries (small blood vessels) inside the posterior pituitary. From here, blood distributes the hormone throughout the body, where it exerts its effect on target cells. Antidiuretic hormone targets kidney cells and reduces urine output. We discuss its action in more detail in Section 40.5. Oxytocin targets cells of smooth muscle in the uterus (womb) and mammary glands. It causes uterine contractions during childbirth and moves milk into milk ducts when a woman nurses a child, events we will discuss in Section 42.11.

Anterior Pituitary Function

The anterior pituitary makes peptide hormones whose secretion is regulated by the hypothalamus (**FIGURE 34.5B**). Most hypothalamic hormones that target cells of the anterior pituitary are releasing hormones, which encourage secretion of hormones. The hypothalamus also makes inhibiting hormones that slow secretion of anterior pituitary hormones.

Table 34.1 Hormones Released From the Human Pituitary Gland

Pituitary Lobe	Secretions	Abbreviation	Main Targets	Main Effects
Posterior Nervous tissue (extension of hypothalamus)	Antidiuretic hormone (vasopressin)	ADH	Kidneys	Induces water conservation as required to maintain extracellular fluid volume and solute concentrations
	Oxytocin	OT	Mammary glands	Induces milk movement into secretory ducts
			Uterus	Induces uterine contractions during childbirth
Anterior Glandular tissue, mostly	Adrenocorticotropic hormone	ACTH	Adrenal glands	Stimulates release of cortisol, an adrenal steroid hormone
	Thyroid-stimulating hormone	TSH	Thyroid gland	Stimulates release of thyroid hormones
	Follicle-stimulating hormone	FSH	Ovaries, testes	In females, stimulates estrogen secretion, egg maturation; in males, helps stimulate sperm formation
	Luteinizing hormone	LH	Ovaries, testes	In females, stimulates progesterone secretion, ovulation, corpus luteum formation; in males, stimulates testosterone secretion, sperm release
	Prolactin	PRL	Mammary glands	Stimulates and sustains milk production
	Growth hormone (somatotropin)	GH	Most cells	Promotes growth in young; induces protein synthesis, cell division; roles in glucose, protein metabolism in adults

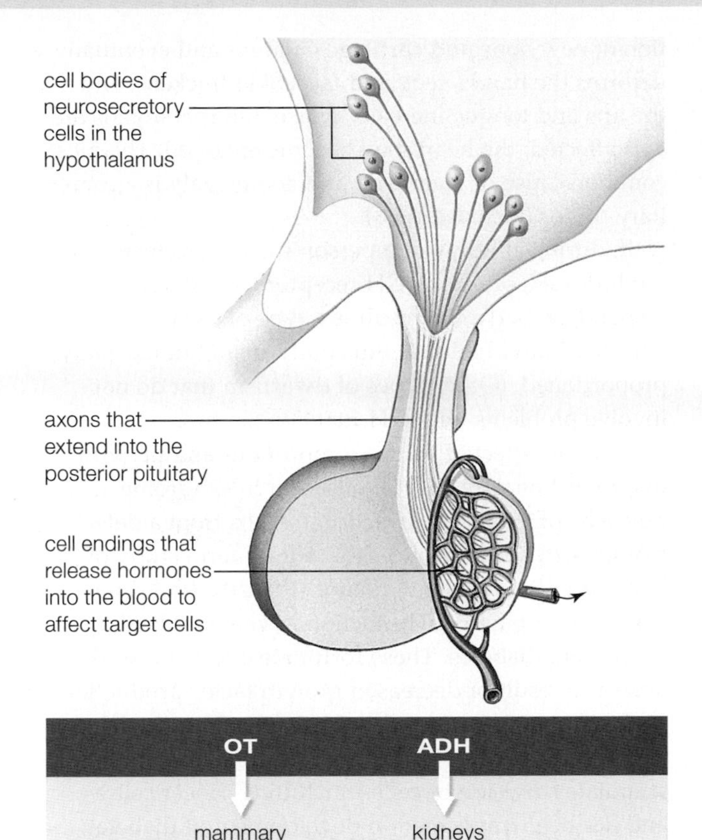

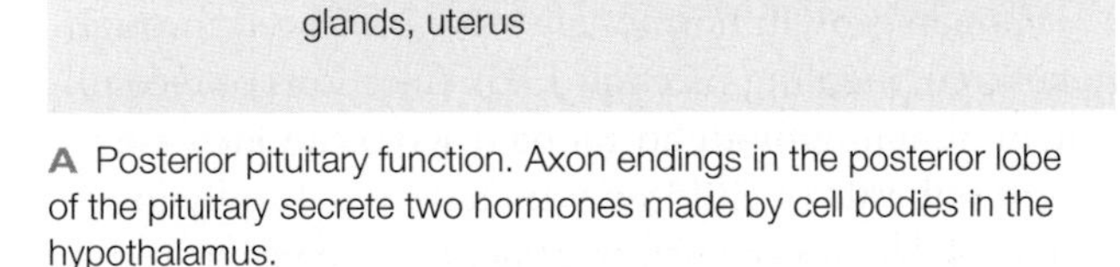

A Posterior pituitary function. Axon endings in the posterior lobe of the pituitary secrete two hormones made by cell bodies in the hypothalamus.

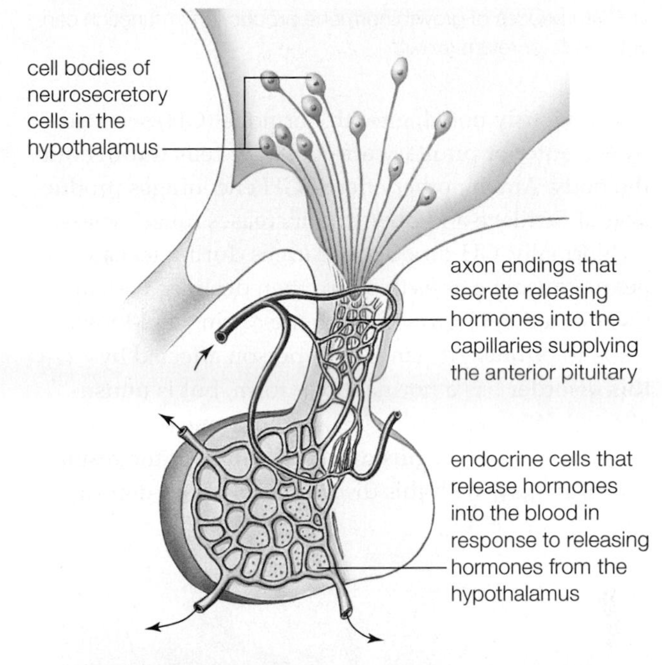

ACTH	TSH	FSH, LH	PRL	GH
adrenal cortex	thyroid gland	ovaries, testes	mammary glands	entire body

B Anterior pituitary function. Endocrine cells in the anterior lobe of the pituitary produce six hormones and secrete them in response to hypothalamic releasing hormones.

FIGURE 34.5 ▸**Animated** Pituitary function. Both lobes of the pituitary interact with the adjacent hypothalamus.

Four hormones produced by the anterior pituitary target other endocrine glands. Adrenocorticotropic hormone (ACTH) stimulates the release of cortisol by adrenal glands. Thyroid-stimulating hormone (TSH) causes the thyroid gland to secrete thyroid hormone. Follicle-stimulating hormone (FSH) and luteinizing hormone (LH) affect sex hormone secretion and production of gametes by gonads (a male's testes or a female's ovaries).

The anterior pituitary also produces two additional hormones. The first, prolactin (PRL), targets cells in the breast's mammary glands, which are exocrine glands. Prolactin contributes to breast development at puberty and governs milk production after a woman gives birth. The second, **growth hormone (GH)**, targets cells throughout the body. It affects metabolism, causing a decrease in fat storage and an increase in synthesis of muscle proteins. It also encourages production of new bone and cartilage.

growth hormone Anterior pituitary hormone that regulates growth and metabolism.
hypothalamus Forebrain region that controls processes related to homeostasis; in concert with the pituitary gland; serves as center for endocrine functions.
neurosecretory cell Specialized neuron that secretes a hormone into the blood in response to an action potential.
pituitary gland Pea-sized endocrine gland in the forebrain that interacts closely with the adjacent hypothalamus.

TAKE-HOME MESSAGE 34.4

How do the hypothalamus and pituitary gland interact?

✔ Some secretory neurons of the hypothalamus make hormones (ADH, OT) that move through axons into the posterior pituitary, which releases them.

✔ Other hypothalamic neurons produce releasers and inhibitors that are carried by the blood into the anterior pituitary. These hormones regulate the secretion of anterior pituitary hormones (ACTH, TSH, LH, FSH, PRL, and GH).

34.5 Growth Hormone Function and Disorders

✔ Disturbances of growth hormone production or function can accelerate or retard growth.

As previously noted, growth hormone (GH) secreted by the anterior pituitary affects target cells throughout the body. Among other effects, GH encourages production of cartilage and bone and increases muscle mass.

Normally, GH production surges during teenage years, causing a growth spurt, then declines with age. Oversecretion of growth hormone during childhood leads to pituitary gigantism. A person affected by this disorder has a normal body form, but is unusually tall. When excessive growth hormone secretion continues into or begins during adulthood, the result is acromegaly. With this disorder, continued deposition of new bone and cartilage enlarges and eventually deforms the hands, feet, and face. Skin thickens, and the lips and tongue increase in size. Internal organs are also affected; the heart may become enlarged. The most common cause of gigantism and acromegaly is a pituitary tumor (**FIGURE 34.6**).

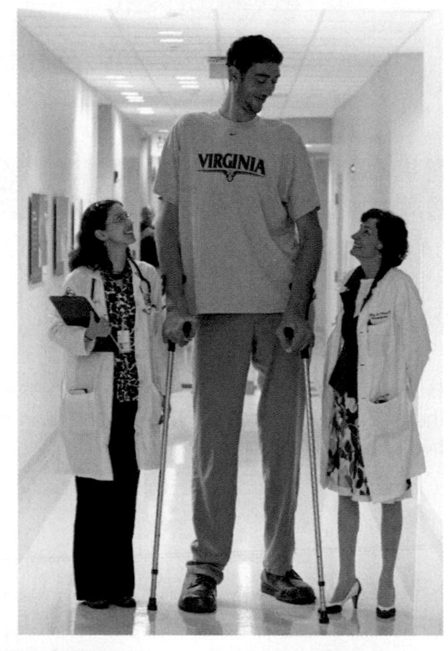

FIGURE 34.6 Sultan Kosen, the tallest living man, with his medical team at the University of Virginia.

Kosen had a pituitary tumor that caused excessive growth hormone secretion. Treatment halted his growth at 8 feet, 3 inches (2.5 meters).

If, during childhood, a person's body produces too little GH, or makes GH receptors that do not respond properly, the result is a type of dwarfism. Affected individuals are unusually small but normally proportioned. (Other types of dwarfism that do not involve problems with GH function are more common. Individuals affected by these conditions end up with disproportionately short limbs.) Laron syndrome, a rare type of inherited dwarfism, results from a defective growth hormone receptor. A long-term study of 99 affected individuals in Equador (**FIGURE 34.7**) suggests that impaired GH function lowers the risk for cancer and diabetes. These fortunate effects most likely arise as a result of decreased growth factor production. In the absence of prompting from GH, the liver does not produce insulin-like growth factor 1. This protein stimulates division of cells—including cancer cells—and has been implicated in development of diabetes.

FIGURE 34.7 A group of Equadorians with Laron syndrome, a heritable from of dwarfism caused by a mutated gene for the GH receptor. Jamie Guevara-Aguirre, at the rear in this photo, has studied the health of this group for more than 22 years.

Human growth hormone is now produced through genetic engineering (Section 15.6). Injections of recombinant human growth hormone (rhGH) can increase the growth rate of children who have a naturally low GH level. However, such treatment is expensive and it remains controversial. Some people object to treating short stature as a defect to be cured.

Injections of rhGH are also used to treat adults who have a low GH level as a result of pituitary or hypothalamic tumors or injury. Injections restore a normal level of GH, and can help affected individuals maintain a healthy bone and muscle mass. Injections of rhGH have also been touted as a way to slow normal aging or boost athletic performance. However, such uses are not approved by regulatory agencies and can have negative side effects, including increased risk of high blood pressure and diabetes.

TAKE-HOME MESSAGE 34.5

What are the effects of too much or too little growth hormone?

✔ Excessive growth hormone causes faster-than-normal bone growth. When the excess occurs during childhood, the result is gigantism. In adults, the result is acromegaly.

✔ A deficiency of GH or malfunction of GH receptors can cause dwarfism.

34.6 Sources and Effects of Other Vertebrate Hormones

✔ A cell in a vertebrate body is a target for a diverse array of hormones from endocrine glands and secretory cells.

The next few sections of this chapter describe effects of the main vertebrate hormones released by endocrine glands other than the pituitary. **TABLE 34.2** provides an overview of this information. In addition to major endocrine glands, vertebrates have hormone-secreting cells in many internal organs. As noted earlier, cells of the small intestine release secretin, a hormone that acts on the pancreas. Other gut hormones affect appetite and digestion. In addition, adipose (fat) tissue makes leptin, a hormone that suppresses appetite. When the oxygen level in blood falls, kidneys secrete erythropoietin, a hormone that stimulates maturation and production of oxygen-transporting red blood cells. Even the heart makes a hormone, atrial natriuretic peptide, which stimulates the kidneys to excrete water and salt.

As you learn about the effects of hormones, keep in mind that most cells have receptors for more than one hormone. The response called up by one hormone may oppose or reinforce that of another. For example, a skeletal muscle fiber has receptors for a variety of hormones, including glucagon, insulin, cortisol, epinephrine, estrogen, testosterone, growth hormone, somatostatin, and thyroid hormone. Thus, blood levels of all of these hormones affect a muscle.

TAKE-HOME MESSAGE 34.6

What are the sources and effects of vertebrate hormones?

✔ Endocrine glands and endocrine cells secrete hormones. Cells in other organs, such as the gut, kidneys, and heart also secrete hormones.

✔ Most cells have receptors for multiple hormones, and the effect of one can be enhanced or opposed by another.

Table 34.2 Vertebrate Hormones Discussed in Sections 34.7 to 34.12

Source	Examples of Secretion(s)	Main Target(s)	Primary Actions
Thyroid	Thyroid hormone	Most cells	Regulates metabolism; has roles in growth, development
	Calcitonin	Bone	Lowers calcium level in blood
Parathyroids	Parathyroid hormone	Bone, kidney	Elevates calcium level in blood
Pancreatic islets	Insulin	Liver, muscle, adipose tissue	Promotes cell uptake of glucose; thus lowers glucose level in blood
	Glucagon	Liver	Promotes glycogen breakdown; raises glucose level in blood
	Somatostatin	Insulin-secreting cells	Inhibits secretion of insulin, glucagon, and some gut hormones
Adrenal cortex	Glucocorticoids (including cortisol)	Most cells	Promote breakdown of glycogen, fats, and proteins as energy sources; thus help raise blood level of glucose
	Mineralocorticoids (including aldosterone)	Kidney	Promote sodium reabsorption (sodium conservation); help control the body's salt–water balance
Adrenal medulla	Epinephrine (adrenaline)	Liver, muscle, adipose tissue	Raises blood level of sugar, fatty acids; increases heart rate and force of contraction
	Norepinephrine	Smooth muscle of blood vessels	Promotes constriction or dilation of certain blood vessels; thus affects distribution of blood volume to different body regions
Gonads			
Testes (in males)	Androgens (including testosterone)	General	Required in sperm formation; development of genitals; maintenance of sexual traits; growth, development
Ovaries (in females)	Estrogens	General	Required for egg maturation and release; preparation of uterine lining for pregnancy and its maintenance in pregnancy; genital development; maintenance of sexual traits; growth, development
	Progesterone	Uterus, breasts	Prepares, maintains uterine lining for pregnancy; stimulates development of breast tissues
Pineal gland	Melatonin	Brain	Influences daily biorhythms, seasonal sexual activity
Thymus	Thymulin	T lymphocytes	Essential for maturation of T lymphocytes (T cells)

34.7 Thyroid and Parathyroid Glands

✔ The thyroid regulates metabolic rate, and the adjacent parathyroids regulate calcium levels.

Feedback Control of Thyroid Function

The human **thyroid gland** lies at the base of the neck, attached to the trachea, or windpipe (**FIGURE 34.8**). It secretes two iodine-containing amines (triiodothyronine and thyroxine) that we refer to collectively as thyroid hormone. Thyroid hormone increases the metabolic activity of cells throughout the body.

The anterior pituitary gland and hypothalamus regulate thyroid hormone secretion by a negative feedback loop (**FIGURE 34.9**). A low blood concentration of thyroid hormone causes the hypothalamus to secrete thyroid-releasing hormone (TRH) ❶. This releasing hormone causes the anterior pituitary to secrete thyroid-stimulating hormone (TSH) ❷. TSH in turn stimulates the secretion of thyroid hormone ❸. When the blood level of thyroid hormone rises, secretion of TRH and TSH declines ❹.

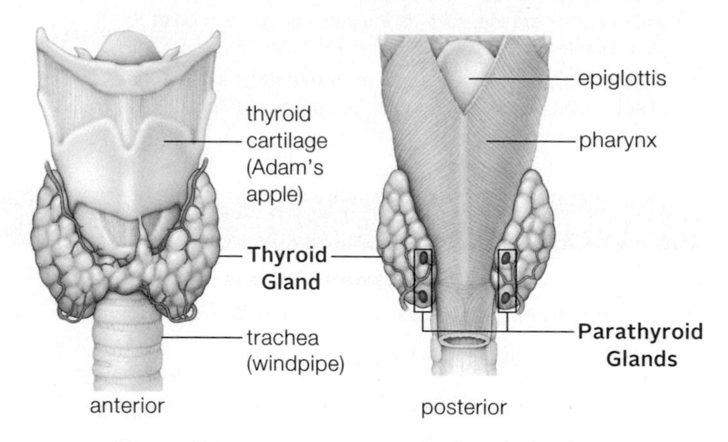

FIGURE 34.8 ▸Animated Location of human thyroid and parathyroid glands.

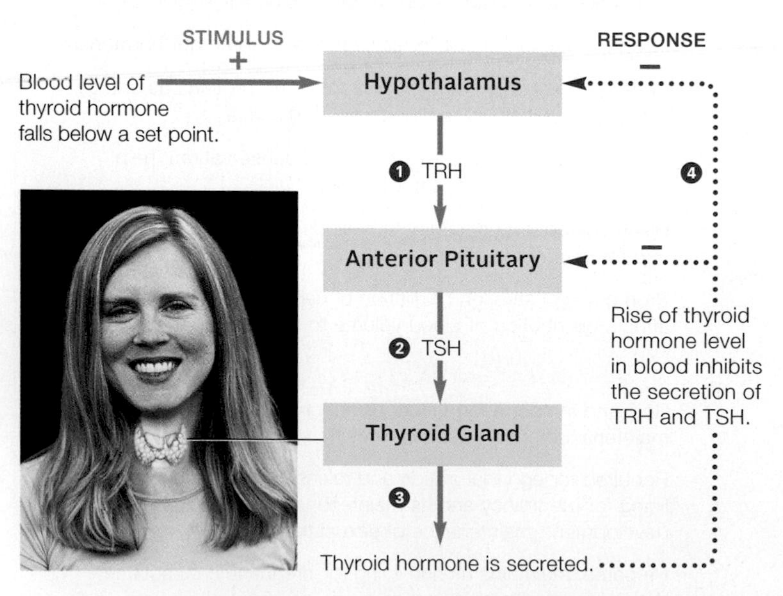

FIGURE 34.9 Negative feedback loop that governs thyroid hormone secretion. The loop involves the hypothalamus and the pituitary's anterior lobe.

FIGURE IT OUT What effect does a high thyroid hormone level have on the hypothalamus?

Answer: It inhibits secretion of TRH.

The thyroid gland also secretes calcitonin. In many animals, this hormone plays an important role in calcium homeostasis. In humans, however, calcitonin secretion occurs mainly during childhood, and adult blood calcium is regulated primarily by the parathyroids. Calcitonin supplements are sometimes used as a treatment for osteoporosis. Human adults normally produce little calcitonin.

Thyroid Disorders

A deficiency in thyroid hormone, a condition called hypothyroidism, results in a reduced metabolic rate. Symptoms include fatigue, depression, increased sensitivity to cold, and weight gain. Affected people often have thyroid enlargement, or goiter (**FIGURE 34.10**).

Thyroid hormone synthesis requires iodine, so adequate iodine intake is essential for health and normal development. A woman's iodine requirement increases during pregnancy and not meeting that requirement raises the risk of miscarriage and of infant mortality. It can also result in cretinism—mental retardation, stunted growth, and deaf-mutism—in her child.

In the United States, use of iodized salt has greatly reduced the incidence of dietary hypothyroidism, so inadequate levels of thyroid hormone most often result from an immune disorder. In some people, the body's white blood cells mistakenly attack the thyroid and destroy its hormone-producing tissue. The result is a type of hypothyroidism known as Hashimoto's disease. Thyroid tumors can also impair thyroid hormone production. Nondietary hypothyroidism is treated by the administration of synthetic thyroid hormone.

An excess of thyroid hormone also has negative health consequences. In Graves' disease, proteins (antibodies) made by white blood cells mimic the effect of thyroid-stimulating hormone. As a result, the thyroid produces an excess of thyroid hormone. Affected people have goiter, as well as anxiety, insomnia, heat intolerance, weight loss, and tremors. Protruding eyes are another symptom. Drugs, surgery, or radiation can be used to reduce thyroid hormone output.

Thyroid Disruptors

Some pollutants impair thyroid function. A sodium–iodide transport protein normally pumps iodide ions

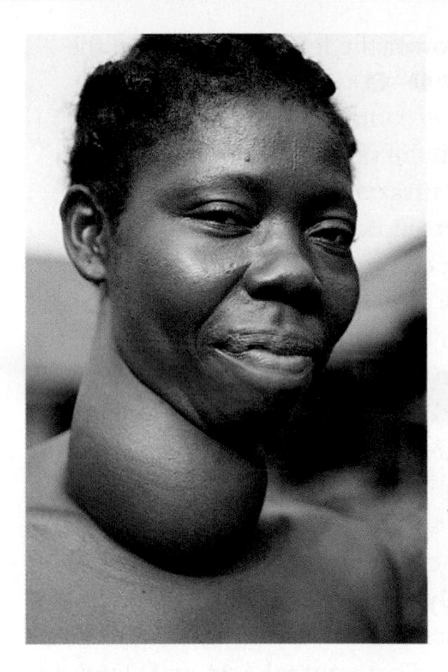

FIGURE 34.10 Goiter caused by a dietary iodine deficiency. The thyroid also enlarges in Graves' disease, an immune disorder that causes overstimulation of the thyroid.

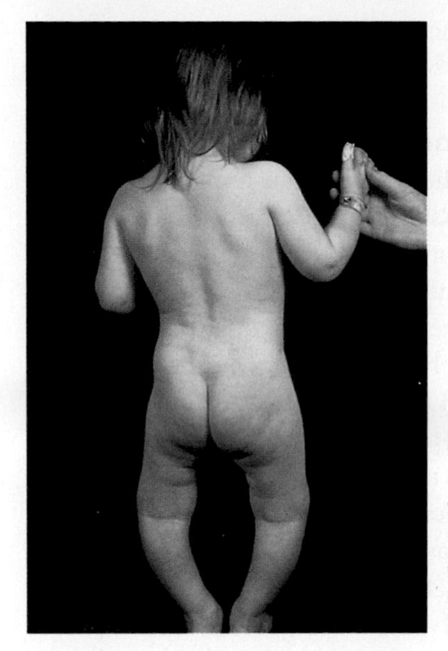

FIGURE 34.11 Child with rickets. Bowed legs result when parathyroid hormone causes existing bone to break down faster than new bone is deposited.

into thyroid cells. However, the transporter binds a chemical pollutant called perchlorate much more strongly than it binds iodide ions. Thus, when perchlorate is present, fewer iodide ions enter thyroid cells, and thyroid hormone production declines. Perchlorate seeps into the ground from facilities that manufacture, test, or dispose of military rockets, fireworks, or other explosives. Most municipal water treatment plants do not remove perchlorate from water, so perchlorate contamination of a source of public drinking water usually forces closure of that water source.

Perchlorate pollution also harms animals. In frogs, a surge in thyroid hormone triggers metamorphosis from a tadpole (the larval form) to an adult. If a tadpole's thyroid tissue is removed, it will keep on growing, but will never undergo metamorphosis. As you might predict, exposure to perchlorate-polluted water can similarly delay or prevent frog metamorphosis.

The Parathyroid Glands

Four **parathyroid glands**, each about the size of a grain of rice, are located on the thyroid's posterior surface (**FIGURE 34.8**). The glands release parathyroid hormone (PTH) in response to a decline in the level of calcium in blood. PTH targets bone cells and kidney cells. In bones, it induces specialized cells to secrete bone-digesting enzymes. Calcium and other minerals released from the bone enter the blood. In the kidneys, PTH stimulates tubule cells to reabsorb more calcium. It also stimulates secretion of enzymes that transform vitamin D to calcitriol, a steroid hormone that encourages cells in the intestinal lining to absorb calcium.

Vitamin D deficiency is the most common cause of the nutritional disorder known as rickets. Without adequate vitamin D, a child does not absorb much calcium from food, so formation of new bone slows. At the same time, the lower-than-normal calcium concentration in the blood triggers PTH secretion. When the concentration of PTH rises, the child's body responds by breaking down existing bones. Bowed legs and deformities in pelvic bones are common symptoms of rickets (**FIGURE 34.11**).

Tumors and other conditions that result in excessive PTH secretion also weaken bone. In addition, they increase the risk of kidney stones, because excess calcium ends up in the blood and urine. By contrast, disorders that reduce PTH output result in a lowered level of calcium in the blood. The resulting seizures and unrelenting muscle contractions can be deadly.

TAKE-HOME MESSAGE 34.7

What are the functions of the thyroid and parathyroid glands?

✔ The thyroid gland secretes an iodine-containing hormone that regulates metabolic rate and is essential to normal development.

✔ The parathyroid glands are the main regulators of blood calcium level in human adults.

parathyroid glands Four small endocrine glands whose hormone product increases the level of calcium in blood.
thyroid gland Endocrine gland at the base of the neck; produces thyroid hormone, which increases metabolism.

CREDITS: (10) Scott Camazine/Science Source; (11) Biophoto Associates/Science Source.

34.8 Pancreatic Hormones

✔ Two pancreatic hormones with opposing effects work together to regulate the level of glucose in the blood.

Regulation of Blood Sugar

The **pancreas**, which lies behind the stomach in the abdominal cavity, (**FIGURE 34.12**), has both exocrine and endocrine functions. The bulk of this organ consists of exocrine cells that secrete digestive enzymes into the small intestine. Scattered among these cells are pancreatic islets, which are clusters of endocrine cells.

Beta cells, the most abundant cells in pancreatic islets, secrete **insulin**—a peptide hormone that causes its target cells to take up and store glucose. After a meal, a rise in the concentration of glucose in the blood stimulates beta cells to release insulin. Insulin's main targets are liver, fat, and skeletal muscle cells. In muscle and adipose cells especially, insulin triggers glucose uptake. In all target cells, it encourages synthesis of fats and proteins and inhibits their breakdown. These cellular responses to insulin lower the level of glucose in the blood (**FIGURE 34.12** ❶–❺).

Pancreatic islets also contain alpha cells. These endocrine cells secrete the peptide hormone **glucagon** when the blood concentration of glucose falls. The glucagon binds to its receptors on liver cells, which respond by activating enzymes that break glycogen into glucose subunits. This response to glucagon raises the level of glucose in blood (**FIGURE 34.12** ❻–❿). By working in opposition, glucagon and insulin maintain the blood's glucose level within a range that keeps cells throughout the body functioning properly.

Diabetes

Diabetes mellitus is a common metabolic disorder. Its name can be loosely translated as "passing honey-sweet water." Diabetics have sweet urine because their liver, fat, and muscle cells do not take up and store glucose as they should. The resulting high blood sugar, or

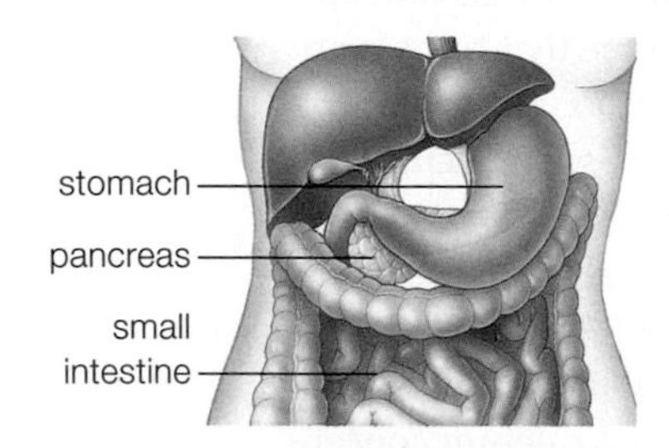

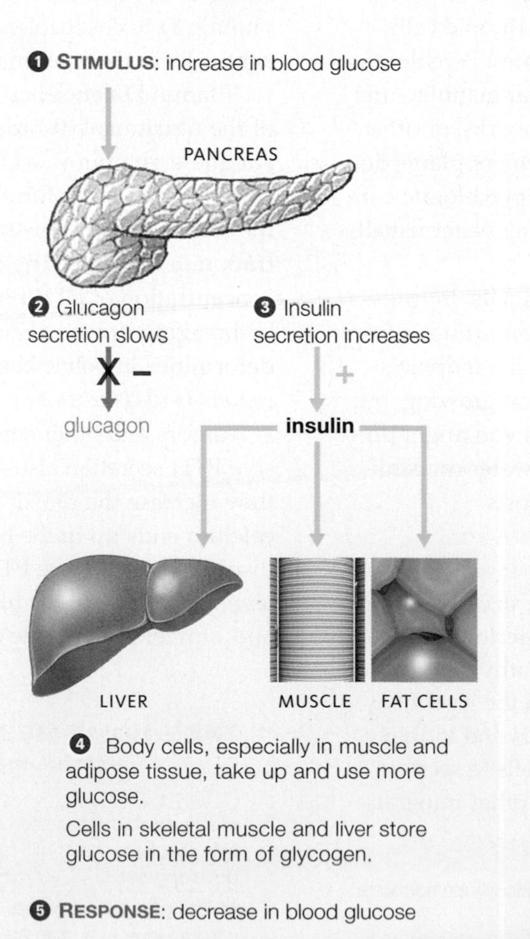

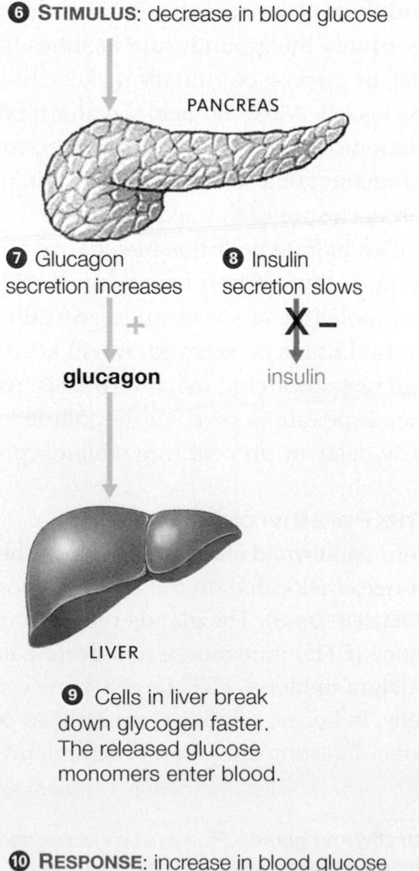

FIGURE 34.12 ▸**Animated** Above, the location of the pancreas. Right, how cells that secrete insulin and glucagon work antagonistically to adjust the level of glucose in the blood.

❶ After a meal, glucose enters blood faster than cells can take it up, so blood glucose increases.

❷ The increase stops pancreatic cells from secreting glucagon, and ❸ stimulates other pancreatic cells to secrete insulin.

❹ In response to insulin, adipose and muscle cells take up and store glucose; cells in the liver and muscle make more glycogen.

❺ As a result, the blood level of glucose declines to its normal level.

❻ Between meals, blood glucose declines as cells take it up and use it for metabolism.

❼ The decrease encourages glucagon secretion, and ❽ slows insulin secretion.

❾ In the liver, glucagon causes cells to break glycogen down into glucose, which enters the blood.

❿ As a result, blood glucose increases to the normal level.

Table 34.3 Some Complications of Diabetes

Eyes	Changes in lens shape and vision; damage to blood vessels in retina; blindness
Skin	Increased susceptibility to bacterial and fungal infections; patches of discoloration; thickening of skin on the back of hands
Digestive system	Gum disease; delayed stomach emptying that causes heartburn, nausea, vomiting
Kidneys	Increased risk of kidney disease and failure
Circulatory system	Increased risk of heart attack, stroke, high blood pressure, and atherosclerosis
Hands and feet	Impaired sensations of pain; formation of calluses, foot ulcers; tissue death that may require amputation

hyperglycemia, disrupts normal metabolism. Cells that do not take up glucose, have to break down proteins and fats to fuel ATP production, and breakdown of these substances yields harmful waste products. At the same time, high blood sugar causes some cells to overdose on glucose and produce other harmful substances. Accumulation of harmful molecules causes the complications associated with diabetes (**TABLE 34.3**).

Type 1 Diabetes There are two main types of diabetes mellitus. Type 1 develops after white blood cells mistakenly identify insulin-secreting beta cells as foreign and destroy them. Symptoms usually appear during childhood and adolescence; thus, this metabolic disorder is known as juvenile-onset diabetes. All affected individuals require injections of insulin, and must monitor their blood sugar concentration carefully. New devices called insulin pumps dispense a continuous supply of insulin (**FIGURE 34.13**).

Type 1 diabetes accounts for only 5 to 10 percent of all reported cases, but it is the most dangerous in the short term. In the absence of a steady supply of glucose, the cells of an affected person use fats and proteins as energy sources. Two outcomes are weight loss and ketone accumulation in the blood and urine. Ketones are normal acidic products of fat breakdown, but too many can alter the acidity and solute levels of body fluids. This condition, called ketosis, can interfere with normal brain function, and extreme cases may lead to coma or death.

glucagon Pancreatic hormone that causes cells to break down glycogen and release glucose.
insulin Pancreatic hormone that causes cells to take up glucose and store it as glycogen.
pancreas Organ that secretes digestive enzymes into the small intestine and hormones into the blood.

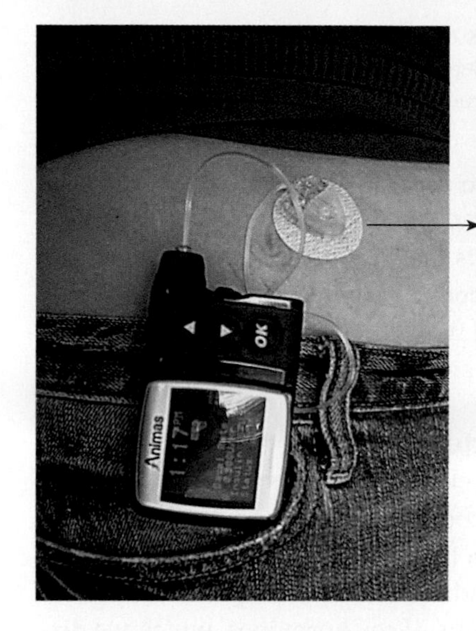

FIGURE 34.13 An insulin pump. The device is programmed to deliver insulin through a hollow tube that projects into the body.

Use of an insulin pump helps smooth out fluctuations in blood sugar, thus lowering the risk of complications that can arise from excessively low or high blood sugar.

Type 2 Diabetes With type 2 diabetes, the more common form of the disorder, insulin levels are normal or even high. However, target cells have an impaired response to the hormone, and blood sugar levels remain elevated. Symptoms typically start to develop in middle age, when insulin production declines.

Diet, exercise, and oral medications can control most cases of type 2 diabetes. Even so, if glucose levels are not lowered, pancreatic beta cells receive continual stimulation. Eventually they may falter, and so will insulin production. When that happens, a type 2 diabetic may require insulin injections.

Worldwide, rates of type 2 diabetes are soaring. By one estimate, more than 150 million people are now affected. Western diets and sedentary lifestyles are contributing factors. The prevention of diabetes and its complications is acknowledged as one of the most pressing public health priorities around the world.

TAKE-HOME MESSAGE 34.8

How do pancreatic hormones maintain the level of glucose in the blood?

✔ Insulin helps cells take up and store more glucose; it lowers the blood level of glucose.

✔ Glucagon triggers breakdown of glycogen; it raises blood glucose.

✔ Diabetes is a metabolic disorder in which the body does not make insulin or the body does not respond to it. As a result, cells do not take up sugar as they should, causing complications throughout the body.

34.9 The Adrenal Glands

✔ An adrenal gland has two functional zones: Its outer cortex secretes steroid hormones. Its inner medulla releases molecules that function as neurotransmitters.

Vertebrates have two **adrenal glands**, one above each kidney (*ad–* means near, and *renal* refers to the kidney). Each adrenal gland is the size of a big grape. Its outer layer is the **adrenal cortex** and its inner portion is the **adrenal medulla**. The two regions are controlled by different mechanisms, and secrete different substances.

The Adrenal Cortex

The adrenal cortex releases steroid hormones. One of these, aldosterone, controls sodium and water reabsorption by kidneys. (Section 40.5 explains this process in detail.) The adrenal cortex also produces and secretes small amounts of sex hormones, which we discuss in Section 34.10. Here we focus on **cortisol**, a hormone that affects metabolism and immune responses.

A negative feedback loop regulates the cortisol level in blood (**FIGURE 34.14**). When the cortisol level decreases, the hypothalamus increases its secretion of CRH (corticotropin-releasing hormone) ❶. CRH stimulates the anterior pituitary to secrete ACTH (adrenocorticotropic hormone) ❷, which in turn causes the adrenal cortex to release cortisol ❸. The blood level of cortisol rises. The rise is detected by the hypothalamus, and it stops secreting CRH, the anterior pituitary stops secreting ACTH, and cortisol secretion slows ❹.

Cortisol helps ensure that adequate glucose is available to the brain by inducing liver cells to break down their store of glycogen, and suppressing uptake of glucose by most cells. Cortisol also induces adipose cells to degrade fats, and skeletal muscles to degrade proteins. The breakdown products of these reactions—fatty acids and amino acids—function as energy sources (Section 7.7).

With injury, illness, or anxiety, the nervous system overrides the negative feedback loop regulating blood cortisol, and the cortisol level in the blood soars. In the short term, this response helps get enough glucose to the brain when food intake is likely to be low. A heightened cortisol level also suppresses inflammatory responses, thus lessening inflammation-related pain.

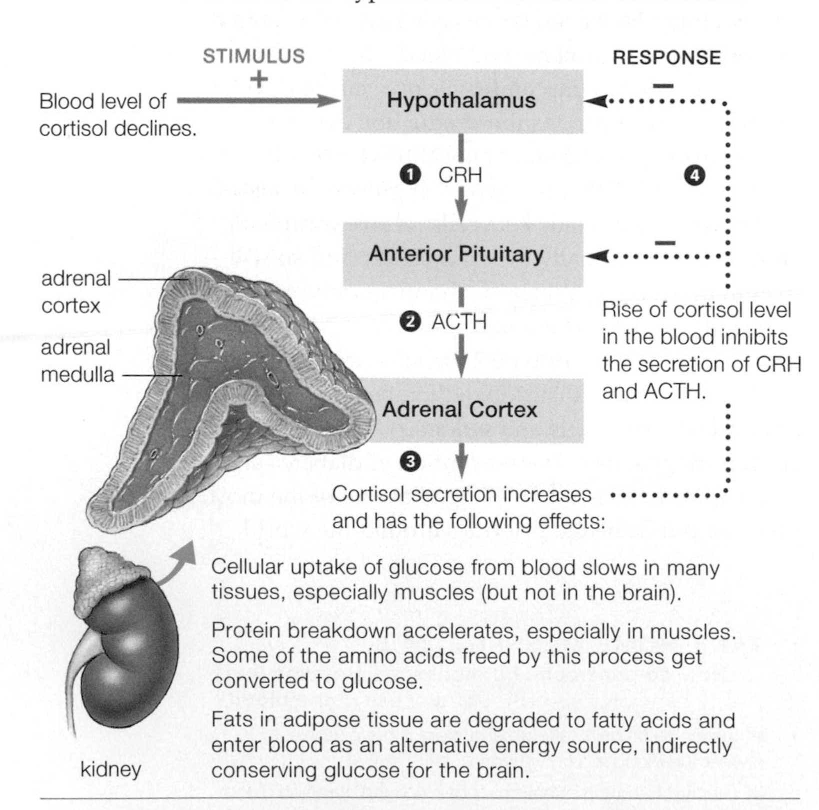

FIGURE 34.14 Structure of the human adrenal gland. An adrenal gland rests on top of each kidney. The diagram shows a negative feedback loop that governs cortisol secretion.

FIGURE IT OUT What effect would a decrease in ACTH have on the rate of fat breakdown in adipose tissue?

Answer: A decrease in ACTH would cause a decrease in cortisol secretion and less fat breakdown.

The Adrenal Medulla

The adrenal medulla contains specialized neurons of the sympathetic division (Section 32.8). Like other sympathetic neurons, those in the adrenal medulla release norepinephrine and epinephrine. However, in this case, the norepinephrine and epinephrine enter blood and function as hormones, rather than acting as neurotransmitters at a synapse. Epinephrine and norepinephrine released into the blood have the same effect on a target organ as direct stimulation by a sympathetic nerve.

Remember that sympathetic stimulation plays a role in the fight–flight response. Epinephrine and norepinephrine dilate the pupils, increase breathing rate, and make the heart beat faster. They prepare the body to deal with an exciting or dangerous situation.

Stress, Elevated Cortisol, and Health

When an animal is frightened or under physical stress, the nervous system triggers increased secretion of cortisol, epinephrine, and norepinephrine. As these hormones find their targets, they divert resources from longer-term tasks and help the body deal with the immediate threat. This stress response is highly adaptive for short periods of time, as when an animal is fleeing from a predator.

What happens when stress is ongoing? For more than twenty years, neurobiologist Robert Sapolsky and his Kenyan colleagues have been studying how olive baboons (*Papio anubis*) interact—specifically how a baboon's social position influences its hormone levels and health. Olive baboons live in large troops with a clearly defined dominance hierarchy. Individuals on top of the hierarchy get first access to food, grooming, and sexual partners. Those at the bottom of the hierarchy must continually relinquish resources to a higher-ranking baboon or face attack (**FIGURE 34.15**). Not surprisingly, the low-ranking baboons tend to have chronically high cortisol levels.

Physiological responses to chronic stress interfere with growth, the immune system, sexual function, and cardiovascular function. Chronically high cortisol levels also harm cells in the hippocampus, a brain region central to memory and learning (Section 32.12).

We also see the impact of long-term elevated cortisol levels in humans affected by Cushing's syndrome, or hypercortisolism. This rare metabolic disorder can be triggered by an adrenal gland tumor, oversecretion of ACTH by the anterior pituitary, or chronic use of the drug cortisone. Doctors often prescribe cortisone to relieve chronic pain, inflammation, or other health problems. The body converts it to cortisol.

Symptoms of hypercortisolism include a puffy, rounded "moon face" (**FIGURE 34.16**) and a fat torso. Blood pressure and blood glucose rise. White blood cell counts decline, so affected people are more prone to

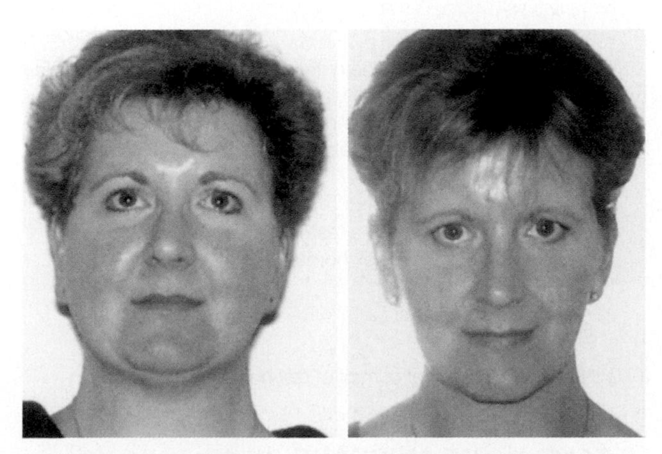

FIGURE 34.16 Cushing syndrome. Left, woman with elevated cortisol levels as a result of an adrenal gland tumor. She has the characteristic puffy moon face. Right, the same woman after removal of the tumor lowered her cortisol to normal levels.

adrenal cortex Outer portion of adrenal gland; secretes aldosterone and cortisol.
adrenal gland Endocrine gland located atop the kidney; secretes aldosterone, cortisol, epinephrine, and norepinephrine.
adrenal medulla Inner portion of adrenal gland; secretes epinephrine and norepinephrine.
cortisol Adrenal cortex hormone that influences metabolism and immunity; secretions rise with stress.

FIGURE 34.15 A dominant baboon (right) raising the stress level—and cortisol level—of a less dominant member of its troop.

infections. Thin skin, decreased bone density, and muscle loss are common. Patients with the highest cortisol level also have the greatest reduction in the size of the hippocampus, and the most impaired memory.

Can status-related social stress affect human health? People who are low in a socioeconomic hierarchy tend to have more health problems than those who are better off. These differences persist even after researchers factor out obvious causes, such as variations in diet and access to health care. By one hypothesis, a heightened cortisol level caused by low social status may be one of the links between poverty and poor health.

Cortisol Deficiency

Tuberculosis and other diseases can damage adrenal glands, and slow or halt cortisol secretion. The result is Addison's disease, or hypocortisolism. In developed countries, this disorder more often arises after autoimmune attacks on the adrenal glands. President John F. Kennedy had this form of the disorder. Symptoms include fatigue, depression, weight loss, and darkening of the skin. If cortisol declines too much, blood sugar and blood pressure fall to life-threatening levels. Addison's disease is treated with synthetic cortisone.

TAKE-HOME MESSAGE 34.9

What are the functions of the hormones secreted by the adrenal glands?

- ✔ The adrenal cortex's main secretions are aldosterone, which affects urine concentration, and cortisol, which affects metabolism and stress responses.
- ✔ The adrenal medulla releases epinephrine and norepinephrine, which prepare the body for excitement or danger.
- ✔ Cortisol secretion is governed by a feedback loop to the hypothalamus and pituitary, but stress breaks that loop and allows the level of cortisol to rise.
- ✔ Long-term cortisol elevation harms health. Insufficient cortisol can be fatal.

CREDITS: (15) Mitch Reardon/Science Source; (16) Permission obtained from Blackwell Publishing © Holt RIG and Hanley NA (2006) *Essential Endocrinology & Diabetes*, edn 5.

34.10 The Gonads

✔ Differences in sex hormone production are the basis for the differences in body form between males and females.

✔ Sex hormones also regulate gamete production.

Gonads are primary reproductive organs, meaning they produce gametes (eggs or sperm). Vertebrate gonads also produce **sex hormones**, steroid hormones that control sexual development and reproduction. **FIGURE 34.17** shows the location of the human gonads. A male's gonads—his testes (singular, testis)—secrete mainly testosterone. A female's gonads—her ovaries—secrete mainly estrogens and progesterone.

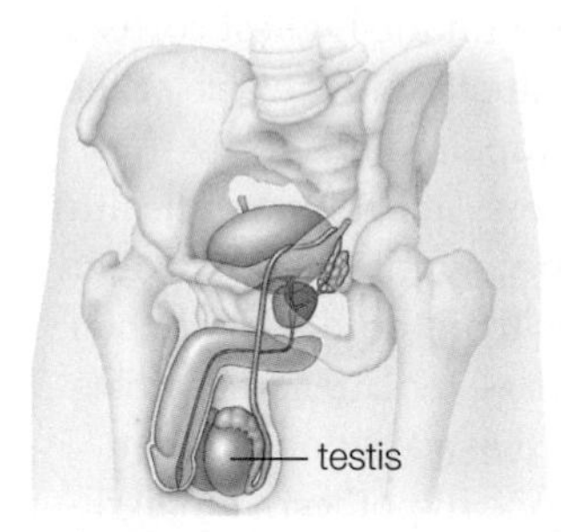

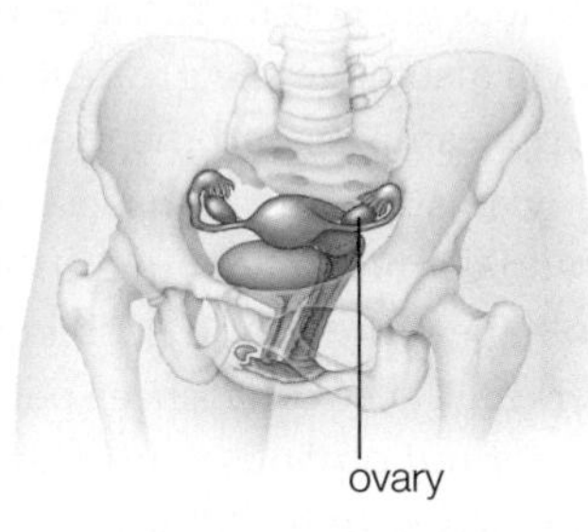

FIGURE 34.17 Human gonads. Testes (left) make sperm and the hormone testosterone. Ovaries (right) make eggs and the hormones estrogen and progesterone.

The hypothalamus and anterior pituitary control sex hormone secretion (**FIGURE 34.18**). In both sexes, the hypothalamus produces gonadotropin-releasing hormone (GnRH). This releasing hormone causes the anterior pituitary to secrete follicle-stimulating hormone (FSH) and luteinizing hormone (LH), which target the gonads and cause them to produce and secrete sex hormones. FSH and LH are sometimes referred to as "gonadotropins." The suffix *-tropin* refers to a substance that has a stimulating effect. In this case, the effect is on the gonads.

As Section 10.4 explained, the presence of an *SRY* gene causes an early human embryo to produce testosterone and develop a male reproductive tract. Embryos that do not make testosterone or do not respond to it develop a female reproductive tract. During childhood, levels of sex hormones are low and, with the exception of their reproductive organs, males and females have similar bodies.

Sex hormone production increases during **puberty**, the period of development when reproductive organs mature and secondary sexual traits appear. **Secondary sexual traits** are traits that differ between the sexes, but do not have a direct role in reproduction. When a girl undergoes puberty, her ovaries increase their estrogen production, causing fat deposition in breasts and on hips, and other female secondary sexual traits. Estrogens and progesterone also regulate egg formation and ready the uterus for pregnancy.

In males, a rise in testosterone output at puberty triggers the onset of sperm production and the development of secondary sexual traits. In humans, these traits include facial hair and a deep voice.

Testes secrete mostly testosterone, but they also make a little bit of estrogen and progesterone. The estrogen is necessary for sperm formation. Similarly, a female's ovaries make mostly estrogen and progesterone, but also a little testosterone. The presence of testosterone contributes to libido—the desire for sex.

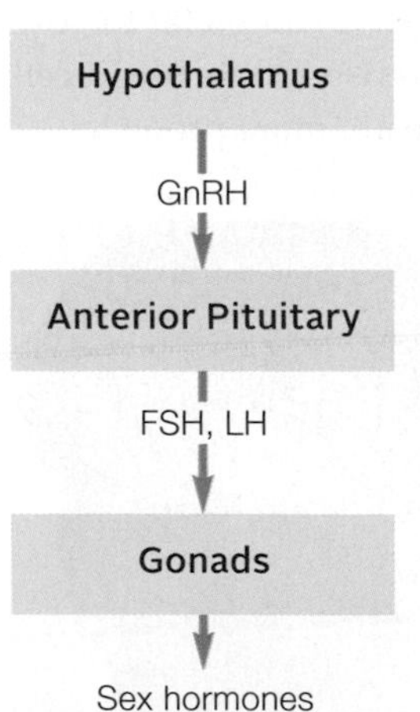

FIGURE 34.18 Central control of sex hormone secretion.

TAKE-HOME MESSAGE 34.10

What are sex hormones?

✔ Sex hormones are steroid hormones that influence the development of sexual traits and reproduction. Production of sex hormones increases at puberty.

✔ In both sexes, gonads secrete sex hormones in response to anterior pituitary hormones (FSH and LH), which are in turn released in response to a hypothalamic releasing hormone.

✔ A male's testes secrete mainly testosterone, and a female's ovaries secrete mainly estrogens and progesterone.

gonads Primary reproductive organs (ovaries or testes); produce gametes and sex hormones.
puberty Period when human reproductive organs mature and begin to function.
secondary sexual traits Traits, such as the distribution of body fat, that differ between the sexes but do not have a direct role in reproduction.
sex hormone A steroid hormone that controls sexual development and reproduction; testosterone in males, estrogens or progesterone in females.

34.11 The Pineal Gland

✔ At night, darkness allows the pineal gland to secrete a hormone that encourages sleep.

The **pineal gland** is a pea-sized, pinecone-shaped gland that secretes the hormone **melatonin**—but only during conditions of low light or darkness. In mammals, including humans, the pineal gland resides deep inside the brain (**FIGURE 34.19**). It receives information about light indirectly, by way of nervous signals from other regions of the brain.

The amount of light varies with the time of day, so melatonin secretion follows a **circadian rhythm**, meaning it varies cyclically over an approximately 24-hour interval. In turn, melatonin secretion influences other circadian rhythms. At night, increased melatonin secretion causes a decline in body temperature and we become sleepy. Just after sunrise, melatonin secretion decreases, body temperature rises, and we awaken. Melatonin's sleep-inducing effects stem from its effects on target cells in the hypothalamus.

Cycles of melatonin secretion that affect sleep–wake rhythms are set by exposure to light, so travelers who fly across multiple time zones are advised to spend some time in the sun to minimize jet lag. The exposure to light can help them reset their internal clock. Taking supplemental melatonin can also help prevent jet lag and alleviate some types of insomnia.

Night shift work and poor sleep habits can disrupt melatonin secretion and increase the risk of cancer. Melatonin normally protects against cancer in two ways. First, it directly inhibits growth of cancer cells. Second, it regulates the secretion of sex hormones that can encourage growth of some cancers.

Melatonin also has seasonal effects. Some people who live at latitudes where daylength fluctuates with the season have seasonal affective disorder (SAD), commonly called the "winter blues." They become depressed in winter, when there is less daylight. SAD arises when a person's rhythm of melatonin secretion gets out of sync with clock time. Exposure to bright light in the morning, a dose of melatonin at an appropriate time, or a combination of the two can diminish or eliminate symptoms of SAD.

Melatonin regulates seasonal changes in many animals. For example, melatonin level influences the timing of seasonal coat color changes in hares and some other mammals that live where winters are cold and snowy (**FIGURE 34.20**). These animals have a dark summer coat, and a white winter one.

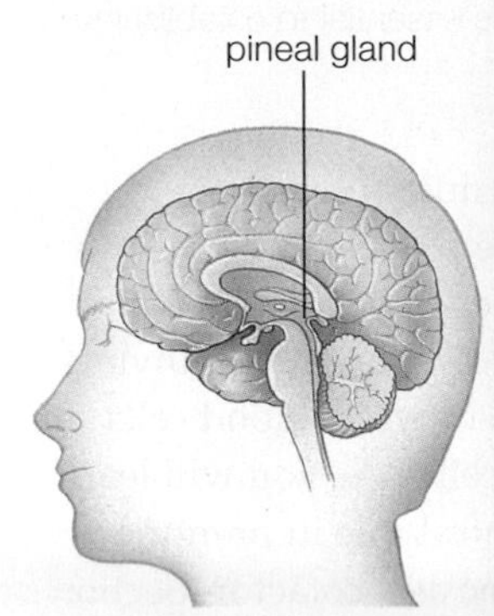

FIGURE 34.19 Location of the human pineal gland, which releases melatonin when the retina is not stimulated by light. Typically, secretion peaks at about 2 A.M., during sleep.

FIGURE 34.20 Hare shedding its dark fall coat as its white winter coat grows in. The white coat helps camouflage the animal against the snow. Melatonin level affects the timing of this change.

TAKE-HOME MESSAGE 34.11

What is the role of melatonin?

✔ Melatonin secreted by the pineal gland during periods of darkness encourages sleepiness and helps set the body's internal clock.

✔ Activities that disrupt nighttime melatonin secretion raise the risk of cancer.

✔ In many animals, seasonal differences in melatonin secretion give rise to seasonal differences in physiology.

circadian rhythm A physiological change that repeats itself on an approximately 24-hour cycle.
melatonin Hormone secreted by the pineal gland under conditions of darkness; affects sleep–wake cycles and protects against cancer.
pineal gland Endocrine gland in brain that secretes melatonin.

34.12 The Thymus

✔ Activity of the thymus early in life is essential to establishing a healthy immune system.

The **thymus** lies beneath the sternum (breastbone) and in front of the heart. It secretes a complex mix of small polypeptides that enhance wound healing and immune function. It also produces a hormone (thymulin) that encourages maturation of white blood cells called T cells (thymus-derived cells). As you will learn in Chapter 37, T cells play a central role in immune defenses. Thymulin requires zinc as a cofactor (Section 5.6); thus a diet deficient in zinc can result in impaired immune function.

The thymus is most active during prenatal development and the first few years of life, when the body's immune defenses are becoming established. Immune function can be severely impaired if the thymus fails to develop (a consequence of a rare genetic disorder), or becomes damaged during infancy. An infant without a functional thymus will die of infection unless he or she receives a transplant of thymus tissue.

The size of the thymus peaks just before puberty. From puberty on, the thymus decreases in size, and more and more of its endocrine cells are replaced by adipose tissue. By age 40, a normal thymus consists largely of fat. However, the endocrine cells that remain will continue to secrete a low but steady supply of hormones for the remainder of the individual's life.

The normal reduction in the activity of the thymus during adulthood does not usually have a dramatic adverse effect because a healthy adult retains many mature T cells that formed at a younger age. However, lowered thymus activity may contribute to the increased incidence of cancer and infectious disease that accompanies aging. Also, the normal reduction in thymus activity is highly problematic for people infected by HIV; the virus kills T cells and without an active thymus, those cells cannot be replaced.

thymus Organ beneath the sternum; produces hormones that enhance immune function and is the site of T cell maturation.

TAKE-HOME MESSAGE 34.12

What is the role of the thymus?

✔ The thymus produces hormones required for maturation of one type of immune cell (T cells).

✔ The thymus is most active during childhood when the immune system is becoming established. During adulthood, the thymus shrinks and secretory cells are replaced by fat.

34.13 Invertebrate Hormones

✔ Animal hormones have a long evolutionary history. Both invertebrates and vertebrate produce hormones.

Evolution of Hormone Diversity

We can trace the evolutionary roots of some vertebrate hormones and hormone receptors back to homologous molecules in invertebrates. For example, receptors for the hormones FSH, LH, and TSH all have a similar structure. The genes that encode these receptors have a similar sequence and all have introns (noncoding DNA) in the same places. The slightly different forms of receptor gene most likely evolved when an ancestral gene was duplicated, then copies of that gene mutated over time (Section 14.5). The ancestral receptor gene apparently evolved very long ago, in the common ancestor of cnidarians and vertebrates. Cnidarians such as sea anemones have a receptor protein gene with a sequence similar to the gene that encodes the vertebrate FSH receptor. Proteins that serve as vertebrate estrogen receptors also have a long history. Like vertebrates, annelids have an estrogen receptor that binds estrogen and functions as a transcription factor. This suggests that annelids, like vertebrates, are adversely affected by chemical pollutants that disrupt estrogen function. Invertebrates do not have the same glands as vertebrates, so they produce homologous hormones in other glands. For example, octopuses produce estrogens and cortisol in a gland near their eye.

Hormones and Molting

Other hormones are unique to invertebrates. For example, arthropods, which include crabs and insects, have a hardened external cuticle that they periodically shed as they grow (Section 24.11). Shedding of the old cuticle is called molting. A soft new cuticle forms beneath the old one before the animal molts. Although details vary among arthropod groups, molting is generally under the control of **ecdysone**, a steroid hormone.

The arthropod molting gland produces and stores ecdysone, then releases it for distribution throughout the body when conditions favor molting. Hormone-secreting neurons inside the brain control ecdysone's release. The neurons respond to internal signals and environmental cues, including light and temperature.

FIGURE 34.21 is an example of the control steps in crabs and other crustaceans. In response to seasonal cues, secretion of a molt-inhibiting hormone declines and ecdysone secretion rises. Ecdysone causes changes in the animal's structure and physiology. The existing

ecdysone Insect hormone with roles in metamorphosis, molting.

Hormones in the Balance (revisited)

Chemicals called phthalates (pronounced THAL-aytes) are commonly used to make plastics more flexible and to stabilize fragrances in scented products. Like atrazine, they are also endocrine disruptors.

Phthalates are not approved for use in human foods, but in 2011 a Taiwanese company illegally added them to a variety of foods and drinks. Children who had ingested these items had a lowered level of TSH (the pituitary hormone that stimulates thyroid hormone production). The reduction in TSH was dose dependent; the more contaminated food and drink a child had taken in, the greater his or her reduction in TSH level. Results of this horrible unplanned experiment are consistent with epidemiological studies that implicate phthalates in disrupted thyroid function. Other studies have found a positive correlation between phthalate exposure and diabetes.

Phthalates also affect sex hormones. They inhibit testosterone synthesis in rats, and exposure of male rats to phthalates during prenatal development interferes with development of their sex organs. Epidemiological studies suggest phthalates may have similar effects in humans. In adult men, a high concentration of phthalates is correlated with a low testosterone level and decreased sperm quality. In women, having a high phthalate level during pregnancy is correlated with an increased likelihood of giving birth to a son with an undescended testicle.

In light of the increasing evidence that exposure to phthalates has negative health effects, some efforts have been made to reduce the use of these chemicals. Phthalates were widely used in pacifiers and teething toys until 2007, when a U.S. law went into effect limiting the phthalate content of products intended for kids under age twelve. Phthalates have also been largely eliminated from cling-wraps and other American-made products designed to come into contact with food.

However, opportunities for phthalate exposure remain. Vinyl, which releases phthalates into the air, is common in furniture, flooring, mattress covers, and shower curtains. Phthalates also remain in artificially scented goods ranging from air fresheners to personal care products.

Pediatrician Sheela Sathyanarayana found that the more scented powders, shampoos, and lotions a mother used on her infant, the higher the level of phthalates in the child's urine. As a result, the American Academy of Pediatrics recommends that parents choose unscented products for use on infants and young children. Similarly, many obstetricians recommend that pregnant women use personal care products that are labeled as organic or free of phthalates. Phthalates do not have to be listed as ingredients if they are a component of a fragrance.

---

cuticle separates from the epidermis and the muscles. Inner layers of the old cuticle break down. At the same time, cells of the epidermis secrete the new cuticle.

The steps in molting differ a bit in insects, which do not have a molt-inhibiting hormone. Rather, stimulation of the insect brain sets in motion a cascade of signals that trigger the production of molt-inducing ecdysone. Chemicals that mimic ecdysone or interfere with its function are sometimes used as insecticides. When such insecticides run off from fields and get into water, they can affect ecdysone signaling pathways in noninsect arthropods, such as crayfish or crabs.

TAKE-HOME MESSAGE 34.13

What types of hormones do invertebrates produce?

- ✔ Some invertebrate hormones are homologous to those in vertebrates.
- ✔ Invertebrates also make hormones with no vertebrate counterpart. Ecdyson, which affects molting in arthropods, is one example.

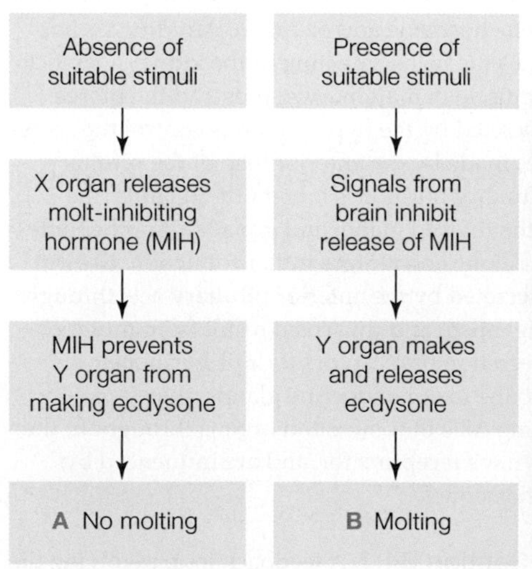

FIGURE 34.21 Control of molting in crabs. Two hormone-secreting organs play a role, the X organ in the eye stalk and the Y organ at the base of the antennae.

(A) In the absence of environmental cues for molting, hormone from the X organ prevents molting. (B) When stimulated by proper environmental cues, the brain sends nervous signals that inhibit X organ activity. With the X organ suppressed, the Y organ releases the ecdysone that stimulates molting.

The photo shows a newly molted blue crab with its old shell. The new shell remains soft for about 12 hours, making it a "soft-shelled crab." During this time, the crab is highly vulnerable to predators, including human seafood lovers.

summary

Section 34.1 **Endocrine disruptors** are synthetic substances that interfere with the endocrine system. The widely used synthetic chemicals atrazine and phthalates are endocrine disruptors, which interfere with the function of sex hormones.

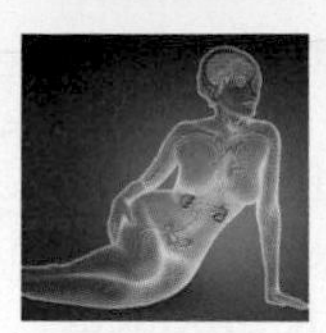

Section 34.2 Neurotransmitters, **local signaling molecules**, and **animal hormones** are signaling molecules that bind to receptors on target cells. Hormones travel through the blood and can convey signals between cells in distant parts of the body. All hormone-secreting glands and cells in a body constitute the animal's **endocrine system**. The endocrine system is closely tied to the nervous system.

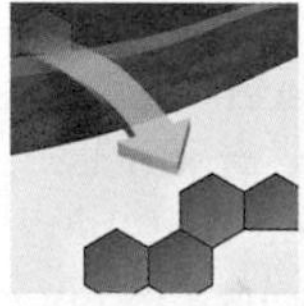

Section 34.3 **Steroid hormones** are lipid soluble and derived from cholesterol. They can enter cells and bind to receptors inside them. **Amino acid–derived hormones** bind to receptors in the cell membrane. Binding triggers the formation of a **second messenger** that elicits changes inside the cell. Only cells with functional receptors for a hormone can respond to it. Variation in the structure of a hormone receptor allow a hormone to elicit different responses in different types of cells. **Releasing hormones** and **inhibiting hormones** target other endocrine glands.

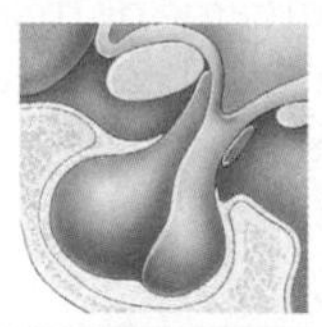

Sections 34.4–34.6 Deep in the forebrain, the **hypothalamus** is structurally and functionally linked with the **pituitary gland**. Axons of **neurosecretory cells** in the hypothalamus extend into the posterior pituitary, where they release antidiuretic hormone and oxytocin. Antidiuretic hormone concentrates the urine by acting in the kidney. Oxytocin targets smooth muscle in mammary glands and the uterus.

Hormones secreted by the hypothalamus control the secretion of hormones made by the anterior lobe of the pituitary. Four anterior pituitary hormones target other glands (the adrenal cortex, the thyroid gland, and gonads). Another anterior pituitary hormone encourages milk production. **Growth hormone (GH)** secreted by the anterior pituitary acts throughout the body. Gigantism and dwarfism result from mutations that affect GH secretion or receptors for this hormone.

In addition to the major endocrine glands, there are hormone-secreting cells in tissues and organs throughout the body. Most cells have receptors for, and are influenced by, many different hormones.

Section 34.7 A feedback loop involving the anterior pituitary and hypothalamus governs secretion of thyroid hormone by the **thyroid gland** at the base of the neck. This iodine-containing hormone increases metabolic rate, and is also is required for normal development. The thyroid also produces calcitonin, which encourages calcium deposition in bone during childhood.

The **parathyroid glands** release a hormone that targets bone and kidney cells, and raises the blood calcium level.

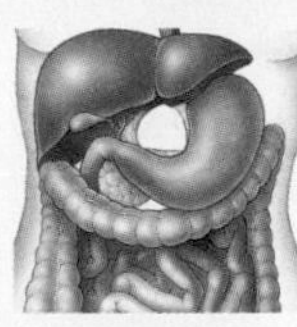

Section 34.8 The **pancreas**, located inside the abdominal cavity, has both exocrine and endocrine functions. Beta cells secrete the hormone **insulin** when the blood glucose level is high. Insulin stimulates uptake of glucose by muscle and liver cells. When the blood glucose level is low, alpha cells secrete **glucagon**, a hormone that causes liver cells to break down glycogen and release glucose. The two hormones work in opposition to keep the blood glucose concentration within an optimal range.

Diabetes occurs when the body does not make insulin (Type 1 diabetes) or its cells do not respond to it (Type 2 diabetes). The resulting disruption of glucose metabolism harms cells throughout the body.

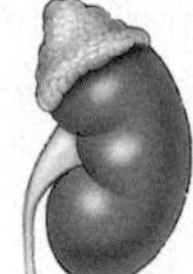

Section 34.9 Vertebrates have an **adrenal gland** atop each kidney. The gland's outer layer, or **adrenal cortex**, secretes aldosterone, which acts in the kidney, and **cortisol**, the stress hormone. The cortisol level in blood is stabilized by a negative feedback loop involving the anterior pituitary and hypothalamus. In times of stress, the nervous system overrides this control and the blood cortisol level soars. Norepinephrine and epinephrine released by neurons of the **adrenal medulla** influence organs as sympathetic stimulation does; they cause a fight–flight response. Long-term elevation of blood cortisol level, as a result of either stress or a disorder, is harmful. A total lack of cortisol is fatal.

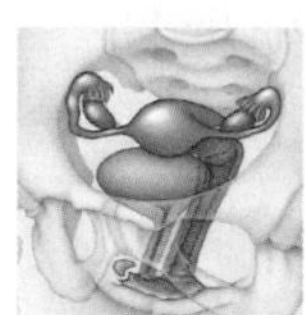

Section 34.10 **Gonads** (ovaries and testes) make gametes and secrete the **sex hormones**. Ovaries secrete mostly estrogens and progesterone. Testes secrete mostly testosterone. Sex hormone output rises at **puberty**, and the increase in the concentration of these hormones encourages development of **secondary sexual traits** such as facial hair in males or rounded breasts in females.

In both sexes, follicle-stimulating hormone (FSH) and luteinizing hormone (LH) from the anterior pituitary govern sex hormone production. LH and FSH secretion are in turn controlled by gonadotropin-releasing hormone from the hypothalamus.

Sections 34.11, 34.12 Light suppresses secretion of **melatonin** by the **pineal gland** in the brain. Melatonin secretion, which occurs in a **circadian rhythm**, causes drowsiness and changes in body temperature. Melatonin also has a protective effect against cancer.

The **thymus** in the chest produces thymulin that helps some white blood cells (T cells) mature. Its function is essential in children, but it is largely inactive in adults over age 40.

Section 34.13 Some invertebrate hormones are homologous to hormones in vertebrates, although they are made in different glands. Other invertebrate hormones such as **ecdysone**, a steroid hormone that regulates molting in arthropods, have no vertebrate counterpart.

data analysis activities

Sperm Counts Down on the Farm Contamination of water by agricultural chemicals affects reproductive function in some animals. Are there effects on humans? Epidemiologist Shanna Swan and her colleagues studied sperm from men in four cities in the United States (**FIGURE 34.22**). The men were partners of women who had become pregnant and were visiting a prenatal clinic, so all were fertile. Of the four cities, Columbia, Missouri, is located in the county with the most farmlands. New York City in New York is in an area with no agriculture.

1. In which cities did researchers record the highest and lowest sperm counts?
2. In which cities did samples show the highest and lowest sperm motility (ability to move)?
3. Aging, smoking, and sexually transmitted diseases adversely affect sperm. Could differences in any of these variables explain the regional differences in sperm count?
4. Do these data support the hypothesis that living near farmlands can adversely affect male reproductive function?

	Location of Clinic			
	Columbia, Missouri	Los Angeles, California	Minneapolis, Minnesota	New York, New York
Average age	30.7	29.8	32.2	36.1
Percent nonsmokers	79.5	70.5	85.8	81.6
Percent with history of STD	11.4	12.9	13.6	15.8
Sperm count (million/ml)	58.7	80.8	98.6	102.9
Percent motile sperm	48.2	54.5	52.1	56.4

FIGURE 34.22 Characteristics of men in four cities. All men were partners of women who visited prenatal health clinics, and so were presumably fertile. STD stands for sexually transmitted disease.

self-quiz

Answers in Appendix VII

1. _______ are signaling molecules that travel through the blood and affect distant cells in the same individual.
   a. Hormones
   b. Neurotransmitters
   c. Pheromones
   d. Local signaling molecules
   e. both a and b
   f. a through d

2. A _______ is synthesized from amino acids and cannot diffuse across the plasma membrane.
   a. steroid hormone
   b. protein hormone
   c. peptide hormone
   d. both b and c

3. Match each pituitary hormone with its target.
   ___ antidiuretic hormone — a. gonads (ovaries, testes)
   ___ oxytocin — b. mammary glands, uterus
   ___ luteinizing hormone — c. kidneys
   ___ growth hormone — d. most body cells

4. Releasing hormones secreted by the hypothalamus cause secretion of hormones by the _______ pituitary lobe.
   a. anterior
   b. posterior

5. In adults, too much _______ can cause acromegaly.
   a. melatonin
   b. cortisol
   c. insulin
   d. growth hormone

6. Function of the _______ declines after puberty.
   a. parathyroid glands
   b. thymus
   c. pancreas
   d. hypothalamus

7. Low blood calcium triggers secretion by _______ .
   a. adrenal glands
   b. ovaries
   c. parathyroid glands
   d. the thyroid gland

8. _______ lowers blood sugar levels; _______ raises it.
   a. Glucagon; insulin
   b. Insulin; glucagon

9. The _______ has endocrine and exocrine functions.
   a. hypothalamus
   b. parathyroid gland
   c. pineal gland
   d. pancreas

10. Secretion of _______ suppresses immune responses.
   a. melatonin
   b. antidiuretic hormone
   c. thyroid hormone
   d. cortisol

11. Exposure to bright light lowers blood _______ levels.
   a. glucagon
   b. melatonin
   c. thyroid hormone
   d. parathyroid hormone

12. True or false? Some heart cells and kidney cells secrete hormones.

13. True or false? Only women make follicle-stimulating hormone (FSH).

14. True or false? All hormones secreted by arthropods such as crabs and insects are also secreted by vertebrates.

15. Match the term listed at left with the most suitable description at right.
   ___ adrenal medulla — a. largely inactive in adults
   ___ thyroid gland — b. a local signaling molecule
   ___ posterior pituitary gland — c. secretes hormones made in the hypothalamus
   ___ pancreatic islets — d. source of epinephrine
   ___ thymus — e. secrete insulin, glucagon
   ___ prostaglandin — f. hormones require iodine

critical thinking

1. Women who were blind from birth almost never get breast cancer, probably as a result of their high level of one hormone. Which hormone is it and why it unusually high in the blind?

2. A diabetic who injects too much insulin can lose consciousness. Explain why injecting excess insulin could impair brain function. Glucagon reverses this effect. Explain how.

35 Structural Support and Movement

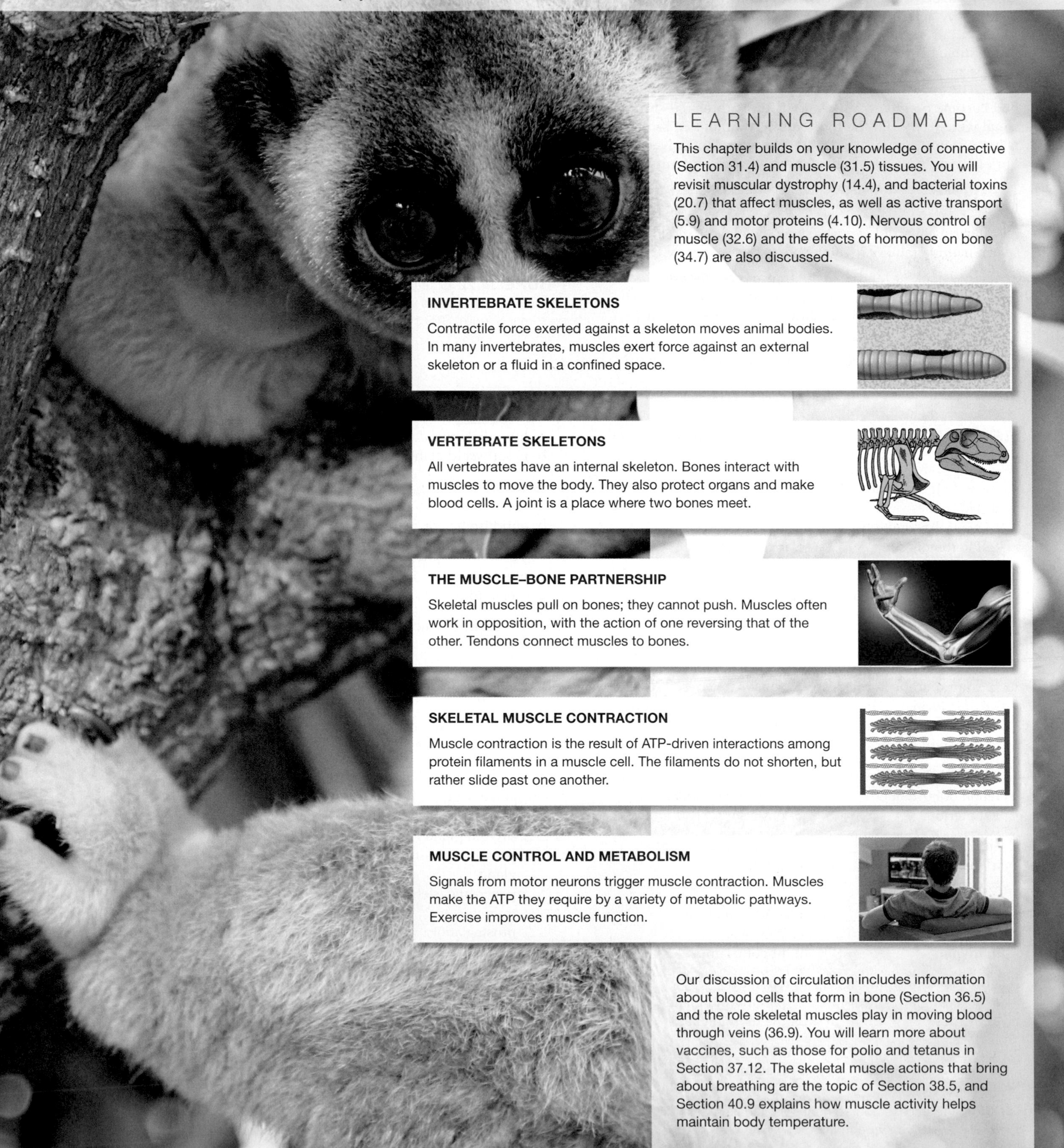

LEARNING ROADMAP

This chapter builds on your knowledge of connective (Section 31.4) and muscle (31.5) tissues. You will revisit muscular dystrophy (14.4), and bacterial toxins (20.7) that affect muscles, as well as active transport (5.9) and motor proteins (4.10). Nervous control of muscle (32.6) and the effects of hormones on bone (34.7) are also discussed.

INVERTEBRATE SKELETONS

Contractile force exerted against a skeleton moves animal bodies. In many invertebrates, muscles exert force against an external skeleton or a fluid in a confined space.

VERTEBRATE SKELETONS

All vertebrates have an internal skeleton. Bones interact with muscles to move the body. They also protect organs and make blood cells. A joint is a place where two bones meet.

THE MUSCLE–BONE PARTNERSHIP

Skeletal muscles pull on bones; they cannot push. Muscles often work in opposition, with the action of one reversing that of the other. Tendons connect muscles to bones.

SKELETAL MUSCLE CONTRACTION

Muscle contraction is the result of ATP-driven interactions among protein filaments in a muscle cell. The filaments do not shorten, but rather slide past one another.

MUSCLE CONTROL AND METABOLISM

Signals from motor neurons trigger muscle contraction. Muscles make the ATP they require by a variety of metabolic pathways. Exercise improves muscle function.

Our discussion of circulation includes information about blood cells that form in bone (Section 36.5) and the role skeletal muscles play in moving blood through veins (36.9). You will learn more about vaccines, such as those for polio and tetanus in Section 37.12. The skeletal muscle actions that bring about breathing are the topic of Section 38.5, and Section 40.9 explains how muscle activity helps maintain body temperature.

35.1 Muscles and Myostatin

The more you use your muscles, the bulkier and more powerful they become. A mature skeletal muscle fiber cannot divide, but it can make more of the proteins involved in muscle contraction. Protein filaments inside a muscle fiber are continually built and broken down. Exercise tilts this process in favor of synthesis, so muscle cells get bigger and the muscle gets stronger. In addition, exercise encourages stem cells in skeletal muscle to divide and differentiate into muscle fibers.

Hormones affect muscle mass. One effect of the sex hormone testosterone is to encourage muscle cells to build more proteins. Men make much more testosterone than women, which is why men tend to be more muscular. Human growth hormone also stimulates synthesis of muscle proteins. Taking synthetic versions of these muscle-building (anabolic) hormones will bulk up muscles and some athletes use synthetic hormones to increase their muscle size. However, most sport organizations consider the use of these drugs to be cheating, and penalize athletes who use them.

Variations in the function of a regulatory protein called myostatin may give some people a natural edge when it comes to bulking up muscles. Myostatin is a regulatory protein that discourages growth of skeletal muscle. If a mutation prevents formation of normal myostatin, the result is myostatin-related muscle hypertrophy. Hypertrophy means "excessive growth." Affected individuals have larger than normal muscles, minimal body fat, and unusual strength. For example, one myostatin-deficient German boy was born with highly muscular arms and thighs. By the time he was 5 years old, he could stand with his arms outstretched, holding a 3-kilogram (more than 5-pound) weight with each hand. This individual is homozygous for a mutation that causes premature termination of the myostatin RNA, thus preventing formation of myostatin. An unusually strong American boy, Liam Hoekstra, has a similar phenotype (**FIGURE 35.1A**). He makes myostatin, but his muscles may not respond to it.

A Liam Hoekstra, shown at age 3, lifting 5-pound (2.3-kilogram) dumbbells. At this time, he had about 40 percent more muscle than an average 3-year-old.

B The bully whippet (right) is homozygous for a mutant myostatin allele, so she does not make functional myostatin and has bulky muscles. A whippet that is homozygous for the wild-type myostatin gene (left) has a comparatively light musculature.

FIGURE 35.1 Muscle hypertrophy.

Myostatin mutations also influence muscle mass and athletic ability in animals. Consider whippets, a type of dog bred for racing. Some whippets, known as bully whippets, are homozygous for a two-base deletion mutation in the myostatin gene. The mutation changes an mRNA codon for cysteine to a stop codon (Section 9.4), so the protein translated from this mRNA is unusually short and nonfunctional. Bully whippets are not usually raced or bred because their heavily muscled body does not match the breeders ideal of how a whippet should look (**FIGURE 35.1B**). The mutant allele is inherited in a pattern of incomplete dominance (Section 13.5), so its effects show up in heterozygous individuals. Whippets heterozygous for the mutation are more likely to win races than normal whippets. Variations in the myostatin gene can affect horses' success on the track as well. Compared to other domestic horses, horses bred to sprint have a higher frequency of a myostatin allele with one particular base pair substitution. Among race horses, individuals homozygous for this mutant allele tend to be the most muscular and the fastest in short distance races.

35.2 Animal Movement

✔ Animals move their bodies and body parts.

✔ Some differences in animal body form reflect adaptation to movement in different environments.

All animals move. During part or all of their life, they are capable of **locomotion**, which is self-propelled movement from place to place. As adults, even sessile animals (those that live fixed in place) usually move some body parts. Consider barnacles, which begin life as free-swimming larvae, then settle and secrete a protective shell. Although adults remain in one place, they still wave their feathery legs to capture food from the water around them. Most animals move on a daily basis to find food and escape from predators.

Animals use diverse mechanisms of locomotion. They swim through water; fly through air; run, walk, or crawl along surfaces; or burrow through soil and sediments. Despite these differences, the same physical law governs all movement: For every action, there is an equal and opposite reaction. For an animal to move in one direction, it must exert a force in the opposite direction. Jet propulsion by squids provides a simple example of this effect (**FIGURE 35.2**). To move, a squid fills an internal cavity (the mantle cavity) with water, then squeezes that water out through a small opening. As the water shoots out in one direction, the squid moves in the opposite direction. By analogy, think of what happens if you inflate a balloon, then release it without tying it shut. As the rubber of the balloon contracts, force exerted by air rushing out makes the balloon shoot the opposite way.

FIGURE 35.2 Jet-propelled squid.

More typically, aquatic animals that swim do so by pushing at water through body movements and movements of fins or flippers. Animals also move along surfaces, both underwater and on land, by pushing against those surfaces.

The effects of friction and gravity increase the amount of energy an animal must expend to move. Water is denser than air, so moving through it produces more friction. Think of how much harder it is to walk through waist-deep water than to walk on land. Aquatic animals that benefit by swimming fast typically have a streamlined body that reduces the effects of friction (**FIGURE 35.3A**). Aquatic animals are generally somewhat buoyant (prone to float), so gravity is less of a constraint for them than it is for land animals.

Snails, snakes, and other animals that creep or slither along surfaces must expend energy to overcome friction between the body and that surface. Snails reduce friction by secreting a lubricating mucus from their large foot. A snake's smooth scales can have a similar effect. By varying the angle of the scales in different parts of its ventral (belly) surface, a snake can minimize friction everywhere except where it pushes off against the ground. In this region, scales are oriented so they act like the tread on a tire; the scales grip the ground so the snake can propel itself forward.

Animals that walk or run minimize friction by reducing the surface area in contact with the ground. Limbs partially or fully elevate their body and only some feet touch the ground at any given time. The cheetah is the fastest land animal (**FIGURE 35.3B**). It can sprint as fast as 65 miles per hour (about 100 kilometers per hour), although it cannot sustain that speed for long. A variety of traits contribute to the cheetah's speed. Compared to other big cats, a cheetah

A Tuna with a streamlined body that minimizes friction with water.

B Cheetah with a flexible backbone and long limbs that lengthen its stride.

FIGURE 35.3 Predators with bodies adapted for speed.

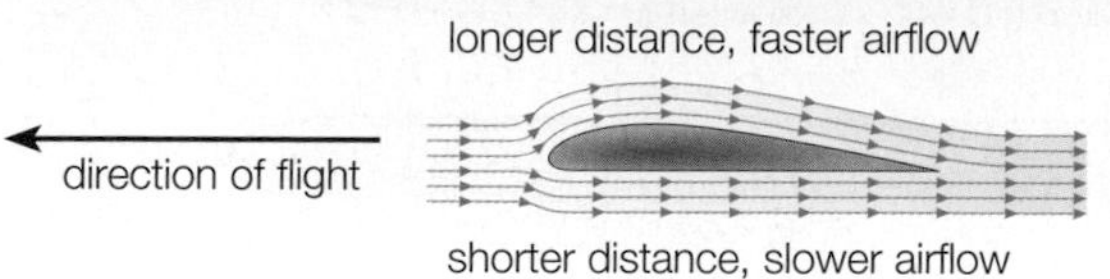

FIGURE 35.4 A soaring gull. Air flows faster over the rounded upper surface of its wings than the flatter lower surface (right). This difference in air velocity between the two surfaces results in an upward force called lift.

has longer legs relative to its body size, a wider range of motion at its hips and shoulders, and a longer, more flexible backbone. Together, these features maximize the length of the cheetah's stride. The cheetah also has a unique sprinting stride: When its back legs land, its backbone becomes bent into a tight C shape. When the animal pushes off again, its spine recoils and helps to force the front of the body forward.

Elastic tissue plays a locomotive role in many animals. Consider kangaroos, which usually move by hopping. Each time a kangaroo lands, elastic material in its hind legs compresses. The compression stores potential energy that is released when this material expands during the next hop.

An ability to fly evolved independently in insects, birds, mammals, and an extinct group of reptiles called pterosaurs. To fly, an animal must overcome gravity to become airborne, and also generate a force to propel itself forward. Wing shape is key to flight. A bird wing typically has the same shape as an airplane wing. In cross-section, its upper surface is convex (bulges outward), while the lower surface is flat or slightly concave (sunken). Consider what happens when the bird is gliding (flying without flapping its wings). As the bird moves forward, air flowing over the upper surface of the wings moves faster than air flowing over the wings' lower surface. This difference in air velocity across the two surfaces creates lift, an upward force, that opposes gravity and keeps the bird aloft (**FIGURE 35.4**). To get off the ground in the first place, the bird has to flap its wings. It spreads its feathers to maximize the size of the wings, then pulls the wings downward. As the air beneath the wings is forced downward, the bird moves in the opposite direction. The bird then folds its wings to minimize their size as it pulls them upward in preparation for another lift-producing downstroke.

Flying, jumping, swimming, and all other forms of locomotion involve interactions between muscles and a skeleton. These interactions, as well as the additional functions of the muscular and skeletal systems, are the focus of the sections that follow.

locomotion Self-propelled movement from place to place.

TAKE-HOME MESSAGE 35.2

How does the form of animals bodies reflect their mode of locomotion?

- ✔ To move, an animal must expend energy to oppose the effects of gravity and friction. The shape of an animal's body can help it minimize this energy cost.
- ✔ Streamlined bodies help aquatic animals reduce friction and move quickly through water. Secreted mucus and smooth scales reduce friction in animals that move along the ground.
- ✔ Elastic tissues that compress and recoil help animals minimize the energy they use to move.
- ✔ The shape of a wing creates lift that allows flight.

35.3 The Vertebrate Endoskeleton

✔ Muscles bring about movement by interacting with a skeleton. There are three types of skeletons.

Invertebrate Skeletons

A system of fluid-filled internal chambers makes up the **hydrostatic skeleton** of soft-bodied invertebrates such as sea anemones and worms. For example, an earthworm's coelom is divided into many fluid-filled chambers, one per segment (Section 24.7). Movement occurs when muscles exert force against these chambers (FIGURE 35.5). Two sets of muscles squeeze the chambers to alter the shape of body segments, much as squeezing a water-filled balloon changes the balloon's shape. Circular muscle rings each earthworm segment; when this muscle contracts, the segment gets longer and narrower. Longitudinal muscle runs the length of each segment; when this muscle contracts, the segment gets shorter and wider. Coordinated changes in the shape of the body segments move the earthworm through soil.

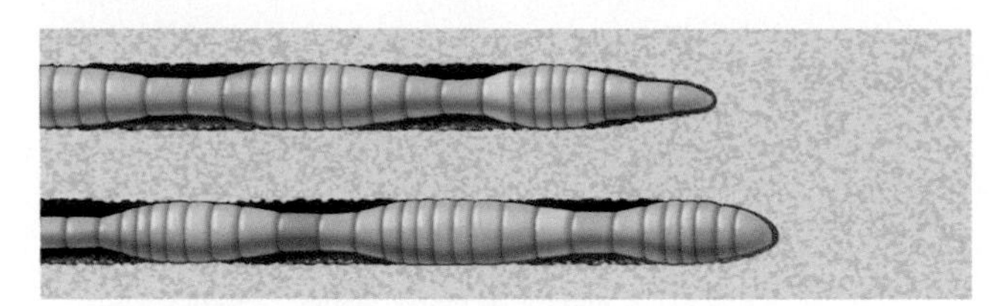

FIGURE 35.5 Hydrostatic skeleton of an earthworm. The worm moves when muscles put pressure on coelomic fluid in individual body segments, causing the segments to change shape.

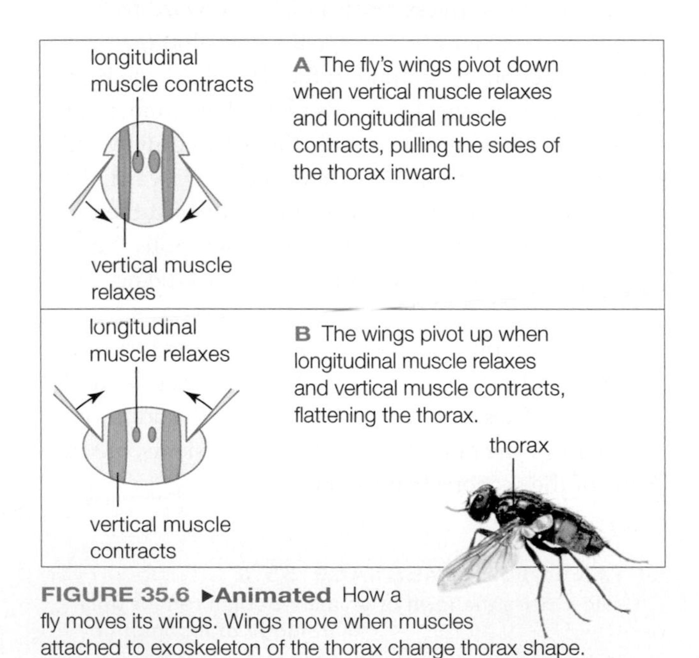

FIGURE 35.6 ▸Animated How a fly moves its wings. Wings move when muscles attached to exoskeleton of the thorax change thorax shape.

An **exoskeleton** is a cuticle, shell, or other hard external body part that receives the force of a muscle contraction. For example, muscles attached to the cuticle of a fly's thorax alter the shape of the thorax, causing the attached wings to flap up and down (FIGURE 35.6). The arthropod exoskeleton has the advantage of also protecting the soft body tissues inside it. However, external skeletons have some drawbacks. An exoskeleton consists of noncellular secreted material, so it cannot grow with the animal. As an arthropod grows, it must periodically molt its old skeleton and replace it with a larger one. Repeatedly producing new exoskeleton uses resources that could otherwise be used for growth or reproduction.

An **endoskeleton** is an internal framework of hard parts. For example, echinoderms such as sea stars have an endoskeleton made of hard, calcium-rich plates.

The Vertebrate Endoskeleton

All vertebrates have an endoskeleton. In sharks and other cartilaginous fishes, it consists of cartilage, a rubbery connective tissue. Other vertebrate skeletons include some cartilage, but consist mostly of bone tissue (Section 31.4).

The term "vertebrate" is a reference to the **vertebral column**, or backbone, a feature common to all members of this group (FIGURE 35.7). The backbone supports the body, serves as an attachment point for muscles, and protects the spinal cord that runs through a canal inside it. Bony segments called **vertebrae** (singular, vertebra) make up the backbone. **Intervertebral disks** of cartilage between vertebrae act as shock absorbers. The vertebral column and bones of the head and rib cage

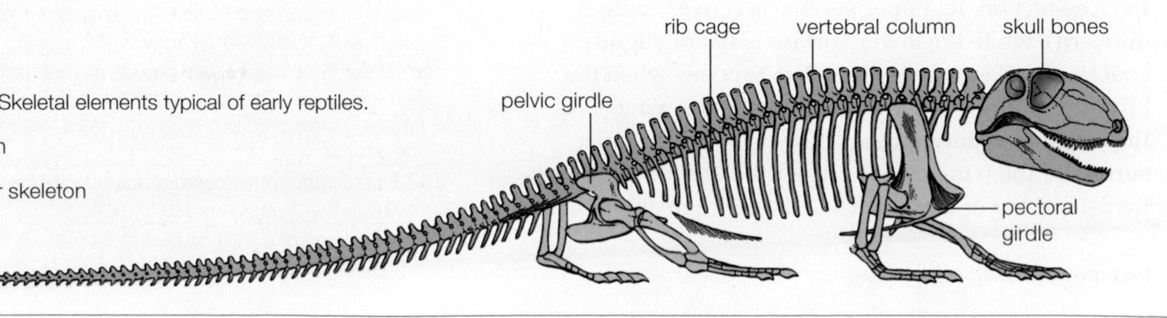

FIGURE 35.7 Skeletal elements typical of early reptiles.

axial skeleton

appendicular skeleton

constitute the **axial skeleton**, the bones that make up the main axis of the body. The **appendicular skeleton** consists of bones of the appendages (limbs or bony fins), and bones that connect the appendages to the axial skeleton (the pectoral girdle at the shoulders and the pelvic girdle at the hip).

You learned earlier how vertebrate skeletons have evolved over time. For example, jaws are derived from the gill supports of ancient jawless fishes (Section 25.4). As another example, bones in the limbs of land vertebrates are homologous to those in the pelvic and pectoral fins of lobe-finned fishes (Section 25.6).

For a closer look at vertebrate skeletal features, consider a human skeleton (**FIGURE 35.8**). The skull's flattened cranial bones fit together as a braincase. Facial bones include cheekbones and other bones around the eyes, the bone that forms the bridge of the nose, and the bones of the jaw.

Ribs and the breastbone, or sternum, form a protective cage around the heart and lungs. Both males and females have twelve pairs of ribs.

The vertebral column extends from the base of the skull to the pelvic girdle. Viewed from the side, our backbone has an S shape that keeps our head and torso centered over our feet when we stand upright.

The scapula (shoulder blade) and clavicle (collarbone) are bones of the pectoral girdle. The clavicle transfers force from the arms to the axial skeleton. The upper arm has a single bone, the humerus. The forearm has two bones, the radius and ulna. Carpals are bones of the wrist, metacarpals are bones of the palm, and phalanges (singular, phalanx) are finger bones.

The pelvic girdle is a basin-shaped ring that encloses the pelvic cavity and supports the weight of the upper body when you stand. It includes bones of the two hips and (at the back) the sacrum and coccyx (tailbone).

Bones of the leg include the femur (thighbone), patella (kneecap), and the tibia and fibula (bones of the lower leg). Tarsals are ankle bones, and metatarsals are bones of the sole of the foot. Like the bones of the fingers, those of the toes are called phalanges.

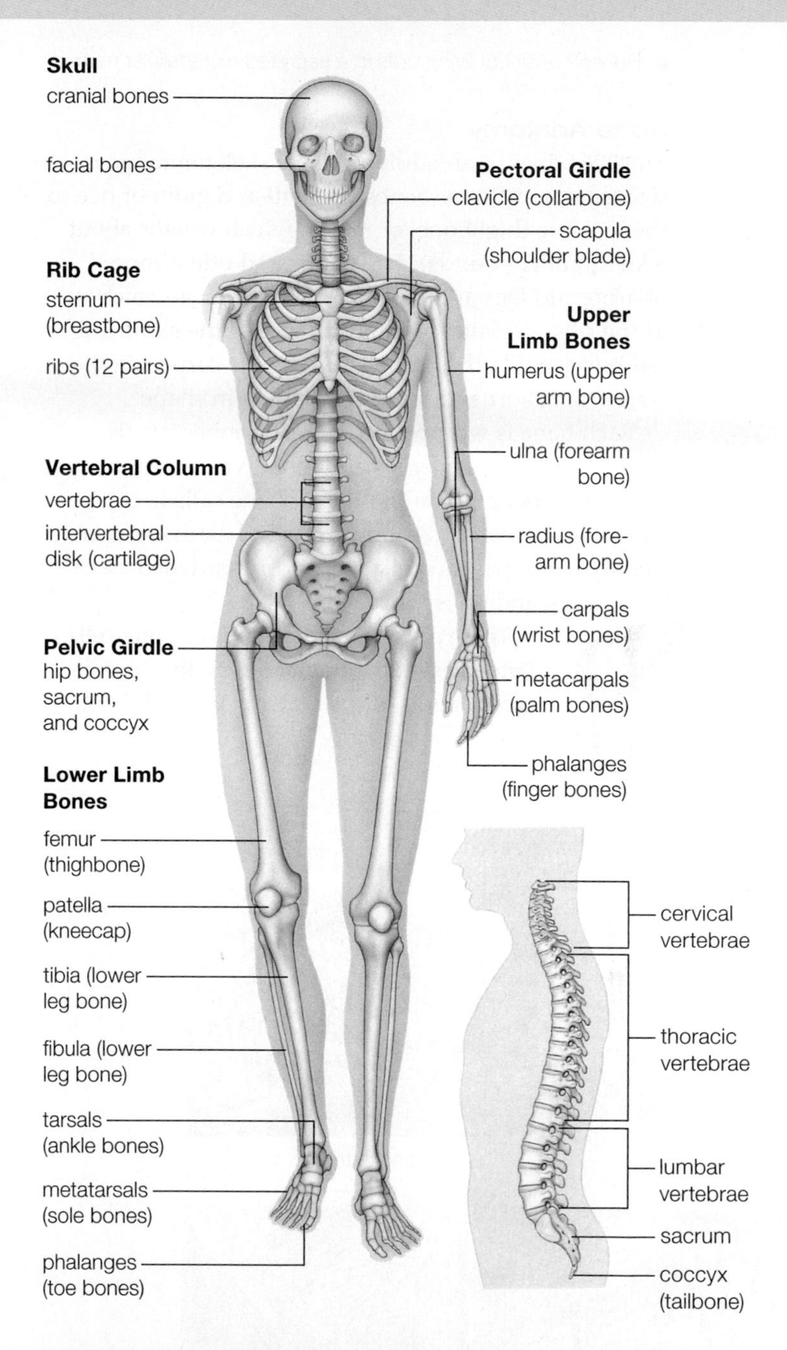

FIGURE 35.8 ▶**Animated** Major bone (tan) and cartilage (light blue) elements of the human skeleton. Lower right, side view of the vertebral column showing its S-shaped curve.

appendicular skeleton Of vertebrates, bones of the limbs or fins and the bones that connect them to the axial skeleton.
axial skeleton Bones of the main body axis; skull, backbone.
endoskeleton Internal skeleton consisting of hard parts.
exoskeleton Of some invertebrates, hard external parts that muscles attach to and move.
hydrostatic skeleton Of soft-bodied invertebrates, a fluid-filled chamber that muscles exert force against, redistributing the fluid.
intervertebral disk Cartilage disk between two vertebrae.
vertebrae Bones of the backbone, or vertebral column.
vertebral column Backbone.

TAKE-HOME MESSAGE 35.3

What type of skeleton do animals have?

✔ In animals with a hydrostatic skeleton, muscle contractions alter the shape of fluid-filled chambers. In those with an exoskeleton, muscles pull on hard external parts. In those with an endoskeleton, muscles pull on hard internal parts.

✔ Vertebrates have an endoskeleton made of cartilage and (in most groups) bone.

35.4 Bone Structure and Function

✔ Bones consist of living cells in a secreted extracellular matrix.

Bone Anatomy

The 206 bones of an adult human's skeleton range in size from middle ear bones as small as a grain of rice to the massive thighbone, or femur, which weighs about a kilogram (2 pounds). The femur and other bones of arms and legs are long bones. Other bones, such as the ribs, sternum, and most bones of the skull, are flat bones. Still other bones, such as the carpals in the wrists, are short and roughly squarish in shape.

Each bone is wrapped in a dense connective tissue sheath that has nerves and blood vessels running through it. Bone tissue consists of bone cells in an extracellular matrix (Section 4.11). The matrix consists mainly of the protein collagen, and is hardened by calcium and phosphorus ions.

There are three main types of bone cells. **Osteoblasts** are bone builders; they secrete components of the matrix. An adult bone has osteoblasts at its surface and in its internal cavities. **Osteocytes** are former osteoblasts that have become surrounded by the matrix they secreted. These are the most abundant bone cells in adults. **Osteoclasts** secrete enzymes that break down the matrix.

A long bone such as a femur includes two types of bone tissue, compact bone and spongy bone (**FIGURE 35.9**). Compact bone forms the outer layer and shaft of the femur. It consists of many functional units called osteons, each having concentric rings of bone tissue, with bone cells in spaces between the rings. Nerves and blood vessels run through a canal in the osteon's center. Spongy bone fills the shaft and knobby ends of long bones. It is strong yet lightweight; many open spaces riddle its hardened matrix.

The cavities inside a bone contain bone marrow. **Red marrow** fills the spaces in spongy bone and is the major site of blood cell formation. **Yellow marrow** fills the central cavity of an adult femur and most other mature long bones. It consists mainly of fat.

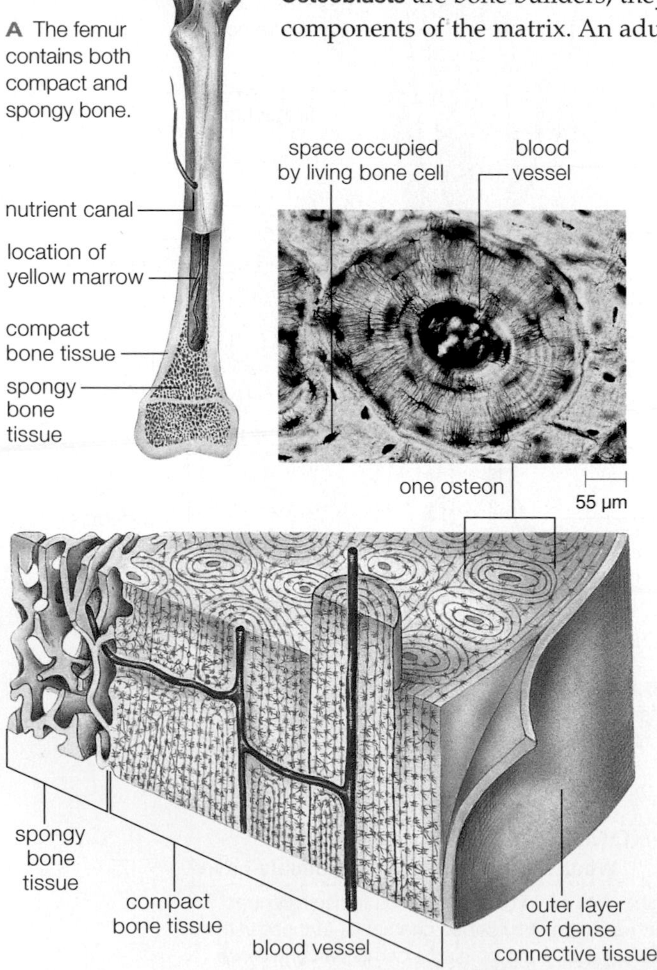

A The femur contains both compact and spongy bone.

B Cross-section through a femur showing osteons. These cylindrical structures consist of concentric layers of compact bone.

FIGURE 35.9 ▶**Animated** Structure of a human adult femur, or thighbone.

Bone Formation and Remodeling

A cartilage skeleton forms in all vertebrate embryos. It remains cartilage in sharks and other cartilaginous fishes. In other vertebrates, osteoblasts move into the embryonic skeleton and convert most of it to bone.

Many bones continue to grow in size until early adulthood. Even in adults, bone remains a dynamic tissue that the body continually remodels. Microscopic fractures caused by normal body movements are repaired. In response to hormonal signals, osteoclasts dissolve portions of the matrix, releasing mineral ions into the blood. Osteoblasts secrete new matrix to replace that broken down by osteoclasts.

Bones and teeth contain most of the body's calcium. Parathyroid hormone, the main regulator of blood calcium, raises the concentration of calcium ions in the blood by encouraging the breakdown of calcium-rich bone matrix. Other hormones also affect bone turnover. Growth hormone and sex hormones encourage bone deposition, whereas cortisol discourages it.

Until an individual is about twenty-four years old, osteoblasts secrete more matrix than osteoclasts break down, so bone mass increases. Bones become denser and stronger. Later in life, bone mass gradually declines as osteoblasts become less active.

Where Bones Meet

A **joint** is an area of contact or near contact between bones. There are three types (**FIGURE 35.10**). At fibrous joints, dense, fibrous connective tissue holds

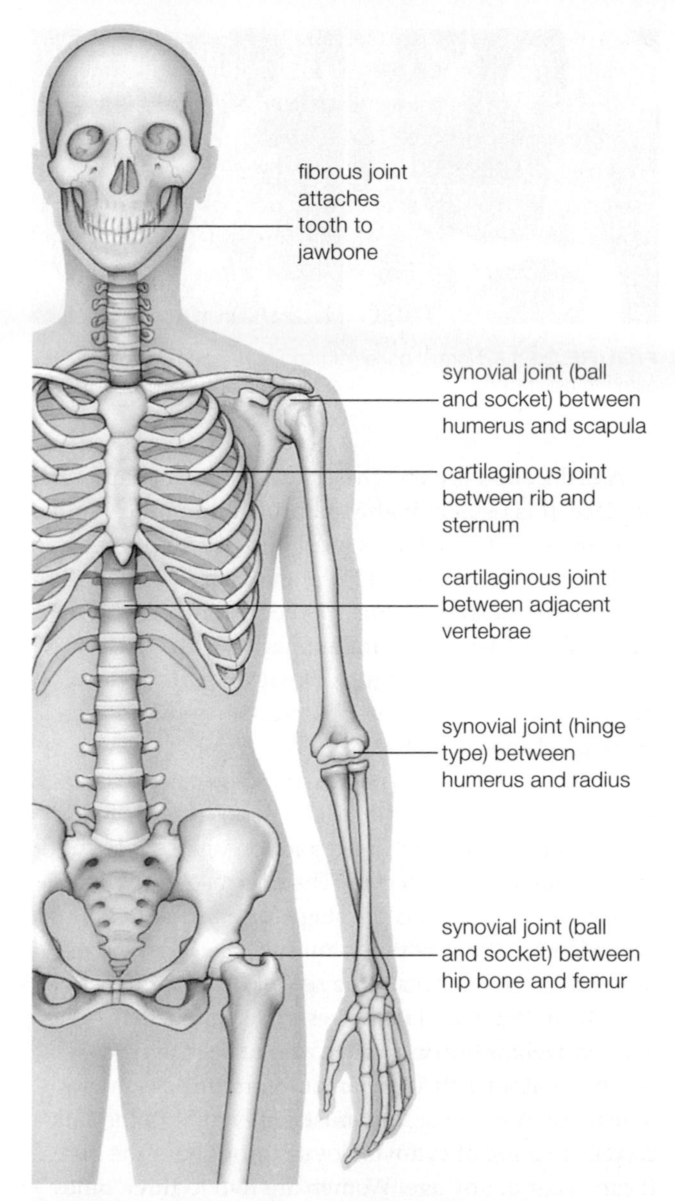

FIGURE 35.10 Examples of the three types of joints.

bones firmly in place. Fibrous joints connect bones of the skull and hold teeth in their sockets in the jaw. Pads or disks of cartilage connect bones at cartilaginous joints. The flexible connection allows just a bit of movement. Cartilaginous joints connect vertebrae to one another and connect some ribs to the sternum.

joint Region where bones meet.
ligament Strap of dense connective tissue that holds bones together at a synovial joint.
osteoblast Bone-forming cell; it secretes a bone's matrix.
osteoclast Bone-digesting cell; it breaks down bone matrix.
osteocyte A mature bone cell; a former osteoblast.
red marrow Bone marrow that makes blood cells.
yellow marrow Bone marrow that is mostly fat; fills cavity in most long bones.

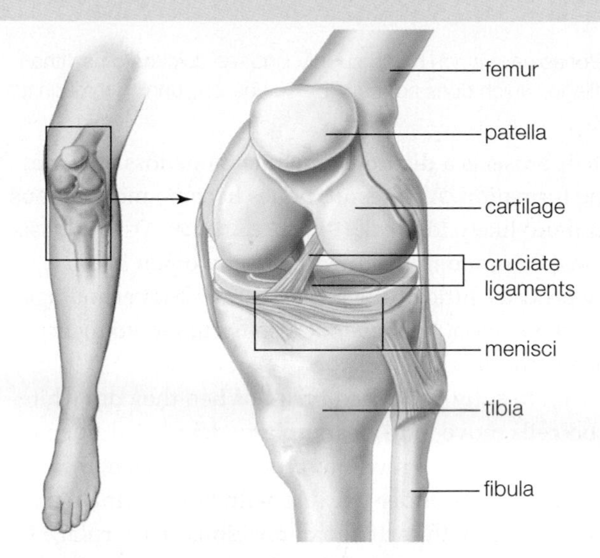

FIGURE 35.11 ▶**Animated** The knee, a hinge-type synovial joint stabilized by ligaments and menisci (wedges of cartilage). A bursa beneath the patella reduces friction with the femur.

Most joints, including the knees, hips, shoulders, wrists, and ankles, are synovial joints. At these joints, the cartilage-covered ends of bones are separated by a small space. Cords of dense, regular connective tissue called **ligaments** hold bones of a synovial joint in place. Some ligaments form a capsule around the joint. The capsule's lining secretes a lubricating synovial fluid.

Synovial joints allow a variety of movements. Ball-and-socket joints at the shoulders and hips provide a wide range of rotational motion. At other synovial joints, including some in the wrists and ankles, bones glide past one another. Joints at the elbows and knees function like a hinged door; they allow the bones to move back and forth in one plane only. **FIGURE 35.11** shows the knee joint. In addition to ligaments, the knee is stabilized by wedges of cartilage called menisci (singular, meniscus). A bursa reduces friction between the patella (kneecap) and femur. A bursa is a fluid-filled sac that reduces friction between parts at a joint.

TAKE-HOME MESSAGE 35.4

What is the structure of bones and joints?

✔ A bone is enclosed by sheath of connective tissues and has an inner cavity containing marrow. Red marrow produces blood cells.

✔ Bone tissue consist of bone cells in a secreted extracellular matrix. A bone is continually remodeled by a hormone-regulated process.

✔ Joints are areas where bones meet and interact. In the most common type, synovial joints, the bones are separated by a small fluid-filled space and are held together by ligaments of fibrous connective tissue.

35.5 Bone and Joint Health

✔ Bones have a rich blood supply, and are quicker to heal than cartilage, which does not have blood vessels running through it.

Osteoporosis is a disorder in which bone loss outpaces bone formation. As a result, bones become more porous and more likely to break (**FIGURE 35.12**). Osteoporosis is most common in postmenopausal women because they produce little of the sex hormones that encourage bone deposition. However, it also occurs in younger women and in men.

Even healthy bones can break. When they do, white blood cells move in to clean up any debris. Additional tiny blood vessels grow into the damaged area, then bone repair gets under way. As with bone formation, repair begins with cartilage deposition. The cartilage fills the gap created by the break, then osteoblasts move in and replace the cartilage with new bone.

To maintain the health of your bones, ensure that your diet provides adequate levels of calcium and of vitamin D, which facilitates calcium absorption from the gut. Get regular exercise to encourage bone renewal. Avoid smoking and excessive consumption of alcohol or caffeine, which can slow bone renewal.

Joints are frequent sites of injury. A sprained ankle, the most common joint injury, occurs when one or more of the ligaments holding bones together at the ankle joint overstretches or tears. The sprain is usually treated immediately with rest, application of ice, compression with an elastic bandage, and elevation of the affected area. After the ankle heals, exercises may help strengthen muscles that stabilize the joint and prevent future sprains.

A tear of the cruciate ligaments in the knee joint may require surgery. "Cruciate" means cross, and these short ligaments cross one another in the center of the joint. They are illustrated in **FIGURE 35.11**. The cruciate ligaments stabilize the knee, and when they are torn completely, bones may shift so the knee gives out when a person tries to stand. A blow to the lower leg, as often occurs in football, can injure a cruciate ligament, but so can a fall or misstep.

Another common knee injury is a torn meniscus. The meniscus is a C-shaped wedge of cartilage that reduces friction between the bones, cushions them, and helps keep them in place. Each knee has two menisci. A minor tear at the edge of the meniscus may heal on its own, but unlike bone, cartilage does not have a rich blood supply and repairs itself very slowly. If a bit of meniscus cartilage gets torn off, it can drift about in the synovial fluid of the joint and end up jammed into a spot where it interferes with normal function.

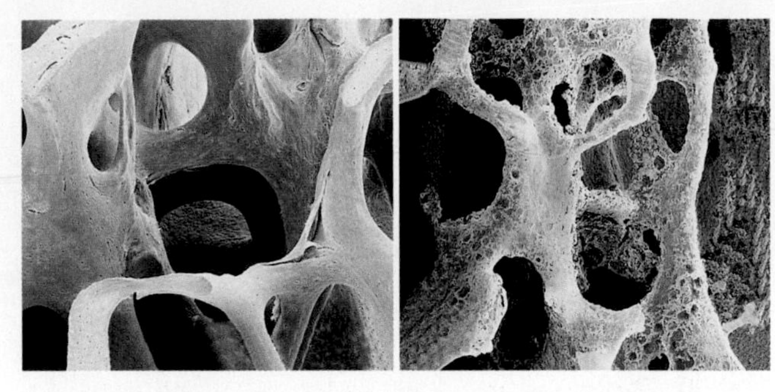

FIGURE 35.12 Normal bone (left) and bone affected by osteoporosis (right). Osteoporosis means "porous bones."

A dislocation occurs when bones of a joint slip out of place. It is usually highly painful and requires immediate treatment. The bones must be put into proper position and immobilized for a time to allow healing.

Arthritis means inflammation of a joint. As you will learn in Chapter 37, inflammation is the body's normal response to an injury. However, with arthritis, inflammation—and the associated pain and swelling—becomes chronic.

The most common type of arthritis is osteoarthritis. It usually appears in older adults, whose cartilage has thinned at a frequently jarred joint or joints. Knees and hips are most often affected. The decrease in rubbery, protective cartilage sets the stage for damage to bones of the joint. An injury to a joint, such as a torn meniscus in the knee, increases the risk of developing osteoarthritis in that joint later. Obesity, which increases the load on weight-bearing joints, also raises the risk.

Rheumatoid arthritis is an autoimmune disorder in which the immune system mistakenly attacks the fluid-secreting lining of synovial joints throughout the body. It can occur at any age. Women are two to three times more likely than men to be affected.

Arthritis can be treated with drugs that relieve pain and minimize inflammation. Joints affected by osteoarthritis can also be replaced with artificial, or prosthetic, joints. Knee and hip replacements are now common and allow a person to resume normal activities.

TAKE-HOME MESSAGE 35.5
What damages bones and joints?

✔ Diet, hormone levels, and other factors affect bone strength. If bones break, new bone is laid down to repair the break.

✔ Shift in bone position or repeated stress can injure joints. The cartilage of joints is very slow to heal.

✔ Arthritis is chronic inflammation of a joint.

CREDIT: (12) Dr P. Motta/Department of Anatomy, University La Sapienza, Rome/SPL/Science Source.

35.6 Skeletal Muscle Function

✔ Skeletal muscles exert their effects by pulling.

Vertebrate skeletal muscles are sometimes referred to as voluntary muscles because they function mainly in intentional movement. However, skeletal muscles also participate in reflexes such as the stretch reflex (Section 32.9). In addition, skeletal muscle plays a role in thermoregulation because muscle activity releases heat.

A sheath of dense connective tissue encloses each vertebrate skeletal muscle and extends beyond it to form a cordlike or straplike **tendon**. Most often, tendons attach each end of a muscle to different bone. Typically one end of a muscle attaches to a bone that remains fixed in place, and the other to a bone that is moved by the muscle's contraction. Consider, for example, the biceps muscle in the upper arm (**FIGURE 35.13**). At one end of this muscle, two tendons attach it to the scapula (shoulder blade). At the opposite end of the muscle, a tendon attaches it to the radius, a bone in the forearm. When the biceps contracts (shortens), the elbow bends and the forearm is pulled toward the shoulder. You can feel the biceps contract if you extend one arm out, palm up, then place your other hand over the muscle and slowly bend your arm at the elbow.

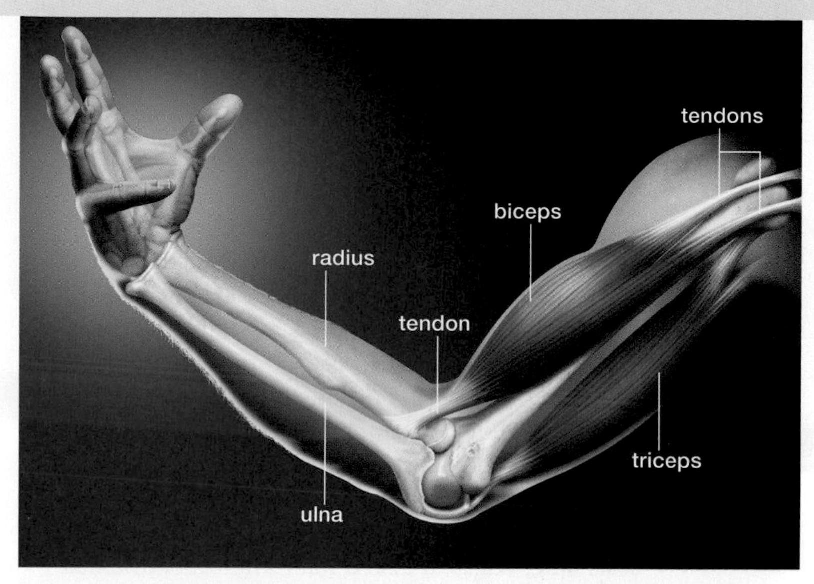

FIGURE 35.13 ▸Animated Opposing muscles of the upper arm. When the biceps contracts and the triceps relaxes, the forearm is pulled toward the upper arm. When the triceps contracts and the biceps relaxes, the arm straightens at the elbow.
FIGURE IT OUT What bone does the biceps pull on?
Answer: The radius

Although the biceps shortens only about a centimeter when it contracts, the forearm moves through a much greater distance. The elbow and many other joints function like a lever, a mechanism in which a rigid structure pivots about a fixed point (the bones are the rigid structures and the joints are the fixed points). Use of a lever allows a small force, such as that exerted by a contracting biceps, to overcome a larger one, such as the gravitational force acting on the forearm.

Keep in mind that muscles can pull but cannot push. To achieve the greatest range of motion, muscles work in opposition: Motion generated by contraction of one muscle is reversed by contraction of another. For example, the triceps in the upper arm opposes the biceps. When the biceps contracts, the triceps relaxes and the forearm is pulled toward the shoulder. Contraction of the triceps coupled with relaxation of the biceps reverses this movement. Other muscles are involved in less conspicuous but equally important movements. For example, muscles associated with the rib cage expand the chest cavity when you inhale.

Most skeletal muscles pull on bones, but some tug on other tissues. Some skeletal muscles pull facial skin to bring about changes in expression. Others attach to and move the eyeball, or open and close eyelids. Skeletal muscle also composes some sphincters. A **sphincter** is a ring of muscle in a tubular organ or at a body opening. The sphincter of skeletal muscle that encircles the urethra (the tube through which urine exits the body) allows voluntary control of urination. Another at the anus allows control over defecation.

Some animals have boneless muscular organs capable of making complex movements. The mammalian tongue is an example. Some tongue muscles have one end attached to the floor of the mouth and one free end. Others are located entirely within the tongue and have no bony attachment at all. They help alter the shape of the tongue by exerting pressure on one another and on connective tissue within the tongue. An elephant's trunk (right) is another example of a boneless muscular organ.

TAKE-HOME MESSAGE 35.6
What are the functions of skeletal muscle?

✔ Most skeletal muscles pull on and move bones. Others pull on skin to alter facial expressions. Rings of skeletal muscle form sphincters that are under voluntary control.

sphincter Ring of muscle in a tubular organ or at a body opening.
tendon Strap of dense connective tissue that connects a skeletal muscle to bone.

35.7 How Muscle Contracts

✔ ATP-fueled movements of protein filaments inside a muscle fiber result in muscle contraction.

Structure of Skeletal Muscle

Properties of skeletal muscle arise from the structure and arrangement of its component fibers (**FIGURE 35.14**). As previously noted, a muscle is enclosed within a sheath of connective tissue ❶. This tissue extends beyond the muscle as a tendon. Additional connective tissue encloses each bundle of muscle fibers within the muscle ❷. Each **skeletal muscle fiber** is a roughly cylindrical cell that runs the length of the muscle and parallels its long axis ❸. A skeletal muscle fiber forms before birth by the fusion of many embryonic cells, so it contains many nuclei, which are positioned at its outer edges. It also contains mitochondria that supply the ATP to necessary for muscle contraction.

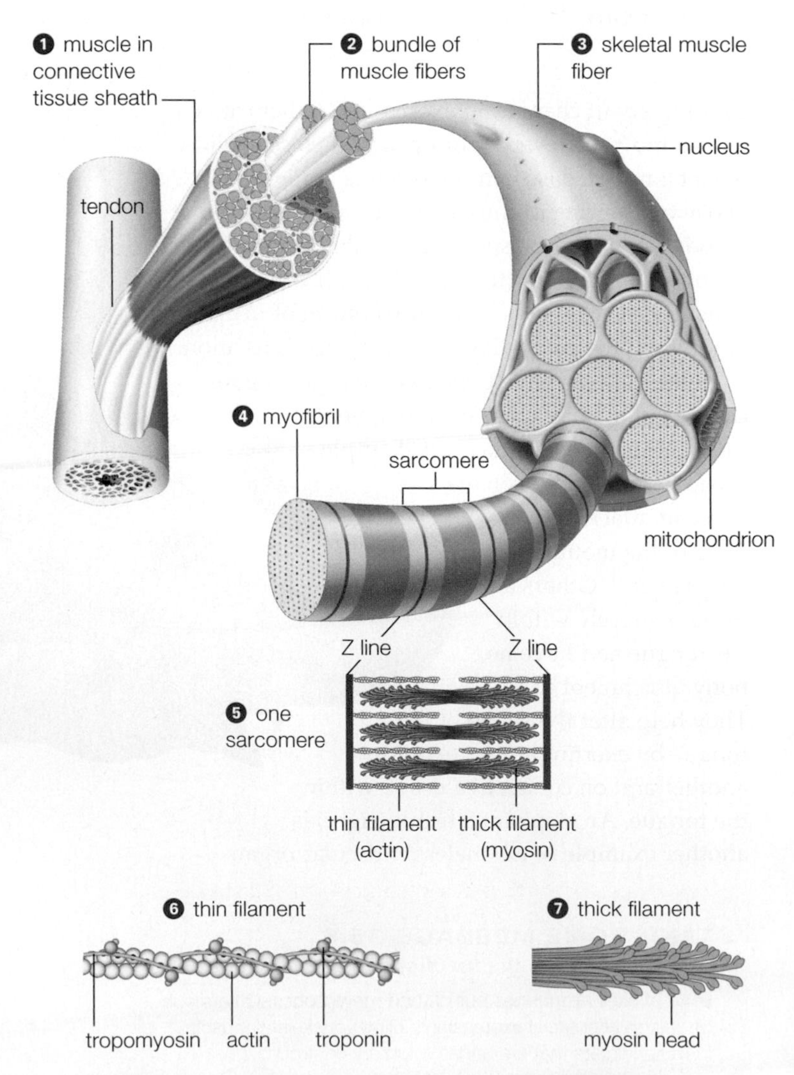

FIGURE 35.14 ▸**Animated** Structure of a skeletal muscle.

The bulk of the muscle fiber's interior is filled with thousands of threadlike structures called **myofibrils** ❹. Each myofibril consists of many identical contractile units, called **sarcomeres,** attached end to end. Each end of a sarcomere is delineated by a Z line, a mesh of cytoskeletal elements that attach the sarcomere to its neighbors. The Z stands for *zwischen*, the German word for "between," and refers to the fact that a sarcomere is the region between two Z lines ❺.

Between a sarcomere's Z lines and perpendicular to them is an array of alternating thin and thick filaments. Thin filaments attach to the Z line and extend inward toward the center of the sarcomere. These filaments consist mainly of the protein actin ❻. Thick filaments reside at the center of the sarcomere, where they are flanked by the free ends of the thin filaments. Thick filaments consist of myosin, a motor protein with a clublike head and a long tail ❼. The myosin head has enzymatic activity; it can bind ATP and break it into ADP and phosphate. It can also bind to the actin of thin filaments.

Muscle fibers, myofibrils, thin filaments, and thick filaments all run parallel with a muscle's long axis. As a result, all sarcomeres in all fibers of a skeletal muscle pull in the same direction—they work together. Skeletal muscle and cardiac muscle appear striated because Z lines and other sarcomere components in all their fibers are aligned. Smooth muscle fibers have sarcomeres, but because the sarcomeres are not aligned, smooth muscle does not have a striped appearance.

The Sliding-Filament Model

The **sliding-filament model** explains how interactions between thick and thin filaments bring about muscle contraction. Neither actin nor myosin filaments change length, and the myosin filaments do not change position. Instead, myosin heads bind to actin filaments and slide them toward the center of a sarcomere. As the actin filaments are pulled inward, the ends of the sarcomere are drawn closer together, and the sarcomere shortens (**FIGURE 35.15A**).

FIGURE 35.15B provides a step-by-step look at events during muscle contraction, starting with the sarcomere in a resting muscle fiber. In a relaxed sarcomere, sites where myosin could bind to actin are blocked. The myosin heads of thick filaments have bound ATP molecules and are in a low-energy conformation ❶. Removing a phosphate group (P*i*) from a bound ATP energizes a myosin head in a manner analogous to stretching a spring ❷. The myosin head remains in this high-energy conformation, with bound

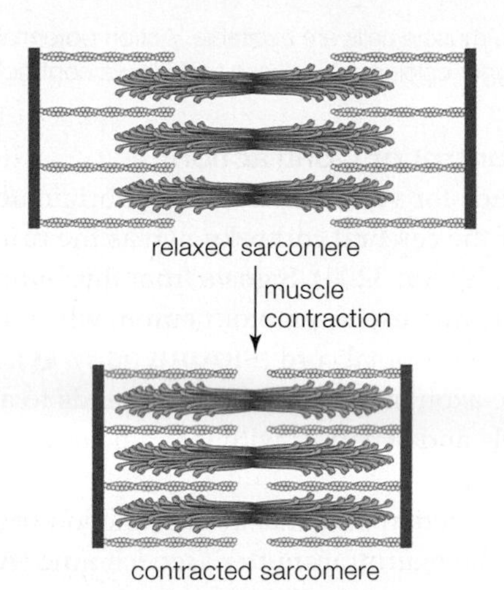

A (Above) During muscle contraction, thin filaments slide inward past thick filaments, reducing the width of the sarcomere.

B Right, molecular mechanism of contraction. For clarity, we show only two myosin heads. A head binds repeatedly to an actin filament and slides it toward the center of the sarcomere.

FIGURE 35.15 ▸Animated The sliding-filament model.

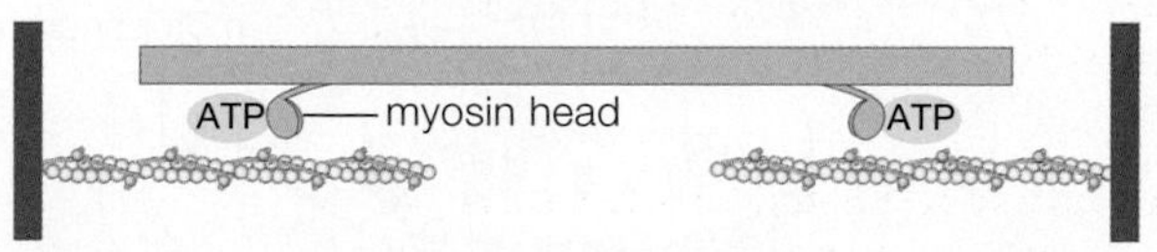

❶ In a relaxed sarcomere, myosin-binding sites on actin are blocked. The myosin heads have bound ATP, and are in their low-energy conformation.

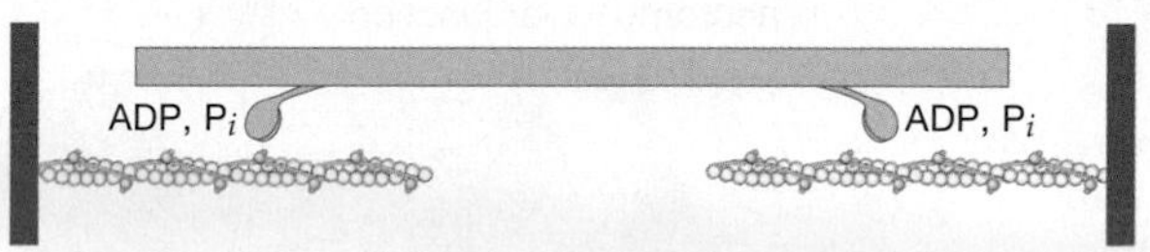

❷ The myosin heads remove a phosphate group from the ATP. Absorbing energy released by breaking the phosphate bond boosts the myosin heads to a high-energy conformation. ADP and phosphate remain bound to each head.

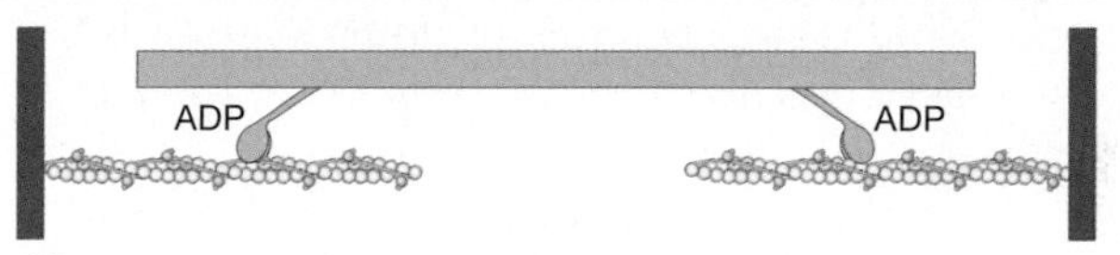

❸ A signal from the nervous system opens up the myosin-binding sites on actin. Each myosin head attaches to actin and releases its bound phosphate group, thus forming a cross-bridge between thick and thin filaments.

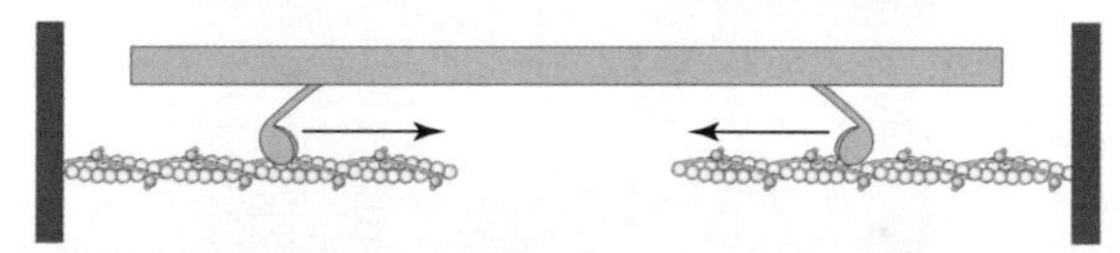

❹ The release of the phosphate group triggers a power stroke, in which the myosin heads snap inward and release ADP. As the heads contract, they pull the attached thin filaments inward.

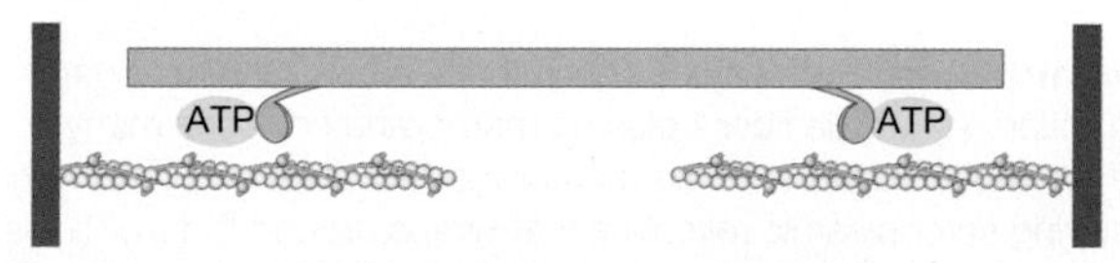

❺ Another ATP binds to each myosin head, causing it to release actin and return to its low-energy conformation.

ADP and phosphate, until a signal from the nervous system excites the muscle (a process we consider in the next section). When this signal arrives, the myosin head releases its bound phosphate group and attaches to actin. This attachment constitutes a cross-bridge between the thin and thick filaments ❸. Attachment is followed by the power stroke ❹. Like a stretched spring returning to its original shape, the myosin head snaps back toward the sarcomere center. As the myosin head moves inward, it pulls the attached thin filament along with it.

During the power stroke, a myosin head releases bound ADP. Afterward, the head can bind to a new molecule of ATP and return to its low-energy conformation ❺. In the process, the myosin head releases its grip on actin, and the cross-bridge is lost. Loss of one cross-bridge does not allow a thin filament to slip backward because other cross-bridges hold it in place. During a contraction, many myosin heads repeatedly bind, move, and release an adjacent thin filament.

skeletal muscle fiber Contractile cell that contains mainly myofibrils and runs the length of a muscle.
myofibrils Of a muscle fiber, threadlike protein structures consisting of contractile units (sarcomeres) arranged end to end.
sarcomere Contractile unit of muscle.
sliding-filament model Explanation of how interactions among actin filaments and myosin filaments bring about muscle contraction.

TAKE-HOME MESSAGE 35.7

How does a muscle contract?

✔ Sarcomeres are lined up end to end in myofibrils that run parallel with muscle fibers. These fibers, in turn, run parallel with the whole muscle. The parallel orientation of skeletal muscle components focuses a muscle's contractile force in a particular direction.

✔ ATP-driven interactions between myosin (thick) and actin (thin) filaments shorten sarcomeres of a muscle cell.

✔ During muscle contraction, the length of actin and myosin filaments does not change, and the myosin filaments do not change position. Sarcomeres shorten because myosin filaments pull neighboring actin filaments inward toward the center of the sarcomere.

35.8 Nervous Control of Muscle Contraction

❶ A signal travels along the axon of a motor neuron, from the spinal cord to a skeletal muscle.

❷ At a neuromuscular junction, arrival of an action potential at the motor neuron's axon terminals causes them to release ACh. The ACh binds to receptors in a muscle fiber's plasma membrane, causing an action potential in that muscle fiber.

❸ Action potentials propagate along a muscle fiber's plasma membrane down to T tubules, then to the sarcoplasmic reticulum, which releases calcium ions. The ions promote interactions of myosin and actin that result in contraction.

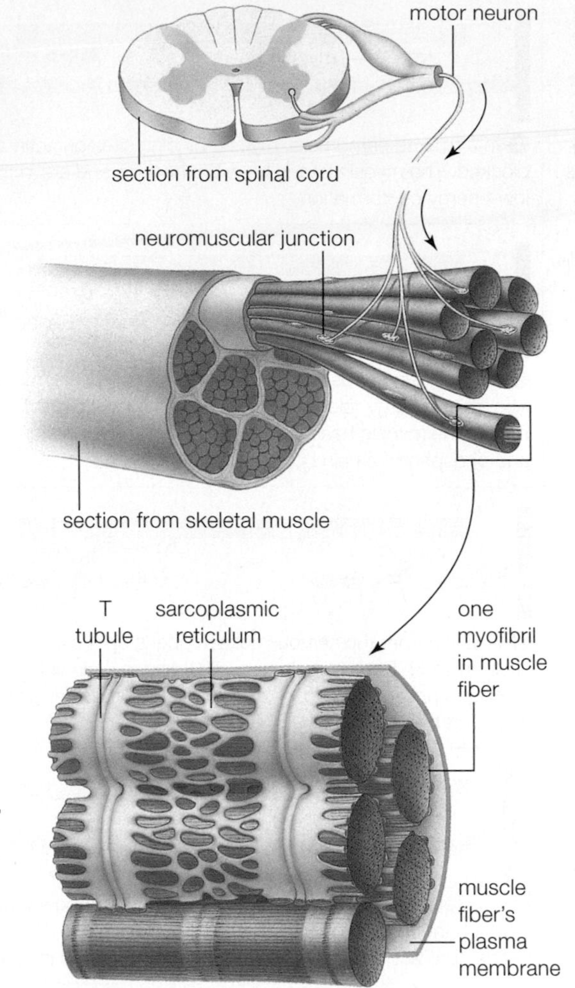

FIGURE 35.16 ▸Animated Pathway by which the nervous system controls skeletal muscle contraction. A muscle fiber's plasma membrane encloses many individual myofibrils. Tubelike extensions of the membrane (T tubules) connect with part of the calcium-storing sarcoplasmic reticulum that wraps around the myofibrils.

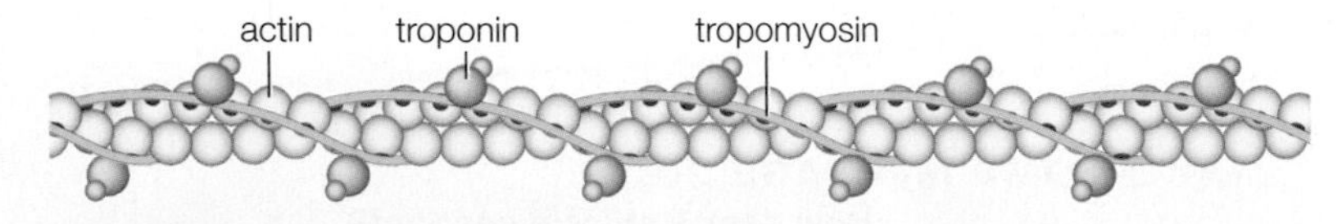

A Resting muscle. Calcium ion (Ca^{++}) concentration is low and tropomyosin covers the myosin-binding sites on actin.

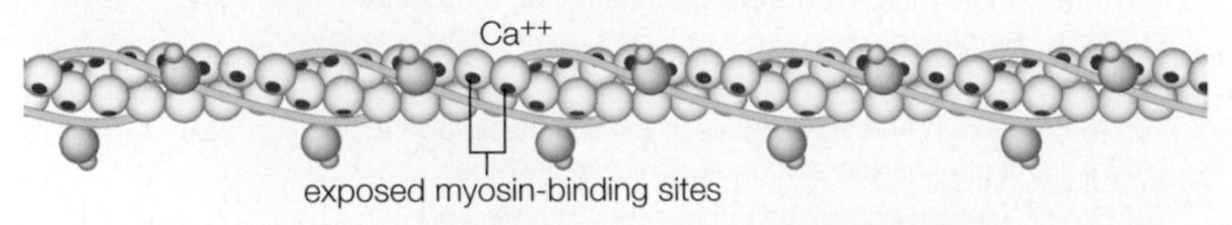

B Excited muscle. Ca^{++} binds to troponin, which shifts and moves tropomyosin, exposing myosin-binding sites on actin.

FIGURE 35.17 ▸Animated Role of calcium in muscle contraction. Configuration of thin filament proteins depends on the Ca^{++} level.

✔ Like neurons, muscle cells are excitable. Action potentials in muscle cells trigger calcium ion release that allows contraction.

Nervous Control of Contraction

Most commands for voluntary movement originate in the portion of the cerebral cortex known as the primary motor cortex (Section 32.11). Signals from this brain region travel to and excite a motor neuron, which has its cell body in the spinal cord (**FIGURE 35.16** ❶). As you know, the axon of a motor neuron extends to a skeletal muscle and synapses with it at a neuromuscular junction ❷.

Arrival of an action potential at this junction triggers the release of the neurotransmitter acetylcholine (ACh).

Like a neuron, a muscle fiber is excitable, meaning it is capable of undergoing an action potential. ACh released at a neuromuscular junction binds to receptors in the muscle fiber's plasma membrane. The binding triggers an action potential that travels along the plasma membrane, then down membranous extensions called T tubules. T tubules convey the action potential to the **sarcoplasmic reticulum**, a specialized smooth endoplasmic reticulum that wraps around myofibrils and stores calcium ions ❸. Arrival of an action potential opens voltage-gated channels in the sarcoplasmic reticulum, allowing calcium ions to follow their gradient and flood out of the organelle. The flow of ions raises the calcium concentration around the myofibrils.

The increase in calcium ion concentration allows actin and myosin filaments to interact by altering the configuration of two proteins: tropomyosin and troponin. These proteins are, along with actin, components of thin filaments. Tropomyosin is a fibrous protein; it forms long polymers that wrap around the actin of a thin filament. Globular molecules of troponin bind to each tropomyosin polymer at intervals along its length.

When a muscle is at rest, tropomyosin polymers block the myosin-binding sites on actin, so actin and myosin cannot interact (**FIGURE 35.17A**). With muscle excitation, binding of calcium ions to troponin causes this protein to change its shape and pull the attached tropomyosin away from actin's myosin-binding sites (**FIGURE 35.17B**). With these binding sites cleared, actin can bind myosin, and the sliding action described in the previous section takes place. After contraction, active transport proteins pump the calcium ions back into the sarcoplasmic reticulum.

Motor Units and Muscle Tension

A motor neuron has many axon endings that synapse on different fibers in a muscle. One motor neuron and

A An 1809 painting showing a wounded soldier as he lay dying of tetanus in a military hospital.

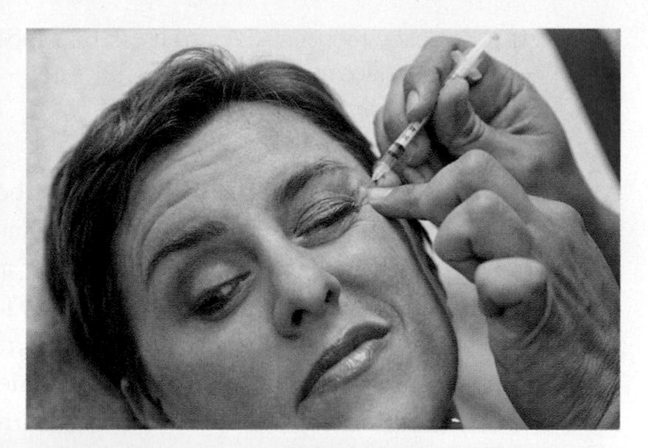

B Botox injections are used to paralyze facial muscles whose contractions cause wrinkles.

FIGURE 35.18 Disruption of muscle function by bacterial (*Clostridium*) toxins. The toxins impair normal signal flow from motor neurons to the muscles that they control.

all of the muscle fibers it synapses with constitute a **motor unit**. Stimulate a motor neuron, and all the muscle fibers on which it synapses will contract simultaneously. The motor neuron cannot make only some of the fibers it controls contract.

The mechanical force generated by a contracting muscle—the **muscle tension**—depends on the number of muscle fibers contracting. Some functions require more muscle tension than others, so the number of muscle fibers controlled by a single motor neuron varies. In motor units that bring about small, fine movements such as those that control eye muscles, one motor neuron synapses with about 5 muscle fibers. By contrast, the biceps of the arm has about 700 muscle fibers per motor unit. Having many fibers contract at once increases the force a motor unit generates.

Disrupted Control of Skeletal Muscle

Disruption of motor neuron signaling impairs muscle function. Consider what happens if *Clostridium tetani* bacteria colonize a wound. Toxin released by the bacteria prevents the release of GABA, a neurotransmitter that normally inhibits motor neuron signaling. As a result, nothing dampens signals calling for muscle contraction, and symptoms of the disease known as tetanus appear. The fists and jaw clench; hence the common name for the disease, lockjaw. The backbone may become locked in an abnormal arch (**FIGURE 35.18A**). Untreated tetanus can be fatal. Vaccines have essentially eliminated the disease in the United States, but the annual global death toll is over 200,000.

motor unit One motor neuron and the muscle fibers it controls.
muscle tension Force exerted by a contracting muscle.
sarcoplasmic reticulum Specialized endoplasmic reticulum in muscle cells; stores and releases calcium ions.

Another *Clostridium* species, *C. botulinum*, makes botulinum toxin. This toxin prevents motor neurons from releasing ACh, so it inhibits muscle contraction. Controlled doses of botulinum toxin are used in some cosmetic procedures. "Botox" is injected into specific facial muscles to inhibit contractions that result in wrinkles (**FIGURE 35.18B**).

Polio is an infectious disease in which a virus infects and destroys motor neurons. Effects range from temporary paralysis, to permanent paralysis, to death. Polio survivors are at risk for postpolio syndrome, a disorder characterized by muscle fatigue and progressive muscle weakness. Thanks to vaccines, no new cases of polio have arisen in the United States since 1979, but sporadic outbreaks continue in developing countries.

Motor neurons are also destroyed in amyotrophic lateral sclerosis (ALS). ALS is sometimes called Lou Gehrig's disease because this famous baseball player died of ALS in the 1930s. Affected people usually die of respiratory failure within a few years of diagnosis, but there are exceptions. Physicist Stephen Hawking has survived more than 40 years since his diagnosis. The cause of ALS is unknown.

TAKE-HOME MESSAGE 35.8

How do nervous signals cause muscle contraction?

✔ A skeletal muscle contracts in response to a signal from a motor neuron. Release of ACh at a neuromuscular junction causes an action potential in the muscle cell.

✔ An action potential results in release of calcium ions, which affect proteins attached to actin. Resulting changes in the shape and location of these proteins open the myosin-binding sites on actin, allowing cross-bridge formation.

35.9 Muscle Metabolism

✔ Muscle contraction requires ATP, which can be supplied by a variety of energy-releasing pathways.

Energy-Releasing Pathways

Muscle contraction requires ATP, but muscle fibers store only a limited amount of it. The fiber has a larger store of creatine phosphate, which can transfer a phosphate to ADP and form ATP (**FIGURE 35.19** ❶). Such phosphate transfers can fuel muscle contraction until other pathways increase ATP output. Some athletes take creatine supplements to increase the amount of creatine phosphate stored in muscle. Research suggests that creatine supplements can enhance performance of tasks that require a quick burst of energy. However, they have no effect on endurance, and the side effects of using such supplements are not fully known.

Aerobic respiration ❷ produces most of the ATP used by skeletal muscle during prolonged, moderate activity. Glucose derived from stored glycogen fuels five to ten minutes of activity, then the muscle fibers rely on glucose and fatty acids delivered by the blood. Fatty acids are the main fuel for activities that last more than half an hour.

Lactate fermentation is a muscle's third source of energy ❸. Some pyruvate is converted to lactate by the fermentation pathway even in resting muscle, but lactate fermentation steps up during exercise. This pathway produces less ATP than aerobic respiration, but has the advantage of operating even when the oxygen level in a muscle is low, as during strenuous exercise.

Types of Muscle Fibers

The primary mechanism of ATP production varies among muscle fibers. Red fibers have many mitochondria and make ATP mainly by aerobic respiration. They are colored red by **myoglobin**, a protein that, like hemoglobin, reversibly binds oxygen. When the blood's oxygen level is high, the gas diffuses into muscle and binds to myoglobin. When the blood oxygen level falls during periods of muscle activity, myoglobin releases oxygen for use in aerobic respiration. White muscle fibers have no myoglobin and few mitochondria. These fibers make ATP mainly by lactate fermentation.

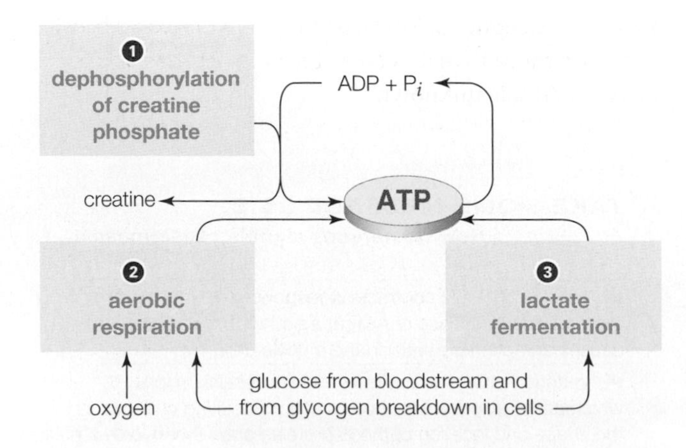

FIGURE 35.19 ▸**Animated** Three metabolic pathways that muscles use to obtain the ATP that fuels their contraction.

FIGURE 35.20 Loris, a slow-moving primate, whose limbs have a large proportion of slow red fibers. Creeping cautiously along branches helps a loris escape the attention of predators.

Muscle fibers can also be subdivided into fast fibers or slow fibers depending on the speed with which their myosin converts ATP to ADP. All white fibers are fast fibers. They contract rapidly but do not produce sustained contractions. The muscles that move your eye consist mainly of white fibers. Red fibers can be either fast or slow. Fast red fibers predominate in the human triceps muscle, which must often react quickly. Muscles involved in maintaining an upright posture, such as those in the back, consist mainly of slow red fibers.

For any given muscle, the relative proportions of fiber types vary among species and reflect the pattern of muscle usage. Limb muscles of cheetahs, which are renowned for their sprinting ability, contain a large proportion of white fibers. By contrast, limb muscles of stealthy, slow-moving lorises (**FIGURE 35.20**) have mostly slow red fibers. Similarly, among human athletes, successful sprinters tend to have a higher-than-average percentage of fast, white fibers in their leg muscles whereas marathoners tend to have a high percentage of slow, red fibers.

Muscles and Myostatin (revisited)

Muscular dystrophies are a class of genetic disorders in which skeletal muscles progressively weaken. With Duchenne muscular dystrophy, symptoms begin to appear in childhood. A mutation of a gene on the X chromosome causes this disorder. This gene encodes dystrophin, a plasma membrane protein of muscle fibers. The altered dystrophin specified by the mutated allele allows foreign material to enter a muscle fiber, causing it to break down.

Muscular dystrophy arises in about 1 in 3,500 males. Like other X-linked disorders (Section 14.4), it rarely causes symptoms in females, who nearly always have a normal version of the gene on their other X chromosome. Affected boys usually begin to show signs of weakness by the time they are three years old, and require a wheelchair when they are in their teens. Most die in their twenties of respiratory failure after the skeletal muscles involved in breathing become affected.

Drug companies hope to develop myostatin inhibitors that can encourage muscle production in people with muscular dystrophy and other muscle-wasting disorders, including cancer. Most myostatin inhibitors currently under investigation binds to and block receptors for myostatin, thus preventing myostatin from exerting its effects. Tests of myostatin inhibitors in animals have shown these drugs increase the differentiation of muscle stem cells into muscle fibers. Clinical trials will determine whether these drugs also improve human muscle function and what, if any, negative side effects they produce.

Effects of Exercise and Inactivity

Engaging in aerobic exercise—low intensity, but long duration—makes skeletal muscles more resistant to fatigue. It increases a muscle's blood supply by boosting growth of new capillaries, increases the number of mitochondria and amount of myoglobin in existing red muscle fibers, and encourages conversion of white fibers to red ones. Engaging in resistance exercise, such as weight lifting, encourages synthesis of additional actin and myosin filaments. The resulting increase in muscle mass allows for stronger contractions.

On the other hand, prolonged sitting can be a hazard to your health (**FIGURE 35.21**). When you sit, your leg muscles decrease their production of lipoprotein lipase (LPL), an enzyme that facilitates uptake of fatty acids and triglycerides from the blood. Decreased LPL activity raises the amount of lipids in the blood—thus increasing risk of cardiovascular disease and diabetes.

There is increasing evidence that the health risks associated with prolonged muscle inactivity persist even if a person also gets regular exercise. In other words, exercising for an hour each morning, although it improves your health in many respects, does not cancel out the negative metabolic effects of sitting in place for hours later in the day.

The best way to prevent health problems associated with a low LPL concentration is to avoid sitting for long intervals. When you must sit to carry out a task, take periodic short breaks to stand or walk around. Any activity that requires your leg muscles to support your weight leads to increased production of LPL.

FIGURE 35.21 Take a break from sitting. Prolonged sitting alters muscle fiber metabolism and may endanger your health.

myoglobin Muscle protein that reversibly binds oxygen.

TAKE-HOME MESSAGE 35.9

What factors affect muscle metabolism?

✔ Muscle contraction requires ATP. When excited, muscle first uses stored ATP, then transfers phosphate from creatine phosphate to ADP to form ATP. With prolonged exercise, aerobic respiration and lactate fermentation provide the ATP.

✔ Exercise increases blood flow to muscles, the number of mitochondria, production of actin and myosin, and the muscle's ability to take up lipids from the blood for use as an energy source.

summary

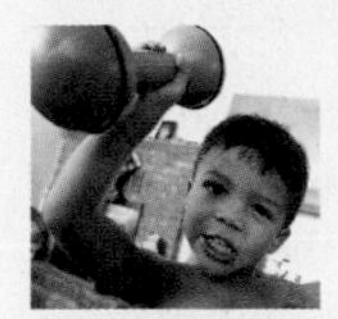

Section 35.1 Muscle fibers do not divide, but they can enlarge by adding proteins. Some hormones encourage this process and myostatin discourages it. Mutations that alter myostatin's effect lead to increased muscle mass and muscle strength.

Section 35.2 Animals are capable of **locomotion** during part or all of their life cycle. Friction and gravity oppose efforts to move. Animals are adapted to use specific forms of locomotion in specific environments.

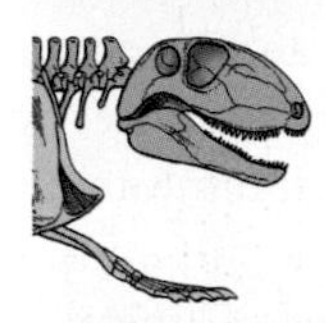

Section 35.3 Soft-bodied invertebrates have a **hydrostatic skeleton**, which consists of chambers full of fluid that muscle contractions redistribute. Arthropods have an **exoskeleton** of secreted hard parts at the body surface. An **endoskeleton** consists of hardened parts inside the body. Echinoderms and vertebrates have an endoskeleton.

All vertebrates have similar skeletal components. The vertebrate skull, vertebral column, and ribs constitute the **axial skeleton**. The **vertebral column** consists of **vertebrae** with **intervertebral disks** between them. Bony fins or limbs and the pectoral and pelvic girdles that attach them to the backbone constitute the **appendicular skeleton**.

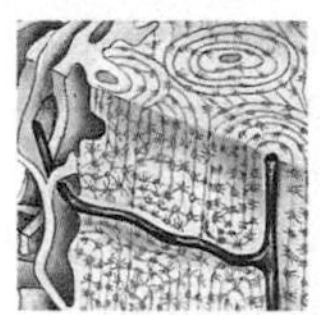

Sections 35.4, 35.5 Bones are organs rich in collagen, calcium, and phosphorus. Some have **red marrow** that makes blood cells; most have **yellow marrow**. In a human embryo, bones develop from a cartilage model. Even in adults, bones are continually remodeled. **Osteoblasts** are cells that synthesize bone, whereas **osteoclasts** break bone down. **Osteocytes** are former osteoblasts enclosed in a matrix of their secretions.

A **joint** is an area of close contact between bones. One or more **ligaments** hold bones together at most joints. Bits of cartilage and fluid-filled bursae cushion joints.

When bones break, cartilage fills the gap, then new bone is deposited. Tears in cartilage are slower to heal.

Section 35.6 Skeletal muscles bring about voluntary movements of body parts, act in reflexes, function in breathing, and generate heat that warms the body. A sheath of connective tissue surrounds each skeletal muscle and extends beyond it as a **tendon**. Most often, the tendon connects the muscle to a bone. A muscle can only exert force in one direction; it can pull but not push. Some skeletal muscles work in pairs to oppose one another's actions. Skeletal muscles also function as **sphincters**.

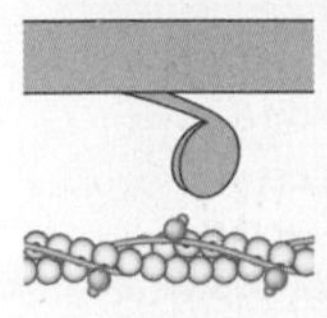

Section 35.7 The internal organization of a skeletal muscle promotes a strong, directional contraction. Many **myofibrils** fill the interior of a **skeletal muscle fiber**. A myofibril consists of **sarcomeres**, units of muscle contraction, lined up along its length. Each sarcomere has parallel arrays of thick and thin filaments.

The **sliding-filament model** describes how ATP-driven sliding of thin filaments past thick filaments shortens the sarcomere and brings about muscle contraction. Thin filaments consist mainly of the globular protein actin. Thick filaments are composed of the motor protein myosin. In an excited muscle, myosin binds to actin and uses energy released by the hydrolysis of ATP to move the thin filament toward the sarcomere center.

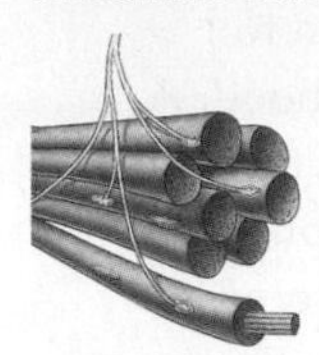

Section 35.8 A motor neuron and all the muscle fibers it controls constitute a **motor unit**. Signals from motor neurons result in action potentials in muscle fibers, which in turn cause the **sarcoplasmic reticulum** to release stored calcium ions. Flow of calcium into the cytoplasm causes two proteins associated with the thin filaments to shift in such a way that actin and myosin heads can interact and bring about a contraction that exerts **muscle tension**.

Diseases such as polio and toxins such as botulinum toxin (Botox) can interfere with motor neuron function and cause skeletal muscle weakness and paralysis.

Section 35.9 Muscle fibers produce the ATP they require for contraction by way of three pathways: dephosphorylation of creatine phosphate, aerobic respiration, and lactate fermentation. Muscle fibers differ in which pathway predominates and how fast they hydrolyze the resulting ATP. Red fibers depend mainly on aerobic respiration. They have many mitochondria and contain **myoglobin**, a protein that can bind and release oxygen. White fibers depend mainly on lactate fermentation and react fast.

Exercise can increase circulation to muscles, increase the number of myofibrils in muscle fibers, and enhance the ability of fibers to burn lipids as fuel. Long periods of uninterrupted sitting have a detrimental effect on the metabolic activity of fibers in leg muscles.

self-quiz

Answers in Appendix VII

1. An endoskeleton consists of ________ .
   a. a fluid in an internal space
   b. hardened plates at the surface of a body
   c. internal hard parts
   d. a fluid that surrounds the body

2. Bones are ________ .
   a. mineral reservoirs
   b. skeletal muscle's partners
   c. sites where blood cells form
   d. all of the above

3. Bones move when ________ muscles contract.
   a. cardiac
   b. skeletal
   c. smooth
   d. all of the above

4. A ligament connects ________ .
   a. bones at a joint
   b. a muscle to a bone
   c. a muscle to a tendon
   d. a tendon to bone

data analysis activities

Building Better Bones Tiffany, shown in **FIGURE 35.22**, was born with multiple fractures in her arms and legs. By age six, she had undergone surgery to correct more than 200 bone fractures. Her fragile, easily broken bones are symptoms of osteogenesis imperfecta (OI), a genetic disorder caused by a mutation in a gene for collagen. As bones develop, collagen forms a scaffold for deposition of mineralized bone tissue. The scaffold forms improperly in children with OI. **FIGURE 35.22** also shows the results of a test of a new drug. Treated children, all less than two years old, were compared to similarly affected children of the same age who were not treated with the drug.

1. An increase in vertebral area during the 12-month period of the study indicates bone growth. How many of the treated children showed such an increase?

2. How many of the untreated children showed an increase in vertebral area?

3. How did the rate of fractures in the two groups compare?

4. Do these results shown support the hypothesis that this drug, which slows bone breakdown, can increase bone growth and reduce fractures in young children with OI?

Treated child	Vertebral area in cm^2 (Initial)	Vertebral area in cm^2 (Final)	Fractures per year
1	14.7	16.7	1
2	15.5	16.9	1
3	6.7	16.5	6
4	7.3	11.8	0
5	13.6	14.6	6
6	9.3	15.6	1
7	15.3	15.9	0
8	9.9	13.0	4
9	10.5	13.4	4
Mean	11.4	14.9	2.6

Control child	Vertebral area in cm^2 (Initial)	Vertebral area in cm^2 (Final)	Fractures per year
1	18.2	13.7	4
2	16.5	12.9	7
3	16.4	11.3	8
4	13.5	7.7	5
5	16.2	16.1	8
6	18.9	17.0	6
Mean	16.6	13.1	6.3

FIGURE 35.22 Results of a clinical trial of a drug treatment for osteogenesis imperfecta (OI), which affects the child shown at right. Nine children with OI received the drug. Six others were untreated controls. Surface area of certain vertebrae was measured before and after treatment. Fractures occurring during the 12 months of the trial were also recorded.

5. Parathyroid hormone stimulates _______ .
   a. bone breakdown
   b. bone deposition
   c. red blood cell formation
   d. all of the above

6. The _______ attaches to the pelvic girdle.
   a. radius
   b. sternum
   c. femur
   d. tibia

7. The _______ is the basic unit of contraction.
   a. osteoblast
   b. sarcomere
   c. twitch
   d. myosin filament

8. In sarcomeres, phosphate-group transfers from ATP activate _______ .
   a. actin b. myosin c. troponin d. Z bands

9. A sarcomere shortens when _______ .
   a. thick filaments shorten
   b. thin filaments shorten
   c. both thick and thin filaments shorten
   d. none of the above

10. Muscle fibers produce ATP by _______ .
    a. aerobic respiration
    b. lactate fermentation
    c. creatine phosphate breakdown
    d. all of the above

11. Injection of *C. botulinum* toxin (Botox) _______ .
    a. causes irreversible muscle contraction
    b. prevents ATP formation by a muscle
    c. prevents release of ACh by motor neurons
    d. all of the above

12. A motor unit is _______ .
    a. a muscle and the bone it moves
    b. two muscles that work in opposition
    c. the amount a muscle shortens during contraction
    d. a motor neuron and the muscle fibers it controls

13. _______ from a motor neuron excites a muscle fiber.
    a. ACh b. GABA c. calcium d. phosphate

14. The sarcoplasmic reticulum stores and releases _______ .
    a. ACh b. GABA c. calcium d. phosphate

15. Match the words with their defining feature.

___ osteoblast	a. conveys excitatory signal
___ myofibrils	b. all in the hands
___ myoglobin	c. site of blood cell production
___ knee	d. binds myosin
___ actin	e. bone-forming cell
___ red marrow	f. donates phosphate to ADP
___ metacarpals	g. binds and releases oxygen
___ T tubule	h. example of a synovial joint
___ creatine phosphate	i. muscle fiber's threadlike parts

critical thinking

1. Continued strenuous activity can cause lactate to accumulate in muscles. After the activity stops, the lactate is converted into pyruvate and used as an energy source. Explain how pyruvate can be used to produce ATP.

2. After death, calcium pumps no longer function and the calcium ion concentration of muscle fiber cytoplasm increases. The result is rigor mortis—a state of postmortem muscle contraction. Explain why this contraction occurs and why it ends only when myosin heads begin to break down.

CREDITS: (22) top, © Cengage Learning; bottom, Courtesy of the family of Tiffany Manning.

36 Circulation

LEARNING ROADMAP

This chapter expands on the discussion of circulatory systems in Section 24.7. You have already been introduced to blood (31.4) and cardiac muscle (31.5). You will draw on your knowledge of hemoglobin (3.2), diffusion and osmosis (5.8), endocytosis (5.10), autonomic nerves (32.8), sickle-cell anemia (9.6, 17.7), and malaria (21.7).

OVERVIEW OF CIRCULATORY SYSTEMS

In an open system, circulatory fluid leaves vessels and flows among tissues. In a closed system, it remains inside vessels and exchanges with tissues occur across vessel walls.

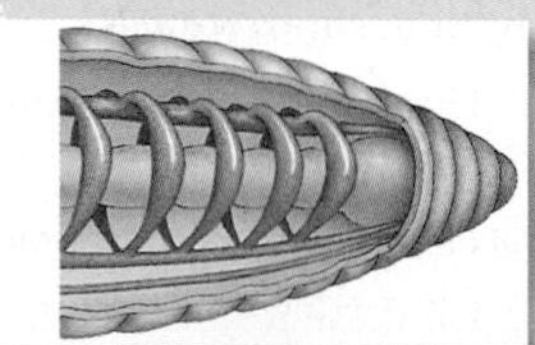

HUMAN CARDIOVASCULAR SYSTEM

A human heart is a muscular pump with four chambers. It pumps blood through two separate circuits: one extends to the lungs and back, the other to tissues throughout the body.

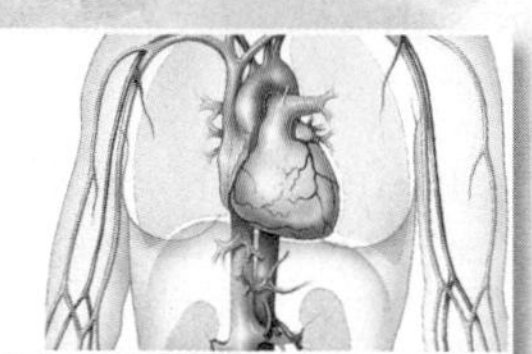

BLOOD AND BLOOD VESSELS

Blood has oxygen-carrying red cells, white cells that fight pathogens, and platelets that aid clotting, all suspended in a fluid plasma. Exchanges with body cells take place at capillaries.

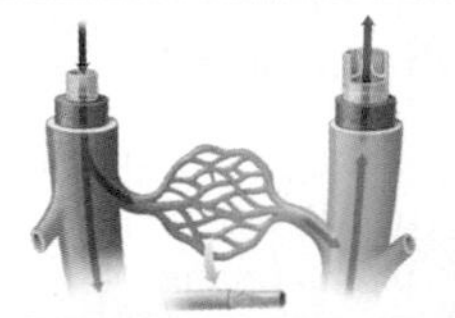

CIRCULATORY DISORDERS

Circulatory function is impaired by abnormal numbers of blood cells, defective blood cells, narrowing of blood vessels, and alteration of the normal heart rhythm.

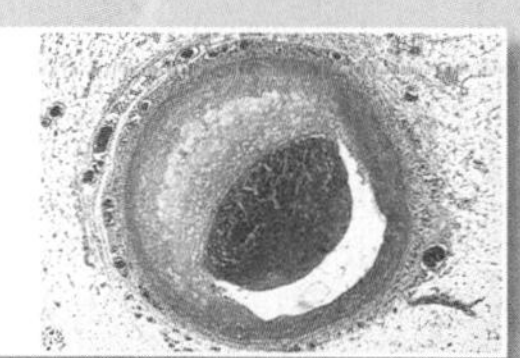

LYMPHATIC CONNECTIONS

Fluid that filters out of blood capillaries returns to blood by way of the lymphatic vascular system. Lymphatic organs help defend against pathogens.

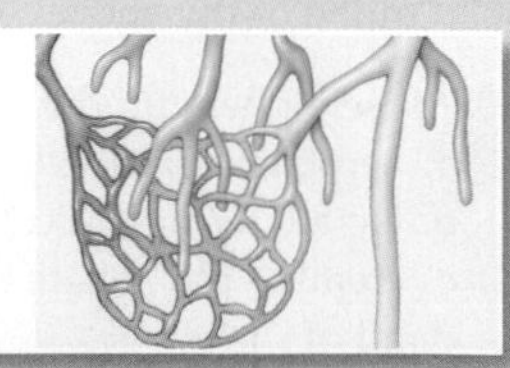

White blood cells act in immunity, the subject of Chapter 37. Red blood cells act in gas exchange (Section 38.7). The role of blood and lymph in the distribution of nutrients is the topic of Section 39.7, and Chapter 40 describes filtration of blood by the kidneys. Section 41.5 explains blood loss during menstruation and Section 41.7 how blood inflates the penis during an erection. Section 42.9 covers exchanges between fetal and maternal bloodstreams.

36.1 A Shocking Save

The heart is the body's most durable muscle. It begins to beat during the first month of human development, and keeps on going for a lifetime. Each heartbeat is set in motion by an electrical signal generated by a natural pacemaker in the heart wall. In some people, this pacemaker malfunctions, causing what is called sudden cardiac arrest. Electrical signaling becomes disrupted, the heart stops beating, and blood flow halts. In the United States, sudden cardiac arrest occurs in more than 300,000 people per year. An inborn heart defect causes most cardiac arrests in people under age 36. In older people, heart disease usually causes the heart to stop functioning.

The chance of surviving sudden cardiac arrest rises by 50 percent when cardiopulmonary resuscitation (CPR) is started within four to six minutes of the arrest. With CPR, a person alternates mouth-to-mouth respiration that substitutes for breathing with chest compressions that keep the victim's blood moving. However, CPR cannot restart the heart. That requires a defibrillator, a device that delivers an electric shock to the chest and resets the natural pacemaker. You have probably seen this procedure depicted in hospital dramas.

Matt Nader (**FIGURE 36.1A**) learned about the importance of CPR and defibrillation when he experienced sudden cardiac arrest during a high school football game. He came off the field after a play, sat on the bench, and felt a burning pain in his chest. His vision blurred, then he passed out. Nader's parents, who are physicians, were at the game and rushed from their seats. They determined that Matt did not have a pulse and, as they examined him, he stopped breathing.

Matt's parents began CPR on their son. At the same time, someone ran to get the school's automated external defibrillator (AED), a device about the size of a laptop computer (**FIGURE 36.1B**). The AED provides simple voice commands describing how to attach electrodes to a person in distress. It then checks for a heartbeat and, if required, shocks the heart.

The AED restarted Nader's heart, saving his life. Cardiologists determined that his sudden cardiac arrest had been caused by a genetic heart defect. To protect him, they implanted a small defibrillator in his body. It will provide a life-saving shock if his heart stops again.

After his recovery, Matt Nader went to testify before the Texas Legislature about his experience and to advocate for wider availability of AEDs in schools. Thanks in part to his efforts, Texas has passed a law requiring that all high schools have an AED available at athletic events and practices.

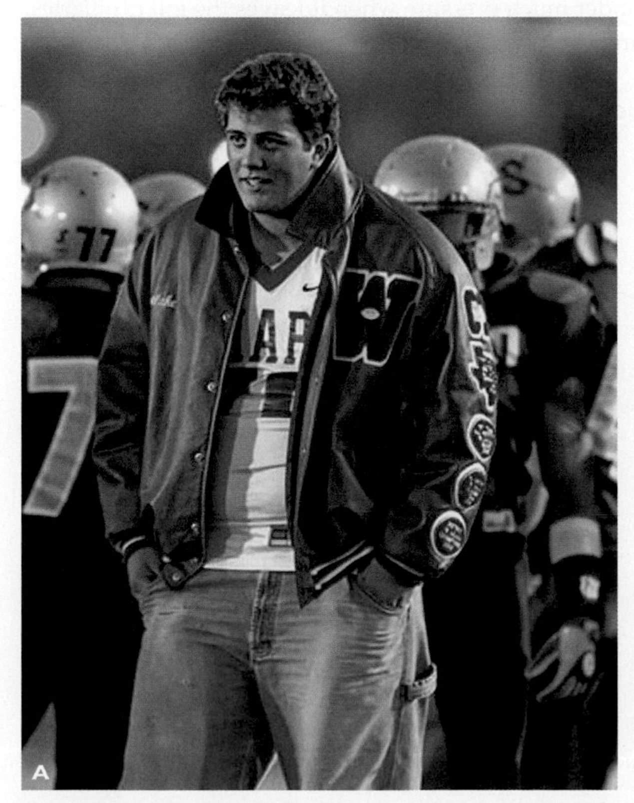

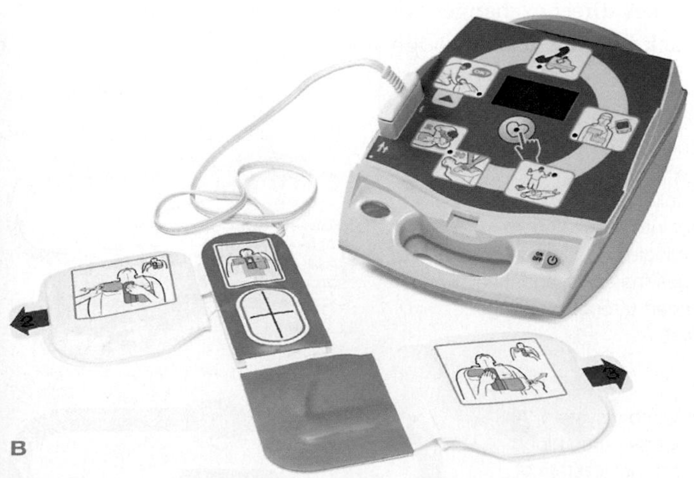

FIGURE 36.1 Surviving sudden cardiac arrest. (**A**) Matt Nader, a talented high school football player, discovered he had a heart defect when his heart stopped during a game. CPR and quick defibrillation saved his life.

(**B**) One type of automated external defibrillator. Such devices are designed to be simple enough to be used by a trained layperson. AEDs are increasingly available in public places, but they only make a difference if someone uses them.

CREDITS: (opposite) National Cancer Institute/Science Source; (1A) Courtesy of the family of Matt Nader; (1B) Courtesy of ZOLL Medical Corporation.

36.2 Circulatory Systems

✔ Most invertebrates and all vertebrates have a circulatory system that moves materials through the body.

All animals must keep their cells supplied with nutrients and oxygen, and all must dispose of cellular wastes. Some invertebrates, including cnidarians and flatworms (Sections 24.5 and 24.6), rely on diffusion alone to accomplish these tasks. In such animals, nutrients and gases reach cells by diffusing across a body surface and then diffusing through the interstitial fluid (the fluid between cells). Wastes diffuse in the opposite direction. Diffusion only works over short distances to move materials quickly, so animals that rely on diffusion to distribute materials have a body plan in which all cells lie close to a body surface. Evolution of circulatory systems allowed more complex body plans.

Open and Closed Circulatory Systems

A **circulatory system** is an organ system that speeds the distribution of materials within an animal body. It includes one or more **hearts** (muscular pumps) that propel fluid through a system of vessels that extends through the body.

Different types of circulatory systems evolved in different animal lineages. Arthropods and most mollusks have an **open circulatory system**. In such systems, a heart or hearts pump fluid called **hemolymph** into open-ended vessels (**FIGURE 36.2A**). Hemolymph leaves the vessels and mixes with the interstitial fluid, where it makes direct exchanges with cells before being drawn back into the heart through pores.

By contrast, annelids, cephalopod mollusks, and all vertebrates have a **closed circulatory system**, in which a heart or hearts pump fluid through a continuous network of vessels (**FIGURE 36.2B**). Fluid pumped through such a system is called **blood**. A closed circulatory system distributes substances faster than an open one. It is "closed" in that blood does not leave blood vessels to bathe tissues. Instead, most exchanges between blood and the cells of other tissues take place across the walls of the smallest-diameter blood vessels, which are called **capillaries**.

Evolution of Vertebrate Circulation

All vertebrates have a closed circulatory system, with a single heart. However, the structure of the heart and the circuits through which blood flows vary among vertebrate groups.

In most fishes, the heart has two chambers, and blood flows in a single circuit (**FIGURE 36.3A**). One chamber, an **atrium** (plural, atria), receives blood. From there, blood enters a **ventricle**, a chamber that pumps blood out of the heart. Pressure exerted by ventricular contractions drives blood through a series of vessels, into capillaries in each gill, then into capillaries in body tissues and organs, and back to the heart. Pressure imparted to blood by ventricular contraction dissipates as blood travels through capillaries, so blood is not under much pressure when it leaves the gill capillaries, and even less as it travels toward the heart.

Adapting to life on land involved coordinated modifications of the respiratory and circulatory systems.

A Open circulatory system. A grasshopper's heart pumps its yellowish hemolymph through a large vessel and out into tissue spaces. Hemolymph mingles with interstitial fluid, exchanges materials, and then reenters the heart through openings in the heart wall.

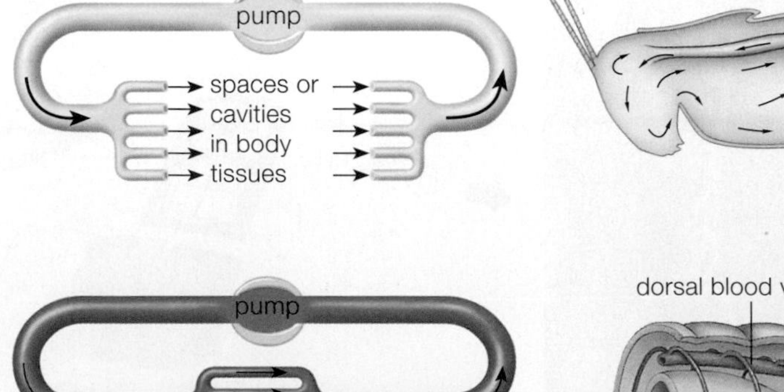

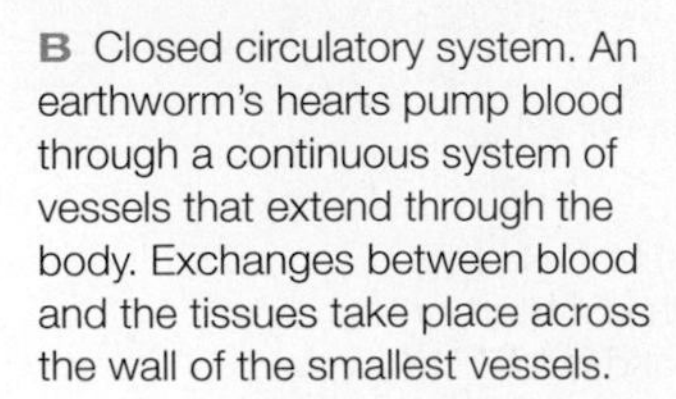

B Closed circulatory system. An earthworm's hearts pump blood through a continuous system of vessels that extend through the body. Exchanges between blood and the tissues take place across the wall of the smallest vessels.

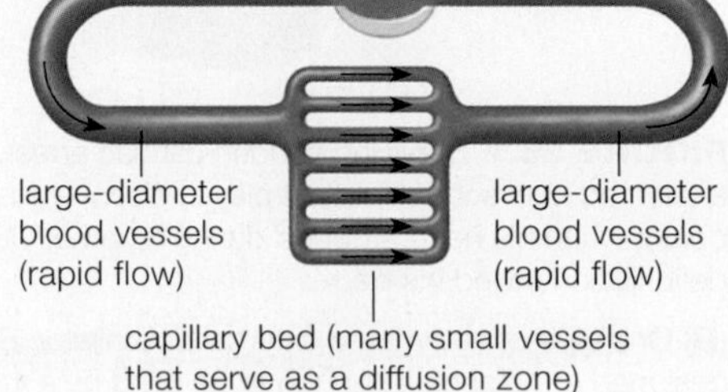

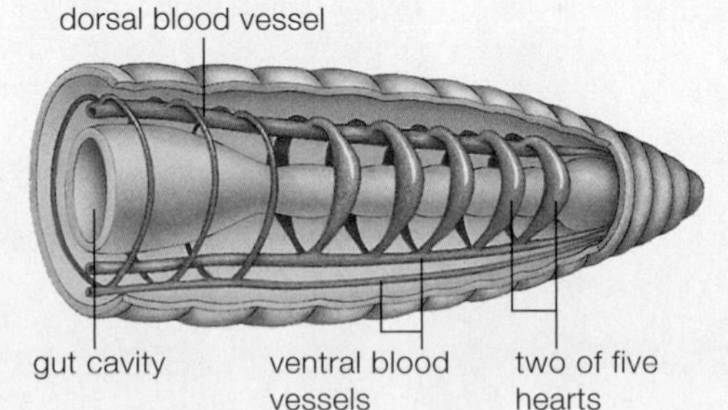

FIGURE 36.2 Comparison of open and closed circulatory systems.

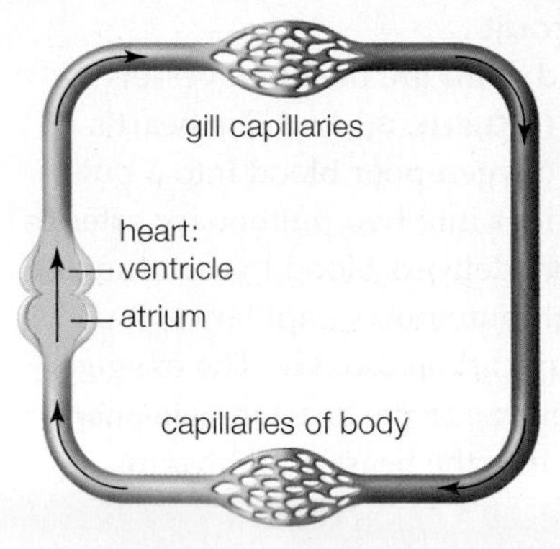

A The fish heart has one atrium and one ventricle. Force imparted by the ventricle's contraction propels blood through a single circuit.

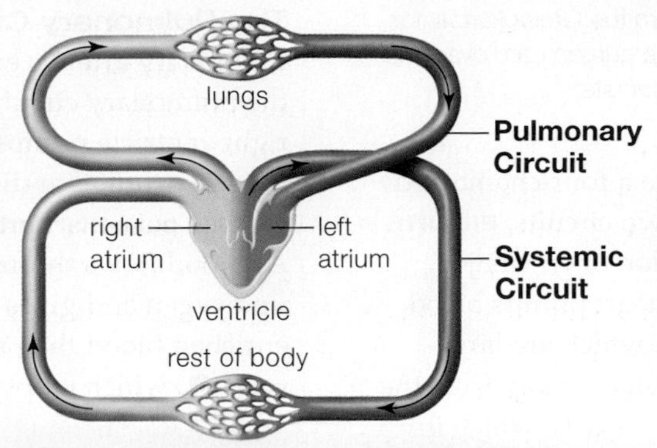

B In amphibians and most reptiles, the heart has three chambers: two atria and one ventricle. Blood flows in two partially separated circuits. Oxygenated blood and oxygen-poor blood mix a bit in the ventricle.

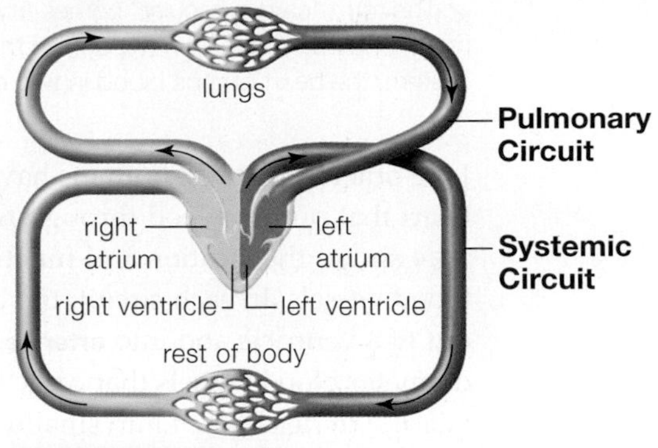

C In birds and mammals, the heart has four chambers: two atria and two ventricles. Oxygenated blood and oxygen-poor blood do not mix.

FIGURE 36.3 ▶**Animated** Variation in vertebrate circulatory systems.

Amphibians and most reptiles have a three-chambered heart, with two atria emptying into one ventricle (**FIGURE 36.3B**). A three-chambered heart speeds blood flow by moving blood through two partially separated circuits. Force of one contraction propels blood through the **pulmonary circuit**—to the lungs, and then back to the heart. (The Latin *pulmo* means lung.) A second contraction sends now-oxygenated blood through the **systemic circuit**. This circuit extends through capillaries in body tissues and returns to the heart.

In birds and mammals, the single ventricle has been divided into two. Their four-chambered heart has two atria and two ventricles (**FIGURE 36.3C**). With two fully separate circuits, only oxygen-rich blood flows to tissues. As an additional advantage, blood pressure can be regulated independently in each circuit. Strong contraction of the heart's left ventricle moves blood quickly through the long systemic circuit. At the same time, the right ventricle can contract more gently, protecting the delicate lung tissue that would be blown apart by higher pressure.

The four-chambered heart of mammals and birds is an example of morphological convergence (Section 18.3). Birds and mammals do not share an ancestor with such a heart. Rather, this trait evolved independently in the two groups. The enhanced blood flow associated with a four-chambered heart supports the high metabolism of these endothermic (heated from within) animals. Endotherms have higher energy needs than comparably sized ectotherms because they must produce heat to maintain their body temperature. Rapid blood flow in an endotherm's body delivers the large amount of oxygen required to keep the aerobic reactions that generate heat going nonstop.

atrium Heart chamber that receives blood.
blood Circulatory fluid of a closed circulatory system.
capillary Small blood vessels; exchanges with interstitial fluid take place across its walls.
circulatory system Organ system consisting of a heart or hearts and fluid-filled vessels that distribute substances through a body.
closed circulatory system Circulatory system in which blood flows through a continuous system of vessels.
heart Muscular organ that pumps fluid through a circulatory system.
hemolymph Fluid pumped through an open circulatory system.
open circulatory system Circulatory system in which hemolymph leaves vessels and flows through spaces in body tissues.
pulmonary circuit Circuit through which blood flows from the heart to the lungs and back.
systemic circuit Circuit through which blood flows from the heart to the body tissues and back.
ventricle Heart chamber that pumps blood out of the heart.

TAKE-HOME MESSAGE 36.2

How do animals distribute substances to cells throughout the body?

✔ Most animals have a circulatory system that speeds the distribution of substances through the body.

✔ Some invertebrates have an open circulatory system; other invertebrates and all vertebrates have a closed circulatory system, in which blood always remains enclosed within the heart or blood vessels.

✔ Fish have a one-circuit circulatory system. All other vertebrates have a short pulmonary circuit that carries blood to and from the lungs, and a longer systemic circuit that moves blood to and from the body's other tissues.

✔ A four-chambered heart evolved independently in birds and mammals. Such a heart allows strong contraction of one ventricle to speed blood through the systemic circuit, while a weaker contraction of the other ventricle protects lung tissue.

36.3 Human Cardiovascular System

✔ The term "cardiovascular" comes from the Greek *kardia* (for heart) and Latin *vasculum* (vessel). In the human cardiovascular system, the heart pumps blood in two circuits.

Like other mammals, humans have a four-chambered heart that pumps blood through two circuits. **FIGURE 36.4** shows the location and function of our major blood vessels. In each circuit, the heart pumps blood out of a ventricle and into **arteries**, which are large-diameter blood vessels that carry blood away from the heart. Arteries branch into smaller vessels, which in turn branch into capillaries inside specific tissues. A network of capillaries in a tissue is called a capillary bed. Blood that passed through a capillary bed returns to the heart by way of a **vein**.

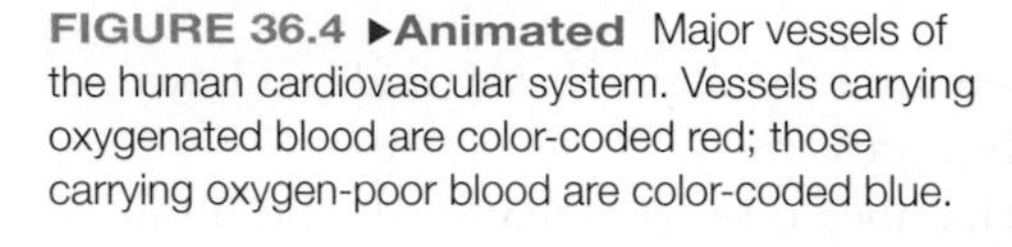

FIGURE 36.4 ▶Animated Major vessels of the human cardiovascular system. Vessels carrying oxygenated blood are color-coded red; those carrying oxygen-poor blood are color-coded blue.

The Pulmonary Circuit

Pulmonary arteries and veins are the main vessels of the pulmonary circuit (**FIGURE 36.5A**). The heart's right ventricle pumps oxygen-poor blood into a pulmonary trunk that divides into two pulmonary arteries ❶. One **pulmonary artery** delivers blood to each lung ❷. As blood flows through pulmonary capillaries, it picks up oxygen and gives up carbon dioxide. The oxygen-enriched blood then returns to the heart in **pulmonary veins** ❸, which empty into the heart's left atrium.

The Systemic Circuit

Oxygenated blood travels from the heart, to body tissues, and back by way of the systemic circuit (**FIGURE 36.5B**). The heart's left ventricle pumps blood into the **aorta**, the body's largest artery ❹. Branches from the aorta convey blood throughout the body. The initial portion of the aorta (the ascending aorta) carries blood toward the head. Carotid arteries, which service the brain, and coronary arteries, which service heart muscle, branch from the ascending aorta. The aorta then turns and descends through the thorax, continuing into the abdomen. Branches from descending portions of the aorta supply most internal organs and the lower limbs. For example, renal arteries deliver blood to the kidneys, and femoral arteries carry blood into each leg.

In the systemic circuit, blood gives up oxygen and picks up carbon dioxide as it flows through capillaries. It then returns to the heart via two large veins ❺. The **superior vena cava** returns blood from the head, neck, upper trunk, and arms. Smaller veins such as the carotid veins and coronary veins drain into the superior vena cava. The longer **inferior vena cava** returns blood from the lower trunk and legs. Renal veins and femoral veins are among the vessels that drain into the inferior vena cava.

In most cases, blood flows through only one capillary bed before returning to the heart. However, blood that passes through the capillaries in the small intestine enters a vein (the hepatic portal vein) that delivers it to a capillary bed in the liver ❻. Journeying through these two capillary beds allows blood to pick up glucose and other substances absorbed from the gut, and deliver them to the liver. The liver stores some of the absorbed glucose as glycogen (Section 3.4). It also breaks down some ingested toxins, including alcohol (Section 5.1).

aorta Largest artery; carries oxygenated blood away from the heart.
artery Large-diameter blood vessel that carries blood away from the heart.
inferior vena cava Vein that delivers blood from the lower body to the heart.
pulmonary artery Vessel that carries blood from the heart to a lung.
pulmonary vein Vessel that carries blood from a lung to the heart.
superior vena cava Vein that delivers blood from the upper body to the heart.
vein Large-diameter vessel that returns blood to the heart.

TAKE-HOME MESSAGE 36.3

What are the two circuits of the human cardiovascular system?

✔ The pulmonary circuit carries oxygen-poor blood from the heart through the pulmonary arteries and to capillaries in the lungs. Pulmonary veins return oxygenated blood to the heart.

✔ The systemic circuit carries oxygenated blood from the heart out the aorta, through branching arteries and to capillaries throughout the body. It returns oxygen-poor blood to the heart by way of veins.

✔ Most blood traveling through the systemic circuit passes through one capillary bed, but blood that flows through capillaries in the intestines also flows through capillaries in the liver.

A **Pulmonary Circuit**

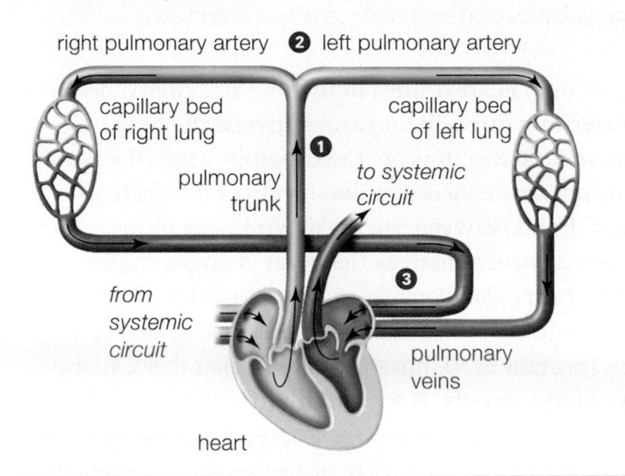

B **Systemic Circuit**

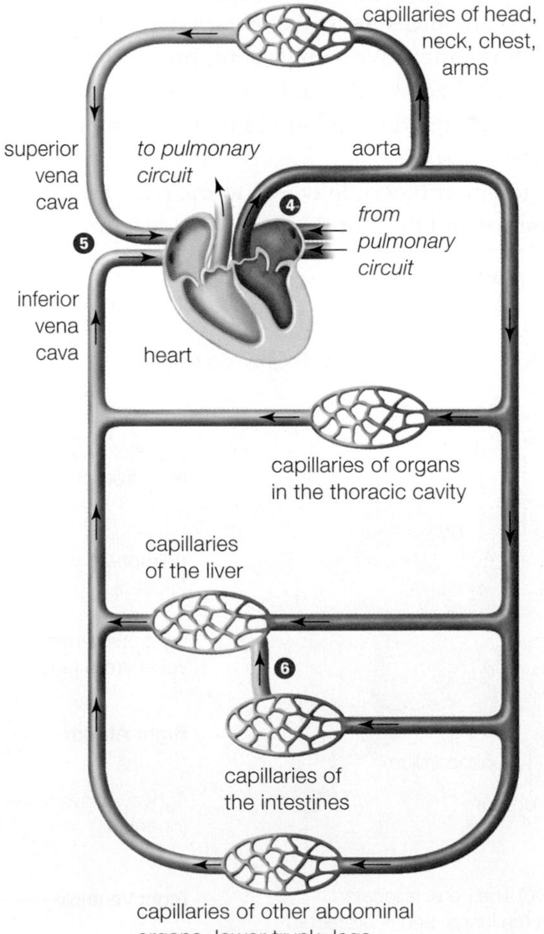

FIGURE 36.5 ▸Animated The two circuits of the human cardiovascular system. Vessels carrying oxygenated blood are shown in red and those carrying deoxygenated blood in blue.

FIGURE IT OUT Which circuit includes the hepatic portal vein, the vein that carries blood from the intestine to the liver?

Answer: The systemic circuit

36.4 The Human Heart

✔ The heart is a durable, muscular pump that contracts in response to its own spontaneous action potentials.

The human heart resides in the thoracic cavity, beneath the sternum (breastbone) and between the lungs (**FIGURE 36.6A**). It is enclosed within a sac (the pericardium) that consists of two layers of connective tissue. Fluid between the sac's two layers reduces the friction between them as the heart changes shape.

The heart wall consists mostly of cardiac muscle cells. Endothelium, a type of simple squamous epithelium (Section 31.3), lines the heart's chambers. It also lines blood vessels. A septum divides the heart into right and left sides (**FIGURE 36.6B**). Each side has an atrium and a ventricle. A pressure-sensitive atrioventricular (AV) valve between the two chambers functions like a one-way door to control blood flow. High fluid pressure forces the valve open. When this pressure declines, the valve swings shut, preventing blood from moving backward. Similar pressure-sensitive valves control the flow of blood into the pulmonary trunk and the aorta.

Oxygen-poor blood delivered to the right atrium by the superior and inferior venae cavae flows through the right AV valve into the right ventricle. The right ventricle pumps this blood through the pulmonary valve into the pulmonary arteries, and through the pulmonary circuit.

Oxygenated blood returns to the left atrium via pulmonary veins. This blood flows through the left AV valve into the left ventricle. The left ventricle pumps the blood through the aortic valve into the aorta. From here, it flows to tissues of the body.

The Cardiac Cycle

The events that occur from the onset of one heartbeat to another are collectively called the **cardiac cycle** (**FIGURE 36.7**). During this cycle, the heart's chambers alternate through **diastole** (relaxation) and **systole** (contraction). First, the relaxed atria expand with blood ❶. Fluid pressure forces AV valves to open and blood to flow into the relaxed ventricles, which expand as the atria contract ❷. Once filled, the ventricles contract. Contraction raises the fluid pressure inside the ventricles and forces the AV valves to close and the aortic and pulmonary valves to open. Blood flows through these valves and out of the ventricles ❸. Now emptied, the ventricles relax while the atria fill again ❹.

FIGURE 36.6 ▸Animated The human heart.

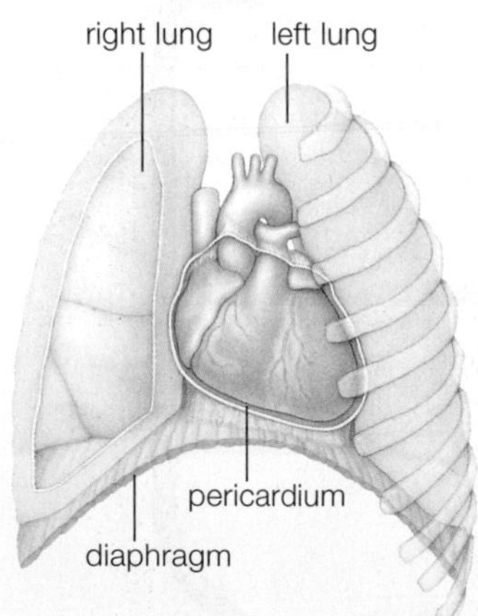

A (Above) The heart is located between the lungs, and is protected by the ribs.

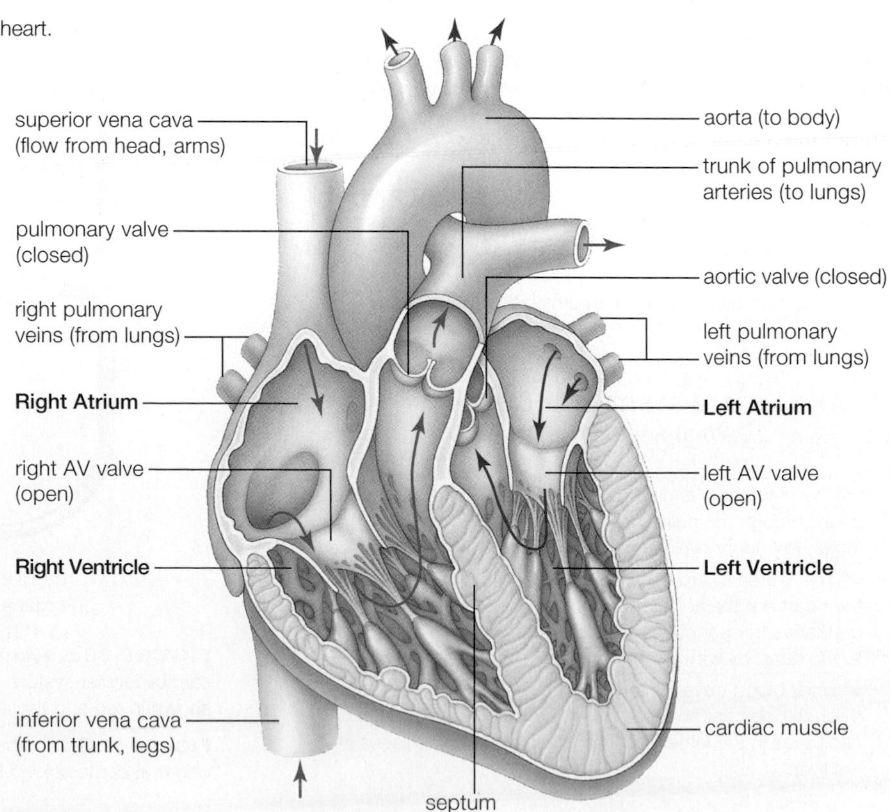

B (Right) Cutaway view of a human heart. Red arrows indicate the flow of oxygenated blood; blue arrows, oxygen-poor blood.

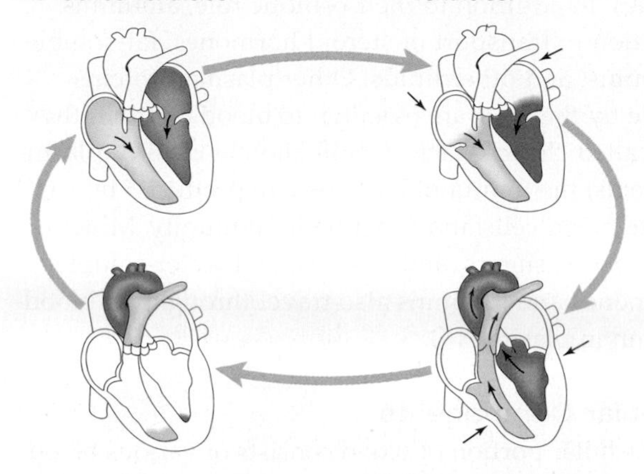

FIGURE 36.7 ▸Animated The cardiac cycle.

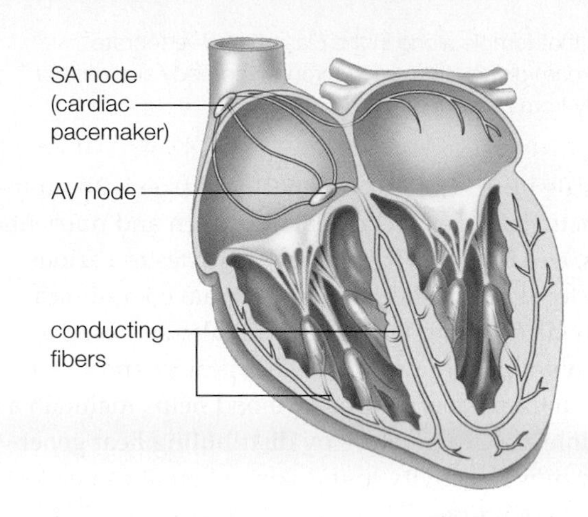

FIGURE 36.8 ▸Animated Cardiac conduction system.

During the cardiac cycle a "lub-dup" sound can be heard through the chest wall. Each "lub" is the heart's AV valves closing. Each "dup" is the heart's aortic and pulmonary valves closing.

Blood circulation is driven entirely by contracting ventricles; atrial contraction only helps fill the ventricles with blood. The structure of the cardiac chambers reflects their different functions. Atria need only generate enough force to squeeze blood into the ventricles, so they have relatively thin walls. By contrast, contraction of ventricles has to produce enough pressure to propel blood through an entire cardiovascular circuit, so ventricle walls are more thickly muscled. The left ventricle, which pumps blood throughout the long systemic circuit, has thicker walls than the right ventricle, which pumps blood only to the lungs and back.

Setting the Pace for Contraction

Like skeletal muscle, cardiac muscle has orderly arrays of sarcomeres that contract by a sliding-filament mechanism. Unlike skeletal muscle, cardiac muscle does not require a signal from the central nervous system to contract. The **sinoatrial (SA) node**, a clump of specialized cells in the wall of the right atrium (**FIGURE 36.8**), serves as the cardiac pacemaker, generating an action potential about seventy times a minute. Signals from the central nervous system only adjust this rate.

Gap junctions (Section 4.11) between adjacent cardiac muscle cells allow action potentials generated by the SA node to spread across the atria. These action potentials cause the atria to contract. During atrial contraction, special noncontractile muscle fibers conduct action potentials to the **atrioventricular (AV) node**. This clump of cells is the only electric bridge to the ventricles; action potentials can only reach the ventricles by way of the AV node. The time it takes for a signal to cross this bridge allows blood from the atria to fill the ventricles before the ventricles contract. From the AV node, the signal travels along conducting fibers in the septum between the heart's left and right halves. The fibers extend to the heart's lowest point and up the ventricle walls. The excitatory signal spreads via gap junctions, causing both ventricles to contract from the bottom up, with a wringing motion.

TAKE-HOME MESSAGE 36.4

How does the structure of the human heart relate to its function?

✔ The four-chambered heart is a muscular pump partitioned into two halves, each with an atrium and a ventricle. Forceful contraction of the ventricles provides the driving force for blood circulation.

✔ The SA node is the cardiac pacemaker. Its spontaneous, rhythmic signals make cardiac muscle cells of the heart wall contract in a coordinated fashion.

atrioventricular (AV) node Clump of cells that serves as the electrical bridge between the atria and ventricles.
cardiac cycle Sequence of contraction and relaxation of heart chambers that occurs with each heartbeat.
diastole Relaxation phase of the cardiac cycle.
sinoatrial (SA) node Cardiac pacemaker; cluster of specialized cells whose spontaneous rhythmic signals trigger contractions.
systole Contractile phase of the cardiac cycle.

36.5 Vertebrate Blood

✔ Cells that tumble along in the plasma of a vertebrate bloodstream distribute oxygen through the body and defend the body from pathogens.

Vertebrate blood is a fluid connective tissue with multiple functions. It delivers essential oxygen and nutrients to cells, and carries cells' metabolic wastes to various organs for disposal. It facilitates internal communications by distributing hormones, and also serves as a highway for cells and proteins that protect and repair tissues. In birds and mammals, blood helps maintain a stable internal temperature by distributing heat generated by muscle activity to the skin, where it can be lost to the surroundings.

A human adult has about 5 liters of blood (a bit more than 5 quarts). Blood is—as the saying goes—thicker than water. Dissolved substances and suspended cells contribute to its greater viscosity. **FIGURE 36.9** shows the components of vertebrate blood.

Plasma

Plasma, the fluid portion of the blood, constitutes about 50 to 60 percent of the blood volume. Plasma is mostly water with dissolved plasma proteins. More than half of these proteins are albumins, which are water-soluble proteins made by the liver. The high albumin content of blood helps create a solute concentration gradient that draws water into capillaries. As you will see in Section 36.8, this osmosis plays an important role in exchanges between blood and tissues. In addition to their osmotic role, albumins function in transport of steroid hormones, fat-soluble vitamins, and other lipids. Other plasma proteins made by the liver are essential to blood clotting; they are called clotting factors. Still another class of plasma proteins, the immunoglobulins, are produced by white blood cells and function in immunity. Mineral ions, gases, sugars, amino acids, and water-soluble hormones and vitamins also travel through the bloodstream in plasma.

FIGURE 36.9 ▶**Animated** Components of human blood.

Cellular Components

The cellular portion of blood consists of various blood cells and platelets. All cellular components of blood descend from stem cells in red bone marrow.

Red Blood Cells **Red blood cells**, or erythrocytes, transport oxygen from lungs to aerobically respiring cells and help move waste carbon dioxide away from them. They are the most abundant cells in blood, accounting for 40 to 50 percent of the blood volume.

In mammals, red blood cells lose their nucleus, mitochondria, and other organelles as they mature. A human red blood cell is a flexible disk with a depression at its center. The cell's flexibility allows it to slip easily through narrow blood vessels, and its thinness facilitates gas exchange.

The interior of a mature red blood cell is filled with hemoglobin, a protein whose structure you learned about in Section 3.6. Most oxygen that enters the blood travels to tissues while bound to the heme group of hemoglobin. In addition to hemoglobin, a mature red blood cell contains sugars, RNAs, and other molecules that sustain it for about 120 days. Ongoing replacements keep the red blood cell count at a fairly stable level. However, during reproductive years, women typically have a lower red blood cell count than men, because women lose some blood during menstruation.

Erythropoietin, a hormone made by the kidney, is necessary for red blood cell production by bone marrow. Erythropoietin produced by genetic engineering is used to treat a low red blood cell count (anemia). It is also abused by athletes who take it to artificially increase their red cell count.

Inherited variations in molecules at the surface of red blood cells are the basis for blood typing. Section 13.5 described the genetic basis of ABO blood types. Variation in another surface protein, called Rh factor, is the basis for Rh blood types.

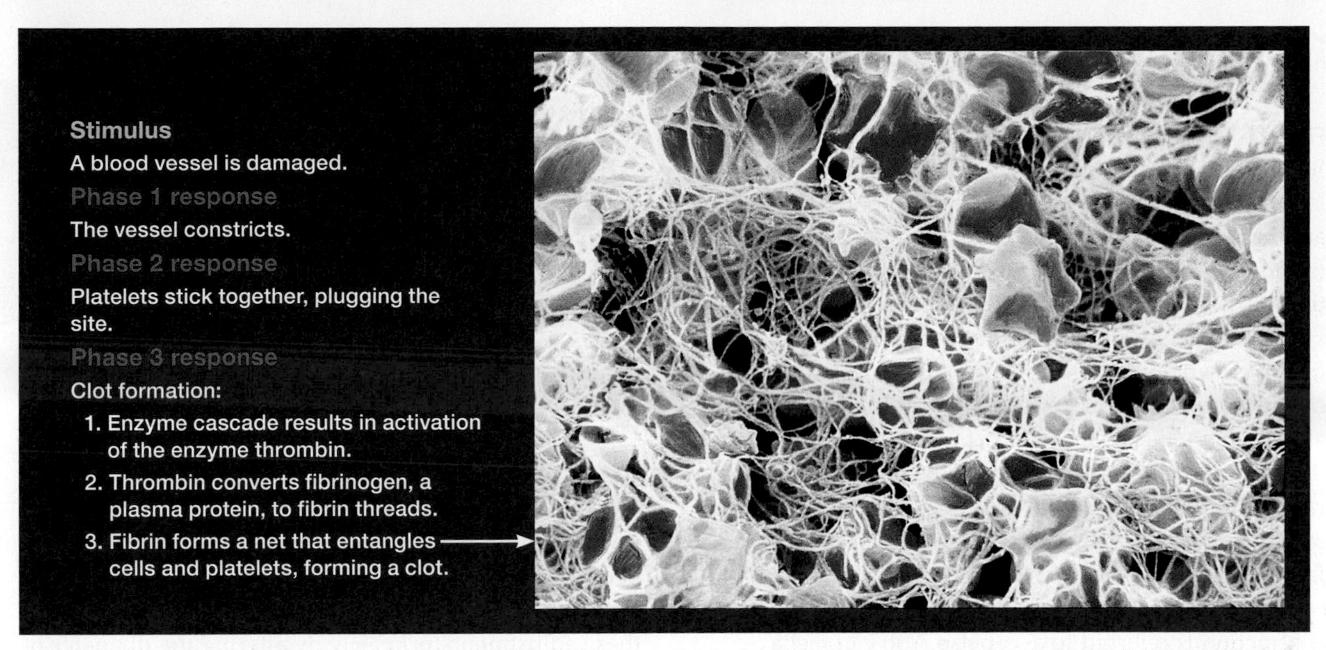

FIGURE 36.10 ▸**Animated** Hemostasis. The micrograph shows the final clotting phase—blood cells and platelets in a fibrin net.

White Blood Cells **White blood cells**, or leukocytes, carry out ongoing housekeeping tasks and function in defense. There are several kinds, and they differ in their size, nuclear shape, and staining traits, as well as function. We discuss the types and roles of white blood cells in detail in the next chapter, but here is a brief preview. Neutrophils, the most abundant white cells, are phagocytes that engulf bacteria and cellular debris. Eosinophils are specialized to attack larger parasites, such as worms. Basophils and mast cells secrete chemicals that have a role in inflammation. Monocytes circulate in the blood for a few days, then move into tissues, where they develop into phagocytic cells known as macrophages. Macrophages interact with lymphocytes to bring about immune responses. The two types of lymphocytes, B lymphocytes and T lymphocytes, protect the body against specific pathogens. B cells secrete the immunoglobulins that are a component of plasma.

Platelets A **platelet** is a membrane-wrapped fragment of cytoplasm that arises when a large cell (a megakaryocyte) breaks up. Hundreds of thousands of platelets circulate in the blood, ready to take part in **hemostasis**. This process stops blood loss from an injured vessel and provides a framework for repairs.

Hemostasis begins when an injured vessel constricts (narrows), reducing blood loss (**FIGURE 36.10**). Platelets adhere to the injured site and release substances that attract more platelets. Plasma proteins convert blood to a gel and form a clot. Clot formation involves a cascade of enzyme reactions. Fibrinogen (a clotting factor) is converted to fibrin by the enzyme thrombin, which circulates in blood as the inactive precursor prothrombin. Prothrombin (also a clotting factor) is activated by an enzyme that is activated by another enzyme, and so on. If a mutation affects any of the enzymes in the cascade of clotting reactions, blood may not clot properly. Such mutations cause the genetic disorder hemophilia. A vitamin K deficiency can impair clotting too, because this vitamin is required as a cofactor for activation of some clotting factors.

TAKE-HOME MESSAGE 36.5

What are the components of human blood and what are their functions?

✔ Blood consists mainly of plasma, a protein-rich fluid that transports wastes, gases, and nutrients.

✔ Blood cells and platelets form in bone marrow and are transported in plasma. Red blood cells contain hemoglobin that carries oxygen from lungs to tissues. White cells help defend the body from pathogens. Platelets are cell fragments that have a role in clotting.

hemostasis Process by which blood clots in response to injury.
plasma Fluid portion of blood.
platelet Cell fragment that helps blood clot.
red blood cell Hemoglobin-filled blood cell that transports oxygen; an erythrocyte.
white blood cell Vertebrate blood cell with a role in housekeeping tasks and defense; a leukocyte.

36.6 Arteries and Arterioles

✔ Arteries and arterioles are large vessels that carry blood from the heart toward capillary beds.

Rapid Transport in Arteries

Blood pumped out of ventricles enters arteries. These large-diameter vessels are ringed by smooth muscle and have an outer covering of highly elastic connective tissue (FIGURE 36.11A). Like other blood vessels and the heart, they are lined by endothelium.

Elastic properties of an artery help keep blood flowing, even when ventricles relax. With each ventricular contraction, pressure exerted by blood forced into an artery causes the artery wall to bulge outward. Then, as the ventricle relaxes, the artery wall springs back like a rubber band that has been stretched. As the artery wall recoils inward, it pushes blood inside the artery a bit farther away from the heart.

The bulging of an artery with each ventricular contraction is referred to as a **pulse**. You can feel a pulse by placing your finger on a pulse point, a body region where an artery runs close to the body surface (FIGURE 36.12).

Adjusting Flow at Arterioles

All blood pumped out of the right ventricle always flows through pulmonary arteries to your lungs. In the systemic circuit, about 20 percent of the blood always flows through the carotid arteries to the brain. The body adjusts the distribution of blood across the rest of the systemic circulation according to its needs. It makes these adjustments primarily by altering the diameter of **arterioles**, which are blood vessels that branch from an artery and deliver blood to capillaries.

Each arteriole (FIGURE 36.11B) is ringed by smooth muscle that responds to commands from the central nervous system. For example, activation of the sympathetic nervous system triggers the fight–flight response (Section 32.8). This response includes **vasodilation** (widening) of arterioles in the limbs, so more blood flows to skeletal muscles. At the same time, **vasoconstriction** (narrowing) of arterioles that deliver blood to the gut decreases their share of the blood supply.

Arterioles also adjust blood flow in response to metabolic activity in nearby tissue. During exercise, skeletal muscle uses up oxygen and releases carbon dioxide and lactic acid. Arterioles delivering blood to the muscle widen in response to these changes.

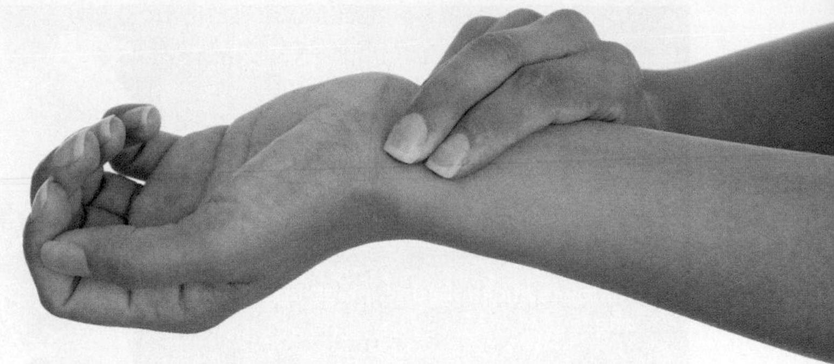

FIGURE 36.12 Checking the pulse in the radial artery, which delivers blood to the hand.

FIGURE IT OUT Which heart chamber exerts the pressure on blood that is felt at this pulse point? Answer: The left ventricle

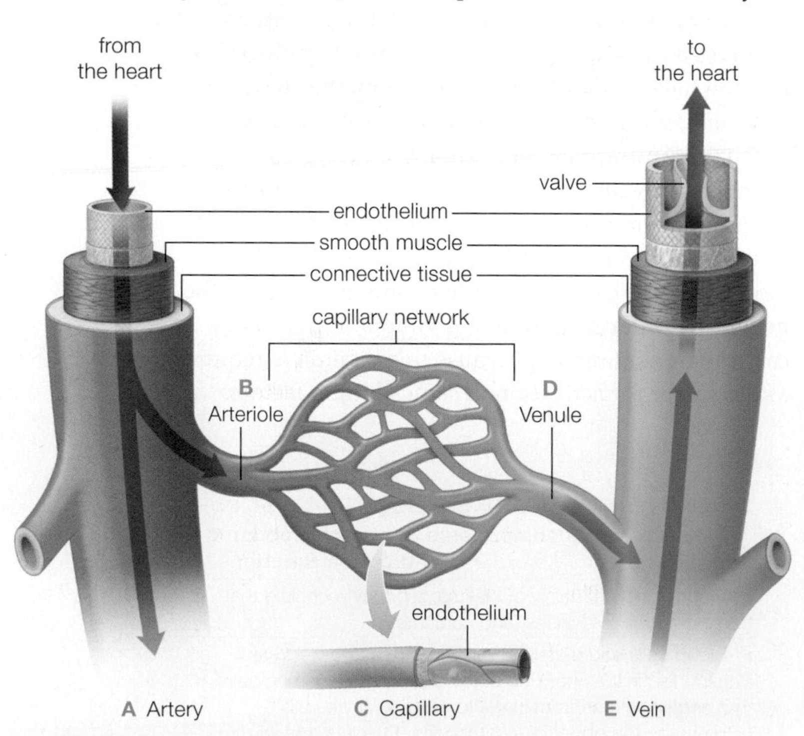

FIGURE 36.11 ▶Animated Structural comparison of human blood vessels and the direction of blood flow through them.

arteriole Blood vessel that conveys blood from an artery to capillaries.
pulse Brief stretching of artery walls that occurs when ventricles contract.
vasoconstriction Narrowing of a blood vessel when smooth muscle that rings it contracts.
vasodilation Widening of a blood vessel when smooth muscle that rings it relaxes.

TAKE-HOME MESSAGE 36.6

How does the structure of arteries and arterioles reflect their function?

✔ Arteries are thick-walled, large-diameter vessels that transport large volumes of blood away from the heart.

✔ Smooth muscle in the wall of arterioles allows adjustments to blood flow in the systemic circuit.

36.7 Blood Pressure

✔ Ventricular contractions are the source of blood pressure, which declines throughout a cardiovascular circuit.

Blood pressure is pressure exerted by blood against the wall of the vessel that encloses it. The right ventricle contracts less forcefully than the left ventricle, so blood entering the pulmonary circuit is under less pressure than blood entering the systemic circuit. In both circuits, blood pressure is highest in arteries, and declines over the course of the circuit (FIGURE 36.13).

Blood pressure is usually measured in the brachial artery of the upper arm (FIGURE 36.14). Two pressures are recorded. **Systolic pressure**, the highest pressure of a cardiac cycle, occurs as contracting ventricles force blood into the arteries. **Diastolic pressure**, the lowest blood pressure of a cardiac cycle, occurs when ventricles are relaxed. We measure blood pressure in millimeters of mercury (mm Hg), a standard unit for describing pressure, and record it as systolic value/diastolic value. Normal blood pressure is about 120/80 mm Hg, or "120 over 80."

Blood pressure depends on the total blood volume, how much blood ventricles pump out (cardiac output), and the degree of arteriole dilation. Sensory receptors in the aorta and in carotid arteries signal the medulla oblongata in the hindbrain when blood pressure rises or falls. In a reflexive response, this brain region calls for changes in cardiac output and arteriole diameter that will restore normal pressure. Vasodilation of arterioles lowers blood pressure; vasoconstriction raises it. This reflex response alters blood pressure over the short term only. Over the longer term, kidneys adjust blood pressure by regulating the amount of fluid lost in urine and thus determining the total blood volume.

Inability to regulate blood pressure can result in hypertension, in which resting blood pressure remains above 140/90. Chronic high blood pressure makes the heart and kidneys work harder, increasing risk of heart disease or kidney failure.

blood pressure Pressure exerted by blood against a vessel wall.
diastolic pressure Blood pressure when ventricles are relaxed.
systolic pressure Blood pressure when ventricles are contracting.

TAKE-HOME MESSAGE 36.7

How is blood pressure recorded and regulated?

✔ Blood pressure is the fluid pressure exerted against a vessel wall. It is recorded as systolic/diastolic pressure.

✔ Adjustments to arteriole diameter, cardiac output, and blood volume regulate blood pressure.

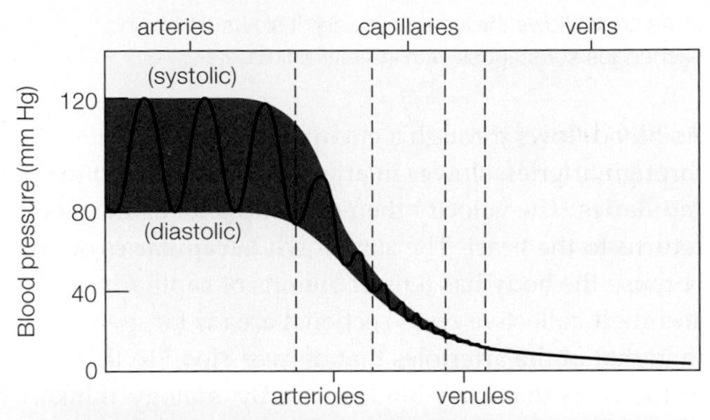

FIGURE 36.13 Plot of changes in fluid pressure as a volume of blood flows through the systemic circuit. Systolic pressure occurs when ventricles contract; diastolic, when ventricles are relaxed.

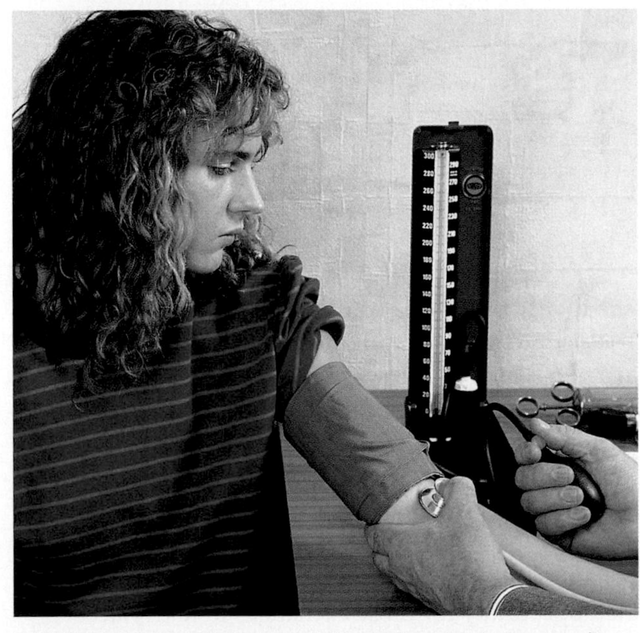

FIGURE 36.14 ▸Animated Measuring blood pressure. A hollow inflatable cuff attached to a pressure gauge is wrapped around the upper arm. A stethoscope is placed over the brachial artery, just below the cuff.

The cuff is inflated with air to a pressure above the highest pressure of the cardiac cycle, when ventricles contract. Above this pressure, you will not hear sounds through the stethoscope, because no blood is flowing through the vessel.

Air in the cuff is slowly released until the stethoscope picks up soft tapping sounds. Blood flowing into the artery under the pressure of the contracting ventricles—the systolic pressure—causes the sounds. When these sounds start, a gauge typically reads about 120 mm Hg. That amount of pressure will force mercury (Hg) to move up 120 millimeters in a glass column of a standardized diameter.

Air is released from the cuff until the sounds stop because blood flows continuously, even when the ventricles are relaxed. The pressure recorded when the sounds first stop is the diastolic pressure, which is usually about 80 mm Hg.

Right, compact monitors are now available that automatically record the systolic/diastolic blood pressure.

36.8 Exchanges at Capillaries

✔ As blood flows through a capillary, it slows down and exchanges substances with interstitial fluid.

As blood flows through a circuit, it moves fastest through arteries, slower in arterioles, and slowest in capillaries. The velocity then picks up a bit as the blood returns to the heart. The slowdown in capillaries occurs because the body has tens of billions of capillaries, and their collective cross-sectional area is far greater than that of the arterioles that deliver blood to them, or the veins that carry blood away. By analogy, think about what happens if a narrow river (representing few larger vessels) delivers water to a wide lake (representing the many capillaries):

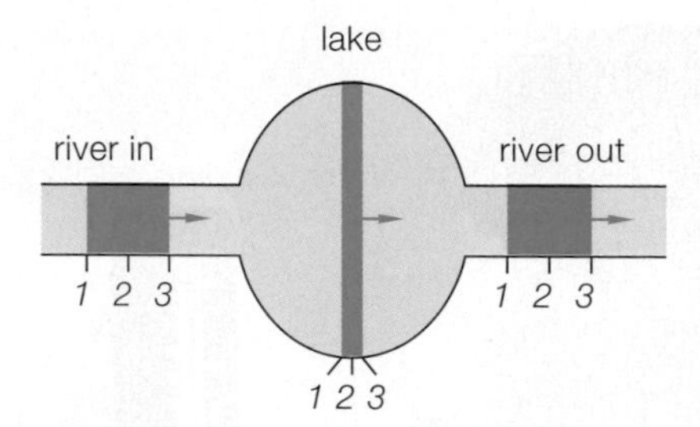

The rate of the flow is constant into and out of the lake: The same volume of water moves from points 1 to 3 in a given interval. However, the velocity of the flow decreases in the lake region. When the volume spreads across a larger cross-sectional area, it flows forward a shorter distance during the specified interval.

Slow flow through small capillaries enhances the rate of exchanges between blood and interstitial fluid. The more time blood spends in a capillary, the greater the opportunity for those exchanges to take place.

Materials move between capillaries and body cells in several ways. The capillary wall consists of a single layer of endothelial cells. In most tissues, spaces between these cells make the capillary wall a bit leaky. This leakiness comes into play mainly at the arterial end of a capillary. Here, pressure exerted by the beating heart forces plasma fluid out through spaces between cells and into the surrounding interstitial fluid (**FIGURE 36.15** ❶). Blood cells and platelets are too big to slip through the spaces between endothelial cells, so they remain in the capillary.

Along the length of the capillary, oxygen released by red blood cells diffuses from blood into interstitial fluid, while nutrients such as glucose are transported in the same direction by membrane proteins ❷. Carbon dioxide (CO_2) diffuses from interstitial fluid into the capillary, and other metabolic wastes are transported into it ❸. Near the venous end of the capillary bed, where blood pressure is lower, water moves by osmosis from interstitial fluid into the protein-rich plasma ❹.

As a result of all the leaking and osmotic movement of fluid, there is a small net outward flow from a capillary bed into interstitial fluid. Fluid lost from the bloodstream by this process is returned by the lymphatic system. Some interstitial fluid enters lymph capillaries ❺, which drain into lymphatic ducts that return fluid to veins near the heart. We consider the lymph vascular system in more detail in Section 36.11.

TAKE-HOME MESSAGE 36.8

How does blood exchange materials with interstitial fluid?

✔ Small molecules cross cells of a capillary by diffusion and larger ones move across by exocytosis.

✔ Fluid rich in oxygen and nutrients also leaks out between cells of the capillary wall.

FIGURE 36.15 Forces affecting capillary exchange.

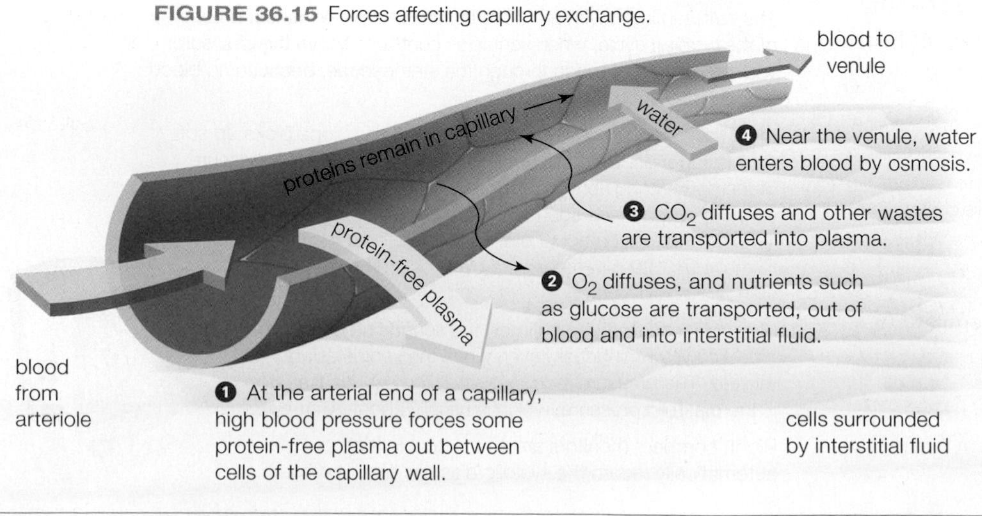

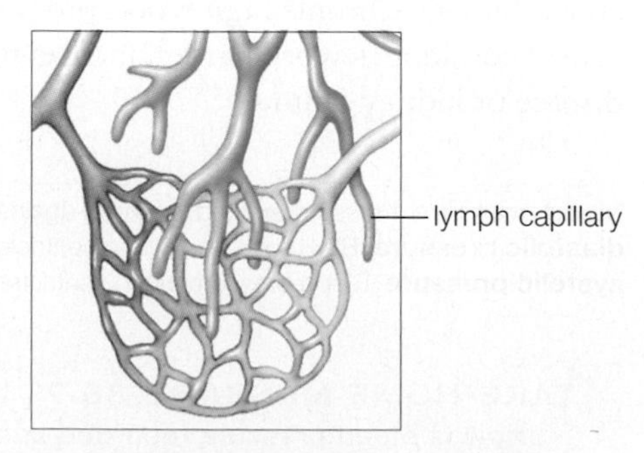

❺ Some fluid leaked by blood capillaries enters neighboring lymph capillaries. This fluid, now called lymph, returns to the bloodstream when large lymph vessels drain into veins at the base of the neck.

36.9 Back to the Heart

✔ Veins are the body's largest blood reservoir.

✔ Skeletal muscle activity helps move blood at low pressure through veins and back to the heart.

Blood from multiple capillaries flows into a thin-walled vessel called a venule, then multiple venules supply each vein. Of all blood vessels, veins hold the greatest volume of blood. When you are at rest, they contain about 60 percent of your total blood volume. Veins also have the lowest blood pressure; by the time blood reaches veins, most of the pressure imparted by ventricular contractions has dissipated.

Several mechanisms help blood at low pressure move back toward the heart. First, veins have one-way valves that prevent backflow. These valves shut when blood starts to reverse direction. For example, valves in the large veins of your leg prevent blood from moving downward in response to gravity when you stand. All vertebrates typically have the same types of arteries and veins, but the number and location of valves in those veins varies (**FIGURE 36.16**).

Smooth muscle in a vein's wall facilitates blood flow too. When this muscle contracts, the vein stiffens so it cannot hold as much blood, pressure on blood in the vein rises, and the blood is forced toward the heart.

Skeletal muscles used in limb movements also help move blood through veins. When these muscles contract, they bulge and press on neighboring veins, squeezing the blood inside them toward the heart (**FIGURE 36.17**). Exercise-induced deep breathing also raises pressure inside veins. The lungs and thoracic cavity expand during inhalation, forcing adjacent organs against veins. The resulting increase in pressure forces blood in a vein forward through a valve.

Sometimes one or more valves in a vein become damaged, causing blood to accumulate in that vein. Damaged valves in the legs can cause varicose veins, which are bulging, twisted veins. Such veins can also arise because of an inherited weakness of the vein wall. Chronic high blood pressure or an occupation that requires prolonged standing increases the risk of having varicose veins.

When blood pools inside a vein, it may clot. A clot that forms in a vessel and remains in place is called a thrombus. A clot or part of a clot that breaks loose and travels through blood vessels to a new location is an embolus. For example, a pulmonary embolus (an embolus in an artery in the lung) usually arises after a blood clot that formed in a vein in the thigh breaks off and travels through the heart into the lung. It is dangerous because it can block blood flow to lung tissue.

FIGURE 36.16 Moving blood against the force of gravity. When a giraffe lowers its head to drink, gravity pulls oxygen-poor blood in its veins toward the brain. Seven valves in each jugular vein prevent backflow through the neck. Humans, who seldom lower their head below their heart, have only one valve in their jugular vein. When a giraffe lifts its head, high blood pressure keeps oxygen-rich blood flowing uphill to the brain. Giraffes have the highest blood pressure of any vertebrate.

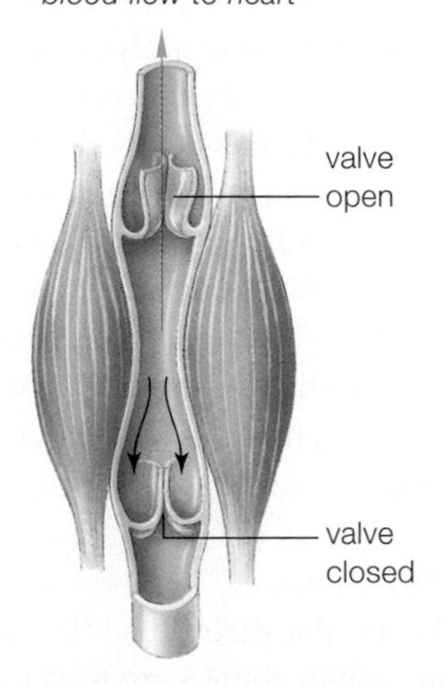

When skeletal muscles contract, they bulge and press on neighboring veins. This puts pressure on the blood in the vein, forcing it forward through the pressure-sensitive valves.

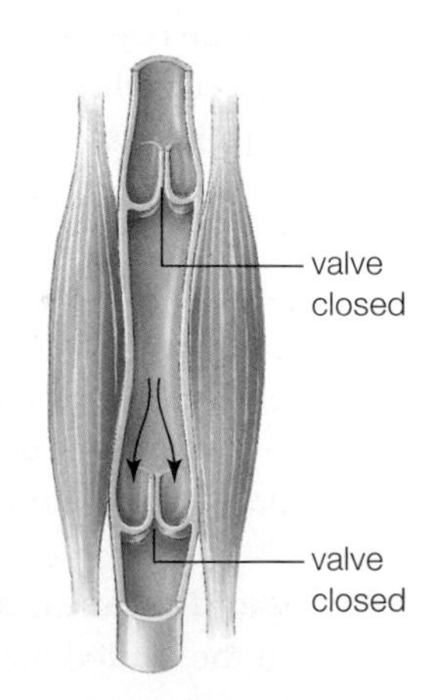

When skeletal muscles relax, the pressure in neighboring veins declines and pressure-sensitive valves shut, preventing blood from moving backward.

FIGURE 36.17 How skeletal muscle activity encourages blood flow through veins.

TAKE-HOME MESSAGE 36.9

What are the functions of veins?

✔ Veins are the body's main blood reservoir. The amount of blood in the veins changes depending on activity level.

✔ Blood pressure in veins is low. One-way valves, activity of skeletal muscle, and respiratory muscle action all help move the blood toward the heart.

36.10 Blood and Cardiovascular Disorders

✔ Altered blood cell quantity or quality can impair health, as can conditions that interfere with normal blood flow.

Altered Blood Cell Count

In anemias, red blood cells are few or somehow impaired, causing oxygen delivery and metabolism to falter. Shortness of breath, fatigue, and chills are common symptoms. Anemia has many causes. It can arise as a result of blood loss from a wound or an infection by a pathogen that kills red blood cells. For example, the protist that causes malaria enters red cells, divides inside them, and then causes the cell to break apart. A diet with too little iron can cause anemia by preventing synthesis of iron-containing heme. Sickle-cell anemia arises from a mutation that causes hemoglobin to change shape at a low oxygen concentration (Section 3.6). Thalassemias occur when mutations disrupt or halt synthesis of a globin chain of hemoglobin (Section 9.6). Disrupted globin synthesis results in cells that are unusually small and misshapen.

If stem cells in bone become cancerous, the result can be overproduction of red blood cells (polycythemia) or white blood cells (leukemia). An excess of red cells increases oxygen delivery, but makes blood more viscous and can raise blood pressure to a dangerous level. Leukemias cause overproduction of abnormally formed white blood cells that do not function properly.

Lymphomas are cancers that originate in B or T lymphocytes. Excessive divisions of the cancerous lymphocytes can produce tumors in lymph nodes and other parts of the lymphatic system.

Cardiovascular Disease

In the United States, cardiovascular disorders kill about a million people every year. Tobacco smoking tops the list of risk factors for these disorders. Other factors include a family history of such disorders, hypertension, a high cholesterol level, diabetes mellitus, and obesity. Regular exercise helps lower the risk of these disorders even when the exercise is not particularly strenuous. Gender is another factor; until about age fifty, males are at greater risk than females.

Atheroclerosis Cardiovascular disease often involves atherosclerosis, in which a buildup of lipids in the arterial wall narrows the lumen, or space inside the vessel. Cholesterol plays a role in this "hardening of the arteries." The human body requires cholesterol to make cell membranes, myelin sheaths, and steroid hormones (Section 3.5). The liver makes enough cholesterol to meet the body's needs, but most of us absorb still more from the food in our gut. Genetics affects how different people's bodies deal with the resulting excess of dietary cholesterol.

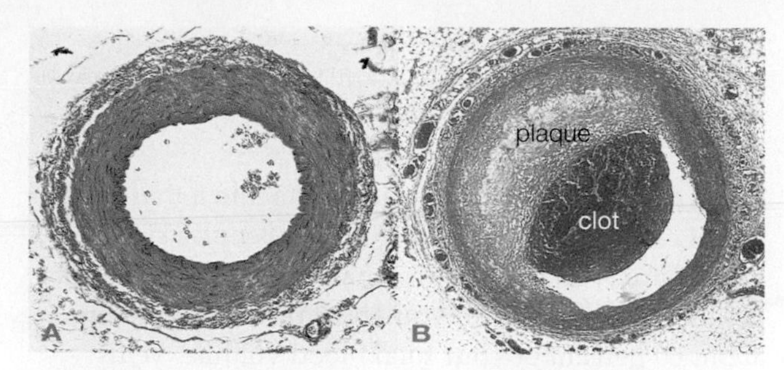

FIGURE 36.18 Sections from (**A**) a normal artery and (**B**) an artery with its interior narrowed by an atherosclerotic plaque. A clot has adhered to the plaque, further narrowing the vessel.

Most cholesterol in blood is bound to protein carriers to form complexes known as low-density lipoproteins, or LDLs. Cells throughout the body, including those lining blood vessels, take up LDLs. A lesser amount of cholesterol is bound up in high-density lipoproteins, or HDLs (Section 3.6). Cells in the liver take up HDLs and use them to produce bile, a substance the liver secretes into the small intestine. Cholesterol incorporated into bile leaves the body in feces.

Having a high LDL level or a low HDL level raises the risk of atherosclerosis. The first sign of trouble is a mass of lipids that builds up in an artery's endothelium (**FIGURE 36.18**). Fibrous connective tissue forms over the mass. The resulting atherosclerotic plaque bulges into the vessel's interior, narrowing its diameter and slowing blood flow. A plaque is hard and can rupture an artery wall, thereby triggering clot formation.

Atherosclerosis raises the risk that a blood vessel in the brain, heart, or other organ will become clogged. Interrupted blood flow to brain tissue causes a stroke, and interrupted blood flow to the heart causes a heart attack. In both cases, if the blockage is not removed fast, irreplaceable heart or brain cells die. Drugs that dissolve clots can restore blood flow and minimize cell death, but only if they are administered within an hour of the onset of an attack. Thus, anyone suspected of having a stroke or heart attack should receive prompt medical attention.

Clogged coronary arteries can be treated with a bypass or angioplasty. With coronary bypass surgery, doctors open a person's chest and use a blood vessel from elsewhere in the body (usually a leg vein) to divert blood around the clogged coronary artery (**FIGURE 36.19A**). In laser angioplasty, laser beams vaporize plaques. In balloon angioplasty, doctors

FIGURE 36.19 Two ways of treating blocked coronary arteries, the main cause of heart attacks in older adults.

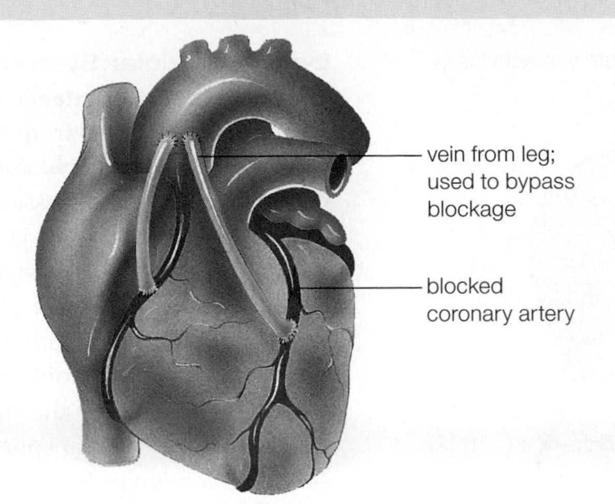

A Coronary bypass surgery. Veins from another part of the body are used to divert blood past the blockages. This illustration shows a "double bypass," in which veins are placed to divert blood around two blocked coronary arteries.

plaque flattened by balloon angioplasty

stent (metal mesh) placed to keep artery open

B Balloon angioplasty and the placement of a stent. After a balloonlike device is inflated in an artery to open it and flatten the plaque, a tube of metal (the stent) is inserted and left in place to keep the artery open.

inflate a small balloon in a blocked artery to flatten the plaques. A wire mesh tube called a stent is then inserted to keep the vessel open (**FIGURE 36.19B**). Angioplasty is also used to open partially blocked carotid arteries. These arteries in the neck supply blood to the brain.

Hypertension Hypertension refers to chronically high blood pressure (above 140/90). Often the cause is unknown. Heredity is a factor, and African Americans have an elevated risk. Diet also plays a role; in some people high salt intake causes water retention that raises blood pressure. Hypertension is sometimes described as a silent killer, because people often are unaware they have it. Hypertension makes the heart work harder than normal, which can cause it to enlarge and to function less efficiently. High blood pressure also increases risk of atherosclerosis.

Rhythms and Arrhythmias Electrocardiograms, or ECGs, record the electrical activity of a beating heart (**FIGURE 36.20**). They can also reveal arrhythmias, which are abnormal heart rhythms caused by malfunction of the SA node.

Bradycardia is a below-average resting cardiac rate. If the resulting slow flow impairs health, implanting an artificial pacemaker can speed the heart rate.

Tachycardia is a faster than normal heart rate. Many people experience palpitations, which are occasional episodes of tachycardia. Palpitations can be brought on by stress, drugs such as caffeine, an overactive thyroid, or an underlying heart problem.

Atrial fibrillation is an arrhythmia in which the atria do not contract normally, but instead quiver. This slows blood flow and increases the risk of clot formation. Often, people who have atrial fibrillation receive anticlotting medication to lower their risk of stroke.

Ventricular contraction is the driving force for blood circulation, so ventricular fibrillation is the most dangerous arrhythmia. Ventricles quiver, and pumping falters or stops, causing loss of consciousness and—if a normal rhythm is not restored—death. A defibrillator can often restore the heart's normal rhythm by resetting the SA node.

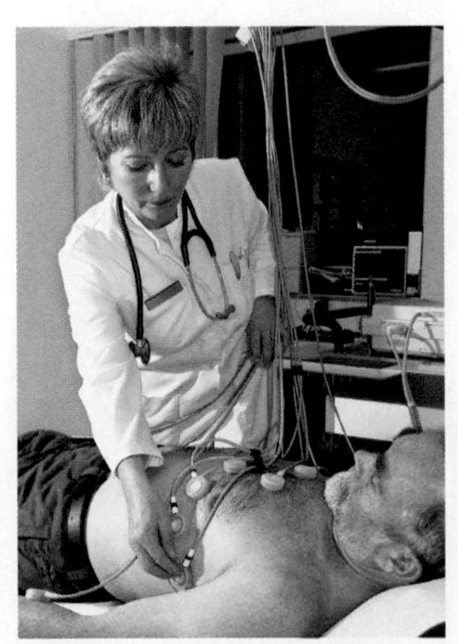

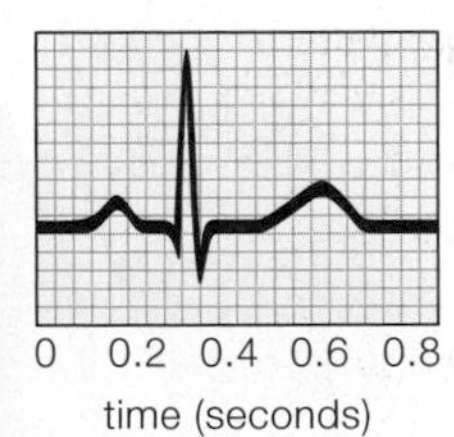

one normal heartbeat

FIGURE 36.20 ▶**Animated** Electrocardiograms. An ECG (above) is a graph showing electrical changes that can be detected by electrodes attached to the skin (left). The changes arise as a result of the activity of the SA node and the transmission of action potentials through the heart.

TAKE-HOME MESSAGE 36.10

What factors impair circulatory function?

✔ Changes to blood cell number or quality can alter blood's ability to carry out its functions.

✔ Atherosclerosis and hypertension raise the risk of heart attack and stroke.

✔ Problems with the cardiac pacemaker cause arrhythmias.

36.11 Interactions With the Lymphatic System

✔ Vessels and organs of the lymphatic system interact closely with the circulatory system.

Tonsils
Defense against bacteria and other foreign agents

Right lymphatic duct
Drains right upper portion of the body

Thymus gland
Site where certain white blood cells acquire means to chemically recognize specific foreign invaders

Thoracic duct
Drains most of the body

Spleen
Major site of antibody production; disposal site for old red blood cells and foreign debris; site of red blood cell formation in the embryo

Some of the lymph vessels
Return excess interstitial fluid and reclaimable solutes to the blood

Some of the lymph nodes
Filter bacteria and many other agents of disease from lymph

Bone marrow
Marrow in some bones is production site for infection-fighting blood cells (as well as red blood cells and platelets)

A

Lymph Vascular System

The **lymph vascular system** consists of vessels that collect water and solutes from the interstitial fluid, then deliver them to the circulatory system. It includes lymph capillaries and vessels (**FIGURE 36.21**). Fluid that moves through these vessels is called **lymph**.

The lymph vascular system serves three functions. First, its vessels return the plasma that leaked out of capillaries and mixed with the interstitial fluid to the circulatory system. Second, lymph vessels deliver fats absorbed from food in the small intestine to the blood. Third, these vessels transport cellular debris, pathogens, and foreign cells to lymph nodes, where white blood cells assess and respond to this material.

Lymph capillaries lie in close proximity to blood capillaries throughout the body (**FIGURE 36.21B**). A lymph capillary begins as a finger-shaped ending in a tissue. Fluid that leaks out of blood capillaries can enter the ending of a lymph capillary. Gaps between the cells in such endings open sporadically as a result of normal body movements.

Lymph capillaries merge to form larger-diameter lymph vessels. Two mechanisms move lymph through these vessels. First, slow wavelike contractions of smooth muscle in the walls of large lymph vessels propel lymph forward. Second, as with veins, the

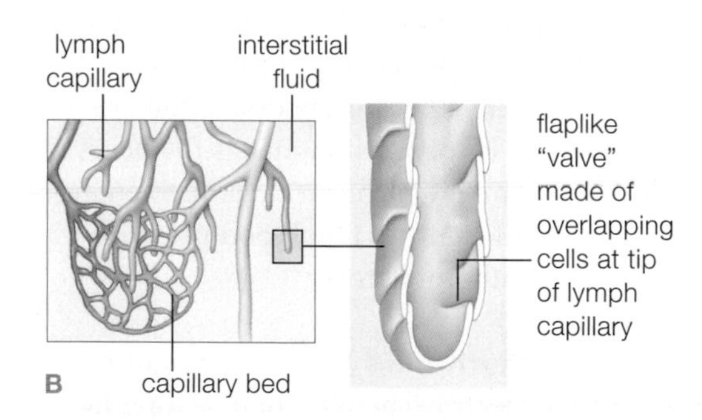

B

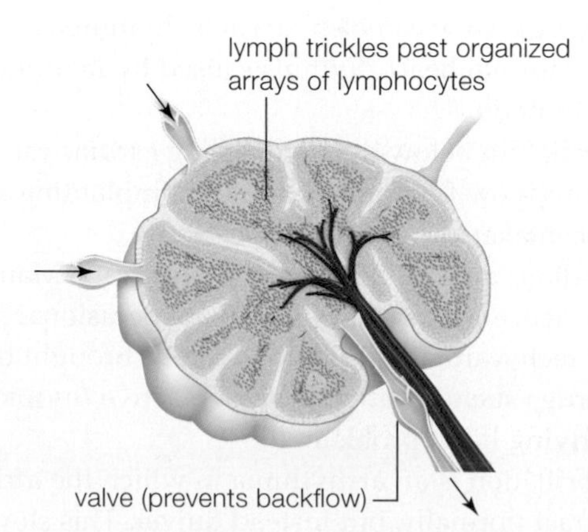

C Lymph node

FIGURE 36.21 ▶**Animated** Components of the lymphatic system.

A Shocking Save (revisited)

Most cardiac arrests do not occur in a hospital, so the presence of a bystander willing to carry out CPR or to use an AED often means the difference between life and death. Sadly, although most cardiac arrests are witnessed, only about 15 percent of victims get CPR before trained personnel arrive.

Most people have an understandable reluctance to engage in mouth-to-mouth contact with a stranger. This can be a problem with traditional CPR, which calls for a rescuer to alternate between exhaling into a victim's mouth to inflate the lungs and providing chest compressions. A new procedure called CCR (cardiocerebral resuscitation) relies on chest compressions alone to move air into and out of the lungs. Proponents of CCR argue that as long as a person's airway is clear, chest compressions will move enough air into and out of a victim's lungs to oxygenate the blood flowing to the person's heart and brain. Early studies indicate that CCR may be as good as or even better than traditional CPR at saving most people who experience a sudden cardiac arrest.

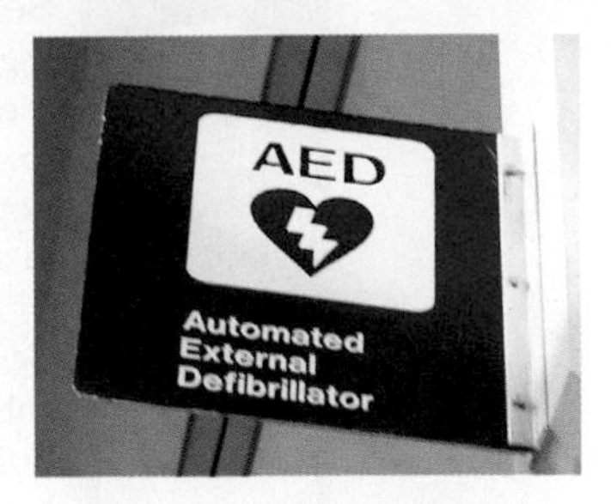

Sign in an airport indicating where an AED is stored.

Another promising development is the rising availability of AEDs. In 2010, approximately 200,000 AEDs were purchased for installation in public places. The use of such devices is already saving lives. A recent study that looked at the outcome of nearly 14,000 out-of-hospital cardiac arrests concluded that, compared to no treatment, bystander use of an AED increased the likelihood of surviving to discharge from the hospital by about 75 percent. That's the good news. The bad news is that most people still do not know what an AED is, where they are available, or how simple it is to use one. The photo at the upper right shows the symbol for an AED in public place.

---

bulging of adjacent skeletal muscles helps move fluid along. Like veins, the lymph vessels have one-way valves that prevent backflow.

The largest lymph vessels converge on collecting ducts that empty into veins near the heart. Each day these ducts deliver about 3 liters of fluid to the blood.

Lymphoid Organs and Tissues

Lymphoid organs and tissues associate with the lymph vascular system; they participate in the body's responses to injury and invasion by pathogens.

Lymph nodes are oval shaped structures located at intervals along lymph vessels. They filter lymph before it returns to the blood (**FIGURE 36.21C**). Each node has a capsule of connective tissue and contains large numbers of lymphocytes and other white blood cells that survey the passing lymph for potential threats.

The white blood cells called T lymphocytes become capable of recognizing and responding to particular pathogens while in the thymus gland. The gland makes hormones necessary for T cell maturation.

The fist-sized **spleen**, which resides in the upper-left portion of the abdomen, is the largest lymphoid organ. During prenatal development it produces red blood cells. After birth, it filters worn-out and damaged red blood cells from the many blood vessels that branch through it. White blood cells in the spleen detect and respond to pathogens in the blood and lymph that filter through it. People can survive without a spleen, but they become more vulnerable to infections.

Tonsils are two patches of lymphoid tissue at the back of the throat. Adenoids are similar tissue clumps at the rear of the nasal cavity. Tonsils and adenoids help the body respond fast to inhaled pathogens. The lining of the intestine and the appendix also include lymphoid tissue.

Mechanisms by which lymphocytes in the lymphoid organs detect and respond to threats are discussed in the next chapter.

lymph Fluid in the lymph vascular system.
lymph node Small mass of lymphatic tissue through which lymph filters; contains many lymphocytes (B and T cells).
lymph vascular system System of vessels that takes up interstitial fluid and carries it (as lymph) to the blood.
spleen Fist-sized lymphoid organ located in the upper abdomen; filters blood and enhances immune function.

TAKE-HOME MESSAGE 36.11

What are the functions of the lymphatic system?

✔ The lymph vascular system consists of tubes that collect and deliver excess water and solutes from interstitial fluid to blood. It also carries absorbed fats to the blood, and delivers disease agents to lymph nodes.

✔ The system's lymphoid organs, including lymph nodes, have specific roles in body defenses.

summary

Section 36.1 When the heart stops pumping, blood flow halts and brain cells begin to die from lack of oxygen. CPR can keep some oxygenated blood moving to cells, but it cannot restart a heart. Reestablishing the normal rhythm requires a shock from a defibrillator.

Section 36.2 A **circulatory system** moves substances through a body by way of a fluid transport medium. Some invertebrates have an **open circulatory system**, in which **hemolymph** pumped out of vessels mixes with interstitial fluid. In other invertebrates and all vertebrates, a **closed circulatory system** confines **blood** inside a **heart** and blood vessels. Exchanges between blood and interstitial fluid take place across walls of **capillaries**.

In fish, a heart with one **atrium** and one **ventricle** pumps blood through a single circuit. A two-circuit system evolved in concert with tetrapod lungs. The **pulmonary circuit** moves blood to lungs and back. The longer **systemic circuit** conveys blood to other body tissues and back.

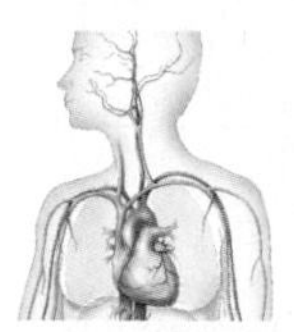

Section 36.3 **Arteries** receive blood from the heart. The **aorta** is the largest artery of the systemic circuit. **Pulmonary arteries** carry oxygen-poor blood to lungs. Exchanges take place across the walls of capillaries, then **veins** carry blood back to the heart. The **superior vena cava** and **inferior vena cava** return blood from the systemic circuit. **Pulmonary veins** return oxygenated blood from the lungs. Most blood flows through one capillary system, but blood in intestinal capillaries also flows through liver capillaries.

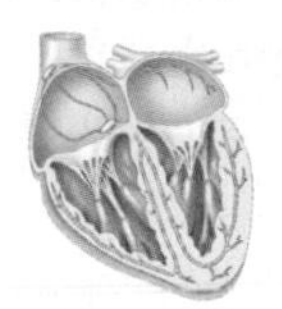

Section 36.4 A human heart has two halves, each with an atrium above a ventricle. Oxygen-poor blood from the body enters the right atrium and moves through a valve into the right ventricle, which pumps blood through a valve into the pulmonary circuit. Oxygen-rich blood enters the heart's left atrium and moves through a valve into the left ventricle, which pumps the blood through a valve into the systemic circuit.

During one **cardiac cycle**, all heart chambers undergo rhythmic relaxation (**diastole**) and contraction (**systole**). Contraction of the ventricles provides the force that propels blood through blood vessels.

The **sinoatrial (SA) node** in the right atrium wall is the cardiac pacemaker. The **atrioventricular (AV) node** serves as an electrical bridge to the ventricles. The delay between atrial contraction in response to SA signaling and ventricular contraction in response to signals arriving via the AV node allows the ventricles to fill fully before they contract.

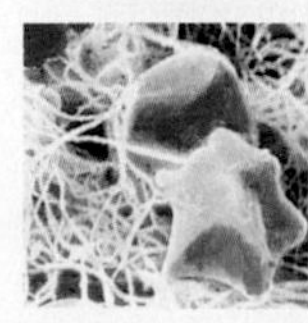

Section 36.5 Blood consists of **plasma**, blood cells, and platelets. Plasma is mostly water. Plasma proteins include albumin and clotting factors, both made by the liver, and immunoglobulins made by white blood cells. Plasma also contains ions, hormones, nutrients, and dissolved gases. Cellular components of the blood are made in red bone marrow. **Red blood cells** contain hemoglobin, a protein that functions in oxygen transport. A hormone (erythropoietin) made by the kidney encourages their synthesis. **White blood cells** help defend the body against damaged cells and pathogens. **Platelets** function in **hemostasis** (clotting).

Sections 36.6–36.9 **Blood pressure** results from the force exerted on blood by ventricular contraction and it declines as blood proceeds through a circuit. It is usually recorded as **systolic pressure** over **diastolic pressure**. A **pulse** is a brief expansion of an artery caused by ventricular contraction. Stretching and recoil of elastic tissue in arteries helps keep blood moving between contractions.

Arterioles are the main site for regulation of blood flow through the systemic circuit. **Vasodilation** of arterioles supplies more blood to a region. **Vasconstriction** decreases inward flow. Adjustments in arteriole diameter alter blood pressure in the short term. Adjustments in urine output affect it over the longer term.

Blood flow slows in capillaries because of their collectively large cross-sectional area. Substances leave a capillary by diffusion, through transport proteins, or in fluid that seeps out between cells. Fluid that seeps out of a capillary at the arterial end is balanced by osmotic uptake of water nearer the vein.

Veins return blood to the heart. Pressure in veins is low, but one-way valves and pressure on veins from skeletal muscles and respiratory movements help move blood forward.

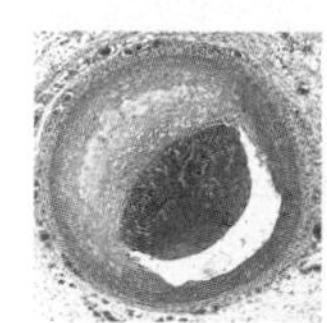

Section 36.10 In a blood disorder, an individual has too many, too few, or abnormal red or white blood cells. Abnormal heart rhythms can slow or halt blood flow. Flow is also impaired when atherosclerosis narrows the interior of a blood vessel.

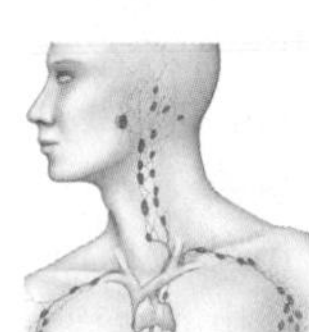

Section 36.11 Some fluid that leaves capillaries enters the **lymph vascular system**. The fluid, now called **lymph**, is filtered by **lymph nodes**. The **spleen** filters the blood and removes any old red blood cells. Tonsils and adenoids are lymphoid organs that help fend off pathogens. The thymus is the site of T cell (T lymphocyte) maturation.

self-quiz

Answers in Appendix VII

1. All vertebrates have ________ .
   a. a closed circulatory system
   b. a two-chambered heart
   c. hemolymph
   d. all of the above

2. In ________, blood flows through two completely separate circuits.
   a. birds b. mammals c. fish d. both a and b

3. The ________ circuit carries blood to and from lungs.
   a. systemic b. pulmonary

data analysis activities

The Stroke Belt Risk of death by stroke is not distributed evenly across the United States. Epidemiologists refer to a swath of states in the Southeast as the "stroke belt" because of the increased incidence of stroke deaths there. By one hypothesis, the high rate of deaths from stroke in this region results largely from a relative lack of access to immediate medical care. Compared to other parts of the country, more stroke-belt residents live in rural settings with few medical services.

FIGURE 36.22 compares the rate of stroke deaths in stroke-belt states (Alabama, Arkansas, Georgia, Mississippi, North Carolina, South Carolina, and Tennessee) with that of New York State. It also breaks down death risk in each region by ethnic group and sex.

1. How does the rate of stroke deaths among blacks living in the stroke-belt compare with whites in the same region?

2. How does the rate of stroke deaths among blacks living in New York compare with whites in the same region?

3. Which group has the higher rate of stroke deaths, blacks living in New York, or whites living in the stroke belt?

4. Do these data support the hypothesis that poor access to care causes the high rate of death by stroke in the stroke belt?

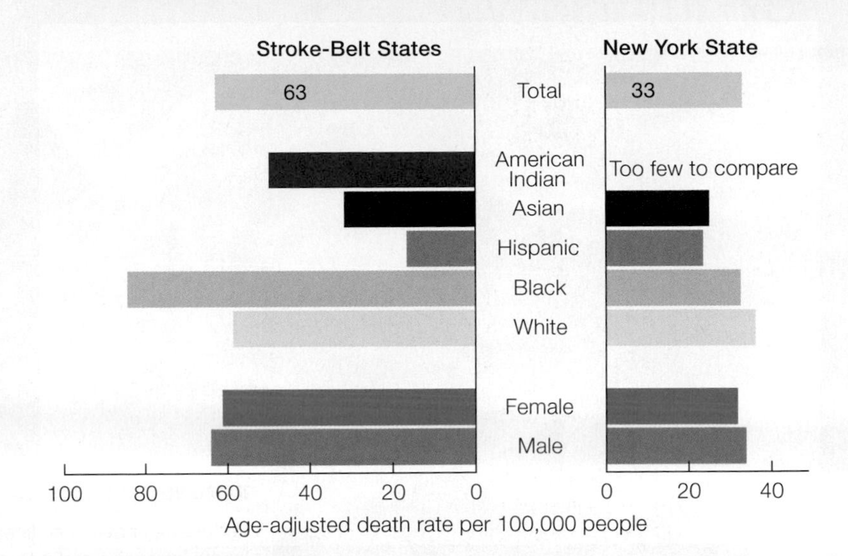

FIGURE 36.22 Comparison of the age-adjusted rate of deaths by stroke in southeastern "stroke-belt" states and in New York State. *Source:* National Vital Statistics System—Mortality (NVSS-M), NCHS, CDC.

4. The plasma protein albumin is made by ________ .
   a. white blood cells
   b. red blood cells
   c. the heart
   d. the liver

5. Platelets function in ________ .
   a. oxygen transport
   b. blood clotting
   c. thermal regulation
   d. both a and b

6. Most oxygen in blood is transported ________ .
   a. in red blood cells
   b. in white blood cells
   c. bound to hemoglobin
   d. both a and c

7. Blood flows directly from the left atrium to ________ .
   a. the aorta
   b. the left ventricle
   c. the right atrium
   d. the pulmonary arteries

8. Contraction of ________ drives the flow of blood through the aorta and pulmonary arteries.
   a. atria
   b. ventricles

9. Blood pressure is highest in the ________ and lowest in the ________ .
   a. arteries; veins
   b. arterioles; venules
   c. veins; arteries
   d. capillaries; arterioles

10. At rest, the largest volume of blood is in ________ .
    a. arteries
    b. capillaries
    c. veins
    d. arterioles

11. Which of the following has the thickest wall?
    a. left atrium
    b. left ventricle
    c. right atrium
    d. right ventricle

12. Lymph nodes filter ________ .
    a. blood
    b. lymph
    c. plasma
    d. all of the above

CENGAGE brain To access course materials, please visit www.cengagebrain.com.

13. Which artery carries oxygen-poor blood?

14. Match the terms with their definition.

___ anemia	a. clot that stays in place
___ hypertension	b. impaired clotting
___ thrombus	c. cancer of bone marrow
___ hemophilia	d. too few red cells
___ stroke	e. high blood pressure
___ atrial fibrillation	f. interrupted blood flow in brain
___ leukemia	g. heart quivers

15. Match the terms with their definition.

___ bone marrow	a. lymphoid organ in abdomen
___ spleen	b. cardiac pacemaker
___ fibrinogen	c. stimulates red cell production
___ HDL	d. largest artery
___ SA node	e. a clotting factor
___ erythropoietin	f. source of all blood cells
___ aorta	g. delivers cholesterol to the liver for use in bile

critical thinking

1. Explain why a blood clot that forms in the leg, then breaks free as an embolus, is more likely to become stuck in a lung than in the brain.

2. Beta blockers are drugs that interfere with stimulation of organs by nerves of the sympathetic system. Would these drugs be useful in treating excessively high blood pressure or excessively low blood pressure?

3. Patients with severe liver disease often have impaired blood clotting. Why?

CREDIT: (22) National Vital Statistics System–Mortality (NVSS-M), NCHS, CDC.

37 Immunity

LEARNING ROADMAP

This chapter revisits bacteria and viruses (Sections 4.4, 20.1–20.4, 20.7) as pathogens; proteins (3.6, 5.7); cell structure and function (4.7, 4.11, 5.8, 5.10, 34.2); cDNA (15.2); genetics (9.3); the internal environment (31.2); epithelia and skin (31.3, 31.8); the circulatory and lymphatic systems (36.5, 36.8, 36.11); immune disorders (15.10, 32.13); and atherosclerosis (36.10).

IMMUNE DEFENSES

Vertebrates have three lines of immune defenses: surface barriers, innate immunity, and adaptive immunity. Leukocytes and signaling molecules function in immune responses.

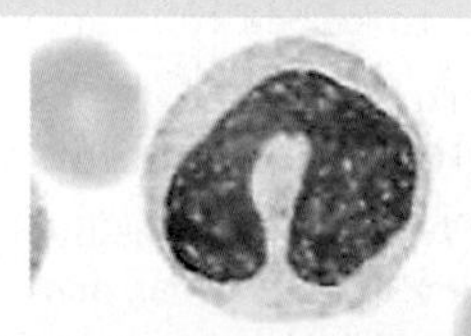

SURFACE BARRIERS

External surfaces of the body come into constant contact with microbial pathogens. Physical, mechanical, and chemical barriers prevent most microbes from entering body tissues.

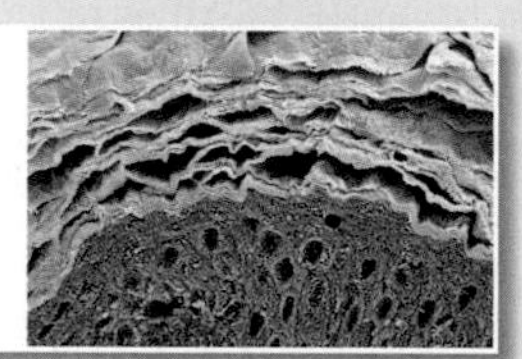

INNATE IMMUNITY

Innate immune responses involve a set of general, immediate defenses. Phagocytic leukocytes, plasma proteins, inflammation, and fever quickly rid the body of most invaders.

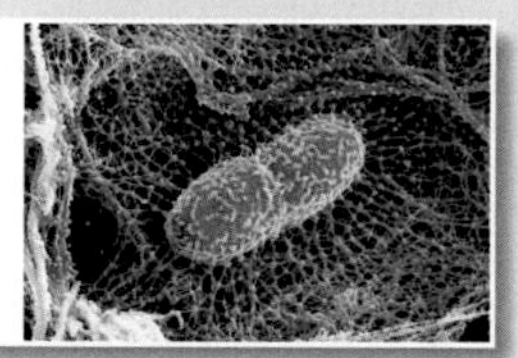

ADAPTIVE IMMUNITY

In an adaptive immune response, leukocytes interact to destroy specific pathogens or altered body cells. Antibodies and other antigen receptors are central to these responses.

IMMUNITY IN OUR LIVES

Vaccines are important in worldwide health programs. Allergies and autoimmune disorders are the result of faulty immune mechanisms. Immune deficiencies can be inherited or acquired.

Inflammatory diseases of the respiratory system will come up again (Section 38.9), as will immunity-enhancing effects of some gut bacteria (39.1), autoimmunity (39.8, 39.11), AIDS and other sexually transmitted diseases (41.9), and apoptosis (42.4). Section 42.11 describes how antibodies pass from mother to child, and Section 48.7 explains why global climate change may exacerbate pollen allergies.

37.1 Frankie's Last Wish

Frankie McCullough had known for a few months that something was not quite right. She had not had an annual checkup in many years; after all, she was only 31 and had always been healthy. She had never doubted her own invincibility until the moment she saw the doctor's face change as he examined her cervix.

The cervix is the lowest part of the uterus, or womb. Cervical cells can become cancerous, but the process is usually slow. The cells pass through several precancerous stages that are detectable by routine Pap tests. Precancerous and even early-stage cancerous cells can be removed from the cervix before they spread to other parts of the body. However, plenty of women like Frankie do not take advantage of regular exams. They do not see a gynecologist unless they have pain or bleeding, which can be symptoms of advanced cervical cancer. Even with treatment, 85 percent of women who end up with this type of cancer will die within five years. Cervical cancer kills thousands of women each year in the United States, and many more in places where routine gynecological testing is not a common practice.

What causes cancer? At least in the case of cervical cancer, we know the answer to that question: Healthy cervical cells are transformed into cancerous ones by infection with human papillomavirus (HPV). HPV is a DNA virus that infects skin and mucous membranes. There are about 100 different types, or strains, of HPV; a few cause warts on the hands or feet, or in the mouth. About 30 others that infect the genital area sometimes cause genital warts, but usually there are no symptoms of infection. Genital HPV is spread very easily by sexual contact: A woman has a 50% chance of being infected with genital HPV within three years of becoming sexually active.

A genital HPV infection usually goes away on its own, but not always. A persistent infection with one of about 10 strains is the main risk factor for cervical cancer (**FIGURE 37.1**). Types 16 and 18 are particularly dangerous: At least one of these two HPV strains is found in most cervical cancers.

In 2006, the U.S. Food and Drug Administration (FDA) approved Gardasil, a vaccine against four types of genital HPV, including types 16 and 18. The vaccine prevents cervical cancer caused by these HPV strains. It is most effective in girls who have not yet become sexually active, because they are least likely to have already become infected with HPV.

The HPV vaccine came too late for Frankie. Despite radiation treatments and chemotherapy, her cervical cancer spread quickly. She died in 2001, leaving a wish for other people: awareness. "If there is one thing I could tell a young woman to convince them to have a yearly exam, it would be not to assume that your youth will protect you. Cancer does not discriminate; it will attack at random, and early detection is the answer." Almost all women newly diagnosed with invasive cervical cancer have not had a Pap test in five years, and many have never had one.

FIGURE 37.1 HPV and cervical cancer. Top, a Pap test reveals HPV-infected cervical cells among normal ones. Infected cells have enlarged, often multiple nuclei surrounded by a clear area. These changes sometimes lead to cervical cancer, which is treatable if detected early enough. Bottom, Frankie McCullough (waving) died of cervical cancer at age 32.

CREDITS: (opposite) Dr. Olivier Schwartz, Institut Pasteur/Science Source; (1) top, Biomedical Imaging Unit, Southampton General Hospital/Science Source; bottom, In memory of Frankie McCullough.

37.2 Integrated Responses to Threats

✔ In vertebrates, the innate and adaptive immune systems work together to combat infection and injury.

✔ Innate immune mechanisms are fast, general responses to tissue damage and invading microorganisms.

✔ Adaptive immune mechanisms can recognize and respond to billions of specific pathogens.

You continually cross paths with a tremendous array of viruses, bacteria, fungi, parasitic worms, and other agents of disease, but you need not lose sleep over this. Humans coevolved with these pathogens, so you have defenses that protect your body from them. The evolution of **immunity**, an organism's capacity to resist and combat infection, began well before multicelled eukaryotes evolved from free-living cells. Mutations in the genes for membrane proteins introduced new molecular patterns that were unique in cells of a given type. As multicellularity evolved, so did mechanisms of identifying the patterns as self, or belonging to one's own body.

By about 1 billion years ago, nonself recognition had also evolved. Cells of all modern multicelled eukaryotes bear a set of receptors that collectively can recognize around 1,000 different nonself cues, which are called pathogen-associated molecular patterns (PAMPs). As their name suggests, PAMPs occur mainly on or in pathogens. They include proteins that make up bacterial flagella and pili (Section 4.4), components of bacterial cell walls, double-stranded RNA unique to some viruses, and so on. A PAMP is an example of **antigen**, which is any molecule or particle recognized by the body as nonself. Binding of a cell's receptors to antigen triggers a set of immediate, general defense responses. In mammals, for example, the binding activates **complement**: a set of proteins that circulate in inactive form throughout the body in blood and tissue fluids. Activated complement proteins mark antigenic particles for uptake by phagocytic white blood cells (Section 5.10), and they can also destroy foreign cells.

Table 37.1 Comparing Features of Innate and Adaptive Immunity

	Innate Immunity	Adaptive Immunity
Response time	Immediate	About a week
How antigen is detected	Fixed set of receptors for pathogen-associated molecular patterns (PAMPs)	Antigen receptors produced by gene recombinations
Specificity	About 1,000 PAMPs	Billions of antigens
Persistence	None	Long-term

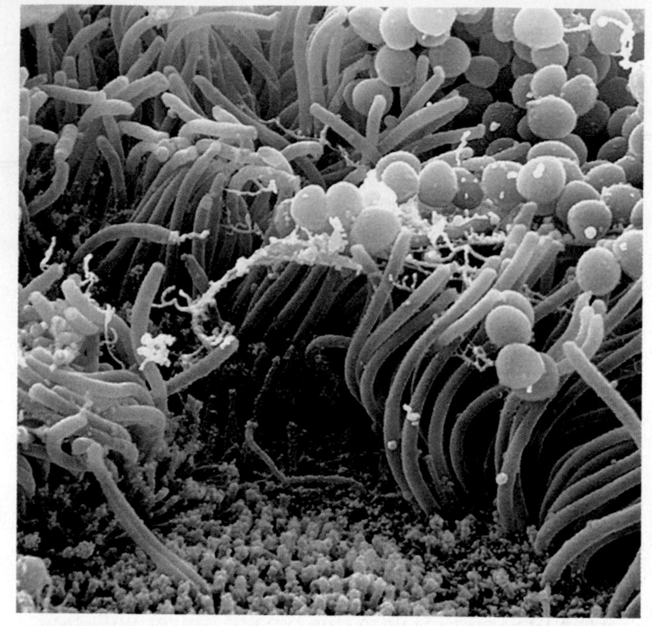

FIGURE 37.2 One physical barrier to infection: Mucus and the mechanical action of cilia can keep pathogens from getting a foothold in nasal epithelia. This micrograph shows *Staphylococcus aureus* bacteria (yellow) stuck in mucus secreted by goblet cells. Cilia on other cells sweep the mucus toward the throat for disposal.

Pattern receptors and the responses they trigger are part of **innate immunity**, a set of immediate, general defenses that help protect all multicelled organisms from infection. Vertebrate animals have an additional set of defenses carried out by interacting cells, tissues, and proteins. This **adaptive immunity** tailors immune defenses to specific pathogens—potentially billions of them—that an individual encounters during its lifetime. TABLE 37.1 compares the two types of immunity.

Three Lines of Defense

The mechanisms of adaptive immunity evolved within the context of innate immunity. The two systems were once thought to operate independently of each other, but we now know they function together. Thus, we describe both systems together in terms of three lines of defense. The first line of defense includes the physical, chemical, and mechanical barriers that prevent most pathogens from entering the internal environment (FIGURE 37.2). The second line of defense, innate immunity, begins after a tissue is damaged, or after antigen is detected inside the body. Its general response mechanisms quickly rid the body of many invaders. An innate immune response also triggers adaptive immunity, which is the third line of defense. Leukocytes (white blood cells, Section 36.5) divide to form huge populations that target a specific antigen and destroy

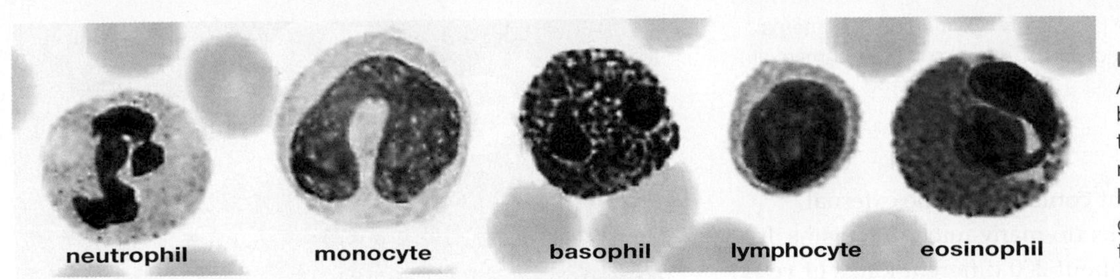

FIGURE 37.3 ▸**Animated** A lineup of leukocytes (white blood cells). These are a few types that circulate in blood. Staining reveals details such as distinctive lobed nuclei and cytoplasmic granules that contain enzymes, toxins, and signaling molecules.

anything bearing it. Some leukocytes that form during an adaptive response persist for years after the infection ends. If the same antigen returns, these memory cells can mount a secondary response.

The Defenders

Leukocytes participate in both innate and adaptive immune responses. Many kinds circulate through the body in blood (**FIGURE 37.3**) and lymph; others populate the lymph nodes, spleen, and other tissues. All communicate with one another by secreting and responding to chemical signaling molecules. These molecules, which include proteins and polypeptides called **cytokines**, allow cells throughout the body to coordinate their activities during an immune response. Interleukins, interferons, and tumor necrosis factors are examples of cytokines that vertebrates have.

Different types of leukocytes are specialized for specific tasks. Those that are phagocytic engulf and digest pathogens, dead cells, or other particles. **Neutrophils**, which circulate in blood, are the most abundant type. Phagocytic **macrophages** that migrate through tissues and tissue fluids develop from monocytes that patrol the blood. **Dendritic cells**, which are also phagocytic, alert the adaptive immune system to the presence of antigen in solid tissue.

Many leukocytes have secretory vesicles called granules. Granules contain cytokines, local signaling molecules, destructive enzymes, and toxins such as hydrogen peroxide. A cell releases the contents of its granules (degranulates) in response to a trigger such as antigen binding. Neutrophils have granules, as do **eosinophils** that target multicelled parasites too big for phagocytosis. **Basophils** and **mast cells** degranulate in response to injury as well as antigen. Mast cells, unlike most other leukocytes, stay anchored in tissues. These cells are often closely associated with nerves, and they also degranulate in response to signaling molecules secreted by cells of the endocrine and nervous systems.

Lymphocytes are a special category of leukocyte with the collective capacity to target billions of specific antigens. **B cells** (B lymphocytes) make antibodies (more about these proteins in Section 37.6). **T cells** (T lymphocytes) play a central role in all adaptive immune responses. Cytotoxic T cells are specialized to kill infected or cancerous body cells. **NK cells** (natural killer cells) are lymphocytes that kill cancerous body cells undetectable by cytotoxic T cells.

adaptive immunity In vertebrates, set of immune defenses that can be tailored to fight specific pathogens as an organism encounters them during its lifetime.
antigen A molecule or particle that the immune system recognizes as nonself. Triggers an immune response.
B cell B lymphocyte. Leukocyte that can make antibodies.
basophil Circulating leukocyte with granules.
complement A set of proteins that circulate in inactive form in blood, and when activated play a role in immune responses.
cytokines Signaling molecules secreted by leukocytes.
dendritic cell Phagocytic leukocyte that alerts the immune system to the presence of antigen in solid tissues.
eosinophil Leukocyte that targets multicelled parasites.
immunity The body's ability to resist and fight infections.
innate immunity In all multicelled organisms, set of immediate, general defenses against infection.
macrophage Phagocytic leukocyte that patrols tissues and tissue fluids.
mast cell Leukocyte with granules that is anchored in many tissues; factor in inflammation.
neutrophil Circulating phagocytic leukocyte.
NK cell Natural killer cell. Leukocyte that can kill cancer cells undetectable by cytotoxic T cells.
T cell T lymphocyte. Leukocyte central to adaptive immunity; some kinds target infected or cancerous body cells.

TAKE-HOME MESSAGE 37.2
What is immunity?

- ✔ Innate immunity of multicelled organisms comprises a system of immediate, general defenses that are triggered by a fixed number of antigens.
- ✔ Vertebrate adaptive immunity is a system of defenses that can specifically target billions of different antigens.
- ✔ White blood cells (leukocytes) are central to both systems; signaling molecules such as cytokines integrate their activities.

37.3 Surface Barriers

✔ A pathogen can cause infection only if it enters the internal environment by penetrating skin or other protective barriers at the body's surfaces.

Normal Flora

Your skin is in constant contact with the external environment, so it picks up many microorganisms. It normally teems with about 200 different kinds of yeast, protozoa, and bacteria. If you showered today, there are probably thousands of microorganisms on every square inch of your external surfaces. If you did not, there may be billions. They tend to flourish in warm, moist areas, such as between the toes. Huge populations inhabit cavities and tubes that open on the body's surface, including the eyes, nose, mouth, and anal and genital openings.

Microorganisms that typically live on human surfaces, including the interior tubes and cavities of the digestive and respiratory tracts, are called **normal flora**. Our surfaces provide them with a stable environment and nutrients. In return, their populations deter more dangerous species from colonizing (and penetrating) body surfaces. Normal flora in the digestive tract help us digest food, and they also make essential nutrients such as vitamins K and B_{12}.

Normal flora are helpful only on body surfaces; many can cause or worsen disease when they invade tissues. Consider a major constituent of normal flora, *Propionibacterium acnes* (left). This bacterium feeds on sebum, a greasy mixture of fats, waxes, and glycerides that lubricates hair and skin. Glands in the skin secrete sebum into hair follicles. During puberty, an increase in sex hormone production triggers an increase in sebum production. Excess sebum combines with dead, shed skin cells and blocks the openings of hair follicles. *P. acnes* can survive on the surface of the skin, but far prefers anaerobic habitats such as the interior of blocked hair follicles. There, the cells multiply to tremendous numbers. Secretions of their flourishing populations leak into internal tissues of the follicles and initiate inflammation (we return to inflammation in Section 37.5). The resulting pustules are called acne.

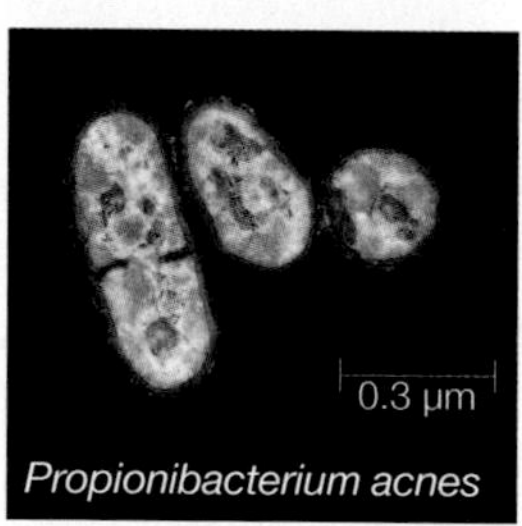

A few of the 400 or so species of normal flora in the mouth are part of **dental plaque**, a thick biofilm of various bacteria and occasional archaea, their extracellular products, and saliva glycoproteins. Plaque sticks tenaciously to teeth (**FIGURE 37.4**). Some of the bacteria in plaque carry out lactate fermentation. The lactate they produce is acidic enough to dissolve minerals that make up the tooth, causing holes called cavities.

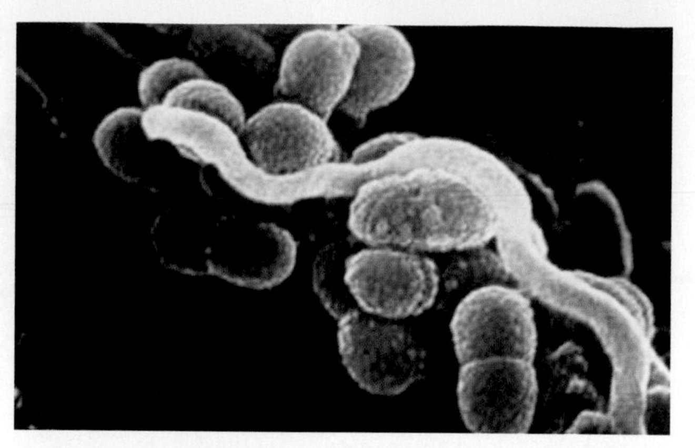

A One of the bacterial species in plaque, *Streptococcus mutans*, is a major contributor to tooth decay and periodontitis.

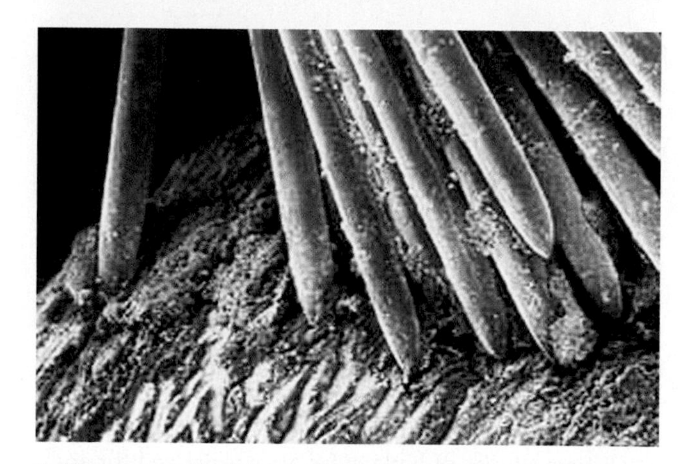

B Micrograph of toothbrush bristles scrubbing plaque on the surface of a tooth.

FIGURE 37.4 Dental plaque.

In young, healthy people, tight junctions (Section 4.11) normally seal gum epithelium to teeth. The tight seal prevents oral microorganisms from entering gum tissue. As we age, the connective tissue beneath the epithelium thins, so the seal between gums and teeth weakens. Deep pockets form, and a nasty collection of anaerobic bacteria and archaea tends to accumulate in them. The microorganisms secrete destructive enzymes and acids that cause inflammation of the surrounding gum, a condition called periodontitis. Periodontal wounds are an open door to the circulatory system and its arteries, and all species of oral bacteria associated with periodontitis are also found in atherosclerotic plaque (Section 36.10). Atherosclerosis is now known to be a disease of inflammation. What role oral microor-

CREDITS: (in text) Kwangshin Kim/Science Source; (4) www.zahnarzt-stuttgart.com.

ganisms play in atherosclerosis is not yet clear, but one thing is certain: They contribute to the inflammation that fuels coronary artery disease.

Other serious illnesses associated with normal flora include pneumonia; ulcers; colitis; whooping cough; meningitis; abscesses of the lung and brain; and cancers of the colon, stomach, and intestine. The bacterial agent of tetanus, *Clostridium tetani*, is considered a normal inhabitant of human intestines. The bacteria responsible for diphtheria, *Corynebacterium diphtheriae*, was normal skin flora before widespread use of the vaccine eradicated the disease. *Staphylococcus aureus* (shown in **FIGURE 37.2**) is a bacterial resident of human skin and linings of the mouth, nose, throat, and intestines; it is also a leading cause of human bacterial disease. A particularly dangerous kind, MRSA (methicillin-resistant *S. aureus*), is now a permanent resident of hospitals worldwide.

Barriers to Infection

In contrast to body surfaces, the blood and tissue fluids of healthy people are typically free of microorganisms (sterile). Surface barriers (**TABLE 37.2**) usually prevent normal flora from entering the body's internal environment. The tough outer layer of skin (Section 31.8) is an example. Microorganisms flourish on skin's waterproof, oily surface, but they rarely penetrate its thick layer of dead cells (**FIGURE 37.5**). The thinner epithelial tissues that line the body's interior tubes and cavities also have surface barriers. Sticky mucus secreted by cells of these linings can trap microorganisms. The mucus contains **lysozyme**, an enzyme that kills bacteria. In the sinuses and respiratory tract, the coordinated beating of cilia sweeps away trapped microorganisms before they have a chance to breach the delicate walls of these structures.

Your mouth is a particularly inviting habitat for microorganisms because it offers plenty of nutrients, warmth, moisture, and surfaces for colonization. Accordingly, it harbors huge populations of normal flora that can resist lysozyme in saliva. Microorganisms that get swallowed are typically killed in the stomach by gastric fluid, a potent brew of protein-digesting enzymes and acid. Any that survive passage to the small intestine are usually killed by salts secreted into intestinal fluid. Hardy cells that reach the large intestine must compete with about 500 resident species specialized to live there and have already established large populations. Any that displace normal flora are typically flushed out by diarrhea.

Lactate produced by *Lactobacillus* bacteria (Section 20.7) helps keep the vaginal pH below the range of tolerance of other bacteria and most fungi. Urination's flushing action usually prevents pathogens from colonizing the urinary tract.

Table 37.2 Examples of Surface Barriers

Mechanical	Mucus; broomlike action of cilia; flushing action of tears, saliva, urine, diarrhea
Chemical	Secretions (sebum, other waxy coatings); low pH of urine, gastric juices, vaginal tract; lysozyme
Physical	Established populations of normal flora; epithelia that line tubes and cavities such as the gut and eye sockets; skin

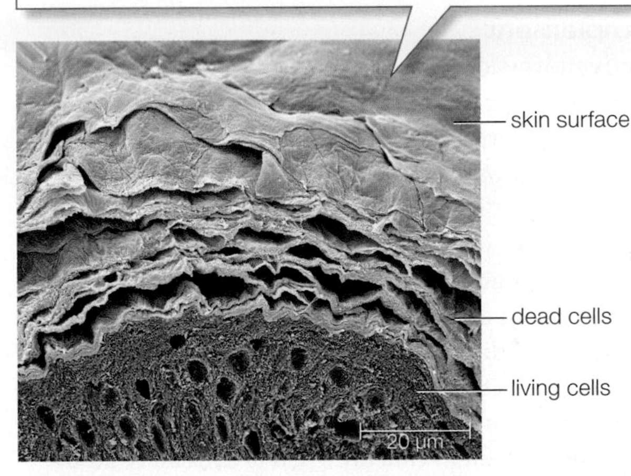

FIGURE 37.5 One surface barrier to infection: the epidermis of human skin. Its thick, waterproof layer of dead cells usually keeps normal skin flora from penetrating internal tissues.

TAKE-HOME MESSAGE 37.3

What prevents microorganisms from entering body tissues or fluids?

✔ Surface barriers usually prevent microorganisms from invading the internal environment.

✔ Skin's tough epidermis is one physical barrier to infection. Resident populations of normal flora deter more dangerous microorganisms from colonizing skin and the linings of internal tubes and cavities.

✔ Lysozyme-containing mucus and other secretions offer more protection, as do the waving action of cilia and the flushing action of urine and diarrhea.

---

dental plaque On teeth, a thick biofilm composed mainly of bacteria, their extracellular products, and saliva proteins.
lysozyme Antibacterial enzyme in body secretions such as saliva and mucus.
normal flora Microorganisms that typically live on human surfaces, including the tubes and cavities of the digestive and respiratory tracts.

✔ Antigen detection and tissue damage trigger complement activation and phagocytosis by leukocytes.

What happens if a pathogen slips by surface defenses and enters the body's internal environment? A second line of defense—the fast-acting, general mechanisms of innate immunity—can keep an invading pathogen from establishing a population in body tissues or fluids. All are general defense mechanisms that normally do not change over an individual's lifetime.

Complement

Inactive forms of about 30 different kinds of complement proteins circulate in the blood and interstitial fluid of vertebrate animals. Complement was named because the proteins were originally identified by their ability to "complement" the action of antibodies in adaptive immune responses (we return to these responses in Sections 37.6 and 37.8).

One type of complement protein can recognize and bind to antibodies clustered on the surface of a cell. The binding initiates a series of reactions that ultimately results in the enzymatic cleavage of another complement protein called C3. C3 is extremely abundant, so its cleavage products accumulate very quickly. Some of the fragments attach directly to cell membranes of invading pathogens as well as any host cell in the vicinity. C3 fragments become enzymatic when they are bound to a membrane, and these activate other complement proteins, which activate other complement proteins, and so on (**FIGURE 37.6A**). The cascading reactions quickly produce huge concentrations of activated complement at the membrane's surface.

Some of the activated complement proteins bind to pathogens, forming a coating that enhances uptake by phagocytic leukocytes. Others assemble into membrane attack complexes—structures that insert themselves into a cell's plasma membrane and form large channels through it (**FIGURE 37.6B,C**). Ions that flow through these channels disturb tonicity and thus trigger cell lysis (Section 20.3).

Activated C3 fragments also diffuse into surrounding tissues, forming a concentration gradient around the site of infection. The gradient recruits phagocytic leukocytes to the site of infection.

Years after complement was named, researchers discovered that some complement proteins can target

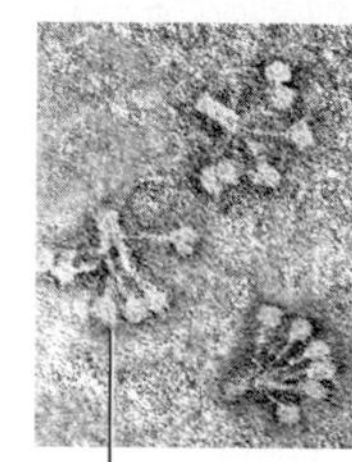

extracellular pathogens independently of antibodies. For example, the complement protein C1Q (shown at left) has six binding sites for simple carbohydrates found on the surface of a variety of pathogens, and it can bind directly to *Candida albicans* (a yeast); HIV and

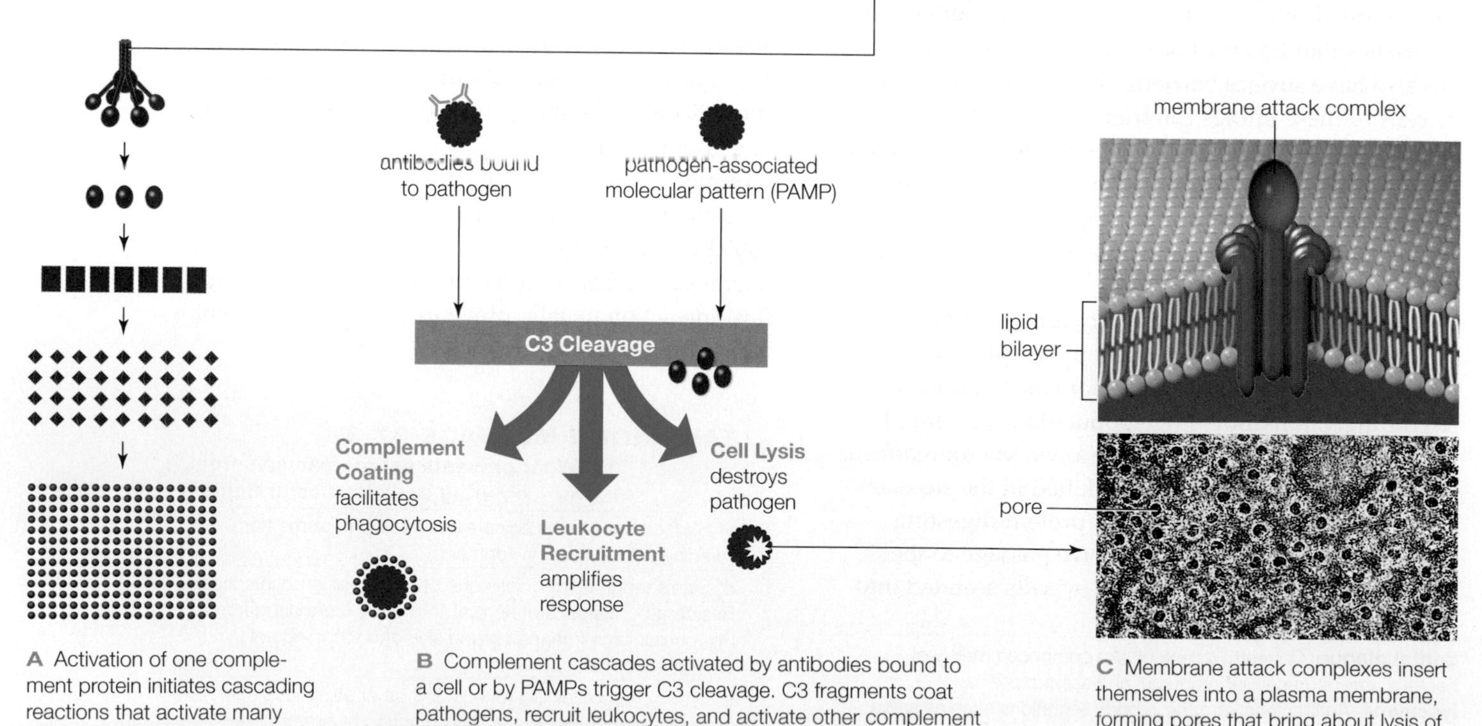

A Activation of one complement protein initiates cascading reactions that activate many other complement proteins.

B Complement cascades activated by antibodies bound to a cell or by PAMPs trigger C3 cleavage. C3 fragments coat pathogens, recruit leukocytes, and activate other complement proteins that assemble as membrane attack complexes.

C Membrane attack complexes insert themselves into a plasma membrane, forming pores that bring about lysis of the cell.

FIGURE 37.6 ▸**Animated** Complement.

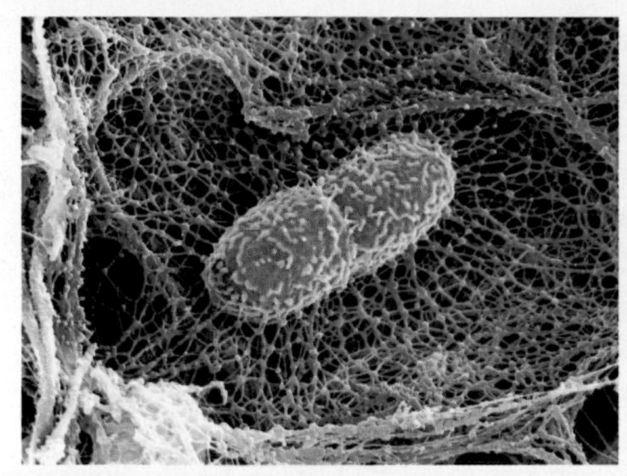

A These *Klebsiella* bacteria (pink) in lung tissue have been ensnared in a neutrophil net.

B Macrophage caught in the process of engulfing tuberculosis bacteria (red).

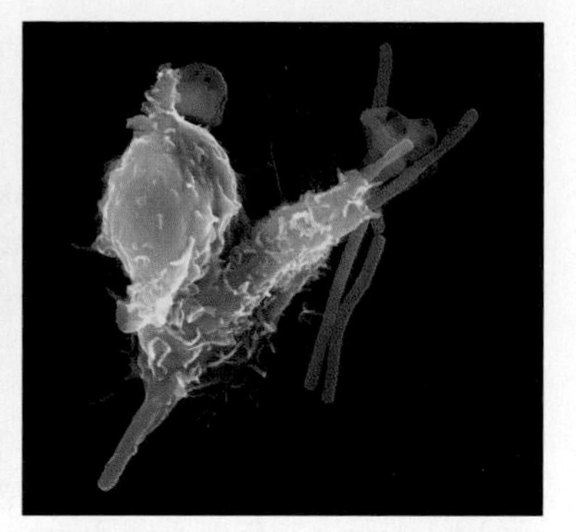

C Dendritic cell engulfing anthrax bacteria (*Bacillus anthracis*, red).

FIGURE 37.7 Phagocytic leukocytes and their role in innate immunity.

influenza virus; *Salmonella* and *Streptococcus* bacteria; and *Leishmania*, a protozoan parasite. When all six of the protein's binding sites are occupied, it activates an enzyme that cleaves C3, thus initiating the complement cascade. Cytoplasmic and mitochondrial proteins that leak out of damaged body cells also trigger a complement cascade.

Normal body cells continuously produce proteins that inactivate complement, thus preventing a complement cascade from spreading too far into healthy tissue. Microorganisms do not make these inhibitory proteins, so they are singled out for destruction.

Phagocytic Leukocytes

Neutrophils, macrophages, and dendritic cells are mobile, and all can follow a chemical trail—a type of movement called **chemotaxis**. For example, these phagocytic cells bear receptors for C3 fragments, and they follow gradients of this activated complement back to an affected tissue, where they engulf cells and particles coated with complement.

Neutrophils Neutrophils are the most abundant leukocyte, but they are also short-lived. Their turnover rate is phenomenal: The bone marrow of an adult human normally produces about 50 billion new neutrophils each day. This abundance is one reason that neutrophils are among the first responders to injury or infection, collecting within minutes at a site of tissue damage. A neutrophil that engulfs a microorganism releases the contents of its granules both into the endocytic vesicle and to the exterior of the cell. Enzymes and toxins released into extracellular fluid destroy all cells in the vicinity—even healthy body cells.

Neutrophils literally explode in response to a certain combination of signaling molecules and complement, in the process ejecting their nuclear DNA and associated proteins along with the contents of their granules. The mixture solidifies into an extracellular matrix that traps pathogens in the vicinity of the secreted antimicrobial compounds (**FIGURE 37.7A**). The net is very effective at killing bacteria.

Macrophages and Dendritic Cells Macrophages in interstitial fluid engulf essentially everything they can except undamaged body cells (**FIGURE 37.7B**). However, these cells are more than just scavengers. Upon engulfing antigen, a macrophage secretes interleukins and other cytokines that alert the immune system to the presence of invading pathogens.

Dendritic cells (**FIGURE 37.7C**) patrol tissues that contact the external environment, such as the lining of respiratory airways. Phagocytosis by dendritic cells plays a critical role in protecting the lungs from pathogens and other harmful particles. However, the main function of dendritic cells is to present antigen to T cells (more on how this works in Section 37.7).

TAKE-HOME MESSAGE 37.4

What happens after antigen is detected inside body tissues or fluids?

✔ Antigen (as well as tissue damage) triggers complement activation.

✔ Activated complement recruits phagocytic leukocytes, coats pathogens in an affected area, and triggers lysis of foreign cells.

✔ Phagocytic leukocytes release cytokines and antimicrobial molecules after engulfing antigen-bearing particles such as complement-coated microorganisms.

chemotaxis Cellular movement in response to a chemical stimulus.

37.5 Inflammation and Fever

✔ Complement activation, antigen detection, or tissue damage can trigger inflammation and fever.

Complement activation and cytokines released by phagocytic leukocytes typically trigger inflammation and fever, which are hallmark processes of an innate immune response.

Inflammation

Inflammation is a fast, local response that simultaneously destroys infected or damaged tissue and jump-starts the healing process (FIGURE 37.8). Inflammation begins when basophils, mast cells, or neutrophils degranulate, releasing the contents of their granules into an affected tissue. Degranulation can occur in response to a number of stimuli, including pattern receptor binding to antigen, for example on the surface of a bacterium ❶. Fragments of C3 complement trigger degranulation when they bind to receptors on leukocyte plasma membranes. Mast cells (FIGURE 37.9A) also degranulate in response to neuromodulators (Section 32.7) released at a site of tissue damage.

A degranulating leukocyte releases cytokines. It also releases prostaglandins and histamines ❷, local signaling molecules that have two effects. First, they cause nearby arterioles to widen, so blood flow to the area increases. The increased flow speeds the arrival of more phagocytic leukocytes, which are attracted to the cytokines. Second, they cause spaces to open up between the cells making up the walls of nearby capillaries. Leukocytes move from blood into tissues by squeezing through these spaces ❸. Lymphocytes also exit capillaries by moving directly through cells of the vessels' walls (FIGURE 37.9B). By the time this occurs, any invading cells have become coated with activated complement ❹, which makes them easy targets for the phagocytic cells ❺.

Symptoms of inflammation include redness and warmth that are outward indications of the area's increased blood flow. The increased permeability of capillary walls allows plasma proteins to escape into interstitial fluid. The fluid becomes hypertonic with respect to blood, and water follows by osmosis. The tissue swells with excess fluid, putting pressure on nerves and thus causing pain.

Inflammation continues as long as its triggers do. After these stimuli subside, for example after invading bacteria have been cleared from an infected tissue, macrophages begin to produce compounds that suppress inflammation and promote tissue repair. If the stimulus persists, inflammation becomes chronic. Chronic inflammation is not a normal condition. It does not benefit the body; rather, it causes or contributes to several diseases, including asthma, Crohn's

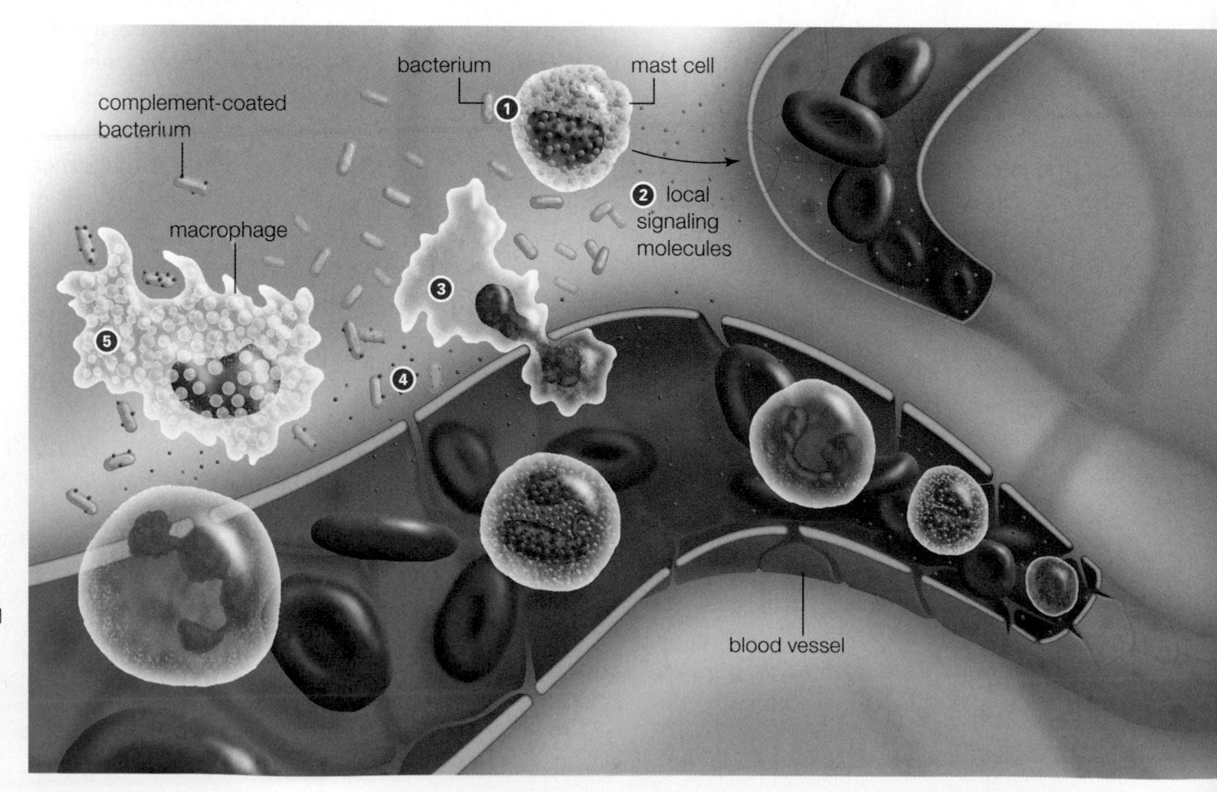

FIGURE 37.8 ▸Animated Example of inflammation as a response to bacterial infection.

❶ Pattern receptors on mast cells in the tissue recognize and bind to bacterial antigen.

❷ The mast cells release signaling molecules (blue dots) that cause arterioles to widen. The resulting increase in blood flow reddens and warms the tissue.

❸ The signaling molecules also increase capillary permeability, which allows phagocytic leukocytes to squeeze through the vessel walls into the tissue. Plasma proteins that leak out of the capillaries cause the tissue to swell with fluid.

❹ Meanwhile, bacterial antigens have triggered complement cascades, and invading bacteria have become coated with complement (purple dots).

❺ Phagocytic leukocytes in the tissue quickly recognize and engulf the complement-coated bacteria.

disease, rheumatoid arthritis, atherosclerosis, diabetes, and cancer.

Fever

Fever is a temporary rise in body temperature above the normal 37°C (98.6°F) that often occurs in response to infection or serious injury. Some cytokines stimulate brain cells to make and release prostaglandins, which act on the hypothalamus (Section 32.10) to raise the body's internal temperature set point. As long as the temperature of the body is below the new set point, the hypothalamus sends out signals that cause blood vessels in the skin to constrict, which reduces heat loss from the skin. The signals also trigger an increase in the rate of heartbeat and respiration, as well as reflexive movements called shivering, or "chills," that increase the metabolic heat output of muscles. These responses all raise the body's internal temperature. If the internal temperature rises too much, sweating and flushing (**FIGURE 37.9C**) quickly lower core temperature to maintain the new set point.

Fever enhances the body's immune defenses by increasing the rate of enzyme activity, thus speeding up the metabolic rate, tissue repair, and the formation and activity of phagocytic leukocytes. In addition, many pathogens multiply more slowly at the higher temperature, so leukocytes can get a head start in the proliferation race against them.

A fever is a sign that the body is fighting something, so it should never be ignored. However, a fever of 40.6°C (105°F) or less does not necessarily require treatment in an otherwise healthy adult. Body temperature usually will not rise above that value, but if it does, immediate hospitalization is recommended. Brain damage or death can occur if the body's core temperature reaches 42°C (107.6°F).

fever An internally induced rise in core body temperature above the normal set point as a response to infection or injury.
inflammation A local response to tissue damage or infection; characterized by redness, warmth, swelling, and pain.

TAKE-HOME MESSAGE 37.5

How do inflammation and fever function in innate immunity?

✔ Inflammation occurs when granular leukocytes release prostaglandins and histamines at a site of tissue damage or infection. These local signaling molecules increase blood flow and attract additional leukocytes to the site.

✔ Fever enhances immune defenses while slowing pathogen growth.

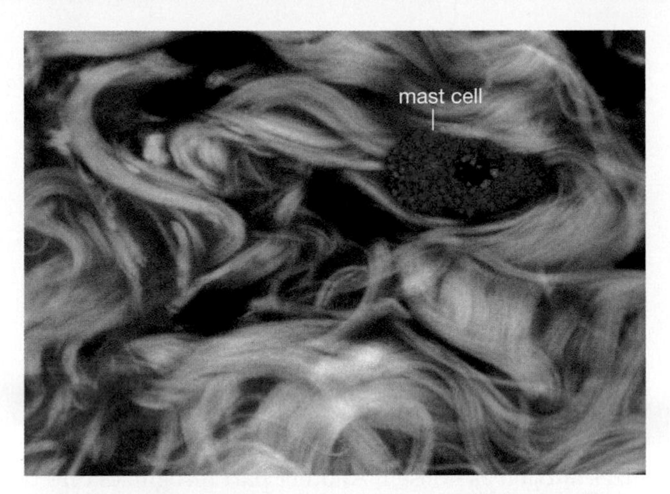

A This human mast cell is embedded in connective tissue of the conjunctiva (Section 33.6). If it degranulates, for example after its pattern receptors bind antigen, the molecules it releases trigger inflammation. Among other effects, these molecules loosen the tightly wound collagen fibers (pink, white) so phagocytic leukocytes can pass through the tissue to reach the affected site.

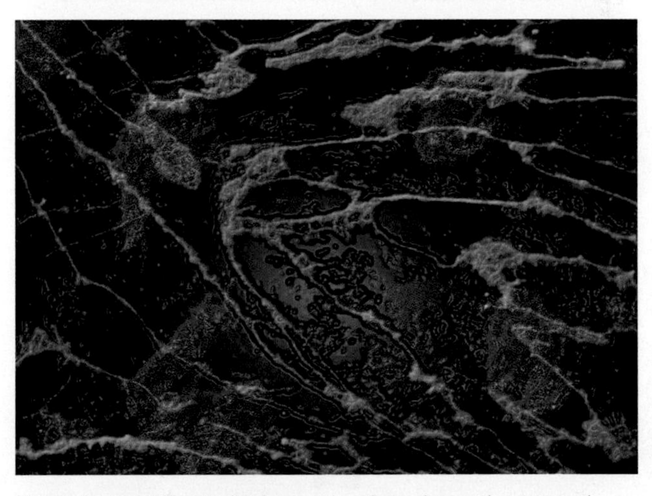

B T cells and NK cells (red) migrating through membranes and cytoplasm of endothelial cells (green). The ability to move directly through other cells allows lymphocytes to exit blood vessels quickly.

C Flushed cheeks are often an outward indication of fever. A raised internal temperature enhances immune defenses and inhibits growth of many pathogens.

FIGURE 37.9 Some mechanisms of innate immunity.

37.6 Antigen Receptors

✔ Antigen receptors give lymphocytes the collective potential to recognize billions of different antigens.

If innate immune mechanisms do not quickly rid the body of an invading pathogen, an infection may become established in internal tissues. By that time, long-lasting mechanisms of adaptive immunity have already begun to target the invaders specifically. These mechanisms are triggered by leukocytes that detect antigen via antigen receptors. Plasma membrane proteins that recognize PAMPs are one type of antigen receptor. Your T cells bear special antigen receptors called **T cell receptors**, or TCRs. Part of a TCR recognizes antigen as nonself. Another part recognizes certain proteins in the plasma membrane of body cells as self. These self-proteins are called **MHC markers** (left) after the major histocompatibility complex genes that encode them. MHC genes have thousands of alleles, so the cells of even closely related individuals rarely bear the same MHC markers.

MHC marker

Antibodies are another type of antigen receptor. **Antibodies** are Y-shaped proteins made and secreted by B cells. Many antibodies circulate in blood, and they can enter interstitial fluid during inflammation, but they do not kill pathogens directly. Instead, they activate complement and facilitate phagocytosis. Antibody binding can also prevent pathogens from attaching to body cells, and can neutralize some toxic molecules.

An antibody molecule consists of four polypeptides: two identical "light" chains and two identical "heavy" chains (**FIGURE 37.10A**). Each chain has a variable and a constant region. When the chains fold up together as an intact antibody, the variable regions form two antigen-binding sites that have a specific distribution of bumps, grooves, and charge. These binding sites are the antigen receptor part of an antibody: They bind only to antigen with a complementary distribution of bumps, grooves, and charge (**FIGURE 37.10B**). Antigen-binding sites vary greatly among antibodies, so they are called hypervariable regions.

An antibody's constant region determines its structural identity, or class: IgG, IgA, IgE, IgM, or IgD (Ig stands for immunoglobulin, another name for antibody). The different classes serve different functions (**TABLE 37.3**). Most of the antibodies circulating in the bloodstream and tissue fluids are IgG, which binds

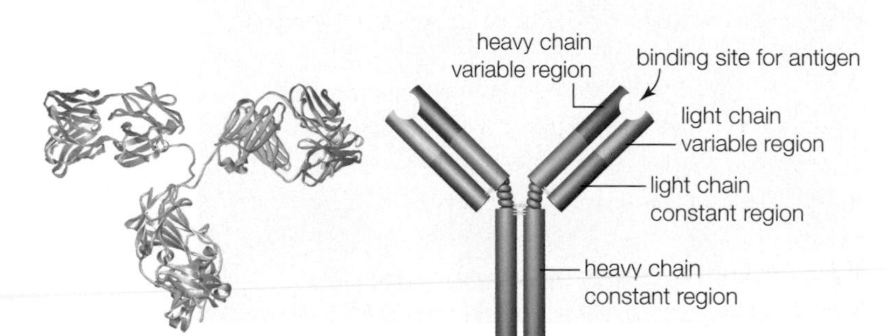

A An antibody molecule consists of four polypeptide chains, each with a variable and a constant region, joined in a Y-shaped configuration. The variable regions fold up as antigen-binding sites.

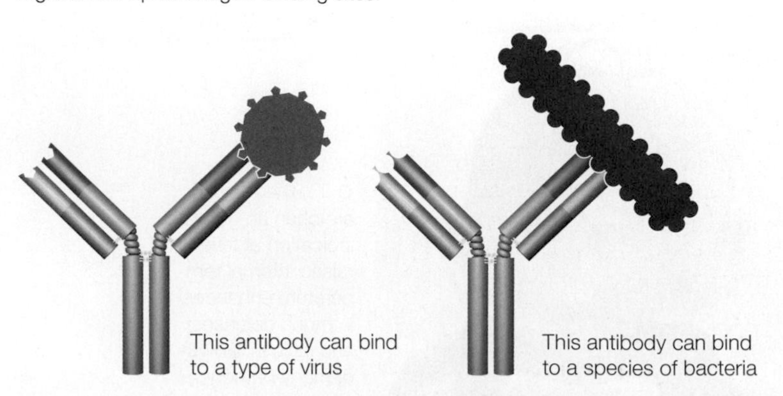

B The antigen-binding sites of each antibody are unique. They bind only to an antigen that has a complementary distribution of bumps, grooves, and charge.

FIGURE 37.10 Antibody structure.

Table 37.3 Antibody Structural Classes

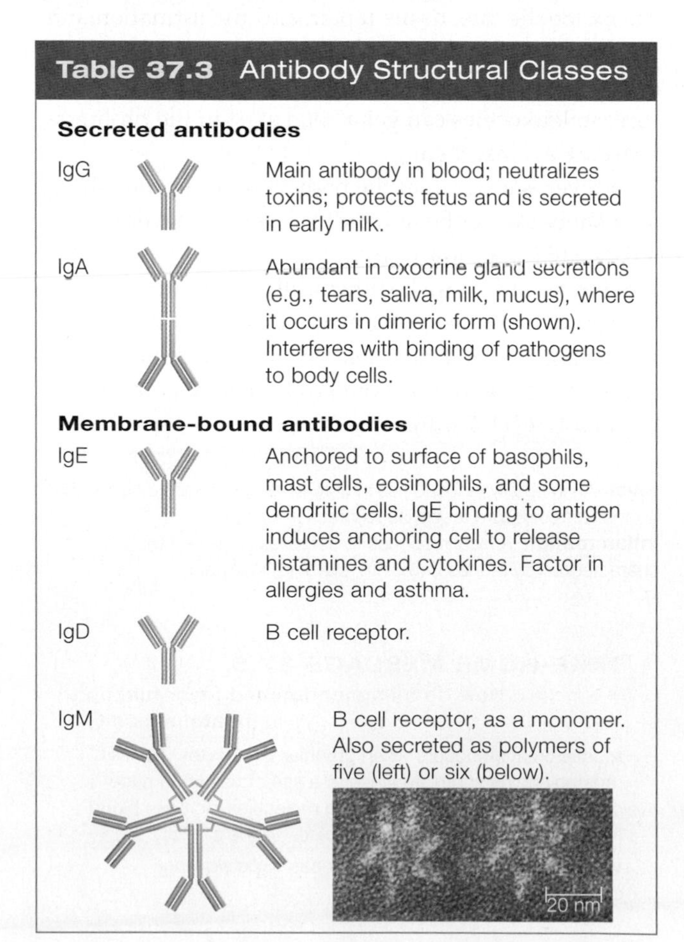

Secreted antibodies	
IgG	Main antibody in blood; neutralizes toxins; protects fetus and is secreted in early milk.
IgA	Abundant in exocrine gland secretions (e.g., tears, saliva, milk, mucus), where it occurs in dimeric form (shown). Interferes with binding of pathogens to body cells.
Membrane-bound antibodies	
IgE	Anchored to surface of basophils, mast cells, eosinophils, and some dendritic cells. IgE binding to antigen induces anchoring cell to release histamines and cytokines. Factor in allergies and asthma.
IgD	B cell receptor.
IgM	B cell receptor, as a monomer. Also secreted as polymers of five (left) or six (below).

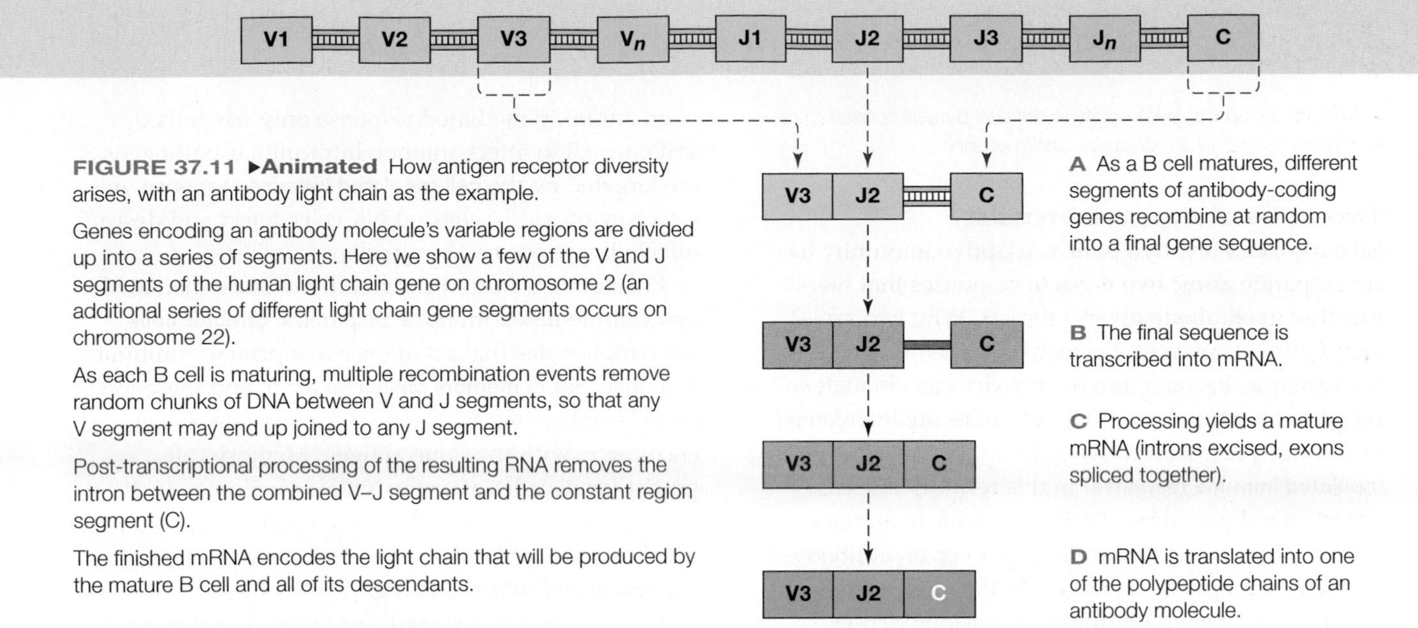

FIGURE 37.11 ▶**Animated** How antigen receptor diversity arises, with an antibody light chain as the example.

Genes encoding an antibody molecule's variable regions are divided up into a series of segments. Here we show a few of the V and J segments of the human light chain gene on chromosome 2 (an additional series of different light chain gene segments occurs on chromosome 22).

As each B cell is maturing, multiple recombination events remove random chunks of DNA between V and J segments, so that any V segment may end up joined to any J segment.

Post-transcriptional processing of the resulting RNA removes the intron between the combined V–J segment and the constant region segment (C).

The finished mRNA encodes the light chain that will be produced by the mature B cell and all of its descendants.

A As a B cell matures, different segments of antibody-coding genes recombine at random into a final gene sequence.

B The final sequence is transcribed into mRNA.

C Processing yields a mature mRNA (introns excised, exons spliced together).

D mRNA is translated into one of the polypeptide chains of an antibody molecule.

pathogens, neutralizes toxins, and activates complement. IgG is the only antibody that can cross the placenta to protect a fetus before its own immune system is active.

IgA is the main antibody in mucus and other exocrine gland secretions (Section 31.3). IgA is secreted as a dimer (two antibodies bound together), which makes the molecule stable enough to patrol harsh environments such as the interior of the digestive tract. There, IgA encounters pathogens before they have a chance to contact body cells. Bound to antigen, IgA triggers mast cells, basophils, macrophages, and NK cells to initiate inflammation.

IgE made and secreted by B cells is taken up by mast cells, basophils, and some types of dendritic cells, and then incorporated into their plasma membranes. Binding of antigen to membrane-bound IgE triggers the anchoring cell to degranulate.

B cell receptors are IgM or IgD antibodies that are not secreted; they stay attached to the B cell's plasma membrane. IgM is also secreted as polymers of five or six antibodies. The polymers are very efficient at binding antigen and activating complement.

Antigen Receptor Diversity

Humans can make billions of unique antigen receptors. This diversity arises because the genes that encode these receptors occur in several segments on different chromosomes, and there are several different versions of each segment (**FIGURE 37.11**). The gene segments become spliced together during B and T cell differentiation, but which version of each segment makes it into the finished antigen receptor gene of a particular cell is random. As a B or T cell differentiates, it ends up with one out of about 2.5 billion potential combinations of gene segments.

Like all other blood cells, lymphocytes form in bone marrow. A new B cell is already making receptors before it even leaves the marrow. The base of each receptor is embedded in the lipid bilayer of the cell's plasma membrane, and the two arms of the "Y" project into the extracellular environment (right). A mature B cell bristles with more than 100,000 receptors. T cells also form in bone marrow, but they mature in the thymus gland, which is part of the endocrine system (Section 34.12). There, they encounter hormones that stimulate them to make receptors.

B cell

B cell receptor

TAKE-HOME MESSAGE 37.6

What are antigen receptors?

✔ Each B cell and T cell makes antigen receptors that can bind a specific antigen. Humans are capable of producing billions of unique antigen receptors.

✔ T cell receptors can discriminate between self (MHC markers) and nonself (antigen).

✔ Antibodies released into the circulatory system activate complement and facilitate phagocytosis.

✔ B cell receptors are antibodies that are not secreted; they remain attached to the B cell's plasma membrane.

antibody Y-shaped antigen receptor protein made only by B cells.
B cell receptor Antigen receptor on the surface of a B cell; an antibody that stays anchored in the B cell's plasma membrane.
MHC markers Self-proteins on the surface of human body cells.
T cell receptor (TCR) Antigen receptor on the surface of a T cell.

37.7 Overview of Adaptive Immunity

✔ Vertebrate adaptive immunity is defined by self/nonself recognition, specificity, diversity, and memory.

Two Arms of Adaptive Immunity

Like a boxer's one-two punch, adaptive immunity has two separate arms: two types of responses that work together to eliminate diverse threats. Why two arms? Not all threats present themselves in the same way. For example, bacteria, fungi, or toxins can circulate in blood or interstitial fluid. These threats are intercepted quickly by phagocytic cells that initiate an **antibody-mediated immune response**. In this response, B cells are triggered to make antibodies specific to antigen detected in extracellular fluid. However, an antibody-mediated immune response is not the most effective way of countering other threats. Consider viruses, bacteria, fungi, and protists that reproduce inside body cells. These intracellular pathogens may be vulnerable to an antibody-mediated response only when they exit one cell to infect another. Intracellular pathogens are targeted by the **cell-mediated immune response**, in which cytotoxic T cells and NK cells detect and destroy infected or cancerous body cells.

Effector cells form during both antibody-mediated and cell-mediated immune responses. **Effector cells** are lymphocytes that act at once in a primary immune response. Some **memory cells** also form, and these long-lived lymphocytes are reserved for possible future encounters with the same antigen. Memory cells can persist for decades after the initial infection ends. If the same antigen enters the body again at a later time, these memory cells carry out a faster, stronger secondary response (FIGURE 37.12).

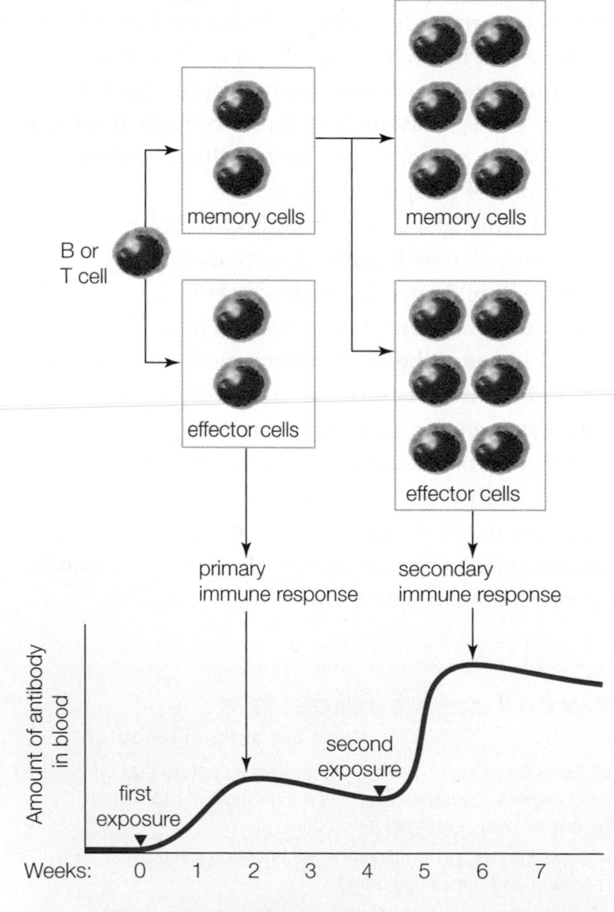

FIGURE 37.12 ▶Animated Primary and secondary immune responses. A first exposure to an antigen triggers a primary immune response in which effector cells fight the infection. Memory cells that also form initiate a faster, stronger secondary response if the antigen enters the body again at a later time.

Lymphocytes and phagocytic leukocytes interact to bring about the four defining characteristics of adaptive immunity:

Self/Nonself Recognition, based on the ability of T cell receptors to recognize self (in the form of MHC markers), and that of all antigen receptors to recognize nonself (in the form of antigen).

Specificity, which means that adaptive immune responses are tailored to combat specific antigens.

Diversity, which refers to the diversity of antigen receptors on a body's collection of lymphocytes. There are potentially billions of different antigen receptors, so an individual has the potential to counter billions of different threats.

Memory, the capacity of the adaptive immune system to "remember" an antigen via memory cells. It takes about a week for B and T cells to respond in force the first time they encounter an antigen. If the same antigen shows up later, the response is faster and stronger.

Antigen Processing

A new lymphocyte is "naive," which means that no antigen has bound to its receptors yet. B cell receptors can bind directly to antigen, but T cell receptors cannot. T cell receptors recognize and bind only to antigen that has been processed by an antigen-presenting cell. Macrophages, B cells, and dendritic cells do the processing (FIGURE 37.13). First, one of these cells engulfs a bacterium or any other particle that bears antigen ❶. A vesicle that contains the antigen-bearing particle forms in the cell's cytoplasm and fuses with a lysosome ❷. Remember from Section 4.7 that lysosomes are vesicles

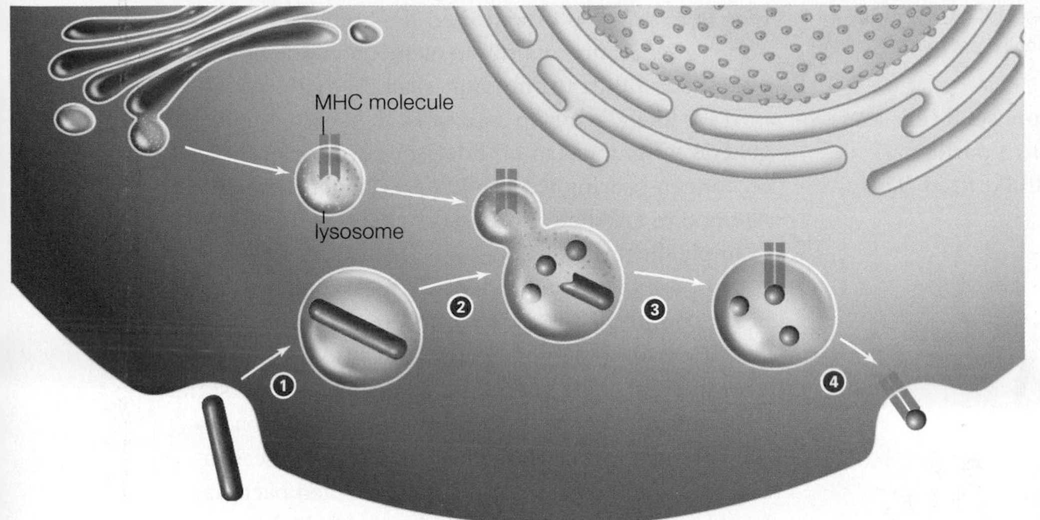

FIGURE 37.13 Antigen processing: What happens when a B cell, macrophage, or dendritic cell engulfs an antigenic particle (here, a bacterium).

❶ An endocytic vesicle forms around a bacterium as it is being engulfed by a phagocytic cell.

❷ The vesicle fuses with a lysosome, which contains enzymes and MHC molecules.

❸ Lysosomal enzymes digest the bacterium to molecular bits. The bits bind to MHC molecules.

❹ The vesicle fuses with the cell's plasma membrane by exocytosis. When it does, the antigen–MHC complex becomes displayed on the cell's surface.

filled with powerful enzymes. These enzymes now proceed to break down the ingested particle into molecular bits. Lysosomes also contain MHC markers that bind to some of the antigen-bearing bits ❸. The resulting antigen–MHC complexes become displayed at the cell's surface when the vesicles fuse with (and become part of) the plasma membrane ❹. The display of MHC markers paired with antigen fragments serves as a call to arms for T cells.

Antigen-bearing particles in blood end up in the spleen; those in solid tissues or interstitial fluid end up in lymph nodes. Phagocytic leukocytes migrate to these organs after engulfing antigen-bearing particles. Particles can also enter these organs directly, in which case they are engulfed by dendritic cells, macrophages, and B cells stationed inside. Either way, the antigen-bearing particles are processed by phagocytic cells, which become antigen-presenting cells.

Every day, billions of T cells filter through each lymph node and the spleen. As they do, they come into close contact with arrays of antigen-presenting cells that have taken up residence in these organs (**FIGURE 37.14**). As you will see shortly, T cells with receptors that recognize and bind to antigen presented by a phagocytic cell stimulate production of effector cells that carry out an immune response. During an infection, the lymph nodes swell because T cells accumulate inside them. When you are ill, you may notice your swollen lymph nodes as tender lumps under the jaw or elsewhere in your body.

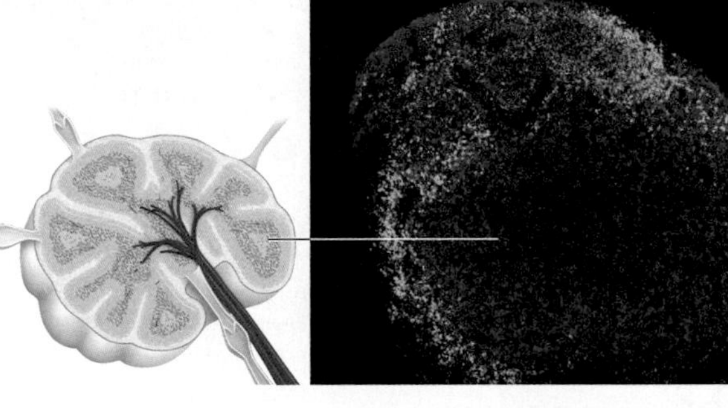

FIGURE 37.14 Lymph is filtered through at least one node before it merges with the bloodstream. The fluorescence micrograph shows T cells (blue) that are passing through a lymph node and interacting with populations of resident B cells (green) and antigen-presenting dendritic cells (red).

antibody-mediated immune response Immune response in which antibodies are produced in response to an antigen.
cell-mediated immune response Immune response in which cytotoxic T cells and NK cells kill infected or cancerous body cells.
effector cell Antigen-sensitized B cell or T cell that forms in an immune response and acts immediately.
memory cell Long-lived, antigen-sensitized B cell or T cell that can act in a secondary immune response.

TAKE-HOME MESSAGE 37.7

What is an immune response?

✔ Interactions between lymphocytes and phagocytic leukocytes bring about vertebrate adaptive immunity, which has four defining characteristics: self/nonself recognition, specificity, diversity, and memory.

✔ The two arms of adaptive immunity work together. Antibody-mediated responses target antigen in blood or interstitial fluid; cell-mediated responses target altered body cells.

✔ Effector and memory cells form during adaptive responses. Memory cells are set aside; if the antigen enters the body again at a later time, these cells initiate a faster, stronger secondary response.

✔ T cell receptors can recognize their specific antigen only in conjunction with MHC markers displayed on the surface of an antigen-presenting cell. The recognition process occurs in lymph nodes and the spleen.

37.8 The Antibody-Mediated Immune Response

✔ In an antibody-mediated immune response, effector B cells form and produce antibodies targeting a specific antigen.

In an antibody-mediated immune response, B cells are triggered to make antibodies specific to a particular pathogen or toxin detected in extracellular fluid. This type of response is often called the humoral response, because it pertains mainly to elements in the blood and other body fluids (from Latin *umor*, body fluid).

If we liken B cells to assassins, then each one has a genetic assignment to liquidate one particular target: an antigen-bearing extracellular pathogen or toxin. Antibodies are their molecular bullets, as the following example illustrates.

Suppose that you accidentally nick your finger. Being opportunists, some *Staphylococcus aureus* cells on your skin immediately enter the cut, invading your internal environment. Complement in interstitial fluid quickly attaches to carbohydrates in the bacterial cell walls, and complement activation cascades begin. Within an hour, complement-coated bacteria tumbling along in lymph vessels reach a lymph node. There, they filter past an army of naive B cells. One of the naive B cells residing in that lymph node makes antigen receptors that recognize a polysaccharide in *S. aureus* cell walls (**FIGURE 37.15** ❶). The B cell's receptors bind to one of the bacteria, and the complement coating stimulates the B cell to engulf it. The B cell is now activated (it is no longer naive).

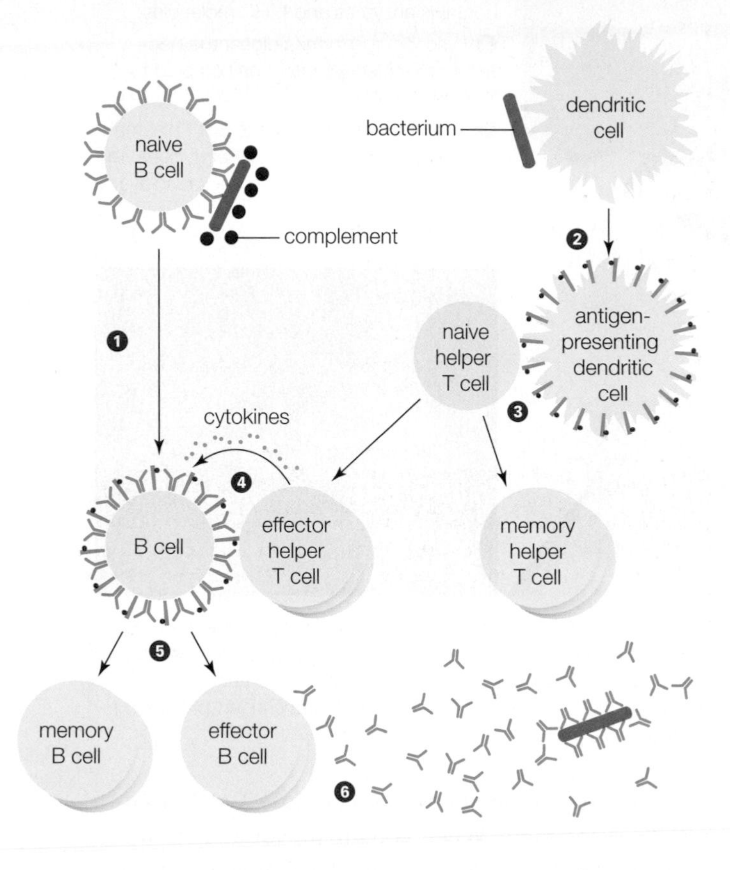

FIGURE 37.15 ▸**Animated** An example of an antibody-mediated immune response.

❶ B cell receptors on a naive B cell bind to an antigen on the surface of a bacterium (red). The bacterium's complement coating (purple dots) triggers the B cell to engulf it. Bacterial fragments bound to MHC markers become displayed at the surface of the B cell.

❷ A dendritic cell engulfs the same kind of bacterium that the B cell encountered. Bacterial fragments bound to MHC markers become displayed at the surface of the dendritic cell.

❸ Antigen–MHC complexes displayed by the dendritic cell are recognized by TCRs on a naive helper T cell. The two cells interact, and then the T cell begins to divide repeatedly by mitosis. Its descendants mature as effector and memory cells.

❹ TCRs on one of the effector cells recognize and bind to the antigen–MHC complexes on the B cell. The binding causes the T cell to secrete cytokines (blue dots).

❺ The cytokines induce the B cell to undergo repeated mitotic divisions. Its many descendants mature as effector B cells and memory B cells.

❻ The effector B cells begin making and secreting huge numbers of antibodies, all of which recognize the same antigen as the original B cell receptor. The new antibodies circulate throughout the body and bind to any remaining bacteria.

Meanwhile, more *S. aureus* cells have been secreting metabolic products into interstitial fluid around your cut. The secretions are attracting phagocytic leukocytes. A dendritic cell engulfs several bacteria, then migrates to the lymph node in your elbow. By the time it gets there, it has digested the bacteria and is displaying their fragments as antigens bound to MHC markers on its surface ❷. In the lymph node, one of your T cells recognizes and binds to the *S. aureus* antigen on the dendritic cell ❸. This T cell is called a helper T cell because it helps other lymphocytes produce antibodies and kill pathogens. The helper T cell and the dendritic cell interact for about 24 hours and then disengage. The helper T cell returns to the circulatory system and begins to divide, and a huge population of identical helper T cells forms. These clones mature as effector and memory cells, each of which has receptors that recognize the same *S. aureus* antigen.

Let's go back to that B cell in the lymph node. By now, it has digested the engulfed bacterium, and it is displaying bits of *S. aureus* bound to MHC molecules on its plasma membrane. The new helper T cells recognize the antigen–MHC complexes displayed by the B cell. One of these helper T cells binds to the B cell. Like long-lost friends, the two cells stay together for a while and communicate ❹. One of the messages that is communicated consists of cytokines secreted by the helper T cell. The cytokines stimulate the B cell to

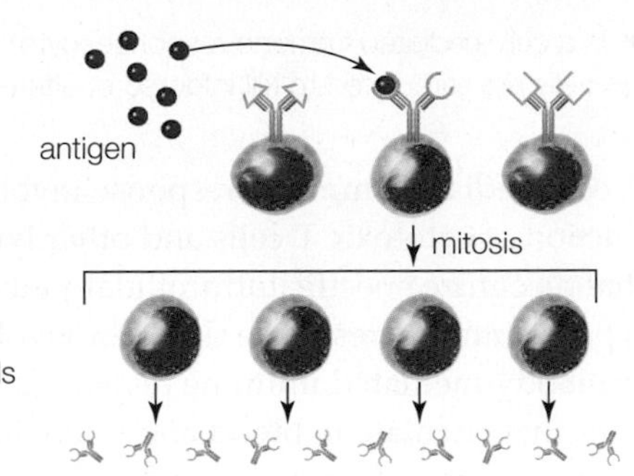

FIGURE 37.16 Clonal selection. Only lymphocytes with receptors that bind to antigen divide and mature. Clonal selection of B cells is shown in this example.

undergo repeated mitotic divisions after the two cells disengage. A huge clonal population of descendant cells forms, and these B cells mature as effector and memory cells ❺. By the theory of clonal selection, the B cell was "selected" because its receptors bound to the antigen. B cells with receptors that did not bind the antigen did not divide to form huge clonal populations (**FIGURE 37.16**).

As the B cells mature into effector and memory cells, they switch antibody classes, which means they start producing antibodies with a different constant region. Instead of making membrane-bound B cell receptors, they now make and secrete IgA, IgG, or IgE. The variable regions of the antibodies are unchanged, so their antigen-binding specificity remains the same: All can recognize and bind to the same *S. aureus* antigen ❻.

Huge numbers of antibodies now circulate throughout the body and attach themselves to any *S. aureus* cells. An antibody coating prevents the bacteria from attaching to body cells and brings them to the attention of phagocytic cells for quick disposal. Antibodies also glue the foreign cells together into clumps, a process called **agglutination**. The clumps are quickly removed from the circulatory system by the liver and spleen.

Antibodies and ABO Blood Typing

As you learned in Section 13.5, a carbohydrate on red blood cell membranes occurs in two forms. This carbohydrate is called H antigen. People with one form of the H antigen have type A blood; people with the other form have type B blood. People with both forms have type AB blood; those with neither are type O.

Early in life, each individual starts making antibodies that recognize molecules foreign to the body, including any nonself form of the H antigen (**TABLE 37.4**). If a blood transfusion becomes necessary, it is especially important to know which H antigens your blood cells carry. A transfusion of incompatible red blood cells can cause a potentially fatal transfusion reaction in which the recipient's antibodies recognize and bind to antigens on the transfused cells. The binding activates complement, which punctures the membranes of the foreign cells, thus releasing a massive amount of hemoglobin that can very quickly cause the kidneys to fail.

Identifying red blood cell surface antigens helps prevent pairing of incompatible transfusion donors and recipients, and also alerts physicians to blood incompatibility problems that may arise during pregnancy. A typical blood typing test involves mixing drops of a patient's blood with antibodies to the different forms of red blood cell antigens. Agglutination occurs when the cells bear antigens recognized by the antibodies (**FIGURE 37.17**).

Table 37.4 Antibodies to H Antigen

ABO Blood Type	H Antigen on Red Blood Cells	Antibodies Present
A	A	anti-B
B	B	anti-A
AB	both A and B	none
O	neither A nor B	anti-A, anti-B

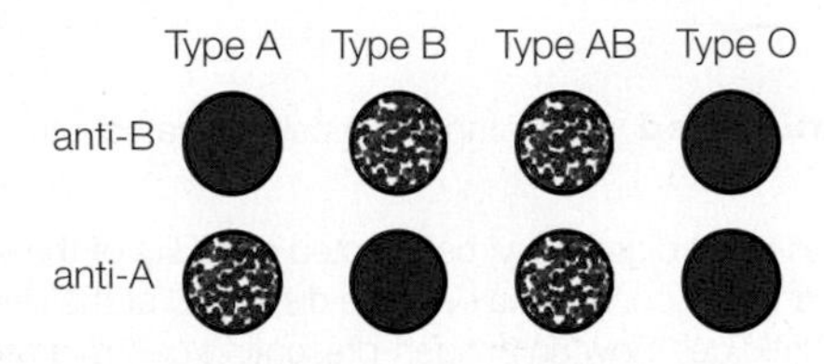

FIGURE 37.17 ABO blood typing test. In such tests, samples of a patient's blood are mixed with antibodies to H antigens. Agglutination (clumping) shows the presence of antigen.

FIGURE IT OUT A person with which blood type can receive a blood transfusion of the other types?

Answer: Type AB

TAKE-HOME MESSAGE 37.8

What happens during an antibody-mediated immune response?

✔ During an antibody-mediated response, B cells produce antibodies that bind to an antigen in blood or interstitial fluid.

✔ Antibody binding can prevent a pathogen from entering body cells, neutralize a toxin, and facilitate the elimination of both from the body.

agglutination The clumping together of foreign cells bound by antibodies; the clumps attract phagocytic cells.

37.9 The Cell-Mediated Immune Response

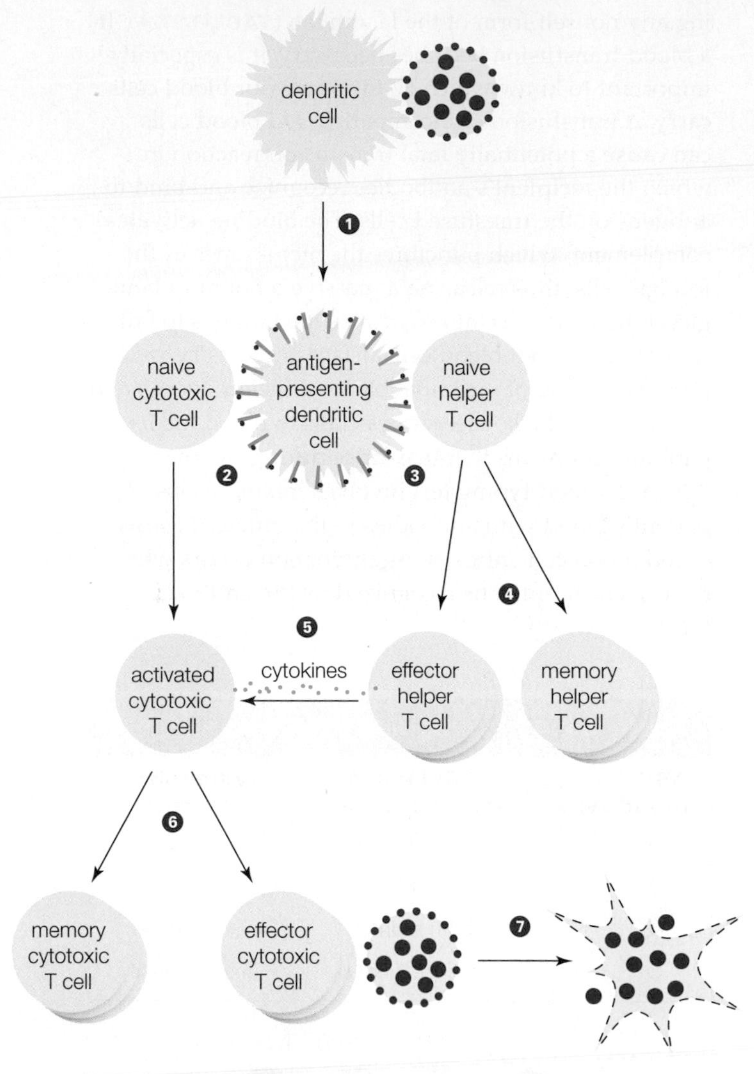

FIGURE 37.18 ▶Animated An example of a cell-mediated immune response.

❶ A dendritic cell engulfs and digests a virus-infected cell. Bits of the virus bind to MHC markers, and the complexes become displayed at the dendritic cell's surface. The dendritic cell, now an antigen-presenting cell, migrates to a lymph node.

❷ Receptors on a naive cytotoxic T cell bind to antigen–MHC complexes displayed by the dendritic cell. The interaction activates the cytotoxic T cell.

❸ Receptors on a naive helper T cell bind to antigen–MHC complexes displayed by the dendritic cell. The interaction activates the helper T cell.

❹ The activated helper T cell divides again and again. Its many descendants mature as effector and memory cells, each with T cell receptors that recognize the same antigen.

❺ The effector cells secrete cytokines.

❻ The cytokines induce the activated cytotoxic T cell to divide again and again. Its many descendants mature as effector and memory cells. Each cell bears T cell receptors that recognize the same antigen.

❼ The new effector cells circulate throughout the body. They kill any body cell that displays the viral antigen–MHC complexes on its surface.

FIGURE IT OUT What do the large red spots represent?

Answer: Viruses

✔ In a cell-mediated immune response, cytotoxic T cells and NK cells are stimulated to kill infected or altered body cells.

A cell-mediated immune response involves the production of cytotoxic T cells and other lymphocytes that recognize specific intracellular pathogens. This type of immune response does not involve antibodies. Antibody-mediated immune responses target pathogens that circulate in blood and interstitial fluid, but they are not as effective against pathogens inside cells.

Ailing body cells typically display molecules that are not found on healthy cells. For example, cancer cells display altered body proteins, and body cells infected with intracellular pathogens display polypeptides of the infecting agent. Both types of cell are killed by lymphocytes that act in cell-mediated responses.

Cytotoxic T Cells: Activation and Action

A cell-mediated immune response often starts in interstitial fluid during inflammation, when a dendritic cell recognizes, engulfs, and digests a sick body cell or the remains of one (**FIGURE 37.18**). The dendritic cell begins to display antigen that was part of the sick cell ❶ as it migrates to the spleen or a lymph node. There, the dendritic cell presents its antigen–MHC complexes to huge populations of naive cytotoxic T cells ❷ and naive helper T cells ❸. (**FIGURE 37.19** shows a T cell inspecting an antigen-presenting dendritic cell.)

Some of the naive cells have T cell receptors that recognize the antigen–MHC complexes presented by the dendritic cell. Within minutes of contact, helper T cells and cytotoxic T cells that recognize these complexes stop their migratory behavior. A tight connection called an immunological synapse forms between the T cell and the antigen-presenting cell (**FIGURE 37.20**). T cell receptors cluster at the point of contact between the two cells, and the connection is stabilized by adhesion proteins that form in a ring around them. The interaction activates the T cells, so they are no longer naive.

Activated helper T cells begin to divide, and their many descendants mature as effector and memory cells ❹. The new effector cells can form immunological synapses with antigen-presenting macrophages. A macrophage that interacts with an activated helper T cell increases its production of pathogen-busting enzymes and toxins, and also secretes more cytokines that attract additional phagocytic leukocytes.

The effector T cells also secrete cytokines ❺. Any cytotoxic T cells that have been activated by interacting with an antigen-presenting cell recognize these cytokines as a signal to divide repeatedly, and their many descendants mature as effector and memory cells ❻. All of the new cytotoxic T cells recognize and bind the same antigen—the one displayed by that first ailing cell.

If B cells are like assassins, then cytotoxic T cells are specialists in cell-to-cell combat. The effector cytotoxic T cells start working immediately. They circulate throughout blood and interstitial fluid, and bind to any other body cell displaying the original antigen together with MHC markers.

A cytotoxic T cell that binds to an ailing body cell (**FIGURE 37.21**) releases protein-digesting enzymes and small molecules called perforins. The perforins assemble into complexes that, like membrane attack complexes, insert themselves into the ailing cell's plasma membrane to form large channels. The channels allow the enzymes to enter the body cell, which then bursts or commits suicide ❼.

As occurs in an antibody-mediated response, memory cells form in a primary cell-mediated response. If the antigen appears in the body again at a later time, these memory cells will mount a faster, stronger secondary response.

The Role of Natural Killer (NK) Cells

In order to kill a body cell, cytotoxic T cells must recognize the MHC molecules on the surface of the cell. However, some infections or cancer can alter a body cell so that it is missing part or all of its MHC markers. NK cells are crucial for fighting such cells because, unlike cytotoxic T cells, NK cells can kill body cells that lack MHC markers. Cytokines secreted by helper T cells also stimulate NK cell division. The resulting populations of NK cells recognize and attack body cells that have antibodies bound to them. They also recognize certain proteins displayed by body cells that are under stress. Stressed body cells with normal MHC markers are not killed; only those with altered or missing MHC markers are destroyed.

Because NK cells can operate early in immune defenses, they are often considered to be part of innate immunity. However, they also have features associated with lymphocytes of adaptive immunity: activation by cytokines, for example, and memory.

TAKE-HOME MESSAGE 37.9

What happens during a cell-mediated immune response?

✔ Cytotoxic T cells and NK cells that form during a cell-mediated immune response kill infected body cells or those that have been altered by cancer.

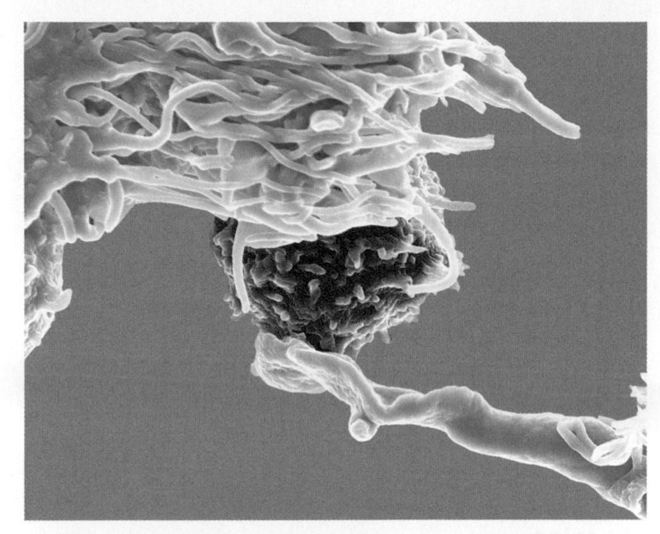

FIGURE 37.19 A human T cell (pink) scanning the surface of a dendritic cell (blue) wrapped around it. The T cell will initiate a cell-mediated immune response if it recognizes antigen–MHC complexes presented by the dendritic cell.

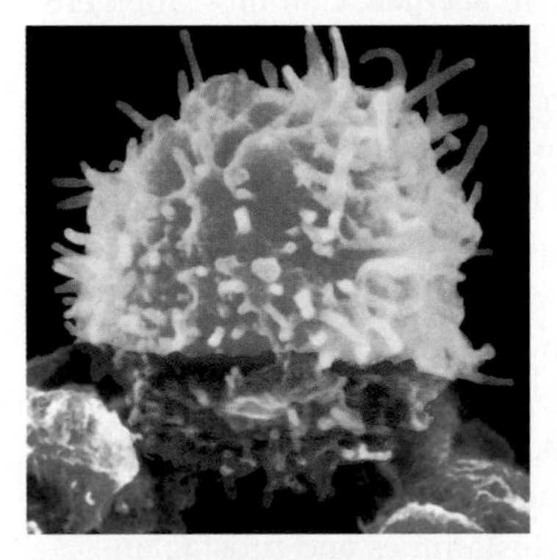

FIGURE 37.20 Zooming in on the point of contact between a T lymphocyte and an antigen-presenting cell. Left, a composite micrograph shows adhesion proteins (red ring) securing bound TCRs (green spot) at the point of contact between the cells. Right, a model shows a TCR (yellow) on a T cell bound to antigen (red) presented in context of an MHC marker (blue).

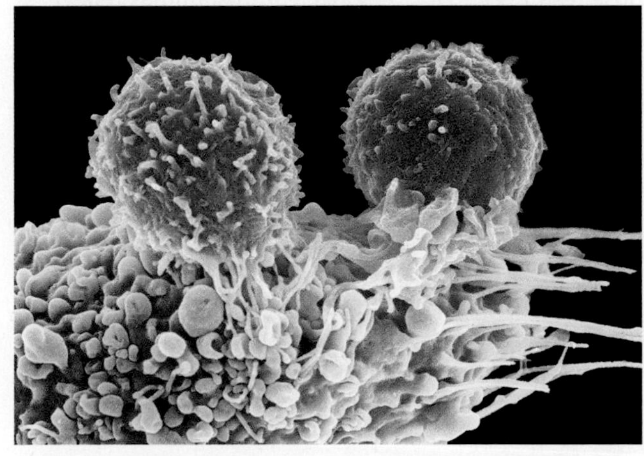

FIGURE 37.21 Cytotoxic T cells (pink) killing a cancer cell.

CREDITS: (19) Dr. Oliver Schwartz, Institute Pasteur/Science Source; (20) left, Courtesy of Dr. Michael Dustin, New York University School of Medicine. From Grakoui A, Bromley SK, Sumen C, Davis MM, Shaw AS, Allen PM, Dustin ML. *The immunological synapse: A molecular machine controlling T cell activation.* Science. 1999; 285:221-7; right, © Cengage Learning; (21) Steve Gschmeissner/Science Source.

37.10 When Immunity Goes Wrong

✔ An allergy is an immune response to something that is ordinarily harmless to most people.

✔ Autoimmune disorders occur when an immune response is misdirected against a person's own healthy body cells.

✔ In immunodeficiency, the immune response is insufficient to protect a person from disease.

Despite built-in quality controls and redundancies in immune system functions, immunity does not always work as well as it should. Its complexity is part of the problem, because there are simply more opportunities for failure to occur in systems with many components. Even a small failure in immune function can have a major effect on health.

A Hay fever is caused by allergy to grass pollen. Symptoms such as sneezing and a runny nose are a result of inflammation of the mucous membranes in respiratory airways.

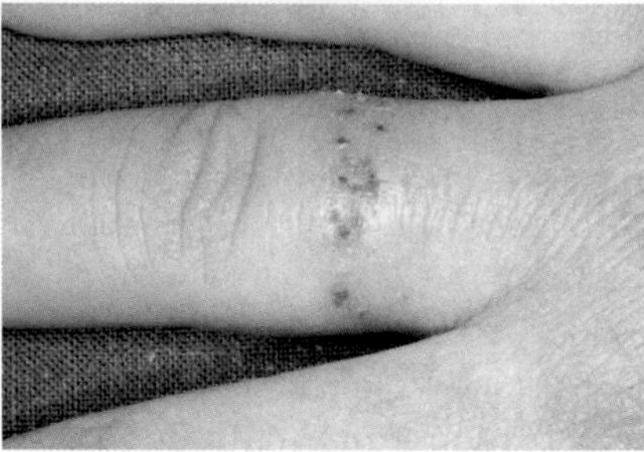

B Direct contact with an allergen can cause a rash—an itchy, irritated patch of skin. In this case, the offending allergen was nickel (a metal) in a ring.

FIGURE 37.22 Examples of allergies.

Allergies

In millions of people, exposure to a normally harmless substance stimulates an immune response. Any substance that is ordinarily harmless yet provokes such responses is called an **allergen**. Common allergens include drugs, foods, pollen, dust mite feces, fungal spores, and venom from bees, wasps, and other insects. Sensitivity to an allergen is an **allergy**. Some people are genetically predisposed to have allergies. Factors such as infections, emotional stress, exercise, and changes in air temperature can trigger or worsen them.

A first exposure to an allergen stimulates B cells to make and secrete IgE, which becomes anchored to mast cells and basophils. Upon a later exposure, antigen binds to the IgE. Binding triggers the anchoring cell to degranulate, and histamines and prostaglandins released by the cells initiate inflammation. If the allergen is detected by mast cells in the lining of the respiratory tract, the resulting inflammation constricts the airways and causes a copious amount of mucus to be secreted; sneezing, stuffed-up sinuses, and a drippy nose result (**FIGURE 37.22A**). Antihistamines relieve these symptoms by dampening the effects of histamines released by the degranulating cells. Other drugs can inhibit mast cell degranulation, thus preventing histamine release.

Skin rashes and other contact allergies do not involve antibodies; they are caused by a cell-mediated response to an allergen (**FIGURE 37.22B**).

Overly Vigorous Responses

Immune defenses that eliminate a threat can also damage body tissues. Thus, multiple mechanisms that limit immune responses are always in play. Consider that some complement proteins can be cleaved spontaneously, even in the absence of infection or tissue damage. Without the fail-safe mechanism of inhibitory molecules that inactivate the complement fragments, complement cascades would occur constantly, with disastrous effects on body tissues.

Acute illnesses arise when mechanisms that limit immune responses fail. Exposure to an allergen sometimes causes a rapid and severe allergic reaction called anaphylactic shock, or anaphylaxis. Huge amounts of inflammatory molecules, including histamines and prostaglandins, are released all at once in all parts of the body. Too much fluid leaks from blood into tissues, causing a sudden and dramatic drop in blood pressure (a reaction called shock). Rapidly swelling tissues constrict the airways and may block them. Anaphylaxis is rare but life-threatening and requires immediate treatment. It may occur at any time, upon exposure to even a tiny amount of allergen. The risks include any prior allergic reaction.

Other types of hyperactive immune responses are not as well understood. Severe episodes of asthma or septic shock occur when too many neutrophils degranulate at once. In a "cytokine storm," too many leukocytes release cytokines at the same time. The cytokine overdose triggers immediate, widespread inflammation that can cause organ failure, with potentially fatal results. Cytokine storm triggered by infection with H5N1 influenza virus (Section 20.4) is a reason that this strain of bird flu has an unusually high mortality rate.

Autoimmune Disorders

People usually do not make antibodies to molecules that occur on their own healthy body cells, in part because the thymus has a built-in quality control mechanism that weeds out T cells with defective receptors. Thymus cells snip small polypeptides from a variety of body proteins and attach them to MHC markers. Maturing T cells that bind too strongly to one of these peptide–MHC complexes have TCRs that recognize a self-protein; those that bind too weakly to the complexes do not recognize MHC markers. Both types of cells die. If this mechanism fails, mature lymphocytes that do not discriminate between self and nonself may be produced. Such lymphocytes can mount an **autoimmune response**, which is an immune response that targets one's own tissues. Autoimmunity is beneficial when a cell-mediated response targets cancer cells, but in most other cases it is not (**TABLE 37.5**).

Antibodies to self-proteins (autoantibodies) may bind to hormone receptors, as in the case of Graves' disease. In this disease, autoantibodies that bind stimulatory receptors on the thyroid gland cause it to produce excess thyroid hormone, which quickens the body's overall metabolic rate (Section 34.7). Antibodies are not part of the feedback loops that normally regulate thyroid hormone production. So, antibody binding continues unchecked, the thyroid continues to release too much hormone, and the metabolic rate spins out of control. Symptoms of Graves' disease include uncontrollable weight loss; rapid, irregular heartbeat; sleeplessness; pronounced mood swings; and bulging eyes.

The neurological disorder called multiple sclerosis occurs when self-reactive T cells attack myelin in the brain and spinal cord (Section 32.13). Symptoms range from weakness and loss of balance to paralysis and blindness. Specific alleles for MHC markers increase susceptibility, but a bacterial or viral infection may trigger the disorder.

Table 37.5 Examples of Autoantibodies Associated With Autoimmune Disorders

Disorder	Autoantibody Target	Affected Area
Crohn's disease	Proteins in neutrophil granules	Gastrointestinal tract
Dermatomyositis	tRNA synthesis enzyme	Muscles, skin
Diabetes mellitus type 1	Islet proteins or insulin	Pancreas
Goodpasture's syndrome	Type IV collagen	Kidney, lung
Graves' disease	TSH receptor	Thyroid
Guillain-Barré syndrome	Lipids of ganglia	Peripheral nervous system
Hashimoto's disease	TH synthesis proteins	Thyroid
Idiopathic thrombocytopenic purpura	Platelet glycoproteins	Blood (platelets)
Lupus erythematosus	DNA, nuclear proteins	Connective tissue
Multiple sclerosis	Myelin proteins	Central nervous system
Myasthenia gravis	Acetylcholine receptors	Neuromuscular junctions
Pemphigus vulgaris	Cadherin	Skin
Pernicious anemia	Parietal cell glycoprotein	Gut epithelium
Polymyositis	tRNA synthesis enzyme	Muscles
Primary biliary cirrhosis	Nuclear pore proteins, mitochondria	Liver
Rheumatoid arthritis	Constant region of IgG	Joints
Scleroderma	Topoisomerase	Arteriole endothelium
Ulcerative colitis	Enzyme in neutrophil granules	Large intestine
Wegener's granulomatosis	Enzyme in neutrophil granules	Blood vessels

Immunodeficiency

Insufficient immune function—immune deficiency—renders an individual vulnerable to infections by opportunistic agents that are typically harmless to those in good health. Primary immune deficiencies, which are present at birth, are the outcome of mutations. Severe combined immunodeficiency (SCID) is an example (Section 15.10). Secondary immune deficiency is the loss of immune function after exposure to a virus or other outside agent. AIDS (acquired immunodeficiency syndrome, described in the next section) is the most common secondary immune deficiency.

allergen A normally harmless substance that provokes an immune response in some people.
allergy Sensitivity to an allergen.
autoimmune response Immune response that inappropriately targets one's own tissues.

TAKE-HOME MESSAGE 37.10

What happens when the immune system does not function as it should?

✔ An allergen is a normally harmless substances that induces an immune response. Sensitivity to an allergen is called an allergy. Severe allergic reactions and other overly vigorous immune responses can be life-threatening.

✔ Autoimmune diseases are caused by inappropriate immune responses to normal body tissues.

✔ Compromised immunity, which can be inherited or triggered by environmental factors, causes an individual to be especially vulnerable to infections.

37.11 HIV and AIDS

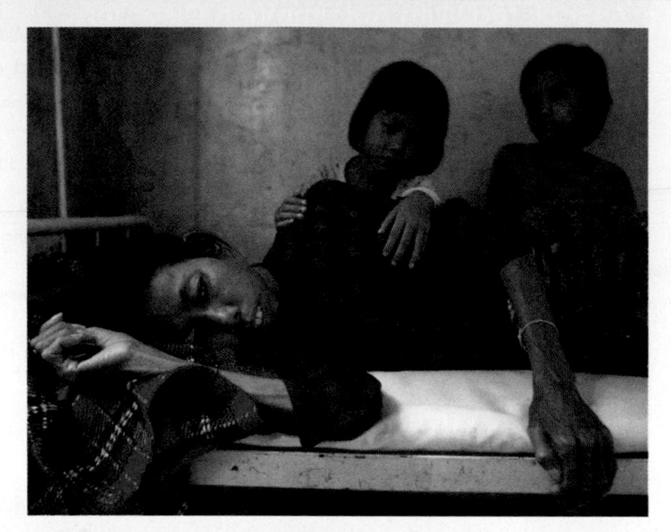

A In Cambodia, a mother lays dying of AIDS in front of her children. This country has the highest incidence of AIDS in Southeast Asia.

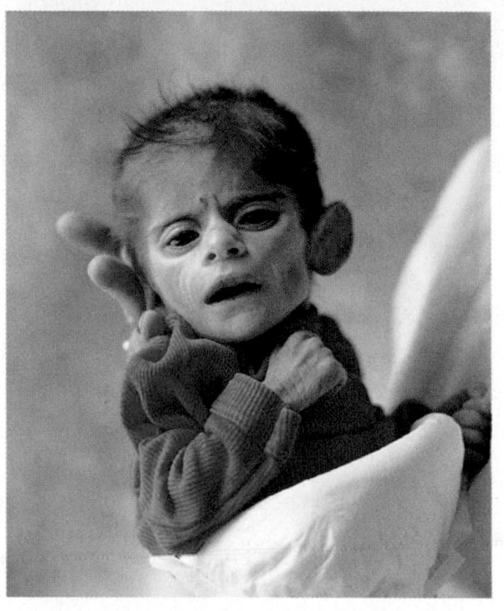

B This baby contracted AIDS from his mother's breast milk.

FIGURE 37.23 Some faces of AIDS.

Table 37.6 Worldwide HIV Incidence

Region	Number Infected	% Adults Infected
Sub-Saharan Africa	22,100,000	4.7
Caribbean Islands	230,000	1.0
Central Asia/East Europe	1,300,000	0.7
North America	1,300,000	0.5
Latin America	1,400,000	0.4
South/Southeast Asia	3,700,000	0.3
Australia/New Zealand	48,000	0.2
Western/Central Europe	860,000	0.2
Middle East/North Africa	250,000	0.1
East Asia	880,000	0.1
Approx. worldwide total	32,100,000	0.8

Source: Joint United Nations Programme HIV/AIDS, 2012 report

✔ AIDS is an outcome of interactions between a virus (HIV) and the human immune system.

The most common secondary immune deficiency is **AIDS**, or acquired immunodeficiency syndrome. AIDS is a syndrome of disorders that occur as a result of infection with HIV, the human immunodeficiency virus (Section 20.1). This virus cripples the immune system, so it makes the body very susceptible to infection by other pathogens and to rare forms of cancer. Worldwide, approximately 32 million individuals are currently infected with HIV (FIGURE 37.23A and TABLE 37.6). AIDS is now considered to be a pandemic by the World Health Organization (WHO).

There is no way to rid the body of HIV, no cure for those already infected. At first, an infected person appears to be in good health, perhaps fighting "a bout of the flu." But symptoms eventually emerge that foreshadow AIDS: fever, many enlarged lymph nodes, chronic fatigue and weight loss, and drenching night sweats. Then, infections caused by normally harmless microorganisms strike. Yeast infections of the mouth, esophagus, and vagina often occur, as well as a form of pneumonia caused by the fungus *Pneumocystis jiroveci*. Gastrointestinal inflammation due to infection by a yeast or virus causes diarrhea. Colored lesions that erupt are evidence of Kaposi's sarcoma, a type of cancer that is common among AIDS patients but rare among the general population. Other cancers are also common, as are infections by cancer-causing viruses such as Epstein–Barr virus. These medical problems are unusual in people with healthy immune systems.

HIV Revisited

HIV mainly infects macrophages, dendritic cells, and helper T cells (FIGURE 37.24). When virus particles enter the body, dendritic cells engulf them. The dendritic cells then migrate to lymph nodes, where they present processed HIV antigen to naive T cells. An army of HIV-neutralizing IgG antibodies and HIV-specific cytotoxic T cells forms.

We have just described a typical adaptive immune response. It rids the body of most—but not all—of the virus. In this first response, HIV infects a few helper T cells in a few lymph nodes. For years or even decades, the IgG antibodies keep the level of HIV in the blood low, and the cytotoxic T cells kill HIV-infected cells.

During this stage, infected people often have no symptoms of AIDS, but they can pass the virus to other people. HIV persists in a few of their helper T cells, in a few lymph nodes. Eventually, the level of virus-neutralizing IgG in the blood plummets, and the production of T cells slows. Why IgG decreases is still a major topic of research, but its effect is certain: The immune system becomes progressively less effective at fighting the virus. The number of virus particles rises, and more and more helper T cells become infected. Lymph nodes begin to swell with infected T cells.

Eventually, the battle tilts as the body makes fewer replacement helper T cells and its immune system is destroyed. Other types of viruses may replicate more quickly than HIV, but the immune system eventually demolishes them. HIV demolishes the immune system. Secondary infections and tumors kill the patient.

Transmission HIV is not transmitted by casual contact; most infections are the result of having unprotected sex with an infected partner. The virus occurs in semen and vaginal secretions, and it can enter a sexual partner through epithelial linings of the penis, vagina, rectum, and mouth. The risk of transmission increases by the type of sexual act; for example, anal sex carries 50 times the risk of oral sex. Infected mothers can transmit HIV to a child during pregnancy, labor, delivery, or breast-feeding (**FIGURE 37.23B**). HIV also travels in tiny amounts of infected blood in syringes shared by intravenous drug abusers, or by hospital patients in less developed countries. Many people have become infected via blood transfusions, but this transmission route is becoming rarer as most blood is now tested prior to use for transfusions.

Testing Most AIDS tests check blood, saliva, or urine for antibodies that bind to HIV antigens. These antibodies are detectable in 99 percent of infected people within three months of exposure to the virus. One test can detect viral RNA eleven days after exposure. Currently, the only reliable tests are performed in clinical settings; home test kits are generally less accurate than clinical tests. A false negative result may cause an infected person to unknowingly transmit the virus.

Treatments Drugs cannot cure AIDS, but they can slow its progress. Of the twenty or so FDA-approved AIDS drugs, most target processes unique to retroviral replication. For example, RNA nucleotide analogs such as AZT interrupt HIV replication when they substitute for normal nucleotides in the viral RNA-to-DNA synthesis process (Sections 15.2 and 20.3). Other drugs such as protease inhibitors affect different parts of the viral replication cycle.

A three-drug "cocktail" of one protease inhibitor plus two reverse transcriptase inhibitors is currently the most successful AIDS therapy, and has changed the typical course of the disease from a short-term death sentence to a long-term, often manageable illness. It also shows promise as a treatment if given immediately after exposure to HIV.

AIDS Acquired immune deficiency syndrome. A secondary immune deficiency that develops as the result of infection by HIV.

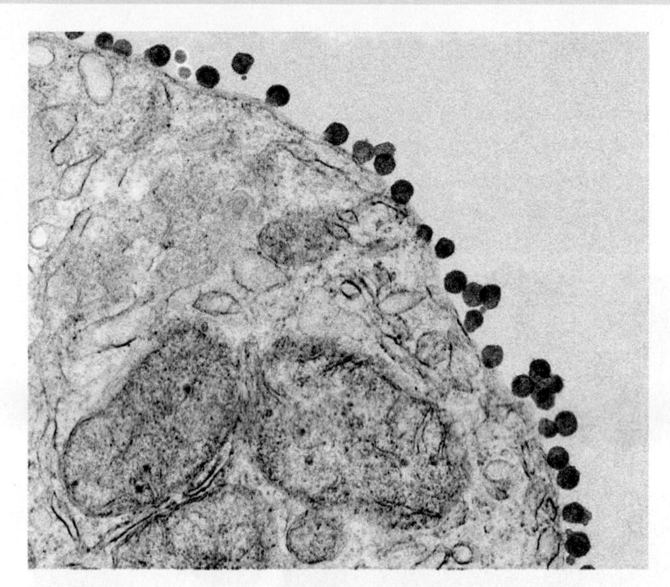

FIGURE 37.24 HIV (red) budding from the surface of a T cell. The virus infects the very cells that would otherwise protect the body from infection.

Prevention Preventive use of a drug that contains two reverse transcriptase inhibitors has recently been shown to greatly reduce the HIV infection rate in high risk populations. However, education is still our best option for halting the global spread of AIDS. In most circumstances, HIV infection is the consequence of a choice: either to have unprotected sex, or to use a shared needle for intravenous drugs. Programs that teach people how to avoid these unsafe behaviors are having an effect on the spread of the virus, but overall, our global battle against AIDS is not being won.

TAKE-HOME MESSAGE 37.11

What is AIDS?

✔ AIDS is a secondary immune deficiency caused by HIV infection. HIV infects lymphocytes and so cripples the human immune system.

37.12 Vaccines

✔ Vaccines are designed to elicit immunity to a disease.

Immunization refers to procedures designed to induce immunity. In active immunization, a preparation that contains antigen—a **vaccine**—is injected or administered orally. The first immunization elicits a primary immune response, just as an infection would. A second immunization, which is called a booster, elicits a secondary immune response for enhanced protection. In passive immunization, one individual receives antibodies that have been purified from the blood of another. Passive immunization offers immediate benefit for someone who has been exposed to a lethal agent such as tetanus, rabies, Ebola virus, or a venom or toxin. Because the antibodies were not made by the recipient's lymphocytes, effector and memory cells do not form, so benefits last only as long as the injected antibodies do.

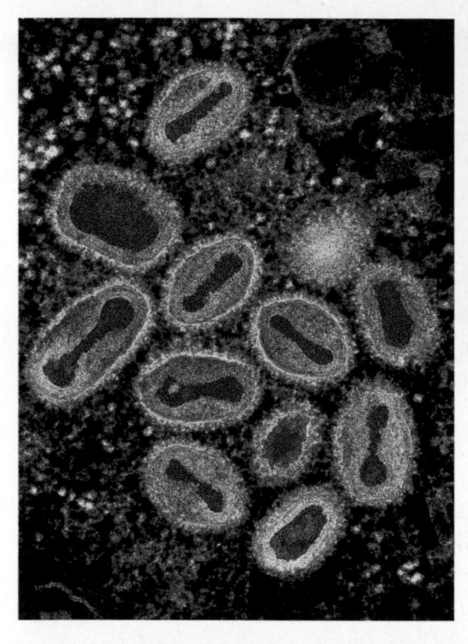

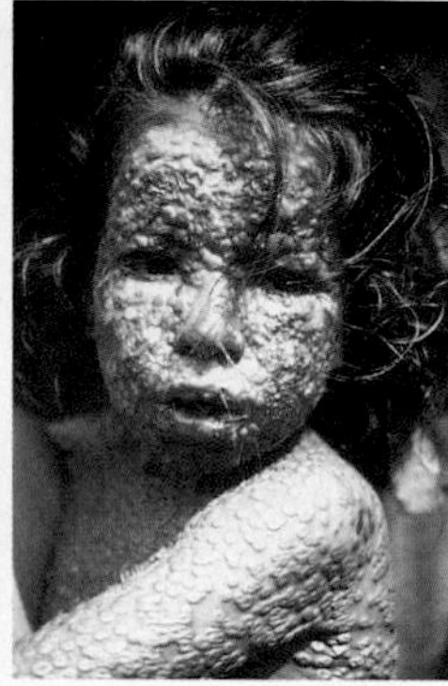

FIGURE 37.25 Smallpox: cause (left) and effect (above). Worldwide use of the smallpox vaccine eradicated naturally occurring cases of smallpox; vaccinations for it ended in 1972.

Table 37.7 Recommended Immunization Schedule for Children

Vaccine	Age of Vaccination
HepB (Hepatitis B)	Birth
HepB boosters	1–2 months, 6–18 months
Rotavirus (RV)	2, 4, and 6 months
DTaP (diphtheria, tetanus, pertussis)	2, 4, and 6 months
DTaP boosters	15–18 months, 4–6 years, 11–12 years
HiB (*Haemophilus influenzae*)	2, 4, and 6 months
HiB booster	12–15 months
PCV13 (Pneumococcus)	2, 4, and 6 months
PCV13 booster	12–15 months
IPV (Inactivated poliovirus)	2 and 4 months
IPV boosters	6–18 months, 4–6 years
Influenza	Yearly, 6 months and older
MMR (measles, mumps, rubella)	12–15 months
MMR booster	4–6 years
Varicella (chicken pox)	12–15 months
Varicella booster	4–6 years
HepA (Hepatitis A, 2 doses)	12–23 months
HPV (Human papillomavirus, 3 doses)	11–12 years
Meningococcal	11–12 years
Meningococcal booster	16 years

Centers for Disease Control and Prevention (CDC), 2014

The first vaccine was developed in the late 1700s, a result of desperate attempts to survive smallpox epidemics that swept repeatedly through cities all over the world. Smallpox is a severe disease that kills up to one-third of the people it infects (**FIGURE 37.25**). Before 1880, no one knew what caused infectious diseases or how to protect anyone from getting them, but there were clues. In the case of smallpox, survivors seldom contracted the disease a second time. They were said to be immune—protected from infection.

At the time, the idea of acquiring immunity to smallpox was extremely appealing. People had been risking their lives on it for over two thousand years by poking into their skin bits of smallpox scabs or threads soaked in pus from smallpox sores. Some survived the crude practices and became immune to smallpox, but many others did not.

By 1774, it was common knowledge that dairymaids usually did not get smallpox after they had contracted cowpox, a mild disease that affects humans as well as cattle. An English farmer, Benjamin Jesty, made this prediction: If people accidentally infected with cowpox become immune to smallpox, then people deliberately infected with it should become immune too. Jesty collected pus from a cowpox sore on a cow's udder, and poked it into the arm of his pregnant wife and two small children. All survived the smallpox epidemic, though they were from that time on subject to derision and rock peltings by neighbors who were convinced they would turn into cows.

In 1796, Edward Jenner, an English physician, injected liquid from a cowpox sore into the arm of a healthy boy. Six weeks later, Jenner injected the boy with liquid from a smallpox sore. Luckily, the boy did not get smallpox. Jenner's experiment showed directly that the agent of cowpox elicits immunity to smallpox.

Jenner named his procedure "vaccination," after the Latin word for cowpox (*vaccinia*). Though it was still controversial, the use of Jenner's vaccine spread quickly through Europe, then to the rest of the world. The last known case of naturally occurring smallpox was in 1977. Use of the vaccine eradicated the disease.

We now know that the cowpox virus is an effective vaccine for smallpox because the antibodies it elicits also recognize smallpox virus antigens. Our increasing knowledge of the immune system has allowed us to develop vaccines for many other infectious diseases (TABLE 37.7). These vaccines have overwhelmingly reduced suffering and deaths, but public confidence is a necessary part of their success. When enough individuals refuse vaccination, outbreaks of preventable and sometimes fatal diseases occur.

Progress on an HIV Vaccine

We still have no vaccine effective against HIV infection. The virus has a very high mutation rate, and the antibodies produced during a normal adaptive response exert selection pressure on it. Thus, a very large number of HIV variants exist. A successful vaccine must efficiently recognize and neutralize all HIV variants, including those that have not yet arisen. Immunization with live, weakened virus is an effective vaccine in chimpanzees, but the risk of infection from the vaccination itself far outweighs its potential benefits in humans. Other types of vaccines have been notoriously ineffective against HIV.

One promising strategy involves reverse engineering HIV antibodies isolated from people with AIDS. These antibodies are being collected and studied in painstaking detail in order to discover the exact parts of the virus they recognize. Those parts of the virus are being used to create new vaccines. Genes encoding the three antibodies are also being inserted into viral vectors for use in gene therapy (Section 15.10), the idea being that the vector will deliver the genes into body cells, which will then start producing antibodies.

immunization Any procedure designed to induce immunity to a specific disease.
vaccine A preparation introduced into the body in order to elicit immunity to a specific antigen.

TAKE-HOME MESSAGE 37.12

How do vaccines work?

✔ Administering a vaccine elicits an immune response in the recipient that protects against future encounters with a disease-causing agent.

✔ Worldwide vaccination programs have greatly reduced suffering and deaths from many diseases. An effective vaccination program requires widespread participation.

Frankie's Last Wish (revisited)

The Gardasil HPV vaccine consists of viral proteins that self-assemble into virus-like particles (VLPs). These proteins are produced by a recombinant yeast, *Saccharomyces cerevisiae*. The yeast carries genes for one surface protein from each of four strains of HPV, so the VLPs carry no viral DNA. Thus, the VLPs are not infectious, but the antigenic proteins they consist of elicit an immune response at least as strong as infection with HPV.

summary

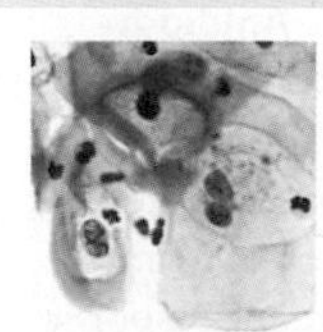

Section 37.1 Screenings, treatments, and vaccines for diseases such as cervical cancer are a direct outcome of our increasing understanding of the way the human body interacts with pathogens.

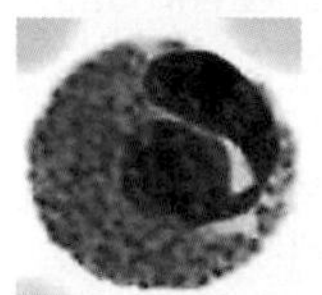

Section 37.2 The body's ability to resist and fight infections is called **immunity**. A microorganism or other antigen-bearing particle that breaches surface barriers triggers **innate immunity**, a set of general defenses that can prevent pathogens from becoming established inside the body. **Adaptive immunity**, which can specifically target billions of different **antigens**, follows. **Complement** and signaling molecules such as **cytokines** help coordinate the activities of white blood cells (leukocytes) such as **dendritic cells**, **mast cells**, **macrophages**, **neutrophils**, **basophils**, and **eosinophils**. Lymphocytes (**B cells**, **T cells**, and **NK cells**) are leukocytes with special roles in immune responses.

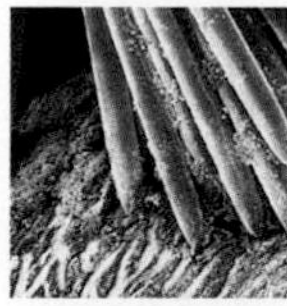

Section 37.3 Most **normal flora** that colonize body surfaces—including the linings of tubes and cavities—do not cause disease unless they penetrate inner tissues. Vertebrates fend off pathogens (such as those that cause **dental plaque**) at body surfaces with physical, mechanical, and chemical barriers (such as **lysozyme**).

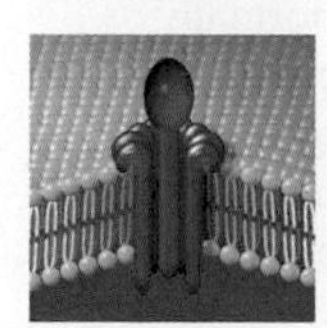

Section 37.4 An innate immune response can be triggered by activated complement or by phagocytic leukocytes that engulf an antigen-bearing particle. Complement is activated by the presence of antigen or tissue damage. Leukocytes follow gradients of activated complement by **chemotaxis**. Activated complement proteins also coat antigenic particles and form complexes that kill invading cells by puncturing their plasma membranes.

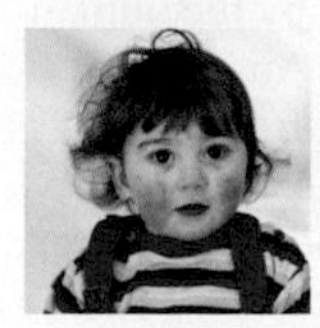

Section 37.5 **Inflammation** begins when leukocytes in infected or damaged tissue release signaling molecules. The molecules attract phagocytic cells and increase blood flow to the affected area, warming and red-

CREDIT: (in text revisited) In memory of Frankie McCullough.

dening it. Plasma proteins that leak from capillaries cause swelling and pain.

Prolonged exposure to an inflammation-provoking stimulus can cause chronic inflammation. **Fever** increases the metabolic rate and can slow pathogen replication.

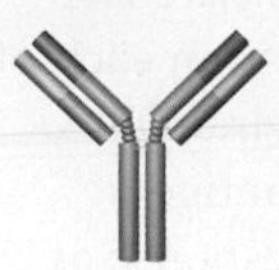

Section 37.6 **T cell receptors** recognize antigen displayed in conjunction with **MHC markers** by an antigen-presenting cell. **B cell receptors** are **antibodies** that have not been released from a B cell. Both kinds of antigen receptors collectively have the ability to recognize billions of specific antigens, a diversity that arises from random recombination of antigen receptor genes.

Section 37.7 B cells and T cells carry out adaptive immune responses. **Antibody-mediated** and **cell-mediated immune responses** work together to rid the body of a specific pathogen. Phagocytic leukocytes present antigen in adaptive responses. **Effector cells** form and target the antigen-bearing particles in a primary response. **Memory cells** that also form are reserved for a later encounter with the same antigen, in which case they trigger a faster, stronger secondary response. Phagocytic leukocytes that engulf, process, and present antigen to T cells in the spleen or lymph nodes are central to all adaptive responses.

Section 37.8 B cells, assisted by T cells, antigen-presenting cells, and signaling molecules, carry out antibody-mediated immune responses. Antibodies that recognize a specific antigen are secreted by B cells during these responses. Antibody binding causes **agglutination** of foreign cells, and tags antigen-bearing particles for phagocytosis. ABO blood typing tests for the presence of H antigen on blood cells by agglutination.

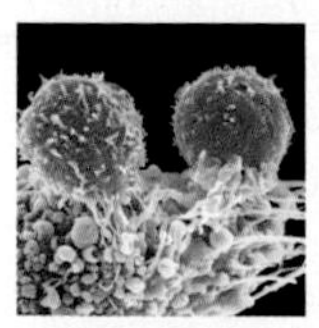

Section 37.9 T cells interact with antigen-presenting cells to carry out cell-mediated immune responses. Cytotoxic T cells and NK cells that form in these responses kill infected body cells or those that have been altered by cancer.

Section 37.10 **Allergens** are normally harmless substances that induce immune responses; sensitivity to an allergen is called **allergy**. A malfunction in the immune system or in its checks and balances can cause dangerous acute illnesses, or chronic and sometimes deadly disorders. Immune deficiency is a reduced capacity to mount an immune response. In an **autoimmune response**, a body's own cells are inappropriately recognized as foreign and attacked.

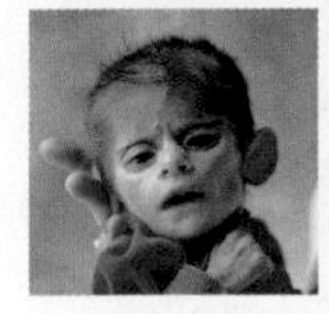

Section 37.11 **AIDS** is caused by the human immunodeficiency virus (HIV). An initial adaptive immune response to infection rids the body of most—but not all—of the virus. The virus infects T cells and other leukocytes, so it eventually cripples adaptive immune responses.

Section 37.12 **Immunization** with **vaccines** designed to elicit immunity to specific disease is an important part of worldwide health programs. Effective vaccination programs require widespread participation.

Self-Quiz

Answers in Appendix VII

1. _______ trigger immune responses.
   a. Cytokines
   b. Lysozymes
   c. Antibodies
   d. Antigens
   e. Histamines
   f. all of the above

2. Which of the following is *not* among the first line of defenses against infection?
   a. skin
   b. acidic gastric fluid
   c. lysozyme in saliva
   d. resident bacterial populations
   e. complement activation
   f. flushing action of diarrhea

3. Which of the following is *not* considered to be part of an innate immune response?
   a. phagocytic cells
   b. fever
   c. histamines
   d. cytokines
   e. inflammation
   f. complement activation
   g. presenting antigen
   h. all take part

4. Which of the following is *not* part of an adaptive immune response?
   a. phagocytic cells
   b. antigen-presenting cells
   c. histamines
   d. cytokines
   e. antigen receptors
   f. complement activation
   g. antibodies
   h. all take part

5. Activated complement proteins _______ .
   a. form attack complexes
   b. promote inflammation
   c. attract phagocytic cells
   d. all of the above

6. Name a defining characteristic of innate immunity.

7. Antibodies are _______ .
   a. antigen receptors
   b. made only by B cells
   c. proteins
   d. all of the above

8. A dendritic cell engulfs a bacterium, then presents bacterial bits on its surface along with a(n) _______ .
   a. MHC marker
   b. antibody
   c. T cell receptor
   d. antigen

9. Antibody-mediated responses are most effective against _______ .
   a. intracellular pathogens
   b. extracellular pathogens
   c. cancerous cells
   d. both a and b
   e. both b and c
   f. a, b, and c

10. Cell-mediated responses are most effective against _______ .
   a. intracellular pathogens
   b. extracellular pathogens
   c. cancerous cells
   d. both a and b
   e. both a and c
   f. a, b, and c

11. Allergies occur when the body responds to _______ .
   a. pathogens
   b. toxins
   c. normally harmless substances
   d. all of the above

data analysis activities

Cervical Cancer Incidence in HPV-Positive Women In 2003, Michelle Khan and her coworkers published their findings on a 10-year study in which they followed cervical cancer incidence and HPV status in 20,514 women. All women who participated in the study were free of cervical cancer when the test began. Pap tests were taken at regular intervals, and the researchers used a DNA probe hybridization test (Section 15.3) to detect specific types of HPV in the women's cervical cells.

The results are shown in **FIGURE 37.26** as a graph of the incidence rate of cervical cancer by HPV type. HPV-positive women are often infected by more than one type, so the data were sorted into groups based on the women's HPV status ranked by type: either positive for HPV16; or negative for HPV16 and positive for HPV18; or negative for HPV16 and HPV18 and positive for any other cancer-causing HPV; or negative for all cancer-causing HPV.

1. At 110 months into the study, what percentage of women who were *not* infected with any type of cancer-causing HPV had cervical cancer? What percentage of women who were infected with HPV16 also had cervical cancer?

2. In which group would women infected with both HPV16 and HPV18 fall?

3. Is it possible to estimate from this graph the overall risk of cervical cancer that is associated with infection of cancer-causing HPV of any type?

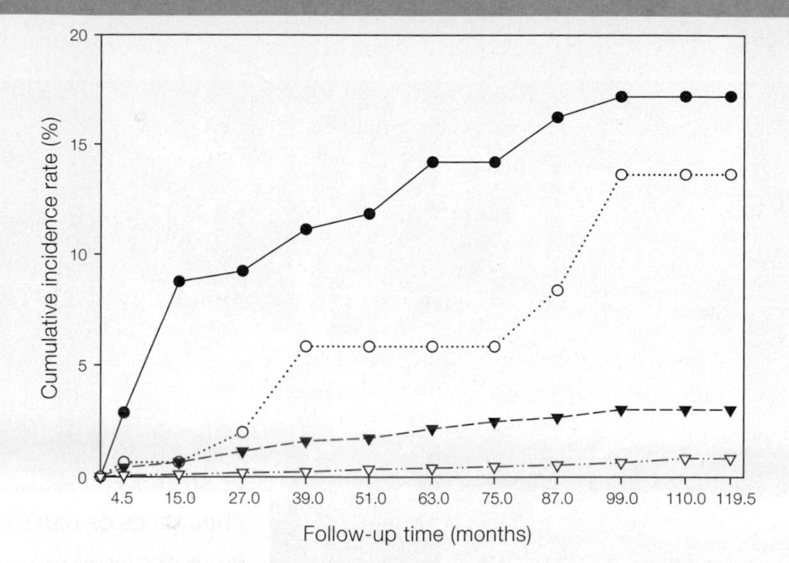

FIGURE 37.26 Cumulative incidence rate of cervical cancer correlated with HPV status in 20,514 women aged 16 years and older.

The data were grouped as follows:

● HPV16 positive

○ HPV16 negative and HPV18 positive

▼ All other cancer-causing HPV types combined

▽ No cancer-causing HPV type was detected.

12. Name a defining characteristic of adaptive immunity.

13. ________ are targets of cytotoxic T cells.
   a. Extracellular virus particles in blood
   b. Virus-infected body cells or tumor cells
   c. Parasitic worms in the liver
   d. Bacterial cells in pus
   e. Pollen grains in nasal mucus

14. Which combination of the following types of antibodies and immune cells is central to hay fever?
   a. IgE and mast cells
   b. IgG and basophils
   c. IgA and lymphocytes
   d. IgM and macrophages

15. Match the immune cell with the function.

___ dendritic cell	a. kills virus-infected cells
___ B cell	b. antigen-presenter
___ helper T cell	c. activates other lymphocytes
___ NK cell	d. kills cells lacking MHC markers
___ cytotoxic T cell	e. makes antibodies

16. Match the immunity concept with the best description.

___ anaphylactic shock	a. recognizes antigen
___ immune memory	b. inadequate immune response
___ autoimmunity	c. general defense mechanism
___ inflammation	d. immune response against one's own body
___ immune deficiency	e. secondary response
___ antigen receptor	f. systemic allergic reaction
___ antigen processing	g. presenting antigen together with MHC molecules

critical thinking

1. A flu shot consists of a vaccine against several strains of influenza virus. This year, you get the shot and "the flu." What happened? (There are at least three explanations.)

2. Monoclonal antibodies are produced by immunizing a mouse with a particular antigen, then removing its spleen. Individual B cells producing mouse antibodies specific for the antigen are isolated from the mouse's spleen and fused with cancerous B cells from a myeloma cell line. The resulting hybrid myeloma cells ("hybridoma" cells) are cloned: Individual cells are grown in tissue culture as separate cell lines. Each line produces and secretes antibodies that recognize the antigen to which the mouse was immunized. These monoclonal antibodies can be purified and used for research or other purposes. Monoclonal antibodies are sometimes used in passive immunization. They are effective, but only in the immediate term. Antibodies produced by one's own immune system can last up to about six months in the bloodstream, but monoclonals delivered in passive immunization often last for less than a week. Why the difference?

3. In what is called transplant rejection, a tissue or organ transplanted from one human into another is immediately destroyed by the recipient's immune system. Which type of leukocyte does the attacking?

CREDIT: (26) Michelle Khan et al, "The Elevated 10-year Risk of Cervical Precancer and Cancer in Women with Human Papillomavirus (HPV) Type 16 or 18 and the Possible Utility of Type Specific HPV Testing in Clinical Practice"; *Journal of the National Cancer Institute*, Vol.97, No. 14, July 20, 2005.

LEARNING ROADMAP

Understanding diffusion (Section 5.8) and aerobic respiration (7.2) will help you understand gas exchange. This chapter revisits the role of red blood cells (36.5) and hemoglobin (3.2) and how evolutionary changes that accompanied the move of vertebrates onto land (25.6) allow respiration in specific environments.

PRINCIPLES OF GAS EXCHANGE

Respiration moves oxygen from air or water in the environment to all metabolically active tissues, and it moves carbon dioxide in the other direction—from tissues to the environment.

GAS EXCHANGE IN INVERTEBRATES

Aquatic invertebrates exchange gases across their body surface or gills. Systems that deliver air to a respiratory surface inside the body adapt other invertebrates to land.

GAS EXCHANGE IN VERTEBRATES

Most vertebrates have either gills or paired lungs. In human lungs, gas exchange occurs in air sacs at the ends of branching airways. Muscle contractions draw air into lungs.

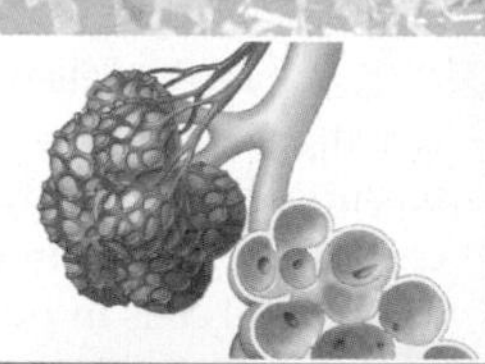

RESPIRATORY ADAPTATIONS

Some animals and humans have adapted to life at high altitudes, where there is less oxygen. Other animals can go without breathing for hours during deep dives.

RESPIRATORY DISEASE AND DISORDERS

Obstructed airways, a faulty respiratory pacemaker, or infectious disease can impair gas exchange. Smoking greatly increases risk of respiratory disease and lung cancer.

The iron needed to build hemoglobin is an example of an essential dietary mineral, a topic we discuss in Section 39.10. The respiratory system plays a role in acid–base balance (Section 40.6) and in the regulation of body temperature (40.9). Section 42.9 explains how a pregnant woman exchanges gases with her developing child.

38.1 Carbon Monoxide—A Stealthy Poison

It was supposed to be a celebration of life. One evening in December 2010, a group of five young men rented a hotel room for a birthday party. The guest of honor, who was turning nineteen, parked in the garage beneath the room. His vehicle had been having battery problems, so he decided to leave it running for a little while. (Running the engine charges the battery.) The next morning, the men were discovered dead by the maid who came to clean their room.

The cause of death was carbon monoxide poisoning, the most common type of fatal inhalation poisoning both in the United States and worldwide. Carbon monoxide (CO), a colorless, odorless gas, is released when organic material burns (**FIGURE 38.1**). Being lighter than air, it rises from its point of release. The young men died when CO in exhaust from the engine idling below them rose into their room.

The incidence of accidental CO poisonings typically increases during the winter, when people close up their homes and start using their wood stoves and oil-, gas-, or kerosene-powered heaters. The incidence of CO poisoning also spikes after a power outage, when people turn to gasoline-powered generators.

Inhaled carbon monoxide exerts its adverse effect by interfering with gas exchange. It binds to hemoglobin to form carboxyhemoglobin (COHb). Hemoglobin has a very high affinity for CO, binding it more than two hundred times more tightly than oxygen. In addition to blocking oxygen-binding sites, binding of CO makes hemoglobin hold more tightly to any oxygen that it does bind. Thus, when CO is present, the blood holds less oxygen and delivers little of that to tissues.

The resulting decline in the rate of oxygen delivery has its most dramatic effects in the brain and heart, the organs with the highest oxygen needs. At a low level, carbon monoxide causes headache, fatigue, irritability, mild nausea, and a shortness of breath. The symptoms are often mistakenly attributed to an infectious disease such as a flu. However, unlike a flu, CO poisoning does not cause a fever. A person with no underlying heart or lung problems will begin to experience symptoms of CO poisoning when the level of COHb in the blood reaches about 15 percent. As this level increases, a person becomes dizzy, light-headed, and confused. Chest pain often occurs because CO binds cardiac myoglobin (the oxygen-storing protein of muscle) even more strongly than it binds hemoglobin. The pain is a signal that heart cells are starved of oxygen. Continued exposure to CO results in seizures and loss of consciousness, followed by coma, cardiac arrest, and death.

FIGURE 38.1 Two common sources of carbon monoxide exposure—automobile exhaust and cigarette smoke. Carbon monoxide is an invisible odorless gas that binds to hemoglobin and impairs normal gas exchange.

A nonsmoker who lives in an area with little air pollution typically has a blood COHb level of 1 to 2 percent. People exposed to air pollution from heavy traffic have heightened blood levels of CO, as do smokers. Smoking cigarettes increases blood COHb by about 5 percent for each pack smoked per day.

Over the short term, a blood COHb level as low as 4 to 6 percent can decrease a healthy, young person's capacity to exercise. Over the longer term, an increased COHb level, such as that seen in smokers, alters the heart's structure. The heart must work overtime to make up for CO's suffocating effects on respiratory function.

38.2 The Nature of Respiration

✔ All animals must supply their cells with oxygen and rid their body of waste carbon dioxide.

✔ A variety of anatomical, behavioral, and physiological traits allow animals to extract oxygen from air or water around them.

Gas Exchanges

In Chapter 7 you learned about aerobic respiration, an energy-releasing pathway that requires oxygen (O_2) and produces carbon dioxide (CO_2) as summarized in the equation below:

$$\underset{\text{glucose}}{C_6H_{12}O_6} + \underset{\text{oxygen}}{6O_2} \longrightarrow \underset{\text{carbon dioxide}}{6CO_2} + \underset{\text{water}}{6H_2O}$$

This chapter focuses on **respiration**, the physiological processes that supply body cells with oxygen from the environment, and deliver waste carbon dioxide from cells to the environment. Respiration depends upon the tendency of gaseous oxygen (O_2) and carbon dioxide (CO_2) to follow their concentration gradients and diffuse between external and internal environments.

Gases enter and leave the animal body by diffusing across a thin, moist layer of cells called the **respiratory surface** (FIGURE 38.2A). A typical respiratory surface is one or two cell layers thick. The respiratory surface is thin because gases diffuse quickly only over very short distances. It must be moist because a gas can only diffuse quickly across a living cell's lipid bilayer when it is in solution.

A second exchange of gases occurs internally, at the plasma membrane of body cells (FIGURE 38.2B). Oxygen diffuses from the interstitial fluid into a cell, and carbon dioxide diffuses in the opposite direction. In invertebrates without a circulatory system, oxygen that crosses the respiratory surface reaches body cells by diffusion. In many invertebrates and all vertebrates, a circulatory system speeds movement of gases between the respiratory surface and body cells.

FIGURE 38.3 Horseshoe crab hemolymph colored blue by the respiratory protein hemocyanin. This copper-containing protein turns blue when it binds oxygen, whereas our respiratory protein, hemoglobin, turns bright red.

Factors Affecting Gas Exchange

Several factors affect how much gas diffuses across an animal's respiratory surface in any given interval.

Size matters: The greater the area of a respiratory surface, the more molecules can cross the surface at once. As a result, the area of a respiratory surface is often surprisingly large relative to the animal's body size. Branchings and foldings can help an extensive respiratory surface fit into a small volume.

Gas concentration gradients across this surface also affect the rate of gas exchange: The steeper the gradient across the membrane, the faster diffusion proceeds. As a result, many animals use muscle movements to keep oxygen-rich air or water from their environment flowing over their respiratory surface. In animals with a circulatory system, this system continuously delivers circulatory fluid (blood or hemolymph) to the respiratory surface and moves it away from that surface after gas exchange has taken place.

Respiratory proteins steepen the oxygen concentration gradient at a respiratory surface of many animals. A **respiratory protein** is a protein that transports oxygen; it has a metal ion or ions that bind O_2 where the O_2 concentration is high, and release it where O_2 concentration is low. **Hemoglobin**, which contains iron, is the main respiratory protein of vertebrates. Hemocyanin, a copper-containing protein, is the respiratory protein in some invertebrates, including octopuses and horseshoe crabs (FIGURE 38.3). When an O_2 molecule binds to a respiratory protein, that O_2 molecule no longer contributes to the O_2 concentration of blood. Thus, respiratory

FIGURE 38.2 Two sites of gas exchange. In some invertebrates, gases diffuse between the two sites. In others, and all vertebrates, a circulatory system facilitates movement of gases.

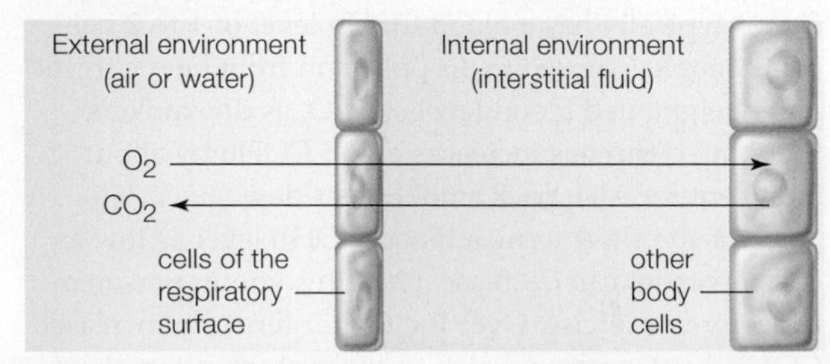

A Cells of the respiratory surface exchange gases with the external and internal environment.

B Other body cells exchange gases with the internal environment.

Respiratory Medium: Water
- High viscosity, harder to move
- Slower diffusion rate for gases
- Lower O_2 concentration
- O_2 concentration varies with flow rate and temperature

Respiratory Medium: Air
- Low viscosity, easier to move
- Higher diffusion rate for gases
- Higher O_2 concentration
- O_2 concentration constant at a given altitude

FIGURE 38.4 Properties of water and air as respiratory media.

proteins lower the effective O_2 concentration in blood and encourage diffusion of O_2 into the blood.

Respiratory Medium—Air or Water?

The respiratory medium, the environmental substance with which an animal exchanges gases, is either water or air (**FIGURE 38.4**). Water is 50 times more viscous (thicker) than air, so moving it over a respiratory surface requires more effort than moving air does. Oxygen also diffuses more slowly through water than through air. In addition, at any given temperature, a volume of water holds much less O_2 than an equivalent volume of air. As a result of these factors, obtaining sufficient O_2 from water requires a greater expenditure of energy than obtaining it from air.

Air is also a more reliable source of O_2. At present, the atmosphere consists of 21 percent O_2 everywhere on Earth. There is less atmosphere at high altitudes than lower ones, but at any given altitude, the proportion of gases is constant. By contrast, the proportion of O_2 in aquatic environments varies widely. Warm water holds less oxygen than cold water, and oxygen dissolves more easily in fast-moving water than still water. Thus, changes in water temperature or flow rate affect oxygen availability. When water temperature rises too high or moving water becomes stagnant, aquatic species that have high oxygen needs can suffocate.

hemoglobin Iron-containing respiratory protein.
respiration Physiological process by which an animal body supplies cells with oxygen and disposes of their waste carbon dioxide.
respiratory protein A protein that reversibly binds oxygen when the oxygen concentration is high and releases it when oxygen concentration is low. Hemoglobin is an example.
respiratory surface Moist surface across which gases are exchanged between animal cells and the external environment.

TAKE-HOME MESSAGE 38.2

What is respiration and what factors influence it?

✔ Respiration supplies cells with oxygen for aerobic respiration and removes carbon dioxide wastes.

✔ Gases are exchanged by diffusion across a respiratory surface: a thin, moist membrane.

✔ The area of a respiratory surface and the steepness of the concentration gradients across it influence the rate of exchange.

✔ Air and water have different properties as respiratory media. Aquatic animals have to expend more energy to obtain oxygen, and the amount of oxygen available can be a limiting factor.

38.3 Invertebrate Respiration

✔ Some invertebrates exchange gases across their entire body surface, but most have specialized organs of gas exchange.

A Jellyfish and other cnidarians have no respiratory organs. All cells in these animals lie close to the body surface and gas exchange takes place by diffusion across that surface.

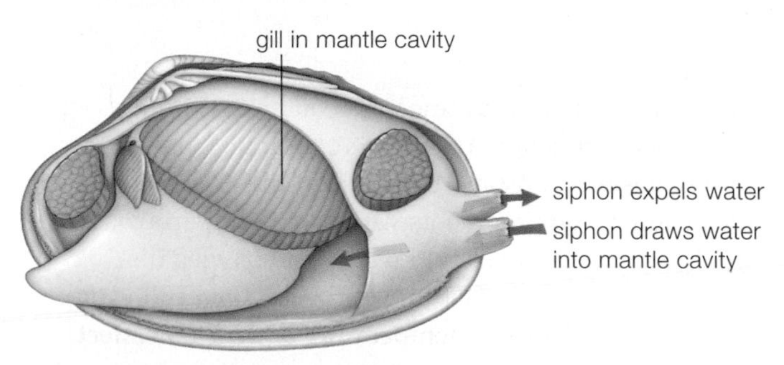

B Clam, a mollusk with internal gills.

C Sea slug, a mollusk with external gills.

FIGURE 38.5 Respiratory surfaces of aquatic invertebrates.

Respiration in Damp or Aquatic Habitats

The oldest invertebrate groups, such as cnidarians and flatworms, have neither respiratory nor circulatory organs (FIGURE 38.5A). These animals live in aquatic or continually damp land environments where their gas exchange needs are met entirely by **integumentary exchange**: diffusion of gases across their outer body surface. Animals that lack both respiratory and circulatory organs are either small and flat or, when larger, have cells arranged in thin layers. In annelids such as earthworms and sludge worms, a closed circulatory system distributes gases that diffuse across the integument. Integumentary exchange also supplements respiratory organs in many gilled invertebrates.

Gills evolved independently in several invertebrate groups, as well as in vertebrates. **Gills** are filamentous or platelike respiratory organs that increase the surface area available for gas exchange. As hemolymph or blood passes through gills, it exchanges gases with the water surrounding the gill.

Gills may be internal or external. Most aquatic mollusks have a gill inside their mantle cavity (FIGURE 38.5B). Sea slugs, which have gills on their body surface (FIGURE 38.5C), are an exception. Many aquatic arthropods such as lobsters and crabs have feathery gills hidden beneath their exoskeleton, where the delicate tissues are protected from damage. These gills evolved as modified branches of limbs.

Respiratory Adaptations to Life on Land

Some land-dwelling invertebrates such as earthworms and nematodes rely solely on integumentary exchange. The need to keep their body surface damp restricts them to moist locations.

Snails and slugs that spend time on land have a lung instead of, or in addition to, a gill. A **lung** is a saclike internal respiratory organ. Inside it, branching tubes deliver air to a respiratory surface supplied by many blood vessels. In snails and slugs, a pore at the side of the body can be opened to allow air into the lung (FIGURE 38.6), or shut to conserve water.

The most successful air-breathing land invertebrates are insects and arachnids, such as spiders. They have a hard integument that helps conserve water but also blocks gas exchange. Insects and some spiders have a **tracheal system** that consists of repeatedly branching, air-filled tubes reinforced with chitin (FIGURE 38.7). Tracheal tubes start at spiracles, which are small openings across the integument. There is usually a pair of spiracles per segment: one on each side of the body. Spiracles can be opened or closed to prevent water

FIGURE 38.6 Land slug, showing the pore that leads to its lung. Compare to **FIGURE 24.19**.

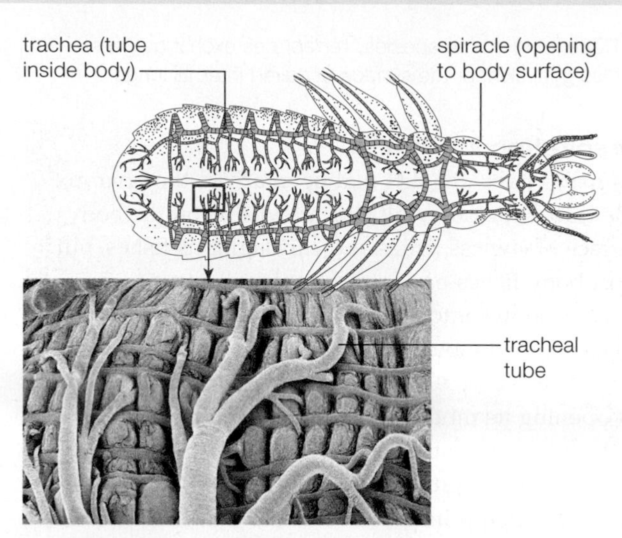

FIGURE 38.7 Insect tracheal system. Air-filled tracheal tubes reinforced by rings of chitin carry air deep inside the body.

loss. Some insecticides kill insects by clogging their spiracles. For example, horticultural oils sprayed on fruit trees kill scale bugs, aphids, and mites by clogging their spiracles and smothering them.

The tips of the finest tracheal branches deliver gases to hemolymph in which gases can dissolve. Oxygen and carbon dioxide diffuse between hemolymph and adjacent cells. Tracheal tubes end right next to cells, so most insects have no need for a respiratory protein such as hemoglobin to carry gases.

Some insects "breathe" by forcing air into and out of tracheal tubes. For example, when a grasshopper's abdominal muscles contract, organs press on the pliable tracheal tubes and force air out of them. When these muscles relax, pressure on tracheal tubes decreases, the tubes widen, and air rushes in.

Spiders and scorpions typically have one or more book lungs in addition to or instead of tracheal tubes. In a book lung, air and hemolymph exchange gases across thin sheets of tissue (**FIGURE 38.8**). The air enters the body through a spiracle. Hemocyanin in a spider's hemolymph picks up oxygen and turns blue-green as it passes through a book lung. It gives up oxygen and becomes colorless in body tissues.

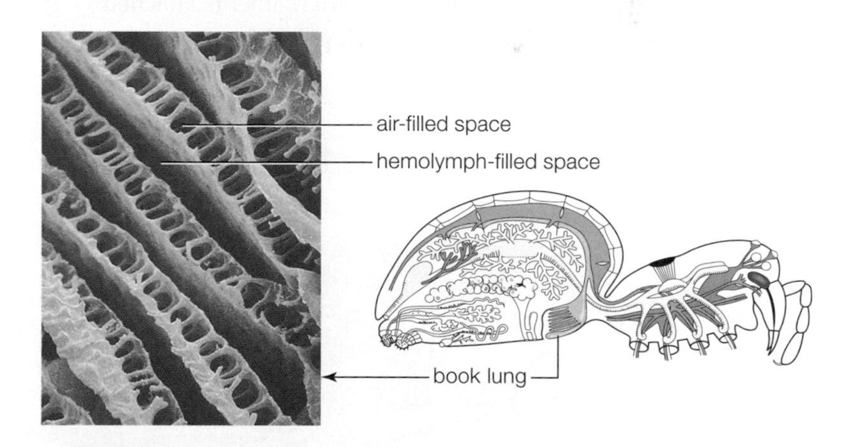

FIGURE 38.8 A spider's book lung. The lung contains many thin sheets of tissue, somewhat like the pages of a book. As hemolymph moves through spaces between the "pages," it exchanges gases with air in adjacent spaces.

TAKE-HOME MESSAGE 38.3

How do invertebrates exchanges gases with their environment?

✔ Some invertebrates do not have respiratory organs; all gas exchange takes place across the body wall. Such integumentary exchange also supplements respiration at gills in many invertebrates.

✔ Gills are filamentous organs that increase the surface area for gas exchange in aquatic habitats.

✔ Some land snails have a lung in their mantle cavity. Land arthropods have tracheal tubes or book lungs, respiratory organs that bring air deep inside their body.

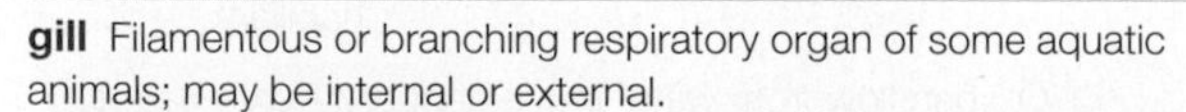

gill Filamentous or branching respiratory organ of some aquatic animals; may be internal or external.
integumentary exchange Gas exchange that occurs across an animal's outer body surface (integument).
lung Internal gas-exchange organ in some air-breathing animals.
tracheal system Of insects and some other land arthropods, tubes that convey gases between the body surface and internal tissues.

38.4 Vertebrate Respiration

✔ Depending on the species, vertebrates exchange gases across gills, skin, or the surface of paired internal lungs.

Respiration in Fishes

All fishes have gill slits that open across the pharynx (the throat region). Gill slits are visible on the body surface of jawless fishes and cartilaginous fishes, but most bony fishes have a moveable gill cover.

Water flows into a fish's mouth, continues into the pharynx, then moves out of the body through the gill slits (**FIGURE 38.9A**). A bony fish sucks water inward by opening its mouth, closing the cover over each gill, and contracting muscles that enlarge the oral cavity. Water is forced out over the gills when the fish closes its mouth, opens its gill covers, and contracts muscles that reduce the size of the oral cavity.

If you could remove the gill covers of a bony fish, you would see that the gills themselves consist of bony gill arches, each with many gill filaments attached (**FIGURE 38.9B**). Inside each gill filament are many capillary beds where gases in water are exchanged with gases in blood.

Water flowing over gills and blood flowing through gill capillaries move in opposite directions (**FIGURE 38.9C**). The result is a **countercurrent exchange**, a mechanism in which two fluids exchange substances while flowing in opposite directions. For the entire length of the capillary, the water next to the capillary holds more oxygen than the blood flowing inside it (**FIGURE 38.9D**). As a result, oxygen continually diffuses from the water into the blood. The blood becomes increasingly oxygenated as it passes through the capillary.

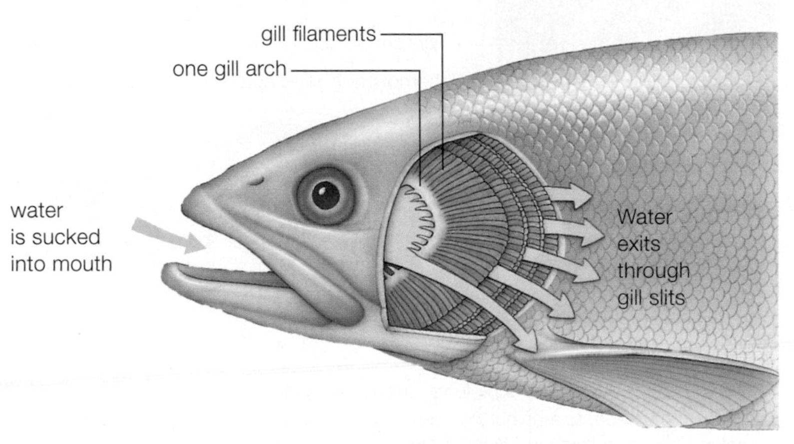

A Bony fish with its gill cover removed. Water flows in through the mouth, over the gills, then out through gill slits. Each gill has bony gill arches with many thin gill filaments attached.

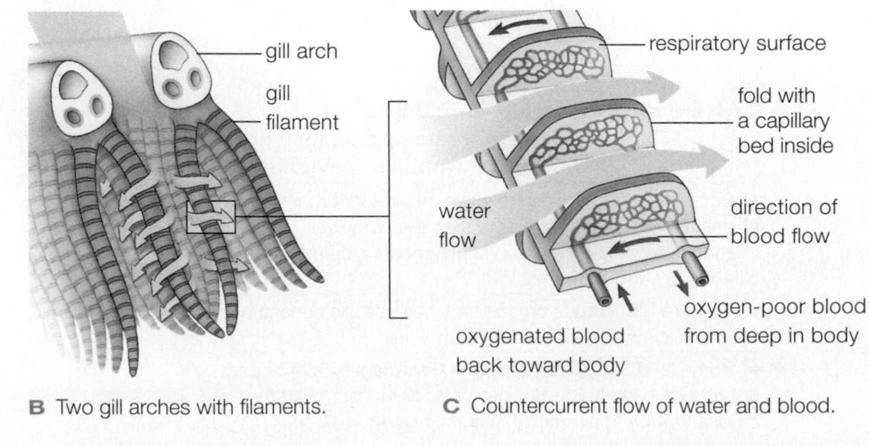

B Two gill arches with filaments.

C Countercurrent flow of water and blood.

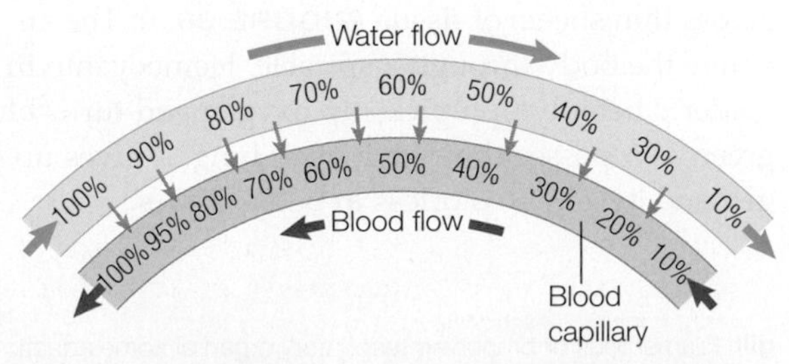

D Oxygen flow from water into a capillary. Percentages indicate the degree of oxygenation of water (blue) and blood (red). All along the capillary, oxygen flows down its concentration gradient from water into blood.

FIGURE 38.9 ▸**Animated** Structure and function of the gills of a bony fish.

Evolution of Paired Lungs

Most tetrapods have paired lungs. The first vertebrate lungs evolved from outpouchings of the gut wall in some bony fishes. Lungs may have helped these fishes survive short trips between ponds. They became increasingly important as aquatic tetrapods began moving onto land (Section 25.6).

Amphibian larvae typically have external gills. Most often, as the animal develops, these gills disappear and are replaced by paired lungs. Amphibians also carry out some gas exchange across their thin-skinned body surface. In all amphibians, most carbon dioxide that forms during aerobic respiration leaves the body across the skin.

All adult frogs have paired lungs. They do not use chest muscles to draw air into their lungs, as you do. Instead, frogs suck in air through their nostrils by lowering the floor of the mouth (**FIGURE 38.10A**). Then they close their nostrils and lift the floor of the mouth and throat. Compression of the oral cavity forces air into the lungs (**FIGURE 38.10B**).

Reptiles (including birds) and mammals are amniotes with a waterproof skin (Section 25.7). Their only

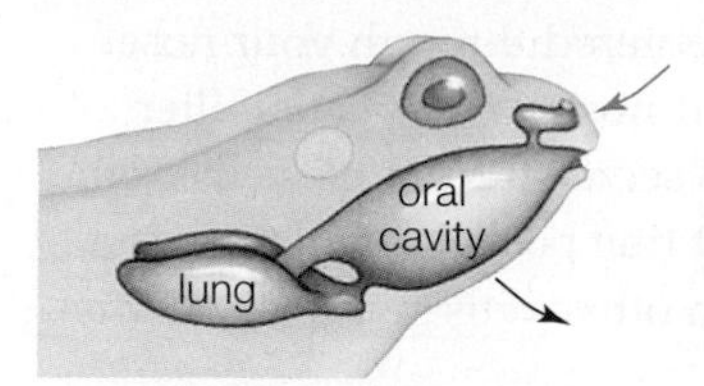

A The frog lowers the floor of its mouth, pulling air into the oral cavity through its nostrils.

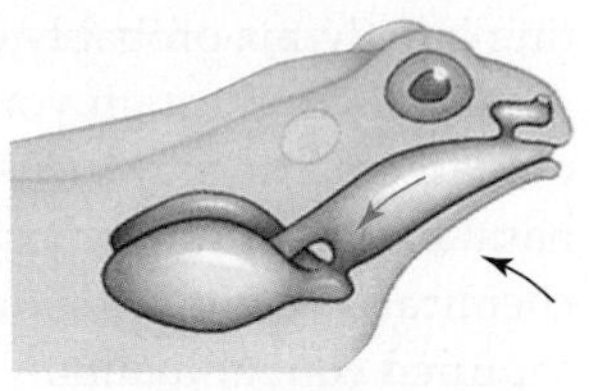

B Closing the nostrils and elevating the floor of the mouth pushes air into lungs.

FIGURE 38.10 ▸Animated How a frog fills its lungs. Black arrows show body wall movements. Blue arrows show air movement. Frogs push air into their lungs, rather than sucking it in as you do.

respiratory surface is the lining of two well-developed lungs. These lungs inflate when muscles increase the size of the thoracic cavity. As the cavity expands, pressure in the lungs declines and air is sucked inward.

In mammals, inhaled air flows through increasingly smaller airways until it reaches tiny sacs (**FIGURE 38.11**). Gas exchange occurs across the walls of these sacs. During exhalation, air retraces its path, flowing out the same way it came in. The lungs do not deflate completely, so a bit of stale air remains behind even after exhalation.

In birds, there are no "dead ends," and there is no stale air inside a lung. Birds have small, inelastic lungs that do not expand and contract when the bird breathes. Instead, large expandable air sacs attached to the lungs inflate and deflate (**FIGURE 38.12**). It takes two breaths to move air through this system. Oxygen-rich air flows through tiny tubes in the lung during inhalations and exhalations. The lining of these tubes serves as the bird's respiratory surface and continual movement of air over this surface increases the efficiency of gas exchange.

We turn next to the human respiratory system. Its operating principles apply to most vertebrates.

countercurrent exchange Exchange of substances between two fluids moving in opposite directions.

TAKE-HOME MESSAGE 38.4

How does the structure of vertebrate respiratory organs affect their function?

✔ Fishes exchange gases with water flowing over their gills. Movement of water and blood in opposite directions enhances this exchange.

✔ Amphibians exchange gases across their skin and push air from their mouth into their lungs. Amniotes suck air into their lungs by expanding the size of their thoracic cavity.

✔ Air flows into and out of mammalian lungs, but it flows continually through bird lungs.

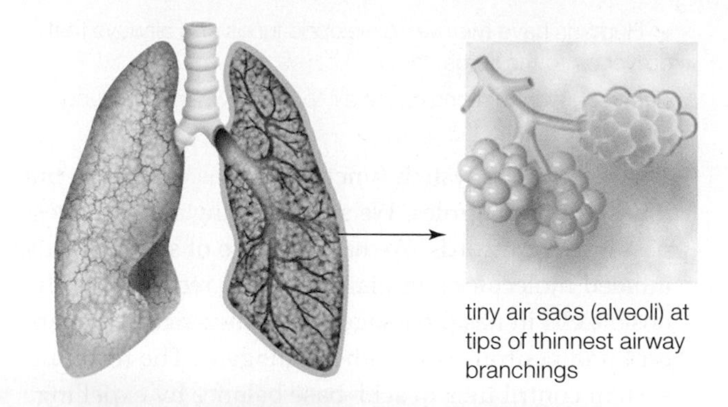

tiny air sacs (alveoli) at tips of thinnest airway branchings

FIGURE 38.11 Mammalian lungs. Expansion of the chest cavity draws air into branching tubes that end at tiny air sacs, where gas exchange occurs.

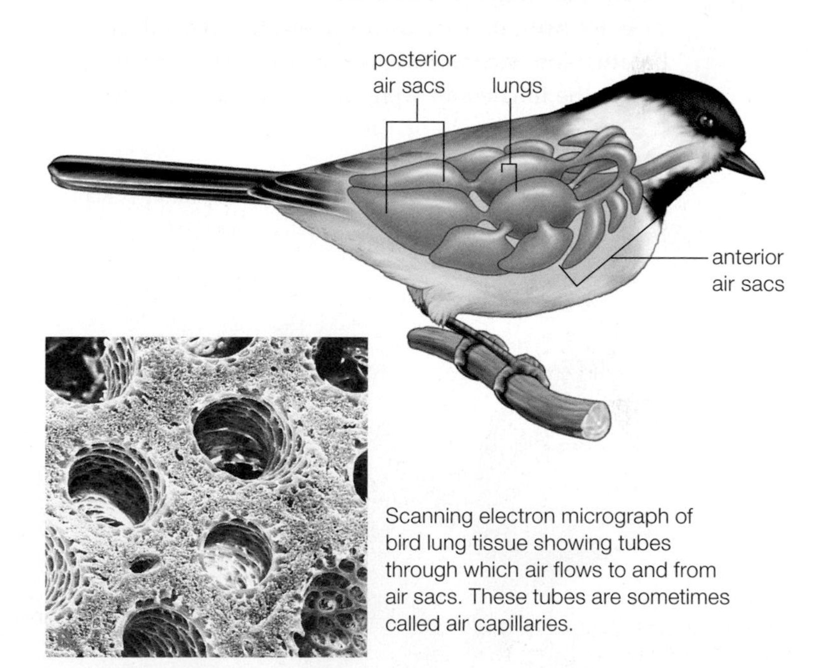

Scanning electron micrograph of bird lung tissue showing tubes through which air flows to and from air sacs. These tubes are sometimes called air capillaries.

FIGURE 38.12 ▸Animated Respiratory system of a bird.

Large air sacs attach to two small, inelastic lungs. Air flows in through many air tubes inside the lung, and into posterior air sacs.

The lining of the tiniest of the air tubes (shown in the micrograph), is the respiratory surface.

It takes two respiratory cycles to move air through the lungs and air sacs of a bird's respiratory system:

Inhalation 1— Muscles expand chest cavity, drawing air in through nostrils. Most of the air flowing in through the trachea goes to lungs and some goes to posterior air sacs.

Exhalation 1— Anterior air sacs empty. Air from the posterior air sacs moves into lungs.

Inhalation 2 — Air in lungs moves to anterior air sacs and is replaced by newly inhaled air.

Exhalation 2 — Air in anterior air sacs moves out of the body and air from posterior sacs flows into the lungs.

38.5 Human Respiratory System

✔ Humans have two well-developed lungs and airways that deliver air to the lungs.

✔ Muscle contractions cause air to be drawn into the lungs.

The respiratory system functions in gas exchange, but it has additional roles. We speak or sing as air moves past our vocal cords. We have a sense of smell because inhaled molecules stimulate olfactory receptors in the nose. Cells in nasal passages and other airways intercept and neutralize airborne pathogens. The respiratory system contributes to acid–base balance by expelling waste CO_2 that can make the blood more acid. Controls over breathing also help maintain body temperature; water evaporating from airways has a cooling effect.

From Airways to Alveoli

The Respiratory Passageways Take a deep breath. Now look at **FIGURE 38.13A** to get an idea of where the inhaled air went. If you are healthy and sitting quietly, air probably entered through your nose. As air moves through your nostrils, tiny hairs filter out large particles. Mucus secreted by some cells of the nasal lining captures most fine particles and airborne chemicals. Waving cilia on other cells sweep away the captured contaminants.

Air from the nostrils enters the nasal cavity, where it is warmed and moistened. From here, it flows next into the **pharynx**, or throat. It continues to the **larynx**, a short airway commonly known as the voice box because it contains a pair of vocal cords (**FIGURE 38.14**). A vocal cord is skeletal muscle with a cover of mucus-secreting epithelium. Contraction of the vocal cords narrows the gap between them, which is called the **glottis**.

When the glottis is wide open, air flows through it silently. When muscle contraction narrows the glottis, flow of air outward through the tighter gap makes vocal cords vibrate so they produce sounds. The tension on the cords and the position of the larynx

FIGURE 38.13 ▸**Animated** The human respiratory system and associated structures.

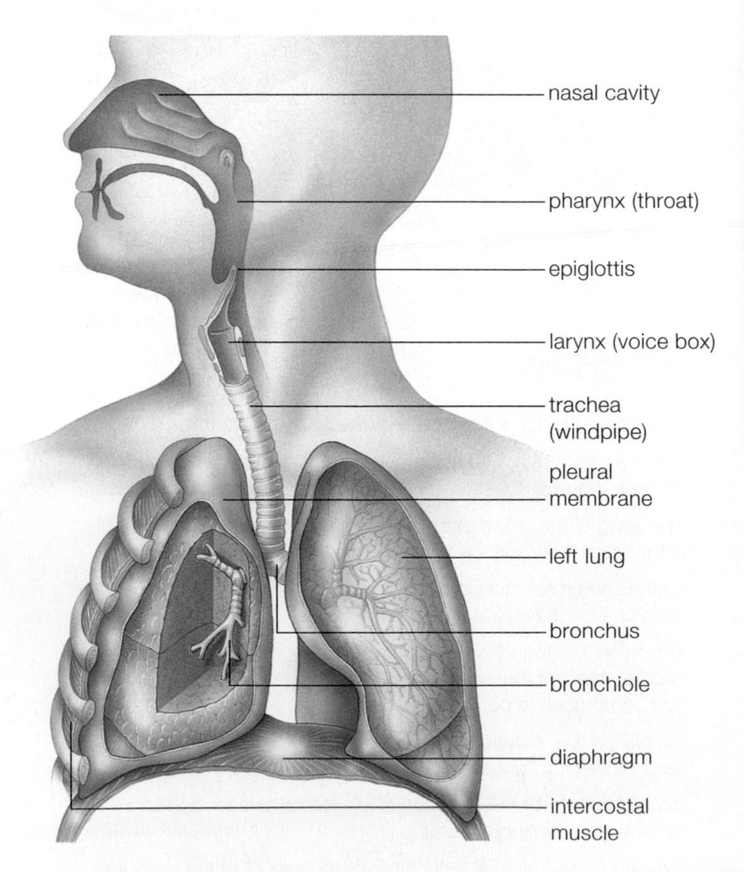

A Airways and respiratory muscles.

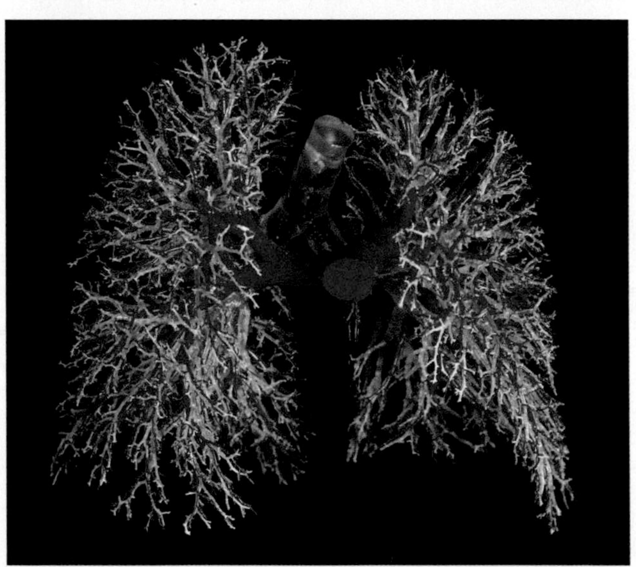

C Cast of airways (white) and blood vessels (red) in human lungs.

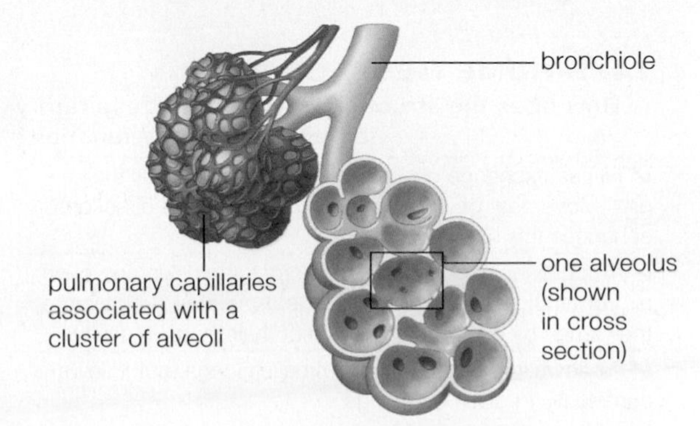

B Close-up view of alveoli and pulmonary capillaries.

determine the sound's pitch. To get a feel for how this works, place one finger on your "Adam's apple," the laryngeal cartilage that sticks out most at the front of your neck. Hum a low note, then a high one. You will feel the vibration of your vocal cords and how laryngeal muscles shift the position of your larynx.

At the entrance to the larynx is a flap of tissue called the **epiglottis**. When the epiglottis points up, air can move into or out of the **trachea**, or windpipe. When you swallow, the glottis closes and the epiglottis flops over to cover the larynx entrance, so food and fluids enter the esophagus. The esophagus connects the pharynx to the stomach.

The trachea branches into two airways, one leading to each lung. Each airway is a **bronchus** (plural, bronchi). Its epithelial lining has many ciliated cells and mucus-secreting cells that fend off respiratory tract infections. Bacteria and airborne particles stick to the mucus. Cilia sweep the mucus toward the throat, where it can be swallowed or expelled by coughing.

The Paired Lungs The lungs are spongy organs that reside in the chest, one on each side of the heart. The rib cage encloses and protects the lungs, and a two-layer-thick pleural membrane covers them. The pleural membrane's outer layer attaches to the wall of the chest cavity and its inner layer attaches to the outer surface of the lungs. A secreted fluid (pleural fluid) fills the small space between the pleural membrane's inner and out layers. This fluid is a lubricant that reduces friction between the membrane's two layers when the lungs expand and contract during respiration.

Air that enters a lung through a bronchus moves through finer and finer branchings of a "bronchial tree." The branches are **bronchioles**. The tips of the finest bronchioles end in respiratory **alveoli** (singular, alveolus), little air sacs where gases are exchanged (**FIGURE 38.13B**).

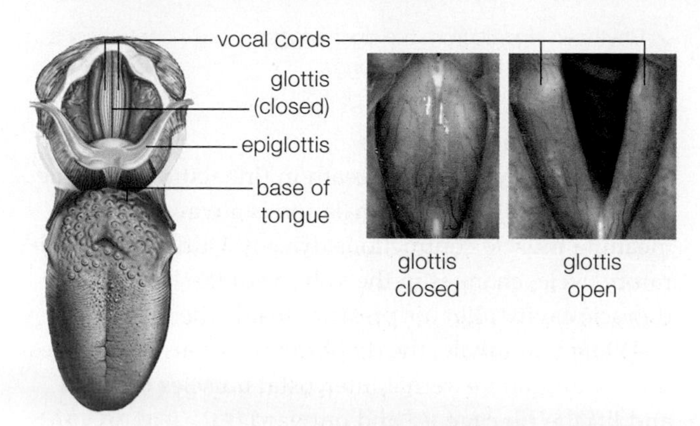

FIGURE 38.14 Vocal cords in the larynx. Action of skeletal muscle in the cords alters the width of the glottis, the gap between the cords.

FIGURE IT OUT Is the glottis open or closed when you are swallowing?

Answer: Closed

Branching blood vessels are associated with the branching airways (**FIGURE 38.13C**). The smallest of these vessels, the pulmonary capillaries, are adjacent to the alveoli. As blood flows through a pulmonary capillary, it exchanges gases with air inside an alveolus. Remember that the pulmonary circuit of the human circulatory system pumps oxygen-poor blood to the lungs and returns oxygen-rich blood to the heart. Systemic circulation then transports oxygen-rich blood to body tissues and returns oxygen-poor, carbon dioxide–rich blood to the heart.

Muscles and Respiration The **diaphragm**, a broad dome-shaped smooth muscle beneath the lungs, partitions the coelom into a thoracic cavity and an abdominal cavity. It is the only smooth muscle that can be voluntarily controlled. You can make your diaphragm contract by deliberately inhaling. The diaphragm and **intercostal muscles**, which are skeletal muscles between the ribs, interact to change the volume of the thoracic cavity during breathing.

TAKE-HOME MESSAGE 38.5

What roles do the components of the human respiratory system play?

✔ In addition to gas exchange, the human respiratory system acts in the sense of smell, voice production, body defenses, acid–base balance, and temperature regulation.

✔ Air enters through the nose or mouth. It flows through the pharynx (throat) and larynx (voice box) to a trachea that branches into two bronchi, one to each lung. Inside each lung, additional branching airways deliver air to alveoli, where gases are exchanged with pulmonary capillaries.

alveolus Tiny sac; in the mammalian lung, site of gas exchange at the tip of a bronchiole.
bronchiole Airway that leads from a bronchus to the alveoli.
bronchus Airway connecting the trachea to a lung.
diaphragm Smooth muscle between the thoracic and abdominal cavities; contracts during inhalation.
epiglottis Tissue flap that covers airway during swallowing to prevent food from entering airways.
glottis Opening formed when vocal cords in the larynx relax.
intercostal muscles The skeletal muscles between the ribs; help change the volume of the thoracic cavity during breathing.
larynx Short airway containing the vocal cords (voice box).
pharynx Tube connecting mouth and digestive tract; in vertebrates it is the throat.
trachea Airway to the lungs; windpipe.

38.6 How We Breathe

✔ Rhythmic signals from the brain trigger muscle contractions that cause air to flow into the lungs.

The Respiratory Cycle

A **respiratory cycle** is one breath in (inhalation) and one breath out (exhalation). Inhalation is always active, meaning muscle contractions drive it. During the respiratory cycle, changes in the volume of the lungs and thoracic cavity alter air pressure inside the lungs.

When you inhale, the diaphragm flattens and moves downward, and external intercostal muscles contract and lift the rib cage up and outward (FIGURE 38.15A). Together, these actions cause the thoracic cavity to expand; as it does, so do the lungs. Pressure in the alveoli falls below atmospheric pressure, and air follows the pressure gradient into the airways.

Exhalation is usually passive. When muscles that cause inhalation relax, the lungs passively recoil and lung volume decreases. This decrease in volume compresses alveoli, so air pressure inside them increases. Air follows this pressure gradient and flows out of the lungs (FIGURE 38.15B).

Exhalation becomes active only during vigorous exercise, or when you consciously expel air. During active exhalation, internal intercostal muscles contract, pulling the thoracic wall inward and downward. At the same time, muscles of the abdominal wall contract, causing intra-abdominal pressure to increase and exerting an upward-directed force on the diaphragm. These actions cause the volume of the thoracic cavity to decrease, so air is forced out of the lungs.

Inward flow of air

Outward flow of air

A **Inhalation.** Diaphragm contracts, moves down. External intercostal muscles contract, lift rib cage upward and outward. Lung volume expands.

B **Exhalation.** Diaphragm, external intercostal muscles return to resting positions. Rib cage moves down. Lungs recoil passively.

FIGURE 38.15 ▸**Animated** Changes in the size of the thoracic cavity during the respiratory cycle. The x-ray images reveal how inhalation and expiration change the lung volume. FIGURE IT OUT What effect does contraction of the diaphragm have on the volume of the thoracic cavity? Answer: It increases the volume.

Respiratory Volumes

Total lung volume, the maximum amount of air that the lungs can hold, averages 5.7 liters in men and 4.2 liters in women. Usually lungs are less than half full. **Vital capacity**, the maximum volume that can move in and out in one cycle, is one measure of lung health. **Tidal volume**—the volume that moves in and out in a normal respiratory cycle—is about 0.5 liter (FIGURE 38.16). Your lungs never fully deflate, so the air inside them always is a mix of freshly inhaled air and stale air left behind during the previous exhalation. Even so, there is plenty of oxygen for gas exchange.

Control of Breathing

Neurons in the medulla oblongata of the brain stem act as the pacemaker for inhalation, initiating an action potential 10–20 times per minute. Nerves deliver the action potentials to the diaphragm and intercostal muscles, these muscles contract, and you inhale. Between signals, the muscles relax and you exhale.

Breathing patterns change with activity level. Active muscle cells produce additional CO_2 that enters blood, where it combines with water to forms carbonic acid (Section 2.6). Chemoreceptors in the walls of carotid arteries and the aorta detect the rise in acidity and signal the brain (FIGURE 38.17). In response, you breathe faster and deeper, so more carbon dioxide is expelled.

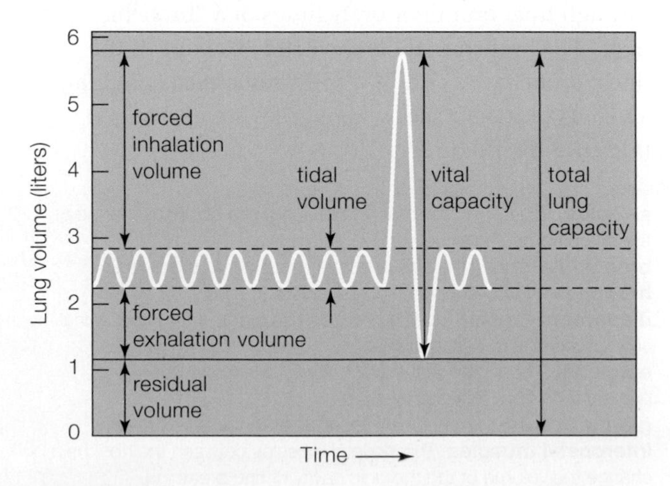

FIGURE 38.16 ▸**Animated** Respiratory volumes. In normal breathing, the tidal volume of air entering and leaving the lungs is only 0.5 liter. Lungs never deflate completely. Even with a forced exhalation, a residual volume of air remains in them.

Chemoreceptors in the artery walls also signal the medulla oblongata when the O_2 concentration in the blood falls to a life-threatening level. This control mechanism usually comes into play only in people with severe lung diseases and at very high altitudes, where there is little oxygen in the air.

Reflexes such as swallowing or coughing can briefly halt breathing. Signals that travel along sympathetic nerves make you breathe faster when you are frightened (Section 32.8). Breathing patterns can also be deliberately altered, as when you hold your breath, or break normal breathing rhythm to talk or sing.

Choking—A Blocked Airway

Choking occurs when food or another object enters and blocks the larynx. A person who is choking cannot move air through the larynx, and so cannot breathe, cough, or speak. If you suspect someone is choking, encourage them to cough. If the person can do so, their airway is not fully obstructed and coughing will probably clear it.

If the person is choking, a rescuer can help clear the airway and keep the person from dying from lack of air. First aid for choking involves two types of maneuvers. Blows to the back can help dislodge the foreign material with little risk of injury to internal organs (**FIGURE 38.18A**). If back blows are ineffective, thrusts to the choker's abdomen (the Heimlich maneuver) can raise intra-abdominal pressure, forcing the diaphragm upward. Raising air pressure inside the lungs by this maneuver can dislodge the object, allowing the victim to resume breathing normally (**FIGURE 38.18B**).

respiratory cycle One inhalation and one exhalation.
tidal volume The volume of air that flows into and out of the lungs during a normal inhalation and exhalation.
vital capacity Amount of air moved in and out of lungs with forced inhalation and exhalation.

TAKE-HOME MESSAGE 38.6
What happens when we breathe?

- ✔ Inhalation is always an active process. Contraction of the diaphragm and external intercostal muscles increase the volume of the thoracic cavity. This reduces air pressure in alveoli below atmospheric pressure, so air moves inward.
- ✔ Exhalation is usually passive. As muscles relax, the thoracic cavity shrinks back down, air pressure in alveoli rises above atmospheric pressure, and air moves out.
- ✔ Only some of the air in the lungs is replaced with each breath. The lungs are never fully emptied of air.
- ✔ The brain controls the rate and depth of breathing.

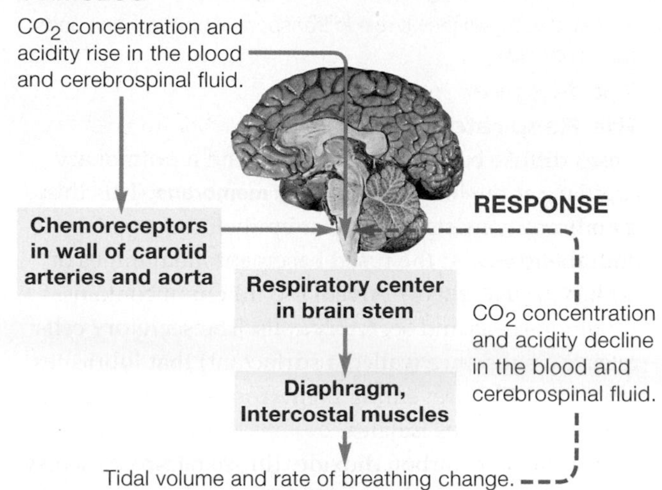

FIGURE 38.17 Respiratory response to an increase in CO_2. When increased CO_2 excites chemoreceptors in arteries and the brain, the brain's respiratory center calls for deeper and more frequent breaths. As a result, more CO_2 is expelled.

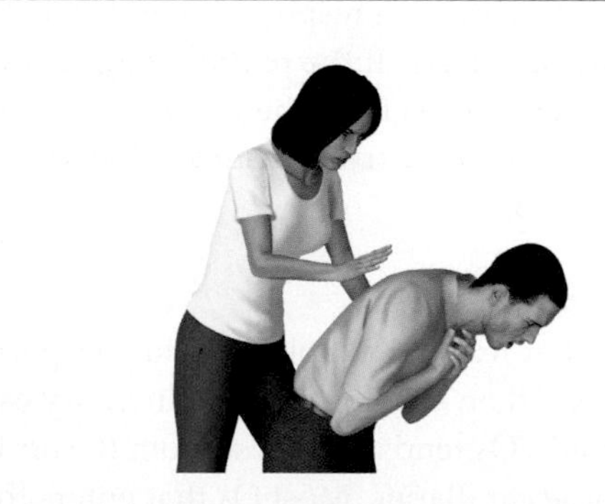

A Back blows. Have the person lean forward. Use the heel of your hand to strike between the shoulder blades.

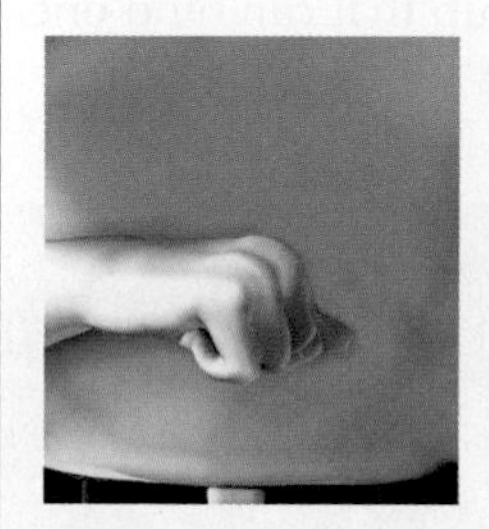
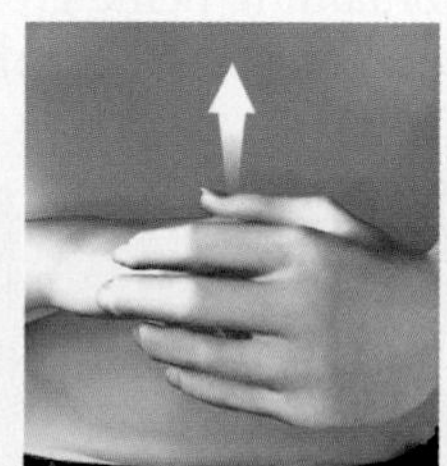

B Abdominal thrusts. Stand behind the person and place one fist below the rib cage, just above the navel, with your thumb facing inward. Cover the fist with your other hand and thrust inward and upward with both fists.

FIGURE 38.18 ▶**Animated** Maneuvers to assist a conscious adult who is choking. The Red Cross recommends that rescuers ask if a victim is choking and wants help. If the person nods, the rescuer should alternate between a series of five back blows and five abdominal thrusts.

38.7 Gas Exchange and Transport

✔ Gases are exchanged by diffusion in alveoli.

✔ Red blood cells play a role in transport of both oxygen and carbon dioxide.

The Respiratory Membrane

Gases diffuse between an alveolus and a pulmonary capillary at the lung's **respiratory membrane**. This thin membrane consists of alveolar epithelium, capillary endothelium, and the fused basement membranes of both (FIGURE 38.19). Alveolar epithelium contains squamous cells and secretory cells. The secretory cells release a substance (called a surfactant) that lubricates alveolar walls, preventing them from sticking together when the alveolus inflates.

Oxygen and carbon dioxide diffuse passively across the respiratory membrane. Which way these gases move depends on their concentration gradients across the membrane, or as we say for gases, partial pressure gradients. The partial pressure of a gas is its contribution to the pressure exerted by a mix of gases. It is measured in millimeters of mercury (mm Hg). Just as a solute tends to diffuse in response to its concentration gradient, so does a gas. If the partial pressure of a gas differs between two regions, the gas will diffuse from the region of higher partial pressure to the region of lower partial pressure.

Oxygen Transport

Inhaled air that reaches alveoli has a higher partial pressure of O_2 than does blood in pulmonary capillaries. As a result, O_2 tends to diffuse from the air into the blood of these capillaries. Most O_2 that enters the blood diffuses into red blood cells, where it binds to hemoglobin. Hemoglobin has four globin subunits, each with an iron-containing heme group that can bind one oxygen molecule (FIGURE 38.20). When O_2 is bound

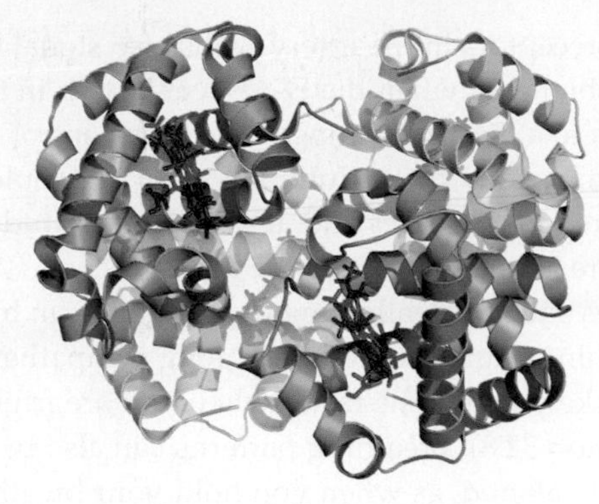

FIGURE 38.20 Structure of hemoglobin, the oxygen-transporting protein of red blood cells. It consists of four globin chains, each associated with an iron-containing heme group, color-coded red.

to one or more of hemoglobin's heme groups, we refer to the molecule as **oxyhemoglobin**.

Heme binds oxygen reversibly, and releases it where the partial pressure of O_2 is lower than that in the alveoli. This occurs in the body tissues serviced by systemic capillaries: In FIGURE 38.21 compare the O_2 partial pressures inside alveoli and at the start of systemic capillaries. Metabolically active tissues also have other characteristics that encourage oxygen release: high temperature, low pH, and high CO_2 partial pressure.

Myoglobin, also an iron-containing respiratory protein, helps cardiac muscle and some skeletal muscles take up oxygen. Structurally, myoglobin resembles a globin chain, but its heme binds oxygen more tightly than globin. The O_2 that hemoglobin gives up near a cardiac muscle cell diffuses into the cell and binds to myoglobin inside it. When the cell requires more oxygen than can be supplied by blood flow, as during periods of intense exercise, O_2 released by myoglobin provides a backup source of this essential gas.

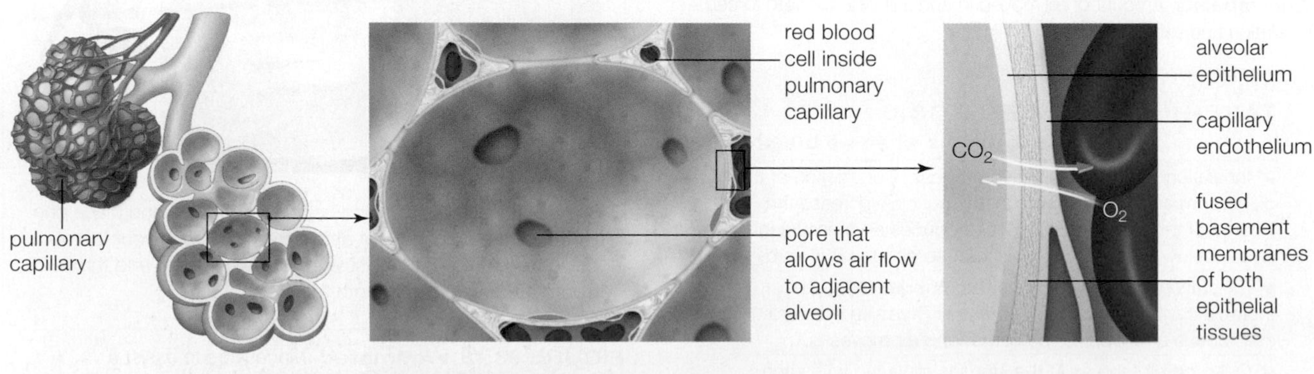

A Surface view of alveoli and associated pulmonary capillaries.

B Cutaway view of one alveolus and adjacent pulmonary capillaries.

C Three components of the respiratory membrane.

FIGURE 38.19 ▶Animated Zooming in on the respiratory membrane in human lungs.

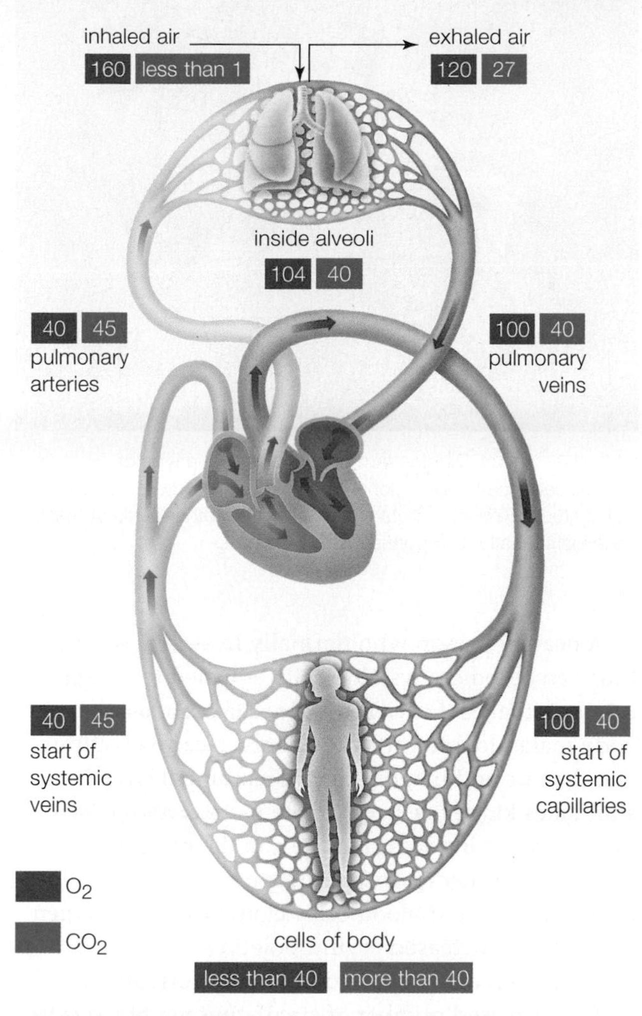

FIGURE 38.21 ▶**Animated** Partial pressures (in mm Hg) for oxygen (red boxes) and carbon dioxide (blue boxes) in the atmosphere, blood, and tissues.

Carbon Dioxide Transport

Carbon dioxide diffuses into the blood from any tissue where its partial pressure is higher than it is in blood. This occurs in the body tissues serviced by systemic capillaries (**FIGURE 38.21**).

Carbon dioxide is transported to the lungs in three forms. About 10 percent remains dissolved in plasma. Another 30 percent reversibly binds with hemoglobin and forms carbaminohemoglobin ($HbCO_2$). However, most CO_2 that diffuses into the plasma is transported as bicarbonate (HCO_3^-). **FIGURE 38.22** shows this

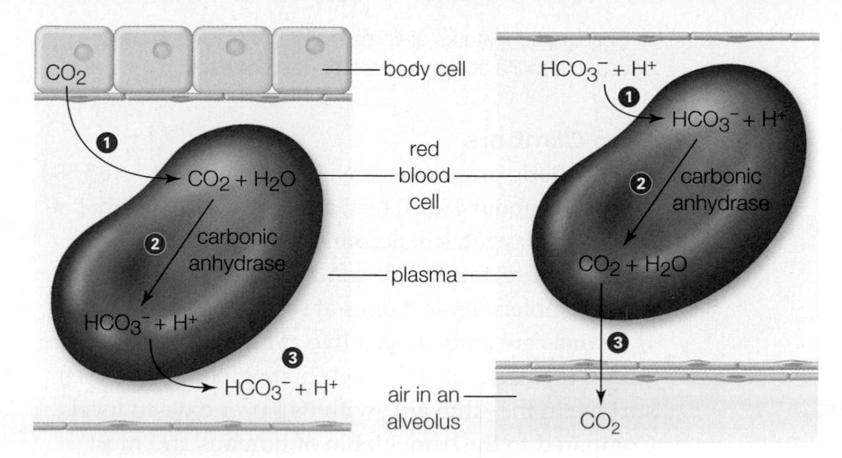

A At a systemic capillary bed:

❶ CO_2 diffuses from a body cell into plasma, then into a red blood cell.

❷ Carbonic anhydrase catalyzes formation of bicarbonate (HCO_3^-).

❸ Bicarbonate diffuses into plasma.

B At a pulmonary capillary bed:

❶ Bicarbonate diffuses from plasma into a red blood cell.

❷ Carbonic anhydrase catalyzes formation of water and CO_2.

❸ CO_2 diffuses across respiratory membrane into air in an alveolus.

FIGURE 38.22 Main mechanism of carbon dioxide transport and exchange. A lesser amount of CO_2 travels to the lungs bound to hemoglobin.

mechanism of transport. Carbon dioxide diffuses from a body cell into a systemic capillary, where it enters a red blood cell. Inside the cell, the enzyme **carbonic anhydrase** catalyzes a reaction between the CO_2 and water. The product of this reaction, carbonic acid (H_2CO_3), dissociates into bicarbonate and H^+ (**FIGURE 38.22A**). Carbonic anhydrase can catalyze the formation of a million molecules of bicarbonate a second. This reaction also occurs spontaneously in plasma, but at a much slower rate.

In pulmonary capillaries, where the partial pressure of CO_2 is relatively low, carbonic anhydrase catalyzes the reverse reaction, forming water and CO_2 from bicarbonate (**FIGURE 38.22B**). The CO_2 diffuses across the respiratory membrane and into the air in an alveolus. It then leaves the body in exhalations.

carbonic anhydrase Enzyme in red blood cells that speeds the reversible conversion of CO_2 into bicarbonate and H^+.
oxyhemoglobin Hemoglobin with oxygen bound to it.
respiratory membrane Membrane consisting of alveolar epithelium, capillary endothelium, and their fused basement membranes; site of gas exchange in lungs.

TAKE-HOME MESSAGE 38.7

How are gases transported in blood?

✔ Most oxygen in blood is bound to hemoglobin, which binds oxygen in alveoli where O_2 partial pressure is high, and releases it in tissues where O_2 partial pressure is lower.

✔ Inside red blood cells, an enzyme speeds the conversion of carbon dioxide to bicarbonate. Most carbon dioxide is transported in blood in the form of bicarbonate.

38.8 Respiratory Adaptations

✔ Specialized features of some respiratory systems adapt organisms to life at high altitude or deep dives.

High Climbers

Atmospheric pressure decreases with altitude. At 5,500 meters, or about 18,000 feet, air pressure is half that at sea level. Oxygen still accounts for 21 percent of the total pressure, but the air contains about half as many oxygen molecules as it does at sea level.

Llamas are animals that live at high altitudes in the Andes (**FIGURE 38.23**). Their hemoglobin helps them survive in the "thin air," with its lower oxygen level. Compared to the hemoglobin of humans and most other mammals, llama hemoglobin binds oxygen more efficiently. Also, the lungs and the heart of a llama are unusually large relative to the animal's body size.

Most people live at lower altitudes where there is plenty of oxygen. When they ascend too fast to high altitudes, delivery of oxygen to their cells plummets. Hypoxia, or cellular oxygen deficiency, is the result.

In an acute compensatory response to hypoxia, the brain signals the heart and respiratory muscles to work harder. People breathe faster and more deeply than usual (they hyperventilate). As a result, CO_2 is exhaled faster than it forms, and ion balances in the cerebrospinal fluid get skewed. Shortness of breath, a pounding heart, dizziness, nausea, and vomiting are symptoms of the resulting altitude sickness.

FIGURE 38.23 Saturation curve for hemoglobin of humans, llamas, and other mammals. **FIGURE IT OUT** At what partial pressure of oxygen do half the heme groups in human blood have oxygen bound?

Answer: 30 mm Hg

FIGURE 38.24 Tibetan farmer with his yaks, a type of domesticated cattle. Directional selection has adapted Tibetans to high altitude. Similarly Tibetan yaks differ from other cattle at many gene loci that affect responses to low oxygen.

A healthy person who normally lives at a low altitude can become physiologically adjusted to a high one, but it takes time. Through **acclimatization**, the body makes long-term adjustments in cardiac output, and the rate and magnitude of breathing. Hypoxia also stimulates kidney cells to secrete more **erythropoietin**. This hormone induces stem cells in the bone marrow to divide repeatedly, and induces descendant cells to differentiate as red blood cells. Under extreme oxygen deprivation, increased erythropoietin secretion can result in a sixfold rise in red blood cell formation.

The increased number of circulating red blood cells improves the oxygen-delivery capacity of blood, but it also puts a strain on the heart. The more blood cells there are, the thicker the blood, and the harder the heart has to work to propel blood through the circulatory system. The heart enlarges, and its stronger-than-normal contractions increase blood pressure, putting a person at risk for the health problems associated with chronic hypertension (Section 36.10).

Typically, members of long-established, high-altitude populations do not experience these ill effects, because natural selection has adapted these populations to their low-oxygen environment. People who live high in the Andes have an unusually high concentration of hemoglobin in their blood, a trait that allows them to maximize the amount of oxygen they acquire from each breath. It also puts them at an increased risk for heart problems. Tibetan plateau dwellers maintain a low hemoglobin level, but take more breaths per minute (**FIGURE 38.24**). Tibetans also make more nitric oxide. This gas encourages blood vessels to widen, increasing blood flow through lung capillaries. Although Andean and Tibetan populations face the same selective pressure (low oxygen), each evolved a different mechanism of countering that pressure. In Andeans, the blood composition was altered; in Tibetans, respiratory rate and blood flow were affected.

Deep-Sea Divers

Water pressure increases with depth. Human divers using tanks of compressed air risk nitrogen narcosis, sometimes called "raptures of the deep." As divers descend, gaseous nitrogen (N_2) dissolves in their interstitial fluid. In neurons, this dissolved nitrogen can disrupt signaling, causing euphoria and drowsiness. Effects increase with increasing depth.

Returning to the surface from a deep dive also has risks. As a diver ascends, the decrease in pressure causes N_2 to move from interstitial fluid into the blood, from which it is eliminated in exhalations. If a diver rises too fast, N_2 bubbles form in the diver's body. The resulting decompression sickness, also known as "the bends," usually begins with joint pain. Bubbles of N_2 slow the flow of blood to organs. If such bubbles form in the brain, heart, or lungs, the result can be fatal.

Humans who train to dive without oxygen tanks can remain submerged for about three minutes. So far, the human free diving record is about 200 meters (700 feet). Other animals can dive much deeper than that (**FIGURE 38.25**). As an air-breathing animal dives deeper and deeper, the weight of more and more water presses on its body. Lungs fully inflated with air would collapse inward under this pressure, so most diving animals move air out of their lungs and into cartilage-reinforced airways before they dive too deep.

Deep dives also require spending long intervals without access to air. The longest dive recorded for a leatherback turtle lasted a little more than an hour. Sperm whales stay submerged for two hours. How does a diving animal, whose lungs are emptied of air and who has no access to the surface, supply its cells with the oxygen they need to survive? First, before the animal dives, it breathes deeply. A sperm whale blows out about 80–90 percent of the air in its lungs with each exhalation; you exhale only about 15 percent. Deep breaths keep oxygen pressure in alveoli high, so more oxygen diffuses into the blood.

Species	Maximum Depth
Sperm whale	2,200 meters
Leatherback turtle	1,200 meters
Southern elephant seal	1,620 meters
Weddell seal	741 meters
Bottlenose dolphin	>600 meters
Emperor penguin	565 meters

FIGURE 38.25 Animal diving records. The photo shows a leatherback turtle.

Second, diving animals can store great amounts of oxygen inside their blood and muscles. They tend to have a large blood volume relative to their body size, a high red blood cell count, and a considerable amount of myoglobin in their muscles. A skeletal muscle of a bottlenose dolphin has about 3.5 times the amount of myoglobin that a comparable skeletal muscle in a dog has. A muscle in a sperm whale has 7 times as much as the dog muscle.

Third, during a dive, oxygen is preferentially distributed to the heart, brain, and other organs that require an uninterrupted supply of ATP. This preferential distribution is made possible by a system of valves that control flow to blood vessels in specific tissues. To make the best use of the limited oxygen supply, metabolic rate and heart rate decrease, as do cellular oxygen uptake and carbon dioxide formation.

Finally, whenever possible, a diving animal makes the most of its oxygen stores by sinking and gliding instead of actively swimming. It conserves energy by avoiding unnecessary movements.

TAKE-HOME MESSAGE 38.8

What are some adaptations that aid respiration in extreme environments?

✔ A big heart and lungs, hemoglobin with an altered oxygen affinity, or additional hemoglobin are adaptations to life at high altitudes, where oxygen is scarce.

✔ Traits such as a high red blood cell count, or a large amount of myoglobin in muscles allow some marine animals to hold their breath during long, deep dives.

acclimatization A body adjusts to a new environment; e.g., after moving from sea level to a high-altitude habitat.

erythropoietin Hormone secreted by kidneys; induces stem cells in bone marrow to give rise to red blood cells.

38.9 Respiratory Diseases and Disorders

FIGURE 38.26 Section from a smoker's lung, showing black carbon deposits.

✔ Genetic disorders, infectious disease, and lifestyle choices can increase the risk of respiratory problems.

Interrupted Breathing

A tumor or damage to the brain stem's medulla oblongata can affect respiratory controls and cause apnea. In this disorder, breathing repeatedly stops and restarts spontaneously, especially during sleep. Sleep apnea also occurs when the tongue, tonsils, or other soft tissue obstructs the upper airways. Breathing may stop for up to several seconds many times each night, preventing deep sleep and causing daytime fatigue. The risk of heart attacks and strokes rises with sleep apnea, because blood pressure soars when breathing stops. Obstructive sleep apnea can be reduced by changes in sleeping position or by wearing a mask that delivers pressurized air. Severe cases require removal of tissue that blocks airways.

Sudden infant death syndrome (SIDS) occurs when an infant does not awaken from an episode of apnea. A defect in the medulla oblongata, the brain's center for control of respiration, is associated with SIDS. Autopsies reveal that infants who died of SIDS typically have fewer receptors for the neurotransmitter serotonin than infants who died of other causes. The low number of serotonin receptors may impair the medulla's normal response to potentially deadly respiratory stress. There are also environmental risk factors. Maternal smoking during pregnancy heightens the risk. Infants who sleep on their back are at lower risk than stomach sleepers.

Lung Diseases and Disorders

Worldwide, about one in three people is currently infected by bacteria that can cause tuberculosis. Most of these people have no symptoms, but about 10 percent of them develop "active TB." They cough up bloody mucus, have chest pain, and find breathing difficult. If untreated, active TB can be fatal. Antibiotics can cure most infections, if taken diligently for at least six months. Unfortunately, multi-drug-resistant strains of *Myobacterium tuberculosis* are increasing in frequency.

Pneumonia is a general term for lung inflammation caused by an infection. Bacteria, viruses, and fungi can cause pneumonia. Typical symptoms include a cough, an aching chest, shortness of breath, and fever. An x-ray reveals lungs filled with fluid and white blood cells instead of air. Treatment and outcome depend on the type of pathogen.

Asthma occurs when an inhaled allergen or irritant triggers inflammation that constricts the airways so breathing becomes difficult. A tendency to have asthma is inherited, but avoiding potential irritants such as cigarette smoke and air pollutants can reduce the frequency of asthma attacks. An acute asthma attack is treated with inhaled drugs that cause dilation of smooth muscle around the airways.

Bronchitis is an inflammation of the ciliated, mucus-producing epithelium of the bronchi. The inflamed cells secrete extra mucus that triggers coughing. Bacteria colonize the mucus, leading to more inflammation, more mucus, and more coughing. Infectious bronchitis most often arises as a result of a viral infection of the upper respiratory tract. Chronic bronchitis most often occurs in tobacco smokers.

Emphysema is a disorder in which the thin, elastic alveolar walls disintegrate. As these walls disappear, the area of the respiratory surface declines. Over time, the lungs become distended and inelastic, causing a constant feeling of being short of breath.

CREDIT: (26) CDC.

Carbon Monoxide—A Stealthy Poison (revisited)

Although carbon monoxide binds tightly to hemoglobin, it does not bind irreversibly. If a person with carbon monoxide poisoning is moved into fresh air, about half of the CO bound to hemoglobin in his or her blood will be replaced by oxygen within four to five hours. The speed at which oxygen replaces carbon monoxide at binding sites can be accelerated by giving a person 100 percent oxygen.

To minimize your carbon monoxide exposure, do not smoke and avoid being in a closed area with people who do. Have wood-burning stove and fireplaces, and fossil fuel–burning heaters checked annually. Remember that any fossil fuel–powered engine releases CO, and should not be allowed to run indoors. If you can smell exhaust from a car, boat, heater, stove, or other device, you are also inhaling CO.

Smoking's Impact

Smoking tobacco has a wide range of negative health effects (**FIGURES 38.26** and **38.27**). It increases the risk of lung infections and many types of cancer. Tobacco smoke contains more than forty carcinogens (cancer-causing chemicals), and more than 80 percent of lung cancers occur in smokers. Once diagnosed with lung cancer, the majority of smokers die within a year. Female smokers are at an especially high risk of developing cancer. On average, their cancers appear earlier than men's, and with lower exposure to tobacco. Tobacco smokers also have a higher risk of cardiovascular disease.

Respiratory effects of marijuana smoke are less well studied than those of tobacco smoke. We do know that marijuana smoke contains carbon monoxide and an assortment of carcinogens, including arsenic and ammonia. Marijuana smokers get an unfiltered dose of these toxins. Nevertheless, the few studies that have been done show no significant impairment of respiratory function or increased risk for lung cancer with marijuana smoking only. On the other hand, people who smoke both marijuana and tobacco have more respiratory problems than tobacco-only smokers.

TAKE-HOME MESSAGE 38.9

What causes common respiratory problems?

✔ Apnea, or interrupted breathing, is caused by tissue obstructing airways or a defective respiratory control center.

✔ Tuberculosis, a bacterial disease, can be fatal, although most people have no symptoms. Pneumonia can be caused by many different pathogens.

✔ In asthma and bronchitis, airways become inflamed and constricted. In emphysema, alveolar sacs become distended and inelastic.

✔ Smoking tobacco increases the risk of lung cancer and other lung disorders.

Risks Associated With Smoking	Reduction in Risks by Quitting
Shortened life expectancy Nonsmokers live about 8.3 years longer than those who smoke two packs a day from their midtwenties on.	Cumulative risk reduction; after 10–15 years, the life expectancy of ex-smokers approaches that of nonsmokers.
Chronic bronchitis, emphysema Smokers have 4–25 times higher risk of dying from these diseases than do nonsmokers.	Greater chance of improving lung function and slowing down rate of deterioration.
Cancer of lungs Cigarette smoking is the major cause.	After 10–15 years, risk approaches that of nonsmokers.
Cancer of mouth 3–10 times greater risk among smokers.	After 10–15 years, risk is reduced to that of nonsmokers.
Cancer of larynx 2.9–17.7 times more frequent among smokers.	After 10 years, risk is reduced to that of nonsmokers.
Cancer of esophagus 2–9 times greater risk of dying from this.	Risk proportional to amount smoked; quitting should reduce it.
Cancer of pancreas 2–5 times greater risk of dying from this.	Risk proportional to amount smoked; quitting should reduce it.
Cancer of bladder 7–10 times greater risk for smokers.	Risk decreases gradually over 7 years to that of nonsmokers.
Cardiovascular disease Cigarette smoking a major contributing factor in heart attacks, strokes, and atherosclerosis.	Risk for heart attack declines rapidly, for stroke declines more gradually, and for atherosclerosis it levels off.
Impact on offspring Women who smoke during pregnancy increase their risk of miscarriage or of an underweight infant.	When smoking stops before fourth month of pregnancy, risk of stillbirth and lower birth weight eliminated.

FIGURE 38.27 Risks incurred by smokers and benefits of quitting.

summary

Section 38.1 Carbon monoxide is the most commonly inhaled poison. It binds to hemoglobin and prevents normal gas transport and exchange. Burning of fossil fuels and other organic material, including tobacco, releases carbon monoxide.

Section 38.2 **Respiration** is a physiological process by which O_2 enters the internal environment and CO_2 leaves it. Both gases diffuse across a **respiratory surface**. Enlarging the respiratory surface, moving air or water past that surface, and having **hemoglobin** or another **respiratory protein** speeds the rate of gas exchange.

Air-breathing animals have a more constant supply of oxygen than aquatic animals and do not have to expend as much energy in respiratory functions.

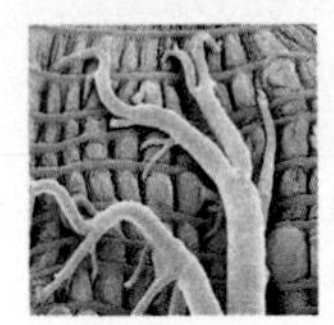

Section 38.3 Some invertebrates do not have special respiratory organs and rely on **integumentary exchange**, diffusion of gases across the body surface. **Gills** enhance respiration in other aquatic invertebrates. On land, **lungs** and **tracheal systems** assist in gas exchange with the air.

Section 38.4 Water flowing over fish gills exchanges gases with blood flowing in the opposite direction inside gill capillaries. This **countercurrent exchange** is highly efficient. Most amphibians have lungs, and also exchange gases across the skin. Reptiles (including birds) and mammals rely on lungs for gas exchange. In birds, air sacs connected to lungs keep air flowing continually past the respiratory surface.

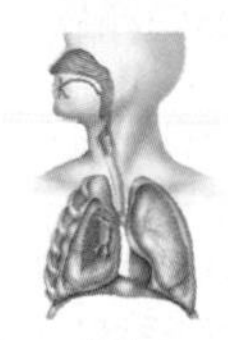

Section 38.5 In humans, air flows through two nasal cavities and a mouth into the **pharynx**, then the **larynx**, and the **trachea** (windpipe). The larynx contains the vocal cords, movements of which alter the size of the opening (the **glottis**) between them. When you swallow, the position of the **epiglottis** at the entrance to the larynx shifts, keeping food out of the trachea. The trachea branches into two **bronchi** that enter the lungs. These two airways branch into **bronchioles**. At the ends of the finest bronchioles are thin-walled **alveoli**, the site of gas exchange. The **diaphragm** at the base of the thoracic cavity and the **intercostal muscles** between ribs are involved in breathing.

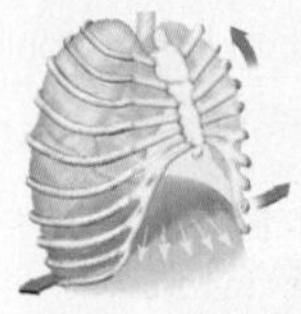

Section 38.6 A **respiratory cycle** is one inhalation and one exhalation. Inhalation is active. As muscle contractions expand the chest cavity, pressure in lungs decreases below atmospheric pressure, and air flows into the lungs. These events are reversed during exhalation, which usually is passive.

Tidal volume is normally far less than **vital capacity**. The lungs never fully deflate. The medulla oblongata in the brain stem adjusts the rate and magnitude of breathing. If a person is choking, a blow to the back or pushing on the abdomen can raise pressure in the lungs and expel the blocking object.

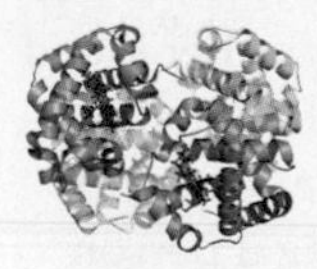

Section 38.7 In human lungs, gases diffuse across a thin **respiratory membrane** that separates the air in alveoli from the blood in pulmonary capillaries. As red blood cells pass through these vessels, hemoglobin binds O_2 to form **oxyhemoglobin**. In systemic capillaries, hemoglobin releases O_2, which diffuses into cells.

Also in systemic capillaries, CO_2 diffuses from cells into the blood. Most CO_2 reacts with water inside red blood cells to form bicarbonate. The enzyme **carbonic anhydrase** catalyzes this reaction, which is reversed in the lungs. There, CO_2 forms and is expelled from the body in exhalations.

Section 38.8 The amount of available oxygen declines with altitude. Physiological changes that occur in response to high altitude are called **acclimatization**. They include altered breathing patterns and an increase in **erythropoietin**, a hormone that stimulates red blood cell formation. Over the longer term, genetic changes can adapt a population to life at a high altitude, as among Tibetans and people living in the Andes.

A variety of adaptive mechanisms allow some turtles and marine mammals to hold their breath for long periods while making deep dives.

Section 38.9 Respiratory disorders include apnea and sudden infant death syndrome (SIDS). Respiratory diseases include tuberculosis, pneumonia, bronchitis, and emphysema. Smoking increases risk of many respiratory problems. It is a leading cause of debilitating disease and deaths.

self-quiz

Answers in Appendix VII

1. Respiratory proteins such as hemocyanin ________ .
   a. contain metal ions
   b. occur only in vertebrates
   c. increase the efficiency of oxygen transport
   d. both a and c

2. In a ________ , air flows continually across the respiratory surface.
   a. fish c. frog
   b. bird d. mammal

3. The tracheal tubes of insects carry ________ to tissues deep inside the body.
   a. hemolymph c. air
   b. blood d. water

4. In human lungs, gas exchange occurs at the ________ .
   a. bronchi c. alveoli
   b. bronchioles d. pleural sacs

data analysis activities

Risks of Radon Radon is a colorless, odorless gas emitted by many rocks and soils. It is formed by the radioactive decay of uranium and is itself radioactive (Section 2.2). There is some radon in the air almost everywhere, but routinely inhaling a lot of it raises the risk of lung cancer. Radon also seems to increase cancer risk far more in smokers than in nonsmokers. **FIGURE 38.28** is an estimate of how radon in homes affects risk of lung cancer mortality. Note that these data show only the death risk for radon-induced cancers. Smokers are also at risk from lung cancers that are caused by tobacco.

1. If 1,000 smokers were exposed to a radon level of 1.3 pCi/L over a lifetime (the average indoor radon level), how many would die of a radon-induced lung cancer?

2. How high would the radon level have to be to cause approximately the same number of cancers among 1,000 nonsmokers?

3. The risk of dying in a car crash is about 7 out of 1,000. Is a smoker in a home with an average radon level (1.3 pCi/L) more likely to die in a car crash or of radon-induced cancer?

Radon level (pCi/L)	Risk of cancer death from lifetime radon exposure	
	Never smoked	Current smokers
20	36 out of 1,000	260 out of 1,000
10	18 out of 1,000	150 out of 1,000
8	15 out of 1,000	120 out of 1,000
4	7 out of 1,000	62 out of 1,000
2	4 out of 1,000	32 out of 1,000
1.3	2 out of 1,000	20 out of 1,000
0.4	>1 out of 1,000	6 out of 1,000

FIGURE 38.28 Estimated risk of lung cancer death as a result of lifetime radon exposure. Radon levels are measured in picocuries per liter (pCi/L). The Environmental Protection Agency considers a radon level above 4 pCi/Liter to be unsafe. For information about testing your home for radon and what to do if the radon level is high, visit the EPA's Radon Information Site at www.epa.gov/radon.

5. Which holds the most dissolved oxygen?
   a. warm, still water
   b. warm, running water
   c. cold, running water
   d. cold, still water

6. When you breathe quietly, inhalation is _______ and exhalation is _______ .
   a. passive; passive
   b. active; active
   c. passive; active
   d. active; passive

7. During inhalation _______ .
   a. the thoracic cavity expands
   b. the diaphragm relaxes
   c. atmospheric pressure declines
   d. both a and c

8. _______ binds to hemoglobin even more strongly than oxygen does.
   a. Carbon dioxide
   b. Carbon monoxide
   c. Oxyhemoglobin
   d. Carbonic anhydrase

9. Most oxygen transported in human blood _______ .
   a. is bound to hemoglobin
   b. combines with carbon to form carbon dioxide
   c. is in the form of bicarbonate
   d. is dissolved in the plasma

10. At high altitudes, _______ .
   a. nitrogen bubbles out of the blood
   b. hemoglobin has fewer oxygen-binding sites
   c. there are fewer O_2 molecules than at low altitudes
   d. both b and c

11. In fish gills, blood and water move _______ .
   a. in the same direction
   b. in opposite directions
   c. through tracheal tubes
   d. in hemolymph

CENGAGE brain To access course materials, please visit www.cengagebrain.com.

12. _______ in arteries sense changes in the acidity of the blood.
   a. Mechanoreceptors
   b. Neurotransmitters
   c. Photoreceptors
   d. Chemoreceptors

13. True or false? Human lungs retain some air even after forced exhalation.

14. The diaphragm is a _______ muscle.
   a. smooth
   b. skeletal
   c. dome-shaped
   d. a and c

15. Match the words with their descriptions.

___ trachea	a. muscle of respiration
___ pharynx	b. gap between vocal cords
___ epiglottis	c. between bronchi and alveoli
___ larynx	d. windpipe
___ bronchus	e. voice box
___ bronchiole	f. keeps food out of airway
___ glottis	g. airway leading to lung
___ diaphragm	h. throat

critical thinking

1. The red blood cell enzyme carbonic anhydrase contains a zinc cofactor. A dietary zinc deficiency does not reduce the number of red blood cells, but it does impair respiratory function by reducing carbon dioxide output. Explain why.

2. A developing fetus gets oxygen from its mother's blood. In an organ called the placenta, fetal capillaries run through and exchange substances with pools of maternal blood. The hemoglobin made by a fetus has different properties than that made after birth. Fetal hemoglobin binds oxygen more strongly at low oxygen levels than does normal hemoglobin. How would fetal hemoglobin's somewhat higher affinity for oxygen benefit a fetus?

39 Digestion and Nutrition

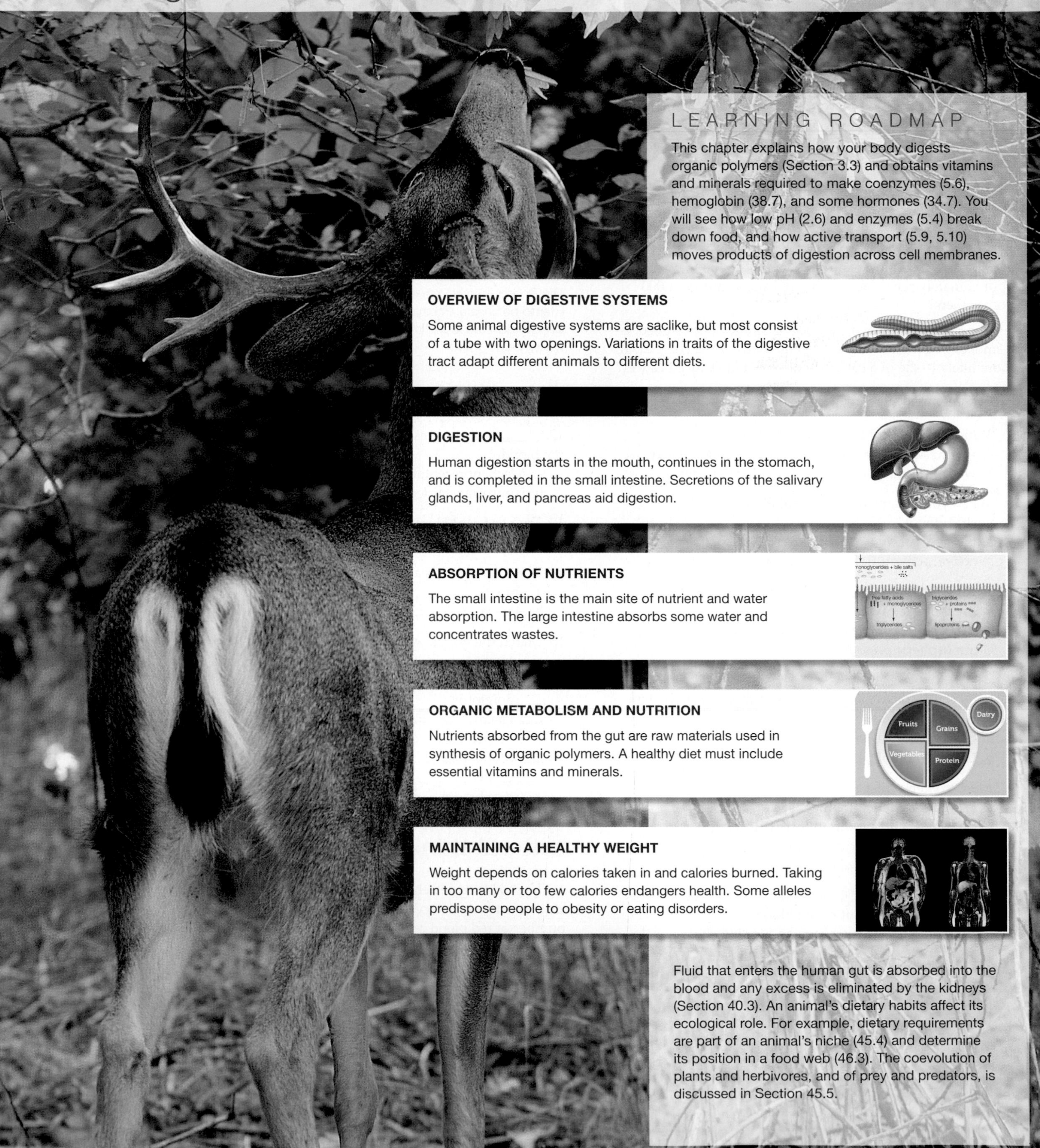

LEARNING ROADMAP

This chapter explains how your body digests organic polymers (Section 3.3) and obtains vitamins and minerals required to make coenzymes (5.6), hemoglobin (38.7), and some hormones (34.7). You will see how low pH (2.6) and enzymes (5.4) break down food, and how active transport (5.9, 5.10) moves products of digestion across cell membranes.

OVERVIEW OF DIGESTIVE SYSTEMS

Some animal digestive systems are saclike, but most consist of a tube with two openings. Variations in traits of the digestive tract adapt different animals to different diets.

DIGESTION

Human digestion starts in the mouth, continues in the stomach, and is completed in the small intestine. Secretions of the salivary glands, liver, and pancreas aid digestion.

ABSORPTION OF NUTRIENTS

The small intestine is the main site of nutrient and water absorption. The large intestine absorbs some water and concentrates wastes.

ORGANIC METABOLISM AND NUTRITION

Nutrients absorbed from the gut are raw materials used in synthesis of organic polymers. A healthy diet must include essential vitamins and minerals.

MAINTAINING A HEALTHY WEIGHT

Weight depends on calories taken in and calories burned. Taking in too many or too few calories endangers health. Some alleles predispose people to obesity or eating disorders.

Fluid that enters the human gut is absorbed into the blood and any excess is eliminated by the kidneys (Section 40.3). An animal's dietary habits affect its ecological role. For example, dietary requirements are part of an animal's niche (45.4) and determine its position in a food web (46.3). The coevolution of plants and herbivores, and of prey and predators, is discussed in Section 45.5.

39.1 Your Microbial "Organ"

Your body is home to 10 trillion or more individual microorganisms, and the vast majority of them reside in your gut. The species composition of this microbial community varies among individuals and among human populations. Researchers have only recently begun to investigate the extent of these variations and their effects on our health, but some interesting connections have already been uncovered.

For example, we now know that one species of bacteria can cause peptic ulcers. An ulcer is a chronically open, inflamed sore at the body surface or in a body lining. Peptic ulcers arise in the stomach or in the part of the small intestine adjacent to the stomach. During the mid-1980s, two Australian physicians proved that most people with peptic ulcers are infected by a spiral-shaped bacterial species called *Helicobacter pylori* (**FIGURE 39.1**). Prior to this discovery, people thought that peptic ulcers were caused by a stressful lifestyle or a diet rich in acidic or spicy foods.

Infection by *H. pylori* is extremely common. Worldwide, more than half of the human population hosts this species. The bacteria live at the surface of the stomach, in mucus secreted by cells of the stomach lining. Most infected people have no symptoms, but 5 to 20 percent of them develop peptic ulcers. Infection by *H. pylori* also increases the risk of stomach cancer sixfold. In 1994, the World Health Organization classified *H. pylori* as the first known bacterial carcinogen.

Treatment with antibiotics can eliminate *H. pylori* from the gut, allow an ulcer to heal, and reduce the risk of gastric cancer. However, clearing *H. pylori* from the gut can also have unintended consequences. Compared to people infected by *H. pylori*, the uninfected are more likely to have gastroesophageal reflux disease (GERD). The main symptom of GERD is a burning pain in the upper abdomen, commonly referred to as heartburn. The pain occurs when acid from the stomach splashes back into the esophagus, the tubular organ between the throat and the stomach. Repeated exposure to acid can damage the tissue of the esophagus and raise the risk of esophageal cancer.

Some gut bacteria also play a role in irritable bowel syndrome (IBS). This noninfectious disorder afflicts about one in five Americans. Affected individuals have frequent abdominal cramps and often feel bloated. Their bowel habits are disrupted, with constipation, diarrhea, or an alternation between these conditions. No single microbial species is thought to cause IBS. Rather, the disorder is associated with a shift in the diversity and proportions of intestinal bacteria. The genera of over- and under-represented bacteria varies with the type of IBS, but several studies have found that most people suffering from the disorder have reduced numbers of *Bifidobacterium* compared with those unaffected by the syndrome.

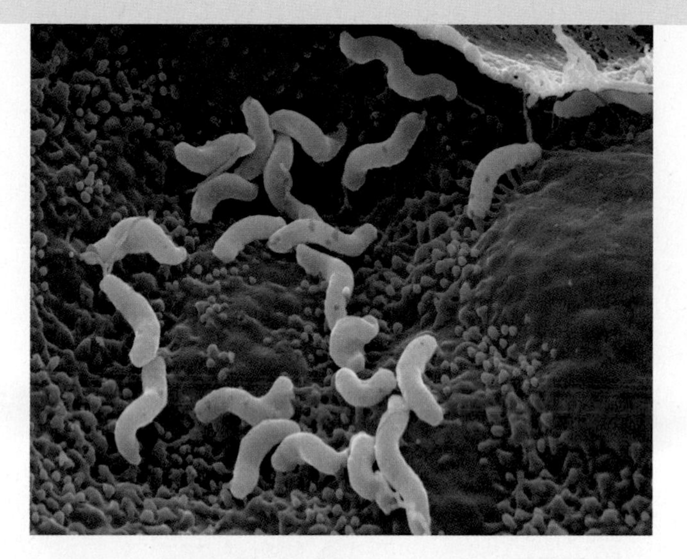

FIGURE 39.1 *Helicobacter pylori* in the human stomach (colorized micrograph). The presence of these bacteria in the stomach raises the risk for peptic ulcers and stomach cancer, but reduces the likelihood of heartburn and cancer of the esophagus.

Additional evidence that a decline in *Bifidobacterium* contributes to IBS comes from studies assessing the effectiveness of a probiotic as a treatment for this disorder. A **probiotic** is a food or supplement that contains living organisms whose presence in the body is thought to enhance health. In controlled clinical studies, IBS patients who receive a probiotic dose of the gut bacteria *Bifidobacterium* had reduced symptoms compared to patients who received *Lactobacillus* bacteria.

There is also mounting evidence that the presence of certain gut microbes exerts influence far beyond the digestive tract. For example, the presence of *H. pylori* in the gut is associated with a lowered risk of immune-related disorders such asthma and allergies. Similarly, having a sizable gut population of *Bifidobacterium* improves immune function.

Given the impact of gut microbes on health, some have described this microbial population as analogous to an organ. Like our body's other organs, this microbial organ consists of cells whose coordinated function and interactions contribute to our health, and whose disruption can result in illness.

probiotic Food or supplement containing living microorganisms whose presence in the body benefits health.

CREDITS: (opposite) Jouan Rius/naturepl.com; (l) Eye of Science/Science Source.

39.2 Animal Digestive Systems

✔ Animals are heterotrophs that typically digest food inside their body, but outside of their cells.

Food Processing Tasks

Being heterotrophs, animals obtain the energy and raw materials they require by breaking down organic compounds made by other organisms. In sponges, individual cells engulf food particles, break them down (digest them), then expel wastes. In other animals, most processing of food and food wastes is extracellular. It occurs in the interior of a hollow sac or tube that opens to the external environment.

Processing food typically involves four tasks:

1. *Ingestion*. Taking food into a digestive organ.
2. *Mechanical and chemical digestion*. Breaking food down into components that can be absorbed into the body. Mechanical digestion smashes food into smaller and smaller fragments. Chemical digestion breaks large polymers into their component subunits.
3. *Absorption*. Movement of nutrient molecules across the wall of a digestive organ and into the body's internal environment.
4. *Elimination*. Expelling from the body any dietary material that was not digested and absorbed.

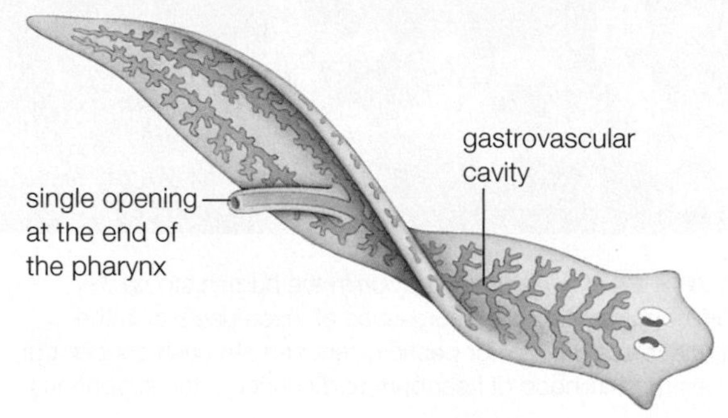

A The saclike gastrovascular cavity of a planarian flatworm has a single opening.

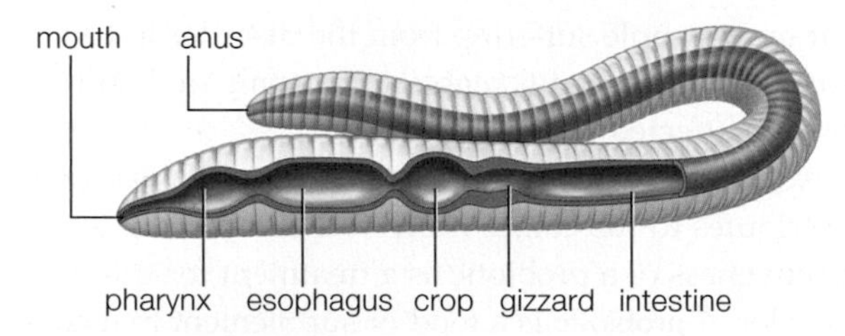

B The complete digestive tract of an earthworm, has two openings (mouth and anus), and specialized regions in between.

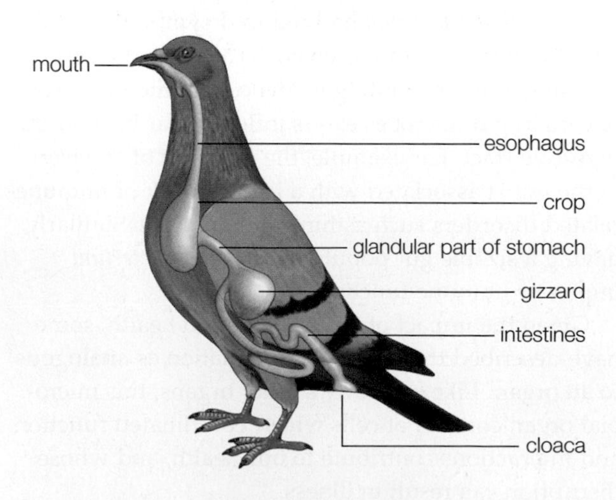

C The complete digestive tract of a bird has two openings (mouth and cloaca), and specialized regions in between.

FIGURE 39.2 ▶**Animated** Examples of digestive systems.

Sac or Tube?

There are two types of digestive systems. Flatworms and cnidarians have the simplest type of system, a saclike **gastrovascular cavity** that functions in both digestion and gas exchange. This cavity has a single opening through which food enters and wastes leave. In flatworms, the opening is at the tip of a muscular pharynx (**FIGURE 39.2A**). In animals that have a gut with a single opening, each load of food must be broken down, its nutrients absorbed, and any wastes eliminated before a new load of food can enter.

Earthworms, most invertebrates, and all vertebrates have a second type of system, a **complete digestive tract**, which is a tubular gut with two openings (**FIGURES 39.2B** and **39.2C**). Food enters a mouth at one end of the tube, and wastes leave through an opening at the other end. Unlike a gastrovascular cavity, a tubular gut has specialized regions. As food travels the length of a complete digestive tract, it passes through regions specialized for food storage, food breakdown, nutrient absorption, and waste compaction. The structure of these regions varies among animal groups.

In a complete digestive tract, a tubular organ called the **esophagus** receives food from the pharynx (throat). In some animals, including earthworms and many birds, part of the esophagus has become enlarged to serve as a food-storing pouch called the **crop**.

The esophagus delivers food to the **stomach**, a muscular digestive organ that chemically and mechanically breaks down food. In earthworms, birds, and other animals that lack teeth, a portion of the stomach may be specialized as a **gizzard**, an organ that grinds up

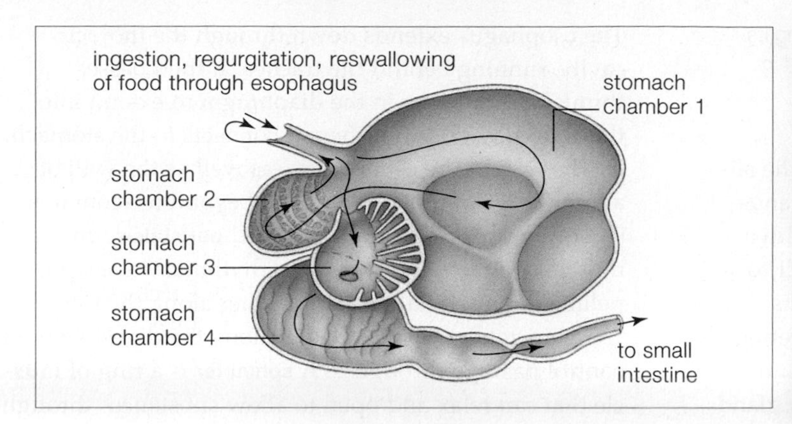

FIGURE 39.3 ▶**Animated** Multiple-chambered stomach of a ruminant such as the mule deer shown at right. Microbes in the first two stomach chambers break down cellulose in plant cell walls. Solids accumulate in the second chamber, forming "cud" that is regurgitated—moved back into the mouth for a second round of chewing. Nutrient-rich fluid moves from the second chamber to the third and fourth chambers, and finally to the intestine.

food and begins the process of mechanical digestion. In the earthworm gizzard, hard bits of dirt ingested by the worm break up food. In birds, the gizzard is a muscular chamber with a hardened lining of glycoprotein. Seeds are difficult to grind up, so seed-eating birds have larger gizzards than those with other diets.

Ruminants such as cattle, goats, sheep, antelope, and deer are hoofed grazers that have multiple stomach chambers (**FIGURE 39.3**). Their digestive system adapts them to a diet of cellulose-rich plant foods. Like other animals, ruminants do not make enzymes that can break down cellulose. However, microbes living in the first two chambers of their stomach do make these enzymes. Solids accumulate in the second chamber, forming a "cud" that is regurgitated—moved back into the mouth for a second round of chewing. Nutrient-rich fluid moves from the second chamber to the third and fourth chambers, and finally to the intestine.

All animals with a complete digestive tract have an **intestine** in which most chemical digestion is carried out and from which nutrients are absorbed. Meat is easier to digest than plant material, so carnivores (animals that eat other animals), typically have a shorter intestine than herbivores (animals that eat plants).

In some animals, digestive wastes exit through an **anus**, an opening that serves this purpose alone. In others, digestive wastes leave the body through a **cloaca**, a multipurpose opening that also releases urinary waste and functions in reproduction. Amphibians, reptiles, and birds have a cloaca, whereas humans and other placental mammals have an anus for digestive waste and a separate opening for urinary waste.

anus Body opening that serves solely as the exit for wastes from a complete digestive tract.
cloaca Body opening that serves as the exit for digestive and urinary waste; also functions in reproduction.
complete digestive tract Tubelike digestive system; food enters through one opening and wastes leave through another.
crop Of birds and some invertebrates, an enlarged portion of the esophagus that stores food.
esophagus Tubular organ that connects the pharynx to the stomach.
gastrovascular cavity Saclike gut; also functions in gas exchange.
gizzard Of birds and some other animals, a digestive chamber lined with a hard material that grinds up food.
intestine Tubular organ in which chemical digestion is completed and from which nutrients are absorbed.
stomach Tubular, muscular organ that functions in chemical and mechanical digestion.

TAKE-HOME MESSAGE 39.2

What is a digestive system and how does its structure reflect its function?

✔ Digestive systems mechanically and chemically degrade food into small molecules that can be absorbed into the internal environment. These systems also expel the undigested residues from the body.

✔ Incomplete digestive systems are a saclike cavity with one opening. Complete digestive systems are a tube with two openings and regional specializations in between.

✔ Structural differences among the digestive tracts of different animal groups are adaptations that allow the animals to exploit different types of foods.

CREDITS: (3) left, Based on A. Romer and T. Parsons, *The Vertebrate Body*, Sixth Edition, Saunders Publishing Company, 1986; right, Jouan Rius/naturepl.com.

39.3 Overview of the Human Digestive System

✔ Humans have a tubular gut. Accessory organs along its length secrete enzymes and other substances that aid in breakdown of food and absorption of nutrients.

The human digestive tract, also referred to as the alimentary canal, extends from the mouth to the anus (**FIGURE 39.4**). If this tube were to be laid out in a straight line, it would extend 6.5 to 9 meters (21 to 30 feet). Extensive coiling and folding allow this lengthy tube to fit inside the body. Mucus-secreting epithelium (mucosa) lines the digestive tract, and accessory digestive organs such as the salivary glands, liver, and pancreas secrete substances into its interior, or **lumen**. These secretions assist in chemical digestion.

Processing of food begins inside the mouth, or oral cavity. Swallowing moves food into the pharynx (throat), from which it proceeds into the esophagus. The esophagus extends down through the thoracic cavity, running behind the trachea. It then passes through an opening in the diaphragm to extend into the abdominal cavity, where it connects to the stomach.

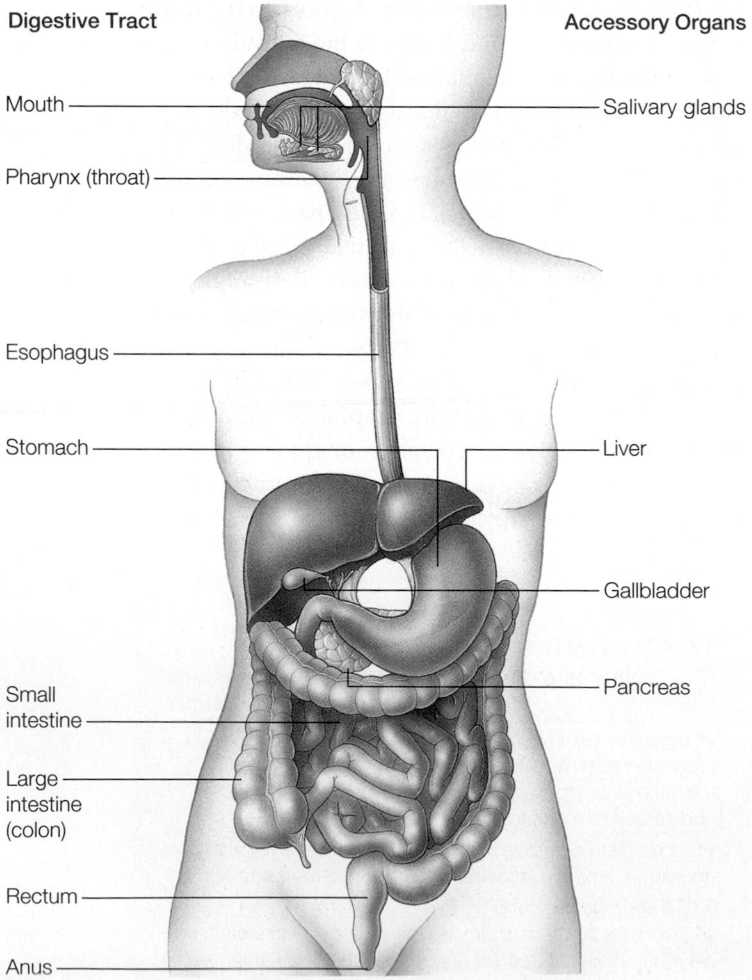

The wall of the esophagus—as well as the wall of all portions of the digestive tract beyond it—contains smooth muscle. By a process called **peristalsis**, this muscle contracts and relaxes in rhythmic waves, propelling any material inside it farther along the tube.

Sphincters at various points along the digestive tract control passage through it. A **sphincter** is a ring of muscle that can relax and open to allow substances through a passageway, or contract to close off the passage. For example, when you swallow, a sphincter in the upper esophagus opens to allow food in, then closes behind it. The sound made during a burp, or belch, arises when this sphincter vibrates as previously swallowed air flows outward through it.

The stomach empties into the intestine, which has two regions that differ in their structure and function. The first region, the **small intestine**, completes the process of chemical digestion and absorbs most nutrients. Secretions from the liver and pancreas assist the small intestine in these tasks. The second region of the intestine, the **large intestine**, absorbs fluid and concentrates waste. The terminal portion of the large intestine, the **rectum**, stores waste until it is expelled from the body through the anus.

Digestive Tract — Mouth; Pharynx (throat); Esophagus; Stomach; Small intestine; Large intestine (colon); Rectum; Anus

Accessory Organs — Salivary glands; Liver; Gallbladder; Pancreas

FIGURE 39.4 ▸**Animated** Components of the human digestive system.

large intestine Organ that receives material from the small intestine, absorbs water, and concentrates and stores waste.
lumen The space within a tubular or hollow organ.
peristalsis Wavelike smooth muscle contractions that propel food through the digestive tract.
rectum Terminal part of the large intestine; stores digestive waste prior to excretion.
small intestine Organ that receives material from the stomach; site of most digestion and absorption.
sphincter Ring of muscle that controls passage through a tubular organ or body opening.

TAKE-HOME MESSAGE 39.3

What type of digestive system do humans have?

✔ Humans have a complete digestive system with a tubular, mucosa-lined gut.

✔ Smooth muscle contractions move material through the digestive tract, and sphincters along the tract control the rate of passage.

✔ Accessory organs positioned adjacent to the gut secrete substances into its interior. These substances aid in digestion of food or absorption of nutrients.

39.4 Chewing and Swallowing

✔ Mechanical digestion, the smashing of food into smaller pieces, begins in the mouth. So does chemical digestion, the enzymatic breakdown of food into molecular subunits.

As a species, humans are omnivores, meaning we are adapted to eat both plant and animal material. Our omnivorous heritage is reflected in our teeth. We have all four types of mammalian teeth, and all are equally well developed (**FIGURE 39.5A**). By contrast, mammals that are carnivorous have enlarged canine teeth for piercing and tearing flesh (**FIGURE 39.5B**), and those that are herbivores have reduced canine teeth (**FIGURE 39.5C**). Premolars of carnivores have a narrow, bladelike surface that helps them shear through meat. Those of herbivores and omnivores like us are broad and flat, the better for grinding plant material.

Each tooth consists mostly of dentin, a bonelike material (**FIGURE 39.6**). Dentin-secreting cells reside in a central pulp cavity, serviced by nerves and blood vessels that extend through the tooth's root. Enamel, the hardest material in the body, covers the tooth's exposed portion, or crown. The crown extends above the gum, and the root is embedded in the bone of the jaw at a fibrous joint.

The tongue is a bundle of membrane-covered skeletal muscle attached to the floor of the mouth. Movements of the tongue help position food where teeth can chop or shred it. Tongue movements also mix food with saliva secreted by salivary glands. **Salivary glands** are exocrine glands that open into the mouth beneath the tongue and on the inner surface of the cheeks next to the upper molars. An enzyme in saliva (salivary amylase) begins the process of chemical digestion by breaking starch into disaccharides (two-sugar units). Like many other reactions of digestion, this is an example of hydrolysis (Section 3.3). Saliva also contains glycoproteins that combine with water to form mucus. The mucus helps small bits of food stick together in easy-to-swallow clumps.

The presence of food at the back of the throat triggers a swallowing reflex. The throat, remember, opens onto both the digestive and respiratory tracts (Section 38.5). When you swallow, the flap of cartilage called the epiglottis flops down to cover the opening to the airway and the vocal cords constrict, so the route between the pharynx and larynx is blocked. This reflex keeps food from accidentally entering the trachea and choking you.

salivary gland Exocrine gland that secretes saliva into the mouth.

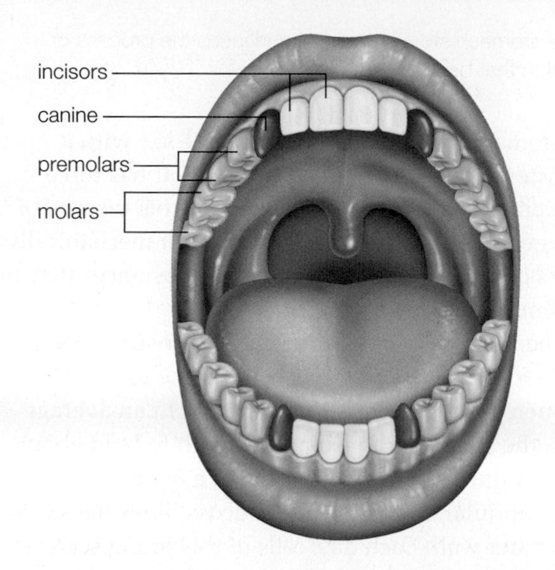

A Adult human mouth. Humans have all four tooth types and all are equally large.

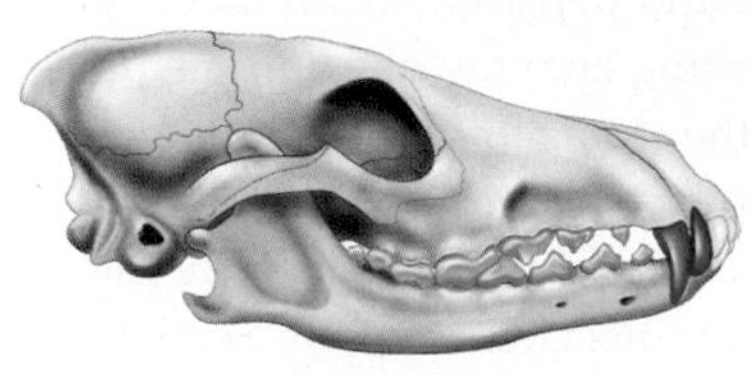

B Carnivore with enlarged canine teeth.

C Herbivore with reduced canines and large broad molars and premolars.

FIGURE 39.5 Tooth form and function.

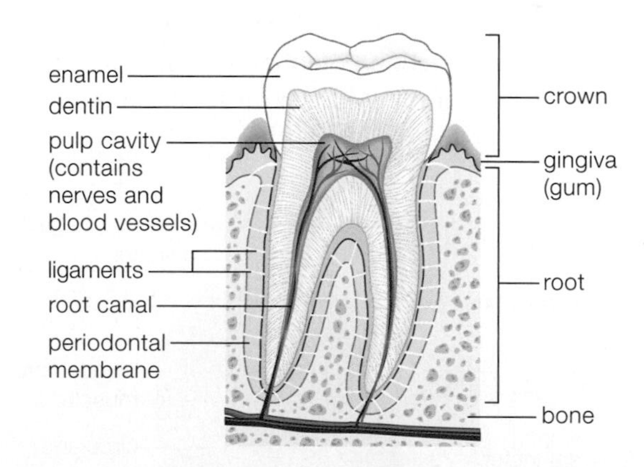

FIGURE 39.6 Structure of a human tooth (molar).

TAKE-HOME MESSAGE 39.4

How is food chewed and swallowed?

✔ Teeth mechanically break food into smaller particles.

✔ The tongue positions food and mixes it with saliva. Enzymes in the saliva begin breakdown of carbohydrates.

✔ Swallowing is a reflex. During swallowing, the epiglottis covers the trachea to prevent food from entering it.

39.5 Food Storage and Digestion in the Stomach

✔ The stomach stores food and continues the process of digestion that began in the mouth.

The stomach is a muscular, stretchable sac with a sphincter at either end (**FIGURE 39.7**). It has three functions. First, it stores food and controls the rate of passage to the small intestine. Second, it mechanically breaks down food. Third, it secretes substances that aid in chemical digestion.

When the stomach is empty, its inner surface is highly folded. As it fills with food, these folds smooth out, increasing the stomach's capacity. In an average adult, the stomach can expand enough to hold about 1 liter of fluid (a little bit more than a quart).

A glandular epithelium, or mucosa, lines the stomach's inner wall. Each day, cells of this lining secrete about 2 liters of **gastric fluid**, a mixture of mucus, hydrochloric acid (HCl), and pepsinogen, an inactive form of the protein-digesting enzyme pepsin. The presence of the acid lowers the stomach's pH to about 2.

When something disrupts the stomach's protective mucus layer, gastric fluid and enzymes can erode the stomach lining, causing an ulcer. As noted in Section 39.1, most ulcers occur after infection by *Helicobacter pylori*. Continual use of nonsteroidal anti-inflammatory drugs such as ibuprofen or aspirin can also cause a stomach ulcer.

Like the heart, the stomach has an internal pacemaker. Spontaneous action potentials generated in the uppermost portion of the stomach cause the smooth muscle in the stomach wall to contract rhythmically about three times a minute. The contractions mix gastric fluid with food to form a semiliquid mass called **chyme**. They also propel a portion of chyme out through the pyloric sphincter and into the first part of the small intestine.

Chemical digestion of proteins begins in the stomach. The acidity of chyme denatures proteins (Section 3.7), so their peptide bonds become exposed. High acidity also converts the protein pepsinogen into pepsin. Pepsin is an enzyme that cleaves the peptide bonds between amino acids, so it chops up the proteins into smaller polypeptides.

The stomach steps up or slows down its secretion of acid depending on when and what you eat. The arrival of food in the stomach, especially protein, triggers endocrine cells in the uppermost region of the stomach to secrete the hormone gastrin into the blood. Gastrin acts on acid-secreting cells of the stomach lining, encouraging them to increase their output. When the stomach is empty, gastrin secretion and acid secretion decline. This prevents excess acidity from damaging the stomach wall.

Conversely, an empty stomach increases its secretion the hormone ghrelin. Ghrelin is a peptide hormone that stimulates the appetite. When people attempt to reduce their weight by eating less, their ghrelin output rises. This is one reason it is difficult to stick to a diet. It may also explain the success of gastric bypass surgery. Such surgery reduces the size of the stomach and the length of the small intestine, so less food can be taken in and fewer nutrients are absorbed. The surgery also removes ghrelin-secreting cells, reducing appetite.

Nervous signals also affect appetite. The wall of the stomach contains stretch receptors. As the stomach fills with food, signals from these neurons travel along the vagus nerve to the brain, causing a feeling of fullness.

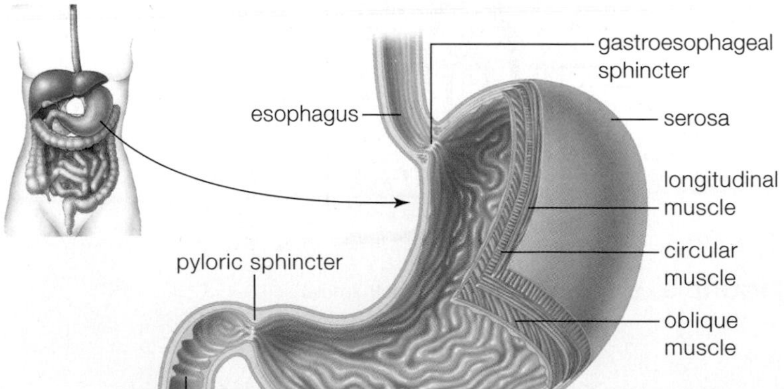

gastroesophageal sphincter
esophagus
serosa
longitudinal muscle
circular muscle
pyloric sphincter
oblique muscle
submucosa
small intestine
mucosa

FIGURE 39.7 ▸**Animated** Location and structure of the stomach. The outermost layer, the serosa, is connective tissue covered by epithelium. Beneath the serosa, three layers of smooth muscle differ in their orientation and direction of contraction. Their coordinated action mixes stomach contents with gastric fluid secreted by the mucosa that lines the stomach's interior. The folds shown on the inner surface become smoothed out when the stomach fills with food.

chyme Mix of food and gastric fluid.
gastric fluid Fluid secreted by the stomach lining; contains digestive enzymes, acid, and mucus.

TAKE-HOME MESSAGE 39.5

What are the functions of the stomach?

✔ The stomach receives food from the esophagus and stretches to store it.

✔ Contractions of stomach muscles break up food and mix it with acidic gastric fluid. They also move the resulting mixture (the chyme) into the small intestine.

✔ Chemical digestion of proteins begins in the stomach.

39.6 Structure of the Small Intestine

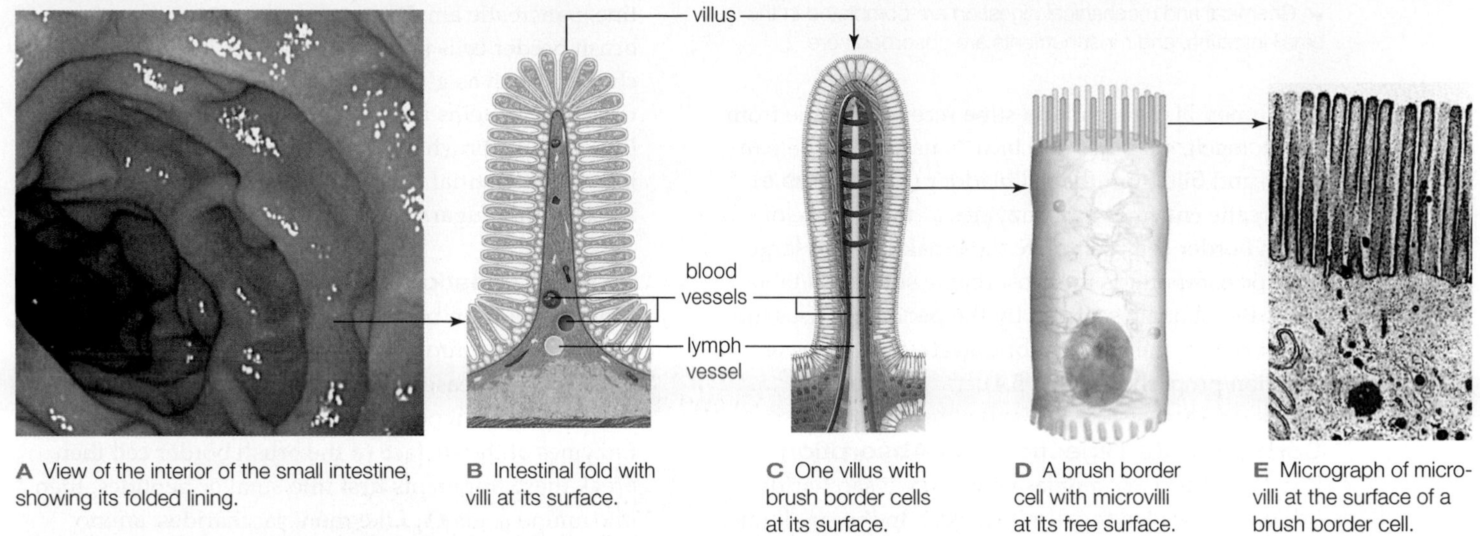

A View of the interior of the small intestine, showing its folded lining.

B Intestinal fold with villi at its surface.

C One villus with brush border cells at its surface.

D A brush border cell with microvilli at its free surface.

E Micrograph of microvilli at the surface of a brush border cell.

FIGURE 39.8 Structure of the small intestine.

FIGURE IT OUT Are microvilli multicelled or smaller than a cell?

Answer: Microvilli are smaller than a cell. Villi are multicellular.

✔ The small intestine has a highly folded lining with many projections that greatly increase its surface area.

Chyme forced out of the stomach through the pyloric sphincter enters the duodenum, the initial portion of the small intestine. The small intestine is "small" only in terms of its diameter—about 2.5 cm (1 inch). It is the longest segment of the gut. Uncoiled, it would extend for 5 to 7 meters (16 to 23 feet).

Most digestion and absorption take place at the surface of the small intestine, which is highly folded (**FIGURE 39.8A**). Unlike the folds of the empty stomach, those of the small intestine are permanent.

The surface of each intestinal fold is covered by **villi** (singular, villus), hairlike multicelled projections about 1 millimeter long (**FIGURE 39.8B,C**). The millions of villi that project from the intestinal lining give it a furry or velvety appearance. Blood vessels and lymph vessels run through the interior of each villus.

Most of the epithelial cells at the surface of a villus have even tinier projections called **microvilli** (singular, microvillus). The 1,700 or so microvilli at the surface of a cell make its outer edge look like a brush. Thus, these cells are sometimes called **brush border cells** (**FIGURE 39.8D,E**). Brush border cells function in both digestion and absorption. Digestive enzymes at the surface of a microvillus break down sugars, polypeptides, and nucleotides. Also at the microvillus surface are many transport proteins that facilitate the movement of nutrients into the microvillus.

Collectively, the many folds and projections of the small intestinal lining increase its surface area by hundreds of times. As a result, the surface area of the small intestine is comparable to that of a tennis court. Having such a large surface area maximizes the number of membrane-embedded digestive enzymes and transport proteins that come into contact with the chyme.

Like the stomach wall, the wall of the small intestine has multiple layers of smooth muscle. The combined action of these muscles mixes the chyme, propels it forward, and forces it up against the wall of the small intestine, thus enhancing digestion and absorption.

TAKE-HOME MESSAGE 39.6

How does the structure of the small intestine affect its function?

✔ The surface of the small intestine is highly folded and each fold has many projections (villi). Brush border cells at the surface of a villus have tiny projections (microvilli) at their surface.

✔ The many folds and projections greatly increase the surface area for the two functions of the small intestine—digestion and absorption.

brush border cell In the lining of the small intestine, an epithelial cell with microvilli at its surface.
microvilli Thin projections from the plasma membrane of some epithelial cells; increase the cell's surface area.
villi Multicelled projections at the surface of each fold in the small intestine.

CREDITS: (8A) Gastrolab/Science Source; (8B–D) After Sherwood and others; (8E) D. W. Fawcett/Science Source.

39.7 Digestion and Absorption in the Small Intestine

✔ Chemical and mechanical digestion are completed in the small intestine, and most nutrients are absorbed here.

The lumen of the small intestine receives chyme from the stomach, enzymes and bicarbonate from the pancreas, and bile from the gallbladder (**FIGURE 39.9**). Pancreatic enzymes and enzymes at the surface of brush border cells complete the breakdown of large organic compounds into absorbable subunits (Table 39.1). Bicarbonate secreted by the pancreas raises the pH of the chyme enough for digestive enzymes to function properly (Section 5.4).

Carbohydrate Digestion and Absorption

Recall that salivary amylase breaks polysaccharides into disaccharides (two-unit sugars). In the small intestine, pancreatic amylase and enzymes at the surface of brush border cells split disaccharides into monosaccharides such as glucose (**FIGURE 39.10** ❶). Active transport proteins move the monosaccharides from the lumen into a brush border cell, then out of that cell and into the interstitial fluid inside the villus ❷. From here, these simple sugars enter the blood.

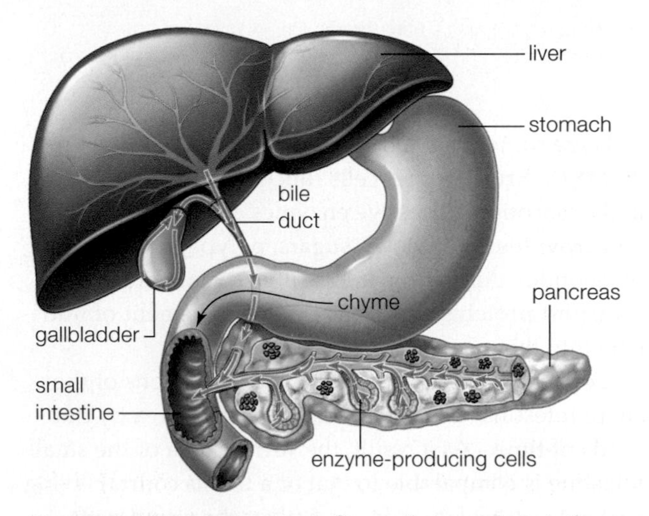

FIGURE 39.9 ▸Animated Organs that contribute to the contents of the small intestine.

Protein Digestion and Absorption

Protein digestion began in the stomach, where pepsin broke proteins into polypeptides. In the small intestine, pancreatic proteases such as trypsin and chymotrypsin cut these polypeptides into smaller fragments. Enzymes at the surface of the brush border cell then break these fragments first into smaller peptides, then into amino acids ❸. Like monosaccharides, amino acids are transported into brush border cells, then into interstitial fluid by membrane proteins ❹. From here, they enter the blood.

Fat Digestion and Absorption

Fat digestion occurs entirely in the small intestine. Here, the effectiveness of pancreatic lipases (fat-digesting enzymes) is enhanced by **bile**, a green liquid containing salts, pigments, cholesterol, and lipids. Bile is made by the **liver**, an organ that also stores glycogen and removes toxins from the blood. The adjacent **gallbladder** stores and concentrates bile. A fatty meal causes the gallbladder to contract, forcing bile out through a short duct into the small intestine.

Bile aids fat digestion by dispersing fat droplets, a process called **emulsification**. Triglycerides from food tend to clump together as fat globules. Muscle contrac-

Table 39.1 Summary of Chemical Digestion

	Enzymes Present	Enzyme Source	Enzyme Substrate	Main Breakdown Products
Carbohydrate Digestion				
Mouth, stomach	Salivary amylase	Salivary glands	Polysaccharides	Disaccharides
Small intestine	Pancreatic amylase	Pancreas	Polysaccharides	Disaccharides
	Disaccharidases	Intestinal lining	Disaccharides	**Monosaccharides*** (such as glucose)
Protein Digestion				
Stomach	Pepsins	Stomach lining	Proteins	Protein fragments
Small intestine	Trypsin, chymotrypsin	Pancreas	Proteins	Protein fragments
	Carboxypeptidase	Pancreas	Protein fragments	**Amino acids***
	Aminopeptidase	Intestinal lining	**Amino acids***	
Lipid Digestion				
Small intestine	Lipase	Pancreas	Triglycerides	**Free fatty acids, monoglycerides***
Nucleic Acid Digestion				
Small intestine	Pancreatic nucleases	Pancreas	DNA, RNA	Nucleotides
	Intestinal nucleases	Intestinal lining	Nucleotides	**Nucleotide bases, monosaccharides***

*** Breakdown products that can be absorbed into the internal environment.**

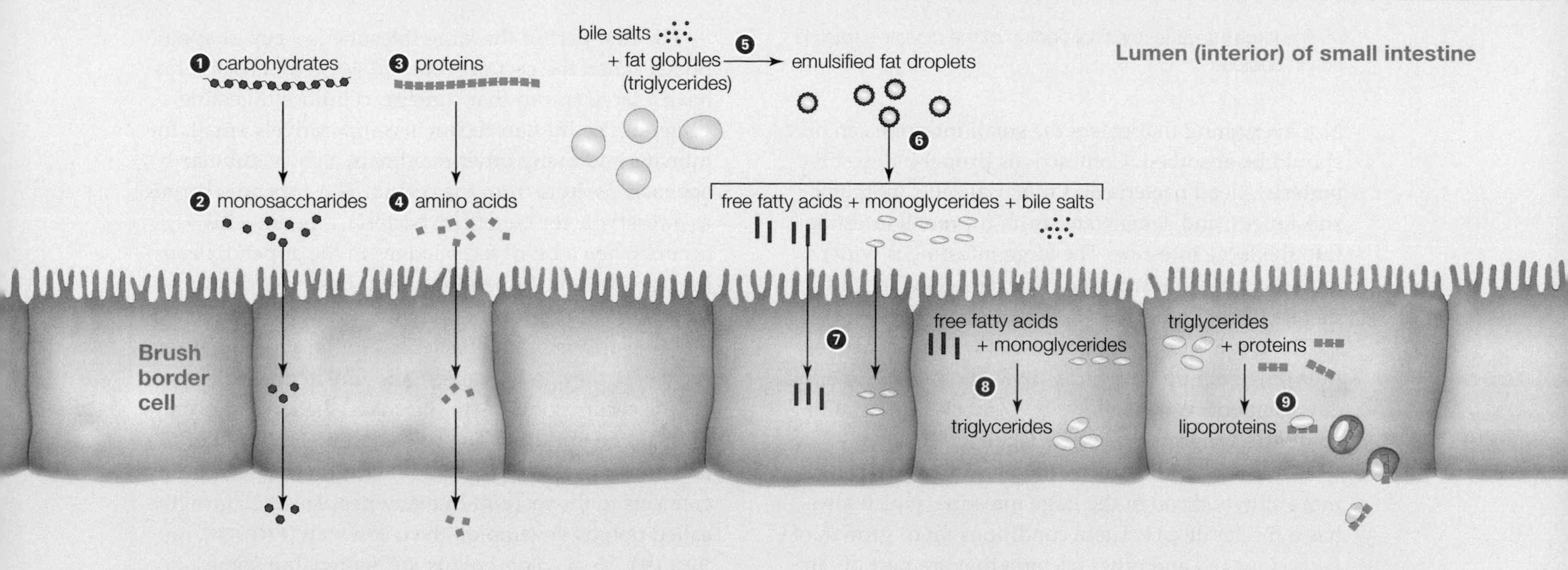

❶ Polysaccharides are broken down to monosaccharides (simple sugars).

❷ Monosaccharides are actively transported into brush border cells, then out into interstitial fluid.

❸ Proteins are broken down to polypeptides, then amino acids.

❹ Amino acids are actively transported into brush border cells, then out into interstitial fluid.

❺ Movements of the intestinal wall break up fat globules into small droplets. Bile salts coat the droplets so they remain separated.

❻ Pancreatic enzymes digest the fat droplets to fatty acids and monoglycerides.

❼ Monoglycerides and fatty acids diffuse across the plasma membrane's lipid bilayer, into brush border cells.

❽ Inside a brush border cell, the products of fat digestion form triglycerides.

❾ Triglycerides associate with proteins to form lipoproteins, which are expelled by exocytosis into interstitial fluid.

FIGURE 39.10 ▶Animated Summary of digestion and absorption in the small intestine.

tions break the globs into smaller droplets, then bile salts coat the droplets so they remain separated ❺. Compared to big globules, many small droplets present a much greater surface area to lipases that break triglycerides into fatty acids and monoglycerides ❻.

Being lipid soluble, fatty acids and monoglycerides produced by fat digestion can enter a villus by diffusing through the lipid bilayer of brush border cells ❼. Inside these cells, the compounds become recombined into triglycerides ❽ that are packaged together with proteins into lipoprotein particles ❾. Lipoproteins enter lymph vessels, which carry them to the blood.

In some people, pebble-like gallstones form in the gallbladder. Most gallstones are harmless, but some block the bile duct. In this case, the gallbladder or gallstones are usually removed surgically.

bile Liquid mixture of salts, pigments, and cholesterol that aids fat emulsification in the small intestine. Produced in the liver, stored and concentrated in the gallbladder, and secreted into the small intestine.
emulsification Dispersion of fat droplets in a water-based fluid.
gallbladder Organ that stores and concentrates bile.
liver Organ that makes bile, stores glycogen, and detoxifies blood.

Water Absorption

Each day, eating and drinking puts 1 to 2 liters of fluid into your small intestine. Secretions from your stomach, accessory glands, and the intestinal lining add 6 to 7 more liters. About 80 percent of the water that enters the small intestine is absorbed here. Transport of salts, sugars, and amino acids across brush border cells creates an osmotic gradient, and water follows that gradient from chyme into the interstitial fluid.

TAKE-HOME MESSAGE 39.7

What are the roles of the small intestine?

✔ Chemical digestion is completed in the small intestine. Enzymes from the pancreas and enzymes embedded in the membrane of brush border cells break large molecules into smaller, absorbable subunits.

✔ Small subunits (monosaccharides, amino acids, fatty acids, and monoglycerides) enter the internal environment when they are absorbed into the interstitial fluid in a villus.

✔ Most fluid that enters the gut is also absorbed across the wall of the small intestine.

39.8 The Large Intestine

✔ The large intestine, the final portion of the digestive tract, is rich in bacteria.

Not everything that enters the small intestine can or should be absorbed. Contractions propel indigestible material, dead bacteria and mucosal cells, inorganic substances, and some water from the small intestine into the large intestine. The large intestine is wider than the small intestine, but shorter—only about 1.5 meters (5 feet) long.

The large intestine concentrates wastes by actively pumping sodium ions across its wall, into the internal environment. Water follows by osmosis. The resulting compacted digestive waste is **feces**.

Compared with other gut regions, material moves more slowly through the large intestine, which also has a moderate pH. These conditions favor growth of *Escherichia coli* and other bacteria that are part of our normal gut flora. *E. coli* makes vitamin B_{12} that we absorb across the lining of the large intestine.

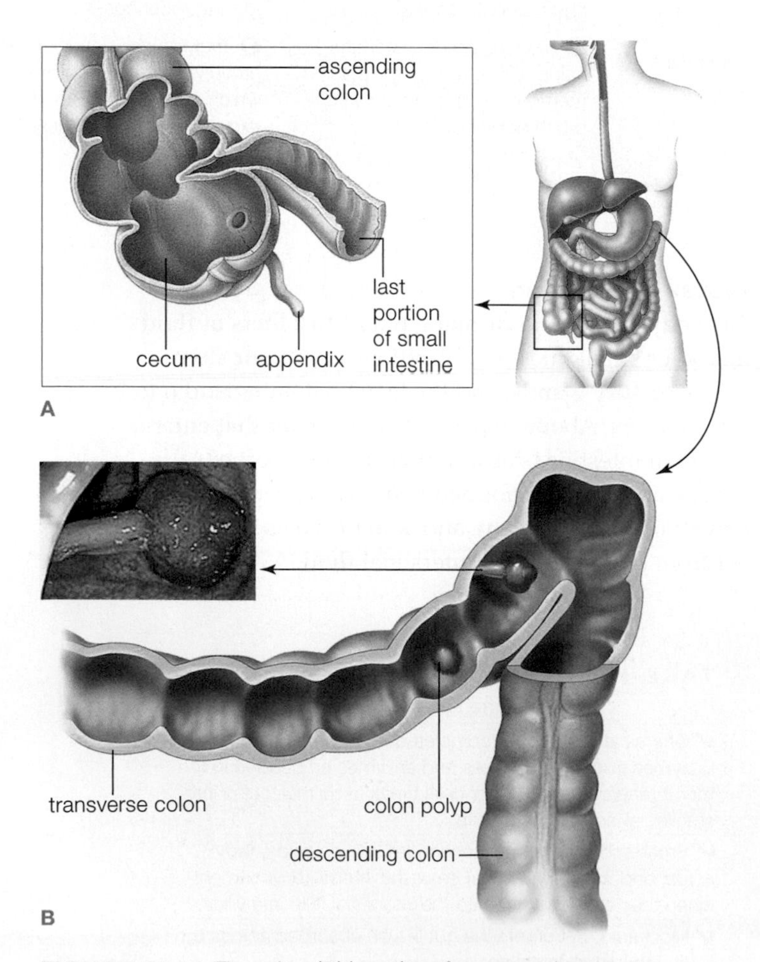

FIGURE 39.11 The colon. (**A**) Location of cecum and appendix. (**B**) Sketch and photo of polyps in the transverse colon.

The first part of the large intestine is a cup-shaped pouch called the cecum (**FIGURE 39.11A**). Herbivores have a large cecum that contains cellulose-digesting bacteria. The human cecum is comparatively small. In humans and many other mammals, a short, tubular **appendix** projects from the cecum. The appendix serves as a reservoir for beneficial bacteria. Appendicitis occurs when a bit of feces lodges in the appendix causing an infection. Appendicitis requires prompt surgical treatment to prevent the inflamed appendix from bursting and releasing bacteria into the abdominal cavity, where they could cause a life-threatening infection.

The cecum connects to the colon, the longest portion of the large intestine. The colon ascends the wall of the abdominal cavity, crosses the cavity, then descends and connects to the rectum. In many people, small growths called polyps develop on the colon wall (**FIGURE 39.11B**). Most colon polyps are benign, but some become cancerous. Colonoscopy, a procedure in which clinicians use a camera to examine the colon, can detect such cancers early, thus increasing the likelihood of curing them before they spread.

After a meal, signals from autonomic nerves cause the colon to contract, propelling feces into the rectum. The resulting stretching of the rectum activates a defecation reflex that opens a sphincter of smooth muscle at the base of the rectum. Voluntary contraction of a sphincter of skeletal muscle at the anus provides control over the timing of defecation (expulsion of feces).

Healthy adults typically defecate once a day, on average. Emotional stress, a diet low in fiber, minimal exercise, dehydration, and some medications can lead to constipation. This means defecation occurs fewer than three times a week, is difficult, and yields small, hardened, dry feces. Occasional constipation usually goes away on its own. A chronic problem should be discussed with a doctor. Infection by a viral, bacterial, or protozoan pathogen can cause an episode of diarrhea, the frequent passing of watery feces. Chronic diarrhea can be a symptom of Crohn's disease or another autoimmune disorder that affects the gut.

appendix Short, tubular projection from the first part of the large intestine; reservoir for beneficial bacteria.
feces Unabsorbed food material and cellular waste that is expelled from the digestive tract.

TAKE-HOME MESSAGE 39.8

What is the function of the large intestine?

✔ The large intestine completes the process of absorption, then concentrates, stores, and eliminates wastes.

39.9 Metabolism of Absorbed Organic Compounds

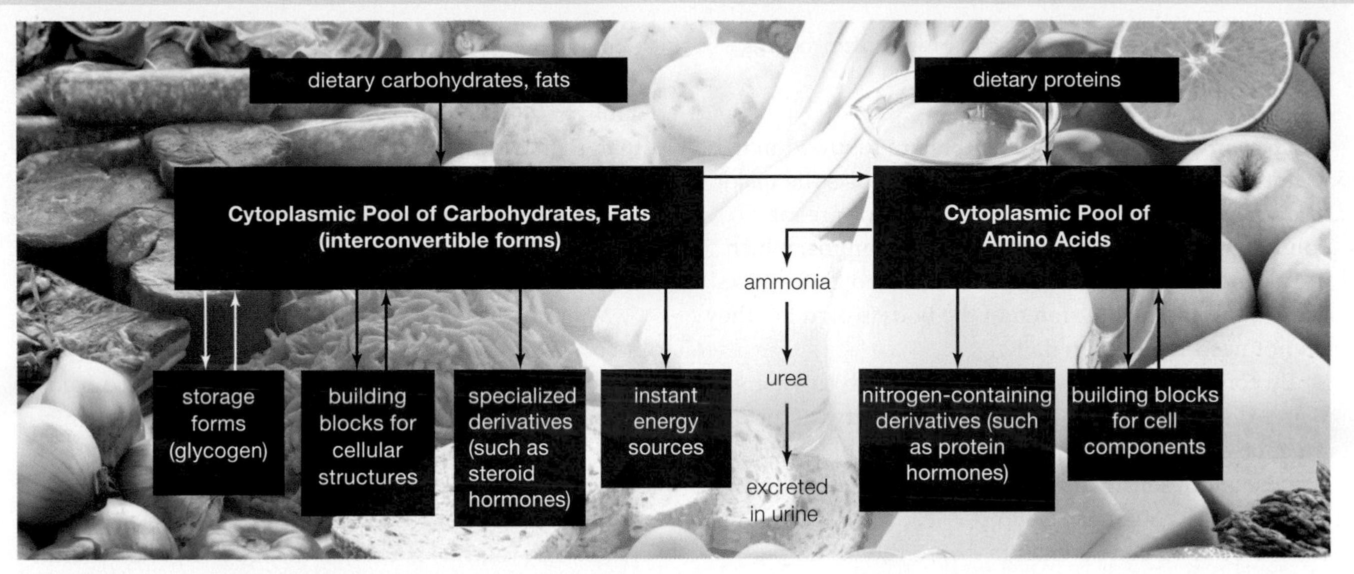

FIGURE 39.12 How cells use dietary carbohydrates, fats, and proteins.

✔ Most absorbed organic compounds are broken down for energy, stored, or used to build larger organic compounds.

Carbohydrates, fats, and proteins are called macronutrients because we require dietary sources of these substances in large amounts. Macronutrients function both as sources of energy and raw materials. As Section 7.7 explained, breakdown products of these molecules serve as reactants in the energy-releasing reactions of aerobic respiration. **FIGURE 39.12** rounds out this picture by illustrating the routes by which the body uses organic compounds obtained from food.

Sugars and starches are hydrolyzed to release glucose, your cells' primary source of energy. When the amount of glucose absorbed exceeds the body's immediate needs, the excess is stored. Blood that flows through capillaries in the small intestine carries glucose-rich blood to the liver, which stores glucose as glycogen. Liver and adipose cells also use glucose to build fats.

In addition to being an energy source, fats are used in cell membranes and to make steroid hormones. They also help you take up fat-soluble vitamins. Your body can remodel carbohydrates to make most of the fatty acids it needs, but some must come from foods. **Essential fatty acids** cannot be synthesized by the body, so these fat components must be obtained from a dietary source such as nuts, seeds, and vegetable oils.

Your body uses amino acids from dietary proteins to build peptides and proteins and to make nucleotides. Most organs do not routinely break down amino acids for a source of energy, but the liver does. Here, the amino group is separated from the amino acid's carbon skeleton, which is then used to fuel the Krebs cycle. The toxic ammonia (NH_3) produced by amino acid breakdown is converted to urea, a somewhat less toxic compound that is excreted in urine.

Essential amino acids are those your body cannot make and must obtain from food. Animal proteins are usually complete, meaning they contain all essential amino acids in the ratio that a human needs. Most plant proteins are incomplete, meaning they lack or contain little of one or more essential amino acids. You can meet your amino acid requirements with a plant-based diet, but doing so requires combining foods. Amino acids missing from one food must be provided by others. As an example, rice and beans together provide all necessary amino acids, but rice alone or beans alone do not. Complementary sources of amino acids do not need to be eaten simultaneously, but can instead be "combined" over the course of a day.

TAKE-HOME MESSAGE 39.9

What happens to macromolecules absorbed from the gut?

✔ Carbohydrates are the body's main sources of energy.

✔ Fats are broken down for energy, and used to make membrane components and steroid hormones.

✔ Proteins provide amino acids for building peptides, proteins, and nucleotides.

essential amino acid Amino acid that the body cannot make and must obtain from food.

essential fatty acid Fatty acid that the body cannot make and must obtain from food.

39.10 Vitamins, Minerals, and Phytochemicals

✔ In addition to macronutrients, normal metabolism requires intake of vitamins and minerals.

Vitamins are organic substances that are required in the diet in very small amounts. TABLE 39.2 lists the major vitamins we need. Vitamins A, D, E, and K are fat soluble. Heat has little effect on these vitamins, which are abundant in both cooked and fresh foods. Because these vitamins can be stored in the body's own fat, they do not need to be eaten daily.

By contrast, water-soluble vitamins such as vitamins B and C are generally not stored in the body, so they must be eaten more frequently. Water-soluble vitamins are also more sensitive to heat, so they are more easily destroyed by cooking.

The body remodels vitamin A into the visual pigment made by rod cells (Section 33.7). In the United States, milk and soy milk products are typically fortified with vitamin A, so a deficiency in this vitamin is rare. In less developed countries, vitamin A deficiency is the most common cause of childhood blindness.

Vitamin D increases calcium uptake from the gut and encourages retention of calcium in bone. It can be obtained from dietary sources, or made in the skin. Vitamin D production by skin requires exposure to sunlight. People with dark skin make less vitamin D than lighter-skinned people, and so are more likely to be deficient in this vitamin.

Vitamin E is an antioxidant (Section 5.6). Being lipid-soluble, it can enter lipid bilayers and it plays an important role in protecting cell membranes.

Vitamin K is a coenzyme that assists enzymes involved in blood clotting. (K denotes the German word *koagulation*.) Absorption of vitamin K produced by bacteria in the large intestine supplements vitamin K extraction from food.

B vitamins are used to build coenzymes that function in a variety of essential pathways. For example, vitamin B_3 (niacin) is used to make NAD, and vitamin B_2 (riboflavin) is used to make FAD. Both of these coenzymes play vital roles in aerobic respiration, as summarized in Section 7.5.

Vitamin C is an antioxidant and it is needed to make collagen, the body's most abundant protein. A deficiency causes scurvy, a disorder in which skin and bones deteriorate and wounds are slow to heal.

Table 39.2 Major Vitamins: Sources, Functions, and Effects of Deficiencies

Vitamin	Main Dietary Sources	Main Functions	Symptoms of Deficiency
Fat-Soluble Vitamins			
A	Orange fruits and vegetables, leafy greens, egg yolk	Component of visual pigments; maintains epithelia	Night blindness, skin problems
D	Fatty fish, egg yolk	Aids uptake and use of calcium	Weak/soft bones, rickets in children
E	Nuts, seeds, vegetable oils helps maintain cell membranes	Antioxidant, aids fat absorption	Muscle weakness, nerve damage
K	Green vegetables	Needed for blood clotting	Impaired blood clotting
Water-Soluble Vitamins			
B_1 (thiamin)	Meats, nuts, legumes	Coenzyme in carbohydrate metabolism	Beri beri (neurological and heart problems)
B_2 (riboflavin)	Meats, eggs, nuts, milk, green vegetables	Component of the coenzyme FAD	Anemia, sores in mouth, sore throat
B_3 (niacin)	Meats, fish, dairy products, nuts, legumes	Component of the coenzyme NAD	Skin and mucous membrane sores, gut pain, diarrhea, psychosis, dementia
B_6	Meats, fish, starchy vegetables, noncitrus fruits	Coenzyme in protein metabolism	Anemia, sores on lips, depression, impaired immune function
B_7 (biotin)	Meats, fish, nuts, legumes, whole grains	Coenzyme in many reactions	Hair loss, dry skin, dry eyes, fatigue, insomnia, depression
B_9 (folic acid)	Meats, fruits, green vegetables, whole grains	Coenzyme in nucleic acid synthesis and amino acid metabolism	Anemia, sores in mouth; deficiency during pregnancy causes neurological birth defects
B_{12}	Meats, seafood, dairy products	Coenzyme in amino acid synthesis	Anemia, fatigue, neurological problems
C (ascorbic acid)	Fruits (especially citrus) and vegetables	Required for collagen synthesis, antioxidant	Scurvy (anemia, bleeding gums, impaired wound healing, swollen joints)

Most people can get all the vitamins they need from a well-balanced diet. Studies have even found an increased death rate among people who take supplemental antioxidants (beta-carotene, vitamin A, and vitamin E). If you take vitamin supplements, be sure the quantities are not excessive.

Minerals are elements that are essential in the diet in small amounts (**TABLE 39.3**). The seven minerals required in the largest quantities are calcium, phosphorus, potassium, sulfur, sodium, chlorine, and magnesium. Calcium and phosphorus, the body's most abundant minerals, are components of teeth and bones. Calcium also plays a role in intercellular signaling, and phosphorus is also used to make nucleic acids. Sodium and potassium are important in nerve function, sulfur is a component of some proteins, chlorine is a component of hydrochloric acid (HCl) in gastric fluid, and magnesium is a cofactor. Deficiencies in any of these seven elements are rare, although calcium levels may be lower in people who eat a solely plant-based diet.

Iodine, another essential mineral, is necessary to produce thyroid hormone, which has roles in development and metabolism (Section 34.7). Iodine is abundant in fish, shellfish, and seaweeds. A deficiency in this mineral causes goiter, which is an enlarged thyroid gland, and disrupts metabolism. In the United States, iodized salt makes deficiency rare.

Iron is an essential component of heme, the chemical group that binds to oxygen in hemoglobin (Section 38.7) and myoglobin. Heme is also a cofactor for many enzymes, including critical components of electron transfer chains. Worldwide, iron is the mineral most commonly deficient in the diet.

Fluorine helps keep teeth strong. In the United States, most public drinking water supplies are fluoridated to help prevent tooth decay.

In addition to vitamins and minerals, a healthy diet should include a variety of **phytochemicals**, also known as phytonutrients. These organic molecules are found in plant foods and they reduce the risk of certain disorders. For example, leafy green vegetables contain the plant pigments lutein and zeaxanthin. A diet low in these phytochemicals increases risk of macular degeneration and blindness (Section 33.8).

mineral With regard to nutrition, an element required in small amounts for normal metabolism.
phytochemical Plant molecules that are not an essential part of the human diet, but may reduce risk of certain disorders; e.g., lutein.
vitamin Organic substance required in small amounts for normal metabolism.

TAKE-HOME MESSAGE 39.10

What roles do vitamins, minerals, and phytonutrients play?

✔ Vitamins are organic molecules with an essential role in metabolism.

✔ Minerals are inorganic substances with an essential role.

✔ Phytochemicals are plant molecules that are not essential but may reduce the risk of certain disorders.

Table 39.3 Major Minerals: Sources, Functions, and Effects of Deficiencies

Mineral	Main Dietary Sources	Main Functions	Symptoms of Deficiency
Calcium	Dairy products, dark green, vegetables	Component of bones and teeth; needed for neuron and muscle action	Stunted growth, fragile bones, nerve impairment
Phosphorus	Whole grains, poultry, red meat	Component of bones, teeth; part of nucleic acids, ATP, phospholipids	Fragile bones, muscle weakness
Sodium	Table salt	Water balance, nerve function	Muscle cramps
Potassium	Meats, fruits, vegetables, grains	Water balance, nerve function	Muscle weakness
Sulfur	Dietary proteins	Component of proteins	Stunted growth
Chlorine	Table salt, vegetables	Acts in water balance, nerve function, formation of stomach acid (HCl)	Muscle cramps, stunted growth, poor appetite
Magnesium	Grains, legumes, nuts dairy products	Coenzyme with role in ATP–ADP cycle; roles in muscle, nerve function	Impaired nerve function
Iodine	Seafood, iodized salt	Component of thyroid hormone	Thyroid and metabolic disorders
Iron	Meats, fish, grains, legumes, green leafy vegetables	Cofactor, component of heme and electron carriers	Iron-deficiency anemia, impaired immune function
Fluorine	Fluoridated water, seafood	Bone, tooth maintenance and amino acid metabolism	Tooth decay

39.11 What Should You Eat?

✔ A healthy diet provides all essential nutrients, as well as fiber, and is low in added salts, sugars, and saturated fats.

Every five years, the United States government issues updated dietary guidelines designed to promote health, prevent disease, and help people maintain a healthy weight. **FIGURE 39.13** shows an example of their recommendations. You can generate your own healthy eating plan by visiting the USDA website: www.choosemyplate.gov.

Fruits, Vegetables, and Whole Grains

Fruits, vegetables, and grains should make up the largest proportion of your diet. These foods provide sugars and starches, your primary sources of energy. Energy stored in food is measured in kilocalories, or as written on food labels, "Calories" (with a capital C). Fruits, vegetables, and grains also provide vitamins, minerals, and dietary fiber.

There are two types of fiber. Soluble fiber consists of polysaccharides that form a gel when mixed with water. Eating foods high in soluble fiber helps lower one's cholesterol level and reduces risk of heart disease.

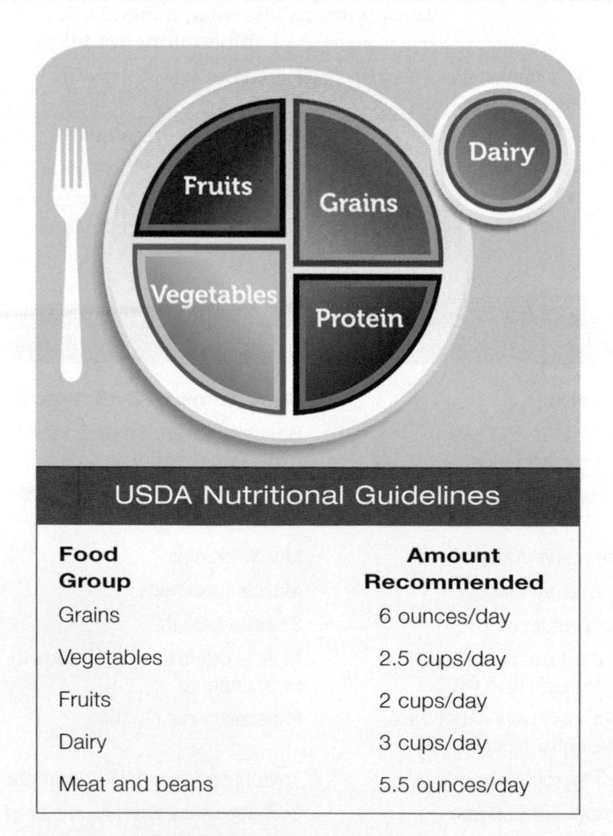

Food Group	Amount Recommended
Grains	6 ounces/day
Vegetables	2.5 cups/day
Fruits	2 cups/day
Dairy	3 cups/day
Meat and beans	5.5 ounces/day

FIGURE 39.13 Example of nutritional guidelines from the United States Department of Agriculture (USDA). These recommendations are for a 20-year-old female, 5 feet 5 inches tall, who weighs 130 pounds and exercises less than 30 minutes a day.

Insoluble fiber such as cellulose does not dissolve. It passes through the human digestive tract more or less intact and its presence helps prevent constipation.

Whole grains provide more vitamins and fiber than their processed counterparts. A whole grain includes all components of a grain seed. For example, whole wheat includes bran (the fiber-rich seed coat) and wheat germ (the protein- and vitamin-rich plant embryo), as well as starchy endosperm. By contrast, white wheat flour is made solely from endosperm.

You may have noticed breads and other grain-based foods labeled as "gluten-free." Gluten is a protein found in wheat and many other grains. Some people have a genetic disorder called celiac disease, in which gluten causes an autoimmune reaction that harms the small intestine. Celiac disease and other types of gluten intolerance are treated by avoiding gluten.

Heart-Healthy Oils

A healthy diet includes fats that provide energy and meet your need for essential fatty acids (**TABLE 39.4**). Essential fatty acids are polyunsaturated (Section 3.5), meaning their tails have two or more double bonds. There are two types of essential fatty acids: omega-3 fatty acids and omega-6 fatty acids. Omega-3 fatty acids, the main fat in oily fish, seem to have special health benefits. Studies suggest that a diet high in omega-3 fatty acids can reduce the risk of cardiovascular disease, lessen the inflammation associated with rheumatoid arthritis, and help diabetics control their blood glucose.

Table 39.4 Main Types of Dietary Lipids

Polyunsaturated Fatty Acids: Liquid at room temperature; essential for health.
- Omega-3 fatty acids
  - Alpha-linolenic acid and its derivatives
  - Sources: Nut oils, vegetable oils, oily fish
- Omega-6 fatty acids
  - Linoleic acid and its derivatives
  - Sources: Nut oils, vegetable oils, meat

Monounsaturated Fatty Acids: Liquid at room temperature. Main dietary source is olive oil. Beneficial in moderation.

Saturated Fatty Acids: Solid at room temperature. Main sources are meat and dairy products, palm and coconut oils. Excessive intake may raise risk of heart disease.

***Trans* Fatty Acids (Hydrogenated Fats):** Solid at room temperature. Manufactured from vegetable oils and used in many processed foods. Excessive intake may raise risk of heart disease.

Oleic acid, the main fat in olive oil, is monounsaturated, which means its carbon tails have only one double bond. Monounsaturated fats have not been shown to have the same benefits as polyunsaturated fats, but can be beneficial if substituted for less healthy fats.

Most animal fats are saturated fats, which are solid at room temperature. Excessive intake of these fats may increase the risk for heart disease, stroke, and some cancers. Coconut oil is rich in saturated fats too, but it has not been shown to have the negative health effects of saturated fats from animal sources.

Trans fats are synthetic fats made from vegetable oils. Their structure makes them even worse for you than saturated fats (Section 3.1). All food labels are now required to show the amounts of *trans* fats, saturated fats, and cholesterol per serving (**FIGURE 39.14**).

Lean Meat and Low-Fat Dairy

Meat (including poultry and fish) is the richest source of protein and iron. Choosing lean meats or fish helps minimize the intake of saturated fats and cholesterol. Eating soybean products such as tofu provides complete protein without harmful fats or cholesterol, as does eating complementary combinations of plant foods such as rice and beans.

Milk and dairy products provide protein, vitamins, and minerals, but whole milk is rich in saturated fats. Low-fat or skim alternatives are better nutritional options. People who are lactose intolerant or who wish to avoid animal products can substitute "milks" made from soybeans, rice, almonds, or other plants.

Minimal Added Salt and Sugar

Prepared foods and drinks are often high in salt and sugar. Salt contains two essential minerals, sodium and chloride, both usually obtained in sufficient quantity from unsalted foods. Ingesting additional salt elevates the body's sodium level, which can raise blood pressure. Food labels show sodium content as a percentage of the recommended daily maximum for a person with normal blood pressure.

Added sugars contribute an astounding 16 percent of the total calories in an average American's diet. Sodas, energy drinks, and sport drinks contribute the greatest proportion of those calories. A typical can of nondiet soda has more than 30 grams of sugar. To avoid the health risks associated with obesity, which we discuss in the section that follows, the American Heart Association recommends that women consume no more than 24 grams of sugar per day and men no more than 36 grams, from all sources.

Nutrition Facts
Serving Size 1 cup (228g)
Servings Per Container 2

Amount Per Serving	
Calories 250	Calories from Fat 110
	% Daily Value*
Total Fat 12g	18%
Saturated Fat 3g	15%
Trans Fat 1.5g	
Cholesterol 30mg	10%
Sodium 470mg	20%
Total Carbohydrate 31g	10%
Dietary Fiber 0g	0%
Sugars 5g	
Protein 5g	
Vitamin A	4%
Vitamin C	2%
Calcium	20%
Iron	4%

* Percent Daily Values are based on a 2,000 calorie diet. Your Daily Values may be higher or lower depending on your calorie needs:

	Calories:	2,000	2,500
Total Fat	Less than	65g	80g
Sat Fat	Less than	20g	25g
Cholesterol	Less than	300mg	300mg
Sodium	Less than	2,400mg	2,400mg
Total Carbohydrate		300g	375g
Dietary Fiber		25g	30g

FIGURE 39.14 How to read a food label. Information on a food label can be used to ensure that you get the nutrients you need without exceeding recommended limits on less healthy substances such as salt and *trans* fats.

FIGURE IT OUT What amount of the fat in a serving of this macaroni and cheese product comes from the least healthy forms of fat (saturated fat and *trans* fat)?

Answer: Of the total fat content in a serving (12 grams), 3 g are saturated fat and 1.5 g are *trans* fat. Thus 4.5 g, or 1/3 the fat, is from unhealthy sources.

TAKE-HOME MESSAGE 39.11

What are the main types of nutrients that humans require and what is the healthiest way to obtain them?

✔ A healthy diet provides energy and all necessary building blocks for assembling essential body components.

✔ Nutritional guidelines are periodically revised in light of new research. Current guidelines call for most calories to come from complex carbohydrates, rather than simple sugars. They also favor fat and protein sources that are low in saturated and *trans* fats.

39.12 Maintaining a Healthy Weight

✔ Maintaining a healthy weight requires balancing energy inputs with energy expenditures.

What Is a Healthy Weight?

Body mass index (BMI) is an indirect measurement of body fat that is based on an individual's height and weight. You can calculate your body mass index with this formula:

$$\text{BMI} = \frac{\text{weight (pounds)} \times 703}{\text{height (inches)}^2}$$

Generally, individuals with a BMI of 25 to 29.9 are said to be overweight. A score of 30 or more indicates **obesity**: an overabundance of fat in adipose tissue that may lead to severe health problems. Conversely, a BMI of 18.5 or lower is considered dangerously underweight. A low BMI can result from anorexia nervosa, an eating disorder in which people restrict their food intake well below a healthy level because of an irrational fear of weight gain.

To maintain your weight, you must balance your caloric intake and your energy output. Energy stored in food is expressed as kilocalories, or Calories (with a capital C). One kilocalorie equals 1,000 calories, which are units of heat energy.

Here is a way to calculate roughly how many kilocalories you should take in daily to maintain a preferred weight. First, multiply the weight (in pounds) by 10 if you are not active physically, by 15 if you are moderately active, and by 20 if you are highly active. Next, subtract one of the following amounts from the multiplication result:

Age:		*Subtract:*
	25–34	0
	35–44	100
	45–54	200
	55–64	300
	Over 65	400

Table 39.5 Energy Expenditures

Activity	Energy expended (Calories/kilogram of body weight/hour)
Sitting quietly	1
Standing	2
Walking slowly	3
Bicycling to class	4–7
Swimming	6–10
Basketball game	8
Running (5 mph)	8

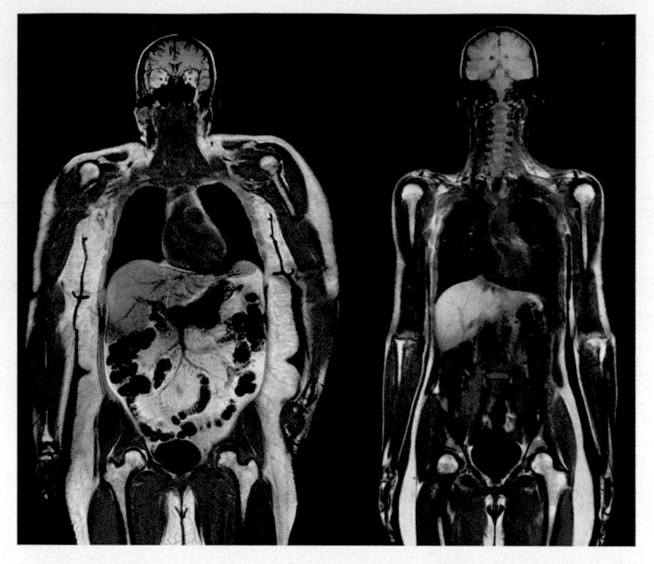

FIGURE 39.15 MRIs (magnetic resonance images) of an obese woman (left) and a woman of normal body weight (right). With obesity, the abdomen fills with fat that squashes internal organs.

For example, if you are 25 years old, highly active, and weigh 120 pounds, you will require 120 × 20 = 2,400 kilocalories daily to maintain weight. Take in more and you gain weight; take in less and you lose it.

Even at rest, you use energy in essential processes such as breathing, generating body heat, fighting off pathogens, moving material though your digestive tract, and circulating your blood. The rate at which you burn energy in these activities is your **basal metabolic rate**. This rate varies among individuals. Big, muscular bodies use more energy than smaller, less muscular ones. Men tend to be more muscular than women of the same weight, so their metabolic rate is higher. In both sexes, metabolic rate slows with age. Thyroid hormone level influences basal metabolic rate too. People with a high level of this hormone have higher metabolic rate than those with a lower level.

Dieting can influence basal metabolic rate. Often, when people eat less, resting metabolic rate declines. This mechanism evolved to promote survival when food became scarce. However, it is a source of great frustration to dieters who have chosen to limit their food intake in the hope of losing weight.

Exercise can help you lose weight because it requires energy. TABLE 39.5 shows energy expenditure during various activities. Exercise that increases the body's muscle mass also helps increase resting metabolic rate.

Why Is Obesity Unhealthy?

Obesity increases the risk for a long list of diseases including high blood pressure, type 2 diabetes, heart

CREDITS: (15) Marty Chobot/National Geographic Creative; (Table 39.5) art, © Cengage Learning; top photo, Yuri Arcurs/Shutterstock; bottom photo, Flashon Studio/Shutterstock.

Your Microbial "Organ" (revisited)

For a probiotic to be effective, it must include beneficial bacteria or spores capable of surviving a trip through the highly acidic stomach, adhering to the colon wall, and growing amidst competing bacteria. Researchers are studying which bacteria meet these requirements and thus are best suited for use in probiotic pills and foods such as yogurt.

An alternative medical procedure, called a fecal transplant, delivers bacteria that are known to do well in the colon exactly where they are needed. During this procedure, a physician takes feces from a healthy donor and uses an endoscope (a long tube with a camera) to insert it deep in the colon of a recipient. The recipient is a person with a chronic debilitating bowel disease, whose colon has been cleansed of bacteria prior to treatment. The goal is to establish a healthy colon microbiota and thus restore the recipient to health. Many patients who have undergone the procedure say it dramatically improved their life, but some physicians worry that the procedure can spread pathogens.

---

disease, stroke, gallbladder disease, osteoarthritis, sleep apnea, and cancers of the uterus, breast, prostate gland, kidney, and colon.

An obese person's internal organs can be squashed by fat (**FIGURE 39.15**), impairing their function. For example, excessive fat in the abdomen impairs the diaphragm's ability to descend downward during inhalation, so breathing becomes more difficult.

Obesity also impairs function at the cellular level. Adipose cells of people who are at a healthy weight hold a moderate amount of triglycerides. In obese people, adipose cells become overstuffed. Like cells stressed in other ways, the overstuffed adipose cells respond by sending out chemical signals that summon up an inflammatory response (Section 37.5). The resulting chronic inflammation harms organs throughout the body and increases the risk of cancer. Overstuffed adipose cells also increase their secretion of chemical messages that interfere with insulin's effects. Recall that insulin encourages cells to take up sugar from the blood (Section 34.8). Type 2 diabetes arises when cells stop responding to insulin.

Causes of Obesity

Geneticists have pinpointed some genes that affect weight. The first to be discovered, the *ob* gene, encodes leptin, a hormone made by adipose cells. Leptin acts in the brain and suppresses appetite. Mutations in the human *ob* gene are very rare, but variations in another obesity-related gene, *fto*, are common. About 16 percent of people of European ancestry are homozygous for an *fto* allele that predisposes them to obesity. Compared to people with two low-risk alleles, those homozygous for the high-risk allele are almost twice as likely to be obese. The function of the protein encoded by the *fto* gene is unknown. We do, however, know that the gene is expressed most strongly in the brain.

Genetics can explain why one person is more likely than another to be overweight, but it cannot explain a national trend toward weight gain. In 1960, only about 15 percent of people in the United States were obese; since then, the obesity rate has more than doubled. Many factors contributed to this increase. The proportion of meals consumed outside the home increased, as did the average portion size of a restaurant meal. Consumption of calorie-rich sodas rose, and physical activity decreased. We spend more time in front of televisions and computers, and fewer of us have jobs that require physical exertion.

Preventing obesity is important. Once a person becomes obese, dieting alone is seldom effective in restoring a normal body weight over the long term. A variety of drugs can reduce weight somewhat, but all have negative side effects and must be taken continually to prevent rebound weight gain. Surgical procedures that reduce stomach volume produce a dramatic sustainable weight loss, but these drastic interventions are expensive and can result in serious complications.

TAKE-HOME MESSAGE 39.12

How does weight affect health?

✔ A person who balances caloric intake with energy expenditures will maintain current weight.

✔ Obesity raises the risk of heart disease, type 2 diabetes, some cancers, and other disorders. These problems may arise because overstuffed adipose cells summon up inflammatory responses in organs throughout the body.

---

basal metabolic rate Rate at which the body uses energy when you are at rest.
body mass index (BMI) Measure that indicates an individual's degree of body fat based on height and weight.
obesity An overabundance of fat, which increases health risks.

summary

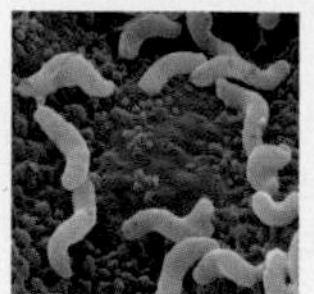

Section 39.1 The human gut contains trillions of living microbes. The numbers and types of these organisms affect digestive function and also influence general health by their effects on the immune system. **Probiotics** are foods that contain living bacteria whose presence in our gut benefits our health.

Section 39.2 A digestive system breaks food down into molecules small enough to be absorbed into the internal environment. It also stores and eliminates unabsorbed materials. Some invertebrates have a **gastrovascular cavity**: a saclike gut with a single opening.

Most animals and all vertebrates have a **complete digestive tract** with a mouth at one end and an **anus** or **cloaca** at the other. An **esophagus** carries food to the **stomach** for digestion, then absorption occurs in the **intestine**. Variations in the structure of vertebrate digestive systems are adaptations to particular diets. For example, a bird's **crop** stores food and its **gizzard** grinds food up. The multiple stomachs of ruminants allow them to digest cellulose-rich plant parts.

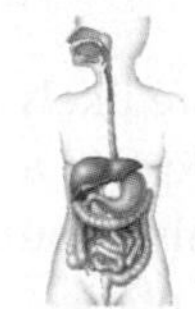

Section 39.3 The human digestive system includes the digestive tract and accessory organs that secrete material into its **lumen**. **Peristalsis** moves food along the tract and **sphincters** regulate this passage. The human intestine has two functional regions: The **small intestine** carries out digestion and most nutrient absorption. The **large intestine** concentrates undigested wastes, which are stored in the **rectum** until eliminated.

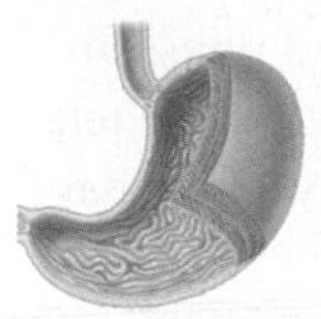

Sections 39.4, 39.5 Digestion starts in the mouth, where teeth break food into bits and it mixes with saliva from **salivary glands**. Saliva contains salivary amylase, which begins the process of starch digestion. Protein digestion begins in the stomach, a muscular sac with a glandular lining that secretes **gastric fluid**. This fluid contains acid, enzymes, and mucus. It mixes with food to form **chyme**.

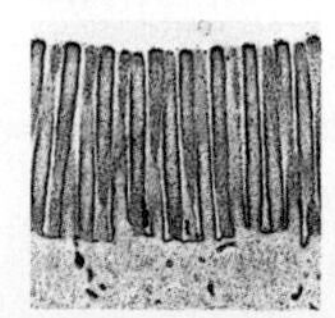

Sections 39.6, 39.7 The small intestine is the longest portion of the gut and has the largest surface area. Its highly folded lining has many **villi** at its surface. Each multicelled villus has a covering of **brush border cells**. These cells have **microvilli** that increase their surface area for digestion and absorption.

Chemical digestion is completed in the small intestine through the action of enzymes from the pancreas, bile from the gallbladder, and enzymes embedded in the plasma membrane of brush border cells.

Carbohydrates are broken into monosaccharides, which are actively transported across brush border cells and enter the blood. Similarly, proteins are broken into amino acids, which are transported across these cells and enter blood.

Bile made in the **liver** and stored and concentrated in the **gallbladder** aids in the **emulsification** of fats. Monoglycerides and fatty acids diffuse into the brush border cells. Here, they recombine as triglycerides, which get a protein coat and are moved by exocytosis into interstitial fluid. The lipoproteins then enter lymph vessels that deliver them to blood.

The small intestine is also the site of most water absorption. Water moves out of the gut by osmosis.

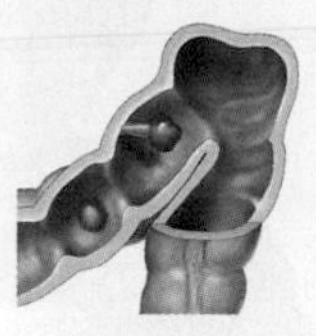

Section 39.8 Additional water and ions are absorbed in the large intestine. The **appendix** is a short extension from the first part of the large intestine (the cecum). The longest portion of the large intestine, the colon, compacts undigested solid wastes as **feces**. Feces are stored in the rectum, a stretchable region just before the anus.

Sections 39.9–39.11 Organic macromolecules (carbohydrates, proteins, and fats) serve as sources of energy and raw materials. Excess sugar is either stored as glycogen in the liver or used to make fat. The body builds most fatty acids and amino acids, but **essential fatty acids** and **essential amino acids** must be obtained from food. The diet must also include **vitamins**, which are small organic molecules, and **minerals**, which are inorganic. **Phytochemicals** promote good health. Plant-based diets can meet all human nutritional needs if foods are combined properly.

Excessive amounts of sugar, salt, and saturated fat are associated with increased health risks, whereas a diet high in fiber and polyunsaturated fats has health benefits.

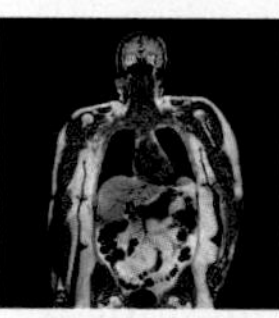

Section 39.12 Maintaining body weight requires balancing energy intake and output. **Body mass index** indicates whether a given height and weight is healthy. **Basal metabolic rate**, the energy expended at rest, varies with age and other factors. With **obesity**, fat deposits press on internal organs and overstuffing of adipose cells leads to chronic inflammation. Obesity has a genetic component.

self-quiz

Answers in Appendix VII

1. A digestive system functions in ________ .
   a. secreting enzymes
   b. absorbing compounds
   c. eliminating wastes
   d. all of the above

2. The enzyme pepsin is activated in the ________ .
   a. mouth
   b. stomach
   c. small intestine
   d. large intestine

3. Most nutrients are absorbed in the ________ .
   a. mouth
   b. stomach
   c. small intestine
   d. large intestine

4. A bird's keratin-lined ________ helps it grind up food.
   a. pharynx
   b. crop
   c. gizzard
   d. gastrovascular cavity

5. Monosaccharides and amino acids absorbed from the gut enter ________ .
   a. blood vessels
   b. lymph vessels
   c. fat droplets
   d. both b and c

data analysis activities

Human Adaptation to a Starchy Diet The human *AMY-1* gene encodes salivary amylase, an enzyme that breaks down starch. The number of copies of this gene varies, and people who have more copies generally make more enzyme. In addition, the average number of *AMY-1* copies differs among cultural groups.

George Perry and his colleagues hypothesized that duplications of the *AMY-1* gene would be selectively advantageous in cultures in which starch is a large part of the diet. To test this hypothesis, the scientists compared the number of copies of the *AMY-1* gene among members of seven cultural groups that differed in their traditional diets. **FIGURE 39.16** shows their results.

1. Starchy tubers are a mainstay of Hadza hunter–gatherers in Africa, whereas fishing sustains Siberia's Yakut. Almost 60 percent of Yakut had fewer than 5 copies of the *AMY1* gene. What percentage of the Hadza had fewer than 5 copies?

2. None of the Mbuti (rain-forest hunter–gathers) had more than 10 copies of *AMY-1*. Did any European Americans?

3. Do these data support the hypothesis that a starchy diet favors duplications of the *AMY-1* gene?

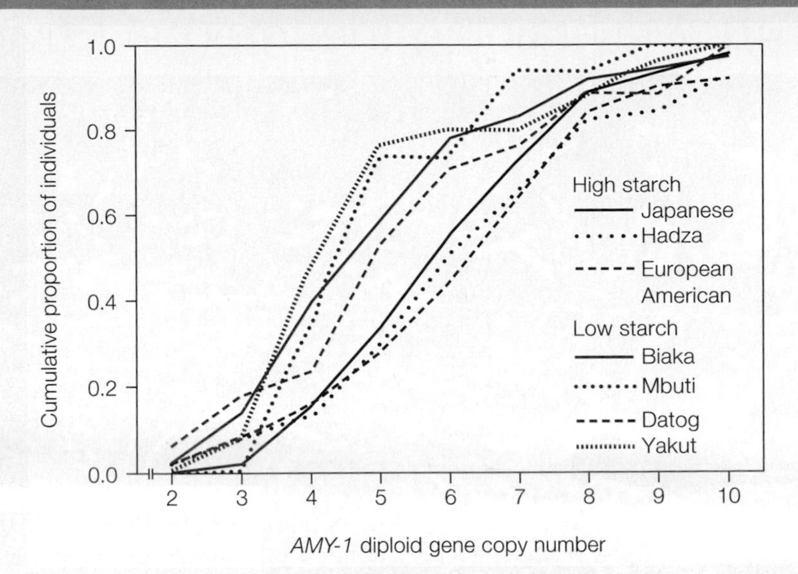

FIGURE 39.16 Number of copies of the *AMY-1* gene among members of cultures with traditional high-starch or low-starch diets. The Hadza, Biaka, Mbuti, and Datog are tribes in Africa. The Yakut live in Siberia.

6. Bacteria in the ________ make essential vitamins.
   a. stomach
   b. small intestine
   c. colon
   d. esophagus

7. The low pH of the ________ aids in protein digestion.
   a. stomach
   b. small intestine
   c. large intestine
   d. esophagus

8. Most water that enters the gut is absorbed across the lining of the ________ .
   a. stomach
   b. small intestine
   c. large intestine
   d. esophagus

9. Iron is the ________ most often deficient in the diet.
   a. phytochemical
   b. vitamin
   c. mineral
   d. enzyme

10. Eating a healthy diet entirely of plant foods requires combining foods to obtain all necessary ________ .
   a. saturated fats
   b. monosaccharides
   c. phytochemicals
   d. amino acids

11. Tiny filaments called ________ increase the surface area of a brush border cell.
   a. villi
   b. microvilli
   c. cilia
   d. flagella

12. ________ is (are) a good source of omega-3 fatty acids that can reduce the risk of heart disease.
   a. Oatmeal
   b. Legumes
   c. Fish
   d. Corn syrup

13. A very low BMI indicates a person is ________ .
   a. obese
   b. overweight
   c. underweight
   d. not getting enough exercise

14. Match each substance with its description.

___ gastrin	a. enzyme that acts in stomach
___ dentin	b. emulsifies fats
___ bicarbonate	c. breaks down polysaccharides
___ bile	d. raises the pH of chyme
___ salivary amylase	e. makes up the bulk of a tooth
___ pepsin	f. hormone made by stomach

15. Match each organ with its digestive function.

___ gallbladder	a. final stop for digestive waste
___ salivary gland	b. makes bile
___ colon	c. compacts undigested residues
___ liver	d. adds enzymes to small intestine
___ esophagus	e. delivers food to the stomach
___ rectum	f. stores, secretes bile
___ stomach	g. secretes gastric fluid
___ pancreas	h. secretes enzyme that begins starch digestion

critical thinking

1. Some people who have gallstones experience pain after they eat, with fatty meals causing the greatest discomfort. Why are fatty meals most likely to trigger gallbladder pain?

2. A python can survive by eating a large meal once or twice a year. When it does eat, microvilli in its small intestine lengthen fourfold and its stomach pH drops from 7 to 1. Explain the benefits of these changes.

3. Although rabbits cannot digest the cellulose in their all-plant diet, bacteria that live in their cecum can. However, to make full use of the nutrients released by bacterial action, a rabbit must eat its feces. Why is this second round of ingestion necessary?

CREDIT: (16) George Perry, et al., Diet and evolution of human amylase gene copy number variation, *Nature Genetics* 39, 1256–1260 (2007).

40 Maintaining the Internal Environment

LEARNING ROADMAP

This chapter's discussion of fluid balance draws on your knowledge of pH (2.6), osmosis (5.8), aerobic respiration (7.2), protein metabolism (39.9), body fluids (31.2), chemoreceptors (33.2) and the adrenal glands (34.9). Its discussion of temperature regulation refers to properties of water (2.5), forms of energy (5.2), and negative feedback controls (31.9).

MAINTAINING THE EXTRACELLULAR FLUID

Animals have organs that maintain the composition and volume of the extracellular fluid within a narrow range. Variations in the structure of these organs adapt animals to different habitats.

THE HUMAN URINARY SYSTEM

Humans have two kidneys that filter blood and make urine. The urine flows through ureters to the bladder, and then out of the body through the urethra.

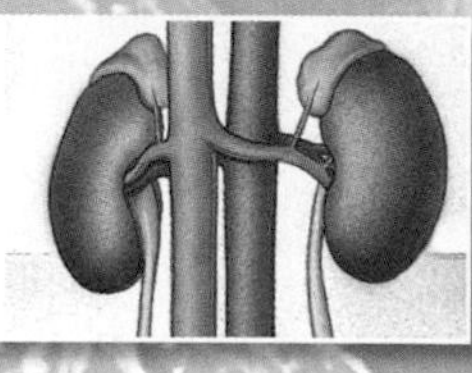

WHAT KIDNEYS DO

Fluid and small solutes from blood enter kidney tubules. Most water and solutes are reabsorbed. Fluid not reabsorbed becomes urine. Hormones adjust urine composition.

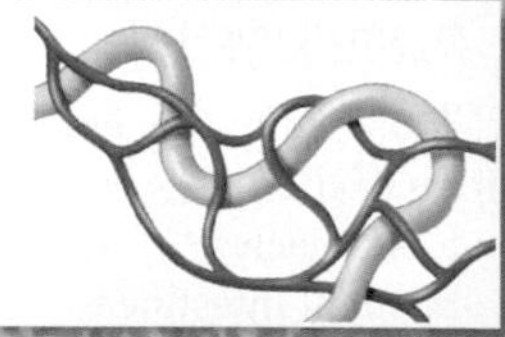

MODES OF THERMOREGULATION

Heat lost to the environment, heat gained from the environment, and heat produced by metabolic activity determine body temperature. Endotherms produce more heat than ectotherms.

RESPONDING TO HEAT AND COLD

Endotherms can stay warm and active even in cold environments by turning up their heat production. Panting and sweating help animals stay cool in hot environments.

The male urethra does double duty, conveying both urine and sperm to the body surface. We discuss its reproductive function in the next chapter. Chapter 41 also explains how urine tests for pregnancy work. The types of environmental conditions an animal can tolerate are an aspect of its ecological niche, which we discuss in Chapter 45.

40.1 Truth in a Test Tube

Light or dark? Clear or cloudy? A lot or a little? Asking about and examining urine is an ancient art (**FIGURE 40.1**). About 3,000 years ago in India, the pioneering healer Susruta reported that some patients produced an excessive amount of sweet-tasting urine that attracted insects. In time, the disorder that causes these symptoms was named diabetes mellitus, which means "passing honey-sweet water." Doctors still diagnose diabetes by testing the sugar level in urine, although they have replaced the taste test with chemical analysis.

Today, physicians routinely check the pH and solute concentrations of urine to monitor their patients' health. Acidic urine suggests metabolic problems. Alkaline urine can indicate an infection. Damaged kidneys will produce urine with an unusually high concentration of the plasma protein albumin. An abundance of certain salts can result from dehydration or trouble with the hormones that control kidney function. Urine tests can also detect chemicals produced by cancer of the prostate gland and of organs of the urinary tract.

A variety of over-the-counter urine tests are available to determine hormone levels. If a woman wants to become pregnant, she can use one test to keep track of the amount of luteinizing hormone, or LH, in her urine. About midway through a menstrual cycle, LH triggers the release of an egg from an ovary. Another over-the-counter urine test can reveal whether she has become pregnant. Still other tests allow older women to check for declining hormone levels in urine, a sign that they are entering menopause.

Not everyone is in a hurry to have their urine tested. Olympic athletes can be stripped of their medals when mandatory urine tests reveal they use prohibited drugs. Major League Baseball players agreed to urine tests only after repeated allegations that certain star players took prohibited synthetic hormones. The National Collegiate Athletic Association (NCAA) tests urine samples from about 3,300 student athletes per year for any performance-enhancing substances as well as for "street drugs."

If you use alcohol, nicotine, marijuana, cocaine, or any other psychoactive drug, your urine tells the tale. After the active ingredient in these drugs enters blood, the liver converts it to other compounds, which we call metabolites. As kidneys filter blood, they add these drug metabolites to the forming urine. Depending on the type of drug and the dosage and frequency of its use, eliminating all drug metabolites from the body can take a few days to nearly a month.

FIGURE 40.1 A seventeenth-century painting depicts a Dutch physician examining a urine specimen. Urine's consistency, color, odor, and—at least in the past—taste supply information about a person's health. Urine forms inside kidneys, and it provides clues to abnormal changes in the volume and composition of blood and interstitial fluid.

It is a tribute to the urinary system that urine is such a remarkable indicator of health, hormonal status, and the types of chemicals we take in. Each day, a pair of fist-sized kidneys filter all of the blood in an adult human body, and they do so more than forty times. When everything goes well, the kidneys rid the body of excess water and excess or harmful solutes, including a variety of metabolites, toxins, hormones, and drugs.

So far in this unit, you have considered several organ systems that keep cells supplied with oxygen, nutrients, water, and other substances. We turn now to the mechanisms that maintain the composition, volume, and temperature of the internal environment.

CREDITS: (opposite) Susumu Nishinaga/Science Source; (1) Alfredo Dagli Orti/The Art Archive/Corbis.

40.2 Regulating Fluid Volume and Composition

✔ All animals constantly acquire and lose water and solutes, yet they must keep the volume and composition of their body fluids stable.

By weight, an animal consists mostly of water, with dissolved salts and other solutes. Fluid outside cells—the extracellular fluid—is the cells' environment. In vertebrates, interstitial fluid (fluid between cells) and plasma (the fluid portion of the blood) constitute the bulk of the extracellular fluid (Section 31.2). To keep the solute composition and volume of extracellular fluid within the ranges cells can tolerate, water and solute gains must equal water and solute losses. An animal loses water and solutes in excretions, exhalations, and secretions. It gains water by eating and drinking. In aquatic animals, water also moves into or out of the body by osmosis across the body surface (Section 5.8).

Metabolic wastes, particularly carbon dioxide and ammonia, affect the composition of the extracellular fluid. Aerobic respiration produces water and carbon dioxide. Carbon dioxide diffuses out across the body surface or leaves with the help of respiratory organs. Protein breakdown produces **ammonia** (**FIGURE 40.2**). In most animals, special organs rid the body of ammonia, other unwanted solutes, and any excess water.

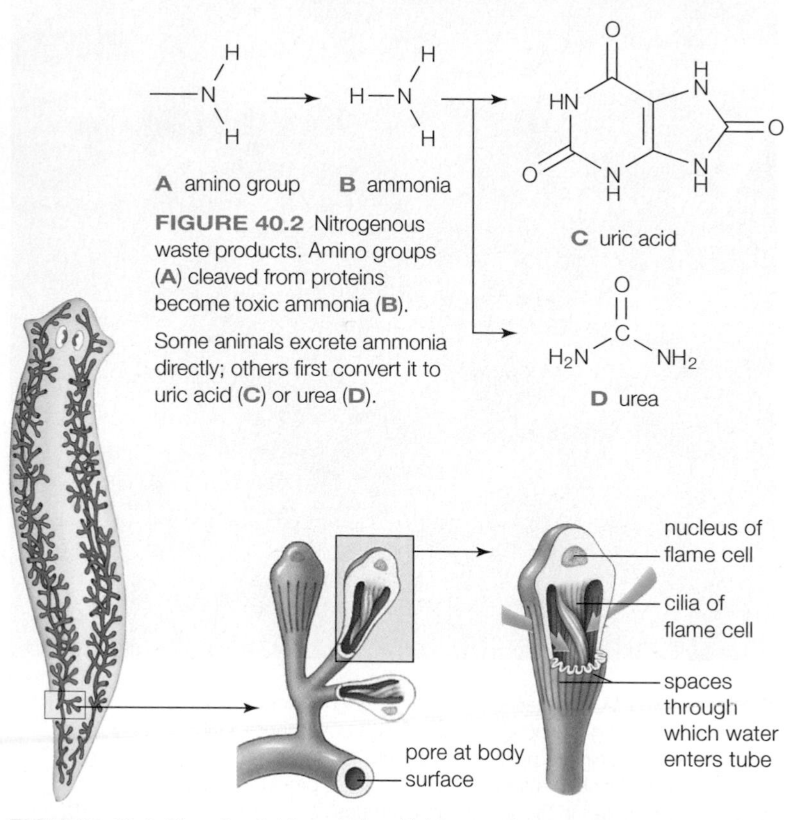

FIGURE 40.2 Nitrogenous waste products. Amino groups (**A**) cleaved from proteins become toxic ammonia (**B**).

Some animals excrete ammonia directly; others first convert it to uric acid (**C**) or urea (**D**).

FIGURE 40.3 Planarian fluid regulation. Fluid enters bulbs at the tips of branching protonephridia. Each bulb has a flame cell and a tubule cell. Beating of flame cell cilia draws fluid in between the cells. The fluid exits the body through a pore.

Fluid Regulation in Invertebrates

Marine invertebrates usually have body fluids with the same solute concentration as seawater. As a result, osmosis produces no net movement of water into or out of the body. By contrast, the body fluids of planarian flatworms and other freshwater invertebrates, have a higher solute concentration than the surrounding water, so water enters these animals by osmosis. Planarians eliminate excess water and waste solutes by means of a pair of branching, tubular excretory organs (protonephridia) that run the length of the body (**FIGURE 40.3**). Along the tubes are bulbs tipped by ciliated flame cells. Cilia movement draws interstitial fluid into the tubes, propels it along, and forces it out of the body through pores at the body surface.

In most animals, a circulatory system interacts with organs that excrete unwanted solutes. Consider the

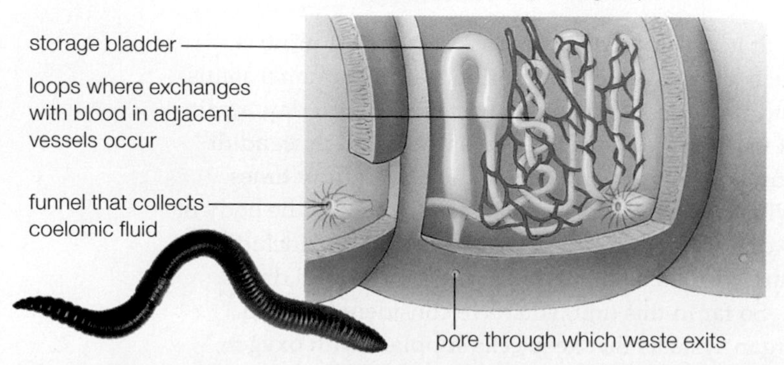

FIGURE 40.4 Earthworm fluid regulation. Coelomic fluid enters a nephridium (green). As fluid travels through it, essential solutes leave this tube and enter adjacent blood vessels (red). Ammonia-rich waste exits the body through a pore.

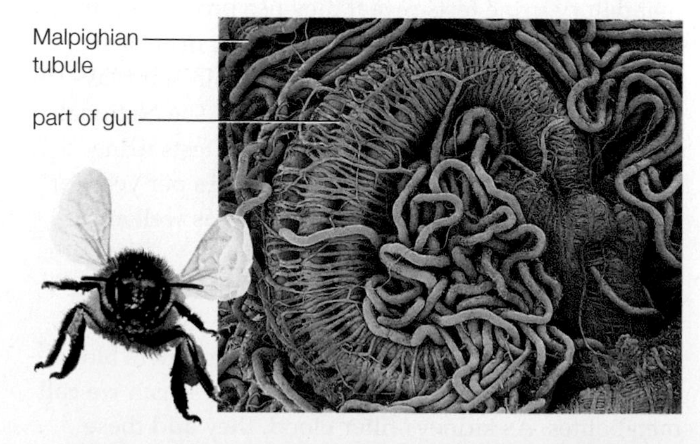

FIGURE 40.5 Insect fluid regulation. A honeybee's Malpighian tubules (gold) are outpouchings of the gut (pink). Uric acid and other waste solutes move from the hemolymph into the tubule interior. The tubules deliver the wastes to the gut for elimination through the anus.

earthworm, a segmented annelid with a fluid-filled body cavity (a coelom) and a closed circulatory system. Most body segments have a pair of tubular excretory organs called nephridia that collect coelomic fluid (**FIGURE 40.4**). As fluid flows through a nephridium, essential solutes and some water exit the nephridium and enter adjacent blood vessels. Ammonia remains in the tube and exits the body through a pore.

Land-dwelling arthropods such as insects do not excrete ammonia. Instead, enzymes in their blood convert ammonia to **uric acid** (**FIGURE 40.2C**). Uric acid and other waste solutes are actively transported into excretory organs called Malpighian tubules, which connect to and empty into the gut (**FIGURE 40.5**). Ammonia can only be excreted when dissolved in water, but uric acid is excreted as crystals mixed with just a tiny bit of water to produce a thick paste.

Fluid Regulation in Vertebrates

Body fluids of bony fishes are less salty than seawater, but saltier than fresh water. A marine bony fish loses water by osmosis across its body surfaces. It replaces this lost water by gulping seawater, then pumping salt out through its gills (**FIGURE 40.6A**). Like other vertebrates, bony fishes have a pair of **kidneys**, organs that filter the blood and produce urine. **Urine** is a mixture of water and soluble wastes. Marine bony fishes produce a small amount of urine. In contrast, a freshwater bony fish produces a large volume of urine because water continually enters its body by osmosis. Solutes lost in urine are offset by solutes absorbed from the gut, and actively transported in across the gills (**FIGURE 40.6B**).

Waterproof skin and water-conserving kidneys adapt amniotes to life on land. Birds and other reptiles convert waste ammonia to uric acid, whereas mammals convert most of it to **urea** (**FIGURE 40.2D**). It takes twenty times more water to excrete a gram of urea than a gram of uric acid. Thus, a typical mammal requires more water than a similarly sized bird or reptile.

Variations in kidney structure adapt mammals to different habitats. Mammals with limited or no access to fresh water tend to have large kidneys for their size and they lose little water in their urine (**FIGURE 40.7**).

ammonia Nitrogen-containing compound that is a waste product of amino acid and nucleic acid breakdown.
kidney Organ of the vertebrate urinary system that filters blood, adjusts its composition, and forms urine.
urea Main nitrogen-containing compound in urine of mammals.
uric acid Main nitrogen-containing compound in the urine of insects, as well as birds and other reptiles.
urine Mixture of water and soluble wastes formed and excreted by the vertebrate urinary system.

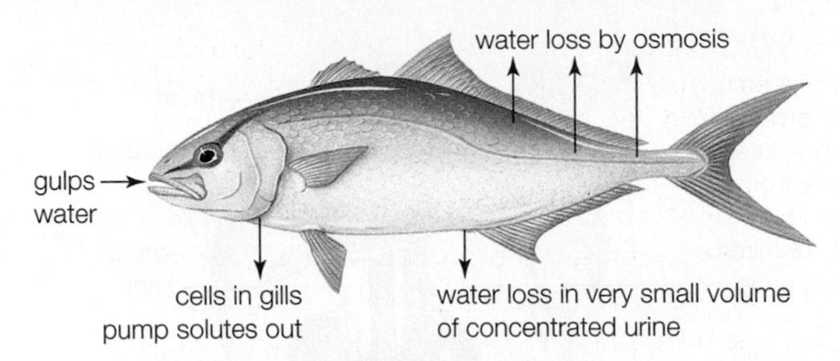

A Marine bony fish with body fluids less salty than the surrounding water; the fish is hypotonic relative to its environment.

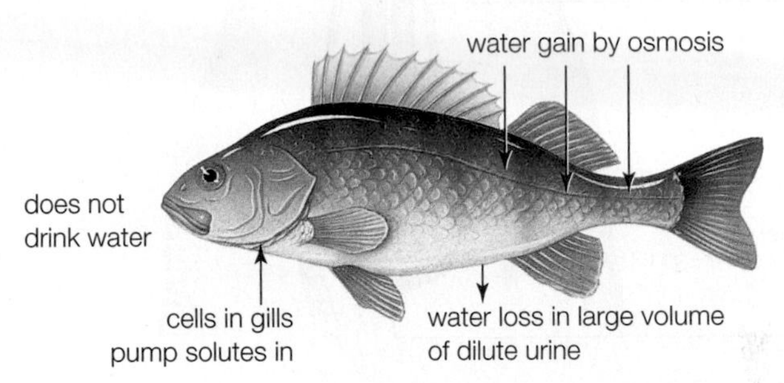

B Freshwater bony fish with body fluids saltier than the surrounding water; the fish is hypertonic relative to its environment.

FIGURE 40.6 Fluid–solute balance in freshwater and marine bony fishes.

	Kangaroo Rat	Human
Daily Water Gain (milliliters):	60 ml	2600 ml
By ingesting solids	10%	33%
By ingesting liquids	0%	54%
By metabolism	90%	13%
Daily Water Loss (milliliters):	60 ml	2600 ml
In urine	23%	58%
In feces	4%	8%
By evaporation	73%	34%

FIGURE 40.7 Comparison of water gains and losses for a desert kangaroo rat and a human. In both species, fluid gains must balance fluid losses.

TAKE-HOME MESSAGE 40.2

How do animals maintain the volume and composition of their body fluid?

✔ Animals must rid their bodies of waste ammonia; many convert it to urea or uric acid before excreting it.

✔ Most animals have organs that interact with a circulatory system to remove wastes from the blood and excrete them.

✔ Invertebrate excretory organs include ammonia-excreting nephridia of earthworms and uric acid–excreting Malpighian tubules of insects.

✔ All vertebrates have kidneys. The volume of urine and type of nitrogen-containing wastes excreted vary among groups.

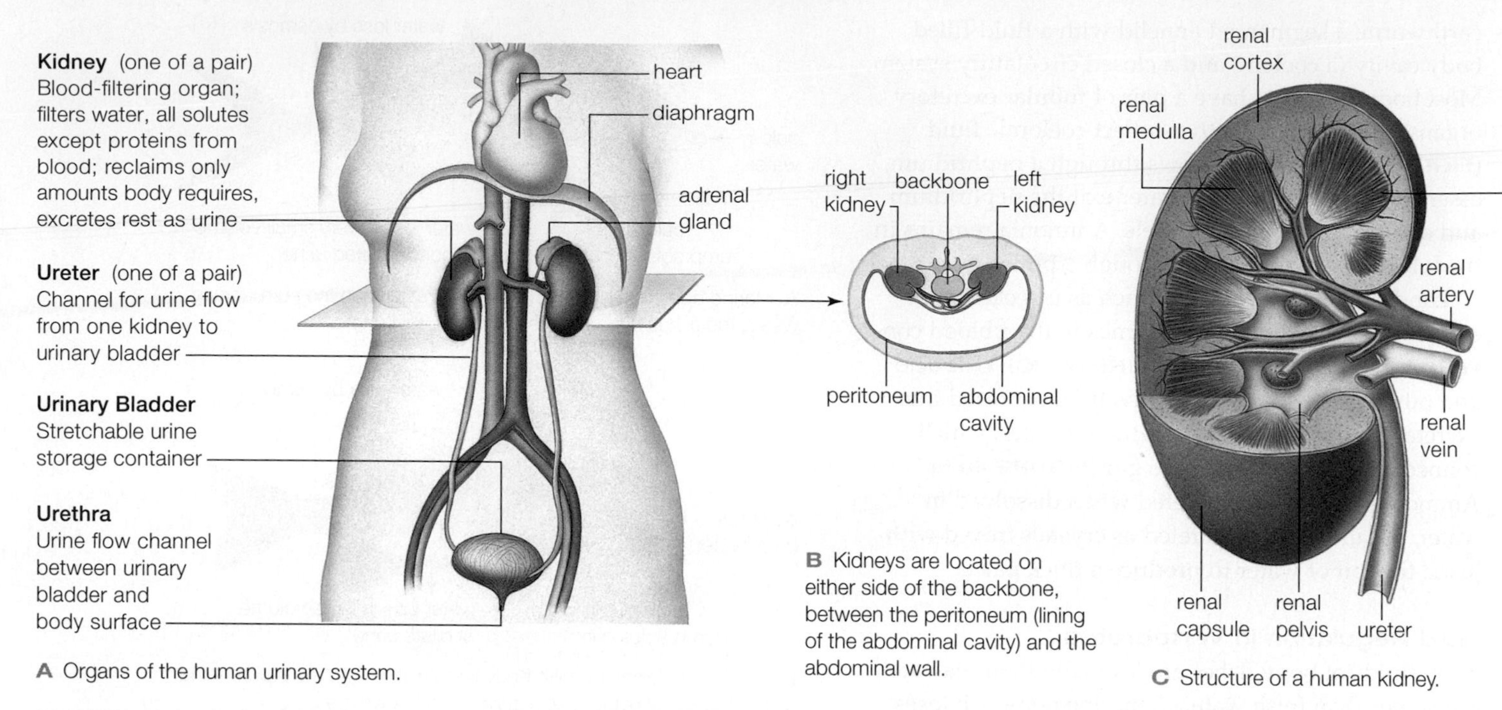

A Organs of the human urinary system.

B Kidneys are located on either side of the backbone, between the peritoneum (lining of the abdominal cavity) and the abdominal wall.

C Structure of a human kidney.

FIGURE 40.8 ▸**Animated** Components of the human urinary system and their functions.

✔ Kidneys filter water and solutes from the blood, adjust the volume and composition of this filtrate, and return most of it to the blood. Fluid not returned to the blood becomes urine.

Organs of the Urinary System

FIGURE 40.8A shows organs of the human urinary system. The bean-shaped, fist-sized kidneys lie just beneath the peritoneum (the membrane that lines the abdominal cavity), and to the left and right of the backbone (**FIGURE 40.8B**). A renal capsule of dense, irregular connective tissue (Section 31.4) is the outermost layer of the kidney. (Renal means "relating to the kidneys.") Inside the renal capsule, kidney tissue is divided into two zones, the outer renal cortex and the inner renal medulla (**FIGURE 40.8C**).

The kidney filters blood and produces urine, which collects in a central cavity called the renal pelvis. From there, the urine flows through a tubular **ureter** to the **urinary bladder**, a hollow, muscular organ that stores the urine. When the bladder is full, stretch receptors signal motor neurons in the spinal cord. In a reflex action, these neurons cause smooth muscle in the bladder wall to contract. At the same time, sphincters encircling the **urethra**, the tube that delivers urine to the body surface, relax. As a result of these two actions, urine flows out of the body. After age two or three, the brain can override the spinal reflex and prevent urine from flowing through the urethra at inconvenient moments.

A male's urethra runs the length of his penis, an organ that conveys both urine and sperm to the body surface. A sphincter cuts off urine flow during times of sexual excitement. In females, the urethra opens onto the body surface near the vagina. The female urethra is a shorter tube, so infectious organisms can more easily reach the urinary bladder. That is one reason women have bladder infections more often than men do.

Tubular Structure of the Kidneys

Each kidney has more than 1 million **nephrons**, each a microscopically small tube (tubule) of cuboidal epithelium and its associated capillaries. The tubule walls are just one cell thick, so substances cross them easily by diffusion or active transport. The nephron is the functional unit of the kidney—the structure that filters blood and produces urine.

Each nephron begins in the cortex, where the tubule wall balloons out and folds back on itself to form a cup-shaped structure called **Bowman's capsule** (**FIGURE 40.9A,B**). Beyond the capsule, the nephron twists a bit and straightens out as a **proximal tubule** (the part of the kidney tubule nearest the beginning of the nephron). After extending down into the renal medulla, the nephron makes a hairpin turn called the **loop of Henle**. The tubule reenters the cortex and twists again as the **distal tubule** (the part of the kidney tubule farthest from the start of the nephron). Distal tubules

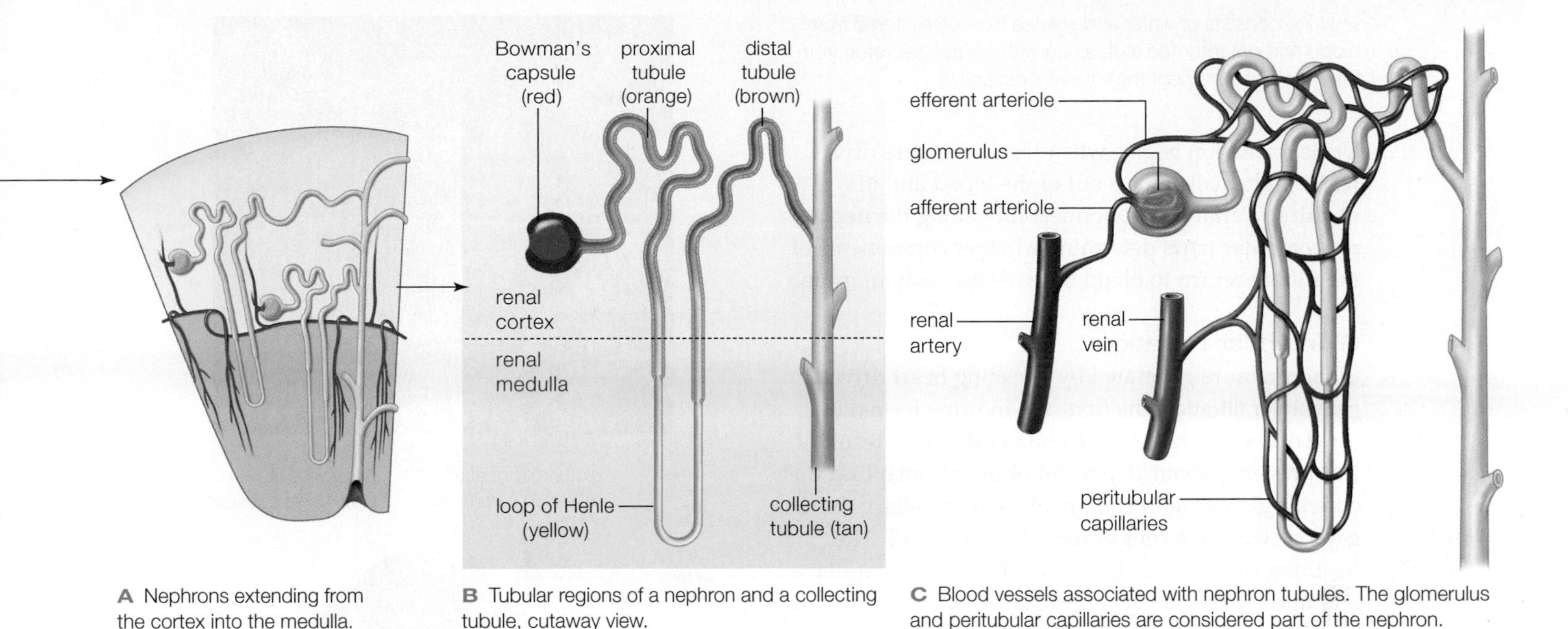

A Nephrons extending from the cortex into the medulla.

B Tubular regions of a nephron and a collecting tubule, cutaway view.

C Blood vessels associated with nephron tubules. The glomerulus and peritubular capillaries are considered part of the nephron.

FIGURE 40.9 ▸**Animated** Orientation and structure of a nephron, the functional unit of the kidney.

of several nephrons drain into a **collecting tubule**. Many collecting tubules extend through the kidney medulla and open into the renal pelvis.

Blood Vessels of the Kidneys

A renal artery carries blood to each kidney. Inside a kidney, the renal artery branches into smaller arterioles called afferent arterioles. Each afferent arteriole leads to a **glomerulus** (plural, glomeruli), a cluster of capillaries in a Bowman's capsule (**FIGURE 40.9C**). Capillaries of a glomerulus are leaky; gaps between the cells in their walls make them about a hundred times more permeable than a typical capillary. As blood flows through a glomerulus, blood pressure forces some fluid out through the gaps and into the Bowman's capsule. *Glomerulus* is the Greek word for filter.

The portion of the blood that is not filtered into a Bowman's capsule continues on into an efferent arteriole. This arteriole branches into **peritubular capillaries**, which thread lacily around the nephron (*peri*–, around). Peritubular capillaries are the site for exchanges between fluid flowing through kidney tubules and the blood. From the peritubular capillaries, blood continues into venules that carry it to the renal vein.

Bowman's capsule In the kidney, the cup-shaped portion of a nephron that encloses the glomerulus and receives filtrate from it.
collecting tubule Kidney tubule that receives filtrate from several nephrons and delivers it to the renal pelvis.
distal tubule Portion of kidney tubule that delivers filtrate to a collecting tubule.
glomerulus In the kidney, a cluster of capillaries enclosed by Bowman's capsule.
loop of Henle U-shaped portion of a kidney tubule; it extends deep into the renal medulla.
nephron Functional unit of the kidney; filters blood and forms urine.
peritubular capillaries Network of capillaries that surrounds and exchanges substances with a kidney tubule.
proximal tubule Portion of kidney tubule that receives filtrate from Bowman's capsule.
ureter Tube that carries urine from a kidney to the bladder.
urethra Tube through which urine expelled from the bladder flows out of the body.
urinary bladder Hollow, muscular organ that stores urine.

TAKE-HOME MESSAGE 40.3

What are the components of the human urinary system?

✔ Two kidneys filter the blood and form urine. Urine flows out of the kidney through ureters, and into a hollow, muscular bladder. When the bladder contracts, urine flows out of the body through the urethra.

✔ A kidney contains microscopic tubules called nephrons that interact with adjacent systems of capillaries. Fluid filtered out of a cluster of capillaries (a glomerulus) enters the first portion of a nephron (Bowman's capsule). As this fluid continues through the tubule and to a collecting duct, it exchanges substances with blood that is flowing through adjacent peritubular capillaries.

40.4 How Urine Forms

✔ Urine consists of water and solutes that were filtered from blood and not returned to it, along with solutes secreted from the blood into the nephron's tubular regions.

Urine formation begins when blood pressure drives water and small solutes out of the blood and into a nephron. Variations in permeability along the nephron's tubular parts determine whether components of the filtrate return to blood or leave the body in urine.

Glomerular Filtration

Blood pressure generated by a beating heart drives **glomerular filtration**, the first step in urine formation (**FIGURE 40.10** and **FIGURE 40.11** ❶). As a result of this pressure, about 20 percent of the plasma fluid that enters a glomerular capillary exits the capillary through gaps between the cells of the capillary's wall. The resulting filtrate—the fluid forced out of the capillary and into the interior of a Bowman's capsule—contains small molecules and ions. Blood cells, platelets, and most plasma proteins are too large to be filtered out of the capillary. They flow on into the efferent arteriole, along with the 80 percent of plasma fluid that did not leave the glomerulus. The cell-free filtrate that enters Bowman's capsule drains into the proximal tubule.

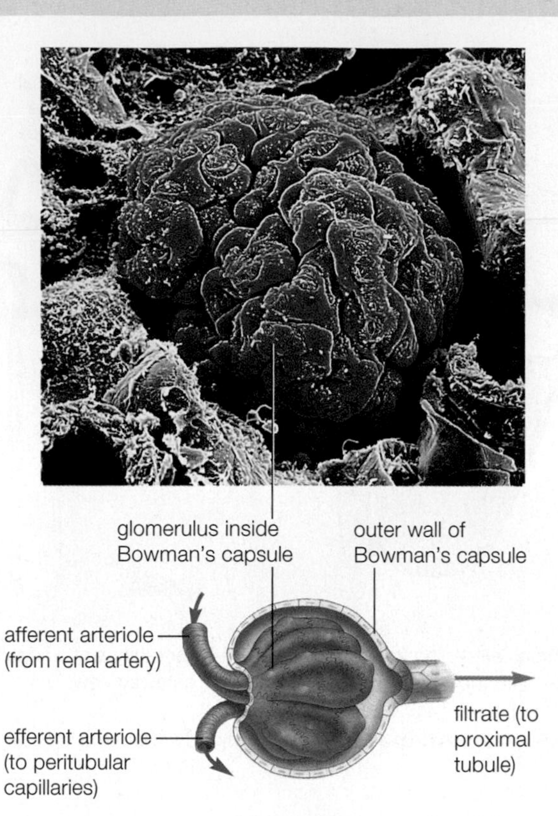

FIGURE 40.10 Glomerular filtration. Pressure exerted on blood by the beating heart forces protein-free plasma out of the glomerular capillaries and into Bowman's capsule.

Tubular Reabsorption

Only a small fraction of the filtrate ends up in urine. **Tubular reabsorption** returns most water and solutes to the blood. Reabsorption begins in the proximal tubule ❷, where active transport proteins move sodium ions (Na^+), chloride ions (Cl^-), potassium ions (K^+), and nutrients such as glucose across the tubule wall and into peritubular capillaries. Water follows the solutes by osmosis, so it moves in the same direction.

Most water and nutrients are reabsorbed from the proximal tubule, but some reabsorption occurs along the entire length of the kidney tubule. About 99 percent of the water that filters into Bowman's capsule returns to blood via tubular reabsorption. All glucose and amino acids that enter Bowman's capsule return to blood the same way, as do most sodium ions, chloride ions, and bicarbonate.

Tubular Secretion

Tubular secretion is the movement of substances from the blood in peritubular capillaries into the filtrate ❸. Membrane proteins in the walls of peritubular capillaries actively transport substances into the interstitial fluid, from which they are similarly transported across the epithelium of the kidney tubule and into the filtrate. Substances that undergo tubular secretion include hydrogen ions (H^+) and breakdown products of foreign organic molecules such as drugs, food additives, and pesticides.

Concentrating the Filtrate

Even the most dilute urine has far more solutes than plasma or the typical interstitial fluid. In the kidney's cortex, the concentration of solutes in the interstitial fluid is similar to that of plasma, but as a tubule descends into the medulla, the interstitial fluid around it becomes increasingly hypertonic. This regional difference in tonicity is essential to the kidney's ability to concentrate the filtrate as urine. For water to leave a kidney tubule by osmosis, the interstitial fluid around that region of tubule must be hypertonic relative to the filtrate inside the tubule (the interstitial fluid must have a higher solute concentration than the filtrate).

The osmotic gradient within the kidney medulla arises because cells in different regions of the nephron wall differ in the types of transport proteins embedded in their plasma membranes. Consider the differences between the two arms of the loop of Henle. The descending arm of the loop of Henle is permeable to

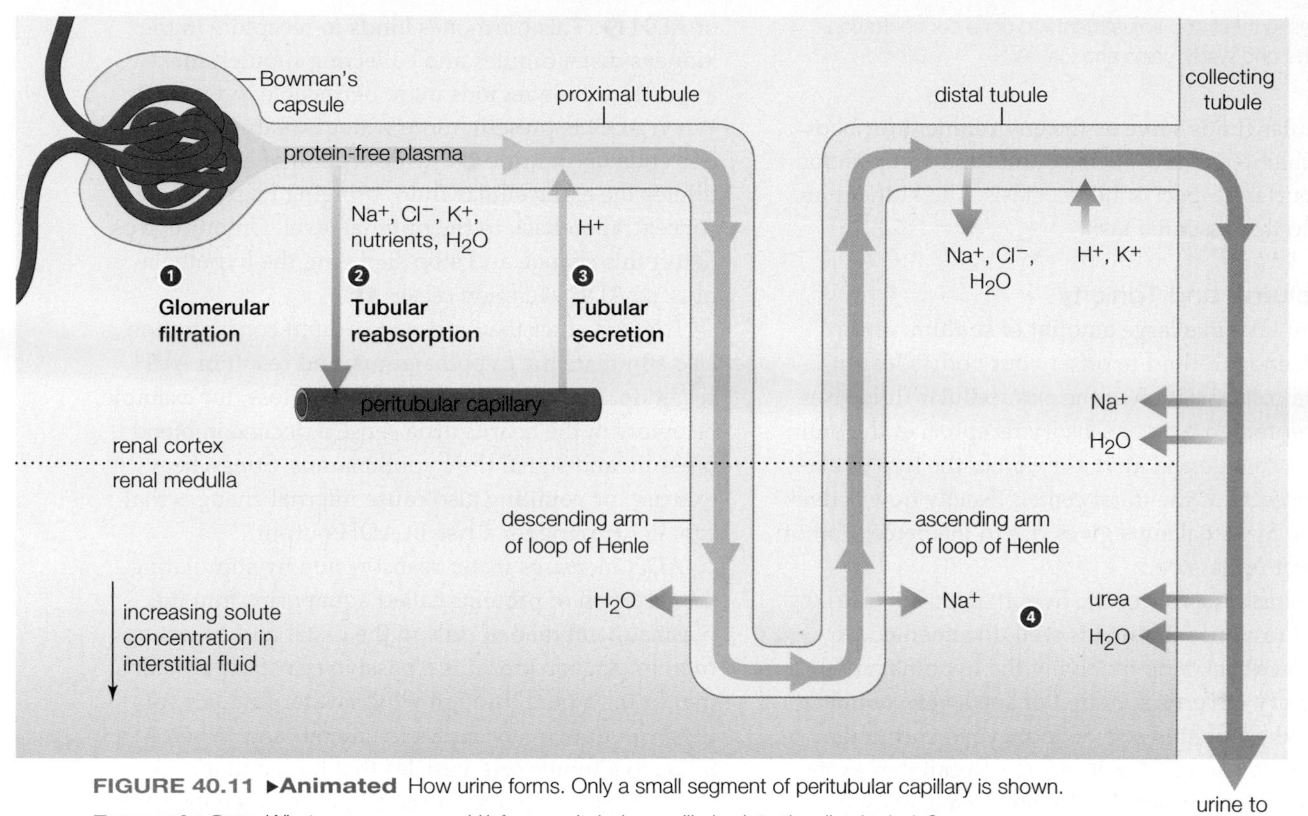

1 Glomerular filtration Cell-free plasma forced out of glomerular capillaries enters Bowman's capsule.

2 Tubular reabsorption Essential ions, nutrients, water, and some urea in the filtrate return to the blood. Green arrows indicate reabsorption.

3 Tubular secretion Wastes and excess ions are moved from the blood into the filtrate. Blue arrows indicate secretion.

4 A high concentration of sodium and urea in the interstitial fluid deep in the medulla draws water out of collecting ducts by osmosis, thus concentrating filtrate as urine.

FIGURE 40.11 ▶Animated How urine forms. Only a small segment of peritubular capillary is shown.

FIGURE IT OUT What process moves H^+ from peritubular capillaries into the distal tubule?

Answer: Tubular secretion

water, but impermeable to sodium ions. The loop of Henle's ascending arm is largely impermeable to water, but its cells actively transport sodium ions from the filtrate into the interstitial fluid. The two arms of the loop of Henle are in close proximity to one another, so as filtrate flows through the loop, the pumping of sodium into the interstitial fluid by the ascending loop causes water to leave the adjacent descending loop by osmosis. As a result, filtrate that flows through the loop of Henle becomes increasing concentrated as it travels through the descending arm, then increasingly dilute as it flows through the ascending arm.

Filtrate loses water once again as it passes through the collecting tubule. Like the loop of Henle, the collecting tubule extends deep into the medulla. In the deepest part of the medulla, urea pumped out of the collecting tubule and sodium pumped out of the ascending loop of Henle make the interstitial fluid hypertonic relative to filtrate in the collecting tubule. Thus, as filtrate descends through this tubule, the increasing solute concentration of the interstitial fluid around the tubule draws water outward by osmosis 4.

The body can adjust how much water is reabsorbed by adjusting the permeability of kidney tubules, a process we discuss in the section that follows.

glomerular filtration First step in urine formation: protein-free plasma forced out of glomerular capillaries by blood pressure enters Bowman's capsule.
tubular reabsorption Substances move from the filtrate inside a kidney tubule into the peritubular capillaries.
tubular secretion Substances move out of peritubular capillaries and into the filtrate in kidney tubules.

TAKE-HOME MESSAGE 40.4

How is urine formed and concentrated?

✔ During glomerular filtration, pressure generated by the beating heart drives water and solutes out of glomerular capillaries and into kidney tubules.

✔ In tubular reabsorption, water, some ions, glucose, and other solutes move out of the filtrate and return to the blood in peritubular capillaries.

✔ In tubular secretion, active transport proteins move solutes such as H^+ from peritubular capillaries into the nephron for excretion.

✔ Differential permeability of the two arms of the loop of Henle sets up a concentration gradient in the interstitial fluid that draws water out of the collecting tubule, concentrating the filtrate as urine.

40.5 Fluid Homeostasis

✔ Changes in thirst and adjustments to urine concentration offset solute and water gains and losses.

Extracellular fluids serve as the environment for body cells, so maintenance of their volume and composition is an important aspect of homeostasis. The kidneys are integral to this essential task.

Fluid Volume and Tonicity

When you take in a large amount of sodium, or do not drink enough fluid to offset your body's loss of water, the concentration of the extracellular fluid rises. Osmoreceptors (a type of sensory receptor) in the brain detect this change and alert a region of the hypothalamus referred to as the thirst center. Exactly how activation of the hypothalamus gives rise to the perception of thirst is not understood.

By contrast, the role of the hypothalamus in the hormonal response to thirst is well documented. A negative feedback loop involving the hypothalamus and pituitary governs secretion of **antidiuretic hormone** (**ADH**), a hormone that acts on kidneys to encourage water reabsorption (**FIGURE 40.12**). Recall that axons of some hormone-producing hypothalamic neurons extend into the posterior pituitary (Section 34.4). When osmoreceptors activated by a rise in sodium signal the hypothalamus ❶, an action potential travels along these axons to the pituitary, where it causes the release of ADH ❷. This hormones binds to receptors in the kidneys distal tubules and collecting tubules, making these tubule regions more permeable to water ❸. When ADH is present, more water is reabsorbed and less ends up in urine ❹. Reabsorption of extra water dilutes the extracellular fluid, bringing its sodium concentration back to the optimal level. Omoreceptors detect this change and stop signaling the hypothalamus, so ADH secretion ceases ❺.

Triggers other than a rise in sodium concentration also stimulate the hypothalamus, and result in ADH secretion. In the event of heavy blood loss, for example, receptors in the heart's atria sense a decline in blood pressure and signal the hypothalamus. Stress, heavy exercise, or vomiting also cause internal changes that can, in turn, trigger a rise in ADH output.

ADH increases water reabsorption by stimulating the insertion of proteins called aquaporins into the plasma membrane of cells in the distal and collecting tubules. An aquaporin is a passive transport protein that forms a pore through which water (but not solutes) can flow freely across the membrane. When ADH binds to a tubule cell, vesicles that hold aquaporin subunits move toward the cells' plasma membrane. As these vesicles fuse with the membrane, the subunits assemble themselves into functional aquaporins. Once in place, aquaporins increase the ability of water to leave the filtrate.

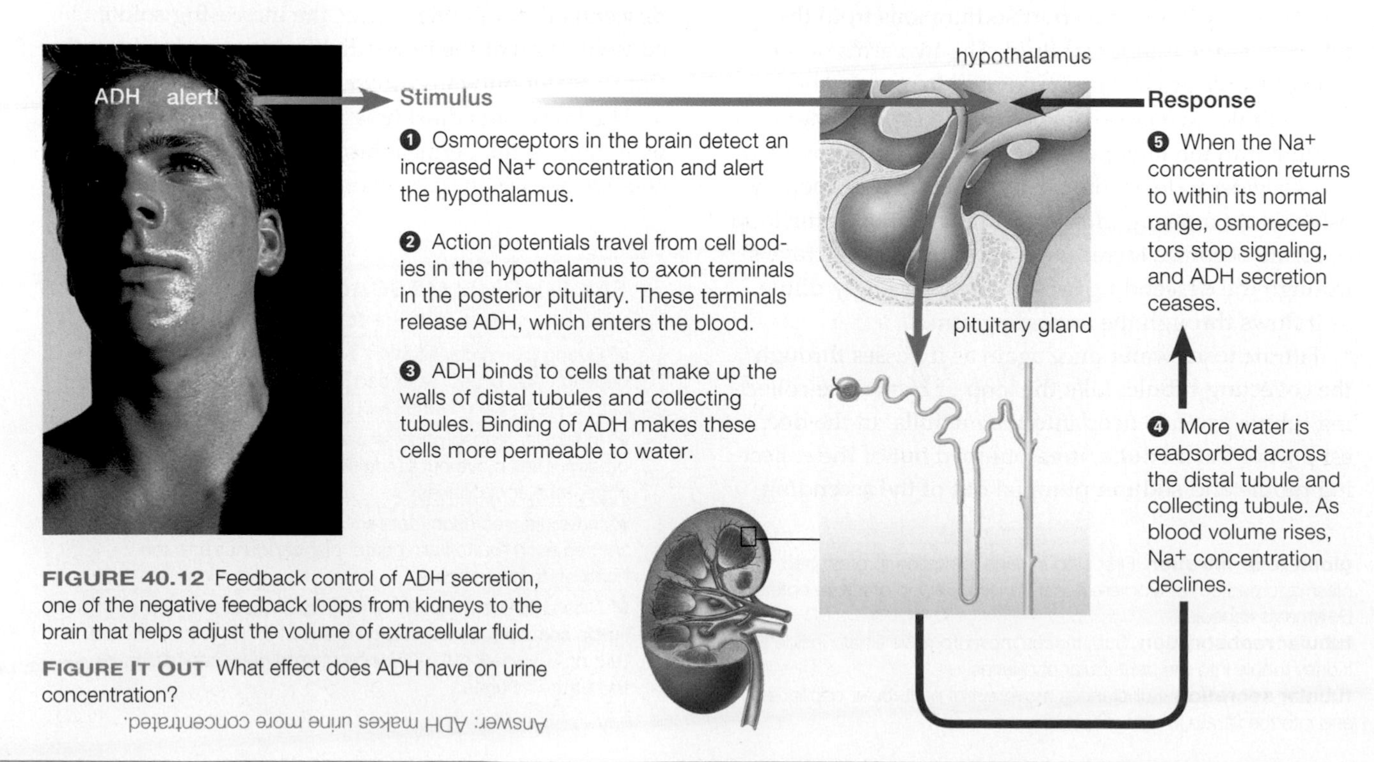

FIGURE 40.12 Feedback control of ADH secretion, one of the negative feedback loops from kidneys to the brain that helps adjust the volume of extracellular fluid.

FIGURE IT OUT What effect does ADH have on urine concentration?

Answer: ADH makes urine more concentrated.

❶ Angiotensinogen made by the liver circulates in the blood. It is converted to angiotensin I by renin, an enzyme released by kidney arterioles when the blood pressure declines.

❷ Another enzyme converts angiotensin I to angiotensin II.

❸ Among its actions, angiotensin II encourages aldosterone secretion by the adrenal cortex.

❹ Aldosterone acts on kidneys to increase Na^+ reabsorption; water follows by osmosis.

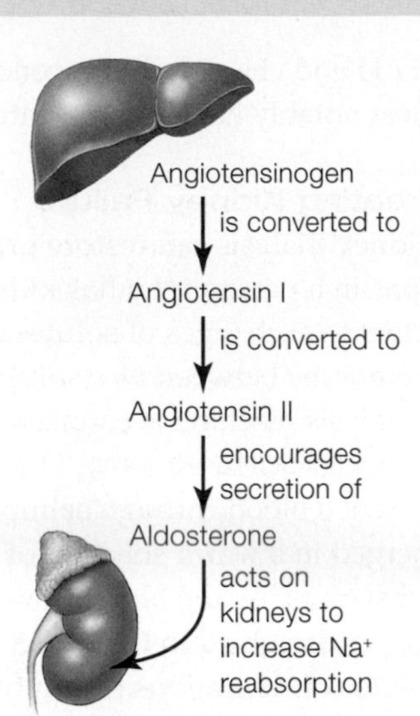

FIGURE 40.13 The renin–angiotensin–aldosterone system. Angiotensin II also encourages secretion of ADH and causes an increase in thirst, thus elevating blood volume and pressure.

A decrease in the volume of the extracellular fluid results in lower blood pressure. This decline in pressure causes cells in the kidney's arterioles to release renin, an enzyme that sets in motion a complex chain of reactions (**FIGURE 40.13**). Renin converts angiotensinogen, a protein secreted by the liver into the blood, into angiotensin I ❶. Another enzyme converts angiotensin I to angiotensin II ❷, which acts in the adrenal glands atop the kidneys. The adrenal cortex responds to angiotensin II by secreting the hormone **aldosterone** into the blood ❸. Aldosterone acts on the collecting tubules, where it encourages sodium reabsorption ❹. Because water follows the sodium by osmosis, aldosterone increases water reabsorption. Retention of additional water raises blood pressure.

Note that both ADH and aldosterone cause urine to become more concentrated, although they do so by different mechanisms.

Atrial natriuretic peptide (ANP) is a hormone that makes urine more dilute. Muscle cells in the heart's atria release ANP when high blood volume causes the atrial walls to stretch. ANP directly inhibits secretion of aldosterone by acting on the adrenal cortex. It also acts indirectly by inhibiting renin release. In addition, ANP increases the glomerular filtration rate, so more fluid enters kidney tubules.

aldosterone Adrenal hormone that makes kidney tubules more permeable to sodium; encourages sodium reabsorption, thus increasing water reabsorption and concentrating the urine.

antidiuretic hormone (ADH) Hormone released in the posterior pituitary; makes kidney tubules more permeable to water; encourages water reabsorption, thus concentrating the urine.

Acid–Base Balance

Metabolic reactions such as protein breakdown and lactate fermentation add hydrogen ions (H^+) to the extracellular fluid. Despite these additions, a healthy body can maintain its H^+ concentration within a tight range, a state known as acid–base balance. Buffer systems and adjustments to the activity of respiratory and urinary systems are essential to this balance.

A buffer system involves substances that reversibly bind and release H^+ or OH^- (Section 2.6). In the body, buffer systems minimize pH changes when acidic or basic molecules enter or leave the extracellular fluid.

The pH of human extracellular fluid usually stays between 7.35 and 7.45. In the absence of a buffer, adding acids to this fluid would decrease its pH. In the presence of bicarbonate, some hydrogen ions combine with bicarbonate to form carbonic acid, which dissociates into carbon dioxide (CO_2) and water:

$$\underset{\text{bicarbonate}}{H^+ + HCO_3^-} \rightleftharpoons \underset{\text{carbonic acid}}{H_2CO_3} \rightleftharpoons CO_2 + H_2O$$

Thus bonded, the hydrogen does not contribute to the pH of the extracellular fluid.

The kidneys contribute to acid–base balance by adjusting the tubular reabsorption of bicarbonate and the tubular secretion of H^+. A decline in pH causes an increase in bicarbonate reabsorption. The bicarbonate moves into peritubular capillaries, where it buffers excess acid. At the same time, tubular secretion of H^+ increases. H^+ leaves the blood and enters the filtrate, where it combines with phosphate or ammonia. The resulting compounds are excreted in the urine.

TAKE-HOME MESSAGE 40.5

How do the kidneys affect the composition of body fluids?

✔ Hormones that act on the kidney alter the amount of water reabsorbed. Antidiuretic hormone concentrates the urine by increasing water reabsorption. Aldosterone increases salt reabsorption, and water follows, so it concentrates the urine. Atrial natriuretic peptide makes urine more dilute by discouraging secretion of aldosterone and increasing the rate of glomerular filtration.

✔ The kidney helps to maintain the pH of the extracellular fluid by adjusting the reabsorption of bicarbonate and secretion of H^+.

40.6 When Kidneys Fail

✔ Kidney failure can be treated with dialysis, but only a kidney transplant can fully restore function.

Causes of Kidney Failure

The vast majority of kidney problems arise as complications of diabetes mellitus or high blood pressure. These disorders damage small blood vessels, including capillaries of nephrons.

High-protein diets increase the amount of urea formed, so such diets force the kidneys to work overtime. Such diets also increase the risk for kidney stones, which are hardened deposits that form when uric acid, calcium, and other wastes settle out of urine and collect in the renal pelvis. Most kidney stones are washed away in urine, but sometimes one lodges in a ureter or the urethra and causes severe pain. A stone that blocks urine flow raises risk of infections and kidney damage.

Kidney function is measured in terms of the rate of filtration through glomerular capillaries. Kidney failure occurs when the filtration rate falls by half. Failure of both kidneys can be fatal because wastes build up in the blood and interstitial fluid. The resulting changes in pH and changes in the concentrations of other ions, most notably Na^+ and K^+, interfere with metabolism.

Treating Kidney Failure

Kidney dialysis can restore proper solute concentrations in a person who has kidney failure. "Dialysis" refers to exchanges of solutes across a semipermeable membrane between two solutions. With hemodialysis, a dialysis machine is connected to a patient's blood vessel (**FIGURE 40.14A**). The machine pumps the patient's blood through semipermeable tubes submerged in a warm solution of salts, glucose, and other substances. As the blood flows through these tubes, wastes dissolved in the blood diffuse out, and blood solute concentrations return to normal levels. Cleansed blood is returned to the patient's body. Typically a person has hemodialysis three times a week at an outpatient dialysis center. By contrast, peritoneal dialysis can be done at home. Each night, dialysis solution is pumped into a patient's abdominal cavity (**FIGURE 40.14B**). Wastes diffuse across the cavity's lining (the peritoneum) into the fluid, which is drained out the following morning. With this procedure, the body lining serves as the dialysis membrane.

Kidney dialysis can keep a person alive through an episode of temporary kidney failure. When kidney damage is permanent, dialysis must be continued for the rest of a person's life, or until a donor kidney becomes available for transplant surgery.

Each year in the United States, about 12,000 people receive a kidney transplant. More than 40,000 others remain on a waiting list because there is a shortage of donated kidneys. The National Kidney Foundation estimates that every day, 17 people die of kidney failure while waiting for a transplant. Most kidneys for transplants come from deceased donors, but the number of living donors is increasing. One kidney is adequate to maintain good health, so the risks to a living donor are mainly related to the surgery—unless the donor's remaining kidney fails.

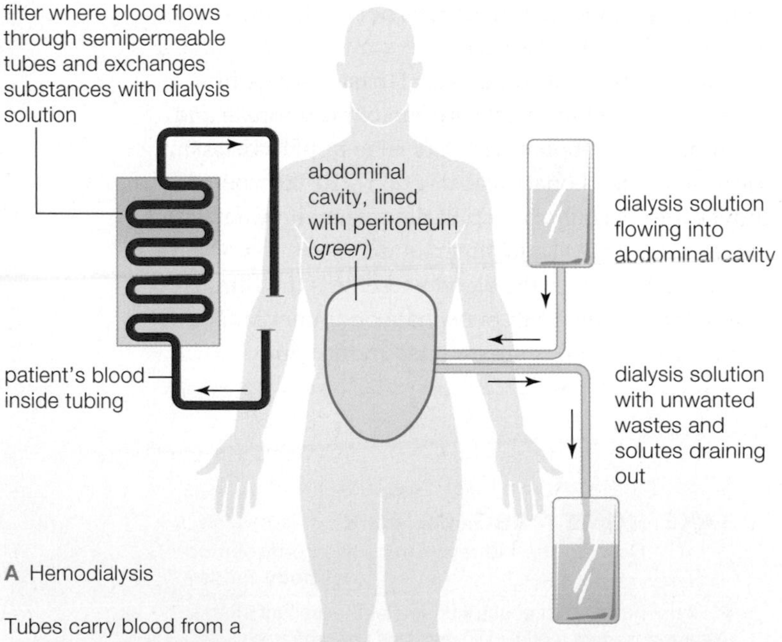

A Hemodialysis

Tubes carry blood from a patient's body through a filter with dialysis solution that contains the proper concentrations of salts. Wastes diffuse from the blood into the solution and cleansed, solute-balanced blood returns to the body.

B Peritoneal dialysis

Dialysis solution is pumped into a patient's abdominal cavity. Wastes diffuse across the lining of the cavity into the solution, which is then drained out.

FIGURE 40.14 ▸Animated Two types of kidney dialysis.

TAKE-HOME MESSAGE 40.6

What causes kidney failure, and how does it affect health?

✔ Kidney failure most often occurs as a complication of diabetes or high blood pressure.

✔ Untreated kidney failure is fatal. Dialysis can keep a person with kidney failure alive, but it must be continued until a person dies or receives a kidney transplant.

40.7 Heat Gains and Losses

✔ Maintaining the body's core temperature is another aspect of homeostasis. Some animals expend more energy than others to keep their body warm.

How the Core Temperature Can Change

Metabolic reactions release heat, and this heat affects an animal's body temperature. Animals also gain heat from, and lose heat to, their surroundings. An animal's internal temperature is stable only when the metabolic heat it produces and the heat it gains from the environment balance heat losses to the environment.

Four processes affect heat gain and loss. **Thermal radiation** is emission of heat from a warm object into the space around it. At rest, a human adult radiates as much heat as a 100-watt incandescent lightbulb. **Conduction** is the transfer of heat within an object or among objects that contact one another. An animal loses heat when it contacts a cooler object, and gains heat when it contacts a warmer one. In **convection**, heated air or water moves away from a hot object, thus carrying heat away from the source. In **evaporation**, a liquid becomes a gas. Energy that powers this process typically comes from the liquid itself in the form of heat, so any remaining liquid cools (Section 2.5). When that liquid is water at a body surface, the cooling decreases body temperature.

Modes of Thermoregulation

Invertebrates, fishes, amphibians, and nonbird reptiles are **ectotherms**, which means "heated from the outside." The body temperature of ectotherms fluctuates with the temperature of the external environment. That is why the chameleon in **FIGURE 40.15** is about the same temperature as its surroundings. Ectotherms typically have a low metabolic rate, and they lack insulating fur, hair, or feathers. They regulate internal temperature by altering their position, rather than their metabolism. A lizard, for example, warms itself by basking on a rock in a sunny spot or cools itself by retreating into a shady burrow.

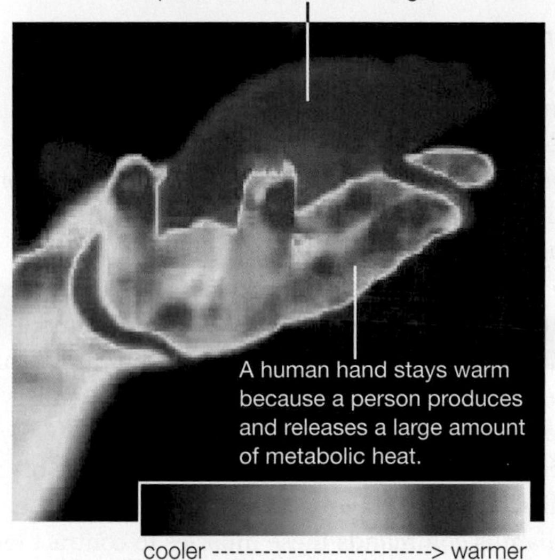

FIGURE 40.15 Endotherm and ectotherm. Photo of a chameleon (an ectothermic lizard) being held by a human (an endotherm). The image was taken using heat-sensitive film.

Most birds and mammals are **endotherms**, which means "heated from within." Compared to ectotherms, endotherms maintain their body temperature within a narrower range and have a higher metabolic rate. For example, a mouse uses thirty times more energy than a lizard of the same body weight. Metabolic heat production allows endotherms to be active in a wider range of temperatures than ectotherms. Fur, hair, or feathers insulate endotherms by minimizing heat transfers.

Some birds and mammals are **heterotherms**, animals that keep their core temperature constant some of the time, but allow it to fluctuate at other times. For example, hummingbirds have a very high metabolic rate when foraging for nectar during the day. At night, the birds' metabolic activity decreases so much that their body may become almost as cool as the surroundings.

conduction Of heat: the transfer of heat within an object or between two objects in contact with one another.
convection Transfer of heat by moving molecules of air or water.
ectotherm Animal whose body temperature varies with that of its environment; it adjusts its internal temperature by altering its behavior.
endotherm Animal that maintains its internal temperature mainly by adjusting its metabolism; for example, a bird or mammal.
evaporation Transition of a liquid to a gas.
heterotherm Animal that sometimes maintains its temperature by producing metabolic heat, and at other times allows its temperature to fluctuate with the environment.
thermal radiation Emission of heat from an object.

TAKE-HOME MESSAGE 40.7

How do animals regulate their body temperature?

✔ Animals can gain heat from the environment, or lose heat to it. They can also generate heat by metabolic reactions.

✔ Fishes, amphibians, and reptiles are ectotherms; their body temperature varies with that of their environment.

✔ Most birds and mammals are endotherms that maintain their body temperature largely by adjusting their production of metabolic heat.

CREDIT: (15) NASA/JPL-Caltech/L. Hermans.

40.8 Adaptations to Heat and Cold

✔ A variety of mechanisms adapt vertebrates to heat, cold, or seasonally fluctuating temperature.

Responses to Cold

Most animals that remain active when it is very cold are endothermic vertebrates. When outside temperature declines, thermoreceptors in their skin send action potentials to the hypothalamus, the brain region responsible for thermoregulation. Signal from the hypothalamus causes vessels that deliver blood to the skin to contract, so less metabolic heat reaches the body surface (**TABLE 40.1**). At the same time, muscle contractions make hair, fur, or feathers "stand up." This response creates a layer of still air next to skin, reducing the amount of heat lost by convection.

Skeletal muscles are a vertebrate's main source of metabolic heat. With prolonged exposure to cold, the hypothalamus commands these muscles to contract ten to twenty times each second. This **shivering response** increases heat production, but it has a high energy cost.

FIGURE 40.16 Woodland jumping mouse (*Napaeozapus insignis*) during its winter hibernation.

Table 40.1 Responses to Heat and Cold

Stimulus	Main Responses	Outcome
Heat stress	Widening of blood vessels in skin; behavioral adjustments; in some species, sweating, panting	Dissipation of heat from body
	Decreased muscle action	Heat production decreases
Cold stress	Narrowing of blood vessels in skin; behavioral adjustments (e.g., minimizing surface parts exposed)	Conservation of body heat
	Increased muscle action; shivering; nonshivering heat production	Heat production increases

Brown adipose tissue helps many types of mammals heat themselves. When body temperature declines, mitochondria in this specialized adipose tissue use the energy released by fatty acid oxidation to produce heat, rather than ATP. The resulting increase in metabolic heat output is called **nonshivering heat production**.

Some animals survive a seasonal cold period through **hibernation**, a state of prolonged inactivity during which the animal's metabolic rate and body temperature decline dramatically. Depending on the species, the animal may reawaken to feed periodically on stored food, or subsist entirely on fat and glycogen accumulated during the prior season. Mammalian hibernators include some bats, bears, and rodents (**FIGURE 40.16**). Many reptiles and amphibians also hibernate during winter, as do some insects.

Responses to Heat

When the temperature of a vertebrate body rises, the hypothalamus triggers responses that decrease core temperature. The animal becomes less active, so the amount of heat produced by the activity of its skeletal muscles declines. Blood delivery to the skin increases, so more metabolic heat escapes into the surroundings. Many vertebrates also reduce their temperature by breathing rapidly (panting), which increases the rate of evaporative cooling from their respiratory tract.

Sweating is the main mechanism of cooling in primates (including humans) and large hoofed mammals. Sweat glands are exocrine glands that release water and solutes through a pore at the skin's surface. Evaporation of sweat cools the skin. Rodents and marsupials cannot sweat, but some cool off by spreading saliva on their skin. Like sweating, this cooling mechanism results in the loss of water, which must be replenished by drinking.

Some desert reptiles and amphibians survive a hot, dry season through **estivation**, a state of prolonged inactivity and reduced metabolism similar to hibernation.

Climate-Related Adaptations in Humans

As humans dispersed out of Africa, the climatic challenges they faced changed. In equatorial Africa, people sweat a lot, so they lose a good deal of sodium. Sodium is an essential nutrient, and was a limited resource for many early populations. Thus, individuals with kidneys that reabsorbed a lot of sodium were at a selective advantage. Later, some humans migrated to colder regions, where they sweated less frequently. In cooler environments, a tendency to reabsorb a large amount of sodium would have been less advantageous. It could

Truth in a Test Tube (revisited)

Solutes and nutrients that the body requires are reabsorbed from the filtrate that enters kidney tubules. Metabolites of drugs also travel in the plasma and are filtered into kidney tubules, but these compounds are not reabsorbed, so they end up in the urine.

How quickly a drug is cleared from the body depends in part on the efficiency of the kidneys, which varies among individuals. Age is one factor; a healthy 35-year-old eliminates drug metabolites from the body about twice as fast as a healthy 85-year-old. There are also inherited differences among people in the rate at which the liver breaks drugs into water-soluble metabolites that kidneys can eliminate.

Some drugs such as marijuana are metabolized to fat-soluble compounds that can be taken up by fat cells, then released over time. As a result, such drugs can be detected in urine longer than drugs that are metabolized to form water-soluble compounds.

---

even be maladaptive, because having a high concentration of sodium in the blood increases risk of hypertension and cardiovascular disorders.

Allele frequencies of some genes involved in sodium reabsorption provide evidence of a climate-related shift in selective pressure. Consider the *ACE* gene, which encodes an enzyme (angiotensin-converting enzyme) needed to make the hormone aldosterone. You learned in Section 40.5 that aldosterone promotes sodium reabsorption from urine. The ancestral allele for the *ACE* gene is most common in African and Middle Eastern populations. The allele that is common in people of European or Asian descent has an insertion mutation, and encodes a less active form of the enzyme.

Distribution of *ACE* alleles in modern populations may be of medical significance. Because people homozygous for the ancestral allele make more aldosterone than those with other genotypes, they reabsorb more sodium. In modern societies, where sodium is abundant, a high capacity for sodium reabsorption may increase risk for cardiovascular disorders.

For humans who settled in cold climates, an ability to produce increased amounts of metabolic heat provided a selective advantage. We see the results of this selection in the distribution of alleles for genes that encode proteins involved in heat production. Alleles that result in an increased capacity for nonshivering heat production tend to be highest in arctic populations (**FIGURE 40.17**).

FIGURE 40.17 Inuit hunter looking for polar bears in the Canadian Arctic. Compared to populations native to warmer regions, the Inuit have higher frequencies of alleles that promote nonshivering heat production by brown adipose tissue.

estivation An animal becomes dormant during a hot, dry season.
hibernation An animal becomes dormant during a cold season.
nonshivering heat production Increase in metabolic heat production that results when brown adipose tissue releases energy as heat, rather than storing it in ATP.
shivering response Rhythmic muscle contractions that generate metabolic heat in response to prolonged exposure to cold.

TAKE-HOME MESSAGE 40.8

What are some animal adaptations to cold and to heat?

- ✔ Some endotherms can remain active when it is cold. Feathers or fur slow their heat loss. Shivering and (in mammals) nonshivering heat production warm the body.
- ✔ Panting and sweating are vertebrate mechanisms of increasing heat loss through evaporative cooling.
- ✔ Some animals become dormant during a cold season (hibernators) or a hot, dry season (estivators).
- ✔ Human populations native to hot climates have a high capacity for sodium reabsorption, a trait that is adaptive where people frequently lose sodium in sweat. Populations native to the Arctic have high frequencies of alleles that promote nonshivering heat production.

CREDITS: (17) Roberta Olenick/Getty Images; (in text revisited) Lawrence Lawry/Science Source.

summary

Section 40.1 The composition of urine provides information about health and about chemicals taken into the body. Hormones and drugs that enter the blood are filtered out by kidneys and end up in the urine.

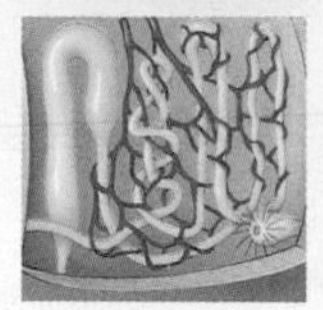

Section 40.2 Plasma and interstitial fluid are the main components of extracellular fluid. Maintaining the volume and composition of this fluid is an essential aspect of homeostasis. Organisms must balance solute and fluid gains with solute and fluid losses. They also must eliminate metabolic wastes such as the **ammonia** produced by the breakdown of proteins and nucleic acids.

Animals in different habitats face different challenges. Those living in water lose or gain water by osmosis. On land, the main challenge is avoiding dehydration. Insects and reptiles (including birds) conserve water by converting ammonia to **uric acid** crystals, which can be eliminated with very little water. Mammals excrete **urea** dissolved in a lot of water. All vertebrates have a pair of **kidneys** that filter the blood and produce **urine**.

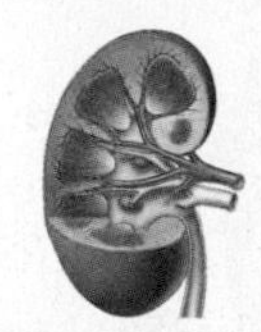

Section 40.3 A human urinary system consists of two kidneys, a pair of **ureters**, a **urinary bladder**, and the **urethra**. Kidney **nephrons** filter blood and form urine. Each nephron starts at **Bowman's capsule** in a kidney's outer region, or renal cortex. The nephron continues as a **proximal tubule**, a **loop of Henle** that descends into and ascends from the renal medulla, and a **distal tubule** that drains into a **collecting tubule**.

Bowman's capsule and capillaries of the **glomerulus** that it cups around serve as a blood-filtering unit. Most filtrate that enters Bowman's capsule is reabsorbed into the **peritubular capillaries** that are adjacent to the kidney tubule. The portion of the filtrate not returned to blood is excreted as urine.

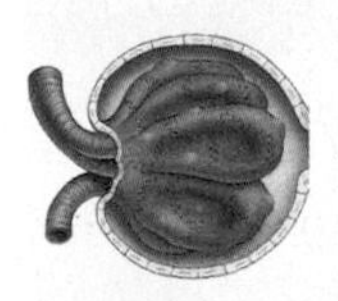

Section 40.4 Urine formation begins when **glomerular filtration** delivers protein-free plasma into a kidney tubule. Most water and solutes are returned to the blood by **tubular reabsorption**. Substances that are not reabsorbed, and substances added to the filtrate by **tubular secretion**, end up in the urine. The amount of water reabsorbed across the distal tubule and collecting tubule can be altered, and affects the concentration of the urine.

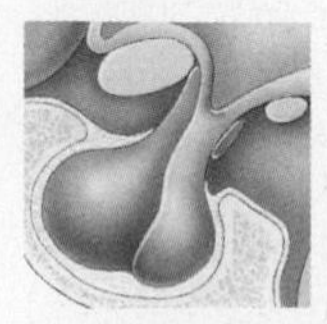

Section 40.5 Hormones regulate the urine's concentration and composition. A region of the hypothalamus regulates thirst. In response to an increase in the solute concentration, the hypothalamus signals the pituitary gland to release **antidiuretic hormone**, a hormone that increases the reabsorption of water through passive transport proteins called aquaporins. **Aldosterone**, a hormone secreted by the adrenal cortex, increases sodium reabsorption. Water follows the sodium by osmosis. Thus, both antidiuretic hormone and aldosterone concentrate the urine.

Atrial natriuretic peptide, a hormone secreted by the heart in response to high blood volume and pressure, slows secretion of aldosterone and thus makes urine more dilute.

The urinary system helps regulate acid–base balance by eliminating H^+ in urine and reabsorbing bicarbonate.

Section 40.6 Kidney stones are hard deposits of material that form in the renal pelvis. Diabetes and hypertension are the most common causes of kidney failure. When both kidneys fail, dialysis or a kidney transplant is required to sustain life.

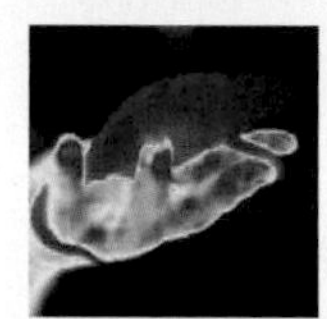

Section 40.7 Animals produce metabolic heat. They also gain or lose heat by **thermal radiation**, **conduction**, and **convection** and lose it by **evaporation. Ectotherms** such as reptiles control core temperature mainly by behavior; their temperature varies with that of the external environment. **Endotherms** (most mammals and birds) regulate temperature by controlling production and loss of metabolic heat. **Heterotherms** control core temperature only part of the time.

Section 40.8 The hypothalamus is the main center for vertebrate temperature control. Increased delivery of blood to the skin, sweating, and panting are responses to heat. Exposure to cold makes blood vessels in the skin constrict, causes hair to stand upright, and elicits a **shivering response**. In mammals, **nonshivering heat production** by brown adipose tissue also provides heat. Some animals enter **hibernation** to survive a cold season or **estivation** to survive a hot, dry season.

Some genetic differences among human populations reflect climate-related adaptations. An increased capacity for urinary reabsorption of sodium (which is lost from the body in sweat) is adaptive in regions with a hot climate. In cold climate regions, an elevated capacity for nonshivering heat production is adaptive.

self-quiz

Answers in Appendix VII

1. An insect's _______ deliver nitrogen-rich waste to its gut.
   a. nephridia
   b. nephrons
   c. Malpighian tubules
   d. contractile vacuoles

2. Body fluids of a marine bony fish have a solute concentration that is _______ its surroundings.
   a. higher than
   b. lower than
   c. equal to

3. Bowman's capsule, the start of the tubular part of a nephron, is located in the _______ .
   a. renal cortex
   b. renal medulla
   c. renal pelvis
   d. renal artery

4. Fluid that enters Bowman's capsule flows directly into the _______ .
   a. renal artery
   b. proximal tubule
   c. distal tubule
   d. loop of Henle

data analysis activities

Pesticide Residues in Urine To carry the USDA's organic label (right), food must be produced without synthetic pesticides that farmers often use on conventionally grown fruits, vegetables, and many grains.

USDA ORGANIC

Chensheng Lu of Emory University used urine testing to see whether eating organic food has a significant effect on the levels of pesticides in children's bodies (**FIGURE 40.18**). Over the course of fifteen days, Lu and his colleagues collected the urine of twenty-three children (aged 3 to 11) and tested it for breakdown products of two synthetic pesticides. The children ate their normal diet of conventionally grown foods for three days, switched to organic versions of the same foods and drinks for five days, then returned to their conventional diet for a week.

Study Phase	No. of Samples	Malathion Metabolite Mean (μg/liter)	Malathion Metabolite Maximum (μg/liter)	Chlorpyrifos Metabolite Mean (μg/liter)	Chlorpyrifos Metabolite Maximum (μg/liter)
1. Conventional	87	2.9	96.5	7.2	31.1
2. Organic	116	0.3	7.4	1.7	17.1
3. Conventional	156	4.4	263.1	5.8	25.3

FIGURE 40.18 Concentrations of metabolites of two pesticides (malathion and chlorpyrifos) in children's urine during the three different phases of the study. The difference in the mean level of metabolites between the organic and conventional phases of the study was statistically significant.

1. During which phase of the experiment did the children's urine contain the lowest level of the malathion metabolite?

2. During which phase of the experiment was the maximum level of the chlorpyrifos metabolite detected?

3. Did switching to an organic diet lower the amount of pesticide residues excreted by the children?

4. Even during the conventional phases of this experiment, the highest pesticide metabolite levels detected were below those considered harmful by the EPA. Given these data, would you spend more to buy organic foods?

5. Blood pressure forces water and small solutes into Bowman's capsule during _______ .
a. glomerular filtration
b. tubular reabsorption
c. tubular secretion
d. both a and c

6. Kidneys return most of the water and small solutes back to blood by way of _______ .
a. glomerular filtration
b. tubular reabsorption
c. tubular secretion
d. both a and b

7. ADH binds to receptors on distal tubules and collecting tubules, making them _______ permeable to _______ .
a. more; water
b. less; water
c. more; sodium
d. less; sodium

8. Increased sodium reabsorption _______ .
a. will make urine more concentrated
b. will make urine more dilute
c. is stimulated by aldosterone
d. both a and c

9. Secretion of H^+ into kidney tubules _______ .
a. makes extracellular fluid less acidic
b. makes urine less acidic
c. can cause acidosis
d. both a and c

10. Match each structure with a function.
___ ureter
___ Bowman's capsule
___ urethra
___ collecting tubule
___ pituitary gland

a. start of nephron
b. delivers urine to body surface
c. carries urine from kidney to bladder
d. secretes ADH
e. target of aldosterone

11. An increased capacity for sodium absorption is adaptive for humans who live in a _______ climate.
a. hot
b. cold
c. rainy
d. temperate

12. The main control center for maintaining mammalian body temperature is in the _______ .
a. anterior pituitary
b. renal cortex
c. adrenal gland
d. hypothalamus

13. Which has a higher metabolic rate?
a. an endotherm
b. an ectotherm

14. How does the human body respond to an increase in core body temperature?
a. Decreased blood flow to the skin
b. Increased sweating
c. Increased activity of brown adipose tissue
d. Increased output of thyroid hormone

15. Which organelle in brown adipose tissue gives this tissue its heightened capacity to produce heat?
a. mitochondria
b. endoplasmic reticulum
c. Golgi bodies
d. ribosomes

critical thinking

1. The kangaroo rat kidney efficiently produces a very small volume of urine (Section 40.2). Compared to a human, its nephrons have a loop of Henle that is proportionally much longer. Explain how a long loop helps the rat conserve water.

2. Compared to closely related species that live in warmer areas, cold dwellers tend to have larger body size and smaller appendages. Think about heat transfers between animals and the habitat, then explain why smaller appendages and larger body size are advantageous in very cold climates.

CREDITS: (17) left, USDA; right, © Cengage Learning.

LEARNING ROADMAP

This chapter expands on mechanisms of reproduction (Sections 11.2,12.1) and gamete formation (12.5). It touches on human sex determination (10.4) and the function of the hypothalamus and pituitary (34.4), sex hormones (34.10), and autonomic nerves (32.8). We conclude with a discussion of HPV (11.6), HIV (20.3), and other sexually transmitted pathogens.

MODES OF ANIMAL REPRODUCTION

Some animals reproduce asexually, but most reproduce sexually. Gametes produced by meiosis in gonads combine in the environment, or in the female's body, depending on the species.

FEMALE REPRODUCTIVE FUNCTION

A human female has ovaries that produce eggs and hormones. Ducts carry eggs toward the uterus, where offspring develop. The vagina receives sperm and also is the birth canal.

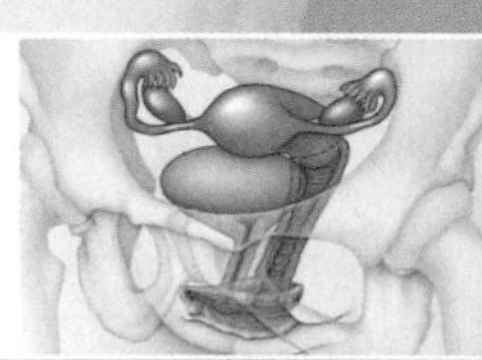

MALE REPRODUCTIVE FUNCTION

A human male has testes that make sperm and secrete the sex hormone testosterone. Sperm mixes with secretions from exocrine glands to form semen, which leaves the body via the urethra.

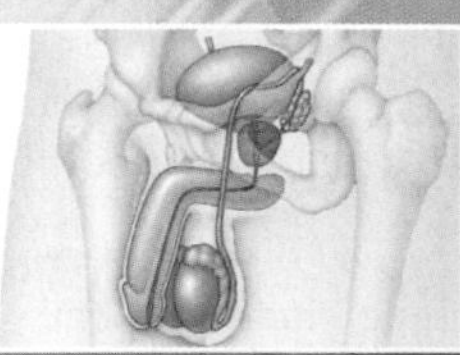

INTERCOURSE AND FERTILIZATION

Sexual intercourse involves an interplay of signals from the nervous and endocrine systems. It can lead to pregnancy, which humans use a variety of methods to prevent.

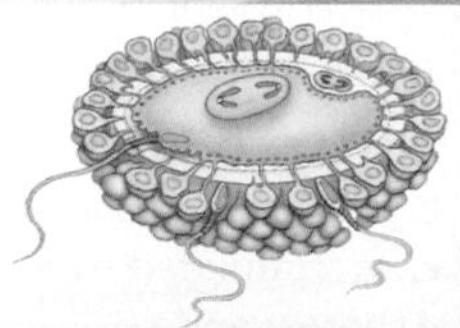

SEXUALLY TRANSMITTED DISEASES

Pathogens are passed between partners by sexual interactions and may be transmitted to offspring during birth. Effects of resulting diseases range from discomfort to death.

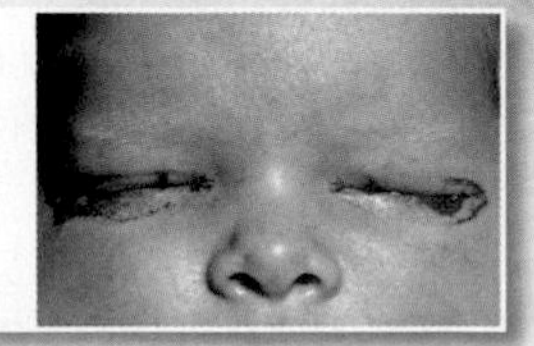

The next chapter continues the story of sexual reproduction with a description of the mechanisms by which sexually reproducing organisms develop. Chapter 43 also touches on animal sexual behavior, courtship, and reproductive success. The evolution of life history traits, such as the number of offspring per reproductive event, is the focus of Section 44.5.

41.1 Assisted Reproduction

***In vitro* fertilization** (IVF) is an assisted reproduction method in which an egg and sperm are combined outside the body. *In vitro* means in glass, and refers to the laboratory glassware in which fertilization takes place. Prior to an IVF procedure, a woman is given hormones to encourage maturation of multiple eggs and to prevent her from releasing them naturally. Then, mature eggs are removed from her body using a hollow needle. The eggs are combined with a partner's or donor's sperm to allow fertilization. After fertilization, each zygote undergoes mitotic divisions, forming a ball of cells (a blastocyst) that can be placed in a woman's uterus (womb) to develop to term.

Louise Brown, the first child conceived by IVF, was born in 1978. At that time, much of the public and many scientists were appalled by the idea of what newspapers referred to as "test tube babies." Scientists worried that this new, unnatural procedure would produce children with psychological and genetic defects. Ethicists warned about the societal implications of manipulating human embryos.

Despite these initial reservations, IVF is now in wide use worldwide. More than 3 million people have been born as a result of this procedure. The first test tube babies have become adults and are now having children of their own. Louise Brown is among them. In 2007, at age 28, she conceived naturally and gave birth to a healthy son.

Research into IVF opened the way to other reproduction technologies. If a man makes sperm, but does not ejaculate properly or does not ejaculate enough sperm to allow fertilization by normal means, intracytoplasmic sperm injection can put his sperm inside his partner's egg (**FIGURE 41.1**). If a woman cannot produce viable eggs, she can obtain donated eggs and use IVF to produce an embryo that she can carry to term. A woman who has normal eggs but cannot sustain a pregnancy herself can have an embryo conceived by IVF implanted in a surrogate mother.

Another recent development is the ability to "bank" eggs—to freeze them and store them for later use. Human sperm have been stored this way for decades, but the Food and Drug Administration did not approve an egg-freezing procedure until 2012. The ability to store eggs after their extraction from a donor has already cut the cost of using donated eggs for IVF. It also has given women who face the loss of fertility as a result of a medical condition or aging a way to retain the option of reproducing later in life.

***in vitro* fertilization** Assisted reproductive technology in which eggs and sperm are united outside the body.

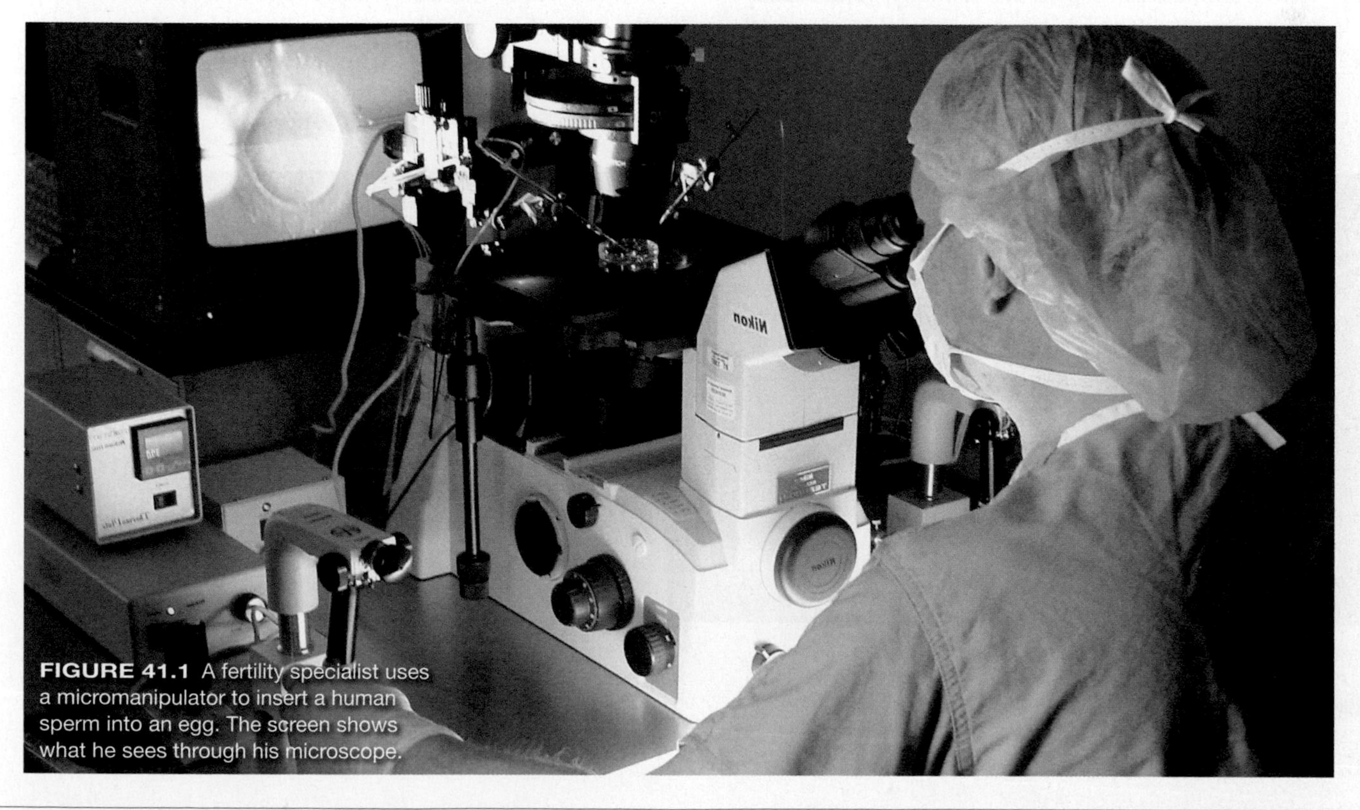

FIGURE 41.1 A fertility specialist uses a micromanipulator to insert a human sperm into an egg. The screen shows what he sees through his microscope.

CREDITS: (opposite) Roger Klein/WaterF/Age Fotostock; (1) © Heidi Speckt, West Virginia University.

41.2 Modes of Animal Reproduction

✔ Sexual reproduction dominates the life cycle of most animals, including many that can also reproduce asexually.

Asexual Versus Sexual Reproduction

With **asexual reproduction**, a single individual produces offspring (Section 11.2). Mutations aside, all offspring are genetic replicas (clones) of the parent and thus identical to one another. Asexual reproduction is advantageous in a stable environment where the gene combination that makes a parent successful is likely to do the same for its offspring.

Invertebrates reproduce by a variety of asexual mechanisms. In some species, a new individual grows on the body of its parent, a process called budding. For example, new hydras bud from existing ones (FIGURE 41.2A). Many corals reproduce by fragmentation, in which a piece of the parent breaks off and develops into a new animal. Some flatworms reproduce by transverse fission: The worm divides in two, leaving one piece headless and the other tailless. Each piece then grows the missing body parts.

Parthenogenesis is a mechanism of asexual reproduction in which female offspring develop from unfertilized eggs. Some invertebrates, fishes, amphibians, lizards, and one bird (the turkey) can produce offspring by parthenogenesis. No mammal is known to reproduce asexually by natural means.

Most animals can reproduce sexually. With **sexual reproduction**, two parents produce haploid gametes that combine at fertilization. An **egg** is a female gamete and a **sperm** is a male gamete. As a result of crossing over and the random assortment of chromosomes during meiosis, each gamete receives a different combination of maternal or paternal alleles (Section 12.4). When an egg and a sperm combine, the result is a genetically unique individual with alleles from both parents.

Producing offspring that differ from both parents and from one another can be advantageous in a changing environment (Section 12.1). By reproducing sexually, a parent increases the likelihood that some of its offspring will have a combination of alleles that suits them to conditions in the changed environment.

Some animals reproduce asexually when conditions are stable and favorable, but switch to sexual reproduction when conditions begin to change. Consider aphids, a type of plant-sucking insect. In early spring, when tender plant parts are plentiful and aphids are not, a female aphid can give birth to several smaller clones of herself every day by parthenogenesis. In autumn, when plants prepare for dormancy, food becomes scarce for aphids and competition for it increases among them. At this time, the aphids switch reproductive modes to produce offspring by sexual means.

Variations on Sexual Reproduction

Sexually reproducing individuals that can make both eggs and sperm are called **hermaphrodites**. Tapeworms and some roundworms are simultaneous hermaphrodites. These worms produce eggs and sperm at the same time, and they can fertilize themselves. Earthworms, land snails, and slugs are simultaneous hermaphrodites too, but they require a partner. Some

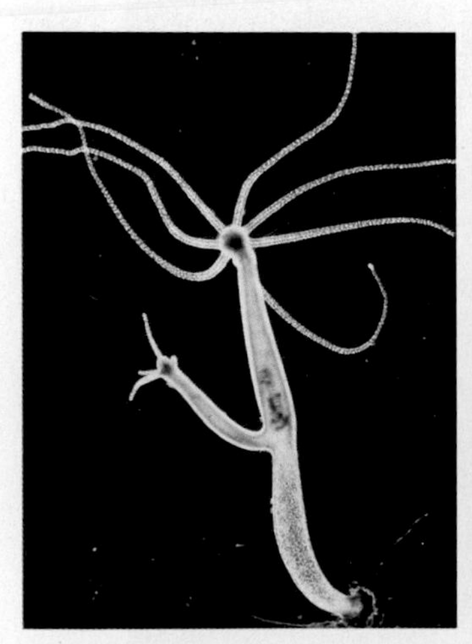

A Asexual reproduction in a hydra. A new individual (left) is budding from its parent.

B Sexual reproduction in hamlets, which are simultaneous hermaphrodites. Each fish lays eggs and also fertilizes its partner's eggs.

C Sexual reproduction in elephants. The male is inserting his penis into the female. Eggs will be fertilized and the offspring will develop inside the mother's body, nourished by nutrients delivered by her bloodstream.

FIGURE 41.2 Examples of animals reproducing.

fishes are sequential hermaphrodites, meaning individuals switch from one sex to another over the course of a lifetime. However, the overwhelming majority of vertebrate species have separate sexes that are fixed for life; each individual remains either male or female.

Most sexually reproducing aquatic invertebrates, bony fishes, and amphibians have **external fertilization**; they release their gametes into water and fertilization occurs there. With **internal fertilization**, sperm fertilize an egg inside the female's body (**FIGURE 41.2C**). Internal fertilization occurs in all cartilaginous fishes and in most land animals, including insects and the amniotes.

After internal fertilization, a female may lay eggs or retain them inside her body while they develop. In all birds and most insects, most development occurs outside the mother's body (**FIGURE 41.3A**). In many sharks, snakes, and lizards embryos develop while enclosed by an egg sac in the mother's body. The eggs hatch inside the mother shortly before she gives birth (**FIGURE 41.3B**).

A developing animal requires nutrients. Embryos of most animals are nourished solely by yolk, a thick fluid rich in proteins and lipids that is deposited in an egg as it forms. The amount of yolk in a egg increases with the time that the egg takes to hatch. Kiwi birds have the longest incubation period of any bird—11 weeks. Their eggs are two-thirds yolk by volume.

By contrast, placental mammals produce almost yolkless eggs and nourish their embryos by means of a placenta (**FIGURE 41.3C**). A **placenta** is an organ that forms during pregnancy and facilitates the exchange of substances between the maternal and embryonic bloodstreams. Placenta-like organs also evolved in several groups of live-bearing fishes and one group of live-bearing lizards (skinks). This independent evolution of similar body parts in different lineages is an example of morphological convergence (Section 18.3).

A Beetle depositing fertilized eggs on a leaf. Yolk in the eggs provides the nutrients that sustain development of the young.

B Snake (adder) giving birth to young that recently hatched in her body. The young were nourished by egg yolk, not by their mother.

C An elk examines her newborn calf. The placenta, the organ that allowed nutrients from her blood to diffuse into the calf's blood, is visible at the left. It is expelled at birth.

FIGURE 41.3 How developing offspring are nourished.

asexual reproduction Reproductive mode by which offspring arise from a single parent only.
egg Female gamete.
external fertilization Sperm and eggs are released into the external environment and meet there.
hermaphrodite Animal that produces both eggs and sperm, either simultaneously or at different times in its life.
internal fertilization Fertilization of eggs inside a female's body.
placenta Organ that factilitates exchanges between the bloodstreams of a developing embryo and its mother.
sexual reproduction Reproductive mode by which offspring arise from two parents and inherit genes from both.
sperm Male gamete.

TAKE-HOME MESSAGE 41.2

How do animals reproduce?

- ✔ Most animals reproduce sexually, but some reproduce only asexually or can switch between asexual and sexual reproduction.
- ✔ External fertilization is typical in aquatic animals. Land-dwelling animals typically fertilize eggs in the female's body.
- ✔ Yolk deposited in an egg as it forms sustains development in most sexually reproducing animal species. In placental mammals, nutrients diffuse from maternal blood into the blood of the developing young.

41.3 Organs of Sexual Reproduction

✔ Most animals have specialized organs that produce gametes, and some have organs that aid in their transfer.

Gonads, Ducts, and Glands

In most animals, gamete production occurs inside special reproductive organs called **gonads.** Eggs are produced in gonads called **ovaries,** and sperm in gonads called **testes** (singular, testis).

The number, structure, and location of gonads varies among animal groups. Jellies typically have gonads on the lining of their gastrovascular cavity (FIGURE 41.4). A sea star has a pair of ovaries or testes in each arm. An octopus has a single testis or ovary near the rear of its head, inside the mantle cavity. A fruit fly has a pair of ovaries or testes inside its abdomen (FIGURE 41.5).

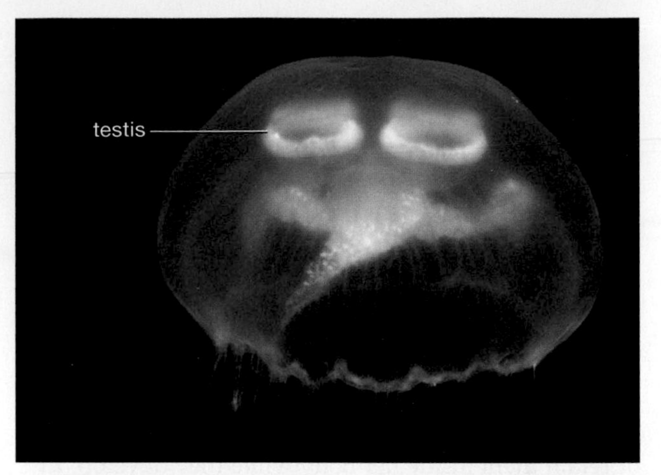

FIGURE 41.4 Moon jelly (*Aurelia*), with four ringlike testes visible through its translucent body.

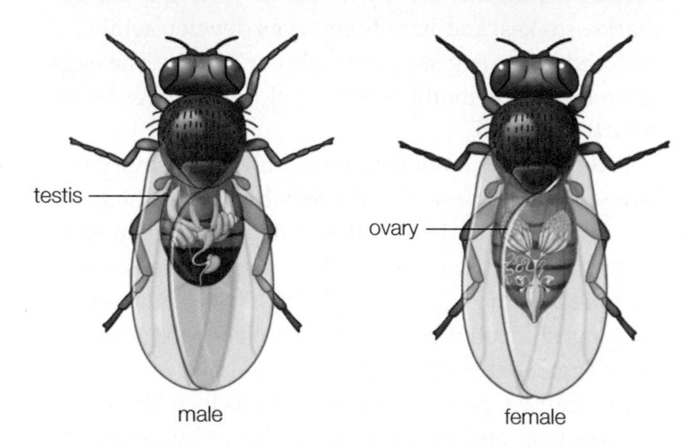

FIGURE 41.5 ▶Animated Fruit fly reproductive system. A series of ducts convey sperm (produced in testes) and eggs (produced in ovaries) out of the body.

Most vertebrates have a pair of either ovaries or testes. Ovaries, when present, are always in the abdominal cavity, as are the testes of fish, reptiles, and birds. In most placental mammals, however, testes descend into a pouch during embryonic development. This pouch suspends the testes outside the abdominal cavity, just beneath the pelvic girdle.

Many animals reproduce only during a specific breeding period. In such species, gonads often shrink or even disappear during the time they are not in use.

For fertilization to occur, gametes must exit from the gonads. In cnidarians such as jellies, the gonads open onto the gastrovascular cavity, from which gametes exit the body. However, in most animals and all vertebrates, newly formed gametes enter into a system of ducts that conveys them to the surface of the body. In live-bearing animals, young develop inside specialized organs derived from such ducts.

In addition to gonads and the ducts that convey gametes, animal reproductive systems often include glands that support gamete development or facilitate fertilization. We delve into the structure and function of the various vertebrate reproductive glands in the sections of this chapter that describe the human reproductive organs.

How Gametes Form

All gonads contain germ cells. These diploid cells give rise to gametes by division. Because germ cells are diploid and eggs and sperm are haploid, the chromosome number must be halved by meiosis (Section 12.2).

Animal sperm production is called **spermatogenesis** (FIGURE 41.6A). It begins with mitosis of a spermatagonium, a diploid germ cell inside the testes. Mitosis and cytoplasmic division of spermatgonia produces primary spermatocytes, which are also diploid cells ❶. Each primary spermatocyte undergoes meiosis I and cytoplasmic division to yield two haploid secondary spermatocytes ❷. A secondary spermatocyte completes meiosis II and undergoes cytoplasmic division to produce two haploid spermatids ❸. The resulting four haploid spermatids then mature into four sperm ❹.

Animal egg production is called **oogenesis** (FIGURE 41.6B). It begins with mitosis and cytoplasmic division of an oogonium, which is a diploid germ cell in an ovary ❶. The resulting diploid cells are the primary oocytes. An **oocyte** is an immature animal egg.

A primary oocyte undergoes meiosis I to produce two haploid nuclei, then undergoes unequal cytoplasmic division to yield two differently sized cells ❷. One cell, the secondary oocyte, has a haploid nucleus and the bulk of the cytoplasm from the primary oocyte. The other cell, a **polar body,** has an identical haploid nucleus, but only a tiny bit of cytoplasm. It has no further role in reproduction and will degenerate.

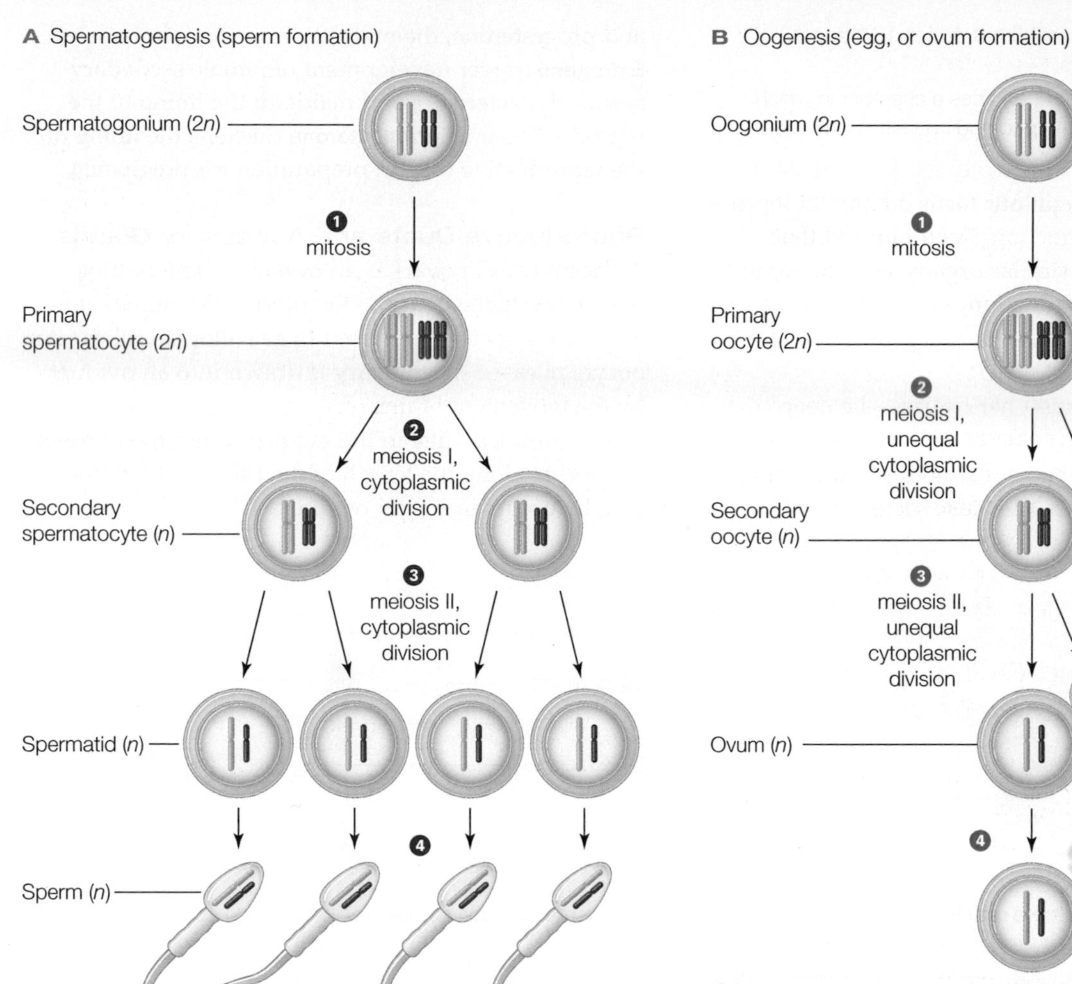

FIGURE 41.6 Gamete formation in animals. **FIGURE IT OUT** How does cytoplasmic division in spermatogenesis differ from that during oogenesis?

Answer: Spermatogenesis involves equal divisions of cytoplasm, and oogenesis involves unequal cytoplasmic divisions.

The secondary oocyte undergoes meiosis II, followed by unequal cytoplasmic division ❸. One cell receives a haploid nucleus with an unduplicated set of chromosomes, along with the bulk of the cytoplasm. This cell is the mature egg, or **ovum** (plural, ova). The other cell receives an identical nucleus, but little cytoplasm. Like the first polar body, this second polar body has no further role and will degenerate ❹. Thus, oogenesis produces only a single ovum from each primary oocyte. An ovum is always much larger than a sperm of the same species.

gonad Animal reproductive organ that produces gametes.
oocyte Immature animal egg.
oogenesis Egg production in an animal.
ovary Egg-producing reproductive organ.
ovum Mature animal egg.
polar body Tiny cell produced by unequal cytoplasmic division during egg production in animals.
spermatogenesis Sperm production in an animal.
testis Sperm-producing reproductive organ.

TAKE-HOME MESSAGE 41.3

What organs play a role in an animal's sexual reproduction?

✔ Sexual reproduction begins with production of gametes (eggs or sperm), which usually occurs inside gonads.

✔ Meiosis produces gametes. During spermatogenesis, one male germ cell gives rise to four sperm. During oogenesis, one female germ cell gives rise to one ovum (egg).

✔ In addition to gonads, most animals have a system of ducts that convey gametes to the body surface, as well as glands that nourish or otherwise support the gametes.

41.4 Reproductive System of Human Females

✔ The reproductive system of human females produces gametes and sex hormones.

✔ The system receives sperm, and has a chamber in which developing offspring are protected and nourished until birth.

With this section, we begin our focus on human reproductive structure and function. Keep in mind that other vertebrates have similar organs, and the same hormones govern their function.

The Female Gonads

A human female's gonads—her ovaries—lie deep inside her pelvic cavity (**FIGURE 41.7A, B**). Ovaries are about the size and shape of almonds. They produce and release oocytes. They also secrete estrogens and progesterone, the main sex hormones in females. **Estrogens** trigger development of female secondary sexual characteristics and maintain the lining of the reproductive tract. **Progesterone** thickens the lining of the reproductive tract in preparation for pregnancy.

Reproductive Ducts and Accessory Glands

Adjacent to each ovary is an **oviduct**, a hollow tube that connects the ovary to the uterus. (Mammalian oviducts are sometimes referred to as Fallopian tubes.) An oocyte released by an ovary is drawn into an oviduct by the movement of fingerlike projections at the oviduct's entrance. Cilia in the oviduct lining then propel the oocyte along the length of the tube. Fertilization usually occurs inside an oviduct.

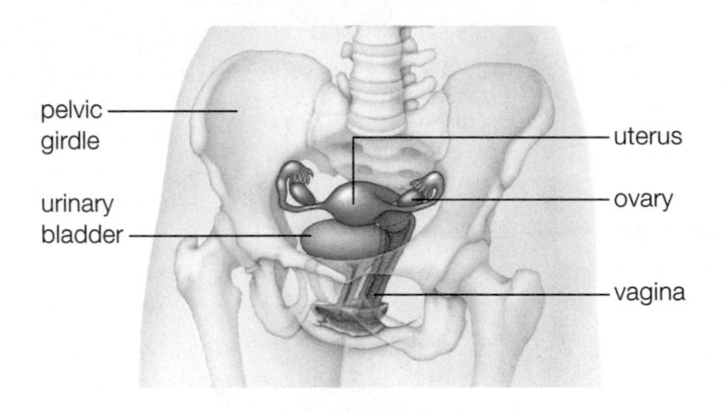

A Location of female reproductive organs.

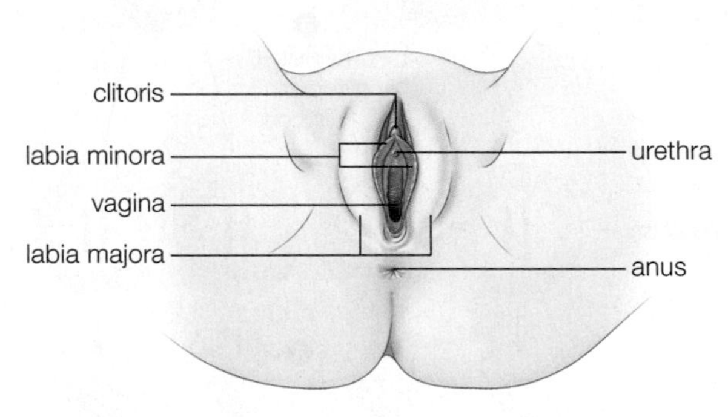

C External sex organs, collectively referred to as the vulva.

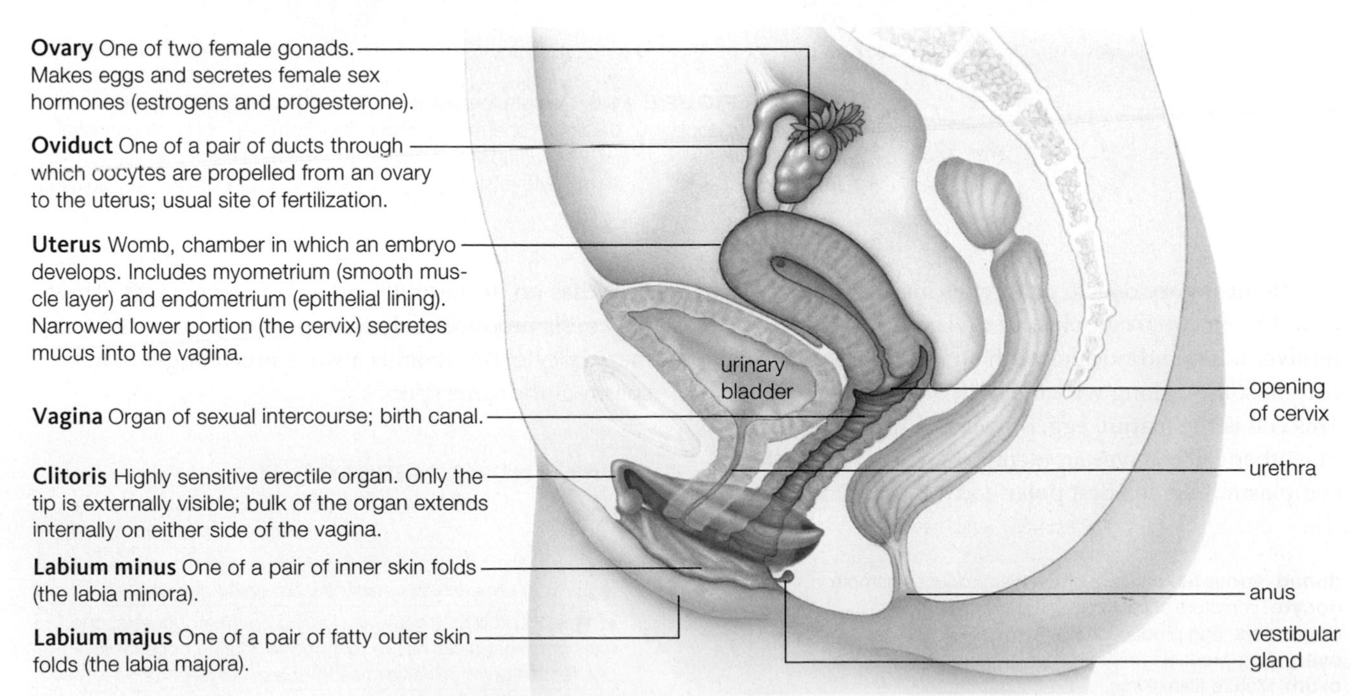

B Human female reproductive organs, shown in longitudinal section.

FIGURE 41.7 ▸Animated Components of the human female reproductive system and their functions.

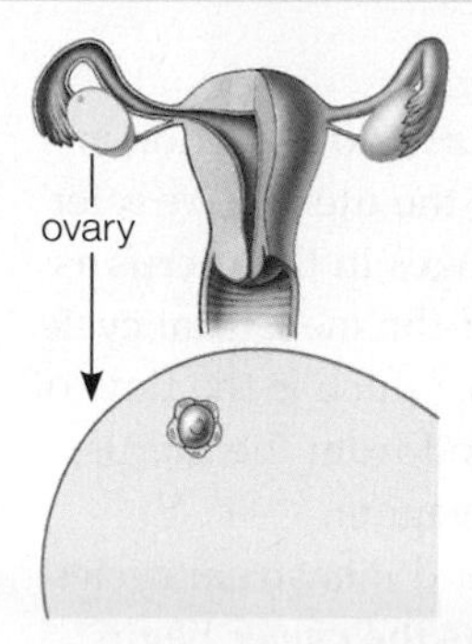

❶ An ovary has many immature follicles, each consisting of a primary oocyte and surrounding follicle cells.

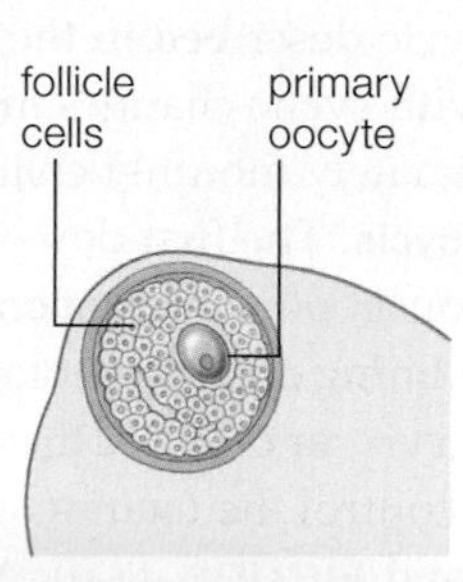

❷ A fluid-filled cavity begins to form in the follicle's cell layer.

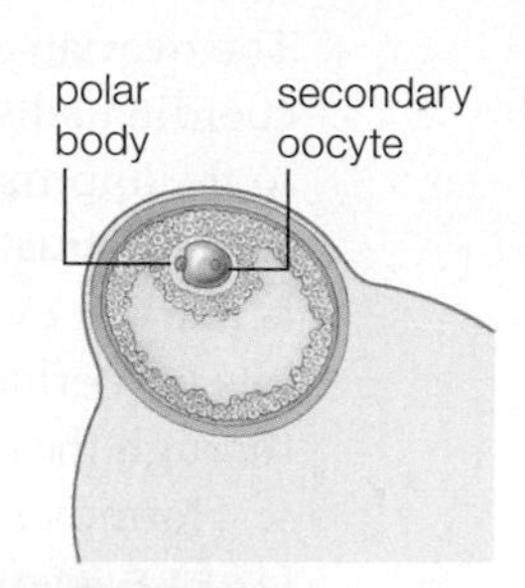

❸ The primary oocyte completes meiosis I and divides unequally, forming a secondary oocyte and a polar body.

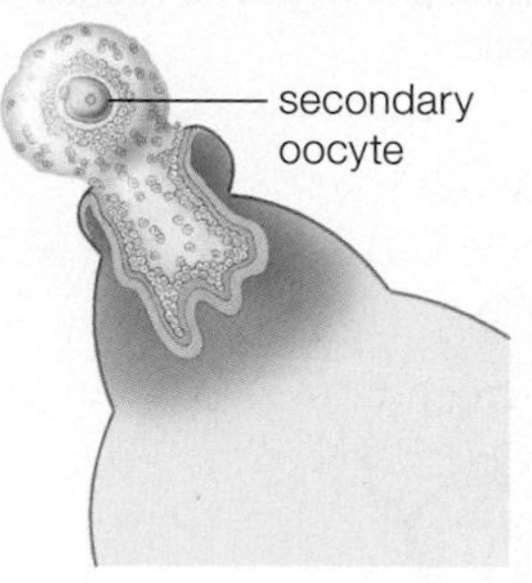

❹ Ovulation. The mature follicle ruptures, releasing a secondary oocyte coated with secreted proteins and follicle cells.

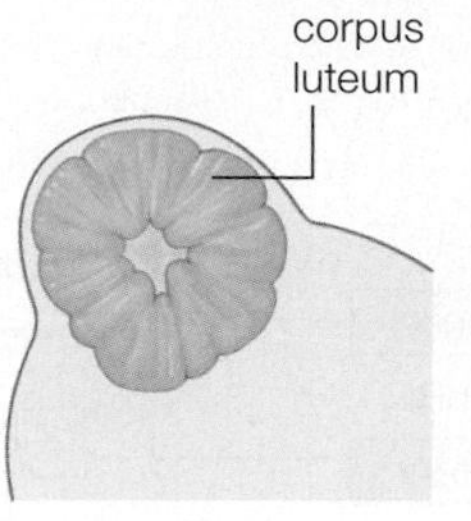

❺ A corpus luteum develops from follicle cells left behind after ovulation.

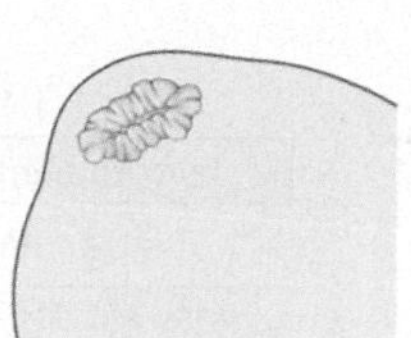

❻ If pregnancy does not occur, the corpus luteum degenerates.

FIGURE 41.8 ▶**Animated** Ovarian cycle. The mature follicle is about the size of a pea and projects from the ovary's surface.

Both oviducts open onto the **uterus**, a hollow, pear-shaped organ above the urinary bladder. A thick layer of smooth muscle (myometrium) makes up the bulk of the uterine wall. The uterine lining (endometrium) consists of glandular epithelium, connective tissues, and blood vessels. The lowest portion of the uterus, a narrowed region called the **cervix**, opens into the vagina. The **vagina**, which extends from the cervix to the body's surface, is the organ of intercourse and the birth canal.

Externally visible organs of the reproductive tract are called genitals. Female genitals include two pairs of liplike skin folds that enclose the openings of the vagina and urethra (FIGURE 41.7C). Adipose tissue fills the thick outer folds, the labia majora. Thin inner folds are the labia minora. The clitoris lies near the anterior junction of the labia minora. It contains erectile tissue and is highly sensitive to tactile stimulation.

Egg Production and Release

In humans, all production of oocytes from germ cells occurs before birth. A girl is born with about 2 million **primary oocytes**—oocytes that entered meiosis but stopped in prophase I, rather than completing meiosis. At puberty (the stage of development when reproductive organs mature) hormonal changes prompt oocytes to mature, one at a time, in an approximately twenty-eight-day ovarian cycle. FIGURE 41.8 shows this cycle. An oocyte and cells around it are an **ovarian follicle** ❶. As the cycle begins, the follicle enlarges and a fluid-filled cavity forms around it ❷. (More than one follicle may begin to develop, but usually only one matures.) The primary oocyte completes meiosis I and undergoes unequal cytoplasmic division to produce a secondary oocyte and a polar body ❸. The **secondary oocyte** has begun meiosis II, but halted in metaphase II. It will not complete meiosis until fertilization.

About two weeks after the follicle began to mature, its wall ruptures and **ovulation** occurs: The secondary oocyte, polar body, and surrounding follicle cells are ejected into the adjacent oviduct ❹.

After ovulation, cells of the ruptured follicle develop into a hormone-secreting **corpus luteum** ❺. (The name means "yellow body" in Latin.) If pregnancy does not occur, the corpus luteum breaks down ❻, and a new follicle will begin to mature.

cervix Narrow part of uterus that connects to the vagina.
corpus luteum Hormone-secreting structure that forms from follicle cells left behind after ovulation.
estrogens Hormones secreted by ovaries; cause development of secondary sexual traits and maintain the reproductive tract's lining.
ovarian follicle Immature animal egg and the surrounding cells.
oviduct Duct that conveys eggs away from an animal ovary.
ovulation Release of a secondary oocyte from an ovary.
primary oocyte Oocyte that has halted in prophase I of meiosis.
progesterone Hormone secreted by ovaries; prepares the reproductive tract for pregnancy.
secondary oocyte Oocyte that has halted in metaphase II of meiosis; released from an ovary at ovulation.
uterus Muscular chamber where offspring develop; womb.
vagina Female organ of intercourse and birth canal.

TAKE-HOME MESSAGE 41.4

What are the functions of female reproductive organs?

✔ A pair of ovaries produce oocytes (eggs). They also make and secrete the hormones estrogen and progesterone.

✔ About once a month, an oocyte released from an ovary enters an oviduct. If fertilization occurs, the resulting embryo develops in the uterus.

✔ The vagina serves as the female organ of intercourse and as the birth canal.

41.5 Female Reproductive Cycles

✔ Cyclic changes in the concentrations of sex hormones govern female reproductive function.

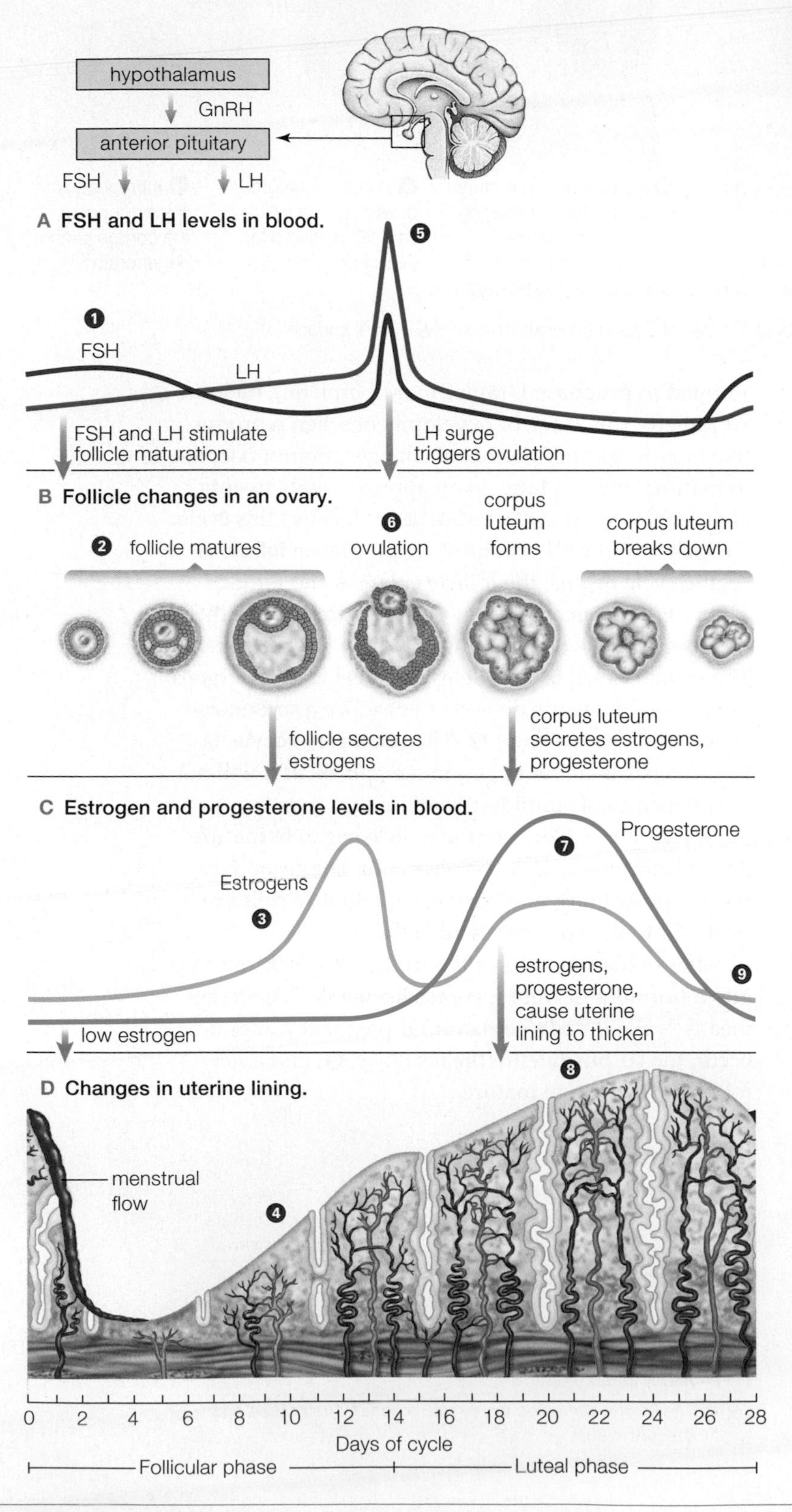

The Menstrual Cycle

The ovarian cycle described in the previous section is coordinated with cyclic changes in the uterus. We refer to the approximately monthly changes in the uterus as the **menstrual cycle**. The first day of the menstrual cycle is marked by onset of **menstruation**, which is the flow of bits of uterine lining and some blood from the uterus, through the cervix, and out of the vagina.

Hormones control the ovarian and menstrual cycles (**TABLE 41.1** and **FIGURE 41.9**). As the cycles begin, secretion of **gonadotropin-releasing hormone** (**GnRH**) by the hypothalamus is stimulating release of FSH and LH from the pituitary ❶. As its name indicates, **follicle-stimulating hormone** (**FSH**) stimulates maturation of an ovarian follicle ❷. The interval of follicle maturation before ovulation is the follicular phase of the cycle. During this time, cells around the oocyte secrete estrogens ❸ that stimulate the endometrium to thicken ❹.

As the level of estrogens increases, a positive feedback loop occurs. Increased estrogen causes the hypothalamus to release more GnRH, so the pituitary releases more FSH, and the follicle grows and produces more estrogens. The rise in estrogens encourages the anterior pituitary to release more **luteinizing hormone** (**LH**), which in females has roles in ovulation and formation of the corpus luteum.

At about the midpoint in the cycle, the high level of estrogens causes a surge in LH secretion ❺. The surge of LH causes the primary oocyte to complete meiosis I and undergo cytoplasmic division. It also causes the follicle to swell and burst. Thus, the midcycle surge of LH is the trigger for ovulation ❻.

The luteal phase of the cycle begins after ovulation. LH stimulates formation of the corpus luteum, which secretes some estrogens and a lot of progesterone ❼. These hormones cause the uterine lining to thicken and encourage blood vessels to grow through it. The uterus is now ready for pregnancy ❽.

FIGURE 41.9 ▸Animated Changes in the ovary and uterus correlated with changing hormone levels. We start with the onset of menstrual flow on day one of a twenty-eight-day menstrual cycle.

(**A,B**) Prompted by GnRH from the hypothalamus, the anterior pituitary secretes FSH and LH, which stimulate a follicle to grow and an oocyte to mature in an ovary. A midcycle surge of LH triggers ovulation and the formation of a corpus luteum. A decline in FSH after ovulation stops more follicles from maturing.

(**C,D**) Early in the cycle, estrogen from a maturing follicle stimulates repair and rebuilding of the endometrium. After ovulation, the corpus luteum secretes some estrogen and more progesterone that primes the uterus for pregnancy.

Table 41.1 Events of a Twenty-Eight-Day Ovarian/Menstrual Cycle

Phase	Events	Day of Cycle
Follicular phase	Menstruation; endometrium breaks down	1–5
	Follicle matures in ovary; endometrium rebuilds	6–13
Ovulation	Oocyte released from ovary	14
Luteal phase	Corpus luteum forms, secretes progesterone; the endometrium thickens and develops	15–28

Secretion of progesterone by the corpus luteum has a negative feedback effect on the hypothalamus and pituitary. A high level of progesterone reduces secretion of FSH and LH, thus preventing maturation of other follicles and inhibiting additional ovulations.

If fertilization does not occur, the level of LH continues to decline, causing the corpus luteum to degenerate. Breakdown of the corpus luteum causes estrogen and progesterone levels to plummet ❾. In the uterus, the decline in estrogens and progesterone causes degeneration of the uterine lining, and menstruation begins. Blood and endometrial tissue flow out of the vagina for three to six days. At the same time, the pituitary increases secretion of FSH and LH once again.

During their reproductive years, many women regularly experience discomfort a week or two before they menstruate. Breasts become tender because estrogens and progesterone cause milk ducts in them to widen. Body tissues swell because premenstrual changes influence aldosterone secretion. (As Section 40.5 explained, aldosterone stimulates reabsorption of sodium and, indirectly, water.) Cycle-associated hormonal changes can also cause depression, irritability, anxiety, headaches, and insomnia. Regular recurrence of these symptoms is known as premenstrual syndrome (PMS). The use of oral contraceptives, which minimize hormone swings, can prevent PMS.

During menstruation, cells in the uterine lining secrete prostaglandins, and these local signaling molecules stimulate contractions of smooth muscle in the uterine wall. Many women do not feel the muscle contractions, but others experience a dull ache or sharp pains commonly known as menstrual cramps. Severe pain and heavy bleeding during menstruation are not normal and may be caused by benign tumors in the uterus called fibroids.

A woman enters **menopause** when all the follicles in her ovaries have either been released during menstrual cycles or have disintegrated as a result of aging. With no follicles left to mature, production of estrogen and progesterone is diminished and menstrual cycles cease. Interestingly, menopause is rare among animals. It is known only in humans and two species of whales.

The Estrous Cycle

All female placental mammals replace their uterine lining on a cyclic basis. However, most do not menstruate—they reabsorb their endometrial lining rather than shedding it. Humans are also unusual in that they can have sex at any time. Most female animals have an **estrous cycle**, which means they are sexually receptive only when they can conceive—a period known as estrus, or "heat." The length of estrous cycles varies among species. In dogs, it is about six months; in mice, it is six days.

In many species, changes in a female's body alert males of her species to the fact she is in estrus. Often a female's labia swell and she produces a discharge containing chemical signals that help attract potential mates. For example, when female dog is heat, her vagina produces a thin discharge containing chemicals that attract male dogs.

TAKE-HOME MESSAGE 41.5

What cyclic changes occur in the ovary and uterus?

- ✔ Every twenty-eight days or so, FSH and LH stimulate maturation of an ovarian follicle.
- ✔ A midcycle surge of LH triggers ovulation—the release of a secondary oocyte into an oviduct.
- ✔ Estrogen secreted by a maturing follicle causes the endometrium to thicken. After ovulation, progesterone from the corpus luteum encourages endometrial thickening.
- ✔ If pregnancy does not occur, the corpus luteum breaks down, progesterone levels drop, the thickened uterine lining is shed, and the cycle begins again.

estrous cycle In female animals, a recurring cyclic variation in sexual receptivity.
gonadotropin-releasing hormone (GnRH) A hypothalamic hormone that induces the pituitary to release hormones (LH and FSH) that act on the gonads.
follicle-stimulating hormone (FSH) Anterior pituitary hormone with roles in ovarian follicle maturation and sperm production.
luteinizing hormone (LH) Anterior pituitary hormone with roles in ovulation, corpus luteum formation, and sperm production.
menopause Permanent cessation of menstrual cycles.
menstrual cycle Reproductive cycle in which the uterus lining thickens and then, if pregnancy does not occur, is shed.
menstruation Flow of shed uterine tissue out of the vagina.

41.6 Reproductive System of Human Males

✔ A human male's reproductive system produces hormones and sperm, which it delivers to a female's reproductive tract.

Male Gonads

A human male's gonads are his testes, also called testicles (**FIGURE 41.10**). They produce sperm and make the male sex hormone **testosterone**. Testes form on the wall of an XY embryo's abdominal cavity. Before birth, testes descend into a **scrotum**, a pouch of loose skin suspended below the pelvic girdle. Inside this pouch, smooth muscle encloses the testes. When a man feels cold, reflexive contractions of this muscle draw his testes closer to his body. When he feels warm, the muscle relaxes, allowing the testes to hang lower so the sperm-making cells inside them do not overheat. These cells function best a bit below normal body temperature.

Reproductive Ducts and Accessory Glands

Sperm form by meiosis in the testes, a process we will discuss shortly. Here we consider the path that sperm travel to the body surface. The journey begins when sperm enter the **epididymis** (plural, epididymides), a coiled duct perched on a testis. The Greek *epi–* means upon and *didymos* means twins. In this context, the "twins" refers to the two testes.

The last region of the epididymis is continuous with the first portion of a vas deferens. In Latin, *vas* means vessel, and *deferens,* to carry away. Each **vas deferens** (plural, vasa deferentia) is a duct that carries sperm away from an epididymis, and to a short ejaculatory duct. Ejaculatory ducts deliver sperm to the urethra, the duct that extends through a male's penis to open at the body surface.

The **penis** is the male organ of intercourse. It has a rounded head (the glans) at the end of a narrower shaft. Nerve endings in the glans make it highly sensitive to touch. Normally, when a man is not sexually excited, a retractable tube of skin called the foreskin covers the glans of his penis. In many cultures, males undergo circumcision, an elective surgical procedure that removes the foreskin. Circumcision reduces the risk of contracting sexually transmitted diseases and of penile cancer, but it is painful, and rare complications require follow-up surgery or impair sexual function.

Three elongated cylinders of spongy tissue fill the interior of the penis. When a male is sexually excited, blood flows into the spongy tissue faster than it flows out. Fluid pressure rises and the normally limp penis becomes stiff and erect.

Sperm stored in the epididymides and first part of the vasa deferentia continue their journey toward the body surface only when a male reaches the peak of sexual excitement and ejaculates. During ejaculation, rhythmic smooth muscle contractions propel sperm

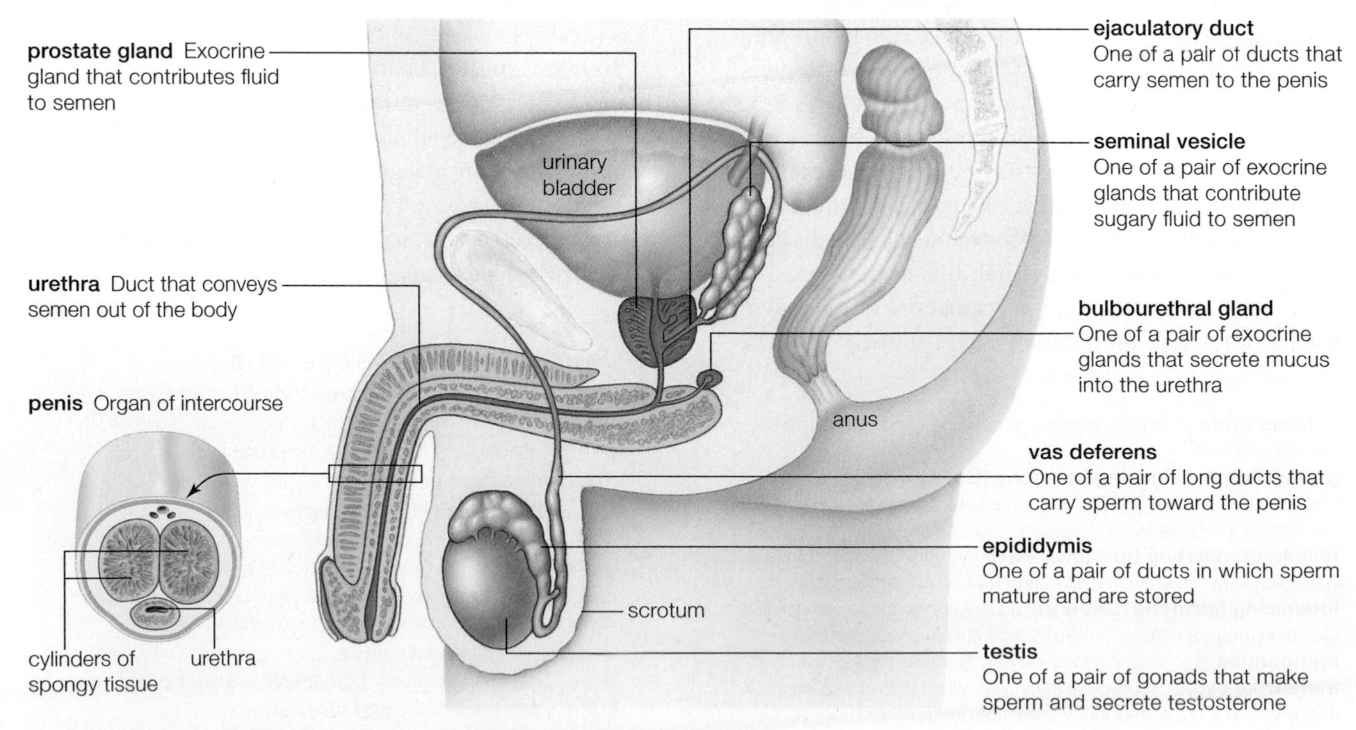

FIGURE 41.10 ▸**Animated** Organs of the human male reproductive system.

and accessory gland secretions out of the body as a thick, white fluid called semen.

Semen is a mix of sperm, proteins, nutrients, ions, and signaling molecules. Sperm constitute less than 5 percent of its volume; the bulk of it consists of secretions from accessory glands. Seminal vesicles, which are exocrine glands near the base of the bladder, secrete fructose-rich fluid into the vasa deferentia. Sperm use the fructose (a sugar) as their energy source. The prostate gland, which encircles the urethra, is the other major contributor to semen volume. Its slightly alkaline secretions help raise the pH of the female reproductive tract, making this passage more hospitable to sperm.

Prior to ejaculation, two pea-sized bulbourethral glands secrete a lubricating mucus into the urethra. This mucus helps clear the urethra of residual urine.

Germ Cells to Sperm Cells

A testis is smaller than a golf ball, yet it contains more than 100 meters of sperm-making **seminiferous tubules** (**FIGURE 41.11**). Diploid male germ cells line the inner wall of each tubule. At puberty, these cells begin to divide by mitosis to produce primary spermatocytes. The spermatocytes undergo meiosis to produce round, haploid cells called spermatids, which then differentiate into specialized cells with a long flagellum—sperm. A seminiferous tubule also contains big, elongated cells called nurse cells (or Sertoli cells) that support developing sperm. The testes' testosterone-secreting cells (Leydig cells) reside between the seminiferous tubules.

Like oogenesis, spermatogenesis is governed by LH and FSH. LH binds to Leydig cells stimulating them to secrete testosterone. FSH targets nurse cells. In combination with testosterone, it causes nurse cells to produce chemicals essential to sperm development.

Newly formed sperm cannot swim and are moved into the epididymis by smooth muscle contractions. Cilia of the epididymis lining push the immature sperm farther along this duct. The sperm mature and become mobile as they move through the epididymis.

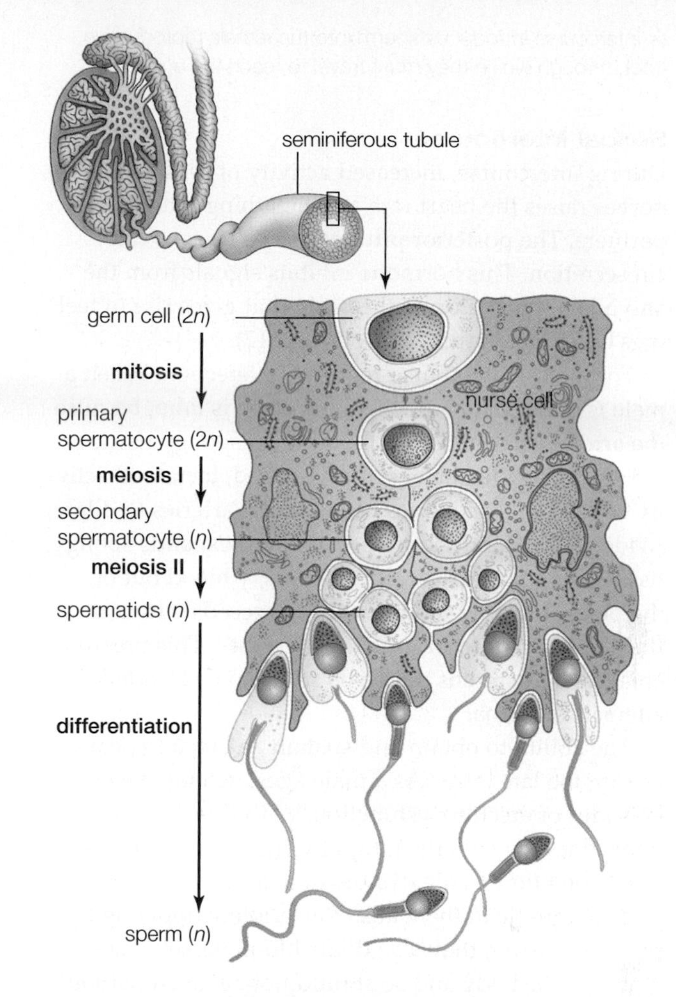

FIGURE 41.11 ▶**Animated** Human spermatogenesis.

Sperm formation takes about 65 days, from start to finish. A sexually mature male makes sperm continually, so on any given day he has millions of them in various stages of development. A sperm stored in the epididymis remains viable for about a month. After that it breaks down and its components are reabsorbed.

TAKE-HOME MESSAGE 41.6

What are the functions of the male reproductive organs?

✔ A pair of testes produce sperm. They also make and secrete the sex hormone testosterone.

✔ Sperm form inside the seminiferous tubules of testes. Their production is governed by LH and FSH.

✔ Ducts convey sperm from a testis to the body surface. Exocrine glands that empty into these ducts provide most of the volume of semen.

✔ The penis contains cylinders of spongy tissue that engorge with blood during sexual excitement.

epididymides A pair of ducts in which sperm formed in testes mature; each empties into a vas deferens.
penis Male organ of intercourse.
scrotum Pouch of skin that encloses a human male's testes.
semen Sperm mixed with secretions from seminal vesicles and the prostate gland.
seminiferous tubules Inside a testis, coiled tubules that contain male germ cells and produce sperm.
testosterone Main hormone produced by testes; required for sperm production and development of male secondary sexual traits.
vas deferens One of a pair of long ducts that convey mature sperm toward the body surface.

41.7 Bringing Gametes Together

✔ Intercourse introduces sperm into the female reproductive tract, through which they must travel to reach the egg.

Sexual Intercourse

During intercourse, increased activity of sympathetic nerves raises the heart rate and breathing rate in both partners. The posterior pituitary steps up its oxytocin secretion. This hormone inhibits signals from the amygdala, the region of the brain that gives rise to feelings of fear and anxiety (Section 32.12).

For males, intercourse requires an erection. When a male is not sexually aroused, his penis is limp, because the arteries that transport blood to its spongy tissue are constricted. When he becomes aroused, increased activity of sympathetic nerves causes these arteries to dilate (widen). The resulting inflow of blood expands spongy tissue and compresses veins that carry blood out of the penis. Inward blood flow now exceeds outward flow, so internal fluid pressure increases. This pressure enlarges and stiffens the penis so it can be inserted into a female's vagina.

The ability to obtain and sustain an erection peaks during the late teens. As a male ages, he may have episodes of erectile dysfunction. With this disorder, the penis does not stiffen enough for intercourse. Drugs prescribed for erectile dysfunction act by relaxing smooth muscle in the walls of arterioles supplying the penis. However, they also dilate blood vessels elsewhere in the body and so should not be taken without a prescription.

When a woman becomes sexually excited, blood flow to the vaginal wall, labia, and clitoris increases. Glands in the cervix secrete mucus, and glands on the labia (the equivalent of a male's bulbourethral glands) produce a lubricating fluid. The vagina itself does not have any glandular tissue. It is moistened by mucus from the cervix and by plasma fluid that seeps out between epithelial cells of the vaginal lining.

Continued mechanical stimulation of the penis or clitoris can lead to orgasm. During orgasm, endorphins flood the brain and evoke feelings of pleasure. At the same time, a surge of oxytocin causes rhythmic contractions of smooth muscle in both the male and female reproductive tract. In males, orgasm is usually accompanied by ejaculation, in which contracting muscles force the semen out of the penis.

The Sperm's Journey

Ejaculation can put 300 million sperm into the vagina. **FIGURE 41.12** shows the structure of a sperm. It is a haploid cell with a "head" that is packed full of DNA

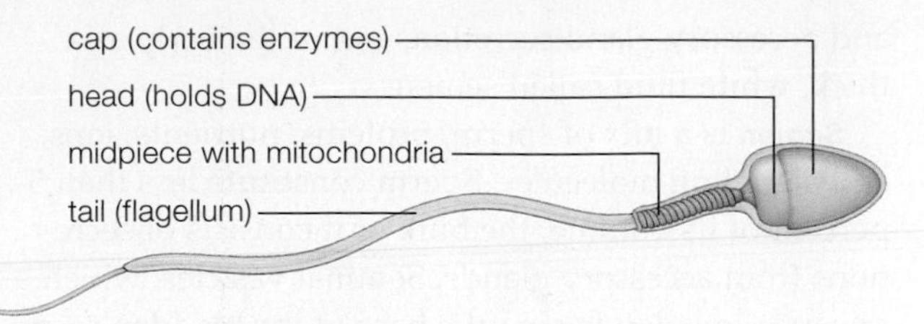

FIGURE 41.12 Structure of a human sperm.

and tipped by an enzyme-containing cap. The enzymes help the sperm penetrate an oocyte. At its other end, the sperm has a flagellum that it uses to swim toward an egg. The sperm does not have ribosomes, endoplasmic reticulum, or Golgi bodies. However, its midsection contains many mitochondria that supply the ATP required for flagellar movement.

While in a male's body, sperm make ATP by breaking down the fructose in seminal vesicle secretions. In a female reproductive tract, they take up carbohydrates from their environment to fuel their movement.

To travel from the vagina into the uterus, a sperm must swim through a canal in the center of the cervix. During parts of the reproductive cycle when a woman's estrogen level is low, a thick plug of acidic mucus bars the passage through this canal. As the woman's estrogen level rises in the prelude to ovulation, the cervix becomes more sperm-friendly. It begins to secrete a thinner, more alkaline mucus that contains glucose, a sugar that sperm can use as fuel. Even so, swimming through cervical mucus is challenging, so only the strongest sperm can pass through it to enter the uterus.

Once sperm are in the uterus, contractions of smooth muscle in the uterine wall help move them, toward the oviducts. Ovulation occurs in one ovary at a time, so only one oviduct leads to an oocyte. However, sperm choose an oviduct randomly, so half of them have no chance of encountering an oocyte.

Fertilization

Fertilization usually happens in the upper part of an oviduct (**FIGURE 41.13** ❶). Of the millions of sperm ejaculated into the vagina, only a few hundred make it this far. Sperm live for three days after ejaculation, so fertilization can occur even if intercourse takes place a few days before ovulation.

A secondary oocyte released at ovulation retains a wrapping of follicle cells. Beneath those cells is the zona pellucida, a coat composed of glycoproteins secreted by the oocyte ❷. Sperm make their way between the follicle cells to the zona pellucida. The plasma membrane of the sperm's head has receptors

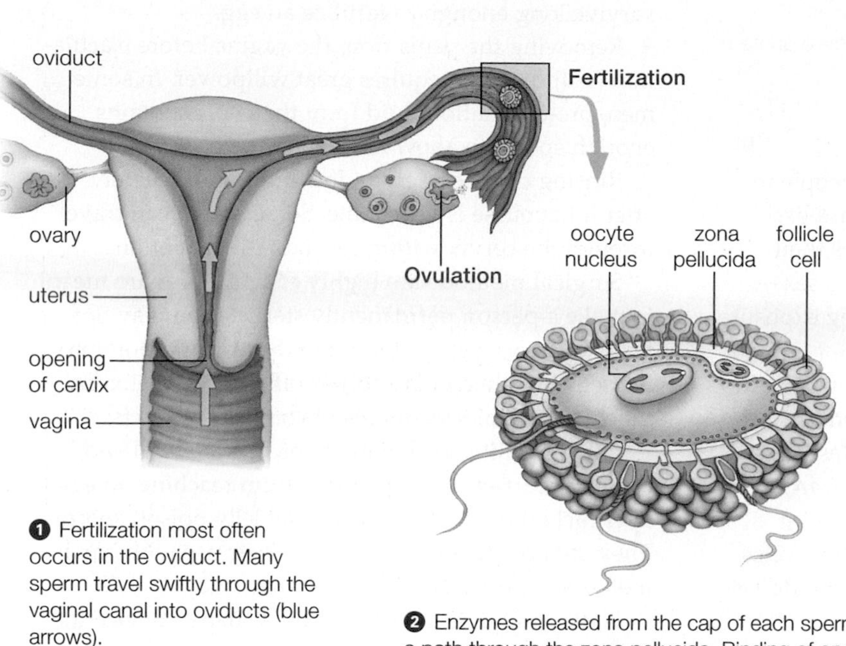

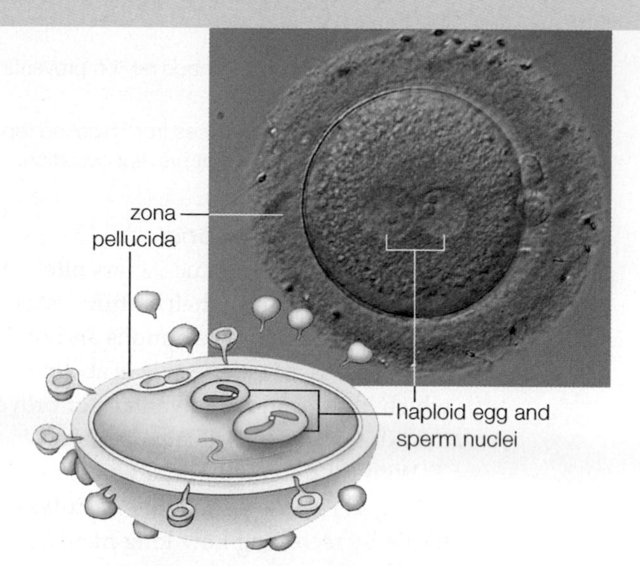

❶ Fertilization most often occurs in the oviduct. Many sperm travel swiftly through the vaginal canal into oviducts (blue arrows).

Inside an oviduct, the sperm surround a secondary oocyte that was released by ovulation.

❷ Enzymes released from the cap of each sperm clear a path through the zona pellucida. Binding of one sperm to the membrane of the secondary oocyte causes the oocyte to release substances that alter the zona pellucida and prevent other sperm from binding.

❸ The sperm is drawn into the oocyte and the oocyte nucleus completes meiosis II. Later, both paternal and maternal nuclear membranes break down and chromosomes become arranged on a bipolar spindle in preparation for the first mitotic division.

FIGURE 41.13 ▸Animated Events in human fertilization. The light micrograph on the right shows a fertilized human oocyte.

FIGURE IT OUT In the micrograph, what are the small cells on the right, just beneath the zona pellucida?

Answer: The polar bodies

that bind species-specific proteins in this coat. This binding triggers release of protein-digesting enzymes from the cap on the sperm's head. The collective effect of enzyme released by many sperm clears a passage through the zona pellucida to the oocyte's plasma membrane. When receptors in this membrane bind a sperm's plasma membrane, the entire sperm is drawn head first into the oocyte.

Typically only one sperm enters the oocyte. Its entry causes the oocyte to secrete proteins that alter the consistency of the zona pellucida. As a result of this alteration, other sperm cannot bind to the zona pellucida or move through it to the oocyte's plasma membrane.

Although it only takes one sperm to fertilize an egg, it takes the presence of many to release enough enzyme to clear a way to the egg plasma membrane. This is why a man who has healthy sperm but a low sperm count can be functionally infertile.

Remember that the secondary oocyte released at ovulation was halted in metaphase of meiosis II. Binding of the sperm membrane to the oocyte membrane causes the oocyte to complete meiosis. Unequal cytoplasmic division follows, producing a single mature egg—an ovum—and the second polar body ❸.

Chromosomes in the egg and sperm nuclei become the genetic material of the zygote (the first cell of the new individual). The sperm also supplies a single centriole. Recall from Section 4.10 that centrioles help microtubules organize themselves. The sperm's centriole replicates to form the pair of centrioles that draw maternal and paternal nuclei together and organize the spindle for the zygote's first mitotic division.

Sperm mitochondria and the flagellum enter the oocyte too, but they are typically broken down. Thus, mitochondria are inherited only from the mother.

TAKE-HOME MESSAGE 41.7

What happens during intercourse and fertilization?

✔ Sexual arousal involves signals from the endocrine and nervous systems.

✔ Ejaculation releases millions of sperm into the vagina. Sperm travel through the uterus toward the oviducts, where fertilization most often occurs.

✔ Penetration of a secondary oocyte by a single sperm causes the oocyte to complete meiosis II, and prevents additional sperm from penetrating the oocyte.

✔ The DNA of the sperm, along with that of the oocyte, become the genetic material of the zygote.

41.8 Contraception and Infertility

✔ Most birth control methods rely on preventing ovulation or blocking fertilization.

✔ Infertility most often arises from blocked reproductive ducts or hormonal disorders that prevent ovulation.

Birth Control Options

Emotional and economic factors often lead people to seek ways to control their fertility. **TABLE 41.2** lists common contraceptive options and compares their effectiveness. Most effective is abstinence—no sex—which has the added advantage of preventing exposure to sexually transmissible pathogens.

With rhythm methods, a woman abstains from sex during her fertile period. She calculates when she is fertile by recording how long menstrual cycles last, checking her temperature daily (**FIGURE 41.14A**), monitoring the thickness of her cervical mucus, or some combination of these methods. However, cycles vary, so miscalculations are frequent and sperm deposited in the vagina up to three days before ovulation can survive long enough to fertilize an egg.

Removing the penis from the vagina before ejaculation (withdrawal) requires great willpower. In some men, pre-ejaculation fluid from the penis contains enough sperm to allow fertilization to occur.

Rinsing out the vagina (douching) immediately after intercourse is unreliable. Some sperm can travel through the cervix within seconds of ejaculation.

Surgical methods are highly effective, but are meant to make a person permanently sterile. Men may opt for a vasectomy. A doctor makes a small incision into the scrotum, then cuts and ties off each vas deferens. A tubal ligation blocks or cuts a woman's oviducts.

Other fertility control methods use physical and chemical barriers to stop sperm from reaching an egg. Spermicidal foam and spermicidal jelly poison sperm. They are not always reliable, but their use with barrier methods reduces the chance of pregnancy.

A diaphragm is a flexible, dome-shaped device that is positioned inside the vagina so it covers the cervix. A diaphragm is relatively effective if it is first fitted by a doctor and used correctly with a spermicide. A cervical cap is a similar but smaller device.

Condoms are thin sheaths worn over the penis or used to line the vagina during intercourse (**FIGURE 41.14B**). Reliable brands can be 95 percent effective when used correctly with a spermicide. Condoms made of latex also offer some protection against sexually transmitted diseases. However, even the best ones can tear or leak, reducing effectiveness.

An intrauterine device, or IUD, is inserted into the uterus by a physician. Some IUDs cause cervical mucus to thicken so sperm cannot swim through it. Others shed copper that interferes with implantation.

Oral contraceptives—birth control pills—are the most common fertility control method in developed countries. "The Pill" is a mixture of synthetic estrogens and progesterone-like hormones that prevents both maturation of oocytes and ovulation (**FIGURE 41.14C**). When taken diligently, oral contraception is highly effective. It also reduces menstrual cramps and lowers the risk of ovarian and uterine cancer. Side effects include nausea, headaches, weight gain, and increased risk of cancer of the breast, cervix, and liver.

Hormones can also be delivered by a birth control patch (an adhesive patch applied to skin), by injections, or by an implant. Injections are effective for several months, whereas an implant such as Implanon is effect for years (**FIGURE 41.14C**). Both methods are quite effective, but may cause sporadic, heavy bleeding.

Table 41.2 Common Methods of Contraception

Method	Mechanism of Action	Pregnancy Rate*
Abstinence	Avoid intercourse entirely	0% per year
Rhythm method	Avoid intercourse when female is fertile	25% per year
Withdrawal	End intercourse before male ejaculates	27% per year
Vasectomy	Cut or close off male's vasa deferentia	>1% per year
Tubal ligation	Cut or close off female's oviducts	>1% per year
Condom	Enclose penis, block sperm entry to vagina	15% per year
Diaphragm, cervical cap	Cover cervix, block sperm entry to uterus	16% per year
Spermicides	Kill sperm	29% per year
Intrauterine device	Prevent sperm entry to uterus or prevent implantation	>1% per year
Oral contraceptives	Prevent ovulation	>1% per year
Hormone patches, implants, or injections	Prevent ovulation	>1% per year
Emergency contraception pill	Prevent ovulation	15–25% per use**

** Percentage of users who get pregnant despite consistent, correct use*
*** Not meant for regular use*

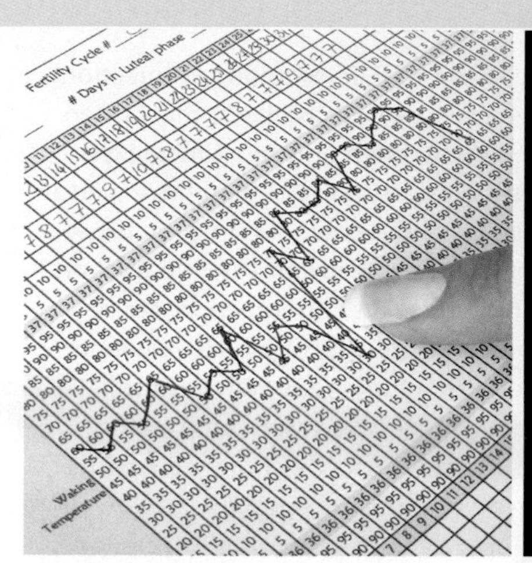

A Abstain from sex when a woman is fertile. Tracking body temperature is one way to estimate when ovulation occurs.

B Use a barrier such as this male condom to prevent sperm from meeting an egg.

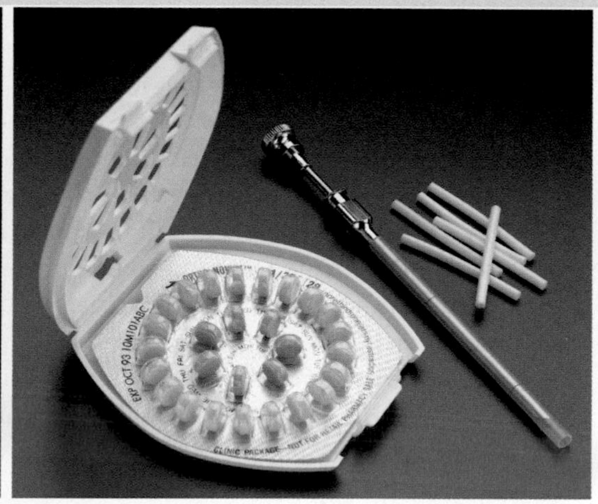

C Use synthetic female hormones to prevent ovulation. Oral contraceptives (left) are taken daily. Implantable contraceptives (right) are injected under the skin.

FIGURE 41.14 Some methods of preventing pregnancy.

If a woman has unprotected sex or a condom breaks, she can turn to emergency contraception. Some so-called morning-after pills are available without a prescription to women over age 17. The most widely used emergency contraceptive pill delivers a large dose of synthetic progesterone that prevents ovulation and interferes with fertilization. It is most effective when taken immediately after intercourse, but has some effect up to three days later. Morning-after pills are not meant for regular use. They cause nausea, vomiting, abdominal pain, headache, and dizziness.

Infertility

About 10 percent of couples in the United States are infertile, which means they do not have a successful pregnancy despite a year of trying. Infertility increases with age. In about half of cases infertility arises in the female partner. The most common cause of female infertility is a hormonal disorder called polycystic ovarian syndrome. With this disorder, a follicle begins to mature, but ovulation does not occur; instead the follicle turns into a cyst (a fluid-filled sac).

Some sexually transmitted diseases can scar the oviducts and prevent sperm from reaching eggs. Endometriosis also interferes with oviduct function. With this disorder, endometrial tissue grows in the regions of the pelvis outside the uterus. Like normal endometrium, the misplaced tissue proliferates and is shed in response to hormones. The result is pain and scarring that can distort oviducts. Endometriosis can also make intercourse painful, thus lessening its frequency and the opportunities for conception.

If fertilization does occur, scarring of oviducts can lead to an embryo implanting in the oviduct, causing a tubal pregnancy that cannot develop to term and threatens the life of the mother. Fibroids, endometriosis, and other uterine problems can interfere with the ability of the embryo to implant.

Male infertility arises when a man makes abnormal sperm or too few sperm, or when sperm are not ejaculated properly. Disorders that lower testosterone can prevent sperm maturation. Some genetic disorders can interfere with sperm motility. In males, as in females, sexually transmitted disease can scar ducts of the reproductive system, causing blockages that keep sperm from leaving the body. Male infertility can also arise when the valve between the bladder and the urethra does not function properly, causing sperm to end up in the bladder rather than leaving the body through the urethra.

TAKE-HOME MESSAGE 41.8

What can prevent pregnancy?

- ✔ Couples can prevent pregnancy by avoiding sex entirely or abstaining when the woman is fertile.
- ✔ Hormones delivered by pills, patches, or injections can prevent a woman from ovulating.
- ✔ Temporary barriers such as condoms or a diaphragm keep sperm and egg apart, as do surgical methods such as a vasectomy or tubal ligation.
- ✔ Infertility can arise from hormonal disorders, blockages of reproductive ducts, and uterine problems that prevent implantation.

CREDITS: (14A) © iStockphoto.com/Ever; (14B) Scott Camazine & Sue Trainor/Science Source; (14C) © SuperStock/SuperStock.

41.9 Sexually Transmitted Diseases

✔ Sex acts transfer body fluids in which some human pathogens travel from one host to another.

Sexually transmitted diseases (STDs) are contagious diseases that spread through sexual contact. Women contract STDs more easily than men, and they have more complications. Women can also infect their offspring during childbirth, so sexually active women who intend to become pregnant should be tested for STDs.

To reduce the risk of STD transmission, physicians recommend use of a latex condom and a lubricant. The condom serves as a barrier to the pathogen and the lubricant helps prevent small abrasions that would make it easier for the pathogen to enter the body.

Trichomoniasis

Trichomoniasis, commonly called "trich," is the most common nonviral STD. It is caused by the flagellated protozoan *Trichomonas vaginalis*, which was shown in **FIGURE 21.6**. Many infected people do not have symptoms, but some infected women have a yellowish discharge, and a sore, itchy vagina. In both sexes, an untreated infection can cause infertility. Some epidemiological studies suggest that, in men, untreated trichomoniasis may increase the risk of benign prostate enlargment and aggressive prostate cancer. A single dose of an antiprotozoal drug can quickly cure the infection. Both partners should be treated.

Bacterial STDs

The most common bacterial STD is chlamydia, which is caused by *Chlamydia trachomatis*. Chlamydias are small bacteria that live as intracellular parasites, relying on their host cell for ATP. In women, an infection of the reproductive tract by *C. trachomatis* most often goes undetected. Some women and most men experience painful urination; most infected men have a clear or yellow discharge from the penis. Left untreated, a *Chlamydia* infection can scar the reproductive tract and lead to infertility in both sexes. An infection can be passed from a mother to child during birth, causing pneumonia and conjunctivitis in the newborn (**FIGURE 41.15A**). Chlamydia can be cured with antibiotics.

Gonorrhea, the second most common bacterial STD is caused by *Neisseria gonorrhoeae*. Men usually develop symptoms within one week of becoming infected; yellow pus oozes from the penis and urination becomes frequent and painful. By contrast, most women have no early symptoms. In both sexes an infection can damage reproductive ducts and cause sterility. A persistent infection can become systemic, meaning the bacteria can spread throughout the body. Systemic gonorrhea most commonly harms the joints and skin, but can also affect the heart and liver. Gonorrhea is treated with antibiotics, but strains resistant to the most widely used antibiotics are increasingly common. After an *N. gonorrhoeae* strain highly resistant to all current antibiotics was isolated from one patient in Japan in 2011, health officials cautioned that a future pandemic of untreatable gonorrhea could be possible.

Syphilis is caused by *Treponema pallidum*, a spiral-shaped bacterium. During sex with an infected partner, these bacteria get onto the genitals or into the cervix, vagina, or oral cavity. They slip into the body through tiny cuts. One to eight weeks later, many *T. pallidum* cells are twisting about inside a flattened, painless

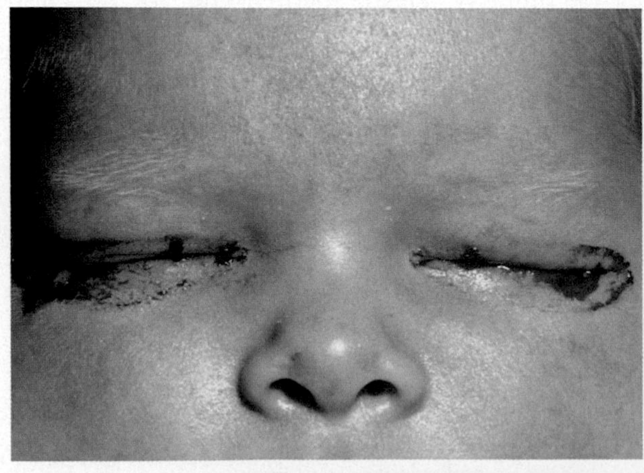

A Passage of pathogens to offspring. This infant's eyes are infected by *Chlamydia* acquired from its mother during birth.

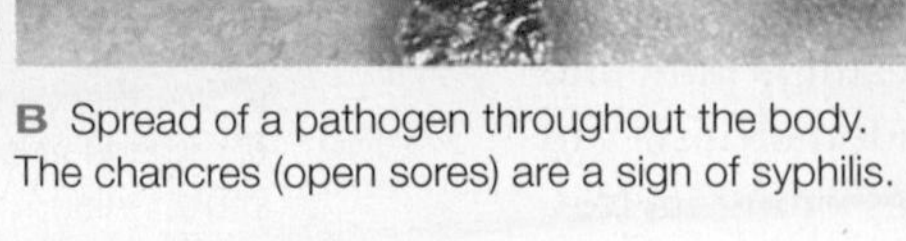

B Spread of a pathogen throughout the body. The chancres (open sores) are a sign of syphilis.

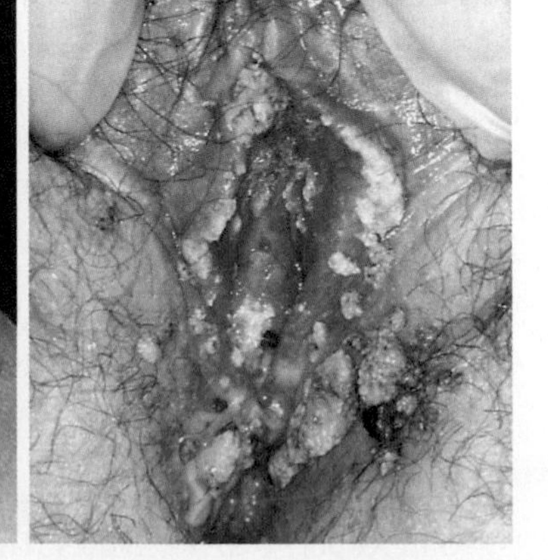

C Warty genitals and increased risk of cancer. The raised white warts on this woman's labia are a sign of HPV.

FIGURE 41.15 A few downsides of unsafe sex.

Assisted Reproduction (revisited)

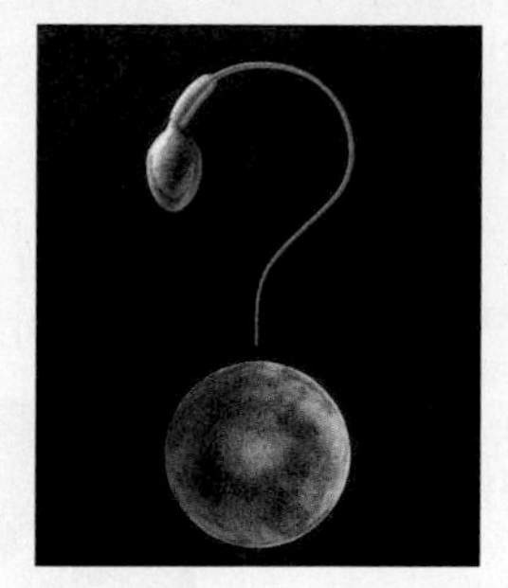

The most common form of assisted reproduction is the use of donor sperm. In the United States, the Federal Drug Administration regulates sperm banks, requiring them to screen donors for general health, test donors for sexually transmitted diseases, take an extensive family history, and test semen for the presence of certain pathogens. It does not, however, mandate any type of genetic screening.

As a result, it is possible for young donors, who are healthy and capable of passing a physical exam, to unkowingly pass on recessive alleles for genetic disorders or alleles for dominant late-onset disorders. In one case, a man who donated sperm as a 23-year-old was later diagnosed with a genetic disorder that weakens the heart and raises the risk of early death. Of the man's 24 children, 22 fathered as a donor and two with his wife, nine have the allele for this disorder. One died at age 2, and two had heart problems by age 15.

To minimize genetic risks, sperm banks now voluntarily screen donors for alleles most commonly associated with genetic disorders. However, even with expensive and extensive testing, they cannot eliminate the possibility of some less common harmful alleles from slipping by.

It's also important to remember that unique allele combinations affecting health can arise whenever people mate, whether with a spouse or a anonymous donor. A shuffling of the genetic cards is integral to sexual reproduction.

---

ulcer (a chancre) at the site of infection. If untreated, the infection can become systemic. Skin chancres appear (**FIGURE 41.14B**) and the liver, bones, and eventually the brain can be damaged. Like gonorrhea, syphilis is treated with antibiotics. As with gonorrhea, the number of antibiotic-resistant cases is on the rise.

Viral STDs

Infection by human papillomaviruses (HPV) is widespread in the United States. Many HPV strains cause bumplike growths called warts, including those common on hands or feet. Other strains of the virus are sexually transmitted and cause genital warts (**FIGURE 41.15C**), and a few of these strains cause cancer (Section 37.1). HPV infection causes nearly all cervical cancers in women and most anal cancers in homosexual men. HPV introduced into the mouth by oral sex raises the risk of mouth and throat cancer.

Vaccinations and screenings are key to the fight against HPV-associated cancers. A vaccine can prevent HPV infection in both males and females, if given before viral exposure. A blood test can determine whether a person has been infected by one of the two strains that most commonly cause cancer, and a Pap test can detect early signs of cervical cancer.

About 45 million Americans have genital herpes caused by herpes simplex virus 2. An initial infection commonly causes small sores at the site where the virus entered the body. The sores heal, but the virus can be reactivated, causing tingling or itching with or without visible sores. Antiviral drugs cannot cure the infection, but they help sores heal faster, and also lessen the likelihood of viral reactivation.

Transmission of the herpes virus from mother to child during birth is rare—the mother must have an active infection at the time of delivery—but it can have serious consequences. If untreated, an infant infected in this manner has a 40 percent chance of death. Those who survive remain infected, and have a heightened risk of health problems including blindness and impaired brain development. Women who become infected for the first time while pregnant are especially likely to have an active infection at delivery.

Infection by HIV (human immunodeficiency virus) can cause AIDS (acquired immunodeficiency syndrome). Sections 20.1 and 20.3 described the HIV virus and Section 37.11 explained how its effects on the immune system can result in AIDS. With regard to sexual transmission, oral sex is least likely to pass on an infection. Unprotected anal sex is 5 times more dangerous than unprotected vaginal sex and 50 times more dangerous than oral sex.

TAKE-HOME MESSAGE 41.9

What are causes and effects of common STDs, and how are they treated?

- ✔ Viruses, bacteria, and protozoans cause common STDs.
- ✔ Effects of infection range from discomfort, to infertility, to systemic problems or cancer.
- ✔ Infected women can pass an STD to their offspring.
- ✔ Bacterial and protozoan pathogens can be killed by drugs. Viral infections cannot be cured with drugs, but antiviral drugs reduce the effects of herpes and HIV infection. An HPV vaccine can prevent infection by this virus.

CREDIT: (in text) Lightspring/Shutterstock.

summary

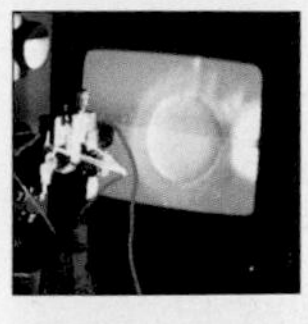

Section 41.1 Millions of people have been born as a result of assisted reproductive methods such as ***in vitro* fertilization** and sperm banking. The variety of reproductive options continues to expand.

Section 41.2 **Asexual reproduction** yields genetically identical copies (clones) of the parent. **Sexual reproduction** produces genetic variety in offspring, which can be advantageous in a changing environment.

Most animals that reproduce sexually have separate sexes, but some are **hermaphrodites** that produce both **eggs** and **sperm**. With **external fertilization**, gametes are released into water. Most animals on land have **internal fertilization**; gametes meet in a female's body. Offspring may develop inside or outside the maternal body. Yolk helps nourish developing young of most animals. In placental mammals, young are sustained by nutrients delivered across the **placenta**.

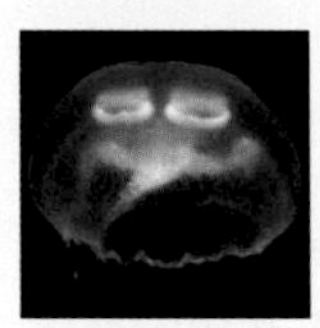

Section 41.3 Gametes form in **gonads**; sperm in **testes** and eggs in **ovaries**. With **spermatogenesis**, meiosis and equal cytoplasmic divisions produce four haploid sperm. With **oogenesis**, meiosis I followed by unequal division produces an **oocyte** and a **polar body**. Meiosis II of this oocyte is followed by another unequal division to yield a second polar body and an **ovum**. In most animals, ducts convey mature gametes to the body surface and glands assist in reproductive function.

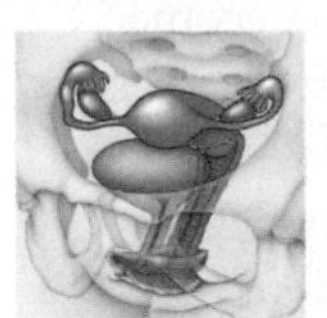

Sections 41.4, 41.5 Human ovaries produce eggs and sex hormones. An **oviduct** conveys an oocyte to the **uterus**. The **cervix** of the uterus opens into the **vagina**, which serves as the organ of intercourse and the birth canal.

From puberty until **menopause**, a woman has an approximately monthly **menstrual cycle**. **Gonadotropin-releasing hormone (GnRH)** causes the pituitary to release **follicle-stimulating hormone (FSH)** and **luteinizing hormone (LH)**. FSH causes an **ovarian follicle** to begin maturing. Follicle cells around a **primary oocyte**, which formed before birth, proliferate and secrete **estrogens** and **progesterone**. A midcycle surge of luteinizing hormone (LH) triggers **ovulation** of the now **secondary oocyte**. After ovulation, the **corpus luteum** secretes progesterone that primes the uterus for pregnancy. When the corpus luteum breaks down, **menstruation** occurs.

Human females do not breed seasonally, unlike female animals that have an **estrous cycle**.

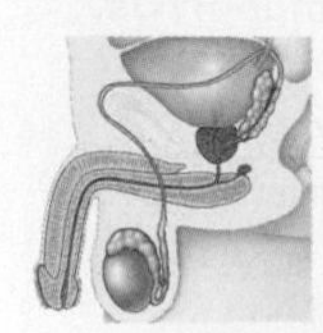

Section 41.6 Human testes reside within a **scrotum**. They produce sperm and **testosterone**. Beginning at puberty, sperm form continually from germ cells inside the testes' **seminiferous tubules**. The sperm mature in an **epididymis** that opens into a **vas deferens**. Secretions from the seminal vesicles and prostate gland join with sperm to form **semen**. Semen is expelled from the body through the urethra that runs through the **penis**. Like oogenesis, spermatogenesis is governed by LH and FSH.

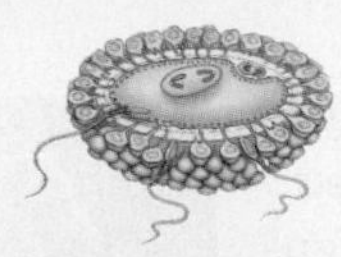

Section 41.7 Hormones and autonomic nerves govern physiological changes during arousal and intercourse. Ejaculation delivers millions of sperm into the vagina. Usually only one penetrates a secondary oocyte, causing it to complete meiosis II. The nucleus of the resulting ovum and that of the sperm supply genetic material of the zygote.

Section 41.8 Humans can choose either to prevent pregnancy by abstaining from sex entirely or to reduce the chance of pregnancy by abstaining during a woman's fertile period, surgically severing reproductive ducts, placing a physical or chemical barrier at the entrance to the uterus, or administering synthetic female sex hormones to prevent ovulation. Infertility can arise as a result of disorders that impair gamete production. Blocked reproductive ducts and uterine problems that interfere with implantation or development also affect fertility.

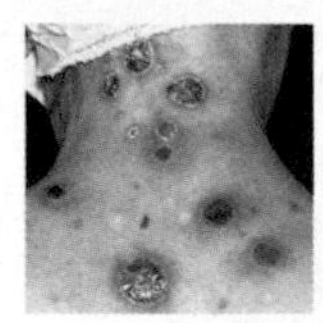

Section 41.9 Sexual intercourse can pass viral, bacterial, and protozoan pathogens between partners. Infected mothers can transmit these pathogens to their offspring. The consequences of a sexually transmitted disease range from mild discomfort to sterility and systemic disease that can be fatal. Protozoan and bacterial STDs can be cured with drugs, but there are no drugs to cure viral STDs.

Self-Quiz

Answers in Appendix VII

1. Sexual reproduction ______ .
   a. requires formation of gametes by meiosis
   b. produces offspring identical in their traits
   c. occurs only in vertebrates
   d. all of the above

2. The ______ is a genetic dead end.
   a. polar body b. oocyte c. ovum d. sperm

3. The cervix is the entrance to the ______ .
   a. oviducts b. vagina c. uterus d. scrotum

4. During a menstrual cycle, a midcycle surge of ______ from the ______ triggers ovulation.
   a. estrogen; follicle cells c. LH; pituitary
   b. progesterone; egg d. FSH; pituitary

5. The corpus luteum develops from ______ and secretes progesterone that causes the lining of the uterus to thicken.
   a. follicle cells c. a primary oocyte
   b. the polar body d. a secondary oocyte

6. Semen contains secretions from the ______ .
   a. adrenal gland c. prostate gland
   b. pituitary gland d. corpus luteum

7. Male germ cells divide by meiosis inside the ______ .
   a. urethra c. prostate gland
   b. seminiferous tubules d. vasa deferentia

data analysis activities

Stuck Sperm Cells lining the epididymis secrete a glycoprotein (beta-defensin) that coats sperm and facilitates their passage through cervical mucus. There are two common alleles for human beta-defensin: a wild-type allele (*wt*) and an allele with a deletion (*del*). To find out if this genetic variation could affect fertility, Gary Cherr and his colleagues tested the ability of sperm from men with different genotypes to pass through a gel barrier. The barrier was designed to simulate cervical mucus. The researchers placed sperm on one side of the gel, then used a video camera to record how many passed into the gel over time. **FIGURE 41.16** illustrates their results.

1. After 4 minutes, how many sperm from men with a wild-type allele were in the camera's view on average?

2. How long did it take for the same number of sperm from men homozygous for the deletion to make it into the view of the camera?

3. Given these results, which genotype or genotypes would you expect to be associated with infertility?

4. The beta-defensin gene is on chromosome 20. What percentage of sperm from a heterozygous man would you expect to carry the *del* allele?

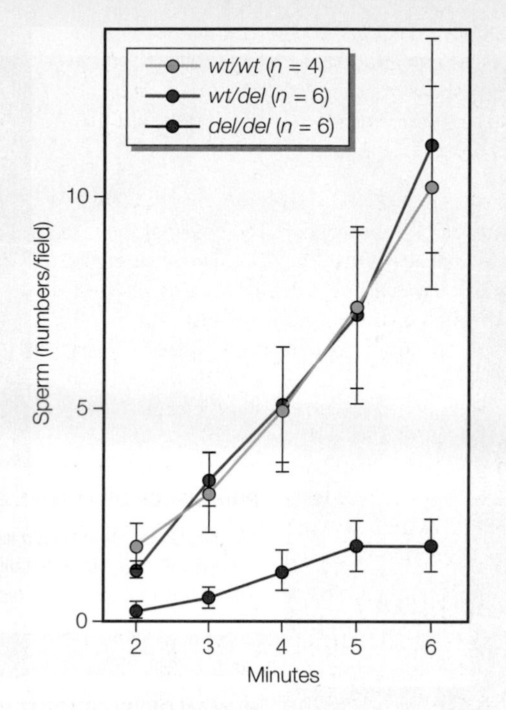

FIGURE 41.16 Effect of beta-defensin genotype on the ability of sperm to penetrate a gel that simulates cervical mucus. Sperm from 16 men was tested: 4 *wt/wt*, 6 *wt/del*, and 6 *del/del*.

Fresh sperm were placed on one side of the gel and a video camera was focused on a region 2.75 mm into the gel. The number of sperm in this region was recorded at one-minute intervals for six minutes. Sperm numbers shown are averages. Bars indicate range.

8. A male attains an erection when ________ .
   a. muscles running the length of the penis contract
   b. Leydig cells release a surge of testosterone
   c. the posterior pituitary releases oxytocin
   d. spongy tissue inside the penis fills with blood

9. Binding of a sperm to an oocyte plasma membrane causes the oocyte to ________ .
   a. enter an oviduct
   b. complete meiosis II
   c. secrete estrogen
   d. expel its nucleus

10. Birth control pills deliver synthetic ________ .
   a. estrogens and progesterone
   b. LH and FSH
   c. testosterone
   d. oxytocin and prostaglandins

11. Sperm in an epididymis pass next into the ________ .
   a. prostate gland
   b. urethra
   c. seminiferous tubules
   d. vas deferens

12. Match each disease with the type of organism that causes it. The choices can be used more than once.

___ chlamydial infection	a. bacteria
___ AIDS	b. protist
___ syphilis	c. virus
___ genital warts	
___ gonorrhea	
___ genital herpes	
___ trichomoniasis	

13. Match each structure with its description.

___ testis	a. conveys sperm out of body
___ epididymis	b. secretes semen components
___ labia majora	c. stores sperm
___ urethra	d. produces testosterone
___ vagina	e. produces estrogens and progesterone
___ ovary	f. usual site of fertilization
___ oviduct	g. lining of uterus
___ prostate gland	h. fat-padded skin folds
___ endometrium	i. birth canal

14. Match each hormone with its source.

___ FSH and LH	a. pituitary gland
___ GnRH	b. ovaries
___ estrogens	c. hypothalamus
___ testosterone	d. testes

Critical Thinking

1. Drugs that interfere with sympathetic nerve signals are often prescribed for men who have high blood pressure. How might such drugs impair sexual performance?

2. Fraternal twins are nonidentical siblings that form when two eggs mature, are released, and are fertilized at the same time. Explain why an unusually high level of FSH raises the likelihood of fraternal twins.

3. Some sperm mitochondria enter an egg during fertilization, but as sperm mature these mitochondria are tagged with a protein (ubiquitin) that marks them for destruction. What organelle do you think carries out this destruction process?

CREDIT: (16) © Cengage Learning, Based on Theodore L. Tollner et.al., A common Mutation in the Defensin *DEFB126* Causes Impaired Sperm Function and Subfertility, *Sci Transl Med* 20 July 2011: Vol.3, Issue 92, p92ra65 *Sci. Transl. Med.* DOI: 10.1126/scitranslmed.3002389.

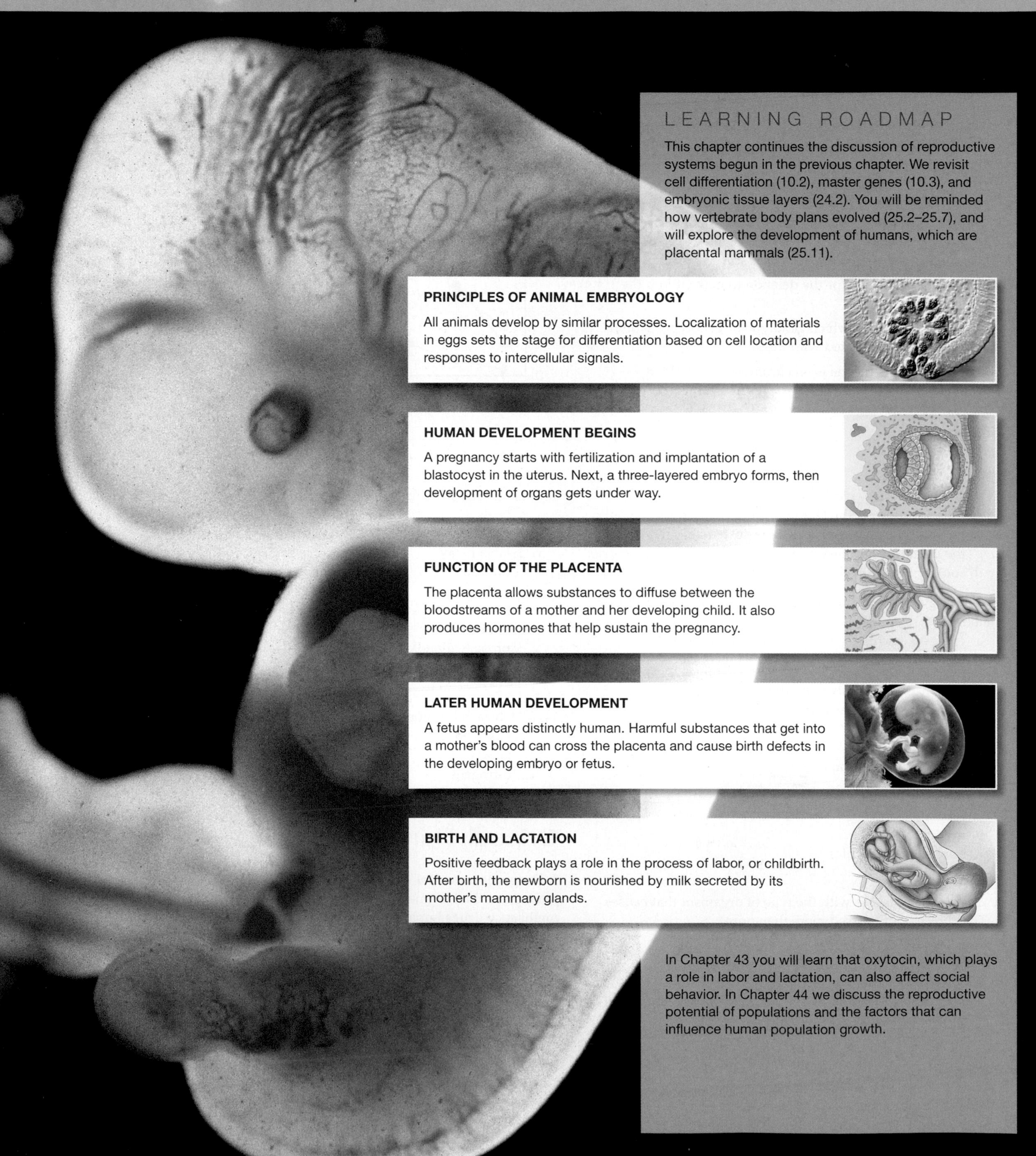

LEARNING ROADMAP

This chapter continues the discussion of reproductive systems begun in the previous chapter. We revisit cell differentiation (10.2), master genes (10.3), and embryonic tissue layers (24.2). You will be reminded how vertebrate body plans evolved (25.2–25.7), and will explore the development of humans, which are placental mammals (25.11).

PRINCIPLES OF ANIMAL EMBRYOLOGY

All animals develop by similar processes. Localization of materials in eggs sets the stage for differentiation based on cell location and responses to intercellular signals.

HUMAN DEVELOPMENT BEGINS

A pregnancy starts with fertilization and implantation of a blastocyst in the uterus. Next, a three-layered embryo forms, then development of organs gets under way.

FUNCTION OF THE PLACENTA

The placenta allows substances to diffuse between the bloodstreams of a mother and her developing child. It also produces hormones that help sustain the pregnancy.

LATER HUMAN DEVELOPMENT

A fetus appears distinctly human. Harmful substances that get into a mother's blood can cross the placenta and cause birth defects in the developing embryo or fetus.

BIRTH AND LACTATION

Positive feedback plays a role in the process of labor, or childbirth. After birth, the newborn is nourished by milk secreted by its mother's mammary glands.

In Chapter 43 you will learn that oxytocin, which plays a role in labor and lactation, can also affect social behavior. In Chapter 44 we discuss the reproductive potential of populations and the factors that can influence human population growth.

42.1 Prenatal Problems

Development of a human infant from a single-celled zygote is a complicated process, so a lot can go wrong along the way. When it does, a pregnancy can end in miscarriage or stillbirth, or a child can be born with a birth defect. With miscarriage, death occurs before 20 weeks. Miscarriages that occur in the first month or two of development often go undetected, so their frequency is not known. The risk of miscarriage for recognized pregnancies is 12 to 15 percent. Most of these occur early in pregnancy. Death of a fetus after 20 weeks is less common, and is called a stillbirth.

Disordered development that is not fatal can result in a birth defect, a structural malformation that is present at birth. According to the Centers for Disease Control, about 3 percent of children born in the United States have some sort of birth defect. A malformed heart, neural tube defects, and cleft lip or palate are common examples (**FIGURE 42.1**).

About half of miscarriages and between 5 and 10 percent of stillbirths result from chromosomal abnormalities (Section 14.6). These abnormalities, and thus the risk of an unsuccessful pregnancy, increase with both paternal and maternal age. Risk of birth defects also rises with parental age.

Other maternal characteristics also influence success of a pregnancy. Very thin women and women who are highly stressed are more likely to miscarry than those who are a bit heavier and happier. On the other hand, obesity is a risk factor for late stillbirth. Having previously given birth to more than five children also raises the risk that a pregnancy will not continue to term.

Some birth defects have an inherited basis, but others result from an environmental factor such as exposure to a teratogen. A **teratogen** is a toxin or infectious agent that interferes with development. Nicotine and alcohol are teratogens; prenatal (before birth) exposure to either raises the risk of cleft lip, as well as other defects. Exposure to the rubella virus during development can cause deafness and heart defects.

Dietary deficiencies cause birth defects too. If a pregnant woman's diet includes too little iodine, her newborn may be affected by cretinism, a disorder that affects brain function and motor skills (Section 34.7). A maternal deficiency in folate (folic acid) can impair neural tube formation. The neural tube develops into the brain and spinal cord.

Although scientists have identified many risk factors for birth defects, most such defects arise from unknown causes. The federally funded National Birth Defects Prevention Study (NBDPS) is designed to further illuminate causes of birth defects. Since 1997, participating researchers have been interviewing and collecting genetic material from women who gave birth to a child with a birth defect, as well as women who gave birth to normal children and can serve as controls. So far, over 35,000 women have participated and the study is ongoing. One recent finding: Work-related maternal pesticide exposure increases risk of a birth defect in which intestines protrude outside the body.

teratogen A toxin or infectious agent that interferes with development and thus raises the risk of birth defects.

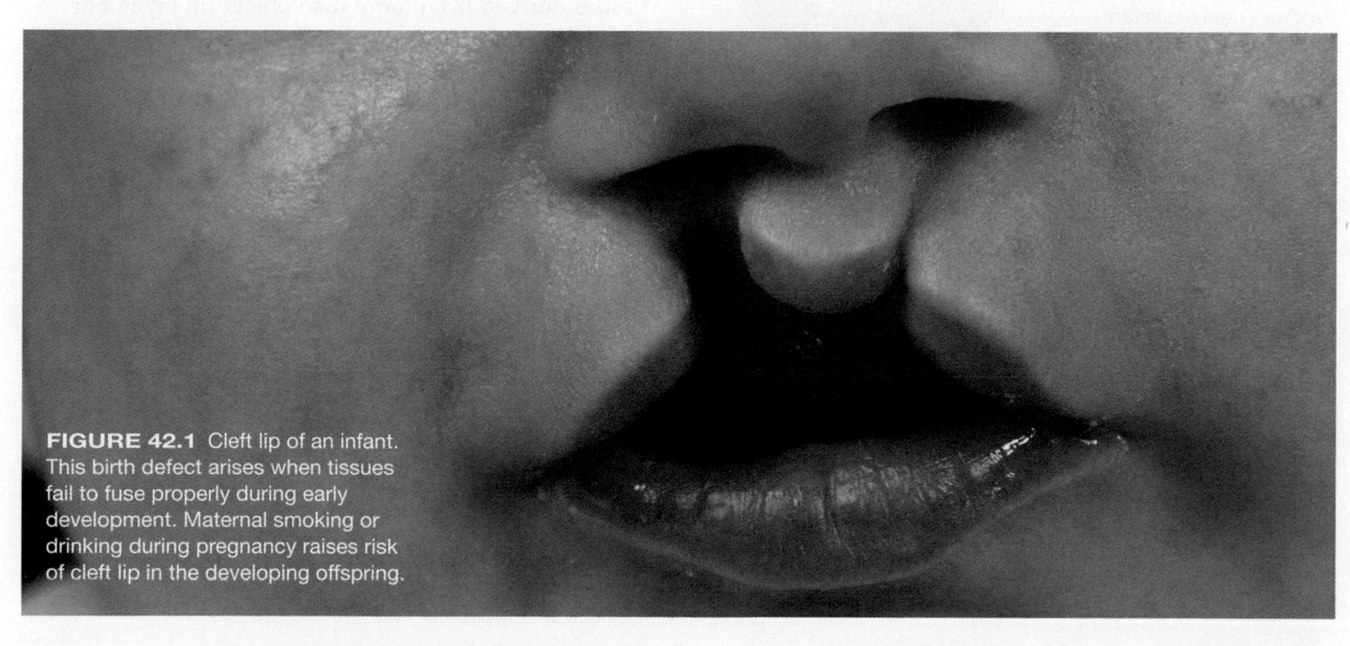

FIGURE 42.1 Cleft lip of an infant. This birth defect arises when tissues fail to fuse properly during early development. Maternal smoking or drinking during pregnancy raises risk of cleft lip in the developing offspring.

CREDITS: (opposite) © Lennart Nilsson/Bonnierforlagen AB; (l) M.A. Ansary/Science Source.

42.2 Stages of Development

✔ All animals pass through the same stages in their developmental journey from a fertilized egg to a multicelled adult.

In all sexually reproducing animals, a new individual begins life as a zygote, the diploid cell that forms at fertilization (Section 12.2). Development from a zygote to an adult typically proceeds through a series of stages (**FIGURE 42.2**). An example of these stages in one vertebrate, the leopard frog, appears in **FIGURE 42.3**. A female frog releases eggs into the water and a male releases sperm onto the eggs. External fertilization produces the zygote. New cells arise when **cleavage** carves up a zygote by repeated mitotic cell divisions ❶. During cleavage, the number of cells increases; however, the zygote's original volume remains unchanged. As a result, cells become more numerous but smaller. The cells that form during cleavage are called blastomeres. They typically become arranged as a **blastula**: a ball of cells that enclose a cavity (a blastocoel) filled with their secretions ❷. Tight junctions hold the cells of the blastula together.

During **gastrulation**, cells move and organize themselves as the layers of the **gastrula** ❸. In most animals and all vertebrates, a gastrula consists of three primary tissue layers, or **germ layers**. The three germ layers give rise to the same types of tissues and organs in all vertebrates (**TABLE 42.1**). The developmental similarity is evidence of a shared ancestry (Section 18.5).

Ectoderm, the outer germ layer, forms first. It gives rise to nervous tissue and to the outer layer of skin or other body covering. **Endoderm**, the inner layer, is the start of the respiratory tract and gut linings. A third layer called **mesoderm** forms between ectoderm and endoderm. This "middle" layer is the source of muscles, connective tissues, and the circulatory system.

Organ formation begins after gastrulation. The neural tube and notochord characteristic of all chordate embryos form early ❹. Many organs incorporate tissues derived from more than one germ layer. For example, the stomach's epithelial lining is derived from endoderm, and the smooth muscle that makes up the stomach wall develops from mesoderm.

In frogs, as in some other animals, a larva (in this case a tadpole) undergoes metamorphosis, a drastic remodeling of tissues into the adult form ❺.

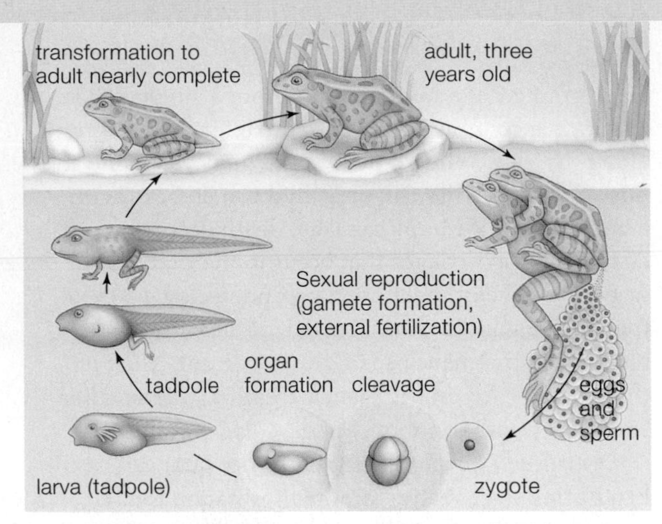

FIGURE 42.3 ▸Animated An example of stages in animal reproduction and development.

Above, overview of reproduction and development in the leopard frog. Opposite page, a closer look at each stage.

Sperm penetrates an egg, the egg and sperm nuclei fuse, and a zygote forms.

Mitotic cell divisions yield a ball of cells, a blastula. Each cell gets a different bit of the egg cytoplasm.

Cell rearrangements and migrations form a gastrula, an early embryo that has primary tissue layers.

Organs form as the result of tissue interactions that cause cells to move, change shape, and commit suicide.

Organs grow in size, take on mature form, and gradually assume specialized functions.

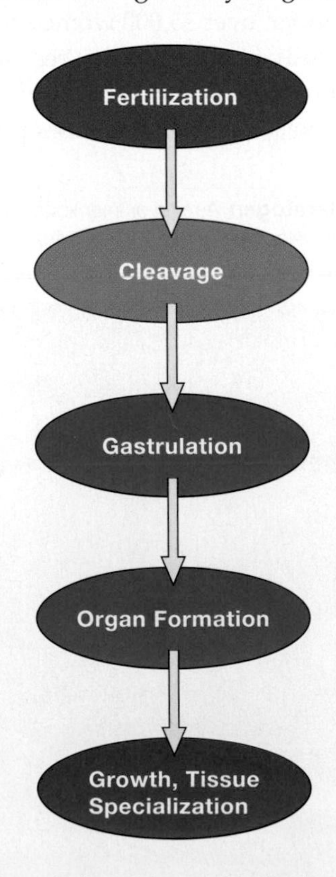

FIGURE 42.2 Stages of development in vertebrates. Fertilization was described in Section 41.7.

Table 42.1 Derivatives of Vertebrate Germ Layers

Germ layer	Derivatives
Ectoderm (outer layer)	Outer layer (epidermis) of skin; nervous tissue
Mesoderm (middle layer)	Connective tissue of skin; skeletal, cardiac, smooth muscle; bone; cartilage; blood vessels; urinary system; peritoneum (coelom lining); gut organs; reproductive tract
Endoderm (inner layer)	Lining of gut and respiratory tract, and organs derived from these linings

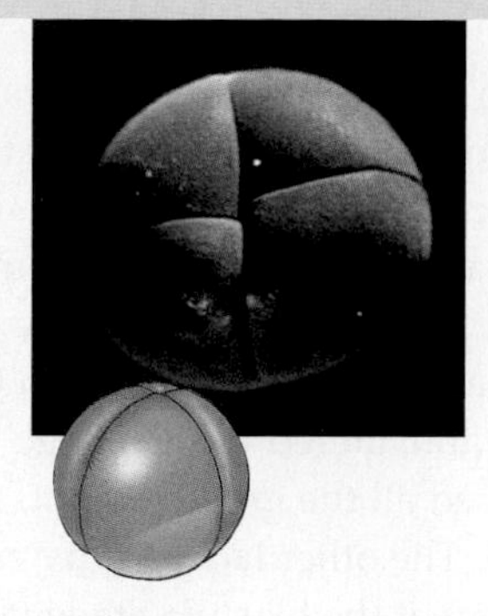

blastocoel

blastula

❶ Here we show the first three divisions of cleavage, a process that carves up a zygote's cytoplasm. In this species, cleavage results in a blastula, a ball of cells with a fluid-filled cavity.

❷ Cleavage is over when the blastula forms.

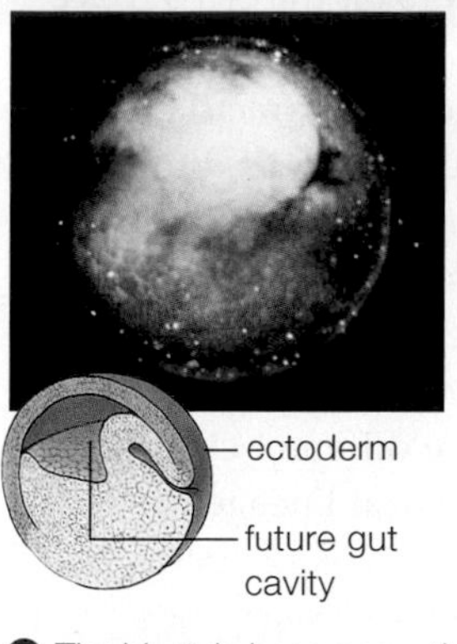

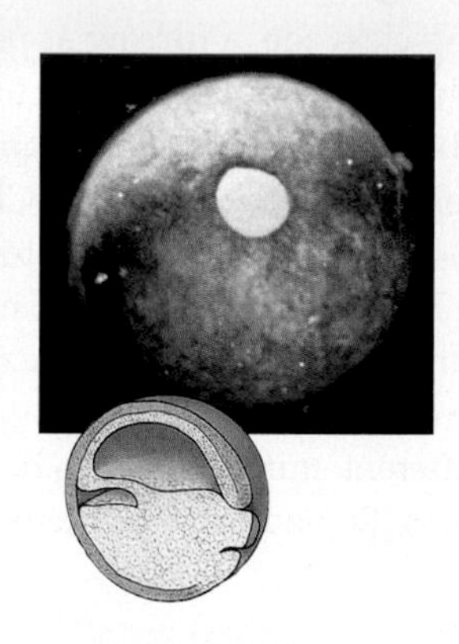

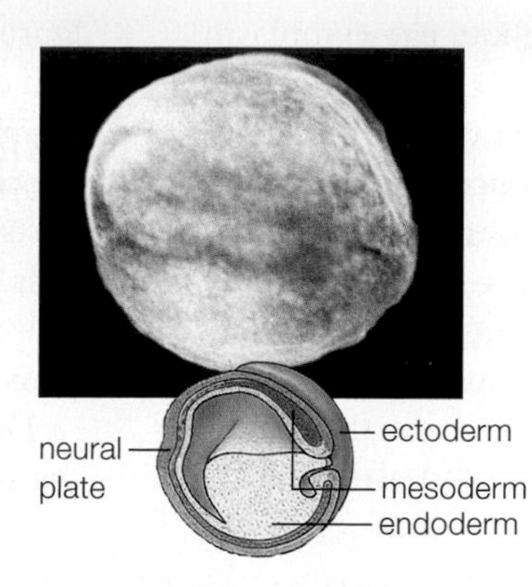

neural tube

notochord

gut cavity

❸ The blastula becomes a three-layered gastrula—a process called gastrulation. At the dorsal lip (a fold of ectoderm above the first opening that appears in the blastula), cells migrate inward and start rearranging themselves.

❹ Organs begin to form as a primitive gut cavity opens up. A neural tube, then a notochord and other organs, form from the primary tissue layers.

Tadpole, a swimming larva with segmented muscles and a notochord extending into a tail.

Limbs grow and the tail is absorbed during metamorphosis to the adult form.

Sexually mature, four-legged adult leopard frog.

❺ The frog's body form changes as it grows and its tissues specialize. The embryo becomes a tadpole, which metamorphoses into an adult.

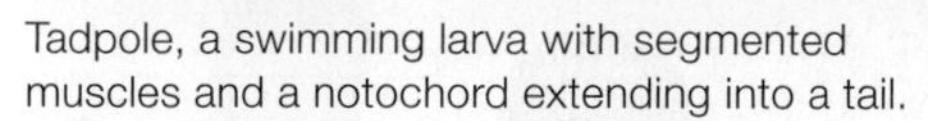

blastula Hollow ball of cells that forms as a result of cleavage.
cleavage In animal developmental, process by which multiple mitotic divisions produce a blastula from a zygote.
ectoderm Outermost primary tissue layer of an animal embryo.
endoderm Innermost primary tissue layer of an animal embryo.
gastrula Three-layered developmental stage formed by gastrulation in an animal.
gastrulation In animal development, process by which cell movements produce a three-layered gastrula.
germ layer One of three primary tissue layers in an early embryo.
mesoderm Middle tissue layer in an animal embryo having three primary tissue layers.

TAKE-HOME MESSAGE 42.2

How does an adult vertebrate develop from a single-celled zygote?

- ✔ A zygote undergoes cleavage, which increases the number of cells. Cleavage ends with formation of a blastula.
- ✔ Rearrangement of blastula cells forms a three-layered gastrula.
- ✔ After gastrulation, organs such as the nerve cord begin forming.
- ✔ Continued growth and tissue specialization produce the adult body.

42.3 From Zygote to Gastrula

✔ Localization of material in the egg cytoplasm affects early development, as do lineage-specific cleavage patterns.

Components of Eggs and Sperm

A sperm consists of paternal DNA and a bit of equipment that helps it swim to and penetrate an egg. The egg has far more cytoplasm. It also has yolk proteins that will nourish the embryo, maternal mRNAs that encode proteins essential to early development, tRNAs and ribosomes to translate the maternal mRNAs, and the proteins required to build mitotic spindles.

Cytoplasmic localization occurs in all oocytes: Instead of being evenly distributed throughout egg cytoplasm, many cytoplasmic components are localized in one region. In a yolk-rich egg, most yolk is at one pole (called the vegetal pole) and the opposite pole (the animal pole) has little. In some amphibian eggs, dark pigment molecules accumulate in the cell cortex, a cytoplasmic region just below the plasma membrane. Pigment is most concentrated close to the animal pole. After fertilization, the cortex rotates to reveal a gray crescent, a lightly pigmented region (**FIGURE 42.4A**).

Early in the 1900s, experiments by Hans Spemann showed that substances essential to development are localized in the gray crescent. In one experiment, he separated the first two blastomeres formed at cleavage. Each had half of the gray crescent and developed normally (**FIGURE 42.4B**). In the next experiment, Spemann altered the cleavage plane. One blastomere received all the gray crescent, and it developed normally. The other lacked gray crescent and halted development at the blastula stage (**FIGURE 42.4C**).

Cleavage—The Start of Multicellularity

During cleavage, a furrow appears on the cell surface and defines the plane of the cut. Beneath the plasma membrane, a ring of microfilaments contracts, dividing the cell in two (Section 11.4). Cleavage puts different parts of the egg cytoplasm into different descendant cells. The types and proportions of substances that cells inherit, as well as their size, depend on where the plane of division occurs.

Different animal lineages have different cleavage patterns. Remember that the bilateral lineage has two

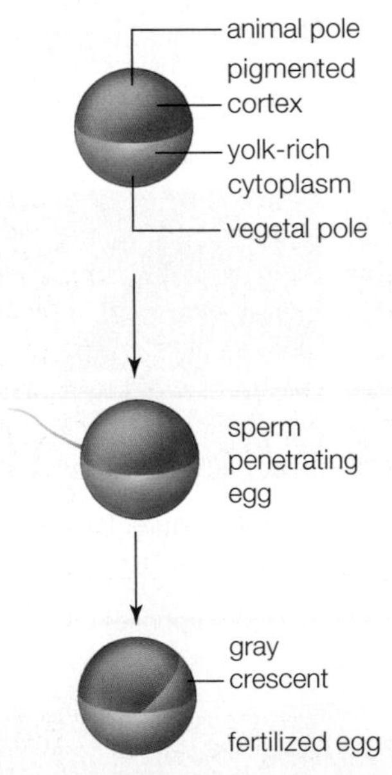

A Many amphibian eggs have a dark pigment concentrated in cytoplasm near the animal pole. At fertilization, the cytoplasm shifts and exposes a gray crescent-shaped region just opposite the sperm's entry point. The first cleavage normally distributes half of the gray crescent to each descendant cell.

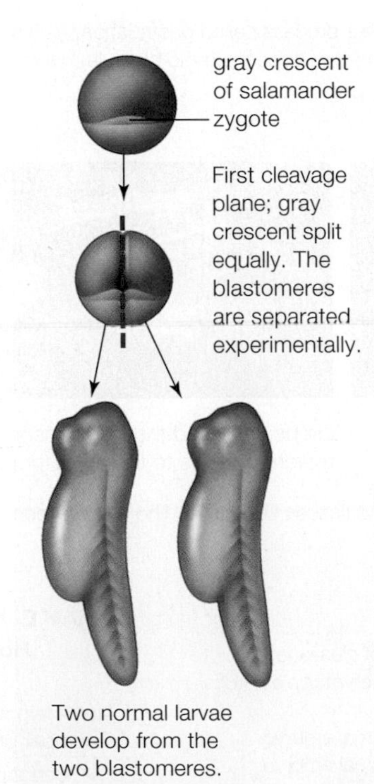

B In one experiment, the first two cells formed by normal cleavage were physically separated from each other. Each cell developed into a normal larva.

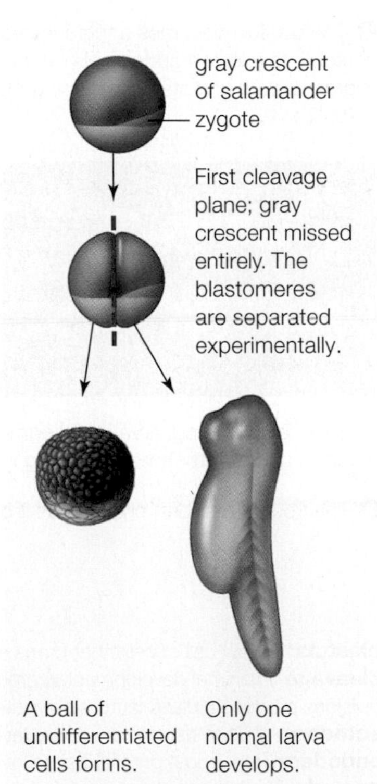

C In another experiment, a zygote was manipulated so one descendant cell received all the gray crescent. This cell developed normally. The other gave rise to an undifferentiated ball of cells.

FIGURE 42.4 ▸Animated Experimental evidence of cytoplasmic localization in an amphibian oocyte.

FIGURE IT OUT Is the gray crescent essential to amphibian development?

Answer: Yes

FIGURE 42.5 Gastrulation in a fruit fly (*Drosophila*). In fruit flies, cleavage is restricted to the outermost region of cytoplasm; the interior is filled with yolk. The series of photographs, all cross-sections, shows sixteen cells (stained gold) migrating inward. The opening through which cells move in will become the fly's mouth. Descendants of the stained cells will form mesoderm. Movements shown in these photos occur during a period of less than 20 minutes.

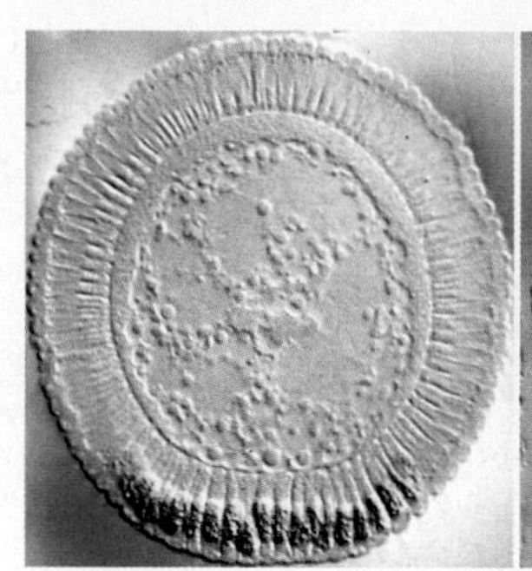

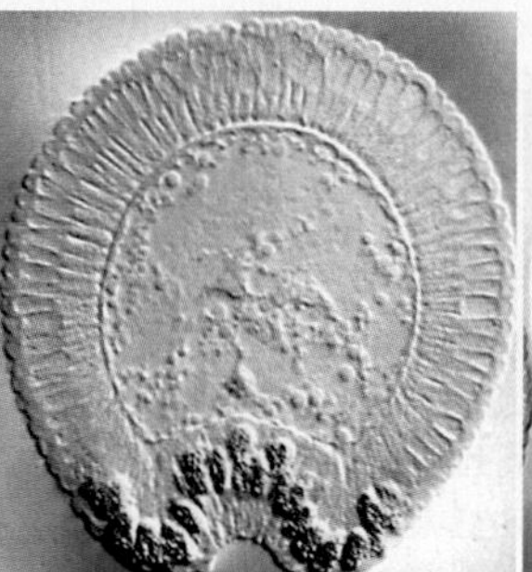

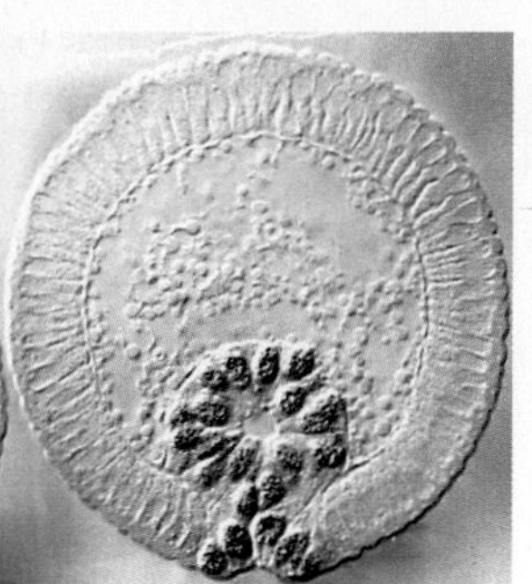

branches—the protostomes and the deuterostomes (Section 24.2). Deuterostomes undergo radial cleavage: the plane of each cell division either parallels the cell's polar axis or is perpendicular to it. In deuterostomes, cells undergo spiral cleavage, in which divisions occur at an oblique angle relative to the polar axis.

The amount of yolk also influences the pattern of division. Insects, frogs, fishes, and birds have yolk-rich eggs. In such eggs, a large volume of yolk slows or blocks some cuts. The result is fewer divisions of the yolky part of the egg than the less yolky part. By comparison, the cuts slice right through the nearly yolkless eggs of sea stars and mammals.

Gastrulation

A hundred to thousands of cells may form at cleavage, depending on the species. Beginning at gastrulation, they migrate about and rearrange themselves (**FIGURE 42.5**). Hilde Mangold, one of Spemann's students, discovered how gastrulation is regulated in vertebrates. Mangold knew that during gastrulation, cells of a salamander blastula move inward through an opening on its surface. Cells in the dorsal (upper) lip of the opening are descended from a zygote's gray crescent, and Mangold suspected that these cells made signals that caused gastrulation. To test her hypothesis, she transplanted dorsal lip material from one embryo to another (**FIGURE 42.6A**). As expected, cells migrated inward at the transplant site, as well as at the usual location (**FIGURE 42.6B**). A salamander larva with two joined sets of body parts developed (**FIGURE 42.6C**). Apparently, the transplanted cells signaled their new neighbors to develop in a novel way. This experiment also explains the results shown in **FIGURE 42.4C**. Without any gray crescent, an embryo does not have dorsal lip cells to signal the start of gastrulation, so development stops short.

cytoplasmic localization Accumulation of different materials in different regions of the egg cytoplasm.

TAKE-HOME MESSAGE 42.3

What are the effects of cytoplasmic localization and cleavage?

✔ Enzymes, mRNAs, yolk, and other materials are localized in specific parts of the cytoplasm of unfertilized eggs. This cytoplasmic localization influences early development.

✔ Cleavage divides a fertilized egg into a number of small cells but does not increase its original volume. The cells—blastomeres—inherit different parcels of cytoplasm that make them behave differently, starting at gastrulation.

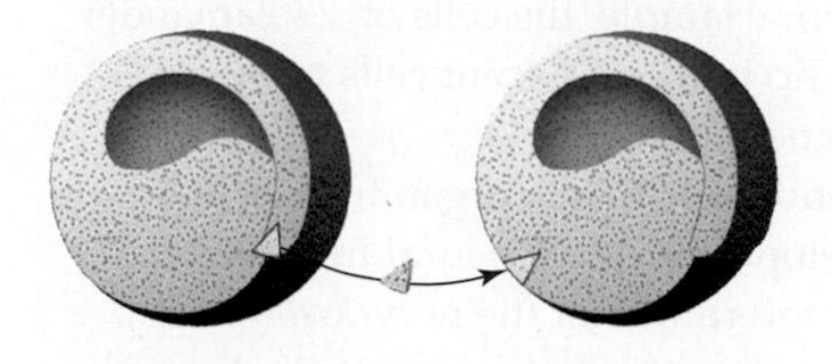

A Dorsal lip excised from donor embryo, grafted to novel site in another embryo.

B Graft induces a second site of inward migration.

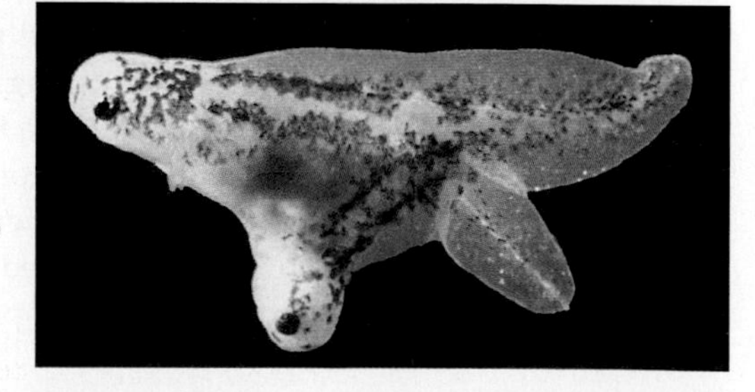

C The embryo develops into a "double" larva, with two heads, two tails, and two bodies joined at the belly.

FIGURE 42.6 ▸Animated Experimental evidence that signals from dorsal lip cells initiate amphibian gastrulation. A dorsal lip region of a salamander embryo was transplanted to a different site in another embryo. A second set of body parts started to form.

42.4 How Specialized Tissues and Organs Form

✔ After gastrulation, cells become specialized as their movement and interaction begin to shape tissues and organs.

Cell Differentiation

All cells in an embryo are descended from the same zygote, so all have the same genes. How then, do the specialized tissue and organs form? From gastrulation onward, different cell lineages express different subsets of genes. This selective gene expression is the key to **differentiation**, the process by which cell lineages become specialized in composition, structure, and function (Section 10.2).

An adult human body has about 200 differentiated cell types. As your eyes developed, cells of one lineage turned on genes for crystallin, a transparent protein that forms the lens in your eyes. No other cells in your body make crystallin.

Keep in mind that a differentiated cell still retains the entire genome. That is why it is possible to clone an adult animal—to create a genetic copy—from one of its differentiated cells (Section 8.7).

Responses to Morphogens

Cell-to-cell communication influences the timing and location of gene expression. Consider **morphogens**, which are signaling molecules that can act over a long distance and affect cells in a concentration-dependent manner. A morphogen is produced in one area of an embryo and diffuses outward, forming a concentration gradient. Cells nearest the source of the morphogen are exposed to a high morphogen concentration and express one set of genes. Cells more distant from the morphogen source are exposed to a lower concentration and turn on different genes.

The bicoid protein of fruit flies is an example of a morphogen. *Bicoid* is a **maternal effect gene**, meaning it is expressed in an immature egg and its product affects development. As an egg matures, bicoid mRNA accumulates at one end—an example of cytoplasmic localization. This mRNA is translated only after fertilization. Then, diffusion of its protein product creates a gradient of this protein along the length of the zygote. This gradient establishes the future fly's front-to-back axis. Where bicoid protein concentration is highest, genes that trigger development of anterior (front end) body structures are expressed. Where it is lowest, genes that trigger development of posterior (back end) body parts are expressed.

The bicoid protein is a transcription factor (Section 10.2) that stimulates the expression of other master genes, which activate other master genes, and so on.

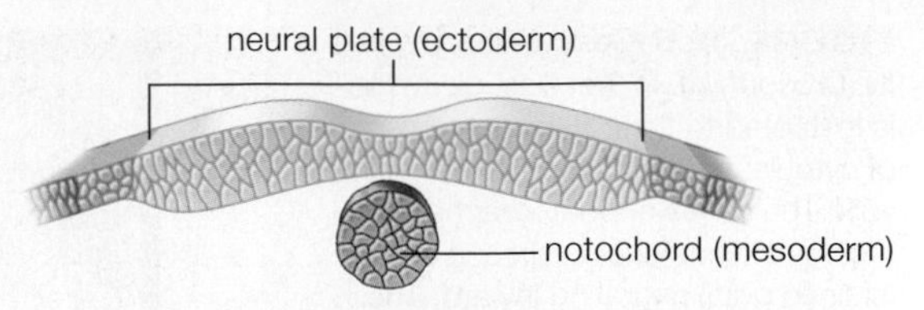

❶ Chemical signals produced by notochord mesoderm induce the ectoderm above it to thicken and form a neural plate.

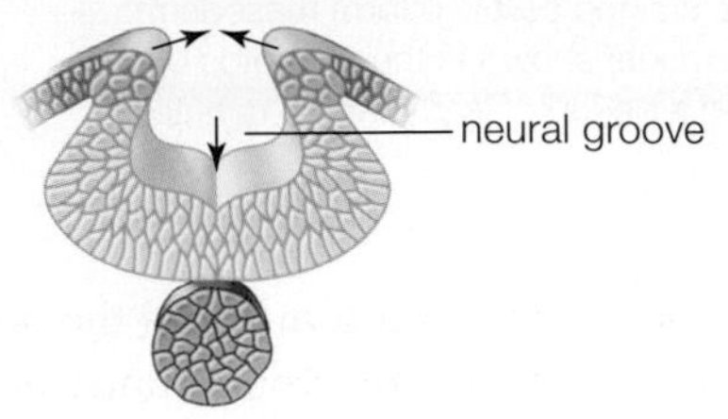

❷ Changes in cell shape cause edges of the neural plate to fold in toward the plate center, forming a neural groove.

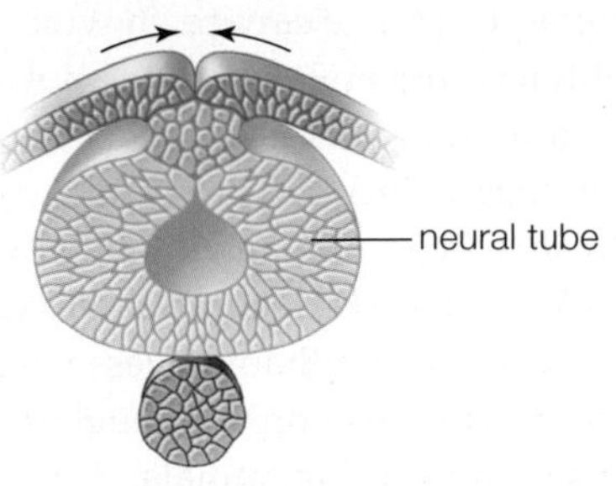

❸ Further folding causes the edges of the neural plate to meet, forming the neural tube.

FIGURE 42.7 ▶**Animated** Neural tube formation in a vertebrate embryo.

This cascade of master gene expression ultimately results in the formation of specific body parts in local regions of the developing embryo (Section 10.3).

Embryonic Induction

Intercellular communication also operates at close range. By the process of **embryonic induction**, cells of one embryonic tissue alter the behavior of cells in an adjacent tissue. For example, the cells of a salamander gastrula's dorsal lip induce adjacent cells to migrate inward and become mesoderm.

After gastrulation, vertebrate organ formation begins with development of the neural tube, which is the embryonic forerunner of the nervous system. Neural tube development is induced by signals from the notochord, which formed earlier from mesoderm (**FIGURE 42.7**). The process begins when ectodermal cells overlying the notochord elongate, forming a thick neural plate ❶. The cell elongation results from the assembly of microtubules inside them. Next, actin fila-

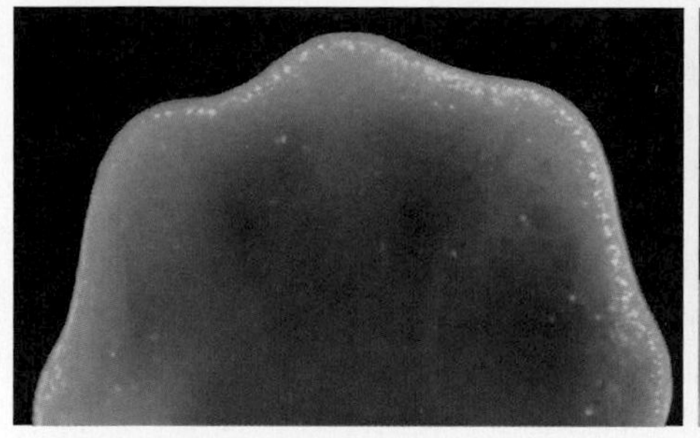
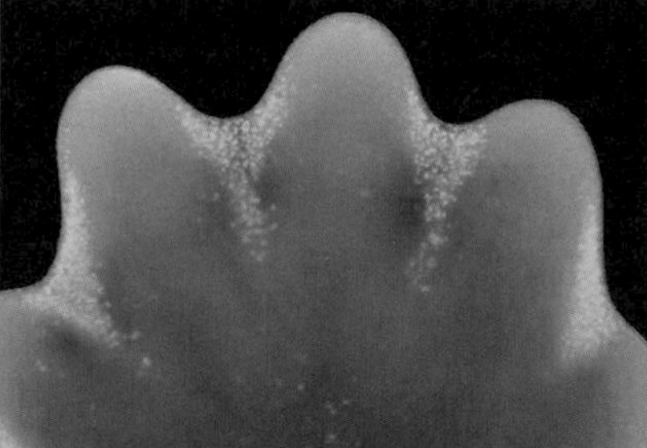
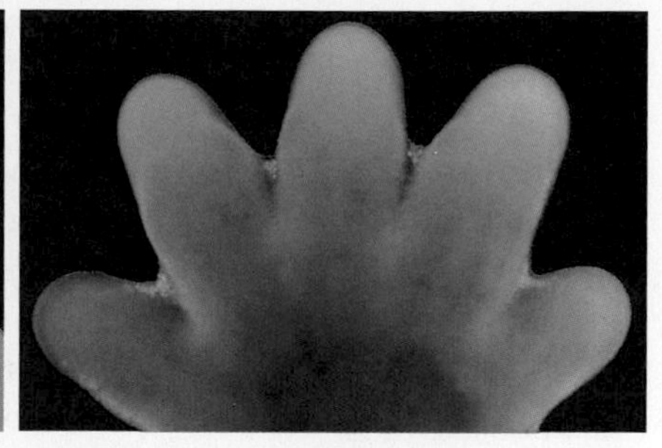

FIGURE 42.8 Apoptosis in the formation of a mouse paw. The yellow stain indicates regions where cells are undergoing apoptosis.

ments in cells at the edges of the neural plate contract, so these cells become wedge-shaped. This change in cell shape causes the edges of the plate to fold inward, forming a U-shaped structure called the neural groove ❷. Eventually the outer edges of the former neural plate meet at the midline, forming the neural tube ❸.

Apoptosis

Programmed cell death, or **apoptosis**, helps sculpt body parts. During apoptosis, cells that are no longer needed self-destruct in response to chemical signals. Depending on the context, an internal or external signal sets in motion a chain of reactions that result in the activation of self-destructive enzymes. Some of these enzymes chop up structural proteins such as cytoskeleton proteins and the histones that organize DNA. Others snip apart nucleic acids. As a result of these activities, the cell dies.

Apoptosis eliminates the webbing between digits of a developing mouse paw (**FIGURE 42.8**) or human hand, and it makes the tail of a tadpole disappear. It also eliminates the tail that develops in the early human embryo.

Cell Migrations

Cell migrations are an essential part of development, For example, ectodermal cells that form at the top of the neural tube (a region called the neural crest) migrate outward to positions throughout the body. Descendants of neural crest cells include the neurons and glial cells of the peripheral nervous system, and the melanocytes in skin.

Cells travel by inching along in an amoeba-like fashion (**FIGURE 42.9**). Assembly of actin microfilaments at one edge of a cell causes that portion of the cell to protrude. Adhesion proteins in the plasma membrane then anchor the advancing portion of the cell to a protein in the membrane of another cell or in the extracellular matrix. Once the front of the cell is anchored, adhesion proteins in the trailing part of the cell release their grip, and the trailing region is drawn forward. The cell may move in response to a concentration gradient of some chemical signal or it may follow a "trail" of molecules that its adhesion proteins bind to. It stops migrating when it reaches a region where its adhesion proteins hold it tightly in place.

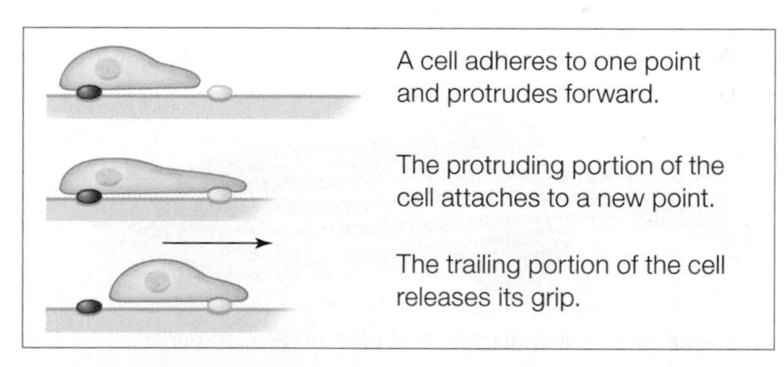

FIGURE 42.9 How cells migrate.

TAKE-HOME MESSAGE 42.4

What developmental processes produce specialized cells, tissues, and organs?

✔ All cells in an embryo have the same genome, but they express different subsets of genes. This selective gene expression is the basis of cell differentiation. It results in cell lineages with characteristic structures and functions.

✔ Cytoplasmic localization results in concentration gradients of signaling proteins called morphogens. Morphogens activate sets of master genes, whose products cause embryonic cells to form tissues and organs in specific places.

✔ Cell migrations, shape changes, and apoptosis help sculpt the form of the body and its parts.

apoptosis Mechanism of programmed cell death.
differentiation Process by which cells become specialized.
embryonic induction Embryonic cells produce signals that alter the behavior of neighboring cells.
maternal effect gene Gene expressed in an egg as it matures; its product influences animal development.
morphogen Chemical encoded by a master gene; diffuses out from its source and affects development.

42.5 An Evolutionary View of Development

✔ Similarities in developmental pathways among animals are evidence of common ancestry.

A General Model for Animal Development

Through studies of animals such as roundworms, fruit flies, fishes, and mice, researchers have come up with a general model for development. The key point of the model is this: Where and when particular genes are expressed determines how an animal body develops.

First, molecules confined to different areas of an unfertilized egg induce localized expression of master genes in the zygote. Products of these master genes diffuse outward, so concentration gradients for these products form along the head-to-tail and front-to-back axes of the developing embryo.

Second, depending on where they fall within these concentration gradients, cells in the embryo activate or suppress other master genes. The products of these genes become distributed in gradients, which affect other genes, and so on.

Third, this positional information affects expression of homeotic genes. As Sections 10.3 and 18.5 explained, all animals have similar homeotic genes. For example, a mouse's *eyeless* gene initiates development of its eyes. Introduce the mouse version of this gene into a fruit fly, and eyes will form in tissues where the introduced gene is expressed.

A Adult zebrafish, an animal used in studies of development.

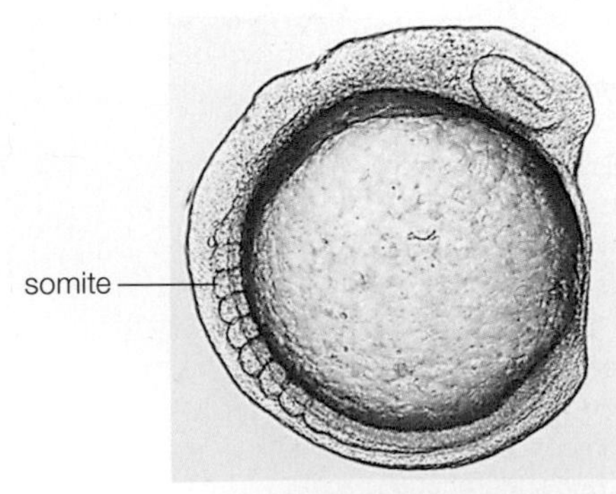

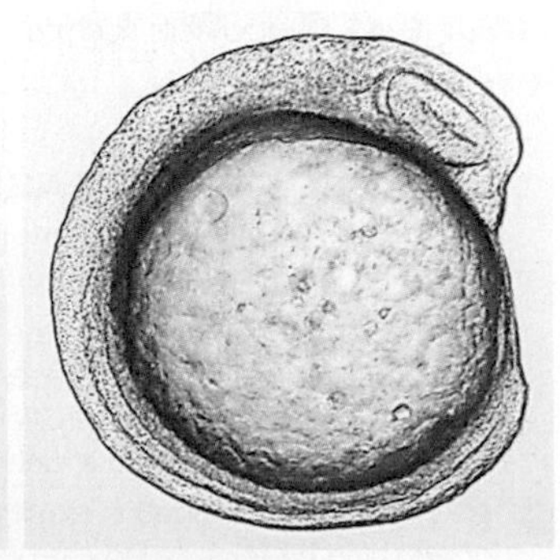

B Normal zebrafish embryo with somites—bumps of mesoderm that give rise to bone and muscle.

C Embryo with a mutation that prevents the formation of somites. It will die early in development.

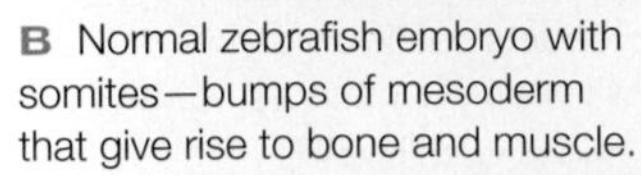

FIGURE 42.10 Lethal effect of a mutation in a zebrafish gene (*fused somites*) that functions in early development.

Developmental Constraints and Modifications

The model just described helps explain why we only see certain types of animal body plans, but it is not the whole story. Body plans are influenced by physical constraints. For example, large body size cannot evolve in an animal without circulatory and respiratory mechanisms that service body cells far from the body surface.

An existing body framework imposes architectural constraints. For example, the ancestors of all modern land vertebrates had a body plan with four limbs. The evolution of wings in birds and bats occurred through modification of existing forelimbs, not by sprouting new limbs. Although it might be advantageous to have both wings and arms, no living or fossil vertebrate with both has been discovered.

Finally, there are phylogenetic constraints on body plans. These constraints are imposed by interactions among genes that regulate development in a lineage. Once master genes evolved, their interactions determined the basic body form. Mutations that dramatically alter these interactions are usually lethal.

Consider the paired bones and skeletal muscles arrayed along a vertebrate's head-to-tail axis. This pattern arises early in development, when the mesoderm on either side of the embryo's neural tube becomes divided into blocks of cells called **somites** (FIGURE 42.10). The somites will later develop into bones and skeletal muscles. A complex pathway involving many genes governs somite formation. Any mutation that disrupts this pathway so that somites do not form is lethal during development. Thus, we do not find vertebrates with an unsegmented body plan, although the number of somites does vary among species.

In short, mutations that affect development led to a variety of forms among animal lineages. These mutations brought about morphological changes through the modification of existing developmental pathways, rather than by blazing entirely new genetic trails.

somites Paired blocks of embryonic mesoderm that give rise to a vertebrate's muscle and bone.

TAKE-HOME MESSAGE 42.5

Why are developmental processes and body plans similar among animal groups?

✔ In all animals, cytoplasmic localization affects expression of sets of master genes shared by most animal groups. The products of these genes cause embryonic cells to form tissues and organs at certain locations.

✔ Once a developmental pathway evolves, drastic changes to genes that govern this pathway are generally lethal.

42.6 Overview of Human Development

✔ Like all animals, humans begin life as a single cell and go through a series of developmental stages.

Chapter 41 introduced the structure and function of human reproductive organs, and explained how an egg and sperm meet at fertilization to form a zygote. The remaining sections of this chapter continue this story with an in-depth look at human development. In this section, we provide an overview of the process and define the stages that we will discuss. Prenatal (before birth) and postnatal (after birth) stages are listed in TABLE 42.2.

It takes about 5 trillion mitotic divisions to go from the single cell of a zygote to the 10 trillion or so cells of an adult human. The process gets under way during a pregnancy that lasts an average of 38 weeks from the time of fertilization.

The first cleavage occurs about 12 to 24 hours after fertilization. It takes about one week for a blastula (called a blastocyst in mammals) to form.

In discussing animals, the term **embryo** generally refers to an individual from the time of first cleavage to the time when it hatches or is born. With regard to humans, the term embryo is usually reserved for the period from 2 weeks to 8 weeks after fertilization. From week nine until birth, the individual is referred to as a **fetus**. During embryonic development, the human body plan is set in place and organs form. During fetal development, the organs mature and begin to function.

We divide the prenatal period into three trimesters. The first trimester includes months one through three; the second trimester, months four through six; the third trimester, months seven through nine.

Births before 36 weeks after fertilization are considered premature. A fetus born earlier than 22 weeks rarely survives because its lungs are not yet fully mature. About half of births that occur before 26 weeks result in some sort of long-term disability.

After birth, the human body continues to grow and its body parts continue to change in proportion. FIGURE 42.11 shows body proportion changes during development. Postnatal growth is most rapid between 13 and 19 years. Sexual maturation occurs at puberty, and bones stop growing shortly thereafter. The brain is the last organ to become fully mature: Portions of it continue to develop until the individual is about 19 to 22 years old.

embryo In animals, a developing individual from first cleavage until hatching or birth; in humans, usually refers to an individual during weeks 2 to 8 of development.

fetus Developing human from about 9 weeks until birth.

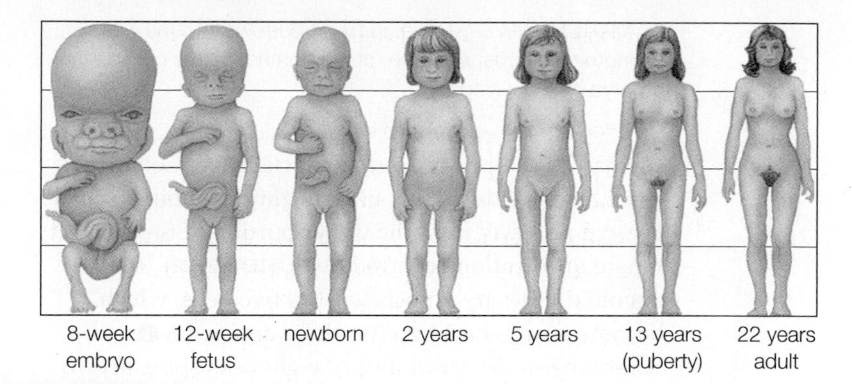

FIGURE 42.11 Observable, proportional changes in prenatal and postnatal periods of human development. Changes in overall physical appearance are slow but noticeable until the teens.

Table 42.2 Stages of Human Development

Prenatal period	
Zygote	Single cell resulting from fusion of sperm nucleus and egg nucleus at fertilization.
Blastocyst (blastula)	Ball of cells with surface layer, fluid-filled cavity, and inner cell mass.
Embryo	All developmental stages from 2 weeks after fertilization until end of eighth week.
Fetus	All developmental stages from ninth week to birth (about 38 weeks after fertilization).
Postnatal period	
Newborn	Individual during the first 2 weeks after birth.
Infant	Individual from 2 weeks to 15 months.
Child	Individual from infancy to about 10 or 12 years.
Pubescent	Individual at puberty, when secondary sexual traits develop. For girls, occurs between 10 and 15 years; for boys, between 11 and 16 years.
Adolescent	Individual from puberty until about 3 or 4 years later; physical, mental, emotional maturation.
Adult	Early adulthood (between 18 and 25 years); bone formation and growth finished. Changes proceed slowly after this.
Old age	Aging processes result in expected tissue deterioration.

TAKE-HOME MESSAGE 42.6

What are the stages of human development?

✔ Like other animals, humans undergo cleavage to form a blastula, which in mammals is called a blastocyst.

✔ Organs form during embryonic development. The fetal period, during which organs enlarge and begin to function, is the longest stage of prenatal development.

✔ Growth and development continues after an individual is born, ending shortly after puberty.

42.7 Early Human Development

✔ After a human blastocyst forms, it burrows into the wall of its mother's uterus. A system of membranes forms outside the embryo as it develops.

Human development starts with cleavage. Cell divisions begin within a day of fertilization, as cilia propel the zygote away from the upper portion of the oviduct where fertilization occurred (**FIGURE 42.12**). The zygote divides by mitosis to form two cells, which become four, which become eight, and so on ❶. Sometimes a cluster of four or eight cells splits in two and each cluster develops independently, resulting in identical twins. More typically, all cells adhere tightly to one another as divisions continue.

By about 5 days after fertilization, the dividing cluster of cells has reached the uterus and formed a **blastocyst**, the mammalian blastula ❷. The blastocyst consists of an outer layer of cells, a cavity (the blastocoel) filled with the cells' fluid secretions, and an inner cell mass. Only the cells of the inner cell mass will give rise to the developing individual. Other cells of the blastocyst give rise to extraembryonic membranes that enclose, protect, and support the embryo.

Up to this point, the developing individual has been nourished by nutrients it absorbed from maternal secretions in the oviduct and uterus. For development to continue, the blastocyst must implant itself in the wall of the uterus, where it can obtain nutrients from the maternal bloodstream. Implantation requires direct contact between the surface of the blastocyst and cells in the lining of the uterus. However, the blastocyst is encased by remnants of the zona pellucida; this protein coat must be shed before implantation can occur.

Once the blastocyst has escaped from its protein coat, it attaches to the lining of the uterus and begins burrowing into it ❸. During implantation, the inner cell mass develops into two flattened layers of cells that are collectively called an embryonic disk ❹. At the same time, extraembryonic membranes typical of amniotes begin to form. One of these membranes, called the **amnion**, will enclose a fluid-filled amniotic cavity between the embryonic disk and the blastocyst surface. Fluid in this cavity (amniotic fluid) acts as a buoyant cradle in which an embryo grows and moves freely, protected from temperature changes and mechanical impacts. As the amnion forms, other cells move around the inner wall of the blastocyst, forming the lining of a yolk sac. In reptiles and birds, this sac holds yolk that nourishes the developing embryo. In humans, it has no

FIGURE 42.12 ▸Animated Early embryonic development.

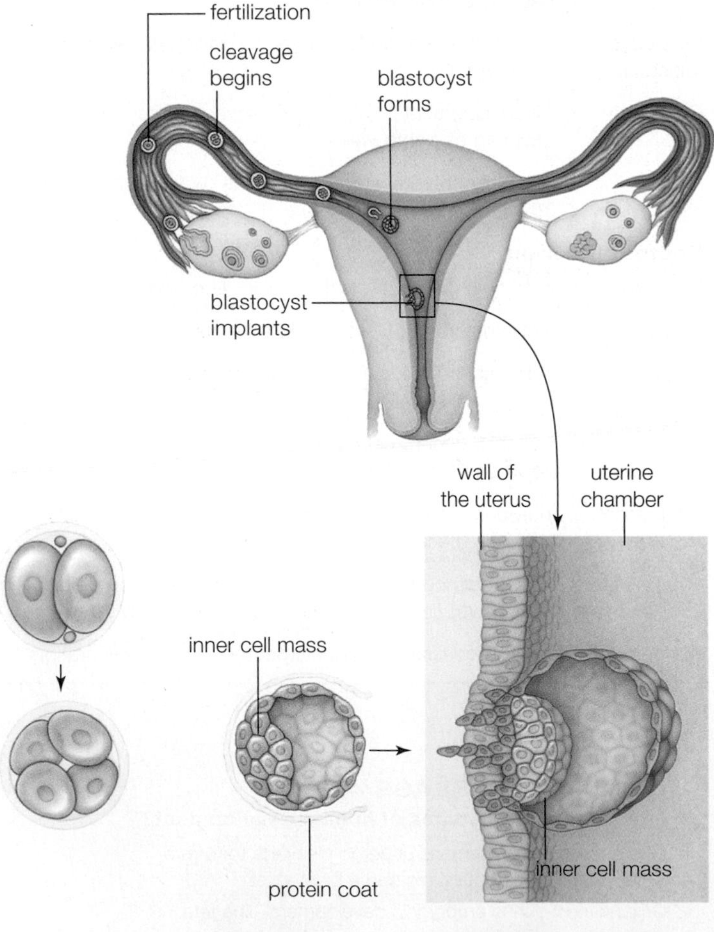

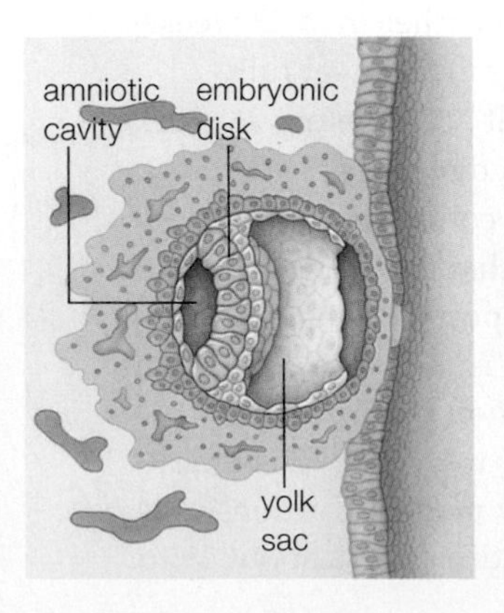

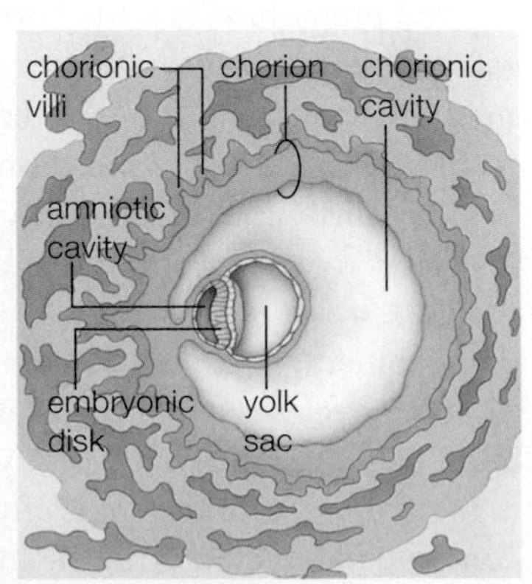

❶ **Days 1–4** Cleavage divides the zygote into an increasing number of smaller cells.

❷ **Days 5–7** The blastocyst forms, expands, then sheds its protein coat.

❸ **Days 8–9** Implantation begins. The blastocyst attaches to the wall of the uterus and begins to burrow into it.

❹ **Days 10–11** The inner cell mass develops into a two-layered embryonic disk. Extraembryonic membranes form.

❺ **Day 14** Chorionic villi (fingerlike extensions of the chorion) begin to grow into blood-filled spaces in the lining of the uterus.

nutritional role. Some cells that form in a human yolk sac give rise to the embryo's first blood cells.

As implantation continues, spaces in the uterine tissue around the blastocyst become filled with blood that seeps from ruptured maternal capillaries. The outermost extraembryonic membrane—the **chorion**—develops many tiny fingerlike projections (chorionic villi) that extend into blood-filled maternal tissues ❺. Later, the chorion will become a major part of the placenta.

The chorion secretes **human chorionic gonadotropin (HCG)**, a hormone that prevents degeneration of the corpus luteum and so maintains the uterine lining and prevents menstruation. By the beginning of the third week, HCG can be detected in a mother's blood or urine. At-home pregnancy tests have a "dipstick" with a region that changes color when exposed to urine that contains HCG. Later in a pregnancy, when the placenta has formed it will also produce HCG.

An outpouching of the yolk sac will become the fourth extraembryonic membrane, the **allantois**. In humans, the allantois forms blood vessels of the umbilical cord, which is the structure that connects the fetus to the placenta.

Gastrulation occurs on about day 15. Cells migrate inward along a depression, called the primitive streak, that forms on the disk's surface ❻. Shortly thereafter, the embryonic disk develops two folds that will merge to form a neural tube, the precursor of the spinal cord and brain ❼. Beneath the neural tube lies the notochord. As development continues, this flexible supporting rod will be replaced by the backbone.

Somites form by the end of the third week ❽. These paired segments of mesoderm will develop into the bones, skeletal muscles of the head and trunk, and overlying dermis of the skin.

Bands of tissue called pharyngeal arches form at the onset of the fourth week ❾. This tissue will later contribute to the pharynx, larynx, and the face, neck, mouth, and nose. In fishes, pharyngeal arches develop into gills, but a human embryo never has gills; it receives oxygen by way of the placenta.

TAKE-HOME MESSAGE 42.7

What occurs during the first two weeks of human development?

✔ Cleavage produces a blastocyst that slips out of its protein coat and implants itself in the mother's uterus.

✔ During implantation, projections from the blastocyst extend into maternal tissues. These and other connections will support the developing embryo.

✔ The blastocyt's inner cell mass will become the embryo. Other blastocyst cells give rise to external membranes.

✔ After the embryonic disk undergoes gastrulation, the neural tube and notochord form. Then, somites and pharyngeal arches develop.

allantois Extraembryonic membrane that, in mammals, becomes part of the umbilical cord.
amnion Extraembryonic membrane that encloses an amniote embryo and amniotic fluid.
blastocyst Mammalian blastula.
chorion Outermost extraembryonic membrane of amniotes; major component of the placenta in placental mammals.
human chorionic gonadotropin (HCG) Hormone first secreted by the blastocyst, and later by the placenta; helps maintain the uterine lining during pregnancy.

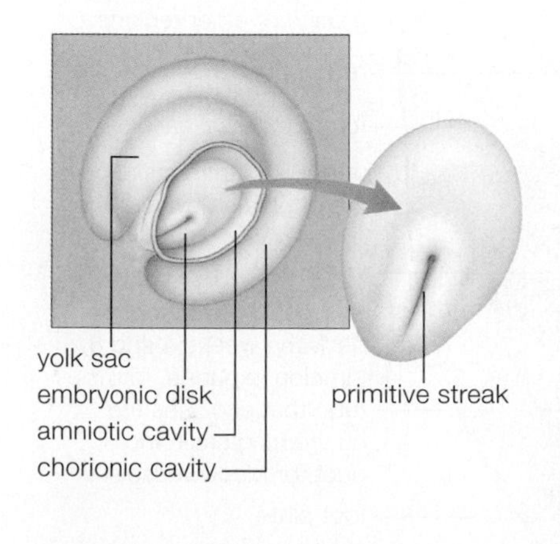

❻ **Day 15** Gastrulation begins at depression called the primitive streak.

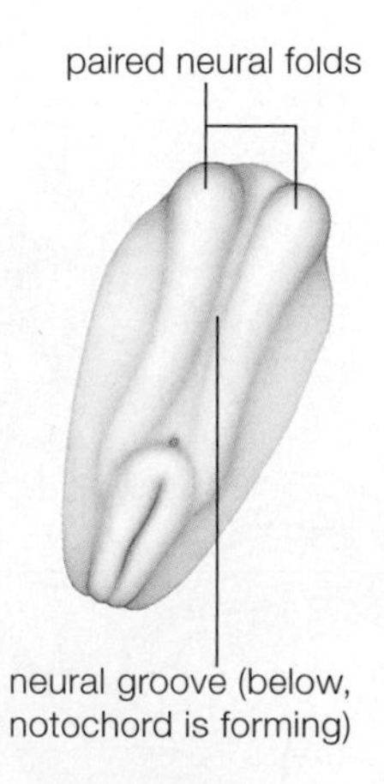

❼ **Day 16** Neural tube begins to form.

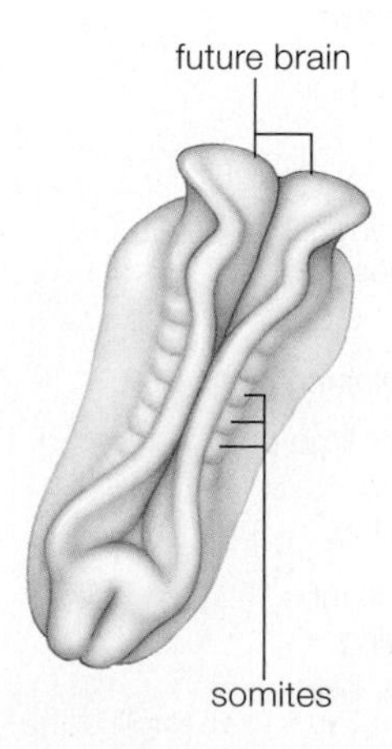

❽ **Days 18–23** Somites develop.

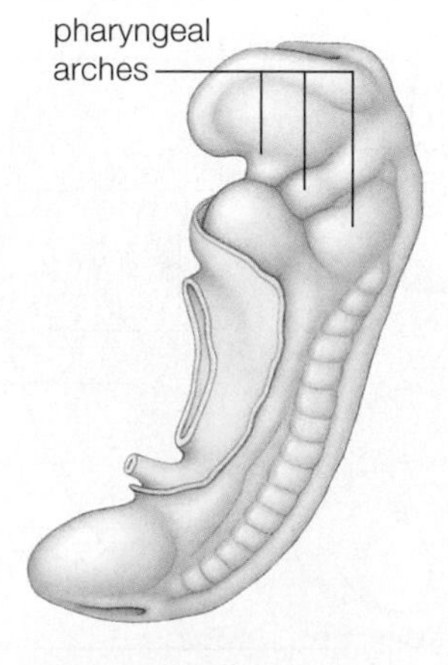

❾ **Days 24–25** Pharyngeal arches appear.

42.8 Emergence of Distinctly Human Features

✔ A human embryo's tail and pharyngeal arches label it as a chordate. The features disappear during fetal development.

When the fourth week ends, the embryo is 500 times the size of a zygote, but still less than 1 centimeter long. Like other chordate embryos, it has a short tail that extends beyond its anus. Growth slows as details of organs begin to fill in. Limbs form and paddles at their ends are sculpted into digits. The umbilical cord and the circulatory system develop. Growth of the head now surpasses that of all other regions (**FIGURE 42.13**). Reproductive organs begin forming.

By the end of the eighth week, all of the organ systems have formed, apoptosis has eliminated the tail, and we define the individual as a human fetus.

In the second trimester, reflexive movements begin as developing nerves and muscles connect. Legs kick, arms wave about, and fingers grasp. The fetus

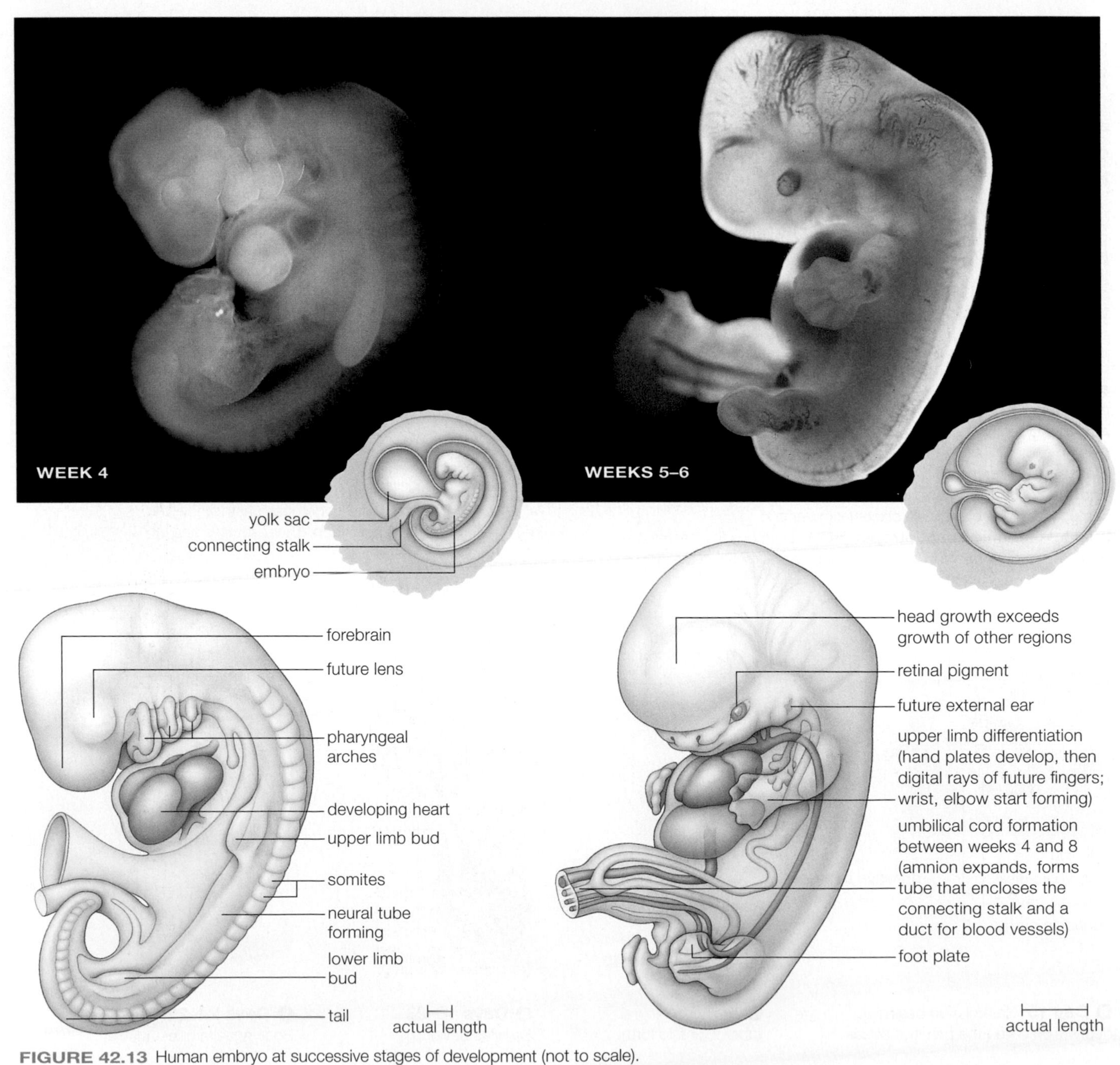

FIGURE 42.13 Human embryo at successive stages of development (not to scale).

frowns, squints, puckers its lips, sucks, and hiccups. When a fetus is five months old, its heartbeat can be heard clearly through a stethoscope positioned on the mother's abdomen. The mother can sense movements of fetal arms and legs. By now, soft fetal hair (lanugo) covers the skin; most will be shed before birth. A thick, cheesy coating (vernix) protects the skin from abrasion.

In the sixth month, eyelids and eyelashes form, and the fetus begins to hear. Eyes open during the seventh month, the start of the final trimester. By this time, all portions of the brain have formed and have begun to function.

TAKE-HOME MESSAGE 42.8

What occurs during the late embryonic and the fetal periods?

✔ The embryo takes on its human appearance by week eight but remains tiny. In the fetal period, organs begin functioning and size increases dramatically.

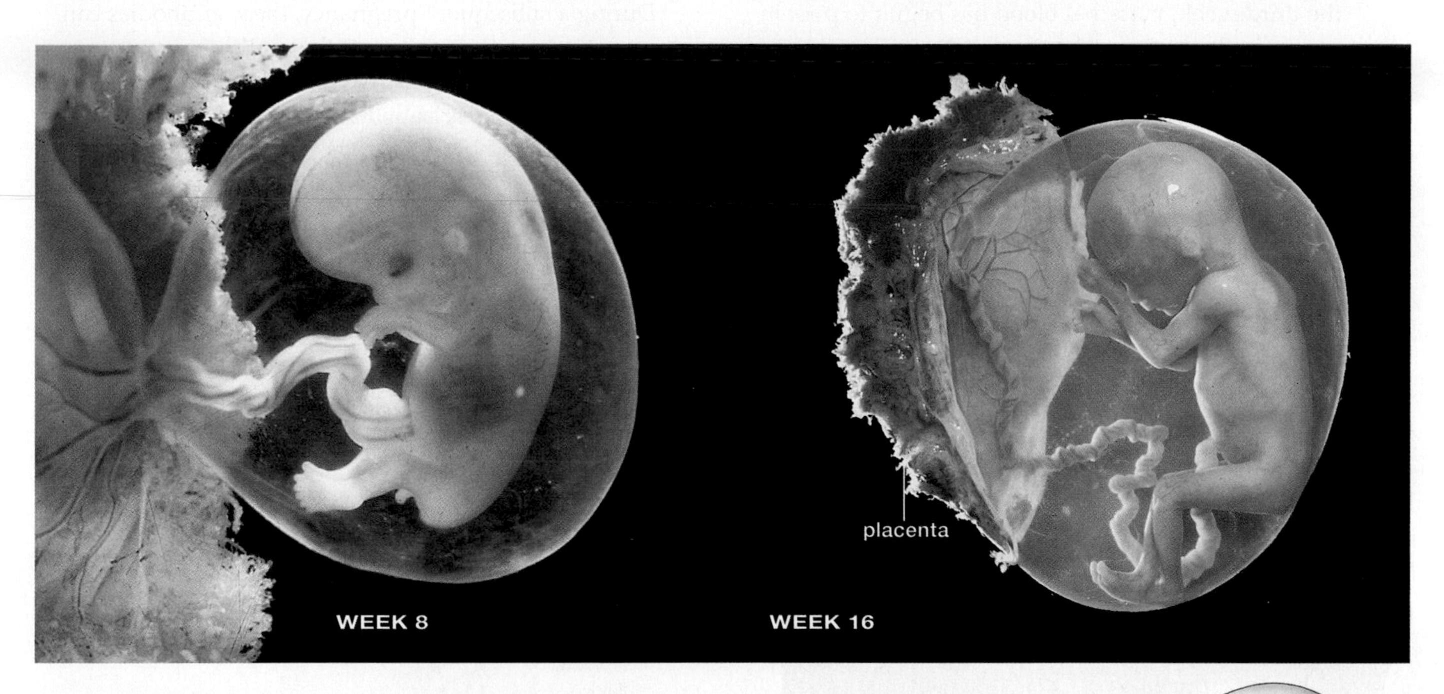

final week of embryonic period; embryo looks distinctly human compared to other vertebrate embryos

upper and lower limbs well formed; fingers and then toes have separated

primordial tissues of all internal, external structures now developed

tail has become stubby

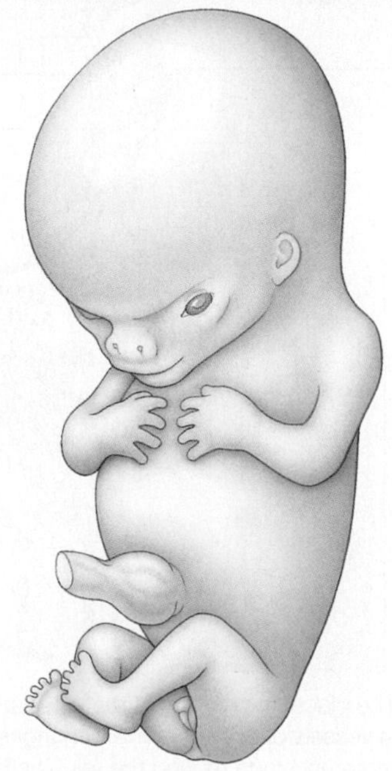

Length: 16 centimeters (6.4 inches)
Weight: 200 grams (7 ounces)

WEEK 29
Length: 27.5 centimeters (11 inches)
Weight: 1,300 grams (46 ounces)

WEEK 38 (full term)
Length: 50 centimeters (20 inches)
Weight: 3,400 grams (7.5 pounds)

During fetal period, length measurement extends from crown to heel (for embryos, it is the longest measurable dimension, as from crown to rump).

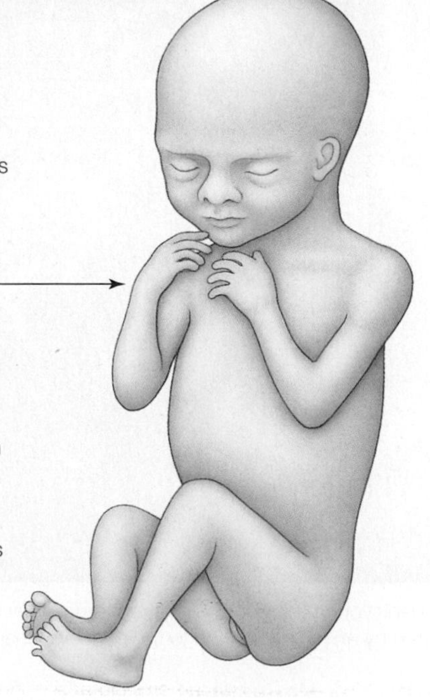

42.9 Structure and Function of the Placenta

✔ Like other placental mammals, a human mother supports her offspring by exchanges across the placenta.

All exchange of materials between an embryo or fetus and its mother takes place by way of the placenta. This pancake-shaped organ consists of uterine lining, extraembryonic membranes, and embryonic blood vessels (**FIGURE 42.14**). At full term, a placenta covers about a quarter of the uterus's inner surface.

The placenta begins forming early in pregnancy. By the third week, maternal blood has begun to pool in spaces in the endometrial tissue. Chorionic villi—tiny fingerlike projections from the chorion—extend into the pools of maternal blood. Embryonic blood vessels extend through a coiled umbilical cord to the placenta, and into the villi.

Maternal and embryonic blood never mix. Instead, substances move between maternal and embryonic bloodstreams by diffusing across the walls of the embryonic vessels in the chorionic villi. Oxygen diffuses from maternal to embryonic blood, and carbon dioxide diffuses the opposite way. Transport proteins assist movement of essential nutrients from the maternal blood into embryonic blood vessels inside the villi.

The placenta also has a hormonal role. From the third month on, it produces HCG, progesterone, and estrogens. These hormones encourage the ongoing maintenance of the uterine lining. Once the placenta takes on this task, the corpus luteum degenerates.

One type of maternal antibody (IgG) is also transported from maternal blood into the fetal bloodstream. This sometimes causes problems if the mother and her child differ in their Rh blood type. An Rh negative woman lacks the Rh antigen on her red blood cells. When she gives birth to her first Rh positive child, exposure to the child's blood during childbirth can cause her to develop antibodies against the Rh antigen. During a subsequent pregnancy, these antibodies can cross the placenta and cause the mother's immune system to attack an Rh positive fetus.

Problems also arise when teratogens travel from mother to child by way of the placenta. The rubella virus crosses the placenta, as do alcohol, nicotine, mercury, and some other dangerous chemicals.

TAKE-HOME MESSAGE 42.9

What is the function of the placenta?

✔ Blood vessels of the embryo's circulatory system extend through the umbilical cord to the placenta, where they run through pools of maternal blood.

✔ Maternal and embryonic blood do not mix; substances diffuse between the maternal and embryonic bloodstreams by crossing vessel walls.

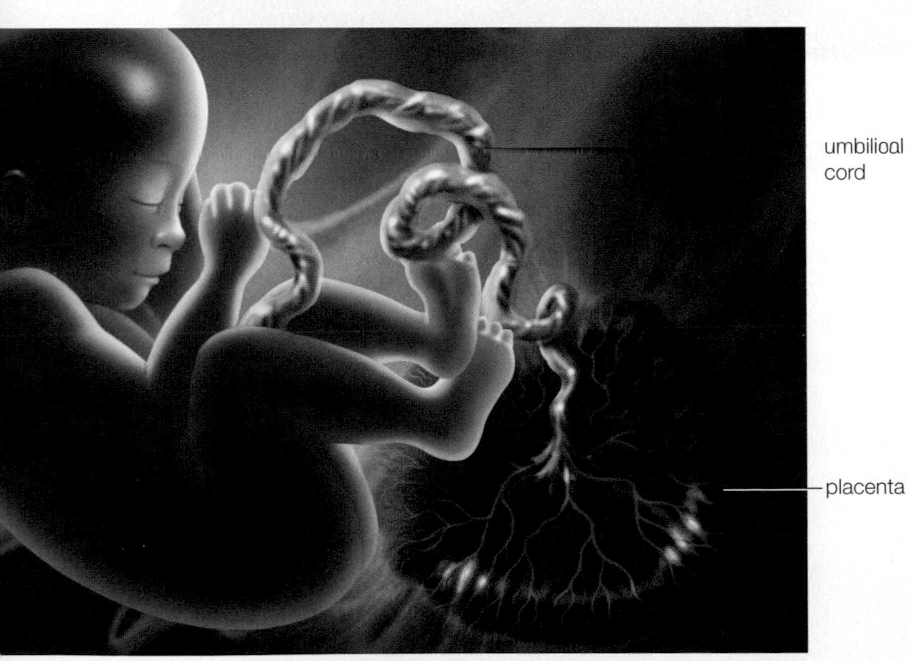

Artist's depiction of the view inside the uterus, showing a fetus connected by an umbilical cord to the pancake-shaped placenta.

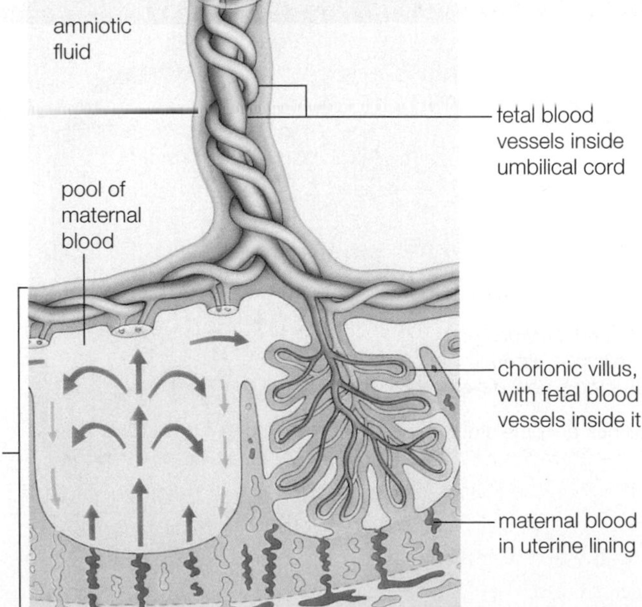

The placenta consists of maternal and fetal tissue. Fetal blood flowing in vessels of chorionic villi exchanges substances by diffusion with maternal blood around the villi. The bloodstreams do not mix.

FIGURE 42.14 ▶**Animated** Life support system of a developing human.

42.10 Labor and Childbirth

✔ Muscular contractions expel a fully developed fetus from its mother's uterus.

A mother's body changes as her fetus nears full term, at about 38 weeks after fertilization. Until the last few weeks, her firm cervix helped prevent the fetus from slipping out of her uterus prematurely. Now the connective tissue of the cervix softens and thins. These changes make the cervix more flexible in preparation for **labor**, the process by which a woman's body expels a fetus from her uterus. Typically, the amnion ruptures right before birth, so amniotic fluid drains out from the vagina. The cervical canal dilates, then contractions of uterine smooth muscle propel the fetus out of the body through the vagina (**FIGURE 42.15**).

A positive feedback mechanism operates during labor. When the fetus nears full term, it typically shifts position so that its head puts pressure on the mother's cervix. Receptors inside the cervix sense the pressure and signal the hypothalamus, which in turn signals the posterior lobe of the pituitary to secrete the hormone **oxytocin**. In a positive feedback loop, oxytocin binds to smooth muscle of the uterus, causing uterine contractions that push the fetus against the cervix. The added pressure triggers more oxytocin secretion, which causes more contractions and more cervical stretching:

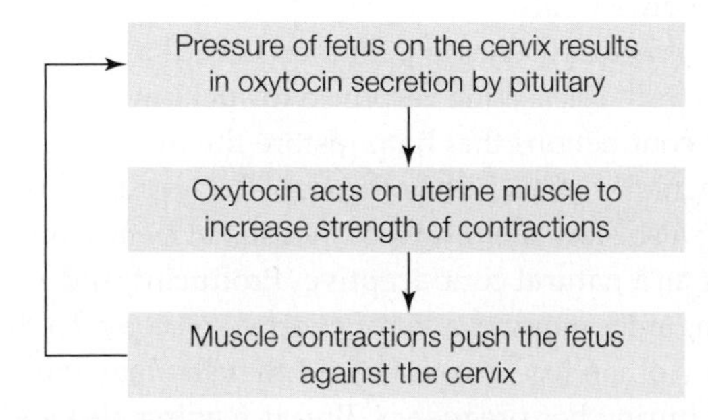

Uterine contractions continue to intensify until they push the fetus out of the mother's body. Synthetic oxytocin is sometimes given to induce labor, or to increase the strength of contractions.

After the newborn emerges, continued muscle contractions expel the placenta from the uterus as the "afterbirth." The umbilical cord that connects the newborn to this mass of expelled tissue is clamped, cut short, and tied. The short stump of cord left in place withers and falls off. The navel marks its former attachment site.

Sometimes, it is difficult, dangerous, or impossible for a child to exit through the vagina. In the United States, about 30 percent of births involve a cesarean section, sometimes called a C-section. During this surgical procedure, a physician cuts the mother's abdominal wall, then opens the uterus and removes the fetus. A variety of conditions can necessitate a surgical delivery. For example, a placenta that grows in the lower portion of the uterus and covers the cervix can make it impossible for the fetus to exit through the vagina. If the placenta dislodges too early or the umbilical cord kinks, blood flow to the fetus is compromised and surgery may be required to ensure its safety. A cesarean section is also used to prevent transmission of a sexually transmitted disease to an infant (Section 41.9).

A Fetus positioned for childbirth; its head is against the mother's cervix, which is dilating (widening).

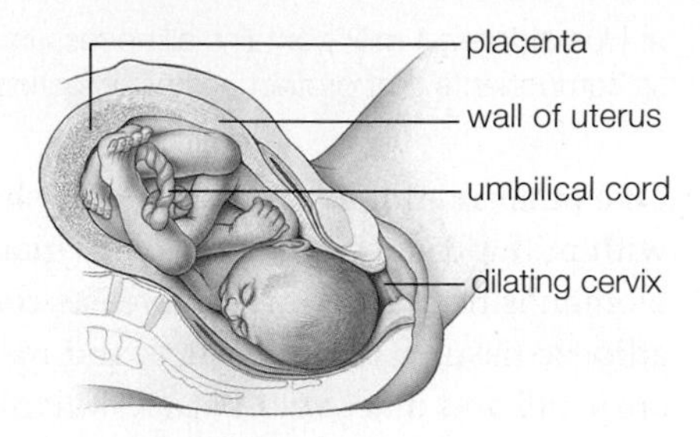

B Muscle contractions stimulated by oxytocin force the fetus out through the vagina.

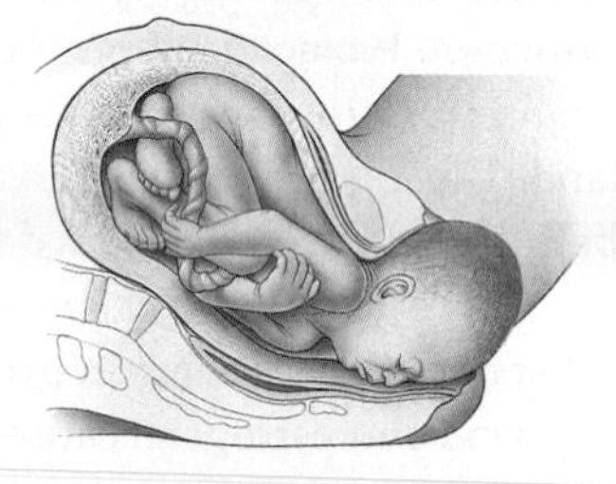

placenta detaching from wall of uterus

umbilical cord

C The placenta detaches from the wall of the uterus and is expelled.

FIGURE 42.15 Expulsion of fetus and afterbirth during normal delivery. Afterbirth consists of the placenta, tissue fluid, and blood.

labor Expulsion of a placental mammal from its mother's uterus by muscle contractions.
oxytocin Posterior pituitary hormone that encourages contraction of smooth muscle of the reproductive tract.

TAKE-HOME MESSAGE 42.10

How is a child born?

✔ During normal childbirth, the hormone oxytocin stimulates increasingly strong contractions of uterine smooth muscle. These contractions propel the child, and then the afterbirth, through the cervix and out of the vagina.

42.11 Milk: Nourishment and Protection

✔ Human breast milk contains all necessary nutrients as well as components that protect against infection.

Like other mammals, humans nourish their newborns with milk produced by mammary glands. When a woman is not pregnant, her breasts consist mainly of adipose tissue. Her milk ducts and mammary glands are small and inactive (**FIGURE 42.16**). In pregnancy, these structures enlarge in preparation for **lactation**, or milk production. **Prolactin**, a hormone secreted by the mother's anterior pituitary, triggers growth of mammary glands during pregnancy.

Just before birth, and for a few days thereafter, mammary glands produce a clear, nutrient-rich fluid called colostrum. After a woman gives birth, a decline in progesterone and estrogens causes milk production to go into high gear. The stimulus of a newborn's suckling results in the release of both prolactin and oxytocin. Prolactin now encourages production of milk proteins. Oxytocin stimulates muscles around the milk glands to contract, forcing milk into the milk ducts.

Human milk is about 88 percent water and contains a variety of ingredients that contribute to an infant's health. Under natural circumstances, human breast milk will be a child's only food during infancy. Thus, it contains all nutrients required for the child's growth and metabolism. The main carbohydrate in milk is lactose (milk sugar), a disaccharide that is broken down by lactase, an enzyme on intestinal villi. Lactose and other carbohydrates serve as a source of energy for an infant, and also as food for bacteria that contribute to its health. In addition to sugars, milk contains fats (mainly triglycerides), easily digested proteins (wheys and casein), and essential vitamins and minerals.

Milk also helps protect an infant from pathogens and encourages development of the immune system. Maternal macrophages (phagocytic white blood cells) delivered by way of milk survive in an infant's throat and gut lining, where they engulf pathogens and produce compounds that encourage maturation of the infant's own immune system. Maternal immunoglobulin A (one type of antibody) in milk provides protection while the infant's immunity gets up to speed. Milk also contains a variety of antimicrobial compounds. One of these, lysozyme, is an enzyme that kills bacteria by destroying their cell walls. Another milk protein, lactoferrin, helps to prevent bacterial, viral, and fungal pathogens from attaching to and infecting human cells.

Children who are breast-fed suffer fewer intestinal and respiratory infections in infancy and are less likely to develop allergies than those who are given formula. Because of milk's role in immune health, the American Academy of Pediatricians recommends that women breast-feed exclusively for the first six months after birth, and continue breast-feeding for the first year.

Nursing women should keep in mind that drugs and pathogens in their body can end up in their milk and affect their child. For example, exposure to nicotine in milk interferes with an infant's ability to sleep, as does exposure to alcohol. Alcohol in milk may also interfere with a child's early motor development. Some viral pathogens, including HIV, can also be transmitted through breast milk.

Breast-feeding benefits a mother as well as her child. The oxytocin released in response to suckling causes uterine contractions that help restore the uterus to its pre-pregnancy state. The hormonal effects of breast-feeding also slow the return of menstrual cycles, thus serving as a natural contraceptive. Producing and secreting milk requires a lot of energy, so breast-feeding helps a woman lose excess weight that she may have gained during her pregnancy. Breast-feeding also lowers a woman's risk of cancers of the reproductive tract and protects against inherited forms of breast cancer that arise in premenopausal women.

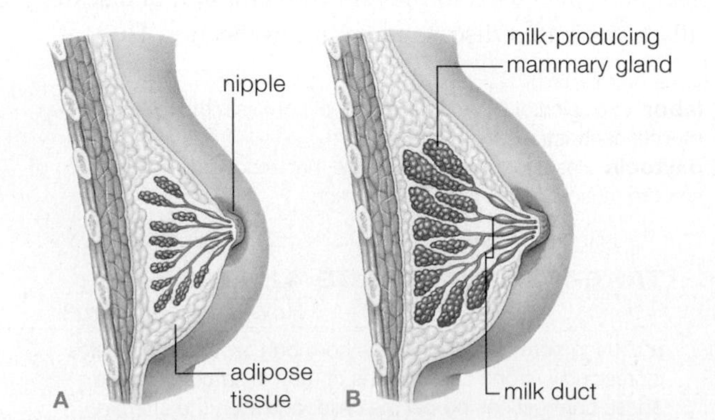

FIGURE 42.16 Cutaway views of (**A**) the breast of a woman who is not pregnant and (**B**) the breast of a lactating woman.

lactation Milk production by a female mammal.
prolactin Anterior pituitary hormone that promotes milk production.

TAKE-HOME MESSAGE 42.11

What occurs during lactation?

✔ Prolactin from the pituitary promotes mammary gland enlargement and milk production. Oxytocin released in response to suckling causes milk secretion.

✔ In addition to nourishing the newborn, breast-feeding protects it from infection. The mother benefits from weight loss, contraceptive effects, and decreased cancer risk.

Prenatal Problems (revisited)

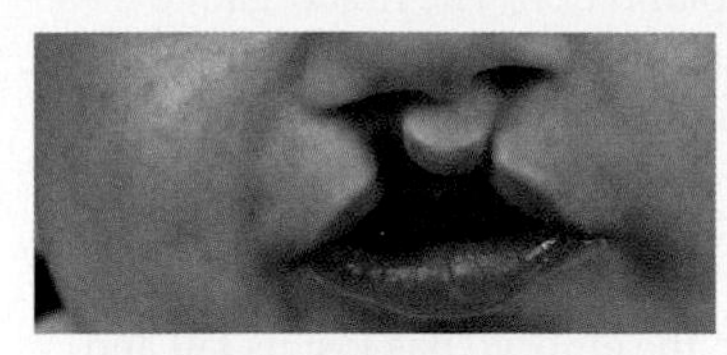

As you now know, human development unfolds in a predictable manner, with specific events occurring at specific times. As a result, we can predict when specific birth defects are likely to arise. FIGURE 42.17 shows when various organs and structures develop and are thus susceptible to the effects of teratogens.

With normal development, fusion of facial tissues during week 7 closes clefts in the developing upper lip and the palate (roof of the mouth). If this fusion does not proceed properly, the child is born with a cleft lip, cleft palate, or both. Each year, about 7,500 children born in the United States have such defects.

As previously noted, a woman's use of alcohol or tobacco during pregnancy increases the risk of cleft lip and palate in her offspring. A maternal deficiency in zinc or folate is another risk factor. Genetics also affects the risk of these birth defects. Consider van der Woude syndrome, which arises from a mutation in a gene for a transcription factor (*IRF6*) that is expressed in the face during early development. The syndrome is governed by autosomal dominant inheritance. People who have a mutant allele at the *IRF6* locus are frequently born with a cleft lip, palate, or both. Differences in frequency of cleft lips among ethnic groups provide additional evidence of the role of genetic factors. Cleft lips occur most often in Native Americans and Asians (1/500), less often in people of European ancestry (1/1000), and least often in those of African descent (1/2500).

Lip and palate clefts are not only a cosmetic concern, they can also interfere with natural feeding by making it impossible for the infant to suck efficiently enough to draw milk from a nipple. Later, as the child matures, they can cause difficulties with speech.

Surgical repair of facial clefts typically begins during a child's first year. Depending on the severity of the defect, additional surgeries may be required as the child grows and develops.

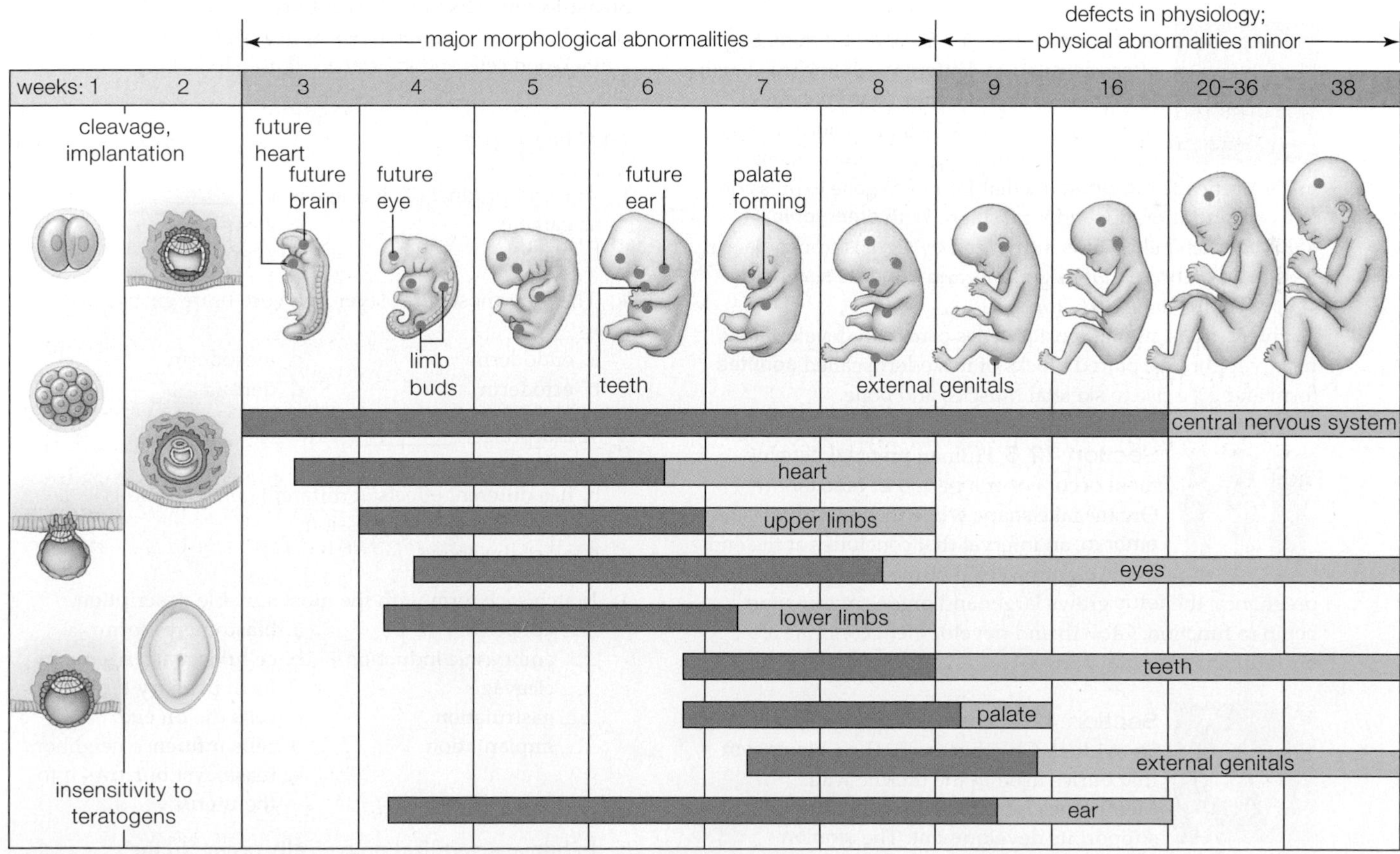

FIGURE 42.17 ▸Animated Teratogen sensitivity. Teratogens are drugs, infectious agents, and environmental factors that cause birth defects. Dark blue signifies the highly sensitive period for an organ or body part; light blue signifies periods of less severe sensitivity. For example, the upper limbs are most sensitive to damage during weeks 4 through 6, and somewhat sensitive during weeks 7 and 8.

FIGURE IT OUT Is teratogen exposure in the 16th week more likely to affect the heart or the genitals?

Answer: Genitals

summary

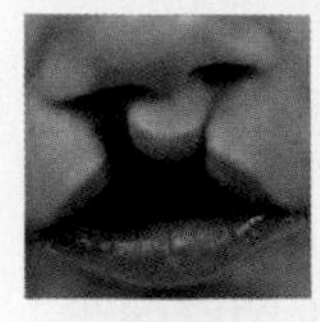

Section 42.1 Problems that arise during development can result in miscarriage, stillbirth, or birth defects. Such problems can have genetic causes or arise after maternal exposure to **teratogens**.

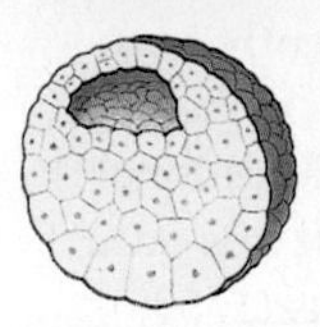

Section 42.2 All sexually reproducing animals have similar stages of development. After fertilization, **cleavage** produces a **blastula**. The blastula undergoes **gastrulation** to produce a **gastrula** with primary tissue layers, or **germ layers**. In vertebrates, three germ layers form: **ectoderm** (outer layer), **mesoderm** (middle layer), and **endoderm** (inner layer).

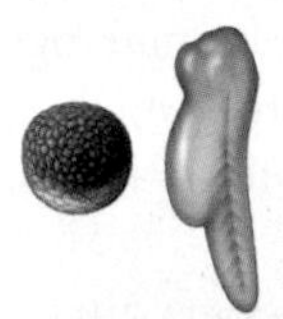

Section 42.3 **Cytoplasmic localization** occurs in all oocytes. Cleavage distributes different portions of the egg cytoplasm into different cells as a blastula forms. Protostomes and deuterostomes differ in the details of their cleavage pattern. The amount of yolk in an egg also influences cleavage. Experimental manipulation of amphibian development demonstrated that material localized in one region of the zygote is essential to gastrulation. Cells derived from this region produce signals that initiate gastrulation.

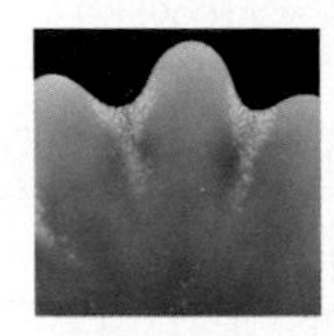

Sections 42.4, 42.5 **Differentiation** begins after gastrulation: Different cell lineages begin to express different genes. **Maternal effect genes** produce mRNAs that are localized in different parts of a zygote. Some of these mRNAs encode **morphogens** that influence gene expression in a concentration-dependent fashion. With **embryonic induction**, cell behavior is influenced by signals from adjacent cells in an embryo. Cell migrations and **apoptosis** (cell suicide) help shape body form.

Interactions among master genes constrain development. In all vertebrates, paired blocks of mesoderm called **somites** form and give rise to skeletal muscles and bone.

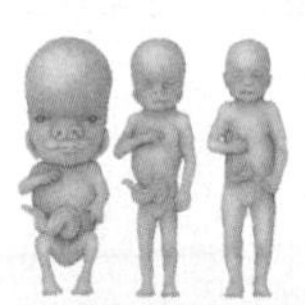

Section 42.6 Human prenatal development occurs over a period of nine months. Organs take shape while the individual is an **embryo**, an interval that concludes at the end of the eighth week. For the remainder of the pregnancy, the **fetus** grows larger and organs mature and begin to function. Growth and development continue after birth (in the postnatal period).

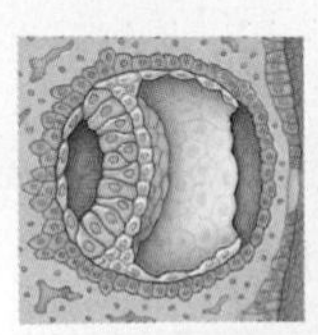

Section 42.7 Human fertilization occurs in an oviduct. Cleavage produces a **blastocyst** that buries itself in the uterine wall. Membranes form outside the blastocyst and support its development. The **amnion** encloses and protects the embryo in a fluid-filled sac. The **chorion** and **allantois** become part of the placenta, which allows exchange of substances between maternal and fetal bloodstreams. After implantation, the chorion produces the hormone **human chorionic gonadotropin (HCG)**.

Gastrulation occurs after implantation. The first organ to form, the neural tube, later becomes the brain and spinal cord. Somites form on either side of the neural tube.

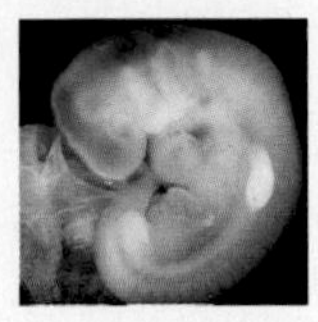

Sections 42.8, 42.9 At four weeks, an embryo has limbs and a tail. By the end of the eighth week, the embryo has lost its tail and pharyngeal arches and has a distinctly human appearance. It continues to grow in size and its organs continue to mature during the fetal period.

The placenta consists of extraembryonic membranes and endometrium. In the placenta, embryonic blood exchanges nutrients and wastes with maternal blood. Teratogens cross the placenta, so a mother's health, nutrition, and lifestyle can affect the growth and development of her future child.

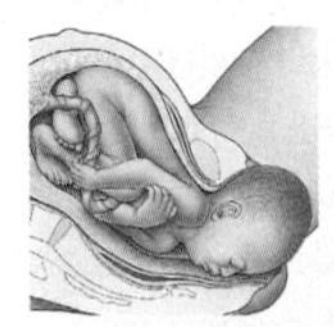

Sections 42.10, 42.11 Hormones typically induce **labor** at about 38 weeks. Positive feedback controls the secretion of **oxytocin**, a hormone that causes contractions that expel the fetus and then the afterbirth. A cesarean section delivers a fetus when vaginal delivery would be impossible or dangerous.

Prolactin regulates milk production by mammary glands and oxytocin encourages milk secretion during **lactation**. Breast-feeding has multiple benefits for both mother and child. Milk contains not only essential nutrients, but also white blood cells and substances that help kill pathogens.

self-quiz

Answers in Appendix VII

1. The end product of cleavage is a ________ .
   a. gamete
   b. blastula
   c. gastrula
   d. zygote

2. The outermost germ layer in a vertebrate gastrula is the ________ .
   a. endoderm
   b. ectoderm
   c. mesoderm
   d. dermis

3. A morphogen ________ .
   a. diffuses through an embryo
   b. has different effects at different concentrations
   c. influences gene expression
   d. all of the above

4. Match each term with the most suitable description.

___ apoptosis	a. blastomeres form
___ embryonic induction	b. cellular rearrangements form primary tissues
___ cleavage	c. cells die on cue
___ gastrulation	d. cells influence neighbors
___ implantation	e. blastocyst burrows into the uterus

5. In humans, fertilization typically occurs in the ________ .
   a. vagina b. uterus c. cervix d. oviduct

6. The ________ , a fluid-filled sac, surrounds and protects a human embryo and keeps it from drying out.
   a. amnion b. allantois c. yolk sac d. chorion

data analysis activities

Birth Defects and Multiple Births A woman who carries multiple offspring at the same time increases the risk of some birth defects. FIGURE 42.18 shows the results of Yiwei Tang's study of birth defects reported in Florida from 1996 to 2000. Tang compared the incidence of various defects among single and multiple births. She calculated the relative risk for each type of defect based on type of birth, and corrected for other differences that might increase risk such as maternal age, income, race, and medical care during pregnancy. A relative risk of less than 1 means that multiple births pose less risk of that defect occurring. A relative risk greater than 1 means multiples are more likely to have a defect.

1. What was the most common type of birth defect in the single-birth group?

2. Was that type of defect more or less common among the multiple-birth newborns than among single births?

3. Tang found that multiples have more than twice the risk of single newborns for one type of defect. Which type?

4. Does a multiple pregnancy increase the relative risk of chromosomal defects in offspring?

	Prevalence of Defect		Relative
	Multiples	Singles	Risk
Total birth defects	358.50	250.54	1.46
Central nervous system defects	40.75	18.89	2.23
Chromosomal defects	15.51	14.20	0.93
Gastrointestinal defects	28.13	23.44	1.27
Genital/urinary defects	72.85	58.16	1.31
Heart defects	189.71	113.89	1.65
Musculoskeletal defects	20.92	25.87	0.92
Fetal alcohol syndrome	4.33	3.63	1.03
Oral defects	19.84	15.48	1.29

FIGURE 42.18 Prevalence, per 10,000 live births, of various types of birth defects among multiple and single births. Relative risk for each defect is given after researchers adjusted for the mother's age, race, previous adverse pregnancy experience, education, Medicaid participation during pregnancy, as well as the infant's sex and number of siblings.

7. The placenta consists of _______ .
   a. embryonic tissue
   b. maternal tissue
   c. paternal tissue
   d. a combination of a and b

8. During the fetal period, _______ .
   a. gastrulation ends
   b. somites form
   c. a tail forms
   d. limb movements begin

9. Human milk contains _______ .
   a. sugars
   b. lysozyme
   c. antibodies
   d. all of the above

10. Match each hormone with its action(s).
   ___ prolactin
   ___ oxytocin
   ___ human chorionic gonadotropin
   a. milk protein production
   b. sustains corpus luteum
   c. causes contraction of uterus, milk ducts

11. Pharyngeal arches of a human embryo will later develop into _______ .
   a. gills
   b. lungs
   c. structures of the head and neck

12. _______ removes webs between developing digits.
   a. Gastrulation
   b. Lactation
   c. Apoptosis
   d. Implantation

13. Most miscarriages _______ .
   a. occur in the first trimester
   c. are caused by exposure to a pathogen
   b. can be prevented by a cesarean section
   d. threaten the life of the mother

14. A pregnancy test detects _______ in the urine.
   a. lysozyme
   b. prolactin
   c. oxytocin
   d. human chorionic gonadotropin

15. Number these events in human development in the correct order.
   ___ gastrulation is completed
   ___ blastocyst forms
   ___ tail is reabsorbed
   ___ implantation occurs
   ___ zygote forms
   ___ neural tube forms
   ___ maternal cervix softens

critical thinking

1. Each year 110,000 people are born with birth defects as a result of prenatal rubella infections. If a woman becomes infected during the first trimester of her pregnancy, her child may be born deaf or blind. Infections later in pregnancy do not increase risk of these effects. Explain why.

2. A nursing mother who has an alcoholic drink will secrete alcohol into her milk for two to three hours afterward. Alcohol also decreases oxytocin production. What effect would decreased oxytocin have on breast-feeding?

3. Tumors that arise from germ cells can contain any tissue type, whereas tumors that arise from somatic cells can only contain a limited number of tissues. Explain why.

43 Animal Behavior

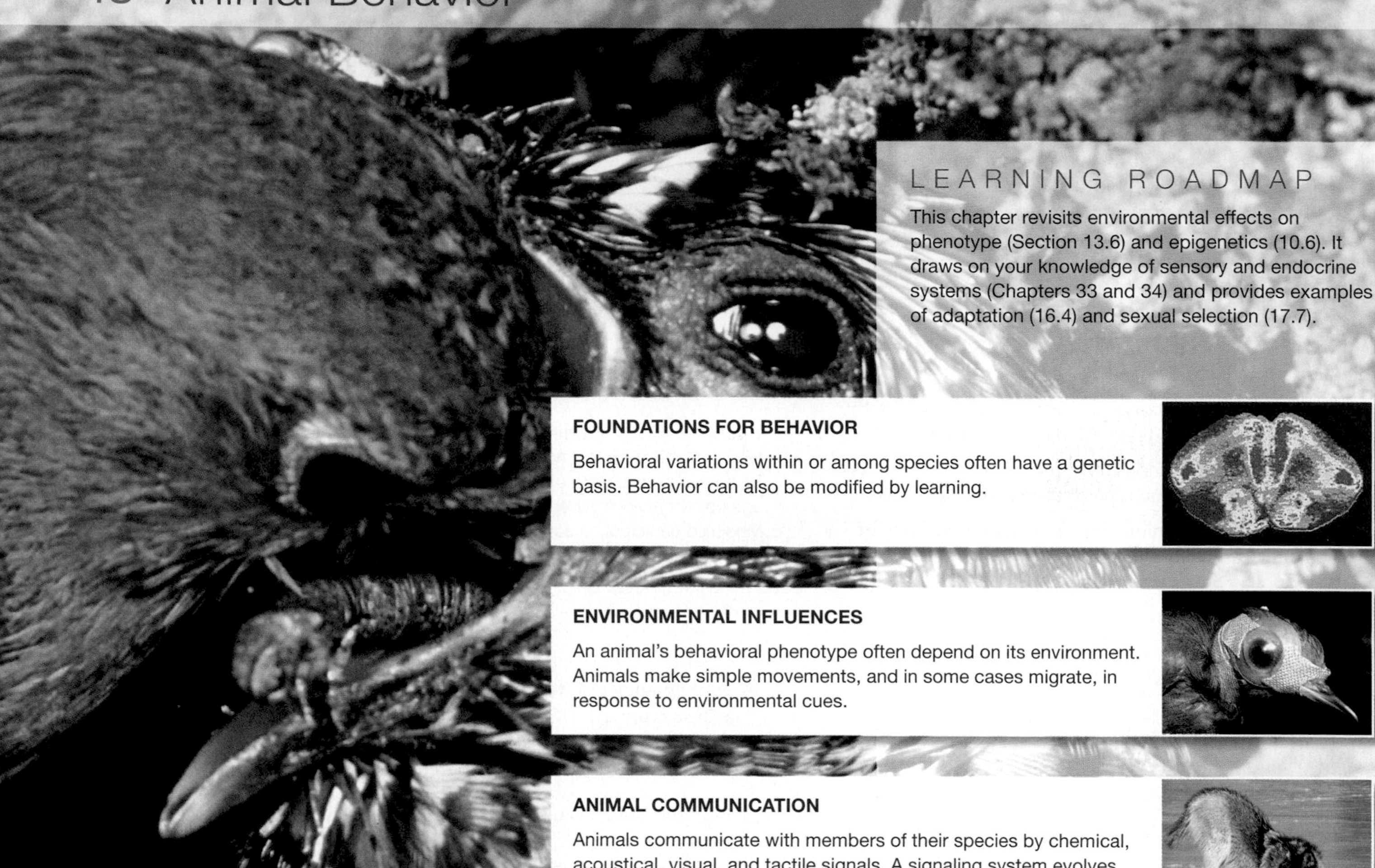

LEARNING ROADMAP

This chapter revisits environmental effects on phenotype (Section 13.6) and epigenetics (10.6). It draws on your knowledge of sensory and endocrine systems (Chapters 33 and 34) and provides examples of adaptation (16.4) and sexual selection (17.7).

FOUNDATIONS FOR BEHAVIOR

Behavioral variations within or among species often have a genetic basis. Behavior can also be modified by learning.

ENVIRONMENTAL INFLUENCES

An animal's behavioral phenotype often depend on its environment. Animals make simple movements, and in some cases migrate, in response to environmental cues.

ANIMAL COMMUNICATION

Animals communicate with members of their species by chemical, acoustical, visual, and tactile signals. A signaling system evolves and persists only if it benefits a signal sender and receiver.

MATING AND PARENTAL CARE

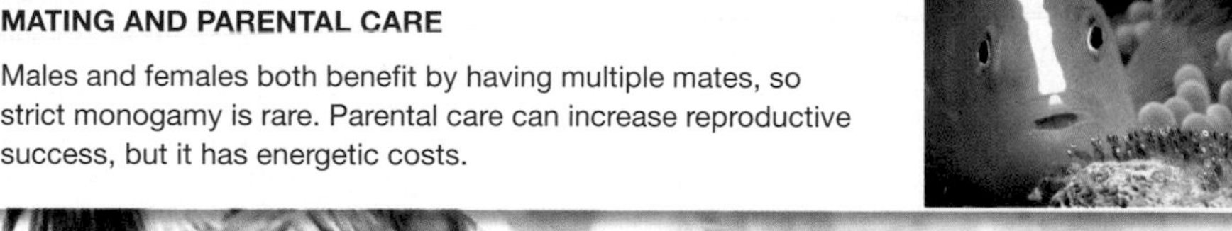

Males and females both benefit by having multiple mates, so strict monogamy is rare. Parental care can increase reproductive success, but it has energetic costs.

SOCIAL BEHAVIOR

Life in social groups has reproductive benefits and costs. Self-sacrificing behavior has evolved among a few kinds of animals that live in large family groups.

In the next chapter you will learn about the effects of behavior on characteristics of populations. Chapter 45 discusses the behavioral interactions between species, such as between predators and their prey. Chapter 48 details some of the ways in which human activities disrupt animal behavior and threaten species with extinction.

43.1 Can You Hear Me Now?

Many animals communicate with other members of their species through sound. They call or otherwise produce sounds that advertise their position to mates or young, scare away potential rivals, or warn one another of dangers. Consider the right whale (**FIGURE 43.1A**), so-called because at one time whalers thought it was the right (correct) whale to hunt. These whales make a variety of sounds. The most frequently used is a whooplike contact call that allows one whale to let others know where it is. Other right whale calls sound like screams, moans, or gunshots.

In the past, whaling drove right whale populations throughout the world to the brink of extinction. The North Atlantic population declined to a low of about 100 animals; today it includes about 400. Unfortunately, this population lives near the eastern coasts of Canada and the United States, where commercial ship traffic is heavy. Potentially fatal collisions with ships are a problem, as is noise pollution.

Marine biologist Susan Parks studies how North Atlantic right whales communicate, the effects of man-made noise on the whales, and ways to minimize any negative effects. By attaching recorders to right whales (**FIGURE 43.1B**), Parks showed that the whales altered their calling in response to the level of shipping noise. When a ship is present, whales call more loudly and their calls become more shrill. The louder the ship noise, the louder the whales call. They behave like humans at a noisy restaurant, shouting at one another in high-pitched voices to be heard above the din. In the sea, as in a restaurant, such vocal alterations improve sound transmission, but vital information can still be drowned out. We do not know if the modifications the whales make to their voices are sufficient to allow their calls to be understood despite the noise.

For humans, chronic exposure to noise is stressful, and the same may be true for whales. It would be impossible to ask tankers to stop their engines for days to see whether noise affects whales, but such a quiet period did occur immediately after the terrorist attacks of September 11, 2001. During this time, another scientist, Rosalind Rollins, was collecting whale feces to test the whales' level of cortisol, a hormone whose secretion increases with stress (Section 34.9). Parks and Rollins pooled data regarding ship-related noise and cortisol secretion and found that the brief decline in shipping noise was accompanied by a significant drop in the whales' cortisol level. In other years, there was no such drop in cortisol after September 11. These data suggest that the normally high noise level related to shipping causes a chronic stress response in right whales. In other mammals, chronic stress dampens the immune response and impairs reproduction. Thus, noise-induced stress may be an important factor in slowing the recovery of this highly endangered species.

A Right whale (*Eubalaena glacialis*) mother and calf.

B Marine biologist Dr. Susan Parks with the recorder she used to investigate the effect of shipping noise on right whale calls. Suction cups attach the device to the whale's body.

FIGURE 43.1 Studying right whale communication.

As this example shows, we can unknowingly interfere with animal signaling. The communication systems of whales and other animals evolved over countless generations in an environment free of human-generated noise, light, and other sensory distractions. Finding ways to meet human needs without unnecessarily disrupting the communication mechanisms of Earth's other species is an ongoing challenge.

CREDITS: (opposite) © AdStock RF/Shutterstock.com; (1A) Richard Du Toit/naturepl.com; (1B) Courtesy of Kelly Slivka/Whale Center of New England, www.whalecenter.org.

43.2 Behavioral Genetics

✔ Much variation in behavior within or among species results from inherited differences. In a few instances, scientists have even pinpointed genes responsible for the variation.

Animals differ in the kinds of stimuli they can perceive, and in their responses to perceived stimuli. Stimulus perception and response are shaped by genes that affect the nervous and endocrine systems.

Genetic Variation Within a Species

One way to investigate the genetic basis of behavior is to examine behavioral and genetic differences among members of a single species. Stevan Arnold's study of snake feeding behavior is an example. There are two populations of garter snakes in the Pacific Northwest. One population lives in coastal forests and preferentially feeds on banana slugs (**FIGURE 43.2**). The other population lives inland, where there are no banana slugs. In this population, fishes and tadpoles are the favored foods. The snakes' difference in food preference is inborn. When offered a banana slug as their first meal, young coastal snakes eat it, but young inland snakes ignore it.

FIGURE 43.2 Coastal garter snake dining on a banana slug. The snake is genetically predisposed to recognize slugs as prey. By contrast, garter snakes from inland regions where there are no banana slugs ignore slugs when experimenters offer them.

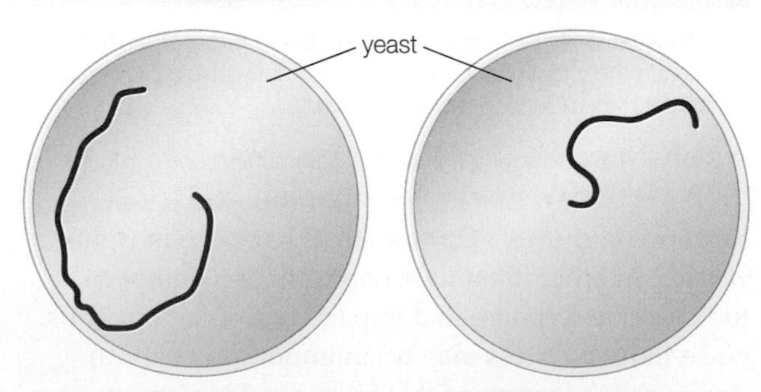

A Rovers (genotype *FF* or *Ff*) move often as they feed. When a rover's movements on a petri dish filled with yeast are traced for 5 minutes, the trail is relatively long.

B Sitters (genotype *ff*) move little as they feed. When a sitter's movements on a petri dish filled with yeast are traced for 5 minutes, the trail is relatively short.

FIGURE 43.3 Genetic polymorphism for foraging behavior in fruit fly larvae. When a larva is placed in the center of a yeast-filled plate, its genotype at the *foraging* locus influences whether it moves a little or a lot while it feeds. Black lines show a representative larva's path.

Arnold hypothesized that snakes from the inland population lack the genetically determined ability to associate the scent of slugs with food. He predicted that if coastal garter snakes were crossed with inland snakes, the resulting offspring would make an intermediate response to slugs. Results from experimental crosses confirmed this prediction. However, we still do not know which gene or genes underlie this difference in behavior.

We know more about the genetic basis of differences in foraging behavior among fruit fly larvae. The fly's wingless, wormlike larvae feed on yeast that grows on decaying fruit. In wild populations, about 70 percent of fruit fly larvae are "rovers," which means they tend to move around a lot as they feed. Rovers often leave one patch of food to seek another (**FIGURE 43.3A**). The remaining 30 percent of larvae are "sitters"; they tend to move little once they find food (**FIGURE 43.3B**). When food is absent, rovers and sitters move the same amount, so both are equally energetic.

The proximate (immediate) cause of the difference in larval behaviors is a difference in alleles of a gene called *foraging*. Flies with a dominant allele of this gene have the rover phenotype. Sitters are homozygous for the recessive allele. The *foraging* gene encodes an enzyme involved in learning about olfactory cues. Rovers make more of the enzyme than sitters.

The ultimate (evolutionary) cause of the behavioral variation in larval foraging behavior is natural selection that arises from competition for food. With limited food, a rover does best when surrounded by sitters, and vice versa. Presumably, when there are lots of larvae of one type, all compete for food in the same way. Thus, frequency-dependent selection (Section 17.7) maintains both alleles of the *foraging* gene.

Genetic Variation Among Species

Researchers also investigate the genetic basis of a behavior by comparing related species. For example, studies of rodents called voles (**FIGURE 43.4**) revealed that inherited differences in the number and distribu-

A Voles are small rodents. Some species form pair bonds, and others are promiscuous.

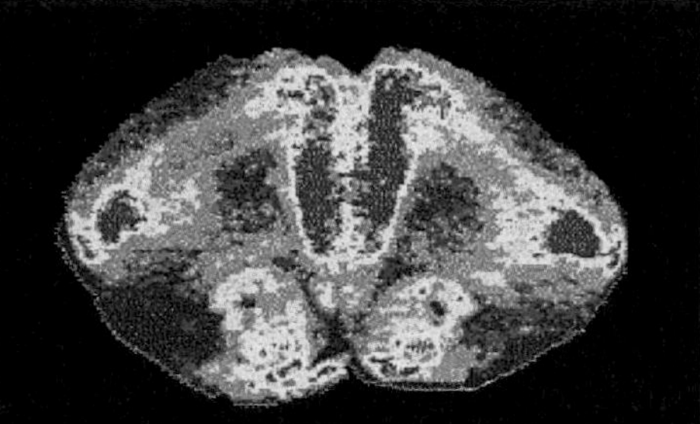

B PET scan of a monogamous prairie vole's brain with many receptors for the hormone oxytocin (red).

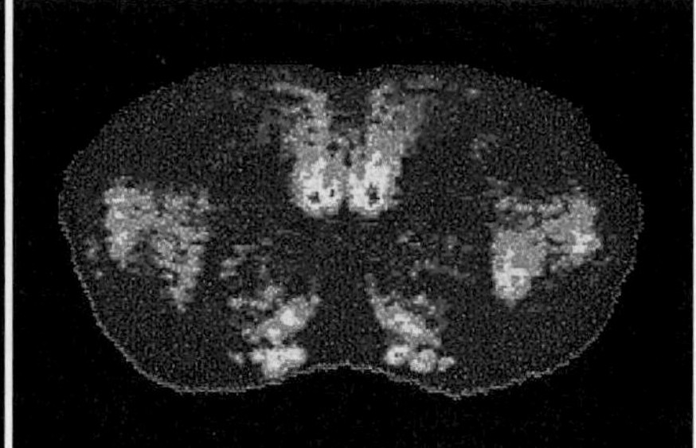

C PET scan of a promiscuous prairie vole's brain with few hormone receptors for oxytocin.

FIGURE 43.4 Differences in the distribution of oxytocin receptors in two species of voles (*Microtus*).

tion of certain hormone receptors influence mating and bonding behavior. Most voles are promiscuous, meaning both males and females have multiple mates and show no special affinity for individuals with whom they have mated. By contrast, prairie voles form a pair bond—a long-term, largely monogamous partnership. This species-specific difference in behavior arises from differences in how the voles respond to two hormones.

The hormone oxytocin is released during sexual intercourse, labor, and lactation in all mammals. In prairie voles, its action is essential to formation and maintenance of the pair bond. Inject the female member of an established prairie vole pair with a chemical that impairs oxytocin function, and she will no longer associate mainly with her partner. All vole species have similar numbers of oxytocin-producing cells. However, compared to members of promiscuous vole species, prairie voles have many more oxytocin receptors in the parts of the brain associated with social learning (**FIGURE 43.4B,C**).

Variations in the distribution of receptors for the hormone aginine vasopressin (AVP) also contribute to differences in the tendency to pair bond. Compared to members of promiscuous vole species, prairie voles have more AVP receptors in their brain. Additional evidence for the importance of AVP in pair bonding behavior comes from an experiment in which the prairie vole AVP receptor gene was transferred into the forebrain of male mice. Mice are naturally promiscuous. However, male mice genetically modified by the prairie vole gene gave up their playboy ways. After the procedure, they preferred a female with whom they had already mated over an unfamiliar female. These results confirmed the role of AVP in fostering monogamy among male rodents.

Human Behavior Genetics

Nearly all human behavioral traits have a polygenic basis and are influenced by the environment. Keep this in mind when you see headlines touting discovery of a gene "for" thrill-seeking behavior, alcoholism, or some other human behavior. Usually, such dramatic headlines are inspired by a study in which a specific allele was found to be associated with a relatively small increase in the tendency to perform a certain behavior.

Insights from animal behavior studies sometimes help researchers understand human behavioral disorders. For example, evidence of oxytocin's role in animal bonding suggests that impaired oxytocin production or reception may contribute to autism. A person with this disorder has trouble making social and emotional attachments. Scientists are collecting data about oxytocin receptor genes of autistic children and their unaffected siblings. The aim of the genetic study is to determine whether inheriting particular alleles of the oxytocin receptor gene raise the likelihood of autism. Oxytocin is also being tested as a treatment for autism. Results of initials study of an oxytocin nasal spray are encouraging. Use of the nasal spray increased activity in brain regions of that govern empathy and reward.

TAKE-HOME MESSAGE 43.2

How do genes influence behavior?

✔ Behavioral differences within a species can arise from allele differences.

✔ Differences between closely related species can also have a genetic basis.

✔ Most human behaviors are complex, polygenic traits. Studies of genetic differences in humans and in animals shed light on human predispositions to particular behaviors.

43.3 Instinct and Learning

✔ Some behaviors are inborn, which means they can be performed without any prior experience.

✔ Most behaviors can be modified as a result of experience.

Instinctive Behavior

All animals are born with the capacity for **instinctive behavior**—an innate response to a specific and usually simple stimulus. For example, a newborn coastal garter snake behaves instinctively when it eats a banana slug in response to the slug's smell.

The life cycle of the cuckoo bird provides several examples of instinct at work. This European bird is a "brood parasite," meaning it lays its eggs in the nest of other birds. A newly hatched cuckoo is blind, but contact with an object beside it stimulates an instinctive response. The hatchling maneuvers the object, which is usually one of its foster parents' eggs, onto its back, then shoves the object out of the nest (**FIGURE 43.5A**). The cuckoo will also shove its foster parents' chicks over the side, if they happen to hatch before it does.

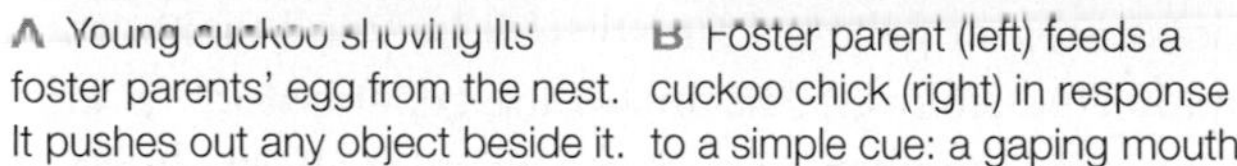

A Young cuckoo shoving its foster parents' egg from the nest. It pushes out any object beside it.

B Foster parent (left) feeds a cuckoo chick (right) in response to a simple cue: a gaping mouth.

FIGURE 43.5 Instinctive behavior.

FIGURE 43.6 Nobel laureate Konrad Lorenz with geese that imprinted on him. The inset photograph shows results of a more typical imprinting episode.

Instinctively shoving objects out of the nest ensures the cuckoo has its foster parents' undivided attention.

Instinctive responses are advantageous only if the triggering stimulus almost always signals the same situation. Doing away with an egg benefits a cuckoo chick because the egg always houses a future competitor for food. However, instinctive responses can open the way to exploitation. Cuckoos often look nothing like the chicks they displace, yet their foster parents feed them all the same (**FIGURE 43.5B**). The cuckoo chick exploits its foster parents' instinctive urge to fill any gaping mouth in their nest with food.

Time-Sensitive Learning

Learned behavior is behavior that is altered by experience. Some instinctive behavior can be modified with learning. A garter snake's initial strikes at prey are instinctive, but the snake learns to avoid dangerous or unpalatable prey. Learning may occur throughout an animal's life, or it may be restricted to a critical period.

Imprinting is a form of instinctive learning that can occur only during a short, genetically determined interval in an animal's life. For example, baby geese learn to follow the large object that bends over them in response to their first peep (**FIGURE 43.6**). With rare exceptions, this object is their mother. When mature, the geese will seek out a sexual partner that is similar to the imprinted object.

A genetic capacity to learn, combined with actual experiences in the environment, shapes most forms of behavior. For example, a male sparrow has an inborn capacity to recognize his species' song when he hears older males singing it. The young male uses these overheard songs as a guide to fill in details of his own song. Males reared alone sing an abnormal, simplified version of their species' song. So do males exposed only to the songs of other species.

The sparrow can only learn his species-specific song during a limited period early in life. To learn to sing normally, he must hear a male "tutor" of his own species during his first 50 or so days of life. Hearing a same-species tutor later will not improve his singing.

Most birds must also practice their song to perfect it. In one experiment, researchers temporarily paralyzed throat muscles of zebra finches who were beginning to sing. After being temporarily unable to practice, these birds never mastered their song. In contrast, temporary throat muscle paralysis in very young birds or adults did not impair later song production. Thus, in zebra finches, there is a critical period for song practice, as well as for song learning.

Conditioned Responses

Nearly all animals are lifelong learners. Most learn to associate certain stimuli with rewards and others with negative consequences.

With **classical conditioning**, an animal's involuntary response to a stimulus becomes associated with another stimulus that is presented at the same time. In the most famous example, Ivan Pavlov rang a bell whenever he fed a dog. Eventually, the dog's reflexive response to food—increased salivation—was elicited by the sound of the bell alone.

With **operant conditioning**, an animal modifies its voluntary behavior in response to consequences of the behavior. This type of learning was first described for behaviors acquired in a laboratory setting. For example, a rat that presses a lever in a laboratory cage and is rewarded with a food pellet becomes more likely to press the lever again. A rat that receives a shock when it enters a particular area of a cage will quickly learn to avoid that area. Similarly, an animal in the wild will learn to repeat behaviors that provide food and to avoid those that cause it discomfort.

Other Types of Learned Behavior

With **habituation**, an animal learns by experience not to respond to a repeated stimulus that has neither positive nor negative effects. For example, pigeons in cities learn not to flee from people who walk past them.

Many animals learn about the landmarks in their environment and form a sort of mental map. This map may be put to use when the animal needs to return home. For example, a fiddler crab foraging up to 10 meters (30 feet) away from its burrow is able to scurry straight home when it perceives a threat.

Animals also learn the details of their social landscape, meaning they learn to recognize mates, offspring, or competitors by appearance, calls, odor, or some combination of cues. For example, two male lobsters usually fight when they meet for the first time (**FIGURE 43.7**). After the fight, they will recognize one another by scent and behave accordingly, with the loser actively avoiding the winner.

FIGURE 43.7 Social learning. Two male lobsters battle at their first meeting. Later, the loser will remember the odor of the winner and avoid him.

FIGURE 43.8 Observational learning. A marmoset opens a container using its teeth. After watching one individual perform this maneuver, other marmosets used the same technique.

With **observational learning**, an animal imitates the behavior of another individual. For example, marmoset monkeys who watched another marmoset open a plastic container and retrieve a treat hidden later mimicked the movements of the animal they observed. Marmosets who watched the demonstrator open the container with its hands imitated this behavior, using their hands in the same way. Those who watched a demonstrator open the box with its teeth attempted to do the same (**FIGURE 43.8**).

TAKE-HOME MESSAGE 43.3

How do instinct and learning shape behavior?

- ✔ Instinctive behavior can initially be performed without any prior experience and it is often elicited by a simple stimulus. Even instinctive behavior may be modified by experience.
- ✔ Certain types of learning can occur only at particular times during an individual's life.
- ✔ Learning alters both voluntary and involuntary behaviors.

---

classical conditioning An animal's involuntary response to a stimulus becomes associated with another stimulus that is presented at the same time.
habituation Learning not to respond to a repeated stimulus.
imprinting Learning that can occur only during a specific interval in an animal's life.
instinctive behavior An innate response to a simple stimulus.
learned behavior Behavior that is modified by experience.
observational learning One animal acquires a new behavior by observing and imitating behavior of another.
operant conditioning A type of learning in which an animal's voluntary behavior is modified by the consequences of that behavior.

43.4 Environmental Effects on Behavioral Traits

✔ Environment influences behavioral phenotypes.

Behavioral Plasticity

Phenotypic plasticity refers to the ability of an individual with a specific genotype to express different phenotypes in different environments. Many animals show **behavioral plasticity**: Their behavioral traits are altered by environmental factors. An ability to learn is one mechanism of behavioral plasticity, but beneficial behavioral changes also arise in other ways.

For example, parasitism induces a change in the feeding preferences of moth larvae commonly called woolly bear caterpillars (**FIGURE 43.9**). These caterpillars are leaf eaters. When healthy, they avoid leaves that have a high concentration of bitter, nonnutritive chemicals called alkaloids. However, caterpillars parasitized by fly larvae prefer to eat alkaloid-rich foods.

Switching off a dietary aversion to alkaloids in the presence of parasites is an adaptive behavior. Alkaloids are toxic to the caterpillars, but they are even more toxic to the caterpillars' parasites. Thus, alkaloid-rich leaves function like a medication for parasitized caterpillars, although one with negative side effects. Because ingested alkaloids harm a caterpillar's parasites more than the caterpillar, a diet high in alkaloids enhances a parasitized caterpillar's likelihood of surviving long enough to reproduce.

As another example, the social environment affects male behavior in some species of cichlid fishes. During the course of its life, a male cichlid can switch back and forth between two social roles. When in a dominant role, a male cichlid courts and breeds with females, and it confronts other males. Dominance is indicated by bright coloration and a dark eye bar (**FIGURE 43.10**).

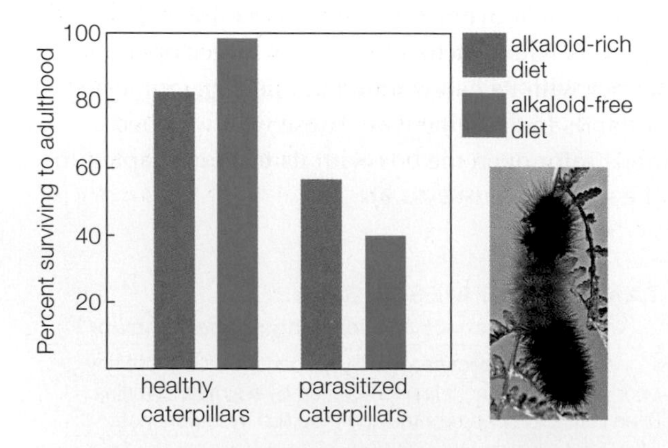

FIGURE 43.9 Effect of dietary alkaloids on parasitized and healthy woolly bear caterpillars (inset photo). In healthy caterpillars, eating alkaloids decreases survival. In parasitized caterpillars, the alkaloid enhances survival by harming parasites.

FIGURE 43.10 Dominant male cichlids with distinctive eye bars face off. Subordinate males flee from such confrontations.

When in a subordinate role, a cichlid flees dominant males, does not court or breed, and has a drab coloration. Whether a male is dominant or subordinate depends on a variety of factors, most importantly whether other dominant males are present. Remove all dominant males, and a subordinate male's brain turns up expression of a regulatory gene. The resulting cascade of changes in gene expression transforms the male's behavioral and physical phenotype from subordinate to dominant within 20 minutes.

Epigenetic Effects

Epigenetic mechanisms cause heritable changes in behavior without altering DNA sequence (Section 10.6). For example, a female rat who received little licking and grooming from her mother early in life has a low tendency to lick and groom her own pups, and her daughters show the same behavior. Researchers have found the lack of tactile stimulation early in a rat's life triggers a change in DNA methylation patterns. Among other effects, the resulting changes in gene expression reduce the number of oxytocin receptors in the brain. The methylated DNA can be passed on for generations, dampening the effects of oxytocin and negatively affecting maternal behavior.

behavioral plasticity The behavioral phenotype associated with a genotype depends on conditions in the animal's environment.

TAKE-HOME MESSAGE 43.4

How does the environment affect behavior?

✔ With behavioral plasticity, environmental factors influence an individual's behavior during its lifetime.

✔ The environment can also cause heritable changes in behavior through epigenetic modifications of DNA.

43.5 Movements and Navigation

✔ All animals move, and some can navigate over extraordinarily long distances.

Taxis and Kinesis

All animals are motile (can move from place to place) during some part of their life cycle. Even simple animals such as planarian flatworms instinctively move toward some stimuli and away from others. An innate directional response to a stimulus is called a **taxis** (plural, taxes). For example, planarians are negatively phototactic (they move away from light) and positively geotactic (they move toward the pull of gravity). A stimulus may also cause an animal to increase or decrease its movements without regard for direction, a response called **kinesis**. For example, planarians are photokinetic, meaning they move more in light than in the dark. The planarian's innate responses to stimuli direct the worm toward a favorable habitat and help it to remain there.

Migration

Most animals move about daily to find food and avoid predators. Some also migrate. During a **migration**, an animal interrupts its daily pattern of activity to travel in a persistent manner toward a new habitat.

Many animal migrations involve seasonal movement to and from a breeding site. For example, birds that nest in the Arctic during the summer typically migrate to a temperate or tropical region to spend the winter. Over the course of a lifetime, a migratory bird may make these trips many times, traveling each year to breed in the region where it hatched. Marine turtles and some whales also make repeated long-distance journeys to breed at the site where their own life began.

For other animals, the migration to a breeding ground is a one-way trip. Consider the Atlantic eel, which spends most of its life in a European or North American river. Then, one fall, it migrates hundreds to thousands of kilometers to the Sargasso Sea, the region of the Atlantic where it hatched 10 to 30 years before. Here, it meets other eels, breeds, and dies. Currents distribute eel larvae throughout the North Atlantic. After the larvae develop into young eels, they swim to a coast and then make their way up a river.

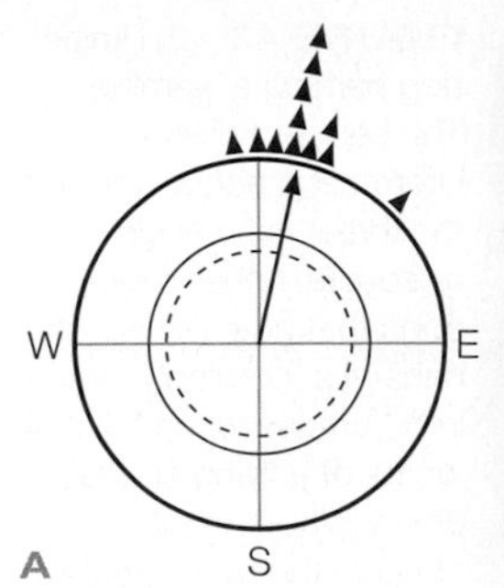

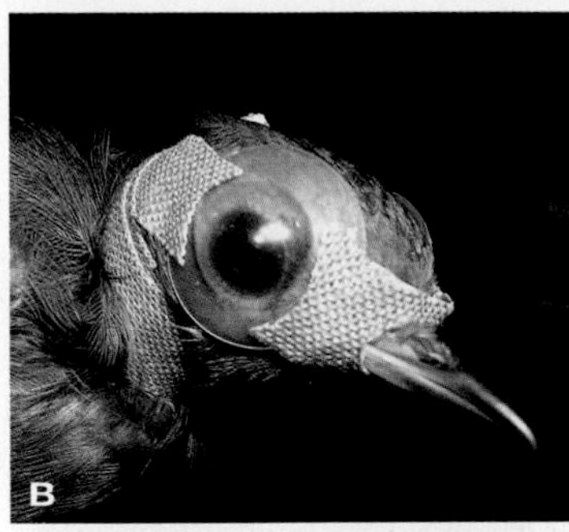

FIGURE 43.11 Vision-based magnetic compass in European robins. (A) Triangles indicate predominant direction of movements made by 12 robins tested indoors in the spring. Arrow indicates average result. (B) A robin with a frosted lens in front of its right eye. Experiments using such lenses have shown that European robins can see magnetic fields.

Researchers have only begun to decipher how migrating animals find their way. Some animals have an innate magnetic compass, meaning they use latitudinal variations in Earth's magnetic field to determine direction. In the spring, a caged European robin will hop repeatedly toward the north, even if it has no view of the outdoors (FIGURE 43.11). Experiments have shown that the bird actually "sees" directional differences in Earth's magnetic field with its eyes.

Animals can also navigate by the sun and stars. Because the position of celestial bodies shifts over time, use of a sun- or star-based compass requires an innate sense of time. A bird that wants to head north must fly to the left of the sun in the morning, and to the right of the sun in the evening.

To find a specific site, an animal must know not only compass directions, but the location of its goal relative to its current location. It does no good to know where north is if you do not also know whether you are north or south of your destination. Localized variations in the Earth's magnetic field provide this information to some species. An animal may also gauge its progress relative to visual or olfactory landmarks. For example, odor cues help guide eels as they travel from the mouth of the river where they spent their adulthood toward the open ocean.

TAKE-HOME MESSAGE 43.5

What factors influence animal movements?

✔ Innate responses to specific stimuli can influence the rate or direction of animal movements.

✔ Some animals migrate between habitats. Navigating to a specific site requires an ability to determine compass direction and a mental map.

kinesis Innate response in which an animal speeds up or slows its movement in reaction to a stimulus.
migration An animal ceases an ongoing pattern of daily activity to move in a persistent manner toward a new habitat.
taxis Innate response in which an animal moves toward or away from a stimulus.

43.6 Communication Signals

FIGURE 43.12 Prairie dog barking a warning. The bark provides information about whether the threat is an eagle or coyote. Other prairie dogs that hear the alarm behave accordingly, diving into burrows when the call warns of a flying predator, or standing erect to observe the movements of a ground predator such as a coyote.

FIGURE 43.13 Courtship display in albatrosses. The display involves a series of movements, some accompanied by calls.

FIGURE 43.14 Threat display of a male collared lizard. This display allows two rival males to assess one another's strengths without engaging in a potentially damaging fight. The display also is called into service to deter potential predators.

✔ A variety of evolved signals facilitate interactions among animals of the same species.

Evolution of Animal Communication

Communication signals are evolved cues that transmit information from one member of a species to another. Use of a communication signal becomes established and persists only if signaling benefits both the signal sender and the signal receiver. If signaling is disadvantageous for either party, then natural selection will tend to favor individuals that do not send or respond to it.

Types of Signals

Chemical signals called **pheromones** convey information among members of a species. There are two categories of pheromone. Signal pheromones cause a rapid shift in the receiver's behavior, whereas priming pheromones cause longer-term responses. Sex attractants that help males and females of many species find each other are signal pheromones. The alarm pheromone a honeybee emits when she perceives a threat to her hive is another. Alarm pheromone causes other honeybees to rush out of the hive and attack a potential intruder. A chemical in the urine of male mice is a priming pheromone; it triggers ovulation in females.

Producing a pheromone requires less energy than calling or gesturing, but it also conveys less information. The chemical is either released or not. By contrast, properties of acoustical, visual, and tactile signals vary continuously, and so convey more information.

Vocal signals help many male vertebrates, including songbirds, whales, frogs, and some fishes, attract prospective mates. Similarly, male crickets attract females by rubbing their legs together to chirp. Male cicadas attract females by using organs on their abdomen to make clicking sounds that can exceed 90 decibels.

Some birds and mammals give alarm calls that inform others of potential threats (**FIGURE 43.12**). In many cases, the calls convey more than simply "Danger!" A prairie dog emits one type of bark when it detects an eagle and another when it sees a coyote.

Evolution shapes the properties of acoustical signals depending on whether or not the sender benefits by revealing its position. Sounds that lure mates are typically easily localized, whereas alarm calls are not.

Visual communication is most widespread in animals that have good eyesight and are active during the day. Bird courtship often involves both visual and acoustical signals. For example, courting albatrosses strike coordinated poses while making distinctive calls (**FIGURE 43.13**). The display assures a prospective

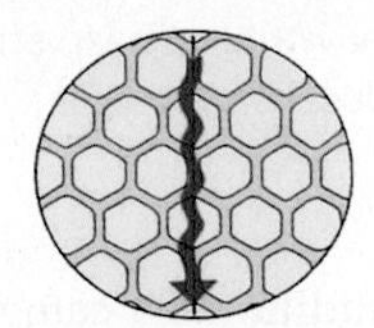

When bee moves straight up comb, recruits fly straight toward the sun.

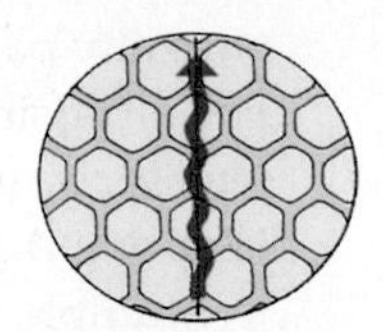

When bee moves straight down comb, recruits fly to source directly away from the sun.

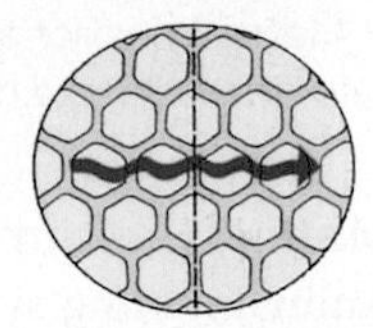

When bee moves to right of vertical, recruits fly at 90° angle to right of the sun.

FIGURE 43.15 ▶**Animated** Honeybee waggle dance, a tactile display. The orientation of the "waggle run" conveys information about the direction of a food source.

mate that the displayer is a member of the correct species and is in good health.

Threat displays advertise the displayer's good health too, but they serve a different purpose (**FIGURE 43.14**). Males of many species compete for access to females. When two potential rivals meet, they often engage in a display that demonstrates their strength and how well armed they are. Most often, the males are not evenly matched, and the weaker individual retreats. Both males benefit by avoiding a fight that could lead to injury.

With tactile displays, information is transmitted by touch. For example, after discovering food, a foraging honeybee worker returns to the hive and dances in the dark, surrounded by a crowd of other workers. If the food is more than 100 meters from the hive, she performs a waggle dance on a vertically orientated honeycomb, moving in a figure eight (**FIGURE 43.15**). During the waggle run portion of this dance, the dancer's orientation provides information about the direction of the food relative to the direction of the sun. For example, if the straight run of her dance proceeds straight up the honeycomb, food is in the same direction as the sun. How fast the dancer moves informs other bees about the distance to the food: the faster her movements, the closer the food.

The same signal may function in more than one context. Dogs and wolves solicit play behavior with a play bow (**FIGURE 43.16**). A play bow informs an animal's prospective playmate that behavior, such as growling, that is normally used in an aggressive context should intead be interpreted as in the context of play behavior.

Eavesdroppers and Counterfeiters

Predators can benefit by intercepting signals sent by their prey. For example, frog-eating bats locate male tungara frogs by listening for their mating calls. This poses a dilemma for the male frogs. Female tungara frogs prefer complex calls, but complex calls are easier for bats to localize. Thus, when bats are near, male frogs call less, and with less flair. The subdued signal is a trade-off between competing pressures to attract a mate and to avoid being eaten.

FIGURE 43.16 A wolf's play bow tells another wolf that its next behavior is meant as play, not aggression.

Other predators lure prey with counterfeit signals. Fireflies are nocturnal beetles that attract mates with flashes of light. When a predatory female firefly sees the flash from a male of the prey species, she flashes back as if she were a female of his own species. If she lures him close enough, she captures and eats him.

communication signal Chemical, acoustical, visual, or tactile cue that is produced by one member of a species and detected and responded to by other members of the same species.
pheromone Chemical that serves as a communication signal between members of an animal species.

TAKE-HOME MESSAGE 43.6

What are the benefits and costs of communication signals?

✔ A communication signal transfers information from one individual to another individual of the same species. Such signals benefit both the signaler and the receiver.

✔ Signals have a potential cost. Some individuals of a different species benefit by intercepting signals or by mimicking them.

43.7 Mates, Offspring, and Reproductive Success

✔ Mating behavior and parental behavior have been shaped by natural selection to maximize reproductive success.

Mating Systems

Animal mating systems were traditionally categorized as promiscuous, polygamous, or monogamous. In recent years, studies that integrate paternity analysis with behavioral observations have shown that species do not always fit cleanly into these categories. Members of a species often vary in their behavior.

Some animals such as prairie voles form a pair bond, which means two individuals mate, preferentially spend time together, and cooperate in rearing offspring. However, participants in a pair bond may also seize opportunities to mate with other individuals. A paternity study of the prairie voles discussed in Section 43.2 found that 7 percent of pair-bonded females produced litters of mixed male parentage; and 15 percent of pair-bonded males sired offspring with females other than their partner. Thus, prairie voles are now described as socially monogamous, but genetically promiscuous. As a species, they form an exclusive social relationship with a single partner. However, the genetic composition of their offspring reflects an occasional tendency to mate with multiple partners.

Even social monogamy is rare in most animals, with the exception of birds. About 90 percent of birds are socially monogamous. Among this group, the vast majority of species that have been examined for paternity patterns are genetically promiscuous. This is not surprising as promiscuity benefits both males and females by increasing the genetic diversity among the offspring they produce.

Multiple matings can also provide another benefit: more offspring. This benefit usually applies most strongly to males. Sperm are energetically inexpensive to produce, so a male's reproductive success is usually limited by access to mates. By contrast, the main limit on a female's reproductive success is her capacity to produce large, yolk-rich eggs or, in mammals, to carry developing young.

When one sex exerts selection pressure on the other, we expect sexual selection to occur (Section 17.7). In many species, females choose among males on the basis of their ability to provide necessary resources. For example, a female hangingfly will only mate with a male while feeding on prey that he has provided as a "nuptial gift" (**FIGURE 43.17A**).

In other species, a male establishes a mating territory, an area that includes resources that females need for reproduction. A **territory** is any area from which an animal or group of animals actively exclude others. A male fiddler crab's territory is a stretch of shoreline in which he excavates a burrow. A male has one oversized claw (**FIGURE 43.17B**) and during mating season, he stands outside his burrow waving it. The claw is used both in the courtship display and in territorial disputes with other males. Females attracted by a male's display

A Valuable gifts. This male hangingfly is dangling a moth as a nuptial gift for a potential mate.

B Showing off a large body part. Male fiddler crabs (top) wave their one enlarged claw to attract a female (bottom).

C Dancing and singing. Male sage grouse gather on a communal display ground called a lek, where they dance, puff out their neck, and make booming calls.

FIGURE 43.17 How to impress a choosy female.

check out the location and size of his burrow before mating with him. Burrow location and size are important because they affect development of the larvae.

In mammals such as elephant seals and elk, the strongest males hold large territories, and mate with all females inside the area. Other males may never get to mate at all unless they sneak into an area under another male's control.

In some birds such as sage grouse, males converge at a communal display ground called a **lek**. Males stamp their feet and emit booming calls by puffing and deflating their large neck pouches for female onlookers (**FIGURE 43.17C**). The most popular males mate with many females.

Parental Care

Parental care requires time and energy that an individual could otherwise invest in reproducing again. It arises only if the genetic benefit of providing care (increased offspring survival now) offsets the cost (reduced opportunity to produce other offspring). If a single parent can successfully rear offspring, individuals who spare themselves this cost by leaving parental duties to their mate will be at a selective advantage.

In about 90 percent of mammals, the female rears young (**FIGURE 43.18A**). In the remaining 10 percent, both sexes participate. No mammal relies solely on male parental care. Female mammals sustain developing young in their body, so they have a greater investment in newborns than males. Female mammals also nurse their young; males typically do not. Two fruit bat species are the exception: Both males and females of these species nurse offspring.

Most fishes provide no parental care, but in those that do, this duty usually falls to the male. Males guard eggs until they hatch (**FIGURE 43.18B**) or, in the case of sea horses, protect them inside their body. The prevalence of male parental care in fishes may be related to sex differences in how fertility changes with age. A female fish's fertility increases dramatically with age, whereas a male's does not. Thus, a female fish who invests resources in care forgoes more future reproduction than a male fish does.

In birds, two-parent care is most common (**FIGURE 43.18C**). Chicks of birds that cooperate in care of their young tend to hatch while in a relatively helpless state. Chicks of sage grouse and other birds in which females alone provide care hatch when more fully developed.

lek Area where male animals perform courtship displays.
territory Region that an animal or group of animals defends.

A Female grizzly bears care for their cub for up to two years.

B A male clownfish guards eggs (foreground) until they hatch.

C A male and a female tern cooperatively caring for their chick.

FIGURE 43.18 Who cares for the young?

TAKE-HOME MESSAGE 43.7

What factors affect mating and parental care?

- ✔ The positive effects of genetic diversity among offspring make monogamy rare.
- ✔ In many species, a few males monopolize mating opportunities either by enticing females to choose them or by defending a territory with many females.
- ✔ Whether parental care occurs and who delivers it depends on the benefits and costs of care to each parent.

CREDITS: (18A) © John Conrad/Corbis; (18B) © Steven Kovacs/SeaPics.com; (18C) © Steve Kaufman/Corbis.

A Selfish herd. Sardines attempt to hide behind one another as predatory sailfish attack.

B When threatened, musk oxen adults (*Ovibos moschatus*) form a ring of horns, often around their young.

FIGURE 43.19 Grouping for defense.

✔ Animals of many species group together during part or all of their lifetime. Such grouping has both benefits and costs.

Benefits of Grouping

Many animals benefit by forming short-term, unstable groups. One benefit is safety in numbers. Whenever animals cluster together, individuals at the margins of a group inadvertently shield others from predators. A **selfish herd** is a temporary aggregation that arises when individual animals move near others to minimize their own risk of predation. Small fish exhibit a selfish herd behavior when under attack by a predator (FIGURE 43.19A). Flocks of birds and herds of ungulates such as zebras show similar behavior.

In a group, multiple individuals can be on the alert for predators. In some cases, an animal that spots a threat will warn others of its approach. Birds, monkeys, meerkats, and prairie dogs are among the animals that make alarm calls in response to a predator. Even in species that do not give alarm calls, individuals benefit from the vigilance of others. Often, when an animal notices another group member beginning to flee, it will do likewise. In what is called the "confusion effect," a predator has a more difficult time choosing one individual to chase among a group of scattering prey.

Not all prey groups flee from predators. Some act together to present a united defense. For example, when threatened by wolves, musk oxen stand back to back, presenting an imposing display of sharp horns (FIGURE 43.19B). Sawfly caterpillars also participate in a group defense against predatory birds. When a group of these caterpillars is approached by a bird, each caterpillar rears up and expels some partly digested eucalyptus leaves. Birds presented with a cluster of caterpillars exuding this slimy, stinky material eat fewer caterpillars than birds offered caterpillars one at a time.

Some endothermic animals, including bats, penguins, and rodents derive an energetic benefit by aggregating when it is cold. By huddling together in a group, each individual reduces the amount of surface area that it exposes to the cold environment. Thus, each animal in the group loses heat more slowly than it would on its own.

Social animals form more permanent multigenerational groups, in which members benefit by cooperating in some tasks. Most often members of a social group are relatives.

Wolves and lions are social animals who cooperate in catching prey. Cooperative hunting allows a group of predators to capture larger or faster prey. Even so, cooperative hunting is often no more efficient than hunting alone. In one study, researchers observed that a solitary lion catches prey about 15 percent of the time. Two lions hunting together catch prey twice as often as a solitary lion, but having to share the spoils of the hunt means the amount of food per lion is the same. When more lions join a hunt, the success rate per lion falls. Wolves have a similar pattern. Among carnivores that hunt cooperatively, hunting success is usually not the major advantage of group living. Individuals hunt together, but also cooperate in fending off scavengers (FIGURE 43.20), caring for one another's young, and defending their territory.

Another benefit of living in a social group is opportunities for imitative learning. Consider how chimpan-

FIGURE 43.20 Cooperating to attain food. Lionesses enjoy the results of a communal hunt, while spotted hyenas work together to try to steal a share.

FIGURE 43.21 A learning opportunity. Chimpanzees (*Pan troglodytes*) using sticks as tools for extracting tasty termites from a nest. This behavior is learned by imitation.

zees learn to make tools. The chimpanzees strip leaves from branches to produce "fishing sticks" that they use to capture termites as food. When a chimpanzee inserts a fishing stick into a termite mound, termites cling to the stick in defense of their nest. The stick is then withdrawn and the termites eaten. Different groups of chimpanzees use different methods of tool-shaping and insect-fishing. In each group, young individuals learn by imitating behavior of others (**FIGURE 43.21**).

Costs of Grouping

In order for a grouping behavior to evolve and persist, the benefits of living in a group must offset the costs. The most obvious cost of group living is increased competition. Individuals that live in groups continually compete with one another for resources, ranging from food to mates. For example, penguins and many other seabirds form dense breeding colonies in which competition for space to nest, nesting material, and food is intense (**FIGURE 43.22**). Given the opportunity, a pair of breeding herring gulls will eat eggs and even the chicks of their neighbors.

Group living can also increase the risk of predation and parasitism. Large groups are more conspicuous to predators than lone individuals, and parasites spread more easily when many hosts are in close proximity.

selfish herd Group that forms when individuals hide behind others to minimize their individual risk of predation.
social animal Animal that lives in a multigenerational group in which members, who are usually relatives, cooperate in some tasks.

TAKE-HOME MESSAGE 43.8

What are the benefits and costs of living in a social group?

✔ Living in a social group can provide benefits, as through improved defenses, shared care of offspring, and greater access to food.

✔ Costs of group living include increased competition and increased vulnerability to infections.

FIGURE 43.22 A colony of nesting penguins. Parasites spread easily through such colonies and the colony's large size makes it easy for predators to locate it.

CREDITS: (20) Gunter Ziesler/Getty Images; (21) © Steve Bloom/stevebloom.com; (22) © A. E. Zuckerman/Tom Stack & Associates.

43.9 Why Sacrifice Yourself?

✔ Extreme cases of sterility and self-sacrifice have evolved in only a few groups of insects and one group of mammals. How are genes of the nonreproducers passed on?

A **eusocial animal** lives in a multigenerational family group with a reproductive division of labor. Permanently sterile workers care cooperatively for the offspring of just a few breeding individuals.

Eusocial Insects

Eusocial insects include some bees, and all ants and termites. In some species, the sterile workers are highly specialized in form and function (**FIGURE 43.23**).

A queen honeybee is the only fertile female in her hive (**FIGURE 43.24A**). She spends her days laying eggs and secreting a pheromone that renders other females in the hive sterile. The queen's sterile female daughters serve as workers; they feed larvae, clean and maintain the hive, and construct honeycomb from waxy secretions. Workers also gather the nectar and pollen that feed the colony. They guard the hive and will sacrifice themselves to repel an intruder.

New queens and males (drones) are produced seasonally. Drones are stingless and subsist on food gathered by their worker sisters. Each day, drones fly out in search of a mate. The occasional lucky one will meet a virgin queen on her one mating flight. The queen mates with many males during her flight, then uses their stored sperm for years. A drone dies after mating.

Like honeybees, termites live in enormous family groups with a queen who specializes in egg production (**FIGURE 43.24B**). Unlike the honeybee hive, a termite mound holds sterile males and females. A king supplies the female with sperm. Winged reproductive termites of both sexes develop seasonally.

Eusocial Mole Rats

Sterility and extreme self-sacrifice are uncommon in vertebrates. The only eusocial mammals known are two species of African mole rat. The best studied is *Heterocephalus glaber*, the naked mole rat. Clans of this nearly hairless rodent live in burrows in dry parts of East Africa. A reproducing female dominates the clan and mates with one to three males (**FIGURE 43.24C**). Nonbreeding members of the clan protect and care for the "queen" and "king" (or kings) and their offspring. Sterile diggers excavate tunnels and chambers. When a digger finds an edible root, it hauls a bit back to the main chamber and chirps. Its chirps recruit others, which help carry food back to the chamber. In this way, the queen, her mates, and her young offspring get fed. Other sterile helpers guard the colony. When a predator appears, they chase and attack it at great risk to themselves.

Evolution of Altruism

A sterile worker in a social insect colony or a naked mole-rat clan engages in **altruistic behavior**—it enhances another individual's reproductive success at its own expense. How did this behavior evolve? According to William Hamilton's **theory of inclusive fitness**, genes for altruism can be favored by natural selection when the altruistic behavior they promote increases the reproductive success of the altruist's close relatives.

It is easy to see the genetic advantage of caring for one's own offspring. They have copies of some of your genes. However, recall that in a sexually reproducing diploid species, each offspring typically inherits half its genes from its mother and half from its father. Thus each individual shares only 50 percent of its genes with each of its parents. It also shares 50 percent of its

A An Australian honeypot ant worker. This sterile female is a living storage container.

B Army ant soldier (*Eciton*), also a sterile female, shows her formidable mandibles.

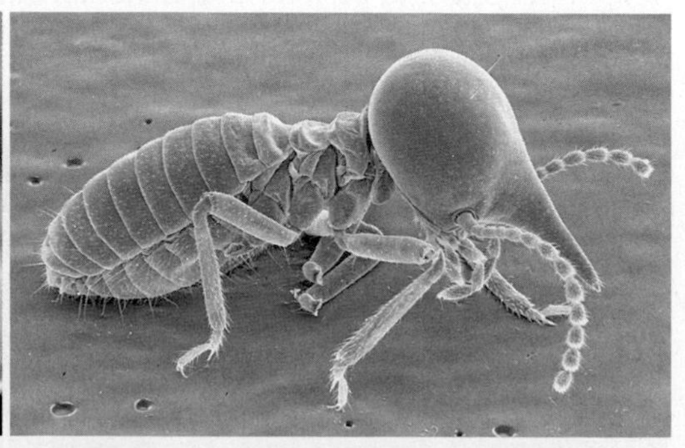

C Eyeless soldier termite (*Nasutitermes*). It shoots gluelike secretions from its nozzle-shaped head. Termite soldiers include both males and females.

FIGURE 43.23 Specialized ways of serving and defending the colony.

Can You Hear Me Now? (revisited)

The noise produced by human activities affects the behavior of many animals that communicate by sound. Like right whales, blue whales that feed off the coast of Southern California "yell" to be heard above shipping noise. However, these whales have an additional problem—sonar signals that miliary vessels use to locate submarines. The sonar signals inhibit blue whale calling, perhaps because they are similar in frequency to the calls of killer whales that prey on blue whales.

Land-dwelling animals alter their calls in response to noise too. Birds that live in cities produce louder, higher-pitched songs than their counterparts in rural environments and may begin singing earlier in the day, when it is less noisy. Frogs and grasshoppers living near heavily used roads also sing more loudly and at a higher pitch than members of the same species that live in quieter environments.

Even when animals are capable of altering their calls, human-induced noise can still drown out their signals. Thus, it is not surprising that some animals appear to suffer a decline in fitness as a result of the noise in their environment. For example, house sparrows who nest in noisy areas produce fewer young than those who nest in a tranquil setting.

---

genes with any of its siblings (brothers and sisters). By sacrificing its own reproductive success to help rear siblings, a sterile worker promotes copies of its "self-sacrifice" genes in these close relatives.

In honeybee and ant colonies, sterile workers assist fertile relatives with whom they share genes. These insects may have an added incentive to make sacrifices. They belong to an order, Hymenoptera, that has a haplodiploid sex determination system: Diploid females develop from fertilized eggs and haploid males from unfertilized ones. As a result, female hymenopterans share 75 percent of their genes with their sisters. By one hypothesis, the unusually high relatedness among sisters may help explain why eusociality has arisen many times in this order. However, high relatedness cannot be the only factor favoring hymenopteran eusocial behavior; most bees and wasps are solitary despite having this type of sex determination.

Inbreeding, which increases genetic similarity among relatives, may have played a role in the evolution of naked mole rat eusociality. However, ecological factors probably also play a role. According to one hypothesis, the mole rat's arid habitat and patchy food supply favors a genetic predisposition to cooperate in digging burrows, searching for food, and fending off competitors. Individuals who strike out on their own are unlikely to have high reproductive success.

A Queen honeybee with her sterile daughters.

B A queen termite dwarfs her offspring and mate. Ovaries fill her enormous abdomen.

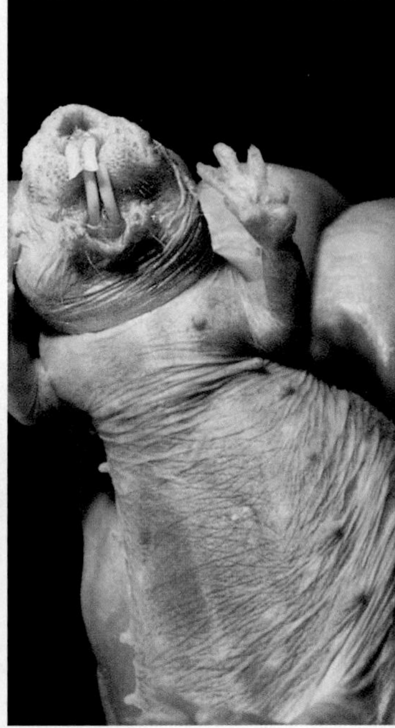

C Naked mole-rat queen.

FIGURE 43.24 Three queens, fertile females in species that have a reproductive division of labor.

altruistic behavior Behavior that benefits others at the expense of the individual.
eusocial animal Animal that lives in a multigenerational family group with a reproductive division of labor.
theory of inclusive fitness Genes associated with altruism can be advantageous if the expense of this behavior to the altruist is outweighed by the reproductive success of relatives.

TAKE-HOME MESSAGE 43.9

How can altruistic behavior be selectively advantageous?

✔ Altruistic behavior may be favored when individuals pass on genes indirectly, by helping relatives survive and reproduce.

✔ Mechanisms that increase relatedness among siblings may encourage the evolution of eusocial behavior.

summary

Section 43.1 Some animals alter their vocal communication in response to noise generated by human activities. Exposure to such noise can stress animals and in some cases lower their reproductive output.

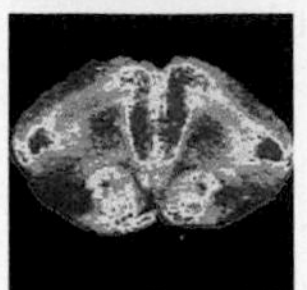

Section 43.2 Behavior starts with genes that influence the development and activity of the nervous and endocrine systems. Studies of behavioral differences within a species or among closely related species can shed light on the proximate and ultimate causes of a behavior.

Section 43.3 **Instinctive behavior** occurs without a prior experience. It is often triggered as a response to a simple stimulus. **Learned behavior** arises in response to experience. **Imprinting** is time-sensitive learning. With **habituation**, an animal learns to disregard certain repeated stimuli. **Classical conditioning** forms an association between two stimuli, and **operant conditioning** forms one between a behavior and its outcome. **Observational learning** involves seeing and imitating a behavior.

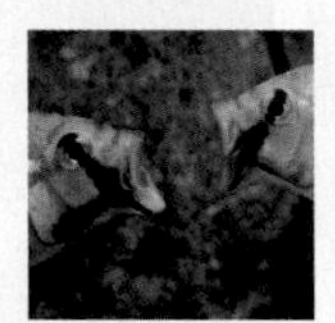

Section 43.4 Behavioral traits can be influenced by an animal's environment. **Behavioral plasticity** allows an animal to respond appropriately to environmental pressures that can vary. Some behaviors that arise in response to a particular environment have an epigenetic effect and are passed on to offspring.

Section 43.5 **Kinesis** (random movement) and **taxis** (directional movement) are instinctual responses to stimuli. During **migration**, an animal interrupts its usual activities to make a journey. Navigation requires the ability to form a mental map and to sense direction.

Section 43.6 **Communication signals** convey information between individuals of the same species. **Pheromones** are chemical communication signals. Signals have been shaped by evolution; for example, courtship calls are easily localized, but alarm calls are not. Predators may take advantage of the communication system of their prey.

Section 43.7 Genetic monogamy, in which a male and female mate only with each other, is extremely rare in animals. Social monogamy is rare in most groups, with the exception of birds. Both males and females benefit by mating with multiple partners, but in general, males are less choosy about mates than females. Depending on the species, females may choose males on the basis of resources they offer, or may assess mate quality by their courtship performances, as in displays at a **lek**. When large numbers of females cluster in a defensible area, males compete with one another to control a mating **territory**.

Parental care has reproductive costs in terms of reduced frequency of reproduction. It is adaptive when benefits to a present set of offspring offset this cost. In mammals, females are usually sole caregivers of their young. When fish provide parental care, the male is usually the caregiver.

Section 43.8 Some animals come together briefly as a **selfish herd** in response to the threat of predation. **Social animals** live in a more permanent, multigenerational group. Grouping allows cooperation in predator detection, defense, and rearing young. Group living also has some costs; it increases the risk of disease and parasitism, and intensifies the competition for resources.

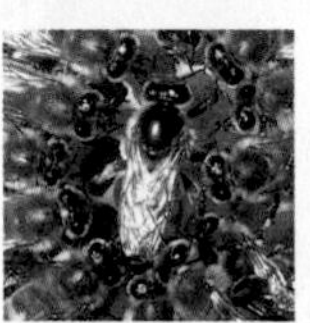

Section 43.9 Ants, termites, and some other insects as well as mole rats are **eusocial animals**: They live in colonies with overlapping generations and have a reproductive division of labor. Most colony members do not reproduce; they assist their relatives instead. The **theory of inclusive fitness** states that **altruistic behavior** can be perpetuated when altruistic individuals help their reproducing relatives. Altruists help perpetuate the genes that lead to their altruism by promoting reproductive success of close relatives who share the same genes.

self-quiz

Answers in Appendix VII

1. Genes affect the behavior of individuals by ________ .
   a. influencing the development of nervous systems
   b. affecting the kinds of hormones in individuals
   c. determining which stimuli can be detected
   d. all of the above

2. Stevan Arnold offered slug meat to newborn garter snakes from different populations to test his hypothesis that the snakes' response to slugs ________ .
   a. was shaped by indirect selection
   b. is an instinctive behavior
   c. is based on pheromones
   d. is an epigenetic effect

3. An animal that navigates by the stars needs ________ .
   a. an ability to sense Earth's magnetic field
   b. pheromone receptors
   c. an internal clock
   d. an acute sense of hearing

4. The honeybee dance language transmits information about distance to food by way of ________ signals.
   a. tactile c. acoustical
   b. chemical d. visual

5. A ________ is a chemical that conveys information between individuals of the same species.
   a. pheromone c. hormone
   b. neurotransmitter d. all of the above

data analysis activities

Alarming Eyespots Section 1.7 described how a peacock butterfly will, when threatened, open its wings to reveal two large eyespots that are hidden when the butterfly is at rest. By one hypothesis, eyespots frighten a predatory bird by mimicking the eyes of the bird's predators. Alternatively, sudden appearance of the spots may act by simply startling the bird. To differentiate between these two possiblities, Martin Olofsson presented peacock butterflies with or without eyespots painted out to domestic chickens and recorded whether the chickens gave an alarm call that is normally given upon sighting a ground predator. **FIGURE 43.25** shows the results.

1. When eyespots were visible, how many birds gave the alarm call? How many did not?
2. How did the number of alarm calls given differ when the butterflies' eyespots were hidden ?
3. Does this data support the hypothesis that butterfly eyespots frighten birds by mimicking their predators?

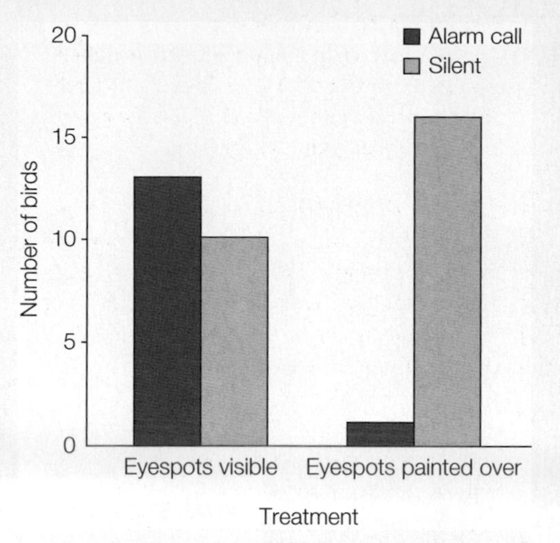

FIGURE 43.25 Response of domestic chickens to the defense display of a peacock butterfly (shown above) with or without eyespots painted over. All chickens were previously unfamiliar with these butterflies.

6. In most _______ , males and females cooperate in care of the young.
   a. mammals
   b. birds
   c. insects
   d. all of the above

7. Generally, living in a social group costs the individual in terms of _______ .
   a. competition for food, other resources
   b. vulnerability to contagious diseases
   c. competition for mates
   d. all of the above

8. Social behavior has evolved because _______ .
   a. social animals are more advanced than solitary ones
   b. under some conditions, the costs of social life to individuals are offset by benefits to the species
   c. under some conditions, the benefits of social life to an individual offset the costs to that individual
   d. under most conditions, social life has no costs to an individual

9. Eusocial insects _______ .
   a. live in extended family groups
   b. include termites, honeybees, and ants
   c. have a reproductive division of labor
   d. all of the above

10. Helping other individuals at a reproductive cost to oneself can be adaptive if those helped are _______ .
   a. members of another species
   b. competitors for mates
   c. close relatives
   d. counterfeit signalers

11. A honeybee worker is a _______ .
   a. fertile female
   b. sterile female
   c. fertile male
   d. sterile male

12. A _______ is a change in the rate of random movement in response to a specific stimulus.
   a. epigenetic trait
   b. taxis
   c. kinesis
   d. migration

13. With _______ , the consequences of a voluntary behavior cause an animal to repeat or avoid that behavior.
   a. classical conditioning
   b. operant conditioning
   c. imprinting
   d. instinct

14. With behavioral plasticity, behavior varies depending on _______ .
   a. the time of day
   b. an individual's age
   c. environmental factors
   d. an individual's genotype

15. Match the terms with their most suitable description.

___ imprinting	a. time-limited learning
___ tactile display	b. requires staying in touch
___ selfish herd	c. communal display ground
___ habituation	d. learning not to respond
___ lek	e. hiding behind others

critical thinking

1. For billions of years, the only bright objects in the night sky were stars or the moon. Night-flying moths used them to navigate in a straight line. Today, the instinct to fly toward bright objects causes moths to exhaust themselves fluttering around streetlights and banging against brightly lit windowpanes. This behavior is not adaptive, so why does it persist?

2. Damaraland mole rats are fur-covered relatives of naked mole rats. In their clans, too, nonbreeding individuals of both sexes cooperatively assist one breeding pair. Even so, breeding individuals in wild Damaraland mole-rat colonies usually are unrelated, and few subordinates move up in the hierarchy to breeding status. Explain why researchers suspect that ecological factors, not genetic ones, were the most important selective force in Damaraland mole-rat altruism.

44 Population Ecology

LEARNING ROADMAP

This chapter considers the structure and growth of populations (Section 1.2). We refer back to sampling error (1.8), directional selection (17.5), gene flow (17.8), evolution of modern humans (26.6), smallpox vaccination (37.12), and animal social behavior (43.8).

VITAL STATISTICS

Populations can be described in terms of their size, density, distribution, and number of individuals in different age categories.

EXPONENTIAL RATES OF GROWTH

A population's size and reproductive base influence its rate of growth. When the population is increasing at a rate proportional to its size, it is undergoing exponential growth.

LIMITS ON INCREASES IN NUMBER

Density-dependent limiting factors result in logistic growth, in which the rate of growth slows as the population nears its carrying capacity.

LIFE HISTORIES

Resource scarcity, disease, and predation can restrict population growth. These limiting factors differ among species and shape their life history patterns.

THE HUMAN POPULATION

Human populations sidestepped limits to growth by way of global expansion into new habitats, cultural interventions, and innovative technologies.

We continue our focus on populations in Chapter 45, with a look at the effect of species interactions on population size. Chapter 48 delves into the way in which expansion of the human population is affecting populations of other species.

44.1 A Honking Mess

Canada geese (*Branta canadensis*) were hunted to near extinction in the late 1800s. In the early 1900s, federal laws and international treaties were put in place to protect them and other migratory birds. In recent decades, the number of geese in the United States has soared. For example, Michigan had about 9,000 of these birds in 1970 and today has more than 300,000.

Canada geese are plant-eating birds, and they often congregate on the grassy lawns of golf courses and parks (**FIGURE 44.1A**). The geese are considered pests because they damage lawns and produce slimy, green feces that soil shoes, stain clothing, and cloud ponds and lakes. Some goose feces also contains bacteria that can sicken people if ingested.

Canada geese pose a more serious problem for air traffic. They are one of the species most commonly involved in collisions with aircraft. Consider what happened to a US Airways flight in 2009: Shortly after the plane took off from New York's LaGuardia airport, both engines failed. Fortunately, the pilot was able to land the plane in the nearby Hudson River, where boats safely unloaded all 155 people aboard (**FIGURE 44.1B**). After the crash, investigators determined that the engine failures occurred because Canada geese were sucked into both engines.

Controlling Canada goose numbers is a challenge, because several different populations spend time in the United States. A **population** is a group of organisms of the same species that live in the same area and interbreed. In the past, nearly all Canada geese seen in the United States were migratory. The geese nested in northern Canada, flew to the United States to spend the winter, then returned to Canada. The common name of the species reflects this tie to Canada.

Most Canada geese still migrate, but some populations have lost this trait. Canada geese breed where they were raised, and nonmigratory geese are generally descendants of birds deliberately introduced to a park or hunting preserve. During the winter, migratory birds often mingle with nonmigratory ones. For example, a bird that breeds in Canada and flies to Virginia for the winter may find itself among geese that have never left Virginia.

Migration is a strenuous and difficult process. Flying hundreds of miles to and from a northern breeding area takes lots of energy and is dangerous. Compared to a migratory bird, one that stays put in a resource-rich area can devote more energy to producing young. If the nonmigrant lives in a suburban or urban area, it also benefits from an unnatural abundance of food (grass) and an equally unnatural lack of predators. Not surprisingly, the biggest increases in Canada geese have been among nonmigratory birds that live where humans are plentiful.

A An Oakland, California park overrun by Canada geese.

B US Airways Flight 1549 floats in New York's Hudson River after collisions with geese incapacitated both of its engines.

FIGURE 44.1 Goose troubles.

In 2006, increasing complaints about Canada geese led the U.S. Fish and Wildlife Service to encourage wildlife managers to look for ways to reduce nonmigratory Canada goose populations, without unduly harming migratory birds. To do so, these biologists are studying which traits characterize each goose population, as well as how populations interact with one another, with other species, and with their physical environment. This sort of information is the focus of the field of population ecology.

population Group of organisms of the same species that live in the same area and interbreed.

CREDITS: (opposite) © Jacques Langevin/Corbis Sygm; (1A) Courtesy of @ Joel Peter; (1B) AP Images/Steven Day.

44.2 Population Demographics

✔ We describe populations in terms of their size, density, distribution, and age structure.

Ecology is the branch of biology that deals with how populations interact with one another and the environment. Ecology is not the same as environmentalism, which is advocacy for protection of the environment. However, environmentalists often cite ecological studies when drawing attention to environmental concerns. Studying populations requires use of **demographics**—statistics that describe a population's traits. A population's demographics include its size, density, distribution, and age structure.

Population Size

Population size is the number of individuals in a population. It is often impractical to count all individuals, so biologists frequently use sampling techniques to estimate population size.

Plot sampling estimates the total number of individuals in an area on the basis of direct counts in a small part of the area. For example, ecologists might investigate the number of grass plants in a grassland or clams in a mudflat by measuring the number of individuals in several 1 meter by 1 meter square plots. The average number of individuals per plot is calculated, then multiplied by the size (in square meters) of the total area. The result is an estimate of the total population size. Plot sampling is most accurate for organisms that are not very mobile, in areas with uniform conditions.

FIGURE 44.2 Florida Key deer marked for a population study.

Scientists use **mark–recapture sampling** to estimate the population size of mobile animals. With this technique, some number of animals in a population are captured, marked (FIGURE 44.2), then released. After allowing a sufficient time to pass for the marked individuals to meld back into the population, the scientists capture animals again. The proportion of marked animals in the second sample is taken to be representative of the proportion marked in the whole population. For example, suppose scientists capture, mark, and release 100 deer in an area. Later, the scientists return and again capture 100 deer. They find 50 of these deer were previously marked, implying that marked deer constitute half of the population. They then infer that the total population is 200 individuals.

Characteristics of individuals in a sample plot or captured group can also be used to infer characteristics of the population as a whole. For example, if half the recaptured deer are of reproductive age, half of the population is assumed to share this trait. This sort of extrapolation is based on the assumption that the individuals in the sample are representative of the general population.

Population Density and Distribution

Population density is the number of individuals per unit area or volume. Examples of population density include the number of dandelions per square meter of lawn, or the number of amoebas per milliliter of pond water. **Population distribution** describes the location of individuals relative to one another. Members of a population may be clumped together, be an equal distance apart, or be distributed randomly.

Clumped Distribution Most populations have a clumped distribution, meaning members of the population are closer to one another than would be predicted by chance alone. A patchy distribution of resources encourages clumping, as when hippopotamuses gather in muddy river shallows (FIGURE 44.3A). Similarly, a cool, damp, north-facing slope may be covered with ferns, whereas an adjacent drier south-facing slope has none. Limited dispersal ability increases the likelihood of a clumped distribution: As the saying goes, the nut does not fall far from the tree. Asexual reproduction also results in clumping. It produces colonies of coral and vast stands of some trees. Finally, as Section 43.8 explained, some animals benefit by grouping together.

Near-Uniform Distribution Competition for limited resources can produce a near-uniform distribution,

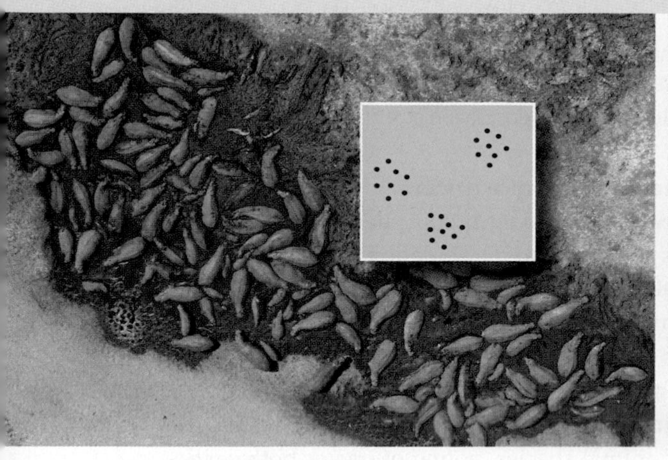

A Clumped distribution of hippopotamuses.

B Near-uniform distribution of nesting seabirds.

C Random distribution of dandelions.

FIGURE 44.3 Population distribution patterns.

with individuals more evenly spaced than would be expected by chance. Creosote bushes in deserts of the American Southwest grow in this pattern. Competition for water among the root systems keeps the plants from growing in close proximity. Similarly, seabirds in breeding colonies often show a near-uniform distribution. Each bird aggressively repels others that get within reach of its beak (**FIGURE 44.3B**).

Random Distribution Members of a population are distributed randomly when resources are uniformly available, and proximity to others neither benefits nor harms individuals. For example, when wind-dispersed dandelion seeds land on the uniform environment of a suburban lawn, dandelion plants grow in a random pattern (**FIGURE 44.3C**).

Age Structure

The **age structure** of a population refers to the number of individuals in various age categories. Individuals are often categorized as pre-reproductive, reproductive, or post-reproductive. Members of the pre-reproductive category have a capacity to produce offspring when mature. Together, pre-reproductive and reproductive individuals constitute a population's **reproductive base**.

Effects of Scale and Timing

The scale of the area sampled and the timing of a study can influence the observed demographics. For example, seabirds are spaced almost uniformly at a nesting site, but the nesting sites are clumped along a shoreline. The birds crowd together during the breeding season, but disperse when breeding is over.

Wildlife managers use demographic information to decide how to manage populations. For example, in considering how to manage Canada geese, wildlife managers began by evaluating the size, density, and distribution of nonmigratory populations. Based on this information, the U.S. Fish and Wildlife Service decided to allow destruction of some eggs and nests, and increased hunting opportunities at times when migratory Canada geese are least likely to be present.

age structure Of a population, the number of individuals in each of several age categories.
demographics Statistics that describe a population.
ecology Study of how populations interact with one another and with their nonliving environment.
mark–recapture sampling Method of estimating population size of mobile animals by marking individuals, releasing them, then checking the proportion of marks among individuals later recaptured.
plot sampling Method of estimating population size of organisms that do not move much by making counts in small plots, and extrapolating from this to the number in the larger area.
population density Number of individuals per unit area.
population distribution Describes whether individuals are clumped, uniformly dispersed, or randomly dispersed in an area.
population size Total number of individuals in a population.
reproductive base Of a population, all individuals who are of reproductive age or younger.

TAKE-HOME MESSAGE 44.2

What are demographics and what factors affect them?

✔ Each population has characteristic demographics, such as its size, density, distribution pattern, and age structure.

✔ Characteristics of the population as a whole are often inferred on the basis of a study of a smaller subsample.

✔ Environmental conditions and interactions among individuals can influence a population's demographics, which often change over time.

44.3 Population Size and Exponential Growth

Starting Size of Population		Net Monthly Increase	New Size of Population
2,000	× r =	800	2,800
2,800	× r =	1,120	3,920
3,920	× r =	1,568	5,488
5,488	× r =	2,195	7,683
7,683	× r =	3,073	10,756
10,756	× r =	4,302	15,058
15,058	× r =	6,023	21,081
21,081	× r =	8,432	29,513
29,513	× r =	11,805	41,318
41,318	× r =	16,527	57,845
57,845	× r =	23,138	80,983
80,983	× r =	32,393	113,376
113,376	× r =	45,350	158,726
158,726	× r =	63,490	222,216
222,216	× r =	88,887	311,103
311,103	× r =	124,441	435,544
435,544	× r =	174,218	609,762
609,762	× r =	243,905	853,667
853,667	× r =	341,467	1,195,134

A Increases in size over time. Note that the net increase becomes larger with each generation.

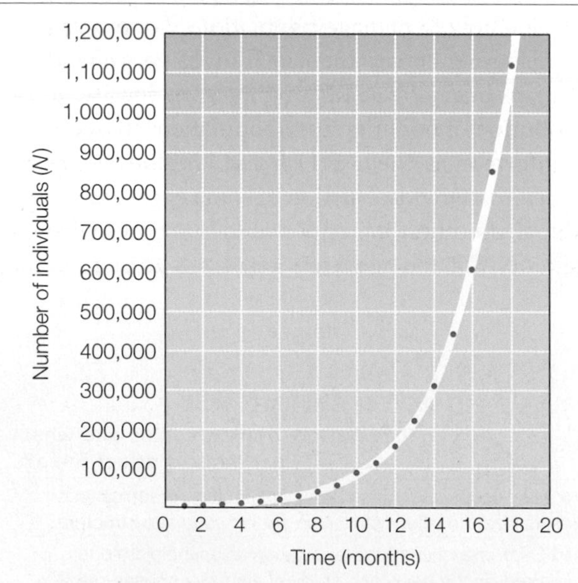

B Graphing numbers over time produces a J-shaped curve.

FIGURE 44.4 ▸**Animated** Exponential growth in a hypothetical population of mice with a per capita rate of growth (*r*) of 0.4 per mouse per month and a population size of 2,000.

✔ Populations fluctuate in size as individuals come and go.

✔ With exponential growth, increases in population size are proportional to the current population size.

Immigration and Emigration

In nature, populations continually change in size. Individuals are added to a population by births and by **immigration**, which is the arrival of new residents that previously belonged to another population. Individuals are removed from a population by deaths and by **emigration**, the departure of individuals who take up permanent residence elsewhere.

In many animal species, young of one or both sexes leave the area where they were born to breed elsewhere. For example, young freshwater turtles typically emigrate from their parental population and become immigrants at another pond some distance away. By contrast, seabirds typically breed where they were born. However, some individuals may emigrate and end up at breeding sites more than a thousand kilometers away. The tendency of individuals to emigrate to a new breeding site is usually related to resource availability and crowding. As resources decline and crowding increases, the likelihood of emigration rises.

Zero to Exponential Growth

If we set aside the effects of immigration and emigration, we can define **zero population growth** as an interval during which the number of births is balanced by an equal number of deaths. As a result, population size remains unchanged, with no net increase or decrease in the number of individuals.

We can measure births and deaths in terms of rates per individual, or per capita. *Capita* means head, as in a head count. Subtract a population's per capita death rate (*d*) from its per capita birth rate (*b*) and you have the **per capita growth rate**, or *r*:

b	–	*d*	=	*r*
(per capita birth rate)		(per capita death rate)		(per capita growth rate)

Imagine 2,000 mice living in the same field. If 1,000 mice are born each month, then the birth rate is 0.5 births per mouse per month (1,000 births/2,000 mice). If 200 mice die one way or another each month, then the death rate is 200/2,000 or 0.1 deaths per mouse per month. Thus, *r* is 0.5 – 0.1, or 0.4 per mouse per month.

As long as *r* remains constant and greater than zero, **exponential growth** will occur, which means that the population's size will increase by the same proportion of its total in every successive time interval. We can calculate population growth (G) for each interval based on the number of individuals (N) and the per capita growth rate:

$$N \times r = G$$

(number of individuals) × (per capita growth rate) = (population growth per unit time)

Applying this equation to our hypothetical population of field mice, shows that after one month, our inital population of 2,000 mice will have increased to 2,800 mice (**FIGURE 44.4A**). In the next month, population size will expand by 1,120 individuals (2,800 × 0.4), bringing the total population size to 3,920. At this growth rate, the number of mice would rise from 2,000 to more than 1 million in under two years! Graphing the increases against time results in a J-shaped curve, which is characteristic of exponential population growth (**FIGURE 44.4B**).

With exponential growth, the number of individuals born into the population increases each generation, although the per capita growth rate stays the same. Exponential population growth is analogous to compound interest in a bank account. The annual interest *rate* stays fixed, yet every year the *amount* of interest paid increases. The annual interest paid into the bank account adds to the balance, so the subsequent interest payment will be based on an increased balance.

Exponential growth occurs in any population in which the birth rate exceeds the death rate—in other words, as long as r is greater than zero. Imagine a bacterium in a culture flask. If this species divides every thirty minutes, then after ten hours (twenty doublings), there will be more than a million bacteria. Curve 1 in **FIGURE 44.5** is a plot of this increase.

Now suppose that 25 percent of the bacteria in our hypothetical population die every thirty minutes. In this scenario, it takes seventeen hours, not ten, for the population to reach 1 million. Thus, deaths slow the rate of increase but do not stop exponential growth (curve 2 in **FIGURE 44.5**).

biotic potential Maximum possible population growth rate under optimal conditions.
emigration Movement of individuals out of a population.
exponential growth A population grows by a fixed percentage in successive time intervals; the size of each increase is determined by the current population size.
immigration Movement of individuals into a population.
per capita growth rate For some interval, the added number of individuals divided by the initial population size.
zero population growth Interval in which births equal deaths.

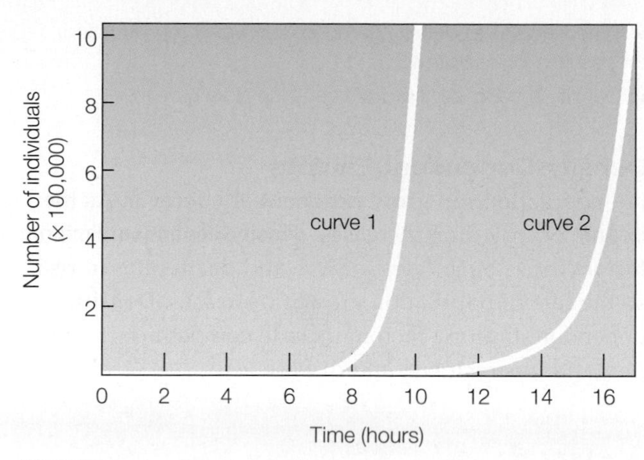

FIGURE 44.5 Effect of deaths on the rate of increase for two hypothetical populations of bacteria. Curve 1 shows population growth for cells that reproduce every half hour with no mortality. Curve 2 shows population growth with 25 percent dying between divisions. Deaths slows the rate of increase, but exponential growth occurs as long as the birth rate exceeds the death rate.

Biotic Potential

The growth rate for a population under ideal conditions is its **biotic potential**. This is a theoretical rate at which the population would grow if shelter, food, and other essential resources were unlimited and there were no predators or pathogens. Factors that affect biotic potential include the age at which reproduction typically begins, how long individuals remain reproductive, and the number of offspring that are produced each time an individual reproduces. These factors are influenced by the environment, and vary by species. Microbes such as bacteria have some of the highest biotic potentials, whereas large-bodied mammals have some of the lowest.

Regardless of the species, populations seldom reach their biotic potential because of the effects of limiting factors. We discuss how limiting factors influence population growth in detail in the next section.

TAKE-HOME MESSAGE 44.3
What causes changes in population size?

- ✔ The size of a population depends on its rates of births, deaths, immigration, and emigration.
- ✔ Subtract the per capita death rate from the per capita birth rate to get r, the per capita growth rate of a population.
- ✔ A population will grow exponentially as long as r is constant and greater than zero. With exponential growth, the number of individuals increases at an ever accelerating rate.
- ✔ The biotic potential of a species is its maximum possible population growth rate under optimal conditions.

✔ When increased density hampers survival or reproduction, the result is logistic growth.

✔ Density-independent factors can also slow growth.

Density-Dependent Factors

No population can grow exponentially forever. As the degree of crowding increases, **density-dependent limiting factors** cause birth rates to slow and death rates to rise, so the rate of population growth decreases. Density-dependent limiting factors include competition, predation, parasitism, and disease.

FIGURE 44.6 Example of a limiting factor. Wood ducks build nests only inside tree hollows of specific dimensions (left). In some places, lack of access to appropriate hollows now limits the size of the wood duck population. Adding artificial nesting boxes (right) to these environments can help increase the size of the duck populations.

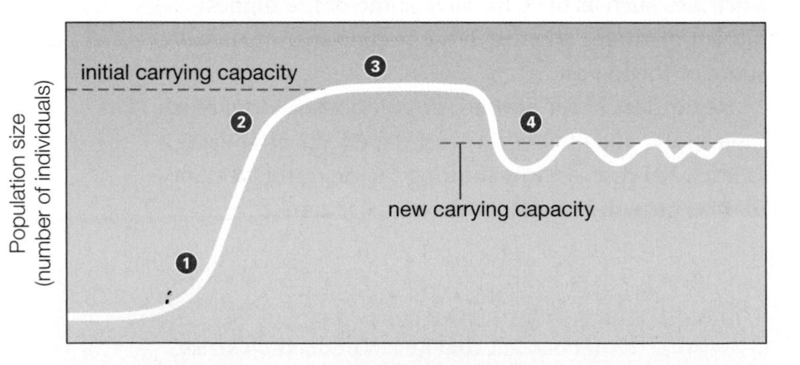

FIGURE 44.7 Logistic growth. Note the initial S-shaped curve.

❶ At low density, the population grows exponentially.

❷ As crowding increases, density-dependent limiting factors slow the rate of growth.

❸ Population size levels off at the carrying capacity for that species.

❹ Any change in the carrying capacity will result in a corresponding shift in population size.

FIGURE IT OUT How does adding nest boxes affect the carrying capacity for wood duck populations?

Answer: It increases carrying capacity.

Any natural area has limited resources. Thus, as the number of individuals in an area increases, so does intraspecific competition: competition among members of the same species. As a result of increased competition, some individuals fail to secure what they need to survive and reproduce. Competition has a detrimental effect even on winners, because energy they use in competition for resources is not available for reproduction. Essential resources for which animals might compete include food, water, hiding places, and nesting sites (FIGURE 44.6). Plants compete for nutrients, water, and access to sunlight.

Parasitism and contagious disease increase with crowding because the closer individuals are to one another, the more easily parasites and pathogens can spread. Predation increases with density too, because predators often concentrate their efforts on the most abundant prey species.

Logistic Growth

Logistic growth occurs when density-dependent factors affect population size over time. Graphing logistic growth results in an S-shaped curve (FIGURE 44.7). When the population is small, density-dependent limiting factors have little effect and the population grows exponentially ❶. Then, as population size and the degree of crowding rise, limiting factors begin to slow growth ❷. Eventually, the population size levels off at the environment's carrying capacity ❸. **Carrying capacity (*K*)** is the maximum number of individuals that a population's environment can support indefinitely.

The equation that describes logistic growth is:

G	$=$	r	$\times$	N	$\times$	$(K - N)/K$
(population growth per unit time)		(maximum per capita population growth rate)		(number of individuals)		(proportion of resources not yet used)

As with the exponential growth equation, G is growth per unit time, r is the per capita growth rate, and N is current population size. K is carrying capacity. The $(K–N)/K$ part of the equation represents the proportion of carrying capacity not yet used. As a population grows, this proportion decreases, so G becomes smaller and smaller. At carrying capacity, the equation becomes $G=rN(0)$; zero individuals are added.

Carrying capacity is species-specific, environment-specific, and can change over time ❹. For example, the carrying capacity for a plant species decreases when nutrients in the soil become depleted. Human activities also affect carrying capacity. For example, human harvest of horseshoe crabs has decreased the carrying

FIGURE 44.8 Overshoot and crash. A reindeer herd introduced to a small island in 1944 increased in size exponentially, then crashed when depletion of the food supply was coupled with an especially cold and snowy winter in 1963–64.

capacity for red knot sandpipers, a type of migratory bird. Horseshoe crab eggs are the sandpipers' main food during their long-distance migration.

Density-Independent Factors

Sometimes, natural disasters or weather-related events affect population size. A volcanic eruption, hurricane, or flood can decrease population size. So can human-caused events such as an oil spill. These events are called **density-independent limiting factors**, because crowding does not influence the likelihood of their occurrence or the magnitude of their effect.

In nature, a population's size is often affected by a combination of density-dependent and density-independent factors. Consider what happened after the 1944 introduction of 29 reindeer to St. Matthew Island, an uninhabited island off the coast of Alaska. When biologist David Klein visited the island in 1957, he found 1,350 well-fed reindeer (**FIGURE 44.8**). Klein returned in 1963 and counted 6,000 reindeer. The population had soared far above the island's carrying capacity. A population can temporarily overshoot an environment's carrying capacity, but the high density cannot be sustained. Klein observed that some effects of density-dependent limiting factors were already apparent. For example, the average body size of the reindeer had decreased. When Klein returned in 1966, only 42 reindeer survived. The single male had abnormal antlers, and was thus unlikely to breed. There were no fawns. Klein figured out that thousands of reindeer had starved to death during the winter of 1963–1964. That winter was unusually harsh, with low temperatures, high winds, and 140 inches of snow. Most reindeer were already in poor condition as a result of increased competition and they starved when deep snow covered their food. A population decline had been expected—a population that exceeds its carrying capacity usually falls back below that capacity—but bad weather magnified the extent of the crash. By the 1980s, there were no reindeer left on the island.

carrying capacity (*K*) Maximum number of individuals of a species that an environment can sustain.
density-dependent limiting factor Factor that increasingly limits population growth as population density increases.
density-independent limiting factor Factor that limits population growth to the same degree regardless of population density.
logistic growth Density-dependent limiting factors cause population growth to slow as population size increases.

TAKE-HOME MESSAGE 44.4

How do environmental factors affect population growth?

✔ Carrying capacity is the maximum number of individuals of a population that can be sustained indefinitely by the resources in a given environment.

✔ With logistic growth, population growth is fastest during times of low density, then it slows as the population approaches carrying capacity.

✔ The effects of density-dependent factors such as disease result in a logistic growth pattern. Density-independent factors such as natural disasters also affect population size.

44.5 Life History Patterns

✔ Natural selection influences traits such as life span, age at maturity, and the number of offspring per breeding event.

Population growth rates are affected by life history traits such as the age at which reproduction begins, how long individuals remain reproductive, number of reproductive events, and the number of offspring produced during each reproductive event.

Quantifying Life History Traits

One way to describe life history traits is to focus on a **cohort**—a group of individuals born during the same interval—from their time of birth until the last one dies. Ecologists often divide a natural population into age classes and record the age-specific birth rates and mortality. The resulting data is summarized in a life table (**TABLE 44.1**). Data in life tables can inform decisions about how changes, such as harvesting a species or altering its environment, will affect a population's numbers. For example, such data were cited in federal court rulings that halted logging in the spotted owl's habitat—old-growth forests of the Pacific Northwest.

Table 44.1 Life Table for an Annual Plant*

Age Interval (days)	Survivorship (number surviving at start of interval)	Number Dying in Interval	Death Rate (number dying/number surviving)	"Birth" Rate in Interval (number of seeds per plant)
0–63	996	328	0.329	0
63–124	668	373	0.558	0
124–184	295	105	0.356	0
184–215	190	14	0.074	0
215–264	176	4	0.023	0
264–278	172	5	0.029	0
278–292	167	8	0.048	0
292–306	159	5	0.031	0.33
306–320	154	7	0.045	3.13
320–334	147	42	0.286	5.42
334–348	105	83	0.790	9.26
348–362	22	22	1.000	4.31
362–	0	0	0	0
		996		

* *Phlox drummondii; data from W. J. Leverich and D. A. Levin, 1979.*

Information about age-specific death rates can also be illustrated by a **survivorship curve**, a plot that shows how many members of a cohort remain alive over time. Ecologists have described three generalized types of curves. A type I curve is convex, indicating survivorship is high until late in life (**FIGURE 44.9A**). Humans and other large mammals that produce one or two young and care for them show this pattern. A diagonal type II curve indicates that the death rate of the population does not vary much with age (**FIGURE 44.9B**). In lizards, small mammals, and large birds, old individuals are about as likely to die of disease or predation as young ones. A type III curve is concave, indicating that the death rate for a population peaks early in life (**FIGURE 44.9C**). Marine animals that release eggs into water have this type of curve, as do plants that release enormous numbers of tiny seeds.

Environmental Effects on Life History

To produce offspring, an individual must invest resources that it could otherwise use to grow and maintain itself. Species differ in the manner in which they distribute parental investment among offspring and over the course of their lifetime. Life history pat-

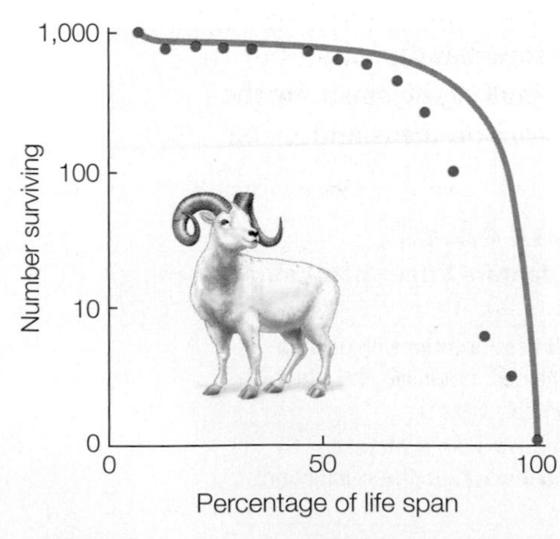

A Type I curve. Mortality is highest very late in life. Data for Dall sheep (*Ovis dalli*).

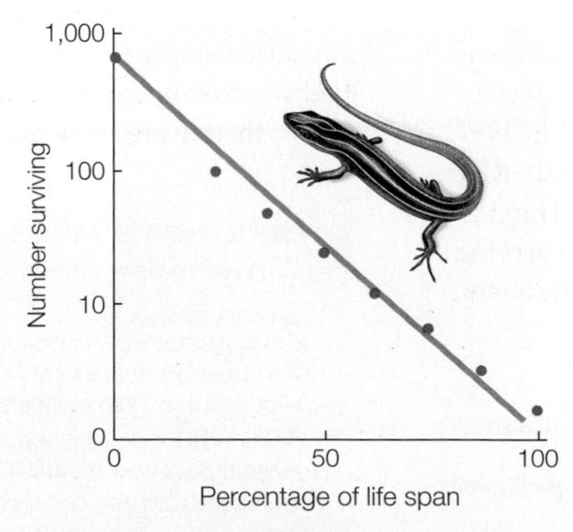

B Type II curve. Mortality does not vary with age. Data for five-lined skink (*Eumeces fasciatus*).

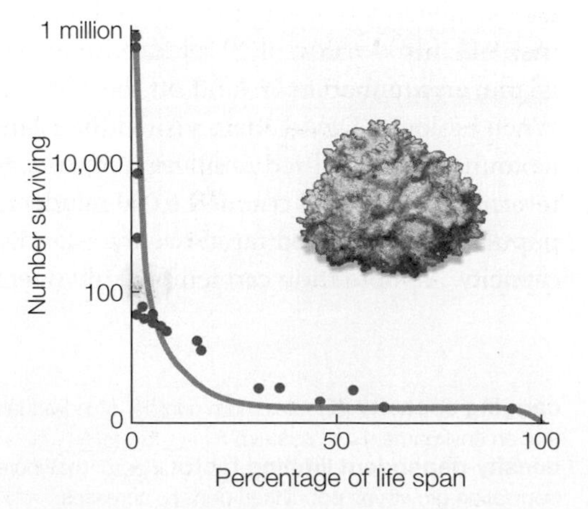

C Type III curve. Mortality is highest early in life. Data for a desert shrub (*Cleome droserifolia*).

FIGURE 44.9 ▶Animated Survivorship curves. Gray lines are theoretical curves. Red dots are data from field studies.

A Fly laying many eggs in a rotting tomato.

Opportunistic life history	Equilibrial life history
shorter development	longer development
early reproduction	later reproduction
fewer breeding episodes, many young per episode	more breeding episodes, few young per episode
less parental investment per young	more parental investment per young
higher early mortality, shorter life span	low early mortality, longer life span
result of *r*-selection	result of *K*-selection

B Elephant with its single calf.

FIGURE 44.10 Two types of life history. Most species have a mix of opportunistic and equilibrial life history traits.

terns vary continuously among species, but ecologists have described two theoretical extremes at either end of this continuum (**FIGURE 44.10**). Both maximize the number of offspring that are produced and survive to adulthood, but they do so under very different environmental conditions.

When a species lives where conditions vary in an unpredictable manner, its populations seldom reach the carrying capacity of their environment. As a result, there is little competition for resources and deaths occur mainly as a result of density-independent factors. Such conditions favor an opportunistic life history, in which individuals produce as many offspring as possible, as quickly as possible. Opportunistic species are said to be subject to ***r*-selection**, because they maximize *r*, the per capita growth rate. They have a short generation time and small body size. Opportunistic species usually have a type III survivorship curve, with mortality heaviest early in life. Weedy plants such as dandelions are an example. They mature within weeks, produce many tiny seeds, then die. Flies are opportunistic animals. A female can lay hundreds of small eggs in a temporary food source (**FIGURE 44.10A**).

When a species lives in a more stable environment, its populations often approach carrying capacity. Under these circumstances, the ability to successfully compete for resources affects reproductive success. Thus, an equilibrial life history, in which parents produce a few, high-quality offspring, is adaptive. Equilibrial species are shaped by ***K*-selection**, in which adaptive traits provide a competitive advantage when population size is near carrying capacity (*K*). Such species tend to have a large body and a long generation time. For example, a female elephant reaches maturity at the age of 10 to 14 years. She then produces only one large calf at a time, and invests in the calf by nursing it after its birth (**FIGURE 44.10B**). Similarly, a coconut palm grows for years before beginning to produce a few coconuts at a time. In both elephants and coconut palms, a mature individual produces young for many years.

Some species have mixes of traits that cannot be explained by *r*-selection or *K*-selection alone. For example, century plants (a type of agave) and bamboo are large and long-lived, but reproduce only once. Atlantic eels and Pacific salmon are unusual among vertebrates in also having a one-shot reproductive strategy. Such a strategy can evolve when opportunities for reproduction are unlikely to be repeated. In the century plant and bamboo, climate conditions that favor reproduction occur only rarely. In the eels and salmon, physiological changes related to migration between fresh water and salt water make a repeat journey impossible.

cohort Group of individuals born during the same time interval.
***K*-selection** Selection favoring traits that allow their bearers to outcompete others for limited resources.
***r*-selection** Selection that favors traits that allow their bearers to produce the most offspring the most quickly.
survivorship curve Graph showing the decline in numbers of a cohort over time.

TAKE-HOME MESSAGE 44.5

How do we describe life history patterns?

✔ Tracking a cohort (a group of individuals) from their birth until the last one dies reveals patterns of reproduction and mortality that can be summarized in a life table.

✔ Survivorship curves are graphs that reveal differences in age-specific survival among species or among populations of the same species.

✔ Species that maximize offspring quantity are said to be *r*-selected, whereas those who maximize offspring quality are *K*-selected.

44.6 Effects of Predation on Life History

✔ Predation can serve as a selection pressure that shapes life history patterns.

A Long-Term Study of Guppies

A long-term study by evolutionary biologists John Endler and David Reznick illustrates the effect of predation on life history traits. Endler and Reznick studied populations of guppies, small fishes native to shallow freshwater streams in the mountains of Trinidad (**FIGURE 44.11**). The scientists focused their attention on a region where many small waterfalls stop guppies in one part of a stream from moving to another. As a result of these natural barriers, each stream holds several populations of guppies that have very little gene flow between them (Section 17.8).

Waterfalls also keep guppy predators from moving from one part of the stream to another. The main guppy predators, killifishes and cichlids, differ in size and prey preferences. The relatively small killifish preys mostly on immature guppies, and ignores the larger adults. Cichlids are bigger fish. They tend to pursue mature guppies and ignore small ones.

Some parts of the streams hold one type of predator but not the other. Thus, different guppy populations face different predation pressures. Reznick and Endler discovered that guppies in regions with the large predator (the cichlid) grow faster and are smaller at maturity than guppies in regions with the small predator (the killifish). Guppies in populations hunted by the larger predator also reproduce earlier, have more offspring at a time, and breed more frequently.

Were these differences in life history traits genetic, or the result of some environmental factor? To find out, the biologists collected guppies from both cichlid- and killifish-dominated streams. They reared the guppies in separate aquariums under identical predator-free conditions. Two generations later, the groups continued to show the differences observed in natural populations. Thus differences between guppies preyed on by different predators must have a genetic basis.

FIGURE 44.11 ▸**Animated** Effects of predation on life history traits in guppies. David Reznik (shown above) and John Endler carried out an experiment using guppies that had evolved in the presence of a predator (pike cichlid) that preferentially eats large guppies. Control group guppies remained in their home pool with pike cichlids. Other guppies were moved to a guppy-free pool with a predator (killifish) that eats small guppies. Data at left show the average age and weight at maturity for descendants of both groups 11 years after the study began.

FIGURE 44.12 Fishermen with a prized catch, a large Atlantic codfish. Both sport fishermen and commercial fishermen preferentially harvested the largest codfish.

Reznick and Endler then made a prediction: If life history traits evolve in response to predation, then these traits will change when a population is exposed to a new predator that favors different prey traits. To test their prediction, they found a stream region with killifish but no guppies or cichlids. Here, they introduced guppies from a site where there were cichlids but no killifish. Thus, at the experimental site, guppies that had previously been preyed upon only by cichlids (which eat big guppies) were now exposed only to killifish (which eat small guppies). The control site was a downstream region where relatives of the transplanted guppies still coexisted with cichlids.

Reznick and Endler revisited the stream over the course of eleven years and thirty-six generations of guppies. Their data showed that guppies at the experimental site evolved. Exposure to a previously unfamiliar predator altered the guppies' rate of growth, age at first reproduction, and other life history traits. By contrast, guppies at the control site showed no such changes. Reznick and Endler concluded that life history traits in guppies can evolve rapidly in response to the selective pressure exerted by predation.

Overfishing of Atlantic Cod

Evolution of life history traits in response to predation is not merely of theoretical interest. It has economic importance. Just as guppies evolved in response to predators, a population of Atlantic codfish (*Gadus morhua*) evolved in response to human fishing. From the mid-1980s to early 1990s, the number of fishing boats targeting the North Atlantic population of codfish increased. As the yearly catch rose, the age of sexual maturity shifted, and fishes that reproduce while young and small became more common. These early-reproducing individuals were at an advantage because fishermen preferentially caught and kept larger fish (**FIGURE 44.12**). Fishing pressure continued to rise until 1992, when declining cod numbers caused the Canadian government to ban cod fishing in some areas. That ban, and later restrictions, came too late to stop the Atlantic cod population from crashing. In some areas, the population declined by 97 percent and still shows no signs of recovery.

Looking back, it is clear that life history changes were an early sign that the North Atlantic cod population was in trouble. Had biologists recognized what was happening, they might have been able to save the fishery and protect the livelihood of fishers and associated workers. Ongoing monitoring of the life history data for other economically important fishes may help prevent similar population crashes in the future.

TAKE-HOME MESSAGE 44.6

How does predation affect life history traits?

✔ When predators prefer large prey, prey individuals who reproduce when still small and young are at a selective advantage. When predators focus on small prey, fast-growing individuals have the selective advantage.

✔ Humans can alter life history traits of a commercially harvested species when they preferentially remove individuals of particular size or age from the population.

44.7 Human Population Growth

✔ The size of the human population is at its highest level ever and is expected to continue to increase.

For most of our history, the human population grew slowly. The growth rate began to increase about 10,000 years ago, and during the past two centuries, it soared (**FIGURE 44.13**). Three trends promoted the increases. First, humans migrated into new habitats and expanded into new climate zones. Second, they developed technologies that increased the carrying capacity of existing habitats. Third, they sidestepped some limiting factors that typically restrain population growth.

Early Innovations and Expansions

Modern humans evolved in Africa by 200,000 years ago, and about 60,000 years ago they began to spread out across the globe (Section 26.6). A large brain and the capacity to master a variety of skills gave humans an unmatched ability to live in a broad range of habitats. Humans learned how to start fires, build shelters, make clothing, manufacture tools, and cooperate in hunts. With the advent of language, knowledge of such skills did not die with the individual.

Domestication of animals and the invention of agriculture by about 10,000 years ago provided a more dependable food supply than traditional hunting and gathering. A pivotal factor was the domestication of wild grasses, including the species ancestral to modern wheat and rice.

From the Industrial Revolution Onward

The rate of human population growth surged during the 1800s as innovations increased the stability of food supplies and decreased the impact of disease for much of the human population.

In the middle of the eighteenth century, people began to use fossil fuel energy to operate machinery. This innovation opened the way to high-yielding mechanized agriculture and improved food distribution systems. Food production was further enhanced in the early 1900s, when chemists discovered a way to convert nitrogen gas to ammonia. Previously this process had been carried out primarily by nitrogen-fixing bacteria. Use of synthetic nitrogen fertilizers dramatically increased crop yields. So did synthetic pesticides that came into widespread use in the mid-1900s.

Disease has historically dampened human population growth. For example, during the mid-1300s, one-third of Europe's population was lost to a pandemic known as the Black Death. Beginning in the mid-1800s,

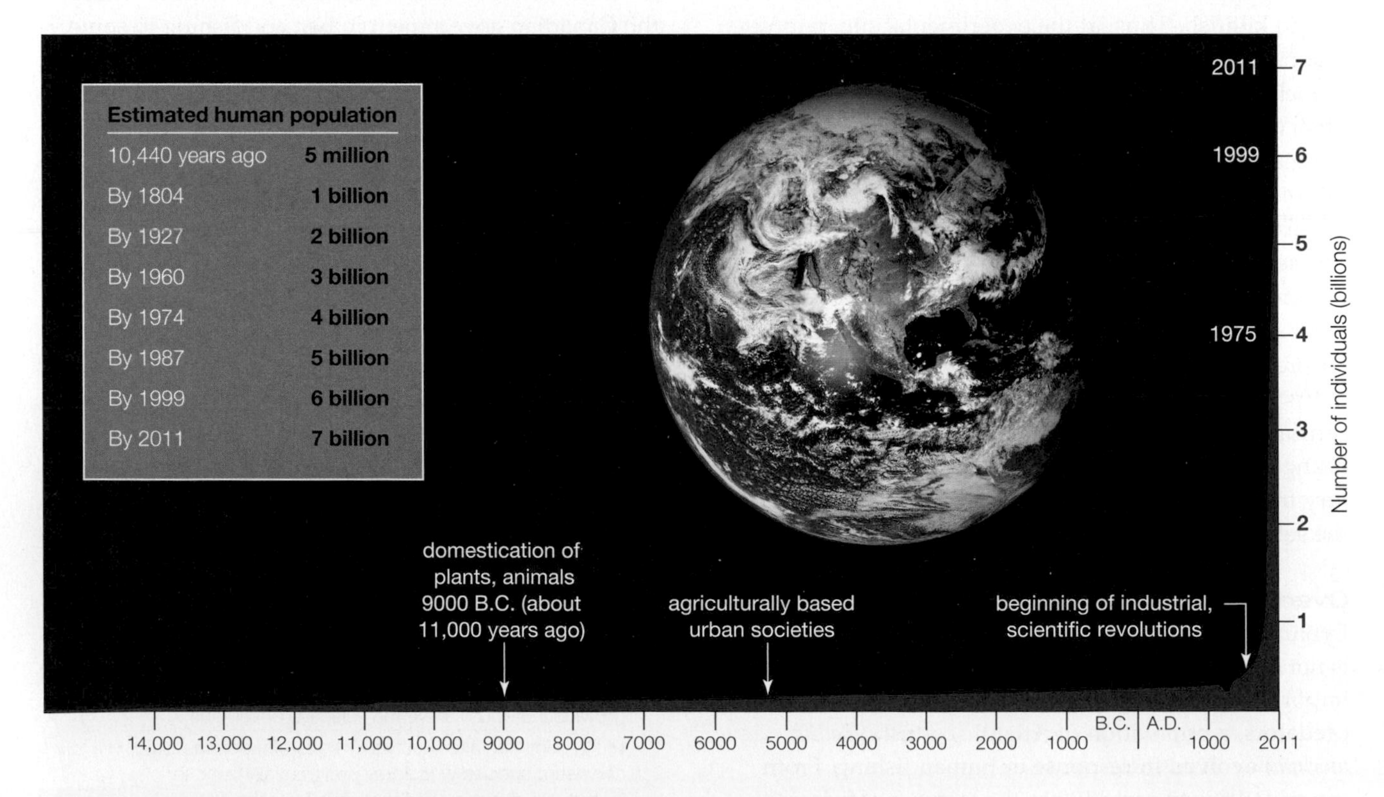

FIGURE 44.13 Growth curve (*red*) for the world human population. To check the current world and U.S. population estimates, visit the U.S. Census website at www.census.gov/popclock.

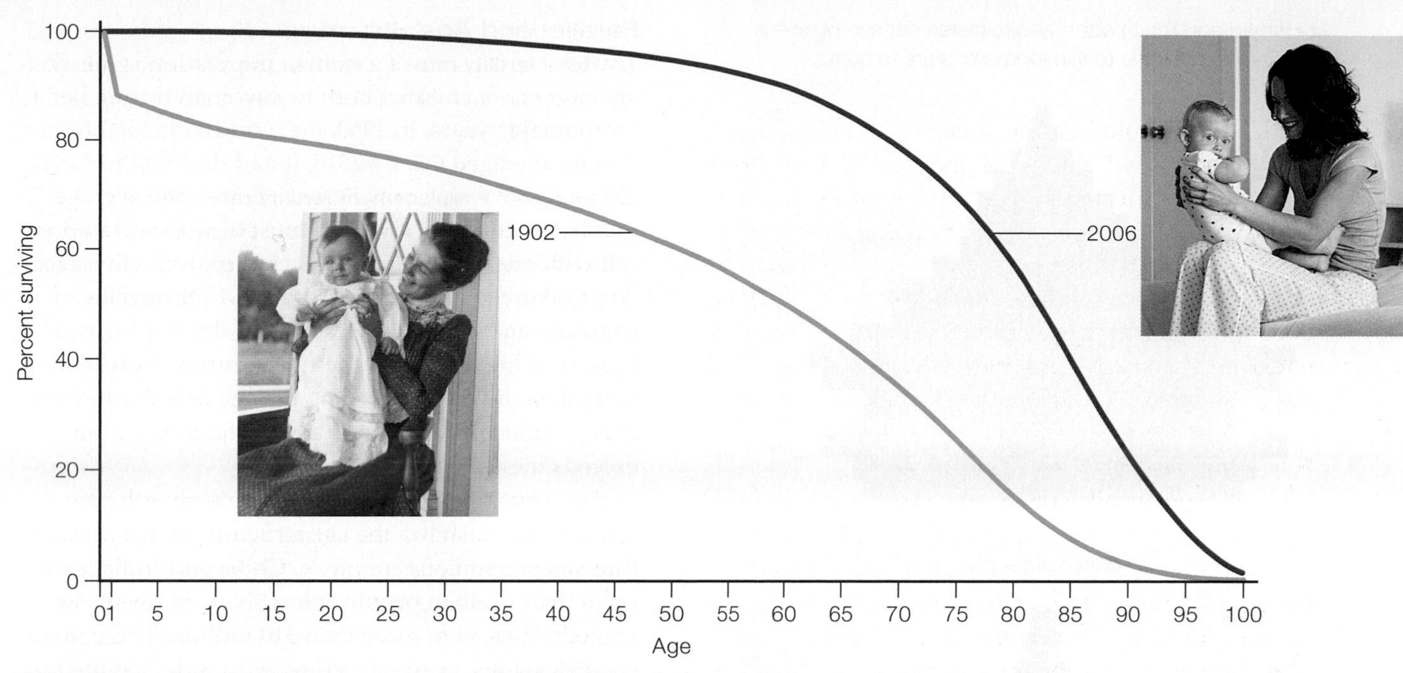

FIGURE 44.14 Survivorship curves for people in the United States in the early 20th and early 21st centuries. Since 1900, infant mortality has declined by more than 90 percent, and mortality during childbirth has declined 99 percent. (CDC/NCHS, National Vital Statistics System and state death registration data)

increased understanding of the link between microorganisms and illness led to improvements in food safety, sanitation, and medicine. People began to pasteurize foods and drinks, heating them to kill harmful bacteria. They also began to protect drinking water. The first modern sewer system was constructed in London, England, in the late 1800s. By diverting wastewater downstream of the city's water source, the system lowered the incidence of deadly waterborne diseases such as cholera and typhoid fever. Beginning in the early 1900s, chlorination and other methods of sterilizing drinking water contributed to a further decline in these diseases in the most industrialized nations.

Advances in sanitation also lowered the death rate associated with medical treatment. In the mid-1800s, Ignaz Semmelweis, a German physician, began urging doctors to wash their hands. His advice was largely ignored until after his death, when Louis Pasteur popularized the idea that unseen organisms cause disease. Acceptance of this concept revolutionized surgery, which had previously been carried out with little regard for cleanliness.

Vaccines helped prevent disease, and antibiotics helped treat them. As Section 37.12 explained, Edward Jenner demonstrated the effectiveness of a vaccine against smallpox in the late 1700s. Antibiotics are a more recent development. Large-scale production of penicillin, the first antibiotic to be widely used, did not begin until the 1940s.

FIGURE 44.14 shows the result of improving nutrition, sanitation, and medicine in the United States: a shift in the survivorship curve. Infant mortality has plummeted and far more people survive to old age.

A worldwide decline in death rates without an equivalent drop in birth rates is responsible for the ongoing explosion in human population size. It took more than 100,000 years for the human population to reach 1 billion in number. Since then, the rate of increase has risen steadily. The population is currently about 7 billion and the United Nations estimates that it will reach 9 billion by 2050.

TAKE-HOME MESSAGE 44.7

What factors contributed to the increase in the human population size?

✔ Innovations such as use of fire and clothing allowed early humans to expand into what would otherwise have been inhospitable habitats.

✔ The invention of agriculture created a stable food supply. More recently, innovations such as chemical fertilizers helped to boost crop yields.

✔ Improved understanding of the causes of disease led to better sanitation and new medical treatments that have increased the average life span.

44.8 Anticipated Growth and Consumption

✔ Human population size—and resource consumption—is expected to continue to rise for many years to come.

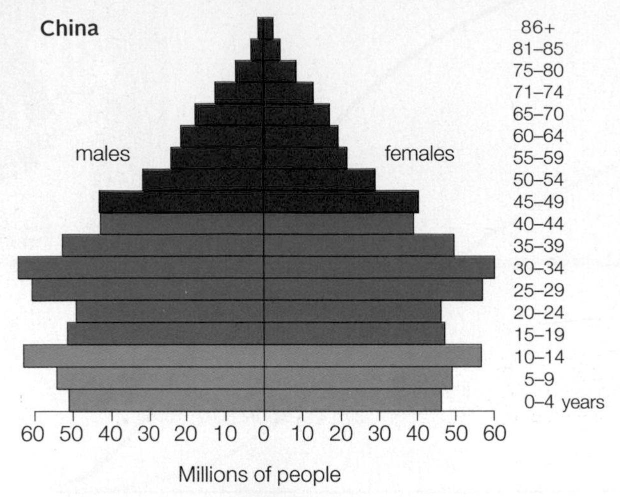

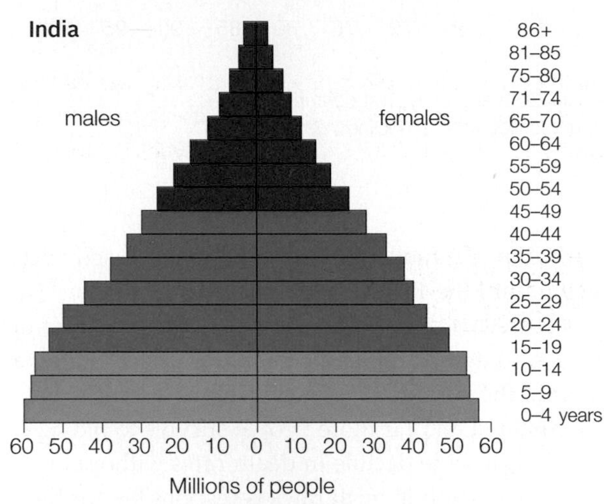

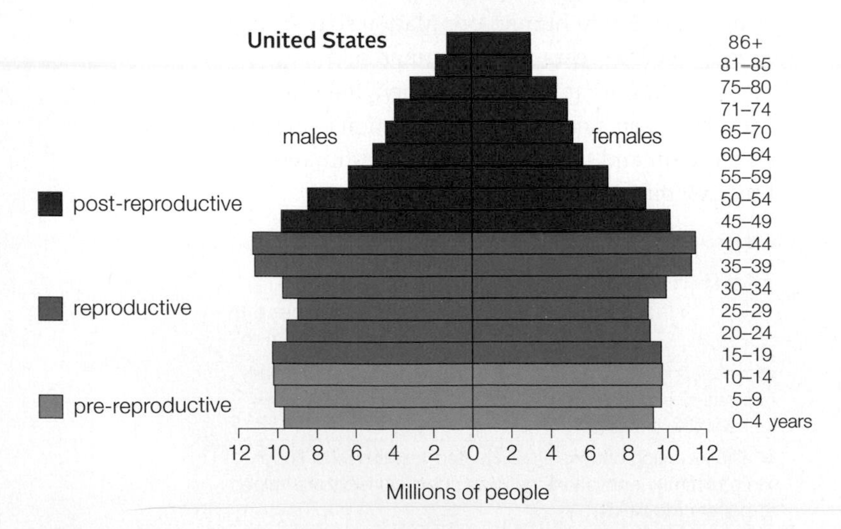

FIGURE 44.15 Age structure for three countries. Green bars represent pre-reproductive individuals. The left side of each chart indicates males; the right side, females. **FIGURE IT OUT** Which country has the largest number of men in the 45 to 49 age group?

Answer: China

Fertility and Age Structure

The **total fertility rate** of a human population is the average number of children born to a woman during her reproductive years. In 1950, the worldwide total fertility rate averaged 6.5. By 2010, it had declined to 2.6. It remains above **replacement fertility rate**—the average number of children a woman must bear to replace herself with one daughter who reaches reproductive age. At present, the replacement rate is 2.1 for developed countries and as high as 2.5 in some developing countries. (It is higher in developing countries because more daughters die before reaching the age of reproduction.) A population grows as long as its total fertility rate exceeds the replacement rate.

Age structure affects a population's growth rate. **FIGURE 44.15** shows the age structure for the world's three most populous countries. China and India have more than 1 billion people apiece. Next in line is the United States, with more than 310 million. These three countries differ in age structure, with India having the greatest proportion of young people. The broader the base of an age structure diagram, the greater the anticipated population growth.

Worldwide, about 1.9 billion people are about to enter their reproductive years. Even if every couple from this time forward has no more than two children, population growth will not slow for sixty years.

A Demographic Transition

Demographic factors vary among countries, with the most highly developed countries having the lowest fertility rates and infant mortality, and the highest life expectancy. The **demographic transition model** describes how changes in population growth often unfold in four stages of economic development (**FIGURE 44.16**). Living conditions are harshest in the preindustrial stage, before technological and medical advances become widespread. Birth and death rates are both high, so the growth rate is low ❶. Next, in the transitional stage, industrialization begins. Food production and health care improve. Death rate drops fast, but birth rate declines more slowly ❷. As a result, population growth rate increases rapidly. India is in this stage.

During the industrial stage, industrialization is in full swing, and birth rate declines. People move to cities, where couples tend to want smaller families. The birth rate moves closer to the death rate, and the population grows less rapidly ❸. Mexico is currently in this stage. In the postindustrial stage, a population's growth rate becomes negative. Birth rate falls below death rate, and population size slowly decreases ❹.

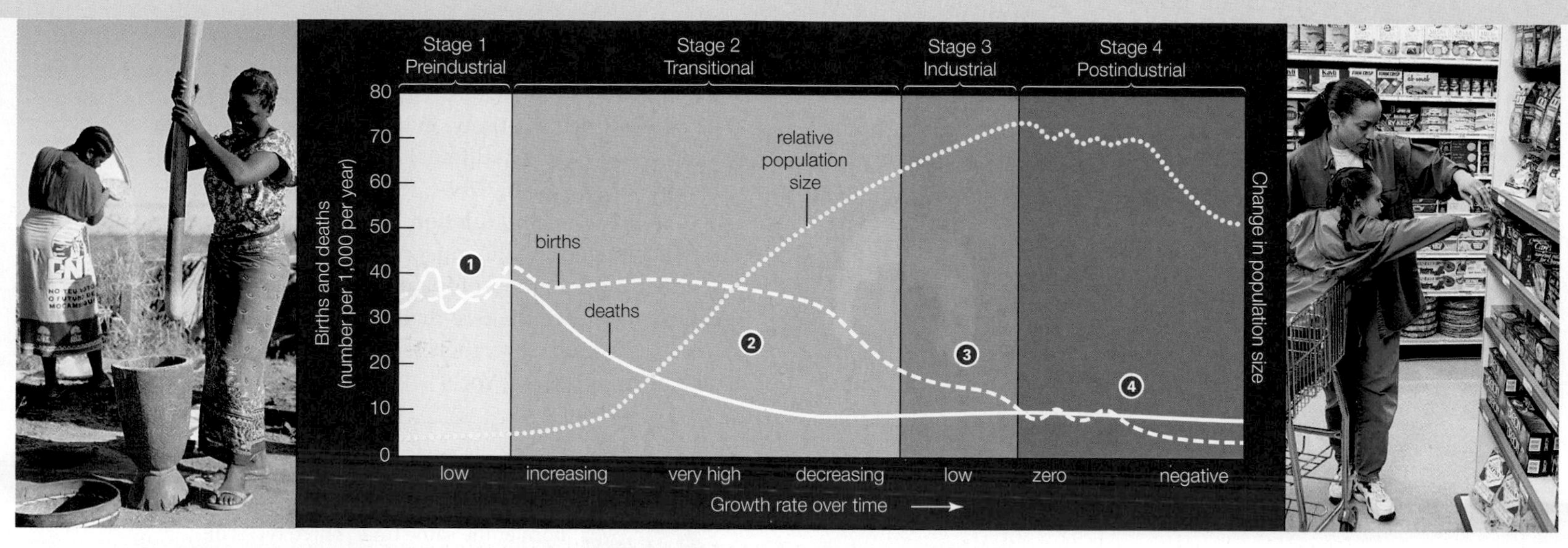

FIGURE 44.16 Demographic transition model for changes in population growth rates and sizes, correlated with long-term changes in economy.

The demographic transition model is based on the social changes that occurred when western Europe and North America industrialized. Whether it can accurately predict changes in modern developing countries remains to be seen. These countries receive aid from the fully industrialized nations, but also must compete with them in a global market.

Resource Consumption

Per capita resource consumption rises with economic and industrial development. Ecological footprint analysis is one method of comparing resource use. An **ecological footprint** is the amount of Earth's surface required to support a particular level of development and consumption in a sustainable fashion. **TABLE 44.2** shows the per capita global footprint data for 2010. The average person in the United States has an ecological footprint nearly three times that of an average world citizen, and about nine times that of an average person living in India.

Billions of people in less developed nations hope to someday enjoy a lifestyle like an average American. Ecological footprint analysis tells us that Earth may not have the resources to make those dreams come true. With current technology, providing everyone alive with an average American's lifestyle would require four times the sustainable resources available on Earth.

Table 44.2 Ecological Footprints*

Country	Hectares per Capita
United States	8.0
Canada	7.0
France	5.0
United Kingdom	4.9
Japan	4.7
Russian Federation	4.4
Mexico	3.0
Brazil	2.9
China	2.2
India	0.9
World Average	2.7

** 2010 data from www.footprintnetwork.org*

TAKE-HOME MESSAGE 44.8

What factors will affect future changes in the human population?

✔ Fertility rates have been declining but remain above replacement level worldwide. Many young people are about to begin reproducing.

✔ Industrialization of less developed countries is predicted to decrease their birth rates over time, as it did for currently industrialized nations.

demographic transition model Model describing changes in birth and death rates that occur as a region becomes industrialized.
ecological footprint Area of Earth's surface required to sustainably support a particular level of development and consumption.
replacement fertility rate Fertility rate at which each woman has, on average, one daughter who survives to reproductive age.
total fertility rate Average number of children the women of a population bear over the course of a lifetime.

A Honking Mess (revisited)

In 2009, investigators from the Federal Aviation Administration were asked to determine why both engines of a passenger plane failed, forcing the plane to land in the Hudson River. The pilot had reported a bird strike, and the investigators found bits of feather, bone, and muscle in the plane's wing flaps and engines (right). Samples of this tissue were sent to the Smithsonian Institute,

which analyzed the DNA. Unique sequences in the DNA identified the tissue in both engines as Canada goose. One engine had female goose DNA. The other had male and female DNA. Researchers were even able to tell which population of geese these unlucky birds belonged to. The mix of hydrogen isotopes varies with latitude, so the isotope mix in a feather provides information about where that feather developed. The birds were migratory; their feathers had developed in Canada, not in New York.

summary

Section 44.1 A **population** is a group of individuals of the same species that live in the same area and interbreed. Canada geese in the United States include migratory populations and resident ones.

Section 44.2 **Ecology** concerns interactions among populations and between populations and the environment. **Demographics** describe populations. **Plot sampling** or **mark–recapture sampling** are used to estimate **population size**. Other demographics include **population density** and **population distribution**, which is most often clumped. **Age structure** is the proportion of individuals in each age category. The size of the **reproductive base** affects population growth.

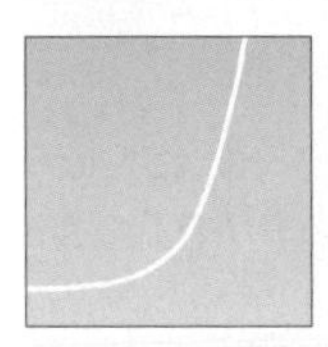

Section 44.3 **Immigration** and **emigration** affect population size. The per capita birth rate minus the per capita death rate is a population's **per capita growth rate** (*r*). **Zero population growth** occurs when birth rate equals death rate.

With **exponential growth**, the population size increases at a fixed rate. In each interval, the number of individuals added is some fixed proportion of the current population size. The time required for a population to double is the doubling time. The maximum possible rate of increase under optimal conditions is a species' **biotic potential**; it is generally not reached.

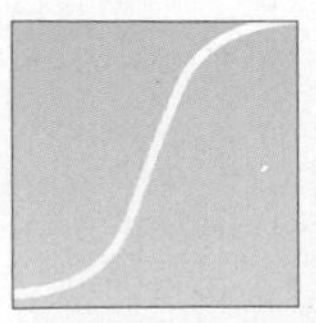

Section 44.4 Limiting factors constrain the growth of natural populations. **Density-independent limiting factors** are conditions or events that are unaffected by crowding. **Logistic growth** occurs because of constraints imposed by **density-dependent limiting factors**. With logistic growth, a population size rises exponentially, then levels off as the population nears **carrying capacity** (***K***).

Sections 44.5, 44.6 The time to maturity, number of reproductive events, number of offspring per event, and life

span are aspects of a life history pattern. A **cohort** is a group of individuals that were born at the same time. Three types of **survivorship curves** are common: a high death rate late in life, a constant rate at all ages, or a high rate early in life. Life histories have a genetic basis and are subject to natural selection. At low population density, ***r*-selection** favors quickly producing as many offspring as possible. At a higher population density, ***K*-selection** favors investing more time and energy in fewer, higher-quality offspring. Predation also affects life histories. Most species have both *r*-selected and *K*-selected traits.

Sections 44.7, 44.8 The human population is currently more than 7 billion. Expansion into new habitats and the invention of agriculture opened the way for early increases in population size. Later, medical and technological innovations lowered death rates, while birth rates continued to increase. The global **total fertility rate** is now declining, but it remains above the **replacement fertility rate**. The **demographic transition model** predicts that economic development may slow population growth. World resource consumption will probably continue to rise because a highly developed nation has a much larger **ecological footprint** than a developing one.

self-quiz

Answers in Appendix VII

1. Most commonly, individuals of a population show a ________ distribution within their habitat.
   a. clumped
   b. random
   c. nearly uniform
   d. none of the above

2. The rate at which population size grows or declines depends on the rate of ________ .
   a. births
   b. deaths
   c. immigration
   d. emigration
   e. a and b
   f. all of the above

3. Suppose 200 fish are marked and released in a pond. The following week, 200 fish are caught and 100 of them have marks. There are about ________ fish in this pond.
   a. 200
   b. 300
   c. 400
   d. 2,000

CREDIT: (in text) National Transportation Safety Board.

data analysis activities

Iguana Decline In 1987, Martin Wikelski began a long-term study of marine iguanas in the Galápagos Islands. He marked iguanas on two islands—Genovesa and Santa Fe—and collected data on how their body size, survival, and reproductive rates varied over time. He found that because iguanas eat algae and have no predators, deaths usually result from food shortages, disease, or old age. In January 2001, an oil tanker ran aground and leaked a small amount of oil into the waters near Santa Fe. FIGURE 44.17 shows the number of marked iguanas that Wikelski and his team counted in their study populations just before the spill and about a year later.

1. Which island had more marked iguanas at the time of the first census?

2. How much did the population size on each island change between the first and second census?

3. Wikelski concluded that changes on Santa Fe were the result of the oil spill, rather than sea temperature or other climate factors common to both islands. How would the census numbers be different from those he observed if an adverse event had affected both islands?

FIGURE 44.17 Shifting numbers of marked marine iguanas on two Galápagos islands. An oil spill occurred near Santa Fe just after the January 2001 census (orange bars). A second census was carried out in December 2001 (green bars).

4. A population of worms is growing exponentially in a compost heap. Thirty days ago there were 300 worms and now there are 600. How many worms will there be thirty days from now, assuming conditions remain constant?
   a. 1,200 b. 1,600 c. 3,200 d. 6,400

5. For a given species, the maximum rate of increase per individual under ideal conditions is its _______ .
   a. biotic potential
   b. carrying capacity
   c. life history pattern
   d. age structure

6. _______ is a density-independent factor that influences population growth.
   a. Resource competition
   b. Infectious disease
   c. Predation
   d. Harsh weather

7. A life history pattern is a set of adaptations that influence an individual's _______ .
   a. longevity
   b. fertility
   c. age at reproductive maturity
   d. all of the above

8. The human population is now about 7 billion. It reached 6 billion in _______ .
   a. 2007 b. 1999 c. 1802 d. 1350

9. Compared to the less developed countries, the highly developed ones have a higher _______ .
   a. death rate
   b. birth rate
   c. total fertility rate
   d. resource consumption rate

10. Species that live in unpredictable habitats are more likely to show traits that are favored by _______ .
   a. *r*-selection b. *K*-selection

11. All members of a cohort are the same _______ .
   a. sex b. size c. age d. weight

CENGAGE brain .com To access course materials, please visit www.cengagebrain.com.

12. The ecological footprint of a person in the United States is about _______ that of a person in India.
   a. half b. twice c. one-ninth d. nine times

13. Match each term with its most suitable description.

___ carrying capacity	a. maximum rate of increase per individual under ideal conditions
___ exponential growth	b. population growth plots out as an S-shaped curve
___ biotic potential	c. maximum number of individuals sustainable by the resources in a given environment
___ limiting factor	d. population growth plots out as a J-shaped curve
___ logistic growth	e. essential resource that restricts population growth when scarce

critical thinking

1. Think back to Section 44.6. When researchers moved guppies from populations preyed on by cichlids to a habitat with killifish, the life histories of the transplanted guppies evolved. They came to resemble those of guppy populations preyed on by killifish. Males became gaudier; some scales formed larger, more colorful spots. How might a decrease in predation pressure on sexually mature fish allow this change?

2. The age structure diagrams for two hypothetical populations are shown at right. Describe the growth rate of each population and discuss the current and future social and economic problems that each is likely to face.

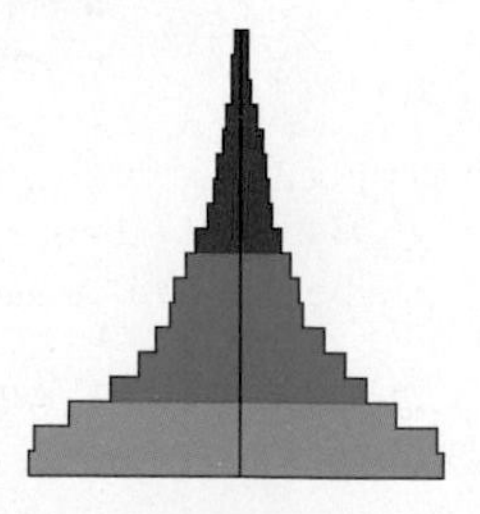

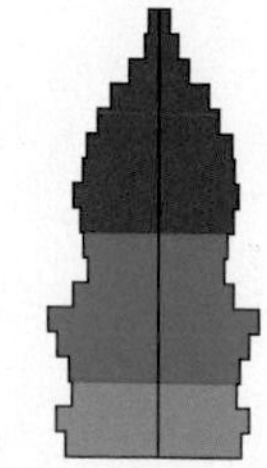

45 Community Ecology

LEARNING ROADMAP

In this chapter, you will see how natural selection (Section 16.4) and coevolution (17.12) shape communities. You will revisit plant–pollinator interactions (22.8), the role of fungi as partners (23.7), and root nodules (28.3). Knowledge of biogeography (16.2) will help you understand how communities in different regions differ.

COMMUNITY CHARACTERISTICS

A community consists of all species in a habitat. A habitat's history, its biological and physical characteristics, and the interactions among species affect community structure.

TYPES OF SPECIES INTERACTIONS

Commensalism, mutualism, competition, predation, and parasitism are types of interspecific interactions. They influence the population size of participating species.

SUCCESSIONAL CHANGES

Community structure changes over time. When a new community forms, early-arriving species often alter the habitat in a way that facilitates their own replacement.

THE ROLE OF DISTURBANCES

Introduction or removal of a species can affect other species within a community. Some species have adapted to a recurring disturbance such as fire.

BIOGEOGRAPHIC PATTERNS

Communities in tropical regions hold the greatest number of species. Models based on studies of islands can be used to predict how many species an area will support.

The next chapter looks at the flow of energy and resources within communities and between communities and the nonliving portion of their environment. We return to global patterns of species richness as we survey biomes in Chapter 47. Chapter 48 addresses the effects of human disturbance on communities.

45.1 Fighting Foreign Fire Ants

Like most ants, red imported fire ants (*Solenopsis invicta*) nest in the ground (**FIGURE 45.1**). Accidentally step on one of their nests, and you will quickly realize your mistake. Fire ants defend their nest by stinging, and their venom causes a burning sensation that gives the ants their common name.

S. invicta is native to South America. The species first arrived in the southeastern United States in the 1930s; the ants traveled as stowaways on a cargo ship. Since then, they have gradually expanded their range across the south, and were accidentally transported to California and New Mexico. More recently, *S. invicta* has been become established in the Caribbean, Australia, New Zealand, and several countries in Asia. Genetic comparisons among *S. invicta* populations indicate that the ants involved in these recent introductions originated in the southeastern United States, rather than South America.

Increased dispersal rates among pest species are an unanticipated side effect of increased global trade and improvements in shipping. Speedier ships make quicker trips, increasing the likelihood that pests hidden away in cargo holds will survive a journey.

Spread of *S. invicta* concerns ecologists because arrival of this species typically has a negative impact on a region's native species. For example, where *S. invicta* becomes prevalent, native ant species decline. In Texas, the *S. invicta*–induced decline of native ant species threatens the Texas horned lizard. Ants are a staple of this lizard's diet, but it cannot eat the red imported fire ants that have largely replaced its normal prey.

S. invicta has also been implicated in population declines of some birds, including bobwhite quails and vireos (a type of songbird). The ants decrease the abundance of insects that the birds would normally feed to their young. They also feed on birds' eggs and on nestlings. Ground-nesting birds are at special risk.

Native plants feel the impact of an *S. invicta* invasion too. Species whose seeds are dispersed by native ants may decline when *S. invicta* replaces these natives. *S. invicta* interferes with pollination by displacing or preying on pollinators such as ground-nesting bees.

Species interactions such as competition between ant species or predation of ants on bird nestlings are one focus of community ecology. A **community** is all the species that live in a region. As you will see, species interactions and disturbances can shift community structure in both small and large ways, some predictable, and others unexpected.

community All species that live in a particular region.

FIGURE 45.1 Red imported fire ants nest mounds in a Texas pasture. The inset photo shows the ants feeding on a quail egg.

45.2 What Factors Shape Community Structure?

✔ Community structure refers to the number and relative abundances of species in a habitat.

Each species lives in a specific type of place, which we call its **habitat**, and all species in a particular habitat constitute a community. Communities often nest one inside another. For example, a community of microbial organisms lives in the termite gut. A termite is part of a larger community of organisms living on a fallen log. The log-dwellers are part of a larger forest community.

Communities that are similar in scale can differ in their species diversity. There are two components to species diversity. The first, species richness, refers to the number of species. The second is species evenness, or the relative abundance of each species. For example, a pond that has five fish species in nearly equal numbers has a higher species diversity than a pond with one abundant fish species and four rare ones.

Community structure is dynamic, which means that the array of species and their relative abundances in a community tend to change over time. Communities change over a long time span as they form and then age. They also change suddenly as a result of natural or human-induced disturbances.

Abiotic (nonbiological) factors such as rainfall, sunlight intensity, and temperature affect community structure. Such factors vary along gradients in latitude, elevation, and—for aquatic habitats—depth.

Species interactions also influence community structure. In some cases, the effect is indirect. For example, when songbirds eat caterpillars, the birds directly reduce the abundance of caterpillars, but they also indirectly benefit the trees that the caterpillars feed on.

We define direct interactions by their effects on both participants (TABLE 45.1). Consider **commensalism**, in which one species benefits and the second is unaffected by the interaction. Epiphytes, which are plants that use the branches or trunk of another plant for structural support, have a commensal relationship with their host (FIGURE 45.2). Having an elevated perch benefits the epiphyte, while the host plant is unaffected.

Species interactions may be fleeting or a long-term relationship. **Symbiosis** means "living together," and in biology this term refers to a relationship in which two species have a prolonged close association that benefits at least one of them. Commensalism, mutualism, and parasitism can be symbiotic relationships, as when commensal, mutualistic, or parasitic microorganisms live inside an animal's gut.

Regardless of whether one species helps or harms another, two species that interact closely for generations can coevolve. As Section 17.12 explained, coevolution is an evolutionary process in which each species acts as a selective agent that shifts the range of variation in the other.

Table 45.1 Direct Two-Species Interactions

Type of Interaction	Effect on Species 1	Effect on Species 2
Commensalism	Beneficial	None
Mutualism	Beneficial	Beneficial
Interspecific competition	Harmful	Harmful
Predation, herbivory, parasitism, parasitoidism	Beneficial	Harmful

FIGURE 45.2 Commensalism. Epiphytic orchid growing on a tree branch. The tree provides orchids with an elevated perch from which they can capture sunlight, and the tree is neither helped nor harmed by the orchid's presence.

TAKE-HOME MESSAGE 45.2

What factors affect community structure?

✔ The types and abundances of species in a community are affected by physical factors such as climate, as well as by species interactions.

✔ A species can be benefited, harmed, or unaffected by its interaction with another species.

commensalism Species interaction that benefits one species and neither helps nor harms the other.
habitat Type of environment in which a species typically lives.
symbiosis One species lives in or on another in a commensal, mutualistic, or parasitic relationship.

45.3 Mutualism

✔ Many species engage in mutually beneficial interactions.

Mutualism is an interspecific interaction that benefits both species. Flowering plants and their pollinators are a familiar example. In some cases, coevolution of two species results in a mutual dependence. For example, there are several species of yucca plant and each is pollinated by a single species of yucca moth, whose larvae develop on that plant species alone (**FIGURE 45.3**). More often, mutualistic relationships are less exclusive. Most flowering plants have more than one pollinator, and most pollinators provide their service to more than one species of plant.

Photosynthetic organisms often supply sugars to their nonphotosynthetic partners, as when plants lure pollinators with nectar. In addition, many plants make sugary fruits that attract seed-dispersing animals. Plants also provide sugars to mycorrhizal fungi and nitrogen-fixing bacteria (Section 28.3). The plants' fungal or bacterial symbionts return the favor by supplying their host with other essential nutrients. Similarly, photosynthetic dinoflagellates provide sugars to corals (Section 21.6), and photosynthetic bacteria and algae feed their fungal partner in a lichen (Section 23.7).

Animals often share ingested nutrients with mutualistic microorganisms that live in their gut. For example, *Escherichia coli* bacteria living in your colon provide you with essential vitamin K. In return, they obtain a steady supply of food and a nice, warm place to live.

Other mutualisms involve protection. For example, an anemonefish and a sea anemone fend off one another's predators (**FIGURE 45.4**). Ants protect bull acacia trees from leaf-eating insects, and in return the tree houses the ant in special hollow thorns and provides them with sugar-rich foods.

From an evolutionary standpoint, mutualism is best considered as a case of reciprocal exploitation. Each participant increases its own fitness by extracting a resource, such as protection or food, from its partner. If taking part in the mutualism has a cost, selection will favor individuals who minimize that cost. Consider nectar production, which is energetically costly for a flower, but serves as a necessary payoff to pollinators. A flower that produces the minimum amount of nectar necessary to keep pollinators coming will be at a selective advantage over one that expends additional energy to produce a more generous serving of nectar.

FIGURE 45.3 An obligate mutualism. Each species of yucca plant (left) has a relationship with one species of yucca moth (right). After a female moth mates, she collects pollen from a yucca flower and places it on the stigma of another flower, then lays her eggs in that flower's ovary. Moth larvae develop in the fruit that develops from the floral ovary. When mature, the larvae gnaw their way out and disperse. Seeds that larvae did not eat give rise to new yucca plants.

FIGURE 45.4 Mutual protection. The stinging tentacles of this sea anemone (*Heteractis magnifica*) protect its partner, a pink anemonefish (*Amphiprion perideraion*) from fish-eating predators. In return, the anemonefish chases away fish that eat sea anemone tentacles. The anemonefish secretes a special mucus that prevents the anemone from stinging it.

TAKE-HOME MESSAGE 45.3

What are the effects of participating in a mutualism?

✔ A mutualism benefits both participants.

✔ In some cases, two species form an exclusive partnership. In others, a species provides benefits to, and receives benefits from, multiple species.

✔ Participating in a mutualism has costs as well as benefits. Selection favors individuals who maximize their benefits, while minimizing their costs.

mutualism Species interaction that benefits both participants.

45.4 Competitive Interactions

✔ Individuals of different species often compete for access to limiting resources.

Interspecific competition, or competition between species, is not usually as intense as competition within a species. The requirements of two species might be similar, but they will nearly always differ more than the requirements of two members of the same species.

The types of resources and environmental conditions an organism requires, and the manner in which it interacts with its living and nonliving environment define its **ecological niche**. Aspects of an animal's niche include the temperature range it can tolerate, the species it eats and is eaten by, and the places it can breed. A description of a flowering plant's niche would include its soil, water, light, and pollinator requirements. The more similar the niches of two species are, the more intensely the species will compete.

Competition takes two forms. With the first, interference competition, one species actively prevents another from accessing some resource. As an example, one species of scavenger will often chase another away from a carcass (**FIGURE 45.5**). As another example, some plants use chemical weapons against potential competition. Aromatic chemicals that ooze from tissues of sagebrush plants, black walnut trees, and eucalyptus trees seep into the soil around these plants. The chemicals prevent other plants from germinating or growing. The second type of competition is exploitative, meaning each species reduces the amount of a resource available to the other simply by using the resource. For example, deer and blue jays both eat acorns. The more acorns birds eat, the fewer there are for deer and vice versa.

FIGURE 45.5 Interspecific interference competition among scavengers. After facing off over a carcass (top), an eagle attacked a fox with its talons (bottom). The fox then retreated, leaving the eagle to exploit the carcass.

Effects of Competition

Species compete most intensely when the supply of a shared resource is the main limiting factor for both (Section 44.4). In the early 1930s, G. F. Gause carried out a series of experiments that led him to describe **competitive exclusion**: Whenever two species require the same limited resource to survive or reproduce, the better competitor will drive the less competitive species to extinction in that habitat. Gause studied interactions between two species of ciliated protists (*Paramecium*) that compete for bacterial prey. He cultured the species separately and together (**FIGURE 45.6**). Within weeks, population growth of one species outpaced the other, which died out.

When competing species do coexist, the presence of each reduces the carrying capacity for the other. For example, the reproductive success of a flowering plant is decreased by competition from other species

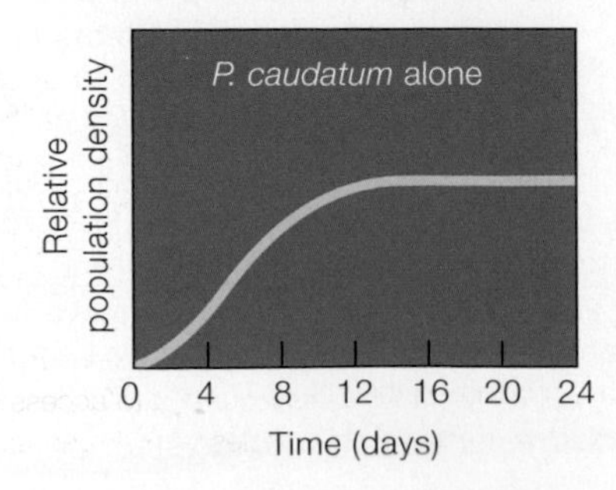

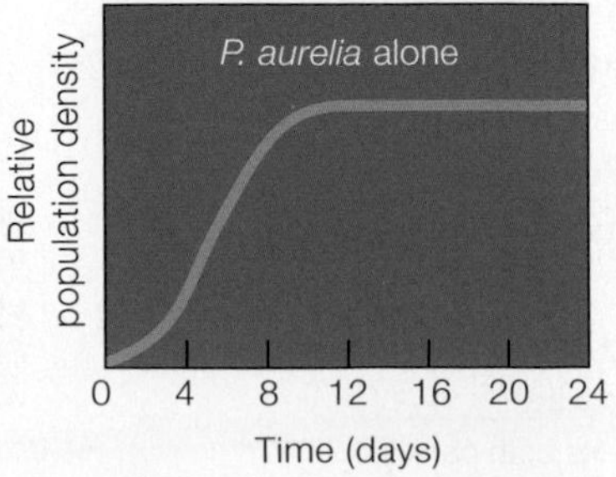

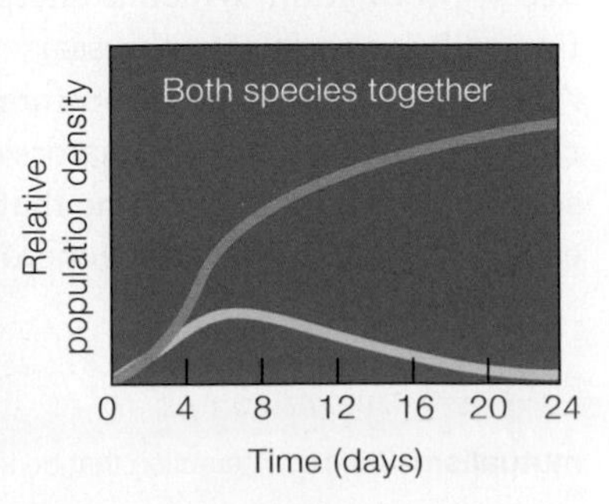

FIGURE 45.6 ▶Animated Competitive exclusion. Growth curves for two *Paramecium* species when grown separately and together.

FIGURE IT OUT Which species is the better competitor?

Answer: *P. aurelia*

that flower at the same time and rely on the same pollinators (**FIGURE 45.7**). Even strikingly different organisms can influence one another through competition. For example, wolf spiders and carnivorous plants called sundews compete for insects in Florida swamps. In experiments, sundews grown in the presence of wolf spiders make fewer flowers than sundews in spider-free enclosures. Presumably, competition for insect prey reduces the energy the plants can devote to flowering.

Mimulus *Lobelia*

FIGURE 45.7 Competing for pollinators. *Mimulus* and *Lobelia* grow together in damp meadows. To test for competition, researchers grew *Mimulus* plants either alone or with *Lobelia.* In mixed plots, pollinator visits to *Lobelia* plants frequently intervened between visits to *Mimulus*. As a result, *Mimulus* in mixed plots produced fewer seeds than those in *Mimulus*-only plots.

Resource Partitioning

Resource partitioning is an evolutionary process by which species become adapted to use a shared limiting resource in a way that minimizes competition. For example, eight species of woodpecker coexist in Oregon forests. All feed on insects and nest in hollow trees, but the details of their foraging behavior and nesting preferences vary. Differences in nesting time also help reduce competitive interactions.

Resource partitioning arises as a result of directional selection on species who share a habitat and compete for a limiting resource. In each species, individuals who differ most from the competing species with regard to resource use will have the least competition and thus leave the most offspring. Over generations, directional selection leads to **character displacement**: The range of variation for one or more traits is shifted in a direction that lessens the intensity of competition for a limiting resource.

Consider the variation in body size among populations of two species of salamander in the genus *Plethodon* (**FIGURE 45.8**). In most parts of their ranges, the two species do not overlap and their body is quite similar. Where the species coexist and compete, their difference in body size becomes more pronounced. One species, *P. cinereus*, is smaller, and the other *P. hoffmani*, is larger. The increased difference in body size may be related to a partitioning of food resources. In regions where the species overlap, *P. cinereus* tends to specialize on smaller insect prey and *P. hoffmani* on larger insects.

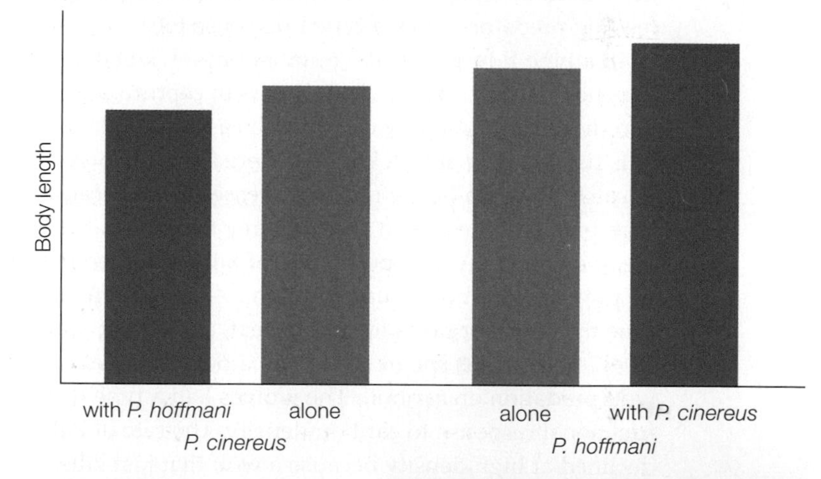

FIGURE 45.8 ▸Animated Possible evidence of character displacement in salamanders (*Plethodon*). Where *P. cinereus* (shown at right) and *P. hoffmani* coexist, their average body lengths (purple bars) differ more than they do in habitats where each species lives alone (orange bars).

TAKE-HOME MESSAGE 45.4

What happens when species compete for resources?

✔ In some interactions, one species actively blocks another's access to a resource. In other interactions, one species is simply better than another at exploiting a shared resource.

✔ When two species compete, selection favors individuals whose needs are least like those of the competing species.

character displacement Outcome of competition between two species; similar traits that result in competition become dissimilar.
competitive exclusion Process whereby two species compete for a limiting resource, and one drives the other to local extinction.
ecological niche The resources and environmental conditions that a species requires.
interspecific competition Competition between two species.
resource partitioning Species adapt to access different portions of a limited resource; allows species with similar needs to coexist.

45.5 Predator–Prey Interactions

✔ The relative abundances of predator and prey populations of a community shift over time in response to species interactions and changing environmental conditions.

Models for Predator–Prey Interactions

Predation is an interspecific interaction in which one species (the predator) captures, kills, and eats another (the prey). Predation removes a prey individual from the population immediately.

Different types of predators have different functional responses, meaning they differ in how changes in prey density affect the rate at which they kill prey. **FIGURE 45.9A** shows the three types of functional responses to increases in prey density.

In a type I response, the proportion of prey killed is constant, so the number killed in any given interval depends solely on prey density. Web-spinning spiders and other passive predators show this type of response. As the number of flies in an area increases, more and more become caught in each spider's web. Suspension-feeding predators show a type I response too.

In a type II response, the number of prey killed depends on the capacity of predators to capture, eat, and digest prey. As prey density increases, the rate of kills rises steeply at first because there are more prey to catch. Eventually, the rate of increase slows, because each predator is exposed to more prey than it can handle at one time. At this point, rate of kills is limited not by prey density, but by handling time—the amount of time the predator takes to subdue, eat, and digest prey.

FIGURE 45.9B shows data from a field study of wolf predation on caribou. The wolves had a type II functional response to caribou density. The rate of kills declined at high density because a wolf that just killed a caribou will not hunt again until it has eaten and digested the first prey.

In a type III response, the number of kills increases slowly until prey density exceeds a certain level, then rises rapidly before leveling off. Three situations can result in a type III response. First, some predators switch among prey, concentrating on the prey species that is most abundant. Second, some predators need to learn how to best capture each prey species; they get more lessons and become more efficient hunters when a particular type of prey is abundant. Third, some prey can escape predation by hiding. Only after prey density rises and some individual prey have no place to hide does the rate of kills increase.

Understanding a predator's functional response helps ecologists predict long-term effects of predation on a prey population. For example, a functional response in which the kill rate rises with prey density can help stabilize prey numbers.

Cyclic Changes in Abundance

Some predator and prey species undergo seemingly coupled cyclic changes in their abundance, with each increase or decline in prey numbers followed by a similar change in predator numbers. The oscillation in populations of Canadian lynx and the snowshoe hare is one example (**FIGURE 45.10A**). Both populations peak about every ten years, with each peak in hare numbers followed by a peak in lynx numbers.

Such cycles are predicted by a model in which prey abundance regulates predator numbers, and predator abundance regulates prey numbers. According to this model, as prey numbers increase, predators find it increasingly easy to locate and catch prey. As a result, the predator's survival and reproduction increases. After predator numbers increase, prey numbers begin to decline as a result of increased mortality and decreased reproduction. Predator numbers continue

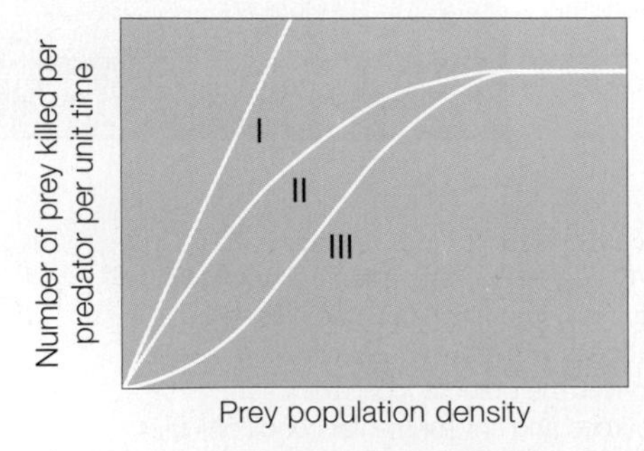

A Three theoretical types of functional response curves.

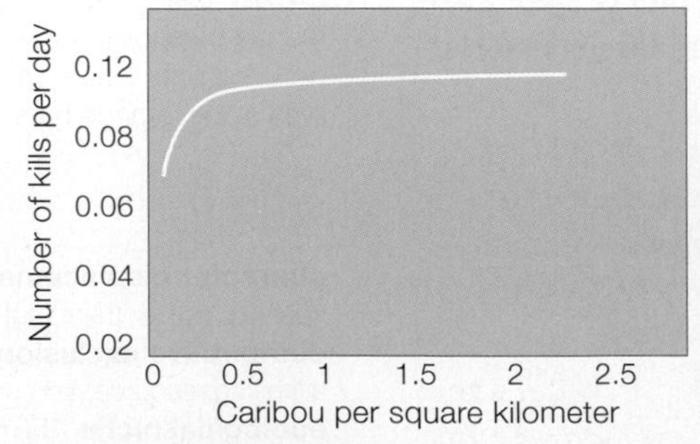

B Example of a type II response from one winter month in Alaska, during which B. W. Dale and his coworkers observed wolf packs (*Canis lupus*) feeding on caribou (*Rangifer tarandus*).

FIGURE 45.9 ▸Animated Functional responses of predators to changes in prey density.

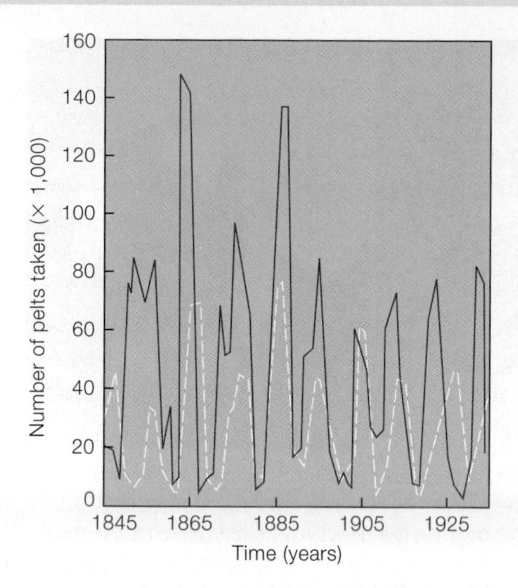

A Abundance of Canadian lynx (dashed line) and snowshoe hares (solid line), based on the numbers of pelts sold by trappers to Hudson's Bay Company during a ninety-year period.

B Canadian lynx pursuing a snowshoe hare.

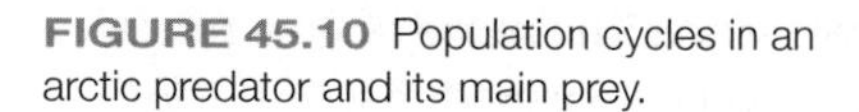

FIGURE 45.10 Population cycles in an arctic predator and its main prey.

to increase for a while, because there are still plenty of prey. Eventually, prey numbers decline to a point where predator reproduction is affected and the predator population declines. After it does, prey numbers begin to rise, and the cycle begins again.

To determine whether this model accurately explains the cause of fluctuations in lynx and hare populations, Charles Krebs and his coworkers tracked hare densities for ten years in the Yukon River Valley of Alaska. They recorded the numbers of wild hares in control plots and in a variety of experimental plots. One type of experimental plot had fencing that protected hares from mammalian predators such as lynx. Another received fertilizer that increased plant growth. Others either provided hares with supplemental food or both excluded mammalian predators and provided supplemental food. Excluding mammalian predators, providing supplemental food, and combining these treatments all significantly increased the average density of hares over that in control plots. However, none of these treatments prevented population cycling. In all plots, hare numbers rose and then declined.

Based on these data, Krebs concluded that the one predator–one prey model did not accurately describe the hare and lynx population cycles. The actual situation is far more complex. For one thing, snowshoe hares have multiple predators. They are eaten by owls and hawks, as well as lynx and other mammals. Food availability also affects hare density. Thus, the observed population oscillations may arise as a result of a three-level interaction between plants, the hares that eat them, and the predators that eat hares.

Ecologists continue to investigate the influence of predation on the density of prey populations. Creating new models and testing the ability of these models to describe what we see in nature is an essential part of this process. Understanding predator–prey dynamics is important because such information allows scientists to make valid decisions about how to best manage threatened predator and prey species.

predation One species captures, kills, and eats another.

TAKE-HOME MESSAGE 45.5

How do predator and prey populations change over time?

- ✔ Predator populations show three general patterns of response to changes in prey density.
- ✔ Population levels of prey may show recurring oscillations.
- ✔ The numbers in predator and prey populations vary in complex ways that reflect the multiple levels of interaction in a community.

45.6 Evolutionary Arms Races

✔ Predators select for better prey defenses, and prey select for more efficient predators.

Coevolution of Predators and Prey

Predator and prey species exert selection pressure on one another. Suppose a mutation gives members of a prey species a more effective defense. Over generations, directional selection will cause this mutation to spread through the prey population. If some members of a predator population have a trait that makes them better at thwarting the improved defense, these predators and their descendants will be at an advantage. Thus, predators exert selection pressure that favors improved prey defense, which in turn exerts selection pressure on predators, and so the process continues over many generations.

You have already learned about some defensive adaptations. Many prey species have hard or sharp parts that make them difficult to eat. Think of a snail's shell or a sea urchin's spines. Cnidarians such as sea anemones and corals have stinging cells on their tentacles. Other prey contain chemicals that taste bad to predators or sicken them. Most defensive toxins in animals are from the plants that they eat. For example, a monarch butterfly caterpillar takes up chemicals from the milkweed plant that it feeds on. A bird that later eats the butterfly will be sickened by these chemicals.

Well-defended prey often have **warning coloration**, a conspicuous color pattern that predators learn to avoid. Monarch butterflies are bright orange, and many stinging wasps and bees share a pattern of black and yellow stripes (**FIGURE 45.11A**). The similar appearance of bees and wasps is an example of **mimicry**, an evolutionary pattern in which one species comes to resemble another. The tendency of well-defended species to have similar coloration is called Müllerian mimicry, after the German naturalist who first described the phenomenon. Müller recognized that well-defended species who share the same type of predator would benefit by having a similar appearance. The more often a predator is stung by a black and yellow striped insect, the less likely it is to attack similar looking insects in the future.

In Batesian mimicry, also named for the scientist who first described it, a species that lacks a defense mimics the appearance of a well-defended species. For example, some stingless insects resemble stinging bees or wasps (**FIGURE 45.11B–D**). These impostors benefit when predators avoid them after a painful encounter with the better-defended species that they mimic.

Stinging is an example of a defensive behavior that can repel a potential predator. Section 1.7 described how eyespots and a hissing sound protect some butterflies from predatory birds. Similarly, a lizard's tail may detach from the body and wiggle a bit as a distraction, allowing the rest of the lizard to escape.

Skunks squirt a foul-smelling, irritating repellent, as do some darkling beetles (**FIGURE 45.12**). Both

FIGURE 45.12 Defense and counter defense. (**A**) *Eleodes* beetles defend themselves by spraying irritating chemicals at predators. (**B**) This defense is ineffective against grasshopper mice, who plunge the chemical-spraying end of the beetle into the ground and devour the insect from the head down.

A Stinging wasp.

B One of its edible mimics.

C Another edible mimic.

D And another edible mimic.

FIGURE 45.11 Examples of mimicry. Edible insect species often resemble toxic or unpalatable species that are not at all closely related. (**A**) A yellow jacket can deliver a painful sting. Nonstinging wasps (**B**), beetles (**C**), and flies (**D**) benefit by having a similar appearance.

CREDITS: (11A–C) Edward S. Ross; (11D) © Nigel Jones; (12) Thomas Eisner, Cornell University.

skunks and beetles announce their intention to spray by assuming a distinctive posture, with their posterior end pointed toward the threat. Some beetles that cannot spray mimic this posture.

Camouflage is a body shape, color pattern, or behavior that allows an individual to blend into its surroundings and avoid detection. Prey benefit when camouflage hides them from predators (FIGURE 45.13A), and predators benefit when it hides them from prey (FIGURE 45.13B,C).

Other predator adaptations include sharp teeth and claws that can pierce protective hard parts. Speedy prey select for faster predators. For example, the cheetah, the fastest land animal, can run 114 kilometers per hour (70 mph). Its preferred prey, Thomson's gazelles, run 80 kilometers per hour (50 mph).

Coevolution of Herbivores and Plants

With **herbivory**, an animal feeds on plants. The number and type of plants in a community can influence the number and type of herbivores present.

Two types of defenses have evolved in response to herbivory. Some plants have adapted to withstand and recover quickly from the loss of their parts. For example, prairie grasses are seldom killed by native grazers such as bison. The grasses have a fast growth rate and store enough resources in their roots to replace the shoots lost to grazers.

Other plants have traits that deter herbivory. Physical deterrents include spines, thorns, and tough leaves that are difficult to chew. Many plants produce secondary metabolites (Section 22.9) that are unpalatable to herbivores or sicken them. Ricin, the toxin made by castor bean plants (Section 9.1), makes all eukaryotic herbivores ill. Caffeine in coffee beans and nicotine in tobacco leaves are defenses against insects.

Capsaicin, the compound that makes some peppers "hot," is an evolved defense against seed-eating mammals. Rodents avoid pepper fruits, leaving them to be eaten by birds. The pepper benefits by deterring rodent seed eaters because rodents chew up and kill seeds, whereas birds excrete the seeds alive and intact.

A Reedlike defensive posture of a bittern. The bird even sways in the wind like a reed.

B Frilly pink body parts of a flower mantis hide it from insect prey attracted to the real flowers.

C Fleshy protrusions give a predatory scorpionfish the appearance of an algae-covered rock. When algae-eating fish come close for a nibble, they end up as prey.

FIGURE 45.13 Camouflage in prey and predators.

When a well-defended plant is abundant, herbivores capable of overcoming the plant's defenses will be at a selective advantage. Such a selection process has given koalas the ability to feed on eucalyptus leaves, which are tough and contain toxic oils. Koalas have evolved specialized teeth to shred the leaves and special liver enzymes that allow them to detoxify the oils.

camouflage Body coloration, patterning, form, or behavior that helps predators or prey blend with the surroundings and possibly escape detection.
herbivory An animal feeds on plant parts.
mimicry A species evolves traits that make it similar in appearance to another species.
warning coloration In many well-defended or unpalatable species, bright colors, patterns, and other signals that predators learn to recognize and avoid.

TAKE-HOME MESSAGE 45.6

How do predation and herbivory influence community structure?

✔ In any community, predators and prey coevolve, as do plants and the herbivores that feed on them.

✔ Defensive adaptations in plants and prey can limit the ability of predators or herbivores to exploit some species in their community.

45.7 Parasites and Parasitoids

✔ Predators have only a brief interaction with prey, but parasites and parasitoids live on or in their hosts.

Parasitism

With **parasitism**, one species (the parasite) benefits by feeding on another (the host), without killing it immediately. Endoparasites such as parasitic roundworms live and feed inside their host (**FIGURE 45.14A**). An ectoparasite such as a tick feeds while attached to a host's external surface (**FIGURE 45.14B**).

Parasitism has evolved in a diverse variety of groups. Bacterial, fungal, protistan, and invertebrate parasites feed on vertebrates. Lampreys (Section 25.3) attach to and feed on other fish. There are even a few parasitic plants that withdraw nutrients from other plants (**FIGURE 45.15**).

Although most parasites do not kill their hosts, parasitism can still decrease the size of a host population. The presence of parasites can weaken a host, making it more vulnerable to predation or less attractive to potential mates. Some parasites can make a host sterile or shift the sex ratio among its offspring.

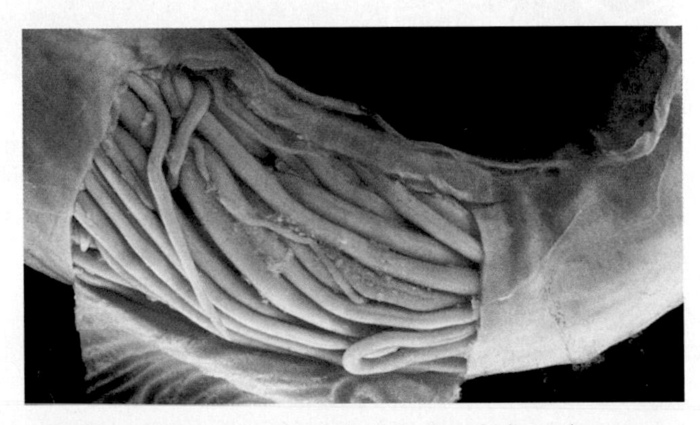

A Endoparasitic roundworms in the intestine of a host pig.

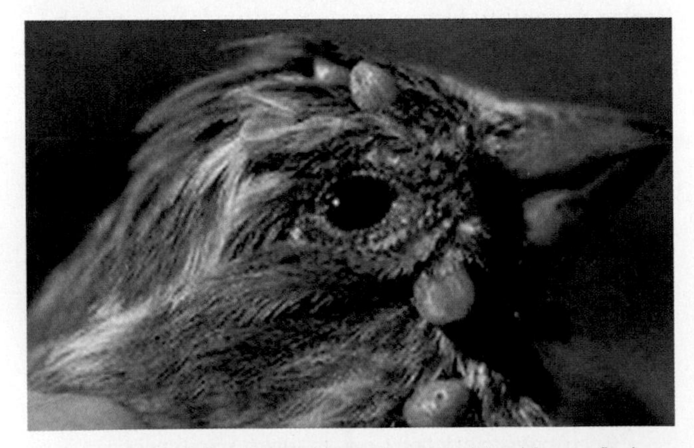

B Ectoparasitic ticks attached to and sucking blood from a finch.

FIGURE 45.14 Parasites of animals.

FIGURE 45.15 Dodder (*Cuscuta*), a parasitic flowering plant with almost no chlorophyll, shown here growing on grass. Modified roots penetrate the host's vascular tissues and absorb water and nutrients from them.

Adaptations to a parasitic lifestyle include traits that allow the parasite to locate hosts and to feed undetected. For example, ticks that feed on mammals or birds move toward a source of heat and carbon dioxide, which may be a potential host. A chemical in tick saliva acts as a local anesthetic, preventing the host from noticing the feeding tick. Endoparasites often have adaptations that help them evade a host's immune defenses.

Among hosts, traits that minimize the negative effects of parasites confer a selective advantage. For example, the allele that causes sickle-cell anemia persists at high levels in some human populations because having one copy of the allele increases the odds of surviving malaria (Section 17.7). Grooming and preening behavior are adaptations that minimize the impact of ectoparasites. Some animals produce chemicals that repel or interfere with parasite activity. For example, one type of seabird (the crested auklet) produces a citrus-scented secretion that repels mosquitoes.

Brood Parasites—Strangers in the Nest

With **brood parasitism**, one species benefits by having another raise its offspring. The European cuckoos described in Section 43.3 are brood parasites, as are North American cowbirds. Not having to invest in parental care allows a female cowbird to produce a large number of eggs, in some cases as many as thirty in a single reproductive season.

When the presence of brood parasites decreases the reproductive rate of the host species, selection favors host individuals that can detect and eject foreign young. Some brood parasites counter this host defense by producing eggs that closely resemble those of the host species. In cuckoos, different subpopulations have

FIGURE 45.16 Brood parasitism. Ants tend a butterfly larva that smells like an ant and imitates sounds made by a queen ant. Ant workers feed such caterpillars more eagerly than their own larvae.

FIGURE 45.17 Biological control agent: a commercially raised parasitoid wasp about to deposit a fertilized egg in an aphid. The wasp larva will devour the aphid from the inside.

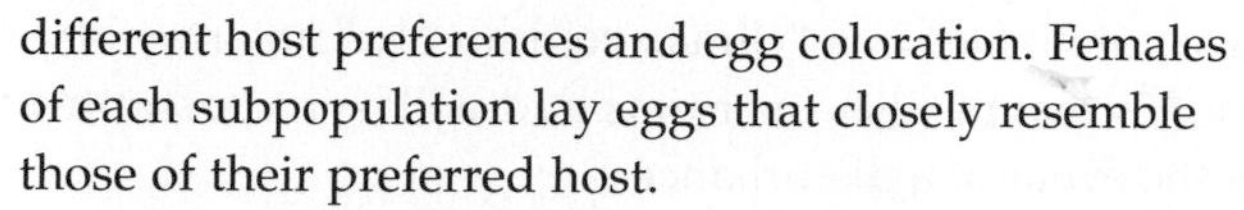

different host preferences and egg coloration. Females of each subpopulation lay eggs that closely resemble those of their preferred host.

Many insects are brood parasites. For example, the Alcon blue butterfly outsources care of its young to ants (**FIGURE 45.16**). Caterpillars of these butterflies smell and sound like ants. Worker ants, fooled by these false cues, carry the caterpillars into their nest, where ants care for them as if they were members of the colony, feeding them and protecting them from predators.

Parasitoids

Parasitoids are insects that lay their eggs in the bodies of other insects. Larvae that hatch from these eggs develop in the host's body, devour its tissue, and eventually kill it. The presence of parasitoids reduces the size of a host population in two ways. First, as the parasitoid larvae grow inside their host, they withdraw nutrients and prevent the host from developing normally and reproducing. Second, the presence of these larvae eventually leads to the death of the host.

Biological Pest Controls

Parasites and parasitoids are commercially raised and released as a form of biological pest control (**FIGURE 45.17**). Biological pest control involves the use of a pest's natural enemies and it has some advantages over pesticides. Most chemical insecticides kill a wide variety of insects, including some that help control pests. Insecticides also have negative effects on human health. By contrast, the parasites and parasitoids used as biological control agents usually target only a limited number of species.

For a species to be an effective biological control, it must be adapted to take advantage of a specific host species and to survive in that species' habitat. The ideal biological control agent excels at finding the target host species, has a population growth rate comparable to the host's, and has offspring that disperse widely.

Of course, introducing a species into a community as a biological control agent always entails some risks. The introduced control agent sometimes attacks nontargeted species in addition to, or instead of, those that they were expected to control. For example, parasitoid wasps were introduced to Hawaii to control stinkbugs that feed on some Hawaiian crops. Instead, the parasitoids decimated the population of koa bugs, Hawaii's largest native bug. Introduced parasitoids have also been implicated in ongoing declines of many native Hawaiian butterfly and moth populations.

brood parasitism One egg-laying species benefits by having another raise its offspring.
parasitism Relationship in which one species withdraws nutrients from another species, without immediately killing it.
parasitoid An insect that lays eggs in another insect, and whose young devour their host from the inside.

TAKE-HOME MESSAGE 45.7

What are the effects of parasites, brood parasites, and parasitoids?

✔ Parasites reduce the reproductive rate of host individuals by withdrawing nutrients from them.

✔ Brood parasites reduce the reproductive rate of hosts by tricking them into caring for young that are not their own.

✔ Parasitoids reduce the number of host organisms by preventing reproduction and eventually killing the host.

45.8 Ecological Succession

✔ Which species are present in a community depends on physical factors such as climate, biotic factors such as which species arrived earlier, and the frequency of disturbances.

Successional Change

The array of species in a community can change over time. Species often alter the habitat in ways that allow other species to come in and replace them. We call this type of change ecological succession.

The process of succession starts with the arrival of **pioneer species**, which are opportunistic colonizers of new or newly vacated habitats. Pioneer species grow and mature fast, and produce many offspring capable of dispersing. Later, other species replace the pioneers. Then the replacements are replaced, and so on.

Primary succession is a process that begins when pioneer species colonize a barren habitat with no soil, such as a new volcanic island or land exposed by the retreat of a glacier (**FIGURE 45.18**). The earliest pioneers to colonize a new habitat are often mosses and lichens (Sections 22.4 and 23.7), which are small, have a brief life cycle, and can tolerate intense sunlight, extreme temperature changes, and little or no soil. Some hardy annual flowering plants with wind-dispersed seeds are also frequent pioneers.

Pioneers help build and improve the soil. In doing so, they often set the stage for their own replacement. For example, many pioneer species partner with nitrogen-fixing bacteria, so they can grow in nitrogen-poor habitats. Seeds of later arrivals find shelter in mats of low-growing pioneer vegetation. Organic wastes and remains that accumulate add volume and nutrients to soil, thereby helping other species to become established. Later successional species often shade and eventually displace earlier ones.

In **secondary succession**, a disturbed area within a community recovers. This process occurs in abandoned agricultural fields and burned forests. Because improved soil is present from the start, secondary succession usually occurs faster than primary succession.

Factors That Influence Succession

The concept of ecological succession was first developed in the late 1800s, and it was hypothesized to be a predictable and directional process. Physical factors such as climate, altitude, and soil type were considered to be the main determinants of which species appeared in what order during succession. In this view, succession culminates in a "climax community," an array of species that persists over time and will be reconstituted in the event of a disturbance.

Ecologists now realize that the species composition of a community changes frequently and unpredictably. Communities do not journey along a well-worn path to a predetermined climax state.

Random events can determine the order in which species arrive in a habitat and thus affect the course of succession. The arrival of a particular species may make it easier or more difficult for others that follow to become established.

Ecologists had an opportunity to investigate these types of species interactions after the 1980 eruption of

A Retreat of a glacier exposes rock and gravel.

B Lichens, mosses, and some annual flowering plants are early pioneers.

C Deciduous trees such as alder, willow, and cottonwood form dense thickets.

D Conifers such as hemlock and spruce move in and overgrow other species.

FIGURE 45.18 ▸Animated One pathway of primary succession in Alaska's Glacier Bay region.

A The 1980 Mount Saint Helens eruption obliterated the community at the base of this Cascade volcano.

B The first pioneer species arrived less than a decade after the eruption.

FIGURE 45.19 A natural laboratory for studies of succession.

Mount Saint Helens leveled about 600 square kilometers (235 square miles) of forest in Washington State (**FIGURE 45.19**). Ecologists recorded the natural pattern of colonization and carried out experiments in plots inside the blast zone. The results showed that the presence of some pioneers helped other, later-arriving plants become established. Other pioneers kept the same late arrivals out.

Disturbances also influence a community's species composition. According to the **intermediate disturbance hypothesis**, species richness is greatest in communities where disturbances are moderate in intensity and frequency (**FIGURE 45.20**). In such habitats, there is enough time for new colonists to arrive and become established but not enough for competitive exclusion to cause extinctions.

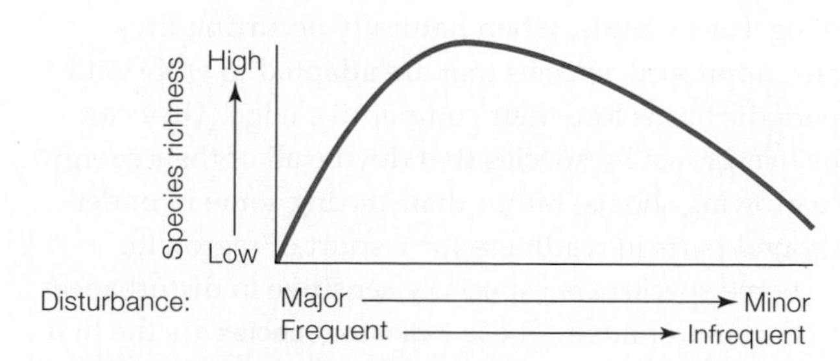

FIGURE 45.20 Graph showing the pattern predicted by the intermediate disturbance hypothesis. Species richness is highest when disturbances are moderate in intensity and frequency.

In short, the modern view of succession holds that three types of factors affect communities: (1) physical factors such as climate, (2) chance events such as the order of species arrival, and (3) the frequency and extent of disturbances. The sequence of species arrivals and the frequency and extent of disturbances vary in unpredictable ways between communities. As a result, it is difficult to predict exactly how the composition of any particular community will change in the future.

intermediate disturbance hypothesis Species richness is greatest in communities where disturbances are moderate in their intensity or frequency.
pioneer species Species that can colonize a new habitat.
primary succession A new community colonizes an area where there is no soil.
secondary succession A new community develops in a disturbed site where the soil that supported a previous community remains.

TAKE-HOME MESSAGE 45.8
What is ecological succession?

✔ Succession is a process in which one array of species replaces another over time. It can occur in a barren habitat (primary succession) or a region where a community previously existed (secondary succession).

✔ Physical factors affect succession, but so do species interactions and disturbances. As a result, the course that succession will take in any particular community is difficult to predict.

CREDITS: (19A) R. Barrick/USGS; (19B) USGS; (20) © Cengage Learning.

45.9 Species Introduction, Loss, and Other Disturbances

✔ The loss or addition of even one species may destabilize the number and abundances of species in a community.

✔ Some species adapted to being disturbed are at a competitive disadvantage if disturbances do not occur.

The Role of Keystone Species

A **keystone species** is a species that has a disproportionately large effect on a community relative to its abundance. Robert Paine coined this term after observing what happened after he removed one species (a sea star) from the rocky intertidal community on the California coast. The sea star, *Pisaster ochraceus*, preys on chitons, limpets, barnacles, and mussels. When Paine experimentally removed *Pisaster* from some plots, mussels took over, crowding out seven other species of invertebrates and reducing the diversity of algae living in the plots (**FIGURE 45.21**). In control plots, where *Pisaster* remained, there was no similar decline in diversity. Paine concluded that this sea star is a keystone species. It normally keeps species diversity in the intertidal zone high by preventing competitive exclusion of other invertebrates and of algae by mussels.

Keystone species need not be predators. For example, the large, herbivorous rodents called beavers can be a keystone species. A beaver cuts down trees by gnawing through their trunk, then uses felled trees to build a dam. Construction of a beaver dam creates a deep pool where a shallow stream would otherwise exist. By altering the physical conditions in a section of the stream, the beaver affects the types of fish and aquatic invertebrates that can live there.

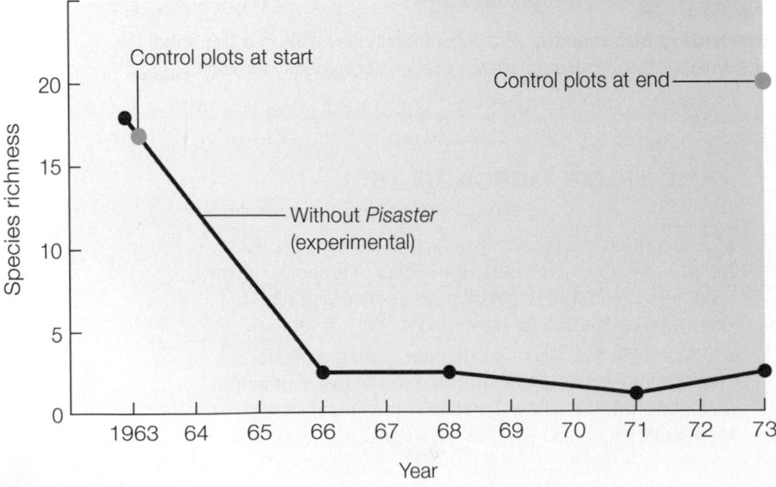

FIGURE 45.21 Evidence that the sea star *Pisaster* (right) is a keystone species. Removal of *Pisaster* from experimental plots resulted in a decline in species richness (brown dots and line). Control plots from which *Pisaster* was not removed showed no comparable decline in species richness (green dots).

FIGURE 45.22 Adapted to disturbance. Some woody shrubs, such as this toyon, resprout from their roots after a fire. In the absence of occasional fire, toyons are outcompeted and displaced by species that grow faster but are less fire resistant.

Adapting to Disturbance

When a community is repeatedly subjected to a particular type of disturbance, species that withstand or benefit from that disturbance are at a selective advantage. For example, in areas subject to periodic fires, some plants produce seeds that require fire to germinate. This ensures that their seedlings come up only after a fire has cleared away most potential competitors. Other plants store resources in the roots, so they can resprout quickly after a fire (**FIGURE 45.22**).

Because different species respond differently to fire, the frequency of fire affects competitive interactions. For example, when naturally occurring fires are suppressed, species that are adapted to withstand periodic burns lose their competitive edge. They can be overgrown by species that devote all of their energy to growing shoots, rather than storing some in underground parts in readiness for a spurt of regrowth.

Some species are especially sensitive to disturbances to the environment. These **indicator species** are the first to do poorly when conditions change, so they can provide an early warning of environmental degradation. For example, a decline in a trout population can be an

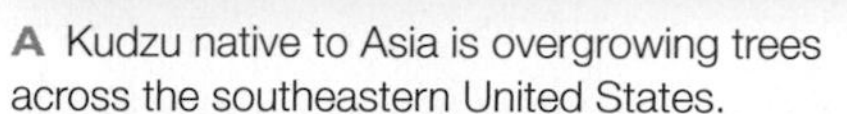

A Kudzu native to Asia is overgrowing trees across the southeastern United States.

B Gypsy moths native to Europe and Asia feed on oaks through much of the United States.

C Nutrias from South America are now in the Gulf States and in the Pacific Northwest.

FIGURE 45.23 Exotic species that are altering natural communities in the United States. To learn more visit the National Invasive Species Information Center online at www.invasivespeciesinfo.gov.

early sign of problems in a stream, because trout are highly sensitive to pollutants and cannot tolerate low oxygen levels.

Species Introductions

The arrival of a new species in a community can also cause dramatic changes. When you hear someone speaking enthusiastically about exotic species, you can safely bet the speaker is not an ecologist. An **exotic species** is a resident of an established community that dispersed from its home range and became established elsewhere. Many species do not do well outside their home range, but an exotic species becomes a permanent member of its new community.

The United states is now home to than 4,500 exotic species. An estimated 25 percent of Florida's plant and animal species are exotics. In Hawaii, 45 percent are exotic. Some species were brought in for use as food crops, to brighten gardens, or to provide textiles. Others, including red imported fire ants, arrived with cargo from distant regions.

One of the most notorious exotic species is a vine called kudzu (*Pueraria lobata*). Native to Asia, it was introduced to the American Southeast as a food for grazers and to control erosion. However, it quickly became an invasive weed. Kudzu overgrows trees, telephone poles, houses, and almost everything else in its path (**FIGURE 45.23A**).

Gypsy moths (*Lymantria dispar*) are an exotic species native to Europe and Asia. They entered the northeastern United States in the mid-1700s and now range into the Southeast, Midwest, and Canada. Gypsy moth caterpillars (**FIGURE 45.23B**) preferentially feed on oaks. Loss of leaves to gypsy moths can weaken trees, making them more susceptible to parasites and disease.

Large semiaquatic rodents called nutrias (*Myocastor coypus*) were imported from South America to be grown for their fur. They were released into the wild in many states. Along the Gulf of Mexico, nutrias now thrive in freshwater marshes (**FIGURE 45.23C**), and their voracious appetite threatens the native vegetation. In addition, their burrowing contributes to marsh erosion and damages levees, increasing risk of flooding.

exotic species A species that evolved in one community and later became established in a different one.
indicator species A species that is especially sensitive to disturbance and can be monitored to assess the health of a habitat.
keystone species A species that has a disproportionately large effect on community structure.

TAKE-HOME MESSAGE 45.9

What types of changes alter community structure?

✔ Removing a keystone species can alter the diversity of species in a community.

✔ A change in the typical frequency of a disturbance can favor some species over others.

✔ Introducing an exotic species can threaten native species.

CREDITS: (23A) Angelina Lax/Science Source; (23B) Photo by Scott Bauer, USDA/ARS; (23C) © Greg Lasley Nature Photography, www.greglasley.net.

45.10 Biogeographic Patterns in Community Structure

✔ The richness and relative abundances of species differ from one habitat or region of the world to another.

Latitudinal Patterns

Biogeography is the scientific study of how species are distributed in the natural world (Section 16.2). Perhaps the most striking pattern of species richness corresponds with distance from the equator. For most major plants and animal groups, the number of species is greatest in the tropics and declines from the equator to the poles (**FIGURE 45.24**). Consider just a few factors that help bring about and maintain this pattern.

First, tropical latitudes intercept more intense sunlight and receive more rainfall, and their growing season is longer. As one outcome, resource availability tends to be greatest and most reliable in the tropics. Thus, the tropics support a degree of specialized interrelationships not possible where species are active for shorter periods.

Second, tropical communities have been established for a long time. Some temperate communities did not start forming until the end of the last ice age. The longer a community has been established, the more time there has been for speciation to occur within it.

Third, species richness may be self-reinforcing. The number of species of trees in tropical forests is much greater than in comparable forests at higher latitudes. Where more plant species compete and coexist, more species of herbivores also coexist, partly because no single herbivore species can overcome all the defenses of all plants. In addition, more predators and parasites can evolve in response to more kinds of prey and hosts. The same principles apply to tropical reefs.

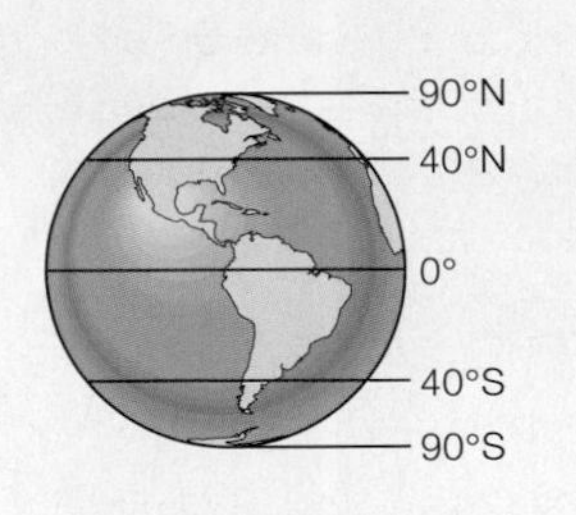

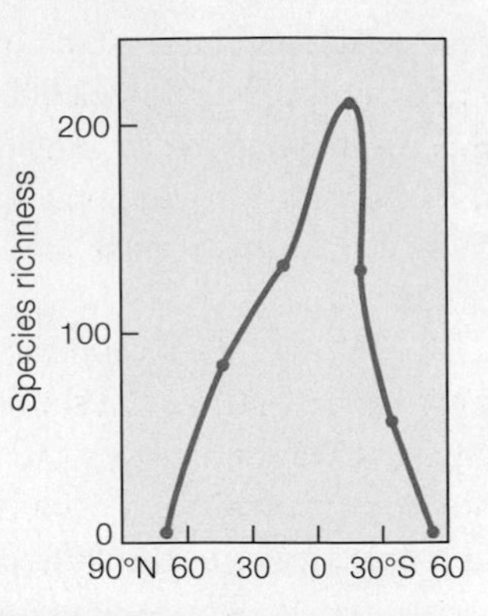

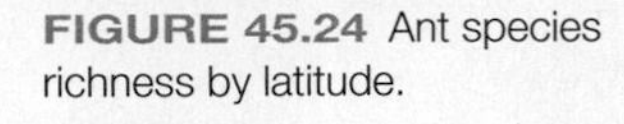

FIGURE 45.24 Ant species richness by latitude.

FIGURE 45.25 Surtsey, a volcanic island, during the time of its formation. The graph below shows the number of vascular plant species found in yearly surveys. Seagulls first began nesting on the island in 1986.

Island Patterns

In the mid-1960s, volcanic eruptions formed a new island 33 kilometers (21 miles) from the coast of Iceland. The island was named Surtsey (**FIGURE 45.25**). Bacteria and fungi were early colonists. The first vascular plant became established on the island in 1965. Mosses appeared two years later and thrived. The first lichens were found five years after that. The rate of arrivals of new vascular plants picked up after a seagull colony became established in 1986.

The number of species on Surtsey will not continue increasing forever. How many species will there be when the number levels off? The **equilibrium model of island biogeography** addresses this question. According to this model, the number of species living on any island reflects a balance between immigration rates for new species and extinction rates for established ones. The distance between an island and a mainland source of colonists affects immigration rates. An island's size affects both immigration rates and extinction rates.

Consider first the **distance effect**: Islands far from a source of colonists receive fewer immigrants than those closer to a source. Most species cannot disperse very far, so they will not turn up far from a mainland.

Species richness is also shaped by the **area effect**: Big islands tend to support more species than small ones for several reasons. First, more colonists will happen upon a larger island simply by virtue of its size. Second, larger islands are more likely to offer a variety of habitats, such as high and low elevations. This variety makes it more probable that a new arrival will find a suitable habitat. Finally, big islands can support larger populations of species than small islands. The larger a population, the less likely it is to become locally extinct as the result of some random event.

Fighting Foreign Fire Ants (revisited)

Invicta means "invincible" in Latin, and the red imported fire ant *S. invicta* lives up to its name. So far, these ants have overcome all control efforts. But scientists have a new weapon. They have imported phorid flies from Brazil for use as a biological control agent (**FIGURE 45.27A**).

The phorid flies are parasitoids and *S. invicta* is their host. The female fly lays an egg in an ant's body. The egg hatches into a larva that grows in the ant and feeds on its soft tissues. Eventually the larva enters the ant's head, and causes it to fall off (**FIGURE 45.27B**). The larva then undergoes metamorphosis to its adult form.

Phorid flies are not expected to kill off all *S. invicta* in affected areas. Rather, the hope is that the flies will reduce the density of the invader's colonies. Ecologists are also exploring other options, such as introducing pathogens that infect *S. invicta* but not native ants.

A Phorid fly attempting to lay its egg in a fire ant.

B Ant that lost its head after a phorid fly larva matured inside it.

FIGURE 45.27 Phorid flies as biological control agents.

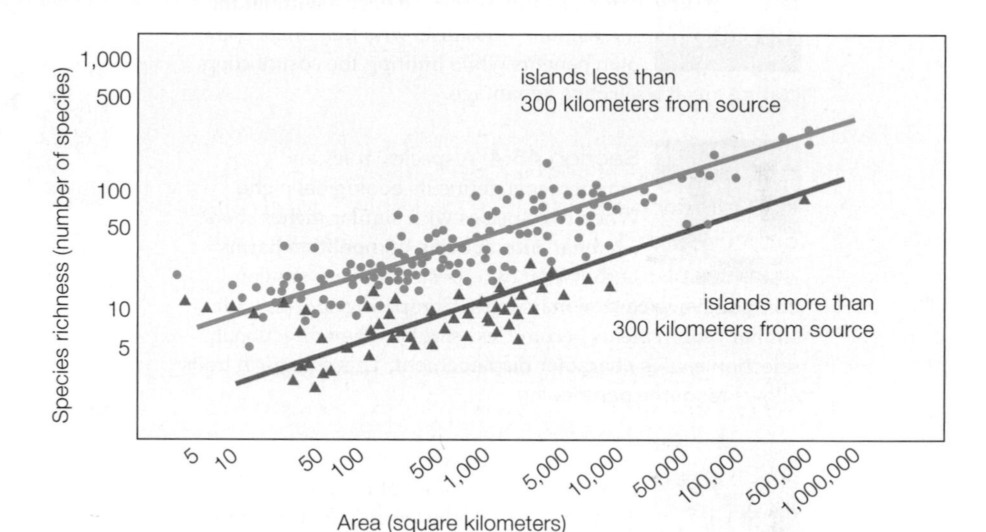

FIGURE 45.26 Island biodiversity patterns.

Distance effect: Species richness on islands of a given size declines as distance from a source of colonists rises. Green circles are values for islands less than 300 kilometers from the colonizing source. Orange triangles are values for islands more than 300 kilometers (190 miles) from a source of colonists.

Area effect: Among islands the same distance from a source of colonists, larger islands tend to support more species than smaller ones.

FIGURE IT OUT Which is likely to have more species, a 100-km^2 island more than 300 km from a colonizing source or a 500-km^2 island less than 300 km from a colonist source?

Answer: The 500-km^2 island

FIGURE 45.26 illustrates how interactions between the distance effect and the area effect can influence the equilibrium number of species on islands.

Robert H. MacArthur and Edward O. Wilson first developed the equilibrium model of island biogeography in the late 1960s. Since then the model has been modified and its use has been expanded to help scientists understand what happens on habitat islands—areas of natural habitat surrounded by a "sea" of habitat that has been disturbed by humans. Many parks and wildlife preserves fit this description. Island-based models can help estimate the size of an area that must be set aside as a protected reserve to ensure survival of a species or a community.

area effect Larger islands have more species than small ones.
distance effect Islands close to a mainland have more species than those farther away.
equilibrium model of island biogeography Model that predicts the number of species on an island based on the island's area and distance from the mainland.

TAKE-HOME MESSAGE 45.10

What are some biogeographic patterns in species richness?

✔ Species richness of communities is highest in the tropics and lowest at the poles. Tropical habitats have the longest growing season, and tropical communities are often older than temperate ones.

✔ When a new island forms, species richness rises over time and then levels off. The size of an island and its distance from a colonizing source influence the number of species it supports when it reaches equilibrium.

summary

Section 45.1 All the species that live in an area constitute a **community**. Introduction of a new species to a community can have negative effects on the species that evolved in that community. Global trade has distributed species such as red fire ants from Brazil to new habitats where they are pests.

Section 45.2 Each species occupies a certain **habitat** characterized by physical features and by the array of species living in it. Species interactions affect community structure. **Commensalism** is an interaction in which one species benefits and the other is unaffected. A **symbiosis** is an interaction in which one species lives in or on another.

Section 45.3 In a **mutualism**, both species benefit from an interaction. Some mutualists cannot complete their life cycle without the interaction. Mutualists who maximize their own benefits while limiting the cost of cooperating are at a selective advantage.

Section 45.4 A species' roles and requirements define its **ecological niche**. When two species with similar niches share a habitat, **interspecific competition** harms both. When two species are very similar, **competitive exclusion** may occur. Competing species with similar requirements become less similar when directional selection causes **character displacement**. This change in traits allows **resource partitioning**.

Sections 45.5, 45.6 **Predation** benefits a predator at the expense of the prey it captures, kills, and eats. Predators and their prey exert selective pressure on one another. Both predators and prey can benefit from **camouflage**. Some well-defended prey have **warning coloration**. With **mimicry**, well-defended species evolve a similar appearance or vulnerable species come to resemble better-defended ones. Plant traits such as thorns or unpalatable secondary metabolites allow plants to escape or survive **herbivory**.

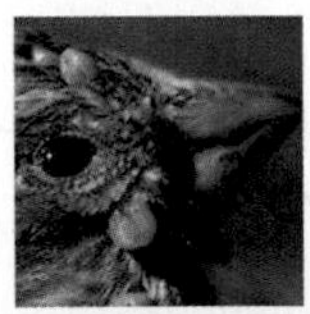

Section 45.7 **Parasitism** involves feeding on a living host. It benefits the parasite at the expense of the host. **Brood parasites** steal parental care from another species. **Parasitoids** are insects that lay eggs in or on a host insect. Parasites and parasitoids are often used in biological control of pest species.

Section 45.8 Ecological succession is the sequential replacement of one array of species by another over time. **Primary succession** happens in newly formed, vacant habitats. **Secondary succession** occurs in disturbed ones. The first species to become established are **pioneer species**. The pioneers may help, hinder, or have no effect on later colonists.

The older idea that all communities eventually reach a predictable climax state has been replaced by models that emphasize the role of chance and disturbances. The **intermediate disturbance hypothesis** holds that disturbances of moderate intensity and frequency maximize species diversity.

Section 45.9 **Keystone species** are especially important in maintaining the composition of a community. The removal of a keystone species or introduction of an **exotic species**—one that evolved in a different community—can alter community structure permanently. An **indicator species** is especially sensitive to environmental change, so its presence or absence can provide early information about an environment's degradation. Some species are adapted to a particular type of periodic disturbance such as wildfires. Such species are at a competitive disadvantage if the disturbance stops occurring.

Section 45.10 Species richness, the number of species in a given area, varies with latitude. The **equilibrium model of island biogeography** predicts the number of species that an island will sustain based on the **area effect** and the **distance effect**. Scientists can use this model to predict the number of species that islands of habitat such as parks can sustain.

self-quiz

Answers in Appendix VII

1. The type of physical environment in which a species typically lives is its ______.
   a. niche
   b. habitat
   c. community
   d. population

2. Which cannot be a symbiosis?
   a. mutualism
   b. parasitism
   c. commensalism
   d. interspecific competition

3. A tick is a(n) ______.
   a. parasitoid
   b. ectoparasite
   c. endoparasite

4. ______ can lead to resource partitioning.
   a. Mutualism
   b. Parasitism
   c. Commensalism
   d. Interspecific competition

5. Match the terms with the most suitable descriptions.

___ mutualism	a. one free-living species feeds on another and usually kills it
___ parasitism	b. two species interact and both benefit by the interaction
___ commensalism	c. two species interact and one benefits while the other is neither helped nor harmed
___ predation	d. one species feeds on another that it lives on or in
___ interspecific competition	e. two species access a resource

Testing Biological Control Ant-decapitating phorid flies are just one of the biological control agents used to battle imported fire ants. Researchers have also enlisted the help of *Thelohania solenopsae*, another natural enemy of the ants. This microsporidian (Section 23.4) is a parasite that infects ants and shrinks the ovaries of the colony's egg-producing female (the queen). As a result, a colony dwindles in numbers and eventually dies out.

Are these biological controls useful against imported fire ants? To find out, USDA scientists treated infested areas with either traditional pesticides or pesticides plus biological controls (both flies and the parasite). The scientists left some plots untreated as controls. **FIGURE 45.28** shows the results.

1. How did population size in the control plots change during the first four months of the study?

2. How did population size in the two types of treated plots change during this same interval?

3. If this study had ended after the first year, would you conclude that biological controls had a major effect?

4. Would your conclusion differ at the end of the time period shown?

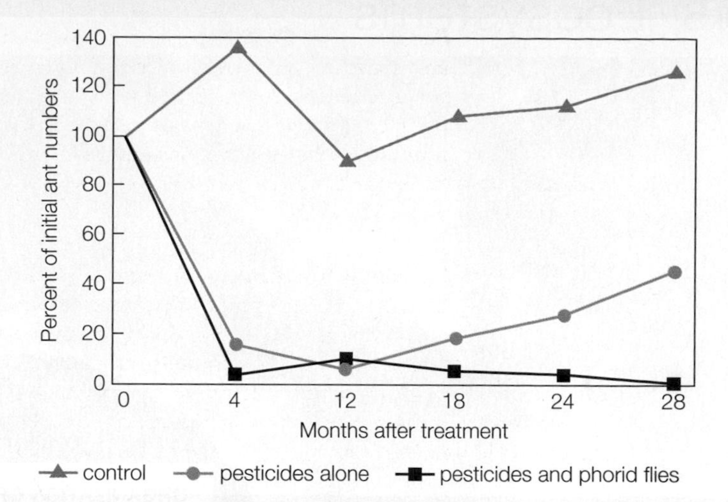

FIGURE 45.28 A comparison of two methods of controlling red imported fire ants. The graph shows the numbers of red imported fire ants over a 28-month period. Orange triangles represent untreated control plots. Green circles are plots treated with pesticides alone. Black squares are plots treated with pesticide and biological control agents (phorid flies along with a microsporidian parasite).

6. Lizards that eat flies they catch on the ground and birds that catch and eat flies in the air are engaged in ______ competition.
   a. exploitative
   b. interference
   c. intraspecific
   d. interspecific
   e. both a and d
   f. both b and c

7. By a currently favored hypothesis, species richness of a community is greatest between physical disturbances of ______ intensity or frequency.
   a. low
   b. intermediate
   c. high
   d. variable

8. With ______, one species evolves to look like another.

9. Growth of a forest in an abandoned corn field is an example of ______.
   a. primary succession
   b. resource partitioning
   c. secondary succession
   d. competitive exclusion

10. Species richness is greatest in communities ______.
   a. near the equator
   b. in temperate regions
   c. near the poles
   d. that recently formed

11. If you remove a species from a community, the population size of its main ______ is likely to increase.
   a. parasite
   b. competitor
   c. predator

12. ______ steal parental care.
   a. Mutualists
   b. Commensalists
   c. Brood parasites
   d. Predators

13. The oldest established land communities are ______.
   a. in the Arctic
   b. in temperate zones
   c. in the tropics
   d. on volcanic islands

CENGAGE brain To access course materials, please visit www.cengagebrain.com.

14. Biological control of pest species ______.
   a. has no side effects
   b. involves mutualists
   c. uses natural enemies
   d. requires use of chemicals

15. Match the terms with the most suitable descriptions.

___ area effect
___ pioneer species
___ indicator species
___ keystone species
___ exotic species
___ resource partitioning

a. native species with large effect
b. first species established in a new habitat
c. more species on large islands than small ones at same distance from the source of colonists
d. species that is especially sensitive to changes in the environment
e. allows competitors to coexist
f. often outcompete, displace native species of established community

critical thinking

1. With antibiotic resistance rising, researchers are looking for ways to reduce use of these drugs. Some cattle once fed antibiotic-laced food now get probiotic feed that can bolster populations of helpful bacteria in the animal's gut. The idea is that if a large population of beneficial bacteria is in place, then harmful bacteria cannot become established or thrive. Which ecological principle is guiding this practice?

2. Flightless birds on islands often have relatives on the mainland that can fly. The island species presumably evolved from fliers that, in the absences of predators, lost their ability to fly. Many island populations of flightless birds are in decline because rats have been introduced to their previously isolated island habitats. Despite current selection pressure in favor of flight, no flightless island bird species has regained the ability to fly. Why is this unlikely to happen?

46 Ecosystems

LEARNING ROADMAP

This chapter builds on prior discussions of energy flow (Sections 1.3 and 5.2), carbon imbalances (6.1), algal blooms (21.6), plant nutrition (28.2), nitrogen fixation (20.7 and 28.3), and mycorrhizal fungi (23.7). Discussions of nutrient cycles draw on your knowledge of plate tectonics (16.7).

ORGANIZATION OF ECOSYSTEMS

An ecosystem consists of a community and its physical environment. There is one-way flow of energy through its components, and a cycling of raw materials among them.

FOOD CHAINS AND WEBS

Food chains are linear sequences of feeding relationships among members of a community. Food chains interconnect as food webs.

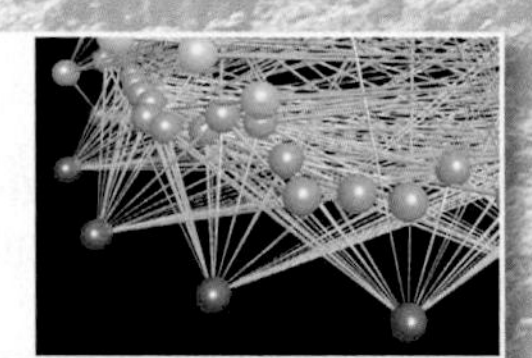

ENERGY FLOW AND PRODUCTIVITY

The amount of energy captured by producers varies among ecosystems. Land ecosystems are typically more productive than marine ones. In many regions, productivity varies seasonally.

BIOGEOCHEMICAL CYCLES

Water, carbon, nitrogen, and phosphorus move from environmental reservoirs, into and through food webs, then back to reservoirs.

DISRUPTED NUTRIENT CYCLES

Human activities such as the use of fossil fuels and synthetic fertilizers have disrupted nutrient cycles, with far-reaching effects on the environment.

The next chapter explains how the differences in climate that influence primary productivity arise. It also describes the types of ecosystems found on land and in aquatic habitats. The final chapter discusses how human activities such as altering nutrient cycles contribute to the ongoing loss of biodiversity.

46.1 Too Much of a Good Thing

All organisms require certain elements to build their bodies and carry out metabolic processes. Phosphorus, for example, is one of the essential elements because it is a component of ATP, phospholipids, nucleic acids, and other biological compounds. Plants meet their phosphorus requirement by taking up phosphates dissolved in soil water. Aquatic producers take up phosphates dissolved in the water around them. Animals obtain phosphorus by eating producers or by eating other animals. Thus, phosphorus taken up from the environment passes from one organism to another. When an organism dies and decays, the phosphorus in its body returns to the environment. As a result, phosphorus moves continually from the environment, through organisms, and back to the environment.

In many habitats, phosphorus is in short supply, so it is the main factor limiting producer growth. This is why fertilizers typically include phosphates. Synthetic phosphate fertilizers are produced by mining rock rich in phosphate and using it to produce phosphoric acid. The acid is then used to make fertilizer.

People spread synthetic phosphate-rich fertilizers to boost plant growth on lawns and agricultural fields. Unfortunately, phosphates do not always stay where they were applied. In may places, the phosphates run off and pollute rivers or lakes, a consequence depicted in **FIGURE 46.1**. Phosphate-containing detergents are a source of phosphate pollution too. Compounds such as sodium triphosphate enhance the cleaning power of laundry detergents and dish detergents. When these phosphate-rich compounds get into aquatic habitats, they have the same effect as phosphate-rich fertilizer.

Addition of phosphate or another nutrient to an aquatic ecosystem is called **eutrophication**. Eutrophication can occur slowly by natural processes, or fast as a result of human actions. Access to phosphorus is usually the main limiting factor for aquatic algae and cyanobacteria, so adding phosphorus to a habitat allows a population explosion of these organisms. The result is an algal bloom (Section 21.6) that clouds water and threatens other aquatic species (**FIGURE 46.2**).

We began using phosphate-rich products when we had no idea of their effects beyond greener lawns and cleaner clothes. Today, we know that daily actions of millions of people can disrupt nutrient cycles that have been in operation since long before humans existed. Our species has a unique capacity to shape the environment to our liking. In doing so, we have become major players in global flows of energy and nutrients even before we fully comprehend how these systems work.

FIGURE 46.1 Poster created by the Washington State Ecology Department to remind people that nutrients from lawn fertilizer can run off and pollute water.

FIGURE 46.2 Effect of phosphate pollution. This aerial photo shows the results of a field experiment in which different combinations of nutrients were added to two artificially separated portions of a lake. Including phosphorus in the mix (lower region) caused an algal bloom that clouded the water.

eutrophication Nutrient enrichment of an aquatic ecosystem.

46.2 The Nature of Ecosystems

✔ In an ecosystem, energy and nutrients from the environment flow among a community of species.

Overview of the Participants

An **ecosystem** is an array of organisms and their physical environment, all interacting through a one-way flow of energy and a cycling of nutrients. It is an open system, because it requires ongoing inputs of energy and nutrients to endure (**FIGURE 46.3**).

Earth's ecosystems are remarkably diverse in their characteristics. In climate, soil type, array of species, and other features, prairies differ from forests, which differ from tundra and deserts. Reefs differ from the open ocean, which differs from streams and lakes. Yet, despite these many differences, all ecosystems are alike in many aspects of their structure and function.

All ecosystems run on energy captured by **primary producers**. These autotrophs, or "self-feeders," obtain energy from a nonliving source—generally sunlight—and use it to build organic compounds from inorganic starting materials. Plants and phytoplankton are the main producers. Chapter 6 explained how they capture energy from the sun and use it in photosynthesis to build sugars from carbon dioxide and water.

Consumers are heterotrophs that obtain energy and carbon by feeding on tissues, wastes, and remains of producers and one another. We can describe consumers by their diets. Herbivores eat plants. Carnivores eat the flesh of animals. Omnivores devour both animal and plant materials. Parasites live inside or on a living host and feed on its tissues. **Detritivores**, such as earthworms and crabs, dine on small particles of organic matter, or detritus. **Decomposers** feed on organic wastes and remains and break them down into inorganic building blocks. The main decomposers are bacteria and fungi.

Energy flows one way—into an ecosystem, through its many living components, then back to the physical environment (Section 5.2). Light energy captured by producers is converted to bond energy in organic molecules, which is then released by metabolic reactions that give off heat. This is a one-way process because heat energy cannot be recycled; producers cannot convert heat into energy stored in chemical bonds.

In contrast, nutrients can be cycled continuously within an ecosystem. The cycle begins when producers take up hydrogen, oxygen, and carbon from inorganic sources, such as the air and water. Producers also take up dissolved nitrogen, phosphorus, and other minerals needed to make organic compounds. Nutrients move from producers into the consumers who eat them. After an organism dies, action of decomposers releases nutrients from the organism's remains into the environment, where they are once again taken up by producers.

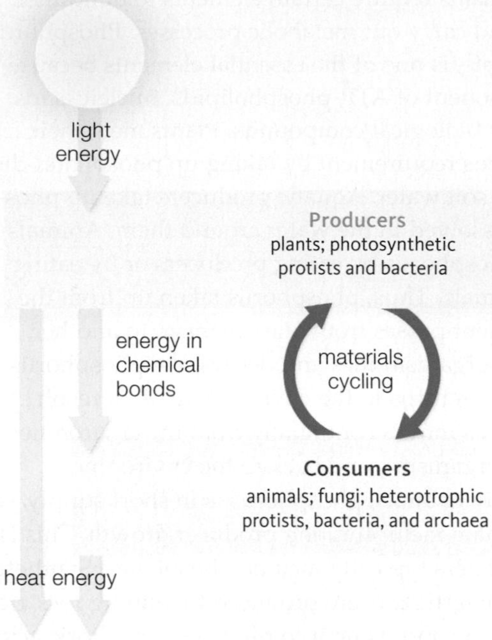

FIGURE 46.3 ▶**Animated** Model for ecosystems on land, in which energy flow starts with autotrophs that capture energy from the sun. Energy flows one way, into and out of the ecosystem. Nutrients get cycled among producers and consumers.

Not all nutrients remain in an ecosystem; typically there are gains and losses. Mineral ions are added to an ecosystem when weathering processes break down rocks, and when winds blow in mineral-rich dust from elsewhere. Leaching and soil erosion can remove minerals (Section 28.2). Gains and losses of each mineral tend to balance out over time in most ecosystems.

Trophic Structure of Ecosystems

All organisms in an ecosystem take part in a hierarchy of feeding relationships called **trophic levels** ("troph" means nourishment). When one organism eats another, energy stored in chemical bonds is transferred from the eaten to the eater. All organisms at the same trophic level in an ecosystem are the same number of transfers away from the energy input into that system.

A **food chain** is a sequence of steps by which some energy captured by primary producers is transferred to organisms at successively higher trophic levels. For example, big bluestem grass and other plants are the major primary producers in a tallgrass prairie (**FIGURE 46.4**). They are at this ecosystem's first trophic level.

FIGURE 46.4 ▶Animated Example of a food chain and corresponding trophic levels in tallgrass prairie, Kansas.

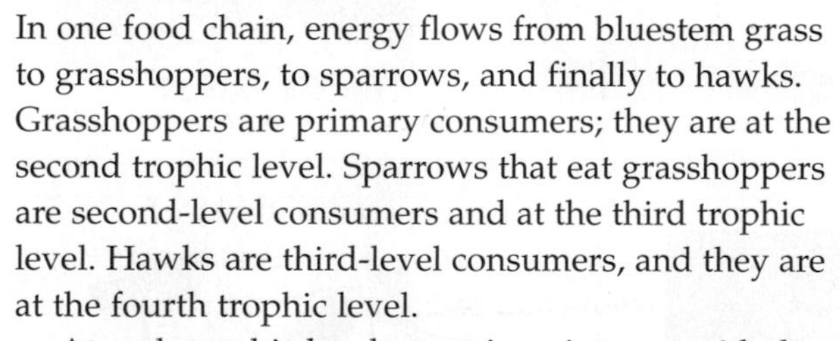

In one food chain, energy flows from bluestem grass to grasshoppers, to sparrows, and finally to hawks. Grasshoppers are primary consumers; they are at the second trophic level. Sparrows that eat grasshoppers are second-level consumers and at the third trophic level. Hawks are third-level consumers, and they are at the fourth trophic level.

At each trophic level, organisms interact with the same sets of predators, prey, or both. Omnivores feed at several levels, so we would partition them among different levels or assign them to a level of their own.

TAKE-HOME MESSAGE 46.2

What is the trophic structure of an ecosystem?

- ✔ An ecosystem includes a community of organisms that interact with their physical environment by a one-way energy flow and a cycling of materials.
- ✔ Autotrophs tap into an environmental energy source and make their own organic compounds from inorganic raw materials. They are the ecosystem's primary producers.
- ✔ Autotrophs are at the first trophic level of a food chain, a linear sequence of feeding relationships that proceeds through one or more levels of heterotrophs, or consumers.

consumer Organism that gets energy and nutrients by feeding on tissues, wastes, or remains of other organisms; a heterotroph.
decomposer Organism that feeds on biological remains and breaks organic material down into its inorganic subunits.
detritivore Consumer that feeds on small bits of organic material.
ecosystem A community interacting with its environment.
food chain Description of who eats whom in one path of energy flow in an ecosystem.
primary producer In an ecosystem, an organism that captures energy from an inorganic source and stores it as biomass; first trophic level.
trophic level Position of an organism in a food chain.

46.3 The Nature of Food Webs

✔ A food web consists of cross-connecting food chains. Its structure reflects environmental constraints and the inefficiency of energy transfers among trophic levels.

An organism that participates in one food chain usually takes part in others as well. Food chains of an ecosystem cross-connect as a **food web**. FIGURE 46.5 shows some participants in an arctic food web.

Ecosystems typically support two types of food webs. In **grazing food webs**, most primary producers are eaten by primary consumers. In **detrital food webs**, most energy in producers flows directly to detritivores.

Detrital food webs tend to predominate in most land ecosystems. For example, in an arctic ecosystem, primary consumers such as voles, lemmings, and hares eat some living plant parts. However, far more plant matter becomes detritus. Bits of dead plant material

human (Inuk)

arctic wolf

arctic fox

Higher Trophic Levels

A sampling of carnivores that feed on herbivores and one another

gyrfalcon

snowy owl

ermine

Second Trophic Level

A sampling of primary consumers (herbivores) that eat plants

vole

arctic hare

lemming

mosquito

flea

Parasitic consumers feed at more than one trophic level.

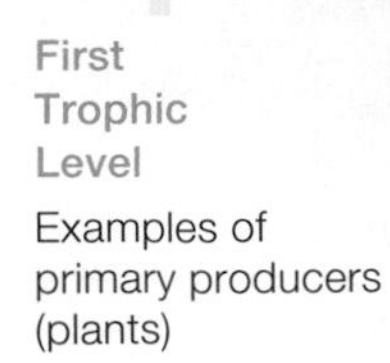

First Trophic Level

Examples of primary producers (plants)

grasses, sedges

purple saxifrage

arctic willow

Detritivores and decomposers (nematodes, annelids, saprobic insects, protists, fungi, bacteria)

FIGURE 46.5 ▶**Animated** A very small sampling of organisms in an arctic food web on land.

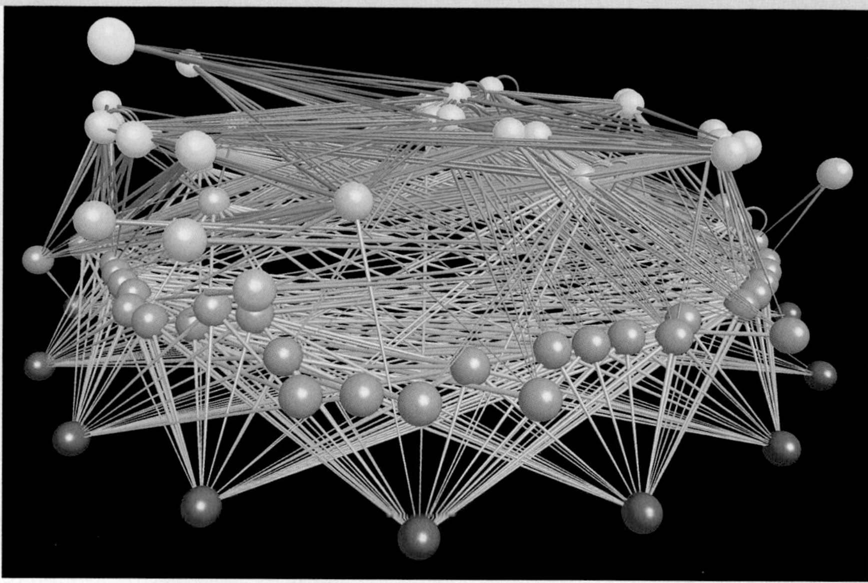

FIGURE 46.6 Computer model (right) for a food web in East River Valley, Colorado (above). Balls signify species. Their colors identify trophic levels, with producers (coded red) at the bottom and predators (yellow) at top. The connecting lines thicken, starting from an eaten species to the eater.

sustain detritivores such as roundworms and decomposers that include soil bacteria and fungi.

Grazing food chains tend to predominate in aquatic ecosystems. Zooplankton (heterotrophic protists and tiny animals that drift or swim) consume most of the phytoplankton. Very little phytoplankton ends up on the ocean floor as detritus.

Detrital food chains and grazing food chains interconnect. For example, many animals at higher trophic levels eat both primary consumers and detritivores. Also, after consumers die, their tissues become food for detritivores and decomposers. Decomposers and detritivores also feed on the wastes from consumers.

How Many Transfers?

When ecologists looked at food webs for a variety of ecosystems, they discovered some common patterns. For example, the energy captured by producers usually passes through no more than four or five trophic levels. Even in ecosystems with many species, the number of transfers is limited. Remember that energy transfers are not very efficient (Section 5.2), so energy losses limit the length of a food chain.

Field studies and computer simulations of aquatic and land food ecosystems reveal additional patterns. Food chains tend to be shortest in habitats where conditions vary widely over time. Chains tend to be longer in stable habitats, such as the ocean depths where weather has no effect. The most complex webs tend to have a large variety of herbivores, as in grasslands. By comparison, food webs with fewer connections tend to have more carnivores.

Diagrams of food webs help ecologists predict how ecosystems will respond to change. Neo Martinez and his colleagues constructed the one shown in **FIGURE 46.6**. By comparing many food webs, they discovered that trophic interactions connect species more closely than previously thought. On average, each species in any food web was two links away from all other species. Ninety-five percent of species were within three links of one another, even in large communities with many species. As Martinez concluded in a paper discussing his findings, "Everything is linked to everything else." He cautioned that extinction of any species in a food web may affect many other species.

detrital food web Food web in which most energy is transferred directly from producers to detritivores.
food web Set of cross-connecting food chains.
grazing food web Food web in which most energy is transferred from producers to grazers (herbivores).

TAKE-HOME MESSAGE 46.3

How does energy flow affect food chains and food webs?

✔ Nearly all ecosystems include both grazing food webs and detrital food webs that interconnect as the system's food web. Tissues of living plants and other producers are the base for grazing food webs. By contrast, remains of producers are the base for detrital food webs.

✔ The cumulative energy losses from energy transfers between trophic levels limit the length of food chains.

✔ Even when an ecosystem has many species, trophic interactions link each species with many others.

CREDITS: (6) left, Courtesy of Dr. Chris Floyd; right, Graphic created by FoodWeb3D program written by Rich Williams courtesy of the Webs on the Web project (www.foodwebs.org).

✔ Primary producers capture energy and take up nutrients, which then move to other trophic levels.

Primary Production

Flow of energy through an ecosystem begins with **primary production**: the rate at which producers capture energy (usually light energy) and convert it into chemical energy. The amount of energy captured by all the producers in an ecosystem is the system's gross primary production. The portion of that energy used for growth and reproduction (rather than for maintenance) is the net primary production of the ecosystem. This is the energy available to the first-level consumers.

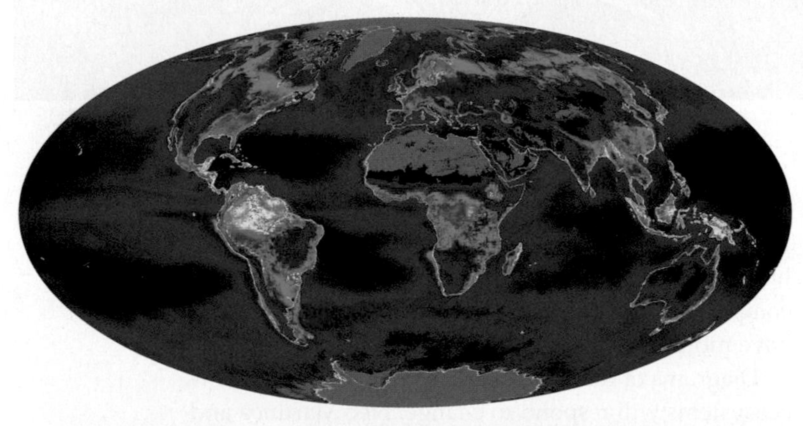

A Summary of annual net primary productivity on land and in the oceans.

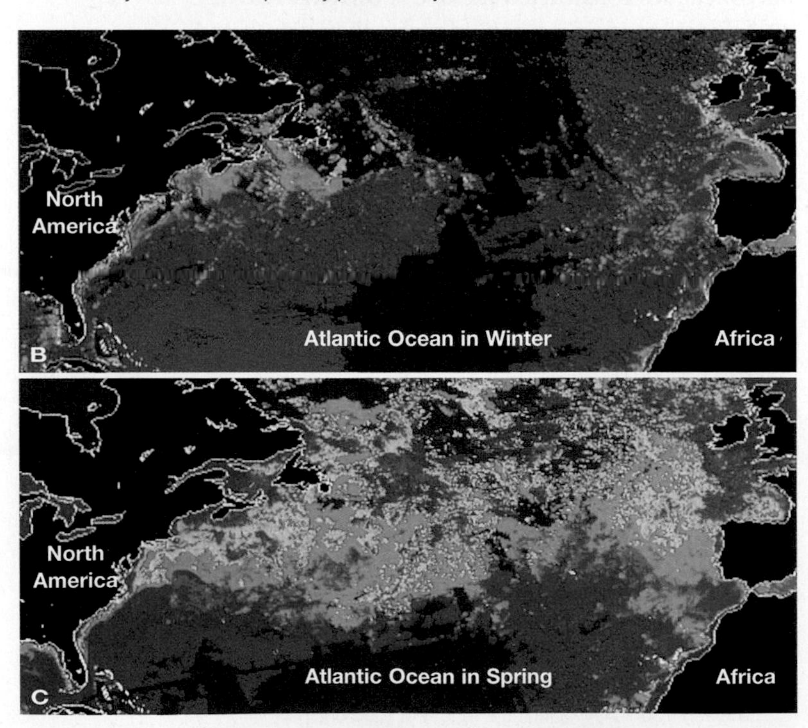

B,C Seasonal changes in net primary productivity of the North Atlantic Ocean.

FIGURE 46.7 Satellite data showing net primary production. Productivity is coded as red (highest) down through orange, yellow, green, blue, and purple (lowest).

Factors that influence the rate of photosynthesis affect primary productivity. As a result, primary production differs among habitats and often varies seasonally (FIGURE 46.7). Per unit area, the net primary production on land tends to be higher than that in the oceans. However, because oceans cover about 70 percent of Earth's surface, marine producers account for nearly half of the global net primary production.

Ecological Pyramids

Ecologists often represent the trophic structure of an ecosystem in the form of ecological pyramids. In such diagrams, primary producers collectively form a base for successive tiers of consumers above them.

A **biomass pyramid** illustrates the dry weight of all organisms at each trophic level. FIGURE 46.8 shows the biomass pyramid for Silver Springs, an aquatic ecosystem in Florida. In this ecosystem, as in most others, primary producers constitute most of the biomass in the pyramid, and top carnivores account for very little. If you visited Silver Springs, you would see a lot of aquatic plants but very few gars (the fish that is the main top predator in this ecosystem). Similarly, if you walked through a prairie, you would see more grams of grass than of hawks.

There are some aquatic ecosystems in which the lowest tier of the biomass pyramid is also the narrowest. In these ecosystems, the main producers are bacteria and single-celled protists—organisms that reproduce fast, rather than investing in building a large body. Because of their quick turnover, a smaller biomass of phytoplankton can support a greater biomass of zooplankton.

An **energy pyramid** illustrates how the amount of usable energy diminishes as it is transferred through an ecosystem. Sunlight energy is captured at the base (the primary producers) and declines with successive levels to the pyramid's tip (the top carnivores). Energy pyramids are always right side up, meaning the pyramid's base is its largest tier. Energy pyramids depict energy flow per unit of water (or land) per unit of time. FIGURE 46.9 shows both the energy pyramid for the Silver Springs ecosystem and the energy flow that this pyramid represents.

biomass pyramid Diagram that depicts the biomass (dry weight) in each of an ecosystem's trophic levels.
energy pyramid Diagram that depicts the energy that enters each of an ecosystem's trophic levels. Lowest tier of the pyramid, representing primary producers, is always the largest.
primary production The rate at which an ecosystem's producers capture and store energy.

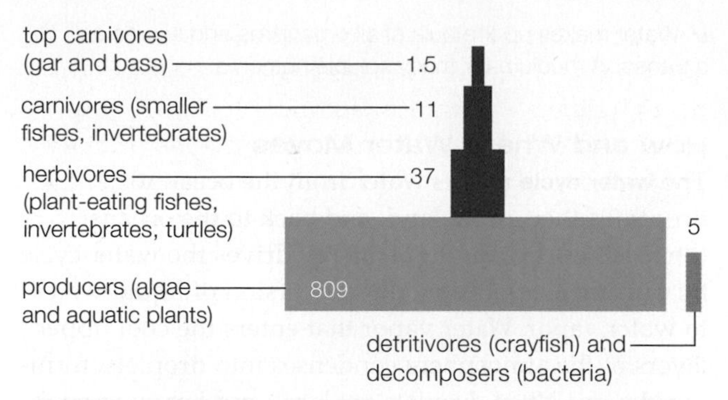

FIGURE 46.8 Biomass (in grams per square meter) for Silver Springs, a freshwater aquatic ecosystem in Florida. In this system, primary producers make up the bulk of the biomass.

Ecological Efficiency

Only 5 to 30 percent of the energy in organisms at one trophic level ends up in the tissues of those at the next level up. Several factors reduce the efficiency of transfers. First, not all energy harvested by consumers is used to build biomass; some is lost as heat. Second, some components of a body may be unavailable to a consumer. The lignin and cellulose that reinforce bodies of most land plants pass largely undigested through some consumers, including humans. Similarly, many animals have some biomass tied up in an internal or external skeleton and hair, feathers, or fur—all of which are difficult for carnivores to digest.

Ecological efficiency is usually higher in aquatic ecosystems than on land, in part because the main aquatic producers—photosynthetic protists—do not make difficult-to-digest lignin as most land plants do. In addition, aquatic ecosystems usually have a higher proportion of ectotherms (such as fishes). This improves ecological efficiency because ectotherms lose less energy as heat than endotherms do.

TAKE-HOME MESSAGE 46.4

How does energy flow through ecosystems?

✔ Primary producers capture energy and convert it into biomass. We measure this process as primary production.

✔ A biomass pyramid depicts dry weight of organisms at each trophic level in an ecosystem. Its largest tier is usually producers, but the pyramid for some aquatic systems is inverted.

✔ An energy pyramid depicts the amount of energy that enters each level. Its largest tier is always at the bottom (producers).

✔ The efficiency of energy transfers tends to be greatest in aquatic systems, where primary producers usually lack lignin and consumers tend to be ectotherms.

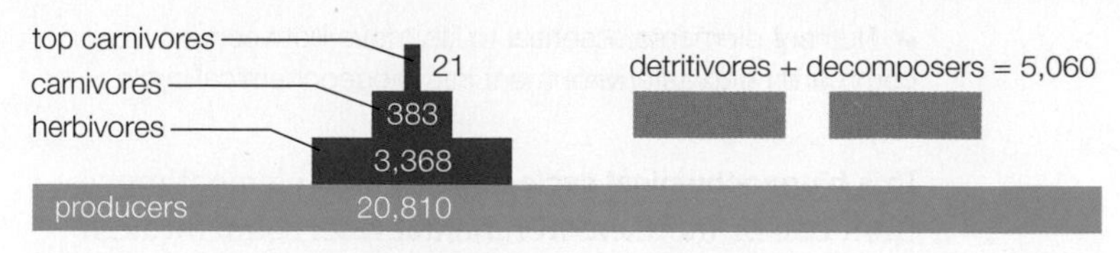

A Energy pyramid for the Silver Springs ecosystem. The width of each tier in the pyramid represents the amount of energy that enters each trophic level annually, as shown in detail below.

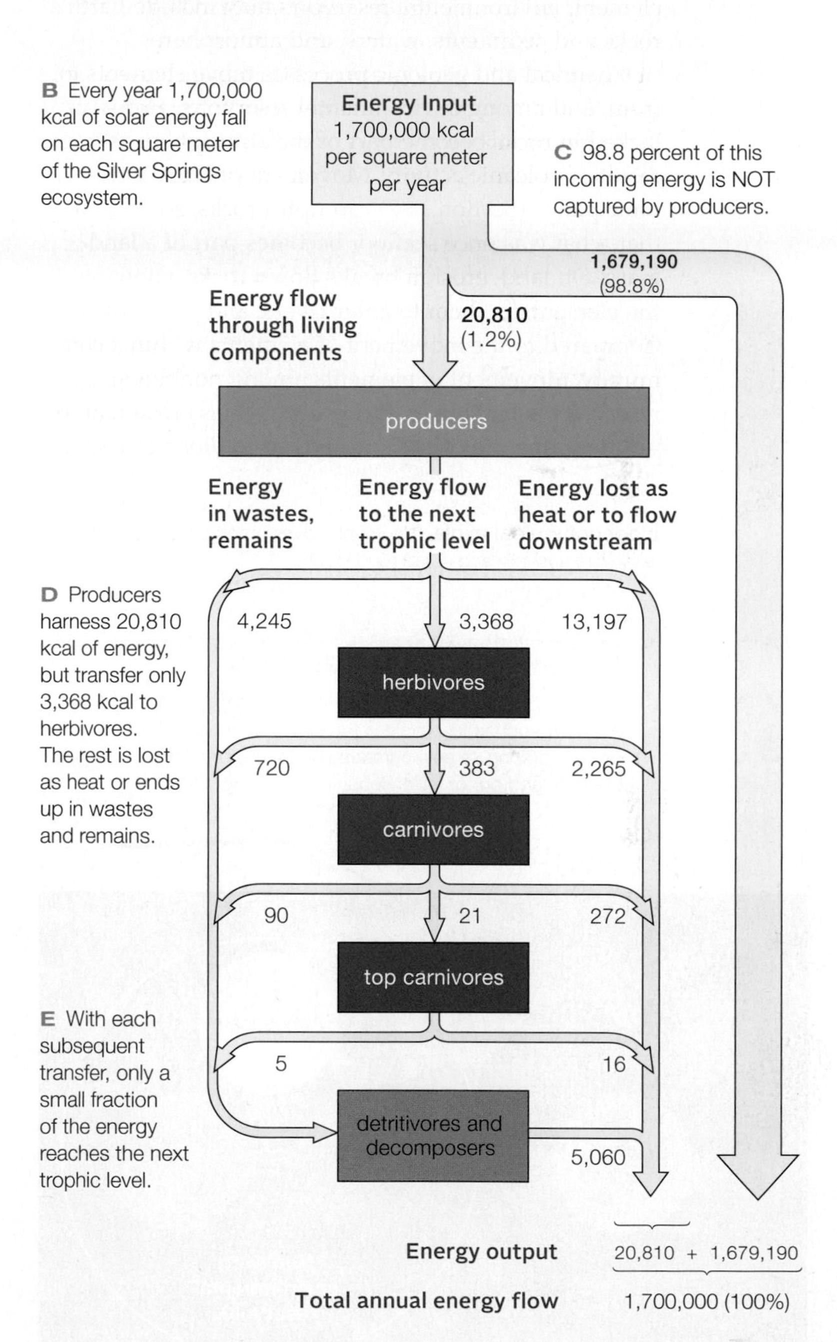

FIGURE 46.9 ▶Animated Annual energy flow in Silver Springs measured in kilocalories (kcal) per square meter per year. **FIGURE IT OUT** What percentage of the energy carnivores received directly from herbivores was later passed on to top carnivores?

Answer: 21/383 × 100 = 5.5 percent

46.5 Biogeochemical Cycles

✔ Nutrient elements essential to life move between a community and its environment in a biogeochemical cycle.

In a **biogeochemical cycle**, an essential element moves from one or more environmental reservoirs, through the living components of an ecosystem, and then back to the reservoirs (**FIGURE 46.10**). Depending on the element, environmental reservoirs may include Earth's rocks and sediments, waters, and atmosphere.

Chemical and geologic processes move elements to, from, and among environmental reservoirs. Elements locked in rocks become part of the atmosphere as a result of volcanic activity. Movement of Earth's tectonic plates (Section 16.7) can uplift rocks, so an area that what was once seafloor becomes part of a landmass. On land, erosion breaks down rocks, allowing the elements in them to enter rivers, and flow to seas. Compared to the movement of elements within a community, movement of elements among nonbiological reservoirs is far slower. Processes such as erosion and uplifting operate over thousands or millions of years.

biogeochemical cycle A nutrient moves among environmental reservoirs and into and out of food webs.

TAKE-HOME MESSAGE 46.5

What is a biogeochemical cycle?

✔ A biogeochemical cycle is the slow movement of a nutrient among its environmental reservoirs and into, through, and out of food webs.

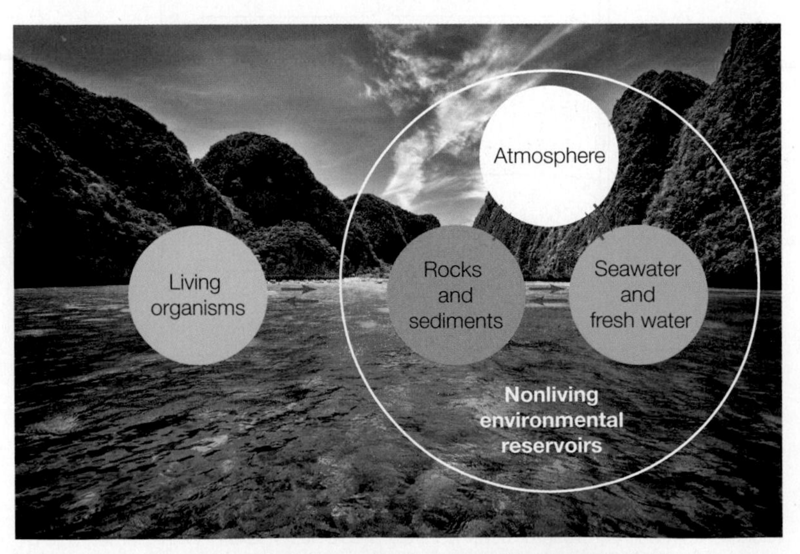

FIGURE 46.10 Generalized biogeochemical cycle. For all nutrients, the cumulative amount in all environmental reservoirs far exceeds the amount in living organisms.

46.6 The Water Cycle

✔ Water makes up the bulk of all organisms and serves as a transport medium for many soluble nutrients.

How and Where Water Moves

The **water cycle** moves water from the ocean to the atmosphere, onto land, and back to the oceans (**FIGURE 46.11**). Sunlight energy drives the water cycle by causing evaporation, the conversion of liquid water to water vapor. Water vapor that enters the cool upper layers of the atmosphere condenses into droplets, forming clouds. When droplets get large and heavy enough, they fall as precipitation (rain, snow, and hail).

Oceans cover about 70 percent of Earth's surface, so most rainfall returns water directly to the oceans. On land, we define a **watershed** as a region in which all precipitation drains into a specific waterway. A watershed may be as small as a valley that feeds a stream, or as large as the 5.9 million square kilometers (2.3 million square miles) of the Amazon River Basin. The Mississippi River Basin watershed includes 41 percent of the continental United States.

Most precipitation that enters a watershed seeps into the ground to become groundwater. **Groundwater** consists of soil water and the water in aquifers. **Soil water** is the water that remains between soil particles. Plant roots tap into soil water as their water source. Soils differ in their water-holding capacity, with clay-rich soils holding the most water and sandy soils the least. Water that drains through soil, often collects in **aquifers**. These natural underground reservoirs consist of porous rock layers that can hold water.

Precipitation that falls on impermeable rock or on saturated soil becomes **runoff**: It flows over the ground into streams. The flow of groundwater and surface water returns water to the oceans.

Movement of water results in the movement of soluble nutrients. Carbon, nitrogen, and phosphorus all have soluble forms that can be moved from place to place by flowing water. As water trickles through soil, it carries nutrient particles from topsoil into deeper soil layers. As a stream flows over limestone, water slowly dissolves the rock and carries carbonates back to the seas where the limestone formed. Flowing water can transport pollutants too; runoff from heavily fertilized lawns and agricultural fields carries dissolved phosphates and nitrates into streams and lakes.

Limited Fresh Water

Although Earth has lots of water, the amount of fresh water available to meet human needs and sustain land ecosystems is severely limited. The vast majority of

Earth's water (97 percent) is seawater, and most fresh water is frozen as ice (**TABLE 46.1**).

Aquifers supply about half of the drinking water in the United States and many of these are in trouble (**FIGURE 46.12**). Overdrafts—removal water from aquifers faster than natural processes replenish it—are common. Overdrawing water from an aquifer lowers the water table (the topmost level at which rock is saturated with water). When the water table falls in an inland area, wells that tap that aquifer can run dry. In coastal areas, overdrawing an aquifer allows saltwater to move in and contaminate the aquifer.

The largest aquifer in the United States, the Ogallala aquifer, stretches from South Dakota to Texas and supplies irrigation water for 27 percent of the nation's crops. For the past thirty years, withdrawals from the Ogallala aquifer have exceeded replenishment by a factor of ten. As a result, the water table has dropped as much as 50 meters (150 feet) in some regions.

Rivers are another source of fresh water. However, water in many rivers is currently over-allocated; the amount of water promised to various stakeholders such as cities and farmers exceeds the amount that currently flows through the river. In such rivers, diversion of water for human use results in lowered or nonexistent flow in some portion of the river. The lack of water can have a devastating effect on biological communities that depend on the river. Rivers convey sediment and nutrients as well as water, so decreased flow alters ecosystems by slowing delivery of these materials to the river's delta, the region where the river approaches the sea.

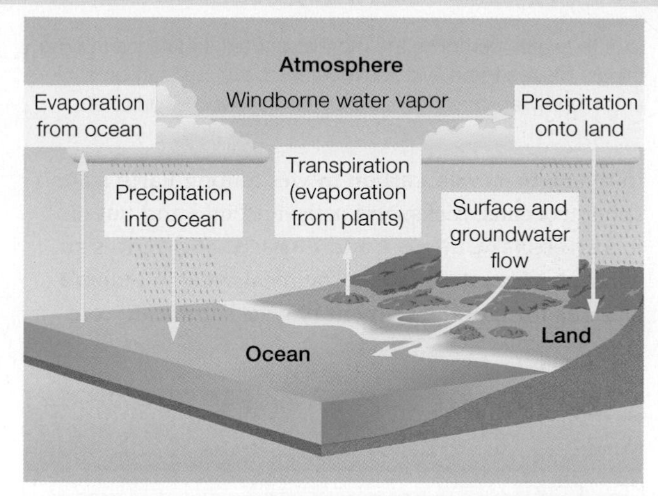

FIGURE 46.11 ▶**Animated** The water cycle. Water moves from the ocean to the atmosphere, land, and back. The arrows identify processes that move water.

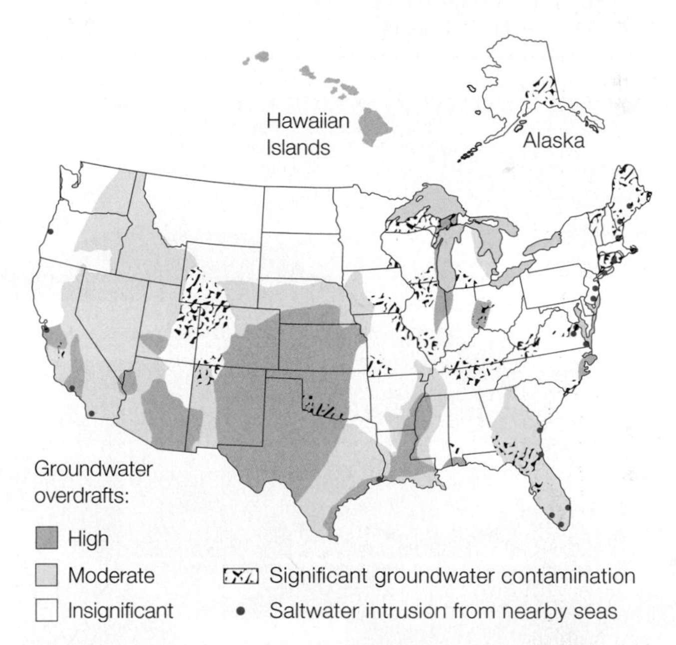

FIGURE 46.12 Groundwater troubles in the United States.

Table 46.1 Environmental Water Reservoirs

Reservoir	Volume (10^3 km^3)
Ocean	1,370,000
Polar ice, glaciers	29,000
Groundwater	4,000
Lakes, rivers	230
Atmosphere (water vapor)	14

aquifer Porous rock layer that holds some groundwater.
groundwater Soil water and water in aquifers.
runoff Water that flows over soil into streams.
soil water Water between soil particles.
water cycle Movement of water among Earth's atmosphere, oceans, and the freshwater reservoirs on land.
watershed Land area that drains into a particular stream or river.

TAKE-HOME MESSAGE 46.6

What is the water cycle and how do human activities affect it?

✔ Water moves slowly from the world ocean—the main reservoir—through the atmosphere, onto land, then back to the ocean.

✔ Fresh water makes up only a tiny portion of the global water supply. Excessive water withdrawals threaten many sources of drinking water.

46.7 The Carbon Cycle

✔ After water, carbon is the most abundant substance in living things. Most carbon is in sedimentary rocks, but carbon can enter food webs as gaseous CO_2 or dissolved bicarbonate.

In the **carbon cycle**, carbon moves among Earth's atmosphere, oceans, rocks, and soils, and into and out of food webs (TABLE 46.2 and FIGURE 46.13). It is an **atmospheric cycle**, a biogeochemical cycle in which a gaseous form of the element plays a significant role.

Table 46.2 Annual Carbon Movement in Gigatons (Billions of Metric Tons)

From atmosphere to plants (carbon fixation)	120
From atmosphere to ocean	107
From ocean to atmosphere	105
From plants to atmosphere	60
From soil to atmosphere	60
From land to ocean in runoff	0.4
Burial in ocean sediments	0.1

The atmosphere holds about 760 gigatons (billion tons) of carbon, mainly in the form of carbon dioxide (CO_2).

Terrestrial Carbon Cycle

Land plants take up CO_2 from the atmosphere and incorporate it into their tissues when they carry out photosynthesis. Plants and other land organisms release CO_2 into the atmosphere; they produce the CO_2 during aerobic respiration.

Soil contains at least 1600 gigatons of carbon, more than twice as much as the atmosphere. Soil carbon consists of humus and living soil organisms. Over time, bacteria and fungi in the soil decompose humus and release carbon dioxide into the atmosphere. The rate of decomposition increases with temperature. In the tropics, decomposition proceeds rapidly, so most carbon is stored in living plants, rather than in soil (FIGURE 46.14). By contrast, in temperate zone forests and grasslands, soil holds more carbon than living plants. Soil is most carbon-rich in the arctic, where

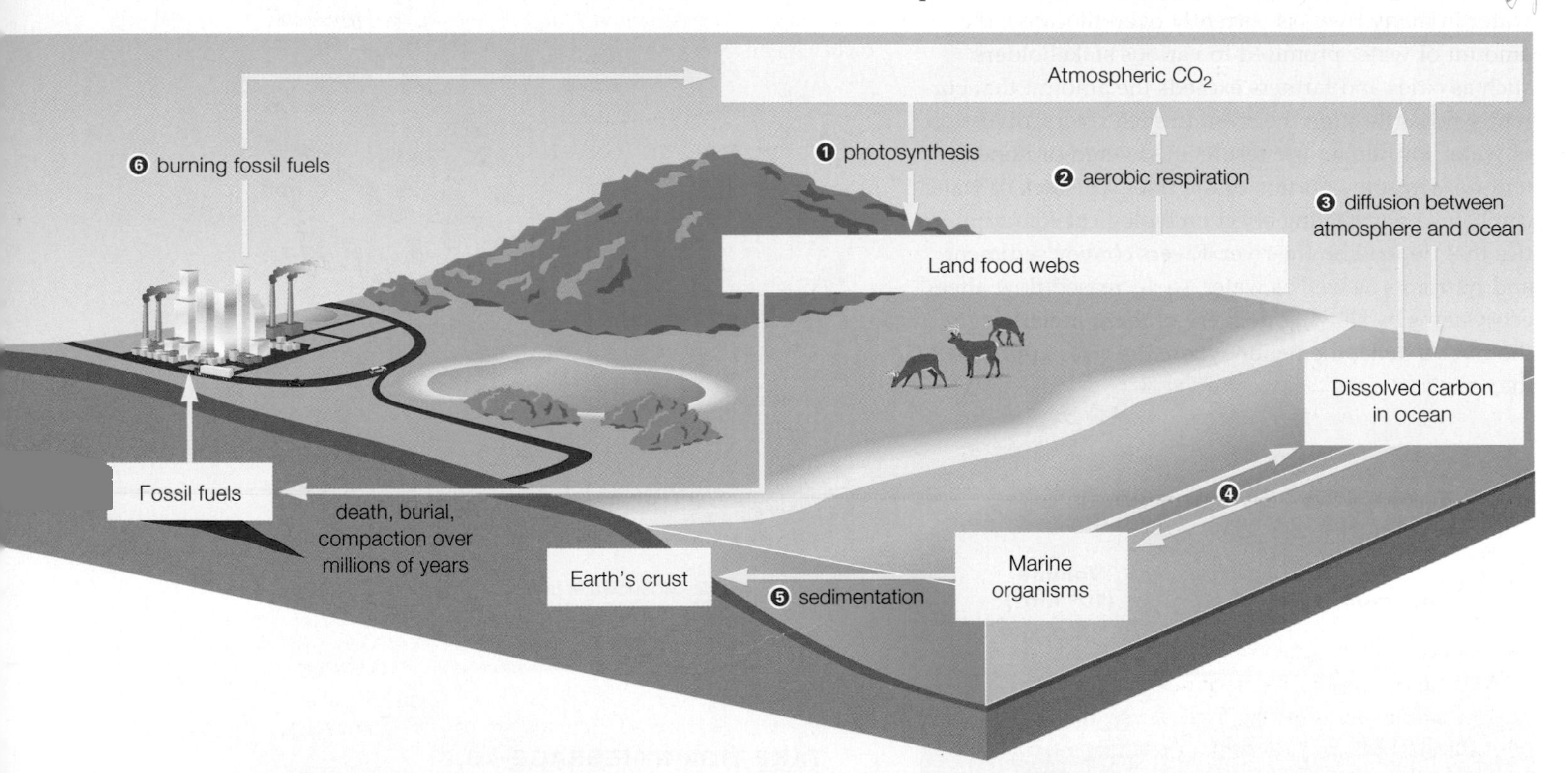

FIGURE 46.13 ▶Animated The carbon cycle. Most carbon is in Earth's rocks, where it is largely unavailable to living organisms.

❶ Carbon enters land food webs when plants take up carbon dioxide from the air for use in photosynthesis.

❷ Carbon returns to the atmosphere as carbon dioxide when plants and other land organisms carry out aerobic respiration.

❸ Carbon diffuses between the atmosphere and the ocean. Bicarbonate forms when carbon dioxide dissolves in seawater.

❹ Marine producers take up bicarbonate for use in photosynthesis, and marine organisms release carbon dioxide from aerobic respiration.

❺ Many marine organisms incorporate carbon into their shells. After they die, these shells become part of the sediments. Over time, the sediments become carbon-rich rocks such as limestone and chalk in Earth's crust.

❻ Burning of fossil fuels derived from the ancient remains of plants puts additional carbon dioxide into the atmosphere.

FIGURE 46.14 Estimates of carbon stored in the plants and soils of different ecosystems.

low temperature hampers decomposition, and in peat-bogs, where acidic, anaerobic conditions do the same (Section 22.4).

Conversion of a forest or grassland to cropland reduces the amount of carbon both above and below ground. Tilling soil (mechanically mixing it) increases the rate at which carbon enters the air because it destroys the hyphae of glomeromycete fungi, a type of mycorrhizal fungi (Section 23.7). These hyphae secrete a gluelike glycoprotein (glomalin) that holds onto organic material and slows its decomposition.

Marine Carbon Cycle

Seawater holds about 40,000 gigatons of carbon, about fifty times as much as the atmosphere. The main form of carbon in seawater is bicarbonate (HCO_3^-), an ion that forms when CO_2 dissolves in water. Marine producers take up bicarbonate and convert it to CO_2 for use in photosynthesis. Marine organisms release CO_2 produced by aerobic respiration into the water. Some marine organisms such as foraminifera, shelled mollusks, and reef-building corals also store carbon in their calcium carbonate–hardened parts.

Marine sediments and sedimentary rocks are Earth's largest carbon reservoir by far, holding more than 65 million gigatons. Limestone and other sedimentary rocks form when the calcium carbonate–rich shells of marine organisms become compacted over millions of years. Such rocks become part of land ecosystems when movements of tectonic plates uplift portions of the seafloor. Carbon in rocks is not available to producers, so this vast store of carbon has little effect on ecosystems. The geological part of the cycle is completed as erosion breaks down rocks, and rivers carry dissolved carbon to the sea.

Carbon in Fossil Fuels

Fossil fuels such as coal, oil, and natural gas hold an estimated 5,000 gigatons of carbon. Deposits of fossil fuels formed over hundreds of millions of years from carbon-rich remains. High pressure and temperature transformed the remains of ancient land plants to coal (Section 22.6). A similar process transformed the remains of plankton to deposits of oil and natural gas. Until recently, the carbon in fossil fuels, like the carbon in rocks, had little impact on ecosystems. Currently, our burning of this fuel adds billions of tons of CO_2 to the atmosphere every year.

atmospheric cycle Biogeochemical cycle in which a gaseous form of an element plays a significant role.
carbon cycle Movement of carbon, mainly between the oceans, atmosphere, and living organisms.

TAKE-HOME MESSAGE 46.7

How does carbon cycle between its main reservoirs?

✔ The largest reservoir is sedimentary rock. Carbon moves into and out of this reservoir over very long time spans and is not available to organisms.

✔ Seawater is the largest reservoir of biologically available carbon. Marine producers take up bicarbonate and convert it to CO_2 for use in photosynthesis.

✔ On land, large amounts of carbon are stored in soil, especially in arctic regions and in peat bogs.

✔ The atmosphere holds less carbon than rocks, seawater, or soil. It serves as the source of CO_2 for land producers. Burning of fossil fuels adds carbon to this reservoir.

46.8 Greenhouse Gases and Climate Change

✔ Concentrations of gases in Earth's atmosphere help determine the temperature near Earth's surface.

The Greenhouse Effect

Carbon dioxide is a **greenhouse gas**, an atmospheric gas whose ability to absorb and reradiate heat energy helps keep Earth warm enough to sustain life. The mechanism by which this occurs is referred to as the greenhouse effect (**FIGURE 46.15**). When radiant energy from the sun reaches Earth's atmosphere, some energy is reflected back into space ❶. However, more energy passes through the atmosphere and warms Earth's surface ❷. When the warmed surface radiates heat energy, greenhouse gases absorb some of that heat, then emit a portion of it back toward Earth ❸. If greenhouse gases did not exist, heat energy emitted by Earth's surface would escape into space, leaving the planet cold and entirely lifeless.

Changing Carbon Dioxide Concentrations

In 1959, researchers began to measure the atmospheric concentration of CO_2 at an observatory near the top of Mauna Loa, Hawaii's highest volcano. The remote site 3,500 meters (11,000 feet) above sea level was chosen because it is almost free of local airborne contamination and is representative of atmospheric conditions for the Northern Hemisphere. For the first time, researchers saw the effects of carbon dioxide fluctuations for the entire hemisphere. The troughs and peaks seen in the line in **FIGURE 46.16A** are annual lows and highs in atmospheric CO_2. The level of CO_2 declines in summer, when the most CO_2 is taken up for photosynthesis. It rises in winter, when photosynthesis declines but aerobic respiration and fermentation continue.

The researchers also noticed a trend: The average annual level of CO_2 is increasing. As additional sites around the world began to monitor atmospheric CO_2, they too detected an ongoing rise.

To put the current increase in perspective, scientists consider historical changes in atmospheric CO_2. Glacial ice provides one window into the past. Such ice forms when snow falls, then is compressed by new snowfalls above it. In some regions, layers of ice have been laid down one atop the next for hundreds of thousands of years. The result is sheets of ice more than a kilometer thick. To determine what conditions were like in the past, scientists use a hollow drill to remove a long ice core. Air bubbles trapped at different depths within the ice provide snapshots of atmospheric conditions at the time that ice formed. So far, analysis of ice cores has provided a history of atmospheric changes that extends back about 800,000 years.

Fossil foraminiferan shells provide information about CO_2 levels in the even more distant past. Foraminifera are single-celled protists that have lived in Earth's oceans for millions of years (Section 21.4). Throughout this time, they have taken up carbon and other elements from seawater and incorporated them into their shells. When atmospheric CO_2 is high, more of it dissolves in the ocean's surface waters, and this increase affects the composition of foraminiferan shells. By studying fossilized shells, scientists can trace how atmospheric CO_2 changed over many millions of years.

Data from glacial ice, foraminiferan fossils, and other sources consistently show that atmospheric CO_2 has risen and fallen many times. These data also show that the current CO_2 concentration is the highest in at least 15 million years.

Changing Climate

Given the greenhouse effect, we would predict that increases in the atmospheric concentration of carbon

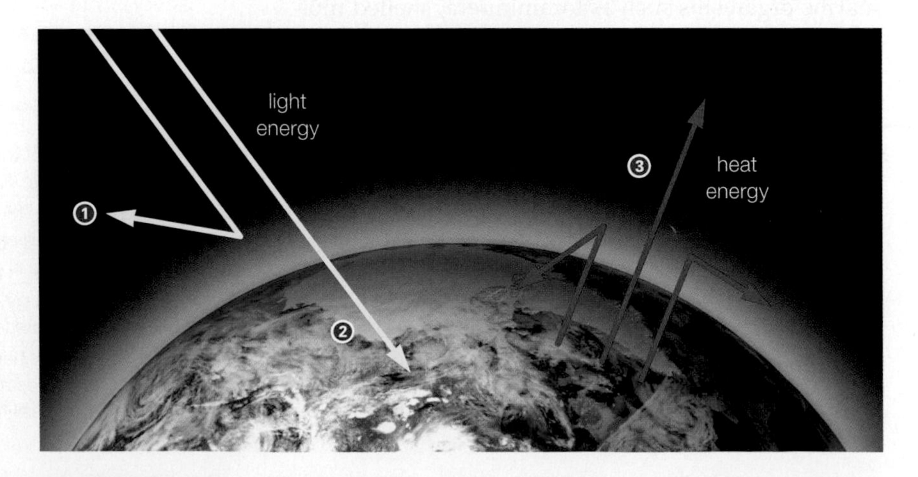

FIGURE 46.15 ▶**Animated** Greenhouse effect.

❶ Earth's atmosphere reflects some sunlight energy back into space.

❷ More light energy reaches and warms Earth's surface.

❸ Earth's warmed surface emits heat energy. Some of this energy escapes through the atmosphere into space. The rest is absorbed and then emitted in all directions by greenhouse gases. Some emitted heat warms Earth's surface and lower atmosphere.

FIGURE IT OUT Do greenhouse gases reflect heat energy toward Earth?

Answer: No. The gases absorb heat energy, then reemit it in all directions.

CREDIT: (15) NASA.

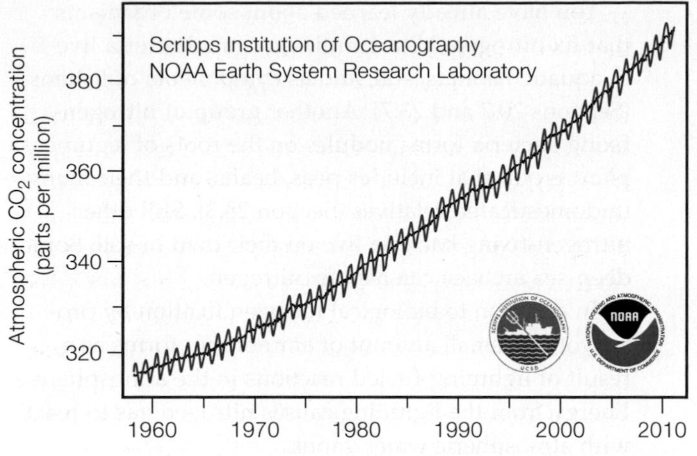

A Atmospheric CO_2 concentration measured at Mauna Loa Observatory. The red line shows seasonal highs and lows. Black is yearly averages.

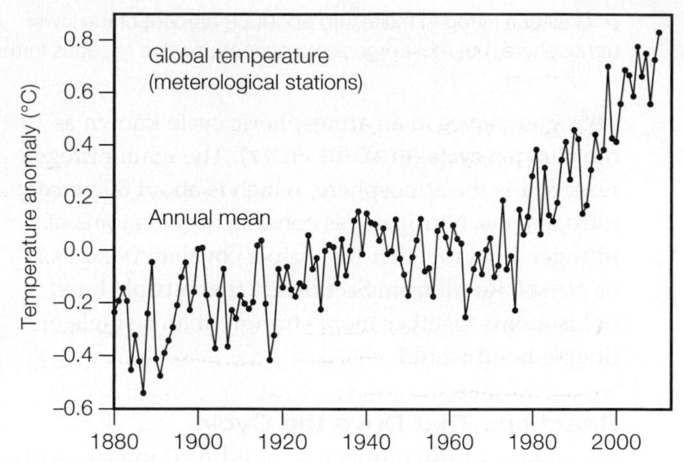

B Global annual mean air temperature based on measurements at meteorological stations. The temperature anomaly (vertical axis) refers to the deviation from the mean temperature of 1951–1980.

FIGURE 46.16 Directly measured changes in atmospheric carbon dioxide and global temperature.

dioxide and other greenhouse gases would raise the temperature of Earth's surface. Evidence from a variety of sources indicate that such a rise in temperature is underway. We are in the midst of **global climate change**, a long-term alteration of Earth's climate. Global warming, an increase in Earth's average surface temperature, is one aspect of this change (**FIGURE 46.16B**).

Earth's climate has varied greatly over its long history. During ice ages, much of the planet was covered by glaciers. Other periods were warmer than the present, and tropical plants and coral reefs thrived at what are now cool latitudes. Scientists can correlate some historical large-scale temperature changes with shifts in Earth's orbit, which varies in a regular fashion over 100,000 years, and Earth's tilt, which varies over 40,000 years. Changes in solar output and volcanic eruptions also influence Earth's temperature. However, most scientists see no evidence that any of these factors have a role in the current temperature rise.

In 2013, the Intergovernmental Panel on Climate Change reviewed the results of many scientific studies related to climate change. The panel included hundreds of scientists from all over the world. After reviewing the data, the panel concluded that it is clear that Earth's climate is warming as a result of human actions, and that limiting this change will require reductions in greenhouse gas emissions.

Temperature of the land and seas affects evaporation, winds, and currents, so many weather patterns will change as temperature rises. For example, warmer temperatures are correlated with extremes in rainfall patterns: periods of drought interrupted by unusually heavy rains. Rising temperature also elevates sea level. These and other effects of global climate change are the focus of Section 48.7.

Looking forward, atmospheric CO_2 is expected to continue to rise as large nations such as China and India become increasingly industrialized. However, efforts are under way to reduce the damage by increasing the efficiency of processes that require fossil fuels, shifting to alternative energy sources that do not release carbon, such as solar and wind power, and developing innovative ways to store carbon dioxide.

TAKE-HOME MESSAGE 46.8

How does carbon dioxide affect climate?

✔ Carbon dioxide is one of the atmospheric gases that absorb heat and emit it toward Earth's surface, thus keeping the planet warm enough for life.

✔ The CO_2 level of the atmosphere is currently rising, most likely as a result of human activity, and global mean temperature is rising with it. Changes in temperature affect other climate factors such as patterns of rainfall.

global climate change A long-term change in Earth's climate.
greenhouse gas Atmospheric gas that absorbs heat emitted by Earth's surface and reemits it, thus keeping the planet warm.

CREDIT: (16) NASA Goddard Institute for Space Studies.

46.9 Nitrogen Cycle

✔ Gaseous nitrogen makes up about 80 percent of the lower atmosphere, but most organisms cannot use this gaseous form.

Nitrogen moves in an atmospheric cycle known as the **nitrogen cycle** (**FIGURE 46.17**). The main nitrogen reservoir is the atmosphere, which is about 80 percent nitrogen gas. Nitrogen gas consists of two atoms of nitrogen held together by a triple covalent bond as N_2, or N≡N. Recall from Section 2.4 that a triple bond holds atoms together more strongly than a single or double bond would.

Reactions That Drive the Cycle

All organisms use nitrogen to build ATP, nucleic acids, and proteins. Photosynthetic organisms also use it to build chlorophyll. Despite the universal need for nitrogen and the abundance of atmospheric nitrogen, no eukaryote can make use of nitrogen gas. Eukaryotes do not have an enzyme that can break the strong bond between the two nitrogen atoms.

Only certain prokaryotes can carry out **nitrogen fixation**; they break the bonds in N_2 and use the nitrogen atoms to form ammonia, which dissolves to form ammonium (NH_4^+) ❶. Biological nitrogen fixation has a high activation energy (Section 5.3); it requires an input of sixteen molecules of ATP to convert one molecule of nitrogen to ammonia.

You have already learned about some organisms that fix nitrogen. Nitrogen-fixing cyanobacteria live in aquatic habitats, soil, and as components of lichens (Sections 20.7 and 23.7). Another group of nitrogen-fixing bacteria forms nodules on the roots of legumes, a plant group that includes peas, beans, and their many undomesticated relatives (Section 28.3). Still other nitrogen-fixing bacteria live on their own in soil. Some deep-sea archaea can also fix nitrogen.

In addition to biological nitrogen fixation by prokaryotes, a small amount of ammonium forms as a result of lightning-fueled reactions in the atmosphere. Energy from the lightning causes nitrogen gas to react with atmospheric water vapor.

Plants take up ammonium from soil water ❷. Animals obtain nitrogen by eating plants or one another. Bacterial and fungal decomposers return the nitrogen in wastes and remains to the soil by a process called **ammonification** ❸.

Nitrification is a two-step process that converts ammonium to nitrates ❹. First, ammonia-oxidizing bacteria and archaea convert ammonium to nitrite (NO_2^-). Other bacteria then take up nitrite and use it in reactions that form nitrates (NO_3^-). Nitrates, like ammonium, are taken up and used by producers ❺. Nitrification is essential to ecosystem health because it prevents ammonium from accumulating to toxic con-

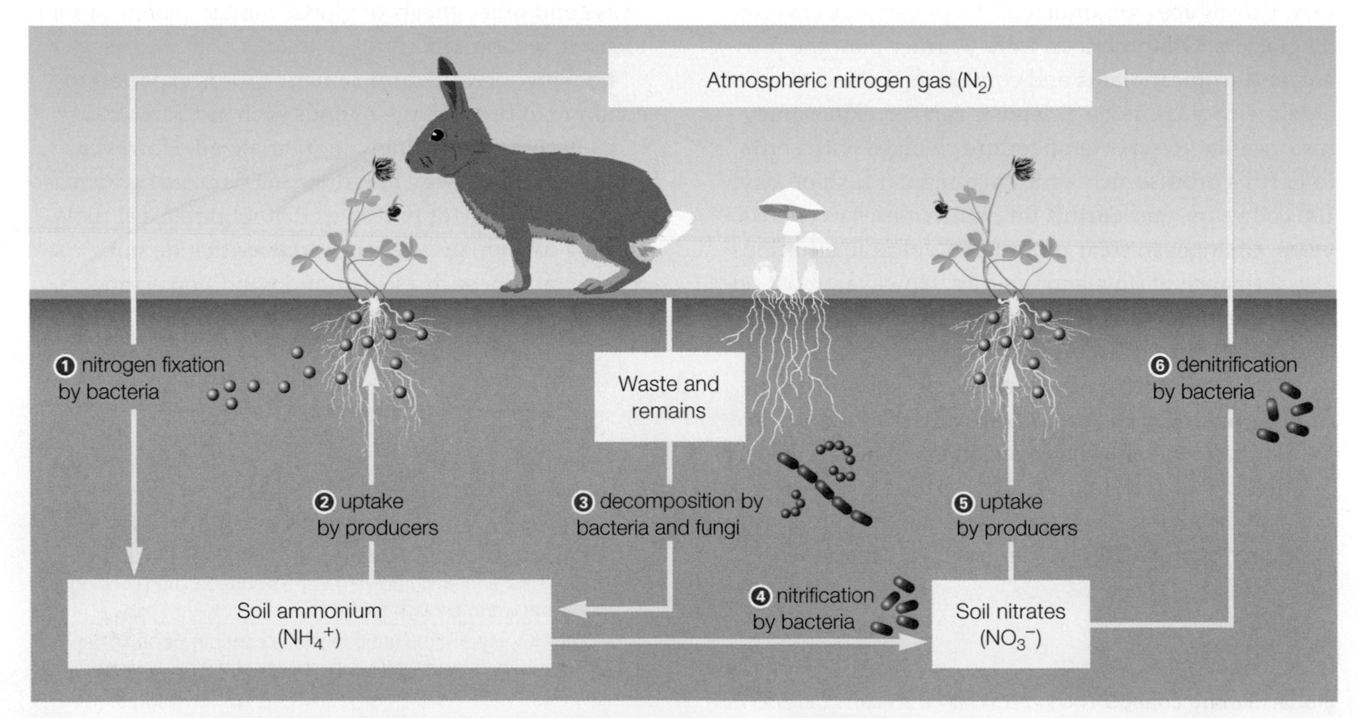

FIGURE 46.17 ▸**Animated** Nitrogen cycle on land. Nitrogen becomes available to plants through the activities of nitrogen-fixing bacteria. Other bacterial species cycle nitrogen to plants. They break down organic wastes to ammonium and nitrates.

centrations. Humans make use of bacteria that carry out this process in sewage treatment plants. Sewage contains large amounts of ammonium formed from urea excreted in urine (Section 40.2).

Denitrification, an anaerobic reaction carried out mainly by bacteria, converts nitrates to nitrogen gas ❻. In ecosystems, denitrification can have important effects on productivity because it results in a decline in the amount of soluble nitrogen available to producers. In sewage treatment plants, denitrifying bacteria are used to remove nitrates from wastewater before the water is released into the environment.

Human Effects on the Cycle

In the early 1900s, German scientists discovered a method of fixing atmospheric nitrogen and producing ammonium on an industrial scale. This process allowed the manufacture of synthetic nitrogen fertilizers that have boosted crop yields (**FIGURE 46.18**). However, use of these fertilizers, along with other human activities, has added large amounts of nitrogen-containing compounds to our air and water. Here we consider two such types of compounds and their effects.

FIGURE 46.18 Spraying synthetic nitrogen fertilizer on a corn field to boost crop yield. The inset photo shows corn grown with adequate nitrogen (left) and in nitrogen-deficient soil (right).

Nitrous Oxide We know from air bubbles in ice cores that the atmospheric concentration of nitrous oxide (N_2O) remained about 270 parts per billion (ppb) for at least a thousand years before the industrial revolution. It is now about 325 ppb and rising. The ongoing increase is the result of burning of fossil fuels, use of synthetic nitrogen fertilizers, and industrial livestock production. Burning fossil fuel releases N_2O directly into the air. Chemical fertilizers and manure from livestock increase atmospheric N_2O by encouraging growth of bacteria that release this gas.

An increased concentration of atmospheric N_2O is a matter of concern for two reasons. First, N_2O is a greenhouse gas, and a highly persistent and effective one. It can remain in the atmosphere for more than 100 years, and it has a warming potential 300 times that of CO_2. Second, N_2O contributes to destruction of the ozone layer. As Section 19.1 explained, ozone high in the atmosphere protects life at Earth's surface from the damaging effects of ultraviolet radiation. We discuss ozone destruction in more detail in Section 48.6.

Nitrates Nitrate from fertilizers and animal manure runs off into streams and lakes or leaches (travels down through the soil) into groundwater. In communities that lack a public sewer system, wastewater from homes also contributes to nitrate pollution. Ingested nitrate inhibits iodine uptake by the thyroid gland and it may increase the risk of thyroid cancer. Drinking nitrate-contaminated water is also correlated with an increased risk for respiratory infections, diabetes, and some cancers. To protect human health, the Environmental Protection Agency (EPA) requires testing of public drinking water sources for excess nitrates.

TAKE-HOME MESSAGE 46.9

How does nitrogen cycle in ecosystems?

- ✔ The atmosphere is the main reservoir for nitrogen, but only nitrogen-fixing prokaryotes can access it.
- ✔ Plants take up ammonium and nitrates from soil. Ammonium is released by nitrogen-fixing bacteria and by fungal and bacterial decomposers. Bacteria and archaea produce nitrates.
- ✔ Nitrogen returns to the atmosphere when denitrifying bacteria convert soluble forms of nitrogen to nitrogen gas.
- ✔ Use of synthetic nitrogen fertilizers and fossil fuels has increased the amount of nitrous oxide (N_2O) in the air and added nitrates to the water.

ammonification Breakdown of nitrogen-containing organic material resulting in the release of ammonia and ammonium ions.
denitrification Conversion of nitrates or nitrites to gaseous forms of nitrogen.
nitrification Conversion of ammonium to nitrates.
nitrogen cycle Movement of nitrogen among the atmosphere, soil, and water, and into and out of food webs.
nitrogen fixation Incorporation of nitrogen from nitrogen gas into ammonia.

46.10 The Phosphorus Cycle

✔ Phosphorus does not occur in gaseous form in habitats that can support life.

✔ Like nitrogen, phosphorus is taken up by plants only in ionized form, and is often a limiting factor on plant growth.

Atoms of phosphorus are highly reactive, so phosphorus does not occur naturally in its elemental form. Most of Earth's phosphorus is bonded to oxygen as phosphate (PO_4^{3-}), an ion that abounds in rocks and sediments. In the **phosphorus cycle**, phosphorus passes quickly through food webs as it moves from land to ocean sediments, then slowly back to land (**FIGURE 46.19**). Because little phosphorus exists in a gaseous form and its major reservoir is sedimentary rock, the phosphorus cycle is described as a **sedimentary cycle**.

Weathering and erosion move phosphates from rocks into soil, lakes, and rivers ❶. Leaching and runoff carry dissolved phosphates to the ocean ❷. Here, most phosphorus comes out of solution and settles as deposits along continental margins ❸. Slow movements of Earth's crust can uplift these deposits onto land ❹, where weathering releases phosphates from rocks once again. Phosphate-rich rocks are mined for use in the industrial production of fertilizer.

All organisms require phosphorus. It is a component of all nucleic acids and phospholipids. The biological portion of the phosphorus cycle begins when producers take up phosphate. Land plants take up dissolved phosphate from the soil water ❺. Land animals get phosphates by eating the plants or one another. Phosphorus returns to the soil in the wastes and remains of organisms ❻. Phosphate-rich droppings from seabird or bat colonies are collected and used as a natural fertilizer. Manure from livestock is also rich in phosphorus.

In the seas, phosphorus enters food webs when producers take up phosphate dissolved in seawater ❼. As on land, wastes and remains continually replenish the phosphorus supply ❽.

phosphorus cycle Movement of phosphorus among Earth's rocks and waters, and into and out of food webs.
sedimentary cycle Biochemical cycle in which the atmosphere plays little role and rocks are the major reservoir.

TAKE-HOME MESSAGE 46.10

How does phosphorus cycle in ecosystems?

✔ Rocks are the main phosphorus reservoir. Weathering puts phosphates into water. Producers take up dissolved phosphates.

✔ Phosphate-rich wastes are a natural fertilizer, and phosphate from rocks can be used to produce fertilizer on an industrial scale.

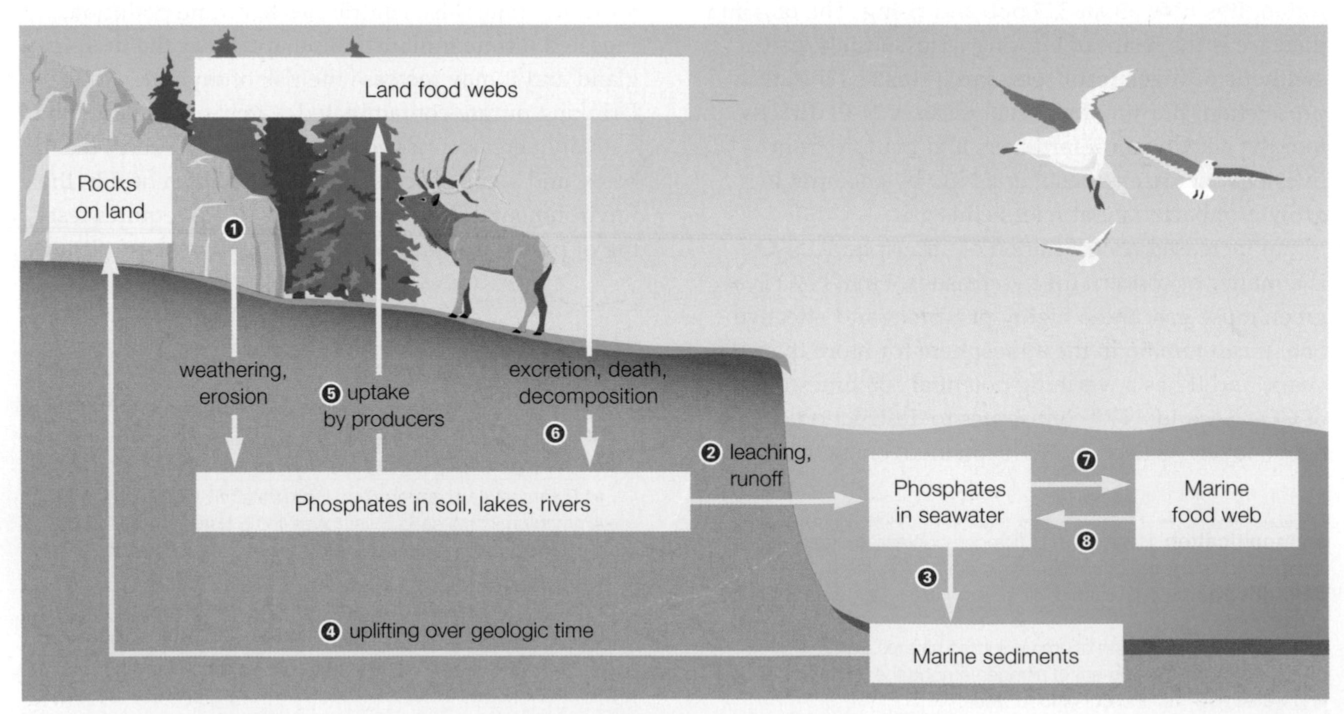

FIGURE 46.19 The phosphorus cycle. In this sedimentary cycle, phosphorus moves mainly in the form of phosphate ions (PO_4^{3-}).

Too Much of a Good Thing (revisited)

Farmers use industrially produced fertilizers because such products are a relatively inexpensive way to enhance crop yield. However, nitrates and phosphates from these fertilizers can pollute waters, both locally and in distant regions. Fertilizer applied to fields in the Midwest not only pollutes local drinking water, it also flows into streams that feed the Mississippi River. The river then delivers its heavy load of phosphates and nitrates to the Gulf of Mexico.

Most excess phosphates and nitrogen entering the Mississippi come from agriculture, but suburbanites and city dwellers also contribute. In some communities, the sewer system receives both sewage and water from storm drains. Heavy rains can overwhelm such a system, causing an overflow that delivers raw sewage into streams. Rain-related overflows can be prevented by a system that keeps sewage and water from storm drains separate. However, such systems typically do not treat water from storm drains before discharging it. Thus, when fertilizer from lawns enters such a system, it is delivered untreated into the river.

Each summer, the excessive nutrient inputs from the Mississippi into the Gulf of Mexico result in formation of a "dead zone," an area of deep water where the oxygen level is too low to support most marine organisms. In 2013, the dead zone encompassed an area a bit larger than the state of Connecticut.

The dead zone forms after eutrophication causes an algal bloom (Section 21.6). The high nutrient level encourages a population explosion of marine algae. When the algae die, their remains drift down to the seafloor. Decomposition of these remains by bacteria then depletes the water of oxygen, making it impossible for most animals to survive. Fishes can swim away from the low-oxygen zone, but less mobile animals suffocate. There are also indirect effects, as when an increase in jellyfishes leads to a decrease in the fish larvae on which the jellies prey. Jellyfishes are among the few animals that thrive in anoxic waters.

summary

Section 46.1 Inorganic substances such as phosphorus and nitrogen serve as essential nutrients for producers, and a deficiency in these substances can limit producer growth. Excessive inputs of nutrients as a result of human activities can alter ecosystem dynamics, as when phosphate from detergents or fertilizers causes **eutrophication**.

Section 46.2 An **ecosystem** consists of an array of organisms along with nonliving components of their environment. There is a one-way flow of energy into and out of an ecosystem, and a cycling of materials among resident species. All ecosystems have inputs and outputs of energy and nutrients.

Sunlight supplies energy to most ecosystems. **Primary producers** convert sunlight energy into chemical bond energy. They also take up the nutrients that they, and all consumers, require. Herbivores, carnivores, omnivores, **decomposers**, and **detritivores** are **consumers**.

Energy moves from organisms at one **trophic level** to organisms at another. Organisms are at the same trophic level if they are an equal number of steps away from the energy input into the ecosystem. A **food chain** shows one path of energy and nutrient flow among organisms. It depicts who eats whom.

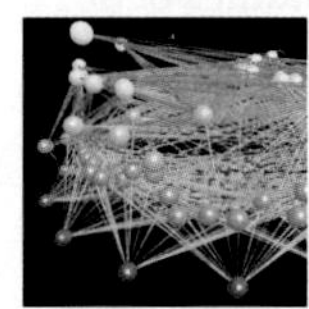

Section 46.3 Food chains interconnect as **food webs**. In a **grazing food web**, most energy captured by producers flows to herbivores. In a **detrital food web**, most energy flows from producers directly to detritivores and decomposers. Both types of food webs interconnect in nearly all ecosystems. The efficiency of energy transfers is low, so most food chains have no more than four or five trophic levels.

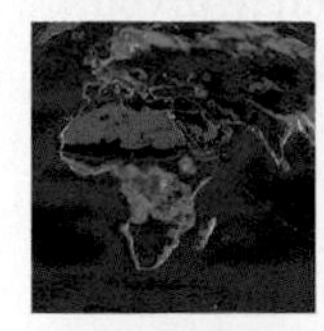

Section 46.4 The **primary production** of an ecosystem is the rate at which producers capture and store energy in their tissues. It varies with climate, seasonal changes, and nutrient availability.

Energy pyramids and **biomass pyramids** depict how energy and organic compounds are distributed among the organisms of an ecosystem. All energy pyramids are largest at their base. If producers get eaten as fast as they reproduce, the biomass of consumers can exceed that of producers, so the biomass pyramid is upside down.

Section 46.5 In a **biogeochemical cycle**, water or some nutrient moves from an environmental reservoir, through organisms, then back to the environment. Chemical and geological processes move the nutrients between their environmental reservoirs.

Section 46.6 In the **water cycle**, evaporation, condensation, and precipitation move water from its main reservoir—oceans—into the atmosphere, onto land, then back to oceans. **Runoff** is water that flows over ground into streams. A **watershed** is an area where all precipitation drains into a specific waterway. Water in **aquifers** and in **soil water** is **groundwater**. Only a small fraction of Earth's water is available as fresh water, and most of that is frozen.

Section 46.7 The **carbon cycle** moves carbon mainly among seawater, the air, soils, and living organisms in an **atmospheric cycle**. The largest carbon reservoir is sedimentary rocks, but living organisms cannot take up carbon from this reservoir. The largest reservoir of biologically available carbon is the ocean. Aquatic producers take up dissolved bicarbonate and convert it to CO_2. Plants take up CO_2 from air. When land organisms die, their wastes and remains contribute to the carbon in soil, which exceeds the amount in the atmosphere. The amount of carbon in soil differs among ecosystems, with the carbon content being highest in regions where decomposition is slowest.

Human use of fossil fuels is moving carbon from a reservoir that was previously inaccessible to living organisms into the air and water.

Section 46.8 The greenhouse effect refers to the ability of **greenhouse gases** to trap heat in the lower atmosphere and thus warm Earth's surface. Carbon dioxide is a greenhouse gas and humans are putting increasing amounts of it into the atmosphere, mainly by burning fossil fuels. Direct measurements of the atmosphere, studies of air bubbles in ice cores, and analysis of the composition of fossil foraminiferan shells indicate that atmospheric CO_2 is currently at its highest level in at least 15 million years. As one would expect, the ongoing rise in carbon dioxide is causing global warming, which is one aspect of an ongoing process of **global climate change**.

Section 46.9 The **nitrogen cycle** is an atmospheric cycle. Air is the main reservoir for N_2, a gaseous form of nitrogen accessible only to nitrogen-fixing prokaryotes. By the process of **nitrogen fixation**, some bacteria and archaea take up N_2 and incorporate it into ammonia that producers can take up and use. Ammonia is also formed by the **ammonification** of organic wastes and remains by bacteria and fungi. Bacteria also carry out **nitrification**, converting ammonium to nitrite and then nitrate, which plants can also take up. Nitrification prevents toxic ammonia excreted in wastes from accumulating. **Denitrification** of nitrate by bacteria converts nitrate in soil or water to a gaseous form, thus allowing it to escape into the air.

Industrial fixation of nitrogen to produce chemical fertilizers, the production of livestock, and the burning of fossil fuels add nitrogen-containing compounds to the air and water. Nitrous oxide is a greenhouse gas that also contributes to destruction of the protective ozone layer. It is formed by burning fossil fuels and by the activity of bacteria in habitats enriched by nitrogen fertilizer. Nitrate is a soluble compound that sometimes pollutes our sources of drinking water. Ingestion of excess nitrate is correlated with a variety of health problems, including cancer. Nitrate enters water as a result of fertilizer use and inadequate treatment of sewage.

Section 46.10 The **phosphorus cycle** is a **sedimentary cycle**. Earth's crust is the largest reservoir of phosphorus, a element that does not occur as a gas in any significant quantity. Producers cannot access the phosphate that is tied up in rocks; they obtain the phosphorus they need by taking up dissolved phosphates. Lack of phosphate often limits plant growth. To overcome this limitation, farmers apply fertilizer. Deposits of bat and bird wastes are mined as a natural phosphate-rich fertilizer, and phosphate-rich rocks are used to produce fertilizer on an industrial scale.

self-quiz

Answers in Appendix VII

1. In most ecosystems, the primary producers use energy from ______ to build organic compounds.
   a. sunlight
   b. heat
   c. breakdown of wastes and remains
   d. breakdown of inorganic substances in the habitat

2. Organisms at the lowest trophic level in a tallgrass prairie are all ______.
   a. two steps away from the original energy input
   b. autotrophs
   c. heterotrophs
   d. both a and b
   e. both a and c

3. All organisms at the top trophic level ______.
   a. capture energy from a nonliving source
   b. obtain carbon from a nonliving source
   c. would be at the top of an energy pyramid
   d. all of the above

4. Primary productivity is affected by ______.
   a. nutrient availability
   b. amount of sunlight
   c. temperature
   d. all of the above

5. Efficiency of energy transfers in aquatic ecosystems is typically higher than in land ecosystems because ______.
   a. aquatic food webs include more endotherms
   b. algae do not make lignin
   c. primary production cannot occur in water
   d. all of the above

6. Most of Earth's fresh water is ______.
   a. in lakes and streams
   b. in aquifers and soil
   c. frozen as ice
   d. in bodies of organisms

7. Earth's largest carbon reservoir is ______.
   a. the atmosphere
   b. sediments and rocks
   c. seawater
   d. living organisms

8. Carbon is released into the atmosphere by ______.
   a. photosynthesis
   b. aerobic respiration
   c. burning fossil fuels
   d. b and c

data analysis activities

Changes in the Air To assess the impact of human activity on the carbon dioxide level in Earth's atmosphere, it helps to take a long view. One useful data set comes from deep core samples of Antarctic ice. The oldest ice core that has been fully analyzed dates back a bit more than 400,000 years. Air bubbles trapped in the ice provide information about the gas content in Earth's atmosphere at the time the ice formed. Combining ice core data with more recent direct measurements of atmospheric carbon dioxide—as in **FIGURE 46.20**—can help scientists put current changes in the atmospheric carbon dioxide into historical perspective.

1. What was the highest carbon dioxide level between 400,000 B.C. and 0 A.D.?

2. During this period, how many times did carbon dioxide reach a level comparable to that measured in 1980?

3. The industrial revolution occurred around 1800. What was the trend in carbon dioxide level in the 800 years prior to this event? What about in the 175 years after it?

4. Was the rise in the carbon dioxide level between 1800 and 1975 larger or smaller than the rise between 1980 and 2013?

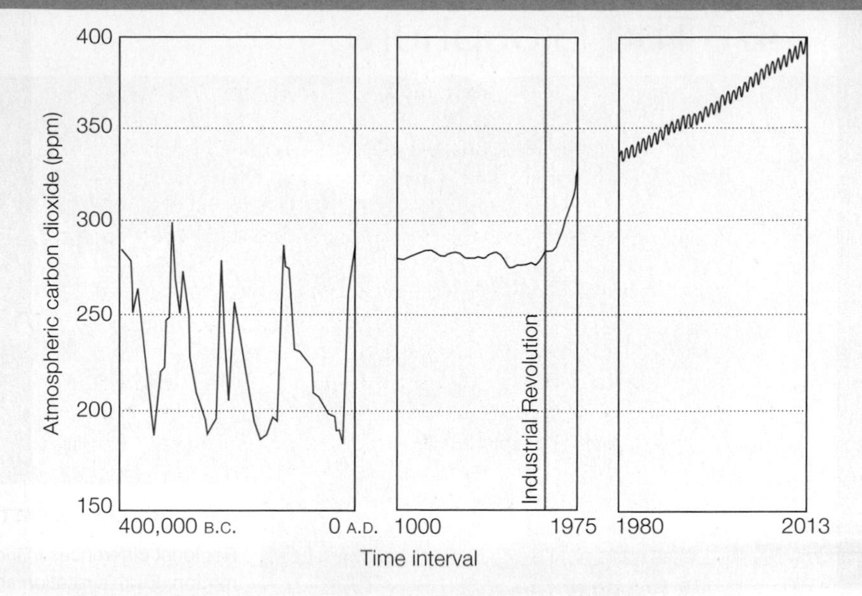

FIGURE 46.20 Changes in atmospheric carbon dioxide levels (in parts per million). Direct measurements began in 1980. Earlier data are based on ice cores.

9. Greenhouse gases _______ .
   a. slow the escape of heat energy from Earth into space
   b. are produced by natural and human activities
   c. are at higher levels than they were 100 years ago
   d. all of the above

10. The _______ cycle is a sedimentary cycle.
   a. phosphorus
   b. carbon
   c. nitrogen
   d. water

11. Earth's largest phosphorus reservoir is _______ .
   a. the atmosphere
   b. the ocean
   c. sedimentary rock
   d. living organisms

12. Plant growth requires uptake of _______ from the soil.
   a. nitrogen
   b. carbon
   c. phosphorus
   d. both a and c
   e. all of the above

13. Nitrogen fixation converts _______ to _______ .
   a. nitrogen gas; ammonia
   b. nitrates; nitrites
   c. ammonia; nitrogen gas
   d. ammonia; nitrates
   e. nitrogen gas; nitrogen oxides

14. Burning fossil fuel releases _______ into the air.
   a. carbon dioxide
   b. nitrous oxide
   c. phosphates
   d. a and b

15. Match each term with its most suitable description.

___ carbon dioxide	a. contains triple bond
___ nitrate	b. mined from sedimentary rock
___ phosphate	c. marine carbon source
___ nitrogen gas	d. soluble form of nitrogen
___ bicarbonate	e. greenhouse gas
___ nitrous oxide	f. greenhouse gas and ozone destroyer

critical thinking

1. Marguerite plants a vegetable garden in Maine. Eduardo plants a similar garden in Florida. Based on climate alone, which garden would you expect to have a higher annual primary productivity? What other factors could affect primary production in each garden?

2. Where does your drinking water come from? An aquifer or an aboveground reservoir? What area is included within your watershed? Visit the Science in Your Watershed site at http://water.usgs.gov/wsc to find out.

3. Scientists study bubbles trapped in ancient glacial ice to determine how concentrations of nitrogen and carbon dioxide gas have changed over time. However, bubbles in glacial ice cannot provide information about changes in phosphorus. Explain why air samples are not useful for this purpose and propose an alternative method to study how the amount of phosphorus in a region has changed over time.

4. Nitrogen-fixing bacteria live throughout the ocean, from its sunlit upper waters to 200 meters (650 feet) beneath its surface. Recall that nitrogen is a limiting factor in many habitats. What effect would an increase in populations of marine nitrogen-fixers have on carbon uptake and primary productivity in those waters?

5. As noted in Section 46.7, mycorrhizal fungi help prevent carbon in the soil from escaping into the atmosphere. These fungi also benefit a host plant by providing it with a share of the phosphorus and nitrogen that their hyphae take up from soil. Most crop plants are capable of forming a relationship with mycorrhizal fungi, if any are present in the soil. Some people have suggested that inoculating soil with fungal spores could help reduce the use of industrially produced fertilizer. What are some potential advantages of using fungi, rather than chemical fertilizers, to enhance plant growth?

47 The Biosphere

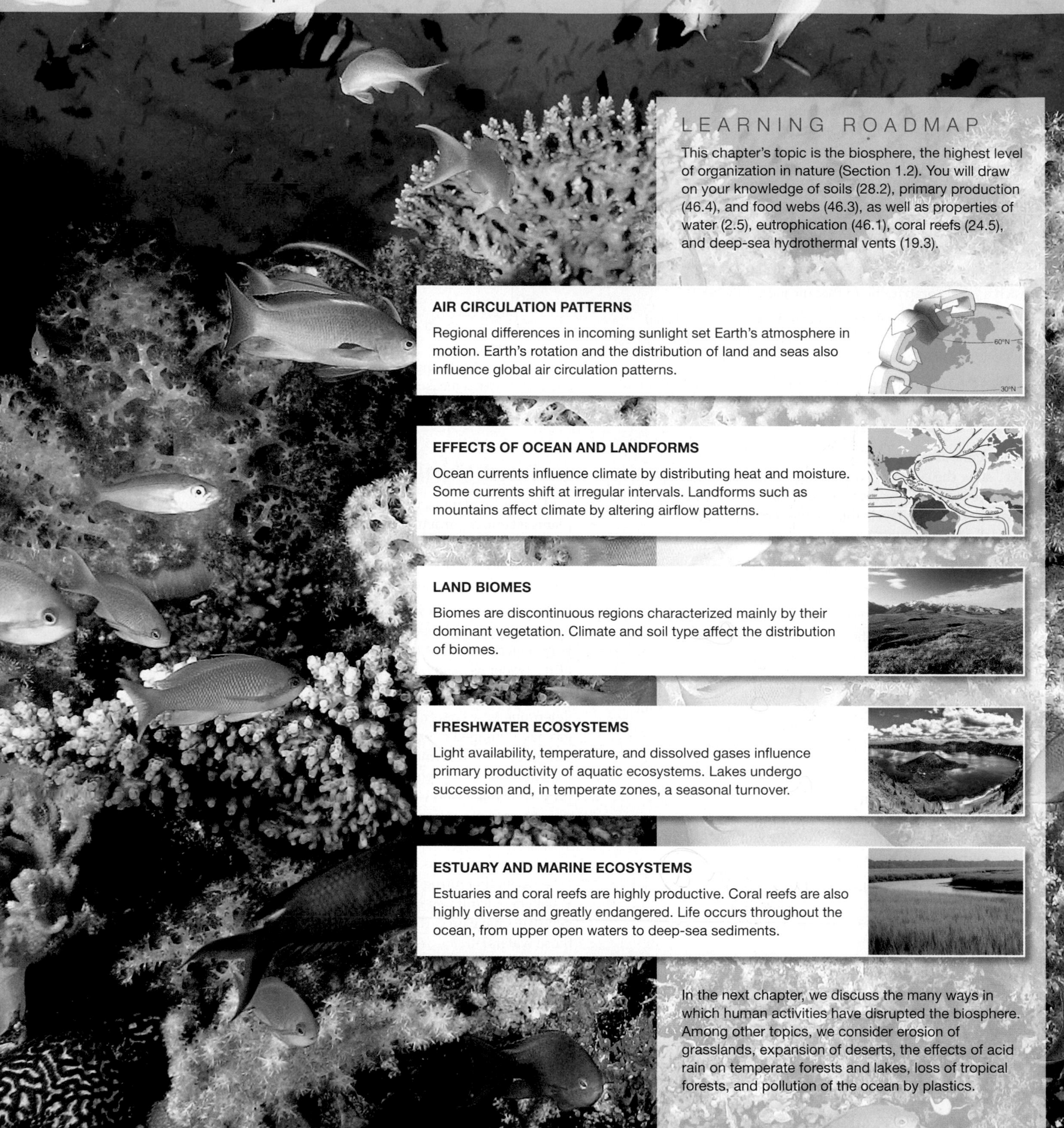

LEARNING ROADMAP

This chapter's topic is the biosphere, the highest level of organization in nature (Section 1.2). You will draw on your knowledge of soils (28.2), primary production (46.4), and food webs (46.3), as well as properties of water (2.5), eutrophication (46.1), coral reefs (24.5), and deep-sea hydrothermal vents (19.3).

AIR CIRCULATION PATTERNS

Regional differences in incoming sunlight set Earth's atmosphere in motion. Earth's rotation and the distribution of land and seas also influence global air circulation patterns.

EFFECTS OF OCEAN AND LANDFORMS

Ocean currents influence climate by distributing heat and moisture. Some currents shift at irregular intervals. Landforms such as mountains affect climate by altering airflow patterns.

LAND BIOMES

Biomes are discontinuous regions characterized mainly by their dominant vegetation. Climate and soil type affect the distribution of biomes.

FRESHWATER ECOSYSTEMS

Light availability, temperature, and dissolved gases influence primary productivity of aquatic ecosystems. Lakes undergo succession and, in temperate zones, a seasonal turnover.

ESTUARY AND MARINE ECOSYSTEMS

Estuaries and coral reefs are highly productive. Coral reefs are also highly diverse and greatly endangered. Life occurs throughout the ocean, from upper open waters to deep-sea sediments.

In the next chapter, we discuss the many ways in which human activities have disrupted the biosphere. Among other topics, we consider erosion of grasslands, expansion of deserts, the effects of acid rain on temperate forests and lakes, loss of tropical forests, and pollution of the ocean by plastics.

47.1 Going With the Flow

Earth's air and seas are in constant circulation, distributing materials on a global scale. Consider what happened after a powerful earthquake that occurred off the northeast coast of Japan triggered a huge tsunami (tidal wave) in March of 2011. Together, the earthquake and tsunami killed more than 20,000 people and decimated coastal cities. Amid this destruction, a nuclear power plant on the shore of the city of Fukushima became damaged and released some radioactive material.

Prevailing winds carried radioisotopes accidentally released into the air at Fukushima eastward. Rain deposited the vast majority of this radioactive material in the Pacific Ocean, but some remained aloft and continued farther east. Radioisotopes from Fukushima were first detected on the west coast of North America about 60 hours after their release, and in Europe about a week later. Within 18 days, some radioisotopes released at Fukushima had circled the globe.

Materials move more slowly in the ocean. Think about the millions of tons of debris that was dragged into the ocean by the tsunami (**FIGURE 47.1**). Most of this material sank, but some stayed afloat and was carried along by surface currents, which also flow predominately east.

Scientists have been monitoring the movement of the floating material and recording when objects that are clearly tsunami debris turn up along the west coast of North America. Several Japanese boats lost during the tsunami have come ashore. One arrived in Washington state in early 2013 carrying five live fishes and a variety of invertebrates native to the tropical Pacific. California received its first verified tsunami debris the same year—a boat belonging to a high school that was destroyed by the tsunami. The boat was cleaned up by American high school students and returned to Japan.

Movement of radioactive water that entered the sea at the Fukushima nuclear plant is also under close scrutiny. This water contains cesium 137 (^{137}Cs), a radioisotope with a half-life of 30 years. Currents are expected to deliver some ^{137}Cs-enriched water to the west coast of North America over a period lasting from 2014 to perhaps as late as 2020. Although this water will certainly contain more ^{137}Cs than normal seawater, scientists think it is unlikely to pose a threat to human health. The water's ^{137}Cs concentration is expected to remain below the level that the Environmental Protection Agency currently allows in drinking water.

FIGURE 47.1 Debris floats in the ocean near the coast of Japan after the 2011 earthquake and tsunami.

CREDITS: (opposite) © John Easley, www.johneasley.com; (1) U.S. Navy photo by Mass Communication Specialist 3rd Class Alexander Tidd/Released.

47.2 Global Air Circulation Patterns

✔ How much solar energy reaches Earth's surface varies from place to place and with the season.

The **biosphere** includes all places on Earth where life exists. The geographical distribution of species within the biosphere depends largely on climate. **Climate** refers to average weather conditions, such as cloud cover, temperature, humidity, and wind speed, over time. Regional climates differ because factors that influence winds and ocean currents vary from place to place. Such factors include the intensity of sunlight, the distribution of landmasses and seas, and elevation.

Seasonal Effects

Each year, Earth rotates around the sun in an elliptical path (**FIGURE 47.2**). Seasonal changes in day length and temperature arise because Earth's axis is not perpendicular to the plane of this ellipse, but rather is tilted about 23 degrees. In June, when the Northern Hemisphere is angled toward the sun, this hemisphere receives more intense sunlight and has longer days than the Southern Hemisphere (**FIGURE 47.2A**). In December, the opposite is true (**FIGURE 47.2C**). Twice a year—on spring and autumn equinoxes—Earth's axis is perpendicular to incoming sunlight. On these days, every place on Earth has 12 hours of daylight and 12 hours of darkness (**FIGURE 47.2B** and **D**).

In each hemisphere, the extent of the seasonal change in daylight increases with latitude. At 25° north or south of the equator, the longest day length is a bit less than 14 hours. By contrast, 60° north or south of the equator, the longest day is nearly 19 hours.

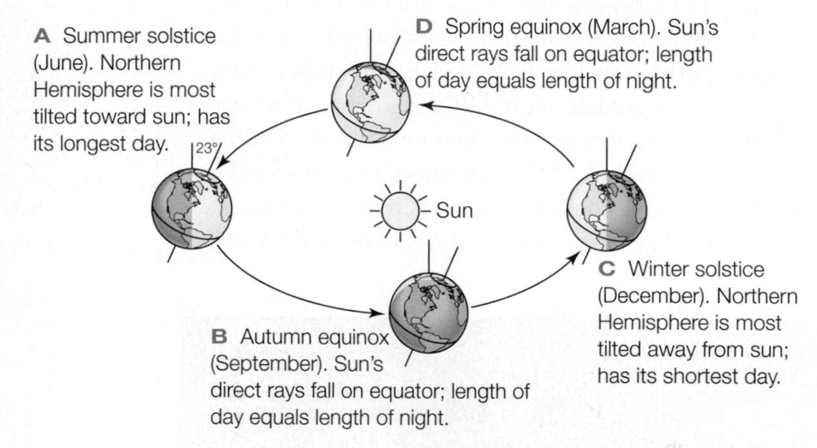

FIGURE 47.2 Earth's tilt and yearly rotation around the sun cause seasonal effects. The 23° tilt of Earth's axis causes the Northern Hemisphere to receive more intense sunlight and have longer days in summer than in winter.

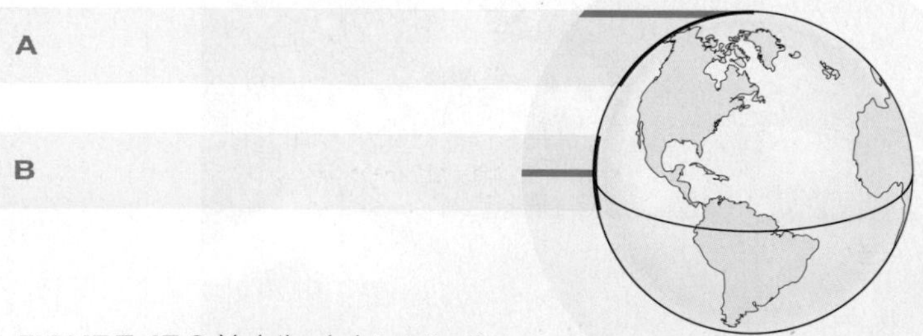

FIGURE 47.3 Variation in intensity of solar radiation with latitude. For simplicity, we depict two equal parcels of incoming radiation on an equinox, a day when incoming rays are perpendicular to Earth's axis.

Rays that fall on high latitudes (**A**) pass through more atmosphere (blue) than those that fall near the equator (**B**). Compare the length of the green lines. Atmosphere is not to scale.

Also, energy in the rays that fall at the high latitude is dispersed over a greater area than energy that falls on the equator. Compare the length of the red lines.

Air Circulation and Rainfall

On any given day, equatorial regions receive more sunlight energy than higher latitudes, so Earth's surface warms more at the equator than at the poles. There are two reasons for this difference in incoming sunlight. First, particles of dust, water vapor, and greenhouse gases absorb some solar radiation or reflect it back into space. Sunlight traveling to high latitudes passes through more atmosphere to reach Earth's surface than light traveling to the equator, so less energy reaches the ground (**FIGURE 47.3A**). Second, energy in an incoming parcel of sunlight is dispersed over a smaller surface area at the equator than at the higher latitudes (**FIGURE 47.3B**).

Knowing about two properties of air can help you understand how regional differences in surface warming give rise to global air circulation and rainfall patterns. First, as air warms, it becomes less dense and rises. Hot air balloonists take advantage of this effect when they ascend by heating the air inside their balloon. Second, warm air can hold more water than cooler air. This is why you can "see your breath" in cold weather. When you exhale, warm air with moisture from your lungs cools, causing the water in that air to condense as tiny droplets.

The global air circulation pattern begins at the equator, where intense sunlight warms the air and causes evaporation from the ocean. As a result, warm, moist air ascends at the equator (**FIGURE 47.4A**). As the air from the equator rises to higher altitudes, it moves north and south and cools, releasing moisture as rain that supports tropical rain forests.

By the time the air has reached 30° north or south of the equator, it has given up most moisture and cooled

Initial Pattern of Air Circulation

D At the poles, cold air sinks and moves toward lower latitudes.

C Air rises again at 60° north and south, where air flowing poleward meets air coming from the poles.

B As the air flows toward higher latitudes, it cools and loses moisture as rain. At around 30° north and south latitude, the air sinks and flows north and south along Earth's surface.

A Warmed by energy from the sun, air at the equator picks up moisture and rises. It reaches a high altitude, and spreads north and south.

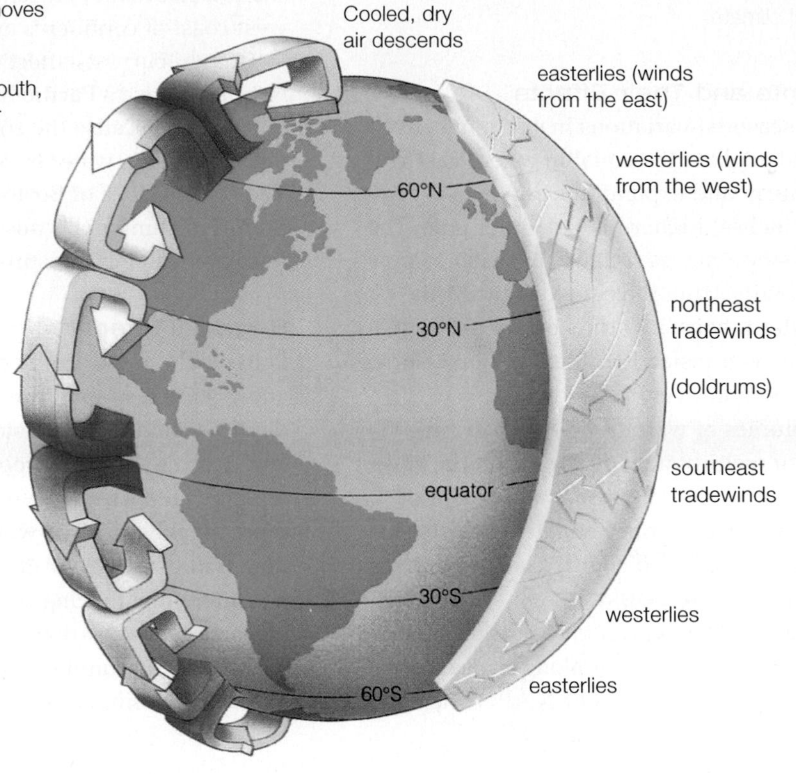

Prevailing Wind Patterns

E Major winds near Earth's surface do not blow directly north and south because of Earth's rotation. Winds deflect to the right of their original direction in the Northern Hemisphere and to the left in the Southern Hemisphere.

FIGURE 47.4 ▸Animated Global air circulation patterns and their effects on climate.

FIGURE IT OUT What is the direction of prevailing winds in the central United States?

Answer: From west to east.

off, so it sinks back toward Earth's surface (**FIGURE 47.4B**). Many of the world's great deserts, including the Sahara, are about 30° from the equator.

As air continues along Earth's surface toward the poles, it again picks up heat and moisture. At a latitude of about 60°, it rises (**FIGURE 47.4C**). The resulting rains support temperate zone forests.

Cold, dry air descends near the poles (**FIGURE 47.4D**). Precipitation is sparse, and polar deserts form.

Surface Wind Patterns

Major wind patterns arise as air in the lower atmosphere moves from latitudes where air is sinking toward those where air is rising. Air masses are not attached to Earth's surface, so Earth spins beneath them, moving fastest at the equator and most slowly at the poles. Thus, as an air mass moves away from the equator, the speed at which the Earth rotates beneath it continually slows. As a result, major winds trace a curved path relative to Earth's surface (**FIGURE 47.4E**). In the Northern Hemisphere, winds curve toward the right of their initial direction; in the Southern Hemisphere, they curve toward the left. For example, between 30° north and 60° north, surface air traveling toward the North Pole is deflected right, or toward the east. Winds are named for the direction from which they blow. Prevailing winds in the United States are westerlies—they blow from west to east.

Winds blow most consistently from one region where air is rising to another such location. Where air actually rises, winds are intermittent, as in the doldrums near the equator.

TAKE-HOME MESSAGE 47.2

How does sunlight affect climate?

✔ Equatorial regions receive more intense sunlight than higher latitudes.

✔ Sunlight drives the rise of moisture-laden air at the equator. The air cools as it moves north and south, releasing rains that support tropical forests. Deserts form where cool, dry air descends. Sunlight energy also drives moisture-laden air aloft at 60° north and south latitude. This air gives up moisture as it flows toward the equator or the pole.

✔ Major surface winds arise as air in the lower atmosphere flows from latitudes where air is sinking toward latitudes where air is rising. These winds trace a curved path relative to Earth's surface because of Earth's rotation.

biosphere All regions of Earth that can support life.
climate Average weather conditions in a region over a long period.

47.3 The Ocean, Landforms, and Climates

✔ Solar heating and the effects of the wind set the ocean's surface water in motion, producing currents that distribute nutrients and affect climate.

Ocean Currents and Their Effects

Latitudinal and seasonal variations in sunlight cause water to heat and cool. At the equator, where vast volumes of water warm and expand, the sea level is about 8 centimeters (3 inches) higher than at either pole. The existence of this slope causes sea surface water to move in response to gravity, from the equator toward the poles. As the water moves, it warms the air above it. At midlatitudes, oceans transfer 10 million billion calories of heat energy to the air every second!

Enormous volumes of water flow as ocean currents. The force of major winds, Earth's rotation, and topography determine the directional movement of these currents. Surface currents circulate clockwise in the Northern Hemisphere and counterclockwise in the Southern Hemisphere (**FIGURE 47.5**).

Swift, deep, and narrow currents of nutrient-poor water flow away from the equator along the east coast of continents. Along the east coast of North America, warm water flows north, as the Gulf Stream. Slower, shallower, broader currents of cold water parallel the west coast of continents and flow toward the equator.

Coastal currents affect coastal climates. Coasts of North America's Pacific Northwest are cool and foggy in summer because the cold California current chills the air, causing water to condense out as droplets. The coastal cities of Boston and Baltimore are hot and humid in summer because the warm Gulf Stream releases heat and moisture into the air over these cities.

Regional Effects

Differences in the ability of water and land to absorb and release heat give rise to coastal breezes. In the daytime, land warms faster than water. As air over the land warms and rises, cooler offshore air moves in to replace it (**FIGURE 47.6A**). After sundown, land cools more quickly than the water, so the breezes reverse direction (**FIGURE 47.6B**).

Differential heating of water and land also causes **monsoons**, which are winds that change direction seasonally. In the summer, the continental interior of Asia heats up, causing air to rise above it. Moist air from

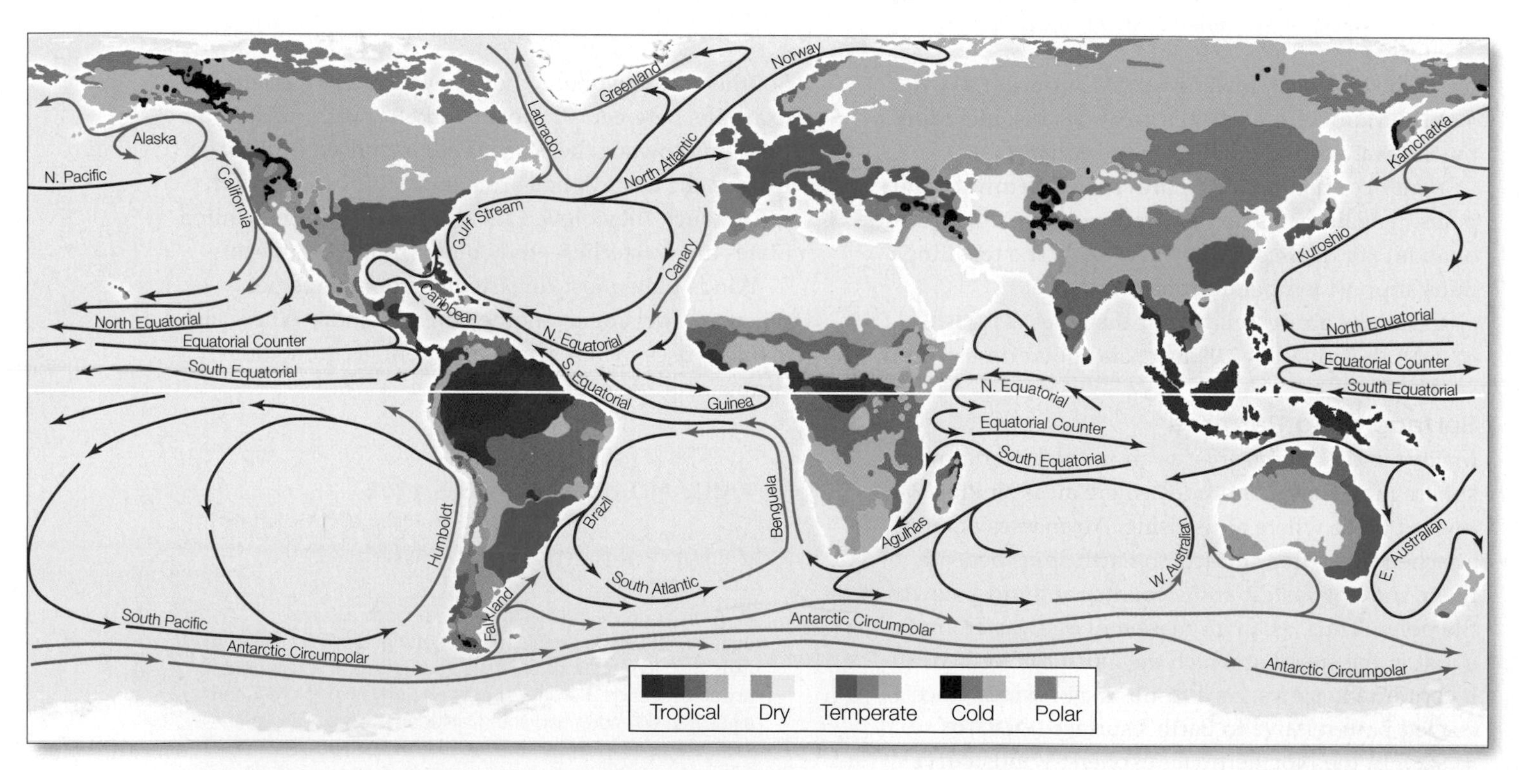

FIGURE 47.5 Major climate zones correlated with surface currents of the world ocean. Warm surface currents start moving from the equator toward the poles, but prevailing winds, Earth's rotation, gravity, the shape of ocean basins, and landforms influence the direction of flow. Water temperatures, which differ with latitude and depth, contribute to the regional differences in air temperature and rainfall.

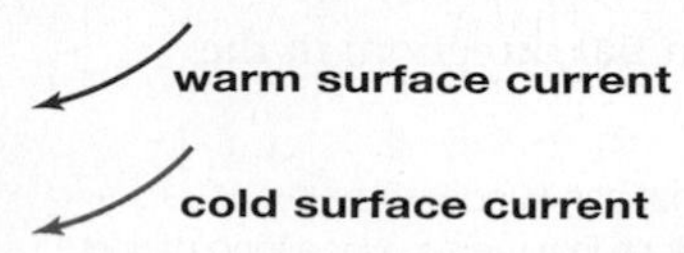

CREDIT: (5) NASA.

over the warm Indian Ocean to the south moves in to replace the rising air, and this north-blowing wind delivers heavy rains. In the winter, the continental interior of Asia is cooler than the ocean. As a result, a cool, dry wind blows from the north toward southern coasts causing a seasonal drought.

Proximity to an ocean moderates climate. Seattle, Washington, has milder winters than Minneapolis, Minnesota, despite Seattle's more northerly latitude. Air over Seattle draws heat from the adjacent Pacific Ocean, a heat source not available to Minneapolis.

Mountains, valleys, and other surface features of the land affect climate too. Suppose you track a warm air mass after it picks up moisture off California's coast. This air moves inland as wind from the west, and piles up against the Sierra Nevada, a high mountain range that parallels the coast. As the air rises in altitude, it cools and loses moisture as rain (**FIGURE 47.7**). The result is a **rain shadow**, a semiarid or arid region of sparse rainfall on the leeward side of high mountains. *Leeward* is the side facing away from the wind. The Himalayas, Andes, Rockies, and other great mountain ranges cast similar rain shadows.

monsoon Wind that reverses direction seasonally.
rain shadow Dry region downwind of a coastal mountain range.

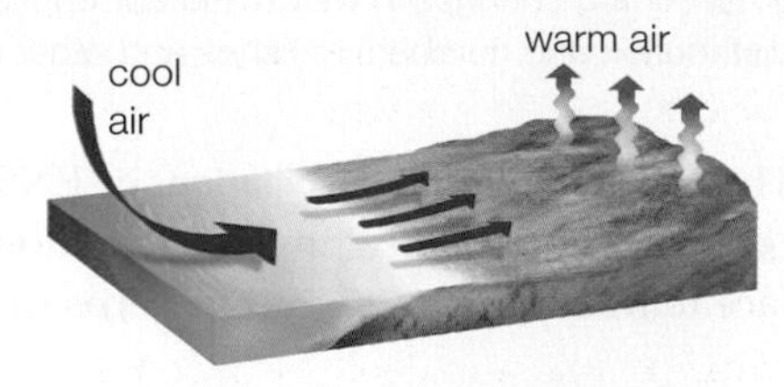

FIGURE 47.6 ▸**Animated** Coastal breezes.

TAKE-HOME MESSAGE 47.3

How do ocean currents arise, and how do they affect regional climates?

✔ Surface ocean currents set in motion by latitudinal differences in solar radiation distribute heat. The currents are affected by winds and by Earth's rotation.

✔ Collective effects of air masses, oceans, and landforms determine regional temperature and annual precipitation.

FIGURE 47.7 ▸**Animated** Rain shadow effect. On the side of mountains facing away from prevailing winds, rainfall is light. White numbers signify elevations, in meters. Black numbers signify annual precipitation, in centimeters, averaged on both sides of the Sierra Nevada, a mountain range.

✔ Recurring changes in winds and currents affect the distribution and numbers of fishes and other marine animals.

The **El Niño Southern Oscillation**, or ENSO, is a naturally occurring, irregularly timed fluctuation in sea surface temperature and wind patterns in the equatorial Pacific. The two extremes of this oscillation are referred to as El Niño and La Niña. Their influence is felt most strongly during the winter and along the western coast of South America, but they affect weather patterns year-round and worldwide (**TABLE 47.1**).

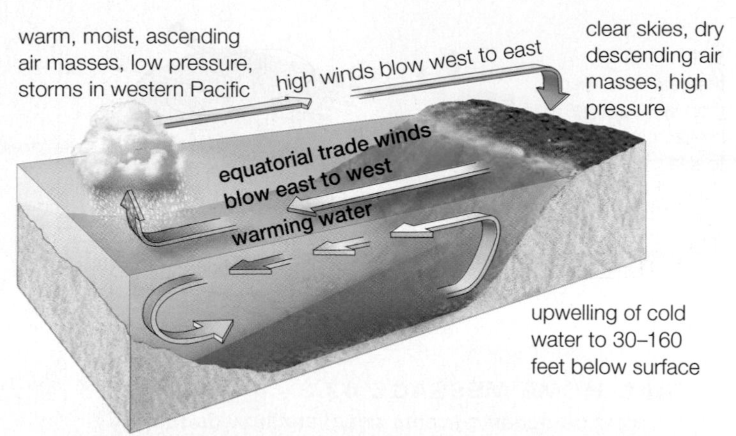

A Conditions during an average year (neither El Niño nor La Niña). Note the upward movement (upwelling) of cool, nutrient-rich water near the coast.

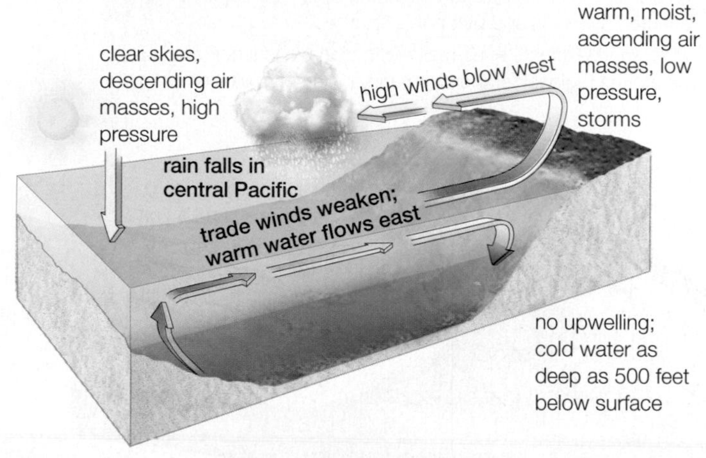

B Conditions during an El Niño. Note the lack of upwelling.

FIGURE 47.8 Effects of El Niño on winds and currents in the equatorial Pacific.

Table 47.1 Effects of El Niño and La Niña

El Niño	La Niña
Western Pacific waters warm	Western Pacific waters cool
Easterly trade winds weaken or reverse	Easterly trade winds strengthen
Less nutrient-rich cold water wells up along South America's west coast	More nutrient-rich cold water wells up along South America's west coast
More rain in western South America	Less rain in western South America
Less rain in Australia	More rain in Australia
Fewer North Atlantic hurricanes	More North Atlantic hurricanes

The term El Niño means "baby boy" and refers to Jesus Christ; it was first used by Peruvian fishermen to describe local weather changes and a shortage of fishes that occurred in some winters and began around Christmas. During an El Niño, unusually warm water flows toward eastern Pacific coasts, displacing the Humboldt Current that would otherwise bring up cooler, nutrient-rich water from the deep (**FIGURE 47.8**). Without this input of nutrients, marine primary producers decline in numbers. The dwindling producer populations and warming water cause a decrease in numbers of small, cold-water fishes, as well as the larger fishes that eat them. This is why Peruvian fishermen catch fewer fishes during an El Niño.

El Niño episodes persist for 6 to 18 months. Often they are followed by a La Niña, in which Pacific waters become cooler than usual. During a La Niña, cold nutrient-rich water flows toward the western coast of South America, phytoplankton populations rebound, and so do populations of fishes that feed on phytoplankton. At other times, waters of the Pacific are not significantly warmer or colder than average.

The most extreme El Niño of the past 100 years occurred during the winter of 1997–1998. Average sea surface temperatures in the eastern Pacific rose by 5°C (9°F) and a plume of warm water extended 9,660 kilometers (6,000 miles) west from the coast of Peru.

The 1997–1998 El Niño/La Niña had extraordinary effects on the primary productivity in the equatorial Pacific. With the massive eastward flow of nutrient-poor warm water, photoautotrophs were almost undetectable in satellite photos that measure primary productivity (**FIGURE 47.9A**). The dwindling populations of producers and the warming water caused decreases in populations of cold-water fishes, as well as the fish-eating animals that feed on them. During the 1997–1998 El Niño, about half of the sea lions on the Galápagos Islands starved to death. California's population of northern fur seals also suffered a sharp decline. Primary productivity rebounded during the La Niña that followed (**FIGURE 47.9B**).

Rainfall patterns shift during an El Niño. During the winter of 1997–1998, torrential rains caused flooding and landslides along eastern Pacific coasts, while Australia and Indonesia suffered from drought-driven crop failures and raging wildfires. An El Niño typically

FIGURE 47.9 Satellite images showing primary production during the 1997–1998 El Niño and the La Niña that followed.

brings cooler, wetter weather to the American Gulf states, and reduces the likelihood of hurricanes.

Outbreaks of some human diseases are more likely to occur during an El Niño. For example, the increased ocean temperature in the Pacific sometimes leads to an increased incidence of cholera, a disease that causes potentially deadly diarrhea. Copepods, a type of small crustacean, serve as a reservoir for cholera-causing bacteria between disease outbreaks. During an El Niño, the rise in the temperature of the ocean's surface results in a rise in the number of cholera-carrying copepods. An El Niño also increases the incidence of malaria in coastal South Asia and Latin America; more rain brings more standing water in which mosquitoes can breed.

Among other duties, the United States National Oceanographic and Atmospheric Administration (NOAA) is charged with studying and predicting El Niño events. This agency collects and analyzes sea temperature data from a system of buoys moored in the tropical Pacific Ocean (**FIGURE 47.10**). The goal of the research is to determine how the El Niño Southern Oscillation affects global weather patterns and the extent of its effects. Such an understanding could help the scientists to develop a method of predicting when an El Niño or La Niña event is likely to occur and which regions are at a heightened risk for flooding, drought, hurricanes, or fires as a result. Predicting and planning for such occurrences could help prevent or minimize their harmful effects. Current data on sea surface temperature, as well as information about the monitoring program and its goals, are available on NOAA's website at www.elnino.noaa.gov.

FIGURE 47.10 Scientists check one of the research buoys that monitor sea temperature in the Pacific. Use of this data can help predict El Niño events.

El Niño Southern Oscillation Naturally occurring, irregularly timed fluctuation in sea surface temperature and wind patterns in the equatorial Pacific; affects weather worldwide.

TAKE-HOME MESSAGE 47.4

What occurs during an El Niño?

✔ During an El Niño event, the equatorial Pacific warms, altering weather patterns. Nutrient flow to coastal waters west of South America slows, causing a decline in primary productivity. Rainfall patterns are altered, and people in some regions have a higher risk of diseases such as cholera.

CREDITS: (9) NASA Goddard Space Flight Center Scientific Visualization Studio; (10) NOAA, LindaStratton.

47.5 Biomes

✔ Similar communities often evolve in widely separated regions as a result of similar environmental factors.

Differences Between Biomes

Biomes are areas of land characterized by their climate and predominant type of vegetation (**FIGURE 47.11**). A biome is discontinuous; most include areas on different continents. For example, the temperate grassland biome includes North American prairie, South African veld, South American pampa, and Eurasian steppe.

The type of biome characteristic of an area depends mainly on rainfall and temperature. Desert biomes get the least annual rainfall, grasslands and shrublands get more, and forests get the most. Deserts occur where temperatures soar highest and tundra where they drop the lowest.

Soils also influence biome distribution. Soils consist of a mixture of mineral particles and varying amounts of humus. Water and air fill spaces between soil particles. Properties of soils vary depending on the types, proportions, and compaction of particles. Deserts have sandy or gravelly, fast-draining soil with little topsoil. Topsoil is deepest in natural grasslands, where it can be more than one meter thick. This is why grasslands are often converted to agricultural uses.

Climate and soils influence primary production, so primary production varies greatly among biomes (**FIGURE 47.12**).

Similarities Within a Biome

Unrelated species living in widely separated parts of a biome often have similar body structures that arose by the process of morphological convergence (Section 18.3). For example, cacti with water-storing stems live in North American deserts and euphorbs with water-storing stems live in African deserts. Cacti and

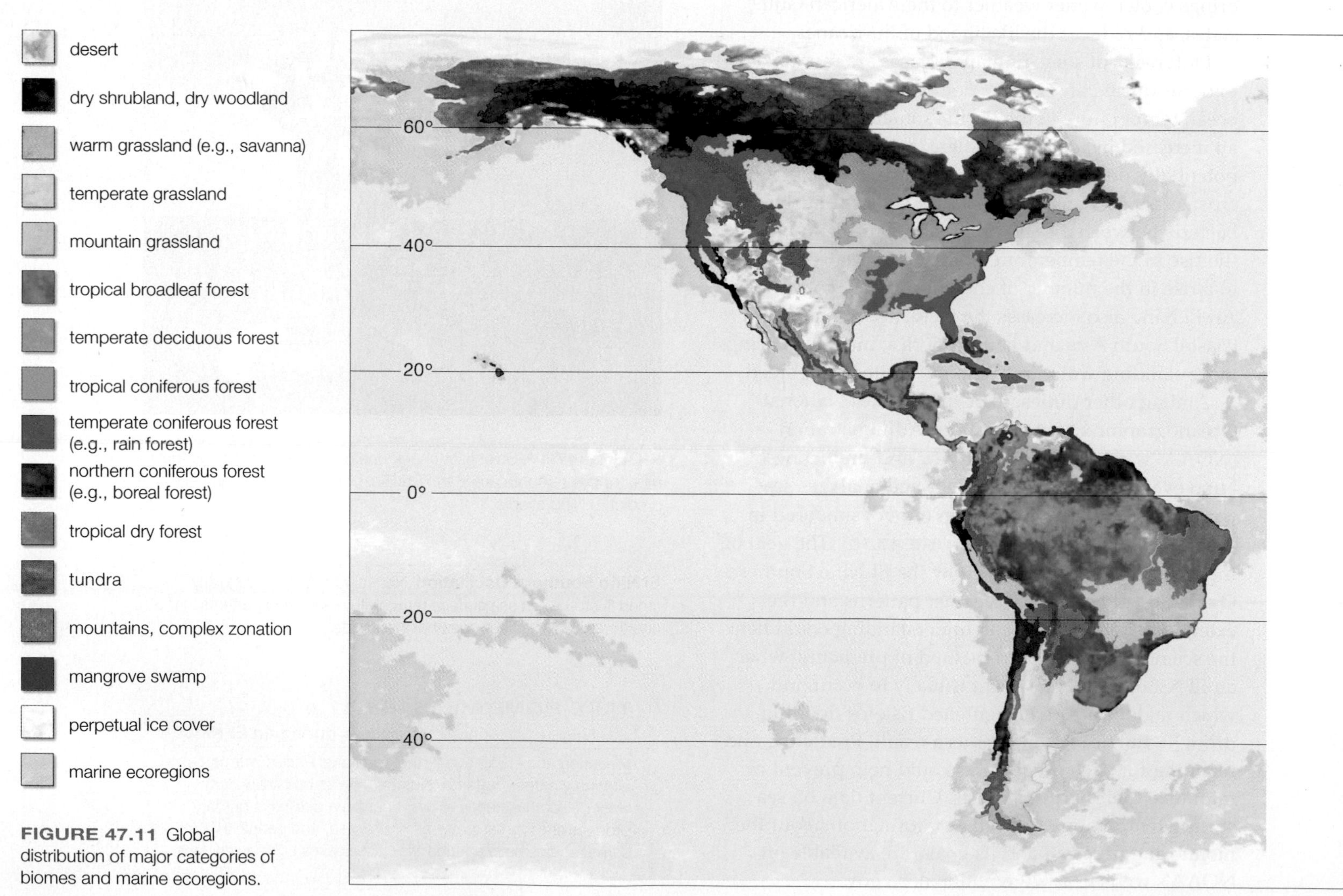

FIGURE 47.11 Global distribution of major categories of biomes and marine ecoregions.

CREDIT: (11) NASA.

euphorbs do not share an ancestor with a water-storing stem. Rather, this feature evolved independently in the two groups as a result of similar selection pressures. Similarly, an ability to carry out C4 photosynthesis evolved independently in grasses growing in warm grasslands on different continents. C4 photosynthesis is more efficient than the more common C3 pathway under hot, dry conditions (Section 6.7).

biome Group of regions that may be widely separated but share a characteristic climate and dominant vegetation.

TAKE-HOME MESSAGE 47.5

What are biomes?

✔ Biomes are vast expanses of land dominated by distinct kinds of plants that support characteristic communities.

✔ The global distribution of biomes is a result of topography, climate, and evolutionary history.

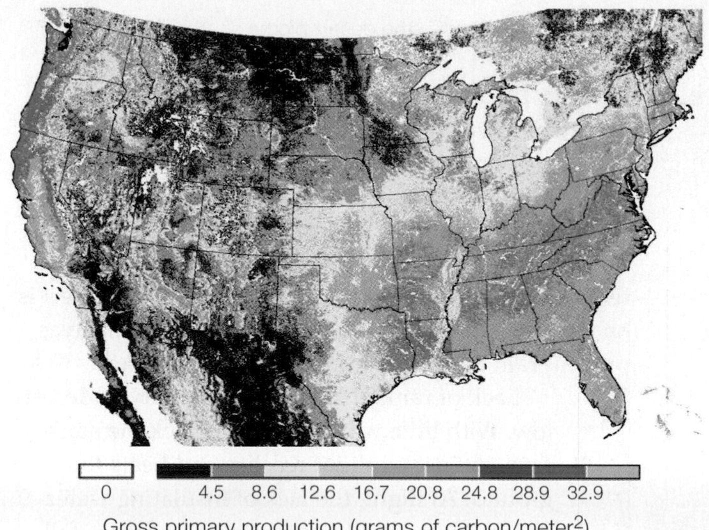

FIGURE 47.12 Remote satellite monitoring of gross primary productivity across the United States. The differences roughly correspond with variations in soil types and moisture.

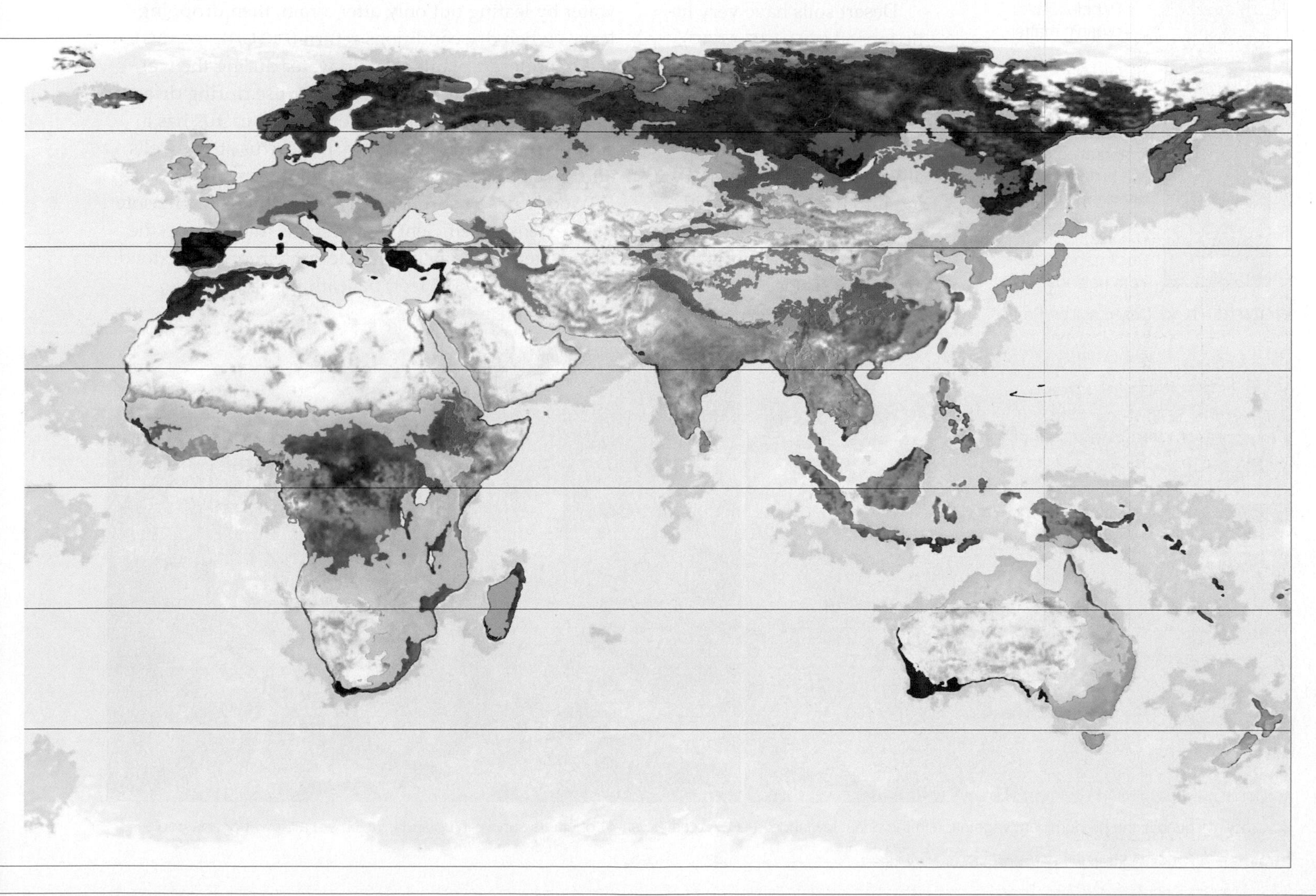

CREDIT: (12) NASA.

47.6 Deserts

✔ Low rainfall shapes the desert biome.

Desert Locations and Conditions

Deserts receive an average of less than 10 centimeters (4 inches) of rain per year. They cover about one-fifth of Earth's land surface and many are located at about 30° north and south latitude, where global air circulation patterns cause dry air to sink. Rain shadows also reduce rainfall. For example, Chile's Atacama Desert is on the leeward side of the Andes, and the Himalayas prevent rain from falling in China's Gobi desert.

Lack of rainfall keeps the humidity in deserts low. With little water vapor to block the sun's rays, intense sunlight reaches and heats the ground. At night, the lack of insulating water vapor in the air allows the temperature to fall fast. As a result, deserts tend to have larger daily temperature shifts than other biomes.

Desert soils have very little topsoil (**FIGURE 47.13**), the layer most important for plant growth. These soils often are somewhat salty, because rain that falls usually evaporates before seeping into the ground. Rapid evaporation allows any salt in rainwater to accumulate at the soil surface.

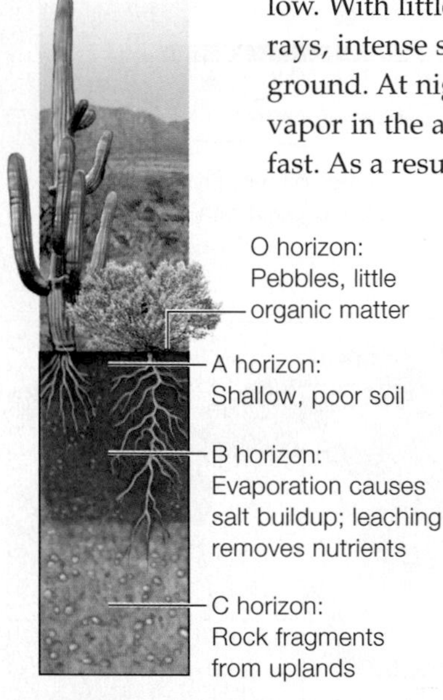

FIGURE 47.13 Desert soil profile.

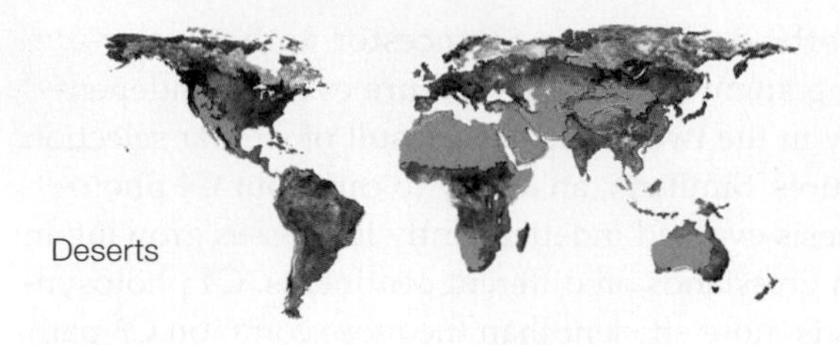

Adaptations to Desert LIfe

Despite their harsh conditions, most deserts support some plant life. Diversity is highest in regions where soil moisture is available in more than one season (**FIGURE 47.14**).

Many desert plants have adaptations that reduce water loss. For example, some have spines or hairs (**FIGURE 47.15A**). In addition to deterring herbivory, these structures reduce evaporation by trapping some water and keeping the humidity around the stomata high. Where rains fall seasonally, some plants conserve water by leafing out only after a rain, then dropping leaves when dry conditions return (**FIGURE 47.15B**).

Other desert plants take up water during the wet season and store it in their body for use during drier times. For example, the stem of a barrel cactus has a spongy pulp that holds water. The cactus stem swells after a rain, then shrinks as the plant uses stored water.

Woody desert shrubs such as mesquite and creosote have extensive, efficient root systems that take up the little water that is available. Mesquite roots can extend up to 60 meters (197 feet) beneath the soil surface.

A Creosote bush is the predominant vegetation in the driest lowlands.

B A greater variety of plants survive in uplands, which are a bit wetter and cooler.

FIGURE 47.14 Vegetation in Arizona's Sonoran Desert.

A Barrel cacti are covered by spines that reduce evaporative water loss. The cactus is a CAM plant.

B Ocotillo, a desert shrub, grows leaves on its stems after a rain, then sheds them when conditions become dry again.

FIGURE 47.15 Perennials adapted to desert conditions.

FIGURE 47.16 Mojave Desert after the rains. Annual poppies sprout, flower, produce seeds, and die within weeks beneath slow-growing perennial cacti.

Alternative carbon-fixing pathways also help desert plants conserve water. Cacti, agaves, and euphorbs are CAM plants (Section 6.7). They open their stomata only at night, when the temperature declines.

Most deserts contain a mix of annuals and perennials (**FIGURE 47.16**). The annuals are adapted to desert life by a rapid life cycle; they sprout and reproduce during the short time that the soil is moist.

Like desert plants, desert animals have adaptations that allow them to conserve water. For example, the highly efficient kidneys of a desert kangaroo rat minimize its water needs (Section 40.2). Most desert animals are not active at the height of the daytime heat (**FIGURE 47.17**).

A The Sonoran desert tortoise spends much of its life inactive. In hot summer months, it ventures out of its burrow only in cool mornings to feed. During the cold winter, when little food is available, it hibernates.

B Lesser long-nosed bats spend spring and summer in the Sonoran Desert. They avoid the daytime heat by resting in caves.

FIGURE 47.17 Two Sonoran Desert animals.

The Crust Community

In many deserts, the soil is covered by a desert crust, a community that can include cyanobacteria, lichens, mosses, and fungi. These organisms secrete organic molecules that glue them and the surrounding soil particles together. The crust benefits members of the larger desert community in important ways. Its cyanobacteria fix nitrogen and make ammonia available to plants. The crust also holds soil particles in place. When the fragile connections within the desert crust are broken, soil can blow away. Negative effects of such disturbance are heightened when windblown soil buries healthy crust in an undisturbed area, killing additional crust organisms and allowing more soil to take flight.

desert Biome with little rain and low humidity; plants that have water-storing and water-conserving adaptations predominate.

TAKE-HOME MESSAGE 47.6

What are deserts?

✔ A desert gets very little rain and has low humidity. There is plenty of sunlight, but the lack of water prevents most plants from surviving here.

✔ The predominant plants in deserts have adaptations that allow them to reduce water lost by transpiration, store water, or access water deep below the soil surface.

✔ Desert animals often spend the day inactive, sheltering from the heat.

✔ Desert soils are held in place by a community of organisms that form a desert crust. Disruption of this crust allows wind to strip away soil.

47.7 Grasslands

✔ Perennial grasses adapted to fire and to grazing are the main plants in grasslands.

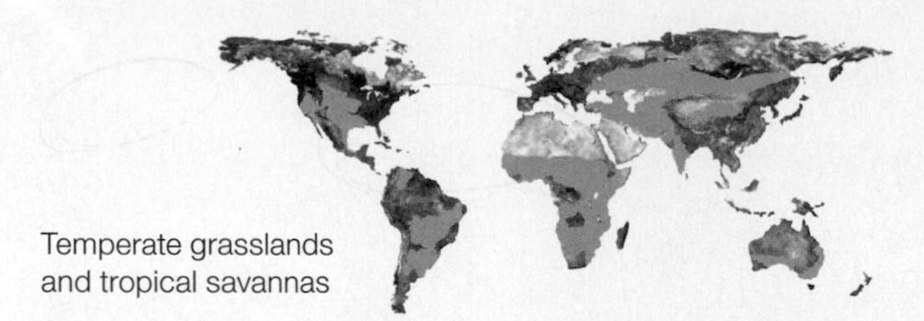
Temperate grasslands and tropical savannas

Grasslands form in the interior of continents between deserts and temperate forests. Their soils are rich, with deep topsoil. Annual rainfall is enough to prevent desert from forming, but not enough to support woodlands. Low-growing grasses and other nonwoody plants tolerate strong winds, sparse and infrequent rain, and intervals of drought. Growth tends to be seasonal. Constant trimming by grazers, along with periodic fires, keeps most shrubs from taking hold.

Temperate Prairies

Temperate grasslands are warm in summer, but cold in winter. Annual rainfall is 25 to 100 centimeters (10 to 40 inches), with rains throughout the year. Grass roots extend profusely through the thick topsoil and help hold it in place, preventing erosion by the constant winds. North America's grasslands are shortgrass and tallgrass prairies (**FIGURE 47.18A,B**).

During the 1930s, much of the shortgrass prairie of the American Great Plains was plowed to grow wheat. Strong winds, a prolonged drought, and unsuitable farming practices turned much of the region into what the newspapers of that time called the Dust Bowl.

Tallgrass prairie has somewhat richer topsoil and slightly more frequent rainfall than shortgrass prairie. Before the arrival of Europeans, it covered about 140 million acres, mostly in Kansas. Nearly all tallgrass prairie has now been converted to cropland. The Tallgrass Prairie National Preserve was created in 1996 to protect the little that remains.

North America's prairies once supported enormous herds of elk, pronghorn antelope, and bison that were prey to wolves. Today, these predators and prey are absent from most of their former range.

Tropical Savannas

Savannas are broad belts of grasslands with a few scattered shrubs and trees. Savannas lie between the tropical forests and hot deserts of Africa, India, and Australia. Temperatures are warm year-round. During the rainy season, 90 to 150 centimeters (35 to 60 inches) of rain falls. Africa's savannas are famous for their abundant wildlife (**FIGURE 47.18C**). Herbivores include giraffes, zebras, elephants, a variety of antelopes, and immense herds of wildebeests. Lions and hyenas are carnivores that eat the grazers.

grassland Biome in the interior of continents where grasses and nonwoody plants adapted to grazing and fire predominate.

TAKE-HOME MESSAGE 47.7
What are grasslands?

✔ Grasslands are biomes dominated by grasses and other nonwoody plants that can withstand fire and grazing.

A Prairie soil profile.

B Bison grazing in a North American prairie.

C Wildebeest grazing on an African savanna.

FIGURE 47.18 Grasslands.

47.8 Dry Shrublands and Woodlands

✔ Regions with cool, rainy winters and hot, dry summers support dry shrublands and woodlands.

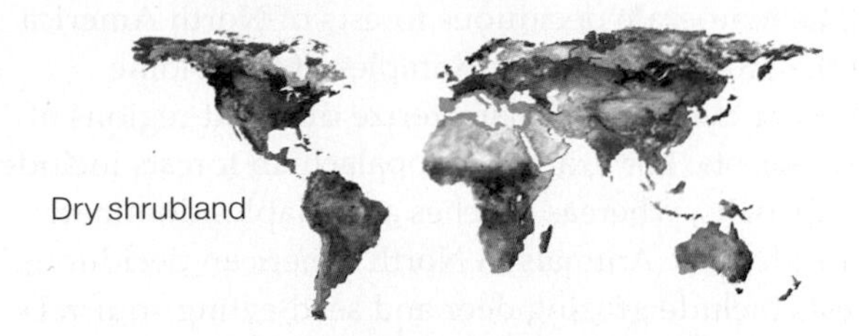

Dry shrubland is a biome dominated by fire-adapted shrubs. It typically occurs along the western coast of continents, between 30 and 40 degrees north or south latitude. Winters are mild and wet, with 25 to 60 centimeters (10 to 24 inches) of rain. Summers are hot and dry. California's dry shrublands, called chaparral, are the state's most extensive biological community (**FIGURE 47.19A**). Dry shrubland also occurs in regions bordering the Mediterranean, as well as Chile, Australia, and South Africa (**FIGURE 47.19B**).

Plants in dry shrublands tend to have small, leathery leaves that help them withstand the summer drought. Many make aromatic oils that help fend off insects, but also make them highly flammable. After a fire, many plant species can resprout from their roots or germinate from fire-resistant seeds.

Dry shrublands grade into **dry woodlands**, where a bit more winter rain allows trees to grow. In a woodland, trees do not shade as much of the ground as they do in a forest, and they tend to be somewhat shorter. Examples of dry woodlands include California's oak woodlands and eucalyptus woodlands of Australia (**FIGURE 47.20**).

dry shrubland Biome dominated by a diverse array of fire-adapted shrubs; occurs in regions with cool, wet winters and a dry summer.
dry woodland Biome dominated by short trees that do not completely shade the ground; occurs in regions with cool, wet winters and a dry summer.

A Oak woodland in California.

B Eucalyptus woodland in Australia.

FIGURE 47.20 Dry woodlands.

TAKE-HOME MESSAGE 47.8

What are dry shrublands and woodlands?

✔ Dry shrublands and woodlands form in areas with a mild, rainy winter and a hot, dry summer.

✔ Dry shrublands are dominated by fire-adapted shrubs; dry woodlands, by low-growing trees that allow a lot of sunlight to reach the ground.

A Chaparral in California.

B Fynbos, a dry shrubland in the Cape region of South Africa.

FIGURE 47.19 Dry shrublands. Evergreen shrubs with small, leathery leaves predominate.

47.9 Broadleaf Forests

✔ Broadleaf trees are the main plants in both temperate and tropical forests.

Semi-Evergreen and Deciduous Forests

Semi-evergreen forests occur in the humid tropics of Southeast Asia and India. They include broadleaf (angiosperm) trees that retain leaves year-round, and deciduous broadleaf trees. A deciduous plant sheds leaves annually, prior to a season when cold or dry conditions would not favor growth. In semi-evergreen forests, deciduous trees shed their leaves at the start of the dry season.

Where less than 2.5 centimeters (1 inch) of rain falls in the dry season, tropical deciduous forests form. Most trees in these forests shed their leaves at the start of the dry season.

Temperate deciduous forests form in the Northern Hemisphere in parts of eastern North America, western and central Europe, and parts of Asia, including Japan. In these regions, 50 to 150 centimeters (about 20 to 60 inches) of precipitation falls throughout the year. Winters are cool and summers are warm.

Growth of temperate deciduous forests is seasonal. Leaves often turn color before dropping in autumn (**FIGURE 47.21**). Winters are cold and trees remain dormant while water is locked in snow and ice. In the spring, when conditions again favor growth, deciduous trees flower and put out new leaves. Also during the spring, leaves that were shed the prior autumn decay to form a rich humus. Rich soil and a somewhat open canopy that lets sunlight through allow shorter understory plants to flourish.

The temperate deciduous forests of North America are the most species-rich examples of this biome. Different tree species characterize different regions of these forests. For example, Appalachian forests include mainly oaks, whereas beeches and maples dominate Ohio's forests. Animals in North American deciduous forests include grazing deer and seed-eating squirrels and chipmunks, as well as omnivores such as raccoons, opossums, and black bears. Native predators such as wolves and mountain lions have been largely eliminated from the forests.

Tropical Rain Forests

Tropical rain forests of evergreen broadleaf trees form between latitudes 10° north and south in equatorial Africa, the East Indies, Southeast Asia, South America, and Central America. Rain that falls throughout the year sums to an annual total of 130 to 200 centimeters (50 to 80 inches).

Regular rains, combined with an average temperature of 25°C (77°F) and little variation in day length, allow photosynthesis to continue year-round. Of all land biomes, tropical forests have the greatest primary production. Per unit area, they remove more carbon from the atmosphere than other forests or grasslands.

Tropical rain forest is the most structurally complex and species-rich biome. The forest has a multilayer structure (**FIGURE 47.22**). Its broadleaf trees can stand

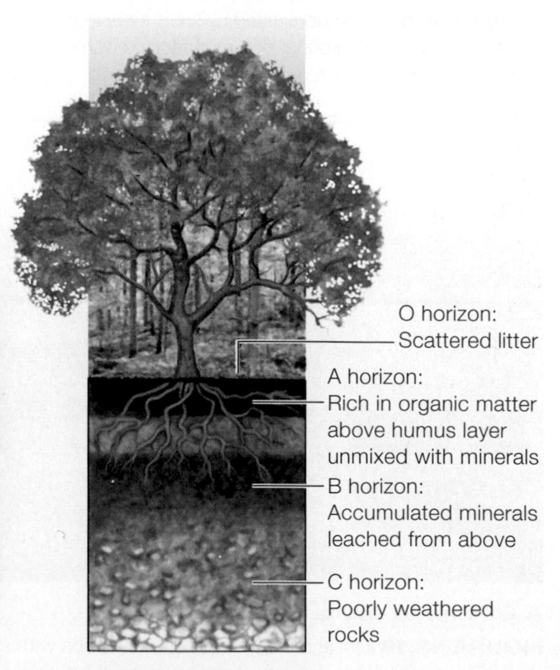

FIGURE 47.21 North American temperate deciduous forest. Forest soil profile at right.

CREDITS: (21) left, © James Randklev/Corbis; (in text, 21 right) © Cengage Learning.

FIGURE 47.22 Tropical rain forest. Forest soil profile at right.

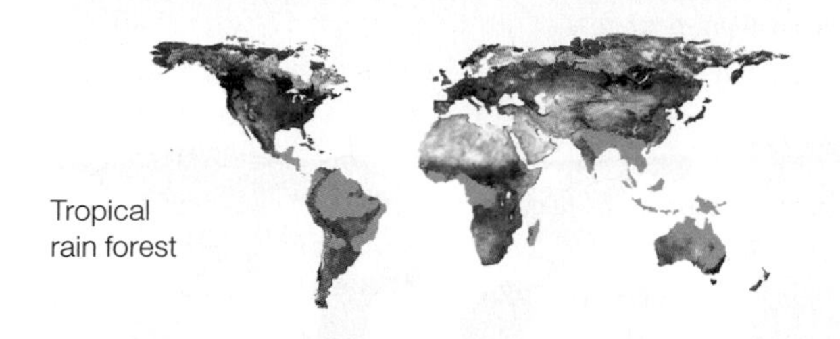

30 meters (100 feet) tall. The trees often form a closed canopy that prevents most sunlight from reaching the forest floor. Vines and epiphytes (plants that grow on another plant, but do not withdraw nutrients from it) thrive in the shade beneath the canopy.

Trees of tropical rain forests shed leaves continually, but decomposition and mineral cycling happen so fast in this warm, moist environment that litter does not accumulate. Soils are highly weathered, heavily leached, and very poor nutrient reservoirs.

Deforestation is an ongoing threat to tropical rain forests. Tropical forests are located in developing countries with fast-growing human populations who look to the forest as a source of lumber, fuel, and potential agricultural land. Deforestation in any region leaves fewer trees to remove carbon dioxide from the atmosphere. In rain forests, it also causes many extinctions. Compared to other land biomes, tropical rain forests have the greatest variety and numbers of insects, as well as the most diverse collection of birds and primates. This great diversity means many species are affected by the loss of any amount of forest. Among the potential losses are species with chemicals that could save human lives. Two chemotherapy drugs, vincristine and vinblastine, were extracted from the rosy periwinkle, a low-growing plant native to Madagascar's rain forests. Today, these drugs help fight leukemia and other cancers. No doubt other similarly valuable species currently live in the rain forests and will go extinct before we learn how they can help us.

temperate deciduous forest Northern Hemisphere biome in which the main plants are broadleaf trees that lose their leaves in fall and become dormant during cold winters.

tropical rain forest Highly productive and species-rich biome in which year-round rains and warmth support continuous growth of evergreen broadleaf trees.

TAKE-HOME MESSAGE 47.9

What are broadleaf forests?

✔ Temperate broadleaf forests grow in the Northern Hemisphere where cold winters prevent year-round growth. Trees lose their leaves in autumn, then remain dormant during the winter.

✔ Year-round warmth and rains support tropical rain forests, the most productive, structurally complex, and species-rich land biome.

47.10 Coniferous Forests

✔ Compared to broadleaf trees, conifers are more tolerant of cold and drought, and can withstand poorer soils. Where these conditions occur, coniferous forests prevail.

Conifers (evergreen trees with seed-bearing cones) are the main plants in coniferous forests. Conifer leaves are typically needle-shaped, with a thick cuticle and stomata that are sunk below the leaf surface. These adaptations help conifers conserve water during drought or periods when the ground is frozen. As a group, conifers tolerate poorer soils and drier habitats than most broadleaf trees. Conifer-dominated forests occur mainly in the Northern Hemisphere.

The most extensive land biome is the coniferous forest that sweeps across northern Asia, Europe, and North America (**FIGURE 47.23A**). It is referred to as **boreal forest**, or taiga, which means "swamp forest" in Russian. The conifers are mainly pine, fir, and spruce. Most rain falls in the summer, and little evaporates into the cool summer air. Winters are long, cold, and dry. Moose are dominant grazers in this biome.

boreal forest Extensive high-latitude forest of the Northern Hemisphere; conifers are the predominant vegetation.

Also in the Northern Hemisphere, montane coniferous forests extend southward through the great mountain ranges (**FIGURE 47.23B**). Spruce and fir dominate at the highest elevations. At lower elevations, the mix becomes firs and pines.

Conifers also dominate temperate lowlands along the Pacific coast from Alaska into northern California. These coniferous forests hold some of the world's tallest trees—Sitka spruce and coast redwoods.

We find other conifer-dominated ecosystems in the eastern United States. About a quarter of New Jersey is pine barrens, a mixed forest of pitch pines and scrub oaks that grow in sandy, acidic soil. Pine forest covers about one-third of the Southeast. Fast-growing loblolly pines dominate these forests and are a major source of lumber and wood pulp. The pines can survive periodic fires that kill most hardwood species. When fires are suppressed, hardwoods outcompete pines.

TAKE-HOME MESSAGE 47.10

What are coniferous forests?

✔ Conifers prevail across the Northern Hemisphere's high-latitude forests, at high elevations, and in temperate regions with nutrient-poor soils.

B Montane coniferous forest near Mount Rainier, Washington.

A Boreal forest (taiga) in Siberia.

FIGURE 47.23 Coniferous forests.

47.11 Tundra

✔ Low-growing, cold-tolerant plants have only a brief growing season on the tundra.

Arctic Tundra

Arctic tundra forms between the polar ice cap and the belts of boreal forests in the Northern Hemisphere. Most is in northern Russia and Canada. It is Earth's youngest biome, having appeared about 10,000 years ago as glaciers retreated at the end of the last ice age.

Conditions in this biome are harsh; snow blankets the ground for as long as nine months of the year. Annual precipitation is usually less than 25 centimeters (10 inches), but cold temperature keeps the snow that does fall from melting. During a brief summer, plants grow fast under the nearly continuous sunlight (**FIGURE 47.24**). Lichens and shallow-rooted, low-growing plants are the producers for food webs that include voles, arctic hares, caribou, arctic foxes, wolves, and brown bears. Enormous numbers of migratory birds nest here in the summer.

FIGURE 47.24 Arctic tundra in the summer. Permafrost underlies the soil.

Only the surface layer of tundra soil thaws during summer. Below that lies **permafrost**, a frozen layer 500 meters (1,600 feet) thick in places. Permafrost acts as a barrier that prevents drainage, so the soil above it remains perpetually waterlogged. The cool, anaerobic conditions in this soil slow decay, so organic remains can build up. Organic matter in permafrost makes the arctic tundra one of Earth's greatest stores of carbon.

As global temperatures rise, the amount of frozen soil that melts each summer is increasing. With warmer temperatures, much of the snow and ice that would otherwise reflect sunlight is disappearing. As a result, newly exposed dark soil absorbs heat from the sun's rays, which encourages more melting.

Alpine Tundra

Alpine tundra occurs at high altitudes throughout the world (**FIGURE 47.25**). Even in the summer, some patches of snow persist in shaded areas, but there is no permafrost. The alpine soil is well drained, but thin and nutrient-poor. As a result, primary productivity is low. Grasses and small-leafed, woody shrubs grow in patches where soil has accumulated to a greater depth. These low-growing plants can withstand the strong winds that discourage the growth of trees.

FIGURE 47.25 Alpine tundra. Low-growing, hardy plants at a high altitude in Washington's Cascade range.

---

alpine tundra Biome of low-growing, wind-tolerant plants adapted to high-altitude conditions.
arctic tundra Highest-latitude Northern Hemisphere biome, where low, cold-tolerant plants survive with only a brief growing season.
permafrost Continually frozen soil layer that lies beneath arctic tundra and prevents water from draining.

TAKE-HOME MESSAGE 47.11

What is tundra?

✔ Arctic tundra prevails at high latitudes, where short, cold summers alternate with long, cold winters.

✔ Alpine tundra prevails in high, cold mountains across all latitudes.

47.12 Freshwater Ecosystems

✔ Freshwater and saltwater provinces cover more of Earth's surface than all land biomes combined. Here we begin our survey of these watery realms.

Lakes

A **lake** is a body of standing fresh water. If it is sufficiently deep, it can be divided into zones that differ in their physical characteristics and species composition (**FIGURE 47.26**). Near shore is the littoral zone, from the Latin *litus* for shore. Here, sunlight penetrates all the way to the lake bottom; aquatic plants and algae that attach to the bottom are the primary producers. The lake's open waters include an upper, well-lit limnetic zone, and—if the lake is deep—a dark profundal zone where light does not penetrate. Primary producers in the limnetic zone can include aquatic plants, green algae, diatoms, and cyanobacteria. These organisms serve as food for rotifers, copepods, and other types of zooplankton. In the profundal zone, where there is not enough light for photosynthesis, consumers feed on organic debris that drifts down from above.

FIGURE 47.26 ▸Animated Lake zonation. A lake's littoral zone extends around the shore to a depth where rooted aquatic plants stop growing. Its limnetic zone is the open waters where light penetrates and photosynthesis occurs. Below that lies the cool, dark profundal zone, where detrital food chains predominate.

Nutrient Content and Succession A lake undergoes succession; it changes over time (Section 45.8). A newly formed lake is oligotrophic: deep, clear, and nutrient-poor, with low primary productivity (**FIGURE 47.27**). Later, as sediments accumulate and plants take root, the lake becomes eutrophic. Eutrophication refers to processes, either natural or artificial, that enrich a body of water with nutrients (Section 46.1).

Seasonal Changes Temperate zone lakes undergo seasonal changes (**FIGURE 47.28**). During winter, a layer of ice forms at the lake surface. Unlike most substances, water is denser as a liquid than as a solid (ice). As water cools, its density increases, until it reaches 4°C (39°F). Below this temperature, additional cooling decreases water's density—which is why ice floats on water (Section 2.5). In an ice-covered lake, water just under the ice is near its freezing point and at its lowest density. The densest (4°C) water resides at the bottom of the lake ❶.

Oligotrophic Lake	Eutrophic Lake
Deep, steeply banked	Shallow with broad littoral
Large deep-water volume relative to surface-water volume	Small deep-water volume relative to surface-water volume
Highly transparent	Limited transparency
Water blue or green	Water green to yellow- or brownish-green
Low nutrient content	High nutrient content
Oxygen abundant through all levels throughout year	Oxygen depleted in deep water during summer
Not much phytoplankton; green algae and diatoms dominant	Abundant, thick masses of phytoplankton; cyanobacteria dominant
Aerobic decomposers favored in profundal zone	Anaerobic decomposers in profundal zone
Low biomass in profundal zone	High biomass in profundal zone

FIGURE 47.27 An oligotrophic lake. Crater Lake in Oregon is a collapsed volcano that filled with snow melt. It began filling about 7,700 years ago. Thus, from a geologic standpoint, it is a young lake.

In spring, the air warms and ice melts. When the resulting meltwater warms to 4°C, it sinks. Wind blowing across the surface of the lake assists in bringing about a **spring overturn**, during which oxygen-rich water at the surface of the lake moves downward while nutrient-rich water from the lake's depths moves up ❷.

In the summer, a lake's waters form three layers that differ in their temperature and oxygen content ❸. The top layer is warm and oxygen-rich. It overlies the **thermocline,** a thin layer where temperature falls rapidly. Beneath the thermocline is the coolest water. The thermocline acts as a barrier that keeps the upper and lower layers from combining. As a result of the thermocline, oxygen from the lake's surface cannot reach the lake's depths, where decomposition is using up oxygen. At the same time, nutrients from those depths cannot escape into surface waters.

In autumn, the upper layer cools and sinks, and the thermocline disappears. During the **fall overturn**, oxygen-rich water moves down while nutrient-rich water moves up ❹.

Overturns influence primary productivity. After a spring overturn, longer day length and an abundance of nutrients support the greatest primary productivity. During the summer, vertical mixing ceases. Nutrients do not move up, and photosynthesis slows. By late summer, nutrient shortages limit growth. Fall overturn brings nutrients to the surface and favors a brief burst of photosynthesis. The burst ends as winter brings shorter days and the amount of sunlight declines.

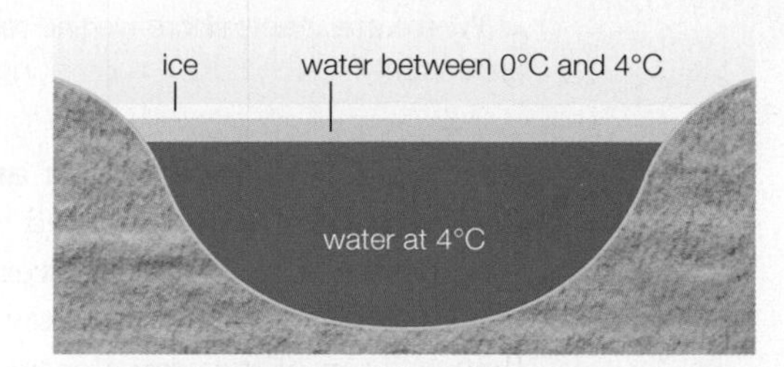

❶ Winter. Ice covers the thin layer of slightly warmer water just below it. Densest (4°C) water is at bottom. Winds do not affect water under the ice, so there is little circulation.

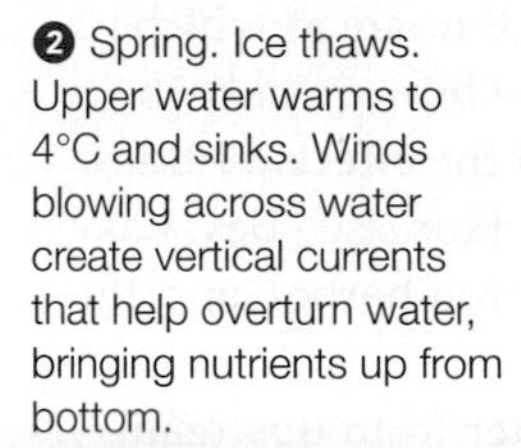
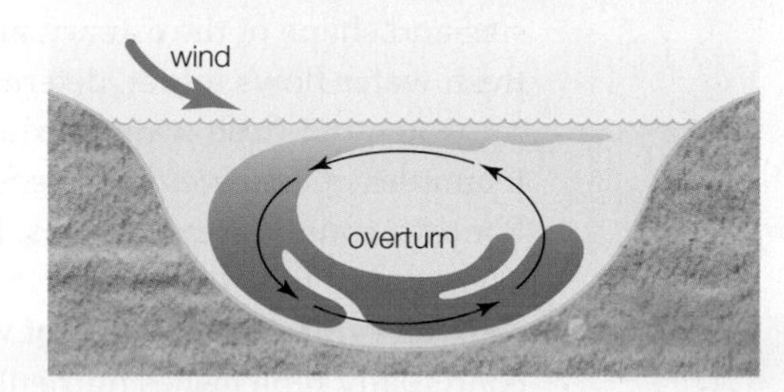

❷ Spring. Ice thaws. Upper water warms to 4°C and sinks. Winds blowing across water create vertical currents that help overturn water, bringing nutrients up from bottom.

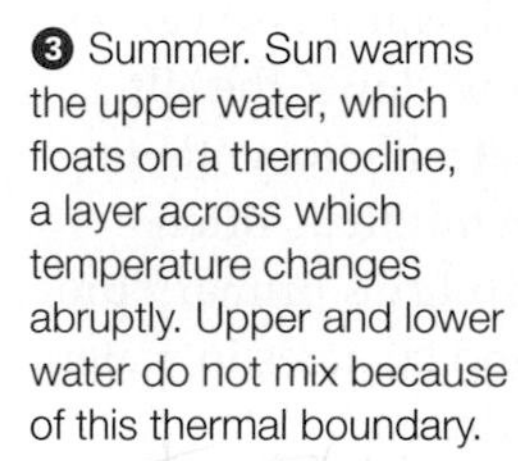
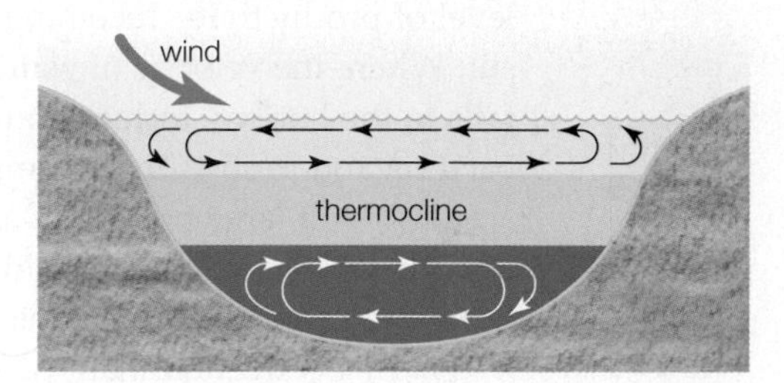

❸ Summer. Sun warms the upper water, which floats on a thermocline, a layer across which temperature changes abruptly. Upper and lower water do not mix because of this thermal boundary.

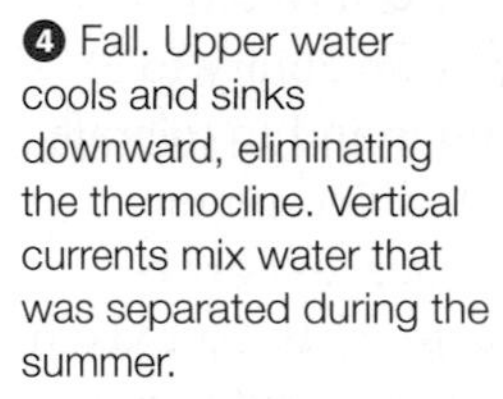
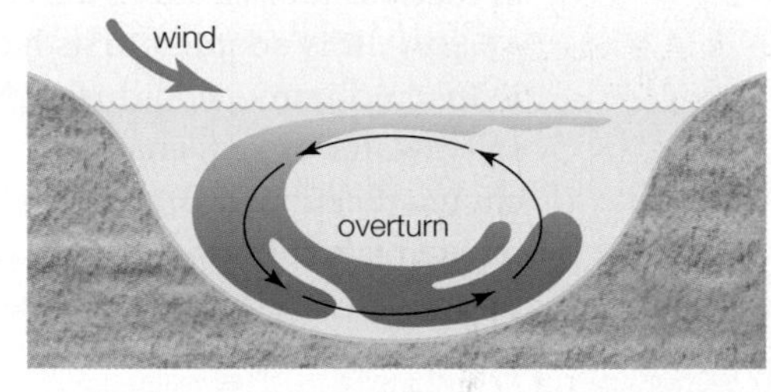

❹ Fall. Upper water cools and sinks downward, eliminating the thermocline. Vertical currents mix water that was separated during the summer.

FIGURE 47.28 Seasonal changes in a temperate zone lake.

Streams and Rivers

Streams are flowing-water ecosystems that begin as freshwater springs or seeps. As they flow downslope, they grow and merge to form rivers. Rainfall, snowmelt, geography, altitude, and shade cast by plants affect flow volume and temperature.

Properties of a stream or river vary along its length. Streambed composition affects solute concentrations, as when limestone rocks dissolve and add calcium. Water that flows rapidly over rocks mixes with air and holds more oxygen than slower-moving, deeper water. Temperature also affects oxygen content. Cold water holds more oxygen than warm water. As a result, different parts of a stream or river support species with different oxygen needs (Section 38.2).

fall overturn During the fall, waters of a temperate zone mix. Upper, oxygenated water cools, gets dense, and sinks; nutrient-rich water from the bottom moves up.
lake A body of standing fresh water.
spring overturn In temperate zone lakes, a downward movement of oxygenated surface water and an upward movement of nutrient-rich water in spring.
thermocline Thermal stratification in a large body of water; a cool midlayer stops vertical mixing between warm surface water above it and cold water below it.

TAKE-HOME MESSAGE 47.12
What factors affect life in freshwater provinces?

✔ Lakes have gradients in light, dissolved oxygen, and nutrients.

✔ Primary productivity varies with a lake's age and, in temperate zones, with the season.

✔ Different conditions along the length of a stream or river create habitat for different organisms.

47.13 Coastal Ecosystems

✔ Where sea meets shore we find regions of high primary productivity.

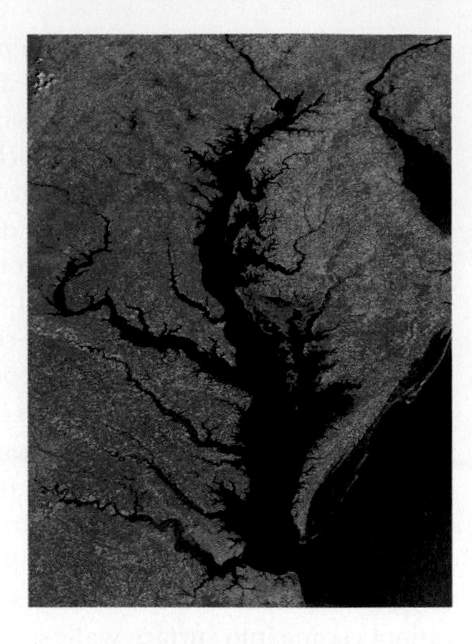

FIGURE 47.29 An aerial view of the Chesapeake Bay, the largest estuary in the United States. Many rivers empty into the bay, which opens to the Atlantic Ocean.

Estuaries—Where Fresh and Saltwater Meet

An **estuary** is a partly enclosed body of water where fresh water from a river or rivers mixes with seawater. Seawater is denser than fresh water, so fresh water floats on top of the seawater where they meet. The size and shape of the estuary, and the rate at which fresh water flows into it, determine how quickly the saltwater and fresh water mix and the effects of tides. Examples of estuaries include San Francisco Bay, Lake Pontchartrain in New Orleans, Boston harbor, and the Chesapeake Bay (**FIGURE 47.29**).

In all estuaries, an influx of water from upstream continually replenishes nutrients and allows a high level of productivity. Incoming fresh water also carries silt. Where the velocity of water flow slows, the silt falls to the bottom, forming mudflats. Photosynthetic bacteria and protists in biofilms on mudflats often account for a large portion of an estuary's primary production. Plants adapted to withstand changes in water level and salinity also serve as producers.

Spartina is the dominant plant in the salt marshes of many estuaries along the Atlantic coast (**FIGURE 47.30A**). It is adapted to its habitat by an ability to withstand immersion during high tides and to tolerate salty, waterlogged, anaerobic soil. Modified parenchyma (Section 27.3) with hollow air spaces allows oxygen taken in by shoots to diffuse to roots. Salt taken up in water by roots is excreted by specialized glands on the leaves. As a result, the leaves are typically covered with salt crystals. The high salt content of *Spartina* and other estuary plants makes them unpalatable to most herbivores, so detrital food webs predominate.

Estuaries and tidal flats of tropical and subtropical latitudes often support nutrient-rich mangrove wetlands (**FIGURE 47.30B**). "Mangrove" is the common term for certain salt-tolerant woody plants that live in sheltered areas along tropical coasts. The plants have prop roots that extend out from their trunk and help the plant stay upright in the soft sediments. Specialized cells at the surface of some exposed roots allow gas exchange with air.

A South Carolina estuary. Cordgrass (*Spartina*) is the dominant plant.

B Florida wetland dominated by red mangroves (*Rhizophora*).

FIGURE 47.30 Coastal wetlands.

FIGURE 47.31 Contrasting coasts. (**A**,**B**) Algae-rich rocky shores where invertebrates abound. (**C**) A sandy shore in Australia shows fewer signs of life. Invertebrates burrow in its sediments.

Estuaries provide important ecological services. Plants in estuaries help slow river flow, thus reducing the risk of flooding. Similarly, they help protect coasts from storm surges. Estuaries' waters serve as nurseries for many species of marine fishes and invertebrates. Migratory birds often stop over in estuaries as they travel from one region to another.

Human activities threaten estuary ecosystems. Pollution from farms and cities flows down rivers and into estuary waters. In the tropics, people have traditionally cut mangroves for firewood. A more recent threat is conversion of mangrove wetlands to shrimp farms. The shrimp mainly end up on dinner plates in the United States, Japan, and western Europe.

Rocky and Sandy Coastlines

Rocky and sandy coastlines support ecosystems of the intertidal zone. As with lakes, an ocean's shoreline is described as its littoral zone. The littoral zone can be divided into three vertical regions that differ in their physical characteristics and diversity. The most obvious difference between the regions is the amount of time they spend underwater. The upper littoral zone, or splash zone, regularly receives ocean spray but is submerged only during the highest of high tides. It gets the most sun, but holds the fewest species. The midlittoral zone is typically covered by water during an average high tide and dry during a low tide. The lower littoral zone, exposed only during the lowest tide of the lunar cycle, has the most diversity.

You can easily see the zonation along a rocky shore (**FIGURE 47.31A,B**). Barnacles live in the midlittoral zone. Algae clinging to rocks are primary producers for the prevailing grazing food web. The primary consumers include a variety of snails.

Zonation is less obvious on sandy shores where detrital food chains start with material washed ashore (**FIGURE 47.31C**). Some crustaceans eat detritus in the upper littoral zone. Nearer to the water, other invertebrates feed as they burrow through the sand.

estuary A highly productive ecosystem where nutrient-rich water from a river mixes with seawater.

TAKE-HOME MESSAGE 47.13

What kinds of ecosystems occur along coastlines?

✔ We find estuaries where rivers empty into seas. The rivers deliver nutrients that foster high productivity.

✔ Mangrove wetlands are common along shorelines in tropical latitudes.

✔ Rocky and sandy shores show zonation, with different zones exposed during different phases of the tidal cycle. Diversity is highest in the zone that is submerged most of the time.

47.14 Coral Reefs

✔ Coral reefs lie just off the coasts of tropical islands and continents.

✔ Coral reefs are the most diverse marine ecosystems.

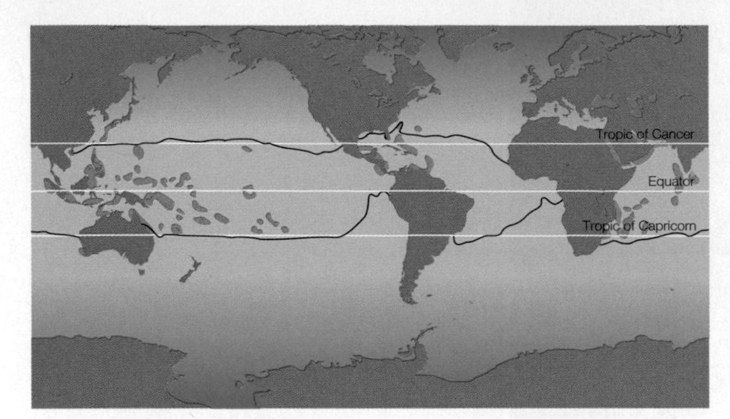

FIGURE 47.32 Distribution of coral reefs. Reef-building corals live in warm seas, here enclosed in dark lines. The greatest diversity of coral species is in the Indo-Pacifiic region.

Coral reefs are wave-resistant formations that consist primarily of calcium carbonate secreted by generations of coral polyps. Reef-forming corals live mainly in shallow, clear, warm waters between latitudes 25° north and 25° south. About 75 percent of all coral reefs are in the Indian and Pacific Oceans (**FIGURE 47.32**). A healthy reef is home to living corals and a huge number of other species (**FIGURE 47.33**). Biologists estimate that about a quarter of all marine fish species are associated with coral reefs.

The largest existing reef, Australia's Great Barrier Reef, parallels Queensland for 2,500 kilometers (1,550 miles), and is the largest example of biological architecture. Scientists estimate it began to form about 600,000 years ago. Today it is a string of reefs, some of them 150 kilometers (95 miles) across (**FIGURE 47.34**). The Great Barrier Reef supports about 500 coral species, 3,000 fish species, 1,000 kinds of mollusks, and 40 kinds of sea snakes.

FIGURE 47.33 A sample of reef biodiversity.

FIGURE 47.34 Satellite photo of Australia's Great Barrier Reef.

A Healthy coral reef near Fiji. The coral gets its color from the pigments of symbiotic dinoflagellates that live in its tissues and supply it with sugars.

B "Bleached" reef near Australia. The coral skeletons shown belong mainly to staghorn coral (*Acropora*), a genus especially likely to undergo coral bleaching.

FIGURE 47.35 ▶**Animated** Healthy and damaged coral reefs.

Photosynthetic dinoflagellates live as mutualistic symbionts inside the tissues of all reef-building corals (Section 24.5). The dinoflagellates are protected within the coral's tissues and provide the coral polyp with oxygen and sugars that it depends on. When stressed, coral polyps expel the dinoflagellates. Because the dinoflagellates give a coral its color, expelling these protists turns the coral white, an event called coral bleaching. When a coral is stressed for more than a short time, the dinoflagellate population in a coral's tissues cannot rebound and the coral dies, leaving only its bleached hard parts behind (**FIGURE 47.35**). The incidence of coral bleaching events has been increasing. Rising sea temperatures and sea level associated with global climate change most likely play a role.

People stress reefs by discharging sewage and other pollutants into coastal waters, by causing erosion that clouds water with sediments, and by destructive fishing practices. Invasive species also threaten reefs. Hawaiian reefs are threatened by exotic algae, including species imported for cultivation during the 1970s.

Assaults on reefs are taking a huge toll. For example, the Indo-Pacific region, the global center for reef diversity, lost about 3,000 square kilometers (1,160 square miles) of living coral reef each year between 1997 and 2003. Reef biodiversity is in danger around the world, from Australia and Southeast Asia to the Hawaiian Islands, the Galápagos Islands, the Gulf of Panama, Kenya, and Florida. The biodiversity on the coral reef offshore from Florida's Key Largo has been reduced by 33 percent since 1970.

TAKE-HOME MESSAGE 47.14

What are coral reefs and how are they threatened?

✔ Coral reefs form by the action of living corals that lay down a calcium carbonate skeleton. Photosynthetic dinoflagellates in the coral's tissues are necessary for the coral's survival.

✔ Rising water temperature, pollutants, fishing, and exotic species contribute to loss of reefs.

✔ Declines in coral reefs will affect the enormous number of fishes and invertebrate species that make their home on or near the reefs.

coral reef Highly diverse marine ecosystem centered around reefs built by living corals that secrete calcium carbonate.

CREDITS: (34) NASA Visible Earth, visibleearth.nasa.gov; (35A) © John Easley, www.johneasley.com; (35B) © Dr. Ray Berkelmans, Australian Institute of Marine Science.

47.15 The Open Ocean

✔ Earth's vast oceans are still largely unexplored. We are only beginning to catalog the diversity they contain.

Pelagic Ecosystems

Like a lake, an ocean shows gradients in light, nutrient availability, temperature, and oxygen concentration. We refer to the open waters of the ocean as the **pelagic province** (FIGURE 47.36). These open waters include the neritic zone—the water over continental shelves—together with the more extensive oceanic zone farther offshore. The neritic zone receives nutrients in runoff from land, and is the zone of greatest productivity.

In the ocean's upper, brightly lit waters, photosynthetic microorganisms are the primary producers, and grazing food chains predominate. Depending on the region, some light may penetrate as far as 1,000 meters (3,000 feet) beneath the sea surface. Below that, organisms live in darkness, and organic material that drifts down from above is the basis of detrital food chains.

Our knowledge of the deep pelagic zone is limited because human divers cannot survive at great depths, where the pressure exerted by the weight of the water above them is very high. Thus, exploration of deeper regions requires submersible vessels. Use of such technology has revealed some extraordinary life forms.

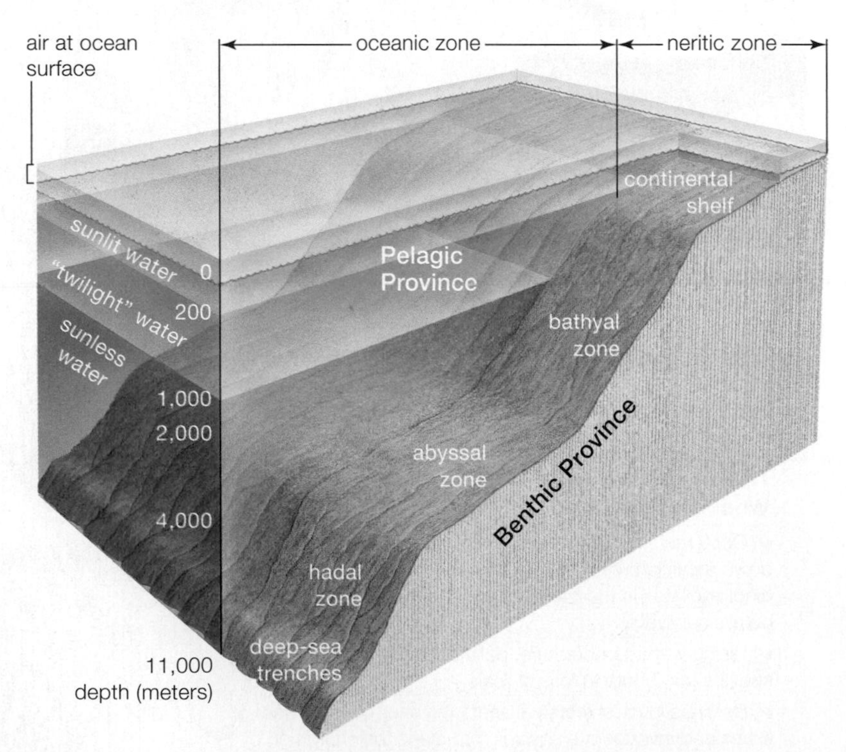

FIGURE 47.36 ▶Animated Oceanic zones. Zone dimensions are not drawn to scale.

FIGURE 47.37 Computer model of seamounts on the seafloor off the coast of Alaska. Patton Seamount, at the rear, stands 3.6 kilometers (about 2 miles) tall, with its peak about 240 meters (800 feet) below the sea surface. Seamounts formed in the same way as volcanic islands; molten rock erupted through the seafloor.

The Seafloor

The **benthic province** is the ocean bottom—its rocks and sediments. Benthic biodiversity is greatest on the margins of continents, or the continental shelves. The benthic province also includes some largely unexplored concentrations of biodiversity on seamounts and at hydrothermal vents.

Seamounts are undersea mountains that stand 1,000 meters or more tall, but are still below the sea surface (FIGURE 47.37). They attract large numbers of fishes and are home to many marine invertebrates. Like islands, seamounts often are home to species that evolved there and are found nowhere else. There are more than 30,000 seamounts, and scientists have just begun to document species that live on them.

The abundance of life at seamounts makes them attractive to commercial fishing vessels. Fishes and other organisms are often harvested by trawling, a fishing technique in which a large net is dragged along the bottom, capturing everything in its path. The process is ecologically devastating; trawled areas are stripped bare of life, and silt stirred up by the giant, weighted nets suffocates filter-feeders in adjacent areas.

At **hydrothermal vents**, hot water rich in dissolved minerals spews out from an opening on the ocean floor. The water is seawater that seeped into cracks in the ocean floor at the margins of tectonic plates and was heated by heat energy from within Earth. Minerals in this water settle out when it mixes with the cold deep-

Going With the Flow (revisited)

Trace amounts of ^{137}Cs-tainted water that reach North America from Japan are not expected to adversely affect life along the continent's west coast. Nevertheless, scientists continue to monitor potentially vulnerable ecosystems for evidence of radioactivity and any possible harm.

One such monitoring program is documenting the level of radiation in kelp forests along the western coast of the United States. Participants collect kelp samples several times a year, then send them to a government laboratory where the samples are tested for radioisotopes from Fukushima. Kelp was chosen because of its important role in the coastal ecosystem and because it tends to take up and concentrate radioactive material, making it easier to detect.

Bits of kelp tested only weeks after the accident at Fukushima showed the presence of iodine 131, a radioisotope produced in nuclear reactors, but rare in nature. Researchers assume the ^{131}I traveled on the winds before falling into the sea and being taken up by the kelp. ^{131}I has a half-life of 8 days, so it is no longer a concern and there is no evidence that exposure to it did the kelp any long-term harm.

By sharing results of their kelp monitoring study online, the scientists hope to put to rest what they consider irrational fears of soaring levels radioactivity along the west coast beaches. Given the diluting effects of trans-ocean travel, the concentration of radioactive material in coastal water and on beaches is not expected to ever climb to a point where it poses a threat to human health.

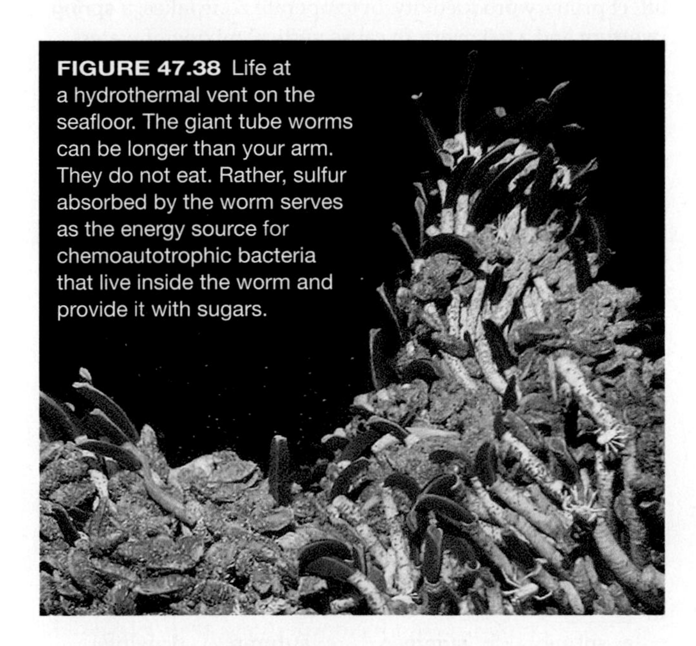

FIGURE 47.38 Life at a hydrothermal vent on the seafloor. The giant tube worms can be longer than your arm. They do not eat. Rather, sulfur absorbed by the worm serves as the energy source for chemoautotrophic bacteria that live inside the worm and provide it with sugars.

sea water. Chemoautotrophic bacteria and archaea that obtain energy by removing electrons from minerals are the main producers. Food webs include diverse invertebrates, including large annelid tube worms (**FIGURE 47.38**).

Life exists even in the deepest sea. A remote-controlled submersible that sampled sediments in the deepest part of the ocean (the Mariana Trench) brought up foraminifera that live 11 kilometers (7 miles) below the surface.

Sediment samples from deep in the Mediterranean Sea turned up another surprise, tiny animals that do not use oxygen (**FIGURE 47.39**). The animals, called loriciferans, live between sand grains and are distant relatives of insects and nematodes. They are the only animals known to live their lives entirely without oxygen. Like some protists, they lack mitochondria, and instead have organelles called hydrogenosomes (Section 21.3).

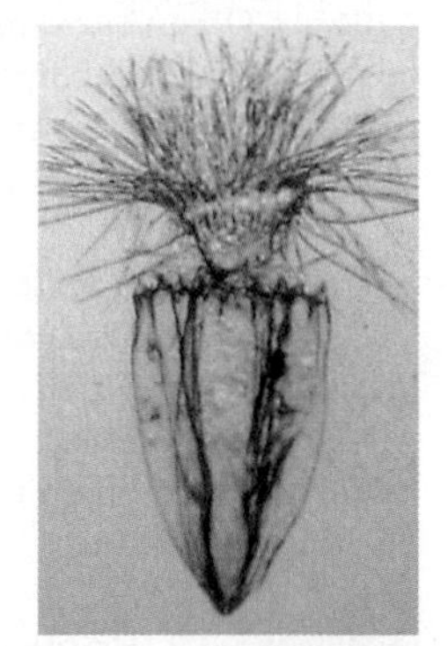

FIGURE 47.39 A deep-sea loriciferan. This tiny animal can live without oxygen.

TAKE-HOME MESSAGE 47.15

What factors affect life in ocean provinces?

- ✔ Oceans have gradients in light, dissolved oxygen, and nutrients. Nearshore and well-lit zones are the most productive and species-rich.
- ✔ On the seafloor, pockets of diversity occur on seamounts and around hydrothermal vents.
- ✔ Animal life exists even in the deepest ocean, and some animals have adapted to life in oxygen-free sediments.

---

benthic province The ocean's sediments and rocks.
hydrothermal vent Rocky, underwater opening where mineral-rich water heated by geothermal energy streams out.
pelagic province The ocean's waters.
seamount An undersea mountain.

CREDITS: (38) NOAA/Photo courtesy of Cindy Van Dover, Duke University Marine Lab; (39) Courtesy of Prof Roberto Danovaro, Polytechnic University of Marche; (in text revisited) U.S. Navy photo by Mass Communication Specialist 3rd Class Alexander Tidd/Released.

summary

Section 47.1 Earth's air and water circulate constantly and distribute energy and materials across the globe. In the aftermath of the 2011 earthquake and tsunami in Japan, these processes distributed radioactive material and debris released by the disaster.

Section 47.2 Global air circulation patterns influence **climate** and the distribution of communities through the **biosphere**. The patterns are set into motion by latitudinal variations in incoming solar radiation. Tropical latitudes receive more sun than higher latitudes. As air heated in the tropics moves toward the poles, its path is influenced by Earth's rotation so that the direction of prevailing winds varies with latitude. Regions where warm air rises and loses moisture have a high annual rainfall; regions where cool air descends are dry.

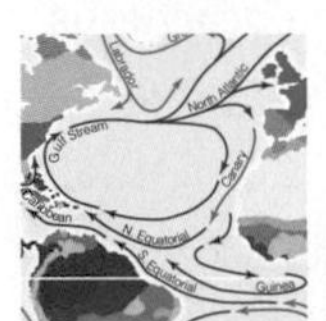

Section 47.3 Latitudinal and seasonal variations in sunlight warm up seawater and create surface currents that are affected by prevailing winds. The currents distribute heat energy worldwide and influence the weather patterns. Ocean currents, air currents, and landforms interact in shaping global temperature zones, as where the presence of coastal mountains causes a **rain shadow** or where **monsoon** rains fall seasonally.

Section 47.4 The **El Niño Southern Oscillation** triggers changes in rainfall and other weather patterns around the world. These changes can harm human health, as when warming waters increase the likelihood of cholera epidemics.

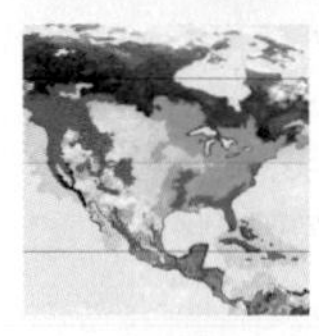

Section 47.5 **Biomes** are discontinuous areas characterized by a particular type of vegetation. Differences in climate, elevation, and soil properties affect the distribution of biomes.

Section 47.6 **Deserts** form around latitudes 30° north and south, where rainfall is sparse and falls seasonally. Desert plants include annuals that grow fast after seasonal rains and perennials that are adapted to withstand a seasonal drought. Desert crust, a community of organisms in the upper soil layer, helps hold soil in place.

Sections 47.7, 47.8 **Grasslands** form in the interior of midlatitude continents. North America's grasslands are prairies. Prairies have a highly fertile soil and most have now been converted to agricultural use. Africa has savannas, which include widely dispersed trees. Both grasslands and savannas support herds of grazing animals.

Southwest coasts of continents have cool, rainy winters and a hot, dry summer. Depending on the amount of rain that falls, they support **dry woodlands** or **dry shrublands** such as California's chaparral. Shrubland plants are adapted to periodic fire.

Sections 47.9–47.11 Broadleaf trees are angiosperms. Most trees in **temperate deciduous forests** shed their leaves all at once just before a cold winter prevents growth. Broadleaf trees in **tropical rain forests** can grow year-round, and these enormously productive forests are home to a large number of species.

Conifers withstand cold and drought better than broadleaf trees and dominate Northern Hemisphere high-latitude **boreal forests.**

Farther north in this hemisphere, **arctic tundra** overlies **permafrost** that stores a large amount of carbon. **Alpine tundra** is a similar biome that occurs at high altitudes.

Section 47.12 Most **lakes**, streams, and other aquatic ecosystems have gradients in the penetration of sunlight, water temperature, and in dissolved gases and nutrients. These characteristics vary over time and affect primary productivity. In temperate zone lakes, a **spring overturn** and a **fall overturn** cause vertical mixing of waters and trigger a burst of productivity. In summer, a **thermocline** prevents upper and lower waters from mixing.

Sections 47.13–47.15 Coastal zones support diverse ecosystems. Among these, the coastal wetlands, **estuaries**, and **coral reefs** are especially productive.

Life persists throughout the ocean. Diversity is highest in sunlit waters at the top of the **pelagic province**, which is the ocean's waters. On the seafloor (in the **benthic province**), diversity is high near deep-sea **hydrothermal vents** and on **seamounts**. Even the deepest ocean sediments support life.

self-quiz

Answers in Appendix VII

1. The Northern Hemisphere is most tilted toward the sun in ________ .
   a. spring b. summer c. autumn d. winter

2. Which latitude will have the most hours of daylight on the summer solstice?
   a. 0° (the equator) c. 45° north
   b. 30° north d. 60° north

3. Warm air ________ and it holds ________ water than cold air.
   a. sinks; less c. sinks; more
   b. rises; less d. rises; more

4. A rain shadow is a reduction in rainfall ________ .
   a. on the inland side of a coastal mountain range
   b. during an El Niño event
   c. that results from global warming

data analysis activities

Sea Temperatures To predict the effect of El Niño or La Niña events in the future, the National Oceanographic and Atmospheric Administration collects information about sea surface temperature (SST) and atmospheric conditions. They compare monthly temperature averages in the eastern equatorial Pacific Ocean to historical data and calculate the difference (the degree of anomaly) to determine if El Niño conditions, La Niña conditions, or neutral conditions are developing. El Niño is a rise in the average SST above 0.5°C. A decline of the same amount is La Niña. **FIGURE 47.40** shows data for 42 years.

1. When did the greatest positive temperature deviation occur during this time period?

2. What type of event, if any, occurred during the winter of 1982–1983? What about the winter of 2001–2002?

3. During a La Niña event, less rain than normal falls in the American West and Southwest. In the time interval shown, what was the longest interval without a La Niña event?

4. What type of conditions were in effect in the fall of 2007 when California suffered severe wildfires?

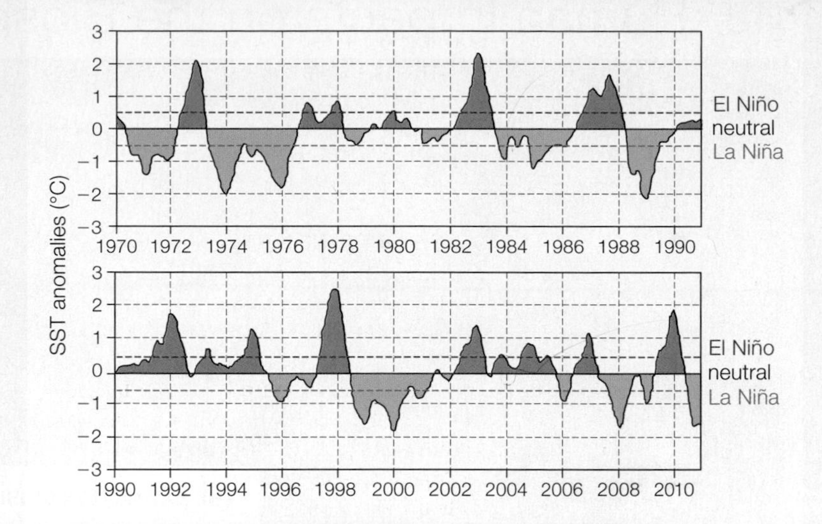

FIGURE 47.40 Sea surface temperature anomalies (differences from the historical mean) in the eastern equatorial Pacific Ocean. A rise above the dashed red line is an El Niño event, a decline below the blue line is La Niña.

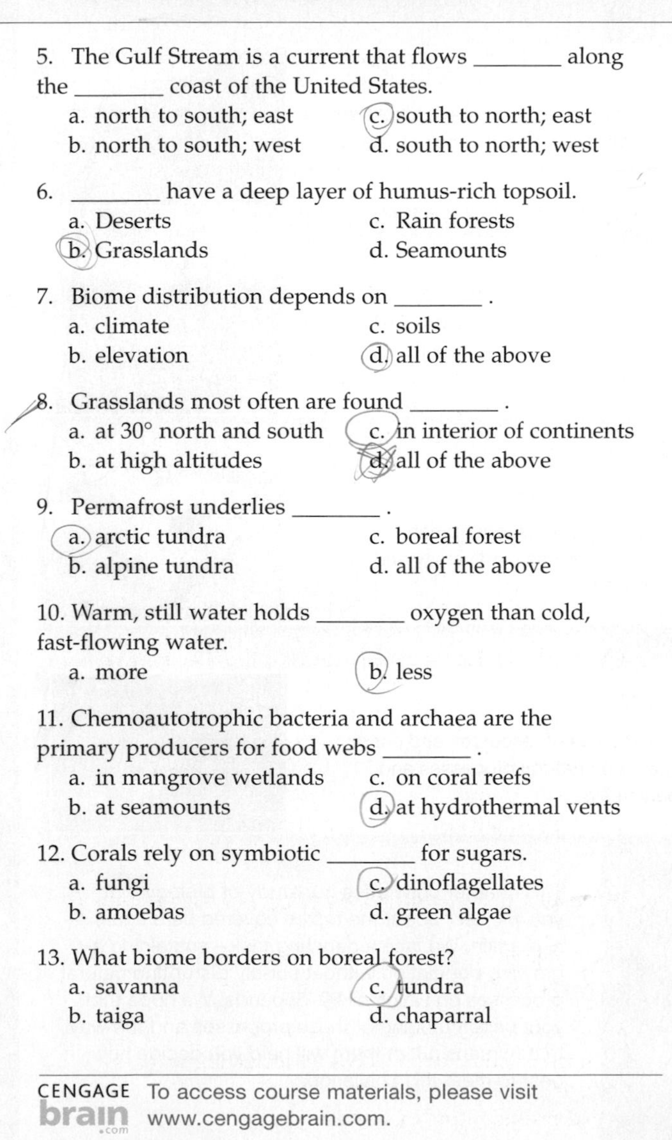

5. The Gulf Stream is a current that flows _______ along the _______ coast of the United States.
   a. north to south; east
   b. north to south; west
   c. south to north; east
   d. south to north; west

6. _______ have a deep layer of humus-rich topsoil.
   a. Deserts
   b. Grasslands
   c. Rain forests
   d. Seamounts

7. Biome distribution depends on _______ .
   a. climate
   b. elevation
   c. soils
   d. all of the above

8. Grasslands most often are found _______ .
   a. at 30° north and south
   b. at high altitudes
   c. in interior of continents
   d. all of the above

9. Permafrost underlies _______ .
   a. arctic tundra
   b. alpine tundra
   c. boreal forest
   d. all of the above

10. Warm, still water holds _______ oxygen than cold, fast-flowing water.
   a. more
   b. less

11. Chemoautotrophic bacteria and archaea are the primary producers for food webs _______ .
   a. in mangrove wetlands
   b. at seamounts
   c. on coral reefs
   d. at hydrothermal vents

12. Corals rely on symbiotic _______ for sugars.
   a. fungi
   b. amoebas
   c. dinoflagellates
   d. green algae

13. What biome borders on boreal forest?
   a. savanna
   b. taiga
   c. tundra
   d. chaparral

14. Unrelated species in geographically separated parts of a biome may resemble one another as a result of _______ .
   a. morphological divergence
   b. morphological convergence
   c. resource partitioning
   d. coevolution

15. Match the terms with the most suitable description.

___ tundra	a. broadleaf forest near equator
___ chaparral	b. partly enclosed by land; where fresh water and seawater mix
___ desert	c. African grassland with trees
___ savanna	d. low-growing plants at high latitudes or elevations
___ estuary	e. dry shrubland
___ boreal forest	f. at latitudes 30° north and south
___ prairie	g. mineral-rich, superheated water supports communities
___ tropical rain forest	h. conifers dominate
___ hydrothermal vents	i. North American grassland

critical thinking

1. London, England, is at the same latitude as Calgary in Canada's province of Alberta. However, the mean January temperature in London is 5.5°C (42°F), whereas in Calgary it is minus 10°C (14°F). Compare the locations of these two cities and suggest a reason for this temperature difference.

2. Increased industrialization in China has environmentalists worried about air quality elsewhere. Are air pollutants emitted in Beijing more likely to end up in eastern Europe or the western United States? Why?

3. The use of off-road recreational vehicles may double in the next twenty years. Enthusiasts would like increased access to government-owned deserts. Some argue that it's the perfect place for off-roaders because "There's nothing there." Explain whether you agree, and why.

CREDIT: (40) Adapted from NOAA.

48 Human Impacts on the Biosphere

LEARNING ROADMAP

In this chapter, we consider how human activities are causing a mass extinction (Section 17.12). In doing so, we reconsider the ecological effects of acid rain (2.6), soil erosion (28.2), aquifer depletion (46.6), and global climate change (46.8).

THE EXTINCTION CRISIS

Human activities have accelerated the rate of extinctions. Habitat loss, degradation, and fragmentation lead to extinctions, as do species introductions and overharvesting.

HARMFUL PRACTICES

Plowing grasslands and cutting down forests can have long-term and long-range effects. Loss of plants allows soil erosion, raises soil temperature, and affects rainfall.

POLLUTANTS

Human activities produce pollutants that harm living organisms, destroy the ozone layer, and cause global climate change.

CONSERVING BIODIVERSITY

All nations have biological wealth that can benefit human populations. Conservation biologists assess biodiversity and investigate the best ways to sustain it.

REDUCING NEGATIVE IMPACTS

Individual actions that minimize the use of resources and energy can help minimize human impacts on natural processes and thereby reduce threats to biodiversity.

This chapter concludes our study of biology, but you will hear about the topics covered here again and again. We face a daunting task—sustaining a growing population without unduly disrupting natural processes on which all life depends. We hope that your understanding of these processes and the way that humans affect them will help you decide how best to meet this challenge.

48.1 A Long Reach

We began this book with the story of biologists who ventured into a remote forest in New Guinea, and the many previously unknown species that they encountered. At the far end of the globe, a U.S. submarine surfaced in Arctic waters and discovered polar bears hunting on the ice-covered sea (**FIGURE 48.1**). The bears were about 270 miles from the North Pole and 500 miles from the nearest land.

Even such seemingly remote regions are no longer beyond the reach of human explorers—and human influence. You already know that increasing levels of greenhouse gases are raising the temperature of Earth's atmosphere and seas. In the Arctic, the warming is causing sea ice to thin and to break up earlier in the spring. This raises the risk that polar bears hunting far from land will become stranded and unable to return to solid ground before the ice thaws.

Polar bears are top predators and their tissues contain a surprisingly high amount of mercury and organic pesticides. These substances entered the water and air far away, in more populated temperate regions. Winds and ocean currents delivered them to the Arctic. Contaminants also travel north inside the bodies of migratory animals such as seabirds that spend their winters in temperate regions and nest in the Arctic.

In places less remote than the Arctic, effects of human populations have a more direct effect. As we cover more and more of the world with our dwellings, factories, and farms, less appropriate habitat remains for other species. We also put species at risk by competing with them for resources, overharvesting them, and introducing nonnative competitors.

It would be presumptuous to think that we alone have had a profound impact on the world of life. As long ago as the Proterozoic, photosynthetic cells were irrevocably changing the course of evolution by enriching the atmosphere with oxygen. Over life's existence, the evolutionary success of some groups has ensured the decline of others. What is new is the increasing pace of change and the capacity of our own species to recognize and affect its role in this increase.

A century ago, Earth's physical and biological resources seemed inexhaustible. Now we know that many practices put into place when humans were largely ignorant of how natural systems operate take a heavy toll on the biosphere. The rate of species extinctions is on the rise and many types of biomes are threatened. These changes, the methods scientists use to document them, and the ways that we can address them are the focus of this chapter.

FIGURE 48.1 Three polar bears investigate an American submarine that surfaced in ice-covered Arctic waters.

CREDITS: (opposite) NASA courtesy of the MODIS Rapid Response Team at Goddard Space Flight Center; (1) U.S. Navy photo by Chief Yeoman Alphanso Braggs.

✔ Extinction is a natural process, but human activities are currently accelerating the rate of extinctions.

Era	Period	mya	Major extinction under way
Cenozoic	Quaternary		With high population growth rates and cultural practices (e.g., agriculture, deforestation), humans become major agents of extinction.
		3	
	Neogene		
		23	
	Paleogene		Major extinction event
		66	
Mesozoic	Cretaceous		Slow recovery after Permian extinction, then adaptive radiations of some marine groups and plants and animals on land. Asteroid impact at K–T boundary, 85% of all species disappear from land and seas.
		145	
	Jurassic		
		201	
	Triassic		Major extinction event
		252	
Paleozoic	Permian		Pangea forms; land area exceeds ocean surface area for first time. Asteroid impact? Major glaciation, colossal lava outpourings, 90%–95% of all species lost.
		299	
	Carboniferous		Major extinction event
		359	
	Devonian		More than 70% of marine groups lost. Reef builders, trilobites, jawless fishes, and placoderms severely affected. Meteorite impact, sea level decline, global cooling?
		419	
	Silurian		Major extinction event
		443	
	Ordovician		Second most devastating extinction in seas; nearly 100 families of marine invertebrates lost.
		485	
	Cambrian		Major extinction event
		541	
	(Precambrian)		Massive glaciation; 79% of all species lost, including most marine microorganisms.

A Dates of the greatest mass extinctions (mya: millions of years ago).

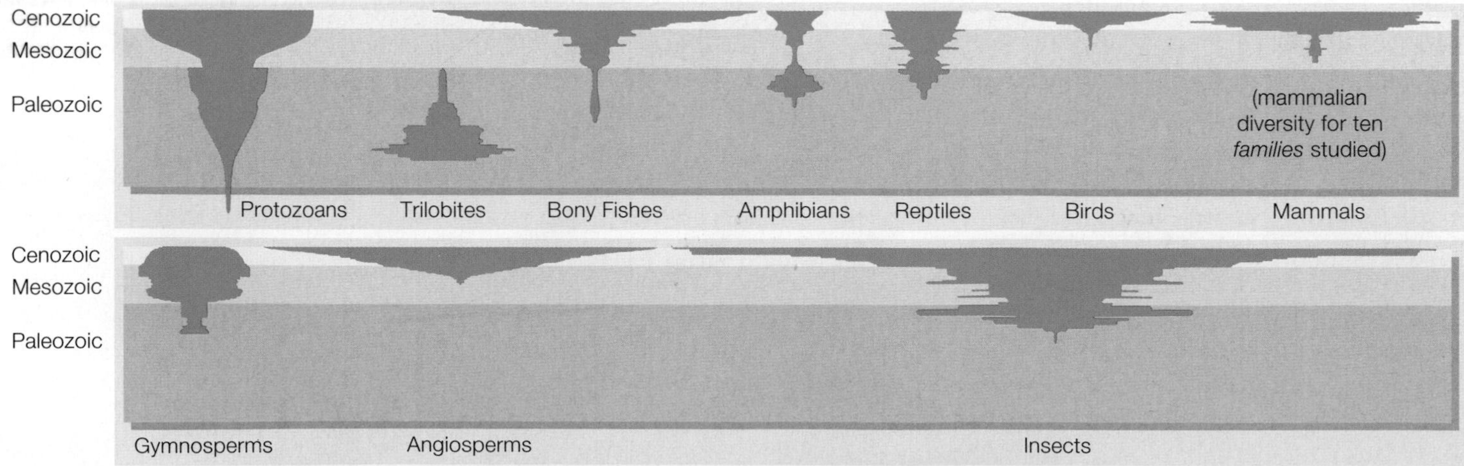

B Changes in diversity for a variety of taxa. Width of each blue shape indicates the number of species in that lineage.

FIGURE 48.2 Extinctions. **FIGURE IT OUT** Which of lineages shown went extinct at the end of the Permian? Answer: Trilobites

Mass Extinction

Extinction, like speciation, is a natural process. Species arise and become extinct on an ongoing basis. Based on several lines of evidence, scientists estimate that 99 percent of all species that have ever lived are now extinct. The rate of extinction picks up dramatically during a **mass extinction**, when a large proportion of Earth's organisms become extinct in a relatively short period of geologic time.

Five great mass extinctions mark the boundaries for the geologic time periods. As you learned in Section 16.1, the mass extinction at the end of the Cretaceous period was most likely caused by an asteroid impact. The greatest mass extinction, in terms of species lost, occurred at the end of the Permian. It took place after increased volcanic activity in what is now Siberia released a large amount of carbon into the atmosphere. It is thought that the resulting warming of Earth's oceans caused the release of methane (natural gas) from frozen deposits deep in the ocean. The ocean became anoxic and methane that entered the air caused explosive conflagrations to occur on land.

Mass extinctions affect all lineages. However, lineages differ in their time of origin, their tendency to branch and give rise to new species, and how long they endure. If we consider the number of species as the measure of success for any lineage, not all lineages are equally successful. **FIGURE 48.2B** illustrates how the number of species changed over time in some major lineages. Expansion of one lineage sometimes occurs at the same time as contraction of another, as when the gymnosperms declined during the adaptive radiation of angiosperms.

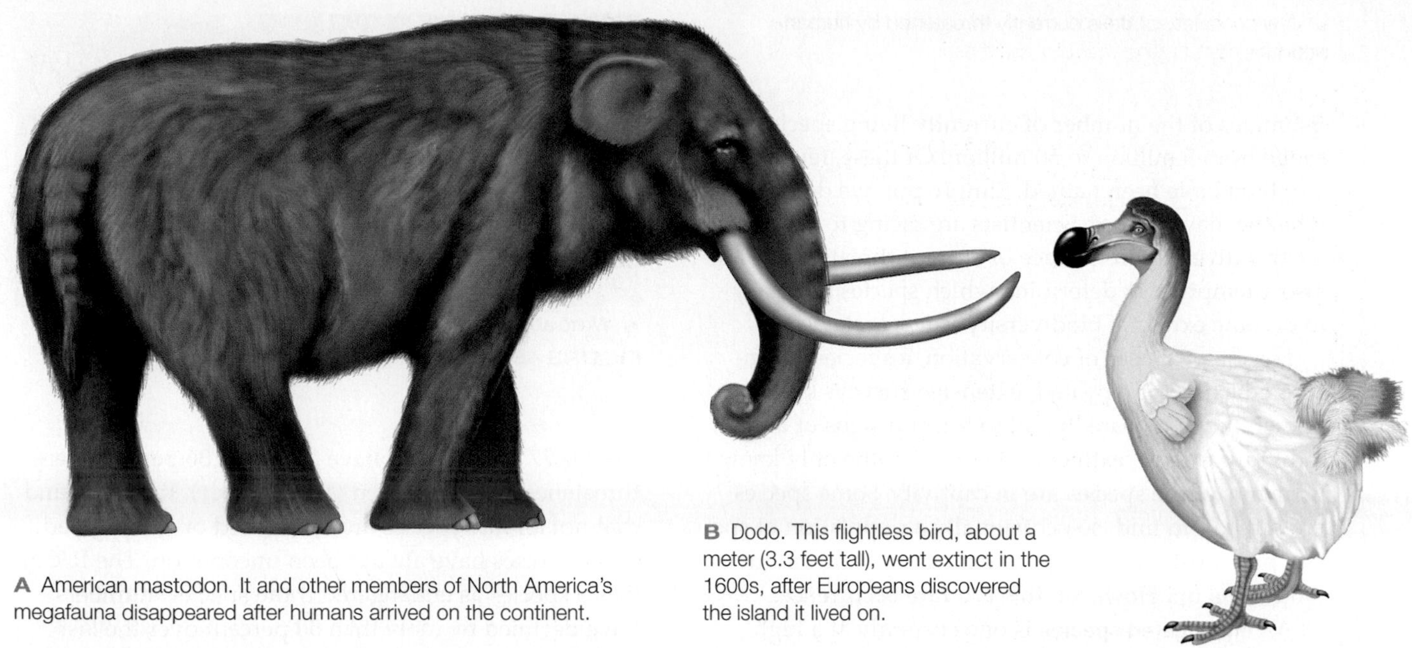

A American mastodon. It and other members of North America's megafauna disappeared after humans arrived on the continent.

B Dodo. This flightless bird, about a meter (3.3 feet tall), went extinct in the 1600s, after Europeans discovered the island it lived on.

FIGURE 48.3 Drawings of two now-extinct animals hunted by humans.

The Sixth Great Mass Extinction

We are presently in the midst of a mass extinction. The current extinction rate is estimated to be 100 to 1,000 times above the typical background rate, putting it on a par with the five major extinction events of the past. Unlike previous extinction events, this one is not the inevitable result of a physical catastrophe such as an eruption or asteroid impact. Humans are the driving force behind the current rise in extinction rate, and our actions will determine the extent of the losses.

Indirect evidence implicates the arrival of humans in the prehistoric decline of many large animals, collectively referred to as megafauna, in Australia and North America. In Australia, the arrival of humans about 40,000 years ago correlates with the onset of extinctions of the continent's largest marsupials, birds, and lizards. In North America, arrival of humans by about 15,000 years ago was followed by the extinction of large herbivores such as camels, giant ground sloths, and mammoths and mastodons (relatives of elephants; **FIGURE 48.3A**). Carnivores such as lions and saber-toothed cats also disappeared.

By one hypothesis, humans directly caused declines in herbivorous species such as mammoths by hunting them. These declines then led to the extinction of their predators. Evidence that humans hunted some megafauna supports this hypothesis. For example, a 13,800-year-old mastodon rib bone found in Washington state had a spear point embedded in it.

However, not all researchers are convinced that hunting was the sole or most important factor in megafaunal extinctions. Demise of these animals has also been attributed to climate change, a comet impact, or effects of human-introduced pathogens. Continued study of sites that date to the time of extinctions will help clarify the relative importance of these factors.

More recent extinctions are more clearly attributable to humans. The World Conservation Union has compiled a list of more than 800 documented extinctions that occurred since 1500. Consider what happened to the dodo (**FIGURE 48.3B**) a big, flightless bird that lived on the island of Mauritius in the Indian Ocean. Dodos were plentiful in 1600, when Dutch sailors first arrived on the island, but 80 or so years later the birds were extinct. Sailors ate some, but destruction of nests and habitat by rats, cats, and pigs introduced by the sailors probably had a greater effect.

mass extinction A large proportion of Earth's organisms become extinct in a relatively short period of geologic time.

TAKE-HOME MESSAGE 48.2

How does the current mass extinction differ from previous ones?

✔ Previous mass extinctions occurred when an event such as an asteroid impact or a series of volcanic eruptions caused a global catastrophe.

✔ The current extinction crisis is the result of human activity.

48.3 Current Diversity and Threats

✔ A wide variety of life is currently threatened by human activities.

Estimates of the number of currently living species range from 5 million to 50 million. Of these, fewer than 2 million have been named. Simply put, we don't know what we have to lose. Scientists are racing to document Earth's diversity in the face of current threats. They are also attempting to determine which species are likely to become extinct if biodiversity is not protected.

For the purposes of conservation, a species is considered extinct if repeated, extensive surveys of its known range repeatedly fail to turn up signs of any individuals. It is "extinct in the wild" if the only known members of the species are in captivity. Some species are difficult to find, so occasionally a population of species previously thought to be extinct or extinct in the wild turns up. However, this is a rare occurrence.

An **endangered species** is one currently at a high risk of extinction in the wild. A **threatened species** is one that is likely to become endangered in the near future. The International Union for Conservation of Nature and Natural Resources (IUCN) reports that of the 48,677 species they have assessed, 36 percent were threatened or endangered (TABLE 48.1). Keep in mind that not all rare species are threatened or endangered. Some species have always been uncommon. The IUCN lists a species as endangered if the species' numbers have declined by more than 50 percent over the last ten years or analysis indicates that there is a more than 20 percent chance that the species will become extinct in the wild within 20 years.

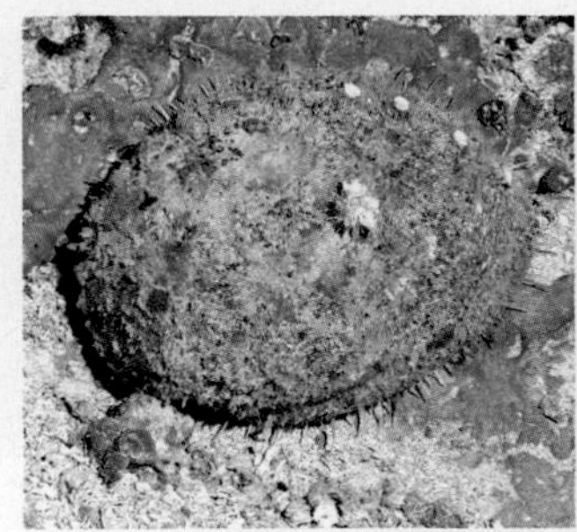

A White abalone.

B Black rhinocerus.

FIGURE 48.4 Two species endangered by overharvesting.

Table 48.1 Global List of Threatened Species*

	Described Species	Evaluated for Threats	Found to Be Threatened
Vertebrates			
Mammals	5,416	4,863	1,094
Birds	9,956	9,956	1,217
Reptiles	8,240	1,385	422
Amphibians	6,199	5,915	1,808
Fishes	30,000	3,119	1,201
Invertebrates			
Insects	959,000	1,255	623
Mollusks	81,000	2,212	978
Crustaceans	40,000	553	460
Corals	2,175	13	5
Others	130,200	83	42
Land Plants			
Mosses	15,000	92	79
Ferns and allies	13,025	211	139
Gymnosperms	980	909	321
Angiosperms	258,650	10,771	7,899
Protists			
Green algae	3,715	2	0
Red algae	5,956	58	9
Brown algae	2,849	15	6
Fungi			
Lichens	10,000	2	2
Mushrooms	16,000	1	1

* *IUCN–WCU Red List, available online at www.iucnredlist.org*

Causes of Species Declines

Overharvesting species is one cause of species decline. Consider the case of the white abalone, a gastropod mollusk native to kelp forests along the coast of California (FIGURE 48.4A). During the 1970s, overharvesting of this species reduced the population to about 1 percent of its original size. In 2001, the abalone became the first invertebrate to be listed as endangered by the United States Fish and Wildlife Service. The white abalone's current population density is too low for effective reproduction in the wild. Abalones release their gametes into the water, a strategy that is effective only when many individuals live within a few meters of one another. In an attempt to boost the species' numbers, some white abalone have been collected for a captive breeding program. If this program succeeds, their offspring will be reintroduced into the wild.

Species are overharvested not only as food, but also for use in traditional medicine, for the pet trade, and for use as ornaments. Some orchids prized by collectors are now nearly extinct in the wild. Elephants are killed for their tusks and rhinos for their horns (FIGURE 48.4B). The majority of elephant tusks end up in China, in the form of decorative carved objects. Rhino horn is used as dagger handles and as a traditional treatment for impotence.

Each species requires a specific habitat, so degradation, fragmentation, or destruction of that habitat also

CREDITS: (Table 48.1) © Cengage Learning; (4A) John Butler, NOAA; (4B) © George Sanker/naturepl.com.

A Texas blind salamander.

B Florida perforate lichen.

C Buffalo clover.

FIGURE 48.5 Three species endangered by habitat loss in the United States.

reduces population numbers. An **endemic species**, a species that lives only in the region where it evolved, is more likely to go extinct than a species with a more widespread distribution. Also, a species with highly specific needs is more likely to go extinct than one that is a generalist. For example, giant pandas are endemic to China's bamboo forests and feed mainly on bamboo. As bamboo forests have disappeared, so have pandas. Their population, which may once have been as high as 100,000 animals, is now reduced to about 1,600 animals in the wild. Habitat degradation threatens Texas blind salamanders too (**FIGURE 48.5A**). The salamander is endemic to the Edwards Aquifer, a series of water-filled, underground limestone formations. Like other aquifers, the salamander's home is threatened by excess withdrawal of water and by pollution. Similarly, a lichen endemic to Florida is endangered by development of its scrubland habitat (**FIGURE 48.5B**).

Introduced species (Section 45.9) pose another threat. Rats that reached islands by stowing away on ships attack and endanger ground-nesting birds. Exotic species often displace natives, as when red imported fire ants eat food that would otherwise feed native ants. Introduced pathogens can decimate species that did not coevolve with them and hence lack defenses.

Decline or loss of one species sometimes endangers others. Consider buffalo clover, a plant that was abundant when many buffalo grazed in the American Midwest (**FIGURE 48.5C**). The clover thrived at the edges of woodlands, where buffalo enriched the soil with their droppings and helped to disperse the clover's seeds. Like most endangered species, buffalo clover faces a number of simultaneous threats. In addition to the loss of buffalo, the clover is threatened by competition from introduced plants, attacks by introduced insects, and development of its habitat for housing.

The Unknown Losses

Endangered species listings have historically focused on vertebrates. Scientists have only recently begun to consider the threats to invertebrates and to plants. Our impact on protists and fungi is largely unknown, and the IUCN does not assess threats to bacteria or archaea.

Microbiologist Tom Curtis is among those making a plea for increased research on microbial ecology and microbial diversity. He argues that we have barely begun to comprehend the vast number of microbial species and to understand their importance. He writes, "I make no apologies for putting microorganisms on a pedestal above all other living things. For if the last blue whale choked to death on the last panda, it would be disastrous but not the end of the world. But if we accidentally poisoned the last two species of ammonia-oxidizers, that would be another matter. It could be happening now and we wouldn't even know." Ammonia-oxidizing bacteria play an essential role in the nitrogen cycle by converting ammonia in wastes and remains to nitrites (Section 46.9).

TAKE-HOME MESSAGE 48.3

How do human activities endanger existing species?

✔ Species decline when humans destroy or fragment natural habitat by converting it to human use, or degrade it through pollution or withdrawal of an essential resource.

✔ Humans also cause declines by overharvesting species and by introducing exotic species that harm native ones.

✔ Most endangered species are affected by multiple threats.

endangered species A species that faces extinction in all or a part of its range.
endemic species A species that remains restricted to the area where it evolved.
threatened species Species likely to become endangered in the near future.

CREDITS: (5A) Joe Fries, U.S. Fish & Wildlife Service; (5B) Joel Sartore/National Geographic Creative; (5C) USDA Forest Service/Photo by Sarena Selbo.

48.4 Harmful Land Use Practices

✔ Human activities have the potential not only to harm individual species, but also to transform entire biomes.

Desertification

Deserts naturally expand and contract over long periods as a result of fluctuations in climate. However, poor agricultural practices that encourage soil erosion can sometimes lead to rapid shifts from grassland or woodland to desert. As the human population increases, more and more people are forced to farm in areas ill-suited to agriculture. Others allow their livestock to overgraze grasslands, killing grass plants. Such practices can result in **desertification**, the conversion of grassland or woodlands to desertlike conditions.

Consider what happened during the mid-1930s, when large portions of prairie on the southern Great Plains were plowed under to plant crops. This plowing exposed the deep prairie topsoil to the force of the region's constant winds. Coupled with a drought, the result was an economic and ecological disaster. Winds carried more than a billion tons of topsoil aloft as sky-darkening dust clouds (FIGURE 48.6), causing the region be known as the Dust Bowl. Tons of displaced soil from the Dust Bowl fell to earth as far away as New York City and Washington, D.C.

Desertification now threatens vast areas. In Africa, the Sahara desert is expanding south into the Sahel region. Ongoing overgrazing of the Sahel strips grasslands of their vegetation and allows winds to erode the soil. Winds carry soil particles aloft and westward, carrying them as far away as the southern United States and the Caribbean (FIGURE 48.7).

In China's northwestern regions, overplowing and overgrazing have expanded the Gobi desert so that dust clouds periodically darken skies above Beijing. Winds carry some of this dust across the Pacific to the United States. In an attempt to hold back the desert, China has planted billions of trees as a "Green Wall."

In a positive feedback cycle, drought encourages desertification, which results in more drought. Plants cannot thrive in a region where the topsoil has blown away. With less transpiration by plants (Section 28.4), less water enters the air, so local rainfall decreases.

The best way to prevent desertification is to avoid farming in areas subject to high winds and periodic drought. If these areas must be used, farming methods that do not repeatedly disturb the soil can minimize the risk of desertification.

Deforestation

The amount of forested land is currently stable or increasing in North America, Europe, and China, but tropical forests continue to disappear at an alarming rate. In Brazil, increases in the export of soybeans and free-range beef have helped make the country the world's seventh-largest economy. However, this economic expansion has come at the expense of the country's woodlands and forests (FIGURE 48.8).

FIGURE 48.6 A giant dust cloud about to descend on a farm in Kansas during the 1930s. A large portion of the southern Great Plains was then known as the Dust Bowl. Drought and poor agricultural practices allowed winds to strip tons of topsoil from the ground and carry it aloft.

FIGURE 48.7 Modern-day dust cloud blowing from Africa's Sahara desert out into the Atlantic Ocean. The Sahara's area is increasing as a result of a long-term drought, overgrazing, and stripping of woodlands for firewood.

FIGURE 48.8 ▶**Animated** Tropical deforestation. Aerial view (left) and ground view (right) of Mato Grosso, Brazil, a region where tropical woodlands and forest have been cleared at an alarming rate, mainly to create pasture for cattle. Cattle producers have been forced to expand into forests by an increase in the industrial production of soybeans and corn in what was formerly prime cattle-rearing territory.

Deforestation has detrimental effects beyond the immediate destruction of forest organisms. Deforestation encourages flooding because rather than being taken up by tree roots, water runs off into streams. Deforestation also raises risk of landslides in hilly areas. Tree roots tend to hold soil in place, so their removal makes waterlogged soil more likely to slide. Deforested areas also lose more nutrients from their soil. **FIGURE 48.9** shows results of an experiment in which scientists deforested a region in New Hampshire and monitored the nutrients in runoff. Deforestation increased the rate of loss of nutrients such as calcium, which are essential for healthy plant growth.

Like desertification, deforestation affects local weather. In hot weather, forests remain cooler than adjacent nonforested areas because trees shade the ground and transpiration causes evaporative cooling. When a forest is cut down, daytime temperatures rise and reduced transpiration results in less rainfall. Once a tropical forest is logged, the resulting nutrient losses and drier, hotter conditions can make it impossible for tree seeds to germinate or for seedlings to survive. Thus, deforestation can be difficult to reverse.

Deforestation also contributes to global climate change in two ways. First, trees that are cut to clear land for other uses are often burned, releasing carbon into the atmosphere. Second, conversion of forest to cropland or pasture decreases the rate at which carbon is taken up by plants.

desertification Conversion of a grassland or woodland to desert.

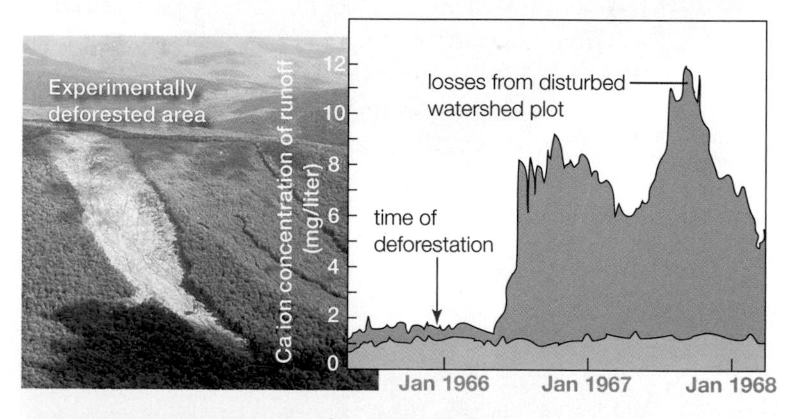

FIGURE 48.9 ▶**Animated** Effect of experimental deforestation on nutrient losses from soil. After deforestation of a portion of forest in the Hubbard Brook watershed, calcium (Ca) levels in runoff increased sixfold (gray). An undisturbed plot in the same forest showed no increase during this time (green).

TAKE-HOME MESSAGE 48.4

What are effects of desertification and deforestation?

✔ Desertification and deforestation not only destroy habitat, they allow increased erosion and can cause changes in local weather patterns.

CREDITS: (8) left, NASA courtesy of the MODIS Rapid Response Team at Goddard Space Flight Center; right, Courtesy of Guido van der Werf, Vrije Universiteit Amsterdam; (9) left, USDA Forest Service, Northeastern Research Station; right, © Cengage Learning.

48.5 Pollutants

✔ Organisms are directly harmed by chemicals and trash we release into the environment.

Pollutants are natural or synthetic substances released into soil, air, or water in greater than natural amounts. Some come from a few easily identifiable sites, or point sources. A factory that discharges pollutants into the air or water is a point source. Pollutants that come from point sources are the easiest to control: Identify the sources, and you can take action there. Dealing with pollution from nonpoint sources is more challenging. Such pollution stems from widespread release of a pollutant. For example, oil that runs off from roads and driveways is a nonpoint source of water pollution.

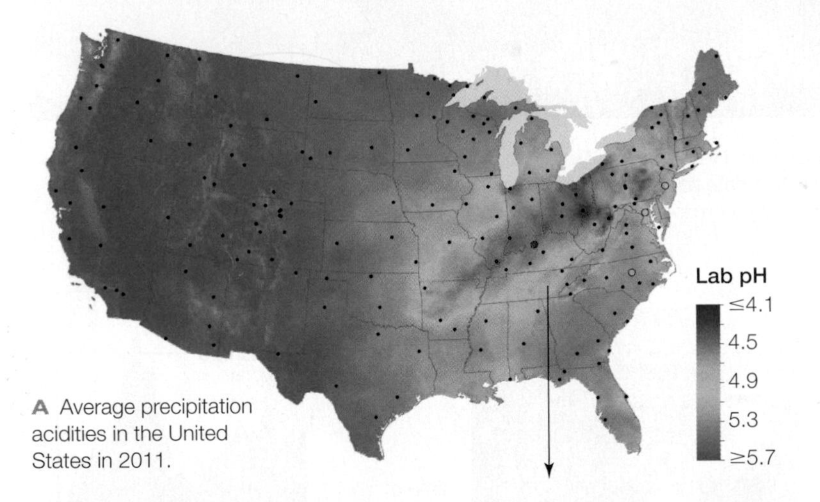

A Average precipitation acidities in the United States in 2011.

B Dying trees in Great Smoky Mountains National Park, where acid rain harms leaves and causes loss of nutrients from soil.

FIGURE 48.10 ▶**Animated** Acid rain. **FIGURE IT OUT** Is rain more acidic on the East Coast or the West Coast?

Answer: The East Coast

Acid Rain

Sulfur dioxides and nitrogen oxides are common air pollutants. Coal-burning power plants release most sulfur dioxides. Combustion of gasoline and oil releases most nitrogen oxides.

In dry weather, airborne sulfur and nitrogen oxides coat dust particles. Dry acid deposition occurs when this coated dust falls to the ground. Wet acid deposition, or **acid rain**, occurs when pollutants combine with water and fall as acidic precipitation. The pH of unpolluted rainwater is 5 or higher (Section 2.6). Acid rain can be ten times more acidic (**FIGURE 48.10A**).

Acid rain that falls on or drains into waterways, ponds, and lakes can harm aquatic organisms. A low pH prevents fish eggs from developing and kills adult fishes. When acid rain falls on a forest, it burns tree leaves and alters the forest's soil. As acidic water drains through the soil, positively charged hydrogen ions displace positively charged nutrient ions such as calcium, causing nutrient loss. The acidity also causes soil particles to release metals such as aluminum that can harm plants. The combination of poor nutrition and exposure to toxic metals weakens trees, making them more susceptible to insects and pathogens, and thus more likely to die. Effects are most pronounced at higher elevations where trees are frequently exposed to clouds of acidic droplets (**FIGURE 48.10B**).

Biological Accumulation and Magnification

Some pollutants can build up inside an organism's body. By the process of **bioaccumulation**, an organism's tissues store a pollutant taken up from the environment, causing the amount in the body to increase over time. The ability of some plants to bioaccumulate toxic substances makes them useful in phytoremediation of polluted soils (Section 28.1).

In animals, hydrophobic chemical pollutants ingested or absorbed across the skin accumulate in fatty tissues. The amount of pollutant in an animal's body increases over time, so long-lived species tend to have a higher amount of fat-soluble pollutants than shorter-lived ones. Within a species, old individuals have a higher pollutant load than younger ones.

Trophic level also affects the pollutant concentration in an organism's tissues. By the process of **biological magnification**, concentration of a chemical increases as the pollutant moves up a food chain. **FIGURE 48.11** provides data documenting biological magnification of DDT (an insecticide) in a salt marsh ecosystem during the 1960s. Notice that the concentration of DDT in ospreys, which are fish-eating birds, was 276,000 times higher than that in the water. As a result of bioaccumulation and biological magnification, even very low environmental concentrations of pollutants can have detrimental effects on a species.

Bioaccumulation and magnification of other pollutants such as methylmercury continue to pose a threat to wildlife and human health (Section 2.1).

Talking Trash

Seven billion people use and discard a lot of stuff. Historically, unwanted material was buried in the ground or dumped out at sea. Trash was out of sight, and also out of mind. We now know that burying garbage can contaminate groundwater, as when lead from discarded batteries seeps into aquifers. We also know that solid waste dumped into oceans harms marine life (**FIGURE 48.12**). In the United States, solid municipal waste can no longer legally be dumped at sea. Nevertheless, plastic constantly enters our coastal waters. Foam cups and containers from fast-food outlets, plastic shopping bags, plastic water bottles, and other litter ends up in storm drains. From there it enters streams and rivers that convey it to the sea.

Ocean currents can carry bits of plastic for thousands of miles. These plastic bits end up accumulating in certain areas of the ocean. Consider the Great Pacific Garbage Patch, a region of the north central Pacific that the media often describes as an "island of trash." In fact, the plastic is not easily visible. Rather, the garbage patch is a region where huge numbers of confetti-like plastic particles swirl slowly around an area as large as the state of Texas. The small bits of plastic absorb and concentrate toxic compounds such as pesticides and industrial chemicals from the seawater around them, making the plastic all the more harmful to marine organisms that mistakenly eat it.

DDT Residues
(In parts per million of wet weight of organism)

Osprey	13.8
Green heron	3.57
Atlantic needlefish	2.07
Summer flounder	1.28
Sheepshead minnow	0.94
Hard clam	0.48
Marsh grass	0.33
Flying insects (mostly flies)	0.30
Mud snail	0.26
Shrimps	0.16
Green alga	0.083
Water	0.00005

FIGURE 48.11 Biological magnification of DDT in an estuary on Long Island, New York, as reported in 1967 by George Woodwell, Charles Wurster, and Peter Isaacson.

FIGURE 48.12 Death by plastic. Scientists found more than 300 pieces of plastic in this albatross chick. One punctured the bird's gut, causing its death. The chick was fed plastic by its parents, who gathered the material from the ocean surface, mistaking it for food.

acid rain Low-pH rain formed when sulfur dioxide and nitrogen oxides mix with water vapor in the atmosphere.
bioaccumulation An organism accumulates increasing amounts of a chemical pollutant in its tissues over the course of its lifetime.
biological magnification A chemical pollutant becomes increasingly concentrated as it moves up through food chains.
pollutant Substance that is released as result of human activities and harms organisms or disrupts natural processes.

TAKE-HOME MESSAGE 48.5

What are some ways that pollutants directly harm living organisms?

✔ Acid rain can make a lake or soil too acidic for life to thrive.

✔ Even small amounts of fat-soluble chemicals can accumulate in tissues and be magnified from one trophic level to the next.

✔ Animals can be harmed when they mistake indigestible trash that enters waterways and oceans for food.

48.6 Ozone Depletion and Pollution

✔ Ozone is said to be "good up high, but bad nearby." It forms a protective sunscreen in Earth's upper atmosphere, but is a harmful pollutant in air near the ground.

Depletion of the Ozone Layer

In the upper atmosphere, between 17 and 27 kilometers (10.5 and 17 miles) above sea level, is a region of high ozone (O_3) concentration referred to as the **ozone layer**. The ozone layer benefits living organisms by absorbing most ultraviolet (UV) radiation from incoming sunlight. UV radiation causes mutations (Section 8.6).

In the mid-1970s, scientists noticed that Earth's ozone layer was thinning. Its thickness had always varied seasonally, but now the average level was declining steadily from year to year. By the mid-1980s, spring ozone thinning over Antarctica was so pronounced that people were referring to the lowest-ozone region as an "ozone hole" (**FIGURE 48.13A**).

Declining ozone quickly became an international concern. With a thinner ozone layer, people would be exposed to more UV radiation, the main cause of skin cancers. Higher UV levels also harm wildlife, which do not have the option of avoiding sunlight. In addition, exposure to higher than normal UV levels affects plants and other producers, slowing the rate of photosynthesis and the release of oxygen into the air.

Chlorofluorocarbons, or CFCs, are major ozone destroyers. These odorless gases were once widely used as propellants in aerosol cans, as coolants, and in solvents and plastic foam. In response to the threat posed by ozone thinning, countries worldwide agreed in 1987 to phase out production of CFCs and other ozone-destroying chemicals. As a result of that agreement (the Montreal Protocol), concentrations of CFCs in the atmosphere are no longer rising dramatically (**FIGURE 48.13B**). However, because CFCs break down slowly, scientists expect them to continue to significantly impair the ozone layer for several decades. Nitrous oxide from fossil fuels is another concern; it also breaks down ozone.

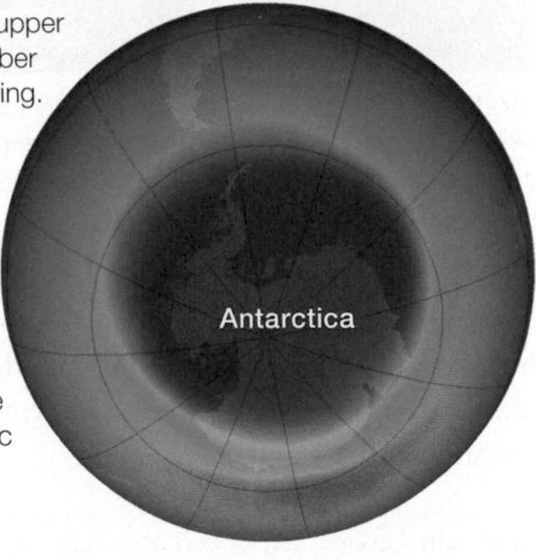

A Ozone levels in the upper atmosphere in September 2012, the Antarctic spring.

Purple indicates the least ozone, with blue, green, and yellow indicating increasingly higher levels.

Check the current status of the ozone hole at NASA's website (http://ozonewatch.gsfc.nasa.gov/).

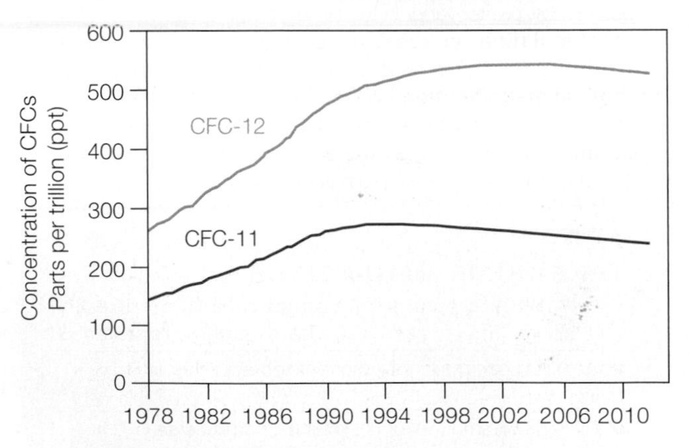

B Concentration of two CFCs in the upper atmosphere. These pollutants destroy ozone. A worldwide ban on CFCs has halted the rise in atmospheric CFC concentration.

FIGURE 48.13 ▶**Animated** Ozone and CFCs.

Near-Ground Ozone Pollution

Near the ground, where there is naturally little ozone, it is considered a pollutant. Ozone irritates the eyes and respiratory tracts of humans and wildlife. It also interferes with plant growth.

Ground-level ozone forms when nitrogen oxides and volatile organic compounds released by burning or evaporating fossil fuels are exposed to sunlight. Warm temperatures speed the reaction. Thus, ground-level ozone tends to vary daily (being higher in the daytime) and seasonally (being higher in the summer).

To help reduce ozone pollution, avoid putting fossil fuels or their combustion products into the air at times that favor ozone production. On hot, sunny, still days, postpone filling a gas tank or using gas-powered appliances until evening, when there is less sunlight to power conversion of pollutants to ozone.

ozone layer High atmospheric layer rich in ozone; prevents most ultraviolet radiation in sunlight from reaching Earth's surface.

TAKE-HOME MESSAGE 48.6

How do human activities affect ozone levels?

✔ Certain synthetic chemicals destroy ozone in the upper atmosphere's protective ozone layer. This layer serves as a protective shield against UV radiation.

✔ Evaporation and burning of fossil fuels increases the amount of ozone in the air near the ground, where ozone is considered a harmful pollutant.

CREDITS: (13A) NASA Ozone Watch; (13B) From www.esrl.noaa.gov.

48.7 Effects of Global Climate Change

✔ Climate change is the most widespread threat to habitats. Among other effects, it melts ice, causing sea level to rise.

Ongoing climate change affects ecosystems throughout the world. Most notably, average temperatures are increasing, especially at temperate and polar latitudes.

A rising temperature raises sea level. Water expands as it is heated, and heat also melts glaciers (**FIGURE 48.14**). Together, thermal expansion and addition of meltwater from glaciers cause sea level to rise. In the past century, sea level has risen about 20 centimeters (8 inches). As a result, some coastal wetlands are disappearing underwater.

The composition of seawater is changing too. In the past 50 years, the upper waters of the Atlantic Ocean have become less salty at the poles, where more meltwater is flowing in. During the same period, subtropical and tropical waters became saltier as warmer ocean temperatures and more intense trade winds encouraged evaporation. Altered rainfall patterns have also promoted salinity changes; annual precipitation has risen at high latitudes and fallen at low ones.

Seawater is also becoming more acidic, because when CO_2 dissolves in a solution, the solution's pH declines. Increased acidity can harm marine life by making less calcium carbonate available. Groups ranging from foraminifera, to corals, to bivalve mollusks require calcium carbonate to build their hard parts.

Warming temperature has widespread effects on biological systems. Many temperate zone species use temperature changes as cues. Warmer than normal springs are causing deciduous trees to leaf out earlier, and spring-blooming plants to flower earlier. Animal migration times and breeding seasons are also shifting. Species arrays in biological communities are changing as warmer temperatures allow some species to expand their range to higher latitudes or elevations. Of course, not all species can move or spread quickly, and warmer temperatures are expected to drive some of these species to extinction. For example, warming of tropical waters is already stressing reef-building corals and increasing the frequency of coral bleaching events.

The warming trend is also expected to affect human health. Deaths from heat stroke are likely to increase, and some infectious diseases will become more widespread. One way of predicting how warmer global temperatures will affect the incidence of disease is to look at what happens during an El Niño event, when sea temperature rises. As Section 47.4 explained, the incidence of cholera and malaria rises during an El Niño. In addition, elevated temperature and CO_2 concentration are expected to increase production of allergy-inducing plant pollen and fungal spores.

A 1941 photo of Muir Glacier in Alaska.

B 2004 photo of the same region.

FIGURE 48.14 Melting glaciers, one sign of a warming world. Water from melting glaciers contributes to rising sea level.

TAKE-HOME MESSAGE 48.7

What are some biological effects of climate change?

✔ Warming water, a rising sea level, and changes in the composition of seawater pose threats to aquatic species.

✔ On land, warming temperatures are altering the timing of flowering and migrations, and the distribution of some species.

✔ Increased warmth will also encourage the spread of some human pathogens and increase the production of allergy-inducing pollen by plants.

CREDITS: (14A) National Snow and Ice Data Center, W. O. Field; (14B) National Snow and Ice Data Center, B. F. Molnia.

48.8 Conservation Biology

✔ Conservation biologists survey and seek ways to protect the world's existing biodiversity.

The Value of Biodiversity

Every nation has several forms of wealth: material wealth, cultural wealth, and biological wealth. Biological wealth is called **biodiversity**. We measure a region's biodiversity at three levels: the genetic diversity within species, species diversity, and ecosystem diversity. Biodiversity is currently declining at all three levels, in all regions.

Conservation biology is the field that addresses these declines by surveying the range of biodiversity, and finding ways to maintain and use biodiversity to benefit human populations. The aim is to conserve biodiversity by encouraging people to value it and use it in ways that do not destroy it.

Why should we protect biodiversity? From a selfish standpoint, doing so is an investment in our future. Healthy ecosystems are essential to our survival. Other organisms produce the oxygen we breathe and the food we eat. They remove waste carbon dioxide from the air and decompose and detoxify wastes. Plants take up rain and hold soil in place, preventing erosion and reducing the risk of flooding. Compounds in wild species can serve as medicines. Wild relatives of crop plants are reservoirs of genetic diversity that plant breeders draw on to protect and improve crops.

Setting Priorities

Protecting biological diversity can pose a challenge. People often oppose environmental protections because they fear that such measures will have adverse economic consequences. However, taking care of the environment can make good economic sense. With a bit of planning, people can both preserve and profit from their biological wealth.

Conservation biologists help us make the difficult choices about which regions should be targeted for protection first. They identify **biodiversity hot spots**, places that have species found nowhere else and are under great threat of destruction. Once identified, hot spots take priority in worldwide conservation efforts.

On a broader scale, conservation biologists define ecoregions, which are land or aquatic regions characterized by climate, geography, and the species found within them. The most widely used ecoregion system was developed by conservation scientists of the World Wildlife Fund. These scientists defined 867 distinctive land ecoregions. **FIGURE 48.15** shows the locations and conservation status of ecoregions that are considered the top priority for conservation efforts.

The Klamath–Siskiyou forest in southwestern Oregon and northwestern California is one of North America's endangered ecoregions. It is home to many rare conifers. Two endangered birds, the marbled murrelet and the northern spotted owl, nest in old-growth

FIGURE 48.15 The location and conservation status of the land ecoregions deemed most important by the World Wildlife Fund.

parts of the forest (**FIGURE 48.16**), and endangered coho salmon breed in streams that run through the forest. Logging threatens all of these species.

By focusing on hot spots and critical ecoregions rather than on individual endangered species, scientists hope to maintain ecosystem processes that naturally sustain biological diversity.

Preservation and Restoration

Worldwide, many ecologically important regions have been protected in ways that benefit local people. The Monteverde Cloud Forest in Costa Rica is one example. During the 1970s, George Powell was studying birds in this forest, which was rapidly being cleared. Powell decided to buy part of the forest as a nature sanctuary. His efforts inspired individuals and conservation groups to donate funds, and much of the forest is now protected as a private nature reserve. The reserve's animals include more than 100 mammal species, 400 bird species, and 120 species of amphibians and reptiles. It is one of the few habitats left for jaguars and ocelots. A tourism industry centered on the reserve provides economic benefits to local people.

Sometimes, an ecosystem is so damaged, or there is so little of it left, that conservation alone is not enough to sustain biodiversity. **Ecological restoration** is work designed to bring about the renewal of a natural ecosystem that has been degraded or destroyed, fully or in part. Restoration work in Louisiana's coastal wetlands is an example. More than 40 percent of the coastal wetlands in the United States are in Louisiana. The marshes are an ecological and economic treasure, but they are troubled. Dams and levees built upstream of the marshes keep back sediments that would normally replenish sediments lost to the sea. Channels cut through the marshes for oil exploration and production have encouraged erosion, and the rising sea level threatens to flood the existing plants. Since the 1940s, Louisiana has lost an area of marshland the size of Rhode Island. Restoration efforts now under way aim to reverse some of those losses (**FIGURE 48.17**).

FIGURE 48.16 The Klamath–Siskiyou forest, one of North America's critical ecoregions. Endangered northern spotted owls (left) are endemic to this coniferous forest.

FIGURE 48.17 Ecological restoration in Louisiana's Sabine National Wildlife Refuge. In places where marshland has become open water, sediments are barged in and marsh grasses are planted on them. Squares are new sediment with marsh grass.

TAKE-HOME MESSAGE 48.8

How do we sustain biodiversity?

- ✔ Biodiversity is the genetic diversity of individuals of a species, the variety of species, and the variety of ecosystems.
- ✔ Conservation biologists identify threatened regions with high biodiversity and prioritize which should be first to receive protection.
- ✔ Through ecological restoration, we re-create or renew a biologically diverse ecosystem that has been destroyed or degraded.

biodiversity Of a region, the genetic variation within its species, variety of species, and variety of ecosystems.
biodiversity hot spot Threatened region with great biodiversity that is considered a high priority for conservation efforts.
conservation biology Field of applied biology that surveys and documents biodiversity, and seeks ways to maintain and use it.
ecological restoration Actively altering an area in an effort to restore or create a functional ecosystem.

CREDITS: (16) background, David Patte, USFWS; inset, USFWS; (17) Diane Borden Bilot, U.S. Fish and Wildlife Service.

48.9 Reducing Negative Impacts

✔ The cumulative effects of individual actions will determine the health of our planet.

Ultimately, the health of life on Earth depends on our ability to live within our limits. The goal is to live sustainably, which means meeting the needs of the present generation without reducing the ability of future generations to meet their own needs.

Promoting sustainability begins with recognizing the environmental consequences of one's lifestyle. People in industrial nations use huge quantities of resources, and the extraction of these resources has negative effects on biodiversity. In the United States, the size of the average family has declined since the 1950s, while the size of the average home has doubled. All of the materials used to build and furnish those larger homes must be extracted from the environment.

For example, an average new home contains about 500 pounds of copper in its wiring and plumbing. Where does that copper come from? Like most other mineral elements used in manufacturing, most copper is mined from the ground (**FIGURE 48.18**). Surface mining strips an area of vegetation and soil, creating an ecological dead zone. It also emits dust into the air, creates mountains of rocky waste, and can contaminate nearby waterways.

FIGURE 48.19 Volunteers restoring the Little Salmon River in Idaho so that salmon can migrate upstream to their breeding site.

Globalization makes it difficult to know the source of the raw materials in products you buy. Resource extraction in developing countries is often carried out under regulations that are less strict or less stringently enforced than those in the United States, so the environmental impact is even greater.

Nonrenewable mineral resources are also used in electronic devices such as phones, computers, televisions, and MP3 players. Reducing consumption by fix-

FIGURE 48.18 Resource extraction. Bingham copper mine near Salt Lake City, Utah. This open pit mine is 4 kilometers (2.5 miles) wide and 1,200 meters (0.75 miles) deep. It is the largest man-made excavation on Earth.

A Long Reach (revisited)

The Arctic is not a separate continent like Antarctica, but rather a region that encompasses the northernmost parts of several continents. Eight countries, including the United States, Canada, and Russia, control parts of the Arctic and have rights to its extensive oil, gas, and mineral deposits. Until recently, ice sheets covered the Arctic Ocean, making it difficult for ships to move to and from the Arctic landmass, but those sheets are breaking up as a result of global climate change (**FIGURE 48.20**). At the same time, ice that covered the Arctic landmass is melting. These changes will make it easier for people to remove minerals and fossil fuels from the Arctic. With the world supply of fuel and minerals dwindling, pressure to exploit Arctic resources is rising. However, conservationists warn that extracting these resources will harm Arctic species such as the polar bear that are already threatened by global climate change.

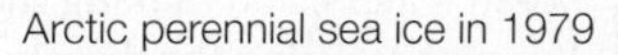

FIGURE 48.20 Declining Arctic perennial ice.

ing existing products is a sustainable resource use, as is recycling. Obtaining nonrenewable materials by recycling reduces the need for extraction of those resources, and it also keep materials out of landfills.

Reducing your energy use is another way to promote sustainability. Fossil fuels such as petroleum, natural gas, and coal supply most of the energy used by developed countries. You already know that burning these nonrenewable fuels contributes to global warming and acid rain. In addition, extracting and transporting these fuels have negative impacts. Oil harms species when it leaks from pipelines, ships, or wells. Similarly, nuclear power plants can have accidents that result in release of dangerous radioactive substances.

Renewable energy sources do not produce greenhouse gases, but they have drawbacks too. For example, dams in rivers of the Pacific Northwest generate renewable hydroelectric power, but they have an adverse effect on the endangered salmon that breed in these rivers. Similarly, wind turbines can harm birds and bats. Panels used to collect solar energy are made using nonrenewable mineral resources, and manufacturing the panels generates pollutants.

To minimize your impact, avoid wasting energy. Look for energy-efficient appliances and lightbulbs. Walking, bicycling, and using public transportation are energy-efficient alternatives to driving. To take a more active role, learn about the threats to ecosystems in your own area. Support efforts to preserve and restore local biodiversity. Many ecological restoration projects are supervised by trained biologists but carried out primarily through the efforts of volunteers (**FIGURE 48.19**). Keep in mind that unthinking actions of billions of individuals are the greatest threat to biodiversity. Each of us may have little impact, but our collective behavior, for good or for bad, will determine the future of the planet.

TAKE-HOME MESSAGE 48.9

What can individuals do to reduce their harmful impact on biodiversity?

✔ Resource extraction and usage have side effects that threaten biodiversity.

✔ You can save energy and other resources by reducing energy consumption and recycling and reusing materials.

CREDIT: (20) NASA.

summary

Section 48.1 Human activities affect even remote places such as the Arctic. Polar bears in the Arctic have pollutants in their bodies and are threatened by thinning sea ice, one effect of global climate change.

Section 48.2 The current rate of species loss is high enough to suggest that a **mass extinction** is under way. Unlike previous mass extinctions, which are attributed to physical causes such as an asteroid impact or a volcanic eruption, this one is caused by human actions and thus is preventable.

Arrival of humans in Australia and North America may have had a role in the extinction of the megafauna on these continents. Many more-recent extinctions certainly resulted from human activity.

Section 48.3 **Endangered species** currently face a high risk of extinction. **Threatened species** are likely to become endangered in the near future. **Endemic species**, which evolved in one place and are present only in that habitat, are highly vulnerable to extinction. Species with highly specialized needs are also especially vulnerable. Our knowledge of existing species is limited and biased toward vertebrates. We know little about the abundance and diversity of microbial species that carry out essential ecosystem processes.

Human activities cause habitat loss, degradation, and fragmentation that can endanger a species. Humans also directly reduce populations by overharvesting. In most cases, a species becomes endangered because of multiple factors. Sometimes, a decline in one species as a result of human activity leads to decline of another species.

Section 48.4 **Desertification** and deforestation dramatically alter a habitat and can even affect local climate. The changes caused by deforestation are especially difficult to reverse.

Section 48.5 Burning fossil fuels, especially coal, releases sulfur dioxide and nitrogen oxides into the atmosphere. These **pollutants** mix with water vapor in the air, then fall to earth as **acid rain**. The resulting increase in the acidity of soils and waters can sicken or kill organisms. With **bioaccumulation**, a pollutant taken up from the environment becomes stored in an organism's tissues, so older animals have a higher pollutant concentration than younger ones. With **biological magnification**, a pollutant increases in concentration as it is passed up a food chain. As a result of bioaccumulation and biological magnification, the tissues of an organism can have a far higher concentration of a pollutant than the environment.

Plastic that washes into or is dumped into oceans is another type of pollutant. It poses a threat to organisms that mistake it for food.

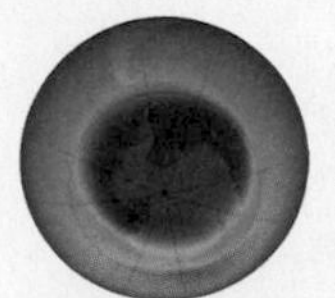

Section 48.6 The **ozone layer** in the upper atmosphere protects against incoming UV radiation. Chemicals called CFCs were banned when they were found to cause thinning of the ozone layer. Near the ground, where the ozone concentration is naturally low, ozone emitted as a result of fossil fuel use is considered a pollutant. It irritates animal respiratory tracts and interferes with photosynthesis by plants.

Section 48.7 Global climate change is raising the sea level, altering the salinity of upper ocean waters, and making the sea more acidic. It is also affecting the distribution and behavior of terrestrial species, allowing some to move into higher elevations or latitudes and altering the timing of migrations and flowering. Global climate change is also expected to have negative effects on human health by increasing heat-related deaths and encouraging outbreaks of some infectious diseases.

Section 48.8 We recognize three levels of **biodiversity**: genetic diversity, species diversity, and ecosystem diversity. All are threatened. The field of **conservation biology** surveys the range of biodiversity, investigates its origins, and identifies ways to maintain and use it in ways that benefit human populations.

Given that resources are limited, biologists attempt to identify **biodiversity hot spots**, regions rich in endemic species and under a high level of threat. Biologists also identify ecoregions, larger regions characterized by their physical characteristics as well as the species in them. The biologists prioritize ecoregions, with the goal of identifying those whose conservation will ensure that a representative sample of all of Earth's current biomes remains intact. When an ecosystem has been totally or partially degraded, **ecological restoration** can help restore biodiversity.

Section 48.9 Individuals can help sustain biodiversity by limiting their energy use and material consumption. Reuse and recycling can help protect species by reducing the use of destructive practices that are required to extract resources.

self-quiz

Answers in Appendix VII

1. True or false? Most species that evolved have already become extinct.

2. Dodos were driven to extinction ________ .
   a. when humans arrived in North America
   b. by overharvesting and introduced species
   c. as a result of global warming
   d. by volcanic eruptions

data analysis activities

Pervasive PCBs Winds carry chemical contaminants produced and released at temperate latitudes to the Arctic, where the chemicals enter food webs. By the process of biological magnification (Section 48.5), top carnivores in arctic food webs—such as polar bears and people—end up with high doses of these chemicals. For example, indigenous arctic people who eat the local wildlife tend to have unusually high levels of hormone-disrupting polychlorinated biphenyls (PCBs) in their bodies. The Arctic Monitoring and Assessment Programme studies the effects of these chemicals on health and reproduction. **FIGURE 48.21** shows the effect of PCBs on the sex ratio at birth in indigenous human populations in the Russian Arctic.

1. Which sex is was most common in offspring of women with less than one microgram per milliliter of PCB in serum?

2. At what PCB concentrations were women more likely to have daughters?

3. In some Greenland villages, nearly all recent newborns are female. Would you expect PCB levels in those villages to be above or under 4 micrograms per milliliter?

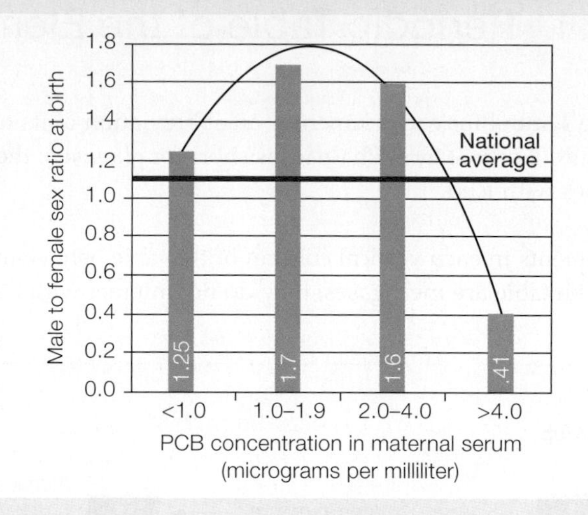

FIGURE 48.21 Effect of maternal PCB concentration on the sex ratio of newborns in indigenous populations in the Russian Arctic. The red line indicates the average sex ratio for births in Russia—1.06 males per female.

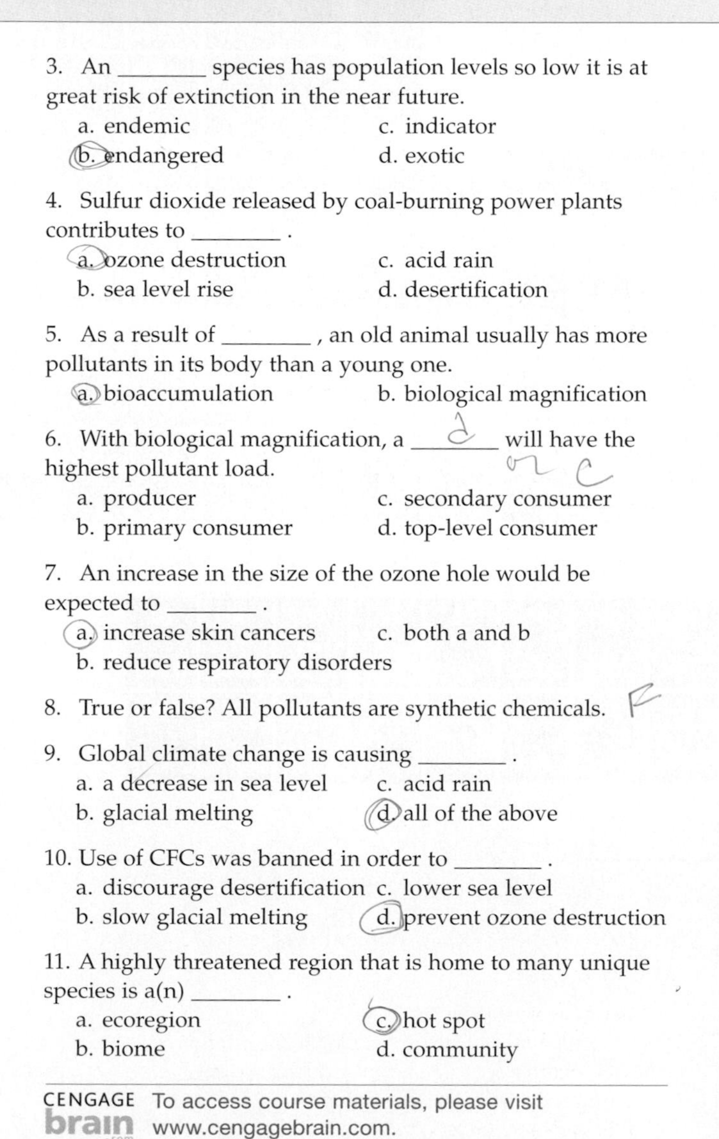

3. An _______ species has population levels so low it is at great risk of extinction in the near future.
   a. endemic
   b. endangered
   c. indicator
   d. exotic

4. Sulfur dioxide released by coal-burning power plants contributes to _______ .
   a. ozone destruction
   b. sea level rise
   c. acid rain
   d. desertification

5. As a result of _______ , an old animal usually has more pollutants in its body than a young one.
   a. bioaccumulation
   b. biological magnification

6. With biological magnification, a _______ will have the highest pollutant load.
   a. producer
   b. primary consumer
   c. secondary consumer
   d. top-level consumer

7. An increase in the size of the ozone hole would be expected to _______ .
   a. increase skin cancers
   b. reduce respiratory disorders
   c. both a and b

8. True or false? All pollutants are synthetic chemicals.

9. Global climate change is causing _______ .
   a. a decrease in sea level
   b. glacial melting
   c. acid rain
   d. all of the above

10. Use of CFCs was banned in order to _______ .
    a. discourage desertification
    b. slow glacial melting
    c. lower sea level
    d. prevent ozone destruction

11. A highly threatened region that is home to many unique species is a(n) _______ .
    a. ecoregion
    b. biome
    c. hot spot
    d. community

CENGAGE brain To access course materials, please visit www.cengagebrain.com.

12. Biodiversity refers to _______ .
    a. genetic diversity
    b. species diversity
    c. ecosystem diversity
    d. all of the above

13. Restoring a marsh that has been damaged by human activities is an example of _______ .
    a. biological magnification
    b. bioaccumulation
    c. ecological restoration
    d. globalization

14. Individuals help sustain biodiversity by _______ .
    a. reducing consumption
    b. reusing materials
    c. recycling materials
    d. all of the above

15. Match the terms with the most suitable description.

___ ozone	a. tree loss alters rainfall pattern and is difficult to reverse
___ biodiversity	b. can increase dust storms
___ acid rain	c. evolved in one region and remains there
___ endemic species	d. coal-burning is a major cause
___ biological magnification	e. results in highest pollution level at top trophic level
___ global climate change	f. good up high, bad nearby
___ deforestation	g. cause of rising sea level
___ desertification	h. genetic, species, and ecosystem diversity

critical thinking

1. Section 34.11 explained how melatonin controls the timing of a hare's coat color change. With this information in mind, explain why increasingly early snow melts in the arctic pose a threat to arctic hares by raising their risk of predation.

2. Two arctic marine mammals that live in the same waters differ in the level of pollutants in their bodies. Bowhead whales have a lower pollutant load than ringed seals. What are some factors that might explain this difference?

Appendix I. Periodic Table of the Elements

The symbol for each element is an abbreviation of its name. Some symbols for elements are abbreviations for their Latin names. For instance, Pb (lead) is short for *plumbum*; the word "plumbing" is related—ancient Romans made their water pipes with lead.

Elements in each vertical column of the table behave in similar ways. For instance, all of the elements in the far right column of the table are inert gases; they do not interact with other atoms. In nature, such elements occur only as solitary atoms.

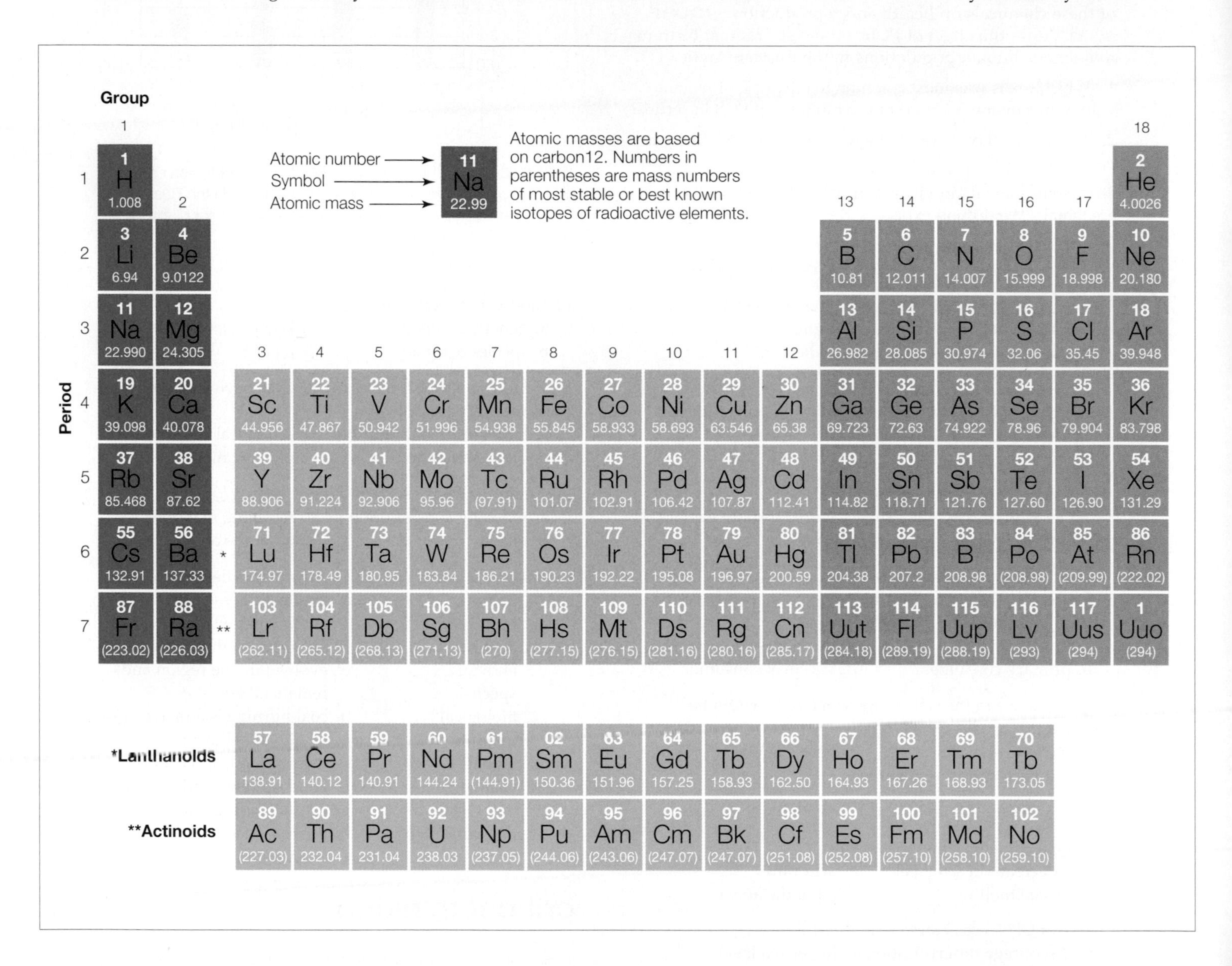

Atomic number → 11; Symbol → Na; Atomic mass → 22.99

Atomic masses are based on carbon12. Numbers in parentheses are mass numbers of most stable or best known isotopes of radioactive elements.

Period \ Group	1	2	3	4	5	6	7	8	9	10	11	12	13	14	15	16	17	18
1	1 H 1.008																	2 He 4.0026
2	3 Li 6.94	4 Be 9.0122											5 B 10.81	6 C 12.011	7 N 14.007	8 O 15.999	9 F 18.998	10 Ne 20.180
3	11 Na 22.990	12 Mg 24.305											13 Al 26.982	14 Si 28.085	15 P 30.974	16 S 32.06	17 Cl 35.45	18 Ar 39.948
4	19 K 39.098	20 Ca 40.078	21 Sc 44.956	22 Ti 47.867	23 V 50.942	24 Cr 51.996	25 Mn 54.938	26 Fe 55.845	27 Co 58.933	28 Ni 58.693	29 Cu 63.546	30 Zn 65.38	31 Ga 69.723	32 Ge 72.63	33 As 74.922	34 Se 78.96	35 Br 79.904	36 Kr 83.798
5	37 Rb 85.468	38 Sr 87.62	39 Y 88.906	40 Zr 91.224	41 Nb 92.906	42 Mo 95.96	43 Tc (97.91)	44 Ru 101.07	45 Rh 102.91	46 Pd 106.42	47 Ag 107.87	48 Cd 112.41	49 In 114.82	50 Sn 118.71	51 Sb 121.76	52 Te 127.60	53 I 126.90	54 Xe 131.29
6	55 Cs 132.91	56 Ba 137.33 *	71 Lu 174.97	72 Hf 178.49	73 Ta 180.95	74 W 183.84	75 Re 186.21	76 Os 190.23	77 Ir 192.22	78 Pt 195.08	79 Au 196.97	80 Hg 200.59	81 Tl 204.38	82 Pb 207.2	83 B 208.98	84 Po (208.98)	85 At (209.99)	86 Rn (222.02)
7	87 Fr (223.02)	88 Ra (226.03) **	103 Lr (262.11)	104 Rf (265.12)	105 Db (268.13)	106 Sg (271.13)	107 Bh (270)	108 Hs (277.15)	109 Mt (276.15)	110 Ds (281.16)	111 Rg (280.16)	112 Cn (285.17)	113 Uut (284.18)	114 Fl (289.19)	115 Uup (288.19)	116 Lv (293)	117 Uus (294)	1 Uuo (294)

*Lanthanoids	57 La 138.91	58 Ce 140.12	59 Pr 140.91	60 Nd 144.24	61 Pm (144.91)	02 Sm 150.36	63 Eu 151.96	64 Gd 157.25	65 Tb 158.93	66 Dy 162.50	67 Ho 164.93	68 Er 167.26	69 Tm 168.93	70 Tb 173.05
**Actinoids	89 Ac (227.03)	90 Th 232.04	91 Pa 231.04	92 U 238.03	93 Np (237.05)	94 Pu (244.06)	95 Am (243.06)	96 Cm (247.07)	97 Bk (247.07)	98 Cf (251.08)	99 Es (252.08)	100 Fm (257.10)	101 Md (258.10)	102 No (259.10)

Appendix II. The Amino Acids

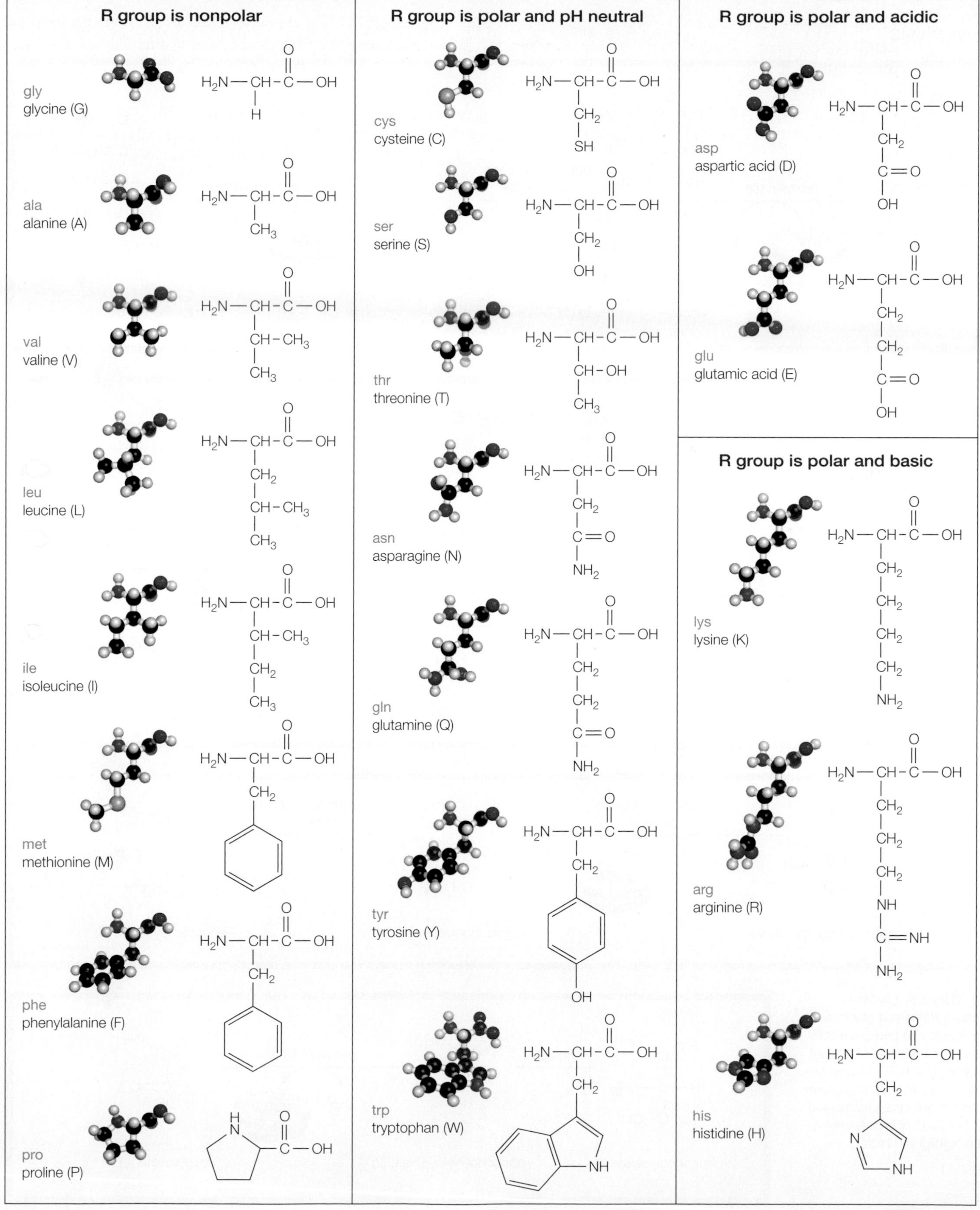

Appendix III. A Closer Look at Some Major Metabolic Pathways

Glycolysis

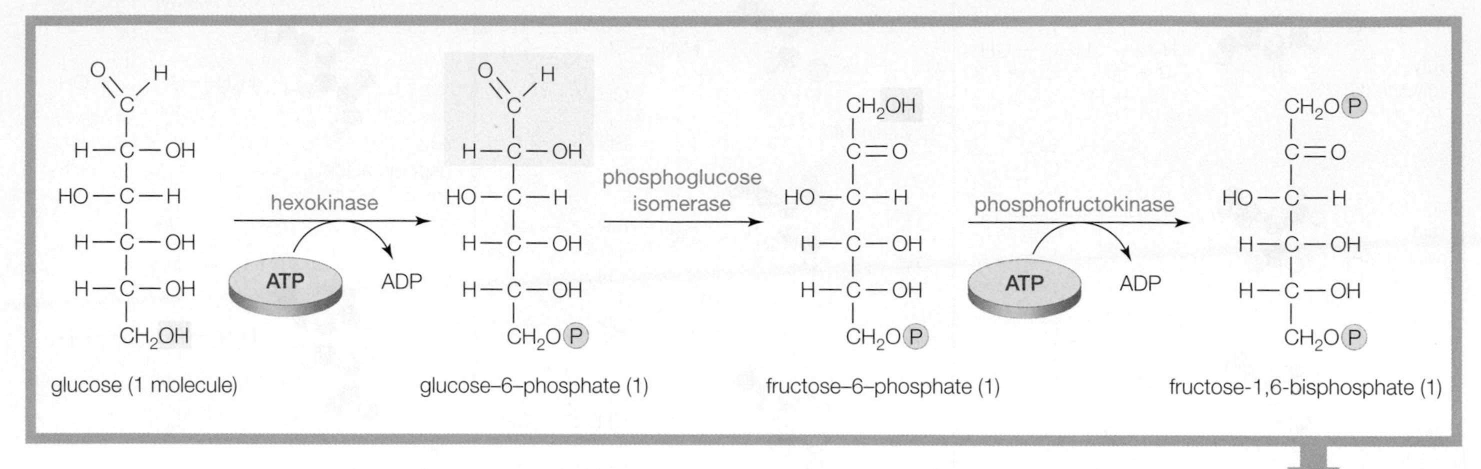

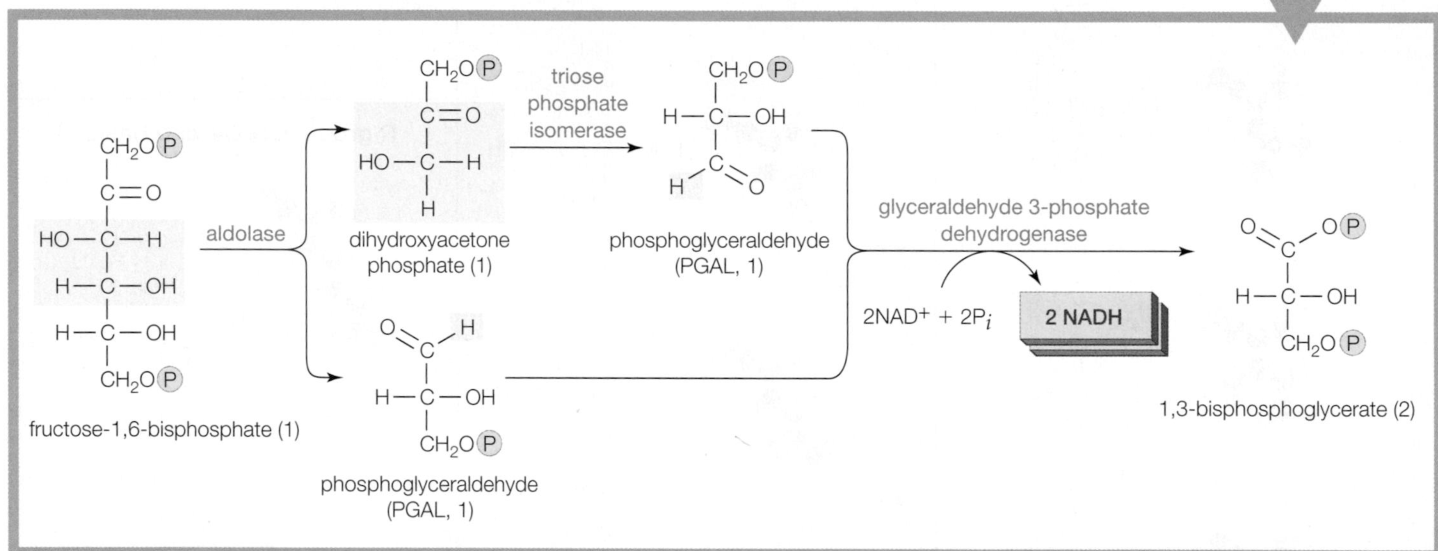

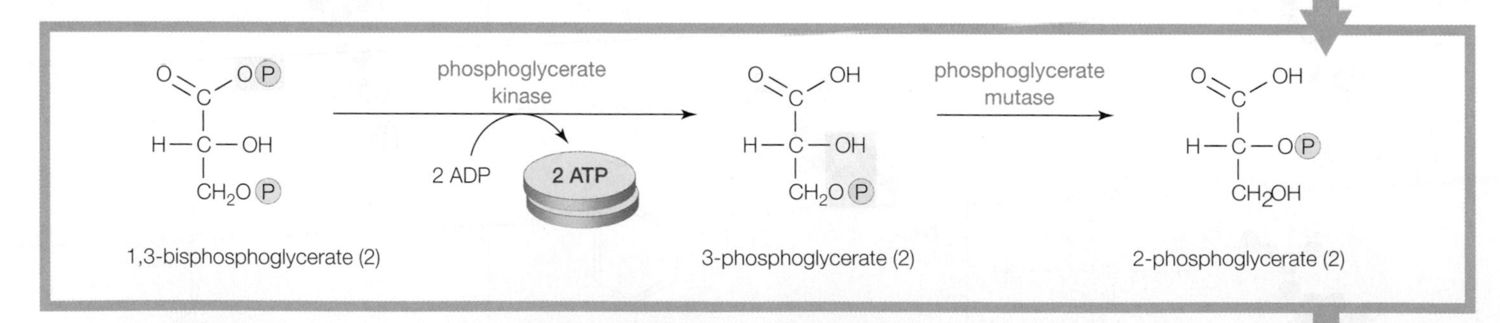

FIGURE A Glycolysis breaks down one glucose molecule into two 3-carbon pyruvate molecules for a net yield of two ATP. Enzyme names are indicated in green; parts of substrate molecules undergoing chemical change are highlighted blue.

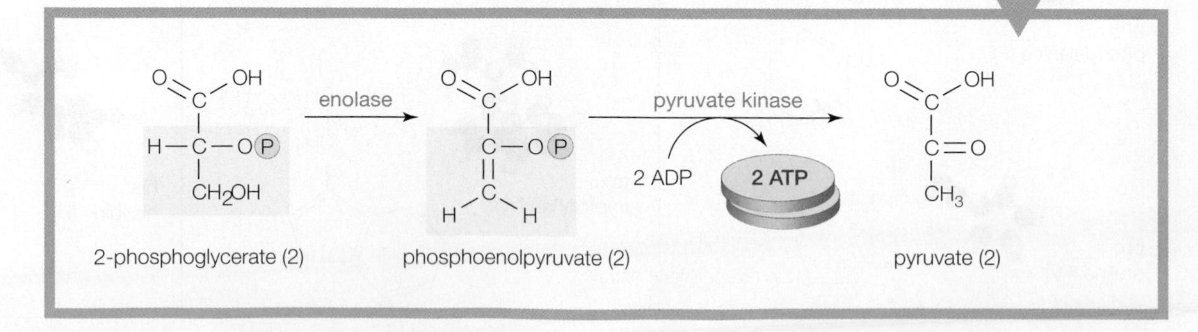

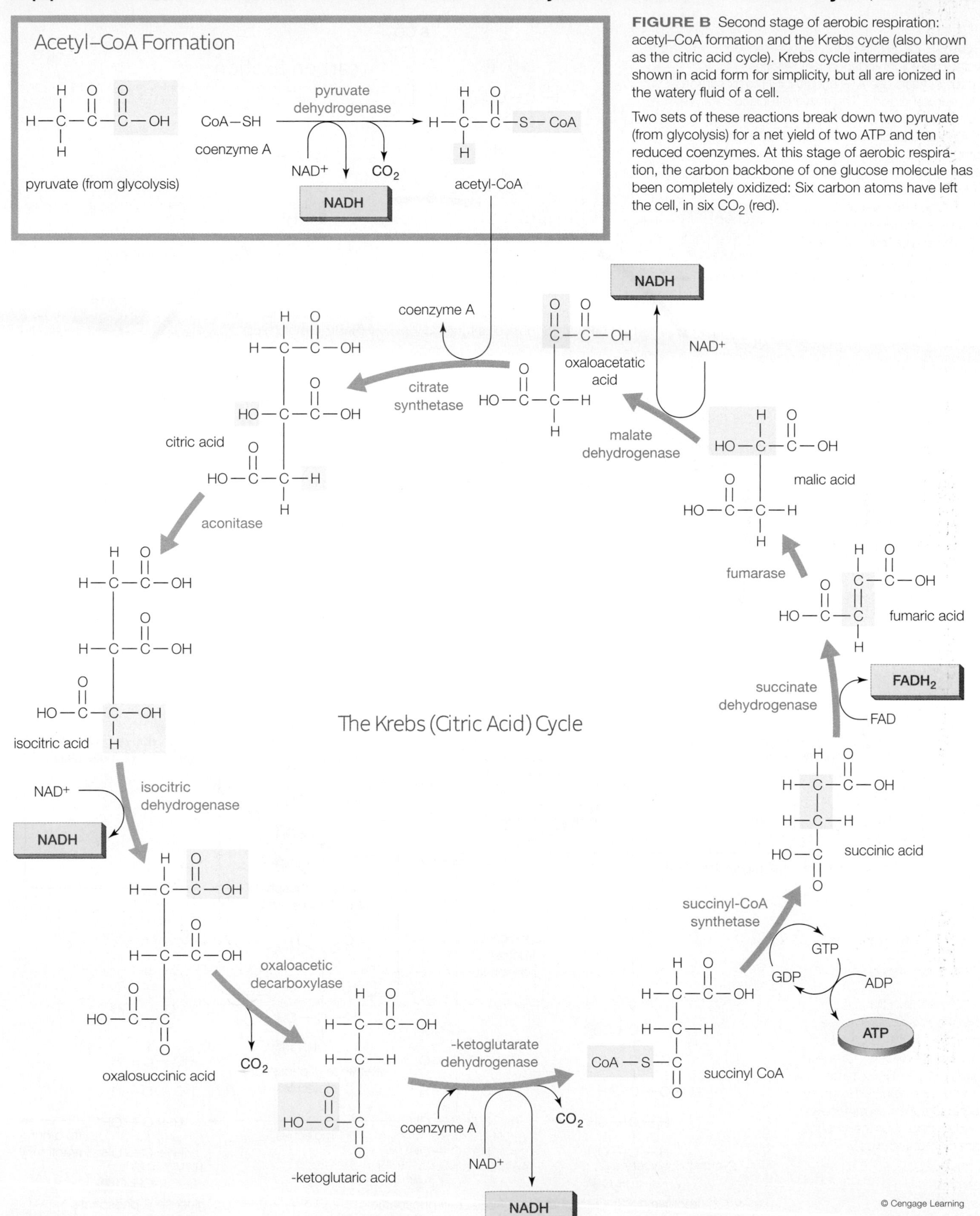

FIGURE B Second stage of aerobic respiration: acetyl–CoA formation and the Krebs cycle (also known as the citric acid cycle). Krebs cycle intermediates are shown in acid form for simplicity, but all are ionized in the watery fluid of a cell.

Two sets of these reactions break down two pyruvate (from glycolysis) for a net yield of two ATP and ten reduced coenzymes. At this stage of aerobic respiration, the carbon backbone of one glucose molecule has been completely oxidized: Six carbon atoms have left the cell, in six CO_2 (red).

FIGURE C Details of the Calvin–Benson cycle. These light-independent reactions of photosynthesis use ATP and NADPH to fix carbon from carbon dioxide. The enzyme rubisco catalyzes the attachment of CO_2 to RuBP. The resulting PGA molecules are converted to PGAL, and the complex series of reactions that follow shuffle carbon atoms among sugar molecules to regenerate RuBP. One molecule of glucose is produced for six CO_2 molecules that enter the reactions. Water and some of the molecular participants are not shown, for clarity.

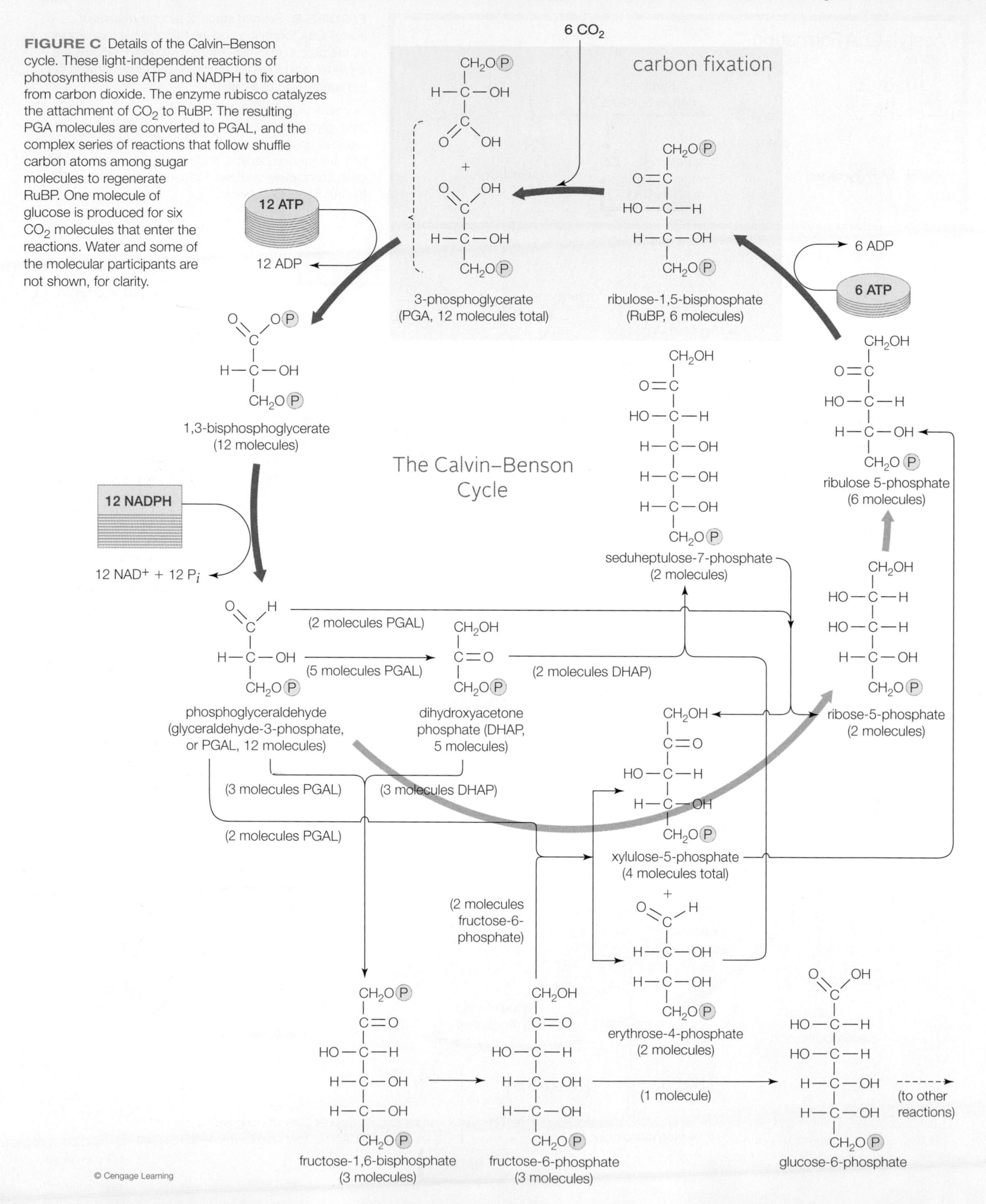

Appendix IV. A Plain English Map of the Human Chromosomes

Haploid set of human chromosomes. The banding patterns characteristic of each type of chromosome appear after staining with a reagent called Giemsa. The locations of some of the 20,065 known genes (as of November, 2005) are indicated. Also shown are locations that, when mutated, cause some of the genetic diseases discussed in the text.

Appendix V. Restless Earth—Life's Changing Geologic Stage

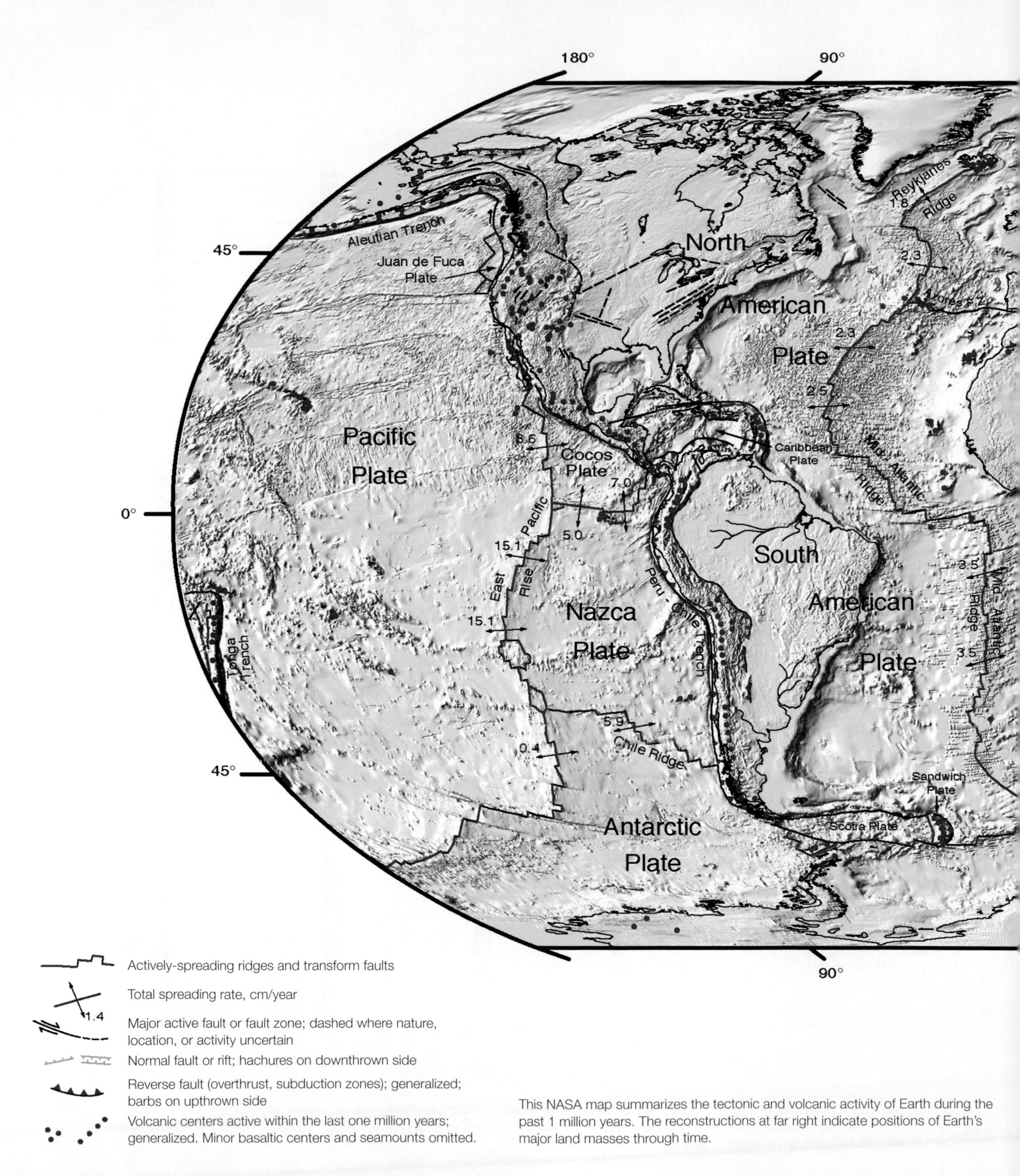

This NASA map summarizes the tectonic and volcanic activity of Earth during the past 1 million years. The reconstructions at far right indicate positions of Earth's major land masses through time.

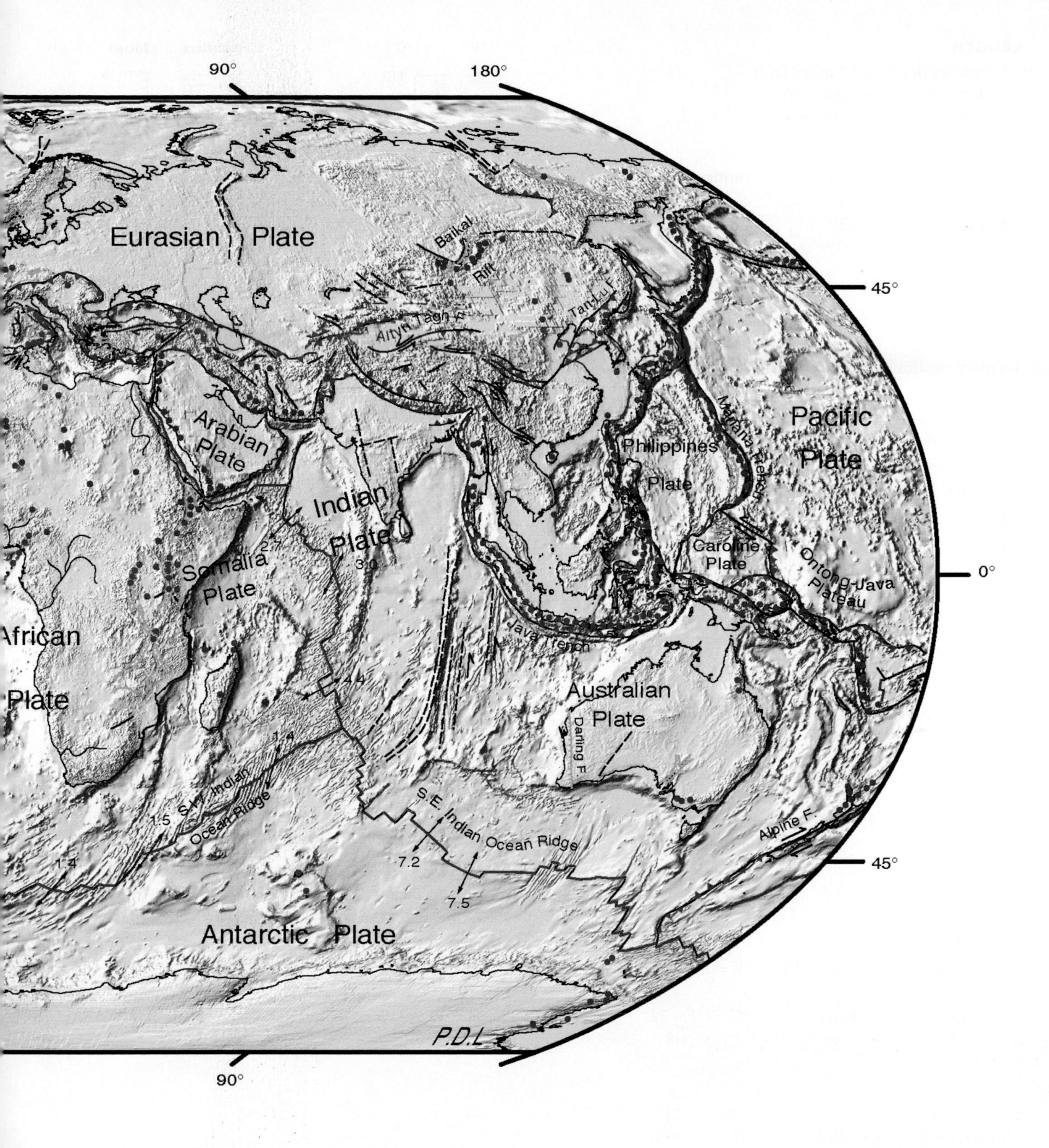
90°
180°
Eurasian Plate
Baikal Rift
Tan-Lu F.
Altyn Tagh F.
45°
Arabian Plate
Pacific Plate
Philippines Plate
Mariana Trench
Indian Plate
2.7
3.0
Somalia Plate
Caroline Plate
Ontong-Java Plateau
0°
African Plate
Java Trench
4.4
Australian Plate
Darling F.
1.4
1.5
S.W. Indian Ocean Ridge
S.E. Indian Ocean Ridge
7.2
7.5
Alpine F.
45°
1.4
Antarctic Plate
P.D.L.
90°

Appendix VI. Units of Measure

LENGTH

1 kilometer (km) = 0.62 miles (mi)
1 meter (m) = 39.37 inches (in)
1 centimeter (cm) = 0.39 inches

To convert	*multiply by*	*to obtain*
inches	2.25	centimeters
feet	30.48	centimeters
centimeters	0.39	inches
millimeters	0.039	inches

AREA

1 square kilometer = 0.386 square miles
1 square meter = 1.196 square yards
1 square centimeter = 0.155 square inches

VOLUME

1 cubic meter = 35.31 cubic feet
1 liter = 1.06 quarts
1 milliliter = 0.034 fluid ounces = 1/5 teaspoon

To convert	*multiply by*	*to obtain*
quarts	0.95	liters
fluid ounces	28.41	milliliters
liters	1.06	quarts
milliliters	0.03	fluid ounces

WEIGHT

1 metric ton (mt) = 2,205 pounds (lb) = 1.1 tons (t)
1 kilogram (kg) = 2.205 pounds (lb)
1 gram (g) = 0.035 ounces (oz)

To convert	*multiply by*	*to obtain*
pounds	0.454	kilograms
pounds	454	grams
ounces	28.35	grams
kilograms	2.205	pounds
grams	0.035	ounces

TEMPERATURE

Celsius (°C) to Fahrenheit (°F): $°F = 1.8\,(°C) + 32$

Fahrenheit (°F) to Celsius: $°C = \frac{(°F - 32)}{1.8}$

	°C	°F
Water boils	100	212
Human body temperature	37	98.6
Water freezes	0	32

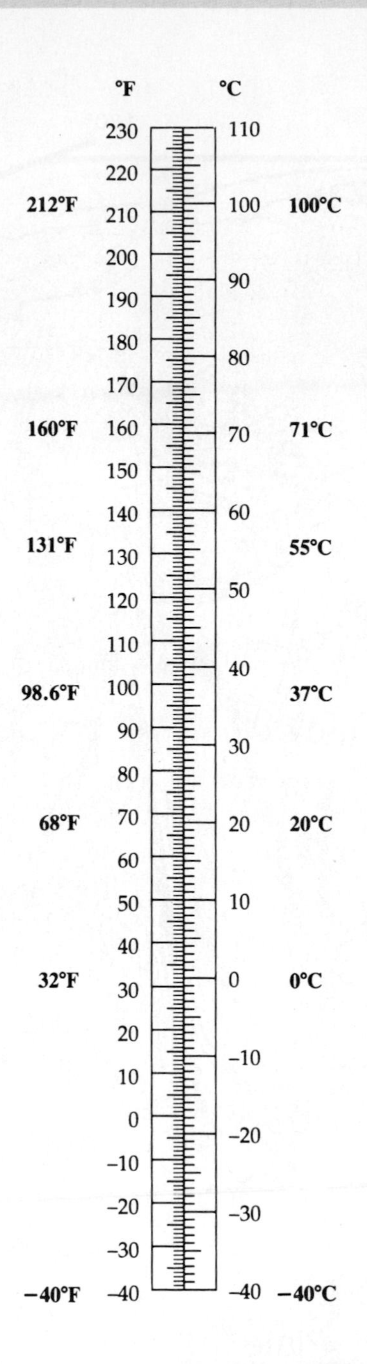

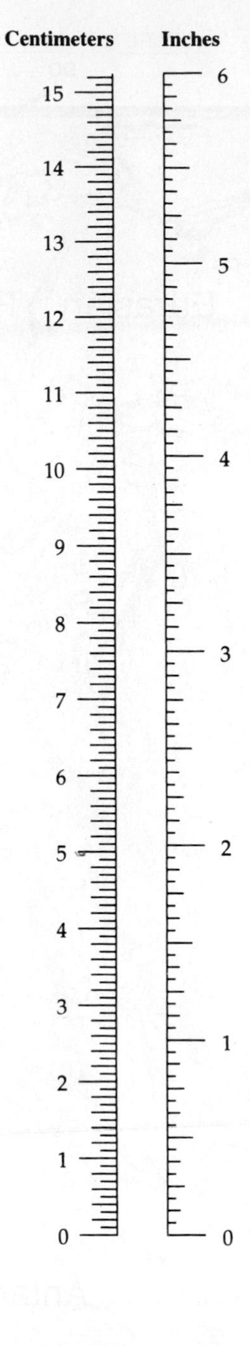

Appendix VII. Answers to Self-Quizzes and Genetics Problems

CHAPTER 1

1. a 1.2
2. c 1.2
3. c 1.3
4. homeostasis 1.3
5. d 1.3
6. a 1.3
7. d 1.3
8. a, d, e 1.2–1.5
9. Animals 1.4
10. a, b 1.2, 1.4
11. domains 1.5
12. b 1.6
13. b 1.8
14. b 1.9
15. c 1.2
e 1.8
b 1.5
d 1.6
a 1.6
f 1.3

CHAPTER 2

1. a 2.3
2. b 2.2
3. d 2.3
4. d 2.2
5. b 2.3
6. a 2.4
7. a 2.4
8. c 2.4
9. c 2.5
10. c 2.5
11. d 2.3, 2.6
12. a 2.6
13. c 2.6
14. b 2.5
15. c 2.5
b 2.2
d 2.5
a 2.2–2.4
f 2.5
e 2.2–2.4

CHAPTER 3

1. c 3.2
2. four 3.2
3. b 3.2, 3.4, 3.5
4. e 3.4, 3.8
5. a 3.4
6. c 3.5
7. False 3.1, 3.5
8. b 3.5
9. e 3.5
10. d 3.6, 3.8
11. d 3.7
12. d 3.8
13. a: amino acid 3.6
b: carbohydrate 3.4
c: polypeptide 3.6
d: fatty acid 3.5
14. c 3.5
e 3.4
f 3.5
d 3.8
a 3.6
b 3.8
15. g 3.6
a 3.5
b 3.6
c 3.5
d 3.8
j 3.5
f 3.8
i 3.6
h 3.4
e 3.4

CHAPTER 4

1. c 4.2
2. any of the following are correct: all organisms consist of one or more cells; or the cell is the smallest unit of life; or each new cell arises from another, preexisting cell; or a cell passes hereditary material to its offspring. 4.2
3. b 4.2
4. False 4.2
5. b 4.2
6. False 4.4
7. c 4.2, 4.4, 4.5
8. a 4.7
9. c, b, d, a 4.7
10. a 4.7
11. False (some cells have a cell wall and/or other ECM) 4.11
12. plasmodesmata 4.11
13. b 4.7, 4.8, 4.9
14. d 4.11
15. h 4.8
g 4.9
b 4.11
f 4.7
e 4.7
a 4.7
d 4.7
c 4.9
i 4.4, 4.10

CHAPTER 5

1. b 5.2
2. c 5.2
3. d 5.2
4. a 5.3
5. c 5.3
6. c 5.4, 5.6
7. temperature, pH, salt, pressure 5.4
8. d 5.2, 5.5, 5.6
9. c 5.5
10. a 5.6
11. more/less 5.8
12. c 5.7–5.9
13. b 5.9
14. a 5.8
15. d 5.10
16. c 5.3
e 5.10
f 5.2
b 5.3
a 5.6
g 5.8
h 5.9
d 5.9
k 5.5
j 5.5
i 5.7

CHAPTER 6

1. weed=autotroph, all others are heterotrophs 6.1
2. a 6.1
3. b 6.1, revisited
4. c 6.2
5. a 6.4
6. d 6.5
7. b 6.4, 6.5
8. b 6.5
9. c 6.5
10. c 6.4, 6.6
11. b 6.6
12. e 6.6
13. a 6.7
14. c 6.7
15. f 6.6
h 6.6
g 6.5
d 6.5
e 6.7
b 6.2, 6.5
a 6.2
c 6.1

CHAPTER 7

1. False 7.2
2. d 7.2, 7.3
3. a 7.2, 7.6
4. c 7.3
5. b 7.2, 7.4, 7.5
6. e 7.4
7. b 7.4
8. c 7.5
9. c 7.6
10. b 7.6
11. c 7.5
12. d 7.7
13. f 7.6
14. b 7.3
c 7.6
a 7.4, 7.6
d 7.5
15. b 7.4
d 7.3
a 7.3, 7.6
c 7.2, 7.5
e 7.2–7.4
f 7.1

CHAPTER 8

1. c 8.3
2. c 8.3
3. b 8.3
4. b 8.3
5. a 8.4
6. b 8.4
7. b 8.4
8. a 8.5
9. d 8.5
10. c 8.5
11. b 8.5
12. d 8.4, 8.5
13. d 8.6
14. d 8.7
15. d 8.2
b 8.1
a 8.3
g 8.4
e 8.5
h 8.5
c 8.4
f 8.6

CHAPTER 9

1. c 9.2
2. b 9.3
3. a 9.2
4. c 9.2
5. a 9.2
6. b 9.2, 9.4
7. c 9.4
8. b 9.3
9. a 9.4
10. a 9.4
11. a 9.3
12. a 9.3, 9.5
13. b 9.5
14. c 9.5
15. c 9.4
b 9.3
e 9.5
a 9.3
f 9.4
d 9.3

CHAPTER 10

1. d 10.2
2. d 10.2, 10.3
3. b 10.2
4. b 10.2
5. h 10.2
6. b 10.2
7. c 10.3
8. c 10.3
9. d 10.3
10. b 10.3
11. b 10.4
12. b 10.4
13. c 10.4
14. b 10.5
15. a 10.4
b 10.2, 10.5
d 10.4
i 10.2
f 10.2
c 10.6
h 10.3
g 10.3
e 10.5

CHAPTER 11

1. d 11.2
2. two 11.2
3. d 11.2
4. e 11.2
5. c 11.2, 11.3
6. c 11.2
7. a 11.2
8. c 11.2
9. a 11.2
10. d 11.4
11. d 11.6
12. a 11.6
13. c 11.4
f 11.3
a 11.6
g 11.4
b 11.4
e 11.6
d 11.3
h 11.5
14. d 11.3
b 11.3
c 11.3
e 11.2
a 11.3
f 11.4

CHAPTER 12

1. b 12.1
2 b 12.2
3. c 12.1, 12.4
4. d 12.2
5. b 12.2
6. a 12.2
7. Sister chromatids are still attached 12.3
8. b 12.2
9. c 12.3
10. b 12.4
11. a 12.4, 12.5
12. b 12.4, 12.5
13. c 12.2, 12.3
d 12.3
a 12.2
f 12.2
e 12.2
b 12.1
g 12.3, 12.4
14. e 12.4

CHAPTER 13

1. b 13.2
2. a 13.2
3. b 13.4
4. b 13.3
5. c 13.3
6. a 13.3
7. b 13.3
8. d 13.4
9. c 13.4
10. False 13.5
11. c 13.5
12. b 13.5
13. d 13.7
14. b 13.4
d 13.3
a 13.2
c 13.2

CHAPTER 14

1. b 14.2
2. b 14.2
3. a 14.2
4. b 14.3
5. False 14.4
6. d 14.3, 14.4 (could be due to both parents carrying an autosomal recessive allele, or the mom carrying an x-linked recessive allele, or a new mutation)
7. d 14.4
8. X from mom, Y from dad 14.4
9. d 14.3
10. Y-linked 14.4
11. d 14.6
12. b 14.6
13. True 14.6
14. c 14.6

15. c	14.6
e	14.5
f	14.6
b	14.5
a	14.2
d	14.5

CHAPTER 15

1. c	15.2
2. a	15.2
3. b	15.2
4. b	15.3
5. c	15.3
6. b	15.4
7. b	15.4
8. d	15.3, 15.5
9. d	15.5
10. c	15.6
11. True	15.7–15.9
12. b	15.8, 15.10
d	15.2, 15.7
e	15.10
f	15.7
h	15.10
13. True	15.10
14. a	15.3
d	15.2
c	15.2
e	15.4
b	15.2
15. c	15.5
f	15.7
d	15.10
b	15.1
a	15.6
e	15.6

CHAPTER 16

1. b	16.2
2. c	16.2
3. d	16.3, 16.4
4. b	16.4
5. b	16.5
6. e	16.5
7. d	16.6
8. Gondwana	16.7
9. 66	16.8
10. a	16.8
11. a	16.4
d	16.2, 16.5
e	16.4
f	16.6
c	16.3
b	16.3
12. a	16.2–16.4
c	16.7
d	16.7
e	16.5
f	16.1

CHAPTER 17

1. a	17.2
2. d	17.2, 17.8
3. a	17.6
b	17.6
4. d	17.7
5. c	17.7
6. b	17.8
7. f	17.2, 17.3, 17.8
8. a	17.9
9. d	17.7, 17.9
10. c	17.11
11. a	17.10
12. c	17.8
d	17.7
e	17.2
b	17.8
a	17.12
f	17.12
13. d	17.2, 17.12

CHAPTER 18

1. c	18.2
2. c	18.2
3. d	18.2
4. c	18.2
5. b	18.2
6. d	18.3
7. d	18.3
8. c	18.4
9. b	18.4
10. b	18.4
11. True	18.4
12. b	18.5
13. e	18.2–18.5
14. True	18.6
15. b	18.2
g	18.2
d	18.5
c	18.3
e	18.4
f	18.3
a	18.2

CHAPTER 19

1. a	19.2
2. d	19.2
3. b	19.3
4. a	19.4
5. b	19.4
6. d	19.4
7. b	19.5
8. d	19.5
9. d	19.6
10. d	19.6
11. endosymbiosis	19.6
12. c	19.5
13. c	19.5
14. b	19.1
15. (1) f	19.2
(2) c	19.4
(3) a	19.5
(4) b	19.6
(5) e	19.6
(6) d	19.6

CHAPTER 20

1. c	20.2
2. b	20.2
3. b	20.3
4. b	20.3
5. d	20.5
6. c	20.5
7. c	20.7
8. d	20.6
9. d	20.4, 20.7
10. c	20.7
11. d	20.5
12. b	20.5
13. d	20.7
14. pandemic	20.4
15. d	20.8
e	20.5
b	20.3
f	20.5
g	20.8
a	20.2
c	20.5

CHAPTER 21

1. c	21.2
2. b	21.3
3. b	21.4, 21.8
4. c	21.6
5. b	21.8
6. cyanobacteria	21.9
7. c	21.2, 21.9
8. a	21.5
9. c	21.10
10. c	21.9
11. d	21.3
12. c	21.10
13. b	21.6
14. e	21.4
c	21.4
a	21.8
b	21.9
d	21.9
15. d	21.3
g	21.7
a	21.6
b	21.8
f	21.8
h	21.9
e	21.9
c	21.10

CHAPTER 22

1. c	22.2
2. d	22.2, 22.3
3. a	22.3
4. False (ferns produce spores inside sori)	22.5
5. a	22.4
6. b	22.3
7. c	22.5
8. a	22.6
9. d	22.5
10. b	22.8
11. True	22.3
12. b	22.9
13. d	22.3
c	22.5
a	22.7
b	22.8
14. c	22.6
h	22.3
a	22.2
b	22.2
e	22.8
f	22.8
d	22.5
g	22.5
15. d	22.4
a	22.5
b	22.6
c	22.6

CHAPTER 23

1. c	23.2
2. a	23.2
3. a	23.2
4. c	23.5
5. d	23.6
6. b	23.2
7. c	23.6
8. d	23.6
9. a	23.3
10. b	23.7
11. c	23.4
12. a	23.7
13. a	23.4, 23.7
14. b	23.5
15. d	23.6
b	23.2
a	23.5
f	23.5
g	23.4
c	23.7
e	23.4, 23.7
i	23.2
h	23.7

CHAPTER 24

1. True	24.2
2. coelom	24.2
3. a	24.2
4. d	24.13
5. a	24.5
6. d	24.7
7. b	24.8
8. c	24.11
9. 2	24.2
10. b	24.8
11. d	24.14
12. d	24.11, 24.15
13. c	24.7
14. f	24.8
h	24.16
d	24.4
b	24.5
c	24.6
a	24.2
g	24.7
e	24.11

CHAPTER 25

1. notochord, dorsal hollow nerve cord, pharynx with gill slits, tail that extends past anus	25.2
2. all four traits	25.2
3. a	25.2
4. a	25.4
5. c	25.5
6. c	25.7
7. f	25.7
8. a	25.3
9. c	25.9
10. c	25.9
11. b	25.1, 25.9
12. a	25.6
13. d	25.5
a	25.5
f	25.10
e	25.6
c	25.2
b	25.11
14. j	25.2
i	25.3
h	25.6
f	25.8
c	25.7
g	25.6
d	25.10
a	25.10
e	25.4
b	25.10
15. a	25.3
e	25.4
d	25.6
c	25.6
b	25.7

CHAPTER 26

1. d	26.2
2. a	26.3
3. c	26.3
4. d	26.2
5. a	26.4
6. c	26.3
7. d	26.2, 26.4
8. c	26.5
9. a	26.6
10. d	26.1
11. b	26.6
12. d	26.6
13. c	26.6
14. c	26.1
d	26.4, 26.5
a	26.4
b	26.6
f	26.5
e	26.6
15. f	26.2
b	26.2
c	26.3
e	26.3
d	26.5
g	26.1
a	26.6

CHAPTER 27

1. b	27.3
2. c	27.3
3. a, b	27.3
4. b	27.4
5. False	27.6
6. c	27.3, 27.7
7. c	27.5
8. a	27.4, 27.6
9. b, c	27.4
10. d	27.6
11. b	27.6
12. a	27.7
13. b	27.8
14. b	27.9
15. c	27.8
d	27.3, 27.5
e	27.8
a	27.4
b	27.4
g	27.2, 27.5
f	27.2, 27.5

CHAPTER 28

1. f	28.2
2. b	28.2
3. e	28.3
4. b	28.3
5. b	28.3
6. c	28.4
7. d	28.4
8. d	28.4
9. c	28.6
10. c	28.5
11. c	28.5
12. a	28.5
13. b	28.6
14. a	28.6
15. c	28.5
a	28.6

Appendix VII. Answers to Self-Quizzes and Genetics Problems (continued)

e 28.6
b 28.3
d 28.4
h 28.4
f 28.6
g 28.4

CHAPTER 29

1. Pollination 29.3
2. a, b, c 29.1 revisited, 29.3
3. b 29.2
4. b 29.2, 29.4
5. b 29.4
6. False 29.3
7. b 29.6
8. a 29.6
9. c 29.6, 29.7
10. It's a pepo 29.7
11. c 29.8
12. d 29.8
13. c 29.9
14. a 29.9
15. c 29.6
f 29.2
g 29.4
e 29.2, 29.4
d 29.8
b 29.4
a 29.4

CHAPTER 30

1. e 30.2
2. b 30.3
3. c 30.6, 30.10
4. a 30.3, 30.4, 30.6
5. d 30.7
6. a 30.8
7. e 30.9
8. c 30.9
9. d 30.10
10. b 30.8
d 30.8
a 30.8
f 30.9
c 30.9
e 30.8
11. b 30.7
d 30.4
a 30.3
e 30.5
c 30.6
f 30.6, 30.10
12. c 30.7
e 30.4
b 30.8
a 30.5
d 30.6
f 30.10

CHAPTER 31

1. a 31.3
2. a 31.3
3. a 31.3
4. b 31.4
5. d 31.4
6. b 31.4
7. b 31.6
8. c 31.5
9. a 31.4
10. b 31.3
11. c 31.3
12. a 31.3
13. b 31.5
14. a 31.9
15. b 31.3
j 31.3
c 31.4
e 31.8
a 31.6
d 31.5
f 31.5
h 31.4
g 31.8
i 31.2

CHAPTER 32

1. a 32.2
2. c 32.3
3. d 32.4
4. False 32.4
5. a 32.5
6. a 32.5, 32.7
7. c 32.6
8. c 32.7
9. a 32.7
10. c 32.8
11. a 32.12, 32.13
12. a 32.12
13. c 32.11
14. a 32.9, 32.10
15. h 32.8
d 32.5
f 32.9
b 32.9
g 32.10
a 32.9
e 32.12
c 32.9

CHAPTER 33

1. b 33.3
2. c 33.2
3. c 33.3
4. e 33.4
5. b 33.3
6. b 33.10
7. b 33.9
8. a 33.3
9. b 33.9
10. b 33.7
11. d 33.6
12. b 33.6
13. c 33.8
14. a 33.6
15. e 33.8
j 33.9
b 33.9
g 33.5
a 33.6
h 33.7
f 33.4
k 33.3
i 33.9
c 33.10
d 33.4

CHAPTER 34

1. a 34.2
2. d 34.3
3. c, b, a, d 34.4
4. a 34.4
5. d 34.5
6. b 34.7
7. c 34.7
8. b 34.8
9. d 34.8
10. d 34.9
11. b 34.11
12. True 34.6
13. False 34.4
14. False 34.13
15. d 34.9
f 34.7
c 34.4
e 34.8
a 34.12
b 34.2

CHAPTER 35

1. c 35.2
2. d 35.4
3. b 35.6
4. a 35.5
5. a 35.5
6. c 35.3
7. b 35.7
8. b 35.7
9. d 35.7
10. d 35.9
11. c 35.8
12. d 35.8
13. a 35.8
14. c 35.9
15. e 35.4
i 35.7
g 35.9
h 35.5
d 35.7
c 35.4
b 35.3
a 35.8
f 35.9

CHAPTER 36

1. a 36.2
2. d 36.2
3. b 36.2
4. d 36.5
5. b 36.5
6. d 36.5
7. b 36.4
8. b 36.4
9. a 36.7
10. c 36.9
11. b 36.4
12. b 36.11
13. pulmonary arteries 36.4
14. d 36.10
e 36.10
a 36.9
b 36.5
f 36.10
g 36.10
c 36.10
15. f 36.5
a 36.11
e 36.5
g 36.10
b 36.4
c 36.5
d 36.3

CHAPTER 37

1. d 37.2
2. e 37.2–37.4
3. g 37.4, 37.5
4. h 37.6–37.9
5. d 37.4, 37.5
6. Immediate, fixed, general, and no antigen memory are all correct. 37.2, 37.4, 37.5
7. d 37.6
8. a 37.6–37.9
9. b 37.7, 37.8
10. e 37.7, 37.9
11. c 37.10
12. Self/nonself discrimination, diversity, specificity, and memory are all correct. 37.7
13. b 37.9
14. a 37.10
15. b 37.7
e 37.6, 37.7
c 37.6, 37.7, 37.9
d 37.9
a 37.9
16. f 37.10
e 37.7
d 37.10
c 37.5
b 37.10, 37.11
a 37.6
g 37.6

CHAPTER 38

1. d 38.2
2. b 38.4
3. c 38.3
4. c 38.5
5. c 38.2
6. d 38.6
7. a 38.6
8. b 38.1
9. a 38.7
10. c 38.8
11. b 38.4
12. d 38.6
13. True 38.6
14. d 38.5
15. d 38.5
h 38.5
f 38.5
e 38.5
g 38.5
c 38.5
b 38.5
a 38.5

CHAPTER 39

1. d 39.2
2. b 39.5
3. c 39.7
4. d 39.2
5. a 39.7
6. c 39.8
7. a 39.5
8. b 39.7
9. c 39.10
10. d 39.9
11. b 39.6
12. c 39.11
13. c 39.12
14. f 39.5
e 39.4
d 39.7
b 39.7
c 39.4
a 39.5
15. f 39.7
h 39.4
c 39.8
b 39.7
e 39.2
a 39.8
g 39.5
d 39.7

CHAPTER 40

1. c 40.2
2. b 40.2
3. a 40.3
4. b 40.3
5. a 40.4
6. b 40.4
7. a 40.5
8. d 40.5
9. a 40.5
10. c 40.3
a 40.3
b 40.3
e 40.5
d 40.5
11. a 40.8
12. d 40.8
13. a 40.7
14. b 40.8
15. a 40.8

CHAPTER 41

1. a 41.2
2. a 41.3
3. c 41.4
4. c 41.5
5. a 41.4
6. c 41.6
7. b 41.7
8. d 41.7
9. b 41.7
10. a 41.8
11. d 41.6
12. a, c, a, c, a, c, b all 41.9
13. d 41.6
c 41.6
h 41.4
a 41.6
i 41.4
e 41.4
f 41.4
b 41.6
g 41.4
14. a 41.4
c 41.4
b 41.5
d 41.3

CHAPTER 42

1. b 42.2
2. b 42.2
3. d 42.4
4. c 42.4
d 42.4
a 42.2
b 42.2
e 42.7
5. d 42.7
6. a 42.7

7. d 42.9
8. d 42.8
9. d 42.11
10. a 42.11
c 42.10, 42.11
b 42.7
11. c 42.7
12. d 42.4
13. a 42.1
14. d 42.7
15. 4 42.7
2 42.7
6 42.8
3 42.7
1 42.7
5 42.7
7 42.10

CHAPTER 43

1. d 43.2
2. b 43.2
3. c 43.5
4. a 43.6
5. a 43.6
6. b 43.7
7. d 43.8
8. c 43.8
9. d 43.9
10. c 43.9
11. b 43.9
12. c 43.5
13. b 43.3
14. c 43.4
15. a 43.3
b 43.6
e 43.8
d 43.3
c 43.7

CHAPTER 44

1. a 44.2
2. f 44.3
3. c 44.2
4. a 44.3
5. a 44.3
6. d 44.4
7. d 44.5
8. b 44.7
9. d 44.8
10. a 44.5
11. c 44.5
12. d 44.8
13. c 44.4
d 44.3
a 44.3
e 44.4
b 44.4

CHAPTER 45

1. b 45.2
2. d 45.2
3. b 45.7
4. d 45.4
5. b 45.3
d 45.7
c 45.2
a 45.7
e 45.4
6. e 45.4
7. b 45.8
8. mimicry 45.6
9. c 45.8
10. a 45.10
11. b 45.4
12. c 45.7
13. c 45.10
14. c 45.7
15. c 45.10
b 45.8
d 45.9
a 45.9
f 45.9
e 45.4

CHAPTER 46

1. a 46.2
2. b 46.2
3. c 46.4
4. d 46.4
5. b 46.4
6. c 46.6
7. b 46.7
8. d 46.7
9. d 46.8
10. a 46.10
11. c 46.10
12. d 46.9, 46.10
13. a 46.9
14. d 46.7, 46.9
15. e 46.8
d 46.9
b 46.10
a 46.9
c 46.7
f 46.9

CHAPTER 47

1. b 47.2
2. d 47.2
3. d 47.2
4. a 47.3
5. c 47.3
6. b 47.7
7. d 47.5
8. c 47.7
9. a 47.11
10. b 47.12
11. d 47.15
12. c 47.14
13. c 47.11
14. b 47.5
15. d 47.11
e 47.8
f 47.6
c 47.7
b 47.13
h 47.10
i 47.7
a 47.9
g 47.15

CHAPTER 48

1. True 48.2
2. b 48.2
3. b 48.3
4. c 48.5
5. a 48.5
6. d 48.5
7. a 48.6
8. False 48.5
9. b 48.7
10. d 48.6
11. c 48.8
12. d 48.8
13. c 48.8
14. d 48.9
15. f 48.6
h 48.8
d 48.5
c 48.3
e 48.5
g 48.7
a 48.4
b 48.4

CHAPTER 13: GENETICS PROBLEMS

1. a. *AB*
   b. *AB*, *aB*
   c. *Ab*, *ab*
   d. *AB*, *Ab*, *aB*, *ab*

2. a. All offspring will be *AaBB*.
   b. 1/4 *AABB* (25% each genotype)
   1/4 *AABb*
   1/4 *AaBB*
   1/4 *AaBb*
   c. 1/4 *AaBb* (25% each genotype)
   1/4 *Aabb*
   1/4 *aaBb*
   1/4 *aabb*
   d. 1/16 *AABB* (6.25% of genotype)
   1/8 *AaBB* (12.5%)
   1/16 *aaBB* (6.25%)
   1/8 *AABb* (12.5%)
   1/4 *AaBb* (25%)
   1/8 *aaBb* (12.5%)
   1/16 *AAbb* (6.25%)
   1/8 *Aabb* (12.5%)
   1/16 *aabb* (6.25%)

3. a. *ABC*
   b. *ABC*, *aBC*
   c. *ABC*, *aBC*, *ABc*, *aBc*
   d. *ABC*, *aBC*, *AbC*, *abC*, *ABc*, *aBc*, *Abc*, *abc*

4. A mating of two M^L cats yields 1/4 *MM*, 1/2 M^LM, and 1/4 M^LM^L. Because M^LM^L is lethal, the probability that any one kitten among the survivors will be heterozygous is 2/3.

5. A mating between a mouse from a true-breeding, white-furred strain and a mouse from a true-breeding, brown-furred strain would provide you with the most direct evidence. Because true-breeding strains typically are homozygous for a trait being studied, all F_1 offspring from this mating should be heterozygous. Record the phenotype of each F_1 mouse, then let them mate with one another. Assuming only one gene locus is involved, these are possible outcomes for the F_1 offspring:
   a. All F_1 mice are brown, and their F_2 offspring segregate: 3 brown : 1 white. *Conclusion*: Brown is dominant to white.
   b. All F_1 mice are white, and their F_2 offspring segregate: 3 white : 1 brown. *Conclusion*: White is dominant to brown.
   c. All F_1 mice are tan, and the F_2 offspring segregate: 1 brown : 2 tan : 1 white. *Conclusion*: The alleles at this locus show incomplete dominance.

6. Yellow is recessive. Because F_1 plants have a green phenotype and must be heterozygous, green must be dominant over the recessive yellow.

7. Possible outcomes of a testcross between an F_1 rose plant heterozygous for height (*Aa*) and a shrubby rose plant:

Gametes F_1 hybrid:

Gametes shrubby plant:	(*A*)	(*a*)
(*a*)	*Aa* climber	*aa* shrubby
(*a*)	*Aa* climber	*aa* shrubby

1:1 possible ratio of genotypes and phenotypes in F_2 generation

8. a. Both parents are heterozygous (*Aa*). Their children may be albino (*aa*) or unaffected (*AA* or *Aa*).
   b. All are homozygous recessive (*aa*).
   c. Homozygous recessive (*aa*) father, and heterozygous (*Aa*) mother. The albino child is *aa*, the unaffected children *Aa*.

9. The data reveal that these genes do not assort independently because the observed ratio is very far from the 9:3:3:1 ratio expected with independent assortment. Instead, the results can be explained if the genes are located close to each other on the same chromosome.

10. a. 1/2 red 1/2 pink 0 white
    b. 0 red All pink 0 white
    c. 1/4 red 1/2 pink 1/4 white
    d. 0 red 1/2 pink 1/2 white

11. Because both parents are heterozygous (*HbA*/*HbS*), the following are the probabilities for each child:
    a. 1/4 *HbS*/*HbS*
    b. 1/4 *HbA*/*HbA*
    c. 1/2 *HbA*/*HbS*

CHAPTER 14: GENETICS PROBLEMS

1. A daughter could develop DMD only if she inherited two mutated alleles, one from each parent, on each of her two X chromosomes. However, males who have the allele are unlikely to father children because they develop the disorder and die early in life.

2. Autosomal recessive. If the allele was inherited in a dominant pattern, individuals in the last generation would all have the phenotype. If it was X-linked, offpsring of the first generation would all have the phenotype.

3. a. A male with an X-linked allele produces two kinds of gametes: one with an X chromosome (and the X-linked allele), and the other with a Y chromosome.
   b. A female homozygous for an X-linked allele produces one type of gamete, which carries the X-linked allele.
   c. A female heterozygous for an X-linked allele produces two types of gametes: one that carries the X-linked allele, and another that carries the partnered allele on the homologous chromosome.

4. 50 percent.

5. a. Anaphase I or anaphase II.
   b. As a result of translocation, chromosome 21 may get attached to the end of chromosome 14. The new individual's chromosome number would still be 46, but his or her somatic cells would have the translocated chromosome 21 in addition to two normal chromosomes 21.

6. A crossover between the two genes during meiosis generates an X chromosome that carries neither mutated allele.

Glossary of Biological Terms

ABA *See* abscisic acid.

abscisic acid (**ABA**) Plant hormone involved in stress responses; inhibits germination. **508**

abscission Process by which plant parts are shed. **508**

acclimatization Physiological adjustment of an individual to a new environment, as occurs after a move from sea level to a high-altitude habitat. **680**

acid Substance that releases hydrogen ions in water. **32**

acid rain Low-pH rain formed when sulfur dioxide or nitrogen oxides mix with water vapor in the atmosphere. **874**

action potential Brief reversal of the voltage difference across the plasma membrane of a neuron or muscle cell. **542**

activation energy Minimum amount of energy required to start a chemical reaction. **80**

activator Regulatory protein that increases the rate of transcription when it binds to a promoter or enhancer. **164**

active site Pocket in an enzyme where substrates bind and a chemical reaction occurs. **82**

active transport Energy-requiring mechanism in which a transport protein pumps a solute across a cell membrane against the solute's concentration gradient. **93**

adaptation (**adaptive trait**) A form of a heritable trait that enhances an individual's fitness in a particular environment. **256**

adaptive immunity In vertebrates, set of immune defenses that can be tailored to fight specific pathogens as an organism encounters them during its lifetime. Characterized by self/nonself recognition, specificity, diversity, and memory. **642**

adaptive radiation Macroevolutionary pattern in which a burst of genetic divergences from a lineage gives rise to many new species. **288**

adaptive trait *See* adaptation.

ADH *See* antidiuretic hormone.

adhering junction Cell junction composed of adhesion proteins that connect to cytoskeletal elements. Fastens animal cells to each other or to basement membrane. **71**

adhesion protein Plasma membrane protein that helps cells stick together in animal tissues. Some types form adhering junctions and tight junctions. **89**

adipose tissue Connective tissue that specializes in fat storage. **524**

adrenal cortex Outer portion of adrenal gland; secretes aldosterone and cortisol. **594**

adrenal gland Endocrine gland located atop the kidney; secretes aldosterone, cortisol, epinephrine, and norepinephrine. **594**

adrenal medulla Inner portion of adrenal gland; secretes epinephrine and norepinephrine. **594**

aerobic Involving or occurring in the presence of oxygen. **117**

aerobic respiration Oxygen-requiring metabolic pathway that breaks down sugars to produce ATP. Includes glycolysis, acetyl–CoA formation, the Krebs cycle, and electron transfer phosphorylation. **118**

age structure Of a population, the number of individuals in each of several age categories. **783**

agglutination The clumping together of foreign cells bound by antibodies; the clumps attract phagocytic cells. Basis of blood typing tests. **655**

AIDS Acquired immune deficiency syndrome. A secondary immune deficiency that develops as the result of infection by the HIV virus. **660**

alcoholic fermentation Anaerobic sugar breakdown pathway that produces ATP, CO_2, and ethanol. **126**

aldosterone Adrenal hormone that makes kidney tubules more permeable to sodium; encourages sodium reabsorption, thus increasing water reabsorption and concentrating the urine. **715**

algal bloom Population explosion of photosynthetic cells in an aquatic habitat. **344**

allantois Extraembryonic membrane that, in mammals, becomes part of the umbilical cord. **753**

allele frequency Abundance of a particular allele among members of a population, expressed as a fraction of the total number of alleles. **271**

alleles Forms of a gene with slightly different DNA sequences; may encode slightly different versions of the gene's product. **190**

allergen A normally harmless substance that provokes an immune response in some people. **658**

allergy Sensitivity to an allergen. **658**

allopatric speciation Speciation pattern in which a physical barrier arises and ends gene flow between populations. **284**

allosteric regulation Control of enzyme activity by a regulatory molecule or ion that binds to a region outside the enzyme's active site. **84**

alpine tundra Biome of low-growing, wind-tolerant plants adapted to high-altitude conditions. **855**

alternation of generations Of land plants and some algae, a life cycle that alternates between a haploid, gamete-producing body and diploid, spore-producing body. **339**

alternative splicing Post-translational RNA modification process in which some exons are removed or joined in different combinations. **153**

altruistic behavior Behavior that benefits others at one's own expense. **776**

alveolate Member of a protist lineage characterized by the presence of small sacs (alveoli) beneath the plasma membrane; a dinoflagellate, ciliate, or apicomplexan. **343**

alveolus (plural **alveoli**) Tiny sac; in the mammalian lung, site of gas exchange at the tip of a bronchiole. **675**

amino acid Small organic compound that is a subunit of proteins. Consists of a carboxyl group, an amine group, and a characteristic side group (R), all typically bonded to the same carbon atom. **46**

amino acid–derived hormone An amine (modified amino acid), peptide, or protein that functions as a hormone. **584**

ammonia Nitrogen-containing compound (NH_3) formed by breakdown of amino acids and nucleic acids. **708**

ammonification Breakdown of nitrogen-containing organic material resulting in the release of ammonia and ammonium ions. **832**

amnion Extraembryonic membrane that encloses an amniote embryo and amniotic fluid. **752**

amniote Vertebrate with eggs that enclose the embryo within waterproof membranes; a reptile, bird, or mammal. **417**

amoeba Single-celled, unwalled protist that uses pseudopods to move and to capture prey. **350**

amoebozoans Lineage of heterotrophic, unwalled protists that live in soils and water; include amoebas and slime molds. **350**

amphibian Tetrapod with a three-chambered heart and scaleless skin; typically develops in water, then lives on land as an air-breathing carnivore. For example, a frog or salamander. **422**

anaerobic Occurring in the absence of oxygen. **117**

analogous structures Similar body structures that evolved independently in different lineages (by morphological convergence). **297**

anaphase Stage of mitosis during which sister chromatids separate and move toward opposite spindle poles. **181**

aneuploid Having too many or too few copies of a particular chromosome. **226**

angiosperm Seed plant that makes flowers and fruits. **368**

animal Multicelled heterotroph that has unwalled cells, develops through a series of stages, and moves about during part or all of its life. **8**, **390**

animal hormone Intercellular signaling molecule secreted by an endocrine gland or cell; travels in the blood to target cells. **582**

annelid A leech or segmented worm with a coelom, complete digestive system, and closed circulatory system. **398**

antenna Of some arthropods, sensory structure on the head that detects touch and odors. **404**

anther Of a flower, pollen-producing portion of a stamen. **484**

anthropoids Primate lineage that includes monkeys, apes, and humans. **437**

antibody Y-shaped antigen receptor protein made only by B cells. Also called immunoglobulin; e.g., IgA, IgG, IgE, IgM, IgD. **650**

antibody-mediated immune response Immune response in which antibodies are produced in response to an antigen. **652**

anticodon In a tRNA, set of three nucleotides that base-pairs with an mRNA codon. **155**

antidiuretic hormone (**ADH**) Hormone released in the posterior pituitary; makes kidney tubules more permeable to water; encourages water reabsorption, thus concentrating the urine. **714**

antigen A molecule or particle that the immune system recognizes as nonself. Its presence in the body triggers an immune response. **642**

antioxidant Substance that prevents oxidation of other molecules. **86**

anus Body opening that serves solely as the exit for wastes from a complete digestive tract. **689**

aorta Largest artery; carries oxygenated blood away from the heart. **624**

apical dominance Effect in which a lengthening shoot tip inhibits the growth of lateral (axillary) buds. **505**

apical meristem Meristem in the tip of a shoot or root; gives rise to primary growth (lengthening) in a plant. *See also* meristem. **460**

apicomplexan Single-celled alveolate protist that lives as a parasite inside animal cells; some cause malaria or toxoplasmosis. **345**

apoptosis Mechanism of cell suicide. **749**

appendicular skeleton Of vertebrates, bones of the limbs or fins and the bones that connect them to the axial skeleton. **607**

appendix Short, tubular projection from the initial segment of the large intestine (the cecum); reservoir for beneficial bacteria. **696**

aquifer Porous rock layer that holds some groundwater. **826**

arachnids Land-dwelling chelicerate arthropods with four pairs of walking legs; spiders, scorpions, mites, and ticks. **405**

archaea Group of single-celled prokaryotic organisms that are more closely related to eukaryotes than to bacteria. **8**, **326**

arctic tundra Highest-latitude Northern Hemisphere biome, where short, cold-tolerant plants survive with only a brief growing season. **855**

area effect The number of species an island can accommodate increases with its size. **814**

arteriole Blood vessel that conveys blood from an artery to capillaries. **630**

artery Large-diameter blood vessel that carries blood away from the heart. **624**

arthropod Invertebrate with jointed legs and a hard exoskeleton that is periodically molted; for example, an insect or crustacean. **404**

asexual reproduction Reproductive mode of eukaryotes by which offspring arise from a single parent only. **178**, **724**

astrobiology The scientific study of life's origin and distribution in the universe. **305**

atmospheric cycle Biogeochemical cycle in which a gaseous form of an element plays a significant role. **828**

atom Fundamental building block of all matter. Consists of varying numbers of protons, neutrons, and electrons. **4**

atomic number Number of protons in the atomic nucleus; determines the element. **24**

ATP Adenosine triphosphate. Nucleotide that consists of an adenine base, a ribose sugar, and three phosphate groups. Functions as a subunit of RNA and as a coenzyme in many reactions. Important energy carrier in cells. **49**

ATP/ADP cycle Process by which cells regenerate ATP. ADP forms when ATP loses a phosphate group, then ATP forms again as ADP gains a phosphate group. **87**

atrioventricular (**AV**) **node** In the heart, a clump of cells that serves as the electrical bridge between the atria and ventricles. **627**

atrium Heart chamber that receives blood. **622**

australopith Informal name for several species of chimpanzee-sized hominins that lived in Africa between 4 million and 1.2 million years ago. **440**

autoimmune response Immune response that inappropriately targets one's own tissues. **659**

autonomic nerves Nerves that relay signals to and from internal organs and to glands. **548**

autosome A chromosome that is the same in males and females. **139**

autotroph Producer. Organism that makes its own food using energy from the environment and carbon from inorganic molecules such as CO_2. **101**

auxin Plant hormone that causes cell enlargement; also has a central role in growth by coordinating the effects of other hormones. Indole-3-acetic acid (IAA) is the most common auxin. **504**

AV node *See* atrioventricular (AV) node.

axial skeleton Bones of the main body axis; skull, backbone. **607**

axon Of a neuron, a cytoplasmic extension that transmits electrical signals along its length and secretes chemical signals at its endings. **540**

B cell B lymphocyte. Leukocyte (white blood cell) that can make antibodies. Central to antibody-mediated immune responses. **643**

B cell receptor Antigen receptor on the surface of a B cell; IgM or IgD antibody that stays anchored in the B cell's plasma membrane. **651**

bacteria (singular **bacterium**) The most diverse and well-known group of prokaryotes. **8**, **326**

bacteriophage Virus that infects bacteria. **135**, **320**

balanced polymorphism Maintenance of two or more alleles of a gene at high frequency in a population. **278**

bark In woody plants, informal term for all living and dead tissues outside of the vascular cambium. **463**

Barr body Inactivated X chromosome in a cell of a female mammal. The other X chromosome is active. **168**

basal body Organelle that develops from a centriole. **69**

basal metabolic rate Rate at which the body uses energy when at rest. **702**

base Substance that accepts hydrogen ions when it dissolves in water. **32**

basement membrane Secreted material that attaches epithelium to an underlying tissue. **522**

base-pair substitution Type of mutation in which a single base pair changes. **158**

basophil Circulating white blood cell that releases the contents of its granules in response to antigen or injury. **643**

behavioral plasticity The behavioral phenotype associated with a genotype depends on conditions in the animal's environment. **768**

bell curve Bell-shaped curve; typically results from graphing frequency versus distribution for a trait that varies continuously. **212**

benthic province The ocean's sediments and rocks. **862**

big bang theory Well-supported hypothesis that the universe originated by a nearly instant distribution of matter through space. **306**

bilateral symmetry Having paired structures arranged so that the right and left halves are mirror images. **390**

bile Liquid mixture of salts, pigments, and cholesterol that aids fat emulsification in the small intestine. Produced in the liver, stored and concentrated in the gallbladder, and secreted into the small intestine. **694**

binary fission Cell reproduction process of bacteria and archaea. **326**

bioaccumulation An organism accumulates increasing amounts of a chemical pollutant in its tissues over the course of its lifetime. **874**

biodiversity Of a region, the genetic variation within its species, variety of species, and variety of ecosystems. **8**, **878**

biodiversity hot spot Threatened region with great biodiversity that is considered a high priority for conservation efforts. **878**

biofilm Community of microorganisms living within a shared mass of secreted slime. **59**

biogeochemical cycle Cycle in which a nutrient moves among environmental reservoirs and into and out of food webs. **826**

biogeography Study of patterns in the geographic distribution of species and communities. **252**

biological magnification A chemical pollutant becomes increasingly concentrated as it moves up through food chains. **874**

biology The scientific study of life. **3**

bioluminescence Production of light by an organism. **344**

biomarker Molecule produced only by a specific type of cell; its presence indicates the presence of that cell type. **311**

biomass pyramid Diagram that depicts the biomass (dry weight) in each of an ecosystem's trophic levels. **824**

biome Group of regions that may be widely separated but share a characteristic climate and dominant vegetation. **846**

biosphere All regions of Earth where organisms live. **5**, **840**

biotic potential Maximum possible population growth rate under optimal conditions. **785**

bipedal Adapted to habitually walking upright. **439**

bird Common name for a member of the only surviving reptile lineage with feathers. **426**

blastocyst Mammalian blastula. **752**

blastula Hollow ball of cells that forms as a result of cleavage in early animal development. **744**

blood Circulatory fluid; in vertebrates it is a fluid connective tissue consisting of plasma, red blood cells, leukocytes (white blood cells), and platelets. **525**, **622**

blood–brain barrier Protective mechanism that controls the composition and concentration of cerebrospinal fluid. **552**

blood pressure Pressure exerted by blood against a vessel wall. **631**

BMI *See* body mass index.

body mass index (BMI) Measure that indicates an individual's degree of body fat based on height and weight. **702**

bone tissue Connective tissue consisting of cells surrounded by a mineral-hardened matrix of their own secretions. **525**

bony fish Common term for a fish whose skeleton includes bone; a ray-finned or lobe-finned fish. **420**

boreal forest Taiga. Extensive high-latitude forest of the Northern Hemisphere; conifers are the predominant vegetation. **854**

bottleneck Reduction in population size so severe that it reduces genetic diversity. **280**

Bowman's capsule In the kidney, the cup-shaped portion of a nephron tubule that encloses a glomerulus and receives filtrate from it. **710**

brain Central control organ in a nervous system. **538**

brain stem The most evolutionarily ancient region of the vertebrate brain; includes the pons, medulla, and midbrain. **552**

bronchiole In a vertebrate lung, one of the many airways that lead from a bronchus to alveoli. **675**

bronchus Large airway that connects the trachea to a lung. **675**

brood parasitism One egg-laying species benefits by having another raise its offspring. **808**

brown alga Multicelled marine protist, such as a kelp, that has the brown accessory pigment (fucoxanthin) in its chloroplasts. **347**

brush border cell In the lining of the small intestine, an epithelial cell with microvilli at its surface. **693**

bryophyte A nonvascular plant with a gametophyte-dominant life cycle; a moss, liverwort, or hornwort. **357**

budding In yeast, a mechanism of asexual reproduction by which a small cell forms on a parent cell, then is released. **380**

buffer Set of chemicals that can keep the pH of a solution stable by alternately donating and accepting ions that contribute to pH. **33**

C3 plant Type of plant that uses only the Calvin–Benson cycle to fix carbon. **110**

C4 plant Type of plant that minimizes photorespiration by fixing carbon twice, in two cell types. **110**

Calvin–Benson cycle Cyclic carbon-fixing pathway that builds sugars from CO_2; the light-independent reactions of photosynthesis. **109**

calyx A flower's outer, protective whorl of sepals. **484**

CAM plant Type of C4 plant that minimizes photorespiration by fixing carbon twice, at different times of day. **111**

camera eye Eye with an adjustable opening and a single lens that focuses light on a retina. **568**

camouflage Body coloration, patterning, form, or behavior that helps predators or prey blend with the surroundings and possibly escape detection. **807**

cancer Disease that occurs when a malignant neoplasm physically and metabolically disrupts body tissues. **185**

capillary Small blood vessel; exchanges with interstitial fluid take place across its walls. **622**

carbohydrate Molecule that consists primarily of carbon, hydrogen, and oxygen atoms in a 1:2:1 ratio. Complex kinds (e.g., cellulose, starch, glycogen) are polymers of simple kinds (sugars). **42**

carbon cycle Movement of carbon, mainly between the oceans, atmosphere, and living organisms. **828**

carbon fixation Process by which carbon from an inorganic source such as carbon dioxide becomes incorporated (fixed) into an organic molecule. **109**

carbonic anhydrase Enzyme in red blood cells that speeds the reversible conversion of CO_2 into bicarbonate and H^+. **679**

cardiac cycle Sequence of contraction and relaxation of heart chambers that occurs with each heartbeat. **626**

cardiac muscle tissue Muscle of the heart wall. **526**

carpel Of flowering plants, a reproductive structure that produces female gametophytes; consists of an ovary, stigma, and often a style. **368**, **484**

carrying capacity (*K*) Maximum number of individuals of a species that an environment can sustain. **786**

cartilage Connective tissue consisting of cells surrounded by a rubbery matrix of their own secretions. **524**

cartilaginous fish Jawed fish with a cartilage skeleton. For example, a shark or ray. **420**

Casparian strip Waxy band between the plasma membranes of abutting root endodermal cells; forms a seal that prevents soil water from seeping through cell walls into the vascular cylinder. **472**

catalysis The acceleration of a chemical reaction by a molecule that is unchanged by participating in the reaction. **82**

catastrophism Now-abandoned hypothesis that catastrophic geologic forces unlike those of the present day shaped Earth's surface. **254**

cDNA Complementary strand of DNA synthesized from an RNA template by the enzyme reverse transcriptase. **235**

cell Smallest unit of life; at minimum, consists of plasma membrane, cytoplasm, and DNA. **4**

cell cortex Reinforcing mesh of cytoskeletal elements under a plasma membrane. **68**

cell cycle The collective series of intervals and events of a cell's life, from the time it forms until it divides. **178**

cell junction Structure that connects a cell to another cell or to extracellular matrix; e.g., tight junction, adhering junction, or gap junction (of animals); plasmodesmata (of plants). **70**

cell-mediated immune response Immune response in which cytotoxic T cells and NK cells kill infected or cancerous body cells. **652**

cell plate A disk-shaped structure that forms during cytokinesis in a plant cell; matures as a cross-wall between the two new nuclei. **182**

cell theory Theory that all organisms consist of one or more cells, which are the basic unit of life; all cells come from division of preexisting cells; and all cells pass hereditary material to offspring. **54**

cellular slime mold Soil-dwelling protist that feeds as solitary cells but congregates under adverse conditions to form a cohesive unit that develops into a fruiting body. **350**

cellulose Tough, insoluble polysaccharide that is the major structural material in plants. **42**

cell wall Rigid but permeable layer of extracellular matrix structure that surrounds the plasma membrane of some cells. **59**

central nervous system Brain and spinal cord. **539**

central vacuole Large fluid-filled organelle in many plant cells. **64**

centriole Barrel-shaped organelle from which microtubules grow. **69**

centromere Of a duplicated eukaryotic chromosome, constricted region where sister chromatids attach to each other. **138**

cephalization Evolutionary process whereby sensory structures and nerve cells accumulate at the from end of a bilateral animal's body. **390**, **538**

cerebellum Hindbrain region responsible for posture and for coordinating voluntary movements. **552**

cerebral cortex Outer gray matter layer of the cerebrum. **554**

cerebrospinal fluid Clear fluid that surrounds the brain and spinal cord and fills cavities within the brain. **550**

cerebrum Forebrain region that acts in language, abstract thought. **552**

cervix Narrow part of uterus that connects to the vagina. **729**

character Quantifiable, heritable characteristic or trait. **294**

character displacement Outcome of competition between two species; similar traits that result in competition become dissimilar. **803**

charge Electrical property. Opposite charges attract, and like charges repel. **24**

chelicerates Arthropod subgroup with specialized feeding structures (chelicerae) and no antennae. Includes horseshoe crabs and arachinds. **405**

chemical bond An attractive force that arises between two atoms when their electrons interact; joins atoms as molecules. *See also* covalent bond, ionic bond. **28**

chemoautotroph Organism that uses carbon dioxide as its carbon source and obtains energy by oxidizing inorganic molecules. **328**

chemoheterotroph Organism that obtains both energy and carbon by breaking down organic compounds. **328**

chemoreceptor Sensory receptor that responds to the presence of a specific chemical. **562**

chemotaxis Cellular movement in response to a chemical stimulus. **647**

chlamydias Bacteria that are intracellular parasites of vertebrates; some cause sexually transmitted disease in humans. **331**

chlorophyll *a* Main photosynthetic pigment in plants. **102**

chloroplast Organelle of photosynthesis in the cells of plants and photosynthetic protists. Has two outer membranes enclosing semifluid stroma. Light-dependent reactions occur at its inner thylakoid membrane; light-independent reactions, in the stroma. **67**

choanoflagellates Heterotrophic protists thought to be the sister group of animals; collared cells strain food from water. **351**

chordate Animal with an embryo that has a notochord, dorsal nerve cord, pharyngeal gill slits, and a tail that extends beyond the anus; a lancelet, tunicate or a vertebrate. **416**

chorion Outermost extraembryonic membrane of amniotes; major component of the placenta in placental mammals. **753**

choroid A blood vessel–rich layer of the middle eye. It is darkened by the brownish pigment melanin to prevent light scattering. **570**

chromatin Collective term for all of the DNA and associated proteins in a cell nucleus. **62**

chromosome A molecule of DNA together with associated proteins; carries part or all of a cell's genetic information. **138**

chromosome number The total number of chromosomes in a cell of a given species. **139**

chyme Mixture of food and gastric fluid that forms in the stomach. **692**

chytrid Fungus that produces flagellated spores. **377**

ciliary muscle A ring-shaped muscle of the eye that encircles the lens and attaches to it by short fibers. **571**

ciliate Single-celled, heterotrophic protist with many cilia. **343**

cilium Short, movable structure that projects from the plasma membrane of some eukaryotic cells. **68**

circadian rhythm A biological activity that is repeated about every 24 hours. **512**, **597**

circulatory system Organ system consisting of a heart or hearts and fluid-filled vessels that distribute substances through a body. **622**

clade A group whose members share one or more defining derived traits. **294**

cladistics Making hypotheses about evolutionary relationships among clades. **295**

cladogram Evolutionary tree diagram that summarizes hypothesized relationships among a group of clades. **295**

classical conditioning An animal's involuntary response to a stimulus becomes associated with another stimulus that is presented at the same time. **767**

cleavage In animal developmental, process by which multiple mitotic divisions produce a blastula from a zygote. **744**

cleavage furrow In a dividing animal cell, the indentation where cytoplasmic division will occur. **182**

climate Average weather conditions in a region over a long period. **840**

cloaca In some vertebrates, body opening that serves as the exit for digestive and urinary waste; also functions in reproduction. **420**, **689**

clone Genetically identical copy of an organism. **133**

cloning vector A DNA molecule that can accept foreign DNA and be replicated inside a host cell. **235**

closed circulatory system Circulatory system in which blood flows to and from a heart or hearts through a continuous series of vessels. **398**, **622**

club fungi Fungi that have septate hyphae and produce spores by meiosis in club-shaped cells. **382**

cnidarian Radially symmetrical invertebrate that has tentacles with stinging cells (cnidocytes). For example, a jelly, sea anemone, or coral. **394**

cnidocyte Stinging cell unique to cnidarians. **394**

coal Fossil fuel formed over millions of years by compaction and heating of plant remains. **364**

cochlea Coiled, fluid-filled structure in the inner ear that holds the sound-detecting organ of Corti. **574**

codominance Effect in which the full and separate phenotypic effects of two alleles are apparent in heterozygous individuals. **208**

codon In an mRNA, a nucleotide base triplet that codes for an amino acid or stop signal during translation. **154**

coelom Of many animals, a body cavity that surrounds the gut and is lined with tissue derived from mesoderm. **391**

coenzyme An organic cofactor; e.g., NAD. **86**

coevolution The joint evolution of two closely interacting species; macroevolutionary pattern in which each species is a selective agent for traits of the other. **288**, **369**

cofactor A coenzyme or metal ion that associates with an enzyme and is necessary for its function. **86**

cohesion Property of a substance that arises from the tendency of its molecules to resist separating from one another. **31**

cohesion–tension theory Explanation of how transpiration creates a tension that pulls a cohesive column of water upward through xylem, from roots to shoots. **474**

cohort Group of individuals born during the same time interval. **788**

coleoptile In monocots, a rigid sheath that protects the plumule (embryonic shoot). **494**

collecting tubule Kidney tubule that receives filtrate from several nephrons and delivers it to the renal pelvis. **711**

collenchyma In plants, simple tissue composed of living cells with unevenly thickened walls; provides flexible support. **452**

colonial organism Organism composed of many integrated cells, each capable of living and reproducing on its own. **338**

commensalism Species interaction that benefits one species and neither helps nor harms the other. **800**

communication signal Chemical, acoustical, visual, or tactile cue that is produced by one member of a species and detected and responded to by other members of the same species. **770**

community All populations of all species in a given area. **5**, **799**

companion cell In phloem, specialized parenchyma cell that provides metabolic support to its partnered sieve element. **478**

comparative morphology The scientific study of similarities and differences in body plans. **253**

competitive exclusion Process whereby two species compete for a limiting resource, and one drives the other to local extinction. **802**

complement A set of proteins that circulate in inactive form in blood; activated complement proteins attract phagocytic leukocytes, coat antigenic particles, and puncture lipid bilayers. **642**

complete digestive tract Tubular gut; food enters through one opening and wastes leave through another. **391**, **688**

compound Molecule that has atoms of more than one element. **28**

compound eye Of some arthropods, an eye that consists of many individual units, each with a lens; excels at detecting movement. **404**, **568**

concentration Amount of solute per unit volume of solution. **30**

condensation Chemical reaction in which an enzyme builds a large molecule from smaller subunits; water also forms. **41**

conduction Of heat: the transfer of heat within an object or between two objects in contact with one another. **717**

cone cell Photoreceptor that provides sharp vision and allows detection of color. **572**

conifer Gymnosperm with nonmotile sperm and woody cones; for example, a pine. **366**

conjugation Mechanism of horizontal gene transfer in which one bacterial or archaeal cell passes a plasmid to another. **327**

conjunctiva Mucous membrane that lines the inner surface of the eyelids and folds back to cover the eye's sclera. **570**

connective tissue Animal tissue with an extensive extracellular matrix; structurally and functionally supports other tissues. **524**

conservation biology Field of applied biology that surveys and documents biodiversity, and seeks ways to maintain and use it. **878**

consumer Organism that gets energy and nutrients by feeding on tissues, wastes, or remains of other organisms; a heterotroph. **6**, **820**

continuous variation Range of small differences in a shared trait. **212**

contractile vacuole In freshwater protists, an organelle that collects and expels excess water. **341**

control group Group of individuals identical to an experimental group except for the independent variable under investigation. **13**

convection Transfer of heat by moving molecules of air or water. **717**

coral reef Highly diverse marine ecosystem centered around reefs built by living corals that secrete calcium carbonate. **860**

cork Tissue that waterproofs, insulates, and protects the surfaces of woody stems and roots. **463**

cork cambium Lateral meristem that produces cork. **463**

cornea Clear, protective covering at the front of the vertebrate eye. **570**

corolla A flower's whorl of petals; forms within sepals and encloses reproductive organs. **484**

corpus luteum Hormone-secreting structure that forms from ovarian follicle cells left behind after ovulation. **729**

cortisol Adrenal cortex hormone that influences metabolism and immunity; secretions rise with stress. **594**

cotyledon Seed leaf; structure that stores nutrients in a flowering plant embryo. Monocots have one; eudicots, two. **451**

countercurrent exchange Exchange of substances between two fluids moving in opposite directions. **672**

covalent bond Type of chemical bond in which two atoms share a pair of electrons. **28**

critical thinking The act of evaluating information before accepting it. **12**

crop Of birds and some invertebrates, an enlarged region of the esophagus that stores food. **688**

crossing over Process by which homologous chromosomes exchange corresponding segments during prophase I of meiosis. **194**

crustaceans Mostly marine arthropod group with two pairs of antennae and a calcium-stiffened exoskeleton. **406**

culture Learned behaviors transmitted between individuals and down through generations. **442**

cuticle Secreted covering at a body surface. **70**, **358**

cyanobacteria Oxygen-producing photosynthetic bacteria. **330**

cycad Tropical or subtropical gymnosperm with flagellated sperm, palmlike leaves, and fleshy seeds. **367**

cytokines Signaling molecules secreted by leukocytes to coordinate their activities during immune responses. **643**

cytokinesis Cytoplasmic division; process in which a eukaryotic cell divides in two after mitosis or meiosis. **182**

cytokinin Plant hormone that promotes cell division in shoot apical meristem and cell differentiation in root apical meristem. Often opposes auxin's effects. **506**

cytoplasm Jellylike mixture of water and solutes enclosed by a cell's plasma membrane. **54**

cytoplasmic localization Accumulation of different materials in different regions of cytoplasm. **746**

cytoskeleton Network of interconnected protein filaments that support, organize, and move eukaryotic cells and their internal structures. *See also* microtubule, microfilament, intermediate filament. **68**

data Experimental results. **13**

decomposer Organism that feeds on wastes and remains; breaks organic material down into its inorganic subunits. **328**, **820**

deductive reasoning Using a general idea to make a conclusion about a specific case. **12**

deletion Mutation in which one or more nucleotides are lost from DNA. **158**

demographics Statistics that describe a population. **782**

demographic transition model Model describing how birth and death rates change as a region becomes industrialized. **794**

denature To unravel the shape of a protein or other large biological molecule. **48**

dendrite Of a motor neuron or interneuron, a cytoplasmic extension that receives chemical signals sent by other neurons and converts them to electrical signals. **540**

dendritic cell Phagocytic white blood cell that alerts the immune system to the presence of antigen in solid tissues. **643**

denitrification Conversion of nitrates or nitrites to gaseous forms of nitrogen. **833**

density-dependent limiting factor Factor that increasingly limits population growth as population density increases. For example, disease or competition. **786**

density-independent limiting factor Factor that limits population growth to the same degree regardless of population density. **787**

dental plaque On teeth, a thick biofilm composed mainly of bacteria, their extracellular products, and saliva proteins. **644**

deoxyribonucleic acid *See* DNA.

dependent variable In an experiment, a variable that is presumably affected by an independent variable being tested. **13**

derived trait A novel trait present in a clade but not in any of the clade's ancestors. **294**

dermal tissues Tissues that cover and protect the plant body *See also* epidermis, periderm. **450**

dermis Deep layer of skin that consists of connective tissue with nerves and blood vessels running through it. **531**

desert Biome with little rain and low humidity; plants with water-storing and water-conserving adaptations predominate. **848**

desertification Conversion of a grassland or woodland to desert. **872**

detrital food web Food web in which most energy is transferred directly from producers to detritivores. **822**

detritivore Consumer that feeds on small bits of organic material. **820**

deuterostomes Lineage of bilateral animals in which the second opening on the embryo surface develops into a mouth. **391**

development Multistep process by which the first cell of a new multicelled organism gives rise to an adult. **7**

diaphragm Smooth muscle between the thoracic and abdominal cavities; contracts during inhalation. **675**

diastole Relaxation phase of the cardiac cycle. **626**

diastolic pressure Blood pressure when ventricles are relaxed. **631**

diatom Single-celled photosynthetic protist with a brown accessory pigment (fucoxanthin) and a two-part silica shell. **346**

differentiation Process by which cells become specialized during development; occurs as different cells in an embryo begin to use different subsets of their DNA. **144**, **748**

diffusion Spontaneous spreading of molecules or ions. **90**

dihybrid cross Cross between two individuals identically heterozygous for two genes; for example, $AaBb \times AaBb$. **206**

dikaryotic Having two genetically different nuclei in a cell ($n + n$). **377**

dinoflagellate Single-celled, aquatic protist with cellulose plates and two flagella; may be heterotrophic or photosynthetic. **344**

dinosaur Reptile lineage that include the ancestors of birds; became extinct at the end of the Cretaceous. **424**

diploid Having two of each type of chromosome characteristic of the species ($2n$). **139**

directional selection Mode of natural selection in which phenotypes at one end of a range of variation are favored. **274**

disaccharide Carbohydrate that is a polymer of two monosaccharides. **42**

disease vector An animal that transmits a pathogen from one host to the next. **324**

disruptive selection Mode of natural selection in which traits at the extremes of a range of variation are adaptive, and intermediate forms are not. **277**

distal tubule Portion of kidney tubule that delivers filtrate to a collecting tubule. **710**

distance effect Islands close to a mainland have more species than those farther away. **814**

DNA Deoxyribonucleic acid. Nucleic acid that carries hereditary information. Consists of two chains of nucleotides twisted into a double helix. **7**, **49**

DNA cloning Set of methods that uses living cells to mass-produce targeted DNA fragments. **235**

DNA library Collection of cells that host different fragments of foreign DNA, often representing an organism's entire genome. **236**

DNA ligase Enzyme that seals gaps in double-stranded DNA. **141**

DNA polymerase DNA replication enzyme. Uses a DNA template to assemble a complementary strand of DNA. **140**

DNA profiling Identifying an individual by analyzing the unique parts of his or her DNA. **240**

DNA replication Process by which a cell duplicates its DNA before it divides. **140**

DNA sequence Order of nucleotides in a strand of DNA. **137**

DNA sequencing *See* sequencing.

dominant Refers to an allele that masks the effect of a recessive allele paired with it in heterozygous individuals. **203**

dormancy Period of temporarily suspended metabolism. **488**

dosage compensation Mechanism in which X chromosome inactivation equalizes gene expression between males and females. **168**

double fertilization Mode of fertilization in flowering plants; one sperm cell fuses with an egg, and a second sperm cell fuses with the endosperm mother cell. **369**, **488**

dry shrubland Biome dominated by a diverse array of fire-adapted shrubs; occurs in regions with cool, wet winters and a dry summer. **851**

dry woodland Biome dominated by short trees that do not completely shade the ground; occurs in regions with cool, wet winters and a dry summer. **851**

duplication Repeated section of a chromosome. **224**

eardrum Membrane that separates the inner and middle ear; vibrates in response to pressure waves (sounds), thus transmitting vibrations to the bones of the middle ear. **574**

ecdysone Insect hormone with roles in metamorphosis, molting. **598**

echinoderms Invertebrate group having hardened plates and spines embedded in the skin or body, and a water–vascular system. **410**

ECM *See* extracellular matrix.

ecological footprint Area of Earth's surface required to sustainably support a particular level of development and consumption. **795**

ecological niche The resources and environmental conditions that a species requires. **802**

ecological restoration Actively altering an area in an effort to restore or create a functional ecosystem. **879**

ecology Study of how populations interact with one another and with their nonliving environment. **782**

ecosystem A community interacting with its environment. **5**, **820**

ectoderm Outermost primary tissue layer of an animal embryo. **744**

ectotherm Animal whose body temperature varies with that of its environment; it adjusts its internal temperature by altering its behavior. **424**, **717**

effector cell Antigen-sensitized B cell or T cell that forms in an immune response and acts immediately. **652**

egg Female gamete. **724**

El Niño Southern Oscillation Naturally occurring, irregularly timed fluctuation in sea surface temperature and wind patterns in the equatorial Pacific; affects weather worldwide. **844**

electron Negatively charged subatomic particle. **24**

electron transfer chain Array of membrane-bound enzymes and other molecules that accept and give up electrons in sequence, thus releasing the energy of the electrons in small, usable steps. **85**

electron transfer phosphorylation Process in which electron flow through electron transfer chains sets up a hydrogen ion gradient that drives ATP formation. **106**

electronegativity Measure of the ability of an atom to pull electrons away from other atoms. **28**

electrophoresis Technique that separates DNA fragments by size. **238**

element A pure substance that consists only of atoms with the same number of protons. **24**

embryo In animals, a developing individual from first cleavage until hatching or birth; in humans, usually refers to an individual during weeks 2 to 8 of development. **751**

embryonic induction Embryonic cells produce signals that alter the behavior of neighboring cells. **748**

embryophytes Clade of land plants; multicelled, photosynthetic species that protect and nourish the embryo on the parental body. **356**

emergent property A characteristic of a system that does not appear in any of the system's component parts. **4**

emerging disease Disease that is relatively new to a species, or has recently expanded its range. **319**

emigration Movement of individuals out of a population. **784**

emulsification Dispersion of fat droplets in a water-based fluid. **694**

endangered species A species that faces extinction in all or a part of its range. **870**

endemic species A species that remains restricted to the area where it evolved. **871**

endergonic Describes a reaction that requires a net input of free energy to proceed. **80**

endocrine disrupter Chemical that interferes with function of the endocrine system. **581**

endocrine gland Ductless gland that secretes hormones into a body fluid. **523**

endocrine system System of hormone-producing glands and secretory cells of a vertebrate body. **582**

endocytosis Process by which a cell takes in a small amount of extracellular fluid (and its contents) by the ballooning inward of the plasma membrane. **94**

endoderm Innermost primary tissue layer of an animal embryo. **744**

endodermis Outer layer of the vascular cylinder in a plant root; sheet of cells just outside the pericycle. **458**

endomembrane system Series of interacting organelles (endoplasmic reticulum, Golgi bodies, vesicles) between nucleus and plasma membrane; produces lipids, proteins. **64**

endoplasmic reticulum (**ER**) Membrane-enclosed organelle that is a continuous system of sacs and tubes extending from the nuclear envelope. Smooth ER makes lipids and breaks down carbohydrates and fatty acids; rough ER modifies polypeptides made by ribosomes on its surface. **64**

endoskeleton Internal skeleton made up of hardened components such as bones. **417**, **606**

endosperm Nutritive tissue in the seeds of flowering plants. **369**, **488**

endospore Spore (resting structure) formed by some soil bacteria; contains a dormant cell and is highly resistant to adverse conditions. **329**

endosymbiont hypothesis Hypothesis that mitochondria and chloroplasts evolved from bacteria that entered and lived inside another cell. **312**

endotherm Animal that maintains its temperature by adjusting its production of metabolic heat; for example, a bird or mammal. **424**, **717**

energy The capacity to do work. **78**

energy pyramid Diagram that depicts the energy that enters each of an ecosystem's trophic levels. Lowest tier of the pyramid, representing primary producers, is always the largest. **824**

enhancer Binding site in DNA for proteins that enhance the rate of transcription. **164**

entropy Measure of how much the energy of a system is dispersed. **78**

enzyme Protein or RNA that speeds up a reaction without being changed by it. **41**

eosinophil Circulating white blood cell with granules; specialized to combat multicelled parasites that are too large for phagocytosis. **643**

epidemic Disease outbreak that occurs in a limited region. **325**

epidermis Outermost tissue layer. Dermal tissue in plants or epidermis in animals. **453**, **530**

epididymides A pair of ducts in which sperm formed in testes mature; each empties into a vas deferens. **732**

epigenetic Refers to heritable changes in gene expression that are not the result of changes in DNA sequence. **172**

epiglottis Tissue flap that covers the trachea during swallowing to prevent food from entering airways. **675**

epiphyte Plant that grows on another plant but does not harm it. **363**

epistasis Polygenic inheritance, in which a trait is influenced by multiple genes. **209**

epithelial tissue Sheetlike animal tissue that covers outer body surfaces and lines internal tubes and cavities. **522**

equilibrium model of island biogeography Model that predicts the number of species on an island based on the island's area and distance from the mainland. **814**

ER *See* endoplasmic reticulum.

erosion *See* soil erosion.

erythropoietin Hormone secreted by kidneys; induces stem cells in bone marrow to give rise to red blood cells. **680**

esophagus Tubular organ that connects the pharynx to the stomach. **688**

essential amino acid Amino acid that the body cannot make and must obtain from food. **697**

essential fatty acid Fatty acid that the body cannot make and must obtain from food. **697**

estivation An animal becomes dormant during a hot, dry season. **718**

estrogens Hormones secreted by ovaries; cause development of secondary sexual traits and maintain the reproductive tract's lining. **728**

estrous cycle In female animals, a recurring cyclic variation in sexual receptivity. **731**

estuary A highly productive ecosystem where nutrient-rich water from a river mixes with seawater. **858**

ethylene Gaseous plant hormone involved in regulating growth and cell expansion. Participates in germination, abscission, ripening, and stress responses. **509**

eudicots Most diverse lineage of angiosperms; members have two seed leaves, branching leaf veins. **370**

eugenics Idea of deliberately improving the genetic qualities of the human race. **247**

euglenoid Flagellated protozoan with multiple mitochondria; may be heterotrophic or have chloroplasts descended from a green alga. **341**

eukaryote Organism whose cells characteristically have a nucleus; a protist, fungus, plant, or animal. **8**

eusocial animal Animal that lives in a multigenerational family group with a reproductive division of labor. **776**

eutrophication Nutrient enrichment of an aquatic ecosystem. **819**

evaporation Transition of a liquid to a gas. **31**, **717**

evolution Change in a line of descent. **254**

evolutionary tree Diagram showing evolutionary connections. **295**

exaptation Evolutionary adaptation of an existing structure for a completely new purpose. **288**

exergonic Describes a reaction that ends with a net release of free energy. **80**

exocrine gland Gland that secretes milk, sweat, saliva, or some other substance through a duct. **523**

exocytosis Process by which a cell expels a vesicle's contents to extracellular fluid. **94**

exon Nucleotide sequence that remains in an RNA after post-transcriptional modification. **153**

exoskeleton Of some invertebrates, hard external parts that muscles attach to and move. **404**, **606**

exotic species A species that evolved in one community and later was introduced into and became established in a different one. **813**

experiment A test designed to support or falsify a prediction. **13**

experimental group In an experiment, a group of individuals who have a certain characteristic or receive a certain treatment as compared with a control group. **13**

exponential growth A population grows by a fixed percentage in successive time intervals; the size of each increase is determined by the current population size. **784**

external fertilization Sperm and eggs are released into the external environment and combine there. **725**

extinct Refers to a species that no longer has living members. **288**

extracellular fluid Of a multicelled organism, body fluid that is not inside cells; serves as the body's internal environment. **520**

extracellular matrix (ECM) Complex mixture of cell secretions; its composition and function vary by cell type. **70**

extreme halophile Organism adapted to life in a highly salty environment. **332**

extreme thermophile Organism adapted to life in a very high-temperature environment. **332**

eye Sensory organ that incorporates a dense array of photoreceptors. **568**

facilitated diffusion Passive transport mechanism in which a solute follows its concentration gradient across a membrane by moving through a transport protein. **92**

fall overturn During the fall, waters of a temperate zone mix. Upper, oxygenated water cools, gets dense, and sinks; nutrient-rich water from the bottom moves up. **857**

fat Lipid that consists of a glycerol molecule with one, two, or three fatty acid tails. Saturated fats have three saturated fatty acid tails. Unsaturated fats have one or more unsaturated fatty acid tails. **44**

fatty acid Organic compound that consists of an acidic carboxyl group "head" and a long hydrocarbon "tail." *See also* saturated fatty acid, unsaturated fatty acid. **44**

feces Unabsorbed food material and cellular waste that is expelled from the digestive tract. **696**

feedback inhibition Regulatory mechanism in which a change that results from some activity decreases or stops the activity. **84**

fermentation A metabolic pathway that breaks down sugars to produce ATP and does not require oxygen. **119**

fertilization Fusion of two gametes to form a zygote; part of sexual reproduction. **191**

fetus Developing human from about 9 weeks until birth. **751**

fever A temporary, internally induced rise in core body temperature above the normal set point as a response to infection or injury. **649**

fibrous root system Root system composed of an extensive mass of similar-sized adventitious roots; typical of monocots. **458**

fight–flight response Response to danger or excitement. Activity of parasympathetic neurons declines, sympathetic signals increase, and adrenal glands secrete epinephrine. **549**

first law of thermodynamics Energy cannot be created or destroyed. **78**

fish General term for a gilled aquatic vertebrate that is not a tetrapod. **418**

fitness Degree of adaptation to an environment, as measured by an individual's relative genetic contribution to future generations. **256**

fixed Refers to an allele for which all members of a population are homozygous. **280**

flagellated protozoan Protist belonging to an entirely or mostly heterotrophic lineage with no cell wall and one or more flagella. **340**

flagellum Long, slender cellular structure used for motility. **59**

flatworm Acoelomate, unsegmented worm; for example, a planarian or tapeworm. **396**

flower Specialized reproductive structure of a flowering plant. **368**, **484**

fluid mosaic Model of a cell membrane as a two-dimensional fluid of mixed composition. **88**

follicle-stimulating hormone (**FSH**) Anterior pituitary hormone with roles in ovarian follicle maturation and sperm production. **730**

food chain Description of who eats whom in one path of energy flow in an ecosystem. **820**

food web Set of cross-connecting food chains. **822**

foraminifera Group of heterotrophic, single-celled protists with a porous calcium carbonate shell and long cytoplasmic extensions. **342**

fossil Physical evidence of an organism that lived in the ancient past. **253**

founder effect After a small group of individuals found a new population, allele frequencies in the new population differ from those in the original population. **280**

fovea Region of the retina where cone cells are most concentrated. **572**

free radical Atom with an unpaired electron; most are highly reactive and can damage biological molecules. **27**

frequency-dependent selection Natural selection in which a trait's adaptive value depends on its frequency in a population. **279**

fruit Mature ovary of a flowering plant, often with accessory parts; encloses a seed or seeds. **368**, **492**

FSH *See* follicle-stimulating hormone.

functional group An atom (other than hydrogen) or a small molecular group bonded to a carbon of an organic compound; imparts a specific chemical property. **40**

fungus Single-cell or multicelled eukaryotic organism with cell walls of chitin; obtains nutrients by extracellular digestion and absorption. **8**, **376**

gallbladder Organ that stores and concentrates bile. **694**

gametangium Gamete-producing organ of a plant. **360**

gamete Mature, haploid reproductive cell; e.g., an egg or a sperm. **190**

gametophyte A haploid, multicelled body that produces gametes in the life cycle of land plants and some algae. **356**

ganglion Cluster of nerve cell bodies. **396**, **538**

gap junction Cell junction that forms a closable channel across the plasma membranes of adjoining animal cells. **71**

gastric fluid Fluid secreted by the stomach lining; contains digestive enzymes, acid, and mucus. **692**

gastrovascular cavity Of some invertebrates, a saclike cavity that functions in digestion and in gas exchange. **391**, **688**

gastrula Three-layered embryo that forms by during early development in most animals. **744**

gastrulation In animal development, process by which cell movements produce a three-layered gastrula. **744**

gene A part of a chromosome that encodes an RNA or protein product in its DNA sequence. **150**

gene expression Process by which the information in a gene guides assembly of an RNA or protein product. **150**

gene flow The movement of alleles into and out of a population. **281**

gene pool All the alleles of all the genes in a population; a pool of genetic resources. **271**

gene therapy Treating a genetic defect or disorder by transferring a normal or modified gene into the affected individual. **246**

genetic code Complete set of sixty-four mRNA codons. **154**

genetic drift Change in allele frequency resulting from chance alone. **280**

genetic engineering Process by which deliberate changes are introduced into an individual's genome. **242**

genetic equilibrium Theoretical state in which an allele's frequency never changes in a population's gene pool. **272**

genetically modified organism (**GMO**) Organism whose genome has been modified by genetic engineering. **242**

genome An organism's complete set of genetic material. **236**

genomics The study of genomes. **240**

genotype The particular set of alleles that is carried by an individual's chromosomes. **203**

genus (plural **genera**) A group of species that share a unique set of traits; also the first part of a species name. **10**

geologic time scale Chronology of Earth's history; correlates geologic and evolutionary events. **264**

germ cell Immature reproductive cell that gives rise to haploid gametes when it divides. **190**

germ layer One of three primary tissue layers in an early animal embryo. **744**

germinate To resume metabolic activity after dormancy. **488**

GH *See* growth hormone.

gibberellin Plant hormone that induces stem elongation and helps seeds break dormancy, among other effects. **507**

gill Respiratory organ of some aquatic animals; may be internal or external. **400**, **670**

ginkgo Deciduous gymnosperm with flagellated sperm, fan-shaped leaves, and fleshy seeds. **367**

gizzard Of birds and some other animals, a digestive chamber lined with a hard material that helps grind up food. **688**

gland cell Secretory epithelial cell. **523**

global climate change A long-term change in Earth's climate. **831**

glomeromycete A type of mycorrhizal fungus in which hyphae branch inside the cell wall of the root cell. **379**

glomerular filtration First step in vertebrate urine formation: protein-free plasma forced out of glomerular capillaries by blood pressure enters Bowman's capsule. **712**

glomerulus In the kidney, a cluster of capillaries enclosed by Bowman's capsule. **711**

glottis Opening between the vocal cords; appears when these cords relax. **674**

glucagon Pancreatic hormone that causes cells to break down glycogen and release glucose. **592**

glycogen Polysaccharide that serves as an energy reservoir in animal cells. **43**

glycolysis Set of reactions in which a six-carbon sugar (such as glucose) is converted to two pyruvate for a net yield of two ATP. **120**

GMO *See* genetically modified organism.

gnetophyte Shrubby or vinelike gymnosperm, with nonmotile sperm; for example, *Ephedra*. **367**

GnRH *See* gonadotropin-releasing hormone.

Golgi body Membrane-enclosed organelle that modifies proteins and lipids, then packages the finished products into vesicles. **65**

gonadotropin-releasing hormone (GnRH) A hypothalamic hormone that induces the pituitary to release hormones (LH and FSH) that act on the gonads. **730**

gonads Primary reproductive organs (ovaries or testes); produce gametes and sex hormones. **596, 726**

Gondwana Supercontinent that existed before Pangea, more than 500 million years ago. **263**

Gram-positive bacteria Lineage of thick-walled bacteria that are colored purple by Gram staining. **331**

grassland Biome in the interior of continents where grasses and nonwoody plants adapted to grazing and fire predominate. **850**

gravitropism Plant growth in a direction influenced by gravity. **510**

gray matter Brain and spinal cord tissue that includes neuron cell bodies, dendrites, and axon terminals, as well as glial cells. **550**

grazing food web Food web in which most energy is transferred from producers to grazers (herbivores). **822**

green alga Common term for one of the single-celled, colonial, or multicelled photosynthetic protists that has chloroplasts containing chlorophylls *a* and *b*; a chlorophyte or charophyte alga. **348**

greenhouse gas Atmospheric gas that absorbs heat emitted by Earth's surface and reemits it, thus keeping the planet warm. **830**

ground tissues Tissues that make up the bulk of the plant body; all plant tissues other than vascular and dermal tissues. **450**

groundwater Soil water and water in aquifers. **826**

growth In multicelled species, an increase in the number, size, and volume of cells. **7**

growth factor Molecule that stimulates mitosis and differentiation. **184**

growth hormone Anterior pituitary hormone that regulates growth and metabolism. **587**

guard cell One of a pair of cells that define a stoma across the epidermis of a leaf or other plant part. **476**

gymnosperm Seed plant that does not make flowers or fruits; for example, a conifer. **366**

habitat Type of environment in which a species typically lives. **800**

habituation Learning not to respond to a repeated stimulus. **767**

half-life Characteristic time it takes for half of a quantity of a radioisotope to decay. **260**

haploid Having one of each type of chromosome characteristic of the species. **190**

HCG *See* human chorionic gonadotropin.

hearing Perception of sound. **574**

heart Muscular organ that pumps blood through a body. **622**

hemoglobin Iron-containing respiratory protein that fills vertebrate red blood cells. **668**

hemolymph Fluid pumped through an open circulatory system. **622**

hemostasis Process by which blood clots in response to injury. **629**

herbivory An animal feeds on plant parts. **807**

hermaphrodite Animal that produces both eggs and sperm, either simultaneously or at different times in its life. **393, 724**

heterotherm Animal that sometimes maintains its temperature by producing metabolic heat, and at other times allows its temperature to fluctuate with the environment. **717**

heterotroph Organism that obtains carbon from organic compounds assembled by other organisms. **101**

heterozygous Having two different alleles of a gene. **203**

hibernation An animal becomes dormant during a cold season. **718**

hippocampus Brain region essential to formation of declarative memories. **555**

histone Type of protein that associates with DNA and structurally organizes eukaryotic chromosomes. **138**

HIV (human immunodeficiency virus) Retrovirus that causes AIDS. **319**

homeostasis Process in which an organism keeps its internal conditions within tolerable ranges by sensing and responding to change. **7**

homeotic gene Type of master gene with a homeodomain; its expression directs formation of a specific body part during development. **166**

hominins Modern humans and their closest extinct relatives. **440**

hominoids Tailless primate lineage that includes apes and humans. **437**

Homo erectus Early human species that arose in Africa by about 2 million years ago; some members of this species left Africa and became established other regions. **442**

Homo habilis Earliest named human species; lived in Africa from about 2.3 to 1.4 million years ago. **442**

Homo neanderthalensis Neanderthals. Extinct hominins that lived in the Middle East, Europe, and Asia; the closest relatives of modern humans. **444**

Homo sapiens Modern humans; only surviving *Homo* species. **444**

homologous chromosomes Chromosomes that have the same length, shape, and genes. In sexual reproducers, one member of a homologous pair is paternal and the other is maternal. **179**

homologous structures Body structures that are similar in different lineages because they evolved in a common ancestor. **296**

homozygous Having identical alleles of a gene. **203**

horizontal gene transfer Transfer of genetic material by a mechanism other than inheritance from a parent or parents. **327**

hormone *See* animal hormone, plant hormone.

human Living or extinct member of the genus *Homo*. **442**

human chorionic gonadotropin (HCG) Hormone first secreted by the blastocyst, and later by the placenta; helps maintain the uterine lining during pregnancy. **753**

human immunodeficiency virus *See* HIV.

humus Decaying organic matter in soil. **470**

hybrid A heterozygous individual. **203**

hydrocarbon Compound or region of one that consists only of carbon and hydrogen atoms. **40**

hydrogen bond Attraction between a covalently bonded hydrogen atom and another atom taking part in a separate covalent bond. **30**

hydrogenosome Organelle that produces ATP and hydrogen gas by an anaerobic pathway; evolved from mitochondria. **340**

hydrolysis Water-requiring chemical reaction in which an enzyme breaks a molecule into smaller subunits. **41**

hydrophilic Describes a substance that dissolves easily in water. **30**

hydrophobic Describes a substance that resists dissolving in water. **30**

hydrostatic skeleton Of soft-bodied invertebrates, a fluid-filled chamber that muscles exert force against, redistributing the fluid. **394**, **606**

hydrothermal vent Underwater opening from which mineral-rich water heated by geothermal energy streams out. **307**, **862**

hypertonic Describes a fluid that has a high solute concentration relative to another fluid separated by a semipermeable membrane. **90**

hypha Component of a fungal mycelium; a filament made up of cells arranged end to end. **376**

hypothalamus Forebrain region that controls processes related to homeostasis; in concert with the pituitary gland; serves as center for endocrine functions. **553**, **586**

hypothesis Testable explanation of a natural phenomenon. **12**

hypotonic Describes a fluid that has a low solute concentration relative to another fluid from which it is separated by a semipermeable membrane. **90**

immigration Movement of individuals into a population. **784**

immunity The body's ability to resist and fight infections. **642**

immunization Any procedure designed to induce immunity to a specific disease. **662**

imprinting Learning that can occur only during a specific interval in an animal's life. **766**

***in vitro* fertilization** Assisted reproductive technology in which eggs and sperm are united outside the body. **723**

inbreeding Mating among close relatives. **281**

incomplete dominance Effect in which one allele is not fully dominant over another, so the heterozygous phenotype is an intermediate blend between the two homozygous phenotypes. **208**

independent variable Variable that is controlled by an experimenter in order to explore its relationship to a dependent variable. **13**

indicator species A species that is especially sensitive to disturbance and can be monitored to assess the health of a habitat. **812**

induced-fit model Substrate binding to an active site improves the fit between the two. **82**

induced pluripotent stem cells (IPSCs) Adult cells that have dedifferentiated so they behave like embryonic stem cells. **533**

inductive reasoning Drawing a conclusion based on observation. **12**

inferior vena cava Vein that delivers blood from the lower body to the heart. **625**

inflammation A local response to tissue damage or infection; characterized by redness, warmth, swelling, and pain. **648**

inheritance Transmission of DNA to offspring. **7**

inhibiting hormone Hormone that deters release of a different hormone by its target endocrine cells. **584**

innate immunity In all multicelled organisms, set of immediate, general defenses against infection. **642**

inner ear Portion of the ear containing the fluid-filled vestibular apparatus and cochlea. **574**

insect Six-legged arthropod with two antennae and two compound eyes. Member of the most diverse class of animals. **408**

insertion Mutation in which one or more nucleotides become inserted into DNA. **158**

instinctive behavior An innate response to a simple stimulus. **766**

insulin Pancreatic hormone that causes cells to take up glucose and store it as glycogen. **592**

integumentary exchange Gas exchange that occurs across an animal's outer body surface (integument). **670**

intercostal muscles The skeletal muscles between the ribs; help change the volume of the thoracic cavity during breathing. **675**

intermediate disturbance hypothesis Species richness is greatest in communities where disturbances are moderate in their intensity or frequency. **811**

intermediate filament Stable cytoskeletal element that structurally supports cell membranes and tissues; also forms external structures such as hair. **68**

internal fertilization Fertilization of eggs inside a female's body. **725**

interneuron Neuron that both receives signals from and sends signals to other neurons. Located mainly in the brain and spinal cord. **540**

interphase In a eukaryotic cell cycle, the interval during which a cell grows, roughly doubles the number of its cytoplasmic components, and replicates its DNA in preparation for division. **178**

interspecific competition Competition between two species. **802**

intestine In a complete digestive system, a tubular organ in which chemical digestion is completed and from which nutrients are absorbed. **689**

interstitial fluid Of a multicelled organism, body fluid in spaces between cells. **520**

intervertebral disk Cartilage disk between two vertebrae. **606**

intron Nucleotide sequence that intervenes between exons and is removed during post-transcriptional modification. **153**

inversion Structural rearrangement of a chromosome in which part of the DNA becomes oriented in the reverse direction. **224**

invertebrate Animal that does not have a backbone. **389**

ion Charged atom. **27**

ionic bond Type of chemical bond in which a strong mutual attraction links ions of opposite charge. **28**

IPSCs *See* induced pluripotent stem cells.

iris Circular muscle that adjusts the shape of the pupil to regulate how much light enters the eye. **570**

iron–sulfur world hypothesis Hypothesis that the metabolic reactions that led to the first cells took place on the porous surface of iron–sulfide-rich rocks at hydrothermal vents. **308**

isotonic Describes two fluids with identical solute concentrations and separated by a semipermeable membrane. **90**

isotopes Forms of an element that differ in the number of neutrons their atoms carry. **24**

joint Region where bones meet. **608**

karyotype Image of an individual's set of chromosomes arranged by size, length, shape, and centromere location. **139**

key innovation An evolutionary adaptation that gives its bearer the opportunity to exploit a particular environment much more efficiently or in a new way. **288**

keystone species A species whose presence has a disproportionately large effect on community structure. **812**

kidney Organ of the vertebrate urinary system that filters blood, adjusts its composition, and forms urine. **709**

kinesis Innate response in which an animal alters its rate of movement in reaction to a stimulus. **769**

kinetic energy The energy of motion. **78**

knockout An experiment in which a gene is deliberately inactivated in a living organism; also, an organism that has a knocked-out gene. **166**

Krebs cycle Cyclic pathway that, along with acetyl–CoA formation, breaks down pyruvate to carbon dioxide in aerobic respiration's second stage. **122**

***K*-selection** Selection that operates in populations near their carrying capacity; favors traits that allow their bearers to outcompete others for limited resources. **789**

labor Expulsion of a placental mammal from its mother's uterus by muscle contractions. **757**

lactate fermentation Anaerobic sugar breakdown pathway that produces ATP and lactate. **126**

lactation Production of milk by a female mammal. **758**

lake A body of standing fresh water. **856**

lancelet Invertebrate chordate with a fishlike shape; retains all the defining embryonic chordate traits into adulthood. **416**

large intestine Organ that receives material from the small intestine, absorbs water, and concentrates and stores waste. Includes the cecum, colon, and rectum. **690**

larva An immature animal with a body form that differs from the adult form. Develops during the life cycle of some animals. **393**

larynx Short airway containing the vocal cords (voice box). **674**

lateral bud Axillary bud; forms in a leaf axil. **461**

lateral meristem Cylindrical sheet of meristem that runs lengthwise through shoots and roots; gives rise to secondary growth (thickening) in a plant. *See also* vascular cambium, cork cambium, meristem. **462**

law of independent assortment During meiosis, members of a pair of genes on homologous chromosomes tend to be distributed into gametes independently of other gene pairs. **206**

law of nature Generalization that describes a consistent natural phenomenon for which there is incomplete scientific explanation. **18**

law of segregation The two members of each pair of genes on homologous chromosomes end up in different gametes during meiosis. **205**

leaching Process by which water moving through soil removes nutrients from it. **471**

leaf vein A vascular bundle in a leaf. **456**

learned behavior Behavior that is modified by experience. **766**

lek Area where male animals congregate and perform courtship displays. **773**

lens Disk-shaped structure that bends light rays so they fall on an eye's photoreceptors. **568**

lethal mutation Mutation that alters phenotype so drastically that it causes death. **270**

LH *See* luteinizing hormone.

lichen Composite organism consisting of a fungus and single-celled algae or cyanobacteria. **384**

ligament Strap of dense connective tissue that holds bones together at a synovial joint. **609**

light-dependent reactions First stage of photosynthesis; convert light energy to chemical energy. **105**

light-independent reactions Second stage of photosynthesis; use ATP and NADPH to assemble sugars from water and CO_2. A noncyclic pathway produces oxygen; a cyclic pathway does not. **105**

lignin Material that stiffens cell walls of vascular plants. **70, 358**

limbic system Group of structures deep in the brain that function in expression of emotion. **555**

lineage Line of descent. **254**

linkage group All of the genes on a chromosome. **207**

lipid A fat, steroid, or wax. **44**

lipid bilayer Double layer of lipids arranged tail-to-tail; structural foundation of cell membranes. **45**

liver Organ that stores glycogen, detoxifies the blood, and produces bile. **694**

loam Soil with roughly equal amounts of sand, silt, and clay. **470**

lobe-finned fish Fish with fleshy fins supported by bones; a coelocanth or lungfish. **421**

local signaling molecule Chemical signal that diffuses through interstitial fluid and affects nearby cells in an animal body. **582**

locomotion Self-propelled movement from place to place. **604**

locus Location of a gene on a chromosome. **203**

logistic growth Density-dependent limiting factors cause population growth to slow as population size increases. **786**

loop of Henle U-shaped portion of a kidney tubule that connects the proximal and distal tubules. **710**

lumen The space within a tubular or hollow organ. **690**

lung Saclike internal respiratory organ in which blood exchanges gases with the air. **400, 670**

luteinizing hormone (LH) Anterior pituitary hormone with roles in ovulation, corpus luteum formation, and sperm production. **730**

lymph Fluid in the lymph vascular system. **636**

lymph node Small mass of lymphatic tissue through which lymph filters; contains many lymphocytes (B and T cells). **637**

lymph vascular system System of vessels that takes up interstitial fluid and carries it (as lymph) to the blood. **636**

lysogenic pathway Bacteriophage replication mechanism in which viral DNA becomes integrated into the host's chromosome and is passed to the host's descendants. **322**

lysosome Enzyme-filled vesicle that breaks down cellular wastes and debris. **64**

lysozyme Antibacterial enzyme in body secretions such as saliva and mucus. **645**

lytic pathway Bacteriophage replication mechanism in which a virus replicates in its host and kills it quickly. **322**

macroevolution Large-scale evolutionary patterns and trends; e.g., adaptive radiation, mass extinction. **288**

macrophage Phagocytic white blood cell that patrols tissues and tissue fluids. **643**

Malpighian tubule Water-conserving excretory organ of insects. **408**

mammal Vertebrate in which females feed young with milk secreted by mammary glands. **428**

mantle Of mollusks, extension of the body wall. **400**

mark–recapture sampling Method of estimating population size of mobile animals by marking individuals, releasing them, then checking the proportion of marks among individuals later recaptured. **782**

marsupial Mammal in which young are born at an early stage and complete development in a pouch on the mother's surface. **429**

mass extinction Macroevolutionary pattern in which a large number of species become extinct in a relatively short period of geologic time. **868**

mass number Of an isotope, the total number of protons and neutrons in the atomic nucleus. **24**

mast cell Leukocyte (white blood cell) with granules that is anchored in many tissues; factor in inflammation. **643**

master gene Gene encoding a product that affects the expression of many other genes. **166**

maternal effect gene Gene expressed in an egg as it matures; its product influences animal development. **748**

mechanoreceptor Sensory receptor that responds to pressure, position, or acceleration. **562**

medulla oblongata Hindbrain region that controls breathing rhythm and reflexes such as coughing and vomiting. **552**

medusa A bell-like cnidarian body fringed with tentacles. **394**

megaspore Of seed plants, haploid spore that forms in an ovule and gives rise to an egg-producing gametophyte. **364, 488**

meiosis Nuclear division process that halves the chromosome number. Basis of sexual reproduction. **190**

melatonin Hormone secreted by the pineal gland under conditions of darkness; affects sleep–wake cycles and protects against cancer. **597**

membrane potential Voltage difference across a cell membrane. **541**

memory cell Long-lived, antigen-sensitized B cell or T cell that can act in a secondary immune response. **652**

meninges The three membranes that enclose the brain and spinal cord. **550**

menopause Permanent cessation of menstrual cycles. **731**

menstrual cycle Reproductive cycle in which the uterus lining thickens and then, if pregnancy does not occur, is shed. **730**

menstruation Flow of shed uterine tissue out of the vagina. **730**

meristem Zone of undifferentiated plant cells; all plant growth arises from divisions of meristem cells. *See also* apical meristem, lateral meristem. **460**

mesoderm Middle tissue layer in an animal embryo having three primary tissue layers. **744**

mesophyll Photosynthetic parenchyma. **452**

messenger RNA (**mRNA**) Type of RNA that carries a protein-building message. **150**

metabolic pathway Series of enzyme-mediated reactions by which cells build, remodel, or break down an organic molecule. **84**

metabolism All of the enzyme-mediated chemical reactions by which cells build and break down organic molecules. **41**

metamorphosis Remodeling of body form during the transition from larva to adult. **404**

metaphase Stage of mitosis at which all chromosomes are aligned midway between spindle poles. **181**

metastasis The process in which malignant cells spread from one part of the body to another. **185**

methanogen Organism that produces methane gas as a metabolic by-product. **333**

MHC markers Self-proteins on the surface of human body cells. **650**

microevolution Change in an allele's frequency in a population. **271**

microfilament Cytoskeletal element composed of actin subunits. Reinforces cell membranes; functions in movement and muscle contraction. **68**

microspore Of seed plants, haploid spore that forms in a pollen sac and gives rise to a pollen grain. **364, 488**

microsporidium Single-celled spore-forming fungus that is an intracellular animal parasite. **379**

microtubule Hollow cytoskeletal element composed of tubulin subunits. Involved in movement of a cell or its parts. **68**

microvilli Thin projections from the plasma membrane of some epithelial cells; they increase the cell's surface area. **522, 693**

middle ear Eardrum and the tiny bones that transfer sound to the inner ear. **574**

migration An animal ceases an ongoing pattern of daily activity to move in a persistent manner toward a new habitat. **769**

mimicry A species evolves traits that make it similar in appearance to another species. **806**

mineral With regard to nutrition, an element required in small amounts for normal metabolism. **698**

mitochondrion Double-membraned organelle that produces ATP by aerobic respiration in eukaryotes. **66**

mitosis Nuclear division mechanism that maintains the chromosome number. Basis of body growth and tissue repair in multicelled eukaryotes; also asexual reproduction in some multicelled eukaryotes and many single-celled ones. **178**

model Analogous system used for testing hypotheses. **13**

molecular clock Technique that uses molecular change to estimate how long ago two lineages diverged. **298**

molecule Two or more atoms joined by chemical bonds. **4**

mollusk Invertebrate with a reduced coelom and a mantle; a chiton, bivalve, gastropod, or cephalopod. **400**

monocots One of two major angiosperm lineages; includes plants such as grasses that have parallel veins and one cotyledon. **370**

monohybrid cross Cross between two individuals identically heterozygous for one gene; for example, $Aa \times Aa$. **204**

monomers Molecules that are subunits of polymers. **41**

monophyletic group An ancestor in which a derived trait evolved, together with all of its descendants. **294**

monosaccharide Simple sugar; carbohydrate that is a monomer of polysaccharides. **42**

monotreme Egg-laying mammal. **428**

monsoon Wind that reverses direction seasonally. **842**

morphogen Chemical encoded by a master gene; diffuses out from its source and affects development in a concentration-dependent manner. **748**

morphological convergence Evolutionary pattern in which similar body parts (analogous structures) evolve independently in different lineages. **297**

morphological divergence Evolutionary pattern in which a body part of an ancestor changes in its descendants. **296**

motor neuron Neuron that controls a muscle or gland. **540**

motor protein Type of energy-using protein that interacts with cytoskeletal elements to move the cell's parts or the whole cell. **68**

motor unit One motor neuron and the muscle fibers it controls. **615**

mRNA *See* messenger RNA.

multicellular organism Organism composed of interdependent cells that vary in their structure and function. **338**

multiple allele system Gene for which three or more alleles persist in a population at relatively high frequency. **208**

muscle tension Force exerted by a contracting muscle. **615**

mutation Permanent change in the DNA sequence of a chromosome. **142**

mutualism Species interaction that benefits both participants. **384, 801**

mycelium Mass of threadlike filaments (hyphae) that make up the body of a multicelled fungus. **376**

mycorrhiza Mutually beneficial partnership between a fungus and a plant root. **379**

myelin Lipid-rich material made by some neuroglia; wraps around axons of some neurons and speeds propagation of action potentials. **548**

myofibrils Of a muscle fiber, threadlike protein structures consisting of contractile units (sarcomeres) arranged end to end. **612**

myoglobin Muscle protein that reversibly binds oxygen. **616**

myriapod Land-dwelling arthropod with two antennae and an elongated body with many segments; a millipede or centipede. **406**

natural killer cell *See* NK cell.

natural selection Differential survival and reproduction of individuals of a population based on differences in shared, heritable traits. Driven by environmental pressures. **256**

nectar Sweet fluid that is exuded by some flowers to attract animal pollinators. **486**

negative feedback A change causes a response that reverses the change; important mechanism of homeostasis. **532**

neoplasm An accumulation of abnormally dividing cells. **184**

nephron Functional unit of the kidney; filters blood and forms urine. **710**

nerve Neuron fibers bundled inside a sheath of connective tissue. **538**

nerve cord Bundle of nerve fibers that runs the length of a body. **396**, **538**

nerve net Of cnidarians, a mesh of interacting neurons with no central control organ. **394**, **538**

nervous tissue Animal tissue composed of neurons and supporting cells (neuroglia); detects stimuli and controls responses to them. **527**

neuroglial cell Cell that supports and assists neurons. **527**

neuromodulator Chemical signal that is released by one neuron and affects the response of other neurons to a neurotransmitter. **547**

neuromuscular junction Synapse between a motor neuron and the muscle it controls. **544**

neuron One of the cells that make up communication lines of a nervous system; transmits electrical signals along its plasma membrane and communicates with other cells through chemical messages. **527**

neurosecretory cell Specialized neuron that secretes a hormone into the blood in response to an action potential. **586**

neurotransmitter Chemical signal released by axon terminals of a neuron; it binds to receptors on a postsynaptic cell. **544**

neutral mutation A mutation that has no effect on survival or reproduction. **270**

neutron Uncharged subatomic particle in the atomic nucleus. **24**

neutrophil Circulating phagocytic white blood cell; most abundant leukocyte. **643**

niche *See* ecological niche.

nitrification Conversion of ammonium to nitrates. **832**

nitrogen cycle Movement of nitrogen among the atmosphere, soil, and water, and into and out of food webs. **832**

nitrogen fixation Incorporation of nitrogen from nitrogen gas into ammonia (NH_3). **330**, **832**

NK cell Natural killer cell. Leukocyte (white blood cell) that can kill cancer cells undetectable by cytotoxic T cells. **643**

node A region of stem where new shoots and roots can form. **454**

nondisjunction Failure of sister chromatids or homologous chromosomes to separate during nuclear division. **226**

nonshivering heat production Increase in metabolic heat production that results when brown adipose tissue releases energy as heat, rather than storing it in ATP. **718**

normal flora Microorganisms that typically live on human surfaces, including the tubes and cavities of the digestive and respiratory tracts. **644**

notochord Stiff rod of connective tissue that runs the length of the body in chordate larvae or embryos. **416**

nuclear envelope A double membrane that constitutes the outer boundary of the nucleus. Nuclear pores in the membrane control the entry and exit of large molecules. **62**

nucleic acid Polymer of nucleotides; DNA or RNA. **49**

nucleic acid hybridization Spontaneous establishment of base-pairing between two nucleic acid strands. **140**

nucleoid Of a bacterium or archaeon, region of cytoplasm where the DNA is concentrated. **59**, **326**

nucleolus In a cell nucleus, a dense, irregularly shaped region where ribosomal subunits are being produced. **63**

nucleoplasm Viscous fluid enclosed by the nuclear envelope. **62**

nucleosome A length of DNA wound twice around a spool of histone proteins. **138**

nucleotide Monomer of nucleic acids; has a ribose or deoxyribose sugar, a nitrogen-containing base, and one, two, or three phosphate groups. E.g., adenine, guanine, cytosine, thymine, uracil. **49**

nucleus Of an atom; core area occupied by protons and neutrons. **24** Of a eukaryotic cell, organelle with a double membrane that holds, protects, and controls access to the cell's DNA. **55**

nutrient Substance that an organism needs for growth and survival but cannot make for itself. **6**

obesity An overabundance of fat, which increases health risks. **702**

observational learning One animal acquires a new behavior by observing and imitating behavior of another. **767**

olfaction The sense of smell. **566**

oncogene Gene that helps transform a normal cell into a tumor cell. **184**

oocyte Immature animal egg. **726**

oogenesis Egg production in an animal. **726**

open circulatory system Circulatory system in which fluid (hemolymph) leaves vessels and mingles with tissue fluid before returning to the heart. **400**, **622**

operant conditioning Type of learning in which an animal's voluntary behavior is modified by the consequences of that behavior. **767**

operator Part of an operon; a DNA binding site for a repressor. **170**

operon Group of genes together with a promoter–operator DNA sequence that controls their transcription. **170**

organ In multicelled organisms, a structure that consists of tissues engaged in a collective task. **4**

organ of Corti An acoustical organ in the cochlea that transduces mechanical energy of pressure waves into action potentials. **574**

organ system In multicelled organisms, set of organs engaged in a collective task that keeps the body functioning properly. **5**

organelle Structure that carries out a specialized metabolic function inside a cell. **55**

organic Describes a compound that consists mainly of carbon and hydrogen atoms. **38**

organism Individual that consists of one or more cells. **4**

organs of equilibrium Sensory organs that respond to body position and motion. **576**

osmosis Diffusion of water across a selectively permeable membrane; occurs in response to a difference in solute concentration between the fluids on either side of the membrane. **90**

osmotic pressure Amount of turgor that prevents osmosis into cytoplasm or other hypertonic fluid. **91**

osteoblast Bone-forming cell; it secretes a bone's matrix. **608**

osteoclast Bone-digesting cell; it breaks down bone matrix. **608**

osteocyte A mature bone cell; a former osteoblast. **608**

outer ear External ear and the air-filled auditory canal. **574**

ovarian follicle Immature animal egg and the surrounding cells. **729**

ovary Egg producing reproductive organ. In flowering plants, the enlarged base of a carpel, inside which one or more ovules form and eggs are fertilized. **368, 484** In animals, the female gonad. **726**

oviduct Duct that conveys eggs away from an animal ovary. **728**

ovulation Release of a secondary oocyte from an ovary. **729**

ovule Of a seed-bearing plant, structure in which a female gametophyte forms. **364, 484**

ovum Mature animal egg. **727**

oxidation–reduction reaction *See* redox reaction.

oxyhemoglobin Hemoglobin with oxygen bound to it. **678**

oxytocin Pituitary hormone with roles in labor, lactation, and bonding behavior. **757**

ozone layer High atmospheric layer rich in ozone; prevents most ultraviolet radiation in sunlight from reaching Earth's surface. **305, 876**

pain Perception of tissue injury. **564**

pain receptor Sensory receptor that responds to tissue damage. **563**

pancreas Organ that secretes digestive enzymes into the small intestine and hormones into the blood. **592**

pandemic Disease outbreak with cases worldwide. **325**

Pangea Supercontinent that formed about 270 million years ago. **262**

parapatric speciation Speciation pattern in which two populations speciate while in contact along a common border. **287**

parasitism Relationship in which one species withdraws nutrients from another species, without immediately killing it. **808**

parasitoid An insect that lays eggs in another insect, and whose young devour their host from the inside. **809**

parasympathetic neurons Neurons of the autonomic system that encourage "housekeeping" tasks in the body. **548**

parathyroid glands Four small endocrine glands whose hormone product increases the level of calcium in blood. **591**

parenchyma In plants, simple tissue composed of living cells that have different functions depending on location; main component of ground tissue. **452**

passive transport Membrane-crossing mechanism that requires no energy input. **92**

pathogen Disease-causing agent. **320**

PCR *See* polymerase chain reaction.

peat bog High-latitude community dominated by *Sphagnum* moss. **360**

pedigree Chart of family connections that shows the appearance of a trait through generations. **218**

pelagic province The ocean's waters. **862**

pellicle Outer layer of plasma membrane and elastic proteins that protects and gives shape to many unwalled, single-celled protists. **340**

penis Male organ of intercourse. **732**

peptide Short chain of amino acids linked by peptide bonds. **46**

peptide bond A bond between the amine group of one amino acid and the carboxyl group of another. Joins amino acids in proteins. **46**

per capita growth rate For some interval, the added number of individuals divided by the initial population size. **784**

pericycle Layer of cells just inside endodermis of a root vascular cylinder; can give rise to lateral roots. **459**

periderm Plant dermal tissue that replaces epidermis during secondary growth of eudicots and gymnosperms. **463**

periodic table Tabular arrangement of all known elements by their atomic number. **24**

peripheral nervous system Nerves that extend through the body and carry signals to and from the central nervous system. **539**

peristalsis Wavelike smooth muscle contractions that propel food through the digestive tract. **690**

peritubular capillaries Network of capillaries that surrounds and exchanges substances with a kidney tubule. **711**

permafrost Continually frozen soil layer that lies beneath arctic tundra and prevents water from draining. **855**

peroxisome Enzyme-filled vesicle that breaks down amino acids, fatty acids, and toxic substances. **64**

petal Unit of a flower's corolla; often showy and conspicuous. **484**

pH Measure of the number of hydrogen ions in a fluid. Decreases with increasing acidity. **32**

phagocytosis "Cell eating"; an endocytic pathway by which a cell engulfs large particles such as microbes or cellular debris. **94**

pharynx Tube connecting mouth and digestive tract; in vertebrates it is the throat. **396, 674**

phenotype An individual's observable traits. **203**

pheromone Chemical that serves as a communication signal among members of an animal species. **567, 770**

phloem Complex tissue of vascular plants; its living sieve elements compose sieve tubes that distribute sugars through the plant body. Each sieve element has an associated companion cell that provides it with metabolic support. **358, 453**

phospholipid A lipid with a phosphate group in its hydrophilic head, and two nonpolar fatty acid tails; main constituent of eukaryotic cell membranes. **45**

phosphorus cycle Movement of phosphorus among Earth's rocks and waters, and into and out of food webs. **834**

phosphorylation A phosphate-group transfer. **87**

photoautotroph Organism that obtains carbon from carbon dioxide and energy from light. **328**

photoheterotroph Organism that obtains its carbon from organic compounds and its energy from light. **328**

photolysis Process by which light energy breaks down a molecule. **106**

photoperiodism Biological response to seasonal changes in the relative lengths of day and night. **512**

photoreceptor Sensory receptor that responds to light. **562**

photorespiration Inefficient sugar-production pathway initiated when rubisco attaches oxygen instead of carbon dioxide to RuBP (ribulose bisphosphate). **110**

photosynthesis Metabolic pathway by which most autotrophs use light energy to make sugars from carbon dioxide and water. **6**

photosystem Cluster of pigments and proteins that converts light energy to chemical energy in photosynthesis. **106**

phototropism Plant growth in a direction influenced by light. **510**

phylogeny Evolutionary history of a species or group of species. **294**

phytochemical Plant molecules that are not an essential part of the human diet, but may reduce risk of certain disorders; e.g., lutein. **699**

phytochrome A light-sensitive pigment that helps set plant circadian rhythms based on length of night. **512**

pigment An organic molecule that can absorb light of certain wavelengths. Reflected light imparts a characteristic color. **102**

pilus Protein filament that projects from the surface of some bacterial and archaeal cells. **59**, **326**

pineal gland Endocrine gland in brain that secretes melatonin. **597**

pinocytosis Endocytic pathway by which fluid and materials in bulk are brought into the cell. **94**

pioneer species Species that can colonize a habitat that lacks an established community. **810**

pituitary gland Pea-sized endocrine gland in the forebrain that interacts closely with the adjacent hypothalamus. **586**

placenta Of placental mammals, organ that forms during pregnancy and allows diffusion of substances between the maternal and embryonic bloodstreams. **429**, **725**

placental mammal Mammal in which a mother and her developing embryo exchange materials by means of an organ called the placenta. **429**

plankton Community of tiny drifting or swimming organisms. **342**

plants Lineage of multicelled, typically photosynthetic eukaryotes adapted to life on land. **8**, **356**

plant hormone Extracellular signaling molecule of plants that exerts its effect at very low concentration. E.g., auxin, gibberellin. **502**

plaque *See* dental plaque.

plasma Fluid portion of blood. **628**

plasma membrane A cell's outermost membrane; controls movement of substances into and out of the cell. **54**

plasmid Of many bacteria and archaea, a small ring of DNA replicated independently of the chromosome. **59**, **327**

plasmodesma Cell junction that forms an open channel between the cytoplasm of adjacent plant cells. **71**

plasmodial slime mold Soil-dwelling protist that feeds as a multinucleated mass. Develops into a fruiting body under adverse conditions. **350**

plastid Double-membraned organelle that functions in photosynthesis, pigmentation, or storage in plants and algal cells; for example, a chloroplast, chromoplast, or amyloplast. **67**

plate tectonics theory Theory that Earth's outer layer of rock is cracked into plates, the slow movement of which conveys continents to new locations over geologic time. **262**

platelet Cell fragment that helps blood clot. **629**

pleiotropy Effect in which a single gene affects multiple traits. **209**

plot sampling Method of estimating population size of organisms that do not move much by making counts in small plots, and extrapolating from this to the number in the larger area. **782**

polar body Tiny cell produced by unequal cytoplasmic division during animal egg production; has no further role in reproduction. **726**

polarity Separation of charge into positive and negative regions. **28**

pollen grain Walled, immature male gametophyte of a seed plant. **359**

pollen sac Of seed plants, sporangium where microspores form and develop into pollen grains. **364**

pollination The arrival of pollen on a receptive stigma of a seed plant. **365**, **486**

pollination vector Environmental agent that moves pollen grains from one plant to another. **486**

pollinator An animal that facilitates pollination by moving pollen from one plant to another. **369**, **483**

pollutant Substance that is released as result of human activities and harms organisms or disrupts natural processes. **874**

polygenic inheritance *See* epistasis.

polymer Molecule that consists of multiple monomers. **41**

polymerase chain reaction (**PCR**) Laboratory method that mass-produces copies of a specific section of DNA. **236**

polyp A pillarlike cnidarian body with a tentacle-ringed mouth at its top. For example, a sea anemone. **394**

polypeptide Long chain of amino acids linked by peptide bonds. **46**

polyploid Having three or more of each type of chromosome characteristic of the species. **226**

polysaccharide Carbohydrate that is a polymer of hundreds or thousands of monosaccharides. **42**

pons Hindbrain region that influences breathing and serves as a functional bridge to the adjacent midbrain. **552**

population A group of organisms of the same species who live in a specific location and breed with one another more often than they breed with members of other populations. **5**, **270**, **781**

population density Number of individuals per unit area. **782**

population distribution Describes whether individuals are clumped, uniformly dispersed, or randomly dispersed in an area. **782**

population size Total number of individuals in a population. **782**

positive feedback A response intensifies the conditions that caused its occurrence. **542**

potential energy Stored energy. **79**

predation One species captures, kills, and eats another. **804**

prediction Statement, based on a hypothesis, about a condition that should exist if the hypothesis is correct. **12**

pressure flow theory Explanation of how a difference in turgor between sieve elements in source and sink regions pushes sugar-rich fluid through a sieve tube. **478**

primary endosymbiosis Evolution of an organelle from bacteria that entered a host cell and lived inside it. **338**

primary growth Lengthening of young shoots and roots; originates at apical meristems. **460**

primary motor cortex Region of brain's frontal lobes that governs voluntary movements. **554**

primary oocyte Immature egg that has halted in prophase I of meiosis. **729**

primary producer In an ecosystem, an organism that captures energy from an inorganic source and stores it as biomass; first trophic level. **820**

primary production The rate at which an ecosystem's producers capture and store energy. **824**

primary somatosensory cortex Region of the brain's parietal lobes that receives sensory information from skin and joints. **554**

primary succession A new community colonizes an area where there is no soil. **810**

primary wall The first cell wall of young plant cells. **70**

primates Mammalian order that includes lemurs, tarsiers, monkeys, apes, and humans. **436**

primer Short, single strand of DNA or RNA that base-pairs with a specific DNA sequence in DNA synthesis. **140**

prion Infectious protein. **48**

probability The chance that a particular outcome of an event will occur; depends on the total number of outcomes possible. **17**

probe Short fragment of DNA designed to hybridize with a nucleotide sequence of interest and labeled with a tracer. **236**

probiotic Food or supplement containing living microorganisms whose presence in the body benefits health. **687**

producer Autotroph. Organism that makes its own food using energy and nonbiological raw materials from the environment. **6**

Glossary of Biological Terms (continued)

product A molecule that is produced by a reaction. **80**

progesterone Hormone secreted by ovaries; prepares the reproductive tract for pregnancy. **728**

prokaryote Informal name for a single-celled organism without a nucleus; a bacterium or archaeon. **8**, **326**

prolactin Anterior pituitary hormone that induces mammary gland enlargement during pregnancy and milk production during lactation. **758**

promoter In DNA, a sequence to which RNA polymerase binds; site where transcription begins. **152**

prophase Stage of mitosis during which chromosomes condense and become attached to a newly forming spindle. **181**

protein Organic molecule that consists of one or more polypeptides. **46**

proteobacteria Most diverse bacterial lineage; includes species that carry out photosynthesis, fix nitrogen or cause human disease. **330**

protist General term for member of one of the eukaryotic lineages that is not a fungus, animal, or plant. **8**, **337**

protocell Membranous sac that contains interacting organic molecules; hypothesized to have formed prior to the earliest life-forms. **309**

proton Positively charged subatomic particle that occurs in the nucleus of all atoms. **24**

proto-oncogene Gene that, by mutation, can become an oncogene. **184**

protostomes Lineage of bilateral animals in which the first opening on the embryo surface develops into a mouth. **391**

proximal tubule Portion of kidney tubule that receives filtrate from Bowman's capsule. **710**

pseudocoelom Incompletely lined body cavity of some animals. **391**

pseudopod A temporary protrusion that helps some eukaryotic cells move and engulf prey. **69**

puberty Period when human reproductive organs mature and begin to function. **596**

pulmonary artery Vessel that carries blood from the heart to a lung. **624**

pulmonary circuit Circuit through which blood flows from the heart to the lungs and back. **623**

pulmonary vein Vessel that carries blood from a lung to the heart. **624**

pulse Brief stretching of artery walls that occurs when ventricles contract. **630**

Punnett square Diagram used to predict the genetic and phenotypic outcomes of a cross. **204**

pupil Adjustable opening that allows light into a camera eye. **570**

pyruvate Three-carbon end product of glycolysis. **120**

radial symmetry Having parts arranged around a central axis, like spokes around a wheel. **390**

radioactive decay Process by which atoms of a radioisotope emit energy and/or subatomic particles when their nucleus spontaneously breaks up. **25**

radioisotope Isotope with an unstable nucleus. **25**

radiolaria Heterotrophic single-celled protists with a porous silica shell and long cytoplasmic extensions. **342**

radiometric dating Method of estimating the age of a rock or a fossil by measuring the content and proportions of a radioisotope and its daughter elements. **260**

radula Of many mollusks, a tonguelike organ hardened with chitin that is used in feeding. **400**

rain shadow Dry region downwind of a coastal mountain range. **843**

ray-finned fish Bony fish with fin supports derived from skin. **420**

reactant A molecule that enters a reaction and is changed by participating in it. **80**

reabsorption *See* tubular reabsorption.

reaction Process of molecular change. **41**

receptor protein Membrane protein that triggers a change in cell activity after binding to a particular substance. **89**

recessive Refers to an allele with an effect that is masked by a dominant allele on the homologous chromosome. **203**

recombinant DNA A DNA molecule that contains genetic material from more than one organism. **234**

rectum Terminal portion of the large intestine; stores digestive waste prior to excretion. **690**

red alga Photosynthetic protist; typically multicelled, with chloroplasts containing red accessory pigments (phycobilins). **348**

red blood cell Hemoglobin-filled blood cell that transports oxygen; an erythrocyte. **628**

red marrow Bone marrow that makes blood cells. **608**

redox reaction Oxidation–reduction reaction in which one molecule accepts electrons (it becomes reduced) from another molecule (which becomes oxidized). Also called electron transfer. **85**

reflex Automatic response that occurs without conscious thought or learning. **550**

releasing hormone Hormone that stimulates release of a different hormone by its target endocrine cells. **584**

replacement fertility rate Fertility rate at which each woman has, on average, one daughter who survives to reproductive age. **794**

repressor Regulatory protein that blocks transcription. **164**

reproduction Processes by which parents produce offspring. *See also* sexual reproduction, asexual reproduction. **7**

reproductive base Of a population, all individuals who are of reproductive age or younger. **783**

reproductive cloning Any of several technologies that produce genetically identical individuals. **144**

reproductive isolation The end of gene flow between populations. **282**

reptile Amniote subgroup that includes lizards, snakes, turtles, crocodilians, and birds. **424**

resource partitioning Species adapt to access different portions of a limited resource; allows species with similar needs to coexist. **803**

respiration Physiological process by which an animal body supplies cells with oxygen and disposes of their waste carbon dioxide. **668**

respiratory cycle One inhalation and one exhalation. **676**

respiratory membrane Membrane consisting of alveolar epithelium, capillary endothelium, and their fused basement membranes; site of gas exchange in lungs. **678**

respiratory protein A protein that reversibly binds oxygen when the oxygen concentration is high and releases it when oxygen concentration is low. Hemoglobin is an example. **668**

respiratory surface Moist surface across which gases are exchanged between animal cells and the external environment. **668**

resting potential Voltage difference across the plasma membrane of an excitable cell that is not receiving stimulation. **541**

restriction enzyme Type of enzyme that cuts DNA at a specific nucleotide sequence. **234**

retina In an eye, a layer densely packed with photoreceptors. **568**

retrovirus Virus whose RNA is used as a template to produce double-stranded viral DNA within a host cell. **323**

reverse transcriptase An enzyme that uses mRNA as a template to make a strand of cDNA. **235**

rhizoid Threadlike structure that anchors a bryophyte. **360**

rhizome Stem that grows horizontally along or under the ground. **362**

ribonucleic acid *See* RNA.

ribosomal RNA (rRNA) RNA that becomes part of ribosomes. **150**

ribosome Organelle of protein synthesis. An intact ribosome has two subunits, each composed of rRNA and proteins. **59**

ribozyme RNA that functions as an enzyme. **308**

RNA Ribonucleic acid. Nucleic acid with roles in gene expression; consists of a single-stranded chain of nucleotides (adenine, guanine, cytosine, and uracil). *See also* messenger RNA, transfer RNA, ribosomal RNA. **49**

RNA polymerase Enzyme that carries out transcription. **152**

RNA world hypothesis Hypothesis that RNA served as the genetic information of early life. **308**

rod cell Photoreceptor that is active in dim light; provides coarse perception of image and detects motion. **572**

root hairs Hairlike, absorptive extensions of a root epidermis cell; form on young roots. **458**

root nodules Swellings of some plant roots that contain mutualistic nitrogen-fixing bacteria. **473**

rotifer Tiny bilateral, pseudocoelomate animal with a ciliated head. **402**

roundworm Unsegmented pseudocoelomate worm with a cuticle that is molted periodically as the animal grows. **403**

rRNA *See* ribosomal RNA.

***r*-selection** Selection that favors traits that allow their bearers to produce the most offspring the most quickly. **789**

rubisco Ribulose bisphosphate carboxylase. Carbon-fixing enzyme of the Calvin–Benson cycle. **109**

runoff Water that flows over soil into streams. **826**

SA node *See* sinoatrial node.

sac fungi Most diverse group of fungi; sexual reproduction produces spores inside a saclike structure (an ascus). **380**

salivary gland Exocrine gland that secretes saliva into the mouth. **691**

salt Ionic compound that releases ions other than H^+ and OH^- when it dissolves in water. **30**

sampling error Difference between results derived from testing an entire group of events or individuals, and results derived from testing a subset of the group. **16**

sarcomere Contractile unit of muscle. **612**

sarcoplasmic reticulum Specialized endoplasmic reticulum in muscle cells; stores and releases calcium ions. **614**

saturated fatty acid Fatty acid with only single bonds linking the carbons in its tail. **44**

scales Flattened structures that grow from and sometimes cover the skin in some groups of vertebrates. **419**

science Systematic study of the observable world. **12**

scientific method Making, testing, and evaluating hypotheses about the natural world. **13**

scientific theory Hypothesis that has not been disproven after many years of rigorous testing. **18**

sclera Fibrous white layer that covers most of the eyeball. **570**

sclerenchyma In plants, simple tissue composed of cells that die when they are mature; their lignin-reinforced cell walls remain and structurally support plant parts. Includes fibers, sclereids. **452**

SCNT *See* somatic cell nuclear transfer.

scrotum Pouch of skin that encloses a human male's testes. **732**

seamount An undersea mountain. **862**

second law of thermodynamics Energy tends to disperse spontaneously. **78**

second messenger Molecule that forms inside a cell when a hormone binds to a receptor in the plasma membrane; sets in motion reactions that alter activity inside the cell. **584**

secondary endosymbiosis Evolution of a chloroplast from a protist that itself contains chloroplasts that arose by primary endosymbiosis. **338**

secondary growth Thickening of older stems and roots; originates at lateral meristems. **462**

secondary metabolite Molecule that is produced by an organism but does not play any known role in its metabolism. Some serve as defense against predation. **371**

secondary oocyte Immature egg that has halted in metaphase II of meiosis; released from an ovary at ovulation. **729**

secondary sexual traits Traits, such as the distribution of body fat, that differ between the sexes but do not have a direct role in reproduction. **596**

secondary succession A new community develops in a disturbed site where the soil that supported a previous community remains. **810**

secondary wall Lignin-reinforced wall that forms inside the primary wall of a plant cell. **70**

sedimentary cycle Biochemical cycle in which the atmosphere plays little role and rocks are the major reservoir. **834**

seed Embryo sporophyte of a seed plant packaged with nutritive tissue inside a protective coat. **359, 491**

seed plant Plant that produces seeds and pollen; an angiosperm or gymnosperm. **357**

selfish herd Group that forms when individuals hide behind others to minimize their individual risk of predation. **774**

semen Sperm mixed with secretions from seminal vesicles and the prostate gland. **733**

semiconservative replication Describes the process of DNA replication, which produces two copies of a DNA molecule: one strand of each copy is new, and the other is conserved (parental). **141**

seminiferous tubules Inside a testis, coiled tubules in which sperm form. **733**

sensation Awareness of a stimulus. **563**

sensory adaptation Slowing or halting of a sensory response to an ongoing stimulus. **563**

sensory neuron Neuron that is activated when its receptor endings detect a specific stimulus, such as light or pressure. **540**

sensory perception The meaning a brain derives from a sensation. **563**

sensory receptor Cell or cell component that detects a specific type of stimulus. **532**

sepal Unit of a flower's calyx; typically photosynthetic and inconspicuous. **484**

sequencing Laboratory method of determining the order of nucleotides in DNA. **238**

sessile animal Animal that lives fixed in place on some surface. **393**

sex chromosome Member of a pair of chromosomes that differs between males and females. **139**

sex hormone A steroid hormone that controls sexual development and reproduction; testosterone in males, estrogens or progesterone in females. **596**

sexual dimorphism Difference in appearance between males and females of a species. **278**

sexual reproduction Reproductive mode by which offspring arise from two parents and inherit genes from both. **189, 724**

sexual selection Mode of natural selection in which some individuals outreproduce others of a population because they are better at securing mates. **278**

shell model Model of electron distribution in an atom. **26**

shivering response in response to prolonged exposure to cold, rhythmic muscle contractions generate metabolic heat. **718**

short tandem repeat In chromosomal DNA, sequences of a few nucleotides repeated multiple times in a row. Used in DNA profiling. **212**

sieve elements Living cells that compose sugar-conducting sieve tubes of phloem. Each sieve tube consists of a stack of sieve elements that meet end to end at sieve plates. **478**

sieve tube Sugar-conducting tube of phloem; consists of stacked sieve elements. **478**

single-nucleotide polymorphism (**SNP**) One-nucleotide DNA sequence variation carried by a measurable percentage of a population. **233**

sink Region of plant tissue where sugars are being used. **478**

sinoatrial (**SA**) **node** Cardiac pacemaker; cluster of specialized cells whose spontaneous rhythmic signals trigger contractions. **627**

sister chromatids The two DNA molecules of a duplicated eukaryotic chromosome, attached at the centromere. **138**

sister groups The two lineages that emerge from a node on a cladogram. **295**

skeletal muscle fiber Contractile cell that contains mainly myofibrils and runs the length of a muscle. **612**

skeletal muscle tissue Muscle that pulls on bones and moves body parts; under voluntary control. **526**

sliding-filament model Explanation of how interactions among actin filaments and myosin filaments bring about muscle contraction. **612**

slime mold *See* cellular slime mold, plasmodial slime mold.

small intestine Organ that receives material from the stomach; site of most digestion and absorption. **690**

smooth muscle tissue Muscle that lines blood vessels and forms the wall of hollow organs. **526**

SNP *See* single-nucleotide polymorphism.

social animal Animal that lives in a multigenerational group in which members, who are usually relatives, cooperate in some tasks. **774**

soil erosion Loss of soil under the force of wind and water. **471**

soil water Water between soil particles. **826**

solute A dissolved substance. **30**

solution Uniform mixture of solute completely dissolved in solvent. **30**

solvent Liquid in which other substances dissolve. **30**

somatic Relating to the body. **190**

somatic cell nuclear transfer (**SCNT**) Reproductive cloning method in which the DNA of an adult donor's body cell is transferred into an unfertilized egg. **144**

somatic nerves Nerves that control skeletal muscle and relay signals from sensory neurons in joints and skin to the central nervous system. **548**

somatic sensations Sensations such as touch and pain that arise when sensory neurons in skin, muscle, or joints are activated. **564**

somatosensory cortex *See* primary somatosensory cortex.

somites Paired blocks of embryonic mesoderm that give rise to a vertebrate's muscle and bone. **750**

sorus Cluster of spore-producing capsules on a fern leaf. **363**

source Region of a plant tissue where sugars are being produced or released from storage. **478**

speciation Evolutionary process in which new species arise. **282**

species Unique type of organism designated by genus name and specific epithet. Of sexual reproducers, often defined as one or more groups of individuals that can potentially interbreed, produce fertile offspring, and do not interbreed with other groups. **10**

specific epithet Second part of a species name. **10**

sperm Male gamete. **724**

spermatogenesis Sperm production in an animal. **726**

sphincter Ring of muscle that controls passage through a tubular organ or body opening. **611**, **690**

spinal cord Portion of central nervous system that connects peripheral nerves with the brain. **550**

spindle Temporary structure that moves chromosomes during nuclear division; consists of microtubules. **181**

spirochetes Lineage of bacteria shaped like a stretched-out spring. **331**

spleen Fist-sized lymphoid organ located in the upper abdomen; filters blood and enhances immune function. **637**

sponge Aquatic invertebrate that has no body symmetry, tissues, or organs; feeds by filtering food from the water. **393**

sporangium Of plants and fungi, a structure in which haploid spores form by meiosis. **360**

sporophyte Diploid, spore-producing body in the life cycle of land plants and some algae. **356**

spring overturn In temperate zone lakes, a downward movement of oxygenated surface water and an upward movement of nutrient-rich water that occurs during spring. **857**

stabilizing selection Mode of natural selection in which an intermediate form of a trait is adaptive, and extreme forms are not. **276**

stamen Floral reproductive organ; consists of a pollen-producing anther and, in most species, a filament. **368**, **484**

starch Polysaccharide that serves as an energy reservoir in plant cells. **42**

stasis Evolutionary pattern in which a lineage changes very little over long spans of time. **288**

statistically significant Refers to a result that is statistically unlikely to have occurred by chance alone. **17**

statolith Amyloplast involved in sensing gravity. **510**

stele Vascular cylinder of a root. **458**

stem cell Cell that can divide to produce new stem cells or differentiate into some or all specific cell types. **519**

steroid Type of lipid with four carbon rings and no fatty acid tails. **45**

steroid hormone Lipid-soluble hormone derived from cholesterol. **584**

stigma Of a flower, the upper part of the carpel. Adapted to receive pollen. **484**

stomach Tubular, muscular organ that functions in chemical and mechanical digestion. **688**

stomata (singular **stoma**) Closable gaps defined by guard cells on plant surfaces; when open, they allow water vapor and gases to diffuse across the epidermis. **110**, **358**

stramenopiles Protist lineage that includes the photosynthetic diatoms and brown algae, as well as the heterotrophic water molds. **346**

strobilus (**strobili**) Of some nonflowering plants such as horsetails and cycads, a cluster of spore-producing structures. **362**

stroma Cytoplasm-like fluid between the thylakoid membrane and the two outer membranes of a chloroplast. Site of light-independent reactions of photosynthesis. **105**

stromatolite Dome-shaped structure composed of layers of bacterial cells, their secretions, and accumulated sediments. **310**

substrate Of an enzyme, a reactant that is specifically acted upon by the enzyme. **82**

substrate-level phosphorylation The formation of ATP by the direct transfer of a phosphate group from a substrate to ADP. **120**

superior vena cava Vein that delivers blood from the upper body to the heart. **625**

surface-to-volume ratio A relationship in which the volume of an object increases with the cube of the diameter, and the surface area increases with the square. Limits cell size. **55**

survivorship curve Graph showing the decline in numbers of a cohort over time. **788**

suspension feeder Aquatic animal that filters food from water around it. **393**

swim bladder Adjustable flotation sac of some bony fish. **420**

symbiosis One species lives in or on another in a commensal, mutualistic, or parasitic relationship. **800**

sympathetic neurons Neurons of the autonomic system that prepare the body for danger or excitement. **548**

sympatric speciation Speciation pattern in which speciation occurs within a population, in the absence of a physical barrier to gene flow. **286**

synapse Region where a neuron's axon terminals transmit chemical signals to another cell. **544**

synaptic integration The summation of excitatory and inhibitory signals by a postsynaptic cell. **545**

systemic acquired resistance In plants, inducible whole-body resistance to a wide range of pathogens and abiotic stressors. **514**

systemic circuit Circuit through which blood flows from the heart to the body tissues and back. **623**

systole Contractile phase of the cardiac cycle. **626**

systolic pressure Blood pressure when ventricles are contracting. **631**

T cell T lymphocyte. Leukocyte (white blood cell) central to adaptive immunity. E.g., helper T cell, cytotoxic T cell. **643**

T cell receptor (TCR) Antigen receptor on the surface of a T cell. **650**

T lymphocyte *See* T cell.

taiga *See* boreal forest.

taproot system An enlarged primary root together with all of the lateral roots that branch from it. Typical of eudicots. **458**

tardigrade Water bear. Tiny coelomate animal with four pairs of legs; in a dormant state, it can survive extremely adverse conditions. **402**

taste receptors Chemoreceptors involved in the sense of taste. **566**

taxis Innate response in which an animal moves toward or away from a stimulus. **769**

taxon (plural **taxa**) A rank of organisms that share a unique set of traits. **10**

taxonomy The science of naming and classifying species. **10**

TCR *See* T cell receptor.

telomere Noncoding, repetitive DNA sequence at the end of chromosomes; protects the coding sequences from degradation. **183**

telophase Stage of mitosis during which chromosomes arrive at opposite spindle poles and decondense, and two new nuclei form. **181**

temperate deciduous forest Northern Hemisphere biome in which the main plants are broadleaf trees that lose their leaves in fall and become dormant during cold winters. **852**

temperature Measure of molecular motion. **31**

tendon Strap of dense connective tissue that connects a skeletal muscle to bone. **611**

teratogen A toxin or infectious agent that interferes with development and thus raises the risk of birth defects. **743**

terminal bud Tip of an actively growing shoot; contains apical meristem. **460**

territory Region that an animal or group of animals defends. **772**

testcross Method of determining genotype of an individual with a dominant phenotype: a cross between the individual and another individual known to be homozygous recessive. **204**

testis Male gonad; sperm-producing reproductive organ. **726**

testosterone Main hormone produced by testes; required for sperm production and development of male secondary sexual traits. **732**

tetrapod Vertebrate with four legs, or a descendant thereof. **417**

thalamus Forebrain region that relays signals to the cerebral cortex. **552**

theory, scientific *See* scientific theory.

theory of inclusive fitness Genes associated with altruism can be advantageous if the expense of this behavior to the altruist is outweighed by the reproductive success of relatives. **776**

theory of uniformity Idea that gradual repetitive processes occurring over long time spans shaped Earth's surface. **255**

therapeutic cloning The use of SCNT to produce human embryos for research purposes. **145**

thermal radiation Emission of heat from an object into the air. **717**

thermocline Thermal stratification in a large body of water; a cool midlayer stops vertical mixing between warm surface water above it and cold water below it. **857**

thermoreceptor Temperature-sensitive sensory receptor. **562**

thigmotropism Plant growth in a direction influenced by contact with a solid object. **511**

threatened species Species likely to become endangered in the near future. **870**

threshold potential Neuron membrane potential at which gated sodium channels open, causing an action potential to occur. **542**

thylakoid membrane A chloroplast's highly folded inner membrane system; forms a continuous compartment in the stroma. Site of light reactions of photosynthesis. **105**

thymus Organ beneath the sternum; produces hormones that enhance immune function and is the site of T cell maturation. **598**

thyroid gland Endocrine gland at the base of the neck; produces thyroid hormone, which increases metabolic rate. **590**

tidal volume The volume of air that flows into and out of the lungs during a normal inhalation and exhalation. **676**

tight junction Cell junction that fastens together the plasma membrane of adjacent animal cells; collectively prevent fluids from leaking between the cells. Composed of adhesion proteins. **71**

tissue In multicelled organisms, specialized cells organized in a pattern that allows them to perform a collective function. **4**

tissue culture propagation Laboratory method in which individual plant cells (typically from meristem) are induced to form embryos. **497**

topsoil Uppermost soil layer; contains the most organic matter and nutrients for plant growth. **471**

total fertility rate Average number of children the women of a population bear over the course of a lifetime. **794**

tracer A substance that can be traced via its detectable component. **25**

trachea Airway to the lungs; windpipe. **675**

tracheal system Of insects and some other land arthropods, tubes that convey gases between the body surface and internal tissues. **670**

tracheid Component of xylem. Tapered cell that dies when mature. Pitted walls that remain interconnect with other tracheids to form water-conducting tubes. **474**

trait An inherited characteristic of an organism or species. **10**

transcription Process by which enzymes assemble an RNA using the nucleotide sequence of a gene as a template. **150**

transcription factor Regulatory protein that influences transcription by binding directly to DNA; e.g., an activator or repressor. **164**

transduction In bacteria and archaea, a mechanism of horizontal gene transfer in which a bacteriophage transfers DNA between cells. **327**

transfer RNA (tRNA) RNA that delivers amino acids to a ribosome during translation. **150**

transformation In bacteria and archaea, a type of horizontal gene transfer in which DNA is taken up from the environment. **327**

transgenic Refers to a genetically modified organism that carries a gene from a different species. **242**

transition state Point during a reaction at which substrate bonds will break and the reaction will run spontaneously. **82**

translation Process by which a polypeptide chain is assembled from amino acids in the order specified by an mRNA. **150**

translocation Of chromosomes, structural change in which a broken piece gets reattached in the wrong location. **224** In plants, movement of organic compounds through phloem. **478**

transpiration Evaporation of water from aboveground plant parts. **475**

transport protein Protein that passively or actively assists specific ions or molecules across a membrane. **89**

transposable element Segment of DNA that can move spontaneously within or between chromosomes. **224**

triglyceride A fat with three fatty acid tails. **44**

tRNA *See* transfer RNA.

trophic level Position of an organism in a food chain. **820**

tropical rain forest Highly productive and species-rich biome in which year-round rains and warmth support continuous growth of evergreen broadleaf trees. **852**

tropism In plants, directional growth response to an environmental stimulus. **510**

trypanosome Parasitic flagellated protist with a single mitochondrion and a membrane-encased flagellum. **341**

tubular reabsorption Substances move from the filtrate inside a kidney tubule into the peritubular capillaries. **712**

tubular secretion Substances move out of peritubular capillaries and into the filtrate in kidney tubules. **712**

tumor A neoplasm that forms a lump. **184**

tunicate Marine invertebrate chordate; a fish-shaped, swimming larva; undergoes metamorphosis to become a barrel-shaped, sessile adult. **416**

turgor Pressure that a fluid exerts against a wall, membrane, or other structure that contains it. **91**

unsaturated fatty acid Fatty acid with one or more carbon–carbon double bonds in its tail. **44**

urea Main nitrogen-containing compound in mammalian urine. **709**

ureter Tube that carries urine from a kidney to the bladder. **710**

urethra Tube through which urine expelled from the bladder flows out of the body. **710**

uric acid Main nitrogen-containing compound excreted by insects, as well as birds and other reptiles. **709**

urinary bladder Hollow, muscular organ that stores urine. **710**

urine Mixture of water and soluble wastes formed and excreted by the vertebrate urinary system. **709**

uterus Muscular chamber where mammalian offspring develop; womb. **729**

vaccine A preparation introduced into the body in order to elicit immunity to a specific antigen. **662**

vacuole A membrane-enclosed organelle filled with fluid; isolates or disposes of waste, debris, or toxic materials. **64**

vagina Female organ of intercourse and birth canal. **729**

variable In an experiment, a characteristic or event that differs among individuals or over time. *See also* dependent variable, independent variable. **13**

vas deferens One of a pair of long ducts that convey mature sperm toward the body surface. **732**

vascular bundle Cylindrical bundle of xylem, phloem, and sclerenchyma fibers that runs through a stem or leaf. **454**

vascular cambium Lateral meristem that produces secondary xylem and phloem. **462**

vascular cylinder Central column of vascular tissue in a plant root. **458**

vascular plant Plant having specialized tissues (xylem and phloem) that transport water and sugar within the plant body. **357**

vascular tissue In vascular plants, a tissue that distributes water or nutrients through the plant body. *See also* xylem, phloem. **358**, **450**

vasoconstriction Narrowing of a blood vessel when smooth muscle that rings it contracts. **630**

vasodilation Widening of a blood vessel when smooth muscle that rings it relaxes. **630**

vector *See* cloning vector, disease vector, pollination vector.

vegetative reproduction Growth of new roots and shoots from extensions or fragments of a parent plant; form of asexual reproduction in plants. **496**

vein Large-diameter vessel that returns blood to the heart. *See also* leaf vein. **624**

ventricle Heart chamber that pumps blood out of the heart. **622**

vernalization Stimulation of flowering in spring by prolonged exposure to low temperature in winter. **513**

vertebrae Bones of the backbone, or vertebral column. **606**

vertebral column Backbone. **606**

vertebrate Animal with a backbone. **416**

vesicle Small, membrane-enclosed organelle; different kinds store, transport, or break down their contents. **64**

vessel element Component of xylem. Type of cell that dies when mature; its pitted wall forms part of a water-conducting vessel. Each vessel consists of a stack of dead vessel elements that meet end to end at perforation plates. **474**

vestibular apparatus System of fluid-filled sacs and canals in the inner ear; contains the organs of equilibrium. **576**

villi Multicelled projections at the surface of each fold in the small intestine. **693**

viral reassortment Two related viruses infect the same individual simultaneously and swap genes. **325**

viroid Small, noncoding, infectious RNA. **321**

virus Noncellular, infectious particle of protein and nucleic acid; replicates only in a host cell. **320**

visceral sensations Sensations that arise when sensory neurons associated with organs inside body cavities are activated. **564**

vision Perception of visual stimuli based on light focused on a retina and image formation in the brain. **568**

visual accommodation Process of adjusting lens shape or position so that light from an object falls on the retina. **571**

vital capacity Amount of air moved in and out of lungs with forced inhalation and exhalation. **676**

vitamin Organic substance required in small amounts for normal metabolism. **698**

vomeronasal organ Pheromone-detecting organ of some vertebrates. **567**

warning coloration In many well-defended or unpalatable species, bright colors, patterns, and other signals that predators learn to recognize and avoid. **806**

water cycle Movement of water among Earth's atmosphere, oceans, and the freshwater reservoirs on land. **826**

water mold Heterotrophic protist that grows as nutrient-absorbing filaments. **347**

watershed Land area that drains into a particular stream or river. **826**

water–vascular system Of echinoderms, a system of fluid-filled tubes and tube feet that function in locomotion. **410**

wavelength Distance between the crests of two successive waves. **102**

wax Water-repellent mixture of lipids with fatty acid tails bonded to long-chain alcohols or carbon rings. **45**

white blood cell Leukocyte. Vertebrate blood cell with a role in housekeeping and defense. **629**

white matter Tissue of brain and spinal cord that consists of myelinated axons. **550**

wood Accumulated secondary xylem. Forms inside the cylinder of vascular cambium in an older plant stem or root. **462**

X chromosome inactivation Developmental shutdown of one of the two X chromosomes in the cells of female mammals. *See also* Barr body. **168**

xenotransplantation Transplantation of an organ from one species into another. **245**

xylem Complex vascular tissue of plants; its dead tracheids and vessel elements form tubes that distribute water and dissolved minerals through the plant body. **358**, **453**

yellow marrow Bone marrow that is mostly fat; fills cavity in most long bones. **608**

zero population growth Interval in which births equal deaths. **784**

zygote Diploid cell formed by fusion of two gametes; the first cell of a new individual. **191**

zygote fungus Fungus that forms a zygospore during sexual reproduction. **378**

Index

Page numbers followed by an f or t indicate figures and tables. ■ indicates human health topics; ■ indicates environmental topics. Bold terms indicate major topics.

O